D1203802

COMPREHENSIVE ORGANIC CHEMISTRY

COMPREHENSIVE ORGANIC CHEMISTRY
EDITORIAL BOARD

Volume 1

Stereochemistry, Hydrocarbons, Halo Compounds, Oxygen Compounds

Edited by J. F. Stoddart, University of Sheffield

Volume 2

Nitrogen Compounds, Carboxylic Acids, Phosphorus Compounds

Edited by I. O. Sutherland, University of Liverpool

Volume 3

Sulphur, Selenium, Silicon, Boron, Organometallic Compounds

Edited by D. Neville Jones, University of Sheffield

Volume 4

Heterocyclic Compounds

Edited by P. G. Sammes, The City University, London

Volume 5

Biological Compounds

Edited by E. Haslam, University of Sheffield

Volume 6

Author, Formula, Subject, Reagent, Reaction Indexes

Edited by C. J. Drayton, Pergamon Press, Oxford

COMPREHENSIVE ORGANIC CHEMISTRY

The Synthesis and Reactions of Organic Compounds

CHAIRMAN AND DEPUTY CHAIRMAN OF THE EDITORIAL BOARD

SIR DEREK BARTON, F.R.S.

AND

W. DAVID OLLIS, F.R.S.

Volume 4 Heterocyclic Compounds

Edited by P. G. SAMMES

THE CITY UNIVERSITY, LONDON

PERGAMON PRESS

OXFORD · NEW YORK · TORONTO · SYDNEY · PARIS · FRANKFURT

U.K.	Pergamon Press Ltd., Headington Hill Hall, Oxford OX3 0BW, England
U.S.A.	Pergamon Press Inc., Maxwell House, Fairview Park, Elmsford, New York 10523, U.S.A.
CANADA	Pergamon of Canada, Suite 104, 150 Consumers Road, Willowdale, Ontario, M2 J1P9, Canada
AUSTRALIA	Pergamon Press (Aust.) Pty. Ltd, P.O. Box 544, Potts Point, N.S.W. 2011, Australia
FRANCE	Pergamon Press SARL, 24 rue des Ecoles, 75240 Paris, Cedex 05, France
FEDERAL REPUBLIC OF GERMANY	Pergamon Press GmbH, 6242 Kronberg-Taunus, Pferdstrasse 1, Federal Republic of Germany

First edition 1979

British Library Cataloguing in Publication Data
Comprehensive organic chemistry
1. Chemistry, Organic
I. Barton, *Sir*, Derek
547 QD251.2 78-40502

ISBN 0-08-021316-2
ISBN 0-08-021319-7 (set)

Printed in Great Britain by A. Wheaton & Co. Ltd., Exeter

Foreword

During more than a century, the development of organic chemistry has been associated with extensive documentation. Vast numbers of textbooks, monographs, and reviews have been published with the objective of summarizing and correlating the results obtained by many thousands of organic chemists working in academic and industrial research laboratories. However, out of this colossal literature there is but a relatively small number of textbooks and multi-volumed works which have become generally accepted as representing real steps forward in the presentation of our subject.

During the classical era of organic chemistry (1820–1940), textbooks which had a profound influence on the teaching of the subject included, for example, works by Armstrong (1874), van't Hoff (1875), Roscoe–Schorlemmer (1878), Richter (1888), Gattermann (1895), van't Hoff–Werner–Eiloart (1898), Meyer–Jacobson (1902), Schmidt–Rule (1926), Karrer (1928), Freudenberg (1933), Richter–Anschütz (1935), and Gilman (1938). These texts provide an opportunity to comment on the relationship between the history of organic chemistry and its associated publications. The *Treatise on Chemistry* by Roscoe and Schorlemmer consisted of three volumes (5343 pages) published in nine parts over the period 1878–1892: the major component was Volume III (6 parts, 3516 pages) which was devoted to organic chemistry. Another instructive example is the important work *Lehrbuch der Organischen Chemie*, produced by Victor Meyer and Paul Jacobson. The increase in size from the edition (1735 pages) published during 1902–1903 to the edition (5115 pages) published over the period 1913–1924 is striking.

Many have expressed concern about the problems of maintaining effective contact with the expanding literature of organic chemistry, but few have allowed themselves to become involved with attempted solutions. The decision to publish Comprehensive Organic Chemistry was not taken lightly. The absence of a work reflecting the current rapid development of modern organic chemistry has been lamented by many eminent chemists, including the late Sir Robert Robinson (1886–1975) who played an important role in the initiation of this project shortly before his death. Comprehensive Organic Chemistry was conceived, designed, and produced in order to meet this deficiency. We realised that the current rate of growth of organic chemistry demanded speedy publication and, furthermore, that its interaction with other subjects including biochemistry, inorganic chemistry, molecular biology, medicinal chemistry, and pharmacology required the collaboration of many authors. The selection of topics to be included in order to justify the work as being comprehensive has not been easy. We recognize that some areas of organic chemistry have not been given the detailed treatment which can be justified, but we have done our best to meet the expectations of the majority of readers. In particular, we have not made a special section for Theoretical Organic Chemistry. This is not because of any lack of appreciation on our part of the importance of Theory. It is because a correct treatment of Theory cannot be made comprehensible in an abbreviated form. It is also because Theory changes with time more rapidly than the facts of the subject. Theory is better treated in our view in specialist monographs. The same arguments apply equally to the fundamental subject of Stereochemistry. Any comments regarding errors and omissions will be appreciated so that they can be dealt with in future editions.

The contents of each volume have been brought together so as to reflect what are judged to be the truly important facets of modern organic chemistry. The information is presented in a concise and logical manner with mechanistic organic chemistry being adopted to provide a constant and correlative theme. The dominating intention of the Editorial Board has been to ensure the publication of a contribution to the literature of organic chemistry which will be genuinely useful and stimulating. Emphasis has therefore been given throughout to the properties and reactions of all the important classes of organic compounds, including the remarkable array of different compounds prepared by synthesis as well as natural products created by biosynthesis. Of course, the study of natural products provided the original foundation stones on which modern synthetic organic chemistry now firmly stands.

As a major presentation of modern organic chemistry, Comprehensive Organic Chemistry will be doubly useful because we have provided, in a separate volume, an extensive index. Not only have the contents of the work been indexed in the ordinary way, but we have also added a substantial number of additional references from the original literature. These do not appear in the text itself. Thus, the reader who wishes to obtain additional information about reactions and reagents mentioned in the text will quickly be able to consult the original literature. The Index volume has been prepared by a team from Pergamon Press.

Our debt to the Authors and to the Volume Editors is considerable. We are very grateful to all our colleagues for the efficient way in which they have tried to meet the challenges (and the deadlines!) which have been presented to them. We hope that the Authors have enjoyed their association with this venture. In a lighter vein, we also trust that their feelings are different from the statement 'this task put system into my soul but not much money into my purse' attributed to Henry Edward Armstrong (1848–1937) after he had written his *Introduction to Organic Chemistry* in 1874.

We are delighted to acknowledge the masterly way in which Robert Maxwell, the Publisher, and the staff at Pergamon Press have supported the Volume Editors and the Authors in our endeavour to produce a work which correctly portrays the relevance and achievements of organic chemists and their contributions to knowledge by research.

D. H. R. Barton
Chairman

W. D. Ollis
Deputy Chairman

Contents

Preface to Volume 4

Organic chemistry is largely comprised of heterocyclic chemistry. As a consequence of the size of the subject, any single volume is bound to omit more systems than it includes. In planning this treatise, the scope of the subject matter, and how best to organize it, were kept very much in mind. The organizational problem was resolved by considering the potential needs of those seeking to use it. Since readers will include those requiring a fresh insight into heterocyclic chemistry, it had to contain details of the more familiar and common systems. Besides such readers, specialists seeking information on more complex and less usual structures of practical importance had also to be included. As a consequence, certain topical areas, such as the chemistry of purine systems (Chapter 17.5) and meso-ionic compounds (Chapter 20.4) are covered in depth. The organizational consequence of such deliberations was very much in the traditional mode, with systems arranged according to the number and type of heteroatoms present and the size and number of rings present. In order to complement other volumes in this treatise, the discussion of fully saturated heterocyclic systems has largely been omitted from Volume 4. For example, cyclic ethers and cyclic amines are mainly dealt with in Chapters 4.4 and 6.1, respectively.

The use of the symbols ly, my, and hy, adopted by some of the authors, refer to low, medium, and high yields, respectively. This symbolism serves as a useful indication of the efficiency of a particular reaction or process.

The limitations on space will inevitably mean that errors of omission have been perpetrated for which I, as editor of this volume, take full responsibility. The contents of individual chapters were arranged by consultation with the authors. It was intended to allow contributors enough freedom, in their presentations, to be able to impart their own individuality and style. This they have achieved, thus making the whole work far more enjoyable to read. I would like to record my thanks to all the contributors for their enthusiastic and efficient help with the preparation of this volume. Without their cooperation the work would still be only a dream, rather than a reality.

London P. G. Sammes

Contributors to Volume 4

Dr. R. T. Brown
Department of Chemistry, University of Manchester

Dr. M. M. Campbell
Department of Chemistry, Heriot-Watt University, Edinburgh

Dr. P. A. Claret
Department of Chemistry, The City University, London

Dr. T. M. Cresp
Department of Chemistry, University College London

Dr. M. R. Grimmett
Department of Chemistry, University of Otago

Dr. D. M. Harrison
Department of Chemistry, The New University of Ulster, Coleraine

Professor A. H. Jackson
Department of Chemistry, University College, Cardiff

Dr. J. A. Joule
Department of Chemistry, University of Manchester

Dr. J. K. Landquist
formerly of ICI Pharmaceuticals Division, Alderley Park, Cheshire

Dr. R. Livingstone
Department of Chemical Sciences, Huddersfield Polytechnic

Dr. E. Lunt
May and Baker Ltd., Dagenham, Essex

Dr. O. Meth-Cohn
Department of Chemistry and Applied Chemistry, University of Salford

Dr. A. E. A. Porter
Department of Chemistry, University of Stirling

Dr. C. A. Ramsden
May and Baker Ltd., Dagenham, Essex

Professor P. G. Sammes
Department of Chemistry, The City University, London

Dr. M. V. Sargent
Department of Organic Chemistry, University of Western Australia, Nedlands

Dr. R. K. Smalley
Department of Chemistry and Applied Chemistry, University of Salford

Dr. D. M. Smith
Department of Chemistry, University of St. Andrews

Professor K. M. Smith
Department of Chemistry, University of California at Davis

Dr. J. Staunton
University Chemical Laboratory, University of Cambridge

Dr. N. D. A. Walshe
Department of Chemistry and Applied Chemistry, University of Salford

Professor D. G. Wibberley
Department of Pharmacy, University of Aston, Birmingham

Contents of Other Volumes

PART 16

NITROGEN SYSTEMS: THE AZINES

16.1

Pyridines

D. M. SMITH
University of St. Andrews

The credit for the discovery of pyridine and its simple alkyl derivatives belongs to the nineteenth-century Scottish chemist, Thomas Anderson. At a meeting of the Royal Society of Edinburgh in April 1846, he presented a paper[1] describing the isolation from coal tar of a new organic base, *picoline**, which was isomeric with aniline. In a subsequent paper[2] he showed that picoline was also a component of the foul-smelling basic oil ('odorine') obtained by dry distillation of deer-horn, and in a third paper,[3] dated April 1851, he described the careful fractionation of this bone-oil, which gave two further bases, *pyridine* and *lutidine**.

The structural relationship between pyridine and benzene was first recognized, apparently independently, by Körner in 1869 and Dewar in 1871,[5] and the formulation of pyridine as azabenzene (**1**) was finally established by Ladenburg[6] in 1888. Proof that picoline and lutidine were pyridine derivatives was achieved by oxidation of the alkyl groups to carboxyl and decarboxylation of the resulting acids (*cf.* Ref. 4).

16.1.1 THE PYRIDINE MOLECULE

Pyridine contains a delocalized 6π-electron system, like benzene, and the molecular dimensions of pyridine (see formula **2**) include C—C and C—H bond lengths extremely close to those of benzene (C—C, 139.7 pm; C—H, 108.4 pm). The resemblance to benzene is also obvious in the ultraviolet[8] and infrared[9] spectra of pyridine. Unlike benzene, however, pyridine has a large dipole moment, 2.26 D†; the polarization of the molecule, which may be represented by the canonical structures (**3**), is reflected in the electron densities at the various ring positions, which may be calculated by molecular-orbital methods, and also in the chemical shifts in both ^1H and ^{13}C n.m.r. spectra.

A typical set of calculated π-electron densities[11] (this particular set being based on Extended Hückel Theory) is given in formula (**4**), and illustrates the general result from all such calculations, namely that the electron densities at the ring positions decrease in the order $1>3>4\approx2$.

The ^1H [12] and ^{13}C [13] chemical shifts‡ of the various atoms in pyridine are shown in formulae (**5**) and (**6**), respectively: the deshielding effect of the hetero-atom is most pronounced at the 2- and 6-positions, and is also noticeable at the 4-position. Interproton coupling constants in simple pyridine derivatives[12] are approximately as follows: $J_{2,3}$, 4.0–6.0; $J_{3,4}$, 6.8–9.2; $J_{2,4}$, 1.0–2.5; $J_{3,5}$, 0.6–1.9; $J_{2,6}$, 0–0.5; $J_{2,5}$, 0–1.8 Hz.

The principal features of the mass spectrum of pyridine itself[14] are intense ions corresponding to $M^{+\cdot}$ and $(M-HCN)^{+\cdot}$. Substituted pyridines may also show $(M-HCN)^{+\cdot}$ ions in their mass spectra, but in the case of alkylpyridines there are competing fragmentation processes which lead to pyridylmethylene (or, more probably, azatropylium) ions (**7**).[14]

* The 'picoline' and 'lutidine' obtained by Anderson were almost certainly mixtures of isomers: for example, Weidel[4] oxidized 'picoline' with potassium permanganate and obtained both pyridine-2- and -3-carboxylic acids.

† For a table of dipole moments of simple pyridine derivatives, see Ref. 10a.

‡ Parts per million downfield from tetramethylsilane: carbon tetrachloride solutions.

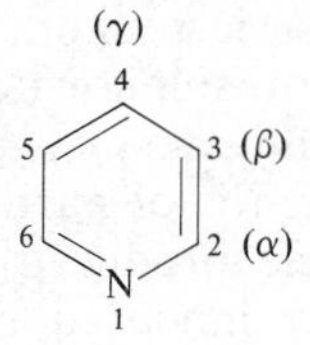

(1)

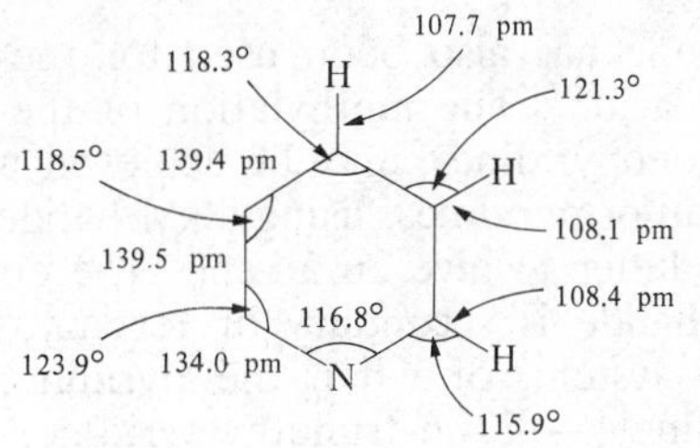

(2)

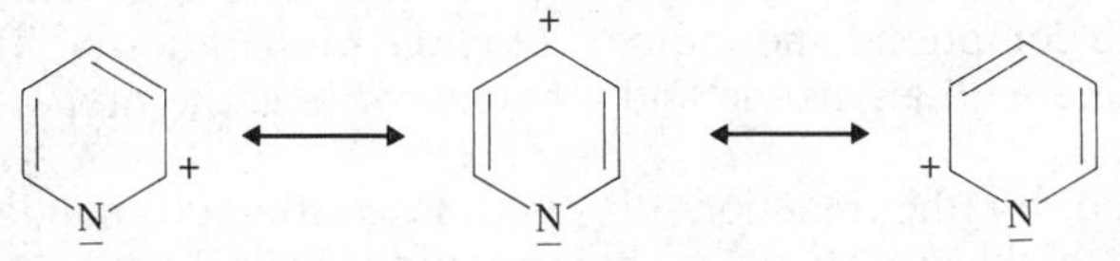

(3)

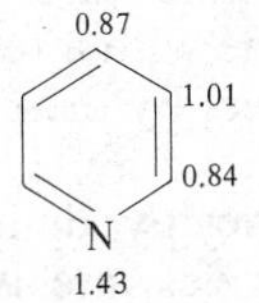

(4)

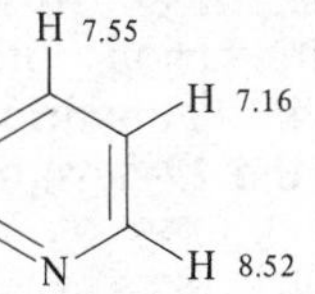

(5)

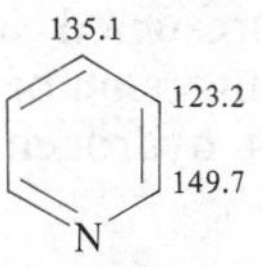

(6)

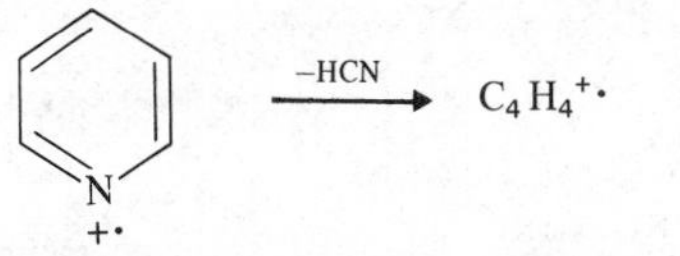

m/e 79 (100%) m/e 52 (76%)

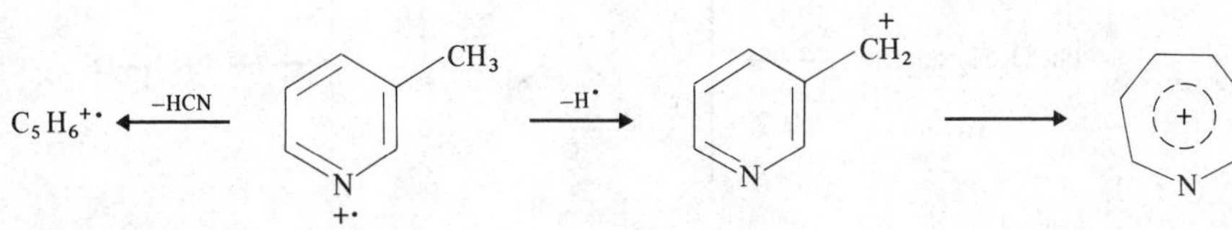

m/e 66 (44%) m/e 93 (100%) m/e 92 (31%) (7)

16.1.2 PYRIDINES AS TERTIARY BASES

16.1.2.1 Protonation

Pyridine and its homologues are weakly basic: pyridine itself has pK_a 5.20 in water at 25 °C, and simple alkylpyridines have pK_a values in the range 5.5–7.5.[15a] Electron-withdrawing substituents in the ring decrease the basicity, especially substituents in the 2- and 6-positions. This is well illustrated in the chlorinated pyridines, the pK_a values of which are as follows: 2-chloro-, 0.72; 3-chloro-, 2.84; 4-chloro-, 3.83; 2,3-dichloro-, −0.85; 2,6-dichloro-, −2.86; pentachloro-, −6.02.

16.1.2.2 Quaternization

Alkyl (and reactive aryl) halides react with pyridines to form quaternary *N*-alkyl- (or *N*-aryl-) pyridinium salts. Methylation, for example, is normally effected by methyl iodide.

Diazomethane has also been used for the methylation of pyridine, in the presence of fluoroboric acid.[16] The methylation of the most weakly basic pyridine derivatives, *e.g.* pentahalogenopyridines, may be achieved using methyl fluorosulphonate.[17]

Alkylation of pyridines, using alkyl halides, may be accompanied by dehydrohalogenation of the latter to give an alkene. The elimination reaction is of particular importance when the halide is secondary or tertiary, when the alkene produced forms part of a conjugated system, or when the pyridine has a sterically hindered hetero-atom (*e.g.* '*sym*-collidine' — 2,4,6-trimethylpyridine) and can thus undergo *N*-protonation more easily than *N*-alkylation.[18]

In some cases, pyridinium salts may be conveniently prepared by the reaction of pyridine with a reactive methylene compound and iodine[19] (*e.g.* the preparation of **8**) and in others by the reaction of pyridine with an alkene and a halogen[20] (*e.g.* the preparation of **9**).

N-Acylpyridinium salts, formed by the reaction of pyridines with acyl halides or anhydrides, are isolable although highly reactive,[21,22] being rapidly hydrolysed even by atmospheric moisture. They undergo nucleophilic attack at the carbonyl carbon atom extremely readily, and are thus very effective acylating agents. They have been detected as intermediates in the pyridine-catalysed hydrolysis of anhydrides[23] (*i.e.* the acylation of water), and are the presumed intermediates in various other acylations in pyridine solution, such as the conversion of acyl chlorides into anhydrides[22,24] or diacyl sulphides[22] (diacylation of water or hydrogen sulphide), or the *O*-acylation of β-keto-esters[25] [*e.g.* (**10**)→(**11**)].

(hy) 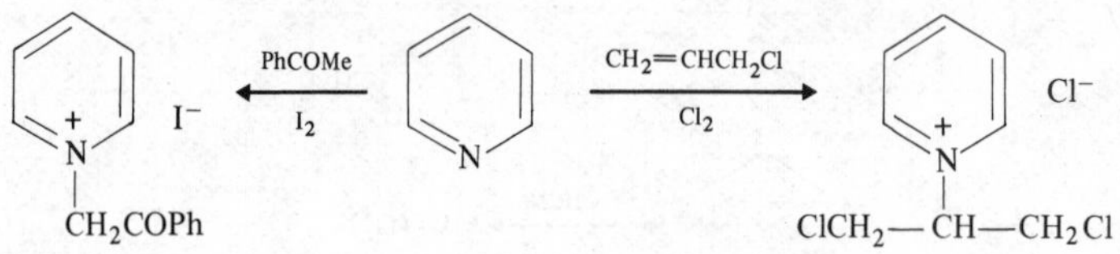 I^- ← PhCOMe, I_2 — pyridine — $CH_2{=}CHCH_2Cl$, Cl_2 → Cl^- (ly)

CH_2COPh (8) $ClCH_2{-}CH{-}CH_2Cl$ (9)

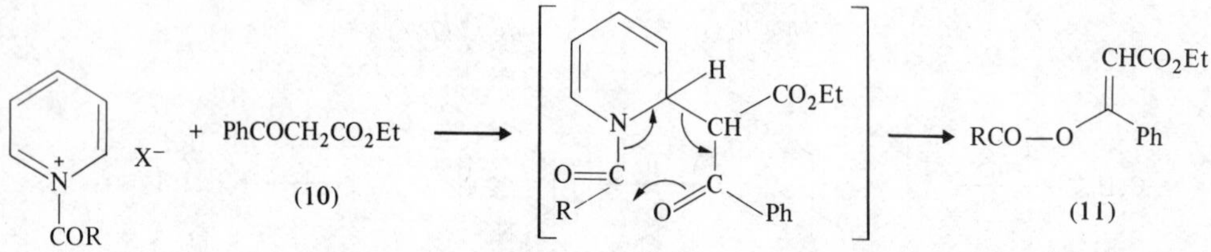

X^- COR + $PhCOCH_2CO_2Et$ (10) → [H, CO_2Et, CH, O=C, R, O, C, Ph] → RCO—O, $CHCO_2Et$, Ph (11)

4-(*N,N*-Dialkylamino)pyridines are evidently much better acylation catalysts than pyridine itself, and have been used to bring about the acylation of sterically hindered alcohols.[26] The increased effectiveness of these pyridine derivatives may be due to their enhanced basicity (the dimethylamino-compound has pK_a 9.71) and to the relative stability of their *N*-acyl derivatives: some of these are stable enough to be stored for several months, and have been used for the *N*-acylation of amino-acids in aqueous alkali.[27]

N-Nitropyridinium salts are, in some cases, effective nitrating agents,[28] pyridine perbromide (*N*-bromopyridinium bromide?) and pyridine hydrobromide perbromide ($C_5H_5\overset{+}{N}H\,Br_3^-$) have both found application as brominating agents,[29] and the pyridine–sulphur trioxide adduct ($C_5H_5\overset{+}{N}{-}SO_3^-$) is a convenient source of sulphur trioxide for the formation of sulphates and sulphonates.[30]

16.1.2.3 *N*-Oxide formation

N-Oxidation of pyridines is normally carried out by reaction with peracids. Peracetic acid, which may be prepared *in situ* from 30% hydrogen peroxide in acetic acid, is the most common reagent,[31] although aromatic peracids (perbenzoic, *m*-chloroperbenzoic, and perphthalic acids, for example) have also been used;[32] the advantage of the latter is that the reactions may be carried out under mild conditions in non-polar solvents such as chloroform. The more weakly basic pyridines require stronger oxidizing agents or reaction conditions: pentahalogenopyridines, for example, are oxidized by peroxytrifluoroacetic acid,[33] or hydrogen peroxide and an organic acid in presence of sulphuric acid.[34]

16.1.2.4 Properties and reactions of pyridine *N*-oxides

The normal representation of pyridine *N*-oxide as (**12a**) takes no account of any back-donation of electrons from the oxygen into the ring (such as may be represented by the canonical forms **12b–d**). That such back-donation is a significant phenomenon is evident from dipole moment measurements[35] and from the ^{1}H [36] and ^{13}C [13] n.m.r. spectra. The dipole moment of pyridine *N*-oxide itself is 4.24 D, much less than the sum of the dipole moment of pyridine and the N^+—O^- bond moment (*ca.* 6.6 D). The ^{1}H and ^{13}C chemical shifts for pyridine *N*-oxide, shown in formulae (**13**) and (**14**) respectively, demonstrate that the deshielding effect at the 2- and 4-positions is less in pyridine *N*-oxide than in pyridine itself. Calculated π-electron densities for the various atoms in pyridine *N*-oxide (a typical set[37] is shown in formula **15**) reveal a higher density at the 2- and 4-positions of pyridine *N*-oxide than those of pyridine.

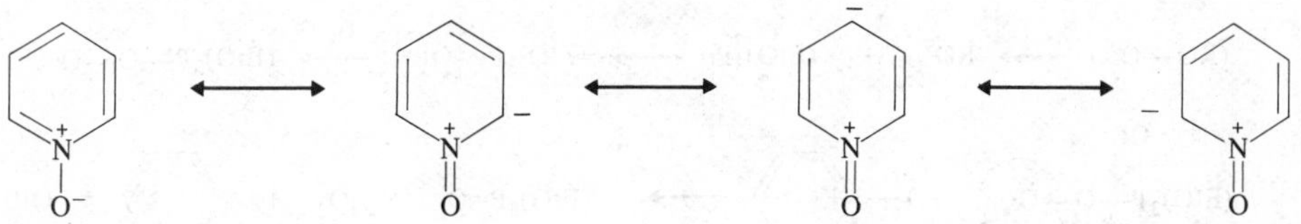

(12a) (12b) (12c) (12d)

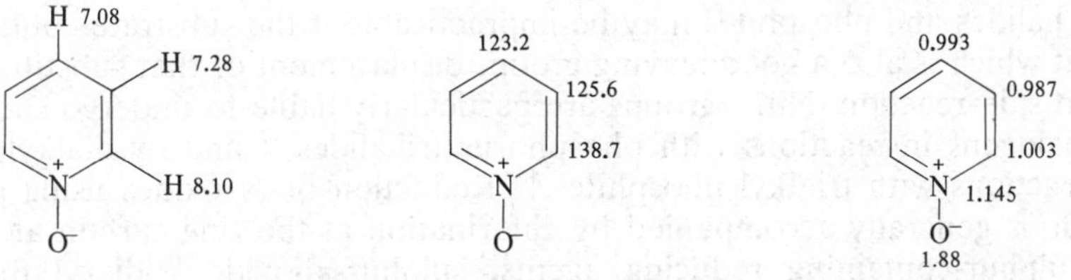

H 7.08, H 7.28, H 8.10 (13)

123.2, 125.6, 138.7 (14)

0.993, 0.987, 1.003, 1.145, 1.88 (15)

The main feature of the mass spectra of pyridine *N*-oxides[38] is an intense $(M-16)^+$ ion.

Pyridine *N*-oxides are weaker bases than the corresponding pyridines: pyridine *N*-oxide itself has pK_a 1.90, and the pK_a values of most simple analogues lie within the range −2 to +3.[39a] Alkylation and acylation of the *N*-oxides takes place on oxygen, and the resulting *N*-alkoxy- and *N*-acyloxy-pyridinium salts are useful synthetic intermediates by virtue of the ease with which they undergo nucleophilic addition–elimination (AE_a) reactions (*cf.* Section 16.1.3.2).

Reduction of pyridine *N*-oxides, to regenerate the parent pyridines, has been effected by a wide variety of reducing agents.[39b,40a] The reagent of choice for any particular case depends on the nature of other substituents in the molecule. For example, catalytic hydrogenation over Raney nickel or palladium–charcoal,[41] or reagents such as iron in acetic acid,[42] ferrous oxalate,[43] or sodium borohydride and aluminium chloride[44] are likely to affect reducible groups in the molecule other than the *N*-oxide.

The most frequently used reducing agents are trivalent phosphorus compounds, especially phosphorus trihalides. Phosphorus trichloride in chloroform reduces a wide variety of *N*-oxides,[39c,40c] and the same is true of triphenylphosphine,[45] although in this case elevated temperatures may be required. The use of phosphite esters for this type of deoxygenation has been less thoroughly explored, but it has been shown[46] that the reaction of pyridine *N*-oxide with triethyl phosphite is facilitated by the presence of oxygen and a peroxide, and so it is possible that, in this case at least, a free-radical mechanism [equation (3)] is operating rather than the more obvious ionic mechanisms [equations (1) and (2)].

$$\text{C}_5\text{H}_5\overset{+}{\text{N}}\text{–O}^- + :\text{PX}_3 \longrightarrow \text{C}_5\text{H}_5\text{N} + \text{X}_3\overset{+}{\text{P}}\text{–O}^- \longleftrightarrow \text{X}_3\text{P=O} \quad (1)$$

$$\text{C}_5\text{H}_5\overset{+}{\text{N}}\text{–O}^- + \text{PX}_3 \longrightarrow \text{C}_5\text{H}_5\overset{+}{\text{N}}\text{–O–}\overset{}{\text{PX}_3} \longrightarrow \text{C}_5\text{H}_5\text{N} + \text{X}_3\text{P=O} \quad (2)$$

$$(\text{RO—OR}) \longrightarrow \text{RO}^{\cdot} + (\text{EtO})_3\text{P} \longrightarrow (\text{EtO})_3\dot{\text{P}}\text{—OR} \xrightarrow{\text{O}_2} (\text{EtO})_3\overset{\text{OR}}{\overset{|}{\text{P}}}\text{—O—O}\cdot$$

$$(\text{EtO})_3\overset{\text{OR}}{\overset{|}{\text{P}}}\text{—O—O}\cdot \;\; \bar{\text{O}}\text{—}\overset{+}{\text{N}}\text{C}_5\text{H}_5 \longrightarrow (\text{EtO})_3\text{P=O} + \text{O}_2 + \text{C}_5\text{H}_5\text{N} + \dot{\text{O}}\text{R} \quad (3)$$

Reduction by a phosphorus halide is clearly impracticable if the pyridine *N*-oxide contains another strongly nucleophilic (*e.g.* amino) substituent, and reductions involving phosphorus halides and phosphites may be impracticable if the substrate contains a 2- or 4-substituent which is also a good leaving group: displacement of that substituent may be an unwanted side-reaction. Nitro-groups are particularly liable to undergo such displacements: by halogens in reactions with phosphorus trihalides,[47] and by dialkylphosphonyl groups in reactions with trialkyl phosphites.[48] Reduction of *N*-oxides using phosphorus pentachloride is generally accompanied by chlorination at the ring carbon atoms.[49]

Among sulphur-containing reducing agents, sulphur dioxide,[50] diaryl disulphides,[51] thiourea,[52] and dimethyl sulphoxide[53] have been used most successfully, although the last-named reagent may affect other substituents in the molecule which are sensitive towards oxidation.

16.1.2.5 Pyridinium ylides

The positively charged hetero-atom in a pyridinium salt has an activating influence on a methylene group attached to it. If this methylene group is also activated by another electron-attracting or electron-accepting group, the pyridinium salt undergoes reactions typical of highly activated methylene compounds such as 1,3-dicarbonyl compounds. The carbanions derived from these pyridinium salts, *e.g.* (**16a**), are generally known as pyridinium ylides.

Pyridinium ylides are formally derived, however, not only by deprotonation of *N*-alkylpyridinium ions,[54] but also by the addition of carbenes to pyridines: in practice the first method is normally used for their preparation, although there are a few instances [*e.g.* the preparation of (**17b**): equation (4)][55] which have involved carbene additions. The dicyanomethylide (**18**) has been obtained[56] by the reaction of pyridine with tetracyano-oxiran [equation (5)], although the general usefulness of this type of reaction has not been demonstrated.

The stability of pyridinium ylides depends on the efficiency with which the negative charge on carbon may be delocalized. Stabilizing groups (R^1 and/or R^2 in formula **16**) which have permitted the isolation of the ylides are electron-acceptors like carbonyl[57] and cyano-functions,[56] and other isolable ylides such as (**17**),[55,58] (**19**),[59] and (**20**)[60] are stabilized by delocalization of the charge into an aromatic system. Electron-accepting substituents in the 4-position of the pyridine ring also assist in stabilizing the ylides: that this should be so is evident from the canonical form (**16c**). As expected, the parent ylide, pyridinium methylide (**21**), is extremely unstable: it is the supposed intermediate in the reactions depicted in equations (6) and (7).[62]

The usefulness of pyridinium ylides as nucleophiles in synthesis, and the methylenic reactivity of *N*-alkylpyridinium salts in general, are the subjects of a series of reviews by Kröhnke.[63] Ylides stabilized by carbonyl functions may be alkylated and acylated,[64] like the anions of 1,3-dicarbonyl compounds, and may take part in the Michael reaction with conjugated enones. Reaction with aldehydes [equation (8)] leads to normal adducts and/or

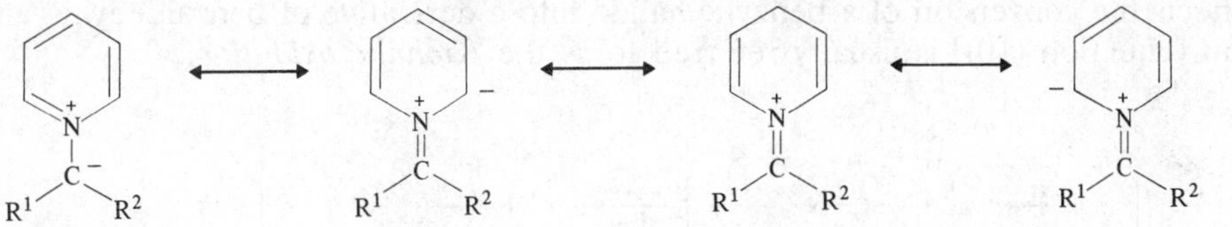

(16a) (16b) (16c) (16d)

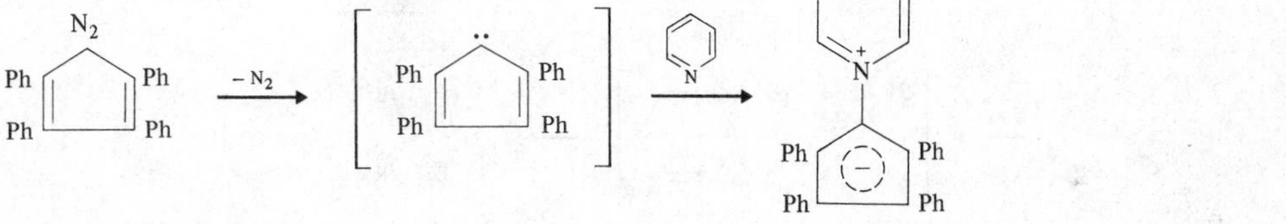

(4)

(17b)

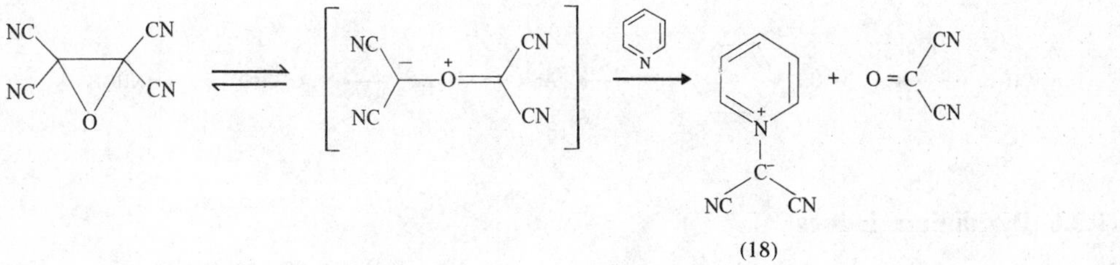

(5)

(18)

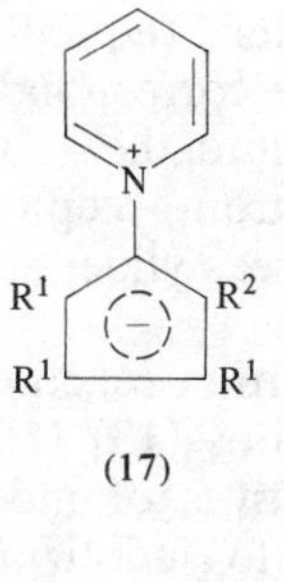

a; $R^1 = R^2 = H$
b; $R^1 = R^2 = Ph$
c; $R^1 = R^2 = Cl$
d; $R^1 = Ph, R^2 = H$

(17) (19) 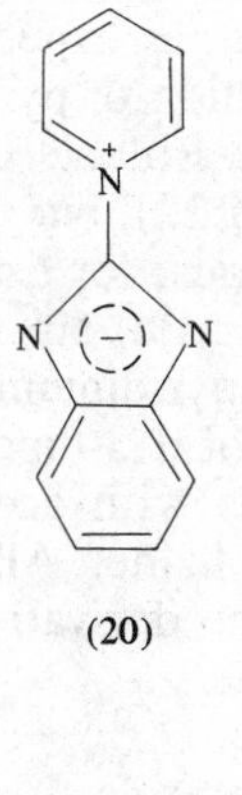 (20)

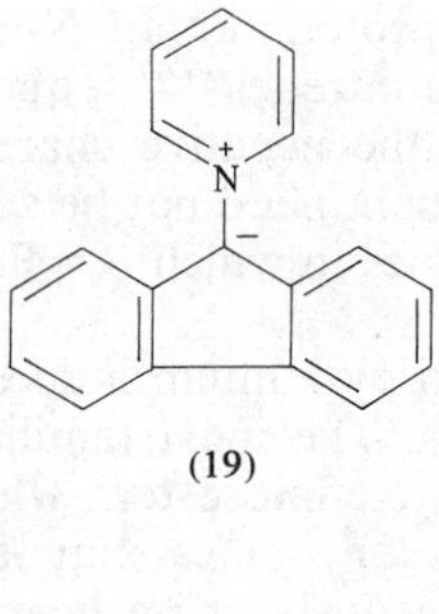

$$\text{Pyridine} + CH_2N_2 \xrightarrow{h\nu} \left[\text{N-}CH_2^-\text{ pyridinium}\right]? \longrightarrow \text{2-methylpyridine} \qquad (6)$$

$$\text{N-}CH_2CO_2^-\text{ pyridinium} \xrightarrow{-CO_2} \left[\text{N-}CH_2^-\text{ pyridinium } \textbf{(21)}\right] \xrightarrow{PhOH} \left[\text{N-}CH_3\text{ pyridinium } \bar{O}Ph\right] \longrightarrow \text{pyridine} + PhOCH_3 \qquad (7)$$

condensation products [unlike the corresponding (Wittig) reaction with phosphonium ylides]. The corresponding reactions with nitrosobenzene derivatives, however [equation (9)], lead to the formation of a nitrone, pyridine being eliminated; a variant of this latter process effects the conversion of a benzylic halide into a derivative of benzaldehyde, and this variant [equation (10)] is usually referred to as the *Kröhnke oxidation.*

$$ArCHO + \text{pyridinium-N-}CHR^1R^2\ X^- \xrightarrow{\bar{O}H} ArCH(OH)\text{—}C(R^1)(R^2)\text{—}\overset{+}{N}C_5H_5\ X^- \xrightarrow[-H_2O]{R^2=H} \left[ArCH{=}C(R^1)\text{—}\overset{+}{N}C_5H_5\ X^-\right] \qquad (8)$$

$$ArNO + \text{pyridinium-N-}CHR^1R^2\ X^- \xrightarrow{\bar{O}H} \left[Ar\text{—}N(O^-)\text{—}C(R^1)(R^2)\text{—}\overset{+}{N}C_5H_5\right] \longrightarrow Ar\overset{+}{N}(O^-){=}C(R^1)R^2 + \text{pyridine} \qquad (9)$$

$$Ar^1CH_2Br \xrightarrow{\text{pyridine}} Ar^1CH_2\overset{+}{N}C_5H_5\ Br^- \xrightarrow{Ar^2NO} Ar^1CH{=}\overset{+}{N}(O^-)\text{—}Ar^2 \xrightarrow[H_2O]{H^+} Ar^1CHO + Ar^2NHOH \qquad (10)$$

16.1.2.6 Pyridinium imines

These, *e.g.* (**22**), are the nitrogen analogues of pyridinium ylides. They may be prepared, as expected, by deprotonation of *N*-aminopyridinium salts[65] [equation (11)] or the reaction of pyridines with nitrenes[66,67] [equation (12)]. Like the corresponding ylides, they are isolable only when the negative charge is delocalizable into the substituent R (formula **22**), but this substituent need not be such a powerful electron-acceptor as in the ylide series: for example, imines in which R = Ph are isolable, whereas ylides such as (**16**; $R^1 = R^2 = Ph$) are not.

Both pyridinium ylides and pyridinium imines undergo cycloaddition reactions characteristic of 1,3-dipolar species. The most familiar of these [equations (13)–(16)] involve reactions with acetylenic or olefinic esters which lead to derivatives of indolizine, or aza-indolizine. Allylic ylides or imines may themselves cyclize to indolizine or aza-indolizine derivatives spontaneously or on heating [equations (17)–(19)].

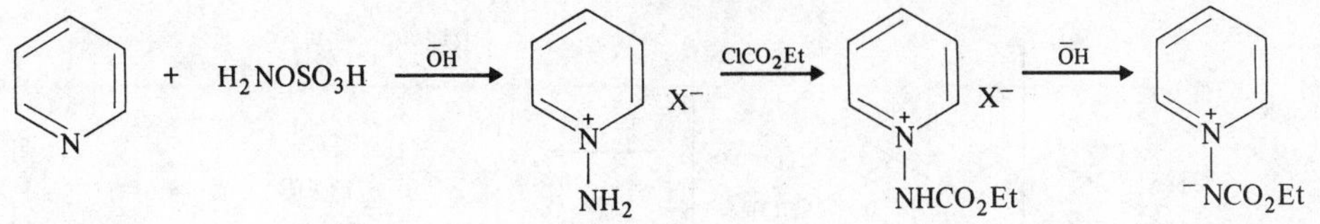

(11)

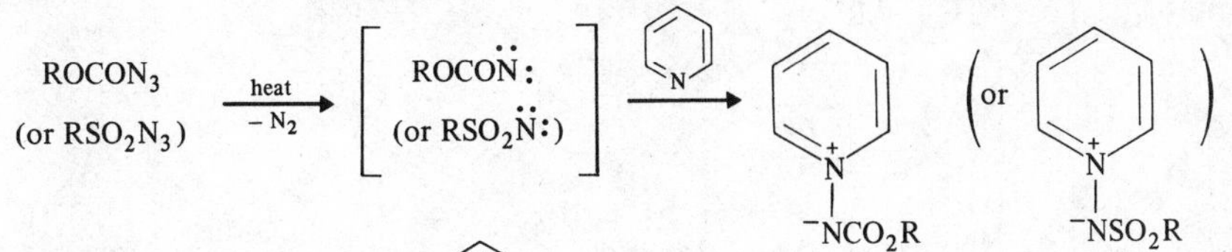

(12)

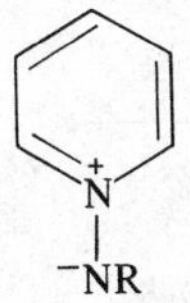

(22)

Equations (16), (17a), and (19a) indicate, however, that the cycloadditions may be complicated by side-reactions. Cleavage of the $N^+—X^-$ bond, as in (17a) and (19a), is a process more familiar in the photochemical reactions of pyridinium ylides and imines (*cf.* Section 16.1.3.4), although it has also been observed[74] in the thermolysis of a simple ylide [equation (20)]. The fine balance which exists between cycloaddition and $N^+—X^-$ cleavage is also evident in the reactions of pyridinium ylides and imines with cyclopropenones

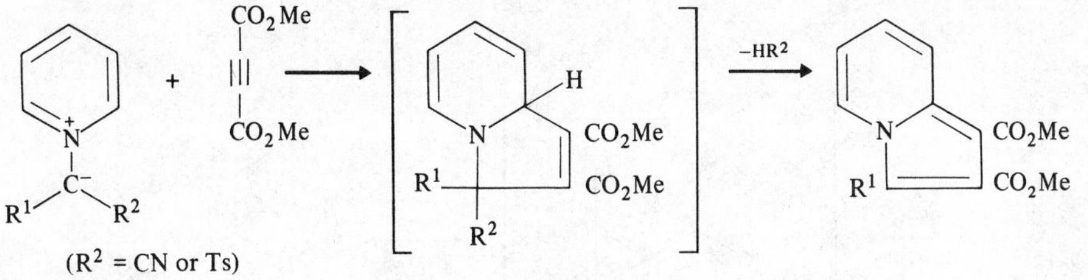

(13)[68]

(R^2 = CN or Ts)

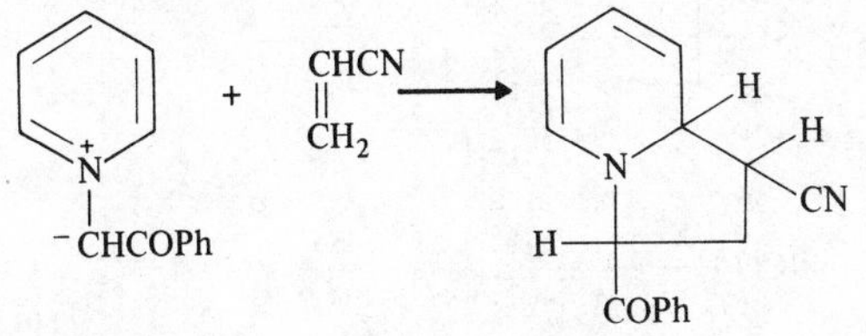

(14)[69]

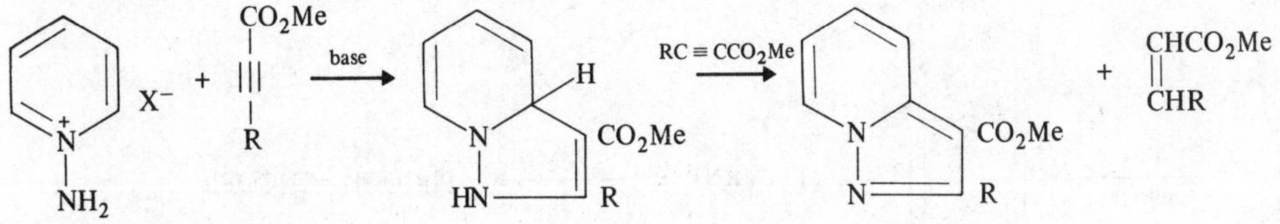

(15)[70]

(R = CO_2Me or H)

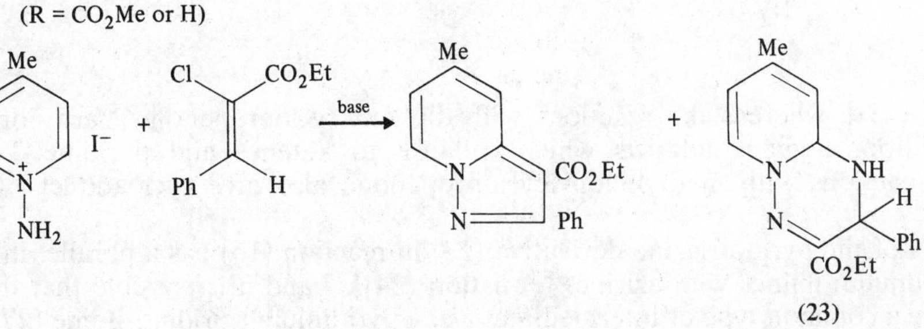

(16)[71]

(23)

(17)[72]

(17a)

(18)[73a]

(19)[73b]

(19a)

heat, *in vacuo* → + [PhCOCH:] → PhCOCH=CHCOPh (20)

[equations (21)–(23)]: whereas the reactions with diphenylcyclopropenone[75] are normal nucleophilic additions, giving adducts which collapse to ketenes and pyridine,[76] the corresponding reactions with methylphenylcyclopropenone also give cycloadducts, (**24**) and/or (**25**).[77]

The formation of the pyridotriazine derivative (**23**) in reaction (16) has a parallel in the reaction of pyridinium imines with azirines [equation (24)],[78] and it is possible that these reactions involve a common type of intermediate, *viz.* a pyridinium aziridino-imine (**27**).

(X = CH or N)

(21)

(22)

(23)

(24)

$(23; R^1 = CO_2Et, R^2 = Ph)$
$(26; R^1 = Ph, R^2 = H)$

16.1.2.7 Other pyridine-containing zwitterions

Pyridinium ylides, in which the negative charge is located on the 2-position in the ring, are reactive intermediates produced by decarboxylation of pyridine-2-carboxylic acid and *N*-alkylpyridine-2-carboxylates.[79] They have been detected by trapping with electrophiles such as aldehydes and diazonium salts. The anions derived from pyridine *N*-oxides by reaction with strong bases may be similarly trapped,[80] as shown overleaf.

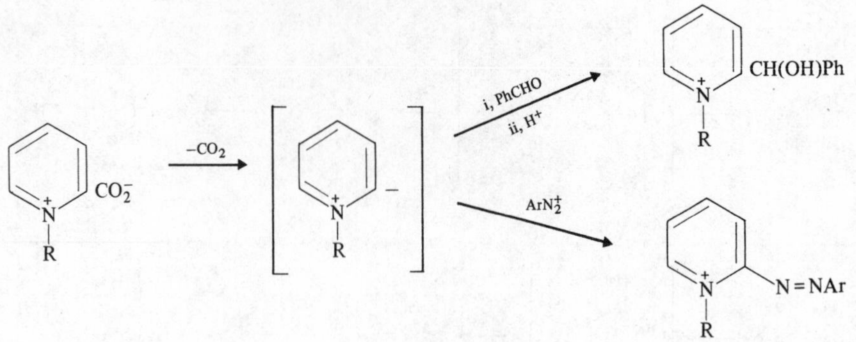

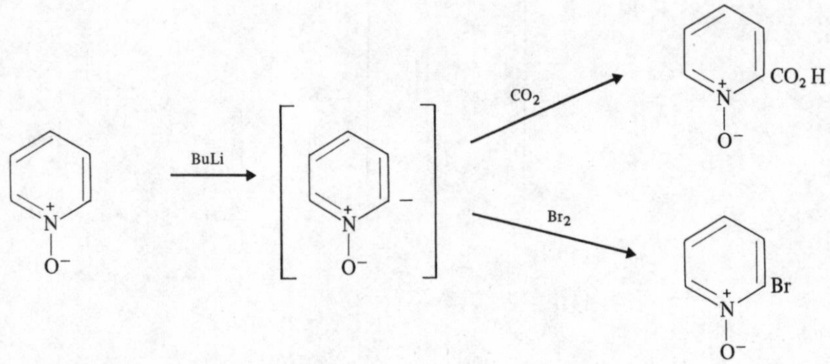

Zwitterionic species also result from the addition of pyridines to epoxides and to suitably activated multiple-bond systems (*e.g.* α,β-unsaturated esters or quinones). Among these reactions, the best-known and most thoroughly investigated is the reaction of pyridine with dimethyl acetylenedicarboxylate (Scheme 1),[81] which formed part of the classic investigation of Diels and Alder in the 1930s. The products vary according to the solvent used: in ether, three derivatives of quinolizine, *viz.* (**28**)–(**30**), are obtained, while in methanol the products, (**31**) and (**32**), are both indolizines.

16.1.3 PYRIDINES AS ANALOGUES OF BENZENE

16.1.3.1 Electrophilic substitution*

The electrophilic substitution of pyridine and its simple derivatives may be accomplished only with extreme difficulty. The partial rate factors for electrophilic substitution of pyridine itself[84a] are of the order of 10^{-6}, *i.e.* similar to those for nitrobenzene; and like nitrobenzene, pyridine undergoes electrophilic substitution at the 3-position. This is in accord with the predictions of simple resonance theory [equation (25)], since in the intermediate cation the electron-attracting hetero-atom does not require to bear a partial positive charge. It is also consistent with electron-density calculations,[11] which show the 3-position to have the highest electron-density of the carbon atoms (formula **4**, p. 5), and with calculations of the stabilities of the intermediate cations by the localization energy method,[83] which predict that (**34**) should be more stable than (**33**) or (**35**).

Since pyridines are basic, however, they may undergo electrophilic attack at the nitrogen prior to reaction at a ring carbon atom. The result is that the species which

* Other accounts of this topic may be found in Refs. 10b, 15b, 82, and 83.

(28)

(29)

(30)

(31)

(32)

SCHEME 1

undergoes electrophilic substitution at carbon is frequently not the free base, but one in which the nitrogen is protonated, quaternized, or co-ordinated to another atom or group. These pyridinium compounds are much less reactive towards electrophiles than the free bases: partial rate factors for electrophilic substitution of pyridinium ions have been estimated[84a,86a] to be as low as 10^{-18} to 10^{-20}.

(25)

(33) (34) (35)

(*i*) *Hydrogen exchange*

Acid-catalysed deuteriation and tritiation of simple pyridines involve attack on the conjugate acids.[87] Pyridones may react as the free bases or the cations,[86f,87,88c,d] and aminopyridines may react as free bases, monocations, or dications,[86d,88b,90] depending on the acidity of the solution.

(*ii*) *Nitration*

This is probably the most extensively studied of all electrophilic substitutions in the pyridine series. Nitration of the more basic pyridines ($pK_a > 1$) usually involves the conjugate acids, whereas that of the weakly basic pyridines ($pK_a < -2.5$) normally occurs on the free bases.[86b] Somewhere in the pK_a range between −2.5 and +1 there is a changeover in mechanism, and also in reactivity: the weak bases are nitrated much more readily than their more strongly basic analogues.

Pyridine cannot be nitrated on carbon using nitronium fluoroborate; such reactions lead only to *N*-nitropyridinium fluoroborate.[89] Successful nitration of pyridine demands extremely vigorous reaction conditions (*e.g.* sodium or potassium nitrate in fuming sulphuric acid at *ca.* 300 °C, or vapour-phase nitration using N_2O_4), and the yields of 3-nitropyridine are very low.[91] Low yields are also obtained in the nitration of 2-methyl-[92a] and 2,4-dimethyl-pyridines,[93] but 2,6-dimethyl- and 2,4,6-trimethyl-pyridines are nitrated in the 3-position (yields > 65%) by potassium nitrate in fuming sulphuric acid at 100 °C.[92] The partial rate factor for this last nitration is 3.5×10^{-12}, and it and the others presumably involve attack on the conjugate acid, since the *N*-methyl salt derived from 2,4,6-trimethylpyridine is also nitrated in good yield under similar conditions.[84b,86a]

The reactions of halogenopyridines with nitrating agents illustrate clearly the difference between nitration of the free base and the conjugate acid. 3-Bromopyridine (pK_a 2.84) is nitrated (as the conjugate acid) only at 270 °C, and in low yield, by potassium nitrate in oleum, and gives the 5-nitro-compound.[94] 2,6-Dichloropyridine (pK_a −2.86), on the other hand, is nitrated (as the free base) at 115 °C by a mixture of concentrated nitric (95%) and sulphuric (90%) acids, and gives almost 50% of the 3-nitro-derivative.[86a]

As in the benzene series, substituents which are strongly electron-donating facilitate electrophilic substitution in pyridines, and direct the incoming electrophile to the *ortho*- and *para*-positions. 2-Alkoxypyridines give 5-nitro-derivatives and 3-alkoxypyridines give 2-nitro-derivatives.[95] Dialkoxypyridines may be dinitrated (in 90% sulphuric acid), despite the combined deactivating influences of the hetero-atom and the first nitro-group; however, the second nitration occurs on the unprotonated mononitro-compound, and it is inhibited by carrying out the reaction in more concentrated acid (which ensures that even the weakly basic mononitro-compound remains protonated).[86a,b]

Hydroxypyridines and pyridones also nitrate easily: 4-pyridone at the 3- (and 5-) position,[88d] 2-pyridone at the 3- or 5-position, depending on the reaction conditions,[96] and 3-hydroxypyridine at the 2-position.[95b,97] 2,4- and 2,6-dihydroxypyridines (4- and 6-hydroxy-2-pyridones) form 3-nitro-derivatives.[95b,98]

The nitration of aminopyridines initially produces nitraminopyridines [equation (26)], which subsequently rearrange to give *o*- or *p*-amino-nitro-compounds.[99] The same is true, apparently, of the nitration of (*N*-alkylamino)pyridines,[100] but (*N,N*-dialkylamino)pyridines undergo 'normal' nitration (*via* the monocations).[101] 3-Aminopyridine derivatives provide some anomalies in this respect: for example, 3-nitraminopyridine gives only azo- and hydroxy-pyridine derivatives on rearrangement,[102] and 3-(*N,N*-dimethylamino)pyridine is nitrated in very low yield.[101] In fact, the only satisfactory nitrations of 3-aminopyridine derivatives appear to be those of the *N*-methyl,[100b] *N*-ethoxycarbonyl,[103] and *N*-methylsulphonyl[104] derivatives, all of which give 2-nitro-compounds as the main products.

Pyridine *N*-oxides are nitrated much more easily than the parent pyridines. The nitration of pyridine *N*-oxide itself, with nitric acid or potassium nitrate in sulphuric

(26)

(27)

(28)

(36)

acid,[105] gives high yields of 4-nitropyridine *N*-oxide, and the partial rate factor for this process[86c] is 4×10^{-6}. Nitration apparently occurs on the unprotonated *N*-oxide,[86c,106] and the higher reactivity and position of substitution may be rationalized in terms of back-donation of electrons from the *N*-oxide oxygen into the ring (*cf.* p. 7).

Although equation (27) adequately represents the mechanism, other mechanisms have been proposed;[106b] these include the formation of an *O*-nitro-cation (**36**) followed by rearrangement of the nitramine type. An intermediate such as (**36**) is almost certainly involved in the nitration of pyridine *N*-oxide with benzoyl nitrate, which gives 3-nitro- and 3,5-dinitro-pyridine *N*-oxides[107] [equation (28)].

Nitration of substituted pyridine *N*-oxides with nitric and sulphuric acids occurs almost invariably at the 4-position, if this is not already occupied by a substituent;[40b] 4-substituted pyridine *N*-oxides are nitrated only with difficulty, by a process involving the conjugate acid. In the presence of strongly electron-donating substituents, however, even nitration of the protonated form becomes relatively easy: thus, for example, 4-hydroxypyridine *N*-oxide[108] (*N*-hydroxy-4-pyridone) and 2,4,6-trimethoxypyridine *N*-oxide[86c] are readily nitrated at the 3-position. The powerful *ortho*- and *para*-directing effects of hydroxy-, alkoxy-, and amino-groups are demonstrated in the nitrations of 2-hydroxy-[109] and 2-(*N,N*-dimethylamino)-pyridine *N*-oxides,[101] each of which gives a 5-nitro-derivative, and in those of 3-hydroxy- and 3,5-dialkoxy-pyridine *N*-oxides, which are nitrated at the 2-position.[86c,97] 2- and 3-monoalkoxypyridine *N*-oxides, however, are nitrated[97,109a] at position 4.

(*iii*) *Halogenation*

In view of the electron-deficient nature of the pyridine ring, it is not surprising that electrophilic halogenation of simple pyridines requires severe reaction conditions.

High-temperature vapour-phase chlorination of pyridine, by molecular chlorine, gives a complex mixture of products. The reaction of chlorine with pyridine in presence of a large excess of aluminium chloride gives 3-chloropyridine in *ca.* 30–35% yield,[110] and this is the only direct electrophilic chlorination procedure which has any real value as a laboratory method [equation (29)].

Bromination of pyridine, on the other hand, may be accomplished relatively simply, by the use of bromine in oleum.[111a] The principal product (*ca.* 90%) is 3-bromopyridine, and small quantities of dibromopyridines are identified as by-products. Bromination of methylpyridines may also be achieved by this method.[111b] These reactions presumably involve electrophilic attack of Br^+ on the pyridine–sulphur trioxide adduct [equation (30)]. Earlier work on bromination of pyridine, in the vapour phase at high temperatures, shows that in these reactions mixtures of brominated products are formed; at 300 °C bromination occurs at the 3- and 5-positions and at 500 °C it occurs at the 2- and 6-positions.[111c]

Direct iodination of pyridine has been relatively little investigated, and no preparatively useful procedure has yet been developed. Pyridine, iodine and oleum at 320 °C produce 3-iodopyridine in only 18% yield.[112]

$$2\ \text{(pyridine-}N\rightarrow AlCl_3) + Cl_2\cdot AlCl_3\ (Cl^+\ AlCl_4^-\ ?) \longrightarrow \text{(3-chloropyridine-}N\rightarrow AlCl_3) + \text{(pyridinium } N^+\text{-H)}\ AlCl_4^- \xrightarrow[\bar{O}H]{H_2O} \text{3-chloropyridine (30–35\%)} + \text{pyridine} \quad (29)$$

$$2\ \text{pyridine} + 2SO_3 + Br_2 \longrightarrow \text{3-bromopyridine} + SO_2 + H_2SO_4 \quad (30)$$

Hydroxy- and amino-pyridines, like phenol and aniline, are very easily halogenated, and even under mild conditions all free positions *ortho* and *para* to the electron-donor group may be substituted (Table 1). Kinetic evidence is available[113] which indicates that 2-aminopyridines undergo bromination as the unprotonated free bases, and there is no reason to suppose that the other members of the series behave differently.

Bromination is the only halogenation reaction of pyridine *N*-oxides on which there is much information. Unlike nitration, which is achieved relatively easily, direct bromination

TABLE 1
Halogenation of Hydroxy- and Amino-pyridines

Substituent	*Position of halogen substituted* Chlorine	Bromine	Iodine
2-hydroxy- (2-pyridone)	5 3,5 [a]	3,5[b]	5 3,5 [c]
3-hydroxy-	2[d]	2 2,6 2,4,6 [d]	2 2,6 2,4,6 [c,d]
4-hydroxy- (4-pyridone)		3,5[e]	3 3,5 [c]
2,4-dihydroxy-		3[f]	
2-amino-	5[g,h]	5 3,5 [g,i]	5[j]
3-amino-	2 2,6 [k]	2 2,6 2,4,6 [k,l]	2,6[m]
4-amino-			3 3,5 [n]
N-oxides			
2-hydroxy-		3 3,5 [b]	
3-hydroxy-		2 2,6 2,4,6 [d]	4,6[d]
4-hydroxy-		3,5[o]	

[a]M. Dohrn and R. Dirksen, U.S. Pat. 1 706 775; *Chem. Abs.*, 1929, **23**, 2189. [b]Ref. 109b. [c] F. W. Broekman and H. J. C. Tendeloo, *Rec. Trav. chim.*, 1962, **81**, 107. [d]Leading references given in Ref. 97. [e]L. Haitinger and A. Lieben, *Monatsh.*, 1885, **6**, 306. [f]C. R. Kolder and H. J. den Hertog, *Rec. Trav. chim.*, 1960, **79**, 474. [g]F. Friedrich and R. Pohloudek-Fabini, *Die Pharmazie*, 1964, **19**, 677. [h]A. E. Tschitschibabin and A. F. Egorov, *Chem. Abs.*, 1929, **23**, 2182. [i]B. A. Fox and T. L. Threlfall, *Org. Synth.*, 1964, **44**, 34. [j]Ref. 130a. [k]Ref. 113. [l]E. Plażek and A. Marcinikov, *Roczniki Chem.*, 1934, **14**, 326. [m]Z. Rodewald and E. Plażek, *ibid.*, 1936, **16**, 130. [n]I. G. Farbenindustrie, *Fr. Pat.* 728 634, *Ger. Pat.* 579 224; *Chem. Abs.*, 1932, **26**, 6071; 1933, **27**, 4880. [o]Ref. 114.

of pyridine *N*-oxide is accomplished only with difficulty. Good yields of the 3-bromo-derivative are obtained, however, by bromination in oleum[115] at 120 °C (this reaction presumably involving attack on the protonated form, or a sulphur trioxide adduct). Bromination in acetic anhydride, in presence of sodium acetate, gives 35% of 3,5-dibromopyridine *N*-oxide,[116] probably by the mechanism shown in equation (31). Reference has already been made (pp. 13–14) to the bromination of pyridine *N*-oxide by deprotonation followed by reaction with bromine, but the yield of 2,6-dibromopyridine *N*-oxide in this reaction is small. The corresponding chlorination is also unsatisfactory.

As expected, the bromination of hydroxypyridine *N*-oxides takes place under relatively mild conditions, and leads to the products indicated in Table 1.

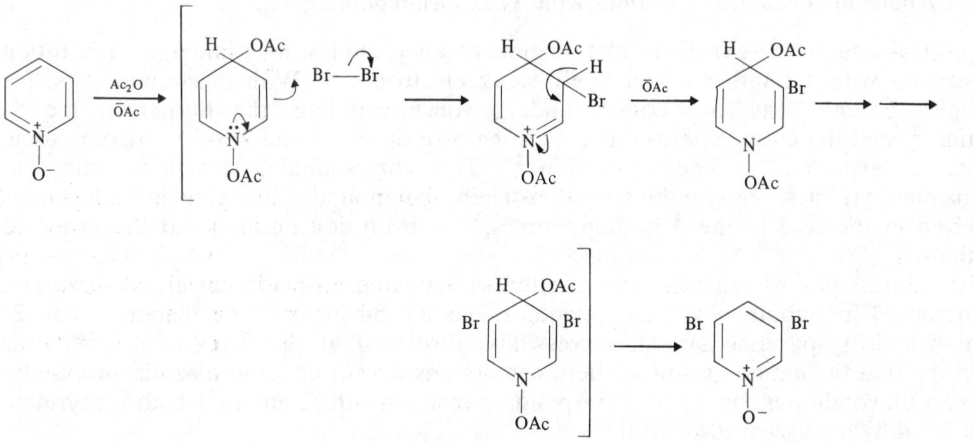

(31)

(*iv*) Sulphonation

Sulphonation of pyridine in oleum, at 220–270 °C in presence of mercuric sulphate, gives pyridine-3-sulphonic acid in high yield;[117] at higher temperatures desulphonation and rearrangement to the 4-sulphonic acid may occur.[117b] The three methylpyridines are similarly monosulphonated at high temperature,[117a] but 2,6-di-t-butylpyridine may be sulphonated very easily,[118a] with sulphur trioxide at −10 °C; at higher temperatures the cyclic sulphone (**37**) is also formed [equation (32)].[118b] The remarkable ease with which this sulphonation occurs is presumably due to steric hindrance around the hetero-atom, which prevents the formation of the normal pyridine–sulphur trioxide adduct.

Hydroxy- and amino-pyridines are sulphonated, as expected, *ortho* or *para* to the hydroxy- or amino-group: 2-aminopyridine[99a] at the 5-position, 3-amino- and 3-hydroxy-pyridines at the 2-position,[119] and 4-aminopyridine and 4-pyridone at the 3-position.[120]

Sulphonation of pyridine *N*-oxide, like bromination, requires severe reaction conditions, and gives the 3-sulphonic acid as the main product.[121a]

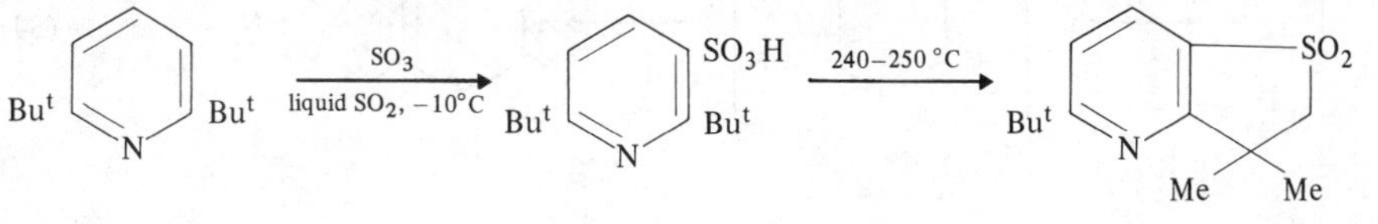

(32)

(37)

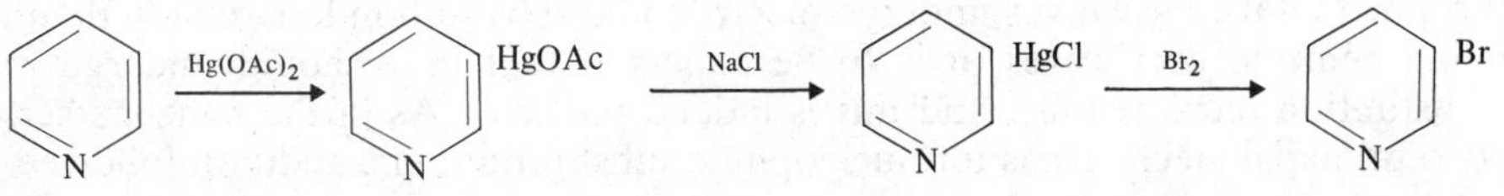

(33)

(38)

(v) Mercuration

The mercuration of pyridine may be carried out by heating pyridine with mercuric acetate at 170–180 °C. Mercuration gives principally 3-pyridylmercuriacetate (**38**), which may be converted into the mercurichloride, and hence into 3-bromopyridine, as shown [equation (33)].[122] Mercuration of methylpyridines[123] and hydroxypyridines[124] occurs at the expected positions, and pyridine *N*-oxide is mercurated mainly at positions 2 and 6.[121b]

(vi) Reactions of 'activated' pyridines with weak electrophiles

Hydroxy- and amino-pyridines, like phenols and arylamines, may undergo substitution by reaction with a number of relatively weak electrophiles. With diazonium salts, for example, 2- and 3-hydroxypyridine undergo diazo-coupling, the former at the 5-position[125] and the latter both at the 2- and the 6-positions;[97] and 2,6-dihydroxypyridine reacts, as expected, at the 3-position.[126] The corresponding reactions with the monoaminopyridines apparently give diazoamino-compounds, but ring substitution is preferred in the case of the diaminopyridines,[127] substitution occurring at the expected positions.

Nitrosation, another characteristic reaction of activated aromatic nuclei, is apparently not observed for mono-substituted pyridines; but 2,6-dihydroxy-, 2,6-diamino-, and 2-amino-6-hydroxy-pyridine are all successfully nitrosated at the 3-position.[126,128] It is surprising that the amino-groups in these compounds do not undergo diazotization under the reaction conditions, since the corresponding reaction with 2-amino-5-hydroxypyridine gives 3,6-dihydroxy-2-nitrosopyridine.[129]

Hydroxypyridines, as their anions, undergo carboxylation by reaction with carbon dioxide, to give the hydroxypyridine-carboxylic acids with the expected orientation.[97,130] 2- and 3-hydroxypyridine may be hydroxylated by potassium persulphate (the Elbs reaction) to give 2,5-dihydroxypyridine,[131] and 3-hydroxypyridine may be hydroxymethylated and aminomethylated (the Mannich reaction),[97] but 2-hydroxypyridine reacts in its pyridone form with formaldehyde, giving *N*-hydroxymethyl-2-pyridone in place of a *C*-hydroxymethylated product.[132]

The thiocyanation of some aminopyridines, by reaction with thiocyanogen, has been described.[133] Where the product contains *ortho*-disposed amino- and thiocyanato-groups, its stability appears to depend on the nucleophilicity of the amino-nitrogen: thus, 2,6-diamino-3-thiocyanatopyridine is isolable, but 3-amino-2-thiocyanatopyridines (with a more nucleophilic amino-group) cyclize spontaneously to thiazolo[5,4-*b*]pyridines [equation (34)].

R, NH_2 —KSCN, Br_2→ [R, NH_2, SCN] ⟶ R, N, S, NH_2 (34)

16.1.3.2 Nucleophilic substitution*

Since pyridine and its derivatives undergo electrophilic substitution less readily than the corresponding benzene derivatives, it is to be expected that they should undergo nucleophilic substitution more readily, and this is indeed the case. As in the benzene series, there are two principal mechanisms for nucleophilic substitution, *viz.* addition followed by

* Other accounts of this topic may be found in Refs. 10c, 15c, 82, 134, and 135.

elimination (*AE*), and elimination followed by addition (*EA*); but in pyridine substitutions there is also a third mechanism, an addition–elimination which leads to '*cine*-substitution', and this is often referred to as the 'abnormal' addition–elimination (AE_a) mechanism.

(*i*) *The addition–elimination* (AE) *mechanism* [*equation* (*35*)]

The position of nucleophilic attack on the pyridine ring is correctly predicted by simple valence-bond theory: stabilization of the anion formed by addition of the nucleophile is best achieved when a partial negative charge may reside on the (electron-attracting) hetero-atom, and the canonical forms for the intermediates in equations (35a) and (35b) show stabilization of this kind. Thus *AE* reactions of pyridines occur most easily at the 2- and 4-positions. Molecular-orbital calculations also indicate that anions arising from nucleophilic addition at the 2- and 4-positions are of lower energy than that arising from addition at the 3-position.[85]

A positively-charged nitrogen in the pyridine ring is more strongly electron-attracting than an uncharged nitrogen, and so pyridinium salts and *N*-oxides are even more reactive than the parent pyridines. Some quantitative comparisons are made in Table 2; they show that halogens at all three positions are considerably more reactive than those of halogenobenzenes, and that the relative reactivities of the 2- and 4-positions vary from one set of conditions to another.

TABLE 2

Substituent Rate Factors[a] for Typical Addition–Elimination Reactions of Halogenopyridines, Halogenopyridinium Salts, and Halogenopyridine *N*-Oxides

Halogeno-compound		*Nucleophile* Methoxide ion[b]	*Piperidine*[c]
Cl, N	2-Cl	2.76×10^{8}	6.11×10^{5}
	3-Cl	9.12×10^{4}	
	4-Cl	7.43×10^{9}	6.73×10^{6}
Cl, $\overset{+}{N}$–Me	2-Cl	1.28×10^{21}	
	3-Cl	2.62×10^{13}	
	4-Cl	4.23×10^{19}	
Cl, $\overset{+}{N}$–O^-	2-Cl	5.30×10^{12}	1.66×10^{9}
	3-Cl	9.67×10^{9}	
	4-Cl	8.33×10^{12}	4.55×10^{8}
cf. Cl, NO_2	2-Cl	2.10×10^{10}	5.90×10^{7}
	3-Cl	5.64×10^{5}	
	4-Cl	7.05×10^{10}	2.51×10^{7}

[a] Relative to chlorobenzene = 1.
[b] M. Liveris and J. Miller, *Austral. J. Chem.*, 1958, **11**, 297; *J. Chem. Soc.*, 1963, 3486.
[c] Data collected by Miller (Ref. 135), p. 247.

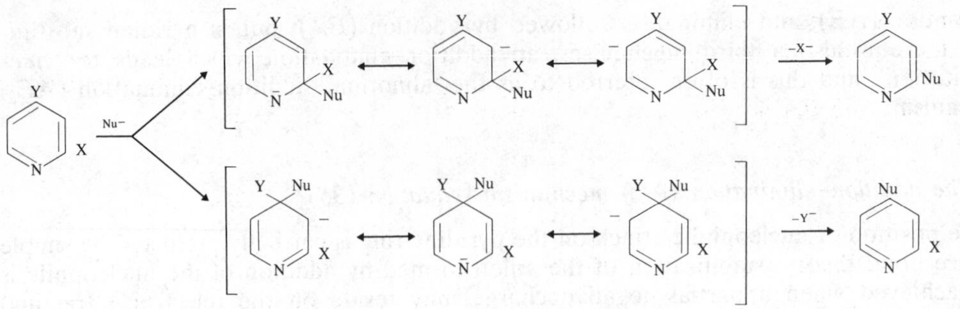

(35a)

(35b)

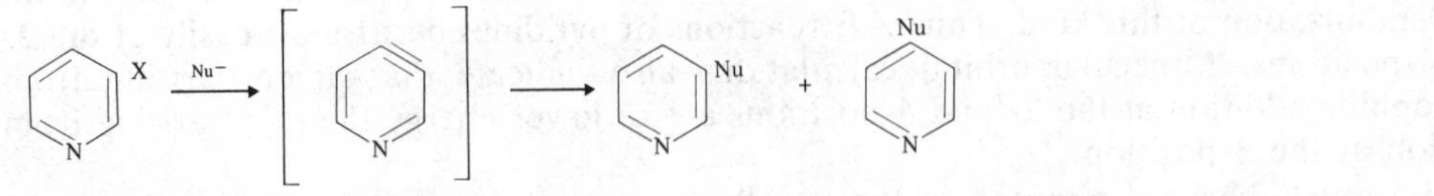

(36)

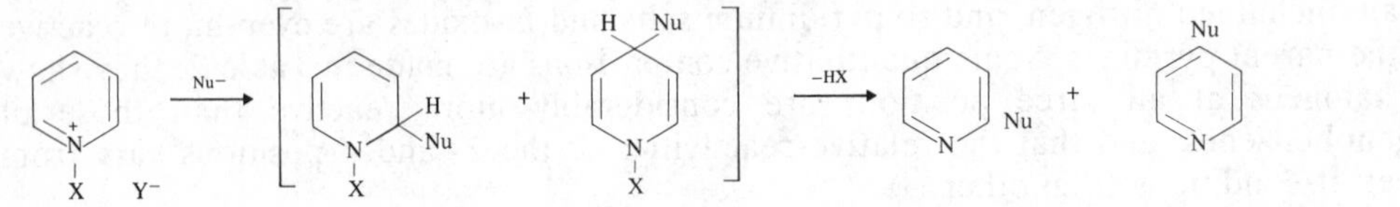

(37)

(ii) *The elimination–addition* (EA) *mechanism* [*equation* (36)]

As in the corresponding reactions in the benzene series, this mechanism is generally observed only when the leaving group is situated in an 'unactivated' (*i.e.* 3-) position, and the nucleophile is strongly basic. The reactive intermediates in these reactions, the dehydropyridines or pyridynes, are discussed in more detail in Section 16.1.5.

(iii) *The abnormal addition–elimination* (AE_a) *mechanism* [*equation* (37)]

Pyridinium salts in which the *N*-substituent is a leaving group (usually alkoxy or acyloxy) react with nucleophiles by addition of the latter at the 2- or 4-position followed by elimination of the *N*-substituent.

(iv) *Competition between* AE *and* EA *mechanisms*

In the 'normal' addition–elimination process, the entering nucleophile becomes attached to the same carbon atom as the leaving group. In the elimination–addition process, however, a pyridyne is formed, and nucleophiles may normally attack either end of the dehydro-bond; in this way two products are formed, one corresponding to direct replacement of the leaving group and one to *cine*-substitution. In highly polarized arynes, however (*cf.* p. 49), the addition of the nucleophile may take place exclusively at one end of the dehydro-bond, and thus give rise to only one product.

In cases where both mechanisms may operate concurrently, it may be difficult to determine the proportion of the reaction taking place by each route. Obviously, if only a *cine*-substitution product is formed, and an AE_a reaction may be ruled out, the mechanism is purely *EA*. In other cases, however, the following criteria may be applied:

1. In a 'pure' *EA* reaction, the isomer ratio of products should be independent of the nature of the leaving group, and should depend only on the polarity of the pyridyne.[136]

2. In a 'pure' *EA* reaction, when the substrate is allowed to react with a mixture of two nucleophiles, $(Nu^1)^-$ and $(Nu^2)^-$ [equation (38)], the ratio of the two direct-substitution products, (**36**) and (**38**), and the ratio of the *cine*-substitution products, (**37**) and (**39**),

should bear a constant relation to the ratio of the concentrations of the two nucleophiles, since the product ratios should depend only on the affinity of the bases for the pyridyne.[137]

Where *AE* and *EA* processes operate concurrently, deviations from the above relationships are observed.

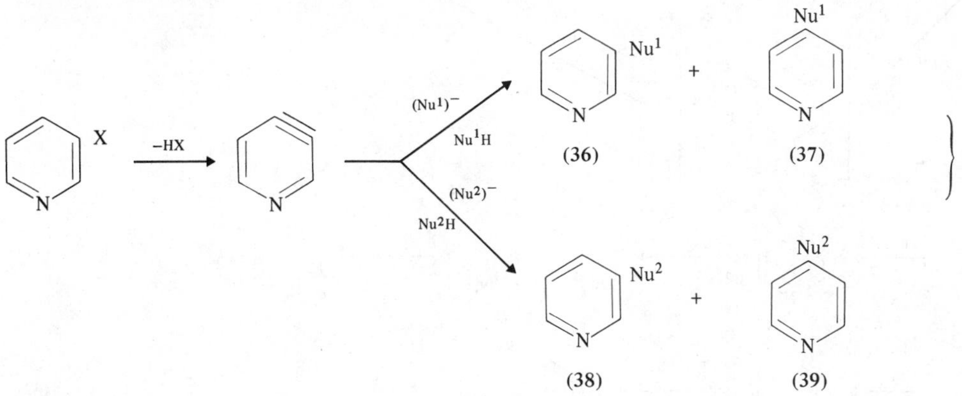

For a 'pure' *EA* reaction, $\frac{[(36)]}{[(38)]} = k\frac{[(Nu^1)^-]}{[(Nu^2)^-]}$ k and k' are independent of the nature of X.

$$\frac{[(37)]}{[(39)]} = k'\frac{[(Nu^1)^-]}{[(Nu^2)^-]}$$

(v) *Typical nucleophilic displacements*

(*a*) *Displacement of hydrogen.* This is a type of substitution for which there is no real parallel in benzene chemistry. Nucleophilic substitution of hydrogen requires the displacement of hydride ions, and so such substitution may be effected only by very strong nucleophiles.

The best-known and most-studied reaction of this type is the Tschitschibabin amination, by which pyridine is converted into 2-aminopyridine by reaction with sodamide. The reaction is widely applicable and has been extensively reviewed.[82,138] 2-Alkylamino-, 2-arylamino-, and 2-hydrazino-pyridines are obtainable by analogous reactions using substituted amide ions.

Diamination of pyridine gives mainly 2,6-diaminopyridine, and substitution occurs at the 4-position only if the 2- and the 6-positions are both blocked. 3-Substituted pyridines are apparently aminated at the 2-position rather than at the 6-position, although the proportion of the 6-amino-compound is increased if the 3-substituent is bulky.[82]

As regards the mechanism of the Tschitschibabin reaction, all the available evidence is consistent with the scheme shown in equation (39). There is no evidence for an *EA* process,[82] although in a few cases, *e.g.* equation (40),[119a] the intermediacy of a pyridyne remains an open question.

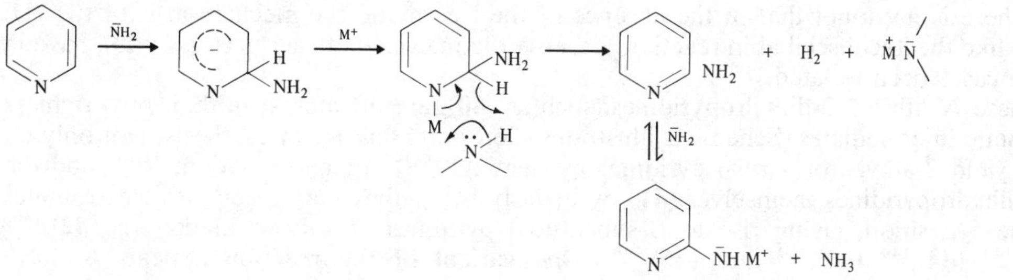

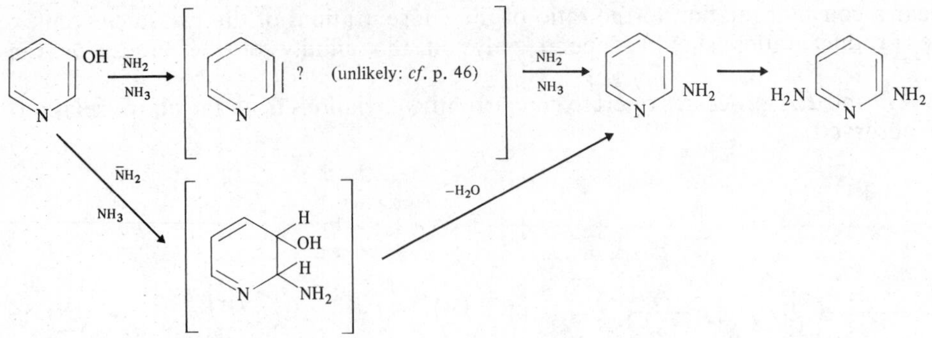

(40)

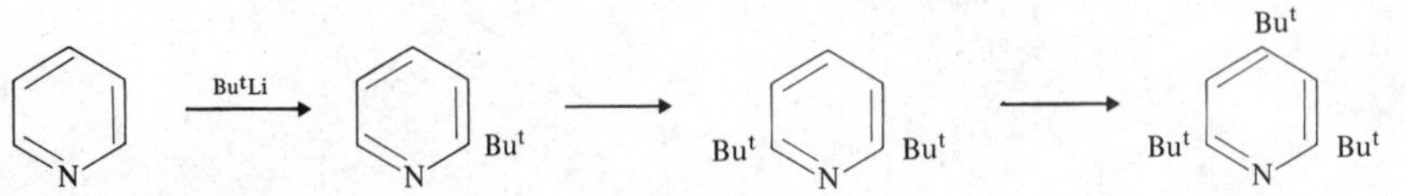

(41)

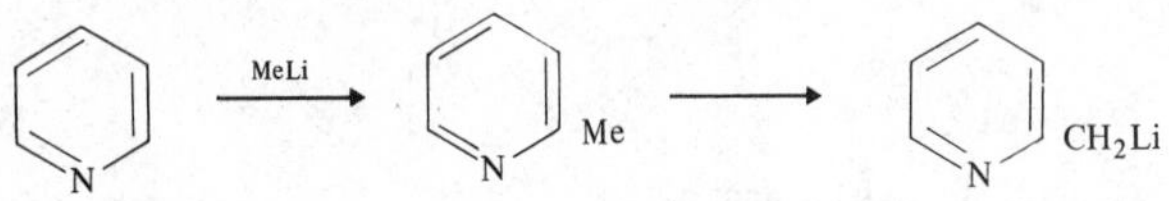

(42)

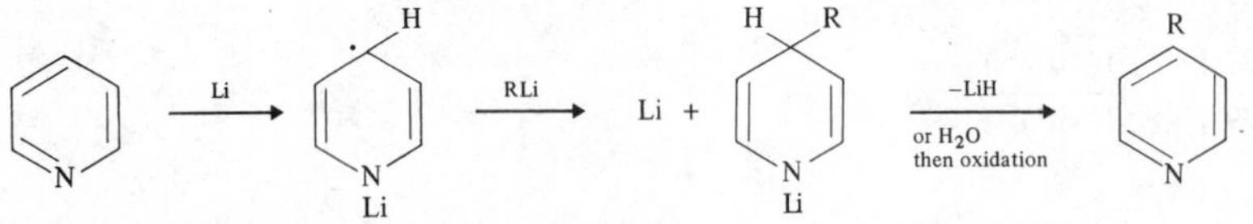

(43)

Substitution of hydrogen by alkyl or aryl groups may take place by the reaction of pyridines with Grignard reagents or organolithium compounds. Like the Tschitschibabin reaction, substitution occurs preferentially at the 2- and/or 6-position. Alkylation using Grignard reagents generally gives only low yields of alkylpyridines,[82] but phenylation gives 2-phenylpyridine in 44% yield.[139] On the other hand, organolithium reagents may be used to make both alkyl- and aryl-pyridines in moderate to good yield.

Where the alkyl group introduced has no α-hydrogen, and so cannot form a carbanion, an excess of the alkyl-lithium reagent may lead to the formation of 2,6-dialkyl- or even 2,4,6-trialkyl-pyridines [equation (41)];[140] but if carbanion formation is possible, this may interfere with further substitution [equation (42)],[141] and in such cases the 2,6-dialkylpyridines are best obtained by the use of a ten-fold excess of the alkyl-lithium reagent.[142] Again like the Tschitschibabin reaction, 3-substituted pyridines react with alkyl- (and aryl-) lithium reagents preferentially at the 2-position, unless the 3-substituent or the attacking nucleophile is bulky.[82,143]

In the presence of the free metal (magnesium or lithium), however, the alkylation of pyridines by alkylmagnesium halides or alkyl-lithium reagents occurs almost exclusively at the 4-position,[144] and the same result may be achieved by direct reaction of the pyridine with the alkyl halide in presence of the metal. It is possible that this is not a straightforward nucleophilic substitution at all [the mechanism in equation (43) has been proposed], but there is no doubt that in the absence of the free metal the mechanism is of the *AE* type, like the Tschitschibabin reaction, because the intermediate adducts, *e.g.* (**40**), have in some cases been isolated.[140b,145]

These *N*-lithio-1,2-dihydropyridines, such as (**40**), are of interest in their own right as synthetic intermediates. Scheme 2 illustrates their potential use in synthesis. Not only do they yield 2-alkyl- (or -aryl-) pyridines by heating[140b,146] or by oxidation,[145,147] and the 1,2-dihydropyridines themselves (**41**) by hydrolysis,[140b] they may react with electrophiles at the 5-position, giving rise to β-substituted pyridines of various kinds, *e.g.* (**42**),[140b] (**43**),[148] (**44**),[149] (**45**),[150] and (**46**).[148b] One variant of the reaction appears to have

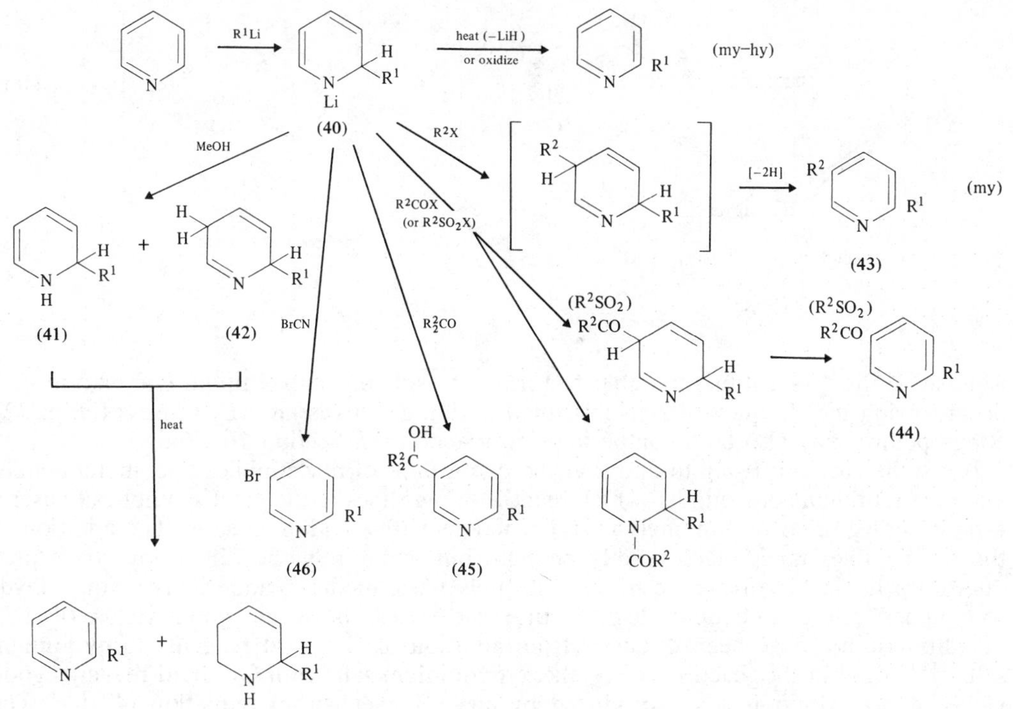

SCHEME 2

particular synthetic potential: the reaction of pyridine with lithium aluminium hydride followed by alkylation [equation (44)]. This gives 3-alkylpyridines in good yield,[151] and complements very well the processes described earlier which lead to 2- and 4-alkylpyridines.

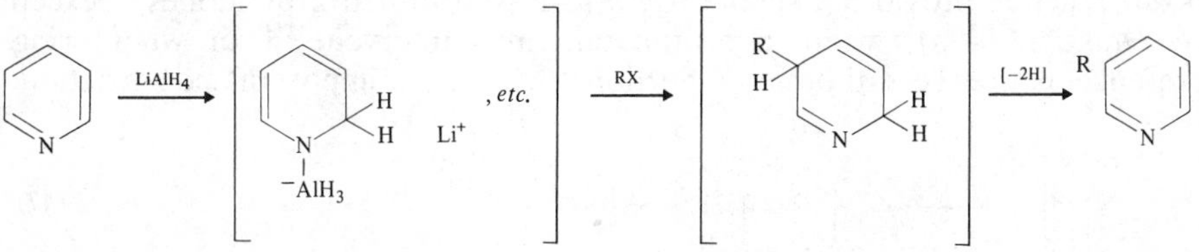

(44)

The intermediate addition compounds such as (**40**) and (**41**), nevertheless, although isolable, are very unstable. Their stability is increased, however, when the pyridine ring carries electron-accepting substituents, especially at the 3- and 5-positions. 3,5-Dicyanopyridine, for example, reacts with Grignard reagents to form two stable dihydropyridine derivatives, (**47**) and (**48**) [equation (45)].[152] 3,5-Dinitropyridines, in which the ring is still more electron-deficient, form adducts of the Meisenheimer type with weaker nucleophiles, *e.g.* methoxide ion [equation (46)].[153]

Pyridinium salts do not generally undergo nucleophilic displacement of hydrogen. Addition of an anionic nucleophile to the pyridinium cation produces a neutral dihydropyridine. If the *N*-substituent is a poor leaving group, the dihydropyridine is often

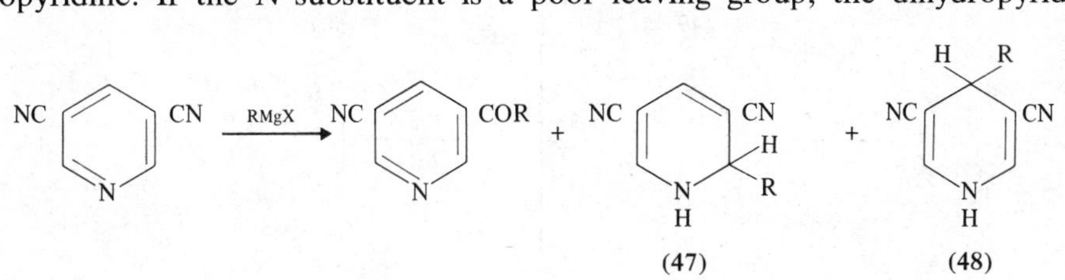

(45)

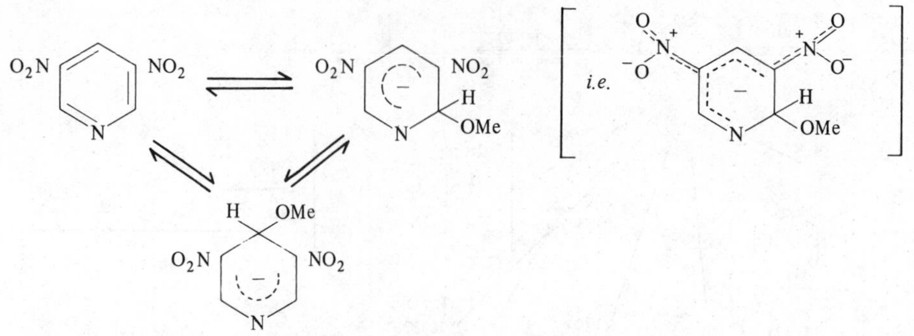

(46)

isolable; if the N-substituent is able to form a reasonably stable anion, elimination may occur, giving the (frequently 2-) substituted pyridine. This is an AE_a process (*cf.* p. 22). Ring-opening may also be a competing side-reaction (*cf.* Section 16.1.6).

It is somewhat surprising to find that the two groups of nucleophiles (*viz.* metal amides and organolithium compounds), which react with pyridines in the most synthetically useful way, do not give correspondingly useful reactions with pyridinium salts. The addition of the nucleophiles takes place readily enough, but apparently the dihydropyridines produced are liable to polymerize in such strongly basic media,[82] and yields of the dihydropyridines are consequently low.[154] In some cases, however, good yields of 1,2-dihydropyridines have been obtained from addition of Grignard reagents to pyridinium salts,[154,155] and in the reactions of N-alkoxypyridinium salts with Grignard reagents good yields of 2-alkylpyridines are produced by an AE_a mechanism [equation (47)].[156] The reaction of cyclopentadienyl anions with pyridinium salts apparently gives 1,4-dihydropyridines, *e.g.* (**49**), which may undergo dehydrogenation to a zwitterionic species, (**50**).[157]

Pyridinium salts, being more reactive towards nucleophiles than pyridines themselves, are able to react with less powerful nucleophiles. Hydroxylation, for example, is readily achieved by reaction of a pyridinium salt with alkaline potassium ferricyanide [equation (48)],[158] and this reaction provides a simple route to N-substituted 2-pyridones,[82] except in cases [*e.g.* equation (48a)] where deprotonation may intervene,[159] or where ring-opening in presence of an excess of base (*cf.* Section 16.1.6) is an important side-reaction.

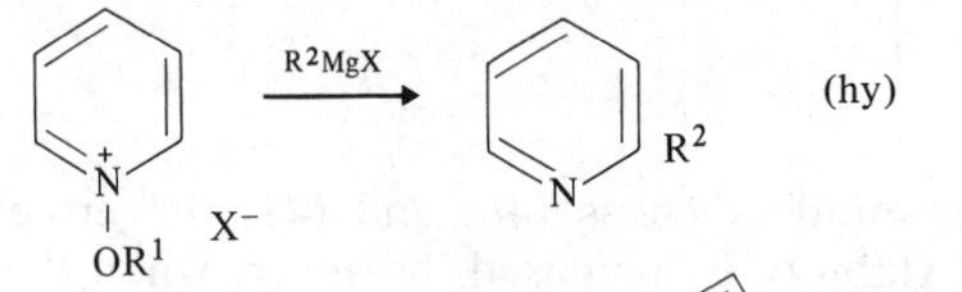

(47)

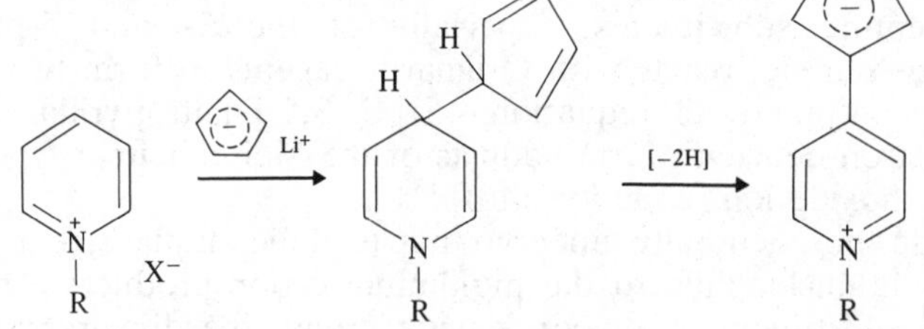

(49) (50)

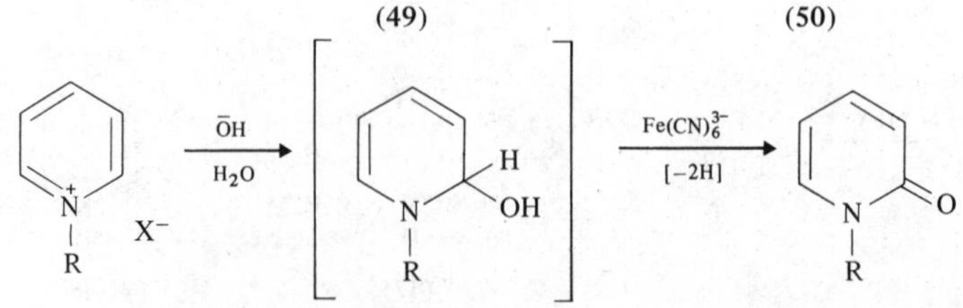

(48)

(48a)

(49)

(50)

(51)

(*cf.* Section 16.1.9)

(52)

(53)

Alkoxide ions attack pyridinium salts at the 2-position, giving dihydropyridines such as (**51**),[160a] but with 'softer' nucleophiles attack at the 4-position is also observed. Enolate ions, for example, add to pyridinium salts at both the 2- and 4-positions[157a,160c,161] [equation (49)] and cyanide ions appear to add preferentially at the 4-position[160b,162] [equation (50)]. 4-Cyano-1,4-dihydropyridines such as (**52**) may be deprotonated by reaction with a second equivalent of cyanide ion, and the resulting carbanion may then attack a second molecule of pyridinium salt [equation (51)]: this process, which bears a formal resemblance to the benzoin condensation, is a potentially useful route to bipyridylium salts.[163]

Among the reactions of pyridine *N*-oxides with nucleophiles, there are few, if any, examples of direct displacement of hydrogen. In general, addition of the nucleophile (at the 2-position, preferentially) is followed by protonation of the *N*-oxide anion, and elimination of water follows [*cf.* equation (47)], or else electrophilic attack on oxygen precedes the nucleophilic addition, and elimination again follows. In either case, the overall reaction is of the AE_a type and is not a direct substitution.

[It must also be noted that *N*-alkoxypyridinium salts may react with nucleophiles in several ways. In addition to the AE_a process and ring-opening (Section 16.1.6), alkylation of the nucleophile, or the formation of an aldehyde — equations (52) and (53), respectively, may occur.[164]]

Typical AE_a reactions of pyridine *N*-oxides are illustrated below. For other examples see references 15d and 165.

Ac_2O ; $\bar{O}Ac$; OAc ; H ; OAc ; OAc ; OAc (Ref. 166)

$NCCH_2CO_2Et$

H ; CN ; CH ; OAc ; CO_2Et ; CN ; CH ; CO_2Et (Ref. 167)

R ; KCN ; PhCOCl ; (R = CF_3,NO_2) ; R ; H ; CN ; OCOPh ; + ; H ; NC ; R ; OCOPh ; R ; CN ; + ; NC ; R (Ref. 168)

RX ; X^- ; OR ; KCN ; CN ; + ; CN (Ref. 169)

SO_2Cl_2 ; Cl ; + ; Cl (Ref. 32b)

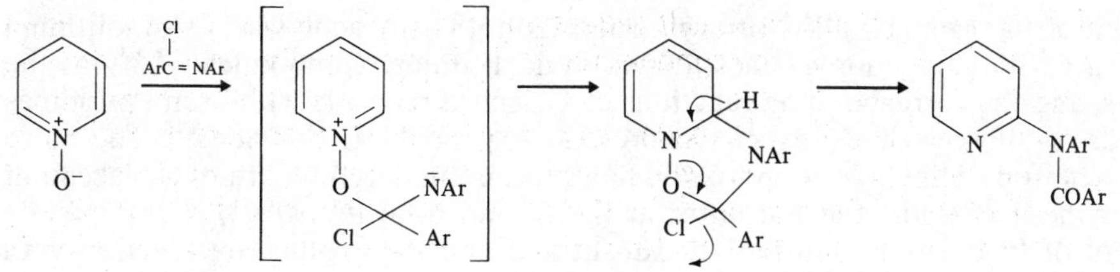

(Ref. 170)

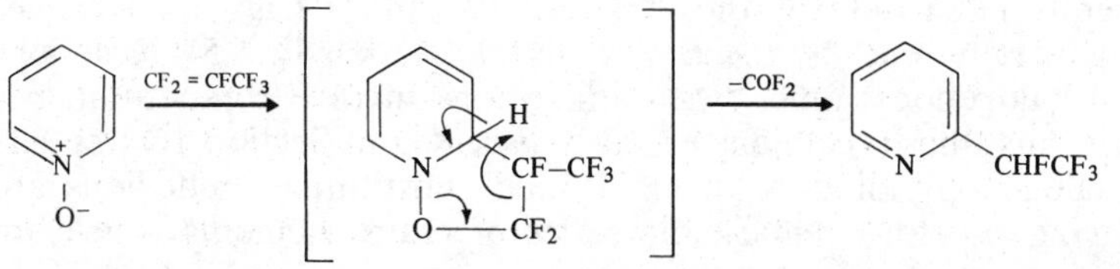

(Ref. 171)

The reaction of pyridine *N*-oxide with Grignard and organolithium reagents is much more complex, however, as shown in Scheme 3.[82,172] Good yields of 2-alkyl- or 2-aryl-pyridines are difficult to obtain by this method. Ring-opening of the initially formed adducts, *e.g.* (**53**), gives 5-substituted pentadienal oximes, *e.g.* (**54**), as the main products, although these may then be recyclized, *e.g.* by acetic anhydride, to the 2-alkyl- (or -aryl-) pyridines. The reaction of pyridine *N*-oxide with acetylide ions similarly gives ring-opened compounds (**55**) and (**56**), in preference to 2-ethynylpyridine.[173]

(*b*) *Displacement of halogens.* Halogenopyridines, like halogenobenzenes, react with nucleophiles by both *AE* and *EA* mechanisms, although the majority of the reactions are of the former type. In the *AE* reaction, the reactivities of halogenopyridines and halogenonitrobenzenes are of approximately the same order (*cf.* Table 2), and 2- and 4-halogeno-substituents in pyridines are thus readily displaced. Protonation, quaternization, or *N*-oxidation of the hetero-atom greatly increases the rate of such halogen displacement. 3-Halogenopyridines undergo substitution less readily, and usually (in the absence of an additional activating group) by the *EA* mechanism.

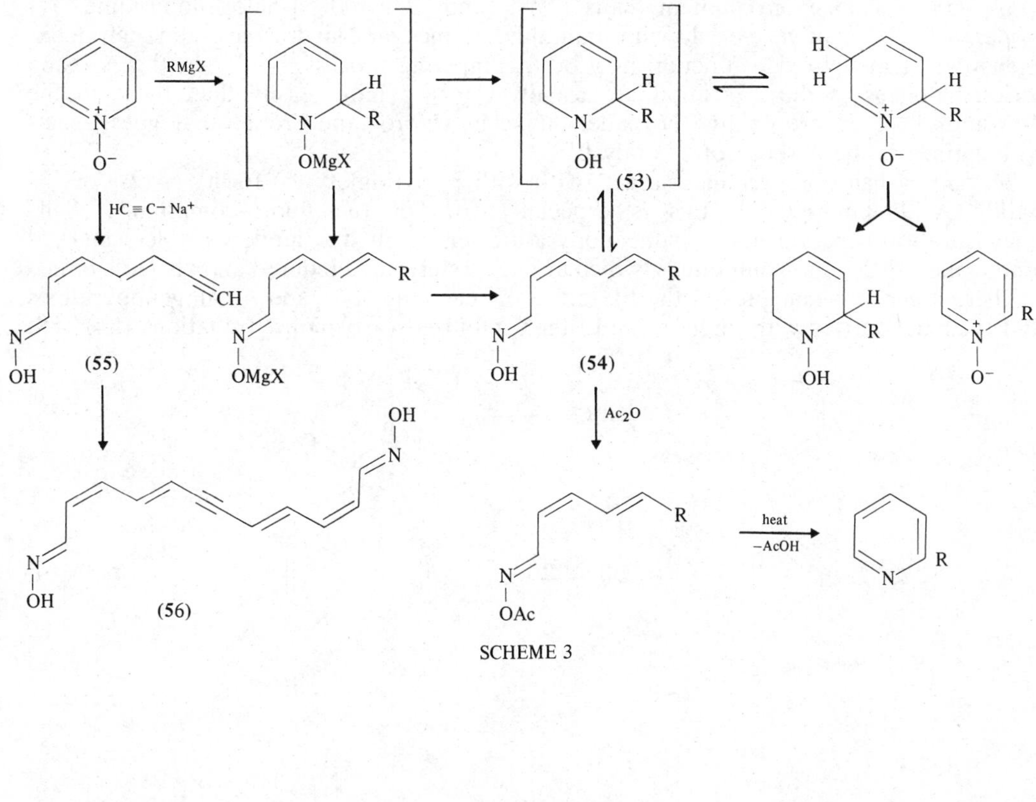

SCHEME 3

Replacement of halogens by alkyl or aryl groups is not easily achieved. Organolithium compounds react with halogenopyridines to give pyridyl-lithium compounds[147b,174] as the main products, and the corresponding reactions of Grignard reagents with bromopyridines give low yields of alkylpyridines, except in presence of cobalt(II) chloride.[175] The more reactive pentachloropyridine and its *N*-oxide, however, do undergo chlorine displacement (the former at the 4-position and the latter at the 2- and 6-positions).[176]

Replacement of halogens by amino- (or substituted amino-) groups, by reaction with the appropriate amide ion, may take place by *AE* or *EA* mechanisms. 2-Halogenopyridines apparently react only by the *AE* route,[177] and 3-halogenopyridines (except 3-fluoropyridine) generally react by the *EA* route (*cf.* Section 16.1.5). Reactions of 3-fluoropyridine, and 4-halogenopyridines, with amide ions involve competition between *AE* and *EA* mechanisms: these reactions are also discussed in Section 16.1.5. The reactions of substituted halogenopyridines with amide and substituted amide ions are more complicated still, giving rise to a remarkable series of rearrangements which are discussed in Section 16.1.6.

In the corresponding reactions of halogenopyridine *N*-oxides with amide ions, it is the 4-halogeno-series where 'normal' *AE* substitution is observed,[178] and the 2-halogeno-series in which competition between *AE* and *EA* processes occurs. This is discussed further in Section 16.1.5.

Replacement of halogens by amino-groups may be accomplished using ammonia or amines if the halogen is 'activated' (*i.e.* at the 2- or 4-position, or at a site *ortho* or *para* to an electron-withdrawing substituent): a considerable amount of kinetic data is available for such reactions.[134,179] Since nucleophilic attack on the ring is assisted when the hetero-atom is positively charged, aminodehalogenations are catalysed by acids,[180] Lewis acids like zinc chloride,[181] and cupric sulphate, the last-named effecting the substitution process even in 'unreactive' 3-halogenopyridines.[112a,182]

Hydrazine[183] and arylamines[181a,184] also react readily with 2- and 4-halogenopyridines, although reactions involving the latter (which are weaker nucleophiles) may require higher temperatures or a catalyst such as copper bronze. Di- and poly-halogenopyridines containing both 2- and 4-halogeno-substituents may undergo mono-aminodehalogenation at either the 2- or the 4-position,[33,185] except for pentafluoropyridine, which reacts exclusively at the 4-position.[186b]

2- and 4-halogenopyridinium salts[181a,187] and 2- and 4-halogenopyridine *N*-oxides[42,105b,188] react very easily with ammonia, amines, and hydrazines, although deoxygenation of the *N*-oxide function may be a competing process.[185a,188a] In the *N*-oxide series, halogens at the 2-position are usually replaced more easily than those in the 4-position.[189] 3-Fluoropyridine *N*-oxide, unlike its chloro- and bromo-analogues, reacts with amines in the absence of a catalyst.[190]

2- and 4-halogenopyridines react with other pyridines to form *N*-(2- or 4-pyridyl)pyridinium salts;[191] this is especially true of the fluoro-compounds. Self-quaternization occurs in the 4-halogenopyridine series, but the tendency is less marked among the 2-halogeno-compounds, which are more sterically hindered and also less basic.

There are many examples in the literature of reactions of 2- and 4-halogenopyridines and their derivatives with enolates and other stabilized carbanions. Equations (54)–(58)

2-chloropyridine + $PhCH_2CN$ $\xrightarrow{NaNH_2}$ 2-pyridyl–CH(CN)Ph (54)[193]

2-bromopyridine + MeCN $\xrightarrow{NaNH_2}$ (2-pyridyl)–CH(CN)–(2-pyridyl) (55)[194]

O_2N, Cl (pyridine) + $EtCH(CO_2Et)_2$ ⟶ O_2N (pyridine) $C(Et)(CO_2Et)_2$ (56)[195]

Me, Br (pyridine N-oxide, $\overset{+}{N}$–O^-) + $Na^+\ \bar{C}H(CO_2Et)_2$ ⟶ Me, N–O, CO_2Et, O $\xrightarrow[H_2O]{\bar{O}H}$ Me, CH_2CO_2H ($\overset{+}{N}$–O^-) (57)[196]

Br (pyridine) + $ArSO_2CH_2R$ $\xrightarrow{NaNH_2}$ (pyridine) $CH(R)SO_2Ar$ (58)[197]

illustrate the diversity of such reactions; further examples are given in Ref. 15e. Halogenopyridines may yield cyanopyridines by reaction with cyanide ions[10d] (even 3-bromopyridine is convertible into 3-cyanopyridine with cuprous cyanide[192]), but this is not the method of choice for the preparation of cyanopyridines.

The reactions of halogenopyridines with oxygen nucleophiles are almost entirely unexceptional. Hydrolysis of 2- and 4-halogenopyridines may be achieved in either aqueous acid or base, the former being an efficient process because protonation of the hetero-atom facilitates nucleophilic attack.[198] For the parent compounds, elevated temperatures may be required,[199] but the more reactive members of the series (including 2-fluoropyridine[200a]) undergo hydrolysis under comparatively mild conditions.[201] Unactivated 3-halogenopyridines are hydrolysed by aqueous hydroxide only at high temperatures and pressures;[202] hydrolysis in the absence of a catalyst is an *EA* reaction, and gives both 3- and 4-hydroxypyridines,[202a] but in presence of cupric sulphate only the 3-hydroxy-compound is obtained.[202b]

The corresponding reactions with alkoxide ions are similar. Substitution of the halogen in 2- and 4-halogenopyridines is relatively easy,[199b] and even unactivated 3-halogenopyridines undergo substitution when heated with alkoxides.[203] In some of these an *EA* mechanism is involved,[203b] but in most cases products of *cine*-substitution are not observed and so it is assumed that the *AE* mechanism operates alone. The reaction of 2-bromopyridines with potassium t-butoxide gives good yields of 2-pyridones,[204] presumably by elimination of isobutene from the intermediate 2-t-butoxypyridines.

2- and 4-halogenopyridinium salts[200b,205] and 2- and 4-halogenopyridine *N*-oxides[105b,188a,206] react, as expected, more easily with oxygen nucleophiles than do the parent halogenopyridines (*cf.* Table 2). In 2,4- and 2,6-dihalogenopyridines, both halogens are displaceable, but one is displaced much more easily than the second.[105a,205b,207] The pentahalogenopyridines undergo hydrolysis first at the 4-position,[33b,186c,208] whereas their *N*-alkylated derivatives react preferentially at the 2-position.[17]

Conversion of 2- and 4-halogenopyridines into pyridine-2- and -4-thiols (2- and 4-thiopyridones) is usually effected with sodium (or potassium) hydrogen sulphide,[209] or with thiourea (followed by hydrolysis),[206a,b,210] and the reactions of halogenopyridines with alkane- or arene-thiolates are exactly analogous to the corresponding reactions with alkoxides and phenoxides.[211] Displacement of halogen by sulphite ions, to give pyridine-2- and -4-sulphonates, is possible,[212] although the yields are extremely variable.

Halogen exchange at the 2- and 4-positions in pyridines is commonly observed, especially if the hetero-atom bears a positive charge.[200b,213] These exchange reactions are not normally of preparative importance, except for the preparation of polyfluorinated pyridines [equation (59)].[186a,208b,214]

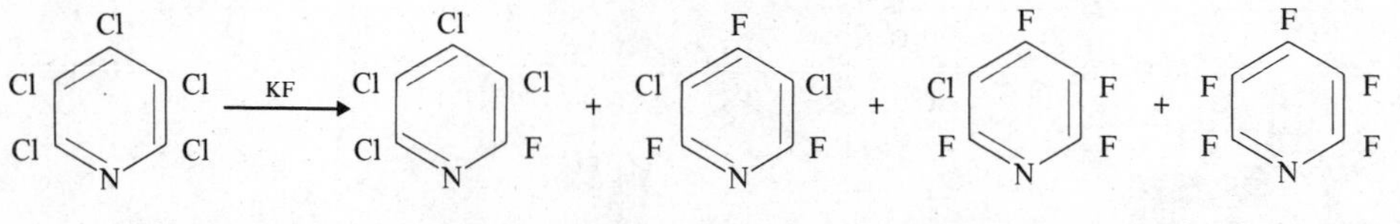

(59)

(*c*) *Displacement of nitrogen-containing groups.* (*A*) *The nitro-group.* By analogy with *o*- and *p*-dinitrobenzene, one might expect 2- and 4-nitropyridines to undergo a facile displacement of nitrite ion by reaction with a variety of nucleophiles. In practice, however, the mobility of the nitro-groups in these compounds depends strongly on the nature of the nucleophile and on any additional activating influences (*e.g.* an *N*-oxide function) in the ring.

Reactions with amines, for example, are surprisingly difficult to accomplish, in the absence of additional activation: this may be provided by a second nitro-group, a carboxyl group, or an *N*-oxide,[215] but if the molecule contains another displaceable group, as in equation (60),[216] this is generally displaced in preference to the nitro-group.

By far the best-known displacement reaction of nitropyridine derivatives is halogenodenitration. 2- and 4-nitropyridines themselves are converted into the corresponding chloro- or bromo-pyridines by reaction with the appropriate hydrogen halide,[217] and the corresponding transformation of the *N*-oxides may be brought about by a variety of reagents, including hydrohalic acids,[218] phosphoryl and sulphuryl chlorides,[105,192] and acyl chlorides.[42,105b,219a] AE_a reactions may also occur under these circumstances, and these lead to the insertion of an additional halogen atom and removal of the *N*-oxide [equations (61) and (62)].[105a,219b]

Hydroxydenitration is a reaction of little preparative significance: a few examples have been recorded[215a,220] but side-reactions [*cf.* equations (63) and (64)] are liable to occur.

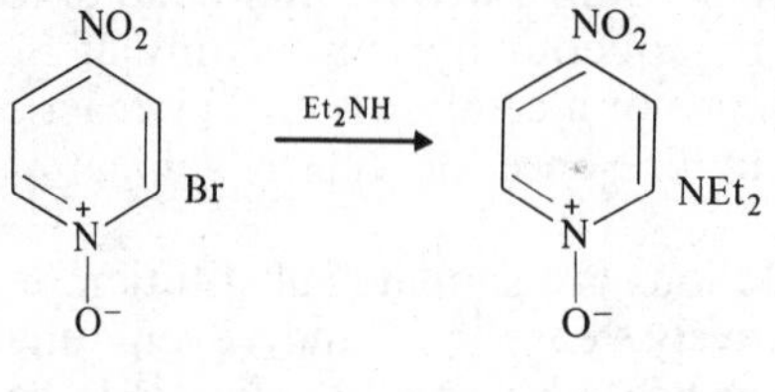

(60)

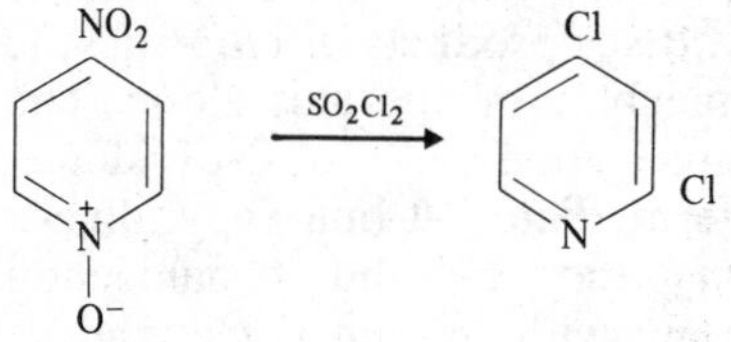

(61)[105a]

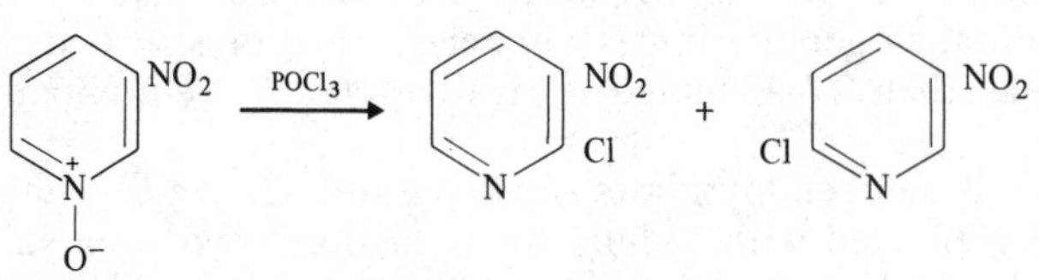

(62)[219b]

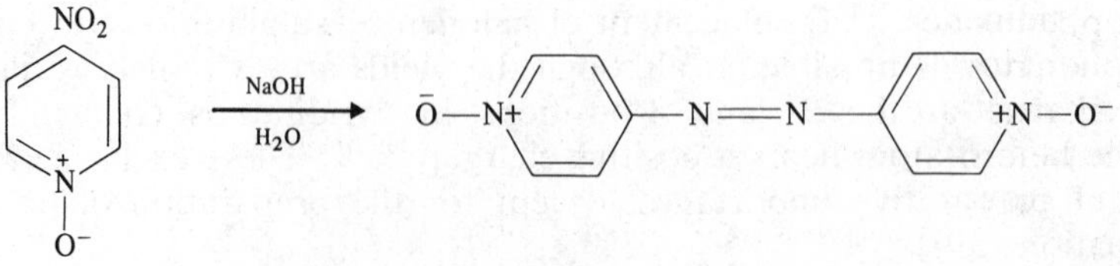

(63)[223]

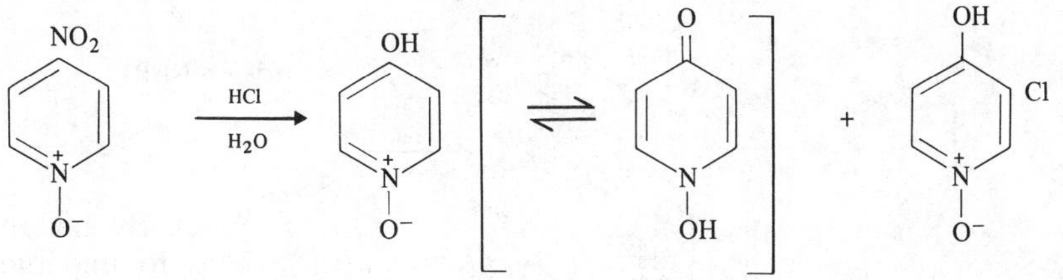

(64)[114]

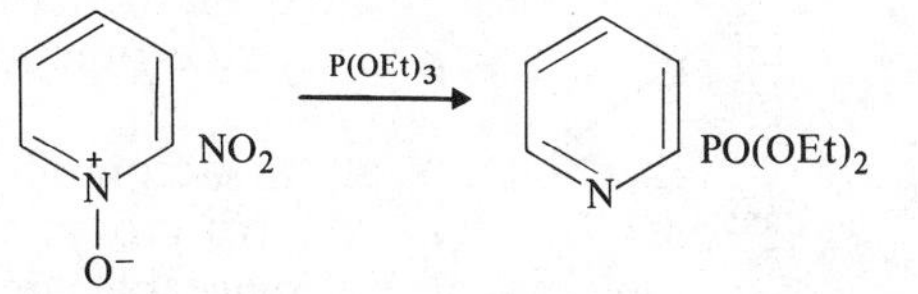

(65)

Replacement of nitro-groups by alkoxy- or aryloxy-groups, however, is well-documented, and kinetic evidence is available to show that 4-nitropyridines and their *N*-oxides are considerably more reactive than the corresponding halogeno-compounds towards alkoxide ions.[221] The reactions of 4-nitropyridine *N*-oxide with thiolate ions are also well-known,[222] and 2-nitropyridine *N*-oxide undergoes substitution as well as deoxygenation on reaction with triethyl phosphite[48] [equation (65)].

(*B*) *The pyridinium group.* 2- and 4-pyridylpyridinium salts, *e.g.* (**56**), readily undergo nucleophilic displacement of pyridine by reaction with a variety of nucleophiles[224] (Scheme 4). The *N,N'*- linked pyridiniopyridone (**57**) and the pyridiniopyridinium dication (**58**) undergo AE_a reactions which are of synthetic utility.[225]

(*C*) *The diazonium group.* Pyridine-2- and -4-diazonium salts display exceptional reactivity towards nucleophiles. Attempts to diazotize 2- and 4-aminopyridines in dilute acid lead directly to hydroxypyridines (pyridones)[226] and, in concentrated hydrohalic acid, directly to halogenopyridines.[191,227] Fluoropyridines are also obtainable directly by diazotization of 2- and 4-aminopyridines in tetrafluoroboric or hexafluorosilicic acid.[228] Diazonium salts derived from 3-aminopyridine and aminopyridine *N*-oxides do not show this exceptional reactivity, and behave like normal arenediazonium salts.

(*d*) *Displacement of oxygen- and sulphur-containing groups.* (*A*) *Hydroxy-, alkoxy-, and aryloxy-groups (and their sulphur analogues).* The relative ease with which 2- and 4-hydroxypyridines are convertible into the corresponding aminopyridines[117b,229] by reaction with ammonia and amines is possibly due to the pyridone tautomers of the starting compounds. On the other hand, the conversions of 2- and 4-hydroxypyridine into 2- and 4-halogeno-pyridines, by reaction with phosphorus tri- or penta-halides, phosphoryl chloride, or thionyl chloride,[199b,230] are reactions of the hydroxy-tautomers, depending on the ease of displacement of a 2- or 4-acyloxy-substituent by halide ions. The conversions of 2- and 4-pyridones into the corresponding thiopyridones by phosphorus pentasulphide[231] may also be regarded as nucleophilic substitution of an *O*-acylated pyridine. 2- and 4-alkoxy-, or aryloxy-, pyridines, their quaternary salts, and their *N*-oxides, may undergo displacement of the alkoxy groups by amines,[232] and alkoxypyridinium salts are even sufficiently reactive to undergo substitution with cyanide ion.[233]

(*B*) *Sulphonyl and sulphonate groups.* Salts of pyridinesulphonic acids undergo the normal nucleophilic substitution reactions of arenesulphonates, *e.g.* fusion with alkali to give hydroxypyridines,[117b,234] and with sodium or potassium cyanide to give cyano-pyridines.[117a,235] However, pyridine-2- and -4-sulphonates undergo certain additional substitution reactions, *e.g.* replacement of the sulphonate by amino-,[117b] hydroxy-,[236] or alkoxy-groups[237] (by reaction, respectively, with amines, aqueous acid, or alkoxides). Alkylsulphonyl groups may also be displaced from the 2- and 4-positions by reaction with alkoxides.[238]

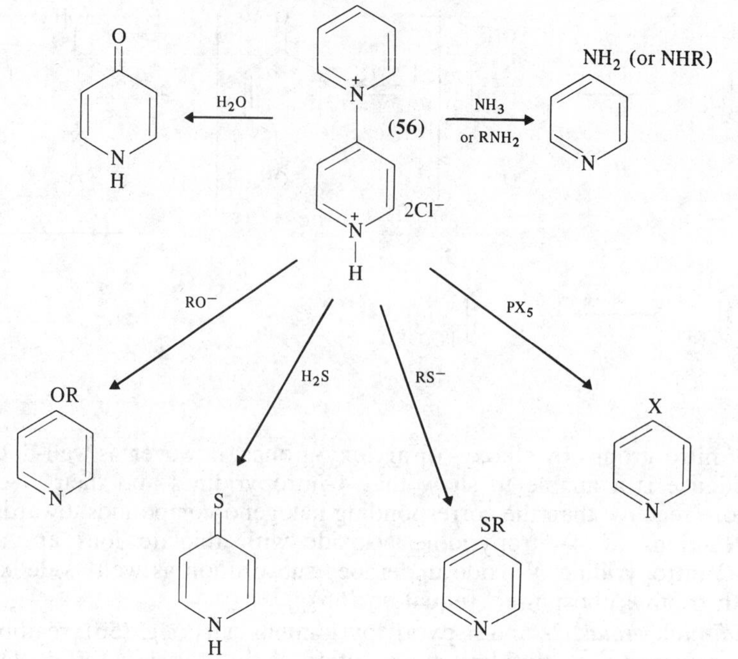

SCHEME 4

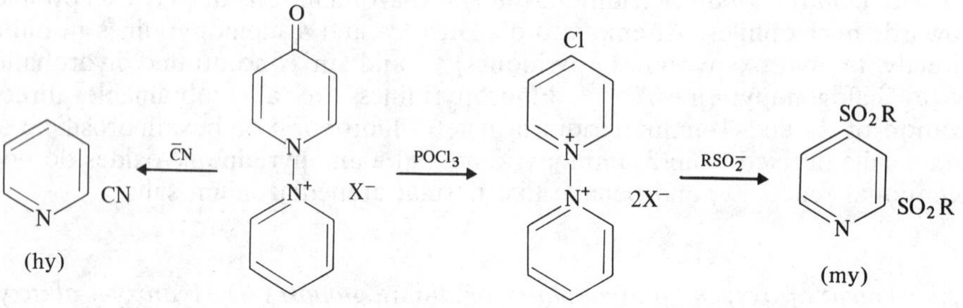

(hy) (57) (58) (my)

(*e*) *Displacement of cyano-groups.* 2- and 4-cyanopyridinium salts are converted into *N*-alkylpyridones by reaction with alkali[239] and 4-cyanopyridinium salts undergo aminodecyanation by reaction with ammonia and amines.[239b,240]

16.1.3.3 Radical substitution

Pyridine, like benzene and other arenes, undergoes substitution of hydrogen (an addition–elimination process) by reaction with a variety of radicals. The principal feature of these substitutions is the predominance of 2-substituted pyridines in the product mixture, and such theoretical calculations as are available[241] agree that the 2-position should be the most reactive towards radical attack. The results collected in Tables 3 and 4 show that, when acid is either initially present, or is generated in the course of the reaction, the predominance of the 2-substituted product is even more striking, and the proportion of 3-substituted product is extremely small. Indeed, attack by radicals at the 3-position of pyridines predominates only if the attacking radical has a 'higher-than-average' amount of electrophilic character, or if the pyridine already bears substituents at the 2- or 4-position.

(i) Halogenation

Reference has already been made (pp. 17–18) to the vapour-phase halogenations of pyridine, which yield, at sufficiently high temperatures, mixtures consisting mainly of 2- (and 2,6-di-) halogenated pyridines.[242] This preponderance of α-substitution is consistent with a radical substitution mechanism, especially since photochemical halogenation of pyridine gives almost exclusively the 2-halogeno-compounds,[243] and it is known that electrophilic halogenation occurs preferentially at the 3-position (pp. 17–18).

(ii) Alkylation[244]

Radical alkylation of pyridines may be achieved in a variety of ways, *e.g.* by thermolysis of dialkyl[245] or diacyl[246] peroxides, or oxidative decarboxylation (electrolytic[247] or chemical[248a]) of carboxylic acids, in the pyridine as solvent, or by thermal decomposition of *N*-alkylpyridinium salts (the Ladenburg rearrangement).[249] Methylation is the alkylation reaction which has been most studied, and the isomer ratios of methylpyridines obtained in such reactions are tabulated below (Table 3). Hydroxymethylation may be achieved

TABLE 3
Radical Substitution of Pyridine

Radical	*Source of radical*	*Isomer ratio* 2-	3-	4-	*Ref.*
(a) *In non-acidic media*					
$CH_3\cdot$	$(CH_3CO_2)_2$	62.7	20.3	17.0	a
	$(Bu^tO)_2$	62.0	22.9	15.1	b
	$Pb(OAc)_4$	62.1	20.5	17.4	a
$C_6H_5\cdot$	$(PhCO_2)_2$	54	32	14	c
	$Pb(OCOPh)_4$	52	32.5	15.5	c
	$PhN{=}NCPh_3$	53	31	16	c
	$PhN_2^+BF_4^-$	51.8	32.4	15.7	d
	$PhN_2^+Cl^-$	51.3	29.8	19.0	e
	PhN(NO)Ac	46	43	11	c
p-$MeC_6H_4\cdot$	$ArN_2^+Cl^-$	55.4	29.3	15.3	e
p-$BrC_6H_4\cdot$	$ArN_2^+Cl^-$	53.3	33.4	13.3	f
p-$MeOC_6H_4\cdot$	$ArN_2^+Cl^-$	57.5	26.3	16.2	f
p-$O_2NC_6H_4\cdot$	$ArN_2^+Cl^-$	44.6	42.7	12.9	e
(b) *In acidic media*					
$CH_3\cdot$	$(Bu^tO)_2+AcOH$	77.9	2.7	19.4	b
	$(Bu^tO)_2+AcOH+HCl$	93.2	0	6.8	b
	$Pb(OAc)_4+AcOH$	76.4	2.9	20.7	b
	Electrolysis of AcOH	73.5	0	26.5	g
$C_6H_5\cdot$	$(PhCO_2)_2+AcOH$	82	18(3- + 4-)		i
	$(PhCO_2)_2+AcOH+HCl$	64.6	4.6	30.8	h
	Electrolysis of BzOH	56	35	9	j
(c) *Pyridinium compounds*					
$CH_3\cdot$	Thermolysis of *N*-methylpyridinium iodide	62.8	6.0	31.2	k
(d) *Pyridine* N-*oxide*					
$C_6H_5\cdot$	PhN=N—NHPh	76.2	7.8	16	l
	$PhN{=}N{-}\overset{+}{N}C_5H_5\ BF_4^-$	87	1	12	l

[a] Ref. 246. [b] Ref. 245. [c] D. H. Hey, C. J. M. Stirling, and G. H. Williams, *J. Chem. Soc.*, 1965, 3963. [d] R. A. Abramovitch and J. G. Saha, *Tetrahedron*, 1965, **21**, 3297. [e] R. A. Abramovitch and J. G. Saha, *J. Chem. Soc.*, 1964, 2175. [f] R. A. Abramovitch and M. Saha, *J. Chem. Soc. (B)*, 1966, 733. [g] Ref. 247. [h] J. M. Bonnier and J. Court, *Compt. rend. (C)*, 1967, **265,** 133. [i] H. J.-M. Dou and B. M. Lynch, *Bull. Soc. chim. France*, 1966, 3815. [j] P. J. Bunyan and D. H. Hey, *J. Chem. Soc.*, 1960, 3787. [k] Ref. 249. [l] R. M. Elofson, F. F. Gadallah, and K. F. Schulz, *J. Org. Chem.*, 1971, **36,** 1526.

TABLE 4

Radical Substitution of Substituted Pyridines[a]

Substituent	Radical	Radical source	Isomer ratio 2-	3-	4-	5-	6-	Ref.
2-CH_3	$C_6H_5\cdot$	$(PhCO_2)_2$		31	15	20	34	b
	$CH_3\cdot$	Thermolysis of methiodide		8.3	32.4	2.1	57.2	c
	$C_6H_5\cdot$	$(PhCO_2)_2$ + AcOH + HCl		8	41	4.5	46	b
2-OMe	$C_6H_5\cdot$	$(PhCO_2)_2$		65	5	7	23	d
2-CN	$C_6H_5\cdot$	$(PhCO_2)_2$		33	7	41.5	18.5	e
3-CH_3	$CH_3\cdot$	$(CH_3CO_2)_2$	55.5		20.5	4.6	19.4	f
	$C_6H_5\cdot$	$(PhCO_2)_2$	41.5		26	9.5	23	b
	$CH_3\cdot$	Thermolysis of methiodide	48.3		23.5	3.6	24.6	c
	$C_6H_5\cdot$	$(PhCO_2)_2$ + AcOH + HCl	41.5		34.5	1.5	22.5	b
3-OMe	$C_6H_5\cdot$	$(PhCO_2)_2$	59		19	8	14	d
3-CN	$C_6H_5\cdot$	$(PhCO_2)_2$	33		25	6	36	e
4-CH_3	$CH_3\cdot$	$(CH_3CO_2)_2$	65.5	34.5				f
	$C_6H_5\cdot$	$(PhCO_2)_2$	54	46				b
	$CH_3\cdot$	Thermolysis of methiodide	96.1	3.9				c
	$C_6H_5\cdot$	$(PhCO_2)_2$ + AcOH + HCl	86	14				b
4-OMe	$C_6H_5\cdot$	$(PhCO_2)_2$	33.8	66.2				d
4-CN	$C_6H_5\cdot$	$(PhCO_2)_2$	24	76				e

[a] The data for methylpyridines take no account of the competing hydrogen abstraction from the methyl group.
[b] J.-M. Bonnier and J. Court, *Compt. rend.* (*C*), 1967, **265,** 133, and references therein.
[c] Ref. 249. [d] R. Dufournet, J. Court, and J.-M. Bonnier, *Bull. Soc. chim. France*, 1974, 1112.
[e] J. Court, S. Vidal, and J.-M. Bonnier, *Bull. Soc. chim. France*, 1972, 3107. [f] Ref. 246.

using methanol and ammonium persulphate,[248b] and cyclohexylation is accomplished by the photolysis of di-t-butyl peroxide in a mixture of cyclohexane and pyridine.[250]

Some data for the methylation of alkylpyridines are also collected above (Table 4). In these reactions, however, abstraction of α-hydrogens from the alkyl groups is a possible side-reaction, and the pyridylalkyl radicals so obtained may themselves undergo alkylation, or else dimerize.[251]

(*iii*) *Arylation*[252]

Homolytic phenylation of pyridine and its derivatives is the most extensively studied of all these radical substitutions: the phenyl radicals may be generated from peroxides, lead tetrabenzoate, or azo- or diazo-precursors (Table 3), and the composition of the isomer mixture obtained remains almost constant, irrespective of the radical source.*

Since the pyridine ring is electron-deficient by virtue of its electron-attracting heteroatom, and reacts more readily with nucleophiles than with electrophiles, it is to be expected that the more nucleophilic character a radical possesses, the more rapidly it should react with pyridine, and the greater should be the degree of substitution at the 2- and 4-positions. This expectation is realized in a study of variously substituted phenyl radicals with pyridine (Table 3a).

Phenylation of substituted pyridines (Table 4) shows an apparent *ortho*-directing effect of substituents when the reactions are carried out in non-acidic media, but the usual preference for α- and γ-substitution in acidic media.

* Radicals generated from *N*-nitrosoacetanilide represent a notable exception.

(iv) Acylation

Homolytic acylation of pyridines (the acyl radicals being generated from aldehydes, a hydroperoxide, and a ferrous salt) gives 2- and 4-acylpyridines;[248c] amidation[248d] and carboxylation[248e] are apparently similar. These reactions are illustrated in Scheme 4a.

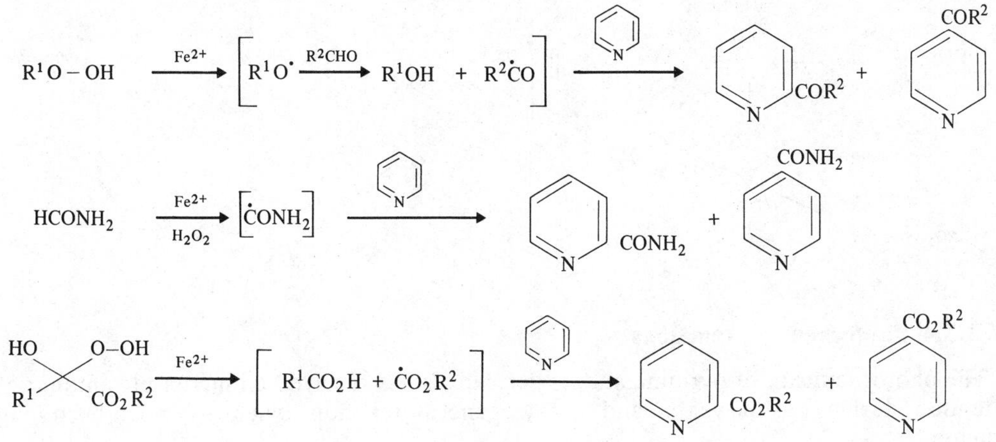

SCHEME 4a

(v) Intramolecular radical substitutions

Pschorr-type cyclizations, involving intramolecular arylation of a pyridine ring, have been used to prepare a number of aza-heterocycles [*cf.* equation (66)], but in cases where the hetero-atom may effect the displacement of the diazonium group, as in equation (67), this displacement normally takes precedence over the arylation. The intramolecular amination of pyridines by nitrenes [equations (68) and (69)] is subject to the same limitation [*cf.* equation (70)].

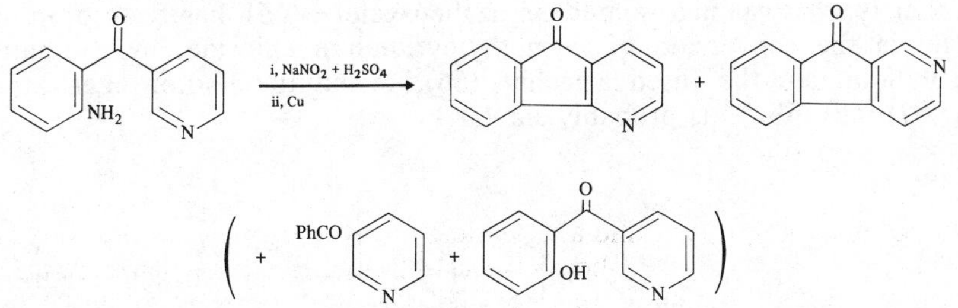

(66)[253]

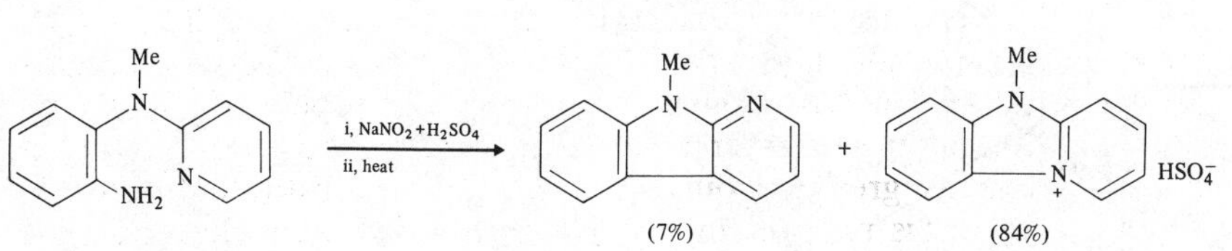

(67)[254]

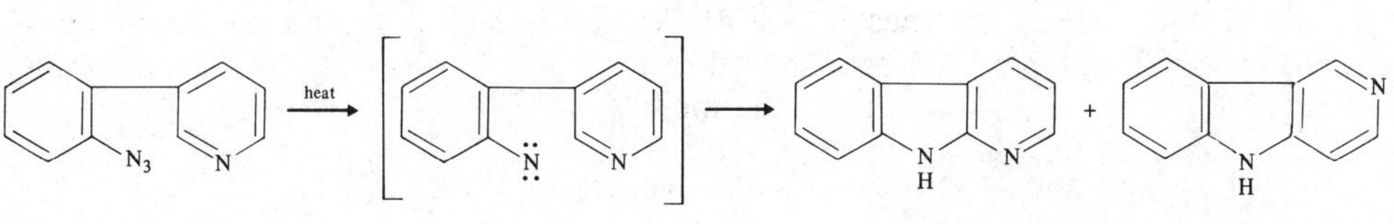

(68)[255]

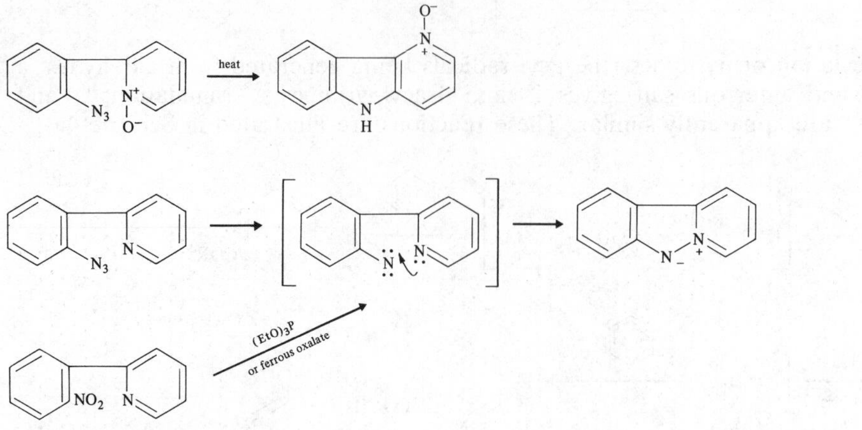

(69)[255]

(70)[255,256]

16.1.3.4 Photochemical reactions

The photochemistry of pyridine and its derivatives has received an increasing amount of attention during recent years, and a few general reaction types are now becoming apparent.

(i) Valence isomerization

Irradiation of pyridine in solution, with radiation of wavelength 253.7 nm, gives the 'Dewar pyridine' (**59**)[257], which is extremely unstable (half-life at 25 °C = 2.5 min). There are, in theory, two 'Dewar pyridines', *viz.* (**59**) and (**60**), but the latter is yet unknown. Derivatives of (**60**), however, have been prepared by irradiation of the substituted pyridines (**61**) and (**62**),[258] and under appropriate reaction conditions the irradiation of (**61**) also gives the azaprismane (**63**).[258a] [This is apparently the only azaprismane to have been isolated, although such compounds have been proposed as intermediates in the photo-isomerizations of methyl- and dimethyl-pyridines:[259] *cf.* equation (71)]. 'Dewar pyridones' (**64**) have also been prepared by photo-isomerization of 2-pyridones.[260]

Yet another type of valence isomer, an azabenzvalene (**65**), has been proposed as the intermediate in the conversion of *N*-methylpyridinium chloride, by irradiation of its aqueous solution, into the fused aziridine (**66**),[261] and the photoisomerization of the zwitterion (**67**) into (**68**)[262] is probably similar.

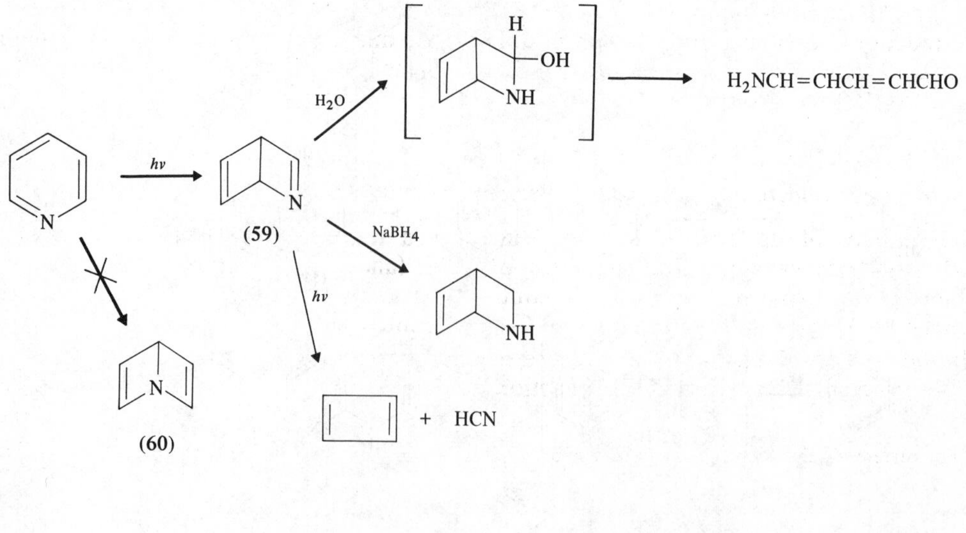

$H_2NCH{=}CHCH{=}CHCHO$

(59)

(60)

+ HCN

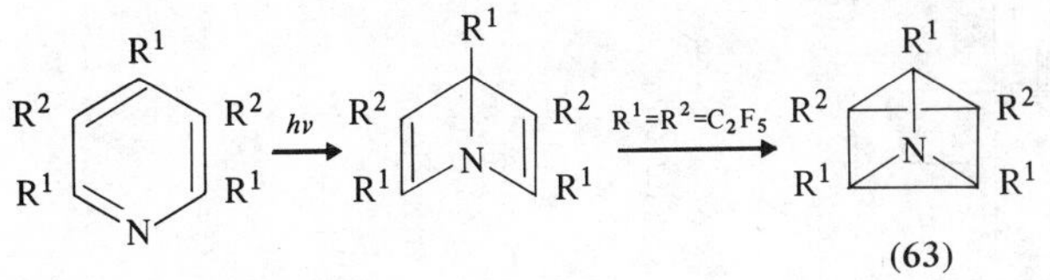

(61; $R^1 = R^2 = C_2F_5$)
(62; R^1 = Me, $R^2 = CF_3$)

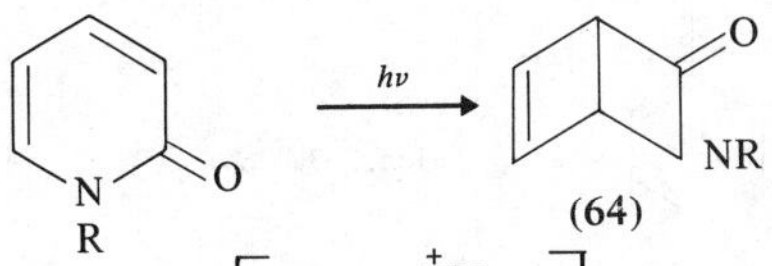

(63)

(64)

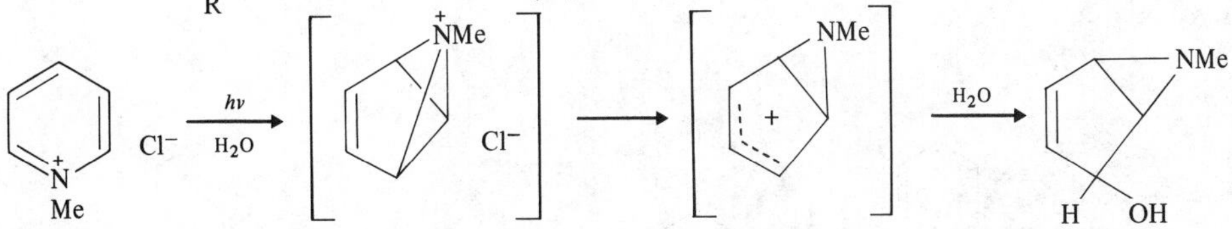

(65)

(66)

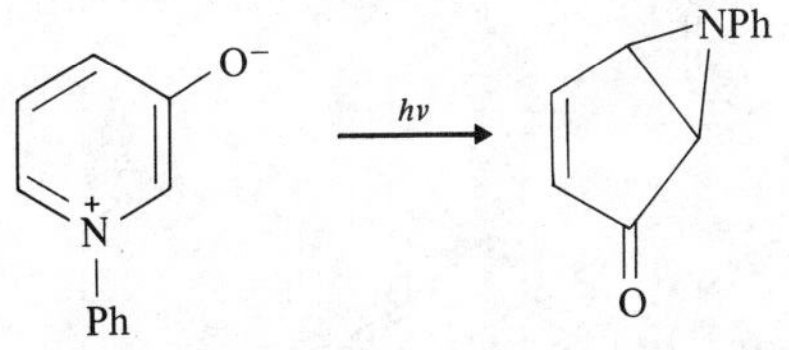

(67)

(68)

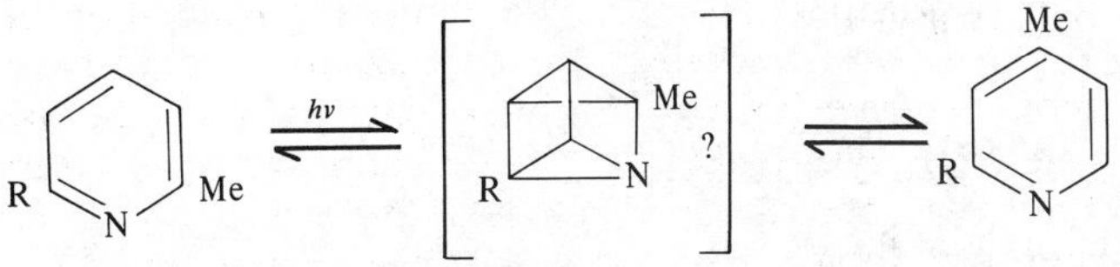

(71)

(*ii*) *Intramolecular arylation*

Just as stilbenes on irradiation yield phenanthrenes, so aza- and diaza-stilbenes yield aza- and diaza-phenanthrenes (Scheme 5). Equations (72) and (73) represent possible side-reactions involving the olefinic double bond, and equations (74) and (75) depict alternative modes of cyclization involving displacement of an *ortho*-halogeno-substituent in one of the aryl groups.

(*iii*) *Cleavage and rearrangement of* N-*oxides,* N-*imines, and* N-*ylides*

Compounds of the type $C_5H_5\overset{+}{N}$—X^- undergo a variety of reactions on irradiation: N—X bond cleavage, giving the parent pyridine (and atomic oxygen, a nitrene, or a carbene); ring expansion and/or ring contraction; and migration of the group X from nitrogen to a ring carbon or to a 2- or 6-substituent.

In the case of the *N*-oxides, all of these reaction types have been observed (Scheme 6)*. In the *N*-imine series, the principal reactions are cleavage to a pyridine and a nitrene, and

* For reviews of this topic, see Refs. 271 and 272.

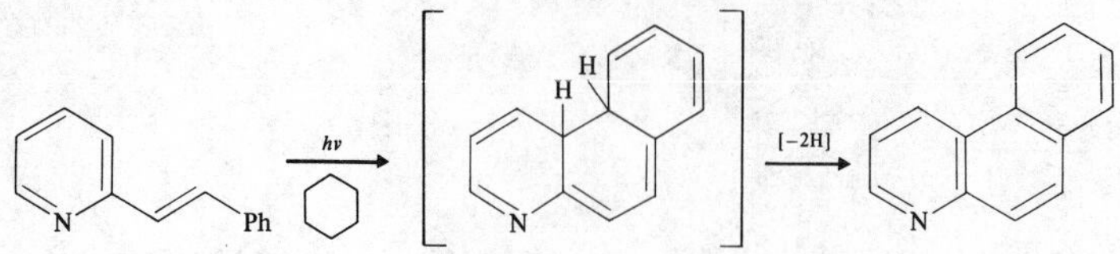

(Ref. 263a, c)

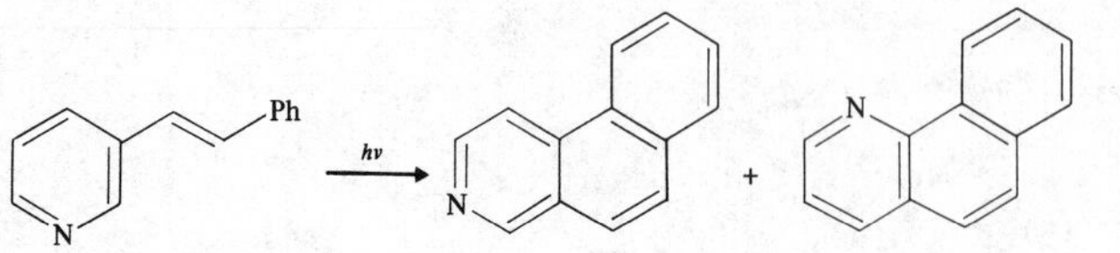

(Ref. 263a, 264)

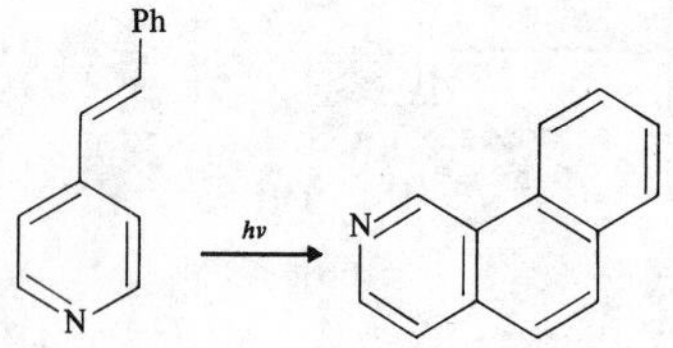

(Ref. 263a, b)

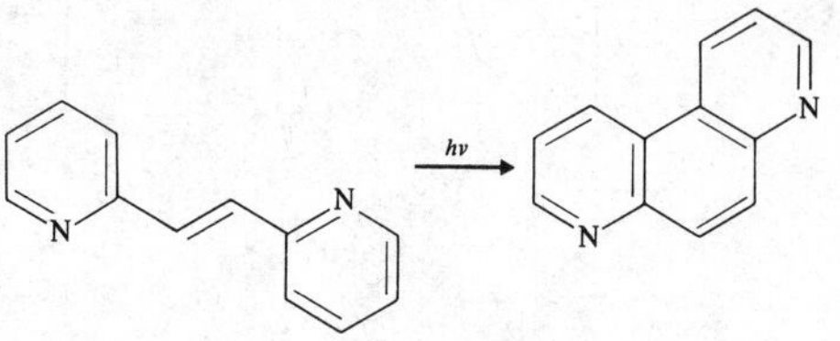

(Ref. 265)

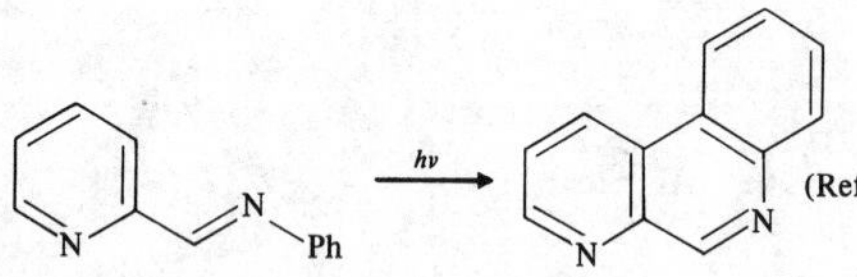

(Ref. 266)

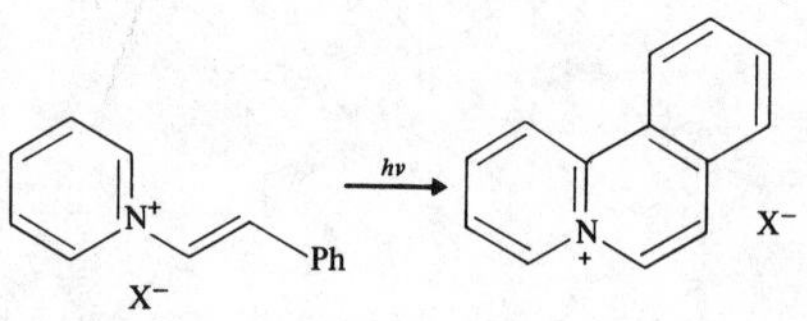

(Ref. 267)

SCHEME 5

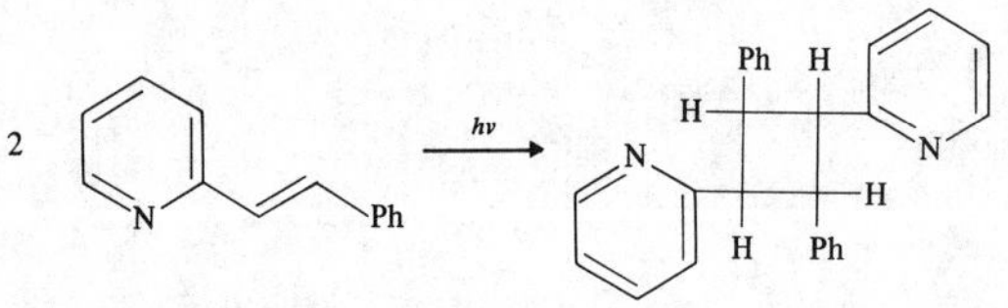

(72)[268]

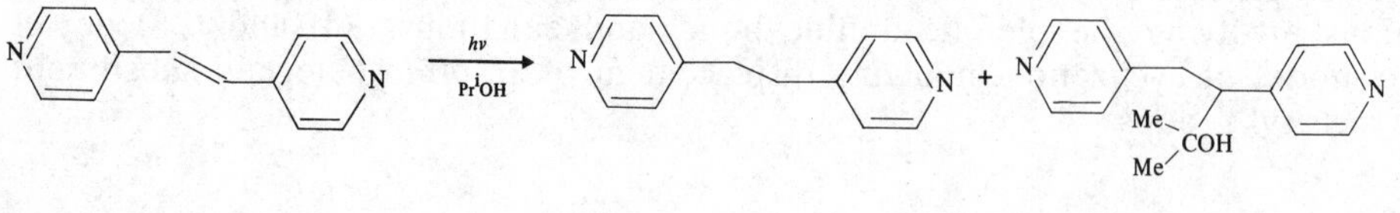

(73)[269]

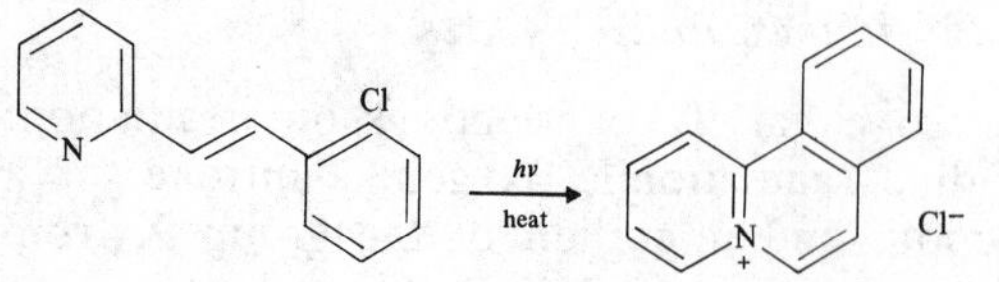

(74)[270a]

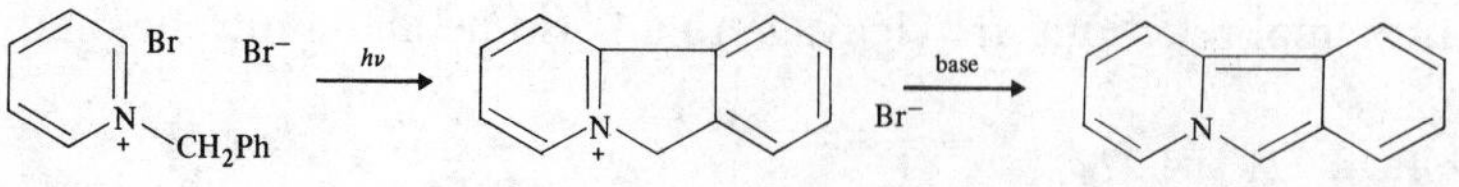

(75)[270b]

R^1=Me, R^2=H; R^1=Me; R^1=H; $h\nu$

SCHEME 6

(hy) (76)[65c,273]

$h\nu$ acetone or benzene; $h\nu$ benzene; C_6H_6; $PhNHCO_2Et$ (ly)

(ly–my) (my) (ly) (77)[274]

$h\nu$ Et_2NH; (hy); ? $PhNH_2$ (hy); Et_2NH; (ly) (78)[275]

$h\nu$ benzene; benzene; (ly) (79)[276]

$h\nu$ benzene; (ly)

ring expansion to a 1,2-diazepine [*e.g.* equations (76)–(78)]. The photochemistry of pyridinium ylides has been relatively little explored, but the photolysis of the dicyanomethylide [equation (79)] leads to cleavage and to ring contraction.

(*iv*) *Other photochemical reactions*

There remain several photo-induced reactions of pyridines which at present may be placed only in a 'miscellaneous' category, and since none of these has yet been shown to have any real preparative significance, the reader must be referred elsewhere* for details of these. Two, however, deserve special mention: the dimerization of 2-pyridones and 2-aminopyridinium salts,[277] which gives the remarkable compounds (**69**) and (**70**) respectively, and the photo-aminoalkylation of pyridine[278] [equation (80)] which provides access to a group of pyridine derivatives not otherwise obtainable.

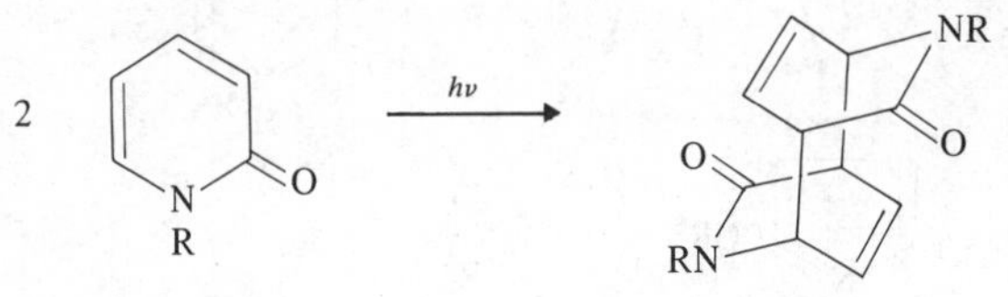

2 (2-pyridone, N-R) $\xrightarrow{h\nu}$ dimer

(**69**)

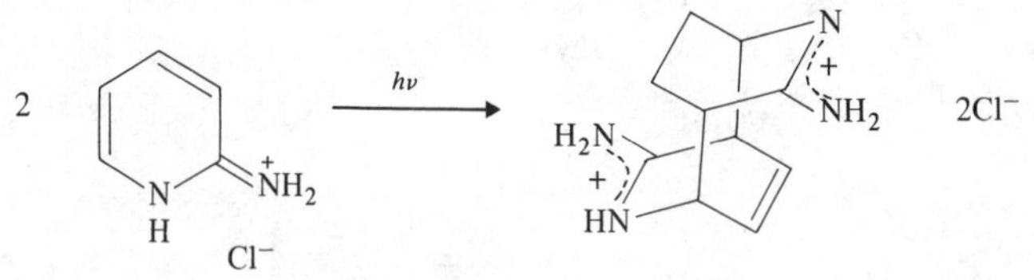

2 (2-aminopyridinium, $\overset{+}{N}H_2$, Cl⁻) $\xrightarrow{h\nu}$ dimer $2Cl^-$

(**70**)

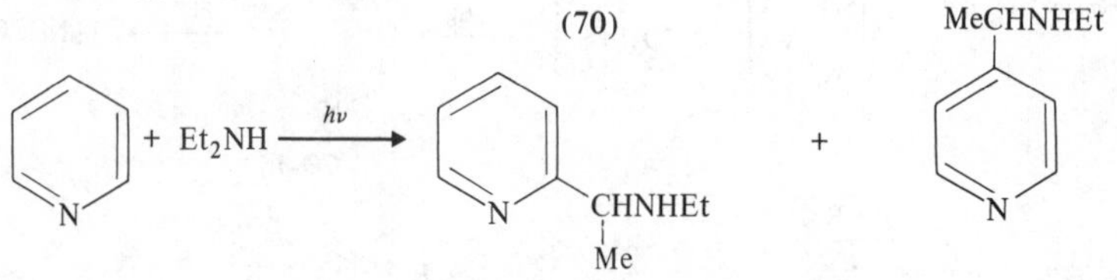

Pyridine + Et_2NH $\xrightarrow{h\nu}$ 2-(CHNHEt, Me)pyridine + 4-(MeCHNHEt)pyridine (80)

16.1.4 REDUCTION OF THE PYRIDINE RING

It is not surprising that pyridines are more susceptible to reduction than benzene derivatives. Since pyridines react readily with nucleophiles, they may be reduced by nucleophilic reducing agents such as complex metal hydrides†, and the more reactive (*i.e.* electron-deficient) pyridine derivatives, *e.g.* pyridinium salts, may be reduced by more weakly nucleophilic reducing agents such as sodium dithionite. These reductions should, in theory, lead to 1,2- and/or 1,4-dihydropyridines [(**71**) and (**72**)], but these may undergo further reduction to a tetrahydropyridine [(**73**) or (**74**)] or even to the hexahydropyridine (usually known as a piperidine: **75**).

Pyridine itself reacts with lithium aluminium hydride to form a complex (**76**) containing both 1,2- and 1,4-dihydropyridine rings;[280] this complex has reducing properties, and has been used for the selective reduction of ketones, but attempts to isolate dihydropyridines from a solution of lithium aluminium hydride in pyridine have proved unsuccessful.[281] However, reduction of pyridine and its simple analogues by lithium aluminium hydride in presence of aluminium chloride[282] gives mixtures of 1,2,3,6-tetrahydropyridines [*e.g.* (**73**)]

* Refs. 10f, 293b.

† For a review, see Ref. 279.

and piperidines. The presence of a Lewis acid in these reactions presumably promotes the further reduction of the initially formed 1,2-dihydropyridine; it has been shown independently that boranes catalyse the reduction of 1,2-dihydropyridines by sodium borohydride[283] and it is possible that the aluminium hydride generated in this reaction functions similarly (*cf.* Ref. 287). The failure to isolate the isomeric tetrahydropyridines [*e.g.* (**74**)] in these reactions may also be tentatively explained on this basis, since these tetrahydropyridines are cyclic enamines, like (**71**), and may themselves undergo further reduction in presence of the Lewis acid.

Sodium borohydride does not apparently react with pyridine and its simple derivatives, but readily reduces those pyridines with more electron-deficient rings (*e.g.* cyanopyridines, pyridinecarboxylic esters, and pyridinium salts). For the reduction of cyanopyridines and pyridinecarboxylic esters, sodium borohydride is preferable to lithium aluminium hydride, since the latter may reduce the substituent as well as the ring; however, it must be noted that 3,5-dicyanopyridines and pyridine-3,5-dicarboxylic esters contain rings so electron-deficient that they may be reduced by either reagent without reduction of the substituents [*cf.* equations (81)–(82)].

The products of these reductions are generally dihydropyridines,[154,279,284] with the ratio of 1,2- to 1,4-dihydropyridine depending on the reducing agent used, and on the reaction conditions. Reduction beyond the dihydropyridine stage is unusual in this series, the few examples recorded being associated with the use of sodium borohydride in protic solvents[285] [equations (81) and (83)].

Complex metal hydride reduction of pyridinium salts, on the other hand, commonly produces tetrahydropyridines along with, or in place of, the mixture of 1,2- and 1,4-dihydropyridines,[154,279,286,287] the relative proportions of the products again depending on the hydride used, and on the reaction conditions (solvent, *etc.*). It is noteworthy, however, that the tetrahydropyridines obtained in these reactions are 1,2,3,6-tetrahydropyridines [*cf.* equation (84)], whereas those produced from the unquaternized pyridines [equations (81) and (83)] are 1,2,3,4-tetrahydropyridines.

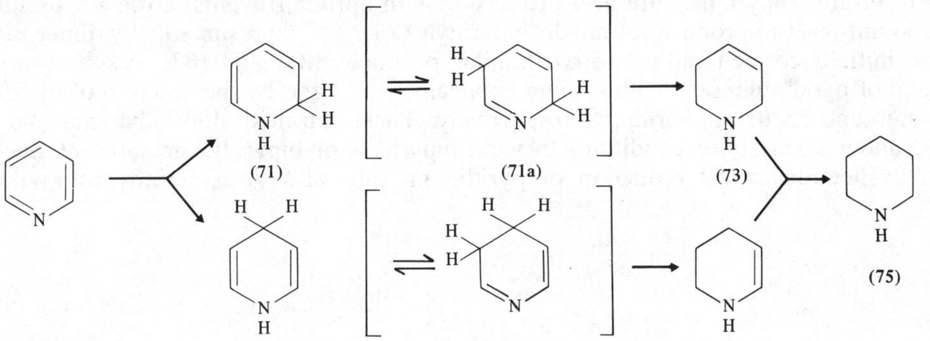

(71) (71a) (73) (75)

(72) (72a) (74)

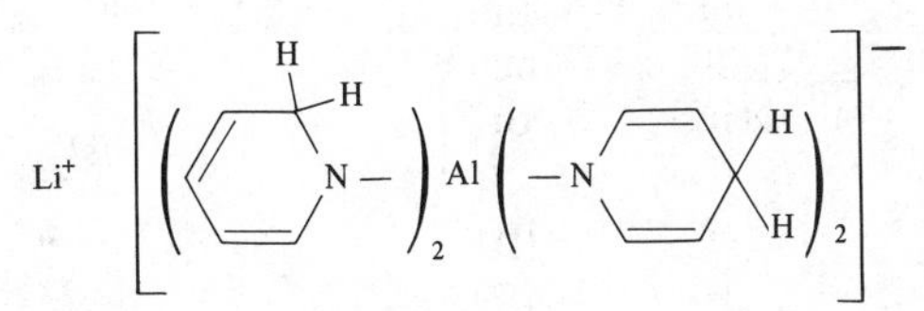

(76)

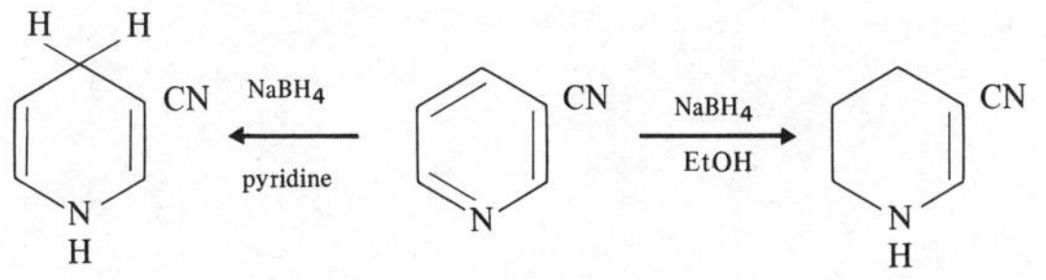

NaBH$_4$, pyridine; NaBH$_4$, EtOH (81)[285a]

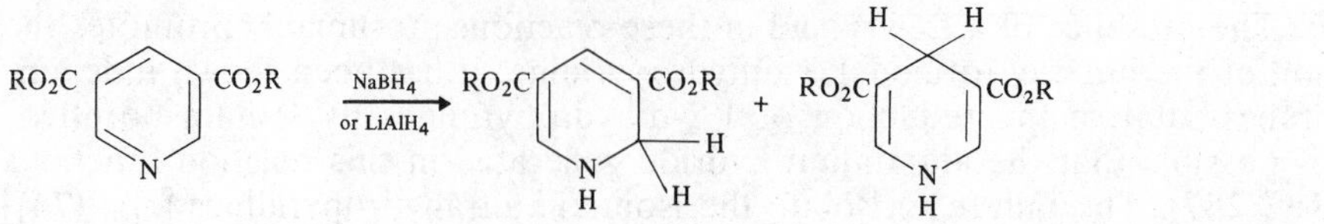

(82)[284]

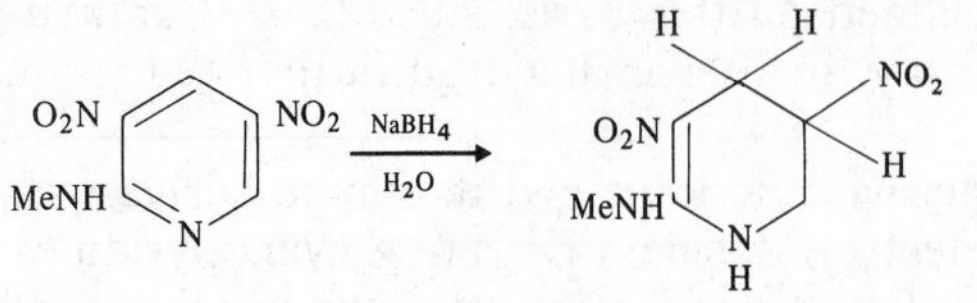

(83)[285b]

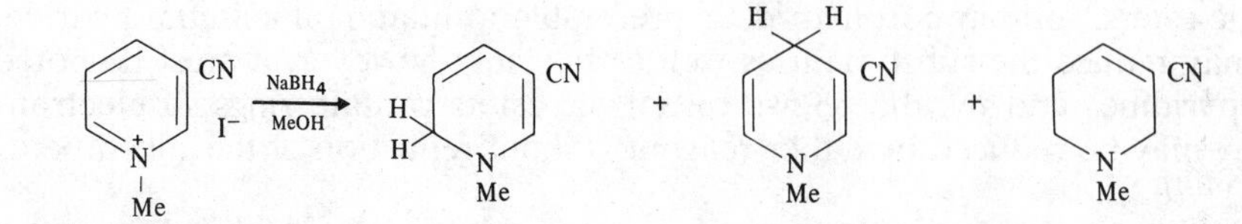

(84)[286]

The interest in sodium dithionite as a reducing agent for pyridinium salts stems largely from the early observation[288] that the coenzyme NAD (nicotinamide adenine dinucleotide, **77**) was reduced to its dihydro-derivative, NADH (**78**), by this reagent. Various model systems related to NAD are reduced under similar conditions, the main product in each case being the 1,4-dihydropyridine,[154] although in some cases the 1,2-dihydro-isomer is obtained as a by-product.[286] The mechanism of dithionite reduction involves the sulphinate (**79**) as intermediate,[154] and this salt on acidification loses sulphur dioxide to give the dihydropyridine.

Reduction of pyridines may also be effected by 'one-electron' processes, for example by a metal such as sodium, or by an electrochemical method. Sodium in ethanol generally reduces pyridines completely, piperidines being the main products, although in some cases 1,2,3,6-tetrahydropyridines are also produced.[287] In aprotic solvents, however, pyridines and sodium react to produce tetrahydrobipyridyls (**80**),[289a,290] presumably by dimerization of the initially formed radical (**80a**). Similar products, (**81**) and (**82**), result from the reaction of pyridinium salts with sodium amalgam,[289b,291] and by the reaction of pyridines with zinc and acetic anhydride,[292] respectively. These tetrahydrobipyridyls may be oxidized under a variety of conditions to yield bipyridyls or bipyridylium salts (*cf.* Section 16.1.9). Electrochemical reduction of pyridinium salts also gives tetrahydrobipyridyls,

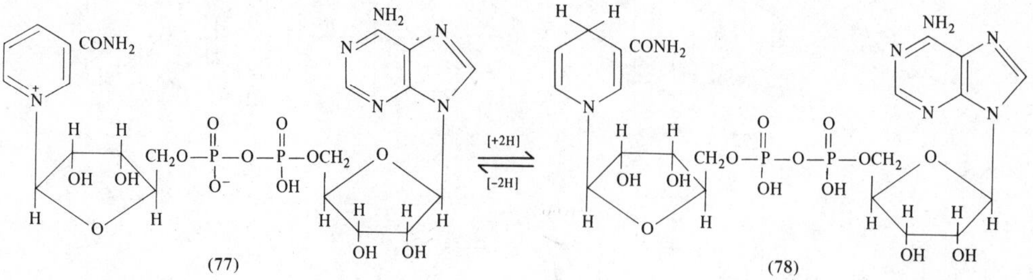

(77)

(78)

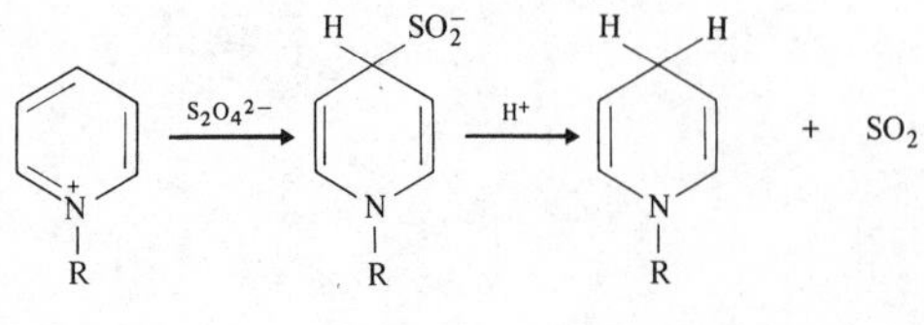

(79)

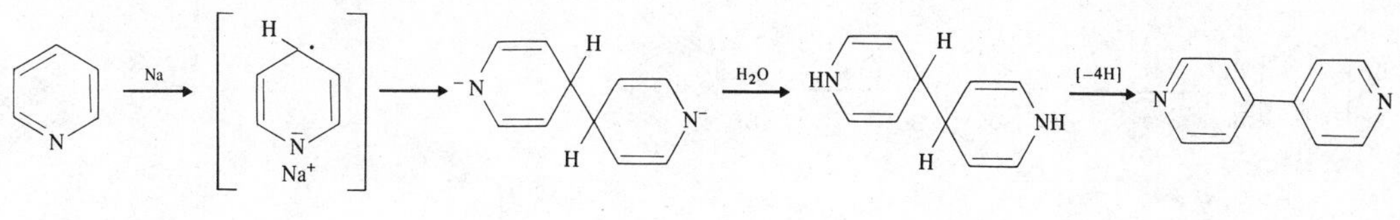

(80a) (80)

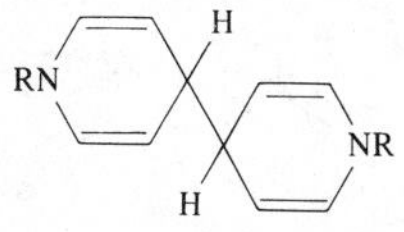

(81) 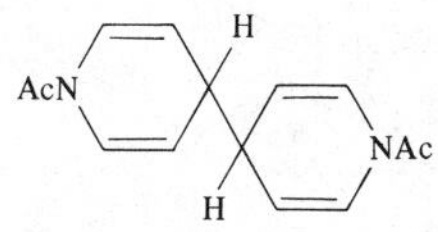 (82)

and/or di- and tetrahydro-pyridines, depending on the substituents in the ring and on the reaction conditions.[154,293a]

The properties and reactions of dihydropyridines[154,293a] and tetrahydropyridines[287,293a] are the subjects of reviews elsewhere.

16.1.5 DEHYDROPYRIDINES (PYRIDYNES)*

16.1.5.1 Generation of pyridynes

There are two possible aza-analogues of dehydrobenzene (benzyne), *viz.* 2,3- and 3,4-dehydropyridine (2,3- and 3,4-pyridyne: **83** and **84**, respectively). They are generated, like benzyne, by a variety of elimination reactions: from 3-halogenopyridines and strong bases,[294] from *ortho*-dihalogenopyridines and lithium amalgam,[294] by lead tetra-acetate oxidation of aminotriazolopyridines,[298] and by thermal decomposition of pyridinediazonium carboxylates[294] (Scheme 7).

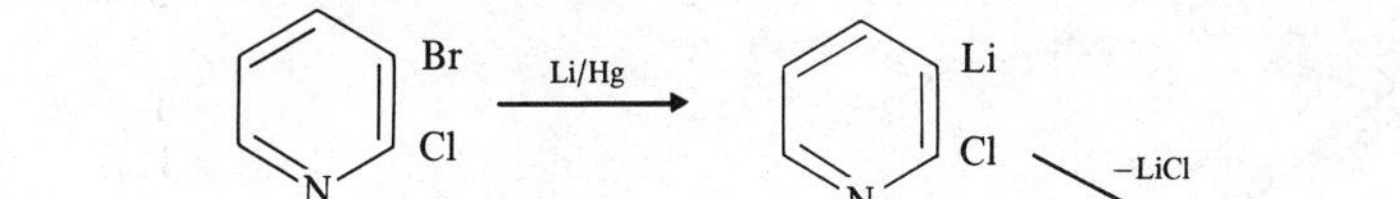

(83)

(84)

SCHEME 7

* For reviews of this topic, see Refs. 137 and 294–297.

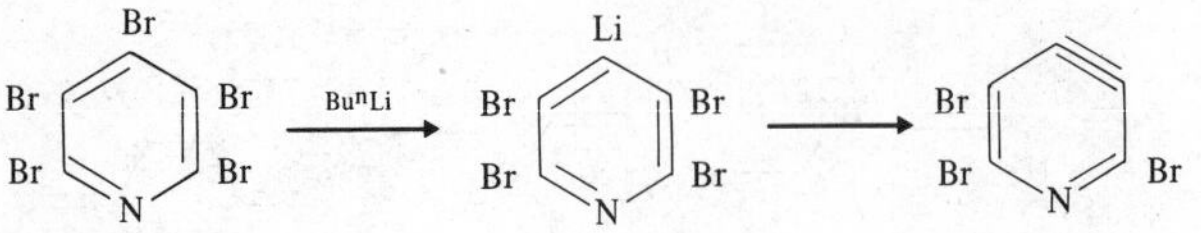

(85)

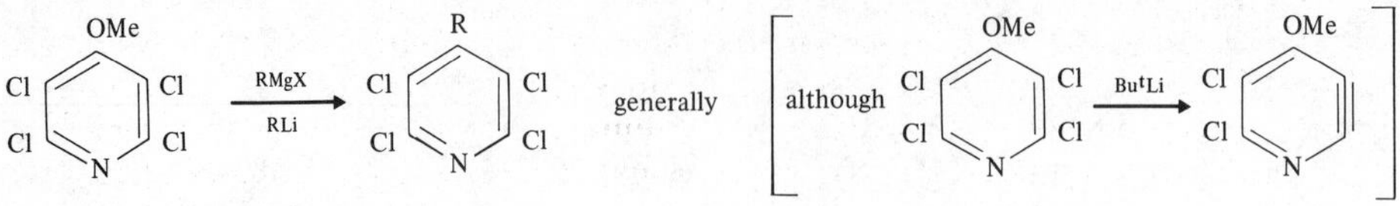

(86)

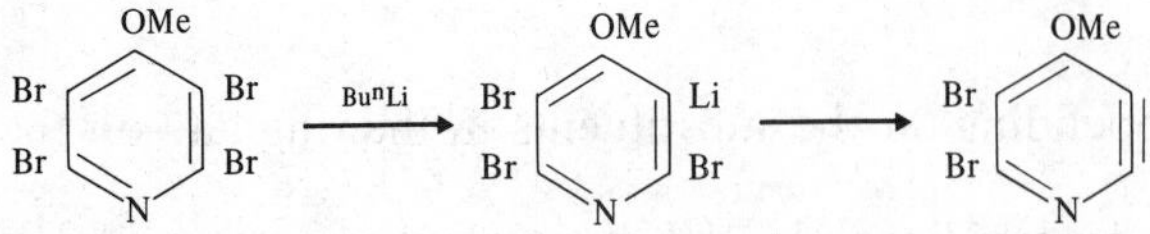

(87)

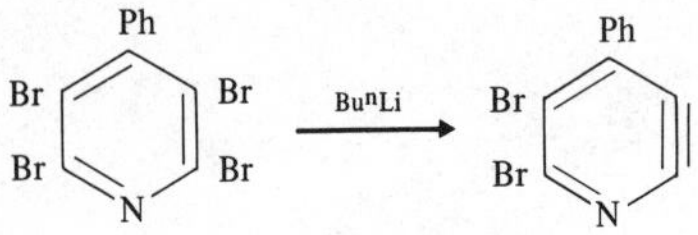

(88)

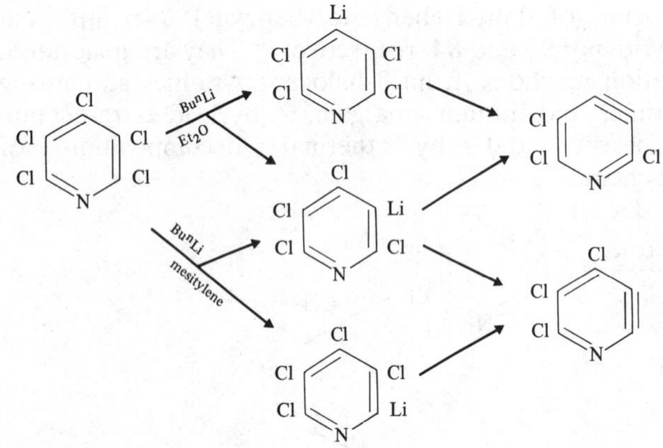

(89)

It is important to note, however, that where a reaction may lead either to a 2,3- or 3,4-pyridyne, only the latter is normally generated. Thus, for example, 3-bromopyridine and potassium amide in liquid ammonia give only 3- and 4-aminopyridine,[294] and 2,5,6-tribromo-3,4-pyridyne is the only dehydropyridine obtained in the reaction of pentabromopyridine with butyl-lithium or with magnesium.[299] Indeed, 2,3-pyridynes have been obtained from 2,3-dihalogenopyridines and organolithium compounds only when the 4-position carries a substituent not readily displaceable as an anion[300a–c,301] [*cf.* equations (85)–(88)]. The only apparent exception to the general rule is provided by pentachloropyridine, which may react with n-butyl-lithium to generate both the 2,3- and the 3,4-pyridyne[300d] [equation (89)].

In terms of simple theory, it seems surprising that 3,4-pyridyne formation is almost invariably preferred to 2,3-pyridyne formation. One might have expected a 2-pyridyl anion, which is the precursor of a 2,3-pyridyne, to be stabilized relative to its 4-pyridyl isomer by the inductive electron-withdrawal of the adjacent heteroatom. Equally, one might have expected that interaction (however feeble) of the lone pair of electrons on nitrogen with the sp^2 orbitals of the aryne would stabilize 2,3-pyridyne relative to its 3,4-isomer (an expectation which is supported by simple Hückel calculations[302]). However, the experimental evidence contradicts this simple theory, not only in respect of pyridyne stability; for it has been shown[303] by deuterium exchange experiments that protons are removed from the 4-position of 3-halogenopyridines much faster than from the 2-position.

Calculations based on Extended Hückel Theory show much better agreement with the experimental results.[304] These show that the order of stability of pyridyl anions is 3>4>2, and that the hetero-atom, far from stabilizing the adjacent anion or aryne, actually *destabilizes* both. This destabilizing effect of the hetero-atom is apparently associated with its lone pair of electrons, because in pyridinium salts and *N*-oxides (where the lone pair is removed but the inductive effect remains), hydrogen–deuterium exchange is fastest at the 2-position,[305] and reactions of 3-halogenopyridine *N*-oxides with strong bases appear to proceed *via* the 2,3-pyridyne *N*-oxide rather than the 3,4-isomer (see below).

16.1.5.2 Trapping of pyridynes

Reagents such as furan and tetraphenylcyclopentadienone ('tetracyclone'), which are familiar trapping agents for benzyne, may also be used as pyridyne traps. The adducts of 3,4-pyridyne, *viz.* (**85**) and (**86**), are obtainable in good yield,[298,306] but the corresponding adducts of 2,3-pyridyne, *viz.* (**87**) and (**88**), are more difficult to isolate;[298,300a,307a] adduct (**87**) is unstable, and is best characterized as the derivative (**89**), obtained by a further Diels–Alder reaction with 2,3-dimethylbuta-1,3-diene.[300a] Furan adducts of substituted 2,3-pyridynes, however, are apparently easier to obtain.[300a–c,301,307] Other known adducts of 3,4-pyridyne include (**90**) and (**91**), in which the arynophiles are 1,3-diphenylnitrilimine and *N*-phenylsydnone, respectively;[306] the remarkable adduct [(**92a**) or (**92b**)] formed by 1,2-addition to cyclopentadiene;[308] and the 3,4-pyridyne dimer [(**93a**) or (**93b**)].[309]

Trihalogeno-3,4-pyridynes do not evidently form stable adducts with furan, but trichloro-3,4-pyridyne forms adduct (**94**) with 1,3-diphenylisobenzofuran,[300d] and both trichloro- and tribromo-3,4-pyridynes form adducts, *e.g.* (**95**)–(**98**), with benzenoid hydrocarbons.[299,300d] An *N*-methylpyrrole adduct, (**99**), of a dichloro-2,3-pyridyne has also been isolated.[300a]

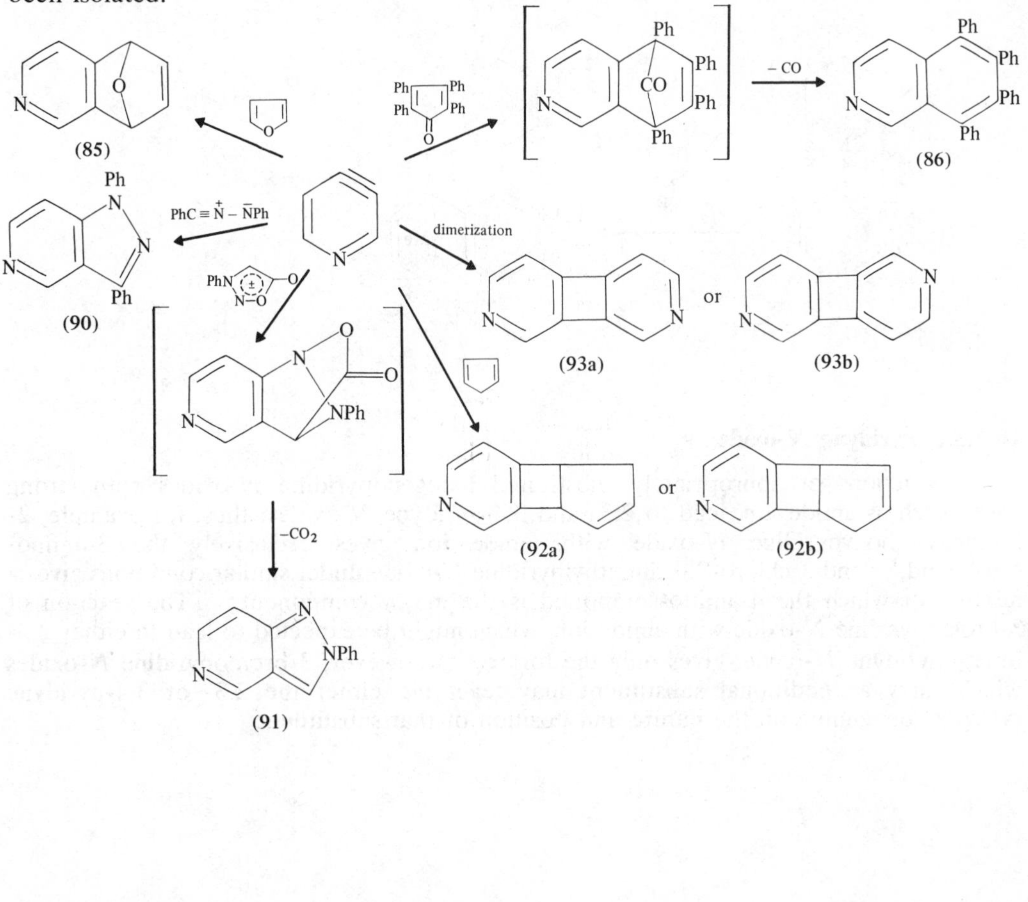

(85) (86) (90) (91) (92a) or (92b) (93a) or (93b)

− CO; dimerization; $PhC\equiv\overset{+}{N}-\overset{-}{N}Ph$; $-CO_2$

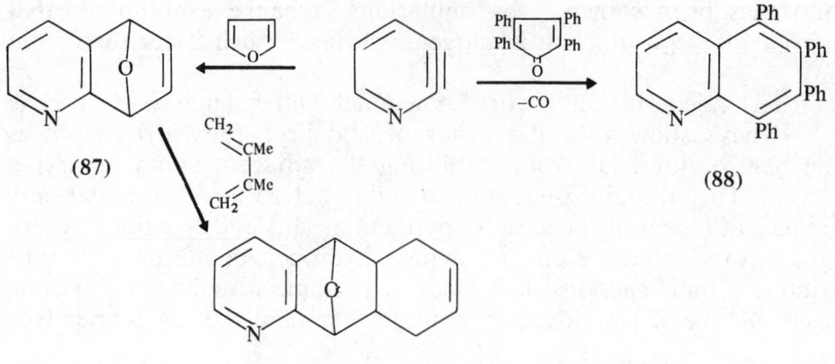

16.1.5.3 Pyridyne *N*-oxides

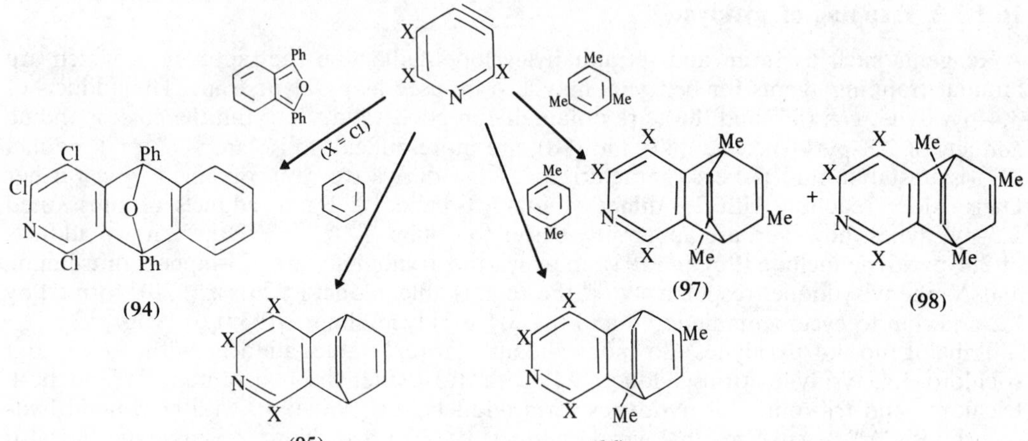

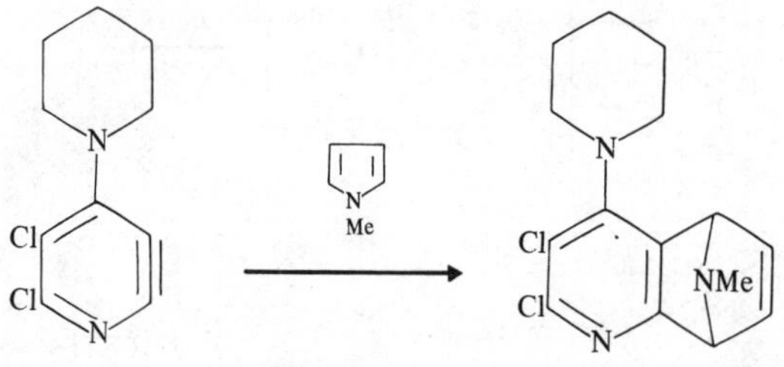

The reactions of appropriately substituted halogenopyridine *N*-oxides with strong bases, such as amide ion, lead to 2,3- and 3,4-pyridyne *N*-oxides: thus, for example, 2-bromo-4-ethoxypyridine *N*-oxide with amide ion gives exclusively the 3-amino-compound,[178] and 3-chloro-2,6-dimethylpyridine *N*-oxide under similar conditions gives a mixture of which the 4-amino-compound is the major component.[310] The reaction of 3-bromopyridine *N*-oxide with amide ion, which might be expected to lead to either 2,3- or 3,4-pyridyne *N*-oxide, gives only the former;[311] however, 3-bromopyridine *N*-oxides which carry an additional substituent may react *via* either the 2,3- or 3,4-pyridyne *N*-oxide, depending on the nature and position of that substituent.

16.1.5.4 Nucleophilic addition to pyridynes and pyridyne *N*-oxides

The addition of nucleophiles to unsymmetrical arynes gives a mixture of two products. In the case of pyridynes and their *N*-oxides, the 'dehydro' bond is heavily polarized, and so the two products are rarely obtained in anything approaching equal proportions. Addition of a nucleophile to a pyridyne produces initially a pyridyl anion, and the orientation of addition is generally such that the more stable anion is produced. It will be recalled that the order of stability of pyridyl anions is 3>4>2, and for the *N*-oxidopyridyl anions the order is 2>3>4.[304,305] This implies that 3,4-pyridynes and their *N*-oxides should undergo nucleophilic addition preferentially at the 4-position (giving 3-pyridyl anions), that 2,3-pyridynes should undergo addition at the 2-position, and that

Addition of $\bar{N}H_2$ (in liquid NH_3) to pyridynes and pyridyne *N*-oxides

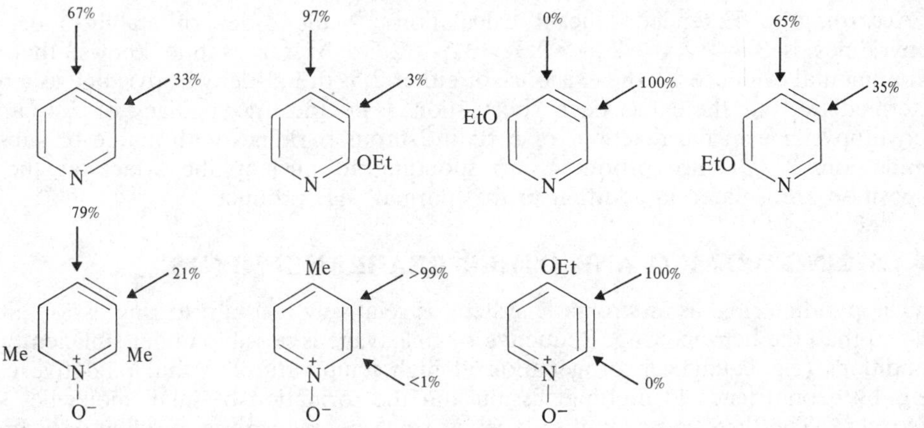

2,3-pyridyne *N*-oxides should undergo addition at the 3-position. In the absence of polar substituents *ortho* to the dehydro-bond, this orientation of addition is indeed observed. The presence of an *ortho*-alkoxy group, however, helps to stabilize an anion on the adjacent centre (presumably by inductive electron-attraction) and directs the incoming nucleophile to the *meta*-position.

TABLE 5
Competitive *AE* and *EA* Processes in Substitution of Halogenopyridines

Halogen	*Nucleophile*	% *AE*	% *EA*
3-F	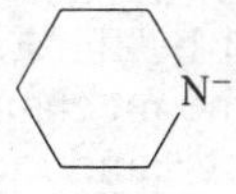 N^- (piperidide ring)	92	8
	$Pr^i_2N^-$	0	100
4-Cl	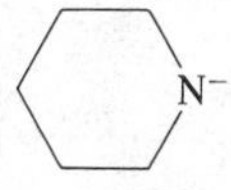 N^- (piperidide ring)	97	3
	Et_2N^-	19	81
	$Pr^i_2N^-$	0	100
4-F	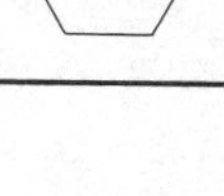 N^- (piperidide ring)	100	0
4-Cl		97	3
4-Br		10	90
4-I		4	96

It is, of course, important to remember that reactions of halogenopyridines and their *N*-oxides with strong bases may not involve dehydropyridine intermediates at all, but may proceed by an *AE* mechanism (*cf.* Section 16.1.3.2, p. 21). In many cases the *AE* and *EA* reactions occur concurrently, and so the isomer ratio of products obtained is not a true reflection of the pyridyne's susceptibility to attack at one end or the other. Data on competitive *AE* and *EA* processes have been collected by Kauffmann and co-workers,[137,177] and some of these are given above (Table 5).

The addition of non-nitrogenous nucleophiles to pyridynes has been recorded in a few instances: all involve 3,4-pyridyne and lead to the expected mixture of 3- and 4-substituted pyridines. The nucleophiles concerned are hydroxide[202a] and methanethiolate ions,[312] and the carbanions derived from pentan-3-one[313] and acetophenone.[314]

16.1.5.5 Other dehydropyridines

According to Extended Hückel calculations,[304] the order of stability for dehydropyridines is 3,4->2,4->2,5->2,3->3,5->2,6-. As far as one knows, there is no experimental evidence for the existence of either 2,5- or 3,5-dehydropyridine as a reactive intermediate, but there has been speculation as to the intermediacy of 2,4- and 2,6-dehydropyridine in the reactions of certain 2-bromopyridines with amide or substituted amide ions,[313,315] since products corresponding to nucleophilic attack at the 4- or 6-position are isolated in addition to the 'normal' *AE* product.

16.1.6 RING-OPENING AND OTHER REARRANGEMENTS

The pyridine ring, as an aromatic system, is relatively resistant to ring fission, although less so than the benzene ring. Reductive ring cleavage is possible under sufficiently severe conditions (*e.g.* catalytic hydrogenation at high temperature),[316] but oxidative cleavage (*e.g.* by ozone) leads to multiple fission and the formation of small molecules such as glyoxal.[317] The three most familiar types of ring-opening among pyridine derivatives are associated with the reactions of pyridinium compounds with nucleophiles (*cf.* Section 16.1.3.2); with the reactions of 2-halogenopyridines with amide or substituted amide ions; and with the generation of pyridylcarbenes and pyridylnitrenes.

The ring-opening of pyridinium compounds by nucleophiles is illustrated in general terms by equation (90). In principle, any nucleophile which has a hydrogen attached to the nucleophilic atom may effect this type of ring-opening, and hydroxide ion,[318] primary amines,[318a,319] and carbanions[320] have all been used for this purpose [*cf.* equations (90)–(93)]. The products of these ring-opening reactions may subsequently undergo recyclization in a different direction, regenerating a pyridine or other heterocyclic ring [*cf.* equations (94)–(96)].

The mechanism of equation (90) clearly cannot apply, however, in the ring-opening of pyridinium salts with secondary amines,[321] and in such reactions the ring-opening of the dihydropyridine intermediate may be initiated by the lone pair on the exocyclic nitrogen [equation (97)]. Even this may not complete the mechanistic picture, for the ring-opening of pyridine *N*-oxides with Grignard reagents and acetylide ions (*cf.* p. 29) cannot proceed by either of the above routes, and such ring-openings are presumably electrocyclic processes [equation (98)].

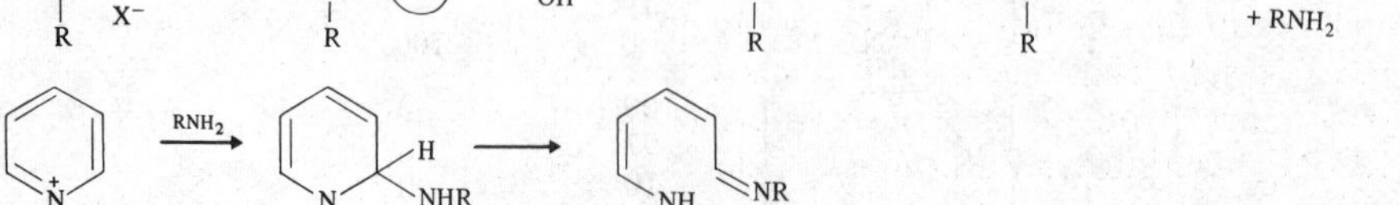

$\bar{O}H$; H_2O; X^-; R; H; O—H; $\bar{O}H$; $\bar{N}$; O; NH; CHO; CHO CHO + RNH_2 (90)

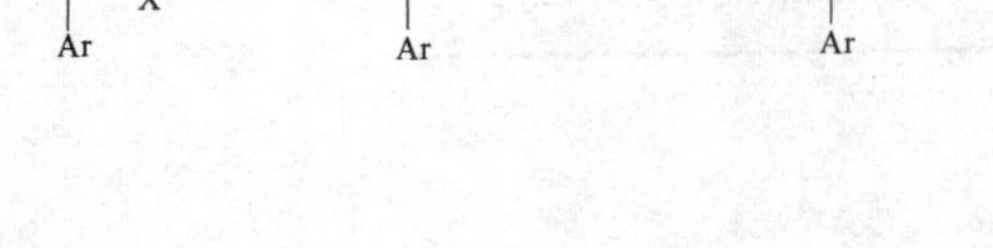

RNH_2; X^-; Ar; H; NHR; NH; NR (91)

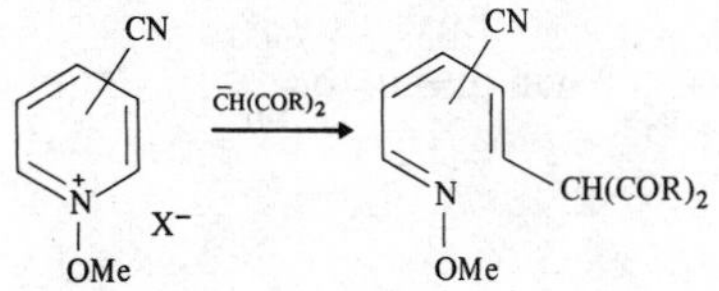

(92)

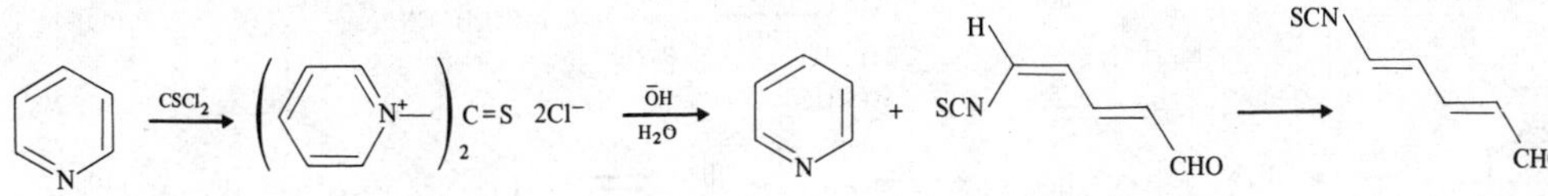

(93)[322]

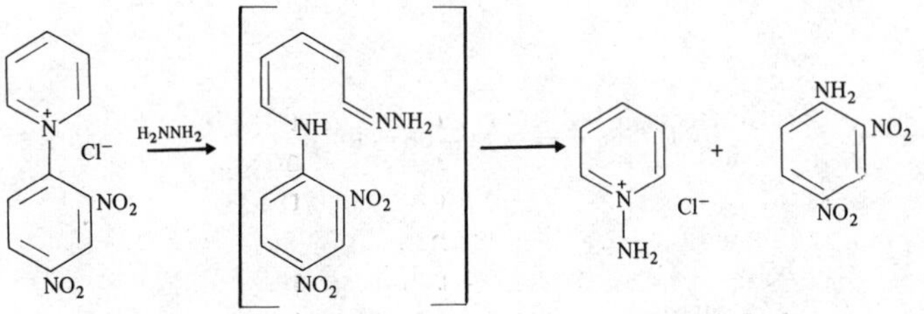

(94)[323]

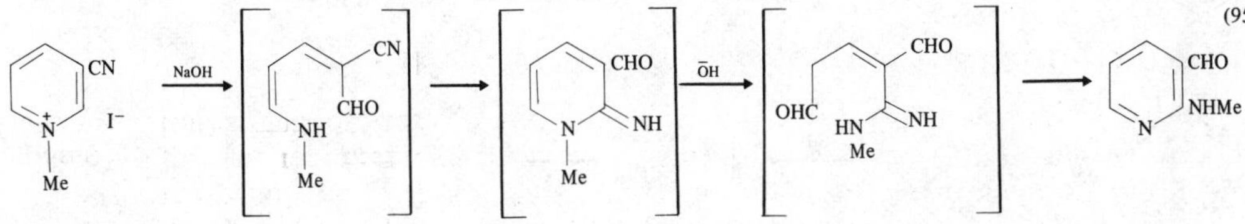

(95)[324]

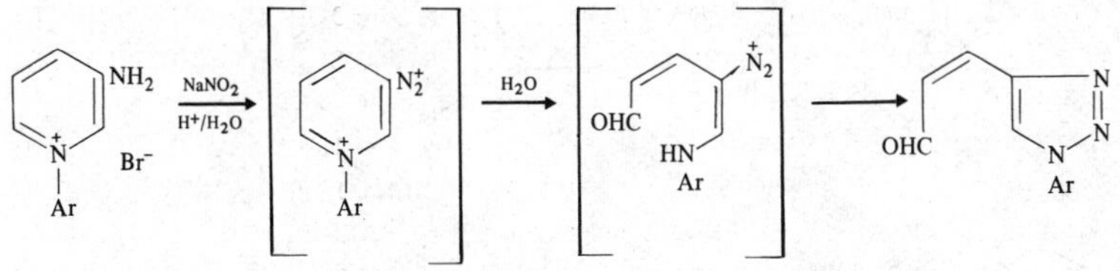

(96)[325]

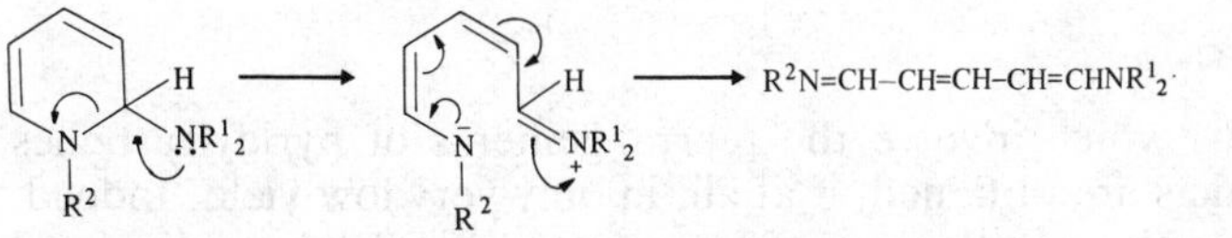

(97)

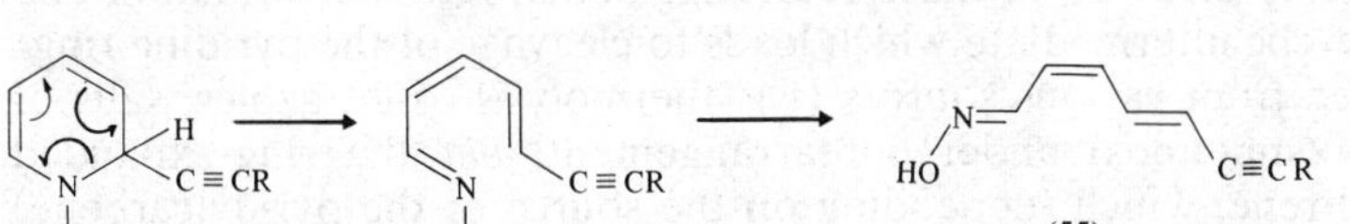

(55)

(98)

The second group of reactions, those involving 2-halogenopyridines and amide ions, lead in some cases to ring-opened products [reactions (99)–(101)], but in others the ring-opened primary product undergoes recyclization to another type of ring system. These ANRORC reactions (Addition of Nucleophile, Ring Opening, Ring Closure) are encountered in several heterocyclic systems other than pyridine; they are exemplified, for the pyridine series, by reactions (102)–(104).

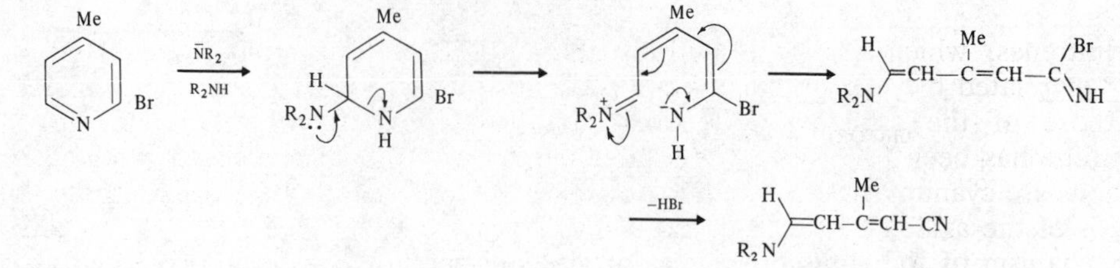

(99)[315]

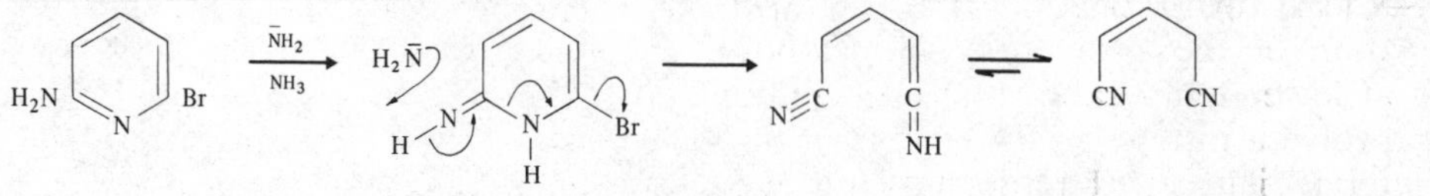

(100)[326]

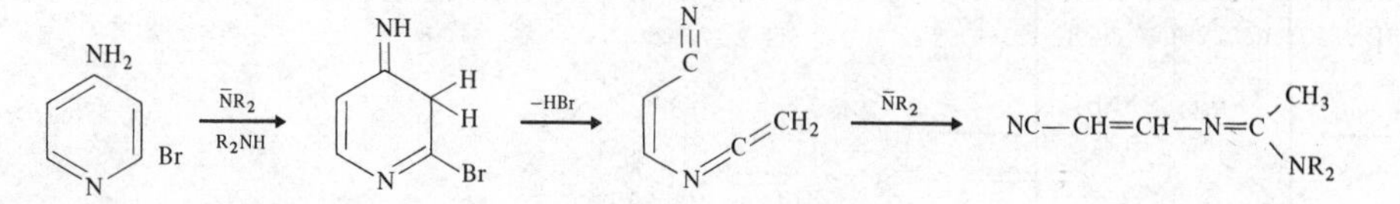

(101)[315]

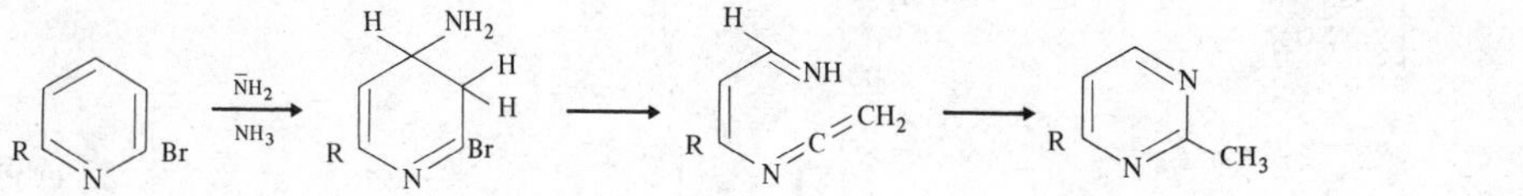

(102)[326]

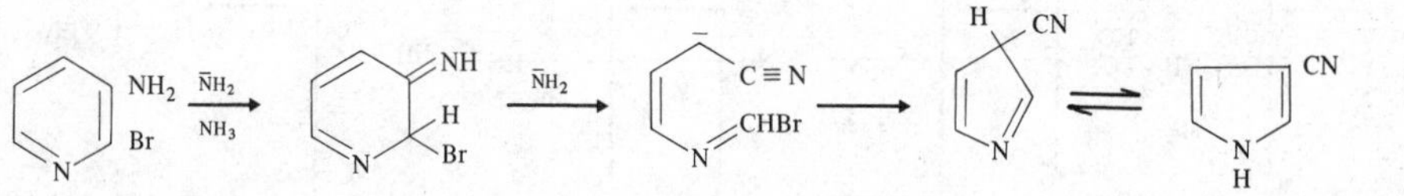

(103)[327]

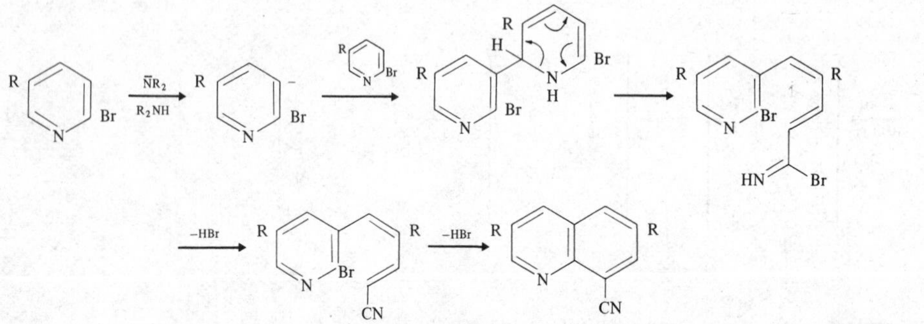

(104)[328]

In the third group of reactions, which involve the rearrangements of pyridylcarbenes and pyridylnitrenes, acyclic products are obtained, if at all, in only very low yield. Indeed, acyclic intermediates are rarely involved in these rearrangements; it is the formation and subsequent collapse of a bicyclic intermediate which leads to cleavage of the pyridine ring.

Pyridylcarbenes, generated from various sources (*e.g.* thermolysis of tetrazoles, salts of tosylhydrazones, or diazo-compounds) undergo rearrangement, *via* the ring-expanded carbenes (**100**), to phenylnitrene, which (depending on the source of the pyridylcarbene) either undergoes ring-contraction to cyanocyclopentadiene (**101**) or else dimerizes to azobenzene.[329,330b] 2-, 3-, and 4-pyridylcarbene all rearrange to phenylnitrene, by a series of ring expansion–ring contraction sequences which lead, effectively, to a 'migration' of the carbene function round the pyridine ring (*cf.* Scheme 8a). (This 'migration' has been demonstrated by examining the rearrangements of pyridylcarbenes which contain an additional methyl group on the ring, for this group may serve as a trap for an adjacent carbene [equations (105) and (106)] and prevent rearrangement to the nitrene[330a]). The cyclizations of phenylpyridylcarbenes (**102**)–(**104**) also illustrate this type of rearrangement.[331]

Pyridylnitrenes, which are generated by thermolysis of triazolopyridines, (**105**) or (**106**),[332b] or related bicyclic systems like (**107**)[332a] or (**108**),[333] undergo reactions analogous to those of the carbenes. The ring-expansion/ring-contraction sequence in 2-pyridylnitrene has been followed by ^{15}N labelling,[332a] and the principal end-products of the reactions are cyanopyrroles. Attempts to generate the *N*-oxide of 2-pyridylnitrene (by thermolysis of the azide) lead to 2-cyano-*N*-hydroxypyrrole.[334]

The mechanism of the ring contraction of aryl nitrenes to these cyano-compounds is complex: at least two distinct pathways have already been recognized[335] (*cf.* Scheme 8b). The formation of the *N*-hydroxypyrrole, however, which differs from the others by occurring at lower temperatures and in solution rather than in the gas phase, probably does not involve a nitrene at all, but rather an acyclic cyano-nitroso-intermediate (**109**); such an intermediate may be intercepted in the corresponding thermolyses of 2-azido-3-halogenopyridine *N*-oxides, and give rise to products such as (**110**) or (**111**).[336]

(100)

(101)

PhN = NPh

PhN: ⟶ products

(label equally distributed)

SCHEME 8a

(105)

(106)

(102) (99%)

(103) (56 : 44)

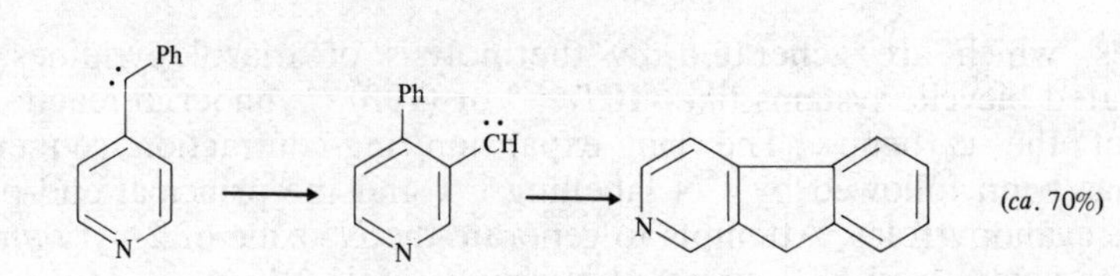

(104) (*ca.* 70%)

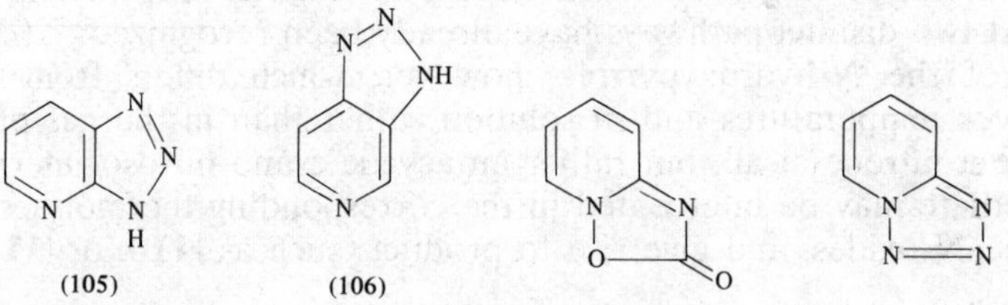

(105) (106) (107) (108)

(label equally distributed between cyano-groups)

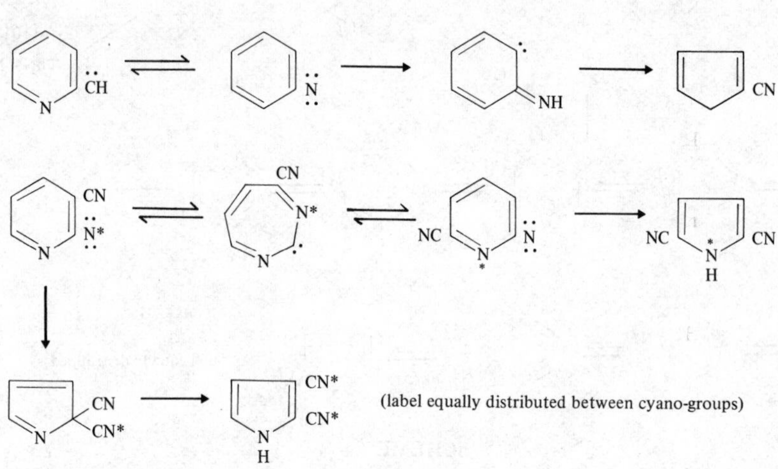

SCHEME 8b

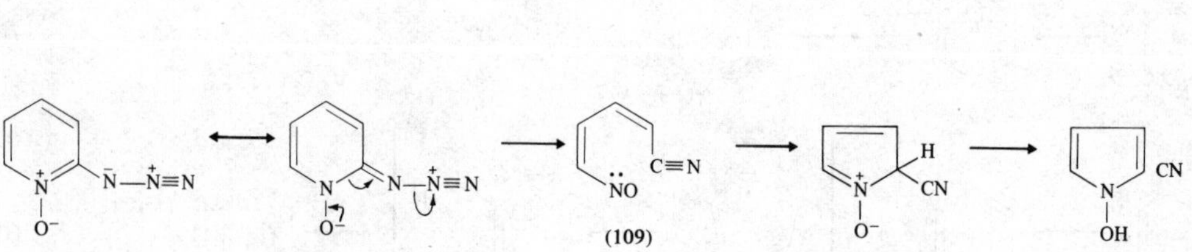

(109)

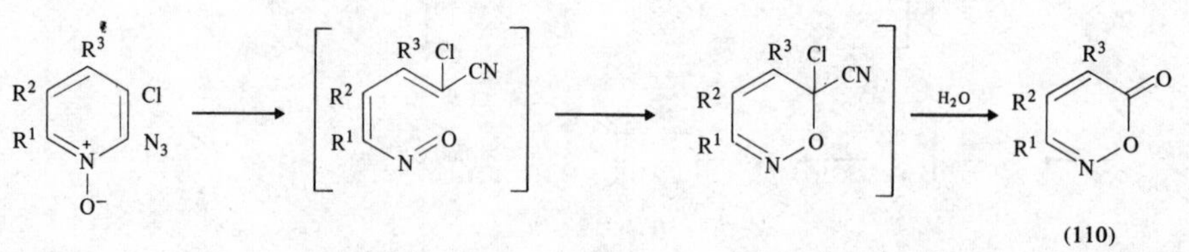

(110)

but

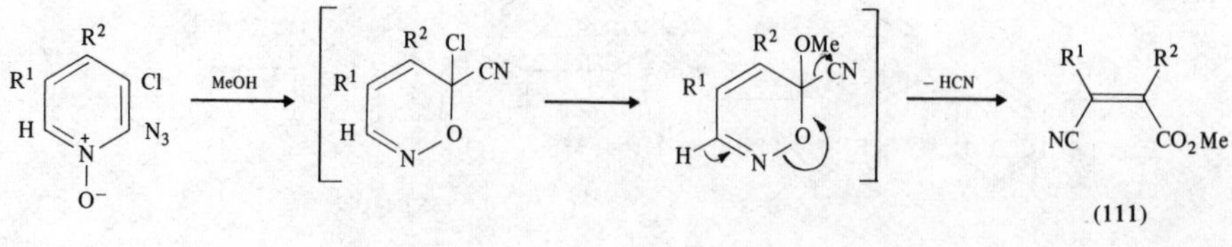

(111)

16.1.7 HYDROXY-, AMINO-, AND MERCAPTO-COMPOUNDS

16.1.7.1 Tautomerism

2- and 4-hydroxypyridines are tautomeric with 2- and 4-pyridones, and spectroscopic evidence leaves no doubt as to the predominance of the pyridone form in polar solvents;[337] the 2- and 4-mercaptopyridine ⇌ 2- and 4-pyridinethione equilibrium also lies in favour of the thione tautomers under these conditions.[337] However, under other conditions, *e.g.* in the gas phase or in non-polar solvents,[338] the position of the equilibrium may be shifted in favour of the hydroxy- or mercapto-forms, and ring substituents may also have a profound effect on the equilibrium position.[339] The *N*-oxides of 2- and 4-hydroxy- and -mercapto-pyridines are also tautomeric, and most of these exist preferentially as *N*-hydroxy-2- and -4-pyridones and -pyridinethiones.[337] 2- and 4-aminopyridines, however, exist as such rather than as pyridineimines, and the same is true of their *N*-oxides.[337]

3-Aminopyridine is an ordinary arylamine, but 3-hydroxypyridine exists to a considerable extent in solution as the pyridinium oxide zwitterion, and 3-mercaptopyridine exists mainly in the zwitterionic form.[337]

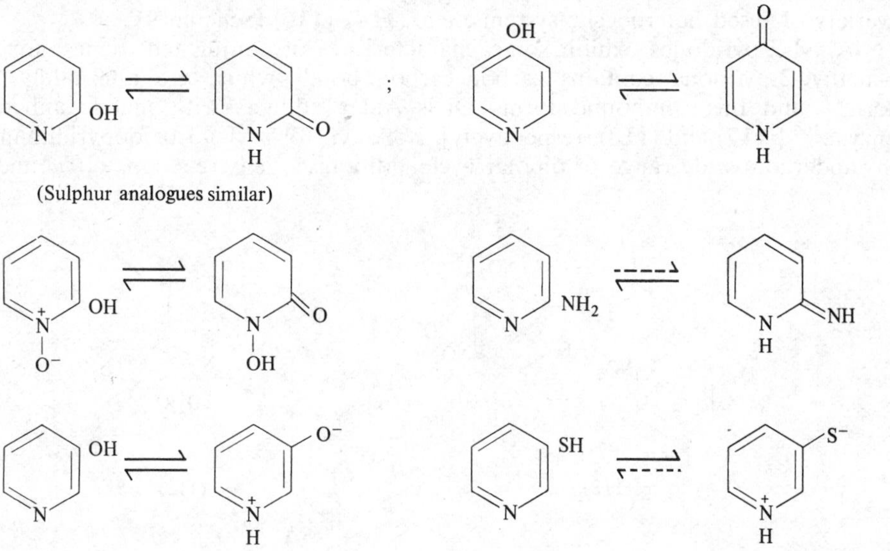

(Sulphur analogues similar)

16.1.7.2 Protonation

Addition of protons to hydroxypyridines or pyridones gives hydroxypyridinium ions,[340] and deprotonation gives anions where the negative charge is located mainly on oxygen.[340a] Thiopyridones also give mercaptopyridinium ions[341a,b] and aminopyridines give aminopyridinium ions.[341a,c]

16.1.7.3 Alkylation and acylation

For hydroxypyridines and pyridones this may occur either at nitrogen or at oxygen, depending on the reagents and conditions used. Alkylation of 2-pyridones generally gives both possible products, although with careful choice of conditions one or other may predominate.[342] Only one of the possible products is normally obtained by alkylation of 3-hydroxypyridines, and methods are available for the preparation of both *N*-[343] and *O*-alkyl[343b,344] derivatives. 4-Pyridones resemble the 2-isomers in giving mixtures of products, but with a marked preference for *N*-alkylation.[199b,345] *N*-Hydroxypyridones alkylate preferentially on the hydroxy-group.[346] Acylation of hydroxypyridines and pyridones is equally complicated: *O*-benzoyl derivatives of all three hydroxypyridines are formed by direct benzoylation,[347] and acetylation gives either *N*- or *O*-acetyl derivatives in the case of 2-pyridone,[348] an *O*-acetyl derivative of 3-hydroxypyridine,[349] an *N*-acetyl derivative of 4-pyridone,[347] and an *N*-acetoxy-compound from *N*-hydroxy-2-pyridone.[350] Both *N*- and *O*-arylsulphonyl derivatives of 2-pyridones are known.[351]

Mercaptopyridines and pyridinethiones are alkylated[352] and acylated[353] almost invariably on sulphur, the only exception (at least, among simple derivatives) being the acetylation of 4-pyridinethione.[354]

Aminopyridines may undergo alkylation on either nitrogen, according to the reaction conditions. The ring nitrogen is the more nucleophilic of the two (the lone pair on the other nitrogen forming part of the conjugated system) and so alkylation in the absence of added base normally occurs at this position.[355] However, 2-amino- and 2-alkylamino-pyridines are alkylated at the exocyclic nitrogen in presence of a strong base.[355b,356] 2-Dimethylaminopyridine is anomalous in undergoing further methylation on the exocyclic nitrogen.[357] Acylation of aminopyridines occurs, apparently exclusively, on the exocyclic nitrogen,[358] and further acylation generally occurs at the same position giving (diacylamino)pyridines,[359] the only notable exceptions being dipicryl derivatives (**112**)[360] and disulphonamido-compounds such as (**113**).[361]

Bifunctional alkylating and/or acylating agents react with 2-aminopyridines to produce a variety of fused heterocyclic systems, *e.g.* (**114**)–(**116**) (Scheme 9).

N-Alkyl-2-pyridones exhibit some characteristics of conjugated dienes: for example, *N*-methyl-2-pyridone contains carbon–carbon bonds which are alternately long and short,[365] and this compound forms Diels–Alder adducts with maleic anhydride and benzyne[366] [(**117**) and (**118**), respectively]. *N*-Alkyl- (or aryl-) 3-oxidopyridinium zwitterions undergo a wide range of dipolar cycloadditions,[367] *e.g.* reactions (107) and (108).

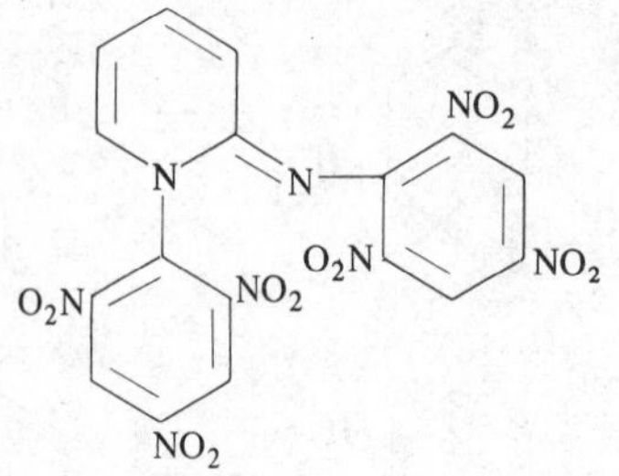

NO_2 N N O_2N NO_2 O_2N NO_2 NO_2

(112)

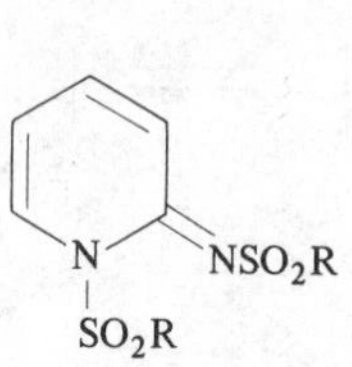

N NSO_2R SO_2R

(113)

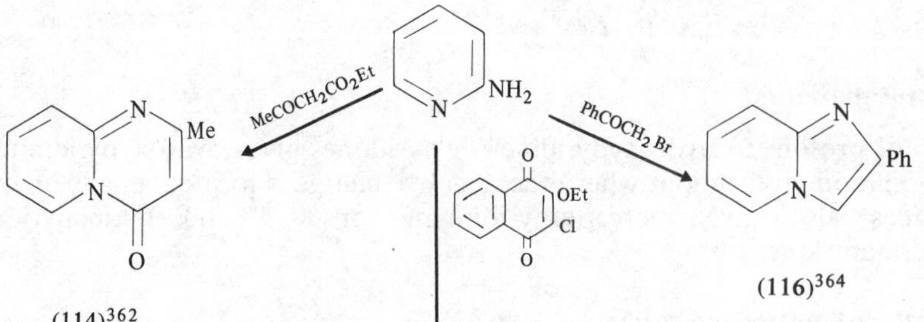

N NH_2 $MeCOCH_2CO_2Et$ $PhCOCH_2Br$ O OEt Cl O

N Me N O

(114)[362]

N N Ph

(116)[364]

O N N O

(115)[363]

SCHEME 9

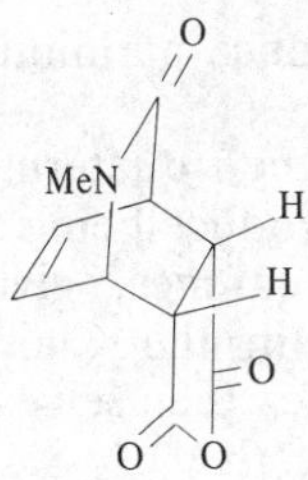

(117) 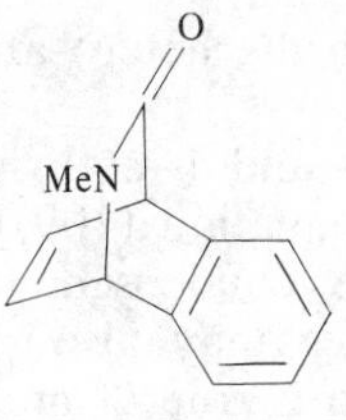 (118) 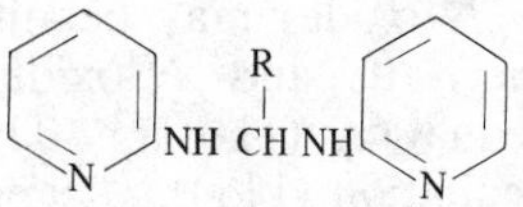 (119)

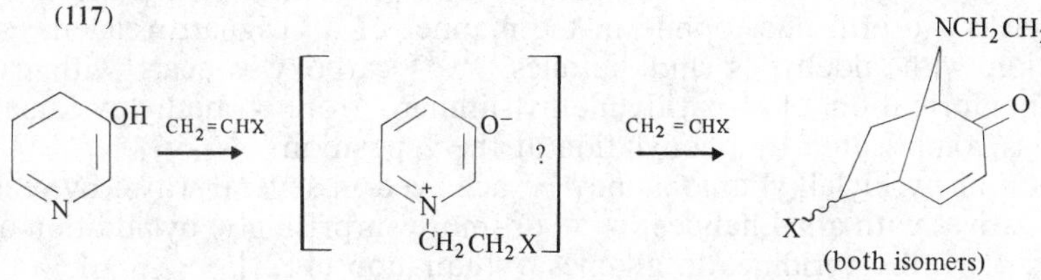

(107)

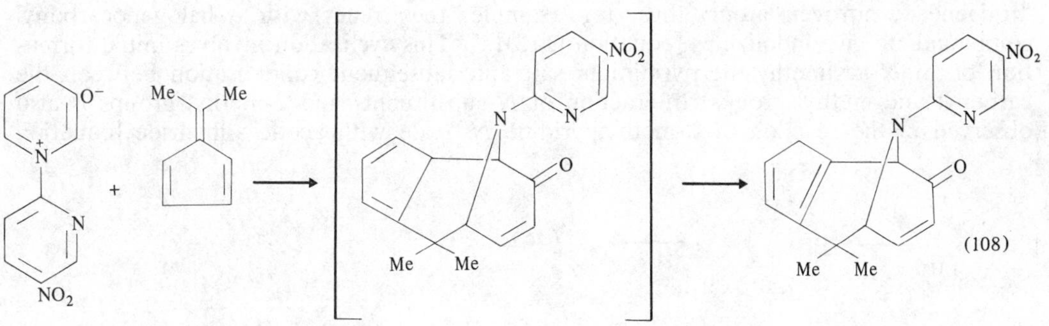

(108)

16.1.7.4 Condensation reactions

Aminopyridines condense normally with nitrosobenzene,[368] but only 3-aminopyridines form stable Schiff bases with aldehydes: 2-aminopyridine and aldehydes give bis-pyridylamino-compounds such as (**119**).[369]

16.1.7.5 Oxidation

For aminopyridines this is discussed further in Section 16.1.11. Oxidation of the sulphur-containing functions is generally similar to that of the corresponding benzene derivatives.[353]

16.1.8 ALKYLPYRIDINES

16.1.8.1 Reactivity of 2- and 4-methylpyridines and related compounds

Methyl and methylene groups attached to the 2- and 4-positions are potential carbanion sources, since deprotonation yields anions which may be stabilized by delocalization as shown below. The stabilization is even more effective when the hetero-atom carries an initial positive charge; thus 2- and 4-methylpyridines, their quaternary salts, and their *N*-oxides all condense with carbonyl compounds. Mild basic catalysts are effective for reactions involving the quaternary salts and *N*-oxides,[370] but the most effective catalysts for condensations of the methylpyridines themselves are acetic anhydride and zinc chloride[371] (presumably because they react with the hetero-atom to produce pyridinium intermediates).

Like other reactive methylene compounds, 2- and 4-methylpyridines, their quaternary salts, and their *N*-oxides may be nitrosated [equation (109)],[372] the product being an oxime, and the salts and *N*-oxides may be condensed with nitrosoarenes, giving azomethines [equation (110)].[373] The *N*-oxides may also participate in the Claisen condensation [equation (111)],[374a] and the same is true of methylpyridines themselves if an additional activating substituent (*e.g.* NO_2) is present in the ring.[374b]

Deprotonation of 2-methylpyridine by an organolithium reagent gives 2-pyridylmethyl-lithium,[141,375] which is a useful nucleophile in the manner of a Grignard reagent, giving alcohols by reaction with aldehydes and ketones,[141,375] carboxylic acids with carbon dioxide,[376] *etc.* The formation of 4-pyridylmethyl-lithium, from 4-methylpyridine and phenyl-lithium,[375] is complicated by phenylation at the 2-position.[150,377]

Further alkylation of pyridylalkyl anions may be achieved in several ways: by reaction of the lithium derivatives with alkyl halides,[377,378] or (more surprisingly) by addition of the sodium derivatives of methylpyridines to alkenes[379] [equation (112)].

2-Alkylpyridines are useful starting materials for the synthesis of bicyclic systems with a 'bridgehead' nitrogen atom: thus, for example, they react with α-halogenocarbonyl compounds to give indolizines [equation (113)].[380] This cyclization involves initial formation of an *N*-acylmethylene-pyridinium salt and subsequent condensation between the carbonyl and methyl groups. Interaction of *N*-substituents and 2-methyl groups is also observed in the reaction of 2-methylpyridine *N*-oxide with acetic anhydride [equation

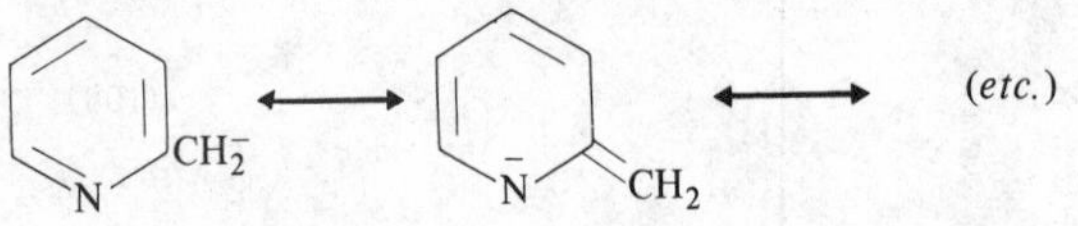

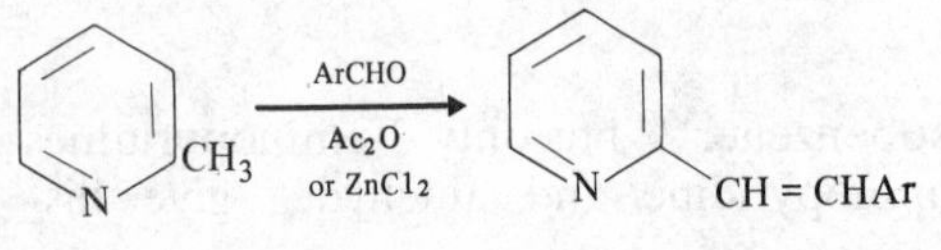

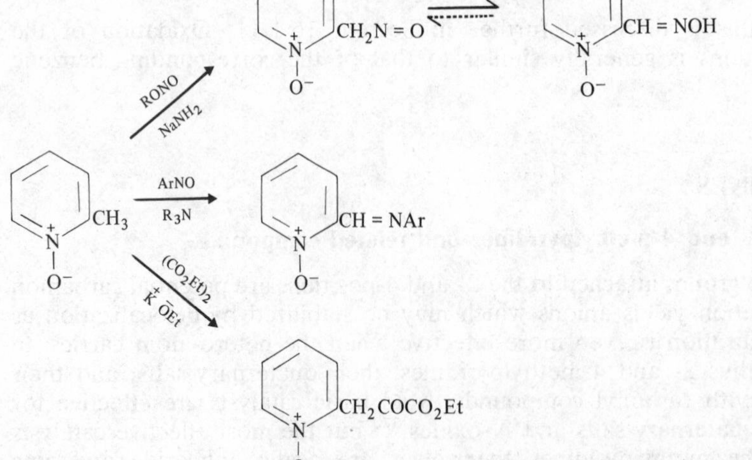

CH_2^- ⟷ $=CH_2$ ⟷ (*etc.*)

CH_3 —ArCHO, Ac_2O or $ZnCl_2$→ $CH=CHAr$

CH_3 (*N*-oxide) —RONO, $NaNH_2$→ $CH_2N=O$ ⇌ $CH=NOH$ (109)

CH_3 (*N*-oxide) —ArNO, R_3N→ $CH=NAr$ (110)

CH_3 (*N*-oxide) —$(CO_2Et)_2$, $\overset{+}{K}\,\overset{-}{O}Et$→ CH_2COCO_2Et (111)

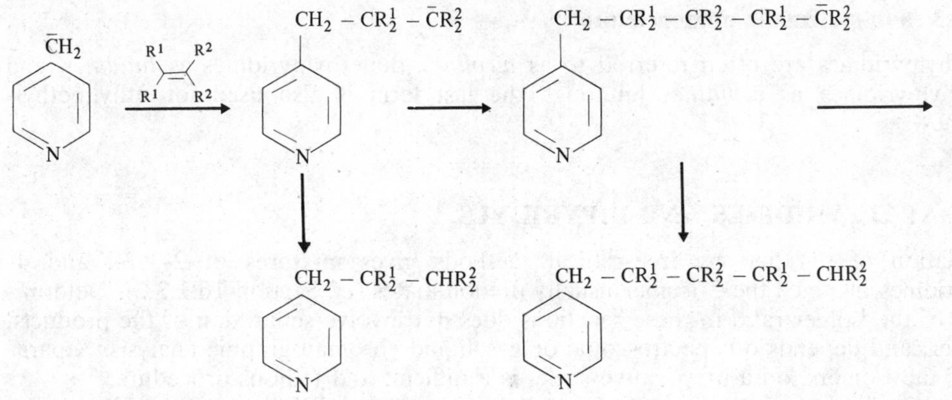

(112)

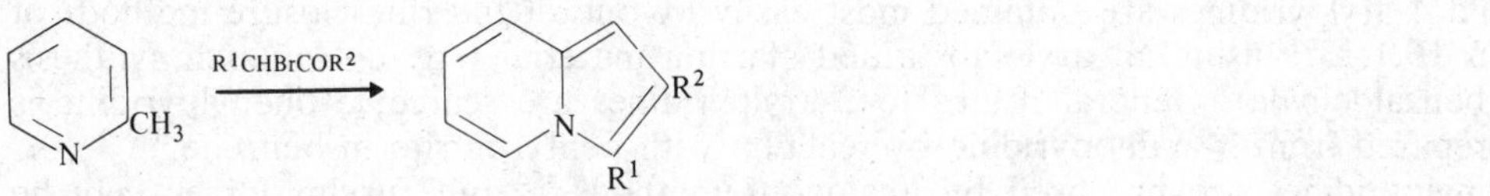

(113)

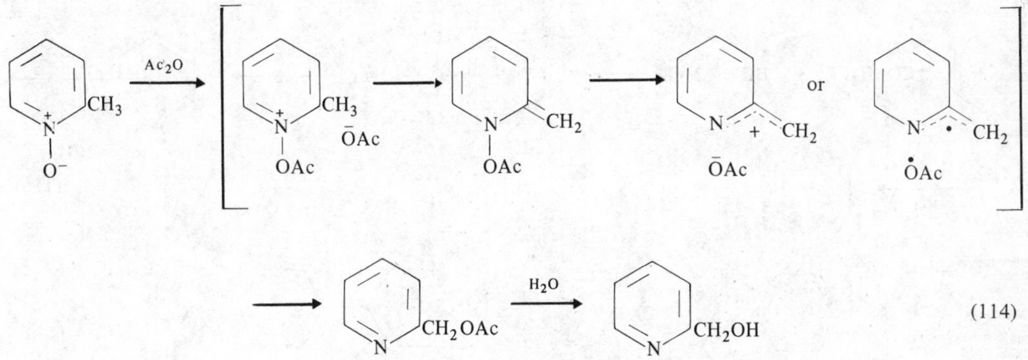

(114)

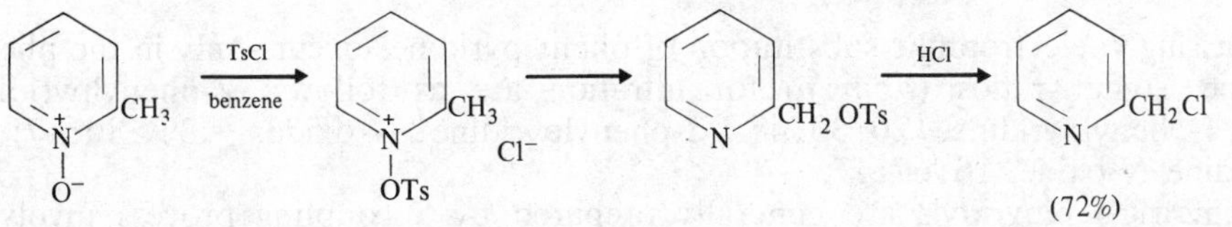

(115)

(114)][381] and toluene-*p*-sulphonyl chloride [equation (115)].[382] The acetic anhydride reaction has been extensively studied, and evidence for both ion-pair and radical mechanisms has been found. 4-Methylpyridine *N*-oxide reacts similarly with acetic anhydride, and both inter- and intra-molecular migrations of the acetoxy-group may occur in this case.[381]

The oxidation of alkylpyridines is considered in Section 16.1.12.

16.1.8.2 Alkenylpyridines

These are prepared by normal olefin syntheses. The vinylpyridines, for example, are normally obtained by dehydration of pyridylethanols.

16.1.8.3 Non-systematic nomenclature

Methylpyridines are often referred to as *picolines*, dimethylpyridines as *lutidines*, and trimethylpyridines as *collidines* (although the last term is also used for ethylmethylpyridines).

16.1.9 ARYLPYRIDINES AND BIPYRIDYLS

Arylation of pyridine by free-radical methods gives mixtures of 2-, 3-, and 4-arylpyridines in which the 2-isomer usually predominates (*cf.* Section 16.1.3.3). Determination of the isomer ratio in these reactions does not involve separation of the products, however, and depends on spectroscopic or gas–liquid chromatographic analysis. Separation of the isomers, on a preparative scale, is a difficult and tedious procedure.

2-Arylpyridines are best prepared from the appropriate aryl-lithium and pyridine (*cf.* p. 24), and 4-arylpyridines are obtained most easily by one of the ring-closure methods of Section 16.1.13.1 using an aryl-substituted starting material (*e.g.* a Hantzsch synthesis using benzaldehyde). General routes to 3-arylpyridines are scarce; 3-phenylpyridine is best prepared from 3-aminopyridine by reaction with pentyl nitrite in benzene.[383]

Thienylpyridines are prepared by analogous methods,[384] but furylpyridines may be obtained by closure of the furan ring on the side-chain of an appropriately substituted pyridine[385] [equation (116)].

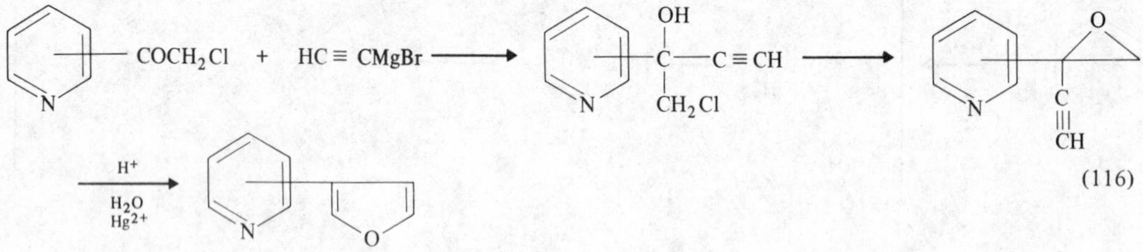

$$\text{Py}{-}COCH_2Cl + HC\equiv CMgBr \longrightarrow \text{Py}{-}C(OH)(CH_2Cl){-}C\equiv CH \longrightarrow \text{Py}{-}C(C\equiv CH)(\text{epoxide}) \xrightarrow[H_2O,\ Hg^{2+}]{H^+} \text{Py}{-}\text{furyl} \quad (116)$$

Me – N⁺ (4,4′-bipyridinium) N⁺ – Me 2X⁻

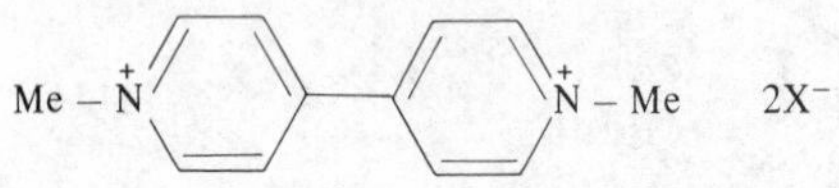

(120) X = halide or $MeSO_4^-$

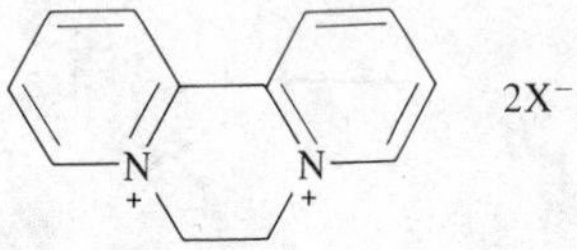

(121) X = halide 2X⁻

Not surprisingly, electrophilic substitution of phenylpyridines occurs only in the phenyl ring.[386] The isomer ratios ($o:m:p$) for nitration are as follows: 2-phenylpyridine, 9:39:52; 4-phenylpyridine, 20:33:47; 2-phenylpyridine *N*-oxide, <3:>80:17; 4-phenylpyridine *N*-oxide, 16:67:17.

The symmetrical bipyridyls are generally prepared by a coupling process involving simple pyridines. 2,2′-Bipyridyl, for example, is obtained from pyridine and Raney nickel,[387] and 3,3′- and 4,4′-bipyridyls are formed by the reaction of pyridine with sodium (*cf.* p. 44) and then with an oxidant.[290] Of the unsymmetrical isomers, 2,4′- and 3,4′-bipyridyls are prepared by Hantzsch syntheses (p. 65) using pyridine-2- and -3-carbaldehydes,[388] and the 2,3′-isomer is the main product of the Gomberg reaction of 3-pyridinediazonium chloride with pyridine.[389] Much of the current interest in bipyridyl chemistry is centred on the herbicides paraquat (**120**) and diquat (**121**) and recent patent literature contains many references to the preparation of these and related compounds.[15g]

16.1.10 HALOGENOPYRIDINES

The available methods for the preparation of halogenopyridines include some which are applicable also to the preparation of halogenobenzenes, *e.g.* electrophilic halogenation of pyridines themselves (p. 17), or replacement of the diazonium group by the Sandmeyer or related process (p. 33).[390] 3-Halogenopyridines may be obtained from pyrroles and dihalogenocarbenes (p. 72), and 3-bromopyridines by mercuration of pyridines followed by bromination (p. 20). 2- and 4-halogenopyridines result from a variety of nucleophilic substitution reactions, especially those involving 2- and 4-nitropyridines (p. 32) and 2- and 4-pyridones (p. 33).

The most important reactions of halogenopyridines are nucleophilic substitution, and elimination to give pyridynes; both of these reaction types have already been discussed (Section 16.1.3.2 and Section 16.1.5, respectively).

Halogenopyridines do not readily form Grignard reagents by direct reaction with magnesium. Pyridyl Grignard reagents[391] and pyridyl-lithium compounds (*cf.* p. 46) are both prepared by reaction of halogenopyridines with alkylmagnesium halides or alkyl-lithium compounds. Halogen–lithium exchange in di- and poly-halogenopyridines occurs most readily at the 3- or 5-position[392] [equation (117)].

(Halogenoalkyl)pyridines are obtainable by the expected routes: direct halogenation of alkylpyridines, the reaction of pyridyl alcohols with hydrogen halides, the addition of hydrogen halides to alkenylpyridines, *etc.* 2- and 4-chloromethylpyridines result from the reaction of 2- and 4-methylpyridine *N*-oxides with toluene-*p*-sulphonyl chloride [equation (115)], and triflurormethylpyridines are obtained from pyridinecarboxylic acids and sulphur tetrafluoride.[393]

Br, Br, N —BuLi→ Li, Br, N

Li, N + Br, N ⇌ Br, N + Li, N (117)

16.1.11 PYRIDINES CONTAINING OTHER NITROGEN SUBSTITUENTS

16.1.11.1 Nitropyridines

Direct nitration of pyridines is a useful preparative method only where the hetero-atom is sterically hindered, or the ring is activated by electron-donating substituents (pp. 16, 17). Nitration of pyridine *N*-oxides followed by reduction, however, is the simplest route to 4-nitropyridines. The nitropyridines are also prepared by oxidation of aminopyridines using a strong peroxy-acid[113,219b,221b,394] (*e.g.* Caro's acid or peroxytrifluoroacetic acid: the latter gives nitropyridine *N*-oxides). This is the usual route to 2-nitropyridines, and is often the method of choice for the 3-nitro-isomers as well.

The nucleophilic substitution reactions of 2- and 4-nitropyridine derivatives have already been described (p. 32). The reduction of nitropyridines is described below.

16.1.11.2 Nitrosopyridines

Direct nitrosation of pyridines containing two electron-donating substituents (p. 20) provides the commonest route to nitrosopyridines. Very little is known about the simpler representatives of the group: a few examples have been prepared by oxidation of the corresponding hydroxylaminopyridines,[91a,395] which are in turn obtained from nitropyridines by reduction.

16.1.11.3 Azopyridines

The symmetrical azopyridines are obtainable from nitropyridines by reduction with, for example, sodium stannite or arsenite,[396] or sodium hydroxide in water;[218] also from aminopyridines by oxidation with hypochlorite.[368,369a] The unsymmetrical azopyridines result from diazo-coupling reactions (pyridinediazonium salts and activated aromatic compounds, or arenediazonium salts and activated pyridines); also from the reaction of aminopyridines with nitrosoarenes.

16.1.11.4 Hydrazino- and azido-pyridines

Hydrazinopyridines are obtained by reduction of nitraminopyridines[99c] or pyridinediazonium salts,[397] or (in the case of the 2- and 4-isomers) by the reaction of halogenopyridines with hydrazine (p. 30). On reaction with nitrous acid the hydrazinopyridines are converted into azidopyridines, and 2- and 4-azidopyridines may also be made by nucleophilic substitution reactions involving azide ions.[398] 2-Azidopyridine itself is tautomeric with tetrazolo[1,5-*a*]pyridine (**108**), the latter predominating in the equilibrium mixture;[399] but 2-azidopyridine *N*-oxides exist as such.[334,336]

16.1.12 PYRIDINES CONTAINING OXIDIZED CARBON SUBSTITUENTS

Alkylpyridines (Section 16.1.8) represent the lowest oxidation level of a carbon substituent attached to the pyridine ring. In theory, therefore, it should be possible to oxidize the alkyl group successively to each of the three higher oxidation levels, *i.e.* to pyridylalkanols, to pyridinecarboxaldehydes and pyridyl ketones, and to pyridinecarboxylic acid derivatives. In practice, however, as with the oxidation of alkylbenzenes, it is often difficult to arrest the oxidation at an intermediate stage, and generally easier to allow complete oxidation (to carboxylic acid or ketone) to occur. The products may then be selectively reduced to the required oxidation level.

16.1.12.1 Pyridylalkanols

These arise by reduction of the corresponding aldehydes, ketones, or esters, or by the action of Grignard reagents or organolithium compounds on these. They also arise from the reactions of pyridyl-lithium[194] or pyridylalkyl-lithium compounds (p. 58) with carbonyl compounds; from the decarboxylation of pyridine-2-carboxylic acid in presence of aldehydes (p. 13); and from the Emmert reaction,[400] in which a ketone is added to pyridine in presence of a metal [usually magnesium or aluminium amalgam: equation (118)]. Finally, they may be obtained from the corresponding halogeno-compounds [equation (115)] or esters [equation (114)] by hydrolysis.

Among the most important pyridylmethanol derivatives are the B_6 or pyridoxine group of vitamins:[401] pyridoxol (**122**), pyridoxal (**123**), and pyridoxamine (**124**), and the related coenzyme codecarboxylase, which is pyridoxal 5-phosphate. The synthesis of these compounds has received a good deal of attention (*cf.* pp. 67, 72).

16.1.12.2 Pyridinecarboxaldehydes

Many of the methods used for the synthesis of benzaldehydes have also been applied to the synthesis of pyridinecarboxaldehydes, but some of these, in the author's experience, are unreliable. Of the oxidative syntheses, the most successful, on a laboratory scale, appear to be ozonolysis of alkenylpyridines,[402] the oxidation of 2- and 4-methylpyridines with a halogen and dimethyl sulphoxide[403a] or with selenium dioxide,[403b] the condensation of a 2-methylpyridine derivative with a nitroso-compound (p. 58), and the reaction of 2-acetoxymethylpyridine *N*-oxide with acetic anhydride[404] [*cf.* equation (114)]. Reduction of carboxylic acid derivatives using complex metal hydrides is capable of giving pyridinecarboxaldehydes in good yield, and esters,[405] acid chlorides,[406] and nitriles[407] all make satisfactory starting points for these reductions.

The reactions of the pyridinecarboxaldehydes generally resemble those of benzaldehydes.

16.1.12.3 Pyridyl ketones

These also resemble the corresponding phenyl ketones, as regards both preparation and reactions. Cyanopyridines and Grignard reagents,[391,408] acylation of pyridyl- (or pyridylalkyl-) lithium compounds (pp. 25, 58),[377,409] and Claisen ester condensations involving pyridinecarboxylates[410] provide standard routes to these compounds, and Friedel–Crafts acylation using pyridinecarboxylic acid chlorides[411] provides a useful alternative route to aryl pyridyl ketones.

16.1.12.4 Pyridinecarboxylic acids and their derivatives

The acids are the end-products of oxidation of most pyridines containing carbon substituents, and pyridinedicarboxylic acids result from the oxidation of fused pyridines such as quinoline and isoquinoline. The pK_a values of the acids indicate that they exist mainly in the zwitterionic form,[412] and this may be responsible, in part at least, for the ease with which they undergo decarboxylation when heated (*cf.* p. 13). Alkylation of pyridinecarboxylate anions gives *N*-alkylpyridinium carboxylates,[413] whereas esterification occurs with alcohols in presence of mineral acid, or with diazoalkanes. Acid chlorides may also be used to obtain the esters, although the chlorides are prone to self-quaternization.

Non-systematic nomenclature. Pyridine-2-, -3-, and -4-carboxylic acids are often referred to as *picolinic*, *nicotinic*, and *isonicotinic* acids, respectively.

16.1.12.5 Cyanopyridines

These are obtained from pyridinecarboxylic acid derivatives by standard routes, *e.g.* by dehydration of the amides;[408a] from pyridinesulphonates and cyanide ion (p. 33); by the Sandmeyer reaction;[397] and from *N*-alkoxypyridinium salts and cyanide ion (p. 28). Their properties are generally as expected for aryl cyanides.

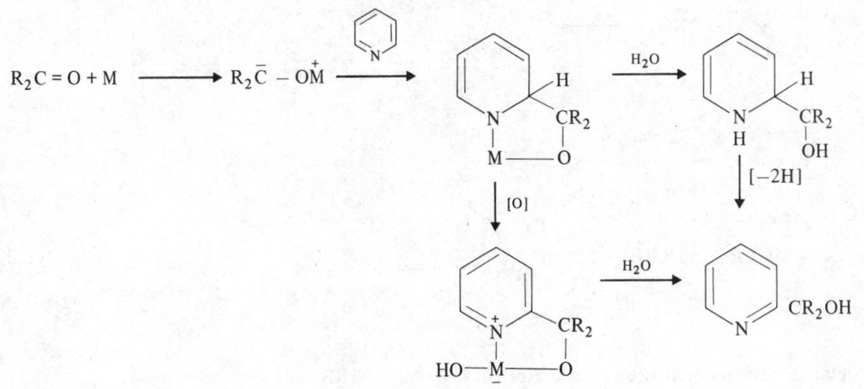

(118)

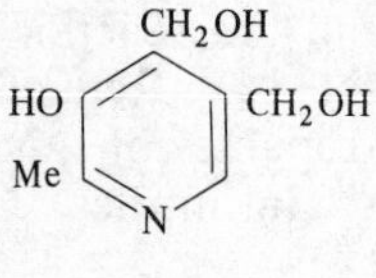

(122) 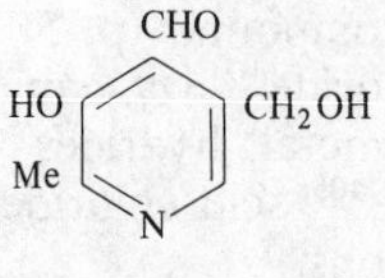 (123) 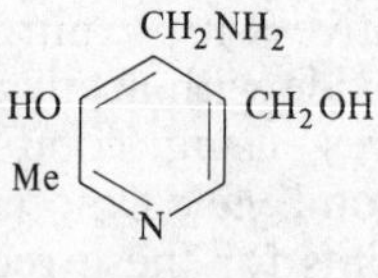 (124) 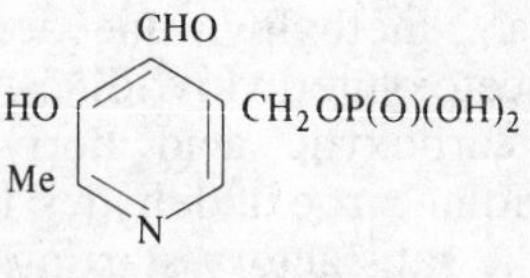 codecarboxylase

16.1.13 PREPARATION OF PYRIDINES*

16.1.13.1 Preparation from acyclic compounds

(i) Ammonia or amines as the source of the hetero-atom

In principle, the synthesis of pyridines is represented in simplest terms by the reaction of a 1,5-dicarbonyl compound, containing 2,3-unsaturation, with ammonia [equation (119a)]. In practice, pent-2-enedial (glutacondialdehyde), as its sodium salt, does undergo cyclization to pyridine by reaction with ammonium acetate,[415] but the usefulness of this as a general method is limited. The corresponding saturated 1,5-dicarbonyl compounds are cyclized to dihydropyridines with ammonia, and oxidation to pyridines may be effected by a variety of methods;[416] alternatively, cyclization using hydroxylamine leads to pyridines directly[417] [equation (119b)]. In other variants of the reaction, 2- and 4-pyridones and 2-aminopyridines may be obtained from reactions of appropriately substituted C_5 compounds and ammonia or amines [equations (120)–(123)].

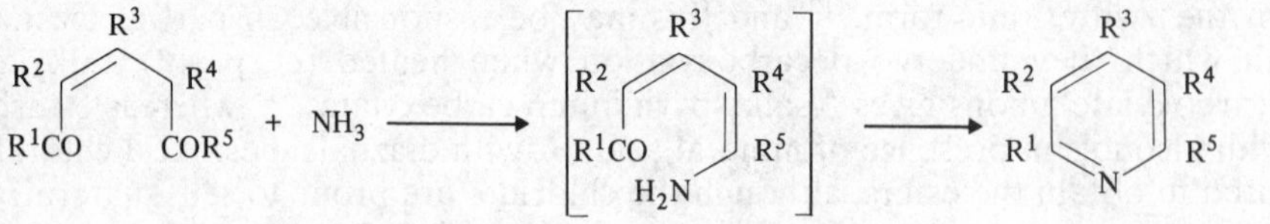

(119a)

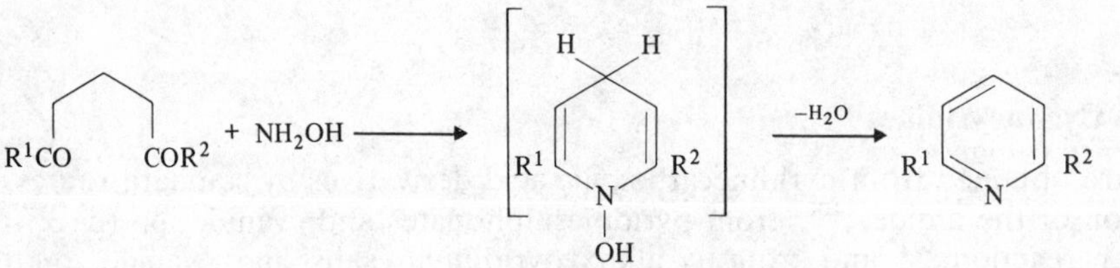

(119b)

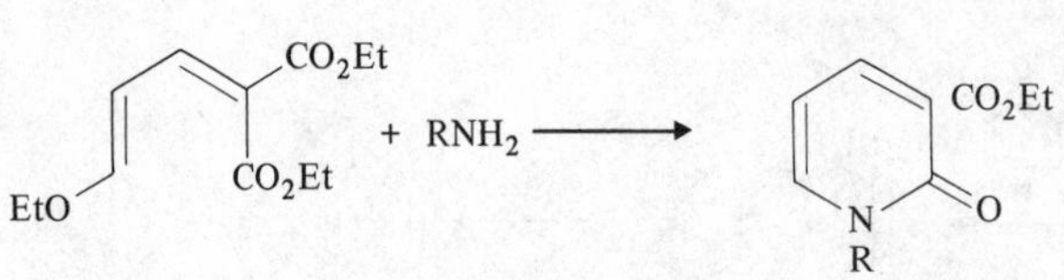

(120)[418]

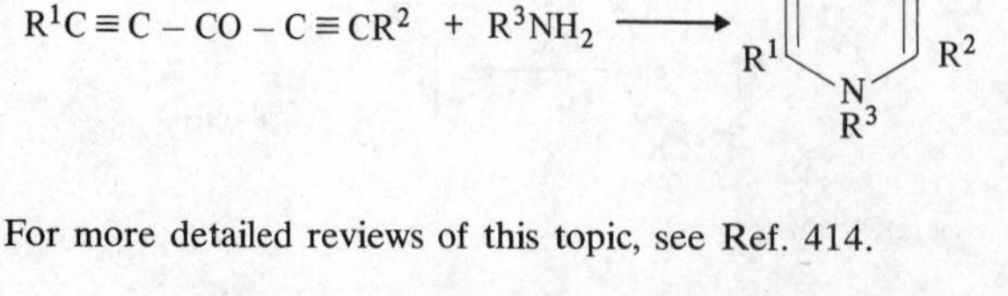

(121)[419]

* For more detailed reviews of this topic, see Ref. 414.

$$^{-}O{-}CH{=}CR^1{-}CO{-}CR^1{=}CH{-}O^{-} + R^2NH_2 \longrightarrow \text{(1-}R^2\text{-3,5-di-}R^1\text{-4-pyridone)} \quad \text{(my)} \qquad (122)^{420}$$

$$R^1C{\equiv}C{-}CR^2{=}CR^3{-}CN + NH_3 \longrightarrow \text{(2-amino-3-}R^3\text{-4-}R^2\text{-6-}R^1\text{-pyridine)} \quad \text{(hy)} \qquad (123)^{421}$$

By far the largest group of useful pyridine syntheses involve the formation of the C_5 unit *in situ*, by one or more base-catalysed condensation reactions. Ammonia or an amine is used as the condensing agent, and in its presence the five-carbon chain undergoes spontaneous cyclization to yield the pyridine (or dihydropyridine).

The C_5 chain is normally constructed from a C_3 and a C_2 unit, or from a C_1 and two C_2 units. The C_3+C_2 synthesis is the more versatile, and is depicted in general terms in Scheme 10. There are essentially two types of C_3+C_2 synthesis: one involves a simple condensation of a 1,3-dicarbonyl compound with a reactive methylene compound [equation (124)], and the second involves Michael addition of a carbanion to a conjugated carbonyl compound [equation (125)]. The intermediacy of the 1,5-dicarbonyl compound, as such, remains an open question: it is perfectly conceivable that ammonia may react with either component at an earlier stage to produce an enamine, and indeed pre-formed enamines may replace either of the components in the reaction [equations (124a,b), (125a)].

There are, obviously, many variants of these general syntheses. In the first type, represented by equation (124), the dicarbonyl compound may be a dialdehyde, a keto-aldehyde, or a diketone. The reactive methylene component is usually doubly activated, and so reactions of this type generally give pyridines with an electron-attracting substituent (R^4) at the 3-position. The use of simple ketones as the C_2 units seldom leads to good yields of pyridines [the self-condensation of acetylacetone with ammonia, as in equation (127),[422] being a notable exception], and the most common forms of this reaction are those shown in equations (124a), (124b), and (126). The use of an enamine in place of a ketone as the C_2 component [*cf.* equation (128)[423]] avoids side-reactions and generally gives higher yields, and the use of an enamine in place of a dicarbonyl compound as the C_3 unit [*cf.* equation (129)[424]] is of particular value when the dicarbonyl compound would be a dialdehyde. The most familiar form of equation (126) is that in which $R^4 = CN$: in this (the Guareschi–Thorpe synthesis) the dicarbonyl compound is condensed with ethyl cyanoacetate and ammonia, or with cyanoacetamide in presence of a base (*e.g.* an alkoxide or an amine), and the product is a 2-pyridone [*cf.* equations (130)[425] and (131)[426]].

In the second type, the C_3 component of equation (125) may be an aldehyde, a ketone, an ester, or an amide, and the unsaturation may be olefinic or acetylenic. Double activation of the C_2 component is not necessary [*cf.* equation (132)[427]], although reactions involving doubly activated methylene groups require milder conditions and generally give better yields. Dihydropyridines are the usual products of such reactions, although spontaneous dehydrogenation may sometimes occur. When the unsaturation is acetylenic, however [equation (133)[428]], pyridines are obtained directly, and the same is true of reaction (134),[429] which illustrates the use of a Mannich base to protect a labile enone, and the use of the pyridinium group as an aid to elimination.

Of all the $2C_2+C_1$ syntheses reported in the literature, the most familiar (and most useful, on a laboratory scale) is the Hantzsch synthesis [equation (135)],[430] in which an aldehyde is condensed with two molecules of a methylenic ketone and ammonia, to give a dihydropyridine, which may readily be oxidized (*e.g.* by nitric acid) to the pyridine itself.

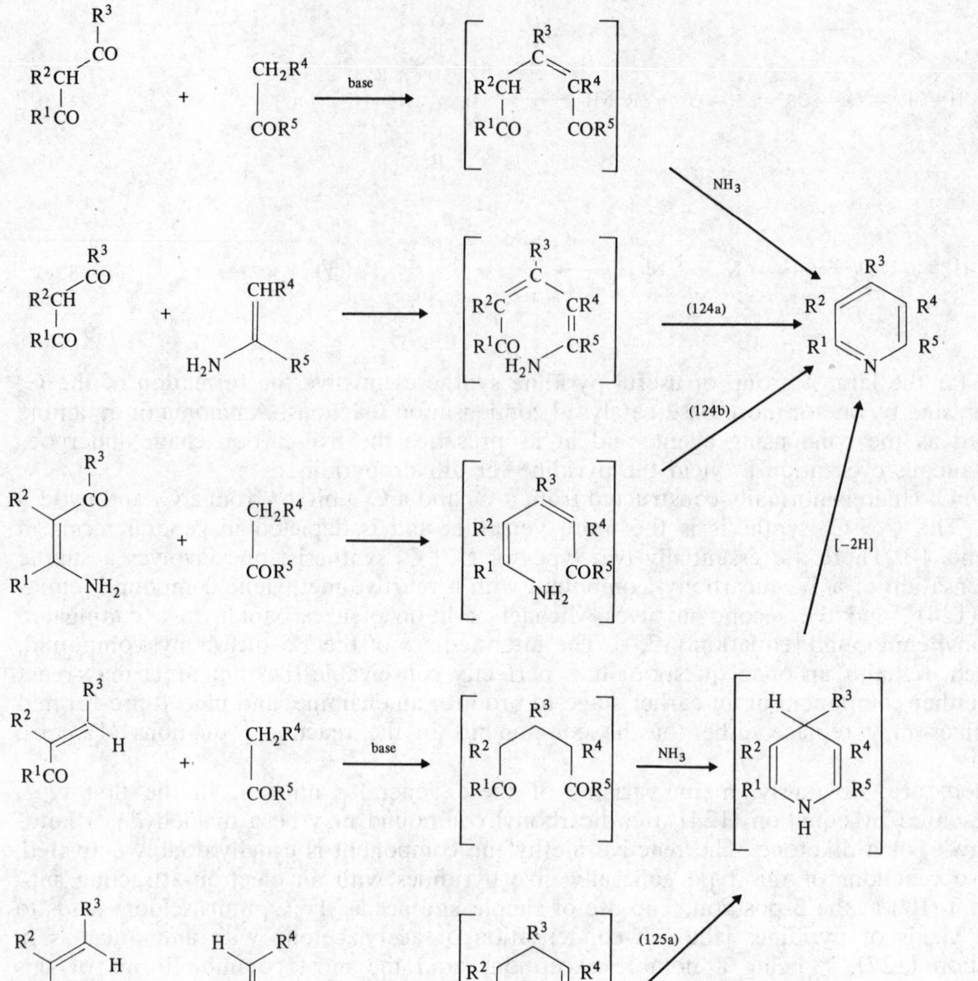

(124)

(125)

(126)

SCHEME 10

The aldehyde provides carbon-4 of the pyridine ring, and the substitution pattern in the product is symmetrical.

On the industrial scale, however, the reactions which have received most attention are the self-condensations of simple aldehydes and ketones with ammonia, which lead to a variety of simple pyridine derivatives. The complexity of these reactions is shown schematically below (Scheme 11) for acetaldehyde: the main products are 2- and 4-methylpyridines, and 5-ethyl-2-methyl- and 3-ethyl-4-methyl-pyridines, and variation of the reaction conditions leads to variation of the product ratio.[431]

$2CH_3COCH_2COCH_3 \xrightarrow{NH_3}$ [pyridine: CH_3, $COCH_3$, CH_3, CH_3] (75%) (127)

$CH_3COCH_2COCO_2Et + CH_3C(NH_2)=CHCO_2Et \longrightarrow$ [pyridine: CO_2Et, CO_2Et, CH_3, CH_3] (90%) (128)

$H_2NCH=CR^1\text{–}CHO + R^2CH_2COR^3 \longrightarrow$ [pyridine: R^1, R^2, R^3] (my–hy) (129)

$CH_3COCH_2CO_2Et + NCCH_2CO_2Et \xrightarrow{NH_3}$ [pyridinone: Me, HO, N–H, O] (hy) (130)

$CH_3COCH_2COCH_2OEt + H_2NCOCH_2CN \xrightarrow[81\%]{\text{piperidine}}$ [pyridinone: CH_2OEt, CN, Me, N–H, O] $\xrightarrow{\text{3 steps}}$ [pyridine: CH_2OEt, H_2N, CH_2NH_2, Me] $\xrightarrow{\text{4 steps}}$ [pyridine: CH_2OH, HO, CH_2OH, Me] pyridoxol (122) (131)

$PhCH=CHCOPh + PhCOCH_3 \xrightarrow[NH_4^+OAc^-,\ heat]{NH_3}$ [pyridine: Ph, Ph, Ph] (ly–my) (132)

$HC\equiv C\text{–}COR^1 + R^2CH=CR^3NH_2 \xrightarrow{heat}$ [pyridine: R^1, R^2, R^3] (hy) (133)

$PhCOCH_2CH_2NMe_2$ [$\longrightarrow PhCOCH=CH_2$] $+ H_2NCOCH_2\overset{+}{N}$[pyridinium] $X^- \longrightarrow$ [pyridinone: Ph, N–H, O] (my) (134)

$R^1CHO + 2R^2COCH_2COR^3 \xrightarrow{NH_3}$ [R^2CO, COR^2, R^1, R^3CO, COR^3] $\longrightarrow$ [1,4-dihydropyridine: R^1, H, R^2CO, COR^2, R^3, R^3, N–H] $\xrightarrow{[-2H]}$ [pyridine: R^1, R^2CO, COR^2, R^3, R^3] (my–hy) (135)

The mechanisms of these self-condensations possibly involve, as a first step, condensation of two molecules of acetaldehyde to give crotonaldehyde (but-2-enal), since crotonaldehyde itself may react with ammonia to give the same four products.[432] The C_8 products may be accounted for as shown in equations (136) and (137), and retro-aldol cleavage of the intermediates (**125**) and (**126**) may lead to the 2- and 4-methylpyridines.

Pyridine itself may be obtained by the condensation of two molecules of acetaldehyde with one of formaldehyde and one of ammonia;[433] from one molecule of crotonaldehyde, one of formaldehyde, and one of ammonia;[434] and from one of acrolein (propenal), one of acetaldehyde, and one of ammonia.[435] Mixtures of products are, of course, obtained in these reactions, but under carefully controlled conditions good yields of pyridine are obtainable in each case.

'Mixed' condensations of aldehydes with ketones and ammonia may be utilized, as for example in the condensation of mesityl oxide (4-methylpent-3-en-2-one) with formaldehyde and ammonia to give 2,4-dimethylpyridine[436] [equation (138)]. However, the

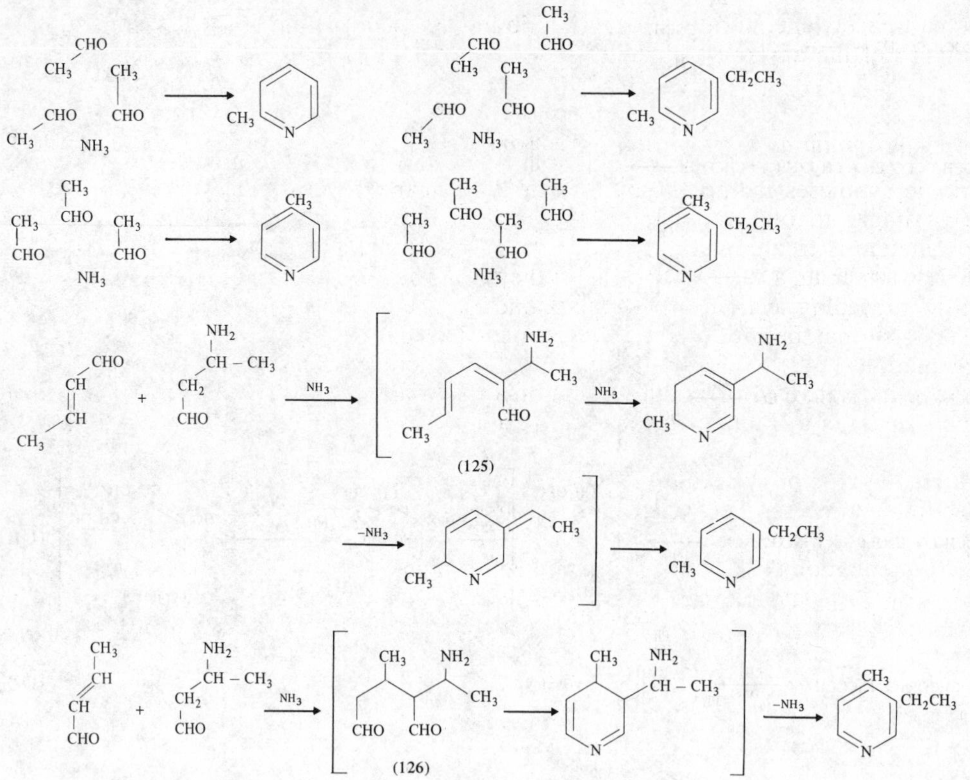

(125)

(136)

(126)

(137)

SCHEME 11

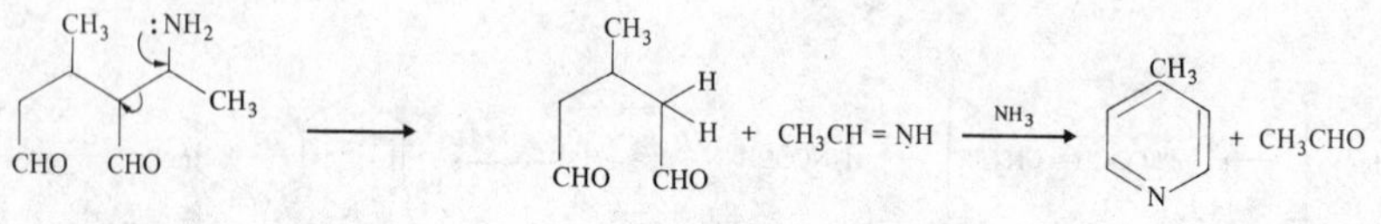

(126)

self-condensation of ketones alone, in presence of ammonia, does not usually give pyridines in useful quantities. One potentially useful laboratory synthesis of pyridines from simple ketones, however, involves the reaction of alkyl phenyl ketones (or benzoins) with hexamethylphosphoramide[437] [equation (139)].

In all of the above reactions involving acetaldehyde and ammonia, the intermediate $CH_3CH{=}NH$ may be involved. This is also formally derivable from the addition of

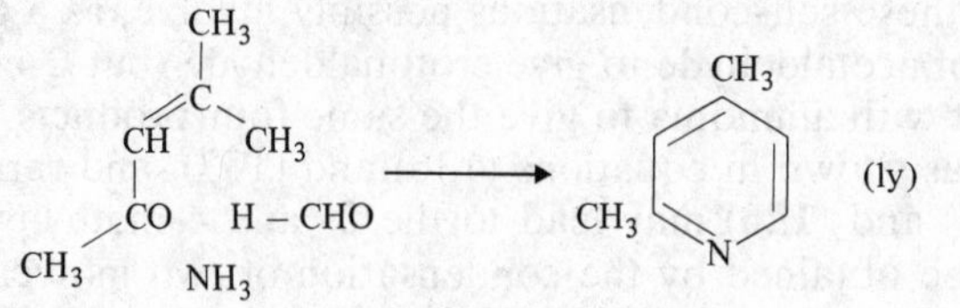

(ly) (138)

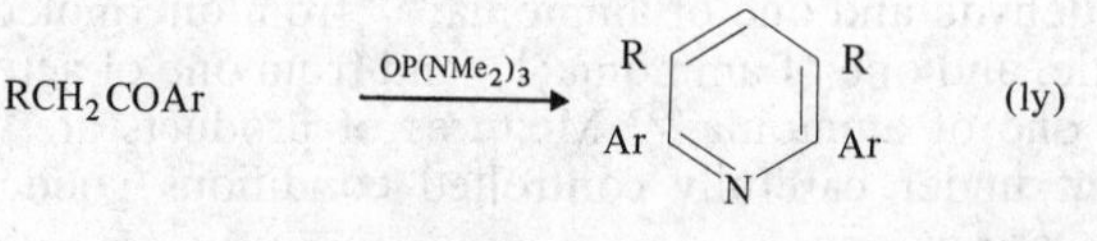

$RCH_2COAr \xrightarrow{OP(NMe_2)_3}$ (ly) (139a)

$ArCH(OH)COAr \xrightarrow{OP(NMe_2)_3}$ (ly) (139b)

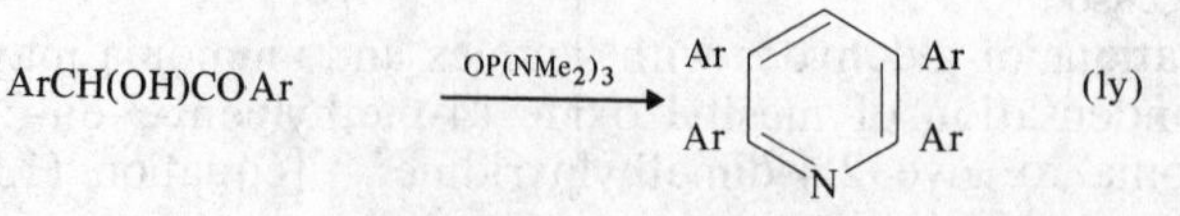

ammonia to acetylene, and these reagents do indeed yield pyridines at high temperatures in presence of the appropriate catalyst.[438]

(*ii*) *A cyano-group as the source of the hetero-atom*

Pyridine syntheses in this group fall into two clearly defined categories. The first of these is similar to others described earlier: a five-carbon chain containing a terminal cyano-function is treated with an acid, HX, giving an imine (or enamine), and this then cyclizes by attacking a carbonyl or a second cyano-group on the other end of the C_5 chain. As in the preceding section, the C_5 compound may be preformed [equations (140)[439] and (141)[440]] or it may be produced *in situ* by a condensation reaction [usually of the C_3+C_2 type: equations (142)[441], (143)[442], and (144)[443]]. Malononitrile undergoes trimerization to a pyridine derivative in presence of ammonia or an amine [equation (145)[444]], and may also participate in a Hantzsch synthesis with aldehydes in presence of alkoxides [equation (146)[445]].

The second type of cyclization in which a cyano-nitrogen provides the hetero-atom is a cycloaddition process. The original preparation of pyridine, by passing acetylene and hydrogen cyanide through a red-hot tube,[446] has been developed into a useful synthetic procedure [equation (147)[447]] by the use of cobalt complexes, and the related Diels–Alder reactions of conjugated dienes with nitriles [equation (148)] has also been developed into a useful laboratory reaction by the use of highly electrophilic nitriles such as RSO_2CN.[448]

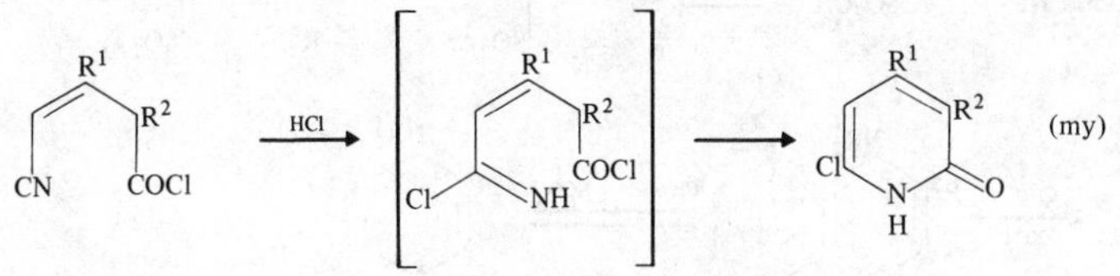

(my) (140)

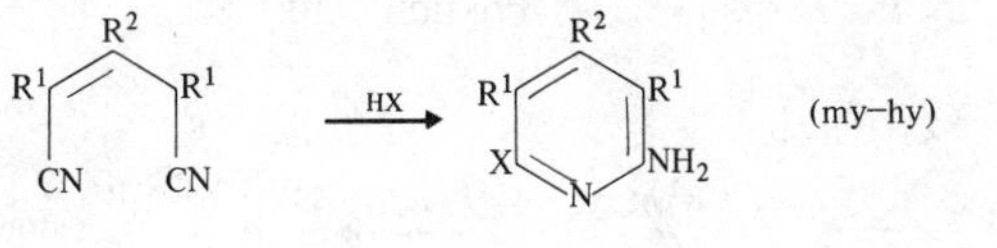

(my–hy) (141)

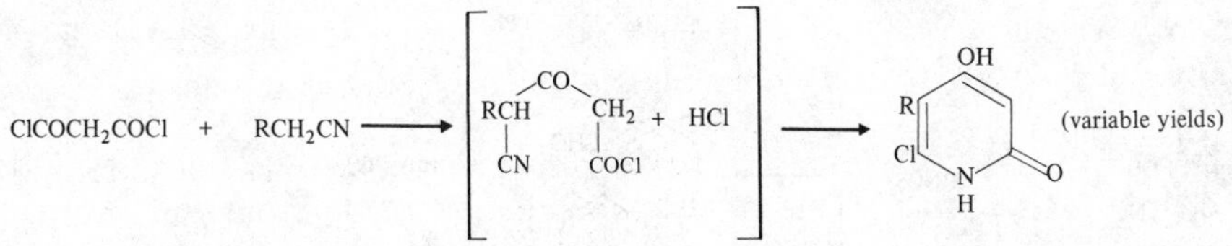

$ClCOCH_2COCl + RCH_2CN \longrightarrow [RCH(CN)COCH_2COCl + HCl] \longrightarrow$ (variable yields) (142)

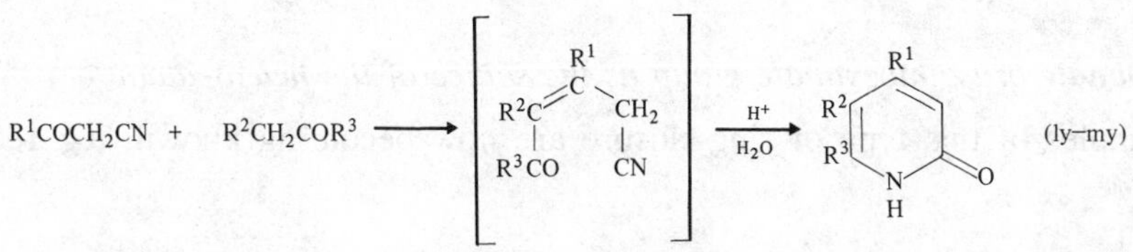

$R^1COCH_2CN + R^2CH_2COR^3 \longrightarrow$ (ly–my) (143)

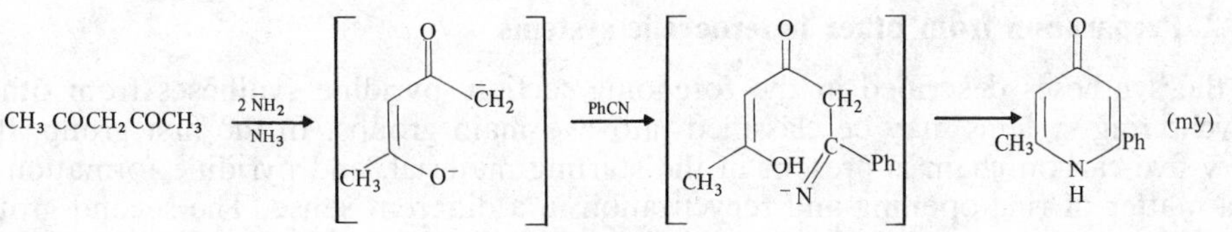

$CH_3COCH_2COCH_3 \xrightarrow[NH_3]{2\,\bar{N}H_2}$ $\xrightarrow{PhCN}$ (my) (144)

$$3CH_2(CN)_2 \xrightarrow{\text{i, NH}_3;\ \text{ii, heat}} \text{(NC, CN, CH}_2\text{CN, H}_2\text{N, NH}_2\text{ pyridine)} \qquad (145a)$$

$$\xrightarrow{Et_2NH} \text{(NH}_2\text{, NC, CN, H}_2\text{N, CH}_2\text{CN pyridine)} \qquad (145b)$$

$$R^1CHO + 2CH_2(CN)_2 \xrightarrow{R^2O^-} [\ldots] \longrightarrow \xrightarrow{[-2H]} \text{(ly–my)} \qquad (146)$$

$$2R^1C\equiv CH + R^2CN \xrightarrow[\text{or Co}^{I}\text{ complex}]{\text{Cobaltocene}} \ldots + \ldots \quad \text{(my–hy)} \qquad (147)$$

$$R^1CH{=}CR^2{-}CR^3{=}CHR^4 + ArSO_2CN \xrightarrow{[-2H]} \ldots + \ldots \quad \text{(ly–my)} \qquad (148)$$

$$R^1_2NC\equiv CR^2 \ldots \xrightarrow[20\,°C]{CH_3CN} \ldots \xrightarrow{R^3CH=CH-NCO} \ldots CONHCH=CHR^3 \quad \text{(hy)} \qquad (149)^{449}$$

$$2R^1C\equiv CR^2 + \pi\text{-}C_5H_5Co(PPh_3)_2 \longrightarrow \ldots \xrightarrow{R^3NCO} \ldots \quad \text{(my–hy)} \qquad (150)^{450}$$

$$\text{CHO–CH=CH–CH=CH–CHO}^{-}\ Na^+ + RNCS \longrightarrow \ldots \quad \text{(my–hy)} \qquad (151)^{451}$$

(*iii*) *An isocyanate or isothiocyanate group as the source of the hetero-atom*

A few examples of this type of ring-closure are now becoming known, *e.g.* reactions (149)–(151).

16.1.13.2 Preparation from other heterocyclic systems

Like the syntheses described in the foregoing section, pyridine syntheses from other heterocyclic ring systems may be classified into two main groups. In the first group, the necessary five-carbon chain is present in the starting material, and pyridine formation is simply a matter of ring-opening and recyclization in a different sense. The second group involves cycloaddition to the heterocycle as the initial step, and the pyridine results from the (often spontaneous) collapse of a bicyclic intermediate.

(*i*) *Ring-opening and recyclization*

(*a*) *Synthesis from furans.* In these syntheses the five carbons of the pyridine are derived from the ring carbons of furan together with the α-carbon of the 2-substituent in the furan. For example, 2-acylfurans react with ammonium salts at moderately high temperatures to give 3-hydroxypyridines [equation (152)[452]] and 2-aminomethylfurans are similarly transformed by treatment with acid and an oxidant [equation (153)[453a]]. The use of the appropriate *N*-substituted reagents in the above sequences usually leads to 3-hydroxypyridinium salts[453b] (but not in every case[454]). The other main pyridine synthesis from furans is illustrated in equation (154): this route involves acid-catalysed ring opening of the cyclic acetal (**127**), and leads to 3-hydroxy-2-pyridones.[453c]

(*b*) *Synthesis from isoxazoles.* A few pyridine syntheses depend on the ring-opening of substituted isoxazoles by catalytic hydrogenation [*e.g.* reaction (155)].[455]

(*c*) *Synthesis from pyrylium salts and pyrones.* The preparation of pyridines from pyrylium salts and ammonia is a useful method for 2,4,6-trisubstituted derivatives,[456] but otherwise the method is little used because of the inaccessibility of the starting compounds. The use of hydroxylamine and hydrazine in this reaction may lead to pyridine *N*-oxides and *N*-aminopyridinium salts, although other modes of recyclization may intervene [equations (156), (157)].[32b,457]

2- and 4-pyrones are converted into pyridones by reaction with ammonia or amines,[458] good yields being generally attainable, although in a few cases recyclization may occur in more than one direction (or may not occur at all).[459]

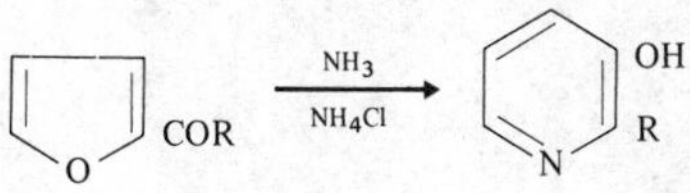

(ly–my) (152)

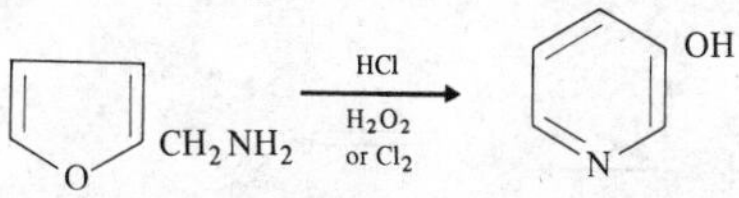

(hy) (153)

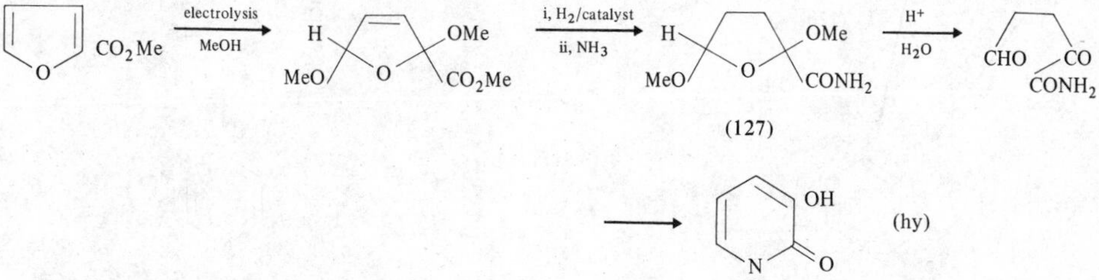

(127)

(hy) (154)

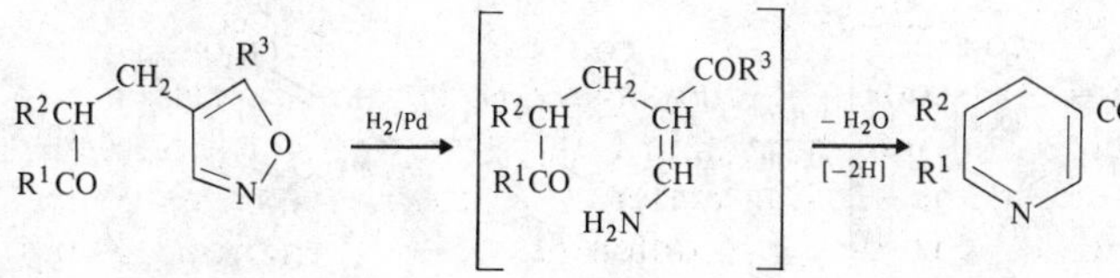

(155)

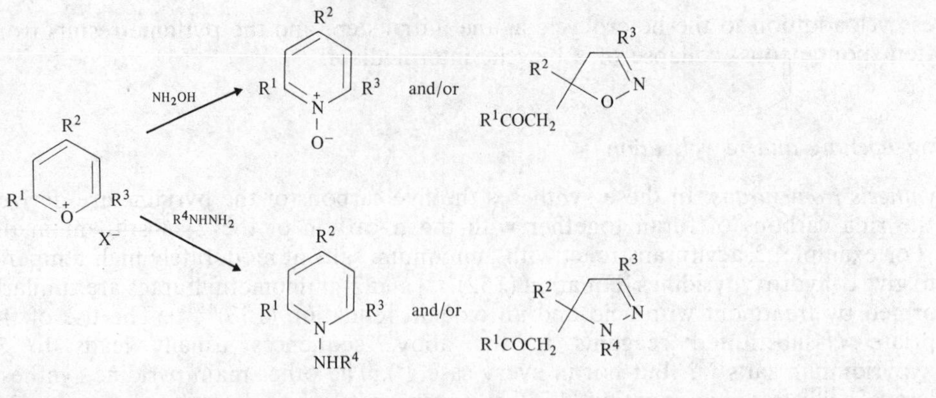

(156)

(157)

(ii) Cycloaddition processes

The best-known of these (although not generally the most useful) is the reaction of pyrroles with dichlorocarbene, which gives 3-chloropyridines [equation (158)].[460] The yield of 3-chloropyridine obtained depends greatly on the source of the carbene[460b,c] and by-products [such as (**128**) and (**129**) in reaction (159)] are liable to be formed in substantial amount.

Most of the cycloaddition processes which yield pyridines are of the Diels–Alder type, as represented by equations (160)–(167). The most studied of these involve oxazoles as the diene components,[461] and reactions of this type provide what are probably the simplest synthetic routes to the pyridoxine vitamins [equation (160)[462]]. The other examples in this series are relatively new reactions which have not yet been fully exploited, but in the pyridazine series, at least [equation (164)], the pyridine synthesis is certainly not a general reaction.

It is also noteworthy that Diels–Alder reactions of certain carbocyclic compounds with nitriles give pyridines: examples of such reactions are given in equations (168) and (169).

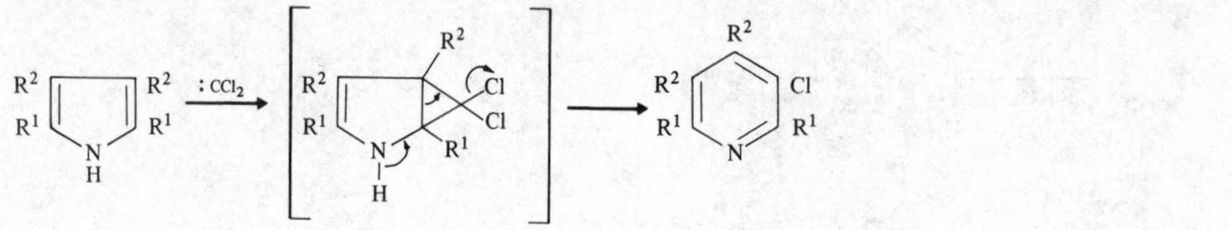

(158)

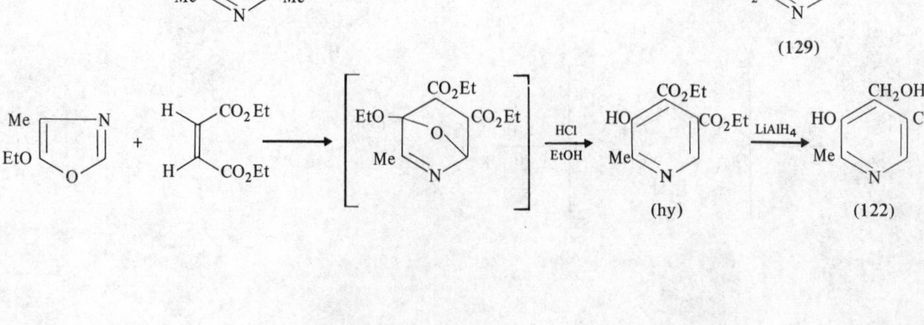

(128)

(159)

(129)

(160)

(hy)

(122)

(my–hy) (161)[463]

(162)[464]

(ly–hy) (163)[465]

(ly–hy) (164)[466a]

(ly–hy) (165)[466b]

(68%) (16%) (15%) (166)[466c]

(60%) (167)[467]

(my–hy) (168)[468]

(169)[469]

16.1.14 NATURAL PRODUCTS CONTAINING THE PYRIDINE RING

Apart from NAD (p. 44) and related nucleotides, and the B_6 (pyridoxine) vitamins (p. 62), the pyridine ring is found in nature in a number of alkaloids. The best known of these are the tobacco alkaloids such as nicotine (**130**) and anabasine (**131**); other simple pyridine alkaloids include ricinine (**132**) and trigonelline (**133**).

N Me N (130) N H N (131)

OMe CN O N Me (132) CO_2^- $\overset{+}{N}$ Me (133)

Acknowledgements

The author wishes to express his thanks to several colleagues and friends for helpful comments and suggestions regarding the contents of this chapter; to the library staff of the Universities of St. Andrews and Glasgow for the facilities made available to him; and most of all to his sister-in-law, Mrs. Maureen Smith, for her expert typing of the manuscript.

References

1. T. Anderson, *Trans. Roy. Soc. Edinburgh*, 1849, **16,** 123; *Annalen*, 1846, **60,** 86.
2. T. Anderson, *Trans. Roy. Soc. Edinburgh*, 1849, **16,** 463; *Annalen*, 1849, **70,** 32.
3. T. Anderson, *Trans. Roy. Soc. Edinburgh*, 1853, **20,** 247; *Annalen*, 1851, **80,** 44.
4. H. Weidel, *Ber.*, 1879, **12,** 1989.
5. *Cf.* L. Dobbin, *J. Chem. Educ.*, 1934, **11,** 596.
6. A. Ladenburg, *Annalen*, 1888, **247,** 1, and references cited therein.
7. B. Bak, L. Hansen-Nygaard, and J. Rastrup-Andersen, *J. Mol. Spectroscopy*, 1958, **2,** 361.
8. W. L. F. Armarego, in 'Physical Methods in Heterocyclic Chemistry', ed. A. R. Katritzky, Academic Press, New York, 1971, vol. 3, pp. 79–87.
9. A. R. Katritzky and A. P. Ambler, in Ref. 8, 1963, vol. 2, pp. 274–292; A. R. Katritzky and P. J. Taylor, *ibid.*, 1971, vol. 4, pp. 375–398.
10. K. Schofield, 'Heteroaromatic Nitrogen Compounds: Pyrroles and Pyridines', Butterworths, London, 1967: (a) p. 124; (b) p. 162; (c) p. 200; (d) p. 227.
11. W. Adam, A. Grimison, and G. Rodriguez, *Tetrahedron*, 1967, **23,** 2513.
12. T. J. Batterham, 'NMR Spectra of Simple Heterocycles', Wiley, New York, 1973, pp. 8–88.
13. F. A. L. Anet and I. Yavari, *J. Org. Chem.*, 1976, **41,** 3589; *cf.* E. Breitmaier and K.-H. Spohn, *Tetrahedron*, 1973, **29,** 1145.
14. Q. N. Porter and A. Baldas, 'Mass Spectrometry of Heterocyclic Compounds', Wiley, New York, 1971, pp. 376–398.
15. D. M. Smith, in 'Rodd's Chemistry of Carbon Compounds', ed. S. Coffey, 2nd Edition, Elsevier, Amsterdam, 1976, vol. 4F: (a) p. 158; (b) p. 79; (c) p. 91; (d) pp. 94, 104; (e) p. 114; (f) p. 135; (g) p. 192.
16. R. Daniels and C. G. Kormendy, *J. Org. Chem.*, 1962, **27,** 1860.
17. E. Ager and H. Suschitzky, *J.C.S. Perkin I*, 1973, 2839.
18. *Cf.* H. C. Brown and A. Cahn, *J. Amer. Chem. Soc.*, 1955, **77,** 1715; W. S. Emerson and T. M. Patrick, *Org. Synth.*, 1958, **38,** 86 (*Coll. Vol.* 4, 980); H. H. Inhoffen, H.-J. Krause, and S. Bork, *Annalen*, 1954, **585,** 132.

19. L. C. King, *J. Amer. Chem. Soc.*, 1944, **66,** 894, 1612; 1948, **70,** 239, 242.
20. F. N. Hayes, H. K. Suzuki, and D. E. Peterson, *J. Amer. Chem. Soc.*, 1950, **72,** 4524; D. H. Wheeler and J. Gross, *J. Amer. Oil Chemists' Soc.*, 1965, **42,** 924; Y. A. Serguchev and E. A. Shilov, *Ukrain. Khim. Zhur.*, 1968, **34,** 969 (*Chem. Abs.*, 1969, **70,** 28320).
21. D. Cook, *Canad. J. Chem.*, 1962, **40,** 2362.
22. H. Adkins and Q. E. Thompson, *J. Amer. Chem. Soc.*, 1949, **71,** 2242.
23. A. R. Butler and V. Gold, *J. Chem. Soc.*, 1961, 4362; A. R. Fersht and W. P. Jencks, *J. Amer. Chem. Soc.*, 1970, **92,** 5432.
24. H. E. Baumgarten, *J. Amer. Chem. Soc.*, 1953, **75,** 1239; C. F. H. Allen *et al.*, *Org. Synth.*, *Coll. Vol.* 3, 28; R. K. Smalley and H. Suschitzky, *J. Chem. Soc.*, 1964, 755.
25. W. E. McEwen *et al.*, *J. Amer. Chem. Soc.*, 1954, **76,** 41, 4540; *J. Org. Chem.*, 1961, **26,** 1684.
26. W. Steglich and G. Höfle, *Angew. Chem. Internat. Edn.*, 1969, **8,** 981; *Synthesis*, 1972, 619.
27. E. Guibé-Jampel and M. Wakselman, *Chem. Comm.*, 1971, 267.
28. C. A. Cupas and R. L. Pearson, *J. Amer. Chem. Soc.*, 1968, **90,** 4742.
29. L. F. Fieser and M. Fieser, 'Reagents for Organic Synthesis', Wiley, New York, 1967, **1,** 966, 967; 1975, **5,** 568.
30. Ref. 29, 1967, **1,** 1127; 1974, **4,** 473.
31. H. S. Mosher, L. Turner, and A. Carlsmith, *Org. Synth.*, 1953, **33,** 79; E. C. Taylor and A. J. Crovetti, *ibid.*, 1956, **36,** 53; 1957, **37,** 63. (*Coll. Vol.* 4, pp. 655, 704, 828).
32. *Cf.* (a) J. Meisenheimer, *Ber.*, 1926, **59,** 1848; B. Bobranski, L. Kochansa and A. Kowalewska, *ibid.*, 1938, **71,** 2385; (b) C. L. Pedersen, N. Harrit, and O. Buchardt, *Acta Chem. Scand.*, 1970, **24,** 3435.
33. S. M. Roberts and H. Suschitzky, *J. Chem. Soc.* (*C*), 1968, 1537; I. Collins and H. Suschitzky, *ibid.*, 1970, 1523.
34. G. E. Chivers and H. Suschitzky, *J. Chem. Soc.* (*C*), 1971, 2867.
35. E. P. Linton, *J. Amer. Chem. Soc.*, 1940, **62,** 1945; A. R. Katritzky, E. W. Randall, and L. E. Sutton, *J. Chem. Soc.*, 1957, 1769; A. N. Sharpe and S. Walker, *ibid.*, 1961, 4522.
36. R. A. Abramovitch and J. B. Davis, *J. Chem. Soc.* (*B*), 1966, 1137; S. Castellano and R. Kostelnik, *J. Amer. Chem. Soc.*, 1968, **90,** 141.
37. R. A. Barnes, *J. Amer. Chem. Soc.*, 1959, **81,** 1935.
38. Ref. 14, p. 384.
39. R. A. Abramovitch and E. M. Smith, in 'Pyridine and its Derivatives', Supplement, ed. R. A. Abramovitch, Wiley, New York, 1974, vol. 2, (a) p. 35; (b) pp. 59–80; (c) p. 69.
40. A. R. Katritzky and J. M. Lagowski, 'Chemistry of the Heterocyclic *N*-Oxides', Academic Press, London and New York, 1971, (a) pp. 166–231; (b) pp. 233–236; (c) pp. 200–202.
41. A. R. Katritzky and A. M. Monro, *J. Chem. Soc.*, 1958, 1263.
42. J. M. Essery and K. Schofield, *J. Chem. Soc.*, 1960, 4953.
43. R. A. Abramovitch and K. A. H. Adams, *Canad. J. Chem.*, 1961, **39,** 2134.
44. H. C. Brown and B. C. Subba Rao, *J. Amer. Chem. Soc.*, 1956, **78,** 2582.
45. E. Howard and W. F. Olszewski, *J. Amer. Chem. Soc.*, 1959, **81,** 1483.
46. T. R. Emerson and C. W. Rees, *J. Chem. Soc.*, 1962, 1917.
47. *Cf.* T. B. Lee and G. A. Swan, *J. Chem. Soc.*, 1956, 771; K. Lewicka and E. Plaźek, *Rec. Trav. chim.*, 1959, **78,** 644.
48. J. I. G. Cadogan, D. J. Sears, and D. M. Smith. *J. Chem. Soc.* (*C*), 1969, 1314.
49. R. C. Duty and G. Lyons, *J. Org. Chem.*, 1972, **37,** 4119.
50. F. A. Daniher and B. E. Hackley, *J. Org. Chem.*, 1966, **31,** 4267.
51. K. Ikura and S. Oae, *Tetrahedron Letters*, 1968, 3791.
52. D. I. Relyea, P. O. Tawney, and A. R. Williams, *J. Org. Chem.*, 1962, **27,** 477.
53. V. J. Traynelis and K. Yamauchi, *Tetrahedron Letters*, 1969, 3619: M. E. C. Biffin *et al.*, *Austral. J. Chem.*, 1974, **27,** 789.
54. For reviews on pyridinium ylides, see A. W. Johnson, 'Ylid Chemistry', Academic Press, London and New York, 1966, pp. 260–265; O. R. Rodig, in 'Pyridine and its Derivatives', Supplement, ed. R. A. Abramovitch, Wiley, New York, 1974, part 1, pp. 353–368.
55. D. Lloyd and M. I. C. Singer, *J. Chem. Soc.* (*C*), 1971, 2939.
56. (a) A. Rieche and P. Dietrich, *Chem. Ber.*, 1963, **96,** 3044; (b) W. J. Linn, O. W. Webster, and R. E. Benson, *J. Amer. Chem. Soc.*, 1965, **87,** 3651; (c) H. Hamberger and R. Huisgen, *Chem. Comm.*, 1971, 1190.
57. *Cf.* F. Kröhnke and co-workers, *Ber.*, 1935, **68,** 1177; 1937, **70,** 538, 543; *Chem. Ber.*, 1962, **95,** 1108, 1118.
58. D. Lloyd and J. S. Sneezum, *Tetrahedron*, 1958, **3,** 334; V. A. Nefedov, *J. Gen. Chem. USSR*, 1966, **36,** 1513; D. Lloyd and M. I. C. Singer, *J. Chem. Soc.* (*C*), 1971, 2941.
59. L. A. Pinck and G. E. Hilbert, *J. Amer. Chem. Soc.*, 1946, **68,** 2011; F. Kröhnke, *Chem. Ber.*, 1950, **83,** 253.
60. G. V. Boyd, *Tetrahedron Letters*, 1966, 3369.
61. W. G. Phillips and K. W. Ratts, *J. Org. Chem.*, 1970, **35,** 3144; J. E. Douglass, M. W. Tabor, and J. E. Spradling, *J. Heterocyclic Chem.*, 1972, **9,** 53.
62. R. Daniels and O. LeRoy Salerni, *Proc. Chem. Soc.*, 1960, 286; H. Wittmann, P. Beutel, and E. Ziegler, *Monatsh.*, 1969, **100,** 1619.

63. F. Kröhnke, *Angew. Chem.*, 1953, **65,** 605; 1962, **74,** 811; 1963, **75,** 181, 317.
64. C. A. Henrick, E. Ritchie, and W. C. Taylor, *Austral. J. Chem.*, 1967, **20,** 2441, 2455.
65. (a) R. Gösl and A. Meuwsen, *Chem. Ber.*, 1959, **92,** 2521; (b) T. Okamoto, M. Hirobe, and A. Ohsawa, *Chem. Pharm. Bull.* (*Japan*), 1966, **14,** 518; (c) A. Balasubramanian, J. M. McIntosh, and V. Snieckus, *J. Org. Chem.*, 1970, **35,** 433; (d) K. Dimroth *et al.*, *Annalen*, 1957, **604,** 221.
66. For leading references, see R. A. Abramovitch and T. Takaya, *J. Org. Chem.*, 1972, **37,** 2022.
67. K. Hafner, D. Zinser, and K. L. Moritz, *Tetrahedron Letters*, 1964, 1733; also Ref. 65(c).
68. W. J. Linn, O. W. Webster, and R. E. Benson, Ref. 56(b); R. A. Abramovitch and V. Alexanian, *J. Org. Chem.*, 1976, **41,** 2144.
69. J. Fröhlich and F. Kröhnke, *Chem. Ber.*, 1971, **104,** 1621.
70. R. Krischke, R. Grashey, and R. Huisgen, *Annalen*, 1977, 498; Y. Tamura *et al.*, *J.C.S. Perkin* 1975, 406, and references cited therein.
71. A. Kakehi and S. Ito, *J. Org. Chem.*, 1974, **39,** 1542.
72. T. Sasaki *et al.*, *J.C.S. Perkin I*, 1973, 2089; Y. Tamura *et al.*, *ibid.*, 2091.
73. Y. Tamura *et al.*, (a) *Tetrahedron*, 1972, **28,** 21; (b) *J.C.S. Perkin I*, 1973, 2580.
74. A. H. Cook *et al.*, *J. Chem. Soc.*, 1941, 502.
75. T. Eicher, E. von Angerer, and A.-M. Hansen, *Annalen*, 1971, **746,** 102; T. Sasaki, K. Kanematsu, and A. Kakehi, *J. Org. Chem.*, 1971, **36,** 2451.
76. A. Kascheres and D. Marchi, *J. Org. Chem.*, 1975, **40,** 2985.
77. A. Kascheres and D. Marchi, *J.C.S. Chem. Comm.*, 1976, 275.
78. A. Kakehi *et al.*, *J. Org. Chem.*, 1976, **41,** 2739.
79. H. Quast and E. Frankenfeld, *Angew. Chem. Internat. Edn.*, 1965, **4,** 691; H. Quast and E. Schmitt, *Annalen*, 1970, **732,** 43; K. W. Ratts, R. K. Howe, and W. G. Phillips, *J. Amer. Chem. Soc.*, 1969, **91,** 6115.
80. R. A. Abramovitch and co-workers, *J. Heterocyclic Chem.*, 1969, **6,** 989; 1972, **9,** 1367; *J. Org. Chem.*, 1972, **37,** 1690, 3584.
81. R. M. Acheson, *Adv. Heterocyclic Chem.*, 1963, **1,** 143.
82. R. A. Abramovitch and J. G. Saha, *Adv. Heterocyclic Chem.*, 1966, **6,** 229.
83. Zh. I. Aksel'rod and V. M. Berezovskii, *Russ. Chem. Rev.*, 1970, **39,** 627.
84. A. R. Katritzky and B. J. Ridgewell, *J. Chem. Soc.*, 1963, (a) 3753; (b) 3882.
85. R. D. Brown and M. L. Heffernan, *Austral. J. Chem.*, 1956, **9,** 83.
86. A. R. Katritzky and co-workers, *J. Chem. Soc.* (*B*), 1967, (a) 1204; (b) 1211; (c) 1213; (d) 1219; (e) 1222; (f) 1226.
87. A. R. Katritzky and C. D. Johnson, *Angew. Chem. Internat. Edn.*, 1967, **6,** 608.
88. A. R. Katritzky and co-workers, *J. Chem. Soc.* (*B*), 1968, (a) 862; (b) 864; (c) 866; (d) 1477.
89. G. A. Olah, J. A. Olah, and N. A. Overchuk, *J. Org. Chem.*, 1965, **30,** 3373.
90. A. R. Katritzky and co-workers, *J. Chem. Soc.* (*B*), 1971, 2363.
91. A. Kirpal and E. Reiter, *Ber.*, 1925, **58,** 699; H. J. den Hertog and J. Overhoff, *Rec. Trav. chim.*, 1930, **49,** 552; P. Schorigin and A. Toptschiew, *Ber.*, 1936, **69,** 1874.
92. (a) E. Plaźek, *Ber.*, 1939, **72,** 577; (b) E. V. Brown and R. H. Neil, *J. Org. Chem.*, 1961, **26,** 3546.
93. L. Achremovicz, T. Batkowski, and Z. Skrowaczewska, *Roczniki Chem.*, 1964, **38,** 1317.
94. E. Plaźek, A. Sorokowska, and D. Tolopka, *Roczniki Chem.*, 1938, **18,** 210.
95. (a) E. Haack, *Ger. Pat.* 568549 (*Chem. Abs.*, 1933, **27,** 2697); (b) A. R. Katritzky, H. O. Tarhan, and S. Tarhan, *J. Chem. Soc.* (*B*), 1970, 114.
96. A. G. Burton, P. J. Halls, and A. R. Katritzky, *J.C.S. Perkin II*, 1972, 1953.
97. K. M. Dyumaev and L. D. Smirnov, *Russ. Chem. Rev.*, 1975, **44,** 823, and references cited therein.
98. F. Kögl, G. M. van der Want, and C. A. Salemink, *Rec. Trav. chim.*, 1948, **67,** 29.
99. (a) F. Korte, *Chem. Ber.*, 1952, **85,** 1012; (b) L. N. Pino and W. S. Zehrung, *J. Amer. Chem. Soc.*, 1955, **77,** 3154; (c) E. Koenigs, M. Mields, and H. Gurlt, *Ber.*, 1924, **57,** 1179.
100. (a) A. E. Tschitschibabin and A. W. Kirssanow, *Ber.*, 1928, **61,** 1223; (b) E. Plaźek, A. Marcinikow, and C. Stammer, *Roczniki Chem.*, 1935, **15,** 365.
101. A. R. Katritzky and co-workers, *J.C.S. Perkin II*, 1972, 1940.
102. W. Czuba, *Roczniki Chem.*, 1960, **34,** 905, 1639, 1647.
103. H. M. Curry and J. P. Mason, *J. Amer. Chem. Soc.*, 1951, **73,** 5043; J. W. Clark-Lewis and M. J. Thompson, *J. Chem. Soc.*, 1957, 442.
104. D. M. Smith and co-workers, unpublished results.
105. (a) H. J. den Hertog and J. Overhoff, *Rec. Trav. chim.*, 1950, **69,** 468; (b) E. Ochiai, *J. Org. Chem.*, 1953, **18,** 534.
106. R. B. Moodie, K. Schofield, and co-workers, *J. Chem. Soc.* (*B*), (a) 1966, 870; (b) 1968, 316.
107. E. Ochiai and C. Kaneko, *Chem. Pharm. Bull.* (*Japan*), 1960, **8,** 28.
108. E. Hayashi, *J. Pharm. Soc. Japan*, 1950, **70,** 142 (*Chem. Abs.*, 1950, **44,** 5880).
109. H. J. den Hertog and M. van Ammers, *Rec. Trav. chim.*, (a) 1955, **74,** 1160; (b) 1956, **75,** 1259.
110. D. E. Pearson *et al.*, *J. Org. Chem.*, 1961, **26,** 789.
111. (a) H. J. den Hertog, L. van der Does, and C. A. Landheer, *Rec. Trav. chim.*, 1962, **81,** 864; (b) van der Does and den Hertog, *ibid.*, 1965, **84,** 951; (c) leading references cited in the above two papers.
112. Z. Rodewald and E. Plaźek, *Ber.*, 1937, **70,** 1159; *Roczniki Chem.*, 1947, **21,** 150.
113. O. von Schickh, A. Binz, and A. Schulz, *Ber.*, 1936, **69,** 2593.

114. H. J. den Hertog and W. P. Combé, *Rec. Trav. chim.*, 1952, **71,** 745.
115. M. van Ammers, H. J. den Hertog, and B. Haase, *Tetrahedron*, 1962, **18,** 227.
116. M. Hamana and M. Yamazaki, *Chem. Pharm. Bull.* (*Japan*), 1961, **9,** 414.
117. (a) S. M. McElvain and M. A. Goese, *J. Amer. Chem. Soc.*, 1943, **65,** 2233; (b) H. J. den Hertog, H. C. van der Plas, and D. J. Buurman, *Rec. Trav. chim.*, 1958, **77,** 963.
118. (a) H. C. van der Plas and H. J. den Hertog, *Rec. Trav. chim.*, 1962, **81,** 841; (b) van der Plas and T. H. Crawford, *J. Org. Chem.*, 1961, **26,** 2611.
119. E. Plaźek, *Roczniki Chem.*, (a) 1936, **16,** 403; (b) 1937, **17,** 97.
120. E. Koenigs and O. Jungfer, *Ber.*, 1924, **57,** 2080.
121. M. van Ammers and H. J. den Hertog, *Rec. Trav. chim.*, (a) 1959, **78,** 586; (b) 1958, **77,** 340; 1962, **81,** 124.
122. (a) N. P. McCleland and R. H. Wilson, *J. Chem. Soc.*, 1932, 1263; C. D. Hurd and C. J. Morrissey, *J. Amer. Chem. Soc.*, 1955, **77,** 4658; (b) M. W. Swaney, M. J. Skeeters, and R. N. Shreve, *Ind. Eng. Chem.*, 1940, **32,** 360.
123. Ref. 122(b); also G. R. Clemo and G. A. Swan, *J. Chem. Soc.*, 1948, 198.
124. T. Takahashi and F. Yoneda, *Chem. Pharm. Bull.* (*Japan*), 1958, **6,** 611.
125. W. H. Mills and S. T. Widdows, *J. Chem. Soc.*, 1908, **93,** 1372.
126. L. Gattermann and A. Skita, *Ber.*, 1916, **49,** 494.
127. I. Ostromislensky, *J. Amer. Chem. Soc.*, 1934, **56,** 1713; J. Barycki and E. Plaźek, *Roczniki Chem.*, 1964, **38,** 553; P. Tomasik *et al.*, *ibid.*, 1970, **44,** 1131.
128. A. I. Titov, *J. Gen. Chem. USSR*, 1938, **8,** 1483.
129. J. A. Moore and F. J. Marascia, *J. Amer. Chem. Soc.*, 1959, **81,** 6049.
130. (a) W. T. Caldwell, F. T. Tyson, and L. Lauer, *J. Amer. Chem. Soc.*, 1944, **66,** 1479; (b) H. Bojarska-Dahlig and P. Nantka-Namirski, *Roczniki Chem.*, 1955, **29,** 1007.
131. E. J. Behrman and B. M. Pitt, *J. Amer. Chem. Soc.*, 1958, **80,** 3717.
132. Cilag Ltd., *Swiss Pat.* 243101; *Chem. Abs.*, 1949, **43,** 7973.
133. J. A. Baker and S. A. Hill, *J. Chem. Soc.*, 1962, 3464; C. A. Okafor, *J. Org. Chem.*, 1973, **38,** 4383.
134. G. Illuminati, *Adv. Heterocyclic Chem.*, 1964, **3,** 285; R. H. Shepherd and J. L. Fedrick, *ibid.*, 1965, **4,** 145.
135. J. Miller, 'Aromatic Nucleophilic Substitution', Elsevier, Amsterdam, 1968, pp. 234–289.
136. R. Huisgen and J. Sauer, *Chem. Ber.*, 1958, **91,** 1453.
137. T. Kauffmann and R. Wirthwein, *Angew. Chem. Internat. Edn.*, 1971, **10,** 20, and references cited therein.
138. F. W. Bergstrom and W. C. Fernelius, *Chem. Rev.*, 1933, **12,** 154; R. Levine and W. C. Fernelius, *ibid.*, 1954, **54,** 537; M. T. Leffler, *Org. Reactions*, 1942, **1,** 91.
139. F. W. Bergstrom and S. H. McAllister, *J. Amer. Chem. Soc.*, 1930, **52,** 2845.
140. (a) F. V. Scalzi and N. F. Golob, *J. Org. Chem.*, 1971, **36,** 2541; (b) R. F. Francis, W. Davis, and J. T. Wisener, *ibid.*, 1974, **39,** 59.
141. K. Ziegler and H. Zeiser, *Annalen*, 1931, **485,** 174; L. A. Walter, *Org. Synth.*, *Coll. Vol.* 3, 757; A. M. Jones, C. A. Russell, and S. Skidmore, *J. Chem. Soc.* (*C*), 1969, 2245.
142. R. F. Francis, J. T. Wisener and J. M. Paul, *Chem. Comm.*, 1971, 1420.
143. R. A. Abramovitch and G. A. Poulton, *J. Chem. Soc.* (*B*), 1969, 901; R. A. Abramovitch, C. S. Giam, and G. A. Poulton, *J. Chem. Soc.* (*C*), 1970, 128.
144. D. Bryce-Smith, P. J. Morris, and B. J. Wakefield, *J.C.S. Perkin I*, 1976, 1977.
145. C. S. Giam and J. L. Stout, *Chem. Comm.*, 1969, 142.
146. K. Ziegler and H. Zeiser, *Ber.*, 1930, **63,** 1847.
147. (a) T. A. Geissman *et al.*, *J. Org. Chem.*, 1946, **11,** 741; (b) H. Gilman and J. T. Edward, *Canad. J. Chem.*, 1953, **31,** 457.
148. C. S. Giam, E. E. Knaus, and co-workers, (a) *Canad. J. Chem.*, 1975, **53,** 2305; (b) *J. Heterocyclic Chem.*, 1976, **13,** 789.
149. C. S. Giam, E. E. Knaus, and F. M. Pasutto, *J. Org. Chem.*, 1974, **39,** 3565.
150. P. Doyle and R. R. J. Yates, *Tetrahedron Letters*, 1970, 3371.
151. C. S. Giam and S. D. Abbott, *J. Amer. Chem. Soc.*, 1971, **93,** 1294.
152. *Cf.* J. Paleček, K. Vondra, and J. Kuthan, *Coll. Czech. Chem. Comm.*, 1969, **34,** 2991.
153. C. A. Fyfe, *Tetrahedron Letters*, 1968, 659; R. Schaal *et al.*, *ibid.*, 1970, 1393; J. Miller *et al.*, *Austral. J. Chem.*, 1970, **23,** 957, 963.
154. See U. Eisner and J. Kuthan, *Chem. Rev.*, 1972, **72,** 1, and references cited therein.
155. L. M. Thiessen, J. A. Lepoivre, and F. C. Alderweireldt, *Tetrahedron Letters*, 1974, 59; R. E. Lyle and D. L. Comins, *J. Org. Chem.*, 1976, **41,** 3250.
156. O. Červinka *et al.*, *Coll. Czech. Chem. Comm.*, 1962, **27,** 567.
157. (a) F. Kröhnke, K. Ellegast, and E. Bertram, *Annalen*, 1956, **600,** 176; (b) D. N. Kursanov and N. K. Baranetskaya, *Bull. Acad. Sci. USSR, Div. Chem. Sci.*, 1961, 1586.
158. R. A. Abramovitch and A. R. Vinutha, *J. Chem. Soc.* (*B*), 1971, 131.
159. *Cf.* H. Decker, *Ber.*, 1905, **38,** 2493; E. Koenigs, K. Köhler, and K. Blindow, *ibid.*, 1925, **58,** 933.
160. J. Kaválek *et al.*, *Coll. Czech. Chem. Comm.*, (a) 1975, **40,** 1166; (b) 1975, **40,** 1932; (c) 1976. **41,** 67.
161. W. von E. Doering and W. E. McEwen, *J. Amer. Chem. Soc.*, 1951, **73,** 2104; T. Severin, H. Lerche, and D. Bätz, *Chem. Ber.*, 1969, **102,** 2163.
162. For examples, and exceptions, see Ref. 154; also R. H. Reuss, N. G. Smith, and L. J. Winters, *J. Org. Chem.*, 1974, **39,** 2027.

163. R. H. Reuss and L. J. Winters, *J. Org. Chem.*, 1973, **38,** 3993; S. Hünig *et al.*, *Annalen*, 1973, 1036.
164. *Cf.* A. R. Katritzky and E. Lunt, *Tetrahedron*, 1969, **25,** 4291.
165. R. A. Abramovitch and I. Shinkai, *Accounts Chem. Res.*, 1976, **9,** 192.
166. J. H. Markgraf *et al.*, *J. Amer. Chem. Soc.*, 1963, **85,** 958.
167. M. Hamana and M. Yamazaki, *Chem. Pharm. Bull. (Japan)*, 1963, **11,** 415.
168. C. Kaneko, *Chem. Pharm. Bull. (Japan)*, 1960, **8,** 286; Y. Kobayashi and I. Kumadaki, *ibid.*, 1969, **17,** 510.
169. W. E. Feely and E. M. Beavers, *J. Amer. Chem. Soc.*, 1959, **81,** 4004; T. Okamoto and H. Tani, *Chem. Pharm. Bull. (Japan)*, 1959, **7,** 130; H. Tani, *ibid.*, 930; M. Ferles and M. Jankovsky, *Coll. Czech. Chem. Comm.*, 1968, **33,** 3848.
170. R. A. Abramovitch and co-workers, *J. Org. Chem.*, 1974, **39,** 1795, 1802; *J. Heterocyclic Chem.*, 1975, **12,** 501.
171. E. A. Mailey and L. R. Ocone, *J. Org. Chem.*, 1968, **33,** 3343; R. E. Banks, R. N. Haszeldine, and J. M. Robinson, *J.C.S. Perkin I*, 1976, 1226.
172. T. J. van Bergen and R. M. Kellogg, *J. Org. Chem.*, 1971, **36,** 1705; P. Schiess *et al.*, *Helv. Chim. Acta*, 1974, **57,** 1676.
173. U. Fritzsche and S. Hünig, *Annalen*, 1974, 1407.
174. H. Gilman and S. M. Spatz, *J. Org. Chem.*, 1951, **16,** 1485.
175. D. I. Davies, J. N. Done, and D. H. Hey, *J. Chem. Soc. (C)*, 1969, 2019.
176. Ya. N. Ivashchenko, S. D. Moshchitskii, and A. K. Eliseeva, *Khim. geterotsikl. Soedinenii*, 1970, 58; F. Binns and H. Suschitzky, *J. Chem. Soc. (C)*, 1971, 1223.
177. (a) M. J. Pieterse and H. J. den Hertog, *Rec. Trav. chim.*, 1961, **80,** 1376; (b) H. J. den Hertog *et al.*, *ibid.*, 1974, **93,** 195; (c) T. Kauffmann, R. Nürnberg, and R. Wirthwein, *Chem. Ber.*, 1969, **102,** 1161.
178. R. J. Martens and H. J. den Hertog, *Rec. Trav. chim.*, (a) 1964, **83,** 621; (b) 1967, **86,** 655.
179. T. O. Bamkole and J. Hirst, *J. Chem. Soc. (B)*, 1969, 848; H. Grube and H. Suhr, *Chem. Ber.*, 1969, **102,** 1570.
180. C. K. Banks, *J. Amer. Chem. Soc.*, 1944, **66,** 1127.
181. (a) O. Fischer, *Ber.*, 1899, **32,** 1297; (b) B. Emmert and W. Dorn, *ibid.*, 1915, **48,** 687.
182. C. Zwart and J. P. Wibaut, *Rec. Trav. chim.*, 1955, **74,** 1062.
183. E. Koenigs, W. Weiss, and A. Zscharn, *Ber.*, 1926, **59,** 316; D. S. Tarbell *et al.*, *J. Amer. Chem. Soc.*, 1948, **70,** 1381.
184. J. P. Wibaut and G. Tilman, *Rec. Trav. chim.*, 1933, **52,** 987.
185. (a) I. Collins, S. M. Roberts, and H. Suschitzky, *J. Chem. Soc. (C)*, 1971, 167; (b) B. J. Wakefield *et al.*, *J.C.S. Perkin I*, 1974, 2310; H. J. den Hertog *et al.*, *Rec. Trav. chim.*, 1950, **69,** 637; S. M. Roberts and H. Suschitzky, *J. Chem. Soc. (C)*, 1968, 2844.
186. R. D. Chambers, J. Hutchinson, and W. K. R. Musgrave, *J. Chem. Soc.*, 1964, (a) 3573; (b) 3736; (c) 5634.
187. A. Michaelis and O. Hillmann, *Annalen*, 1907, **354,** 91.
188. A. R. Katritzky, *J. Chem. Soc.*, (a) 1956, 2404; (b) 1957, 191.
189. G. Coppens *et al.*, *Bull. Soc. chim. Belges*, 1961, **70,** 480.
190. M. Bellas and H. Suschitzky, *J. Chem. Soc.*, 1963, 4007.
191. J. P. Wibaut and W. J. Holmes-Kamminga, *Bull. Soc. chim. France*, 1958, 424.
192. R. Jujo, *J. Pharm. Soc. Japan*, 1946, **66,** 21; *Chem. Abs.*, 1951, **45,** 6200.
193. L. Pannizzon, *Helv. Chim. Acta*, 1944, **27,** 1748.
194. N. Sperber *et al.*, *J. Amer. Chem. Soc.*, 1951, **73,** 3856.
195. W. Gruber, *Canad. J. Chem.*, 1953, **31,** 1181.
196. R. Adams and W. Reifschneider, *J. Amer. Chem. Soc.*, 1957, **79,** 2236.
197. T. R. Lewis and S. Archer, *J. Amer. Chem. Soc.*, 1951, **73,** 2109.
198. J. D. Reinheimer *et al.*, *J. Amer. Chem. Soc.*, 1962, **84,** 2770.
199. (a) C. Räth, *U.S. Pat.* 1778784; *Brit. Pat.* 288628; *Chem. Abs.*, 1931, **25,** 116; 1929, **23,** 607; (b) L. Haitinger and A. Lieben, *Monatsh.*, 1885, **6,** 279.
200. H. L. Bradlow and C. A. VanderWerf, *J. Org. Chem.*, (a) 1949, **14,** 509; (b) 1951, **16,** 1143.
201. E. Koenigs and A. Fulde, *Ber.*, 1927, **60,** 2106; R. R. Bishop, E. A. S. Cavell, and N. B. Chapman, *J. Chem. Soc.*, 1952, 437.
202. (a) J. A. Zoltewicz and A. A. Sale, *J. Org. Chem.*, 1971, **36,** 1455; (b) H. Maier-Bode, *Ber.*, 1936, **69,** 1534.
203 (a) J. A. Zoltewicz and A. A. Sale, *J. Org. Chem.*, 1970, **35,** 3462; (b) W. J. van Zoest and H. J. den Hertog, *Rec. Trav. chim.*, 1974, **93,** 166.
204. G. R. Newkome *et al.*, *Synthesis*, 1974, 707.
205. (a) J. P. Wibaut, B. W. Speekman, and H. M. van Wagtendonk, *Rec. Trav. chim.*, 1939, **58,** 1100; (b) J. P. Wibaut, P. W. Haayman, and J. van Dijk, *ibid.*, 1940, **59,** 202.
206. (a) E. Shaw *et al.*, *J. Amer. Chem. Soc.*, 1950, **72,** 4362; (b) M. Hamana and M. Yamazaki, *J. Pharm. Soc. Japan*, 1961, **81,** 612 (*Chem. Abs.*, 1961, **55,** 24742); E. Profft and G. Schulz, *Arch. Pharm.*, 1961, **294,** 292.
207. *Cf.* M. Forchiassin, G. Illuminati, and G. Sleiter, *J. Heterocyclic Chem.*, 1969, **6,** 879.
208. R. E. Banks *et al.*, *J. Chem. Soc.*, 1965, (a) 575, (b) 594.
209. W. Marckwald, W. Klemm, and H. Trabert, *Ber.*, 1900, **33,** 1556; J. R. Thirtle, *J. Amer. Chem. Soc.*, 1946, **68,** 342.

210. M. A. Phillips and H. Shapiro, *J. Chem. Soc.*, 1942, 584; M. P. V. Boarland and J. F. W. McOmie, *ibid.*, 1951, 1218; B. Iddon *et al.*, *J.C.S. Perkin I*, 1974, 2300.
211. See, for example, R. Adams and A. Ferretti, *J. Amer. Chem. Soc.*, 1959, **81,** 4927; L. Bauer and A. Gardella, *J. Org. Chem.*, 1963, **28,** 1320; J. Bratt and H. Suschitzky, *J.C.S. Perkin I*, 1973, 1689.
212. R. F. Evans and H. C. Brown, *J. Org. Chem.*, 1962, **27,** 1329; R. N. Haszeldine and co-workers, *J. Chem. Soc.* (*C*), 1969, 1660.
213. R. Graf, *Ber.*, 1931, **64,** 21; *J. prakt. Chem.*, 1932, **133,** 36; 1937, **148,** 13.
214. J. Hamer *et al.*, *Rec. Trav. chim.*, 1962, **81,** 1058; M. M. Boudakian, *J. Heterocyclic Chem.*, 1968, **5,** 683.
215. (a) W. Czuba and E. Plaźek, *Rec. Trav. chim.*, 1958, **77,** 92; (b) G. M. Badger and R. P. Rao, *Austral. J. Chem.*, 1964, **17,** 1399; (c) R. M. Johnson, *J. Chem. Soc.* (*B*), 1966, 1062.
216. Z. Talik, *Roczniki Chem.*, 1961, **35,** 475.
217. H. J. den Hertog *et al.*, *Rec. Trav. chim.*, 1949, **68,** 275; 1950, **69,** 1281.
218. H. J. den Hertog and W. P. Combé, *Rec. Trav. chim.*, 1951, **70,** 581.
219. E. C. Taylor and J. S. Driscoll, (a) *J. Amer. Chem. Soc.*, 1960, **82,** 3141; (b) *J. Org. Chem.*, 1960, **25,** 1716.
220. H. J. den Hertog, F. W. Broekman, and W. P. Combé, *Rec. Trav. chim.*, 1951, **70,** 105; H. J. den Hertog and J. de Bruyn, *ibid.*, p. 182.
221. (a) A. Dondoni, A. Mangini, and G. Mossa, *J. Heterocyclic Chem.*, 1969, **6,** 143; (b) R. M. Johnson, *J. Chem. Soc.* (*B*), 1966, 1058.
222. T. Takahashi, I. Yamashita, and J. Iwai, *J. Pharm. Soc. Japan*, 1958, **78,** 943 (*Chem. Abs.*, 1958, **52,** 20144); T. Okamoto and M. Itoh, *Chem. Pharm. Bull.* (*Japan*), 1963, **11,** 785.
223. E. Ochiai and M. Katada, *J. Pharm. Soc. Japan*, 1943, **63,** 186 (*Chem. Abs.*, 1951, **45,** 5151).
224. E. Koenigs and H. Greiner, *Ber.*, 1931, **64,** 1049; D. Jerchel, H. Fischer, and K. Thomas, *Chem. Ber.*, 1956, **89,** 2921; D. Jerchel and L. Jacob, *ibid.*, 1958, **91,** 1266; J. P. Wibaut, S. Herzberg, and J. Schlatmann, *Rec. Trav. chim.*, 1954, **73,** 140.
225. M. P. Sammes, Ho King Wah, and A. R. Katritzky, *J.C.S. Perkin I*, 1977, 327.
226. R. Camps, *Arch. Pharm.*, 1902, **240,** 345; E. Kalatzis, *J. Chem. Soc.* (*B*), 1967, 273, 277.
227. W. Marckwald, *Ber.*, 1894, **27,** 1317; C. F. H. Allen and J. R. Thirtle, *Org. Synth.*, *Coll. Vol. 3*, 136; T. Talik and E. Plaźek, *Roczniki Chem.*, 1955, **29,** 1019; 1959, **33,** 387, 1343.
228. R. D. Beaty and W. K. R. Musgrave, *J. Chem. Soc.*, 1952, 875; P. B. Desai, *J.C.S. Perkin I*, 1973, 1865, and references cited therein.
229. H. Vorbrüggen, *Synthesis*, 1973, 301.
230. H. von Pechmann and O. Baltzer, *Ber.*, 1891, **24,** 3144; F. H. Case, *J. Amer. Chem. Soc.*, 1946, **68,** 2574; A. H. Berrie, G. T. Newbold, and F. S. Spring, *J. Chem. Soc.*, 1952, 2042; K. Heyns and G. Vogelsang, *Chem. Ber.*, 1954, **87,** 13.
231. H. King and L. L. Ware, *J. Chem. Soc.*, 1939, 873; J. Renault, *Bull. Soc. chim. France*, 1953, 1001.
232. *Cf.* A. F. Vompe *et al.*, *Tetrahedron*, 1958, **2,** 361; T. Itai, *Chem. Abs.*, 1955, **49,** 327.
233. H. Tani, *J. Pharm. Soc. Japan*, 1960, **80,** 1418 (*Chem. Abs.*, 1961, **55,** 6477).
234. L. Marion and W. F. Cockburn, *J. Amer. Chem. Soc.*, 1949, **71,** 3402.
235. E. Ochiai and I. Suzuki, *Pharm. Bull.* (*Japan*), 1954, **2,** 247; Y. Suzuki, *ibid.*, 1957, **5,** 13, 78; *J. Pharm. Soc. Japan*, 1961, **81,** 917 (*Chem. Abs.*, 1961, **55,** 27305).
236. *Cf.* U. Schmidt and G. Giesselmann, *Chem. Ber.*, 1960, **93,** 1590.
237. *Cf.* P. Blumbergs *et al.*, *J. Org. Chem.*, 1969, **34,** 4065.
238. G. B. Barlin and W. V. Brown, *J. Chem. Soc.* (*B*), 1967, 648.
239. (a) R. I. Ellin, *J. Amer. Chem. Soc.*, 1958, **80,** 6588; (b) E. Posiomek, *J. Org. Chem.*, 1963, **28,** 590.
240. J. K. Landquist, *J.C.S. Perkin I*, 1976, 454.
241. *Cf.* R. D. Brown, *J. Chem. Soc.*, 1956, 272; J. Paleček, V. Skala, and J. Kuthan, *Coll. Czech. Chem. Comm.*, 1969, **34,** 1110.
242. J. P. Wibaut and J. R. Nicolai, *Rec. Trav. chim.*, 1939, **58,** 709; H. J. den Hertog and J. P. Wibaut, *ibid.*, 1932, **51,** 381.
243. M. M. Boudakian and co-workers, *J. Heterocyclic Chem.*, 1967, **4,** 375, 377.
244. For a review, see H. J.-M. Dou, G. Vernin, and J. Metzger, *Bull. Soc. chim. France*, 1971, 4593.
245. K. C. Bass and P. Nababsing, *J. Chem. Soc.* (*C*), 1970, 2169.
246. R. A. Abramovitch and K. Kenaschuk, *Canad. J. Chem.*, 1967, **45,** 509.
247. S. Goldschmidt and M. Minsinger, *Chem. Ber.*, 1954, **87,** 956.
248. F. Minisci and co-workers, (a) *Tetrahedron*, 1971, **27,** 3573; (b) *ibid.*, p. 3655; (c) *J.C.S. Perkin II*, 1972, 2034; (d) *Tetrahedron Letters*, 1970, 15; (e) *ibid.*, 1973, 645.
249. P. A. Claret and G. H. Williams, *J. Chem. Soc.* (*C*), 1969, 146.
250. H. J.-M. Dou, G. Vernin, and J. Metzger, *Compt. rend.* (*C*), 1971, **272,** 854; *Bull. Soc. chim. France*, 1971, 3553.
251. *Cf.* K. K. Chiu and H. H. Huang, *J. Chem. Soc.* (*C*), 1969, 2758.
252. For a review of earlier work, see G. Vernin, H. J.-M. Dou, and J. Metzger, *Bull. Soc. chim. France*, 1972, 1173.
253. R. A. Abramovitch and G. Tertzakian, *Canad. J. Chem.*, 1965, **43,** 940.
254. R. A. Abramovitch, D. H. Hey, and R. D. Mulley, *J. Chem. Soc.*, 1954, 4263.
255. R. A. Abramovitch and K. A. H. Adams, *Canad. J. Chem.*, 1961, **39,** 2516.
256. J. I. G. Cadogan and co-workers, *J. Chem. Soc.*, 1965, 4831.
257. K. E. Wilzbach and D. J. Rausch, *J. Amer. Chem. Soc.*, 1970, **92,** 2178; *cf.* O. L. Chapman, C. L. McIntosh and J. Pacansky, *ibid.*, 1973, **95,** 614.

258. (a) M. G. Barlow, J. G. Dingwall, and R. N. Haszeldine, *J.C.S. Perkin I*, 1973, 1542; (b) Y. Kobayashi *et al.*, *Chem. Pharm. Bull.* (*Japan*), 1976, **24,** 2219.
259. S. Caplain and A. Lablache-Combier, *Chem. Comm.*, 1970, 1247; *cf.* T. J. van Bergen and R. M. Kellogg, *J. Amer. Chem. Soc.*, 1972, **94,** 8451.
260. R. C. De Selms and W. R. Schleigh, *Tetrahedron Letters*, 1972, 3563, and references cited therein; H. Furrer, *Chem. Ber.*, 1972, **105,** 2780; E. Ager, G. E. Chivers, and H. Suschitzky, *J.C.S. Perkin I*, 1973, 1125.
261. L. Kaplan, J. W. Pavlik, and K. E. Wilzbach, *J. Amer. Chem. Soc.*, 1972, **94,** 3283.
262. A. R. Katritzky and H. Wilde, *J.C.S. Chem. Comm.*, 1975, 770.
263. C. E. Loader and C. J. Timmons, (a) *J. Chem. Soc.* (*C*), 1966, 1078; (b) *ibid.*, 1967, 1457; (c) P. L. Kumler and R. A. Dybas, *J. Org. Chem.*, 1970, **35,** 125, 3825.
264. G. Galiazzo, P. Bortolus, and G. Cauzzo, *Tetrahedron Letters*, 1966, 3717.
265. H.-H. Perkampus and G. Kassebeer, *Annalen*, 1966, **696,** 1.
266. H.-H. Perkampus and B. Behjati, *J. Heterocyclic Chem.*, 1974, **11,** 511.
267. R. E. Doolittle and C. K. Bradsher, *Chem. and Ind.* (*London*), 1965, 1631.
268. J. L. R. Williams, *J. Org. Chem.*, 1960, **25,** 1839; J. L. R. Williams, S. K. Webster, and J. A. Van Allan, *ibid.*, 1961, **26,** 4893.
269. D. G. Whitten and Y. J. Lee, *J. Amer. Chem. Soc.*, 1970, **92,** 415.
270. A. Fozard and C. K. Bradsher, *J. Org. Chem.*, (a) 1966, **31,** 2346; (b) 1967, **32,** 2966.
271. G. G. Spence, E. C. Taylor, and O. Buchardt, *Chem. Rev.*, 1970, **70,** 231.
272. F. Bellamy and J. Streith, *Heterocycles*, 1976, **4,** 1391.
273. T. Sasaki *et al.*, *J. Org. Chem.*, 1970, **35,** 426.
274. R. A. Abramovitch and T. Takaya, *J. Org. Chem.*, 1973, **38,** 3311.
275. V. Snieckus and G. Kan, *Chem. Comm.*, 1970, 172.
276. J. Streith *et al.*, *Bull. Soc. chim. France*, 1969, 948.
277. L. A. Paquette and G. Slomp, *J. Amer. Chem. Soc.*, 1963, **85,** 765; E. C. Taylor and R. O. Kan, *ibid.*, p. 776; M. Laing, *Proc. Chem. Soc.*, 1964, 343.
278. D. Bryce-Smith, A. Gilbert, and S. Krestonosich, *Tetrahedron Letters*, 1977, 385.
279. R. E. Lyle and P. S. Anderson, *Adv. Heterocyclic Chem.*, 1966, **6,** 45.
280. P. T. Lansbury and J. O. Peterson, *J. Amer. Chem. Soc.*, 1963, **85,** 2236.
281. F. Bohlmann, *Chem. Ber.*, 1952, **85,** 390.
282. M. Ferles, *Chem. Abs.*, 1961, **55,** 24740.
283. F. Liberatore, V. Carelli, and M. Cardellini, *Tetrahedron Letters*, 1968, 4735.
284. E. Booker and U. Eisner, *J.C.S. Perkin I*, 1975, 929.
285. (a) S. Yamada and co-workers, *Tetrahedron Letters*, 1969, 3101; (b) A. Signor, A. R. Katritzky, and co-workers, *J. Chem. Soc.* (*B*), 1970, 1567.
286. R. M. Acheson and G. Paglietti, *J.C.S. Perkin I*, 1976, 45.
287. M. Ferles and J. Pliml, *Adv. Heterocyclic Chem.*, 1970, **12,** 43.
288. O. Warburg, W. Christian, and A. Griese, *Biochem. Z.*, 1935, **282,** 157.
289. B. Emmert, *Ber.*, (a) 1917, **50,** 31; (b) 1919, **52,** 1351.
290. C. R. Smith, *J. Amer. Chem. Soc.*, 1924, **46,** 414.
291. J. E. Colchester and J. H. Entwistle, *Brit. Pat.* 1077367; *U.S. Pat.* 3478042 (*Chem. Abs.*, 1968, **68,** 87175; 1970, **72,** 31627.
292. A. T. Nielsen *et al.*, *J. Org. Chem.*, 1964, **29,** 2175.
293. (a) R. E. Lyle, in 'Pyridine and its Derivatives', Supplement, ed. R. A. Abramovitch, Wiley, New York, 1974, vol. 1, p. 137; (b) R. A. Abramovitch and G. M. Singer, *ibid.*, p. 91.
294. H. J. den Hertog and H. C. van der Plas, *Adv. Heterocyclic Chem.*, 1965, **4,** 122, and references cited thorein.
295. T. Kauffmann, *Angew. Chem. Internat. Edn.*, 1965, **4,** 543.
296. R. W. Hoffmann, 'Dehydrobenzene and Cycloalkynes', Verlag Chemie, Weinheim, and Academic Press, New York, 1967, Chapter 6.
297. H. J. den Hertog and H. C. van der Plas, in 'The Chemistry of Acetylenes', ed. H. G. Viehe, Marcel Dekker, New York, 1969, Chapter 17.
298. G. W. J. Fleet and I. Fleming, *J. Chem. Soc.* (*C*), 1969, 1758.
299. D. J. Berry and B. J. Wakefield, *J. Chem. Soc.* (*C*), 1969, 2342.
300. B. J. Wakefield and co-workers, *J. Chem. Soc.* (*C*), (a) 1969, 1973; (b) 1969, 2376; (c) 1971, 1227; (d) 1968, 2727.
301. H. Heaney and co-workers, *J. Chem. Soc.* (*C*), 1969, 1908.
302. H. L. Jones and D. L. Beveridge, *Tetrahedron Letters*, 1964, 1577.
303. J. A. Zoltewicz and C. L. Smith, *Tetrahedron*, 1969, **25,** 4331.
304. W. Adam, A. Grimison, and R. Hoffmann, *J. Amer. Chem. Soc.*, 1969, **91,** 2590.
305. R. A. Abramovitch, G. M. Singer, and A. R. Vinutha, *Chem. Comm.*, 1967, 55; J. A. Zoltewicz and G. M. Kauffmann, *J. Org. Chem.*, 1969, **34,** 1405.
306. T. Sasaki, K. Kanematsu, and M. Uchide, *Bull. Chem. Soc. Japan*, 1971, **44,** 858.
307. (a) M. Mallet, G. Quéguiner, and P. Pastour, *Compt. rend.* (*C*), 1972, **274,** 719; (b) H. N. M. van der Lans and H. J. den Hertog, *Rec. Trav. chim.*, 1968, **87,** 549.
308. T. Kauffmann *et al.*, *Angew. Chem. Internat. Edn.*, 1964, **3,** 650.

309. J. M. Kramer and R. S. Berry, *J. Amer. Chem. Soc.*, 1971, **93,** 1303.
310. T. Kato and T. Niitsuma, *Chem. Pharm. Bull. (Japan)*, 1965, **13,** 963.
311. R. Peereboom and H. J. den Hertog, *Rec. Trav. chim.*, 1974, **93,** 281.
312. J. A. Zoltewicz and C. Nisi, *J. Org. Chem.*, 1969, **34,** 765.
313. H. Boer and H. J. den Hertog, *Tetrahedron Letters*, 1969, 1943.
314. R. Levine and W. W. Leake, *Science*, 1955, **121,** 780.
315. H. N. M. van der Lans, H. J. den Hertog, and A. van Veldhuizen, *Tetrahedron Letters*, 1971, 1875.
316. J. I. Jones, *J. Chem. Soc.*, 1950, 1392.
317. W. Shive, E. G. Ballweber, and W. W. Ackermann, *J. Amer. Chem. Soc.*, 1946, **68,** 2144; F. L. J. Sixma, *Rec. Trav. chim.*, 1952, **71,** 1124.
318. (a) T. Zincke, *Annalen*, 1904, **330,** 361; 1904, **333,** 296; (b) W. König, *J. prakt. Chem.*, 1905, **70,** 19.
319. W. König, *J. prakt. Chem.*, 1904, **69,** 105; R. Oda and S. Mita, *Bull. Chem. Soc. Japan*, 1963, **36,** 103.
320. J. Schnekenburger, D. Heber, and E. Heber-Brunschweiger, *Annalen*, 1976, 1799.
321. T. Zincke and W. Würker, *Annalen*, 1905, **338,** 107; S. S. Malhotra and M. C. Whiting, *J. Chem. Soc.*, 1960, 3812; J. Kavalek *et al.*, *Coll. Czech. Chem. Comm.*, 1974, **39,** 2047, 2056.
322. F. T. Boyle and R. Hull, *J.C.S. Perkin I*, 1974, 1541.
323. Y. Tamura *et al.*, *J. Heterocyclic Chem.*, 1972, **9,** 865.
324. J. H. Blanch and K. Fretheim, *J. Chem. Soc. (C)*, 1971, 1892.
325. W. König *et al.*, *J. prakt. Chem.*, 1965, **30,** 96.
326. J. W. Streef and H. J. den Hertog, *Rec. Trav. chim.*, 1969, **88,** 1391.
327. H. J. den Hertog and co-workers, *Tetrahedron Letters*, 1966, 4325.
328. H. N. M. van der Lans and H. J. den Hertog, *Tetrahedron Letters*, 1973, 1887.
329. W. D. Crow, M. N. Paddon-Row, and D. S. Sutherland, *Tetrahedron Letters*, 1972, 2239, and references cited therein.
330. W. D. Crow and co-workers, *Austral. J. Chem.*, 1975, **28,** (a) 1741, (b) 1763.
331. C. Mayor and C. Wentrup, *J. Amer. Chem. Soc.*, 1975, **97,** 7467.
332. W. D. Crow and C. Wentrup, (a) *Chem. Comm.*, 1969, 1387; (b) *Tetrahedron*, 1970, **26,** 3965.
333. R. F. C. Brown and R. J. Smith, *Austral. J. Chem.*, 1972, **25,** 607.
334. R. A. Abramovitch and B. W. Cue, *J. Amer. Chem. Soc.*, 1976, **98,** 1478.
335. C. Thétaz and C. Wentrup, *J. Amer. Chem. Soc.*, 1976, **98,** 1258; R. Harder and C. Wentrup, *ibid.*, 1259.
336. R. A. Abramovitch and co-workers, *J. Heterocyclic Chem.*, 1976, **13,** 415.
337. A. R. Katritzky and J. M. Lagowski, *Adv. Heterocyclic Chem.*, 1963, **1,** 339.
338. P. Beak *et al.*, *J. Amer. Chem. Soc.*, 1976, **98,** 171; J. Frank and A. R. Katritzky, *J.C.S. Perkin II*, 1976, 1428.
339. *Cf.* A. Gordon, A. R. Katritzky, and S. K. Roy, *J. Chem. Soc. (B)*, 1968, 556; E. Spinner and G. B. Yeoh, *ibid.*, 1971, 279.
340. (a) E. Spinner and J. C. B. White, *J. Chem. Soc. (B)*, 1966, 996; (b) J. P. Shoffner, L. Bauer, and C. L. Bell, *J. Heterocyclic Chem.*, 1970, **7,** 479.
341. (a) A. R. Katritzky and R. E. Reavill, *J. Chem. Soc.*, 1965, 3825; (b) E. Spinner, *ibid.*, 1962, 3127; (c) E. Spinner, *ibid.*, 1962, 3119.
342. N. Kornblum and G. P. Coffey, *J. Org. Chem.*, 1966, **31,** 3447, 3449; H. Tieckelmann *et al.*, *ibid.*, 1967, **32,** 4040.
343. (a) S. L. Shapiro, K. Weinberg, and L. Freedman, *J. Amer. Chem. Soc.*, 1959, **81,** 5140; (b) D. A. Prins, *Rec. Trav. chim.*, 1957, **76,** 58.
344. N. Clauson-Kaas and co-workers, *Acta Chem. Scand.*, 1969, **23,** 1791.
345. H. Meyer, *Monatsh.*, 1905, **26,** 1311.
346. J. N. Gardner and A. R. Katritzky, *J. Chem. Soc.*, 1957, 4375.
347. (a) C. J. Cavallito and T. H. Haskell, *J. Amer. Chem. Soc.*, 1944, **66,** 1166; (b) P. A. Singgih and M. J. Janssen, *Tetrahedron Letters*, 1971, 4223; I. Fleming and D. Philippides, *J. Chem. Soc. (C)*, 1970, 2426.
348. A. McKillop, M. J. Zelesko, and E. C. Taylor, *Tetrahedron Letters*, 1968, 4945.
349. H. Bojarska-Dahlig and T. Urbanski, *Chem. Abs.*, 1954, **48,** 1337.
350. E. C. Taylor, F. Kienzle, and A. McKillop, *J. Org. Chem.*, 1970, **35,** 1672.
351. M. Hamer and E. P. Lira, *J. Heterocyclic Chem.*, 1972, **9,** 215.
352. See, for example, Refs. 209, 231; also A. Albert and G. B. Barlin, *J. Chem. Soc.*, 1959, 2384; D. J. Fry and J. D. Kendall, *ibid.*, 1951, 1716; R. A. Jones and A. R. Katritzky, *ibid.*, 1958, 3610.
353. H. L. Yale, in 'Pyridine and its Derivatives', Supplement, Wiley, New York, 1975, part 4, p. 189.
354. A. M. Comrie, *J. Chem. Soc.*, 1963, 688.
355. (a) A. E. Tschitschibabin, R. A. Konowalowa, and A. A. Konowalowa, *Ber.*, 1921, **54,** 814; T. M. Sharp, *J. Chem. Soc.*, 1939, 1855; (b) A. E. Tschitschibabin and E. D. Ossetrowa, *Ber.*, 1925, **58,** 1708; E. Koenigs, H. Friedrich, and H. Jurany, *ibid.*, 2571.
356. A. E. Tschitschibabin and I. L. Knunjanz, *Ber.*, 1928, **61,** 427.
357. R. Frampton, C. D. Johnson, and A. R. Katritzky, *Annalen*, 1971, **749,** 12.
358. R. A. Jones and A. R. Katritzky, *J. Chem. Soc.*, 1959, 1317; 1961, 378.
359. A. W. Johnson, T. J. King, and J. R. Turner, *J. Chem. Soc.*, 1960, 1509; P. Grammaticakis, *Bull. Soc. chim. France*, 1959, 480; P. A. Lyon and C. B. Reese, *J.C.S. Perkin I*, 1974, 2645.
360. B. W. Harris, M. D. Coburn, and E. P. Papadopoulos, *J. Heterocyclic Chem.*, 1976, **13,** 845.
361. H. Dorn, G. Hilgetag, and co-workers, *Chem. Ber.*, 1964, **97,** 695; *Rec. Trav. chim.*, 1964, **83,** 189.

362. M. Shur and S. S. Israelstam, *J. Org. Chem.*, 1968, **33,** 3015.
363. W. L. Mosby, *J. Org. Chem.*, 1961, **26,** 1316.
364. R. Adams and J. S. Dix, *J. Amer. Chem. Soc.*, 1958, **80,** 4618.
365. I. Nan Hsu and B. M. Craven, *Acta Cryst.* (*B*), 1974, **30,** 998.
366. H. Tomisawa and H. Hongo, *Tetrahedron Letters*, 1969, 2465; C. L. Bell *et al.*, *J. Heterocyclic Chem.*, 1968, **5,** 859.
367. N. Dennis, A. R. Katritzky, and Y. Takeuchi, *Angew. Chem. Internat. Edn.*, 1976, **15,** 1.
368. N. Campbell, A. W. Henderson, and D. Taylor, *J. Chem. Soc.*, 1953, 1281.
369. (a) *Cf.* A. Kirpal and E. Reiter, *Ber.*, 1927, **60,** 664; (b) R. Tiollais, G. Bouget, and H. Bouget, *Compt. rend.* (*C*), 1962, **254,** 2597.
370. A. P. Phillips, *J. Org. Chem.*, 1947, **12,** 333; D. Jerchel and H. E. Heck, *Annalen*, 1958, **613,** 171; other references cited in these papers.
371. B. D. Shaw, *J. Chem. Soc.*, 1924, **125,** 1930; B. D. Shaw and E. A. Wagstaff, *ibid.*, 1933, 77; C. E. Kaslow and R. D. Stayner, *J. Amer. Chem. Soc.*, 1945, **67,** 1716.
372. T. Kato and Y. Goto, *Chem. Pharm. Bull.* (*Japan*), 1963, **11,** 461.
373. A. Kauffmann and L. G. Valette, *Ber.*, 1912, **45,** 1736; *cf.* G. R. Bedford, A. R. Katritzky, and H. M. Wuest, *J. Chem. Soc.*, 1963, 4600.
374. (a) R. Adams and S. Miyano, *J. Amer. Chem. Soc.*, 1954, **76,** 3168; (b) M. H. Fisher and A. R. Matzuk, *J. Heterocyclic Chem.*, 1969, **6,** 775.
375. O. F. Beumel, W. N. Smith, and B. Rybalka, *Synthesis*, 1974, 43.
376. R. B. Woodward and E. C. Kornfeld, *Org. Synth.*, *Coll. Vol.* 3, 413.
377. J. P. Wibaut and J. W. Hey, *Rec. Trav. chim.*, 1953, **72,** 513, 522.
378. C. Osuch and R. Levine, *J. Amer. Chem. Soc.*, 1956, **78,** 1723.
379. H. Pines and co-workers, *J. Org. Chem.*, 1969, **34,** 2113, 2119; 1972, **37,** 27799; other papers cited in these.
380. E. T. Borrows and D. O. Holland, *Chem. Rev.*, 1948, **42,** 615.
381. For leading references, see T. Cohen and G. L. Deets, *J. Amer. Chem. Soc.*, 1972, **94,** 932; V. J. Traynelis and A. I. Gallagher, *J. Org. Chem.*, 1970, **35,** 2792; R. Bodalski and A. R. Katritzky, *J. Chem. Soc.* (*B*), 1968, 831.
382. J. F. Vozza, *J. Org. Chem.* 1962, **27,** 3856.
383. J. I. G. Cadogan, *J. Chem. Soc.*, 1962, 4257.
384. H. Wynberg, T. J. van Bergen, and R. M. Kellogg, *J. Org. Chem.*, 1969, **34,** 3175.
385. P. Ribereau *et al.*, *Compt. rend.* (*C*), 1975, **280,** 293.
386. A. R. Hands and A. R. Katritzky, *J. Chem. Soc.*, 1958, 1754; A. R. Katritzky and M Kingsland, *ibid.* (*B*), 1968, 862; F. De Sarlo and J. H. Ridd, *ibid.* (*B*), 1971, 712.
387. W. H. F. Sasse, *Org. Synth.*, 1966, **46,** 5.
388. F. Kuffner and F. Straberger, *Monatsh.*, 1957, **88,** 793.
389. R. L. Frank and J. V. Crawford, *Bull. Soc. chim. France*, 1958, 419.
390. T. Talik, Z. Talik, and H. Ban-Oganowska, *Synthesis*, 1974, 293; J. I. G. Cadogan, D. A. Roy, and D. M. Smith, *J. Chem. Soc.* (*C*), 1966, 1249; A. Roe and G. F. Hawkins, *J. Amer. Chem. Soc.*, 1947, **69,** 2443.
391. J. P. Wibaut and L. G. Heeringa, *Rec. Trav. chim.*, 1955, **74,** 1003, and references cited therein.
392. W. E. Parham and R. M. Piccirilli, *J. Org. Chem.*, 1977, **42,** 277.
393. M. S. Raasch, *J. Org. Chem.*, 1962, **27,** 1406; Y. Kobayashi and E. Chinen, *Chem. Pharm. Bull.* (*Japan*), 1967, **15,** 1896.
394. A. Kirpal and W. Böhm, *Ber.*, 1932, **65,** 680.
395. E. Ochiai and H. Mitarashi, *Chem. Pharm. Bull.* (*Japan*), 1963, **11,** 1084, 1586.
396. H. J. den Hertog, C. H. Henkens, and J. van Roon, *Rec. Trav. chim.*, 1952, **71,** 1145.
397. C. Räth, *Annalen*, 1931, **486,** 95.
398. J. H. Boyer *et al.*, *J. Amer. Chem. Soc.*, 1953, **75,** 5298.
399. M. Tišler, *Synthesis*, 1973, 123, and references cited therein.
400. R. A. Abramovitch and A. R. Vinutha, *J. Chem. Soc.* (*C*), 1969, 2104.
401. For a review, see S. A. Harris, E. E. Harris, and R. W. Burg, 'Kirk-Othmer Encyclopaedia of Chemical Technology', 2nd Edition, Interscience, New York, 1968, vol. 16, p. 806.
402. E. P. Hart, *J. Chem. Soc.*, 1952, 4540; R. H. Callighan and M. H. Wilt, *J. Org. Chem.*, 1961, **26,** 4912.
403. (a) B. E. Hackley and co-workers, *J. Org. Chem.*, 1970, **35,** 841; (b) T. Slebodzinski *et al.*, *Przemysl Chem.*, 1969, **48,** 90 (*Chem. Abs.*, 1969, **71,** 38751); T. Banas and Z. Skrowaczewska, *Roczniki Chem.*, 1969, **43,** 739.
404. V. Boekelheide and W. J. Linn, *J. Amer. Chem. Soc.*, 1954, **76,** 1286.
405. V. M. Mićović and M. Lj. Mihailović, *Rec. Trav. chim.*, 1952, **71,** 970; H. Rutner and P. E. Spoerri, *J. Org. Chem.*, 1963, **28,** 1898; G. Quéguiner and P. Pastour, *Bull. Soc. chim. France*, 1968, 4117; 1969, 3678.
406. H. C. Brown and B. C. Subba Rao, *J. Amer. Chem. Soc.*, 1958, **80,** 5377; L. I. Zakharkin, D. N. Maslin, and V. V. Gavrilenko, *J. Org. Chem. USSR*, 1966, **2,** 2153.
407. G. Hesse and R. Schrödel, *Annalen*, 1957, **607,** 24.
408. (a) F. B. LaForge, *J. Amer. Chem. Soc.*, 1928, **50,** 2477, 2484; (b) P. C. Teague, A. R. Ballentine, and G. L. Rushton, *ibid.*, 1953, **75,** 3429.
409. J. P. Wibaut *et al.*, *Rec. Trav. chim.*, 1951, **70,** 962, 989, 1054.
410. C. F. Koelsch, *J. Org. Chem.*, 1945, **10,** 34, and references cited therein.
411. F. J. Villani, M. S. King, and D. Papa, *J. Org. Chem.*, 1952, **17,** 249; F. J. Villani and M. S. King. *Org. Synth.*, 1957, **37,** 6.

412. For a discussion of this and related spectroscopic evidence, see H. P. Stephenson and H. Sponer, *J. Amer. Chem. Soc.*, 1957, **79,** 2050.
413. H. Meyer, *Ber.*, 1903, **36,** 616.
414. F. Brody and P. R. Ruby, in 'Pyridine and its Derivatives', ed. E. Klingsberg, Interscience, New York, 1960, part 1, p. 99; N. S. Boodman *et al.*, *ibid.*, Supplement, ed. R. A. Abramovitch, Wiley, New York, 1974, part 1, p. 183.
415. P. Baumgarten, *Ber.*, 1924, **57,** 1622; Z. Arnold, *Experientia*, 1959, **15,** 415.
416. I.C.I. Ltd., *Brit. Pats.* 1077573; 1099642; 1102261 (*Chem. Abs.*, 1968, **68,** 87175; **69,** 27263, 27266); *Belg. Pat.* 669512 (*Chem. Abs.*, 1966, **65,** 7153).
417. E. Knoevenagel, *Annalen*, 1894, **281,** 25; Yu. I. Chumakov and V. P. Sherstyuk, *Tetrahedron Letters*, 1965, 129.
418. T. B. Windholz, L. H. Peterson, and G. J. Kent, *J. Org. Chem.*, 1963, **28,** 1443.
419. M. Chauvelier, *Bull. Soc. chim. France*, 1949, 611; 1954, 734.
420. I. E. El-Kholy, M. M. Mishrikey, and R. F. Atmeh, *J. Heterocyclic Chem.*, 1973, **10,** 665.
421. F. Ya. Perveev and N. V. Koshmina, *J. Org. Chem. USSR*, 1968, **4,** 167; 1969, **5,** 1302; *cf.* J. Liebscher and H. Hartmann, *Z. Chem.*, 1973, **13,** 342.
422. A. Dornow and A. Machens, *Chem. Ber.*, 1947, **80,** 502.
423. O. Mumm and O. Böhme, *Ber.*, 1921, **54,** 726.
424. E. Breitmaier, S. Gassenmann, and E. Bayer, *Tetrahedron*, 1970, **26,** 5907; G. Bouchon, K.-H. Spohn, and E. Breitmaier, *Chem. Ber.*, 1973, **106,** 1736.
425. A. Dornow and E. Neuse, *Arch. Pharm.*, 1955, **288,** 174.
426. S. A. Harris and K. Folkers, *J. Amer. Chem. Soc.*, 1939, **61,** 1245.
427. R. L. Frank and R. P. Seven, *J. Amer. Chem. Soc.*, 1949, **71,** 2629.
428. F. Bohlmann and D. Rahtz, *Chem. Ber.*, 1957, **90,** 2265.
429. J. Thesing and A. Müller, *Chem. Ber.*, 1957, **90,** 711.
430. F. W. Bergstrom, *Chem. Rev.*, 1944, **35,** 94; see also J. A. Berson and E. Brown, *J. Amer. Chem. Soc.*, 1955, **77,** 444; T. Channat and U. Eisner, *J.C.S. Perkin I*, 1975, 926.
431. R. L. Frank and co-workers, *J. Amer. Chem. Soc.*, 1946, **68,** 1368; A. E. Tschitschibabin and co-workers, *J. prakt. Chem.*, 1924, **107,** 132; *Bull. Soc. chim. France*, 1936, **3,** 772; 1937, **4,** 1826.
432. J. D. Butler, *J. Catalysis*, 1966, **6,** 26.
433. *Cf.* F. E. Cislak and W. R. Wheeler, *U.S. Pat.* 2807618 (*Chem. Abs.*, 1958, **52,** 2932).
434. J. Veitch, *Brit. Pat.* 1135854 (*Chem. Abs.*, 1969, **70,** 47310).
435. K. R. Hargrave, *Brit. Pat.* 963887 (*Chem. Abs.*, 1964, **61,** 9472); A. Tamono and M. Doya, *Brit. Pat.* 1141526 (*Chem. Abs.*, 1969, **70,** 106396).
436. L. E. Cooper and K. R. Hargrave, *Brit. Pat.* 968946 (*Chem. Abs.*, 1964, **61,** 13289).
437. R. S. Monson and A. Baraze, *Tetrahedron*, 1975, **31,** 1145; *J. Org. Chem.*, 1975, **40,** 1672.
438. *Cf.* W. Reppe and co-workers, *Annalen*, 1956, **601,** 128.
439. G. Simchen and G. Entenmann, *Angew. Chem. Internat. Edn.*, 1973, **12,** 119.
440. E. L. Little *et al.*, *J. Amer. Chem. Soc.*, 1958, **80,** 2833; F. Johnson *et al.*, *J. Org. Chem.*, 1962, **27,** 2473.
441. T. Stensrud, E. Bernatek, and M. Johnsgaard, *Acta Chem. Scand.*, 1971, **25,** 523, and references therein.
442. C. R. Hauser and C. J. Eby, *J. Amer. Chem. Soc.*, 1957, **79,** 728; J. F. M. Wajon and J. F. Arens, *Rec. Trav. chim.*, 1957, **76,** 65.
443. C. Kashima *et al.*, *Bull. Chem. Soc. Japan*, 1969, **42,** 2389.
444. J. W. Ducker and M. J. Gunter, *Austral. J. Chem.*, 1973, **26,** 2567.
445. A. S. Alvarez-Insua, M. Lora-Tamayo, and J. L. Soto, *J. Heterocyclic Chem.*, 1970, **7,** 1305.
446. W. Ramsay, *Phil. Mag.*, 1876, (v), **2,** 269.
447. H. Bönnemann, R. Brinkmann, and H. Schenkluhn, *Synthesis*, 1974, 575; Y. Wakatsuki and H. Yamazaki, *ibid.*, 1976, 26.
448. J. C. Jagt and A. M. van Leusen, *Rec. Trav. chim.*, 1973, **92,** 1343.
449. R. Fuks, *Tetrahedron*, 1970, **26,** 2161; *cf.* J. H. MacMillan and S. S. Washburne, *J. Org. Chem.*, 1973, **38,** 2982.
450. P. Hong and H. Yamazaki, *Synthesis*, 1977, 50.
451. J. Becher and E. G. Frandsen, *Acta Chem. Scand.* (*B*), 1976, **30,** 863; *Tetrahedron*, 1977, **33,** 341.
452. K. Aso and co-workers, *Chem. Abs.*, 1940, **34,** 431, 3273, 6278; W. Gruber, *Canad. J. Chem.*, 1953, **31,** 564, and references therein.
453. N. Clauson-Kaas and co-workers, *Acta Chem. Scand.*, (a) 1956, **10,** 1603; 1967, **21,** 1104; (b) 1969, **23,** 1785; (c) 1955, **9,** 17, 23, 30; 1960, **14,** 938.
454. See, for example, K. G. Lewis and C. E. Mulquiney, *Chem. and Ind.* (*London*), 1968, 1249.
455. G. Stork and co-workers, *J. Org. Chem.*, 1971, **36,** 2784.
456. W. Dilthey and H. Dierichs, *J. prakt. Chem.*, 1935, **144,** 1.
457. A. T. Balaban, *Tetrahedron*, 1968, **24,** 5059; 1970, **26,** 739.
458. See, for example, K. Heyns and G. Vogelsang, *Chem. Ber.*, 1954, **87,** 1337; J. Fried and R. C. Elderfield, *J. Org. Chem.*, 1941, **6,** 566.
459. *Cf.* J. A. Van Allan *et al.*, *J. Heterocyclic Chem.*, 1971, **8,** 919.
460. (a) E. R. Alexander, A. B. Herrick, and T. M. Roder, *J. Amer. Chem. Soc.*, 1950, **72,** 2760; (b) F. S. Baker *et al.*, *Chem. and Ind.* (*London*), 1969, 1344; (c) R. L. Jones and C. W. Rees, *J. Chem. Soc.* (*C*), 1969, 2249, 2255.
461. For a review, see M. Ya. Karpeiskii and V. L. Florent'ev, *Russ. Chem. Rev.*, 1969, **38,** 540.

462. R. A. Firestone, E. E. Harris, and W. Reuter, *Tetrahedron*, 1967, **23,** 943.
463. Takeda Chemical Industries, *Fr. Pat.* 1400843 (*Chem. Abs.*, 1965, **63,** 9922).
464. T. Jaworski and S. Kwiatkowski, *Roczniki Chem.*, 1970, **44,** 555.
465. E. Buschmann and W. Steglich, *Angew. Chem. Internat. Edn.*, 1974, **13,** 484.
466. H. Neunhoeffer and G. Werner, *Annalen*, (a) 1973, 437; (b) 1974, 1190; (c) 1972, **761,** 39.
467. H. Neunhoeffer and H.-W. Frühauf, *Annalen*, 1972, **758,** 120.
468. W. Dilthey *et al.*, *Ber.*, 1935, **68,** 1159; T. Jaworski and co-workers, *Roczniki Chem.*, 1953, **27,** 468; 1967, **41,** 1521.
469. R. G. Pews, E. B. Nyquist, and F. P. Corson, *J. Org. Chem.*, 1970, **35,** 4096.

16.2

Diazines and Benzodiazines

A. E. A. PORTER
University of Stirling

16.2.1 INTRODUCTION

The diazines are a group of compounds formally derived from benzene by the replacement of two of the ring carbon atoms by nitrogen. Clearly three isomeric diazines are possible with the nitrogen atoms in a 1,2-, 1,3-, or a 1,4-relationship, giving rise to the pyridazines (**1**), the pyrimidines (**2**), and the pyrazines (**3**), respectively.

(1)  (2) 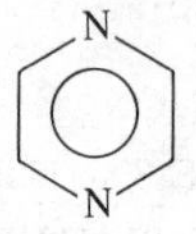 (3)

No naturally occurring pyridazines have yet been reported and indeed this comes as no surprise because of the paucity of chemical compounds containing two nitrogen atoms bonded to one another in nature. A number of pyridazines have found commercial use, in particular maleic hydrazide (**4**) is used as a selective plant-growth regulator.

Many pyrazine derivatives occur naturally. Methoxypyrazines such as (**5**) occur as natural flavouring materials in sweet corn, peas, and many other leguminous plants. The antibiotic aspergillic acid (**6**) is a derivative of 1-hydroxy-2-pyrazinone. Many pyrazines are used as medicinal agents and are typified by the synthetic pyrazinamide (**7**), which has found extensive use as a tuberculostat.

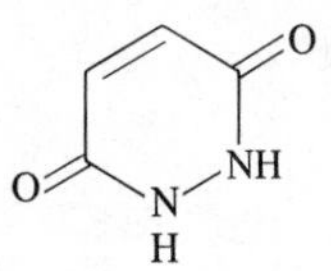

(4) 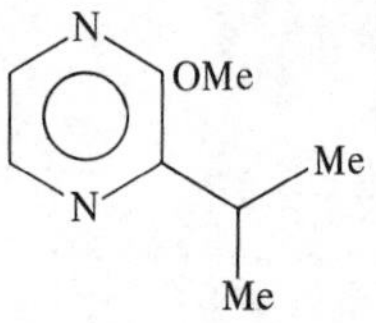 (5) 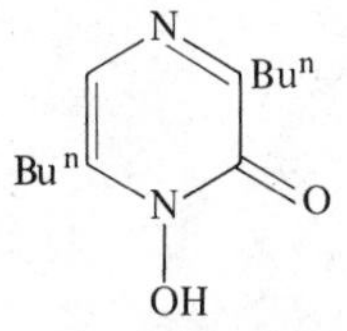 (6) 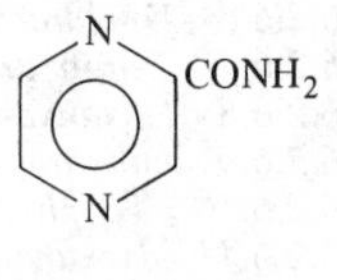 (7)

Pyrimidines are by far the most ubiquitous members of the diazine family with uracil (**8**) and thymine (**9**) being constituents of RNA and DNA, respectively, and with cytosine (**10**) being present in both. The pyrimidine skeleton is also present in many natural products, such as vitamin B_1 (thiamine), and many synthetic compounds, such as barbituric acid (**11**) and hypnotics such as veronal (**12**).

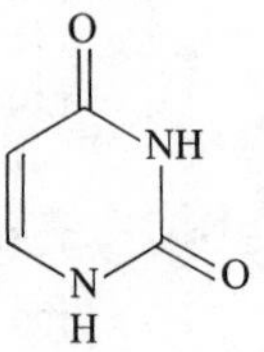

(8) 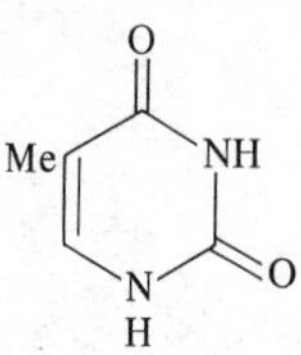 (9) 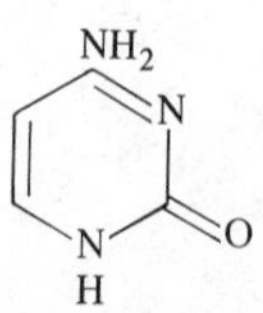 (10) 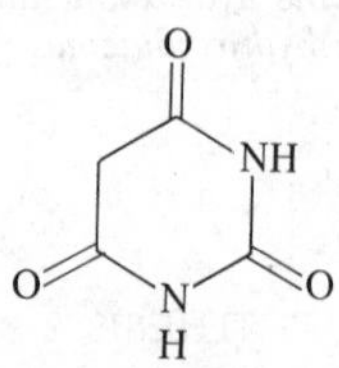 (11) 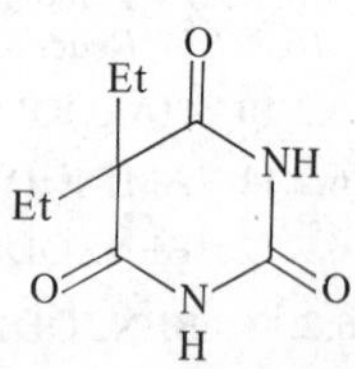 (12)

A comparison of the physical properties of the parent diazines is shown in Table 1.

TABLE 1
Physical Properties of the Diazines

	Pyridazine	*Pyrimidine*	*Pyrazine*
M.p.	−6.4 °C	22.5 °C	54 °C
B.p.	207 °C	124 °C	121 °C
pK_a[a]	2.33	1.30	0.65
Dipole moment[b]	3.94 D	2.10 D	0
λ_{max}	246, 340 nm	243, 298 nm	260, 328 nm

[a] A. Albert, R. J. Goldacre, and J. Phillips, *J. Chem. Soc.*, 1948, 2240.
[b] W. C. Schneider, *J. Amer. Chem. Soc.*, 1948, **70,** 627.

16.2.2 PYRIDAZINE SYNTHESIS

Most syntheses of the pyridazine ring are based on the addition of hydrazine or its derivatives to a suitably 1,4-disubstituted carbon chain.[1,2] The condensation of hydrazine derivatives with 1,4-dicarbonylethenes,[3] maleic anhydride derivatives,[4] 4-aldehydo- or 4-keto-acids,[5] butenolides,[6] 2-bromo- or 2-hydroxy-1,4-dicarbonyl compounds,[7] and dihydrofuran derivatives[8] results in the formation of pyridazine rings.

The major advantage in the use of maleic anhydride (Scheme 1) and its derivatives, or 1,4-dicarbonylethenes (Scheme 2), lies in the fact that the resultant pyridazine requires no

SCHEME 1

further oxidation although, depending on reaction conditions, the former route may result in numerous by-products, including linear hydrazides or derivatives of *N*-aminomaleimide.

SCHEME 2

1,4-Diketones react with hydrazine and its derivatives to yield 1,4- (**13**) or 4,5-dihydropyridazines (**14**) which are frequently unstable, undergoing rapid oxidation or disproportionation to the corresponding pyridazines on distillation.

1,4-Keto- or aldehydo-acids also undergo facile condensation with hydrazine and its derivatives to produce dihydropyridazines (Scheme 3), which may be oxidized with bromine in acetic acid to the corresponding pyridazinones.

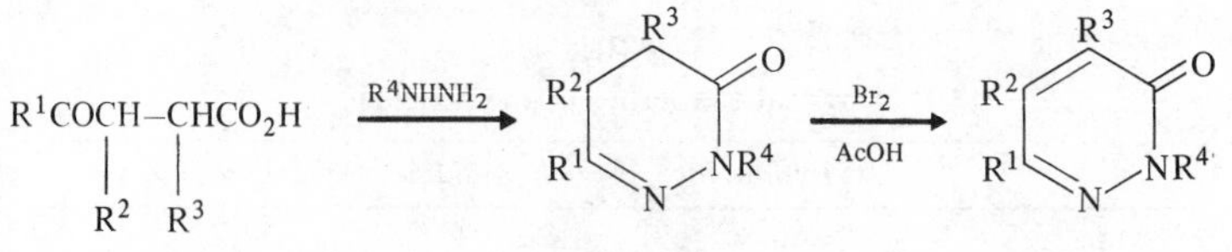

SCHEME 3

A useful synthesis of 3(2*H*)-pyridazinones (**15**) involves the reaction of α-diketones with hydrazine derivatives in the presence of an ester containing an active methylene group (Scheme 4).[9] Two mechanistic pathways have been suggested to rationalize this reaction: the formation of a monohydrazone of the diketone followed by condensation with the ester, or the formation of the acid hydrazide followed by condensation with the ketone.

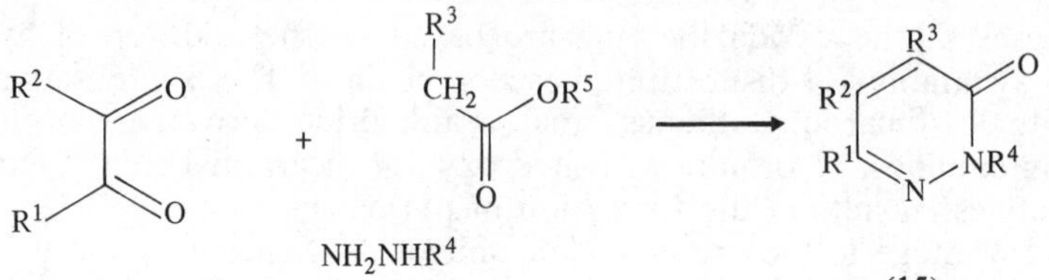

SCHEME 4

Two other methods are commonly used in the construction of the pyridazine ring, both of which involve the use of [4+2] cycloaddition reactions. Thus reaction of butadiene and its derivatives with azodicarboxylic esters results in the formation of tetrahydropyridazines (Scheme 5) which, after decarboxylation, may be oxidized to the corresponding

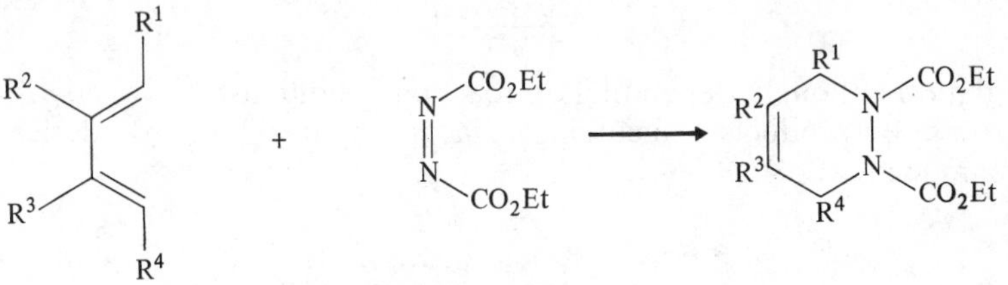

SCHEME 5

pyridazines. Alternatively, cycloaddition reactions of tetrazines with acetylenes (ethynes) or olefins (ethenes) results in the formation of pyridazines or dihydropyridazines, respectively (Scheme 6). If the olefin contains a substituent X which is readily lost as HX, then

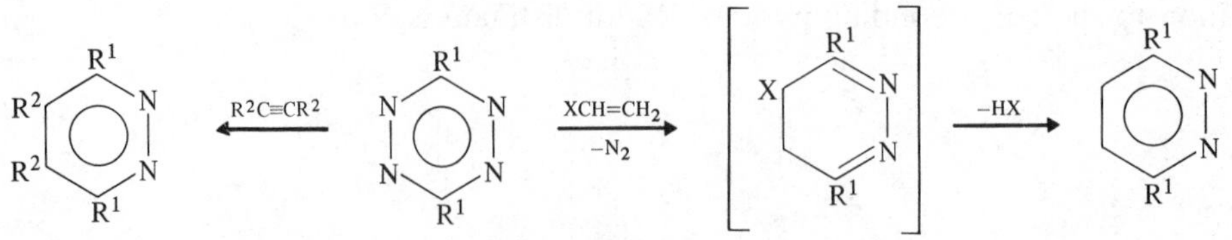

SCHEME 6

the reaction ultimately leads to pyridazine derivatives. The reaction with electron-rich olefins is extremely facile, the deep-red tetrazine being decolourized instantly with the concomitant evolution of nitrogen.

16.2.3 PYRIMIDINE SYNTHESIS

The biological importance of pyrimidine derivatives has prompted considerable attention to the synthesis of substituted pyrimidines and many detailed reviews have appeared.[10,11] Synthetic strategies have involved four main routes based on the condensation of two fragments, as illustrated by (**16**)–(**19**). Of these strategies, that illustrated by (**16**),

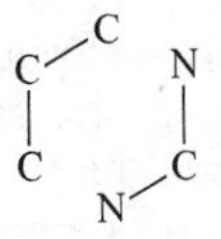

(16) 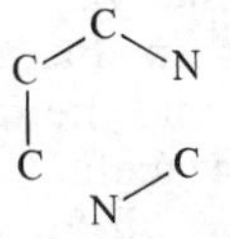 (17) 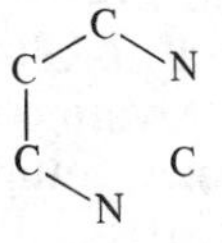 (18) 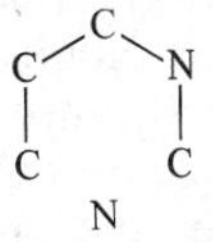 (19)

i.e. the condensation of a three-carbon unit with an N—C—N fragment, appears to be the most widely used, offering direct entry into the pyrimidine nucleus. This approach has been called 'the common synthesis'[10] because of its general applicability to the synthesis of a whole range of pyrimidine derivatives. The great versatility in this synthesis rests with the fact that one or both of the groups of the three-carbon fragment may be present as an aldehyde, ketone, ester, or nitrile group. β-Dialdehydes (or their equivalent), β-ketoaldehydes, β-ketoesters, malonic ester, β-aldehydo- or β-keto-nitriles, and many other combinations of these groups or their masked derivatives, may be used. The nitrogen-containing fragment may be an amidine, urea, thiourea, or guanidine, and acetylacetone serves as an excellent illustrative example in that it readily undergoes reaction with formamidine,[12] guanidine,[13] urea,[14] or thiourea[15] to produce the corresponding 4,6-dimethylpyrimidines (Scheme 7).

An important limitation of this reaction is that in practice not all possible combinations of reactants produce the expected products, as typified by the failure of malondialdehyde to undergo condensation with formamidine to produce pyrimidine. It is often possible, however, to achieve transformations of this type by modification of the reaction conditions and/or reactants. Thus 1,1,3,3-tetraethoxypropane, a readily available precursor of malondialdehyde, reacts with formamide over an alumina catalyst at 200 °C to produce pyrimidine in 70% yield.

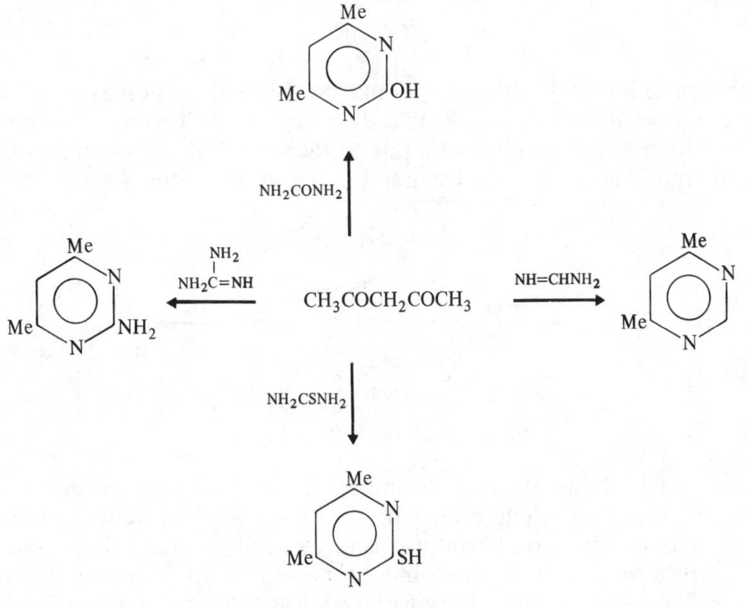

$CH_3COCH_2COCH_3$ — NH_2CONH_2; $NH_2C(NH_2)=NH$; $NH=CHNH_2$; NH_2CSNH_2

SCHEME 7

β-Dialdehydes and β-aldehydoketones are useful in that unsubstituted and 4-substituted pyrimidines may be prepared by suitable condensation reactions.

An additional advantage in the use of β-dicarbonyl compounds rests in the acidity of the α-protons: this permits the simple introduction of new substituents and adds a new dimension to the functionalization of the pyrimidine ring. For example, nitromalondialdehyde and phenylmalondialdehyde readily undergo condensation with urea to produce 2-hydroxy-5-nitropyrimidine and 2-hydroxy-5-phenylpyrimidine, respectively.

The use of β-aldehydo- and β-keto-esters in the synthesis of pyrimidines is of considerable importance in that thymine, uracil, and their derivatives are best synthesized by this method. Attempts to react ethyl formylacetate with urea have met with little success, but when the formyl acid is generated *in situ* by the oxidative decarboxylation of malic acid and reacted with urea, then modest yields[16] of uracil may be obtained (Scheme 8). Thymine has also been synthesized by an analogous route.[17] In contrast, formylacetic

CO_2H–CH_2–CHOH–CO_2H $\xrightarrow{H_2SO_4}$ [CO_2H–CH_2–CHO] $\xrightarrow{NH_2CONH_2}$ (uracil: O, NH, NH, O)

SCHEME 8

ester reacts readily with guanidine to form isocytosine derivatives[18] (Scheme 9). Substituted β-aldehydoesters are widely used in the synthesis of 5-substituted thymine derivatives.

(OEt, OH, O; NH_2, NH_2, NH) $\longrightarrow$ (OH, N, NH_2, N)

SCHEME 9

β-Ketoesters usually react with exceptional facility with amidines, urea, *etc.*, and the reaction more obviously takes place in two discrete steps. Ethyl acetoacetate reacts with urea to form an intermediate ureido-compound (Scheme 10), which may subsequently be induced to undergo ring closure on treatment with sodium ethoxide.

(OEt, O, Me, O) $\xrightarrow{NH_2CONH_2}$ (O, OEt, NH_2, Me, N, H, O) $\xrightarrow[EtONa]{EtOH}$ (O, NH, Me, N, H, O)

SCHEME 10

Probably the most frequently used example of 'the common synthesis' involves the reaction of derivatives of malonic ester with an N—C—N fragment such as urea, which leads directly to derivatives of barbituric acid as well as other important substituted pyrimidines. Barbituric acid (**11**) was first prepared in 1878 by the condensation of malonic acid with urea in the presence of phosphoryl chloride, but these conditions have now been superseded[19] by the ethoxide method. Substituted malonic ester derivatives also react under these conditions to yield a number of important barbiturate drugs: phenobarbitone (**20**; R^1 = Et, R^2 = Ph), nembutal (**20**; R^1 = Et, R^2 = 2-methylbutyl), amytal (**20**; R^1 = Et, R^2 = isopentyl), and seconal (**20**; R^1 = allyl, R^2 = 1-methylbutyl) have all been prepared by this route.

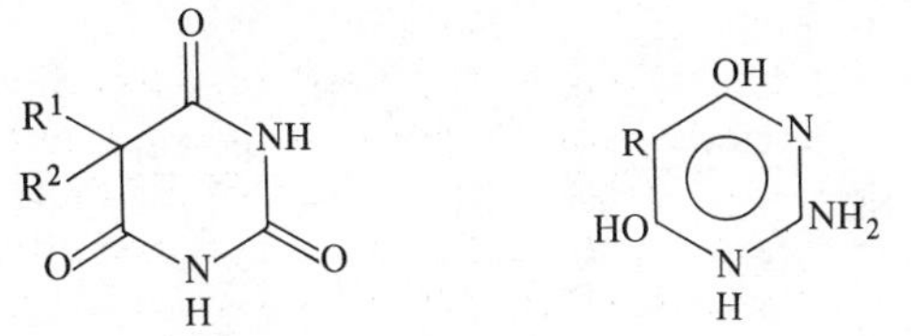

(20) (21)

Thiourea reacts in a similar way, although conditions under which the reactions are carried out vary. Typically, alkyl and dialkyl malonates seem to react more rapidly with thiourea than with urea. The facility with which guanidine reacts with derivatives of malonic ester is reflected in the relatively short reaction periods required to form 2-amino-4,6-dihydroxypyrimidines (**21**).

An important modification of the common synthesis may be seen in the use of a three-carbon fragment containing a nitrile group. Typical examples of such fragments are β-aldehydonitriles and cyanoacetic acid derivatives. The principal advantage in the use of nitriles rests in the fact that direct introduction of amino-substituents into the 4- and/or 6-positions of the pyrimidine ring may be achieved. In its simplest form, cyanoacetaldehyde (as its diethyl acetal) reacts with urea in a two step process to yield (Scheme 11) 4-amino-2-hydroxypyrimidine (cytosine). [^{15}N]Cytosine has been prepared by this method.

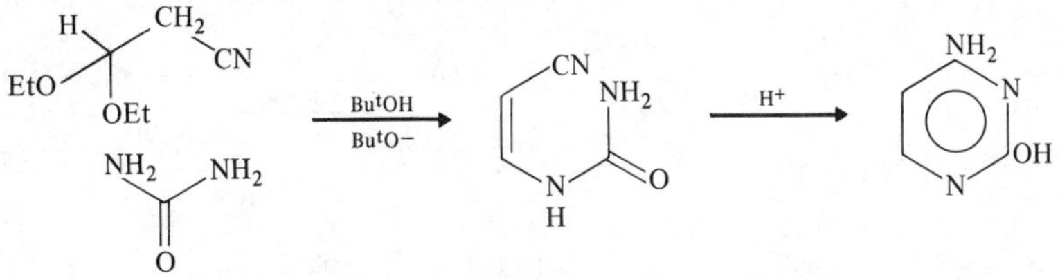

SCHEME 11

Ethyl ethoxymethylenecyanoacetate has been used as an aldehydonitrile but in principle it may function as an ester-nitrile, an ester-aldehyde or an aldehydonitrile and products are largely determined by reaction conditions. With amidines the condensation proceeds by way of an isolable intermediate which may undergo two different cyclizations (Scheme 12) and much use has been made of this reaction in the synthesis of 5-substituted cytosine derivatives.

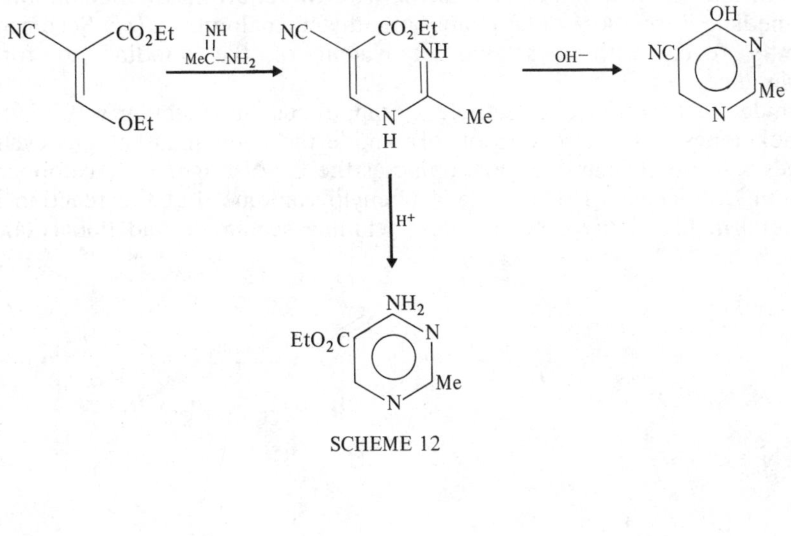

SCHEME 12

β-Dinitriles such as malonitrile exhibit anomalous behaviour. Condensation of malonitrile with amidines, such as formamidine or benzamidine, results in the formation of 4-amino-5-cyanopyrimidines and the reaction is believed to involve the intermediate formation (Scheme 13) of aminomethylenemalonitrile.

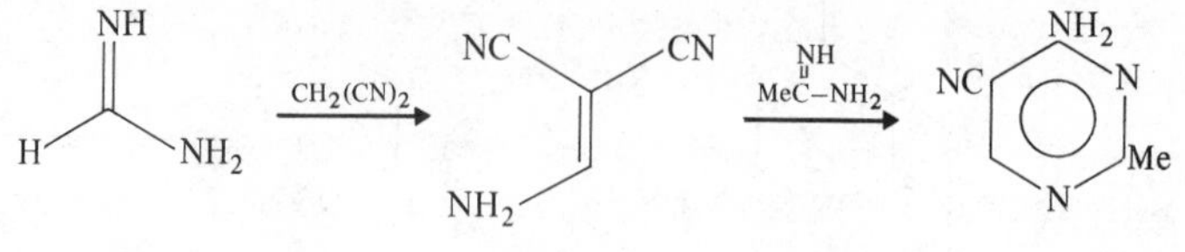

SCHEME 13

Other main methods of synthesis of the pyrimidines are based on the strategies indicated by (**17**), (**18**), and (**19**), although a variety of less general methods exist. Although the common synthesis (**16**) is more general than syntheses involving constructions as illustrated by (**17**), (**18**), and (**19**), these methods are complementary and should not be regarded as less useful.

Type (**17**) syntheses are centred around the condensation of a β-cyano- or β-acylenamine (the N—C—C—C fragment, readily available by the direct amination of ethoxymethylenemalonic ester derivatives) with a suitable C—N-containing fragment.

Many C—N-containing fragments have found use, for example ethyl β-aminocrotonate undergoes an extremely facile reaction with phenyl isocyanate or methyl isocyanate[20] to form an intermediate ureido-derivative (Scheme 14) which undergoes cyclization on treatment with base.

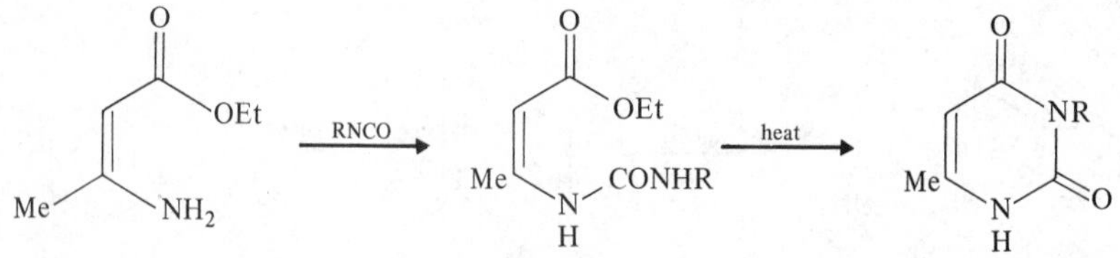

SCHEME 14

The analogous procedure with saturated β-aminoketones results in the formation of dihydropyrimidines, which may be oxidized to the corresponding pyrimidines by a bromination–dehydrobromination sequence. Other C—N-containing fragments which have found use include imidates,[21] imidoyl halides,[22] amides and thioamides,[23] but the full potential of these reactions does not appear to have been developed.

In many cases the acylenamine is not isolated but generated *in situ* during the course of the reaction: thus the reaction of malonitrile with two equivalents of an amidine results in the intermediate formation of a β-aminomethylenemalonitrile (*cf.* Scheme 13) which subsequently reacts with a second equivalent of the amidine to form the 4-aminopyrimidine.

Formamide reacts with active methylene compounds[10] in a similar way to form isolable β-enaminoketones. With an excess of formamide the β-enaminoketones cyclize to form pyrimidines with the formamide functioning as the C—N fragment. Acetophenone reacts with formamide (Scheme 15) to form 4-phenylpyrimidine, but the reaction is far from being general in that many other amides including acetamide and thioacetamide fail to

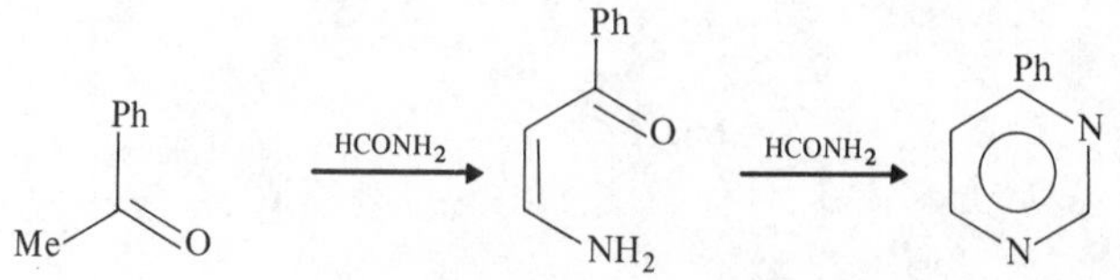

SCHEME 15

undergo similar reactions. Furthermore, the activation of the methylene group, *e.g.* by a phenyl substituent, is essential since simple ketones or nitriles fail to react. A closely related reaction (Scheme 16) involves the reaction of β-diketones with an excess of formamide and this also proceeds by way of an isolable β-acylenamine.[24]

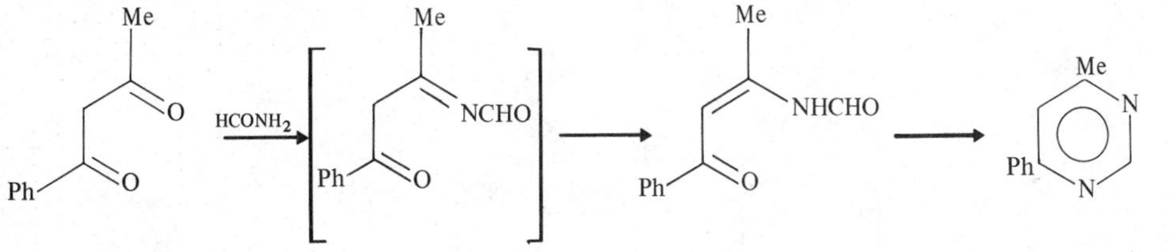

SCHEME 16

Syntheses of the type (**18**) category involve in principle the condensation of a one-carbon fragment with a derivative of 1,3-diaminopropane, but more usually the 1,3-diamino-compound is at a higher level of oxidation and derivatives of malondiamide are commonly used. A typical example of this class involves the reaction of malondiamide with an ester such as malonic ester[25] to give a derivative of 4,6-dihydroxypyrimidine (Scheme 17). The reaction may also be carried out using formate esters,[26] formamide,

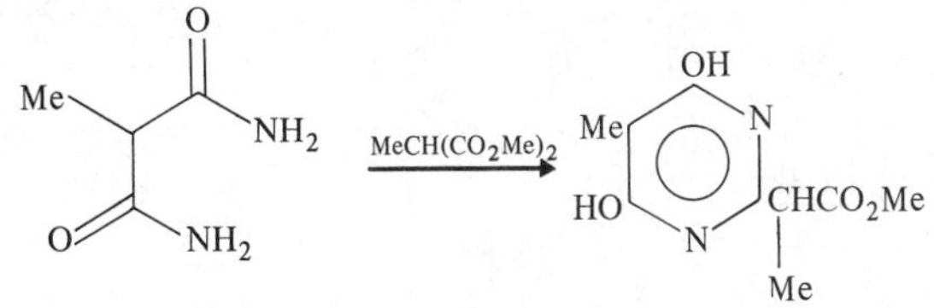

SCHEME 17

acetamide, CS_2,[27] and carbonates.[28] Although the reaction is not directly comparable, the malondiamide may be replaced by malonyl chloride, *e.g.* diethyl malonyl chloride reacts with urethane to yield diethylbarbituric acid (Scheme 18). Malondiamidines[29] behave in a similar way to the diamides, giving rise to 4,6-diaminopyrimidines on condensation with formate esters.

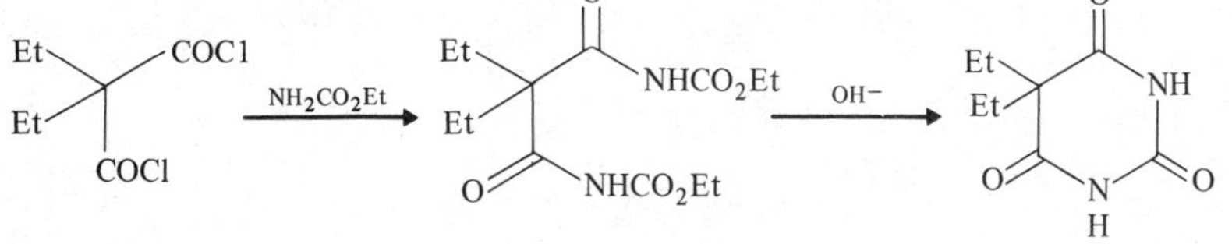

SCHEME 18

Syntheses of type (**19**) are not very common but they have found some application in the synthesis of uracil and thymine derivatives and the synthesis due to Shaw [30] is perhaps one of the best illustrative examples. This synthesis depends on whether the 5-position of the pyrimidine ring is to be substituted or free from substituents. The condensation of cyanoacetic acid with *N*-methylurethane in acetic anhydride (Scheme 19) leads to the amide (**22**), which may be induced to undergo condensation with formaldehyde or ethyl orthoformate to produce the enol ether (**23**). Reaction of (**23**) with an amine followed by base-catalysed cyclization yields[31] the 5-cyanopyrimidine (**24**).

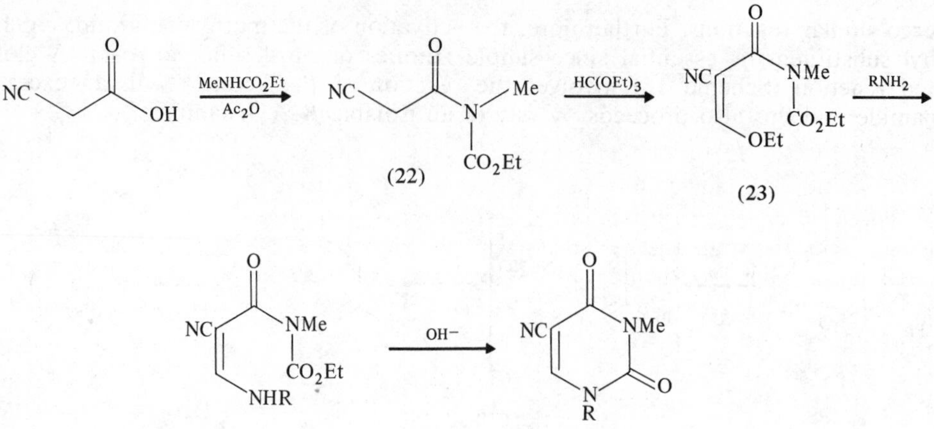

SCHEME 19

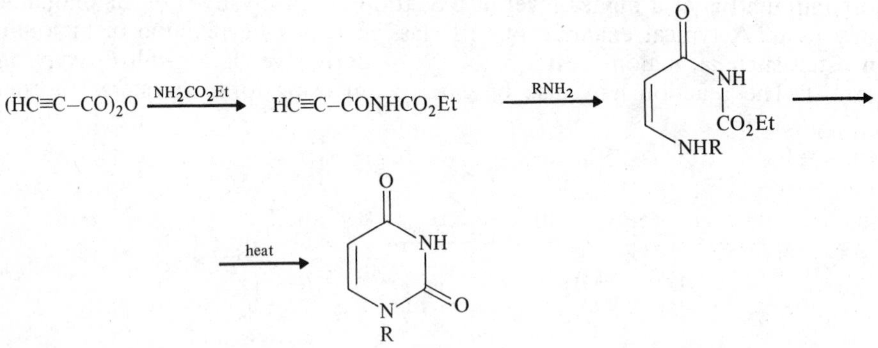

SCHEME 20

The procedure for the synthesis of pyrimidines unsubstituted at position-5 utilizes propiolic anhydride as a starting point (Scheme 20). The reaction is of particular importance since, when the amine RNH_2 is tribenzylribosamine, the nucleoside uridine (**25**) can be produced.[30]

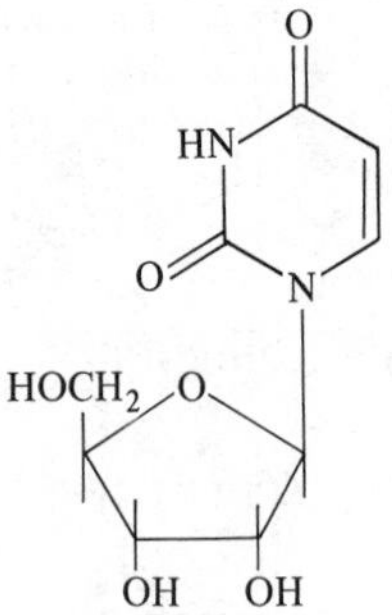

(25)

16.2.4 PYRAZINE SYNTHESIS

Cheeseman and Werstiuk[32] have classified pyrazine syntheses into six main categories (**26**)–(**31**), of which those indicated by (**26**) and (**27**) are the most important.

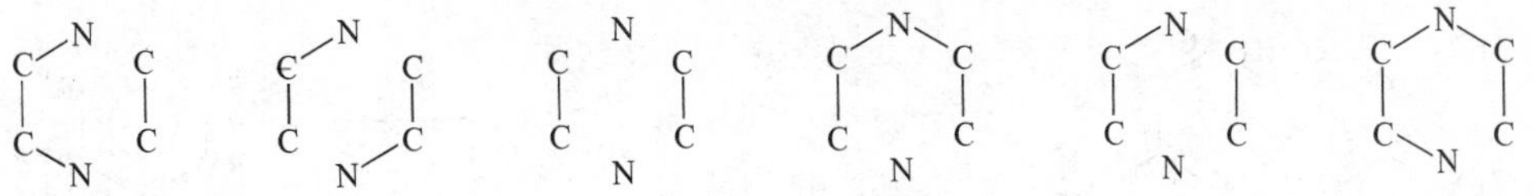

(26) (27) (28) (29) (30) (31)

Syntheses of type (**26**) represent one of the classical approaches to the synthesis of pyrazines and their annulated analogues, and the reaction usually proceeds in two discrete steps. Thus 1,2-diaminoethane undergoes condensation with 1,2-diketones to yield an intermediate dihydropyrazine (Scheme 21). These dihydropyrazines are usually

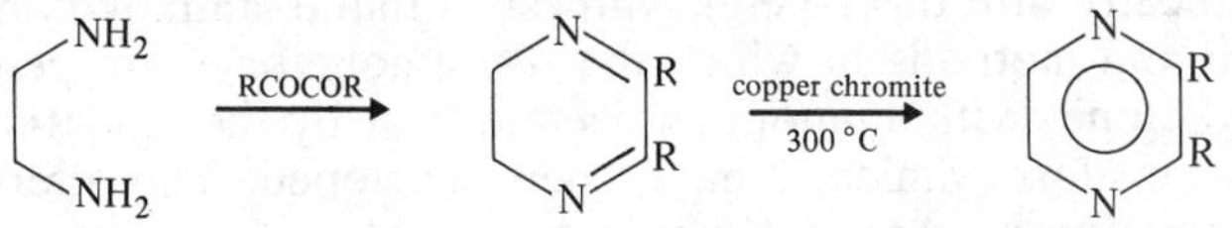

SCHEME 21

very resistant to oxidation in the absence of some activating substituent. Typically, oxidation to the corresponding pyrazines is carried out over copper chromite at 300 °C. This method appears to offer a general method for the synthesis of alkyl, aryl, and aralkyl pyrazines, *e.g.* reaction of 1,2-diaminopropane with 2,3-dioxobutane yields 5,6-dihydro-2,3,5-trimethylpyrazine, which has been oxidized to the corresponding pyrazine under somewhat milder conditions using an autoxidative procedure over KOH pellets.[33]

A simple variant of this method has been successfully applied to the synthesis of some functionalized derivatives. Hydroxypyrazines, the immediate precursors of some important flavouring methoxypyrazines,[34] have been prepared by the condensation of α-amino-acid amides with α-diketones (Scheme 22); by the simple expedient of using the corresponding α-amino-acid amidines, *e.g.* aminoacetamidine, aminopyrazines may be derived.

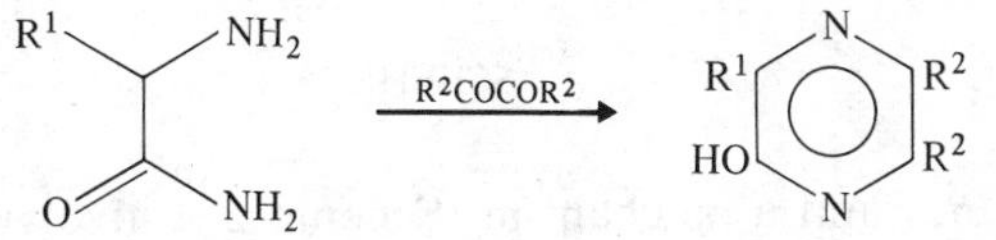

SCHEME 22

Pyrazinecarboxylic acids have also been prepared by an analogous route involving the reaction of a 1,2-diamino-acid with an α-diketone (Scheme 23) followed by aerial oxidation of the intermediate 5,6-dihydropyrazine.[35]

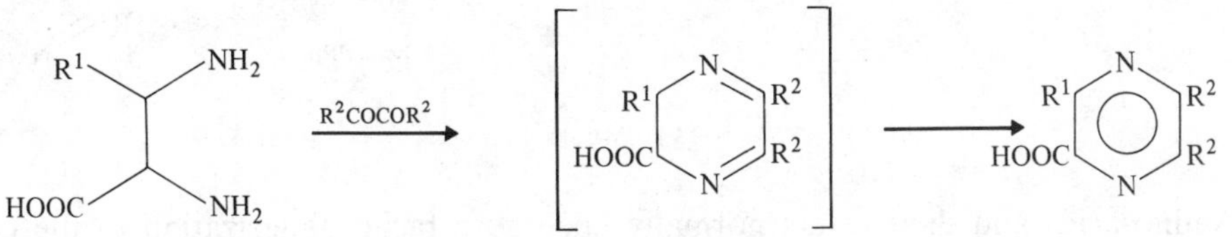

SCHEME 23

Syntheses of type (**27**) generally involve the dimerization of α-acylamino-compounds, *e.g.* the self condensation of two molecules of an α-aminoketone. Again the reaction usually proceeds in two discrete steps (Scheme 24) with the intermediate formation of a 2,5-dihydropyrazine which is subsequently oxidized. One essential difference between 5,6-dihydro- and 2,5-dihydro-pyrazines lies in the facility with which oxidation to the

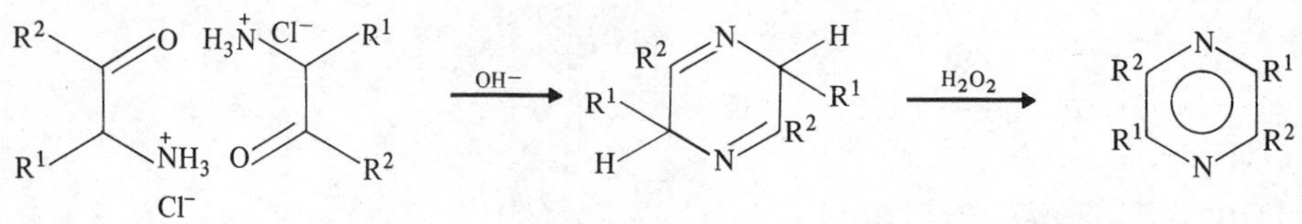

SCHEME 24

corresponding pyrazine occurs, with the latter undergoing facile oxidation even with such weak oxidizing agents as molecular iodine,[36] although more usually hydrogen peroxide or mercuric chloride are used.

The principal difficulty with this type of synthesis is that α-aminoketones themselves are difficult to prepare and methods in which the α-aminoketones are generated *in situ* by heating α-hydroxyketones with ammonium acetate,[37] or by the oxidation of ketones with ammoniacal potassium ferricyanide,[38] have been developed. The alternative method of generating α-aminoketones from α-bromoketones does give rise to 2,5-disubstituted pyrazines, although they are usually contaminated, to a varying extent, with the isomeric

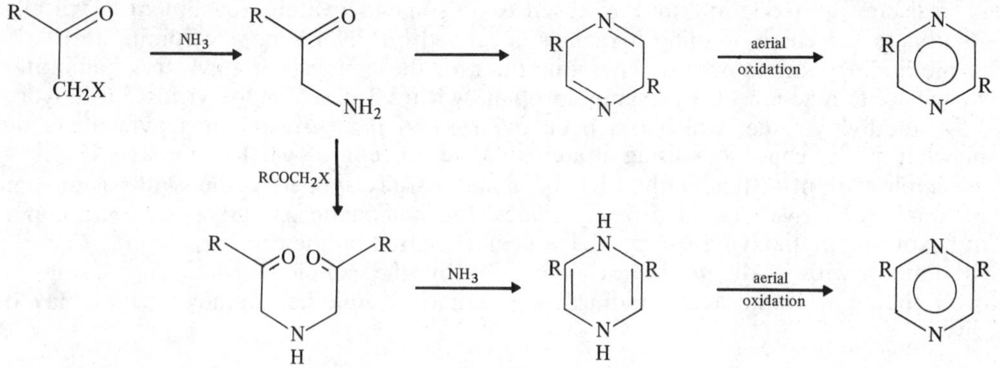

SCHEME 25

2,6-disubstituted pyrazines and a mechanism (Scheme 25) involving the formation of an α-aminoketone and dialkylamine is thought probable.[39] A recent modification of this reaction[40] involves the reaction of α-haloketones with sodium azide to generate the α-azidoketone. On reaction with triphenylphosphine the α-azidoketone generates specifically 2,5-dihydropyrazines (Scheme 26).

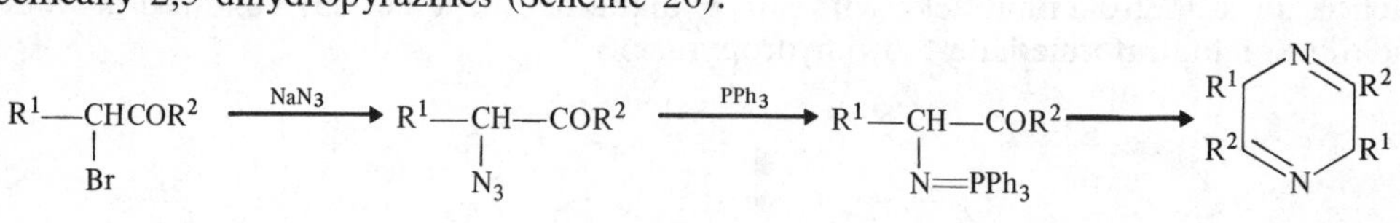

SCHEME 26

α-Amino-acids and their esters normally undergo a facile dimerization to the corresponding 2,5-dioxopiperazines, and treatment of the latter with trialkyloxonium fluoroborate generates 2,5-dialkoxy-3,6-dihydropyrazines which are easily oxidized with DDQ to the corresponding 2,5-dialkoxypyrazines[41] (Scheme 27). An interesting extrusion reaction has been observed with 3,6-dihydropyrazines (**32**): thus thermolysis of (**32**; $R^1 = CH_2Ph$) in the absence of air results in the elimination of toluene with the formation of a monobenzyldiethoxypyrazine. Although the full potential of this reaction has not been explored, it could offer a useful route to the synthesis of a number of otherwise inaccessible pyrazines.

$R^1—CH(NH_2)—COOR^2$ —heat→ (2,5-dioxopiperazine) —$Et_3O^+ BF_4^-$→ (32) —DDQ→ ; (32) —heat→ + R^1H

SCHEME 27

2,5-Dioxopiperazines also serve as a useful starting point for the synthesis of halogenated pyrazines, and Spring and Baxter have found[42] that mixtures of monochloro- and dichloro-pyrazines are formed on treatment of the dioxopiperazines with phosphoryl chloride. The precise ratio of products depends markedly on the substitution of the starting dioxopiperazine.

The methods of pyrazine synthesis indicated by types (**28**), (**29**), (**30**), and (**31**) are beyond the scope of this work and readers are referred to Ref. 32 for fuller information.

16.2.5 GENERAL CHEMISTRY OF THE DIAZINES

16.2.5.1 Electrophilic substitution

The known reluctance of pyridine to take part in electrophilic substitution reactions would suggest that the introduction of a second azomethine nitrogen into the ring would have the effect of rendering the diazines even less reactive towards electrophiles and, with a few exceptions, this is the case for unsubstituted diazines. Chlorination of 2-methylpyrazine with molecular chlorine in carbon tetrachloride results in the formation of a good yield of 2-chloro-3-methylpyrazine, but although this reaction formally corresponds to electrophilic substitution, it is believed to proceed by an addition–elimination mechanism. Similarly, pyrimidine reacts with bromine at 160 °C to form 5-bromopyrimidine in good yield, but again the mechanism of this reaction probably involves the intermediacy of a perbromide (*cf.* pyridine) rather than conventional electrophilic substitution.

When the diazine ring is activated by one or more electron donating (+*M*) substituents, electrophilic substitution becomes a more serious proposition. In general, when two activating substituents are present the ring becomes comparable in reactivity to benzene, and when three activating substituents are present electrophilic substitution is facile with partial rate factors approaching those of activated benzene rings such as phenol.

Electrophilic reagents almost invariably attack the pyrimidine ring at position-5, which is the carbon atom least depleted in electronic charge. Given the presence of suitable activating groups, nitrosation, nitration, diazo-coupling, halogenation, and sulphonation all occur under the appropriate conditions. For example, nitration of 2,4-dihydroxy-6-methylpyrimidine (**33**) may be effected using nitric acid in acetic acid at 20 °C to give the 5-nitro-derivative (**34**) in good yield.

(33) (34)

Diazo-coupling is an extremely useful reaction for the synthesis of 5-aminopyrimidines in that both the coupling reaction and subsequent reductions may be effected under mild conditions.

The positions of the activating substituents are critical in many cases: thus whilst 4,6-dihydroxypyrimidine and 4,6-diaminopyrimidine undergo both nitration and nitrosation, 2,4-dihydroxypyrimidine will not undergo nitrosation and 2,4-diaminopyrimidine will undergo neither nitration nor nitrosation.

The behaviour of the pyrazines and the pyridazines towards electrophiles is not as clearly documented as that of the pyrimidines. Like pyrimidine, pyridazine itself is extremely resistant to attack and simple phenyl-substituted pyridazines undergo substitution in the phenyl ring in preference to the pyridazine ring. However, like pyrimidine when suitable activating substituents are present, substitution will occur as seen in the nitration[2] of 4-amino-3,6-dimethoxypyridazine (**35**) to 4-amino-3,6-dimethoxy-5-nitropyridazine (**36**).

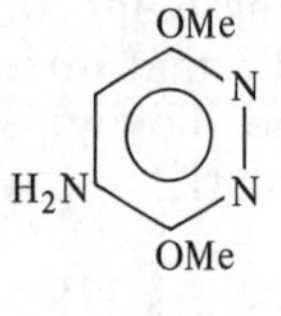

OMe, N, N, H_2N, OMe (35)   OMe, O_2N, N, N, H_2N, OMe (36)

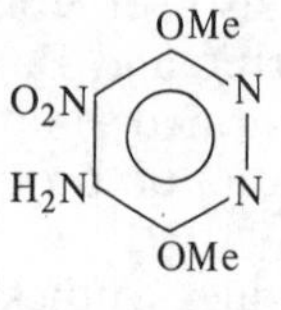

A number of electrophilic substitution reactions are in evidence in the pyrazine series, although the reactions are largely centered around halogenation. Chlorination of 2-amino-3-methoxycarbonylpyrazine results in ring chlorination at the 5-position (Scheme 28), but the amino-group is also chlorinated, a result which has also been observed in pyridazine chemistry. The ease with which the *N*-chloroamine is reduced with sodium bisulphite means that this side reaction does not represent a serious problem.

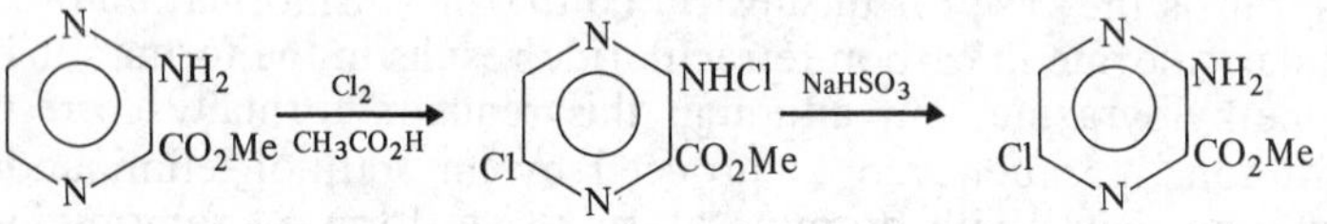

N, NH_2, CO_2Me, N $\xrightarrow[CH_3CO_2H]{Cl_2}$ Cl, N, NHCl, CO_2Me, N $\xrightarrow{NaHSO_3}$ Cl, N, NH_2, CO_2Me, N

SCHEME 28

16.2.5.2 Nucleophilic substitution

In contrast to the pyridines, little work has been carried out on direct amination reactions of the diazines, probably because aminodiazines are more conveniently prepared by alternative methods. Pyrazine undergoes direct amination with $NaNH_2/NH_3$[43] to yield 2-aminopyrazine, but no yield has been quoted for this reaction. 4-Methylpyrimidine undergoes reaction with sodamide in refluxing decalin[44] and although some 2-amino-4-methylpyrimidine is formed, the yield is low and some polyamination is observed. To date no details of the direct amination of pyridazines have been reported.

In general, most nucleophilic substitution reactions of the diazines involve the displacement of some substituent other than hydrogen, and usually a halogen atom is involved. This stems from the fact that halodiazines are readily available from the corresponding oxodiazines which, in turn, are primary products of synthesis. Ring-opening reactions are also common, especially with the pyrimidines. Many examples are given in Section 17.5.13.

The reactivity of halogen atoms attached to the diazine ring is greatly influenced by the nature of the halogen atom itself, its position in the ring, the nucleophile, and the influence of other substituents. Halodiazines are more reactive than similarly activated

2-halopyridines: thus 2-chloropyridine is a stable substance whereas 2-chloropyridazine undergoes spontaneous decomposition at 0 °C.

In an examination of the reactivity of chlorodiazines with the *p*-nitrophenoxide ion, it has been established[45] that the order of reactivity of the chlorodiazines is: 2-chloropyrimidine > 4-chloropyrimidine > 4-chloropyridazine ≈ 3-chloropyridazine ≈ chloropyrazine. Although this order of reactivity serves as a good 'rule of thumb', it is not inviolate and the reverse order of reactivity is usually observed for 2- and 4-chloropyrimidines.[46]

2-Fluoropyrazine undergoes a facile reaction with sodium azide[47] to give 2-azidopyrazine (**37**), which exists in dynamic equilibrium with tetrazolo[1,5-*a*]pyrazine (**38**), the position of the equilibrium being solvent dependent. The reaction is very rapid and consistent with the general observation that fluorine substituents are more labile during nucleophilic substitution reactions in aromatic systems (*cf.* Sanger's reagent) than are other halogen atoms.

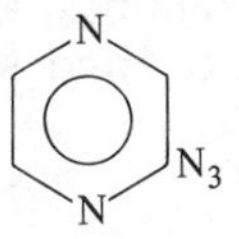

(37) 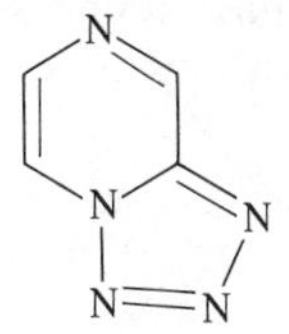 (38) 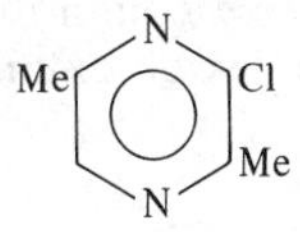 (39)

Chloropyrazine undergoes all of the expected displacement reactions with NH_3, $MeNH_2$, Me_2NH, NaOMe, and NaSH, although alkyl substituents greatly reduce the facility with which displacement occurs. Thus although 2-chloro-3,6-dimethylpyrazine (**39**) readily reacts with NaOEt, it fails to undergo aminolysis even under drastic conditions.

Dihalogenated diazines usually show appreciable differences in the reactivity of the two halogen atoms, a halogen atom located *para* to a ring nitrogen atom being most easily displaced. Halogens located *ortho* to a ring nitrogen atom, although relatively reactive, usually require more drastic conditions for displacement. 3,4,6-Trichloropyridazine (**40**) reacts with methanolic sodium methoxide at 25 °C to yield 3,6-dichloro-4-methoxypyrazine (**41**) in good yield.[48] 3,6-Dichloropyridazine (**42**) reacts with hot aqueous sodium hydroxide to yield 6-chloro-3(2*H*)-pyridazinone (**43**), but the second halogen is remarkably resistant to displacement. When both of the halogen atoms are in a similar environment, substituent effects may dictate which halogen atom is displaced more easily, as illustrated by the aminolysis of 3,6-dichloro-4-methylpyridazine (**44**) which results[49] in the formation of (**45**) and (**46**) in a ratio of 10 : 1.

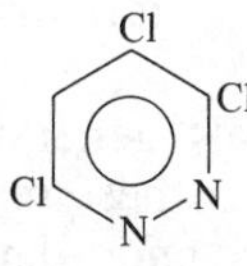

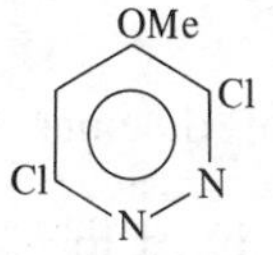

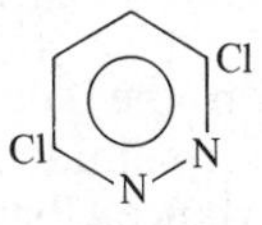

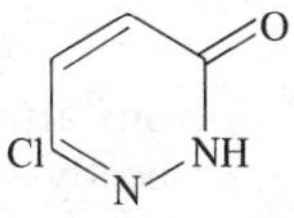

(40) (41) (42)

(43) 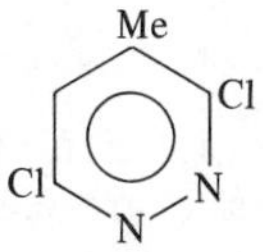 (44) 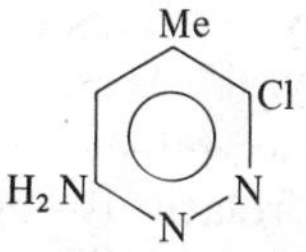 (45) 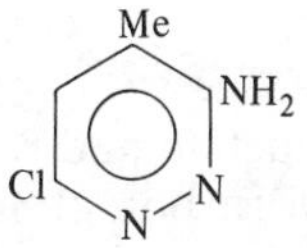 (46) 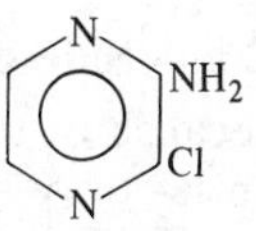 (47)

Nitro groups and other substituents which generally aid nucleophilic substitution have a profound effect on the reactivity of the ring–halogen bond. This influence has been used extensively, as illustrated by the synthesis (Scheme 29) of 4-dimethylaminopteridine.[50]

Dihalogenopyrazines usually lose one of the halogen atoms fairly readily and the remaining halogen becomes resistant to displacement. Thus 2,3-dichloropyrazine on heating with ammonia under pressure[51] yields 2-amino-3-chloropyrazine (**47**).

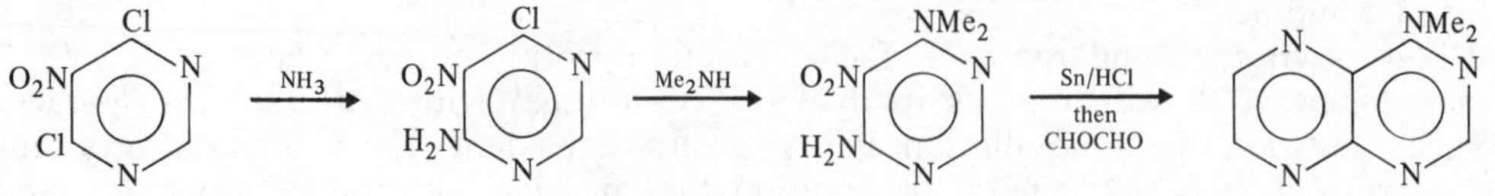

SCHEME 29

Although many other nucleophiles are available, most reported substitution reactions are limited to a few common nucleophiles such as hydrazine, amines, alkoxides, sodium azide, and, to a lesser extent, cyanides, cyanates, thiocyanates, and sodium hydrogen sulphide.

16.2.5.3 Reaction with radicals

Homolytic aromatic substitution of the diazines is a little used phenomenon and information on the relative reactivity of a given position of the diazine ring is scarce. Positions α and γ to the ring nitrogen atoms appear to be the most reactive.[52] Phenylation of pyridazine using phenyl radicals generated by the thermal decomposition of *N*-nitrosoacetanilide[53] results in the formation of 4-phenylpyridazine (**48**) in low yield. Similarly, the arylation of pyrimidine with aryl radicals generated from *p*-nitrobenzene-diazonium salts leads to the formation of (**49**) and (**50**) in a ratio of 3 : 2, albeit in low overall yield.

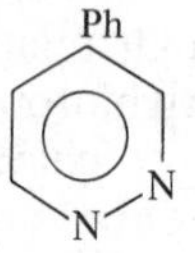

(48)  (49) 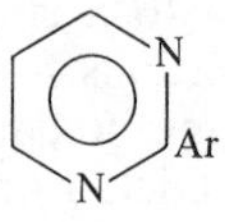 (50) 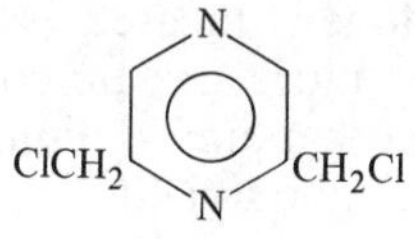 (51)

Recent work on the substitution of diazines with radicals generated by hydrogen abstraction from formamide[54] and its derivatives[55] suggests that these reactions may be of preparative importance. Pyrazine is claimed to give 2-carboxamido-derivatives in yields greater than 80% (based on recovered starting material). The full potential of this reaction does not seem to have been exploited.

Alkyldiazines behave normally in that radical substitution into the side chain occurs readily. Treatment of 2-methylpyrazine with *N*-chlorosuccinimide in the presence of a radical initiator results in the formation of 2-chloromethylpyrazine, and chlorination of 2,6-dimethylpyrazine with molecular chlorine under photochemical conditions results in the formation[56] of the bischloromethyl compound (**51**).

16.2.5.4 Side-chain reactivity

Because of the inductive and resonance effects of the nitrogen atoms, protons attached to an α- or γ-methyl carbon atom are relatively acidic and are easily removed by strong bases. The resultant carbanions undergo typical condensation and alkylation reactions. 2-Methylpyrazine reacts with sodamide in liquid ammonia to form the corresponding anion (Scheme 30), which has been alkylated[57] with simple alkyl bromides. By the simple expedient of using a two-molar excess of 2-methylpyrazine, di- and tri-alkylations are minimized and yields of the desired monoalkylated pyrazines are high. Pyrazinyl ketones have been prepared by an analogous route involving the condensation of the anion with esters.[58]

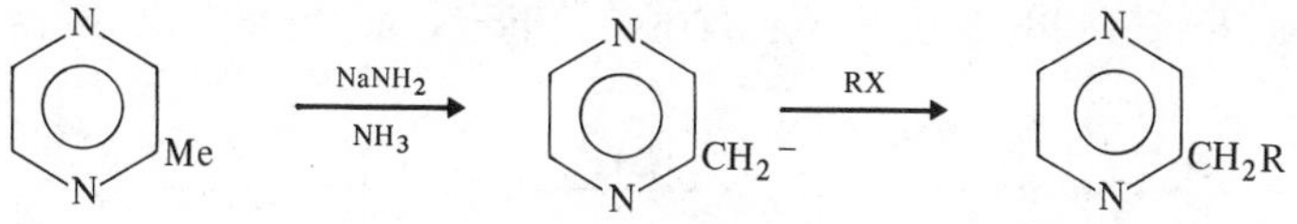

SCHEME 30

Condensation reactions are frequently aided by the use of Lewis acid catalysts, *e.g.* 2,5-dimethylpyrimidine undergoes condensation with benzaldehyde in the presence of zinc chloride[59] to form the corresponding benzylidenepyrimidine (Scheme 31). Some

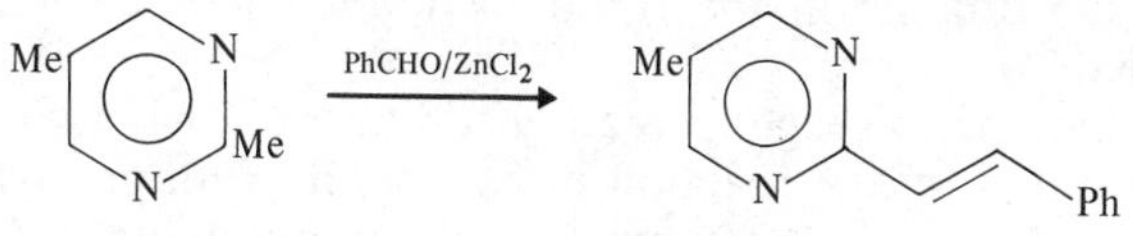

SCHEME 31

selectivity may be observed during anion formation, as seen in the reaction of 2,4,6-trimethylpyrimidine with phenyl-lithium. Alkylation of the intermediate anion with methyl iodide leads to the formation of 2,6-dimethyl-4-ethylpyrimidine (**52**), resulting from the exclusive alkylation of the 4-carbanion.

4-Methylpyridazine readily undergoes condensation with chloral in the presence of pyridine to yield (**53**).

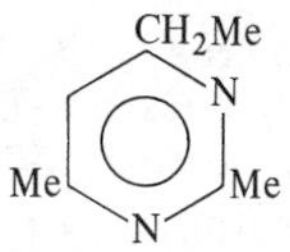

(52) 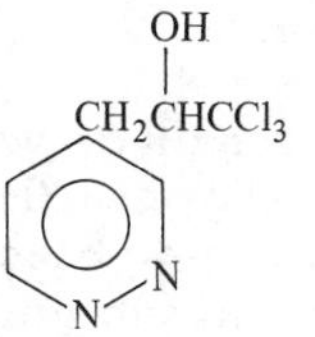 (53)

16.2.5.5 Reactions with organometallic reagents

The introduction of hetero-atoms into an aromatic ring results in an increase in electron density at the site occupied by the hetero-atom which, in turn, results in greater localization of the double bonds and organometallic reagents tend to attack a —C=N— bond (*cf.* the addition of organometallic reagents to a carbonyl group), or undergo conjugate addition to a —C=C—C=N— group, the type of addition being governed by the nature of the organometallic reagent and the solvent employed during the reaction. For example, the reaction of pyridazine with phenylmagnesium bromide results in the formation of 4-phenylpyridazine[60] in low yield. The reaction is believed to proceed by way of an intermediate 4,5-dihydropyridazine (Scheme 32) which undergoes spontaneous

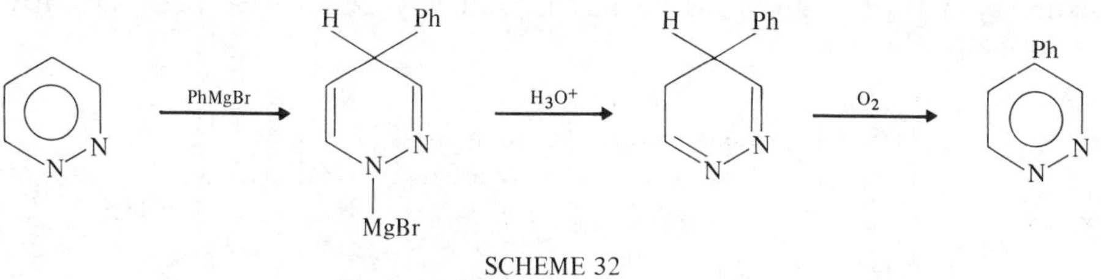

SCHEME 32

aerial oxidation during work up. Butylmagnesium halides behave in a similar way and usually give higher yields of the 4-butylpyridazine.

In contrast, with organolithium reagents, 3-substituted pyridazines are formed, often in yields of preparative value. Butyl-lithium reacts with pyridazine in ether solution (Scheme 33) to generate 3-butyl-2,3-dihydropyridazine, which on oxidation with potassium permanganate in acetone yields 3-butylpyridazine. When ether/tetrahydrofuran mixtures are

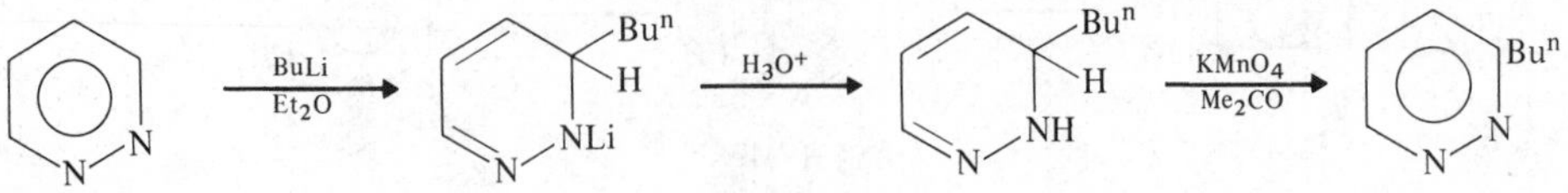

SCHEME 33

used as a solvent for the reaction, mixtures of 3- and 4-butylpyridazines are obtained. Symmetrically 3,6-disubstituted pyridazines yield exclusively 4-substituted derivatives, but as expected if the substituents in the 3,6-dialkylpyridazine are different, two different products are obtained.[61]

Grignard and organolithium reagents readily add to the 3,4-bond of pyrimidine at room temperature, as illustrated by the reaction of pyrimidine with phenylmagnesium bromide (Scheme 34) to yield the stable 4-phenyl-3,4-dihydropyrimidine after hydrolysis of the intermediate magnesium complex. Oxidation of the dihydropyrimidine is readily accomplished with $KMnO_4$ in acetone.[62]

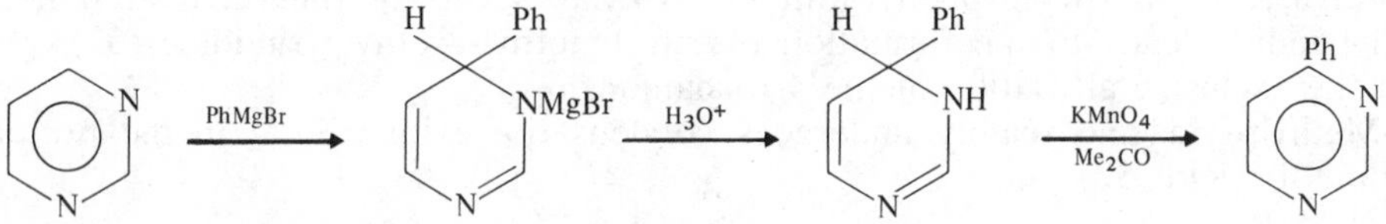

SCHEME 34

The reaction of pyrimidines with organolithium reagents, as with the analogous pyridazines, is more facile than the reaction with Grignard reagents and a number of 4-substituted pyrimidines have been prepared by this method. When the 4-position of the pyrimidine ring is blocked, substitution occurs at positions-2 or -6 with equal ease,[63] as evidenced by the formation of 4-methyl-2-phenyl- and 4-methyl-6-phenyl-pyrimidines during the reaction of phenyl-lithium with 4-methylpyrimidine. In principle, starting from pyrimidine, three different alkyl or aryl residues may be separately introduced into the ring, although to date this possibility appears to have been ignored.

Pyrimidine differs from pyridazine and pyrazine in that the 5-position of the ring is neither α or γ to one of the ring nitrogen atoms, and it is to be expected that substituents at this site should enjoy reactivity more closely aligned with substituted benzene derivatives. Indeed, if 5-bromopyrimidine is treated with butyl-lithium at −100 °C the 5-lithio-derivative is formed. At higher temperatures (−80 °C) the reaction is complicated by addition reactions of the butyl-lithium to the pyrimidine ring.

In the pyrazine ring all of the —C═N— bonds are equivalent and the addition of organometallic reagents follows the general trends observed with the other diazines. Treatment of acetonylpyrazine (**54**) with phenyl-lithium results[64] in the formation of 2-phenyl-6-acetonylpyrazine (**55**), indicating that addition, rather than anion formation at the acidic acylmethylene group, is favoured. Methylpyrazines have been similarly alkylated with methyl-lithium.[65]

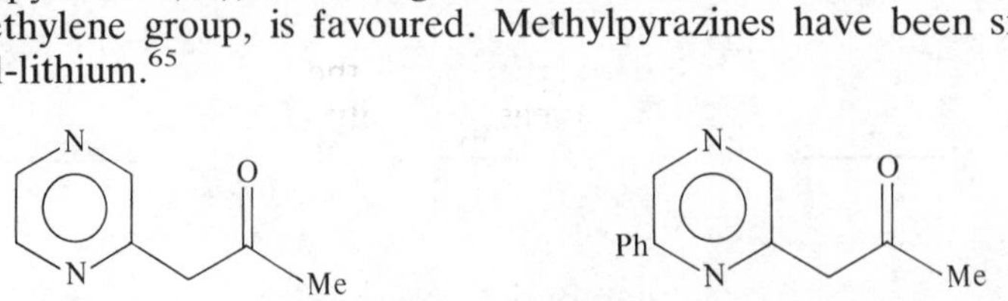

(54) (55)

In favourable circumstances, functionalized diazines may react with Grignard reagents in a more conventional sense. 2-Acetylpyrazine has been prepared[66] (Scheme 35) by the reaction of methylmagnesium iodide with 2-cyanopyrazine and, under normal Grignard conditions, addition of the organometallic reagent occurs preferentially at the cyano group.

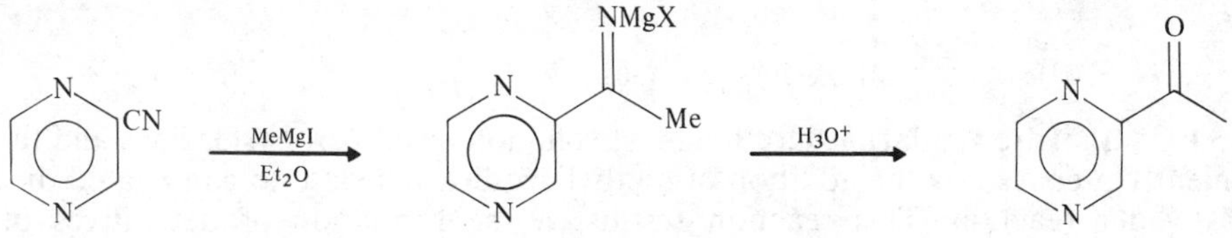

SCHEME 35

16.2.6 REDUCED DIAZINES

16.2.6.1 Introduction

The chemistry of reduced diazines is an ill-defined area and very few generalizations are possible, making a detailed description of each of the reduction products necessary. Wherever possible, sources of further reference will be indicated in the following brief summary.

16.2.6.2 Dihydropyridazines

Of the six possible dihydropyridazines, only the 1,2- (**56**), 1,4- (**57**), and 4,5-dihydropyridazines (**58**) are known.[1]

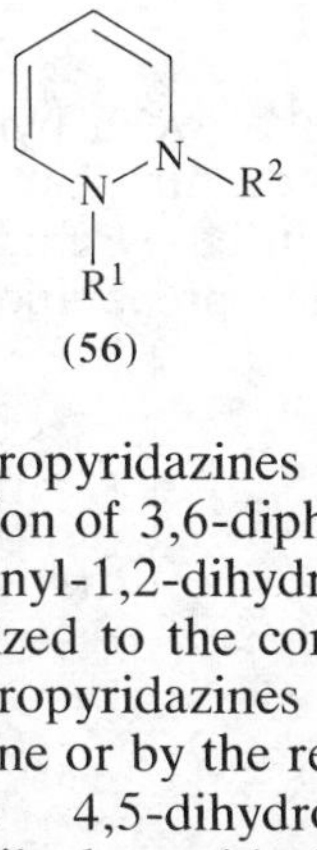

(56) 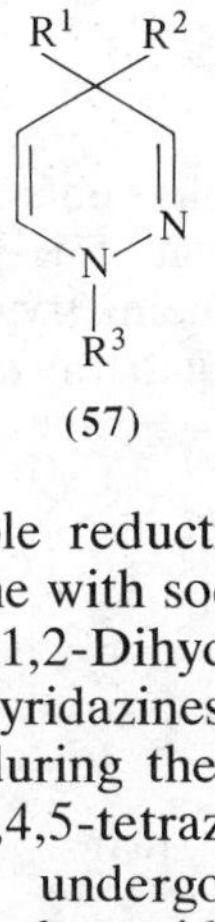 (57) 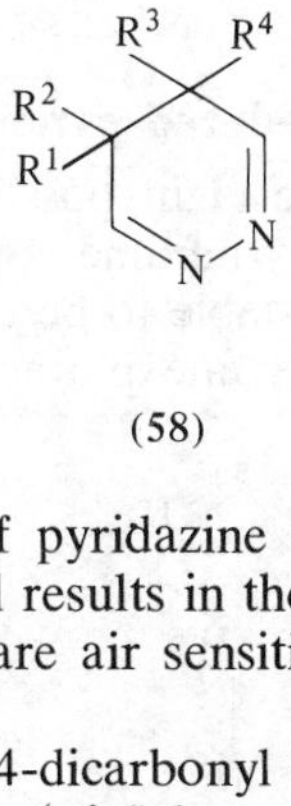 (58)

1,2-Dihydropyridazines are the simple reduction products of pyridazine derivatives. Thus reduction of 3,6-diphenylpyridazine with sodium in ethanol results in the formation of 3,6-diphenyl-1,2-dihydropyridazine. 1,2-Dihydropyridazines are air sensitive and are rapidly oxidized to the corresponding pyridazines.

1,4-Dihydropyridazines are formed during the reaction of 1,4-dicarbonyl compounds with hydrazine or by the reaction of 1,2,4,5-tetrazines with olefins (*cf.* Scheme 6) with the intermediate 4,5-dihydropyridazines undergoing isomerization to the more stable 1,4-dihydropyridazines. 4,5-Dihydropyridazines would appear to be the first-formed products during the reaction of 1,4-diketones with hydrazine; however, under normal reaction conditions the corresponding 1,4-dihydropyrazines result, presumably because of the increased thermodynamic stability of the 1,4-dihydropyridazine ring.

4,5-Dihydropyridazines are usually observed as products of the addition of organometallic reagents to pyridazines. They appear to be stable under neutral conditions and require vigorous oxidation to convert them to the corresponding pyrazines. Substitution of the ring generally results in increased stabilization.

16.2.6.3 Tetrahydropyridazines

Of the four possible tetrahydropyridazines,[1] only three are known; 1,2,3,4-tetrahydropyridazine (**59**) has not been described to date.

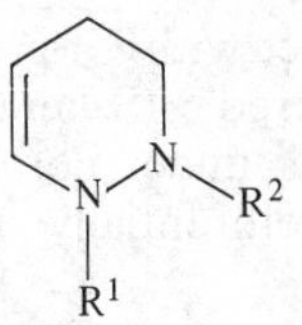

(59) 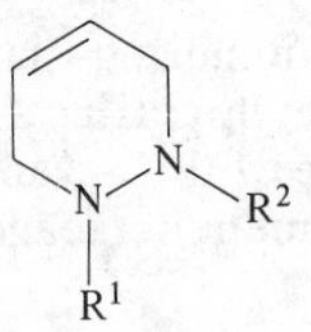 (60) 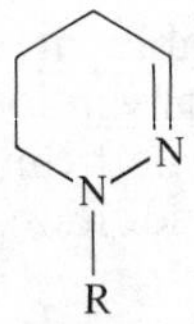 (61)

1,2,3,6-Tetrahydropyridazines are the most common tetrahydropyridazines and they are conveniently prepared by the addition of diethyl azodicarboxylate to conjugated dienes in a Diels–Alder reaction. This reaction results in the formation of derivatives of (**60**) bearing ethoxycarbonyl groups at positions-1 and -2 of the pyridazine ring. These may be removed in a stepwise manner to give 1,2,3,6-tetrahydropyridazines which are unsubstituted on the ring nitrogen atoms. This reaction is sometimes complicated by the isomerization of the 1,2,3,6-tetrahydropyridazines to the isomeric 1,4,5,6-tetrahydropyridazines (**61**).

1,4,5,6-Tetrahydropyridazines are usually prepared by the reduction of 4,5-dihydro-3(2*H*)-pyridazinones with lithium aluminium hydride, although with an excess of LAH total reduction to the hexahydropyridazines may be observed.

16.2.6.4 Hexahydropyridazines

Hexahydropyridazines are usually prepared by catalytic hydrogenation of the double bond in the readily accessible 1,2,3,6-tetrahydropyridazines, although reduction with sodium in ethanol has sometimes been employed.

16.2.6.5 Reduced pyrimidines

There are eight possible partially hydrogenated pyrimidines between pyrimidine and hexahydropyrimidine, (**62**)–(**69**), although it is by no means certain that they are all sufficiently stable to be isolated.[10] The chemistry of partially reduced pyrimidines appears to be largely unexplored and there are gaps in methods for their preparation.

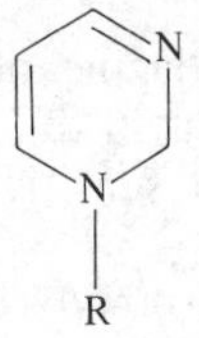

(62) 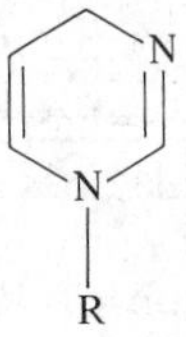 (63) 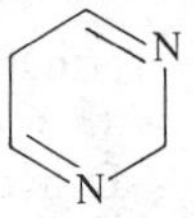 (64) 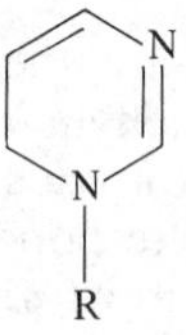 (65)  (66)

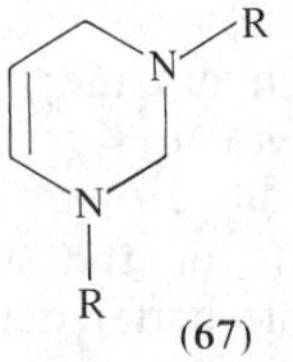

(67) 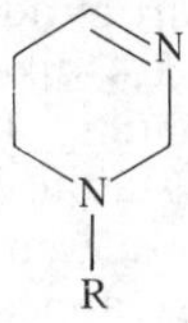 (68) 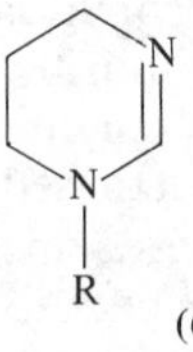 (69)

16.2.6.6 Dihydropyrimidines

Probably the most common dihydropyrimidines are 5,6-dihydropyrimidines. These compounds are usually synthesized directly by the reaction of urea with an α,β-unsaturated carbonyl compound. This type of reaction is typified by the reaction of acrylic

SCHEME 36

esters[67] with urea to yield 5,6-dihydrouracil (Scheme 36). Hofmann degradation of succindiamide derivatives[68] also results in the formation of 5,6-dihydropyrimidines, presumably through the intermediacy of a monoisocyanate (Scheme 37).

SCHEME 37

Direct reduction of pyrimidines to the corresponding 5,6-dihydro-derivatives is possible, as in the reduction of uridine by catalytic hydrogenation over Adams catalyst[69] to form 5,6-dihydrouridine. It is of interest that 5,6-dihydropyrimidine nucleosides are more susceptible to hydrolytic cleavage than the pyrimidine nucleosides.

1,2-, 1,4-, and 1,6-dihydropyrimidines have been observed during the addition of organometallic reagents to substituted pyrimidines, but are otherwise uncommon in pyrimidine chemistry. 2,5-Dihydropyrimidines have been synthesized by direct methods or by the application of reductive techniques such as Raney nickel desulphurization. The direct methods usually involve the condensation of derivatives of malondiamide with carbonyl compounds, as illustrated by the synthesis of 5,5-diethyl-4,6-dioxo-2-phenyl-2,5-dihydropyrimidine (**70**) from benzaldehyde and diethylmalondiamide.[70] Virtually any

(70) (71) (72) (73)

aldehyde or ketone may be employed in this reaction and in some instances formamide may be used to prepare the 2,5-dihydropyrimidine free from substituents at position-2,[71] although this reaction formally involves the intermediacy (Scheme 38) of a pyrimidine which is subsequently reduced by formamide.

SCHEME 38

(74)

Desulphurization of monothiopyrimidines with Raney nickel is a relatively facile process and the 2,5-dihydropyrimidine (**74**) has been prepared[72] by the desulphurization of (**71**).

16.2.6.7 Tetrahydropyrimidines

Catalytic reduction of pyrimidines is generally more likely to result in the formation of tetrahydropyrimidines than dihydropyrimidines. Furthermore, the resultant tetrahydro-derivatives usually have the double bond in the 1,2(2,3)-position.

In 1922, Mannich and Heilner[73] reported that hydrogenation of 1,2-dihydro-1-methyl-2-oxo-4-phenylpyrimidine (**72**; R = Me) over palladium gives mainly 1,4,5,6-tetrahydro-2-hydroxy-1-methyl-4-phenylpyrimidine (**73**) as the major product. More usually, phenyl-substituted pyrimidines undergo reduction of the phenyl ring during such catalytic reductions.[74]

More conventionally, tetrahydropyrimidines are synthesized directly from derivatives of 1,3-diaminopropane. When a 1,3-diamine is acylated or diacylated,[75] ring closure to a 1,4,5,6-tetrahydropyrimidine may be effected under a variety of conditions. The reaction may also be brought about without the prior acylation merely by heating the 1,3-diamine with a suitable carbonyl compound.[76] Esters, carbonates, formaldehyde, CS_2, phosgene, thiophosgene, and formamide all serve as suitable one-carbon fragments.

A few tetrahydropyrimidines have been prepared by electrolytic reduction, *e.g.* barbituric acid[77] and related compounds yield trimethyleneureas amongst other products, although this method appears to be of little preparative value.

16.2.6.8 Hexahydropyrimidines

Most hexahydropyrimidines are synthesized by the direct condensation of a 1,3-diamine with an aldehyde or ketone. They frequently exist in equilibrium[78] with an open-chain system (Scheme 39) with the position of the equilibrium being dependent on factors such

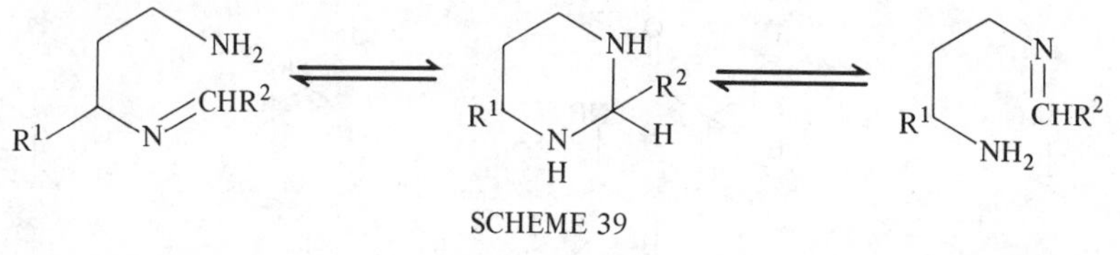

SCHEME 39

as the degree of substitution of the ring and solvent. Catalytic reduction of pyrimidines to hexahydropyrimidines is rare since under normal hydrogenation conditions hexahydropyrimidines undergo reductive ring cleavage.

16.2.6.9 Dihydropyrazines

Derivatives of 1,2- (**75**), 1,4- (**76**), 5,6- (**77**), and 2,5-dihydropyrazines (**78**) are known, but the establishment of a precise structure for a given dihydropyrazine is often difficult

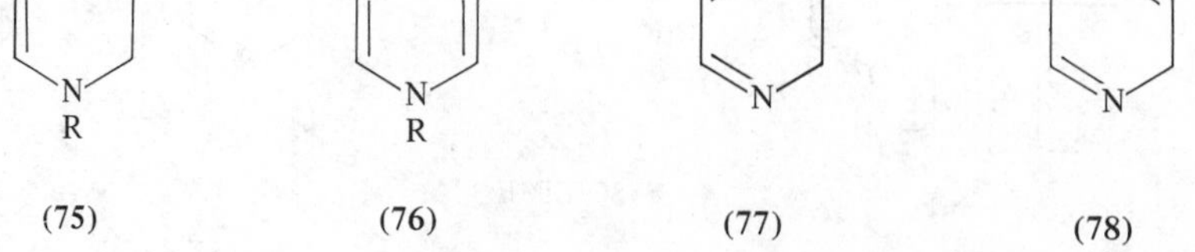

(75) (76) (77) (78)

because of the extreme lability of dihydropyrazines under mildly acidic or basic conditions. Generally ring substitution helps to limit isomerization within the dihydropyrazine series.

1,2-Dihydropyrazines are formed during the addition of organometallic reagents to pyrazines. They are unstable, undergoing rapid aerial oxidation to the corresponding pyrazines. 1,4-Dihydropyrazines are also generally unstable. The bistrimethylsilyl derivative (**79**),[79] formed during the reduction of pyrazine with lithium in the presence of trimethylsilyl chloride, spontaneously decomposes in air. When strongly electron-withdrawing ethoxycarbonyl groups are present the 1,4-dihydropyrazine structure becomes more stable and compound (**80**) is a stable solid which may be prepared by the reduction of the parent pyrazine with sodium dithionite.[80]

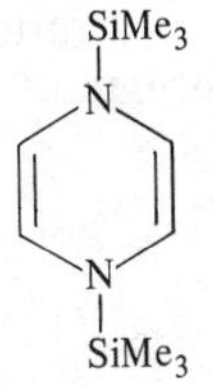

(79) 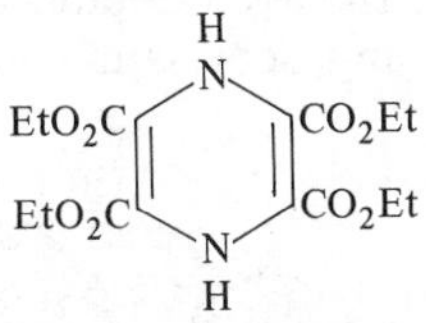 (80)

5,6-Dihydropyrazines are common products of direct synthesis (Scheme 21) and are relatively stable, requiring heating at 300 °C over a copper chromite catalyst to bring about dehydrogenation to the corresponding pyrazines.

2,5-Dihydropyrazines are common and are usually formed by the dimerization of α-aminoketones. Oxidation usually occurs more readily than with 5,6-dihydropyrazines, although isomerization to 1,2- and 1,4-dihydropyrazines may occur. 2,5-Dioxopiperazines, the dimerization products of α-amino-acids, are formally dihydropyrazines and the *O*-alkyl derivatives have been oxidized to pyrazines.

16.2.6.10 Tetrahydropyrazines

Tetrahydropyrazines are uncommon and are usually prepared synthetically rather than by direct reduction methods. Stevens and co-workers[81] have prepared 2-phenyl-3,3-dimethyl-3,4,5,6-tetrahydropyrazine directly from ethylenediamine (Scheme 40).

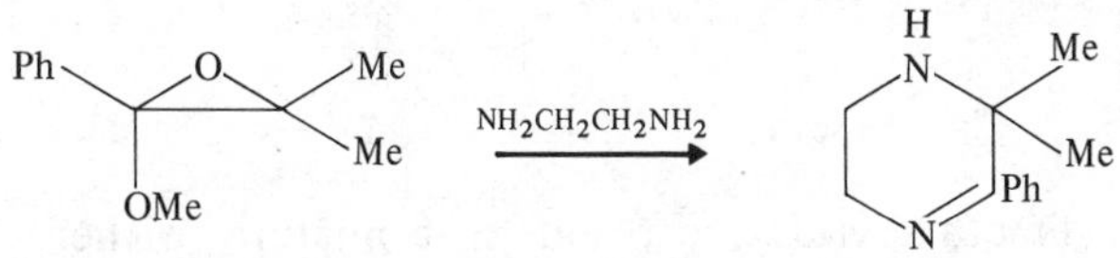

$NH_2CH_2CH_2NH_2$

SCHEME 40

16.2.6.11 Hexahydropyrazines

Hexahydropyrazine (piperazine) is conveniently prepared by the reduction of pyrazine with sodium in ethanol, although, commercially, the reaction of ethylenediamine with ethylene oxide represents a more practical synthesis. The bifunctional nature of piperazine is highly desirable and this compound has found extensive use in medicinal chemistry.

16.2.7 DIAZINE *N*-OXIDES

16.2.7.1 Synthesis

The diazine *N*-oxides are generally prepared by methods analogous to their monoazine counterparts, and typical oxidizing agents such as H_2O_2/AcOH, *m*-chloroperbenzoic acid, monoperoxyphthalic acid, permaleic acid, and peroxytrifluoroacetic acid may be used. In the case of the parent diazines, pyridazine and pyrazine may be oxidized directly, but pyrimidine tends to undergo decomposition during normal oxidizing conditions, resulting in low yields of the *N*-oxide.[82]

Most derivatives of the diazines are readily oxidized to the corresponding *N*-oxides directly, and once the pyrimidine ring is substituted, *e.g.* by alkyl or alkoxy groups, oxidation occurs readily and the *N*-oxides are stable under the oxidizing conditions.

When unsymmetrical diazines are subjected to oxidation, mixed products are observed. 2-Methylpyrazine on oxidation with one equivalent of peracetic acid (generated *in situ* from H_2O_2/AcOH) results in mixtures of 2-methylpyrazine 1-oxide (**81**) and 2-methylpyrazine 4-oxide (**82**), which may be separated by fractional crystallization.[83] 3-Methylpyridazine similarly gives a mixture[84] of the *N*-oxides (**83**) and (**84**) in a ratio of 1 : 3, and 4-methylpyridazine yields the 1-oxide and the 2-oxide in a ratio of 1 : 4, the structures of the product *N*-oxides having been confirmed by unambiguous synthesis.[85]

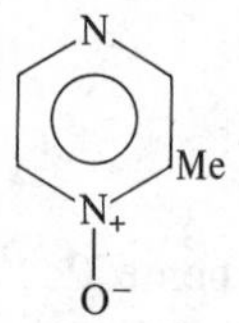

(81) 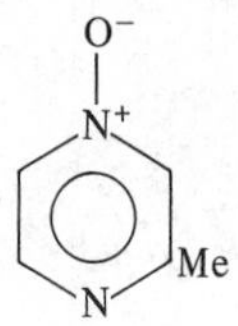 (82) 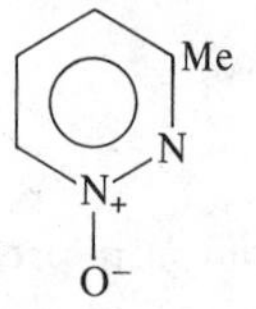 (83) 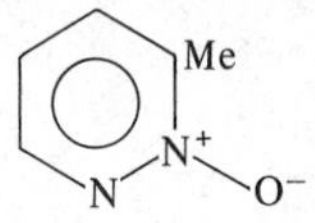 (84)

Alkyl-substituted pyrimidines are also stable to peracetic acid and 4-methylpyrimidine yields mixtures of the 3-oxide (**85**) and the 1-oxide (**86**) with the former predominating.[86] These results on the oxidation of simple diazines support the existence of a weak short range $+I$ effect of a methyl substituent which results in the preferential *N*-oxidation of an adjacent ring-nitrogen atom. This is in keeping with the observation that substituents generally exhibit a profound effect on which of the ring nitrogen atoms is oxidized.

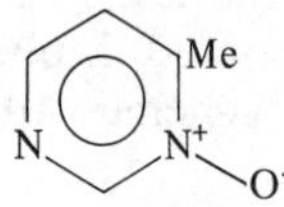

(85) 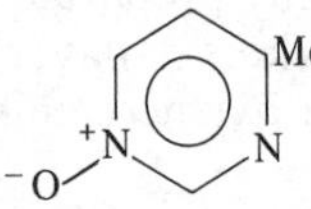 (86) 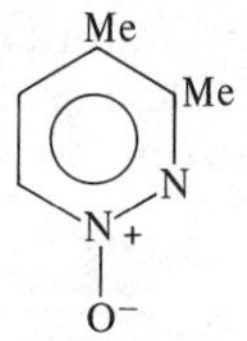 (87) 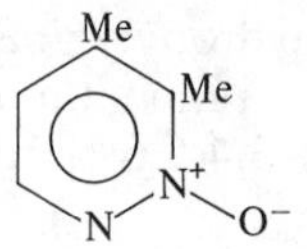 (88)

Oxidation of 3,4-dimethylpyridazine results in a mixture of the 1-oxide (**87**) and the 2-oxide (**88**) in a ratio of 2 : 1.[87] 6-Chloro-3,4-dimethylpyridazine (**89**) gives 83% of the 2-oxide (**90**) and only 0·7% of the 1-oxide (**91**), whereas 3,4-dimethyl-6-methoxypyridazine (**92**) gives exclusively the 2-oxide (**93**) and these orientation effects appear to be partly steric and partly electronic in nature.

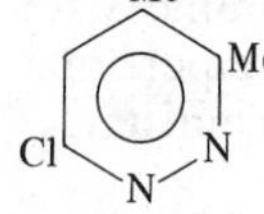

(89) 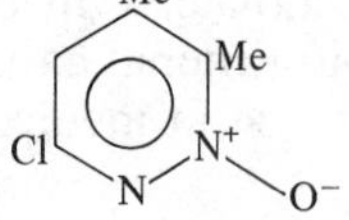 (90) 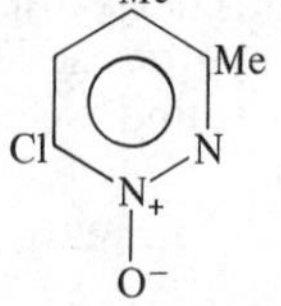 (91) 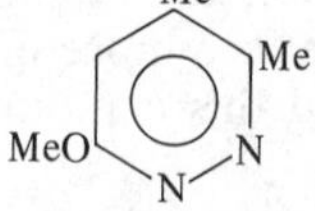 (92) 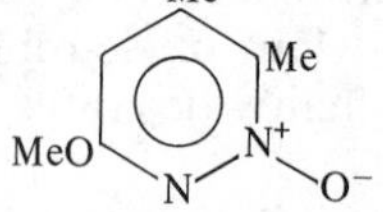 (93)

Hayashi *et al.*[88] have shown that *N*-oxidation of monoalkoxydiazines, where the alkoxy-group is located on a carbon atom adjacent to a ring nitrogen atom, always takes place at the nitrogen atom remote from the alkoxy-group, probably as a consequence of steric hindrance. This is illustrated by oxidations such as (**94**)→(**95**).[89]

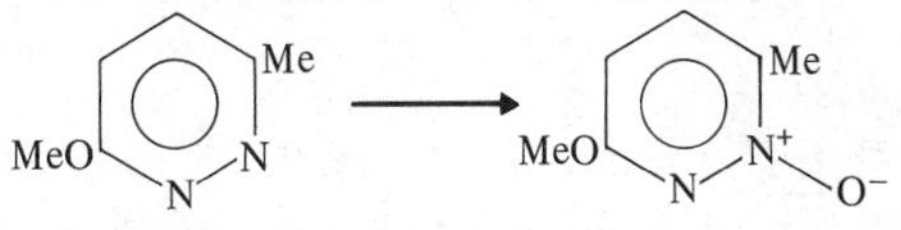

(**94**) (**95**)

When substituents with large negative inductive effects, such as a halogen atom, are present in the ring, oxidation usually occurs on the nitrogen atom remote from the halogen. Thus oxidation of 2-chloro-3,6-dimethylpyrazine (**96**) results in the preferential formation[90] of the 4-oxide (**97**). 3-Chloro-5-methylpyridazine behaves similarly.[85]

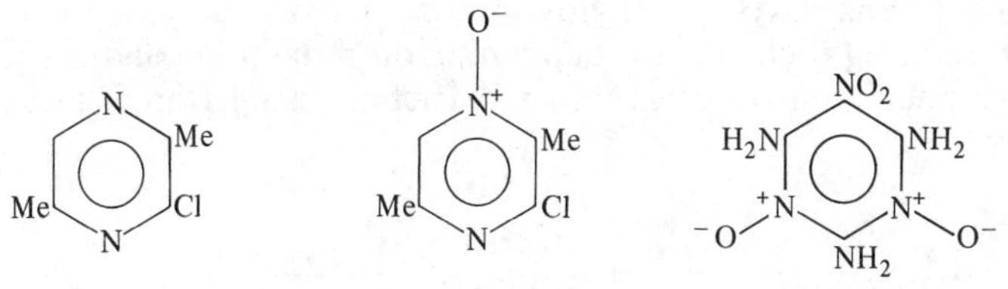

(**96**) (**97**) (**98**)

Of the three diazines only pyrazine and its derivatives may be oxidized directly to the di-*N*-oxide, and the oxidant of choice appears to be peroxytrifluoroacetic acid.[90] Yields of pyrazine di-*N*-oxides are usually good. An isolated report on the synthesis of the pyrimidine di-*N*-oxide (**98**) by direct oxidation[91] has appeared but there is no evidence that this procedure is general. A more general route to pyrimidine di-*N*-oxides has recently been published[92] which involves the oxidation of 1-hydroxy-1,2,5,6-tetrahydropyrimidine 1-oxides (Scheme 41) with MnO_2.

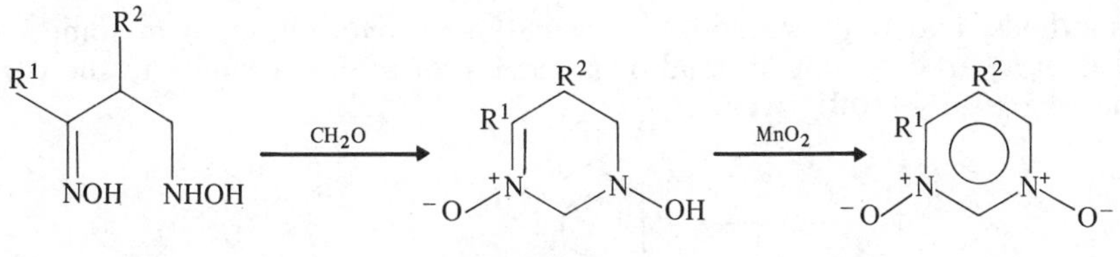

SCHEME 41

Until recently, pyridazine di-*N*-oxides were unknown, and they have never been prepared by direct methods. Spyroudis and Vargvoglis[93] have developed a route based on the phenyliodine bistrifluoroacetate oxidation of the bis-oximes of 1,2-diacylethylenes (Scheme 42), which produces the di-*N*-oxides in modest yields.

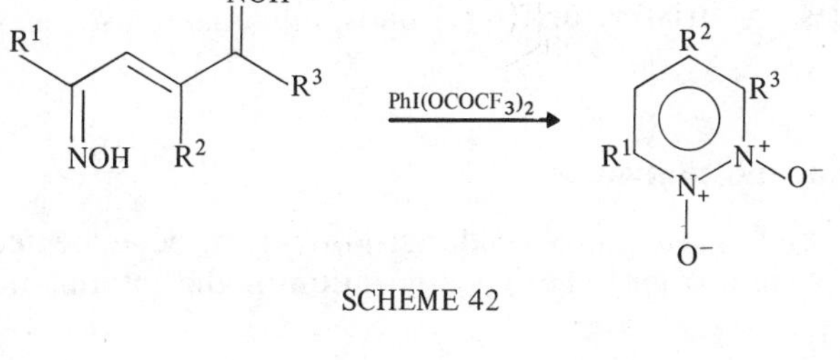

SCHEME 42

16.2.7.2 Reactions of diazine *N*-oxides

Diazine *N*-oxides generally undergo all of the common reactions of their monoazine counterparts, and although it is assumed that the reactions are similar, little mechanistic work has been carried out to establish the justification of this assumption.

16.2.7.3 Deoxygenation

One of the main characteristics of the diazine *N*-oxides is their stability. Thus whilst aliphatic *N*-oxides are easily reduced to tertiary amines with H_2SO_3, at room temperature, diazine *N*-oxides tend to be much more stable and examples of simple reductive deoxygenation are rare. In fact this stability of diazine *N*-oxides imposed a barrier to the use of the *N*-oxides in synthetic manipulation of the diazine nucleus until suitable deoxygenation techniques were developed.

One of the most common techniques for the removal of the oxygen atom involves catalytic hydrogenation, and it is of some interest to note that considerable selectivity may be achieved. Thus 4-benzyloxy-6-methylpyrimidine 1-oxide may be debenzylated or the *N*-oxide function reduced (Scheme 43) depending on reduction conditions.[94] Both catalytic reduction over palladium on charcoal or reduction using Raney nickel appear to be

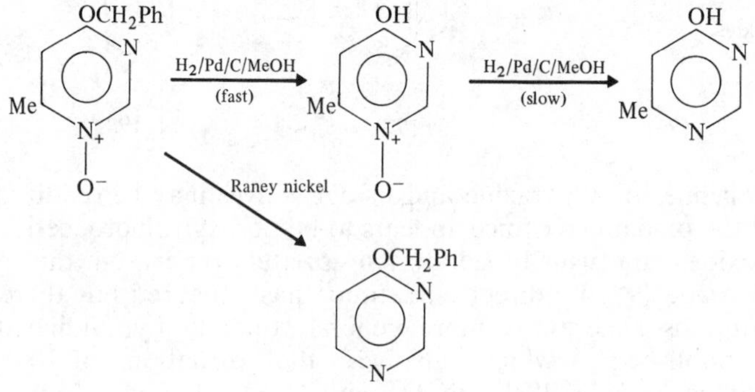

H_2/Pd/C/MeOH (fast); H_2/Pd/C/MeOH (slow); Raney nickel

SCHEME 43

general methods, and both methods are usually selective, although in some cases hydrogenolysis of a carbon–halogen bond or reduction of a nitro-group[95] to the corresponding amine [*e.g.* (**99**)→(**100**)] occurs.

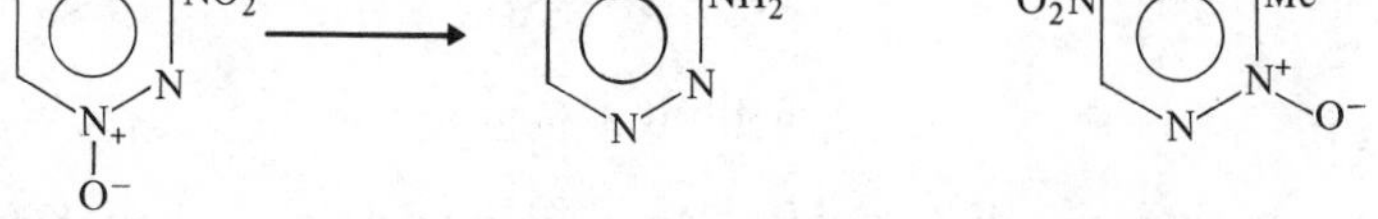

(99) (100) (101)

Treatment of a chloroform solution of *N*-oxides with phosphorus trichloride or tribromide results in selective deoxygenation with the concomitant formation of phosphoryl halides.[96] Clearly the presence of active hydrogen atoms will complicate the use of phosphorus trihalides and trialkyl or triaryl phosphites have also been successfully used for deoxygenation.

16.2.7.4 Electrophilic substitution

The positions α and γ to an *N*-oxide group might be expected to be prone to electrophilic attack owing to mesomeric stabilization of the intermediate σ-complex and this has been found to be the case. Nitration of pyridazine 1-oxide[97] and its 3-, or 6-, mono- or di-substituted derivatives with mixed acids yields the corresponding 4-nitropyridazine 1-oxides. The reaction appears to be sensitive to substituent effects, as evidenced by the fact that 3-methylpyridazine 1-oxide could not be nitrated even under forcing conditions,[98] whereas 3-methylpyridazine 2-oxide is nitrated to give 3-methyl-5-nitropyridazine 2-oxide (**101**) in excellent yield.

To date, electrophilic substitution of the diazine mono-*N*-oxides appears to be largely restricted to a consideration of pyridazine with few examples in the pyrazine or pyrimidine series. Clearly, in view of the ease with which the *N*-oxide may be removed, it offers an attractive route to substitution of the diazine ring.

An alternative method of nitration has been introduced which tends to alter the position of substitution within the ring relative to the *N*-oxide group. Treatment of pyridazine 1-oxide with silver nitrate in the presence of an acid chloride results in the formation of 3-nitropyridazine 1-oxide.[99] The proposed mechanism of this reaction involves the formation of an acyl nitrate, which dissociates to produce a nitronium ion (Scheme 44) that functions as the electrophile.

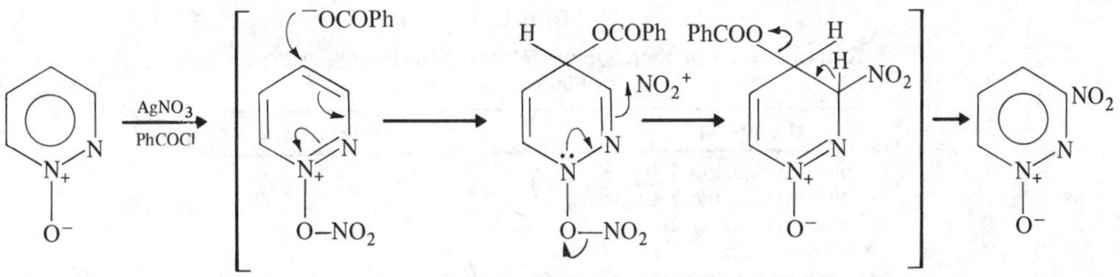

SCHEME 44

Direct halogenation of pyridazine *N*-oxides has also been reported. Thus 3-hydroxypyridazine 1-oxide is readily halogenated at the 4- and 6-positions using chlorine or bromine in acetic acid.[100]

16.2.7.5 Nucleophilic substitution

The diazine *N*-oxides show similarities in their reactions to the pyridine *N*-oxides, with all of the diazine rings being activated towards nucleophilic substitution, especially if the positive charge on nitrogen is enhanced by the intermediate formation of a complex with an electron acceptor. Two distinct types of nucleophilic attack are observed: those in which the N—O bond is retained in the product and those in which the N—O bond is cleft during the reaction.

Ring substituents show enhanced reactivity towards nucleophilic substitution, with substituents α to the *N*-oxide generally showing greater reactivity than those in β-positions. 2-Chloropyrazine 1-oxide (**102**) is readily converted into 2-hydroxypyrazine 1-oxide (**103**) [1-hydroxy-2-oxo-(1*H*)-pyrazine (**104**)] on treatment with dilute aqueous sodium hydroxide,[101] whereas 2-chloropyrazine is stable under the same conditions. 2-Chloropyrazine 4-oxide (**105**) is also stable under these conditions.

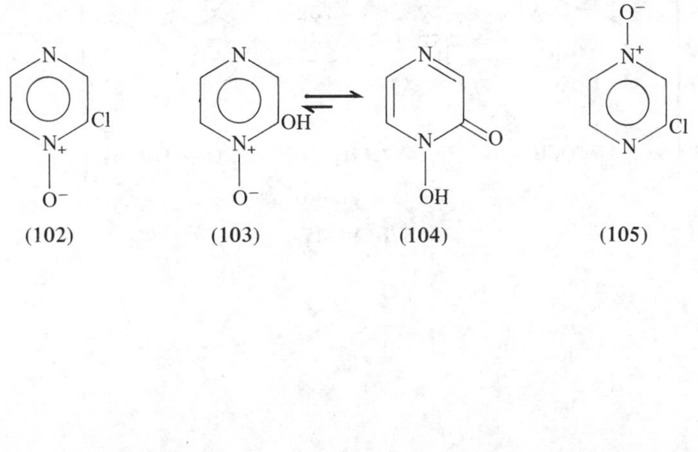

(102) (103) (104) (105)

The activating influence of the *N*-oxide function is further illustrated by the fact that the reaction of 2-chloro-3-methylpyrazine 4-oxide (**106**) with thiourea to yield the thione (**107**) occurs at an appreciably greater rate[102] than with the parent diazine (**108**). A detailed examination[103] of the reactivity of chloropyridazine 1-oxides has appeared and

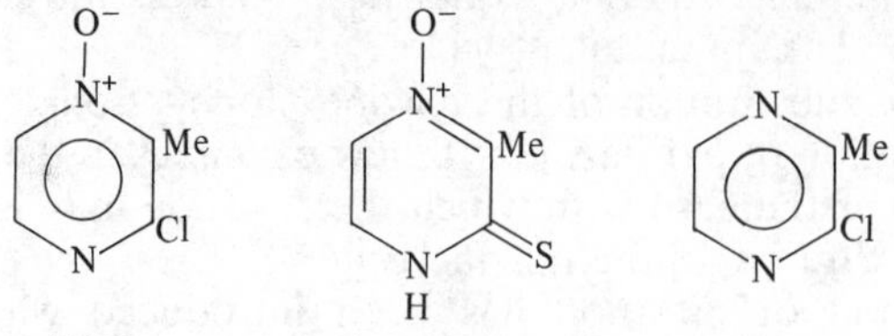

(106) (107) (108)

the relative rates for the reaction with piperidine at 50 °C are shown in Table 2. These data clearly indicate that the combined effect of a *meta* *N*-oxide substituent and an *ortho* or *para* nitrogen is greater than a *meta* nitrogen and a *para* *N*-oxide group, on the activation of the halogen atom.

TABLE 2
Relative Rates of Reaction of Chloropyridazine *N*-Oxides with Piperidine

Compound	*Relative rate*
3-Chloropyridazine 1-oxide	41
4-Chloropyridazine 1-oxide	18
5-Chloropyridazine 1-oxide	5.6
6-Chloropyridazine 1-oxide	1

With 3,6-dichloropyridazine 1-oxides the 3-halogen appears to be more susceptible to displacement,[104] but mixtures usually result during displacement reactions. For example, 3,6-dichloropyridazine 1-oxide (**109**) reacts with sodium ethoxide to yield 6-chloro-3-ethoxypyridazine 1-oxide (**110**) and 3-chloro-6-ethoxypyridazine 1-oxide (**111**) in a ratio of 6.5 : 1. Other factors, such as the size of the nucleophile, appear to play an important role in determining the ratio of products.

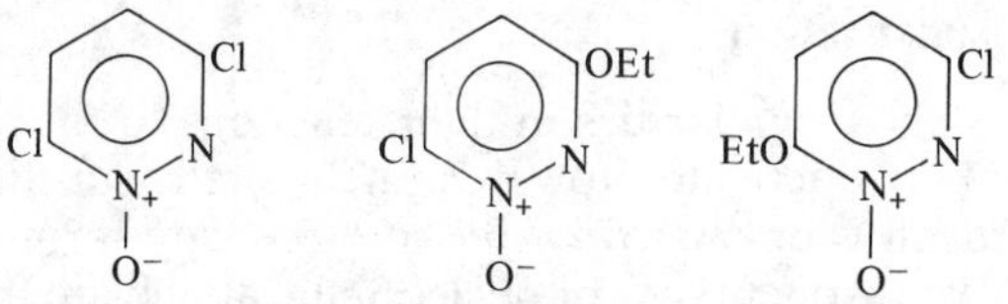

(109) (110) (111)

Nitro-groups may also be displaced from nitrodiazine *N*-oxides and displacements from the 3-, 4-, 5-, or 6-positions of pyridazine *N*-oxides have been reported. On treatment of nitropyridazine 1-oxides with acetyl chloride,[105] or with 15–20% HCl solution, the nitro-substituent is replaced by chlorine and a probable mechanism for this reaction is shown in

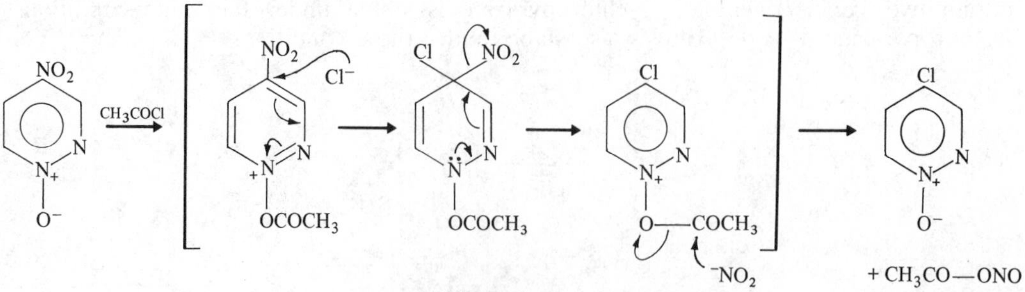

CH_3COCl $^-NO_2$ $+ CH_3CO—ONO$

SCHEME 45

Scheme 45. Support for this mechanism comes from the observation that when methyl substituents are adjacent to the *N*-oxide, nitrosation (presumably by NO^+ generated from the acetyl nitrite) occurs and oximes or nitriles result (Scheme 46).

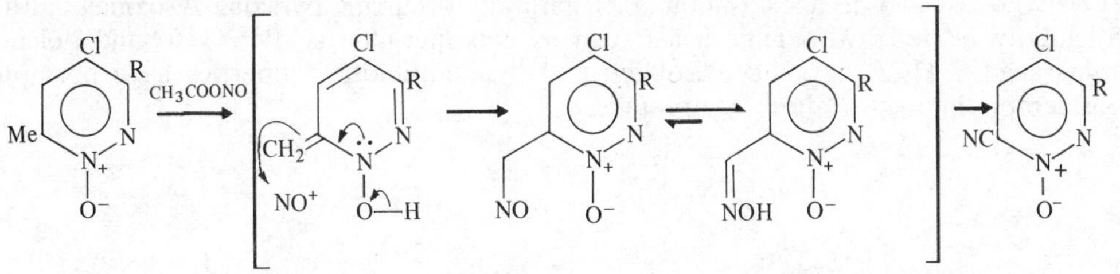

SCHEME 46

Nucleophilic reactions in which the N—O bond is cleft are also very common amongst the reactions of the diazine *N*-oxides. Treatment of pyrazine 1-oxide with phosphoryl chloride results in the formation of 2-chloropyrazine.[106] Similarly, treatment of pyrazine 1-oxide with acetic anhydride results in the formation of 2-acetoxypyrazine and pyrimidine *N*-oxide yields 4-acetoxypyrimidine.[107] When an alkyl substituent is present α to the *N*-oxide, substitution into the alkyl group results. 2-Methylpyrazine 1-oxide (**112**) and 4,6-dimethylpyrimidine 1-oxide (**113**) yield 2-acetoxymethylpyrazine (**114**) and 4-acetoxymethyl-6-methylpyrimidine (**115**), respectively. Analogous substitution reactions have been observed with phosphoryl halides.

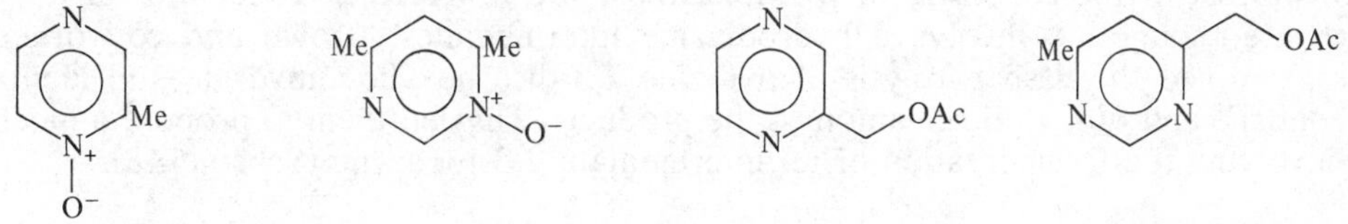

(112) (113) (114) (115)

Reissert addition reactions appear to be fairly common, although studies have largely centred around derivatives of the pyrimidine *N*-oxides. Typically the reactions are carried out by treating the pyrimidine mono-*N*-oxide with sodium or potassium cyanide and benzoyl chloride under alkaline conditions to form 2-cyanopyrimidines, as illustrated[108] by the reaction of 4-methoxypyrimidine 1-oxide, which yields 2-cyano-4-methoxypyrimidine. A mechanism (Scheme 47) analogous to that postulated for quinoline

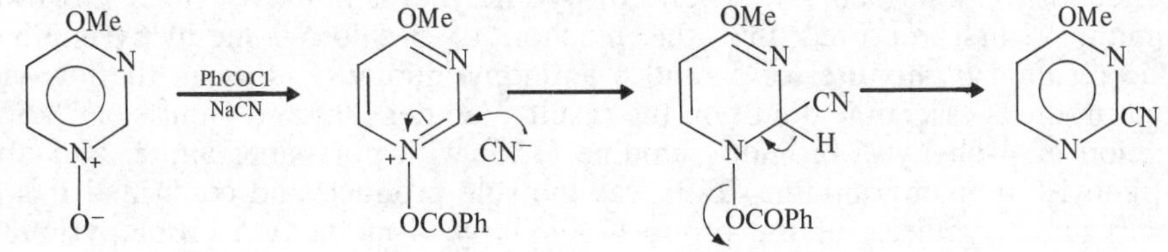

SCHEME 47

N-oxides appears probable. Numerous other examples of this reaction have appeared, with the newly introduced cyano-group favouring entry at the 2-position of the pyrimidine ring.[109]

If potassium carbonate is used in place of potassium cyanide, then 2-hydroxypyrimidines result,[110] although clearly more direct methods for the synthesis of 2-hydroxypyrimidines are available.

It has already been pointed out that alkyl groups attached to a diazine nucleus readily suffer proton abstraction with strong bases and the resultant carbanions undergo normal condensation reactions; this tendency is enhanced with the diazine *N*-oxides. 2-Methylpyrazine 1,4-dioxides undergo condensation with benzaldehyde in the presence of sodium hydroxide to form[111] styrylpyrazine di-*N*-oxides.

Over the last two decades a number of naturally occurring pyrazine *N*-oxides and the structurally related hydroxamic acids such as neoaspergillic acid[112] (**116**) and pulcherriminic acid[113] (**117**) have been isolated and their antibiotic properties have prompted considerable interest in these compounds.

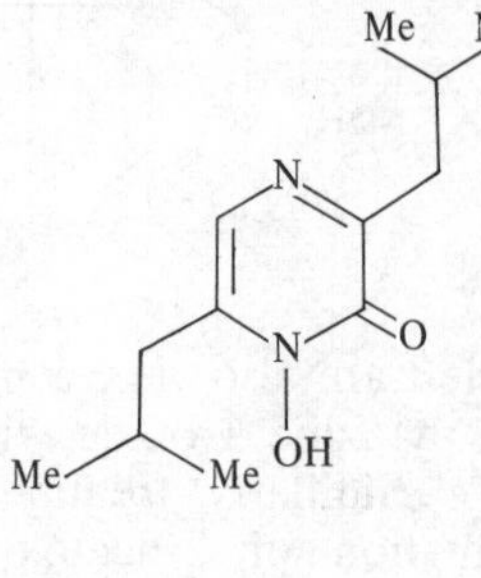

(116) 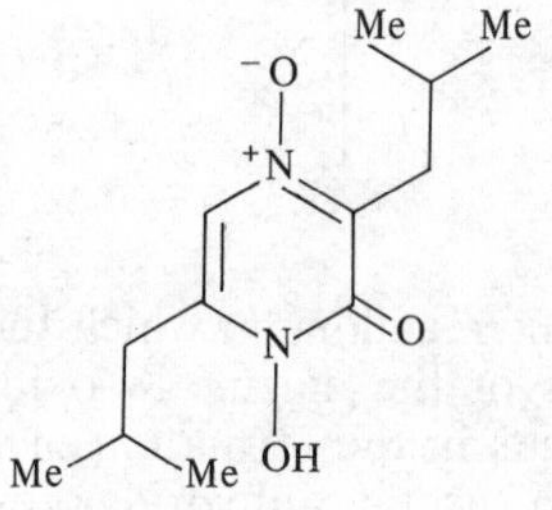 (117)

16.2.8 DIAZYNES

In contrast to the chemistry of pyridine, there are few recorded reactions of diazines which are thought to involve dehydrodiazine intermediates. Brown and co-workers[114] have examined the flash pyrolysis of pyrazine-2,3-dicarboxylic anhydride and identified maleonitrile and fumaronitrile amongst the products. This led them to propose a mechanism involving the fragmentation of an intermediate 2,3-pyrazyne (Scheme 48).

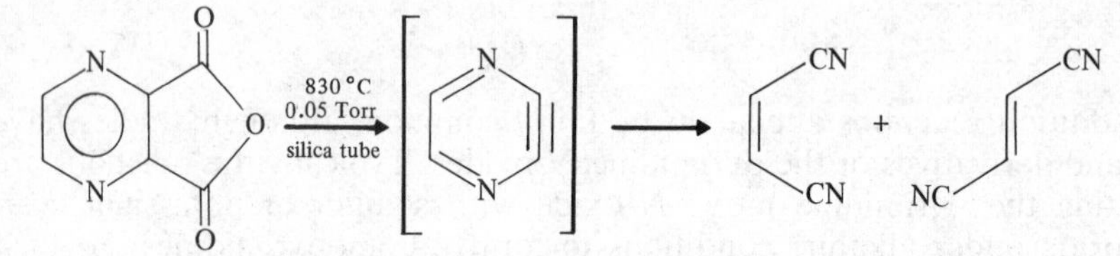

830 °C, 0.05 Torr, silica tube

SCHEME 48

Evidence for the intermediacy of 4,5(5,6)-pyrimidynes is more convincing. Schwan and Tieckelmann[115] first reported that the reaction of 5-chloro-2-methylpyrimidine with sodamide resulted in mixtures of 5- and 6-aminopyrimidines, although the low yields of aminated products cast some doubt on the result. Van der Plas and Geurtsen[116] examined the reaction of 4-phenyl-5-bromopyrimidine (**118**) with potassium amide and observed that 4-phenyl-6-aminopyrimidine (**119**) was the sole product, and confirmed this finding by unambiguous synthesis of the product. Similarly, 4-methoxy-5-bromopyrimidine on reaction with potassium amide resulted in the formation of 4-methoxy-6-aminopyrimidine and these results, along with supporting observations,[117] support the intermediacy of a 5,6-dehydropyrimidine (**120**).

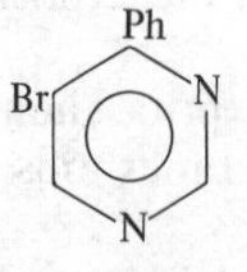

(118) 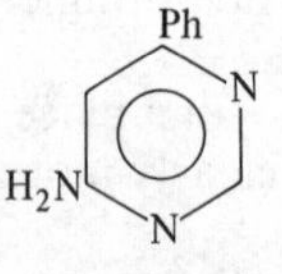 (119) 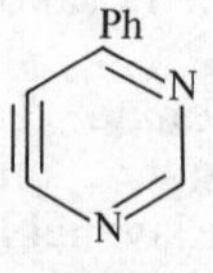 (120)

4,5-Pyridazyne derivatives are uncommon. Kauffmann and Risberg,[118] on examination of the aminolysis of 4-chloro-1-methyl-2-phenyl-pyridazine-3,6-dione (**121**) with piperidine, observed mixtures of the 4-piperidino- (**122**) and 5-piperidino-derivatives (**123**) and the pyridazyne (**124**) has been invoked as an intermediate. Clearly, an alternative mechanism involving Michael addition of the amine followed by elimination of HCl might be expected to furnish a similar product mixture, casting some doubt on the intermediacy of (**124**).

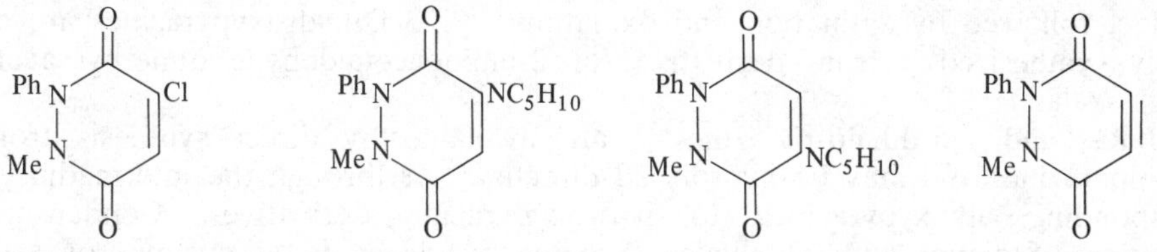

(121) (122) (123) (124)

16.2.9 HYDROXYDIAZINES

The hydroxydiazines are an important class of compounds in that they are generally available by direct synthesis and serve as a useful starting point for functional modification. Hydroxypyridazines represent a typical example of these principles, as illustrated by the synthesis of 3,6-dihydroxypyridazine (Scheme 1) by the reaction of maleic anhydride with hydrazine, and its conversion to dihalopyridazines by phosphoryl halides. A related approach[119] using derivatives of β-formylacrylic acids results in the formation of monohydroxypyridazines.

The synthesis of hydroxypyrimidines has already been encountered. Generally the use of urea as a one-carbon fragment results in the formation of 2-hydroxypyrimidines, whereas the reaction of a β-formyl ester with an amidine leads to 4-hydroxypyrimidines, and a β-diester with an amidine yields 4,6-dihydroxypyrimidines. Careful choice of the starting materials thus permits the synthesis of mono-, di-, or tri-hydroxypyrimidines at will. Hydrolysis of halopyrimidines has also led to hydroxypyrimidines, but the reaction is of little value in most instances since the halopyrimidines are themselves derived from hydroxypyrimidines.

5-Hydroxypyrimidines (**125**) represent an anomaly amongst the hydroxydiazines in that they are the only hydroxydiazines in which the hydroxyl group is not α or γ to the ring nitrogen atoms. 5-Hydroxypyrimidines have been prepared by treatment[120] of 5-bromopyrimidine with $Ba(OH)_2$ in the presence of copper powder at 140 °C. Alternative procedures have centred around the synthesis of simple derivatives of 5-hydroxypyrimidines which can be modified by routine transformations (*e.g.* Scheme 49).

Hydroxypyrazines are generally prepared by direct synthesis, from aminopyrazines, by the decarboxylation of hydroxypyrazinecarboxylic acids or by the hydrolysis of halopyrazines. Perhaps the most direct synthesis of monohydroxypyrazines involves the condensation of an α-amino-acid amide with an α-diketone (Scheme 22). A direct

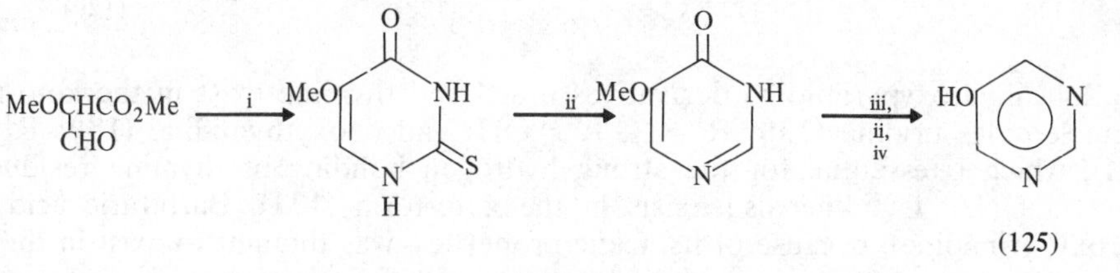

MeOCHCO$_2$Me (CHO) →(i) → (ii) → (iii, ii, iv) (125)

i, NH_2CSNH_2; ii, Raney nickel; iii, P_4S_{10}; iv, H_3O^+.

SCHEME 49

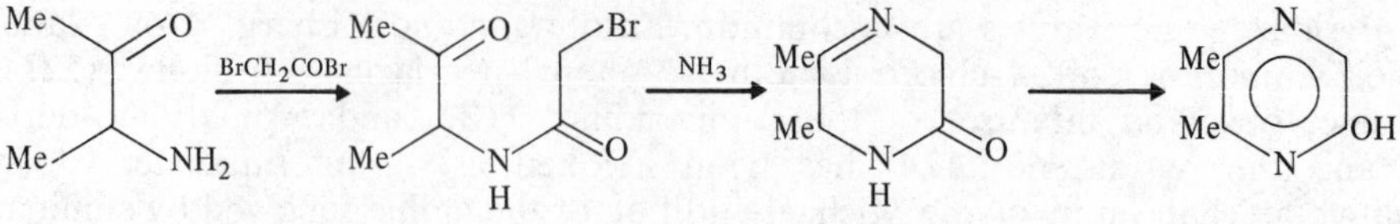

SCHEME 50

synthesis (Scheme 50) involves the condensation of an α-aminoketone with bromoacetyl bromide, followed by amination and oxidation.[121] 2,3-Dihydroxypyrazines are conventionally synthesized[122] from derivatives of 2-aminoacetaldehyde dimethyl acetal and diethyl oxalate.

2-Halo- and 2,5-dihalo-pyrazines[123] are available by direct synthesis from 2,5-dioxopiperazines and may be hydrolysed directly[124] or through the intermediacy of the corresponding alkoxypyrazines to hydroxypyrazine derivatives. Condensation of piperazine-2,5-diones with aldehydes (Scheme 51) leads to alkylidene- or arylidene-piperazine-2,5-diones, which on treatment with aqueous base at 100 °C yield the corresponding dihydroxypyrazines.[125]

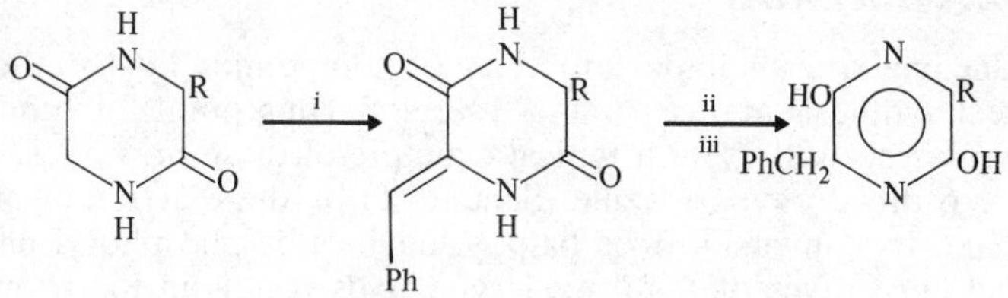

i, $PhCHO/Ac_2O/AcONa$; ii, OH^-/100 °C; iii, H_3O^+.

SCHEME 51

16.2.9.1 Tautomerism in the hydroxydiazines

Although it is convenient to refer to hydroxydiazines, the hydroxydiazines show an overwhelming preference for the keto-form. In the case of 2-hydroxypyrimidine (**126**) the predominant tautomer is pyrimidin-2(1*H*)-one (**127**), which is readily identified by the presence of N—H and C—O stretches in the i.r. spectrum and by comparison of its u.v. spectrum with those of the fixed model systems (**128**) and (**129**).

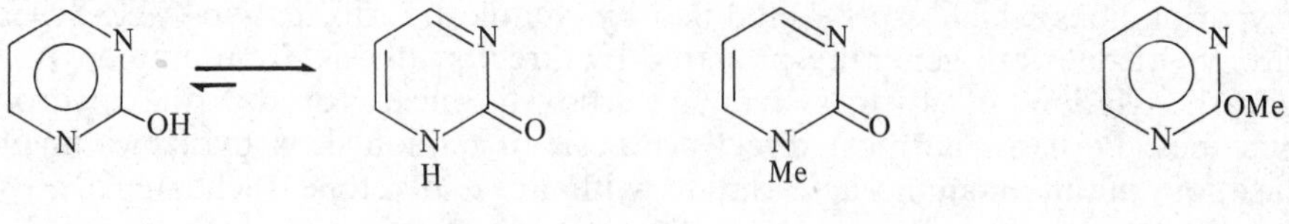

(126) (127) (128) (129)

The 2,4-dihydroxypyrimidine derivatives uracil and thymine exist in the keto-form in their nucleosides uridine (**130**; $R^1 = H$, $R^2 = OH$) and deoxythymidine (**130**; $R^1 = CH_3$, $R^2 = H$), which is essential for the strong hydrogen bonding of thymine residues with adenine in DNA. Cytidine also exists in the keto-form (**131**). Barbituric acid (2,4,6-trihydroxypyrimidine), because of its acidic properties, was thought to exist in the dioxo-monohydroxy-form, but X-ray evidence suggests that the trioxo-form is preferred.[126]

As previously stated, 5-hydroxypyrimidine is the exception amongst the diazines in that the hydroxyl group is neither *ortho* nor *para* to the ring nitrogen atoms and there is no

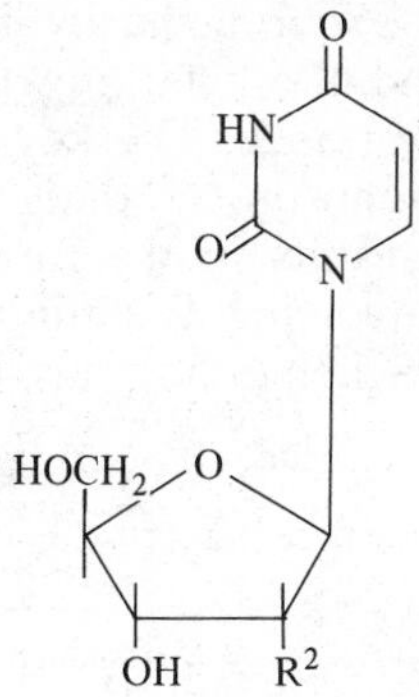

(130) 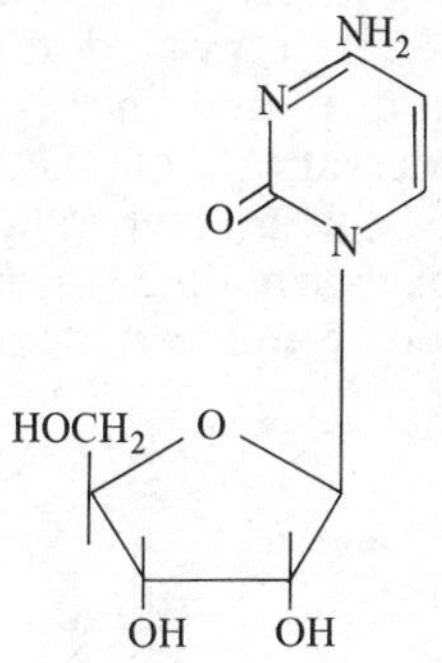 (131)

possibility of hydroxyl–oxo tautomerism. The hydroxyl group is truly phenolic in character (*cf.* 3-hydroxypyridine). Like their pyrimidine counterparts, pyridazines exist predominantly in the oxo-form[127] in the solid state and in solution. Dihydroxypyridazines, in particular maleic hydrazide, have been investigated, and spectroscopic data suggest that they exist in the monohydroxy-mono-oxo-form (**132**).

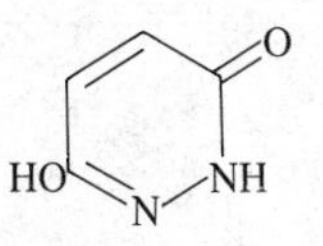

(132) 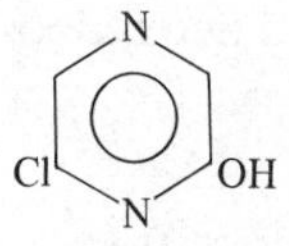 (133) 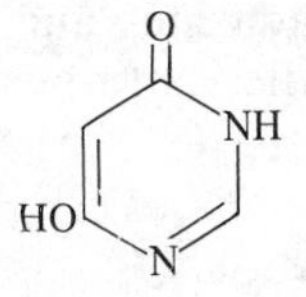 (134) 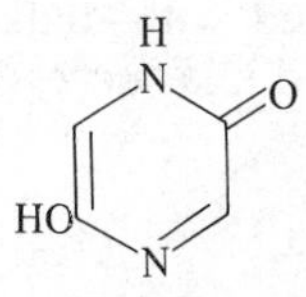 (135)

In the case of pyrazines with hydroxyl substituents the oxo-tautomer is generally favoured, although substituents play a dominant role in determining the position of the tautomeric equilibrium, and Godwin[128] has shown from u.v. spectral data that 2-chloro-6-hydroxypyrazine (**133**) exists as the hydroxyl rather than the oxo-tautomer. Little confirmatory evidence is available on equilibria in 2,3-dihydroxypyrazines, but precedent suggests that the dioxo- or the mono-oxy-monohydroxy-forms should be favoured. 2,5-Dihydroxypyrazines, like 4,6-dihydroxypyrimidines, are unable to exist in the bis-amide forms, as once one of the hydroxyl groups tautomerizes to an oxo group [(**134**) and (**135**)], the second hydroxyl group cannot revert to the amide form owing to the cross conjugation of the imidic acid grouping.

16.2.9.2 Reactions of the hydroxydiazines

Alkylation represents one of the most important reactions of the hydroxydiazines, particularly in the pyrimidine series. Alkylations are usually performed using alkyl halides, alkaline dimethyl sulphate, or diazomethane. Generally reaction of the hydroxydiazines with alkyl halides or dimethyl sulphate favours *N*-alkylation, whereas diazomethane usually results in mixed *O*- and *N*-alkylation in varying proportions. Maleic hydrazide reacts with one equivalent of diazomethane to form the *O*-methyl derivative and with an excess of diazomethane the *O,N*-dimethyl derivative is formed.[129]

Alkoxydiazines are more usually prepared from the halodiazines by reaction with alkoxides, although more recently McMinn[130] has observed that trialkyloxonium salts specifically *O*-alkylate hydroxypyrimidines; similar results have been obtained with hydroxypyrazines.[131]

Alkylation of 2-hydroxypyrimidine with diazomethane leads to a 4 : 1 mixture of *N*-methyl- and *O*-methyl–pyrimidines,[10] and methyl iodide generally leads to selective

N-alkylation in an entirely analogous way to the pyridazines. Since many naturally occurring pyrimidines are alkylated at N-1, a method of alkylation which is highly specific for the *N*-alkylation of pyrimidines is of great importance. The key to this problem was found in the observation of Hilbert and Johnson[132] that treatment of 2,5-dialkoxypyrimidines with methyl iodide at 20 °C results in the formation of 1-methyl-2(1*H*)-oxo-4-methoxypyrimidine. A mechanism involving the initial quaternization of N-1 followed by iodide-induced demethylation (Scheme 52) has been confirmed. At

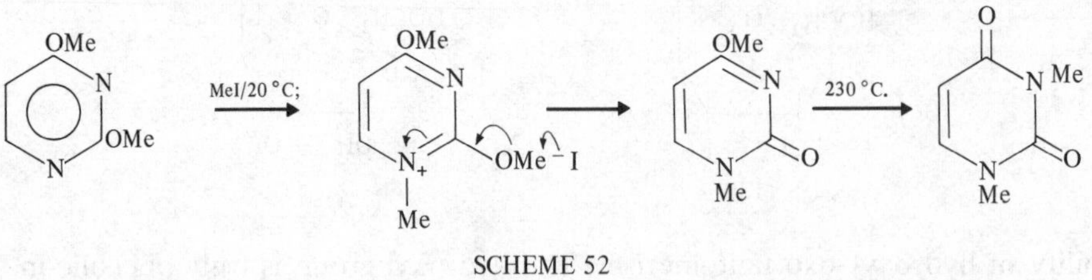

SCHEME 52

ambient temperatures the reaction stops at the mono-*N*-alkylated pyrimidine, but at temperatures above 230 °C the di-*N*-alkylated pyrimidine results. By the simple expedient of using triacetoxyribofuranosyl bromide it has proved possible to synthesize uridine (**130**) and cytidine (**131**) in good overall yields. O→N Rearrangements of this type have also been observed with dialkoxypyridazines and application of the Hilbert–Johnson reaction has resulted in a successful synthesis of ribosides and deoxyribosides of maleic hydrazide.[133]

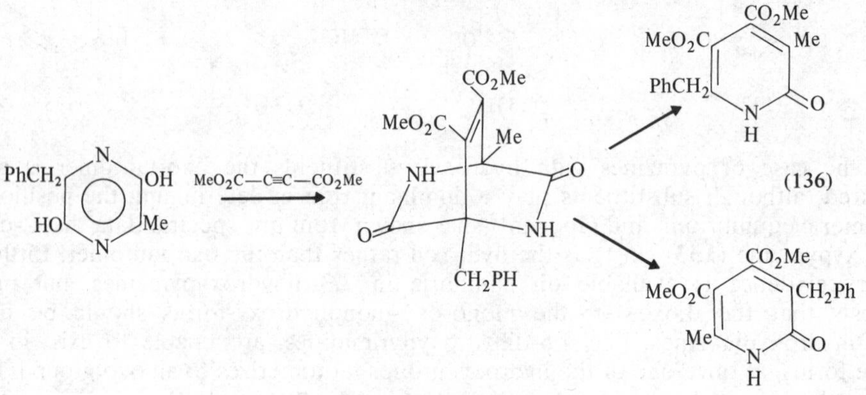

(136)

(137)

SCHEME 53

2,5-Dihydroxypyrazines and 4,6-dihydroxypyrimidines have been shown to undergo cycloaddition reactions with electron-deficient and strained olefins,[134] supporting the existence of tautomeric structures such as (**134**) and (**135**). The reaction of 3-benzyl-6-methyl-2,5-dihydroxypyrazine (Scheme 53) with dimethyl acetylenedicarboxylate results in the formation of a bicyclic adduct which undergoes a retro-Diels–Alder reaction, eliminating isocyanic acid to produce the isomeric pyridones (**136**) and (**137**).

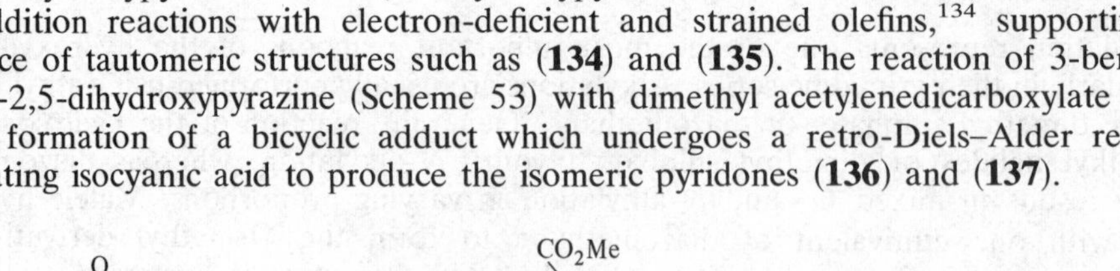

SCHEME 54

In the case of the corresponding 4,6-dihydroxypyrimidines the intermediate bicyclic adduct was not isolable, but yielded the pyridone directly[135] and because of the symmetry of the intermediate adduct (Scheme 54) a single pyridone is formed by the elimination of the isocyanic acid bridge. These reactions have been extended[136] to derivatives such as (**138**) which undergo intramolecular [4+2] cycloaddition reactions to furnish stable tricyclic adducts, *e.g.* (**139**), that undergo facile hydrolysis to cyclopentane derivatives (**140**).

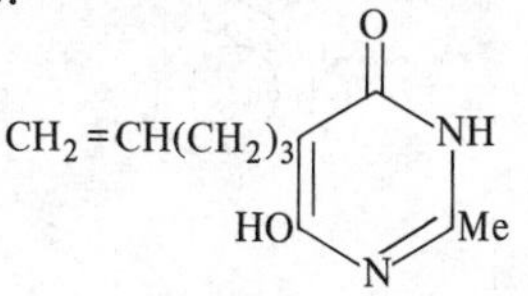

(138) 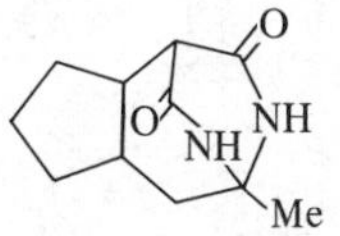 (139) 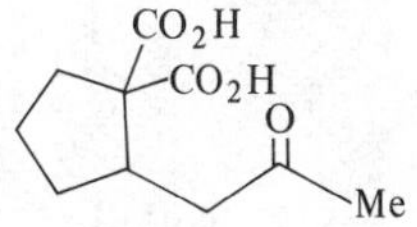 (140)

16.2.10 HALODIAZINES

Halogenated diazines play an important role in diazine chemistry since, by nucleophilic displacement of the halogens, numerous otherwise inaccessible diazines become available.

There are several methods of introducing halogen into the diazine nucleus, but by far the most common method involves the halogenation of hydroxydiazines using phosphoryl halides, phosphorus trihalides and, less commonly, phosphorus pentahalides. In some instances it is possible to introduce halogen atoms directly during synthesis, as illustrated by the synthesis of 4,5-dichloro-3-hydroxypyridazine (Scheme 1, $R^1 = R^2 = Cl$) and the direct synthesis of 5-halogenopyrimidine derivatives[137] from halomalondialdehydes, although such reactions rely on the stability of the resulting halo-compounds under the reaction conditions employed.

Halogenation with phosphoryl halides is simple in most cases, with short reflux periods converting the hydroxydiazine directly into the corresponding halodiazine. For example, maleic hydrazide is readily converted[138] into 3,6-dichloropyridazine during three hours at reflux and 2-chloropyrazine[139] has been prepared under similar conditions. 2-Chloro-, 4-chloro-, and 2,4-dichloro-pyrimidines have also been prepared using phosphoryl halides, but reaction times are usually longer and more forcing conditions are necessary.[10]

In some instances oxidative halogenation has been observed, as in the conversion of 2,5-dioxopiperazines to a mixture of mono- and di-chloropyrazines. This unusual reaction probably proceeds by way of the intermediacy of the bisimidoyl halide (Scheme 55). The

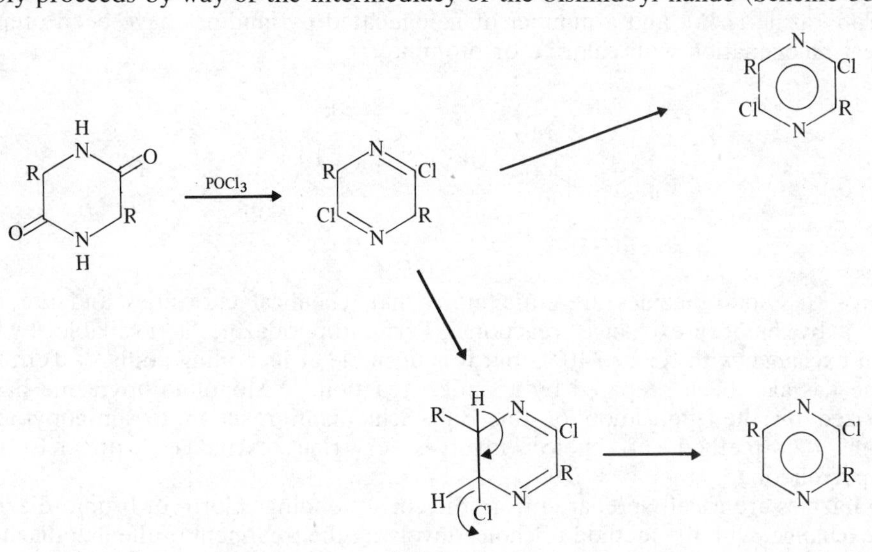

SCHEME 55

product ratio is remarkably susceptible to variation of substituents: thus 3,6-dimethylpiperazine-2,5-dione gives largely the monochloropyrazine, whereas 3,6-dibenzylpiperazine-2,5-dione gives predominantly the 2,5-dichloropyrazine.[140] Hexahydropyridazine-3,6-dione (**141**) similarly undergoes oxidative bromination on treatment with an excess of $POBr_3$,[141] giving 3,6-dibromopyridazine (**142**).

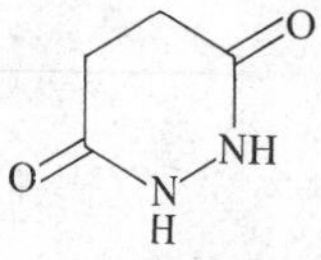

(141) 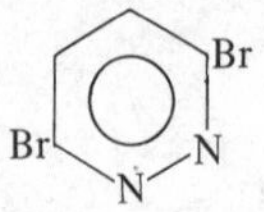 (142) 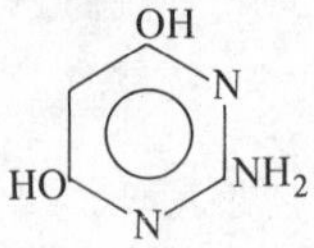 (143) 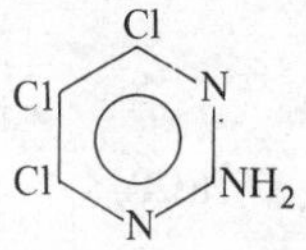 (144)

Frequently, particularly in the pyrimidine series, conversion of hydroxydiazines to the corresponding halo-compounds is difficult and a number of simple modifications have been introduced to cut down reaction times. The use of phosphoryl halides in conjunction with high-boiling bases such as *N,N*-dimethylaniline or *N,N*-diethylaniline has been shown to increase the rates of some halogenation reactions.[10] Phosphorus pentahalides are generally much more powerful halogenating agents and they are often used in conjunction with phosphoryl halides to effect halogenations that are particularly difficult or which proceed at relatively slow rates. An example of such an application may be seen in the chlorination of 2-hydroxypyrimidine, which with phosphoryl chloride alone takes 12 hours at reflux to complete, whereas with added phosphorus pentachloride the reaction is complete within one hour.[142] Caution should be exercised when using phosphorus pentahalides as further halogenation of the diazine can occur, as observed during the bromination of 2-hydroxypyrazine with $POBr_3/PBr_5$ mixtures. Aside from the expected 2-bromopyrazine, 2,6-dibromopyrazine[143] is also formed in significant quantities. 2-Amino-4,6-dihydroxypyrimidine (**143**) with mixtures of $POCl_3/PCl_5$ similarly gives 2-amino-4,5,6-trichloropyrimidine (**144**).

High-temperature halogenation with PCl_5 has been utilized in the synthesis of polychlorodiazines and Chambers *et al.*[144] have prepared tetrachloropyrazine by heating pyrazine with PCl_5 in sealed tubes at 320 °C.

Direct halogenation of the diazines is not commonplace, but a number of examples of halogenation of diazines containing substituents which activate the ring towards electrophilic attack have been reported. 2-Amino-3-methoxycarbonylpyrazine (**145**) undergoes direct chlorination[32] with molecular chlorine to produce the corresponding 5,6-dichloropyrazine (**146**), and a number of halogenated pyrimidines have been prepared[10] by direct halogenation with chlorine or bromine.

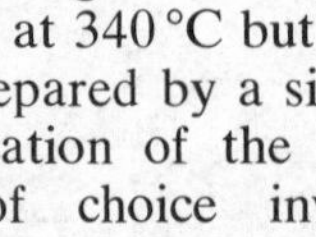

(145) 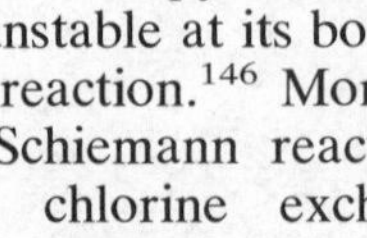 (146)

Fluoro- and iodo-diazines are little more than chemical curiosities and are usually prepared by halogen-exchange reactions. Perfluoropyridazine is available by direct halogen exchange with KF at 340 °C but it is unstable at its boiling point.[145] Tetrafluoropyrazine has also been prepared by a similar reaction.[146] Monofluoropyrazine has been synthesized by the application of the Balz–Schiemann reaction to aminopyrazine,[147] although the method of choice involves chlorine exchange with KF in *N*-methylpyrrolidone.[128]

Iododiazines are usually prepared from the corresponding chloro- or bromo-diazines by direct exchange, with the method of choice involving the treatment of the halodiazine with acetone saturated with sodium iodide containing a catalytic amount of HI.[148]

16.2.11 AMINODIAZINES

There are several methods[1,10,32] which have been applied to the synthesis of aminodiazines. In the pyrimidine series the most common method involves direct introduction of the amine group during synthesis. Utilizing guanidine, 2-aminopyrimidines are directly available, and 4-amino-, 6-amino-, and 4,6-diamino-pyrimidines are directly available by the use of nitriles and dinitriles as the three-carbon fragment in a 'common synthesis' approach.

Aminopyridazines and aminopyrazines are not usually synthesized directly but are usually prepared by transformation of functional groups within the diazine ring; one of the most common methods for the introduction of an amino-group involves the replacement of halogen atoms by reaction with ammonia or amines (*cf.* Section 16.2.5.2). The reaction proceeds most easily in the case of fluorodiazines, *e.g.* fluoropyrazine on stirring with aqueous ammonia for[149] three days yields 2-aminopyrazine in 70% yield, whereas reaction of chloropyrazine or 3-chloropyridazine requires heating the chlorodiazine under pressure with ammonia.[139,150]

The introduction of a second amino-group by halogen displacement generally requires more drastic conditions. Thus 3,6-dichloropyridazine on reaction with ammonia at 100 °C for six hours yields 3-amino-6-chloropyridazine,[150] and only low yields of 3,6-diaminopyridazine have been achieved even at high temperatures and pressures.[151]

Direct amination has only been satisfactorily applied in the pyrazine series and appears to be of limited value. The application of the Hofmann reaction, however, does represent a good general method for the introduction of amino-groups into the diazine nucleus and has been applied in the pyrazine,[152] pyridazine,[153] and pyrimidine series.[154] It appears to be the method of choice for the synthesis of 5-aminopyrimidines, which are not easily accessible by other methods.

U.v. and i.r. spectroscopic evidence suggests that, in contrast to the hydroxydiazines which exist predominantly in the oxo-form, the aminodiazines exist almost exclusively as the amino rather than the corresponding imino-tautomers.[155]

Amino-substituents in the diazine ring exhibit similar properties to the aminopyridines. Thus the amino-group is readily acylated and generally activates the ring towards electrophilic substitution. Diazotization is readily accomplished under normal conditions but the stability of the resultant diazonium salts varies markedly. Generally, when the amino-substituent is α or γ to the ring nitrogen atoms, hydrolysis to the corresponding oxo-compounds occurs, *e.g.* the diazonium salt of aminopyrazine decomposes during the diazotization reaction at 0 °C to form the pyrazinone in 30% yield.[156] Diazonium salts of 5-aminopyrimidines are generally stable[10] and have been coupled to β-naphthol, or transformed to 5-iodopyrimidines by reaction with potassium iodide.

16.2.12 BENZODIAZINES

The benzodiazines, as the name implies, are diazine rings to which a benzene ring is fused. Because there are two possible sites of fusion of a benzene ring on to the pyridazine nucleus, a total of four isomeric benzodiazines are possible: cinnoline (benzo[*c*]pyridazine) (**147**), phthalazine (benzo[*d*]pyridazine) (**148**), quinazoline (**149**), and quinoxaline (**150**). Some of the physical properties of the benzodiazines are listed in Table 3.

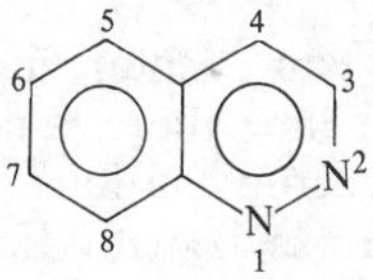

(147) 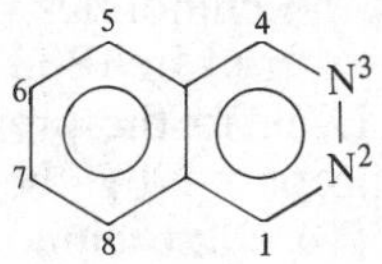 (148) 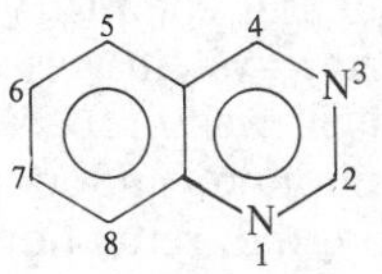 (149) 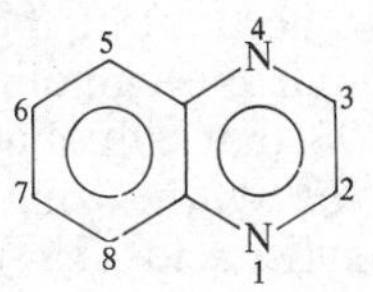 (150)

TABLE 3
Physical Properties of the Benzodiazines

	Cinnoline	*Phthalazine*	*Quinazoline*	*Quinoxaline*
M.p. (°C)	40–41	90–91	48	31
B.p. (°C)	$114_{0.35}$	189_{29}	241_{764}	220–223
pK_a[a]	2.3	3.5	3.5(1.9)[c]	0.7
μ (D)[b]	4.14	4.88	2.2	0.51
λ_{max} (nm)	390, 322, 308, 286, 275	360, 290, 259	370, 308, 270, 220	340, 312, 232

[a] A. Albert, 'Heterocyclic Chemistry', Athlone Press, London, 1959.
[b] R. D. Brown and B. A. W. Coller, *Theor. Chim. Acta*, 1967, **7,** 259.
[c] pK_a measurements of quinazoline taken by normal methods give a pK_a value of 3.5; however, this is a composite figure based on a series of hydration equilibria.

No cinnolines or phthalazines occur naturally and this is to be expected (*cf.* the pyridazines). Cinnoline itself is toxic and shows antibacterial activity against *E. coli,* and many cinnoline and phthalazine derivatives have found use in medicinal chemistry. A limited number of alkaloids of the vasicine, evodiamine, and febrifugine type containing the quinazoline ring exist naturally, and some are physiologically active although generally the low therapeutic index of these materials is not sufficient to warrant clinical application. The most active derivative of quinazoline known at the present time is the 2-iminoperhydroquinazoline tetrodotoxin (**151**), which is a powerful neurotoxin. Many synthetic quinazolines are now marketed successfully, *e.g.* 3,4-dihydro-2-methyl-4-oxo-3-*o*-tolylquinazoline (**152**) is used as a hypnotic and 7-chloro-2-ethyl-1,2,3,4-tetrahydro-4-oxo-6-sulphamoylquinazoline (**153**) functions as an oral diuretic.

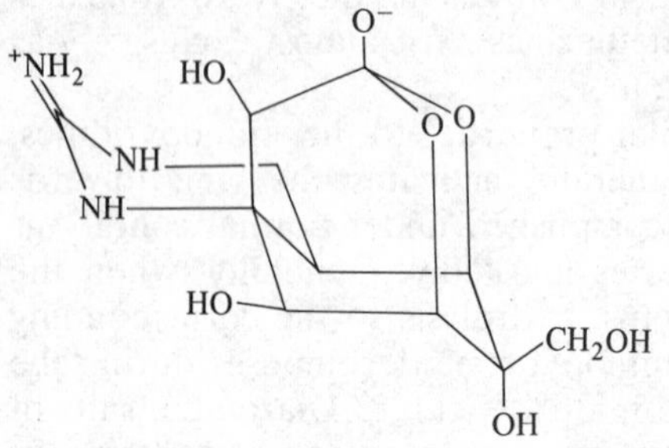

(151) 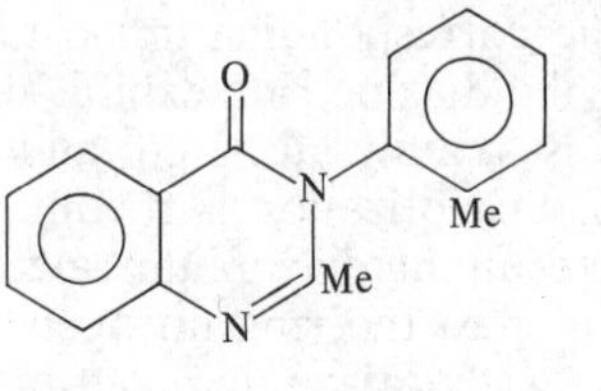 (152) Methaqualone 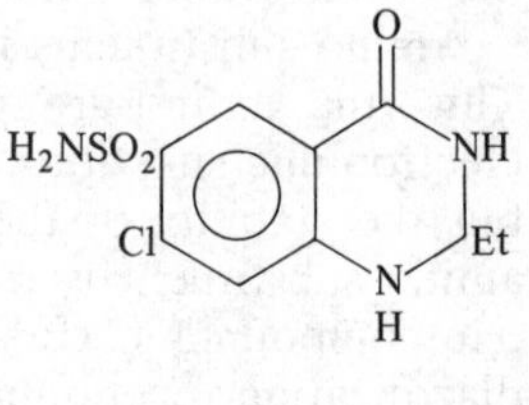 (153) Quinethazone

Several quinoxalines occur naturally and the peptide antibiotic echinomycin[157] is a peptide derivative of quinoxaline-2-carboxylic acid.

16.2.13 BENZODIAZINE SYNTHESIS

16.2.13.1 Cinnolines

The cinnolines have been rather less studied than the other benzodiazines and this is due in part to the difficulties encountered in the preparation of the cinnoline ring system.[158] However, in recent years two major reviews have appeared[159,160] which present a concise account of the general chemistry of the cinnolines.

Although the cinnoline ring system was discovered in 1883 by von Richter, it was not until 1897 that cinnoline itself was prepared. Even to the present time there is no direct method of synthesizing cinnoline, and it is prepared by the decarboxylation[161] of the 4-carboxylic acid (**154**), by the reduction of 4-chlorocinnoline (**155**) with Fe/H_2SO_4 followed by oxidation of the intermediate 1,4-dihydrocinnoline (**156**) with HgO, or by the reduction of 4-hydroxycinnoline with $LiAlH_4$.

COOH Cl

N N N N N N H

(154) (155) (156)

The original von Richter synthesis involves the intramolecular cyclization of the diazonium salt of *o*-aminophenylpropriolic acid (Scheme 56) and results in the formation of 4-hydroxycinnoline-3-carboxylic acid. The von Richter reaction has been extended[162] and represents a general method for the synthesis of 4-hydroxycinnolines.

$$\text{C}\equiv\text{C—COOH},\ \text{NH}_2 \xrightarrow[\text{HCl}]{\text{HNO}_2} \text{C}\equiv\text{C—COOH},\ \text{N}_2^+ \longrightarrow \text{OH},\ \text{CO}_2\text{H},\ \text{N—N}$$

SCHEME 56

The Widman–Stoermer synthesis is the name applied to the cyclization of the diazonium salts of *o*-aminostyrene derivatives (Scheme 57). Within certain limitations the synthesis is general; R^1 can be an alkyl, aryl, or heteroaryl group but not hydrogen, and

$$\text{R}^1,\ \text{CHR}^2,\ \text{NH}_2 \xrightarrow[\text{HCl}]{\text{HNO}_2} \text{R}^1,\ \text{H},\ \text{R}^2,\ +,\ \text{N}{=}\text{N} \longrightarrow \text{R}^1,\ \text{R}^2,\ \text{N—N}$$

SCHEME 57

R^2 can be hydrogen, alkyl, or aryl. Clearly the method cannot be applied to the synthesis of cinnoline, and all cinnolines prepared by this method will contain substituents at positions-3 and/or -4.

A related reaction, the Borsche synthesis, utilizes *o*-aminoacetophenone derivatives, and a mechanism (Scheme 58) involving the cyclization of the enol form of acetophenone

$$\text{O},\ \text{CH}_2\text{R},\ \text{NH}_2 \xrightarrow[\text{HCl}]{\text{HNO}_2} \text{O—H},\ \text{CHR},\ \text{N}_2^+ \longrightarrow \text{OH},\ \text{R},\ \text{N—N}$$

SCHEME 58

appears probable. Although the von Richter, Widman–Stoermer, and Borsche reactions enjoy different names, the similarity of mechanism in the cyclization step is obvious.

Two approaches to the synthesis of cinnolines which are conceptually different from the diazonium salt cyclization utilize a preformed nitrogen–nitrogen bond in the form of phenylhydrazine as the starting point. The reaction of phenylhydrazine with diethyl mesoxalate results in the formation of the corresponding hydrazone (Scheme 59) which, on saponification, reaction with thionyl chloride, and treatment of the resultant acid chloride with a Friedel–Crafts catalyst, results in the formation of 4-hydroxycinnoline-3-carboxylic acid.[163]

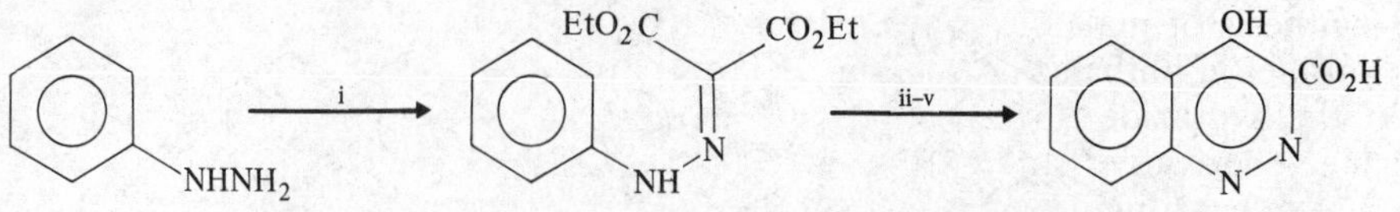

i, $EtOCOCOCO_2Et$; ii, OEt^-; iii, H_3O^+; iv, $SOCl_2$; v, $TiCl_4$.

SCHEME 59

The second method involves the treatment of a preformed benzaldehyde phenylhydrazone with oxalyl chloride followed by Friedel–Crafts cyclization of the *N*-benzylideneamino-*N*-phenyloxamoyl chloride to *N*-benzylideneaminoisatin. Treatment of the isatin (Scheme 60) with hot aqueous KOH gives cinnoline derivatives in yields ranging from 75–85%.

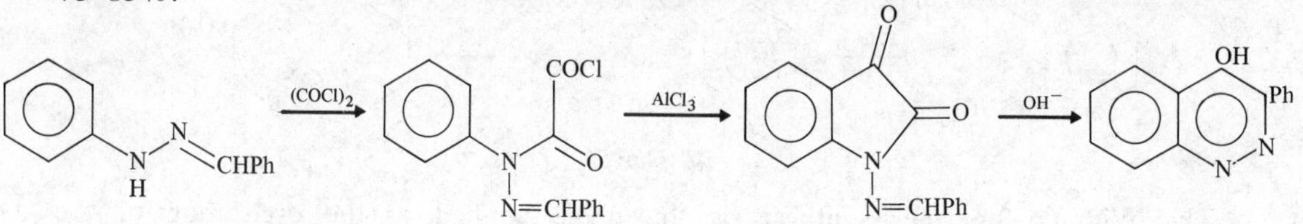

SCHEME 60

16.2.13.2 Phthalazines

Phthalazines are readily synthesized by a variety of methods which have their parallel in pyridazine synthesis. The first synthesis of phthalazine was due to Gabriel and Pinkus and utilizes[164] the reaction of aqueous hydrazine with $\alpha,\alpha,\alpha',\alpha'$-tetrachloro-*o*-xylene (**157**) under pressure at 150 °C for two hours, although this method has been modified and phthalazine is currently manufactured by treating (**157**) with hydrazine sulphate in 90% sulphuric acid. Phthalazine has also been prepared[165] in high yield by the reaction of phthalaldehyde (**158**) with alcoholic hydrazine hydrate, and this method offers a general synthesis of phthalazine derivatives unsubstituted in the heterocyclic ring.

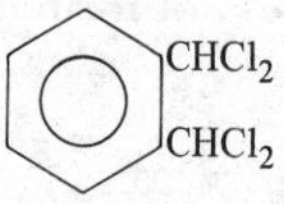

(157) 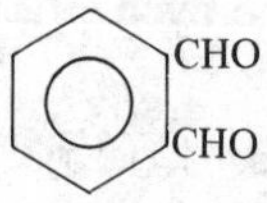 (158) 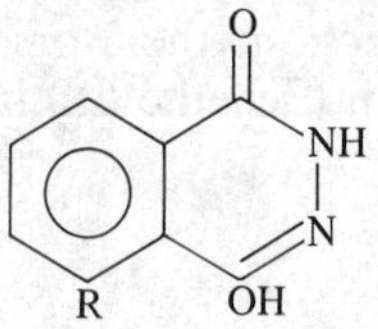 (159)

1,4-Dialkyl-, -diaryl-, and -aralkyl-phthalazines nave been synthesized by the reaction of 1,2-diacylbenzene derivatives (*cf.* 1,2-diacylethylenes in the pyridazine series) (Scheme 61) with hydrazine.

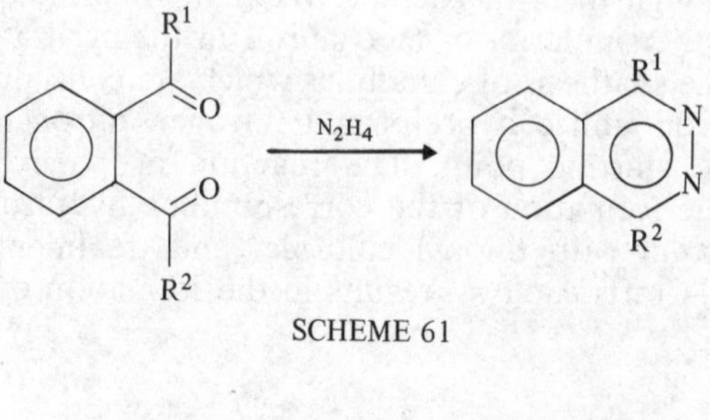

SCHEME 61

The synthesis of maleic hydrazide by the reaction of maleic anhydride with hydrazine has its parallel in phthalazine chemistry. Thus phthalic anhydride reacts with hydrazine to yield phthalhydrazide (**159**; R=H) in good yield. Phthalhydrazide is a common phthalazine derivative and is invariably the product of hydrazinolysis of common phthalimido-derivatives used during the protection of amino-groups during the Gabriel synthesis and in peptide synthesis. Its ease of synthesis makes it an ideal starting point for the synthesis of phthalazine derivatives, particularly those in which the benzene ring is unsubstituted.

5-Aminophthalhydrazide (**159**; $R=NH_2$) is known as 'luminol' on account of its strong chemiluminescence when oxidized with alkaline potassium ferricyanide.

An approach to 1,4-disubstituted phthalazines involves the reaction of benzyne with 1,4-disubstituted tetrazines (Scheme 62).[166] This approach at first sight appears to be general, although in practice the problems in making the starting tetrazines severely limit its application.

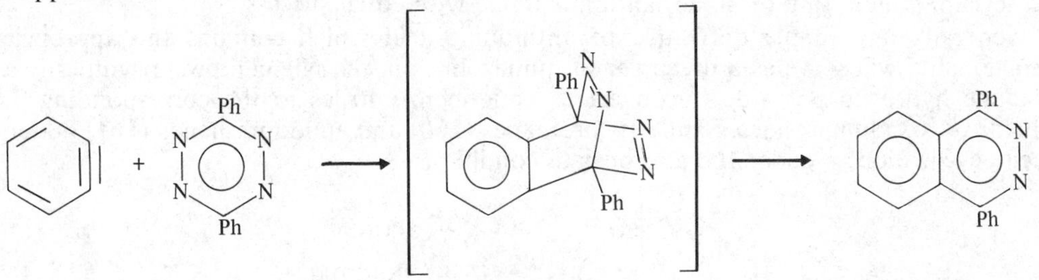

SCHEME 62

16.2.13.3 Quinazolines

Syntheses of quinazolines which may be applied to quinazoline itself are rare and most syntheses of quinazoline rely on modification of a preformed derivative. It may be prepared directly by a modification of the method due to Reidel, which involves the reductive cyclization (Scheme 63) of bisformamido-derivatives of *o*-nitrobenzaldehyde.

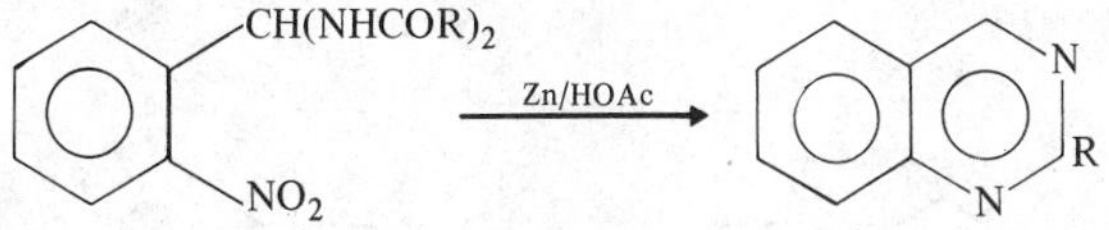

SCHEME 63

The reaction may be applied to the synthesis of 2-substituted quinazolines or quinazolines with substituents in the aromatic ring, but because derivatives of *o*-nitroacetophenone do not condense with aliphatic amides, the 4-position of the quinazoline ring is invariably unsubstituted.[167]

Brown[168] has listed a number of methods for the synthesis of quinazolines and the more general of the synthetic methods use anthranilic acid or a simple derivative as a starting point. In 1895 Niementowski observed that fusion of anthranilic acid with formamide resulted in the formation of 3,4-dihydro-4-oxoquinazoline. He recognized the potential of this reaction and studied the reaction of anthranilic acid with higher amides under similar conditions. Although the reaction with formamide occurs readily at 120–130 °C, the reaction with higher amides requires progressively more forcing conditions and reaction times increase dramatically. The reaction is further complicated by decarboxylation of the anthranilic acid, which occurs to a significant extent at elevated temperatures. Meyer and Wagner[169] subjected the reaction to a detailed mechanistic examination and established that it proceeds in discrete steps (Scheme 64), and the steps indicated several possibilities

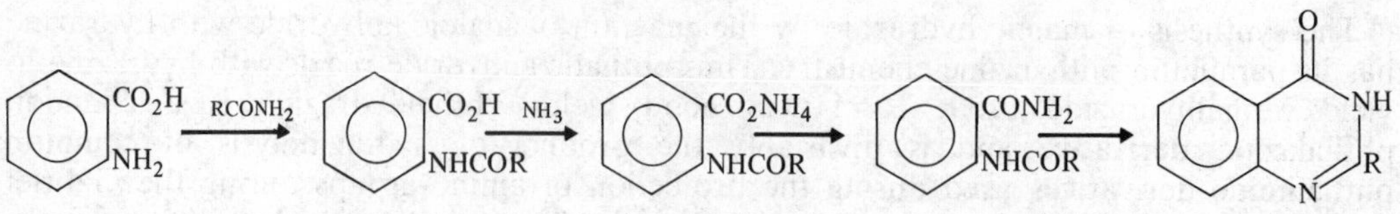

SCHEME 64

in terms of yield improvement in the reaction, and alternative starting materials for the synthesis.

The yields in the synthesis with higher amides may be improved by an initial fusion at lower temperatures to bring about the formation of the acylanthranilic acid, followed by fusion at higher temperatures in order to effect dehydration to the amide and dehydrative ring closure. Alternatively, anthranilic acid may be replaced by its ammonium salt, *N*-acylammonium salt, or an acylanthranilamide with equal success.

Generally any simple derivative of anthranilic acid which contains the appropriate functionality will serve as a precursor of quinazolines in the Niementowski synthesis, *e.g.* alkaline hydrogen peroxide[170] converts *o*-amidobenzonitriles to the corresponding 3,4-dihydro-4-oxoquinazolines. Similarly, urethanes (**160**) and amidourethanes (**161**) undergo facile cyclization[171] under the appropriate conditions.

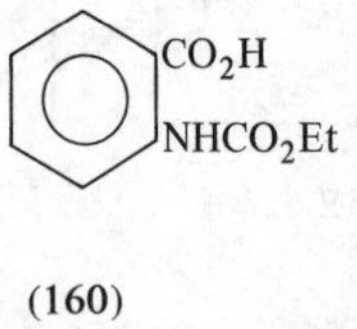

(**160**) 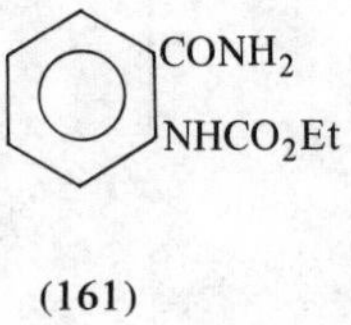 (**161**)

When anthranilic acid and its derivatives serve as the starting point in quinazoline synthesis, 4-oxo-derivatives always result, but by the simple expedient of starting from *o*-aminobenzaldehyde or *o*-aminoacetophenone Bischler[168] was able to synthesize quinazoline derivatives (Scheme 65) with alkyl, aryl, or hydrogen substituents at the 4-position. The reaction is of general applicability and has been used[172] in the preparation of alkylquinazolines bearing methoxy-substituents in the benzene ring.

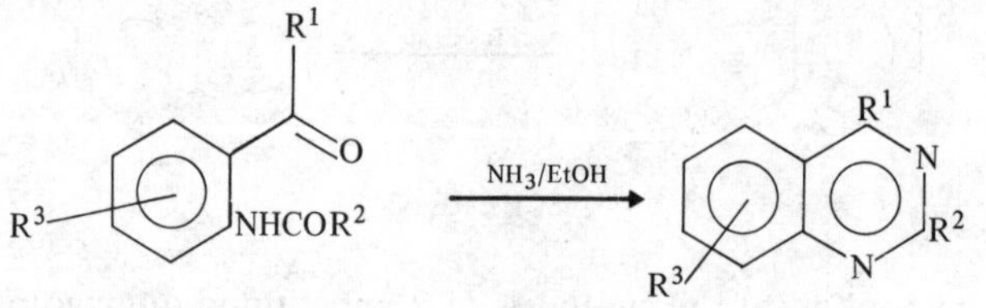

SCHEME 65

16.2.13.4 Quinoxalines

The classical synthesis of quinoxalines involves the condensation of *o*-phenylenediamine with glyoxal, an α-ketoaldehyde, or an α-diketone (Scheme 66) and the reaction is so facile that it is not only valuable in a preparative sense but also as a method of

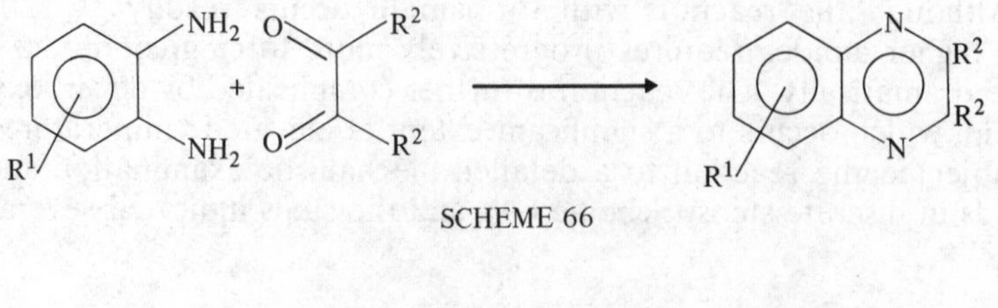

SCHEME 66

characterizing diketones. The method is general and quinoxaline itself may be prepared by the reaction of glyoxal with *o*-phenylenediamine,[173] although in certain instances a number of hindered α-diketones have failed to react.[159]

The arylglyoxal (**162**) reacts with *o*-phenylenediamine to form the bis-Schiff base (**163**) and quinoxaline formation was not observed. When difficult reactions of this type are encountered it is often possible to induce the reaction to occur by the use of dimethylaniline as a solvent.

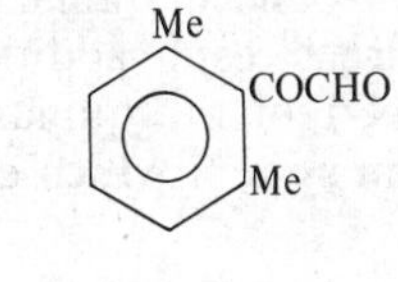

Me COCHO Me

(162)

N=CH—CO— Me Me N CH Me CO Me

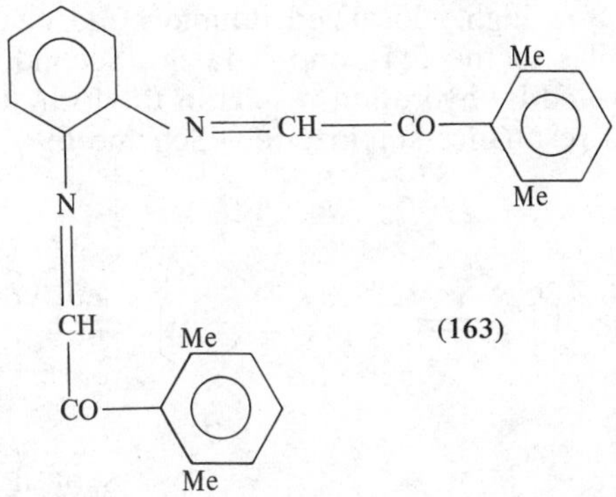

(163)

A number of simple variations of the dialdehyde/diamine reaction appear to work well. Thus replacement of the dialdehyde with an α-halogenoketone results in the formation of 2-substituted quinoxalines, and 2-phenylquinoxaline has been prepared in this way from phenacyl chloride (Scheme 67). Similarly, glyoxylic and pyruvic acids and their esters react to give 2-hydroxyquinoxalines in high yields.[174]

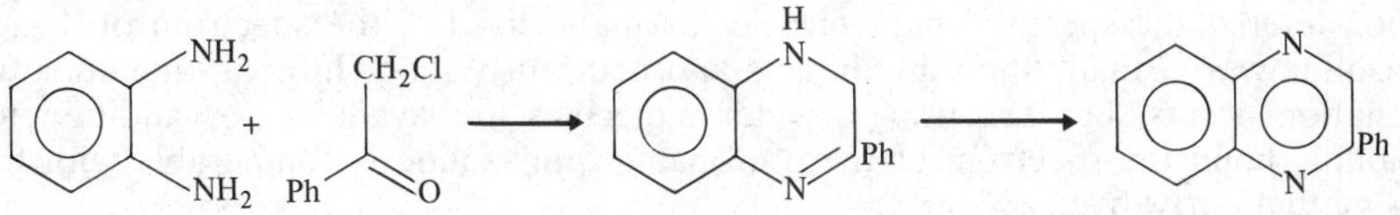

NH_2 + CH_2Cl NH_2 Ph O → H N Ph N → N Ph N

SCHEME 67

A widely applicable and unambiguous synthesis of 2-hydroxyquinoxalines[175] was introduced by Horner *et al.* in 1953. The method is based on the reduction of *o*-nitro-*N*-phenylglycine derivatives which are directly available by the reaction of *o*-chloronitrobenzene derivatives with glycine. The reduction of the dinitrophenylglycine (**164**) results in an intramolecular cyclization (Scheme 68) to a 1,2,3,4-tetrahydro-2-oxoquinoxaline which on subsequent permanganate oxidation yields the 2-hydroxyquinoxaline.

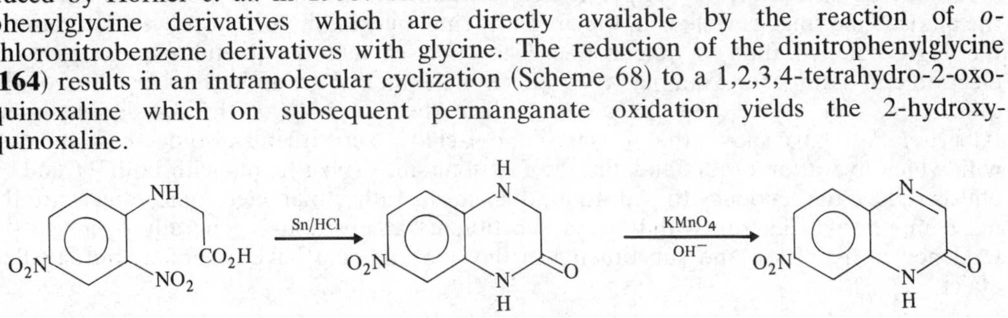

NH O_2N CO_2H NO_2 (164) —Sn/HCl→ H N O_2N N H O —$KMnO_4$, OH^-→ N O_2N N H O

SCHEME 68

16.2.14 GENERAL CHEMISTRY OF THE BENZODIAZINES

The chemical reactivity of the benzodiazines follows predictable trends, but, as might be expected, specific behaviour of a given member of the series may have no counterpart in the other members of the series. This statement is adequately illustrated with reference to quinazoline chemistry. All of the benzodiazines are relatively stable, but on heating quinazoline with aqueous acid or aqueous alkali it undergoes decomposition to benzaldehyde and formic acid. All of the benzodiazines are basic and undergo protonation in aqueous acid, but the cation of quinazoline differs from the cations of the other benzodiazines in that it undergoes reversible addition of water across the 3,4-double bond, and it has been suggested that this reactivity is due to two factors. Firstly, the effect of fusion of a benzene ring on to a pyrimidine ring results in the 3,4-bond in quinazolines developing a highly localized double bond character, and as a direct result addition of nucleophiles to the 3,4-bond is facile. Secondly, protonation at N-1 of the quinazoline ring followed by hydration results in the formation of an amidinium system which enjoys enhanced resonance stabilization (Scheme 69).

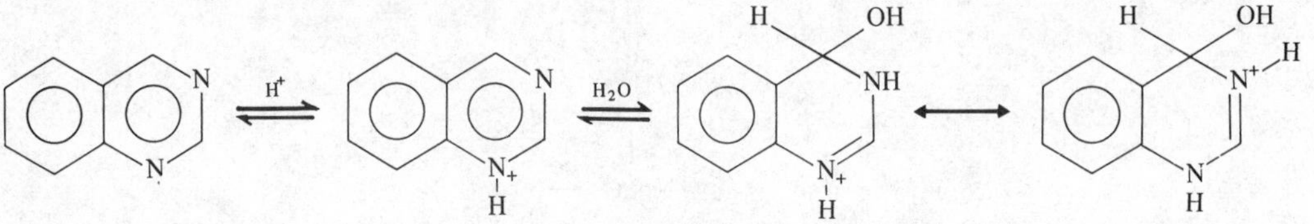

SCHEME 69

This reversible addition of water has been described as covalent hydration and generally dictates the chemistry under aqueous, acidic conditions. The evidence supporting the existence of hydration is strong, and stems from the observations by Albert *et al.*[176] that quinazoline is an order of magnitude more basic than 4-methylquinazoline, whereas, had normal trends been followed, 4-methylquinazoline would be expected to be a stronger base. U.v. spectral data indicated that protonation of the quinazoline ring in aqueous solution altered the spectral characteristics dramatically, but the spectrum of 4-methylquinazoline was virtually the same in acid and neutral solution. In non-aqueous solvents, protonation occurs, but because no water molecules are available for addition to the 3,4-double bond the spectrum of the protonated quinazoline is comparable with that of the 4-methyl derivative.

Further support for the existence of the hydrated form of quinazoline comes from the observation that mild oxidation of quinazoline in aqueous acid gives rise to 3,4-dihydro-4-oxoquinazoline, as would be expected from the oxidation of an α-carbinolamine.

As may be inferred from the previous discussion, substituents at positions-2 and -4 of the quinazoline ring, as well as in the aromatic ring, play an important role in determining the degree of hydration of quinazoline. The effect of the substituents may be steric or electronic in nature; for example, a $+I$ effect of the methyl group in the 4-position of the quinazoline ring would be expected to reduce the reactivity of the 3,4-double bond. Albert *et al.*[177] have shown that 4-cyano- and 4-chloro-substituents also decrease the ease with which hydration occurs and cite the fact that since substituents with both $+I$ and $-I$ effects reduce the tendency to hydration, then steric, rather than electronic, effects are the most important. Electron-withdrawing substituents at position-2 generally decrease the tendency to hydration and substituents in the benzene ring have a similar, but smaller, effect.

16.2.14.1 Electrophilic substitution

In general, electrophilic substitution in the benzodiazine series has not been extensively examined but certain trends would be expected. In the case of the corresponding diazines, unless activating substituents are present, electrophilic substitution does not occur. The benzodiazines, by virtue of the benzene ring, have alternative positions available and substitution can, and does, occur in the benzene ring, albeit under forcing conditions.

Little experimental work has been carried out on the electrophilic substitution of cinnolines, presumably owing to the fact that substituted cinnolines are more easily prepared by direct synthesis. Molecular orbital calculations[178] suggest that the relative order of reactivities at the cinnoline ring positions towards electrophilic substitution are $5=8>6=7>3\gg4$. This suggests that positions-5 and -8 should be the most reactive and this has been confirmed experimentally in the case of nitration[179] in sulphuric acid, which results in the formation of 5-nitrocinnoline (**165**) and 8-nitrocinnoline (**166**) in 33% and 28% yields, respectively. In 80% sulphuric acid the species undergoing nitration is the 2-cinnolinium cation (**167**) rather than cinnoline; the cation is nitrated some 287 times more slowly than the isoquinolinium cation!

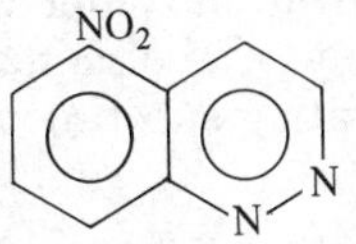

NO_2 N N NO_2 N N N $\overset{+}{N}H$

(165) 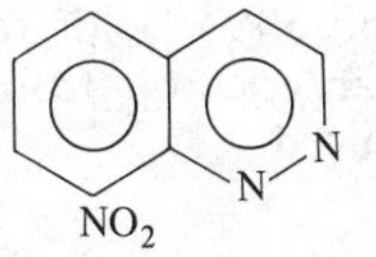 (166) 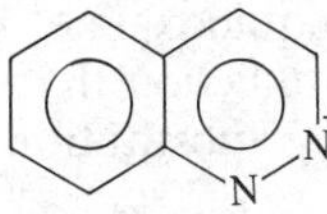 (167)

The 5- and 8-positions of the phthalazine ring are equivalent and are expected to be the most reactive sites towards substitution. Nitration of phthalazine with potassium nitrate in concentrated sulphuric acid results in the formation of 5-nitrophthalazine (**168**) as the main product with 5-nitro-4(3*H*)-phthalazinone (**169**) as a by-product.[180]

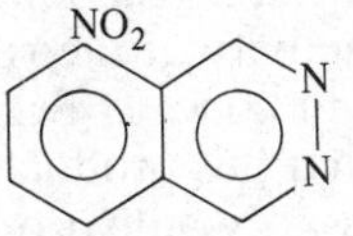

NO_2 N N O NO_2 NH N

(168) 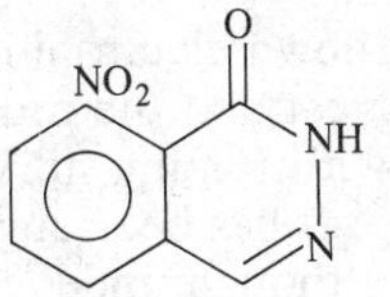 (169)

The symmetry of the quinoxaline ring renders the 5- and 8-positions equivalent and electron localization energy calculations[181] indicate that substitution should occur preferentially at position-5(8). Nitration is only effected under forcing conditions: using concentrated nitric acid and oleum at 90 °C for 24 hours a 1.5% yield of 5-nitroquinoxaline (**170**) is obtained. The major product is 5,6-dinitroquinoxaline (**171**), formed in 24% yield. This is a curious result since the predicted major dinitration product is 5,7-dinitroquinoxaline.

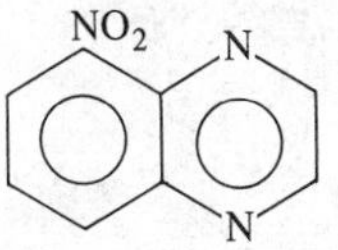

NO_2 N N NO_2 O_2N N N

(170) 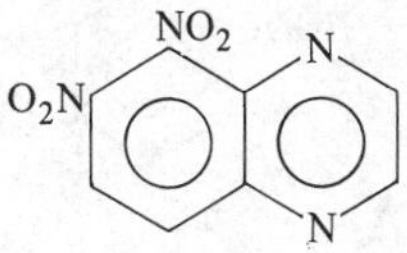 (171)

When activating substituents are present in the benzene ring, nitration becomes more facile and usually occurs in accordance with predictions based on simple valence bond theory.

When activating substituents are present in the hetero-ring the situation becomes more complex and can vary with reaction conditions, as illustrated by the nitration of 2-hydroxyquinoxaline. Nitration in acetic acid results in the formation of 7-nitro-1,2-dihydro-2-oxoquinoxaline (**172**), whereas nitration with mixed acids results in the formation of 6-nitro-1,2-dihydro-2-oxoquinoxaline (**173**) and the different products are probably a reflection of the species which is being nitrated, the neutral 2-hydroxyquinoxaline in acetic acid, and the protonated quinoxaline (**174**) in mixed acids.

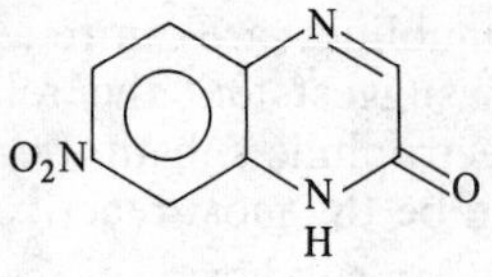

(172) 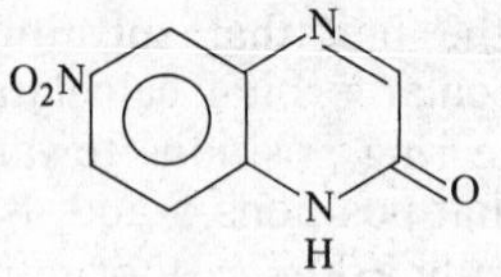 (173) 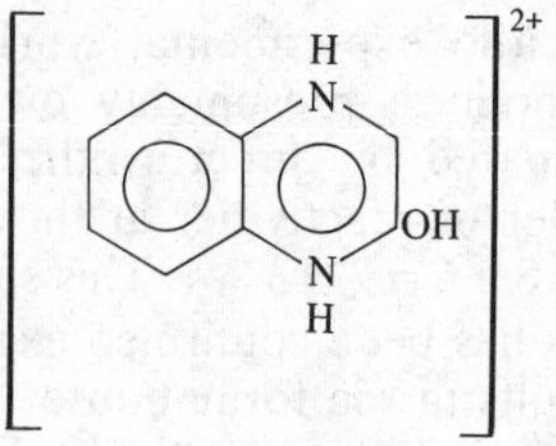 (174)

Attempts to brominate 2-methylquinoxaline with bromine in acetic acid[182] failed to result in ring bromination and a mixture of 2-bromomethylquinoxaline (**175**) and 2-dibromomethylquinoxaline, in 27 and 33% yields respectively, was obtained. This result supports the prediction that, in the absence of strongly activating substituents, side-chain rather than ring substitution is favoured.

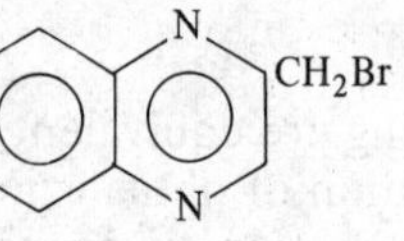

(175) 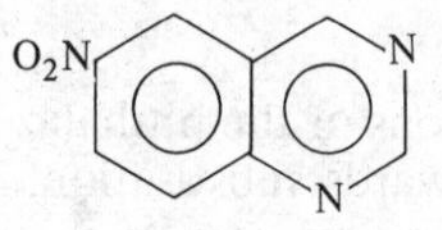 (176)

Nitration is the only known electrophilic substitution reaction of quinazoline. Theoretical considerations[183] suggest that the order of reactivity of the ring positions is $8>6>5>7\gg4>2$, but nitration with fuming nitric acid and conc. H_2SO_4 gave a 56% yield of 6-nitroquinazoline (**176**).[184] It has been suggested that this product might be due to nitration of the hydrated cation, although under the necessarily anhydrous conditions this appears unlikely. The species undergoing nitration is probably the mono- or di-cation.[185]

The initial stages of permanganate oxidation of aromatic substrates are believed to be analogous to electrophilic substitution, and alkaline permanganate oxidizes the benzodiazines to the corresponding diazinecarboxylic acids. Thus cinnoline and phthalazine give pyridazine-3,4- (**177**) and pyridazine-4,5-dicarboxylic acids (**178**), respectively. Quinazoline gives largely pyrimidine-4,5-dicarboxylic acid (**179**) along with minor amounts of 3,4-dihydro-4-oxoquinazoline, the latter product being due to the oxidation of hydrated quinazoline, and quinoxaline yields pyrazine-2,3-dicarboxylic acid (**180**).

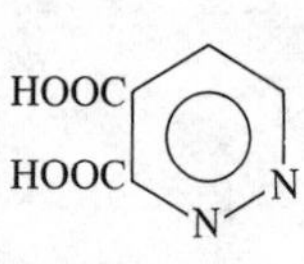

(177) 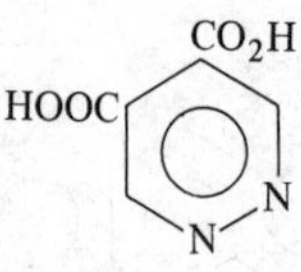 (178) 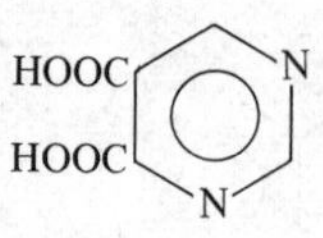 (179) 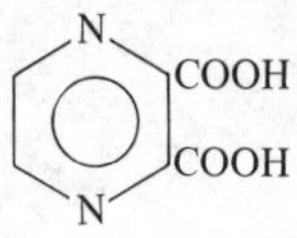 (180)

16.2.14.2 Nucleophilic substitution

Direct nucleophilic substitution reactions of the benzodiazines, such as amination, have not been reported and this is almost certainly due to the fact that substituted benzodiazines are more easily accessible by other methods. Quinazoline may be directly aminated with sodamide in the 4-position in accord with expectations,[186] and, similarly, 4-hydrazinoquinazoline was prepared by the reaction of quinazoline with hydrazine, but in view of the known reactivity (see above) of the 3,4-bond of quinazoline these reactions may involve an addition/oxidation sequence (Scheme 70) rather than direct nucleophilic substitution.

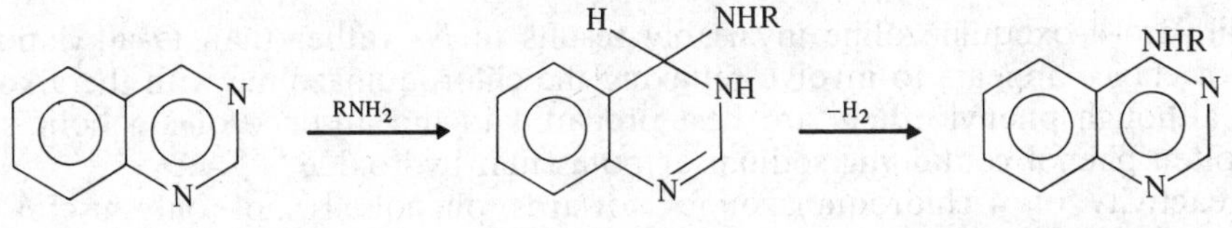

SCHEME 70

As in the case of the diazines, nucleophilic substitution reactions are more usually confined to the displacement of a halogen atom. The kinetics of substitution of a number of monochlorobenzodiazines with the chlorine atom bonded to the heterocyclic ring have been investigated[187] (Table 4) and although rate data for 4-chloroquinazoline and 3-chlorocinnoline are notably absent, the rates of reaction of 2-chloroquinazoline and 4-chloroquinazoline with piperidine are 4.79×10^{-4} and 3.1 l. mol^{-1} s^{-1}, respectively. This clearly puts a difference in reactivity of 6500 between 2- and 4-chloroquinazoline in favour of the 4-isomer. No rate data are available for 3-chlorocinnolines; however, in 3,4-dichlorocinnoline the 4-chlorine is the most labile during substitution reactions, and the relative reactivities of the chlorobenzodiazines may be listed as 4-chloroquinazoline > 2-chloroquinoxaline > 4-chlorocinnoline > 2-chloroquinazoline > 1-chlorophthalazine ≈ 3-chlorocinnoline, with some degree of confidence.

TABLE 4
Rate Constants for the Reaction of Ethoxide with Halobenzodiazines

Halobenzodiazine	*Rate constant* (l mol^{-1} s^{-1}) *at* 20 °C
2-Chloroquinazoline	3.00×10^{-3}
2-Chloroquinoxaline	8.35×10^{-3}
1-Chlorophthalazine	1.86×10^{-3}
4-Chlorocinnoline	4.74×10^{-3}

The effect of fusing a benzene ring on to a halodiazine is to increase the reactivity of the halogen atom towards nucleophilic substitution, presumably through the extra resonance stabilization of the transition state leading to products. The rate constant for the reaction of4-chloroquinazoline with piperidine at 20 °C is 3.1 l mol^{-1} s^{-1} compared with a rate constant of 1.5×10^{-3} l mol^{-1} s^{-1} for 4-chloropyrimidine, confirming a difference in reactivity of 2000. However, the difference in reactivity between 2-chloroquinazoline and 2-chloropyrimidine is only in the order of 1.4 ($4.79 \times 10^{-4} : 3.34 \times 10^{-4}$) and clearly points to the enhanced stability of resonance forms such as (**181**)–(**182**) relative to (**183**)–(**184**).

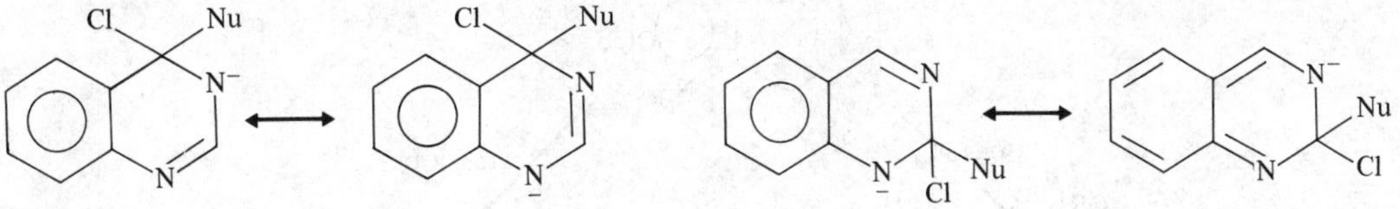

(181) (182) (183) (184)

4-Chloroquinazoline reacts spontaneously with methanol to give 4-methoxyquinazoline hydrochloride and the alcoholysis appears to be autocatalysed by the liberated hydrochloric acid. The reaction with ethanol is fairly rapid in the presence of acid, but it is of interest to note that 4-chloroquinazoline may be crystallized from ethanol to which a trace of sodium hydroxide has been added. Hydrolysis of 4-chloroquinazoline in boiling water is complete in 15 minutes and a mechanism involving covalent hydration appears probable.[188]

The displacement of the halogen atom in 4-chloroquinazolines by alkoxide ions offers the best synthetic method of preparation of 4-quinazolinyl ethers, since direct alkylation of 3,4-dihydro-4-oxoquinazoline invariably results in *N*- rather than *O*-alkylation. The method of choice appears to involve refluxing the chloroquinazoline with the alkoxide in alcohol, although phenyl ethers are best prepared in the absence of alcoholic solvents using molten phenol containing sodium or potassium hydroxide.

The reactivity of 4-chloroquinazolines towards phenols is not only useful in the synthesis of aryloxyquinazolines but also serves as an efficient method of transposing phenols to the corresponding anilines in high overall yield.[189] Phenoxides react with 4-chloroquinazolines in dimethylformamide to yield 4-aryloxyquinazolines, which on thermolysis at *ca.* 300 °C undergo a Chapman rearrangement to the corresponding 3-aryl derivative (Scheme 71). On aqueous hydrolysis the intermediate 3-aryl-3,4-dihydro-4-oxoquinazoline decomposes to produce the arylamine and phenylbenzoxazinone. This

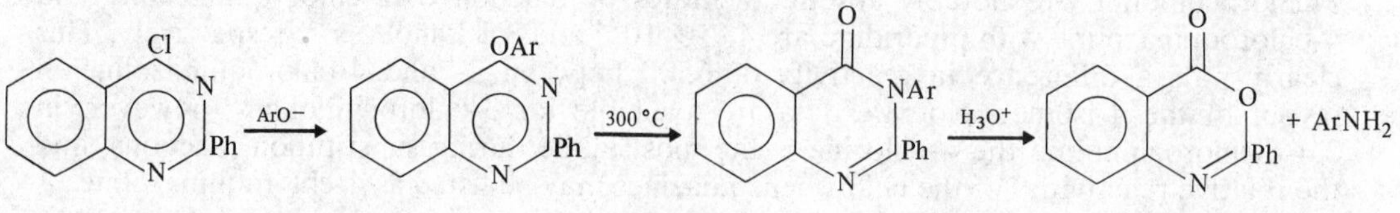

SCHEME 71

transformation has the added advantage that the chloroquinazoline may be regenerated in two simple steps from the by-product.

The remarkable difference in reactivity[168] between the 2- and 4-positions in 2,4-dichloroquinazoline makes selective substitution possible. Thus reaction with one equivalent of alkoxide or with an excess of alkylamine results in the formation of the 4-alkoxy-2-chloro- and 4-amino-2-chloro-quinazolines, respectively. Arylamines in general show somewhat different behaviour (Scheme 72) in that the reaction does not stop at monoamination.

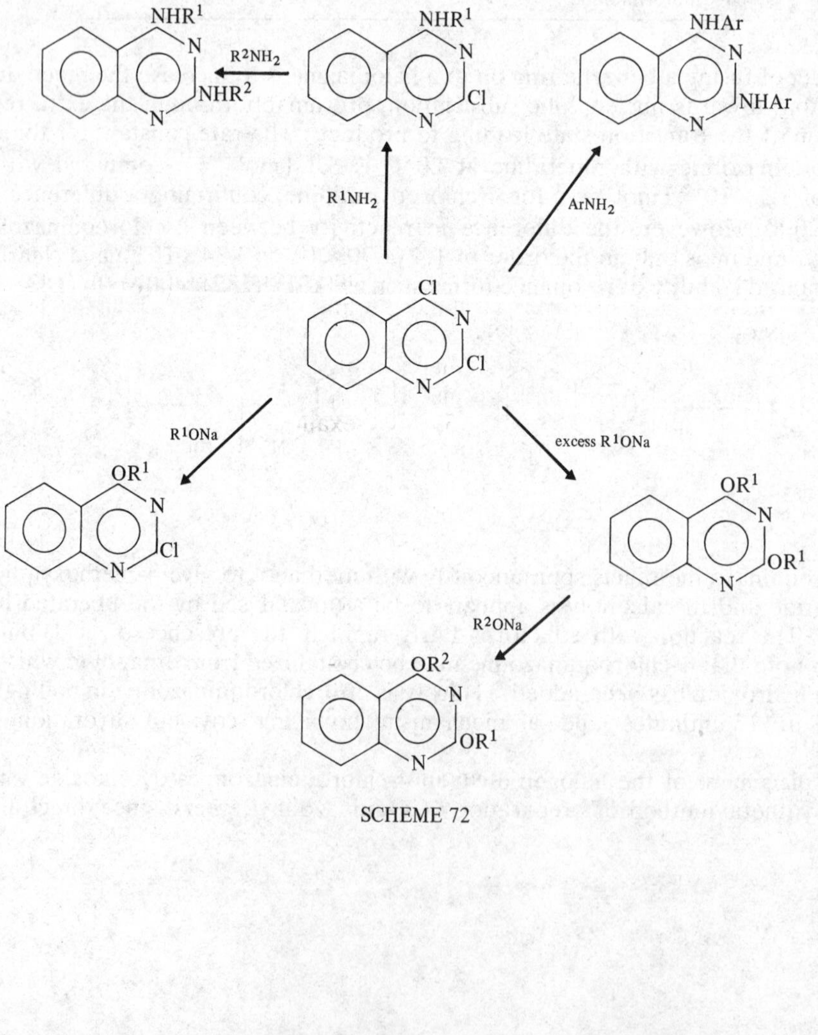

SCHEME 72

2-Chloroquinazolines also participate in substitution reactions and the 2-amino-, 2-alkylamino-, and 2-hydrazino-derivatives may be prepared by the direct reaction with alcoholic ammonia, alkylamines, and hydrazine, respectively. The reaction of 2-chloroquinazoline with triethylamine is of interest in that 2-diethylaminoquinazoline results, probably with loss of ethyl chloride. Other examples of this behaviour have been observed in quinazoline chemistry.

Nucleophilic substitution in the phthalazine series centres around the displacement of halogen atoms from halophthalazines. 1-Halophthalazines readily react with refluxing solutions of alkoxides in the appropriate alcohol to yield the 1-alkoxyphthalazines. Similarly, the reaction of phenols in the presence of K_2CO_3 at 100 °C results in the formation of phenoxyphthalazines.

Aminophthalazines are directly available by the reaction of 1-halophthalazines, 1-alkoxyphthalazines, or methylthiophthalazines with ammonia or amines.

By analogy, 1,4-dichlorophthalazines react with sodium alkoxides to give 1,4-dialkoxyphthalazines,[190] although the reaction may be controlled to give 1-alkoxy-4-chlorophthalazines. 1,4-Dialkoxyphthalazines have served as a starting point for the synthesis of 1-alkoxy-4-hydroxylaminophthalazines, which show useful antipyretic, anti-inflammatory, antihypertensive, and bronchodilatory effects and function as respiratory stimulants. When 2,5-dialkoxyphthalazines are stirred with an alcoholic solution of buffered hydroxylamine hydrochloride, one of the alkoxy-groups is selectively displaced (Scheme 73).

OEt, N, N, OEt —i→ NHOH, N, N, OEt

i, $NH_2OH{\cdot}HCl/CH_3CO_2Na/EtOH/20$ °C.

SCHEME 73

1,4-Dichlorophthalazines react with ammonia or alkylamines in a stepwise manner to give the 1-amino- or 1-alkylamino-4-chlorophthalazine, but reaction with aromatic amines generally results in 1,4-diarylaminophthalazines unless stoichiometric quantities of the amine are used. More conveniently, 1,4-diaminophthalazines may be prepared from dialkoxyphthalazines by high-temperature reaction with the appropriate amine. Thus refluxing 1,4-diphenoxyphthalazine with morpholine results in the formation of 1,4-dimorpholinophthalazine.

Aside from the enhanced reactivity of halogen atoms in the heterocyclic ring, halogen atoms bonded to the benzene ring have also been shown to possess enhanced reactivity. Chambers *et al.*[191] have prepared 1,4,5,6,7,8-hexafluorophthalazine and studied the nucleophilic substitution with methoxide at different temperatures. All fluorine atoms are displaced (Scheme 74) progressively, although no rate data are available.

Halide, methanesulphonyl ($MeSO_2$—), methanesulphinyl (MeSO—) , nitrile, and mercapto groups in the 4-position of the cinnoline ring are labile and readily undergo substitution reactions, but, since many of the more important cinnoline derivatives are directly available from 4-chlorocinnoline, some of its more important transformations are shown in Scheme 75.

3-Halocinnolines are generally much less reactive than the isomeric 4-halocinnolines, although substitution will occur when sufficiently stringent conditions are employed. 3-Bromocinnoline reacts with aqueous ammonia at 130–140 °C in the presence of copper sulphate to give 3-aminocinnoline, and with sodium methoxide in methanol at 120 °C in sealed tubes to yield the 3-methoxy-compound. As with dihaloquinazolines, 3,4-dihalocinnolines will react in a stepwise manner and different substituents may be introduced in this way.

i, NaOMe / − 15 °C; ii, NaOMe/0 °C; iii, NaOMe/45 °C; iv, NaOMe/ 85 °C / sealed tube.

SCHEME 74

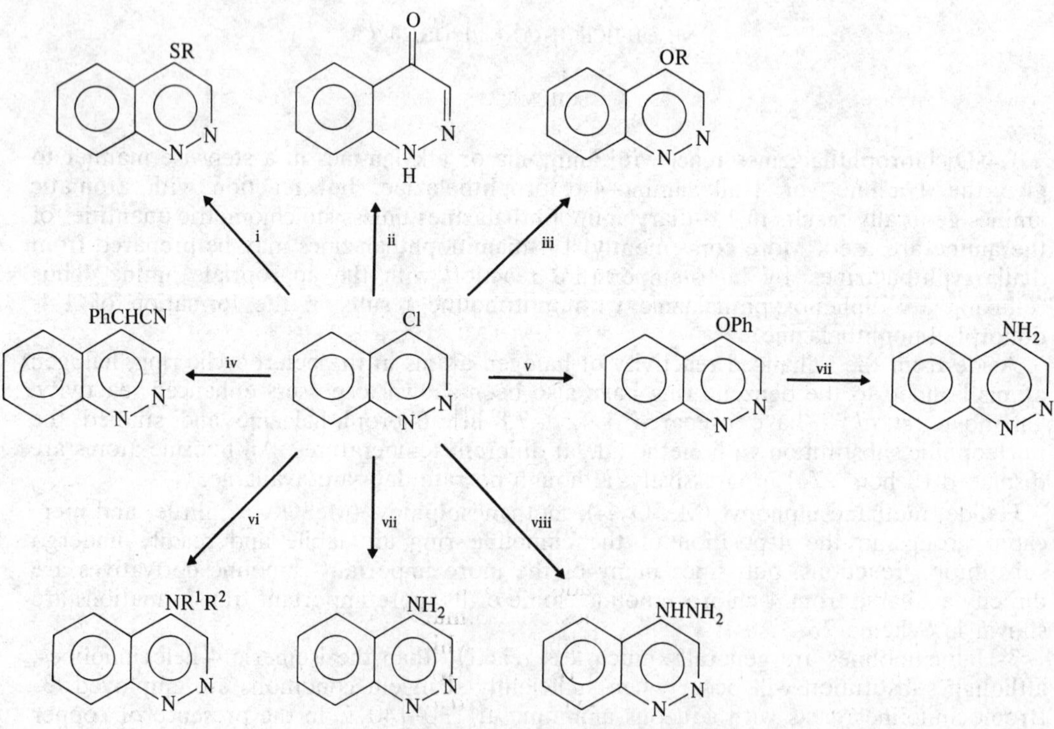

i, RSNa; ii, H_2O; iii, RONa; iv, $PhCH_2CN/NaNH_2/C_6H_6$; v, PhONa; vi, R^1R^2NH; vii, NH_4OH; viii, N_2H_4.

SCHEME 75

16.2.14.3 Reactions of the benzodiazines with organometallic reagents and other nucleophiles

The anomalous hydration of quinazoline has already been discussed and if the 3,4-bond is highly localized as suggested then addition of nucleophiles other than water might also be expected. This prediction is observed and common anionic reagents add readily, *e.g.* sodium bisulphite, hydrogen cyanide, acetophenone, butan-2-one, and cyclohexanone (Scheme 76) all add across the 3,4-double bond to form stable addition products.

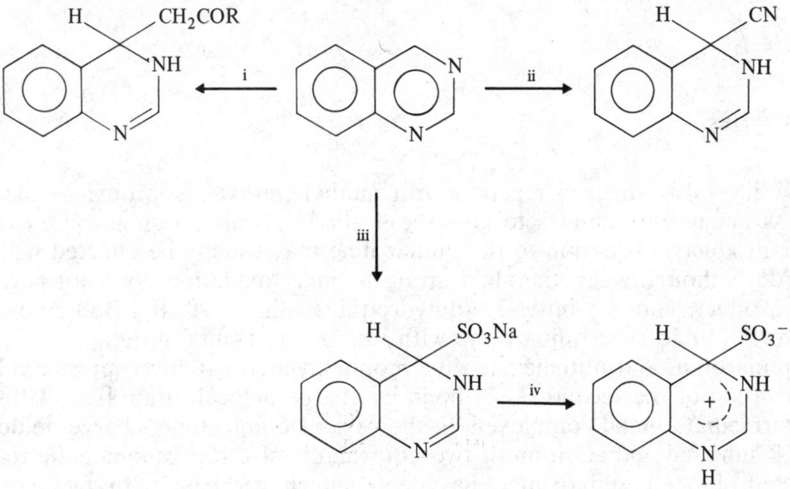

i, $RCOCH_3/OH^-$; ii, HCN; iii, $NaHSO_3$; iv, H_3O^+.

SCHEME 76

Quinoxaline, to a lesser extent, also undergoes addition reactions and addition compounds with sodium bisulphite and HCN are known. Cinnolines and phthalazines appear to be stable under these conditions.

All of the benzodiazines undergo addition reactions with organometallic reagents and although examples in the cinnoline series are not well documented,[160] 1,4-addition is predicted. The reaction of ethyl-3-phenyl-4-cinnoline-4-carboxylate (**185**) is anomalous in that the 1-phenyl-1,4-dihydrocinnoline (**186**) and the 1,4-dihydrocinnoline (**187**) are the only products, formed in low yield, during the reaction of (**185**) with phenyl-lithium.

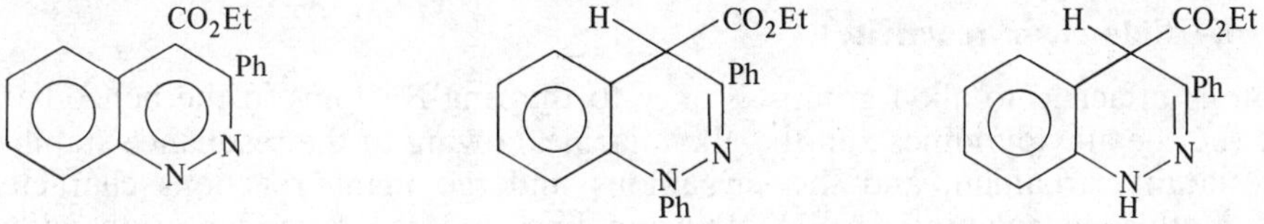

(185) (186) (187)

Phthalazine readily undergoes[192] 1,2-addition reactions with Grignard reagents to yield 1-alkyl- or 1-aryl-1,2-dihydrophthalazines. Phenylmagnesium bromide adds to the C—N bond to give 1-phenyl-1,2-dihydrophthalazine, which is unstable under the reaction conditions, and undergoes facile aerial oxidation to 1-phenylphthalazine in an overall yield of 44%. In an analogous way, alkyl Grignard reagents react but the resultant 1,2-dihydrophthalazines are generally more stable, although they may be oxidized with potassium ferricyanide to the corresponding phthalazines. Thus 3-dimethylaminopropyl-magnesium chloride on reaction with phthalazine gives 1-(3-dimethylaminopropyl)-1,2-dihydrophthalazine (**188**), which may be oxidized[193] to the phthalazine (**189**) with

$K_3Fe(CN)_6$ in 55% yield. When 1-alkyl- or 1-aryl-phthalazines are used, addition generally occurs at the 3,4-bond.

Alkyl- and aryl-lithium derivatives also react with phthalazine, but generally the yields of alkylated phthalazines are lower than those obtained with the corresponding Grignard reagents.

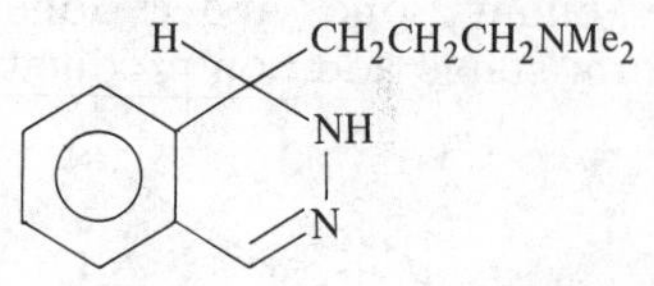

(188) 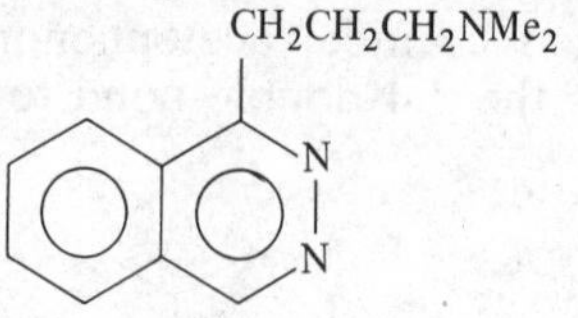 (189) 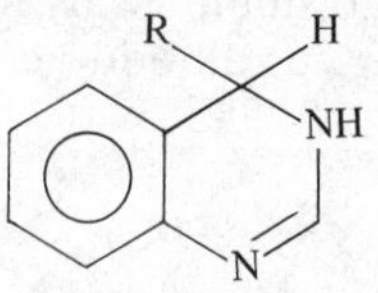 (190)

Quinazolines also undergo reaction with methyl-, ethyl-, isopropyl-, benzyl-, t-butyl- and phenyl-magnesium halides to give the 4-alkyl-3,4-dihydroquinazolines (**190**), which are stable products. Oxidation to the quinazoline may usually be effected with potassium ferricyanide, although when the alkyl group is bulky oxidation does not always give the expected product. Thus 4-t-butyl-3,4-dihydroquinazoline (**190**; $R = Bu^t$) on oxidation with alkaline ferricyanide gave quinazoline with loss of the t-butyl group.

Both quinazoline and phthalazine give monoalkylated products, presumably owing to the deactivation of the second C—N bond by charge delocalization [see (**191**) and (**192**)] in the intermediate metal complexes. In the case of quinoxaline, charge delocalization of this type is minimal and addition of two equivalents of the organometallic reagent might be expected. Indeed, quinoxaline has been shown to react with two equivalents of Grignard reagent, as illustrated by the formation of 2,3-diallyl-1,2,3,4-tetrahydroquinoxaline (**193**) during the reaction with allylmagnesium bromide.

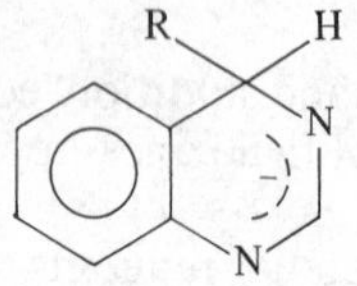

(191) 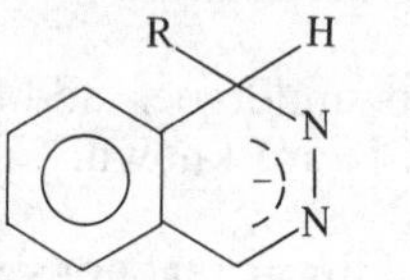 (192) 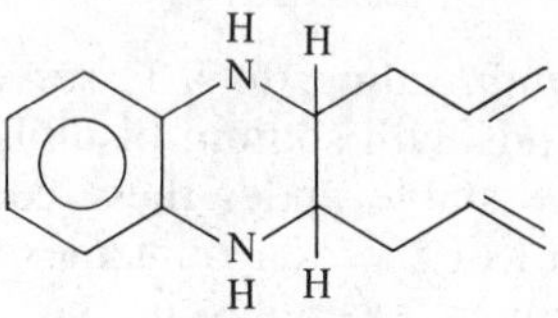 (193)

16.2.14.4 Side-chain reactivity

Protons attached to alkyl groups α or γ to the ring-N atoms in the benzodiazines are acidic (*cf.* the alkylpyridines and the alkyldiazines) owing to the resonance stabilization of the resultant carbanion, and the carbanions undergo many reactions characteristic of active methylene compounds. 4-Methylcinnoline undergoes condensation with benzaldehyde to yield 4-styrylcinnoline (**194**) in good yield, although many benzaldehyde derivatives such as salicaldehyde or *p*-nitrobenzaldehyde fail to react or give rise to polymeric products. Zinc chloride, which has been used to catalyse this type of reaction, has little effect on product yield in the cinnoline series.

On treatment of 4-methylcinnoline with ethyl nitrite under acidic conditions, nitrosation occurs, resulting in the formation of the oxime (**195**) in high yield. 3-Methylcinnoline, however, undergoes neither condensation reactions with aldehydes nor reaction with ethyl nitrite, attesting to the difference in acidity of protons in the 3-methyl group.

1-Methylphthalazines, 4-methylquinazolines, and 2-methylquinoxalines, like 4-methylcinnoline, all undergo condensation reactions with aromatic aldehydes, but in general the low yield of product from these reactions makes them a chemical curiosity rather than

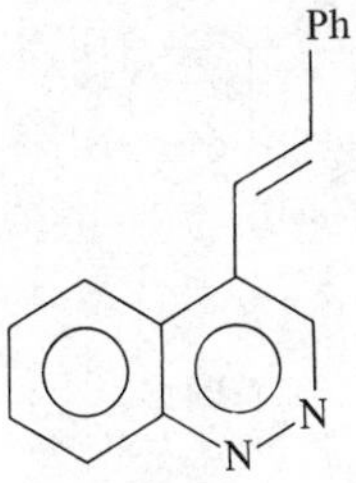

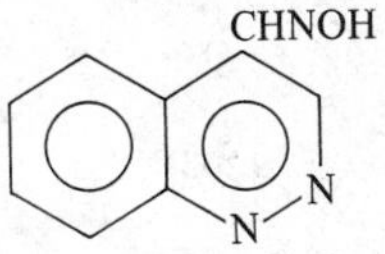

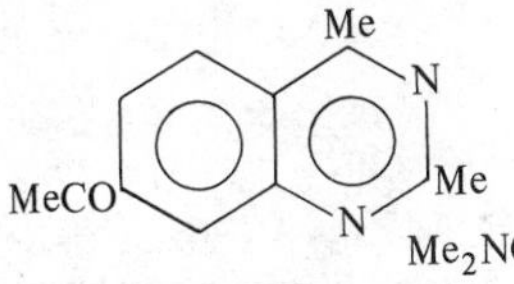

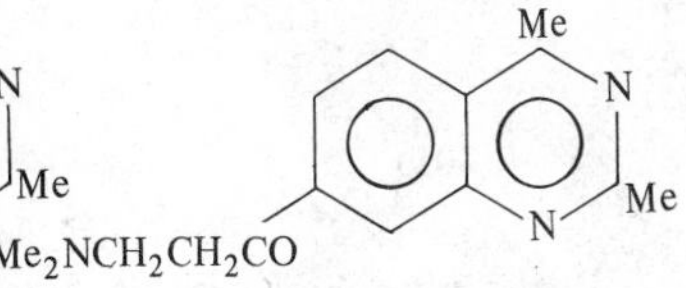

(194) (195) (196) (197)

preparative methods for the synthesis of alkenyl derivatives. Siegel and Cristensen[194] have carried out definitive experiments which clearly indicate the relative reactivity of alkyl groups located at positions-2 and -4 in the quinazoline ring. The Mannich reaction (Scheme 77) of 4-methyl- and 2,4-dimethyl-quinazoline gave exclusively the 4-substituted

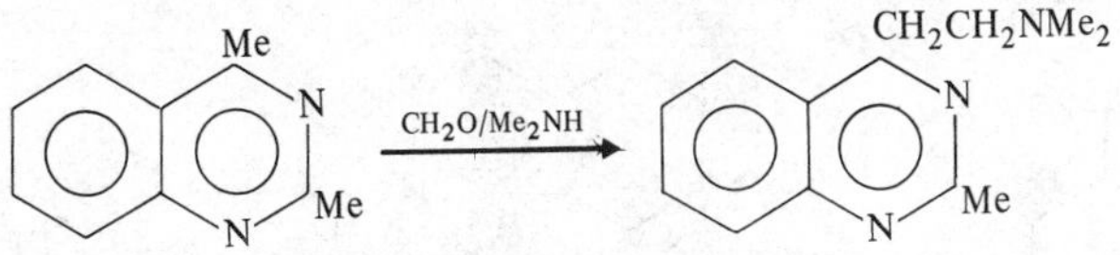

CH_2O/Me_2NH

SCHEME 77

products when morpholine or dimethylamine were used as bases. The reactivity of the 4-methyl group, however, is lower than that of an acetyl group, as 7-acetyl-2,4-dimethylquinazoline (**196**) gave 2,4-dimethyl-7-(3′-dimethylaminopropionyl)quinazoline (**197**) under similar reaction conditions.

1-Methylphthalazine has also been shown to react with aqueous formaldehyde. Thus heating an aqueous mixture under reflux for four hours results in 1-(2-hydroxyethyl)phthalazine (**198**), whereas refluxing in pyridine for 16 hours yields the 2 : 1 condensation product (**199**). Most of the alkylbenzodiazines undergo reaction with a pyridine/iodine mixture to give quaternary salts such as (**200**), clearly confirming a general trend in reactivity.

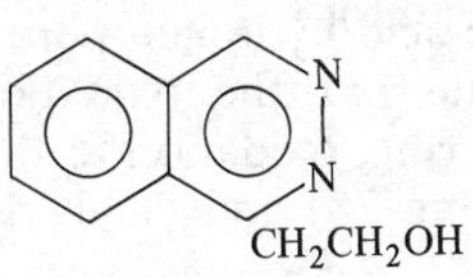

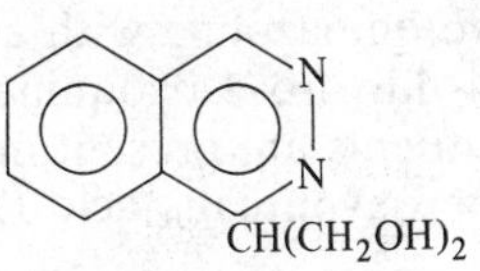

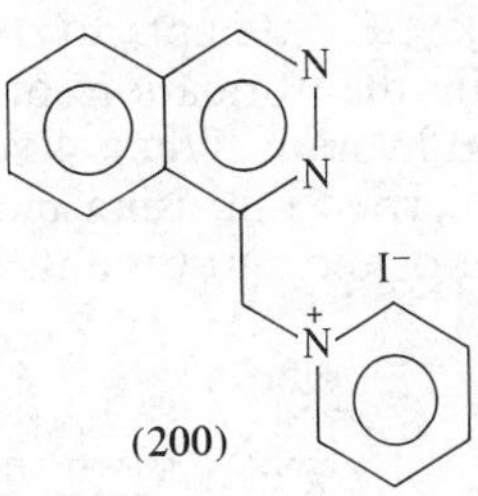

(198) (199) (200)

16.2.14.5 *N*-Oxides

The first known synthesis of cinnoline *N*-oxides is that due to Atkinson and Simpson,[195] who established that treatment of a number of 3,4-disubstituted cinnolines with hydrogen peroxide/acetic acid resulted in high yields of the mono-*N*-oxides, which were formulated as cinnoline 1-oxides. Subsequent work has shown that when there is a bulky substituent in the 3-position, 1-oxide formation is favoured. Where no such substituent is present, 2-oxide formation is preferred. Cinnoline itself on oxidation gives a mixture of the 1-oxide (**201**), the 2-oxide (**202**), and the 1,2-dioxide (**203**) in a ratio of 25.9 : 49.2 : 0.3. The di-*N*-oxide is directly available by further oxidation of the mono-*N*-oxides.

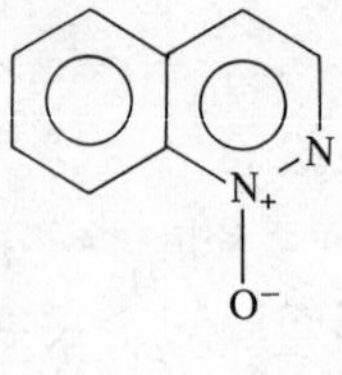

(201) 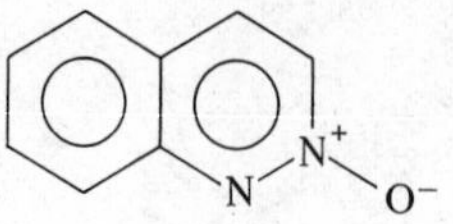 (202) 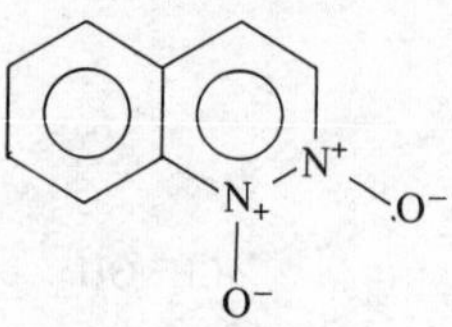 (203) 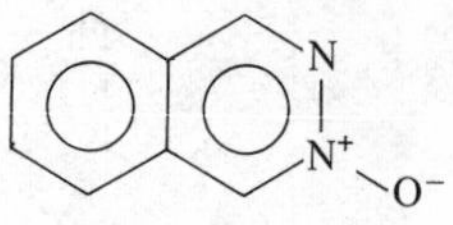 (204)

Phthalazine *N*-oxides have been prepared by direct oxidation using monoperoxyphthalic acid. Thus phthalazine at 0 °C yields phthalazine 2-oxide (**204**) in high yield. With unsymmetrical phthalazines, generally both mono-*N*-oxides are produced although the precise ratio of the isomers is dependent on steric and electronic effects of the substituents. 1-Isopropyl-4-phenylphthalazine gives exclusively the 3-oxide (**205**), whereas 1-methyl-4-phenylphthalazine preferentially forms the 2-oxide (**206**). No phthalazine 2,3-dioxides have been described.

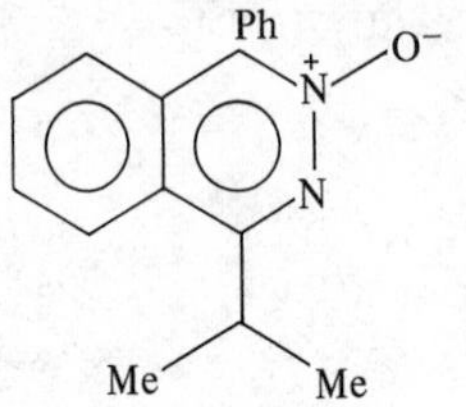

(205) 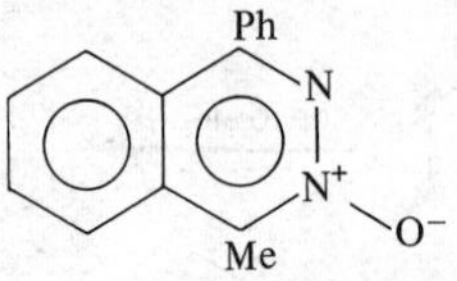 (206) 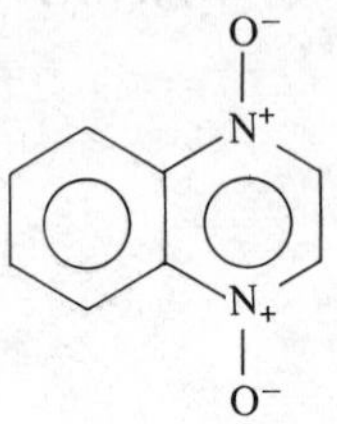 (207)

Quinoxalines are more easily converted to their *N*-oxides than any other member of the benzodiazine series (*cf.* the diazines). During the oxidation the use of peracetic acid rather than acetic acid/hydrogen peroxide is recommended, since the latter reagent may oxidize quinoxaline to quinoxaline-2,3-dione. Generally one equivalent of peracid generates the mono-*N*-oxide and an excess results in the formation of the di-*N*-oxide (**207**). The effect of substituents in the 2- or 3-positions is entirely similar to that observed with other members of the series.

Quinazoline 1-oxide is not known, although a number of 4-substituted derivatives have been prepared. Attempts to oxidize quinazoline with acetic acid/hydrogen peroxide have resulted in the formation of 3,4-dihydro-4-oxoquinazoline *via* the oxidation of the protonated hydrate. When 4-substituents are present in the ring, oxidation to the 1-oxide has been carried out using monoperoxyphthalic acid (Scheme 78) and yields depend on the nature of the substituent.

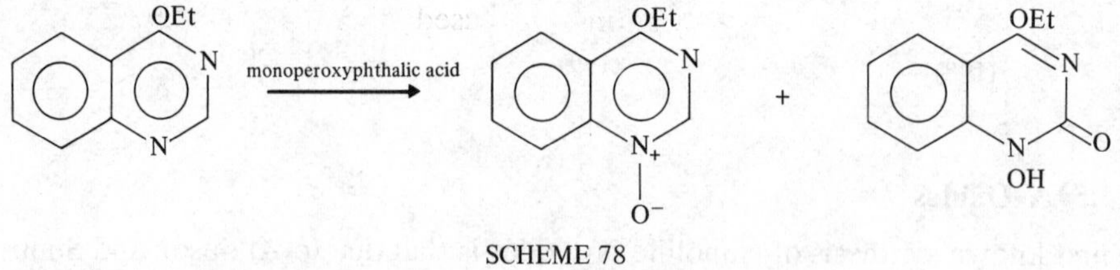

monoperoxyphthalic acid

SCHEME 78

Quinazoline 3-oxide has not been prepared by direct oxidation but may be prepared by the condensation of ethyl orthoformate (Scheme 79; R = H) with the oximes of *o*-aminobenzaldehydes or -ketones.

The benzodiazine *N*-oxides generally undergo all of the reactions previously encountered in the diazine series. Thus deoxygenation may be accomplished with phosphorus trihalides, and Reissert addition is observed, as is activation of the heterocyclic ring

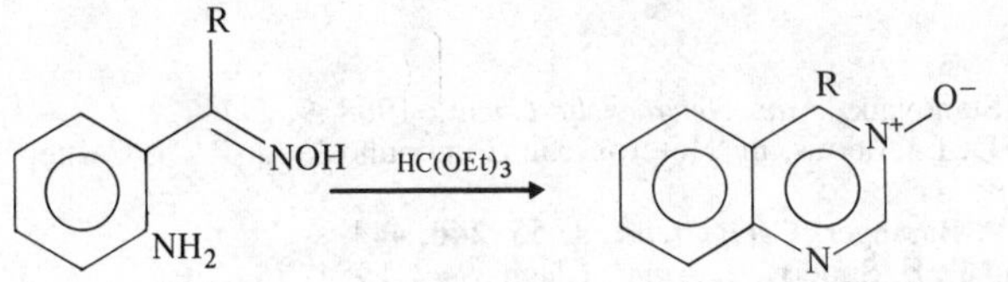

SCHEME 79

towards electrophilic substitution; reductive acetylation and reductive chlorination with acetic anhydride and phosphoryl chloride respectively, and substitution into an α-methyl group, are common reactions.

Quinazoline 3-oxides are anomalous in their behaviour, and readily undergo hydrolysis under strongly basic conditions, quinazoline itself yielding the oxime of *o*-aminobenzaldehyde. Because of its tendency to exist in the hydrated form (**208**), quinazoline 3-oxide is readily oxidized to 3,4-dihydro-3-hydroxy-4-oxoquinazoline (**209**). Nucleophiles such as HCN or hydrazine readily add to the 3,4-bond to form unstable derivatives (**210**), which undergo spontaneous loss of water with the concomitant formation of 4-substituted quinazolines.

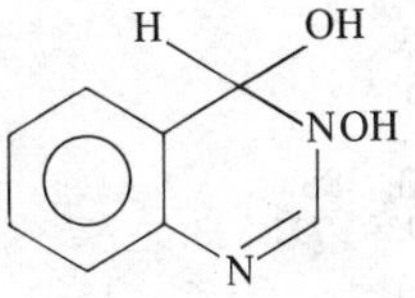

(208) 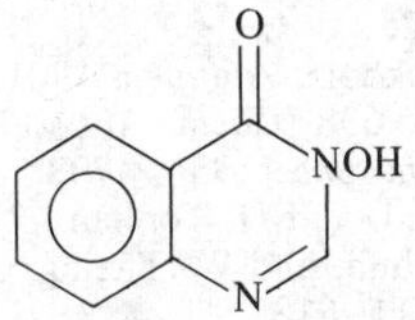 (209) 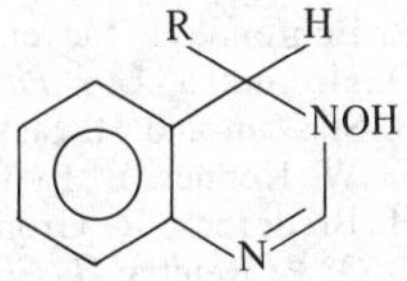 (210)

Under the conditions of normal Reissert addition, ring fission to the aminobenzaldehyde oxime (**211**) occurs and reaction with acetic anhydride results in a similar fragmentation to *o*-isocyanobenzonitrile (**212**).

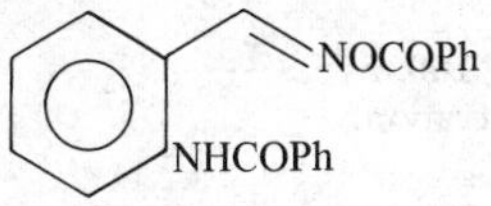

(211) 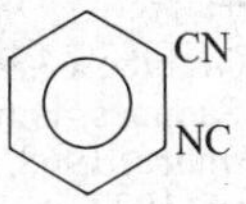 (212) 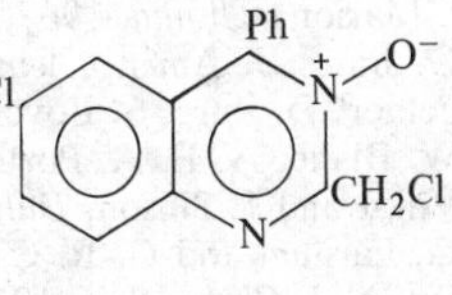 (213)

In an attempt to effect simple substitution reactions of 6-chloro-2-chloromethyl-4-phenylquinazoline 3-oxide (**213**), Sternbach and co-workers[196] made an important discovery. With dialkylamines the 2-dialkylaminomethyl compounds are formed in accord with predictions, but when primary alkylamines are used a ring expansion occurs, *e.g.* with

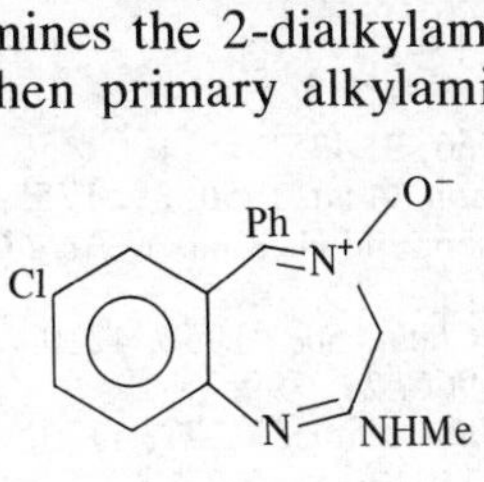

(214) 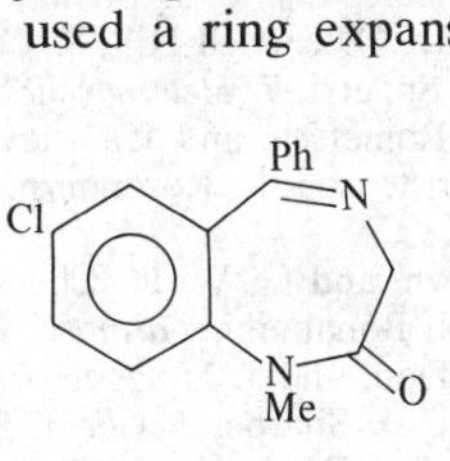 (215)

methylamine the product of solvolysis is 7-chloro-2-methylamino-5-phenyl-3*H*-1,4-benzodiazepin 4-oxide (**214**). Compound (**214**) and the related (**215**) are currently marketed under the well-known trade names librium and valium, respectively, and are probably amongst the most widely used tranquilizers currently in production.

References

1. M. Tisler and B. Stanovnik, *Adv. Heterocyclic Chem.*, 1968, **9,** 211.
2. J. D. Mason and D. L. Aldous, in 'Heterocyclic Compounds', ed. R. N. Castle, Wiley, New York, 1973, vol. 28.
3. J. Levisalles and P. Baranger, *Compt. rend.*, 1955, **240,** 444.
4. R. H. Mizzoni and P. E. Spoerri, *J. Amer. Chem. Soc.*, 1954, **76,** 2201.
5. M. Amorosa and L. Lipparini, *Ann. Chim.* (*Rome*), 1959, **49,** 322.
6. J. A. Giles and J. N. Schumacher, *Tetrahedron*, 1961, **14,** 246.
7. I. Kumashiro, *Nippon Kagaku Zasshi*, 1961, **82,** 928 (*Chem. Abs.* 1962, **57,** 11 183).
8. G. W. H. Cheeseman, *J. Chem. Soc.*, 1960, 242.
9. P. Schmidt and J. Druey, *Helv. Chim. Acta*, 1954, **37,** 134.
10. G. W. Kenner and A. Todd, in 'Heterocyclic Compounds', vol. 6, ed. R. C. Elderfield, Wiley, New York, 1957.
11. D. J. Brown, in 'The Chemistry of Heterocyclic Compounds', ed. A. Weissberger, Interscience, New York, 1962, vol. 16.
12. R. R. Hunt, J. F. M. McOmie, and E. R. Sayer, *J. Chem. Soc.*, 1959, 525.
13. C. A. C. Haley and P. Maitland, *J. Chem. Soc.*, 1951, 3155.
14. W. J. Hale, *J. Amer. Chem. Soc.*, 1914, **36,** 104.
15. R. R. Hunt, J. F. M. McOmie, and E. R. Sayer, *J. Chem. Soc.*, 1959, 527.
16. A. von Meyer, *J. prakt. Chem.*, 1889, **39,** 156.
17. H. L. Wheeler and D. F. McFarland, *Amer. Chem. J.*, 1910, **43,** 19.
18. R. O. Roblin, J. H. Williams, P. S. Winnek, and J. P. English, *J. Amer. Chem. Soc.*, 1940, **62,** 2002.
19. W. Traube, *Annalen*, 1923, **432,** 226.
20. R. Behrend, H. Meyer, and O. Buckholz, *Annalen*, 1901, **304,** 200.
21. O. Hromatka, *U.S. Pat.* 1941, 2 235 638 (*Chem. Abs.*, 1942, **35,** 4041).
22. B. Staskun and H. Stephen, *J. Chem. Soc.*, 1956, 4708.
23. G. W. Kenner, B. Lythgoe, A. Todd, and H. Topham, *J. Chem. Soc.*, 1943, 388.
24. H. Bredereck, R. Grompper, H. Schuh, and W. Theilig, *Angew. Chem.*, 1959, **71,** 573.
25. F. G. P. Remfry, *J. Chem. Soc.*, 1911, 610.
26. R. Hull, *J. Chem. Soc.*, 1951, 2214.
27. Bayer, *Ger. Pat.* 1906, 168 407.
28. Bayer, *Ger. Pat.* 1905, 163 136.
29. G. W. Kenner, B. Lythgoe, A. Todd, and H. Topham, *J. Chem. Soc.*, 1943, 574.
30. G. Shaw and R. N. Warrener, *J. Chem. Soc.*, 1958, 153.
31. G. Shaw and R. N. Warrener, *J. Chem. Soc.*, 1959, 50.
32. G. W. H. Cheeseman and E. S. G. Werstiuk, *Adv. Heterocyclic Chem.*, 1972, **14,** 99.
33. J. P. Marion, *Chimia*, 1967, **21,** 510.
34. R. G. Jones, *J. Amer. Chem. Soc.*, 1949, **71,** 78.
35. E. Felder, D. Pitre, S. Boveri, and E. B. Grabitz, *Chem. Ber.*, 1967, **100,** 555.
36. K. W. Blake, A. E. A. Porter, and P. G. Sammes, unpublished observation.
37. N. Vinot and J. Pinson, *Bull. Soc. chim. France*, 1968, 4970.
38. D. G. Farnum and G. R. Carlson, *Synthesis*, 1972, 191.
39. F. Tutun, *J. Chem. Soc.*, 1910, 2495.
40. E. Zbiral and J. Stroh, *Annalen*, 1969, **727,** 231.
41. K. W. Blake, A. E. A. Porter, and P. G. Sammes, *J. C. S. Perkin I*, 1972, 2494.
42. R. A. Baxter and F. S. Spring, *J. Chem. Soc.*, 1947, 1147.
43. M. L. Crossley and J. P. English, *U.S. Pat.* 1946, 2 394 963 (*Chem. Abs.*, 1946, **40,** 3143).
44. E. Ochiai and M. Yanai, *J. Pharm. Soc. Japan*, 1939, **59,** 18.
45. T. L. Chan and J. Miller, *Austral. J. Chem.*, 1967, **20,** 1595.
46. N. B. Chapman and D. Q. Russell-Hill, *J. Chem. Soc.*, 1956, 1563.
47. H. Rutner and P. E. Spoerri, *J. Heterocyclic Chem.*, 1966, **3,** 435.
48. K. Eichenberger, R. Rometsch, and J. Druey, *Helv. Chim. Acta*, 1956, **39,** 1755.
49. S. Linholter, A. B. Kristensen, R. Rosenoern, S. E. Nelsen, and H. Kaaber, *Acta Chem. Scand.*, 1961, **15,** 1660.
50. A. Albert, D. J. Brown, and G. W. H. Cheeseman, *J. Chem. Soc.*, 1952, 4219.
51. G. Palamidessi and N. Bonanomi, *Farmaco. Ed. Sci.*, 1966, **21,** 799.
52. G. Vernin, H. J. M. Dou, and J. Metzger, *Bull. Soc. chim. France*, 1972, 1173.
53. C. M. Atkinson and C. J. Sharpe, *J. Chem. Soc.*, 1959, 3040.
54. F. Minisci, G. P. Gardini, R. Galli, and F. Bertini, *Tetrahedron Letters*, 1970, 15.
55. G. P. Gardini, F. Minisci, G. Palla, A. Arnone, and R. Galli, *Tetrahedron Letters*, 1971, 59.
56. A. Hirschberg and P. E. Spoerri, *J. Org. Chem.*, 1961, **26,** 2356.
57. J. D. Behun and R. Levine, *J. Org. Chem.*, 1961, **26,** 3379.
58. M. R. Kamal and R. Levine, *J. Org. Chem.*, 1962, **27,** 1355.
59. A. Holland, *Chem. and Ind.* (*London*), 1954, 786.
60. R. L. Letsinger and R. Lasco, *J. Org. Chem.*, 1956, **21,** 812.
61. I. Crossland and L. K. Ramussen, *Acta Chem. Scand.*, 1965, **19,** 1652.

62. H. Bredereck, R. Gompper, and H. Herlinger, *Angew. Chem.*, 1958, **70,** 571.
63. M. D. Potter and T. Henshall, *J. Chem. Soc.*, 1956, 2000.
64. S. K. Chakrabartty and R. Levine, *J. Heterocyclic Chem.*, 1967, **4,** 109.
65. G. P. Rizzi, *J. Org. Chem.*, 1968, **33,** 1333.
66. V. K. Smith, P. River, and S. Kushner, *U.S. Pat.*, 1954, 2 677 686.
67. E. Fischer and H. Roeder, *Ber.*, 1901, **34,** 3751.
68. S. Gabriel, *Ber.*, 1905, **38,** 630.
69. P. A. Lavene and F. B. La Forge, *Ber.*, 1912, **45,** 608.
70. H. Burrows and C. A. Keane, *J. Chem. Soc.*, 1907, 269.
71. W. R. Boon, H. C. Carrington, N. Greenhalgh, and C. H. Vasey, *J. Chem. Soc.*, 1954, 3263.
72. A. Einhorn and H. Von Diesbach, *Ber.*, 1907, **40,** 4902.
73. C. Mannich and G. Heilner, *Ber.*, 1922, **55,** 365.
74. K. Folkers and T. B. Johnson, *J. Amer. Chem. Soc.*, 1933, **55,** 3361.
75. H. Hoffmann, *Ber.*, 1888, **21,** 2332.
76. S. R. Aspinall, *J. Amer. Chem. Soc.*, 1940, **62,** 2160.
77. J. Tafel and H. B. Thomson, *Ber.*, 1908, **40,** 4489.
78. G. E. K. Branch, *J. Amer. Chem. Soc.*, 1916, **38,** 2466.
79. R. A. Sulzbach and A. F. M. Iqbal, *Angew. Chem. Internat. Edn.*, 1971, **10,** 127.
80. H. I. X. Mager and W. Behrends, *Rec. Trav. chim.*, 1960, **79,** 282.
81. C. L. Stevens, K. G. Taylor, and M. E. Munk, *J. Org. Chem.*, 1964, **29,** 3574.
82. T. Kato, Y. Yamanaka, and T. Shibata, *J. Pharm. Soc. Japan*, 1967, **87,** 1096.
83. M. Asai, *J. Pharm. Soc. Japan*, 1959, **79,** 1273.
84. M. Ogata and H. Kano, *Chem. Pharm. Bull.* (*Japan*), 1963, **11,** 29.
85. M. Ogata and H. Kano, *Chem. Pharm. Bull.* (*Japan*), 1963, **11,** 35.
86. M. Ogata, H. Watanabe, K. Tori, and H. Kano, *Tetrahedron Letters*, 1964, 19.
87. T. Nakagome, *Chem. Pharm. Bull.* (*Japan*), 1963, **11,** 721.
88. E. Hayashi, T. Higashino, C. Iijima, Y. Kono, and T. Doihara, *Yakugaku Zasshi*, 1962, **82,** 584.
89. H. Igeta, *Chem. Pharm. Bull.* (*Japan*), 1959, **7,** 938.
90. K. W. Blake and P. G. Sammes, *J. Chem. Soc.* (*C*), 1970, 1070.
91. T. J. Delia, D. E. Portlock, and D. L. Venton, *J. Heterocyclic Chem.*, 1968, **5,** 449.
92. A. Ya. Tikhonov and L. B. Volodarsky, *Tetrahedron Letters*, 1975, 2721.
93. S. Spyroudis and A. Vargvoglis, *Synthesis*, 1976, 837.
94. H. Yamanaka, *Chem. Pharm. Bull.* (*Japan*), 1959, **7,** 158.
95. T. Itai and S. Natsume, *Chem. Pharm. Bull.* (*Japan*), 1963, **11,** 342.
96. M. Hamana, *Yakugaku Zasshi*, 1951, **71,** 263.
97. T. Nakagome, *Yakugaku Zasshi*, 1963, **82,** 253.
98. M. Ogata and H. Kono, *Chem. Pharm. Bull.* (*Japan*), 1963, **11,** 29.
99. E. Ochiai and C. Kaneko, *Chem. Pharm. Bull.* (*Japan*), 1957, **5,** 56.
100. H. Igeta, M. Yamada, Y. Yoshioka, and Y. Kawazoe, *Chem. Pharm. Bull.* (*Japan*), 1967, **15,** 1411.
101. G. Palamidessi and L. Bernardi, *Gazzetta*, 1963, **93,** 339.
102. J. Cullen and D. Harrison, *J. Chem. Soc.* (*C*), 1966, 495.
103. S. Sako and T. Itai, *Chem. Pharm. Bull.* (*Japan*), 1966, **14,** 269.
104. S. Sako, *Chem. Pharm. Bull.* (*Japan*), 1962, **10,** 956.
105. M. Ogata, *Chem. Pharm. Bull.* (*Japan*), 1963, **11,** 1511.
106. B. Klein, N. E. Hetman, and M. E. O'Donnell, *J. Org. Chem.*, 1963, **28,** 1682.
107. H. Bredereck, R. Grompper, and H. Herlinger, *Chem. Ber.*, 1958, **91,** 2832.
108. R. Huisgen, *Angew. Chem.*, 1965, **75,** 628.
109. E. Ochiai and Y. Yamanaka, *Chem. Pharm. Bull.* (*Japan*), 1955, **3,** 175.
110. R. B. Augier and W. V. Corrau, *J. Org. Chem.*, 1961, **26,** 1891.
111. A. S. Elina and I. S. Musatova, *Khim. geterotsikl. Soedinenii*, 1967, 419 (*Chem. Abs.*, 1969, **70,** 87 747).
112. R. G. Micetich and J. C. Macdonald, *J. Chem. Soc.*, 1964, 1507.
113. A. J. Klyver, J. P. van der Walt, and A. J. van Triet, *Proc. Nat. Acad. Sci. U.S.A.*, 1953, **39,** 583.
114. R. F. C. Brown, W. D. Crow, and R. K. Solly, *Chem. and Ind.* (*London*), 1966, 343.
115. T. J. Schwan and H. Tiekelmann, *J. Org. Chem.*, 1964, **29,** 941.
116. H. C. van der Plas and G. Geurtsen, *Tetrahedron Letters*, 1964, 2093.
117. T. Kauffmann, J. Hansen, K. Udluft, and R. Wirthwein, *Angew. Chem.*, 1964, **76,** 590.
118. T. Kauffmann and A. Risberg, *Tetrahedron Letters*, 1963, 1459.
119. D. T. Moury, *J. Amer. Chem. Soc.*, 1953, **75,** 1909.
120. P. A. Lavene and F. B. La Forge, *Ber.*, 1912, **45,** 608.
121. Y. A. Tota and R. Elderfield, *J. Org. Chem.*, 1942, **7,** 313.
122. G. Palamidessi and L. Panizzi, *Fr. Pat.* 1964, 1 372 807 (*Chem. Abs.*, 1965, **62,** 1674).
123. G. Karmas and P. E. Spoerri, *J. Amer. Chem. Soc.*, 1957, **78,** 680.
124. G. W. H. Cheeseman, personal communication.
125. K. W. Blake and P. G. Sammes, *J. Chem. Soc.* (*C*), 1970, 980.
126. G. A. Jeffery, S. Ghose, and J. O. Warwicker, *Acta Cryst.*, 1961, **14,** 881.
127. P. Cucka, *Acta Cryst.*, 1963, **16,** 318.
128. R. A. Godwin, Ph.D. Thesis, London University, 1970.

129. F. Arndt, *Angew. Chem.*, 1949, **61,** 397.
130. D. G. McMinn, *Synthesis*, 1976, 824.
131. R. O. Cain and A. E. A. Porter, unpublished observation.
132. G. E. Hilbert and T. B. Johnson, *J. Amer. Chem. Soc.*, 1930, **52,** 2001.
133. J. Pliml and F. Sorm, *Coll. Czech. Chem. Comm.*, 1965, **30,** 3744.
134. A. E. A. Porter and P. G. Sammes, *Chem. Comm.*, 1970, 1103.
135. P. J. Machin, A. E. A. Porter, and P. G. Sammes, ***J.C.S. Perkin I,*** 1973, 404.
136. P. G. Sammes and R. A. Watt, *J. C. S. Chem. Comm.*, 1975, 502.
137. R. O. Roblin, P. S. Winnek, and J. P. English, *J. Amer. Chem. Soc.*, 1942, **64,** 567.
138. P. Coad, R. A. Coad, S. Clough, J. Hyepock, R. Salisbury, and C. Wilkins, *J. Org. Chem.*, 1963, **28,** 218.
139. G. W. H. Cheeseman, *J. Chem. Soc.*, 1960, 242.
140. K. W. Blake, Ph.D. Thesis, London University, 1970.
141. C. Pedrali and A. Montegani, *J. Org. Chem.*, 1958, **23,** 778.
142. T. Matsukawa and M. Ohta, *J. Pharm. Soc. Japan*, 1949, **69,** 491.
143. A. E. Erikson and P. E. Spoerri, *J. Amer. Chem. Soc.*, 1946, **68,** 400.
144. R. D. Chambers, J. A. H. MacBride, and W. K. R. Musgrave, *Chem. and Ind.* (*London*), 1966, 1721.
145. R. D. Chambers, J. A. H. MacBride, and W. K. R. Musgrave, *Chem. and Ind.* (*London*), 1966, 904.
146. C. G. Allison, R. D. Chambers, J. A. H. MacBride, and W. K. R. Musgrave, *J. Chem. Soc.* (*C*), 1970, 1203.
147. H. Rutner and P. E. Spoerri, *J. Heterocyclic Chem.*, 1965, **2,** 492.
148. A. Hirschberg and P. E. Spoerri, *J. Org. Chem.*, 1961, **26,** 1907.
149. H. Rutner and P. E. Spoerri, *J. Heterocyclic Chem.*, 1966, **3,** 435.
150. G. W. Anderson, H. E. Faith, H. W. Marson, P. S. Winnek, and R. O. Roblin, *J. Amer. Chem. Soc.*, 1942, **64,** 2902.
151. T. V. Gortinskaya and M. N. Shchukina, *Zhur. obshchei Khim.*, 1960, **30,** 1518.
152. B. Camarino and G. Palamidessi, *Gazzetta*, 1960, **90,** 1807.
153. C. M. Atkinson and R. E. Rodway, *J. Chem. Soc.*, 1959, 1.
154. R. Urban and O. Schnider, *Helv. Chim. Acta*, 1958, **41,** 1806.
155. A. R. Katritzky and J. M. Lagowski, *Adv. Heterocyclic Chem.*, 1962, **1,** 131.
156. R. A. Baxter, G. T. Newbold, and F. S. Spring, *J. Chem. Soc.*, 1947, 370.
157. A. Dell, D. H. Williams, H. R. Morris, G. A. Smith, J. Feeney, and G. C. K. Roberts, *J. Amer. Chem. Soc.*, 1975, **97,** 2497.
158. T. L. Jacobs, in 'Heterocyclic Chemistry', vol. 6, ed. R. C. Elderfield, Interscience, London and New York, 1957.
159. J. C. E. Simpson, in 'The Chemistry of Heterocyclic Compounds', vol. 5, ed. A. Weissberger, Interscience, New York, 1953.
160. G. M. Singerman, in 'The Chemistry of Heterocyclic Compounds', vol. 27, ed R. N. Castle, Interscience, New York, 1972.
161. J. S. Morley, *J. Chem. Soc.*, 1951, 1971.
162. K. Schofield and J. C. E. Simpson, *J. Chem. Soc.*, 1945, 512.
163. H. J. Barber, K. Washbourne, W. R. Wragg, and E. Lunt, *J. Chem. Soc.*, 1961, 2828.
164. S. Gabriel and G. Pinkus, *Ber.*, 1893, **26,** 2210.
165. A. Hirsch and D. Orphanos, *J. Heterocyclic Chem.*, 1965, **2,** 206.
166. J. Sauer and G. Heinrichs, *Tetrahedron Letters*, 1966, 4979.
167. W. L. F. Armarego, *J. Chem. Soc.*, 1962, 561.
168. W. L. F. Armarego, in 'The Chemistry of Heterocyclic Compounds', vol. 24, ed. D. J. Brown, Interscience, New York, 1967.
169. J. F. Meyer and E. C. Wagner, *J. Org. Chem.*, 1943, **8,** 239.
170. E. C. Taylor, R. J. Knopf, and A. L. Borror, *J. Amer. Chem. Soc.*, 1960, **82,** 3152.
171. R. J. Grout and M. W. Partridge, *J. Chem. Soc.*, 1960, 3546.
172. A. Albert and A. Hampton, *J. Chem. Soc.*, 1954, 505.
173. J. C. Cavagnol and F. Y. Wiselogle, *J. Amer. Chem. Soc.*, 1947, **69,** 795.
174. G. W. H. Cheeseman, *Adv. Heterocyclic Chem.*, 1963, **2,** 203.
175. L. Horner, U. Schwenk, and E. Junghanus, *Annalen*, 1953, **579,** 212.
176. A. Albert, D. J. Brown, and D. Wood, *J. Chem. Soc.*, 1954, 3832.
177. A. Albert, W. L. F. Armarego, and E. Spinner, *J. Chem. Soc.*, 1961, 2689.
178. M. J. S. Dewar and P. M. Maitlis, *J. Chem. Soc.*, 1957, 2521.
179. J. S. Morley, *J. Chem. Soc.*, 1951, 1971.
180. S. Kanahara, *Yakugaku Zasshi*, 1964, **84,** 489.
181. M. J. S. Dewar and P. M. Maitlis, *J. Chem. Soc.*, 1957, 2518.
182. A. S. Elina, *J. Gen. Chem. USSR*, 1959, **29,** 2728.
183. M. J. S. Dewar and P. M. Maitlis, *J. Chem. Soc.*, 1957, 2521.
184. R. C. Elderfield, T. A. Williamson, W. J. Gensler, and C. B. Kremer, *J. Org. Chem.*, 1947, **12,** 405.
185. J. H. Ridd, in 'Physical Methods In Heterocyclic Chemistry', vol. I, ed. A. R. Katritzky, Academic Press, London and New York, 1963.
186. T. Higashino, *J. Pharm. Soc. Japan*, 1960, **80,** 1563.
187. N. B. Chapman and D. Q. Russell-Hill, *J. Chem. Soc.*, 1956, 1563.

188. J. R. Keneford, J. S. Morley, J. C. E. Simpson, and P. H. Wright, *J. Chem. Soc.*, 1950, 1104.
189. R. A. Scherrer and H. R. Beatty, *J. Org. Chem.*, 1972, **37,** 1681.
190. B. G. Murray, D. G. Parsons, and A. F. Turner, *Brit. Pat.* 1968, 1 133 406.
191. R. D. Chambers, J. A. H. MacBride, W. K. R. Musgrave, and I. S. Reilly, *Tetrahedron Letters*, 1970, 55.
192. A. Mustafa, A. R. Harhash, and A. A. S. Saleh, *J. Amer. Chem. Soc.*, 1960, **82,** 2735.
193. A. Marxer, *Helv. Chim. Acta*, 1966, **49,** 572.
194. J. Siegel and B. E. Christensen, *J. Amer. Chem. Soc.*, 1951, **73,** 5777.
195. C. M. Atkinson and J. C. E. Simpson, *J. Chem. Soc.*, 1947, 1469.
196. L. H. Sternbach and F. Reeder, *J. Org. Chem.*, 1961, **26,** 1111.

16.3
Triazines and Tetrazines

A. E. A. PORTER
University of Stirling

16.3.1 TRIAZINES

Six-membered aromatic rings containing three nitrogen atoms are known as triazines, and there are three possible arrangements of the nitrogen atoms in the ring as seen in the 1,2,3-triazines (**1**), the 1,2,4-triazines (**2**), and the 1,3,5-triazines (**3**). It is of interest to note that whilst 1,3,5-triazines are amongst the oldest known compounds, the first authenticated 1,2,3-triazine derivative was not prepared until as recently as 1960. 1,3,5-Triazines are sometimes referred to as the *sym* (symmetrical) isomer and 1,2,4-triazines as the *asym* (asymmetrical) form; 1,2,3-triazines (often present in fused systems) are often referred to as the *vic* (vicinal) isomers.

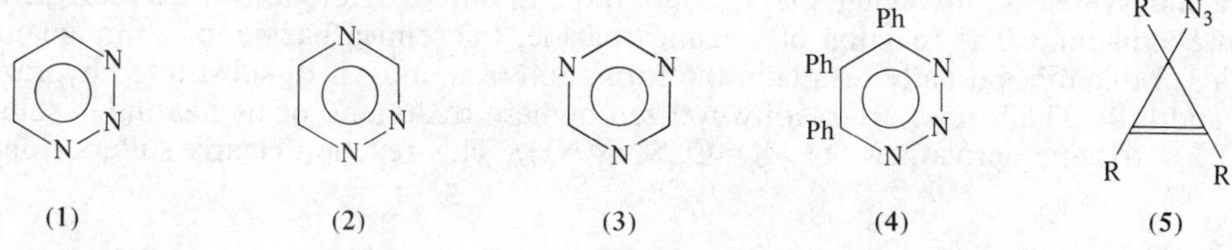

(1) (2) (3) (4) (5)

Of the three parent triazines, only 1,3,5-triazine is known. It exists as a colourless crystalline compound melting at 85 °C and boiling at 114 °C. It shows characteristic u.v. maxima at 272 (π–π^*) and 320 (n – π^*) nm.

16.3.1.1 1,2,3-Triazines

1,2,3-Triazine has not been prepared and until 1960 there were no proven examples of monocyclic 1,2,3-triazines. The known propensity of fused 1,2,3-triazines to undergo degradation under mild conditions indicates that if this ring system is amenable to practical synthesis then relatively mild conditions must be employed.

The first derivative of 1,2,3-triazine to be prepared, 4,5,6-triphenyl-1,2,3-triazine (**4**) was described[1] by Chandros and Smolinsky;[2] the method of synthesis involved the thermolysis of 1,2,3-triphenylcyclopropenyl azide (**5**; R = Ph), and this reaction has been shown to be general, culminating in the synthesis of several alkyl- and aryl-1,2,3-triazines.

The prospect of heating organic azides is somewhat daunting to most organic chemists; however, the reaction appears to occur under extremely mild conditions, as evidenced by the conversion of 1,2,3-trimethylcyclopropenyl azide (**5**; R = Me) to 4,5,6-trimethyl-1,2,3-triazine, which occurs on allowing the azide to stand at room temperature for two days. The final proof of the 1,2,3-triazine structure has come from a complete X-ray crystallographic analysis of tri(*p*-methoxyphenyl)-1,2,3-triazine.[3]

Little is known about the general chemistry of the monocyclic 1,2,3-triazines and the only reactions which have been reported are those of hydrolysis, which results in the formation of 1,3-diketones, thermolysis, and photolysis. The products of thermolysis are to a degree dependent on substituents. Thus 4,5,6-triphenyl-1,2,3-triazine (**4**) leads to diphenylindanone imine (**6**) and nitrogen, whereas 4,5,6-trimethyl-1,2,3-triazine yields but-2-yne and acetonitrile. Photolysis usually favours the latter pathway to yield acetylenes, nitriles, and molecular nitrogen.

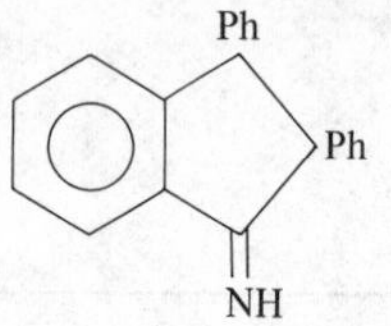

Ph Ph NH

(6)

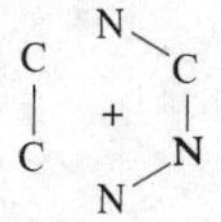

C C N C N N +

(7)

N C C C N N +

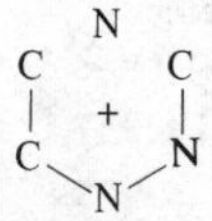

(8)

16.3.1.2 1,2,4-Triazines

The parent of this class has never been prepared, although in part this has probably been due to lack of interest rather than the complexity of the problem, and a number of simple possibilities for the conversion of readily available 1,2,4-triazine derivatives to 1,2,4-triazine exist.[4] Quantum mechanical calculations predict the 1,2,4-triazine ring to be stable with a resonance energy of 75 kJ mol^{-1}.

Two basic strategies have been used to construct the 1,2,4-triazine ring as illustrated by (**7**) and (**8**). The first of these is not conceptually dissimilar from the common approach to pyrazine synthesis, involving the condensation of an α-diketone with a derivative of ethylenediamine. The reaction of a semicarbazide, thiosemicarbazide, or aminoguanidine with 1,2-diketones usually results in the formation of mono- or di-substituted hydrazones (**9**) and (**10**). The former are readily cyclized by base treatment, or by heating in solution, to 1,2,4-triazine derivatives (**11**; X = O, S, or NH). This reaction clearly suffers from the

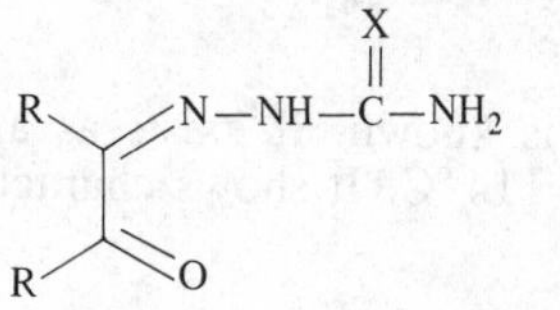

R R N—NH—C(=X)—NH$_2$ O

(9)

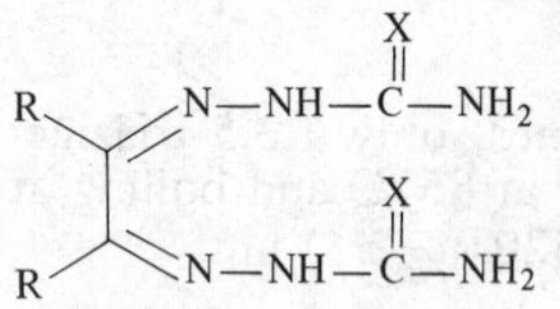

R R N—NH—C(=X)—NH$_2$ N—NH—C(=X)—NH$_2$

(10)

R N N XH N

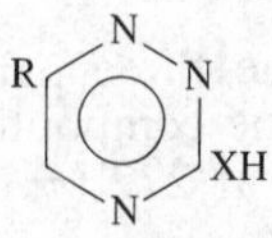

(11)

limitation that if the starting 1,2-diketone contains different alkyl or aryl residues, then two products may arise from the condensation. In some instances it is possible to control which isomer is produced, as seen in the reaction of phenylglyoxal, which under acidic conditions gives rise to the 6-phenyl-1,2,4-triazine and under basic conditions (Scheme 1) the isomeric 5-phenyl-1,2,4-triazine.

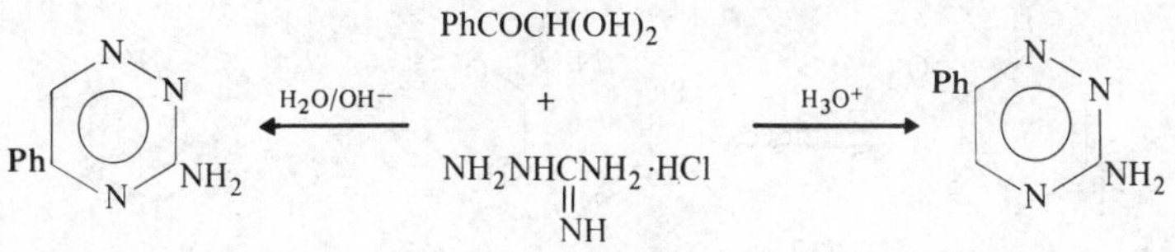

SCHEME 1

The second synthetic strategy involves the reaction of an α-diketone with an acid hydrazide to form an intermediate hydrazone (Scheme 2), which on subsequent treatment with ammonia undergoes cyclization to the triazine in a manner reminiscent of the final ring closure of the Hantzsch pyridine synthesis. When the 1,2-diketone is unsymmetrical,

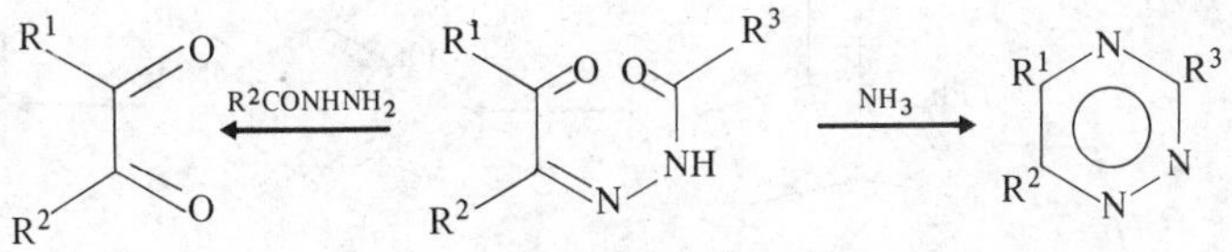

SCHEME 2

i.e. $R^1 \neq R^2$, the most reactive carbonyl group initially forms the hydrazone and determines the substitution pattern of the triazine, although, in a situation where there is only a marginal difference in the reactivity of the carbonyl groups, mixed products are to be expected.

In an important modification of this reaction, Saraswathi and Srinivasan[5] have used α-haloketones in place of α-diketones, which appears to give greater control over the selectivity of the reaction. Reaction of the α-haloketone with two equivalents of the acid hydrazide (Scheme 3) initially results in the formation of the hydrazone, which undergoes

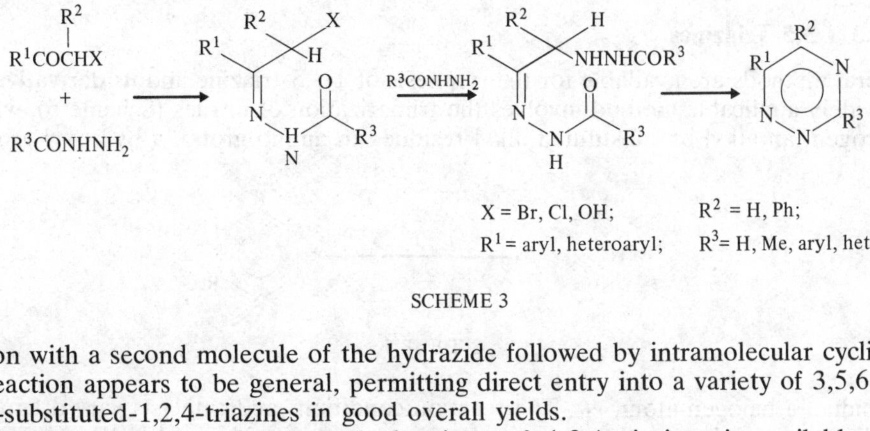

X = Br, Cl, OH; R^2 = H, Ph;
R^1 = aryl, heteroaryl; R^3 = H, Me, aryl, heteroaryl;

SCHEME 3

reaction with a second molecule of the hydrazide followed by intramolecular cyclization. The reaction appears to be general, permitting direct entry into a variety of 3,5,6-, 3,6-, and 6-substituted-1,2,4-triazines in good overall yields.

Little information on the general chemistry of 1,2,4-triazines is available as the synthesis of a given derivative usually involves direct ring synthesis from acyclic precursors rather than post-cyclizational modification of the ring. The 3-hydroxy- and 3-mercapto-1,2,4-triazines (**11**) are thought to exist in the amide and thioamide forms and alkylation occurs specifically on N-2. 5,6-Diphenyl-3,4-dihydro-3-oxo-1,2,4-triazine (**12**) appears to undergo most of the simple functional group manipulations expected: treatment with phosphoryl chloride yields the 3-chloro-derivative (**13**), which may be transformed into alkoxy- or amino-derivatives under standard conditions.

In 1969 Neunhoffer *et al.*[6] established that 1,2,4-triazines undergo facile addition reactions with electron-rich olefins. Thus reaction of 5-phenyl-1,2,4-triazine with 1-dimethylamino-1-ethoxyethylene resulted in the formation of pyridine derivatives, and by

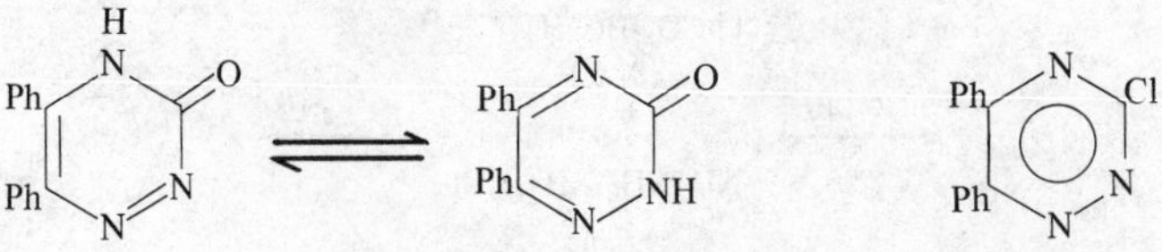

(12) (13)

analogy with the reaction of 1,2,4,5-tetrazines under similar conditions they proposed a mechanism involving an initial Diels–Alder reaction (Scheme 4). The reaction with electron-rich acetylenes follows a different course, giving rise to pyrimidine derivatives, and two alternative mechanisms[7] have been proposed.

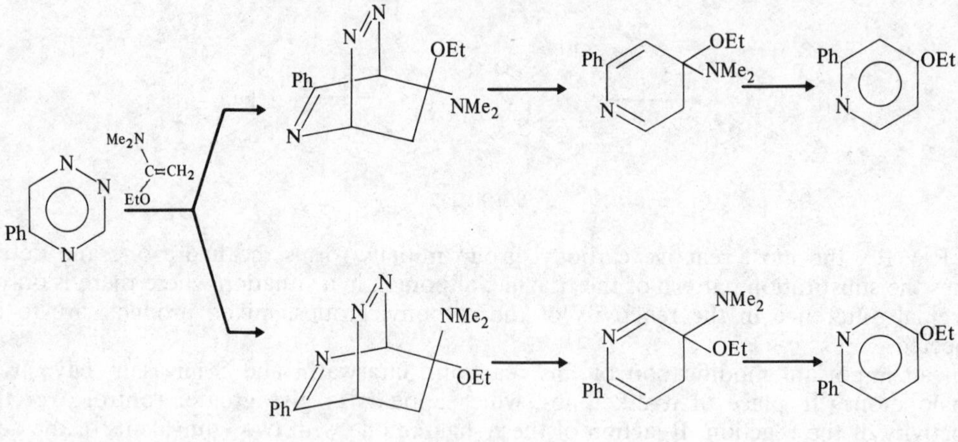

SCHEME 4

16.3.1.3 1,3,5-Triazines

Several methods are available for the synthesis of 1,3,5-triazine and its derivatives. The most widely applicable method involves the trimerization of nitriles (Scheme 5), where R is hydrogen, an alkyl or substituted alkyl residue, an amino group, a hydroxyl group, an

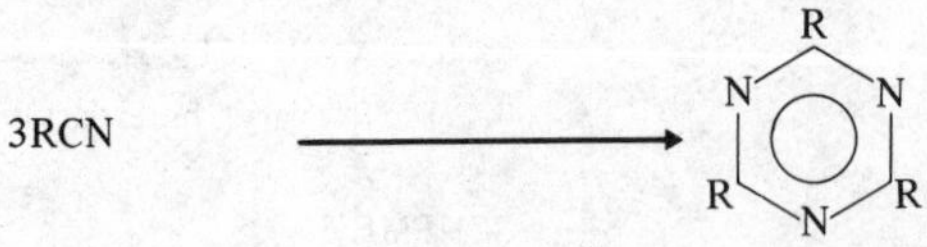

3RCN ⟶

SCHEME 5

aryl residue, a halogen atom, *etc.* The reaction conditions and catalysts vary dramatically depending on R and more than one mechanism is probably involved in the trimerization reaction.

Aromatic and aliphatic nitriles may be trimerized by the application of high pressure alone, although heat is generally required. More frequently some form of catalyst is used, and catalysts ranging from anhydrous hydrogen chloride, sodium metal, Grignard reagents, chlorosulphonic acid, and triphenylmethylsodium have found application.

1,3,5-Triazine itself has been prepared by hydrogen chloride catalysed trimerization of hydrogen cyanide. The reaction appears to be complex and proceeds through the intermediacy[8] of the so-called sesquihydrochloride of hydrogen cyanide, 2HCN·3HCl, which is formed by the reaction of HCl with HCN. Overall yields of 1,3,5-triazine are as high as 60%.

Cyanic acid (formed *in situ* by the reaction of potassium cyanate with sulphuric acid) undergoes spontaneous trimerization to yield cyanuric acid (**14**), which is formally 2,4,6-trihydroxy-1,3,5-triazine (**15**), although X-ray evidence confirms the triketo-structure (**14**). The monohydroxy-form (**16**) is probably present to a small extent in solution, which accounts for the acidity of cyanuric acid (*cf.* barbituric acid). Cyanogen chloride and cyanamide on polymerization yield cyanuric chloride (**17**) and melamine (**18**), respectively.

The principal problem with all of the syntheses involving the direct polymerization of a nitrile is that the triazine produced is symmetrically substituted, *viz.* 2,4,6-trisubstituted. If different substituents are required at the 2-, 4-, or 6-positions, then the 'trimerization method' is not strictly applicable.

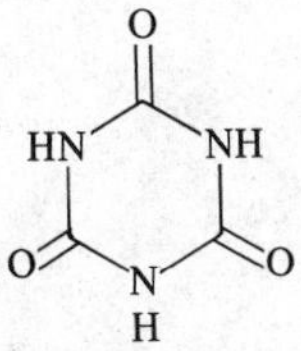

(14) 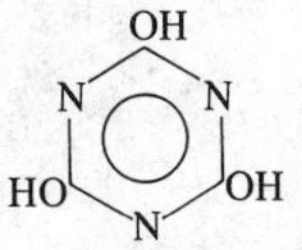 (15) 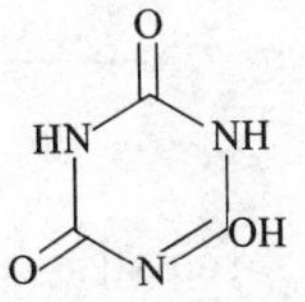 (16) 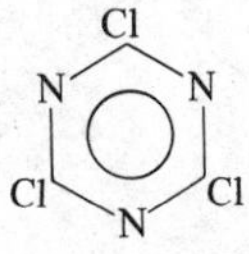 (17) 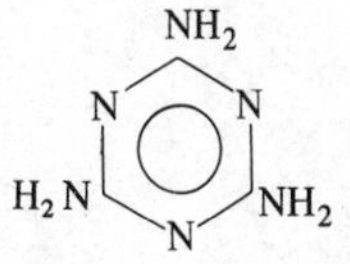 (18)

An attempt to circumvent this problem by the polymerization of mixed nitriles has met with some limited success. Thus trichloroacetonitrile and acetonitrile may be copolymerized at −15 °C, in the presence of hydrogen chloride, to yield 2-methyl-4,6-bistrichloromethyl-1,3,5-triazine (**19**), but the reaction appears to be limited in application.

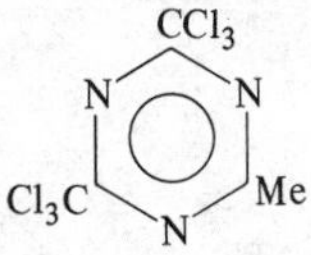

(19) 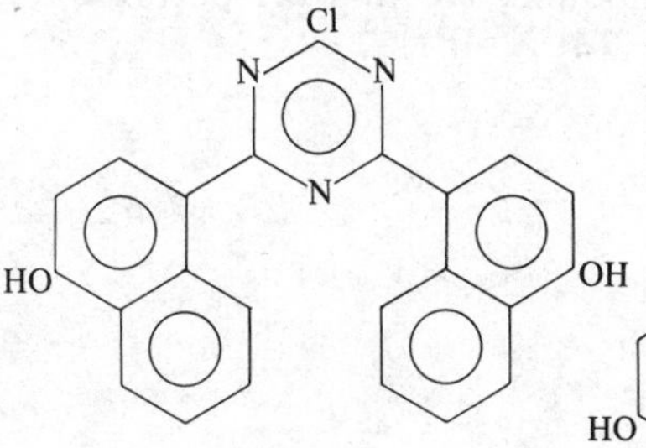 (20) 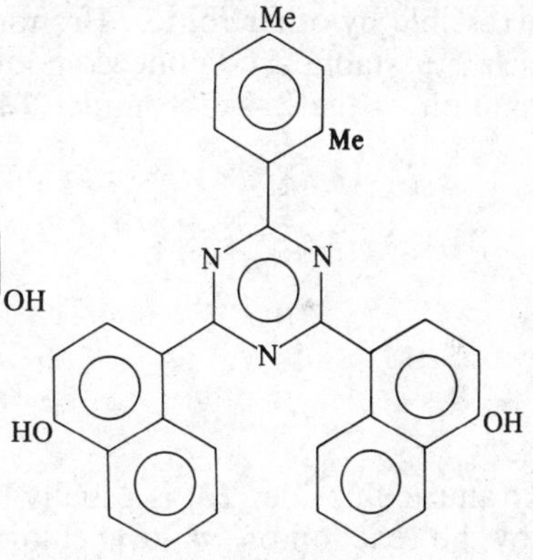 (21)

Cyanuric chloride (**17**) reacts with activated aromatic substrates under Friedel–Crafts conditions, and unsymmetrically substituted 1,3,5-triazines have been prepared in this way. Treatment of cyanuric chloride with 2-equivalents of β-naphthol in the presence of $AlCl_3$ gave the diarylated triazine (**20**), which on subsequent reaction with m-xylene in the presence of $AlCl_3$ gave (**21**).

In 1960 it was established that triazines react with amidines under conditions which permit the successive introduction of alkyl groups into the triazine ring (Scheme 6). This reaction appears to offer a unique method for the successive introduction of differing alkyl residues into the triazine ring and may be applicable to the synthesis of amino and other substituted triazines.

(*i*) *Electrophilic substitution*

The facility with which electrophilic substitution occurs in π-deficient heterocyclic species generally decreases on increasing the number of hetero-atoms in the ring and for

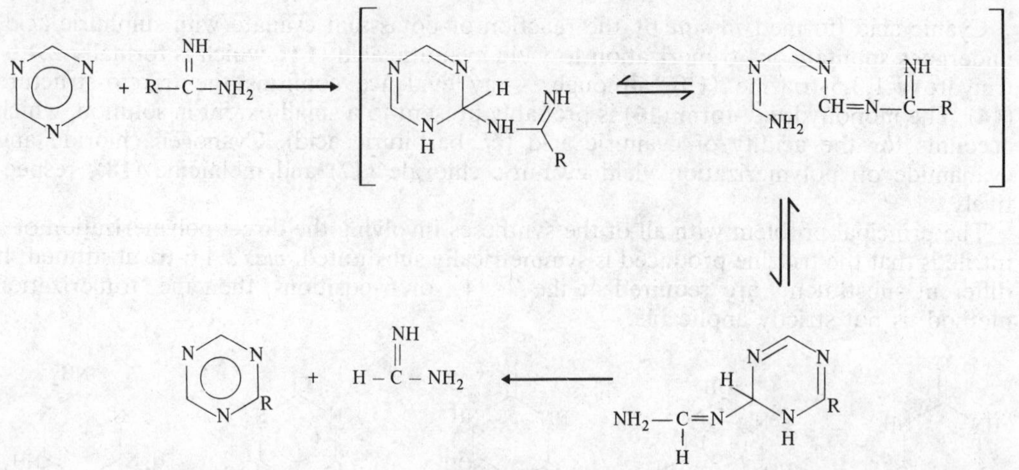

SCHEME 6

this reason the triazines would be expected to be extremely inert. The only reported attempts at substitution of 1,3,5-triazines by electrophiles relate to halogenation reactions. The reaction of chlorine with 1,3,5-triazine in carbon tetrachloride results in the formation of an insoluble hygroscopic solid believed to be the perchloride. At 140–200 °C in a sealed tube, chlorine in CCl_4 results in the formation of cyanuric chloride (**17**) and 2,4-dichloro-1,3,5-triazine (**22**; X = Cl) in 25 and 4% yields, respectively. This method is only of importance in the preparation of 2,4-dichloro-1,3,5-triazines, which are relatively inaccessible by other routes. Bromine in CCl_4 results in the formation of the perbromide, which is a stable crystalline compound. Heating the perbromide at 150 °C results in the formation of the 2,4-dibromide (**22**; X = Br).

PhNH—N=CH—N=NPh

(22) (23) (24)

Cyanuric fluoride (**23**) is usually prepared by the reaction of cyanuric chloride with SF_4 or by the reaction of cyanuric chloride with HF at −78 °C and then at 0 °C rather than by direct substitution of the triazine ring.

(*ii*) *Nucleophilic substitution*

Nucleophiles attack the 1,2,3-triazine ring with great ease; even a 10% solution in water[9] at 25 °C is almost completely hydrolysed to ammonium formate after 10 minutes. The ring is readily cleft with primary amines or sodamide, which yield *N,N′*-disubstituted formamidines and sodium cyanamide, respectively. Phenylhydrazine reacts to yield *N,N*-diphenylformazan (**24**).

Cyanuric chloride (**17**) is by far the most important derivative of 1,3,5-triazine in terms of nucleophilic substitution reactions. The halogen atoms are displaced under normal substitution conditions with the appropriate nucleophile and, as expected, the ease with which they are displaced progressively decreases. Cyanuric chloride reacts with alcohols at 30 °C in the presence of sodium bicarbonate to produce the monoalkoxydichlorotriazine (Scheme 7). When the temperature is increased to 70 °C, the dialkoxy-derivative is formed. The third chlorine atom is generally resistant to displacement and requires alkoxides in order to produce the trialkoxytriazine.

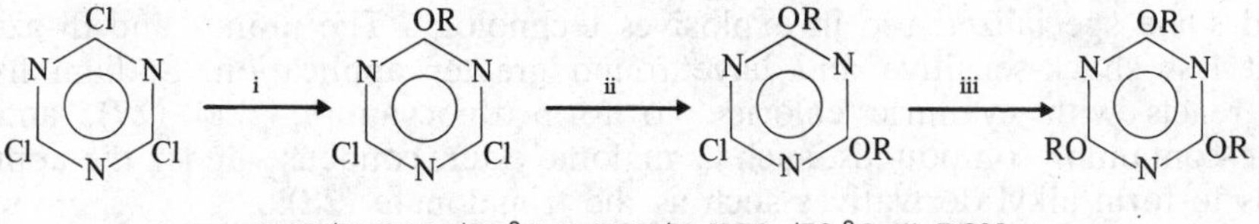

i, $ROH/NaHCO_3$/30 °C; ii, $ROH/NaHCO_3$/70 °C; iii, RONa.

SCHEME 7

During the course of the substitution reactions the alcohol is usually used in excess, but when the alcohol (primary, secondary, or tertiary) is heated somewhat below the boiling point with an excess of cyanuric chloride the corresponding alkyl chlorides are produced in high yields.[10] This reaction is of preparative value in organic synthesis.

The reaction of cyanuric chloride with amines also occurs in a stepwise manner; with dialkylamines the first chlorine atom is easily displaced at subambient temperatures. On warming the mixture of amine and monoaminated triazine to 50 °C the second chlorine atom is displaced and reaction at 100 °C results in the formation of 2,4,6-trisdialkylamino-triazine.

Cyanuric chloride is used extensively in the dyestuffs industry, where it displays two main benefits. It is possible to introduce separately two different dyestuffs with different colouring properties into the triazine ring. The use of cyanuric chloride is not merely a convenient procedure for the combination of azo and other dye units to form more complex dyes, but special properties have been claimed for cyanurated dyes. The dyes once produced enjoy an improved purity of shade and colouring properties and usually show a higher affinity for cellulose fibres. These principles are adequately illustrated in the production of the dyestuff Chlorantine Fast Green (**25**). A blue and a yellow component are joined by means of a triazine ring to give a green shade and the remaining chlorine atom is displaced with aniline to produce the final dyestuff.

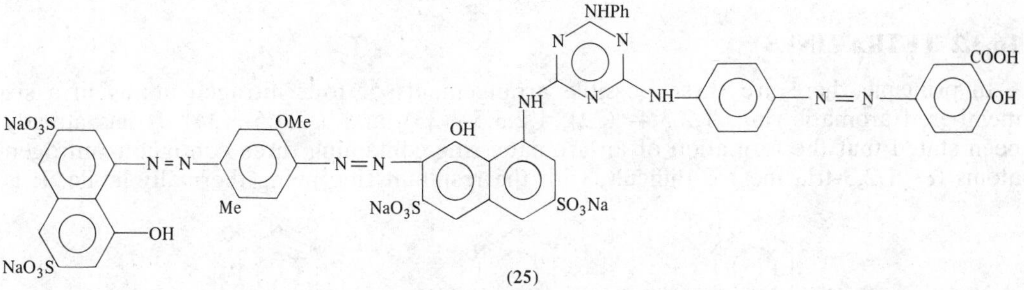

(25)

More recently the third carbon–chlorine bond has been left intact and the disubstituted derivative of cyanuric acid then retains an active site, which becomes a focus of attack by hydroxyl or amino-groups present in the fibre which is being dyed (*e.g.* cotton fibres). Treatment of the fibre with the dyestuff under controlled conditions of pH and temperature then binds the dyestuff chemically to the fibre, giving rise to truly 'fast' colours which do not fade and possess other highly desirable properties.

Cyanuric chloride reacts with many of the more common nucleophiles. Cyanuric triazide (**26**) is formed directly by the reaction of cyanuric chloride with sodium azide. The azide is one of the most shock-sensitive compounds known and may detonate spontaneously or on heating. Although it is too sensitive to use in conventional detonation caps, it

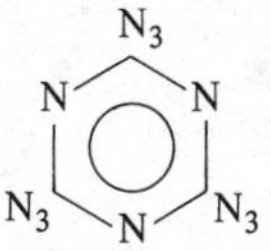

(26) 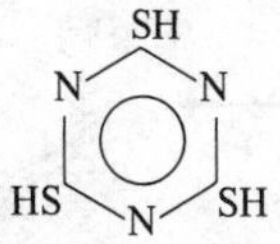 (27) 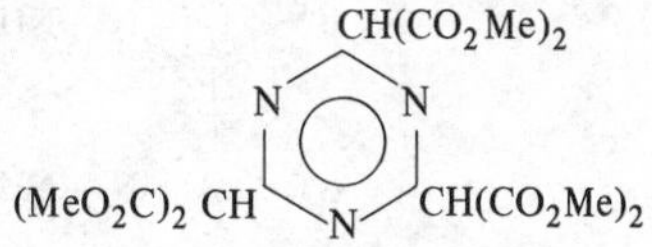 (28)

has found some specialized use in explosives technology. The mono- and di-azides are somewhat less shock-sensitive and have found greater application. Sodium hydrogen sulphide reacts with cyanuric chloride to form thiocyanuric acid (**27**), and active methylene-containing compounds, such as malonic ester, condense under the appropriate conditions to form alkyl derivatives such as the trimalonate (**28**).

Although *N*-alkylmelamine derivatives are usually prepared by the reaction of cyanuric chloride with amines, melamine (**18**) is more conveniently prepared by direct synthesis from dicyandiamide on an industrial scale. Melamine is used commercially for the synthesis of melamine/formaldehyde thermosetting resins. The characteristics of the resins depend on the polymerization conditions, but hydroxymethylmelamines such as (**29**) are probably produced as initial products.

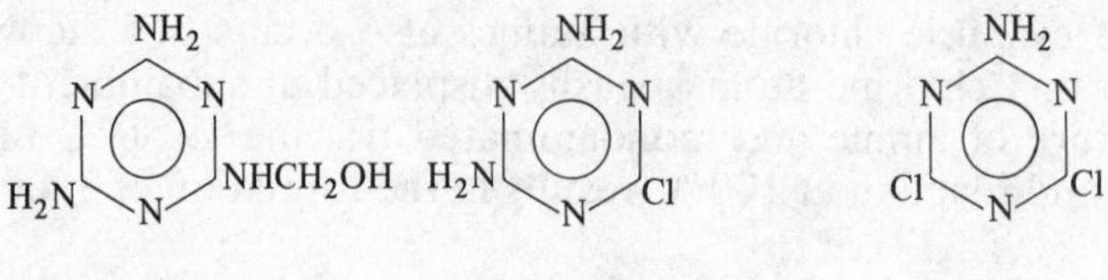

(29) (30) (31)

The reaction of ammonia with cyanuric chloride gives excellent yields of melamine. The reaction is highly exothermic and requires a temperature of 100 °C in order to effect complete substitution. At 50 °C two of the chlorine atoms are displaced and at 0 °C only one reacts, giving the monochlorodiamino- (**30**) and dichloromonoamino-triazine (**31**), respectively. These products are potentially useful in that reduction with HI/PI_3 results in the formation of monoamino- and diamino-1,3,5-triazines. Alternatively, substitution of the remaining halogen atoms with other nucleophiles may lead to unsymmetrically substituted 1,3,5-triazines.

16.3.2 TETRAZINES

In principle there are three possible arrangements of four nitrogen atoms in a six membered aromatic ring: 1,2,3,4- (**32**), 1,2,3,5- (**33**), and 1,2,4,5- (**34**). It has already been stated that the formation of an aromatic ring containing three contiguous nitrogen atoms (*cf.* 1,2,3-triazines) is difficult, with the resultant ring being thermally labile. It is

(32) (33) (34)

therefore to be expected that rings containing four contiguous nitrogen atoms would be exceptionally labile and difficult, if not impossible, to prepare. Suffice it to say that to the present there are no known 1,2,3,4-tetrazines whose structures have been established unambiguously. There are many reports of dihydro-1,2,3,4-tetrazines, although structural proof in this series is sadly lacking. The principal reaction employed during the synthesis of the dihydro-1,2,3,4-tetrazine ring was first described in 1888 by H. von Pechmann[4] and involves the oxidation of bishydrazones of α-diketones (Scheme 8). Most subsequent

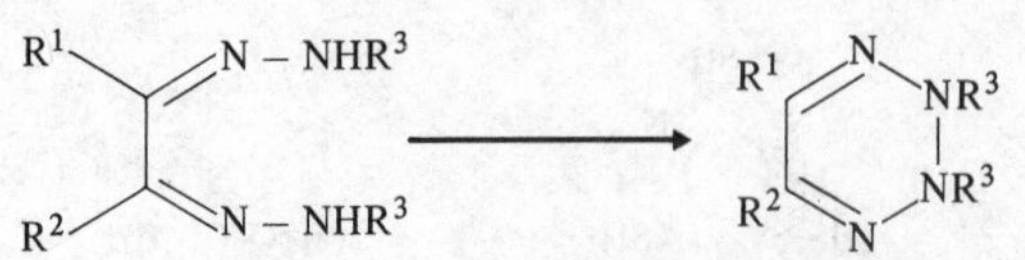

SCHEME 8

syntheses are based on this reaction and vary only in the choice of the oxidant. The resultant 2,3-dihydro-1,2,3,4-tetrazines vary in properties dramatically and there seems some doubt as to whether they have the claimed cyclic structures or are the bis-diazo-compounds (**35**).

No record of the synthesis of a compound containing the 1,2,3,5-tetrazine ring has yet appeared, and in view of the fact that the 1,2,3-triazine ring has only surfaced in recent years this should come as no surprise. In contrast, 1,2,4,5-tetrazines are a well-established

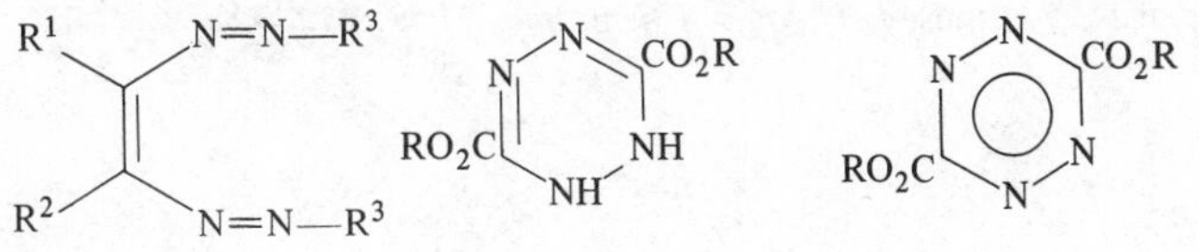

(35) (36) (37)

class of compounds and the parent of the series, 1,2,4,5-tetrazine (**34**), was prepared as early as 1900. Tetrazines are generally prepared by the oxidation of 1,2-dihydro-tetrazines, which are available by a variety of methods. Dimerization of diazoacetic ester in the presence of strong base results in the formation of the dihydrodicarboxylic ester (**36**), which may be oxidized to the tetrazine (**37**) with pentyl nitrite, bromine, air, ferric chloride, hydrogen peroxide or, more commonly, nitrous acid. 1,2-Dihydro-1,2,4,5-tetrazines are also formed from the reaction of nitriles, thioamides, imidates, and bisimidoyl halides with hydrazine. In general the reactions proceed in low yield and other compounds are frequently the principal products.

1,2,4,5-Tetrazines are weakly basic compounds which are deep red or violet in colour, in contrast to the dihydrotetrazines which are virtually colourless. 1,2,4,5-Tetrazine may be prepared by the decarboxylation of the dicarboxylic acid (**37**; R = H), albeit in low yield. It exists as deep purple rods which melt at 99 °C and is unstable in air. It may be stored in sealed tubes under its own vapour, but a report of an explosive decomposition during sublimation has appeared. It is extremely sensitive to acid and treatment with dilute hydrochloric acid yields hydrazine hydrochloride, nitrogen, and formic acid. Base treatment also brings about decomposition.

No details of electrophilic substitution reactions of 1,2,4,5-tetrazines have appeared, and in view of the apparent instability of the unsubstituted molecule this comes as no surprise. One of the most important and characteristic reactions of tetrazines involves their reactions with olefins (ethenes) and acetylenes (ethynes). Olefins react spontaneously with 1,2,4,5-tetrazines to form bicyclic adducts (Scheme 9) which on heating undergo loss

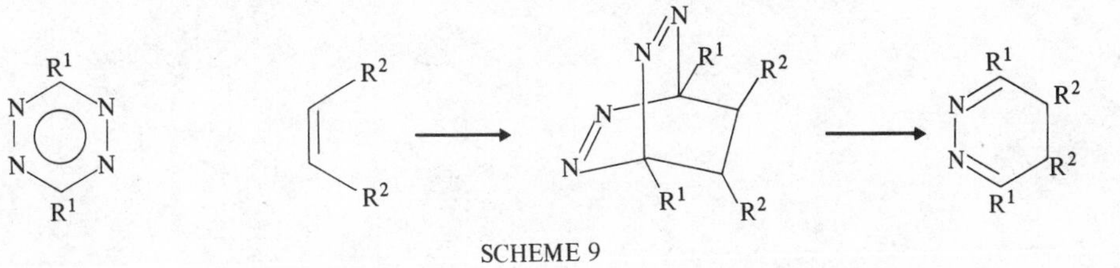

SCHEME 9

of nitrogen to form dihydropyridazines.[11] The reaction is conveniently followed by the disappearance of the characteristic red colouration of the tetrazine, and this reaction has been suggested as a simple chemical test for olefins. The reaction with acetylenes is somewhat slower and thermolysis of the intermediate adduct leads to pyridazine derivatives.

References

1. R. J. Kobylecki and A. McKillop, *Adv. Heterocyclic Chem.*, 1976, **19**, 216.
2. E. A. Chandros and G. Smolinsky, *Tetrahedron Letters*, 1960, 19.
3. O. Oeser and L. Schiele, *Chem. Ber.*, 1972, **105**, 3695.

4. J. G. Erikson, P. F. Wiley, and V. P. Wystrach, in 'The Chemistry of Heterocyclic Compounds', vol. 10, ed. A. Weissberger, Interscience, New York, 1956.
5. T. V. Saraswathi and V. R. Srinivasan, *Tetrahedron*, 1977, **33,** 1043.
6. H. Neunhoffer, H. W. Fruhauf, and I. Mitt, *Tetrahedron Letters*, 1969, 3151.
7. H. Neunhoffer and W. Fruhauf, *Tetrahedron Letters*, 1970, 3355.
8. E. M. Smolin and L. Rapoport, in 'The Chemistry of Heterocyclic Compounds', vol. 13, ed. A. Weissberger, Interscience, New York, 1959.
9. C. Grundmann and A. Kreutzberger, *J. Amer. Chem. Soc.*, 1954, **76,** 5646.
10. S. R. Sandler, *J. Org. Chem.*, 1970, **35,** 3967.
11. R. A. Carboni and R. V. Lindsey, *J. Amer. Chem. Soc.*, 1959, **81,** 4342.

16.4
Quinolines

P. A. CLARET
The City University, London

16.4.1 OCCURRENCE AND ISOLATION OF QUINOLINE AND ALKYLQUINOLINES

16.4.1.1 The quinoline ring system

Quinoline (1-azanaphthalene or benzo[*b*]pyridine) (**1**) is one of the three possible azanaphthalenes; the other two, isoquinoline (2-azanaphthalene or benzo[*c*]pyridine) (**2**) and the quinolizinium cation (4a-azanaphthalene or benzo[*a*]pyridine) (**3**) are discussed in the following sections.

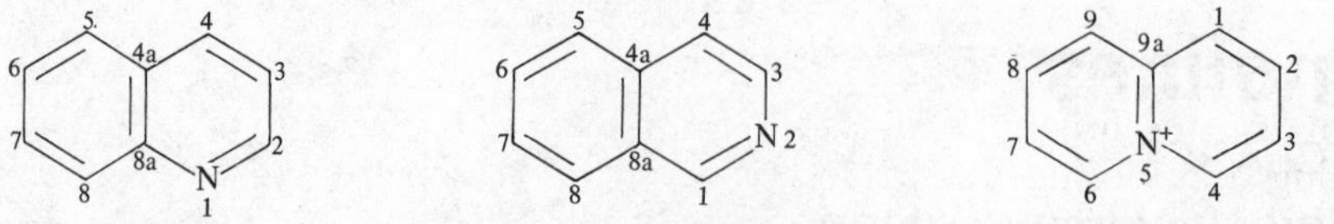

(1) Quinoline (2) Isoquinoline (3) Quinolizinium cation

Note that quaternary carbons 8a and 4a in quinoline and isoquinoline are sometimes numbered 9 and 10, respectively

The bicyclic ring structure related to naphthalene, first suggested by Körner, was established on the basis of the constitution, syntheses, and chemical properties of the base itself, particularly the Friedlander synthesis, degradative oxidation, including ozonolysis, and 'ring opening' reactions together with the consistency of the consequent isomerism and properties of simple quinoline derivatives. More recently this structure has been amply confirmed by X-ray crystallographic analysis and spectroscopic techniques, particularly the electronic (u.v.) spectrum and ^{1}H and ^{13}C n.m.r.; all of these topics are discussed in more detail below.

16.4.1.2 Occurrence and isolation

Runge[3] (1834) described the isolation, from coal tar, of crude quinoline, presumably also containing isoquinoline and alkyl derivatives of both bases. Later, Gerhardt[4,5] (1842) isolated quinoline from the alkaloids cinchonine (**4**) and quinine (**5**) by vigorous distillation with alkali. Distillation of quinine also gives 6-methoxyquinoline. Quinoline has also been obtained from crude petroleum, shale oil, and tobacco by destructive distillation, and by direct synthesis, as described below.

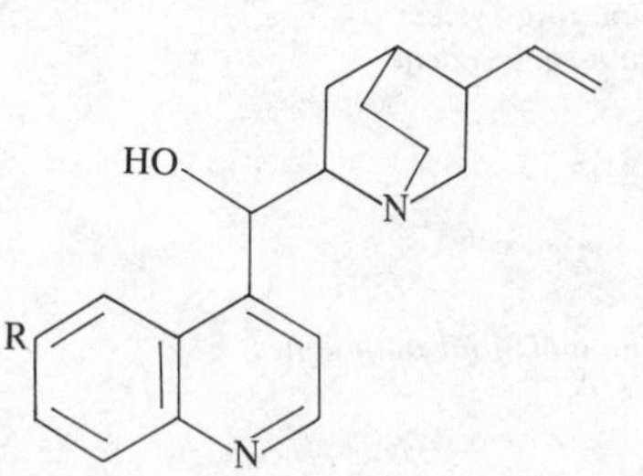

(4) R = H, Cinchonine; (5) R = OMe, Quinine

A series of papers by J. R. Bailey and co-workers in *J. Amer. Chem. Soc.* from 1930–1941 gives details of the separation of quinoline and many of the mono-, di-, tri-, and tetra-alkylquinolines from Californian crude oil. Distillates with a boiling range of about 230–240 °C, for quinoline, or higher (up to 300 °C), for alkylquinolines, were extracted with sulphuric acid (30%) and the bases liberated by the addition of alkali (20% sodium hydroxide). Further purification consisted of redissolving in acid and washing the extracts with ether or xylenes before treatment with alkali. Alternatively, the bases were converted into addition compounds with cresols which were separated and decomposed with sodium hydroxide. Final purification was achieved by fractional distillation at atmospheric or reduced pressures.

The chemistry of quinoline and its compounds has been reviewed.[1,2]

16.4.2 SYNTHESES OF QUINOLINES

16.4.2.1 General discussion

Quinolines are usually synthesized from the readily available aromatic amines by completing and closing the heterocyclic ring. A number of approaches are possible, depending upon which of the bonds (a)–(d) in (**6**) completes the ring closure and the method of introduction of any missing carbons, of which (**7**)–(**10**) are the most common. Thus the Skraup and Doebner–von Miller syntheses, and their many variations, and the Combes, Conrad–Limpach, and Knorr syntheses of quinolones all use a three-carbon intermediate, method (**7**), with ring closure of bond (d). The Friedlander and Pfitzinger syntheses are examples of method (**8**) in which a two-carbon unit is used to complete the heterocyclic ring with the final ring closure usually at bond (c), but variations, particularly when N is part of a NO_2 group, are known in which the final ring closure is to nitrogen, bond (a). Cyclization of cinnamanilides and derivatives of *o*-aminocinnamic acid are examples of methods (**9**) and (**10**), respectively.

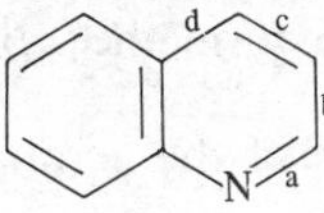

(**6**) 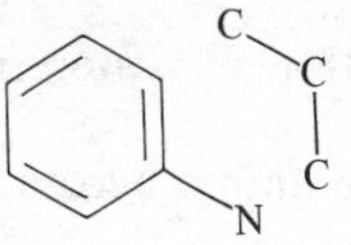 (**7**) 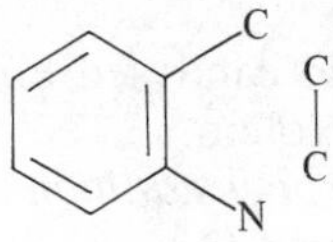 (**8**)

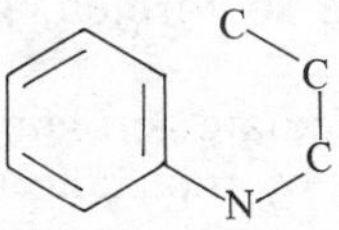

(**9**) 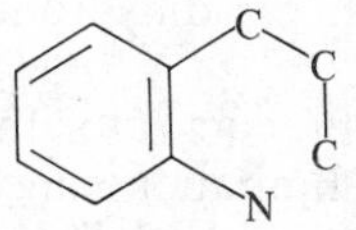 (**10**)

The first reported syntheses of quinoline were by Koenigs, who passed the vapours of ethylaniline, and other alkylanilines, over heated litharge (1879) and, in a second method, heated the addition product of aniline and acrolein (1880). These early syntheses were quickly followed by those of Skraup (later in 1880) and of Doebner and von Miller (1881), both of which have been widely used and are closely related mechanistically to Koenigs' acrolein synthesis.

16.4.2.2 The Skraup synthesis[6,7]

An aniline derivative having a vacant *ortho*-position is heated with glycerol, sulphuric acid, and an oxidizing agent, which may conveniently be the nitro-compound corresponding to the aromatic amine. Other oxidizing agents such as arsenic pentoxide, iodine, ferric salts, or *m*-nitrobenzenesulphonic acid can replace the aromatic nitro-compound. The inorganic oxidizing agents usually reduce tar formation and give cleaner products. Ferrous sulphate[8] or boric acid[9] have been recommended as additives to moderate the exothermic reaction, which may become violent.

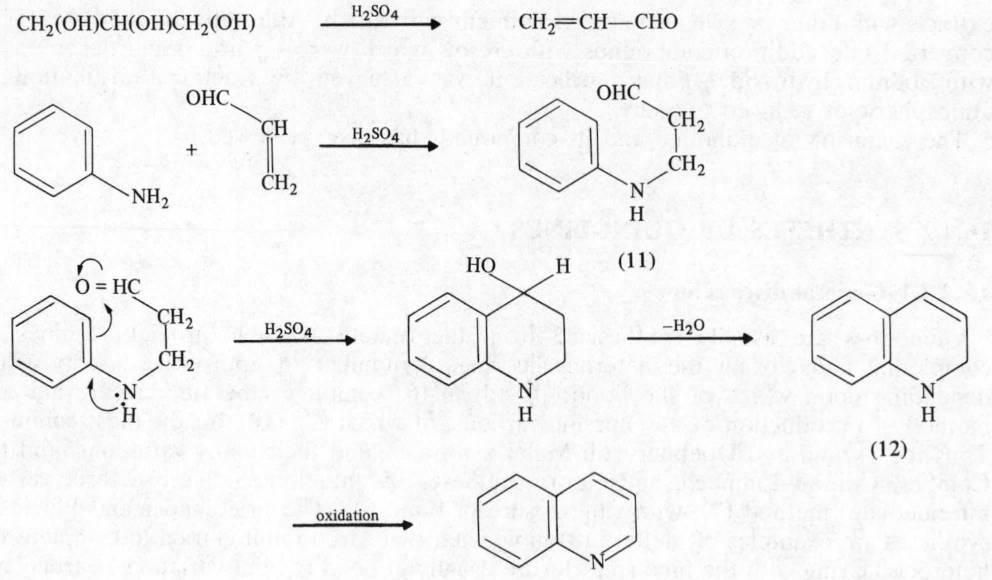
CH₂(OH)CH(OH)CH₂(OH)
H₂SO₄
CH₂=CH—CHO
OHC
CH
CH₂
NH₂
H₂SO₄
OHC
CH₂
CH₂
N
H
(11)
O = HC
CH₂
CH₂
N
H
H₂SO₄
HO
H
N
H
−H₂O
N
H
(12)
oxidation
N

SCHEME 1

The superficially complicated reaction can be explained as a sequence of familiar steps (Scheme 1):

(i) An acid-catalysed elimination of water from glycerol to form the $\alpha\beta$-unsaturated aldehyde, acrolein.

(ii) A *Michael addition* of the amine across the vinyl group of the acrolein to give 3-anilinopropanal (**11**).

(iii) Ring closure results from electrophilic attack by the carbonyl carbon at the ring position *ortho* to the amino-group, followed by elimination of water to extend the conjugated system.

(iv) Oxidation of the 1,2-dihydroquinoline (**12**), so formed, gives the fully aromatic heterocycle.

With glycerol as the precursor for the key intermediate acrolein, the heterocyclic ring is unsubstituted, but this limitation can be overcome by replacing glycerol with preformed $\alpha\beta$-unsaturated carbonyl intermediates having appropriate substituents (equation 1). Thus crotonaldehyde (**13a**) gives 2-methylquinoline (quinaldine) (**14a**) and methyl vinyl ketone

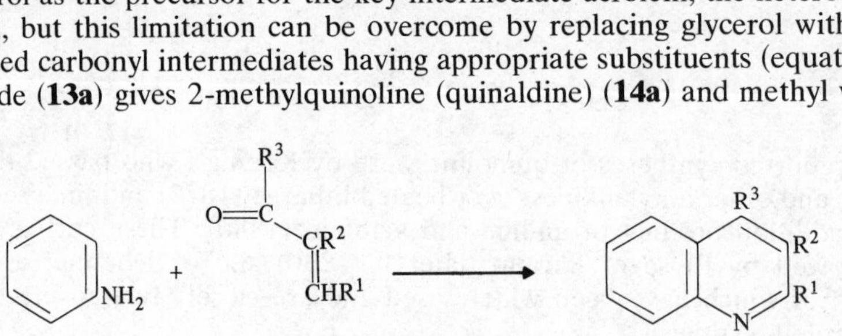
R³
O=C
CR²
CHR¹
NH₂
R³
R²
R¹
N

(1)

(13) (14)

a, R^1 = Me, R^2 = R^3 = H; b, R^1 = R^2 = H, R^3 = Me

(**13b**) gives 4-methylquinoline (lepidine) (**14b**). The use of glycerol has the advantage of maintaining a steady low concentration of acrolein, which reacts rapidly with the amine and minimizes losses by polymerization (which causes poor yields of quinoline when preformed acrolein is used). Less reactive intermediates such as those indicated above may be used satisfactorily. It is also possible that the *Michael addition*, stage (ii), could involve the aldehyde in the form of its Schiff base, in which case (equation 2) ring closure would occur with elimination of amine instead of water. That the Schiff base (**19**) does not

(2)

(3)

i, Michael addition; ii + iii, ring closure, $-PhNH_2$; iv, oxidation

SCHEME 2

undergo ring closure directly (*i.e.* before the Michael addition) as originally proposed by Skraup (equation 3) is shown by the formation of 2-methylquinoline (**18**) from crotonaldehyde *via* (**15**)–(**17**) (Scheme 2) rather than 4-methylquinoline (**20**).

Substituted anilines give quinoline derivatives with substituents in the carbocyclic ring. Anilines substituted in the *ortho*-position having only one site for ring closure give 8-substituted quinolines. The symmetrical *para*-substituted anilines undergo ring closure at either *ortho*-position to give 6-substituted quinolines, but *meta*-substituted anilines can give both 5- and 7-substituted quinolines. There have been conflicting reports on the orientation of ring closure, which can vary slightly with the reaction conditions used. A study by Palmer[10] using gas chromatographic and spectroscopic methods of product analysis indicates that a deactivating and *meta*-directing substituent (to electrophilic attack) such as NO_2 ($-I, -M, -E$) in position-3 of the amine favours ring closure at position-2, giving the 5-nitroquinoline as the major product. Strongly activating and

TABLE 1
Orientation of Quinolines Formed from 3-Substituted Anilines by the Skraup Reaction[a]

Substituent	*% Composition of quinolines* 5-	7-
OMe	22	78
OEt	19	81
NMe_2	25	75
Et	25	60
CF_3	42	58
F	25	75
Cl	45.5	54.5
Br	46	54
I	49	51
NO_2	78	22

[a] See Ref. 10 for details of % sulphuric acid (70% approx.) and reflux temperatures (140 °C approx.).

ortho/para directing groups such as (OMe) (+*M*, +*E*, −*I*) favour ring closure at position-6 of the amine (*para* to the substituent) to give mainly the 7-substituted quinoline. The halogens (deactivating but *ortho/para* directing) (−*I*, +*M*, +*E*) and weaker activating groups such as alkyl (+*I*, +*H*) give mixtures of intermediate composition (see Table 1). Electron-withdrawing substituents on the amine (*e.g.* NO_2, Cl) moderate the final electrophilic ring closure, whereas electron-donating groups (*e.g.* OMe) facilitate ring closure but may result in reduced yields owing to tar formation. A procedure in which the whole reaction is completed rapidly in about 90 seconds has been described by Elderfield[11] and co-workers which improves the yields under these conditions, and also when labile substituents are present on the amine.

16.4.2.3 The Doebner–von Miller synthesis[12]

In this synthesis, which is closely related to the *Skraup synthesis*, an aromatic amine together with an aldehyde is heated with hydrochloric acid. Doebner and von Miller originally (1881) used glycol from which acetaldehyde can be formed by elimination of water, but other aldehyde precursors, such as paraldehyde or acetals, may be used in place of the aldehyde itself. Zinc chloride has also been used as a condensing agent with, or instead of, hydrochloric acid. The most likely mechanism was originally considered to involve the formation of an $\alpha\beta$-unsaturated carbonyl intermediate (crotonaldehyde from acetaldehyde) by an acid-catalysed aldol condensation and elimination of water. A Michael condensation with the aromatic amine followed by ring closure and oxidation, as in the Skraup reaction, would then complete the synthesis.[13a] However, experiments carried out in a deuterated solvent medium[13b] have shown that no significant amount of deuterium is incorporated at position-3 of the quinoline ring, which would result if a Michael addition involving an $\alpha\beta$-unsaturated ketone was involved. It is therefore more likely that the quinoline ring is formed as a result of the self-condensation of two molecules of the Schiff base as shown in Scheme 3, where the nucleophilic molecule is depicted in the enamine form and the electrophilic molecule in the ketimine form. The intermediates (**16**) and (**17**) are common to the Skraup synthesis (Scheme 2).

However, crotonaldehyde may well be formed, even if it does not participate in the main ring-closure process, as the secondary amines *N*-ethylaniline and *N*-butylaniline are

PhNH CH CH$_2$ + CH—Me N H$^+$ → i PhNH CH CH$_2$ CH—Me N H → ii PhNH H Me N H

→ iii PhNH Me N H → iv Me N H → v Me N

(16) (17) (18)

i and ii, H$^+$; iii, −H$^+$; iv, −PhNH$_2$; v, oxidation (−2H), see discussion

SCHEME 3

also formed during the reaction[13a] and are believed to be produced as a result of the oxidation of the intermediate dihydroquinoline (**17**) by the Schiff bases of acetaldehyde and crotonaldehyde. No tetrahydroquinoline derivative was present in the secondary amine fraction (which is normally separated by treatment with nitrous acid), thus showing that simple disproportionation of the dihydroquinoline is not significant. The addition of added oxidizing agents, as in the Skraup synthesis, does not appear to improve the yields in the Doebner–von Miller synthesis.

In general the aldehyde R^1CH_2CHO will give the quinoline (**21**), and a mixture of two aldehydes R^1CH_2CHO and R^2CH_2CHO (both with α-hydrogen atoms) will give the four quinolines (**21**)–(**24**). If one of the aldehydes is formaldehyde the product from the mixed reaction is simplified as described below.

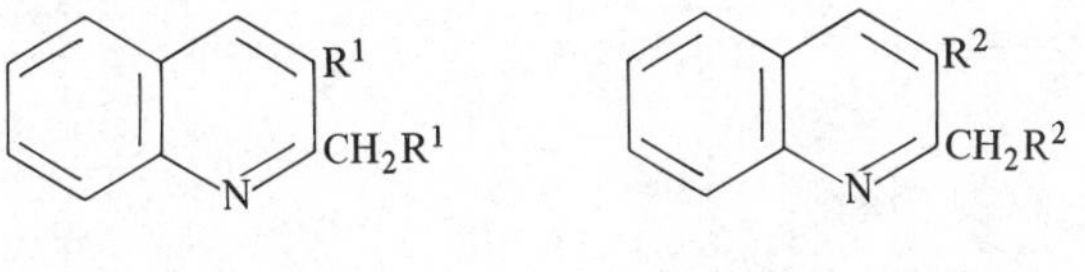

(21) (22)

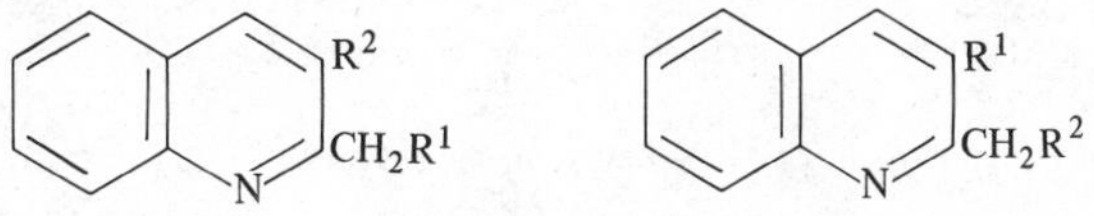

(23) (24)

Further modifications are known in which one or both molecules of aldehyde in the *Doebner–von Miller synthesis* are replaced by other compounds. In the *Beyer synthesis*[14] one mole of a methyl ketone with one mole of aldehyde gives a 2,4-disubstituted quinoline (equation 4, $R^2 = H$). This can be extended[16] by the use of higher ketones to

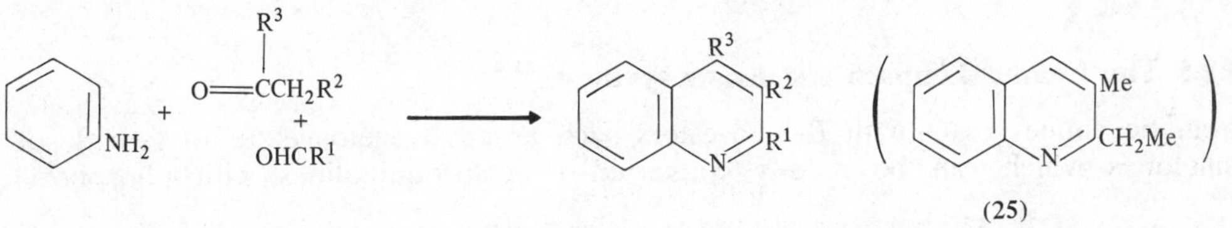

(4)

(25)

give substituents at position-3, and by the use of formaldehyde (or dimethoxymethane), with its restricted role in the *aldol condensation* (no α-hydrogen atom), substituents at position-2 can be eliminated. Quinolines with substituents at position-3 only of the heterocyclic ring, which are the most difficult to prepare by the Skraup and Doebner–von Miller syntheses, can also be obtained,[17] (equation 4, $R^1 = R^3 = H$).

Whenever mixtures of carbonyl compounds are used, products from alternative aldol condensations may also be formed. Thus in the preparation of 3-methylquinoline (equation 4, $R^1 = R^3 = H$, $R^2 = Me$), some of the 2-ethyl homologue (**25**) is formed as well, as a result of a normal Doebner–von Miller reaction with the propanal.[17]

The use of two moles of a ketone, first suggested by Riehm, who heated the anil of the ketone with hydrochloric acid, gives a 2,2′-dialkyl-1,2-dihydroquinoline (**26**) as an intermediate, which then eliminates an alkane to form a quinoline.[15,18] Equation 5 illustrates the reaction with acetone.

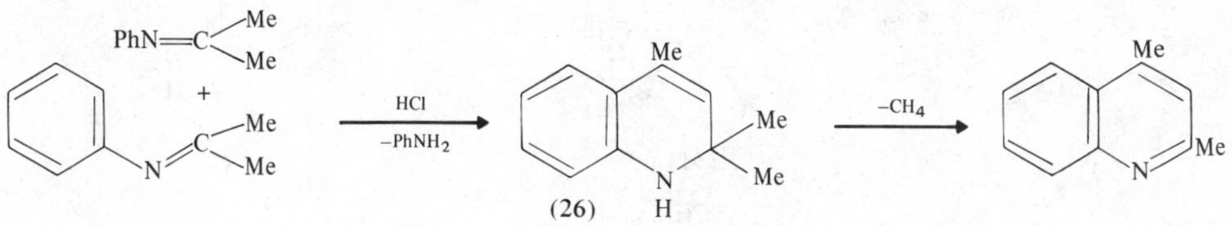

(5)

(26)

16.4.2.4 The Combes synthesis[19]

When the mono-anils of 1,3-diketones are heated with sulphuric acid, 2,4-disubstituted quinolines are formed. Heptane-3,5-dione and aniline give 2,4-diethylquinoline (**27**) (equation 6). The mono-anils of 2-nitro- or 2-cyano-malonodialdehyde give the corresponding 3-substituted quinolines (**28**).[20] Similarly, when 2-substituted malonic esters are refluxed with an aromatic amine in diphenyl ether under nitrogen, 3-substituted 4-hydroxy-2-quinolones are formed (**28a**) (equations 7 and 7a).[20a]

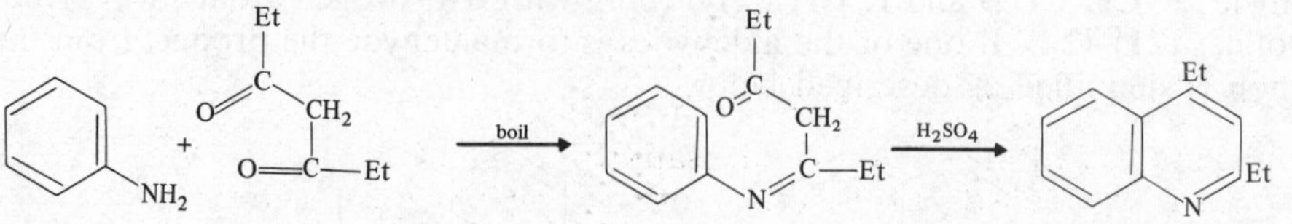

(6)

(27)

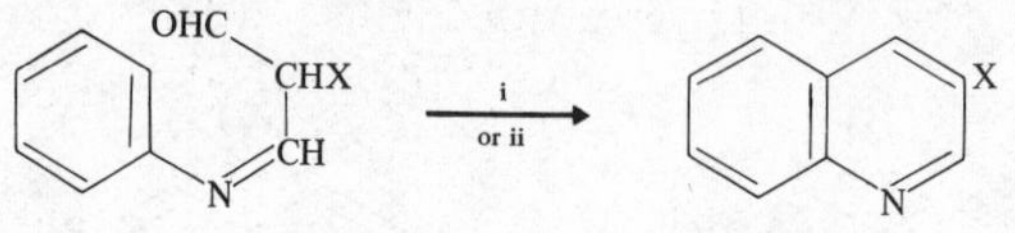

(7)

(28)

i, X = CN, $ZnCl_2$; ii, X = NO_2, HCl

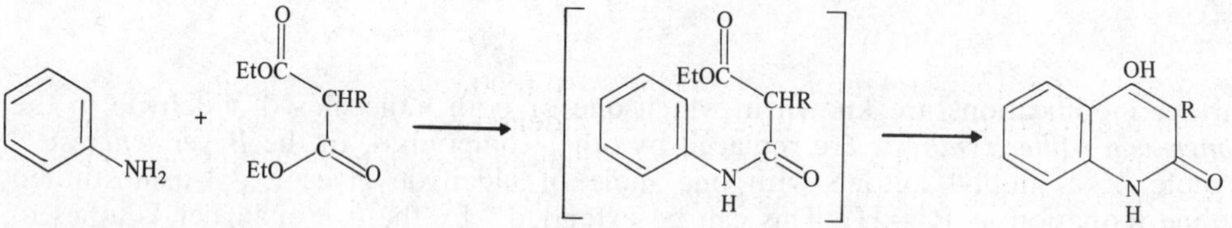

(7a)

(28a)

16.4.2.5 The Conrad–Limpach and Knorr syntheses[21,22]

Aromatic amines react with β-keto-esters such as ethyl acetoacetate to give 2- or 4-quinolones which can be readily converted to chloroquinolines with phosphoryl

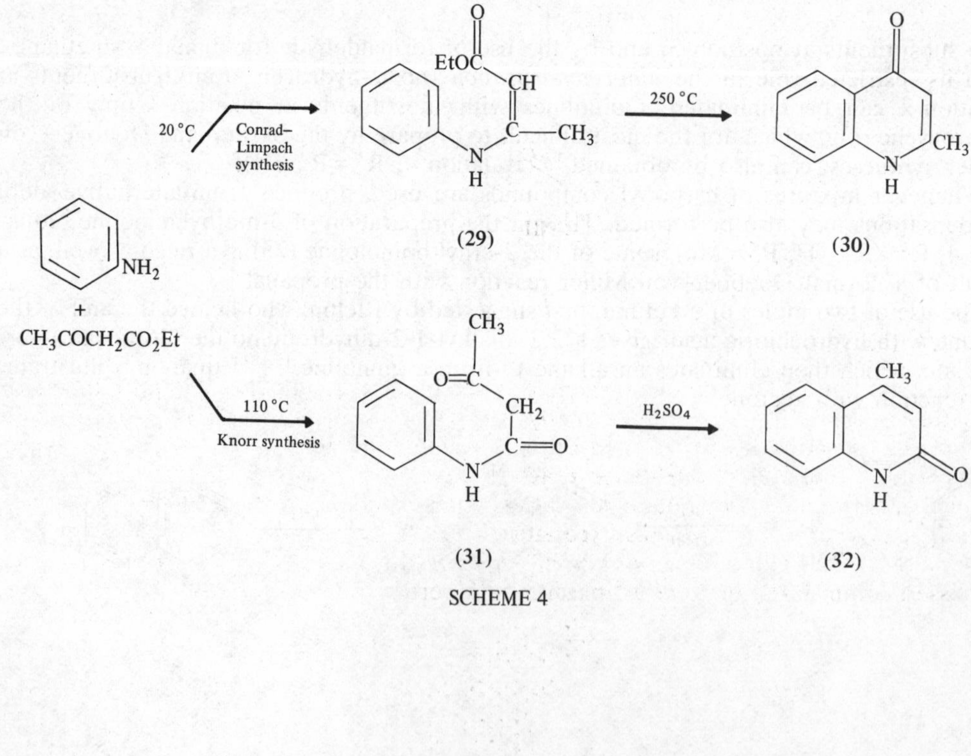

(29) (30)

(31) (32)

SCHEME 4

chloride[28] and then reduced to quinolines. At low temperatures (20 °C) the amine condenses at the more reactive keto-group (kinetic control) to give an anilinocrotonic ester (**29**) which, on heating, cyclizes to the 4-quinolone (**30**) (*Conrad–Limpach synthesis*). At higher temperatures (110–140 °C) the initial product is the thermodynamically favoured anilide of acetoacetic acid (**31**), formed by condensation of the amine at the ester carbonyl. This anilide undergoes ring closure on heating alone or with sulphuric acid to give the 2-quinolone (**32**) (*Knorr synthesis*) (Scheme 4).

16.4.2.6 The Friedlander synthesis[23]

Quinolines are formed by the condensation of *o*-aminobenzaldehydes or *o*-aminoacetophenones with aldehydes, ketones, or other compounds containing an active hydrogen function such as $—CH_2—CO—$ or $—CH_2—CN$. Acetaldehyde and *o*-aminobenzaldehyde (equation 8, $R^1 = R^2 = R^3 = R^4 = H$) give quinoline itself.

R⁴ CO, NH₂, R¹ + H₂C—R³, O=C—R² —NaOH or piperidine→ [intermediate] → quinoline (R¹, R², R³, R⁴) (8)

The first stage is the formation of the Schiff base followed by ring closure of carbons-3 and -4 of the quinoline ring by a Knoevenagel condensation. The method suffers from the disadvantage that amino-aldehydes are bifunctional and prone to self-condensation. This can be overcome either by the use of the Pfitzinger modification described below, or by the use of the less-reactive nitro-compounds in place of the amines. The Knoevenagel condensation then occurs first and ring closure to the nitrogen can be achieved by reduction of the nitro-group. This is illustrated by the formation of 2-aminoquinoline from *o*-nitrobenzaldehyde and acrylonitrile (equation 9).

CHO, NO₂ + CH₃—CN —NaOEt→ CH=CH—CN, NO₂ —Sn/HCl→ CH=CH—CN, NH₂ —NaOH→ 2-aminoquinoline (NH₂) (9)

16.4.2.7 The Pfitzinger synthesis[24]

In this method the *o*-aminobenzaldehyde used by Friedlander is replaced by the anion of isatinic acid (isatic acid, *o*-aminobenzoylformic acid) formed by alkaline cleavage of the lactam ring of isatin, or a substituted isatin. Condensation with a keto-methylene moiety, or its oxime, as in the Friedlander reaction, gives a quinoline-4-carboxylic acid (cinchoninic acid) which may be decarboxylated by heating alone to its melting point, or with calcium oxide or copper powder (equation 10). This method has been widely used (especially by Buu-Hoi and co-workers)[25] for the synthesis of quinoline derivatives likely to possess antimalarial or tumour-inhibiting properties.

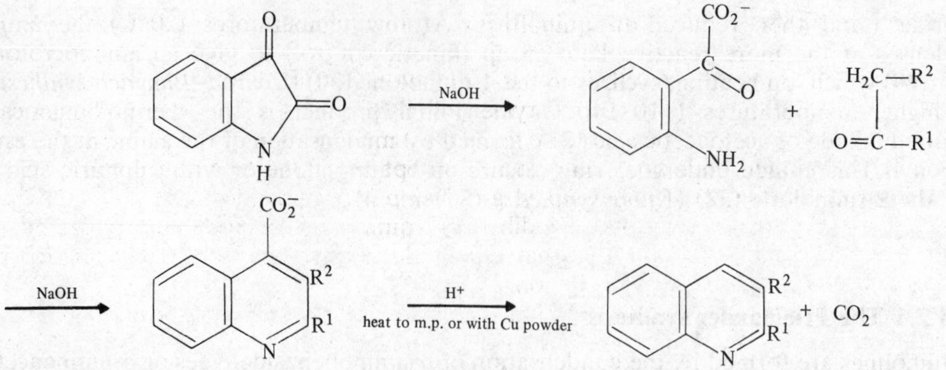

(10)

16.4.2.8 Cyclization of *o*-amino-derivatives of cinnamic acid

The *o*-amino-derivatives of *allo*-cinnamic acid (*i.e.* the *cis*-isomer) readily undergo ring closure to give 2-quinolones. The *o*-amino-derivatives of *trans*-cinnamic acid, obtained by reduction of the nitro-compounds, on treatment with acetic anhydride or sulphuric acid give bimolecular anhydrides which on heating also give 2-quinolones following rearrangement to the *cis*-isomer (equation 11).

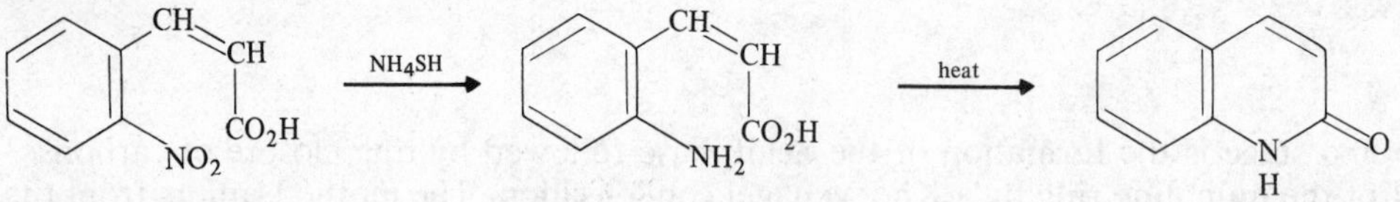

(11)

This reaction[26] was one of the earliest syntheses of the quinoline ring system. Later workers have used esters of cinnamic acid and alternative reducing agents have included ferrous hydroxide and tin or zinc with hydrochloric acid.

16.4.2.9 Cyclization of cinnamanilides

When heated with polyphosphoric acid,[27,28] cinnamanilides give 4-aryl-3,4-dihydro-2-quinolones (**33**); with aluminium trichloride, 2-quinolones (**34**) are formed with elimination of the aryl group as a hydrocarbon (Scheme 5).

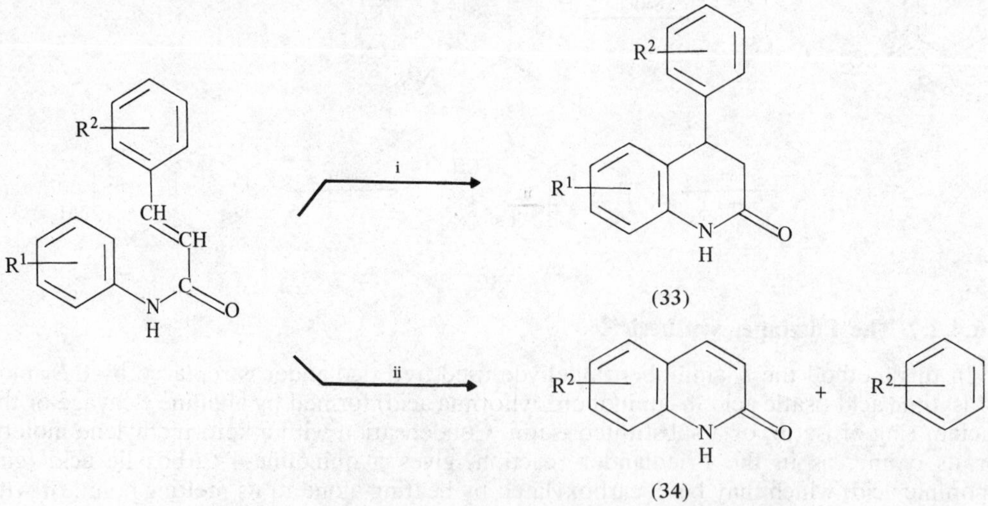

(33)

(34)

i, Polyphosphoric acid (100–190 °C, 10–30 min); ii, $AlCl_3$ (3 mol, 100 °C, 1 h)

SCHEME 5

With polyphosphoric acid, route (i), the reaction proceeds with a variety of *para*-substituents in the N-aryl ring ($R^1 = H$, *p*-Br, *p*-Me, *p*-OMe, *p*-Cl) and presumably *meta*-substituents as well, but when R^1 is an *ortho*-substituent ring closure is hindered. Electron-withdrawing groups in the C-aryl ring (*e.g.* $R^2 = p\text{-}NO_2$) can prevent ring closure and strongly electron-releasing groups (*e.g.* *p*-OMe) may cause loss of the aryl group, as in route (ii), unless the reaction temperature is kept as low as possible. The order of increasing ease of elimination of the aryl group is *p*-chlorophenyl < phenyl < *p*-tolyl < *p*-anisyl. The stability of the 4-aryl-3,4-dihydro-2-quinolone once formed, when heated under reaction conditions, indicates that elimination occurs as part of the ring-closure process and not in a subsequent reaction.

By route (ii) three moles of aluminium chloride were found to be necessary to give optimum yields and this was attributed to the coordination of two moles of the reagent, one to the amide nitrogen and one to the carbonyl oxygen. Cinnamanilides gave 2-quinolones with loss of the C-aryl group as the hydrocarbon, but the *N*-arylamides of β-methylcrotonic acid gave 4,4-dimethyl-3,4-dihydro-2-quinolone (**34a**) without loss of the methyl group (equation 12).

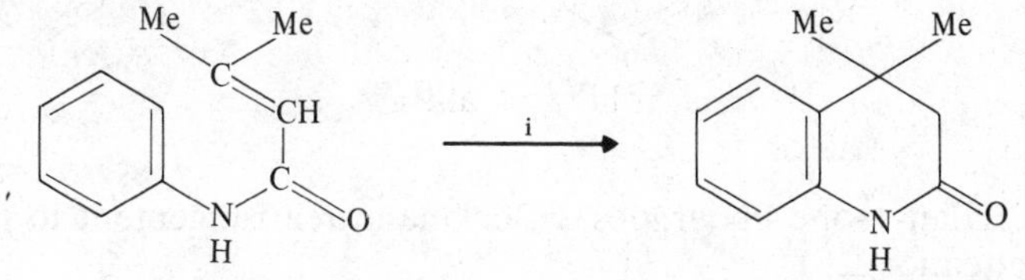

(12)

(34a)

i, $AlCl_3$, 3 mol, 100 °C, 1 h

16.4.2.10 Ring-enlargement methods

Quinoline derivatives can be prepared from indoles by ring enlargement using alkyl halides[29] (equations 13 and 14), with chloroform and sodium ethoxide[30] (dichlorocarbene) (equation 15), and from isatin by treatment with diazomethane,[31] a reaction characteristic

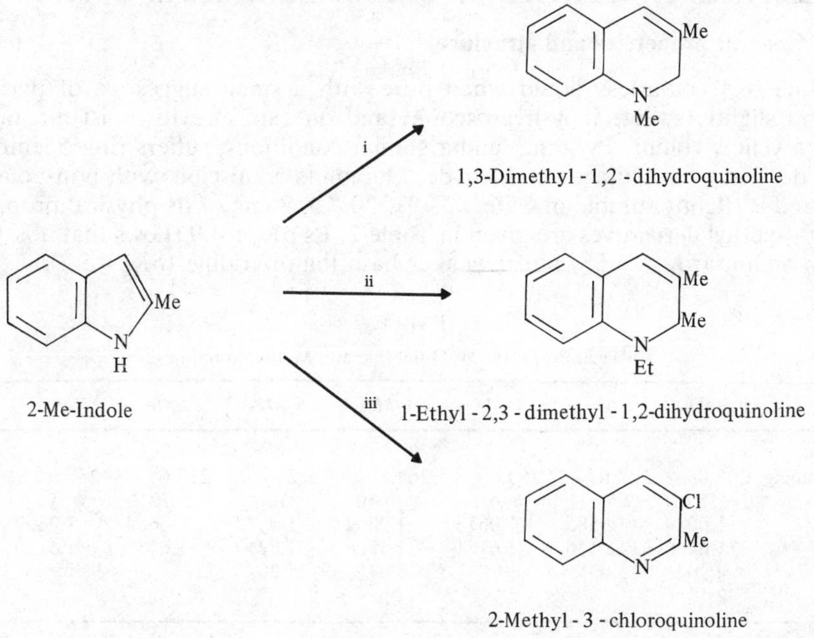

(13)

1,3-Dimethyl-1,2-dihydroquinoline

2-Me-Indole

(14)

1-Ethyl-2,3-dimethyl-1,2-dihydroquinoline

(15)

2-Methyl-3-chloroquinoline

i, 2 MeI/100 °C/80% yield; ii, 2 EtI; iii, $CHCl_3$ + NaOEt (: CCl_2)

$$\text{Isatin} \xrightarrow{\text{i}} \text{3-Hydroxy-2-quinolone} \qquad (16)$$

i, CH_2N_2 $(-N_2)$

of cyclic ketones, (equation 16). The mechanism for the reactions 13 and 14 has not been established but is consistent with the following ring enlargement of the quaternary salt (equation 17):

$$\xrightarrow{\text{2 EtI}} \quad I^- \quad \xrightarrow{\text{heat}} \quad + \text{HI} \qquad (17)$$

The oxime of indan-1-one undergoes a Beckmann rearrangement to give 3,4-dihydro-2-quinolone[32] (equation 18).

$$\xrightarrow{PCl_5} \qquad (18)$$

16.4.3 PHYSICAL PROPERTIES OF QUINOLINE AND ITS HOMOLOGUES

16.4.3.1 General properties and structure

Quinoline is a colourless liquid when pure with a smell suggestive of pyridine, but milder and slightly sweet. It is hygroscopic, and on exposure to moist air and light it develops a yellow colour. Pyridine, under similar conditions, suffers ring opening to give products derived from glutaconic aldehyde. Quinoline is miscible with non-polar organic solvents and is slightly soluble in water (0.7%, 20 °C). Some of its physical properties and those of its methyl derivatives are given in Table 2. Its $pK_a = 4.9$ shows that it is a stronger base than aniline ($pK_a = 4.58$) but a weaker base than pyridine ($pK_a = 5.2$).

TABLE 2

Physical Properties of Quinoline and Methylquinolines

	Quinoline	*2-Me-*	*3-Me-*	*4-Me-*	*5-Me-*	*6-Me-*	*7-Me-*	*8-Me-*
B.p.[a] (°C/760 mmHg)	238.05	247.6	259.6	264.2	262.7	258.6	257.6	247.8
M.p.[a] (°C)	−15.6	−2 to −1	16 to 17	9 to 10	19	−22	39	—
d_4^{20} [a]	1.0929	1.0585	1.0673	1.0862	1.0832	1.0654	1.0609	1.0719
n_D^{20} [a]	1.6268	1.6116	1.6171	1.6206	1.6219	1.6157	1.6150	1.6164
pK_a [a]	4.90	5.83		5.67	5.20			
μ (D)[b]	2.18							

[a] 'Handbook of Chemistry and Physics', ed. R. C. Weast, 1976, 57th edition, Chemical Rubber Co., Cleveland, Ohio. [b] A. D. Buckingham, *J. Chem. Soc.*, 1956, 1405.

Körner's representation of quinoline as 1-azanaphthalene invites further comparisons of the valence bond structures and the electronic spectra of quinoline and naphthalene. As with naphthalene, various classical bond structures are possible representing favourable (low-energy) electron distributions (**35**), (**36**), and (**37**). The introduction of the electronegative nitrogen atom with its non-bonded electron pair enables a number of additional dipolar structures, (**38**)–(**43**), to be considered.

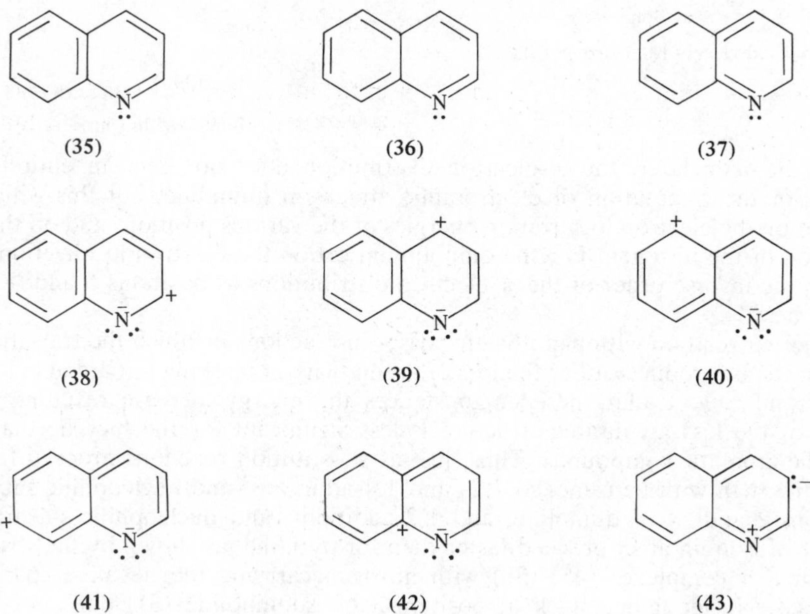

As with naphthalene, structure (**35**) with two benzenoid rings is more favourable energetically than structures (**36**) or (**37**), each of which have one ring with a quinonoid bond distribution. Of the dipolar structures, those in which the nitrogen carries the negative formal charge, (**38**)–(**42**), will be inductively favoured compared with the four possible structures, such as (**43**), in which the nitrogen is formally positive. These latter structures have the added disadvantage that the linear *sp* hybridized orbitals required of the nitrogen would not overlap efficiently with the carbon sp^2 orbitals as part of a six-membered ring (*cf.* benzyne structure). Furthermore, of these charged structures, only (**38**) and (**39**) retain the benzenoid character of the carbocyclic ring. On this basis quinoline can be considered as a mesomeric compound in which the significant contributing structures are (**35**)–(**39**), all of which retain at least one benzenoid ring, with structure (**35**), having two benzenoid rings, as the most important.

This interpretation implies a degree of bond fixation favouring structure (**35**) and an unsymmetrical distribution of π-electrons, with the nitrogen atom rich at the expense of the carbons, particularly C-2 and C-4 and to a lesser extent C-5, C-7, and C-8a. Further evidence, both physical and chemical, is available which is consistent with this description of the structure of quinoline including (a) the dipole moment of 2.18 D; (b) the stabilization energy of 226 kJ mol^{-1}, which compares with a value of 292 kJ mol^{-1} for naphthalene measured on the basis of the difference between ΔH calculated from the bond energies of structure (**35**) and the value determined experimentally; (c) a comparison of the spin–spin coupling constants measured from the n.m.r. spectra of quinoline derivatives as discussed below; (d) the pattern of ring degradation by ozonolysis and the orientation of coupling reactions of hydroxy-derivatives of quinoline, also discussed in detail below; (e) the calculated π-bond orders for quinoline[33a] (**44a**) and the bond lengths (pm) measured by X-ray crystallography of copper 8-hydroxyquinolinate[33b] (**44b**).

0.748 0.534 0.531 0.749 0.580 0.578 0.749 0.537 0.527 N 0.743

(44a)

π - Bond orders

(ethylene = 1.0; benzene = 0.667)

136.5 135.7 142.4 139.7 138.1 N 131.8 O Cu

(44b)

Bond lengths (pm)

As will be seen later, the π-electron distribution does not give an entirely correct indication of the orientation of electrophilic attack on quinoline, but this will be more dependent on the electron localization energies of the various positions and on the state of protonation of the nitrogen. For the quinolinium cation the electron localization energies are not in the inverse order of the π-electron distributions as positions-5 and -8 have the lowest values.[34]

A further correlation with naphthalene arises in reactions in which the transition states, intermediates, or products suffer the loss of aromaticity in one ring but are able to retain it in the second ring. Under these circumstances the energy increase resulting from the adoption of the less favourable structure is less significant for the bicyclic than for the monocyclic aromatic compounds. Thus, just as substitution reactions proceed faster with naphthalene than with benzene, so 1,2- and 1,4-additions, and nucleophilic substitutions at positions-2 and -4 of quinoline, and 1,2-additions, and nucleophilic substitutions at position-1 of isoquinoline, proceed faster than for pyridine, as shown by the structures of the principal σ-complexes (**45**)–(**50**) with nitrogen carrying the negative charge in the heterocycles. Nucleophilic attack at position-3 of isoquinoline (**51**) does not have this advantage as the carbocyclic ring has the quinonoid structure. Nucleophilic substitution at position-3 of quinoline has the added disadvantage that the nitrogen cannot carry the formal negative charge (**52**).

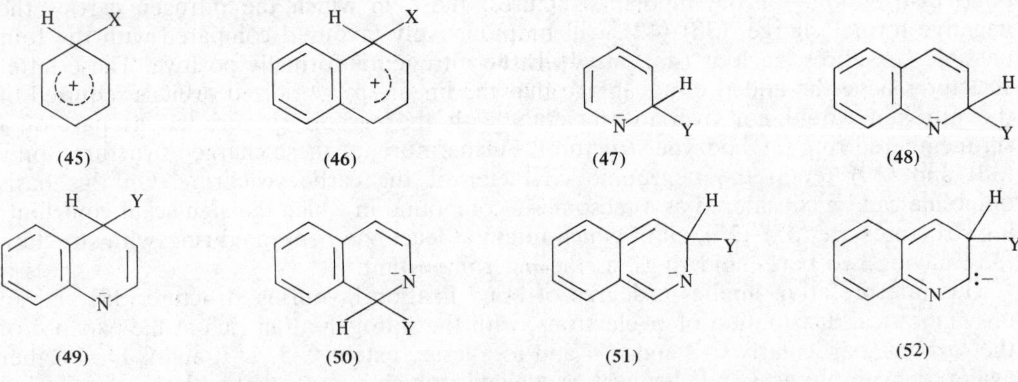

The relative stabilities of the Reissert addition compounds of quinoline and isoquinoline and the deprotonated methylene bases from 2- and 4-methylquinolinium salts and 1-methylisoquinolinium salts compared with corresponding pyridine compounds (see below) can also be explained on this basis.

16.4.3.2 Spectroscopic properties

The spectroscopic properties of quinoline and its derivatives have been reviewed.[35] The following points are of particular interest in connection with the structure and chemistry of quinoline. The close similarity of the absorption bands associated with $\pi \rightarrow \pi^*$ transitions in the u.v. spectrum of quinoline with those of naphthalene confirm the similarities in structure first suggested by Körner and discussed above. In the quinoline spectrum the absorptions due to the $n \rightarrow \pi^*$ transitions of the non-bonded nitrogen electrons overlap, and are masked by, the stronger $\pi \rightarrow \pi^*$ transitions and do not show as a separate band as in the case of pyridine when compared with benzene (Table 3).

TABLE 3
Ultraviolet Spectra (main peaks only)

Compound	*Solvent*	*Ref.*	λ_{max} (nm) (log ε)
Quinoline	Water (pH 6.3)	a	226(4.36) 275(3.51), 299(3.46), 312(3.52)
Isoquinoline	Hexane	c(i)	216(4.91) 266(3.61), 306(3.35), 318(3.56)
Quinolizinium (iodide)	Water	b	226(4.25) 272(3.42), 283(3.47), 325.5(4.23)
Naphthalene	Hexane	c(ii)	221(5.07) 275(3.75), 297(2.47), 311(2.40)
Pyridine	Hexane	c(iii)	198(3.78) 256(3.28), 276(2.2) shoulder
Benzene	Hexane	c(iv)	204(3.94) 254(2.40)

[a] Ref. 35a, Vol. III, p. 121. [b] A. Richards and T. S. Stevens, *J. Chem. Soc.*, 1958, 3067. [c] D.M.S., U. V. Atlas, Butterworths, London, 1966: (i) spectrum H2/1; (ii) spectrum E1/1; (iii) spectrum G5/1; (iv) spectrum D1/1.

The values of the ^1H and ^{13}C chemical shifts in the n.m.r. spectra of quinoline (Table 4) are consistent with the electron densities predicted (and calculated) for the proposed structure. Thus H-2, H-4, and H-8 all resonate downfield as a result of deshielding by the nitrogen and a similar situation is apparent for C-2, C-8a, C-4, and C-8. Both H-3 and C-3 resonate at the highest field strengths, consistent with the highest electron density at this position.

TABLE 4
^1H and ^{13}C Chemical Shifts for Quinoline Positions (δ p.p.m. downfield from TMS)[a]

Nucleus	*Solvent*	*1*	*2*	*3*	*4*	*4a*	*5*	*6*	*7*	*8*	*8a*
^1H	Neat liquid	—	9.52	7.09	7.83	—	7.58	7.35	7.60	9.43	—
^1H	CCl_4	—	8.81	7.26	8.00	—	7.68	7.43	7.61	8.05	—
^{13}C	CCl_4	—	150.9	121.7	136.1	128.9[b]	128.5	127.0	129.9	130.5	149.3[b]

[a] All data from Ref. 35b, Vol. IV, p. 219 except b.
[b] P. A. Claret and A. G. Osborne, *Spectroscopy Letters*, 1975, **8**, 385.

Further confirmation of bond fixation, or reduced double bond character of the 2,3- and 6,7-bonds of quinoline, is given by a comparison of the vicinal and benzylic proton coupling constants and the ^{13}C–^1H interactions. As shown in Table 5, $J_{2,3}$ and $J_{6,7}$ are significantly lower than the other *ortho*-proton coupling constants and the related ^3J coupling constants between methyl ^{13}C and *ortho*-protons in the methylquinolines show a

TABLE 5
^1H–^1H Coupling Constants (Hz) for Liquid Quinoline[a]

o-*Coupling (three bond)*		m-*Coupling (four bond)*		*Long-range coupling*	
$J_{2,3}$	4.3	$J_{2,4}$	1.8	$J_{4,8}$	0.9
$J_{3,4}$	8.3	$J_{5,7}$	1.6	$J_{5,8}$	0.3
$J_{5,6}$	8.2	$J_{6,8}$	1.1	$J_{4,5}$[b]	0.4
$J_{6,7}$	6.8				
$J_{7,8}$	8.3				

[a] All data from Ref. 35b, Vol. IV, p. 219 except b. [b] P. A. Claret and A. G. Osborne, *Spectroscopy Letters*, 1973, **6**, 103.

TABLE 6

Methyl Carbon (^{13}C)–Aromatic Proton Three-bond Coupling Constants (Hz) for Methylquinolines[a]

$J_{2',3}$	2	$J_{3',4}$	5	$J_{5',6}$	5
$J_{3',2}$	3	$J_{4',3}$	5	$J_{6',5}$	5
$J_{6',7}$	4	$J_{7',8}$	—		
$J_{7',6}$	4	$J_{8',7}$	5		

[a] Ref. 37. $CDCl_3$ solvent. The position of the methyl carbon is given first.

similar trend (Table 6). A comparison of the various benzylic coupling constants in methylquinolines, which are believed to depend on the π-character of the aromatic bond,[36] is even more significant as there is practically no measurable interaction across the 2,3- or 6,7-bonds, as shown in Table 7. Furthermore, a comparison of the methyl proton signals of 2-chloro-4-methylquinoline (**53**) and 4-chloro-2-methylquinoline (**54**) shows that the methyl proton signal for compound (**53**) is a doublet, as a result of benzylic

TABLE 7

Methyl Proton–Ring Proton (four-bond) Benzylic Coupling Constants (Hz) for Methylquinolines[a]

Coupling constant		*Ref.*	*Coupling constant*		*Ref.*
$J_{2',3}$	~0	b	$J_{3',4}$	0.95	b
$J_{3',2}$	~0	b	$J_{4',3}$	0.95	b
$J_{6',7}$	~0	c	$J_{5',6}$	0.8	c
$J_{7',6}$	~0	c	$J_{6',5}$	0.8	c
			$J_{7',8}$	0.8	c
			$J_{8',7}$	0.8	c

[a] CCl_4 solvent. The position of the protons of the methyl group are given first. [b] Ref. 36. [c] Ref. 37.

coupling with H-3, but for compound (**54**) it is a singlet. In the same spectra H-3 appears usually as a poorly resolved quartet for (**53**) and as a sharp singlet for (**54**).[37] The one-bond ring-carbon–ring-proton (J_{CH}) coupling constants for positions 4, 5, 6, 7, and 8 are all about 161 Hz, but the coupling constants for bonds nearer to the heterocyclic nitrogen atom are significantly larger: position-2 = 178 Hz, position-3 = 165 Hz.

Me H N Cl   Cl H N Me

(53)   (54)

16.4.4 CHEMICAL PROPERTIES OF QUINOLINE AND ALKYLQUINOLINES

16.4.4.1 Salt formation

As expected from their chemical structure and pK_a values, quinoline and the alkylquinolines are readily protonated on the ring nitrogen to form stable salts, many of which, particularly the dichromates and picrates, are sparingly soluble in water and may be used to purify the bases by crystallization. With alkyl, acyl, and aroyl halides, and dimethyl sulphate, they form quaternary salts, which on treatment with alkali give initially the quaternary hydroxides (equation 19). In solution, however, nucleophilic attack by the

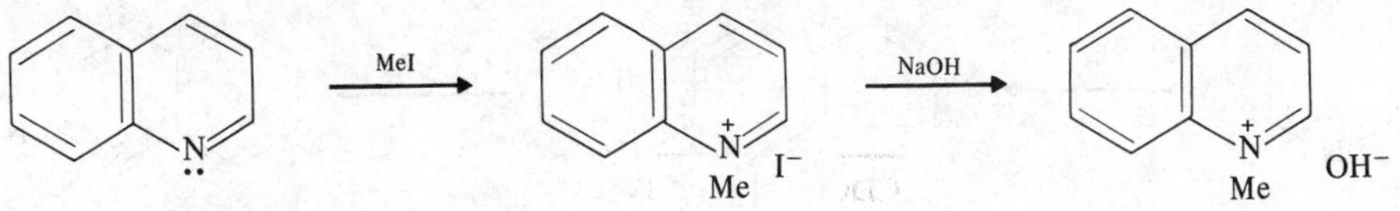

MeI NaOH (19)

$\overset{+}{N}$ I^- Me $\overset{+}{N}$ OH^- Me

hydroxyl ion at position-2 results in the formation of an equilibrium mixture of the ionized form with the un-ionized carbinolamine base and possibly also with the tautomeric *N*-alkyl-*o*-aminocinnamaldehyde (equation 20).

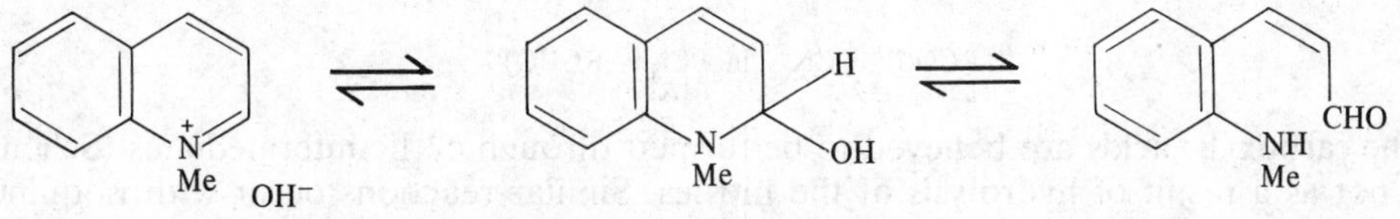

(20)

$\overset{+}{N}$ Me OH^- ⇌ N Me H OH ⇌ CHO NH Me

Evidence for the equilibria is provided by the fall in electrical conductivity after the addition of one equivalent of alkali to a quinolinium salt and the formation of a phenylhydrazone by the product. Oxidation of the hydroxide with potassium ferricyanide gives the 2-quinolone, which could be formed either from the carbinolamine or from the *o*-aminocinnamaldehyde followed by ring closure of the amino-acid to the lactam (equation 21). With alcohols the carbinolamines form ethers, and dimeric ethers can be formed by self-condensation (equation 22).

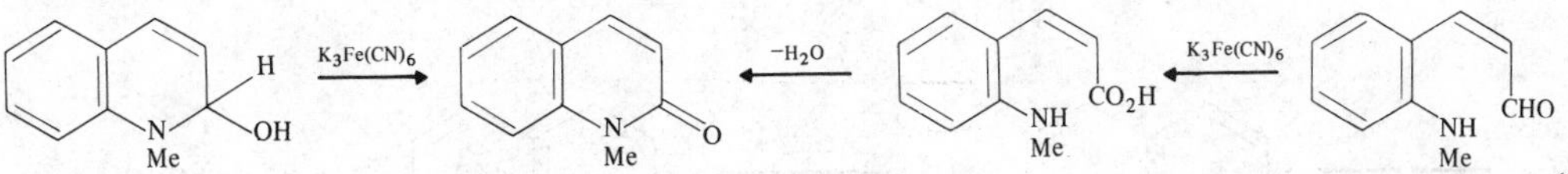

$K_3Fe(CN)_6$ $-H_2O$ $K_3Fe(CN)_6$

H OH N Me; N Me O; CO_2H NH Me; CHO NH Me

(21)

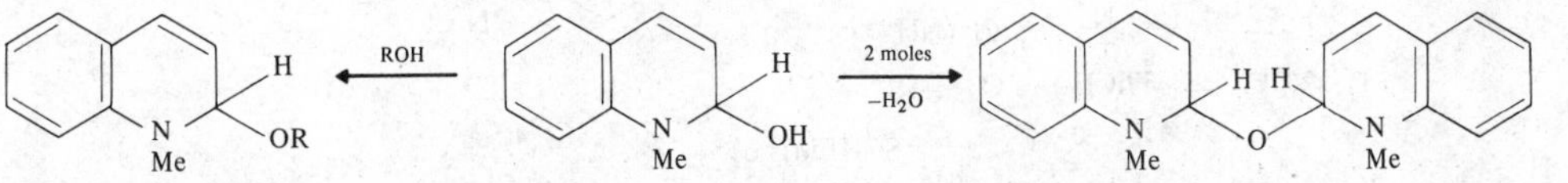

ROH 2 moles $-H_2O$ (22)

H OR N Me; H OH N Me; H H N Me O N Me

When heated in sealed tubes at 300 °C, 1-alkylquinolinium iodides undergo the Ladenburg rearrangement in which the 1-alkyl substituent becomes partly attached to the ring carbons and partly converted into an alkane or alkene. The reaction product from 1-methylquinolinium iodide has been shown[38b] to contain 2-, 4-, 5-, and 8-methylquinolines, identified by gas chromatography, together with an unresolved mixture believed to consist of the 6- and 7-methylquinolines and possibly containing a small amount of the 3-isomer. The mechanism[38a] is presumably similar to that of the corresponding pyridinium salts, which has been shown to involve a 7π-electron radical intermediate.[39]

Addition of cyanide ion to a 1-alkylquinolinium cation occurs at position-4 and provides a route to 4-cyanoquinolines and the 4-carboxylic acids (cinchoninic acids) (equation 23). The action of cyanide ion on 1-acyl- or 1-aroyl-quinolinium salts, on the

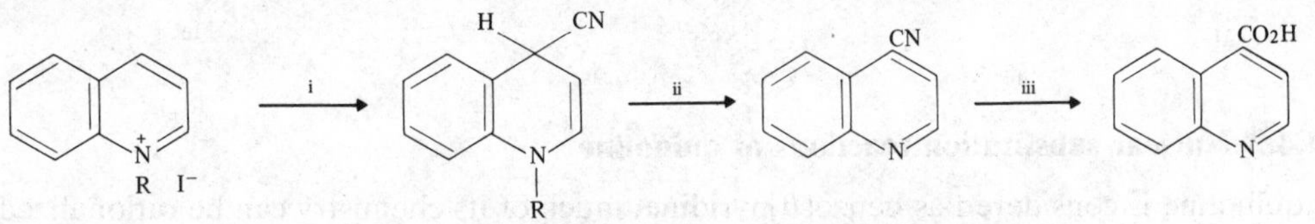

i ii iii (23)

$\overset{+}{N}$ R I^-; H CN N R; CN N; CO_2H N

i, CN^-; ii, (a) I_2 in pyridine, (b) heat (−RI); iii, H^+

other hand, results in addition at position-2 to give 1-acyl- or 1-aroyl-2-cyano-1,2-dihydroquinolines (**55**), known as Reissert compounds.[40] Reissert compounds can be converted into 2-carboxylic acids with concentrated hydrochloric acid or into 2-nitriles with phosphorus pentachloride (equation 24).

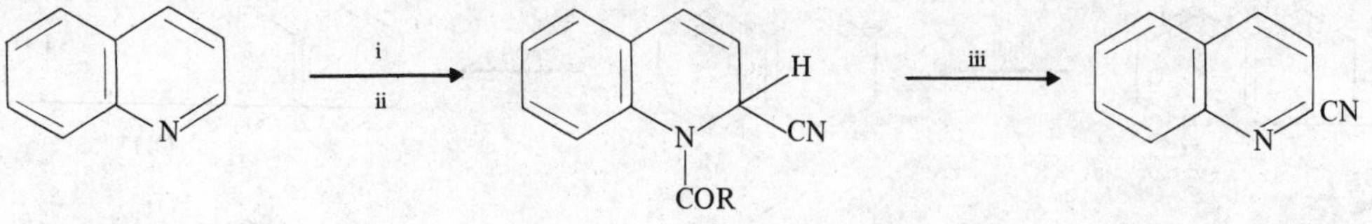

(24)

(55)

R = aryl or alkyl

i, RCOCl; ii, CN^-; iii, PCl_5 (−RCHO)

The carboxylic acids are believed to be formed through cyclic intermediates (Scheme 6) and not as a result of hydrolysis of the nitriles. Similar reactions occur with isoquinoline and phenanthridine but not with pyridine, which lacks the stabilizing influence of the

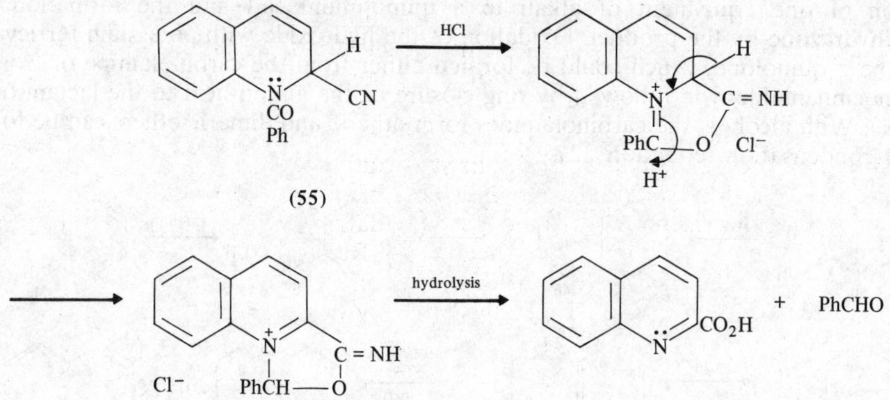

(55)

SCHEME 6

carbocyclic ring (see Section 16.4.4.2 below). The Reissert reaction (Scheme 6) has also been developed as a method for the conversion of acyl halides RCOCl to aldehydes RCHO.

Alkyl groups at positions-2 and -4 of the quinolinium ring readily lose an α-hydrogen to a base to give reactive tertiary amines with exocyclic double bonds; these undergo addition reactions as shown in equation 25.

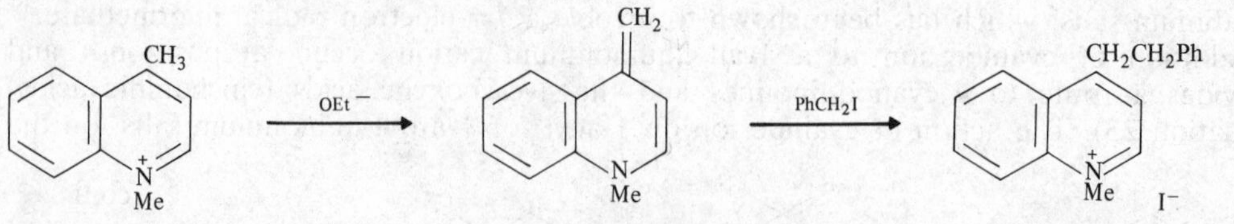

(25)

16.4.4.2 Nuclear substitution reactions of quinoline

If quinoline is considered as benzo[*b*]pyridine, much of its chemistry can be rationalized in terms of that of benzene (or naphthalene) and pyridine. The electron-rich nitrogen atom of the free base is the main centre of attack by electrophiles, but when this is protonated or otherwise quaternized, electrophilic substitution occurs preferentially at the carbons of the carbocyclic ring. As expected, if the transition states are similar to the σ-complexes, positions-5 and -8 are favoured kinetically as the corresponding intermediates can be represented by two canonical structures which still preserve the aromatic character of the heterocyclic ring, *e.g.* (**56**) and (**57**), whereas the intermediates resulting from electrophilic attack at positions-6 and -7 have only one, *e.g.* (**58**).

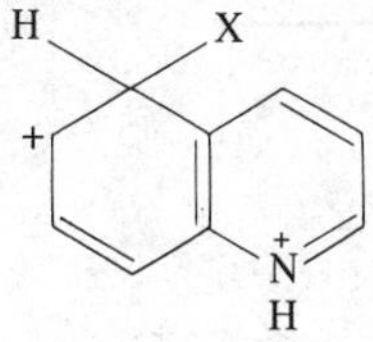

(56) 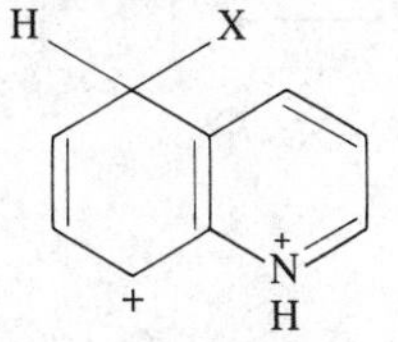 (57) 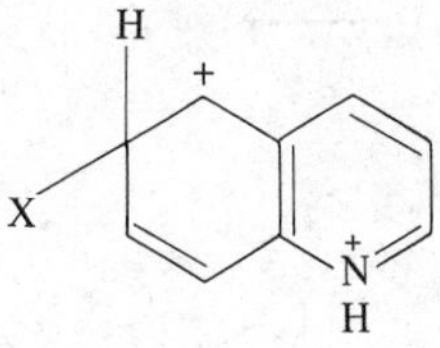 (58)

This is also consistent with the calculated values of the electron delocalization energies for electrophilic attack on the quinolinium cation.[34] As a π-deficient heterocycle, quinoline is less reactive towards electrophiles than benzene, particularly in acid solution. Nucleophilic attack, on the other hand, is favoured especially at the electron-deficient positions-2 and -4 in the heterocyclic ring and the quinolinium ion is even more reactive. The course of homolytic substitution is more difficult to predict as it depends on the reactivity of the radical and the state of protonation of the substrate, but the more nucleophilic radicals in acid media attack positions-2 and -4 specifically.

In non-acidic media, addition compounds may be formed which lead to electrophilic substitution at position-3 or to nucleophilic substitution at positions-2 and -4. Oxidation as an electrophilic process requires stronger conditions than for benzene and in general the heterocyclic ring is more resistant to degradation than the carbocyclic ring, but substituent effects may reverse this tendency. Reduction of the electron-deficient heterocyclic ring occurs readily, but again substituents can reverse the reactivities of the two ring systems. These principles are illustrated by the following specific examples.

(*i*) *Nitration*

Nitric acid and sulphuric acid at 0 °C give a mixture of 5- and 8-nitroquinolines in almost equal proportions, and kinetic studies[41] have shown that the reacting species is the quinolinium cation. Although more reactive than pyridine, the partial rate factor for positions-5 and -8 is only 4×10^{-7}. More vigorous conditions give a mixture of 5,7- and 6,8-dinitroquinolines with the second nitro-group entering *meta* to the first (equation 26).

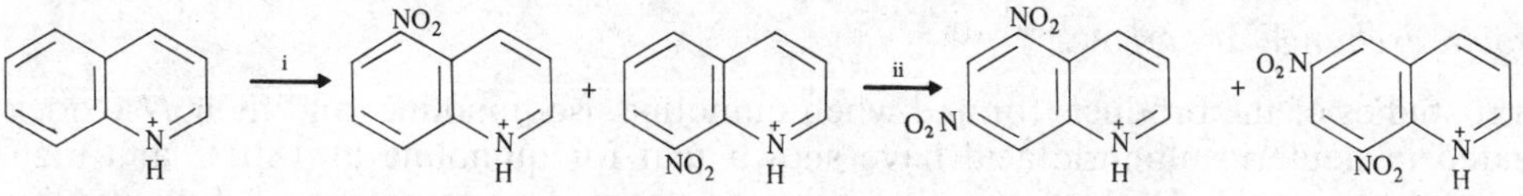

(26)

i, HNO_3, H_2SO_4, 0 °C; ii, HNO_3, H_2SO_4, heat

Further nitration of 6- and 7-nitroquinolines by heating the mononitro-compounds in sealed tubes with the calculated amount of potassium nitrate in sulphuric acid at 130–140 °C gives mixtures of 5,6- and 6,8-dinitroquinolines from 6-nitroquinoline and 5,7- and 7,8-dinitroquinolines from 7-nitroquinoline. Here the higher reactivities of positions-5 and -8 enable them to compete successfully against the directing influence of the first nitro-group.

With acetic anhydride as solvent, or with dinitrogen tetroxide as nitrating agent, the main nitration product is 3-nitroquinoline, but in very low yield (~6%). Some 6- and 8-nitroquinolines are also formed but, significantly, none of the 5-isomer. This is consistent with the intermediate formation of an addition product,[35e] such as that shown in Scheme 7. Key intermediates such as (**59**) are also possible for substitution at positions-6 and -8 but not at positions-5 or -7.

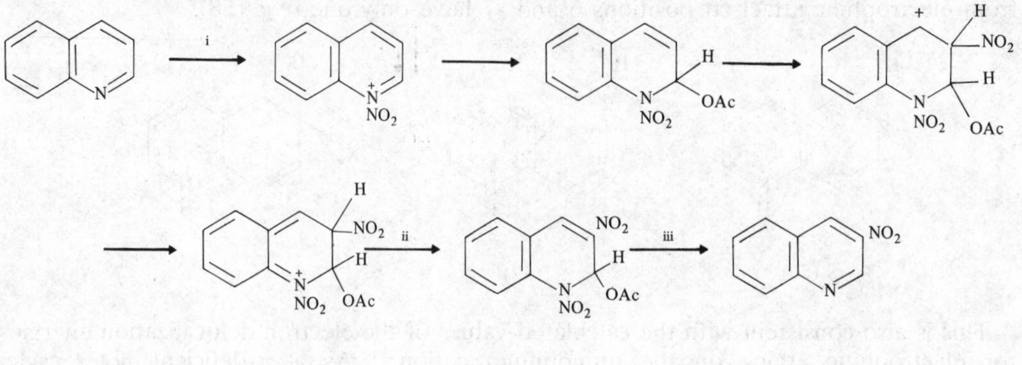

(59)

i, $HNO_3 + Ac_2O$; ii, $-H^+$; iii, $-NO_2^+$, $-OAc^-$

SCHEME 7

Nitrato-derivatives of transition metals have been used for the nitration of quinoline[42] and tetranitratotitanium(IV), $Ti(NO_3)_4$, in carbon tetrachloride gave 3-nitroquinoline (87%) and 8-nitroquinoline (10%), whereas tetranitratozirconium(IV), $Zr(NO_3)_4$, gave 7-nitroquinoline (90%).

The nitration of 5- and 7-methylquinolines[43] provides a further illustration of the directing influence of substituents and also of 'bond fixation'. The 5-methylquinoline is nitrated at positions-6 and -8, both of which are conjugated (*ortho* and *para*) with the methyl group, but 7-methylquinoline is nitrated only at position-8, presumably because of the lack of conjugation across the 6,7-bond to the other *ortho*-position.

The reduced reactivity of even the carbocyclic ring in quinoline compared with benzene, and the increased electron-attracting power of the charged quinolinium system, was shown[41a] by the nitration of 2-phenylquinoline and 1-methyl-2-phenylquinolinium sulphate with nitric acid under identical conditions. Only the phenyl rings were nitrated, and whereas the quaternary salt gave 97% of the *m*-nitrophenyl product, 2-phenylquinoline, which is incompletely protonated under these conditions, gave 2-*m*-nitrophenylquinoline (40%) and 2-*p*-nitrophenylquinoline (60%). 4-Phenylquinoline behaves similarly but also gives 5% of the *o*-nitrophenylquinoline.

(*ii*) *Proton exchange at carbons*[43a]

N.m.r. studies of the products formed when quinoline, isoquinoline, and their *N*-oxides are heated in deuteriosulphuric acid have shown that for quinoline at 180 °C and high acidities (50–90% acid, $D_0 = -4$ to -9), exchange occurs first at position-8 followed by positions-5 and -6. Under these conditions no exchange was observed at other positions over prolonged periods. At 245 °C and lower acidities ($D_0 = 0.3$ to -4.8), exchange occurred first at position-2, followed at increasing acidities by positions-3, -8, -5, and finally -6 and -7. No exchange was observed at position-4. Reaction rates at positions-2 and -3 decreased with increasing acidity, but the rates at the other positions increased. Since increasing acidity would be expected to reduce the activities of the basic species but increase the activities of electrophilic species, the results are considered to be consistent with exchange at positions-8, -5, -6, and -7 by direct electrophilic attack on the

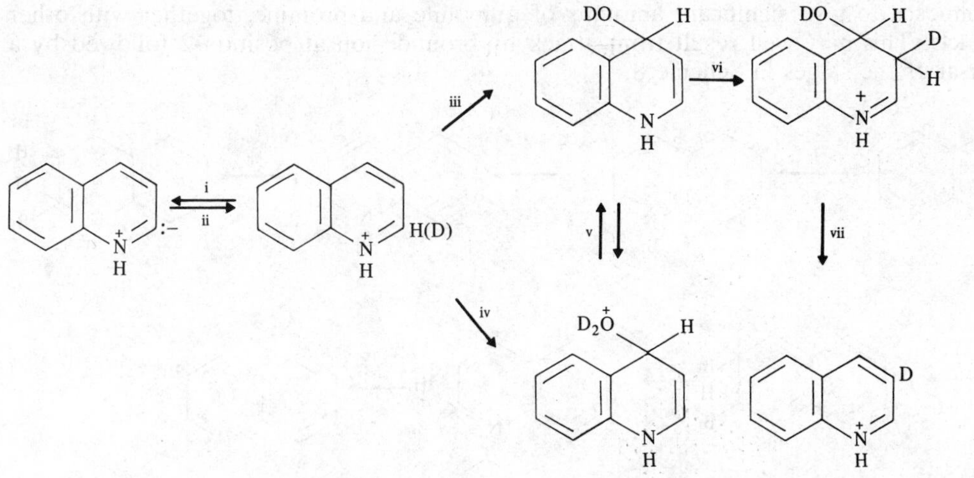

(27)

Conditions: 245 °C, 40% D_2SO_4. i, + OH^- (OD^-), – H_2O (HOD); ii, + H^+(D^+); iii, + OD^- (slow); iv, + D_2O, (slow); v, fast; vi, + D^+ (fast); vii, –HOD (fast)

quinolinium cation, at position-2 following deprotonation by a base, and at position-3, (at least up to $D_0 = -2$) by an addition–elimination sequence (equation 27). The overall pattern is then similar to that for other electrophiles such as NO_2^+ (see above) and Br^+ (see below).

(*iii*) *Sulphonation*

Sulphuric acid at 220 °C gives mainly the 8-sulphonic acid with some of the 5-isomer. At 300 °C the thermodynamically favoured 6-sulphonic acid is the main product. Under these conditions both the 5- and the 8-isomers undergo rearrangement to the 6-isomer (equation 28). This behaviour parallels the relationship between the α- and β-naphthalene sulphonic acids.

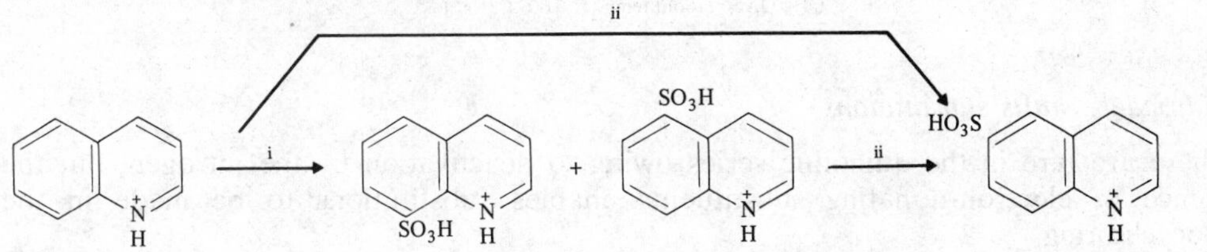

(28)

i, H_2SO_4, 220 °C, or 35% oleum, 90 °C; ii, H_2SO_4, 300 °C

(*iv*) *Halogenation*

Bromine and silver sulphate in sulphuric acid solution[44] give good yields of the 5- and 8-bromoquinolines in almost equal proportions by the normal electrophilic mechanism; chlorine and iodine behave in a similar manner.[45] Bromine at 300 °C gives 3-bromoquinoline[46] in 25% yield and at 450–500 °C 2-bromoquinoline in 50–60% yield, in this case by a homolytic mechanism. Bromine in carbon tetrachloride and quinoline gives a good yield of 3-bromoquinoline by an addition–elimination mechanism similar to that for nitration in acetic anhydride[35e] (Scheme 8). Sulphuryl chloride, under similar conditions, gives 3-chloroquinoline. When heated strongly the hydrobromide of 3-bromoquinoline decomposes to give significant amounts of quinoline and bromine, together with other products. This may well result from attack by bromide ion at position-2 followed by a reversal of the stages in Scheme 8.

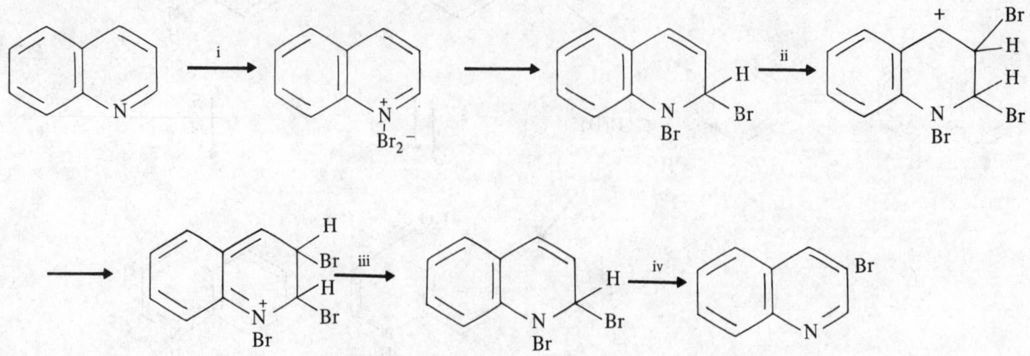

i, Br_2, CCl_4, pyridine; ii, Br^+; iii, $-H^+$; iv, $-Br_2$

SCHEME 8

(*v*) *Mercuration*

This is an example of electrophilic substitution under non-acidic conditions, and mercuric acetate readily gives the quaternary 1-mercuriacetate. At 160 °C further substitution occurs and treatment with sodium chloride gives the 3- and 8-mercurichlorides (equation 29).

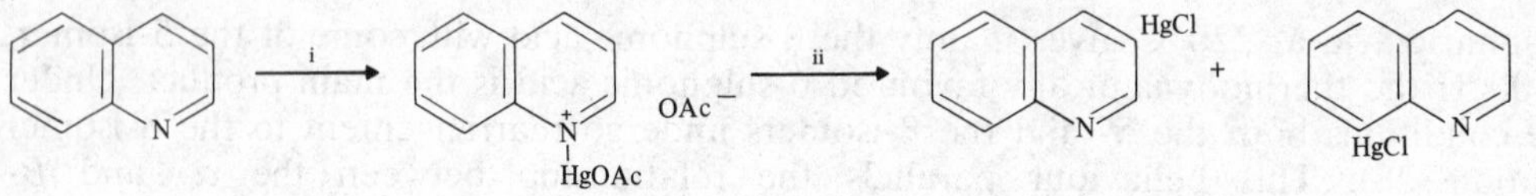

(29)

i, $Hg(OAc)_2$, room temp.; ii, 160 °C, NaCl

(*vi*) *Friedel–Crafts substitutions*

These are rare in the quinoline series owing to deactivation by the nitrogen, but the presence of electron-donating substituents enables substitutions to be made in the carbocyclic ring.[47]

(*vii*) *Amination*

Amination by direct nucleophilic attack by the amide ion (Chichibabin reaction[48]) occurs preferentially at position-2. This position and position-4 are more reactive than the corresponding positions in pyridine as a result of stabilization of the addition product by the second aromatic ring [see (**60**)]. With potassium amide in the presence of potassium nitrate as a hydride ion acceptor, 10% of 4-aminoquinoline is formed, and if position-2 is blocked (*e.g.* by phenyl) high yields of the 4-amino-product are formed. The use of barium amide in liquid ammonia under pressure at room temperature gives the highest yield of 2-aminoquinoline (80%) and the addition of a soluble barium salt improves the yields with other cations. In addition to liquid ammonia, xylene, toluene, and dimethylaniline have been used as solvents (equation 30).

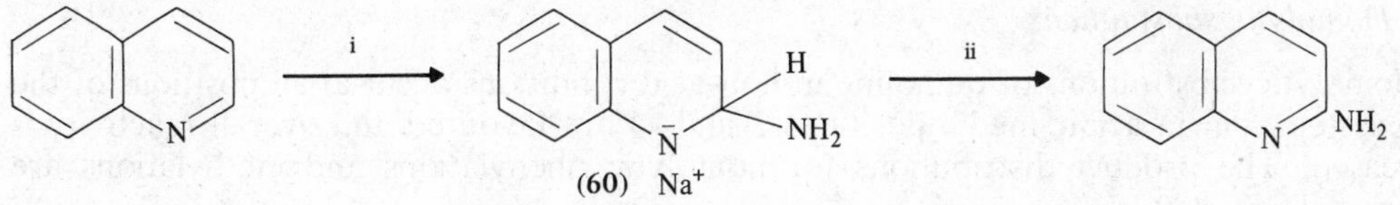

(30)

(60) Na^+

i, $NaNH_2$, xylene, 100 °C; ii, $-H^-$; yield 40%

(*viii*) *Alkylation and arylation*

Alkylation and arylation by organolithium compounds or Grignard reagents occur by a similar nucleophilic mechanism and give mainly 2-substituted quinolines. As with amination, the final stage involves the loss of hydride ion so that a mild oxidizing agent such as nitrobenzene is an advantage (equation 31). Treatment of quinoline with methyl

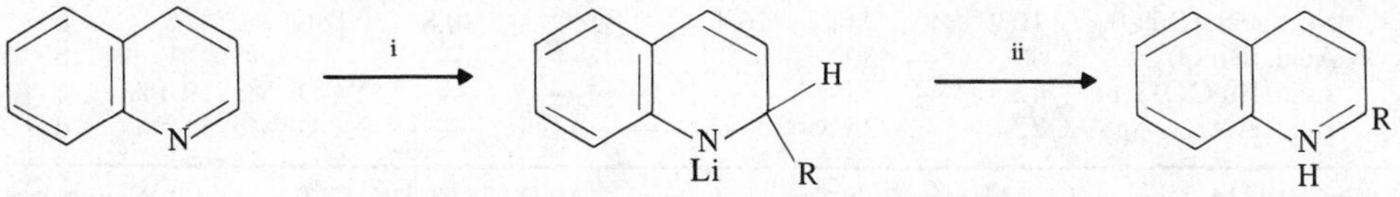

(31)

i, LiR, benzene; ii, H_2O, $PhNO_2$ $(-H^-)$; R = alkyl or aryl

sulphinylmethylpotassium[48a] in DMSO results in alkylation at position-4 in high yield (90%), a reaction which does not occur with pyridine (equation 32).

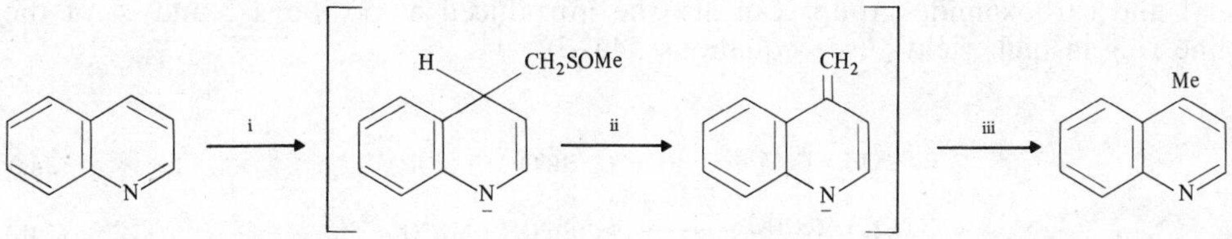

(32)

i, $MeSO\bar{C}H_2\ \overset{+}{K}$, DMSO, 70 °C (+ $MeSO\bar{C}H_2$); ii, $-$HOSMe; iii, + H^+; yield 90%

(*ix*) *Hydroxylation*

Hydroxylation to 2-quinolone can be accomplished by heating with potassium hydroxide (equation 33a) or by the action of hypochlorites (equation 33b). Both reactions proceed through a 1,2-addition product; in the latter case initial attack could be by either Cl^+ or OH^-.

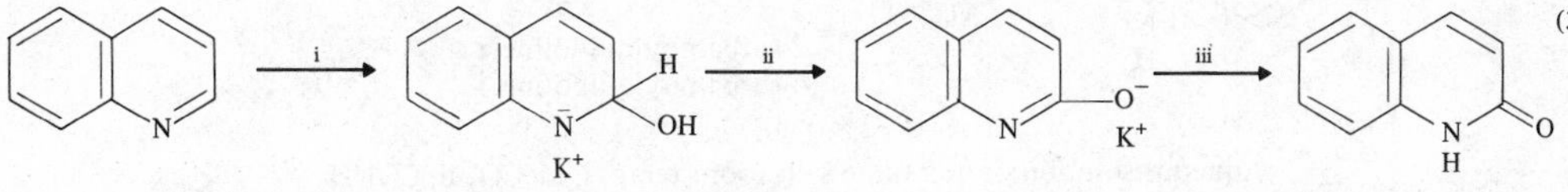

(33a)

i, KOH, 250 °C; ii, $-H_2$; iii, H^+

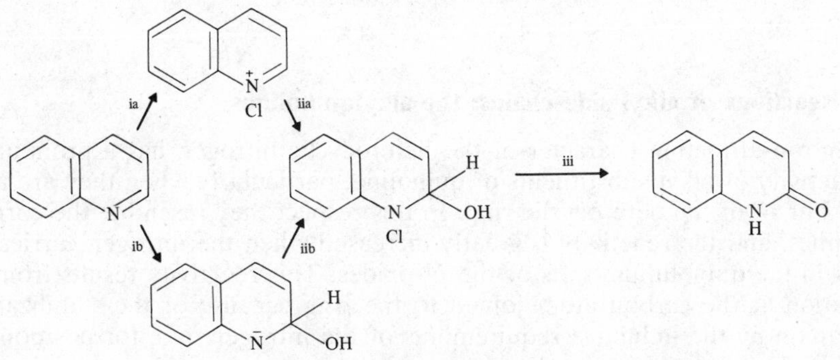

(33b)

ia, NaOCl (Cl^+); ib, NaOCl (OH^-); iia, OH^-; iib, Cl^+; iii, NaOH ($-$HCl)

(*x*) *Homolytic substitutions*

Homolytic substitutions of quinoline in non-acid conditions occur at all positions of the ring system, but in acidic media positions-2 and -4 are favoured and overall reactivity is increased. The product distributions for homolytic phenylations and methylations are given in Table 8.

TABLE 8
Homolytic Substitutions of Quinoline

Reagent	*Conditions and radical source*	*Position and % product* 2	3	4	5	6	7	8	*Notes*	*Ref.*
Ph·	Non-acid, $(PhCO_2)_2$	11	6	18	22	5.5	5.5	32		a
Ph·	Acid, $(PhCO_2)_2$	35	*	38	8	*	*	13	* Total 6%	a
Me·	Non-acid, $(Bu^tO)_2$	10.2	†	24.2	16.1	†	†	30.8	† Total 18.7%	b
Me·	Acid, $(Bu^tO)_2$	49	—	50	—	—	—	—		b
Me·	Acid, Bu^tOOH, Fe^{2+}	42.5	—	38.4	—	—	—	—	2,4-Di-Me, 19.1%	c
Me·	Acid, AcOH, Ag^{2+}	23	—	25.5	—	—	—	—	2,4-Di-Me, 51.5%	d

[a] G. Vernin, H. J. M. Dou, and J. Metzzer, *Bull. Soc. chim. France*, 1971, 2612. [b] K. C. Bass and P. Nababsing, *J. Chem. Soc.* (*C*), 1969, 388; 1970, 2169. [c] T. Caronna, G. P. Gardini, and F. Minisci, *Chem. Comm.*, 1969, 201. [d] See Ref. 49.

Acetyl and carboxamide groups can also be introduced at positions-2 and -4 of the quinoline ring in high yield[49] (see equations 34a–d).

$$Bu^tOOH + Fe^{2+} \longrightarrow Bu^tO\cdot + Fe(OH)^{2+} \qquad (34a)$$

$$Bu^tO\cdot + MeCHO \longrightarrow Bu^tOH + Me\dot{C}O \qquad (34b)$$

$$Bu^tO\cdot + HCONH_2 \longrightarrow Bu^tOH + \dot{C}ONH_2 \qquad (34c)$$

Quinolinium ($\overset{+}{N}$–H) —i→ 2-acetylquinoline, 4-acetylquinoline } 15.4% + 2,4-diacetyquinoline 61.6% (34d)

—ii→ 2-carbamoylquinoline, 4-carbamoylquinoline } 92%

Aqueous solution of quinoline salt, room temp. i, $Me\dot{C}O$; ii, $\dot{C}ONH_2$

(Ref. 49, pp. 150 and 166)

16.4.4.3 Reactions of alkyl side-chains: the alkylquinolines

The electron-attracting character of the heterocyclic nitrogen has a pronounced effect on the reactivity of alkyl substituents of quinoline, particularly when they are attached to positions-2 or -4 of the heterocyclic ring. In this respect they resemble the corresponding alkylpyridines, and the reactivity is greatly increased when the nitrogen carries a positive charge, as in the quinolinium salts or the *N*-oxides. This reactivity results from the facile deprotonation at the carbon atom joined to the ring because of the stabilization of the resulting anion by the inductive requirements of the nitrogen and, for positions-2 and -4 (but not -3), further stabilization by mesomerism as shown in (**61**).

$$\left[\text{quinolin-2-yl}\text{-}\bar{C}H_2 \longleftrightarrow \text{quinolin-}\bar{N}\text{-}2\text{-ylidene}{=}CH_2 \right]$$

(61)

A typical reaction is condensation with aldehydes (equations 35a and 35b). The free bases require an acid catalyst, such as hydrogen chloride or zinc chloride, but the quaternary salts condense readily with a trace of piperidine. Benzaldehydes give the

$$\text{4-}CH_3\text{-quinoline} + MeOC_6H_4CHO \xrightarrow[\text{ii}]{\text{i}} \text{4-}(CH{=}CHC_6H_4OMe)\text{-quinoline} \quad (35a)$$

(62)

i, HCl, 160 °C, 4 h; ii, NH_4OH; yield 29%

$$\text{2-}CH_3\text{-}N^+\text{-Me quinolinium } I^- + Me_2NC_6H_4CHO \xrightarrow{\text{i}} \text{2-}(CH{=}CHC_6H_4NMe_2)\text{-}N^+\text{-Me quinolinium } I^- \quad (35b)$$

i, piperidine

(63)

benzylidene derivatives (styrylquinolines) (**62**) and (**63**) and, in special cases, *e.g.* with formaldehyde, the intermediate carbinol (**64**) can be isolated (see also Section 16.4.6.1). Ease of deprotonation is in the order 4>2>3 but, possibly owing to steric effects, condensation usually occurs most readily at position-2 (equation 36). Self-condensation of these quinolinium salts to give the important synthetic cyanine dyes is discussed in Section 16.4.6.2.

$$\text{2,4-}(CH_3)_2\text{-quinoline} \xrightarrow{\text{i}} \text{4-}CH_3\text{-2-}CH_2CH_2OH\text{-quinoline} \xrightarrow{\text{ii}} \text{4-}CH_3\text{-2-}CH(CH_2OH)_2\text{-quinoline} \quad (36)$$

(64)

i, 1 mol HCHO; ii, 1 mol HCHO

The deprotonated 2- and 4-alkylquinolinium salts can be formulated as methylene bases (**66**). These are highly reactive (*cf.* enamines) but, unlike the corresponding pyridine compounds, they can usually be isolated. From a study of the reactions of the methylene base (**66**) of benzo[*f*]quinolinium ethiodide (**65**), Mills and Raper[50] proposed the mechanism shown in Scheme 9 for the condensation of alkylquinolinium salts with aldehydes.

The necessity for an acidic condensing agent for the free bases suggests that coordination to nitrogen may be necessary to enable a methylene base to be formed, which could then follow the sequence suggested in Scheme 9. Equation 37 shows the possible function of zinc chloride. The carbanions, formed by deprotonation of the alkylquinolines at the

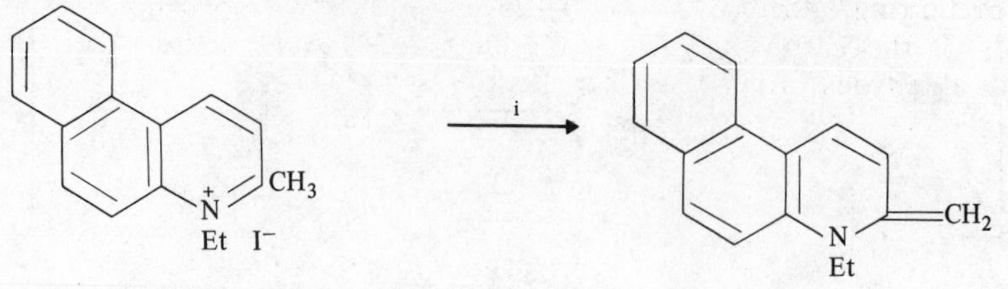

(65) $\xrightarrow{\text{i}}$ (66) $\xrightarrow{\text{ii}}$ $\xrightarrow{\text{iii}}$

CH$_3$, $\overset{+}{N}$, Et I$^-$; N, Et, =CH$_2$; N, Et, =CHCHC$_6$H$_4$NMe$_2$, OH; $\overset{+}{N}$, Et I$^-$, CH = CHC$_6$H$_4$NMe$_2$

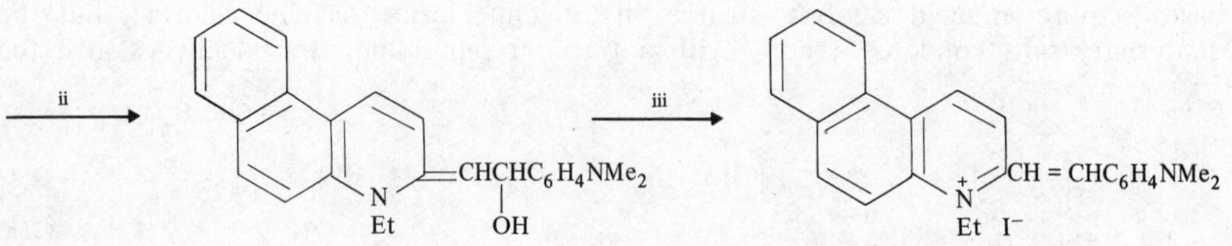

i, Piperidine (–HI); ii, $Me_2NC_6H_4CHO$; iii, + HI (from piperidinium iodide)

SCHEME 9

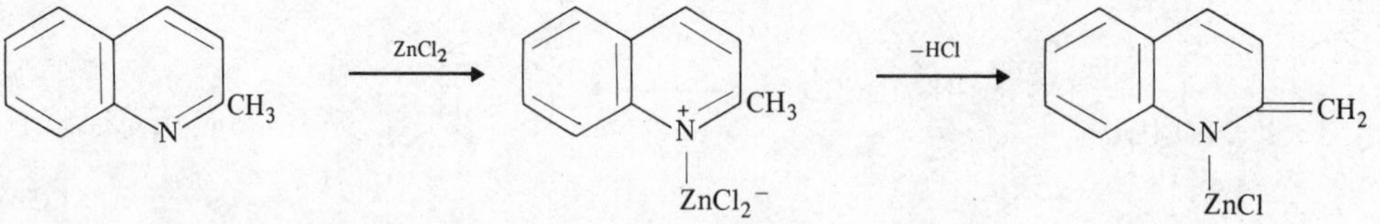

$\xrightarrow{ZnCl_2}$ $\xrightarrow{-HCl}$ (37)

N, CH$_3$; $\overset{+}{N}$, CH$_3$, ZnCl$_2$$^-$; N, =CH$_2$, ZnCl

carbon atom next to the ring with amide ion, participate in Claisen condensations and Michael additions; they can also be alkylated by alkyl halides (equations 38 and 39).

N, CH$_3$ + PhCO$_2$Et $\xrightarrow{\text{i}}$ N, CH$_2$COPh (38)

i, KNH_2 in liq. NH_3

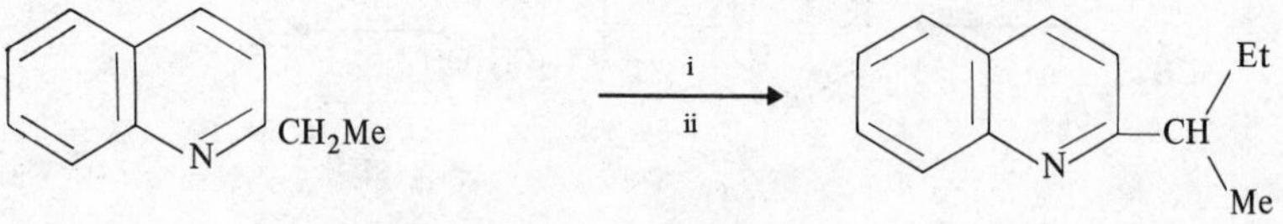

N, CH$_2$Me $\xrightarrow[\text{ii}]{\text{i}}$ N, CH(Et)(Me) (39)

i, KNH_2 in liq. NH_3; ii, EtBr

Although oxidation of quinoline and alkylquinolines with potassium permanganate or hydrogen peroxide usually results in degradation of the ring, as described below, it is possible to oxidize the alkyl side-chains to hydroxycarbonyl (carboxyl) with chromic acid (CrO_3) or, with better results, potassium dichromate and sulphuric acid.[51] An intermediate carbonyl state can sometimes be isolated (*e.g.* acetyl from ethyl groups) and methyl groups can be oxidized to formyl (aldehyde) with selenium dioxide. The ease with which oxidation to carboxylic acids occurs depends upon the position of alkylation and follows the sequence 8 > (5, 6, 7, or 3) > 4 > 2, but methyl groups in the 2- or 4-positions are the most easily converted to aldehyde by selenium dioxide,[52] followed by substituents in the carbocyclic ring,[53] with the 3-methyl group least reactive[54] (equations 40 and 41). Higher yields of the carboxylic acids are obtained by oxidation of the condensation products with aldehydes rather than the alkylquinolines themselves.

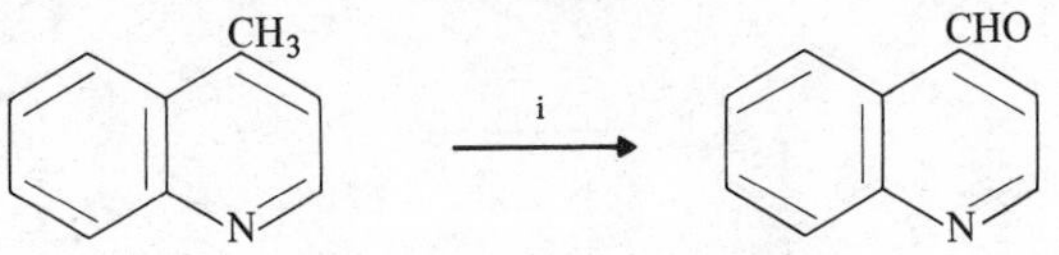

(40)

i, SeO_2, 135°C, 1 h, xylene; yield 61%

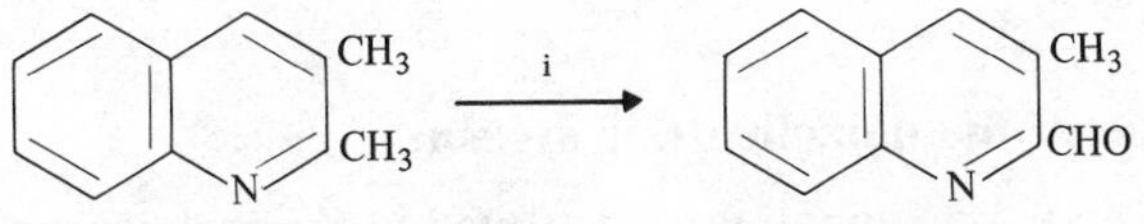

(41)

i, SeO_2, 130–135 °C, 1.5 h, xylene; yield 50%

16.4.4.4 Oxidation of the quinoline ring system

The course of degradative oxidation of quinoline and its derivatives is complex and either ring may be opened.[55] Quinoline[56] itself and alkylquinolines substituted in the carbocyclic ring give pyridine 2,3-dicarboxylic acid (quinolinic acid) (equation 42). Lepidine (4-methylquinoline) behaves similarly to give 4-methylquinolinic acid.

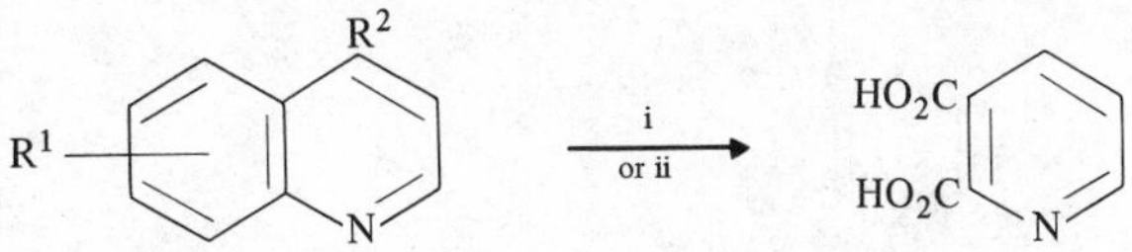

(42)

i, alkaline $KMnO_4$; ii, $H_2O_2 + Cu^{2+}$; R^1 = 5 -, 6 -, 7 -, or 8 - alkyl or H;
R^2 = Me or H

Quinolines with alkyl or aryl substituents at position-2 suffer ring opening of the heterocyclic ring[57] on oxidation, preferably with acid permanganate, to give acyl- or aroyl-anthranilic acids (equation 43). An alkyl substituent at position-3 induces the disruption of both rings on oxidation[55] to give oxalic acid, carbon dioxide, and ammonia.

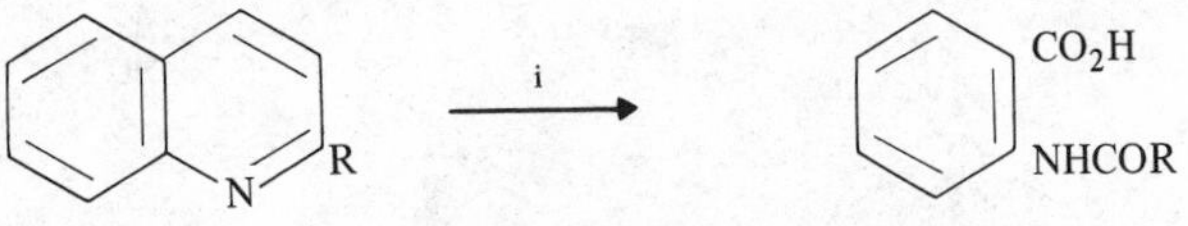

(43)

i, acid $KMnO_4$; R = alkyl or aryl

The products of ozonolysis are consistent with the proposed structure of quinoline and also with the variations in double-bond character within the ring.[58] Ozone attacks the 5,6- and 7,8-bonds rapidly and the 3,4- and 8a,4a-bonds more slowly. Reduction of the appropriate ozonides of quinoline (**67**) gave glyoxal and pyridine-2,3-dicarbaldehyde, of 5,8-dimethylquinoline (**68**) gave glyoxal and 2,3-diacetylpyridine, and of 2,3-dimethylpyridine (**69**) gave glyoxal and diacetyl, indicating ring fission as indicated. With peracetic

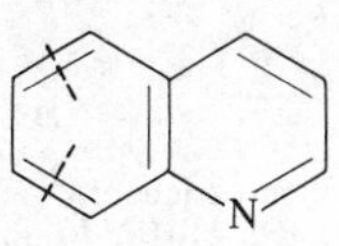

(67) 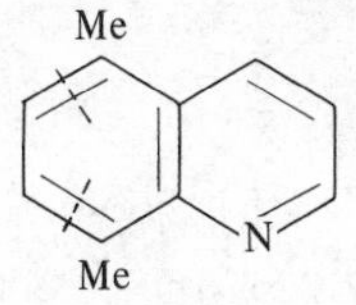 (68) 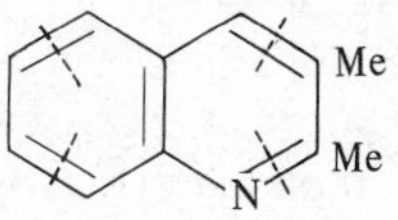 (69)

acid and other percarboxylic acids, quinoline, in common with pyridine and other tertiary amines, is converted into the 1-oxide (equation 44), which is an important synthetic intermediate discussed in more detail below.

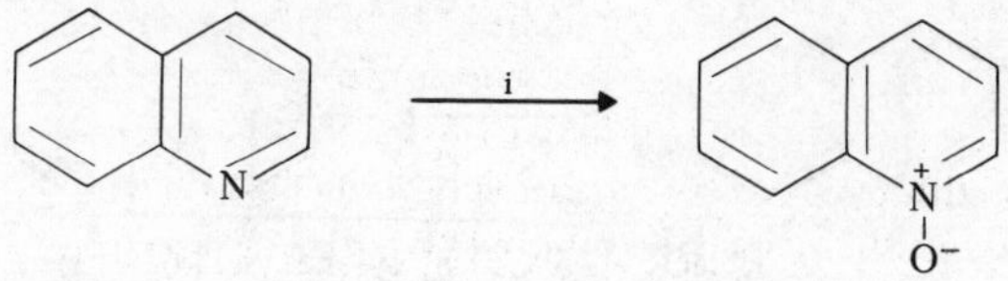

(44)

i, $CH_3CO_2H + H_2O_2$, 65 °C; yield 95%

16.4.4.5 Reduction of the quinoline ring system

Quinoline itself and alkylquinolines substituted in the carbocyclic ring are preferentially reduced in the heterocyclic ring. Normally two, four, or ten hydrogens are added, although derivatives of octahydroquinoline have been reported. With sodium in liquid ammonia, lithium aluminium hydride, or diethylaluminium hydride, quinoline gives high yields of 1,2-dihydroquinoline, and the 1,4-isomer has been prepared[59] by reduction with lithium in liquid ammonia at −33 °C. The two compounds are readily distinguished by their u.v. spectra and n.m.r. coupling constants, *viz.* 1,2-dihydroquinoline, $\lambda_{max} = 345$ nm, $J_{3,4} = 10$ Hz; 1,4-dihydroquinoline, $\lambda_{max} = 298$ nm, $J_{2,3} = 7.6–8.2$ Hz. The 5,6- and 5,8-dihydroquinolines have also been prepared.

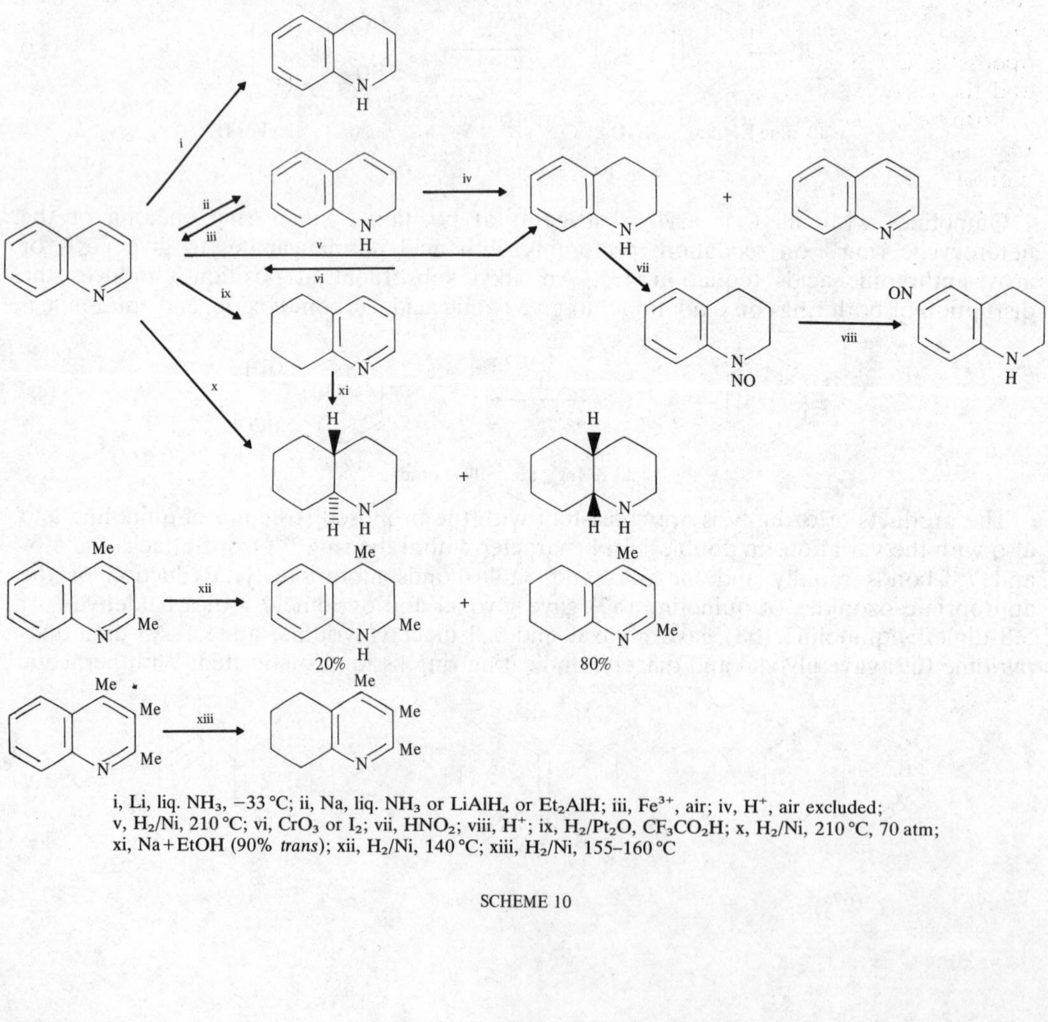

i, Li, liq. NH_3, −33 °C; ii, Na, liq. NH_3 or $LiAlH_4$ or Et_2AlH; iii, Fe^{3+}, air; iv, H^+, air excluded; v, H_2/Ni, 210 °C; vi, CrO_3 or I_2; vii, HNO_2; viii, H^+; ix, H_2/Pt_2O, CF_3CO_2H; x, H_2/Ni, 210 °C, 70 atm; xi, Na + EtOH (90% *trans*); xii, H_2/Ni, 140 °C; xiii, H_2/Ni, 155–160 °C

SCHEME 10

In acid solution and in the absence of air, 1,2-dihydroquinoline disproportionates to quinoline and 1,2,3,4-tetrahydroquinoline. The tetrahydroquinoline can be prepared directly by reduction of quinoline with sodium and alcohol, tin and hydrochloric acid, or by catalytic hydrogenation;[59] it is readily oxidized back to quinoline by chromic acid or even by mild oxidizing agents such as iodine. More vigorous reduction with hydrogen iodide and phosphorus, or by catalytic hydrogenation, gives a mixture of *cis*- and *trans*-decahydroquinolines containing from 35% to 80% of the *trans*-isomer depending on the reaction conditions. Reduction of the carbocyclic ring to give 5,6,7,8-tetrahydroquinolines becomes increasingly significant as alkyl substituents are introduced into the heterocyclic ring,[60] and 2,3,4-trimethylquinoline is converted completely to the 5,6,7,8-tetrahydro-derivative. Quinoline itself can be reduced in the carbocyclic ring[61] using a platinum oxide catalyst in trifluoroacetic acid. Further reduction with sodium and alcohol then gives *trans*-decahydroquinoline in 90% yield. 1,2,3,4-Tetrahydroquinoline behaves chemically as an aromatic secondary amine and its *N*-nitroso-derivative undergoes an acid-catalysed rearrangement to 6-nitroso-1,2,3,4-tetrahydroquinoline (Scheme 10).

16.4.4.6 Ring-opening reactions

Both the carbocyclic and heterocyclic rings of quinoline can be opened by oxidation as described in Section 16.4.4.4 above. The formation of the *o*-aminocinnamaldehyde tautomer of the carbinolamine base from 1-methylquinolinium hydroxide constitutes an opening of the heterocyclic ring (see Section 16.5.4.1 above). Opening of the heterocyclic ring also results from the Hofmann and Emdé exhaustive methylation reactions, although success in the first stage by the Hofmann method requires distillation under reduced pressure to remove the 2-allyldimethylaniline[62] (**70**). The Hofmann reaction (Scheme 11) opens the 2,3-bond and the Emdé procedure opens the 1,8a-bond between the nitrogen and the carbocyclic ring.

With benzoyl chloride and phosphorus pentoxide under the conditions of the von Braun reaction,[63] 1,2,3,4-tetrahydroquinoline undergoes ring contraction to the isomeric 2-methyl-2,3-dihydroindole (**71**); Scheme 12 is a suggested, but probably oversimplified, reaction sequence.

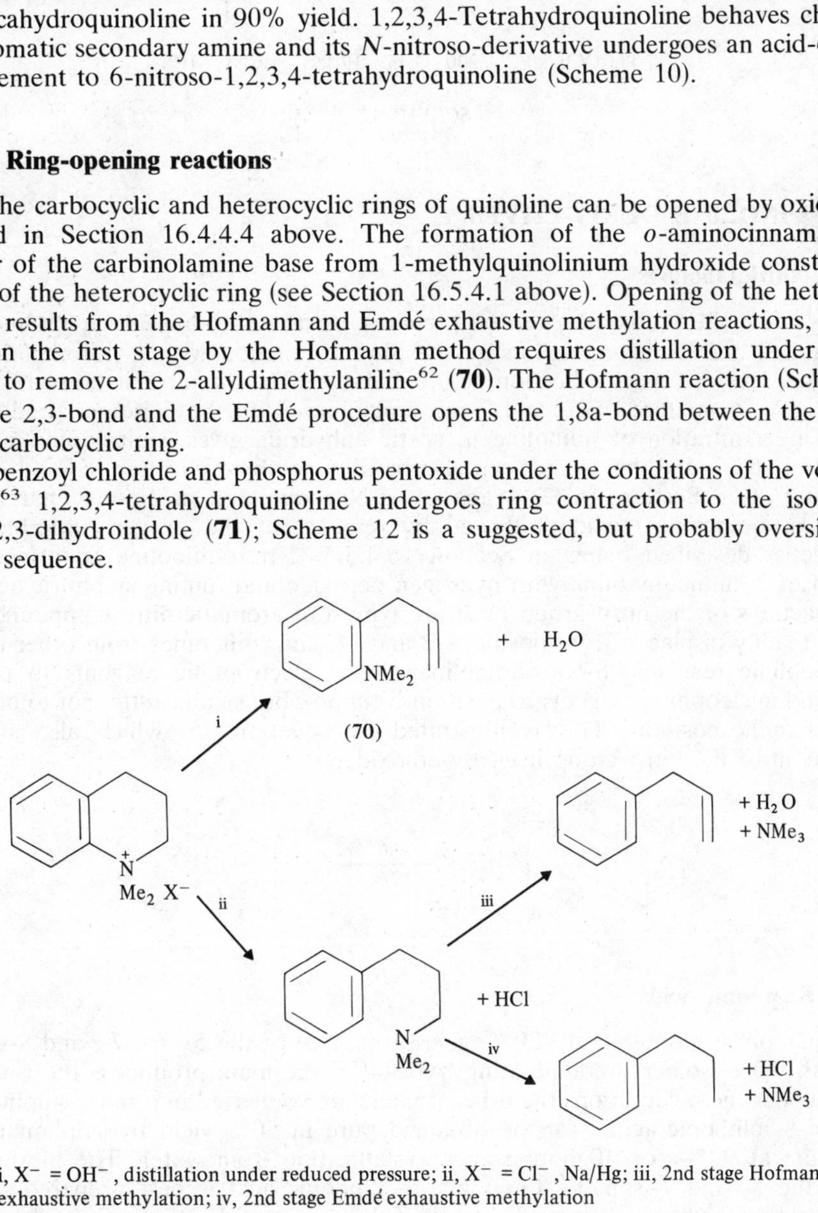

i, X^- = OH^-, distillation under reduced pressure; ii, X^- = Cl^-, Na/Hg; iii, 2nd stage Hofmann exhaustive methylation; iv, 2nd stage Emdé exhaustive methylation

SCHEME 11

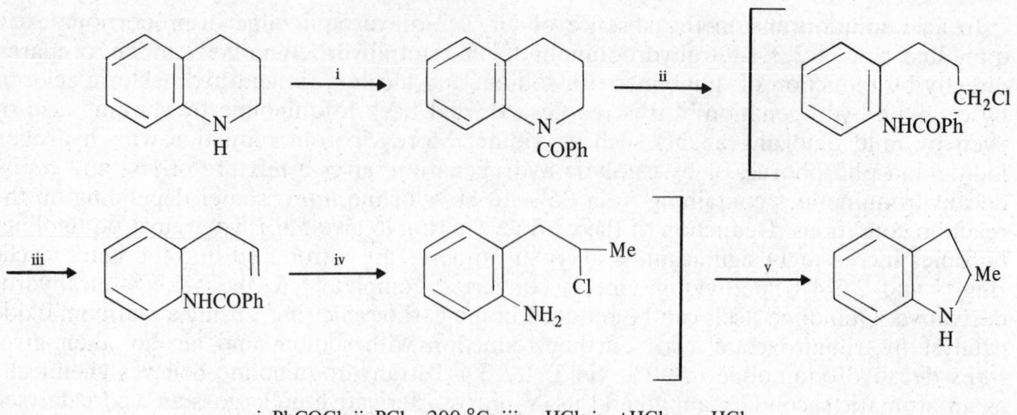

i, $PhCOCl$; ii, PCl_5, 200 °C; iii, –HCl; iv, +HCl; v, –HCl

SCHEME 12

16.4.5 QUINOLINE DERIVATIVES

16.4.5.1 Nitroquinolines

The 5- and 8-nitroquinolines can be prepared by direct nitration of quinoline, and 8-nitroquinoline and 6-nitroquinoline by the Skraup reaction with *o*-nitroaniline and *p*-nitroaniline, respectively (see Section 16.4.2.2 above). The Skraup reaction with *m*-nitroaniline[10] gives a mixture of 5- and 7-nitroquinolines with the 5-isomer predominating (78%). Direct nitration of quinoline in acetic anhydride gives 3-nitroquinoline, but in poor yield (~6%), and this isomer can also be obtained by cyclization of 2-nitromalonodialdehyde[20] (see Section 16.4.2.4 above). 4-Nitroquinoline is best prepared by the nitration of quinoline 1-oxide followed by removal of the oxygen with phosphorus trichloride as described below in Section 16.4.5.5; 2-nitroquinoline is formed by the oxidation of 2-aminoquinoline with hydrogen peroxide and fuming sulphuric acid.

The reactions of the nitro-group itself are typical of aromatic nitro-compounds except that it is readily displaced from positions-2 and -4, and sometimes from other positions, by nucleophilic reagents. 6-Nitroquinoline directs electrophilic reagents to position-8 (*meta*), and nucleophilic reagents to position-5 (*ortho*), but significantly, not to position-7, the other *ortho*-position. This is illustrated by equation 45, which also shows the displacement of the nitro-group itself by ethoxide.

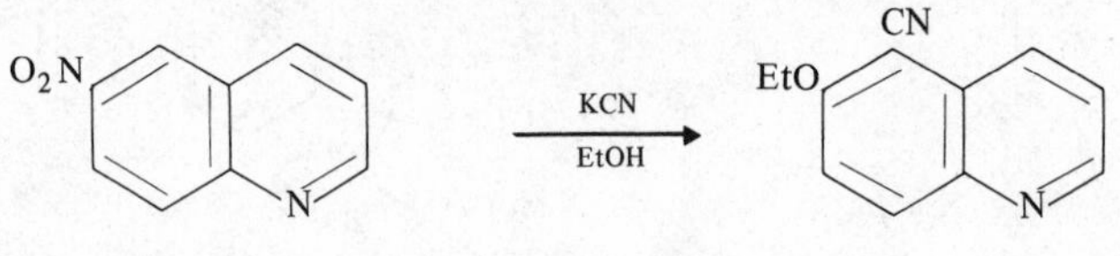

O_2N — KCN / EtOH → CN, EtO (45)

16.4.5.2 Sulphonic acids

Sulphonation of quinoline at 220 °C gives a mixture of the 5-, 6-, 7-, and 8-sulphonic acids, with the 8-isomer predominating. At 300 °C the main product is the 6-sulphonic acid and under these conditions the other isomers are converted into the 6-sulphonic acid. Quinoline-8-sulphonic acid[64] can be obtained pure in 50% yield by sulphonation with 35% oleum at 90 °C for 40 hours, and crystallization from water. The mother liquor contains the 5- and 7-isomers. Quinoline-6-sulphonic acid is best prepared from sulphanilic acid by the Skraup reaction, and the 5-sulphonic acid is the major product of this reaction with metanilic acid, but with orthanilic acid the sulphonic acid group is displaced.

The sulphonic acid group can be introduced into positions-2 and -4 by nucleophilic displacement of chlorine[65] (equation 46) and can itself be displaced by nucleophiles under more vigorous conditions. In other respects the reactions of quinolinesulphonic acids are typical of aromatic sulphonic acids in general.

(46)

i, $NaHSO_3$, (35% solution), boiling

16.4.5.3 Halogen compounds

Methods for the preparation of 2-, 3-, 5-, and 8-bromoquinolines and 3-chloroquinoline by direct substitution have been described above (Section 16.4.4.2). Other derivatives with halogens substituted in the carbocyclic ring are usually prepared from the appropriately substituted anilines by the methods discussed in Section 16.4.2 or from the amino-compounds by the Sandmeyer or Gattermann reactions. Probably the best method for the preparation of 2- and 4-chloroquinolines is by the action of phosphoryl chloride, or phosphoryl chloride and phosphorus pentachloride, on the 2- or 4-quinolones, which are usually readily available by the Conrad–Limpach or Knorr syntheses (Section 16.4.2.5), from *o*-aminocinnamic acid derivatives (Section 16.4.2.8), or from cinnamanilides (Section 16.4.2.9). The hydrochloride of quinoline *N*-oxide gives a mixture of the 2- and 4-chloroquinoline *N*-oxides, containing 62% of the 4-chloro-isomer, when treated with sulphuryl chloride or phosphoryl chloride.

Halogens substituted at the 3-position of quinoline and in the carbocyclic ring have properties generally typical of halogens attached to aromatic compounds except that they are more easily replaced by nucleophiles such as OEt^- or RNH_2. Halogens at the electron-deficient 2- and 4-positions are even more readily displaced, as in the preparation of quinoline-4-sulphonic acid (equation 46). In general, temperatures of 50–100 °C are sufficient for the 2- and 4-halogenoquinolines, whereas halogens at other positions require temperatures of 150–200 °C for nucleophilic displacement.

16.4.5.4 Hydroxyquinolines (quinolinols and quinolones)

The hydroxyquinolines fall into two main groups: those substituted in the carbocyclic ring or at position-3, the quinolinols, which have phenolic properties, and those substituted at positions-2 and -4, the quinolones, which, in spite of their name, have amido-properties. In all cases tautomerism can occur between the *O*-protonated and *N*-protonated forms.[66a] The quinolinols form an equilibrium mixture of the hydroxy-form (**72**) and the *N*-protonated zwitterionic form (**73**), as shown for 3-hydroxyquinoline (equation 47a). For the equilibrium as written, $pK_T = -1.08$, indicating an approximately

(47a)

(**72**) (**73**)

i, $pK_T = -1.08$

tenfold preponderance of the hydroxy-form.[67] The 2- and 4-isomers, however, as in the pyridones, exist almost entirely in the *N*-protonated form stabilized by mesomerism as shown for 2-quinolone ('carbostyril') by the hybrid structure (**76**) ↔ (**77**) in equation 47b.

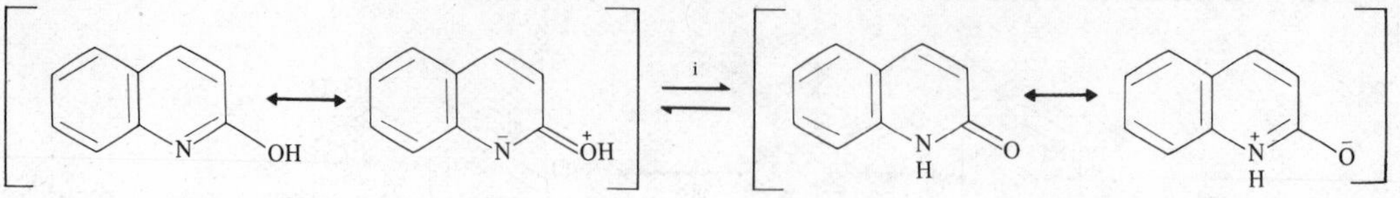

(74) (75) (76) (77) (47b)

i, $pK_T = 3.88$

The *N*-protonated (quinolone) tautomers have the advantage that the dipolar canonical structures (**77**) carry the negative charge on oxygen, the more electronegative element, whereas the reverse is the case for the *O*-protonated tautomers (**75**). Structure (**77**) is probably the best single representation of 2-quinolone. In this case $pK_T = 3.88$, showing a preponderance of the *N*-protonated form by a factor of over 7500 : 1.[67] A similar situation exists for the 4-quinolone ($pK_T = 4.19$) which, by the principle of vinylogy, retains its amido-properties.[67] Clearly for 3-hydroxyquinoline (and also for the 6- and 8-isomers), mesomeric stabilization of the two tautomeric forms is not possible, as structures corresponding to (**75**) and (**76**) cannot be formulated.

These conclusions, based on pK_a measurements, are well supported by other physical properties,[66a] notably by comparisons between the u.v. and i.r. spectra of the oxygenated compounds themselves and their *O*- and *N*-methyl ethers. 4-Quinolone-3-carboxylic acid, however, is reported to exist mainly in the hydroxy-form. Chemical evidence for the structures of these compounds is frequently discussed, including their reactivity to particular reagents such as ferric chloride or diazomethane, and the structures of any reaction products formed, notably, for example, whether alkylation occurs on oxygen or nitrogen. However, such evidence, in the case of tautomeric systems, is usually unreliable and frequently misleading, as pointed out by Katritzky and Lagowski.[66b]

The 2- and 4-quinolones can usually be prepared directly by the Conrad–Limpach or Knorr syntheses or from *o*-aminocinnamic acids or cinnamanilides, as outlined in Section 16.4.2. 2-Quinolones can also be prepared directly from quinolines by hydroxylation with potassium hydroxide or hypochlorous acid (see Section 16.4.4.2). The 6- and 7-quinolinols are usually prepared by the Skraup reaction from the appropriate aminophenols or anisidines. The latter give the methyl ethers which can be hydrolysed with hydrogen bromide. The 7-isomer, however, is likely to be only the minor product from *meta*-substituted anilines. The 3-, 5-, and 8-quinolinols can be obtained by diazotization of the aminoquinolines and in the case of the 5-isomer a higher yield has been reported using a reverse Bucherer reaction,[68] as shown in equation 48. 8-Hydroxyquinoline can also be prepared from the 8-sulphonic acid by alkali fusion.

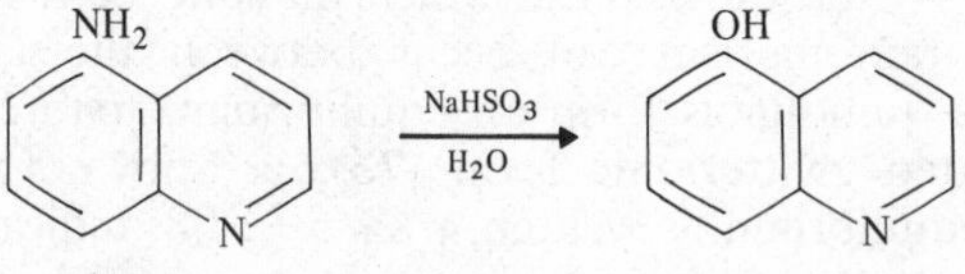

NH_2 — $NaHSO_3$, H_2O → OH (48)

The 'aromatic' quinolinols (namely the 3-, 5-, 6-, 7-, and 8-isomers), as their name suggests, have typical phenolic properties, *e.g.* they give violet colours with ferric chloride, couple with diazonium cations, and participate in Reimer–Tiemann and Bucherer reactions; their acetate esters usually undergo the Fries rearrangement with aluminium chloride to give acetyl derivatives. As a result of 'bond fixation' the 6- and 7-quinolinols behave like β-naphthol and couple at the 5- and 8-positions, respectively; the Reimer–Tiemann reaction is also directed to these positions. Scheme 13 illustrates these reactions for 6-hydroxyquinoline.

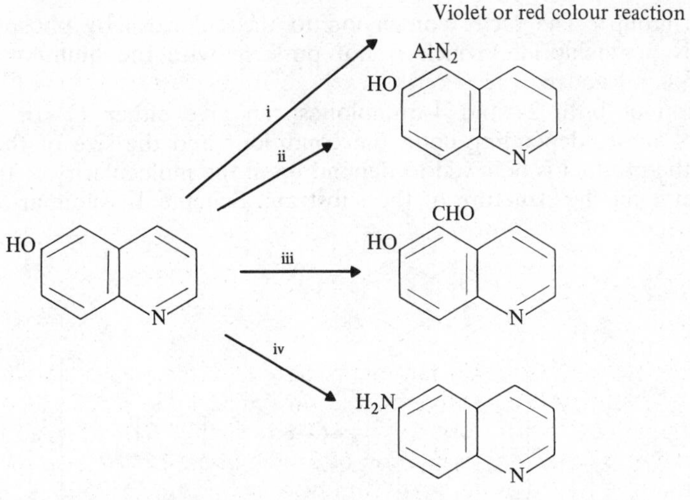

Violet or red colour reaction

i, $FeCl_3$ soln. (fresh); ii, $[ArN_2]^+$; iii, $CHCl_3$ + NaOH; iv, $NaHSO_3 + NH_3$

SCHEME 13

8-Hydroxyquinoline or 'oxine' shows a broad OH stretching absorption in the i.r. spectrum at 3412 cm^{-1} as a result of hydrogen bonding to nitrogen [see (**78**)]. Also as a result of the proximity of the hydroxyl group to the heterocyclic nitrogen it forms insoluble chelate compounds with a great variety of metal ions, including Cu^{2+}, Bi^{3+}, Mn^{2+}, Fe^{3+}, Al^{3+}, and Ni^{2+}. These compounds are suitable for the determination of the metals gravimetrically or in some cases volumetrically by bromination of the 'oxine' complex with excess of bromine (to the 5,7-dibromo-derivative), and back titration of the bromine. The hydrogen of the hydroxyl group is displaced and the metal is linked to both the oxygen and nitrogen. Four-covalent metal complexes therefore require two molecules of 'oxine' for each atom of metal [see (**79**)] and six-covalent complexes require three molecules. The use of derivatives of 8-hydroxyquinoline for their bactericidal and fungicidal properties is discussed in Section 16.4.6.1.

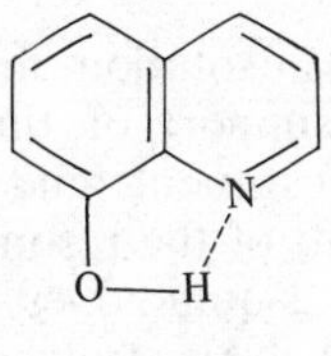

(78)

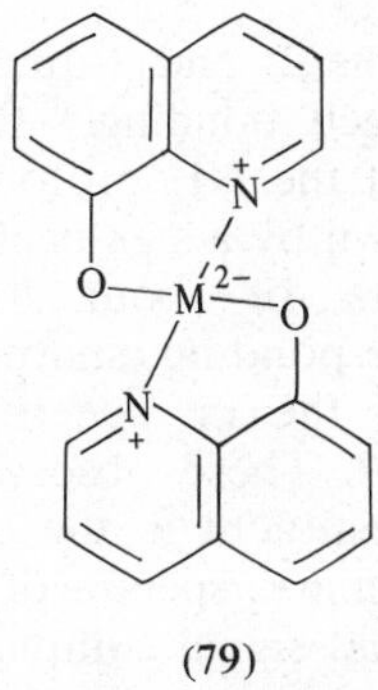

(79)

The phenolic properties of the 2- and 4-quinolones are less pronounced, although ferric chloride reagent does give a reddish colour and the 4-isomer is reported to give a poor yield of the 3-aldehyde in the Reimer–Tiemann reaction. The 2- and 4-alkoxy (ether) and acyloxy (ester) derivatives are more reactive to dealkylation or hydrolysis than the corresponding substituents at other positions. 2-Quinolone, called carbostyril, is more easily hydrogenated to the 1,2-dihydro-product than quinoline itself. Substituent effects are also shown by the lack of reactivity of the 4-methyl group in 4-methyl-2-quinolone and the formation of the 6-nitro-2-quinolone on nitration (*i.e.* electrophilic attack *para* to the amido-group). The facile conversion to the chlorides by phosphoryl chloride and phosphorus pentachloride, which is not possible with the quinolinols, is discussed in Section 16.4.5.3 above.

Alkylation of both 2- and 4-quinolones can give either *O*- or *N*-alkyl ethers, or mixtures of these, depending upon the conditions and the size of the alkyl group. The nature of the product is believed to depend upon the molecularity of the reaction (S_N1 or S_N2) rather than the structure of the substrate. Scheme 14 summarizes the reactions of 2-quinolone.

i, CH_2N_2; ii, MeI + KOH/EtOH; iii, EtI + KOH/EtOH; iv, EtI + Ag^+

SCHEME 14

The structures of the 2- and 4-quinolones, even in solution, may involve cyclic dimers formed by π-hydrogen bonding.[69] Evidence in support of this is provided[70] by the observed shielding of the NH protons by an 8-substituent (and to a lesser extent by a 5-substituent) as shown by a significant up-field shift of the resonance of the NH protons in the n.m.r. spectra, of about 2.5 p.p.m. for 2-quinolones and 1.4 p.p.m. for 4-quinolones. A corresponding sharpening of the NH stretching absorptions in the $3000\ cm^{-1}$ region of the i.r. spectrum of a quinolone is also apparent when an 8-substituent is present. These observations provide a useful method for detecting the presence of an 8-substituent in the quinolones. Further evidence of a similar nature is provided by the ^{13}C n.m.r. spectra of the quinolones. Thus C-8 can be recognized by its characteristic high-field signal compared with the other ring carbons,[71a] and the carbon

(49)

i, u.v. irradiation; ii, O_3 + AcOH; iii, H_2O_2; iv, HCl

NCH_3 resonance shows a downfield shift of about 7 p.p.m. when an 8-substituent is present.[71b] Photolysis of 2-quinolone in ethanol gives a dimer which, on the basis of its u.v., i.r., and n.m.r. spectra and degradative oxidation to *cis-trans-cis*-cyclobutanetetracarboxylic acid, has been identified as the head-to-head dimer across the 3,4-double bond[74b] (equation 49).

16.4.5.5 Quinoline 1-oxide[72]

Oxidation of quinoline with peracetic acid gives the important intermediate quinoline 1-oxide. As with pyridine 1-oxide the importance stems from its versatility in substitution reactions, particularly in the formation of 4-substituted derivatives, owing to the ability of the oxygen to act as either an electron donor or acceptor, as shown by (**80**)–(**82**), where *

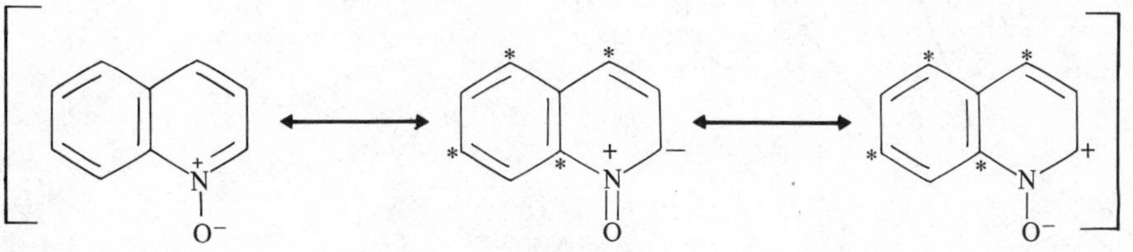

(80) (81) (82)

indicates alternative sites for the formal charge on the ring. Protonation occurs on the oxygen and with alkyl and acyl halides it forms quaternary salts (Scheme 15).

Electrophilic substitution can occur either in the carbocyclic ring, usually at positions-8 and -5 as for quinoline, or at position-4 in the heterocyclic ring. The former results from attack on the cation and the latter from attack on the free base. Nitration[73] is particularly interesting as, at low temperatures (20 °C), the 8-nitro 1-oxide is the main product, together with the 5-nitro-isomer, but at 65–70 °C high yields of the 4-nitroquinoline 1-oxide are formed. Nitration with benzoyl nitrate or by the action of metallic nitrates in

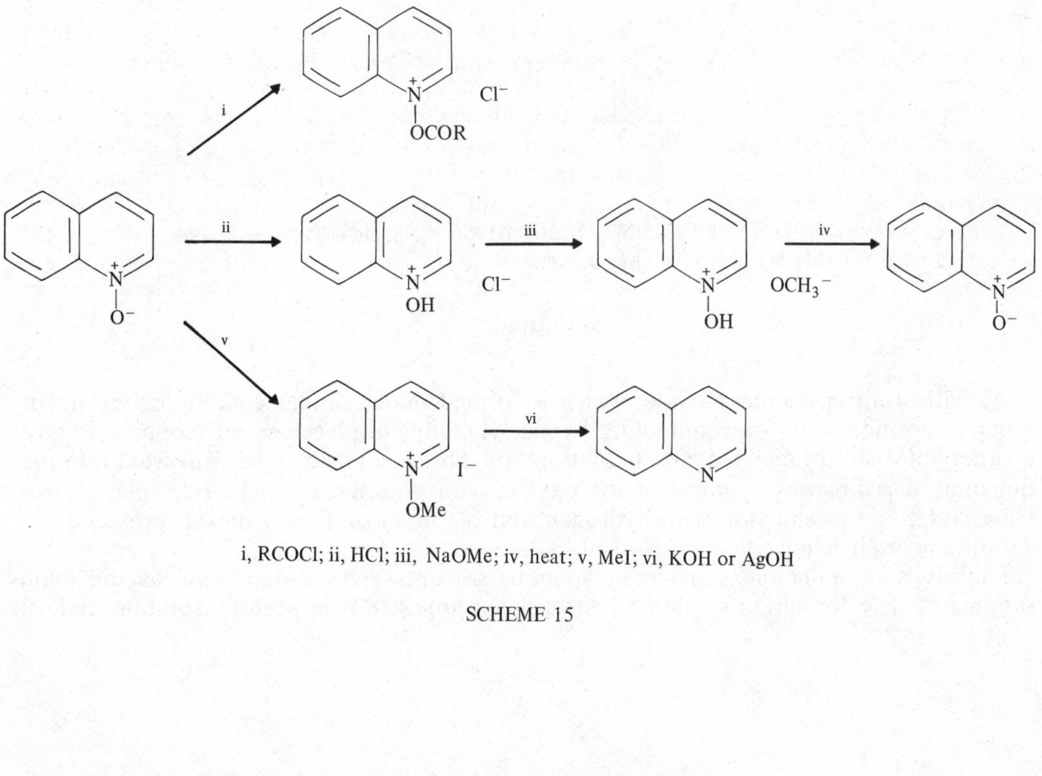

i, RCOCl; ii, HCl; iii, NaOMe; iv, heat; v, MeI; vi, KOH or AgOH

SCHEME 15

dimethyl sulphoxide on 1-methoxyquinolinium methosulphate gives 3-nitroquinoline 1-oxide by nucleophilic attack. As with quinoline, bromination in acetic anhydride gives the 3- and 6-bromo-substitution products by a similar addition–elimination mechanism (see Section 16.4.4.2 and Scheme 8). These substitutions are summarized in Scheme 16.

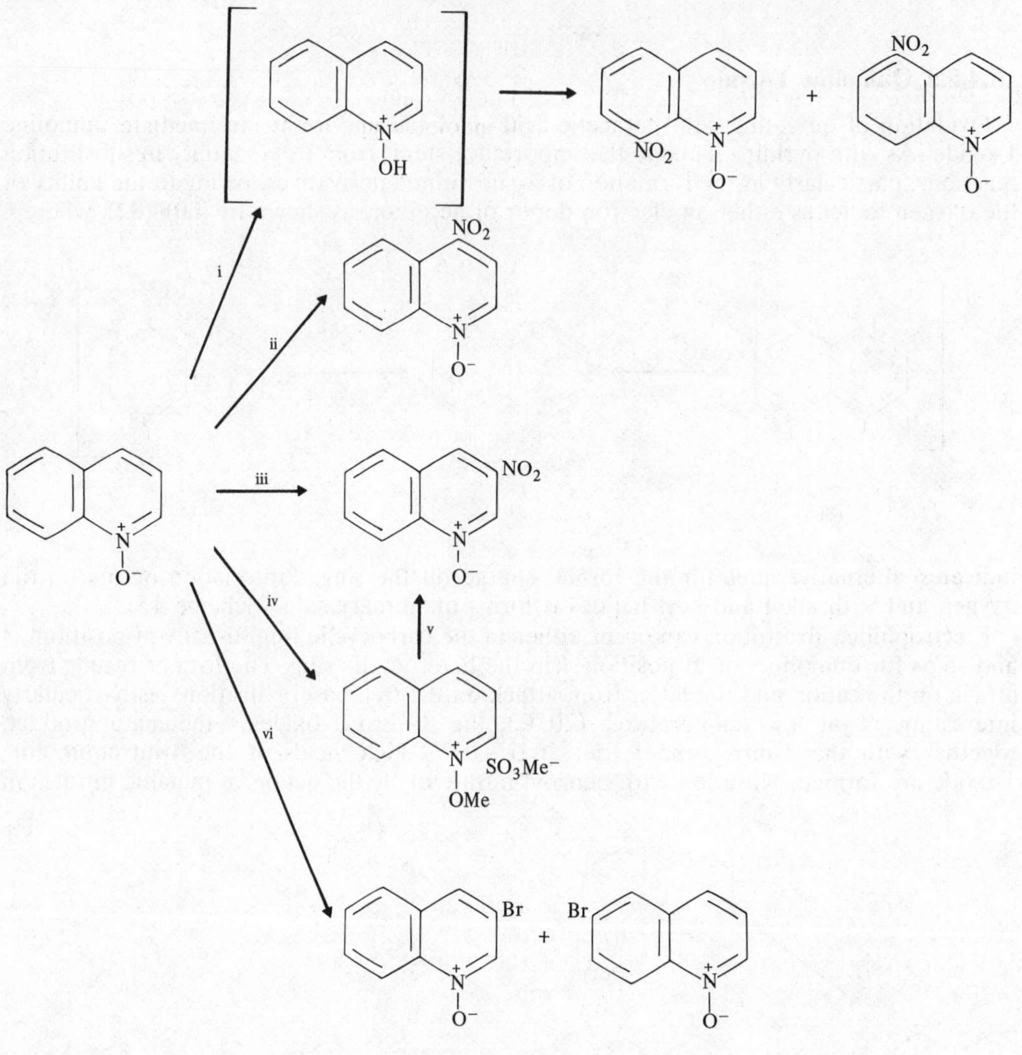

i, HNO_3, H_2SO_4, 0–20 °C; ii, HNO_3, H_2SO_4, 65–70 °C; iii, $PhCOONO_2$; iv, Me_2SO_4; v, KNO_3, Me_2SO, 150 °C; vi, Br_2, AC_2O

SCHEME 16

As with 4-nitropyridine 1-oxide, which is formed under similar conditions, the nitro-group at position-4 of 4-nitroquinoline 1-oxide is readily displaced by nucleophiles to give a variety of 4-substituted 1-oxides (equation 50). These can usually be converted into the quinoline derivative by removal of the oxygen with phosphorus trichloride, phosphorus tribromide, or by reduction with hydrogen and platinum or Raney nickel, provided the substituent itself is not affected or displaced.

Photolysis of quinoline 1-oxide in aqueous solvents gives 2-quinolone as the main product.[74a] The formation of stable benzo[*f*]oxazepines (**83**) in acetone solution and, in

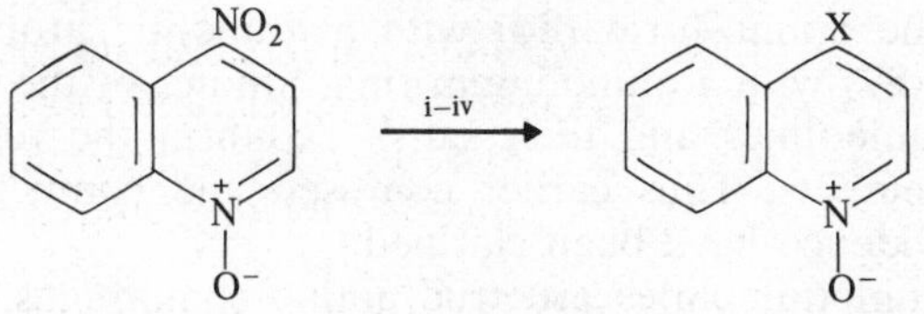

(50)

i, NaOR (X = OR); ii, NaSR (X = SR); iii, AcCl (X = Cl); iv, R_2NH (X = R_2N)

the absence of water, from 2-cyano- and 2-phenyl-quinoline 1-oxide and in other cases, on addition of water, 1-acyl derivatives of indole, suggests the following mechanism involving oxaziridine intermediates (Scheme 17).

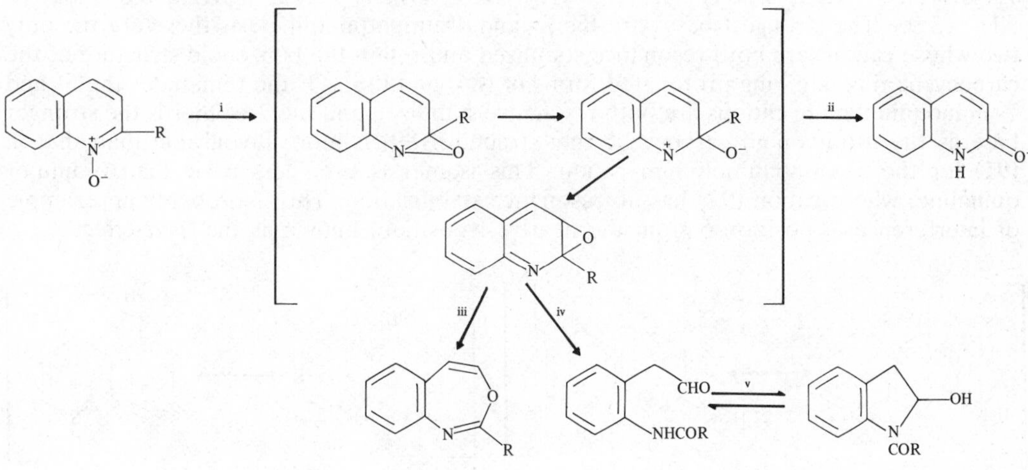

(83)

i, photolysis; ii, aqueous solution, R = H; iii, acetone solution, R = CN or Ph; iv, H_2O; v, tautomeric equilibrium

SCHEME 17

16.4.5.6 Aminoquinolines

The possibilities for tautomerism in the aminoquinolines[66c] are similar to those in the hydroxyquinolines and quinolones, but in this case the equilibrium between the amino- and imino-forms is strongly in favour of the amino-structure, even for the 2- and 4-aminoquinolines. The equilibrium for 4-aminoquinoline (equation 51) shows[75] a pK_T of

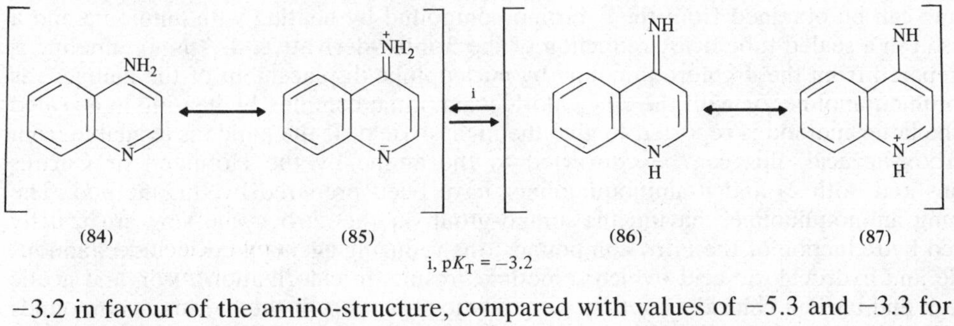

(84) (85) (86) (87) (51)

i, $pK_T = -3.2$

-3.2 in favour of the amino-structure, compared with values of -5.3 and -3.3 for the 2- and 4-aminopyridines.

The position of the equilibrium is reversed compared with the corresponding quinolones because, in this case, the imino dipolar form (**87**) does not have an advantage over its amino counterpart (**85**) as a result of atomic electronegativities and, of the fully aromatic structures, the amino-form (**84**) with a quinoline aromatic ring has a lower energy than structure (**87**) with a quinolinium ring. Studies of the i.r. and u.v. absorption spectra of the aminoquinolines and their salts[66c] confirm the results above, which are based on pK measurements. Thus earlier confused and conflicting assignments based largely on chemical evidence have been clarified.

Although all the aminoquinolines are true amino-compounds, the 2- and 4-isomers have properties which distinguish them from both the 3-isomer and isomers substituted in the carbocyclic ring, which have properties similar to aniline or the naphthylamines. All the aminoquinolines are preferentially protonated on the ring nitrogen, as shown by the bathochromic shift of the u.v. absorption band in acid solution, and have the following 1st pK_a values which should be compared with the pK_a = 4.94 for quinoline itself: 2-NH_2 = 7.34, 3-NH_2 = 4.95, 4-NH_2 = 9.17, 5-NH_2 = 5.51, 6-NH_2 = 5.62, 7-NH_2 = 6.65, and 8-NH_2 = 3.93. The strongest bases are the 2- and 4-aminoquinolines, as these are the only two whose cations are both resonance stabilized and retain the benzenoid structure of the carbocyclic ring, *e.g.* ring protonated forms of (**84**) and (**85**). Of the remainder the 5- and 7-aminoquinolinium cations are both resonance stabilized and the 7-isomer is the stronger base as the *p*-quinonoid carbocyclic ring structure (**89**) is more favourable than that in (**91**) for the 5-aminoquinolinium cation. This isomer is even less basic than 6-aminoquinoline, whose cation (**92**) has no resonance stabilization. This is probably an example of interference at position-5 from the nearby 4-position, known as the '*peri*-effect'.

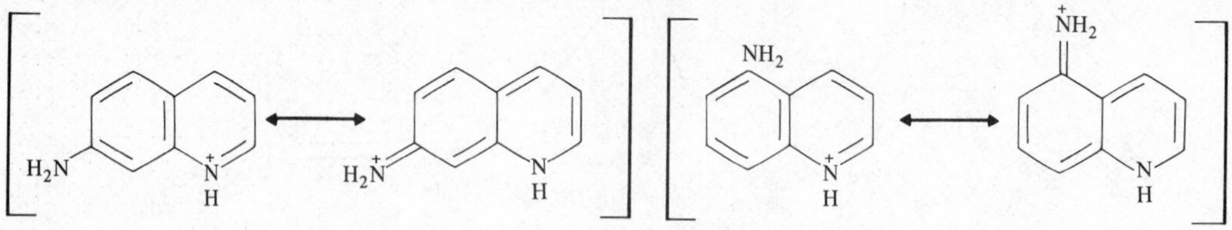

H_2N … $\overset{+}{N}$ H ⟷ $H_2\overset{+}{N}$ … N H NH_2 … $\overset{+}{N}$ H ⟷ $\overset{+}{N}H_2$ … N H

(88) (89) (90) (91)

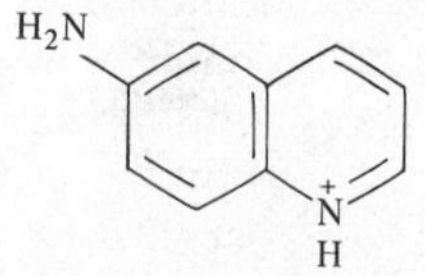

H_2N … $\overset{+}{N}$ H

(92)

Quinolines with an amino-group at position-2 are usually prepared from the parent quinoline by direct amination (Chichibabin[48] reaction; Section 16.4.4.2). 3-Aminoquinoline can be obtained from the 3-bromo-compound by heating with ammonia and a cupric salt in a sealed tube or by reduction of the 3-nitro-derivative. 4-Aminoquinoline is best prepared from the 4-chloroquinoline by nucleophilic displacement of the halogen, as for 3-bromoquinoline, or with the zinc chloride–ammonia complex by heating in a sealed tube; the latter method is reported to give the highest yield. If the amide is available from the carboxylic acid, this can be converted to the amine by the Hofmann or Curtius reactions and both 2- and 4-aminoquinolines have been prepared by this method. The remaining aminoquinolines having the amino-group in the carbocyclic ring are usually prepared by reduction of the nitro-compounds; the reducing agents used include stannous chloride and hydrochloric acid (which sometimes results in chlorination), iron and acetic acid, iron and barium chloride solution, or catalytic reduction. If the hydroxy-compounds are available, the Bucherer reaction with sodium hydrogen sulphite and ammonia may be used.

All the aminoquinolines except the 2- and 4-isomers can be described as 'aromatic' in the sense that they behave like the naphthylamines. They form stable diazonium salts and the diazonium group can be replaced in the normal manner by the Sandmeyer, Gattermann, and Balz–Schiemann reactions. The Gattermann reaction is generally more successful than the Sandmeyer reaction which, for example, fails to give an 8-cyano-derivative. The diazonium salts can be reduced to hydrazines and some coupling reactions with other diazonium ions are successful. The aminoquinolines take part in Skraup and Doebner–von Miller reactions to give phenanthrolines in which 6-aminoquinoline undergoes ring closure only to position-5. The 2- and 4-aminoquinolines, on the other hand, particularly the 4-isomer, are not easily diazotized and 2-aminoquinoline with alkali is converted to 2-quinolone. On nitration the main products are the nitramines together with some ring-nitro compounds.[76] The nitramines undergo the usual acid-catalysed rearrangement to ring nitro-compounds (Scheme 18).

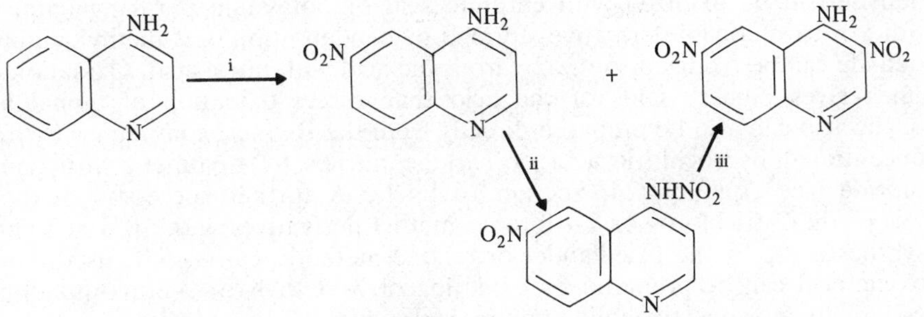

NH_2 $\xrightarrow{i}$ O_2N NH_2 + O_2N NH_2 NO_2

ii O_2N $NHNO_2$ iii

i, HNO_3, H_2SO_4, −7 to 2 °C (isolated from crude product); ii, HNO_3, H_2SO_4; iii, H^+

SCHEME 18

16.4.5.7 Quinolinecarboxylic acids

These acids and their simple derivatives (*e.g.* esters, amides, acyl chlorides, and nitriles) behave like their homocyclic aromatic counterparts. The 2- and 4-derivatives, quinaldinic acid and cinchoninic acid respectively, having the hydroxycarbonyl group attached to electron-deficient carbon atoms, have slightly modified properties; also the properties of the 2- and 8-carboxylic acids are affected by the proximity of the heterocyclic nitrogen as indicated below. The acids themselves exist in the zwitterionic forms [*e.g.* (**92a**)] and, with the exception of the 8-isomer, are slightly stronger acids than benzoic acid ($pK_a = 5.27$), as shown by the following pK_a values in 50% methanol[77] (isomer positions in parenthesis): (2-) 4.96, (3-) 4.62, (4-) 4.53, (5-) 4.81, (6-) 4.98, (7-) 4.97, and (8-) 7.20. The pK_a values of the 2- and 8-carboxylic acids are increased by hydrogen bonding to nitrogen [see (**92b**) and (**92c**)].

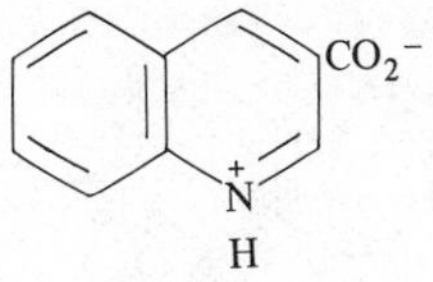

CO_2^- $\overset{+}{N}$ H (92a)

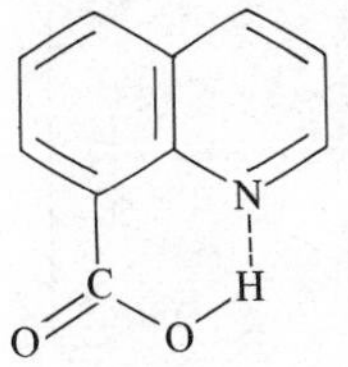

N H C O O (92b)

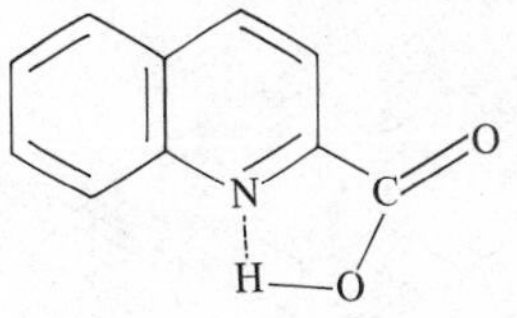

N C O H O (92c)

The esters of all the quinolinecarboxylic acids participate in the Claisen ester condensation with substituted acetic esters (RCH_2CO_2Et) to give β-keto-esters which when hydrolysed under mild conditions give the quinolyl alkyl ketone (equation 52). In view of the failure of the Friedel–Crafts acylation, unless powerful activating groups are present, this represents an important method of preparing the ketones.

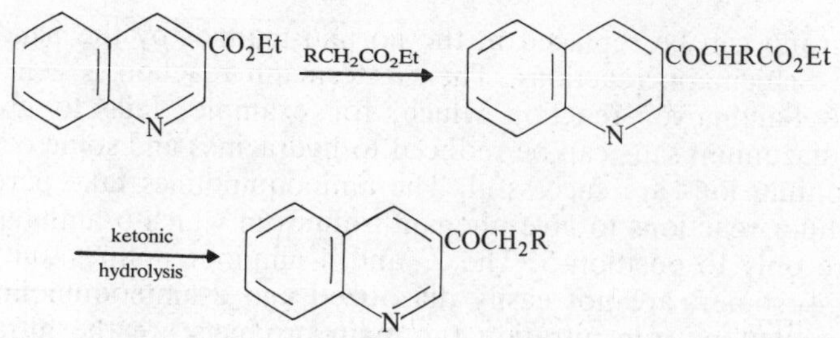

(52)

Quinoline-2-carboxylic acid can be prepared by oxidation of the condensation products of 2-methylquinoline with aldehydes. Thus 2-styrylquinoline formed by condensation with benzaldehyde can be oxidized with chromic acid or potassium permanganate, and the mono-, di-, or tri-methylol derivatives formed by condensation of 2-methylquinoline with formaldehyde can be oxidized to the 2-carboxylic acid with nitric acid. Oxidation of these compounds gives higher yields of the acid than direct oxidation of quinaldine. The 2-carboxylic acid can also be prepared directly from the Reissert compounds by treatment with concentrated hydrochloric acid or *via* the nitriles by treatment with phosphorus pentachloride (see equation 24, Section 16.4.4.1). A further successful method is by hydrolysis of the 2-trichloro- or 2-tribromo-methyl derivatives with 10% sulphuric acid; direct synthesis, *e.g.* by the Friedlander or related methods, can also be used. Quinoline-3-carboxylic acid can be prepared by oxidation of 3-methyl- or 3-ethyl-quinoline or by hydrolysis (with aqueous/alcoholic sodium hydroxide) of the nitrile prepared from 3-bromoquinoline and anhydrous cuprous cyanide. A further method is by carbonation (with carbon dioxide) of 3-quinolyl-lithium, formed from 3-bromoquinoline and butyl-lithium. Quinoline-4-carboxylic acid can be prepared by oxidation of 4-methylquinoline (lepidine) or, preferably, its condensation products with aldehydes, as described for the 2-carboxylic acid. A further method is from 4-cyanoquinoline, prepared by direct addition of cyanide to a quinolinium cation, followed by oxidation and hydrolysis (Section 16.4.4.1, equation 23), and direct synthesis by the Pfitzinger method can be used for substituted cinchoninic acids for which the appropriate isatin and keto-methylene derivatives are available.

The carboxylic acids having the hydroxycarbonyl group in the carbocyclic ring can be synthesized by standard methods.[78] Oxidation of the appropriate methylquinoline, either directly or after conversion to the aldehyde with selenium dioxide,[77] has been used to prepare all the carbocyclic-substituted quinolinecarboxylic acids. The nitriles, obtained either by the Sandmeyer reaction from the aminoquinolines (for the 5- and 8-carboxylic acids), or from the sulphonic acids by fusion with potassium cyanide (for the 6- and 8-carboxylic acids), have been successfully hydrolysed to the acids. Direct synthesis by the Skraup reaction gives good yields of the 5-carboxylic acid from *m*-aminobenzoic acid (and

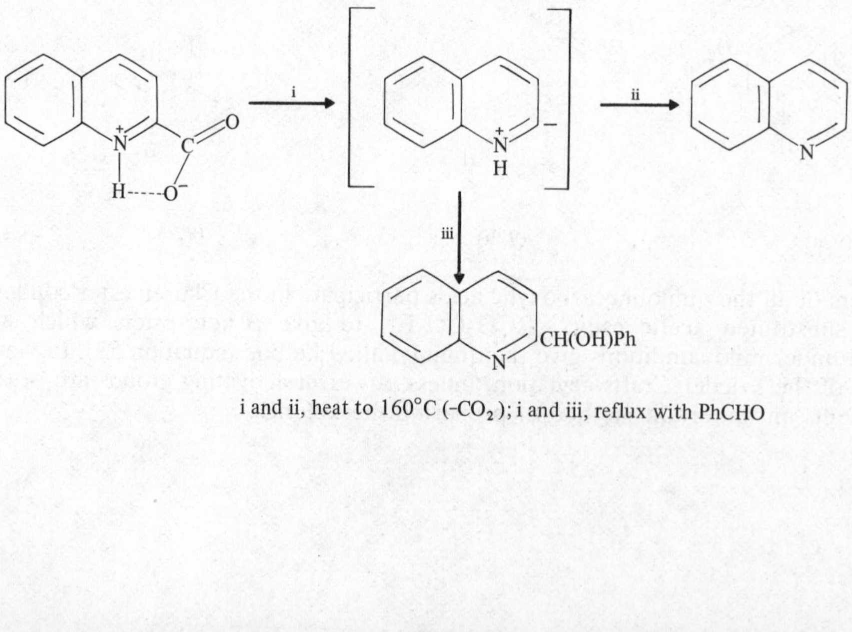

(53)

i and ii, heat to 160°C ($-CO_2$); i and iii, reflux with PhCHO i, KOH, H_2O, fuse ($-H_2$); ii, $+H^+$ (54)

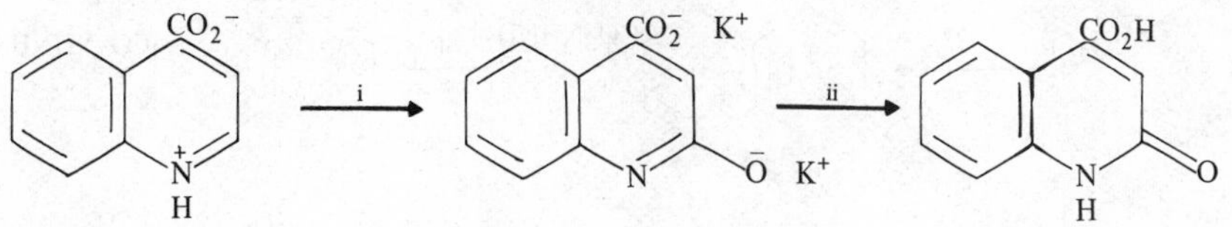

the 7-isomer, if formed, is not isolated)[77] and of the 8-carboxylic acid from *o*-aminobenzoic acid. The yield of the 6-carboxylic acid by this method using *p*-aminobenzoic acid, however, is poor.

When the hydroxycarbonyl group is attached to the heterocyclic ring, decarboxylation is possible and is particularly facile for the 2-isomer, quinaldinic acid. The reaction proceeds through an intermediate ylide which can be trapped if a reactive carbonyl compound is present[79] (equation 53). The 4-carboxylic acid, although stable to fused aqueous potassium hydroxide which converts it to the 4-hydroxycarbonyl-2-quinolone (equation 54), can be decarboxylated by heating with copper powder. Other substituted cinchoninic acids decarboxylate on heating to their melting points[25] (see the Pfitzinger synthesis, Section 16.4.2.7 and equation 10).

16.4.5.8 Quinolinecarbaldehydes

All seven of the quinolinecarbaldehydes are known and the usual method of preparation is by oxidation of the appropriate methyl derivative with selenium dioxide[52–54] (see Section 16.4.4.3, equations 40 and 41). Many of the methods commonly used for the synthesis of aromatic and other aldehydes such as the Gattermann and Gattermann–Koch syntheses and the Rosenmund reduction do not give good yields of quinoline aldehydes, but other methods which have been successful include the Reimer–Tiemann reaction with quinolinols (see Section 16.4.5.4, Scheme 13), the hydrolysis of mono- or di-bromomethylquinolines, obtained from the tribromomethylquinolines by reduction (equation 55), and the method of McFadyan and Stevens[80] which involves hydrolysis of the benzenesulphonyl hydrazide of a quinolinecarboxylic acid (equation 56).

The properties of all the aldehydes, with very few exceptions, can be described as typical of aromatic (*i.e.* tertiary) aldehydes. They participate, for example, in the Perkin

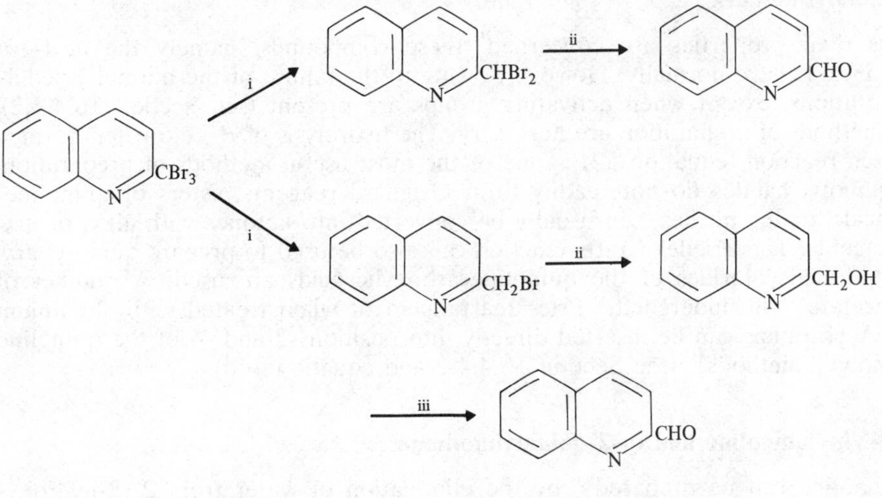

(55)

i, $SnCl_2$ or $TiCl_3$ (calc. amount);
ii, $AgNO_3$, H_2O ; iii, HNO_3

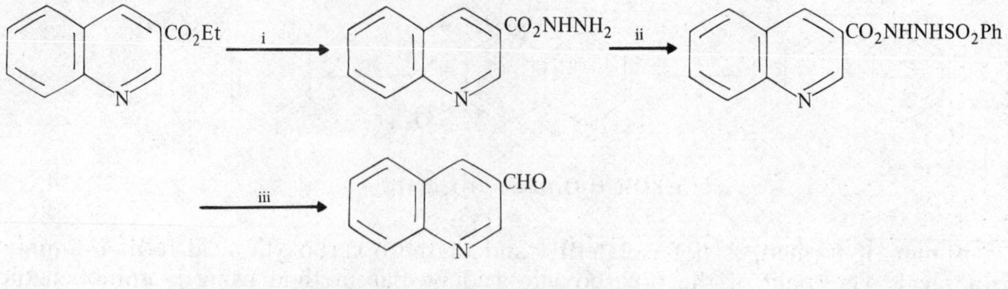

(56)

i, NH_2NH_2; ii, $PhSO_2Cl$; iii, Na_2CO_3, ethylene glycol, 150 °C

reaction, the Cannizzaro reaction, and Knoevenagel condensations with active methylene compounds. They can be reduced by the Wolff–Kishner method, they form Schiff bases with amines and react normally with Grignard reagents to give the carbinols. Oxidation with hydrogen peroxide gives high yields of the acids. The 2- and 4-quinolinecarbaldehydes, however, do not undergo the normal benzoin condensation with potassium cyanide. The 4-isomer appears to give a normal acyloin which is then reduced by further molecules of the aldehyde as in a crossed-Cannizzaro reaction (equation 57). Various products have been obtained by the action of potassium cyanide on quinoline-2-aldehyde.[81]

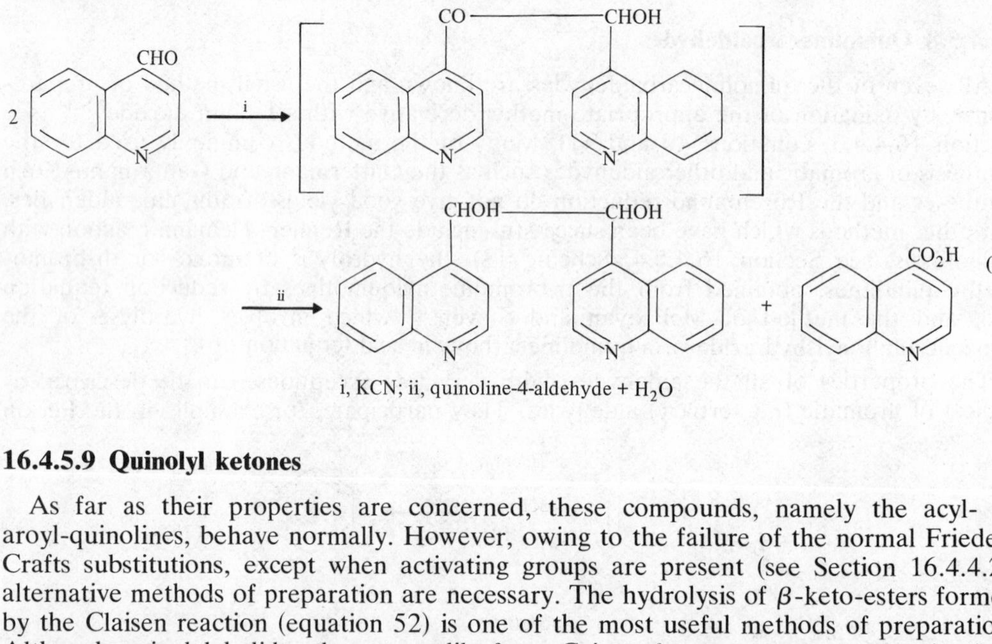

(57)

i, KCN; ii, quinoline-4-aldehyde + H_2O

16.4.5.9 Quinolyl ketones

As far as their properties are concerned, these compounds, namely the acyl- or aroyl-quinolines, behave normally. However, owing to the failure of the normal Friedel–Crafts substitutions, except when activating groups are present (see Section 16.4.4.2), alternative methods of preparation are necessary. The hydrolysis of β-keto-esters formed by the Claisen reaction (equation 52) is one of the most useful methods of preparation. Although quinolyl halides do not readily form Grignard reagents, esters of quinolinecarboxylic acids, or the nitriles, can usually be converted into ketones with alkyl or aryl Grignard reagents. The Friedel–Crafts reaction can also be used to prepare quinolyl aryl ketones if the acid chlorides of the quinolinecarboxylic acids are used. A number of quinolinol acetates will undergo the Fries rearrangement when treated with aluminium trichloride. Acyl groups can be inserted directly into positions-2 and -4 of the quinoline ring by homolytic methods[49] (see Section 16.4.4.2 and equation 34d).

16.4.5.10 2-Vinylquinoline and di-2-quinolylmethane

2-Vinylquinoline can be prepared[81] by the elimination of water from 2-(2-hydroxyethyl)quinoline, the primary condensation product of formaldehyde with 2-methylquinoline (equation 58). It polymerizes even more readily than styrene (unless stabilized

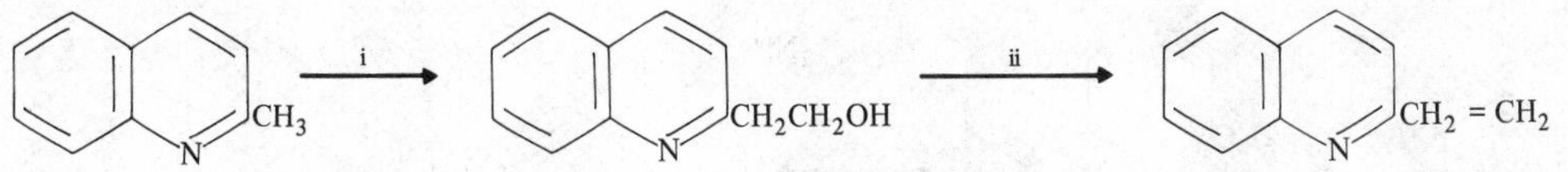

(58)

i, CH_2O, $EtOH/H_2O$, heat 24 h; ii, NaOH + PhNH-β-$C_{10}H_7$, reduced pressure

with a radical scavenger such as hydroquinone), and is used for the manufacture of commercial polymers and copolymers.

Di-2-quinolylmethane, together with some tri-2-quinolylmethane, is formed[82] when 2-chloroquinoline is heated with 2-methylquinoline, preferably in a sealed tube, for one hour at about 200 °C. The diquinolylmethane can be separated from the tarry reaction product as the dipicrate, from which the free base is liberated with sodium hydroxide. The solution in organic solvents is colourless in the cold but the solute turns red on warming, or on melting the solid. This results from isomerism of the colourless form (**93**) to the red form (**94**), a structure related to that of the cyanine dyes (see Section 16.4.6.2). On treatment with one mole of dimethyl sulphate, the monomethyl derivative (**95**) so formed is unable to tautomerize and remains as deep-red crystals with a metallic lustre (see Scheme 19).

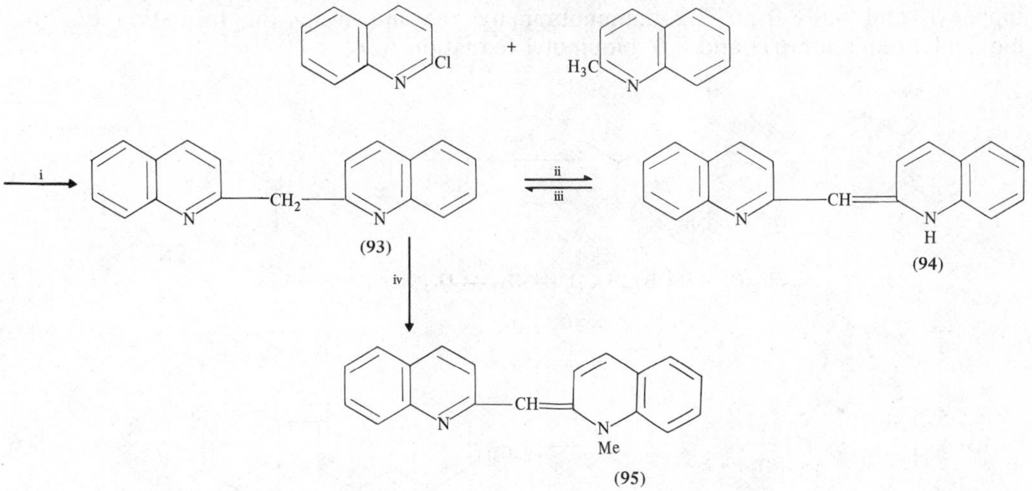

(93) (94) (95)

i, 24 h, sealed tube, 200 °C; ii, warm, or trace of acid; iii, alkali; iv, Me_2SO_4

SCHEME 19

16.4.5.11 Biquinolyls[83]

A large number of the 28 possible biquinolyls have been prepared, either by ring-closure reactions (Skraup or Friedlander syntheses) or by pyrolysis methods, when the structures are less certain unless confirmed by other evidence. The following are interesting examples.

2,2′-Biquinolyl is formed (15% yield) when quinoline is heated with nickel on alumina at 325 °C. The structure has been confirmed by a Friedlander synthesis from *o*-aminobenzaldehyde and diacetyl (equation 59). 2,3′-Biquinolyl has been obtained in 30–40% yield by heating quinoline with sodium. This structure has been confirmed by a number of

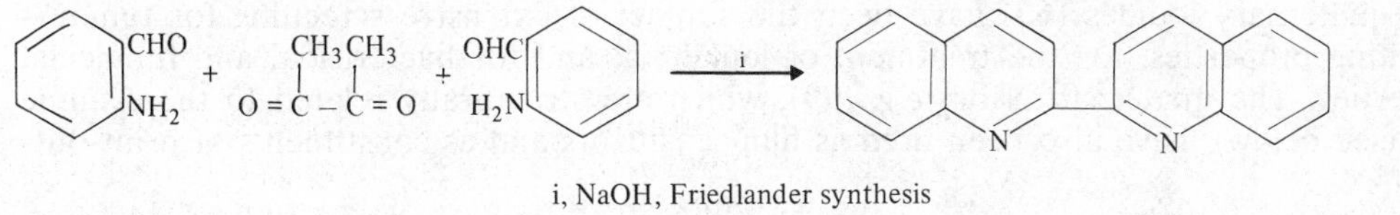

(59)

i, NaOH, Friedlander synthesis

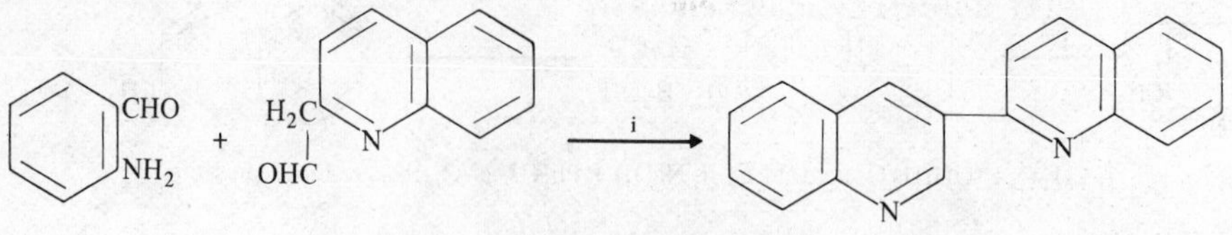

(60)

i, NaOH, Friedlander synthesis

independent investigations, including a Friedlander synthesis from *o*-aminobenzaldehyde and 2-quinolylacetaldehyde (equation 60). This reaction contrasts with the action of pyridine, which under similar conditions gives 2,2′- and 4,4′-bipyridyls. 5,5′- and 7,7′-biquinolyls have been made by reduction of the bromo-derivatives with hydrazine and palladium on calcium carbonate. The Skraup reaction on an appropriate aminophenylquinoline, or a double Skraup reaction on a diaminobiphenyl, has been used to prepare the following biquinolyls: 2,6′- from 2-*p*-aminophenylquinoline, 4,6′- from 4-*p*-aminophenylquinoline, and mixtures of 2,5′- and 2,7′- from 2-*m*-aminophenylquinoline, 4,5′- and 4,7′- from 4-*m*-aminophenylquinoline, 6,6′- from benzidine, 6,8′- from 2,4′-diaminobiphenyl, and 8,8′- from 2,2′-diaminobiphenyl, as shown by the formation of 2,6′-biquinolyl (equation 61) and 6,6′-biquinolyl (equation 62).

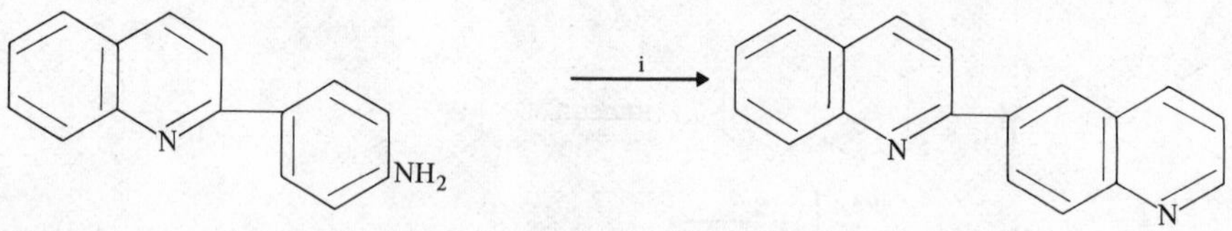

(61)

i, $CH_2(OH)CH(OH)CH_2OH$, H_2SO_4, As_2O_5,Skraup synthesis

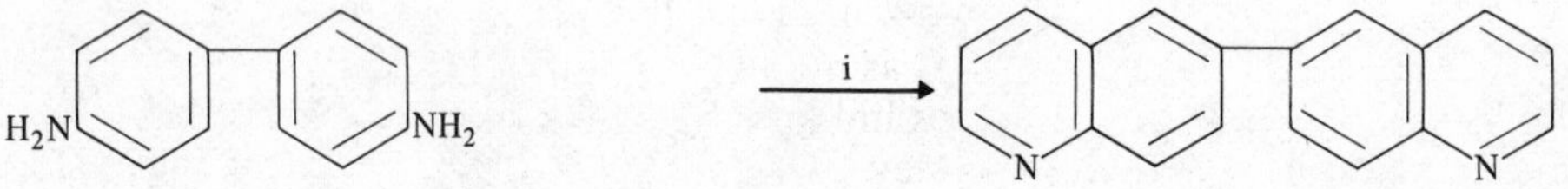

(62)

i,Skraup synthesis

16.4.6 NATURAL AND SYNTHETIC COMPOUNDS OF QUINOLINE

16.4.6.1 8-Hydroxyquinolines and styrylquinolines

Derivatives of 8-hydroxyquinoline have bactericidal and fungicidal properties and the halogenated compounds (*e.g.* 5-chloro-7-iodo-8-hydroxyquinoline, 'Vioform') are amoebicides. The bactericidal properties have been shown to be associated with the presence of the iron chelate complex.[84] The Cu^{II} derivative has been used as a protection against mildew.

Styrylquinolines (formed from 2- or 4-methylquinolines or their quaternary iodides by condensation with benzaldehydes) and particularly *p*-dimethylaminostyrylquinolines and their quaternary iodides (**63**) have been the subject of extensive screening for tumour-inhibiting properties, for the treatment of leukemia, and for bactericidal and fungicidal properties. The quaternary salts, *e.g.* (**99**), which are structurally related to the cyanine dyes (see below), have also been used as film sensitizors and as constituents of print-out-papers.

16.4.6.2 Cyanine dyes and Quinoline Yellow

The cyanines are an important group of dyes used for improving the colour sensitivity of photographic emulsions. Their chemistry and uses have been reviewed by Hamer.[85,86]

Owing to their instability to light they are not used as fabric dyes, but they contain one of the few organic structures capable of absorbing light in the visible and infrared regions and transferring the energy to a photographic emulsion as a 'latent image' of activated silver halide crystals. In this way the sensitivity of a silver bromide emulsion can be extended from its natural position in the ultraviolet and blue regions of the spectrum ($\lambda = 350–530$ nm) to the green, red, and near-infrared regions (up to $\lambda = 900–1000$ nm). With suitable combinations of these dyes the sensitivity can be balanced throughout the visible spectrum to give 'panchromatic' emulsions. For example, the dye cyanine (**96**) (a green sensitizer) is formed by base-catalysed condensation from equimolar quantities of the quaternary ethiodides of quinoline and 4-methylquinoline, and the dye pinacyanol (**97**) (a red sensitizer) is formed similarly from two moles of 1-ethyl-2-methylquinolinium iodide and ethyl orthoformate (equations 63 and 64).

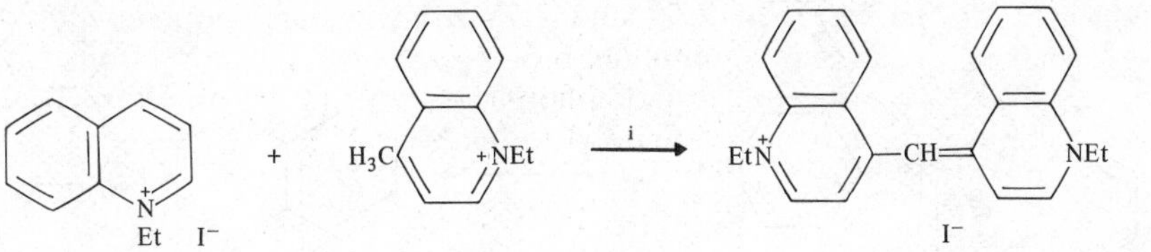

Et I⁻ + H_3C +NEt —i→ EtN — CH= — NEt (63)

I⁻

(**96**) Cyanine

i, KOH (alcoholic)

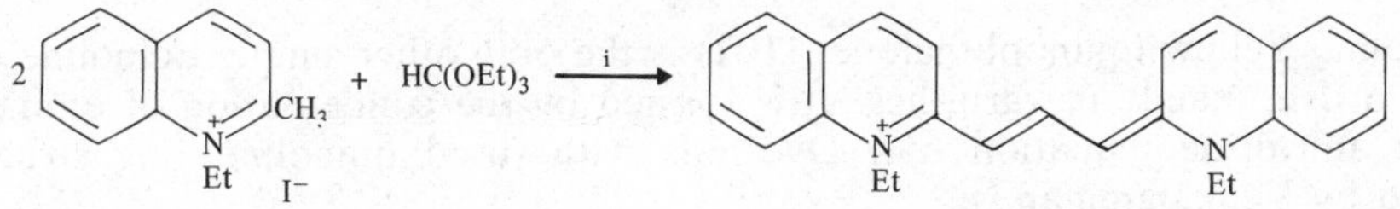

2 N Et I⁻ CH_3 + $HC(OEt)_3$ —i→ N Et — N Et (64)

(**97**) Pinacyanol

i, pyridine (dry)

Cyanine and pinacyanol can be classified as mono- and tri-methincyanines, respectively, according to the number of —CH= groups linking the quinoline nuclei. Pentamethin- and heptamethin-cyanines are also known; they give sensitivity at longer wavelengths, but they are less stable. An odd number of methine connecting groups is necessary as the photosensitizing properties depend upon the delocalization of the cationic charge over the conjugated system between the two nitrogen atoms. The quaternary 2- and 4-*p*-dimethylaminostyrylquinolines discussed above have structurally related chromophores, see (**98**) and (**99**).

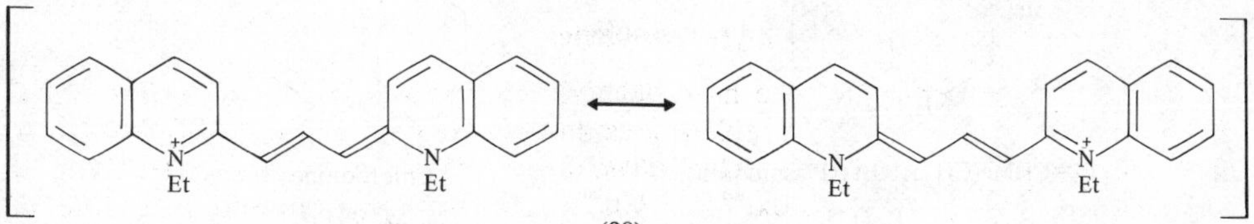

[N⁺ Et — N Et ⟷ N Et — N⁺ Et] I⁻

(**98**)

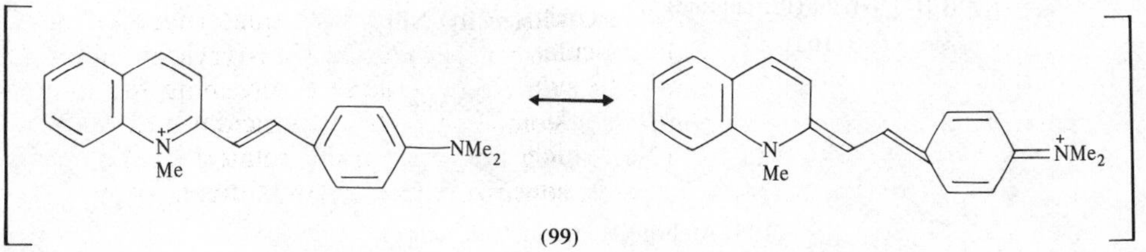

[N⁺ Me — NMe_2 ⟷ N Me — =$\overset{+}{N}Me_2$] I⁻

(**99**)

In addition, one or more of the methine groups can be replaced by the azomethine nitrogen (—N=) to give the azocyanines, and other heterocyclic bases, particularly thiazole (giving thiazolocyanines) and benzothiazole (giving thiacyanines), can replace quinoline. The points of linkage of the bases with the polymethine chain may vary, provided that the essential conjugation is preserved, and the products are associated with further unsystematic trivial names. Thus monomethincyanines with quinoline bases may have 2,2′-linkages (pseudocyanines), 2,4′-linkages (isocyanines), or 4,4′-linkages (simple cyanines). Trimethincyanines with 2,2′-linkages are known as carbocyanines and the corresponding penta- and hepta-methincyanines are dicarbocyanines and tricarbocyanines, respectively. The apocyanines, which are not important photographically, have no methine groups, but instead they have one reduced ring and 3,2′- or 3,4′-links to preserve the conjugation: see (**100**). A systematic nomenclature has also been devised in which, for example, pinacyanol, a carbocyanine (**97**), would be named as bis-2-(1-ethylquinolyl)trimethincyanine iodide.

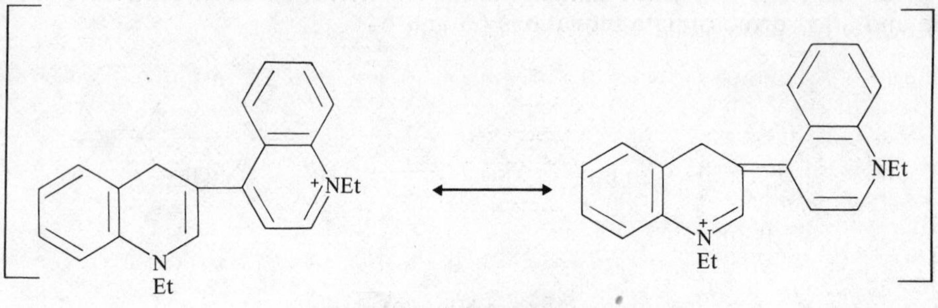

$^+$NEt ⟷ NEt I^- N Et $\overset{+}{N}$ Et

(100) Diethyl erythroapocyanine

Quinoline Yellow (quinophthalone) (**101**) is the only other simple quinoline derivative used as a dye, mainly in varnishes. It is formed by the condensation of quinaldine and phthalic anhydride (equation 65). Dyestuffs with fused quinoline ring structures are reviewed by Venkataraman.[87]

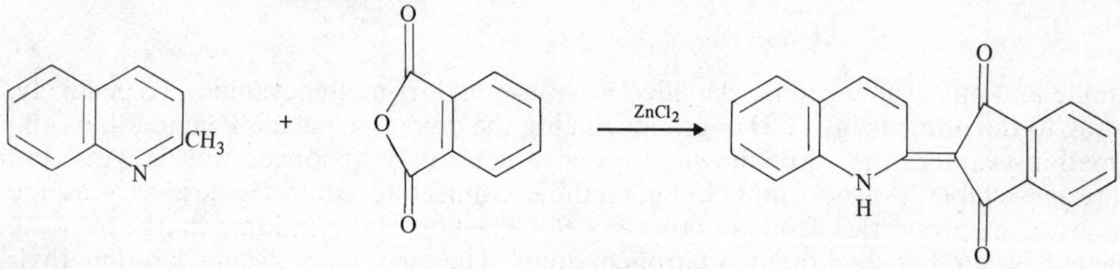

CH_3 N + O O O $\xrightarrow{ZnCl_2}$ O O N H (65)

(101)

MeO NHR N

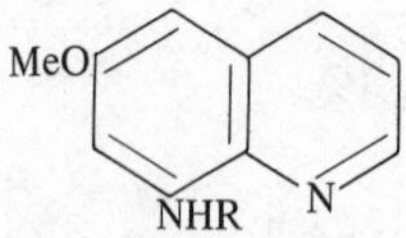

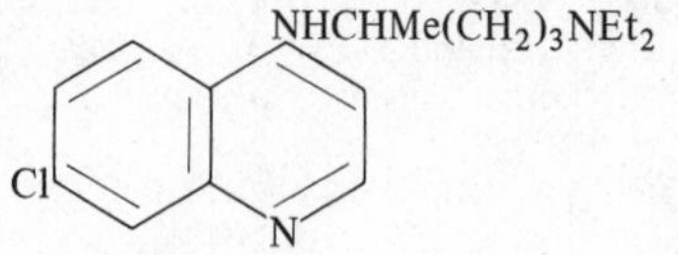

R = $CHMe(CH_2)_3NEt_2$ (Plasmoquin)

(102)

R = $(CH_2)_5NHPr^i$ (Pentaquine)

(103)

$NHCHMe(CH_2)_3NEt_2$ Cl N

Chloroquine

(104)

$NHCHMe(CH_2)_3NEt_2$ OMe Cl N

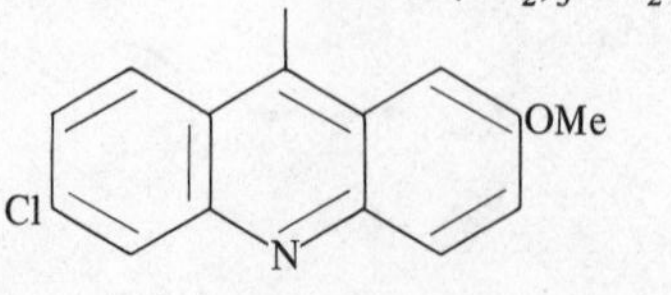

(105) Atebrin (Mepacrin, Quinacrin)

MeO, NH_2, NO_2 —i→ MeO, NO_2, N —ii→ MeO, NH_2, N —iii→ MeO, N, $NHCHMe(CH_2)_3NEt_2$ (66)

i, Skraup synthesis; ii, Sn/HCl; iii, $MeCHCl(CH_2)_3NEt_2$

16.4.6.3 The quinoline alkaloids[90,91] and synthetic antimalarials

The use of Cinchona bark for the treatment of malaria was introduced into Europe from Peru in the 17th century. The active principles are the Cinchona alkaloids,[90a,91a] which are quinoline bases and include cinchonine (**4**) and quinine (**5**) (Section 16.4.1.1). The structure of quinine was first reported by Rabe[88] in 1908 and a complete synthesis was achieved by Woodward and Doering[89] in 1945. The isolation of quinoline from cinchonine and quinine by Gerhardt was followed by an intensive search for derivatives of this base with antimalarial or therapeutic properties. The most successful synthetic antimalarial drugs include Plasmoquin (**102**), Pentaquine (**103**), chloroquine (**104**), and the acridine derivative Atebrin (Mepacrin, Quinacrin) (**105**). These drugs have now been largely replaced in clinical use owing to undesirable side effects. The synthesis of Plasmoquin is outlined above (equation 66).

References

1. (a) R. H. F. Manske, *Chem. Rev.*, 1942, **30,** 113; (b) F. W. Bergstrom, *Chem. Rev.*, 1944, **35,** 77; (c) R. C. Elderfield, 'Heterocyclic Compounds', Wiley, New York, Chapman and Hall, London, 1952, vol. 4, chapter 1.
2. N. Campbell, in 'Rodds Chemistry of Carbon Compounds', ed. S. Coffey, Elsevier, Amsterdam, 2nd edn, 1976, vol. 4F p. 231.
3. F. F. Runge, *Ann. Phys. Chem.*, 1834, **31** (2), 65.
4. Ch. Gerhardt, *Ann. Chim. Phys.*, 1842, **7** (3), 251.
5. Ch. Gerhardt, *Annalen*, 1842, **44,** 279.
6. Zd. H. Skraup, *Ber.*, 1880, **13,** 2086.
7. R. H. F. Manske and M. Kulka, 'The Skraup Synthesis of Quinolines', in 'Organic Reactions', ed. R. Adams, Wiley, New York, 1953, vol. 7, p. 59.
8. H. T. Clarke and A. W. Davis, *Org. Synth., Coll. Vol. I*, 478.
9. E. W. Cohn, *J. Amer. Chem. Soc.*, 1930, **52,** 3685.
10. M. H. Palmer, *J. Chem. Soc.*, 1962, 3645.
11. R. C. Elderfield, *J. Amer. Chem. Soc.*, 1946, **68,** 1584.
12. O. Doebner and W. von Miller, *Ber.*, 1881, **14,** 2812.
13a. W. H. Mills, J. E. G. Harris, and H. Lambourne, *J. Chem. Soc.*, 1921, 119 (1294).
13b. T. P. Forrest, G. A. Dauphinee, and W. F. Miles, *Canad. J. Chem.*, 1969, **47,** 2121.
14. C. Beyer, *J. prakt. Chem.*, 1886, **33,** 393.
15. W. R. Vaughan, *Org. Synth., Coll. Vol. III*, 329.
16. P. A. Claret and A. G. Osborne, *Org. Prep. Proced. Internat.*, 1974, **6,** 149.
17. P. A. Claret and A. G. Osborne, *Chem. and Ind.* (*London*), 1970, 401.
18. J. Craig, *J. Amer. Chem. Soc.*, 1938, **60,** 1485.
19. A. Combes, *Bull. Soc. chim. France*, 1888, **49,** 89.
20. J. S. Morley and J. C. E. Simpson, *J. Chem. Soc.*, 1948, 2024.
20a. E. A. Clarke and M. F. Grundon, *J. Chem. Soc.*, 1964, 438 and references therein.
21. M. Conrad and L. Limpach, *Ber.*, 1888, **21,** 523.
22. L. Knorr, *Annalen.*, 1886, **236,** 69.
23. P. Friedlander, *Ber.*, 1882, **15,** 2572.

24. W. Pfitzinger, *J. prakt. Chem.*, 1886, **33** (2), 100.
25. N. P. Buu-Hoi, P. Jacquignon, and T. B. Loc, *J. Chem. Soc.*, 1958, 738.
26. L. Chiozza, *Annalen.*, 1852, **83,** 117.
27. R. T. Conley and W. N. Knopka, *J. Org. Chem.*, 1964, **29,** 496.
28. K. M. Johnston, R. M. Luker, and G. H. Williams, *J. Chem. Soc.*, 1972, 1648 and references therein.
29. E. Fischer and A. Steehe, *Annalen*, 1887, **242,** 353.
30. G. Magnanini, *Ber.*, 1887, **20,** 2608; 1888, **21,** 1940.
31. G. Heller, *Ber.*, 1919, **52,** 741.
32. F. S. Kipping, *J. Chem. Soc.*, 1894, **65,** 489.
33. (a) A. T. Amos and G. G. Hall, *Mol. Phys.*, 1961, **4,** 25; (b) G. J. Palenik, *Acta Cryst.*, 1964, **17,** 687.
34. R. D. Brown and R. D. Harcourt, *J. Chem. Soc.*, 1959, 3451.
35. 'Physical Methods in Heterocyclic Chemistry', ed. A. R. Katritzky, Academic Press, New York and London, vols. I and II, 1963; vols. III and IV, 1971: (a) 'Ultra violet spectroscopy', by S. F. Mason (vol. II) and W. L. F. Armarego (vol. III); (b) 'Nuclear Magnetic Resonance Spectroscopy', by R. F. M. White, (vol. II) and R. F. M. White and H. Williams (vol. IV); (c) 'Infrared spectroscopy', by A. R. Katritzky and P. J. Taylor (vol. IV); (d) 'Mass Spectroscopy', by G. Spiteller (vol. III); (e) J. H. Ridd (vol. I) p. 109.
36. H. Rottendorf and S. Sternhell, *Austral. J. Chem.*, 1964, **17,** 1315.
37. P. A. Claret and A. G. Osborne, *Org. Magn. Resonance*, 1976, **8,** 147, and previously unpublished results.
38. P. A. Claret and G. H. Williams, (a) *J. Chem. Soc. (C)*, 1969, 146; (b) unpublished results.
39. J. Kuthan, N. V. Koshmina, J. Palacek, and V. Skala, *Coll. Czech. Chem. Comm.*, 1970, **35,** 2787.
40. F. D. Popp, *Adv. Heterocyclic Chem.*, 1968, **9,** 1.
41. M. W. Austin and J. H. Ridd, *J. Chem. Soc.*, 1963, 4204.
41a. R. J. W. Le Fevre and F. C. Matur, *J. Chem. Soc.*, 1930, 2236.
42. R. G. Coombes and L. W. Russell, *J. C. S. Perkin I*, 1974, 1751.
43. P. A. Claret and A. G. Osborne, *Tetrahedron*, 1977, **33,** 1765.
43a. U. Bressel, A. R. Katritzky, and J. R. Lee, *J. Chem. Soc. (B)*, 1971, 4.
44. P. B. D. De la Mare, M. Kiamuddin, and J. H. Ridd, *Chem. and Ind. (London)*, 1958, 361.
45. M. Kiamuddin and M. E. Haque, *Chem. and Ind. (London)*, 1964, 1753.
46. H. E. Jansen and J. P. Wibaut, *Rec. Trav. chim.*, 1937, **56,** 699.
47. W. Borsche and H. Groth, *Annalen*, 1941, **549,** 238.
48. M. T. Leffler, *Org. Reactions*, 1942, **1,** 91.
48a. G. A. Russell and S. A. Weiner, *J. Org. Chem.*, 1966, **31,** 248.
49. F. Minisci and O. Porta, *Adv. Heterocyclic Chem.*, 1974, **16,** 123.
50. W. H. Mills and R. Raper, *J. Chem. Soc.*, 1925, **127,** 2466.
51. R. A. Glenn and J. R. Bailey, *J. Amer. Chem. Soc.*, 1941, **63,** 639.
52. C. E. Kwartler and H. G. Lindwall, *J. Amer. Chem. Soc.*, 1937, **59,** 524.
53. V. M. Rodionov and M. A. Berkengeim, *J. Gen. Chem. (U.S.S.R.)*, 1944, **14,** 330 (*Chem. Abs.*, 1944, **39,** 40769).
54. M. Seyhan, *Chem. Ber.*, 1952, **85,** 425.
55. W. von Miller, *Ber.*, 1891, **24,** 1919.
56. S. Hoogewerff and W. A. van Dorp, *Ber.*, 1879, **12,** 747.
57. O. Doebner and W. von Miller, *Ber.*, 1882, **15,** 3077.
58. J. P. Wibaut and H. Boer, *Rec. Trav. chim.*, 1955, **74,** 241.
59. A. J. Birch and P. G. Lehman, *Tetrahedron Letters*, 1974, 2395.
60. J. von Braun, *Ber.*, 1922, **55,** 3779; 1923, **56,** 1338; 1924, **57,** 382.
61. F. W. Vierhapper and E. L. Eliel, *J. Org. Chem.*, 1975, **40,** 2729.
62. D. A. Archer, H. Booth and P. C. Crisp, *J. Chem. Soc.*, 1964, 249.
63. J. von Braun and A. Steindorf, *Ber.*, 1904, **37,** 4723.
64. G. E. McCasland, *J. Org. Chem.*, 1946, **11,** 277.
65. J. Walker, *J. Chem. Soc.*, 1947, 1552.
66. A. R. Katritzky and J. M. Lagowski, *Adv. Heterocyclic Chem.*, (a) 1963, **1,** 347 and 1976, **1** (suppl.), 87, 152; (b) **1,** 320; (c) **1,** 404–407.
67. S. F. Mason, *J. Chem. Soc.*, 1958, 674.
68. E. B. Hartshorn and S. L. Baird, Jr., *J. Amer. Chem. Soc.*, 1946, **68,** 1562.
69. J. C. Petersen, *J. Phys. Chem.*, 1971, **75,** 1129.
70. P. A. Claret and A. G. Osborne, *Spectroscopy Letters*, 1976, **9,** 157.
71. P. A. Claret and A. G. Osborne, (a) *Spectroscopy Letters*, 1976, **9,** 167; (b) unpublished results.
72. A. R. Katritzky and J. M. Lagowski, 'The Chemistry of the Heterocyclic N-oxides', Academic Press, New York and London, 1971.
73. E. Ochiai, *J. Org. Chem.*, 1953, **18,** 534.
74. (a) O. Buchardt, P. L. Kamler, and C. Lohse, *Acta Chem. Scand.*, 1969, **23,** 159; (b) O. Buchardt, *ibid.*, 1964, **18,** 1389.
75. S. J. Angyal and C. L. Angyal, *J. Chem. Soc.*, 1952, 1461.
76. J. C. E. Simpson and P. H. Wright, *J. Chem. Soc.*, 1948, 2023.
77. R. C. Elderfield and Siegel, *J. Amer. Chem. Soc.*, 1951, **73,** 5622.
78. R. C. Elderfield, Ref. 1(c), p. 225.
79. B. Brown and D. Ll. Hammick, *J. Chem. Soc.*, 1949, 659.

80. J. S. McFadyen and T. S. Stevens, *J. Chem. Soc.*, 1936, 584.
81. C. A. Buehler and J. O. Harris, *J. Amer. Chem. Soc.*, 1950, **72,** 5015.
82. G. B. Backmann and D. D. Micucci, *J. Amer. Chem. Soc.*, 1948, **70,** 2381.
83. Ref. 1(c), p. 316.
84. A. Albert, M. I. Gibson, and S. D. Rubbo, *Brit. J. Exp. Pathol.*, 1953, **34,** 119 (and previous papers).
85. F. M. Hamer, *Quart. Rev.*, 1950, **4,** 327.
86. F. M. Hamer, 'The Cyanine Dyes and Related Compounds', Interscience, New York, 1963.
87. K. Venkataraman, 'Synthetic Dyes', Academic Press, New York and London, 1952, vol. 2, chapter 38.
88. P. Rabe, *Ber.*, 1908, **41,** 62.
89. R. B. Woodward and W. E. Doering, *J. Amer. Chem. Soc.*, 1945, **67,** 860.
90. 'Chemistry of the Alkaloids', ed. S. W. Pelletier, Van Nostrand Reinhold, New York, 1970: (a) W. Solomon, 'The Cinchona Alkaloids', chapter 11, p. 301; (b) P. J. Scheuer, 'The Furoquinoline Alkaloids', chapter 12, p. 355.
91. 'The Alkaloids', ed. R. H. F. Manske, Academic Press, New York and London: (a) M. R. Uskokovic and G. Grethe, 'Cinchona Alkaloids', 1973, vol. 14, p. 181; (b) H. T. Openshaw, 'Quinoline Alkaloids other than Cinchona Alkaloids', 1967, chapter 9, p. 223.

16.5

Isoquinolines

P. A. CLARET
The City University, London

16.5.1 OCCURRENCE AND ISOLATION OF ISOQUINOLINE AND ITS HOMOLOGUES

16.5.1.1 The isoquinoline ring system

Isoquinoline[1] (2-azanaphthalene or benzo[*b*]pyridine) (**1**) is one of the three possible azanaphthalenes (see Section 16.4.1.1) and is usually represented and numbered in the same way as quinoline. As might be expected from the similarities in their structures, there is a close relationship between the physical and chemical properties of isoquinoline and those of quinoline, modified only as a result of the different position of the nitrogen atom relative to the carbocyclic ring. The following account will deal with most of the chemistry of isoquinoline and its simple derivatives by comparison with that of quinoline and its derivatives, concentrating in detail only on the points of difference between the two series.

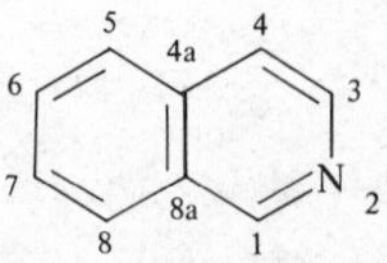

(1)

As with quinoline, the evidence in favour of the bicyclic structure of isoquinoline is based on its constitution, methods of synthesis, and chemical properties, especially degradation by oxidation, ozonolysis, and 'ring opening'. It has been confirmed by the observed pattern of the isomerism of its derivatives and particularly by physical methods, including X-ray crystallography, ultraviolet spectroscopy, and ^{1}H and ^{13}C n.m.r. spectroscopy, as described in more detail below.

16.5.1.2 Occurrence and isolation of isoquinoline and alkylisoquinolines

Isoquinoline was first isolated in 1885 by Hoogewerff and van Dorp from the 'quinoline' fraction of coal tar bases as its sparingly soluble bisulphate salt. Isoquinoline and its 1-methyl, 3-methyl, and 1,3-dimethyl derivatives have also been obtained from the bases isolated from the distillation products of crude petroleum and shale oil. Other methods of separation, from the complex mixtures of bases obtained from these sources, include fractional precipitation by successive additions of 10% sodium hydroxide solution to a solution of the mixed hydrochlorides and extraction with chloroform. The fractions so obtained have increasing basic strengths and the stronger basic properties of isoquinoline compared with those of quinoline enable the separation to be achieved. Azeotropic distillation (*e.g.* with ethylene glycol) has also been used.

Apart from its formation during the pyrolysis of coal and the destructive distillation of shale and crude petroleum, the isoquinoline structure occurs naturally in the opium alkaloids, usually, but not always, with the heterocyclic ring in a reduced (dihydro or tetrahydro) form, and with a substituent at position-1 (see Section 16.5.6).

16.5.2 SYNTHESES OF ISOQUINOLINE COMPOUNDS

16.5.2.1 General discussion

Although there is a close parallel between the chemical and physical properties of quinoline and isoquinoline, this is not the case with the usual methods of synthesis as the difference in the ring structure necessitates a different choice of starting materials. In the most widely used methods the final ring-closure completes the heterocyclic ring by intramolecular electrophilic attack at a ring position of an aromatic intermediate. The 'reagent' is a positive carbon centre suitably placed on a side-chain *ortho* to the position of ring closure. Thus in the Bischler–Napieralski, Pictet–Gams, and Pictet–Spengler syntheses outlined below, the intermediates can be represented by (**3**) and ring closure is at bond (a) in (**2**). In the Pomeranz–Fritsch synthesis the intermediates can be represented by (**4**) and ring closure is at bond (e). The use of isocoumarins or homophthalic acid derivatives as starting compounds provides examples of ring closure to nitrogen at bonds

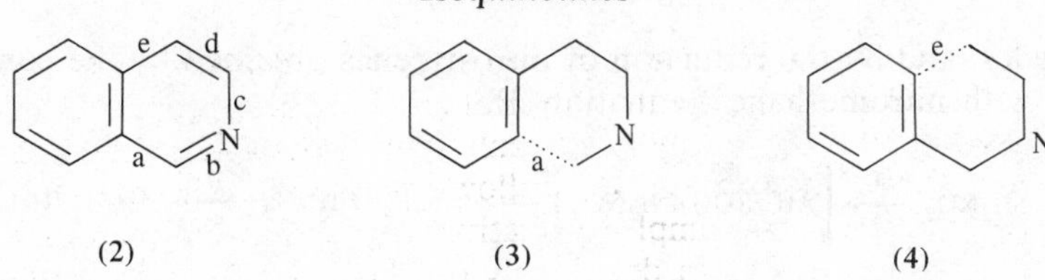

(2) (3) (4)

(b) and (c), and the rearrangements of phthalimido-acetic esters result in the closure of bond (d).

16.5.2.2 The Bischler–Napieralski synthesis[1c,2a]

This synthesis was originated by Bischler and Napieralski in 1893, and has since been subjected to many improvements. It results in the formation of 3,4-dihydroisoquinolines by ring closure of *N*-(2-phenylethyl)amides. The amides are usually formed by the acylation or aroylation of 2-phenylethylamines. The ring-closure stage is an acid-catalysed, intramolecular, electrophilic, aromatic substitution. Successful reagents include phosphorus pentoxide, phosphorus pentachloride, and phosphoryl chloride in boiling xylene or decalin[3] (equation 1), phosphorus pentoxide in pyridine,[4] and polyphosphoric acid.[5] Final conversion to the aromatic isoquinoline is usually accomplished by dehydrogenation with palladized charcoal.

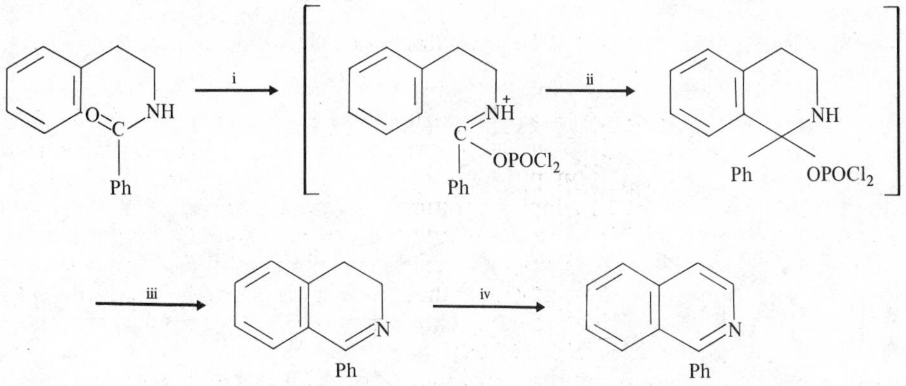

(1)

i, P_2O_5, $POCl_3$, xylene, reflux 1 h (+ $POCl_3$ – Cl^-);
ii, $-H^+$; iii, $-POCl_2(OH)$; iv, Pd/C, 160 °C.

Substituents attached to the aromatic ring of the phenylethylamide exert a predictable influence on the electrophilic ring-closure stage. Thus an electron donor, such as methoxy, improves both the yield and rate of reaction and is particularly effective at position-3 of the amide when it directs ring-closure exclusively to the *para*-position, giving a 6-methoxy-3,4-dihydroisoquinoline. Ring closure to the *ortho*-position to give the 8-methoxy-isomer does not occur, probably owing to steric interactions. Electron-withdrawing groups such as nitro or halogens have the reverse effect and yields of dihydroisoquinolines with these substituents in the carbocyclic ring are poor by this method,[6] but better results can be achieved using the Pomeranz–Fritsch synthesis. Isoquinolines unsubstituted at position-1 require amides of formic acid (*N*-formylamines) as precursors, but yields are poor in the ring-closure stage unless the benzene ring contains electron-releasing groups such as alkoxy. Higher reaction temperatures with decalin as a solvent also give improved yields.

The necessary 2-phenylethylamines can be prepared by a number of methods, *e.g.* from β-arylethyl bromides by the Gabriel phthalimide synthesis, by the reduction of oximes of arylacetaldehydes, and by the reduction of nitrostyrenes obtained by the condensation of arylaldehydes with nitromethane[7] (equation 2).

$$ArCHO + CH_3NO_2 \xrightarrow{i} [ArCH(OH)CH_2NO_2] \xrightarrow{ii} ArCH=CHNO_2 \xrightarrow{iii} ArCH_2CH_2NO_2 \quad (2)$$

i, OEt^-; ii, $-H_2O$; iii, H_2/Pt (1 mol).

16.5.2.3 The Pictet–Gams synthesis[2a,8]

In this modification of the Bischler–Napieralski synthesis, introduced by Pictet and Gams in 1909, the acyl or aroyl derivative of a 2-hydroxy- or 2-methoxy-phenylethylamine is used for the cyclization and produces the fully aromatic isoquinoline directly. The intermediate can be synthesized from the readily available aryl alkyl ketones by nitrosation of the alkyl group and reduction of the isonitroso (oxime) tautomer to the amine. Then either the original keto-group is reduced followed by acylation (or aroylation) of the amino function, or the reduction stage can follow the acylation (see Scheme 1).

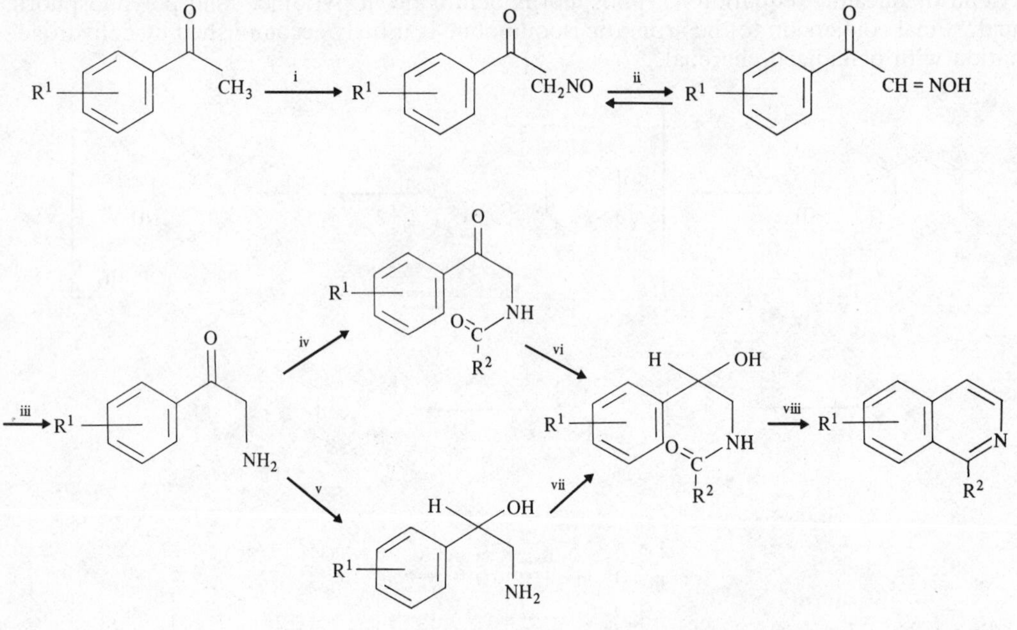

R^1 = electron-donating substituent or H; R^2 = alkyl or aryl (or H if R is electron donating).

i, HNO_2; ii, tautomeric equilibrium; iii, sodium dithionite, $Na_2S_2O_6$, or H_2/Pt;

iv, R^2COCl; v, vi, Na/Hg, H_2O, 50 °C; vii, R^2COCl; viii, P_2O_5, $POCl_3$, toluene, reflux.

SCHEME 1

In the subsequent ring closure, styryl amides have been isolated from the reaction mixture in the early stages although they are eventually converted into isoquinolines. They have been shown to arise by elimination of water in the course of the intermediate formation of a 4,5-substituted oxazoline[11] (Scheme 2). The use of this reaction for the first synthesis of the alkaloid papaverine[8] (**5**) is shown (equation 3) although a number of alternative methods have since been devised.

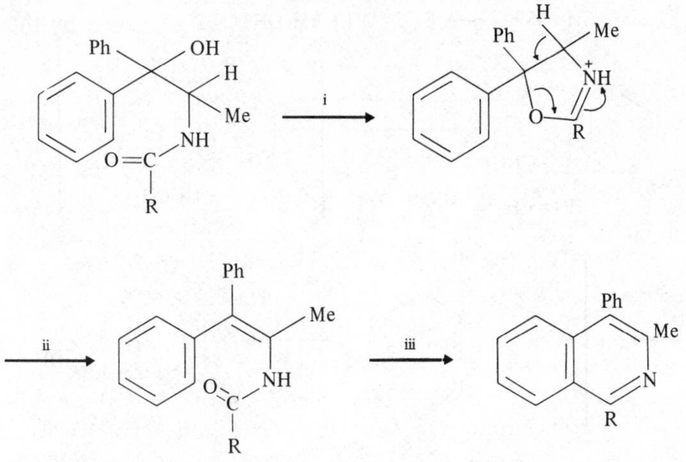

i, P_2O_5, PhMe or decalin, boil ($-OH^-$); ii, $-H^+$; iii, $-H_2O$.

SCHEME 2

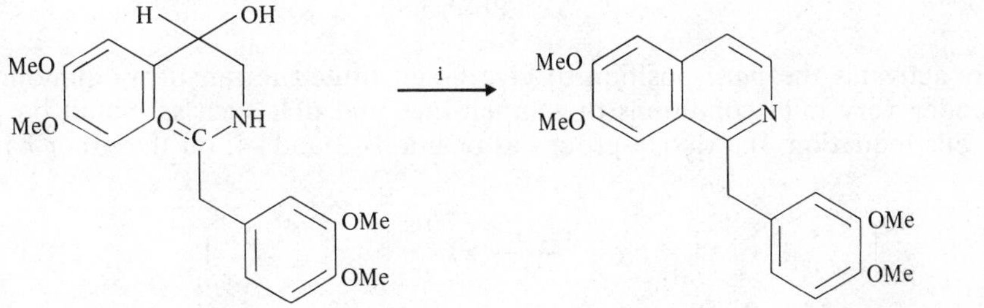

(3)

(5) Papaverine

i, P_2O_5, $POCl_3$, toluene reflux.

16.5.2.4 The Pictet–Spengler synthesis[2b]

A suitable intermediate of type (**3**) can be readily prepared by the condensation of a 2-phenylethylamine with an aldehyde to form an imine; the ring closure can then be induced by an acid as in the Bischler–Napieralski synthesis. In this case the ring closure is the result of a Mannich reaction and a tetrahydroisoquinoline is formed, as the imine is in a lower oxidation state than the amide in the Bischler–Napieralski reaction, whereas the hydroxyamide in the Pictet–Gams modification is in a higher oxidation state and therefore gives the fully aromatic isoquinoline. The tetrahydroisoquinolines, like the dihydro-compounds, can be dehydrogenated by heating with palladized charcoal.

Pictet and Spengler[9] condensed formaldehyde, in the form of its dimethyl acetal, with 2-phenylethylamine and concentrated hydrochloric acid in a single-stage process to give tetrahydroisoquinoline (Scheme 3). In general, when formaldehyde is used, isoquinolines are formed which are unsubstituted at position-1, and which are difficult to prepare by the Bischler–Napieralski and Pictet–Gams syntheses. Some later workers have preferred to use a two-stage process in which the imine is isolated first and then cyclized by the addition of acid. Secondary amines give 2-substituted tetrahydroisoquinolines.

As with the two previous syntheses, the ring-closure stage is sensitive to substituent effects. An alkoxy- or hydroxy-group at position-3 of the aromatic ring of the imine

$CH_2(OMe)_2 \xrightarrow{i} CH_2=\overset{+}{O}H + 2MeOH$

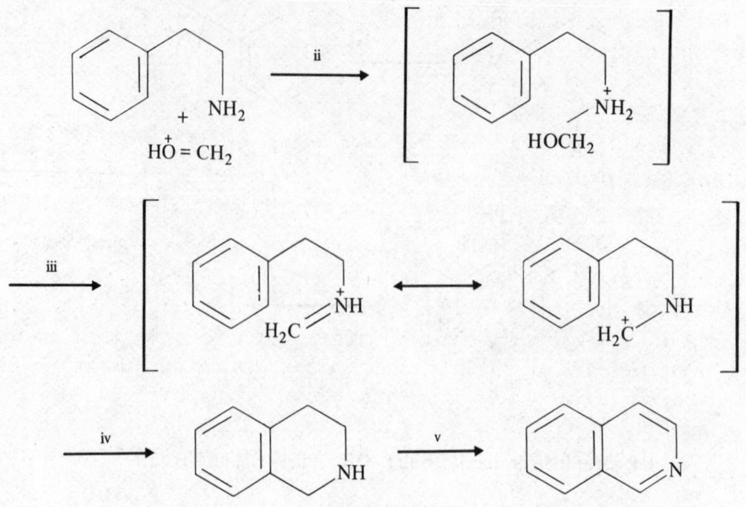

i and ii, hydrolysis of acetal ($+H^+$); iii, $-H_2O$; iv, $-H^+$; v, Pd/C, heat.

SCHEME 3

strongly activates the *para*-position to give 6-substituted tetrahydroisoquinolines in high yield under very mild conditions of temperature and pH, such as would be possible in living cells (equation 4). Alkoxy-groups at positions-2 and -4, on the other hand, appear

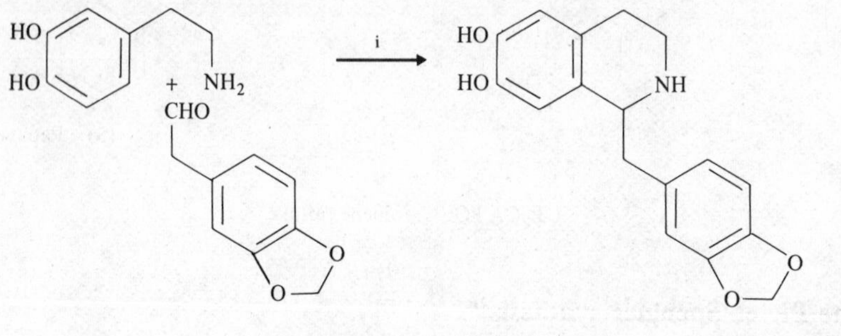

(4)

i, pH = 6–7, room temp.

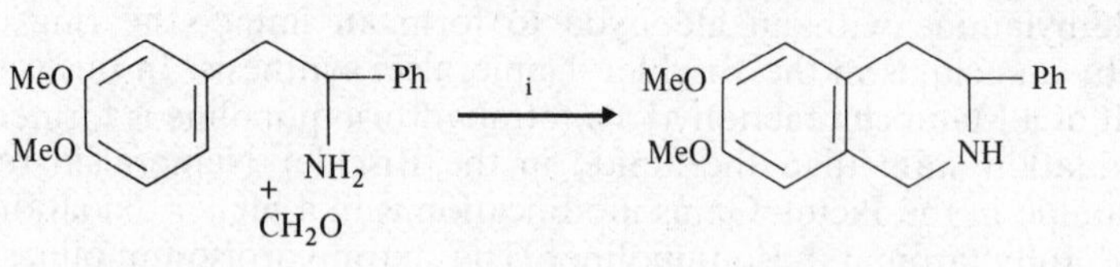

(5)

i, H^+, good yield

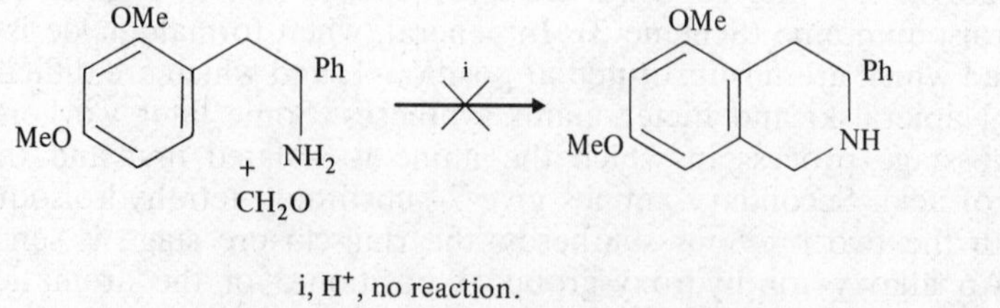

(6)

i, H^+, no reaction.

to prevent ring-closure,[10] presumably as a result of their negative inductive effect (−*I*) (equations 5 and 6). Ring-closure has not been reported *ortho* to alkoxy-groups, but 7,8-dihydroxytetrahydroisoquinolines together with the 6,7-isomers are formed during the ring closure of 3,4-dihydroxy-imines.

16.5.2.5 The Pomeranz–Fritsch synthesis

The condensation of benzaldehyde or a substituted benzaldehyde with the acetal of aminoacetaldehyde gives a Schiff base (aldimine) with the skeleton structure (**4**) which can be cyclized with acids to an isoquinoline. The acetal group is necessary to protect the bifunctional aminoaldehyde from polymerization by self condensation; it is sensitive to acid hydrolysis and does not prevent the final ring closure. As a result of the electrophilic nature of the ring closure the substituent effects discussed in the previous syntheses are operative. Electron-releasing groups on the aromatic ring are favourable and, with 3-substituted benzaldehydes, ring-closure is normally to the *para*-position rather than to the *ortho*-position. Thus the 3-substituted aldehydes give 7-substituted (rather than 5-substituted) isoquinolines, which are the ones difficult to prepare by the previous methods (equation 7).[13] As indicated previously, halogen substituents in the aromatic ring, although deactivating, do not prevent ring-closure and *m*-chloro- and *m*-bromo-benzaldehydes give mixtures of the 7- and 5-substituted isoquinolines; *m*-hydroxy-benzaldehyde gives mainly 7-hydroxyisoquinoline (64%) with about 5% of the 5-isomer.

EtO OEt

HO CHO Cl —i→ HO Cl N —ii→ HO Cl N (7)

i, $H_2NCH_2CH(OEt)_2$, C_6H_6 ($-H_2O$); ii, (a) H_2SO_4 (76%), 8 °C, 40 h, (−2EtOH), (b) NH_4OH; yield 64%, Ref. 13.

The poor yields obtained in the formation of ketimines, as opposed to aldimines, in the first stage of this synthesis makes it unsuitable for the preparation of 1-substituted isoquinolines, but this limitation can be overcome by preparing an isomeric imine from a substituted benzylamine and the hemiacetal of glyoxal.[14] The latter can be obtained from the acetal of acrolein by oxidation and cleavage of the glycol so formed (equations 8 and 9).

$$CH_2=CH_2CH(OEt)_2 \xrightarrow{i} CH_2(OH)CH_2(OH)CH(OEt)_2 \xrightarrow{ii} O=CHCH(OEt)_2 \quad (8)$$

i, $KMnO_4$; ii, $Pb(OAc)_4$

EtO OEt

MeO CH(Me) NH_2 —i→ MeO N Me —ii→ MeO N Me (9)

i, O=$CHCH(OEt)_2$, piperidine, toluene, reflux ($-H_2O$); ii, (a) H_2SO_4 (76%), HCl, 10–20 °C, 5½ days (−2EtOH), (b) NaOH; yield 50%, Ref. 14.

16.5.2.6 Syntheses of isoquinolones (isocarbostyrils)

The following syntheses give hydroxyisoquinolines, or their isoquinolone tautomers in the case of the 1- and 3-isomers. Isocoumarin derivatives, *e.g.* (**6**), with aqueous or alcoholic ammonia or primary amines are converted into isocarbostyril derivatives (**7**) (equation 10). When the diammonium salt of homophthalic acid (**8**) is heated, it is converted into homophthalimide (**9**), a tautomeric form of 3-hydroxyisocarbostyril (**10**) (equation 11). α-Phthalimido-derivatives of the esters of aliphatic acids (**11**) with sodium ethoxide undergo a rearrangement to 3-substituted 4-hydroxyisocarbostyrils (**12**) (equation 12). Ethyl α-phthalimidoacetate (**11**; $R^1 = H$) gave 3-ethoxycarbonyl-4-hydroxyisocarbostyril (**12**; $R^2 = CO_2Et$), but if R^1 = alkyl or aryl it is retained with loss of the ester group ($R^1 = R^2$).

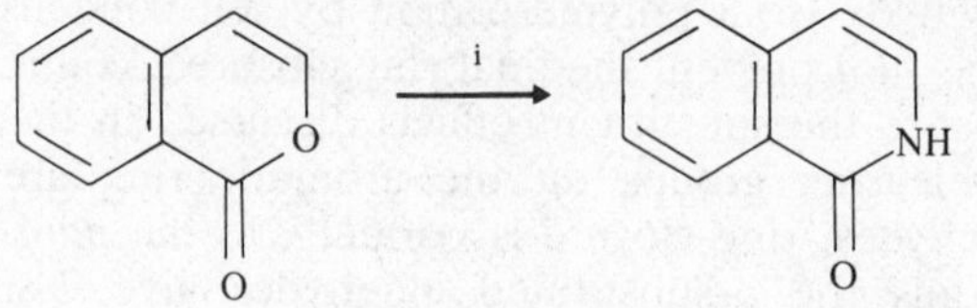

(6) → (7) (10)

i, NH_3, H_2O or EtOH

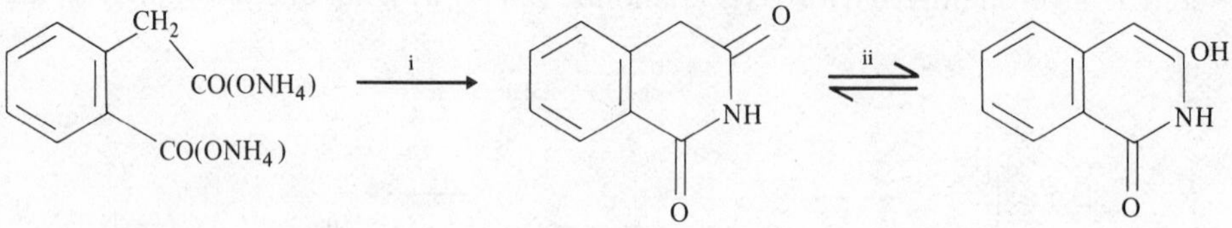

(8) → (9) ⇌ (10) (11)

i, heat; ii, tautomeric equilibrium

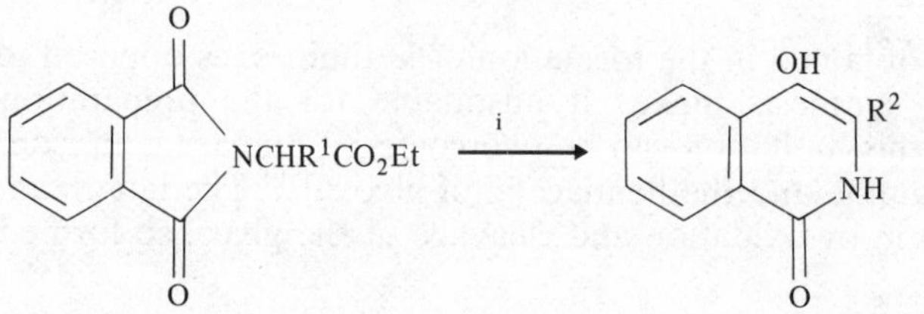

(11) → (12) (12)

i, OEt^-

16.5.2.7 Syntheses of isoquinolines *via* the Beckmann rearrangement

The oxime of cinnamaldehyde (**13**; $R^1 = R^2 = H$) when submitted to the conditions for the Beckmann rearrangement with phosphorus pentoxide gives isoquinoline, presumably by ring-closure of the *N*-styrylformamide which would normally result from the Beckmann rearrangement (equation 13). The oxime of α-chlorocinnamaldehyde gives 3-chloroisoquinoline, and other examples have been reported of the formation of isoquinolines from oximes *via* the Beckmann rearrangement, *e.g.* equation 13 for $R^1 = H$, $R^2 = Me$, but the reaction is not of general application.

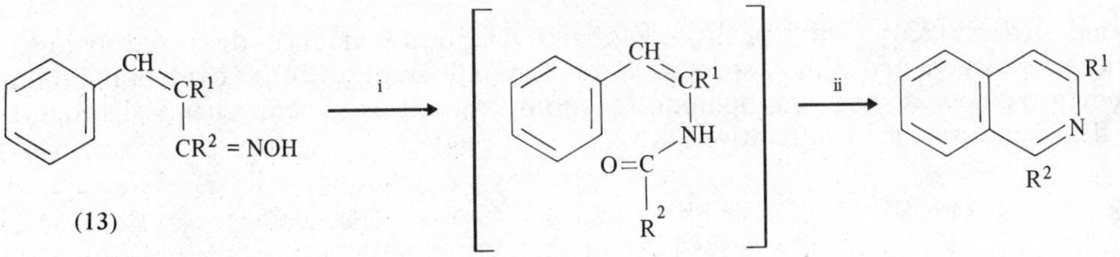

(13)

i, P_2O_5, Beckmann rearrangement; ii, $-H_2O$.

16.5.3 PHYSICAL PROPERTIES OF THE ISOQUINOLINE BASES

16.5.3.1 General properties and structure

Although familiar as a liquid, isoquinoline, when pure, is a colourless solid: m.p. 26.5 °C, b.p. 243 °C. It has a smell suggestive of almonds or benzaldehyde. It is only slightly soluble in water, less so than quinoline, and typically becomes yellow on exposure to moist air and light. The separation of the nitrogen from the carbocyclic ring results in an increase in basic strength ($pK_a = 5.14$) compared with quinoline ($pK_a = 4.9$); it has a higher dipole moment ($\mu = 2.60$ D) owing to greater charge separation in the contributing dipolar valence bond structures (**14**)–(**22**).

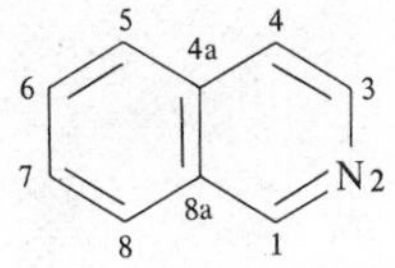

(14) 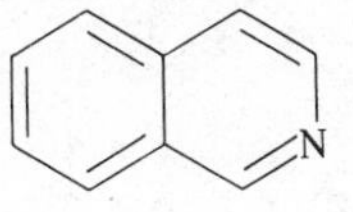 (15) 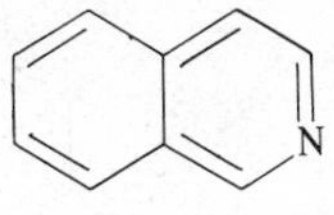 (16)

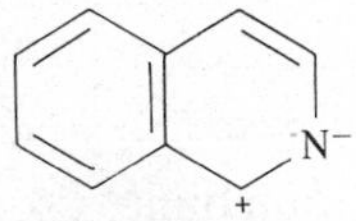

(17) 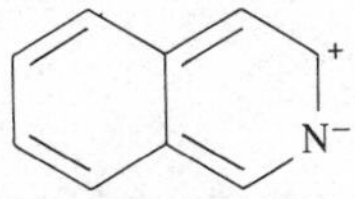 (18) 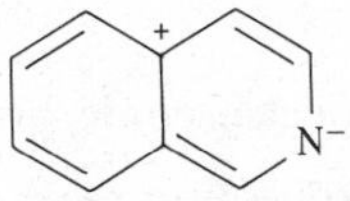 (19)

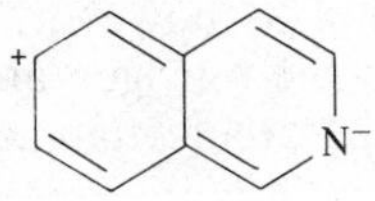

(20) 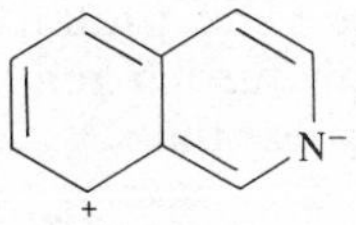 (21) 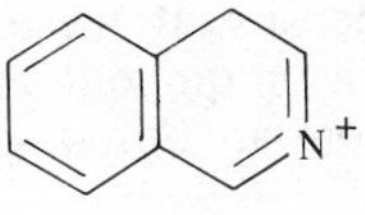 (22)

For reasons similar to those in the discussion of quinoline, structure (**14**) is the most favoured resonance form and, of the dipolar structures with the formal negative charge on nitrogen, only (**17**) retains the aromatic character of the carbocyclic ring; however, the high dipole moment is indicative of a significant contribution of structures such as (**20**). Structures such as (**22**) with positive charge on nitrogen are unfavoured. This picture suggests a degree of 'bond fixation' favouring structure (**14**) with C-1 more reactive to nucleophilic attack than C-3 and the other positions. This is consistent with calculated

π-bond orders (**23**)[15] and spectroscopic and chemical evidence discussed below, in particular the u.v. and n.m.r. spectra, X-ray crystallography (**24**),[16] oxidation products including ozonolysis, and ring-opening reactions, as well as general chemical properties and the isomerism of its derivatives.

(23)

π-Bond orders for isoquinoline (ethylene = 1.0, benzene = 0.667)[15]

(24)

Bond lengths (pm)[16]

A comparison with quinoline and pyridine would indicate that positions-1 and -3 should be compared with α-positions in pyridine and position-4 with β-positions. There is no site equivalent to the γ-position of pyridine or the 4-position of quinoline. As already noted, the 3-position in isoquinoline is much less reactive than the 1-position since transition states and intermediate complexes resulting from attack at position-3 do not have the advantage of stabilization by a fully aromatic carbocyclic ring, *e.g.* for attack by nucleophile **:B** see (**25**) and (**26**), and as with quinoline this restriction will apply also to the stabilities of Reissert compounds and methylene bases.

(25)

(26)

16.5.3.2 Spectroscopic properties[17]

The relationships between the ultraviolet spectra of quinoline, isoquinoline, and naphthalene have been discussed in Section 16.4.3.2 and are shown in Table 3 of that section. The ^1H and ^{13}C n.m.r. spectra of isoquinoline are analysed in Tables 1 and 2. Both protons and carbons at positions-1 and -3 next to the nitrogen resonate further downfield than the corresponding nuclei at other positions, consistent with the low electron density at these positions. The proton at position-8, however, is not as far downfield as in quinoline as it is now further removed from the nitrogen atom. The *ortho* coupling constant $J_{6,7}$ is significantly less than $J_{5,6}$ and $J_{7,8}$, the other *ortho* couplings in the

TABLE 1

^1H and ^{13}C Chemical Shifts for Isoquinoline[a]

		1	2	3	4	4*a*	5	6	7	8	8*a*
$^1H^b$	Neat	9.45	—	8.77	7.48	—	7.57	7.96	7.37	7.74	—
$^1H^b$	CCl_4	9.13	—	8.45	7.50	—	7.71	7.57	7.49	7.86	—
$^{13}C^c$		153.3	—	144.0	121.0	136.2	127.0	130.7	127.7	128.1	129.3

[a] δ (p.p.m. downfield from TMS). [b] P. J. Black and M. L. Hefferman, *Austral. J. Chem.*, 1966, **19,** 1287. [c] R. J. Pugmire, D. M. Grant, M. J. Robbins, and R. K. Robbins, *J. Amer. Chem. Soc.*, 1969, **91,** 638.

TABLE 2

^1H–^1H Coupling Constants (Hz) for Liquid Isoquinoline[a]

Three-bond o-coupling		*Four-bond m-coupling*		*Long range five-bond couplings*	
$J_{3,4}$	5.8	$J_{1,3}$	~0	$J_{1,5}$	<0.5
$J_{5,6}$	8.6	$J_{5,7}$	0.9	$J_{4,8}$	~0.8
$J_{6,7}$	7.0	$J_{6,8}$	1.1	$J_{5,8}$	0.8
$J_{7,8}$	8.4				
		Four-bond benzylic coupling[b]			
		$J_{CH_3(3),4}$	~0.45		

[a] P. J. Black and M. L. Hefferman, *Austral. J. Chem.*, 1966, **19,** 1287, except b. [b] A. G. Osborne, City University, London, previously unpublished.

carbocyclic ring, in line with its lack of π-character. The corresponding 2,3-bond cannot be assessed because of the position of the nitrogen. The only available figure for benzylic coupling constants $J_{CH_3(3),4}$, though smaller than in quinoline, shows that the 3,4-bond possesses π-character. The ring-carbon–ring-proton coupling constants are similar in value to those in quinoline and are significantly higher on the positions next to the nitrogen.

16.5.4 CHEMICAL PROPERTIES OF ISOQUINOLINE

16.5.4.1 Salt formation

Isoquinoline is a stronger base than quinoline and about the same strength as pyridine. It therefore readily forms isoquinolinium salts by protonation of the nitrogen, and 2-substituted quaternary salts with appropriate alkyl, acyl, and aroyl halides and dimethyl sulphate. In general the reactivity at position-1 is similar to that of quinoline at position-2 but, for the reasons given above, position-3 is unreactive although next to nitrogen. Addition of alkali to the salts gives the ionized form of the hydroxide, which forms a carbinol base by attack of hydroxyl at position-1. This is oxidized by potassium ferricyanide to the 1(2*H*)-isoquinolone. The 2-benzoylisoquinolinium salt with cyanide ion gives a Reissert compound, again only at position-1. Scheme 4 gives examples of these reactions.

16.5.4.2 Nuclear substitution reactions

Calculations of the electron localization energy[18] of the isoquinolinium cation indicate that positions-5 and -8 should be favoured for electrophilic attack on the protonated base. Of these, substitution at position-5 would proceed through a more favourable σ-complex. In less acidic media with some of the base unprotonated, the nitrogen would be the favoured position for electrophilic attack, and 1,2-addition could lead to electrophilic substitution at position-4, corresponding to the formation of 3-substituted quinolines. Nucleophilic attack should be limited to position-1 and lead either to 1,2-addition or, by loss of hydride (oxidation), to nucleophilic substitution at position-1. The quaternary isoquinolinium ion is further activated towards nucleophilic reagents and 1,2-addition products are usually formed. The following examples illustrate these principles.

(*i*) *Nitration*

Nitric acid and sulphuric acid at room temperatures give a mixture of 5- and 8-nitroisoquinolines containing about 90% of the 5-isomer.[19] At higher temperatures the

i, ii, KOH; iii, $K_3Fe(CN)_6$; iv, PhCOCl; v, KCN; vi, PCl_5 (−PhCHO); vii, HCl; viii, −HCl; ix, HCl; x, hydrolysis

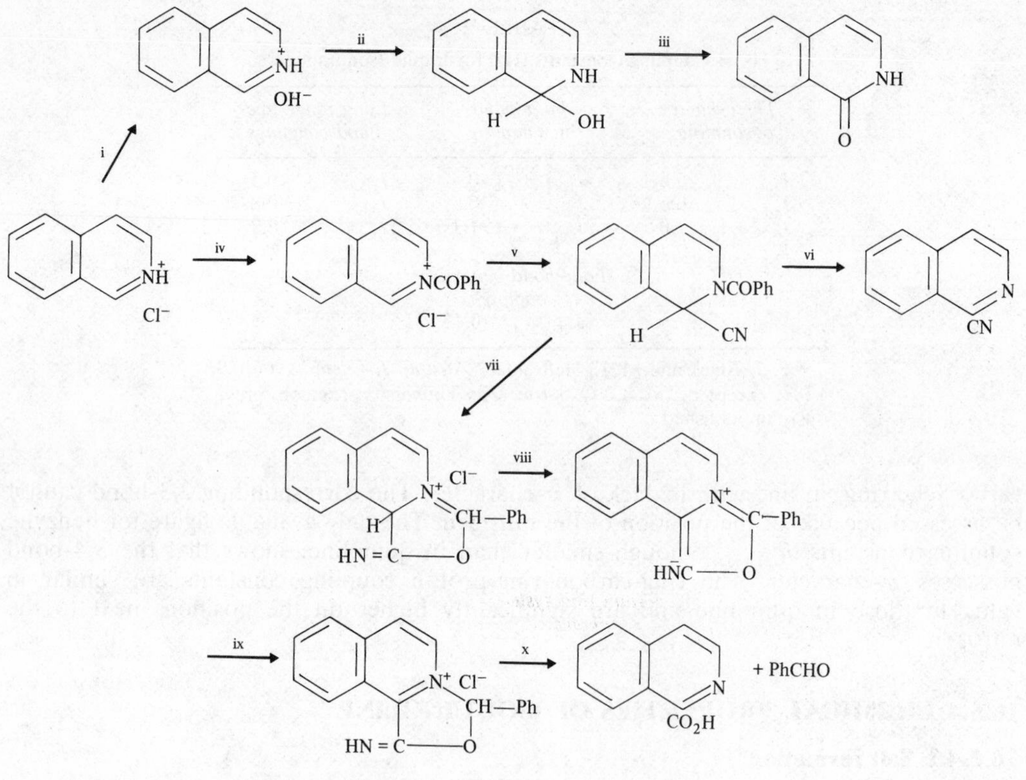

SCHEME 4

proportion of the 8-isomer is slightly increased. In 81.3% sulphuric acid at 25 °C it is the cation[20] that is nitrated at the rate of 13.9×10^{-2} l mol^{-1} min^{-1}. Nitration of 1- and 4-benzylisoquinolines gives the 4′-nitrobenzyl derivatives with no nitration of the isoquinoline ring, but 1-phenylisoquinoline is reported to give 1-*m*-nitrophenyl-5-nitroisoquinoline (equations 14a, b, and c).

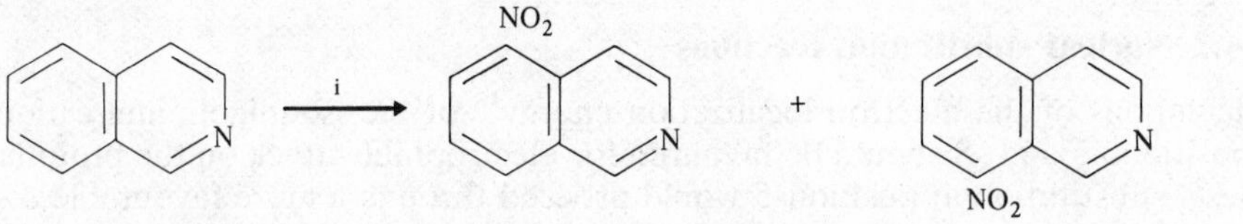

(14a)

i, HNO_3, H_2SO_4(81.3%), 25 °C; isomer ratio 5-NO_2 : 8-NO_2 = 9 : 1.

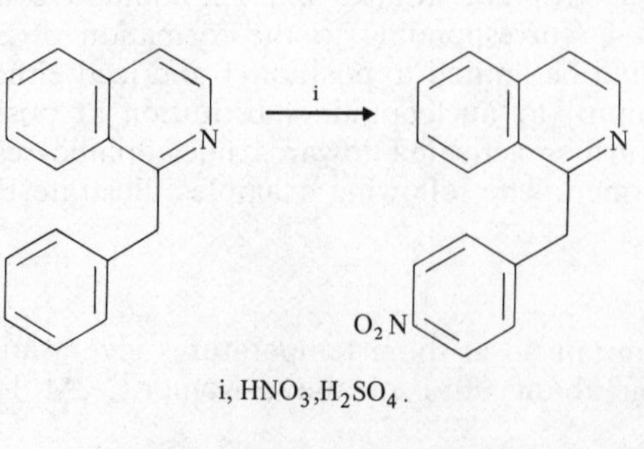

(14b)

i, HNO_3,H_2SO_4.

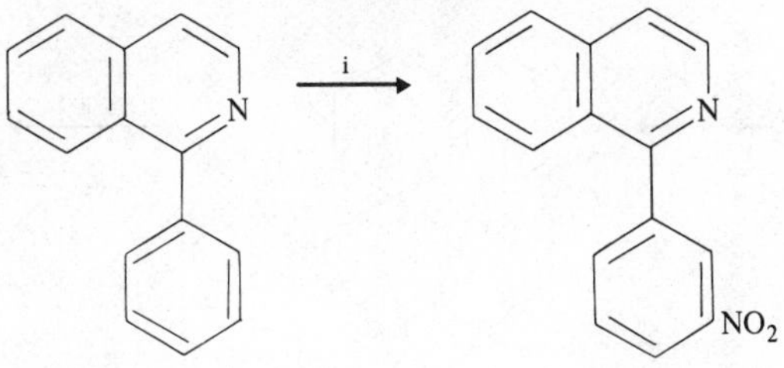

(14c)

i, HNO_3, H_2SO_4.

(*ii*) *Sulphonation*

With oleum of strength 40% or higher SO_3 content, and temperatures up to 180 °C, isoquinoline-5-sulphonic acid is the main product. At higher temperatures the formation of the 8-sulphonic acid in addition to the 5-isomer has been established.

(*iii*) *Halogenation*

Bromination may take place by direct electrophilic attack at position-5 with an aluminium chloride catalyst. By heating isoquinoline hydrobromide with bromine, or with bromine in sulphur monochloride, substitution is at position-4 by the addition–elimination mechanism. Iodine monochloride acts as an iodinating agent, but halogen derivatives other than bromo are usually prepared from the amino-compounds (see Scheme 5).

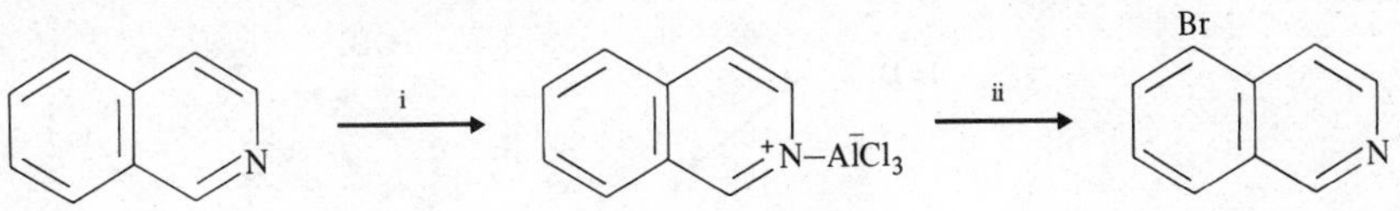

i, $AlCl_3$; ii, Br_2

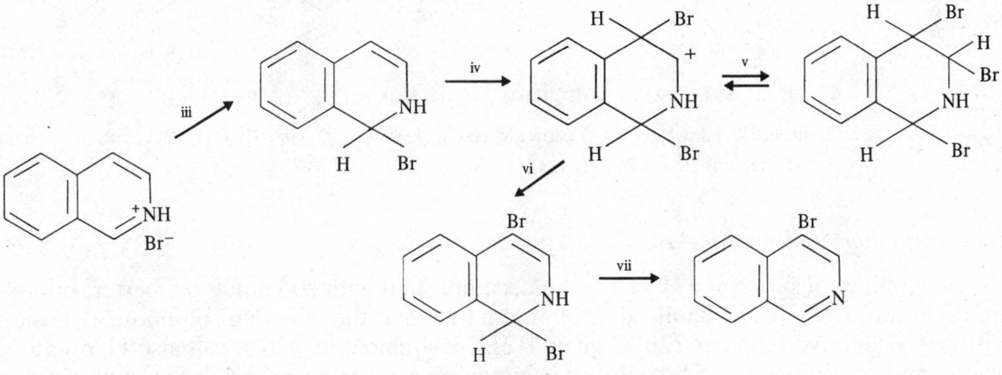

iii, heat; iv, Br^+; v, $\pm Br^-$; vi, $-H^+$; vii $-HBr$.

SCHEME 5

(*iv*) *Mercuration*

When heated with mercuric acetate the substituent enters position-4 by the addition–elimination mechanism, and may be converted to the mercuri-chloride derivative by treatment with sodium chloride solution (equation 15).

(15)

i, $Hg(OAc)_2$, 160 °C; ii, via further addition–elimination; iii, NaCl solution.

(*v*) *Proton exchange*

In deuteriosulphuric acid at 245 °C and lower acidities (40% acid), exchange occurs at positions-1 and -4 by mechanisms believed to be similar to those for positions-2 and -3, respectively, for quinoline (see Section 16.4.4.2 and Ref. 43a, Section 16.4) (equation 16). At 180 °C and higher acidities (90% acid), exchange is mainly at position-5 by direct attack on the cation. No exchange was observed at position-3.

(16)

245 °C, 40% D_2SO_4. i, $+OH^-$ $(OD)^-$, $-H_2O$ (HOD); ii, $+H^+$ (D^+);
iii, $+OD^-$ (slow); iv, $+D_2O$ (slow); v, (fast); vi, $+D^+$ (fast); vii, –HOD (fast);
(see Section 16.4.4.2, Ref. 43a).

(*vi*) *Amination and hydroxylation*

Nucleophilic substitution at C-1 can be accomplished with sodamide or, better, potassamide in an inert solvent or in liquid ammonia to give the 1-aminoisoquinoline. Fusion with potassium hydroxide at 220 °C gives 1(2*H*)-isoquinolone, which can also be obtained under mild conditions in 75% yield by using an enzyme from rabbit liver.[21] Substitution has not been observed at C-3 even when position-1 is blocked, *e.g.* with phenyl (equations 17 and 18).

(17)

i, KNH_2, liq. NH_3, $-H_2$; ii, H^+.

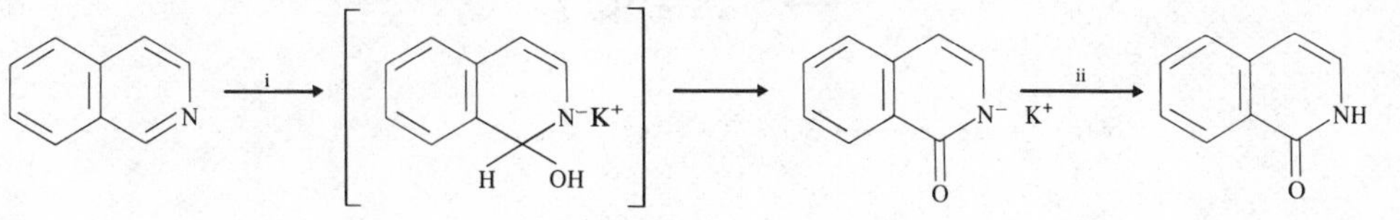

(18)

i, KOH, 220 °C, $-H_2$; ii, H^+.

(*vii*) *Alkylation and arylation*

In view of the failure of the electrophilic Friedel–Crafts reactions with all but the strongly activated quinolines and isoquinolines, these substitutions must be carried out by nucleophilic or homolytic processes. Nucleophilic substitution is achieved as with quinoline using organolithium compounds, Grignard reagents or, for methylation, methyl sulphinylmethylpotassium. Substitution occurs only at C-1 and stable 1-alkyl-1,2-dihydro-isoquinolines can be isolated by hydrolysis of the lithium complex (equations 19, 20, and 21).

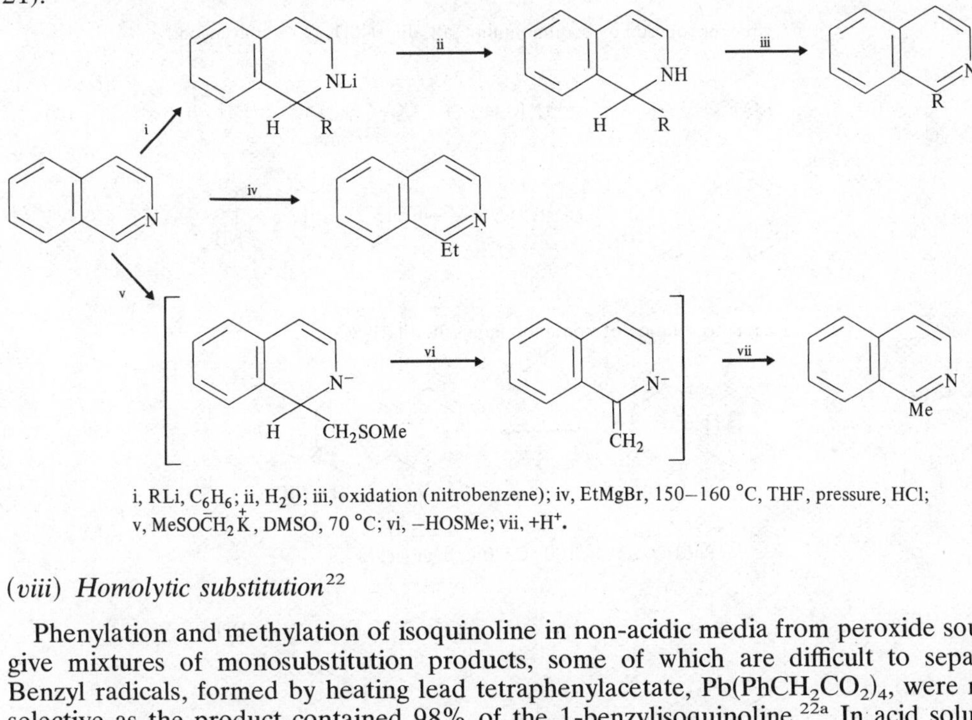

(19)

(20)

(21)

i, RLi, C_6H_6; ii, H_2O; iii, oxidation (nitrobenzene); iv, EtMgBr, 150–160 °C, THF, pressure, HCl; v, $MeSO\bar{C}H_2\overset{+}{K}$, DMSO, 70 °C; vi, –HOSMe; vii, $+H^+$.

(*viii*) *Homolytic substitution*[22]

Phenylation and methylation of isoquinoline in non-acidic media from peroxide sources give mixtures of monosubstitution products, some of which are difficult to separate. Benzyl radicals, formed by heating lead tetraphenylacetate, $Pb(PhCH_2CO_2)_4$, were more selective as the product contained 98% of the 1-benzylisoquinoline.[22a] In acid solution, substitution by all radicals is limited to the 1-position. High yields can be obtained by the method of Minisci and co-workers[22] using radicals, including Me·, Et·, CH_3CO·, and $CONH_2$·, generated in aqueous solutions of the protonated heterocyclic base by redox systems or peroxides (see equations 22 and 23).

16.5.4.3 Reactions of the alkyl side-chain: alkylisoquinolines

The discussion of the particular reactivity of alkyl groups at positions-2 and -4 of quinoline (Section 16.4.4.3) applies also to 1-alkylisoquinolines and, to a much lesser degree, to 3-alkylisoquinolines. The alkyl groups at other positions have the reactivity normally associated with attachment to an aromatic nucleus. Both 1- and 3-methyl-isoquinolines condense with benzaldehyde;[23,24] the difference in reactivity is apparent from the different conditions required and the yields of styrylisoquinolines formed (equations 24 and 25).

$$S_2O_8^{2-} + Ag^+ \longrightarrow SO_4^{\cdot-} + SO_4^{2-} + Ag^{2+}$$

$$SO_4^{\cdot-} + Ag^+ \longrightarrow SO_4^{2-} + Ag^{2+}$$

$$EtCO_2H + Ag^{2+} \longrightarrow EtCO_2\cdot + Ag^+ + H^+$$

$$EtCO_2\cdot \longrightarrow Et\cdot + CO_2$$

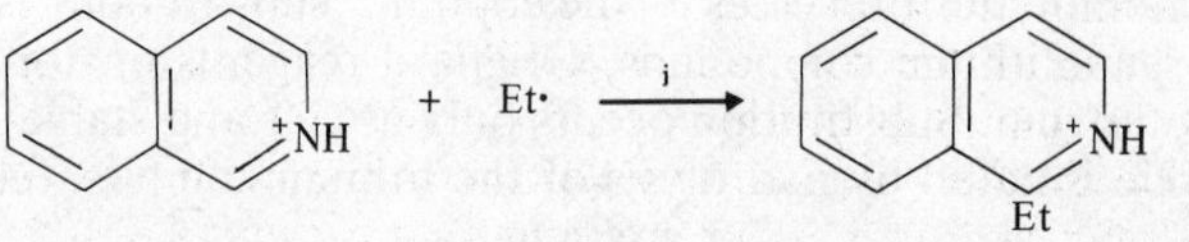

(22)

i, aqueous solution of isoquinolinium salt, dil. H_2SO_4, 33% conversion.

$$RCOCO_2H + Ag^{2+} \longrightarrow RCO\cdot + CO_2 + Ag^+ + H^+$$

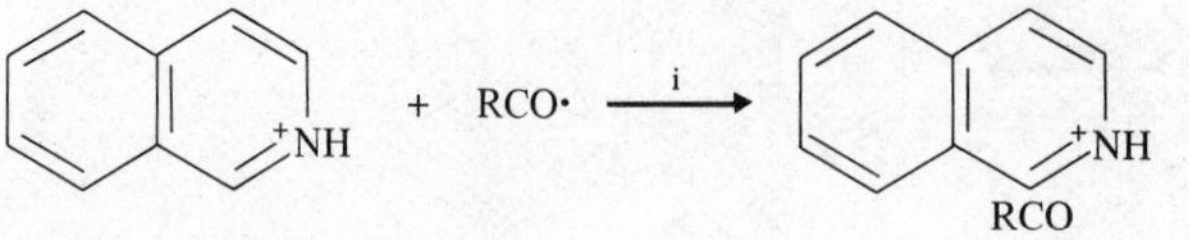

(23)

i, aqueous solution of isoquinolinium salt, dil. H_2SO_4.[22]

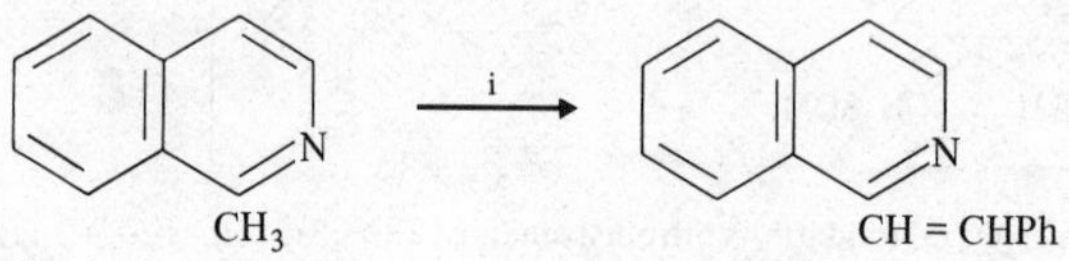

(24)

i, PhCHO, $ZnCl_2$, 100 °C, 20 h, high yield.

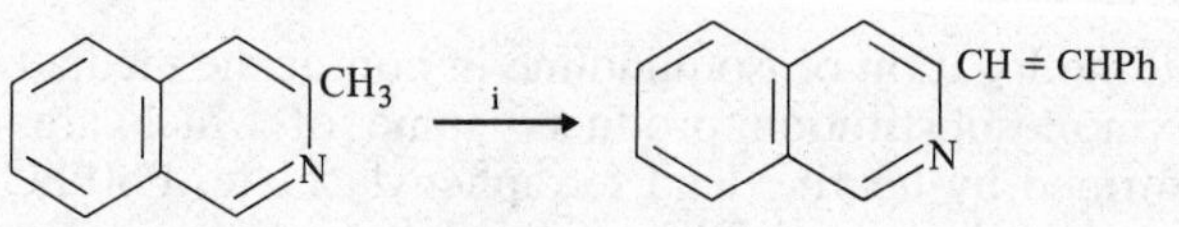

(25)

i, PhCHO, $ZnCl_2$, 160–165 °C, prolonged heating, yield 3.8%.

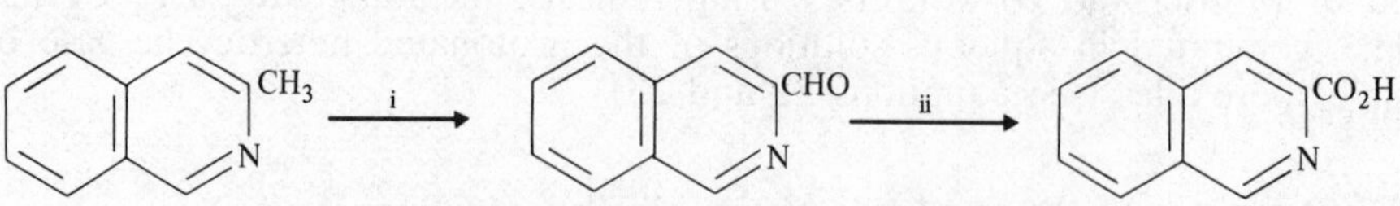

(26)

i, SeO_2; ii, H_2O_2

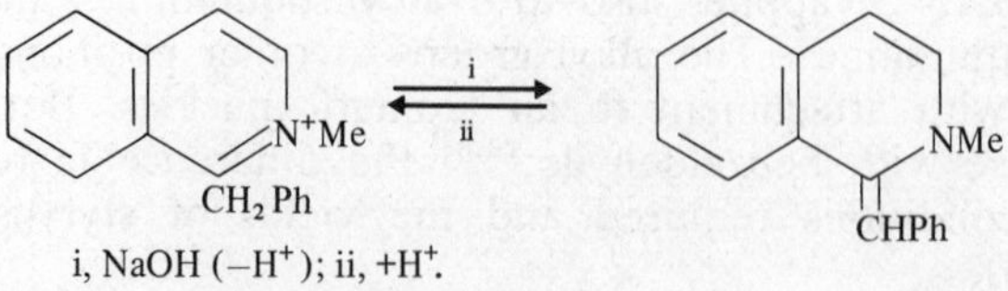

(27)

i, NaOH ($-H^+$); ii, $+H^+$.

Methyl groups at positions-1 and -3, and possibly other positions, can be oxidized by selenium dioxide[25] to formyl and subsequently to hydroxycarbonyl (carboxylic acid), *e.g.* by hydrogen peroxide *via* the peracid (equation 26). With alkyl or benzyl groups at position-1, isoquinolinium salts form stable, ether-soluble, methylene bases with alkali which revert to the quaternary form with acids (equation 27).

16.5.4.4 Oxidation

Degradative oxidation of isoquinoline under appropriate conditions provides evidence of the presence of both six-membered rings and establishes the points of fusion. It provided important evidence for the original structural assignment. Neutral potassium permanganate gives phthalimide as the main product, but alkaline potassium permanganate gives a mixture of phthalic acid and pyridine-3,4-dicarboxylic acid (cinchomeronic acid)[26] (equation 28). The susceptibility of the two rings to degradation is finely balanced. If the carbocyclic ring has an electron-attracting substituent, *e.g.* $-NO_2$, it is protected and oxidation opens the heterocyclic ring and *vice-versa*, so that better yields of pyridine-3,4-dicarboxylic acid can be made from isoquinoline by nitration, and reduction of the mixture of 5- and 8-nitro-compounds to amino-compounds before oxidation.

O NH O ← i — N — ii → CO_2H CO_2H + HO_2C HO_2C N

iii ↓ (28)

NO_2 CO_2H CO_2H ← iv — NO_2 N — v → NH_2 N — vi → HO_2C HO_2C N

i, $KMnO_4$ (neutral); ii, $KMnO_4$ (alkaline); iii, HNO_3/H_2SO_4; iv, $KMnO_4$ (alkaline); v, Sn, HCl; vi, $KMnO_4$ (alkaline), high yield.

As with quinoline, the products of ozonolysis indicate that the 3,4-, 5,6-, and 7,8-bonds are attacked preferentially, thus providing chemical evidence of their greater double-bond character and a degree of bond fixation. With peracetic acid and other organic peracids, isoquinoline 2-oxide is formed (see Section 16.5.5.4). Degradative oxidation can often be used to distinguish between quinoline and isoquinoline compounds, as the formation of anthranilic acid or its derivatives can only result from a quinoline compound whereas phthalimide or phthalic acid indicates an isoquinoline compound.

16.5.4.5 Reduction of the isoquinoline ring: decahydro-, tetrahydro-, and dihydro-isoquinolines

Isoquinoline is more resistant to reduction than quinoline, which has the advantage that substituents such as nitro, vinyl, and halogen can be selectively reduced more easily without affecting the ring structure.

(i) Decahydroisoquinolines

Vigorous hydrogenation at 450 °C causes the complete disruption of the isoquinoline ring. However, in acetic acid solution containing mineral acid, isoquinoline can be reduced[27] quantitatively by hydrogen and a platinum catalyst to a mixture of *cis*- (**27**) and *trans*-decahydroisoquinolines (**28**) in the ratio 4 : 1. From this mixture the *cis*-form was isolated by fractional crystallization of the picrate salts, and the *trans*-form was isolated by selective dehydrogenation of the *cis*-form over palladium when it remained together with the less strongly basic 5,6,7,8-tetrahydroisoquinoline (**29**) and isoquinoline.

(*ii*) *Tetrahydroisoquinolines*

The 5,6,7,8-isomer cannot be prepared by direct hydrogenation as the heterocyclic ring is reduced preferentially, but it has been prepared as described above by partial dehydrogenation of decahydroisoquinolines.[27] The 1,2,3,4-tetrahydroisoquinolines (**30**) are of

H NH H (27) Ratio 4 : 1 H NH H (28)

iii

i ii

N + isoquinoline + (28)

(29)

ii or ix

iv

N NH (30)

ii iv v

vi

N (32)

+ NMe I⁻ vii NMe (31) viii NMe

i, H_2/Pt; ii, Pd/C; iii, Pd; iv, Na/EtOH or Sn/HCl or H_2/Pt or H_2/Ni;
v, partial oxidation (special cases); vi, MeI; vii, $LiAlH_4$ or Na, liq. NH_3; viii, Sn/HCl; ix, I_2.

SCHEME 6

interest as they are formed in the Pictet–Spengler synthesis. They can be prepared by reduction of isoquinolines with sodium and alcohol, tin and hydrochloric acid, and by controlled hydrogenation with platinum, Raney nickel, or copper chromite catalysts. They can be converted to isoquinolines by dehydrogenation by heating with palladium charcoal[28] or by chemical oxidation with iodine. Unlike the corresponding tetrahydroquinoline, which is an aromatic secondary amine, 1,2,3,4-tetrahydroisoquinoline is a more strongly basic aliphatic secondary amine.

(*iii*) *Dihydroisoquinolines*

1-Substituted 1,2-dihydroisoquinolines (**31**) are formed as the products of addition of lithium alkyls to isoquinoline (equation 19), and in the Reissert reaction (Scheme 4). They are unstable (reactive) compounds which polymerize and are readily autoxidized in air. They can be prepared by the reduction of quaternary isoquinolines with lithium aluminium hydride[29] or with sodium and liquid ammonia. With tin and hydrochloric acid they are reduced to 1,2,3,4-tetrahydroisoquinolines. The 3,4-dihydroisoquinolines (**32**) are of interest as they are formed by the Bischler–Napieralski reaction. They are stable to acids and many can be distilled, although the 1-phenyl isomer disproportionates[30] on distillation at 340–345 °C, and the 1-benzyl isomer with potassium hydroxide at 200 °C gives toluene and isoquinoline. In general the 3,4-dihydroisoquinolines may be converted to isoquinolines by chemical oxidation, or better by dehydrogenation by palladium on charcoal.[23] Reduction with zinc or tin and dilute acid gives 1,2,3,4-tetrahydroisoquinoline. In particular cases they have been prepared by partial oxidation of the tetrahydroisoquinolines (Scheme 6).

16.5.4.6 Ring-opening reactions

Ring opening by oxidation and reduction processes has been described above (Sections 16.5.4.4 and 16.5.4.5). Exhaustive methylation by the Hofmann method results in the opening of the 2,3-bond, and the Emdé method opens the 1,2-bond as shown in Scheme 7, starting from 2,2′-dimethyl-1,2,3,4-tetrahydroisoquinolinium hydroxide (Hofmann) or chloride (Emdé).

$\overset{+}{N}Me_2$ X^- —i→ NMe_2 —ii→ CH_2OH

$\overset{+}{N}Me_2$ X^- —iii→ NMe_2 —iv→

i, X = OH, heat; ii, MeI, anion exchange ($I^- \rightarrow OH^-$), heat;
iii, X = Cl, Na/Hg; iv, MeI, anion exchange ($I^- \rightarrow Cl^-$), Na/Hg.

SCHEME 7

Quaternization of isoquinoline with electron-attracting substituents such as 2,4-dinitrophenyl, *e.g.* (**33**), leads to ring opening of the orange carbinol base (**34**) to give the violet-red open-chain aldehyde tautomer (**35**), confirmed by i.r. analysis which shows an OH absorption for (**34**) and an NH absorption for (**35**). When the quaternary salt is boiled with aniline, ring opening gives the anil (**36**). This, after an exchange of amine function on the vinyl side-chain, with loss of the 2,4-dinitrophenyl group, can then cyclize again to give 2-phenylisoquinolinium chloride (**37**), which cannot be prepared by direct quaternization (Scheme 8).

$\overset{+}{N}C_6H_3(NO_2)_2$ Cl^- (33) —i→ $NC_6H_3(NO_2)_2$, H, OH (34) —ii→ $NHC_6H_3(NO_2)_2$, CHO (35)

(33) —iii→ [$NC_6H_3(NO_2)_2$, H, NHPh → $NHC_6H_3(NO_2)_2$, CH = NPh (36) → NHPh, CH = NPh] —iv→ $\overset{+}{N}Ph$ Cl^- (37)

i, NaOH, ii, heat; iii, $PhNH_2$, boil ($+2PhNH_2$, $-NH_2C_6H_3(NO_2)_2$, $-HCl$); iv, $-PhNH_2 + HCl$.

SCHEME 8

16.5.5 ISOQUINOLINE DERIVATIVES

16.5.5.1 Nitro-compounds and sulphonic acids

The substituent groups of the nitroisoquinolines and isoquinolinesulphonic acids retain their normal reactivity. Nitro-groups can be selectively reduced to amino-groups with iron and dilute acetic acid, with tin and hydrochloric acid, or by catalytic hydrogenation. Although nitro-substituents reduce the basicity of the isoquinoline, the mononitro-compounds are soluble in acids and can be quaternized; dinitro-compounds, however, do not form stable salts and do not react with alkyl halides.

The 5-sulphonic acid can be converted to the hydroxy-compound by alkali fusion but is resistant to removal by acid hydrolysis, owing to the electron deficiency of the isoquinoline ring.

16.5.5.2 Halogen compounds

In addition to preparation by direct substitution, described above, many halogen derivatives of isoquinolines can be obtained from the corresponding amino-compounds *via* the diazonium salts, and 1-bromoisoquinoline is formed by the action of phosphorus tribromide on 1(2*H*)-isoquinolone. Bromine attached to the carbocyclic ring or at position-4 can be replaced by the cyano-group on heating with cuprous bromide. The

TABLE 3
Rate Constants[31] for the Displacement of Chlorine from Chloroisoquinolines and Chloroquinolines by Reaction with Sodium Ethoxide in Ethanol at 20 °C

Chloro-compound	*Rate constant* k (l mol^{-1} s^{-1})
1-Chloroisoquinoline	6.9×10^{-7}
4-Chloroquinoline	6.5×10^{-7}
2-Chloroquinoline	6.3×10^{-7}
3-Chloroisoquinoline	1.2×10^{-11}

most significant feature in the reactivity of the halogen compounds is the contrast between the reactivities of halogen substituents at positions-1 and -3 of isoquinoline with nucleophiles[31] such as alkoxides or amines. As shown by the rate constants in Table 3 for the chloro-compounds,[31] 1-chloroisoquinoline resembles 2- and 4-chloroquinolines and is more reactive than the 3-isomer by a factor of approximately 50 000. Chlorine at positions-1 and -3 can, however, be replaced successively[33] (equation 29), and the increased reactivity towards nucleophiles resulting from the formation of the isoquinolinium cation is shown by the displacement of chlorine by iodine when the 1-chloro-compound is quaternized with methyl iodide[34] (equation 30).

(29)

i, $NH_2C_6H_4Cl$-*p*, heat

(30)

i, MeI, 100 °C, sealed tube, 2 days

16.5.5.3 Hydroxyisoquinolines (isoquinolinols and isoquinolones)

The direct synthesis of some isoquinolones and the formation of 1(2*H*)-isoquinolone (isocarbostyril) by nucleophilic substitution of isoquinoline has been discussed above and the preparation of isocarbostyril from isoquinoline 2-oxide is described in Section 16.5.5.4 below. The hydroxyisoquinolines exist in solution as equilibrium mixtures of the isoquinolinol and isoquinolone tautomers.[35] The 4-, 5-, 6-, 7-, and 8-isomers have typical phenolic properties consistent with the predominance of the hydroxy-isomers over the *N*-protonated amido or zwitterionic forms, as shown for 4-hydroxyisoquinoline (**38**) and (**39**). The latter (and also the 5- and 7-isomers) cannot be formulated as isoquinolones such as (**41**), and all the phenolic isomers are usually represented as isoquinolinols. 1-Hydroxyisoquinoline, on the other hand, exists almost completely in the amido (isoquinolone) form as the hybrid [(**40**)$\leftrightarrow$(**41**)] pK_T=4.85.[35c] In these respects the isoquinoline compounds are analogous to the corresponding quinoline derivatives. The equilibrium between the two tautomeric forms of the 3-isomer however, is finely balanced[36a] ($pK_T \approx 0$), although the substituent is α to the nitrogen. This is related to 'bond fixation' and comparisons of its u.v. spectrum with those of compounds of fixed structure indicate that the colourless isoquinolinol form (**42**) predominates in non-hydroxylic solvents and the yellow isoquinolone form [(**43**) ↔ (**44**)] in water.[37] Here the advantages of the amido-structure are offset by loss of benzenoid character in the carbocyclic ring. In ethanol/chloroform solvent mixtures the concentrations of the two tautomers are comparable. The hydroxy-groups of isoquinolinols (or of reagents used in the syntheses of isoquinolines) are frequently protected by conversion to the benzyl ether, which will withstand conditions such as those required for the Bischler–Napieralski reaction but which can be removed subsequently by reduction.

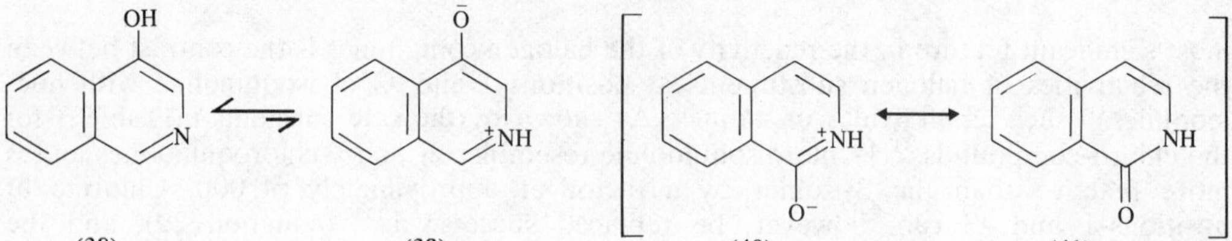

(38) (39) (40) (41)

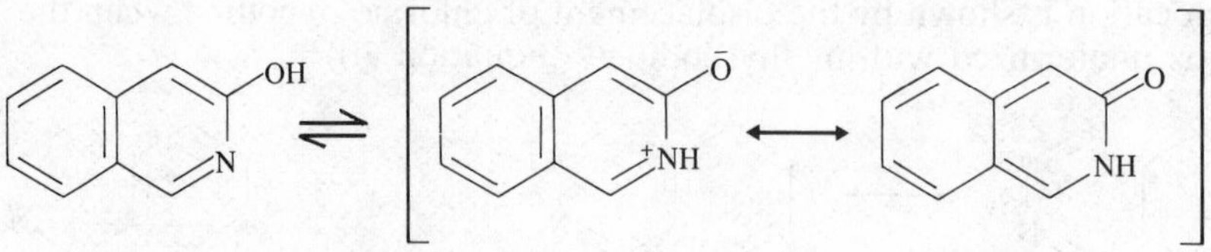

(42) (43) (44)

16.5.5.4 Isoquinoline 2-oxide

Isoquinoline reacts with organic peracids (*e.g.* acetic acid and hydrogen peroxide) to give the *N*-oxide, isoquinoline 2-oxide (**45**), a reaction typical of tertiary amines. The particular importance of these derivatives in the chemistry of pyridine and quinoline stems largely from the activation of the 4-position of each of these compounds to both electrophilic and nucleophilic reagents, to give products not readily obtained from the original heterocycles. This is not possible with isoquinoline *N*-oxide as the corresponding position (γ to N) is involved in ring fusion and so the compound is less valuable synthetically, although a number of rearrangements to positions-1 and -4 (but not -3) are known. When heated with acetic anhydride followed by hydrolysis of the primary product, isoquinoline 2-oxide gives 1-isoquinolone (60%)[38] (equation 31). With toluene-*p*-sulphonyl chloride a 4-toluene-*p*-sulphonate ester is formed which on hydrolysis gives 4-hydroxyisoquinoline in which the oxygen possibly originates from the *N*-oxide as shown by the use of ^{18}O labelled toluene-*p*-sulphonyl chloride[39] (equation 32).

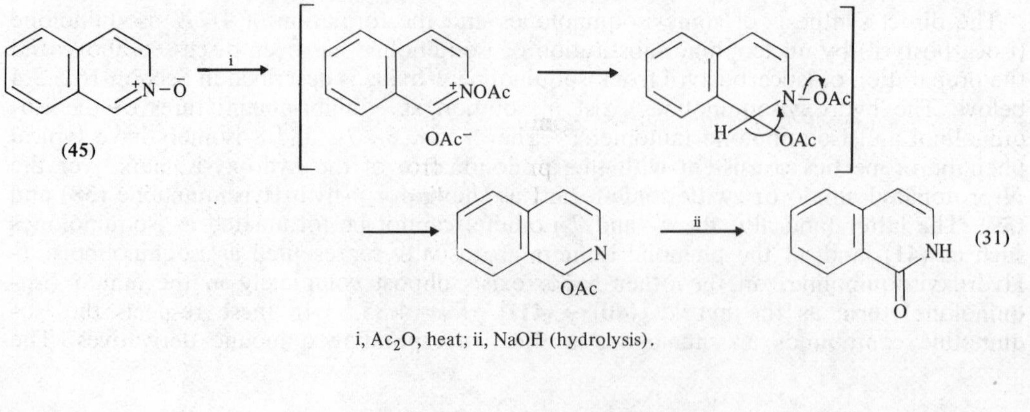

(45) (31)

i, Ac_2O, heat; ii, NaOH (hydrolysis).

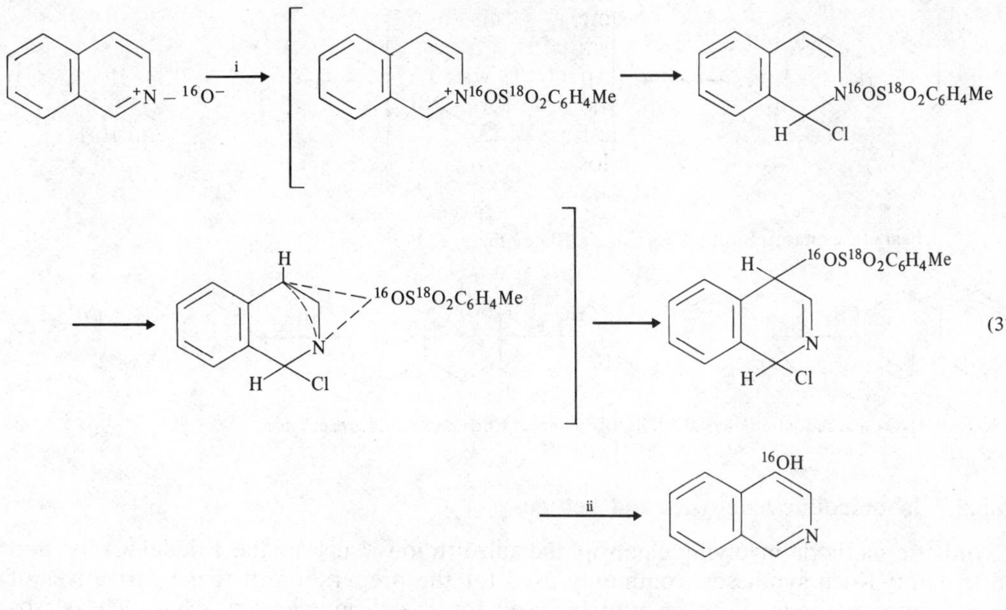

i, p-$MeC_6H_4S^{18}O_2Cl$; ii, NaOH, hydrolysis.

(32)

16.5.5.5 Aminoisoquinolines

For the reasons discussed in connection with the aminoquinolines, all the amino-isoquinolines exist predominantly[35d,36b] in the amino-form (**46**) rather than the imine form (**47**); thus for 1-aminoisoquinoline (**46**), $pK_T = 3.8$ (equation 33).[36c] 1-Aminoisoquinoline is also the strongest base, $pK_a = 7.62$,[36c] and 3-aminoisoquinoline is the weakest base ($pK_a = 5.0$). All seven isomers are known; 1-aminoisoquinoline can be prepared by direct amination (Chichibabin reaction), 3-aminoisoquinoline can be prepared from 3-methyl-isoquinoline (equation 35); the 4-amino-isomer can be prepared by displacement of bromine from 4-bromoisoquinoline with alcoholic ammonia and by reduction of the 4-nitro-compound. The 5-, 6-, 7-, and 8-amino-compounds can also be obtained by reduction of the nitro-derivatives or from the isoquinolinols by the Bucherer reaction.

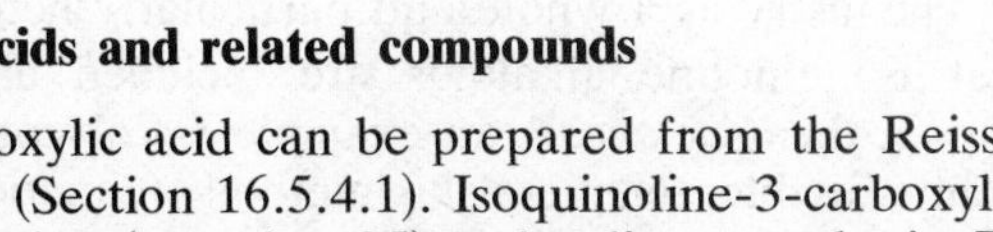

(46) (47) (33)

i, $pK_T = 3.8$.

16.5.5.6 Carboxylic acids and related compounds

Isoquinoline-1-carboxylic acid can be prepared from the Reissert derivative by treatment with strong acid (Section 16.5.4.1). Isoquinoline-3-carboxylic acid can be prepared from 3-methylisoquinoline (equation 35) or by direct synthesis. The remaining acids are usually prepared by hydrolysis of the nitriles. When the hydroxycarbonyl group is attached to the heterocyclic ring, the acids can be decarboxylated by heating the free acid or its silver salt. In boiling benzaldehyde the intermediate ylide can be trapped as the alcohol (**48**) (equation 34).

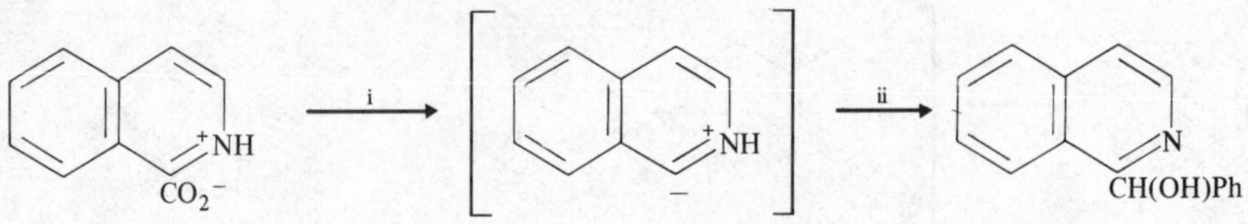

(34)

(48)

i, heat alone or as Ag^+ salt ($-CO_2$); ii, PhCHO, reflux.

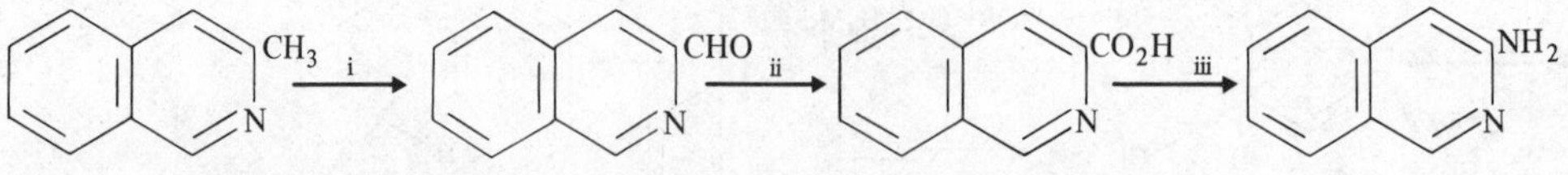

(35)

i, SeO_2; ii, H_2O_2 and reduction of peracid; iii, Hofmann or Curtius–Schmidt reactions.

16.5.5.7 Isoquinoline aldehydes and ketones

Synthetic methods involving electrophilic substitutions, such as the Friedel–Crafts and Gattermann–Koch syntheses, commonly used for the preparation of these derivatives of normal aromatic compounds, cannot be used for π-deficient heterocycles such as isoquinoline. As with quinoline, however, the 3-carbaldehyde has been prepared[40] by oxidation of 3-methylisoquinoline with selenium dioxide. The aldehyde was further oxidized with hydrogen peroxide to the peracid from which the 3-carboxylic acid was obtained (equation 35). Acylisoquinolines have been prepared by the ketonic hydrolysis of the keto-esters formed by the Claisen ester condensation[41] between esters of isoquinolinecarboxylic acids and ethyl acetate. The Friedel–Crafts reaction can also be used to prepare isoquinolyl aryl ketones provided the acid chloride of the isoquinoline function is employed. 1-Acylisoquinolines can be prepared by direct homolytic substitution (see Section 16.5.4.2; equation 23).

16.5.6 ISOQUINOLINE ALKALOIDS[42]

Isoquinoline compounds do not have many well known uses in the chemical or industrial fields, but the ring system is a structural component of a large number of naturally occurring plant alkaloids. The family of isoquinoline alkaloids is subdivided into many groups, and in addition the ring structure frequently forms a part of more complicated systems in alkaloids not normally included in the isoquinoline family, *e.g.* morphine and codeine (**62a**) and (**62b**). The phenylethylamine structural unit common to the isoquinoline alkaloids is present in the aromatic amino-acids phenylalanine and tyrosine, which are primary precursors in their biosynthesis.[43] This subject has been investigated by many research groups including those of Winterstein and Trier, Robinson, and Barton. The isolation and characterization of the alkaloids has constituted one of the great achievements of organic chemistry. The following brief account of some of the isoquinoline alkaloids is intended to show how these investigations have influenced the development of organic chemistry as a whole and particularly heterocyclic chemistry.

Some of the simplest isoquinoline alkaloids are included among the tetrahydroisoquinoline group found in the *Cactaceae*, an example is calycotomine (**49**) which has been synthesized by the Bischler–Napieralski method (equation 36). The benzylisoquinoline and pavine group includes the fully aromatic alkaloid papaverine (**5**), a powerful coronary vasodilator, the synthesis of which by the Pictet–Gams method has been given (equation 3, p. 209). In the aporphine group, which includes over fifty bases, further ring-closure gives as the common structural feature the tetracyclic unit present in apomorphine (**50**), which has been synthesized[44] by the method shown in equation (37).

MeO, MeO, NH_2 + O=C(Cl)CO_2Et →(i) MeO, MeO, NH, O, CO_2Et →(ii) MeO, MeO, NH, CH_2OH

(36)

(49) Calycotomine

i, –HCl; ii (a) P_2O_5 (Bischler–Napieralski), (b) $LiAlH_4$.

$\overset{+}{N}$Me + NO_2, CH_3 →(i) NMe, NO_2 →(ii) NMe, NH_2

(37)

→(iii) NMe

(**50**) Apomorphine

i, NaOEt; ii, H_2/Pt; iii, Pschorr synthesis.

The dibenzopyrrocoline alkaloids contain the reduced pyrrocoline (indolizine) ring system as well as the isoquinoline structure. Robinson and Schöpf and co-workers,[45] independently, synthesized the dibenzopyrrocoline derivative (**52**) by the oxidation of the benzylisoquinoline alkaloid laudanosoline (**51**). The intention was to synthesize the compound (**53**) with the apomorphine ring system, but ring closure occurred from carbon to nitrogen instead of to position-8 of the carbocyclic ring (equation 38). It was not until 20 years later that two alkaloids with this structure were characterized after isolation from the Australian shrub *Cryptocaria bowiei* (Lauracaea), *e.g.* cryptaustoline (**54**).

The protoberberine alkaloids are a group of over 35 compounds with the basic tetracyclic structure present in tetrahydropalmatine (**55**), which has been synthesized[46] (equation 39). As well as the isoquinoline structure this group contains the reduced quinolizine ring system. Some of the members of this group have the nitrogen in the quaternary state to give a dihydrodibenzoquinolizinium ring system; an example is berberine itself (**56**).

An interesting example of the protopine group is cryptopine (**57**). It has a 10-membered ring which shows spectroscopic evidence of transannular interaction between the nitrogen and the carbonyl group (the C=O absorption is at 1675 cm^{-1}, the amide frequency, instead of at about 1693 cm^{-1} for an aryl ketone). On protonation (equation 40) the carbonyl absorption disappears altogether, indicating the formation of the tetracyclic structure (**58**). The benzophenanthridine group of alkaloids includes chelidonine (**59**) and other members which are completely aromatic in structure. Corydaldine (**60**) is an example of the small group of isoquinolone alkaloids which are believed to be formed in the plant by oxidative degradation of more complex benzylisoquinolines. Emetine (**61**),

(53)

(51) Laudanosoline

[O]

$^{-}OH^{-}$

(52)

(38)

X^-

(54) Cryptaustoline

i

ii

Br^-

(39)

iii

(55) Tetrahydropalmatine

(56) Berberine

i, SeO_2; ii (a) NH_2OH, (b) 2,3-dimethoxybenzyl bromide, (c) H^+; (iii) H_2/Pt.

(40)

(57) Cryptopine

(58)

(59) Chelidonine

(60) Corydaldine

(61) Emetine

(62a) R = H, Morphine
(62b) R = Me, Codeine

with a hexahydrobenzo[*a*]quinolizine ring system, is a member of the epicacuanha group of alkaloids, and morphine (**62a**) and codeine (**62b**), which are phenanthrene derivatives, are both in the morphine group. All three are examples of alkaloids in other families which also contain the isoquinoline ring system.

References

1. (a) R. H. F. Manske, *Chem. Rev.*, 1942, **30,** 145; (b) F. W. Bergstrom, *ibid.*, 1944, **35,** 77; (c) W. J. Gensler, in 'Heterocyclic Compounds', ed. R. C. Elderfield, Wiley, New York, 1952, chapter 2, p. 344; (d) S. F. Dyke, in 'Rodd's Chemistry of Carbon Compounds', ed. S. Coffey, Elsevier, Amsterdam, 2nd edn., 1976, Vol. 4F, p. 357.
2. W. M. Whaley and T. R. Govindachari, in 'Organic Reactions', ed. R. Adams, Wiley, New York, 1951, vol. 6, (a) p. 74, (b) p. 151.
3. W. M. Whaley and W. H. Hartung, *J. Org. Chem.*, 1949, **14,** 650.
4. N. Itoh and S. Sugasawa, *Tetrahedron*, 1957, **1,** 45.
5. H. R. Snyder and F. X. Werber, *J. Amer. Chem. Soc.*, 1950, **72,** 2962.
6. A. McCoubrey and D. W. Mathieson, *J. Chem. Soc.*, 1949, 696.
7. R. F. Nystrom and W. G. Brown, *J. Amer. Chem. Soc.*, 1948, **70,** 3738.
8. A. Pictet and A. Gams, *Ber.*, 1909, **42,** 2943.
9. A. Pictet and T. Spengler, *Ber.*, 1911, **44,** 2030.
10. J. S. Buck and W. S. Ide, *J. Amer. Chem. Soc.*, 1934, **56,** 1769; 1937, **59,** 726.
11. P. Gams, A. O. Fitton, J. R. Frost, M. M. Zakaria, and G. Andrew, *J. C. S. Chem. Comm.*, 1973, 889.
12. W. J. Gensler, in Ref. 2, p. 191.

13. R. H. F. Manske and M. Kulka, *Canad. J. Res.*, 1949, **27B,** 161.
14. E. Schlittler and J. Muller, *Helv. Chim. Acta*, 1948, **31,** 914.
15. A. T. Amos and G. G. Hall, *Mol. Phys.*, 1961, **4,** 25.
16. F. Genet, *Bull. Soc. Fr. Mineral Cryst.*, 1965, **88,** 463.
17. See chapter 16.4, Ref. 35.
18. R. D. Brown and R. D. Harcourt, *Tetrahedron*, 1960, **8,** 23.
19. M. J. S. Dewar and P. M. Maitlis, *J. Chem. Soc.*, 1957, 2521.
20. R. B. Moodie, K. Schofield and M. J. Williamson, *Chem. and Ind.* (*London*), 1963, 1283.
21. W. E. Knox and W. I. Grossmann, *J. Amer. Chem. Soc.*, 1948, **70,** 2172.
22. F. Minisci and O. Porta, *Adv. Heterocyclic Chem.*, 1974, **16,** 123.
22a. K. C. Bass and P. Nababsing, *J. Chem. Soc.* (*C*), 1969, 388.
23. W. H. Mills and J. L. B. Smith, *J. Chem. Soc.*, 1922, 2724.
24. H. Erlenmeyer, H. Baumann, and E. Sorkin, *Helv. Chim. Acta*, 1948, **31,** 1978.
25. C. E. Teague, Jr. and A. Roe, *J. Amer. Chem. Soc.*, 1951, **73,** 688.
26. S. Hoogewerff and W. A. Van Dorp, *Rec. Trav. chim.*, 1885, **4,** 285.
27. B. Witkop, *J. Amer. Chem. Soc.*, 1948, **70,** 2617.
28. E. Spath and A. Burger, *Ber.*, 1927, **60,** 704.
29. H. Schmid and P. Harrer, *Helv. Chim. Acta*, 1949, **32,** 960.
30. W. Huckel and G. Graner, *Chem. Ber.*, 1957, **90,** 2017.
31. G. Illuminati, in Ref. 22, 1964, vol. 3, p. 285.
32. N. B. Chapman and D. Q. Russell-Hill, *J. Chem. Soc.*, 1956, 1563.
33. R. D. Haworth and S. Robinson, *J. Chem. Soc.*, 1948, 777.
34. N. I. Fisher and F. M. Hamer, *J. Chem. Soc.*, 1934, 1905.
35. A. R. Katritzky and J. M. Lagowski, in Ref. 22, 1963, vol. 1, (a) p. 311, (b) p. 335, (c) p. 350, (d) p. 404.
36. J. Elquero, C. Marzin, A. R. Katritzky, and P. Linda, *Adv. Heterocyclic Chem. Suppl.* 1, 1976, (a) 99, (b) 152, (c) 153.
37. D. A. Evans, G. F. Smith, and M. A. Wahid, *J. Chem. Soc.* (*B*), 1967, 590.
38. M. M. Robison and B. L. Robison, *J. Org. Chem.*, 1956, **21,** 1337; *J. Amer. Chem. Soc.*, 1958, **80,** 3443.
39. S. Oae, T. Kitao, and Y. Kitaoka, *Tetrahedron*, 1963, **19,** 827.
40. C. E. Teague and A. Roe, *J. Amer. Chem. Soc.*, 1951, **73,** 688.
41. J. J. Padbury and H. G. Lindwall, *J. Amer. Chem. Soc.*, 1945, **67,** 1268.
42. M. Shamma, in 'Chemistry of the Alkaloids', ed. S. W. Pelletier, Van Nostrand Reinhold, New York, 1970, chapter 3, p. 31.
43. I. D. Spenser, in Ref. 42, chapter 21, p. 688.
44. J. A. Weisbach, C. Burns, E. Macko, and B. Douglas, *J. Medicin. Chem.*, 1963, **6,** 91.
45. (a) R. Robinson and S. Sugasawa, *J. Chem. Soc.*, 1932, 789; (b) C. Schopf and K. Thierfelder, *Annalen*, 1932, **497,** 22.
46. C. K. Bradsher and N. L. Dutta, *J. Amer. Chem. Soc.*, 1960, **82,** 1145.

16.6

Quinolizines and Quinolizinium Salts

P. A. CLARET

The City University, London

16.6.1 GENERAL INTRODUCTION

16.6.1.1 The quinolizine and quinolizinium ring systems

The quinolizinium cation[1,13] is of interest as one of the three aromatic azanaphthalene ring systems, isoelectronic with naphthalene, quinoline, and isoquinoline; see Section 16.4.1.1. Because the nitrogen atom is sited at the angular position of the ring, the fully aromatic structure can only exist as the mesomeric cation [(**1**) ↔ (**2**) ↔ (**3**)].

(1) (2) (3)

There are three possible structures for the non-aromatic parent quinolizines depending on the location of the additional hydrogen atom, namely 2*H*-quinolizine (**4**), 4*H*-quinolizine (**5**), and 9a*H*-quinolizine (**6**). None of these has been isolated in stable form, although there is some evidence that the 4*H*-isomer at least may have a transient existence[2] and a number of synthetic and naturally occurring derivatives are known. The aromatic quinolizinium cation [(**1**) ↔ (**2**) ↔ (**3**)] has been synthesized[3a] and also the fully hydrogenated cycloaliphatic tertiary amine quinolizidine (**7**) (see Section 16.6.2.3). In earlier works the quinolizines were known as 'pyridocolines' and the quinolizinium salts as 'dehydroquinolizinium salts'.

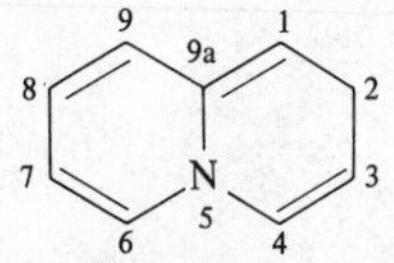

(4) 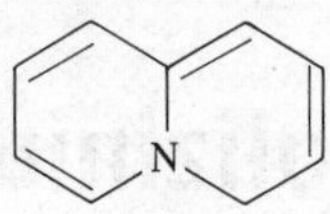 (5) 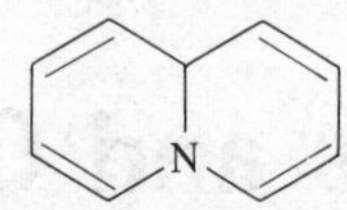 (6) 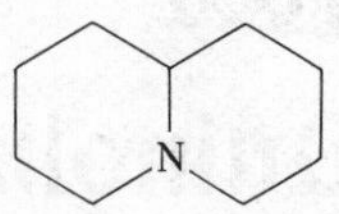 (7)

16.6.1.2 Occurrence of quinolizines and quinolizinium salts

The quinolizine ring system occurs in a number of alkaloids,[4a] frequently in the fully reduced quinolizidine form as in the Lupins, a family of the *Leguminosae* including gorse, broom, and laburnum. Examples are the simple quinolizidine derivative lupinine (**8**) [and (**13**)], present in yellow lupin seeds, cytisine (**9**), a tricyclic poisonous alkaloid of the ornamental shrub *Cytisus laburnum*, and anagyrine (**10**), one of the tetracyclic quinolizidine alkaloids from lupins which is closely related to other members of this group. Supervirine (**11**), present in *Gelsemium sempervirens*, contains the aromatic quinolizinium ring structure.

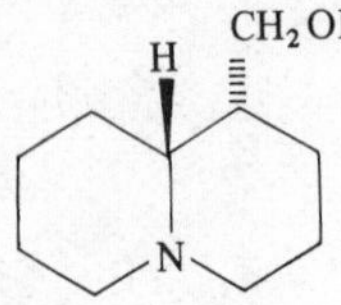

(**8**) Lupinine

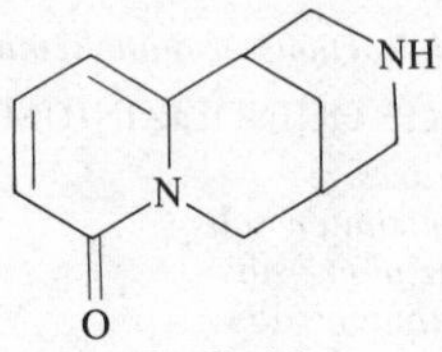

(**9**) Cytisine

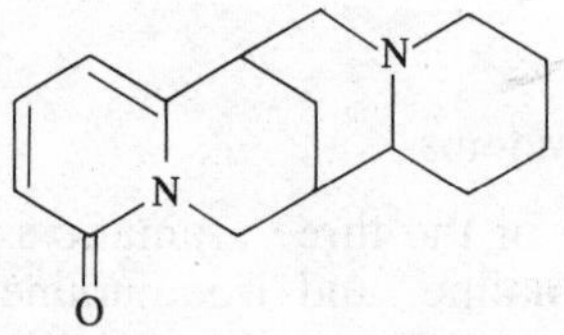

(**10**) Anagyrine

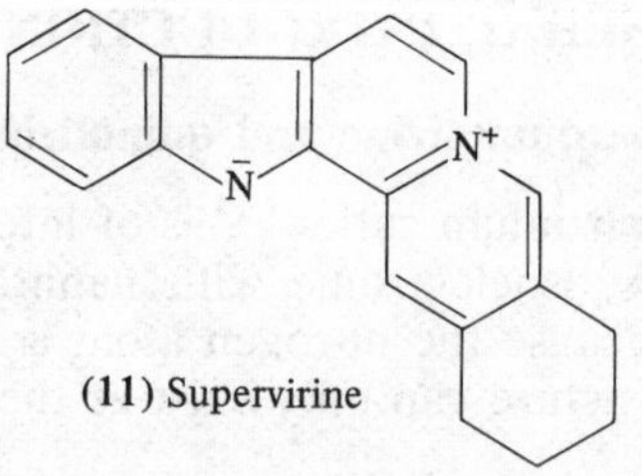

(**11**) Supervirine

As a simple example of the elegant and painstaking chemistry involved in the study of these compounds, the structure of lupinine was determined by chemical degradation, and by total synthesis [from the substituted malonic ester (**12**)] of (±)-lupinine and its diastereomer (±)-epilupinine (Scheme 1).

(±)-Lupinine (**13**) epimerizes to epilupinine (**14**) when its solution in benzene is refluxed with sodium. This suggests that the primary alcohol group in lupinine is in the structurally less favourable axial position, which is confirmed by the evidence of hydrogen bonding in its i.r. spectrum [see (**13**)] and the ready cyclization of its toluene-*p*-sulphonyl ester to (**15**) (Scheme 2).

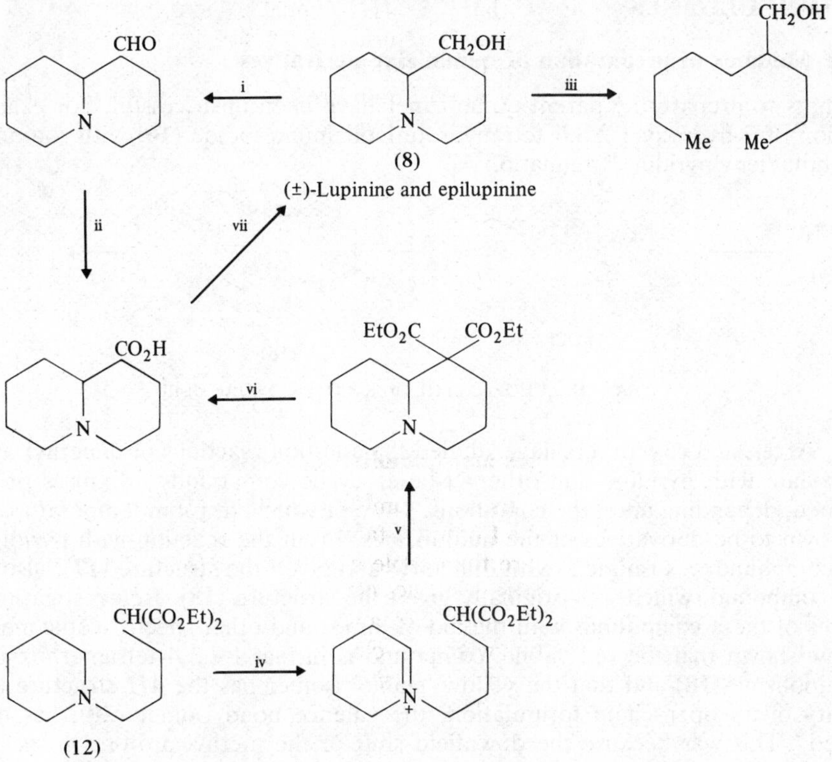

CHO CH$_2$OH CH$_2$OH
i iii
N N Me Me
(8)
(±)-Lupinine and epilupinine
ii vii
EtO$_2$C CO$_2$Et
CO$_2$H
vi
N N
v
CH(CO$_2$Et)$_2$ CH(CO$_2$Et)$_2$
iv
N N$^+$
(12)

i, ii, oxidation; iii, three-stage Hofmann exhaustive methylation and hydrogenation; iv, Hg(OAc)$_2$ ($-H^-$); v, OH^-; vi, hydrolysis, decarboxylation; vii, $LiAlH_4$.

SCHEME 1

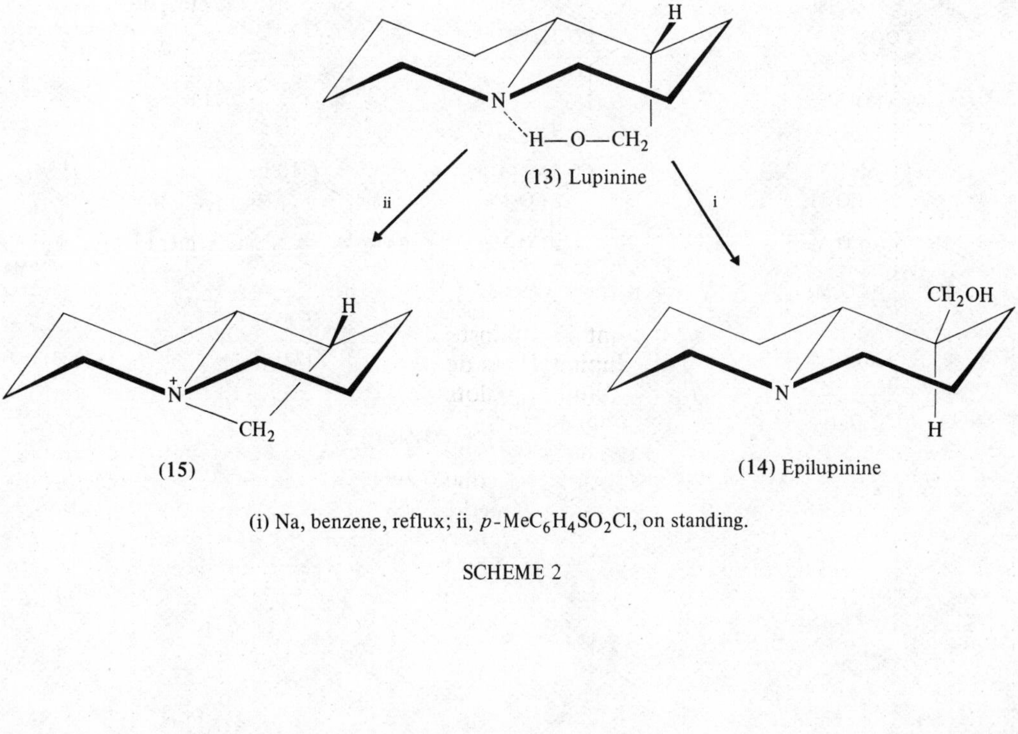

H
N
H—O—CH$_2$
(13) Lupinine
ii i
H CH$_2$OH
N$^+$ N
CH$_2$ H
(15) (14) Epilupinine

(i) Na, benzene, reflux; ii, p-$MeC_6H_4SO_2Cl$, on standing.

SCHEME 2

16.6.2 QUINOLIZINES

16.6.2.1 Methods of preparation of quinolizine derivatives

Attempts to prepare the parent quinolizines have been unsuccessful. For example, the distillation of 2-hydroxy-1,2,3,4-tetrahydroquinolizinium iodide (**16**) with sodium acetate gave 2-butadienylpyridine[3a] (equation 1).

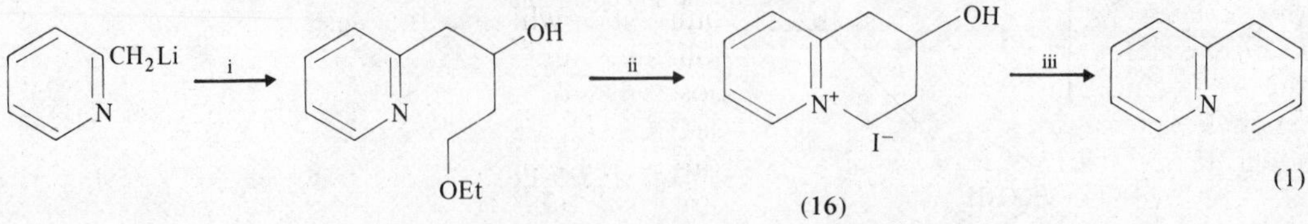

(16) (1)

i, $EtO(CH_2)_2CHO$; ii, (a) HI, (b) K_2CO_3; iii, NaOAc, distil

Diels, Alder, and co-workers have studied the addition reactions of dimethyl acetylenedicarboxylate with pyridine and other N-heterocyclic compounds. Mixtures of products are formed, depending upon the conditions, some of which, at room temperature in ether, were shown to be derivatives of the quinolizines. From the reaction with pyridine, a red 'labile' compound was isolated (with difficulty) and given the structure (**17**); also a yellow 'stable' compound, which was originally given the structure (**18**). Reinvestigations of the structures of these compounds with the aid of n.m.r. and other spectroscopic methods[6,7,8] have now shown that the red 'labile' compound is in fact 1,2,3,4-tetramethoxycarbonyl-9a*H*-quinolizine (**18**) and that the yellow 'stable' isomer has the 4*H*-structure (**19**). The possibility of an open-chain formulation, the valence bond isomer (**20**), has also been suggested.[8] This was because the downfield shift of the methyl protons in the 7-methyl analogue of the 'stable' isomer from 3-methylpyridine (**21**) indicated that they were attached to an aromatic pyridine ring, as in the model compound (**22**), rather than to a non-aromatic structure such as (**23**). However, resonance with the ester groups [see Section 16.6.2.2, (**34**)] could account for an aromatic ring current.[5]

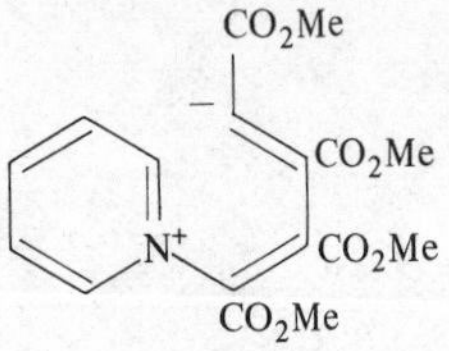

(17) 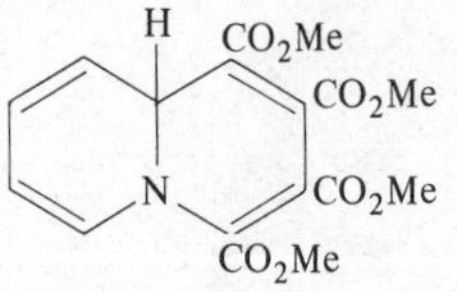 (18) 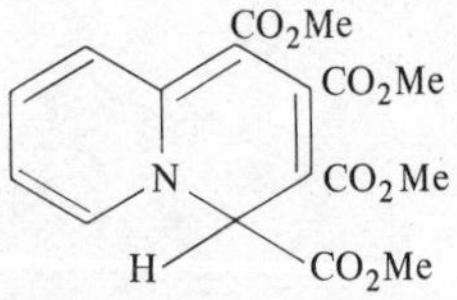 (19)

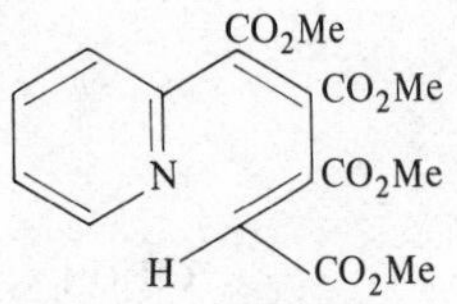

(20) 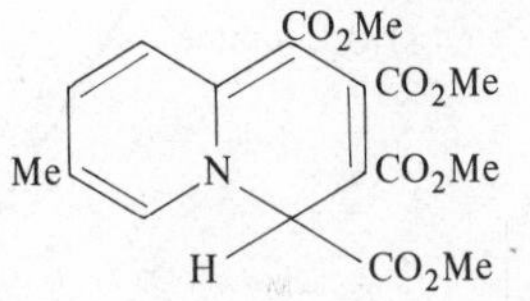 (21) 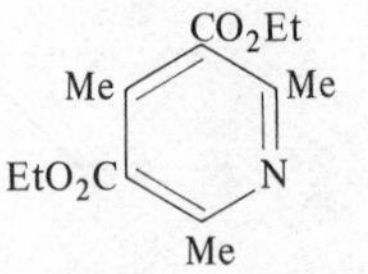 (22) 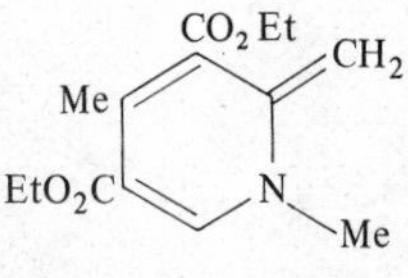 (23)

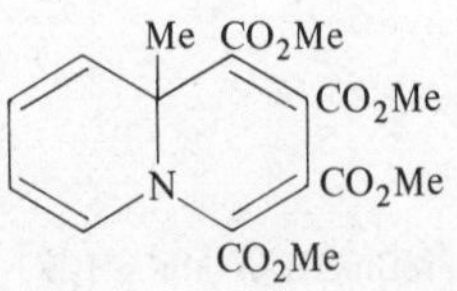

(24)

If 2-methylpyridine is used in place of pyridine, the adduct isolated is the 9a-methyl compound (**24**), which is unable to tautomerize, suggesting that the 9a*H*-isomer is formed first and isomerizes to the more favourable 4*H*-isomer.

Diels and Alder also isolated similar products from the reaction between dimethyl acetylenedicarboxylate and both quinoline and isoquinoline. They suggested structures analogous to (**17**) and (**18**) for the 'labile' and 'stable' isomers, respectively, formed from quinoline. Reinvestigations[9,10] of the structures of these compounds have shown that, as in the reaction with pyridine, the 'labile' compound (which in this case is predictably more readily isolated) is the 9a*H*-benzo[*c*]quinolizine ester (**25**), and the more stable isomer has the 4*H*-quinolizine structure (**26**). The possibility of an open-chain formulation (**27**) has again been considered,[9] but in this case there is further evidence[10] in favour of the quinolizine structure (**26**) as in acid solution there is a hypsochromic shift of the u.v. absorptions, also shown by the pyridine analogue, (**19**) or (**20**). The open-chain structure (**27**) would be expected to show a bathochromic shift to longer wavelengths on protonation of the nitrogen as it is a quinoline derivative (see Section 16.4.5.6). The hypsochromic shift is compatible with protonation at C-3 [see (**37**)].

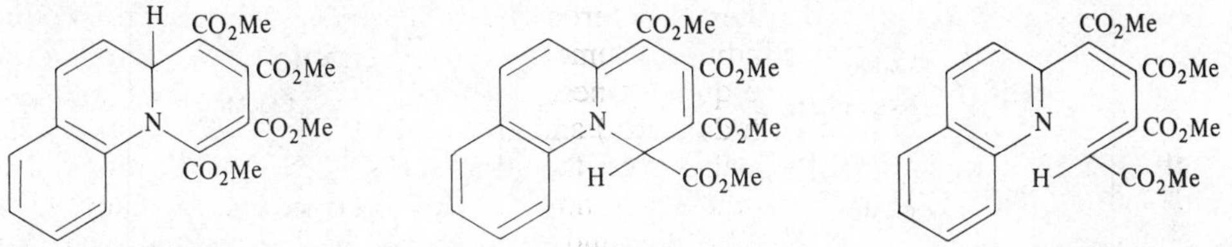

(25) (26) (27)

The reaction of dimethyl acetylenedicarboxylate with isoquinoline[11] in ether gives the tetramethyl ester of 11b*H*-benzo[*a*]quinolizine-1,2,3,4-tetracarboxylic acid (**28**). Although in this case more stable, it is the equivalent of the 'labile' 9a*H*-quinolizines formed from pyridine and quinoline. It can be converted into the 4*H*-isomer (**29**) by boiling in xylene or with a sulphuric acid–acetic acid mixture. On treatment with bromine the 4*H*-isomer (**29**) is oxidized to the quinolizinium perbromide (**30**), but the 9a*H*-isomer (**28**) can be oxidized firstly to the quinolizone (**31**) and finally to the pyridine derivative (**32**) (see Scheme 3).

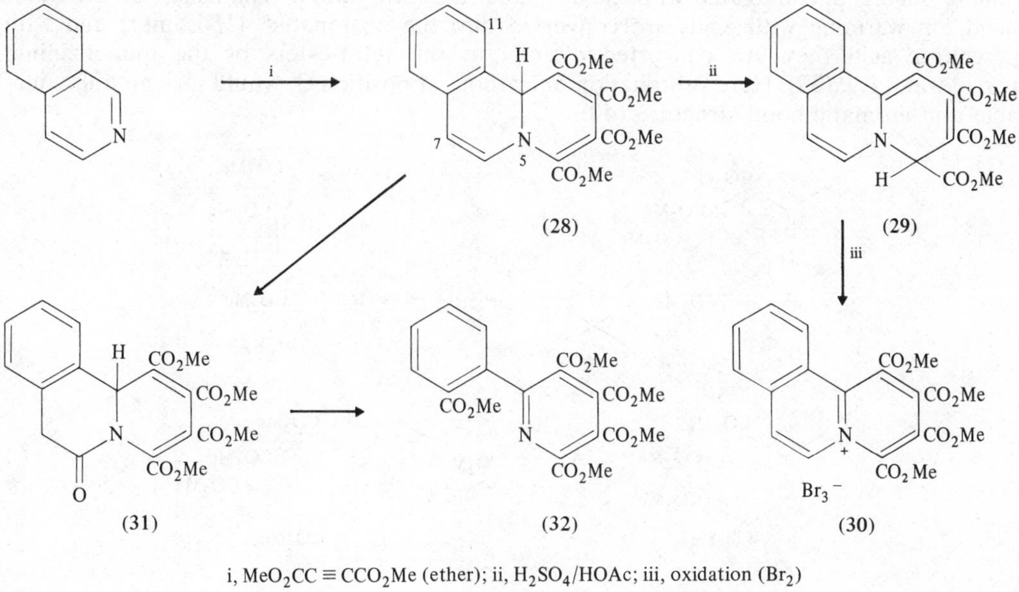

(28) (29) (31) (32) (30)

i, $MeO_2CC{\equiv}CCO_2Me$ (ether); ii, H_2SO_4/HOAc; iii, oxidation (Br_2)

SCHEME 3

16.6.2.2 Properties of quinolizine derivatives

The greater stability of the quinolizine tetracarboxylic esters (**18**) and (**19**) compared with that of the unsubstituted parent compounds can be explained as the result of resonance involving the ester groups at positions-1 and -3, conjugated with the nitrogen atom, such as [(**33**) ↔ (**34**)] and [(**35**) ↔ (**36**)], in which the 4*H*-isomer also has the advantage of the contribution of structure (**34**) containing a benzenoid ring.

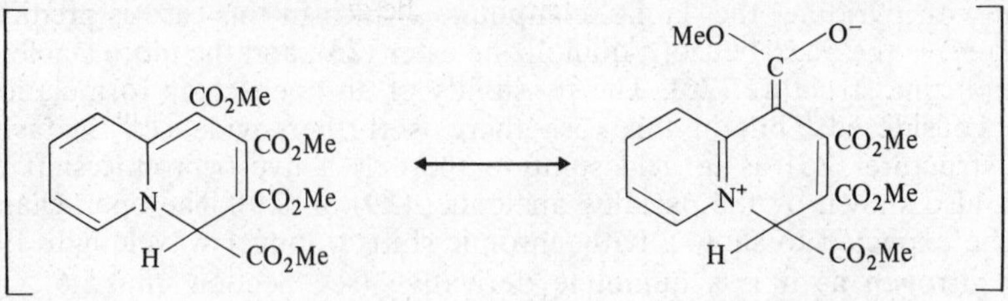

(33) (34)

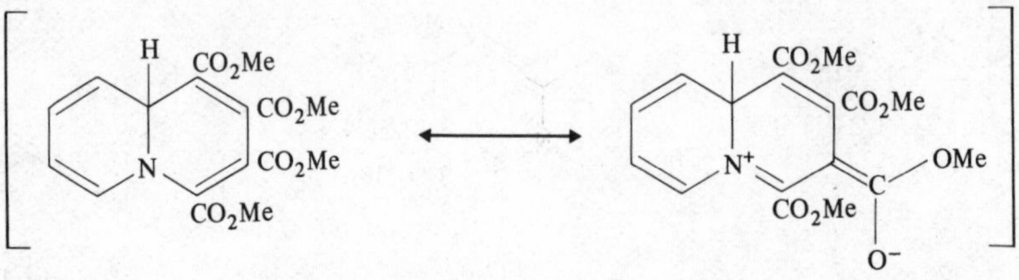

(35) (36)

The 4*H*-quinolizine ester (**19**) is a mono-acid base which forms a stable salt with perchloric acid in methanol. Unless there is a substituent (*e.g.* Me) at position-9, protonation occurs on carbon at position-3. This is evident from the u.v. spectrum of the salt, which is similar to that of a 3,4-dihydroquinolizine having the additional double bond conjugated with the aromatic heterocyclic ring (**37**). 4*H*-Quinolizine esters with a substituent at position-9 are protonated at position-1; their spectra show that the additional double bond is unconjugated with the aromatic ring (**38**). 9a*H*-Quinolizines, on the other hand, on warming with acids are converted into the protonated 4*H*-isomers and with perchloric acid they are converted directly to the tetra-esters of the quinolizinium perchlorate, *e.g.* (**39**). Here protonation on carbon, at position-3, would give an unfavourable non-aromatic bond structure (**40**).

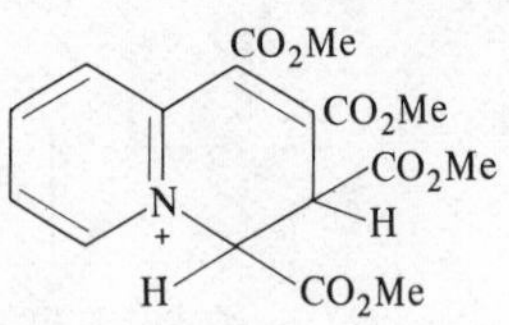

(37) 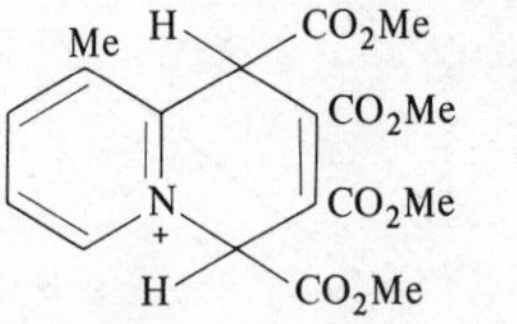 (38)

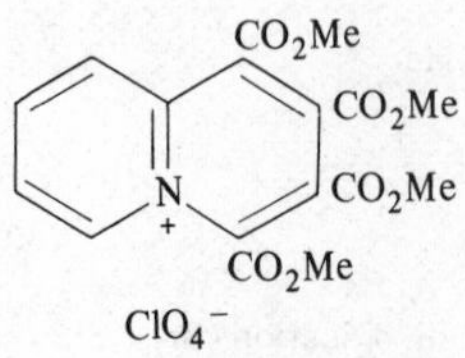

(39) 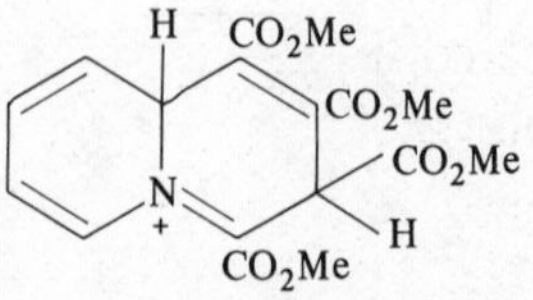 (40)

On warming with concentrated alkali the labile 9a*H*-adducts are degraded to the original heterocyclic base, but the stable 4*H*-adducts give the 2-methylsubstituted base. Thus (**18**) gives pyridine but (**19**) gives 2-methylpyridine.

The 2- and 4-quinolizones, which are the conjugate bases of the corresponding 2- and 4-hydroxyquinolizinium cations, are also derivatives of 2*H*- and 4*H*-quinolizine. They exist as resonance structures with contributions from the dipolar forms 2-quinolizone [(**41**) ↔ (**42**)] and 4-quinolizone [(**43**) ↔ (**44**)] (equations 2 and 3). The deprotonated 1- and 3-hydroxyquinolizinium salts are restricted to the zwitterionic structures (**45**) and (**46**) (equations 4 and 5).

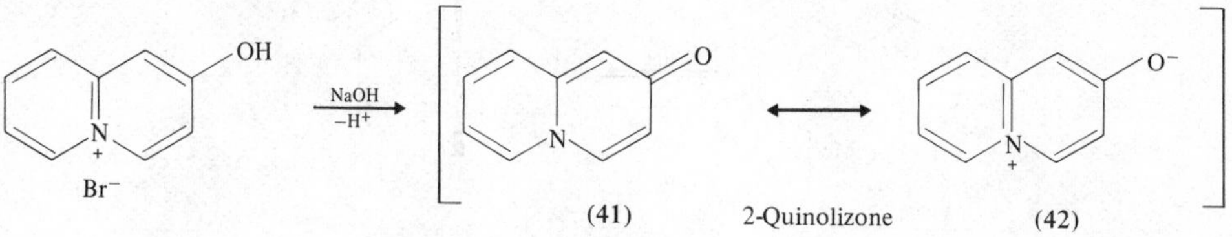

OH; Br^-; NaOH, $-H^+$; O; (41) 2-Quinolizone (42); O^- (2)

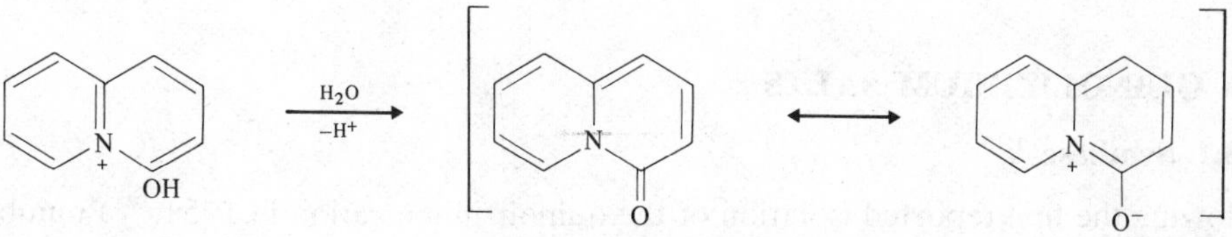

OH; H_2O, $-H^+$; O; (43) 4-Quinolizone (44); O^- (3)

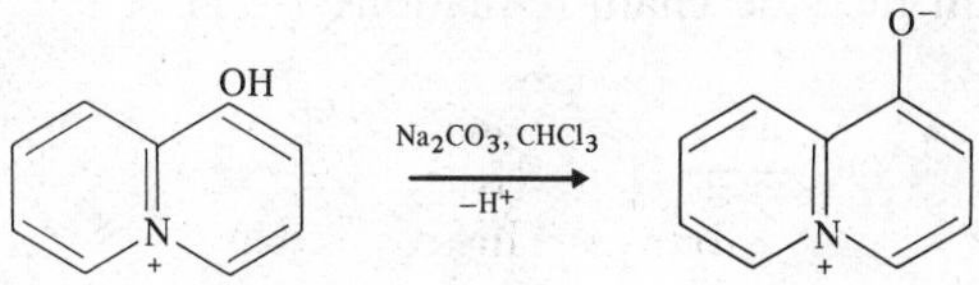

OH; Na_2CO_3, $CHCl_3$, $-H^+$; O^-; (45) (4)

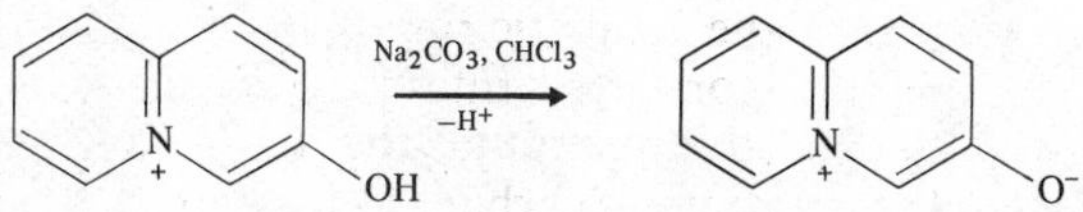

OH; Na_2CO_3, $CHCl_3$, $-H^+$; O^-; (46) (5)

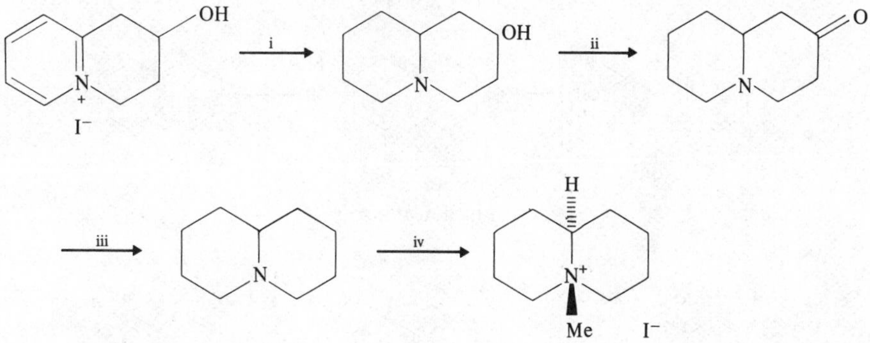

OH; I^-; i; OH; ii; O

iii; iv; H; N^+; Me I^-; (7); (47)

i, H_2/Ni; ii, oxidation; iii, reduction (Wolf–Kishner); iv, MeI

SCHEME 4

16.6.2.3 Quinolizidine

This fully hydrogenated octahydroquinolizine (**7**) has been prepared from 2-hydroxy-1,2,3,4-tetrahydroquinolizinium bromide (equation 7),[3b] by the series of reactions shown in Scheme 4. It has the properties expected of a tertiary cycloaliphatic amine. Protonation gives the quinolizidinium cation and direct methylation leads to the *trans*-4-methyl-quinolizidinium salt (**47**). The corresponding *cis*-isomer (**48**) may be synthesized by cyclization of the piperidine derivative (**49**) (equation 6).

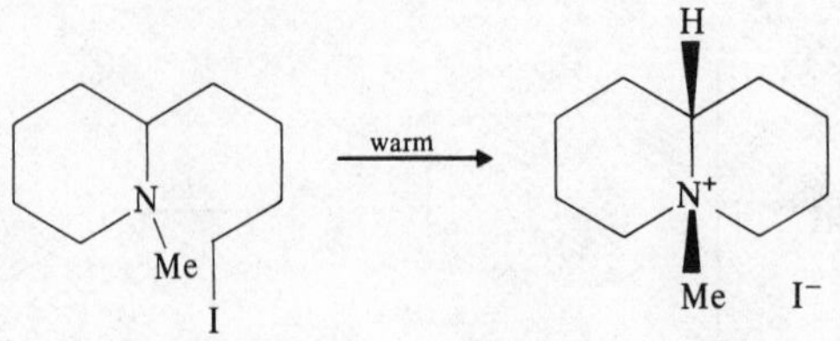

(**49**) —warm→ (**48**) (6)

16.6.3 QUINOLIZINIUM SALTS

16.6.3.1 Syntheses[1,13]

Following the first reported isolation of the quinolizinium cation in 1951,[14] a number of syntheses have been developed from 2-butylpyridine derivatives where the final ring closure is achieved by attack of the heterocyclic nitrogen on an appropriately placed electron-deficient carbon in the side chain (equations 7–11).

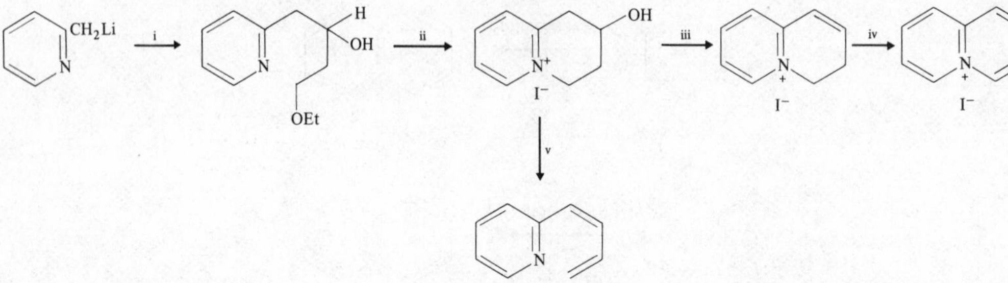

(7)

i, $EtO(CH_2)_2CHO$; ii (a) HI, (b) K_2CO_3; (iii) Ac_2O, H_2SO_4(trace); iv, Pd/C; v, NaOAc[15]

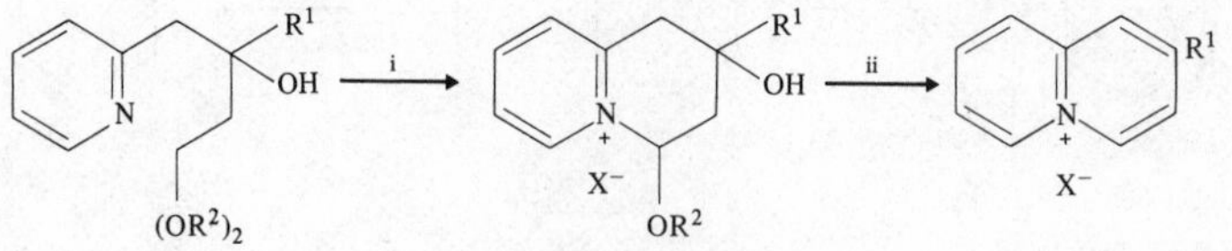

(8)

i, H^+; ii, Ac_2O/H_2SO_4[16]

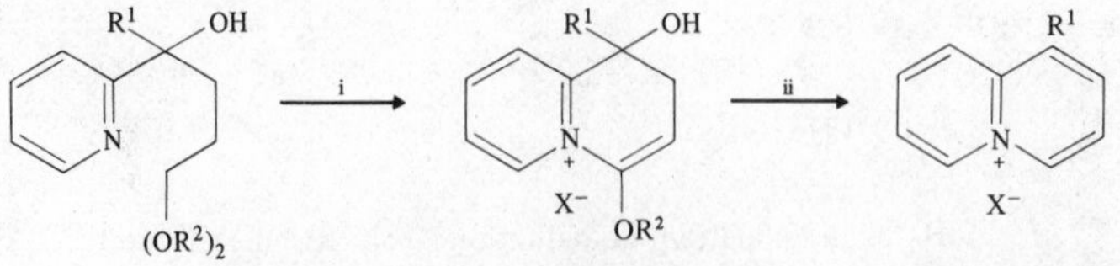

(9)

i, H^+; ii, Ac_2O/H_2SO_4[17a]

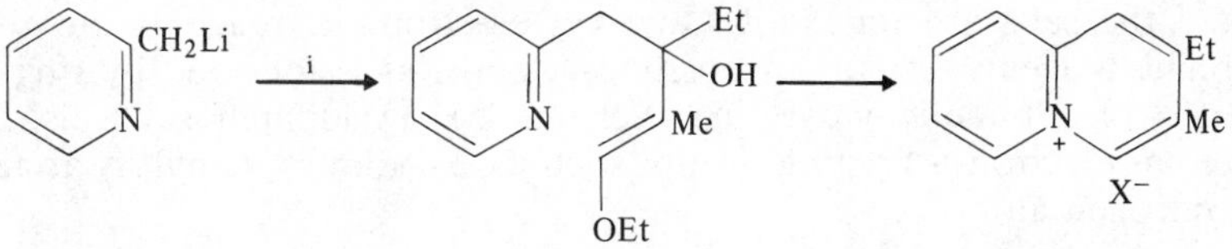

(10)

i, $EtCOC(Me)=CH(OEt)$; ii, picric acid, ethanol; X^- = picrate[18]

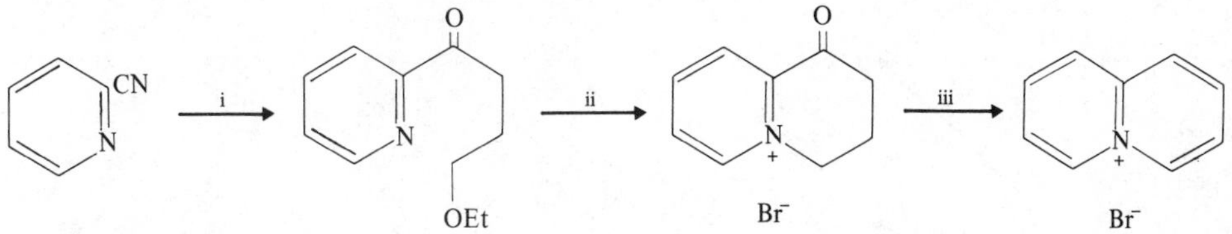

(11)

i, $EtO(CH_2)_3MgBr$; ii, warm; iii, Ac_2O, boil[17b]

The method of Glover and Jones[17b] (equation 11) is convenient and versatile and it has been used to prepare both alkyl- and aryl-quinolizinium salts substituted at any of the possible ring positions and also, starting from 2-cyanoquinoline or from 1- or 3-cyano-isoquinoline, the three possible benzoquinolizinium ring systems can be obtained. 1-Cyanoisoquinoline, for example, gives a benzo[*a*]quinolizinium salt (equation 12).

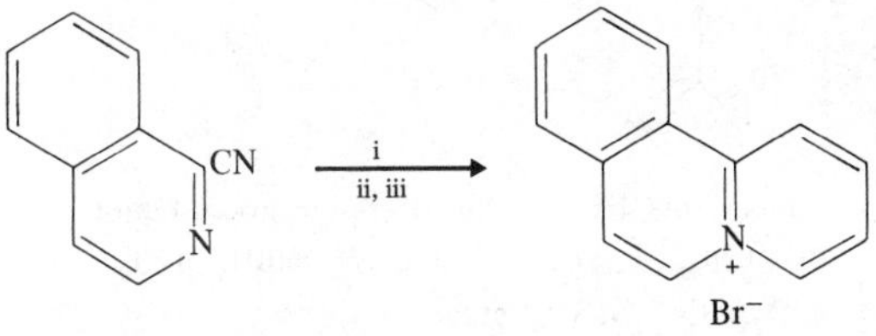

(12)

i, ii, iii, as equation (11)

16.6.3.2 Properties and reactions of quinolizinium salts

The quinolizinium salts are water-soluble crystalline solids, which may be difficult to isolate from aqueous solutions containing other soluble salts; they usually melt without decomposition. The u.v. spectrum of the quinolizinium cation is closely similar to those of naphthalene, quinoline, and isoquinoline (see Table 3, Section 16.4.3.2). In their chemical properties the salts resemble quaternary pyridinium salts, with positions-1 and -3 corresponding to β-positions of the pyridinium system and positions-4 and -2 to the α- and γ-positions, respectively. The correspondence can be extended to the properties of the simple derivatives. The electron localization energies have been calculated by Acheson and Goodall[19] for electrophilic and nucleophilic attack [(**50**) and (**51**)].

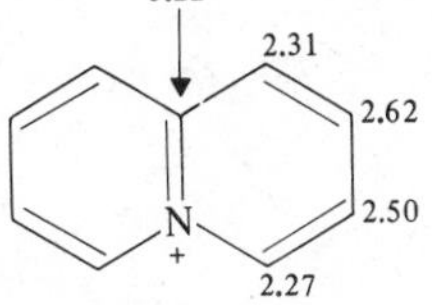

(50) Electrophilic attack (A_E)

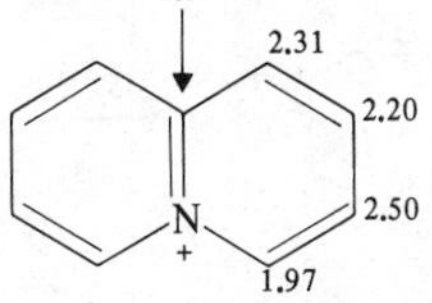

(51) Nucleophilic attack (A_N)

Localization energies A ($-\beta$) for quinolizinium cation
Coulomb parameter $h = 2$; resonance parameter $k = 1$[19]

As expected the salts are unreactive towards electrophilic reagents but react readily with nucleophilic reagents, usually at position-4, and this is followed by ring opening to give derivatives of 2-butadienylpyridine (Scheme 5). Pyridinium salts also suffer ring opening when an electron-attracting group, such as 2,4-dinitrophenyl, is attached to the heterocyclic nitrogen atom.

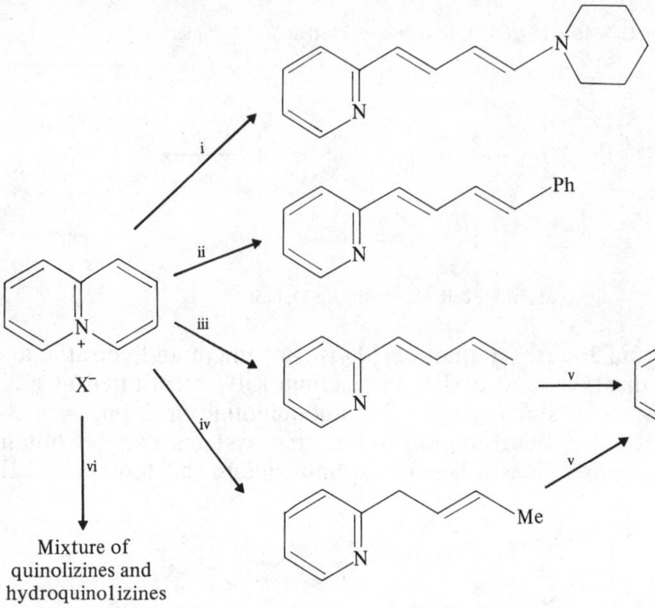

Mixture of quinolizines and hydroquinolizines

i, Piperidine, heat; ii, PhMgBr, ether, room temp.; iii, $LiAlH_4$, THF, room temp.; iv, $NaBH_4$, THF, room temp.; v, H_2/Pd; vi, $NaBH_4$, ethanol

SCHEME 5

16.6.4 DERIVATIVES OF QUINOLIZINIUM SALTS

16.6.4.1 Methyl derivatives

As with most of the derivatives of quinolizinium salts, the methyl derivatives are usually obtained by direct synthesis, rather than by substitution reactions [see Section 16.6.2.1]. The isomers with the methyl group at positions-2, -4, -6 or -8 (*i.e.* α or γ to the nitrogen) have a reactivity similar to that of the 2- and 4-methylpyridines; they are readily deprotonated and form condensation products with aldehydes and aromatic nitroso-compounds. 2-Methylquinolizinium iodide with *p*-dimethylaminobenzaldehyde gives 2-(*p*-dimethylaminostyryl)quinolizinium iodide (**52**) (equation 13).

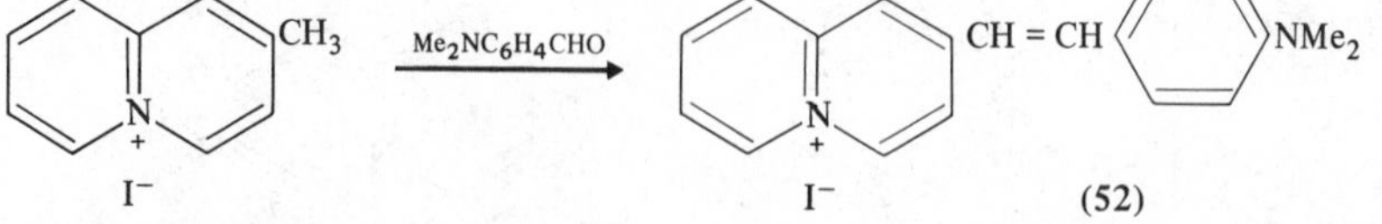

CH_3 $\xrightarrow{Me_2NC_6H_4CHO}$ CH = CH NMe_2 (13)

I^- I^- (52)

16.6.4.2 Hydroxyquinolizinium salts

The hydroxyquinolizinium salts may be compared with the hydroxy-derivatives of quaternary pyridinium salts. The 1- and 3-isomers have phenolic properties and on deprotonation are restricted to the zwitterionic structure. The 2- and 4-hydroxy-compounds on deprotonation form quinolizones which are derivatives of the quinolizines; they have amido-properties and are mesomeric structures analogous to the pyridones (see equations 2–5). Examples of all four monohydroxyquinolizinium salts have been synthesized (equations 14–19).

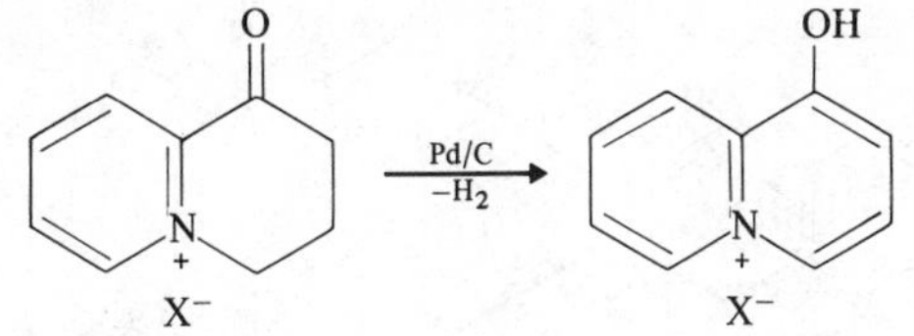

Pd/C, $-H_2$ (14)[20]

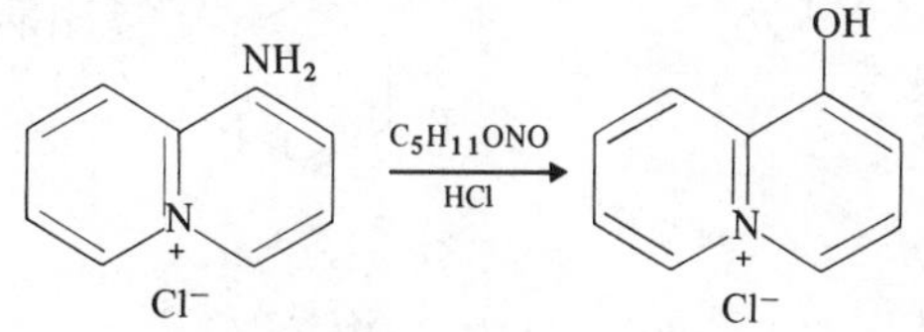

$C_5H_{11}ONO$, HCl (15)[20]

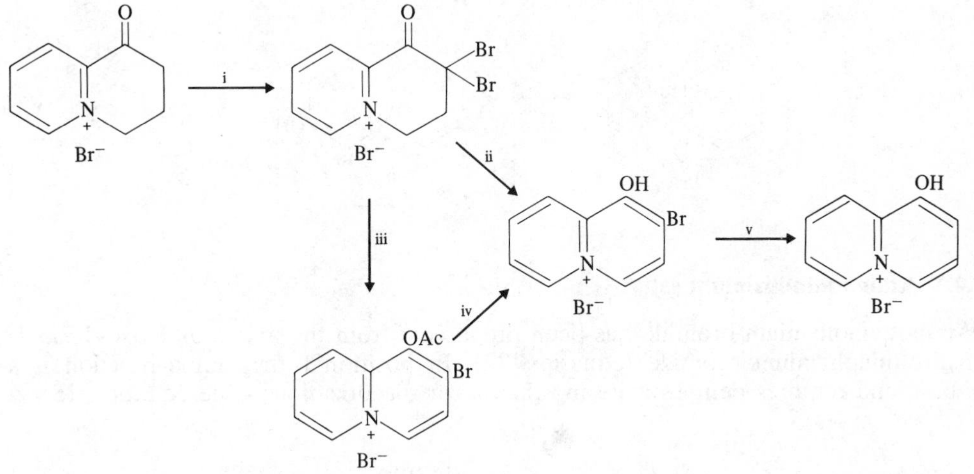

(16)

i, Br_2; ii, 160 °C; iii, Ac_2O, heat; iv, H^+, H_2O; v, H_2, Pd/C[21a]

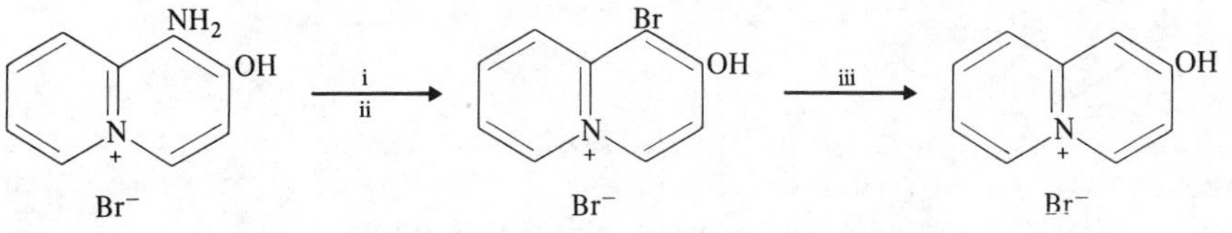

(17)

i, HNO_2, H^+; ii, Br^-; iii, H_2, Pd/C ($-HBr$)[21b]

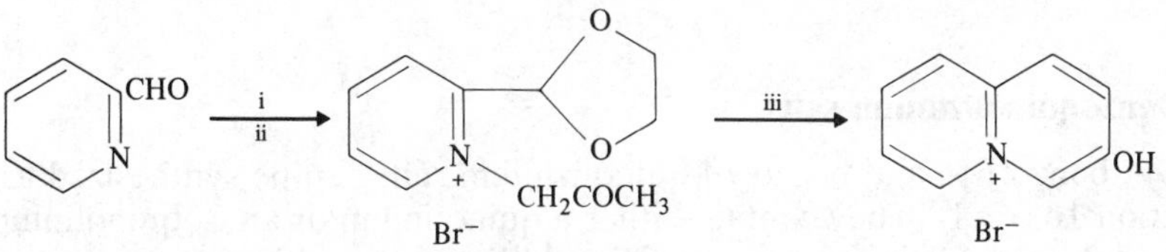

(18)

i, $HOCH_2CH_2OH$, H^+; ii, $BrCH_2COCH_3$; iii, HBr[22]

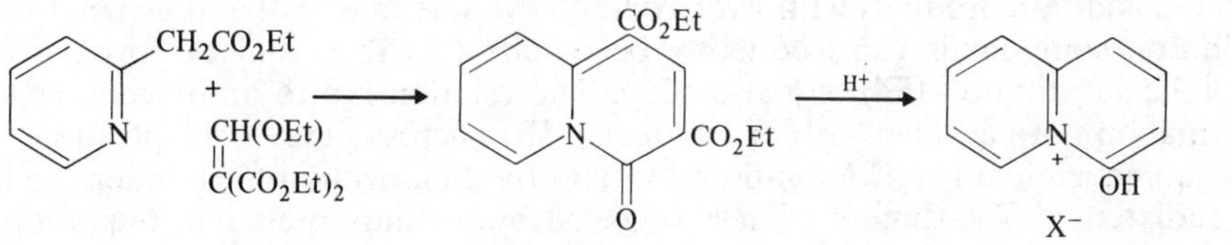

H^+ (19)[23]

The phenolic properties of the 1- and 3-hydroxyquinolizinium salts are illustrated by their substitution reactions with bromine and diazonium salts (equations 20 and 21).

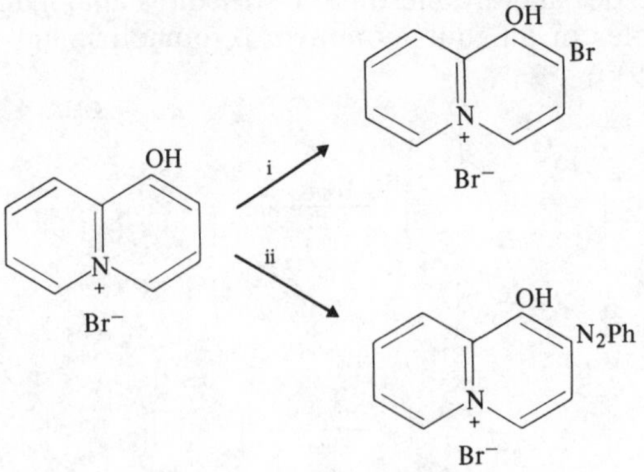

(20)

i, Br_2, HBr; ii, $PhN_2^+Cl^-$, H^+ [21a, 24]

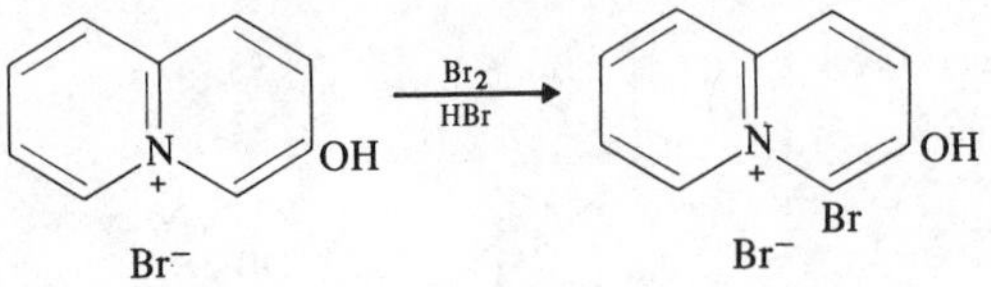

(21)[24]

16.6.4.3 Aminoquinolizinium salts

1-Aminoquinolizinium bromide has been prepared[20] from the oxime of 1-oxo-1,2,3,4-tetrahydroquinolizinium bromide (equation 22). In position-1 the amino-function is a weak base and requires pentyl nitrite in ethanol for diazotization[20] (see equation 15).

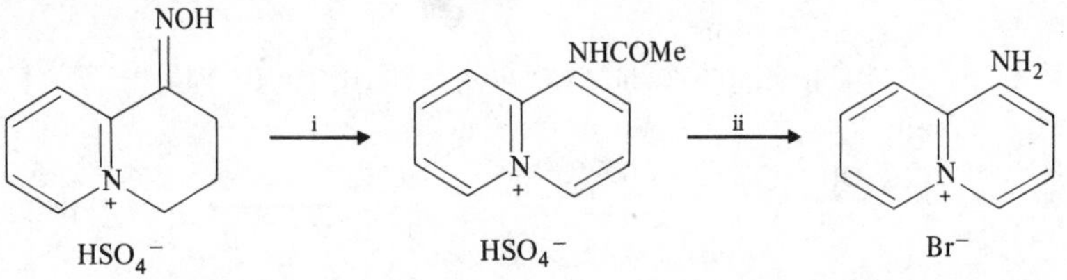

(22)

i, Ac_2O, H_2SO_4; ii, hydrolysis (HBr)

16.6.4.4 Benzoquinolizinium salts

Benzo[*a*]-, benzo[*b*]-, and benzo[*c*]-quinolizinium salts can be synthesized as described above (Section 16.6.3.1). They contain either a quinolinium or an isoquinolinium function in addition to the quinolizinium system. This additional aromatic ring gives added stability to the adducts formed by the reaction of nucleophiles.[25] Thus benzo[*b*]quinolizinium salts form stable addition products with hydroxyl and cyanide ions (**53**) and Grignard reagents. The addition compounds can frequently be oxidized to the aromatic form, resulting in nucleophilic substitution (**54**) (equation 23). The relationship to anthracene is shown by the formation of a Diels–Alder adduct (**55**) across the 6,11-positions of the benzo[*b*]quinolizinium ring[26] (equation 24), and by dimerization (**56**) under the influence of u.v. radiation.[27] The dimerization is reversed by heating in ethanol (equation 25).

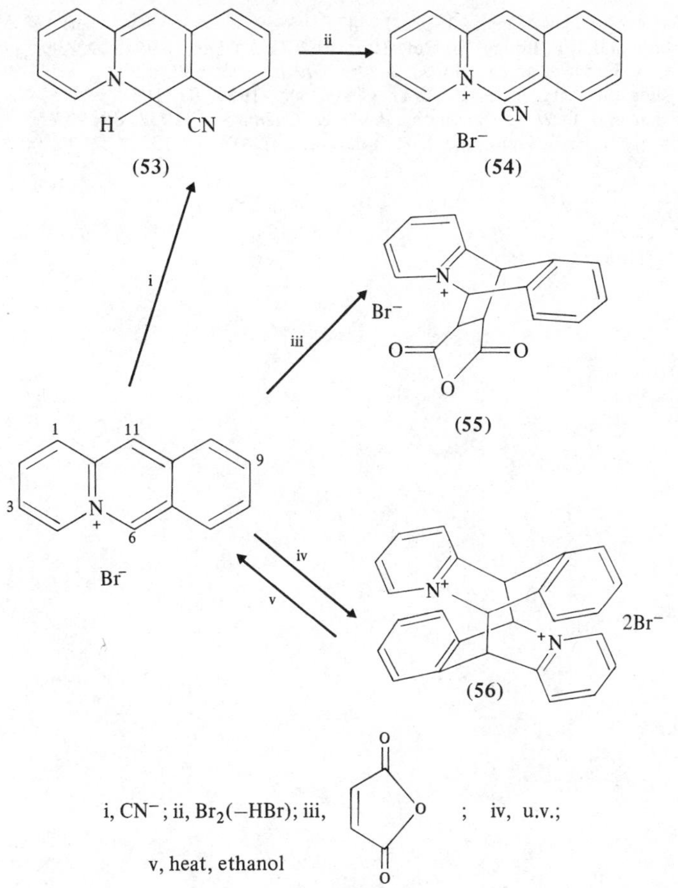

(23)

(24)

(25)

i, CN^-; ii, Br_2(−HBr); iii, [maleic anhydride]; iv, u.v.;

v, heat, ethanol

References

1. B. S. Thyagarajan, (a) *Chem. Rev.*, 1954, **54,** 1019; (b) *Adv. Heterocyclic Chem.*, 1965, **5,** 291.
2. T. Miyadera and Y. Kishida, *Tetrahedron Letters*, 1965, 905.
3. (a) V. Boekelheide and W. G. Gall, *J. Amer. Chem. Soc.*, 1954, **76,** 1832; (b) V. Boekelheide and J. P. Lodge, *ibid.*, 1951, **73,** 3681.
4. (a) R. K. Hill, in 'Chemistry of the Alkaloids', ed. S. W. Pelletier, van Nostrand Reinhold, New York, 1970, p. 414; (b) E. E. van Temelen and R. L. Foltz, *J. Amer. Chem. Soc.*, 1960, **82,** 502.
5. R. M. Acheson, Ref. 1(b), 1963, vol. 1, p. 143.
6. L. M. Jackman, A. W. Johnson, and J. C. Tebby, *J. Chem. Soc.*, 1960, 1579.
7. R. M. Acheson, G. A. Taylor, P. Higham, and R. E. Richards, *J. Chem. Soc.*, 1960, 1691.
8. J. A. Elvidge and L. M. Jackman, *J. Chem. Soc.*, 1961, 859.
9. E. E. van Tamelen, P. E. Aldrich, P. Bender, and G. Miller, *Proc. Chem. Soc.*, 1959, 309.
10. R. M. Acheson, N. J. Earl, P. Higham, R. E. Richards, G. A. Taylor, and J. M. Vernon, *Proc. Chem. Soc.*, 1960, 281.
11. R. M. Acheson and F. Hole, *J. Chem. Soc.*, 1962, 748.
12. B. S. Thyagarajan, Ref. 1(b).
13. Gurnos Jones, in 'The Organic Chemistry of Nitrogen', ed. N. V. Sidgwick, revised by I. T. Millar and H. D. Springall, Clarendon Press, Oxford 1966, chapter 27, p. 762.
14. A. G. Beaman, Ph.D. Thesis, Harvard, 1951.
15. V. Boekelheide and W. G. Gall, Ref. 3(a).
16. A. N. Nesmeyanov and M. I. Ribinskaya, *Doklady Akad. Nauk S.S.S.R.*, 1957, **116,** 63.
17. (a) E. E. Glover and G. Jones, *J. Chem. Soc.*, 1959, 1686; (b) *ibid.*, 1958, 3021.
18. A. Richards and T. S. Stevens, *J. Chem. Soc.*, 1958, 3067.
19. R. M. Acheson and D. M. Goodall, *J. Chem. Soc.*, 1964, 3225.
20. A. R. Collicut and G. Jones, *J. Chem. Soc.*, 1960, 4101.
21. (a) A. Fozzard and G. Jones, *J. Chem. Soc.*, 1963, 2203; (b) *ibid.*, 1964, 2760.

22. E. Schraufstatter, *Angew. Chem.*, 1962. **74,** 874.
23. V. Boekelheide and J. P. Lodge, (a) Ref. 3(b); (b) *J. Org. Chem.*, 1954, **19,** 499.
24. P. A. Duke, A. Fozzard, and G. Jones, *J. Org. Chem.*, 1965, **30,** 526.
25. C. K. Bradsher and J. H. Jones, *J. Amer. Chem. Soc.*, 1959, **81,** 1938.
26. C. K. Bradsher and T. W. G. Solomons, *J. Amer. Chem. Soc.*, 1958, **80,** 933.
27. C. K. Bradsher, L. E. Beavers, and J. H. Jones, *J. Org. Chem.*, 1957, **22,** 1740.

16.7

Mono- and Polyaza-anthracenes and -phenanthrenes, Naphthyridines, and Polyazanaphthalenes

D. G. WIBBERLEY
University of Aston, Birmingham

16.7.1 INTRODUCTION

The scope of this chapter is the discussion of those heteroaromatic ring systems represented in Scheme 1, where N may be 0, 1, 2, 3, or 4 N atoms. N-Bridgehead compounds, quinolines, isoquinolines, quinazolines, quinoxalines, cinnolines, phthalazines, and their benzo-derivatives are excluded, having been discussed in other chapters.

Some 60 polyazanaphthalenes and well over 1000 polyaza-anthracenes and -phenanthrenes are theoretically possible and this theoretical diversity is paralleled by a

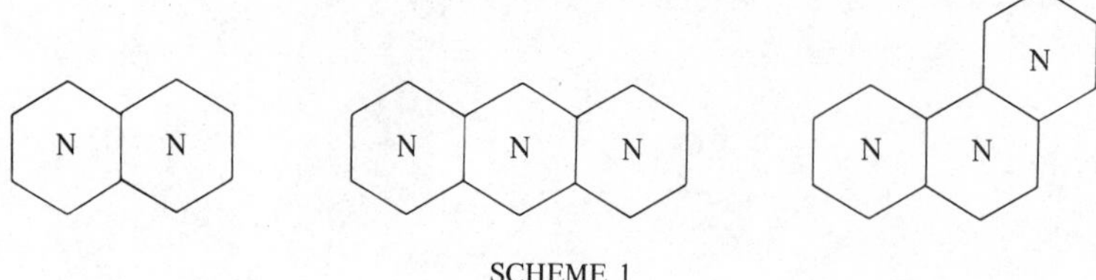

SCHEME 1

broad practical approach to the subject. Thus in a typical recent year such as that covered by the 1975 *Chemical Abstracts*, out of more than 200 substantial papers in this area, 51 different ring systems were mentioned, many of these in only one publication. Undoubtedly the main spur for this broad approach is the continued search for compounds of therapeutic importance. In each of the four most thoroughly researched ring systems of recent years, the pteridines (**1**), the 1,8-naphthyridines (**2**), the pyrido[2,3-*d*]pyrimidines (**3**), and the pyrimido[5,4-*d*]pyrimidines (**4**), at least one drug is now marketed (*cf.* Section 16.7.6) and very large numbers of derivatives of these and related ring systems are synthesized annually for chemo-therapeutic and pharmacological evaluation.

(1) (2) (3) (4)

Throughout this chapter the nomenclature and numbering of specific compounds is in accordance with the recommendations of the International Union of Pure and Applied Chemistry and with the further advice of the Chemical Society Publications. There is much to be said, however, for the use of 'aza-nomenclature' in preference to the numbers and letters currently used to show the fusion faces of simpler heterocycles. Thus 1,3,8-triazanaphthalene (**3**) and 1,3,5,7-tetra-azanaphthalene (**4**) are much more readily designated by a wider group of chemists than the accepted names and the difficulties are further confounded in tricyclic systems. For this reason the aza-nomenclature will be used as it is in the chapter heading, for broad general comment.

In view of the very large number of ring systems under discussion, it would obviously be inappropriate to discuss named systems separately. The preparations and properties are therefore treated with no division according to type of compound but with separate discussion of, for example, the nature and mechanisms of the bond-forming process, or the mechanisms of substitution, or ring-opening reactions.

During the last 15 years there have been major reviews of the naphthyridines,[1] pyridopyrimidines,[2] pteridines,[3] acridines,[4] and phenanthridines.[5] In addition, there have been less extensive, or less accessible, reviews of diazabicyclic ring systems,[6] 1,5-naphthyridines,[7] azapteridines,[8] benzonaphthyridines,[9] riboflavine analogues,[10] and on the diuretic activity of tri- and tetra-azanaphthalenes.[11] A broader coverage of heteroaromatic literature of both five- and six-membered polycyclic compounds has been given for the years 1970–1971[12] and 1972–1973,[13] and the recent *Specialist Periodical Reports* and the *Annual Reports* of the Chemical Society make reference to important new syntheses or reactions.

16.7.2 SYNTHETIC ROUTES

The synthetic routes discussed in this chapter are classified according to the position, nature, and mechanism of bond formation in the cyclization step. Very many of the syntheses under consideration involve the interaction of a difunctional reagent in which two bond-forming reactions are effected in the same operation. In those cases an arbitrary placing is made into that bond-forming classification which appears to be the more difficult.

16.7.2.1 Electrophilic attack at a ring-C atom

This is an extremely versatile type of synthesis, applicable to most of the heterocyclic systems under consideration but particularly those with *b*- or *c*-fused pyridines. The antecedents are in those well-known reactions which often carry a name and which have been used for many years for the synthesis of naphthalenes, quinolines, and isoquinolines. Structural requirements for the intermediate are an electrophilic group, placed four atoms away from an aromatic ring and a vacant *ortho*-position; the precursor of this intermediate is an aromatic amine or aminomethyl derivative containing a vacant *ortho*-position. Thus Skraup, Doebner, Conrad–Limpach and Pomeranz–Fritzch reaction-types have been extended to the synthesis of important ring systems such as the four naphthyridines,[1,14] pyrido[2,3-*d*]pyrimidines,[2,15] acridines,[16] and phenanthridines.[17] Recent examples of the synthesis of less-common or novel ring systems which also illustrate the variety of electrophilic groupings which may be involved are the pyrimido[4,5-*b*]quinolines[18] and pyrimido[5,4-*b*]quinolines[19] which were obtained from suitable acids, the pyrimido[4,5-*c*]isoquinolines[20] from amides, the benzo[*b*][1,8]naphthyridines,[21] from nitriles and a pyrimido[5,4-*g*]pteridine[22] from a nitro-compound. A suitable choice of reagent, catalyst, and temperature for the two-stage reaction from the aromatic amine can give good yields without the necessity of isolating the intermediates. Thus a convenient synthesis of a pyrido[2,3-*f*]quinoline[23] from a *m*-phenylenediamine is effected by a double Conrad–Limpach type of reaction with ethyl acetoacetate.

Et O C=O CO$_2$Et H N O HN O N H → Et O O$^-$ H N H CO$_2$Et HN + N H O → H N O CO$_2$Et HN N H O

(5) (6)

Constraining factors are a strongly π-deficient parent ring system which renders either the original amino-group or the *ortho*-position unreactive towards attacking electrophiles, an unsymmetrical electrophile which can theoretically lead to the formation of two products, and a ring N atom at the alternative *ortho*-position which can yield N-bridgehead compounds. Thus the formation of a pyrido[3,2-*d*]pyrimidine (**6**) from a 4-ethoxycarbonylvinylaminopyrimidine required the presence of electron-donating groups in the pyrimidine ring (**5**),[24] and, because of the lack of basicity of the 4-amino-group, 4-aminoquinoline does not undergo Conrad–Limpach or Doebner cyclizations. It does, however, react with the more electrophilic ethoxymethylenemalonic ester (EMME), a valuable reagent in all these syntheses, and the intermediate 4-ethoxycarbonylvinylamino-quinoline cyclizes at the electron-rich 3-position to yield a benzo[*h*][1,6]naphthyridine.[25] In general an sp^2-hybridized ring-nitrogen atom is a better nucleophile than a ring-carbon atom, and in those reactions where both cyclizations are possible, N-bridgehead compounds are formed more easily. Such compounds, however, are more susceptible to both hydrolysis and thermal ring-opening and so catalyst and temperature can be used to vary the nature of the product. Thus 1,8-naphthyridines may be prepared by thermal re-arrangement of pyrido[1,2-*a*]pyrimidines,[26] and anthyridinones or pyrimido[4,5-*b*]-[1,8]naphthyridines may be prepared from 2-amino-1,8-naphthyridine.

A large number of potentially biologically important pyrido[2,3-*d*]pyrimidines have been synthesized from 4-aminopyrimidines bearing oxo- or amino-substituents in 2- and/or 6-positions. In those reactions where an unsymmetrical bifunctional electrophile, such as a keto-aldehyde, an unsymmetrical 1,3-diketone, an acylacetate, or pyruvate is involved, the compound isolated is that one expected by attack of the more reactive carbonyl function at the 5-position of the pyrimidine ring rather than at the exocyclic amino group.[2]

16.7.2.2 Nucleophilic attack at a ring-C atom

In view of the greater susceptibility of polyaza-heteroaromatic compounds to nucleophilic attack at ring C atoms, it is, at first sight, somewhat surprising that so little use has been made of this type of synthetic method. 1,2,3,4-Tetrahydro-1,8-naphthyridines have been prepared in good yield by treatment of 3-(3-pyridyl)propylamines with sodium in toluene, but the difficulties in the preparation of suitable starting materials have inhibited the extension of this type of route to other ring systems.[28] More commonly one additional bond-forming step is also involved in the formation of the new heterocycle. Thus pyrimido[5,4-*e*]-*asym*-triazines are obtained from *asym*-triazines and guanidine,[29] and a methoxy-group is nucleophilically displaced in the synthesis of pyrido[4,3-*e*]-*asym*-triazine.[30] A number of cyclizations involve attack at a ring C=O grouping by an exocyclic amino-group: thus pyridazinoquinoxalines (**8**) are obtained by treatment of the hydrazone (7) with boiling acetic acid.[31]

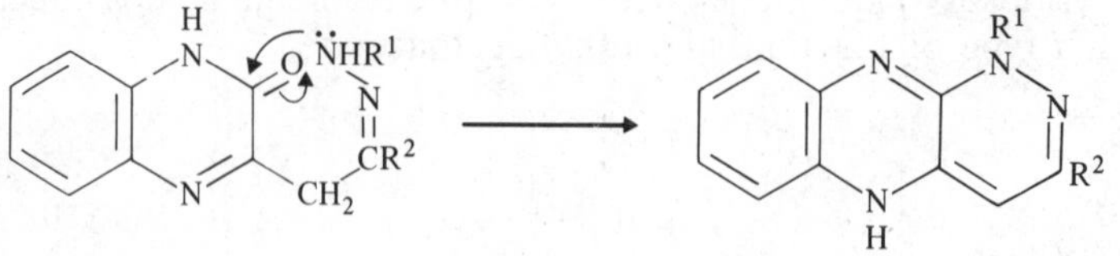

(7) (8)

16.7.2.3 Amine–carbonyl addition reactions

This is the most versatile route for the synthesis of polyaza-heteroaromatic compounds and has been used at one time or another for the synthesis of all the well known bi- or tri-cyclic ring systems. It involves the intramolecular addition of an amino or substituted amino group to a carbonyl-containing group such as an aldehyde acid, ester, or amide, with subsequent elimination of a small molecule such as water, alcohol, or ammonia. In those cases where no additional reagent is employed or where an additional reagent such as ammonia or hydrazine is caused to react with a di-carbonyl compound, the mechanism is unequivocally of this type. Similarly, in the well-known synthesis of fused pyrazines by reaction of *ortho*-diamines with α,β-dicarbonyl compounds, there can be no doubt that amine–carbonyl addition–elimination reactions are involved in the cyclization stage. On the other hand, a number of other ring systems are formed by reactions which involve both amine–carbonyl and other C—C or N—N bond-forming reactions. Although intermediates are seldom isolated, it is presumed, for the purposes of classification, that the N—C bond-forming step does not occur in the final cyclization stage. Thus all the pyrido[*d*]pyrimidine ring systems can be prepared by this type of reaction[2] and often the readily available amino-pyridinecarboxylic acids are used as starting material. The final cyclization can be at either of the two pyridine ring N atoms, *e.g.* 2-amino-4,6-dimethylnicotinic acid is converted into the corresponding 2-ureidonicotinic acid by treatment with potassium cyanate, and the latter is cyclized to 5,7-dimethylpyrido[2,3-*d*]pyrimidine-2,4-(1*H*,3*H*)-dione by the action of heat.[32] Cyclizations at the other N-atom are even more common, *e.g.* 4-amidonicotinamides (**10**) readily yield the corresponding pyrido[4,3-*d*]-pyrimidine-4(3*H*)-ones (**12**).[33] The diamines may be prepared from pyrido[4,3-*d*][1,3]-oxazin-4-ones (**9**), which are themselves derived from 4-aminonicotinic acid and the appropriate anhydride, or from the 4-amidonicotinic esters (**11**) which are obtained in two

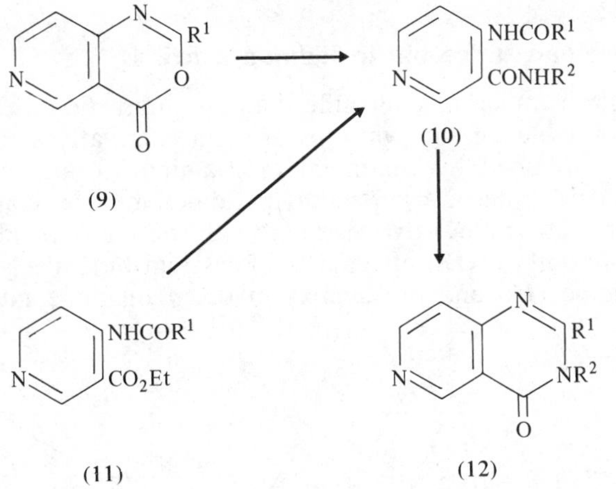

(9) (10) (11) (12)

steps from the same 4-aminonicotinic acid.[33] The ease of cyclization of these (**10**) and similar pyridine diamides varies with the nucleophilicities of the amino-groups and the electrophilicities of the carbonyls. In some cases isolation of the diamide (**10**) is not necessary, but in others high temperature and a base or Lewis-acid catalyst are required. Other O-containing ring systems such as the pyrans and cyclic anhydrides yield fused pyridine and pyrazine rings on treatment with amines or hydrazine and the problems of the high reactivity of amino-substituents can be avoided by their *in situ* preparation from suitable *ortho*-NO_2 compounds. In the latter case, a suitable reducing agent such as sodium borohydride in the presence of palladium on charcoal can give good yields of cyclic hydroxamic acids.[34]

The insertion of a bi-functional electrophile between two suitable amino-groups is a valuable route for fused pyrimidines and triazines. Pyrimido[5,4-*e*]-*asym*-triazines (**14**), for example, may be prepared from pyrimidines (**13**)[35] or from triazines (**15**)[36] by reaction with orthoformic esters.

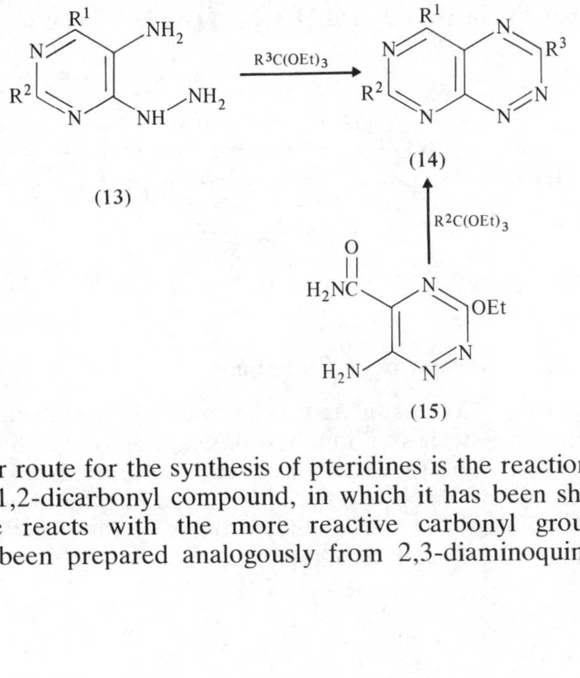

(13) $R^3C(OEt)_3$ (14) $R^2C(OEt)_3$ (15)

The most popular route for the synthesis of pteridines is the reaction of a 4,5-diamino-pyrimidine with a 1,2-dicarbonyl compound, in which it has been shown that the more nucleophilic amine reacts with the more reactive carbonyl group.[37] Pyrazino[2,3-*b*]quinolines have been prepared analogously from 2,3-diaminoquinoxalines and, in a closely related type of reaction, hydrazine and an *ortho*-diester or di-aldehyde yield pyrimido[4,5-*d*]pyridazines,[38] pyrido[3,4-*d*]pyridazines,[39] pyridazino[4,5-*b*]-quinoxalines,[40] and pyridazino[4,5-*b*]quinolines.[41]

16.7.2.4 Amine–imine and amine–nitrile addition reactions

This type of reaction is much less versatile than the interaction of amino and carbonyl groups, but can be valuable for the synthesis of amino-derivatives. Thus a 2,4-diamino-pteridine *N*-oxide is obtained by treatment of 2-amino-3-cyanopyrazine 1-oxide and guanidine,[42] and 2- or 4-cyano-3-cyanomethylpyridines may be converted into amino-ethoxy-1,7- or -2,6-naphthyridines by treatment with sodium alkoxides.[43] The dimerization of 2-methylnicotinonitrile (**18**) effected by potassium butoxide yields a 6-pyridyl-5-amino-1,5-naphthyridine (**16**) and presumably involves an imine–nitrile addition at the cyclization stage (**17**).[44]

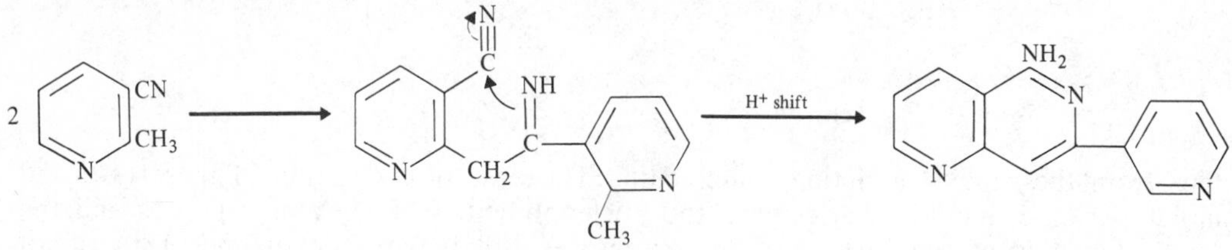

2 CN CH3 N → N C NH C CH2 N CH3 —H+ shift→ NH2 N N N

(16) (17) (18)

16.7.2.5 Carbanion–nitroso and related interactions

An unequivocal route to the pteridines which appears to give scope for other ring systems is the reaction of phosphonate anions of esters [O=P(OEt)$_2\bar{C}$(R)CO_2Et] with 4-amino-5-nitrosopyrimidines,[45] and a similar double activation of a methylene group has been used in another synthesis of pteridines from 4-amino-5-nitrosopyrimidines and *N*-acylmethylpyrimidine salts.[46] Nitrative and nitrosative cyclizations fall into this category: thus convenient routes to derivatives of the powerful poison toxoflavin[47] involve the nitrosation of hydrazones (**19**) with sodium nitrite in acetic acid. Closely similar reactions, but under nitrating conditions (KNO_3 and H_2SO_4 in HOAc), yield fervenulin 4-oxides.[48]

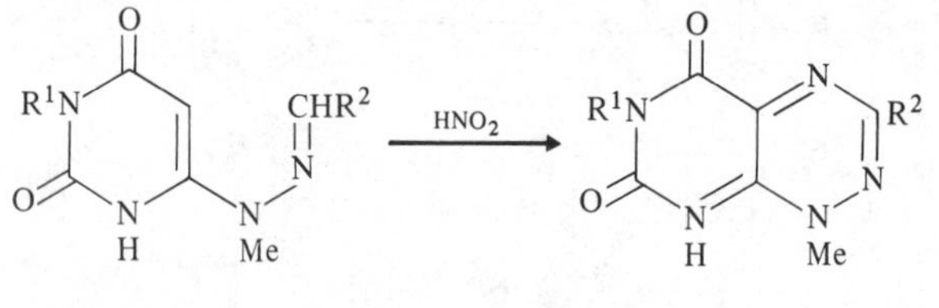

O R1N CHR2 N O N H N Me —HNO2→ O R1N N R2 N O N H N Me

(19) (20)

16.7.2.6 Exocyclic carbon–carbon bond formation

The extensions of the Dieckmann and Friedlander reactions involve this type of mechanism. The former has been used in a new preparation of 2,3-dihydrobenzo[*h*][1,6]-naphthyridin-4(1*H*)-ones,[49] and the latter has been used fairly extensively for the synthesis of pyridopyrimidines[2] and naphthyridines.[1] The type of catalyst necessary for carbanion formation has been shown to be dependent on the nature of the carbanion in the synthesis of both 1,8- and 1,6-naphthyridines.[50]

16.7.2.7 Nitrogen–nitrogen bond formation

This type of route could be alternatively called amine–diazonium salt–nitroso or –nitro group interactions. One of the two common routes for the synthesis of benzo[1,2,3]triazines involves the diazotization of anthranilamide and related compounds and this has been extended, for example, to the preparation of pyrido[2,3-]-[51] and [3,2-*e*]-*sym*-triazines[52] (**22**) by diazotization of *o*-aminopyridinecarboxamides (**21**). Nitro-group interactions are involved in the conversion of 4-guanidino-3-nitropyridines to pyrido-[4,3,*e*]-*asym*-triazines,[53] and nitroso in the synthesis of a pyrimidino[4,5-*e*]-*asym*-triazine.[54]

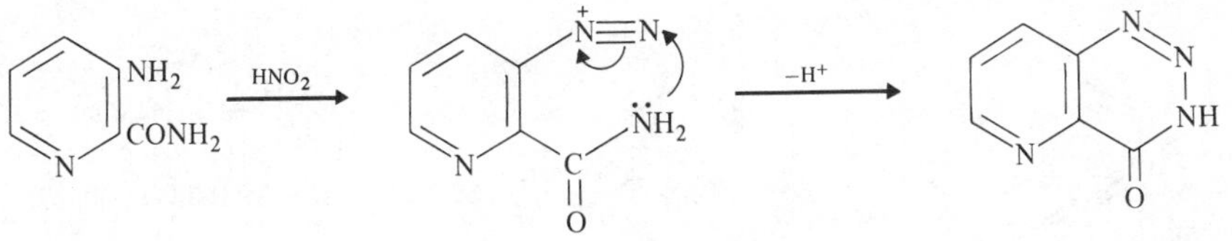

(21) (22)

16.7.2.8 Oxidation of di- and tetra-hydro-compounds

Several useful routes to the pyrido-[55] and pyrimidino-triazines[56] and to the pteridines[57] yield dihydro-derivatives in the cyclization step. The latter are then readily oxidized to the fully aromatic compound under relatively mild conditions with reagents such as Ag_2O, $KMnO_4$, H_2O_2, $K_3Fe(CN)_6$, oxygen from the air, and, particularly successfully, activated MnO_2. A related alternative reaction is the bromination–dehydrobromination of dihydro-pyrido[2,3-*d*]pyrimidines[58] to yield strongly antibacterial analogues of nalidixic acid.

16.7.2.9 Miscellaneous cyclization reactions

Isolated examples of the synthesis of bi- or tri-cyclic ring systems from purely aliphatic starting materials are not easy to classify mechanistically. For example, tetra-acetylethylene is a new intermediate for both pyridazino[4,5-*d*]pyridazines and pyrrolo-[3,4-*d*]pyridazines.[59] An ynamine–isocyanate non-concerted 1,4-dipolar addition reaction is postulated in the formation of a benzo[*h*]-1,6-naphthyridine (**22b**) from the 4-quinolyl-isocyanate (**22a**).[60]

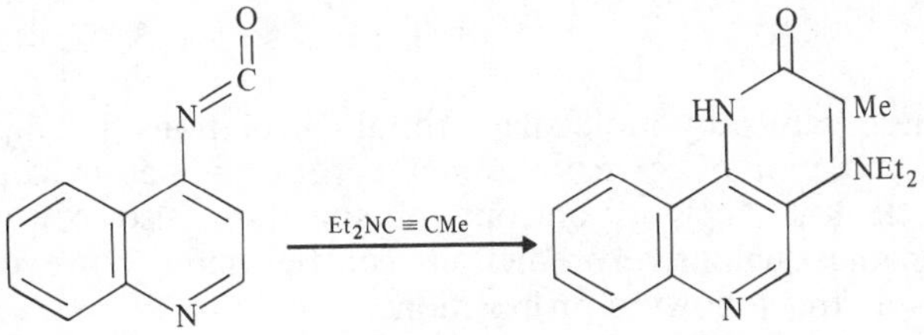

(22a) (22b)

An aryne intermediate is presumably involved in a one-stage cyclization of *o*-chloro-benzylideneaniline to phenanthridine in excellent yield,[61] and a concerted cyclization with HCN loss is suggested to operate in an unusual conversion of fused pyrimidines into fused pyridines.[62]

16.7.3 PHYSICAL PROPERTIES

The physical properties of the polyaza-naphthalenes, -anthracenes, and -phenanthrenes are quite well documented and, at least in terms of contemporary qualitative electronic mechanisms, generally explicable. As in other fields of organic chemistry, structural proof of the identity of new compounds rests largely on spectroscopic data.

16.7.3.1 pK_a Values

The pK_a values of polyaza-heteroaromatic compounds and their substituted derivatives have proved useful in structure determination, studies of covalent hydration, and biological structure–activity work. pK_a Values for a number of mono-, di-, and tri-azanaphthalenes are shown in Scheme 2. No satisfactory method has yet been found for

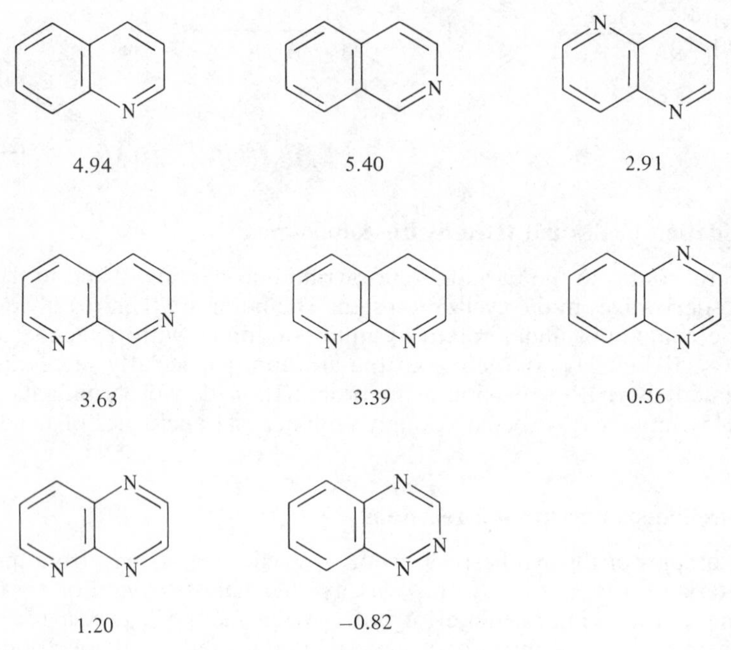

pK_a values (no hydration)

SCHEME 2

the derivation of these values by molecular orbital calculations, owing, no doubt, to the serious difficulties in selection of correct parameters for ring N atoms and their α-C atoms and because other factors such as entropy of solvation also control the energy of protonation. Reasonable qualitative predictions can be made, however, on the basis of electronic structures as the following propositions.

(i) π-Deficient N-heteroaromatic compounds are invariably considerably weaker bases than the corresponding N-heteroparaffinic compounds because the lone-pair of electrons is in an sp^2-hybridized orbital, rather than the sp^3 of the latter, and this tighter orbital reduces the availability of the lone-pair for protonation.

(ii) Annulation of a benzene ring has little effect.

(iii) A ring N-atom in a 2-position tends to be more readily protonated than one in a 1-position.

(iv) Addition of further ring N-atoms has a base-weakening effect by virtue of both inductive ($-I$) and mesomeric ($-M$) electron-attraction forces.

(v) A *para* ring N-atom has a very pronounced base-weakening effect, possibly because of the low contribution of any $—\overset{+}{N}—$ form to the resonance hybrid of the protonated form.

(vi) A moderate base-strengthening effect appears in compounds with two N-atoms in *peri*-positions (1,8-diazanaphthalenes), possibly because H-bonded structures tend to stabilize the protonated structure.

(vii) Substituent groups exert their expected inductive and mesomeric effects.

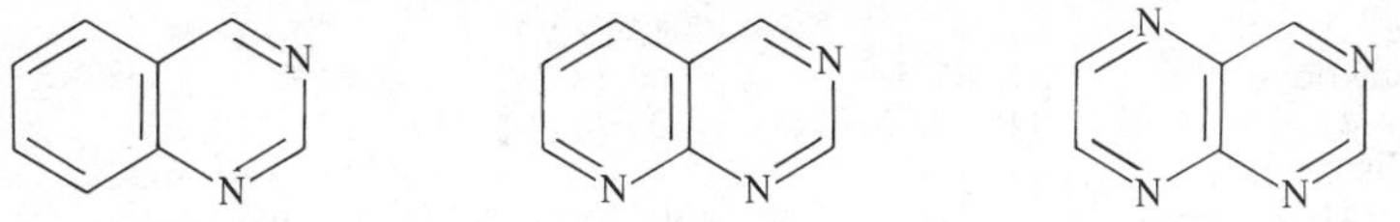

3.51 3.85 4.12

Apparent pK_a values (hydration involved)

SCHEME 3

On the above basis the pK_a values recorded in Scheme 2 may be readily explained. In certain other compounds, however, pK_a values obtained by ordinary potentiometric titration or by spectrophotometric determination appear abnormal. Thus all three values in Scheme 3 are much higher than expected and that of pteridine especially so in view of the value for pyrido[2,3-*b*]pyrazine recorded in Scheme 2. The explanation is that the pK_a values, as determined, are actually composite values, arising from equilibrium between stable hydrated cations and a stable neutral anhydrous species.[63] When pK_a values are determined by rapid titration techniques, the values for the anhydrous bases are found to be ~1.5 for quinazoline and ~2.0 for pteridine and values for all three covalent hydrates decrease in the expected manner (Scheme 4) with increasing ring N-atoms.

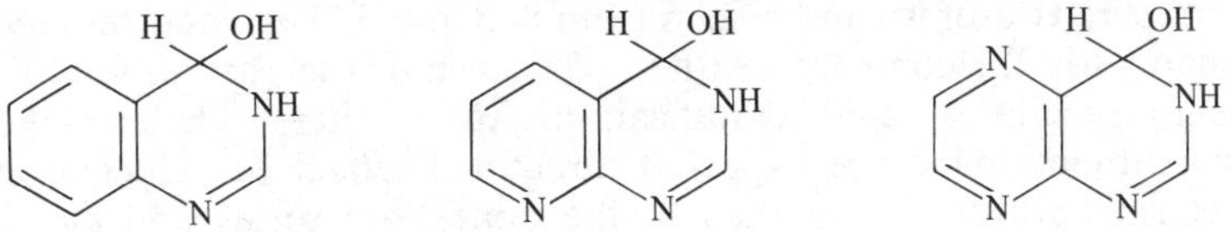

7.77 6.56 4.79

pK_a Values of covalent hydrates

SCHEME 4

16.7.3.2 Spectral properties used in structural assignments

Ultraviolet spectra of all the common di- and tri-azanaphthalenes and many aza-phenanthrenes and -anthracenes have been determined and are well documented in the literature. Theoretical studies are reasonably well correlated in this field and the review by Mason[64] gives an excellent comprehensive treatment. In general terms the lowest energy $\pi \rightarrow \pi^*$ bands of the azanaphthalenes do not vary greatly in position, but the $n \rightarrow \pi^*$ bands move progressively to longer wavelengths with increasing aza-substitution. Successive annulations shift the $\pi \rightarrow \pi^*$ bands of three azines to lower frequencies approximately twice as fast as the $n \rightarrow \pi^*$ bands and the latter absorptions tend to become submerged in the strong $\pi \rightarrow \pi^*$ absorptions (Table 1). In *ortho*-diazines the atomic lone-pair orbitals on the adjacent N atoms overlap, giving rise to a bonding and an anti-bonding lone-pair molecular orbital. As a consequence, *ortho*-diazines invariably absorb at longer wavelengths than isomeric diazines. The results of these effects and the decreasing fine-structure observed in increasing aza-substituted compounds are well illustrated in Mason's review and in a useful chapter by Albert on the subject.[65]

TABLE 1

Ultraviolet Spectra of Some Azanaphthalenes

Compound	$n \rightarrow \pi^*$ $\lambda_{max}(\log \varepsilon)$	$\pi \rightarrow \pi^*$ $\lambda_{max}(\log \varepsilon)$	$\lambda_{max}(\log \varepsilon)$	$\lambda_{max}(\log \varepsilon)$
Quinoline		313 (3.73)	270 (3.59)	225 (4.48)
1,8-Naphthyridine	330 (2.60)	308 (3.76)	254 (3.61)	
Pyrido[3,2-*d*]-pyrimidine	345 (2.07)	305 (3.80)	251 (3.54)	
Pyrido[2,3-*d*]-pyrimidine		304 (3.64)	244 (3.48)	
Pteridine	387 (1.92)	301 (3.87)	235 (3.46)	210 (4.04)

Experimentally determined values have been used in structural assignments, studies of covalent hydration, and tautomerism studies of naphthyridines,[1] pyridopyrimidines,[2] pteridines,[3] pyrimido[5,4-*e*]-*asym*-triazines, and acridines,[4] but, as in other fields, other spectroscopic techniques are now more generally used.

Infrared spectroscopy has been of particular value in distinguishing between oxo- and hydroxy-tautomers and typical examples are seen in naphthyridines, pyridopyrimidines, and pteridines. Armarego has presented a comprehensive study of the i.r. spectra of the six 1,*x*-naphthyridines, the four pyridopyrimidines, the two pyridopyrazines, and pyrazino[2,3-*b*]pyrazine.[66] In many cases correlations with particular vibrational modes can be made by comparison with naphthalene, and a similar exercise by Mason[67] gives the picture for pteridine and 16 monosubstituted derivatives.

^{1}H n.m.r. spectra have been of considerable value for proof of cyclization reactions, isomer differentiation, and determination of positions of substitution in very many polyaza-heteroaromatic systems. The powerful deshielding effect of a ring N-atom on both *ortho*- and *peri*-H-atoms can be seen from Scheme 5. This, and the invariably larger coupling constants for $J_{5,6}$ compared with $J_{6,7}$ because of the partial bond fixation in 5,6-positions, usually facilitates rapid identification. As in other π-deficient heteroaromatic series, ring substituents have the expected directional effect but electron-donating substituents affect ring protons more than in the benzenoid series and electron-attracting

Chemical shifts for ring protons in di-, tri-, and tetra-azanaphthalenes

SCHEME 5

substituents less. In fused pyrimidine ring systems generally, a signal from H-4 undergoes a greater downfield shift than one from H-2 when the solvent polarity is increased (*e.g.* $CDCl_3 \rightarrow Me_2CO \rightarrow$ DMSO).[2,68,69]

Typical of the value of n.m.r. spectra is the use to distinguish between pyrido[3,2-*d*]pyrimidines and isomeric N-bridgehead pyrimido[1,2-*a*]pyrimidines,[70] and the demonstration that 7-phenacyl- and 6-phenacyl-isoxanthopterins exist as their methine tautomers.[71] ^1H n.m.r. spectroscopy has been used to examine deuterium exchange of both methyl[72] and ring[73] protons in lumazine derivatives, and the nature of covalent adducts is invariably proved by this type of spectroscopy.[74,75]

^{13}C n.m.r. spectra have been used less frequently for structural assignments, but here again the potential applicability is considerable. Thus the extent, position, and proportion of pteridine covalent hydrates is readily found by ^{13}C n.m.r. spectroscopy.[76]

The mass spectra of very many polyaza-heteroaromatic compounds are now routinely recorded in the literature and those of naphthyridines, several benzonaphthyridines and pyridopyrimidines, pyrido[2,3-*d*]pyridazines, quinoxalino[2,3-*d*]pyridazines, pteridines, and pyrimido[5,4-*e*]-*asym*-triazines have been described in some detail and fragmentation pathways proposed.

In general these ring systems are less stable the greater the number of ring N atoms and of exocyclic substituents. The parent compounds, alkyl, or aryl derivatives are relatively stable with prominent molecular ions and peaks visible for $m/2e$ ions. In these cases HCN or RCN loss is generally the most important fragmentation, followed by further HCN or RCN and C_2H_2 loss for fused benzenes, pyridines, or pyrazines. Pteridine itself fragments by successive losses of HCN and deuterium labelling has shown that 75% of the HCN comes from C^4N^3 and 19% from C^2N^1.[77] Fused 1,2,4-triazines and benzo[*c*]cinnolines show initial N_2 loss, but in fused pyridazines bond fixation inhibits N_2 loss and HCN loss dominates.[78] Fragmentation of 5,7-disubstituted pyrimido[5,4-*e*]-*asym*-triazines is dominated by consecutive losses of N_2 and HCN from the triazine ring, and comparison of the spectra with those of the corresponding pteridines showed that the fused triazine ring fragmented much more readily than the fused pyrazine.[78]

The mass spectra of quinazolin-4(3*H*)-ones, pyridopyrimidin-4(3*H*)-ones, and pteridin-4(3*H*)-ones are closely similar, with strong molecular ions and important losses of CO followed by HCN; fused pyrimidine-2,4(1*H*, 3*H*)-diones show loss of HNCO.[2]

X-ray diffraction methods are seldom used for proof of structure but examples do occur for certain naphthyridines,[1] a pyrimido[4,5-*d*]pyridazine,[79] a pyrimido[5,4-*d*]-pyrimidine,[80] a pyrido[2,3-*d*]pyridazine,[81] and a new flavin compound, roseoflavin.[82] A recent redetermination of the crystal structure of pteridine by X-ray diffraction studies has led to significant shifts in the observed bond distances and angles from previously recorded values,[83] and the present molecular geometry is more in accord with results of molecular orbital calculations.

16.7.4 CHEMICAL REACTIONS

16.7.4.1 Electrophilic substitution reactions

Electrophilic substitution reactions are normally understood to be those reactions involving attack of an electrophile at a ring-C atom with subsequent replacement of an attached H atom or other group with a new substituent. As such they are difficult reactions and uncommon in all fused π-deficient heteroaromatic ring systems. In a broader sense, however, we can also regard as an electrophilic substitution the replacement of a ring-N lone-pair of electrons by a new grouping as a result of attack by an electrophile. Some account has already been given of proton attack (Section 16.7.3.1) and the principles given there are also commonly applicable to quaternization[84] and *N*-oxide formation,[85] both of which topics have been well-covered in review articles. Also feasible with fused pyridine-ring systems are complex formation at the ring-N atom with halogens,

Lewis acids, and metals. The conversion of N—H to N—alkyl on treatment of 2- or 4-oxo-derivatives with alkyl halides is a common substitution reaction referred to in Section 16.7.4.6.

Electrophilic substitution reactions at ring-C atoms become increasingly difficult with increasing numbers of ring-N atoms, so it is not surprising that few such reactions are reported in the literature. An unusual *N*-methylation which results in the conversion of an isoalloxazine (**23**) to an alloxazine (**24**) is claimed to occur because the N-1 position of isoalloxazines is a hard nucleophilic centre.[86]

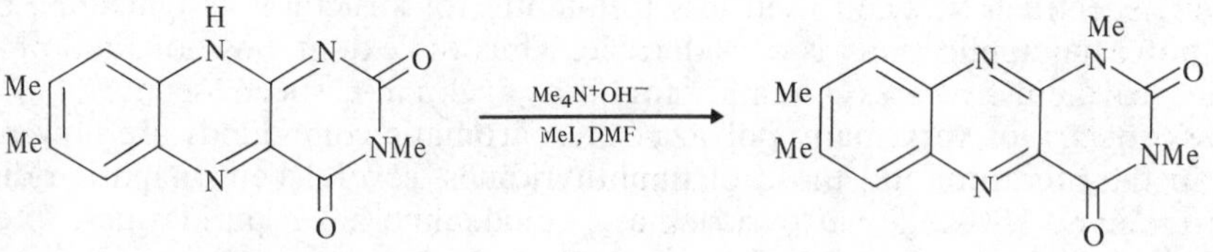

$Me_4N^+OH^-$, MeI, DMF

(23) (24)

The Kress method (treatment of hydrohalides with bromine in nitrobenzene) has been shown to be superior to the Eisch method (formation of a heterocycle–bromine complex which is decomposed by pyridine) for the bromination of both 1,7- and 1,8-naphthyridine.[87] In the former case, 3- and 5-bromo- and 3,5-dibromo-derivatives are produced, and in the latter, 3-bromo- and 3,6-dibromo-derivatives are formed, in agreement with theoretical calculations of ground-state π-electron distribution. Electrophilic substitution reactions of triazanaphthalenes are extremely rare, but one report[88] describes the successful 3-bromination of pyrido[2,3-*d*]pyridazine as a result of pyrolysis of the heterocycle–bromine adduct at 170 °C. Electron-donating substituents will obviously facilitate these reactions, and rare examples of nitration are found in the conversion of 1,5-naphthyridin-2- and -4-ones and 1,6-naphthyridin-2- and -5-ones to 3-nitro-derivatives.[1]

16.7.4.2 Nucleophilic substitution reactions

An extremely comprehensive review of the reactivity of azines with nucleophiles was produced by Shepherd and Fedrick in 1965.[89] Within this review the relative reactivity at different ring-positions of all the major polyaza-bicyclic compounds is deduced from known kinetic, experimental, and theoretical data, and comparisons made with published work in this area. By application of the principles given by these authors, the ease of nucleophilic substitution can be postulated for any position of any polyaza-heteroaromatic compound and the effect of substituents, cationization of ring N-atom, or variation of reaction conditions are all considered. In addition, the various possible mechanisms are reported and discussed.

In practical terms, nucleophilic substitution reactions are second only to ring syntheses in importance as a technique for the synthesis of specific compounds. By far the most important type of substitution reaction, mentioned annually in well over half of the publications in this area, is that of the displacement of a chloro, or less commonly bromo, alkoxy, or thio, substituent by a primary, secondary, or tertiary amine. Many of the derivatives thus produced are of considerable biological importance and reference is made (in Section 16.7.6) to the importance of piperidinyl, piperazinyl, and morpholino-derivatives which are invariably produced by treatment of chloro-compounds with the appropriate heterocyclic base. A less common amination procedure is the use of phosphoramines to convert oxo- or hydroxy-derivatives directly into amines without the usual intervention of the chloro-substituent. Thus 4-hydroxypteridines give moderate yields of 4-amino-derivatives.[90]

Another unusual substitution is that of the first reported replacement of a trifluoroacetyl group by fluorine on treatment with BF_3–Et_2O,[91] and methylation in dimethyl sulphoxide[92] has been shown to involve initial attack by $CH_3SO_2CH_2^-$ ions.

By the correct choice of nucleophile and a knowledge of the order of reactivity of the various ring positions, selective replacement of identical substituents can be achieved. The treatment of 2,4-dibromo-1,8-naphthyridine with toluene-*p*-sulphonylhydrazine followed by sodium carbonate, for example, yields 4-bromo-1,8-naphthyridine,[93] and the 10-chloro-group remains unaffected on treatment of 2,4,10-trichloropyrimido[5,4-*b*]-quinoline with sodium methoxide, amines, or sodium hydrosulphide.[94] An increasing number of ring-N atoms renders these aza-heterocycles liable to ring-opening rather than substitution reactions, and covalent hydration or similar nucleophilic addition reactions are very common in fused pyrimidines and pyrazines (see later). An example of the complexity produced by these competing reactions is the formation of the pteridine dimer (**25**), which results from carbanion formation from one molecule of a 6,7-dimethyl-pteridine followed by nucleophilic attack and covalent hydration of another molecule.[95]

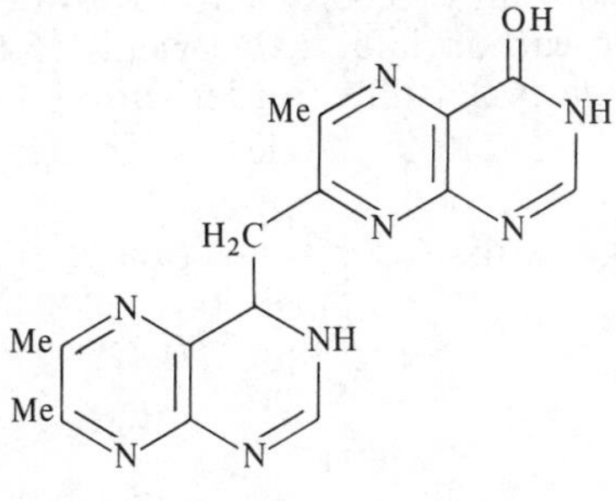

(25)

As in the case of monocyclic azines, the action of $POCl_3$ on oxides can yield chloro-deoxygenated derivatives and the site of chlorination may be the adjacent ring. Thus the 1-*N*-oxide of pyrido[2,3-*d*]pyrimidin-2,4(1*H*,3*H*)-dione, which exists in the tautomeric cyclic hydroxamic acid form (**26**), yields 6-chloropyrido[2,3-*d*]pyrimidin-2,4(1*H*,3*H*)-dione (**29**), presumably by the mechanism shown.[96] A thorough investigation of mechanisms of nucleophilic substitution reactions will probably reveal increasing numbers of reactions not following the conventionally accepted S_N(Ar2) routes. For example, it has recently been shown that aminations of 8-chloro-1,7- and 2-chloro-1,8-naphthyridines occur by $S_N(AE_n)$ and $S_N(ANROC)$ mechanisms.[97]

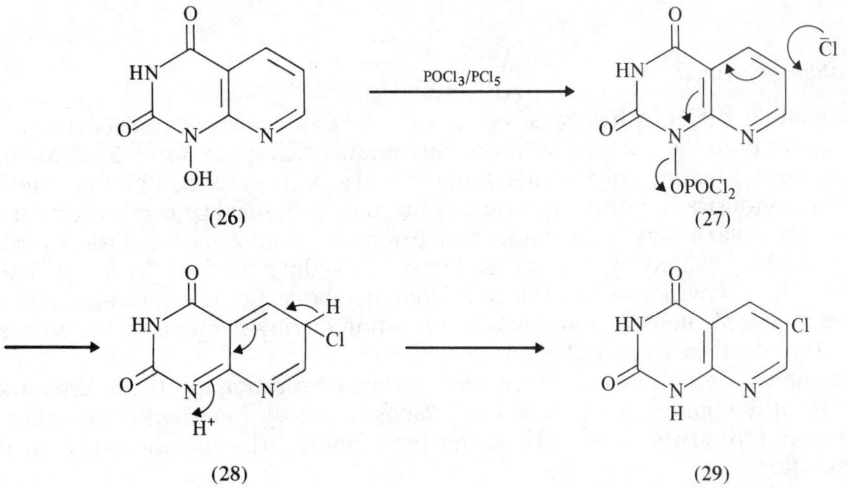

$POCl_3/PCl_5$

(26) (27)

(28) (29)

16.7.4.3 Addition reactions

The initial adduct formed in the nucleophilic substitution reactions just discussed (16.7.4.2) may, in the case of di- or tri-cyclic ring systems, be quite stable at normal temperatures and pressures. Best known of such adducts are the covalent hydrates, but an increasing number of polyaza-heteroaromatic compounds are now being shown to be susceptible to attack by a variety of nucleophiles. Factors which determine the position and ease of the reaction are the nature of the nucleophile, the positions of lowest electron density, and the nature of the adjacent ring or substituent groups capable of stabilizing the adduct structure.

(i) Hydrogenations

The pteridine ring system is one which may be attacked at either the pyrimidine or the pyrazine ring. Thus folic acid is hydrogenated in the pyrazine ring by $NaBH_4$, and the method can be adapted to yield 5-alkyl-5,6,7,8-tetrahydro-derivatives; catalytic reduction of the pteridin-7(8*H*)-one-6-carboxylate (**31**) yields 5,6-dihydro-compounds (**30**), whereas sodium borohydride gives 3,4-dihydro-derivatives (**32**).[98]

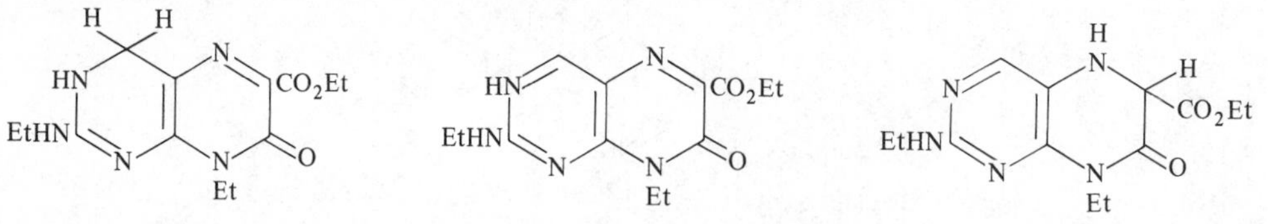

(30) (31) (32)

If two electron-donating groups are present in the pyrimidine ring, however, as in the case of pterins and lumazines, attack either by, borohydride or by catalytic hydrogenation yields 7,8-dihydro- or easily oxidizable 5,6,7,8-tetrahydro-derivatives.[99]

In a similar manner, the reduction of either the pyridine or the pyrazine ring in a series of pyrido[2,3-*b*]pyrazines was determined by the nature of substituent groups.[100] Electrochemical reduction of the same ring system gave 1,4-dihydro-derivatives which were too unstable to be isolated and which were transposed into 1,2- and 3,4-dihydro- and 5,6,7,8-tetrahydro-derivatives. Substituent groups can of course be affected and in the latter example a 3-bromo-substituent was lost in reductions under alkaline conditions. Similarly, the reduction of 9-formylacridines with $LiAlH_4$ yielded 9-hydroxymethylacridanes.[101]

Certain pyrido[3,4-*d*]pyrimidin-4(3*H*)-ones are similarly reduced at both the exocyclic group and the ring 1,2-C=N group to yield 1,2,3,4-tetrahydro-derivatives.[102]

(ii) Covalent hydration

This is another subject which was well reviewed in *Advances in Heterocyclic Chemistry* in 1965, where both the qualitative[103] and quantitative[104] aspects were treated. Since that date many other examples of the phenomenon have been described and new techniques evolved for demonstration of its occurrence, but the theoretical principles remain unchallenged. In particular it has been shown that resonance stabilization is just as essential as electron deficiency caused by extra ring-N atoms in stabilizing these hydrates. This is the reason why fused pyrimidines (**33**) by virtue of the amidine-type resonance are so susceptible to the phenomenon and why the position of ring-N atoms in the adjacent ring (**34**) can also affect the extent of hydration.

New examples of ring systems undergoing covalent hydration are the internal diquaternary 1,8-naphthyridinium salt (**35**) at the 1,2-position, 5-trifluoromethylpyrimido[5,4-*e*]-*asym*-triazines (**36**) at the 5,6-position, and pyrazino[2,3-*b*]quinoxalines (**37**) at the 1,2- and 3,4-positions.

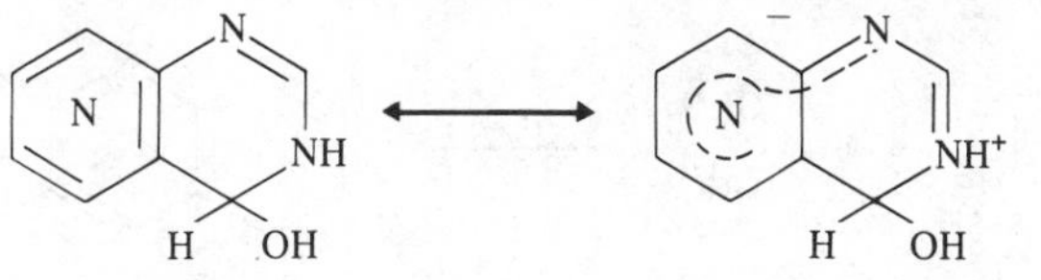

(33) (34)

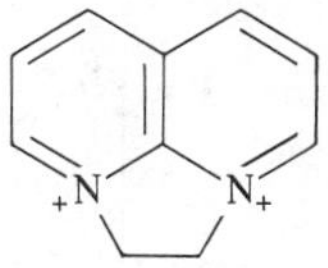

(35) 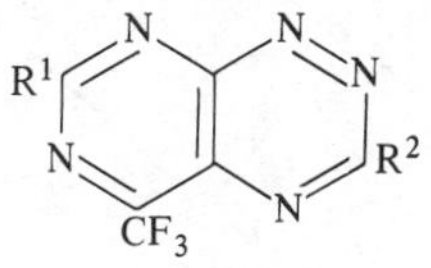 (36)

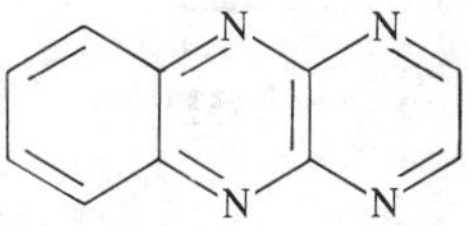

(37)

In addition to u.v. spectra, ionization constants, chemical oxidation, and ^{1}H n.m.r. spectroscopy, ^{13}C n.m.r. spectroscopy and mass spectrometry have more recently been used to determine the position and extent of covalent hydration.

(*iii*) *Other addition reactions which do not modify the basic ring structure*

Certain other nucleophiles can form covalent adducts, as does water, with similar diazines and the same factors affect ease of formation and stability. For example, pyrimido[5,4-*e*]-*asym*-triazines readily add amines and alcohols across the 5,6-bond, and transetherification of 5-alkoxy-derivatives of the same ring system by treatment with silver oxide in a boiling alcohol[105] is similarly suggested to involve covalent alcoholates. This ability to form amine adducts has been used in unusual preparations of 5-alkylamino-pyrimido[5,4-*e*]-*asym*-triazines from either unsubstituted or 5-alkyl derivatives by treatment with amines and one or two aerial oxidation steps.

Sodium hydrogen sulphite, dimedone, barbituric acid, and thiobarbituric acid all add to the 5,6- and 7,8-positions of pteridin-4-one, which is unaffected by a weaker nucleophile, water. Addition of sodium hydrogen sulphite also occurs across 5,6- and 7,8-bonds of pteridine-2,4-dione (lumazine) and 4-aminopteridine and across the 7,8-double bond of 2-aminopteridine-4,6-dione (xanthopterin).

Sodium hydrogen sulphite is a powerful nucleophile, reacting even with pyridine. The additional stabilization of fused rings is seen by the fact that quinoline is more reactive than pyridine, and acridine more reactive than quinoline, to this reagent and in the latter case sodium 9,10-dihydroacridine-9-sulphonate is produced from aqueous solutions at room temperature.

Carbanion attack occurs at the 3,4-bond in both pyrido-[3,2-*d*]- and -[3,4-*d*]-pyrimidines. The dicovalent hydrate (**38**) of pyrazino[2,3-*b*]quinoxaline (**40**) is converted into a series of analogous alcoholates (**39**) on treatment with the appropriate alcohol in the presence of toluene-*p*-sulphonic acid.[106] The disproportionation of pyrimido[4,5-*b*]-quinolinium salts (**41**) at pH 9.0[107] into dihydro- (**42**) and dihydro-oxo-derivatives (**43**) has been shown by deuterium-labelling studies to involve H-transfer from the 5-position of one molecule as its pseudo-base anion to the 5-position of another molecule in its cation form.

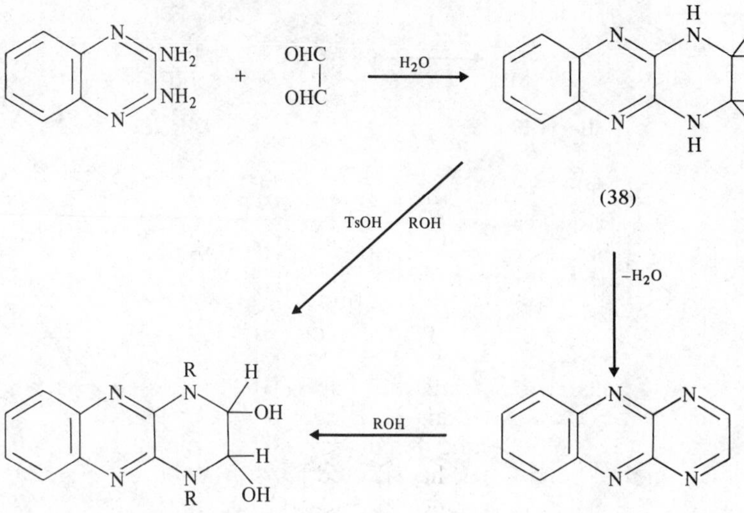

(38)

(39)

(40)

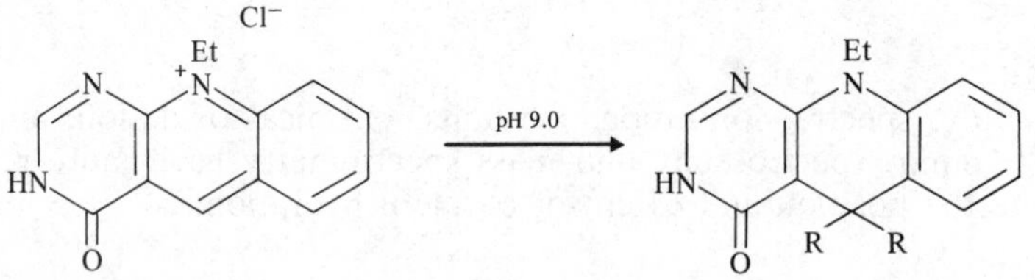

(41)

(42) R = H

(43) R,R = O

16.7.4.4 Ring-modifying reactions

Heteroaromatic rings with more than one N atom are relatively easily opened and the product may either recyclize to the same ring or to a new ring system or retain its open-chain structure. The first category of ring–chain isomerization is more common in five-membered and N-bridgehead compounds, but the tautomerism of protonated covalent hydrates (*e.g.* **44**) and open-chain aldehydo-amidines (*e.g.* **45**) falls into this category.

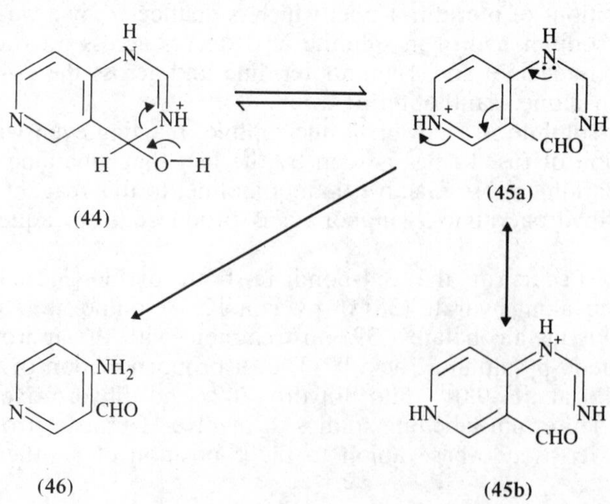

(44)

(45a)

(46)

(45b)

The Dimroth reaction is not too common in this type of compound, but the simplest example in the pteridine series involves the conversion of the anion of 2-amino-3-methylpteridin-4(3*H*)-one into 2-methylaminopteridin-4(3*H*)-one and the rate constants are in agreement with formation of the intermediate pyrazine.

(*i*) *Ring-modifying reactions which yield ring-opened products*

The ease with which aza-heteroaromatic compounds ring-open is dependent on the strength of the ring-opening reagent (generally a nucleophile), the number and position of ring-N atoms in the ring under consideration, and the presence of adjacent rings or substituents which stabilize ring-opened products. A good example is pyrido[4,3-*d*]pyrimidine, which decomposes in 4 minutes at pH 2 to yield 4-aminonicotinaldehyde (**46**). This is the least stable of the pyridopyrimidines because of the position of the pyridine ring-N atom which stabilizes the open-chain amidine (**45**).[108]

Albert has discussed the ring-opening reactions of the simpler six-membered aza heteroaromatic compounds,[103] and Clark has studied more complicated systems, particularly the pteridines and other fused pyrimidines and pyrazines of similar structure.[109]

Typical recent examples of reagents used to effect ring-opening are sodium hydroxide, ammonia, amines, hydroxylamines, methoxyamine, hydrazine and substituted hydrazines, sodium borohydride, lithium aluminium hydride, resorcinol, and β-naphthol.

Fused pyrimidines, under most circumstances, are the least stable of the mono-, di-, and tri-azines and although electron-donating substituents render ring-opening less easy, fused pyrimidin-4(3*H*)-ones and -2,4-(1*H*,3*H*)-diones are still unstable, particularly to powerful nucleophiles such as hydrazine.

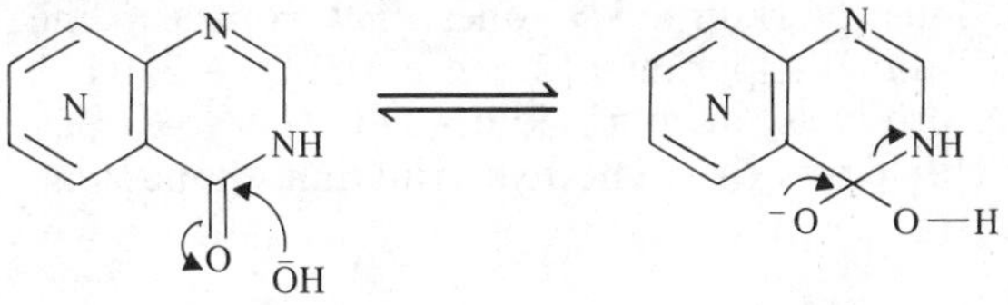

(47) (48)

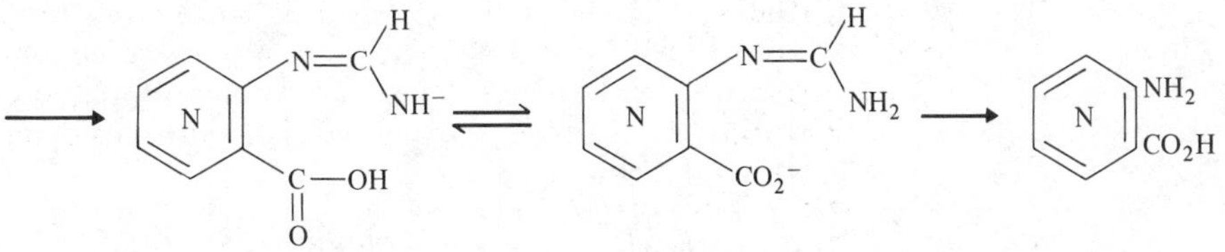

(49)

The mechanism for the ring-opening of such fused pyrimidin-4(3*H*)-ones (**47**) under alkaline conditions is presumably as indicated, (**48**) $\rightarrow$ (**49**), although a cleavage at the 1,2-position is also possible. The ease of ring-opening of pyrido-[2,3-*d*]- and -[4,3-*d*]-pyrimidine-4(3*H*)-thiones (*e.g.* **50**) and of 4-mercaptopteridine[110] on treatment with chloroacetic acid and sodium carbonate has been shown to occur at the 2,3-bond and to be in part due to the ease of departure of the 4-carboxymethylthiolate anion and in part to the stabilization of the initial anion (**51**), formed by attack of the base at position-2 by the suitable positioning of other ring-N atoms (**52**). Similar ring-opening at the 2,3-position, presumably initiated by attack at the 2-position, occurs in the treatment of pyrido[2,3-*d*]pyrimidine 3-oxide with acetic anhydride or alkali hydroxide. The very mild conditions required to convert 3-arylpyrido[3,4-*d*]pyrimidin-4(3*H*)-ones into 3-aminopyridines on treatment with $LiAlH_4$ is claimed to be due to the stabilization of an intermediate anion by delocalization of the charge from N to O or an aryl ring.[102]

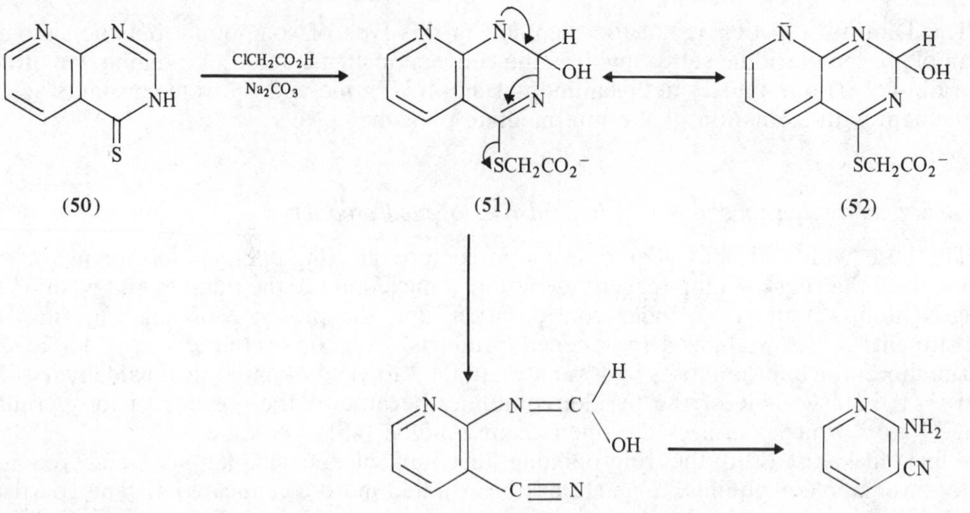

(50) (51) (52)

Nucleophiles containing an amino-group next to an electronegative atom are useful in facilitating the less-common ring-opening of pteridin-4(3*H*)-ones in the pyrazine ring, and intermediates are postulated such as (**53**) in which the 6-position is subjected to attack by a further molecule of the nucleophile.[111] Similar treatment of pyrimido[5,4-*e*]-*asym*-triazin-5(6*H*)-ones (**54**) with amines or hydrazines yields 6-amino-*asym*-triazine-5-carboxamides and -5-carbohydrazines (**56**), whereas hydroxylamine yielded 6-hydroxy-derivatives (**57**) presumably *via* ring-cleavage (**55**) and recyclization.[112] No products were formed resulting from cleavage of the triazine ring, but reaction conditions were milder than those for analogous pteridin-4(3*H*)-ones. By contrast, the reaction conditions required to cleave 1,3-dimethylpyrimido[4,5-*d*]pyrimidine-2,4(1*H*,3*H*)-diones (**58**) are more vigorous.[113] The products are pyrimidine-2,4(1*H*,3*H*)-diones (**59**) even with hydrazine ($R=NH_2$), which cleaves the dimethylpyrimidine dione ring of the lumazines.

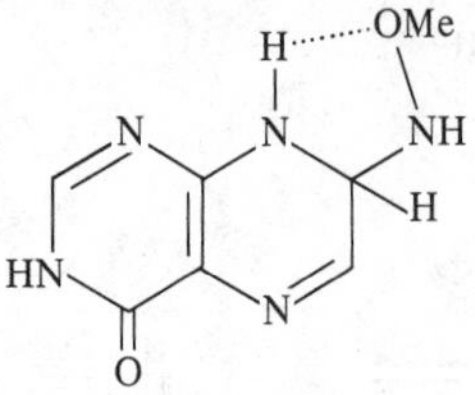

(53)

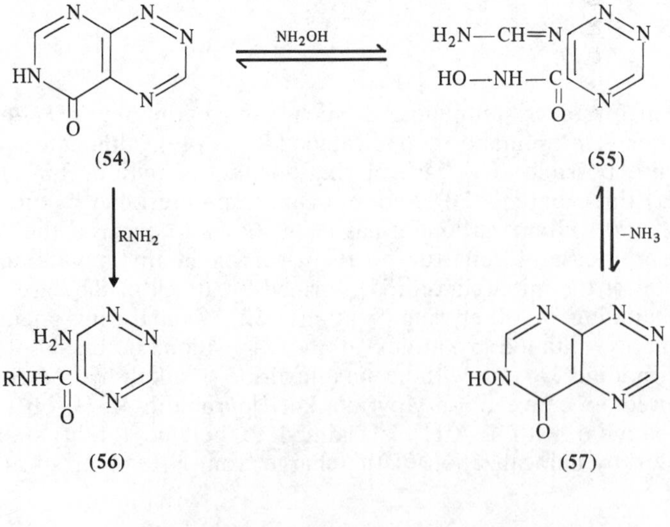

(54) (55)

(56) (57)

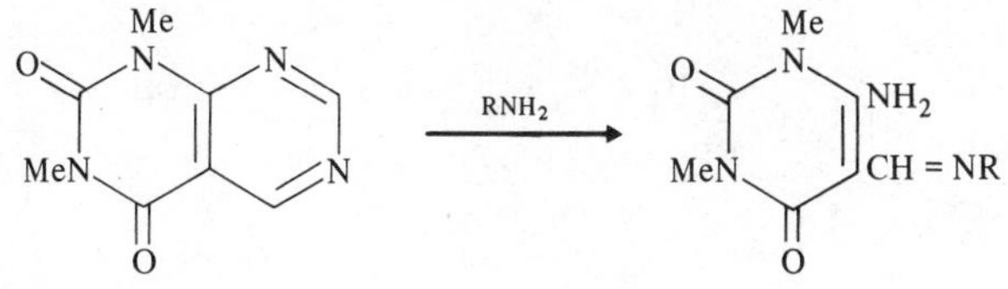

(58) (59)

(*ii*) *Ring-modifying reactions which yield a new ring system*

Van der Plas has recently reviewed[114] the ring transformations of heterocycles and has indicated their very considerable extent. Within this series, however, this type of reaction has been little explored and some of the relatively few examples have been unexpected. Thus treatment of pyrido[3,2-*c*]pyridazine with Zn–Hg in acetic acid yielded pyrrolo[3,2-*b*]pyridine,[115] and two groups of workers have shown that pyrido[2,3-*d*]pyrimidine 3-oxide can be converted into 1,8-naphthyridines on treatment with certain carbanions;[115,116] a reactive ring-contraction of pyrimido[5,4-*e*]-*asym*-triazin-4(3*H*)-one to 9-aminohypoxanthine is effected by sodium hydrosulphite in acetic acid. Not surprisingly, the pteridines have been investigated in most detail, in many cases for the purpose of understanding biologically occurring ring transformations. Thus it has been shown that lumazines can be converted into riboflavines by a chemical synthesis which closely simulates the known enzymatic route,[117] and hydrolysis of 10-alkyl-3-methyl-isoalloxazines (flavins) (**60**) and their 5-oxides with benzyltrimethylammonium hydroxide in dimethylformamide in the dark yielded the corresponding spirohydantoins (**62**), whereas the same reaction under irradiation brought about elimination of the 10-alkyl group.[118] A ring-contraction of a pteridine to a purine occurs on treatment of a 6-methylpteridine 5-oxide with acetic anhydride,[114] and a rearrangement of a pteridine (**63**) to a pyrazino[1,2-*a*]-*sym*-triazine (**64**) on autoxidation. A series of bicyclic pyrimidine derivatives with fused benzene, pyridine, or another pyrimidine ring underwent ring-opening with subsequent closure to yield fused pyridine derivatives on treatment with an active methylene reagent such as malononitrile or dimedone and, in the latter case, tricyclic structures resulted.

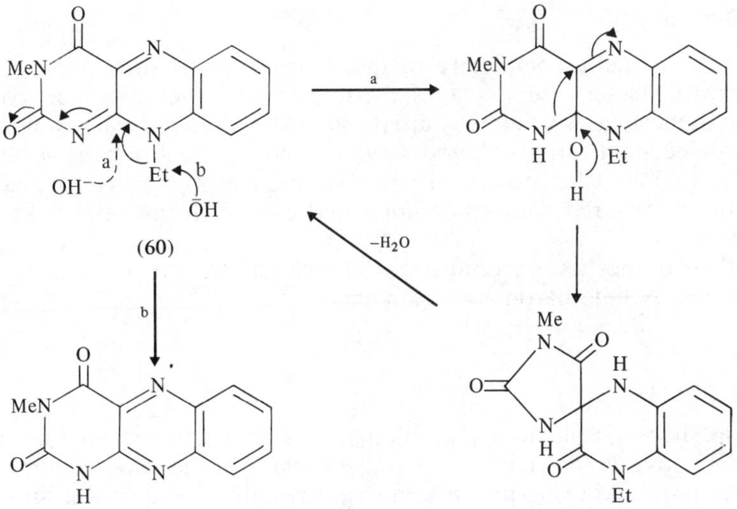

(60)

(61) (62)

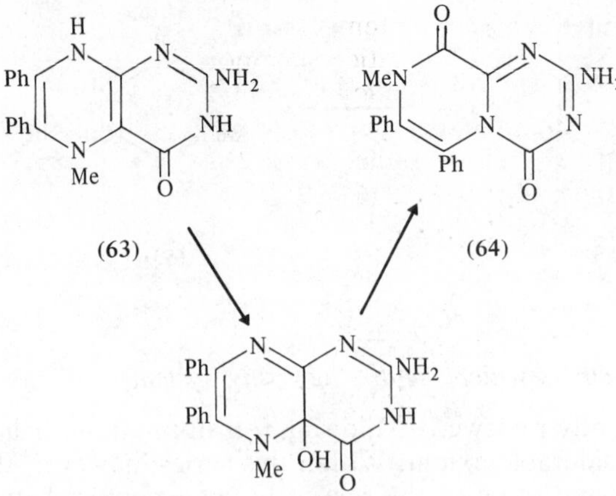

(63)

(64)

16.7.4.5 Photochemical reactions

Photochemically induced reactions are rarely used in six-membered polyaza-heteroaromatic compounds. A dealkylation of 10-alkylisoalloxazines initiated by such irradiation has already been mentioned[118] and a photoreduction of pteridin-7(8*H*)-one to the 5,6-dihydro-derivative occurs on irradiation on moist cellulose. Exposure of solutions of acridine to sunlight yields 9,10-dihydroacridine in dilute, and tetrahydro-9,9′-biacridyl in concentrated, solutions. A ring-contraction to pyrrolopyridines occurs on treating amino-derivatives of 1,6-, 1,7-, and 1,8-naphthyridines with nitrous acid in the presence of u.v. light.[1]

16.7.4.6 Properties of derivatives

The properties of substituted derivatives are in close agreement with those expected from a knowledge of the monocyclic analogues. Ring annulation often enhances an existing effect, as does the introduction of additional ring-N-atoms at positions *ortho* and *para* to substituent groups.

(*i*) *Hydroxy-derivatives*

These can exist virtually completely in the hydroxy-form when present in a fused benzene ring or, to a lesser extent, when *meta* to a ring-N atom, or almost completely in the amide form in most cases when they are positioned α or γ to a ring-N atom. Thus the proportion of amide : enol form for pyrid-4-one, quinol-4-one and acrid-4-one increases from $2.2\times10^3:1$ to $2.4\times10^4:1$ to $1.0\times10^7:1$. Where 2- or 4-one structures are possible, the latter are often favoured and both anionic and cationic forms tend to exist substantially in Kekulé forms.

In general, their properties are predictable and closely similar to those of the simpler hydroxy-monocyclic N-heteroaromatic compounds.

(*ii*) *Amino-compounds*

These exist as the amino-tautomer no matter what ring position they occupy; 2- and 4-substituted derivatives appreciably increase the basicity of the parent compounds owing to the resonance possible in ring-protonated structures, but the amino-groups themselves are very weakly basic. As in the case with monocyclic analogues, 2- and 4-substituted amino-compounds are difficult to diazotize and readily yield hydroxy-compounds on treatment with nitrous acid at a moderate temperature. The introduction of both hydroxy- and amino-groups renders N-heteroaromatic rings more susceptible to electrophilic substitution and less susceptible to nucleophilic substitution or ring-opening reactions.

An α-amino substituent offers the possibility of annulation to N-bridgehead compounds by reaction with a bifunctional electrophile. Thus 2-amino-1,8-naphthyridines (**65**) can yield an imidazo[1,2-*a*][1,8]naphthyridine (**66**).[119]

(65) $\xrightarrow{BrCH_2COR^3}$ (66)

(*iii*) *Halogeno-derivatives*

These are important as intermediates in the conversion of hydroxy-compounds to other substituted derivatives. Their preparations, therefore, have been studied with the aim of obtaining optimum yields. For many chlorodehydroxylations, phosphoryl chloride alone is adequate, but for the less reactive compounds conversion is only effected when phosphorus pentachloride or tertiary amines are added.

(*iv*) *Alkyl and aryl derivatives*

These tend to confer stability on N-heteroaromatic ring systems. Methyl groups α or γ to a ring-N atom show the type of activity well exemplified by α- and γ-methylpyridines and quinolines, with additional enhancement when more than one ring-N atom is involved. Thus they react with halogens or halogenosuccinimide to form halogenomethyl derivatives, with selenium dioxide to form aldehydes, with $KMnO_4$ to form acids, with aldehydes to form styryl derivatives and, when more than one ring-N atom is present, with esters in Claisen-type condensation reactions. Where methyl groups are positioned α to a sufficiently basic ring-N atom, Tschitschibabin-type reactions are possible with bromocarbonyl compounds to form N-bridgehead pyrrolo[1,2-*a*] ring systems.

(*v*) *Carboxylic acids*

These, when positioned α or γ to a ring-N atom readily lose CO_2 on heating. Suitably positioned they may be involved in intramolecular attack at nucleophilic centres such as a ring N atom, an exocyclic amino group or even a reactive aromatic ring C atom.

(*vi*) *Other substituted derivatives*

Electron-attracting substituents such as nitro, sulphonic and carboxylic acids, and trifluoromethyl groups are difficult to introduce other than by ring-closure methods and when present will render their attached rings even more susceptible to nucleophilic addition and substitution reactions. Azido-groups situated *ortho* to a ring-N atom often allow the possibility of ring–chain tautomerism, *e.g.* interconvertible forms of 3-azidopyrido[2,3-*e*]-*asym*-triazine (**67**) and pyrido[2,3-*e*]tetrazolo[5,1-*c*]-*asym*-triazine (**68**) have been isolated.[120]

N-Oxides are potentially very valuable derivatives, but their preparation and reactions remain largely an unexplored subject in these ring systems, although one systematic study

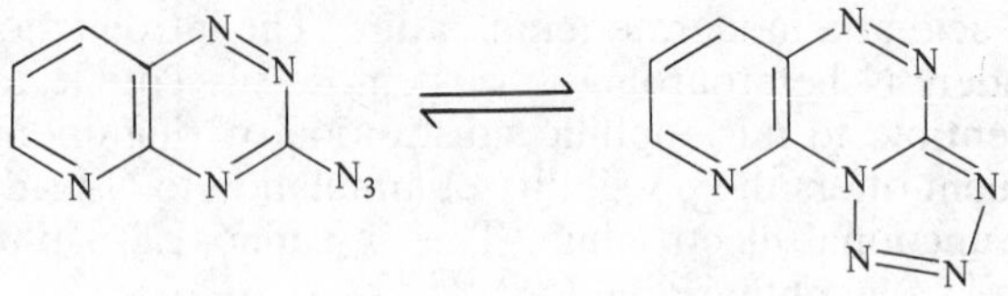

(67) (68)

of the synthesis of naphthyridine *N*-oxide has appeared.[121] Typical of the possible scope which they offer is the rearrangement of the pterin 8-oxide (**69**) to the trifluoroacetyl derivative (**70**) of xanthopterin in a convenient five-step synthesis of the latter from aliphatic starting materials.[122] An interesting investigation of the position of *N*-oxidation in the pteridines has demonstrated that orientation is linked to the nature of an N-1 substituent, and that the conversion of a pteridine into its *N*-5-oxide results in a decrease of basicity of 2 pK_a units whereas conversion into a *N*-8-oxide results in a decrease of 1 pK_a unit.[123]

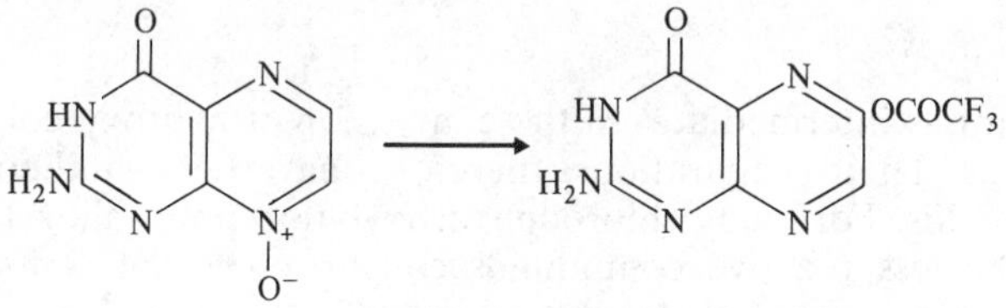

(69) (70)

16.7.5 NATURAL PRODUCTS

The pteridine ring system is present in many essential plant and animal compounds and a very considerable literature covers the chemical, biological and physical properties of compounds such as folic acid and its hydrogenated derivatives, and the pteridine pigments leucopterin, xanthopterin, and isoxanthopterin. Alloxazine and the vitamin riboflavin are benzopteridines with a similar wide coverage in the literature. Much of the recent work in this field has been concerned with unravelling the details of biochemical pathways involving these compounds and the mechanisms of enzymatic formation of riboflavins and the pteridines themselves. Two natural products with the pyrimido[5,4-*e*]triazine ring system are the poisonous antibiotics toxoflavin and fervenulin and several workers have attempted structural modifications of these two compounds in the hope of retaining the anti-bacterial activity but reducing the toxicity. A number of minor alkaloids possess naphthyridine, benzonaphthyridine, or other diaza-anthracene and -phenanthrene structures, often as their hydrogenated derivatives.

16.7.6 SYNTHETIC PRODUCTS OF TECHNOLOGICAL IMPORTANCE

The stimulus for the synthesis of many of these polyaza-heteroaromatic compounds has been the search for new compounds of therapeutic value[12,13] and biological properties are recorded in most of the patent literature in this field. Drugs currently marketed containing these ring systems include the potent gram-negative antibacterial agent nalidixic acid (**71**), the powerful diuretic triamterene (**72**), one of the best of all anti-cancer drugs, methotrexate (**73**), the platelet inhibitor dipyridamole (**74**), the well-known acridine antibacterials proflavine and acriflavine, the analeptic tacrine, the useful anti-malarial mepacrine, and the phenanthridine trypanocide, ethidium.

The large number of publications which refer to biological activity makes a critical summary difficult, particularly since much of the patent literature reports even moderate

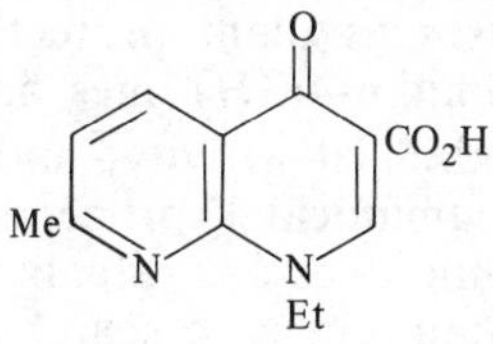

(71)

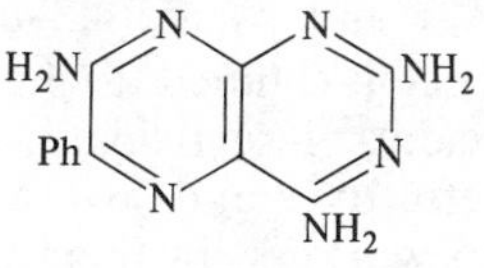

(72)

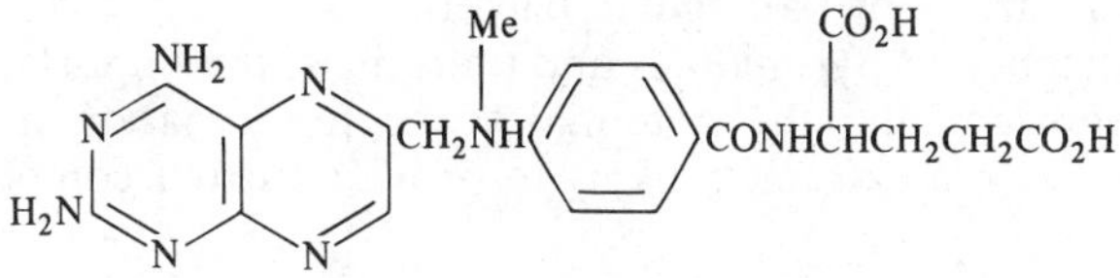

(73)

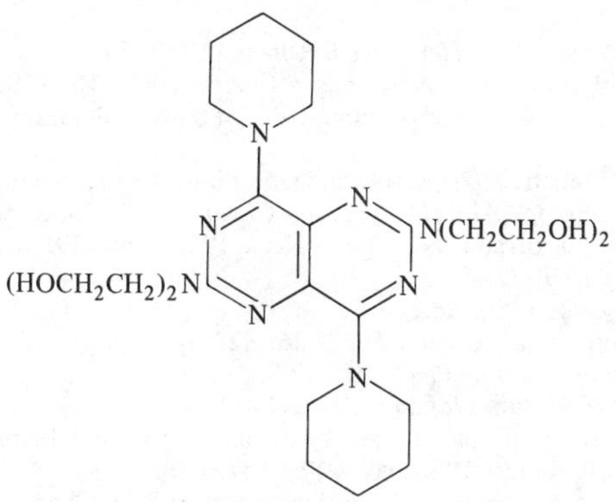

(74)

activity, to ensure protection for a series of compounds. More critical animal and, ultimately, clinical trials reduce to a very small percentage the number of compounds likely to be valuable as drugs. Nevertheless, the repeated publication of activity in a particular ring system in the last five or six years is as clear an indication as we have of potential success. In this category, anti-bacterial activity has been reported on many occasions for 1,8-naphthyridines other than nalidixic acid, for pyrido[2,3-*d*]pyrimidines and less frequently for pteridines other than the marketed ones, and for pyrimido[5,4-*d*]-*vic*-triazines. A large number of patents, particularly from Japan, record anti-inflammatory, analgesic, and CNS depressant properties for pyrido[2,3-*d*]pyrimidines. Diuretic activity is common in many tri- and tetra-azanaphthalenes and Japanese workers have determined structure–activity relationships. Pyrimido[4,5-*d*]-, pyridazino[3,4-*d*]-, and pyrido[3,4-*d*]-pyridazines and certain of their dihydro-derivatives were active.

Morpholino-groups disubstituted in the pyridazine ring, phenyl groups on the other hetero-ring, and alkyl or alkoxy-groups at the 8-position of pyrido[3,4-*d*]pyridazines enhanced the activity. Diuretic activity has also been claimed by other workers for pteridines other than triamterene, particularly those containing amino and amide substituents and for similarly substituted pyrido[2,3-*d*]- and pyrido[3,4-*d*]-pyrimidines and 1,8-naphthyridines.

A number of pyrimido[5,4-*d*]pyrimidines other than dipyridamole have also shown coronary dilating or broncho-dilatory effects, and occasional claims for anti-cancer activity have been made for pteridines other than folic acid analogues, for some pyrimido[5,4-*e*]-*asym*-triazines, and acridine *N*-oxides.

Occasional reference is found to possible uses as plant protection agents: pyrido-[2,3-*d*]-, pyrido[3,4-*d*]-, and pyrido[4,3-*d*]-pyrimidin-4(3*H*)-ones and pyrimido[4,5-*d*]-pyridinones have all been claimed to have herbicidal activity; and pyrido[3,2-*d*]-*vic*-triazines show acaricidal, insecticidal, and nematocidal properties. A number of naphthyridines are useful dyes and several acridine dyes are valuable fluorescent stains.

Hitchings and his co-workers, in their very comprehensive search for the competitive inhibitors of folic acid containing a fused 2,4-diaminopyrimidine, showed that some pyrido[3,2-*d*]pyrimidines and a very large number of pyrido[2,3-*d*]pyrimidines were highly active against a variety of pathogenic bacteria.[2]

The biological importance of pyrimidines and pteridines, the important drugs and active compounds so far reported, and the potential scope for variation in ring structure all combine to make this an area extremely likely to provide further compounds of biological importance in the future.

References

1. W. W. Paudler and T. J. Kress, *Adv. Heterocyclic Chem.*, 1970, **11,** 123.
2. W. J. Irwin and D. G. Wibberley, *Adv. Heterocyclic Chem.*, 1969, **10,** 149.
3. 'Proceedings of the 3rd International Symposium, 1962', ed. W. Pfleiderer and E. C. Taylor, Pergamon, Oxford, 1964.
4. R. M. Acheson, in 'The Chemistry of Heterocyclic Compounds', Interscience, New York, 1973, vol. 9.
5. B. R. T. Keene and P. Tissington, *Adv. Heterocyclic Chem.*, 1971, **13,** 315.
6. M. V. Rubtsor and E. S. Nikitskaya, *Russ. Chem. Rev.*, 1965, **34,** 439.
7. J. Pomorski, *Wiad. Chem.*, 1970, **24,** 773.
8. F. Yoneda, *Kayaka No Ryoiki*, 1970, **24,** 1077.
9. W. Sliwa and J. Mlochowski, *Wiad. Chem.*, 1972, **26,** 823.
10. N. A. Polyakova *et al.*, *Vitam.-Vitam. Prep.*, 1973, 70.
11. S. Yurugi *et al.*, *Chem. Pharm. Bull.* (*Japan*), 1976, **24,** 2057.
12. D. G. Wibberley, in 'MTP International Review of Science, Organic Chemistry Series One, Heterocyclic Compounds', Butterworths, London, 1973, vol. 4, p. 167.
13. D. G. Wibberley, in 'International Review of Science, Organic Chemistry Series Two, Heterocyclic Compounds', Butterworths, London, 1975, vol. 4, p. 181.
14. L. Achremowicz and J. Mlochowski, *Roczniki Chem.*, 1973, **47,** 1383; J. Pomorski, *Roczniki Chem.*, 1974, **48,** 321.
15. T. Shono and J. Matsumoto, *Chem. Pharm. Bull.* (*Japan*), 1971, **19,** 1426; H. Ogura and M. Sakaguchi, *Chem. Pharm. Bull.* (*Japan*), 1973, **21,** 2014.
16. A. Albert, 'The Acridines', Arnold, London, 1966.
17. Allen, 'Six-membered Heterocyclic Nitrogen Compounds with Three Condensed Rings', Interscience, New York and London, 1958.
18. D. H. Kim and A. A. Santilli, *J. Heterocyclic Chem.*, 1975, **12,** 181.
19. E. M. Levine and T. J. Bardos, *J. Heterocyclic Chem.*, 1972, **9,** 91.
20. T. Koyana *et al.*, *Yakugaka Zasshi*, 1973, **93,** 330.
21. A. I. Mikhalev and M. E. Konshia, *Khim. geterotsikl. Soedinenii*, 1974, **7,** 968.
22. Y. Maki *et al.*, *Tetrahedron Letters*, 1971, 4271.
23. R. A. Henry *et al.*, *J. Heterocyclic Chem.*, 1976, **13,** 375.
24. W. J. Irwin and D. G. Wibberley, *J. Chem. Soc.* (*C*), 1967, 1745.
25. K. A. Roseman *et al.*, *J. Medicin. Chem.*, 1970, **13,** 230.
26. I. Hermecz *et al.*, *J.C.S. Perkin I*, 1977, 789.
27. S. Carboni *et al.*, *Gazzetta*, 1967, **97,** 1274; 1968, **98,** 1174; 1969, **99,** 677.
28. E. M. Hawes and D. G. Wibberley, *J. Chem. Soc.* (*C*), 1966, 315.
29. T. Sugimoto and S. Matsuura, *Bull. Chem. Soc.* (*Japan*), 1975, **48,** 725.
30. E. Gy. T. Gyogyszervegyeszeti Gyar, *Ger. Offen.*, 1973, 2 237 074.
31. Y. S. Andreichikov *et al.*, *Khim. geterotsikl. Soedinenii*, 1974, **12,** 1690.
32. A. Dornow and D. Wille, *Chem. Ber.*, 1965, **98,** 1505.
33. A. G. Ismail and D. G. Wibberley, *J. Chem. Soc.* (*C*), 1967, 2613.
34. J. B. Bapat, D. St. C. Black, and R. F. C. Brown, *Adv. Heterocyclic Chem.*, 1970, **11,** 199.
35. C. Temple *et al.*, *J. Org. Chem.*, 1971, **36,** 2974; K. J. M. Andrews and B. P. Tong (Hoffman-La Roche), *U.S. Pat.*, 1974, 3 813 393.
36. D. J. Brown and R. K. Lynn, *Austral. J. Chem.*, 1974, **27,** 1781.
37. J. Weinstock, *U.S. Pat.*, 1970, 3 505 329.
38. S. Yurugi and S. Kikuchi, *Ger. Offen.*, 1971, 2 046 577.
39. L. Matsuara, *Jap. Pat.*, 1970, 7 033 185.

40. T. G. Koksharova *et al.*, *Khim. geterotsikl. Soedinenii*, 1972, **2,** 274.
41. A. Godard *et al.*, *Bull. Soc. chim. France*, 1972, **4,** 1588.
42. M. Chaykovsky *et al.*, *J. Heterocyclic Chem.*, 1973, **10,** 425.
43. F. Alhaique *et al.*, *Tetrahedron Letters*, 1975, 173.
44. J. Mayer and M. H. Sherlock, *U.S. Pat.*, 1973, 3 928 367.
45. E. C. Taylor and B. E. Evans. *J. Chem. Soc.* (*C*), 1971, 189; R. D. Youssefyeh and A. Kalums, *Chem. Comm.*, 1970, 1371.
46. W. W. Parish and H. S. Broadbent, *J. Heterocyclic Chem.*, 1971, **8,** 527.
47. F. Yoneda *et al.*, *J. Heterocyclic Chem.*, 1974, **11,** 83.
48. F. Sakuma *et al.*, *Chem. Pharm. Bull.* (*Japan*), 1976, **24,** 338.
49. F. S. Mikhailitsyn *et al.*, *Khim. geterotsikl. Soedinenii*, 1975, **12,** 1660.
50. E. M. Hawes and D. K. J. Gorecki, *J. Heterocyclic Chem.*, 1972, **9,** 703.
51. D. Harrison and A. C. B. Smith, *J. Chem. Soc.*, 1960, 2157.
52. T. Nakagome *et al.*, *Ger. Offen.*, 1973, 2 166 355.
53. A. Lewis and R. G. Shepherd, *J. Heterocyclic Chem.*, 1971, **8,** 47.
54. F. Yoneda *et al.*, *J. Heterocyclic Chem.*, 1970, **7,** 1443.
55. A. Lewis and R. G. Shepherd, *J. Heterocyclic Chem.*, 1971, **8,** 41.
56. D. J. Brown and T. Sugimoto, *Austral. J. Chem.*, 1971, **24,** 633.
57. A. Albert and K. Ohla, *J. Chem. Soc.* (*C*), 1970, 1540.
58. M. Pesson *et al.*, *Compt. rend.*, 1974, **278,** 717.
59. G. Adembri *et al.*, *J. Chem. Soc.* (*C*), 1970, 1536.
60. N. D. Heindel *et al.*, *J. Heterocyclic Chem.*, 1971, **8,** 1047.
61. S. Kessar *et al.*, *Tetrahedron*, 1973, **29,** 167, 177, 419.
62. A. Albert and W. Pendergast, *J.C.S. Perkin I*, 1973, 794; A. Albert and H. Mizuno, *ibid.*, 1973, 1615.
63. A. Albert, 'Heterocyclic Chemistry', Wiley Interscience, 2nd edition, New York, 1967, p. 127.
64. S. F. Mason, in 'Physical Methods in Heterocyclic Chemistry', ed. A. R. Katritzky, Academic Press, New York, 1963, vol. II, chapter 7.
65. A. Albert, 'Heterocyclic Chemistry', Wiley Interscience, 2nd edn., New York, 1967, chapter 11.
66. W. L. F. Armarego *et al.*, *Spectrochim. Acta*, 1964, **22,** 117.
67. S. F. Mason, *J. Chem. Soc.*, 2336.
68. W. L. F. Armarego and T. J. Batterham, *J. Chem. Soc.* (*B*), 1966, 750.
69. R. M. White, in 'Physical Methods in Heterocyclic Chemistry', Academic Press, New York, 1963, vol. II, chapter 9.
70. B. S. Hurlbert and B. F. Valenti, *J. Medicin. Chem.*, 1968, **11,** 708.
71. Y. Iwanami and T. Seki, *Bull. Chem. Soc. Japan*, 1972, **45,** 2829.
72. J. M. McCandless and R. Stewart, *Canad. J. Chem.*, 1970, **48,** 263.
73. R. L. Beach and G. W. E. Plant, *Biochemistry*, 1970, **9,** 760.
74. J. Clark and F. S. Yates, *J. Chem. Soc.* (*C*), 1971, 2475.
75. D. J. Brown and T. Sugimoto, *J. Chem. Soc.* (*C*), 1971, 2616.
76. U. Ewers *et al.*, *Chem. Biol. Pteridines, Proc. 5th Internat. Symp.*, 1975, 687.
77. T. Goto *et al.*, *J. Org. Chem.*, 1965, **30,** 1844.
78. J. Clark, *Org. Mass Spectrometry*, 1972, **6,** 467.
79. K. Shinozaki *et al.*, *Takeda Kenkyasho Ho*, 1972, **31,** 432.
80. H. M. Berman *et al.*, *Tetrahedron Letters*, 1973, 3099.
81. S. H. Simonsen *et al.*, *J. Heterocyclic Chem.*, 1975, **12,** 1329.
82. R. Miura *et al.*, *J.C.S. Chem. Comm.*, 1973, 703.
83. C. D. Shirrell and D. E. Williams, *J.C.S. Perkin II*, 1975, 40.
84. G. F. Duffin, *Adv. Heterocyclic Chem.*, 1964, **3,** 2.
85. A. R. Katritzky and J. M. Lagowski, 'Chemistry of the Heterocyclic N-oxides', Academic Press, London, 1971.
86. L. Hevesi, *Tetrahedron Letters*, 1976, 1389.
87. H. C. van der Plas and M. Wozniak, *J. Heterocyclic Chem.*, 1976, **13,** 961.
88. D. B. Paul and H. J. Rodda, *Austral. J. Chem.*, 1969, **22,** 1745.
89. R. G. Shepherd and J. L. Fedrick, *Adv. Heterocyclic Chem.*, 1964, **3,** 150.
90. G. R. Gapski and J. M. Whiteley, *Chem. Biol. Pteridines, Proc. 5th Internat. Symp.*, 1975, 627.
91. T. Novinson *et al.*, *Tetrahedron Letters*, 1973, 3149.
92. G. Kovacs *et al.*, *Hung. Teljes.*, 1971, 1161.
93. W. Czuba and M. Wozniak, *Synthesis*, 1974, 809.
94. N. E. Britikova *et al.*, *Khim. geterotsikl. Soedinenii*, 1974, **4,** 554.
95. A. Albert and H. Mizuno, *J.C.S. Perkin I*, 1973, 1974.
96. Tzoong-Chyh Lee and G. Salemnick, *J. Org. Chem.*, 1975, **40,** 3608.
97. H. C. Van der Plas *et al.*, *Tetrahedron Letters*, 1976, 2087.
98. E. C. Taylor *et al.*, *J. Org. Chem.*, 1971, **36,** 4012.
99. W. Pfleiderer, *Annalen*, 1971, 747.
100. N. Vinot and P. Maitti, *Bull. Soc. chim. France*, 1973, 3100.
101. C. Tashiro, *Japan Kokai*, 1973, 73, 52 776.
102. I. R. Gelling and D. G. Wibberley, *J. Chem. Soc.* (*C*), 1971, 780.

103. A. Albert and W. L. F. Armarego, *Adv. Heterocyclic Chem.*, 1964, **4,** 1.
104. D. D. Perrin, *Adv. Heterocyclic Chem.*, 1964, **4,** 43.
105. D. J. Brown and T. Sagimoto, *J. Chem. Soc.* (*C*), 1970, 2661.
106. A. N. Hughes and D. Kleemola, *J. Heterocyclic Chem.*, 1976, **13,** 13.
107. J. Clark and B. Parvisi, *J.C.S. Perkin I*, 1967, 131.
108. W. L. F. Armarego, *J. Chem. Soc.*, 1962, 4094.
109. J. Clark, *Lect. Heterocyclic Chem.*, 1976, **3,** 79 (*Chem. Abstr.*, 1976, **85,** 159 928).
110. E. C. Taylor *et al.*, *J. Amer. Chem. Soc.*, 1960, **82,** 6058.
111. J. Clark and G. Neath, *J. Chem. Soc.* (*C*), 1969, 1297.
112. J. Clark and C. Smith, *J. Chem. Soc.* (*C*), 1972, 247.
113. J. Clark and M. S. Morton, *J. Chem. Soc.* (*C*), 1974, 1812.
114. H. C. Van der Plas, 'Ring Transformations of Heterocycles', Academic Press, London and New York, 1973.
115. V. G. Kartsev *et al.*, *Khim. geterotsikl. Soedinenii*, 1975, **10,** 1431.
116. T. Higashimo and E. Hayashi, *Chem. Pharm. Bull.* (*Japan*), 1973, **21,** 2643.
117. T. Paterson and H. C. S. Wood, *J.C.S. Perkin I*, 1972, 1051.
118. F. Yoneda *et al.*, *J.C.S. Chem. Comm.*, 1977, 175.
119. J. F. Harper and D. G. Wibberley, *J. Chem. Soc.* (*C*), 1971, 2985.
120. A. Messmer *et al.*, *J. Heterocyclic Chem.*, 1973, **10,** 575.
121. W. W. Paudler *et al.*, *J. Heterocyclic Chem.*, 1972, **9,** 1001.
122. E. C. Taylor and P. A. Jacobi, *J. Amer. Chem. Soc.*, 1973, **95,** 4455.
123. W. Pfleiderer and W. Hutzenlaub, *Chem. Ber.*, 1973, **106,** 3149.

PART 17

NITROGEN SYSTEMS: THE AZOLES AND OTHER SYSTEMS

17.1
Pyrroles

A. H. JACKSON
University College, Cardiff

17.1.1 INTRODUCTION

The five-membered ring system of pyrrole (**1**) is one of the most ubiquitous throughout the plant and animal kingdoms because of its involvement as a sub-unit of haem and the chlorophylls; the biosynthetically related vitamin B_{12} is also a tetrapyrrole, as are the animal and plant bile pigments. Monopyrrolic natural products include porphobilinogen, the precursor of all natural pyrrole pigments based on the porphyrin or corrin nuclei, and a number of antibiotics, *etc.*, including the tripyrrolic prodigiosins which have an entirely different biosynthetic origin. The indole ring system, in which a pyrrole ring is fused to a benzene ring, is also widespread in Nature and is found in the amino-acid tryptophan, many alkaloids, and indigo.

Pyrrole itself was first obtained by Runge in 1834, and isolated in pure form some 20 years later by the dry distillation of horn and hoof, and was later synthesized by heating the ammonium salt of mucic acid (see below). Its constitution and structural formula (**1**) were established soon afterwards. However, some of the greatest contributions to our knowledge of pyrroles, and especially their syntheses, were made by Hans Fischer and his school at Munich in the course of their studies of the structures and synthesis of haem and chlorophyll in the period from 1910 to 1944; much of this work was summarized in the well-known book by Fischer and Orth[1] in 1934 and in the companion volumes[2,3] on the natural pigments in 1937; more recent general reviews of pyrrole chemistry have been given by Corwin,[4] Stevens,[5] Baltazzi and Krimen,[6] and in a book by Schofield;[7] two monographs by Gossauer[8] and by Jones and Bean[9] are the most complete and exhaustive accounts yet to be published.

17.1.2 STRUCTURE AND PHYSICAL PROPERTIES[10a]

Pyrrole is a colourless liquid, b.p. 129 °C, with an odour somewhat reminiscent of chloroform, which slowly turns brown on exposure to air. It is slightly hygroscopic, and is slightly miscible in water, but freely soluble in most organic solvents. Its structure (**1**) was first suggested by Baeyer in 1870, on the basis of chromic acid oxidation to maleimide (**2**), and its formation by zinc dust distillation of succinimide (**3**).

β' β α' α N H   4 3 5 2 N 1 H

(1)

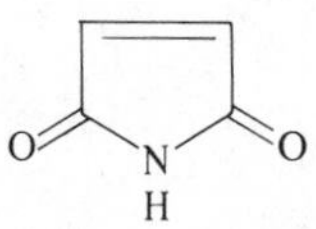

O N H O   O N H O   N H H   H H N

(2) 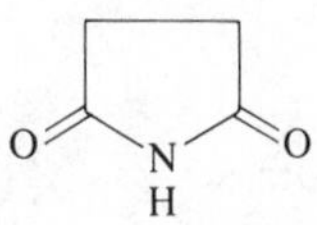 (3) 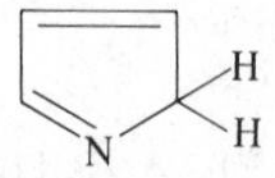 (4) 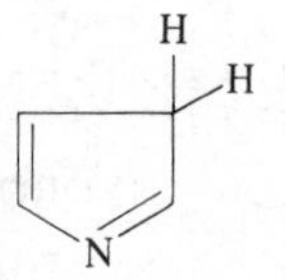 (5)

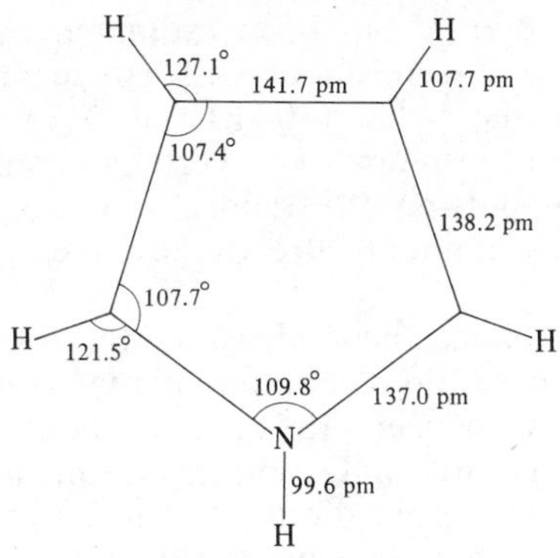

FIGURE 1

The accepted IUPAC numbering system for pyrrole is shown in structure (**1**), together with the older nomenclature in which the carbon atoms adjacent to nitrogen were referred to as α and α' and the other two carbons as β and β'. In the earlier literature the α- and β-*pyrrolenine* tautomers (**4**) and (**5**) of pyrrole were often invoked to explain the course of certain reactions; however, modern physical methods have failed to show their existence, although the protonated forms clearly exist in acid solution and derivatives with two substituents on the saturated carbon atoms can be prepared (see below).

Pyrrole itself is completely planar and the bond lengths and bond angles shown in Figure 1 were obtained originally by electron diffraction and later confirmed by microwave spectroscopy. These data are in accord with the recognition that pyrrole is an aromatic compound having a monocyclic system of 6π-electrons (two from each of the double bonds and two from the lone pair on nitrogen), *i.e.* in agreement with the Hückel $(4n+2)$ rule. The α—β bonds are slightly longer, the β—β bond is slightly shorter, and the N—α bond considerably shorter, than the corresponding bonds in cyclopentadiene (134, 147 and 151 pm, respectively).

Calculations of the electronic structure of pyrrole have been carried out by a variety of molecular orbital methods, *e.g.* simple HMO, MINDO, PPP-SCF, SCF-LCAO, CNDO/2, and LCGTO-SCF, and these have been summarized by Gossauer.[8a] The calculated π-electron densities on nitrogen are in the 1.6–1.8 range, but for both α- and β-carbons are 1.0–1.1. The calculated π-bond orders for the N—C, C_α—C_β, and C_β—$C_{\beta'}$ bonds are in the ranges 0.37–0.49, 0.75–0.87, and 0.45–0.6, respectively. The electronic structure of the pyrrolate anion (formed by removal of the proton from nitrogen with strong base) has also been calculated. The data obtained also agree well with the experimental findings that in basic media pyrrole has a considerable tendency to undergo alkylation, acylation, or Michael reactions on nitrogen, whereas in neutral, or acidic, media, substitution on carbon is the normal rule.

The alternative valence-bond description of the pyrrole molecule suggests that it can be regarded as a resonance hybrid of the five mesomeric forms (**6a–e**). The uncharged form (**6a**) is the major contributor and the two forms (**6d,e**), with a negative charge on the β or β' carbon atoms, are least important. Similar resonance structures may be written for other-five-membered heterocyclic aromatics such as thiophen and furan, and in all cases only one uncharged form can be written, in contrast to the two for benzene.

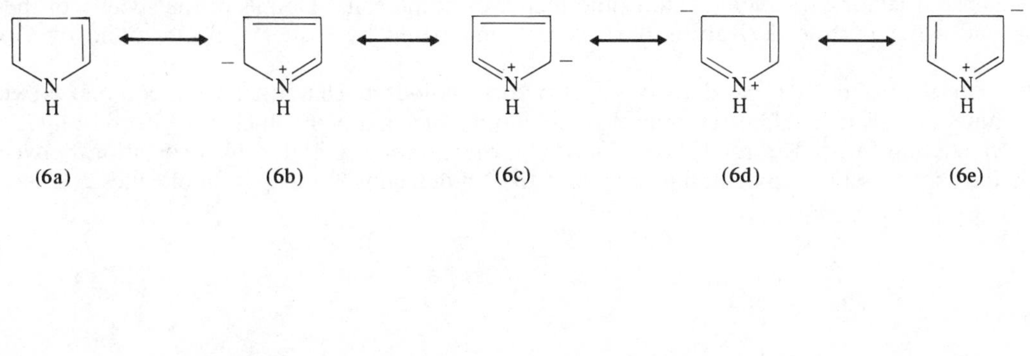

(**6a**) (**6b**) (**6c**) (**6d**) (**6e**)

The resonance energy of pyrrole has been estimated by several authors, and values ranging from 62 to 129 kJ mol^{-1} were obtained.[8b] The mean value of about 100 kJ mol^{-1} is in good agreement with the value (103 kJ mol^{-1}) calculated by the valence-bond method,[10b] and although quite substantial, is nevertheless only two-thirds that of benzene. Other physical evidence supporting the delocalized aromatic structure of pyrrole is provided by bond length measurements already described (Figure 1), and also by dipole moment measurements.

The dipole moment of pyrrole has been measured by several different methods, and the precise value obtained depends on the experimental conditions;[8c] thus in non-polar solvents such as cyclohexane or benzene the value is about 1.80 D at 25 °C. On the other hand, in the neat liquid it is about 1.55 D, and in solvents to which it can hydrogen-bond it may be considerably higher, *e.g.* in dioxan about 2.1 D, and in triethylamine, about 3.0 D. However, the dipole is always headed into the ring away from nitrogen, whereas in the saturated analogue, pyrrolidine, the dipole (1.57 D) points towards the nitrogen; this evidence clearly indicates partial delocalization of the lone pair of electrons on nitrogen over the ring system as a whole. Wheland and Pauling's early estimate[11] of the dipole moment based on MO π-electron density calculations was 1.96 D; since then many other workers have also attempted to calculate the dipole moment of pyrrole using a variety of sophisticated MO methods.[8d] The dipole moments of a wide range of substitute pyrroles have also been determined experimentally and the results have been summarized by Gossauer.[8e]

17.1.3 SPECTRA

17.1.3.1 Electronic spectra

The longest wavelength absorption bond of pyrrole in the ultraviolet region is at 207.5 nm (log ε 3.88) in cyclohexane, and 208 nm (log ε 3.86) in 95% ethanol. It is not clear whether this is due to an n–π^* transition; calculations have been carried out by a variety of MO methods.[8f] Alkyl substituents, whether on nitrogen or carbon, generally give rise to a small hypsochromic shift (*ca.* 3–8 nm) as anticipated. On the other hand, negative substituents on carbon such as acyl, alkoxycarbonyl, carboxyl, and cyano effect a pronounced bathochromic shift to 260–300 nm, with a corresponding increase in intensity ($\varepsilon > 10\,000$); furthermore, the α-substituted pyrroles usually absorb at some 15–20 nm to longer wavelengths than the corresponding β-substituted pyrroles because the latter are effectively cross-conjugated (see structures **7a** and **7b**). Aryl substitution also results in

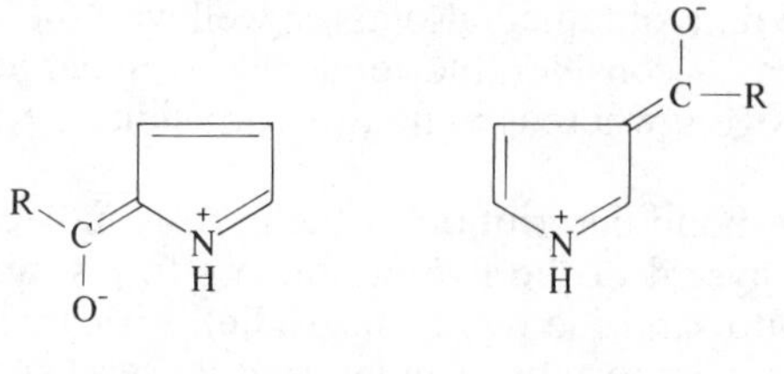

(7a) (7b)

substantial long wavelength shifts and increases in intensity. Details of the spectra of these and other pyrrole derivatives have been summarized by Eisner,[12] Scott,[13] and by Gossauer.[8]

Pyrrole and its alkyl derivatives also form coloured charge-transfer complexes with quinones, picric acid, tetracyanoethylene, iodine, *etc.*, many of which have been isolated in crystalline form. The ready formation of these complexes is due to association between the 'π-excessive'[14] pyrrole nucleus, and the 'π-deficient'[14] acceptor molecules.

17.1.3.2 Infrared spectra

A detailed discussion of the infrared spectra of pyrrole and its simple derivatives is given by Gossauer[8g] and the most interesting feature is the N—H absorption. The N—H stretching frequency assigned to the non-associated pyrrole molecule is 3496 cm^{-1} (in CCl_4), the intensity of which is higher than that of the N—H bond in saturated secondary amines. There is a clear relationship between the acidity of the N—H hydrogen atom and the i.r. stretching frequency, and the effects of substituents at both the α- and β-positions has been shown to be approximately additive. Intramolecular hydrogen bonding occurs[15,16] between the NH group and α-carbonyl or α-alkoxycarbonyl groups, as shown in structures (**8a**) and (**8b**) and the N—H stretching frequency may be lowered by as much as 45 cm^{-1}. Evidence has also been obtained for intermolecular hydrogen bonding between the NH and carbonyl groups. Likewise, the carbonyl frequencies are also lowered slightly (20–30 cm^{-1}) by intramolecular hydrogen bonding. The carbonyl stretching frequencies also clearly show the effects of conjugation with the pyrrole nucleus (see structures **7a** and **7b**), the α-carbonyl frequencies (*e.g.* 1732–1710 cm^{-1} in esters) being some 20 cm^{-1} lower than the corresponding β-carbonyl frequencies (*e.g.* 1711–1701 cm^{-1} in esters).

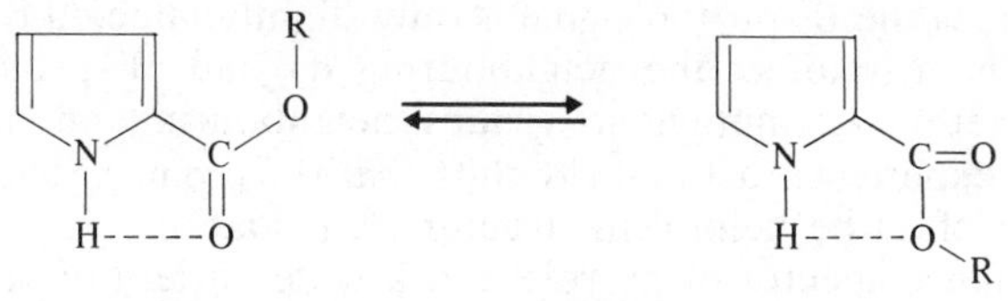

(8a) (8b)

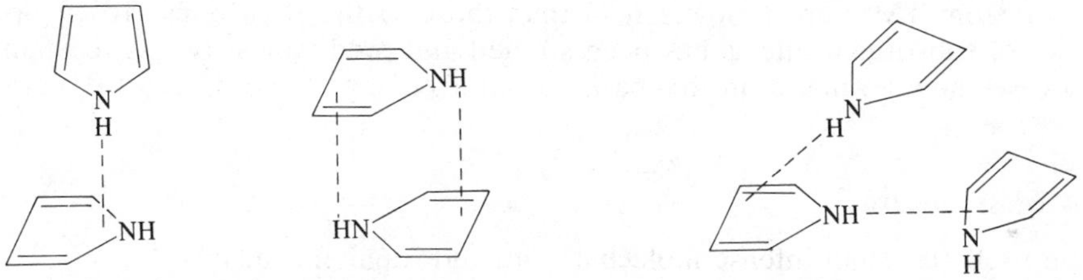

(9a) (9b) (9c)

Apart from the effects discussed above, there is clear evidence from i.r. spectra and from a variety of other physical techniques that association occurs between pyrrole molecules owing to hydrogen bonding of the NH group of one pyrrole with the π-electron system of another. Several models have been proposed,[8g] *e.g.* structures (**9a–c**), and experimental values for the association constant (0.14–0.40 l mol^{-1}) and heat of formation (6 kJ mol^{-1}) of the dimer have been obtained.

17.1.3.3 Nuclear magnetic resonance spectra

In its ^{1}H n.m.r. spectrum pyrrole shows two multiplets centred at 3.67τ (α-H) and 3.86 (β-H), and a broad signal at *ca.* 2.75τ (for the NH proton);[17] the CH signals sharpen up considerably on substitution of the NH for deuterium, and in presence of the potassium salt of pyrrole all the signals become very sharp. As with other aromatic systems, the signals vary somewhat with concentration and temperature owing to intermolecular association; the signals are also affected by association with aromatic or polar solvents. Coupling constants are shown in Table 1, and ^{14}N–H, ^{15}N–H, and ^{13}C–H coupling constants have also been determined.[8h]

The low-field shifts of the ring protons are in accord with the induced ring current expected of an aromatic ring system, although not so great as in benzene. On this basis, Elvidge[18] has suggested that pyrrole has 59% of the aromatic character of benzene, and

TABLE 1
Coupling Constants (Hz) of Ring Protons in Pyrroles

$J_{1,2}$	2.3–3.2	$J_{2,4}$	1.3–1.9
$J_{1,3}$	2.2–3.0	$J_{2,5}$	1.9–2.2
		$J_{2,3}$	2.3–3.2
		$J_{3,4}$	2.8–4.1

this is qualitatively in agreement with the lower value of the resonance energy of pyrrole compared with benzene. However, this simple view has been disputed by other workers,[19] and is also discussed more extensively by Gossauer[8i] and by Maroni.[20]

Alkyl substituents in the pyrrole nucleus cause neighbouring protons to move 0.2–0.3 p.p.m. to higher field, and those on the more distant carbon atoms 0.1–0.2 p.p.m. upfield. Electronegative substituents on nitrogen (*e.g.* carbonyl) give rise to a marked paramagnetic shift of the α-protons (*ca.* 0.5 p.p.m.) and a smaller low-field shift (*ca.* 0.1 p.p.m.) to the β-protons.[21] A negative group on the α-carbon atom causes a low-field shift of the neighbouring β-protons signal (by 0.7–0.8 p.p.m.) and also of the α'-proton resonance (*ca.* 0.4 p.p.m.), whereas the β'-proton signal is only slightly affected (*ca.* 0.1 p.p.m.). Negative β-substituents similarly affect the neighbouring α- and β'-proton resonances.[22,23] The resonances of methyl or methylene groups neighbouring an acyl, or other negative, substituent also experience a low-field shift (*ca.* 0.3 p.p.m.). These shifts are in general additive and can often be helpful in structural elucidation.

The ^{13}C resonance spectra of pyrrole and a wide variety of substituted pyrroles[8j,24,25] have been determined and the data in the literature is expanding rapidly with the widespread advent of Fourier transform instruments. The α-carbon resonances of pyrrole (118 p.p.m. from TMS) are at lower field than those of the β-carbons (107.9 p.p.m.).

A series of substituent effects has been studied and predictions[2] of ^{13}C resonances can in many cases now be made on this basis (to an accuracy of the order of 0.5 p.p.m.).

17.1.3.4 Mass spectra

Pyrrole itself shows an intense molecular ion, and fragments at *m/e* 41, 40, 39, and 28 have been attributed to the ions shown in the accompanying fragmentation (Scheme 1). *N*-Alkylpyrroles either fragment[26] with loss of the alkyl substituent, or by loss of a hydrogen atom (Scheme 2); whether the ion resulting from the latter process undergoes ring enlargement to a pyridinium species is unknown (*cf.* the corresponding transformations of benzylic cations to tropylium ions). Both α- and β-*C*-alkylpyrroles undergo benzylic-type fragmentation at the bond β to the ring (Scheme 3) and as in the *N*-alkyl series the resulting ion may also be a pyridinium species. The fragmentation patterns of a

SCHEME 1

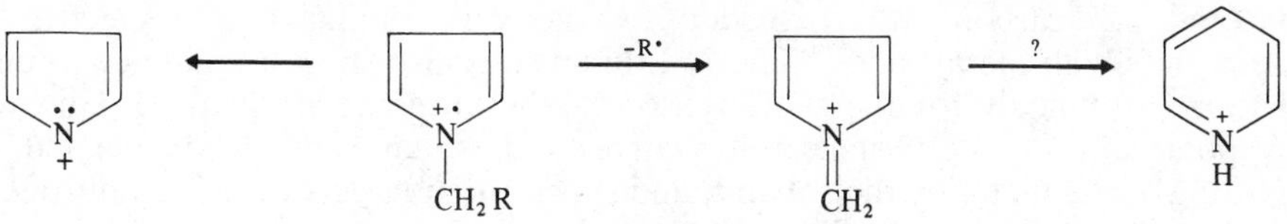

SCHEME 2

wide variety of acylpyrroles and pyrrolecarboxylic ester derivatives used in porphyrin synthesis have also been studied;[8k,27] in general the fragmentations observed are unexceptional and follow now well-accepted pathways.[27]

The ionization potential of pyrrole and several derivatives have been measured by photoionization mass spectrometry and other methods, values for the first and second ionization of pyrrole being about 8.20 and 8.90 eV, respectively.[cf.81] Molecular orbital calculations have also been performed by a variety of methods and a number of these calculations give good agreement with the experimental results.

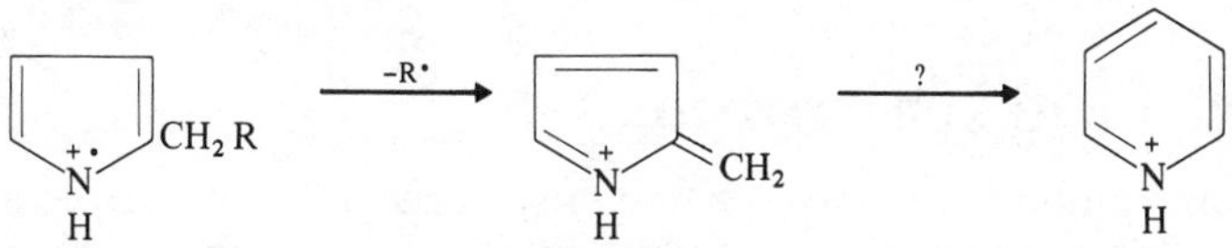

SCHEME 3

17.1.4 PYRROLE METAL DERIVATIVES

The weakly acidic pyrrole NH group has a pK_a of about 17.5,[28] but this is lowered quite dramatically by the presence of electron-withdrawing groups in the nucleus.[8m] Thus two alkoxycarbonyl groups reduce the pK_a to about 13 and α-nitropyrrole has a pK_a of 10.6; 2,4-dinitropyrrole has a pK_a of 6.15, and 2,5-dinitropyrrole of 3.6, showing the more effective conjugation of α-electron-withdrawing substituents with the lone pair of electrons on nitrogen. Halo-substituents also have a marked effect on increasing the acidity of pyrroles.

The sodium and potassium salts of pyrrole may be prepared by treatment with the metal, or metal amide, in liquid ammonia, or by direct reaction with the metal in an inert solvent. The lithium salt may be prepared by reaction with lithium in liquid ammonia, or with butyl- or phenyl-lithium in ethereal solution. *N*-Substituted pyrroles undergo metallation with butyl- or phenyl-lithium on the α-position.

Pyrroles also react very readily with Grignard reagents in ether to afford the *N*-pyrrylmagnesium derivatives (**10**). These are of considerable synthetic utility because alkylation and acylation occur mainly on carbon, whereas the alkali-metal derivatives undergo alkylation and acylation on nitrogen. This difference in reactivity is attributed to coordination of the magnesium ion with the pyrrolyl anion, whereas the alkali metal salts (**11**) are probably completely dissociated. Proof that the magnesium is associated with

(10) (11)  (12) (13)

nitrogen (and not carbon) was provided by the very similar n.m.r. spectra of the magnesium and alkali-metal salts.[29] The *N*-lithium salts also show a tendency to undergo *C*-alkylation, presumably for the same reason as the Grignard derivatives. However, in the more polar solvents the Grignard derivatives and lithium salts are dissociated, owing to complexing of the metal by the solvent, and substitution largely occurs on nitrogen (*e.g.* in hexamethylphosphoramide[30]). The nature of the alkylating agent is also important, the more reactive alkyl halides (*e.g.* allyl halides) effecting rapid *C*-substitution.

Other pyrrole *N*-metal complexes have also been prepared, *e.g.* those with cadmium or zinc, and these have occasionally been utilized in the synthesis of *C*-alkyl- or *C*-acyl-pyrroles. 1-Methyl-2-copper(I) complexes[30] have been prepared from the corresponding lithium salts, and *C*-mercury(II) complexes may be formed by direct reaction of pyrrole with mercury salts.

Several transition metal π-complexes of pyrrole have been synthesized in recent years,[31] *e.g.* (**12**) and (**13**). The former is a monoaza-analogue of ferrocene, but the bispyrrolyl (diaza) analogue is, so far, unknown. Bispyrrolyltitanocene analogues are very unstable and highly reactive liquids.[32]

17.1.5 ELECTROPHILIC SUBSTITUTION

Electrophilic substitution is the most characteristic reaction[20] of pyrrole and most of its simple derivatives, whereas they are virtually inert to nucleophiles. Indeed, pyrrole is often compared with phenol, or aniline, in its reactivity towards electrophiles and oxygen, or oxidizing agents. Because of their 'π-excessive' character,[14] pyrrole and its simple alkyl derivatives are polymerized by many of the conventional acidic electrophiles, although they may be stabilized sufficiently by electron-withdrawing substituents, such as alkoxy-carbonyl groups, so that nitration and sulphonation can be carried out successfully. Another important feature of the chemistry of pyrroles is their amphoteric character; for example, although dilute aqueous solutions of pyrrole are neutral, the proton on nitrogen is not only sufficiently acidic to be removed by strong bases (see preceding section) but the nucleus is protonated in acidic media.

Pyrrole, itself, undergoes electrophilic substitution reactions predominantly on the α-carbon atoms, although if these positions are blocked substitution will occur at the β-positions; substitution on nitrogen (except for reversible protonation reactions) normally only occurs in basic media *via* the anion. The preference for α-substitution may be explained qualitatively by considering the Wheland intermediates for α- and β-substitution (see Scheme 4); there are three resonance forms for the intermediate for α-substitution and the positive charge is delocalized over four atoms, whereas there are only two resonance forms for the intermediate σ-complex for β-substitution and this system is cross-conjugated. In both cases the pyrrole ring is effectively behaving like an enamine. The energy profiles for reactions at the α- and β-positions are shown in Figure

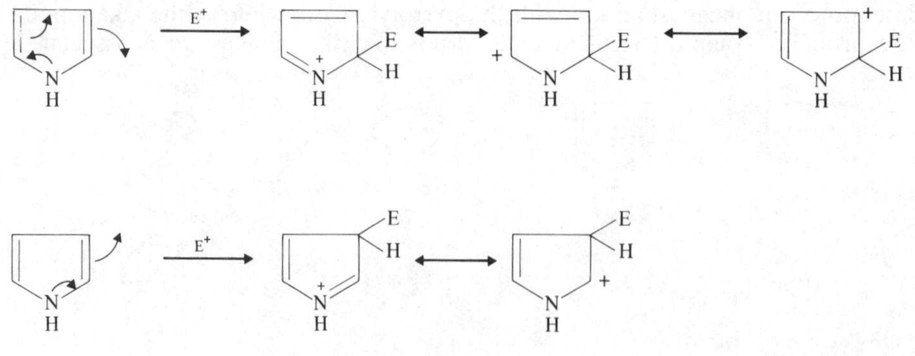

SCHEME 4

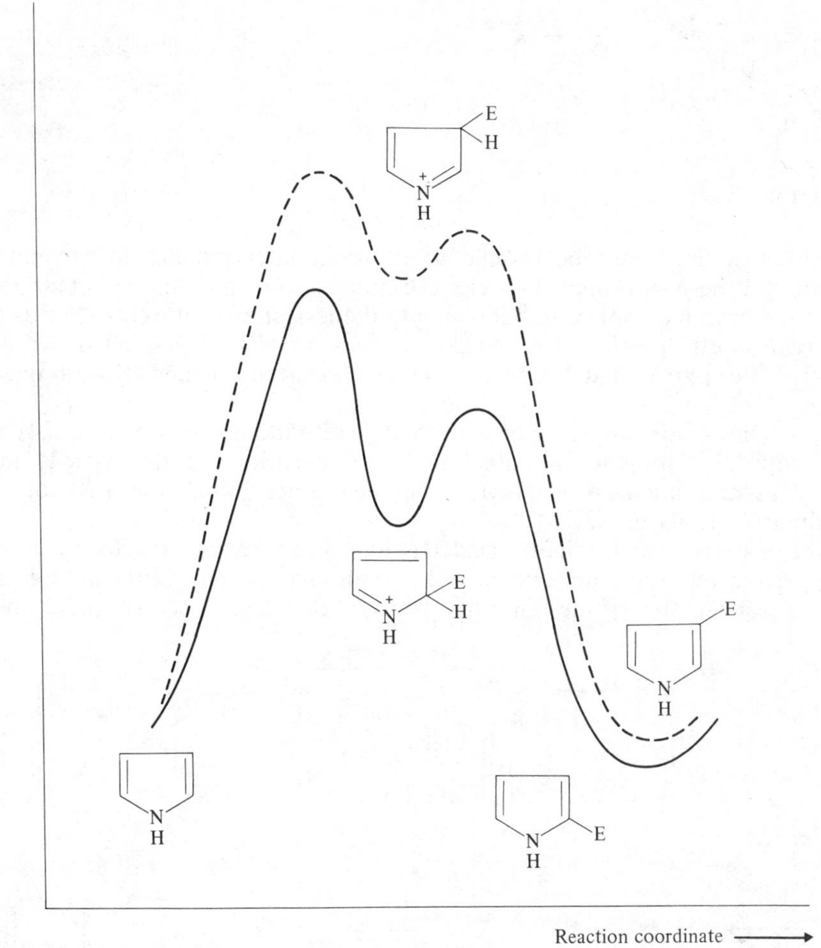

FIGURE 2

2, and the greater stabilization of the σ-complex involved in α-substitution leads to a lower activation energy. The presence of an electron-withdrawing group in the pyrrole nucleus may, however, alter the situation. Thus an α-alkoxycarbonyl group may cause an incoming electrophile to enter the corresponding β′-position[33] (rather than the α′-position), *i.e.* in a pseudo-'*meta*'-position to the electron-withdrawing substituent (see below).

17.1.5.1 Protonation

The proton attached to nitrogen undergoes very rapid exchange in alkali, or acid, and even at neutral pH; the latter can readily be demonstrated by addition of D_2O to a deuteriochloroform solution of pyrrole in an n.m.r. tube and observing the immediate disappearance of the N—H resonance in the n.m.r. spectrum. Exchange of the protons attached to carbon, however, only occurs under more acidic conditions, the α-protons exchanging at about twice the rate of the β-protons (*cf.* Figure 2). In aprotic solvents, complete protonation at the α-position may occur; and indeed if hydrogen chloride gas is passed into an ether solution of krypto-pyrrole, a crystalline hydrochloride (**17**) is formed.[17] Similar salts have also been prepared from other alkylpyrroles and the structures clearly demonstrated by n.m.r. and u.v. spectral measurements.[17,34,35] This is clearly the thermodynamically most stable salt, and the β-protonated salts (**15**) have not been

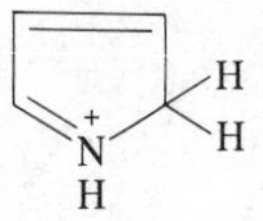

(14) 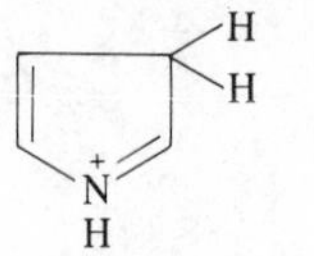 (15) 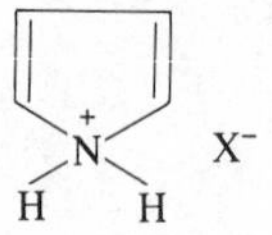 (16) 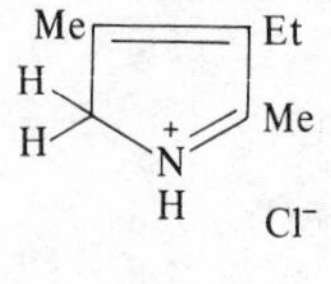 (17)

observed although they must be formed as transient intermediates to account for β-proton exchange; the *N*-protonated species (**16**) must also be an intermediate in the rapid exchange of the proton or nitrogen, but owing to the loss of aromatic character its lifetime is very short. *N*-Methylpyrrole also undergoes exchange of both α- and β-hydrogens in acidic solution, but pyrrole aldehydes and ketones, *etc.*, tend to undergo protonation on oxygen.

The pK_a of the conjugate acid of pyrrole (corresponding to structure **14**) is −3.8 (compared with 11.3 for the saturated analogue, pyrrolidine); the basicity increases considerably as the number of alkyl groups increases[35] and the pK_a of 2,3,4,5-tetramethylpyrrole is about +3.7.

Treatment of pyrrole with mineral acid has long been known to afford a mixture of polymers ('pyrrole-red'), and under controlled conditions a trimer (**19**) can be isolated in low yield.[36] The formation of the latter had been rationalized as shown in Scheme 5, *via*

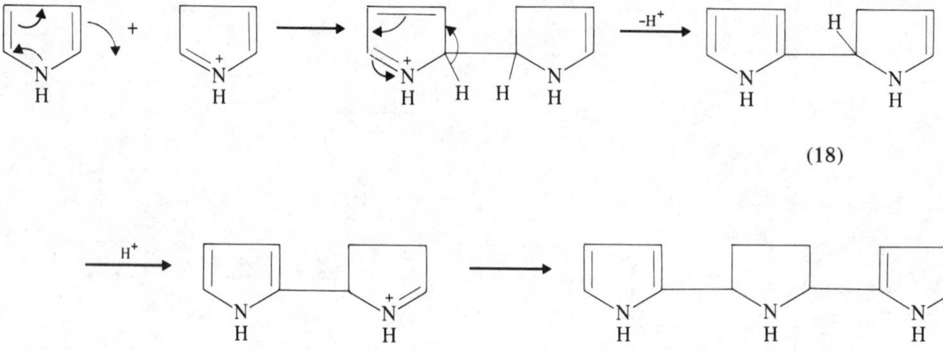

(18)

(19)

SCHEME 5

initial interaction of unprotonated pyrrole with the β-protonated form.[37,38] The dimer (**18**) of pyrrole itself has not been isolated, but dimers are readily formed from α-mono- and α,β-di-alkylpyrroles;[38] alkylpyrrole trimers are not formed, presumably in part because of steric hindrance, and perhaps also because of the lower electrophilicity of the dimer salts owing to electron release by the alkyl group(s).

17.1.5.2 Nitration

The usual nitrating reagents used with benzene cause extensive decomposition of pyrrole, but at lower temperatures acetyl nitrate (prepared from nitric acid and acetic anhydride) affords α-nitropyrrole (**20**) in moderate yield, together with smaller amounts of the β-nitropyrrole (**21**). Kinetic studies have shown that the rate of nitration at C-2 is about four times that at C-3, in accord with the product distribution.[39] *N*-Alkylpyrroles give larger amounts of the β-nitropyrroles, whilst 2-methylpyrrole (which is more reactive) gives a mixture of 2-methyl-5-nitro- and 2-methyl-3-nitro-pyrroles in which the former predominates.

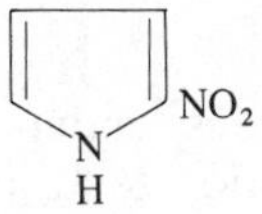

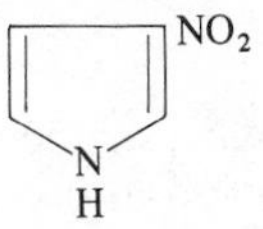

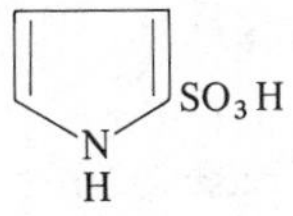

(20) (21) (22)

17.1.5.3 Sulphonation

Sulphuric acid itself causes complete destruction of pyrrole, but the 2-sulphonic acid (**22**) can be prepared in excellent yield by heating with the pyridine–sulphur trioxide complex at 100 °C. If both α-positions are blocked, sulphonation occurs at a β-position.

17.1.5.4 Halogenation

Tetrahalopyrroles are formed very rapidly by reaction with the halogens themselves, and it is difficult to prepare the monohalo-derivatives except under very special conditions. These halopyrroles are not very stable compounds and decompose rapidly in air and light; the tetraiodopyrrole has even been used as an antiseptic. α-Alkylpyrroles tend to undergo free radical substitution by halogens, sulphuryl chloride, or t-butyl hypochlorite at the methylene group next to the nucleus as well as nuclear substitution, and more complex products are obtained; in the presence of electron-withdrawing groups, however, nuclear substitution may occur cleanly, and the reactions have some synthetic utility as the halogen may be used as a protecting group.

Nuclear carboxylic acid, and acyl groups, can often be displaced by bromine or iodine, and the latter can subsequently be hydrogenolysed catalytically, thus providing a useful synthetic route to the corresponding unsubstituted pyrroles. Two examples are shown (**23**→**25** and **26**→**27**).

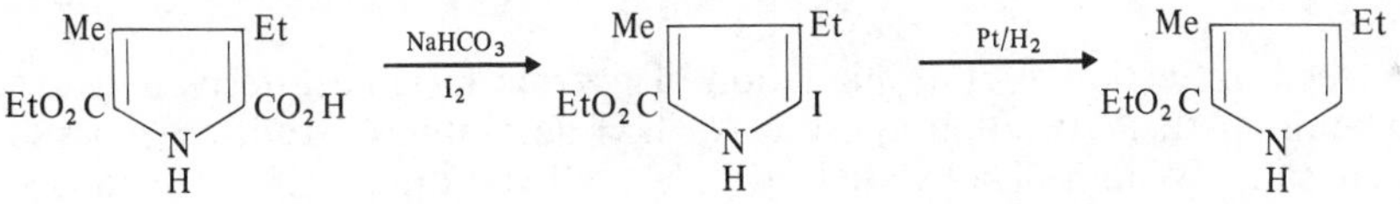

(23) (24) (25)

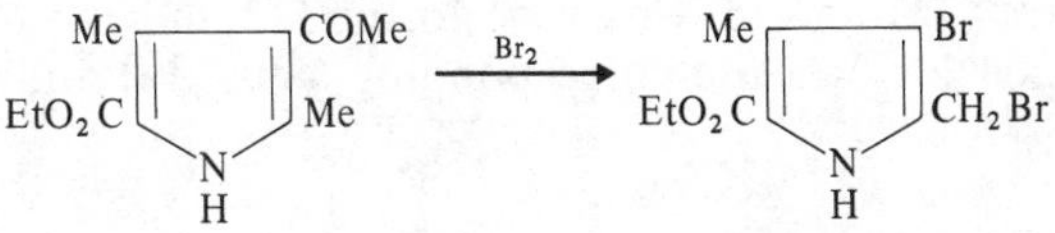

(26) (27)

Knorr's pyrrole (**28**) will even undergo reaction[43] with six moles of sulphuryl chloride to form the hexachloropyrroline (**29**), but the monochloromethyl and monobromomethyl derivatives (**30**) can be prepared under milder conditions.

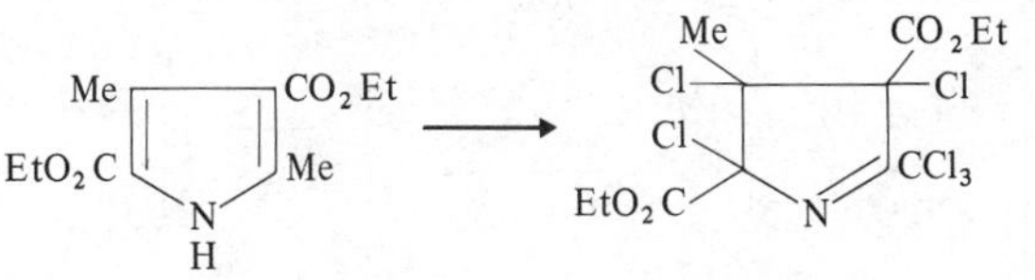

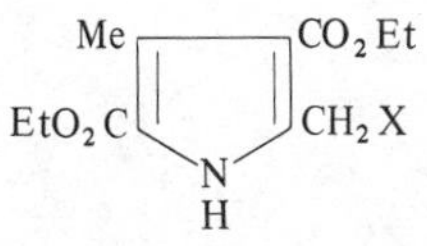

(28) (29) (30) X = Cl, Br

17.1.5.5 Alkylation

Pyrroles will react directly with allylic or benzylic haldes under mild conditions in presence of weak bases, but polyalkylation and polymerization tend to occur unless most of the positions are already blocked. Less-reactive alkyl halides will also react directly with pyrrole at elevated temperatures, *e.g.* methyl iodide and pyrrole react at 100–150 °C to give polymethylated pyrroles (Scheme 6) and polymeric material.

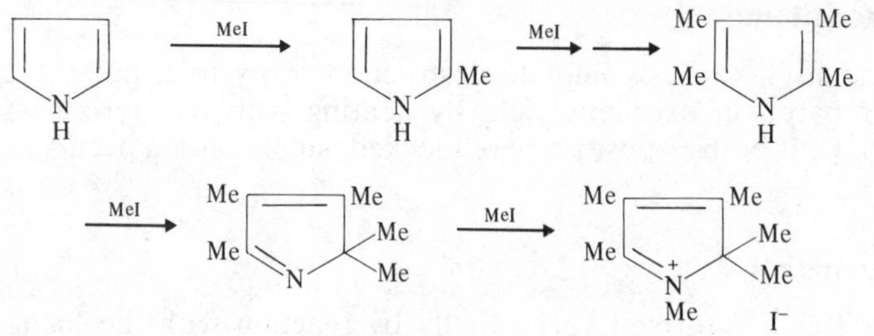

SCHEME 6

Friedel–Crafts isopropylation of pyrrole-2-carboxylic ester affords the related '*meta*'-substitution product (**31**), as well as the 4,5-di-isopropyl derivative (**32**);[33] the latter may well be formed by a second substitution at the 4-position followed by rearrangement. Friedel–Crafts alkylation of pyrrole and alkylpyrroles, however, tends to lead to polyalkylation and polymerization and is not very useful preparatively.

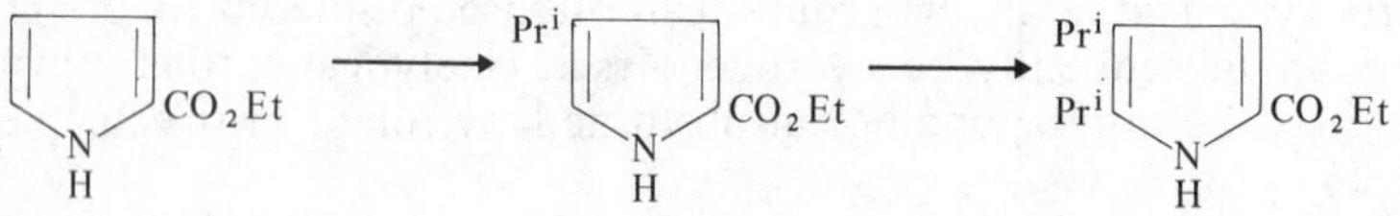

(31) (32)

As described in Section 17.1.4, alkylation of pyrrole Grignard derivatives leads mainly to *C*-alkylation at the 2-position and the *N*-alkyl derivatives are best prepared from the alkali-metal salts. With polyalkylated pyrroles, alkylation of the Grignard derivatives occurs at both the 2- and 3-positions, although the former predominates. In those cases where there is already a substituent at the 2- (or 3-) position, the weakly basic non-aromatic *pyrrolenines* (**33**) and (**35**) may be formed,[44] and these readily undergo further alkylation on nitrogen to form the corresponding salts (**34**) and (**36**).

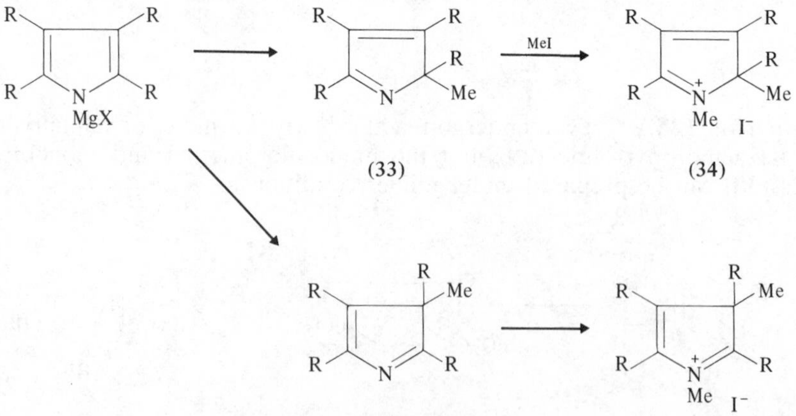

(33) (34)

(35) (36)

17.1.5.6 Acylation

N-Acylation of pyrroles is readily carried out by reaction of the alkali-metal salts with acid chlorides, and *N*-acetylpyrrole is also formed in low yield by direct acetylation with acetic anhydride in presence of sodium acetate. Acylation of pyrrole Grignard reagents affords predominantly the *C*-acyl products with smaller amounts of *N*-acyl derivatives.

Friedel–Crafts acylation of pyrroles carrying negative substituents can be carried out in presence of catalysts such as aluminium chloride, tin(IV) chloride, boron trifluoride in carbon disulphide, dichloromethane, nitromethane, and similar solvents. Many examples of intramolecular *C*-alkylation have also been described. Pyrrole, especially alkylpyrroles, can be acylated directly with acid anhydrides in the absence of a catalyst; pyrrole itself gives 2-acetyl- and 2,5-diacetyl-pyrroles on heating with acetic anhydride at temperatures over 100 °C, and additional alkyl groups facilitate reaction. Trifluoroacetic anhydride, or trichloroacetyl chloride, react directly with pyrrole to give the 2-trihaloacetyl derivatives (**37**);[45] the latter can readily be hydrolysed in alkali to afford pyrrolecarboxylic acid (**38**) in excellent yield. Acetyl trifluoroacetate and pyrrole give 2-acetylpyrrole in good yield. Formylation of the less-reactive pyrrole-2-carboxylic esters with alkyl dichloromethyl ethers can be achieved[46] under Friedel–Crafts-type conditions to afford the 4-formylpyrrole-2-carboxylic esters (**39**), *i.e.* '*meta*' substitution.

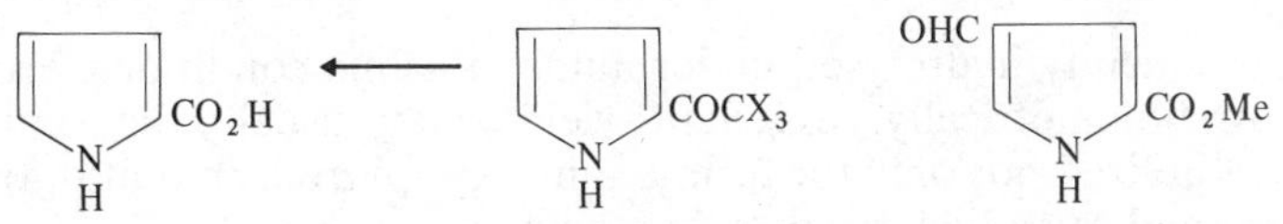

CO_2H ← $COCX_3$ OHC CO_2Me

(38) (37) X = Cl, F (39)

Direct formylation of pyrrole can also be achieved by the Gattermann reaction[47] using anhydrous hydrogen cyanide and dry hydrogen chloride gas in ether or chloroform solution. The intermediate aldimine hydrochloride, formed as shown in Scheme 7, is hydrolysed under mildly basic conditions to the 2-formylpyrrole. The closely related Houben–Hoesch reaction, using aromatic or aliphatic nitriles in presence of hydrogen chloride, affords the 2-acylpyrroles (see Scheme 7). Substituted pyrroles can be formylated and acylated in a similar fashion, the alkylpyrroles reacting more readily, but those

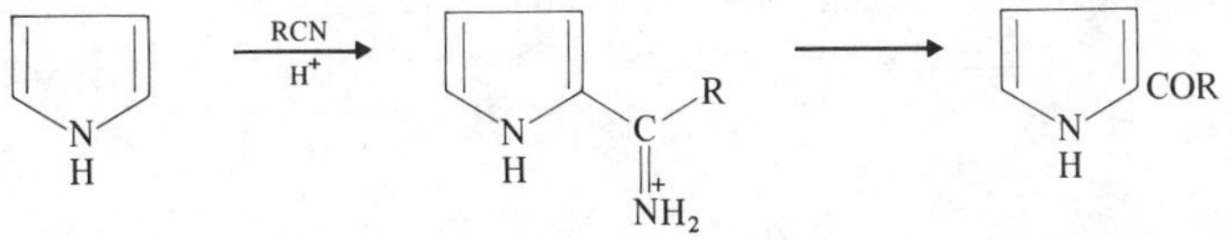

RCN, H^+ → $C(R)=\overset{+}{N}H_2$ → COR

R = H, alkyl or aryl

SCHEME 7

with electron-withdrawing substituents may require catalysis, *e.g.* with zinc chloride or boron trifluoride. Alternatives to the direct use of hydrogen cyanide are zinc cyanide[48] and *sym*-triazine.[49] More recently, triethyl orthoformate in the presence of trifluoroacetic acid has been developed[50] as a less hazardous alternative to the Gattermann reaction for formylation (see Scheme 8). The Reimer–Tiemann reaction, with pyrrole and chloroform

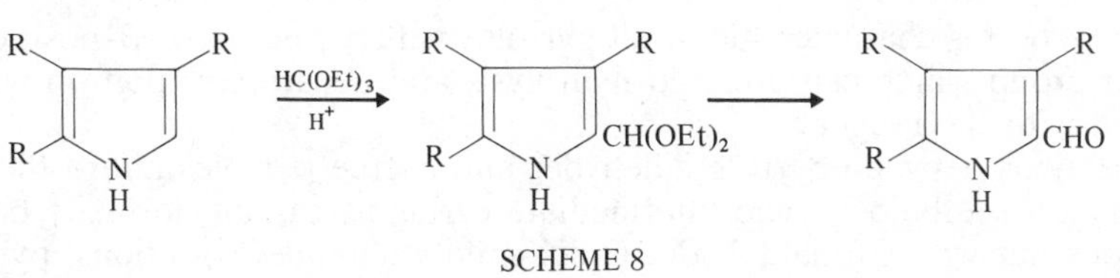

$HC(OEt)_3$, H^+ → $CH(OEt)_2$ → CHO

SCHEME 8

in strongly alkaline medium, also affords 2-formylpyrrole, but other ring-expansion reactions also occur owing to formation of dichlorocarbene (see below).

Currently, the most general useful method for formylation[52] and acylation[53] of pyrroles is the Vilsmeier–Haak procedure (which may also be used for other reactive aromatics such as thiophen, indole, and even polymethoxybenzenes). The pyrrole is treated with the phosphoryl chloride complex of an *N,N*-dialkylamide (Scheme 9) and the intermediate

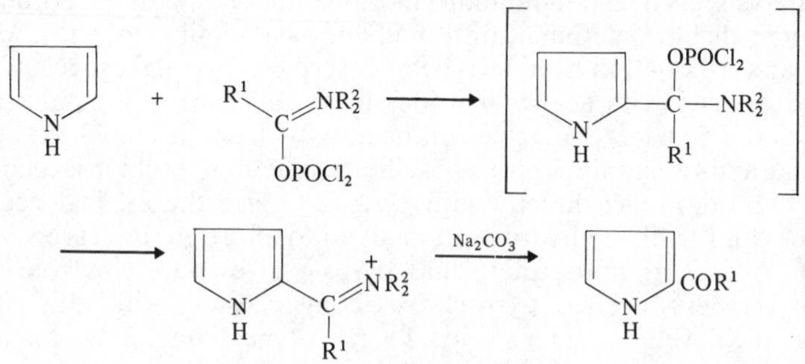

SCHEME 9

imine salt is subsequently hydrolysed under mildly alkaline conditions. Each stage can be followed spectrophotometrically, excellent yields may usually be obtained, and the method has been utilized not only for formyl- and acyl-pyrroles but also for aroylpyrroles, dipyrryl ketones, and pyrrolylpyrrolines (see below).

17.1.5.7 Reactions with aldehydes and ketones

The classical Ehrlich's colour reaction of pyrroles with *p*-dimethylaminobenzaldehyde takes place under mildly acidic conditions and gives purple coloured products, *e.g.* **(40)** (Scheme 10); an intermediate carbinol (39) is formed which then undergoes elimination of

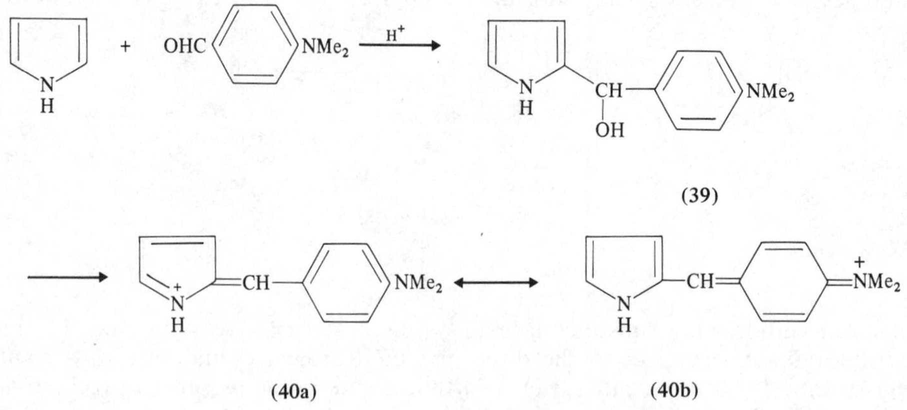

SCHEME 10

water. The reaction is characteristic of all pyrroles with a free α- or β-position or with a nuclear ester group which can undergo hydrolysis and decarboxylation on warming with the acid used with the reagent.

Similar reactions between pyrrole aldehydes and α-free pyrroles afford the orange-red dipyrromethenes (see below). The intermediate carbinols cannot normally be isolated in these reactions but with formaldehyde, under mildly alkaline conditions, pyrrole affords

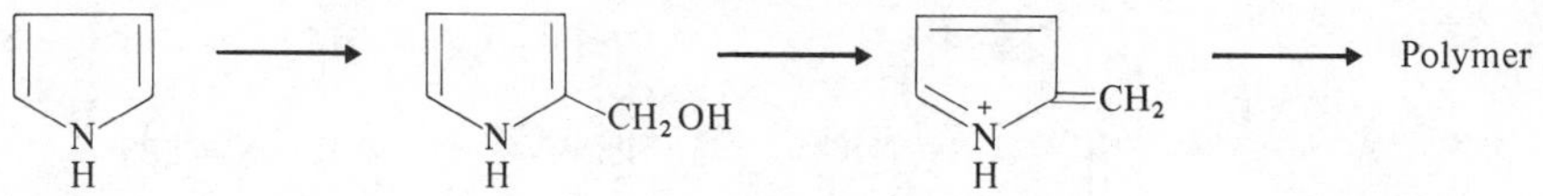

SCHEME 11

N-hydroxymethyl or 2,5-bis(hydroxymethyl) derivatives; under acidic conditions, however, formaldehyde and other aliphatic aldehydes condense with pyrrole to form polymers owing to self-condensation of the methylene pyrrolenines formed by elimination of water (Scheme 11); under some conditions small amounts of porphin may be obtained.[53] This reaction is reminiscent of the polymerization of phenols with aldehydes. Symmetrical dipyrromethanes may be formed, however, if electron-withdrawing substituents are present in the pyrrole ring (see below). The acid-catalysed condensation[54] of acetone with pyrrole affords a stable cyclic tetramer (**41**) (a porphyrinogen), polymer formation probably being inhibited because of the steric effects of the methyl groups in forcing the pyrrole rings into a coplanar conformation and hence assisting ring closure of the open-chain tetrapyrrolic precursor of the porphyrinogen. Pyrrole and its 3,4-alkyl derivatives react with aryl aldehydes in acidic solution in presence of air to give *meso*-tetra-arylporphyrins, *e.g.* (**42**) in 10–20% yield; 3,4-dialkylpyrroles will also react with aliphatic aldehydes and formaldehyde to form porphyrins in presence of air.

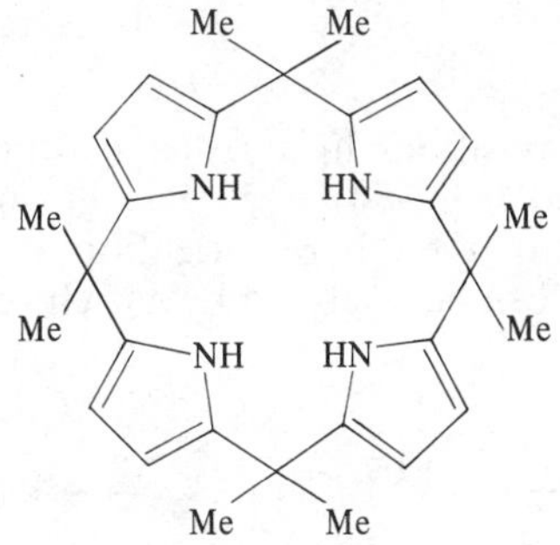

(41) 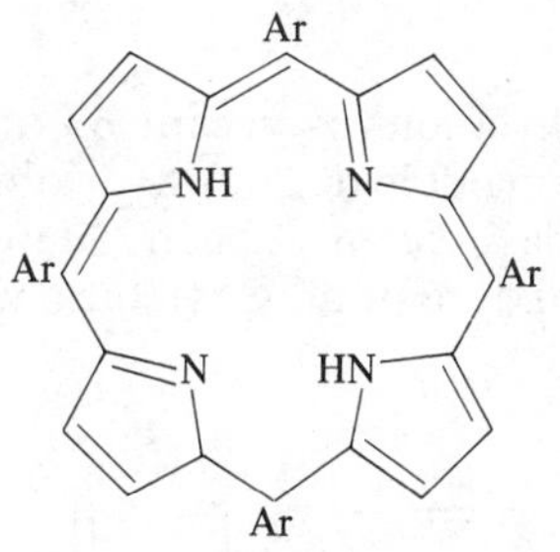 (42)

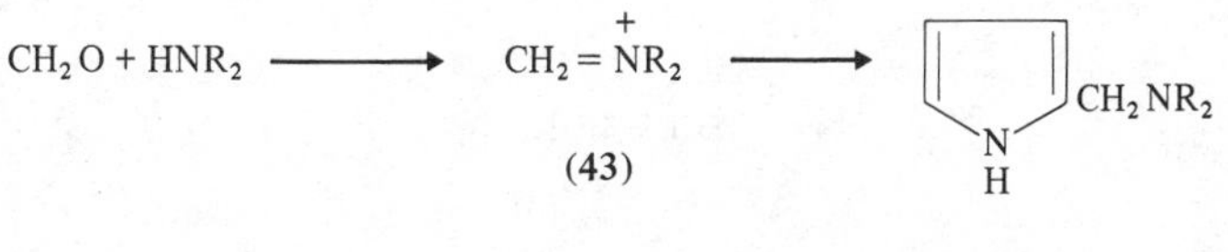

$$CH_2O + HNR_2 \longrightarrow CH_2{=}\overset{+}{N}R_2 \longrightarrow \text{(44)}$$

(43)

(44)

The formation of pyrrole Mannich bases (**44**) from pyrroles, formaldehyde, and a secondary amine is a preparatively useful extension[56] of this type of reaction involving an imine intermediate; another example is the synthesis of the bipyrrole moiety of prodigiosin (see below).

17.1.5.8 Other electrophilic substitution reactions

Apart from the main reactions summarized above, the majority of which are of preparative importance, a number of other electrophilic substitution reactions of pyrroles are of interest preparatively and diagnostically.

Acrylic esters, acylonitrile, and maleic anhydride undergo Michael-type addition reactions with some α-free pyrroles in the presence of acidic catalysts or bases such as sodium ethoxide or benzyltrimethylammonium hydroxide to form *C*-alkylated pyrroles.[57] Pyrrole,

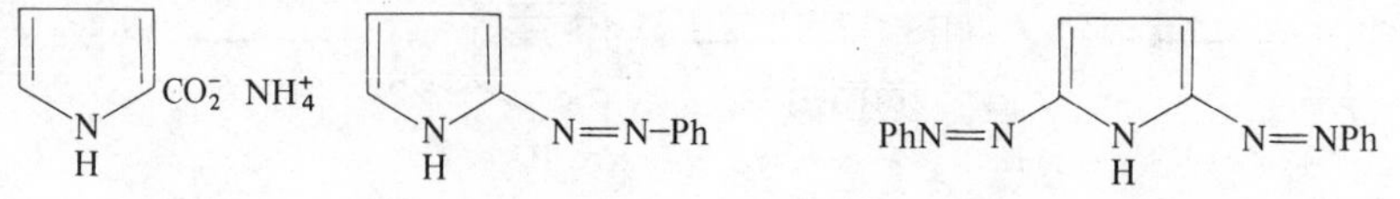

(45) (46) (47)

like phenol, can be carboxylated directly by heating with aqueous ammonium carbonate under pressure to form the ammonium salt (**45**) of pyrrole-2-carboxylic acid. Similarly, rapid azo-coupling occurs with pyrroles having unsubstituted α- or β-positions at neutral or slightly acidic pH, *e.g.* pyrrole forms the monoazo-derivative (**46**), but under basic conditions the bis derivative (**47**) is formed. Nitrosation of pyrrole and alkylpyrroles may result in ring-opening or oxidation of the ring or the alkyl groups, but in the presence of an electron-withdrawing group (as in **48**) a mononitrosopyrrole, *e.g.* (**49**), may be obtained in moderate yield. Isothiocyanates and isocyanates also react with pyrrole and alkylpyrroles with vacant positions to form thioamides and amides, respectively.

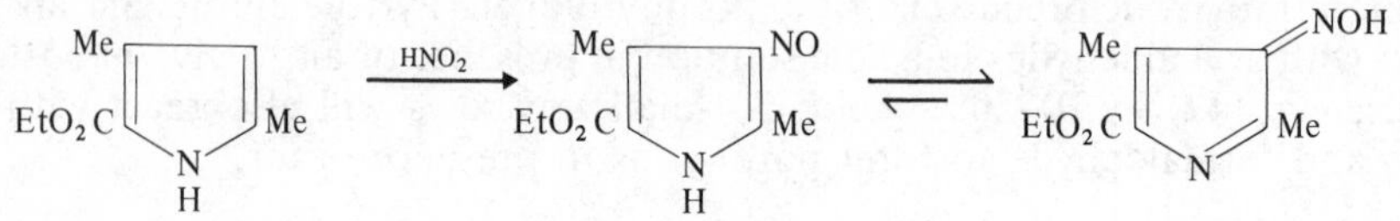

(48) (49a) (49b)

'Reductive alkylation' of vacant α- and β-positions in pyrroles can be carried out by heating with formaldehyde and hydriodic acid, *e.g.* (**50**)→(**51**) as shown in Scheme 12. This reaction has recently been adapted for the characterization of porphyrins by reduction to a mixture of alkylpyrroles, which can be analysed by gas chromatography and mass spectrometry.[58]

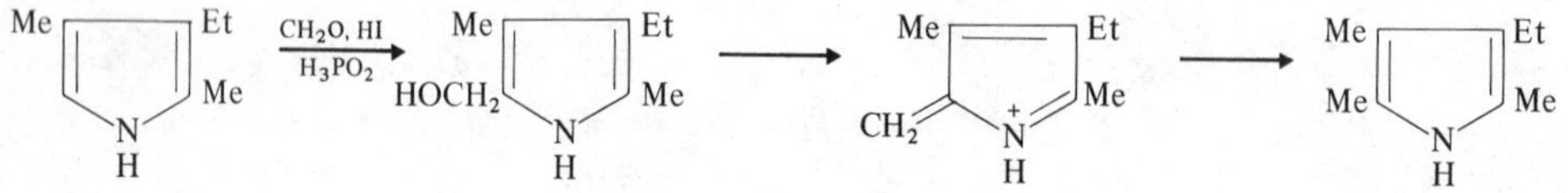

(50) (51)

SCHEME 12

17.1.5.9 Summary of substituent effects

Treibs[59] has recently elaborated a series of substitution rules for electrophilic substitution in pyrroles, which may be summarized as follows:

(i) an electron-withdrawing substituent in the α-position tends to direct substitution into the β'- (and α'-) positions;

(ii) an electron-releasing substituent in the α-position directs substitution to the neighbouring β-position or to the α'-position;

(iii) an electron-withdrawing β-substituent will undergo substitution in the α'-position;

(iv) an electron-releasing β-substituent tends to direct substitution into the neighbouring α-position.

Whether substitution occurs as predicted depends, of course, also upon whether or not the appropriate position is unsubstituted, and to the electron-releasing or -withdrawing capabilities of the substituents already present in relation to the normal tendency to α-substitution (*cf.* Scheme 4). The relative reactivities of pyrrole and other heterocyclic systems have been reviewed by Marino.[20]

17.1.6 REACTIONS WITH CARBENES[60] AND NITRENES

Treatment of pyrrole with chloroform and strong bases gives a mixture of the 2-formylpyrrole (**53**) and 3-chloropyridine (**54**). The mechanism of the reaction probably involves primary formation of dichlorocarbene, which inserts into the 2,3-double bond of pyrrole,[61] and the intermediate (**52**) then rearranges as shown. Isolation of dichloromethylpyrrolenines has been observed in a number of cases, *e.g.* (**55**) from 2,5-dimethylpyrrole; however, the latter does not rearrange to a chloropyridine under similar conditions[62] although with very strong bases such as butyl-lithium rearrangement occurs[63] as shown to give both 2- and 3-chlorodimethylpyridines (**56**) and (**57**). Improved yields of pyridine derivatives can be obtained by using carbenes derived from dichloromethane and n-butyl-lithium, or from pyrolysis of sodium trichloroacetate.[62]

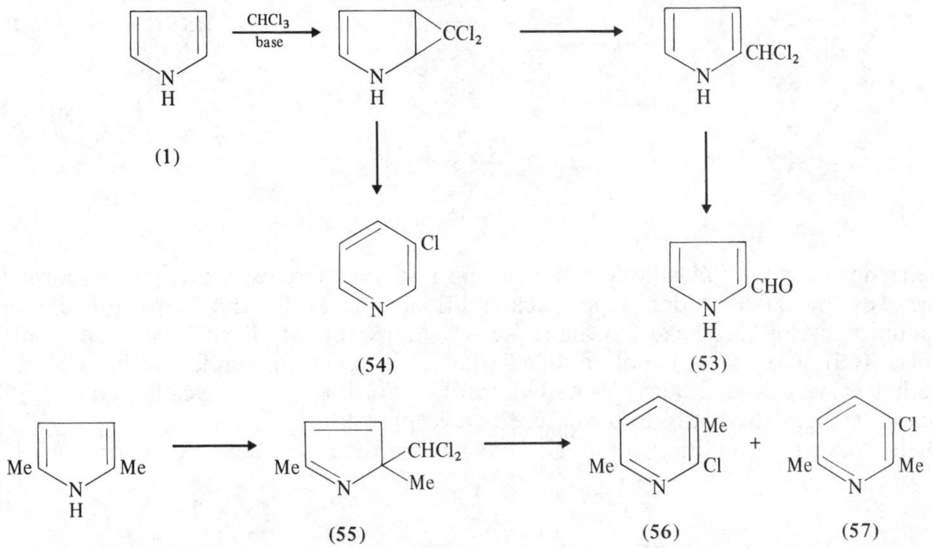

(1) —$CHCl_3$, base→ [CCl_2 adduct] → [$CHCl_2$] ; (54) Cl; (53) CHO

Me, Me, N, H → Me, $CHCl_2$, Me, N (55) → Me, Me, Cl, N (56) + Me, Cl, Me, N (57)

Carbenes derived from copper- or light-catalysed decomposition[64] of diazoacetic ester react with pyrroles to give α-substituted pyrrolacetic esters, or the β-products if the α-positions are blocked. *N*-Alkoxycarbonyl-substituted pyrroles, on the other hand, afford 2,3-homopyrroles with diazomethane or diazoacetic ester and copper(I) salts, *e.g.* (**58**), whilst the bis-homopyrrole derivatives (**59**) may also be obtained from pyrrole.[65]

Ethoxycarbonylnitrene (derived pyrolytically from azidoformic ester) affords 2-amino-*N*-ethoxycarbonylpyrrole (**61**),[66] and the mechanism suggested (Scheme 13) involves 2,3-addition to form the homoazopyrrole (**60**) followed by rearrangement to a 2,5-adduct and ring opening; alternatively, the 2,5-adduct may be formed directly.

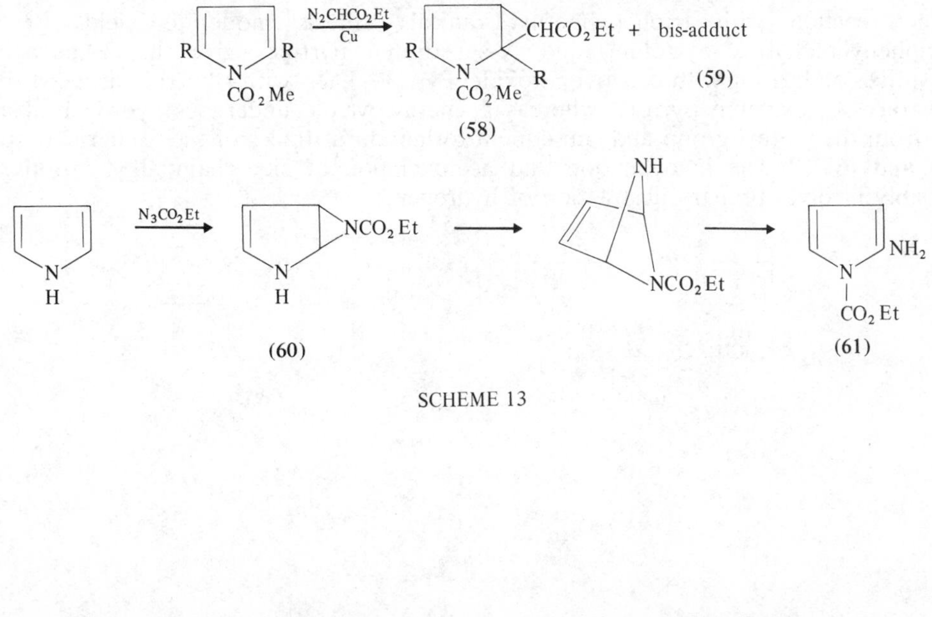

R, R, N, CO_2Me —N_2CHCO_2Et, Cu→ R, $CHCO_2Et$, R, N, CO_2Me (58) + bis-adduct (59)

N, H —N_3CO_2Et→ NCO_2Et, N, H (60) → NH, NCO_2Et → NH_2, N, CO_2Et (61)

SCHEME 13

17.1.7 DIELS–ALDER REACTIONS

Pyrrole itself does not readily undergo Diels–Alder reactions, presumably because of its aromatic character due to the delocalization of the lone pair of electrons on nitrogen with the diene system, and typical dienophiles such as maleic acid, acetylenedicarboxylic ester, and azodicarboxylic ester undergo Michael-type additions (see above). However, *N*-methoxycarbonylpyrrole reacts with dimethyl acetylenedicarboxylic ester to give 1,3,4-trimethoxycarbonylpyrrole (**63**),[67] presumably by the Diels–Alder, retro-Diels–Alder process shown; the intermediate (**62**) has been isolated and other aspects of its chemistry studied. *N*-Methyl- and *N*-benzyl-pyrrole give mainly Michael-type adducts with acetylenedicarboxylic ester,[68] but small amounts of other by-products are also formed and these have been attributed to reactions of Diels–Alder adducts.

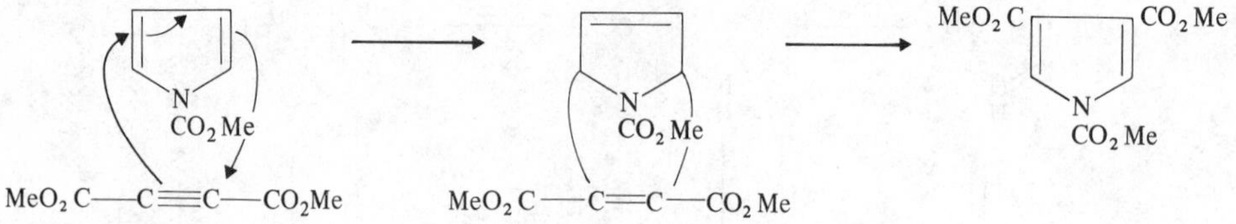

N
CO_2Me
$MeO_2C—C≡C—CO_2Me$
→
N
CO_2Me
$MeO_2C—C=C—CO_2Me$
(62)
→
MeO_2C CO_2Me
N
CO_2Me
(63)

Benzynes react with *N*-alkoxycarbonyl- and *N*-alkyl-pyrroles to give azanorbornadiene derivatives by Diels–Alder type cycloaddition, *e.g.* (**64**); the powerful dienophile hexafluorobicyclo[2,2,0]hexa-2,5-diene adds[69] to pyrrole itself to give mono- and di-adducts (**65**) and (**66**). Dipolarophiles may also react in similar fashion, *e.g.* 2,2-dimethylcyclopropanone and *N*-methylpyrrole afford the azabicycloheptanone (**67**),[70] which rearranges thermally to a mixture of two pyrroles.

Diels–Alder type additions of singlet oxygen are discussed below (Section 17.1.11).

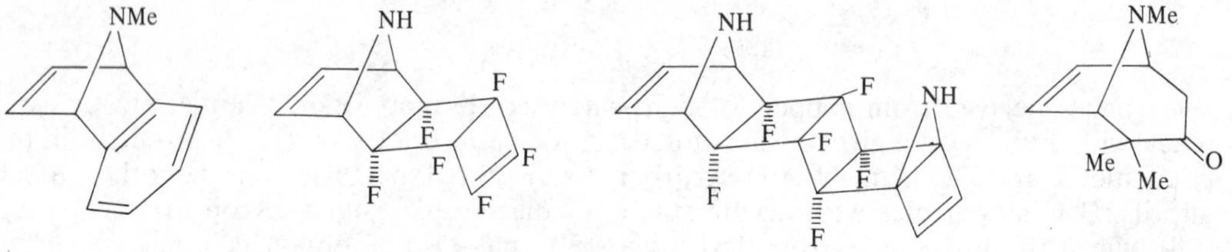

NMe
(64)

NH F F F F F F
(65)

NH F F F NH F F F
(66)

NMe Me Me O
(67)

17.1.8 FREE RADICAL REACTIONS

The reaction with triphenylmethyl radicals affords moderate yields of 2,4-di(triphenylmethyl)-Δ^3-pyrroline, and *N*-substituted pyrroles give the 2-benzoyloxy-derivatives on heating with benzoyl peroxide. Pyrrole reacts with t-butyl hydroperoxide to give traces of pyrrolinylpyrrole, whereas *N*-methylpyrrole undergoes hydrogen abstraction from the methyl group and subsequent formation of traces of the 'dimeric' products (**68**) and (**69**).[71] The halogenation and acetoxylation of side-chain alkyl groups also probably involves primary abstraction of hydrogen.

N N
CH_2——CH_2
(68)

CH_2—N
N
CH_3
(69)

Relatively stable coloured radicals have been obtained by lead dioxide or polarographic oxidation of tetraphenyl- and pentaphenyl-pyrroles, and radical reactions are almost certainly involved in the autoxidative polymerization of alkylpyrroles.

17.1.9 RING-OPENING REACTIONS

Whereas the furan ring is very readily opened under acidic conditions and that of thiophen is relatively stable, the pyrrole ring is intermediate in reactivity. Thus treatment of pyrrole with hydroxylamine in presence of half an equivalent of hydrochloric acid gives the 1,4-dioxime (**70**) in moderate yield, and 2,5-dimethylpyrrole also reacts in a similar fashion. The mechanism of reaction may involve initial protonation to form the β-pyrrolenine salt which then undergoes nucleophilic addition of hydroxylamine, *etc.*, and finally ring-opening (Scheme 14). Alternatively, hydrolysis to the 1,4-diketone might occur before oxime formation. A similar reaction occurs with 2,4-dinitrophenylhydrazine.

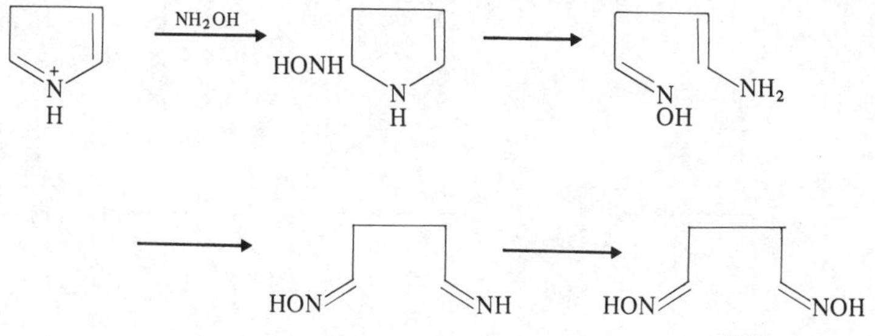

NH_2OH HONH N H N OH NH_2 HON NH HON NOH

(70)

SCHEME 14

2,5-Dimethylpyrrole undergoes self-condensation on heating in acetic acid solution and the isoindole (**71**) formed was characterized[71a] by reduction with zinc or tin and acid to a mixture of stereoisomers of 1,3,4,5-tetramethylindoline (**72**). Again it seems likely that the reaction is initiated by protonation of the pyrrole followed either by hydrolysis to

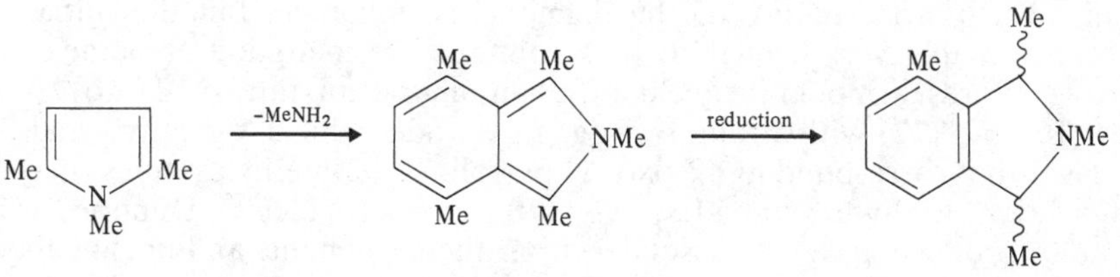

Me Me N Me $-MeNH_2$ Me Me NMe Me Me reduction Me Me NMe Me

(71) (72)

acetonylacetone and subsequent reaction with the dimethylpyrrole, or by direct reaction of the β-pyrrolenine with the dimethylpyrrole.

If pyrrole is heated in a sealed tube with sodium bisulphite, a disodium salt is formed and this has been formulated as (**73**); the product readily undergoes ring opening under alkaline or acidic conditions.

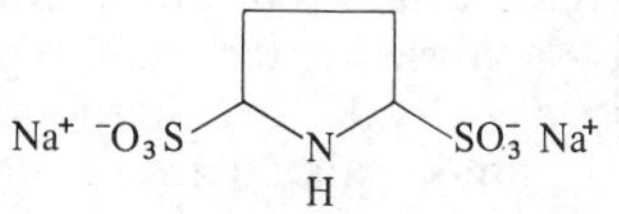

$Na^+ \; ^-O_3S$ N H $SO_3^- \; Na^+$

(73)

17.1.10 REDUCTION

Catalytic hydrogenation of pyrrole to pyrrolidine can be achieved at moderate temperatures and pressures over a variety of catalysts including platinum, palladium, and other noble metals, or Raney nickel, and virtually quantitative yields are obtained. Rhodium on alumina can be used even as low as 40 °C. The presence of α- or β-alkoxycarbonyl groups tends to make the ring more resistant to hydrogenation, so that side-chain acyl groups can be selectively reduced; however, *N*-alkoxycarbonyl groups facilitate reduction of the ring, presumably because the lone pair of electrons is less readily available for delocalization to form the aromatic sextet.

Zinc and hydrochloric acid treatment of pyrrole and *N*-methylpyrrole leads to formation[72] of the Δ^3-pyrrolines (**74**) and pyrrolidines (**75**), presumably *via* reduction of the α- and β-pyrrolenine salts, respectively (Scheme 15).

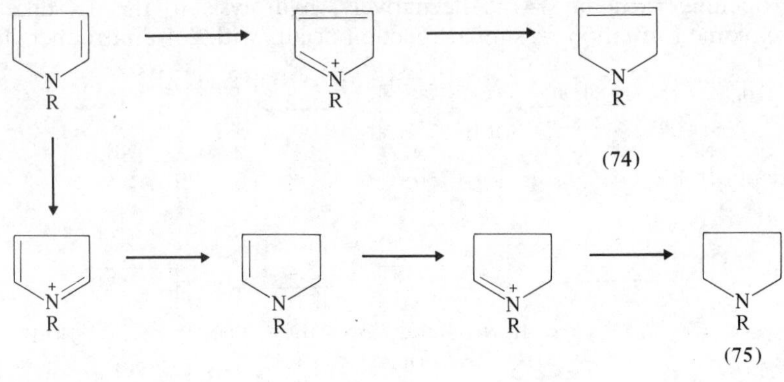

SCHEME 15

17.1.11 OXIDATION

Ozonolysis of pyrroles at low temperatures affords low yields of aldehydes which have been attributed to 2,3- and 2,5-addition reactions.[73] Similar products are also formed by reaction with permanganate.

On standing in air and light, pyrrole autoxidizes to form reddish brown, and eventually black, pigments ('pyrrole black'),[74] and the rate of oxidation is increased by alkyl substituents. The precise nature of the pigments is unknown, but the initial oxidation products are probably very similar to those obtained by peroxide or peracid oxidation. Thus pyrrole affords a moderate yield of a tautomeric mixture of 2-oxo-Δ^3- and -Δ^4-pyrrolines (**76**) and (**77**) with strong hydrogen peroxide, whilst *N*-methyl- and 3-methylpyrroles give the corresponding 2-oxo-Δ^3-pyrrolines; 2-methylpyrroles may undergo further oxidation to hydroperoxides, *e.g.* (**78**), or peroxides.[75] Dimeric[76] (**79a**) and trimeric[77] (**80**) products have been isolated from these pigments and presumably arise by nucleophilic addition of pyrrole to the initial 2-oxo-Δ^4-pyrrolines; similar dimeric products may also be formed in the autoxidation of alkylpyrroles, *e.g.* (**79b**), and *N*-methylpyrrole. Preparatively the best route to the 2-oxo-Δ^3-pyrrolines is by oxidation with warm hydrogen peroxide in pyridine,[78] *e.g.* (**81**)→(**82**). Perbenzoic acid oxidizes *N*-methylpyrrole to *N*-methylsuccinimide in excellent yield,[79] whilst chromic acid oxidation[80] of pyrroles and alkylpyrroles (**81**) leads to maleic imides (**83**) (with loss of the α-alkyl groups).

The last reaction is of considerable analytical importance[80] as porphyrins and bile pigments can also be oxidized to maleic imides in this manner and hence the nature of the side-chains can be determined by analysis of the maleic imides formed.[81] Treatment of α-alkylpyrroles with nitrous acid also leads to oxidation of both the ring and α-alkyl group, *e.g.* (**84**)→(**85**). However, unlike alkylbenzene derivatives the alkyl groups of alkylpyrroles cannot be oxidized selectively by any reagents, unless the ring is stabilized by electron-withdrawing substituents.

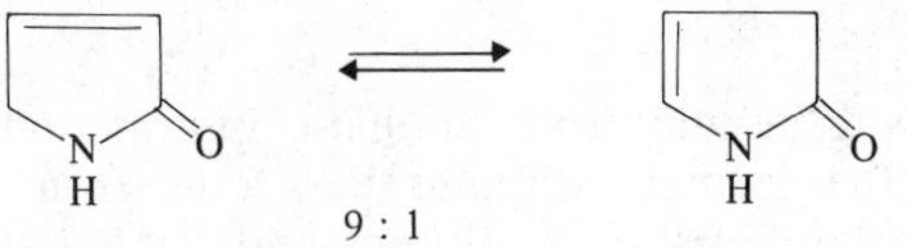

(76) 9 : 1 (77) 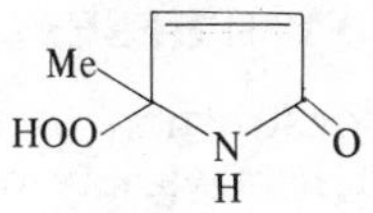 (78)

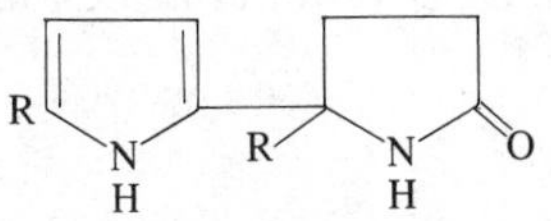

(79) a; R = H
b; R = Me

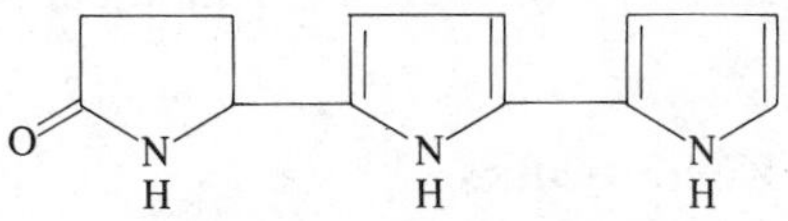

(80)

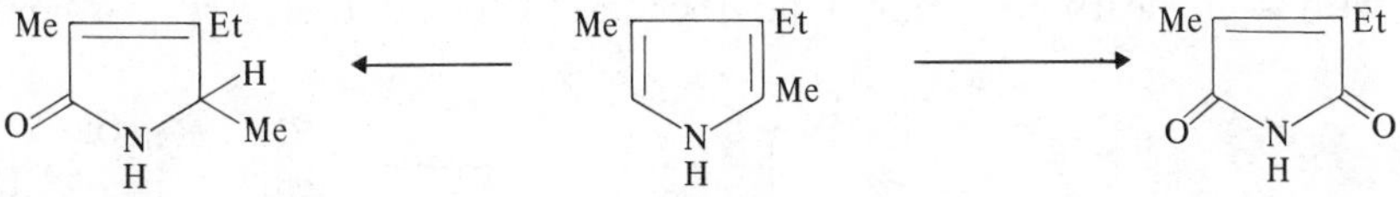

(82) (81) (83)

Sensitized photo-oxidation[82] of pyrrole affords 5-hydroxy-2-pyrrolone (**87a**), probably *via* singlet oxygen addition in a Diels–Alder fashion (Scheme 16) to form an intermediate cyclic peroxide (**86**). Allylpyrroles also undergo photo-oxidation in a similar fashion[83] with formation of the hydroxypyrrolones, *e.g.* (**87b**); 3,4-dialkylpyrroles may also undergo further oxidation to maleimides (**88**). The course of these oxidation reactions has been of considerable interest in recent years in relation to the mechanisms of photo-oxidation of bile pigments *in vivo*. The photo-oxidation of a range of mono-, di-, tri-, tetra-, and penta-phenylpyrroles has also been extensively studied[84,85] (*cf.* Scheme 17).

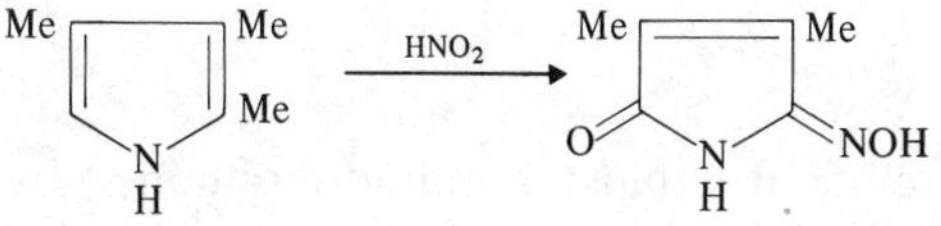

(84) HNO_2 (85)

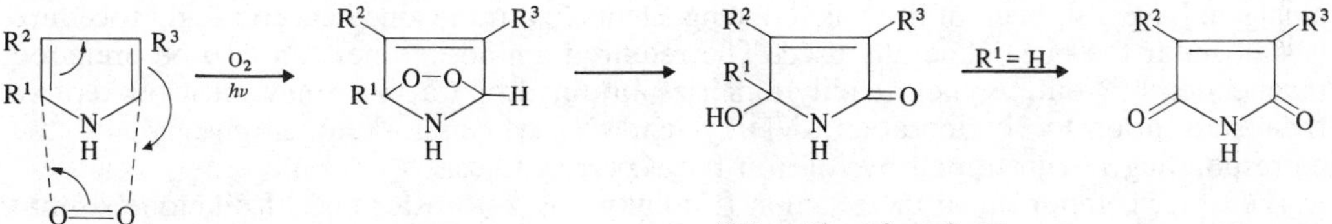

(86) (87) a; $R^1 = R^2 = R^3 = H$
b; $R^1 = R^3 = Me$, $R^2 = Et$
(88)

SCHEME 16

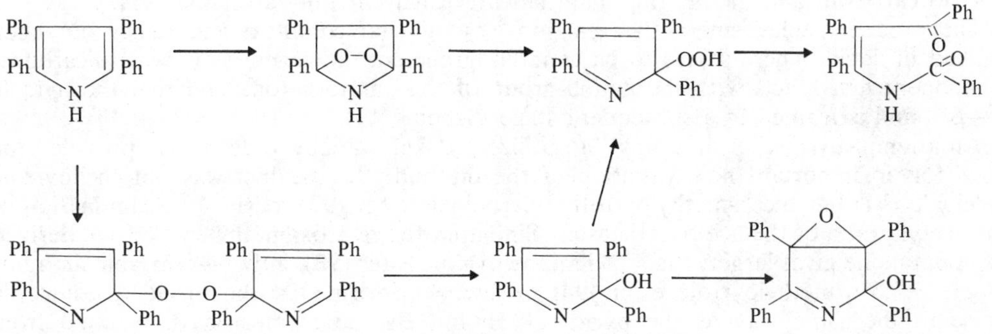

SCHEME 17

17.1.12 SYNTHESES

The construction of the pyrrole ring system from aliphatic intermediates has been achieved in a variety of ways, and the greatest stimulus has come from the need to prepare intermediate pyrroles for the synthesis of porphyrins and related compounds of biological importance. The most important synthetic methods lead to substituted pyrroles, the side-chains of which are later modified, or degraded, as required (Section 17.1.13).

17.1.12.1 Knorr synthesis

This method, and variations, constitute the most important and widely used series of pyrrole ring syntheses. The original method involved the condensation of an α-amino-ketone with a β-dicarbonyl compound (Scheme 18) and the classic example[86] is the

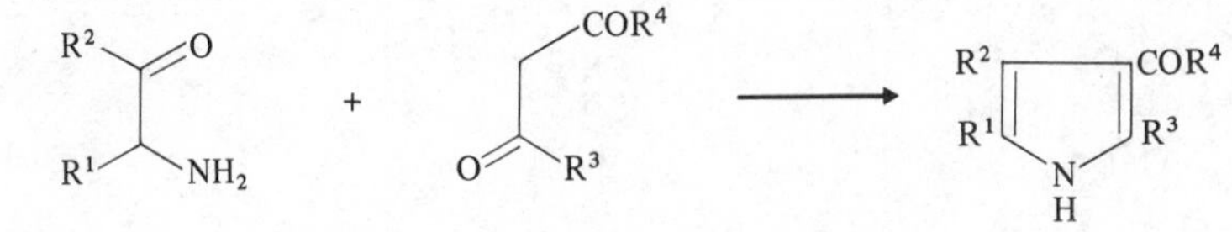

SCHEME 18

formation of 'Knorr's pyrrole' (**89**) by reductive condensation of ethyl oximinoaceto-acetate with ethyl acetoacetate, using zinc in glacial acetic acid.

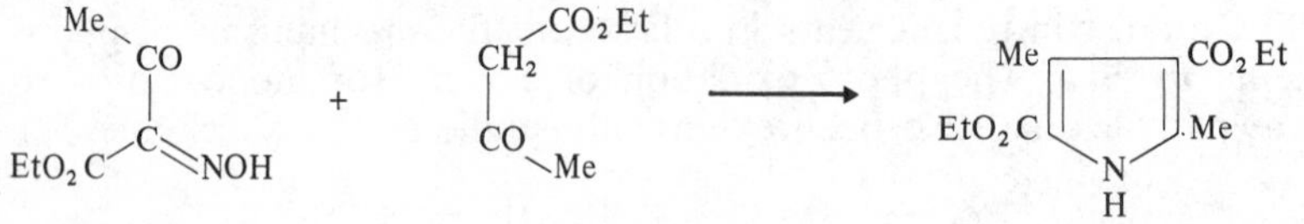

(89)

The amino-ketone component is most commonly obtained by *in situ* reduction of an oximino-derivative, obtained by nitrosation of a ketone or β-dicarbonyl compound (with sodium nitrite, or pentyl nitrite in glacial acetic acid); sodium dithionite and sodium amalgam have also been utilized as reducing agents,[87] but the zinc in acetic acid procedure is still by far the most generally used. The required amino-ketones can also be prepared independently[88] but, as they readily dimerize to dihydropyrazines, they must be kept as their salts prior to condensation; *N*-alkyl and *N*-aryl amino-ketones give rise to the corresponding *N*-substituted pyrroles in the Knorr synthesis.[89]

The other component in the reaction is usually a β-keto-ester or a β-diketone so that the methylene group is sufficiently reactive to undergo condensation with the keto-group of the α-amino ketone at a faster rate than the latter self-condenses. The yields in the Knorr synthesis are generally only of the order of 40–60%, but the reactions are relatively easy to carry out and the starting materials are usually readily available.

The precise mechanism of the reaction is somewhat complex, and it has not been studied in detail. There seems to be general agreement, however, that the β-dicarbonyl component first reacts with the amino-group of the amino-ketone and that the pyrrole β—β' bond is formed in a subsequent stage (Scheme 19).

An attempt by Fischer and Fink[90] to utilize a β-keto-aldehyde derivative provided the basis for an important new variation of the method; the product was not the pyrrole aldehyde (**91**) but the 5-methylpyrrole-2-carboxylic ester (**92**), derived by elimination of the acetyl group of the acetoacetic ester. Similarly, the hydroxymethylene ketone derived from butanone gives largely the 4,5-dimethylpyrrole ester (**93**), although variable amounts of the 3,4,5-trimethylpyrrole ester (**94**) may also be formed;[91] the former presumably arose in the same way as the pyrrole (**92**), but the latter must have resulted from elimination of the hydroxymethylene (or formyl) group.

SCHEME 19

(91) (92) (93) (94) (95)

(96) a; $R^1 = CO_2Et$
b; $R^1 = CN$

SCHEME 20

A further development was the use of amino-malonic esters by Kleinspehn,[92] and these on condensation with β-diketones (**95**) gave pyrrole esters (**96**) (Scheme 20); the amino-malonic esters can be generated by nitrosation and zinc/acetic acid reduction in the usual manner. α-Cyanopyrroles may also be prepared in a similar manner starting from ethyl cyanoacetate (or even malononitrile). More recently it has been shown that oximino-β-keto-esters will also reductively condense with alkyl-β-diketones (**97a**) in this way[93,94] to form trialkylpyrroles (**98**), although acetylacetone (**97b**) itself reacts in the normal Knorr fashion to give the β-acetylpyrrole (**99**). There was originally some ambiguity concerning which acetyl group had been cleaved in the reaction; however, this was resolved both by use of a different keto-ester[93] (showing that the acyl group was lost, and the same product obtained) and by ^{14}C labelling experiments.[95] The two different reaction pathways have been attributed to the effect of the alkyl group in reducing the acidity of the β-diketone component, and thus rendering it less likely to undergo condensation with the carbonyl group in the amino-keto-ester.

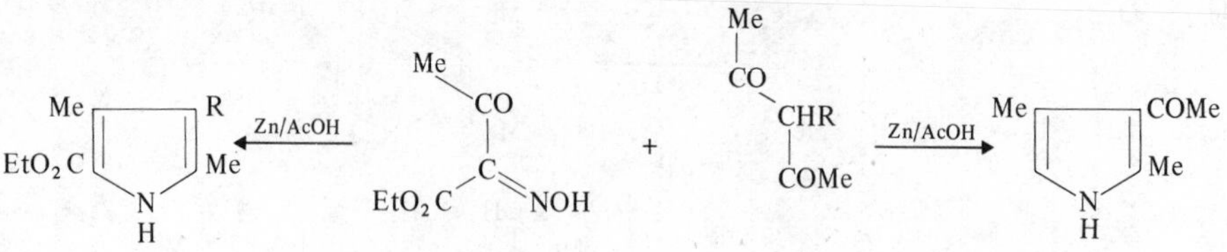

(98) (97) a; R = alkyl b; R = H (99)

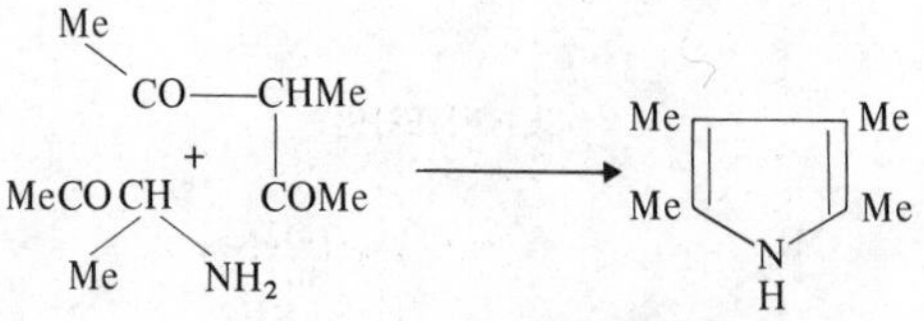

(100)

Further applications of this new procedure made it one of the most versatile pyrrole syntheses and even tetra-alkylpyrroles may be prepared directly in this way, *e.g.* (**100**).[96]

The amino-ketone moiety may be derived not only from malonates or β-keto-esters, but also from β-diketones or β-ketoamides, the products being α-acylpyrroles or pyrrole amides, respectively. Furthermore, unsymmetrical diketones may also be utilized[97,98] and the reaction is usually regiospecific, *e.g.* (**101**)→(**102**). A variant of this method, due to Plieninger,[99] is to acylate the enamine (**103**) derived from amino-malonic ester and a β-keto-ester and to cyclize the product to an α-methyl-β'-alkylpyrrole (**104**) as shown.

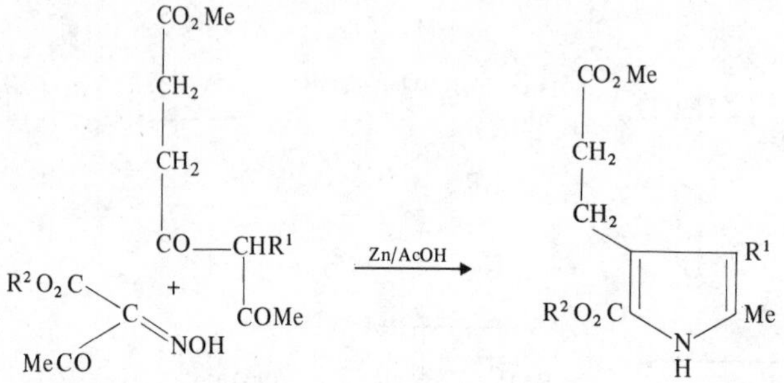

(101) a; R^1 = Me b; $R^1 = CH_2CO_2Me$ (102)

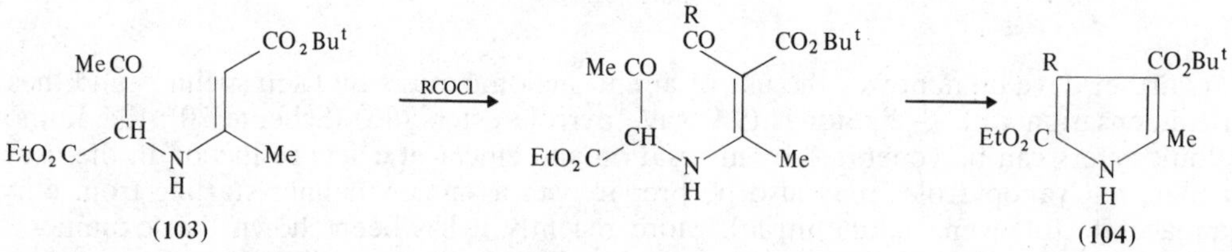

(103) (104)

17.1.12.2 Hantzsch synthesis

This method[100] involves the condensation of an α-haloketone with a β-keto-ester in the presence of ammonia (or an amine) and is closely akin to the Feist–Benary synthesis[101] of furans. The probable reaction pathway is shown in Scheme 21, the intermediate amino-crotonic ester (**105**) undergoing β-alkylation in the usual manner expected of an enamine. The Hantzsch synthesis has been somewhat overshadowed by the Knorr synthesis, but is useful for the preparation of 2,5-dialkyl- and 2,4,5-trialkyl-pyrrole-3-carboxylic esters.[102]

SCHEME 21

The scope of the synthesis has recently been extended by McDonald[103] by use of α-halo-derivatives of other aldehydes, and he has shown that in certain cases benzyl and t-butyl 4-carboxylates are accessible.

The Feist method[104] is formally analogous to the Hantzsch synthesis, and involves condensation of acyloins with aminocrotonic ester in acetic acid in presence of zinc chloride (Scheme 22), whilst Grob[105] has shown that α-chloroacrylonitriles will condense

SCHEME 22

with aminocrotonic esters to form cyanopyrrolines which undergo fragmentation of hydrogen cyanide on heating to form pyrroles (Scheme 23). Nitropropene has also been utilized in a similar process.[106]

SCHEME 23

17.1.12.3 Paal–Knorr synthesis

This method[107] involves the condensation of a 1,4-diketone with ammonia, or a primary amine, and in general gives pyrroles in very good yield (Scheme 24); a large number of examples of its use have been summarized by Gossauer,[8n] but the method is somewhat limited by the availability of appropriate γ-diketones. Use of hydroxylamine or hydrazine

SCHEME 24

instead of an amine leads to *N*-hydroxy- or *N*-amino-pyrroles.[108,109] The original preparation of pyrrole itself by dry distillation of the ammonium salt (**106**) of mucic acid is probably also an example of the Paal–Knorr synthesis, and it has since been extended to *N*-alkylpyrroles.[109]

(**106**)

17.1.12.4 Other syntheses

Several other methods have also been devised for the formation of pyrrole derivatives,[8] but most have had only limited application. For example, dipolar addition of azirines (or of nitrile ylides) to acetylenes has been exploited by Huisgen, *e.g.* (**107**)→(**108**); oxazolium oxides, thiazolium oxides, or imidazolium imines will also serve as dipolarophiles in similar reactions to form pyrroles.[110,111]

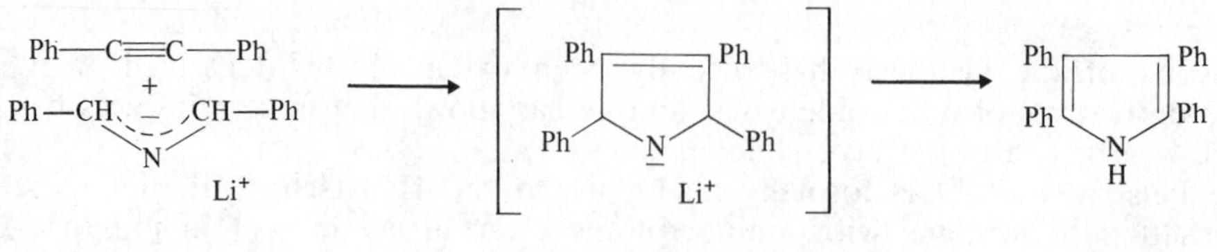

(107) (108)

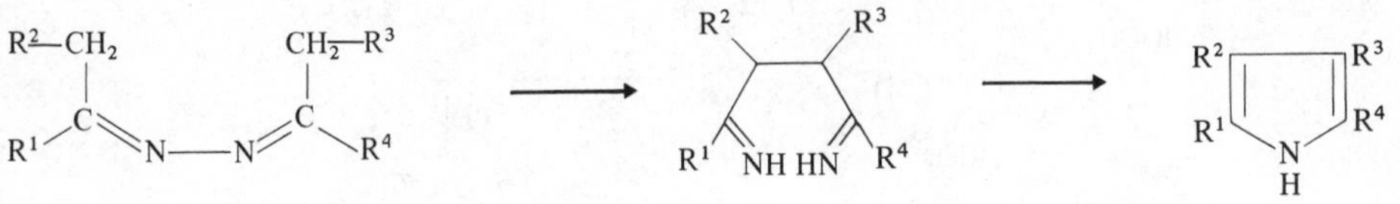

(109) (110)

The Piloty synthesis[112] in which dialkylketazins (**109**) undergo thermal rearrangement in presence of zinc chloride to form pyrroles (**110**) is an interesting method because of its direct analogy with the Fischer indole synthesis. Pyrroles have also been synthesized from furans by heating with ammonia or primary amines at high temperature in the gas phase.[113] α-Diketones (**111**) will condense with suitably activated dialkylamines (**112**) to

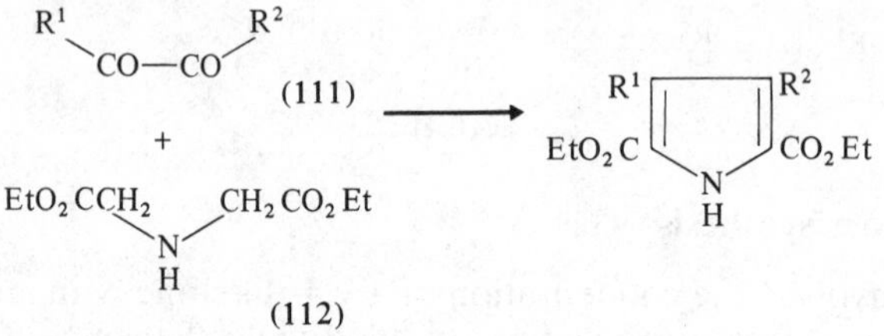

(111) (112)

form pyrroles;[114] a similar procedure has also been utilized in furan and thiophen syntheses. α-Free pyrrole-α'-carboxylic esters (**115**) may be synthesized[115] from tosyl glycine esters (**113**) and α,β-unsaturated ketones (**114**), as shown in Scheme 25.

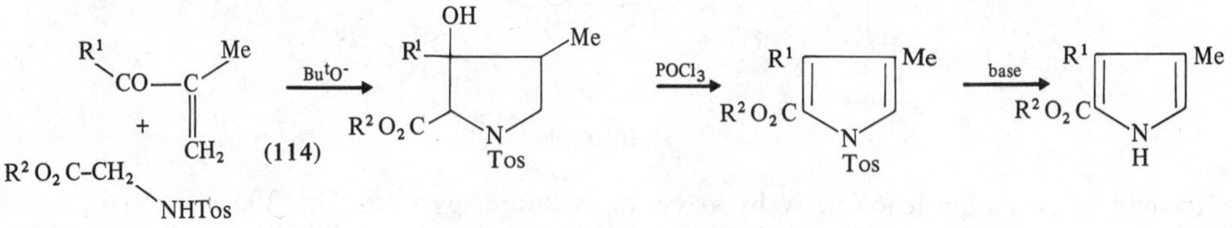

(113) (114) (115)

SCHEME 25

17.1.13 REACTIONS OF PYRROLE DERIVATIVES

The reactivities of various substituents and side-chain derivatives of pyrrole show considerable variations from those of the corresponding benzenoid derivatives, owing to the electron-releasing character of the pyrrole nucleus.

17.1.13.1 α-Alkylpyrroles

These compounds (**116**) readily undergo halogenation and acetoxylation at room temperature, or even below, to form mono-, di-, and even tri-substituted derivatives (**117a–c**). These can be utilized synthetically to prepare pyrrole aldehydes (**118**), carboxylic acids (**119a**), amides (**119b**), or esters (**119c**) (Scheme 26) and in the preparation of dipyrromethanes and dipyrrylmethenes (see below).

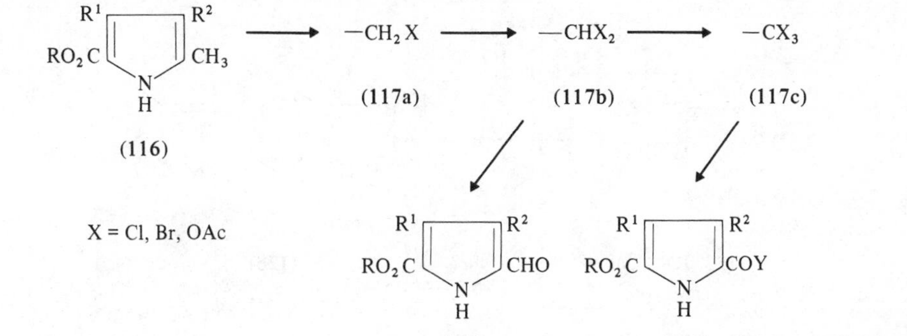

(116) → $-CH_2X$ (117a) → $-CHX_2$ (117b) → $-CX_3$ (117c)

X = Cl, Br, OAc

(118) (119) Y = OH, OR^3, NR_2

SCHEME 26

17.1.13.2 Pyrrylcarbinols

Although relatively stable in neutral or alkaline media, these compounds (**120**) are very unstable in acid, because of the readiness with which they form methylenepyrrolenine carbonium ions (**121**); if other positions are unsubstituted, polymers are formed, but otherwise dipyrrylmethanes (**122**) are produced. Lithium aluminium hydride, or diborane, reduction[116] results in formation of the corresponding alkylpyrrole (**116**), presumably by hydride addition to an intermediate methylenepyrrolenine, *cf.* (**121**). The methiodides (**125**) of the related Mannich bases (**124**), which may be formed[117] from the haloalkylpyrroles (**117a**) or by direct reaction of a pyrrole with formaldehyde and a secondary amine,[118] will undergo reaction with nucleophiles such as cyanide or acetamidomalonic esters in presence of base to give excellent yields of substituted pyrroles,[119] *e.g.* (**125**) and (**126**), respectively.

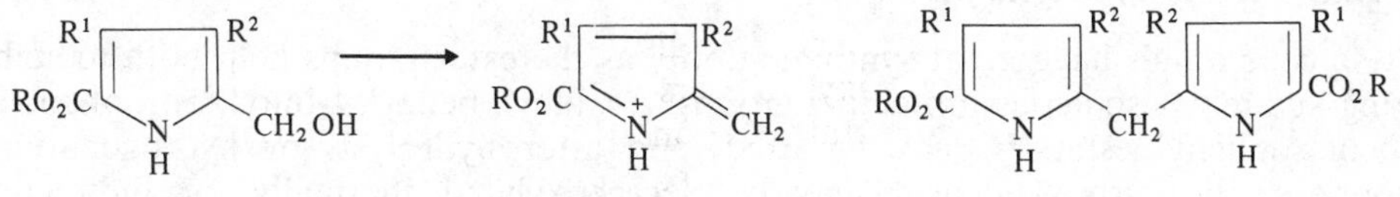

(120) → (121) (122)

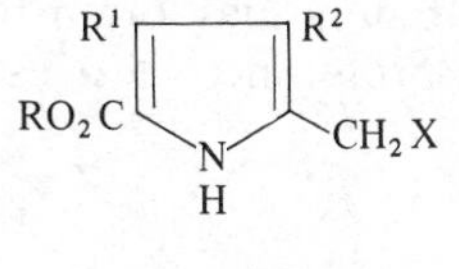

(123) X = NMe_2 (125) X = CN

(124) X = $\overset{+}{N}Me_3$ (126) X = $C(CO_2Et)_2NHCOMe$

17.1.13.3 Pyrrole aldehydes and ketones

In general, these compounds (**127**) are less reactive than the corresponding benzenoid derivatives and in some respects may be regarded as vinylogous amides, *e.g.* they undergo protonation or oxidation. The aldehydes do not undergo the Cannizzaro or Perkin reactions but will condense with a variety of compounds containing active methylene groups, *e.g.* cyanoacetic ester, malononitrile, nitromethane, malonic ester, *etc.*[80] They will also react with other unsubstituted pyrroles under acidic conditions to form dipyrrylmethenes (**128**). The carbonyl group is reduced to methylene by lithium aluminium hydride, by diborane,[116] or by the Wolf-Kishner method, and this is a useful method for the synthesis of alkylpyrroles.

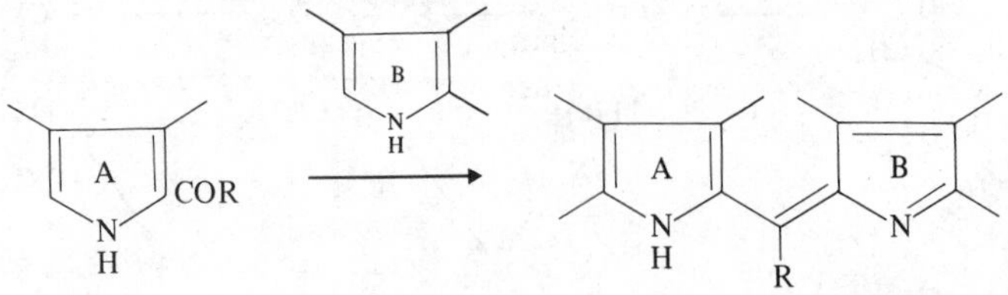

(127) R = H, alkyl (128)

17.1.13.4 Pyrrolecarboxylic acids

These derivatives (**129**) are considerably less stable than benzoic acid, and often readily decarboxylate on heating, although much higher temperatures are required[120] if one, or more, additional electron-withdrawing substituents are present (*e.g.* **130**). However, in these cases decarboxylation may be assisted by heating in ethanolamine,[121] or by iodination followed by catalytic reduction[122] [*cf.* (**23**)→(**25**)]. Carboxyl groups may also be displaced directly in some cases by nitration, bromination, or even by diazo-coupling.

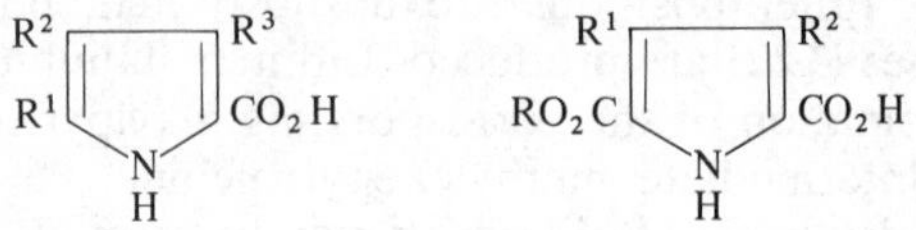

(129) R^1, R^2, R^3 = alkyl (130)

17.1.13.5 Pyrrolecarboxylic esters

These compounds have great synthetic utility as the ester groups help both to stabilize the ring and may also act as protecting groups, *e.g.* alkyl, benzyl, t-butyl, nitrophenyl, and pentachlorophenyl esters[123] may be used, and after hydrolysis or hydrogenation, as appropriate, the carboxylic acid may be decarboxylated thermally, or indirectly, as indicated above. t-Butyl esters may often be de-esterified and decarboxylated in one step by treatment with cold trifluoroacetic acid. β-Esters may often be hydrolysed preferentially by warm concentrated sulphuric acid, in preference to α-esters.[1]

17.1.14 HYDROXYPYRROLES

α-Hydroxypyrroles (**131**) exist predominantly in the tautomeric pyrrolin-2-one forms with the double bond in the Δ^3- or Δ^4-position.[78,123,124] In the unsubstituted compound, its 4-alkyl (**132**), and 3-acyl derivatives (**133**), the Δ^3-tautomer predominates, as shown by n.m.r. spectroscopic measurements; on the other hand, if acyl or ester substituents are present in the 4-position, then the Δ^4-isomer (**134**) is favoured. They may by synthesized either directly by oxidation of pyrroles unsubstituted in one, or both, α-positions with hydrogen peroxide in pyridine[78] (see above) or by ring synthesis. Examples of the latter include the reduction and cyclization of cyanohydrins (**135**) derived from β-keto-esters[124]

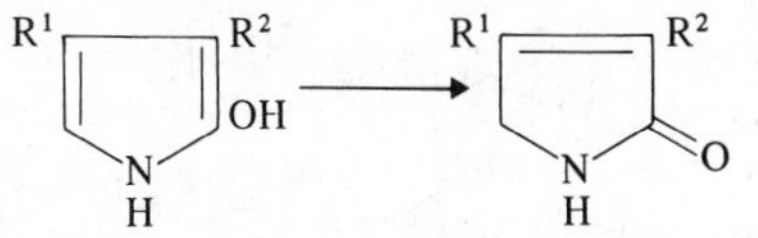

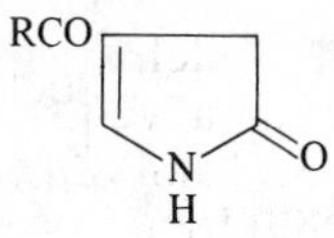

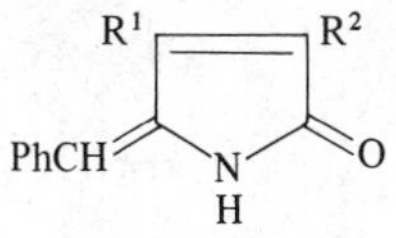

(131) (132) R^1 = alkyl, R^2 = H (134) (137)
(133) R^1 = H, R^2 = acyl

(Scheme 27) and the cyclization of acylsuccinic esters (**136**) with ammonia or primary amines[125] (Scheme 28). Other syntheses are summarized by Gossauer.[8p]

The 5-unsubstituted Δ^3-pyrrolin-2-ones react readily with aldehydes and even ketones to give orange products,[124] *e.g.* (**137**). Treatment with acetic anhydride[123] in presence of

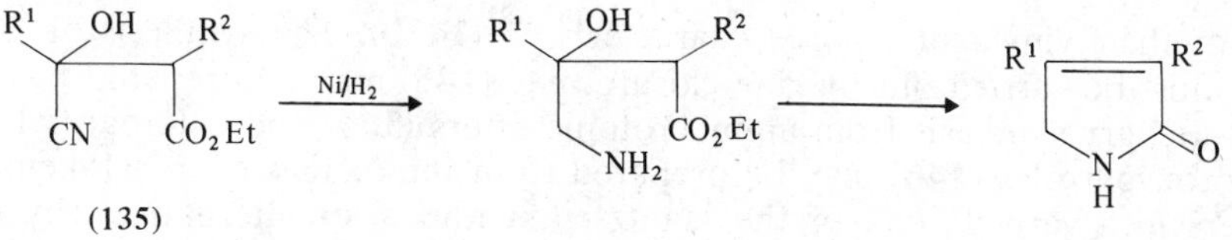

(135)

SCHEME 27

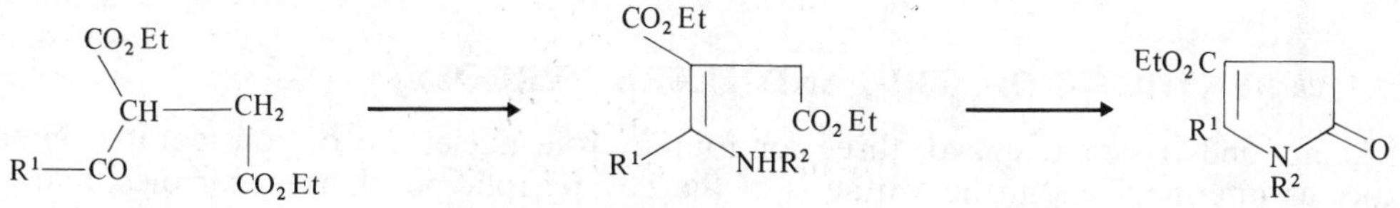

(136)

SCHEME 28

pyridine, or concentrated sulphuric acid, affords the acetoxypyrroles (**138**); the corresponding alkoxypyrroles (**139**) may be prepared by *O*-alkylation[126,127] with diazomethane, dimethyl or diethyl sulphate, or Meerwein's reagent (triethyloxonium fluoroborate). Diborane reduction of the latter results[127] in cleavage of the alkoxy-group and affords a useful method for preparing α,α'-diunsubstituted pyrroles (**140**).

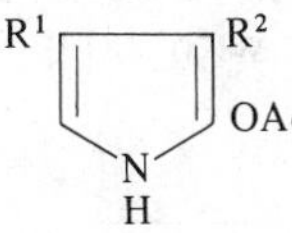

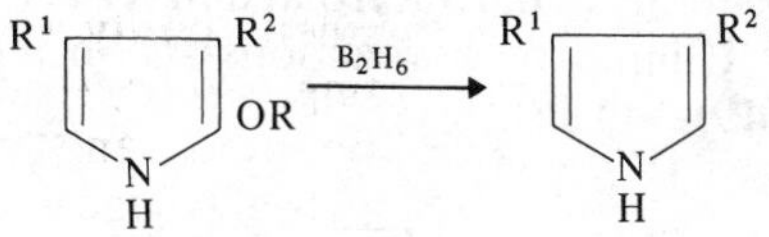

(138) (139) R = Me, Et (140)

Like the α-analogues, the simple alkyl β-'hydroxy'-pyrroles (**141**) also exist largely in the 3-keto-form, *e.g.* (**142**), although the presence of an ester substituent in the 2-position favours[128] the 3-hydroxy-tautomer, *e.g.* (**143**). The β-oxygen functions can be alkylated[128] and acylated[129] in the same way as with the α-analogues, and even the β-hydroxypyrroles which exist largely in the keto-form (**142a**) do not show the normal properties of a ketone

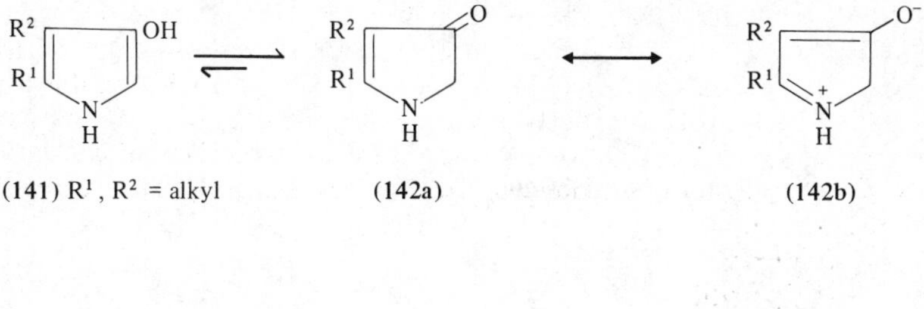

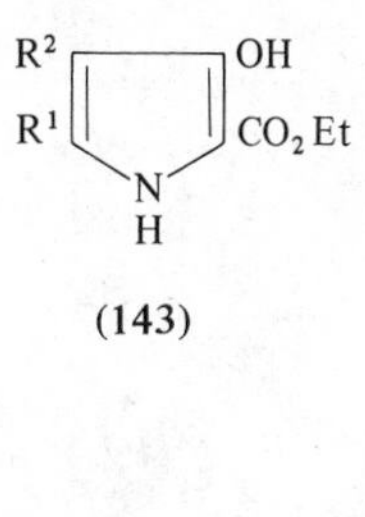

(141) R^1, R^2 = alkyl (142a) (142b) (143)

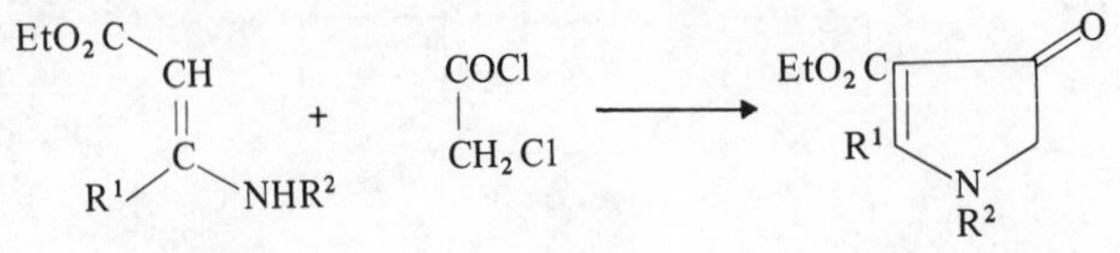

(144) (145)

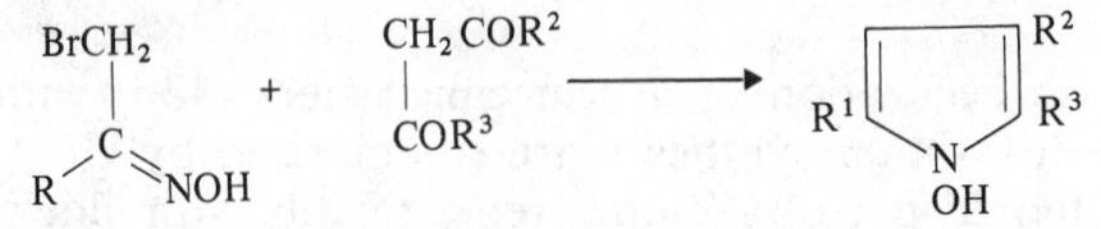

(146)

because of their vinylogous amide character[130,98] (**142b**). The synthesis of β-'hydroxy'-pyrroles must be carried out by ring closure, *e.g.* (**145**) may be prepared by variation of the Feist–Benary synthesis from aminocrotonic esters (**144**) and chloroacetyl chloride.[98]

N-Hydroxypyrroles (**146**) may be prepared from the oximes of α-haloketones and β-keto-esters in a variant[108,131] of the Hantzsch synthesis, or alternatively by Paal–Knorr synthesis from hydroxylamine and 1,4-diketones. They are reduced to the parent pyrroles by zinc and acetic acid.[108]

17.1.15 SYNTHETIC DI-, TRI-, AND TETRA-PYRROLES

Compounds containing two, three, or four pyrrole nuclei are of considerable importance as intermediates in the synthesis of the tetrapyrrolic porphyrins, corroles, and bile pigments,[132,133] and also of the tripyrrolic prodigiosins. The pyrrole nuclei are joined by one carbon fragment in the porphyrins and bile pigments, but a direct pyrrole–pyrrole linkage also occurs in the corroles and prodigiosins. Biosynthetic studies have also recently provided a considerable impetus to synthetic work in this area.[134]

17.1.15.1 Bipyrroles

Four different types are known,[8,135] 1,1′-, 2,2′-, 2,3′-, and 3,3′-bipyrroles. The 2,2′-bipyrroles are the most important and the parent compound (**148**) may be prepared by Vilsmeier condensation[136] of butyrolactam (**147**) (Scheme 29). α-Bromo- or α-iodopyrroles undergo Ullman-type reactions[137] on heating with copper powder to form bipyrroles, *e.g.* (**149**)→(**150**).

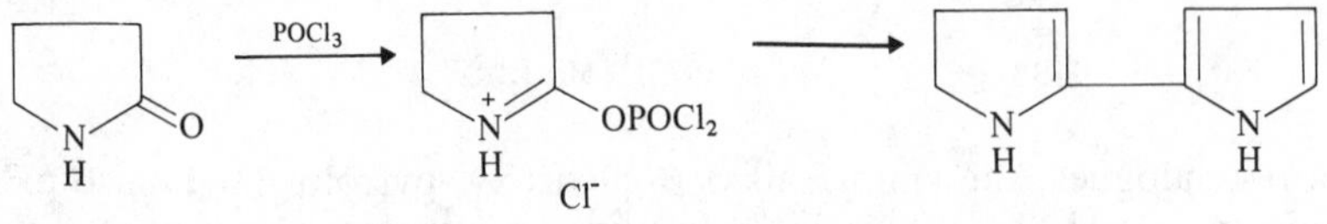

(147)

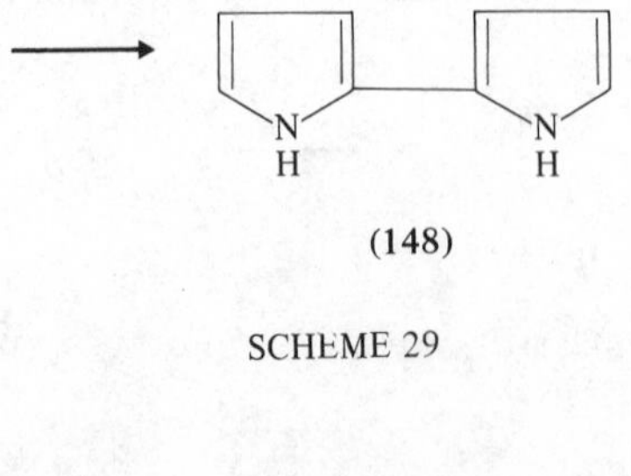

(148)

SCHEME 29

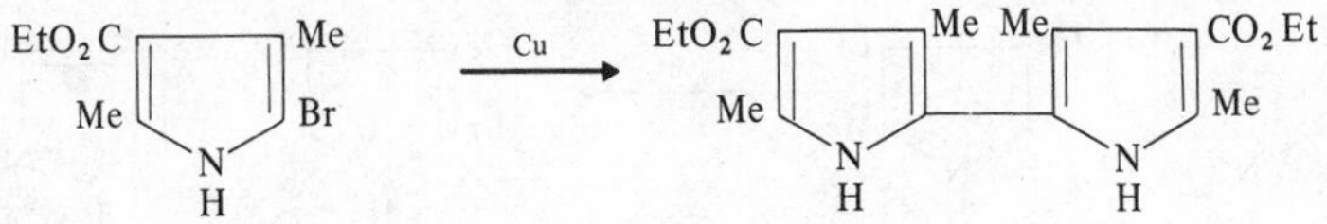

(149) (150)

17.1.15.2 Dipyrrylmethanes

As with the bipyrroles, different types of dipyrrylmethanes are known, but again the most important class are the 2,2′-dipyrrylmethanes. The parent compound and simple alkyl derivatives are not very stable,[1] and are sensitive to aerial oxidation and to acids. However, the presence of one or more electron-withdrawing groups results in a considerable increase in stability. Symmetrical dipyrrylmethanes (**152**) may be synthesized by acid-catalysed self-condensation[138,139] of α-halomethyl-, α-acetoxymethyl- or α-methoxymethyl-pyrroles (**151**), whilst unsymmetrical pyrromethanes (**154**) are available[140,141] by condensation of these pyrroles (**151**) with α-unsubstituted pyrroles (**153a**).

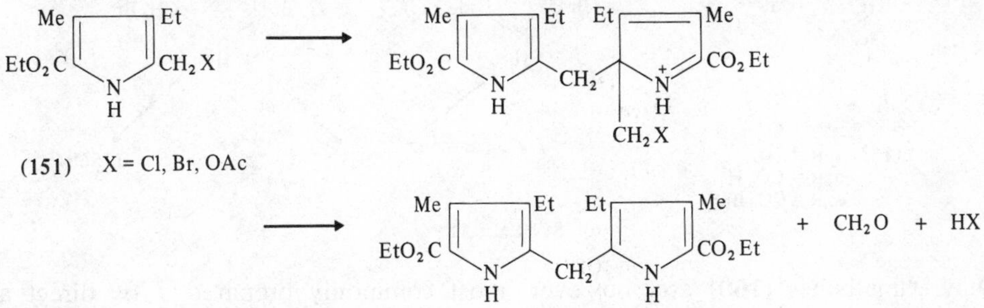

(151) X = Cl, Br, OAc

(152)

The latter reaction has been extended by the use of α-carboxypyrroles (**153b**) in which the carboxyl group is displaced during the course of the reaction (Scheme 30). In a further extension[140] of this type of synthesis the pyridinium methyl analogues (**151**) may be condensed with the lithium or sodium salts (**153c**) of the carboxylic acids (**153b**), and

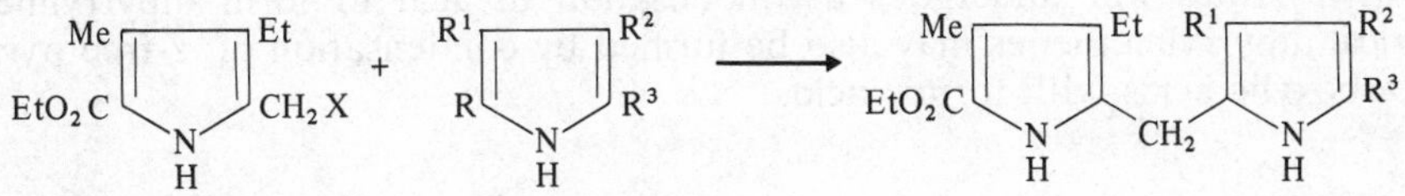

(151) X = Cl, Br, OAc, $\overset{+}{N}C_5H_5$ (153) a; R = H b; R = CO_2H c; R = CO_2^- (154)

SCHEME 30

other amino-methyl analogues of (**151**) have also been utilized in similar syntheses.[141] In some cases the use of acid or Friedel–Crafts catalysts may facilitate the reaction.

Symmetrical dipyrrylmethanes may in some cases also be prepared by direct condensation of formaldehyde with an α-free pyrrole in which the other three positions are blocked by substituents; if aldehydes other than formaldehyde are employed, *meso*-substituted pyrromethanes are formed.

17.1.15.3 Dipyrrylmethenes

Like the bipyrroles and dipyrrylmethanes, the 2,2′-dipyrrylmethenes are the most important isomers. They may be prepared by direct oxidation of dipyrrylmethanes, *e.g.* with bromine or sulphuryl chloride; α-methyl groups may also undergo substitution

(155) (156) a; X = H
b; X = Br

simultaneously, and nuclear carboxyl groups will be displaced, *e.g.* (**155**)→(**156a**). Dipyrrylmethene (**156a**) may also be prepared directly[142] by bromination of kryptopyrrole (**157a**) (Scheme 31); the precise sequence of steps in the reaction is unknown, but the bromomethene (**156b**) may also be formed. Pyrrolecarboxylic acids, *e.g.* (**157b**), may also be used, and with t-butyl esters, *e.g.* (**157c**), products of type (**156b**) may be formed exclusively, the ester group being removed either directly by the bromine or by hydrogen bromide generated in the reaction.[143]

(157) a; R = H
b; R = CO_2H
c; R = CO_2Bu^t

SCHEME 31

Dipyrrylmethenes (**160**) are, however, most commonly prepared[cf.2] by direct acid-catalysed condensation of a pyrrole aldehyde (**158**) with an α-unsubstituted pyrrole (**159a**) or the corresponding carboxylic acid (**159b**). A possible complication[144] in this synthesis is that the initially formed dipyrrylmethene (**160**) may react further with the α-free pyrrole (**159a**) by nucleophilic addition of the methine bridge to form a tripyrrylmethane (**161**); the latter may then revert to the initial dipyrrylmethene (**160**), or alternatively cleave to give the symmetrical product (**162**) (Scheme 32).

α-Formylpyrroles will undergo self-condensation in acid to form dipyrrylmethenes; symmetrical dipyrrylmethenes may also be formed by condensation of α-free pyrroles or pyrrolecarboxylic acids with formic acid.

(158) a; R = H
b; R = CO_2H
(159) (160)

(162) (161)

SCHEME 32

Dipyrrylmethenes are bright orange in colour (λ_{max} 440–450 nm) and form deep red salts (λ_{max} 500 nm) with acids and chelate with metal ions. Their structures may be represented as a resonance hybride of two forms (**163a**) and (**163b**). The central methine carbon is susceptible to nucleophilic attack, *e.g.* by alcohols, water, cyanoacetic acid,[145] malononitrile,[146] Grignard reagents,[147] *etc.*, and dipyrrylmethenes are readily reduced to the corresponding dipyrrylmethanes. Indeed, dipyrrylmethene-5,5′-dicarboxylic esters are almost unknown because of the ease with which hydration of the central methene linkage occurs[148] to form the *meso*-hydroxydipyrrylmethane.

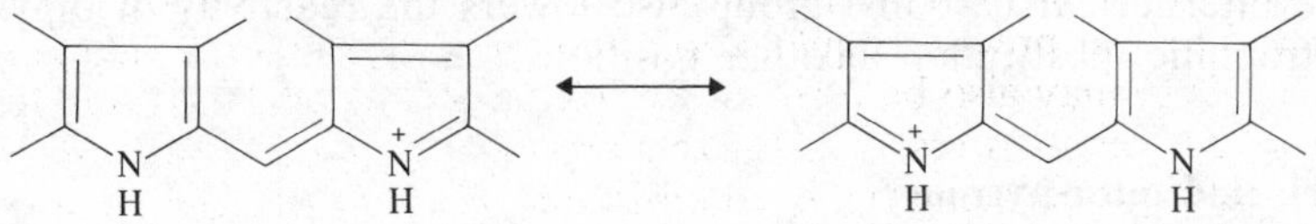

(**163a**) (**163b**)

17.1.15.4 Dipyrryl ketones

Various types of dipyrryl ketones are known, but as with the preceding dipyrryl compounds discussion will be confined to the 2,2′-isomers, because of their importance as intermediates in porphyrin synthesis. The original methods of preparation involved the reaction of pyrrole Grignard reagents with pyrrole acid chlorides, or Friedel–Crafts type reactions,[cf.1] and 2,2′-dipyrryl ketone (**165**) itself was prepared by reaction of phosgene with the pyrrole Grignard reagent (**164**). However, a variation of the Vilsmeier–Haak

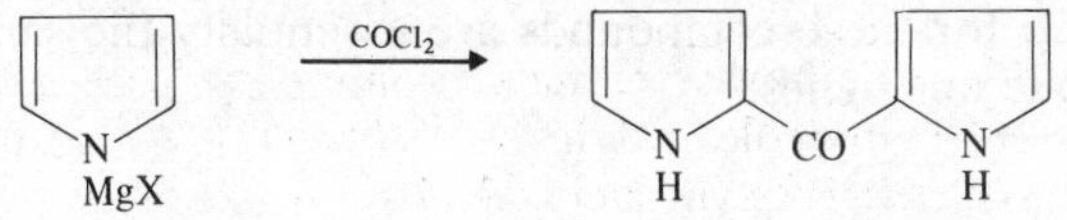

$COCl_2$

(**164**) (**165**)

method is now the most widely used method[149] and involves the coupling of a phosphoryl chloride activated pyrrole amide (**166**) with an α-free pyrrole (**167**) (Scheme 33). In some instances dipyrrylmethanes may also be oxidized to dipyrryl ketones.[150]

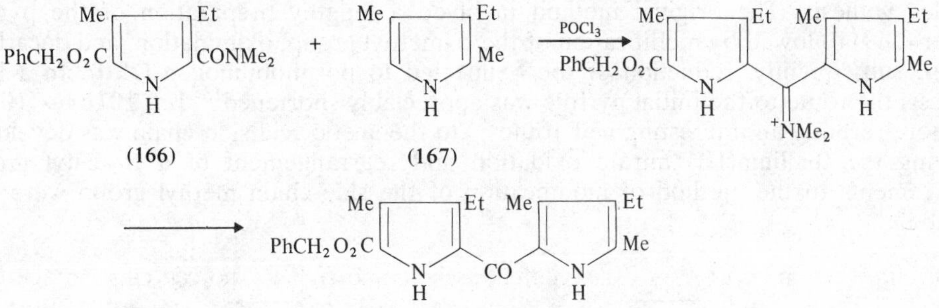

(**166**) (**167**)

SCHEME 33

The carbonyl group in dipyrryl ketones is very unreactive towards the usual ketone reagents,[1] although azines may be formed on heating to high temperatures with hydrazine. Furthermore, they are only reduced very slowly by borohydrides, whereas diborane effects rapid reduction to dipyrrylmethanes, probably *via* an intermediate dipyrrylmethene.[149] This type of reactivity is attributed to the vinylogous amide character of dipyrryl ketones, as expressed in the dipolar resonance forms (**168b**) and (**168c**). This is supported by the low-frequency carbonyl band (1550 cm^{-1}) in the infrared, and the relatively high degree

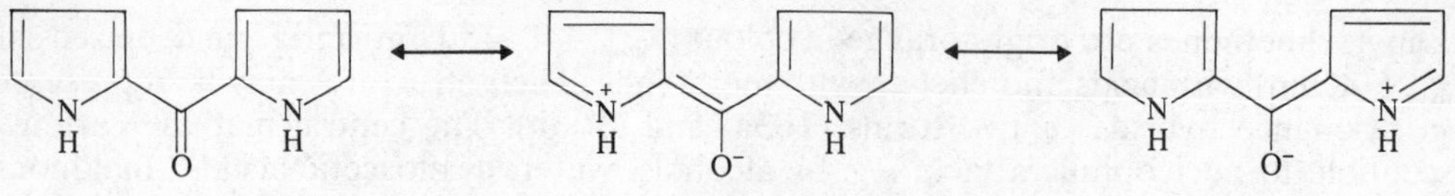

(168a) (168b) (168c)

of conjugation shown by the principal absorption band (*ca.* 360 nm) in the ultraviolet region; in acid, protonation occurs on oxygen with formation of an orange salt (λ_{max} *ca.* 420 nm). The internuclear carbonyl group also lowers the reactivity of dipyrryl ketones towards electrophilic substitution at other positions.

17.1.15.5 Tri- and tetra-pyrroles

Before 1950 the classical Fischer synthesis[2] from dipyrrylmethenes was virtually the only general method for porphyrin synthesis. Since then there has been a growing interest in the rational synthesis[132] of porphyrins *via* di- and tetra-pyrroles. Further interest has also been stimulated by the desire to synthesize a range of macrocycles related to the corrin nucleus of vitamin B_{12} which contain a direct pyrrole–pyrrole linkage, and also by studies of possible biosynthetic intermediates[141] *en route* to porphyrins and corrins. These syntheses have been discussed in detail elsewhere[132] (see also Section 17.2), but suffice it to say that tripyrrenes, and the tetrapyrrolic bilenes, biladienes, bilatrienes, and *a*- and *b*-oxobilanes have all been utilized in porphyrin synthesis; likewise, corrole derivatives have been synthesized either by forming the direct pyrrole–pyrrole linkage in the final ring-closure step, or from a tetrapyrrole with a central pyrrole–pyrrole linkage.[132] The synthetic methods used for these compounds are essentially the same as those described above for the dipyrrolic analogues.

17.1.16 PORPHOBILINOGEN

The most important natural monopyrrole is porphobilinogen (**170**), the biosynthetic precursor of porphyrins, chlorophylls, and vitamin B_{12} (see Section 30.2). It was originally isolated[151] in crystalline form from the urine of a porphyric patient in 1952, and its structure determined[152] by Cookson and Rimington in 1954. Several syntheses have now been described, and Scheme 34 illustrates an approach based on the use of the Knorr pyrrole synthesis. The original method involved a lengthy preparation of the pyrrole triester (**169**) followed by modification of the α-methyl group, oximination, and decarboxylation; subsequently, reduction of the oxime led to porphobilinogen (**170**). In a later synthesis the route to the initial pyrrole was appreciably shortened[154] [*cf.* **101b**)→(**102b**)] and more recently an interesting new route[155] to the acetic acid side-chain was developed involving the thallium(III) nitrate oxidation and rearrangement of a β-acetyl group; improvements to the method of modification of the side-chain methyl group were also described.[155]

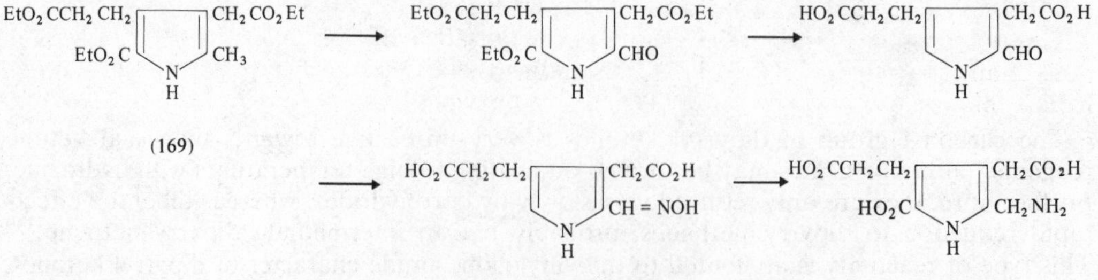

(169) (170)

SCHEME 34

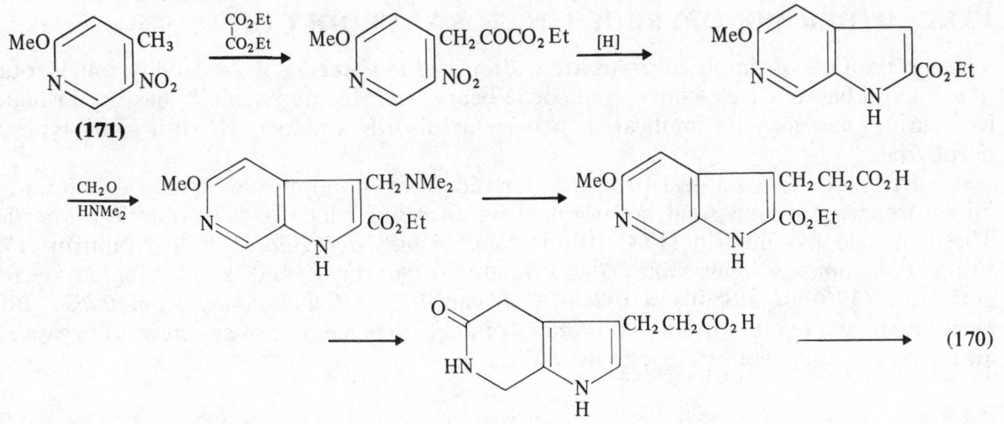

(171)

(172)

SCHEME 35

An entirely different approach was adopted by Rapoport[156] (Scheme 35) starting from the 5-nitropyridine (**171**), and this method has been extended and improved[157,158] and has proved to be particularly useful for synthesizing labelled compounds in connection with biosynthetic studies; the relatively stable lactam (**172**) of porphobilinogen is the primary product of this synthesis and it may be hydrolysed to porphobilinogen as required. More recently a novel route to porphobilinogen (Scheme 36) has been reported by Evans.[159] It

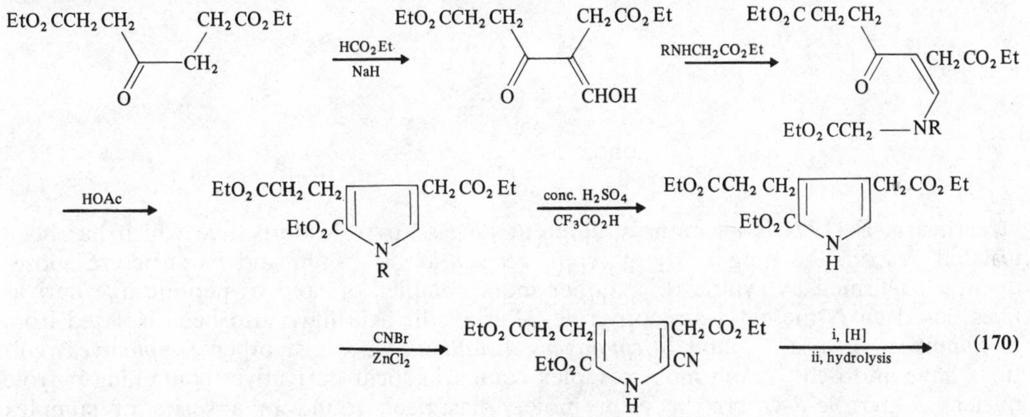

SCHEME 36

is interesting to note that the biosynthesis of porphobilinogen from aminolaevulinic acid (**173**), which involves a Knorr-type condensation, has also been imitated in the laboratory but in low yield, although enzymic catalysis has been used successfully.

Porphobilinogen is a colourless crystalline material when first prepared but slowly turns pink in air, and rapidly polymerizes[160] in acid or alkaline solution in the presence of air to form a mixture of isomeric uroporphyrins. The enzymic condensation to form the type III, and type I, isomers is discussed in Section 30.2.

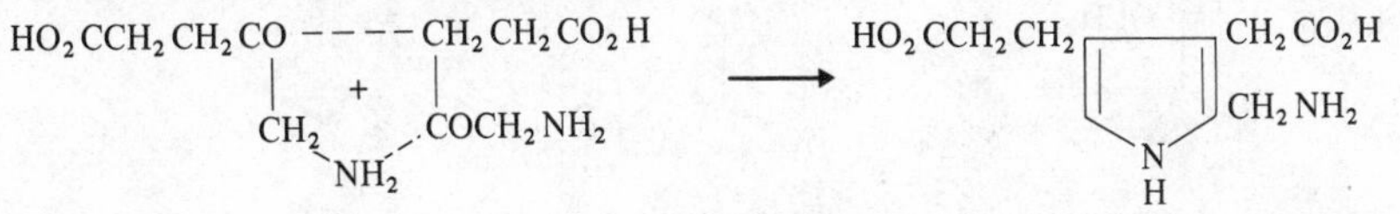

(173) (170)

17.1.17 OTHER MONOPYRROLIC NATURAL PRODUCTS

Trace amounts of simple pyrroles are widespread in nature, *e.g.* 2-acetylpyrrole is found in tea and tobacco leaves, coffee, and cocoa beans, *etc.* Haemopyrrole[161] has been isolated from urine, and may be implicated in nervous disorders associated with some types of porphyrias.

A variety of other natural products derived from monopyrroles have been found in micro-organisms, plants, and high organisms, many of which have antibiotic properties. These include pyoluteorin (**174**) (from *Pseudomonas aeruginosa*),[162] pyrrolnitrin (**175**) (from *Pseudomonas pyrrocinia*[163] and related organisms), and a brominated pyrrole derivative (**176**) isolated from a marine organism[164] (*Pseudomonas bromoutilis*). Both pyrrolnitrin and pyoluteorin and a number of analogues have been synthesized because of an interest in their antibiotic activity.[162,163]

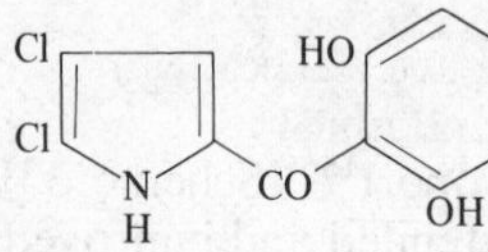

(174) 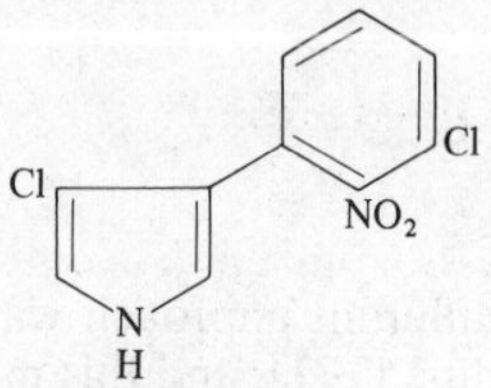 (175)

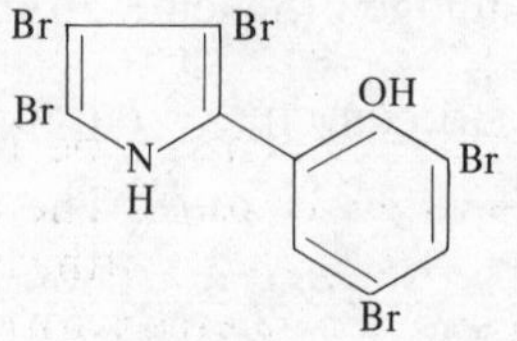

(176)

Verucarin-E (**177**) is an entirely different type of pyrrole derivative which has been isolated[165] from the fungus *Myrothecium verrucaria* by Tamm and its structure subsequently confirmed by synthesis.[166] Other more complex di- and tri-peptide-like derivatives based on *N*-methyl-4-aminopyrrole-2-carboxylic acid have also been isolated from *Streptomyces netropsis*[167] and *Streptomyces distallicius*,[168] whilst other *Streptomyces* cultures have afforded[169] even more complex coumarin sugar derivatives containing pyrrole nuclei. A pyrrole-2-carboxylic ester moiety has been found in a series of complex alkaloids isolated from the leguminous plants *Ryania speciosa*,[170] *Virgilia oroboides*,[171] and *Readea membranaceae*,[172] and bachrotoxin (**178**), a highly toxic steroidal alkaloid secretion[173] of the South American frog *Phyllobates*, also contain a pendant pyrrole ester unit.

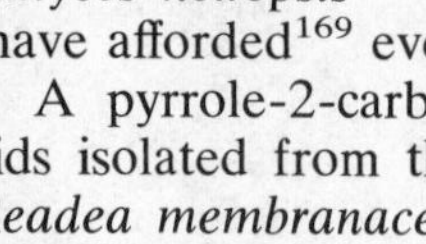

(177) 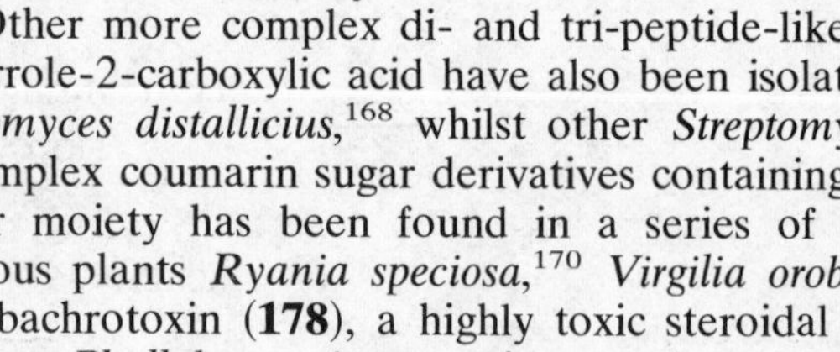 (178)

The 4-methylpyrrole-2-carboxylic ester[174] (**179**) is a trail-marking secretion of the leaf-cutting ant *Atta taxana*, and ketopyrroles (**180**) have been reported as sex-pheromones in the butterfly family *Danainae.*[175,176] Oroidin (**181**) is a complex di-bromopyrrole derivative isolated from the sponge *Agelas oroides*,[177] whilst a series of long-chain 3-alkylpyrrole-2-aldehydes have been obtained from the sponge *Oscapella lobularis*,[178] and pyrrolyl polyenes from the fungus *Wallemia sebi.*[179]

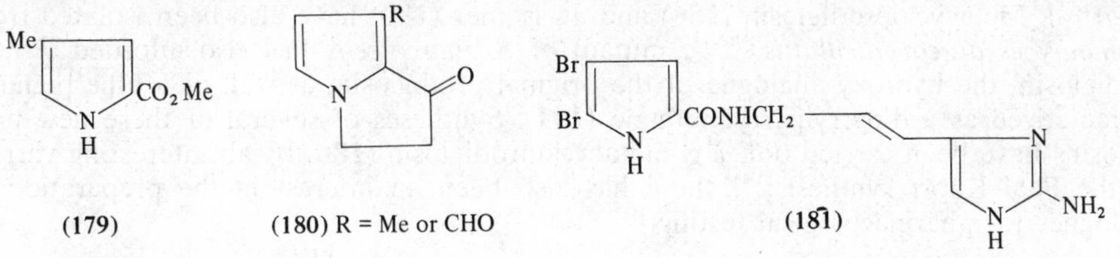

(179) (180) R = Me or CHO (181)

17.1.18 PRODIGIOSINS

The prodigiosins are an interesting series of antibacterial and antifungal orange-red pigments derived from the basic pyrryl-dipyrrylmethene unit (**182**), now given the trivial

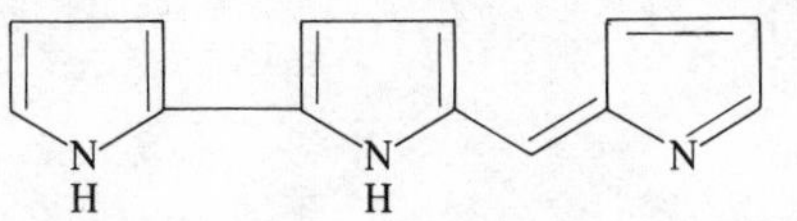

(182)

name 'prodigiosin'. The first of these pigments to be isolated was the prodigiosin (**184**) from *Serratia marcescens* (or *Bacillus prodigiosus*). The tentative structure assigned on the basis of chemical degradations by Wrede and Rotthaas[180] was later confirmed, both by isolation of a biosynthetic precursor,[181] the methoxyformylbipyrryl (**183**) from a mutant of *S. marcescens*, and by total synthesis[182] (Scheme 37); the basic tripyrrolic nucleus (**182**) has now also been synthesized.[183]

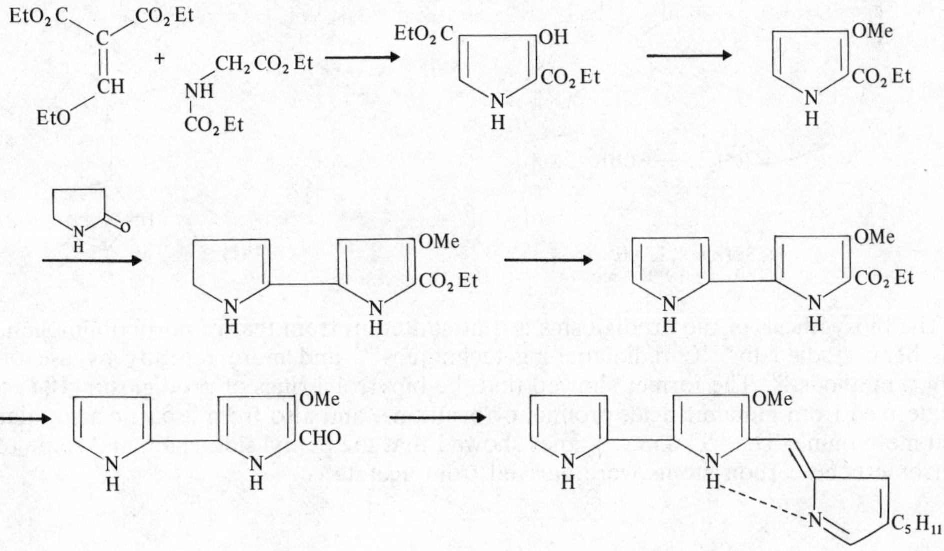

(183) (184)

SCHEME 37

Since the original discovery of prodigiosin (**184**), a variety of other prodigiosins have now been isolated, which differ from one another mainly in the substitution pattern in the terminal pyrrole ring of the dipyrrylmethene unit, *e.g.* undecylprodigiosin[184] (**185**) and metacycloprodigiosin[185] (**186**) (from *Streptomyces longisporus ruber*), nonylprodigiosin[186] (**187**) and a 2,10-nonamethyleneprodigiosin[187] (**188**) (from *Actinomadura madural*[186]), and a 2,10-(1′-methyldecamethylene)prodigiosin[188] (**189**) (from *Actinomadura pelletieri*[187]). Metacycloprodigiosin (**186**) and an isomer (**190**) have also been isolated from *Actinomyces aureoverticillatus.*[190] A mutant of *S. marcescens* has also afforded[191] norprodigiosin, the hydroxy-analogue of the original prodigiosin, as well as a blue pigment characterized as a dipyrrylpyrrylmethene (**191**). Syntheses of several of these new prodigiosins have been carried out, *e.g.* metacycloprodigiosin (**186**) by an interesting variant on the Paal–Knorr synthesis;[191] there has also been an interest in the preparation of analogues for pharmacological testing.[192]

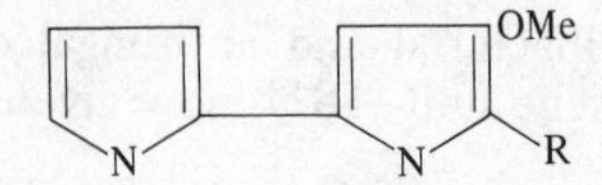

OMe N H N H R

(186) R = Et N

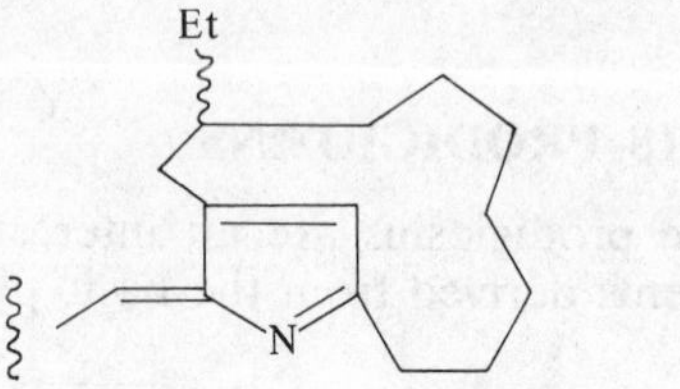

(185) R = $C_{11}H_{23}$ N

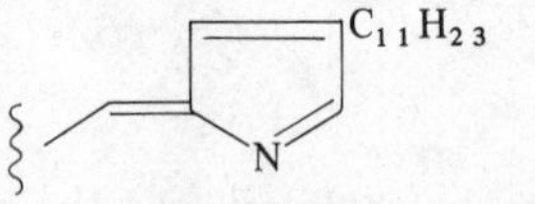

(187) R = C_9H_{19} N

(190) R = Me CH$(CH_2)_5$Me Me N

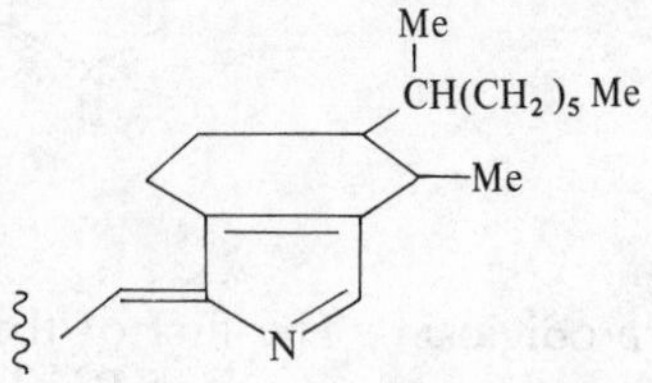

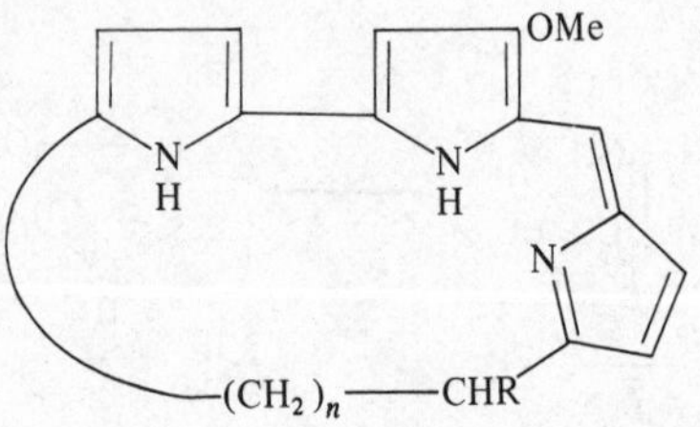

OMe N H N H N $(CH_2)_n$ CHR

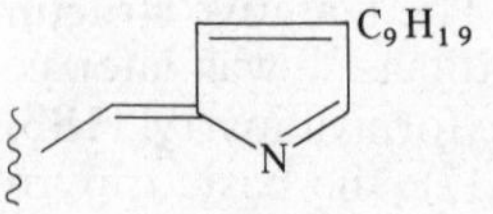

(188) $n = 8$, R = H
(189) $n = 10$, R = Me

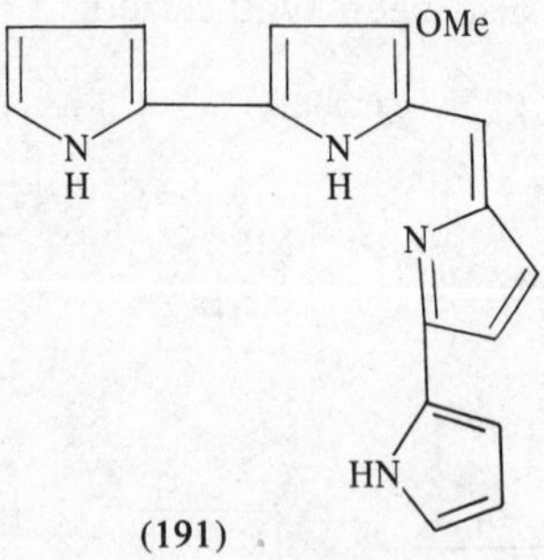

OMe N H N H N HN

(191)

The biosynthesis of the prodigiosins is quite different from that of porphobilinogen, and has been studied by ^{14}C radiolabelling techniques[194] and more recently by use of ^{13}C n.m.r. methods.[195] The former showed that the bipyrrolyl rings of prodigiosin (**184**) could be derived from glutamic acid, proline, or ornithine, and also from aspartic acid, alanine, and methionine. The ^{13}C n.m.r. studies showed that the pentyl side-chain and some of the pyrromethene carbon atoms were derived from acetate.

17.1.19 BILE PIGMENTS

Bile pigments are open-chain tetrapyrroles which arise in nature as a result of the oxidative degradation of haem, and their chemistry and mode of formation have been reviewed in several books and articles.[2,196–202] In man and the higher animals they are essentially waste products, but in some lower animals and in algae and plants they have an important biological function. The primary product of ring-opening of haem is the blue-green pigment biliverdin-IXα (**192**); this trivial name is based upon its derivation by ring-opening of the IX-isomer of protoporphyrin at the α-position. Biliverdin-IXα then undergoes reduction[201] by biliverdin reductase (with NADPH as co-factor) to the orange-yellow pigment bilirubin-IXα (**193**); the yellow pigmentation of the skin observed in jaundice is due to bilirubin, and ox-gallstones provide a convenient natural source for large quantities of bilirubin. In normal metabolism, however, bilirubin is excreted with the bile into the duodenum as a soluble glucuronide 'conjugate', but hydrolysis back to free bilirubin takes place in the lower intestine. Further reduction by bacterial organisms then leads to a complex mixture of products known as urobilins, *e.g.* (**194**), and urobilinogens, the most highly reduced being the colourless stercobilinogen (**195**).

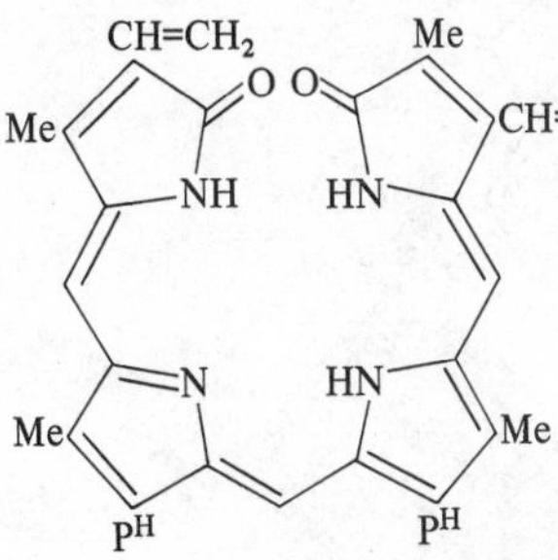

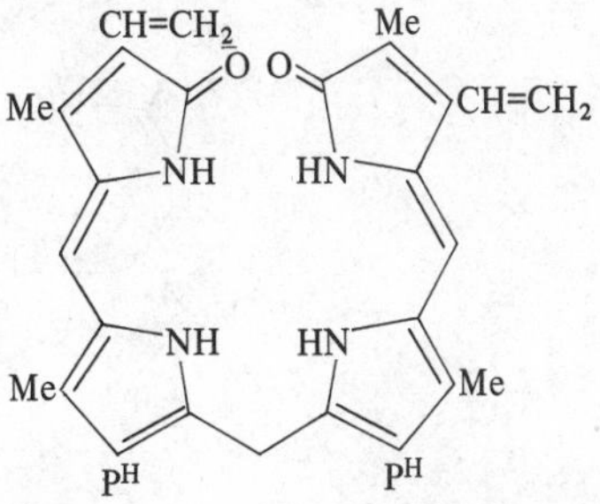

$P^H = CH_2CH_2CO_2H$

(**192**) (**193**)

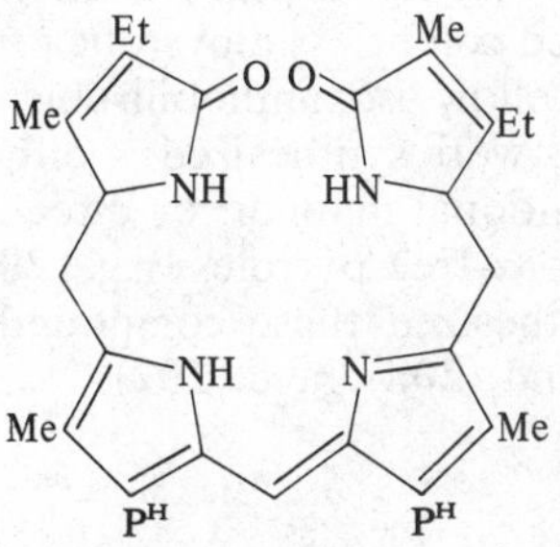

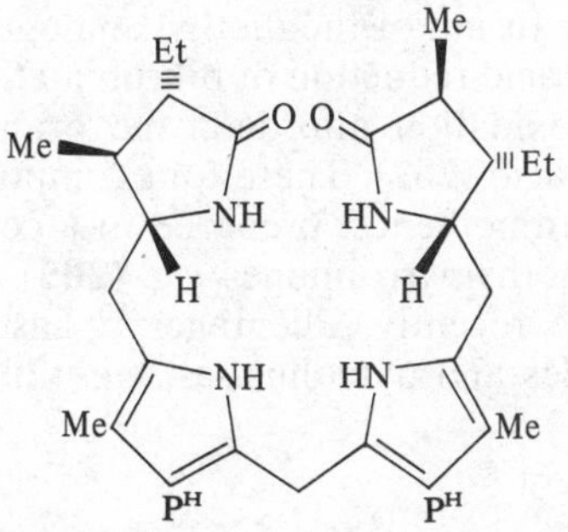

(**194**) (**195**)

Whilst the vast majority of bile pigments are derived from cleavage of haem at the α-position, the green integumental pigment of the caterpillar of the cabbage white butterfly has been shown[81,203] to be biliverdin-IX, and recently related pigments have been found in other lepidoptera.[204]

The pigments of red and blue-green algae are chromoproteins derived from bile pigments. Phycoerythrobilin[205–208] (**196a**), for example, may be obtained by cleavage of the protein under mild conditions, and phycocyanobilin[208,209] (**197**) may be obtained from phycocyanin, the photosynthetic pigment of blue-green algae. Phytochrome, which occurs in all higher plants as well as in algae, is an important photoreceptor controlling growth and development in plants, and the time of flowering; it has been difficult to obtain in sufficient amounts for full characterization but structure (**198**) has been proposed.[208] It exists in two forms, P_r (λ_{max} 724 nm) and P_{fr} (λ_{max} 665 nm) which undergo photochemical interconversion, probably due to a *cis–trans* isomerization process. Aplysioviolin[81,210] (**196b**) is a purple defensive secretion produced by the sea-hare *Aplysia*, which is thought to be derived from the biliprotein of red algae eaten by the animal as part of its food.

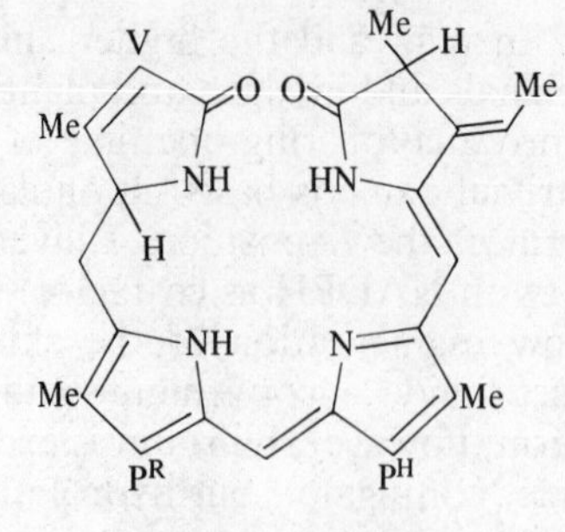

(196) a; R = H
b; R = Me

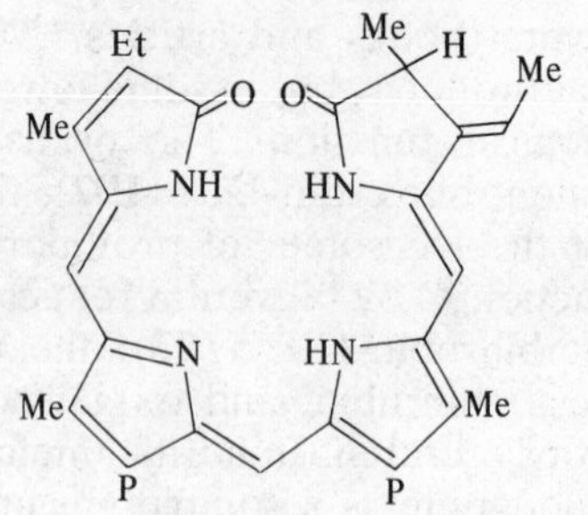

(197)

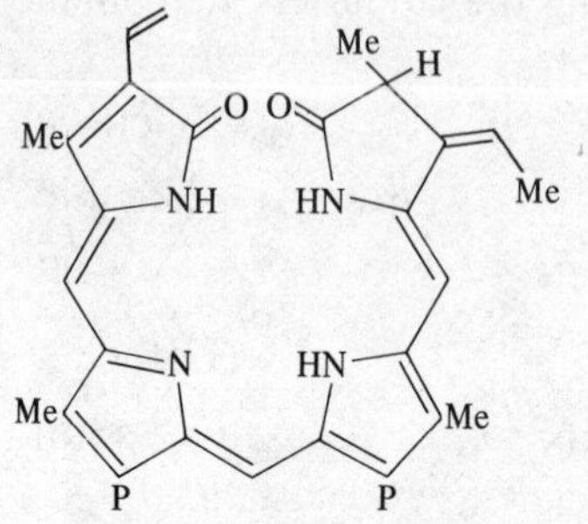

(198)

The structure of bilirubin (**193**) was originally deduced by Fischer[2] in 1931 following the isolation[211] of two pigments, isoneoxanthobilirubic acid (**199**) and neoxanthobilirubic acid (**200**), after fusion of the diethyl analogue of bilirubin with resorcinol. Mild hydrogen iodide in acetic acid reduction of bilirubin afforded colourless dipyrrolic products which on permanganate oxidation afforded[2] the orange-yellow isoxanthobilirubic acid (**201**) and xanthobilirubic acid (**202**). These four compounds were synthesized[212] either by hydrolysis of α-bromopyrromethenes, prepared in a conventional manner, or directly by condensation of bromomethylpyrrolinones *e.g.* (**205**), with α-free pyrroles, *e.g.* (**206**), followed by oxidation. More recently, Plieninger[213] has synthesized these compounds directly from pyrrole aldehydes and pyrrolinones, *e.g.* (**207**) and (**208**) gave (**202**).

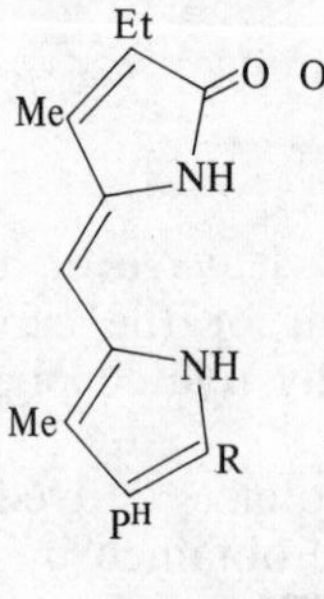

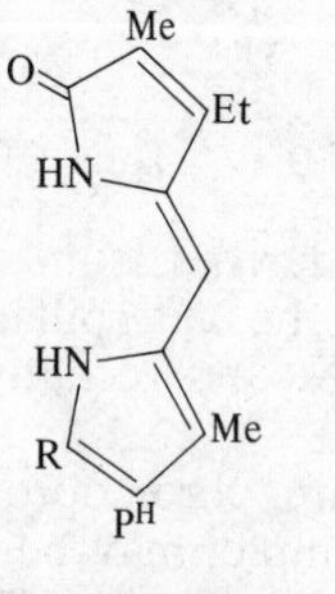

(199) R = H (200) R = H
(201) R = Me (202) R = Me
(203) R = CHO (204) R = CHO

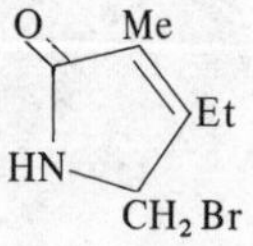

(205)

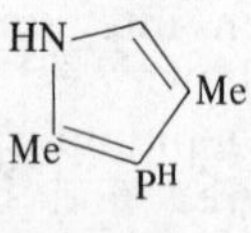

(206)

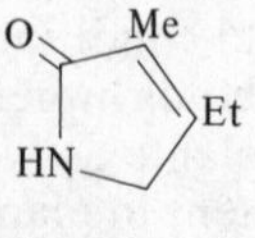

(207)

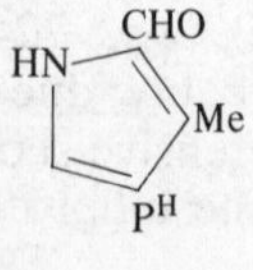

(208)

meso-Biliverdin, the diethyl analogue of (**192**), has been synthesized[214] by condensation of formylisoneoxanthobilirubic acid (**203**) with neoxanthobilirubic acid (**200**), and *meso*-bilirubin was also obtained[214] by coupling the hydroxymethyl analogue of (**203**) with (**200**). These syntheses could also be carried out in the alternative manner, *e.g.* (**204**) and (**199**) give *meso*-biliverdin. Fischer later synthesized[215] biliverdin (**192**) and bilirubin (**193**) by similar methods using urethanes as protecting groups for the potential vinyl side-chains; the latter were prepared by Schmidt degradation of propionic acid side-chains, and in the final stages of the synthesis were degraded to vinyl groups by hydrolysis and exhaustive methylation (Scheme 38). In more recent studies, vinylpyrrolinones have been prepared[216] and have enabled more direct syntheses to be carried out. Further modifications have led to syntheses of the reduced systems present in the urobilins,[132] stercobilins,[132,217] and the algal bile pigments.[132,218]

$$\text{pyr—}CH_2CH_2CO_2H \longrightarrow \text{—}CH_2CH_2NHCO_2Et \longrightarrow \text{—}CH_2CH_2NH_2$$

$$\longrightarrow \text{—}CH_2CH_2\overset{+}{N}Me_3 \longrightarrow \text{—}CH{=}CH_2$$

SCHEME 38

Much of the structural work on the algal and plant bile pigments has relied on oxidative degradation. Permanganate oxidation[219] affords pyrrolecarboxylic acids derived from the central rings of the bile pigments, but Rüdiger[81] has now shown that chromic acid oxidation (used originally by Fischer for degradation of porphyrins[2]) is a superior process. Bile pigments are rapidly oxidized under much milder conditions than porphyrins, *e.g.* at pH 1.7 (with sodium dichromate in 1% sodium bisulphate) biliverdin (184) affords methylvinylmaleimide (**209**) from the terminal ring and the 2,5-diformylpyrrole (**210**) from the central rings; in more acidic media (*e.g.* with chromic oxide in 1 M sulphuric acid) the latter is further oxidized to haematinic acid (**211**). Rüdiger has applied these techniques[81] extensively in determining the structures of the algal pigments, aplysiovilin, and the caterpillar pigment, the degradation products being isolated by thin-layer chromatography; the reactions can also be carried out directly on the biliproteins themselves in order to distinguish the linkages to the proteins. More recent studies[220] have shown that gas chromatography–mass spectrometry can also be used for identification of maleimides on a micro-scale, and this method has proved useful in confirming that a urobilinoid pigment contained ethyl, rather than vinyl, groups in both terminal rings.[221]

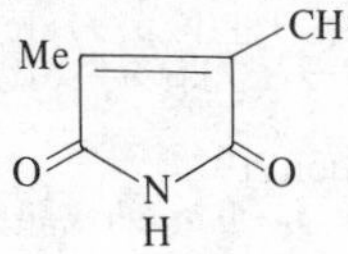

(209) 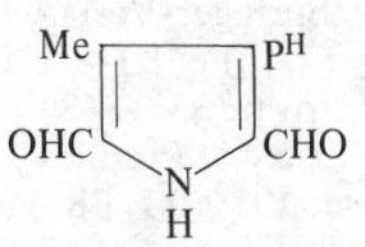 (210) 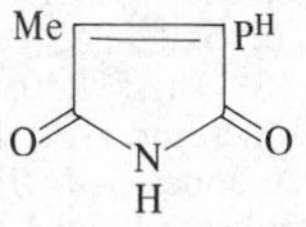 (211)

The broad outlines of mammalian bile pigment formation in nature have already been indicated; the plant and algal pigments are thought to be formed in a similar manner, as indicated by radiolabelling studies carried out in algae with aminolaevulinic acid.[222] However, photochemical ring-opening of a magnesium (rather than an iron) complex has also been suggested. The precise mechanism of the ring-opening process is still a matter of speculation and several comprehensive reviews of the present situation have been published recently.[196,199–202]

In conclusion it is appropriate to mention several aspects of the chemistry of bilirubin which have attracted attention. For example, in acidic media it undergoes a reversible cleavage reaction at the central methine bridge with formation of a mixture of three isomers.[223] Cleavage to give a mixture of two azo-pigments (**212**) and (**213**) also occurs on

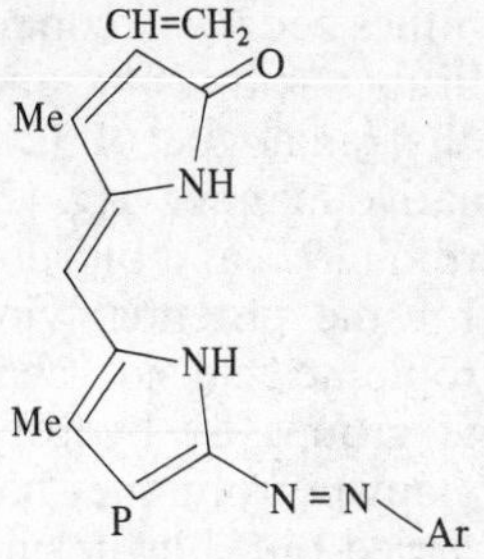

(212) 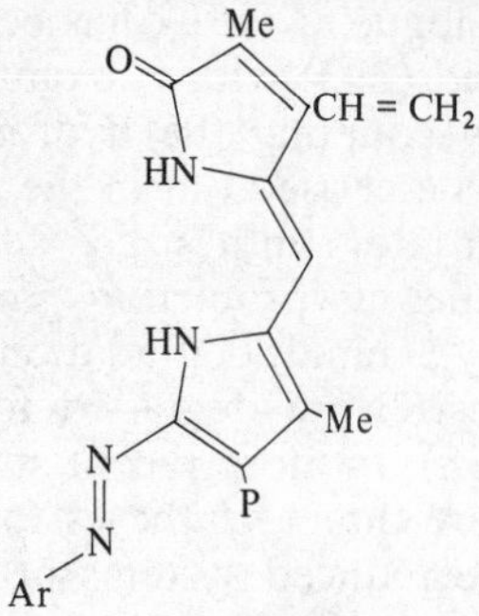 (213) 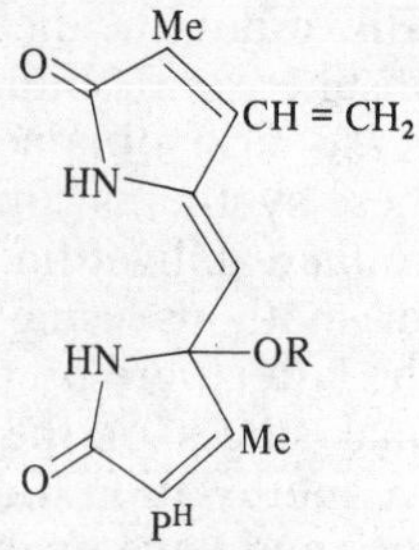 (214) R = H, Me

treatment with arenediazonium salts,[198] and this forms the basis of the Van den Bergh reaction which has long been used for the quantitative estimation of bilirubin. The formation of coloured pigments from bilirubin on treatment with nitrous acid (the Gmelin reaction) has also been known for many years,[2] but the precise nature of the products is still confused. Photo-oxidation[224,225] of bilirubin affords a variety of products, including maleimides and various so-called 'propentdyopents', *e.g.* (**214**); biliverdin may also be formed but it inhibits the oxidation, whereas bilirubin sensitizes its own destruction through formation of singlet oxygen;[224] the mechanism of these processes has been the subject of intense study in recent years owing to the widespread use of phototherapy for treatment of hyperbilirubinaemia in new-born infants. The X-ray crystal structure of bilirubin has also been determined recently, and shows that it has a so-called 'ridge-tile' conformation.

References

1. H. Fischer and H. Orth, 'Die Chemie des Pyrrols', vol. I, Akademische Verlag, Leipzig, 1934.
2. H. Fischer and H. Orth, 'Die Chemie des Pyrrols', vol. II (i), Akademische Verlag, Leipzig, 1937.
3. H. Fischer and H. Stern, 'Die Chemie des Pyrrols', vol. II (ii), Akademische Verlag, Leipzig, 1937.
4. A. H. Corwin, in 'Heterocyclic Compounds, vol. I, ed. R. C. Elderfield, Wiley, New York, 1950.
5. T. S. Stevens, in 'Chemistry of Carbon Compounds' ed. E. H. Rodd, Elsevier, Amsterdam, 1957, vol. 4, p. 28.
6. E. Baltazzi and L. I. Krimen, *Chem. Rev.*, 1963, **63,** 511.
7. K. Schofield, 'Heteroaromatic Nitrogen Compounds: Pyrrole and Pyridine', Butterworths, London, 1967.
8. A. Gossauer, 'Die Chemie der Pyrrole', Springer-Verlag, Berlin, 1974 (a) p. 18, (b) p. 11, (c) p. 21, (d) p. 27, 28, (e) p. 22–25, (f) p. 49, (g) p. 60–77, (h) p. 83–85, (i) p. 93, (j) p. 89–92, (k) p. 94–101, (l) p. 30–34, (m) p. 130, (n) p. 240, (o) p. 301–303, (p) p. 339–343.
9. R. A. Jones and G. P. Bean, 'The Chemistry of Pyrroles', Academic, London, 1977.
10. (a) R. A. Jones, *Adv. Heterocyclic Chem.*, 1970, **11,** 383; (b) M. Simonetta, *J. Chim. phys.*, 1952, **49,** 68.
11. G. W. Wheland and L. Pauling, *J. Amer. Chem. Soc.*, 1935, **57,** 2086.
12. U. Eisner and P. H. Grove, *J. Chem. Soc.*, 1958, 922.
13. A. I. Scott, 'Interpretation of the Ultra-violet Spectra of Natural Products', Pergamon Press, Oxford, 1964.
14. A. Albert, 'Heterocyclic Chemistry', University of London, London, 1959.
15. R. J. Abraham, E. Bullock, and S. Mitra, *Canad. J. Chem.*, 1959, **37,** 1859.
16. *Cf.* R. A. Jones and A. G. Moritz, *Spectrochim. Acta*, 1965, **21,** 295 and R. W. Guy and R. A. Jones, *Austral. J. Chem.*, 1966, **19,** 107.
17. R. J. Abraham and H. J. Bernstein, *Canad. J. Chem.*, 1969, **37,** 1056.
18. J. A. Elvidge and L. M. Jackman, *J. Chem. Soc.*, 1961, 859; J. A. Elvidge, *Chem. Comm.*, 1965, 160.
19. H. A. P. de Jongh and H. Wynberg, *Tetrahedron*, 1965, **21,** 515; D. W. Davies, *Chem. Comm.*, 1965, 258; R. J. Abraham and W. A. Thomas, *J. Chem. Soc.* (*B*), 1966, 127.
20. G. Marino, *Adv. Heterocyclic Chem.*, 1971, **13,** 235.
21. R. A. Jones, T. Mc. L. Spotswood, and P. Chenychit, *Tetrahedron*, 1967, **23,** 4469.
22. R. J. Abraham and H. J. Bernstein, *Canad. J. Chem.*, 1971, **39,** 905.
23. S. Gronowitz, A.-B. Hörnfeldt, B. Gestblom, and R. A. Hoffman, *Arkiv Kemi*, 1961, **18,** 133.
24. E. Lippmaa, M. Mägi, S. S. Novikon, L. I. Kmelnitski, A. S. Prihodko, O. V. Lebedev, and L. V. Epishina, *Org. Magn. Resonance*, 1972, **4,** 153, 197.
25. R. J. Abraham, R. D. Lapper, K. M. Smith, and J. F. Unsworth, *J. C. S. Perkin II*, 1974, 1004.

26. H. Budzikiewicz, C. Djerassi, and D. H. Williams, 'Mass Spectrometry of Organic Compounds', Holden-Day, San Francisco, 1967, p. 596.
27. H. Budzikiewicz, C. Djerassi, A. H. Jackson, G. W. Kenner, D. J. Newman, and J. M. Wilson, *J. Chem. Soc.*, 1964, 1949.
28. G. Yagil, *Tetrahedron*, 1967, **23,** 2855.
29. M. G. Reinecke, H. W. Johnson, Jr., and J. F. Sebastian, *J. Amer. Chem. Soc.*, 1963, **85,** 2859.
30. N. Gjøs and S. Gronowitz, *Acta Chem. Scand.*, 1971, **25,** 2596.
31. F. Seel and V. Sperbe, *J. Organometallic Chem.*, 1968, **14,** 405.
32. D. C. Bradley and K. J. Chivers, *J. Chem. Soc.* (*A*), 1968, 1967.
33. H. J. Anderson and L. C. Hopkins, *Canad. J. Chem.*, 1966, **44,** 1831; J. K. Groves, H. J. Anderson and H. Nagy, *Canad. J. Chem.*, 1971, **49,** 2427.
34. G. P. Bean, *Chem. Comm.*, 1971, 421.
35. Y. Chiang and E. B. Whipple, *J. Amer. Chem. Soc.*, 1963, **85,** 2763.
36. A. Pieroni and A. Moggi, *Gazzetta*, 1923, **53,** 126.
37. H. A. Potts and G. F. Smith, *J. Chem. Soc.*, 1957, 4018.
38. G. F. Smith, *Adv. Heterocyclic Chem.*, 1963, **2,** 287.
39. A. R. Cooksey, K. J. Morgan, and D. P. Morrey, *Tetrahedron*, 1970, **26,** 5101.
40. A. P. Terentev and J. A. Shadkina, *Comp. Rev. Acad. Sci. U.R.S.S.*, 1947, **55,** 227.
41. A. H. Kleinspehn and G. G. Kleinspehn, *J. Amer. Chem. Soc.*, 1954, **76,** 5641; 1953, **75,** 5295.
42. H. Fischer and H. Scheyer, *Annalen*, 1923, **434,** 237.
43. J. H. Mathewson, *J. Org. Chem.*, 1963, **28,** 2153; H. Booth, A. W. Johnson, and R. A. Langdale-Smith, *J. Chem. Soc.*, 1963, 650.
44. J. L. Wong and M. H. Ritchie, *Chem. Comm.*, 1970, 142; J. L. Wong, M. H. Ritchie, and C. M. Gladstone, *ibid.*, 1971, 1093.
45. S. Clementi and G. Marino, *Tetrahedron*, 1969, **25,** 4599; *J. C. S. Perkin II*, 1972, 71; *Gazzetta*, 1970, **100,** 556.
46. P. E. Sonnet, *J. Medicin. Chem.*, 1971, **15,** 97.
47. H. Fischer and W. Zerweck, *Ber.*, 1922, **55,** 1942.
48. R. A. Nicolaus, L. Margoni, and L. Caglioti, *Ann. Chim.* (*Rome*), 1956, **46,** 293.
49. A. Kreutzberger, *Arch. Pharm.*, 1969, **302,** 828.
50. P. S. Clezy, C. J. R. Fookes, and A. J. Liepa, *Austral. J. Chem.*, 1972, **25,** 1979.
51. G. F. Smith, *J. Chem. Soc.*, 1954, 3842; R. M. Silverstein, E. E. Ryskiewicz, C. Willard, and R. C. Koehler, *J. Org. Chem.*, 1955, **20,** 668.
52. W. C. Anthony, *J. Org. Chem.*, 1960, **25,** 2049; P. A. Burbridge, G. L. Collier, A. H. Jackson, and G. W. Kenner, *J. Chem. Soc.* (*B*), 1967, 930.
53. P. Rothemund, *J. Amer. Chem. Soc.*, 1936, **58,** 625.
54. A. H. Corwin, A. B. Chivers, and C. B. Stone, *J. Org. Chem.*, 1964, **29,** 3702.
55. A. D. Adler, L. Sklar, F. R. Lonzo, J. D. Finarelli, and M. G. Finarelli, *J. Heterocyclic Chem.*, 1968, **5,** 669; G. H. Barnett, M. F. Hudson, and K. M. Smith, *J. C. S. Perkin I*, 1975, 1401.
56. H. Fischer and C. Nenitzescu, *Annalen*, 1925, **443,** 113; U. Eisner, A. Lichtarowicz, and R. P. Linstead, *J. Chem. Soc.*, 1957, 733.
57. A. Treibs and K.-H. Michl, *Annalen*, 1954, **589,** 163; V. Carelli, M. Cardellini, and F. Marlachi, *Ann. Chim.* (*Rome*), 1963, **53,** 309.
58. B. V. Gregorovitch, K. S. Y. Liang, D. M. Clugston, and S. F. MacDonald, *Canad. J. Chem.*, 1968, **46,** 3291; M. S. Stoll, G. H. Elder, D. E. Games, D. S. Millington, and A. H. Jackson, *Biochem. J.*, 1973, **131,** 429.
59. A. Treibs and G. Fritz, *Annalen*, 1958, **611,** 162.
60. C. W. Rees and C. E. Smith, *Adv. Heterocyclic Chem.*, 1964, **3,** 57.
61. J. Hine and J. M. Van der Veen, *J. Amer. Chem. Soc.*, 1959, **81,** 6446.
62. R. L. Jones and C. W. Rees, *J. Chem. Soc.* (*C*), 1969, 2249, 2255.
63. A. Gambacorta, R. Nicoletti, and M. L. Forcellese, *Tetrahedron*, 1971, **27,** 985.
64. G. M. Badger, J. A. Elix, and G. E. Lewis, *Austral. J. Chem.*, 1967, **20,** 1777.
65. F. W. Fowler, *Angew. Chem.*, 1971, **83,** 147.
66. K. Haffner and W. Kaiser, *Tetrahedron Letters*, 1964, 2185.
67. R. M. Acheson and J. A. Vernon, *J. Chem. Soc.*, 1961, 457; 1963, 1008.
68. L. Mandell and W. A. Blanchard, *J. Amer. Chem. Soc.*, 1957, **79,** 6198.
69. M. G. Barlow, R. N. Haszeldine, and R. Hubbard, *J. Chem. Soc.* (*C*), 1971, 90.
70. N. J. Turro, S. S. Edelson, J. R. Williams, T. R. Darling, and W. B. Hammond, *J. Amer. Chem. Soc.*, 1969, **91,** 2283.
71. R. G. Gritter and R. G. Chriss, *J. Org. Chem.*, 1964, **29,** 1163.
71a. C. O. Bender and R. Bonnett, *J. Chem. Soc.* (*C*), 1968, 3036; R. Bonnett and J. D. White, *J. Chem. Soc.*, 1963, 1648.
72. C. B. Hudson and A. V. Robertson, *Tetrahedron Letters.*, 1967, 4015.
73. J. P. Wibaut, *Konink. Ned. Akad. Wetenschap. Proc. Ser. B*, 1965, **68,** 117 (*Chem. Abs.*, 1965, **63,** 11 306).
74. P. Pratesi, *Gazzetta*, 1937, **67,** 188, 199.
75. G. P. Gardini and V. Bocchi, *Gazzetta*, 1972, **102,** 91.
76. L. Chierici and G. P. Gardini, *Tetrahedron*, 1966, **22,** 53.

77. V. Bocchi, L. Chierici, and G. P. Gardini, *Tetrahedron*, 1967, **23,** 737.
78. J. H. Atkinson, R. S. Atkinson, and A. W. Johnson, *J. Chem. Soc.*, 1964, 5999.
79. I. Nabih and E. Helmy, *J. Pharm. Soc.*, 1967, **56,** 649.
80. G. Plancher and C. Ravenna, *Atti Acad. Lincei*, 1905, **14I,** 214; *cf.* also Ref. 2.
81. W. Rüdiger, in 'Porphyrins and Related Compounds', ed. T. W. Goodwin, Academic Press, London, 1969; and in L. Zechmeister, *Fortschr. Chem. org. Naturstoffe*, 1971, **29,** 60.
82. P. De Mayo and S. T. Reid, *Chem. and Ind.* (*London*), 1962, 1576.
83. D. A. Lightner, G. S. Bisacchi, and R. D. Norris, *J. Amer. Chem. Soc.*, 1976, **98,** 802 and refs. therein.
84. Ch. Dufraisse, G. Rio, and A. Raujou, *Compt. rend.* (*C*), 1967, **265,** 310.
85. H. H. Wasserman and A. Liberles, *J. Amer. Chem. Soc.*, 1960, **82,** 2086; H. H. Wasserman and A. H. Miller, *Chem. Comm.*, 1969, 199.
86. H. Fischer, *Org. Synth.*, 1935, **15,** 17; 1937, **17,** 96.
87. W. Küster, *Z. physiol. Chem.*, 1922, **121,** 135; A. Treibs, R. Schmidt, and W. Zinsmeister, *Chem. Ber.*, 1957, **90,** 79.
88. O. Piloty and P. Hirsch, *Annalen*, 1913, **395,** 63; *cf.* also Ref. 1.
89. G. K. Almström, *Annalen*, 1913, **400,** 131; 1915, **409,** 291.
90. H. Fischer and E. Fink, *Z. physiol. Chem.*, 1944, **280,** 123; 1948, **283,** 152.
91. P. J. Crook, A. H. Jackson, and G. W. Kenner, *Annalen*, 1971, **748,** 26; P. S. Clezy and A. W. Nichol, *Austral. J. Chem.*, 1965, **18,** 1977; M. W. Roomi and S. F. MacDonald, *Canad. J. Chem.*, 1970, **48,** 1689.
92. G. G. Kleinspehn, *J. Amer. Chem. Soc.*, 1955, **77,** 1546.
93. E. Bullock, A. W. Johnson, E. Markham, and K. B. Shaw, *J. Chem. Soc.*, 1958, 1430.
94. G. G. Kleinspehn and A. H. Corwin, *J. Org. Chem.*, 1960, **25,** 1048.
95. J. W. Harbuck and H. Rapoport, *J. Org. Chem.*, 1971, **36,** 853.
96. A. W. Johnson, E. Markham, R. Price, and K. B. Shaw, *J. Chem. Soc.*, 1958, 4254.
97. H. Plieninger, P. Hess, and J. Ruppert, *Chem. Ber.*, 1968, **101,** 240.
98. A. Treibs and A. Ohorodnik, *Annalen*, 1958, **611,** 139, 149.
99. H. Plieninger and H. Husseini, *Synthesis*, 1970, 587.
100. A. Hantzsch, *Ber.*, 1890, **23,** 1474.
101. F. Feist, B. Widmer, and R. Dubusc, *Ber.*, 1902, **35,** 1545; E. Benary, *Ber.*, 1911, **44,** 493.
102. G. Karschun, *Ber.*, 1904, **37,** 2183, 2196.
103. M. W. Roomi and S. F. MacDonald, *Canad. J. Chem.*, 1970, **48,** 1689.
104. F. Feist and E. Stenger, *Ber.*, 1902, **35,** 1558, D. M. McKinnon, *Canad. J. Chem.*, 1965, **43,** 2628.
105. C. A. Grob and K. Camenisch, *Helv. Chim. Acta*, 1953, **36,** 49.
106. C. A. Grob and H. P. Schad, *Helv. Chim. Acta*, 1955, **38,** 1121.
107. *Cf.* H. S. Broadbent, W. S. Burham, R. K. Olsen, and R. M. Sheeley, *J. Heterocyclic Chem.*, 1968, **5,** 757.
108. R. Ramasseul and A. Rassat, *Bull. Soc. chim. France*, 1970, 4330.
109. W. Keller–Schierlein, M. L. Mihailovic, and V. Prelog, *Helv. Chim. Acta*, 1958, **41,** 220.
110. H. Gotthardt, R. Huisgen, and H. O. Bayer, *J. Amer. Chem. Soc.*, 1970, **92,** 4340.
111. R. Huisgen, H. Gotthardt, H. O. Bayer, and F. C. Schaefer, *Chem. Ber.*, 1970, **103,** 2611; K. T. Potts and D. N. Roy, *Chem. Comm.*, 1968, 1061.
112. O. Piloty, *Ber.*, 1910, **43,** 489; G. M. Robinson and R. Robinson, *J. Chem. Soc.*, 1918, **113,** 639; R. Baumes, R. Jacquier, and G. Tarrago, *Bull. Soc. chim. France*, 1974, 1147.
113. Ju. K. Jurjew, *Zhur. obshchei Khim.*, 1938, **8,** 1934.
114. K. Dimroth and V. Pintschorino, *Annalen*, 1961, **639,** 102; M. Friedman, *J. Org. Chem.*, 1965, **30,** 859.
115. W. G. Terry, A. H. Jackson, G. W. Kenner, and G. Kornis, *J. Chem. Soc.*, 1965, 4389.
116. K. M. Biswas and A. H. Jackson, *Tetrahedron*, 1968, **24,** 1145.
117. H. Fischer and C. Nenitzescu, *Annalen*, 1925, **443,** 113.
118. W. Herz, K. Dittmer, and S. J. Cristol, *J. Amer. Chem. Soc.*, 1947, **69,** 1698.
119. W. Herz and R. L. Settine, *J. Org. Chem.*, 1959, **24,** 201.
120. D. M. MacDonald and S. F. MacDonald, *Canad. J. Chem.*, 1955, **33,** 573; S. F. MacDonald, *J. Chem. Soc.*, 1952, 4176.
121. E. J. Chu and T. C. Chu, *J. Org. Chem.*, 1954, **19,** 266.
122. A. H. Corwin and G. G. Kleinspehn, *J. Amer. Chem. Soc.*, 1953, **75,** 2098; 1954, **76,** 6541.
123. C. A. Grob and A. P. Ankli, *Helv. Chim. Acta*, 1949, **32,** 2010, 2023.
124. H. Plieninger and M. Decker, *Annalen*, 1956, **598,** 198.
125. M. Pesson, D. Humbert, M. Dursin, and H. Tecker, *Compt. rend.* (*C*), 1971, **272,** 478.
126. H. Plieninger, H. Bauer, A. R. Katritzky, and U. Lerch, *Annalen*, 1962, **654,** 165.
127. H. Plieninger, H. Bauer, W. Bühler, J. Kurze, and U. Lerch, *Annalen*, 1964, **680,** 69.
128. R. Chong and P. S. Clezy, *Austral. J. Chem.*, 1967, **20,** 935.
129. J. Davoll, *J. Chem. Soc.*, 1953, 3802.
130. H. Bauer, *Annalen*, 1970, **736,** 1.
131. V. Spiro and I. Fabra, *Ann. Chim.* (*Rome*), 1960, **50,** 1635.
132. A. H. Jackson and K. M. Smith, in 'The Total Synthesis of Natural Products', ed. J. ApSimon, Wiley, New York, 1974, p. 143.
133. K. M. Smith (ed.), 'Porphyrins and Metalloporphyrins', Elsevier, Amsterdam, 1975.
134. A. R. Battersby and E. McDonald, in Ref. 133, p. 61.
135. M. Farnier, S. Soth, and P. Fournari, *Canad. J. Chem.*, 1976, **54,** 1083.

136. H. Rapoport and G. Castagnoli, Jr., *J. Amer. Chem. Soc.*, 1962, **84,** 2178; H. Rapoport and J. Bordmer, *J. Org. Chem.*, 1964, **29,** 2727.
137. R. Grigg, A. W. Johnson, and J. W. F. Wasley, *J. Chem. Soc.*, 1963, 359; R. Grigg and A. W. Johnson, *J. Chem. Soc.*, 1964, 3315; D. Dolphin, R. Grigg, A. W. Johnson, and J. Leng, *J. Chem. Soc.*, 1965, 1460.
138. A. W. Johnson, I. T. Kay, E. Markham, R. Price, and K. B. Shaw, *J. Chem. Soc.*, 1959, 3416.
139. J. Ellis, A. H. Jackson, A. C. Jain, and G. W. Kenner, *J. Chem. Soc.*, 1964, 1935; F. Morsingh and S. F. MacDonald, *J. Amer. Chem. Soc.*, 1960, **82,** 4377.
140. A. Hayes, G. W. Kenner, and N. R. Williams, *J. Chem. Soc.*, 1958, 3779; A. H. Jackson, G. W. Kenner, and D. Warburton, *J. Chem. Soc.*, 1965, 1328.
141. *E.g.* J. A. S. Cavaleiro, A. M. Rocha da Gonzalves, G. W. Kenner, and K. M. Smith, *J.C.S. Perkin I*, 1973, 2471; A. R. Battersby, J. F. Beck, and E. McDonald, *ibid.*, 1974, 160; A. Valasinas, E. S. Levy, and B. Frydman, *J. Org. Chem.*, 1974, **39,** 2872; A. Valasinas and B. Frydman, *ibid.*, 1976, **41,** 2991, and earlier references cited therein.
142. H. Fischer, E. Baumann, and H. J. Riedl, *Annalen*, 1929, **475,** 205.
143. A. H. Jackson, G. W. Kenner, and J. Wass, *J.C.S. Perkin I*, 1972, 1475.
144. A. H. Corwin and K. W. Doak, *J. Amer. Chem. Soc.*, 1955, **77,** 464; J. H. Paden, A. H. Corwin, and W. A. Bailey, Jr., *ibid.*, 1940, **62,** 418.
145. A. C. Jain and G. W. Kenner, *J. Chem. Soc.*, 1959, 185.
146. P. Bamfield, A. W. Johnson, and J. Leng, *J. Chem. Soc.*, 1965, 7001.
147. H. Booth, A. W. Johnson, F. Johnson, and R. A. Langdale-Smith, *J. Chem. Soc.*, 1963, 650.
148. K. J. Brunnings and A. H. Corwin, *J. Amer. Chem. Soc.*, 1944, **66,** 337.
149. J. A. Ballantine, A. H. Jackson, G. W. Kenner, and G. McGillivray, *Tetrahedron, Suppl. No. 7*, 1966, 241.
150. P. S. Clezy and A. J. Liepa, *Austral. J. Chem.*, 1970, **23,** 2461.
151. R. G. Westall, *Nature*, 1952, **170,** 614.
152. G. H. Cookson and C. H. Rimington, *Biochem. J.*, 1954, **57,** 476.
153. A. H. Jackson and S. F. MacDonald, *Canad. J. Chem.*, 1957, **35,** 715; G. P. Arsenault and S. F. MacDonald, *ibid.*, 1961, **39,** 2043.
154. H. Plieninger, P. Hess, and J. Ruppert, *Chem. Ber.*, 1968, **101,** 240.
155. G. W. Kenner, K. M. Smith, and J. F. Unsworth, *J. C. S. Chem. Comm.*, 1973, 43.
156. B. Frydman, S. Reil, M. E. Despuy, and H. Rapoport, *J. Amer. Chem. Soc.*, 1969, **91,** 2338; 1970, **92,** 1810.
157. A. R. Battersby, D. A. Evans, K. H. Gibson, E. McDonald, and L. Nixon, *J.C.S. Perkin I*, 1973, 1546.
158. B. Frydman, G. Buldain, and J. C. Repetto, *J. Org. Chem.*, 1973, **38,** 1824.
159. M. L. Jones, C. Froussious, and D. A. Evans, *J. C. S. Chem. Comm.*, 1976, 472.
160. *Cf.* D. Mauzerall, *J. Amer. Chem. Soc.*, 1960, **82,** 2601, 2605.
161. D. G. Irvine, W. Bayne, and J. R. Majer, *J. Chromatog.*, 1970, **48,** 334.
162. A. J. Birch, P. Hodge, R. W. Richards, R. Takeda, and T. R. Watson, *J. Chem. Soc.*, 1964, 2641; G. R. Birchall, C. G. Hughes, and A. H. Rees, *Tetrahedron Letters*, 1970, 4879; D. G. Durham, and A. H. Rees, *Chem. and Ind.* (*London*), 1974, 343.
163. H. Imanaka, M. Kousaka, G. Tamura, and K. Arima, *J. Antibiot.* (*Tokyo*) *Ser. A.*, 1965, **18,** 207; H. Nakano, S. Umio, K. Kariyone, K. Tauaka, T. Kishimoto, H. Noguchi, I. Ueda, K. Nakamura, and Y. Morimoto, *Tetrahedron Letters*, 1966, 737; E. E. Garcia, L. E. Benjamin, and R. I. Fryer, *J. Heterocyclic Chem.*, 1974, **11,** 275.
164. P. R. Burkholder, R. M. Pfister, and F. M. Leitz, *Appl. Microbiol.*, 1966, **14,** 649; F. M. Lovell, *J. Amer. Chem. Soc.*, 1966, **88,** 4510; S. Hanessian and J. S. Kaltenbronn, *ibid.*, 1966, **88,** 4509.
165. E. Härri, W. Loeffler, H. P. Stigg, H. Stähelin, Ch. Stoll, Ch. Tamm, and D. Wiesinger, *Helv. Chim. Acta*, 1962, **45,** 839; E. Fetz and Ch. Tamm, *ibid.*, 1966, **49,** 349.
166. P. Pfäffli and Ch. Tamm, *Helv. Chim. Acta*, 1969, **52,** 1911, 1921; J. K. Groves, N. E. Cundaswam, and H. J. Anderson, *Canad. J. Chem.*, 1973, **51,** 1089; A. Gossauer and K. Suhl, *Helv. Chim. Acta*, 1976, **59,** 1698.
167. M. Julia and N. P. Joseph, *Bull. Soc. chim. France*, 1967, 4348.
168. S. Penco, S. Redaelli, and F. Arcamone, *Gazzetta*, 1967, **97,** 1110.
169. A. E. Karr, *Ger. Pat.* 1 905 328 (1969); *Chem. Abs.*, 1970, **72,** 11 385.
170. R. B. Kelly, D. J. Whittingham and K. Wiesner, *Canad. J. Chem.*, 1951, **29,** 965.
171. G. C. Garrans and J. Harley-Mason, *J. Chem. Soc.*, 1964, 2202.
172. A. H. Manchanda, J. Nabney, and D. W. Young, *J. Chem. Soc.* (*C*), 1968, 615.
173. E. X. Albuquerque, J. W. Daly, and B. Witkop, *Science*, 1971, **172,** 995.
174. J. H. Tumlinson, R. M. Silverstein, J. C. Moser, R. G. Brownlee, and J. M. Ruth, *Nature*, 1971, **234,** 348.
175. J. A. Edgar, C. C. J. Culvenor, and L. W. Smith, *Experientia*, 1971, **27,** 761.
176. J. Meinwald and H. C. J. Ottenheym, *Tetrahedron*, 1971, **27,** 3307.
177. S. Forenza, L. Minale, R. Riccio, and E. Fattorusso, *Chem. Comm.*, 1971, 1129.
178. G. Cirmino, S. Destefano, and L. Minale, *Experientia*, 1975, **31,** 1387.
179. Y. Badar, W. J. S. Lockley, T. P. Toube, B. C. L. Weedon, and L. R. G. Valadon, *J.C.S. Perkin I*, 1973, 1416.
180. F. Wrede and O. Hettche, *Ber.*, 1929, **62,** 2678; F. Wrede and A. Rothhaas, *Z. physiol. Chem.*, 1933, **219,** 267.
181. H. H. Wasserman, J. E. McKeon, L. Smith, and P. Forgione, *J. Amer. Chem. Soc.*, 1960, **82,** 506; *Tetrahedron*, 1966, Suppl. 8, 647.

182. H. Rapoport and K. G. Holden, *J. Amer. Chem. Soc.*, 1962, **84,** 635.
183. W. R. Hearn, M. K. Elson, R. H. Williams, and J. Medina-Castro, *J. Org. Chem.*, 1970, **35,** 142.
184. H. H. Wasserman, G. C. Rogers, Jr., and D. D. Keith, *Chem. Comm.*, 1966, 825.
185. H. H. Wasserman, G. C. Rogers, Jr., and D. D. Keith, *J. Amer. Chem. Soc.*, 1969, **91,** 1263.
186. N. N. Gerber, *Appl. Microbiol.*, 1969, **18,** 1.
187. N. N. Gerber, *Tetrahedron Letters*, 1970, 809.
188. N. N. Gerber, *J. Antibiot.*, 1974, **24,** 636.
189. N. N. Gerber *J. Heterocyclic Chem.*, 1973, **10,** 925.
190. Yu. M. Khokhlova, L. N. Sergeeva, N. S. Vulfson, V. I. Zaretskii, V. G. Zaikin, V. I. Sheichenko, and A. S. Khokhlov, *Khim. prirod. Soedinenii*, 1968, **4,** 307.
191. W. R. Hearn, R. E. Worthington, R. C. Burgus, and R. H. Williams, *Biochem. Biophys. Res. Comm.*, 1964, **17,** 517; B. S. Deol, J. R. Alden, J. L. Still, A. V. Robertson, and J. Winkler, *Austral. J. Chem.*, 1974, **27,** 2657.
192. H. H. Wasserman, D. D. Keith, and J. Nadelson, *Tetrahedron*, 1976, **32,** 1867.
193. E. Campaigne and G. M. Shutski, *J. Heterocyclic Chem.*, 1976, **13,** 497.
194. G. S. Marks and L. Bogorad, *Proc. Nat. Acad. Sci. U.S.A.*, 1960, **46,** 25; D. M. Shrimpton, G. S. Marks, and L. Bogorad, *Biochim. Biophys. Acta*, 1963, **71,** 408; R. P. Williams, C. L. Gott, and S. M. H. Qadri, *J. Bacteriol.*, 1971, **106,** 444; S. M. H. Qadri and R. P. Williams, *Biochim. Biophys. Acta*, 1971, **230,** 181.
195. R. J. Cushley, D. R. Anderson, S. R. Lipsky, R. J. Sykes, and H. H. Wasserman, *J. Amer. Chem. Soc.*, 1971, **93,** 6284; R. J. Cushley, R. J. Sykes, C. K. Shaw, and H. H. Wasserman, *Canad. J. Chem.*, 1975, **53,** 148.
196. R. Lemberg and J. W. Legge, 'Haematin Compounds and Bile Pigments', Interscience, New York, 1949; R. Lemberg, *Rev. Pure Appl. Chem. (Australia)*, 1956, **6,** 1.
197. C. H. Gray, 'The Bile Pigments', Methuen, London, 1953.
198. T. K. With, 'Bile Pigments', Academic Press, New York, 1968.
199. A. H. Jackson, in 'Iron in Biochemistry and Medicine', ed. A. Jacobs and M. Worwood, Academic Press, London, 1974, p. 145.
200. P. O'Carra, in Ref. 133, p. 123.
201. R. Schmid and A. F. McDonagh, *Ann. N.Y. Acad. Sci.*, 1975, **244,** 533.
202. M. F. Hudson and K. M. Smith, *Chem. Soc. Rev.*, 1975, **4,** 363.
203. W. Rüdiger, W. Klose, M. Vuillaume, and M. Barbier, *Experientia*, 1968, **24,** 1000.
204. M. Choussier and M. Barbier, *Helv. Chim. Acta*, 1975, **58,** 2651.
205. C. O'hEocha and P. O'Carra, *Phytochemistry*, 1966, **5,** 993.
206. H. L. Crespi, L. J. Boucher, G. Norman, J. J. Katz, and R. C. Dougherty, *J. Amer. Chem. Soc.*, 1967, **89,** 3642.
207. W. Rüdiger, P. O'Carra, and C. O'hEocha, *Nature*, 1967, **215,** 1477.
208. D. J. Chapman, H. W. Siegelman, and W. J. Cole, in 'Porphyrins and Related Compounds' ed. T. W. Goodwin, Academic Press, London, 1968, p. 107.
209. D. J. Chapman, H. W. Siegelman, and W. J. Cole, *J. Amer. Chem. Soc.*, 1967, **89,** 3643.
210. W. Rüdiger, *Z. physiol. Chem.*, 1967, **348,** 1554.
211. W. Siedel and H. Fischer, *Z. physiol. Chem.*, 1933, **214,** 145.
212. W. Siedel and H. Fischer, *Z. physiol. Chem.*, 1933, **214,** 146, 163; W. Siedel, *Z. physiol. Chem.*, 1935, **231,** 181, 197.
213. H. Plieninger and U. Lerch, *Annalen*, 1966, **698,** 196.
214. W. Siedel, *Z. physiol. Chem.*, 1935, **237,** 8; 1937, **245,** 257.
215. H. Fischer and H. Plieninger, *Z. physiol. Chem.*, 1942, **274,** 231.
216. H. Plieninger and R. Steinstrasser, *Annalen*, 1969, **723,** 149; H. Plieninger, K.-H. Heutschel, and R.-D. Kohler, *ibid.*, 1974, 1522.
217. H. Plieninger, K. Ehl, and A. Tapia, *Annalen*, 1970, **736,** 62.
218. A. Gossauer and W. Hirsch, *Annalen*, 1974, 1496.
219. R. A. Nicolaus, *Rass. Med. Speri.*, 1960, **7,** Suppl. 2.
220. A. H. Jackson, D. S. Millington, and D. E. Games, *Adv. Mass Spectrometry*, 1974, **6,** 215.
221. D. E. Games, A. H. Jackson, and G. Cooper, *Trans. Biochem. Soc.*, 1976, **4,** 204.
222. R. F. Troxler and L. Bogorad, *Plant Physiol.*, 1966, **41,** 491; R. F. Troxler and R. Lester, *Biochemistry*, 1967, **6,** 3840.
223. A. F. McDonagh and F. Assisi, *Biochem. J.*, 1972, **129,** 797.
224. A. F. McDonagh, *Biochem. Biophys. Res. Comm.*, 1971, **44,** 1306.
225. D. A. Lightner, *Photochem. Photobiol.*, 1974, **19,** 457; D. A. Lightner and Y. T. Park, *Tetrahedron Letters*, 1976, 2209; D. A. Lightner and C. S. Pak, *Experientia*, 1976, **32,** 1107; R. Bonnett and J. C. M. Stewart, *J. C. S. Perkin I*, 1975, 224.

17.2
Porphyrins, Corrins, and Phthalocyanines

K. M. SMITH,
University of California, Davis

17.2.1 PORPHYRINS AND CHLORINS

17.2.1.1 Structure and nomenclature

The bare tetrapyrrolic nucleus (**1**) is known as 'porphin', and the porphyrins are formally derived from this by substitution of some or all of the peripheral positions with a variety of side-chains. Porphyrinogens (**2**), which feature predominantly in biosynthetic pathways to the pyrrole pigments (Chapter 30.2), consist of four pyrrole rings linked together to form a colourless macrocycle by four methylene groups. Abstraction of six hydrogen atoms from porphyrinogen (**2**) affords the porphyrin nucleus which, being highly conjugated, is often brilliantly coloured; thus, it is incorrect to refer to the four individual subunits in the porphin nucleus (**1**) as 'pyrrole' rings. The chlorin nucleus (**3**) is a dihydroporphyrin. It is usually drawn[1] with ring D reduced, and in the natural products (chlorophylls) in which this nucleus occurs, the two additional hydrogens are added in a *trans* fashion.

Tetrahydroporphyrins (**4**) are also known to occur naturally, as in the bacteriochlorophyll-a molecule. However, the reduced rings are always opposite, and tetrahydroporphyrins (**5**) in which the adjacent subunits are reduced, though chemically accessible, have never been found in natural pigments.

Two systems for numeration of the porphyrin ring are currently in use. The IUPAC system is shown in structure (**6**), and this system, which was designed to achieve consistency between porphyrins and corrins, will be used throughout this section. The

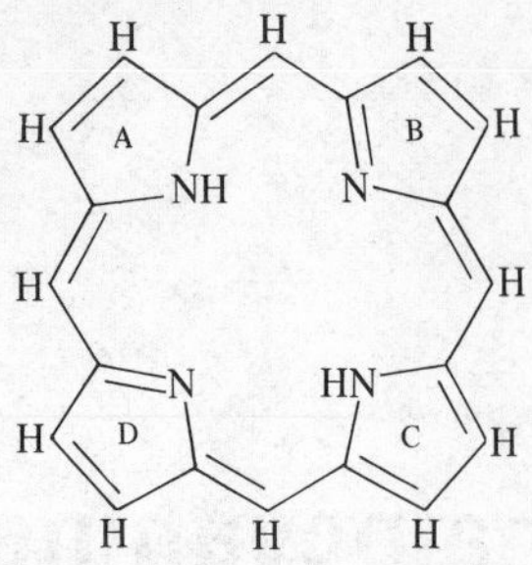

(1) 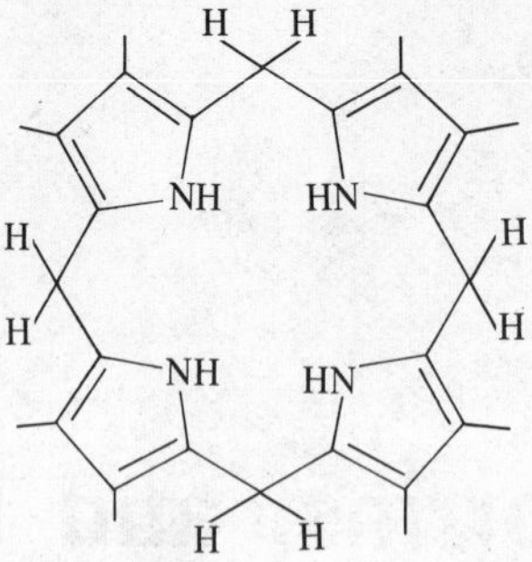 (2) 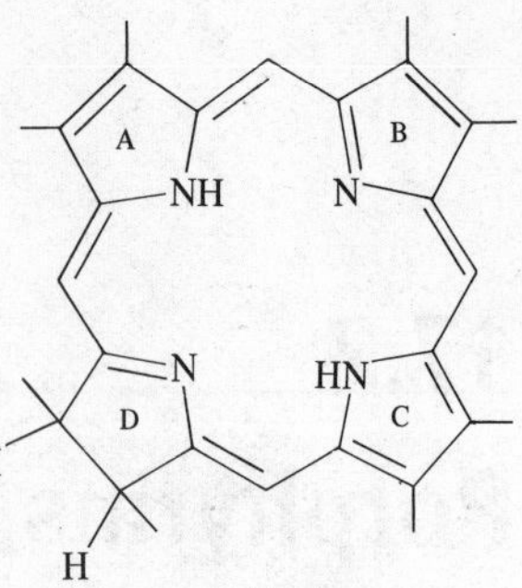 (3)

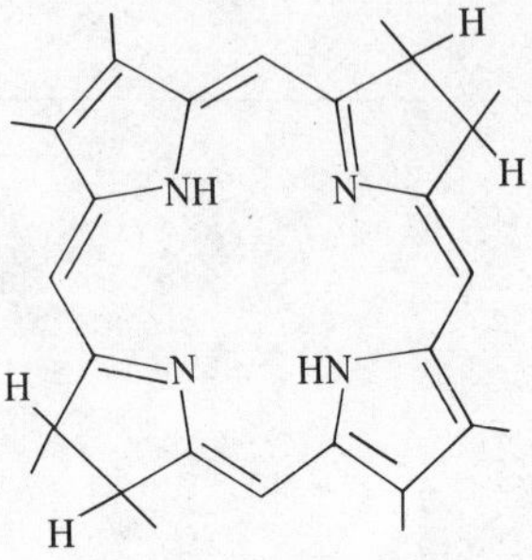

(4) 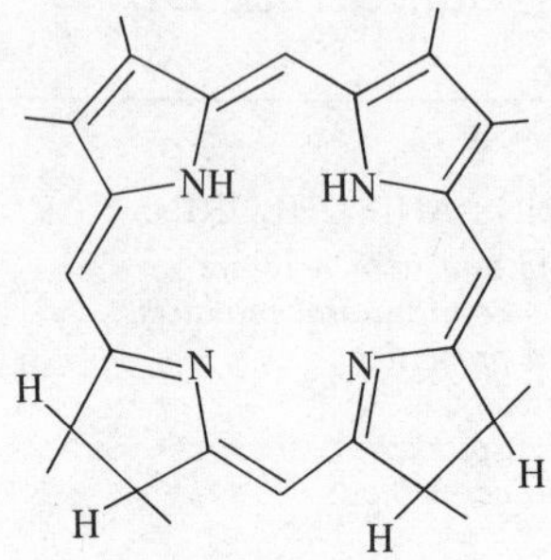 (5)

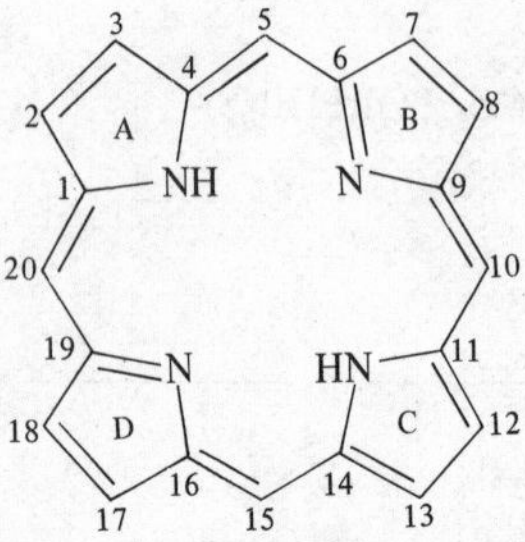

(6) 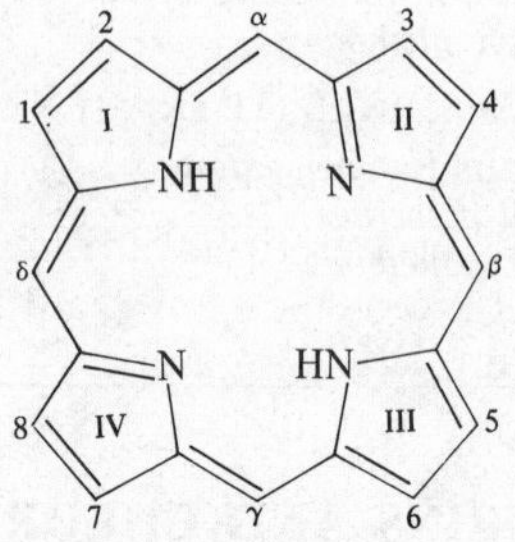 (7)

major disadvantage of the IUPAC recommended system is that it may divorce contemporary research from the monumental body of early work which used the Fischer system (**7**).[1] There can be few fields of chemical endeavour possessing such a rich history which is still utilized daily in current research.

The four inter-subunit carbons (5, 10, 15, 20) are known as the *meso* carbons. The eight remaining carbons able to accept substituents are known as the peripheral positions. If the eight peripheral substituents are of two types (*e.g.* Me, Et), and if one of each is present on every subunit, then four isomers are possible.[1] In the Me, Et case these are called aetioporphyrins, and the 'type' isomers are shown in structures (**8**)–(**11**). If the ethyl group is replaced by propionic acid the coproporphyrin type isomers are produced; furthermore, if the methyls are replaced by acetic acids, and the ethyls by propionic acid substituents, then the uroporphyrins result. When three different types of substituent are present (four of one kind, and two pairs of others) then fifteen type isomers are possible. The biologically significant porphyrins with only two types of substituent always have the type-III orientation, and with three substituents, the type-IX orientation (which is derived from type-III by modification of the 3 and 8 side-chains) is the natural arrangement. A fuller discussion of porphyrin isomerism and structure can be found in a recent monograph.[2]

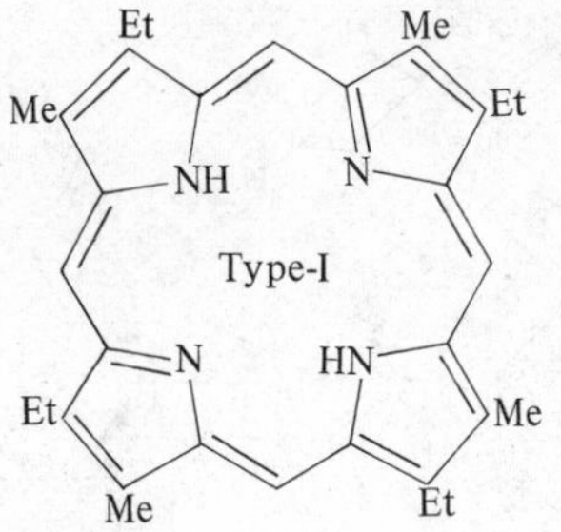

(8) 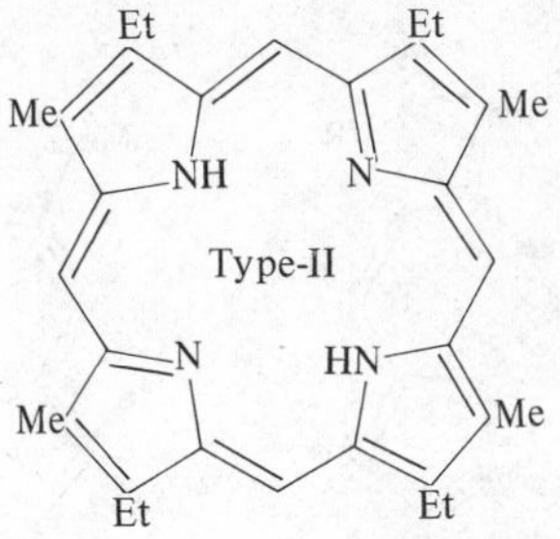 (9)

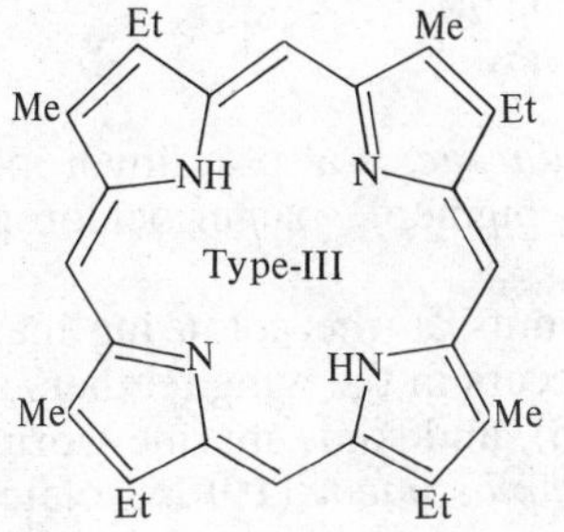

(10) 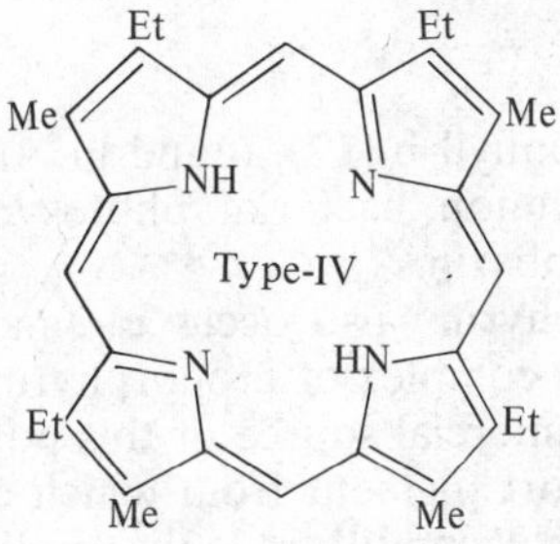 (11)

17.2.1.2 Occurrence in natural products

Porphyrins and chlorins are the most ubiquitous pigments to be found in Nature. Haem (**12**), the iron(II) complex of protoporphyrin-IX, is the prosthetic group of myoglobin and haemoglobin; (**12**), or a peripherally modified form, is found in the cytochromes[3] and in enzymes such as peroxidases and catalases. (Discussion of the mode of action of these systems can be found in Chapter 24.2.) A large variety of chlorophylls is known; with only one exception, they all possess the dihydroporphyrin (**3**) chromophore. The most abundant photosynthetic pigments are chlorophylls-a (**13**) and -b (**14**), which normally co-occur in a ratio of about 3:1. Chlorophyll-d (**15**) is found in some species of *Rhodophyceae*, and green sulphur bacteria such as *Chlorobium thiosulphatophilum*, *C. limicola*, and *Chloropseudomonas ethylicum* produce complex mixtures of pigments called *Chlorobium* chlorophylls. The exceptions to the dihydroporphyrin nucleus are the chlorophylls-c (**16**), which are a mixture of 8-vinyl- and 8-ethyl-porphyrins.

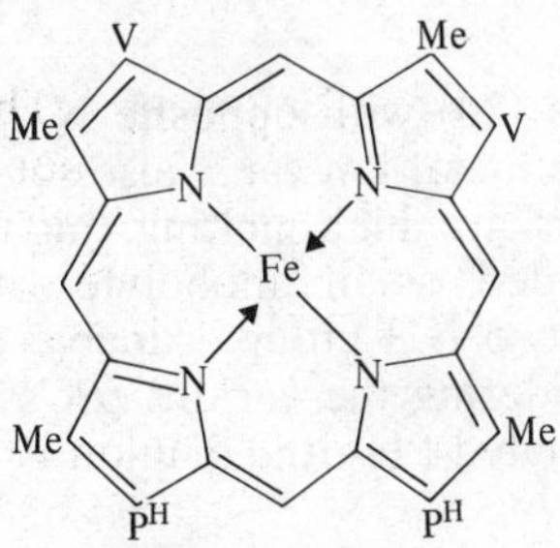

(12)

V = $CH{=}CH_2$
P^R = $CH_2CH_2CO_2R$
Ac = $COCH_3$

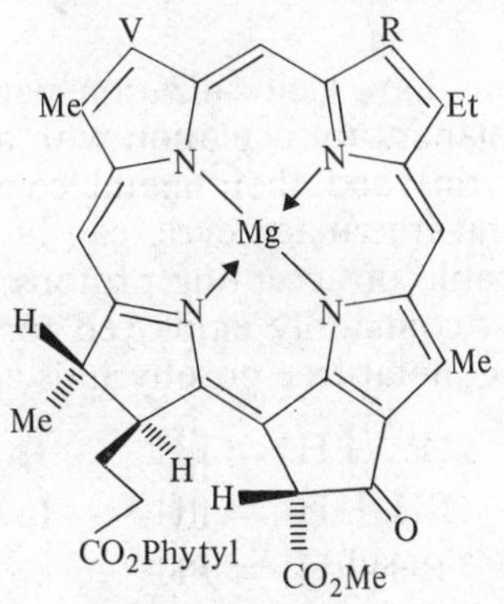

(13) R = Me
(14) R = CHO

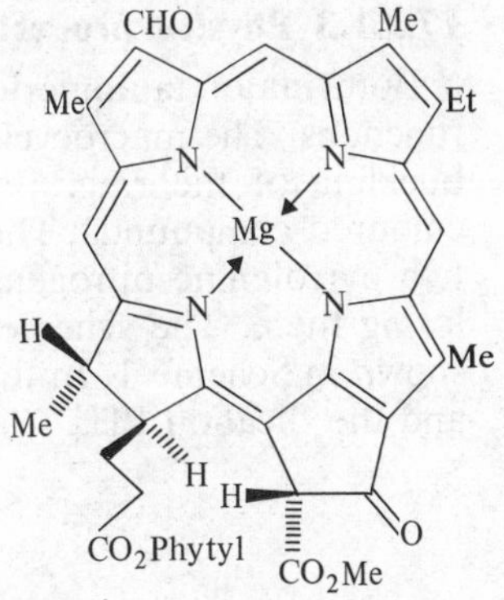

(15)

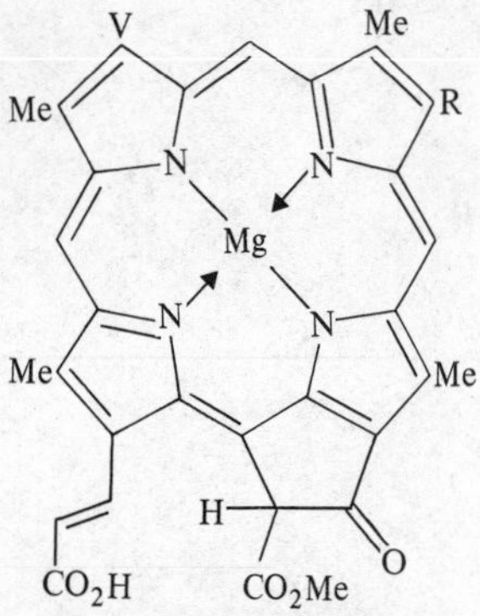

(16) R = Et, V

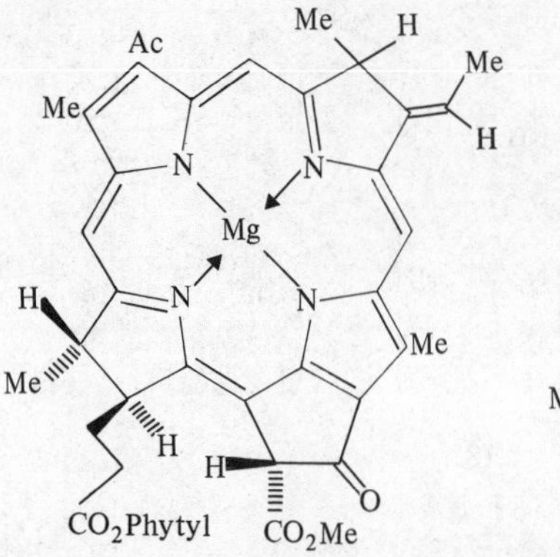

(17)

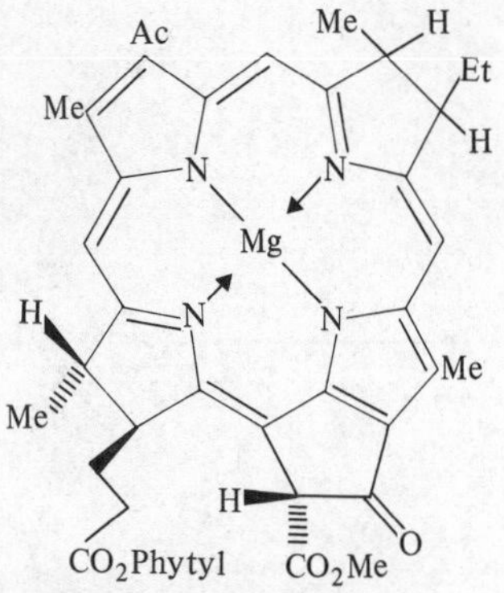

(18)

Bacteriochlorophyll-b (**17**), found in *Athiorhodaceae,* is also a dihydroporphyrin, whereas the other common bacterial photosynthetic pigment, bacteriochlorophyll-a (**18**), is a tetrahydroporphyrin.

Metalloporphyrins also occur in small amounts as the colouring matter in egg shells. The copper(II) complex of uroporphyrin-III occurs in the wing feathers of *Turacus indicus* (the main commercial source of this porphyrin), and some marine worms possess a green oxygen transport pigment from which *Spirographis haem* (**19**) is isolated.

Under normal conditions only small amounts of metal-free porphyrins are found in Nature. Reasonable amounts of some porphyrins can be obtained from root nodules of leguminous plants, as well as in the Harderian gland of rodents, in yeast mutants, in some bird shells, and in the faeces of carnivores and herbivores. Under abnormal conditions, for example during porphyrias,[4] metal-free porphyrins can be produced and deposited in quite large amounts.

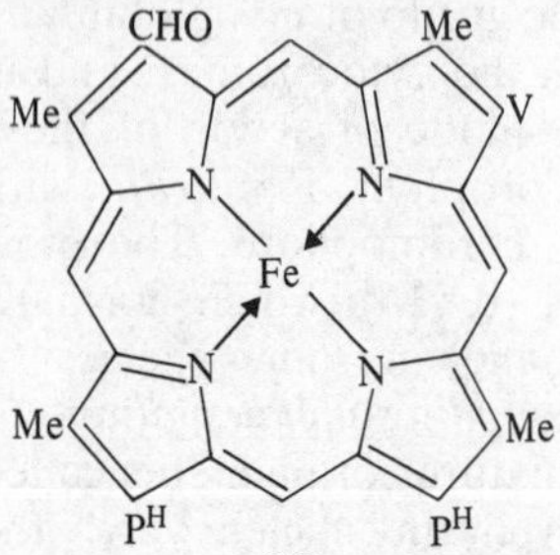

(19)

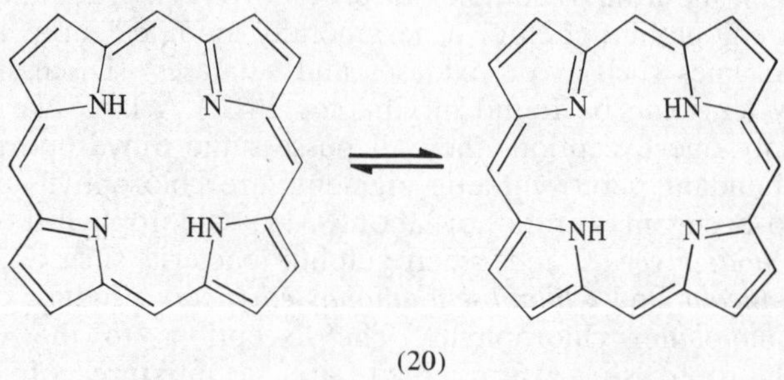

(20)

17.2.1.3 Physical properties

Both major tautomeric forms have delocalization pathways (**20**) with opposite N—H functions. The macrocycle is planar, but chelation with a large metal ion can cause some buckling of the ring.[5] Porphyrins and their metal complexes are high-melting, highly coloured compounds. The metal-free macrocycle can be regarded[6] as an ampholyte with two pyrrolenine nitrogens capable of accepting protons, and two NH groups capable of losing them. The scheme most commonly employed for considering the various p*K*'s is shown in Scheme 1. In this, the metal-free porphyrin is abbreviated PH_2, the dianion P^{2-}, and the dication PH_4^{2+}.

$$P^{2-} + H^+ \rightleftarrows PH^- \quad (pK_1)$$
$$PH^- + H^+ \rightleftarrows PH_2 \quad (pK_2)$$
$$PH_2 + H^+ \rightleftarrows PH_3^+ \quad (pK_3)$$
$$PH_3^+ + H^+ \rightleftarrows PH_4^{2+} \quad (pK_4)$$

SCHEME 1

Porphyrins behave as very weak acids, strong bases such as alkoxides being required to allow spectrophotometric observation of the dianion. Both pK_1 and pK_2 (for aetioporphyrin) have been estimated to be approximately +16. pK_3 for octa-alkylporphyrins is about 5.5 (measured, for solubility reasons, in 2.5% sodium dodecyl sulphate), but spectroscopic observation of the monocation is usually difficult owing to the resonance stabilization available in the symmetrical dication. Generally, pK_3 and pK_4 differ by only two or three units. Chlorins tend to be weaker bases than their porphyrin counterparts.

17.2.1.4 Spectroscopy

The brilliant red colour of many porphyrins and derivatives is manifested in their electronic absorption spectra. However, the major feature of these spectra, the so-called Soret band found around 400 nm and possessing an extinction coefficient sometimes as large as 400 000, is largely unseen by the naked eye. What is seen is a series of satellite peaks of extinction 15 000 or less lying between 500 and 650 nm. These satellite peaks are known to vary with the peripheral substituents, and the experienced researcher can accurately interpret subtle changes in hue in terms of structural features. The intense Soret band is a characteristic of the 18π-electron delocalization pathway present in the porphyrin nucleus; when the ring is broken open, or if the conjugated pathway is interrupted in some other way, the Soret band disappears almost completely. Metal-free chlorins (**3**) are usually green; in this case the Soret band is still present, but by far the major satellite band is found between 600 and 660 nm with molar extinction coefficient often around 70 000. Chelation of porphyrins with metal ions usually results in an increase of the Soret absorption, and simplification of the satellite peaks to give only two. These can vary in extinction and wavelength depending upon the metal ion involved. Diprotonation also causes increased Soret extinction and simplification of the satellite bands in a similar fashion to metal complex formation. No such simplification of the satellite bands in chlorins is found upon chelation; usually the long wavelength band suffers a hypsochromic shift as the Soret increases in extinction.

The aromaticity of the porphyrin macrocycle has been extensively investigated using nuclear magnetic resonance spectroscopy.[2] The ring current due to the large delocalization pathway in the porphyrin nucleus has been used to investigate aggregation and a variety of other phenomena. Because of deshielding, the *meso* protons appear at about $0\ \tau$ (*cf.* benzene at $2.8\ \tau$) while the shielded N—*H* resonances occur between 12 and $15\ \tau$. Measurements of chemical shifts in proton and carbon-13 n.m.r. spectra have been complicated by the concentration dependence of the chemical shifts, due largely to aggregation of the molecules into layers in solution.[2] When one porphyrin molecule closely approaches another in solution the ring current of one has the effect of causing an upfield shift of protons or carbons in the substituents of another. An analysis of these upfield shifts can be used to determine the geometrical structure of these dimers or higher aggregates in solution. By far the major use for n.m.r. spectroscopy in porphyrin work is for identification of side-chains and for determining whether or not a porphyrinic product is isomerically pure. In this connection the use of shift reagents has been shown to be particularly advantageous.[7] Paramagnetic n.m.r. spectra of haems and haemoproteins have also been measured. If the low-spin cyanoferrihaems or haemoproteins[8] are used, then the lone electron causes dramatic shifts of the resonances associated with the porphyrin, which are spread far away from those peaks due to solvent or protein residues. Moreover, the shifts of the peaks can be directly related to spin densities in the haem, and hence are sensitive to minute perturbations of the type which occur under physiological conditions when the haemoprotein is carrying out its biological role.

Mass spectrometry provides an easy method for the accurate determination of molecular weights. In the absence of very labile substituents the base peak is invariably the molecular ion. It is a little surprising that the macrocycle suffers no detectable fragmentation; instead it acts as an inert support for the substituents which can therefore be studied in great detail.[9] The highly aromatic nature of the porphyrin macrocycle is confirmed by the abundant series of doubly charged ions, some of which have intensity as large as 20% of the base peak. Triply charged ions have also been observed in some spectra.

Countless original papers dealing with single-crystal X-ray structures of metal-free and metal-containing porphyrins have been published.[5] Only rarely is this technique used in order to establish the identity of a compound. More often it is the case that the stereochemistry of a known compound of particular interest is investigated. Importantly, X-rays were used to establish the delocalization pathway shown in structures (**20**), and thus eliminate an alternative hypothesis.

Electron paramagnetic resonance (e.p.r.) has been employed in the study of paramagnetic complexes, or of the one-electron oxidation products from diamagnetic metalloporphyrins; this technique has yielded important information on the electronic structure of the macrocycle, as well as being a key technique for investigation of the phenomenon of photosynthesis and other natural phenomena in which radicals or cation-radicals participate.

Owing to the assignment problem, infrared and Raman spectroscopy have not reached their full potential. However, as synthetic methods develop it is anticipated that labelling of groups (usually with deuterium) will enable peaks to be assigned, and so these techniques will achieve more prominence. Mössbauer spectroscopy has achieved limited success, particularly in the study of haems (iron complexes), where it can give detailed information on the electronic configurations of diamagnetic and paramagnetic states. It is limited, however, in that a nuclide having a low-lying isomeric nuclear energy must be used as an absorbing nucleus, and the excited state of that same nuclear isomer must be populated as a result of a nuclear reaction or decay. In the metalloporphyrin series therefore, only ^{57}Fe, ^{119}Sn, and ^{129}I have been investigated.

17.2.1.5 Chemical properties

Perhaps the most well-known chemical property of porphyrin systems is the relative ease with which metal complexes are formed; complexes with most metals in the Periodic Table have been prepared and characterized.[2,10] The facility of metal complex formation is underscored by the fact that most pyrrole pigments with physiological functions of any significance occur as metal complexes (*e.g.* Fe in oxygen storage and transport pigments, and in cytochromes, catalases and peroxidases; Mg in chlorophylls; Co in vitamin B_{12}).

As befits a highly aromatic system, porphyrin systems and particularly metalloporphyrins, undergo a wide variety of electrophilic substitutions.[2,11] Early workers had concentrated on metal-free porphyrins, and since many electrophilic substitutions utilize acidic conditions (*e.g.* nitration, sulphonation) they had experienced difficulty owing to the fact that the porphyrin is readily diprotonated to give the dication which, as might be expected, is non-nucleophilic. However, as early as 1929, Fischer had realized that di-acetylation of deuteroporphyrin-IX (see Section 17.2.1.6) needed to be carried out on the iron(III) complex. Electronegative porphyrin ligands are obtained by complexation with divalent central metal ions (Mg > Zn > Cu > Ni > Pd) (this can be correlated with the ease of removal of electrons from the porphyrin ring by electrochemical methods, *i.e.* with the metalloporphyrin oxidation potential, *vide infra*). Such metal complexes usually, but not always, substitute at the *meso* positions. For example, magnesium porphin (**21**) is brominated (*N*-bromosuccinimide) exclusively at the 5-, 10-, 15-, and 20-positions to give (**22**), and mono-*meso*-formylation of the copper complex (**23**) is accomplished using the Vilsmeier method ($POCl_3$/DMF) to give (**24**). On the other hand, some specificity is exhibited by the analogous chlorins, which tend to undergo substitution at the 15- and 20-positions adjacent to the partially reduced ring. Deuteration has usually been used to measure the relative susceptibilities of various metalloporphyrins towards electrophilic reactions; this has the advantage that, within the confines of the deuterium isotope effect, introduction of one substituent does not greatly affect the further reactivity of the compound studied. Nitrations of most porphyrins are usually accomplished in very high yield using nitric and acetic acids at 0 °C; up to three nitro-groups can be introduced under controlled conditions, and the dinitro-products are usually a mixture of 5,10- and 5,15-disubstituted compounds. Chlorins, which are sensitive to oxidation, are usually nitrated under electrophilic conditions using nitronium tetrafluoroborate. Chlorination of porphyrin free bases is accomplished with hydrogen peroxide and hydrochloric acid, and usually affords the *meso*-tetrachloro-derivative. Mono- and dichloro-products are obtained with difficulty under highly controlled two phase conditions, the disubstituted product usually being the 5,15-dichloro-derivative only. As with nitration, chlorins tend to be chlorinated at the 15- and 20-positions. Copper porphyrins also react with thiocyanogen to give the *meso*-thiocyanatoporphyrin which can be hydrolysed to the mercapto-derivative. Reactions between copper porphyrins and carbenes are known, but they tend to give mixtures of products owing to addition at *meso* positions as well as β–β' double bonds. Nitrenes also react with metal-free porphyrins to give insertion products and *meso*-substitution compounds.

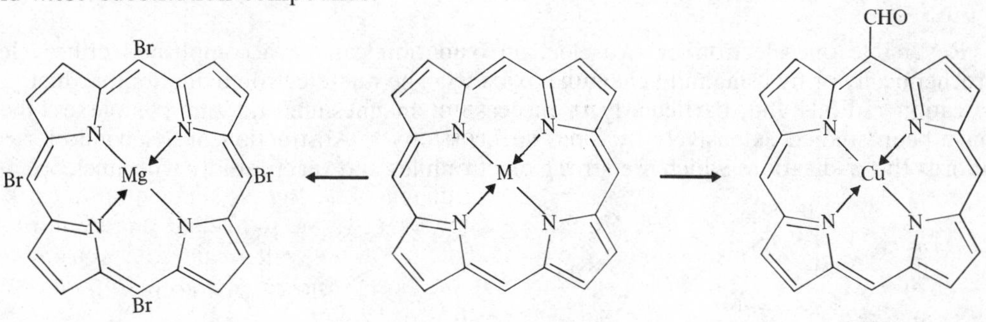

(22)

(21) M = Mg
(23) M = Cu

(24)

Irreversible oxidation of the porphyrin nucleus can be accomplished with either chromium trioxide or potassium permanganate. The products from the CrO_3 reaction are maleimides (Scheme 2) and this procedure has been used extensively for structure elucidation in porphyrins and for tracking down the fate of isotopic labels in biosynthetic studies.[2] In the absence of side-chains which are liable themselves to oxidation, it is possible to establish the identities of the substituents on the subunits of any porphyrin; the total structure, however, cannot be deduced using this method because of the various structural permutations which are possible. Vinyl groups (*e.g.* in protoporphyrin-IX) are degraded by CrO_3; thus, in Shemin's early biosynthetic work, the protoporphyrin-IX (**25**) was first hydrogenated to give mesoporphyrin-IX (**26**). CrO_3 oxidation (Scheme 3) then furnished ethylmethylmaleimide and haematinic acid. Permanganate oxidation affords pyrrole-2,5-dicarboxylic acids (Scheme 4) which can be readily identified by paper chromatography. As with CrO_3 oxidation, vinyl and formyl substituents are oxidized to the corresponding carboxylic acid group. In the permanganate method, the *meso* carbon atoms are retained in the 2- and 5-carboxylic acid function, unlike the CrO_3 technique, where they are eliminated as carbon dioxide.

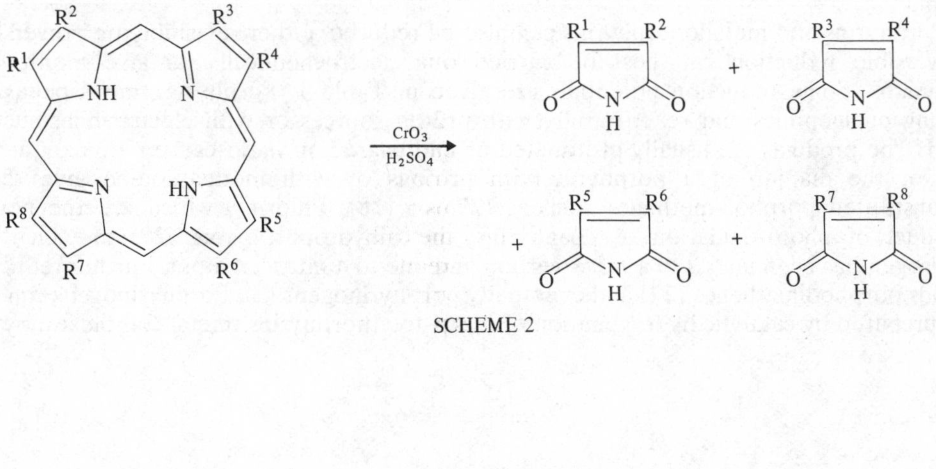

SCHEME 2

(25) $\xrightarrow[(h\nu)]{H_2/Pd}$ (26) $\xrightarrow[H_2SO_4]{CrO_3}$ Ethylmethylmaleimide (A,B); Haematinic acid (C,D)

SCHEME 3

Reversible one-electron or two-electron oxidation can be accomplished either electrochemically or by using mild chemical oxidants. The one-electron oxidation products are π-cation radicals and, particularly in the case of magnesium and zinc complexes, these have been studied extensively by e.p.r. spectroscopy.[12] Abstraction of a second electron affords the π-dications which are strong electrophiles and react rapidly with nucleophiles.

$\xrightarrow{KMnO_4}$

SCHEME 4

π-Cation radicals of metalloporphyrins also react with some types of nucleophile. The ease of oxidation depends upon the central metal ion; Table 1 shows the oxidation potentials for removal of one ($E_{\frac{1}{2}}^{1}$) and two ($E_{\frac{1}{2}}^{2}$) electrons from a series of metal complexes of octaethylporphyrin.[13] The effect of the metal ion is as expected from theoretical treatments of Gouterman,[14] and the oxidation potentials appear in most cases to parallel susceptibilities towards electrophilic substitution. These one- and two-electron abstractions are usually perfectly reversible. In a few cases, oxidation occurs initially at the central metal ion (see Table 1), and methods for deciding whether the point of oxidation is the porphyrin ligand or the metal ion have been outlined.[2,13] The π-cation radicals of some metalloporphyrins are so stable that they can be crystallized and stored in bottles for long periods. On the other hand, π-dications have never been isolated in a pure, solid form.

Porphyrins and metalloporphyrins can also be reduced, either reversibly or irreversibly. Reversible reduction can best be carried out electrochemically to give mono- and di-anions. Some reduction potentials are given in Table 1. Porphyrin anions behave as strong nucleophiles and react rapidly with proton sources or with electrophiles such as MeI. The products are usually protonated or methylated on *meso*-carbon atoms; quenching of the dianion of a porphyrin with protons or with methyl iodide gives 5,15-disubstituted porphodimethenes such as (**27**) or (**28**). Phlorins, which are the normal product of photo-oxidation of porphyrins, are dihydroporphyrins (**29**) in which one hydrogen has been added to a *meso*-carbon and one to a nitrogen atom. Further reduction yields porphodimethenes (**27**) and eventually porphyrinogens (**2**). Porphyrinogens can also be prepared by catalytic hydrogenation of metal-free porphyrins (metal complexes, by and

TABLE 1

Half-Wave Potentials ($E_{\frac{1}{2}}$) (in volts *vs.* sce) of Octaethylporphyrins, Using Cyclic Voltammetry with a Platinum Button Electrode[a]

	Ligand oxidations		*Metal*	*Ligand reductions*	
	$E^2_{\frac{1}{2}}$	$E^1_{\frac{1}{2}}$	$E^1_{\frac{1}{2}}$	$E^1_{\frac{1}{2}}$	$E^2_{\frac{1}{2}}$
H_2OEP[b]	1.30	0.81		−1.46	−1.86
H_4OEP^{2+}		1.65			
AgOEP		1.10	0.44 III ⇌ II	−1.29	
Al(OH)OEP	1.28	0.95		−1.31	
CaOEP	0.86	0.50		−1.68	
CdOEP	1.04	0.55		−1.52	
CoOEP		1.00	Irreversible III ⇌ II −1.05 II ⇌ I		
Cr(OH)OEP	1.22	0.99	0.79 IV ⇌ III −1.14 III ⇌ II	−1.35	
CuOEP	1.19	0.79		−1.46	
Fe(OH)OEP[c]	1.24	1.00	−0.24 III ⇌ II	−1.33	
Ga(OH)OEP	1.32	1.01		−1.34	−1.80
$Ge(OH)_2OEP$	1.36	1.09		−1.31	
In(OH)OEP	1.36	1.08		−1.19	−1.59
MgOEP	0.77	0.54		−1.68	
Mn(OH)OEP	1.40	1.12	−0.42 III ⇌ II	−1.61	
MoO(OH)OEP		1.43	−0.21 V ⇌ IV	−1.30	−1.72
NiOEP		0.73		−1.50	
PbOEP		0.65	0.90 II ⇌ IV	−1.30	
PdOEP		0.82		−1.53	
Sb(OH)OEP		1.40		−1.07	
Sc(OH)OEP	1.03	0.70		−1.54	
$Si(OH)_2OEP$	1.19	0.92		−1.35	
$Sn(OH)_2OEP$[c]		1.40		−0.90	−1.30
Tl(Cl)OEP	1.32	1.03		−1.21	−1.69
Tl(OH)OEP	1.31	1.00		−1.24	
VOOEP	1.25	0.96		−1.25	−1.72
ZnOEP	1.02	0.63		−1.61	

[a] J.-H. Fuhrhop, K. M. Kadish, and D. G. Davis, *J. Amer. Chem. Soc.*, 1973, **95**, 5140; [b] OEP = octaethylporphyrin dianion; [c] μ-Oxo-dimer.

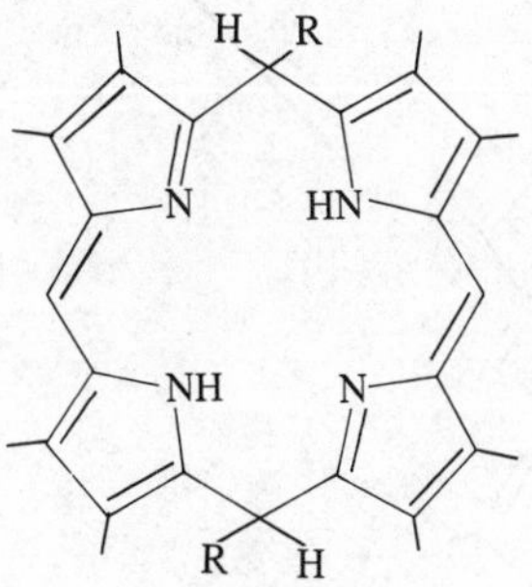

(27) R = H
(28) R = Me

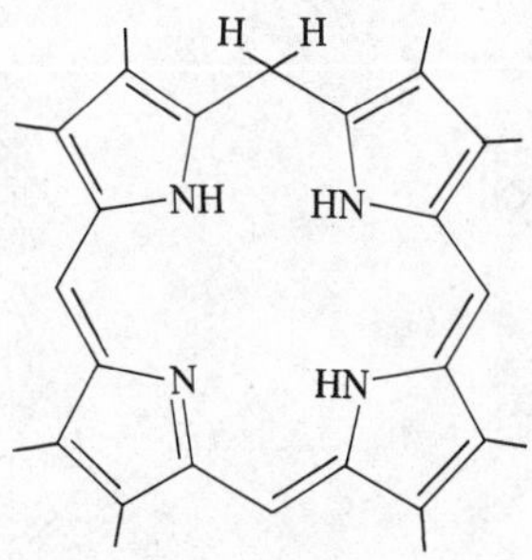

(29)

i, HI; ii, Paraformaldehyde added to HI reagent.

SCHEME 5

large, do not give porphyrinogens unless the central metal ion is reduced out of the macrocycle — an example of this is Tl^{III}, which is readily reduced to Tl^{I}, an ion which is too large to fit into the porphyrin central cavity). When other reducible groups (*e.g.* vinyl) are present as substituents, porphyrins can be reduced to porphyrinogens using sodium amalgam. Treatment of porphyrinogens with oxygen (usually in the presence of light), with iodine, or with high-potential quinones (*e.g.* 2,3-dichloro-5,6-dicyanobenzoquinone) accomplishes smooth conversion back to the porphyrin state.

Irreversible reduction of porphyrins using hydriodic acid has been used (like CrO_3 and $KMnO_4$ oxidation) for structure elucidation. Using hydriodic acid alone, the procedure usually gives homologous mixtures of unstable (2-unsubstituted or 2,5-diunsubstituted) pyrroles which are poorly suited to separation and identification (Scheme 5A). Bearing in mind the fact that each porphyrin subunit can furnish up to four different pyrrolic products, the results can often be impossible to untangle. However, if the reaction is carried out in the presence of formaldehyde, then the unsubstituted positions suffer 'reductive *C*-methylation'[15] and only one perfectly stable pyrrole (Scheme 5B) is obtained from each subunit.

SCHEME 6

There are relatively few examples of nucleophilic substitution of porphyrins or metalloporphyrins. The most common examples are nucleophilic attack upon porphyrin π-cation radicals or π-dications (which are obtained by abstraction of one or two electrons respectively from the porphyrin π-system). Likewise, there are relatively few fully substantiated examples of radical attack upon porphyrins. Porphyrins have been shown to react with nitrogen dioxide to give *meso*-nitroporphyrin metal complexes in high yield, but this reaction is believed to involve oxidation of the metalloporphyrin (by nitrogen dioxide) to give the π-cation radical, followed by further attack by a molecule of nitrogen dioxide and loss of a proton (Scheme 6).

Porphyrin and metalloporphyrin photochemistry has been studied extensively,[16] partly because a magnesium dihydroporphyrin is the prosthetic group in the vitally important phenomenon of photosynthesis from which all of our food originates. Particularly facile reactions are photo-oxidation and photoreduction, but in the absence of a full discussion[16] of excited states and electron distribution in the porphyrin and metalloporphyrin macrocycle, further discussion is inappropriate.

One of the most important chemical transformations of the porphyrin ring system is that of cleavage to give open-chain tetrapyrrolic substances. A blood corpuscle serves, on average, a useful life of about 120 days; after this time, in response to an unknown trigger, the prosthetic haem group (**12**) of the haemoglobin or myoglobin is degraded to give initially biliverdin-IXα, and then bilirubin-IXα and other reduction and re-oxidation products, which are excreted in the faeces. (*N.B.* The -IXα nomenclature is due to Fischer and indicates that a type-IX porphyrin has been ring-opened by rupture at the 'α' (5) carbon atom). Only the iron atom is kept for reutilization. Interestingly, almost all natural bile pigments have the -IXα substituent orientation, the only exception to this rule being the tegumental pigment of the caterpillar of the cabbage white butterfly, which is biliverdin-IXγ. Two major theories of haem catabolism are still under consideration, but the most recent work appears to indicate that the process is enzymatic. (For further discussion of bile pigments see Section 17.1.) It is generally accepted that an intermediate

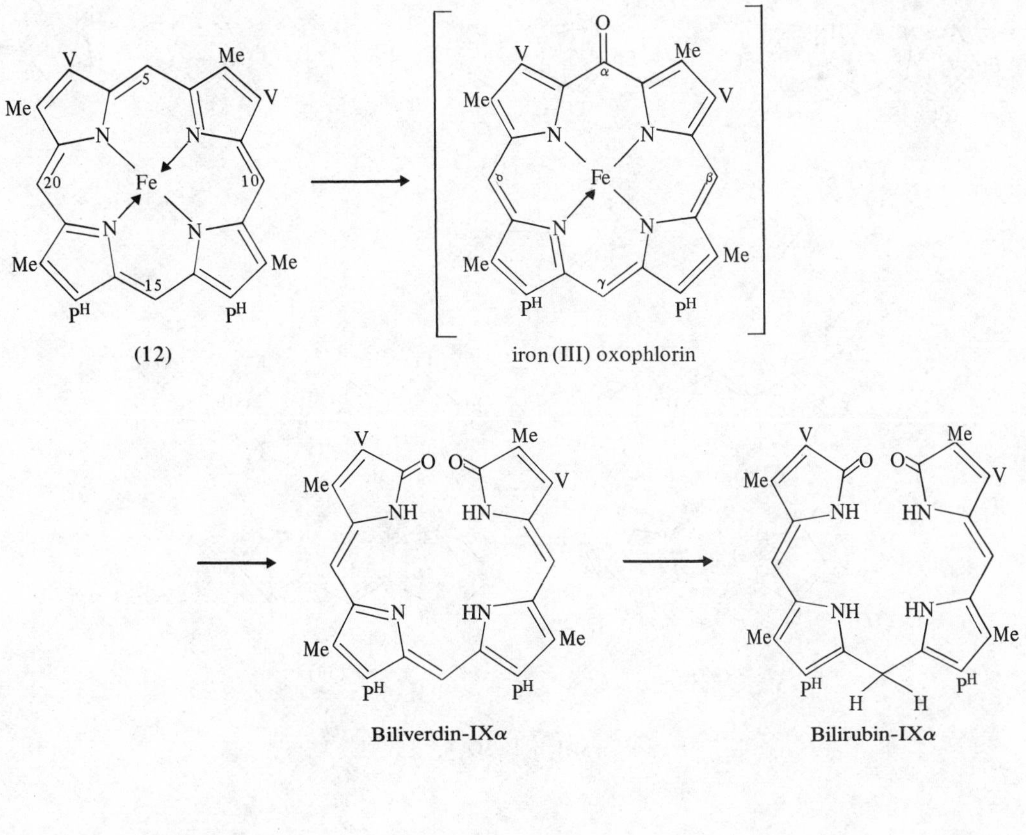

(12)

iron (III) oxophlorin

Biliverdin-IXα

Bilirubin-IXα

in haem catabolism is an iron oxophlorin, in which the carbonyl group is sited at the 5 (α) position; this carbon is known to be expelled as carbon monoxide. Chemical oxidation of porphyrins and metalloporphyrins to give oxophlorins is readily accomplished, but there is no obvious *meso*-bridge specificity if an unsymmetrical porphyrin is used. A chemical analogue of the conversion of haems into biliverdins is also known; this is called 'coupled oxidation' and involves oxidation (usually with oxygen or hydrogen peroxide) of an iron(II) porphyrin in the presence of mild reducing agents such as ascorbic acid or hydrazine. The reducing agent keeps the haem in the iron(II) state (and also reduces oxygen to hydrogen peroxide if the former reagent is used). Use of haem (**12**) affords approximately equal amounts of all four possible biliverdins (-IXα, -IXβ, -IXγ, and -IXδ). On the other hand, if myoglobin is subjected to coupled oxidation, only biliverdin-IXα is obtained; haemoglobin gives a mixture consisting of -IXα and -IXβ. If the haemoproteins are denatured and then subjected to coupled oxidation, the random mixture of all four biliverdins then results, indicating that the intact protein may have some type of directional influence on the site of haem rupture.

Porphyrins, *e.g.* octaethylporphyrin (**30**), can be methylated on the central nitrogens with comparative ease.[17] Treatment with methyl iodide affords the *N*-methyl derivative (**31**) together with the *N,N'*-dimethyl iodide (**32**) with *trans* methyl groups. Methylation with methyl fluorosulphonate gives a *N,N',N''*-trimethyl derivative (**33**) together with a *cis-N,N''*-dimethyl compound (**34**), which is presumably obtained by decomposition of (**33**). The methyl groups are sterically very hindered, and as a result the porphyrin macrocycle is heavily buckled.

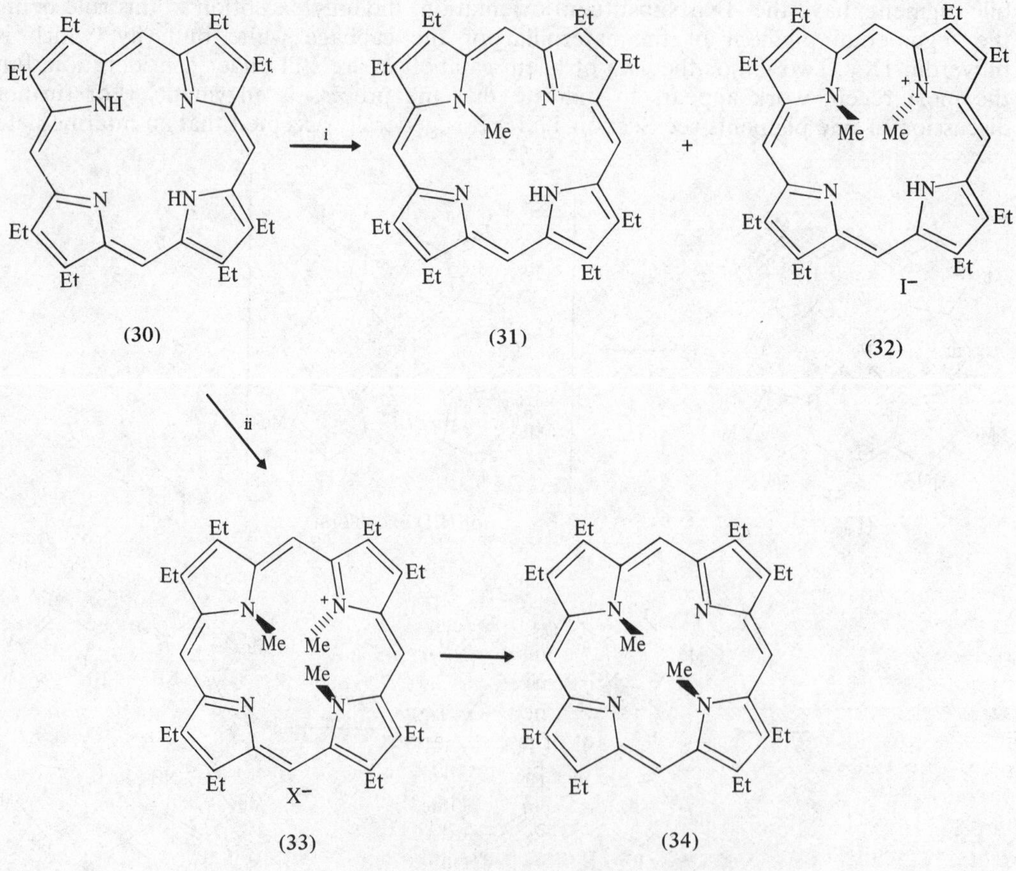

(30) (31) (32)

(33) (34)

i, MeI; ii, $MeSO_3F$.

17.2.1.6 Preparative methods[18]

(*i*) *From polymerization of monopyrroles*

This often highly efficient approach can only be used unambiguously if the pyrrole 3- and 4-substituents are identical. Otherwise, a mixture of all four possible type isomers results; for example, if the pyrrole 3-substituent were to be methyl and the 4-substituent ethyl, then the aetioporphyrins (**8**)-(**11**) would be obtained, and the proportions of each would be statistically predictable as 1/8 Type-I; 1/8 Type-II; 1/2 Type-III; 1/4 Type-IV. Of course, in the enzymatic tetramerization of porphobilinogen to uroporphyrinogen-III (Section 30.2), only the natural type-III isomer is produced owing to the controlling effect of the enzymes porphobilinogen deaminase and uroporphyrinogen-III cosynthetase.

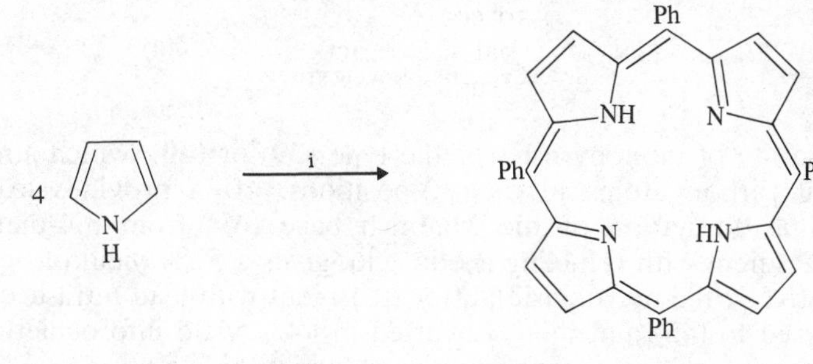

(35)

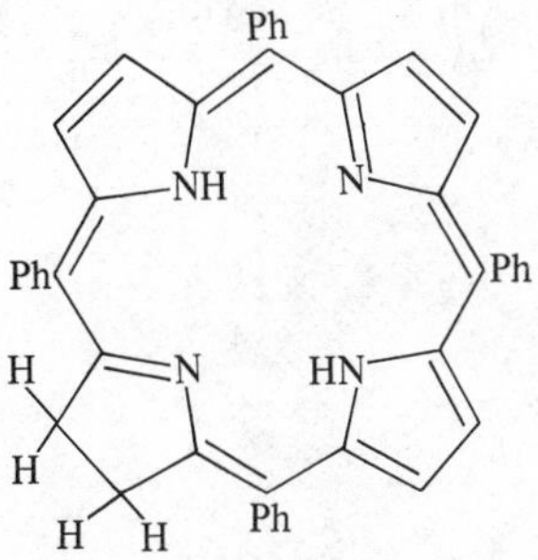

(36)

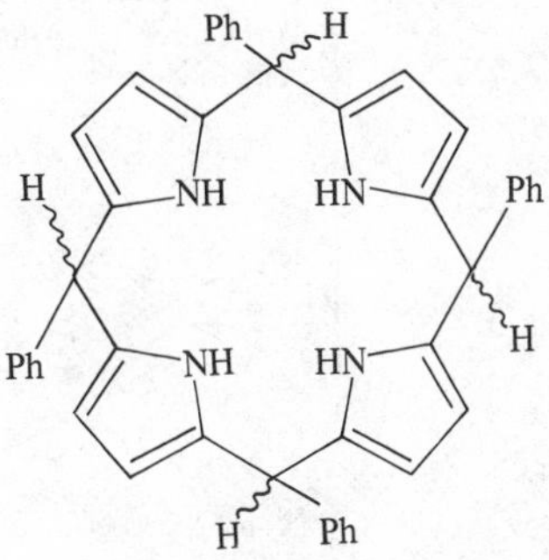

(37)

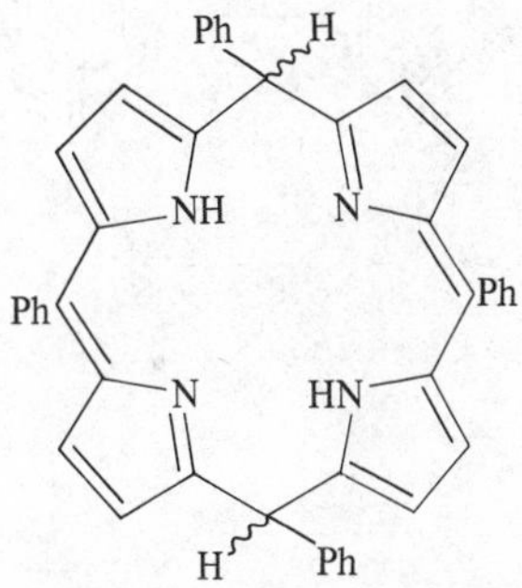

(38)

i, 4PhCHO in $CH_3CH_2CO_2H$

Several kinds of chemical polymerization of monopyrroles to porphyrins have been described, and these vary only in the nature of the monopyrrole 2- and 5-substituents. Treatment of pyrroles with unsubstituted 2- and 5-positions with agents which are able to supply the four *meso* carbons will lead to porphyrins. For example, treatment of pyrrole itself with formic acid gives a very low yield of porphin (**1**), but perhaps the best known application of this principle is the Rothemund reaction, from which the ubiquitous *meso*-tetraphenylporphyrin (**35**) is obtained. The optimum conditions[19] for this reaction involve heating equimolar proportions of pyrrole and benzaldehyde in refluxing propionic acid for about 30 minutes, and a yield approximating 20% of the porphyrin (**35**) results. By most standards of organic chemistry this yield might be regarded as mediocre, but bearing in mind the complexity of the product, and the fact that it can be prepared on the kilogram scale from the simplest of starting materials, this procedure represents one of the successes in this field. Indeed, on account of this very reaction (which can be operated even by theoretical chemists with no reduction in yield!) *meso*-tetraphenylporphyrin (**35**) has been the porphyrin of choice for many of the kinetic and other physical studies carried out on porphyrins in the past 20 years. The product filtered from the cooled propionic acid reaction mixture is usually contaminated (to an extent of 2–10%) with the chlorin (**36**), and this is best removed by brief treatment with 2,3-dichloro-5,6-dicyanobenzoquinone.[20] The mechanism of the synthesis has been investigated[21] and the porphyrinogen (**37**) and porphodimethene (**38**) have been shown to be intermediates.

(39) (40)

X is a good leaving group

Self-condensation of monopyrroles of the type (**39**) or (**40**), which already possess the future bridging carbon atoms in their 2-positions, often provides excellent yields of porphyrins. Thus, formation of the Mannich base (**41**) from 3,4-diethylpyrrole (**42**), followed by treatment with refluxing acetic acid, gives a 52% yield of octaethylporphyrin (**30**).[22] Alternatively, the pyrrole (**43**), after treatment with lead tetra-acetate to give (**44**), can be hydrolysed to (**45**) and then converted in 44% yield into octaethylporphyrin (**30**) by heating in acetic acid containing potassium ferricyanide.[23]

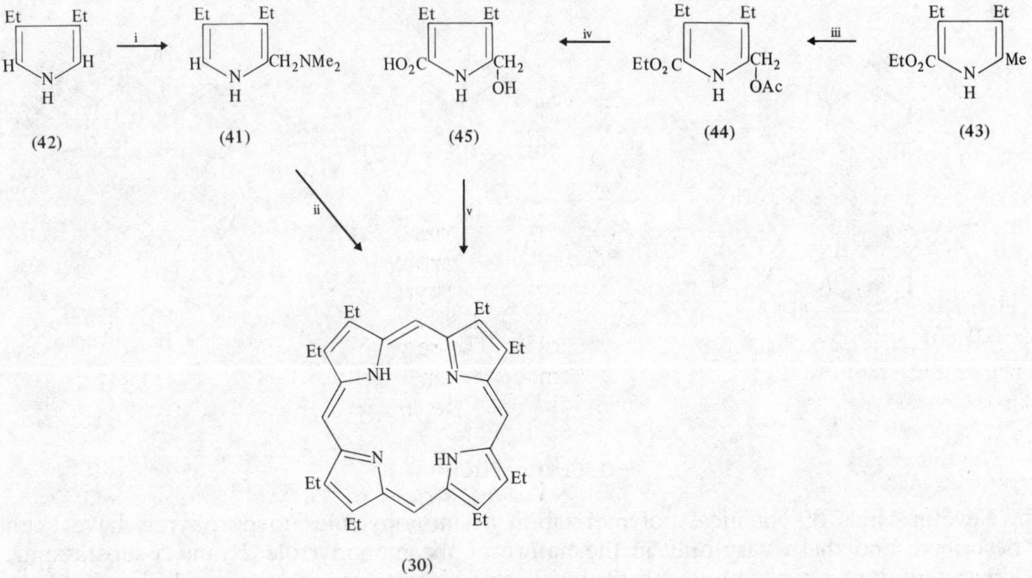

i, CH_2O, $HNMe_2$; ii, heat in AcOH; iii, $Pb(OAc)_4$; iv, OH^-; v, heat, AcOH, K_3FeCN_6

(ii) From dipyrrolic intermediates

Such syntheses can be subdivided into two types, namely those employing dipyrromethenes and those from dipyrromethanes. Most of Fischer's early work relied[1] heavily on the use of pyrromethenes; his method is exemplified in the synthesis of deuteroporphyrin-IX (**46**) from which haemin chloride (**47**) was subsequently synthesized[24] (Scheme 7). Condensation of the unsymmetrically substituted dipyrromethene (**48**) with the symmetrical one (**49**) (symmetrical about the interpyrrolic carbon atom) in an

i, heat, succinic acid; ii, $FeSO_4$, O_2; iii, Ac_2O, $SnCl_4$; iv, removal of iron; v, KOH, EtOH; vi, heat in vacuum; vii, $FeSO_4$, O_2, Cl^-.

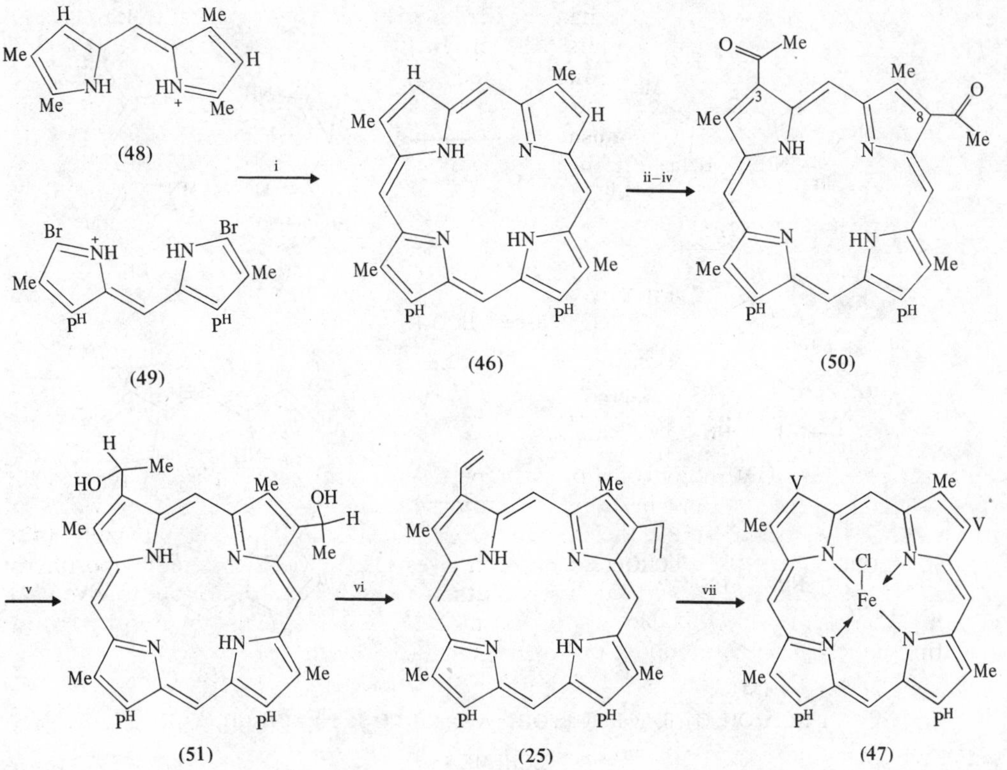

SCHEME 7

organic acid melt gave a low yield of deuteroporphyrin-IX (**46**). Friedel–Crafts acetylation of the iron(III) complex gave 3,8-diacetyldeuteroporphyrin-IX (**50**) (after removal of the metal ion), which was reduced to give haematoporphyrin-IX (**51**) and then dehydrated to afford protoporphyrin-IX (**25**). Insertion of iron(III) using ferrous acetate and oxygen then gave, after anion exchange with chloride, the required haemin chloride (**47**). Apart from representing a tremendous achievement for early synthetic strategy, this synthesis also served to confirm Küster's proposal for the macroring system present in the porphyrins.

The symmetry restrictions inherent in this type of '2+2' approach should be stressed. Only if one of the two dipyrromethenes is symmetrically substituted does one product result. However, by judicious choice of substituent arrays Fischer was able to employ this type of synthesis on many occasions, even when both dipyrromethenes were unsymmetrically substituted. This required great skill in separation of the resulting porphyrin mixtures (before the advent of chromatography), but a whole host of new porphyrins were synthesized and fully characterized. The method can also be applied to self-condensation of single dipyrromethenes (*e.g.* **52**). Treatment of this compound in refluxing formic acid affords a good yield of aetioporphyrin-I (**8**).

Similar symmetry restrictions apply to porphyrin syntheses proceeding through dipyrromethanes. (For synthetic approaches to both dipyrromethenes and dipyrromethanes, see Section 17.1.) The usefulness of these intermediates was first pointed out by MacDonald,[25] and employed most notably in the synthesis of uroporphyrin-III (**53**) (Scheme 8). Thus, condensation of the 5,5′-diunsubstituted dipyrromethane (**54**) with the 5,5′-diformyldipyrromethane (**55**) in the presence of hydriodic acid gave yields of porphyrin (**53**) approaching 65%. Self-condensation of a 5-formyl-5′-unsubstituted dipyrromethane

i, heat, HCO_2H.

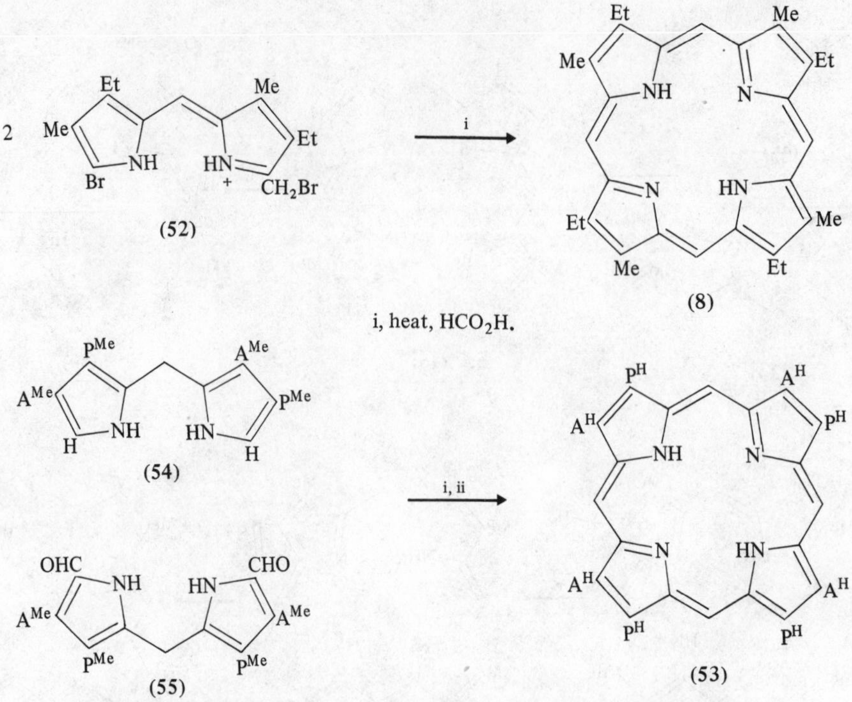

i, HI, AcOH, O_2; ii, hydrolysis, OH^-; $A^R = CH_2CO_2R$; $P^R = CH_2CH_2CO_2R$.

SCHEME 8

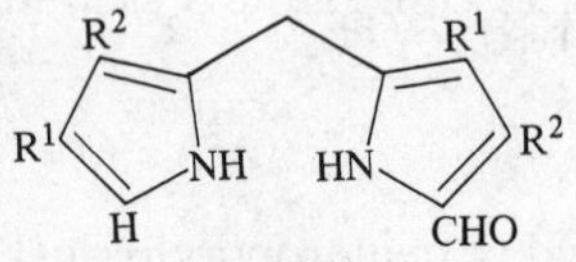

(56)

(*e.g.* **56**) can also be employed to afford regularly substituted porphyrins. The symmetry limitations of this general type of synthesis were overcome ingeniously in Woodward's total synthesis of chlorophyll-a[26] (**13**). The porphyrin (**57**) (Scheme 9) was constructed from a 5,5′-diunsubstituted dipyrromethane (**58**) and another (**59**) bearing the future 5- and 15-*meso*-carbons, and, in principle, two modes of condensation were possible. However, ambiguity was removed by prior formation of a Schiff's base (**60**) in which the two components were fixed in space in the required orientation.

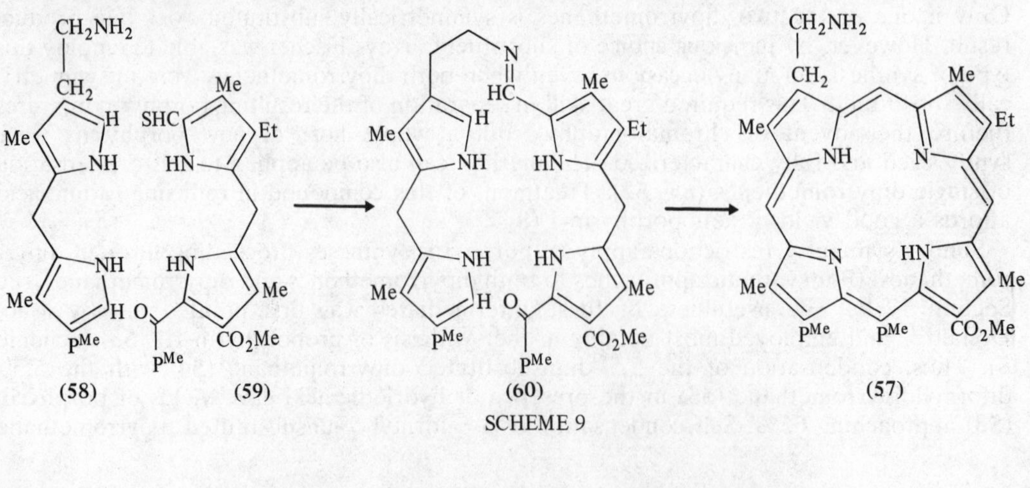

SCHEME 9

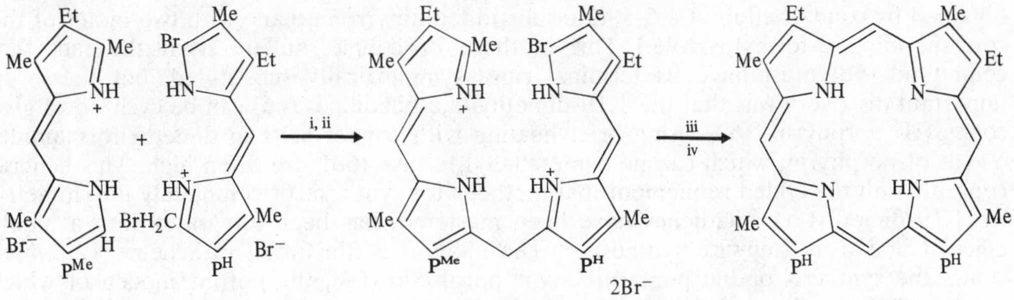

i, $SnCl_4$; ii, HBr, MeOH; iii, heat; iv, hydrolysis, OH^-

SCHEME 10

(*iii*) *From open-chain tetrapyrrolic intermediates*

Most modern developments in porphyrin synthesis[18] have employed the concept that if an open-chained tetrapyrrolic intermediate can be constructed with the correct sequence of substituents, and furthermore it can be cyclized to porphyrin without any redistribution of the individual pyrrole rings, then all former symmetry restrictions will be eliminated.

Johnson and his co-workers were able to develop a highly successful route[27] which is basically a two step version of Fischer's dipyrromethene route. The method is exemplified in the synthesis of mesoporphyrin-IX (**26**) (Scheme 10), in which dipyrromethene (**61**) is alkylated in the presence of stannic chloride with the dipyrromethene (**62**) to give a high yield of the a,c-biladiene (**63**), initially as its tin(IV) salt. Heating in *ortho*-dichlorobenzene affords the required porphyrin (**26**), again in high yield.

Yet another method for preparation of a,c-biladienes[28] is shown in Scheme 11. In this method, treatment of a 5,5′-diformyldipyrromethane (**64**) with two moles of a 2-unsubstituted pyrrole (**65**) gives an a,c-biladiene (**66**). (The same biladiene could also be

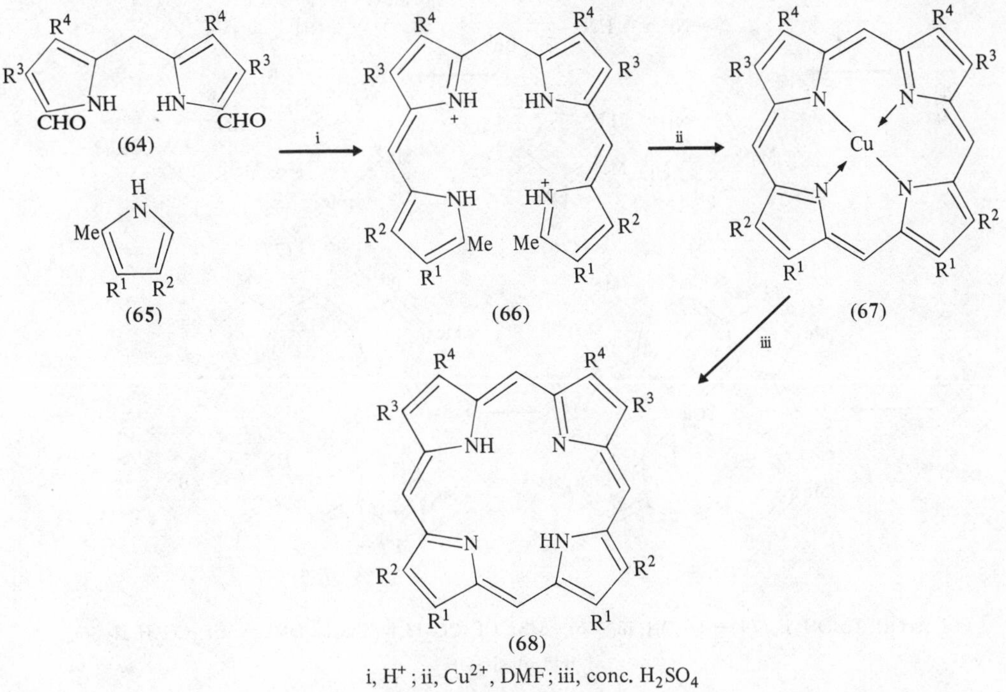

i, H^+; ii, Cu^{2+}, DMF; iii, conc. H_2SO_4

SCHEME 11

obtained by condensation of a 5,5′-diunsubstituted dipyrromethane with two moles of the corresponding 2-formylpyrrole.) This method, of course, suffers from the fact that compound (**66**) must have its terminal rings symmetrically substituted, but the more important discovery was that the 1,19-dimethyl-a,c-biladienes (**66**) can be cyclized to give copper(II) porphyrins (**67**) using brief heating with copper salts in dimethylformamide. Yields of porphyrin, which can be demetallated to give (**68**), are often high. This general concept, with the added refinement that methods for synthesis of completely unsymmetrical 1,19-dimethyl-a,c-biladienes have been mastered, has been integrated into a highly efficient and truly stepwise synthesis.[29] The method is illustrated in Scheme 12, which shows the synthesis of the pentacarboxylic porphyrin (**69**), the porphyrinogen of which (Section 30.2) has been shown[30] to be an intermediate in normal haem metabolism. Thus, the differentially protected and unsymmetrically substituted dipyrromethane (**70**) was synthesized by condensation of pyrroles (**71**) and (**72**) in acetic acid containing a catalytic quantity of toluene-*p*-sulphonic acid hydrate.[31] Catalytic hydrogenolysis gave the dipyrromethanecarboxylic acid (**73**) which was condensed, under strictly controlled acidic conditions, with the 2-formylpyrrole (**74**) to give the tripyrrole salt (**75**) still bearing an intact t-butyl ester. Further condensation of (**75**) with another 2-formylpyrrole (**76**) in trifluoroacetic acid containing hydrobromic acid affords a high yield of the a,c-biladiene dihydrobromide (**77**). Cyclization, using copper(II) chloride in DMF, gave the copper porphyrin (**78**), which was demetallated with 10% trifluoroacetic acid in concentrated

i, AcOH, TosOH; ii, (74) + TosOH; iii, (76) + HBr, CF_3CO_2H; iv, $CuCl_2$, DMF; v, CF_3CO_2H, H_2SO_4; vi, hydrolysis, OH^-

SCHEME 12

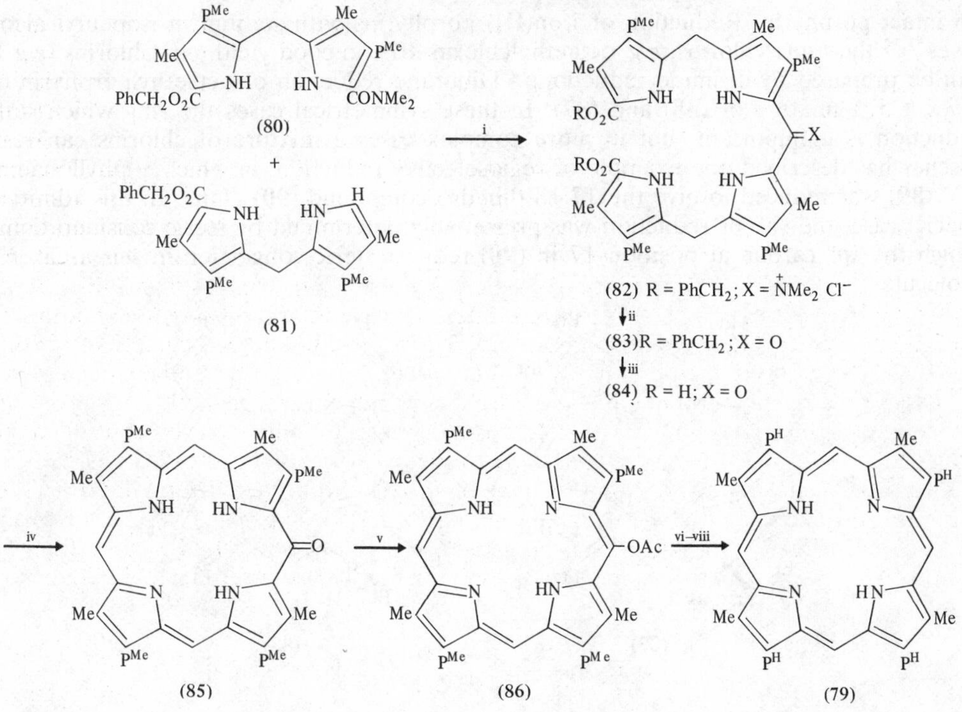

(80) + (81) $\xrightarrow{\text{i}}$

(82) R = $PhCH_2$; X = $\overset{+}{N}Me_2$ Cl^-

↓ii

(83) R = $PhCH_2$; X = O

↓iii

(84) R = H; X = O

$\xrightarrow{\text{iv}}$ (85) $\xrightarrow{\text{v}}$ (86) $\xrightarrow{\text{vi–viii}}$ (79)

i, $POCl_3$; ii, aq. Na_2CO_3; iii, H_2, Pd/C; iv, $HC(OMe)_3$, CCl_3CO_2H; v, Ac_2O, pyridine; vi, H_2, Pd/C; vii, DDQ; viii, hydrolysis, OH^-

SCHEME 13

sulphuric acid, and then de-esterified to give the required porphyrin-pentacarboxylic acid (**69**). Labels (^{13}C or ^{14}C) are readily incorporated into this procedure, as indicated in Scheme 12, by carrying out the Vilsmeier formylation (Section 17.1) for preparation of the formylpyrrole (**76**) using labelled DMF.

One of the earliest porphyrin syntheses using open-chain tetrapyrroles involved b-oxobilanes as intermediates.[32] Moreover, the first macrocyclic products obtained in this synthesis (Scheme 13) are oxophlorins, which are important intermediates in the process of haem catabolism (see Section 17.2.1.5). The approach is illustrated in the synthesis of coproporphyrin-III (**79**), which occasionally occurs naturally, but is best known since its porphyrinogen is a key intermediate in haem and chlorophyll metabolism (Section 30.2). In a type of Vilsmeier reaction, the 5-dimethylcarbamoyldipyrromethane (**80**) was condensed with the 5-unsubstituted dipyrromethane (**81**) to give the tetrapyrrolic imine salt (**82**). Hydrolysis gave the b-oxobilane (**83**), which was debenzylated using catalytic hydrogenation to afford the dicarboxylic acid (**84**). Cyclization using trimethylorthoformate (as a one-carbon unit) in presence of trichloroacetic acid gave the oxophlorin (**85**) after aerial oxidation. The oxygen function was eliminated by treatment with acetic anhydride in pyridine, which furnished the *meso*-acetoxyporphyrin (**86**), followed by catalytic hydrogenation to the porphyrinogen, and then re-oxidation to the *meso*-unsubstituted porphyrin tetramethyl ester using 2,3-dichloro-5,6-dicyanobenzoquinone. Alkaline hydrolysis then gave the porphyrin (**79**).

(*iv*) *Syntheses of chlorins*

Owing to the ease with which chlorins are oxidized to porphyrins, no direct synthesis of chlorins has yet been described. Instead, chlorins are usually prepared by reduction of an intact porphyrin. Reduction of iron(III) porphyrins with sodium in isopentyl alcohol gives[33,34] the *trans*-chlorins, *e.g.* octaethylchlorin (**87**), in good yield. *cis*-Chlorins (*e.g.* **88**) can be prepared by di-imide reduction.[34] Diborane reduction of octaethylporphyrin (**30**) gives a 5:1 mixture of (**88**) and (**87**). In these symmetrical cases the ring which suffers reduction is unimportant, but in more complex cases a mixture of chlorins can result. Fischer has described one example of regioselective reduction, in which γ-phyllohaemin-XV (**89**) was reduced to give the 17,18-dihydro-compound (**90**) alone. In this admittedly special case, the site of reduction was presumably determined by steric considerations in which the sp^3 carbon at position-17 in (**90**) reduces steric congestion in this area of the molecule.

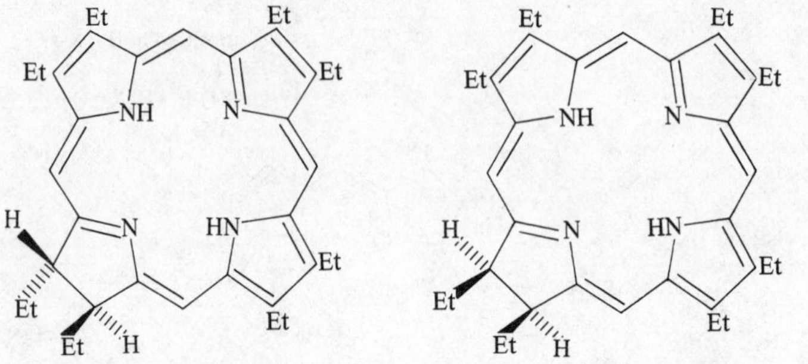

(87)

(88)

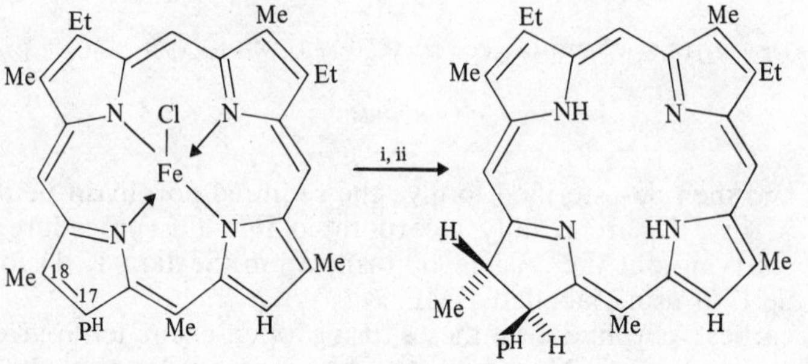

(89)

(90)

i, Na, isopentyl alcohol; ii, demetallation

17.2.2 CORRINS

17.2.2.1 Structure and nomenclature

The corrin ring, with its numbering system, is shown in structure (**91**). The key natural product containing the corrin skeleton is vitamin B_{12} (**92**), which is an artefact from the coenzyme (**93**). The coenzyme has a 5′-deoxyadenosyl group attached directly to cobalt through a cobalt–carbon bond, and during isolation this group is displaced by cyanide to give vitamin B_{12} (**92**), also known as cyanocobalamin.[35–37] The corrin part is a remarkable type of modified porphyrin system in which the 'type-III' substituent pattern (**94**) (*cf.* uroporphyrinogen-III (**95**), one of its biosynthetic precursors; see Section 30.2) can still be discerned.

The structure elucidation of vitamin B_{12} constituted an impressive contribution from X-ray crystallographic analysis.[38] Nevertheless, much chemistry had been carried out on the system, but this will be referred to only briefly. Acid hydrolysis of vitamin B_{12} affords

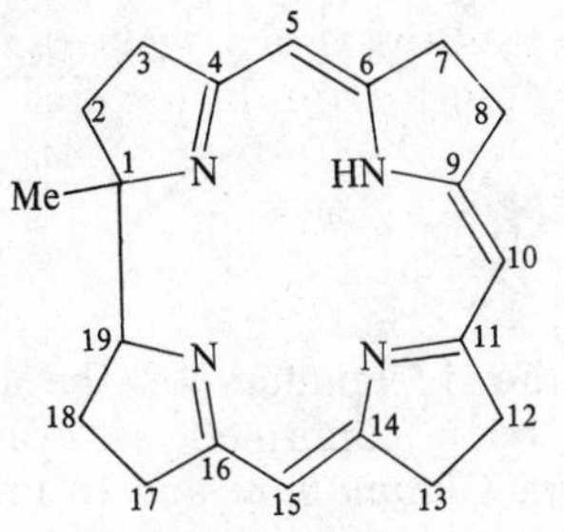

(91)

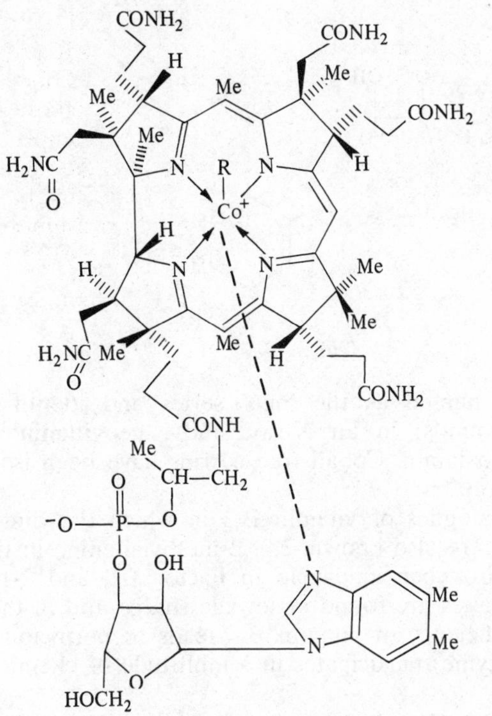

(92) R = CN

(93) R =

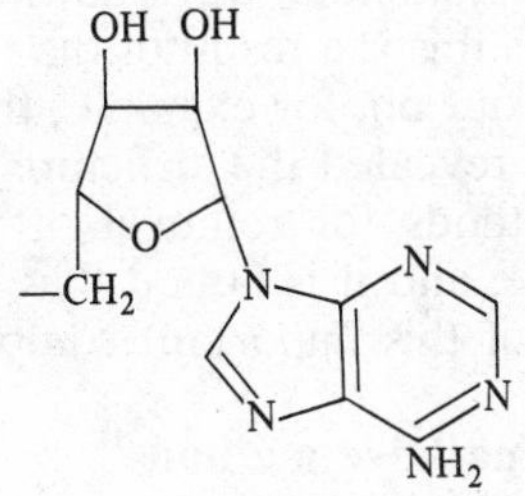

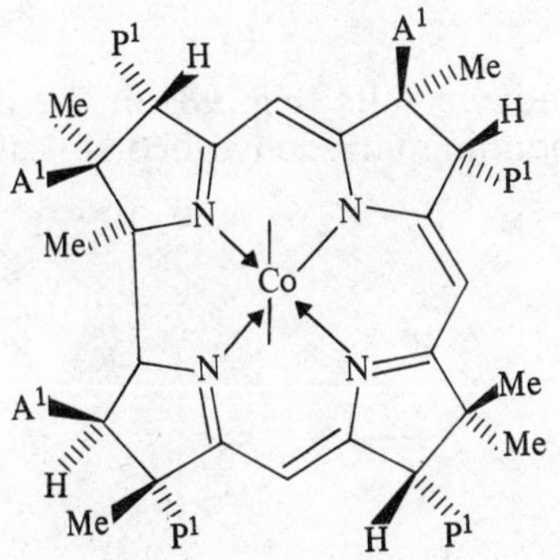

(94)

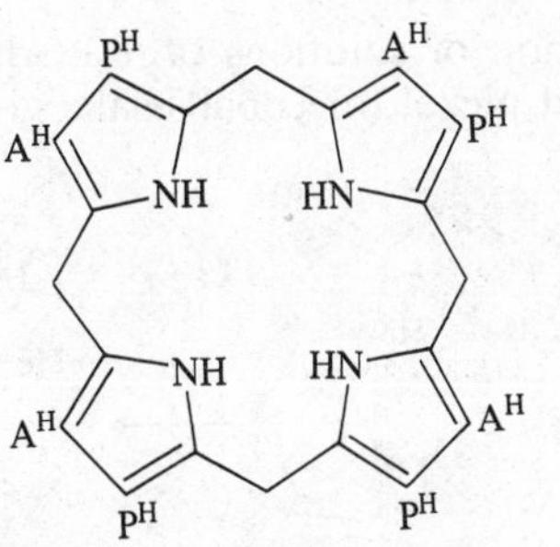

(95)

$A^1 = CH_2CONH_2$; $P^1 = CH_2CH_2CONH_2$

the nucleotide (**96**) and D-1-aminopropan-2-ol. Mild hydrolysis gives aetiocobalamin in which the aminopropanol is still attached to the macrocycle, but in which the nucleotide is cleaved. Alkaline hydrolysis gives the lactam (**97**). Treatment of vitamin B_{12} with strong acids (such as trifluoroacetic acid) gives a second series of pigments (along with the normal acid hydrolysis products), and these are slightly darker[39] than the normal series. They have

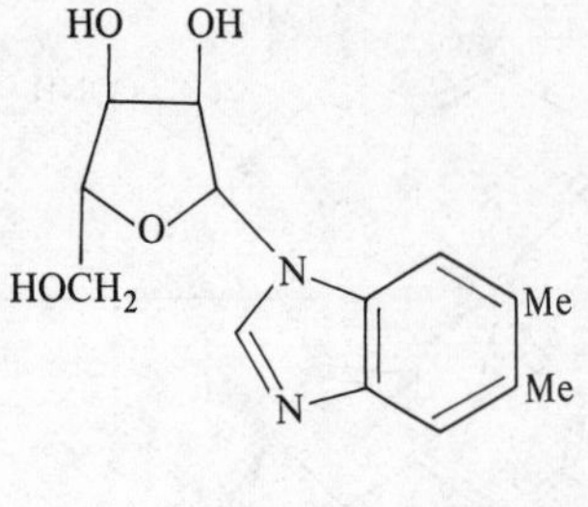

(96)

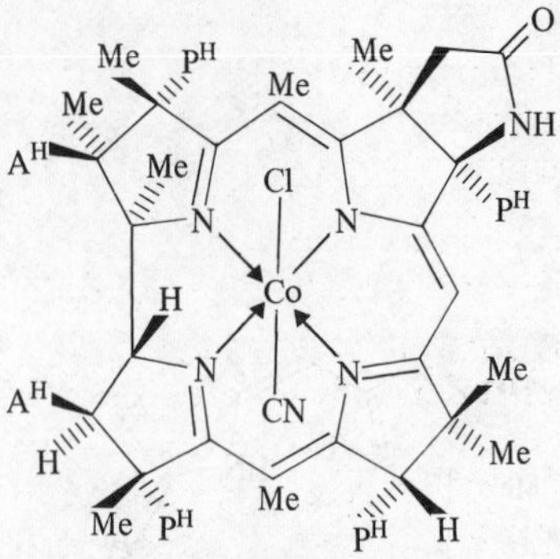

(97)

been named as the 'neo' series and identified[40] as the 13-epimers of the natural compounds; in an X-ray study, neovitamin B_{12} has been confirmed as cyano-13-epicobalamin. Cobalt-free corrins have been isolated from *Chromatium* and from *Streptomyces*.[41]

Analogues of vitamin B_{12} in which the dimethylbenzimidazole is replaced by other bases are also known, *e.g.* 2-methyladenine in Factor A, adenine in pseudo-vitamin B_{12}, 5-hydroxybenzimidazole in Factor III, and 2-methylhypoxanthine in Factor H. These analogues are found in sewage sludge and in the digestive tracts of animals.

Deficiency of vitamin B_{12} leads to pernicious anaemia, and the vitamin itself (or its coenzyme) participates in a multitude of chemical transformations (see Section 24.3).

17.2.2.2 Chemical transformations of the corrin nucleus

Very little systematic work on the chemistry of the corrin nucleus has been carried out. In some ways this is a result of the fact that traditional chemical degradation studies of the type carried out on, for example, the chlorophylls, were cut short by the X-ray investigation[38] which revealed the structure of the vitamin. Much work is presently attempting to discover methods for removal of the cobalt atom, which is tenaciously bound to the chromophore, and it is hoped that as a result of these studies we shall learn more about the nucleus of this important compound.

17.2.2.3 Preparative methods[18]

(i) From pyrrolic intermediates

Aeration of solutions of 1,19-dimethyl-a,c-biladiene salts (*e.g.* **98**) in the presence of base and nickel (or cobalt) salts gives the corresponding tetradehydrocorrin salts (*e.g.* **99**)

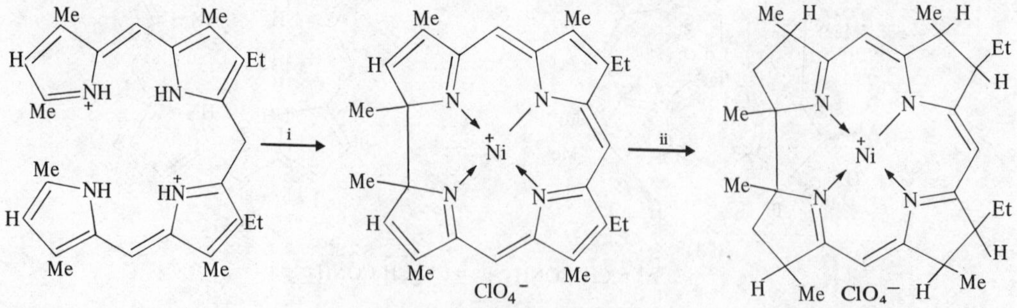

(98) (99) (100)

i, O_2, piperidine, Ni^{2+}; ii, H_2, Raney Ni, 160 °C, 100 atm

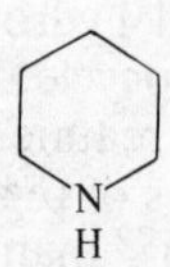

SCHEME 14

(Scheme 14).[42] Catalytic hydrogenation affords[43] nickel corrins (*e.g.* **100**); crystalline nickel(II) and cobalt(II) complexes of 1,19-dimethylcorrin have also been obtained by less vigorous reductions of appropriate metal tetradehydrocorrin salts lacking any peripheral substituents.[44]

(*ii*) *From non-pyrrolic precursors*

(*a*) *Eschenmoser's approach.*[45] The target corrin in Eschenmoser's early work was the model (**101**). This was to be approached by coupling of the two halves (**102**) and (**103**),

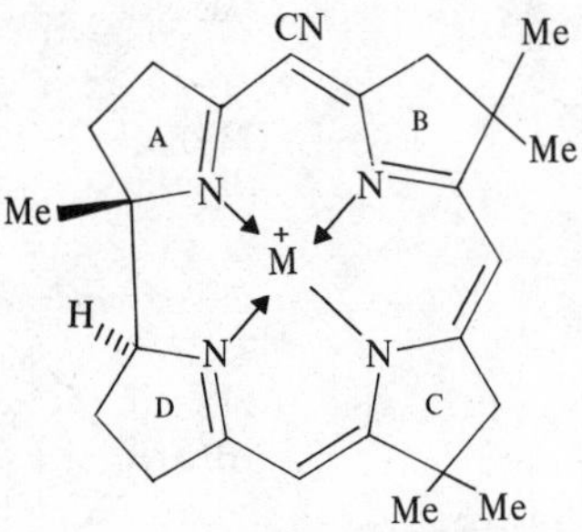

(101) 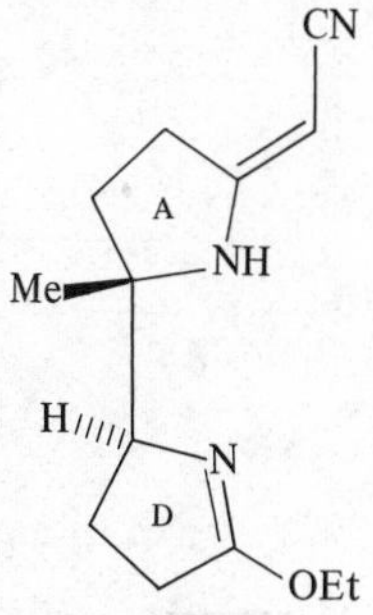 (102) 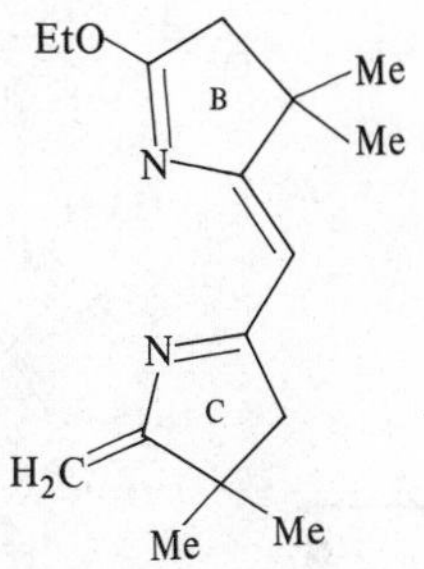 (103)

and the successful syntheses of these building blocks are shown in Schemes 15 and 16. Thus, Diels–Alder cycloaddition of isoprene and tetramethyl ethylenetetracarboxylate gave the tetra-ester (**104**), which was cleaved using sodium in liquid ammonia to give

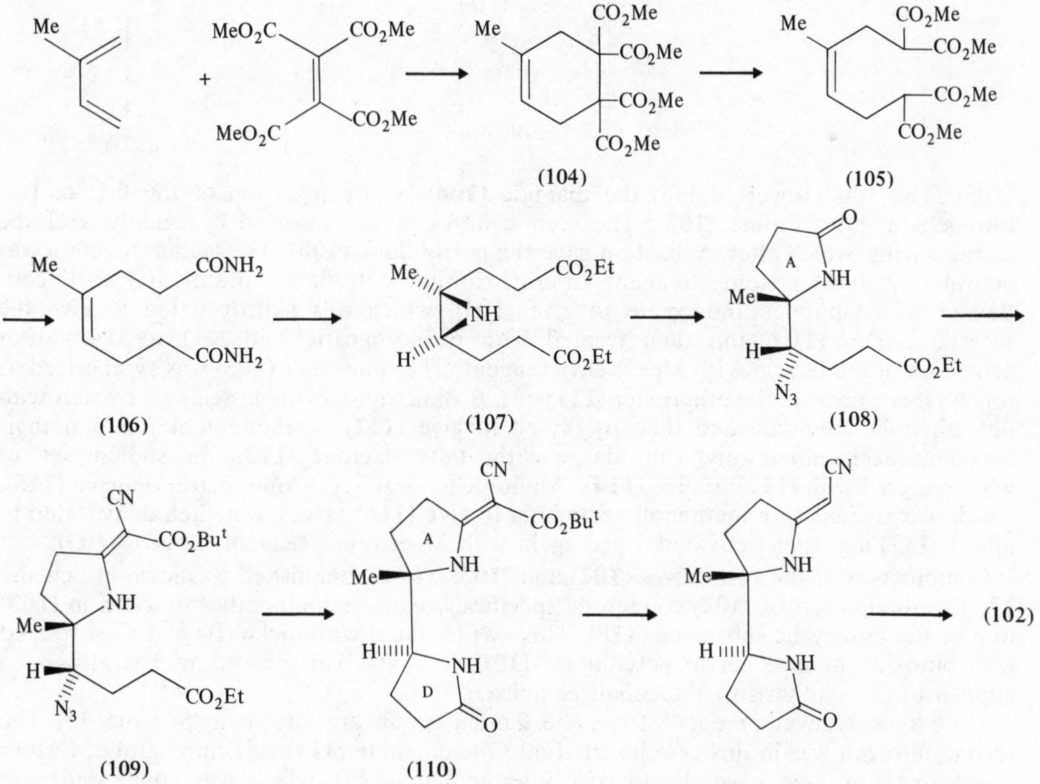

(104) (105) (106) (107) (108) (109) (110) (102)

SCHEME 15

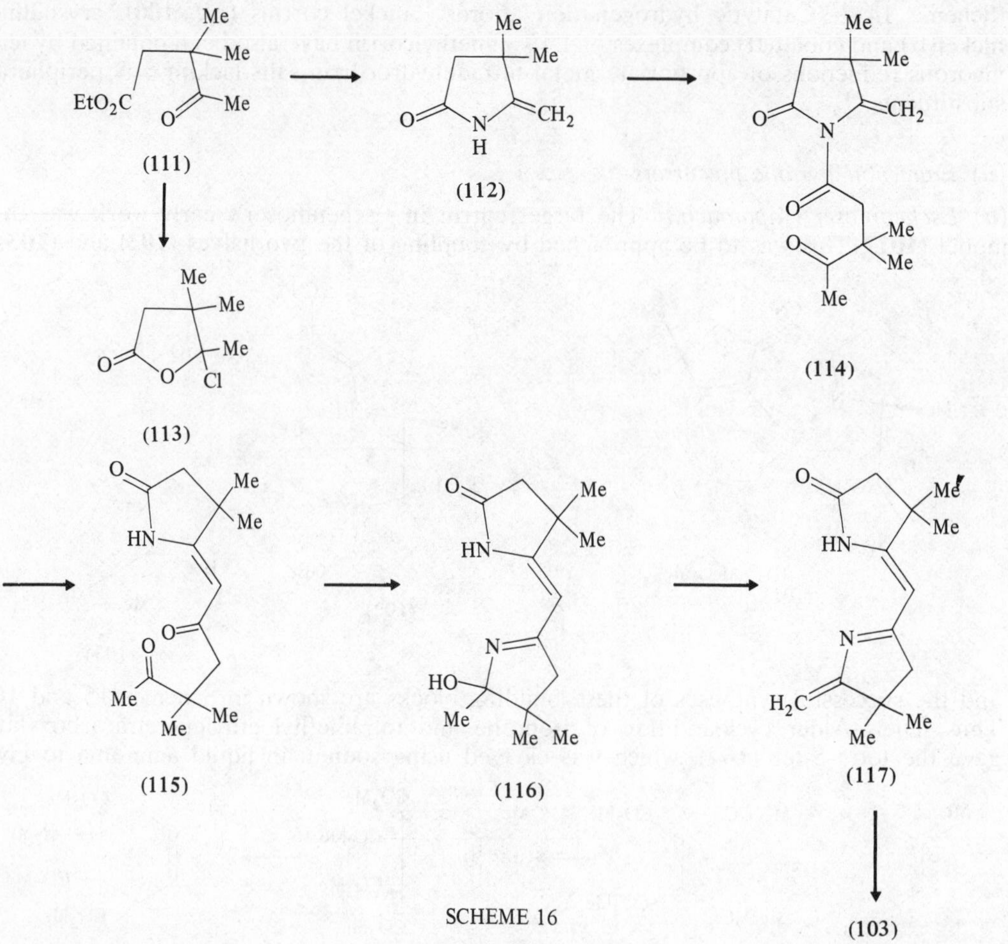

SCHEME 16

(**105**). This was converted into the diamide (**106**) before insertion of the first of two nitrogens as an aziridine (**107**). The second nitrogen was inserted by azidolysis of the aziridine ring, which after cyclization gave the pyrrolidone (**108**). The lactam function was activated with Meerwein's reagent (triethyloxonium tetrafluoroborate) and then condensed with t-butyl cyanoacetate to give (**109**), which was hydrogenated to give the bicyclic lactam (**110**) and then treated with trifluoroacetic acid to give (**102**) after activation of the lactam with Meerwein's reagent. The other half (**103**) was synthesized as follows (Scheme 16). The ethyl ester (**111**) of β,β-dimethyl-laevulinic acid was treated with hot ethanolic ammonia and then pyrolysed to give (**112**). Treatment of β,β-dimethyl-laevulinic acid with thionyl chloride gave the butyrolactone (**113**), the sodium salt of which reacted with (**112**) to give (**114**). A photochemical acyl group migration gave (**115**), which was treated with methanolic ammonia to give (**116**), which was then dehydrated to afford (**117**) and then activated, once again with Meerwein's reagent, to give (**103**).

Combination of the two halves (**102**) and (**103**) was accomplished as shown in Scheme 17. The sodium salt of (**102**) condensed specifically with the iminoether function in (**103**) to give the tetracyclic substance (**118**). This was chelated with nickel(II) and then treated with butoxide to give corrin perchlorate (**101**; M = Ni). This procedure has also been applied to the synthesis of the cobalt complex.

More recent developments[46] from the Zurich group are shown in Scheme 18. The second nitrogen was in this case inserted into the enamide (**119**) as a nitro-group; further transformations then gave the bicyclic intermediate (**120**), which was condensed with (**103**) to give the dicyanocorrin-cobalt(III) complex (**121**).

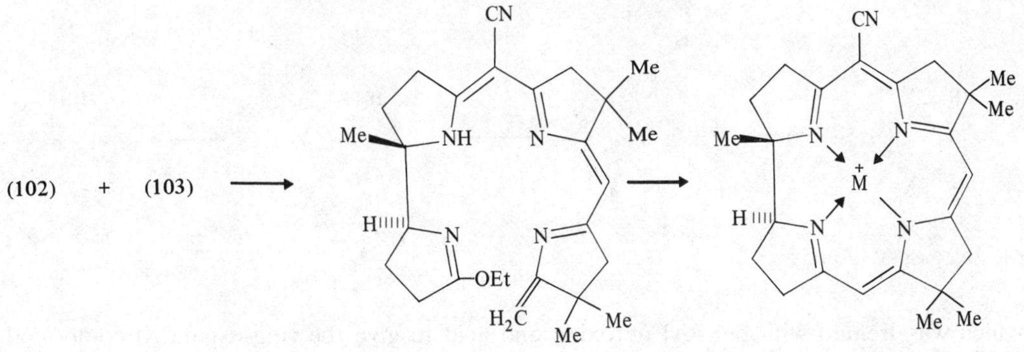

(102) + (103) ⟶ (118) ⟶ (101) M = Ni

SCHEME 17

In the two approaches so far discussed for synthesis of the corrin nucleus, a metal ion has been used as a template around which the two halves have been fused. Using the principle of 'sulphide contraction' (Scheme 19), a method for synthesis of metal-free corrins was devised.[47] Thus, the sodium salt of the precorrin (**122**) was transformed into the thiolactam (**123**). Loose complexation with zinc(II) ions caused cyclization to (**124**),

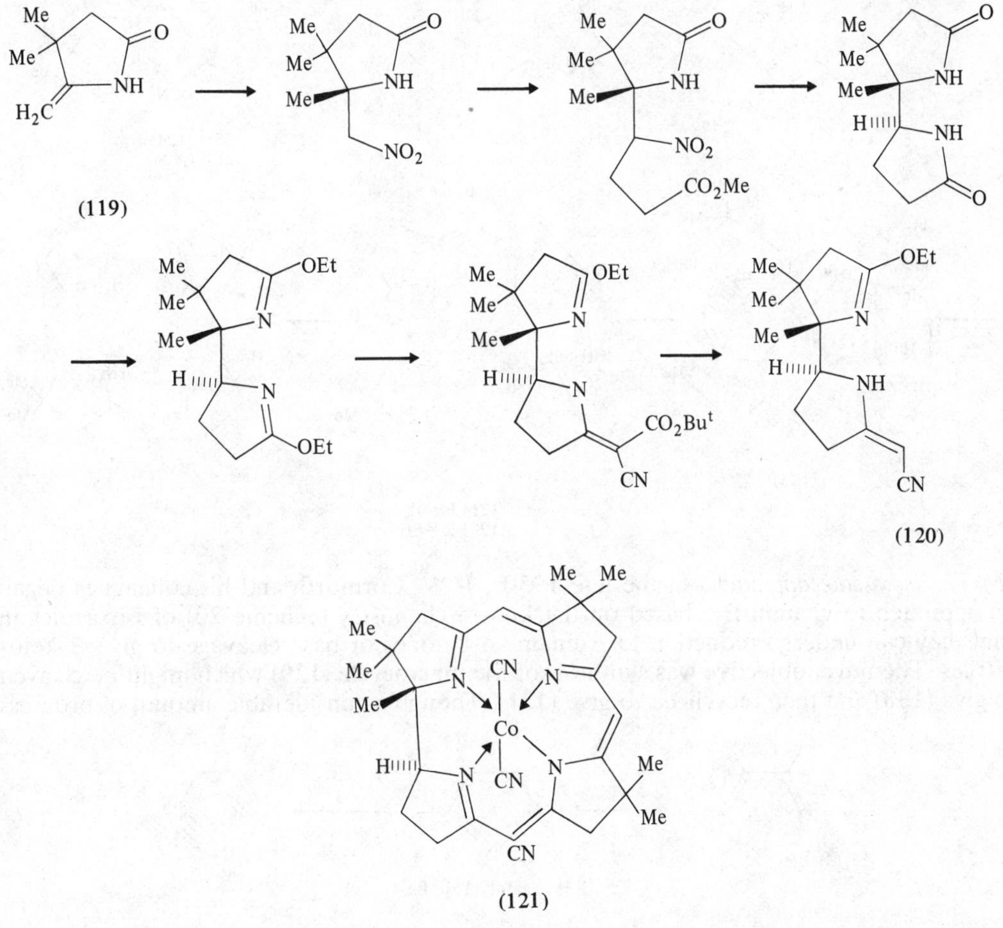

(119) ⟶ ⟶ ⟶ ⟶ ⟶ ⟶ (120)

(121)

SCHEME 18

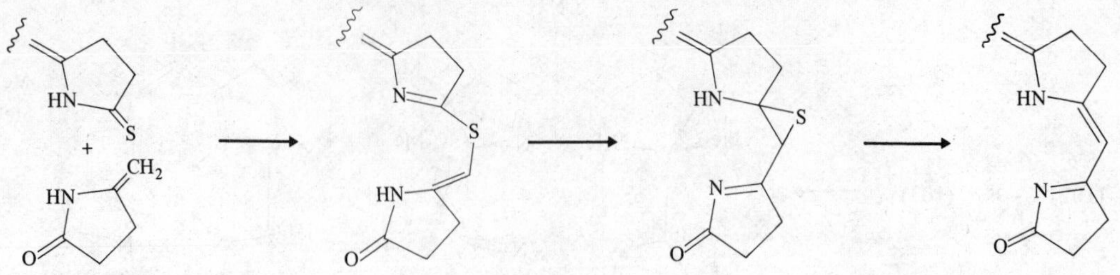

SCHEME 19

which was treated with benzoyl peroxide and acid to give the ring-expanded compound (**125**). Contraction with trifluoroacetic acid in DMF gave the corrins (**126**) and (**127**), and the major of these [(**127**)] was desulphurized with triphenylphosphine in the presence of acid to give (**126**); demetallation with trifluoroacetic acid gave the required metal-free corrin (**128**), a source of a whole variety of metal complexes.

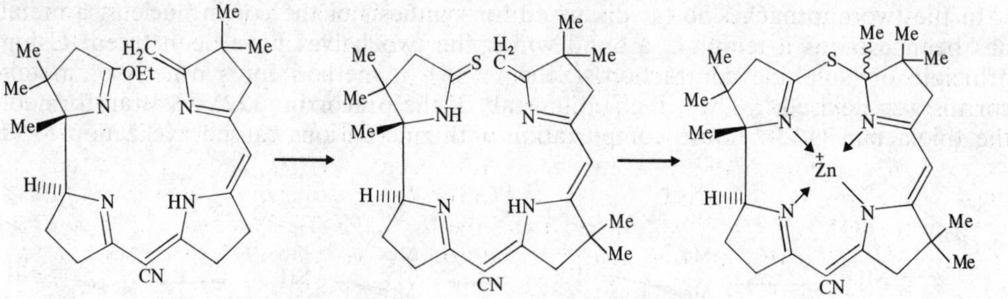

(122) (123) (124)

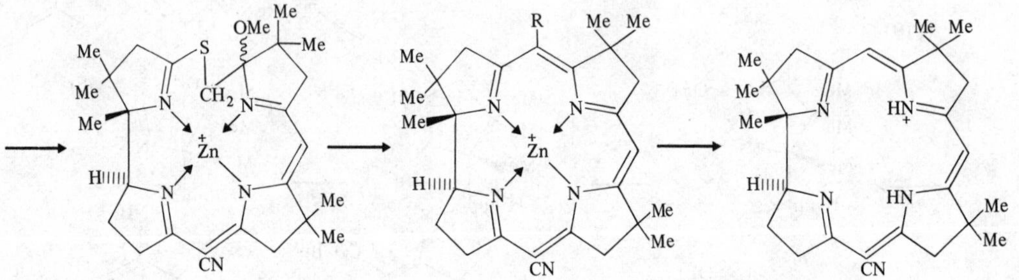

(125) (126) R = H (127) R = SH (128)

(*b*) *The isoxazole approach.* In the late 1950s, J. W. Cornforth and his colleagues began an approach to vitamin B_{12} based on the known chemistry (Scheme 20) of isoxazoles in that they can undergo reduction to β-enamino-ketones or base cleavage to give β-keto-nitriles. The novel objective was synthesis of the macrocycle (**129**) which might be cleaved to give (**130**) and then recyclized to give (**131**). Though a considerable amount of progress

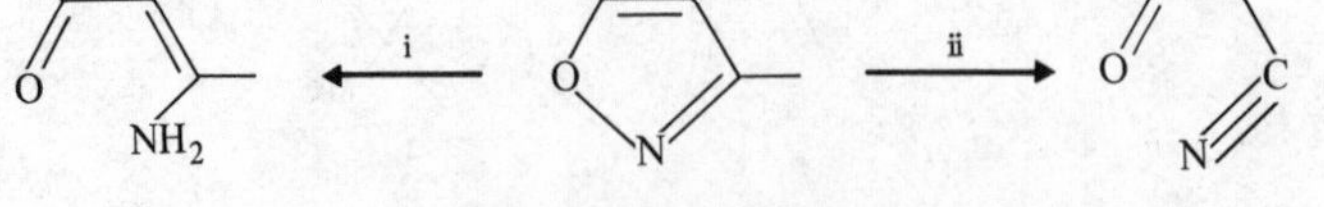

i, H_2, catalyst; ii, base

SCHEME 20

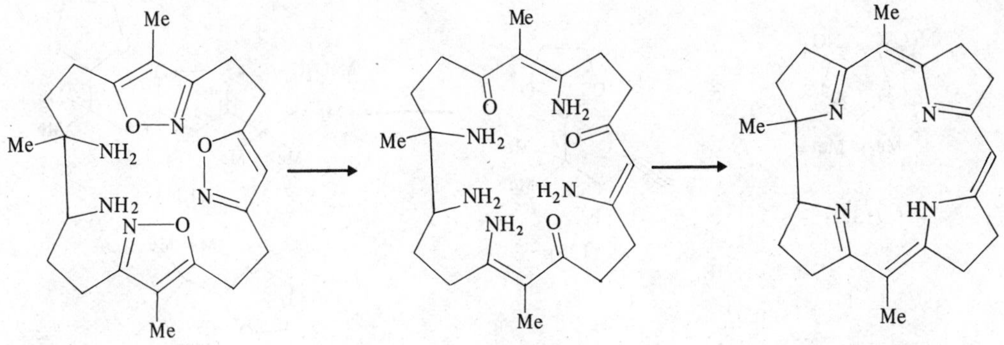

(129) (130) (131)

in this direction was achieved,[18] the isoxazole route was independently conceived by Stevens and, in a model series, carried to a successful conclusion.[48,49] In this work, the isoxazoles were prepared by one of the three routes shown in Scheme 21; either primary nitro-compounds were dehydrated with phosphoryl chloride or phenyl isocyanate, or else aldoximes were dehydrogenated with lead tetra-acetate (*syn*) or with *N*-bromosuccinimide (*syn* and *anti*). Scheme 22 shows Stevens' tri-isoxazole route to metal-free corrins. The nitrile oxides are unstable and were generated *in situ* in the presence of an appropriate acetylene. Thus, nitro-compound (**132**) was converted into its nitrile oxide in the presence of acetylene (**133**) and gave the mono-isoxazole (**134**); the acetal protection was removed with acid and the resulting aldehyde was converted into its oxime (**135**). Dehydrogenation and cycloaddition with another molecule of acetylene (**133**) gave the bis-isoxazole (**136**); repetition of this sequence of reactions gave the

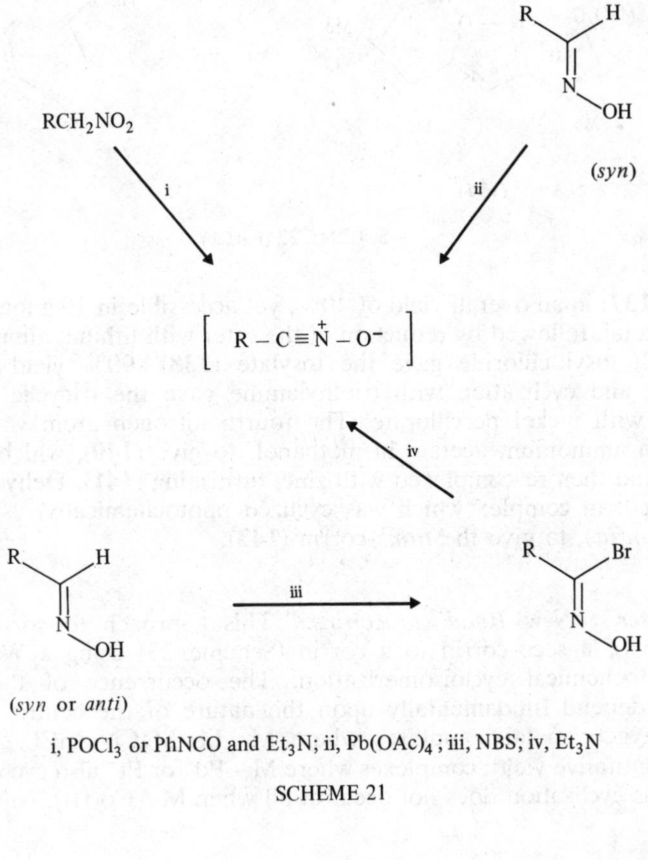

RCH_2NO_2 —i→ $[R-C\equiv\overset{+}{N}-O^-]$ ←ii— RCH=N–OH (*syn*)

RCH=N–OH (*syn* or *anti*) —iii→ RC(Br)=N–OH —iv→ $[R-C\equiv\overset{+}{N}-O^-]$

i, $POCl_3$ or PhNCO and Et_3N; ii, $Pb(OAc)_4$; iii, NBS; iv, Et_3N

SCHEME 21

SCHEME 22 (Part 1)

tri-isoxazole (**137**) in an overall yield of 40%, yet accessible in 10 g lots. Protection of the ketone as its ketal, followed by reduction of the ester with lithium aluminium hydride and treatment with tosyl chloride gave the tosylate (**138**) (90% yield overall). Catalytic hydrogenation and cyclization with triethylamine gave the tricyclic ligand (**139**) after complexation with nickel perchlorate. The fourth nitrogen atom was incorporated by treatment with ammonium acetate in methanol, to give (**140**), which was demetallated with cyanide and then re-complexed with zinc, furnishing (**141**). Dehydration gave (**142**), the key seco-corrin complex which was cyclized photochemically, using Eschenmoser's method (*vide infra*), to give the *trans*-corrin (**143**).

(*c*) *Eschenmoser's 'New Road' to corrins.*[50] This approach to corrins depends upon the cyclization of a seco-corrin to a corrin (Scheme 23) using a Woodward–Hoffman 'allowed' photochemical cycloisomerization. The occurrence of this cyclization was shown[50,51] to depend fundamentally upon the nature of the central metal ion. In the absence of oxygen, metal complexes where M = Li, MgCl, ZnCl, or CdCl cyclize in essentially quantitative yield; complexes where M = Pd^+ or Pt^+ also cyclize, but much more slowly, whereas cyclization does not occur at all when M = Co(III), Ni(II), or Cu(II).

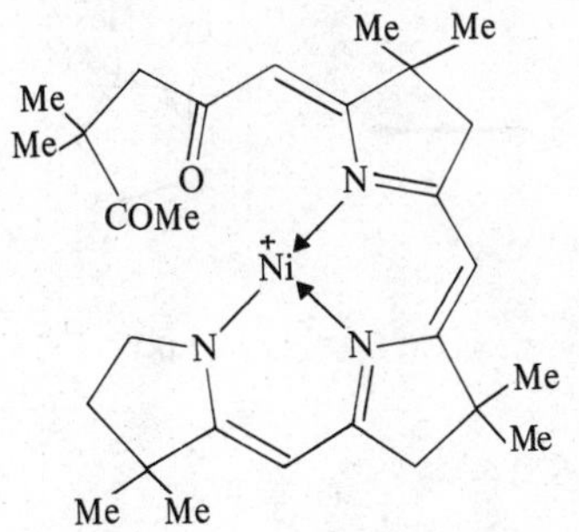

ClO_4^-

(139)

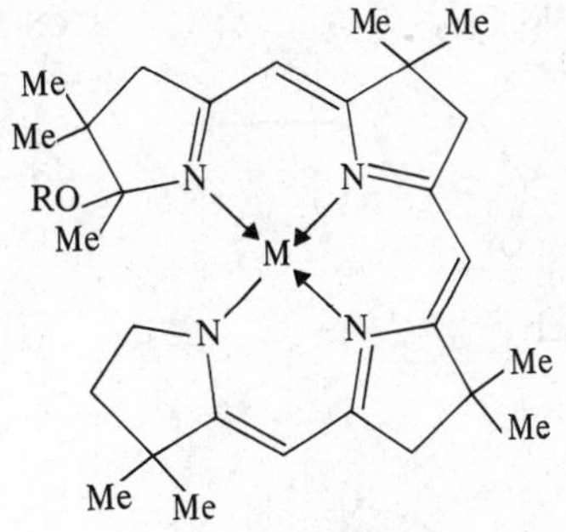

(R = H, OAc)

(140) M = Ni^+ ClO_4^-
(141) M = Zn^+ Cl^-

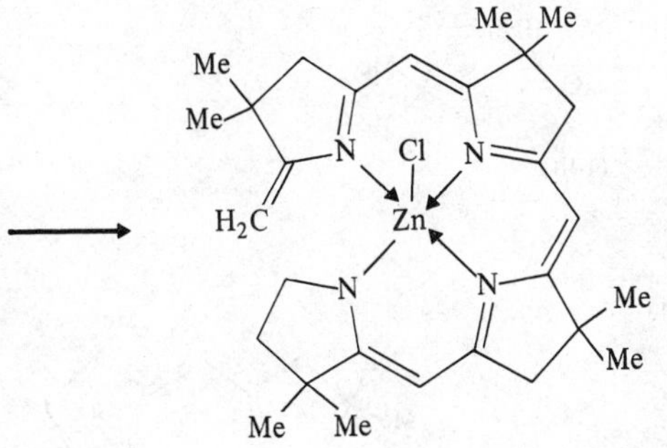

(142)

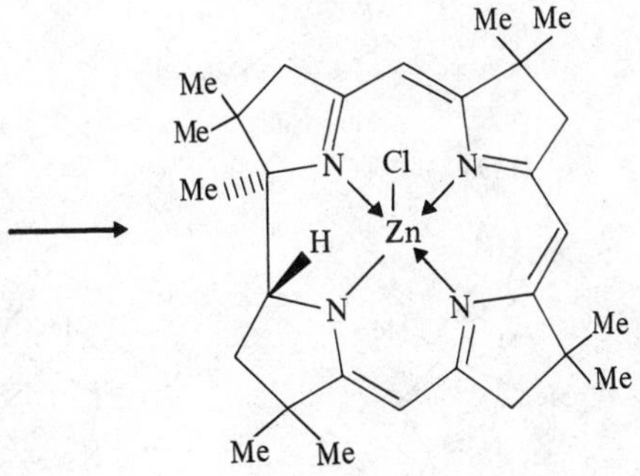

(143)

SCHEME 22 (Part 2)

The synthetic approach to the model corrin ligand is shown in Scheme 24; enamide (**144**) was treated with potassium cyanide to give (**145**), which upon treatment with phosphorus pentasulphide gave the thiolactam (**146**). Oxidation with benzoyl peroxide gave the disulphide (**147**), which, in the presence of enamide (**144**), afforded the bicyclo-thio-bridged compound (**148**). Sulphur extrusion (Scheme 19) using triphenylphosphine gave the vinylogous amidine (**149**), which was deprotected to give the exocyclic methylene compound (**150**) using butoxide. Repetition of this series of transformations gave the

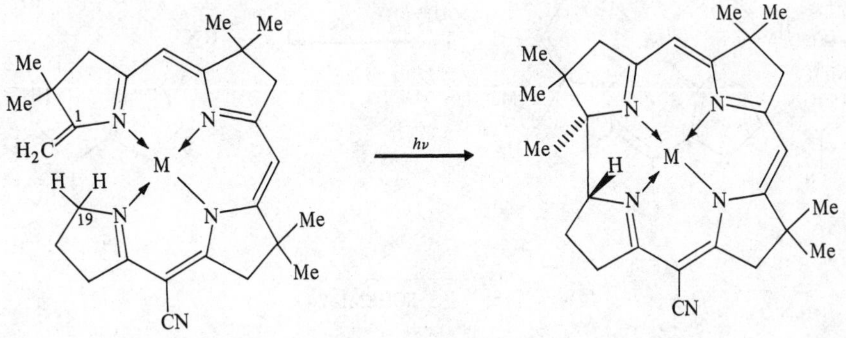

M = Li, MgCl, ZnCl, CdCl, Pd^+, Pt^+

SCHEME 23

(144) (145) (146)

(147) (148)

(149) (150) (151)

(152) (153)

SCHEME 24

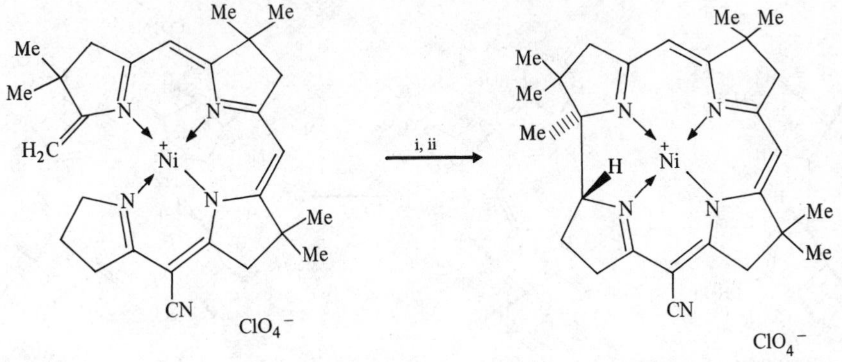

(153) (154)

i, +1.22 V; ii, −0.30 V

SCHEME 25

tricyclic compound (**151**). Alkylation of the silver complex of (**151**) with Meerwein's reagent (formation of the silver complex prevented *N*-alkylation) followed by treatment with enamine (**152**) gave the tetracyclic ligand, isolated as its nickel complex (**153**). The nickel ion was removed using cyanide to give access to the free ligand, from which the other metal complexes required for study of the photocyclization (Scheme 23) were obtained.

Very recently the Zurich group have also demonstrated[51] (Scheme 25) that A/D seco-corrin complexes can be cyclized using electrochemical techniques. Preparative electrochemical oxidation of the nickel(II) seco-corrin (**153**) (which does *not* cyclize under photochemical conditions) in dry solvents consumed one electron equivalent; subsequent reduction led to a mixture from which the corrin complex (**154**) was isolated in 18% yield. This work has been extended to the electrochemical cyclization of a Δ^{18}-dehydro-1,19-seco-corrin complex (**155**), prepared[51] from the methoxy-compound (**156**) by treatment with butoxide (Scheme 26). Electrochemical reduction in the presence of acid consumed about two electron equivalents and gave smoothly the corrin complex (**157**) in 50% yield.

(*d*) *Total syntheses of vitamin B_{12}. The Woodward–Eschenmoser approach.* Full discussion of this landmark in organic chemistry is not possible in the space available here

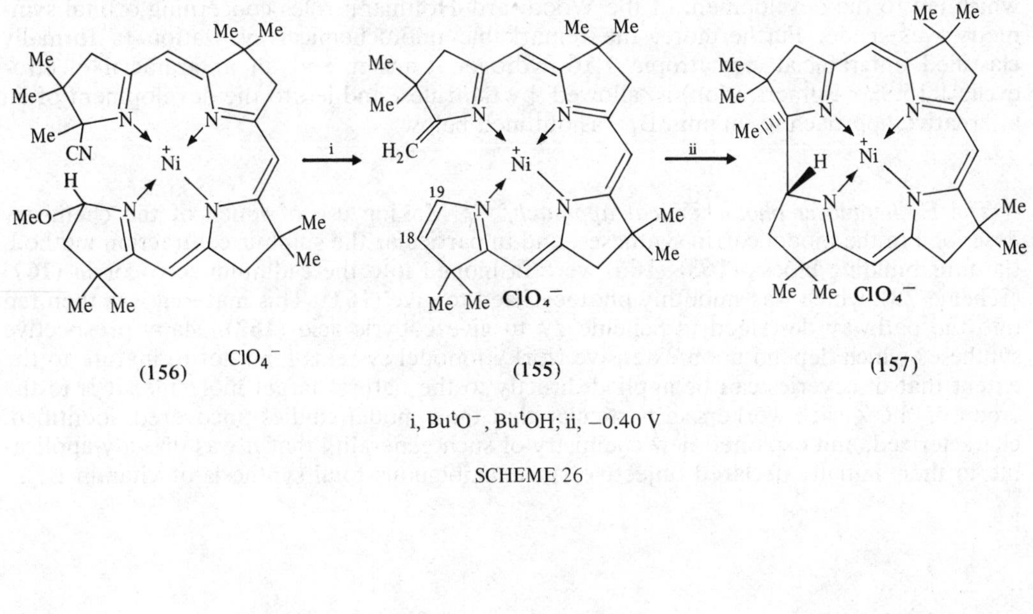

(156) (155) (157)

i, Bu^tO^-, Bu^tOH; ii, −0.40 V

SCHEME 26

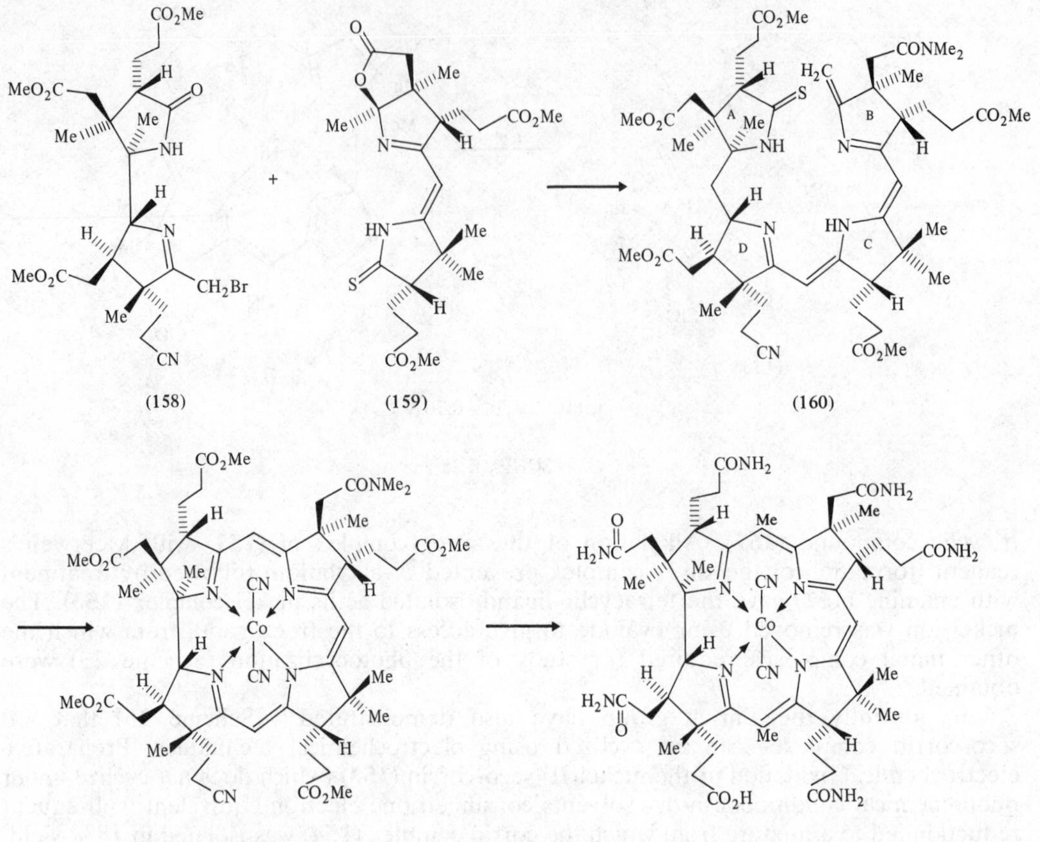

(158) (159) (160)

(161) SCHEME 27 (162)

(see, however, Refs. 18, 50, 52). Briefly, in a series of inimitably conceived reactions, the A–D component (**158**) and the B–C component (**159**) were synthesized from building blocks obtainable in any chemistry laboratory. These were joined (Scheme 27) to give the A-D-C-B molecule (**160**), and then cyclized to give the corrin (**161**). Further transformations, including 'methylation' at the 5- and 15-positions gave cobyric acid (**162**), which was then finally converted into vitamin B_{12} (**92**) itself. It should not go without mention that, during exploratory work on an eventually abandoned route to (**158**), the key observation which led to the development of the Woodward–Hoffmann rules concerning orbital symmetry was made. Furthermore, the remarkable photochemical cyclization (a formally classified antarafacial sigmatropic 1,16-hydrogen transfer and an antarafacial electrocyclic 1,15-π,σ isomerization) is 'allowed' by the rules, and led to the development of an alternative approach to vitamin B_{12}, as outlined below.

The Eschenmoser photochemical approach.[50,51] Making use of much of the chemistry described in the model corrin syntheses, and in particular the sulphur contraction method, the four building blocks (**163**)–(**166**) were fashioned into the cadmium seco-corrin (**167**) (Scheme 28), which was smoothly photocyclized to give (**161**). This material was then fed into the pathway described in Scheme 27 to give cobyric acid (**162**). Many prospective syntheses which depend upon extensive work in model systems tend not to mature to the extent that discoveries can be applied directly to the natural target molecule. It is to the credit of the Zurich workers in particular that their model studies uncovered, identified, characterized, and exploited new chemistry of such generality that it was directly applicable to their initially declared objective, an unambiguous total synthesis of vitamin B_{12}.

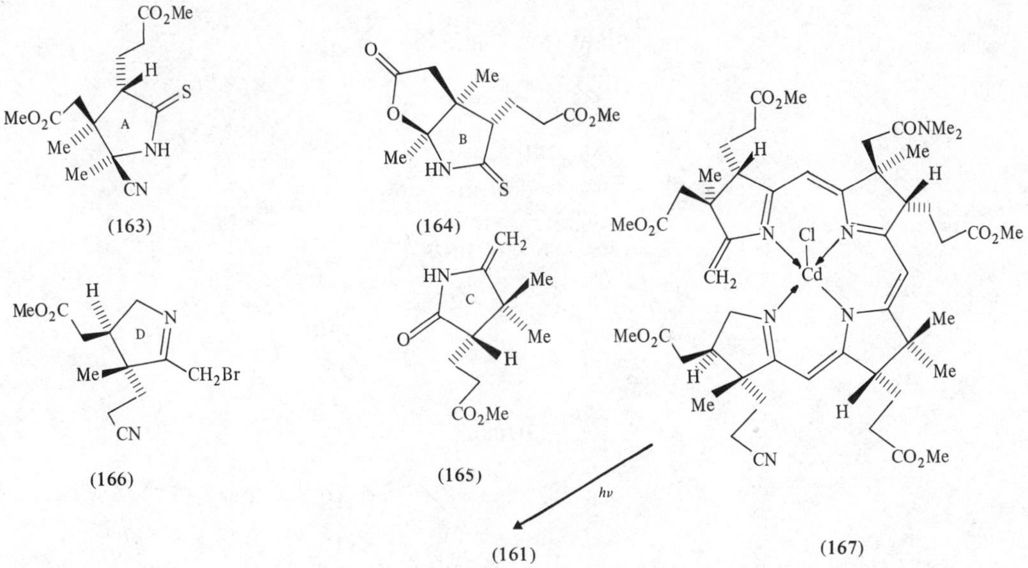

(163) (164) (166) (165) (161) (167)

SCHEME 28

17.2.3 PHTHALOCYANINES (TETRABENZOTETRA-AZAPORPHINS)[53]

17.2.3.1 Structure and nomenclature

Phthalocyanines are tetra-azaporphins (**168**); the numbering system is as shown. X-ray studies[54] showed phthalocyanine and its metal complexes to be planar and tetradentate (*cf.* porphyrins).

17.2.3.2 Chemical properties

As with porphyrins, the two central NH's are exchangeable, and metal ions can be readily incorporated, up to 40 metallic derivatives having been reported.[55] Of the metal complexes, copper(II) phthalocyanine is by far the most important, owing to its use as a pigment in the dyestuffs industry. It can be halogenated to give pigments of slightly different shade containing 15 or 16 halogen atoms (in the benzene rings). Up to eight chloromethyl groups can be introduced using dichlorodimethyl ether in sulphuric acid; functionalization of this type allows the pigments, after further modification, to be applied to fibres. Copper phthalocyanine can be sulphonated and chlorosulphonated to give useful products which are used extensively by the dyestuffs industry.

The nucleus can be degraded with nitric acid, so nitrophthalocyanines are prepared directly from nitrophthalic acid derivatives. From these, reduction affords aminophthalocyanines which can be diazotized and coupled as with other dyestuffs.

17.2.3.3 Preparative methods

During the manufacture of phthalimide from phthalic anhydride and ammonia in a reactor vessel, a blue impurity was observed, and this was shown by Linstead[56] to be iron phthalocyanine, the iron being provided from the contaminated reactor vessel. Having established the structure of this novel macrocycle, and rationalized its formation, more direct methods for phthalocyanine formation were devised. Thus, heating of 2-cyanobenzamide (**169**) with magnesium gave the magnesium complex (**170**), which

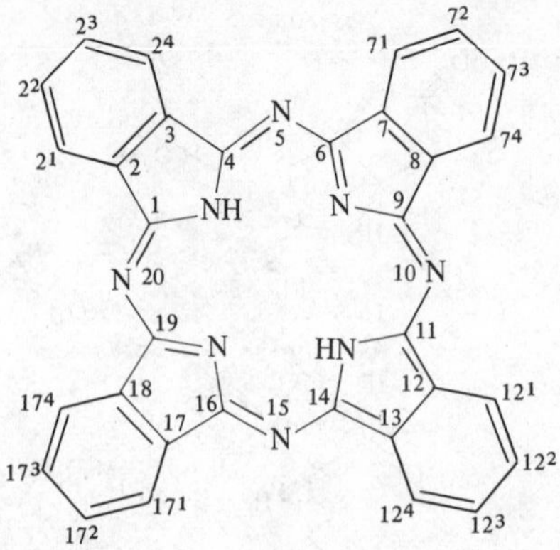

(168)

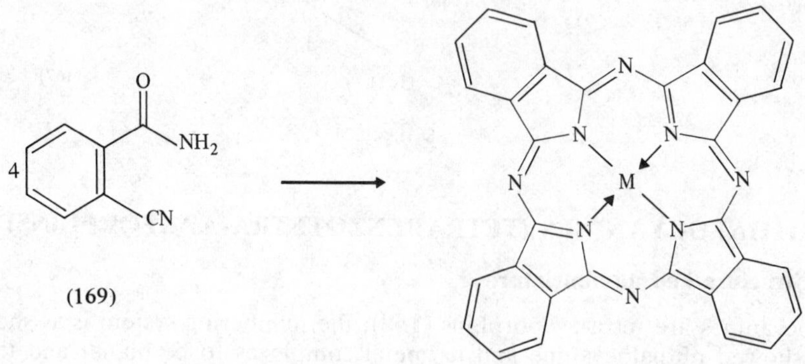

(169)

(170) M = Mg
(171) M = Cu

was demetallated to give the free phthalocyanine (**168**). The metal-free compound could also be prepared directly by substitution of antimony for magnesium in the reaction. Copper phthalocyanine (**171**) was prepared by reacting phthalonitrile with cuprous chloride, and a variety of other metal complexes are accessible in this way, *e.g.* dilithium and disodium salts by use of the appropriate metal pentoxide.

References

1. H. Fischer and H. Orth, 'Die Chemie des Pyrrols', Akademische Verlag, Leipzig, vols. II(i) and II(ii), 1937 and 1940.
2. 'Porphyrins and Metalloporphyrins', ed. K. M. Smith, Elsevier, Amsterdam, 1975.
3. R. Lemberg and J. Barrett, 'Cytochromes', Academic Press, London, 1973.
4. G. Dean, 'The Porphyrias', Pitman Medical Press, London, 1963.
5. J. L. Hoard, in 'Porphyrins and Metalloporphyrins', ed. K. M. Smith, Elsevier, Amsterdam, 1975, p. 317.
6. J. N. Phillips, *Rev. Pure Appl. Chem.*, 1960, **10,** 35.
7. M. S. Stoll, G. H. Elder, D. E. Games, P. O'Hanlon, D. S. Millington, and A. H. Jackson, *Biochem. J.*, 1973, **131,** 429; A. R. Battersby, E. Hunt, E. McDonald, J. B. Paine III, and J. Saunders, *J. C. S. Perkin I*, 1976, 1008.
8. J. S. Morrow and F. R. N. Gurd, *CRC Crit. Rev. Biochem.*, 1975, **3,** 221.
9. K. M. Smith, in 'Porphyrins and Metalloporphyrins', ed. K. M. Smith, Elsevier, Amsterdam, 1975, p. 381.
10. J. E. Falk, 'Porphyrins and Metalloporphyrins', Elsevier, Amsterdam, 1964.
11. H. H. Inhoffen, J. W. Buchler, and P. Jäger, *Fortschr. Chem. org. Naturstoffe*, 1968, **26,** 284.
12. J. Subramanian, in 'Porphyrins and Metalloporphyrins', ed. K. M. Smith, Elsevier, Amsterdam, 1975, p. 555.
13. J.-H. Fuhrhop, K. M. Kadish, and D. G. Davis, *J. Amer. Chem. Soc.*, 1973, **95,** 5140.

14. M. Zerner and M. Gouterman, *Theor. Chim. Acta*, 1966 **4,** 44.
15. R. A. Chapman, M. W. Roomi, T. C. Morton, P. T. Krajcarski, and S. F. MacDonald, *Canad. J. Chem.*, 1971, **49,** 3544.
16. F. R. Hopf and D. G. Whitten, in 'Porphyrins and Metalloporphyrins', ed. K. M. Smith, Elsevier, Amsterdam, 1975, p. 667.
17. A. H. Jackson and G. R. Dearden, *Ann. New York Acad. Sci.*, 1973, **206,** 151.
18. A. H. Jackson and K. M. Smith, in 'Total Synthesis of Natural Products', ed. J. W. ApSimon, Wiley, New York, 1973, vol. 1, p. 143.
19. A. D. Adler, F. R. Longo, J. D. Finarelli, J. Goldmacher, J. Assour, and L. Korsakoff, *J. Org. Chem.*, 1967, **32,** 476.
20. G. H. Barnett, M. F. Hudson, and K. M. Smith, *J. C. S. Perkin I*, 1975, 1401.
21. D. Dolphin, *J. Heterocyclic Chem.*, 1970, **7,** 275.
22. H. W. Whitlock and R. Hanauer, *J. Org. Chem.*, 1968, **33,** 2169; U. Eisner, A. Lichtarowicz, and R. P. Linstead, *J. Chem. Soc.*, 1957, 733.
23. H. H. Inhoffen, J.-H. Fuhrhop, H. Voigt, and H. Brockmann Jr., *Annalen*, 1966, **695,** 133; W. Siedel and F. Winkler, *ibid.*, 1943, **554,** 162.
24. H. Fischer and K. Zeile, *Annalen*, 1929, **468,** 98.
25. G. P. Arsenault, E. Bullock, and S. F. MacDonald, *J. Amer. Chem. Soc.*, 1960, **82,** 4384.
26. R. B. Woodward, *Angew. Chem.*, 1960, **72,** 651; *Pure Appl. Chem.*, 1961, **2,** 383.
27. R. L. N. Harris, A. W. Johnson, and I. T. Kay, *J. Chem. Soc.* (*C*), 1966, 22.
28. A. W. Johnson and I. T. Kay, *J. Chem. Soc.*, 1961, 2418.
29. J. A. P. Baptista de Almeida, G. W. Kenner, J. Rimmer, and K. M. Smith, *Tetrahedron*, 1976, **32,** 1793.
30. A. H. Jackson, H. A. Sancovich, A. M. Ferramola, N. Evans, D. E. Games, S. A. Matlin, G. H. Elder, and S. G. Smith, *Phil. Trans. Roy. Soc. Ser. B*, 1976, **273,** 191.
31. J. A. S. Cavaleiro, A. M. d'A. Rocha Gonsalves, G. W. Kenner, and K. M. Smith, *J. C. S. Perkin I*, 1973, 2471.
32. A. H. Jackson, G. W. Kenner, G. McGillivray, and K. M. Smith, *J. Chem. Soc* (*C*), 1968, 294.
33. H. Fischer, K. Platz, J. H. Helberger, and H. Niemer, *Annalen*, 1930, **479,** 41.
34. H. Whitlock, R. Hanauer, M. Y. Oester, and B. K. Bower, *J. Amer. Chem. Soc.*, 1969, **91,** 7485.
35. R. Bonnett, *Chem. Rev.*, 1963, **63,** 573.
36. S. F. Dyke, in 'The Chemistry of the Vitamins', in 'The Chemistry of Natural Products', vol. VI, Interscience, New York, 1965, p. 107.
37. T. C. Stadtman, *Science*, 1971, **171,** 859.
38. D. Crowfoot-Hodgkin, J. Kamper, J. Lindsey, M. McKay, J. Pickworth, J. H. Robertson, C. B. Shoemaker, J. G. White, R. J. Prosen, and K. N. Trueblood, *Proc. Roy. Soc. Ser. A*, 1957, **242,** 288.
39. W. Friedrich and K. Bernauer, *Z. Naturforsch.*, 1954, **9B,** 685.
40. R. Bonnett, J. M. Godfrey, and V. B. Math, *J. Chem. Soc.* (*C*), 1971, 3736.
41. J. I. Toohey, *Proc. Nat. Acad. Sci. USA*, 1965, **54,** 934; K. Sato, S. Shimzu, and S. Fukui, *Biochem. Biophys. Res. Comm.*, 1970, **39,** 170.
42. D. Dolphin, R. L. N. Harris, J. Huppatz, A. W. Johnson, and I. T. Kay, *J. Chem. Soc.* (*C*), 1966, 30; R. Grigg, A. W. Johnson, R. Kenyon, V. B. Math, and K. Richardson, *ibid.*, 1969, 176.
43. I. D. Dicker, R. Grigg, A. W. Johnson, H. Pinnock, and P. van den Broek, *J. Chem. Soc.* (*C*), 1971, 536.
44. A. W. Johnson and W. R. Overend, *J. C. S. Perkin I*, 1972, 2681.
45. E. Bertele, H. Boos, J. D. Dunitz, F. Elsinger, A. Eschenmoser, I. Felner, H. P. Gribi, H. Gschwend, E. F. Meyer, M. Pesaro, and R. Scheffold, *Angew. Chem.*, 1964, **76,** 393.
46. I. Felner, A. Fischli, A. Wick, M. Pesaro, D. Bormann, E. L. Winnacker, and A. Eschenmoser, *Angew. Chem.*, 1967, **79,** 863.
47. A. Fischli and A. Eschenmoser, *Angew. Chem.*, 1967, **79,** 865.
48. R. V. Stevens, C. G. Christensen, R. M. Cory, and E. Thorsett, *J. Amer. Chem. Soc.*, 1975, **97,** 5940.
49. R. V. Stevens, *Tetrahedron*, 1976, **32,** 1599.
50. A. Eschenmoser, *Quart. Rev.*, 1970, **24,** 366.
51. A. Eschenmoser, *Chem. Soc. Rev.*, 1976, **5,** 377.
52. R. B. Woodward, *Pure Appl. Chem.*, 1968, **17,** 519; 1971, **25,** 283; 1973, **33,** 145; A. Eschenmoser, 23rd IUPAC Congress, Boston, *Pure Appl. Chem. Suppl.*, 1971, **2,** 69; *Naturwiss.*, 1974, **61,** 513.
53. F. H. Moser and A. L. Thomas, 'Phthalocyanine Compounds', Reinhold, New York, 1963.
54. J. M. Robertson, *J. Chem. Soc.*, 1935, 615; 1936, 1195.
55. A. B. P. Lever, *Adv. Inorg. Chem. Radiochem.*, 1965, **7,** 28.
56. R. P. Linstead, *J. Chem. Soc.*, 1934, 1016; G. T. Byrne, R. P. Linstead, and A. R. Lowe, *ibid.*, 1934, 1017; R. P. Linstead and A. R. Lowe, *ibid.*, 1934, 1022.

17.3

Diazoles, Triazoles, Tetrazoles, and Their Benzo-analogues

M. R. GRIMMETT
University of Otago

17.3.1 INTRODUCTION

These compounds may be regarded as being formally derived from pyrrole (or indole) by the replacement of one or more of the annular carbons by nitrogen atoms. Thus pyrazole (**1**), imidazole (**2**), 1,2,3-triazole (**3**), 1,2,4-triazole (**4**), tetrazole (**5**), indazole (**6**), benzimidazole (**7**), and benzotriazole (**8**) comprise the parent compounds of the class.

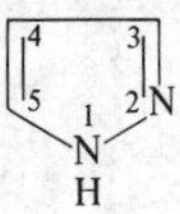

(1) 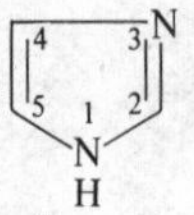 (2) 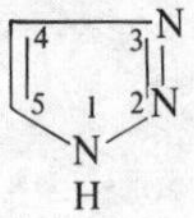 (3) 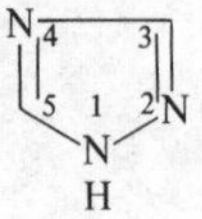 (4) 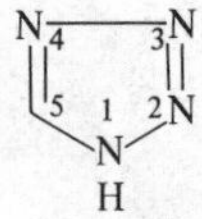 (5)

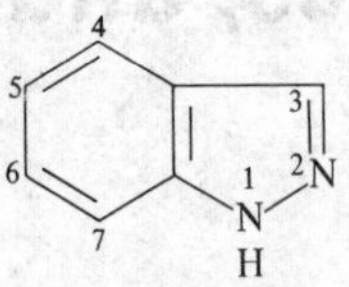

(6) 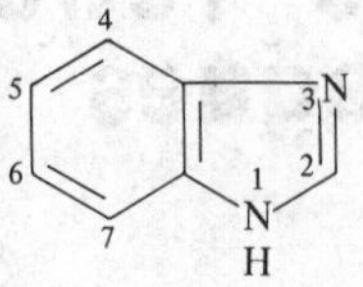 (7) 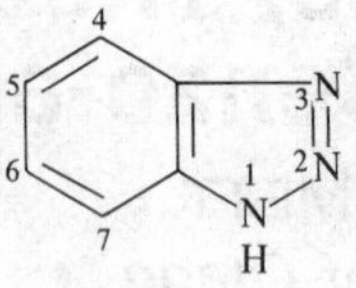 (8)

Interest in many of these compounds has been stimulated because of their applications in industry and agriculture, and because of their biological and analytical importance. Traditional interest in pyrazoles stemmed from their drug, dye, and anaesthetic uses. More recently there have been applications of reduced pyrazoles (pyrazolines) as chemical bleaching agents, as luminescent and fluorescent substances, and in the cinematographic film industry. Pyrazoles, themselves, can be used as antioxidants in fuels, but their major practical applications have been in the medical and agricultural fields where derivatives have such widely-differing activities as bacteriostats and bacteriocides, insecticides, fungicides, sedatives, anticarcinogens, and psychopharmacological agents. Orisul (**9**) is an example of a pyrazole sulphonamide. The imidazole nucleus is found in a number of naturally-occurring compounds such as histamine (**10**), histidine (**11**), allantoin (**12**), and pilocarpine (**13**). Histamine, the decarboxylation product of the essential amino-acid histidine, is implicated in the development of allergies, and in consequence has generated intense interest in antihistamine drugs. Allantoin is the end product of nitrogen metabolism in some animals; pilocarpine is an imidazole alkaloid. A number of nitroimidazoles have found medical applications as bacteriostats. The benzimidazole nucleus appears in vitamin B_{12}.

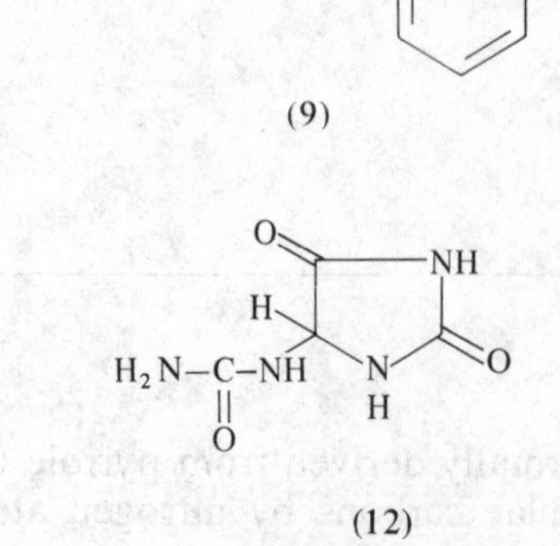

(9) 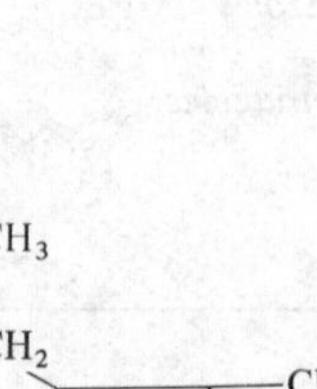 (10) 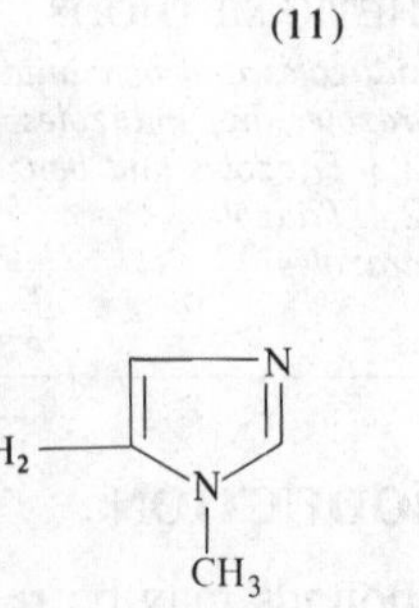 (11)

(12) (13)

In recent years, 1,2,3-triazoles have proved valuable as light-stabilizers and as optical brightening agents. They are precursors of azapurines which are potential carcinostatic agents. 1,2,4-Triazoles have found applications as herbicides, plasticizers, adhesives, and in the photographic industry. Tetrazolium salts have been of importance as antibacterial agents which are subject to enzymatic reduction to formazans which stain the nuclei of bacterial cells. Pentamethylenetetrazole (Metrazole) is a powerful central nervous system stimulant used to counteract barbiturate over-doses. A number of tetrazoles and their metal salts are of value as propellants and primers. This reflects their instability.

A number of reviews and monographs dealing with the polyazoles are listed for those who wish to delve more deeply into the field. References prior to 1966 may be found in the article by Katritzky and Weeds;[1] more recent references are listed separately.[2–7]

17.3.2 NOMENCLATURE AND STRUCTURE

Knorr first described and correctly formulated pyrazoles in 1883, introducing the name to describe the relationship to pyrrole. The systematic name is 1,2-diazole. The dimensions (pm) of the planar molecule are illustrated in Figure 1 with the bond lengths being compared with those for pyrrole and pyridine[2] (values in parenthesis). It is possible to see close similarities with both parent molecules. Furthermore, both C—N bonds and both C—C bonds are equal within the experimental error. The situation of the two annular nitrogen atoms at adjacent ring positions of pyrazole and indazole considerably modifies the properties compared with imidazole and benzimidazole.

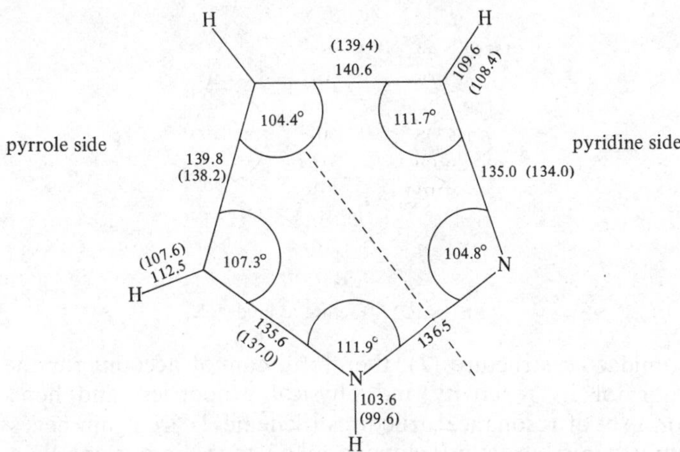

FIGURE 1 Structure of pyrazole

Pyrazole exhibits aromatic character with properties resembling both pyridine and pyrrole. In fact the molecule can be regarded as 'overlapping' pyrrole and pyridine in order that predictions concerning the reactivities of the ring carbons can be made. Position-4 should be the most activated for electrophilic attack. The structure of pyrazole is best represented by a mesomeric structure **(14)** or by a set of resonance structures

pyridine β / pyrrole β (4); pyridine α / pyrrole γ (3); pyrrole α / pyridine γ (5); 1 N–H; 2 N

(1a) (14)

SCHEME 1

(Scheme 1). Tautomerism in *N*-unsubstituted pyrazoles (see Section 17.3.5) makes the 3- and 5-positions equivalent. Except in discussions which relate to tautomerism, 3- (or 5-) substituted pyrazoles will be designated as 3-substituted pyrazoles.

Imidazole, previously known as glyoxaline or iminazole, is systematically named 1,3-diazole. X-ray studies point to a planar molecule with dimensions (pm) as illustrated[8] in Figure 2. The molecule exhibits, as does pyrazole, the aromaticity associated with six π-electrons; one from each carbon, one from the 'pyridine' nitrogen, and two from the 'pyrrole' nitrogen. This leaves the pyridine-type nitrogen with an unshared electron pair and confers on it basic and nucleophilic properties. A similar situation exists with benzimidazole, which can be envisaged as two overlapping sextets. The aromaticity of these compounds can be inferred from the appearance of the proton magnetic resonance spectrum in which the proton signals all are at low field.

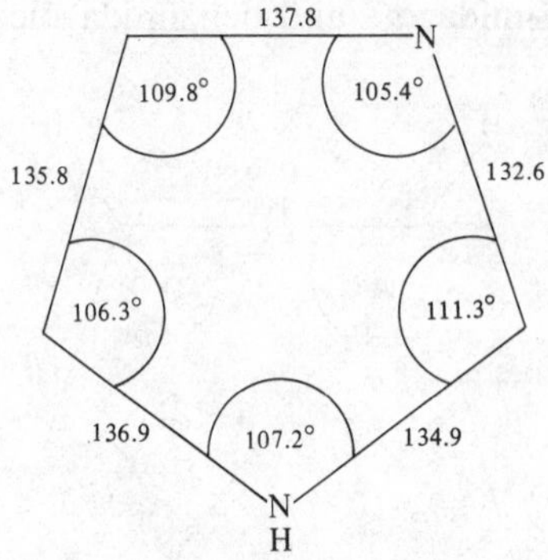

FIGURE 2 Structure of imidazole

The classical imidazole structure (**2**), therefore, cannot account for the dimensions of the molecule, nor for its reactivity and physical properties, and hence a mesomeric structure (**15**), or a set of resonance structures (Scheme 2) are again necessary to describe the ring. Certainly 'ionic' structures appear to be more important than in the case of benzene, but, although dipole moment measurements indicate considerable polarization of the imidazole ring, the extent of this is less than would be required to yield an ionic structure. As will be seen later (Section 17.3.3), these dipole moment measurements are complicated by the propensity for imidazole to self-associate in some solvents.

(15)

A number of attempts to apply quantum-mechanical calculations to imidazole have resulted in the publication of π-electron densities which are contradictory. Molecular orbital theoretical studies have more successfully computed dipole moments, some spectroscopic properties, and pK_a values.[2,4,6]

SCHEME 2

As with imidazole and pyrazole, the triazole rings are planar 6π-electron aromatic systems with distortions of the π-system induced by the annular nitrogens. Two possible combinations of three nitrogens and two carbons account for the 1,2,3- (**3**), and 1,2,4-triazoles (**4**). The former (**3**) were originally called *vic*- (or vicinal) triazoles, and the latter (**4**) *sym*- (or symmetrical) triazoles.

Three possible classes of 1,2,3-triazoles are named 1*H*-1,2,3-triazole (**16**), 2*H*-1,2,3-triazole (**17**), and 4*H*-1,2,3-triazole (**18**). Furthermore, triazoles unsubstituted on nitrogen can be regarded as either 1*H*- or 2*H*- depending on which is the preferred tautomer. Such compounds are still listed as *vic*-triazoles in *Chemical Abstracts.* Although (**16**) and (**17**) are fully aromatic systems, (**18**) is not and this is reflected in its rarity. The term 'osotriazole' refers to derivatives of 2*H*-1,2,3-triazole, particularly those derived from osazones. The nomenclature of benzotriazoles as 1*H*- or 2*H*-benzotriazoles follows.

(16) (17) (18)

1,2,4-Triazoles can be regarded as 1*H*- (**19**) or 4*H*- (**20**) compounds. A benz-analogue is not possible without a nitrogen atom at a ring junction.

(19) (20)

Although measurements have not yet been applied to 1,2,3-triazole, X-ray studies show that in the crystalline state, and in the vapour phase, 1*H*-1,2,4-triazole exists in a planar form (Figure 3).

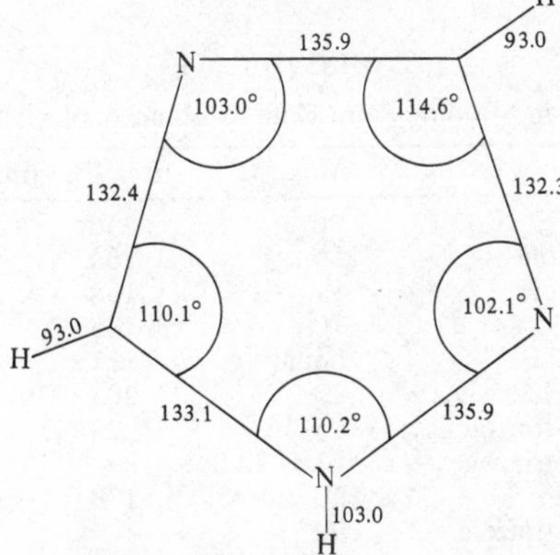

FIGURE 3 Structure of 1,2,4-triazole

(21) ⇌ (22)

Tetrazoles can be regarded as 1*H*- (**21**) or 2*H*-tetrazoles (**22**). The molecules are planar and exhibit aromatic properties with the sextet comprising one electron from carbon, two from the 'pyrrole' nitrogen, and one each from the three 'pyridine' nitrogens. The calculated resonance energy is high (230–260 kJ mol^{-1}, *cf.* imidazole, 60 kJ mol^{-1}, pyrazole, 122 kJ mol^{-1}). Although no X-ray study of the parent molecule has been reported, some data are available for the anion (Figure 4).

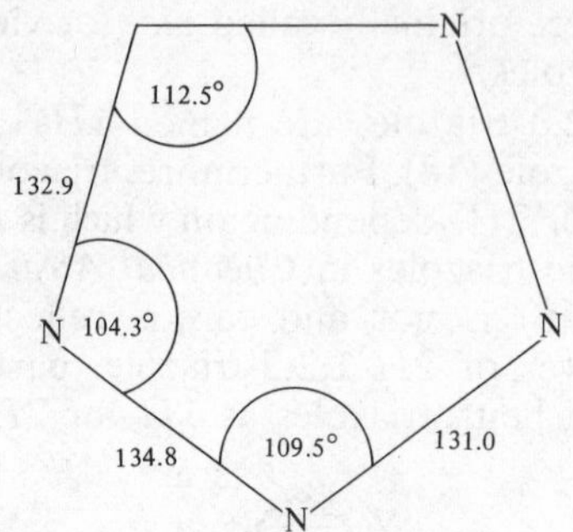

FIGURE 4 Structure of the tetrazole anion

17.3.3 GENERAL AND PHYSICAL PROPERTIES

The introduction of one or more 'pyridine' nitrogen atoms to pyrrole drastically affects the properties. Water solubility is increased and intermolecular hydrogen bonding becomes possible. When the molecules are unsubstituted on the pyrrole nitrogen they generally exist as crystalline compounds with relatively high melting and boiling points (see Table 1). Such compounds are usually soluble in polar solvents, but only sparingly soluble in non-polar solvents. The solubility in solvents such as water is a function of the capacity of the azoles to hydrogen bond with water, a property which is most marked with imidazole. The solubility characteristics are reversed and there is a corresponding decrease in melting and boiling points when a substituent is introduced at N-1. An apparent exception exists with 1-methyl-1,2,3-triazole, but the most striking fact is the much higher boiling point than pyrrole. Although in the vapour state imidazole and pyrazole are monomeric, in non-polar solvents association through hydrogen bonding is common. Early attempts to determine the molar mass of imidazole by cryoscopic or ebulliometric

TABLE 1

Melting and Boiling Point Data for some Azoles

Compound	*M.p.* (°C)	*B.p.* (°C) (mmHg)
Imidazole	90	256
4-Methylimidazole	56	264
1-Methylimidazole	−6	198
Pyrazole	70	185
1-Methylpyrazole	liquid	127
1,2,3-Triazole	23	203 (739)
1-Methyl-1,2,3-triazole	15–16	228 (752)
2-Methyl-1,2,3-triazole	21.5–22.0	89–90 (714)
1,2,4-Triazole	20	178
4-Methyl-1,2,4-triazole	90	—
Pyrrole	—	130

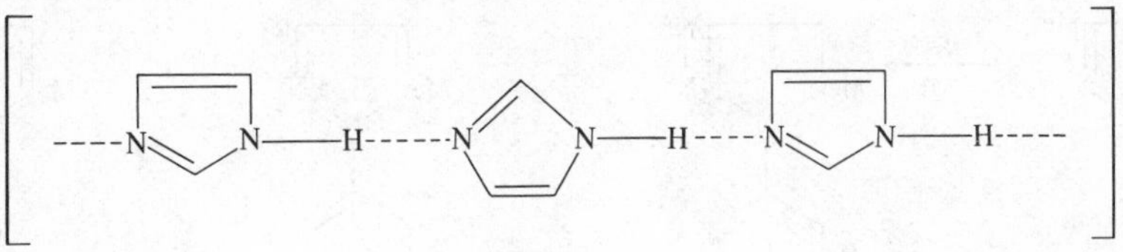

(23)

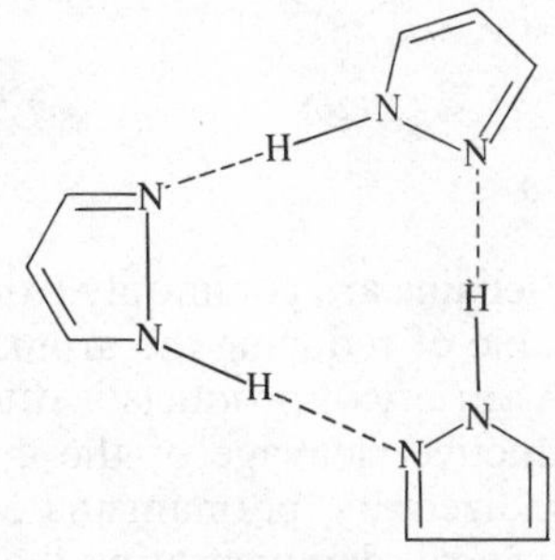

(24)

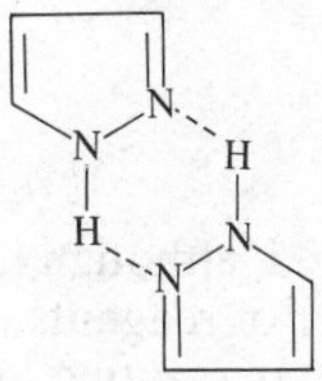

(25)

methods in solvents such as benzene gave anomalously high results as a consequence of the propensity of the azole to associate as linear polymers of up to 20 molecules (**23**). Pyrazole, in contrast, forms a strainless trimer (**24**) and a similar dimer (**25**). Polar solvents and dioxan are able to disrupt this association, as does substitution of the 'pyrrole' hydrogen by an alkyl or aryl group. While 1- and 2-substituted benzotriazoles are mononuclear, the unsubstituted compounds form linear associates. Tetrazoles follow the general trends with the melting and boiling points of 2-substituted derivatives being generally lower than for the 1-substituted isomers.

The intermolecular hydrogen bonding in covalent solvents has been a problem also in dipole moment determinations, and accurate values can only be obtained in high dilutions or in cases where the association is disrupted. A selection of values is shown in Table 2. In benzene, pyrazole shows a dipole moment of 5.24×10^{-30}C m, presumably because of the formation of dimeric species in that solvent. Dioxan can disrupt the intermolecular hydrogen bonding so that the value of the dipole moment in dioxan approaches that of 1-methylpyrazole. The value obtained for 1,2,3-triazole has been interpreted as indicating that the predominant form is 2*H*-1,2,3-triazole since it is closer to that of pyrazole than imidazole. Comparison with the isomeric *N*-ethyltetrazoles shows tetrazole to be predominantly 1*H*-1,2,3,4-tetrazole.

With a few notable exceptions azoles show remarkable stability to acids, bases, oxidizing agents, and heat. Few examples of the oxidative cleavage of the pyrazole ring

TABLE 2

Dipole Moments of Azoles (10^{-30} C m) (in benzene)

Pyrazole	7.39*
1-Methylpyrazole	7.64
Imidazole	12.81
1-Methylimidazole	12.11
1,2,3-Triazole	5.90
1-Phenyl-1*H*-1,2,3-triazole	13.59
2-Phenyl-2*H*-1,2,3-triazole	3.23
1,2,4-Triazole	10.57*
1-Phenyl-1*H*-1,2,4-triazole	9.61
4-Phenyl-4*H*-1,2,4-triazole	18.78
Tetrazole	17.05*
1-Ethyl-1*H*-1,2,3,4-tetrazole	18.21
2-Ethyl-2*H*-1,2,3,4-tetrazole	8.84

* In dioxan.

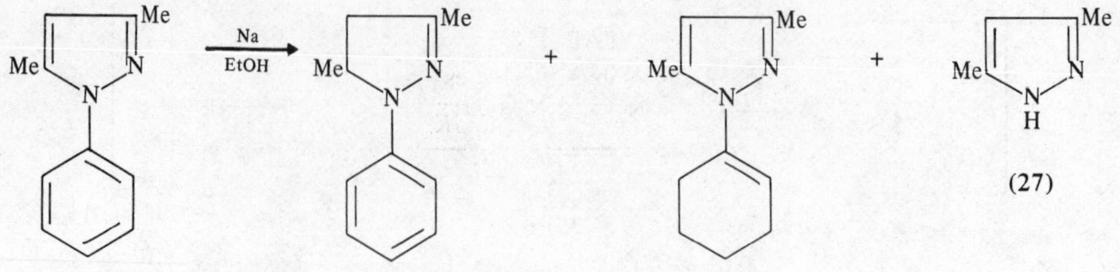

(26)

(27)

SCHEME 3

have been reported, although hydrocarbon side-chains are commonly oxidized to carboxyl groups. A number of reagents, though, are capable of reducing the aromatic compound to the corresponding pyrazoline or pyrazolidine. A variety of products is often obtained. The 3,5-dimethylpyrazole (**27**) results from the reductive cleavage of the cyclohexenyl compound (**26**) (Scheme 3). Benzimidazole is oxidized by permanganate, dichromate, or hydrogen peroxide to imidazole-4,5-dicarboxylic acid, demonstrating the great stability of the imidazole ring. While many triazoles and tetrazoles can be distilled without decomposition, the introduction of a third (or fourth) nitrogen to the five-membered ring certainly makes the compounds less stable. The nuclei are generally stable to oxidation, but benzotriazole explodes at 160 °C (2 mm) and 1,2,3-triazole halides explode violently above 260 °C. Tetrazole is explosive when heated above its melting point, and nitrogen side-chains may decrease thermal stability even further, *e.g.* aqueous solutions containing more than about 2% of tetrazolediazonium cation detonate at 0 °C. However, the considerable stability to chemical reagents of the tetrazole ring system is evidenced in the degradation of 5-cyano-2-phenyltetrazole (**28**) to tetrazole (Scheme 4).

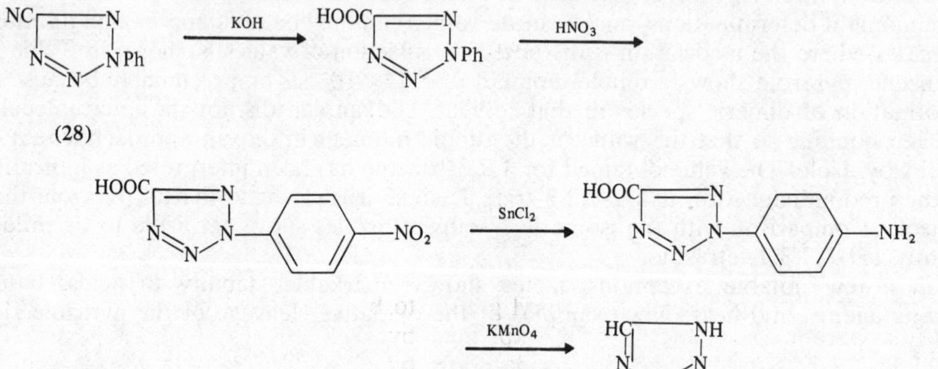

(28)

SCHEME 4

17.3.4 IONIZATION CONSTANTS

The azoles can act as both acids and bases as long as the NH proton is not substituted. Table 3 lists some representative pK_a values.

The amphoteric properties may be deduced from examination of the mesomeric or resonance structures for the compounds. Imidazole is a surprisingly strong base, presumably because of the stability of its symmetrical mesomeric cation (**29**) (Scheme 5). Imidazole forms stable salts with a wide range of organic and inorganic acids. Pyrazole salts, on the other hand, are often readily hydrolysed and frequently hygroscopic. The basic strength of pyrazole is considerably less than that of imidazole, and this fact illustrates a common characteristic of bases in which two nitrogen atoms are adjacent. As

TABLE 3
pK_a Values for Azoles

Compound	*Basic* pK_a	*Acid* pK_a
Pyridine	5.2	—
Pyrrole	—	16.5
Pyrazole	2.5	—
Indazole	~1.3	~14
Imidazole	6.95	14.52
4-Methylimidazole	7.61	—
2,4-Dimethylimidazole	~8.5	—
4-Phenylimidazole	6.10	—
4-Bromoimidazole	3.60	12.32
4-Nitroimidazole	−0.05	9.30
Benzimidazole	5.5	13.2
1,2,3-Triazole	1.17	9.4
1-Methyl-1,2,3-triazole	1.25	—
2-Methyl-1,2,3-triazole	<1	—
1,2,4-Triazole	2.2	~10
1-Methyl-1,2,4-triazole	3.2	—
4-Methyl-1,2,4-triazole	3.4	—
Benzotriazole	—	8.2
Tetrazole	—	4.8
1-Methyltetrazole	4.3(?)	—
2-Methyltetrazole	3.6(?)	—

further annular nitrogen atoms are introduced the basic strength decreases and the acidic nature becomes more important. Both of the triazoles are comparable in acidity with phenol, but less basic even than pyrazole. The effect is most marked in 1,2,3-triazole in which the base-weakening effect of the second pyridine nitrogen atom is felt equally by both of the other nitrogen atoms. It is of interest to note that while 1-methyl-1H-1,2,3-triazole is slightly more basic than the parent, 2-methyl-2H-1,2,3-triazole is a weaker

(29)

SCHEME 5

base. One would normally expect N-alkyl groups to be base-strengthening, but it seems that the apparent discrepancy can be explained by the ability of 1-methyl-1H-1,2,3-triazole and 1,2,3-triazole to form imidazolium-type cations (**30**) and (**31**), while 2-methyl-2H-1,2,3-triazole must give a pyrazolium-type cation (**32**). The triazoles form a number of characteristic metal salts, some of which can only be obtained in anhydrous conditions.

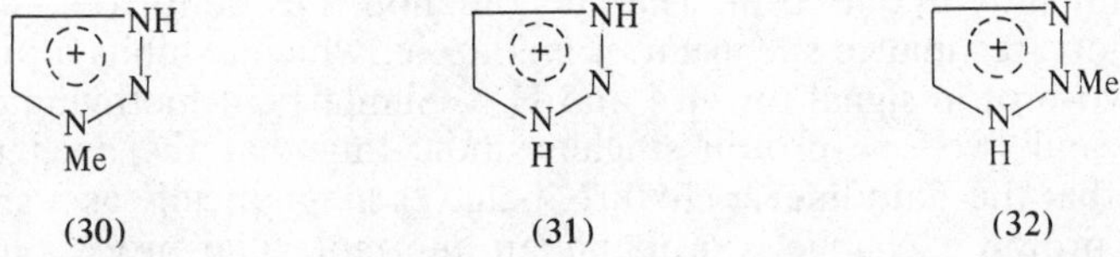

(30) (31) (32)

The basic strength of tetrazole has not been measured, but it would be expected to be very low because of the competition of the four annular nitrogens for the available electronic charge. Certainly the pyridine type nitrogens make the NH considerably acidic, with the result that the metal salts of tetrazole are relatively stable. The sodium and barium salts can even be prepared in water, but this is not surprising when it is realized that as an acid tetrazole is comparable with acetic acid. The basic pK_a values listed for 1- and 2-methyltetrazoles seem suspiciously high (comparable with aniline) and merit re-examination.

The effects of ring substituents on the pK_a values follow normal electronic considerations. A fused benzene ring is base-weakening and acid-strengthening, and substituents on such a benzene ring have entirely predictable effects (*e.g.* 5-chlorobenzotriazole, pK_a 7.7; 4,5,6,7-tetrachlorobenzotriazole, pK_a 5.48). Although the acid strengths of imidazoles are less marked than the basic strengths, the imidazole anion is again able to distribute the charge equally over the two nitrogen atoms. The weakly acidic natures of the compounds are evidenced by their reactions with alkali metals, alkalies, and other strong bases. Some metal salts (*e.g.* silver salts) are formed readily, but the reports of copper and zinc salts in the earlier literature probably refer to coordination complexes through the pyridine nitrogen. Such complexes, common to all of the azoles, are being increasingly studied because of their biological significance. In haemoglobin, Fe(II) is octahedrally coordinated with the four nitrogens of haem, with a histidine unit, and with either oxygen or a water molecule. Other examples exist where metal ions such as Cd(II), Zn(II), and Mn(II) are coordinated with the imidazole ring of histidine.

One major consequence of the basicity of imidazole is that, at physiological pH of about 7.4, substantial quantities of both the free base and protonated imidazolium species are present in the histidine unit of a protein. This means that histidine can act as either a proton-acceptor or a proton-donor according to the demands of its immediate environment. Histidine units in a number of enzymes (*e.g.* ribonuclease, aldolase, some proteases) may have this role. The buffering action of histidine in the haemoglobin–oxyhaemoglobin system is a further consequence.[7] The comment has been made[7] that the imidazole groups of histidine units in a polypeptide are the strongest bases present in any quantity at physiological pH; furthermore, the imidazolium cations are the strongest acids present in substantial amount (with variations in pK_a induced by the local environment).

17.3.5 TAUTOMERISM

Throughout the series there are many examples of tautomerism. A pyrrole-type hydrogen is particularly mobile, and other examples of the isomerism are common in azole *N*-oxides, aminoazoles, and hydroxyazoles (and the corresponding thiols). In some cases the situation has been confused by invalid experimental results, particularly those which draw conclusions from chemical reactivity — a notoriously misleading criterion for tautomeric character.

17.3.5.1 Imidazoles

When imidazole has an unsubstituted NH group there is the possibility of tautomerism, with consequent equivalence of the 4- and 5-positions in the ring. Often the 4- and 5-tautomers cannot be isolated as separate compounds, although they may enter chemical reactions predominantly in one form. The phenomenon is evident from an examination of the proton magnetic resonance spectrum of imidazole, which exhibits a one-proton signal for H-2 and a two-proton signal for H-4 and H-5. Similarly, 4-methylimidazole* displays a methyl singlet and two one-proton singlets, indicating that the exchange of the NH proton is so fast that the 'tautomeric mixture' behaves magnetically as a single compound (Scheme 6). The proton exchange is probably an intermolecular process involving two or more imidazole molecules, or perhaps one imidazole molecule and two hydroxylic solvent molecules. That the tautomerism in imidazoles is due to prototropy can be shown when imidazole is dissolved in concentrated sulphuric acid. Proton loss from nitrogen is retarded

* The compound formerly designated as 4(5)-methylimidazole is now usually referred to as 4-methylimidazole; the existence of the other tautomer is understood.

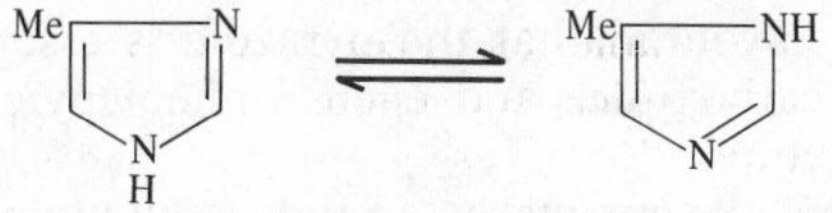

SCHEME 6

and spin–spin coupling involving protons on nitrogen becomes observable, showing a symmetrical imidazolium cation (**29**). This retardation of the exchange rate by acid has been used to demonstrate that electron-withdrawing substituents appear to favour the 4-substituted tautomer.[2,4] When 2-deutero-4(5)-bromoimidazole (**33**) is dissolved in D_2SO_4, and 1,2-dideutero-4(5)-bromoimidazole (**34**) in H_2SO_4, the n.m.r. spectra reveal that the 1,4-structures are preferred. The larger coupling constant is assumed to infer coupling between adjacent protons. Even this result, though, must be interpreted with caution since it does not take into account the possibility that one tautomer may protonate at a faster rate than the other. It is not improbable in some reactions that a minor tautomer may react preferentially (Scheme 7).

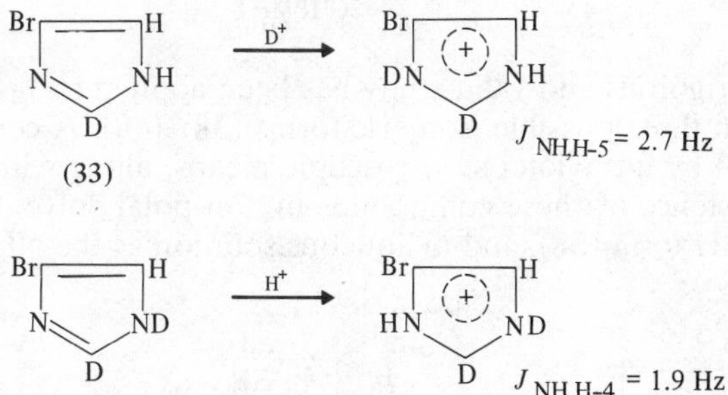

(33) $J_{NH,H\text{-}5} = 2.7$ Hz

(34) $J_{NH,H\text{-}4} = 1.9$ Hz

SCHEME 7

A more reliable procedure for studying tautomerism utilizes 'fixed model' compounds. A comparison of the pK_a values for 4(5)-nitroimidazole (−0.05), 1-methyl-4-nitroimidazole (−0.53), and 1-methyl-5-nitroimidazole (2.13) suggests that the structure (**35**) (X = NO_2, Y = H) predominates, and that in the cation the two protons are no longer equivalent. (The greater basic character of 1-methyl-5-nitroimidazole is a result of the basicity being caused by the heterocyclic nitrogen atom most remote from the substituent.) Tautomeric equilibrium constants (K_T) indicate the extent by which the 4-tautomer predominates.[9] The Hammett correlation of imidazole pK_a values with σ_m constants also leads to the conclusion that an electron-attracting substituent favours the 4-position and an electron-releasing group favours the 5-position.

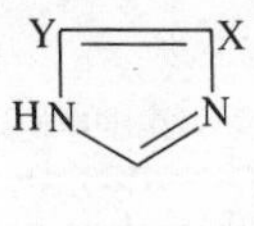

(35) 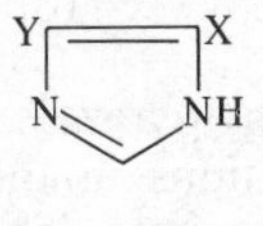 (36)

Y	X	K_T = (35)/(36)
H	NO_2	500
H	Cl	447
Cl	NO_2	186

The introduction of a substituent other than hydrogen on the NH group prevents any possibility of tautomerism; thus 1-methyl-4- and 1-methyl-5-nitroimidazoles act as completely separate compounds.

17.3.5.2 Pyrazoles

In *N*-unsubstituted pyrazoles the 3- and 5-positions are equivalent. As with imidazoles, though, substituents can modify the tautomeric equilibrium to the extent that one tautomer may be so highly favoured that the pyrazole is essentially homogeneous, *e.g.* methyl 3-phenylpyrazole-5-carboxylate and even 3-phenylpyrazole appear to approach 100% of the tautomeric mixture.

It is useful to include here the results of a study of tautomerism based on chemical activity. When 3(5)-methylpyrazole is ozonized, the yields of products (glyoxal and pyruvaldehyde) isolated are suggested to provide evidence for the existence of 3-methylpyrazole (**37**) as the major tautomer (Scheme 8). These results, however, can be explained equally well by a greater reactivity of the 3-methyl tautomer with ozone, but any conclusions drawn from such a study must be regarded with extreme suspicion.

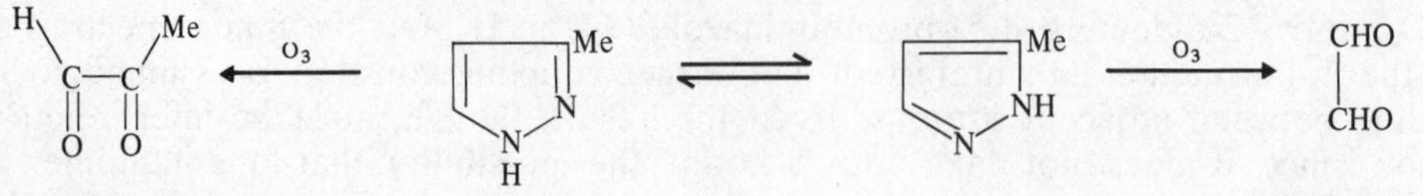

(0.18 mol) $\xleftarrow{O_3}$ (37) $\rightleftharpoons$ $\xrightarrow{O_3}$ CHO–CHO (0.02 mol)

SCHEME 8

A much more rigorous and valid study has been applied to 1-substituted 5-pyrazolones, which can exist in three possible isomeric forms (**38**)–(**40**). A comparison[10] with the 'fixed models' (**41**)–(**43**) by ultraviolet spectroscopic means, along with other physical evidence, points to the existence of these compounds in non-polar solvents as mainly the CH form (**39**) with some NH form (**38**), and in aqueous solution as the NH form (**38**) and some OH form (**40**).

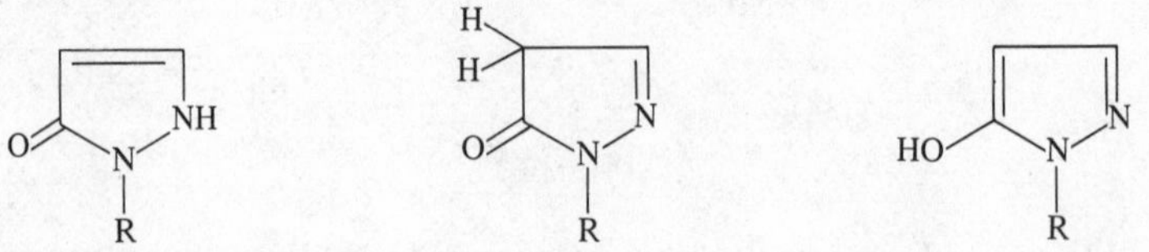

(38) (NH form) (39) (CH form) (40) (OH form)

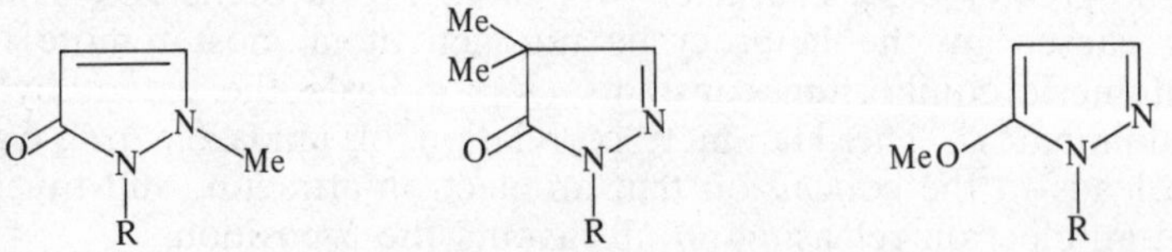

(41) (42) (43)

Substitution of the NH of 3(5)-methylpyrazoles gives mixtures of products (both 1,3- and 1,5-). One of the crucial orientations in the pyrazole series has been that which distinguishes 1,3- from 1,5-dimethylpyrazole. The unequivocal synthesis (Scheme 9) of 1,3-dimethylpyrazole from 1,3-dimethylpyrazolin-5-one (**44**), together with n.m.r. spectroscopic methods, were able to solve the problem. Chemical shift data for the two

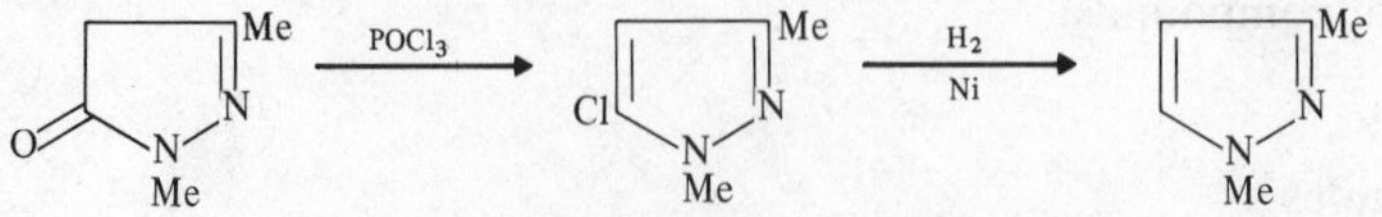

(44) $\xrightarrow{POCl_3}$ $\xrightarrow{H_2/Ni}$

SCHEME 9

dimethyl isomers show that in the 5-methyl compound the signal for H-3 (7.27 p.p.m.) is at lower field, and CH_3-5 (2.19 p.p.m.) is at higher field, than the corresponding H-5 (7.13 p.p.m.) and CH_3-3 (2.22 p.p.m.) protons in 1,3-dimethylpyrazole.

17.3.5.3 Triazoles

In dilute solution, 1,2,3-triazole appears to contain both the 1*H*- and 2*H*-forms, but in more concentrated solutions the molecules become associated by hydrogen bonding through the 1*H*-forms. Various physical methods have been brought to bear on the problem. The dipole moment for 1,2,3-triazole has a value closer to that of pyrazole than imidazole (see Table 2), suggesting that the predominant tautomer is 1*H*-1,2,3-triazole. The ^{1}H n.m.r. spectrum of 1,2,3-triazole has signals for H-4 and H-5 as a singlet at δ 7.90 ($CDCl_3$) which shifts to 7.96 in the anion and to 8.69 (CF_3CO_2H) in the cationic species. Furthermore, in deuteroacetone the spectrum shows a temperature dependence which provides evidence for the existence of both the 1*H*- and 2*H*-tautomers in solution at low temperatures. At 23 °C the spectrum displays two small doublets at δ 8.12 and 7.70, besides the major singlet at 7.83. At −90 °C these peaks have shifted so that the two doublets appear at 8.32 and 7.83, and the singlet at 8.02. Since the intensities are approximately equal, the changes can be ascribed to a reversible solvation of the triazole which prevents the rapid exchange of protons within aggregates of triazole molecules.[5]

A similar temperature study of 1,2,4-triazole shows the presence of the 1*H*-tautomer at −34 °C. The same tautomer is indicated in the vapour phase by microwave spectroscopy.

17.3.5.4 Tetrazole

Both the 1*H*- and 2*H*-forms of tetrazole are possible. Comparison of the dipole moment of tetrazole with moments of the 1- and 2-ethyltetrazoles (Table 2) show that the 1*H*-form is preferred for the parent base, a conclusion which is supported by molecular orbital calculations. ^{14}N n.m.r. studies, though, suggest that the 2*H*-form is predominant. Alkylation of tetrazole gives rise to 1- and 2-alkyl isomers, but this cannot provide evidence for the predominance of either tautomer.

It is of interest to note that 5-aminotetrazoles, which are theoretically capable of existence as imino-tautomers, have been shown to exist mainly as the amines (Scheme 10). This appears to be a general property of all monoaminoazoles.

SCHEME 10

17.3.6 ELECTROPHILIC SUBSTITUTION

Whereas pyrrole is very reactive towards electrophilic reagents, the introduction of a pyridine-type nitrogen reduces the availability of electrons. The azoles should therefore react with electrophiles in the manner of a pyrrole with one or more electron-attracting substituents in the ring. In consequence, rather severe conditions may be required for some substitution reactions, and weak electrophiles (*e.g.* Friedel–Crafts acylation or alkylation) may not react. The situation is further complicated by the possibility of electrophilic attack at one of the pyridine nitrogen atoms. In addition, since some electrophiles are generated in strongly acid media, the substrate may be a highly deactivated azolium cation.

17.3.6.1 Imidazoles and benzimidazoles

It is of advantage to treat electrophilic attack separately at a ring nitrogen and at carbon atoms. In theory the former can involve a number of mechanistic possibilities (Scheme 11). Examples of the (unlikely) electrophilic attack upon the conjugate acid (S_E2 cA) are lacking, while an S_E2 mechanism would require the disruption of the aromatic sextet in

S_E2 S_E2' S_E2 cA S_E2 cB

SCHEME 11

the transition state. The classical method of imidazole alkylation provides an example of the S_E2' mechanism whereby the base and an alkyl halide (or sulphate) are heated together in a sealed tube (Scheme 12). Yields are not particularly good since quaternary salt formation is difficult to suppress, and the hydracid salt of the imidazole is also formed.

SCHEME 12

When the substrate becomes the imidazole anion, generated in basic conditions (*e.g.* with sodium in liquid ammonia, or in sodium ethoxide), the much more favourable S_E2 cB pathway leads to high yields of 1-substituted imidazoles uncontaminated by quaternary salt.

The alkylation of unsymmetrical imidazoles and benzimidazoles proceeds in a complex manner[6] and it is probably premature at this stage to attempt to reach any firm conclusions. As an example, though, the methylation of 4-nitroimidazole in neutral medium gives mainly 1-methyl-5-nitroimidazole (**45**), but in basic solution (where the imidazole anion reacts) the major product is 1-methyl-4-nitroimidazole (**46**) (Scheme 13).

(45) + (46) $\left(\frac{46}{45} = \frac{1}{350}\right)$

(46) + (45) $\left(\frac{46}{45} = 9\right)$

i, Me_2SO_4; ii, Na, liq. NH_3.

SCHEME 13

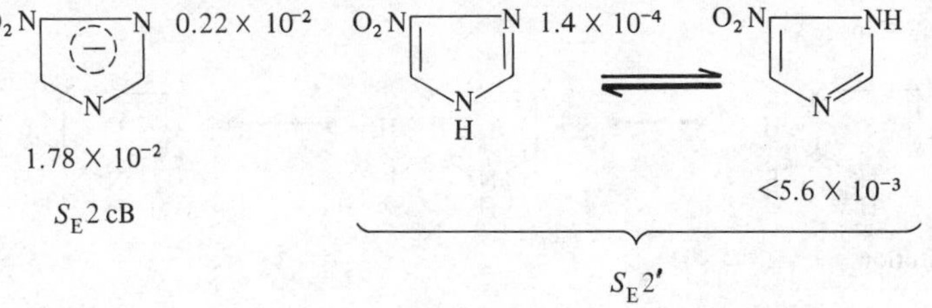

O_2N (anion) 1.78×10^{-2}, $S_E2\,cB$ 0.22×10^{-2} O_2N (NH) 1.4×10^{-4} ⇌ O_2N (NH) $<5.6 \times 10^{-3}$ S_E2'

FIGURE 5

Examination of the structures of the heterocyclic substrates involved permits prediction of the observed experimental results. Rate coefficients (1 mol^{-1} s^{-1}) for the methylation of 4-nitroimidazole with dimethyl sulphate under S_E2 cB and S_E2' conditions are quoted in Figure 5.[11] When 4-bromo- and 4-phenyl-imidazoles are methylated under S_E2' conditions, the ratios of 1,4- : 1,5-products are respectively 1 : 34 and 5 : 1. In all three cases discussed the electron-withdrawing nature of the substituent makes the 4-substituted compound the predominant tautomer. Even though the minor tautomer may react more rapidly (as is the case), one would expect the 1-methyl-5-substituted product to predominate. Only in 4-phenylimidazole is this prediction reversed, presumably for steric reasons.

The *N*-acylation of imidazoles is achieved by reaction with an acid halide in an anhydrous solvent. The products, known as azolides, are readily hydrolysed (see Section 17.3.11).

The quaternization of 1-alkylimidazoles is also a result of electrophilic attack at the unsubstituted nitrogen atom. Even in the presence of electron-withdrawing groups a powerful alkylation agent is still capable of forming a quaternary salt. Thus, 1-benzoyl-4-phenylimidazole can be quaternized with triethyloxonium tetrafluoroborate (Scheme 14).

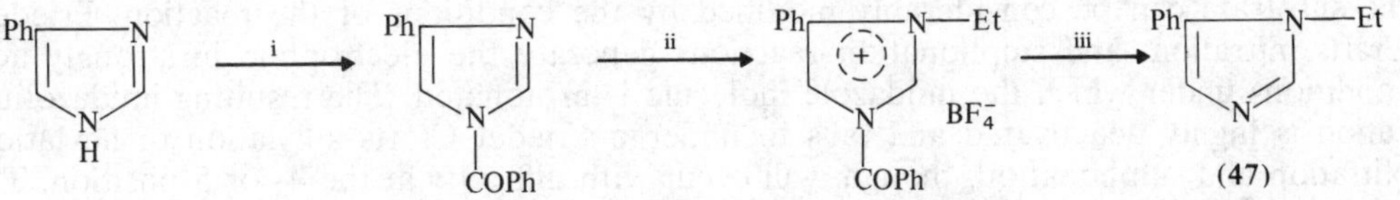

Ph–imidazole (NH) —i→ Ph–imidazole (N–COPh) —ii→ Ph–imidazolium (N–Et, N–COPh) BF_4^- —iii→ Ph–imidazole (N–Et) **(47)**

i, PhCOCl, dry benzene; ii. $Et_3O^+BF_4^-$, iii, EtOH.

SCHEME 14

Ethanolysis removes the aroyl group to provide a synthetic route to 1-ethyl-5-phenylimidazole (**47**), the isomer which is not formed to any extent on direct ethylation of 4-phenylimidazole in which steric factors limit attack at the nitrogen adjacent to the phenyl substituent.

Electrophilic attack at ring-carbon atoms is rather complicated in imidazoles and attempts to correlate reaction mechanisms, electron-density calculations, and experimental results have met with only limited success. Recent SCF calculations[12] predict orders of electrophilic substitution for imidazole as 5>2>4 and benzimidazole as 7>6>5>4. These results do not take into account the tautomeric equivalence of the 4- and 5-positions of imidazole, but they are in accord with the observed orientation of attack in benzimidazole (see **2a**).

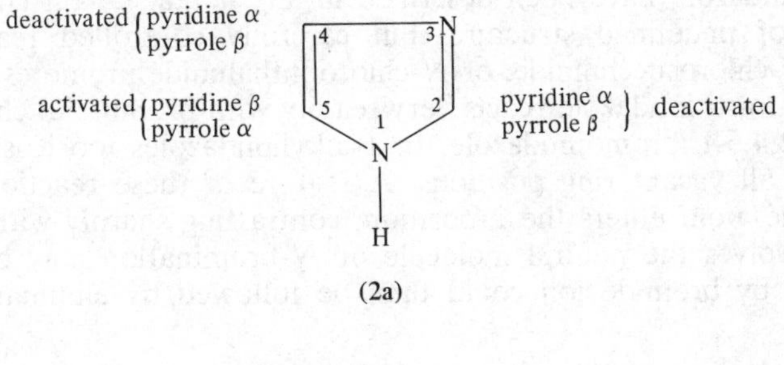

deactivated {pyridine α, pyrrole β} (4); activated {pyridine β, pyrrole α} (5); (2) pyridine α, pyrrole β} deactivated; N3; N1–H

(2a)

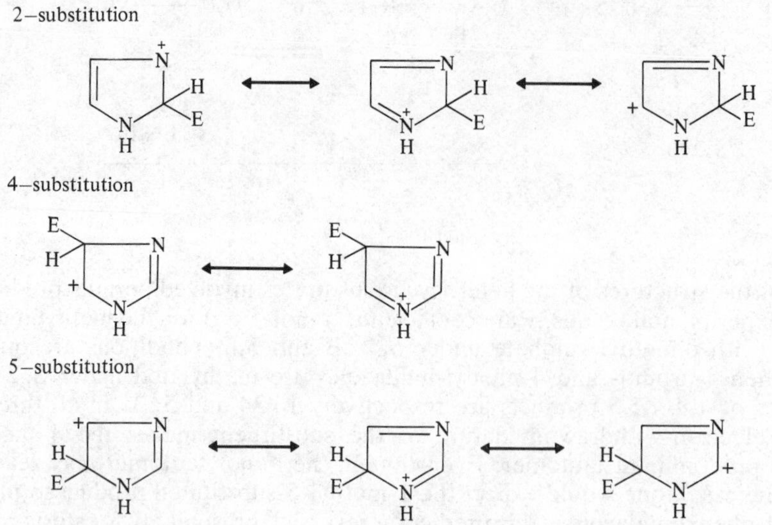

2–substitution

4–substitution

5–substitution

SCHEME 15

If imidazole can be considered as a combination of pyrrole and pyridine, the most activated ring-carbon to electrophilic attack will be C-5. Resonance structures for the isolated molecule suggest that all ring-carbons may be susceptible to electrophiles, while consideration of the stabilities of the expected σ-intermediates for electrophilic (Scheme 15) attack suggests that 5-substitution is most likely.

Separation of the reactions into acidic, neutral, and basic offers some advantages since the substrate can be considerably modified by the conditions of the reaction. Friedel–Crafts, nitration, and sulphonation reactions generate the electrophile in strongly acid conditions under which the imidazole molecule is protonated. The resulting imidazolium cation is highly deactivated and fails to undergo Friedel–Crafts alkylation or acylation. Nitration and sulphonation, though, will occur with difficulty at the 4- or 5-position. The reaction is accompanied by ring-opening and (presumably) some oxidation. Imidazole nitrates at about 10^{-10} times the rate of nitration of benzene but about 10^9 times faster than pyridine (Scheme 16). Dinitration occurs only with extreme difficulty and never in the 2-position.

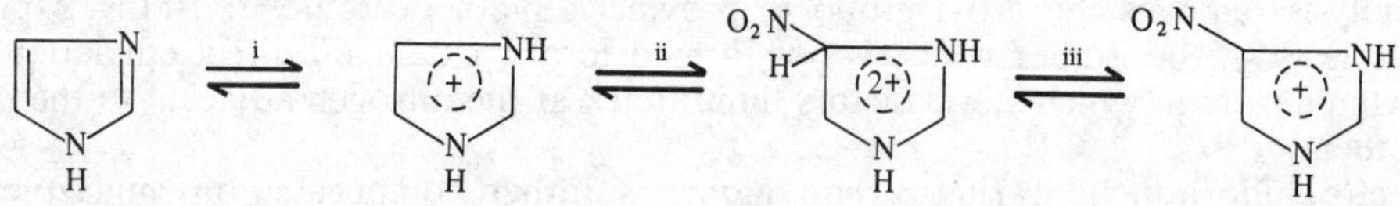

i, H^+, ii, NO_2^+, slow; iii, $-H^+$, fast.

SCHEME 16

Halogenation is an even more complex reaction with considerable variations, depending on the reagent, substrate, and reaction conditions employed. Only the bromination and iodination of imidazole have been described in any detail. Direct chlorination seems to give products of undefined structure, but carefully controlled reaction with sodium hypochlorite, *N*-chlorosuccinimide, or *N*-chlorophthalimide promotes chlorination at C-4 and C-5. In contrast, imidazole reacts very readily with bromine in chloroform, water, or ether to yield 2,4,5-tribromoimidazole. In 1-alkylimidazoles too it is difficult to prevent bromination of all vacant ring positions. A feature of these reactions is the ease with which a bromine atom enters the 2-position, contrasting sharply with nitration. Perhaps bromination involves the neutral molecule or *N*-bromination may be implicated. Nucleophilic attack by bromide ion could then be followed by elimination of HBr.[13] The

i, Br_2, CS_2, heat; ii, –HBr; iii, BrCN; iv, –HCN.

SCHEME 17

second hypothesis is supported by the remarkable reaction between imidazole and cyanogen bromide, which forms hydrogen cyanide and 2-bromoimidazole (Scheme 17).

Direct monobromination of imidazole and 1-methylimidazole has been accomplished using the reagent 2,4,4,6-tetrabromocyclohexa-2,5-dienone.

The bulk of the mechanistic work in the area relates to iodination. In aqueous alkali, iodine converts imidazole to 2,4,5-tri-iodoimidazole with some 2,4-di-iodoimidazole. That 1-substituted imidazoles fail to react lends support to the theory that it is the imidazole anion which reacts. In fact the anion is about 10^8 times as reactive as the neutral molecule. Iodination of 2,4,5-trideuteroimidazole shows a large kinetic isotope effect ($k_H/k_D = 4.5$), present also for 4,5-dideuteroimidazole, but absent for 2-deuteroimidazole. This demonstrates that loss of a proton from C-4 is rate determining in monoiodination (Scheme 18). The iodoimidazole anion is then rapidly iodinated again.

SCHEME 18

Deuterium exchange in imidazole takes place in neutral or basic reaction conditions. The proton attached to nitrogen exchanges rapidly in D_2O, while the proton at C-2 is exchanged at 37 °C with $t_{\frac{1}{2}} \sim 700$ minutes. At 150 °C the reaction is virtually complete in two hours with or without added base, and in acid conditions it does not occur. 1-Methylimidazole can undergo 2-deuteration in D_2O with the rate of the reverse reaction being essentially independent of pH as long as the conditions do not become acid, when the rate drops rapidly to zero. A mechanism involving an ylide (**48**) derived from the conjugate acid is believed to be involved (Scheme 19). The deuteration of imidazole itself does not appear to involve the imidazole anion either, but rather depends on two parallel processes: rate-determining proton abstraction from the imidazolium cation by D_2O and by OD^- to give the ylide at C-2 (analogous to **48**), followed by deuteration there. An additional path involving proton abstraction from the imidazole neutral molecule accounts for the 4-substitution pD profile.

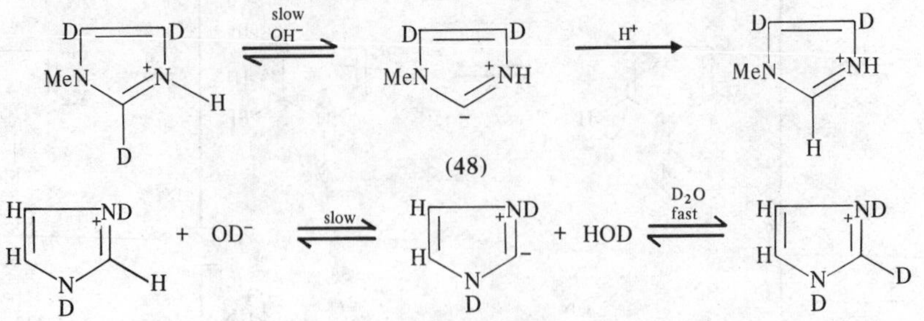

(48)

SCHEME 19

In alkaline medium, imidazoles can undergo diazo-coupling. The reaction has been known for many years and is the basis of the Pauly test for imidazoles whereby the diazonium salt derived from an aromatic amine (*e.g.* sulphanilic acid) reacts in basic solution with an imidazole to produce a red, orange, or yellow dye. The test is particularly sensitive and is capable of detecting microgram quantities of imidazoles as long as the compound has a free NH group and at least one ring-carbon unsubstituted. Some electron-withdrawing substituents (*e.g.* ethoxycarbonyl) hinder the reaction and benzimidazoles fail to couple. The observation that *N*-substituted imidazoles will not react coupled with kinetic studies has demonstrated that, over the pH range 7–11, coupling at C-2 involves the imidazole anion (Scheme 20). It is a bimolecular reaction with a negligible

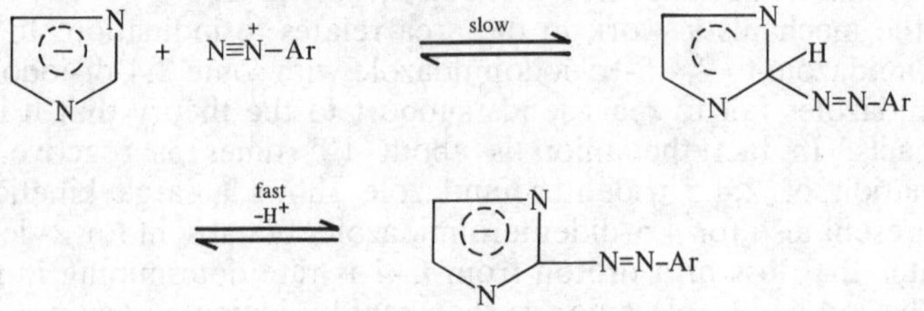

SCHEME 20

deuterium isotope effect for 2,4,5-trideuteroimidazole. The reaction, then, differs in some important respects from iodination. The reason for preferential 2-substitution has not been convincingly explained.

Other electrophilic substitution reactions (*e.g.* carbonyl reactions, metallation) have been discussed elsewhere.[2] In benzimidazoles the hetero-ring is highly deactivated to electrophilic substitution, with the result that the few reactions which do take place occur in the benzene ring, generally at positions-5 or -6. Nucleophilic attack is much more likely with benzimidazole.

17.3.6.2 Pyrazoles and indazoles

As with the 1,3-diazoles, pyrazoles are subject to electrophilic substitution, either on a ring carbon or at the pyridine nitrogen. Reaction conditions can again considerably modify product orientation.

Reactions at the pyridine nitrogen parallel, to some extent, those of imidazoles with alkylation, acylation, and (sometimes) nitration occurring. The last-named reaction is of particular interest, for, when pyrazoles with free NH groups are nitrated with nitronium acetate (acetic anhydride–nitric acid mixture), it is possible to isolate *N*-nitropyrazoles. This reaction does not usually occur with imidazoles, which are much more basic than pyrazoles and tend to form nitrate salts under the reaction conditions used. If, however, the basic nature of imidazole is decreased by the introduction of an electron-withdrawing

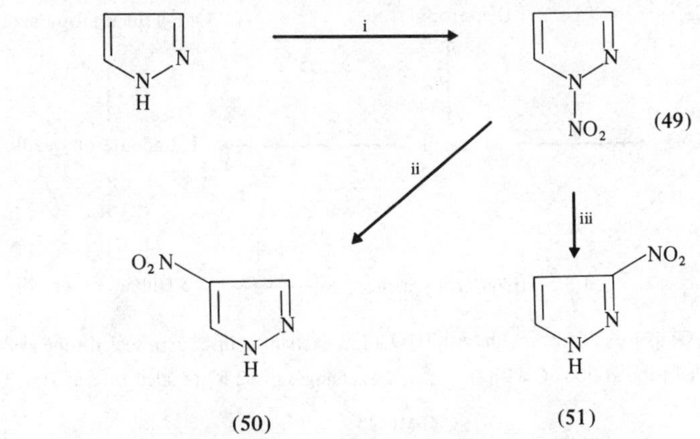

(49)

(50) (51)

i, NO_2^+ $MeCO_2^-$; ii, conc. H_2SO_4; iii, 130–140 °C, 10 h.

SCHEME 21

group (*e.g.* 4-nitro), then *N*-nitro-products can be isolated. The weakly basic pyrazoles and indazoles react much more universally. The *N*-nitropyrazole (**49**) products are of particular interest, for when they are treated with concentrated sulphuric acid they are converted to 4-nitropyrazoles (**50**), by dissociation followed by normal electrophilic nitration at C-4 by NO_2^+ ion. If, on the other hand, they are heated in a high-boiling solvent, 3-nitropyrazole (**51**) is formed through what has been classed as a thermally-induced [1,5]-sigmatropic rearrangement (Scheme 21). Such a thermal rearrangement of 5-methyl-1-nitropyrazole (**52**) produces 93% of 3-methyl-4-nitropyrazole (**53**) and 7% of 3-methyl-5-nitropyrazole (**54**). Since the isomerization to the adjacent carbon is blocked by the 5-methyl group, the formation of (**54**) must result from an intermediate rearrangement to 3-methyl-1-nitropyrazole (Scheme 22). The major importance of these reactions lies in their synthetic utility as an approach to pyrazoles nitrated at carbon atoms normally only subject to nucleophilic attack.[14] The thermal rearrangement is common to 2-nitro-2*H*-indazoles and to 1-nitro-1*H*-1,2,4-triazoles.

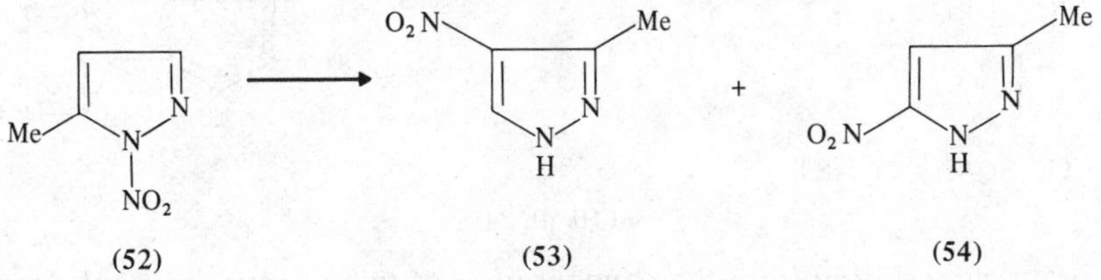

(52) (53) (54)

SCHEME 22

Indazole can also be alkylated and acylated but gives mixtures of 1- and 2-substituted products. Fortunately, separation of the alkylated products is simplified by the customary higher-boiling points of 2-alkyl-2*H*-indazoles. 2-Acyl-2*H*-indazoles are less stable than the 1-isomers into which they often isomerize on heating or long standing.

Electrophilic attack at a pyrazole ring-carbon usually takes place most readily at carbon-4, the position of highest π-electron density.[12] Certainly the annular nitrogen atoms reduce electron density at the 3- and 5-positions, particularly so in acidic media in which the pyrazolium cation exists. The reactivity of the 4-position in pyrazole is intermediate between the reactivities of benzene and phenol, but the pyrazolium cation is less reactive even than benzene. Thus, reactions with the weaker electrophiles such as in Friedel–Crafts, Vilsmeier, and diazo-coupling reactions are virtually absent. The Vilsmeier–Haack formylation of 1,3- and 1,5-dimethylpyrazoles in the presence of phosphoryl chloride and aluminium chloride, however, is reported to give satisfactory yields.

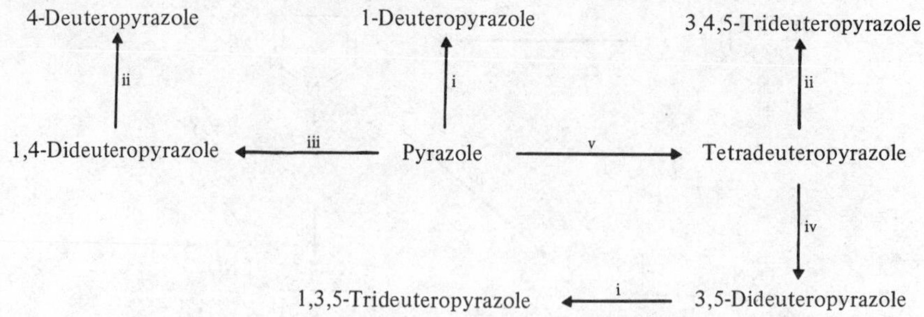

i, 3 exchanges in 3 h with D_2O; ii, 1 exchange in 3 h with H_2O; iii, 2 exchanges in 12 h in sealed tube at 200 °C with D_2O; iv, 2 exchanges in 12 h in sealed tube at 200 °C with H_2O; v, 2 exchanges in 12 h in sealed tube at 200 °C with 1 M NaOD.

SCHEME 23

Deuterium exchange reactions with pyrazole have been examined in considerable detail (Scheme 23). It can be seen that the order of ease of replacement is H-1 > H-4 ≫ H-3 or H-5, which only exchange in the presence of base. There are marked differences from the behaviour of imidazole, but discussion in detail lies outside the scope of this text (see Ref. 2, p. 38).

The halogenation of pyrazoles is a particularly facile reaction leading, with a variety of reagents, to the 4-halopyrazole. Under more forcing conditions a dihalogenated product, or one which also involves side-chain halogenation, can be obtained (Scheme 24). In

i, Cl_2, HOAc; ii, Cl_2, CCl_4.

(55)

SCHEME 24

addition, products resulting from the condensation of two or more pyrazole units (**55**) are often formed. Many early workers noted the presence of orange-red 'perbromides' (**56**), formed by the action of excess bromine on pyrazoles. These compounds are unstable and readily lose HBr to give 1-bromopyrazoles (**57**) (Scheme 25). The bromine atom at N-1 is rather mobile, conferring properties similar to *N*-bromosuccinimide, and it may be that 1-bromopyrazoles are implicated in the process of *C*-bromination as a source of bromonium ions.

(56) (57)

SCHEME 25

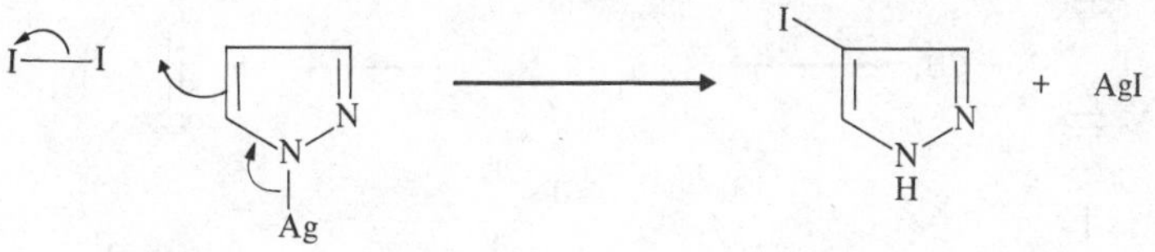

(58)

SCHEME 26

Pyrazole does not react directly with iodine, but silver pyrazole is converted to 4-iodopyrazole (**58**) (Scheme 26). Again *N*-iodopyrazoles are often formed as stable compounds, which rearrange on heating to *C*-iodo-products (Scheme 27).

Metallation reactions of pyrazoles are of synthetic importance. Butyl- or phenyl-lithium give firstly an *N*-lithiopyrazole, and then a second equivalent of reagent causes *C*-metallation. On carboxylation some pyrazole-3-carboxylic acid is formed. The same reagents react with 1-methylpyrazole at carbon-5.

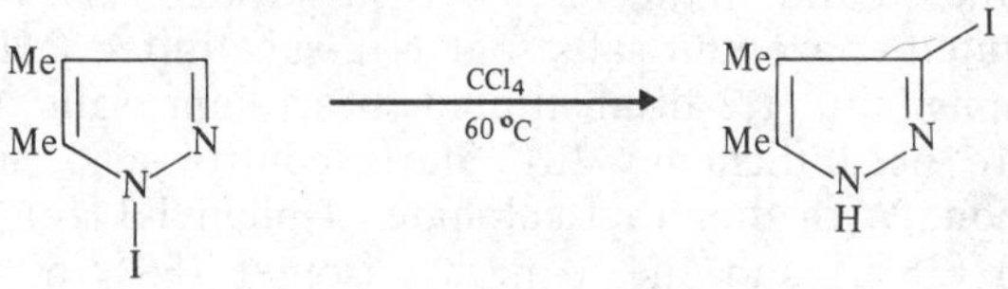

SCHEME 27

The nitration of pyrazole in concentrated sulphuric acid leads quantitatively to 4-nitropyrazole *via* electrophilic attack by NO_2^+ on the pyrazolium cation. That the pathway is of low reactivity is attested to by a partial rate factor of 2.1×10^{-10}.

Recent SCF calculations[12] of π-electron distributions for indazoles predict orders of electrophilic substitution as follows: 1*H*-indazole, 7>5; 2*H*-indazole, 3>5>7>4. These are in accord with the observed orientation of substitution. Halogenation commonly gives 3,5-dihalo-products unless carefully controlled. In acidic medium though, indazoles mono- or di-nitrate in the 5- and 6-positions, while sulphonation gives rise to the 7-sulphonic acid. In acetic anhydride (conditions more nearly neutral) nitration leads to 3-nitro- and 3,5-dinitro-indazoles.

When 1-phenylpyrazoles are nitrated the products isolated display a marked dependence on the reagent used. Whereas nitration in acetic anhydride generally leads to 4-nitration in the pyrazole ring, in sulphuric acid the nitro-group enters the *para*-position of the phenyl ring. Kinetic studies[15] demonstrate that the nitration in sulphuric acid proceeds by nitration of the protonated 1-phenylpyrazolium species in which the hetero-ring is highly deactivated. The nitration process in acetic anhydride is not fully understood. It may involve nitration of the free base species but there are some anomalies. In particular, 5-methyl-1-phenylpyrazole, a compound in which steric effects of the 5-methyl group force the phenyl ring out of coplanarity with the pyrazole nucleus, nitrates in acetic anhydride to yield a 3 : 1 mixture of 3-nitro- and 4-nitro-products.

17.3.6.3 Triazoles

Once again electrophilic substitutions can take place at a ring-nitrogen or a ring-carbon atom.

Both 1,2,3-triazoles and 1-substituted 1*H*-1,2,3-triazoles can be alkylated by a variety of reagents which include alkyl halides, dimethyl sulphate, diazoalkenes, or by the Mannich reaction. Unsubstituted 1,2,3-triazoles appear to substitute selectively on N-1 when methylated with methyl iodide as their silver or thallium salts, or in alkaline solution. Diazomethane gives a high proportion of 2-methyl-2*H*-1,2,3-triazole, though,

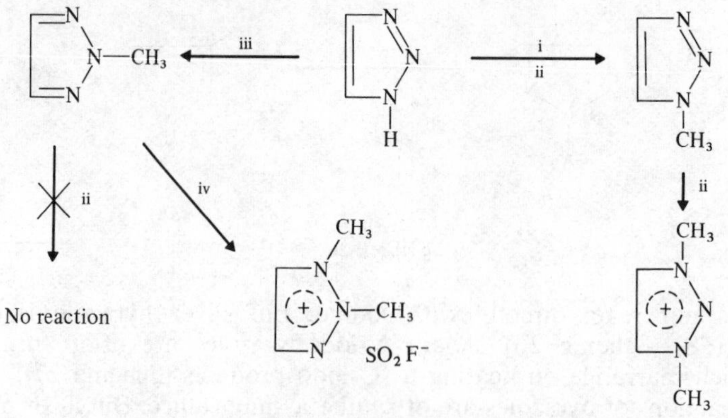

No reaction

i, Ag salt; ii, CH_3I; iii, CH_2N_2; iv, CH_3OSO_2F.

SCHEME 28

when it reacts with the neutral molecule. N-1-Substituted 1*H*-1,2,3-triazoles tend to form the 1,3-disubstituted triazolium salts, but N-2-substituted 2*H*-1,2,3-triazoles resist alkylation, only forming the 1,2-disubstituted triazolium salts when very powerful methylating agents are used (Scheme 28). Steric control also manifests itself in the orientation of alkylation. With dimethyl sulphate, 4-phenyl-1*H*-1,2,3-triazole gives the 2-methyl isomer (**58**) (38%) and the 1-methyl isomer (**59**) (62%), but none of the hindered 1-methyl-5-phenyl compound (**60**) (Scheme 29).

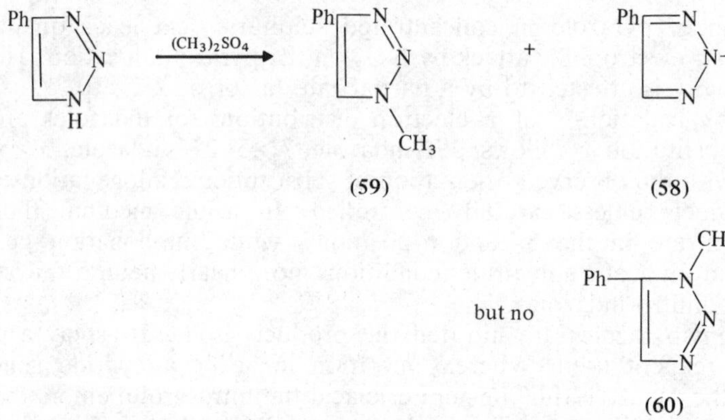

$(CH_3)_2SO_4$

(**59**) (**58**)

but no

(**60**)

SCHEME 29

In basic media 1,2,4-triazoles give 1-substituted derivatives, and early reports of 4-alkylated products must be viewed with some suspicion. Substituents may modify the direction of attack and matters can become highly complex with hydroxy (or '-one') derivatives which can alkylate on nitrogen, oxygen, or carbon.

Quaternizing alkylation of 1-alkyl-1*H*-1,2,4-triazoles occurs at N-4. The proof of structure of a quaternary salt of this series can be obtained as illustrated on alkali treatment, which degrades the ring to methylamine and methylhydrazine. Had quaternization taken place at N-2 some *sym*-dimethylhydrazine would have been formed. Quaternization of 4-substituted 4*H*-1,2,4-triazoles occurs at N-1 or N-2 (Scheme 30). When there is no NH group present, as in 1-phenyl-1*H*-1,2,4-triazole, it is possible to achieve *C*-alkylation by using butyl-lithium and methyl iodide at low temperatures.

N-Acylation reactions occur in the expected fashion, but is is only recently that the use of nuclear magnetic resonance has allowed unequivocal determination of the orientation of substitution. Apparently acetylation of 1,2,3-triazoles in the cold gives the 1-acetyl-1*H*-compounds, which rearrange on heating above 100 °C to the 2-acetyl-2*H*-isomers.

$MeNH_2 + MeNHNH_2$

i, MeI; ii, 50% aq. KOH.

SCHEME 30

With unsymmetrical triazoles, 1-acetyl-5-substituted compounds (**61**) are the minor products (Scheme 31). At equilibrium (~100 °C) the mixtures of isomers obtained consist of about 90% 2-acetyl and $\not>$10% of 1-acetyl, with some exceptions. Since it seems evident that 1,2,3-triazole has the 1*H*-form as the dominant tautomer, the reaction with acetyl chloride in benzene at 0 °C to give mainly the 1-acetyl product is readily accounted for in terms of an S_N2' reaction. The rearrangement to the 2-acetyl compound can take place either by heating or on treatment with base. An alternative approach to 1-acetyl-1,2,3-triazoles is by reaction of the 2-trimethylsilyl derivatives with acetyl chloride. The mechanism is presumably analogous to an S_E2' process. Acylation of 1,2,4-triazoles gives the 1-acyl products.

MeCOCl

(61)

SCHEME 31

There is some evidence that *N*-halogenation occurs in triazoles, perhaps giving rise to intermediates of *C*-halogenation.

Electrophilic substitutions at ring-carbon atoms of triazoles are not particularly well known. SCF calculations[12] of π-electron distributions for the ground states of triazoles predict initial electrophilic attack at C-5 in 1,2,3-triazole, support the resistance of 1,2,4-triazole to substitution by electrophiles, and predict initial attack at C-4 and C-7 of 1*H*-benzotriazole, as is observed experimentally. The observed inactivity of 1,2,4-triazole (and of tetrazoles) may be inferred from the very low-field chemical shifts of the ring protons in the proton magnetic resonance spectra (see Table 4). These results point to low electron density at the ring-carbon atoms, although the use of such data as an index must be treated with caution since reaction conditions can considerably modify the molecules.

TABLE 4
Some n.m.r. Data for Azoles

Compound	δ (p.p.m.)($CDCl_3$)
Pyrazole	H-3(5) 7.61; H-4* 6.31
Imidazole	H-4(5)* 7.25; H-2 7.86
1,2,3-Triazole	H-4(5)* 7.90
1,2,4-Triazole	H-3(5) 8.30
1-Methyltetrazole	H-5 9.25

* Usual position for electrophilic attack.

The attempted chlorination of 1,2,3-triazole and its 2-methyl derivative gave only the hydrochloride salts, but 1-methyl-1*H*-1,2,3-triazole yields 4-chloro-1-methyl-1*H*-1,2,3-triazole. Bromination is more facile giving, with 1,2,3-triazole, either the 4,5-dibromo-compound or with excess hypobromite the 1,4,5-tribromo-derivative. Acidic electrophiles react only with difficulty. Thus nitration in sulphuric acid of 2-phenyl-2*H*-1,2,3-triazole gives initially the *p*-nitrophenyl compound; further nitration results in some substitution in the heterocyclic ring. Bromination with silver sulphate and concentrated sulphuric acid exhibits a similar preference for the phenyl ring. Diazo-coupling occurs only in the presence of activating substituents.

In benzotriazoles, electrophilic substitution reactions take place in the 4- or 7-positions, an orientation which contrasts with the behaviour of benzimidazole and indazole, which substitute in the 5- or 6-positions.

Attempts to nitrate 1,2,4-triazole generally have only given the nitrate salt and one is left with the impression that the ring carbons of 1,2,4-triazole are particularly inert to electrophilic attack.

17.3.6.4 Tetrazoles

The high degree of deactivation engendered by the annular nitrogens makes electrophilic substitution reactions at carbon-5 a difficult proposition. Attack at nitrogen, though, is not prevented.

Most alkylations of tetrazoles have been carried out in basic medium in which the conjugate base of tetrazole is the probable reactive species. Although two isomeric alkyltetrazoles are usually formed (Scheme 32), the experimental data at present available

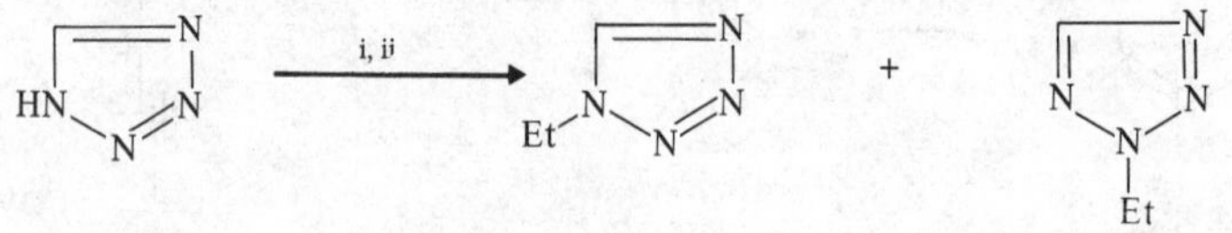

i, Ag salt; ii, EtI, C_6H_6.

SCHEME 32

are inadequate to reveal any general trends. A more detailed study of the alkylation of 5-aminotetrazole (as the anionic species) showed that methylation at N-1 predominates over methylation at N-2, and a small proportion of alkylation of the exocyclic amino-group also occurs.

When a 1- or 2-substituted tetrazole is quaternized the incoming group usually attaches at N-4, although there are exceptions known.

In contrast to the orientation observed in alkylation, tetrazoles commonly acylate to give 2-acyl-2*H*-1,2,3,4-tetrazoles. These compounds are very reactive acylating agents and undergo ring modifications to form oxadiazoles (see Scheme 47).

While reaction of 1-phenyltetrazole with bromine or mercuric acetate leads to substitution at carbon-5, such reactions are rare. It is of interest to note that 5-phenyltetrazole nitrates to form 5-(*m*-nitrophenyl)tetrazole. Arylazoles are normally nitrated (in acid conditions) in the *para*-position of a phenyl ring[15] unless there are deactivating groups present, *e.g.* 2-phenylimidazole-4,5-dicarboxylic acid gives mainly *meta*-nitration. The result with 5-phenyltetrazole confirms the considerable combined electron-withdrawing powers of the four annular nitrogen atoms. In contrast, 1- and 2-phenyltetrazoles nitrate in the *para*-position of the benzene ring.

17.3.7 REACTIONS INVOLVING NUCLEOPHILIC ATTACK

Unless electron-withdrawing groups are present the diazoles are not particularly susceptible to nucleophilic attack, but introduction of one or two more nitrogen atoms to the ring increases the likelihood of such reactions. Quaternary salts of the azoles are generally susceptible to nucleophilic attack, which may result in substitution or in ring-opening.

17.3.7.1 Imidazoles and benzimidazoles

Imidazole reacts only sluggishly with nucleophiles in the absence of electron-withdrawing substituents. In benzimidazole the azole ring is sufficiently electron-deficient and such reactions as the halogenation of benzimidazol-2-ones and the Chichibabin reaction occur readily (the latter will not occur with imidazoles) (Scheme 33). Bulky groups at N-1 can hinder the latter reaction.

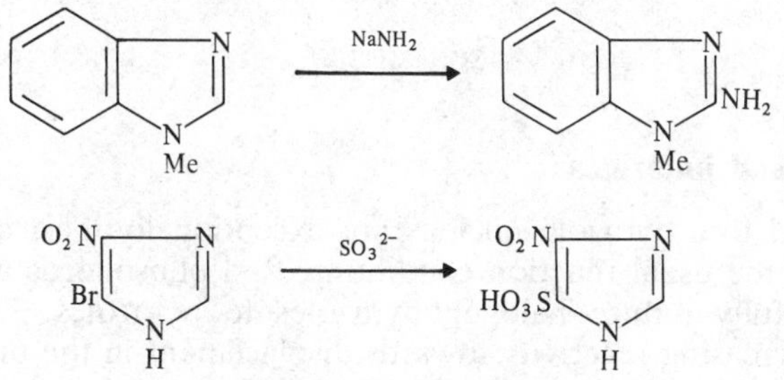

SCHEME 33

Replacement of halogen atoms can take place if there is a strong nucleophile available and an electron-withdrawing group in the ring. Although 2-bromo-1-methylimidazole will not react with piperidine at 200 °C, both 2-bromo-1-methyl-4-nitro- and 5-bromo-1-methyl-4-nitro-imidazoles are quite reactive. When the reagent is made more nucleophilic (lithium piperidide in dry ether), 5-bromo-1-methylimidazole is found to undergo 25% conversion to 1-methyl-5-piperidinoimidazole by conventional nucleophilic substitution. A further 16% of the product is 1-methyl-2-piperidinoimidazole (**62**) (Scheme 34), which is considered to arise by an addition–elimination process.[16]

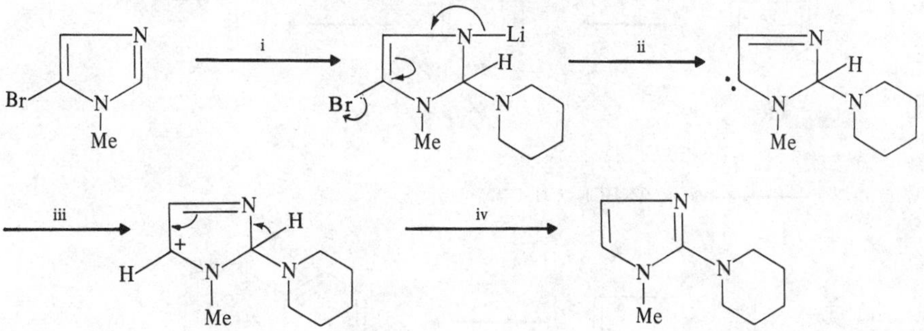

i, $Li^+ \bar{N}C_5H_{10}$; ii, –LiBr; iii, H^+; iv, $-H^+$.

SCHEME 34

Often, nucleophilic attack results in ring fission. In particular, aminoimidazoles are unstable owing to hydrolytic cleavage and quaternary imidazolium salts decompose to primary amines under the influence of alkali. When the classical Schotten–Baumann technique is used to prepare benzoylimidazoles, the ring breaks to yield a 1,2-dibenzamidoethene (Scheme 35). This reaction is a consequence of attack of OH^- at carbon-2 of imidazolium, and has long been known as the Bamberger degradation. The susceptibility of carbon-2 of imidazolium to nucleophilic attack is reflected in the acidity of a proton attached to it. Such a proton very readily exchanges in D_2O solution.

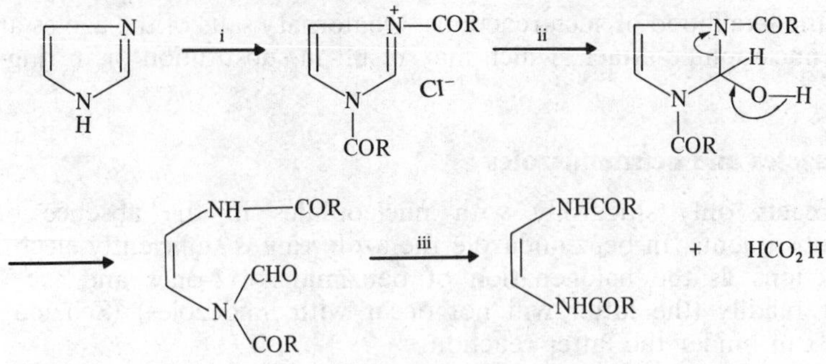

i, RCOCl, NaOH; ii, OH^-; iii, H_2O.

SCHEME 35

17.3.7.2 Pyrazoles and indazoles

Halogens attached to a pyrazole nucleus are exceptionally inert and will not undergo displacement under the usual reaction conditions. Red phosphorus and hydriodic acid at 140–160 °C successfully reduce halogenopyrazoles to pyrazoles. In 1-phenylpyrazoles, halogens display decreasing reactivity towards displacement in the order 5->4->3-, and the halide activity increases in the series Cl<Br<I. The presence of electron-withdrawing groups markedly assists nucleophilic displacements of halogen. Pyrazole quaternary salts are particularly reactive in nucleophilic reactions, and in these compounds 3- and 5-halogen substituents can be readily displaced by a variety of nucleophiles including hydroxyl, alkoxyl, amino, alkylamino, thioalkyl, and cyanide. In contrast to the replacement of halogen, direct amination of pyrazoles (involving hydrogen replacement) has not been observed. In fact sodamide can cause ring-opening (as can alkalies) or removal of an *N*-aryl group. The ring-opening can also occur when sodamide reacts with indazoles (Scheme 36). Nucleophilic replacement of hydrogen by halogen is unknown in pyrazoles,

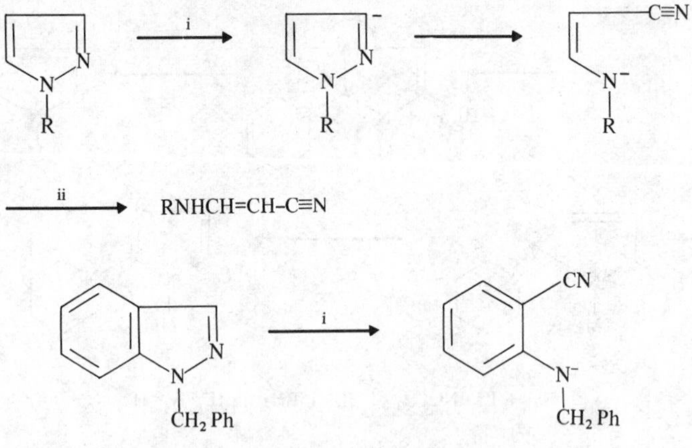

RNHCH=CH-C≡N

i, NH_2^-; ii, H^+.

SCHEME 36

but the standard transformation of amino to halogen *via* the diazonium salt is quite common. Under copper catalysis iodine atoms at carbons-3, -4, and -5 of 1-methylpyrazole can be replaced by acetylenic groups. Many chloro- and bromo-pyrazoles have been prepared from hydroxypyrazoles or pyrazolones on reaction with phosphorus halides. The substitution occurs best in the 5- and 3-positions.

17.3.7.3 Triazoles

Although, in common with the other azoles, direct nucleophilic amination has not been reported, replacement of a halogen by amino is not uncommon in triazoles. In 1-methyl-1*H*-1,2,3-triazoles a 5-chloro-group, but not a 4-chloro-group, can be displaced (Scheme 37). If a bimolecular, addition–elimination mechanism is involved in this process, then it can be seen that the intermediate for 5-substitution is more favourable than that for

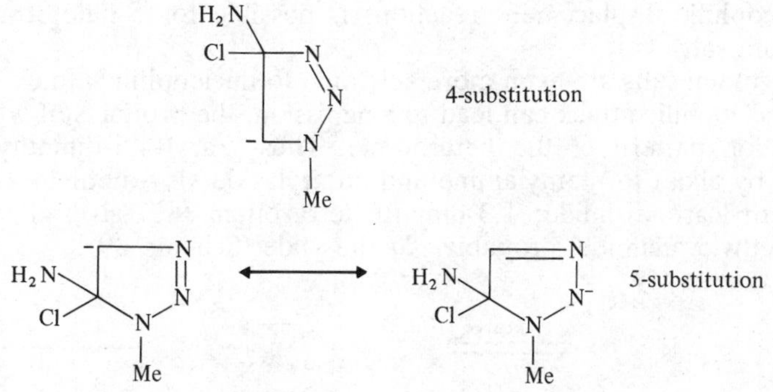

4-substitution

5-substitution

SCHEME 37

4-substitution. Under appropriate conditions the reaction product may be that resulting from a Dimroth rearrangement. Replacement of halogen by other nucleophiles such as thiol and cyanide are known. Similarly, nucleophilic substitution reactions of 1,2,4-triazolediazonium salts take place normally, but even in the triazoles some activation is often needed, either by incorporation of an electron-withdrawing substituent, or by using a triazolium salt.

Interesting combinations of nucleophilic hydroxylation and electrophilic bromination have been observed in the reactions of 1,2,3-triazolium salts with alkali. When 4-bromo-1,3-dimethyl-1,2,3-triazolium tosylate (**63**) is treated with 1 *M* sodium hydroxide at 100 °C the equilibrium shown is set up rapidly. Evidently (**63**) is converted to an ylide which is further brominated by (**63**). Subsequently (**63**) and the dibromo-compound undergo very much slower nucleophilic substitution to give 1,3-dimethyl-1,2,3-triazole 4-oxides (**64**), where R = H and R = Br, respectively (Scheme 38). Ring-cleavage as a

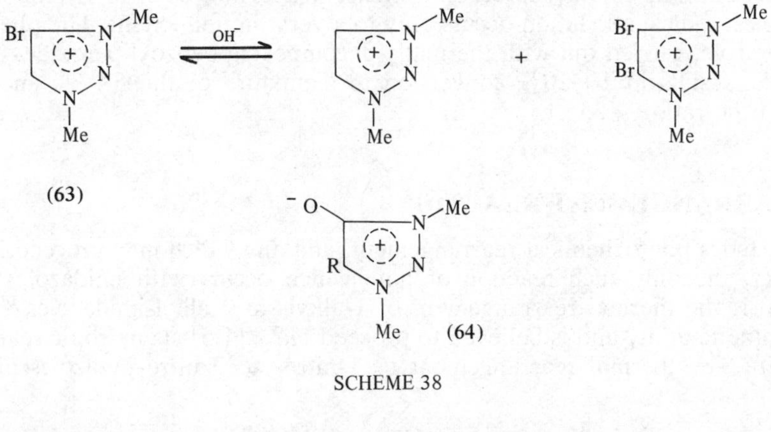

(63)

(64)

SCHEME 38

result of nucleophilic attack, usually by alkali, has been described earlier (17.3.6.3). Hydroxide ion reacts with 5-halo-1-methyl-3-nitro-1*H*-1,2,4-triazoles by a second-order substitution of halogen or nitro to give the respective oxotriazolines in a 30 : 1 ratio. Regardless of the nature of substituents, the 5-position is more reactive than the 3-position. With alkoxide as the nucleophile the product ratio becomes 300–700 : 1.

17.3.7.4 Tetrazoles

With three pyridine-type nitrogens, nucleophilic reactions take place rather more readily with tetrazoles. With piperidine, 5-bromo-1-methyl-1*H*-1,2,3,4-tetrazole and 5-bromo-2-methyl-2*H*-1,2,3,4-tetrazole react much faster than analogous bromo-1,2,4- and -1,2,3-triazoles, bromoimidazoles, or bromopyrazoles. In addition, the 5-bromo-1-methyltetrazole is much more susceptible to nucleophilic attack than the 2-methyl isomer. A temperature difference of 120 °C is required to achieve comparable rates. A wide range of nucleophilic displacement reactions is possible for 5-halotetrazoles and tetrazole-5-diazonium salts.

Tetrazolium salts are even more activated to nucleophilic attack. As with other azolium salts, nucleophilic attack can lead to ring fission, the products of which are defined by the substitution pattern of the heterocycle. Thus 1-aryl-4,5-dimethyltetrazolium iodide is cleaved by alkali to methylamine and an arylazide; 1,4-diethyltetrazolium gives nitrogen and diethylcarbodi-imide; 1,3-dimethyltetrazolium (**65**) gives diazomethane and potassium methylcyanamide, probably *via* the ylide (Scheme 39).

(65) $\xrightleftharpoons{KOH}$ [ylide] $\longrightarrow$ MeN=N-N(Me)C≡N $\xrightarrow{KOH}$ CH_2N_2 + $(MeN\text{-}C{\equiv}N)^-K^+$

17.3.8 RADICAL SUBSTITUTION

Knowledge of radical substitution in the azoles is scant. Studies in the imidazole series have been limited to alkylation and arylation. When methyl (or other alkyl) radicals are generated by the silver-catalysed decarboxylation of acetic (or higher alkyl carboxylic) acid with ammonium peroxydisulphate, they can be introduced at C-2 of imidazoles. Yields of greater than 50% are difficult to obtain. Radical phenylation too occurs with only 10–20% conversion, and positions of substitution vary with variation in reaction conditions. The imidazolium cation is mainly substituted at C-2 (as in radical alkylation), while the neutral molecule is subject to a higher proportion of 5-phenylation.

In pyrazoles, radical alkylation occurs only to a very limited extent. The phenylation of 1-methylpyrazole, carried out with thermally-decomposing benzoyl peroxide, or *N*-nitrosoacetanilide, results in 10–20% conversion to a mixture of the 5-, 3-, and 4-phenylpyrazoles in the ratio 94 : 5 : 1.

17.3.9 REARRANGEMENT REACTIONS

If one excludes photochemical rearrangement, and any which may proceed in the mass spectrometer, the only such reaction of note which occurs with imidazoles (and benzimidazoles) is the thermal rearrangement of 1-alkyl- to 2-alkyl-imidazoles.[17] The reaction has synthetic utility and is believed to proceed *via* a [1,5]-sigmatropic rearrangement (Scheme 40). The thermal rearrangement of 1-nitro- to 3-nitro-pyrazoles (17.3.6.2) is

[1,5]-H

SCHEME 40

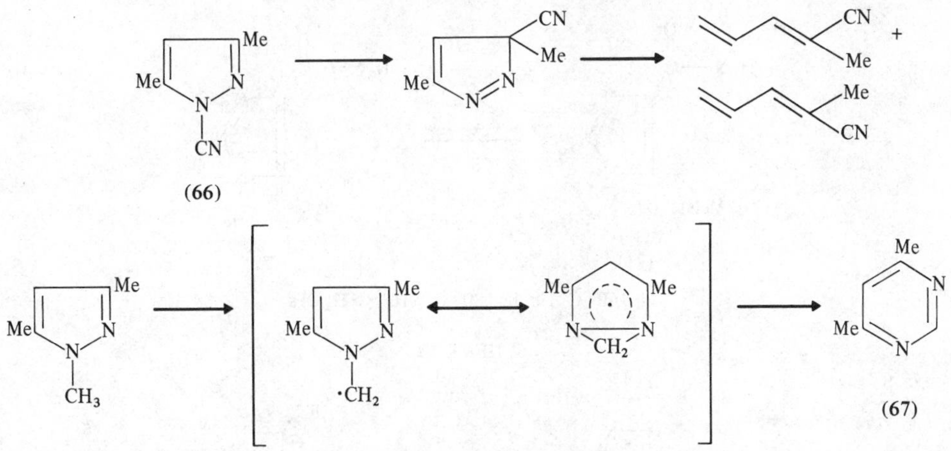

(66)

(67)

SCHEME 41

analogous. Whereas pyrazole is very stable to heat, under severe conditions (650 °C, 0.05 mmHg) 1-cyano-3,5-dimethylpyrazole (**66**) forms alkenes following a similar rearrangement. At 800 °C/0.05 mmHg, 1,3,5-trimethylpyrazole reacts in a different way to form the pyrimidine (**67**) (Scheme 41).

A widely-occurring rearrangement reaction of heteroaromatic chemistry is the Dimroth rearrangement, first recognized in the triazole series. When 1-substituted 5-amino-1,2,3-triazoles (**68**) are heated, an equilibrium is established (in which a diazoimine (**69**) intermediate is implicated) with a 5-substituted amino-1,2,3-triazole (**70**) (Scheme 42).

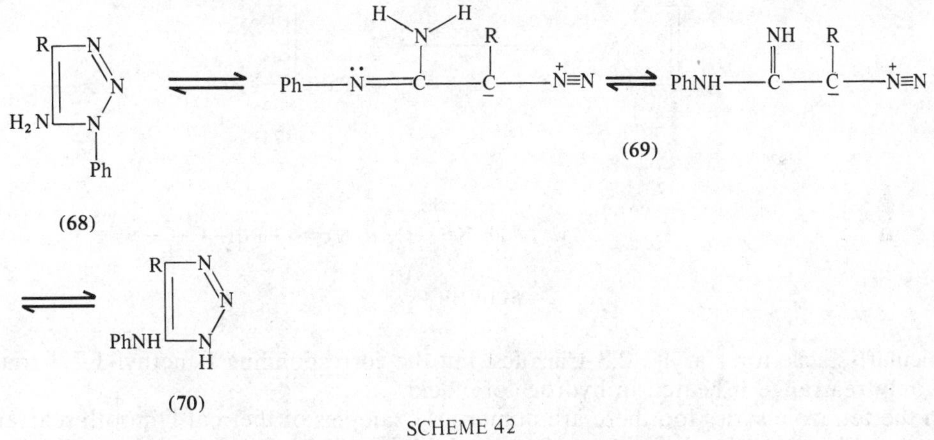

(68) (69) (70)

SCHEME 42

Although the equilibrium is established thermally, its position can be influenced by the solvent basicity. In general terms, the more basic the solvent, the greater the proportion of the more acidic NH triazole (**70**) which exists in the equilibrium mixture. Electron-attracting groups and bulky groups favour the tautomer which has them on the exocyclic nitrogen; alkyl groups favour the ring nitrogen. For example, 5-methylamino-1,2,3-triazole-4-carboxamide (**71**) is converted entirely into 5-amino-1-methyl-1,2,3-triazole-4-carboxamide (**72**) when heated at 160 °C for an hour, but even at 180 °C in basic medium the reverse reaction will not take place (Scheme 43). A benzyl group behaves similarly in neutral conditions, but in basic solution reverts to the ring-substituted compound (Scheme 44).

5-Mercapto-1,2,3-triazoles (**73**) and 5-amino-1,2,3-thiadiazoles (**74**) can be interconverted *via* a Dimroth rearrangement (Scheme 45). If the latter are heated in basic solvents they are converted to triazoles; the reverse occurs in acidic media. The reaction is

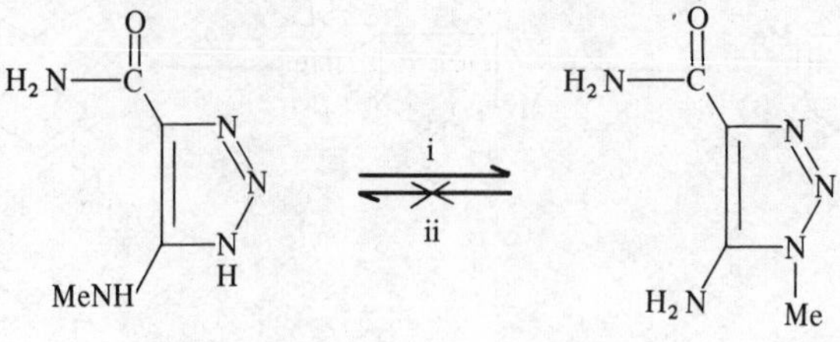

(71) (72)

i, 160 °C, 1 h; ii, 180 °C, NH_3, C_2H_5OH.

SCHEME 43

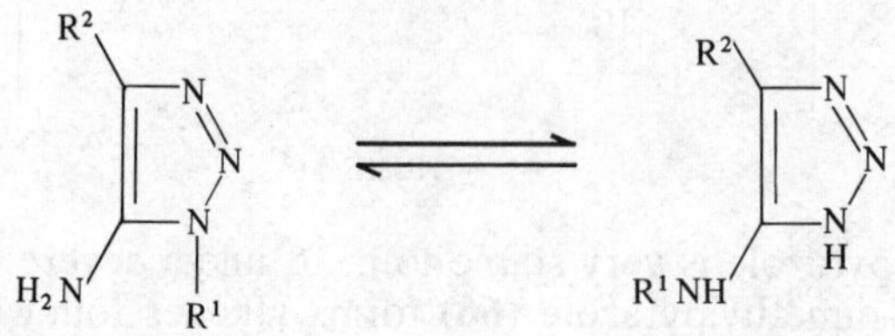

Favoured when

In neutral and basic solution (R^1 = Me)
In neutral solution (R^1 = CH_2Ph)

Favoured when

In basic solution (R^1 = Ar, CH_2Ph)

SCHEME 44

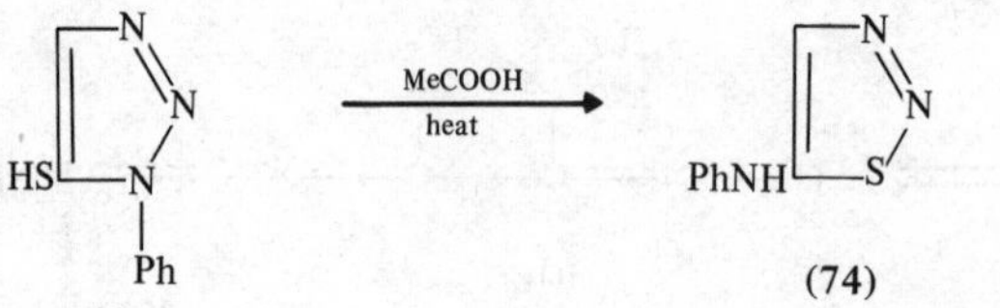

(73) (74)

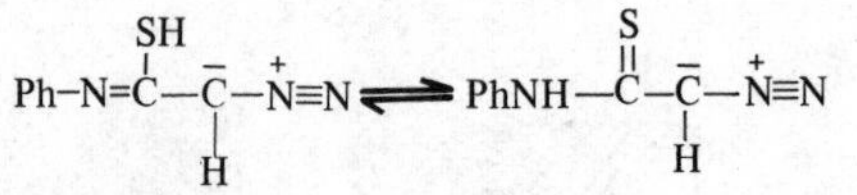

SCHEME 45

particularly facile for 1-aryl-1,2,3-triazoles, but the corresponding 1-methyl-1,2,3-triazole will only rearrange if heated in hydrochloric acid.

In the tetrazole series too there are numerous examples of thermal Dimroth rearrangements. Such a transformation was first noticed in 1953 when the intermediacy of a *C*-azidoformamidine ('guanyl azide') (**75**) was recognized (Scheme 46). Substituent and solvent effects parallel those in the triazoles.

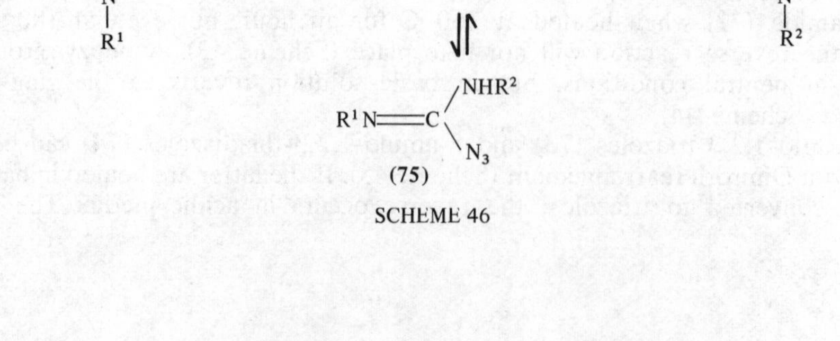

(75)

SCHEME 46

A similar transformation of tetrazoles is their conversion into oxadiazoles on boiling with acetic anhydride or acyl halides (brief heating with acetic anhydride can sometimes also promote a Dimroth rearrangement). The reaction is general for a variety of 5-alkyl-, -aryl-, or -amino-tetrazoles and proceeds through ring-opening of the initially formed 2-acyl-2*H*-1,2,3,4-tetrazole (**76**) to give an intermediate, which loses nitrogen to become an acyl nitrilimine (**77**). Cyclization of this gives the oxadiazole (**78**) (Scheme 47). The

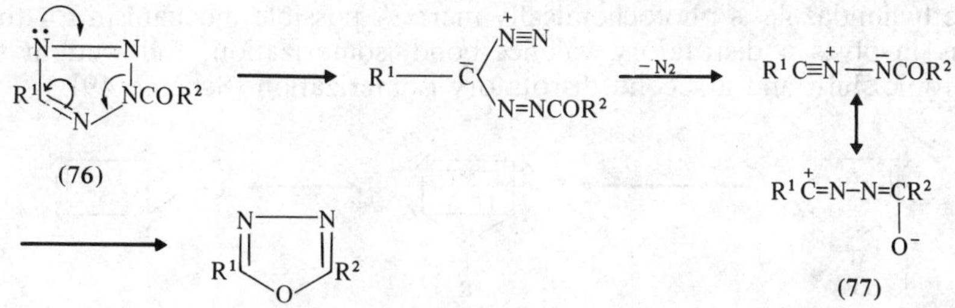

(76) (77) (78)

SCHEME 47

ring-opening has been shown to be the rate-determining step, and electron-releasing groups at C-5 and electron-attracting groups in the acyl function facilitate the reaction.

In principle 1,5-disubstituted tetrazoles can exist either as the cyclic tetrazoles (**79**) or as acyclic imide azides (azidomethines) (**80**). This constitutes an example of ring–chain tautomerism. Spectroscopic studies allow experimental distinction between these two

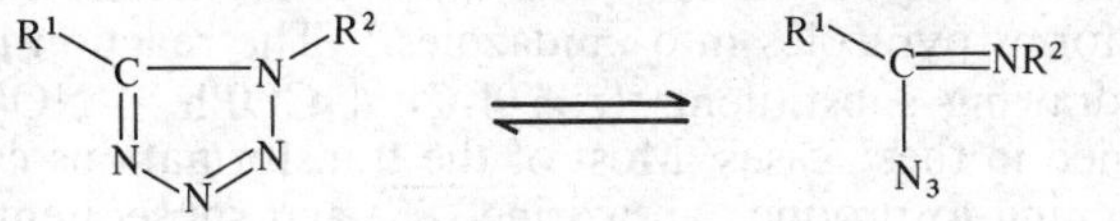

(79) (80)

forms, and as a general rule it appears that compounds with an aryl group on carbon and substituted on nitrogen by hydroxyl, methoxyl, or benzoyl are imide azides; compounds with aryl substituents on carbon and hydrogen, alkyl, or aryl groups on nitrogen are tetrazoles. That is, a strongly electron-withdrawing group on nitrogen stabilizes the acyclic form. The introduction of such an acceptor group at N-1 of 5-aminotetrazole induces

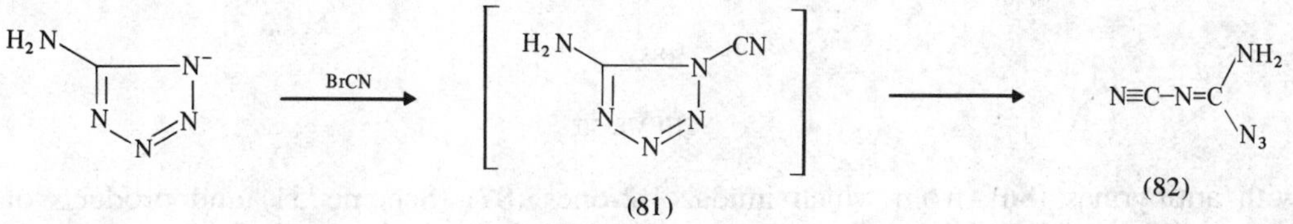

(81) (82)

SCHEME 48

ring-opening. A particular example of this is in the reaction of an alkali-metal salt of 5-aminotetrazole with cyanogen bromide. The intermediate 5-amino-1-cyanotetrazole (**81**) readily opens to form a cyanoguanyl azide (**82**) (Scheme 48). A further example of this phenomenon is provided by 4-azidopyrimidines (**83**), which exist as the tetrazolo isomers (**84**) in some solvents.

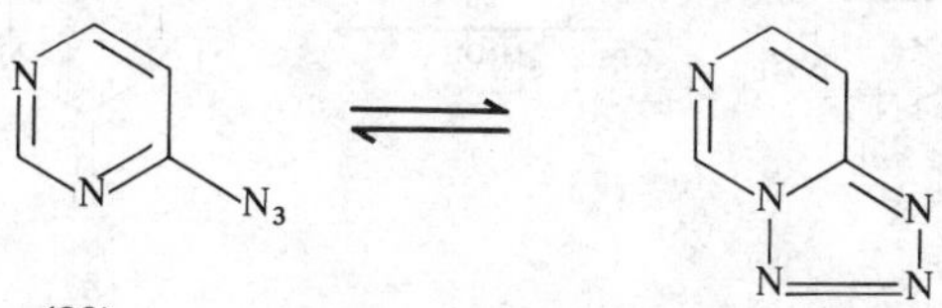

(83) (84)

17.3.10 PHOTOCHEMISTRY

Although imidazoles cannot be converted photochemically to pyrazoles, the reverse is true. Alkylimidazoles, though, are isomerized to other alkylimidazoles. In t-butanol, 1,4-dimethyl- and 1,2-dimethyl-imidazoles are interconvertible, and 1,4,5-trimethylimidazole gives the 1,2,5-isomer in a variety of solvents. In cyclohexene, however, 1,2,5-trimethylimidazole gives less than 5% of the 1,4,5-isomer, being largely destroyed. Furthermore, 1,2,4-trimethylimidazole is photochemically inert. A possible mechanism for these interconversions involves a disrotatory valence-bond isomerization, followed in turn by a 1,3-sigmatropic shift, and a second disrotatory isomerization (Scheme 49).

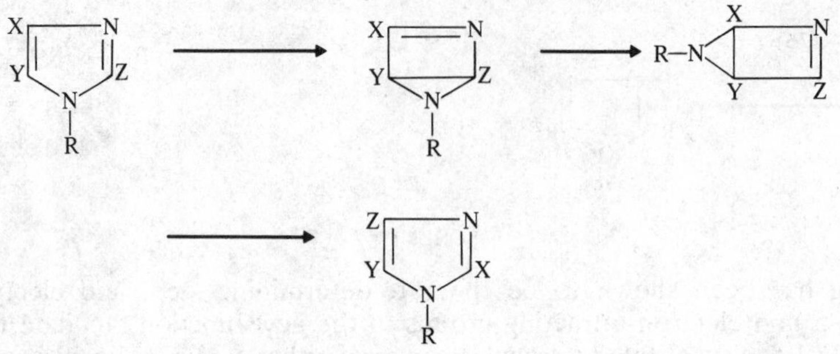

SCHEME 49

As mentioned above, one of the most important photochemical reactions of the azoles is that which transforms pyrazoles into imidazoles.[18] The reaction is fairly general except that electron-withdrawing substituents (*e.g.* 4-Cl, 1-COPh, 4-NO_2) inhibit the reaction and resins are formed in these cases. Most of the transformations can be accounted for in terms of a ring-opening to produce an azirine (**85**) and subsequent ring closure (Scheme 50). When allowance is made for tautomerism in *N*-unsubstituted pyrazoles, the consequence is an interchange of N-2 and C-3 of the pyrazole. Similar rearrangements also occur

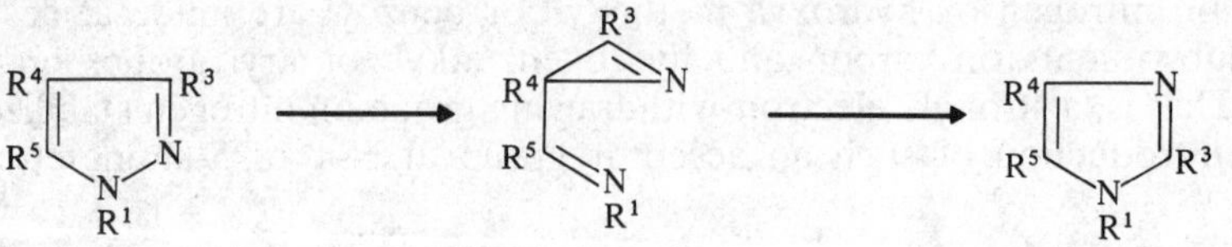

(85)

SCHEME 50

with antipyrines (**86**), from which imidazol-2-ones (**87**) (Scheme 51) and products of ring-opening result. While 1-alkylindazoles afford 2-alkylaminobenzonitriles by photolytic ring cleavage, 2-alkylindazoles rearrange in good yield to 1-alkylbenzimidazoles (Scheme 52). This reaction is surprisingly similar to the antipyrine photolysis.

There are several examples known of photochemical extrusion of nitrogen from the 1,2,3-triazole ring. When nitrogen is lost from a 1*H*-triazole the intermediate formed can

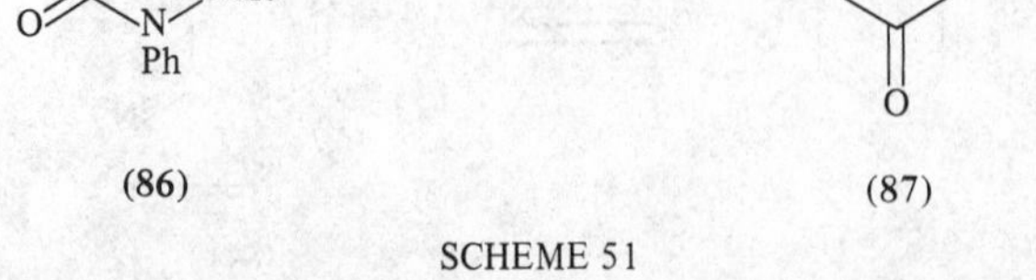

(86) (87)

SCHEME 51

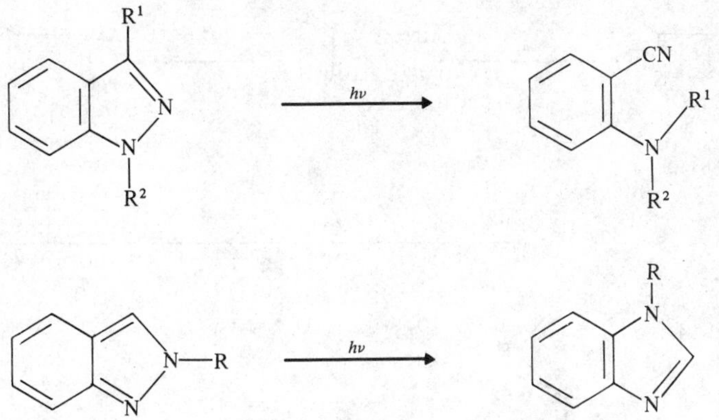

SCHEME 52

be written as an imino-carbene, a zwitterion, or a diradical (Scheme 53). Subsequent reactions of this intermediate which have been observed include: (i) a photochemical Wolff rearrangement, which implies that the intermediate may have a singlet carbene structure; (ii) ring closure involving the substituent R^1 (where R^1 = phenyl, benzoyl, or

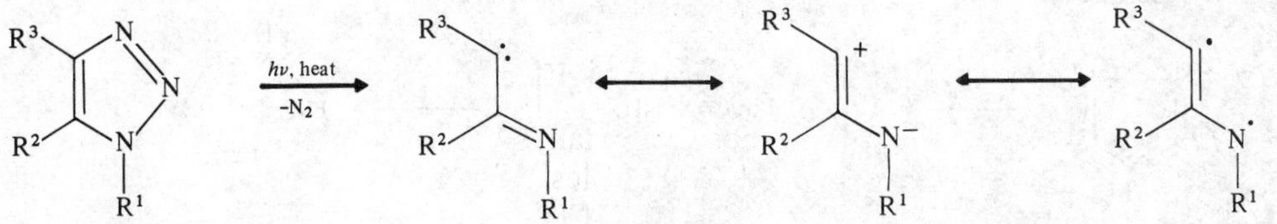

SCHEME 53

2-pyrimidinyl); (iii) ring closure to form a 1*H*-azirine, followed by rearrangement to a 2*H*-azirine isomer (Scheme 54). The last reaction seems to be a more favourable process than a Wolff rearrangement since no ketenimines are formed.

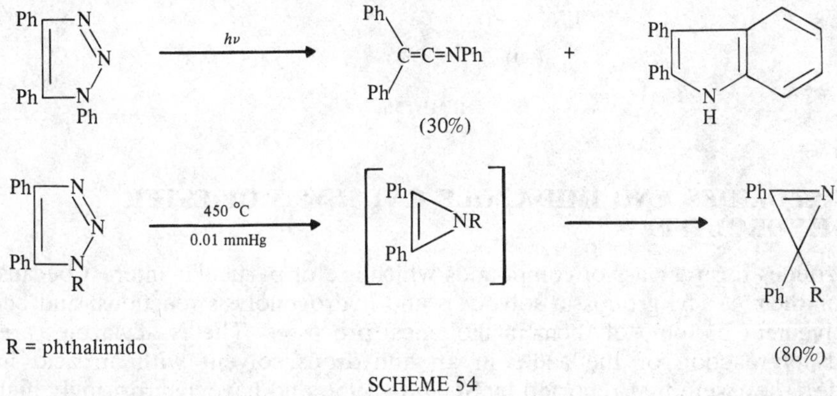

R = phthalimido

SCHEME 54

Although 2,4,5-triphenyl-2*H*-1,2,3-triazole has been reported to be photochemically stable, other 2*H*-1,2,3-triazoles may not be since 2,4,5-trimethyl-2*H*-1,2,3-triazole, when irradiated in ether, gives acetonitrile (22%) and a number of other products.

Mesoionic 4-aryl-1-phenyl-1,2,4-triazole-4-ones (**88**) photolyse to give aryl isocyanates, azo-compounds, and benzimidazoles. The last-named probably form from a Ph—N=C(R)N̈ fragment (Scheme 55).

The photolysis of 5-phenyltetrazole has been studied in some depth. In dioxan or tetrahydrofuran the products are nitrogen and 3,6-diphenyl-1,2-dihydro-1,2,4,5-tetrazine

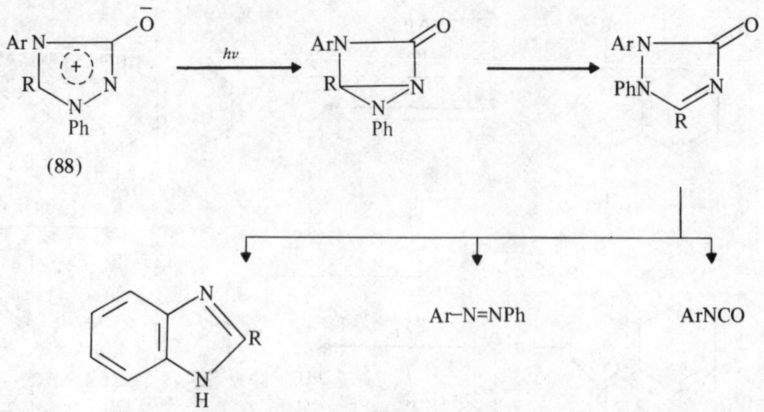

(88)

SCHEME 55

(this product is converted into 3,6-diphenyl-1,2,4,5-tetrazine (**89**) by oxidation during work-up, and into 3,5-diphenyl-1,2,4-triazole (**90**) and benzonitrile as secondary products) (Scheme 56). In the formation of the nitrilimine intermediate it can be shown by ^{15}N isotope studies that the nitrogen lost comes from the 3- and 4-positions of the original 5-phenyltetrazole.

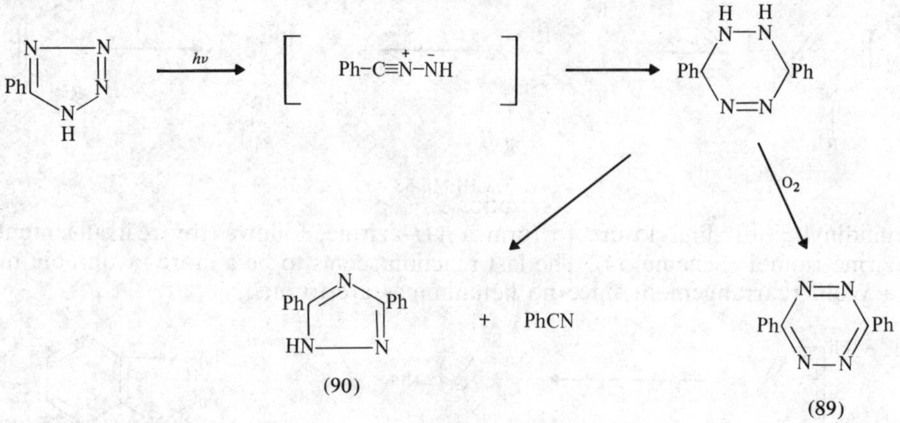

(90)

(89)

SCHEME 56

17.3.11 AZOLIDES AND IMIDAZOLE CATALYSIS OF ESTER HYDROLYSIS

The azolides form a class of compounds which are of particular interest because of the lability of their *N*-acyl groups in solvolysis and hydrogenolysis reactions, and because of the involvement of some of them in biological processes. The *N*-acylazoles are usually prepared by reaction of the azole in an anhydrous solvent with an acid halide or anhydride. They were first reported by Staab in 1956 and have a surprisingly high degree of reactivity in nucleophilic reactions, in which they compare well with acid halides and anhydrides with respect to reactivity. Detailed surveys of the chemistry of these compounds have appeared elsewhere.[2,19] The importance of the azolides has been extended to their use in peptide synthesis in which *N*-formyl- and *N*-benzyloxycarbonyl-imidazoles are employed as formylating and benzyloxycarbonylating agents.

When the hydrolysis rates of a series of azolides (see Table 5) are compared, it can be seen that an increase in the number of ring nitrogens increases the reactivity, and in the isomeric azolides, those with the greater number of adjacent nitrogens display lower reactivity (pyrazolides are hydrolysed more slowly than imidazolides; 1,2,3-triazolides

TABLE 5
The Hydrolysis of Azolides[a]

Azolide	$k' \times 10^5 (s^{-1})$	$\tau_{\frac{1}{2}}$(min)
1-Acetylpyrrole	$\rightarrow 0$	$\rightarrow \infty$
1-Acetylpyrazole	1.27	908
1-Acetylimidazole	28.1	41
2-Acetyl-1,2,3-triazole	43.5	26.6
1-Acetyl-1,2,4-triazole	180	6.4
2-Acetyltetrazole	>2000	>0.5

[a] Values of k' and the half-lives ($\tau_{\frac{1}{2}}$) for the hydrolysis of *N*-acetylazoles in conductivity water at 25 °C and pH 7.0.[19]

react more slowly than 1,2,4-triazolides). In comparison with other amides the electron pair of the amide nitrogen in the azolides is involved in the π-system, and this leads to a rather positive carbonyl carbon atom and consequent susceptibility to nucleophilic attack. The infrared spectra of *N*-acylazoles in general have carbonyl stretching bands at frequencies higher (*e.g.* 1-acetylimidazole, 1747 cm^{-1}) than those usually associated with aliphatic saturated ketones (1705–1725 cm^{-1}), and much higher than those for disubstituted amides (1650 cm^{-1}). This illustrates the minor importance of ring nitrogen–acyl group conjugation (**91**), a consequence of participation of the nitrogen atom in the aromatic structure of the ring. The increase in reaction rate caused by extra ring nitrogens

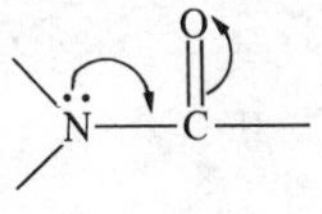

(91)

is a function of the increased delocalization of the unshared electron pair of the amide nitrogen, and the effect of adjacent annular atoms comes about because these cannot assume the same electron densities as those separated by a carbon atom.

The *N*-acylazoles are valuable synthetic intermediates for the preparation of acids, esters, amides, aldehydes, ketones, hydrazides, acyl halides, *etc.* (Scheme 57).

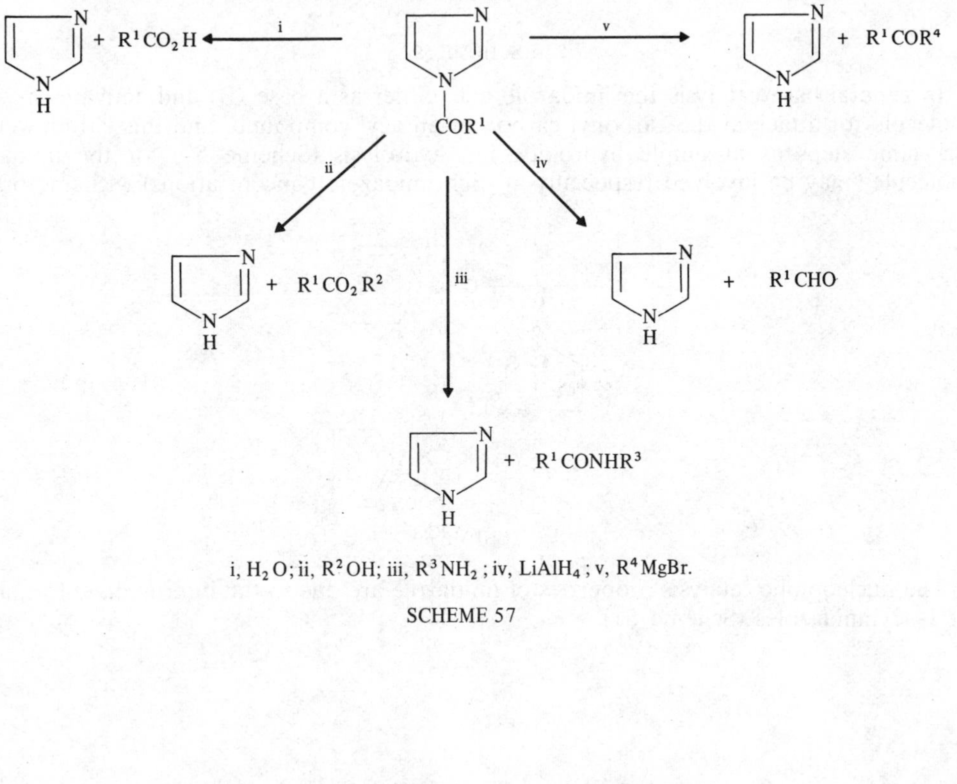

i, H_2O; ii, R^2OH; iii, R^3NH_2; iv, $LiAlH_4$; v, R^4MgBr.

SCHEME 57

Kinetic studies of some of these nucleophilic reactions show that aminolysis and alcoholysis occur by a bimolecular addition–elimination process, as does the neutral hydrolysis of the imidazolides of aryl carboxylic acids. Sterically hindered aliphatic carboxylic acid azolides appear to be hydrolysed in water by an S_N1 process, but some workers prefer an S_N2 process for the hydrolysis of 1-acetylimidazole (*cf.* Scheme 58).

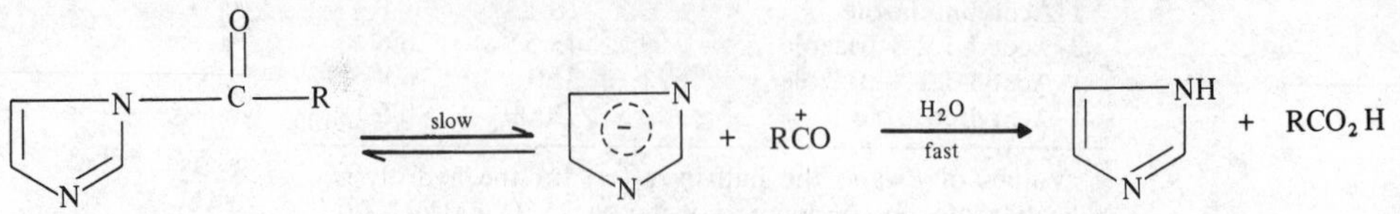

slow
$\overset{+}{RCO}$
H_2O
fast
+ RCO_2H

SCHEME 58

1-Acetylimidazole can be made enzymically, probably by the sequence:

Acetyl phosphate + coenzyme A ⇌ acetyl coenzyme A + phosphate
Acetyl coenzyme A + imidazole ⇌ 1-acetylimidazole + coenzyme A

Since the imidazolyl group of a histidine residue may be implicated in the mode of action of such hydrolytic enzymes as trypsin and chymotrypsin, much effort has been directed into the study of imidazoles as catalysts of ester hydrolysis. Two pathways have been found to be involved: *general base catalysis* and *nucleophilic catalysis*.

More specifically

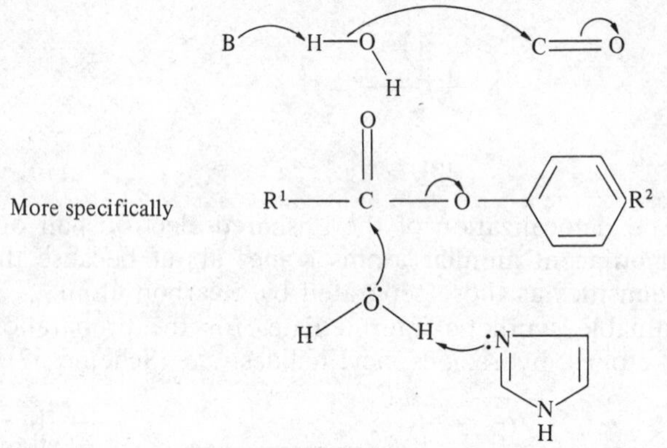

SCHEME 59

In general base catalysis the imidazole acts either as a base (B) and activates a water molecule for attack at the carbonyl carbon of an acyl compound, and this is followed by the same steps as in simple hydroxide ion hydrolysis (Scheme 59), or the imidazole molecules may be involved (especially at high imidazole concentrations) (Scheme 60).

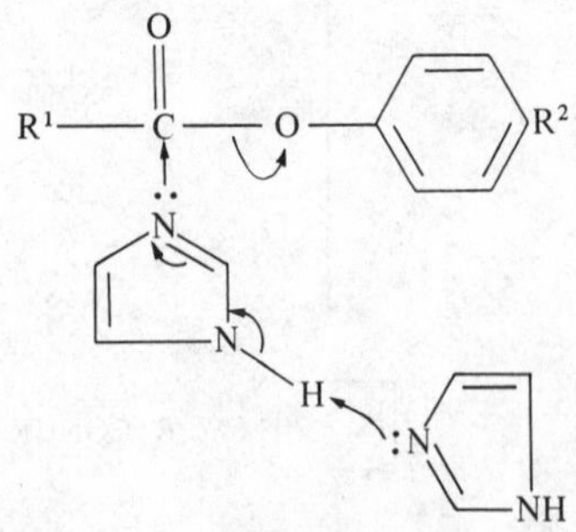

SCHEME 60

The nucleophilic catalysis properties of imidazole are due to the intermediate formation of 1-acylimidazoles (Scheme 61).

SCHEME 61

In the hydrolysis of esters with good leaving groups (*e.g.* *p*-nitrophenyl acetate) the imidazole neutral molecule is the effective catalyst, and the more basic the imidazole is, the more effective it is. Where the esters have poorer leaving groups (*e.g.* *p*-cresol acetate) the imidazole anion is involved and general base catalysis predominates. Thus, for imidazoles with $pK_a > 4$, catalysis by the anion is the main reaction. Imidazole is a much more effective nucleophile than amines in this type of reaction since it is a tertiary amine with minimal steric hindrance and with the facility to delocalize the positive charge resulting from the nucleophilic addition to the carbonyl carbon. Furthermore the 1-acylimidazole is highly reactive (see Table 5) in aqueous solution compared with other amides.

The transacylation reaction between a carboxylic acid and an *N,N′*-carbonyldi-imidazole (**92**) provides an important route to *N*-acylimidazoles. The diazolide of carbonic acid (**92**) exhibits very high reactivity towards nucleophilic reagents because of the electron-attraction from both sides of the carbonyl function. Although water hydrolyses *N,N′*-carbonyldi-imidazole at room temperature with vigorous evolution of carbon dioxide, the compound is crystalline and much more easily handled than phosgene, which has similar properties (Scheme 62). Variation of the nucleophile allows the transacylation reaction to be applied to the synthesis of esters, amides, hydrazides, hydroxamic acids, peptides, anhydrides, acyl halides, esters of peracids, *C*-acyltriphenylphosphine-alkylenes, and diacyl peroxides. The use of the reagent in peptide synthesis in particular gives good yields and the mild reaction conditions usually (but not always) minimize racemization.

(92)

i, RCO_2H; ii, imidazole.

SCHEME 62

17.3.12 AZOLE QUATERNARY SALTS

The quaternization of 1-substituted imidazoles and pyrazoles (see Section 17.3.5) occurs on the unsubstituted nitrogen. This can be demonstrated by alkaline fusion of the quaternary salt (Scheme 63). Such a reaction results in complete cleavage of the ring with the formation of 1,2-dialkylhydrazine — a product which defines the quaternary salt as the

$$\text{1,2-dimethylpyrazolium } I^- \xrightarrow[200\,^\circ C]{OH^-} \text{MeNHNHMe}$$

SCHEME 63

compound alkylated on both nitrogens. Rates of ethylation with ethyl iodide of 1-substituted imidazoles is in the expected order 1-CH_3>1-CH_2Ph>1-Ph. As could be deduced, the quaternization of imidazoles which have an electron-withdrawing substituent is more difficult.

Most quaternary salts are hygroscopic, soluble in polar solvents, and rather insoluble in non-polar solvents. Some are crystalline, but many are viscous liquids.

The quaternization process renders the ring much more susceptible to nucleophilic attack, *viz.* alkaline cleavage and nucleophilic displacements (Section 17.3.6). Much of the chemical interest in these compounds comes from their dealkylation reactions, which can be induced either thermally or with basic reagents. The thermal reaction presumably involves a nucleophilic substitution process (either S_N2 or S_N1) which is largely governed by steric factors. Usually the smaller alkyl group appears as alkyl halide, but the reaction (Scheme 64) is not simple as is seen by the order of cleavage of groups from imidazolium salts: allyl (2.8)>methyl (1.0)>benzyl (0.56)>ethyl (0.14)>butyl, propyl (0.04–0.09)> isopropyl (0.01–0.03) (relative rates compared with methyl in parenthesis).

$$\text{1-}R^1\text{-3-}R^2\text{-imidazolium } I^- \xrightarrow{\text{heat}} \text{1-}R^1\text{-imidazole} + R^2I$$
$$\text{1-}R^1\text{-3-}R^2\text{-imidazolium } I^- \xrightarrow{\text{heat}} \text{1-}R^2\text{-imidazole} + R^1I$$

SCHEME 64

Phenyl and vinyl groups do not cleave — they are conjugated with the hetero-ring. In addition, electron-withdrawing groups weaken an adjacent nitrogen–alkyl bond, while variation of the anion also affects product distribution. One complication of this reaction is the transalkylation process. When 1-alkylpyrazoles (or imidazoles) are heated in an open vessel at 150 °C with benzyl chloride the alkyl group is replaced by benzyl. Both decompositions of the quaternary salt (**93**) formed *in situ* are reversible, but the more volatile alkyl halide is removed from the reaction mixture, allowing the 1-benzyl-3,5-dimethylpyrazole to accumulate (Scheme 65). The reaction is of considerable synthetic

$$PhCH_2Cl + \text{1-R-3,5-dimethylpyrazole} \rightleftharpoons \text{1-R-2-}CH_2Ph\text{-3,5-dimethylpyrazolium } Cl^- \; (93) \rightleftharpoons \text{1-}CH_2Ph\text{-3,5-dimethylpyrazole} + RCl$$

(93)

R = Me, Et, Pr

SCHEME 65

importance since it provides a method of removal of an *N*-methyl group (the benzyl group which replaces it can be removed by reductive or oxidative procedures).

A further consequence of quaternary salt formation is that some azolium salts can lose a proton to form an ylide. For example, H-2 in imidazolium salts undergoes very rapid deuterium exchange *via* an ylide (**94**). The labile nature of this proton can be attributed to

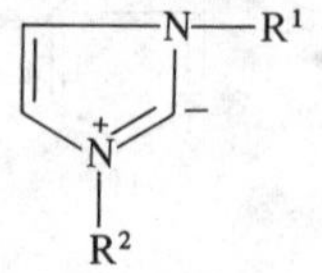

(94)

several factors: (i) the high *s*-character of the C—H bond (*cf.* acidity of acetylene); (ii) a favourable inductive effect (the reason HCN is a stronger acid than acetylene); (iii) resonance stabilization of the ylide. Pyrazolium salts exchange ring protons very much more slowly for they cannot stabilize the ylide as readily as imidazolium. In the latter the incipient carbanion is located between two adjacent nitrogen atoms which can share the positive charge (Scheme 66).

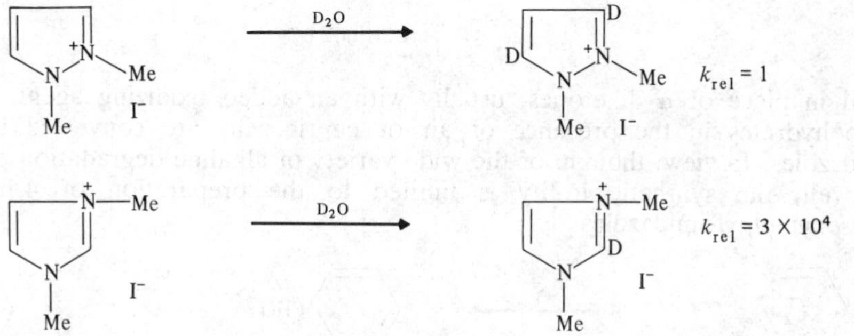

$k_{rel} = 1$

$k_{rel} = 3 \times 10^4$

SCHEME 66

17.3.13 SYNTHETIC METHODS

17.3.13.1 Imidazoles and benzimidazoles

One immediate problem is the absence of any one general method of wide application for imidazoles and benzimidazoles.[4] Scheme 67 illustrates some generally useful sequences.

The first method (i) involves formation of the 1,5-, 2,3-, 3,4- (and sometimes the 1,2-) bonds, and includes most of the early methods of imidazole synthesis. All of these reaction types are deficient in some way, such as the difficulty of synthesis of starting materials, low yields, and, more often than not, the formation of mixtures of products.

One method utilizes the reaction of an α-dicarbonyl compound with ammonia. Benzil reacts with aqueous ammonia to form 2,4,5-triphenylimidazole (**95**) (Scheme 68), formerly known as lophine. It seems likely that ammonia induces cleavage of the bond between the two carbonyl functions, producing benzaldehyde and ammonium benzoate (which is often dehydrated under the reaction conditions to benzamide). Condensation between benzil, benzaldehyde, and two molecules of ammonia then yields lophine. Under similar conditions, unsymmetrical dicarbonyl compounds give rise to mixtures of imidazoles. Pyruvaldehyde presumably undergoes alkaline cleavage to form acetaldehyde and formaldehyde since its products with ammonia are 4-methyl- and 2,4-dimethylimidazoles in about equal quantities. Acyloins (α-hydroxycarbonyl compounds) can be

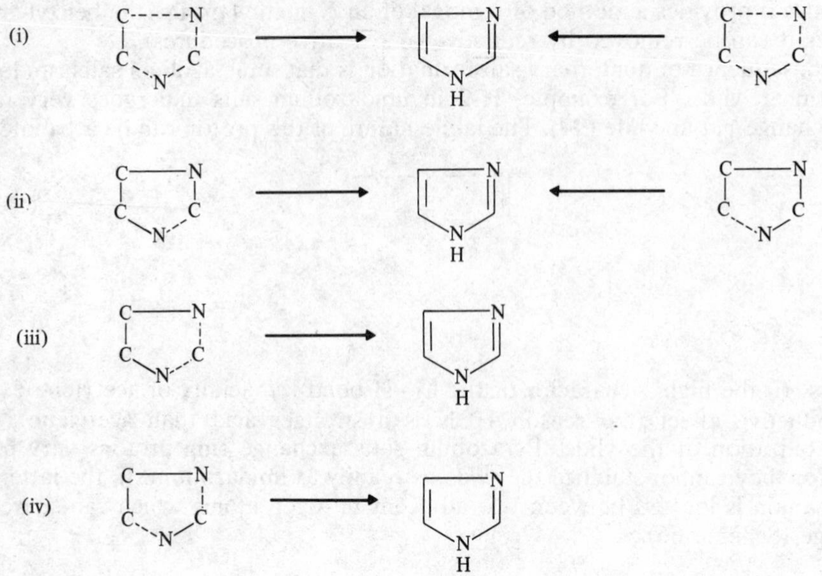

(i)

(ii)

(iii)

(iv)

SCHEME 67

used in place of α-diketones, usually with an added oxidizing agent. Thus, reducing carbohydrates in the presence of air or cupric salts are converted by ammonia to imidazoles. In view, though, of the wide variety of alkaline degradation products formed as well, the synthetic utility is limited to the preparation of 4-methyl- and 4-hydroxymethyl-imidazoles.[20]

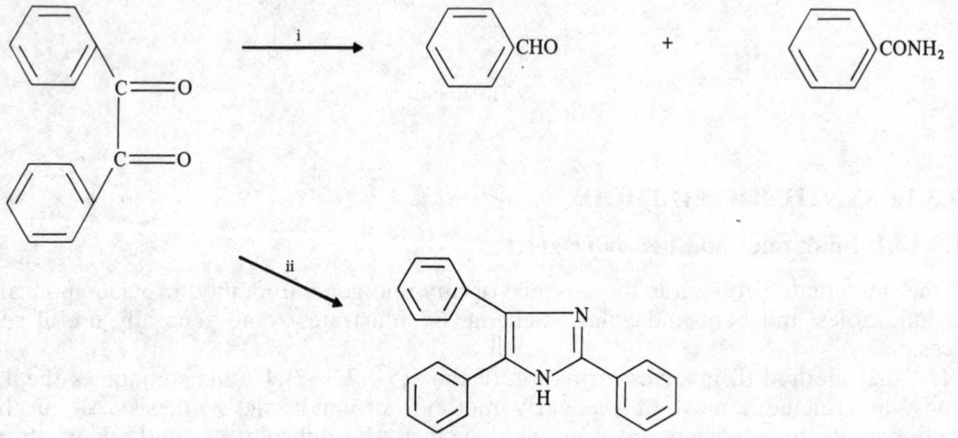

(95)

i, NH_3, H_2O; ii, $2NH_3$, PhCHO.

SCHEME 68

Probably the most useful advance in recent years has come in the 'formamide synthesis' developed by Bredereck and Theilig.[21] The interaction of α-diketones, α-hydroxy-, α-halogeno-, or α-amino-ketones, α-ketol-esters, or (under reducing conditions) α-oximinoketones with formamide usually results in high yields (40–90%) of imidazoles. Imidazole itself is prepared in 60% yield from the reaction of bromoacetaldehyde (as its glycol acetal), formamide, and ammonia at 180 °C (Scheme 69).

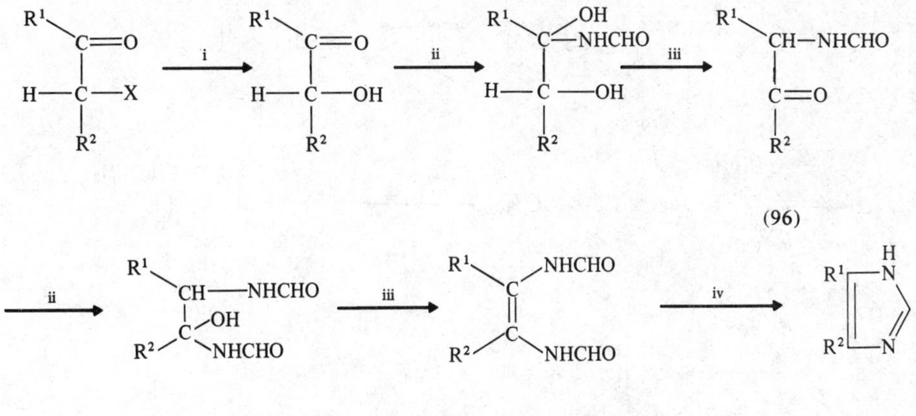

i, OH^-; ii, $HCONH_2$; iii, $-H_2O$; iv, $-HCO_2H$.

SCHEME 69

The α-formamidoketone (**96**) can be isolated and reacts with formamide to produce the imidazole. Oxazoles will form in preference in these reactions unless a large excess of formamide or ammonia is incorporated. Unfortunately, amides other than formamide (and occasionally acetamide) do not enter the reaction sequence, and so the reaction is limited mainly to the synthesis of 2-unsubstituted imidazoles. The difficulty of obtaining suitably substituted α-diketones or acyloins has been partially overcome by the reaction of an aldehyde with propane-1,3-dithiol. The stable dithiane anion formed can be converted to either of the desired starting materials.

In method (ii) the 1,2- and 1,5-bonds are made. The old Wallach synthesis ring-closed an *N,N'*-disubstituted oxamide with phosphorus pentachloride to give a chlorine-containing compound (Scheme 70) which is reducible by hydriodic acid to a 1-substituted imidazole.

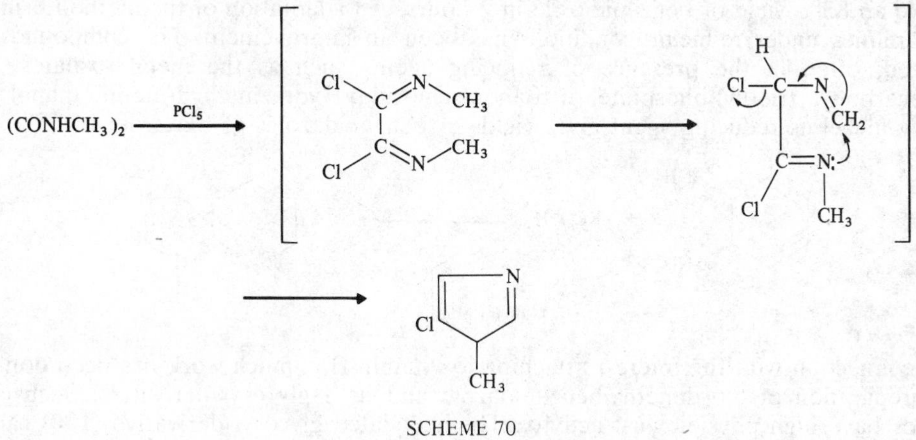

SCHEME 70

Ring-closure of formylglycine amidines (**97**) induced by heat or phosphoryl chloride leads to 5-aminoimidazoles, while imidates (**98**) condense with aminoacetaldehyde dimethyl acetal to form an amidine hydrochloride (**99**), which cyclizes to an imidazole (Scheme 71).

The cyclization of *N*-haloamidines with sodium ethoxide to form benzimidazoles is believed to involve a nitrene intermediate (Scheme 72).

The formation of the 1,2- and 2,3-bonds [method (iii)] utilizes an alkylene (or aryl) diamine with the appropriate carboxylic acid, alcohol, or aldehyde, generally at high temperatures in the presence of a dehydrogenating agent such as Pt/Al_2O_3. The reaction

R—NH—C(=NH)—CH_2NHCHO (97) →(i) 1-R-2-amino-imidazole (H_2N, N, R)

R—C(=NH·HCl)—OCH_3 (98) + $NH_2CH_2(OCH_3)_2$ → R—C(=NH)—NH—CH_2—$CH(OCH_3)_2$ (99)

→(ii) 2-R-imidazole (N, R, N–H)

i, 150 °C, 30 min or $POCl_3$, 90 °C, 30 min; ii, aq. HCl.

SCHEME 71

R—C(=N—Ph)—NHBr →($C_2H_5O^-$) [R—C(=N—Ph)—N̈:] → 2-R-benzimidazole (N, R, N–H)

SCHEME 72

is complicated experimentally as a method for preparing imidazoles, but occurs readily for benzimidazoles (Scheme 73). In fact 1,2-diaminobenzene reacts with formic acid at 100 °C to give an 83% yield of benzimidazole in 2 hours. An adaptation of the method using *o*-nitroanilines under reducing conditions has been an improvement. The compounds are cyclized, often in the presence of reducing agents such as the metal oxalates, iron pentacarbonyl, triethyl phosphite, or titanous chloride. Hydrazine hydrate in ethanol with Raney nickel as reducing agent gives yields of benzimidazoles of the order of 65%.

1,2-(NH_2)$_2$C$_6$H$_4$ + RCO_2H → 2-R-benzimidazole (N, R, N–H)

SCHEME 73

In connection with the interest attaching to vitamin B_{12}, much work has been done on the preparation of 5,6-dimethylbenzimidazole and its 1-glycosyl derivative. Such compounds have high physiological activity. The acetylated glycosyl derivative (**100**) can be prepared by the reactions shown (Scheme 74).

Even ketones will react with *o*-phenylenediamine. The primary reaction products, though, are benzimidazolines (or in the case of a diaryl ketone, a Schiff base) which thermally eliminate a hydrocarbon molecule to form the fully aromatic product.

The last major reaction sequence (iv) involves formation of the 1,5- and 2,3-bonds, and is exemplified by the Marckwald synthesis in which α-aminoketones react with cyanates, thiocyanates, or isothiocyanates. The products, 3*H*-imidazol-2-ones or 3*H*-imidazole-2-thiones, may then be converted to imidazoles by oxidation or dehydrogenation. As with other methods, the α-aminocarbonyl compounds have proved accessible only with difficulty, but are often prepared by reduction of α-amino-acids. The oxidative removal of the

i, RBr; ii, $2H_2$; iii, CS_2, $Ba(OH)_2$; iv, $HC(OEt)_3$; v, Raney Ni.

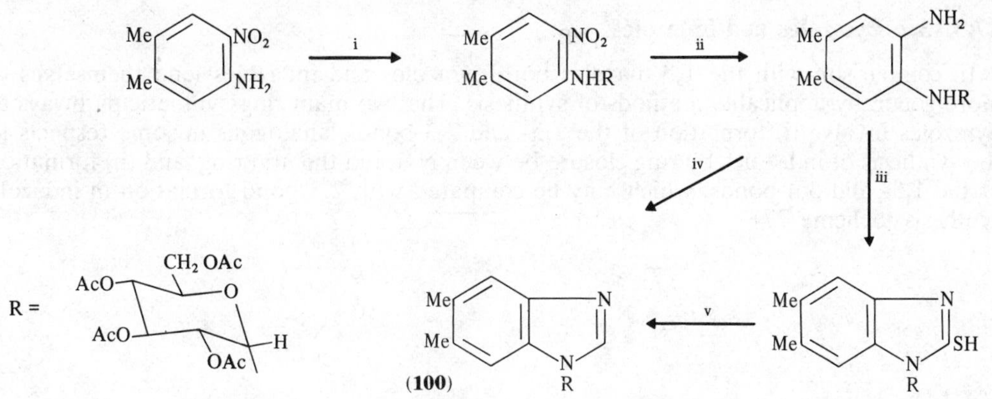

SCHEME 74

thione function is believed to proceed through the unstable sulphinic acid, which loses sulphur dioxide with great facility. The use of Raney nickel is now generally preferred (Scheme 75).

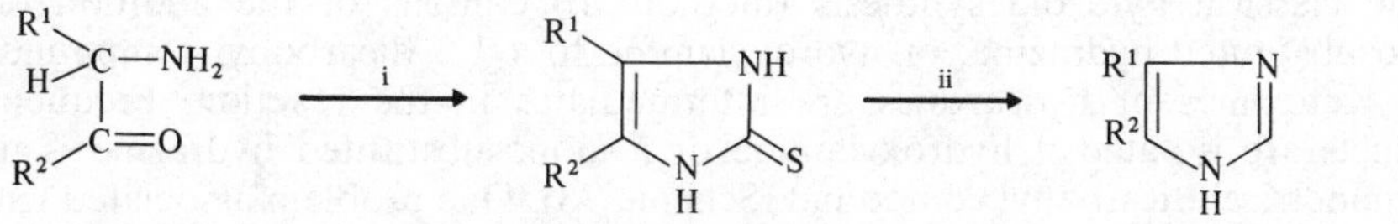

i, NCS^-; ii, Raney Ni or HNO_3.

SCHEME 75

A further type (iv) reaction leads to the synthesis of imidazole *N*-oxides (inaccessible by direct oxidation of the heterocycle). This reaction utilizes the interaction of an aldoxime or aldimine with an isonitrosoketone. The 1-hydroxyimidazole product (tautomeric with the *N*-oxide) predominates in non-polar solvents, but increasing amounts of the *N*-oxide form appear as the solvent polarity increases (Scheme 76).

Other more specific methods of imidazole synthesis have been detailed elsewhere.[4,6]

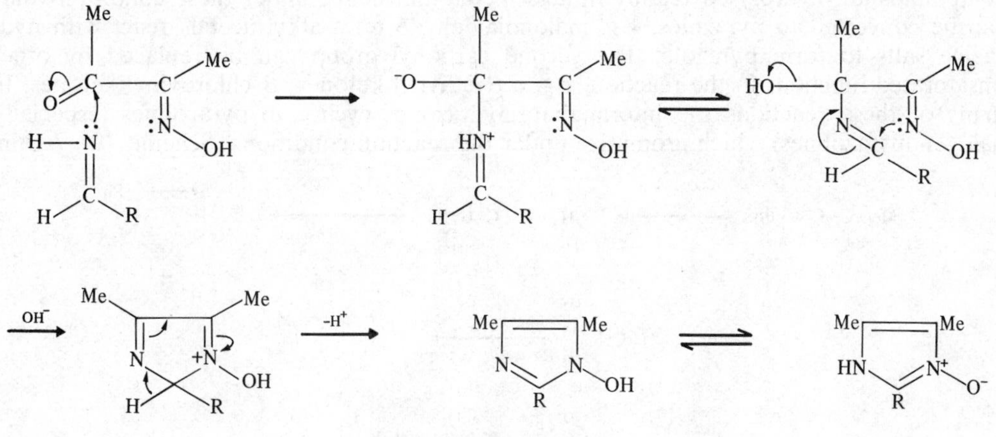

SCHEME 76

17.3.13.2 Pyrazoles and indazoles[3]

In comparison with the 1,3-diazoles both pyrazoles and indazoles lend themselves to more generally-applicable methods of synthesis. The two main ring-synthetic pathways to pyrazoles involve (i) formation of the 1,5- and 2,3-bonds, analogous in some respects to the synthesis of indazoles by ring closure between N-1 and the aryl ring, and (ii) formation of the 1,5- and 3,4-bonds, which may be compared with 2,3-bond formation in indazole synthesis (Scheme 77).

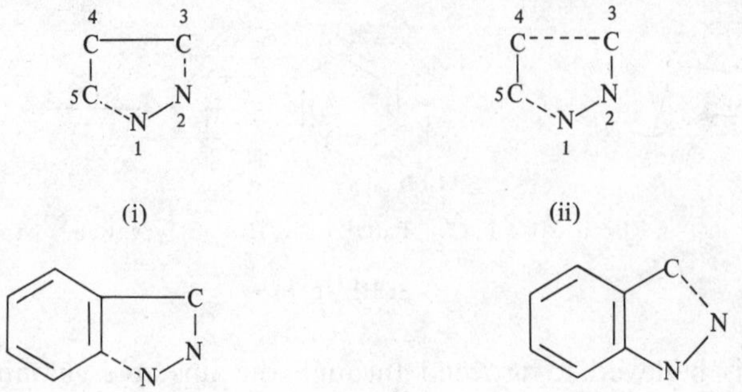

(i) (ii)

SCHEME 77

The classical pyrazole synthesis (method (i)) consists of the addition of hydrazine, monosubstituted hydrazine, or hydroxylamine to a 1,3-dicarbonyl compound (or precursor). Ketoximes or hydrazones are intermediates in the reaction. Frequently isomeric products are isolated if hydroxylamine or a monosubstituted hydrazine is used with an unsymmetrical dicarbonyl compound (Scheme 78). One problem associated with this mode

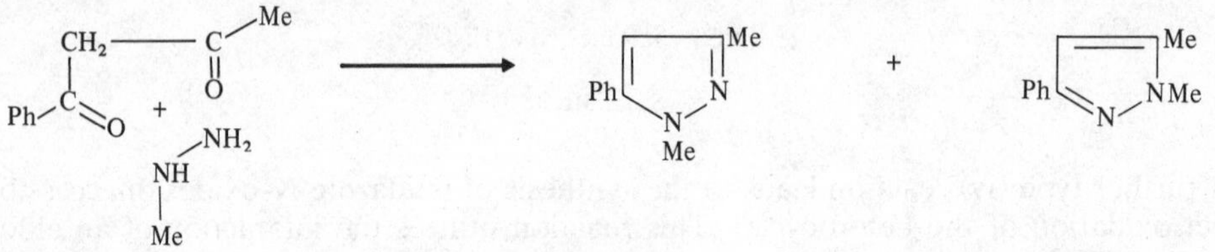

SCHEME 78

of synthesis is that the dicarbonyl compounds are commonly made by Claisen condensations, which may give mixtures of products. Occasionally such mixtures have not been recognized, resulting in confusing data on some pyrazoles. Acetals of 1,3-dicarbonyl compounds are hydrolysed readily in acid media and hence under these conditions they can be converted to pyrazoles, *e.g.* malondialdehyde tetra-alkyl acetals react with hydrazine salts to form pyrazole. The second carbonyl group can be replaced by other unsaturated functions in the reaction, *e.g.* α,β-ethynyl ketones, β-chlorovinyl ketones. In many of these reactions the intermediate hydrazones cyclize to pyrazolines (especially halogenopyrazolines) which aromatize under the reaction conditions (Scheme 79). Again,

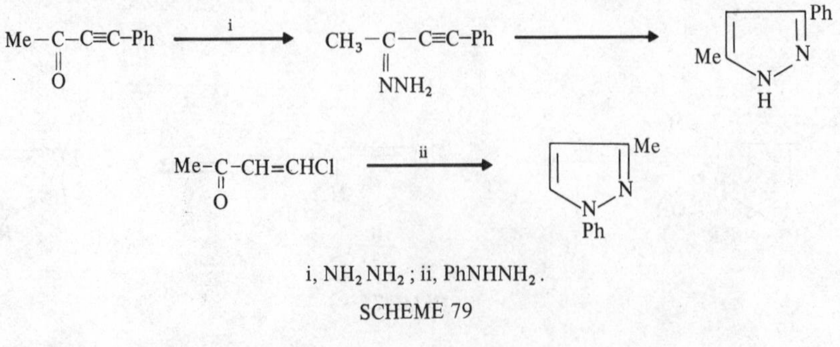

i, NH_2NH_2 ; ii, $PhNHNH_2$.

SCHEME 79

these methods often result in the formation of mixtures of isomers since there can be direct competition between the process leading to hydrazone function and that in which preliminary Michael-type addition of the hydrazine to the electron-deficient unsaturated linkage occurs. The preferred reaction is usually difficult to predict because the natures of the molecules involved, the solvent, and the temperature can all affect the reaction sequence.

Analogous syntheses of indazoles are from hydrazones of *ortho*-substituted aromatic aldehydes and ketones. When the *ortho*-substituent is halogen, the cyclization is accomplished at high temperatures in the presence of a carbonate, which facilitates elimination of hydrogen halide. Other *ortho*-substituents which can be eliminated on cyclization include alkoxyl and nitro (Scheme 80).

CH=NNHCOPh, O_2N, Cl → (K_2CO_3, heat) → O_2N, N, N, COPh

SCHEME 80

Type (ii) syntheses of pyrazoles comprise the 1,3-dipolar addition of a diazoalkane to an acetylene which has its triple bond activated by an electron-withdrawing substituent. The diazoalkane can be considered as a compound which displays both electrophilic and nucleophilic reactivity at the 1- and 3-positions. The methyl ester of phenylacetylenecarboxylic acid (methyl phenylpropiolate) reacts with methyl diazoacetate to form the two possible isomeric pyrazoles (**101**, **102**) in about equal quantities (Scheme 81). The place of acetylenes in these reactions may be taken by an alkene substituted by a group capable of elimination during the reaction. Thus, aliphatic diazo-compounds and β-chlorovinyl ketones give 3-acylpyrazoles *via* the unstable 4-chloropyrazolines.

(a) H–C≡C–CHO + $^-N{=}N{-}^+CH_2$ → (Et_2O, 0 °C) → CHO, N, N–H

(b) Ph–C≡C–CO_2Me + N_2CHCO_2Me → Ph, CO_2Me, MeO_2C, N, N–H **(101)** + MeO_2C, Ph, MeO_2C, N, N–H **(102)**

SCHEME 81

The most frequently used route to the indazole ring system (Ref. 3, p. 294) consists of the diazotization of suitably substituted anilines which have a hydrocarbon group in the *ortho*-position (Scheme 82). The *o*-toluidine starting materials appear to react best when they have nitro-substituents. When these are absent, alkaline medium best promotes the cyclization, which is presumably initiated by proton abstraction from the methyl group. Cyclization can also take place with diazotized *o*-aminoketones or *o*-aminonitriles when treatment with a reducing agent follows. Somewhat similar reactions occur when *N*-nitroso-*o*-toluidines are heated, and from azo-compounds. If a methyl group is situated *ortho* to the azo-linkage, cyclization is facile on heating with a nitroso-compound and sodium carbonate (Scheme 83). A nitro-group *para* to the methyl assists the reaction, which leads to a 2-substituted indazole (**103**).

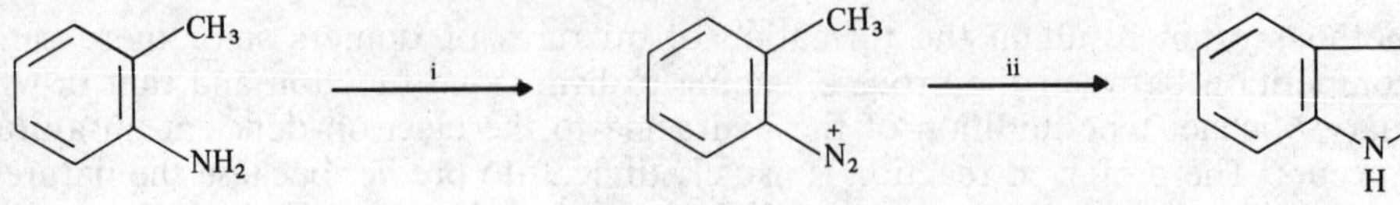

(92%)

i, HNO_2 ; ii, $Me_4\overset{+}{N}\,OAc^-$

SCHEME 82

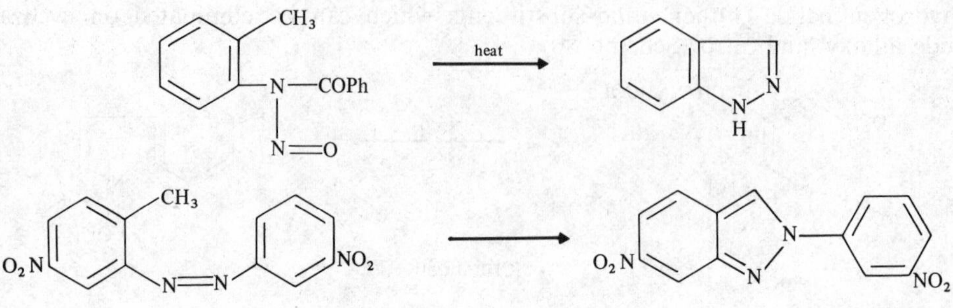

(103)

SCHEME 83

Methods of synthesis which form the N—N bond are of importance in indazole chemistry, but not in the formation of pyrazoles. The reduction of *o*-nitrobenzylanilines provides an example of the synthesis of 2-arylindazoles (Scheme 84).

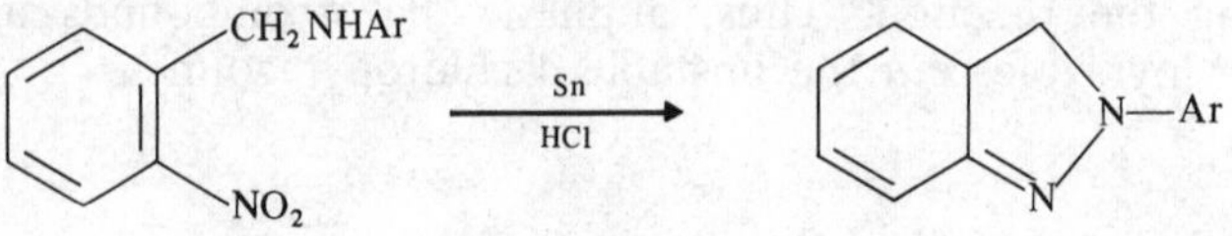

SCHEME 84

Other heterocyclic compounds can sometimes act as sources of pyrazoles. The aromatization of pyrazolines has already been mentioned. Sydnones react with unsaturated compounds to yield pyrazoles, and γ-pyrones react with hydrazines to open the ring and form pyrazoles in a general reaction which works for chromones, flavones, and flavonones, as well as their sulphur analogues, *e.g.* Scheme 85.

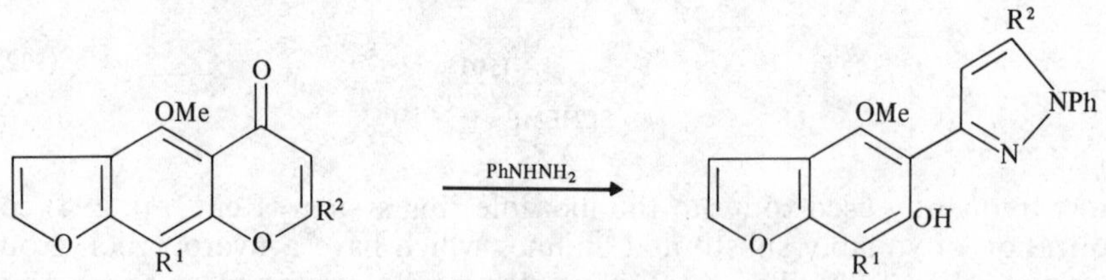

SCHEME 85

17.3.13.3 1,2,3-Triazoles and benzotriazoles[5,22]

The methods of ring formation can again be divided into a number of main classes in spite of some classification difficulties: (i) formation of the 1,2-bond; (ii) formation of the 1,5- and 3,4-bonds; (iii) formation of the 1,2- and 4,5-bonds (Scheme 86).

(i) (ii) (iii)

SCHEME 86

The N—N bond is formed [method (i)] particularly in the synthesis of 2*H*-1,2,3-triazoles from α-diketone bis-arylhydrazones. This very general reaction has been employed extensively for the preparation of sugar 'osotriazoles' from osazones; these osotriazoles are capable of ready degradation to simple 2*H*-1,2,3-triazoles, and this was the original method by which von Pechmann in 1888 prepared the first mononuclear triazoles. A possible mechanism is given in Scheme 87. The reaction requires an oxidizing agent

SCHEME 87

such as copper(II) salts, manganese dioxide, iodine, mercuric oxide, or acid dichromate. An analogous route is provided by the intramolecular elimination of water between adjacent hydrazone and oxime functions (Scheme 88). If, however, an oxidant is incorporated, the product is a 1,2,3-triazole *N*-oxide. When bis-hydrazones of α-diketones are oxidized, 1-amino-1*H*-1,2,3-triazoles are formed. Unfortunately, unless the oxidation is

SCHEME 88

carefully controlled the hydrazone may be completely oxidized to yield an acetylenic product. Partial avoidance of this problem can be achieved by using toluene-*p*-sulphonylhydrazones (**104**), which cyclize either in acid or base (Scheme 89). However, two isomeric products are sometimes formed, and in some instances base-catalysed cyclization still results in alkyne formation.

1*H*-Triazoles and unsubstituted 1,2,3-triazoles can be approached by ring-closure of α-diazoimines and α-diazoamides. The former are available *in situ* by reaction of α-diazoketones with amines (Scheme 90).

Benzotriazoles are commonly prepared by ring-closure of *o*-aminodiazonium salts, *o*-aminoazo-compounds, or, under oxidizing conditions, *o*-aminohydrazines. The corresponding *o*-nitro-compounds can be used if a reducing agent is incorporated (Scheme 91).

Method (ii) is exemplified by the reactions of azides with a wide variety of functionalized carbon compounds such as acetylenes, active methylene compounds, alkoxides, α-acylphosphorus ylides, enamines, enol ethers, and suitably substituted alkenes. The

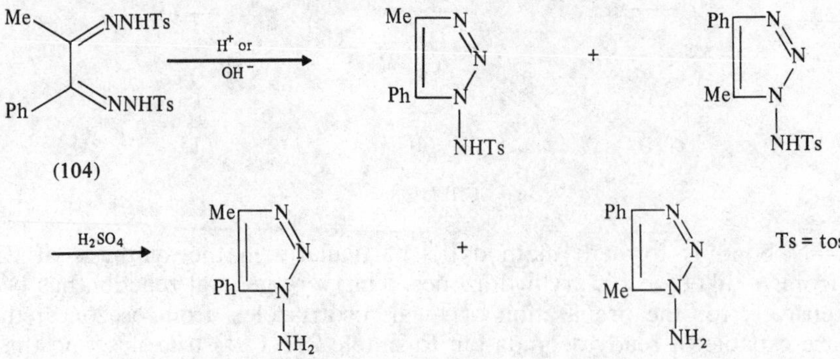

SCHEME 89

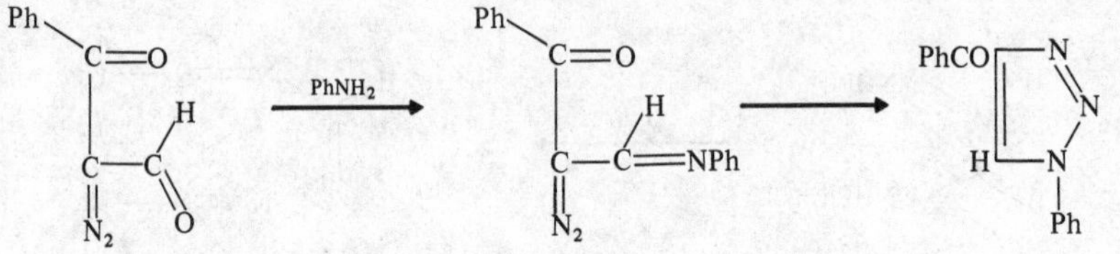

Ts = tosyl

SCHEME 90

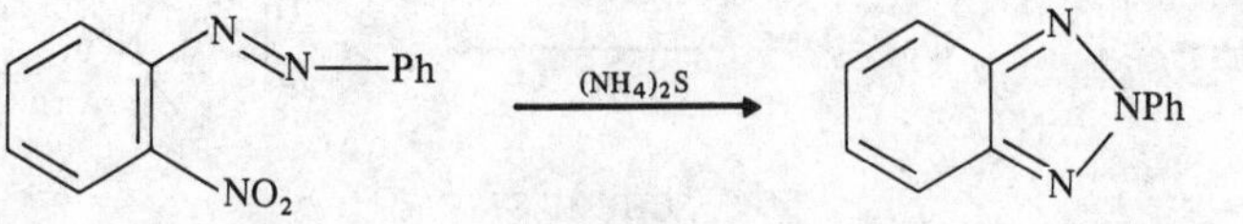

SCHEME 91

variety of azides which react include alkyl and aryl azides, trimethylsilyl azide, sodium azide, sulphonyl azides, and alkoxycarbonyl azides.

The accepted mechanism for reaction of an acetylene with an azide is a concerted 1,3-dipolar addition, which suffers from a number of preparative disadvantages. Firstly, many reactions require elevated temperatures which can lead to decomposition (often explosive) of the azide. A second deficiency is that unsymmetrical alkynes commonly give rise to two isomeric products (Scheme 92). It has been possible, though, to make some

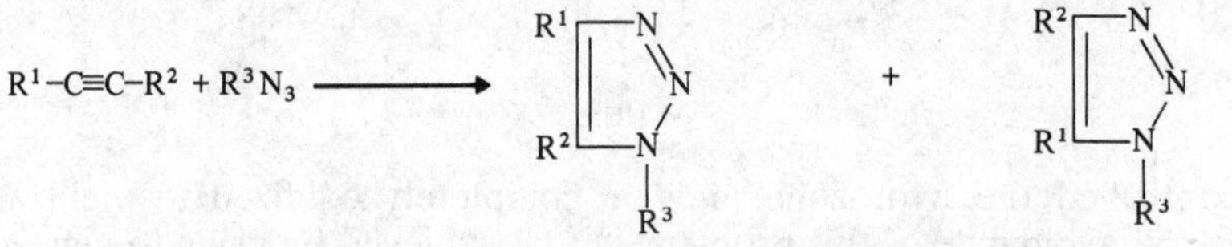

$$R^1{-}C{\equiv}C{-}R^2 + R^3N_3 \longrightarrow$$

SCHEME 92

generalizations as to the regiospecificity of the reaction: electron-withdrawing groups tend to appear in the 4-position, and electron-releasing groups in the 5-position; bulky groups usually occupy the less hindered 4-position. Furthermore, it is sometimes possible to accomplish regiospecific synthesis of 5-substituted triazoles from terminal acetylenes by using the Grignard derivative or the acetylide under mild conditions. The reaction mechanism which applies in these latter instances may not be of the thermal 1,3-dipolar type, but, rather, may involve nucleophilic attack of the acetylene anion on the terminal azide nitrogen, followed by 1,5-anionic cyclization to the triazolyl anion (**105**) (Scheme 93). Again, reactions which involve hydrazoic acid (or sodium azide) are unlikely to be 1,3-dipolar additions and probably follow a similar pathway to that below.

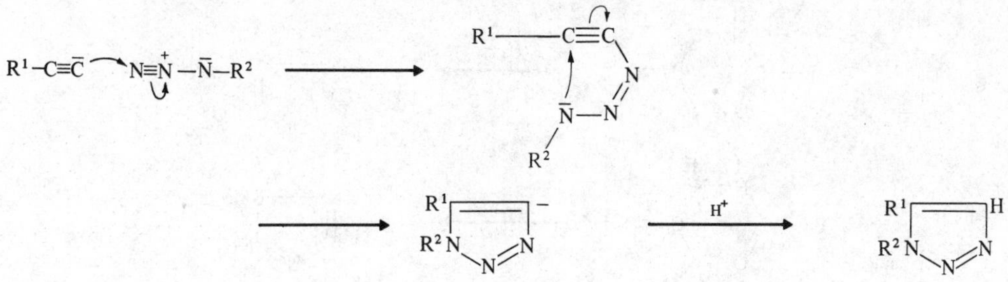

(105)

SCHEME 93

Active methylene compounds react in a base-catalysed condensation with azides to form 1*H*-1,2,3-triazoles. This is the most useful route to compounds bearing 5-hydroxy- or -amino-groups, and aryl or carbonyl functions at C-4. The general mechanism can be regarded as proceeding through the triazoline (**106**) which aromatizes (Scheme 94).

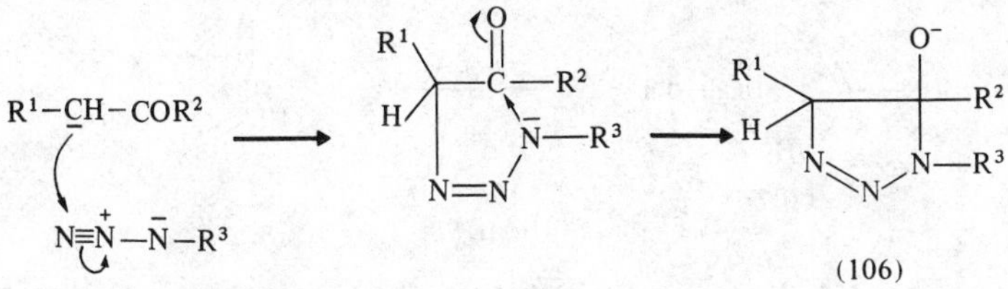

(106)

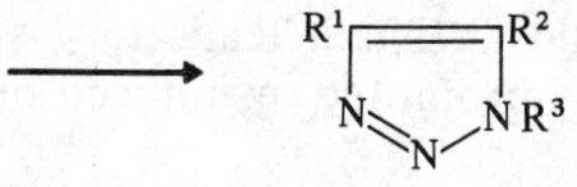

SCHEME 94

The formation of the 1,2- and 4,5-bonds [method (iii)] provides another example of 1,3-dipolar addition. Diazoalkanes can react with activated nitriles such as cyanogen, cyanogen halides, and cyanic acid esters (Scheme 95). With less activated nitriles, catalysts such as aluminium alkyls are necessary.

Synthetic routes to 1,2,3-triazoles which involve rearrangement or degradation of other heterocyclic systems have been discussed elsewhere.[5,22]

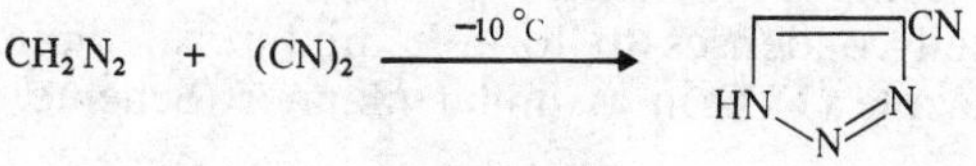

$$CH_2N_2 + (CN)_2 \xrightarrow{-10\,^{\circ}C}$$

SCHEME 95

17.3.13.4 1,2,4-Triazoles[22]

Major methods of synthesis include: (i) formation of the 3,4-bond; (ii) formation of the 3,4- and 4,5-bonds; (iii) formation of the 2,3- and 3,4-bonds; (iv) formation of the 1,5- and 2,3-bonds; and (v) formation of the 1,5- and 3,4-bonds (Scheme 96).

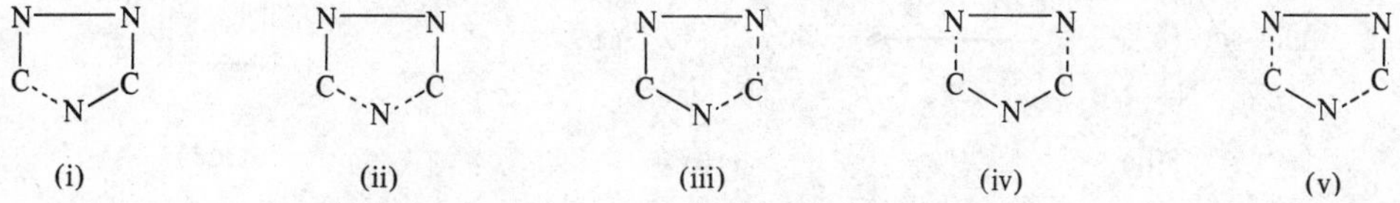

(i) (ii) (iii) (iv) (v)

SCHEME 96

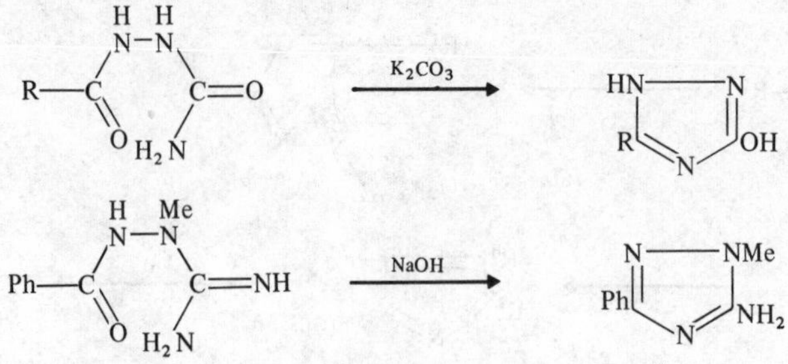

SCHEME 97

One of the most common routes [method (i)] is the intramolecular condensation in alkaline medium of acyl derivatives of semicarbazides, thiosemicarbazides, or aminoguanidines (Scheme 97). In acid conditions, acylthiosemicarbazides tend to be transformed into thiadiazoles rather than into triazoles, but there are some examples of successful syntheses. The oxidative cyclization of benzaldehyde thiosemicarbazone (**107**) gives rise to a mercaptotriazole (Scheme 98). Sometimes the ring-closures can be induced thermally.

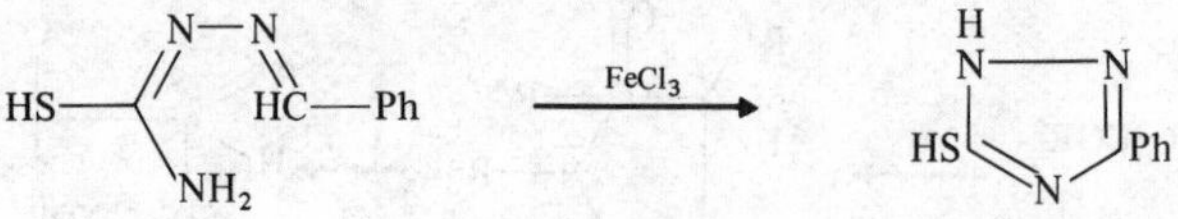

(107)

SCHEME 98

An example of method (ii) is provided by the intermolecular condensation of diacylhydrazines with amines. The value of this method is that it leads to 4-substituted 4*H*-1,2,4-triazoles. Again cyclization may be thermally promoted or induced by a dehydrating agent (Scheme 99).

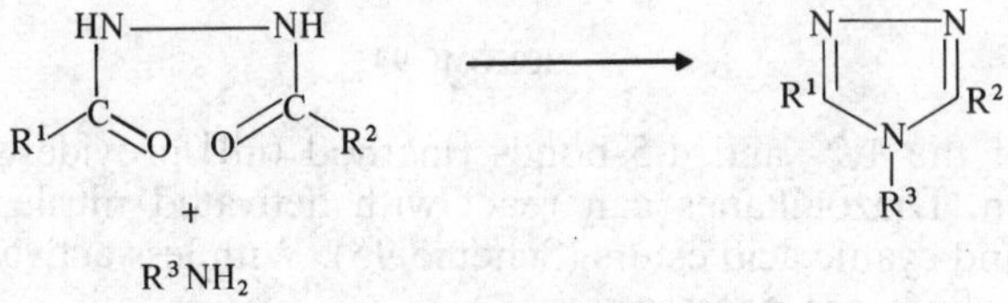

SCHEME 99

The 2,3- and 3,4-bonds form [method (iii)] when amidrazones (**108**) or hydrazidines cyclize with carbonyl compounds such as acid chlorides or anhydrides. Phenylbenzimidrazone (**109**) self-condenses to form a mixture of diphenyltetrazine (**110**) and 3,4,5-triphenyl-1,2,4-triazole (**111**) in a similar manner (Scheme 100).

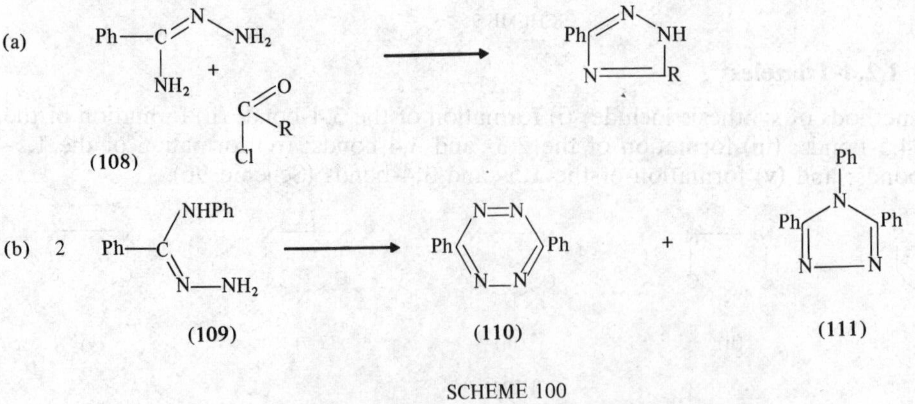

(a) (108)

(b) 2 (109) (110) (111)

SCHEME 100

SCHEME 101

The Einhorn–Brunner reaction of hydrazines with diacyl amines in mildly acid conditions gives rise to 1,2,4-triazoles by forming the 1,5- and 2,3-bonds [method (iv)] (Scheme 101). The reaction seems very general and has the advantage that unsymmetrical diacylamines do not produce mixtures of isomeric triazoles. In addition, the corresponding thioacyl or imino-compounds will take part in the reaction.

The Pellizzari reaction of hydrazides and amides at high temperature in the absence of solvent produces 1,2,4-triazoles by formation of the 1,5- and 3,4-bonds [method (v)]. Complications include concurrent dehydration of the amide and sometimes aryl interchange between amide and hydrazide (Scheme 102).

SCHEME 102

Besides the ring-synthetic methods there are several molecular rearrangements which can result in 1,2,4-triazole formation. Ring contractions of tetrazines and the pyrolysis of 5-substituted tetrazoles are examples.

17.3.13.5 Tetrazoles[23]

Major synthetic pathways can be classified as: (i) formation of the 1,2-bond (or 1,2- and 4,5-bonds); (ii) formation of the 2,3-bond; (iii) formation of the 2,3- and 4,5-bonds; and (iv) formation of the 2,3- and 3,4-bonds (Scheme 103).

SCHEME 103

Either hydrazoic acid or azides add to compounds with carbon–nitrogen unsaturation [*e.g.* cyanides, cyanates, thiocyanates, and carbodi-imides — method (i)]. Presumably the intermediates formed are imide azides which cyclize to the tetrazole (Scheme 104) (imide azide–tetrazole tautomerism has been discussed in Section 17.3.9). Although yields of 5-alkyltetrazoles are high, the reaction may require several days to reach completion. The problem of handling hydrazoic acid solutions can be overcome in a number of ways: by generating it *in situ* from sodium azide, or by the use of ammonium azide. A general

$$\mathrm{RCN} + \mathrm{HN_3} \longrightarrow \mathrm{R{-}C{=}NH} \ (\mathrm{N_3}) \rightleftharpoons \text{tetrazole}$$

SCHEME 104

$$\mathrm{Ph{-}C(OEt){=}NH}\ \xrightarrow{HN_3}\ \text{5-phenyltetrazole}$$

(112)

SCHEME 105

synthesis of 1-monosubstituted or 1,5-disubstituted tetrazoles involves reaction between imide chlorides and azides. Initially there is nucleophilic displacement of chlorine by azide to form an imide azide. Groups other than chlorine can also be displaced, *e.g.* ethyl benzimidate (**112**) loses its ethoxy-group *en route* to 5-phenyltetrazole. (Scheme 105).

Among the most valuable synthetic approaches to tetrazoles is the Schmidt reaction between ketones and hydrazoic acid in the presence of strong acids. When one mole of hydrogen azide reacts with one mole of a carbonyl compound an *N*-substituted amide is formed; two or more moles of hydrogen azide result in tetrazole formation (Scheme 106).

$$R_2C{=}O + H^+ \longrightarrow R_2\overset{+}{C}OH$$

$$R_2\overset{+}{C}OH + HN_3 \longrightarrow R_2C(OH){-}HN{-}\overset{+}{N}{\equiv}N \longrightarrow R_2C{=}N{-}\overset{+}{N}{\equiv}N + H_2O$$

$$R_2C{=}N{-}\overset{+}{N}{\equiv}N \longrightarrow R\overset{+}{C}{=}NR + N_2$$

$$R\overset{+}{C}{=}NR + HN_3 \longrightarrow \text{1,5-disubstituted tetrazole (R, NR)} + H^+$$

SCHEME 106

In effect, the reaction is analogous to reactions of type (i). The acid catalyst converts the carbonyl compound to a carbonium ion, which combines readily with one molecule of hydrogen azide. The intermediate dehydrates and rearranges to an imino-carbonium ion, with simultaneous loss of nitrogen. A futher molecule of hydrogen azide then combines with the imino-carbonium ion.

Such ions are also generated in the course of the Beckmann rearrangement of oximes, and when nitriles are exposed to the effects of acids. Thus oximes and nitriles can be sources of tetrazoles when treated with hydrogen azide. In Scheme 107, cyanide migration occurs at the time of nitrogen elimination.

$$RC{\equiv}N + H^+ \longrightarrow R\overset{+}{C}{=}NH$$

$$R\overset{+}{C}{=}NH + HN_3 \longrightarrow R{-}C({=}NH){-}HN{-}\overset{+}{N}{\equiv}N \longrightarrow RNH{-}\overset{+}{C}{=}NH + N_2$$

$$RNH{-}\overset{+}{C}{=}NH + HN_3 \longrightarrow \text{5-amino-1-R-tetrazole } (H_2N, NR) + H^+$$

SCHEME 107

Oxidative ring-closure of formazans provides an example of type (ii) synthesis in which the tetrazole ring is formed by closure of the 2,3-bond. Formazans are made by coupling diazonium salts with aldehyde arylhydrazones (or sugar osazones). A variety of mild oxidizing agents convert the formazans (**113**) to 2,3-disubstituted tetrazolium salts (**114**) (Scheme 108).

$$PhCH{=}NNHPh + \overset{-}{Cl}\overset{+}{N_2}Ph \longrightarrow Ph{-}C(N{=}N{-}Ph){=}N{-}NHPh$$

(113)

$\longrightarrow$ (114) Cl^-

(114)

SCHEME 108

Both the 2,3- and 4,5-bonds are made [method (iii)] when acylhydrazines react with diazonium compounds in alkaline medium to form 1,5-disubstituted tetrazoles. Intermediate in the reaction are tetrazenes (**115**), which can be isolated, but which are conveniently cyclized *in situ* by strong base (Scheme 109). If symmetrical diacylhydrazine is used, one of the acyl groups is eliminated by hydrolysis during the course of the reaction.

$$MeCONHNH_2 + Cl^-\overset{+}{N_2}Ph \xrightarrow{Na_2CO_3} MeCONHNHN{=}NPh \xrightarrow{NaOH}$$

(115)

SCHEME 109

Finally, the 2,3- and 3,4-bonds may be formed [method (iv)] in the hydrazidine–nitrite reactions. One of the most convenient methods of tetrazole synthesis employs the action of nitrous acid on a hydrazidine (Scheme 110). The method is applicable to a wide range of 5-substituted tetrazoles as well as fused-ring compounds. Thus, 5-aminotetrazole can be prepared from aminoguanidine, and 2-hydrazinoquinoline is converted to a tetrazoloquinoline.

$$H_2N{-}C({=}NH){-}NHNH_2 + HNO_2 \longrightarrow H_2N{-}C({=}NH){-}N_3 \longrightarrow$$

$$\xrightarrow{N_2H_4} \quad NHNH_2 \quad \xrightarrow{HNO_2}$$

SCHEME 110

References

1. A. R. Katritzky and S. M. Weeds, *Adv. Heterocycl. Chem.*, 1966, **7,** 250.
2. K. Schofield, M. R. Grimmett, and B. R. T. Keene, 'Heteroaromatic Nitrogen Compounds, The Azoles', Cambridge University Press, Cambridge, 1976.
3. R. Fusco, in 'The Chemistry of Heterocyclic Compounds, Pyrazoles, Pyrazolines, Pyrazolidines, Indazoles, and Condensed Rings', ed. R. H. Wiley, Interscience, New York, 1967.
4. M. R. Grimmett, *Adv. Heterocycl. Chem.*, 1970, **12,** 104.
5. T. L. Gilchrist and G. E. Gymer, *Adv. Heterocycl. Chem.*, 1974, **16,** 33.
6. A. F. Pozharskii, A. D. Garnovskii, and A. M. Simonov, *Russ. Chem. Rev.*, 1966, **35,** 122.
7. C. A. Matuszak and A. J. Matuszak, *J. Chem. Educ.*, 1976, **53,** 280.

8. S. Martinez-Carrera, *Acta Cryst.*, 1966, **20,** 783; G. J. Visser and A. Vos, *Acta Cryst.*, 1971, **B27,** 1802.
9. G. G. Gallo, C. R. Pasqualucci, P. Radaelli, and G. C. Lancini, *J. Org. Chem.*, 1964, **29,** 862.
10. J. Elguero, R. Jacquier, and G. Tarrago, *Bull. Soc. chim. France*, 1967, 3772, 3780; A. R. Katritzky and F. W. Maine, *Tetrahedron*, 1964, **20,** 299, 315; G. A. Newman and P. J. S. Pauwels, *J. Chem. Soc.* (*C*), 1970, 1842; *Tetrahedron*, 1970, **26,** 1571; A. R. Katritzky, *Chimia*, 1970, **24,** 134.
11. A. Grimison, J. H. Ridd, and B. V. Smith, *J. Chem. Soc.*, 1960, 1352, 1357; J. H. Ridd and B. V. Smith, *ibid.*, 1960, 1363.
12. M. Kamiya, *Bull Chem. Soc. Japan*, 1970, **43,** 3344.
13. J. A. Joule and G. F. Smith, 'Heterocyclic Chemistry', van Nostrand Reinhold, London, 1972, p. 308.
14. J. W. A. M. Janssen, H. J. Koeners, C. G. Kruse, and C. L. Habraken, *J. Org. Chem.*, 1973, **38,** 1777.
15. M. R. Grimmett, S. R. Hartshorn, K. Schofield, and J. B. Weston, *J. C. S. Perkin II*, 1972, 1654.
16. D. A. de Bie, H. C. van der Plas, and G. Guertsen, *Rec. Trav. chim.*, 1971, **90,** 594.
17. C. G. Begg, M. R. Grimmett, and P. D. Wethey, *Austral. J. Chem.*, 1973, **26,** 2435.
18. H. Tiefenthaler, H. Dörscheln, W. Göth, and H. Schmid, *Helv. Chim. Acta*, 1967, **50,** 2244; P. Beak and W. Messer, *Tetrahedron*, 1969, **25,** 3287.
19. H. A. Staab, *Angew Chem. Internat. Edn.*, 1962, **1,** 351.
20. M. R. Grimmett, *Rev. Pure Appl. Chem.*, 1965, **15,** 101.
21. G. Bredereck, R. Gompper, H. G. Schuh, and G. Theilig, *Angew. Chem.*, 1959, **71,** 753; and in 'Newer Methods of Preparative Organic Chemistry' ed. W. Foerst, Academic Press, New York, 1964, vol. 3. p. 241.
22. J. H. Boyer, in 'Heterocyclic Compounds' ed. R. C. Elderfield, Wiley, New York, 1961, vol. 7. p. 384.
23. F. R. Benson, in 'Heterocyclic Compounds' ed. R. C. Elderfield, Wiley, New York, 1967, vol. 8. p. 1.

17.4
Indoles and Related Systems

R. T. BROWN AND J. A. JOULE*
University of Manchester,
and
P. G. SAMMES†
The City University, London

* Indoles.
† Carbazoles, isoindoles, indolizines.

For many decades the chemistry of indoles has formed the basis of much research effort, which continues unabated to the present day. In large measure this was due to the varied biological activity of many natural indolic compounds, which created an interest in their structure elucidation and synthesis, and also in the preparation of synthetic analogues as potential chemotherapeutics. In recent years mechanistic studies have increased in importance as chemists have sought to clarify their theoretical understanding of indole reactivity, and the full range of modern spectroscopic techniques has been deployed on indolic compounds. For extensive and detailed coverage of the literature prior to 1970, reference should be made to the excellent monographs by Sundberg[1] and by authors in the 'Heterocyclic Compounds' series.[2]

17.4.1 GENERAL PROPERTIES

17.4.1.1 Structure

Indole (**1**) comprises a benzene ring fused to the 2- and 3- positions of a pyrrole nucleus. Numbering of the atoms begins with the nitrogen and proceeds around the ring as indicated in (**1**); positions-2 and -3 are often referred to as α and β, respectively.

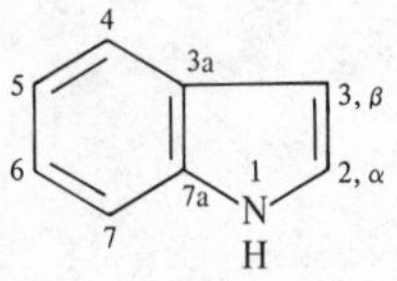

4 3a 5 3, β 6 1 2, α 7a N 7 H

(1)

137 137 149 134 142 139 138 139 140 N 140

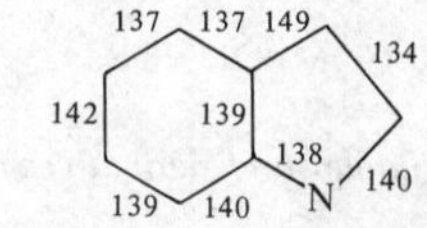

Measured bond lengths (pm)

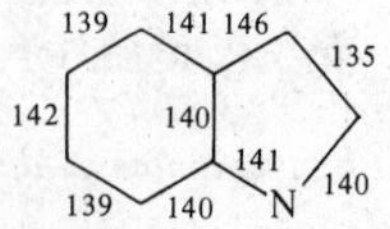

139 141 146 135 142 140 141 139 140 N 140

Calculated bond lengths (pm)

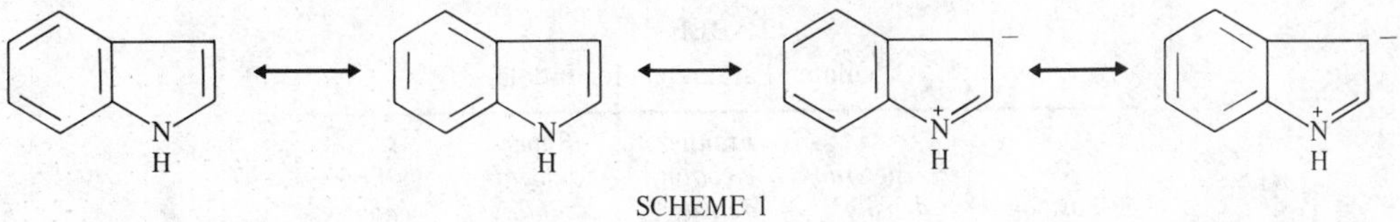

SCHEME 1

Indole is planar with a conjugated system of 10 π-electrons — two from the nitrogen and eight from the carbons — and hence is a heteroaromatic molecule. In simple resonance terms indole is a hybrid of the main canonical forms given in Scheme 1, with lesser contributions from structures with the negative charge on the indole α-position and the benzene ring. Thus the nitrogen atom carries a fractional positive charge.

Although experimental data are lacking for indole itself, an X-ray structure determination on the 3-methylindole/trinitrobenzene complex confirmed the planarity of the indole system.[3] In addition, values for the bond lengths were obtained which agreed well with those calculated for indole by the self-consistent field molecular orbital (SCFMO) method.[4] Both theory and X-ray measurements indicated a shortening of the N—C bonds relative to an N—C single bond, as expected from resonance conjugation of the nitrogen electron pair.

Classically, the aromaticity of indole was established by its chemical behaviour in undergoing substitution rather than addition reactions, and has been confirmed by the more recent criterion of a paramagnetic ring current effect in its proton magnetic resonance spectra. A resonance energy of *ca.* 200 kJ mol^{-1} has been estimated.[4,5]

17.4.1.2 Reactivity

Like pyrrole (Chapter 17.1) indole is a π-excessive molecule, *i.e.* the π-electron density on the carbon atoms is greater than in benzene, and this is reflected in the formation of π-complexes with acceptors such as 1,3,5-trinitrobenzene. Therefore, like pyrrole, indole is subject to ready electrophilic substitution and resistant to nucleophilic substitution. The major difference is in the preferred site for electrophilic attack — C-2 in pyrrole and C-3 in indole — which is directly attributable to the presence of a benzene ring in the latter. Addition of an electrophile E^+ to C-2 of indole would generate a non-benzenoid 2*H*-indolium cation (**2**), whereas attack at C-3 affords the more stable 3*H*-indolium cation (**3**)

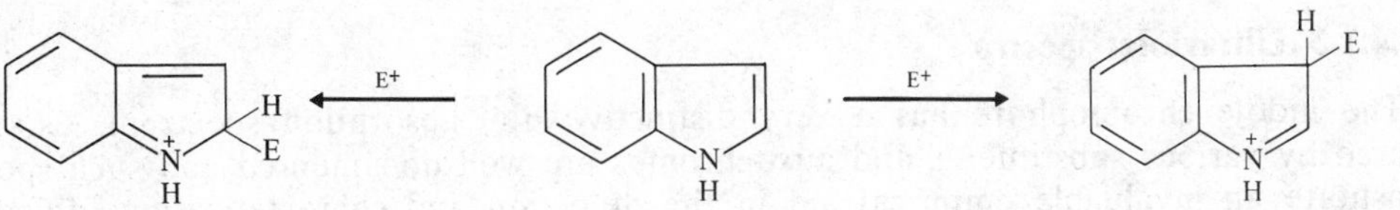

(2) (1) (3)

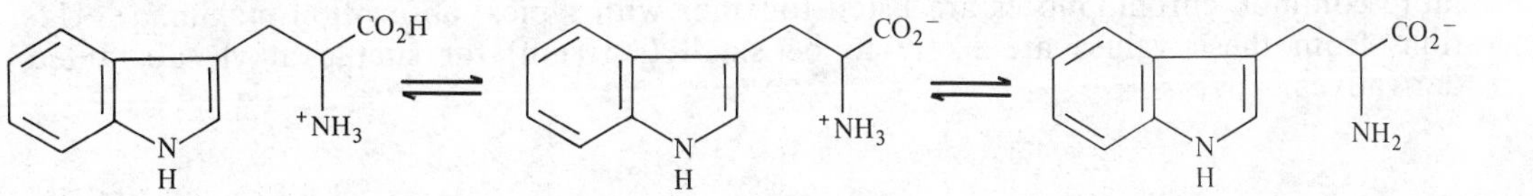

(5) (6) (7)

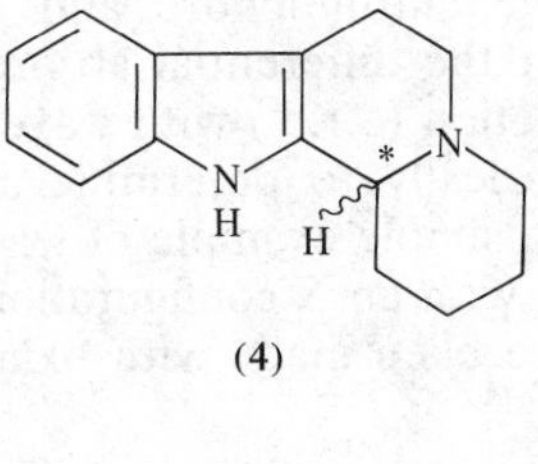

(4)

TABLE 1
Calculated Parameters for Indole

Position	*π-Electron density*[a]	*Frontier electron density*[b]	*Super-delocal-izability*[c]	*Free valence*[d]
N	+0.27	0.278	—	—
2	−0.02	0.219	1.148	0.440
3	−0.11	0.595	1.238	0.457
3a	−0.05	0.005	—	0.107
4	+0.01	0.345	1.032	0.451
5	−0.05	0.058	0.911	0.402
6	0.00	0.217	0.966	0.408
7	−0.04	0.233	0.978	0.441
7a	−0.01	0.046	—	0.138

[a] See Ref. 6. [b] See Ref. 8. [c] See Ref. 9. [d] A. Streitwieser, Jr. and J. I. Brauman, 'Supplemental Tables of Molecular Orbital Calculations', Pergamon, New York, 1965, vol. 2, p. 318.

in which the aromaticity of the benzene ring is retained. This qualitative argument has been supported by various theoretical MO calculations including π-electron density,[6] localization energy,[7] frontier electron density,[8] and super-delocalizability[9] (see Table 1), which all indicate C-3 as the site for electrophilic substitution.

However, that π-electron density in the ground state molecule makes the correct prediction may be fortuitous, since it assumes that the transition state resembles the starting material more than the intermediate. Probably a more reliable parameter is one such as super-delocalizability, which gives more weight to the relative energies of transition states.

Free radical reactions of indoles follow a different pattern to heterolytic reactions, giving complex mixtures including products of substitution in the benzene ring. Such behaviour seems to be best accommodated by the atomic free-valence parameter, which reflects the changes in π-electron density on substitution at a particular position.[10] For indole these suggest that there is little difference in reactivity between the six available carbon atoms (see Table 1).

17.4.1.3 Ultraviolet spectra

The indole chromophore has a very distinctive u.v. absorption spectrum. Changes caused by various substituents and auxochromes are well documented and such spectra constitute an invaluable empirical aid in the detection and characterization of indolic natural products, particularly the several hundred alkaloids.[11] In Table 2 some of the more common chromophores are listed together with typical absorption maxima; deviations from these values are likely to be small (<10 nm) for simple alkyl substituted derivatives.

17.4.1.4 C.d. and o.r.d. spectra

Interactions between the indole chromophore and adjacent chiral centres can be detected by the variation in either the differential absorption of left and right circularly polarized light (c.d.) or optical rotation (o.r.d.) with wavelength. The sign of the resulting Cotton effects can be used empirically to determine absolute configurations, as with tetrahydro-β-carboline alkaloids, a simple example of which is (**4**), where a positive effect in the 280 nm region is correlated with an *S* configuration at C* and a negative one with an *R*.[12a,b] Similar assignments have been made with oxindole derivatives.[12c]

TABLE 2

Typical u.v. Maxima for Common Indolic Chromophores

Chromophore	λ_{max} nm($\varepsilon \times 10^{-3}$) *in alcohol*			
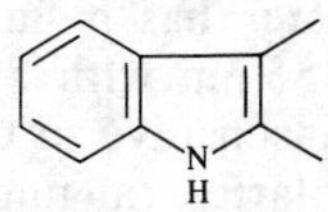	226 (27.8)	282 (6.2)	290 (5.5)	
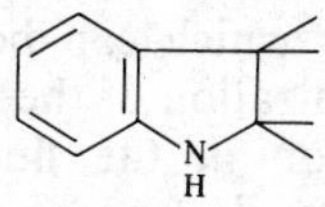	210 (32)	254 (25)	306 (5.5)	
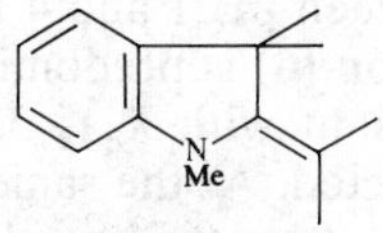	293 (45)			
	219 (22.4)	258 (5.1)		
	H^+ 275			
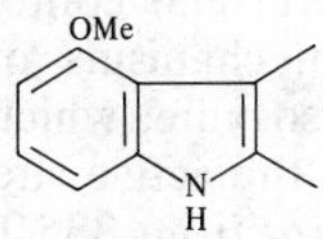	227 (35)	273 (7.6)	293 (7.4)	
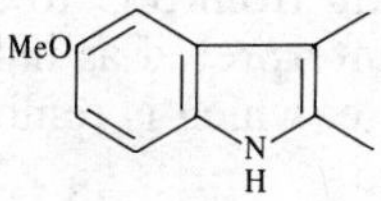	226 (24.2)	285 (7.8)		
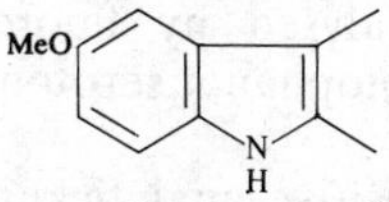	229 (33)	271 (4.5)	300 (5.1)	
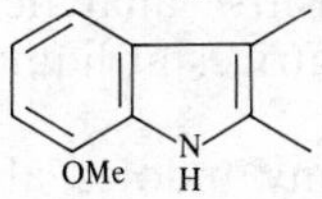	226 (42)	271 (7.1)		
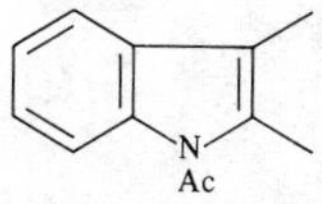	255 (10)	281 (4.2)	290 (4.3)	
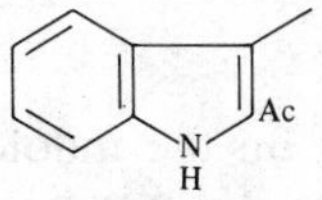	238 (15)	312 (21)		
	251 (17.8)	270 (13.6)	297 (9.6)	
	HO^- 252 (7.5)	275 (27.5)	315 (11)	
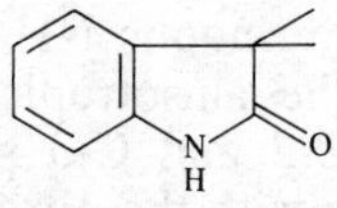	210 (28.2)	252 (7.4)	280 (1.4)	
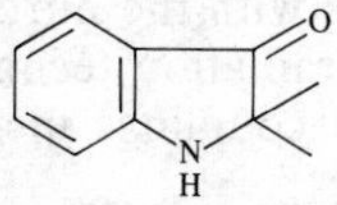	230 (23.5)	251 (6.4)	400 (3.3)	
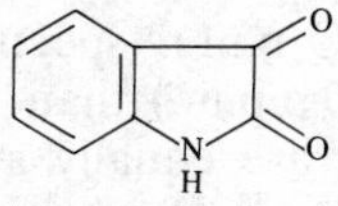	243 (22.3)	249 (18.6)	298 (3.0)	422 (0.7)

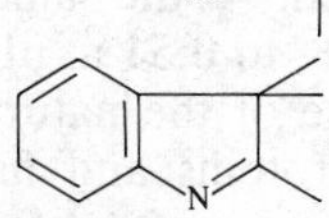

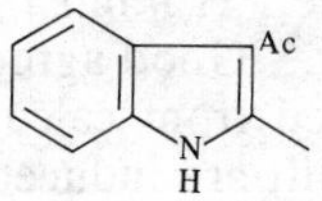

17.4.1.5 Fluorescence and related spectra

The highly characteristic fluorescence spectra exhibited by indoles form the basis of several methods of identification and analysis.[13] These are particularly useful for biological molecules since they are inherently more sensitive than u.v. techniques and can often detect sub-microgram quantities of material. Indole itself in water has a fluorescence excitation maximum at 287 nm and an emission maximum at 355 nm with a quantum efficiency of 46%.[14] Although the former maximum is relatively insensitive to the nature of the solvent, the latter shows marked changes with lessening polarity, ranging down to 297 nm in cyclohexane. Substituents can also have large effects which can be usefully correlated with structural features. For instance, changes in the ionization of the α-amino-acid function in tryptophan with pH are mirrored by variation in the fluorescence spectrum.[14] No fluorescence is detectable below pH 1, but between pH 1 and 4 the cation (**5**) has a maximum at 345 nm; from pH 4.1 to 8.6 the zwitterion (**6**) is predominant with λ_{max} 352 nm, and from pH 10.7 to 11.0 the anion (**7**) is present with λ_{max} 359 nm; in highly alkaline solution a dianion with λ_{max} 420 nm can be detected. At the same time the quantum yield increases from 0.085 at pH 1.5–2, to 0.20 at pH 4–8, to 0.51 at pH 10.9.[14]

Since the indolic fluorescence of tryptophan is the most intense of the natural amino-acids, it is largely responsible for that associated with most proteins and finds many applications in biology and medicine, *e.g.* as a probe for structural and conformational changes in proteins, antibody affinities in immunology, and mechanisms of enzyme action.[13b,c,15] One particular example is the hydrolytic enzyme lysozyme, which contains six tryptophan residues including three apparently associated with the active site. Binding of substrate leads to a blue shift of 10 nm in the emission spectrum from 335 to 325 nm accompanied by an increase in quantum yield. Such behaviour is interpreted as indicating an interaction between carboxyl and indole groups at the active site, which is removed by binding to the substrate.[16]

Many other important indolic metabolites can also be analysed by fluorometric methods, including, for example, tryptamine, 5-hydroxytryptophan, serotonin (5-hydroxytryptamine), indoleacetic acid, and kynurenine.

The phosphorescence of indoles can equally well serve as a sensitive analytical method except for the drawback of being applicable only to rigid glass matrices at liquid nitrogen temperatures.[13]

As discussed later (p. 440), on oxidation in basic media many indoles also exhibit chemiluminescence.

17.4.1.6 Nuclear magnetic resonance spectra

Since the magnetic environment varies markedly with position on the indole nucleus, the chemical shifts of hydrogens in the proton magnetic resonance (^{1}H n.m.r.) spectra of indole (see Table 3) and its derivatives are relatively easy to assign.[17] Good agreement has been obtained between observed and theoretical values estimated from calculated ring currents and π-electron densities.[17b] Some changes in chemical shift are induced by polar solvents such as acetone, which forms a hydrogen bond with the indolic NH. The C-2 proton is consequently deshielded owing to its proximity to the anisotropic carbonyl group.[18a] Benzene also causes changes of $\pm$0.5 p.p.m. in the C-2 and C-3 protons in indoles, attributed to preferential association of solvent molecules with the nitrogen.[18b] A thorough investigation of the spin–spin interactions by double irradiation techniques has revealed an unusually complex series of vicinal and long-range couplings in indole, as shown in Table 3.

In the past few years technical improvements have made ^{13}C n.m.r. spectroscopy a routine tool. It has the advantage of a much greater range (600 p.p.m.) than ^{1}H n.m.r. (20 p.p.m.) while the principles of shielding and chemical shifts are equally applicable. Tetrahydrocarbazole (**8**) is given as a relatively simple example.[19a] However, the ^{13}C

TABLE 3

Proton position	$\tau(CCl_4)$ (p.p.m.)	$\tau([^2H_6]Me_2CO)$ (p.p.m.)	*Spin–spin coupling,*[a] J(Hz)
N			1,2: 2.5; 1,3: 2.0; 1,4: 0.8
2	3.26	2.72	2,3: 3.1
3	3.58	3.54	
4	2.44	2.40	4,5: 7.8; 4,6: 1.2; 4,7: 0.9
5	3.00	3.03	5,6: 7.1; 5,7: 1.3
6	2.95	2.93	
7	2.53	2.54	6,7: 8.1

[a] P. J. Black and M. L. Hefferman, *Austral. J. Chem.*, 1965, **18,** 353.

n.m.r. spectra of more complex molecules such as indole alkaloids (*q.v.*) are also readily interpreted,[19b] and together with ^{1}H n.m.r. provide a powerful technique for structure determination.

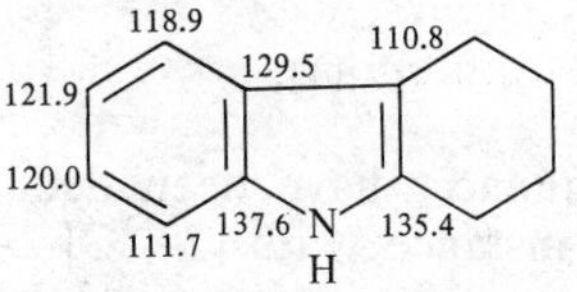

(8) ^{13}C n.m.r. assignments (δ from Me_4Si) for tetrahydrocarbazole

17.4.1.7 Mass spectrometry

The fragmentation behaviour of simple indoles on electron impact has been well investigated.[20a,b] Because of the stabilizing influence of the aromatic ring the molecular ions are always large, and the common breakdown is by expulsion of HCN. Alkyl substituents at the 2- and 3-positions readily undergo β-cleavage with loss of a hydrogen or alkyl fragment, perhaps with the formation of a quinolinium ion (see Scheme 2).

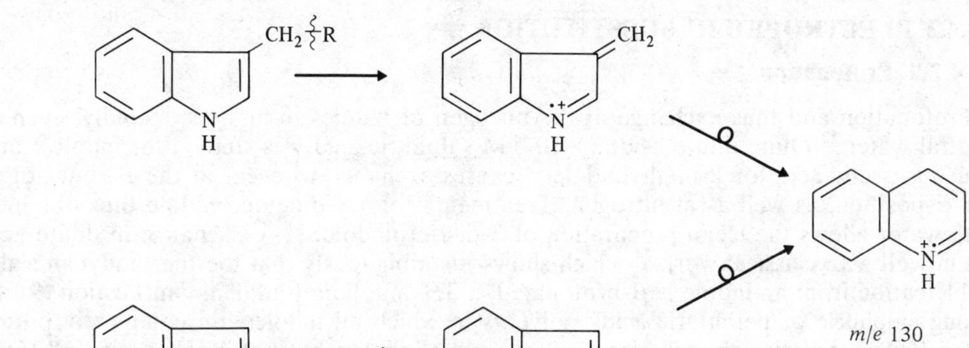

SCHEME 2

(a) ■ = ^{13}C, * = ^{12}C
(b) ■ = ^{12}C, * = ^{13}C

m/e 131

(a) *m/e* 103
(b) *m/e* 104

SCHEME 3

Details of the fragmentation sequences have been deduced from metastable ions and labelling experiments. Thus, for instance, with [2-^{13}C]-2-methylindole the predominant ion formed by loss of HCN from *m/e* 130 was *m/e* 103 rather than *m/e* 104, *i.e.* C-2 was expelled with the HCN, but when the methyl group was labelled the major ion was at *m/e* 104, showing retention of the ^{13}C atom (Scheme 3). These results demonstrated that the preferred mode of formation of the quinolinium intermediate involved migration of C-3 rather than N-1.[20b]

However, the major application of mass spectrometry for indoles has been in the structure elucidation of natural products, particularly the indole alkaloids (*q.v.*). Again the indole nucleus stabilizes certain fragments preferentially and a body of empirical knowledge has been built up which can be used to obtain structural information.[21]

17.4.2 ELECTROPHILIC SUBSTITUTION

17.4.2.1 Protonation

Protonation and thus exchange at the nitrogen of indoles occurs very rapidly, even in neutral water.[22] Quite dilute acid, 5×10^{-3} M sulphuric acid was studied for indole[22] and aqueous acetic acid for 2-methylindole,[23] causes exchange to occur at the β- but not at the α-position, as well as at nitrogen. Treatment[22] of 1,3-dideuteroindole thus obtained with water allows the clean preparation of 3-deuteroindole. H-3 Exchange in dilute acid fits in well with classical work,[24] which shows unambiguously that the thermodynamically stable cation from an indole is β-protonated, a 3*H*-indolium (indoleninium) cation (**9**). In strong sulphuric or perchloric acids, solvents in which all indoles are completely protonated (trifluoroacetic acid will also convert[25] most indoles, but not indole itself, fully into their salts), u.v. and n.m.r. evidence[21a] clearly showed that the principal cation present, to the level of sensitivity of the spectroscopic methods employed, had a proton added at C-3. Thus, for example, the u.v. absorption corresponded closely with that of model 3,3-dialkylindoleninium cations, absorption of which is distinctly and characteristically different (Section 17.4.1.3) from that of an indole, or of a styrene type such as would have been expected from an *N*-protonated species. The n.m.r. evidence for β-protonation included, for example, for the cation from 3-methylindole in 18 M sulphuric acid, the observation of a methyl doublet resulting from coupling with the added C-3 proton. Although 3-protonated indoles were thus demonstrated to greatly predominate at equilibrium, it was shown[24a] that both 1- and 2-protonation were also occurring in strong acid solution. Thus slow exchange of the α-proton was observed and, in some cases, *N*-protonated cations (**11**) could be trapped out[24a] as solid salts by irreversible precipitation with ether.

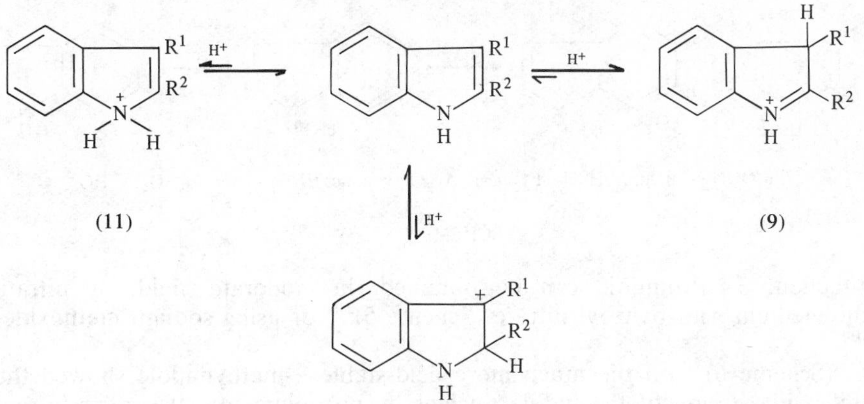

(11) (9)

(10)

It was possible to make a quantitative comparison[24b] of the basicity of simple indoles and these measurements too were entirely consistent with the concept of preferred β-protonation. Thus 2-methylindole (pK_a −0.28) and 1-methylindole (pK_a −2.32) are stronger bases than indole (pK_a −3.63), illustrating the hyperconjugative stabilizing effect of, in particular, the α-methyl on the immonium system of the salt. In contrast, 3-methylindole (pK_a −4.55) is a weaker base than the parent heterocycle since the 3-methyl group has a base-weakening effect ascribable to a decrease in the electron density at the β-position in the indole, and also that in its cation the methyl group cannot hyperconjugate with the unsaturated system. Interestingly the *N*-hydrogen of 3-methylindole was exchanged[26] more rapidly than that of indole in dioxan–10^{-4} M sulphuric acid. Further, the relative rate of exchange at C-2 in this indole is increased and this can be accounted for[24a] by recognizing that the 3-methyl group *can* exert a hyperconjugative stabilizing effect on the 2-protonated cation (**10**).

Protonated indoles are involved in oligomerization reactions (see p. 433) and in certain electrophilic nitrations (see p. 420). Indole can be reversibly converted[27] into a bisulphite addition complex, presumably by way of an equilibrium concentration of β-protonated cation, under conditions which lead to the crystallization of the complex (Scheme 4).

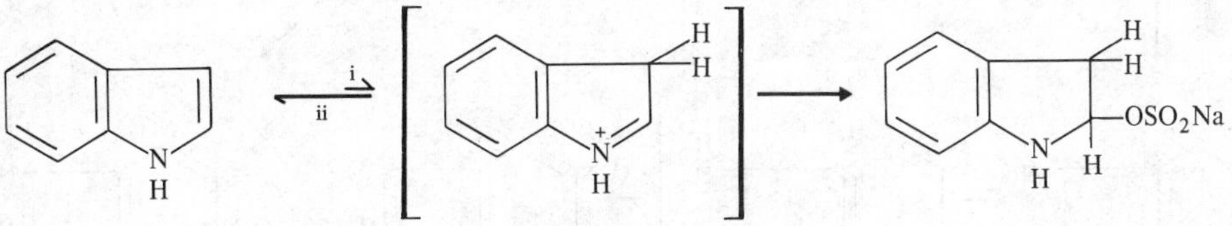

i, aq.$NaHSO_3$, Et_2O; ii, H_2O

SCHEME 4

17.4.2.2 Nitration

Attempted nitration of indole itself under the usual strongly acidic conditions leads[28] only to uncharacterizable products which probably result from acid-catalysed polymerization (see p. 433) and/or oxidation of the sensitive unsubstituted hetero-ring. The bisulphite adduct of indole (Scheme 4) can be nitrated,[27] however, after *N*-acetylation, and thus yields 5- and 7-nitroindoles after deprotection (Scheme 5). 1-Acetyl-2,3-dihydroindole (1-acetylindoline) also gives a 5-nitro-derivative, which in turn can be further nitrated to give 1-acetyl-5,7-dinitroindoline.[29] Such indolines can be dehydrogenated to the corresponding fully aromatic heterocycles after hydrolytic removal of the

i, $PhCOONO_2$; ii, aq.$NaHSO_3$, Et_2O; iii, Ac_2O; iv, conc. HNO_3–conc. H_2SO_4; v, HO^-

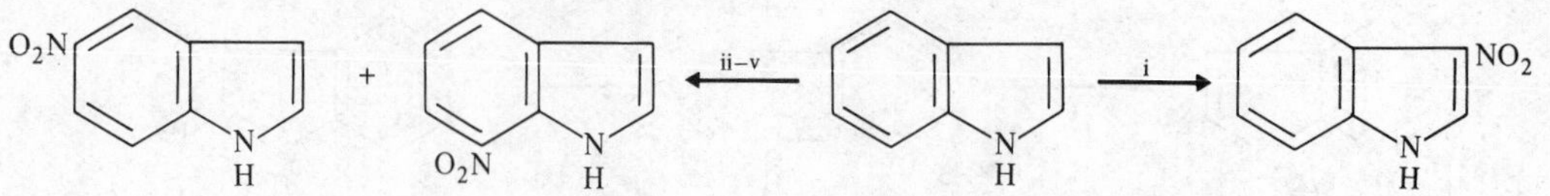

SCHEME 5

amide function. 3-Nitroindole can be obtained, in moderate yield, by nitration in non-acidic medium with benzoyl nitrate (Scheme 5)[30a] or using sodium methoxide–ethyl nitrate.[30b]

Studies (Scheme 6)[31] on the much more acid-stable 2-methylindole showed that, for electrophilic substitution of the indole nucleus by nitronium ion, the intrinsic reactivity order is 3>6>4. Thus in nitric acid or in nitric acid–acetic acid, 3-substitution is rapidly followed by C-4 (minor) and C-6 (major) nitration. Since a relatively high temperature (90 °C) is necessary to initiate nitration under these conditions, yields are poor due to competing oxidations. No 3-monosubstituted product could be obtained from reaction with nitric acid, but 2-methyl-3-nitroindole *is* almost certainly the initial product for when it was separately prepared by nitration with benzoyl nitrate it was shown to be smoothly nitrated by nitric acid, even at room temperature, to a mixture of 2-methyl-3,4- and -3,6-dinitroindoles. It is important to realize that this nitration of 2-methyl-3-nitroindole is not plagued with oxidation difficulties because the hetero-ring nitro-group stabilizes the ring to which it is attached. Both dinitroindoles could be futher nitrated to 2-methyl-3,4,6-trinitroindole.

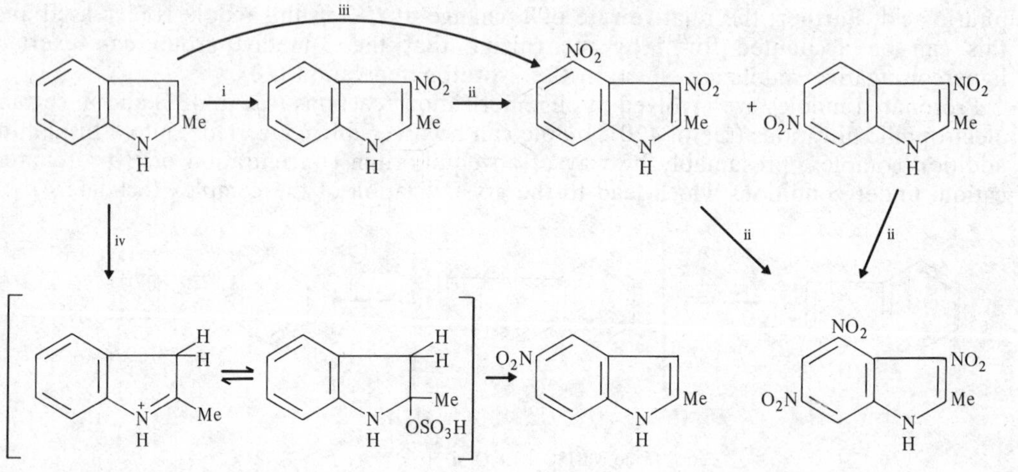

(12)

i, $PhCOONO_2$; ii, conc. HNO_3, room temp.; iii, conc. HNO_3, 90 °C; iv, conc. HNO_3–conc. H_2SO_4

SCHEME 6

The nitration of 2-substituted indoles in concentrated sulphuric acid–concentrated nitric acid[28,31] is smooth, but takes a different course to that discussed above. It is known that indoles are completely protonated in such strongly acidic media; clearly then, attack cannot occur at the indole β-position now occupied with an added proton and so mono-nitration occurs on the homocyclic ring, rather surprisingly at C-5. The in-doleninium salt is generally considered to be the substrate undergoing electrophilic attack. This interpretation is supported by the significant model reactions (Scheme 7)[31,32] in which

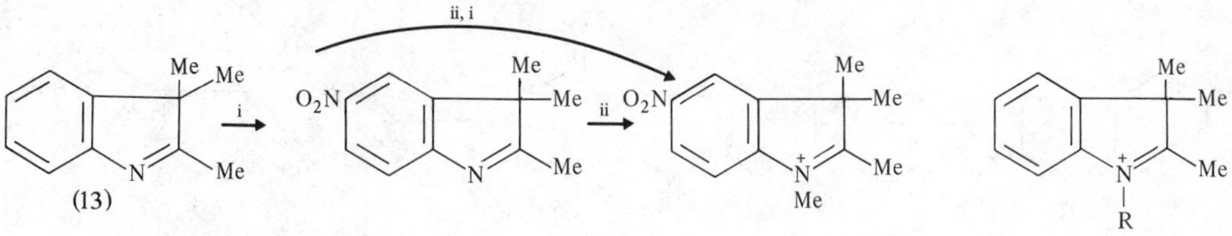

ii, i

Me Me (13) i → O_2N Me Me Me ii → O_2N Me Me Me N Me Me Me Me N R

i, conc. HNO_3–conc. H_2SO_4; ii, MeI

(14) a; R = H
b; R = Me

SCHEME 7

the 3,3-dialkylindolenine (**13**), which would of course be present in strong acid as the proto-indoleninium salt (**14a**), or its metho-salt (**14b**), likewise give 5-nitro-derivatives.

Thus the evidence suggests that an indoleninium cation is the species undergoing electrophilic substitution at C-5 even though the positively charged nitrogen might have been expected to be *meta*-directing and lead to the formation of a 4- or 6-nitroindole. Nitration of indoline[33] in strong acid does indeed give a 6-nitroindoline product, in accord with this line of reasoning. One explanation[34] for this apparent anomaly is that nitration is actually occurring not on the indoleninium salt but on a small equilibrium concentration of the much more nucleophilic neutral adduct (**12**) (Scheme 6) formed from the salt by C-2 addition of anion, such as bisulphate.

17.4.2.3 Nitrosation

The reaction of indole with nitrous acid is complex, but from the tarry product two bis-indoles (**15**) and (**16**) can be isolated (Scheme 8). One can understand their formation as involving the anticipated β-nitrosation, followed by a sequence in which a second

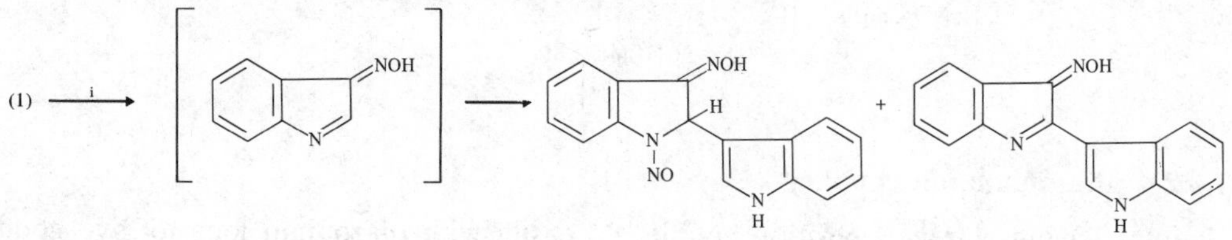

(1) —i→ [NOH N] → NOH H N NO N H + NOH N N H

(16) (15)

i, $NaNO_2$, aq. AcOH

SCHEME 8

indole adds to the oximino-3*H*-indole tautomer of the substitution product and then either oxidation to (**15**) or *N*-nitrosation to (**16**) completes the sequence. It is relevant to this rationalization that both the authentic 3-nitroso-derivative of indole, preparable by a base-catalysed nitrosation[36] (see p. 438) and the 3-nitroso-derivative of 2-methylindole, which *can* be obtained by acidic nitrosation[37] as well as by a base-catalysed process,[31] exist predominantly in the oximino-3*H*-indolic tautomeric form. That the 2-methylindole nitrosation is free from complications is rationalized by the very much reduced electrophilicity of the methyl-substituted α-position in the 2-methyl-3-oximino-3*H*-indole (**17**).

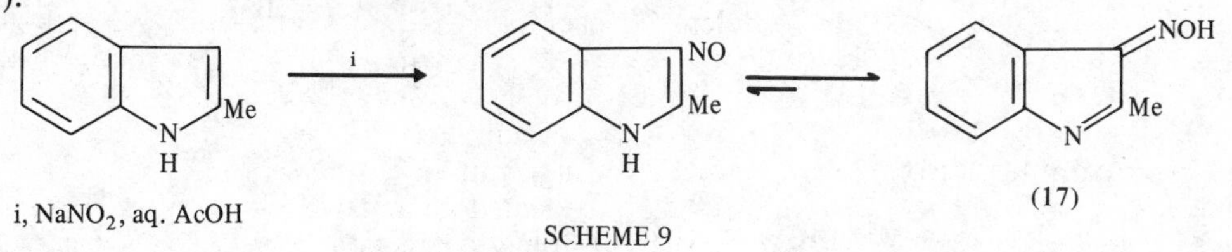

Me N H —i→ NO Me N H ⇌ NOH Me N

(17)

i, $NaNO_2$, aq. AcOH

SCHEME 9

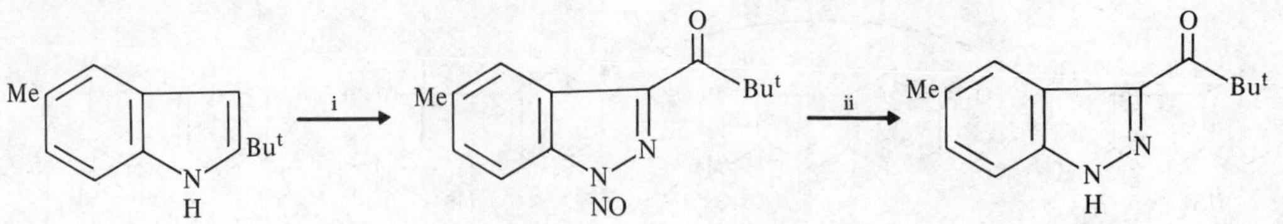

i, excess $NaNO_2$, aq. AcOH; ii, AcOH, heat

SCHEME 10

3-Methylindole reacts with nitrous acid to give[35] an *N*-nitroso-derivative; this is consistent with recent mechanistic studies[31] which suggested that the *C*-nitrosation of 2-methylindole may in fact proceed *via* an *N*-nitroso-intermediate. The observed rearrangement of 2-t-alkylindoles to 3-acylindazoles (Scheme 10)[38] during reaction with an excess of nitrous acid has also been rationalized in terms of a sequence in which *N*-nitrosation occurs first, though that these same indoles give 3-nitroso-derivatives on reaction with one mole of nitrous acid does not seem consistent with such an explanation.

A different type of complication arises when 1-aryl- or -alkyl-2-methylindoles are nitrosated,[39] for here side-chain substitution occurs as well as β-nitrosation (Scheme 11). No such side-chain substitution was reported[38] for 2-alkylindoles having *N*-hydrogen.

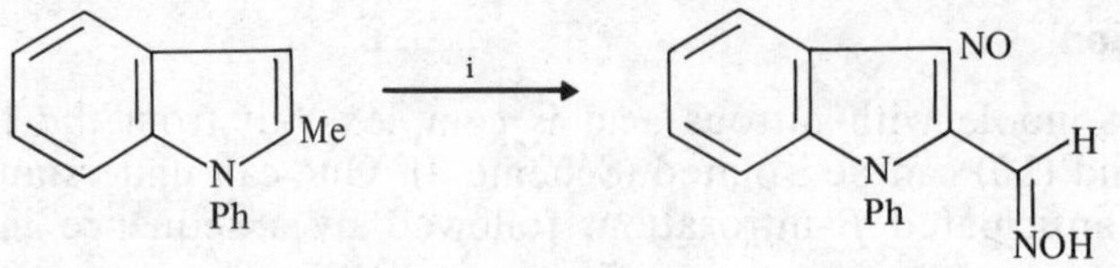

i, $NaNO_2$, aq. AcOH

SCHEME 11

17.4.2.4 Diazonium ion coupling

Indoles having a free β-position couple[40,41] easily with diazonium ions to give good yields of β-substituted products (Scheme 12), the tautomeric form of which has not been determined. Mechanistic studies[41] have shown that reaction can occur either on the neutral indole, or, about 10^8 times faster, on a small equilibrium concentration of indolyl anion (see p. 436), depending on pH. A recent study[42] of the reactions of 4-nitro-2-methoxyphenyl diazonium cation with alkylindoles revealed interesting product types.

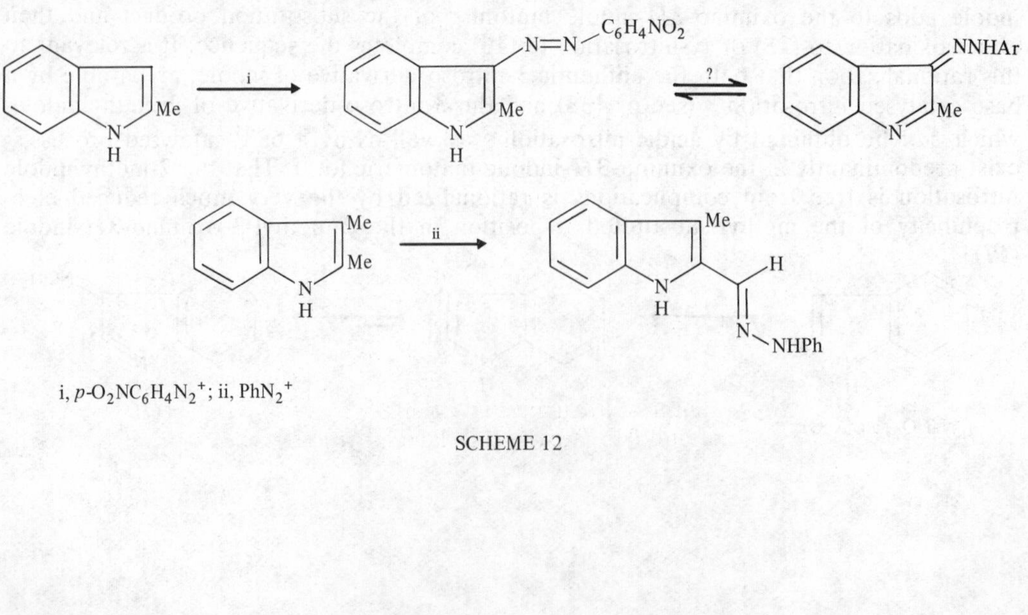

i, p-$O_2NC_6H_4N_2^+$; ii, PhN_2^+

SCHEME 12

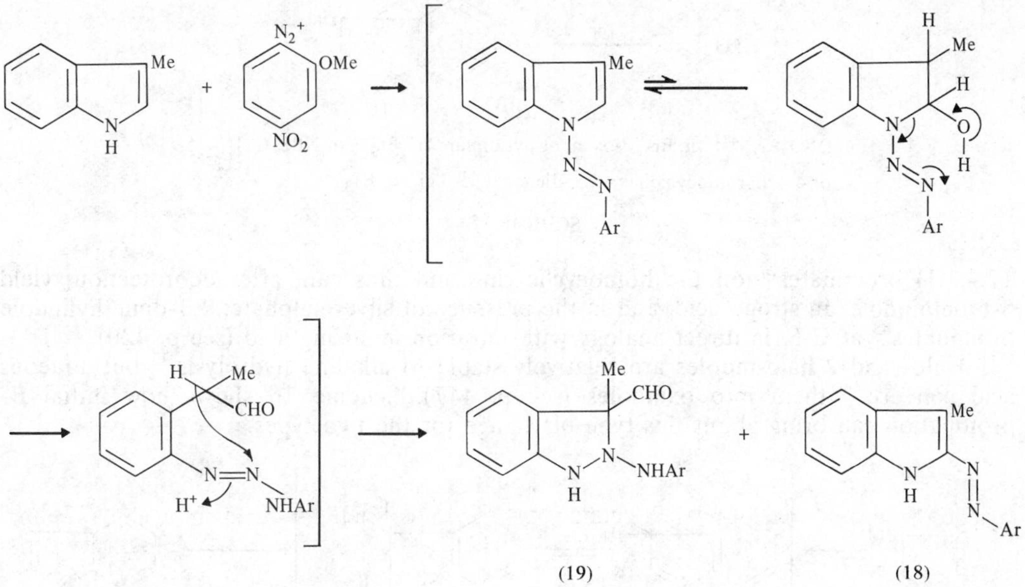
Me
N_2^+
OMe
NO_2
N
H
Ar
NHAr
CHO
H^+
(19)
(18)

SCHEME 13

Whereas 2-methylindole gave a normal β-azo-derivative, α-methyl substitution was observed with 2,3-dimethylindole (Scheme 12). 3-Methylindole gave a mixture of products, (**18**) and (**19**), with (**18**) predominating from reaction at pH 3 and (**19**) from reaction at pH 6–7. It seems that the latter must result from initial attack at nitrogen, though in contrast to nitrosation, no evidence for initial N-coupling was found in mechanistic studies.[41] Scheme 13 outlines a possible sequence for the formation of (**19**).

17.4.2.5 Sulphonation

There seem to be no recorded examples of sulphonation of the homocyclic ring of indoles in strong acid, such as might have been expected by analogy with nitration at very low pH (Section 17.4.2.2); it is possible, however, to sulphonate[43] in the heterocyclic ring (Scheme 14), at C-3 if unsubstituted or at C-2 if the β-position carries an alkyl group. The pyridine–sulphur trioxide complex is utilized to minimize problems associated with acidity.

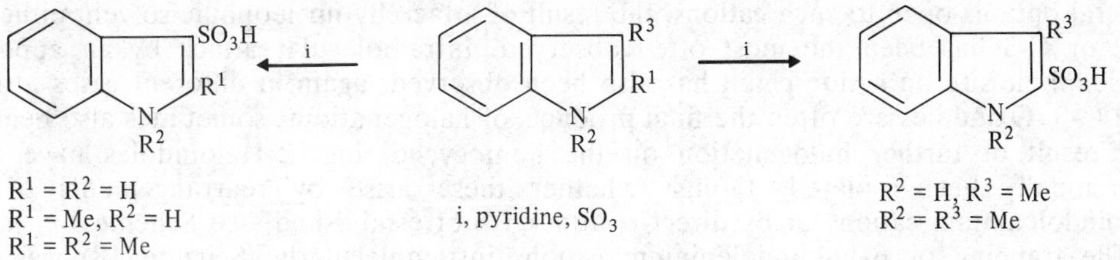
SO_3H
R^1
R^2
R^3
i
$R^1 = R^2 = H$
$R^1 = Me, R^2 = H$
$R^1 = R^2 = Me$
i, pyridine, SO_3
$R^2 = H, R^3 = Me$
$R^2 = R^3 = Me$

SCHEME 14

17.4.2.6 Halogenation

3-Chloro-,[44a] -bromo-,[44b] and -iodo-indoles[44c] can be prepared from the parent heterocycle by electrophilic substitution under appropriately mild conditions (Scheme 15). The principal precaution which must be taken is the avoidance of conditions which would lead to the build-up of appreciable quantities of hydrogen halide in the presence of acid-sensitive indole or haloindole. The *N*-acetylated indole bisulphite adduct (Section

(1) $\xrightarrow[\text{iii}]{\text{i, ii or}}$ 3-Hal-indole

i, SO_2Cl_2;[44a] ii, Br_2, dioxan, or pyridinium H^+ Br_3^-, or NBS, CCl_4, or 4,4-dibromocyclohexa-2,5-dienone;[44b] iii, aq. KI_3[44c]

SCHEME 15

17.4.1.1) brominates[27] on the homocyclic ring and thus can, after deprotection, yield 5-bromoindole. In strong acid, and in the presence of silver sulphate, 2,3-dimethylindole brominates[45] at C-5, in direct analogy with nitration in strong acid (see p. 420).

3-Halo- and 2-halo-indoles are relatively stable to alkaline hydrolysis,[46] but aqueous acid converts[44] them into oxindoles (see p. 447). Scheme 16 shows how initial β-protonation can bring about this type of change for the two types.

i, H_2O, H^+

SCHEME 16

It is believed that 3-substituted indoles still react with electrophilic halogen fastest at a β-position to give 3-haloindoleninium cationic intermediates; because of the possibilities open to such cations, such as the solvolytic modification discussed above, for example, the results from halogenations of 3-substituted indoles are complex and precise products depend critically on the conditions used. All the observed product types can be rationalized, however, in terms of β-haloindoleninium cationic intermediates. Of the several options open to such cations, the result of attack by nucleophilic solvent either at C-2 or C-3 has been the most often observed. Intramolecular attack by an apposite nucleophilic site on a side chain has also been observed, again in different cases at C-2 and C-3. Oxindoles are often the final products of halogenations, sometimes also bearing the result of further halogenation on the homocyclic ring. 2-Haloindoles have also occasionally been isolated, though whether these arise by rearrangements of 3-haloindoleninium cations or by direct α-attack is not established (see Scheme 17[47]).

The trapping of β-haloindoleninium cations intramolecularly (Scheme 18) was observed[48] when indole-3-propionic acid was treated with *N*-bromosuccinimide–acetic acid. Yet a further product type resulting from intramolecular trapping of an intermediate salt was observed[49] when *N*-acetyltryptamine (or methyl *N*-acetyltryptophanate) was brominated with *N*-bromosuccinimide. Nucleophilic addition of the aliphatic nitrogen then hydrogen halide elimination led to the interesting pyrrolo[2,3-*b*]indole (**20**).

C-2 Side-chain halogenation has been observed, for example, in the bromination of 1-acetyl-2,3-dimethylindole[50a] and the iodination of 1,2,3-trimethylindole.[50b] It seems likely that these reactions involve an iodo-enamine type intermediate, as in the reaction of 1,2,3-trimethylindole with iodine and pyridine (Scheme 19).

Me Br N H + Br Me Br N H

Me N H

ix → Br H Br N H O

i ii iii iv vi, vii vi, viii

Me Br N H

H Me N H O Br Me N H O Me N H $\overset{+}{N}$ v → H Me N H O

i, NBS, glacial AcOH; [47c] ii, NBS, dry Bu^{t}OH; [47c] iii, 2 × NBS, dry Bu^{t}OH; [47c] iv, NBS, pyridine; [47a,b] v, aq. HCl; vi, – 60 °C, Br_2, Et_2O; [47a] vii, 10% aq. HCl; viii, piperidine; ix, 2 × NBS, aq. AcOH [47c]

SCHEME 17

CO_2H N H i → Br O O N H O

N H NHAc ii → [Br $\overset{+}{N}$ H NHAc ⇌ Br N H H N Ac] →

N H N Ac

(20)

i, NBS, AcOH; ii, NBS, pH 9

SCHEME 18

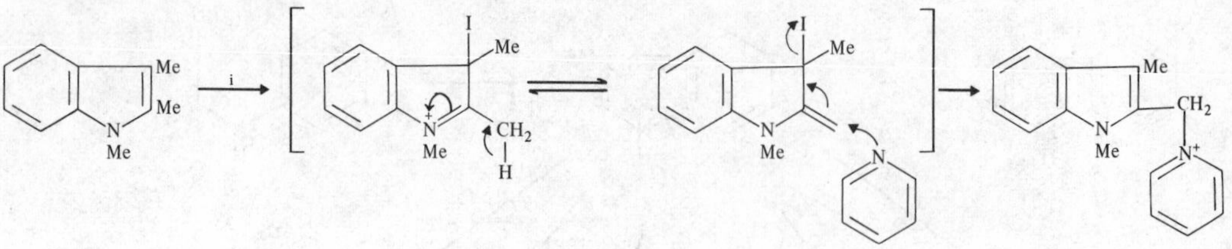

i, I_2, pyridine

SCHEME 19

N,N-Dichlorourethane, another source of positive halogen, has been used for the chlorination of indoles;[51] thus indole and both indole-2- and -3-carboxylic acids all give 3,3,5-trichloro-oxindole with this reagent (see Scheme 20).

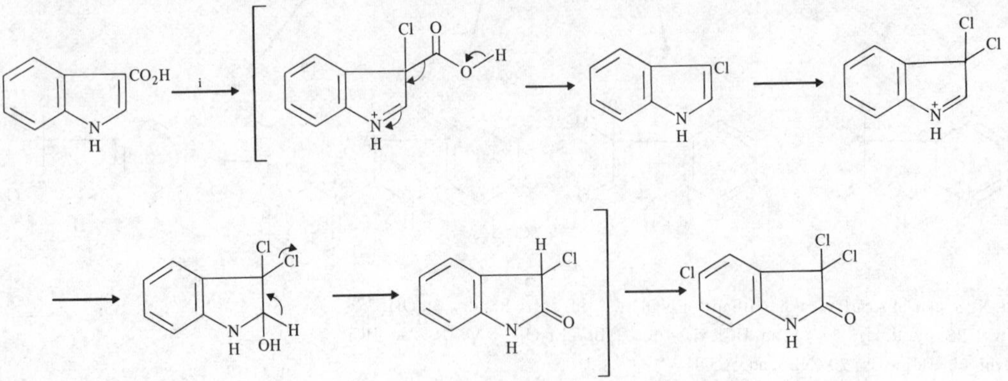

i, Cl_2NCONH_2

SCHEME 20

17.4.2.7 Alkylation

The involvement of the indole nitrogen lone pair in the hetero-ring aromatic sextet means that, unlike simple secondary amines, reaction with alkyl halides does not occur at nitrogen. Methyl iodide does not react at all with indole below about 100 °C. Under forcing conditions, however, alkylation[52] does take place, but initially at carbon and not at nitrogen. A series of alkylations, deprotonations, and a C-3→C-2 methyl shift occur and the final product is the indoleninium salt (**21**). It seems likely that the sequence shown in Scheme 21, in which the 2-methyl group is introduced by migration from C-3, rather than by direct α-attack on (**22**), represents the actual course of events (see also below). It was

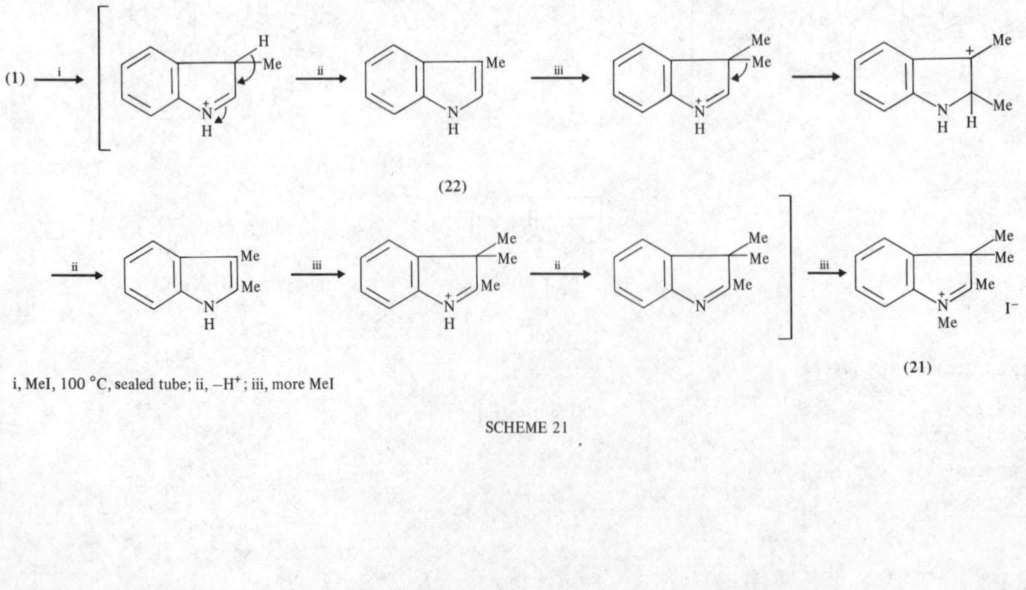

(22)

(21)

i, MeI, 100 °C, sealed tube; ii, $-H^+$; iii, more MeI

SCHEME 21

recognized very early that 1,2-migrations of the Wagner–Meerwein type, such as that which introduces the α-methyl in Scheme 21, could occur within an indoleninium framework and the process is referred to as the Plancher rearrangement.[52b] It can be illustrated unambiguously by the separate preparation of an indolenine (or oxidation level equivalent) and then its acid-catalysed rearrangement. Scheme 22[53] shows three such examples. In the first[53a] a thermodynamically controlled equilibrium is set up between two

i, 48% HBr, heat; ii, 5% H_2SO_4, heat; iii, 150 °C, H_3PO_4

SCHEME 22

indolenine isomers by a double (*i.e.* C-2→C-3 as well as C-3→C-2) migration. The second process is an example[53b] of a general phenomenon whereby the formation of a more stable 2,3-disubstituted indole from a 3,3-disubstituted-2-unsubstituted indolenine is catalysed by acid and rendered irreversible by a final proton loss. The final example[53c] shows how an indoleninium system can be generated from a carbinolamine. Migratory aptitudes in the Plancher rearrangement have been established[54] as $PhCH_2 >$ allyl $> Pr^i >$ $Pr^n >$ Et $>$ Me.

A particularly neat demonstration[55a] of the intermediacy and rearrangement of 3,3-dialkylindoleninium cations in the overall α-alkylation of 3-substituted indoles was provided in the boron trifluoride catalysed closure of indol-3-ylbutan-1-ol to tetrahydrocarbazole (Scheme 23). Tritium labelling of the indolylic carbon revealed an almost complete randomization of carbon atoms 1 and 4 in the product. This demonstrated that virtually all of the product had arisen *not* by direct α-attack but by β-attack and subsequent rearrangement of the symmetrical spiro-indoleninium cation (**23**).

There is, however, now clear evidence[55b,56] that α-alkylation in indoles can occur by a direct process. For example, an analogous study[55b] of 6-methoxyindol-3-ylbutan-1-ol

(23)

R = H, ~ 50 : 50 label * at C-1 and C-4
R = OMe, 37% label ⊛ at C-4

(24)

SCHEME 23

closure to 7-methoxy-1,2,3,4-tetrahydrocarbazole showed that 25% of the product was formed by direct electrophilic alkylation of the α-position. The effect of the 6-methoxyl is to specifically enhance the nucleophilicity at C-2 [see resonance contributor (**24**)] with respect to C-3 and thus relatively to accelerate direct α-attack. A complementary study[56] (Scheme 24) used more reactive carbon electrophiles: allylating agents. Indole itself, for example, reacts[57] with dimethylallyl bromide in aqueous acetic acid–sodium acetate to give 3-dimethylallylindole smoothly even at room temperature. Methallylation of 3,3-dimethylallylindole gave a mixture of (**25**) and (**26**). About one half of (**25**), the product which must arise either by direct α-attack or by β-attack and then a C-3 → C-2 shift, was shown to result from the former pathway. This conclusion was drawn from a comparison with the product composition obtained by the acid-catalysed rearrangement of indolenine (**27**). 3-Methylindole gives[57] only 2-dimethylallyl-3-methylindole; this result probably reflects *some* direct α-attack but also the very much greater migratory aptitude of the dimethylallyl substituent in a 3-dimethylallylated intermediate.

(1) —i→ ; ii, iii ; (27) ; v ; iv ; (25) ; 1 : 1 ; 1 : 9 ; + ; (26)

i, $Me_2C{=}CHCH_2Br$, aq. AcOH, NaOAc; ii, MeMgI; iii, $MeCH{=}CHCH_2Br$; iv, $MeCH{=}CHCH_2Br$, aq. AcOH, NaOAc; v, CF_3CO_2H

SCHEME 24

The rate of solvolysis of indol-3-ylethyl tosylates is greater[58] than that of comparable homoaromatic tosylates. This is supported by labelling studies and interpreted as meaning that a three-membered spiro-cyclic indoleninium cation is involved. Exactly analogous intermediates can be invoked to explain the ring expansions[59] of tetrahydrocarbazolyl-methanol tosylates (Scheme 25) on solvolysis.

Simple benzyl halides are not reactive enough to alkylate indoles, but the more reactive *o*- and *p*-hydroxybenzyl halides will react.[60a] When 2-hydroxy-3,5-dimethylbenzyl chloride was allowed to react with 3-methylindole, the intriguing tetracycle (**28**) was obtained;[60b] one may envisage its formation by an intramolecular trapping of indoleninium intermediate by nucleophilic phenolic oxygen (Scheme 26).

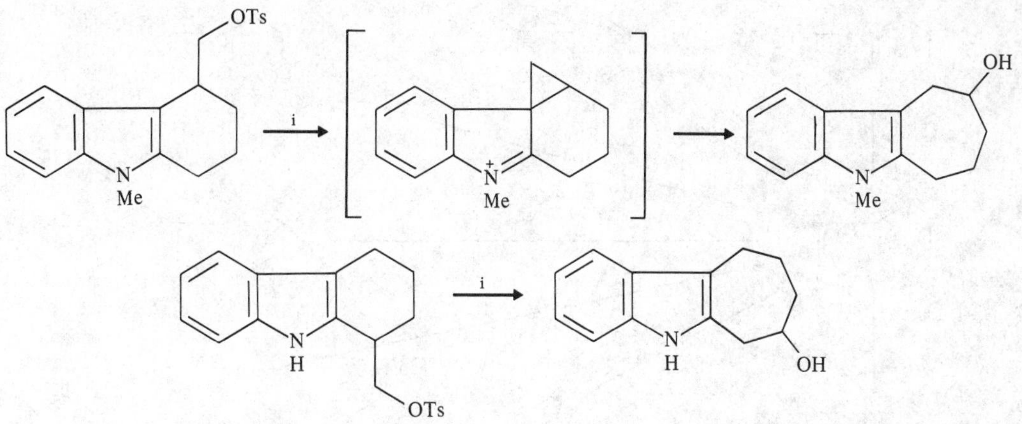

i, aq. Me_2CO

SCHEME 25

Polyphosphate ester has been shown[61] to convert 2,3-disubstituted indoles efficiently into 3-ethylindolenines. α-Chloroketones alkylate[62] indoles in acetic acid–phosphoric acid. Indole is reported[63] to alkylate with ethyleneiminium borofluoride to yield tryptamine. Olefins having carbonyl, cyano, nitro, or 2- or 4-pyridyl groups in conjugation are

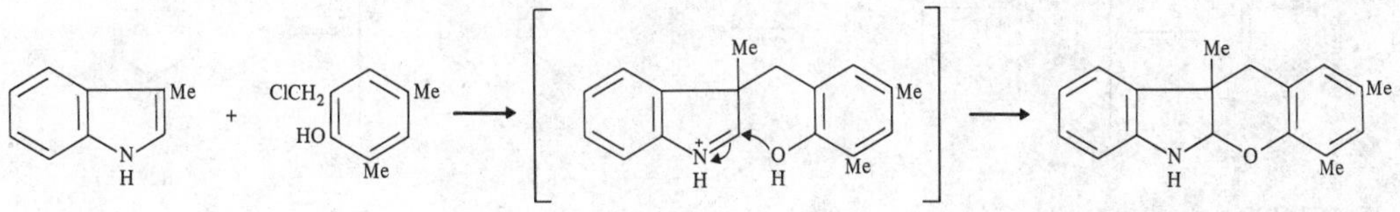

(28)

SCHEME 26

sufficiently electrophilic to alkylate indoles at the β-position. Acids are used to catalyse these conjugate additions in some cases. A study[64] of indole reactivity with nitroalkenes (in the absence of acid) showed 2-methyl- > 3-methyl- > 1,2-dimethyl- > 2-phenyl- > indole > 1-methyl- > 2,3-dimethyl-indole (no reaction).

Scheme 27 shows typical examples[64,65] of the types of substitution product which can be obtained in this way. Particularly noteworthy is the synthesis of *N*-acetyltryptophan and of nitroethylindoles, which can be easily transformed by reduction into tryptamines.

The formation (Scheme 28) of the tricyclic ketone (**29**) by reaction of 1,3-dimethylindole with mesityl oxide has mechanistic significance.[66] Its formation is viewed as involving β-attack to generate an indoleninium cation, which is in turn trapped by intramolecular enol addition to the immonium system at C-2. Though a similar result might be expected from reaction between methyl vinyl ketone and 3-methylindole, only the α-substitution product was isolated;[65a] an explanation may be that any tricyclic material which is formed, by a route like that in Scheme 28, can more easily revert to starting materials, by a sequence (available in this case) initiated by proton loss from nitrogen. This would then allow the more stable α-substituted indole to build up.

17.4.2.8 Reaction with carbonyl carbon

Two types of product can result from the acid-catalysed condensation of ketones or aldehydes with indoles (Scheme 29); these are ones in which either one or two moles of the indole combine with the carbonyl component. In the strongly acidic media necessary for reaction, both aldehydes and ketones condense with indoles having a free hetero-ring

R^1 = Me or H
R^2 = Me, H, or Ph

i, CH_2=$CHNO_2$;[64] ii, CH_2=$CHCO_2H$, AcOH;[65a] iii, CH_2=$C(NHAc)CO_2H$, AcOH, Ac_2O;[65b] iv, α-pyridyl-CH=CH_2, AcOH;[65c] v, ClCH=CHCOPh;[65d] vi, CH_2=CHCOMe, AcOH, Ac_2O,[65e]

SCHEME 27

(29)

i, Me_2C=CHCOMe, EtOH, HCl

SCHEME 28

(30) (31) (32) (33)

R = Me or H

i, Me_2CO, H_2SO_4;[68] ii, CH_2O, HCl;[70] iii, p-$Me_2NC_6H_4CHO$, EtOH, HCl; iv, Me_2CO, H^+.[72]

SCHEME 29

carbon position to give carbinols, but these are not isolated for acid-catalysed dehydration then occurs very rapidly. In some cases, from indoles with a free β-position, crystalline 3-alkylideneindoleninium salts have been obtained,[67] most derived from reaction with aromatic aldehydes, though the simplest so far obtained[68] is **(30)** from acetone and 2-methylindole. Highly coloured cations **(31)** are formed in solution from α- or β-unsubstituted indoles on reaction with *p*-dimethylaminobenzaldehyde in acidic ethanol: this forms the basis of the Ehrlich test for such indoles.[69]

The more usual outcome from indole/carbonyl compound interaction, however, is the isolation of a di-indolylmethane [*e.g.* **(32)**[70]]; these are formed by the nucleophilic addition of a second mole of the indole to a 3-alkylideneindoleninium cation intermediate (Scheme 29).

3-Methylindole also gives di-indolylmethanes, reaction occurring at the α-position.[71] Indeed, condensation at both hetero-ring positions has been observed in the interaction of acetone and indole itself, giving the pentacyclic product **(33)**.[72] Cyclic ketones react[73] with 1,2-dimethylindole in acetic acid–phosphoric acid to give 3-cycloalkenylindoles.

Condensation of indoles with dicarbonyl compounds can be complex.[74] One of the simpler cases[75] is the reaction of indole with hexane-2,5-dione in which both carbonyl groups react at the two available hetero-ring positions and 1,4-dimethylcarbazole is the result of the loss of two moles of water (Scheme 30).

Me; R = H; i; R; R = Me; ii; Me Me; N H; Me; N H; N H; N H; Ph

R=Me or H

i, $MeCOCH_2CH_2COMe$; ii, PhCHO, H^+

SCHEME 30

17.4.2.9 Reaction with immonium

The Mannich reaction proceeds easily with indoles and many examples are known. In its simplest form it involves formaldehyde–dimethylamine and (often) acetic acid, which together generate $CH_2{=}\overset{+}{N}Me_2$ as the attacking electrophile. The usual product type is a β-alkylaminoindole (Scheme 31[76]), but indoles with a blocked β-position can give *N*- or α-substituted products. Indole itself (Scheme 31) will give[76a] an *N*-substitution product but only under mild conditions with formaldehyde and dimethylamine in the absence of acid. The *N*-dimethylaminomethylindole is easily isomerized, presumably by reversion and re-attack at C-3, to give the thermodynamically more stable *C*-substitution product, 3-dimethylaminomethylindole (gramine).[76b] A study[77] of 3-propylindole also showed *N*-substitution to occur under mild conditions and that, in this case too, isomerization to a *C*-substituted product, necessarily the α-dimethylaminomethylindole, took place after

iii; N; N H; i; (1); ii; N; Me_2N; NMe_2; N H

i, CH_2O, AcOH, piperidine;[76c] ii, CH_2O, Me_2NH, 0 °C;[76a] iii, CH_2O, Me_2NH; AcOH.[76b]

SCHEME 31

more vigorous treatment. 2,3-Disubstituted indoles[78] gave *N*-dimethylaminomethyl products and 1,2,3-trialkylindoles can be substituted[79] on the α-side chain.

These findings raise the question as to whether all intermolecular indole Mannich substitutions proceed *via* *N*-dimethylaminomethyl intermediates (see also p. 421). This question is as yet unanswered although probably initial N-attack is *not* a necessary prerequisite for *C*-substitution, for there are a very large number of examples (Scheme 32[80] shows two) of the formation of tetrahydro-β-carboline derivatives by intramolecular

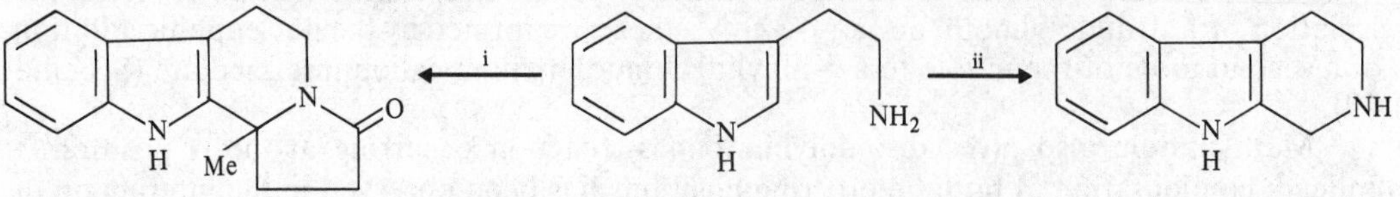

i, $MeCOCH_2CH_2CO_2H$;[80b] ii, CH_2O, H_2SO_4 [80a]

SCHEME 32

Mannich cyclization of tryptamine or tryptamine derivatives where steric factors preclude intramolecular attack at nitrogen. These cyclizations have been extensively studied because of their relevance for the synthesis of indole alkaloids. Several methods have been developed for the generation of requisite immonium systems other than by classical interaction of carbonyl, amine, and acid and two[81] of these are shown in Scheme 33.

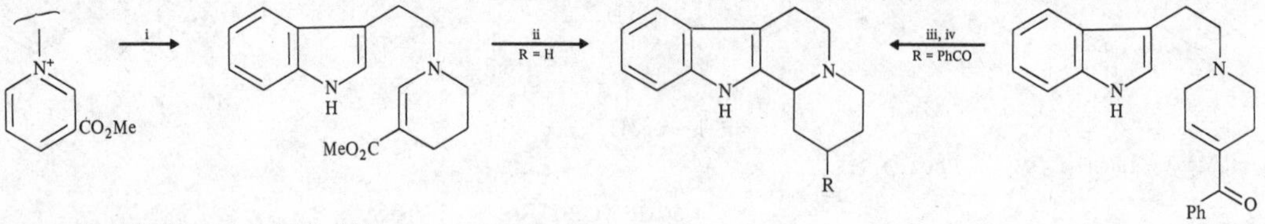

i, H_2, Pd/C; ii, H^+; iii, NaOMe, MeOH; iv, AcOH

SCHEME 33

A further mechanistic consideration relates to the question as to whether the observed α-substitutions are direct or go *via* 3,3-disubstituted indoleninium cations followed by migration. Certainly,[82] given aminoalkyl and alkyl groups at the β-position of such an intermediate, the former would migrate to the exclusion of the other, thus giving the observed α-Mannich product. One[83a] recent experiment[83b] (Scheme 34), which appears to represent the trapping of a 3-alkyl-3-aminoalkyl (indoleninium) cation, is the reaction of methyl tryptophanate with excess of formaldehyde in the presence of hydrogen and palladium, which led to variable yields of (**34**).

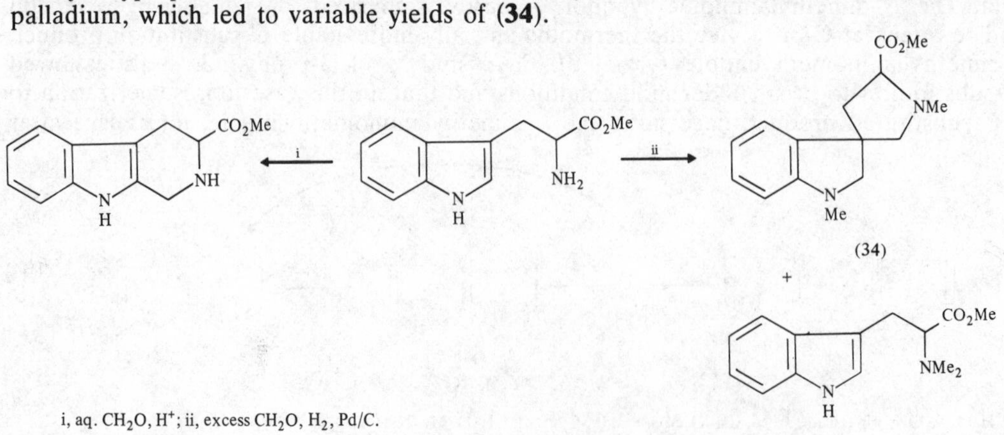

(34)

i, aq. CH_2O, H^+; ii, excess CH_2O, H_2, Pd/C.

SCHEME 34

The alternative view of α-Mannich substitutions rests on the known reversibility,[84] even at carbon, of amino-alkylation. In this case a sequence of fast β-attack *and* fast reversal reactions can be envisaged, the build up of α-substitution product proceeding by a slower, direct substitution. Certainly direct α-alkylation has been demonstrated (see p. 427) and also, as in alkylation, a 6-hydroxyl group has been shown[85] to accelerate the rate of formation of α-Mannich product, presumably therefore at least partially by direct attack.

At an appropriate concentration and at a pH where indole is incompletely protonated, a dimer and a trimer as well as higher oligomers[86b] are formed.[86a] This sensitivity to acids, particularly of those indoles having no group at C-2, must always be borne in mind in designing conditions for indole reactions. The mechanism for the formation of oligomers involves protonated indole acting as an electrophilic immonium system which Mannich-substitutes unprotonated indole; Scheme 35 shows how the formation of dimer (**35**) and trimer (**36**) are envisaged. 3-Methylindole gives a dimer by substitution at an α-position. 2-Methylindole does not dimerize, the protonated form being insufficiently electrophilic at the hindered α-carbon.

(1) ⇌ [...] (35) ... NH_2 + ... (1) ... NH_2 ... (36)

SCHEME 35

17.4.2.10 Acylation

Formylation[87] of indoles by the Vilsmeier–Haack procedure (Scheme 36) is very efficient and gives 3-formylindoles; *N*-formylation[88] occurs with β-substituted indoles but the normal, strongly alkaline work-up immediately hydrolyses these back to starting

i,iii ← (1) → ii [NMe_2] → iii CHO

NH → iv N

i, Me_2NCOMe, $POCl_3$; ii, DMF, $POCl_3$; iii, NaOH; iv, $POCl_3$

SCHEME 36

indole. Even deactivated indoles, like 2-ethoxycarbonylindole, β-formylate[89] easily by this method. The use of dimethylamides of acids other than formic allows the method to be extended[90] for the preparation of 3-ketones. Intramolecular Vilsmeier reactions[91] with N_b-acyltryptamines gives dihydro-β-carbolines, ring formation stabilizing the imine bond.

One of the few mechanistic studies[88] of indole reactivity was carried out employing the Vilsmeier reaction: it was shown that in order of reactivity 2-methyl- > 1-methyl- > indole > 2,3-dimethyl- [95% reaction goes at N] ≈ 3-methylindole (at N). The work also provided an estimate of relative $\beta:\alpha$ reactivity of 460.

The β-position of the indole nucleus is sufficiently nucleophilic to react with acid anhydrides at elevated temperatures without the necessity for a Lewis acid catalyst. Thus indole can be β-acetylated[92] with acetic anhydride (the presence of acetic acid helps to suppress *N*-acetylation) to a mixture of 1-acetyl- and 1,3-diacetyl-indoles. *N*-Acyl groups on indoles are extremely easily hydrolysed by mild base. *N*-Acetylindole, which results from reaction with acetic anhydride–sodium acetate, presumably by way of the indolyl anion (see p. 436), is insufficiently reactive to ring acylate. 3-Methyl- and 1,3-dimethyl-indoles can be α-acetylated[93] with acetic anhydride, but this requires the use of boron trifluoride as Lewis acid catalyst. Intramolecular acylations at the indole α-position[93,94] have been catalysed by polyphosphoric acid (Scheme 37).

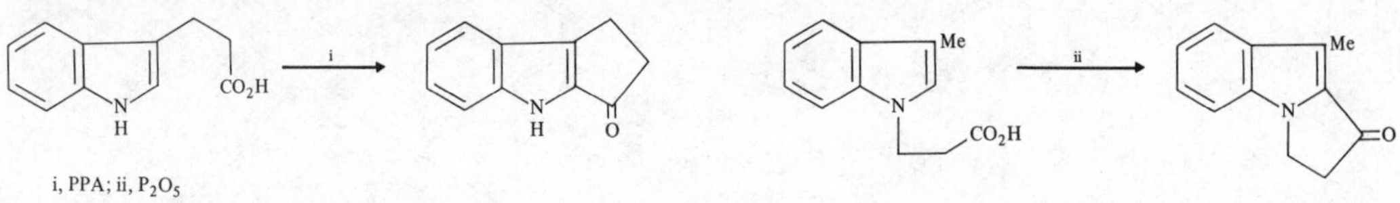

i, PPA; ii, P_2O_5

SCHEME 37

Trifluoroacetic anhydride acylates[95] indole at C-3 in dimethylformamide solution, but mainly at nitrogen in dichloromethane. Mild alkaline hydrolysis of 3-trifluoroacetylindoles is a good way of preparing indole-3-carboxylic acids.

Oxalyl chloride reacts[96] smoothly with β-unsubstituted indoles at room temperature and thus provides, by subsequent esterification and reduction, a route[97] to tryptophols or, by reaction with ammonia or an amine and then reduction, a useful method[96b] for the synthesis of tryptamines (Scheme 38).

N-Acetyl- or *N*-methyl-2,3-dimethylindoles acetylate at C-6[98] in the presence of aluminium chloride.

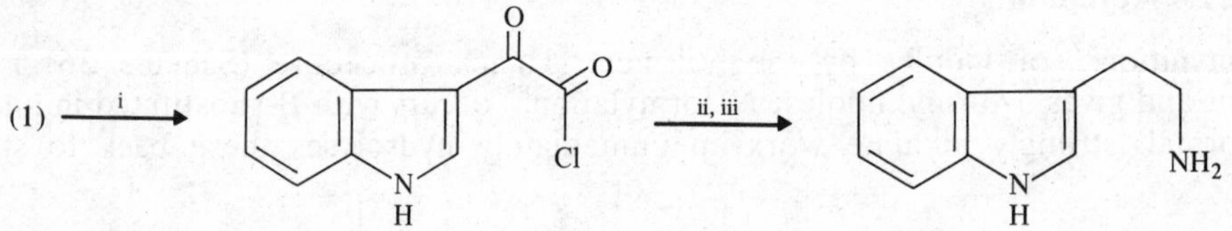

i, ClCOCOCl; ii, NH_3; iii, $LiAlH_4$

SCHEME 38

17.4.3 NUCLEOPHILIC ATTACK AND DEPROTONATION

17.4.3.1 Acidity and structure of salts

The most acidic proton on the indole nucleus is the *N*-hydrogen. With a pK_a of 16.97[99a] it is only slightly more acidic than that of an alcohol, but indoles like alcohols can be easily converted into salts by strong bases such as sodium hydride, alkyl Grignard reagents or, in most cases, potassium t-butoxide. The deprotonated species is known as an indolyl anion. Indoles carrying appropriate acidifying substituents, 2- or 3-acyl groups for example, are considerably stronger acids (3-formylindole has a pK_a of 12.36[99a] while that of 2-formylindole is 14.00[99b]) and require correspondingly weaker conditions for deprotonation, aqueous or alcoholic sodium hydroxide being sufficient. It is interesting to note the greater acidifying effect of the extended conjugation possible in the 3-acylindolyl anion, represented by canonical form (**38b**), and not open to the α-isomer.

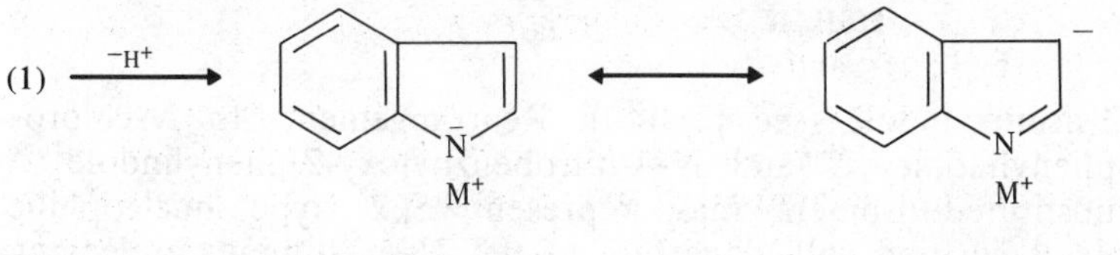

(1) $\xrightarrow{-H^+}$

(37) a; M = Na, K, or Li

b; M = MgBr or MgI

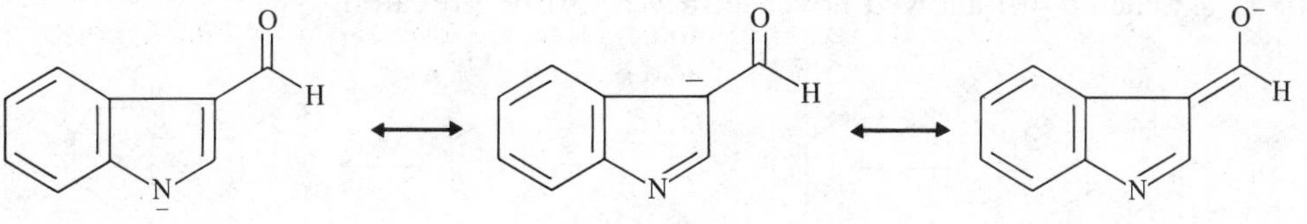

(38a)

(38b)

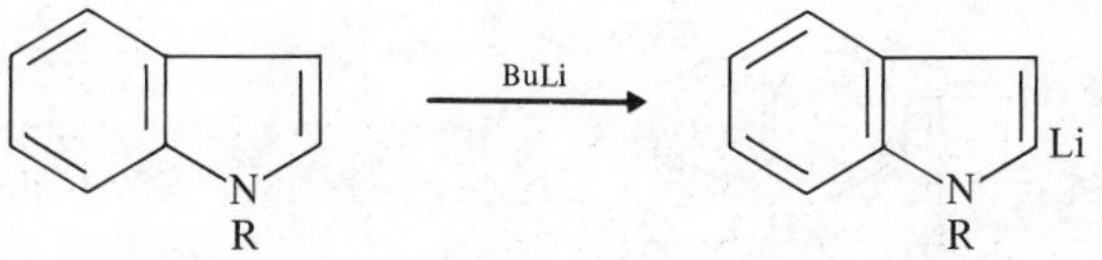

BuLi

R = Me or $PhSO_2$

Because of the differing reactivity of the ambident indolyl anion in association with alkali metal cations (**37a**) as against magnesium in halomagnesylindoles (indolyl Grignard reagents) (**37b**) (see p. 437), there was for a long time a controversy over their structures. N.m.r. studies[100] have revealed that both types are essentially ionic in character and both have metal associated with nitrogen, but they differ in the tightness of that association. Indolyl Grignard species are strongly associated in all solvents examined except HMPT, while alkali metal salts are looser ion pairs. It was shown,[100] for example, that rapid exchange of *N*-hydrogen occurs in a mixture of indole and a sodioindole, whereas the reverse is true for indole and indolylmagnesyl bromide except in HMPT solution.

N-Methyl-[101a] and *N*-benzenesulphonyl-indole[101b] are deprotonated by an alkyllithium at the α-carbon. An instructive comparison[101c] of the effects of these two nitrogen substituents is seen in the deprotonation of 5-methoxy-1-methylindole and 5-methoxy-1-benzenesulphonylindole. In the former case competitive deprotonation led to lithiation at the α-position *and* at both homocyclic ring positions *ortho* to the methoxyl group, whereas in the latter case the *N*-benzenesulphonyl substituent increases the acidity of the α-proton sufficiently that only deprotonation at C-2 was observed.

17.4.3.2 Nucleophilic substitution

Examples of nucleophilic addition or substitution on a neutral indole nucleus are very rare. Phenyl Grignard reagent adds to 1-methyl-3-benzoylindole (Scheme 39);[102a] comparable 1,4-conjugate addition to 2-acylindoles apparently does not occur.[102b] The silver-ion-assisted displacement of iodine in 3-iodoindole by acetate is a convenient method[103]

PhMgBr

SCHEME 39

for the preparation of 3-acetoxyindole (see p. 449). Rearrangements to *N*-chloroindole,[104a] *N*-tosyloxy-2-phenylindole,[104b] and *N*-4-nitrobenzoyloxy-2-phenylindole[104c] to their respective 3-substituted isomers may represent S_N2' type nucleophilic displacements at the indole β-position with departure of the *N*-substituents as leaving groups. The first clear-cut examples of the operation and synthetic potential of such a process have recently been provided:[105] intramolecular alcoholate substitution of *N*-benzenesulphonyl-2-acylindoles with departure of the *N*-substituent as benzenesulphinate (*e.g.* Scheme 40) allowed novel tetracycles to be prepared.

i, NaOH, H_2O, EtOH

SCHEME 40

17.4.3.3 Reactions of deprotonated indoles

The synthetic utility of deprotonated indoles lies in their reactivity with electrophiles and thus in the overall alkylation, acylation, carboxylation, *etc.* of the starting indole. There is a preparatively useful difference between indolyl Grignard reagents and other salts. At the simplest level one can summarize this difference by saying that indolyl Grignard reagents react with electrophiles mainly at carbon, the indole β-carbon, whereas alkali-metal salts react mainly at nitrogen. This difference in reactivity is attributed[100,106] to the closer association of the nitrogen and magnesium and the looser, more ion-separated character of the sodium, potassium, and lithium derivatives. Parallels can be drawn with the ambident reactivity of enolate anions, in association with different counter ions. This view is substantiated by the reaction of the indolyl Grignard reagent in HMPT where, as the solvent promotes separation of the metal and the nitrogen, its 'typical' reactivity is completely inverted[106] and it reacts with methyl iodide only at nitrogen. There is a slight tendency for alkali-metal salts to react[106] partially at the β-carbon, more for lithium than for sodium than for potassium, but this can be eliminated in analogous fashion, by carrying out the reaction in a dipolar ionizing solvent. In addition to the differences discussed above, there is a further aspect to the reactivity of indolyl anions which can affect the N/C-3 product ratio: this is the nature of the electrophile.[107] Thus halides which tend to react by an S_N1 process give more β-attack than do, for example, primary alkyl halides.

The reaction of indolyl Grignard reagent with water regenerates indole. A study[108] using D_2O revealed the process to be complex, however, N and/or C-3 deuteration occurring and depending, in a rather strange way, on the quantity of D_2O added. 3-Deuteroindole, which might have been expected to be easily available in this way, could be prepared[109] by exchange *via* an indolyl anion, in aqueous alkaline solution, though it is more usually prepared by acid-catalysed exchange (see p. 418).

N-Alkylation of indoles was traditionally best achieved[110a] using a sodium salt prepared by reaction with sodamide in liquid ammonia. Recently much more convenient methods have been evolved for high yielding *N*-alkylations. For example, all of the following conditions give more than a 90% yield in the conversion of indole into *N*-methylindole: potassium hydroxide in DMSO,[110b] sodium hydride in HMPT,[110c], thallous ethoxide in DMF,[110a] or phase-transfer catalysis[110e,f] with mixtures of benzene, 50% aqueous sodium hydroxide, and a quaternary salt such as tetra-n-butylammonium hydroxide.[110e] Simpler conditions still suffice for the alkylation of more acidic indoles; 3-formylindole, for example, *N*-alkylates[111] with sodium carbonate as base.

A large variety[112] of C-3 alkylations, acylations, *etc.* have been achieved using the indolyl Grignard reagent (see Scheme 41[113]).

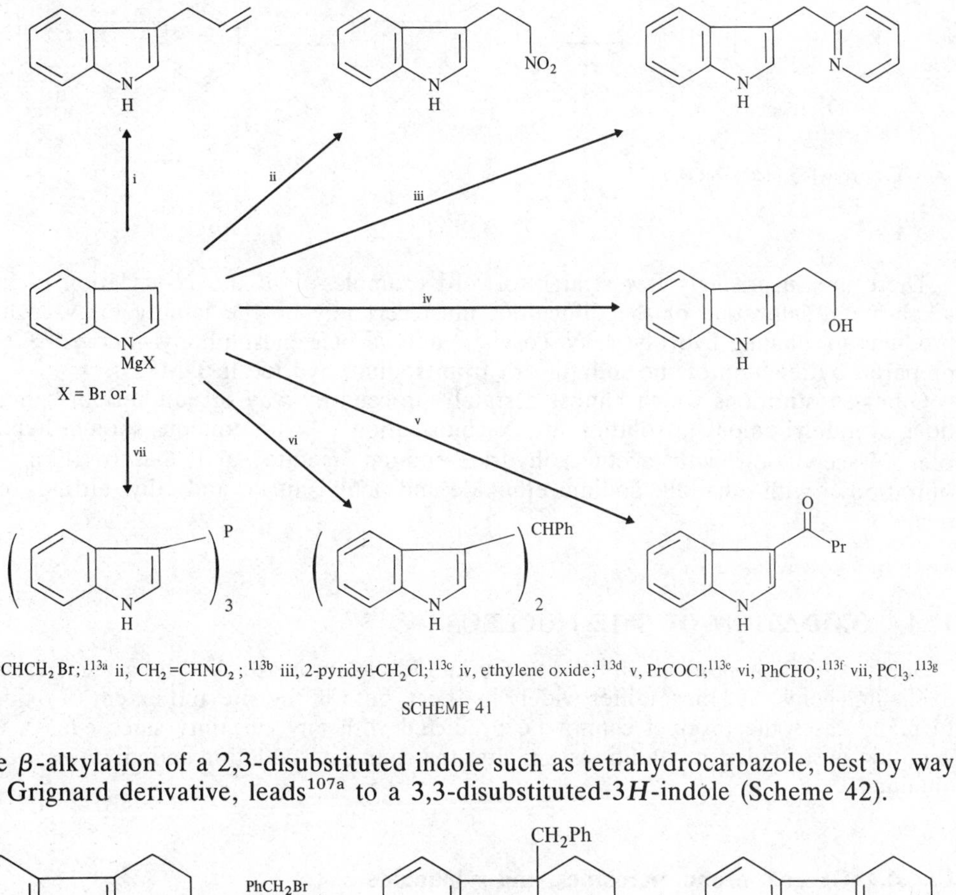

i, $CH_2{=}CHCH_2Br$;[113a] ii, $CH_2{=}CHNO_2$;[113b] iii, 2-pyridyl-CH_2Cl;[113c] iv, ethylene oxide;[113d] v, PrCOCl;[113e] vi, PhCHO;[113f] vii, PCl_3.[113g]

SCHEME 41

The β-alkylation of a 2,3-disubstituted indole such as tetrahydrocarbazole, best by way of its Grignard derivative, leads[107a] to a 3,3-disubstituted-3*H*-indole (Scheme 42).

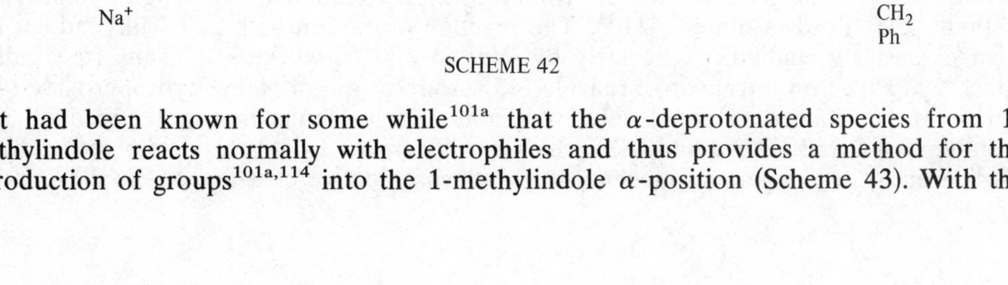

SCHEME 42

It had been known for some while[101a] that the α-deprotonated species from 1-methylindole reacts normally with electrophiles and thus provides a method for the introduction of groups[101a,114] into the 1-methylindole α-position (Scheme 43). With the

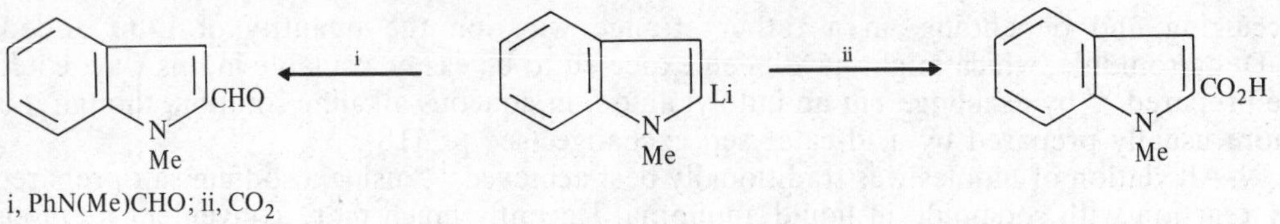

i, PhN(Me)CHO; ii, CO_2

SCHEME 43

introduction[115] of *p*-toluenesulphonyl and benzenesulphonyl[101b] groups as removable protecting groups for nitrogen in indole chemistry and the finding[101b] that 1-benzenesulphonylindole could likewise be deprotonated and substituted at the α-carbon, a unique method[101b] became available for the introduction of substituents into the α-position of an *N*-unsubstituted indole (*e.g.* Scheme 44). The protecting group can normally be removed by alkaline hydrolysis and is particularly easily cleaved (during work-up) from 2-acyl-1-benzenesulphonylindole products.

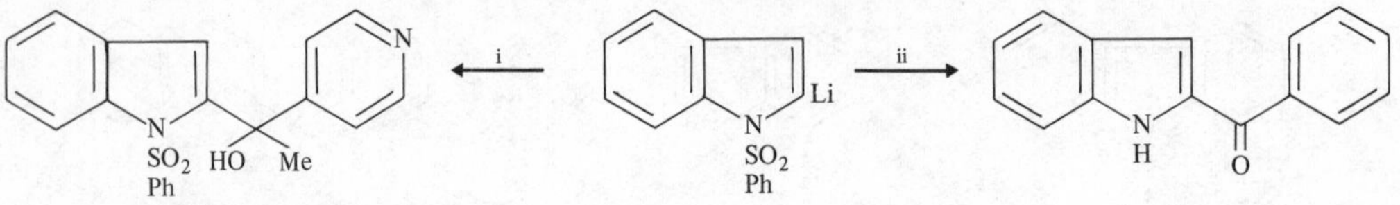

i, 4-pyridyl-COMe; ii, $PhCO_2Et$.

SCHEME 44

There are surprisingly few straightforward examples[116] of the *N*-acylation of indolyl alkali-metal salts; one of the difficulties must certainly be the lability of *N*-acylindole products to alkaline hydrolysis. *N*-Tosyl-[115] and *N*-benzenesulphonyl-indoles[101b] can be prepared by reaction of the sodium salt from sodium hydride in DMF.

Other substitutions which almost certainly proceed by way of equilibrium concentrations of indolyl anions in solution are *N*-chlorination[104a] with aqueous sodium hypochlorite, *N*-acetylation with acetic anhydride–sodium acetate, and β-nitrosation[117] and -nitration[118] with ethanolic sodium ethoxide and pentyl nitrite and ethyl nitrate, respectively.

17.4.4 OXIDATION OF THE NUCLEUS

Because of its high electron density the indole system is attacked by a great variety of oxidizing agents, but these differ widely in effect, both in the site and extent of oxidation. In many cases the product cannot be predicted with any certainty since even a minor variation in structure or conditions can alter the course of oxidation and mixtures are often obtained.

17.4.4.1 Oxygen, ozone, peroxides, and sulphoxide

In the presence of air and light, indole itself is autoxidized to indoxyl (**39**), which reacts further to give indigo (**40**) and a 'trimer' (**41**) by oxidative coupling. Similarly, 2-methylindole affords a dimer (**42**).[119] The reaction of oxygen with 2,3-dialkylindoles has been extensively studied, particularly by Witkop and co-workers.[120] Thus free radical attack by oxygen on tetrahydrocarbazole (**43**) produces an indolenyl hydroperoxide (**44**), which can be isolated in good yield from non-polar solvents. In a polar medium it is unstable and forms a dioxetan (**45**) which readily opens to a keto-lactam (**46**). This is considered to be the general mechanism by which oxidative cleavage of the 2,3-bond in

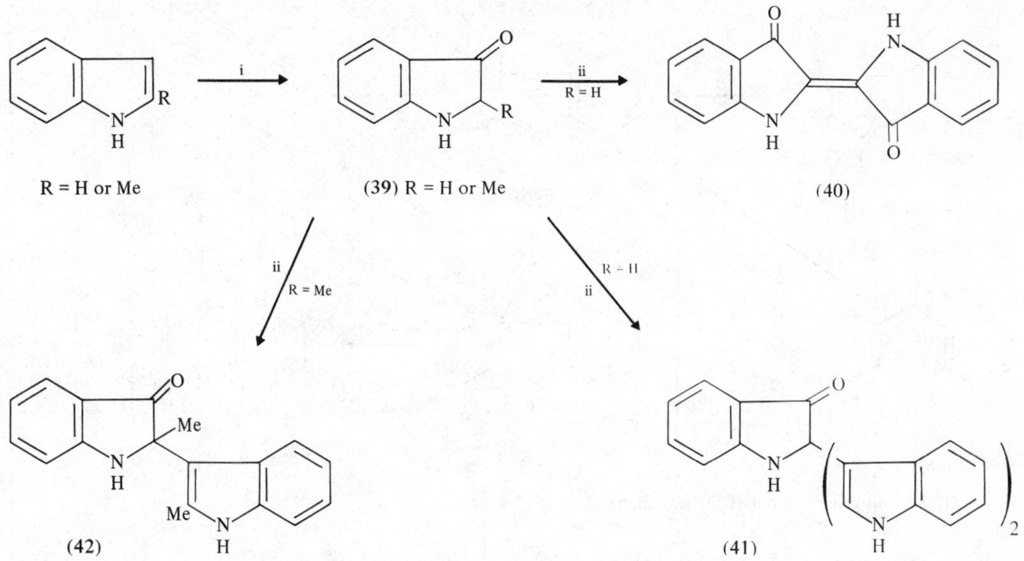

R = H or Me

(39) R = H or Me

(40)

(42)

(41)

i, O_2, *hν*; ii, [O]

indoles occurs. For instance, the metabolism of tryptophan *in vivo* to formylkynurenine (**47**) requires oxygen and a specific enzyme, and it is likely that a 3-hydroperoxide is again involved.

In addition to the expected keto-amides (**48**), autoxidation of 2,3-diethyl- and 2,3-dimethyl-indoles gave 2-acylindoles (**49**).[121] One of several mechanisms proposed for the latter is outlined in Scheme 45, involving tautomerism of the 3-hydroperoxyindolenine (**50**) to (**51**) and allylic rearrangement to an indolic hydroperoxide (**52**) which decomposes to (**49**).

When oxygen reacted with 2-isopropyl-3-methylindole in hexane solution, the initially formed 3-hydroperoxide (**53**) underwent further oxidation to 3-methyldioxindole (**54**) and the acylamide (**55**) as indicated in Scheme 46.[122a] The two products are formed by competing fragmentations of (**56**) triggered by fission of the peroxide and involving two equivalent hydroxyl groups. Autoxidation of 2-benzyl-3-phenylindole afforded an analogous dioxindole (**57**), but oxygen and a catalyst oxidized[122b] the methylene group to give 2-benzoyl-3-phenylindole.

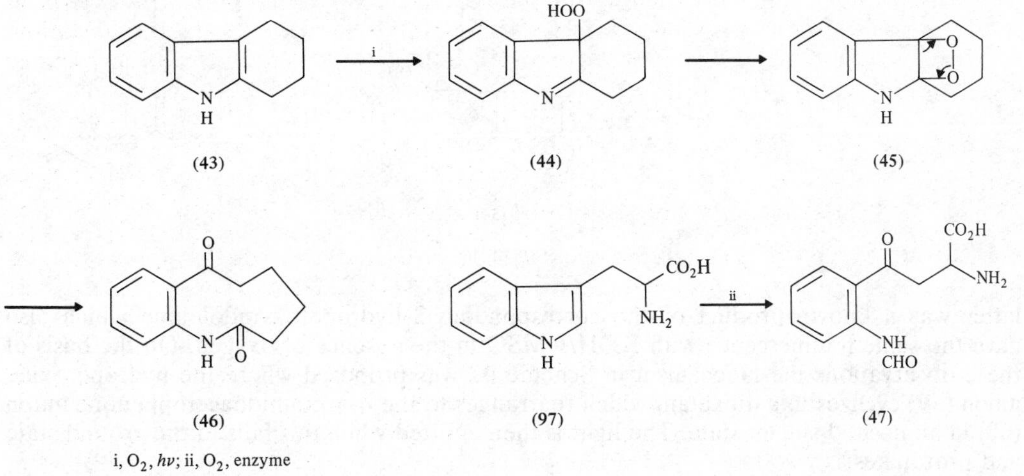

(43)

(44)

(45)

(46)

(97)

(47)

i, O_2, *hν*; ii, O_2, enzyme

R = H or Me

(48) (50) (49) (51) (52)

i, O_2, $h\nu$, petroleum ether; ii, H_2O, heat; iii, heat

SCHEME 45

Yet another mode of transformation was shown by 2,3-dimethyl- and 2-methyl-3-ethylindole, which both gave coupled products (**58**), presumably formed by attack of unreacted indole on an intermediate 3-hydroperoxide.[123]

Many indoles exhibit chemiluminescene when exposed to oxygen in alkaline solution in an aprotic medium. Under these conditions, 2,3-dimethylindole emitted light at 520 nm which correlated with the fluorescence spectrum of *o*-acetamidoacetophenone.[124] The

(53) (56) (56) (54) (55)

SCHEME 46

latter was a known product of the corresponding 3-hydroperoxyindolenine which also gave the same luminescence with KOH/DMSO in the absence of oxygen. On the basis of these observations the mechanism in Scheme 47 was proposed where the hydroperoxide anion (**59**) cyclizes to a dioxetan, which rearranges to the *o*-acetamidoacetophenone anion (**60**) in an excited singlet state. The light is then emitted when this falls to the ground state and protonates.

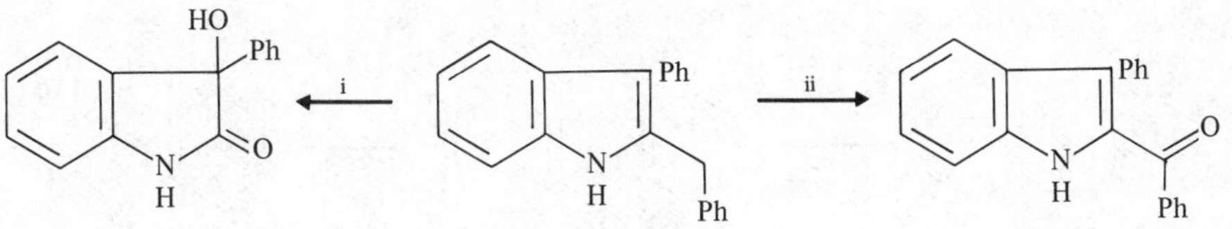

(57)

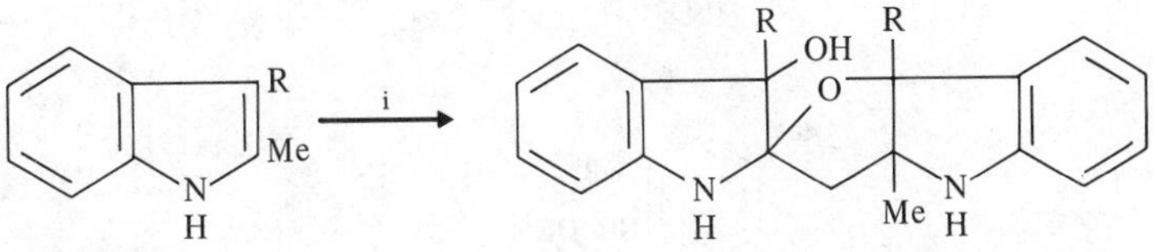

R = Me or Et

(58) R = Me or Et

i, O_2, $h\nu$; ii, Pt, O_2

Irradiation of (**61**) in the presence of oxygen and a sensitizer leads *inter alia* to insertion of an oxygen atom into the 3,3α-bond to give (**62**) and (**63**).[125]

In general, ozonolysis of simple 2- and 3-substituted indoles cleaves the 2,3-bond to *o*-acylaminophenyl ketones and aldehydes (**64**).[126] Hydrogen peroxide and peracids lead to the same coupled products as autoxidation with indole and 2-methylindole; 3-methylindole is oxidized to 3-methyloxindole by peracetic acid, but cleaved to *o*-formamidoacetophenone by perbenzoic acid.[122] A more general reagent for 2,3-bond cleavage is hydrogen peroxide with ammonium molybdate.[127] Treatment of 1-methylindoles with dibenzoyl peroxide affords the corresponding 2- and 3-benzoyloxyindoles and in 62–81% yield.[128]

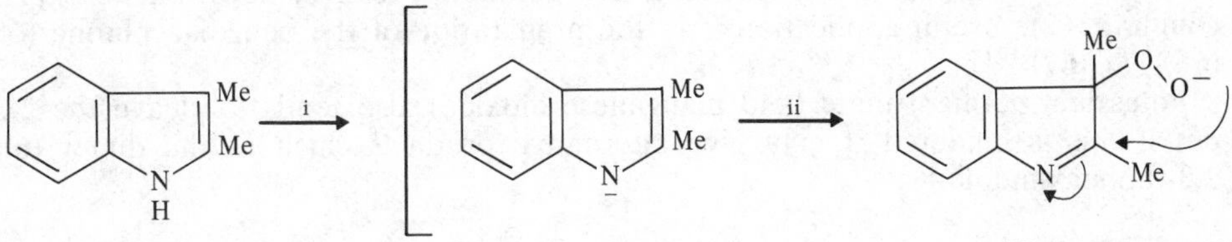

(59)

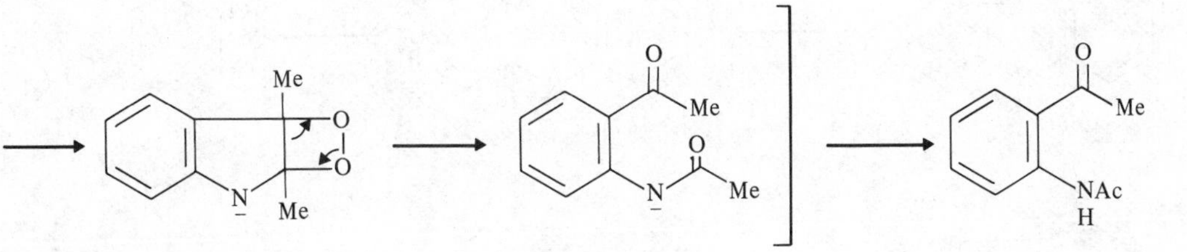

(60)

i, KOH; ii, O_2; iii, H^+

SCHEME 47

An excellent new general procedure for oxidizing 3-substituted indoles to oxindole in high yield uses dimethyl sulphoxide in concentrated hydrochloric acid.[129a] A likely mechanism involves nucleophilic attack by the sulphoxide on a 3*H*-indolium cation followed by loss of dimethyl sulphide (Scheme 48). Recently a method for oxidizing 2,3-dialkylindoles to the 3-acyl derivatives in high yield with DDQ has been reported.[129b]

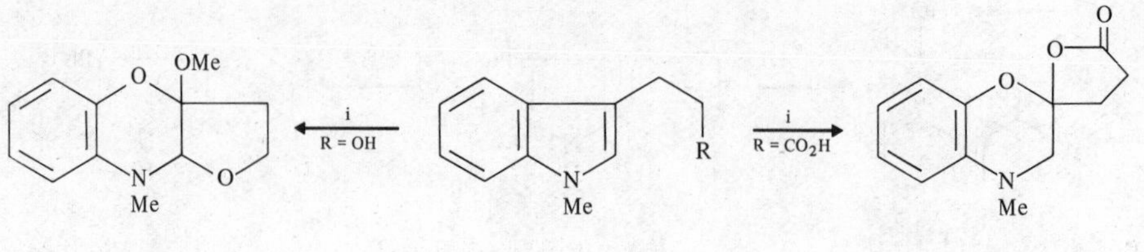

(62) (61) (63)

i, O_2, sensitizer, MeOH, –70 °C

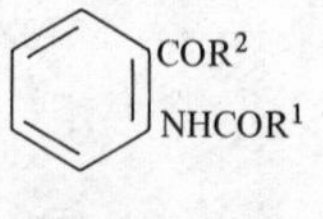

(64)

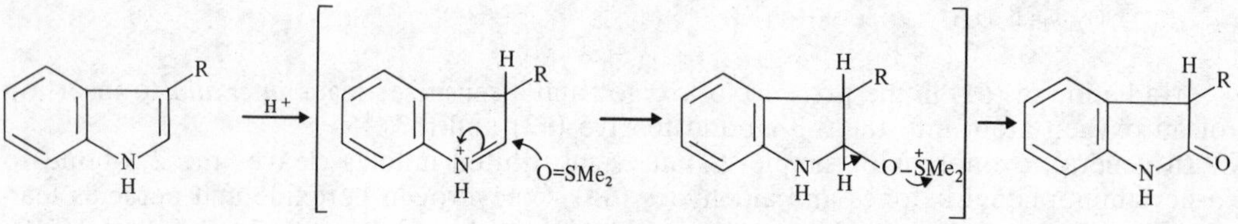

SCHEME 48

17.4.4.2 Inorganic oxidizing agents

Although chromic acid often cleaves indoles to *o*-acylamino-anthranilic acids and -ketones, the yields tend to be low unless the benzene ring is stabilized by an electron-withdrawing group.[130] Other products are oxindoles, isatins, and dimers from indoxyl coupling. One useful application is in the preparation of the benzdiazepinone (**65**) from (**66**) *via* (**67**).[131]

Potassium permanganate (and manganese dioxide) also tends to cleave the 2,3-bond, but if this is hindered it may give alternative products, such as the dimer (**68**) from 2,3-diphenylindole.[132]

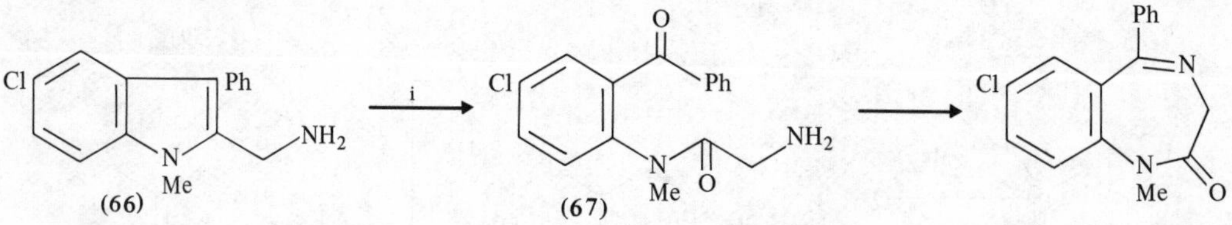

(66) (67) (65)

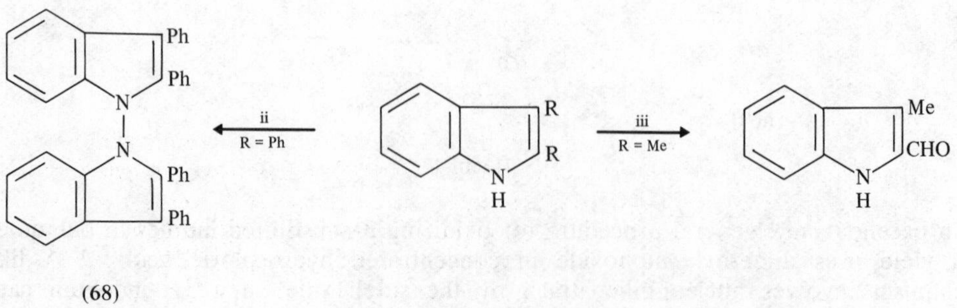

(68)

i, CrO_3, AcOH; ii, $KMnO_4$; iii, HIO_4

A much milder and more specific reagent for the cleavage is sodium periodate, which gives high yields of keto-amides from substituted indoles, *e.g.* 85% in the case of 2,3-dimethylindole.[133] On the other hand, reaction with periodic acid takes a different course; oxidation of the α-methylene group affords 2-formyl-3-methylindole in low yield. The former reagent has been used in the conversion[134] of tetrahydro-β-carbolines (**69**) into γ-quinolones (**70**) *via* keto-lactams (**71**).

Lead tetra-acetate can also oxidize α-methylene groups to carbonyls or *gem*-diacetates, but its most extensive use has been in transformations of tetrahydro-β-carboline alkaloids (Scheme 49[135]). The initial product is a β-acetoxyindolenine (**72**), which can eliminate acetic acid to give (**73**); the latter on heating with an excess of lead tetra-acetate forms the carbolinium system (**74**). Alternatively, (**72**) can rearrange in acid solution to an oxindole (**75**), and in alkali to an indolinone (**76**).

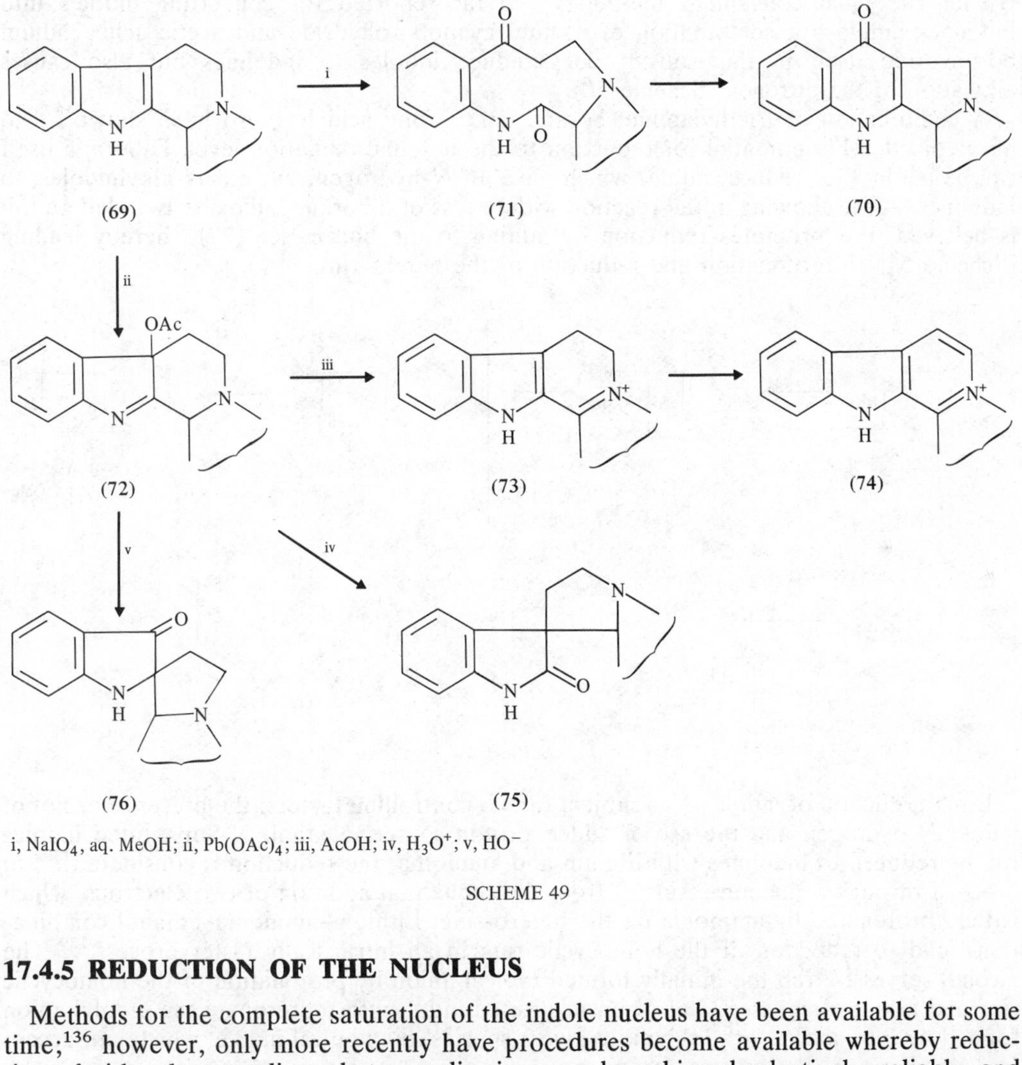

(69) (71) (70) (72) (73) (74) (76) (75)

i, $NaIO_4$, aq. MeOH; ii, $Pb(OAc)_4$; iii, AcOH; iv, H_3O^+; v, HO^-

SCHEME 49

17.4.5 REDUCTION OF THE NUCLEUS

Methods for the complete saturation of the indole nucleus have been available for some time;[136] however, only more recently have procedures become available whereby reduction of either homocyclic or heterocyclic rings can be achieved selectively, reliably, and smoothly. Most of the methods available for reduction of the hetero-ring employ an acidic medium and almost certainly proceed by reduction of an equilibrium concentration of 3*H*-indolium cation.[137a] One of the difficulties attending the use of acids with indoles is the possibility of oligomerization (see p. 433); in this context combinations of zinc dust and phosphoric acid[137b] and platinum, hydrogen, ethanol, and fluoroboric acid[137a]

i, H_2, Pt, AcOH; ii, Zn, H_3PO_4 or H_2, Pt, EtOH, HBF_4 or $NaCNBH_3$, AcOH or $Et_3N{\cdot}BH_3$, 6N HCl, heat; iii, $NaBH_4$, AcOH

SCHEME 50

minimize this difficulty and provide means for reduction to 2,3-dihydroindoles (indolines). By far the most convenient method[137c] so far reported for converting indoles into indolines employs a combination of sodium cyanoborohydride and acetic acid; sodium borohydride itself in this solvent does reduce indoles to indolines but also causes alkylation of the nitrogen (Scheme 50).

A combination of triethylamine, borane, and strong acid has also been shown[137d] to have considerable potential for reduction to the indoline oxidation level. Diborane itself can be made[138] to reduce indoles which have an *N*-hydrogen, but not *N*-alkylindoles, to indolines. After allowing initial reaction with excess of diborane, alkoxide is added and, it is believed, this promotes reduction by adding to the boron, see **(77)**, thereby leading (Scheme 51) to protonation and reduction of the hetero-ring.

(77)

i, excess B_2H_6; ii, NaOMe, MeOH

SCHEME 51

Birch reduction of indoles[139] is subject to two controlling factors: the presence or not of indole *N*-hydrogen and the use of added proton source, alcohol. *N*-Substituted indoles can be reduced to indolines with lithium and ammonia; the reduction is considered[139a] to proceed through a dianion resulting from the sequential addition of two electrons, which is then protonated by ammonia on the hetero-ring. Lithium–ammonia–ethanol combinations lead to reduction of the homocyclic ring in an intrinsically faster process.[139] The alcohol serves to trap the initially formed radical anion by protonation of the homocyclic ring and also, in the case[139b] of *N*-unsubstituted indoles, to reprotonate the indolyl anion formed by interaction with lithium, to give a species, the neutral indole, which *can* accept

i, Li, NH_3; ii, Li, NH_3, EtOH

an electron. Thus by this method 5-methoxy-1-methylindole cleanly gave[139a] 4,7-dihydro-5-methoxy-1-methylindole, and indole itself[139b] gave a mixture of 4,7-dihydro- and 4,5,6,7-tetrahydro-indole.

17.4.6 FREE RADICAL ATTACK

Remarkably little is known of the reactivity of indoles to free radical attack and rationalization on the basis of currently known facts is difficult. Interaction[140] of indole with benzyl radicals led, very inefficiently, to a mixture of 1- and 3-mono- and 1,3- and 2,3-di-benzylated indoles. Hydroxylation[141] with hydrogen peroxide–ferrous sulphate, on the other hand, gave very poor yields[141a] of products, 4-, 5-, 6-, and 7-hydroxyindoles, resulting from attack in the homocyclic ring; the 4-HO isomer was the major product in another study.[141b] Somewhat more efficient were the reactions[142] ensuing when benzoyl peroxide was decomposed in the presence of 1-methylindole and 1,3-dimethylindole, though here again substitution at different positions was observed. The former gave 3-benzoyloxy-1-methylindole while the latter was substituted at the 2-position.

17.4.7 REACTIONS WITH CARBENES

Of the small number of examples of carbenoid attack on indoles, by far the most studied[143] is that between an indole and a halocarbene which results, depending on substitution and detailed reaction conditions, in a 3-formylated product (the result of classical Reimer–Tiemann reaction, possible only for a 3-unsubstituted indole), a 3-halomethyl-3*H*-indole (from 2,3-disubstituted indoles), or a combination of these different types. There is no unanimous view[143c,144] as to the details of these processes, but an entirely plausible sequence[143c] involving a halocyclopropyl intermediate [such as (**78**)] explains all types of product, in its ability to ring open by two modes. Both the position of halogen in ring-expanded products[145] from dichlorocarbene and the position of radio-tracer[143c] in ring-expanded products from reaction between ^{14}CHBr and 3-methylindole are in accord with such a sequence. Scheme 52 shows how the processes would operate for 5-methoxyindole.[145]

Diazoesters and diazoketo-esters generate carbenes which react[146] with indoles having a free β-position to give products of overall substitution at that position. It is not known whether these reactions proceed *via* a cyclopropyl intermediate.

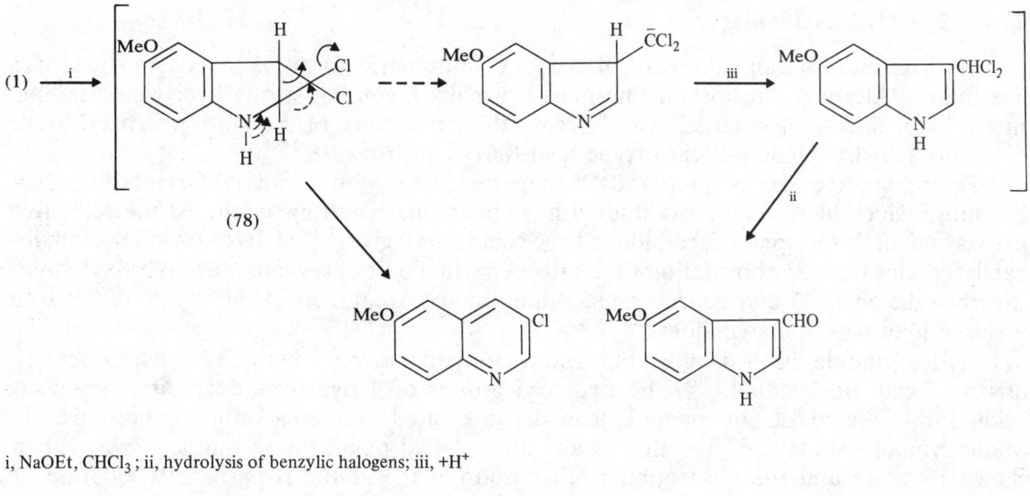

i, NaOEt, $CHCl_3$; ii, hydrolysis of benzylic halogens; iii, $+H^+$

SCHEME 52

17.4.8 OXYINDOLES

17.4.8.1 Homocyclic oxyindoles

Indoles, such as the important vasoconstrictor and neurotransmitter substance serotonin (**79**), in which a hydroxyl substituent is attached to the homocyclic ring, show normal phenolic properties. An interesting internal comparison of the susceptibilities to electrophilic attack of the indole α-position with those *ortho* to phenolic hydroxyl emerges

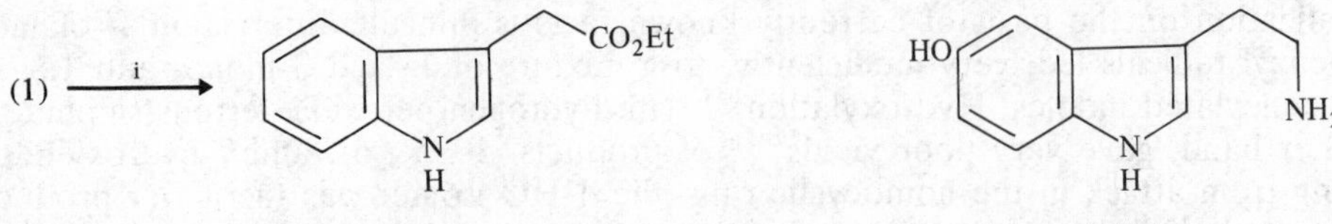

(1) $\xrightarrow{\text{i}}$ [3-indolylacetic ester, CO_2Et]; (79) [serotonin: HO, NH_2, N–H]

i, N_2CHCO_2Et, Cu, heat

from deuteration and aminomethylation studies[147] on such indoles. In aqueous alkaline solution at 60 °C, 5- and 6-hydroxyindoles exchange on *o*-hydrogen, at C-4 and C-7 respectively, the former faster and the latter somewhat slower than their indole β-hydrogen, which exchanges much faster than any other.[147a] All four of these indoles undergo[147b] the Mannich reaction under neutral conditions in the phenolic ring, rather than in the hetero-ring, in each case giving only one product by attack at C-5 (4-hydroxyindole), C-4 (5-hydroxyindole), C-7 (6-hydroxyindole), and C-6 (7-hydroxyindole), respectively.

Homocyclic indolic quinones have been obtained[148] (Scheme 53) by oxidation of hydroxyindoles with potassium nitrosodisulphonate (Fremy's salt) in cases where the hetero-ring carries a substituent as protecting group.

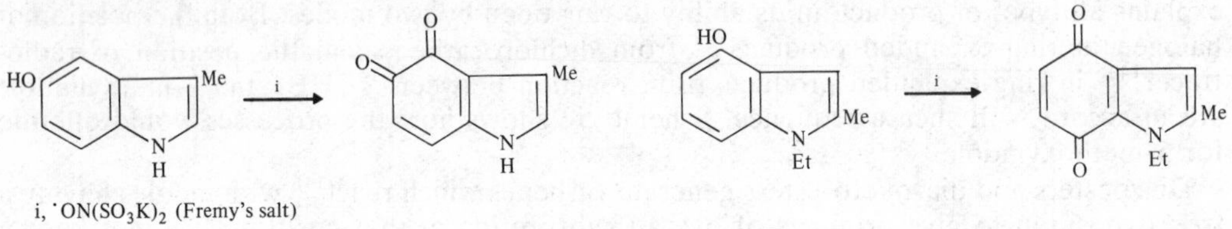

i, $\cdot ON(SO_3K)_2$ (Fremy's salt)

SCHEME 53

17.4.8.2 1-Hydroxyindoles

1-Hydroxyindoles can adopt an alternative tautomeric form, *viz.* the *N*-oxide of a 3*H*-indole. Although the former tautomer is usually favoured, strong hydrogen bonding, inter-[149a] or intra-molecular,[149b] can increase the importance of the more polarized form; 5-bromo-1-hydroxyindole-2-carboxylic acid (**80**) is such a case.[149b]

1-Hydroxyindoles can be prepared[149a] by peracid oxidation of indolyl Grignard species, but those which have been most studied have been made by ring synthesis: the reductive cyclization of 2-nitrophenylacetaldehyde (Scheme 54) gives[150a] 1-hydroxyindole, alkali-catalysed closure–decarboxylation of 2-nitrobenzylmalonic acid leads to 1-hydroxyindole-2-carboxylic acid,[150b] and concentrated sulphuric acid treatment of the oxime of benzoin yields 1-hydroxy-2-phenylindole.[150c]

1-Hydroxyindole itself is unstable and readily dimerizes[149a] to (**81**); its *O*-acetate however, can, be handled.[150a] The hydroxyl groups of 1-hydroxyindoles are somewhat acidic and like those of phenols can be alkylated under alkaline conditions. 1-Methoxyindole obtained in this way and 1-hydroxy-2-phenylindole have been shown[150a,151] to undergo electrophilic substitution at the indole β-position. Oxidation of 1-hydroxyindoles to isatogens (see p. 456) can be achieved[152] with 4-nitroperbenzoic acid.

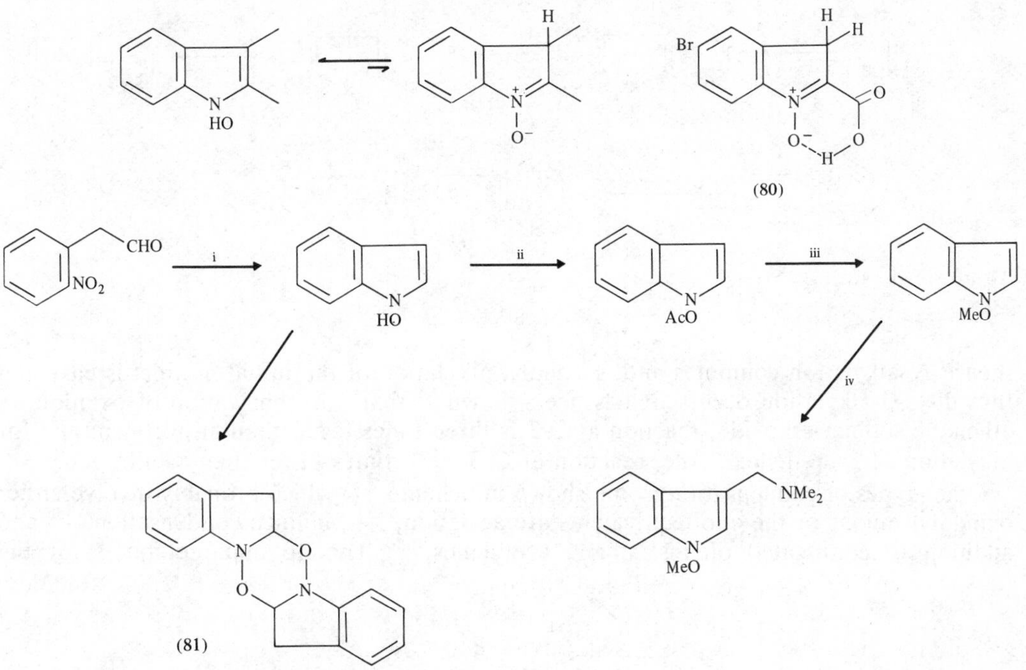

Br

(80)

i, Zn, NH_4Cl; ii, Ac_2O; iii, MeI, NaOMe; iv, CH_2O, Me_2NH, AcOH

(81)

SCHEME 54

Transformation of the hydroxyl of 1-hydroxyindoles into an ester and hence into a leaving group leads to an interesting type of reactivity (see also p. 436). For example, the immediate products of *O*-tosylation[153a] or *O*-4-nitrobenzoylation[153b] of 1-hydroxy-2-phenylindole thermally rearrange to 3-substituted esters. It is possible that these rearrangements involve β-nucleophilic attack by an equilibrium concentration of tosylate or 4-nitrobenzoate with departure of the *N*-substituent as a leaving group followed by tautomerism (Scheme 55).

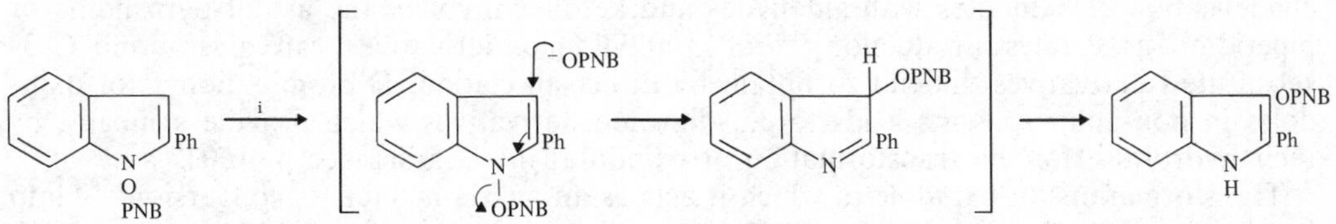

i, MeOH, heat; PNB = *p*-nitrobenzoyl

SCHEME 55

17.4.8.3 2-Hydroxyindoles (oxindoles)

2-Hydroxyindole exists in the carbonyl tautomeric form and is commonly known as oxindole, though indolin-2-one is strictly correct. In many of its reactions oxindole behaves simply as a lactam; however, abstraction of proton from the carbon α to carbonyl, *i.e.* the indole β-position, is much easier than is generally true for amides, for the resulting anion derives some stabilization from an indolic resonance contributor (Scheme 56). The alkylation of such anions might be expected to lead to both *O*- and *C*-3-alkylation, but in practice only the latter is observed. There are, however, other complicating features in such base-catalysed alkylations: firstly, *N*-deprotonation and

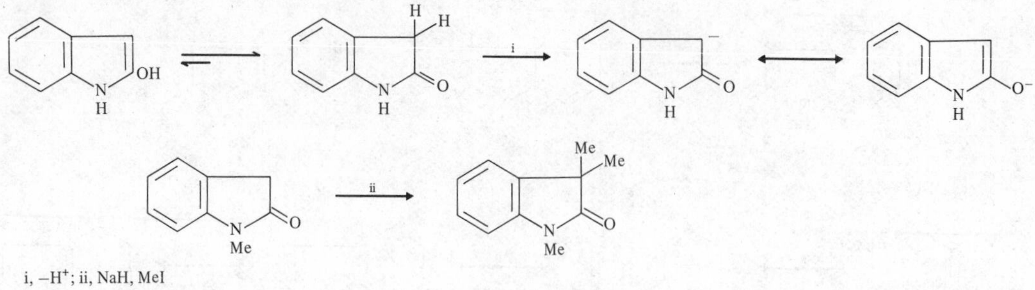

i, $-H^+$; ii, NaH, MeI

SCHEME 56

thence *N*-alkylation competes and, secondly, alkylation of the initial product is easy and thus di-*C*-3-alkylation occurs. It has been shown[154] that, for benzylation of oxindole in ethanolic sodium ethoxide, reaction at C-3 is three times faster than at nitrogen, and for alkylation of 3-methyloxindole, reaction at C-3 is 30 times faster than at nitrogen.

Other types of condensation (some shown in Scheme 57) which probably involve either oxindolyl anions or the enol equivalents are acylation,[155a] Mannich condensation,[155b] and addition to conjugated olefins[155c] and acetylenes.[155d] The optimal method[156] for the

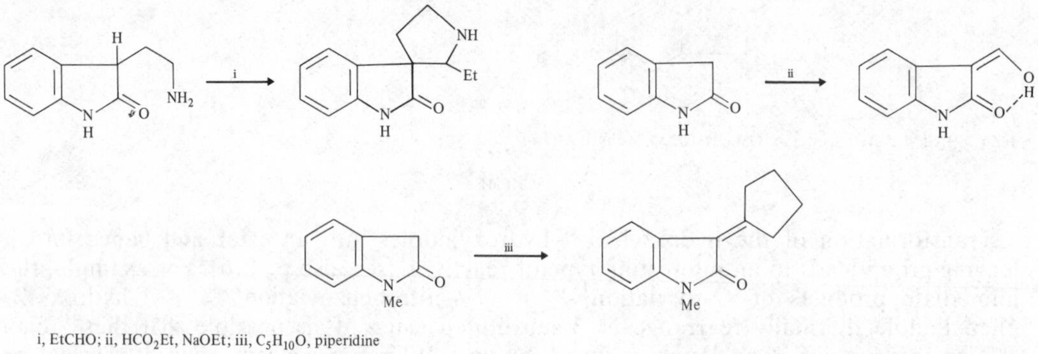

i, EtCHO; ii, HCO_2Et, NaOEt; iii, $C_5H_{10}O$, piperidine

SCHEME 57

condensation of oxindoles with aldehydes and ketones involves the use of pyrrolidine or piperidine as catalyst; reduction[156a] of 3-alkylidene derivatives can give mono-*C*-3-substituted derivatives difficult to obtain by direct alkylation. Dibromination[157] of oxindoles in non-polar solvents leads to 3,3-dibromo-derivatives which provide a means, by their hydrolysis, for the transformation of oxindoles into isatins (see p. 450).

Transformations of oxindole in which it acts as an amide include its conversion[158a] into 2-chloroindole with phosphorus oxychloride and into thio-oxindole[158b] by reaction with a limited quantity of phosphorus pentasulphide. *O*-Alkylation of oxindole can be achieved[159a] by reaction with Meerwein's reagent. The product exists,[159b] in solution, as a mixture of 2-ethoxyindole and 2-ethoxy-3*H*-indole.

17.4.8.4 3-Hydroxyindoles (indoxyls)

Studies on the position of tautomeric equilibrium between 3-hydroxyindole and its carbonyl counterpart, indolin-3-one (**82a**), trivially referred to as indoxyl, have been hampered by its instability. It seems[160a] that the major tautomer in polar solvents is the latter and its stable *N*-acetyl derivative (**82b**) exists in the keto-form.[160b] Indoxyl is very easily autoxidized to indigo (**40**), the dyestuff of the ancient world, *via* what is probably a radical-coupling process[161] proceeding through leucoindigo (**84**).

(82) a; R = H
b; R = Ac

(84)

(40)

(83)

i, O_2; ii, $Na_2S_2O_4$; iii, MeI, NaOH; iv, Ac_2O; v, H_2O, Na_2SO_3; vi, Ac_2O, H_2O; vii, PhCHO, piperidine; viii, aq. NaOH or aq. HCl, N_2; ix, PhCHO

SCHEME 58

Like oxindole, deprotonation α to carbonyl, *i.e.* the indole α-position, is particularly easy since the anion derives stabilization from an indolic resonance contributor. 3-Acetoxyindole (**83**)[162] ('indoxyl acetate') is more easily handled than indoxyl itself and can conveniently serve as a precursor for it in condensation[163] reactions (Scheme 58).

Reduction of indoxyls with lithium aluminium hydride leads to 3-hydroxy-2,3-dihydroindoles, which on loss of water give rise to indoles. This can occur either by loss of a *C*-2-hydrogen (a process which has been utilized[164] in the preparation of 2-deuteroindole) or by migration[165] of a substituent from C-2, if no hydrogen is present at that carbon (Scheme 59).

i, DCl; ii, $LiAlH_4$; iii, H^+

SCHEME 59

17.4.8.5 Dioxyindoles

The *ortho*-quinone type indole with the two carbonyl groups in the hetero-ring is known as isatin; no alternative tautomeric dicarbonyl form is possible. Isatins are usually prepared by ring synthesis (see p. 465). They can be transformed into oxindoles by reduction of the ketonic carbonyl either with hydrazine and then sodium ethoxide[166a] or by catalytic hydrogenation in acetic acid.[166b] Grignard addition followed by lithium aluminium hydride reduction and elimination of water leads to 3-substituted indoles; many other nucleophilic reagents also react with isatin at the very electrophilic C-3: a couple of examples are shown in Scheme 60.[167] Isatogens (3*H*-indol-3-one 1-oxides),

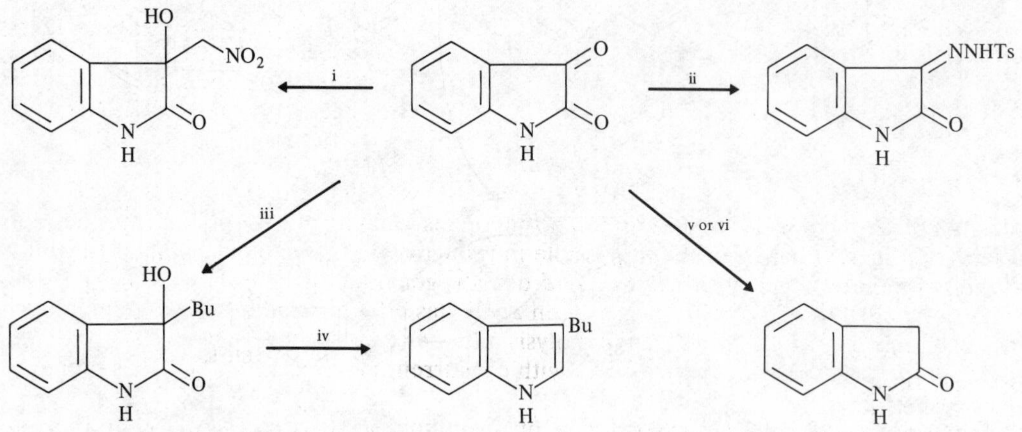

i, $MeNO_2$, Et_2NH; ii, $TsNHNH_2$; iii, BuMgBr; iv, $LiAlH_4$; v, H_2, Pd/C,AcOH, 60 °C; vi, N_2H_4 then NaOEt

SCHEME 60

which are usually prepared by base-catalysed ring closure of 2-nitrophenylalkenes or alkynes, undergo an intriguing series of transformations, initiated by nucleophilic addition at C-2. The synthesis[168a] (Scheme 61) of 2-phenylisatogen is typical, as is the reaction[168b] of this isatogen with aqueous acid.

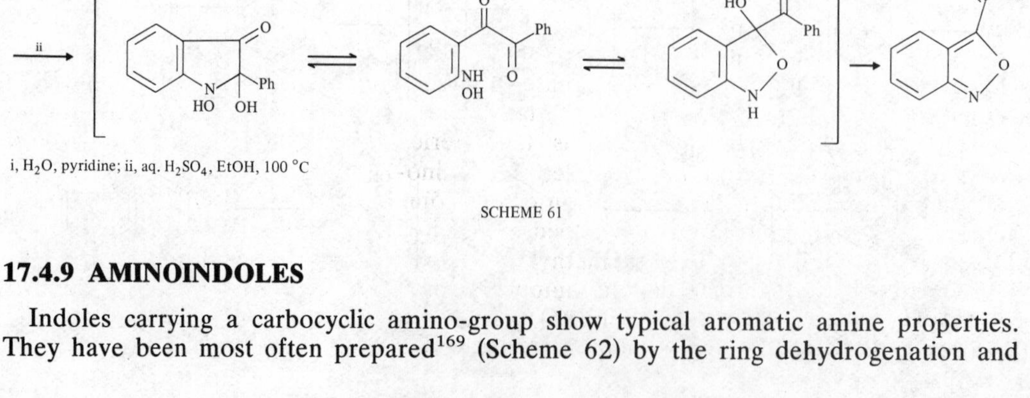

an isatogen

i, H_2O, pyridine; ii, aq. H_2SO_4, EtOH, 100 °C

SCHEME 61

17.4.9 AMINOINDOLES

Indoles carrying a carbocyclic amino-group show typical aromatic amine properties. They have been most often prepared[169] (Scheme 62) by the ring dehydrogenation and

i, H_2, Ni; ii, H_2, Ni, NaOH; iii, Ni, NaOH

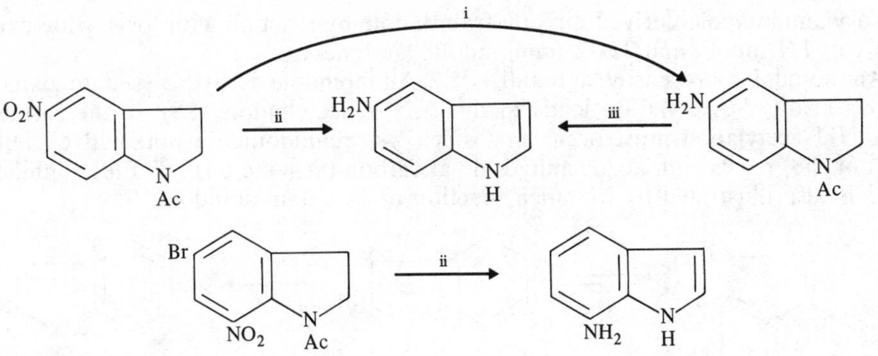

SCHEME 62

substituent reduction of nitro- or amino-indolines obtained in turn *via* nitration of *N*-acylindolines. It has been found possible to reductively remove blocking halogen whilst at the same time reducing nitro-group *and* dehydrogenating the hetero-ring. The replacement of halogen by the amino-group *via* an aryne has been achieved[170] at both indole and indoline oxidation levels. The ammonolysis of a 4- or 6-bromo-substituent from the corresponding indole-2-carboxylic acids with concurrent loss of carbon dioxide is preparatively useful.[171]

The best method[172] for the synthesis of 1-aminoindoles involves *N*-amination with hydroxylamine sulphonate; this method can be used for 1-aminoindole itself. Substituted derivatives have been obtained by the reduction[173] of *N*-nitroso-derivatives with zinc–acetic acid or lithium aluminium hydride. The 1-amino group has normal properties, condensing for example with carbonyl compounds; after condensation with unsaturated aldehydes, Lewis acid catalysis can lead to pyridazino[2,3-*a*]indoles (Scheme 63).[174]

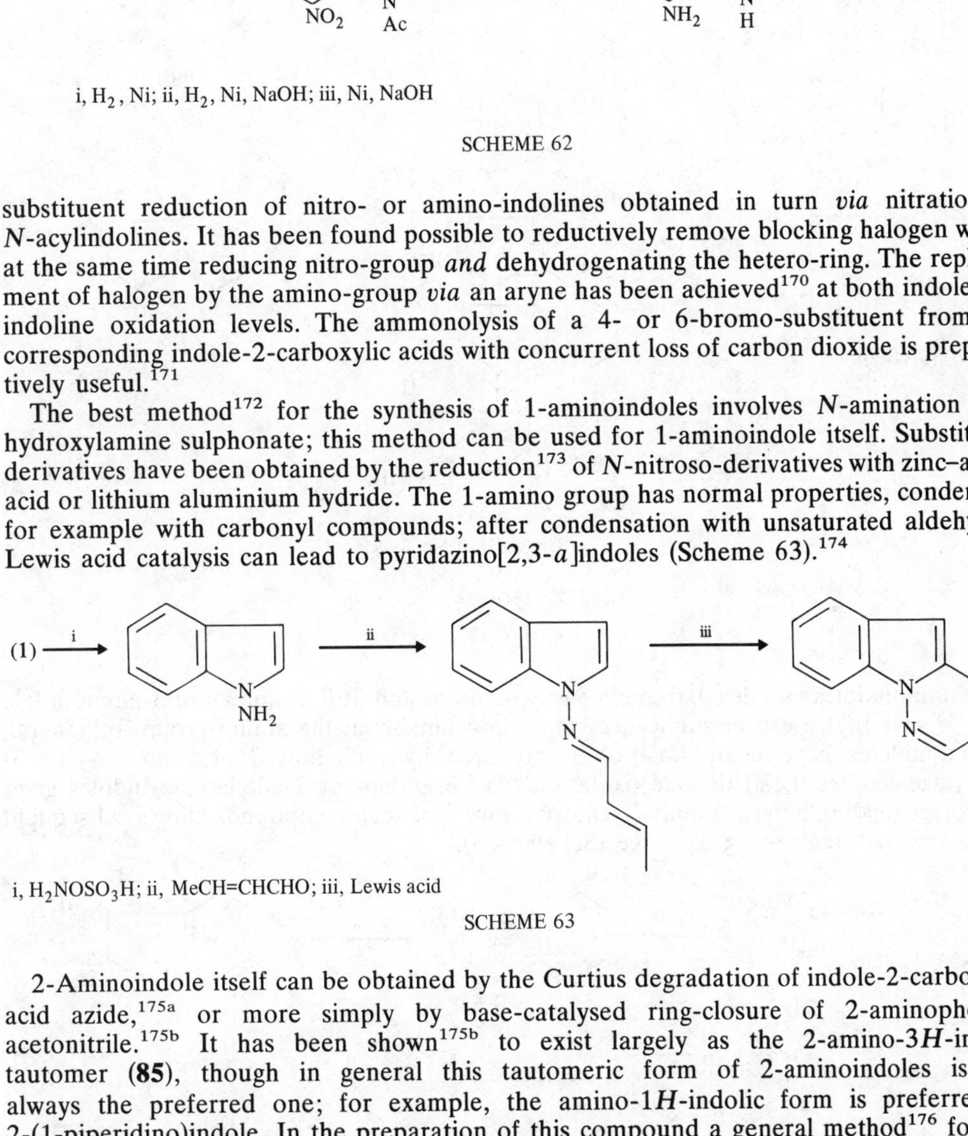

i, H_2NOSO_3H; ii, MeCH=CHCHO; iii, Lewis acid

SCHEME 63

2-Aminoindole itself can be obtained by the Curtius degradation of indole-2-carboxylic acid azide,[175a] or more simply by base-catalysed ring-closure of 2-aminophenylacetonitrile.[175b] It has been shown[175b] to exist largely as the 2-amino-3*H*-indole tautomer (**85**), though in general this tautomeric form of 2-aminoindoles is not always the preferred one; for example, the amino-1*H*-indolic form is preferred in 2-(1-piperidino)indole. In the preparation of this compound a general method[176] for the synthesis of 2-aminoindoles was developed in which thio-oxindoles react with amines.

Reaction of 2-aminoindole with methyl iodide results in *N*-1-alkylation.[175b] In the product (**86**), the 2-amino-1*H*-indole tautomer is preferred over the alternative 2-imino possibility. Interestingly, the pK_a of (**86**) is higher (9.60) than that of 2-aminoindole (8.15), though N-1 and C-3 protonation respectively lead to analogous cations (**87**).

2-Tosylaminoindole derived directly from indole by reaction with tosyl azide exists as a mixture of 1*H*-indolic and 3*H*-2-iminoindolic tautomers.[177]

2-Aminoindoles are easily autoxidized. 2-Aminoindole is hydrolysed to oxindole by aqueous acid.[175a] Acetylation leads to the *N,N'*-diacetylindole (**88**), in the formation of which *N*-1-acetylation must occur first since 2-acetamidoindole obtained by mild hydrolysis of (**88**) reacts with acetic anhydride at carbon (Scheme 64).[175b] Electrophilic attack at C-3 is also illustrated by Mannich reaction of 2-acetamidoindole.[175a]

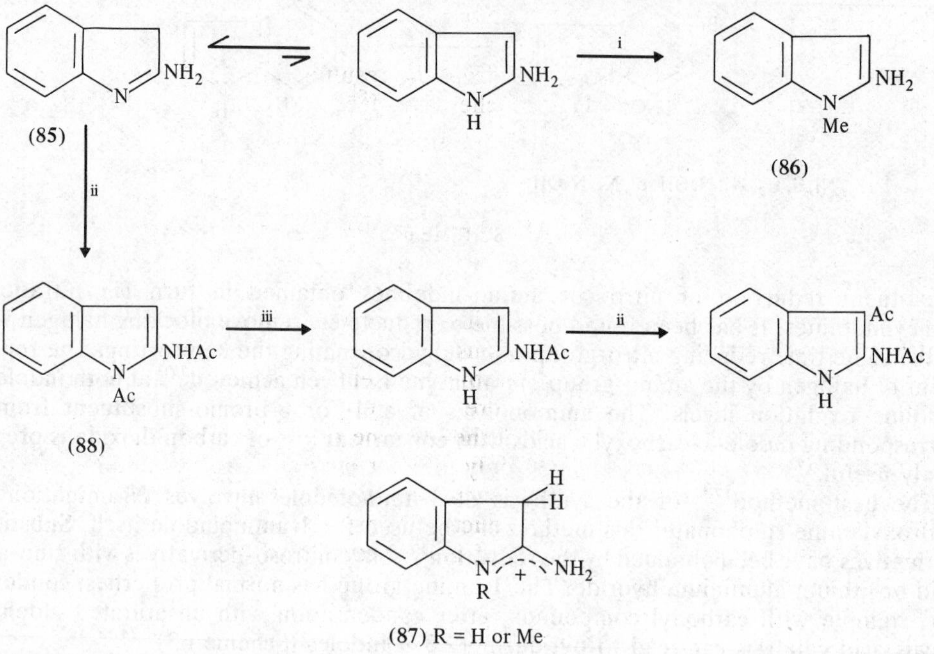

(85) (86) (88) (87) R = H or Me

i, MeI; ii, Ac_2O; iii, aq. $NaHCO_3$

SCHEME 64

3-Aminoindoles are also extremely easily oxidized and little is known of 3-aminoindole itself[178] save that it can be mono-acetylated, presumably at the amino-group. In general, 3-aminoindoles have been most often prepared by reduction[179] of 3-nitroso- or 3-phenylazo-indoles. Lead dioxide oxidation[180] of 3-amino- or 3-alkylamino-indoles gives the corresponding imines. Imino-*N*-chlorination[181] of such compounds allows subsequent base-catalysed hetero-ring cleavage (Scheme 65).

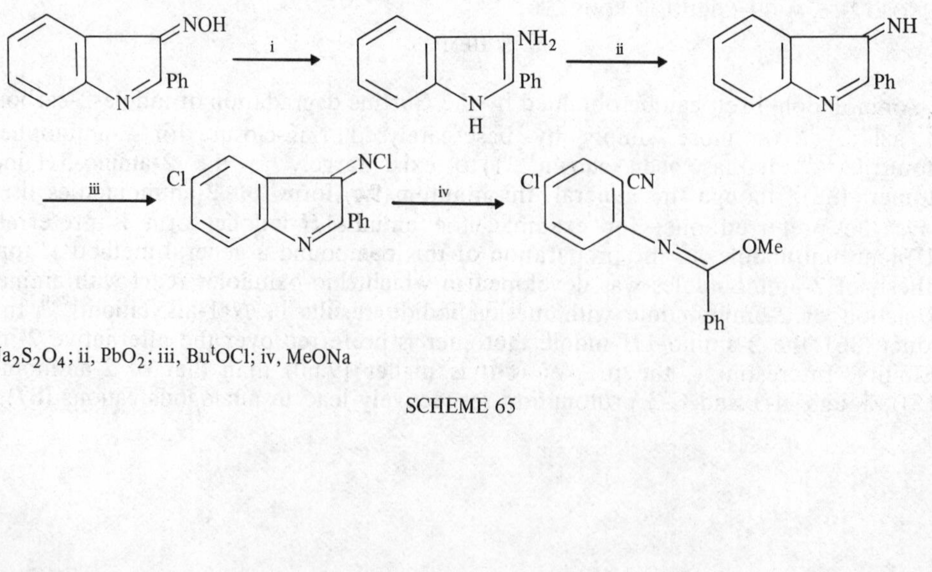

i, $Na_2S_2O_4$; ii, PbO_2; iii, Bu^tOCl; iv, MeONa

SCHEME 65

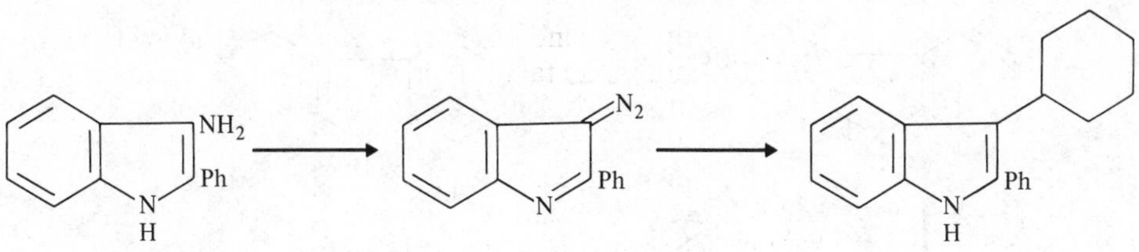

i, HNO_2; ii, *hv*, cyclohexane

SCHEME 66

3-Aminoindoles can be diazotized and from the resulting salts, by loss of proton, stable 3-diazo-3*H*-indoles are obtained, photochemical decomposition of which leads to carbenes (Scheme 66).[182]

17.4.10 REACTIONS OF INDOLYL—C—X COMPOUNDS

17.4.10.1 General

The electron-releasing indole nucleus will interact with any potential electron-deficient centre attached to the α- and β-positions (Scheme 67), and will thus accelerate the departure of leaving groups in indolylcarbinyl systems. Assistance of this kind is much more effective in C-3 carbinyl (**89**), where the intermediate (**90**) (or the corresponding neutral species) retains the aromaticity of the benzene ring, than in a C-2 system involving an intermediate in which nitrogen can only interact *via* a non-benzenoid canonical form (**91**). A reaction proceeding through such intermediates is completed by conjugate addition of a nucleophile Y^- to give products (**92**) or (**93**). Analogous reaction sequences have been used extensively in synthesis of indolic compounds.

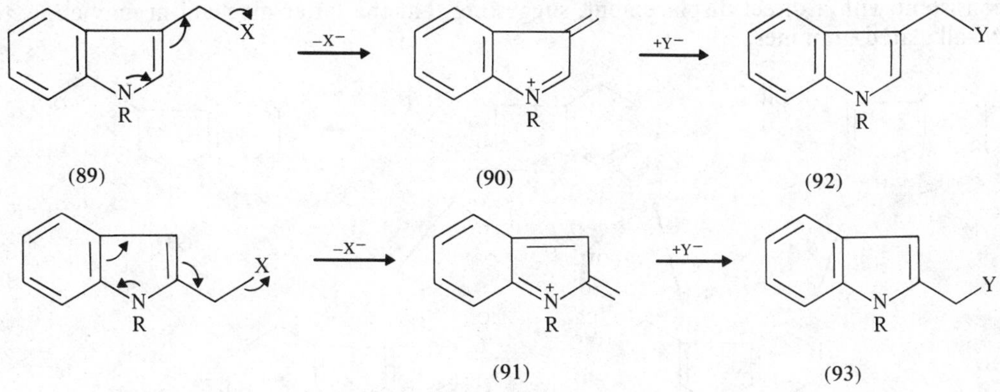

(89) (90) (92)

(91) (93)

SCHEME 67

17.4.10.2 Gramine (3-dimethylaminomethylindole) and derivatives

Owing to its ready preparation (p. 431) and the ease with which the dimethylamino-group can be replaced by nucleophiles, gramine (**94**) is an important intermediate for synthesizing C-3 substituted indoles[183] (Scheme 68). Particularly useful are alkylations of various carbon nucleophiles with gramine or its quaternary salts (methiodide, methosulphate, *N*-oxide), which are even more reactive (Scheme 69). Displacement by cyanide ion gives 3-indolylacetonitrile, which affords 3-indolylacetic acid (IAA) (**95**) by hydrolysis or tryptamine (**96**) by reduction. Other important synthetic routes use diethyl acetamidomalonate or ethyl nitroacetate to give compounds which are readily converted into tryptophan (**97**).

SCHEME 68

Such reactions could occur either by an elimination–addition mechanism or by direct nucleophilic displacement. An investigation by Albright and Snyder[184a] with an optically active gramine analogue showed that only racemic products were obtained on reaction with diethyl acetamidomalonate, malonate, and piperidine. These results indicated the former mechanism for gramine, and were supported by a rate study on the reaction of a thiophenoxide with gramine methiodide.[184b] However, the rate of reaction of the corresponding *N*-1-methyl derivative with the sulphide anion was much slower and more consistent with a direct displacement, suggesting that the latter mechanism may apply to N_a-alkylated gramines.

SCHEME 69

A neat synthesis[185] of tryptamines proceeds *via* a chloromethylgramine (**98**), prepared from chloroacetaldehyde and dimethylamine or piperidine. The intermediate is presumably an aziridine which undergoes reductive C—N cleavage with sodium borohydride (Scheme 70).

Cl NR$_2$ $\overset{+}{N}R_2$ Cl^- $NaBH_4$ N H NR$_2$

(98)

SCHEME 70

17.4.10.3 Isogramine (2-dimethylaminomethylindole) and derivatives

Isogramine has not been so extensively used as the C-3 isomer, partly because it is not so readily available by synthesis (*e.g.* Ref. 185a). Some displacement reactions with cyanide and diethyl acetamidomalonate have been carried out on its methiodide to give the corresponding α-substituted products in fair to good yield.[186b,c] The rate of reaction in the α-series is much slower, as explained above.

However, the isogramine system is in fact very common in a substituted form as the ubiquitous tetrahydro-β-carboline unit in several hundred indole alkaloids (*q.v.*) and synthetic analogues. If the basic nitrogen atom is assisted in acting as a leaving group by protonation, alkylation, or acylation, then cleavages analogous to those above can occur.

KCN Me CN

(99) (100)

During synthesis of the indole alkaloid tetrahydrosecodine the system (**100**) was derived by a cyanide displacement[187a,b] on (**99**). Examples of analogous cleavages include a nucleophilic displacement by acetate after an intramolecular acylation,[187c] or reduction of (**101**) to (**102**) with alkali metals in liquid ammonia[188] (Scheme 71).

i, n = 1,2 or ii, n = 2 Me $(CH_2)_n$

(101) (102)

i, Li, NH_3, ii, $LiAlH_4$

SCHEME 71

17.4.10.4 3-Indolylcarbinols

Depending on the reducing agent, reduction of carbonyl groups attached to C-3 either affords an alcohol or leads to hydrogenolysis (Scheme 72). Thus with sodium borohydride

i, $NaBH_4$; ii, $LiAlH_4$; iii, B_2H_6

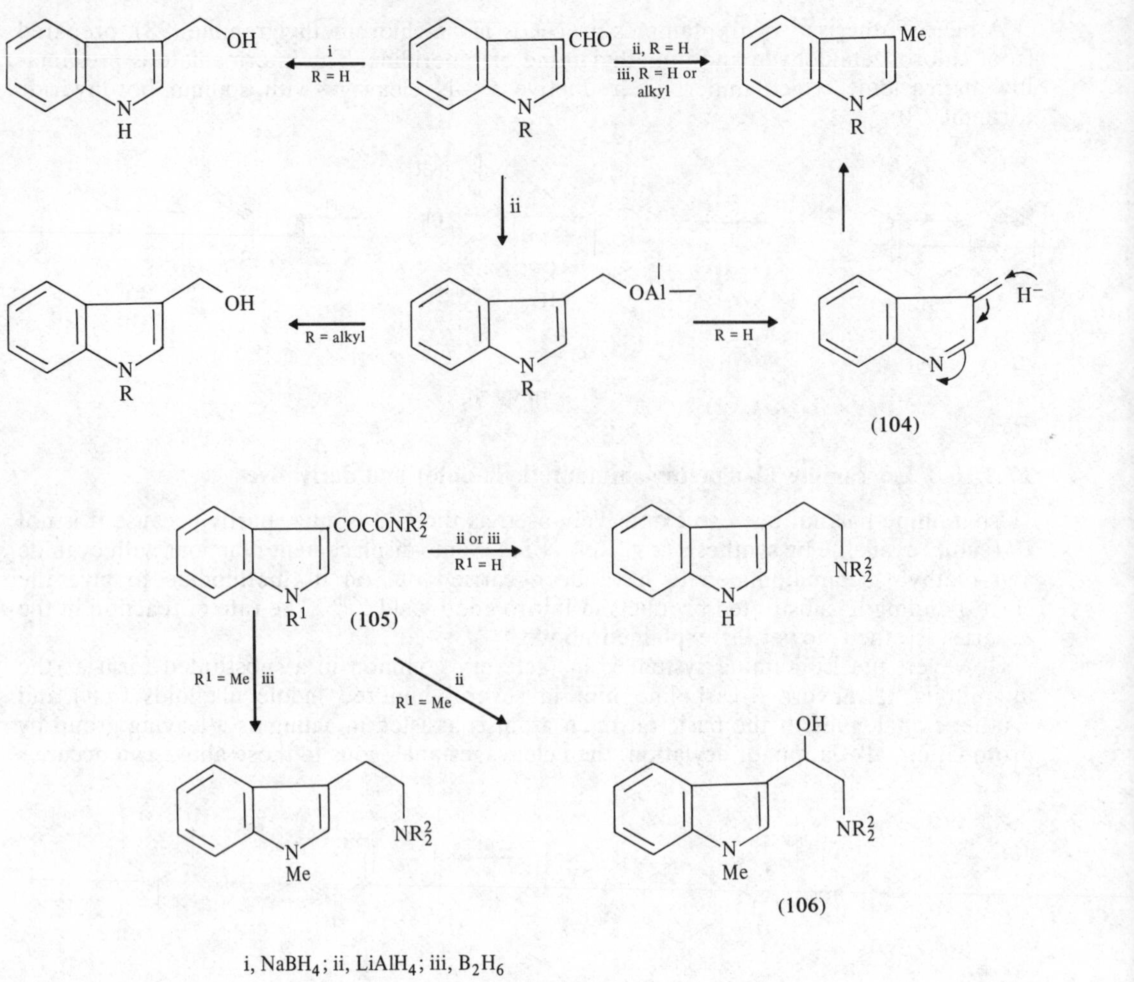

SCHEME 72

at room temperature, 3-formylindole is reduced to 3-hydroxymethylindole. With the more forceful lithium aluminium hydride, 3-methylindole is produced, but if an alkyl group is present on the nitrogen, reduction stops at the alcohol stage.[189] The explanation would appear to be that displacement of the aluminate complex occurs only when a proton can be lost from nitrogen to give 3-methyleneindolenine (**104**), which is subsequently reduced. However, with diborane the corresponding borate complex can eliminate even when the nitrogen is alkylated and 3-methylindoles are produced in both cases. These considerations are also valid in the reduction of indole-3-glyoxamides (**105**), an important procedure for making tryptamines. Lithium aluminium hydride usually reduces both carbonyl functions to methylene groups to give the desired ethanamine side chains, but if an N-1 alkyl group is present the reaction often stops at the β-hydroxytryptamine (**106**) stage. Again, diborane can achieve complete reduction.[189c]

Readiness to form 3-methyleneindolenines is the reason why 3-hydroxymethylindoles and the related 3-methoxy-compound react with cyanide anion and piperidine to afford 3-cyano- and 3-piperidinomethyl derivatives.[184a,189b,190] Likewise the production of di-indolylmethanes (**107**) from indolylmethanols under mild conditions[189a,b] is attributable to formation of an intermediate which undergoes nucleophilic attack by a molecule of the starting material to give (**108**) (Scheme 73). Loss of formaldehyde in a retro-aldol manner then affords the di-indolylmethane.

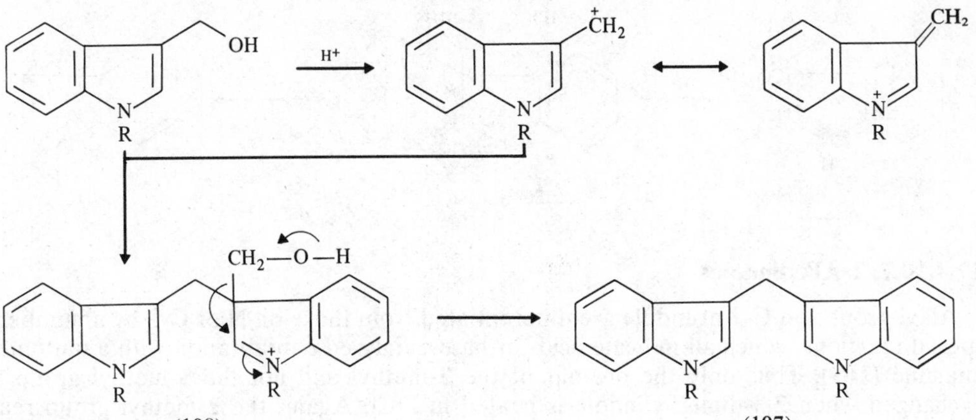

SCHEME 73

17.4.10.5 2-Indolylcarbinols

A 2-hydroxyalkyl group appears to be more stable than a C-3 carbinol, presumably because elimination generates a less favourable intermediate. With diborane, 2-acetyl-3-methylindole afforded the alcohol in addition to 2-ethyl-3-methylindole,[189c] indicating a much slower breakdown of the borate complex. Again, 2-hydroxymethylindoles do not tend to form di-indolylmethanes. However, nucleophilic displacement of a hydroxyl group can occur under rather more forcing conditions than required for 3-indolylcarbinols. Two alkaloid examples[191] both involve intramolecular reaction.

17.4.10.6 Indole-2- and -3-carboxylic acids

The distinctive feature of the 2- and 3-carboxylic acids is the ease with which they undergo decarboxylation. For the former this is an important step in a procedure for synthesizing tryptamines.[192a] The rate of decarboxylation for indole-2-carboxylic acids in mineral acids is increased by electron-releasing groups and decreased by electron-withdrawing substituents. A mechanism consistent with this behaviour involves protonation at C-3 and loss of CO_2 to give an ylide as indicated in Scheme 74. Other effective catalysts are copper salts and quinoline, which is also compatible with an ylide mechanism.[192b]

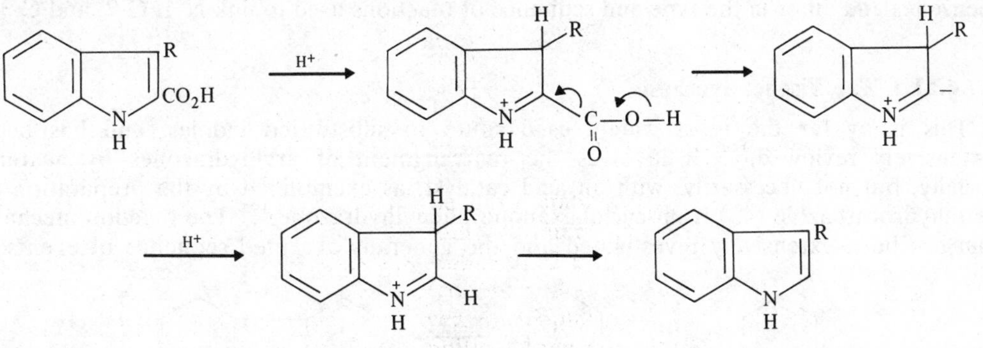

SCHEME 74

Indole-3-carboxylic acid can be decarboxylated by heating in either acidic or alkaline solution.[192c] Protonation at C-3 would afford the imine equivalent of a β-keto-acid (Scheme 75).

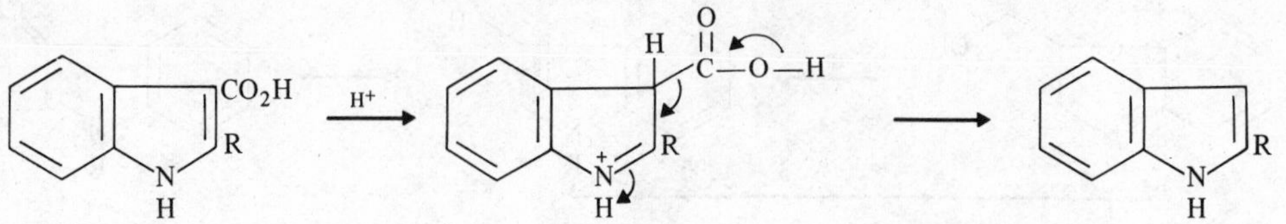

SCHEME 75

17.4.10.7 2-Alkylindoles

Alkyl groups on C-2 of indole are differentiated from those on N or C-3 by a number of special reactions, which all indicate acid- or base-catalysed equilibration with a tautomeric enamine (**109**). Thus only the protons of the 2-methyl and not the 3-methyl group are exchanged when 2,3-dimethylindole is heated in DCl. Again, the 2-methyl group reacts exclusively with *N*-bromosuccinimide[193a] and in the Mannich reaction[193b] (Scheme 76).

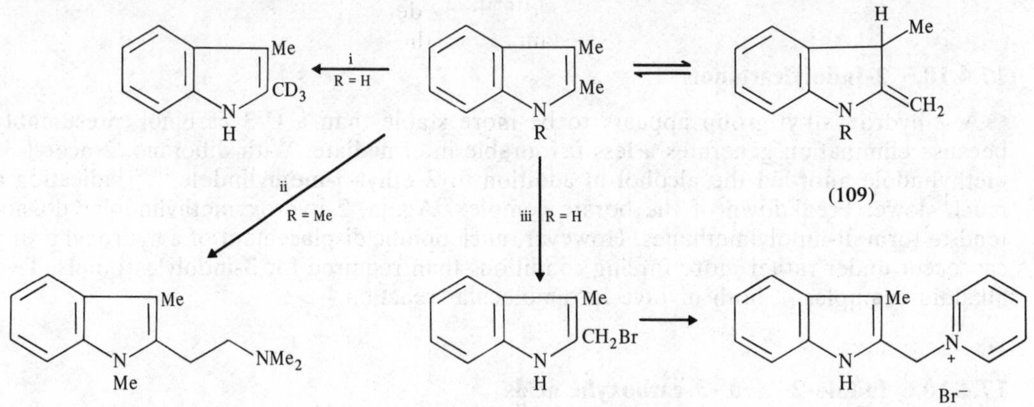

(109)

i, 3M DCl, 100 °C; ii, CH_2O, Me_2NH, AcOH, 90 °C; iii, NBS, py

SCHEME 76

17.4.11 SYNTHESES OF THE INDOLE SKELETON

Of the very large number of reactions used to synthesize indoles, there are relatively few which can be considered as general methods, the others being either variants of these or restricted to the particular compounds. Essentially all approaches utilize substituted benzenes and differ in the type and sequence of reactions used to link N-1, C-2, and C-3.

17.4.11.1 The Fischer synthesis

This is by far the most widely used route to substituted indoles, and has been extensively reviewed.[194] It involves the rearrangment of arylhydrazones by heating, usually, but not necessarily, with an acid catalyst, as exemplified by the preparation of tetrahydrocarbazole (43) from cyclohexanone phenylhydrazone.[195] The reaction mechanism has been extensively investigated and the generally accepted sequence of events is

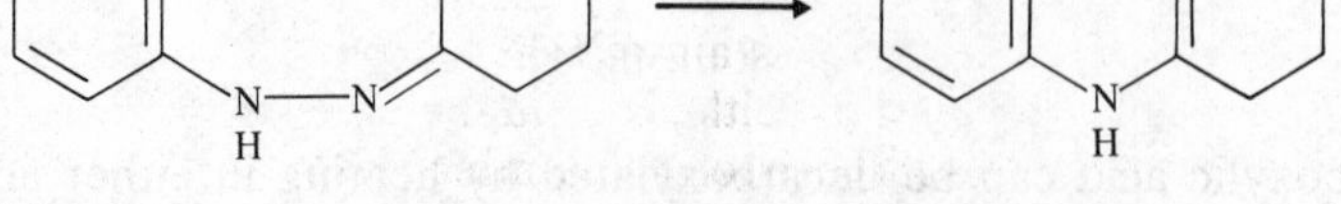

i, AcOH, heat, ii, heat

(43)

SCHEME 77

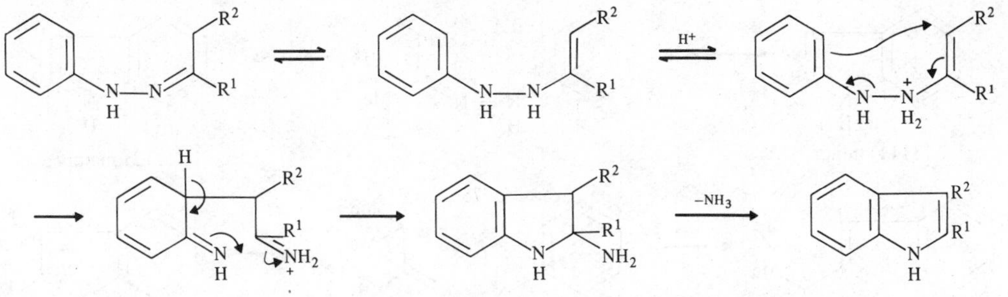

SCHEME 78

outlined in Scheme 78. The crucial step is formation of the 3—3a bond, which probably occurs by an electrocyclic process analogous to the *o*-Claisen rearrangement of allyl phenyl ether.[196] Evidence for such a stage in the sequence was obtained by Bajwa and Brown,[197] who isolated the salt (**110**) where aromatization is prevented by *o*-methyl groups. Previous work with ^{15}N labels had already demonstrated that N-2 of phenylhydrazones was lost as ammonia, in accordance with the mechanism shown.[198]

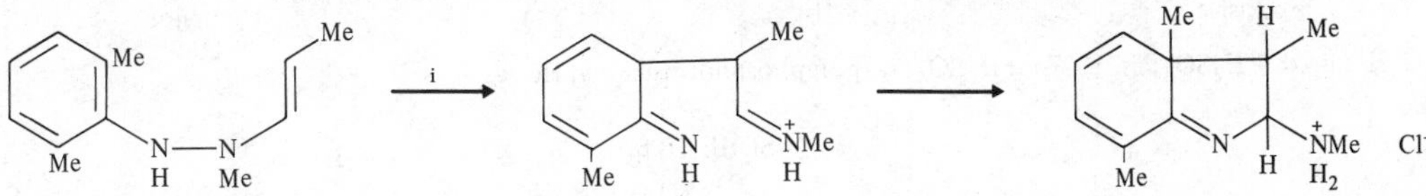

(110)

i, HCl, PhH, heat

SCHEME 79

Experimental conditions employed vary greatly, since cyclization depends upon the nature of the arylhydrazone, the catalyst, and temperature. In general, electron-withdrawing substituents on the benzene ring retard the reaction. The traditional catalysts are hydrogen chloride or zinc chloride, but many other proton or Lewis acids have been used such as formic, acetic, sulphuric, and (poly)phosphoric acids, boron trifluoride, cation exchange resins, and polyphosphate ester.

A general problem arises with phenylhydrazones of unsymmetrical ketones, which can give two alternative products. In some cases both are obtained in comparable amounts, in others one predominates, and the factors involved are complex and not completely understood. However, some empirical predictions can be made from the alkyl groups on the ketone, *e.g.* methyl alkyl ketones usually give 3-alkyl-2-methylindoles (**112**) rather than (**111**).

Again, the direction of closure can be greatly influenced by the nature and concentration of the catalyst and the solvent[199] (Scheme 80). Cyclization of benzyl methyl ketone phenylhydrazone with hydrochloric acid gave exclusively 2-methyl-3-phenylindole, whereas with polyphosphoric acid the alternative 2-benzylindole was the major product; with the derivative of isopropyl methyl ketone, dilute sulphuric acid afforded the 3*H*-indole (**114**) in preference to the indole (**113**), but in concentrated acid a high yield of the latter was obtained.

An important extension of the method is the Japp–Klingemann reaction, which provides access to many phenylhydrazones unobtainable by direct methods. The scope of the reaction has been reviewed by Phillips.[200a] It involves coupling of an arenediazonium salt to an activated carbanion derived from an $\alpha\beta$-dicarbonyl system or equivalent, as in Scheme 81. Depending on whether R^2 is alkyl or hydrogen, base-promoted β-dicarbonyl cleavage gives the arylhydrazone of either an α-keto-ester or an α-diketone. One example of its use is in a neat synthesis of tryptamines from piperid-2-one-3-carboxylic acids *via* (**115**).[200b] An extension of the reaction to enamines is also feasible, as with the preparation of (**116**)[200c] (Scheme 82).

(111) minor

(112) major

(113) major (hy)

(114) major (hy)

exclusive

major

i, H^+; ii, 78% H_2SO_4; iii, 10% aq. H_2SO_4; iv, polyphosphoric acid; v, HCl

SCHEME 80

ArN_2^+ + $^-C(COMe)(CO_2R^2)-CH_2R^1$ → $Ar-N=N-C(COMe)(CO_2R^2)-CH_2R^1$

R^2 = alkyl → $Ar-NH-N=C(CO_2R^2)-CH_2R^1$

R^2 = H, $-CO_2$ → $Ar-NH-N=C(COMe)-CH_2R^1$

SCHEME 81

(115)

(116)

i, PhN_2^+; ii, HO^-; iii, H^+, heat

SCHEME 82

17.4.11.2 The Bischler synthesis

α-Arylamino-ketones or -aldehydes are readily prepared from α-halo-precursors and arylamines, and on heating with an acid catalyst cyclize to indoles with loss of water (Scheme 83). The scope of this reaction has been extensively reviewed.[201] Where R^2 and

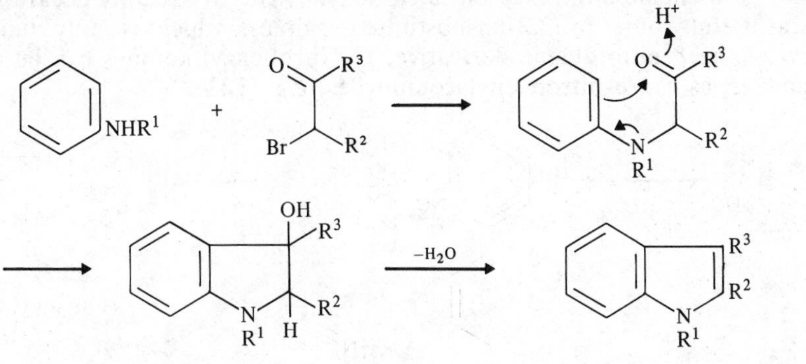

SCHEME 83

R^3 are the same, good yields are often obtained, but difficulties may arise when they are different, since partial or complete rearrangement occurs to afford mixtures, *e.g.* (**118**) and (**119**) from (**117**).[202] An *N*-alkyl group apparently prevents this rearrangement in many cases.

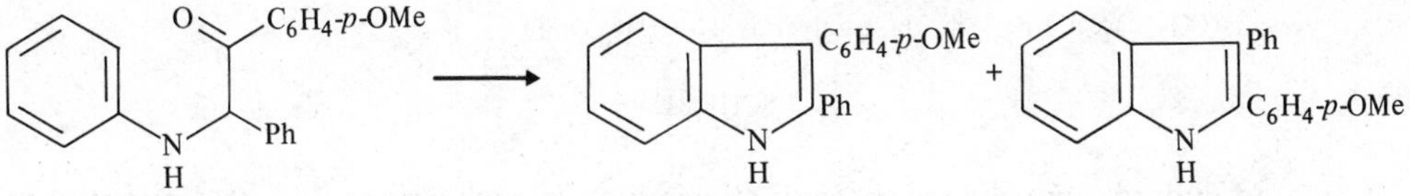

(117) major (118) minor (119)

SCHEME 84

17.4.11.3 The Madelung synthesis

Treatment of *o*-alkylanilides of carboxylic acids with strong bases ($NaNH_2$, $KOBu^t$, *etc.*) at high temperatures (200–400 °C) results in the formation of indoles (Scheme 85). Owing to the drastic conditions necessary to generate the carbanion required to cyclize to the amide carbonyl, the reaction is limited to indoles with stable alkyl or aryl substituents.[203]

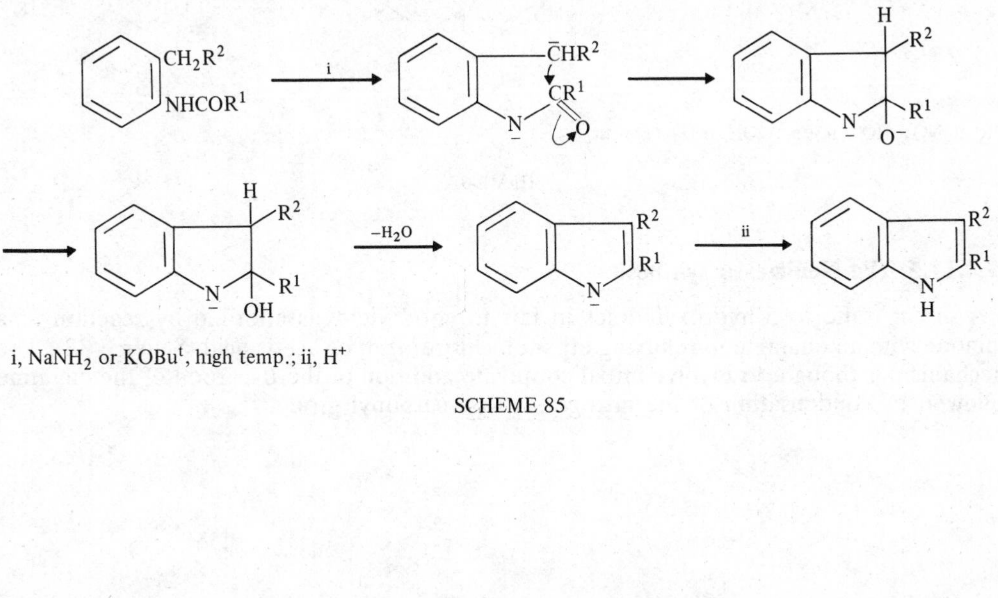

i, $NaNH_2$ or $KOBu^t$, high temp.; ii, H^+

SCHEME 85

17.4.11.4 The Reissert and related syntheses

Base-catalysed condensation of oxalic acid esters with the activated methyl group of *o*-nitrotoluenes affords *o*-nitrophenylpyruvates, which can be reductively cyclized to indole-2-carboxylic acids or esters. An example is the preparation of ethyl indole-2-carboxylate[204a] (Scheme 86). Since the 2-carboxylic acids are readily decarboxylated, this provides a useful route to 2,3-unsubstituted indoles, which is only limited by the availability of the *o*-nitrotoluene derivative. *O*-Nitrobenzyl ketones can be cyclized in a similar manner, as can *o*-nitrophenylacetonitriles, *e.g.* **(121)**.[204b]

Me, NO_2 —i, ii→ CO_2Et, O, NO_2 —iii→ CO_2Et, N, H

(120) (hy)

Cl, NO_2 —iv, v→ $CONMe_2$, CN, NO_2 —iii→ $CONMe_2$, N, H

(121)

i, $(CO_2Et)_2$; ii, EtO^-; iii, H_2, catalyst; iv, $NCCH_2CONMe_2$; v, base

SCHEME 86

A related and widely used synthesis is the reductive cyclization of *o*-,*β*-dinitrostyrenes, which are readily available from base-catalysed condensation of *o*-nitrobenzaldehydes and nitroalkanes (Scheme 87) or occasionally by nitration of a *β*-nitrostyrene. A variety of reducing agents has been employed, most commonly iron and acetic acid or catalytic hydrogenation.[205]

CHO, NO_2 —i→ NO_2, NO_2, R —ii→ NH_2, NH_2, R

→ NH, NH_2, R → R, NH_2, N, H → R, N, H

my-hy

i, RCH_2NO_2, HO^-; ii, Fe, AcOH or H_2, catalyst

SCHEME 87

17.4.11.5 The Nenitzescu synthesis

A direct route to 5-hydroxyindoles in fair to good yields is afforded by reaction of a quinone with an enamine in refluxing ethanol, chloroform, or acetic acid (Scheme 88). The mechanism is thought to involve initial conjugate addition of the *β*-carbon of the enamine followed by condensation of the nitrogen with a carbonyl group.[206]

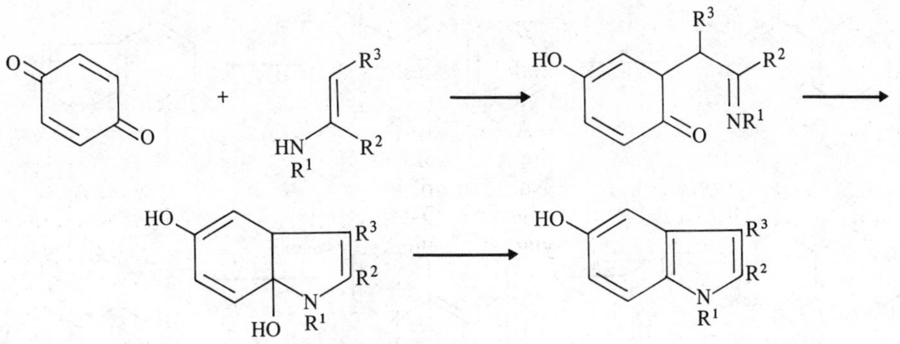

SCHEME 88

17.4.11.6 Miscellaneous syntheses

A new general and viable method for preparing indoles from anilines has been described.[207] Reaction of an *N*-chloroaniline with α-thiomethyl ketones or aldehydes (**122**; R = H or alkyl) followed by treatment with mild base and reduction gives indoles with a variety of substituents (Scheme 89). A variation in the procedure by using acid and reduction provides a route to oxindoles.

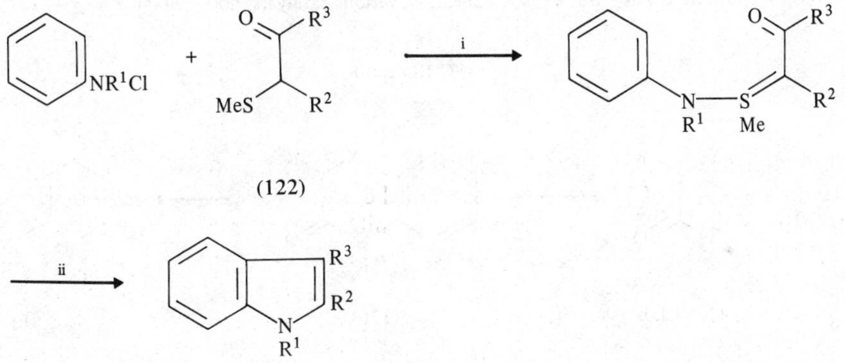

(122)

i, Et_3N; ii, Raney Ni or $LiAlH_4$

SCHEME 89

Several attempts have been made to find novel routes to indoles by using pyrolytic methods to generate reactive intermediates, *e.g.* *o*-styrylnitrenes from the corresponding azides[208] and ketones from Meldrum's acid derivatives (**123**).[209] Various catalysts are also effective at high temperatures in converting ethylaniline[210a] and 2-(*o*-nitrophenyl)ethanol[210b] to indole (Scheme 90).

17.4.11.7 Syntheses of oxindoles

Excluding the oxidative transformation of indoles discussed above, the most general route to oxindoles involves the Friedel–Crafts cyclization of anilides of α-halocarboxylic acids (**126**)[211] (Scheme 91). Other reactions of general utility are the reduction of *o*-nitrophenylacetic acid derivatives (**127**)[211b] and the base-catalysed cyclization of *N*-acylphenylhydrazides (**128**).[212]

(my–hy)

(123)

(hy)

(124)

(125)

i, heat; ii, flash vacuum pyrolysis, 500 °C; iii, Cu, heat; iv, various catalysts, 500–750 °C

SCHEME 90

(X = Cl, Br)

(126)

(hy)

(127)

(128)

(my)

i, $AlCl_3$; ii, H_2, catalyst; iii, CaH_2, heat

SCHEME 91

17.4.11.8 Syntheses of indoxyls (indolin-3-ones)

The best method of preparing these labile compounds is *via* the enol acetate, which can be obtained from anthranilic acid derivatives (**129**) by sodium acetate catalysed condensation in the presence of acetic anhydride (Scheme 92). Liberation of the free indolinone is carried out in the absence of air owing to the very ready oxidative coupling.[213]

i, $ClCH_2CO_2H$, NaOH, heat; ii, Ac_2O, NaOAc, heat; iii, NaOAc, H_2O, N_2

SCHEME 92

17.4.11.9 Syntheses of isatins (indoline-2,3-diones)

Isatins, important intermediates for the synthesis of indolic and other compounds, are usually made by reaction of an aniline with chloral and hydroxylamine to give an oximinoanilide (**130**), which is cyclizèd with sulphuric acid (Scheme 93).[214] An alternative method involves an intramolecular Friedel–Crafts reaction on a monoanilide from oxalyl chloride.

i, Cl_3CCHO, H_2NOH, HCl; ii, conc. H_2SO_4; iii, $(COCl)_2$; iv, $AlCl_3$

SCHEME 93

17.4.12 INDOLE ALKALOIDS

The indole alkaloids comprise a heterogeneous collection of natural products containing an indolic or closely related nucleus whose occurrence is restricted (with a few exceptions) to higher plants and some fungi. By traditional definition they are nitrogenous bases, but the current classification embraces neutral and acidic compounds, and over 1000 are now known, with a wide range of simple to complex structures.[11,215] Consequently, a comprehensive coverage is not feasible except in specialized multi-volume compendia such as the Manske series[216] and the *Specialist Periodical Reports.*[217] However, mainly as a result of biosynthetic studies, the majority of indole alkaloids can be grouped in a relatively small number of structural types derived from the ubiquitous tryptophan or tryptamine and their derivatives, and a few alkaloids can be used as illustrative examples. Again the field has provided a stimulating and productive challenge to synthetic chemists which has been recently reviewed[218] and will not be discussed in any detail.

17.4.12.1 Non-tryptamine alkaloids

The simplest basic alkaloids are gramine, 3-methylaminomethyl-, and 3-aminomethylindoles which have been isolated from sprouting barley (*Hordeum vulgare*) and derived biosynthetically from tryptophan. The 20 or so carbazole derivatives from the Indian curry plant, *Murraya koenigii*, and other Rutaceous species are less easily classified but apparently comprise an indole combined with one, two, or three isoprene units, *e.g.* murrayanine (**131**), girinimbine (**132**), mahanimbine (**133**), and isomahanimbine (**134**).[219] Their structures were deduced by a combination of chemical reactions and spectroscopy, and confirmed by synthesis. For example, (**133**) was obtained from 3-hydroxy-4-methylcarbazole and citral in the presence of pyridine (Scheme 94). The isomeric structure (**134**) for isomahanimbine was assigned from the n.m.r. doublets at τ 3.3 and 2.3 for the H-3,4 AB system and a broad singlet at τ 2.12 for H-5.

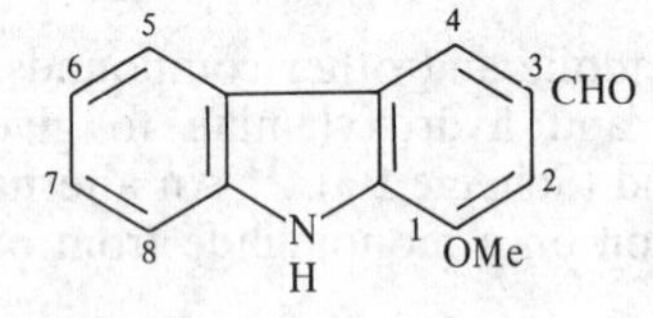

(131)

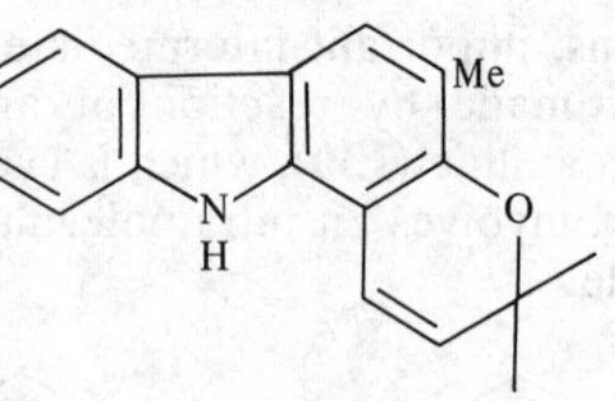

(132)

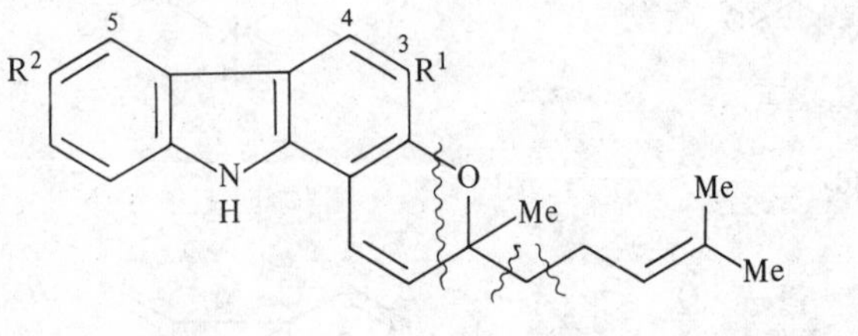

	R^1	R^2
(133)	Me	H
(134)	H	Me

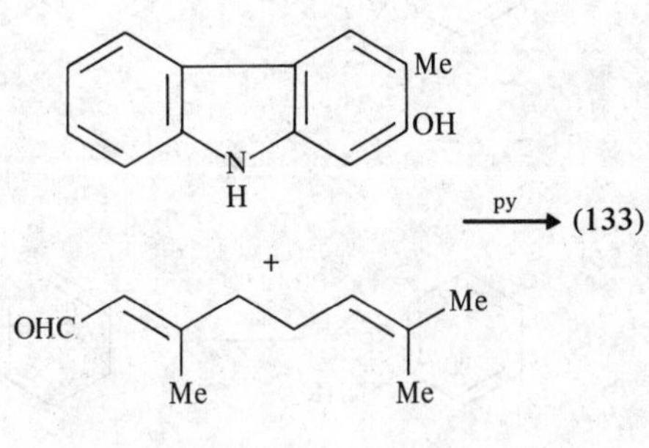

SCHEME 94

17.4.12.2 Non-isoprenoid tryptamine and tryptophan alkaloids

Several simple tryptamine derivatives have marked physiological effects. *N,N*-Dimethyltryptamine is widely distributed in plants, some of which are used as narcotic snuff by American Indian tribes, and the corresponding 4-oxygenated derivative, psilocybin (**135**), is the psychotropically active principle of the hallucinogenic Mexican mushroom, *Psilocybe mexicana.*[220] A major constituent of the Calabar bean (*Physostigma venenosum*), used as an ordeal poison in West Africa, is physostigmine (**136**), which has found application in opthalmology.[221] The structures of this and most of its congeners were elucidated many years ago by chemical methods and synthesis, but the relative and absolute configuration of (−)-physostigmine was not established until comparatively recently.[222] On the basis of relative stabilities the B/C junction had been regarded as *cis*, a reasonable assumption that was confirmed by nuclear Overhauser effect (n.O.e.) measurements in the ^{1}H n.m.r. spectrum. Irradiation of the C-3a or N-8 methyl group signals caused a 15% enhancement in the intensity of the H-8a absorption, whereas no effect was observed when the N-1 methyl was irradiated. It was concluded that the N-8 methyl must be *cis* to H-8a, and that the aromatic nitrogen does not undergo rapid inversion. The absolute stereochemistry followed from two independent degradations to molecules of known chirality: *S*-(+)-3-ethyl-*N*,3-dimethyloxindole and the amino-acid (**138**) (Scheme 95). Correlations with other related alkaloids have been made by o.r.d. comparisons, *e.g.* geneserine (**139**). Until recently geneserine was thought to be merely the *N*-8-oxide of

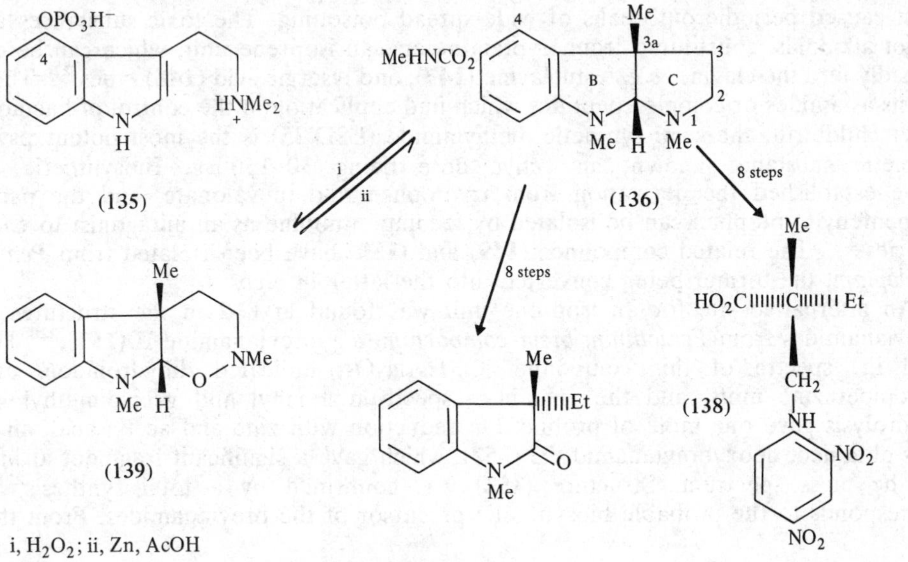

(135) (136) (138) (139)

i, H_2O_2; ii, Zn, AcOH

SCHEME 95

physostigmine on the basis of chemical interconversion. It has now been assigned structure (**139**) since the mass spectrum lacked the $M-16$, -17, and -18 peaks characteristic of *N*-oxides, and there was no significant change in the n.m.r. chemical shift of the C-2 methylene protons relative to physostigmine.[223a] Again, the B/C ring junction was found to be *cis* from n.O.e. measurements.[223b]

The β-carboline skeleton is obtained by condensation of tryptamine or tryptophan with an aldehyde or acid, and some of the simpler (non-terpenoid) alkaloids derived in this way are harman (**140**), 3-carboxyharman (**141**), and harmaline (**142**). More complex examples are typified by the pyridine alkaloid (**143**) from *Nauclea diderrichii*, elaeocarpidine (**144**) from *Elaeocarpus densiflorus*, perlolyrine (**145**) from the perennial rye-grass *Lolium perenne*, and evodiamine (**146**) from the Chinese plant *Evodia rutaecarpia.*[225]

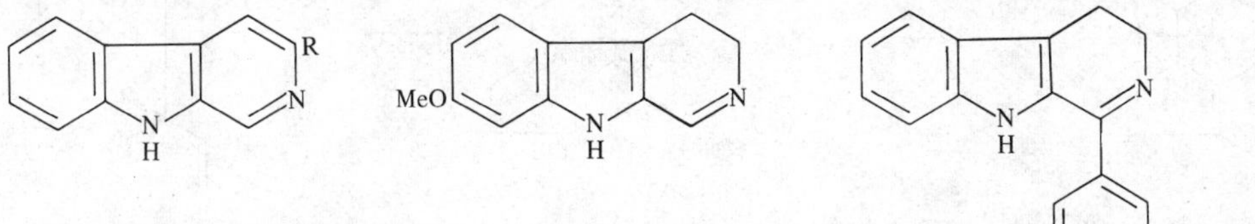

(140) R = H

(141) R = CO_2H

(142)

(143)

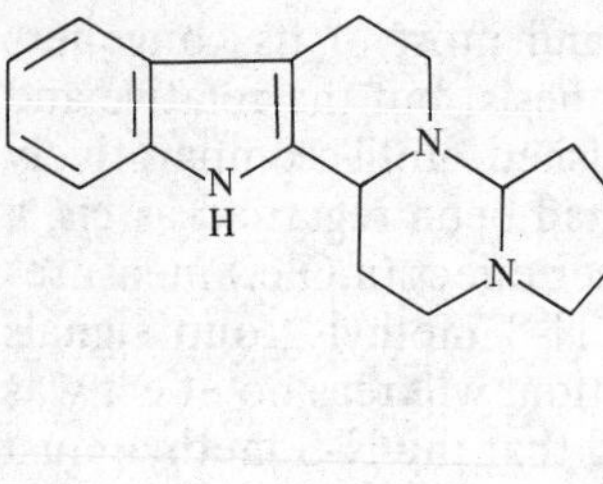

(144) 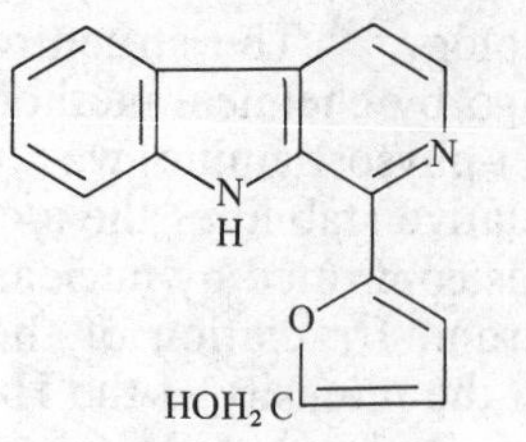 (145) 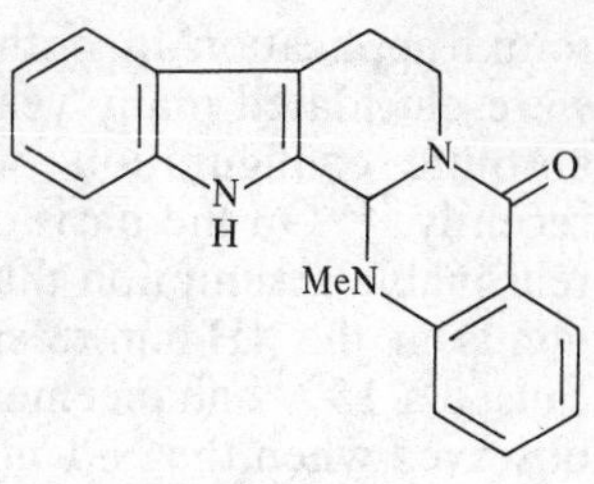 (146)

17.4.12.3 Isoprenoid tryptamine and tryptophan alkaloids

Infection of ears of rye by the fungus *Claviceps purpurea* produces ergot, which in the past caused periodic outbreaks of wide-spread poisoning. The toxic substances are the ergot alkaloids, constituted from tryptamine and an isoprene unit, which can be divided broadly into the clavine, *e.g.* festuclavine (**147**), and lysergic acid (**148**) types.[226] The latter exists as amides or complex peptides which find application in the control of haemorrhage after childbirth; the semi-synthetic diethylamide (LSD 25) is the most potent psychotomimetic substance known, an active dose being 30–150 μg. Biosynthetic studies have established the derivation from tryptophan and mevalonate, and the parent 4-isopentenyltryptophan can be isolated by feeding ethionine as an antagonist to *Claviceps* species.[226c] The related compounds (**149**) and (**150**) have been isolated from *Penicillium cyclopium*, the former being converted into the latter *in vivo.*[227]

An alternative site for an isoprene unit was found at C-2 in the structures of the brevianamides from *Penicillium brevi-compactum*, *e.g.* brevianamide E (**151**).[228] The u.v. and i.r. spectra of this compound, $C_{21}H_{15}N_3O_3$, indicated dihydroindole and diketopiperazine units, and the ^{1}H n.m.r. spectrum a vinyl and *gem*-dimethyl groups. Hydrolysis gave one mole of proline. On reduction with zinc and acetic acid, an indole was obtained, deoxybrevianamide E (**152**), which gave a significant fragment at *m/e* 198 in the mass spectrum. Structure (**152**) was confirmed by a total synthesis[228b] and corresponds to the probable biosynthetic precursor of the brevianamides. From this and

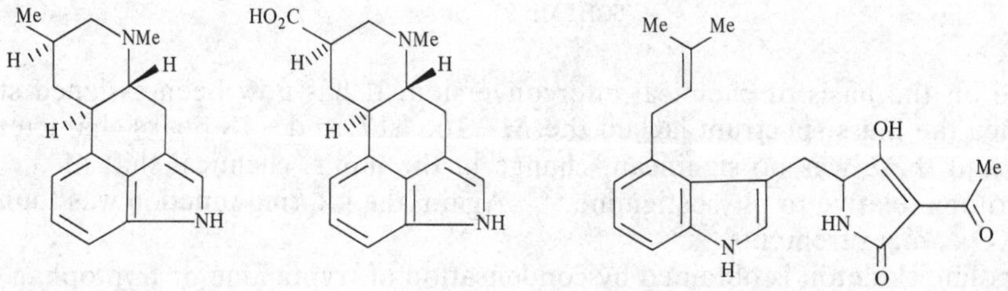

(147) (148) (149)

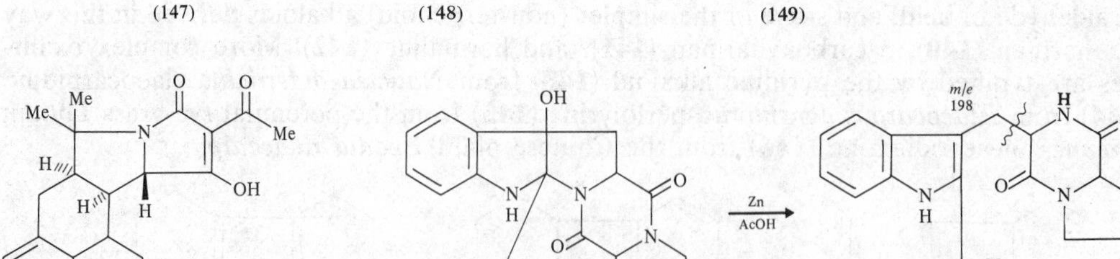

(150) (151) (152)

SCHEME 96

other evidence, structure (**151**) was assigned to brevianamide E. A relative of the brevianamides is the indoxyl austamide, whose structure (**153**) was obtained with the aid of a detailed ^{1}H n.m.r. analysis.[228c]

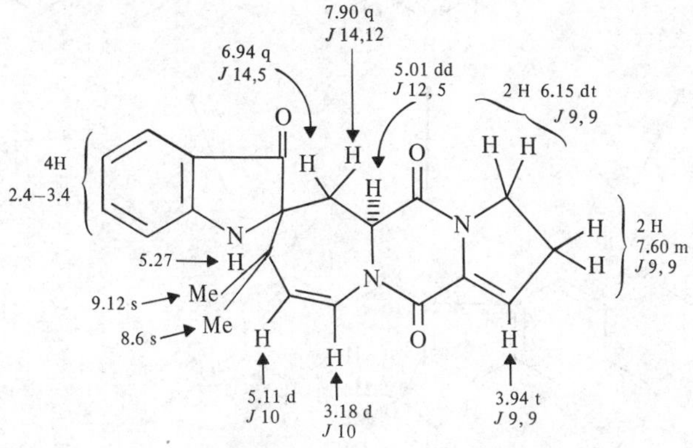

(153) N.m.r. assignments (τ) for austamide

A further elaboration is an additional isoprene unit at C-6, as in neoechinulin (**154**), isolated from *Aspergillus amstelodami*, which provides an example of a classical structure determination (Scheme 97).[229] Ozonolysis of neoechinulin, $C_{23}H_{25}N_3O_3$, gave formaldehyde and acetone from the two isopentenyl units; oxidation with alkaline permanganate gave 2-aminoterephthalic acid and established the location of the 6-alkenyl group. Catalytic hydrogenation gave an octahydro-derivative (**155**) by reduction of three C═C and one C═O bonds. Subsequent oxidation to an *N*-acylanthranilic acid, and hydrolysis to 2,2-dimethylbutyric acid, showed that the second isoprene unit was attached to C-2 of the indole *via* the carbon carrying the *gem*-dimethyl groups. Basic hydrolysis of octahydroechinulin gave 2,6-di-isopentyltryptophan, whereas alkali cleaved neoechinulin to a mixture of an indole (**157**) and a 3-formylindole (**156**). The final piece of evidence was for the dehydrotryptophan unit, which came from the isolation of tetraketopiperazine from the oxidation products, and completed the structure proof.

17.4.12.4 Monterpenoid tryptamine and tryptophan alkaloids

The common feature of the majority of indole alkaloids is the combination of tryptamine or occasionally tryptophan with a C_9 or C_{10} unit of monoterpenoid origin. Although there is an endless variation in complexity among the several hundred structures,[11] most can be arranged into three basic groups (Scheme 98) depending on the skeleton of the C_9/C_{10} unit: (i) *Corynanthé–Strychnos* type (**158**), *e.g.* ajmalicine (**161**) and strychnine (**162**); (ii) the *Aspidosperma* type (**159**), *e.g.* tabersonine (**163**); the *Iboga* type (**160**), *e.g.* catharanthine (**164**). Where a C_9 unit is involved, it is invariably the carbon atom (C-22) indicated by the broken line which is lost. A common numbering system for all carbon and nitrogen atoms is now used for these alkaloids, based on that originally used for yohimbine,[231] since the *Aspidosperma* and *Iboga* systems are formed by rearrangement of the *Corynanthé* skeleton. In denoting stereochemistry the α/β convention for substituents below and above the plane respectively is followed as in steroids.

The biosynthetic precursors of the majority of monoterpenoid alkaloids are tryptamine and secologanin (**165**), which condense to produce two diastereoisomers, vincoside (**166**) (H-3β) and strictosidine (H-3α) (Scheme 99). After the crucial step of cyclizing N-4 to C-21, the *Corynanthé*-type alkaloids can be derived by rational sequences. The problems of structure elucidation are compounded by the number of stereoisomers that may be produced. Corynantheine (**167**), for example, has three chiral centres, but since H-15 always retains the same configuration as in secologanin, only four stereoisomers need to

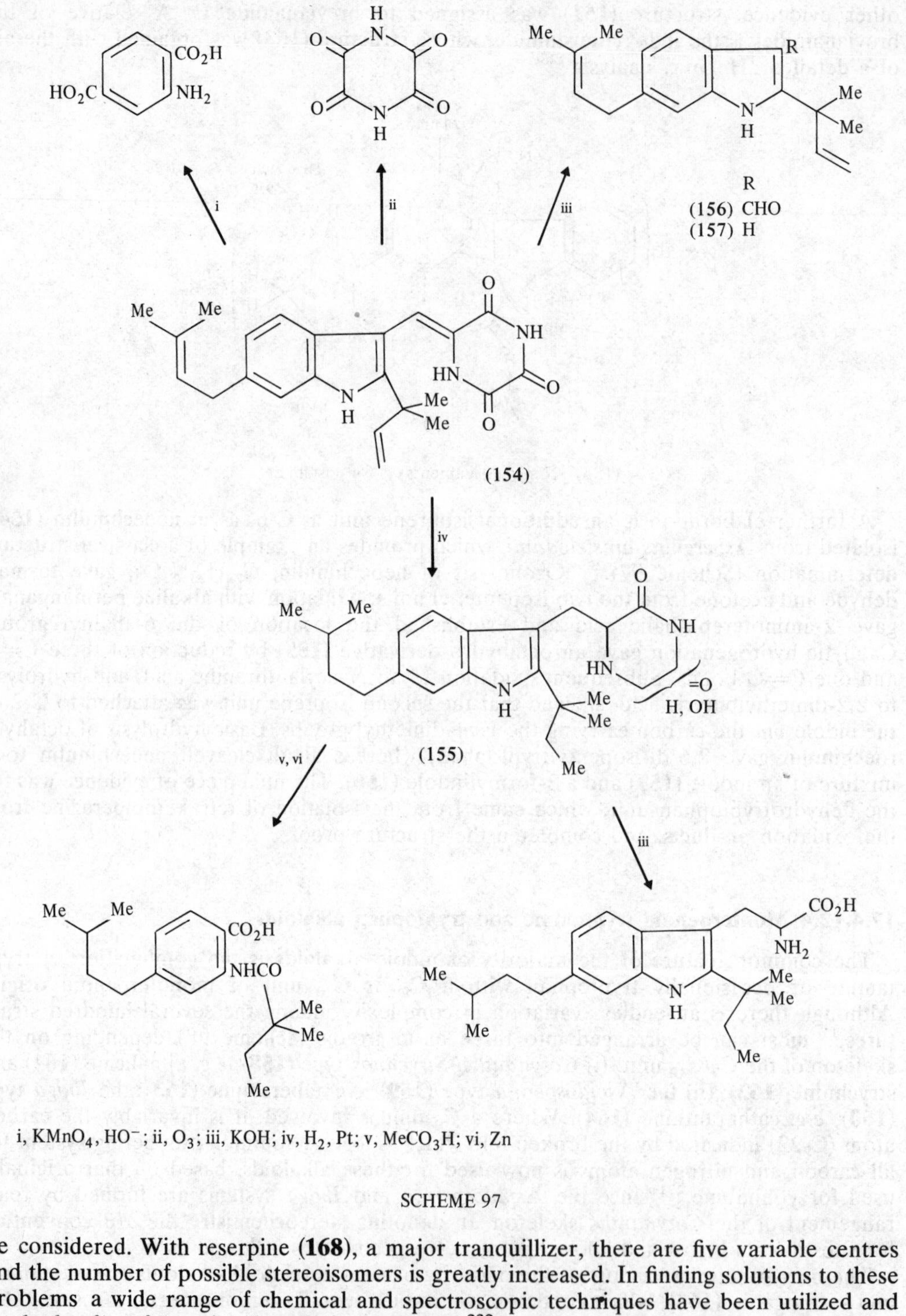

(156) R = CHO
(157) R = H

(154)

(155)

i, $KMnO_4$, HO^-; ii, O_3; iii, KOH; iv, H_2, Pt; v, $MeCO_3H$; vi, Zn

SCHEME 97

be considered. With reserpine (**168**), a major tranquillizer, there are five variable centres and the number of possible stereoisomers is greatly increased. In finding solutions to these problems a wide range of chemical and spectroscopic techniques have been utilized and methods of conformational analysis developed.[232]

A relatively simple example is adirubine (**169**), one of the small number of *Corynanthé* alkaloids derived from tryptophan rather than tryptamine (Scheme 100).[233] Many essential features of the structure were indicated by the mass spectrum, where the indicated cleavages gave ions characteristic of the tetracyclic *Corynanthé* skeleton,[21] in addition to

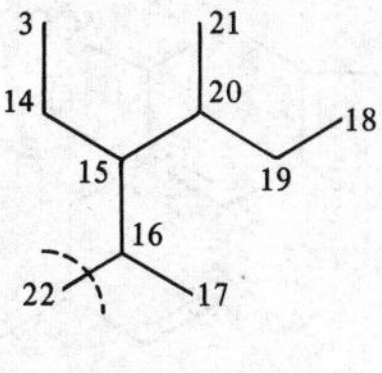

(158) 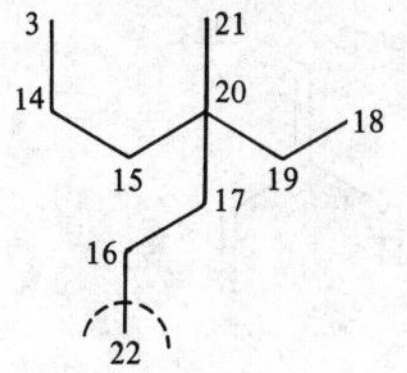 (159) 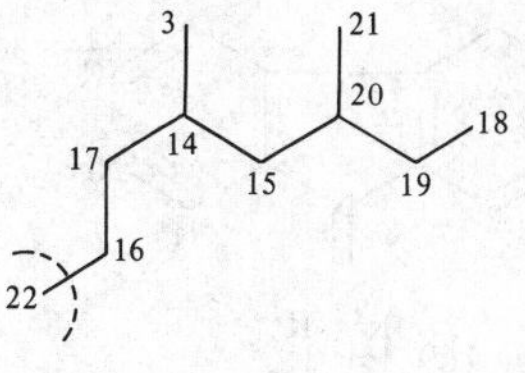 (160)

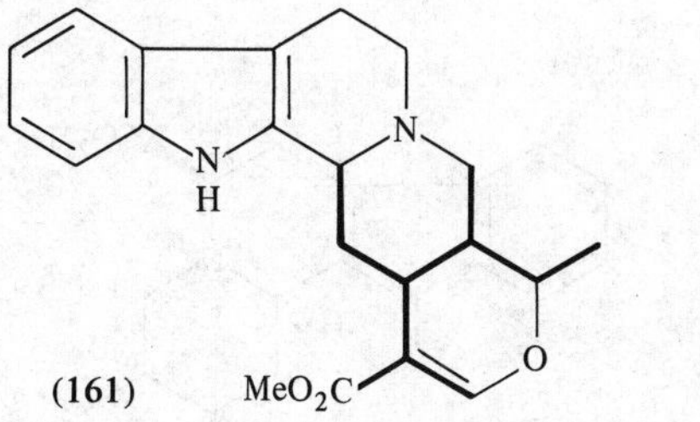

(161) 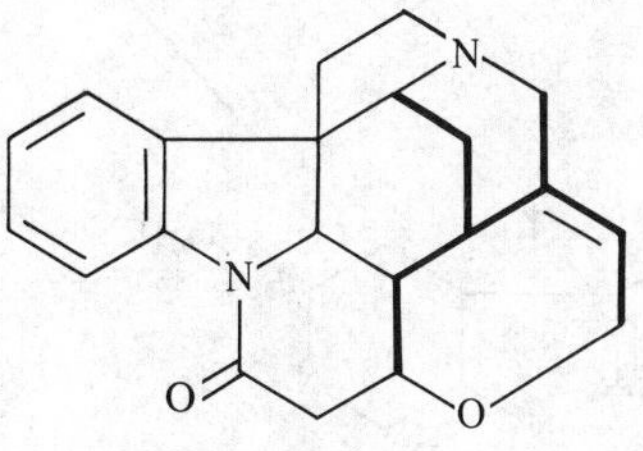 (162)

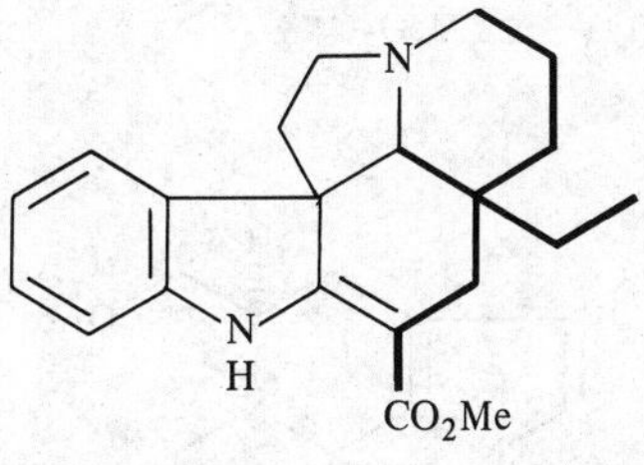

(163) 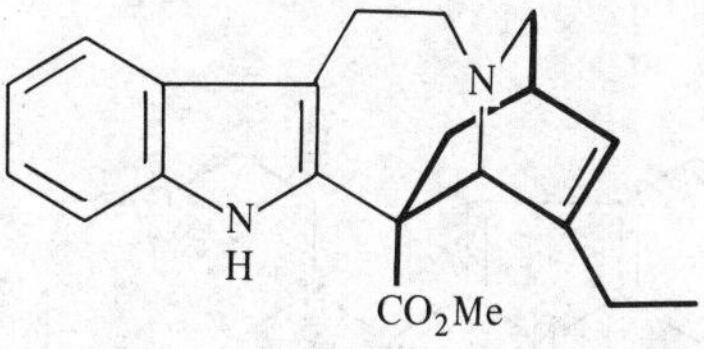 (164)

SCHEME 98

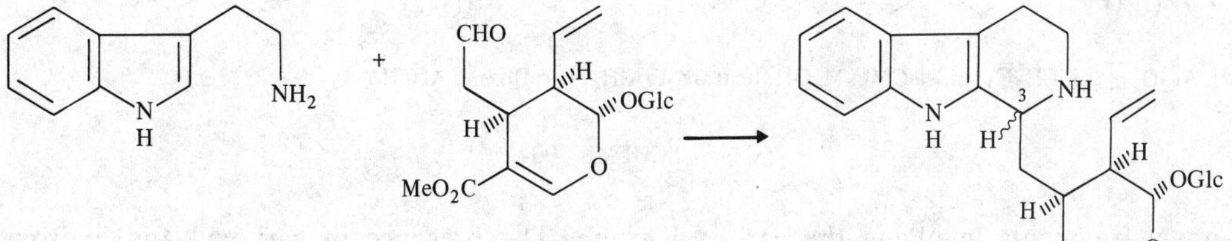

(165) (166)

SCHEME 99

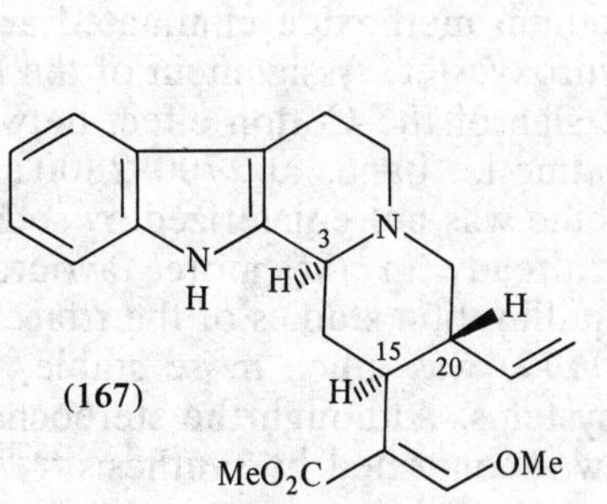

(167) 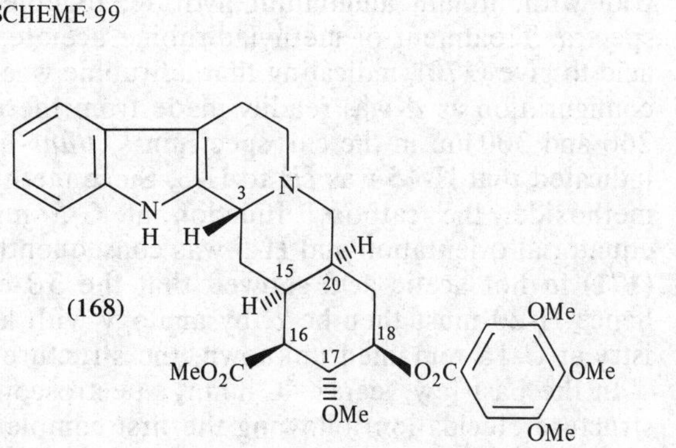 (168)

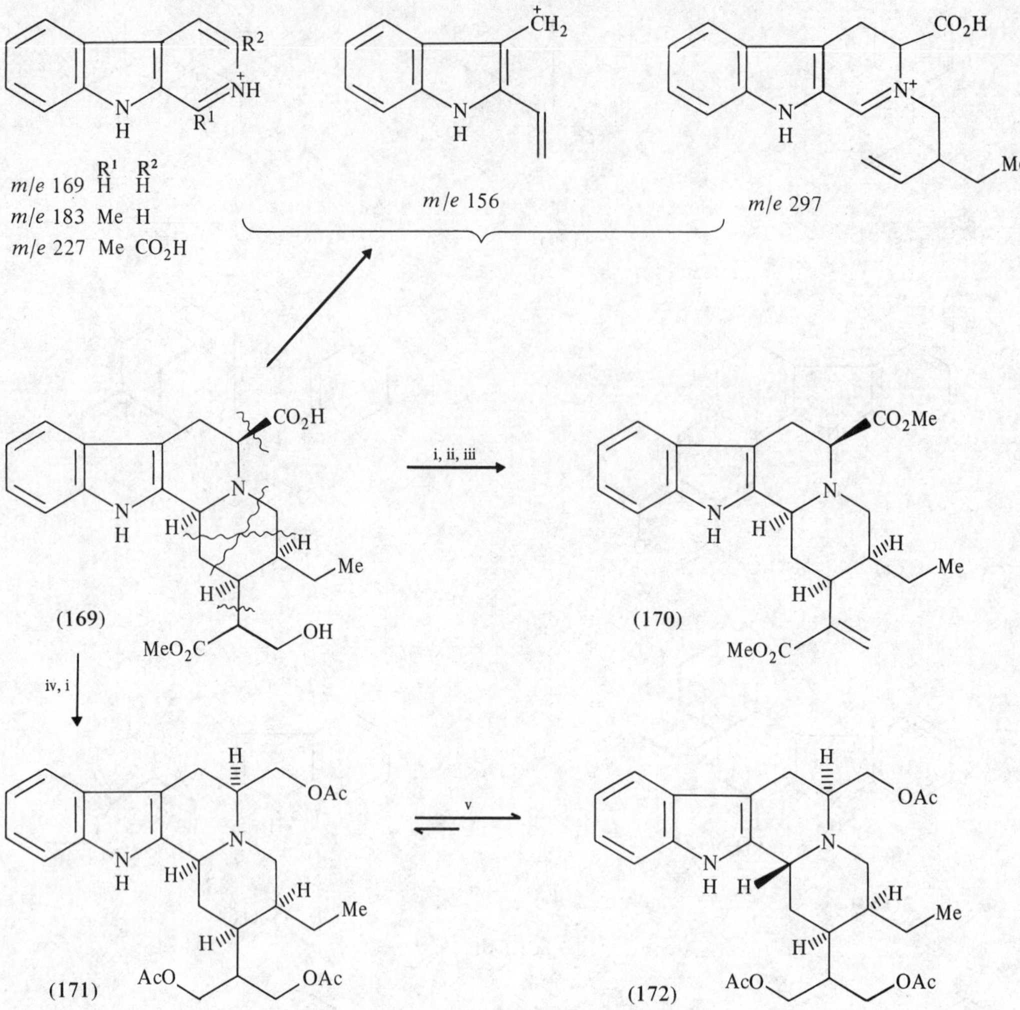

	R^1	R^2
m/e 169	H	H
m/e 183	Me	H
m/e 227	Me	CO_2H

m/e 156 *m/e* 297

(169) (170) (171) (172)

i, Ac_2O, py; ii, CH_2N_2; iii, NaOMe, MeOH, heat; iv, $LiAlH_4$; v, refluxing AcOH

SCHEME 100

novel fragments involving the carboxyl group. The presence of carboxyl and hydroxyl groups was confirmed by formation of methyl and acetate esters and by reduction to a triol with lithium aluminium hydride, together with appropriate n.m.r., i.r., and mass spectra. Treatment of methyladirubine acetate with sodium methoxide eliminated acetic acid to give (**170**), indicating that adirubine was a β-hydroxyester. Assignment of the H-3 configuration as α was readily made from the positive sign of the Cotton effect between 260 and 300 nm in the c.d. spectrum;[12] *trans*-quinolizidine i.r. bands at 2700–2800 cm^{-1} indicated that H-15 was *cis* to H-3. Since methyladirubine was not epimerized by sodium methoxide, the carboxyl function at C-5 must be already in the more favourable equatorial orientation and H-5 was consequently α. Equilibration studies of the triacetate (**171**) in hot acetic acid showed that the 3β-epimer (**172**) was much more stable, and hence H-20 must then be α by analogy with known systems. Although the stereochemistry at C-16 remained unknown, the structure (**169**) was confirmed by synthesis.[233d]

In the past few years, ^{13}C n.m.r. spectroscopy has made an increasing contribution to structure elucidation following the first complete analysis of a complex alkaloid with the

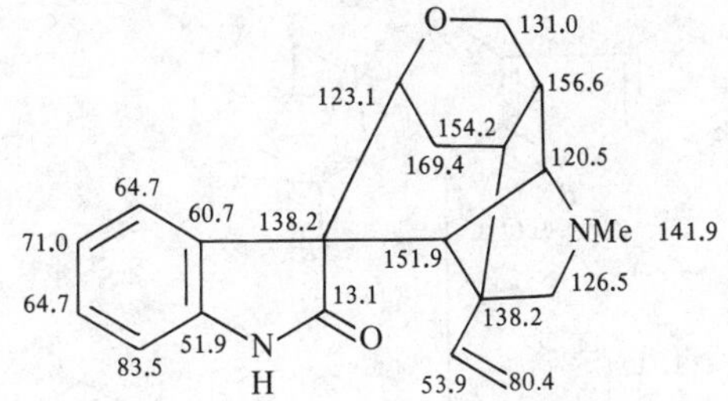

(173) ^{13}C n.m.r. assignments (δ from CS_2) for gelsemine

oxindole gelsemine (**173**) by Wenkert and his group.[234a] A particular example is the revision of the structure of the hexacyclic *Aspidosperma* alkaloid vindolinine to (**174**) when ^{13}C n.m.r. revealed one more quaternary carbon and one less methine carbon than required by the previously accepted structure (**175**).[234b] The stereochemistry was subsequently determined by correlation with the known (−)-vincadifformine (**176**) (Scheme 101). The numerous *Aspidosperma* and related alkaloids have been extensively reviewed.[235]

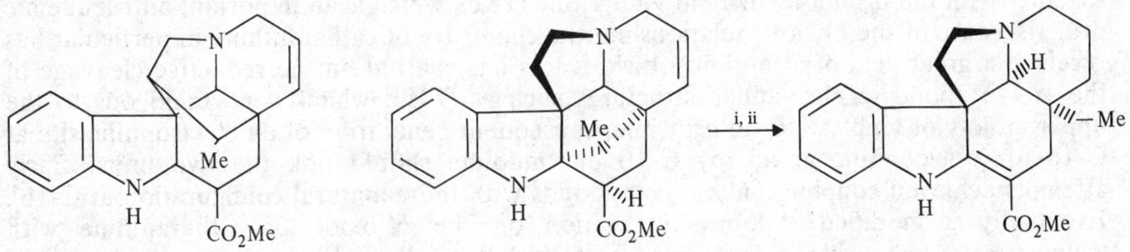

(175) (174) (176)

i, I_2; H_2, catalyst

SCHEME 101

Unlike the *Corynanthé* alkaloids, the *Aspidosperma* and *Iboga* type both exist as two enantiomeric series, the fixed chirality at C-15 being lost through cleavage of the 15–16 bond on rearrangement during biosynthesis. Although this had long been apparent for the *Aspidosperma* alkaloids, *e.g.* (+)- and (−)-vincadifformine, only recently was it demonstrated for the relatively small group of *Iboga* alkaloids from a study of the c.d. spectra.[236] Most belong to the (−)-ibogamine (**177**) series,[237] whereas (+)-catharanthine (**178**) is a member of the enantiomeric series.

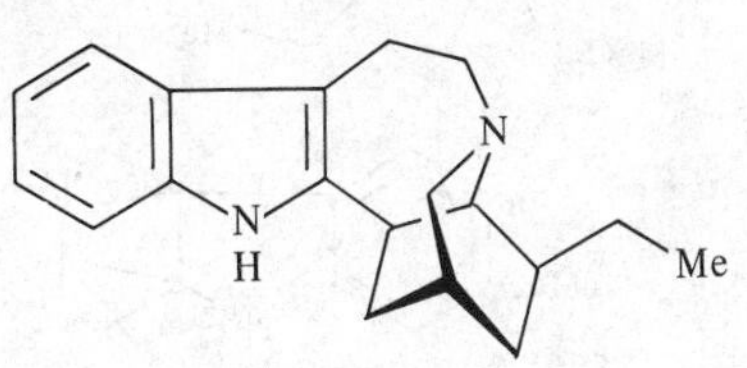

(177) 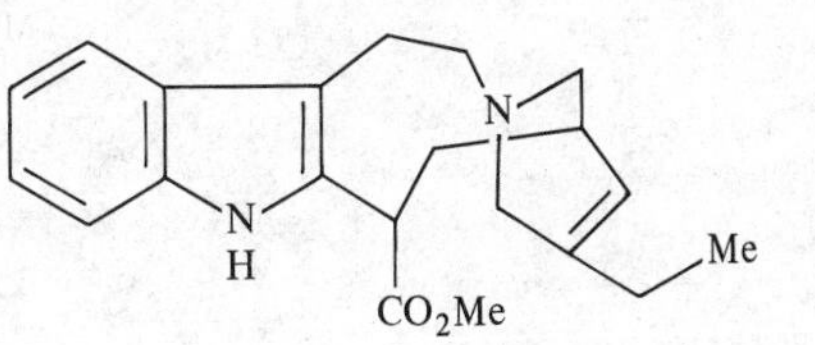 (179)

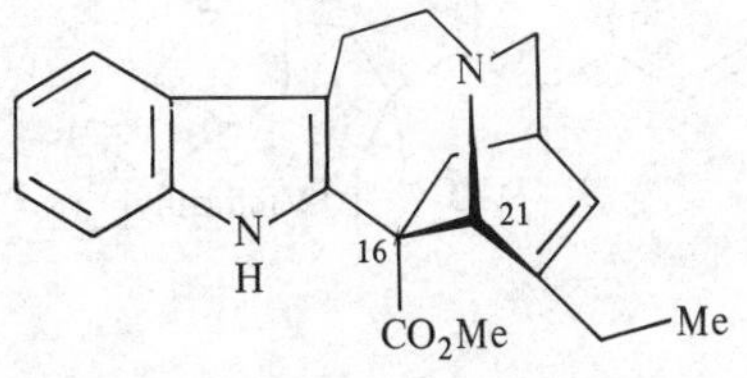

(178) 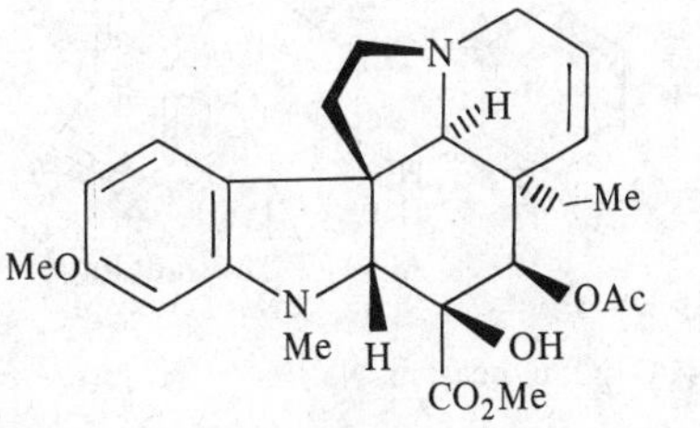 (180)

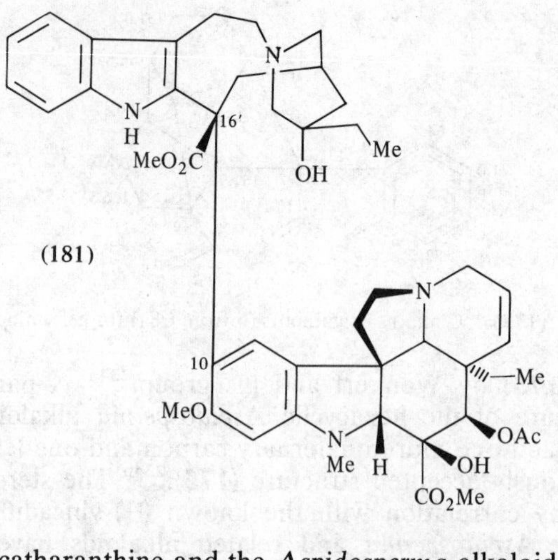

(181)

In *Vinca rosea* both catharanthine and the *Aspidosperma* alkaloid vindoline (**180**) occur together with the bisindole alkaloid vinblastine (**181**), which is an important anti-leukemic drug. Because of the obvious relationship, the chemistry of catharanthine in particular has received a great deal of attention. Much of this has centred on the reductive cleavage of the 16—21 bond to cleavamine structures such as (**179**), which were analogous to the upper moiety of vinblastine. In principle, subsequent generation of an electrophilic site at C-16 for nucleophilic attack by C-10 of vindoline should link the two units. Early attempts achieved coupling only to compounds with the unnatural configuration at C-16′. Eventually a modified Polonovski reaction on the *N*-oxide of catharanthine with trifluoroacetic anhydride in the presence of vindoline followed by reduction with sodium borohydride afforded a bisindole with the natural configuration (Scheme 102).[238]

Many other bisindole alkaloids are known, some of which are also important drugs, such as the muscle relaxant curare alkaloids. A comprehensive review of their chemistry and pharmacology has been published.[239]

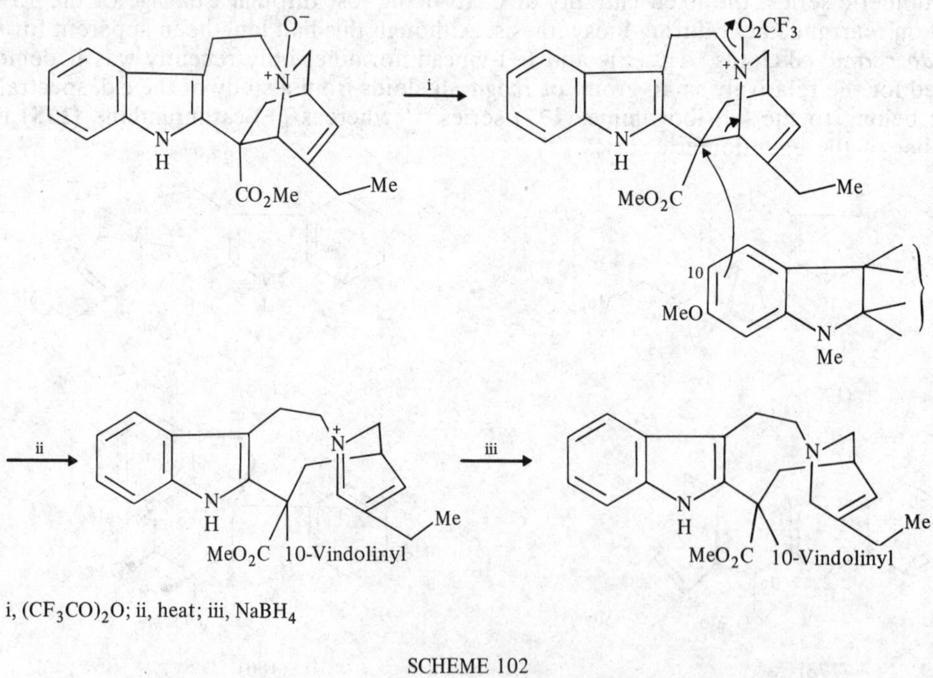

i, $(CF_3CO)_2O$; ii, heat; iii, $NaBH_4$

SCHEME 102

17.4.13 CARBAZOLES

17.4.13.1 Properties of carbazole

Carbazole (**182**), or dibenzopyrrole, can be obtained from the anthracene fraction of coal tar, whence it is completely purified only with difficulty. Total synthesis affords pure material, m.p. 245–246 °C, as a non-fluorescent, colourless compound. As a side product

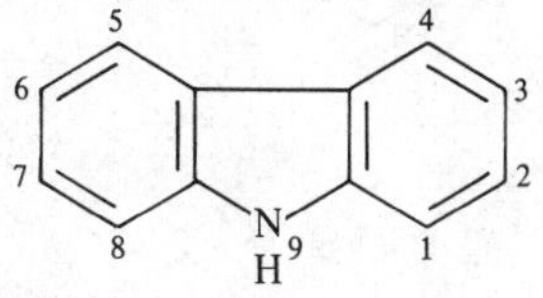

(182) 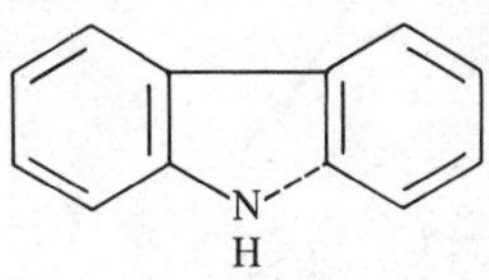 (183) 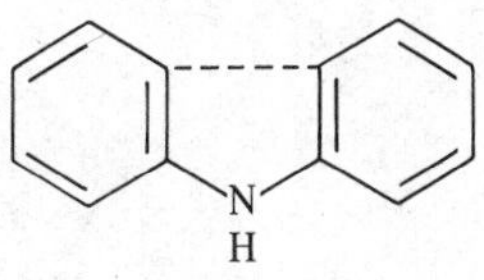 (184)

from coal tar it was originally investigated as an alternative feedstock for the production of dyes. Currently these are of minor importance and its main use is as a source of *N*-vinylcarbazole, which is readily polymerized to give polymers with good heat-resistant properties.[240]

Carbazole has an estimated resonance energy of *ca.* 305 kJ mol^{-1}, which compares with the value of 293 kJ mol^{-1} for biphenyl. The presence of the extra benzene ring, compared with indole, imparts further protection to the pyrrole ring. As a consequence, carbazole is less reactive towards electrophilic attack than indole. The N—H bond is more acidic than that in diphenylamine and it reacts with bases or Grignard reagents to give salts. The u.v. spectrum of carbazole shows similarities to that of phenanthrene.

17.4.13.2 Syntheses

Preparative routes to carbazoles follow one of two main paths, both involving the formation of the central ring as in (**183**) and (**184**). The Fischer indole synthesis can be modified to give carbazoles (the Borsche synthesis) (reaction 1). The intermediate tetrahydrocarbazoles, *e.g.* (**185**), available from this reaction are of interest as precursors and

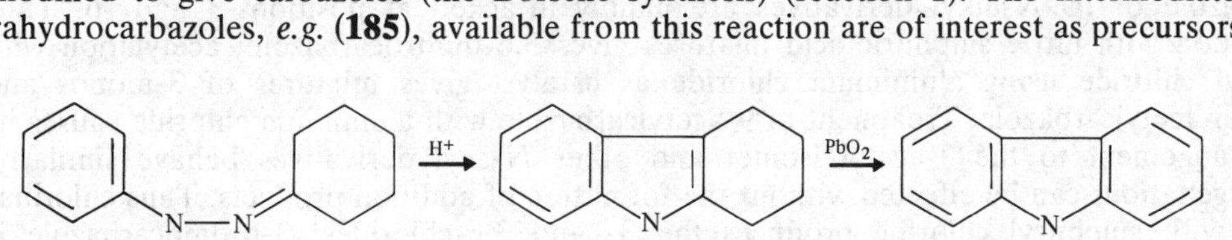

H^+ → (185) PbO_2 → (1)

models for a variety of indole alkaloids (see Section 17.4.12). Other dehydrogenating agents include sulphur and selenium. This route involves final cyclization by formation of the aryl–nitrogen link, as in (**183**). Another synthesis, in which a similar path is followed, commences with 2-nitrobiphenyls. Deoxygenation of these with triethyl phosphite yields the corresponding carbazole.

Carbon–carbon bond formation, see (**184**), is achieved in the Graebe–Ullmann synthesis (Scheme 103), which commences with 2-aminodiphenylamines. Diazotization produces an intermediate benzotriazole (**186**) that loses nitrogen on heating, with formation of the carbazole. If *N*-alkyldiphenylamines are used the reaction proceeds directly to the carbazole. Carbazoles can also be prepared from phenothiazines by the extrusion of sulphur, *e.g.* with copper, or by condensation of indoles with 1,4-diketones (reaction 2).

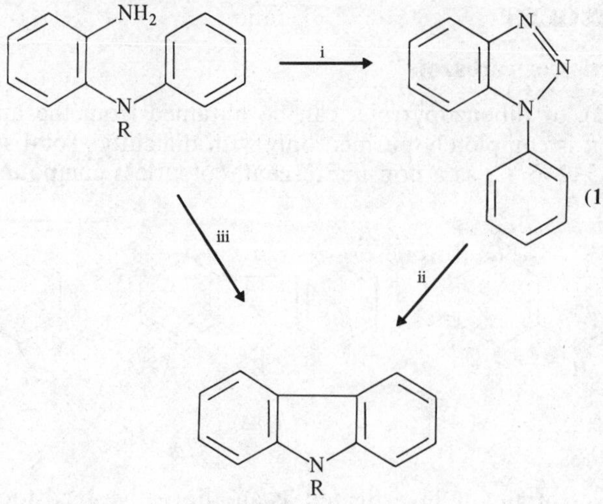

i, NO^+, R = H; ii, NO^+, R = alkyl; iii, heat

SCHEME 103

+ $MeCOCH_2CH_2COMe$ → (2)

17.4.13.3 Reactions

Salts of carbazole generally react directly on the nitrogen atom. Thus magnesium salts, prepared by the interaction of carbazole with Grignard reagents, react with carbon dioxide, ethyl chloroformate, or acid chlorides to yield the *N*-substituted derivatives. Alkali-metal salts can also be alkylated on the nitrogen atom.

Electrophilic substitution depends on the nature of the nitrogen substituent. The free carbazole or its *N*-alkyl derivatives are mainly attacked at positions-3 and -6. Thus nitration with nitric–sulphuric acid mixtures gives 3,6-dinitrocarbazole; acetylation with acetyl chloride using aluminium chloride as catalyst gives mixtures of 3-mono- and 3,6-di-acetylcarbazole. Treatment of 9-acetylcarbazole with aluminium chloride causes its rearrangement to the 3-acetyl isomer and other *N*-acyl derivatives behave similarly. Halogenations can be effected without the formation of addition products. Thus chlorination with sulphuryl chloride produces the 3- and 3,6-chlorides; 3-bromocarbazole is formed by reaction of carbazole with *N*-bromosuccinimide in carbon tetrachloride in the presence of dibenzoyl peroxide as catalyst; 3-iodocarbazole is formed from carbazole by reaction with potassium iodide and potassium iodate in acetic acid.

Carbazole is not reactive enough to couple directly with diazonium salts. It is claimed that vigorous reduction with tin and hydrochloric acid yields the hexahydro-derivative (**187**) and selective reduction to the tetrahydro-state (**185**) can be achieved by catalytic

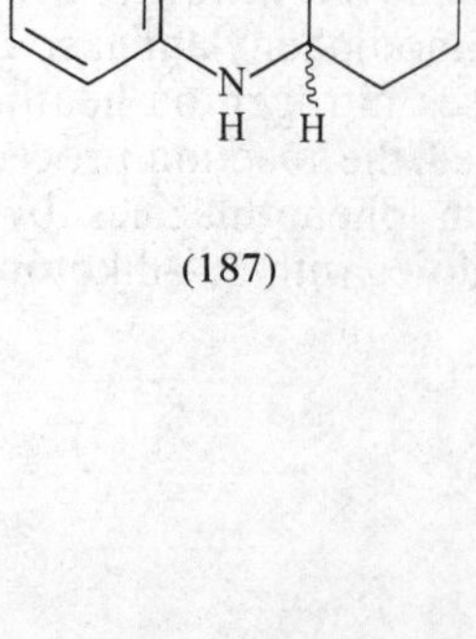

(187) 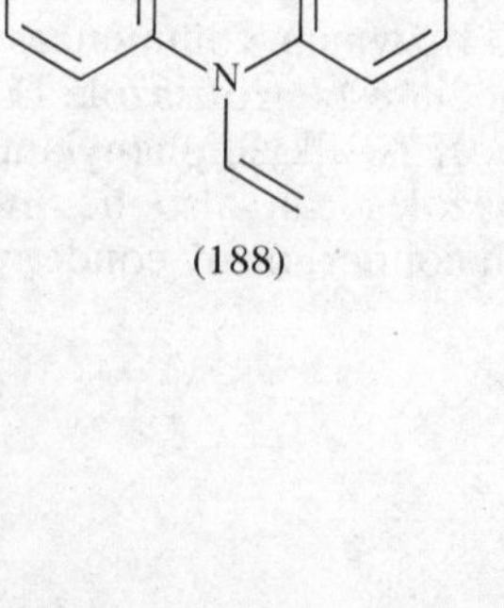 (188)

hydrogenation over copper chromite. Oxidation of the parent system tends to give mixtures.

Under the influence of base, the nitrogen atom of carbazole can participate in addition reactions. Thus, sodium hydroxide catalyses the addition of carbazole to acetylene, producing *N*-vinylcarbazole (**188**). Michael-type additions of carbazole, for example to acrylonitrile, also occur under the influence of base catalysts.

N-Vinylcarbazole is an important monomer which can either be polymerized under acidic or free radical conditions or co-polymerized with other monomers to produce a variety of plastics. These tend to be brittle, but they generally possess good resistance to heat and they therefore find applications in the electronics and electrical industries as insulators.

17.4.14 ISOINDOLES

17.4.14.1 Properties

Isoindole (**189**) is isomeric with indole. The parent compound and 2-unsubstituted derivatives have the opportunity to tautomerize with the 1*H*-isomer, *i.e.* isoindolenine (**190**) (or 1*H*-isoindole). Compared with indole it is much more unstable, rapidly undergoing oxidation in air to form mixtures and polymers. Its detailed characterization and isolation was not accomplished until 1972,[241] although it had previously been made in

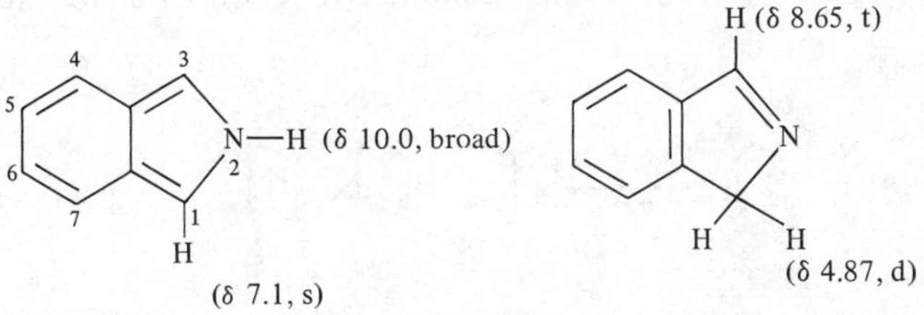

4 3 5 6 7 1 2 N—H (δ 10.0, broad) H (δ 7.1, s)

(189)

H (δ 8.65, t) N H H (δ 4.87, d)

(190)

dilute solutions. It can be obtained pure and in nearly quantitative yield by the flash pyrolysis of, for example, 7-azabenzonorbornene at 600 °C (reaction 3), but the product is obtained as a mixture with the isoindolenine isomer. The mixture of isomers is

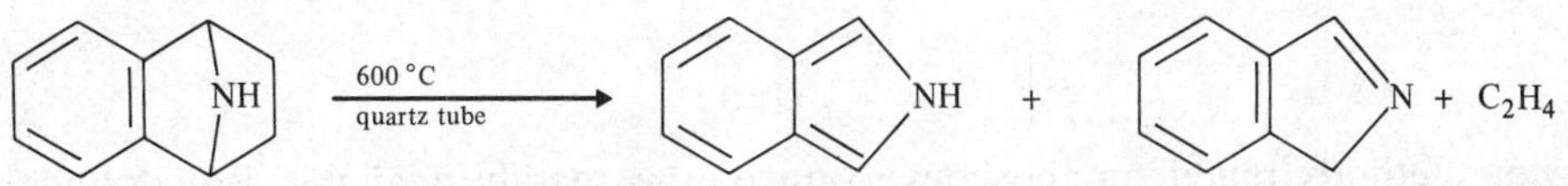

NH —(600 °C, quartz tube)→ NH + N + C_2H_4 (3)

readily analysed by ^{1}H n.m.r. spectroscopy. The mixture so obtained is *not* an equilibrium mixture since, on leaving it at room temperature, the concentration of the isoindolenine isomer decreases with the complete formation of isoindole. Thus isoindole is thermodynamically more stable than its isoindolenine isomer at room temperature. Furthermore, the isomerization occurs without the incorporation of deuterium into positions-1 and -3 when carried out in the presence of a deuterated solvent such as heavy water. The N—H group of the product does exchange with deuterium. This result suggests the occurrence of a 1,5-sigmatropic rearrangement in the isomerization step.

The greater stability of the isoindole form (**189**) over that of the Schiff's base (**190**) is ascribed to the lower π-electron energy of the former, it behaving as a 10π-electron system.[242] The isoindole system can also be considered to have *ortho*-quinoidal character across the fused benzene ring. In order to explain both its relative stability as well as its comparative reactivity, it has been argued that the endothermic contribution to the activation energy on disruption of the system is more than offset by the exothermic contribution due to formation of the benzene ring, as, for example, in cycloadditions across the 1,3-positions. In other words it behaves as a hyper-reactive system! Photoelectron spectroscopy has shown that the system has a low resonance energy compared with naphthalene as well as a close, low-lying excited state.[243]

As might be anticipated, the position of equilibrium between the isoindole and isoindolenium isomers can be shifted by varying the substitution pattern, for example:

unsubstituted isoindole	*ca.* 100%
1-phenyl	91%
1-*p*-methoxyphenyl	69%
1-*p*-dimethylaminophenyl	50%

In general, substitution aids the isolation of these systems. 2-Methylisoindole, for example, cannot undergo tautomerism, can be isolated as a solid with m.p. 90–91 °C, but does not react with methyl iodide, illustrating the absence of amine properties from the nitrogen atom.

17.4.14.2 Syntheses

The ingenuity of chemists is amply illustrated in the range of synthetic methods developed for the preparation of isoindoles. Earlier methods have been reviewed.[244] Flash pyrolysis methods for the formation of isoindoles have been developed by using better leaving groups. In this manner the tetrazine adduct (**191**) rapidly eliminates both nitrogen and the pyridazine (**192**) to produce the isoindole (reaction 4) at low temperatures.[245]

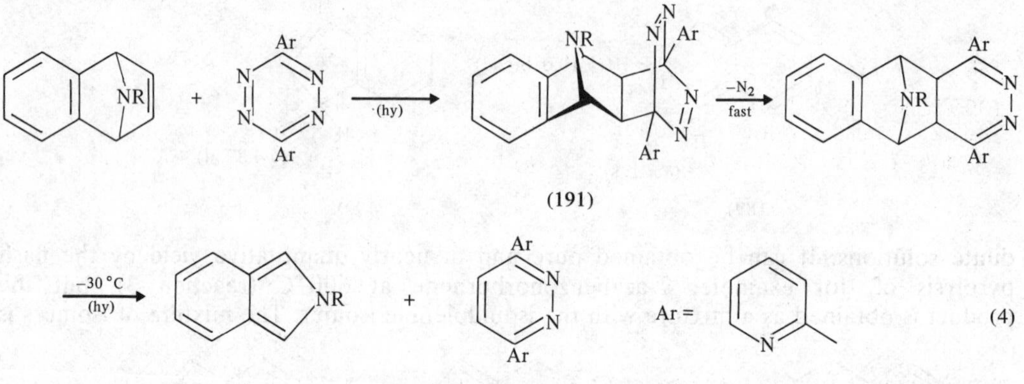

NR + Ar-tetrazine-Ar —(hy)→ (191) —$-N_2$, fast→ —$-30\,°C$ (hy)→ isoindole NR + (192)   Ar = 2-pyridyl   (4)

Earlier methods relied on the conversion of the readily available isoindolines, using either a sequential oxidation or elimination process. Since isoindoles are themselves readily oxidized, elimination has been the preferred process. The first clear indication of a simple isoindole derivative was obtained in 1951 by Wittig,[246] who reacted the isoindolinium salt (**193**) with phenyl-lithium to produce 2-methylisoindole. Under similar conditions the spiro-derivative (**194**) gave the isoindole (**195**) in low yield. The benzylmethyl salt (**196**) gave a variety of products depending on the reaction conditions. At lower temperatures a Sommelet rearrangement product (**197**) formed, whilst at moderate temperature a Stevens rearrangement product (**198**) was preferred. Only at higher

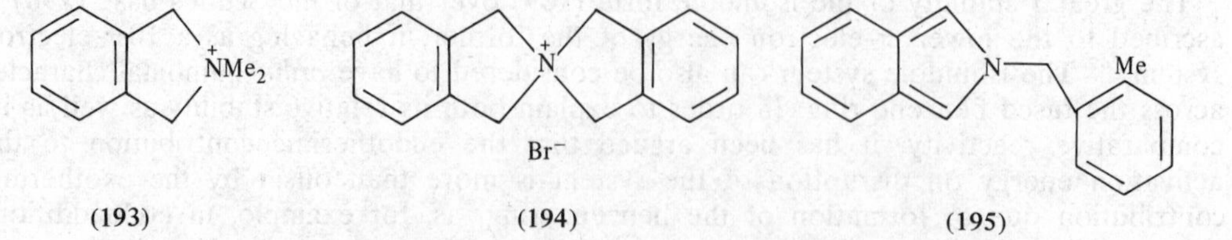

$\overset{+}{N}Me_2$ (193) $\overset{+}{N}$ Br^- (194)   N–CH(Me)Ph (195)

NMe

(197)

Me

−30 °C

Me

$\overset{+}{N}$

CH_2Ph

Base 80–120 °C

NMe

(198)

(196)

> 180 °C

NMe

SCHEME 104

temperatures was 2-methylisoindole formed (Scheme 104). Other *N*-substituted isoindolines can also produce isoindoles. 2-Benzyloxyisoindoline (**199**) produced a mixture of isoindole and isoindolenine on treatment with base, whilst the *N*-oxide (**200**) undergoes a

$NOCH_2Ph$

(199)

Me

$\overset{+}{N}$

O^-

(200)

rearrangement followed by elimination to the isoindole (in medium yield) on treatment with acetic anhydride at low temperature.[247] A related elimination reaction, which is only suitable for the preparation of substituted isoindoles, employs *N*-toluene-*p*-sulphonyl derivatives (reaction 5).[248]

Ph

NSO_2 Tol $\xrightarrow{K^+ \ ^-OBu^t}$

Ph

Ph

NH (my) (5)

Ph

An alternative approach to substituted isoindoles commences with phthalimidines by reaction with Grignard reagents (reaction 6). Reduction of *N*-methylphthalimidine with lithium aluminium hydride can likewise be made to generate 2-methylisoindole.[249]

NMe $\xrightarrow[\text{(anisole)}]{\text{EtMgBr, 70 °C}}$ NMe (my) (6)

O

Et

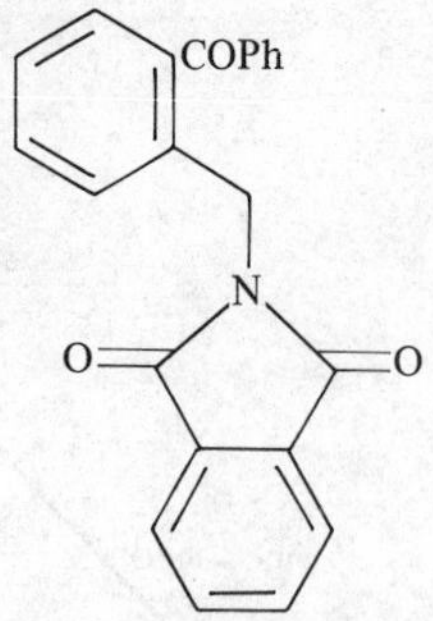

(201)

The intramolecular condensation of aminomethylaryl systems bearing *ortho*-carbonyl functions also generates the isoindole unit. For example, the phthalimide derivative (**201**) reacts with hydrazine to liberate the amine, which preferentially condenses with the ketone group to produce, after equilibration, 1-phenylisoindole. A modified Leuckart reaction, using ammonium formate on 1,2-dibenzoylbenzene, produces 1,3-diphenylisoindole by a condensation, partial reduction, and further condensation sequence.[250] The latter method is related to an earlier process which involves the reaction of ammonia with a simple 1,4-diketone (*c.f.* the Paal–Knorr reaction) under the influence of an acid catalyst. The pyrrole initially formed (reaction 7) reacts further with more of the

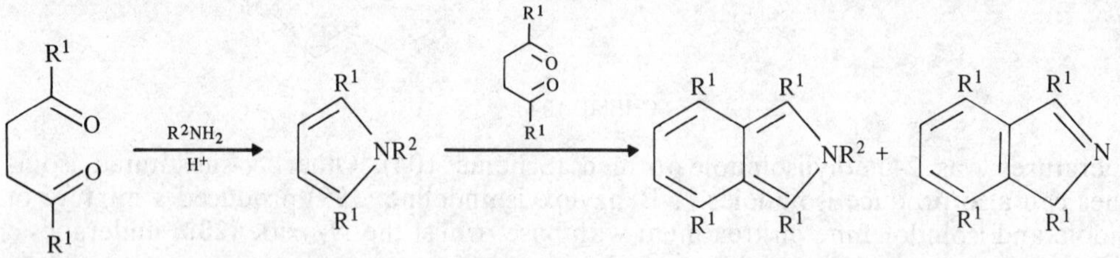

(7)

diketone, with loss of water, to give mixtures of isoindoles and isoindolenines.[251] Alternatively, substituted pyrroles can be reacted directly with appropriately substituted diketones, thus extending the scope of the reaction, or the pyrroles themselves can be used as a source of the dicarbonyl system, by hydrolysis *in situ* (reaction 8).[252]

Other methods used for the synthesis of isoindoles include the preparation and reaction of the cyclohexadiene (**202**) (reaction 9) and the process illustrated in reaction 10, which

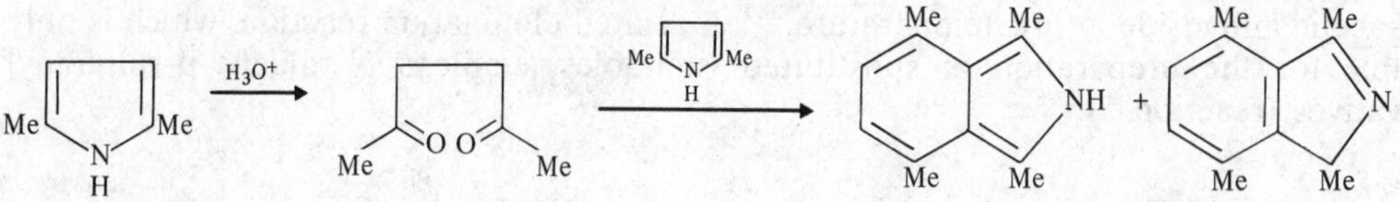

(8)

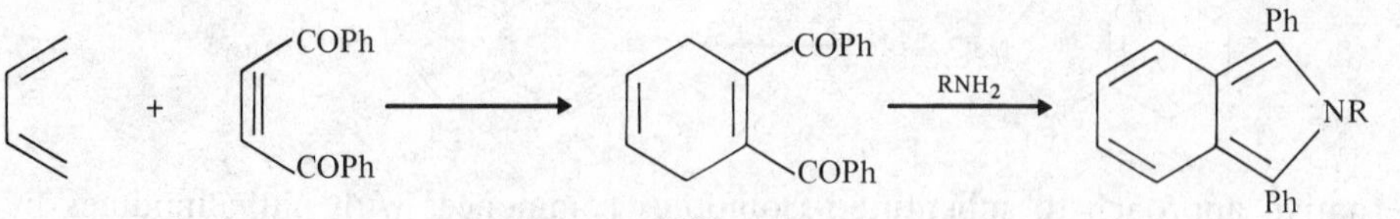

(202)

(9)

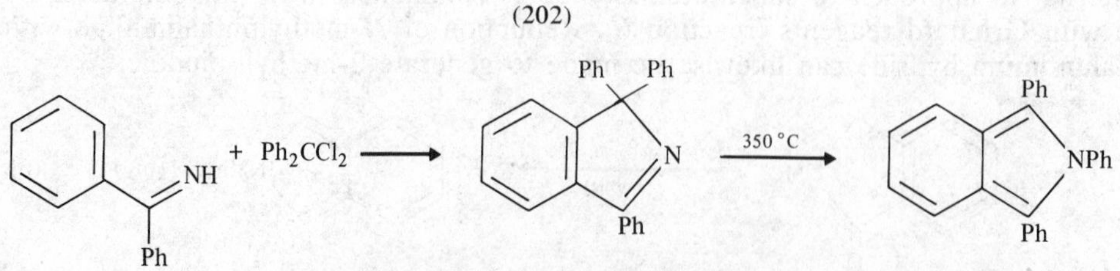

(10)

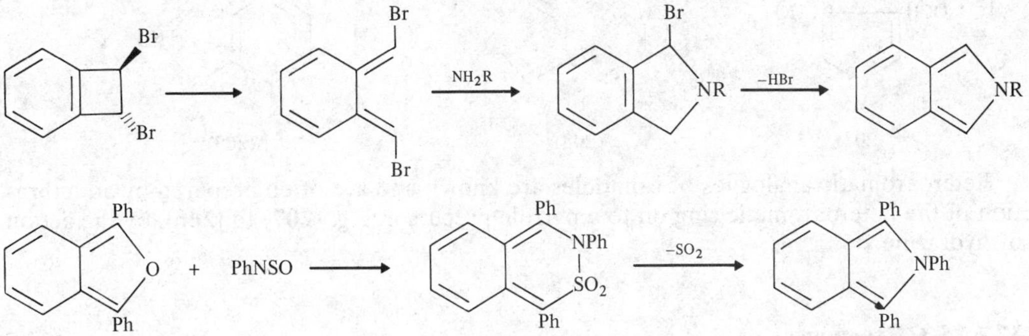

SCHEME 105

capitalizes on the ability of phenyl groups to participate in 1,5-sigmatropic rearrangements.[253] Two methods which capitalize on the reactivity of *ortho*-quinonedimethides are illustrated in Scheme 105.[244]

An important ring-contraction takes place when benzodiazepines of the type (**203**) are treated with base and the process presumably takes place through intermediate formation of a tricyclic ß-lactam system (*e.g.* **204**) (Scheme 106) followed by its rapid collapse to the isoindole (**205**).

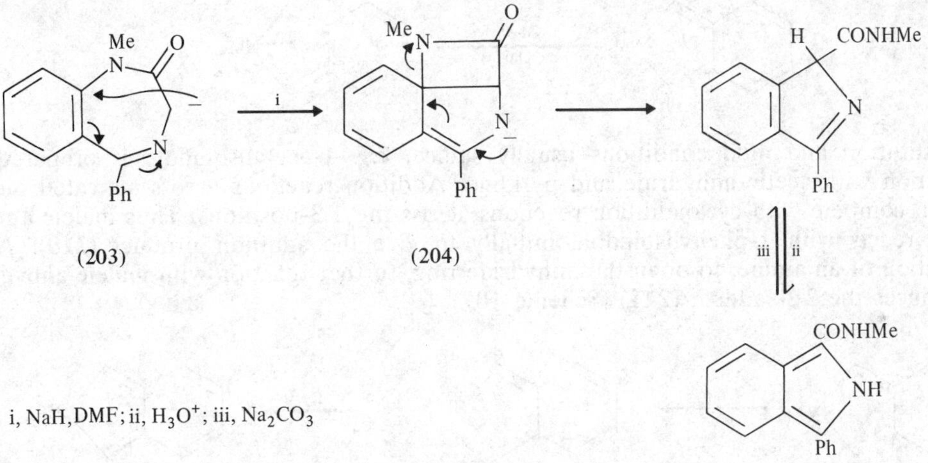

(203) (204) (205)

i, NaH,DMF; ii, H_3O^+; iii, Na_2CO_3

SCHEME 106

Intramolecular cyclizations to aryne intermediates can lead to the formation of isoindoles, as in the example illustrated (reaction 11).[254] This particular example relies on *in situ* oxidation of the initially formed isoindolinine and mixtures of the reduced and aromatic forms are obtained. A more efficient method is to build-in a good leaving group, as in the case of the aminoacetonitrile derivative (**206**) (reaction 12).

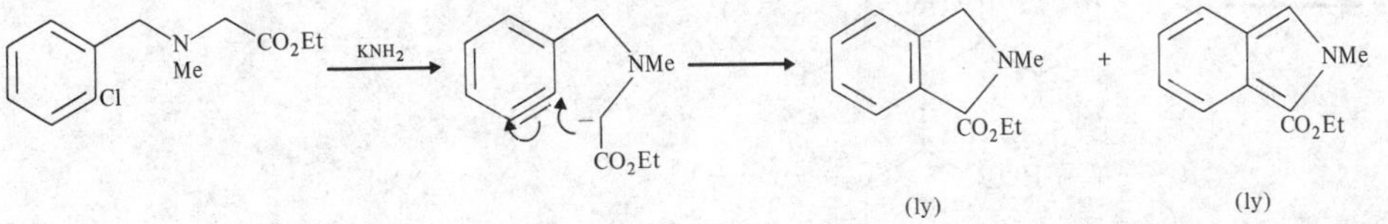

(11)

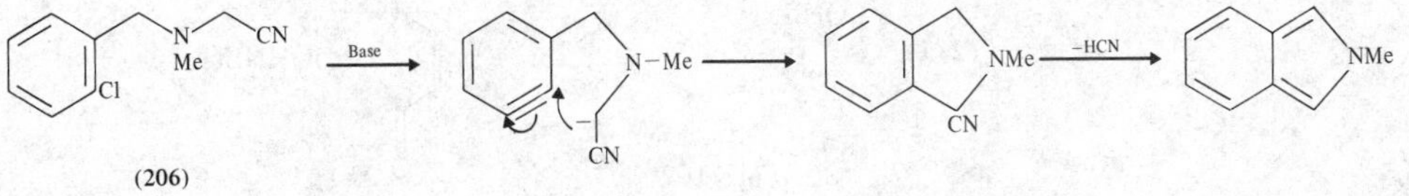

(12)

(206)

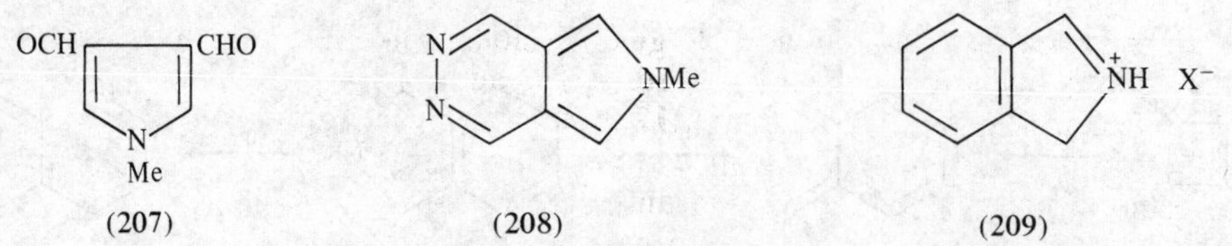

OCH CHO N Me (207) — N N NMe (208) — NH X⁻ (209)

Heteroaromatic analogues of isoindoles are known and are often prepared by adumbration of the heteroaromatic ring on to a pyrrole precursor, *e.g.* (**207**) to (**208**), by the action of hydrazine.

17.4.14.3 Reactions

The isoindole system undergoes electrophilic substitution reactions, preferentially at position-1, and acids give isoindolenium salts (**209**). They give a positive Ehrlich test with *p*-dimethylaminobenzaldehyde, provided either positions-1 or -3 are vacant. They also participate in the Mannich reaction (equation 13). The system is sensitive to electrophilic

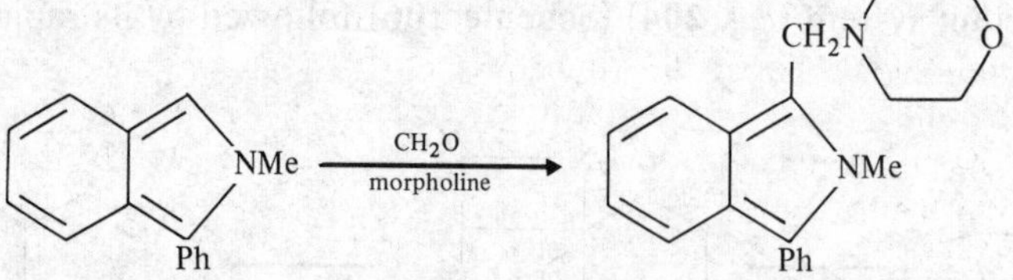

$$\text{NMe, Ph} \xrightarrow[\text{morpholine}]{CH_2O} \text{CH}_2\text{N(morpholino), NMe, Ph} \quad (13)$$

substitution and mild conditions usually suffice, *e.g.* 1-acetylisoindole is prepared by reaction with acetic anhydride and pyridine. Addition reactions across activated olefins often compete with cycloaddition reactions across the 1,3-positions. Thus maleic anhydride reacts with 1-phenylisoindole initially to give the addition product (**210**). After addition of an amine, to open the anhydride ring, further reaction with maleic anhydride produces the 2:1 adduct (**211**) (Scheme 107).

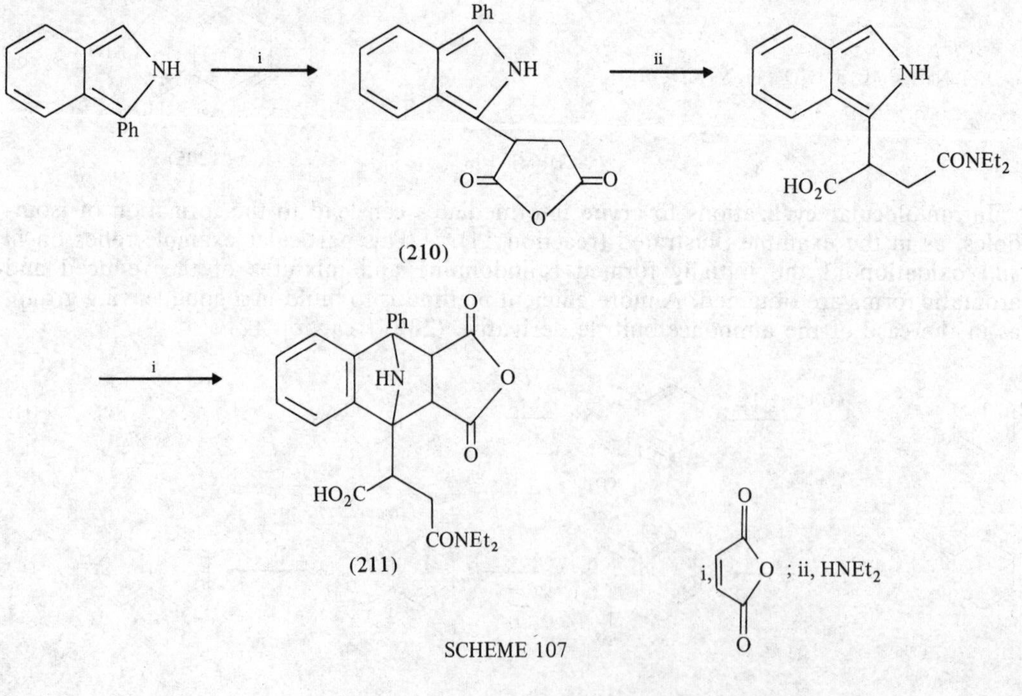

NH Ph —i→ Ph NH O O O (210) —ii→ NH HO_2C $CONEt_2$

—i→ Ph HN O O O HO_2C $CONEt_2$ (211)

i, maleic anhydride; ii, $HNEt_2$

SCHEME 107

1,3-Disubstituted isoindoles readily give cycloaddition products across the 1,3-positions. Photo-oxidation can also afford cyclic adducts, 1,2,3-triphenylisoindole, for example, yielding the cyclic peroxide (**212**). On warming this can be made to liberate oxygen, reforming the isoindole, whilst mild reduction liberates 1,2-dibenzoylbenzene. Presumably because of steric hindrance between the methyl groups 1,3,4,7-tetramethylisoindole is photo-oxidized to give the hydroperoxide (**213**).

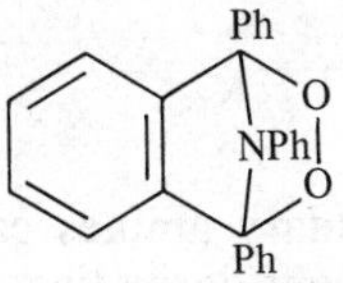

(212) 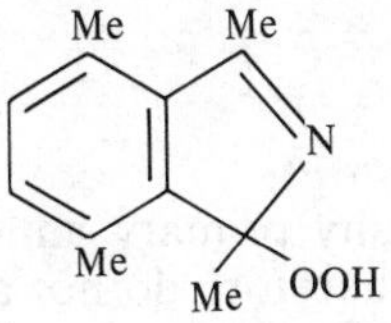 (213) 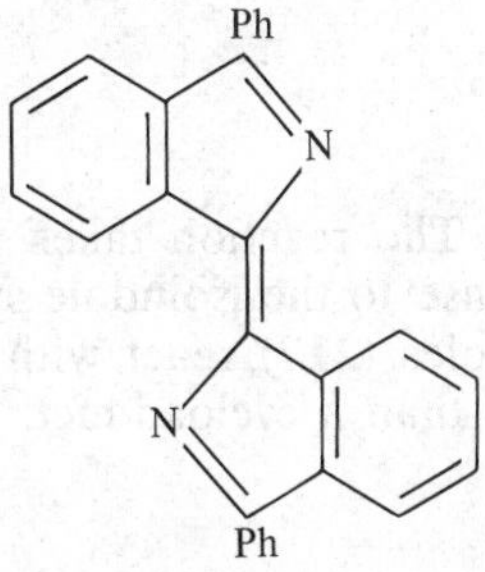 (214)

Isoindoles are sensitive to air and often form mixtures of products or polymers. Treatment of 1-phenylisoindole with air produces, for example, the dimer (**214**). Isoindoles are also susceptible to reduction, initially producing the corresponding isoindoline, effective reagents being Raney nickel, zinc–acetic acid, tin–hydrochloric acid, *etc.* High-pressure reductions, using Raney nickel, generate perhydroisoindoles.

Just as in pyrroles and indoles, 2*H*-isoindoles are acidic and metal salts can be prepared. The salts produced by reaction with Grignard reagents show enhanced reactivity towards electrophiles.

17.4.14.4 Derivatives

The most widely known derivatives of isoindole are the phthalocyanines (see Section 17.2.3). Phthalocyanine (**215**), the parent of this class of dyes, is obtained by heating either phthalonitrile or 2-cyanobenzamide with magnesium oxide. This reaction affords the magnesium complex (**216**), which can be freed from the metal by treatment with cold concentrated sulphuric acid. These dyes are very stable, being attacked only by very strong oxidizing agents.

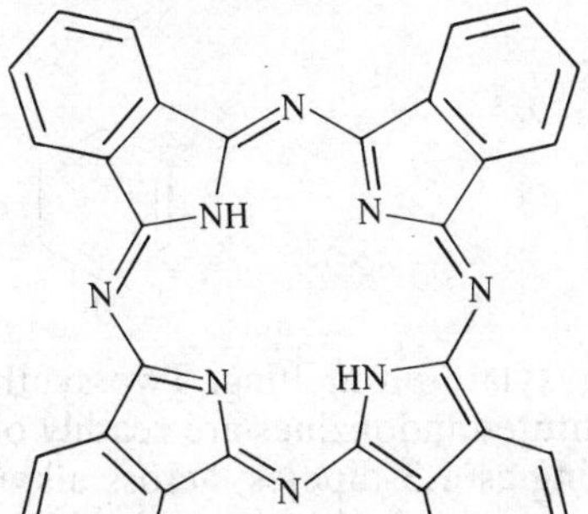

(215) 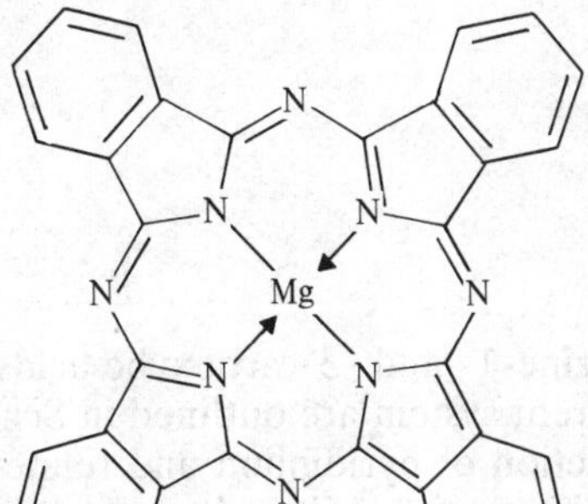 (216)

The most sensitive method currently available for the detection of amino-acids and peptides involves the preparation of intensely fluorescent derivatives by a reaction involving *o*-phthalaldehyde and a thiol, originally 2-mercaptoethanol.[255] The structure of the fluorescent derivative has been shown to be a 1-alkylthio-2-substituted isoindole, *e.g.*

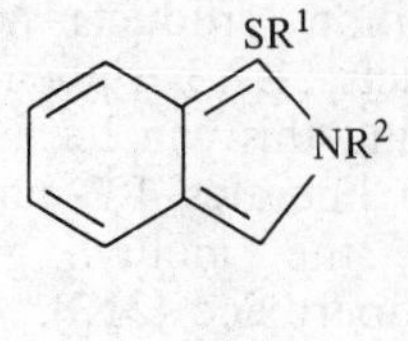

SR^1 NR^2 (217) SR^1 NR^2 MeO_2C $CHCO_2Me$ (218)

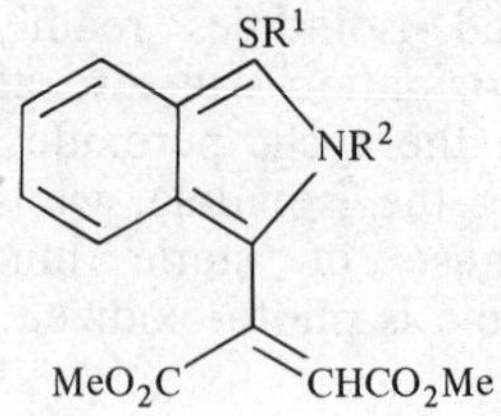

(**217**). The reaction takes place with any primary amine, but secondary amines cannot condense to the isoindole system and therefore do not afford fluorescent derivatives. The isoindoles (**217**) react with dimethyl butynedioate to give a 1:1 addition product (**218**) rather than a cycloadduct.[256]

17.4.15 INDOLIZINES

Indolizine (**219**), formerly called pyrrocoline, is a solid, m.p. 74 °C, exhibiting a blue fluorescence and possessing an odour resembling that of naphthalene. It is isomeric with indole and can be represented by a variety of resonance structures, the most important of which is the structure (**220**), which shows the least charge separation of these alternative

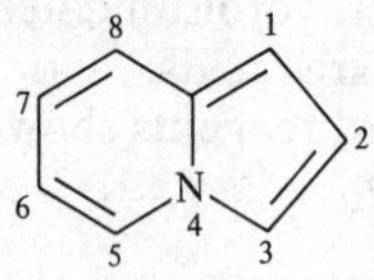

(219) 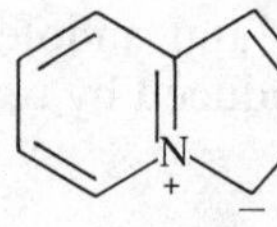 (220)

resonance forms. Indolizines behave as very weak bases which protonate preferentially at position-3, at which position electrophilic substitution is also favoured (reaction 14).

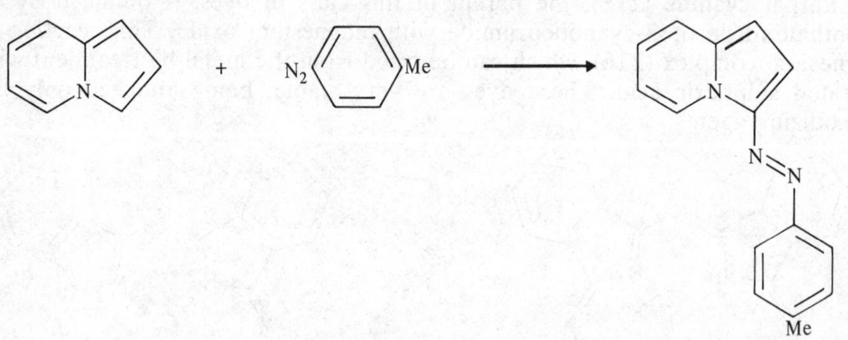

(14)

Indolizine-1- and -3-carboxylic acids readily decarboxylate on melting. Two syntheses of the parent system are outlined in Scheme 108. Substituted indolizines are readily obtained by reaction of pyridinium and related ylides, behaving as 1,3-dipoles, across alkynes and alkenes (reaction 15). 2-Acylpyridines may also be condensed with aromatic aldehydes to give 1-hydroxy-3-arylindolizines; if the latter reaction is carried out in the presence of ammonium ions (Scheme 109), the corresponding 1-amino-system can also be prepared.[259]

Diphenylcyclopropenone reacts with pyridine to give a 2:1 cycloadduct, identified as the indolizine (**221**) (reaction 16).[260] Pyridinium-*N*-allyl ylides, *e.g.* (**222**), also cyclize to produce the dihydroindolizines which readily oxidize to the aromatic system (reaction 17); an interesting extension of this 1,5-cyclization is illustrated in Scheme 110.[261]

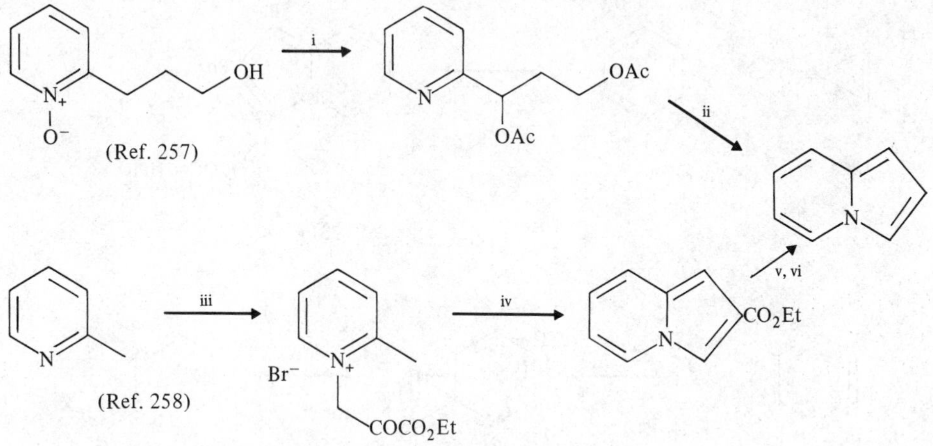

(Ref. 257)

(Ref. 258)

i, Ac_2O; ii, heat; iii, $BrCH_2COCO_2Et$; iv, $NaNHCO_3$; v, NaOH; vi, heat, CaO

SCHEME 108

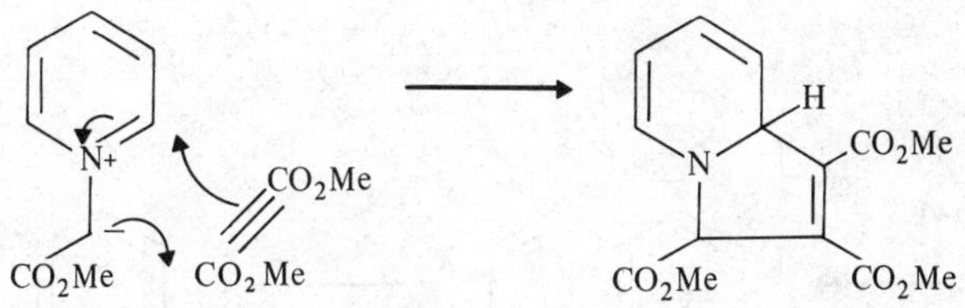

(15)

Indolizines are often used as precursors of certain cyclazines, tricyclic aromatic systems containing a central nitrogen atom. Thus cycl[3.2.2]azine (**223**) has a 10π-electron structure and conforms to Hückel's rule for aromaticity. Its preparation from 5-methylindolizine is presented (Scheme 111). In accord with localization energy calculations, it undergoes electrophilic substitution mainly at position-1.[262]

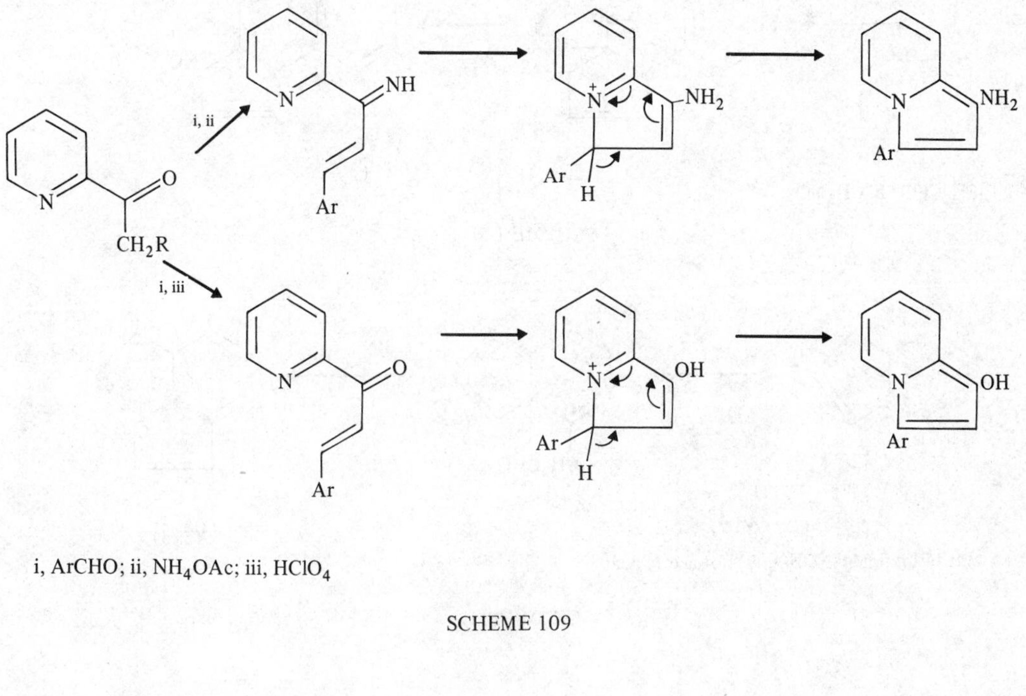

i, ArCHO; ii, NH_4OAc; iii, $HClO_4$

SCHEME 109

2 Ph, Ph, O + N → Ph, Ph, N, OCOC=CHPh, Ph (16)

(221)

+N, –, X → N, H, X →[O] N, X (17)

(222)

N →i N+, COR →ii +N, –, NO_2, R → N, H, NO_2, R → N, NO_2, R

i, $ClCH_2COR$; ii, CH_3NO_2, base

SCHEME 110

N, CH_3 →i, ii N, CH_2CHO →iii 7, 7a, 1, 6, 2, N, 5, 4a, 2a, 4, 3

(223)

i, Bu^nLi; ii, Me_2NCHO; iii, hot acetic acid

SCHEME 111

References

1. R. J. Sundberg, 'The Chemistry of Indoles', Academic Press, New York, 1970.
2. 'The Chemistry of Heterocyclic Compounds', ed. W. J. Houlihan, Wiley-Interscience, New York, 1972, vol. 25.
3. A. W. Hanson, *Acta Cryst.*, 1964, **17,** 559.
4. M. J. S. Dewar and G. J. Gleicher, *J. Chem. Phys.*, 1966, **44,** 759.
5. G. W. Wheland, 'Resonance in Organic Chemistry', Wiley, New York, 1955, p. 99.
6. J. E. Bloor and D. L. Breen, *J. Amer. Chem. Soc.*, 1967, **89,** 6835.
7. R. D. Brown and B. A. W. Coller, *Austral. J. Chem.*, 1959, **12,** 152.
8. K. Fukui, T. Yonezawa, C. Nagata, and H. Shingu, *J. Chem. Phys.*, 1954, **22,** 1433.
9. P. Millié, J. P. Malrieu, J. Benaim, J. Y. Lallemand, and M. Julia, *J. Medicin. Chem.*, 1968, **11,** 207.
10. L. Salem, 'Molecular Orbital Theory of Conjugated Systems', Benjamin, New York, 1966.
11. (a) M. Hesse, 'Indolalkaloide in Tabellen', Springer-Verlag, Berlin, 1968; (b) B. Gabetta and G. Mustich, 'Spectral Data of Indole Alkaloids', Inverni della Beffa, Milan, 1975.
12. (a) W. Klyne, R. J. Swan, N. J. Dastoor, A. A. Gorman, and H. Schmid, *Helv. Chim. Acta*, 1966, **50,** 115; (b) C. M. Lee, W. Trager, and A. H. Beckett, *Tetrahedron*, 1967, **23,** 375.
13. (a) D. M. Hercules, 'Fluorescence and Phosphorescence Analysis', Interscience, New York, 1966; (b) S. Udenfriend, 'Fluorescence Assay in Biology and Medicine', Academic Press, New York, 1962, vol. I; (c) *idem.*, *ibid.*, 1969, vol. II.
14. J. W. Bridges and R. T. Williams, *Biochem. J.*, 1968, **107,** 225.
15. H. N. Eisen, *Methods Med. Res.*, 1964, **10,** 115.
16. (a) M. Shinitzky, V. Grisaro, D. M. Chipman and H. Sharon, *Arch. Biochem. Biophys.*, 1966, **115,** 232; (b) S. S. Lehrer and G. D. Fassman, *Biochem. Biophys. Res. Comm.*, 1966, **23,** 133.
17. (a) S. P. Hiremath and R. S. Hasmane, *Adv. Heterocyclic Chem.*, 1973, **15,** 278; (b) R. B. Herman, *Internat. J. Quantum Chem.*, 1968, **2,** 165; (c) P. J. Black, R. D. Brown and M. L. Hefferman, *Austral. J. Chem.*, 1967, **20,** 1325.
18. (a) R. V. Jardine and R. K. Brown, *Canad. J. Chem.*, 1963, **41,** 2067; (b) J. Ronayne and D. H. Williams, *J. Chem. Soc. (B)*, 1967, 805.
19. (a) R. H. Levin, J.-Y. Lallemand and J. D. Roberts, *J. Org. Chem.*, 1973, **38,** 1983; (b) E. Wenkert, J. S. Bindra, C.-J. Chang, D. W. Cochran, and F. M. Schell, *Accounts Chem. Res.*, 1974, **7,** 46.
20. (a) H. Budzikiewicz, C. Djerassi, and D. H. Williams, 'Interpretation of Mass Spectra of Organic Compounds', Holden-Day, San Francisco, 1964; (b) M. Marx and C. Djerassi, *J. Amer. Chem. Soc.*, 1968, **90,** 679.
21. H. Budzikiewicz, C. Djerassi, and D. H. Williams, 'Structure Elucidation by Mass Spectrometry', Holden-Day, San Francisco, 1964, vol. I.
22. R. L. Hinman and C. P. Bauman, *J. Org. Chem.*, 1964, **29,** 2437.
23. B. C. Challis and E. M. Millar, *J.C.S. Perkin II*, 1972, 1116.
24. (a) R. L. Hinman and E. B. Whipple, *J. Amer. Chem. Soc.*, 1962, **84,** 2534; (b) R. L. Hinman and J. Lang, *ibid.*, 1964, **86,** 3796.
25. A. H. Jackson and A. E. Smith, *J. Chem. Soc.*, 1964, 5510.
26. R. L. Hinman and J. Land, *Tetrahedron Letters*, 1960, 12.
27. J. Thesing, G. Semler, and G. Mohr, *Chem. Ber.*, 1962, **95,** 2205.
28. W. E. Noland, L. R. Smith, and D. C. Johnson, *J. Org. Chem.*, 1963, **28,** 2262.
29. W. E. Noland and K. R. Rush, *J. Org. Chem.*, 1964, **29,** 947.
30. (a) G. Berti, A. Da Settimo, and E. Nannipieri, *J. Chem. Soc. (C)*, 1968, 2145; (b) A. Angelico and G. Verlardi, *Atti Accad. Lincei*, 1904, **13,** i, 241; *Gazzetta*, 1904, **34,** ii, 57.
31. W. E. Noland, L. R. Smith, and K. R. Rush, *J. Org. Chem.*, 1965, **30,** 3457.
32. K. Brown and A. R. Katritzky, *Tetrahedron Letters*, 1964, 803.
33. R. Ikan, E. Hoggmann, E. D. Bergmann, and A. Galun, *Israel J. Chem.*, 1964, **2,** 37; T. Kinoshita, H. Inoue, and F. Imoto, *Nippon Kagaku Zasshi*, 1957, **78,** 1372 (*Chem. Abs.*, 1960, **54,** 491).
34. G. F. Smith and R. Atittulla, unpublished work.
35. H. F. Hodson and G. F. Smith, *J. Chem. Soc.*, 1957, 3546.
36. A. Angeli and G. Marchetti, *Atti Accad. Lincei*, 1907, **16,** i, 381.
37. B. C. Challis and A. J. Lawson, *J.C.S. Perkin II*, 1973, 918.
38. F. Piozzi and M. Dubini, *Gazzetta*, 1959, **89,** 638; F. Piozzi and A. Umani-Ronchi, *ibid.*, 1964, **94,** 1248.
39. F. G. Mann and R. C. Howarth, *J. Chem. Soc.*, 1944, 670.
40. W. Madelung and O. Wilhelm, *Ber.*, 1924, **57,** 234.
41. J. H. Binks and J. H. Ridd, *J. Chem. Soc.*, 1957, 2398; B. C. Challis and H. S. Rzepa, *J.C.S. Perkin II*, 1975, 1209.
42. T. F. Spande and G. G. Gleaver, *J. Amer. Chem. Soc.*, 1973, **95,** 3400.
43. G. F. Smith and D. A. Taylor, *Tetrahedron*, 1973, **29,** 669.
44. (a) J. C. Powers, *J. Org. Chem.*, 1966, **31,** 2627; (b) L. A. Yanovskaya, *Doklady Akad. Nauk S.S.S.R.*, 1950, **71,** 693 (*Chem. Abs.*, 1950, **44,** 8354); K. Piers, C. Meimaroglou, R. V. Jardine, and R. K. Brown, *Canad. J. Chem.*, 1963, **41,** 2399; M.-M. Canet and J.-P. Boca, *Bull. Soc. chim. France*, 1967, 1294; V. Calo, F. Ciminale, L. Lopez, F. Noso, and P. E. Todesco, *J.C.S. Perkin I*, 1972, 2567; (c) R. D. Arnold, W. M. Nutter, and W. L. Stepp, *J. Org. Chem.*, 1959, **24,** 117.

45. A. N. Kost, L. G. Yadin, V. A. Budylin, and N. G. Yaryshev, *Khim. Geterotsikl. Soedinenii Akad. Nauk Latv.*, 1965, 632 (*Chem. Abs.*, 1966, **64,** 3457).
46. R. Weissgerber, *Ber.*, 1913, **46,** 651; R. L. Hinman and C. P. Bauman, *J. Org. Chem.*, 1964, **29,** 1206.
47. (a) T. Hino, M. Nakagawa, T. Wakatsuki, and S. Akaboshi, *Chem. Pharm. Bull.* (*Japan*), 1967, **15,** 1800; (b) T. Hino, M. Nakagawa, T. Wakatsuki, K. Ogawa, and S. Yamada, *Tetrahedron*, 1967, **23,** 1441; T. Kobayashi and N. Inokuchi, *ibid.*, 1964, **20,** 2055; (c) R. L. Hinman and C. P. Nauman, *J. Org. Chem.*, 1964, **29,** 2437.
48. W. B. Lawson, A. Patchornik, and B. Witkop, *J. Amer. Chem. Soc.*, 1960, **82,** 5918.
49. M. Chuo, T. F. Spande, and B. Witkop, *J. Amer. Chem. Soc.*, 1970, **92,** 343.
50. (a) S. G. P. Plant and M. L. Tomlinson, *J. Chem. Soc.*, 1933, 955; (b) E. Lippmann, K. Richter, and M. Muhlstadt, *Z. Chem.*, 1965, **5,** 186.
51. T. A. Foglia and D. Swern, *J. Org. Chem.*, 1968, **33,** 4440; J. M. Muchowski, *Canad. J. Chem.*, 1970, **48,** 422.
52. (a) H. Brunner, *Monatsh.*, 1896, **17,** 253, 479; 1897, **18,** 95, 527; 1900, **21,** 156; 1906, **27,** 1183; (b) G. Ciamician and G. Plancher, *Ber.*, 1896, **29,** 2475; G. Plancher, *Gazzetta*, 1898, **28,** ii, 374, 418; 1900, **30,** ii, 564; *Atti Accad. Lincei*, 1902, **11,** ii, 182.
53. (a) F. J. Evans, G. G. Lyle, J. Watkins, and R. E. Lyle, *J. Org. Chem.*, 1962, **27,** 1553; (b) R. Robinson and H. Suginome, *J. Chem. Soc.*, 1932, 298; (c) B. Witkop and J. B. Patrick, *J. Amer. Chem. Soc.*, 1953, **75,** 2572.
54. A. H. Jackson and P. Smith, *Tetrahedron*, 1968, **24,** 2227.
55. (a) A. H. Jackson and P. Smith, *Tetrahedron*, 1968, **24,** 6119; (b) R. Iyer, A. H. Jackson, P. V. R. Shannon, and B. Naidoo, *J.C.S. Perkin II*, 1973, 873.
56. G. Casnati, A. Dossera, and A. Pochini, *Tetrahedron Letters*, 1972, 5277.
57. G. Casnati, M. Franciani, A. Guareschi, and A. Pochini, *Tetrahedron Letters*, 1969, 2485.
58. W. D. Closson, S. A. Roman, G. T. Kwiatkowski, and D. A. Corwin, *Tetrahedron Letters*, 1966, 2271; M. Julia, H. Sliwa, and J. Lenzi, *Bull. Soc. chim. France*, 1966, 3359.
59. M. Julia and J. Lenzi, *Bull. Soc. chim. France*, 1971, 4084.
60. (a) M. Wakselman, G. Decodts, and M. Vilkas, *Compt. rend.* (*C*), 1968, **266,** 1089; (b) M. Wakselman, G. Decodts, and M. Vilkas, *ibid.*, 1968, **267,** 1063.
61. Y. Kanaoka, K. Miyashiba, and O. Yonemitsu, *Tetrahedron*, 1969, **25,** 2485.
62. K. Freter, *J. Org. Chem.*, 1972, **37,** 2010.
63. E. Pfeil and V. Hander, *Angew. Chem. Internat. Edn.*, 1967, **6,** 178.
64. W. E. Noland and R. F. Lange, *J. Amer. Chem. Soc.*, 1959, **81,** 1203.
65. (a) S. Ghosal, *J. Indian Chem. Soc.*, 1964, **41,** 496; (b) H. R. Snyder and D. S. Matteson, *J. Amer. Chem. Soc.*, 1955, **77,** 1257; (c) A. P. Gray and W. L. Archer, *ibid.*, 1957, **79,** 3554; (d) M. Colonna and L. Marchetti, *Gazzetta*, 1965, **95,** 868; (e) J. Szmuszkovicz, *J. Amer. Chem. Soc.*, 1957, **79,** 2819; J. R. Merchant and S. S. Salgar, *J. Indian Chem. Soc.*, 1963, **40,** 83.
66. B. Robinson and G. F. Smith, *J. Chem. Soc.*, 1960, 4574.
67. R. Walther and J. Clemen, *J. prakt. Chem.*, 1930, **61,** 256; K. A. Schellenberg and G. W. McClean, *J. Amer. Chem. Soc.*, 1966, **88,** 1077.
68. A. H. Cook and J. R. Majer, *J. Chem. Soc.*, 1944, 486.
69. For a study of the limitations of the Ehrlich test, see D. Kumfer, *Analyt. Biochem.*, 1964, **8,** 75.
70. E. Leete, *J. Amer. Chem. Soc.*, 1959, **81,** 6023; J. Thesing, *Chem. Ber.*, 1954, **87,** 692.
71. W. E. Noland and D. N. Robinson, *Tetrahedron*, 1958, **3,** 68.
72. W. E. Noland and M. R. Venkiteswaran, *J. Org. Chem.*, 1961, **26,** 4263.
73. K. Freter, *J. Org. Chem.*, 1975, **40,** 2525.
74. See, for example, W. E. Noland, M. R. Venkitswaran, and C. G. Richards, *J. Org. Chem.*, 1961, **26,** 4241.
75. R. Robinson and J. E. Saxton, *J. Chem. Soc.*, 1952, 976.
76. (a) S. Swaminathan and K. Narasimhan, *Chem. Ber.*, 1966, **99,** 889; (b) H. Kuhn and O. Stein, *ibid.*, 1937, **70,** 567; (c) W. J. Brehm and H. Lindwall, *J. Org. Chem.*, 1950, **15,** 685.
77. J. Wolinsky and J. E. Sundeen, *Tetrahedron*, 1970, **26,** 5427.
78. S. Swaminathan and S. Ranganathan, *J. Org. Chem.*, 1957, **22,** 70; S. Swaminathan, S. Ranganathan, and S. Sulochana, *ibid.*, 1958, **23,** 707.
79. J. Thesing and G. Semler, *Annalen*, 1964, **680,** 52.
80. (a) E. Späth and E. Lederer, *Ber.*, 1930, **63,** 2102; (b) S. Wawzonel and J. D. Nordstrom, *J. Medicin. Chem.*, 1965, **8,** 265.
81. (a) E. Wenkert, *Accounts Chem. Res.*, 1968, **1,** 78; (b) M. S. Allen, A. J. Gaskell, and J. A. Joule, *J. Chem. Soc.* (*C*), 1971, 736.
82. A. H. Jackson and A. E. Smith, *Tetrahedron*, 1968, **24,** 403.
83. (a) J. R. Williams and L. R. Unger, *J.C.S. Chem. Comm.*, 1976, 1605; (b) for other incidental evidence, see R. B. Woodward, M. P. Cava, W. D. Ollis, A. Hunger, H. V. Daeniker, and K. Schenker, *Tetrahedron*, 1963, **19,** 247; G. Büchi, K. E. Matsumoto, and H. Nishimura, *J. Amer. Chem. Soc.*, 1971, **93,** 3299; E. Wenkert, K. G. Dave, C. T. Gnewuch, and P. W. Sprague, *J. Amer. Chem. Soc.*, 1968, **90,** 5251.
84. A. Ebnother, P. Niklaus, and R. Suess, *Helv. Chim. Acta*, 1969, **52,** 629; G. F. Smith and J. T. Wrobel, *J. Chem. Soc.*, 1960, 792.
85. J. B. Hester, *J. Org. Chem.*, 1964, **29,** 2864.
86. (a) G. F. Smith, *Adv. Heterocyclic Chem.*, 1963, **2,** 287; (b) J. Youmans, *J. Heterocyclic Chem.*, 1976, **13,** 949.

87. G. F. Smith, *J. Chem. Soc.*, 1954, 3842.
88. S. Clementi, P. Linda, and G. Marino, *J.C.S. Chem. Comm.*, 1972, 427.
89. T. Nogrady and L. Morris, *Canad. J. Chem.*, 1969, **47,** 1999.
90. W. C. Anthony, *J. Org. Chem.*, 1960, **25,** 2049.
91. E. Späth and E. Lederer, *Ber.*, 1930, **63,** 2102.
92. J. E. Saxton, *J. Chem. Soc.*, 1952, 3592; G. Hart, D. R. Liljegren, and K. T. Potts, *ibid.*, 1961, 4267.
93. K. Ishizumi, T. Shioire, and S. Yamada, *Chem. Pharm. Bull. (Japan)*, 1967, **15,** 863.
94. P. Hoechst and E. Roeder, *Arch. Pharm.*, 1975, **308,** 779 (*Chem. Abs.*, 1976, **84,** 30 941).
95. A. Cipiciani, S. Clementi, P. Linda, G. Savelli, and G. V. Sebastiani, *Tetrahedron*, 1976, **32,** 2595.
96. (a) M. S. Kharash, S. S. Kane, and H. C. Brown, *J. Amer. Chem. Soc.*, 1940, **62,** 2242; (b) M. E. Speeter and W. C. Anthony, *ibid.*, 1954, **76,** 6208.
97. R. C. Elderfield and B. Fisher, *J. Org. Chem.*, 1958, **23,** 949.
98. W. Borsche and H. Groth, *Annalen*, 1941, **549,** 238; N. N. Suvorov and N. P. Sorokina, *J. Gen. Chem. USSR*, 1960, **30,** 2036.
99. (a) G. Yagil, *Tetrahedron*, 1967, **23,** 2855; (b) W. J. Scott, W. J. Bover, K. Bratin, and P. Zuman, *J. Org. Chem.*, 1976, **41,** 1952.
100. M. G. Reinecke, J. F. Sebastian, H. W. Johnson, and C. Pyun, *J. Org. Chem.*, 1971, **36,** 3091.
101. (a) D. A. Shirley and P. A. Roussel, *J. Amer. Chem. Soc.*, 1953, **75,** 375; (b) R. J. Sundberg and H. F. Russell, *J. Org. Chem.*, 1973, **38,** 3324; (c) R. J. Sundberg and R. L. Parton, *ibid.*, 1976, **41,** 163.
102. J. Szmuszkowicz, *J. Org. Chem.*, 1962, **27,** 511; (b) *Belg. Pat.* 637 355 (*Chem. Abs.*, 1965, **62,** 7731).
103. R. D. Arnold, W. M. Nutter, and W. L. Stepp, *J. Org. Chem.*, 1959, **24,** 117.
104. (a) M. de Rosa, *J.C.S. Chem. Comm.*, 1975, 482; (b) R. J. Sundberg, *J. Org. Chem.*, 1965, **30,** 3604; (c) P. G. Gassman, G. A. Campbell, and G. Mehta, *Tetrahedron*, 1972, **28,** 2749.
105. M. M. Cooper, G. J. Hignett, R. F. Newton, J. A. Joule, M. Harris, and J. D. Hinchley, *J.C.S. Chem. Comm.*, 1977, 432.
106. M. G. Reinecke, J. F. Sebastian, H. W. Johnson, and C. Pyun, *J. Org. Chem.*, 1972, **37,** 3066.
107. (a) M. Nakazaki, *Bull. Chem. Soc. Japan*, 1961, **34,** 334; (b) B. Cardillo, G. Casnati, A. Pochini, and A. Ricca, *Tetrahedron*, 1967, **23,** 3771.
108. J. C. Powers, W. P. Meyer, and T. G. Pearsons, *J. Amer. Chem. Soc.*, 1967, **89,** 5812.
109. B. C. Challis and E. M. Millar, *J.C.S. Perkin II*, 1972, 1111.
110. (a) K. T. Potts and J. E. Saxton, *Org. Synth.*, 1960, **40,** 68; (b) H. Heaney and S. V. Ley, *Org. Synth.*, 1974, **54,** 58; (c) G. M. Rubottom and J. C. Chabala, *Synthesis*, 1972, 566; (d) A. Banerji and J. Banerji, *Indian J. Chem.*, 1975, **13,** 945; (e) A. Barco, S. Benetti, G. P. Pollini, and P. G. Baraldi, *Synthesis*, 1976, 124; (f) V. Bocchi, G. Casnati, A. Dossena, and F. Villani, *ibid.*, 414.
111. A. H. Jackson and A. E. Smith, *J. Chem. Soc.*, 1964, 5510.
112. R. E. Heacock and S. Kasparek, *Adv. Heterocyclic Chem.*, 1969, **10,** 43.
113. (a) R. M. Acheson and A. R. Hands, *J. Chem. Soc.*, 1961, 744; (b) W. E. Noland and P. J. Hartmann, *J. Amer. Chem. Soc.*, 1954, **76,** 3227; (c) J. I. DeGraw, J. G. Kennedy, and W. A. Skinner, *J. Heterocyclic Chem.*, 1966, **3,** 67; (d) H. R. Snyder and F. Pilgrim, *J. Amer. Chem. Soc.*, 1948, **70,** 1962; (e) J. Szmuszkowicz, *ibid.*, 1960, **82,** 1180; (f) R. Majima and M. Kotake, *Ber.*, 1922, **55B,** 3865; (g) Q. Mingoia, *Gazzetta*, 1930, **66,** 144.
114. A. Alemany, M. Bernabe, E. F. Alvarez, M. L.-Tamayo and O. Nieto, *Bull. Soc. chim. France*, 1967, 780.
115. R. E. Nowman, D. D. Evans, and R. J. Islip, *Chem. and Ind. (London)*, 1971, 33.
116. T. Shen, *U.S. Pat.* 3 242 185 (*Chem. Abs.*, 1966, **64,** 17 555).
117. A. Angeli and G. Marchetti, *Atti Accad. Lincei*, 1907, **16,** i, 381.
118. F. Angelico and G. Verladi, *Atti Accad. Lincei*, 1904, **13,** i, 241.
119. (a) B. Witkop and J. B. Patrick, *J. Amer. Chem. Soc.*, 1951, **73,** 713; (b) B. Witkop, *Annalen*, 1947, **558,** 98.
120. B. Witkop and J. B. Patrick, *J. Amer. Chem. Soc.*, 1951, **73,** 2196 and refs. therein.
121. (a) E. Leete, *J. Amer. Chem. Soc.*, 1961, **83,** 3645; (b) W. I. Taylor, *Proc. Chem. Soc.*, 1962, 247.
122. (a) H. H. Wasserman and M. B. Floyd, *Tetrahedron Letters*, 1963, 2009; (b) F. Y. Chen and E. Leete, *ibid.*, 2013.
123. S. McLean and G. I. Dimitrienko, *Canad. J. Chem.*, 1971, **49,** 3642.
124. (a) N. Sugiyama and M. Akutagawa, *Bull. Chem. Soc. Japan*, 1967, **40,** 240; (b) N. Sugiyama, H. Yamamoto, and Y. Omote, *ibid.*, 1968, **41,** 1917.
125. I. Saito, M. Irnuta, S. Matsugo, and T. Matsuura, *J. Amer. Chem. Soc.*, 1975, **97,** 7191.
126. (a) D. W. Ockenden and K. Schofield, *J. Chem. Soc.*, 1953, 3440; (b) B. Witkop, *Annalen*, 1944, **556,** 103; (c) H. Uchino, D. Higashi, and M. Uchida, *Chem. Abs.*, 1970, **72,** 121 288.
127. T. Suchiro and A. Nakagawa, *Bull. Chem. Soc. Japan*, 1967, **40,** 2919.
128. Y. Kanaoka, M. Aiura, and S. Hariya, *J. Org. Chem.*, 1971, **36,** 458.
129. (a) W. E. Savige and A. Fontana, *J.C.S. Chem. Comm.*, 1976, 599; (b) Y. Oikawa and O. Yonemitsu, *J. Org. Chem.*, 1977, **42,** 1213.
130. W. E. Noland, L. R. Smith, and K. R. Rush, *J. Org. Chem.*, 1965, **30,** 3457.
131. H. Yamamoto, S. Inaba, T. Hirohashi, and K. Ishizumi, *Chem. Ber.*, 1968, **101,** 4245.
132. V. Dave, *Canad. J. Chem.*, 1972, **50,** 3397.
133. (a) L. J. Dolby and D. L. Booth, *J. Amer. Chem. Soc.*, 1966, **88,** 1049; (b) L. J. Dolby and G. W. Gribble, *J. Org. Chem.*, 1967, **32,** 1391.
134. C. R. Hutchinson, G. J. O'Loughlin, R. T. Brown, and S. B. Fraser, *J.C.S. Chem. Comm.*, 1974, 928.

135. N. Finch, G. W. Gemenden, I. H. Hsu, and W. I. Taylor, *J. Amer. Chem. Soc.*, 1963, **85,** 1520.
136. L. Willstäter and D. Jacquet, *Ber.*, 1918, **51,** 777; H.-J. Teuber and G. Schmitt, *ibid.*, 1969, **102,** 713.
137. (a) A. Smith and J. H. P. Utley, *Chem. Comm.*, 1965, 427; (b) L. J. Dolby and G. W. Gribble, *J. Heterocyclic Chem.*, 1966, 124; (c) G. W. Gribble, P. D. Lord, J. Skotnicki, S. E. Dietz, J. T. Eaton, and J. L. Johnson, *J. Amer. Chem. Soc.*, 1974, **96,** 7812; (d) J. G. Berger, *Synthesis*, 1974, 508.
138. S. A. Monti and R. R. Schmidt, *Tetrahedron*, 1971, **27,** 3331.
139. (a) W. A. Remers, G. J. Gibbs, C. Pidack, and M. J. Weiss, *J. Amer. Chem. Soc.*, 1967, **89,** 5513; (b) W. A. Remers, G. J. Gibbs, C. Pidacks, and M. J. Weiss, *J. Org. Chem.*, 1971, **36,** 279.
140. J. Hutton and W. A. Waters, *J. Chem. Soc.*, 1965, 4253.
141. (a) E. Eich and H. Rochelmeyer, *Pharm. Acta Helv.*, 1966, **41,** 109; (b) M. Julia and F. Ricalens, *Compt. rend. (C)*, 1972, **275,** 613.
142. Y. Kanaoka, M. Aiura, and S. Hariya, *J. Org. Chem.*, 1971, **36,** 458.
143. (a) C. W. Rees and C. E. Smithen, *Chem. and Ind. (London)*, 1962, 1033; (b) B. Robinson, *Tetrahedron Letters*, 1962, 139; (c) H. E. Dobbs, *Tetrahedron*, 1968, **24,** 491.
144. C. W. Rees and C. E. Smithen, *Adv. Heterocyclic Chem.*, 1964, **3,** 57.
145. R. H. Marchant and D. G. Harvey, *J. Chem. Soc.*, 1951, 1808.
146. R. W. Jackson and R. H. Manske, *Canad. J. Res.*, 1935, **13,** 170; J. Ratusky and F. Sorm, *Chem. listy*, 1957, **51,** 1091.
147. (a) J. W. Daly and B. Witkop, *J. Amer. Chem. Soc.*, 1967, **89,** 1032; (b) F. Troxler, G. Bormann, and F. Seeman, *Helv. Chim. Acta*, 1968, **51,** 1203.
148. H.-J. Teuber and G. Thaler, *Chem. Ber.*, 1956, **89,** 489; W. A. Remmers and M. J. Weiss, *J. Amer. Chem. Soc.*, 1966, **88,** 804.
149. (a) M.-M. Canet and J.-P. Boca, *Bull. Soc. chim. France*, 1967, 1296; (b) R. M. Acheson, C. J. Q. Brookes, D. P. Dearnalay, and B. Quest, *J. Chem. Soc. (C)*, 1968, 504.
150. (a) R. M. Acheson, D. M. Littlewood, and H. E. Rosenberg, *J.C.S. Chem. Comm.*, 1974, 671; (b) A. Reissert, *Ber.*, 1896, **29,** 639; (c) E. Fischer and H. Hütz, *ibid.*, 1895, **28,** 585.
151. M. Colonna and P. Bruni, *Gazzetta*, 1964, **94,** 1448.
152. C. C. Bond and M. Hooper, *Synthesis*, 1974, 443.
153. (a) R. J. Sundberg, *J. Org. Chem.*, 1965, **30,** 3604; (b) P. G. Gassman, G. A. Campbell, and G. Mehta, *Tetrahedron*, 1972, **28,** 2749.
154. I. Gruda, *Canad. J. Chem.*, 1972, **50,** 18.
155. (a) E. Wenkert, N. K. Bhattacharyya, N. K. Reid, and T. S. Stevens, *J. Amer. Chem. Soc.*, 1956, **78,** 797; (b) A. B. A. Jansen and C. G. Richards, *Tetrahedron*, 1965, **21,** 1327; (c) P. L. Julian and H. C. Printy, *J. Amer. Chem. Soc.*, 1953, **75,** 5301; (d) J. A. Ballantine, R. J. S. Beer, and A. Robertson, *J. Chem. Soc.*, 1958, 4779.
156. (a) R. W. Daisley and J. Walker, *J. Chem. Soc. (C)*, 1971, 1375; (b) G. N. Walker, R. T. Smith, and B. N. Weaver, *J. Medicin. Chem.*, 1965, **8,** 626.
157. E. Giovannini and P. Portmann, *Helv. Chim. Acta*, 1948, **31,** 1375.
158. (a) J. C. Powers, *J. Org. Chem.*, 1966, **31,** 2627; (b) T. Hino, K. Tsuneoka, M. Nakagawa, and S. Akaboshi, *Chem. Pharm. Bull. (Japan)*, 1969, **17,** 550.
159. (a) H. Plieninger and H. Bauer, *Angew. Chem.*, 1961, **73,** 433; (b) J. Harley-Mason and T. J. Leeney, *Proc. Chem. Soc.*, 1964, 368.
160. (a) K. Parry and G. F. Smith, unpublished observations; (b) S. J. Holt, A. E. Kellie, D. G. O'Sullivan, and P. W. Sadler, *J. Chem. Soc.*, 1958, 1217.
161. G. A. Russell and G. Kaupp, *J. Amer. Chem. Soc.*, 1969, **91,** 3851.
162. A. Etienne, *Bull. Soc. chim. France*, 1948, 651.
163. (a) M. Hooper, and W. N. Pickethly, *J.C.S. Perkin I*, 1972, 1607; (b) W. I. O'Sullivan and E. J. Rothery, *Chem. and Ind. (London)*, 1972, 849.
164. G. W. Kirby and S. W. Shah, *Chem. Comm.*, 1965, 381.
165. B. Witkop, *J. Amer. Chem. Soc.*, 1950, **72,** 614.
166. (a) A. H. Jackson, *Chem. and Ind. (London)*, 1965, 1652; (b) J. M. Muchowski, *Canad. J. Chem.*, 1969, **47,** 857.
167. M. P. Cava, R. L. Little, and D. R. Napier, *J. Amer. Chem. Soc.*, 1958, **80,** 2257; G. Tacconi and S. Pietra, *Farmaco (Pavia), Ed. Sci.*, 1963, **18,** 409.
168. (a) F. Krohnke and M. M.-Delius, *Chem. Ber.*, 1951, **84,** 932; (b) J. L. Pinkus, G. G. Woodyard, and T. Cohen, *J. Org. Chem.*, 1965, **30,** 1104.
169. H. E. Johnson and D. C. Crosby, *J. Org. Chem.*, 1963, **28,** 2794.
170. M. Julia, Y. Huang, and J. Igolen, *Compt. rend.*, 1967, **265,** 110.
171. H. Pleninger, *Chem. Ber.*, 1955, **88,** 370.
172. M. Somei and M. Natsume, *Tetrahedron Letters*, 1974, 370.
173. D. I. Haddlesey, P. A. Mayor, and S. S. Szinai, *J. Chem. Soc.*, 1964, 5269.
174. M. Somei and M. Natsume, *Tetrahedron Letters*, 1974, 3605.
175. (a) H. Rinderknecht, H. Koechlin, and C. Niemann, *J. Org. Chem.*, 1953, **18,** 971; (b) J. Kebrle and K. Hoffmann, *Helv. Chim. Acta*, 1956, **31,** 116.
176. T. Hino, M. Nakagawa, T. Hashizume, N. Yamaji, Y. Niwa, K. Tsuneoka, and S. Akabashi, *Tetrahedron*, 1971, **27,** 775.
177. A. S. Bailey, M. C. Chumm, and J. J. Wedgewood, *Tetrahedron Letters*, 1968, 5953.

178. W. Madelung, *Annalen*, 1914, **405,** 92.
179. R. J. Richman and A. Hassner, *J. Org. Chem.*, 1968, **33,** 2548.
180. J. Schmitt, C. Perrin, M. Langlois, and M. Suquet, *Bull. Soc. chim. France*, 1969, 1227.
181. J. Schmitt, M. Langlois, C. Callet, and C. Perrin, *Bull. Soc. chim. France*, 1969, 2008.
182. U. Simon, O. Süs, and L. Horner, *Annalen*, 1966, **697,** 17.
183. J. H. Brewster and E. L. Eliel, *Org. Reactions*, 1953, **7,** 99.
184. (a) J. D. Albright and W. R. Snyder, *J. Amer. Chem. Soc.*, 1959, **81,** 2239; (b) E. Baciochi and A. Schiroli, *J. Chem. Soc. (B)*, 1969, 401.
185. M. Julia, J. Bagot, and O. Siffert, *Bull. Soc. chim. France*, 1973, 1424.
186. (a) S. Siddapa and G. A. Bhat, *J. Chem. Soc. (C)*, 1971, 178; (b) W. Schindler, *Helv. Chim. Acta*, 1957, **40,** 2156; (c) W. R. Snyder and P. L. Cook, *J. Amer. Chem. Soc.*, 1956, **78,** 969.
187. (a) J. Harley-Mason and Atta-ur-Rahman, *Chem. Comm.*, 1967, 708; (b) R. T. Brown, G. F. Smith, K. S. J. Stapleford, and D. A. Taylor, *Chem. Comm.*, 1970, 190; (c) L. J. Dolby and S. Sakai, *Tetrahedron*, 1967, **23,** 1.
188. (a) D. Herbst, R. Rees, G. A. Hughes and H. Smith, *J. Medicin. Chem.*, 1966, **9,** 864; (b) L. J. Dolby and D. L. Booth, *J. Org. Chem.*, 1965, **30,** 1550.
189. (a) J. Thesing, *Chem. Ber.*, 1954, **87,** 692; (b) E. Leete, *J. Amer. Chem. Soc.*, 1959, **81,** 6023; (c) K. M. Biswas and A. H. Jackson, *Tetrahedron*, 1968, **24,** 1145, (d) R. Littell and G. R. Allen, Jr., *J. Org. Chem.*, 1973, **38,** 1504.
190. C. Runbi and G. Ordando, *Chem. Abs.*, 1955, **49,** 3940.
191. (a) G. Grete, H. L. Lee, and M. R. Uskoković, *Helv. Chim. Acta*, 1976, **59,** 2268; (b) G. Büchi, R. E. Manning, and S. A. Monti, *J. Amer. Chem. Soc.*, 1964, **86,** 4631.
192. (a) R. A. Abramovitch, *J. Chem. Soč.*, 1956, 4593; (b) E. Piers and R. K. Brown, *Canad. J. Chem.*, 1962, **40,** 559; (c) J. C. Powers, *Tetrahedron Letters*, 1965, 655.
193. (a) H. Sakakibara and T. Kobayashi, *Tetrahedron*, 1966, **22,** 2475; (b) J. Thesing and G. Semler, *Annalen*, 1964, **680,** 52.
194. B. Robinson, *Chem. Rev.*, 1963, **63,** 373; 1969, **69,** 227.
195. A. H. Kelly, D. H. McLeod, and J. Parrick, *Canad. J. Chem.*, 1965, **43,** 296.
196. R. B. Carlin and E. E. Fisher, *J. Amer. Chem. Soc.*, 1948, **70,** 3421.
197. G. S. Bajwa and R. K. Brown, *Canad. J. Chem.*, 1970, **48,** 2293; 1969, **47,** 785.
198. (a) C. F. H. Allen and C. V. Wilson, *J. Amer. Chem. Soc.*, 1943, **65,** 611; (b) K. Clusius and H. R. Weisser, *Helv. Chim. Acta*, 1952, **35,** 400.
199. (a) N. P. Buu-Hoi, P. Jacquignon, and O. Perin-Roussel, *Bull. Soc. chim. France*, 1965, **78,** 5854; (b) H. Illy and C. Funderbunk, *J. Org. Chem.*, 1968, **33,** 4283.
200. (a) R. R. Phillips, *Org. Reactions*, 1959, **10,** 143; (b) R. A. Abramovitch and D. Shapiro, *J. Chem. Soc.*, 1956, 4589, 4593; (c) A. Jackson, A. J. Gaskell, N. D. V. Wilson, and J. A. Joule, *Chem. Comm.*, 1968, 364.
201. (a) P. L. Julian, E. W. Meyer, and H. C. Printy, in 'Heterocyclic Chemistry', ed. R. C. Elderfield, Wiley, New York, 1953, vol. 3., p. 32; (b) W. A. Remers, in 'Heterocyclic Compounds', ed. W. J. Houlihan, Wiley-Interscience, New York, 1972, vol. 25, p. 317.
202. J. K. Landquist and C. J. Marsden, *Chem. and Ind.* (*London*), 1966, 1032.
203. (a) C. F. H. Allen and J. Van Allan, *Org. Synth., Coll. Vol.* 3, 1955, 592; (b) Ref. 201b, p. 385; (c) R. L. Augustine, A. J. Gustavsen, S. F. Wanat, I. C. Pattison, and K. S. Houghton, *J. Org. Chem.*, 1973, **38,** 3004.
204. (a) W. E. Noland and F. J. Baude, *Org. Synth.*, 1963, **43,** 40; (b) J. Bourdais and C. Germain, *Tetrahedron Letters*, 1970, 195.
205. (a) I. Baxter and G. A. Swan, *J. Chem. Soc. (C)*, 1968, 468; (b) P. J. Mulligan and S. La Berge, *J. Medicin. Chem.*, 1970, **13,** 1248.
206. R. Littell, G. O. Morton, and G. R. Allen, Jr., *J. Amer. Chem. Soc.*, 1970, **92,** 3740.
207. (a) P. G. Gassman, T. J. Van Bergen, D. P. Gilbert, and B. W. Cue, *J. Amer. Chem. Soc.*, 1974, **96,** 5495, 5508, 5512; (b) P. G. Gassman, D. P. Gilbert, and T. J. Van Bergen, *J.C.S. Chem. Comm.*, 1974, 201.
208. R. J. Sundberg, H. J. Russell, W. V. Ligon, Jr., and L.-S. Lin, *J. Org. Chem.*, 1972, **37,** 719.
209. G. J. Baxter, R. F. C. Brown, and G. L. McMullen, *Austral. J. Chem.*, 1974, **27,** 2605.
210. (a) W. E. Erne, G. A. Mills, and R. K. Smith, *Chem. Abs.*, 1961, **55,** 4553; (b) J. Bakke, H. Heikman, and E. B. Hellgren, *Acta Chem. Scand. (B)*, 1974, **28,** 393.
211. (a) W. C. Sumpter, *Chem. Rev.*, 1945, **37,** 443; (b) A. H. Beckett, R. W. Daisley, and J. Walker, *Tetrahedron*, 1968, **24,** 6093.
212. A. S. Endler and E. I. Becket, *Org. Synth.*, 1957, **37,** 60.
213. (a) H. C. F. Su and K. C. Tsou, *J. Amer. Chem. Soc.*, 1960, **82,** 1187; (b) A. Hassner and M. J. Haddadin, *J. Org. Chem.*, 1963, **28,** 224.
214. (a) C. S. Marvel and G. S. Hieirs, *Org. Synth. Coll. Vol. 1*, 1952, 327; (b) F. D. Popp, *Adv. Heterocyclic Chem.*, 1975, **18,** 1.
215. B. Gabetta, *Fitoterapia*, 1973, **44,** 3.
216. 'The Alkaloids', ed. R. H. F. Manske, Academic Press, New York, 1960 onwards, vols. I–XVI.
217. 'The Alkaloids', ed. J. E. Saxton, (Specialist Periodical Reports), The Chemical Society, London, 1971 onwards, vols. 1–7.

218. J. P. Kutney, in 'MTP International Review of Science', Organic Chemistry Series I, Butterworth, London, 1973, vol. 9, p. 27.
219. Ref. 216, 1971, vol. XIII, chapter 6.
220. A. Hofmann, R. Heim, A. Brack, H. Kobel, A. Frey, H. Ott, T. Petrzilka, and F. Troxler, *Helv. Chim. Acta*, 1959, **42,** 1557.
221. (a) Ref. 216, vol. II, chapter 13; (b) *ibid.*, vol. VII, chapter 10; (c) *ibid.*, vol. VIII, chapter 2; (d) *ibid.*, vol. X, chapter 5; (e) *ibid.*, vol. XIII, chapter 4.
222. (a) G. R. Newkome and N. S. Bhacca, *Chem. Comm.*, 1969, 385; (b) R. K. Hill and G. R. Newkome, *Tetrahedron*, 1969, **25,** 1249; (c) R. B. Longmore and B. Robinson, *Chem. and Ind.* (*London*), 1969, 622.
223. (a) C. Hootele, *Tetrahedron letters*, 1969, 2713; (b) B. Robinson and D. Moorcroft, *J. Chem. Soc.* (*C*), 1970, 2077.
224. (a) Ref. 216, vol. VIII, chapter 3; (b) G. W. A. Slywka and R. A. Locock, *Tetrahedron Letters*, 1969, 4635.
225. (a) Ref. 216, vol. VIII, chapter 4; (b) S. R. Johns, J. A. Lamberton, and A. A. Soumis, *Austral. J. Chem.*, 1969, **22,** 801; (c) J. A. Jeffreys, *J. Chem. Soc.* (*C*), 1970, 1091.
226. (a) Ref. 216, 1965, vol. VIII, chapter 21; (b) D. Gröger, *Fortschr. Chem. Forsch.*, 1966, **6,** 159; (c) S. Augurell and J.-E. Lindren, *Tetrahedron Letters*, 1968, 5127.
227. C. W. Holzapfel, R. D. Hutchinson, and D. C. Wilkins, *Tetrahedron*, 1970, **26,** 5239.
228. (a) A. J. Birch and J. J. Wright, *Tetrahedron*, 1970, **26,** 2329; (b) R. Ritchie and J. E. Saxton, *J.C.S. Chem. Comm.*, 1975, 611; (c) P. S. Steyn, *Tetrahedron Letters*, 1971, 3331.
229. M. Barbetta, G. Casnati, A. Pochini, and A. Selva, *Tetrahedron Letters*, 1969, 4457.
230. Ref. 216, vol. I, chapter 7; vol. II, chapters, 13, 15; vol. III, chapter 16; vol. VII, chapter 10; vol. VIII, chapters 5–15, 17–20, 22; vol. X, chapter 11; vol. XI, chapters 1–9; vol. XII, chapter 3; vol. XIV, chapters 2–5.
231. J. Le Men and W. I. Taylor, *Experientia*, 1965, **21,** 508.
232. Ref. 216, vol. VIII, chapter 13; vol. XIV, chapter 3.
233. (a) R. T. Brown, C. L. Chapple, and G. K. Lee, *J.C.S. Chem. Comm.*, 1972, 1007; (b) R. T. Brown and A. A. Charalambides, *Phytochemistry*, 1975, 2527; (c) R. T. Brown and D. M. Duckworth, *J.C.S. Chem. Comm.*, 1976, 530; (d) E. E. van Tamelen and C. Dorschel, *ibid.*, 529.
234. (a) E. Wenkert, C.-J. Chang, A. O. Clouse, and D. W. Cochran, *Chem. Comm.*, 1970, 961; (b) A. Ahond, M.-M. Janot, N. Langlois, G. Lukais, P. Potier, P. Rasoanaivo, M. Sangaré, N. Neuss, M. Plat, J. Le Men, E. W. Hagaman, and E. Wenkert, *J. Amer. Chem. Soc.*, 1974, **96,** 633.
235. Ref. 216, vol. VIII, chapters 11, 14; vol. XI, chapter 4.
236. K. Blaha, Z. Kobikova, and J. Trojanek, *Tetrahedron Letters*, 1972, 2763.
237. Ref. 216, vol. VIII, chapter 9; vol. XI, chapter 9.
238. (a) P. Potier, N. Langlois, Y. Langlois, and F. Guéritte, *J.C.S. Chem. Comm.*, 1975, 670. (b) J. P. Kutney, *Heterocycles*, 1975, **3,** 205.
239. A. A. Gorman, M. Hesse, H. Schmid, P. G. Waser, and W. H. Hopf, in 'The Alkaloids', ed. J. E. Saxton (Specialist Periodical Reports), The Chemical Society, London, 1971, vol. 1., p. 200.
240. For reviews on carbazole chemistry, see W. Freudenberg, in 'Heterocyclic Compounds', ed. R. C. Elderfield, Wiley, New York, 1952, vol. 3, p. 291; N. Campbell and B. M. Barclay, *Chem. Rev.*, 1947, **40,** 359.
241. R. Bonnett and R. F. C. Brown, *J.C.S. Chem. Comm.*, 1972, 393; J. E. Bornstein, D. E. Remy, and J. E. Shields, *ibid.*, 1972, 1149; R. Bonnett, R. F. C. Brown, and R. G. Smith, *J.C.S. Perkin I*, 1973, 1432.
242. E. Chacko, J. Bornstein, and D. J. Sardella, *Tetrahedron Letters*, 1977, 1095.
243. M. H. Palmer and S. M. F. Kennedy, *J.C.S. Perkin II*, 1976, 81; W. Rettig and J. Wirz, *Helv. Chim. Acta*, 1976, **59,** 1054.
244. J. D. White and M. E. Mann, *Adv. Heterocyclic Chem.*, 1969, **10,** 113.
245. G. M. Priestley and R. N. Warrener, *Tetrahedron Letters*, 1972, 4295.
246. G. Wittig, H. Tenhaeff, W. Schoeh, and G. Koenig, *Annalen*, 1951, **572,** 1.
247. R. Kreher and J. Seubert, *Angew. Chem. Internat. Edn.*, 1966, **5,** 967.
248. J. C. Emmett, D. F. Veber, and W. Lwowski, *J.C.S. Chem. Comm.*, 1965, 272.
249. R. Kreher and J. Seubert, *Tetrahedron Letters*, 1966, 3015.
250. J. C. Emmett and W. Lwowski, *Tetrahedron*, 1966, **22,** 1011.
251. H. Fletcher, *Tetrahedron*, 1966, **22,** 2481.
252. C. O. Bender and R. Bonnett, *J.C.S. Chem. Comm.*, 1966, 198.
253. W. Theilacker, H. J. Blum, W. Heitmann, H. Kalenda, and H. J. Meyer, *Angew. Chem.*, 1964, **673,** 96.
254. I. Ahmed, G. W. H. Cheeseman, B. Jaques, and R. G. Wallace, *Tetrahedron*, 1977, **33,** 2255; B. Jaques and R. G. Wallace, *J.C.S. Chem. Comm.*, 1972, 397.
255. J. R. Benson and P. E. Hare, *Proc. Nat. Acad. Sci. USA*, 1975, **72,** 619.
256. S. S. Simons and D. F. Johnson, *J.C.S. Chem. Comm.*, 1977, 374.
257. V. Boekelheide and W. Feely, *J. Org. Chem.*, 1957, **22,** 589.
258. D. R. Bragg and D. G. Wibberley, *J. Chem. Soc.*, 1963, 3277.
259. J. Frohlich and F. Kröhnke, *Chem. Ber.*, 1971, **104,** 1629, 1645.
260. J. W. Lown and K. Matsumoto, *Canad. J. Chem.*, 1971, **49,** 1165.
261. W. Kiel and F. Kröhnke, *Chem. Ber.*, 1972, **105,** 3709.
262. R. J. Windgassen, W. H. Saunders, and V. Boekelheide, *J. Amer. Chem. Soc.*, 1959, **81,** 1459; V. Boekelheide, F. Gerson, E. Heilbronner, and D. Mueche, *Helv. Chim. Acta*, 1963, **46,** 1951.

17.5

Purines and Analogues

E. LUNT

May & Baker Ltd., Dagenham

17.5.1 OCCURRENCE AND IMPORTANCE OF PURINES

The chemistry of the purine, or imidazo[4,5-*d*]pyrimidine (**1**), ring system has been one of the most widely investigated. The eccentric numbering shown, of historical derivation,[1] has been retained in accordance with current general practice. This is due, of course, to the occurrence in the building blocks of the human and animal organism, the nucleic acids, of the bases adenine (**2**; R=H) and guanine (**3**; R=H) in the form of their 9-phosphorylated sugar derivatives, the nucleotides (see Chapter 22.2), and their metabolites such as uric acid (**4**), found in urine and gallstones. Guanine itself occurs in guano, the excretion of seabirds used commercially as a fertilizer, whilst the beverage alkaloids theobromine, theophylline, and caffeine are *N*-methyloxopurine derivatives. The coenzyme NAD, nicotinamide adenine dinucleotide, is also very important (see Chapter 22.2).

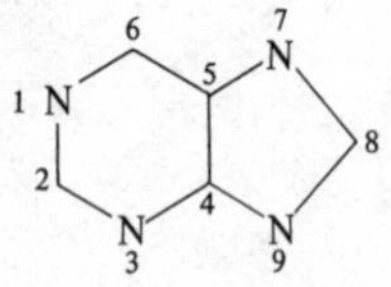

(1) 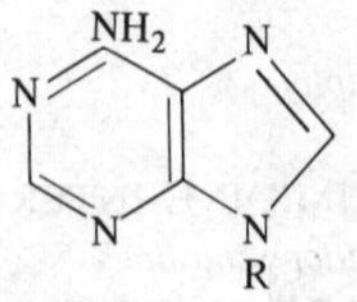 (2) 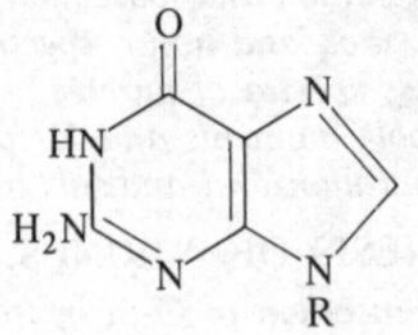 (3)

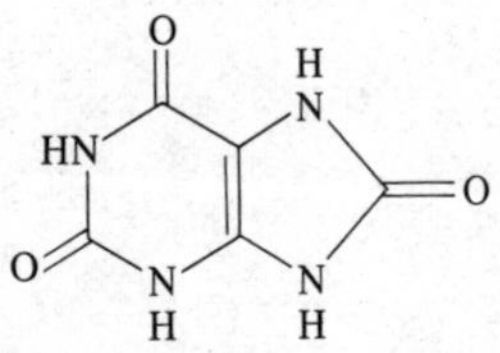

(4) 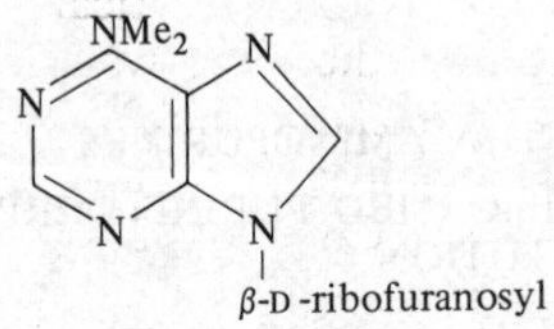 (5) β-D-ribofuranosyl (6) β-D-psicofuranosyl

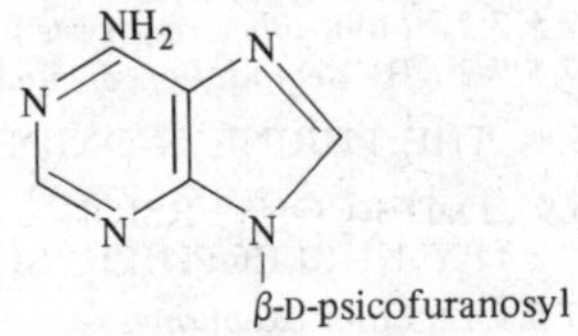

In recent years, new naturally-occurring purines have been recognized, including the antibiotics puromycin (**5**) and angustmycin C (**6**), the anticholesteraemic eritadenine (**7**) (from the fungus *Lentinus edodes*), the marine shellfish toxin saxitoxin (**8**) and its relatives, and the fused imidazopurine derivatives wye base and wyosine (**9a, b**), isolated from bacterial tRNA anti-codon fractions.

During the past few years, many purines have been claimed to have a multitude of useful biological and other activities, but the main areas of activity, apart from the anti-tumour/anti-leukemia field (6-mercaptopurines), include immunosuppressives (where azathioprine (**10**) represents a considerable advance), hypocholesteraemics (analogues of eritadenine), antiviral (Vidarabine, **11**), anti-allergics and bronchodilators (including phosphodiesterase inhibitors), cardiac stimulants, and coronary or other vasodilators, diuretics and hypotensives, xanthine oxidase inhibitors (for gout treatment), and cytokinins (plant growth regulators, analogues of those occurring naturally).

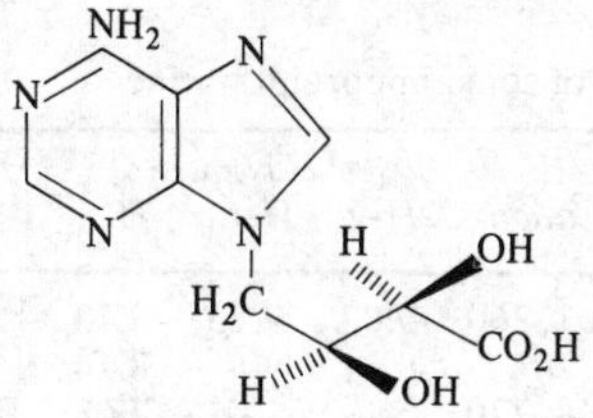

(7)

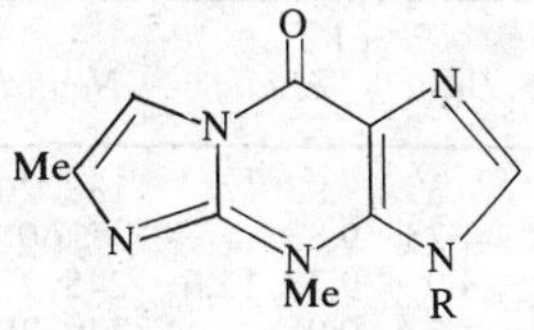

(9) a; R = H
b; R = β-D-ribofuranosyl

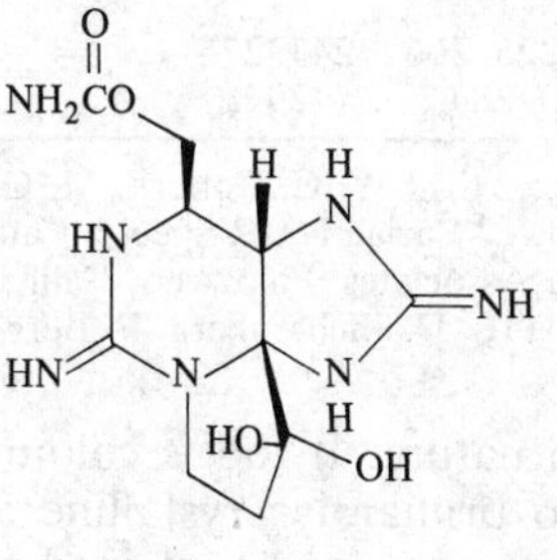

(8)

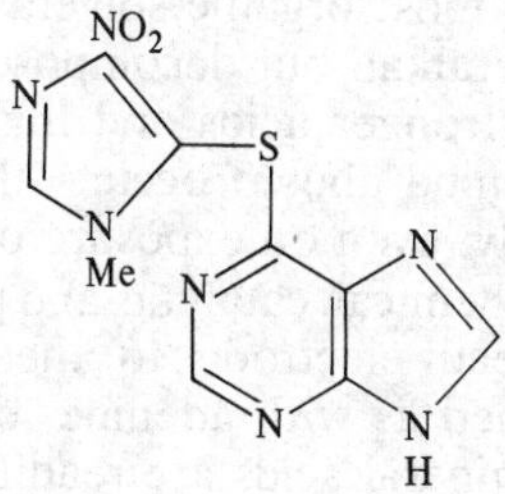

(10)

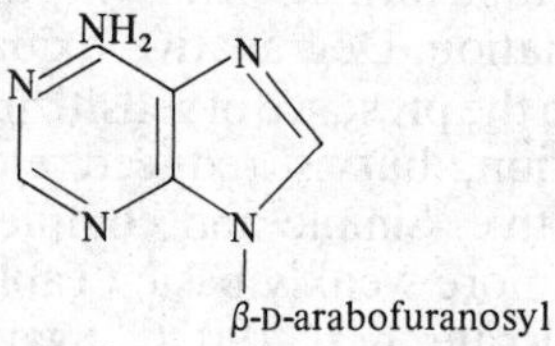

(11) β-D-arabofuranosyl

Comprehensive review articles on the purine system have been quite rare over the years, and the best for general organic chemistry is the comprehensive account by Lister,[2] whilst physical-organic and spectral data and analytical methods are to be found in the work edited by Zorbach and Tipson.[3] Other reviews cover electronic aspects of purine tautomerism,[4] synthesis of purines from uric acid,[5] and nucleophilic and electrophilic substitution,[6] as well as many aspects of physical-organic chemistry in the proceedings of a symposium.[7] For nucleoside reviews, see Chapter 22.2.

17.5.2 ADENINE AND GUANINE

Adenine is present in all living matter. It is a colourless, crystalline substance having m.p. 352–354 °C. It is soluble in cold water (40 parts), soluble in alcohol, but poorly soluble in many other organic solvents. It is stable to hot alkali and towards boiling dilute hydrochloric acid. It is, however, broken down by 6N acid at 150 °C to 4-amino-5-imidazolecarboxamide, and complete degradation occurs with HCl at higher temperatures, with hot concentrated sulphuric acid, or in oxidizing conditions with acid potassium dichromate or potassium permanganate. It is unchanged by hydrogenation but is reduced at a dropping mercury electrode in acid solution. Adenine, being an ampholyte (Table 1), forms salts with acids and bases. The picrate (m.p. 296 °C) is convenient for identification and gravimetric analysis. Simple metal salts are formed with monovalent cations, but with higher-valent cations many complex salts are formed, particularly involving cupric, cobalt, and nickel ions (see Section 17.5.4). Spectral characteristics of adenine are given in Table 1.

TABLE 1

Spectroscopic and other physical properties of some important purines

	pK_a[a]		*U.v.* λ_{max} (nm)[a]			^1H *N.m.r.* (δ)			
	Basic	*Acidic*	*Neutral*	*Anion*	*Cation*	*H-2*	*H-6*	*H-8*	*Solvent*
Purine	2.39	9.93	188, 263	219, 271	201, 260	9.03	9.23	8.73	[$^2H_6$]DMSO[b]
Adenine	4.25	9.83	260	267	262	8.11	—	8.14	[$^2H_6$]DMSO[c]
Guanine	3.0	9.32, 12.6	245, 274	243, 273	248, 270	—	—	7.68	D_2O/NaOD[d]
Isoguanine	4.5	9.0	240, 286	235, 284	230, 284	—	—	—	—
Hypoxanthine	1.98	8.94, 12.1	249	258	248	8.12	—	7.97	[$^2H_6$]DMSO[e]
						7.87	—	8.08	D_2O/NaOD[d]
Xanthine	—	7.7, 11.94	225, 266	241, 276	—	—	—	7.85	9:1 [$^2H_6$]DMSO/ D_2O[f]
Uric acid	—	5.4, 10.6	284	292	—	—	—	—	

[a] From tables in Ref. 3; [b] Ref. 3, p. 313; [c] W. C. Coburn, M. C. Thorpe, J. A. Montgomery, and K. Hewson, *J. Org. Chem.*, 1965, **30,** 1114; [d] Varian NMR Spectra Catalogue, ed. N. S. Bhacca, D. P. Hollis, L. F. Johnson, and J. A. Pier, Varian Associates, Palo Alto, Calif., 1963; [e] J. A. Montgomery and H. J. Thomas, *J. Org. Chem.*, 1966, **31,** 1411; [f] D. Lichtenberg, F. Bergmann, and Z. Neiman, *J. Chem. Soc.* (*C*), 1971, 1676.

Guanine is also ubiquitous in nature. It too is colourless and crystalline, with no m.p. <350 °C. In fact, guanine is so brilliantly crystalline as to have been used in artificial pearling agents. It is insoluble in water, ethanol, and most organic solvents, but readily soluble in acids and bases. It is stable to boiling dilute alkali but decomposed by heating with acids. Hot dilute acid gives xanthine, but with stronger acids and higher temperatures, first formation of the usual imidazole (see adenine above) occurs, then complete degradation. Degradative oxidation occurs quite readily, even on exposure of solutions to light in the presence of sensitizing dyes. It is resistant to chemical, catalytic, and polarographic reduction, but is reduced electrochemically at a lead electrode to the 1,6-dihydro-derivative. Simple and complex salts are readily formed as with adenine, but guanine is much more weakly basic (Table 1) and its salts with mineral acids are readily hydrolysed. The picrate, m.p. 190 °C, again serves as a good reference. Further physical and spectral data are given in Table 1. For more detailed information a short review on guanine is available.[8]

17.5.3 NOMENCLATURE OF PURINES

The literature nomenclature of purines presents a very confused picture, with a bewildering array of names based on purine itself, on the above alkaloids and xanthine, uric acid, adenine, and guanine, or on the systematic nomenclature using the imidazopyridine ring system. This last has not been considered for use in this review as it leads to numbering difficulties. Throughout, a system based either on the current Chemical Society method, based on the purine system, or on the xanthine, theobromine, adenine, guanine, uric acid, *etc.* nomenclature has been used in accordance with the needs of convenience or clarity.

Substituents have been attached in an additive manner, except for oxo- and thio-derivatives, and structures have been drawn in the aromatic form, again except for the latter, where the oxo- or thioxo-tautomer is preferred, with the hydrogen arbitrarily placed on N-1, N-3, N-7, or N-9 as appropriate, except where work on exact tautomers is being discussed.

17.5.4 PHYSICAL PROPERTIES OF PURINES

Purine and its alkyl and aryl analogues are fairly stable crystalline solids of medium melting-points and moderate to high solubility, but with the introduction of amino- and/or oxo-groups, the naturally-occurring purines (adenine, guanine, *etc.*) become very crystalline, high melting and relatively insoluble, a fact which has hindered physicochemical investigations. Guanine is so brilliantly crystalline that it has been used in artificial pearling compositions (see above). The physical properties and spectral data for a few important purines are given in Table 1.

17.5.4.1 Molecular orbital calculations

Because of the importance of purines in the chemistry of the nucleic acids, many and varied calculations of chemical reactivity, tautomerism, spectral characteristics, energies of ionization and of complex formation, *etc.* have been made using a wide variety of theoretical chemical techniques. The earlier attempts (using simple Hückel, frontier orbital theory, charge densities, and localization energies) showed a considerable measure of disagreement with each other and with experimental findings, but in recent years more sophisticated calculations have been made using such techniques as the SCFMO (Pariser–Pople–Parr) method,[9] MINDO,[10] and extended Hückel, CNDO/2, SCF/CI, and MCNDO methods.

Increasingly good agreement between prediction and experiment has been found, especially for tautomeric equilibria[4,11] and protonation sites, dipole moments,[7,9] u.v. spectra,[12] and emission (fluorescence and phosphorescence) spectra.

The main difficulty lies in the prediction of which state the purine molecule will have attained at the point of reaction. This is, of course, especially difficult for the very important problem of hydrogen-bonding, which is of crucial importance in nucleic acid structure, and many calculations have been devoted to the interactions of the naturally occurring base pairs, adenine–thymine and guanine–cytosine, in various physical states, both in isolation and *in situ* in the Watson–Crick double-helical configuration.[13] Unnatural base pairs which might be involved in mutagenic or oncogenic processes, or in information transmission, have also been studied. One calculation appeared to show that the base pairs would be most stable when in the 'half-protonated' form,[14] but the whole picture is still far from clear. A comparison of the different calculation methods for adenine and guanine is available.[15] A short review is also extant.[7]

17.5.4.2 Tautomerism of purines

Another area for which extensive experimental and theoretical data are available is that of the exact tautomeric forms in which the various important purine molecules exist.

The problem is that, in the various methods of obtaining experimental estimates which have been used, the physical state of the purine is not self consistent, which has led to a lack of agreement on the correct structure of the major tautomer for some molecules.

For instance, crystal forces may adversely influence X-ray crystallographic measurements in the solid state, whilst in solution the main methods used, u.v. spectroscopy, potentiometric titration to obtain pK_a values, and n.m.r. spectroscopy, all operate at different concentration levels. There is also the problem of possible crucial differences between the ground and excited states. There seems to be a general consensus on the main tautomers of adenine (**2**; R = H) and guanine (**12, 13**), and on the general principles that amino-groups in purines exist in the —NH_2 form, thioxo-groups (with a few possible

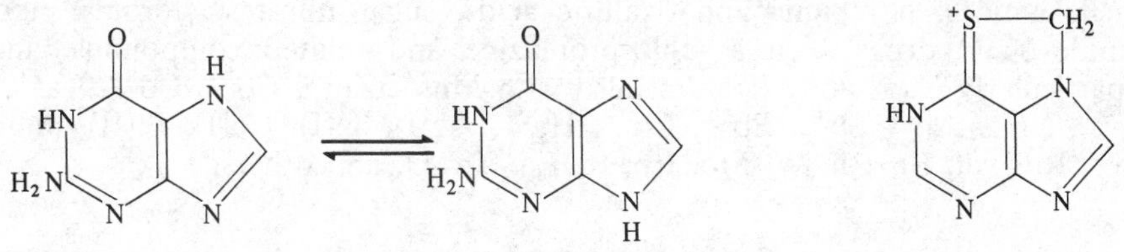

(12) (13) (14)

exceptions) and oxo-groups in the C=O (amide) form, and that the imidazole proton resides in most purines on N-9, although in purine itself N-7 seems to be favoured, from dipole moment studies. Some reviews are available,[4,7] as well as a section in a recent book on heterocyclic tautomerism (Ref. 458, pp. 502 *et seq.*).

17.5.4.3 Protonation and ionization of purines

The ionization constants for purines have been determined both by titrimetric and u.v. spectroscopic methods.[3,7] Some examples are quoted in Table 1.

17.5.4.4 I.r., u.v., and n.m.r. spectra

I.r. spectra have been of relatively little use in the chemical identification of purines, but have been of great value in the studies of hydrogen-bonding.[2,3,7]

U.v. spectra have been of great use for pK_a determination, tautomerism studies, and identification and purity determination of individual purines.[3,16]

In recent years, detailed n.m.r. measurements, especially by Bergmann's group, have begun to provide much useful information, especially in the fields of tautomerism and hydrogen-bonding, and, in the establishment of the structure of the nucleosides, in particular, the anomeric nature of the glycosidic linkages (see Chapter 22.2).[3,7]

17.5.4.5 Mass spectra of purines

Mass spectral measurements,[3] often in conjunction with g.l.c. separation of purines in the form of their pertrimethylsilylated derivatives, have shown fragmentation patterns characteristic of nitrogen heterocyclic molecules, *i.e.* loss of HCN, MeCN, and, in the case of oxo-derivatives, C=O fragments. In some 7-methyl-6-thio-derivatives, an unusual $[M-1]^+$ ion is attributed to bridged five-membered structures (**14**) between the sulphur and adjacent methyl groups.[17]

17.5.4.6 Dipole moments and X-ray crystallography

The use of dipole moments has been severely limited by the insolubility of many purines in suitable solvents. They have provided some useful information on tautomeric structures, however, and could become more important as techniques develop.

X-ray crystallography has been of great use in the establishment of the correct structures of various natural products, in the study of hydrogen-bonding and protonation, and in the structural study of the complexes formed by purines with themselves (stacking), and with a vast number of types of organic compound (including other purines and pyrimidines (thymine, cytosine, uracils, barbituric acids, *etc.*), many aromatic molecules, amino-acids (especially aromatic), other organic acids and amides, indoles, quinoline *N*-oxides, high potential quinones, saturated and unsaturated lactones, ε-caprolactam and other cyclic amides, parabanic and oxaluric acids, sugar nitrates, steroids, riboflavin, nicotinamides, and drugs such as chlorpromazine and related compounds, and with metals, particularly Cu^{2+} and Ni^{2+}, but also with Mn^{2+}, Zn^{2+}, Co^{2+}, Co^{3+}, Ca^{2+}, Mg^{2+}, Be^{2+}, Fe^{3+}, Cr^{3+}, Al^{3+}, Sb^{5+}, Pb^{2+}, Cd^{2+}, Hg^{2+}, Ag(I), Tl(I), Cu(I), Pt(II), and Pd(II) complexes, Ru, Au, Bi, Rh, W, Mo, Re, Ir, Eu, Th, U, and others.

17.5.4.7 Separation and analysis of purines

Amongst techniques used for the analytical and preparative separation of purines are paper, column (sometimes on copper- and silver-impregnated substrates), gas–liquid, ion-exchange, thin-layer, and high-performance liquid chromatography, paper electrophoresis, and gel-filtration on Sephadex G-10.

17.5.5 SYNTHESIS OF PURINES FROM PYRIMIDINE INTERMEDIATES

17.5.5.1 By insertion of C-8, or from 5-acylaminopyrimidines

(i) 8-Unsubstituted Purines

The most widely used purine synthesis is of this type, the so-called Traube synthesis, after its discoverer,[18] who cyclized the triaminopyrimidine (**15**) with formic acid to obtain guanine (**3**; R = H), the reaction occurring *via* the 5-acylamido-compound (**16**), which is often isolated and cyclized in a second stage, *e.g.* with aqueous alkali.

Difficulties may be encountered through replacement of labile halo- or alkoxy-groups, or loss of sugar substituents in attempted nucleoside preparations, whilst with thiopyrimidines (**17**) the alternative ring closure to the thiazolopyrimidines (**18**) may occur.[19] Low temperature and high pH favour thiopurine formation. A modification which avoids this difficulty is the cyclization of the 5-formylamino-4-aminopyrimidine directly using P_2S_5/pyridine, which favours the formation of the 6-thiopurine (hy)[20] and has received extensive application in the literature. Many other reagents have since been used in place of formic acid in the preparation of 8-unsubstituted purine derivatives, such as mixtures of formic acid with sodium formate or acetic anhydride, dithioformic acid (*via* the corresponding 5-thioformyl derivative), of value for glycosyl derivatives (see Chapter 22.2) because of the milder cyclization conditions, ethyl orthoformate, diethoxymethyl acetate, formamide, dimethylformamide/phosphorus oxychloride (the Vilsmeier–Haack reagent), and formamidine. Formic acid remains the choice in many cases, however, as illustrated by a recent synthesis of hypoxanthine in 60% overall yield from thiourea, the sulphur being removed in the cyclization step after oxidation to SO_2H.[21]

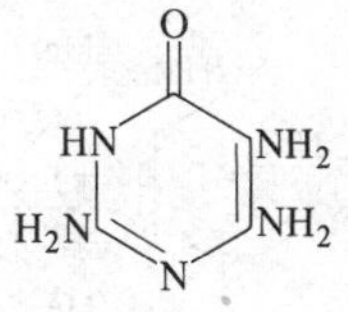

(15) 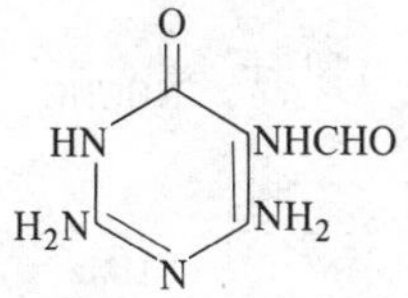 (16)

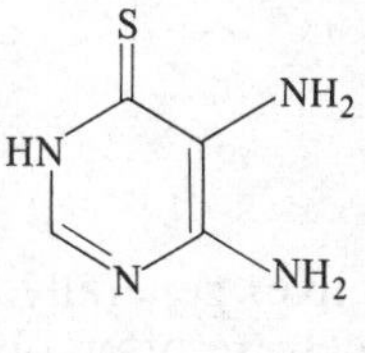

(17) 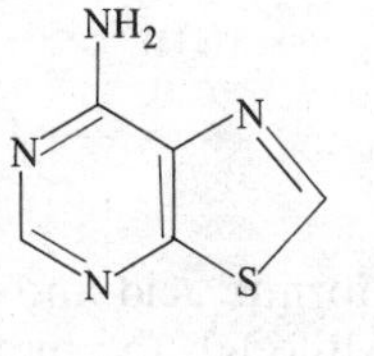 (18) 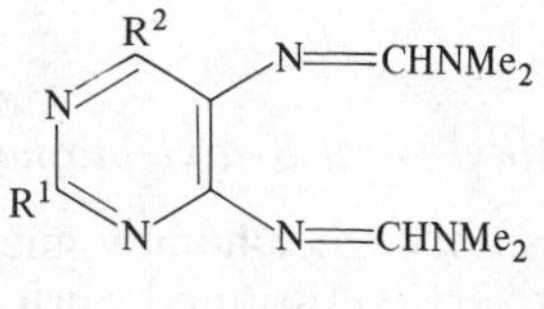 (19)

The orthoformate is usually used with an acid catalyst such as HCl or acetic anhydride, when the cyclization often occurs in one stage without isolation of the intermediate. Diethoxymethyl acetate, the product of the above reagents, has also been used, *e.g.* in a synthesis of eritadenine analogues. Formamide, which was first introduced as solvent for the formic acid cyclization, has latterly been used as a one-step (hy) cyclizing agent in its own right, especially in the presence of a trace of acid catalyst, such as HCl, H_2SO_4, or (industrially) a pyrimidine sulphamic acid, but it is not applicable to halo- or glycosyl-substituted pyrimidines.

The modified Vilsmeier–Haack procedure, involving cyclization of the intermediate azomethine (**19**) at lower temperatures *in situ*,[22] has the advantage that halopyrimidines may be converted to halopurines. Good results are also obtained with amino- and hydroxy-purines. *sym*-Triazine, an *in situ* source of formamidine, has also been used to generate the one-carbon unit.

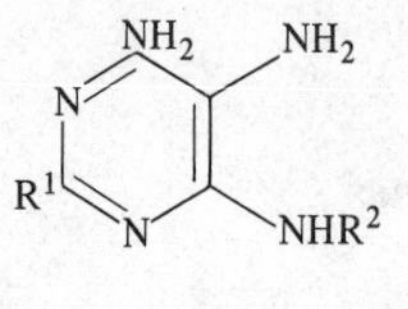

(20) 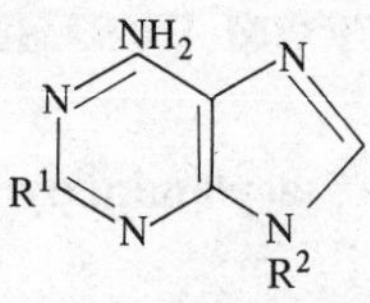 (21) 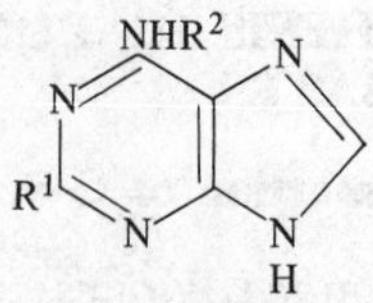 (22)

In general, in all these types of cyclization, when a 4,5,6-triaminopyrimidine carrying an alkyl substituent on one of the 4,6-amino-groups is used, *e.g.* (**20**), the direction of cyclization almost always favours (>75%) the formation of the 9-substituted adenine derivative (**21**) over the 6-substituted aminopurine (**22**), reflecting the enhanced basicity of the alkylated nitrogen. Intermediate unsubstituted 5-acylaminopyrimidines may also be alkylated at the 5-nitrogen *in situ*, in the alkaline solution, and then cyclized to the 7-substituted purine, *e.g.* (**23**) → (**24**).[23] With other polyalkylaminopyrimidines, cyclization may occur very readily, *e.g.* rapid formation of the purinium salt (**26**) from 5-formamido-4,6-bis(methylamino)pyrimidine (**25**) (Scheme 1).[24]

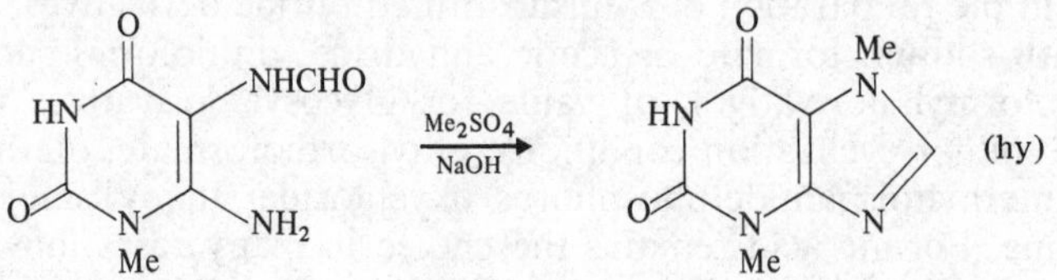

(23) $\xrightarrow[\text{NaOH}]{Me_2SO_4}$ (24) (hy)

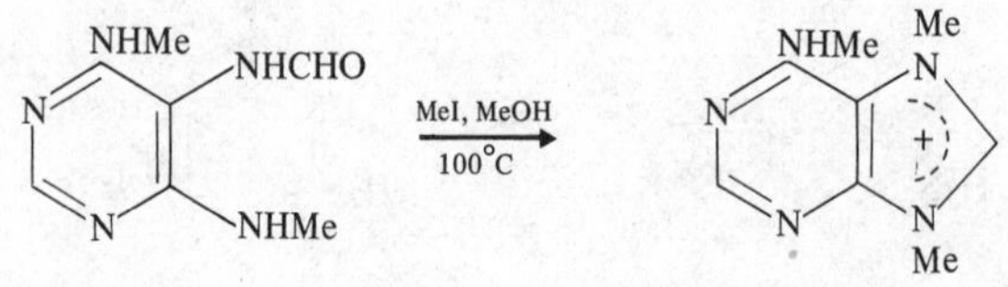

(25) $\xrightarrow[100^\circ C]{\text{MeI, MeOH}}$ (26)

SCHEME 1

(ii) 8-Alkyl- and 8-aryl-purines

The use of higher homologues of formic acid and their derivatives generally affords 8-alkyl- or -aryl-substituted purines. Although the methyl analogues are often obtained by preforming the 5-amidopyrimidine and subsequent cyclization, acetic anhydride has been extensively used for 8-methylpurines, especially for the 9-alkyl derivatives. With oxo-pyrimidines the reaction usually stops at the acetamido stage. Thio- or halo-pyrimidines may give side-reactions. Higher acids or anhydrides (up to propionic) give on heating the acylamidopyrimidines, which may be cyclized in one of the usual ways. Trifluoroacetic acid, or better, trifluoroacetic anhydride, gives 8-CF_3 purines directly. Other acids which have been used, with subsequent cyclization, include various substituted acetic acids such as phenyl- and phenoxy-acetic acids, glycollic acid, mandelic acid, crotonic and cinnamic acids, and isatoic anhydride. The use of benzoic acids with polyphosphoric acid as the cyclizing agent to give 8-arylpurines has been noteworthy,[25] phosphorus oxychloride also being used.[26] Dicarboxylic acids normally lead to 8,8′-bispurines, whilst oxalic acid at moderate temperatures gives the oxalylamino-derivative, cyclized to the purine-8-carboxylic acid with alkali.[27] Higher temperatures lead to oxo-pteridines *via* the Isay reaction.[28] Glycine and its derivatives give 8-aminomethyl derivatives, *e.g.* (**27**). Acetimidate salts have also recently been used.[29]

Alternatively, the 5-acylamino-derivative may be prepared using the acid chloride (especially aroyl chlorides), anhydride, or lactone (including sugar lactones) and subsequently cyclized to the purine using alkaline treatment, or, for 8-arylpurines, by fusion of the sodium salt, by use of polyphosphoric acid, or with freshly distilled phosphoryl halides; the latter reagent also converts any oxo-groups present into halo-groups.

The use of the aldehyde plus an oxidizing agent (ferric chloride, nitrobenzene, air+Pd catalyst) to give 8-substituted purines has not been particularly successful, concomitant alkylation being observed, especially with formaldehyde. The use of an aldehyde bisulphite adduct, however, is more promising.[31] Glyoxal gives 8,8′-bis-7,8-dihydropurines, instead of the expected pteridines.[32]

Acetylacetone in acetic acid gives the 8-methylpurine with loss of acetone, dibenzoylmethane giving the 8-phenyl derivatives.[33]

The use of amidines with diaminopyrimidines has been largely limited to the use of acetamidine or benzamidine (and heterocyclic amidines) for 8-methyl and 8-phenyl (heterocyclyl) purines,[34] respectively. The amidine hydrochloride is usually fused with the pyrimidine and anhydrous sodium acetate. Alkyl, aryl, amino-, oxo-, and thio-purines have been successfully prepared using these methods.

(*iii*) *8-Oxo- and 8-thio-purines*

The reagent originally used for cyclization of diaminopyrimidines to the 8-oxo-compounds (uric acid derivatives) was ethyl chloroformate, but this has now largely been superseded by one-step procedures involving urea fusion, or use of phosgene in alkaline solution, although benzyl chloroformate has more recently been utilized, especially for 9-alkyl derivatives, the intermediate urethane, *e.g.* (**28**), being cyclized by fusion.

The urea fusion method has been successfully used with a variety of pyrimidines, (alkyl, aryl, thio-, alkylthio-, amino-, and even some halo-derivatives) yielding the appropriate 8-oxopurines. As the use of *N,N*-dimethylurea gives only 7,9-unsubstituted 8-oxopurines, the mechanism must involve the loss of both the urea nitrogens as ammonia. Confirmation has come from the use of ^{15}N-labelled reagents.[35] With phosgene the reaction is carried out in alkaline solution, a one-step reaction occurring at room temperature in favourable cases. 4,5-Diaminouracil (**29**) does not give uric acid itself under these conditions, but does so on treatment of its per(trimethylsilyl) derivative (**30**) with triethylamine and phosgene in toluene.[36]

Thiophosgene has had very limited use for preparing the corresponding 8-thioxo-analogues, which are most usually prepared by the use of carbon disulphide, either in pyridine or sodium hydroxide–ethanol solution. The reaction is believed to proceed *via* dithiocarbamates of type (**31**). Dimethylformamide has also been found useful as a

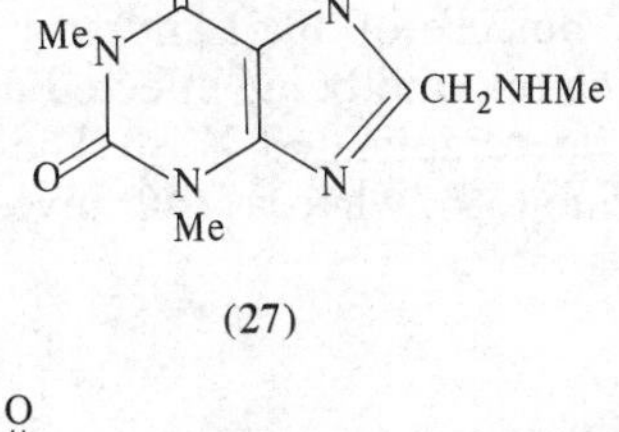

(27) 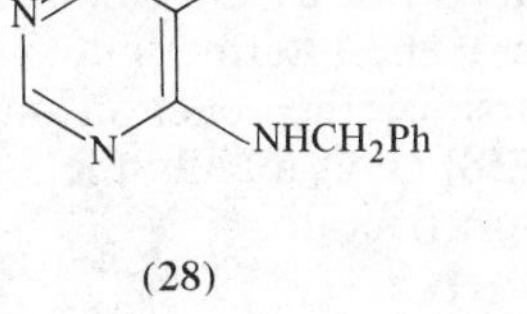 (28)

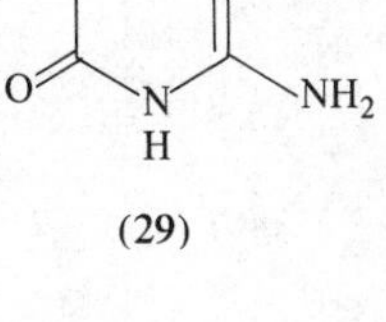

(29) 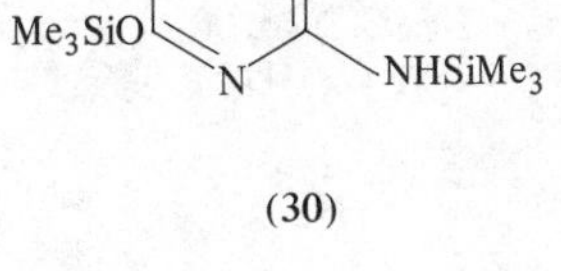 (30) 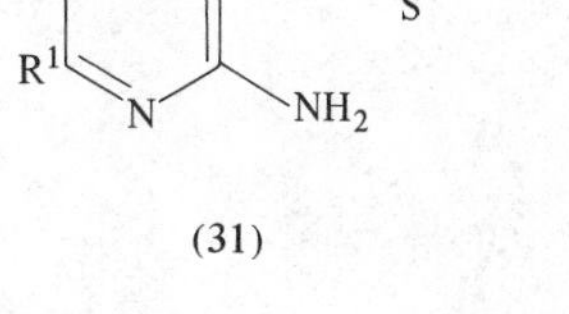 (31)

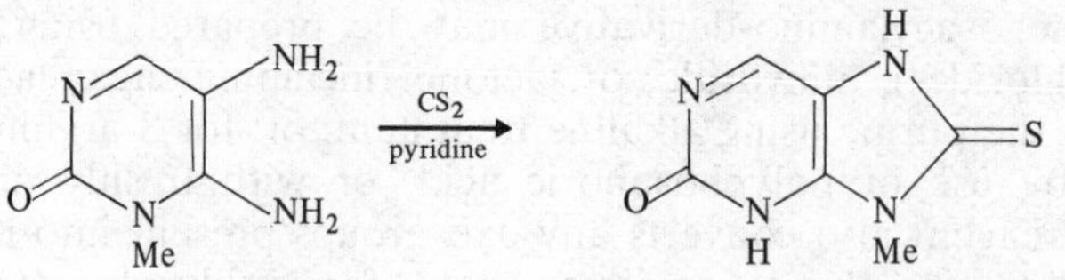

(32) (33)

SCHEME 2

solvent. The conditions are usually mild and even glycosylpurines can be readily prepared. Here again difficulties are encountered using halo- or alkylthio-pyrimidines owing to their prior substitution by thio-groups, with subsequent cyclization to thiazolopyrimidines. Dimroth rearrangement can also be a problem; for instance (**32**) gives (**33**), not the expected 3-methyl analogue, with CS_2/pyridine (Scheme 2).[37]

Thiourea may be used in a fusion reaction similar to that with urea, but more forcing conditions are required, leading to increased danger of side-reactions. The reaction is

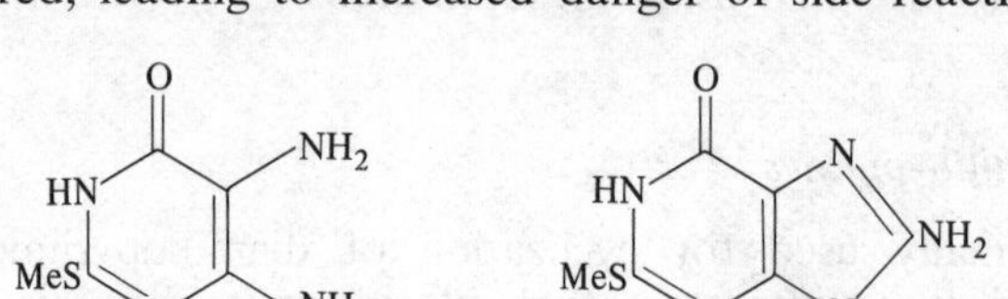

(34) (35)

believed to occur *via* guanidine thiocyanate, formed by thiourea rearrangement, which leads in some cases to the formation of 8-aminopurines, *e.g.* (**34**) gives (**35**). *N*,*N*-Dimethylthiourea requires a lower temperature, but here *S*-methylation may occur by migration, *e.g.* in the formation of (**37**) from (**36**) (Scheme 3).[38]

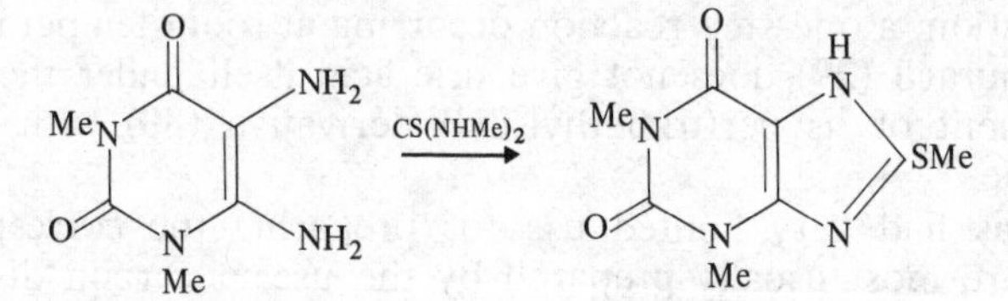

(36) (37)

SCHEME 3

Potassium cyanate is not a convenient reagent, but use of alkyl and aryl isocyanates is possible, cyclization of the intermediate 5-ureido-compound being effected using hydrochloric acid. The course of the reaction may also be complicated here. The 5-substituted ureido-pyrimidine (**38**) gives mainly the 9-arylpurine (**39**), whereas (**40**) gives a 7-alkyl (or -aryl) 8-oxopurine on fusion.

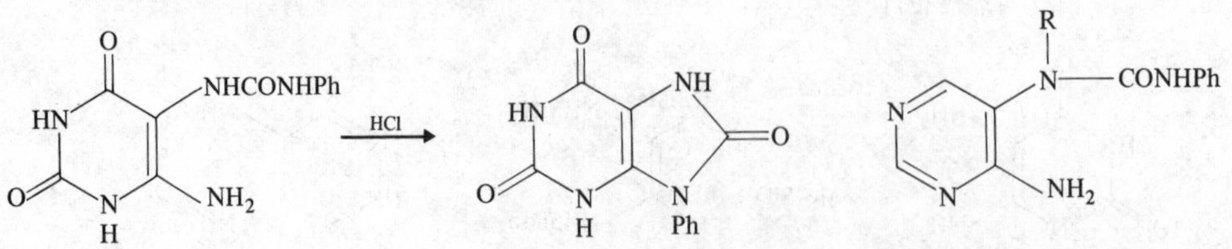

(38) (39) (40)

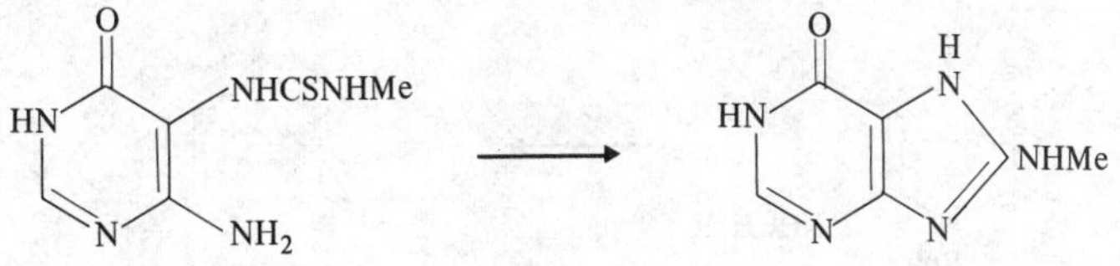

(41) (42)

Isothiocyanates have been more widely used than their oxygen analogues, especially in the theophylline series. Here again the reaction may give either 9-substituted or 9-unsubstituted 8-thio-derivatives. If sulphur-removing reagents, such as mercuric oxide, are added, H_2S is eliminated, leading to 8-alkylaminopurines, *e.g.* (**41**) gives (**42**).[39]

(iv) 8-Amino- and 8-alkoxy-purines

The preparation of 8-aminopurines by cyclization reactions with guanidines has not proved fruitful and only a few examples have been noted, including those of 2-oxo- and 6-oxo-4,5-diaminodihydropyrimidines, whilst guanidine thiocyanate converted the methylthio-derivative (**43**) to (**44**) (Scheme 4). Many close analogues, however, have failed to give 8-aminopurines, so the reaction remains unpredictable.

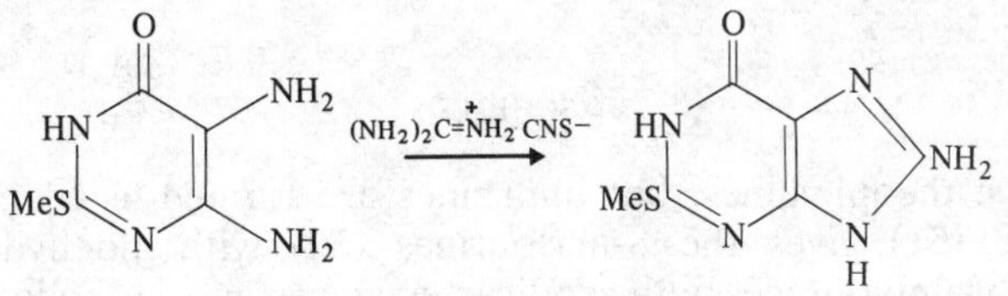

(43) (44)

SCHEME 4

The use of tetraethyl orthocarbonate to give 8-ethoxypurines, *e.g.* (**45**) → (**46**), has also been noted (Scheme 5).[40]

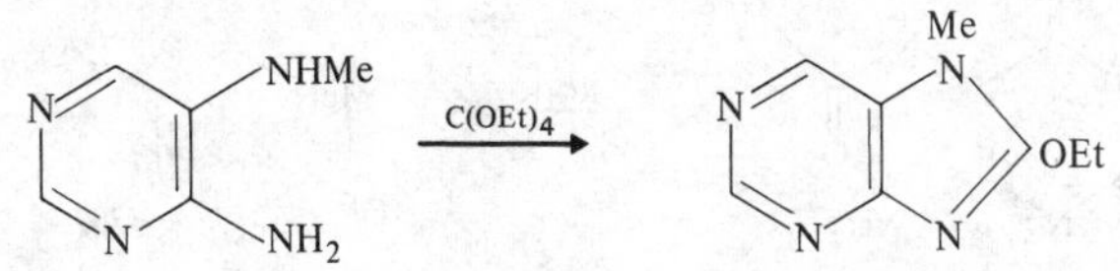

(45) (46)

SCHEME 5

17.5.5.2 Syntheses from other 4,5-substituted pyrimidine derivatives

The *in situ* reductive cyclization of 5-nitro-, 5-nitroso-, and 5-phenylazo-4-amino-pyrimidines in formic acid are really only extensions of the Traube synthesis above. A new reaction involving cyclization of 5-nitrosopyrimidines has received considerable attention recently, especially in the theophylline series. By it, 5-nitroso-4-RCH_2-substituted aminouracils, *e.g.* (**47**), easily prepared *in situ* using pentyl nitrite, cyclize in hot xylene or butanol,[41] or with ethanolic HCl, to give 8-R-substituted purines, *e.g.* (**48**). This reaction has recently been used in the establishment of the structures of a series of isomeric unsymmetrically 1,3-disubstituted uracil derivatives.[42]

With 4-s-alkylaminopyrimidines, 8,8-dialkyl non-classical structures are formed, which rearrange to 7,8-disubstituted purines.

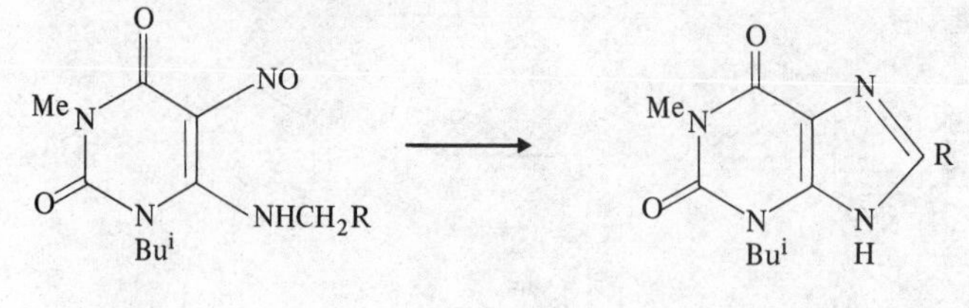

(47) (48)

The use of aldehydes alone with the nitroso-compound gives mixtures of the 8-substituted purines and their 7-oxides, but the presence of formic acid or rongalite (hy), or use of the aldehyde hydrazone[43] or *N,N*-dimethylhydrazone,[44] avoids this.

A recent (hy) synthesis of 8-dimethylaminotheophylline involved cyclization of the nitrosopyrimidine with $POCl_3$/DMF,[45] whilst the uracil (**49**) gives 7-aryltheophyllines (**50**) with nitrosobenzenes in acetic anhydride (Scheme 6).[46] 5-Phenylazouracils under these

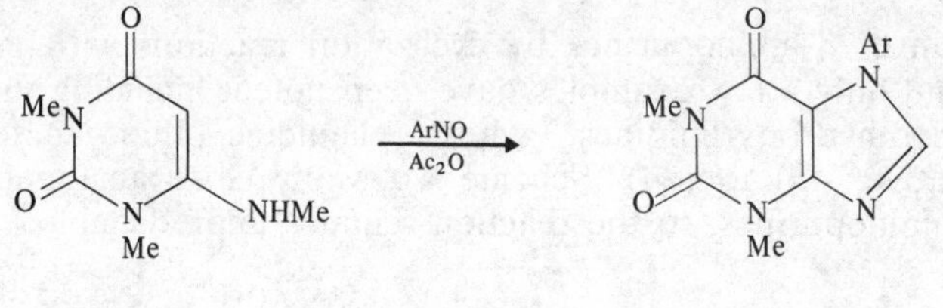

(49) (50)

SCHEME 6

conditions also give theophyllines, but lumazines are formed as by-products, whilst the urazolyl derivative (**51**) gives the 8-arylpurines (**52**) with aldehydes.[47] Recently the cyclization of 6-alkylaminouracils with azodicarboxylic ester, *via* the intermediate (**53**), has been described,[48] whilst benzylidenetriphenylphosphorane gives the 8-aryltheophylline directly.[49]

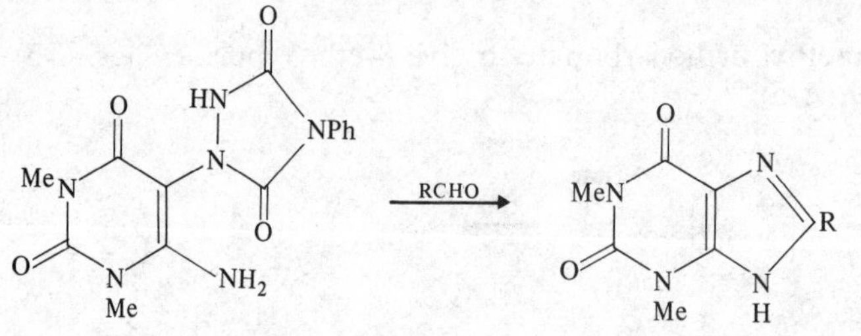

(51) (52)

5-Aminopyrimidine-4-carboxamide derivatives have been cyclized to 8-oxopurines under the conditions of the Hofmann rearrangement, presumably *via* the intermediate isocyanate (**54**).[50] The hydrazides (**55**) likewise undergo the Curtius reaction, leading to the same products.

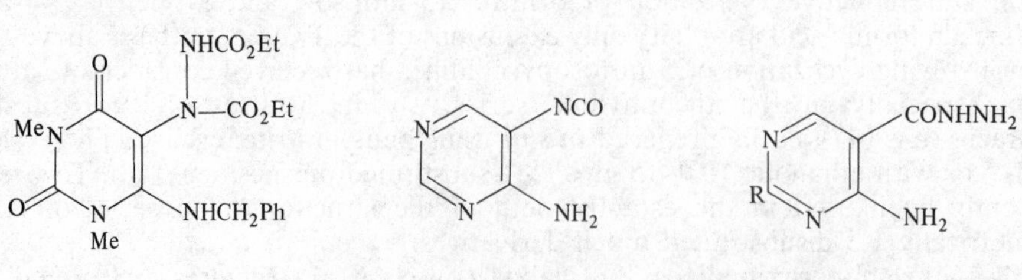

(53) (54) (55)

17.5.6 SYNTHESIS OF PURINES FROM IMIDAZOLE INTERMEDIATES

Among the earliest syntheses of purines from imidazole intermediates were those involving the Hofmann degradation of imidazole-4,5-dicarboxamides (**56**),[51] or the Lössen degradation of the corresponding dihydroxamic acids (**57**), which gives 1-hydroxyxanthines. Ring-substituted amides give the 9-substituted purines, but in the Lössen case the 7-substituted derivatives are obtained.[52]

More recently a number of routes have been described from imidazole intermediates of the types (**58**), (**59**), and (**60**) which parallel those believed to be involved in biological and prebiotic synthesis (Section 17.5.7).

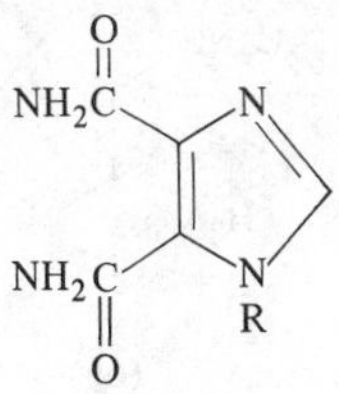

(56) 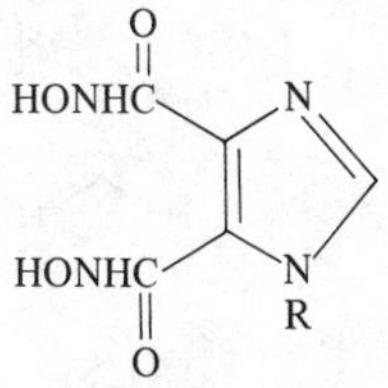 (57) 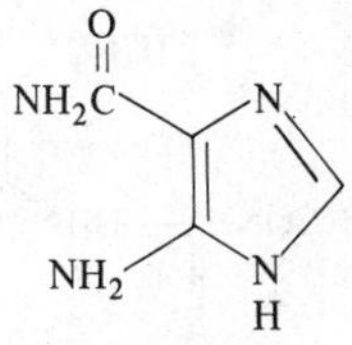 (58)

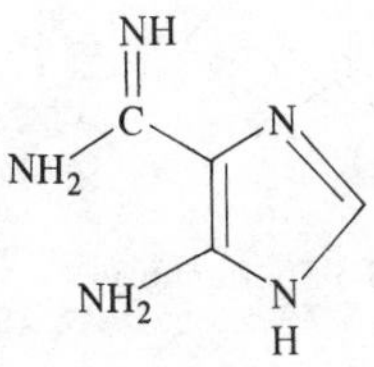

(59) 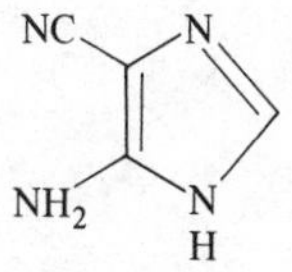 (60) 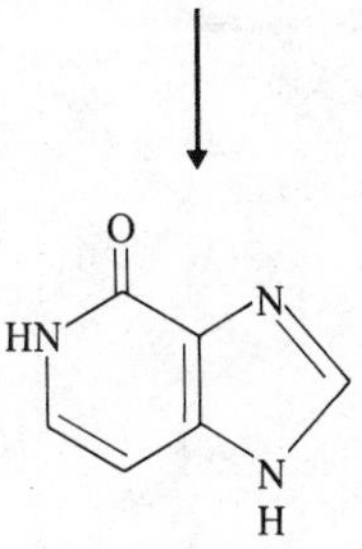 (61)

The amide (**58**), which leads to 6-oxopurines, has been widely employed, being formylated with formic acid/acetic anhydride, followed by ring closure with dilute alkali, to give (**61**); even $NaHCO_3$ is sufficient, providing especially mild conditions for nucleoside synthesis (see Section 22.2.1). Many other 7-, 8-, or 9-substituted purines have been prepared in this way. Other conditions for the synthesis of 2-unsubstituted or substituted purin-6-ones involve sodium ethoxide in ethanol, or a one-step synthesis with the latter and carboxylic or orthocarboxylic esters, or diethoxymethyl acetate. The last has recently been utilized in a synthesis of the C-2 glycoside (**62**).[53]

Other one-step syntheses involve the use of amides such as formamide or trifluoracetamide as solvent and reactant.

The insertion of a C=O group has been performed with aqueous KCNO, by fusion with urea, or, more successfully, especially with labile intermediates, using phosgene or one of its equivalents, such as *S*-phenylthiochloroformate,[54] cyclization being effected thermally, or with a base. Even urethane has recently been employed in an 8-alkylpurine synthesis, whilst another variant used polyhalo-compounds such as chloroform, carbon tetrachloride, or polychloro-ethylenes and -ethanes, with sodium alkoxides.[55]

Related cyclizations of 4-benzoylimidazoles, *e.g.* (**63**), lead to 6-phenylpurines.[56]

A 2-C=S group has usually been inserted using CS_2 in pyridine, or thio-acid derivatives such as potassium ethyl xanthate; thiourea has been rarely used. The use of alkyl or acyl isothiocyanates (cyclized in pyridine, or with barium hydroxide) also gives the 2-thiopurine *via* thioureido-intermediates, *e.g.* (**64**).

The ester (**65**) has been less widely employed, as it requires a two-step synthesis (*e.g.* diethoxymethyl acetate, followed by methanol/ammonia) with most reagents, but cyclization using cyanate or thiocyanate ion, or alkyl isothiocyanates, urea, and formamide have all been reported.

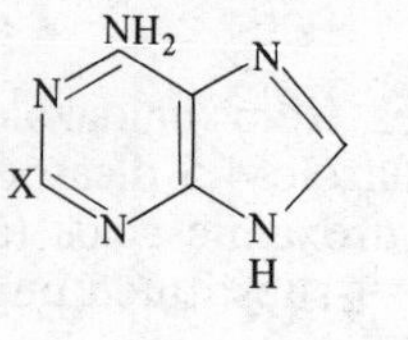

(62) X = β-D-ribofuranosyl 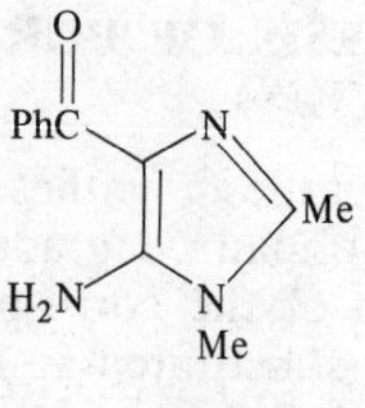 (63)

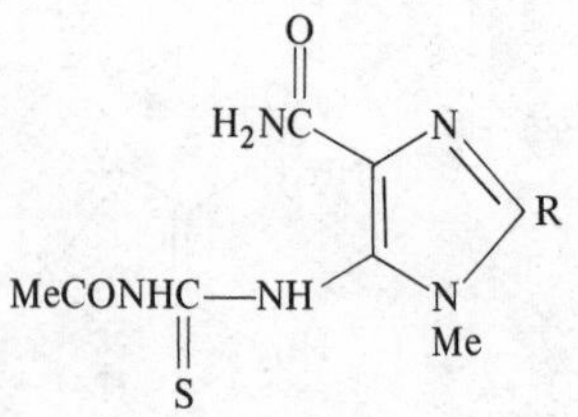

(64) 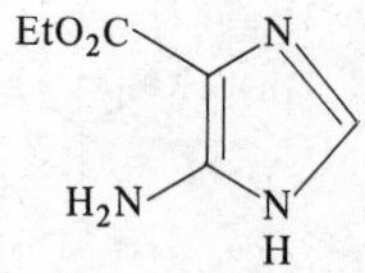 (65) 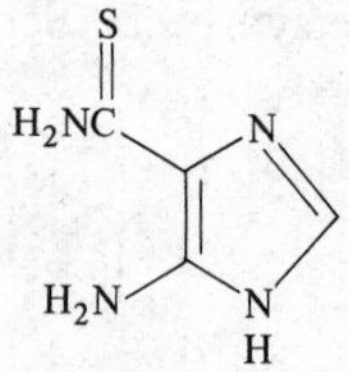 (66)

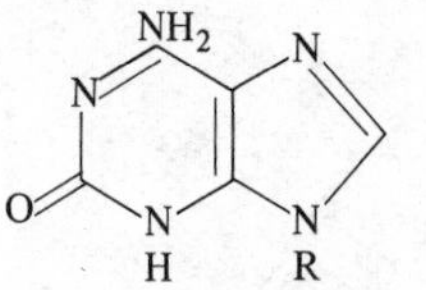

(67) 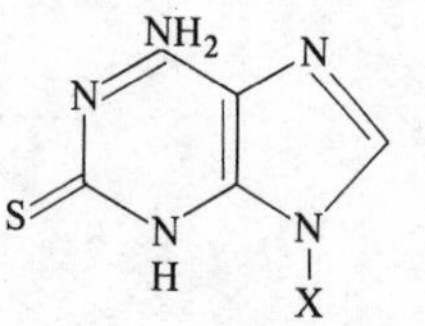 (68) X = β-D-ribofuranosyl

The corresponding 6-thiopurines have been prepared in analogous ways starting from the thioamide (**66**). Reagents, which seem to have been used to prepare mainly 2-unsubstituted 6-thio- or 2-oxo-6-thio-purines, include formic acid/acetic anhydride (followed by sodium bicarbonate), formamide, ethyl formate/sodium ethoxide, or urea.

The most widely-used imidazole intermediates for the preparation of the 6-aminopurines have been the amidine (**59**) and the nitrile (**60**). Adenine and its derivatives arise from the treatment of the amidine with formic acid/acetic anhydride, followed by aqueous alkali, the cyclization step being particularly rapid. 2-Substituted derivatives arise from the use of amides, especially trifluoracetamide, or (lower alkyl) from the use of triethyl orthoesters, or of aldehydes in oxidizing conditions.[57] The use of urea, or of phosgene in aqueous alkali, gives the important isoguanine derivatives (**67**), whilst recently the 2-thioadenosines, *e.g.* (**68**), have been prepared using carbon disulphide, alkyl isothiocyanates, or alkyl dithiocarbonates in DMF in the presence of potassium carbonate.

The nitrile (**60**) gives the adenine analogues on short refluxing of its acyl derivative (from orthoformate/acetic anhydride treatment) with methanolic ammonia. Alkylamines give 1-alkyladenines, whilst hydrazine leads to the 1-amino-derivative. Ethyl orthoformate in glycerol apparently leads directly to the purine, possibly *via* an imidate intermediate. Attempted utilization of the imidate (**69**) led to dimerization to the imidazolyladenosine (**70**).[58] Other 2-unsubstituted adenines have been prepared using formamide, formamidine acetate, formic acid/ammonium formate, or diethoxymethyl acetate/ammonia, whilst the sugar 1-imidate derivative (*e.g.* **71**), led to C-8 nucleosides.[59] Other 2-substituted adenines are obtained using nitriles followed by ammonia in methanol, whilst fusion with urea gives isoguanine derivatives, as with the amidine above. Guanidine either alone,[60] or in xylene,[61] gives 2,6-diaminopurines, whilst with CS_2/pyridine 6-thiopurines are obtained. Intermediates such as (**72**) are believed to be involved.[62] Triethyl orthoformate followed by sodium sulphide also gives 6-thiopurines, presumably *via* the thioamide, *e.g.* (**73**).[53,63]

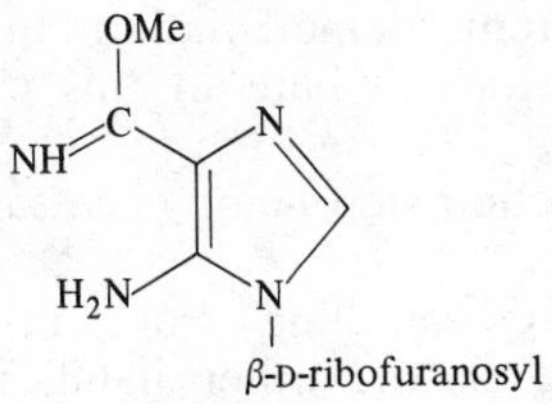

(69)

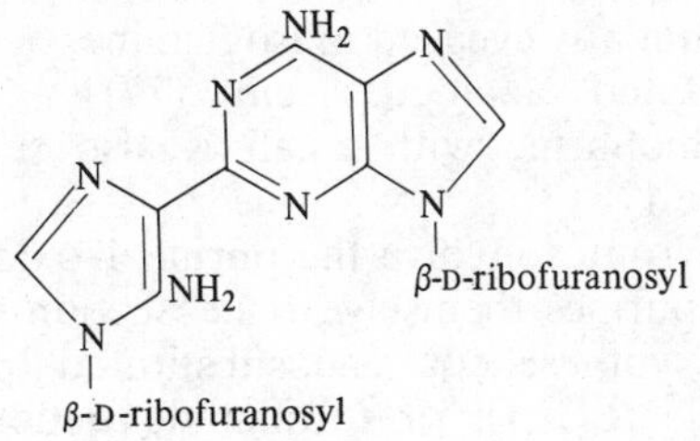

(70)

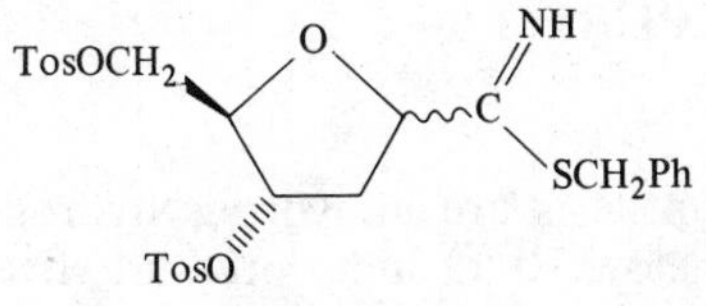

(71)

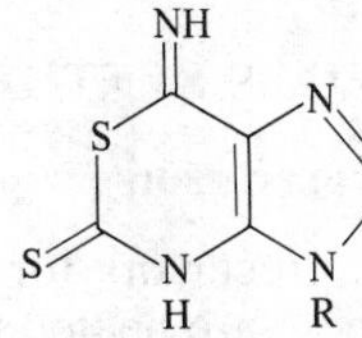

(72)

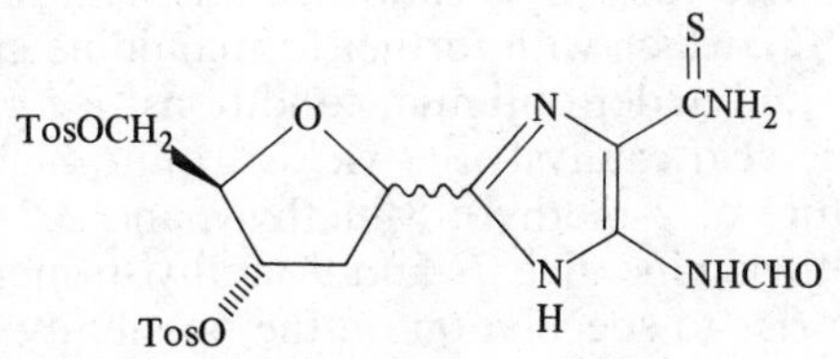

(73)

Other 5-cyano-4-substituted aminoimidazoles which have been cyclized include the urethane (**74**) (formic acid) and the ethoxymethylene derivatives (**75**) (ammonia/methanol).[64]

The imidazoles for these cyclizations can be either built up by well-known imidazole syntheses (see Chapter 17.3), isolated from fermentation mixtures involving bacteria such as *B. subtilis* (mutants), *S. olivaceous*, *etc.*, *en route* to the purines of which they are the biosynthetic precursors, or by degradation of existing purines (or nucleosides formed in the above bacterial fermentations). The last method can lead to the interconversion of one purine to another, exemplified by the degradation of adenine to the imidazole (**76**), which

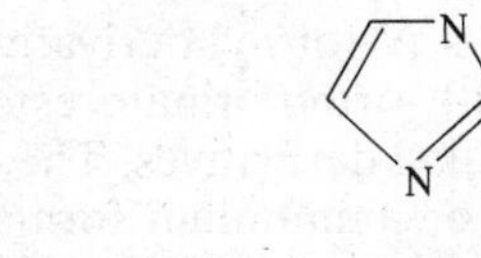

R = Tetraacetyl-β-D-glucopyranosyl

(74)

(75)

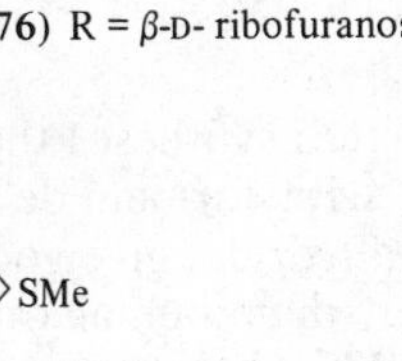

(76) R = β-D- ribofuranosyl

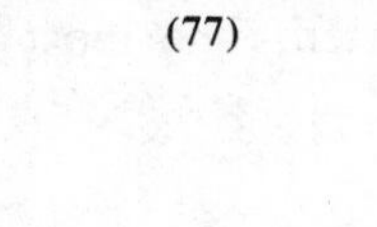

R = β-D-ribofuranosyl 3′, 5′-cyclic phosphate

(77)

(78)

is subsequently cyclized to isoguanine, or the conversion of 1,N^6-etheno-cyclic-AMP to its 2-substituted analogues *via* (**77**).[65] The apparent demethylation of 1-methyl-8-methylthiopurine with alkali is also related, the intermediate in this case being the aldehyde (**78**).[66]

Other routes involve the purine 1-oxides, a reduction step being necessary to regenerate the purines themselves (see Section 17.5.8).

The synthesis of 8-unsubstituted purines, *e.g.* xanthine from uric acid, with formamide/$POCl_3$, or formamide under pressure, on the other hand, involves prior opening of the imidazole ring, followed by recyclization to the new purine.

17.5.7 MISCELLANEOUS SYNTHESES OF PURINES

17.5.7.1 From acyclic precursors

The simplest acyclic precursors for purine synthesis are anhydrous hydrogen cyanide and ammonia. When these are heated together below 70 °C under pressure, five molecules of HCN combine in a series of steps involving addition and subsequent elimination of ammonia. The trimer of HCN (**79**) is first formed, followed by addition of two moles of ammonia to give the diamidine (**80**). Cyclization of this with formamidine (from HCN + NH_3) gives the imidazole (**81**), which with further formamidine gives adenine. A variety of conditions have been used, and under optimum conditions (*e.g.* HCN with ammonium salt in liquid ammonia + organic acid catalyst) the yield is quite high. From acetamidine and ammonia at 120 °C a mixture of 2-methyl-, 8-methyl-, and 2,8-dimethyl-adenine results, whilst liquid HCN and methylamine give 7- and 9-methyl-6-methylaminopurines.

These results have given rise to speculation on the possibility of similar reactions being involved in the prebiotic synthesis of purines in 'primitive Earth' conditions, in support of which are the discovery of purines in hydrolysates of the Murchison chondrite meteorite,[67] and the formation of purines when a methane/ammonia/water mixture is exposed to a high-frequency electric discharge.[68]

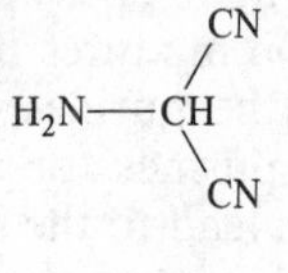

H_2N—CH(CN)—CN

(79)

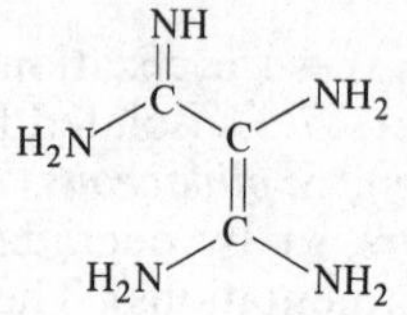

H_2N—C(=NH)—C(NH$_2$)=C(NH$_2$)—NH_2

(80)

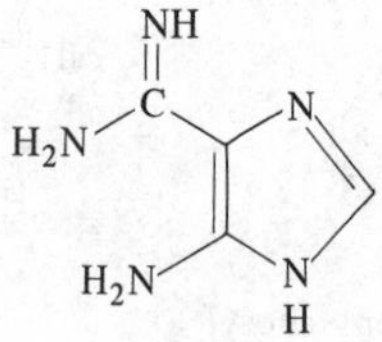

(81)

H_2N(NC)C=C(CN)NH_2

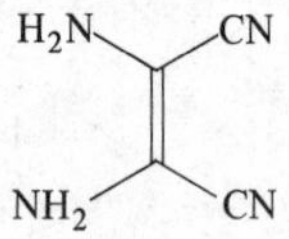

(82)

In related syntheses the aminomalonamide amidine (**80**), or its *N*-formyl derivative, on heating with formamide in acetonitrile or DMF, or better with orthoformate esters in acetic anhydride, give good yields of hypoxanthine or its 2,8-dialkyl derivatives. The HCN tetramer, diaminomaleonitrile (**82**), with formamidine acetate or ammonium formate at 130–140 °C gives low yields of adenine, and the photodecomposition of azidoacetonitriles to purines probably involves similar intermediates.[69] *N*-Cyanomethylphthalimide and related aminoacetonitriles with formamidine acetate or trisformamidomethane/formamide give moderate yields of adenines and 7-substituted adenines, whilst the formation of purine from aminoacetonitrile itself has been reported. The most successful one-step syntheses of this type involve the use of formamide. Poor to moderate yields of adenine are obtained alone or with ammonia, even at 180 °C under pressure, but higher yields result when phosphorus oxychloride is used at 120 °C, especially when an oxidizing agent such as picric acid or nitric acid/urea[70] are added. Formamide with alkyl halides or formates, or even alcohols at 170–200 °C, is claimed to give mixtures of alkylpurines. Ethyl orthoformate, or formamidine and ammonia, give only low yields even at 200 °C.

The above syntheses are believed to involve primary imidazole intermediates, but another related series of syntheses, starting from various derivatives of malonodinitrile or its close relatives, have been shown to go primarily *via* pyrimidine intermediates, followed by terminal imidazole ring-closure.[71] These syntheses involve the pressure hydrogenation (Pd/C or Raney nickel catalyst) of mixtures of the oximinomalonodinitrile or its potassium salt, or related arylazonitriles, esters, or amides with ammonia and formamide.[71,72] In favourable conditions moderate yields of adenine are obtained. Chemical reductions are also possible, and a good yield is claimed from the oximinodinitrile with sodium dithionite/ammonia at 170–190 °C. The use of guanidine carbonate or amidines with oximinomalononitrile gives 2,6-diaminopurines, or 2-substituted adenines, respectively.

Ethyl acetamidocyanoacetate with ethyl orthoformate–ammonium acetate gives a high yield of hypoxanthine, whilst phenylazocyanoacetylurea gives xanthine and guanine.[72] The use of the corresponding *N'*-methylurea gives 3-methylxanthine.

17.5.7.2 From other ring systems

The aminofurazancarboxamidoxime (**83**), obtained from malononitrile with nitrous acid followed by hydroxylamine, gives a moderate yield of adenine on reduction with formic acid and Raney nickel or zinc dust.[73] A related compound, the 7-aminofurazano[3,4-*d*]-pyrimidine (**85**), obtained on lead tetra-acetate oxidation of the 2-substituted 4,6-diamino-5-nitrosopyrimidine (**84**), on treatment with alkylamines followed by acylation and reductive cleavage gives 2-, 8-, and/or 9-substituted adenines.[74]

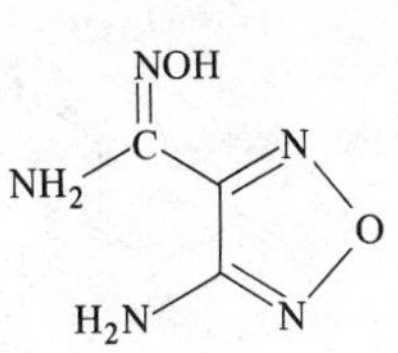

NOH; NH$_2$; C; N; O; H$_2$N; N

(83)

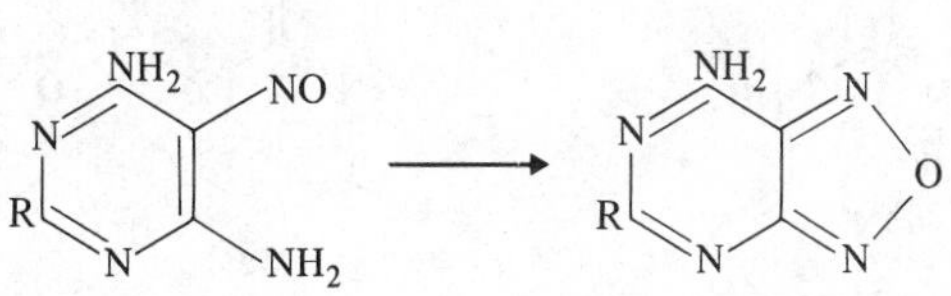

NH$_2$; NO; N; R; N; NH$_2$

(84)

NH$_2$; N; N; O; R; N; N

(85)

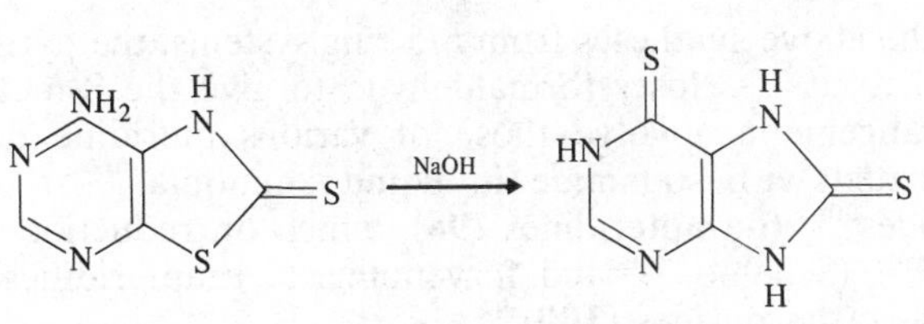

NH$_2$; H; N; N; S; N; S; NaOH

(86)

S; H; N; HN; S; N; N; H

(87)

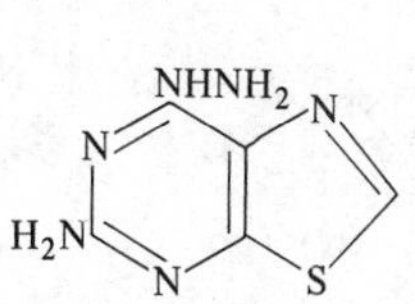

NHNH$_2$; N; N; H$_2$N; N; S

(88)

S; N; HN; H$_2$N; N; N; NH$_2$

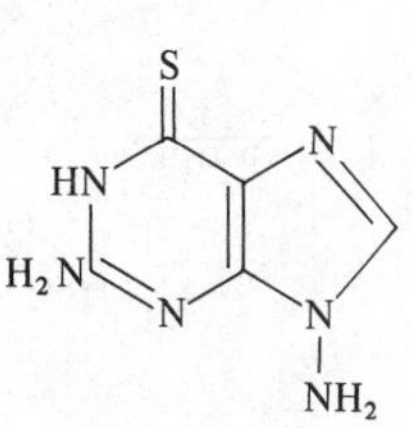

(89)

The 2-thio-7-aminothiazolo[5,4-*d*]pyrimidines (**86**), obtained initially on reaction of 4,5-diamino-6-chloropyrimidines with carbon disulphide, rearrange in hot dilute sodium hydroxide solution to give 6,8-dithiopurines (**87**). The corresponding 2-amino-derivatives give 8-amino-6-thiopurines *via* ring-opening and subsequent recyclization. An attempt to prepare a 7-azapteridine *via* the thiazolo[5,4-*d*]pyrimidine (**88**) gave instead the 9-aminopurine (**89**).[75] Some 2-substituted oxazolo[5,4-*d*]pyrimidines (**90**) on treatment with amines give 9-substituted purines (**91**). With aqueous alkali, or on heating in DMF or excess of amine, the 7-amino-oxazolo[5,4-*d*]pyrimidines, *e.g.* (**92**), (from 4-cyano-5-ethoxymethyleneamino-oxazoles with ammonia or amines) rearrange to the hypoxanthines (**93**) or their 9-substituted derivatives.[76] With a 5-alkoxy-substituent the reaction takes a different course, giving 3-substituted or 3,8-disubstituted xanthines, *e.g.* (**94**).[77]

The chlorotetrazolo[1,5-*c*]pyrimidine (**95**) with refluxing diethoxymethyl acetate gives the 6-chloro-8-ethoxypurine *via* azide–tetrazole tautomerism, followed by loss of N_2 to give a nitrene intermediate, and cyclization.[78]

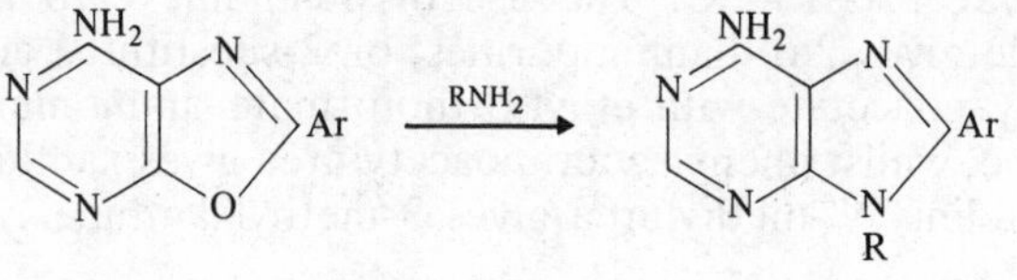

(90) (91)

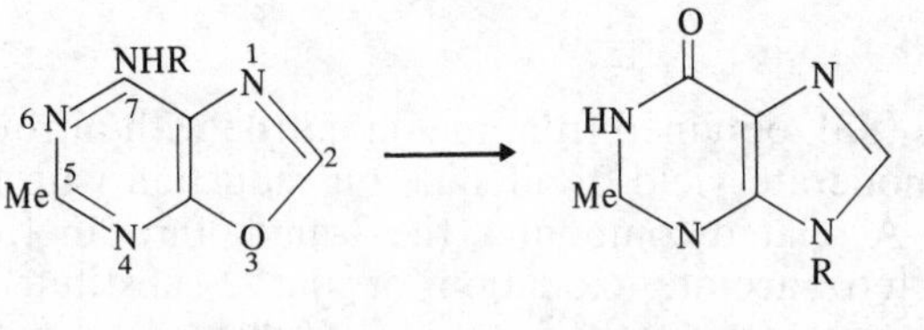

(92) (93) (94)

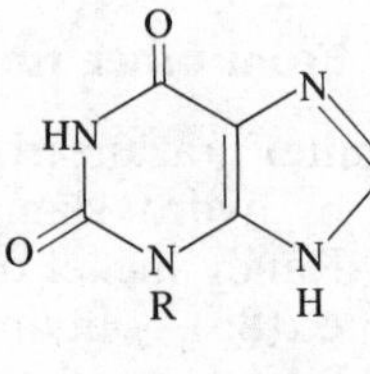

R = Me, Et, CH_2Ph, *etc.*

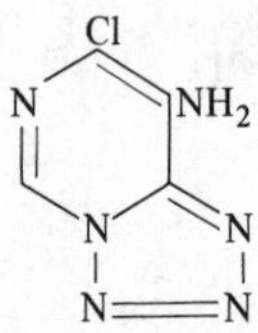

(95) 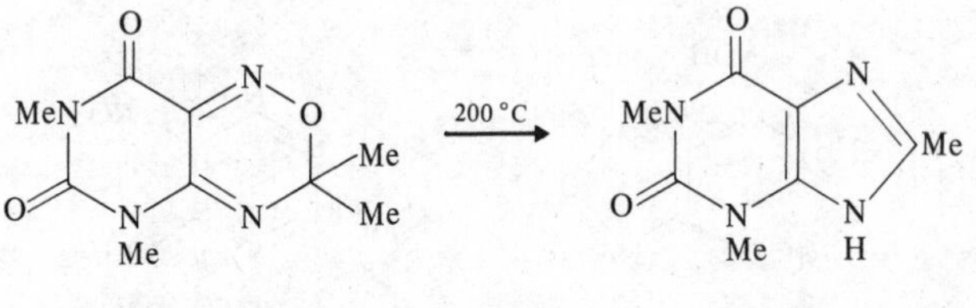 (96) (97)

In addition to the above syntheses from 6/5 ring systems, the pyrimido[5,4-*c*]oxadiazine (**96**) when heated to 200 °C loses formaldehyde to give the 8-methylpurine (**97**). Other 6/6 system rearrangements involve those of various pteridine derivatives, which give purines when treated with sodamide in liquid ammonia,[79] or on refluxing in acetic anhydride (*N*-oxides[80]), 6-azapteridines (**98**), which on reduction give the 8-substituted theophyllines (**99**)[81] (Scheme 7), and fervenulins (7-azapteridines) which on treatment with formamide give the purines (**100**).[82]

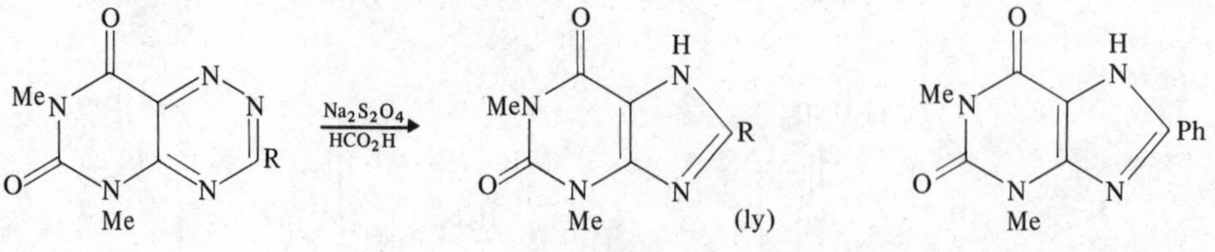

(98) (99) (100)

SCHEME 7

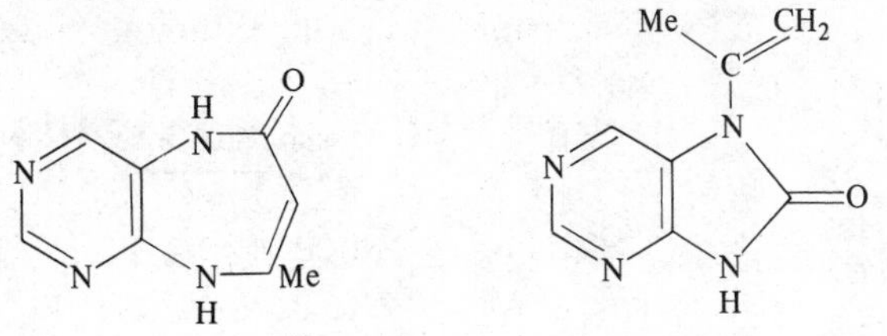

(101) (102)

Finally, the (6/7) diazepinones (**101**) with alkoxide give (**102**),[83] whilst the (5/5) tricyanoimidazo[1,5-*b*]-*sym*-triazole, obtained from cyanogen and cyanide ion, on methylation and alkaline hydrolysis/decarboxylation gave the amide (**103**), which rearranged to 3-methylguanine (**104**) (Scheme 8).[84]

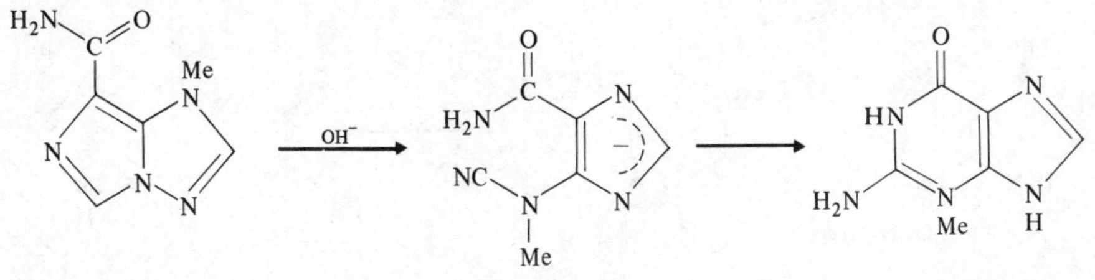

(103) (104)

SCHEME 8

17.5.7.3 By degradation of nucleosides

An increasing source of purine analogues is the hydrolysis of nucleosides and nucleotides, either obtained from natural sources or, more recently, isolated by modern separative techniques from fermentation reactions cultured with a wide variety of mutant species of *Bacillus*, *Brevibacterium*, *Corynebacterium*, *Arthrobacter*, *etc.* (see Chapter 22.2).

Although alkaline conditions, for example pH 10 at >100 °C,[85] have been successfully used in the above hydrolyses, acid conditions are usually employed to avoid ring-opening of the bases. Those recently recommended include hot hydrochloric acid (preferably 0.1 N at 80 °C),[86] with or without a thallium sulphate catalyst, hot 72% perchloric acid,[87] anhydrous hydrogen fluoride (preferably with a trace of HCl catalyst),[88] cation exchange resins,[89] or chloromethylenedimethylammonium chloride in DMF.[90]

17.6.8 THE PURINE *N*-OXIDES AND *N*-AMINOPURINES

Purine *N*-oxides have become of great interest in recent years because of their powerful oncogenic nature, in particular of the 3- and 7-oxides and their derivatives. The 1-oxides are not active in this respect.

The synthesis of the 1-oxides is generally by oxidation of the corresponding purines with peracids, or by simple transformations of those readily formed in this way, such as adenine 1-oxide. Cyclization of imidazole intermediates such as (**105**)[91] has also been used. The oncogenic 3-oxides and their *N*-alkyl derivatives[92] have been extensively investigated by Brown and his co-workers.[7,93] Guanine with pertrifluoroacetic acid gives the 3-oxide, converted on acid hydrolysis to xanthine 3-oxide. The oncogenic 7-oxides[84] have usually been prepared by cyclization of nitropyrimidines with aldehydes[95] or anils, or simply by oxidative cyclization of 4-alkylaminouracils, *e.g.* the formation of (**106**) (Scheme 9).[96] 9-Oxides have recently been prepared by cyclization of imidazole *N*-oxide intermediates, *e.g.* (**107**).[97]

(105) R = β-D-ribofuranosyl

KNO$_3$, HOAc, H_2SO_4, 70 °C

(106)

SCHEME 9

The *N*-oxides and their derivatives undergo several interesting rearrangements. In the presence of acylating agents (*e.g.* acetic acid or acetic anhydride) the *N*-oxygen equivalent is transferred either to an adjacent carbon atom, as for (**108**) to (**109**),[98] to one remote, as

(107) (108) Ac_2O (109)

in the purine system (**110**) to (**113**), or to an exocyclic alkyl group. The mechanism for these reactions has been extensively investigated[7] for the xanthine 3-oxides and shown to involve the mesomeric nitrenium and carbonium ions (**111**) and (**112**).

In the presence of other nucleophiles such as nitrite, chloride, pyridine, or methionine (MeS^-), 8-substituted xanthines are formed.[7] A similar pyridinium salt is formed at the 2-position using pyridine with hypoxanthine 1-oxide, and gives guanine (**3**) on alkaline hydrolysis. These 2- or 8-substitution processes are related to those observed with *N*-oxides and phosphorus halides, and to certain rearrangements of *N*-alkoxypurines observed during alkylations of purine oxides. By concomitant homolytic fission, a radical

(110) (111) (112) (113)

is formed, which undergoes reduction or polymerization, and may be responsible for the biological activity. However, the attack at the 8-position by tryptophan and tyrosine, and by methionine, paralleling the behaviour of other carcinogenic compounds such as *N*-acetoxy-2-acetamidofluorene,[7,100] suggests other pathways.

Recently *N*-methyl migrations in methylated *N*-oxides, such as (**114**) to (**115**), have been observed by Japanese workers,[101] and the kinetics and mechanism have been extensively studied. Alkylation of added nucleophiles may also occur.[102]

N-Aminopurines have recently been prepared by amination with hydroxylamine-*O*-sulphonic acid[103] or its equivalents, the Sheradsky[103,104] and Krause reagents.[105] They are also available by synthetic routes involving cyclizations of hydrazinopyrimidines, *e.g.* (**116**) to (**117**), and by a one-step cyclization (my) from formamide at 160–200 °C in the absence of oxygen.

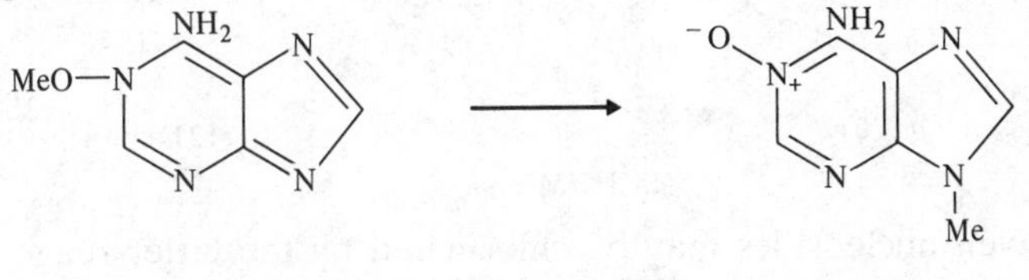

(**114**) (**115**)

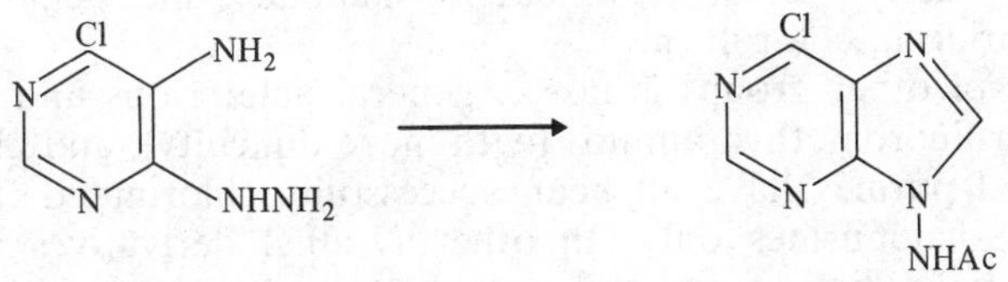

(**116**) (**117**)

17.5.9 INTERCONVERSION OF PREFORMED PURINE DERIVATIVES BY NUCLEOPHILIC SUBSTITUTION

Apart from the preparations from various precursors outlined in the previous sections, the other main source of any desired purine derivatives is by interconversion starting from those initially prepared.

These interconversions may be *via* alkylation or other forms of electrophilic substitution (Section 17.5.10), or by rearrangement (Section 17.5.13), but are most often performed by nucleophilic interconversions of oxo-, halo-, thio- (or alkylthio-), and amino-purines.

17.5.9.1 From oxopurines

By far the most important methods of interconversion starting from oxopurines begin with their prior transformation to the corresponding halo-derivatives, *e.g.* (**118**) to (**119**), of which the chloro-derivatives are by far the most important,[107] owing both to the ease of replacement of the chlorine atom by a wide variety of other groups and to the convenient differential reactivity of chlorine groups in the 2-, 6-, and 8-positions of the purine ring.

This transformation is normally carried out using a phosphorus halide, such as phosphorus oxychloride, or a mixture of this with phosphorus pentachloride, either alone or in an inert solvent such as toluene. This latter mixture may cause nuclear chlorination as a side-reaction (see Section 17.5.10). Tertiary aliphatic or aromatic amines (such as dimethylaniline) are often used as catalysts and/or acid scavengers, although in this case some side reactions, involving replacements of halides by amine fragments, *e.g.* (**120**) to (**121**) (Scheme 9a), or complex dealkylative rearrangements of alkyl derivatives,[108] have

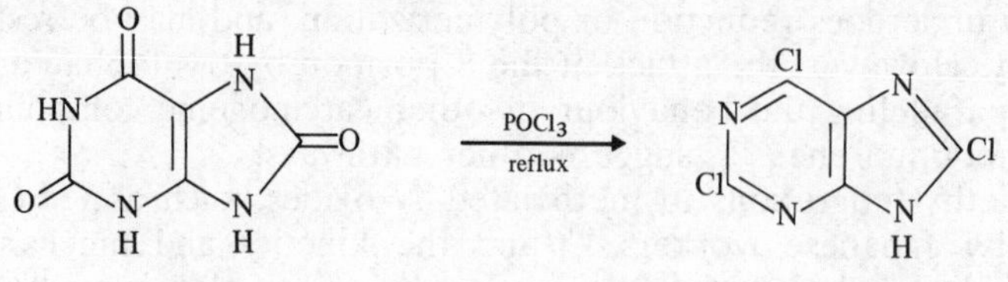

(118) $\xrightarrow[\text{reflux}]{POCl_3}$ (119)

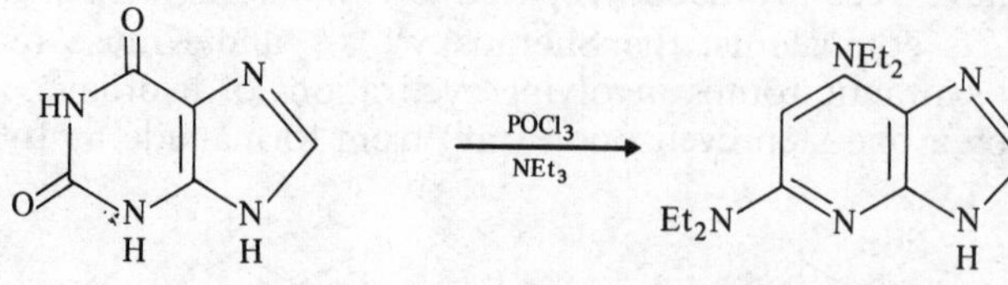

(120) $\xrightarrow[NEt_3]{POCl_3}$ (121)

SCHEME 9a

been reported. Even nucleosides may be chlorinated by the latter reagent, provided the sugar hydroxy-groups are protected as acetoxy- or benzoyloxy-groups. Other reagents used include thionyl chloride/dimethylformamide and chloromethylenedimethyl-ammonium chloride in chloroform.

The presence of other groups is not in general deleterious and alkyl (including some *N*-alkyl), aryl, trifluoromethyl, amino- (with more difficulty), and alkylthio- (but not free thio-) substituted purines have all been successfully chlorinated. Methylated uric acids give 2,6-dioxo-8-halopurines only. In other *N*-alkyl derivatives where the oxo-group cannot exist in the hydroxy-form, the methyl groups are often removed and the dealky-lated halo-compound results, *e.g.* (**122**) gives (**123**). These demethylations are believed to occur by successive chlorinations of the methyl group prior to its displacement. Oxo-groups cannot be easily converted to fluoro-groups, whilst 8-bromopurines are more usually prepared by electrophilic substitution.

The other valuable transformation of 6-oxopurines is their conversion to the corres-ponding thio-derivatives by phosphorus pentasulphide in inert solvents such as pyridine. 2- and 8-oxo-groups are usually inert, and the preparation of polythio-compounds is usually carried out from mixed 6-oxo-thiopurines. Again, aminopurines react more slowly.

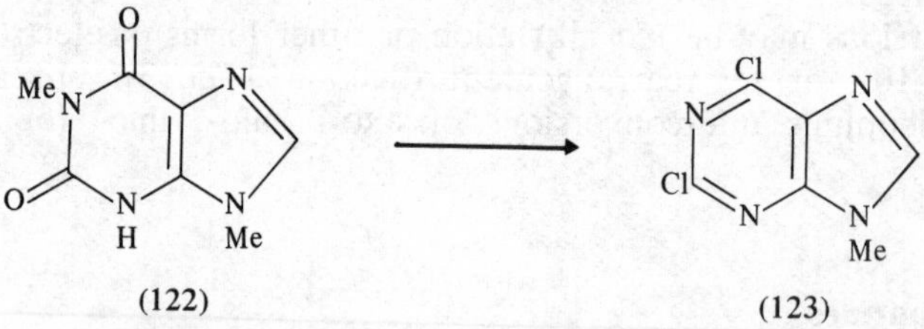

(122) → (123)

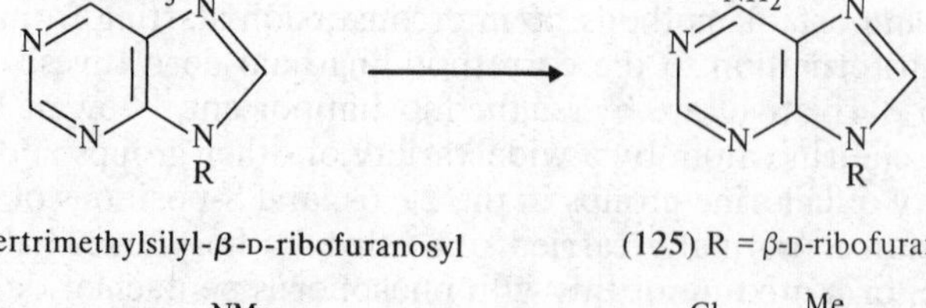

(**124**) R = pertrimethylsilyl-β-D-ribofuranosyl → (125) R = β-D-ribofuranosyl

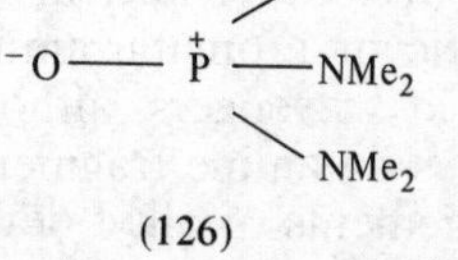

(126) 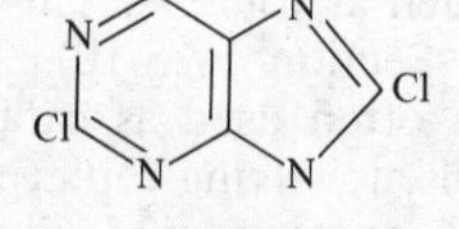 (127)

Conversions of oxo-groups to amino-derivatives *via* mercury-catalysed reactions of *O*-trimethylsilyl derivatives, *e.g.* (**124**) → (**125**), have been reported,[109] and hypoxanthines have been converted directly to aminopurines (hy) using phosphoramides such as (**126**) at 225–230 °C.[110]

17.5.9.2 From halopurines

The overriding importance of the chloropurines mentioned above is due both to the ease of replacement of the chlorine atom by a wide variety of other groups, and to the convenient differential reactivity of chlorine groups in the 2-, 6-, and 8-positions of the purine ring. In *N*-unsubstituted purines this order of reactivity to hydroxide ion or amines has been shown in kinetic studies to be $6>2>8$. This is different from the order predicted by the Fukui FMO superdelocalizability theory, which has been so successful with methylthiopurines,[6] but this is probably due to the halopurines reacting as anions in these solutions. In the case of 7- and 9-alkyl-substituted purines, such as (**127**), where the anion is not formed, the initial attack is at C-8, in agreement with that found for the methylthiopurines,[6] which also react as neutral species.

Halogen atoms, especially in the 6- and 8-positions, have been replaced by hydrogen, leading to the corresponding unsubstituted purines, formerly by use of hydriodic acid/phosphorus reagents but now usually by catalytic hydrogenation with Pd catalysts in the presence of an acid scavenger, such as magnesium oxide. The use of deuterium or tritium in place of hydrogen leads to the corresponding labelled derivatives,[112] which are also prepared by base-catalysed hydrogen exchange at the 8-position, which proceeds *via* a proton abstraction mechanism.

Replacement of chlorine by hydroxyl proceeds with ease in alkaline solution only if other activating groups are present in the molecule; ring-opening of the imidazole ring following hydroxyl attack at C-8 can complicate the reaction, especially when 7- or 9-alkyl groups are present.

Aminochloropurines are particularly inert towards sodium hydroxide; 2-amino-6-chloropurine, for instance, being recovered largely unchanged from boiling 0.1 N sodium hydroxide solution.[113]

Hydrolyses with hydrochloric acid are preferable in many cases, especially for 7- or 9-alkyl derivatives. Refluxing dilute acid is often sufficient, but 30% hydrochloric acid has been used in difficult cases, although these strong acid conditions may cause rearrangements (see Section 17.5.13). A different order of reactivity of polyhalopurines may result from cation formation.

Replacement of chlorine with alkoxy-groups proceeds normally by heating with alkali-metal alkoxides. N^7,N^9-Diunsubstituted derivatives may be fairly resistant to attack owing to Coulombic repulsion by their anions. A recent paper uses antimony chloride followed by ethanol.[114]

Chlorine can also be replaced by alkyl- or aryl-thio-groups, using the alkali-metal mercaptides; a large number of 6-*S*-substituted purines having been prepared in this way, especially following the use of 6-mercaptopurine as a powerful anti-tumour agent, including the immunosuppressive drug azathioprine (**10**) itself.

The replacement of chlorine with a thio-group may be carried out successfully with sodium or potassium hydrosulphide solutions, usually prepared *in situ* in ethanol, but a method often employed is heating with thiourea[115] or thioacetamide[116] in ethanol, the intermediate thiouronium salts undergoing either spontaneous or alkali-induced decomposition. Side reactions may occur but often lead to easily separated non-acidic by-products.[117] Recently, similar reactions using the corresponding selenium intermediates have become important.[118]

The most widely used reaction of chloropurines has undoubtedly been that with amines, leading to a wide variety of purines of synthetic and, particularly, biological importance such as kinetin (**128**), (*E*)- and (*Z*)-zeatin (**129**), *N*-benzyladenine (**130**), the base of puromycin (**131**), and other similar substituted adenines.

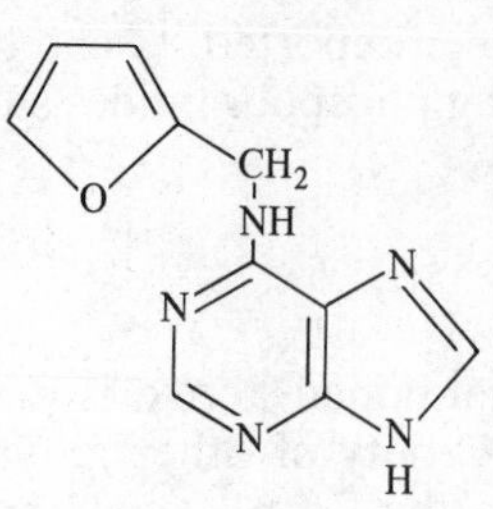

(128)

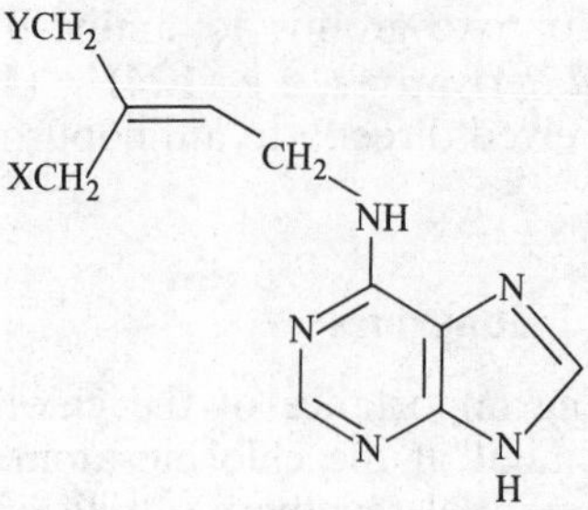

(129) a; X = H Y = OH
b; X = OH Y = H

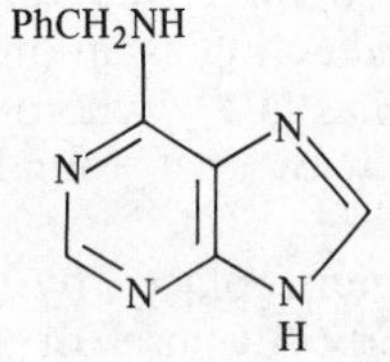

(130)

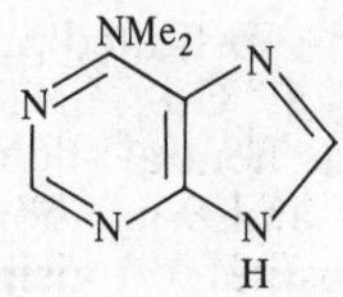

(131)

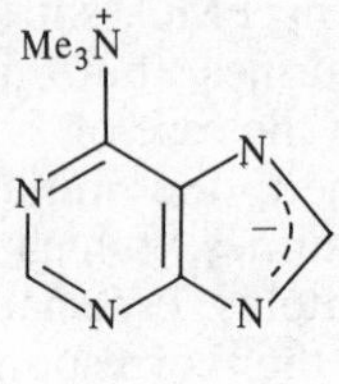

(132)

The use of ammonia itself to give adenine and its congeners may require high temperatures and pressure conditions, especially in the introduction of a second amino-group in dichloro-derivatives, but concentrated aqueous ammonia or ethanolic ammonia at 100 °C may suffice if activating (*e.g.* alkyl) groups are present. Stepwise replacement of multiple chlorine atoms is usually possible by careful selection of conditions.

Primary and secondary amines usually give replacement without difficulty at atmospheric pressure, usually below 100 °C. For example, puromycin base (**131**) is formed with methanolic dimethylamine from the corresponding halopurine.[119]

Tertiary amines, such as trimethylamine or pyridine, react in the cold to give the quaternary ammonium derivatives, which give stable betaines (**132**). These may themselves be used in nucleophilic displacement reactions (see below).

Among other groups which have been inserted into purines by replacement of chlorine are fluoro (using silver fluoride in toluene or xylene), hydroxylamino and alkoxyamino, hydrazino and substituted hydrazino, azido (a convenient route to aminopurines by reduction),[120] sulphonic acid (using aqueous sodium sulphite), thiocyanato, diethyl malonate (followed by hydrolysis/decarboxylation to give alkyl purines, also prepared by use of Wittig reagents[121]), and aryl (using aryl-lithiums).[122]

17.5.9.3 From thiopurines

The replacement of a thio-group by hydrogen has been one of the most widely used purine reactions, being often used for final removal of sulphur after other transformations have been conveniently effected due to the presence of the thio- or alkylthio-group. The usual means is by Raney nickel in aqueous, aqueous alkaline, or ethanolic solution. The reaction is effective (hy) in the presence of alkyl, hydroxy-, alkoxy-, or amino-groups, only halothiopurines proving unsuccessful. In the case of *N*-alkylated purines, reduction to dihydro-derivatives can be effected.

8-Thio- and in some cases 2-thio-groups may also be removed oxidatively, either with nitrous acid, nitric acid, acid hydrogen peroxide (*via* the 8-SO_2H derivative),[123] or iodine/sodium iodide, depending on the other groups present. Oxidative conversion of a thio-group to chloro was formerly carried out using chlorine in ethanol or ethanolic HCl. This has recently been shown to be dangerous and acetonitrile or hydrochloric acid are proposed as alternative solvents.[124] The alkylthio-derivatives (readily obtained from the thio-compounds by alkylation at low temperature), are more suitable for this reaction, especially the methylthio-derivatives. The preparation of bromopurines by a modification of these procedures often provides the best route[125] to these compounds and 8-iodopurines have been obtained using iodine. Thiopurines are not, in general, suitable for conversion direct to oxo- and amino-derivatives, as the forcing conditions required bring about ring-opening and/or Dimroth-type rearrangements. The methylthio-derivatives, prepared as above, or specifically using dialkyl disulphide–Bu_3P, are suitable, however, and their reactions have been widely studied in recent years. Kinetic studies have shown the order of replacement to be $8>6>2$, as predicted by FMO theory, except where steric considerations intervene,[6] and recently reactions with methoxide have been studied.[127]

Practical examples of the replacement of —SMe groups with amines, especially in the 6-position, abound in the literature. A recent report recommends the use of mercuric chloride in the reaction of weak nucleophiles such as anilines with methylthiopurines.[128] Furthermore, mild oxidation with hydrogen peroxide/acetic acid, chlorine/aqueous methanol, or $KMnO_4$ gives the methanesulphonyl derivatives, whose reactivity with nucleophiles is equal to or greater than that of the corresponding chloro-compounds.[129]

The replacement of a thio-group by a seleno-group, using hydrogen selenide in aqueous solution, has become important, especially for 6-seleno-analogues of the thiopurine drugs. Deselenation with Raney nickel, too, is much easier than desulphurization. A recent paper reports preparation of mixed 6-purinyl disulphides, *e.g.* (**133**), using the reagent (**134**).[130]

BuS—S

N N N N H

(133)

BuS—S—C(=NH)—NH_2

(134)

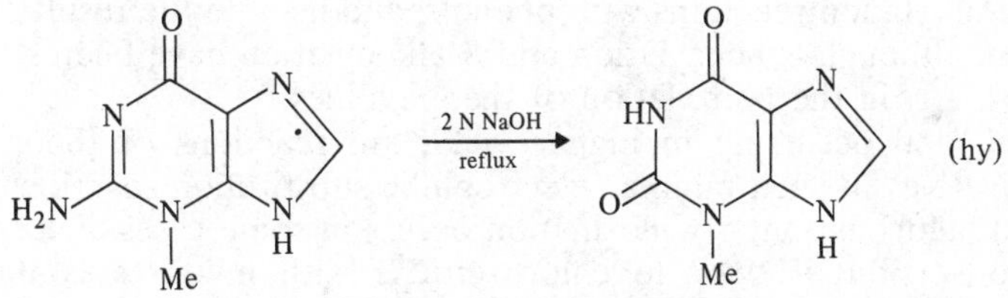

(135) H$_2$N, N, N, N, N, H, Me, O —— 2 N NaOH, reflux ——> (136) (hy)

SCHEME 8b

17.5.9.4 From aminopurines

The important displacement reactions of aminopurines are few, because in so many cases the aminopurine is the 'end of the line'. The conversion of amino to oxo is in many cases effected by warming with nitrous acid, especially in the case of adenine and its simple alkyl derivatives. When other oxo-groups are already present, however, the amino-group is inert to nitrous acid. Recently, nitrosyl chloride/pyridine in DMF has been described as a powerful new reagent for this transformation.[131] Isoguanines, which are inert to nitrous acid, may be converted to oxo-derivatives by heating with strong mineral acids. *N*-Alkylation activates amino-groups to hydroxide-ion substitution; for example, (**135**) gives (**136**) (Scheme 8b).

The modified Schiemann reaction[132] of amines leads to fluoropurines, using sodium nitrite/48% HBF_4, in low to moderate yields. The corresponding reaction using concentrated HCl does not seem to be of much use, but in the nucleoside series a good yield of (**137**) was obtained.

(137) R = β-D-ribofuranosyl

Finally, the 6-trimethylammoniopurines mentioned above, although they have a tendency to undergo alkyl migration[133] to 9-alkyl-6-dimethylaminopurines, possess great facility for nucleophilic displacement[134] which has only recently begun to be explored. The facile production of 6-fluoropurine by the reaction of the quaternized product with KHF_2 in ethanol at 50 °C,[135] or in DMF,[136] is evidence of the great potential of this method, as fluoropurines are very reactive in nucleophilic substitutions.[136]

A very large number of the displacements described above may also be carried out on the corresponding nucleosides, excluding those, of course, where the conditions (*e.g.* strongly acidic or basic media) would lead to fission of the glycosidic linkage. The presence of the nucleoside appears to have no consistent effect on the ease of the nucleophilic substitutions. Activation, especially of chloro-groups, is achieved on conversion of purines to their *N*-oxides, but ring-opening side-reactions also become more facile (see Section 17.5.13).

17.5.10 INTERCONVERSION OF PREFORMED PURINE DERIVATIVES BY ELECTROPHILIC SUBSTITUTION

17.5.10.1 At carbon

Electrophilic substitution at purine carbon occurs only in those derivatives with many activating substituents, such as xanthines and methylated xanthines, and then only at the 8-position. Alkylation has generally been observed with groups, such as but-2-enyl,[137] which give *C*-alkylation reactions with phenols, and may be the result of a rearrangement from nitrogen, although 8-benzylation and 8-alkenylation have been reported to occur as side-reactions, *e.g.* in the benzylation of theophyllines.[138]

More usual, and occurring in higher yield, are reactions of theophylline and theobromine derivatives in such familiar electrophilic substitution reactions as nitration, azo-coupling, and halogenation. 8-Chlorination occurs in some cases as a side reaction in the attempted conversion of oxo- to chloro-groups with mixtures containing phosphorus pentachloride, but is best performed with gaseous chlorine in an inert solvent such as chloroform or acetic acid. Bromination[139] has been widely studied, and occurs readily with adenine, guanine, hypoxanthine, xanthine, and methylated xanthines, giving the 8-bromo-analogues, which provide convenient precursors for many 8-amino-derivatives of these compounds.

Azo-coupling occurs only with the most activated substrates such as xanthine and its derivatives, whilst nitration, although failing with xanthine itself, gives good yields with methylxanthine derivatives, the preferred medium being acetic acid. An apparent displacement of an 8-thio-group by a nitro-group[140] is believed to occur *via* prior oxidative removal of the sulphur followed by nitration. The 8-nitro-groups themselves may undergo nucleophilic displacement by such reagents as hydrochloric and hydrobromic acids, or by pyridine, giving the salt (**138**), by a dimethylamino-group using toluene-*p*-sulphonyl chloride in DMF, or by fluorine using liquid hydrogen fluoride. Finally, purine itself gives 8-thiopurine (hy) on fusion with sulphur at 245 °C.[141]

17.5.10.2 At oxygen

O-Alkylation is very rare in the purine field, the best-known examples being in the alkylation of *N*-methyl derivatives of uric acid with diazomethane, or of their silver salts with methyl iodide. Even here the picture is not consistent, as uric acid itself with diazomethane gives 1,3,7,9-tetramethyluric acid, although this, of course, could result from the rearrangement of initially formed *O*-alkyl derivatives. The complex mixture of isomers formed has recently been studied by g.l.c.–m.s.[142,143]

O-Acylation is also very rare in the purine field, uric acid again providing an example: (**139**) is formed from 1,7-dimethyluric acid with refluxing acetic anhydride.

Trimethylsilylation, or triethylsilylation, is useful for derivatizing oxopurines for g.l.c. analysis, *e.g.* xanthine gives (**140**).

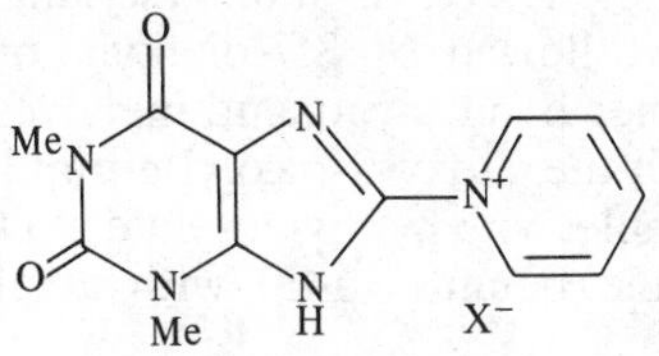

(138) 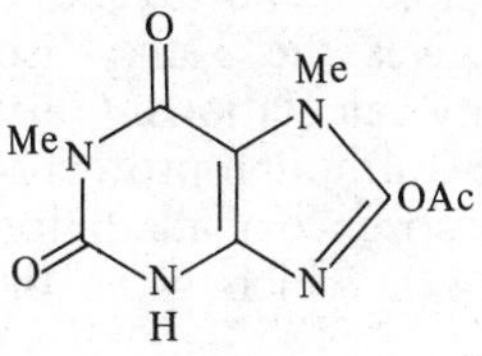 (139) 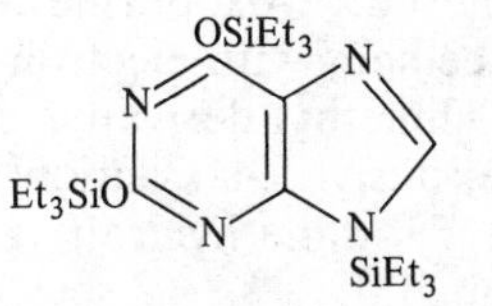 (140)

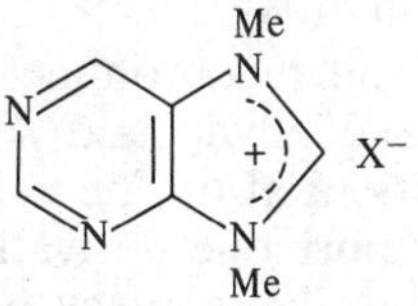

(141) 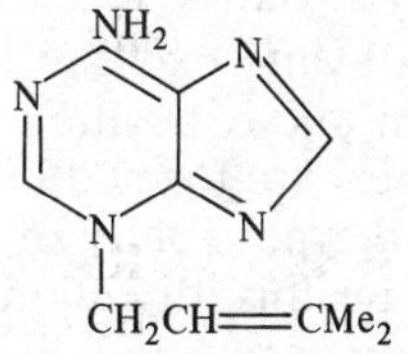 (142) 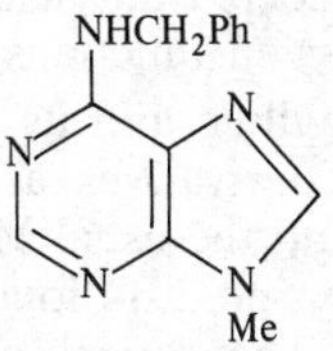 (143)

17.5.10.3 At nitrogen

Purine itself gives only 9-alkyl derivatives, *e.g.* with dimethyl sulphate/alkali, diazomethane, or using the silver or thallium salt with methyl iodide in DMF. An excess of methyl iodide in DMF gives the 7,9-dimethylpurinium salt (**141**), whilst 6-methylpurine gives 6,9-dimethylpurine as the main product.

With other purines different patterns are observed, depending on whether the anion (sodium salt) or neutral base in aprotic solvent is used for alkylation.

With adenine anions a mixture of 3- and 9-alkyl derivatives is observed. With few exceptions[144] the 9-isomer predominates, as in the synthesis of eritadenine (**7**) and its analogues. With the free adenine in DMF the 3-alkylation product predominates, *e.g.* in the preparation of the natural product, triacanthine (**142**), or using neat butyrolactone.[145] Other neat alkylating agents such as ethylene carbonate,[146] trimethyl phosphate,[147] or epichlorohydrin[148] gave similar results.

N^6-Benzoylation or -acetylation do not seem to affect the picture, although some anomalies have been observed,[149] and the derivatives are of course well-known to give good yields of 9-glycosides (see Chapter 22.2). At high temperatures, alkyl migrations may occur (*cf.* Section 17.5.13). A selective methylation of N^6-benzyladenine using dimethyloxosulphonium methylide gave (hy) 9-methyl-N^6-benzyladenine (**143**).[150] Reactions with acrylonitrile, or ethyl acrylate and sodium methoxide, give 9-adducts, as does dihydropyran in acid- or mercuric chloride-catalysed reactions.

In contrast, alkylation of guanine is much more complex. Methyl sulphate and alkali gives a mixture of 9-, 7-, and 3-methyl derivatives, together with a little 2-methylamino-4-oxopurine,* and most monoalkylguanines are best prepared by indirect methods, although fusion with tetramethylammonium hydroxide at 260 °C is said to give only the 9-methyl derivative. When held in place in the Watson–Crick double helix in nucleic

* *Cf.* Ref. 151.

acids, both *in vitro* and *in vivo*, guanine appears to be alkylated predominantly in the 7-position by such biological alkylating agents as *N*-nitroso-*N*-ethylurea,[152] 1-phenyl-3,3-dimethyltriazine,[153] ethyl mesylate,[152] *etc.* Similar results are obtained using certain cobalt complexes, which attach to the 9-position of xanthines, thus causing alkylation at N-7,[155] and the use of thallium salts has also been recommended for 9-substitution.[156] The use of directing groups in nucleoside synthesis is, of course, well-known (see Chapter 22.2).

A convenient permethylation for g.l.c. derivatization may be performed with an excess of NaOMe/MeI in DMSO.

Mono-oxopurines, for example, tend to be dialkylated, but if monoalkylation is achieved, 2-oxopurine anion gives the 3-alkyl isomer, followed by 3,7-dialkylation, in agreement with electron-density calculations. On the other hand, hypoxanthine gives the 1,7-dimethyl derivative directly, although protection with the 1-oxy-2-picolyl group leads to exclusive 7-alkylation. In the absence of alkali, for example, with methyl tosylate in DMF, the 7,9-dimethyl salt, *e.g.* (**144**), which gives the stable betaine (**145**) with alkali, is formed.

Xanthine anions, on the other hand, are alkylated in the order N-3, N-7, N-1, in the order of decreasing acidity, methylation giving first theobromine then caffeine. Blocking of the oxygen functions by trimethylsilylation leads to alkylation at N-7.[157]

Theophylline and its analogues alkylate exclusively at N-7, and theobromines at N-1. Mannich derivatives are formed in these positions with formaldehyde and secondary amines, and the usual Michael adducts are formed in these positions also. The polyalkylation of purine-2,6-diones leads to the possibility of mesoionic purinone derivatives, *e.g.* (**146**).[158] Theoretical possibilities in this direction have been the subject of recent reviews.[159,160] The complex alkylation of purine-6,8-diones has recently been fully investigated.[161]

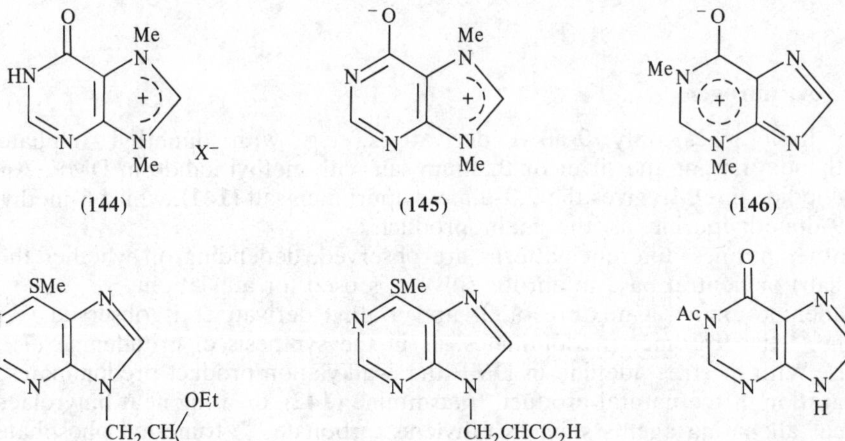

(144) (145) (146)

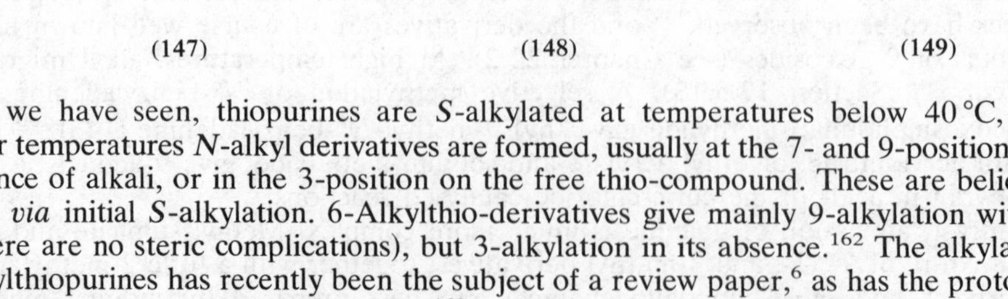

(147) (148) (149)

As we have seen, thiopurines are *S*-alkylated at temperatures below 40 °C, but at higher temperatures *N*-alkyl derivatives are formed, usually at the 7- and 9-positions in the presence of alkali, or in the 3-position on the free thio-compound. These are believed to result *via* initial *S*-alkylation. 6-Alkylthio-derivatives give mainly 9-alkylation with base (if there are no steric complications), but 3-alkylation in its absence.[162] The alkylation of methylthiopurines has recently been the subject of a review paper,[6] as has the protonation and alkylation of 8-methylthiopurines.[163]

An interesting recent application involves the 9-alkylation of 6-methylthio and other purine anions with bromoacetal to give, *e.g.* (**147**), followed by hydrolysis and Strecker synthesis to the alanine derivatives (**148**).[164]

Eritadenines have also been prepared *via* 6-methylthio-derivatives,[165] whilst use of the 9-Michael adducts in a synthesis of NAD analogues has been reported.[166]

Finally, chloropurines are usually alkylated mainly at position-9 with alkali (a little 7-derivative being formed in some cases). This reaction has been widely used recently in the preparation of eritadenines from 6-chloropurines,[165] by subsequent amination of the chloro-derivatives. In absence of alkali, mixtures of 7- and 9-substituted products are obtained, in which the former may predominate. The use of the thallium salt gives exclusively 9-alkylation.[156] With 6-chloropurines, Michael additions occur at the 9-position, as does the vinylation reaction (hy) with vinyl acetate/mercuric chloride. This last reaction has been widely used in preparations of purine-containing polymers, the chloro-groups being converted to the desired 6-substituted derivatives.

In recent years many alkylations have been performed using the above general methods, particularly *en route* to 9-substituted purines, congeners of nucleosides, and other natural products such as eritadenine.

2- and 6-amino purines are acylated (or diacylated)[167] on N-6 (N-2) using anhydrides, or acid chlorides in pyridine, and this has been widely used for directing alkylation and glycosidation (see Chapter 22.2). Isocyanates also react at N-6 to give ureido-derivatives,[168,169] as do chloroformates, although here, 9-substituted derivatives may predominate, depending on the conditions.[168,169]

Ring *N*-acylation is best known for oxopurines. Hypoxanthine with acetic anhydride/DMF gives 1-acetylhypoxanthine (**149**); xanthine gives 7-acylxanthines.[170] Uric acid is acylated in the 7-position, or if this is blocked by an alkyl group, as in 7-methyl- and 1,7-dimethyl-uric acids, 8-acetoxy-derivatives, *e.g.* (**139**), are formed. Theobromine and theophylline are acylated at the 1- and 7-position respectively and in recent years several acyl or acylbis derivatives[171] have been prepared as potential pro-drugs.

Recently, using the new reagent trifluoromethyl methanethiolsulphonate, purine has been converted to its 9-methylsulphonyl derivative. 6- and 8-thiopurines gave 7- or 9-derivatives.[172]

17.5.11 FUSED PURINE SYSTEMS

The N^1,N^6-cyclic intermediate (**159**) mentioned in Section 17.5.13 is one member of a large class of interesting systems in which five- or six-membered rings are fused across adjacent or 1,3-linked fragments of the purine system. Among the rings so attached in recent years are included (various reduction states) imidazo[*a*], imidazo[*b*], imidazo[*e*], imidazo[*f*], imidazo[*i*], 1,2,4-oxadiazolo[*f*], 1,2,4-oxadazolo[*i*], 1,3-oxazino[*f*], oxazolo[*f*], pyrimido[*e*], pyrimido[*f*], pyrimido[*i*], pyrrolo[*f*], tetrazolo[*i*], 1,3-thiazino[*f*], thiazolo[*e*], thiazolo[*f*], 1,3,4-triazepino[*f*], 1,2,4-triazino[*f*], and 1,2,4-triazolo[*i*].

The fluorescent N^1,N^2-etheno-bridged system occurs naturally in the wye bases (**9**), constituents of tRNA anti-codons,[173] and similar compounds are also obtained from adenine and α-haloaldehydes.

Many of the sulphur-containing rings have been desulphurized with Raney nickel to give new purine derivatives, both for synthetic and structural identification purposes.

17.5.12 RADICAL REACTIONS OF PURINES

The radical reactions of purines which lead to useful products are not great in number. Extensive oxidation occurs in the presence of oxygen, owing to attack of hydroxyl radicals, generated photochemically or chemically with Fenton's reagent. Attack is at C-6 (if free), followed by *C*-8-oxidation to 6,8-dioxopurines or uric acid derivatives,[174] in agreement with calculations.[175]

Further oxidation, especially under attack by high-energy radiation, occurs at the C-4—C-5 bridge to give derivatives such as (**150**), together with alloxan and parabanic acid

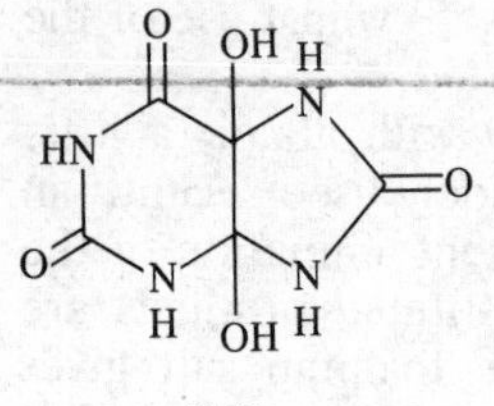

(150)

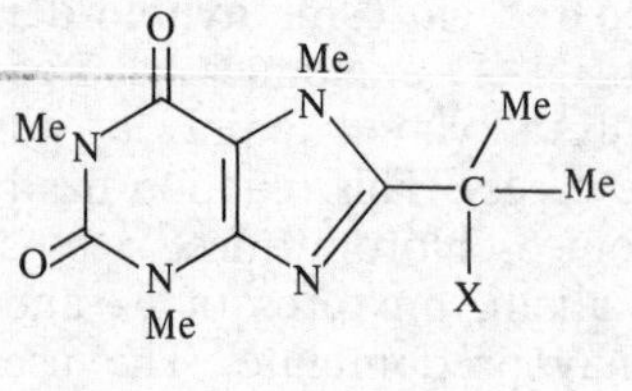

(151) X = NH_2 or OH

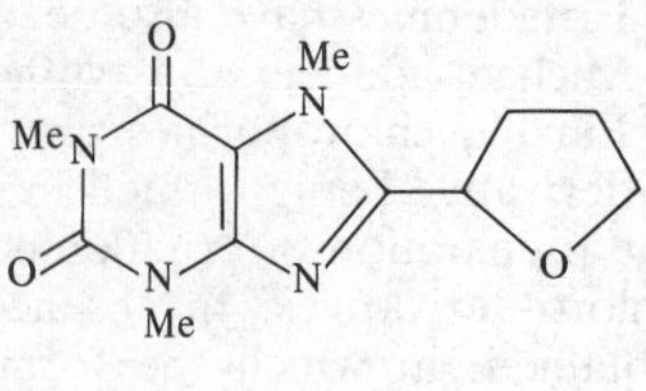

(152)

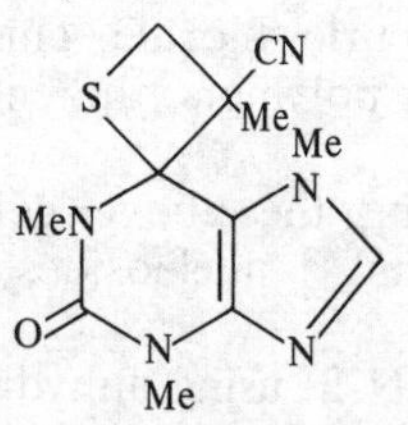

(153)

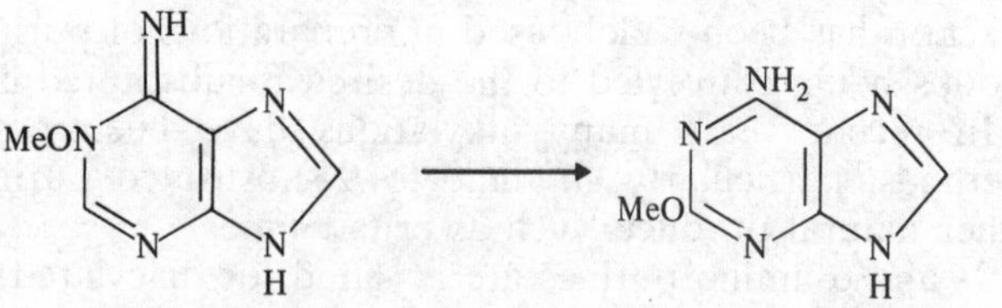

(154)

(155)

derivatives. In the absence of oxygen, substitution reactions at the 6- and 8-positions, of radicals derived from alcohols,[176] amines,[177,178] and ethers,[179] give rise to 6- or 8-alkylpurines or to products of the type (**151**) or (**152**), which have been described recently, as have photodealkylation reactions of 8-hydroxyalkylpurines,[180] and adducts from 8-substituted caffeines. t-Butyl hydroperoxide/ferrous sulphate/sulphuric acid in DMF gives (my–hy) 2- or 8-methyl derivatives from several purines.[181] Irradiation of 6-thiopurines in the presence of acrylonitrile derivatives gives the interesting spiro-adducts (**153**).[182] Irradiation of 1-methoxyadenine or adenine 1-oxide gives rise to deoxygenation to adenine, or conversion to 2-substituted analogues (**154**) to (**155**). 1-Hydroxyisoguanine gives an 8-hydroxy-derivative, while 6-methylpurine 1-oxides undergo rearrangement, reduction, and ring-opening reactions.[183] Radicals are also believed to be involved in the oncogenic behaviour of *N*-hydroxyxanthines (see Section 17.5.8). Reviews on some aspects of the subject are available.[7,184,185]

17.5.13 REARRANGEMENT AND RING-OPENING REACTIONS OF PURINES

The principal rearrangement involving the purine-ring system is the Dimroth rearrangement, also known for isolated pyrimidine systems. In its simplest form this involves an alkali-induced ring-opening reaction of a 1-alkylpurine or nucleoside with formation of the imidazole intermediate (**156**), followed by rotation about the N-1—C-6 bond and recyclization to give the exocyclic alkylamino-derivative (**157**) (Scheme 10). The reaction can also occur in the N^1-unsubstituted purine and has been observed with the help of ^{15}N labelled intermediates.[186,187] Many examples are known, involving a wide variety of substituent groups, including nucleosides in which the reaction is enhanced.[188] It can also be invoked to explain many otherwise puzzling rearrangements such as the conversion (**158**) to (**161**), which is believed to occur *via* cyclization to the imidazo[2,1-*i*]purine (**159**), followed by hydrolysis to (**160**) and final isomerization[189] (Scheme 11), as well as many of those occurring during nucleophilic transformations or attempted Traube-type pyrimidine cyclizations. In forcing alkaline conditions, further hydrolysis to 5-aminoimidazoles occurs. The formation of caffeidine (**162**) from caffeine is a well-known example. Similar rearrangements occur with 6-aminopurine 1-oxides (**163**) and 1-alkoxypurinium salts, giving rise to N^6-hydroxylamino- and alkoxyamino-purines (**164**). The ring-opening reactions are facilitated in this way, especially in nucleosides, and the reaction has recently been widely used with an *O*-benzyl substituent[190] to furnish imidazole intermediates for cyclization to new purine derivatives, as described in Section 17.5.6, the oxygen functions being removed at a suitable intermediate stage.

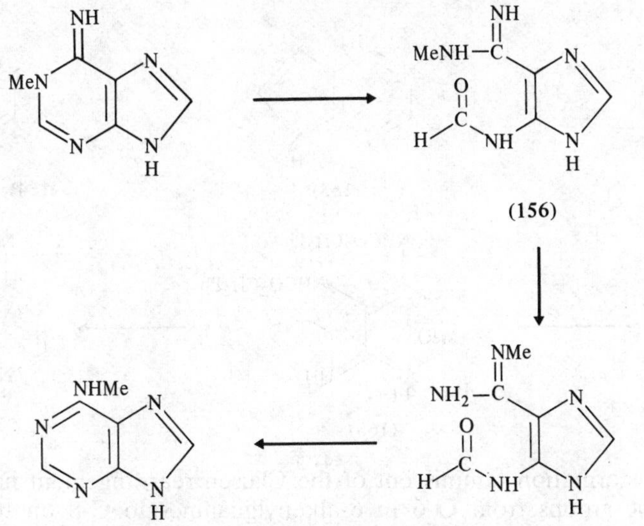

(156)

(157)

SCHEME 10

7- and 9-mono- or 7,9-di-*N*-alkylated oxopurines undergo degradation with alkali with rupture of the imidazole ring, leading to 5-formylalkylamino-4-aminopyrimidines, as met with in the discussion on the Traube synthesis (Section 17.5.5). These reactions are especially favoured in *N*-alkylated oxopurines because of the prevention of stabilization by anion formation. An opening of the imidazole ring in adenines by diethyl pyrocarbonate has recently been described.[191]

Another rearrangement involving opening of the imidazole ring occurs on treatment of 3-ethyl- (or 3-benzyl-) adenine with benzyl chloroformate and sodium bicarbonate to give the 8-hydroxypurine (**166**) *via* the pyrimidine (**165**).[192]

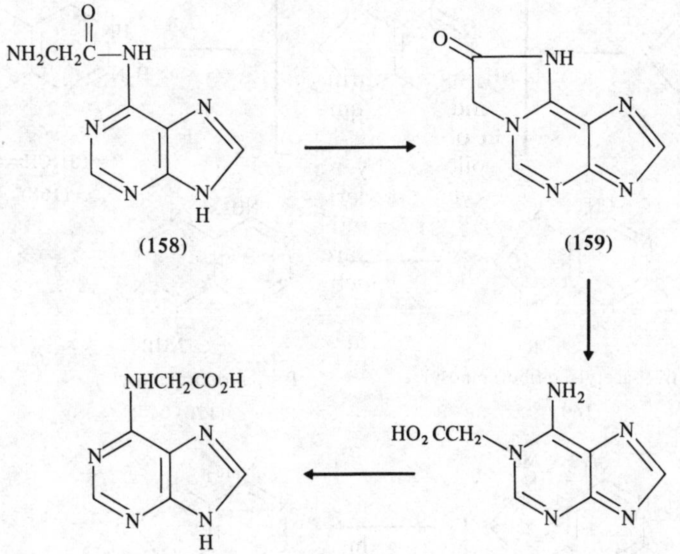

(158)

(159)

(161)

(160)

SCHEME 11

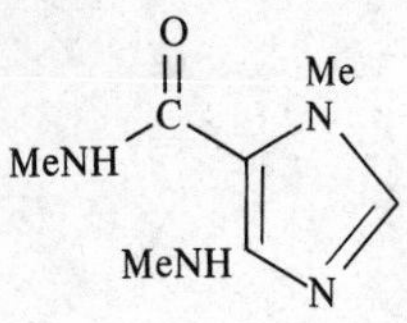

(162) 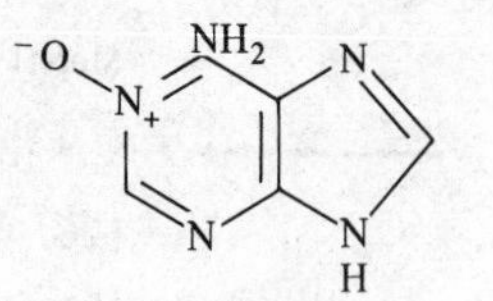 (163) 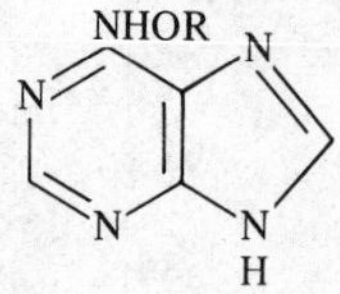 (164) R = H, Me, *etc.*

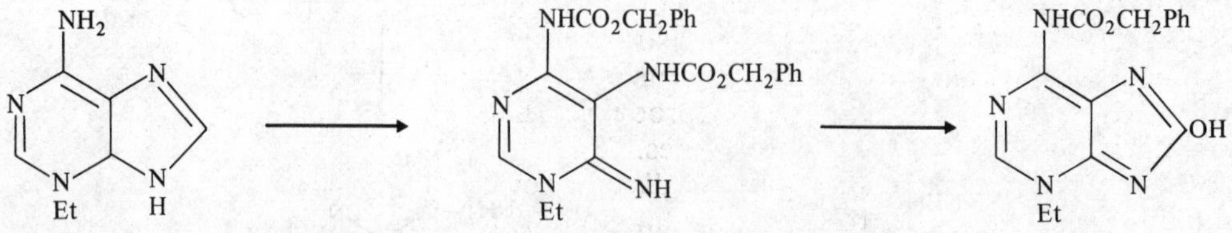

(165) (166)

Recently, a transformation reminiscent of the Claisen rearrangement has involved the migration of allylic groups from O-6 in 6-alkenylguanines to C-8 on treatment of 2-amino-6-chloropurines with allylic alkoxides,[193] (**167**) → (**168**), believed to proceed *via* a double [3,3]-sigmatropic shift involving C-5. If C-8 is blocked, the shift is to N-3 and N-7.[194]

Ring-expansions of purines to pteridines have been reviewed,[7] and the formation of the pyrimido[4,5-*d*]pyrimidine (**170**) on deacetylation of (**169**) described.[195]

Many alkylated purines undergo rearrangements involving true migration of the alkyl groups (as distinct from the apparent migration of the Dimroth rearrangement) under forcing conditions. For example, 3-benzyl-N^2-acetylguanine hydrobromide on heating in dimethylacetamide gives a mixture of the 7- and 9-benzyl isomers, and similar rearrangements of N^6-acyladenines, and of quaternary salts (halides) such as (**171**) → (**172**), are well-known, as are similar migrations of glyosidic groups (Chapter 22.2). In the case of the quaternary salt the perchlorate was found to be stable.[196]

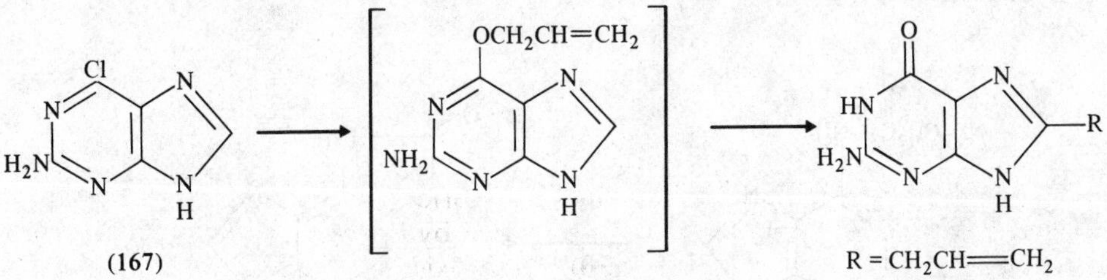

(167) R = $CH_2CH{=}CH_2$ (168)

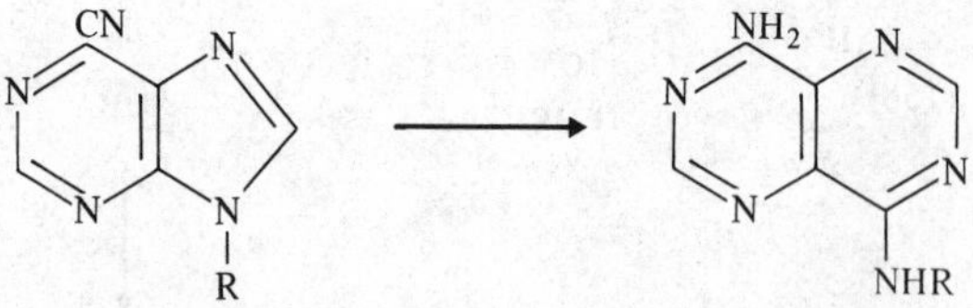

R = tri-*O*-acetyl-β-D-ribofuranosyl (169)

R = β-D-ribofuranosyl (170)

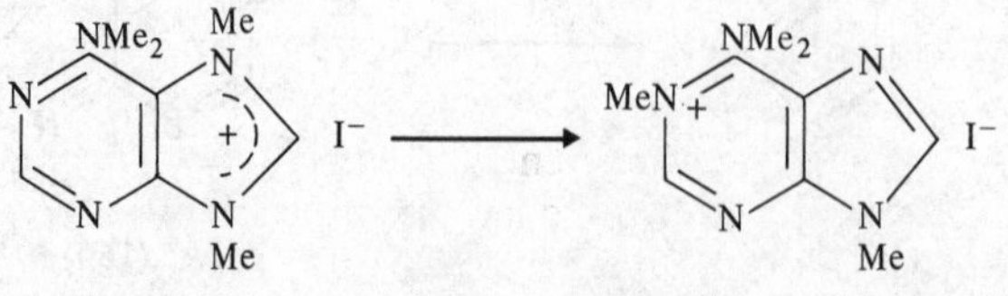

(171) (172)

17.5.14 REDUCTION AND OXIDATION OF PURINES

Although 1,6-dihydropurines have long been recognized as the primary products of electrolytic reduction of purines at a mercury or lead cathode, preparation is usually by catalytic reduction or by chemical means.

The first method is only really suitable for 6-unsubstituted purines since hypoxanthine, (methylated) xanthines, adenine, and guanine are not reduced. The products are unstable to air oxidation (reversion to purines), and to hydrolysis to 5-amino-4-amino methyl-imidazoles, and are best isolated as acetyl derivatives.

N-Alkylated purines, especially with charged rings, such as (**173**), are readily reduced by sodium borohydride, the reduction occurring preferentially in the charged ring to give dihydro-derivatives.[197,198] The initial products of Raney nickel desulphurization of thiopurines are also usually dihydropurines.

Although purines in general do not display covalent hydration properties, various additions across the 1,6-double bond leading to dihydro-derivatives have been observed, among them those of barbituric acids,[199] resorcinol, and other aromatics,[200] and potassium hydrogen sulphite[201] which gives adducts of the type (**174**) and (**175**).

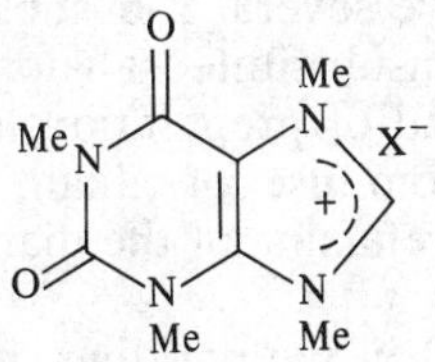

(173) 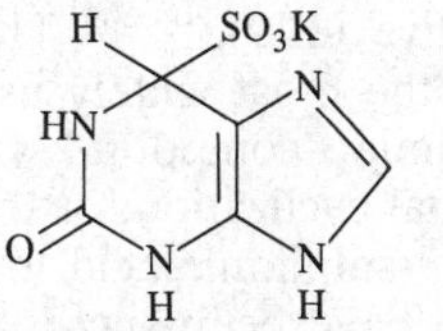 (174)

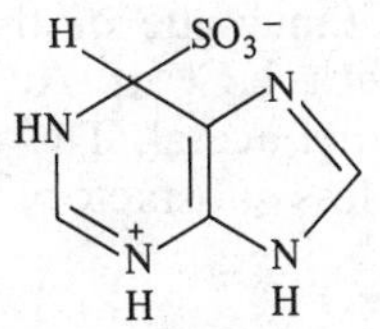

(175) 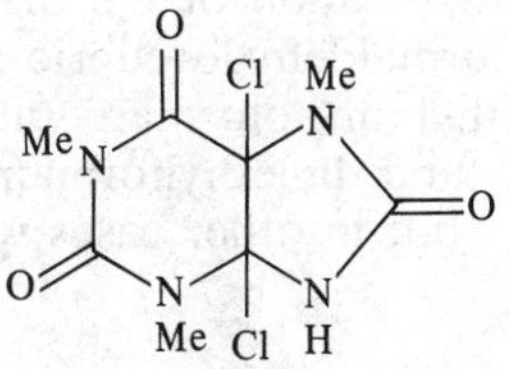 (176)

The main pathway of oxidation of purines is by prior conversion to oxygen functions at the 6- and 8-positions to give uric acids, followed by oxidative attack at the 4,5-double bond to give 4,5-dihydroxyuric acid (**150**). Final oxidative degradation gives alloxan or parabanic acid derivatives and urea.

This pathway appears to be common to electrochemical, biological, radiation-induced, and chemical oxidation. When the latter involves initial 4,5-addition of chlorine to, *e.g.* (**176**), much interesting chemistry ensues (see Ref. 2, pp. 240 *et seq.*).

17.5.15 GENERAL POINTS ON AZA- AND DEAZA-PURINES

Discussion of monoaza- (2-, 4-, 5-, 6-, and 8-) and monodeaza- (1-, 3-, 7-, and 9-) purines will be confined here largely to questions of nomenclature, natural occurrence, and synthesis with only incidental, and not exhaustive, details of the chemistry of these compounds.

As far as nomenclature is concerned, the sometimes bewiiaering array of names and numbering methods used in the literature and abstracts will be replaced by the current systematic name and numbering as given by the Patterson–Capell–Walker Ring Index,[202] or that currently accepted in the *Chemical Abstracts Index Guide* (Vol. 76 onwards).

Most synthetic aza- and deaza-purines have been prepared as analogues of the natural purines, with potential anti-metabolite activity, but notable success in this direction has only been achieved with the 2-aza-, 7-deaza-, and 8-aza-analogues, although synthetic difficulties have limited work on the 4-aza- and 6-aza-analogues.

1- and 3-deaza-analogues seem to resemble the corresponding imidazoles in their biological activity, and some success has been achieved in fields where the latter have proved useful (see Sections 17.5.16 and 17.5.18).

17.5.16 1-DEAZAPURINES

1-Deazapurines are derivatives of the imidazo[4,5-*b*]pyridine system, the numbering being as shown in (**177**). As far as is known they do not occur naturally, unlike the 3-isomers (Section 17.5.18). In general they are fairly strong bases and weaker acids than benzimidazole, the pK_a's of the parent compound being 3.95 and 11.08.

The earliest synthesis of a 1-deazapurine was by Chichibabin, who cyclized 2,3-diaminopyridine (and its 5-chloro-derivative) with refluxing acetic anhydride to give the 2-methylimidazo[4,5-*b*]pyridine derivative (**178**) (my).[203] Little further work was done for many years until the use of 98% formic acid gave several 2-unsubstituted analogues *via* the 3-formyl derivative (**179**).[204,205] The fully unsubstituted analogue soon followed.[206] This has since been the most widely used method of preparation, the cyclization conditions for the formamido-compound varying from use of alkali, dry distillation with magnesium, or thermal cyclization, or by simple refluxing of the diamine in formic acid or formic acid/toluene-*p*-sulphonic acid in butanol.

Substituents which have been successfully used in the pyridine ring include halogen in various positions, amino, acetamido, ethoxycarbonyl, alkoxy, hydroxyalkyl, aryl, nitro (hy), carbamoyl and sulphamoyl, and thioalkyl.

Other methods of cyclization of 2,3-diaminopyridines to give 2-unsubstituted derivatives are by use of formaldehyde–cupric acetate[207] (analogue of the Wiedenhagen imidazole synthesis), ethyl orthoformate (alone or with HCl or Ac_2O), diethoxymethyl acetate, DMF/$POCl_3$, and dimethylformamide dimethyl acetal. The use of dithioformate was recommended,[207] but in other cases was found less satisfactory.[208]

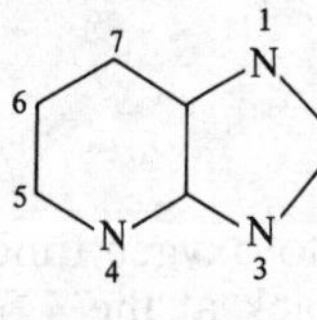

(**177**) 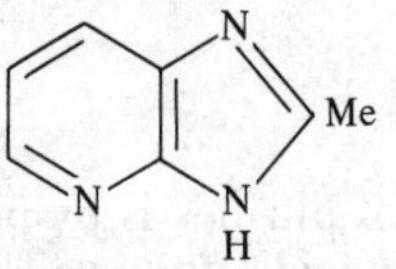 (**178**) 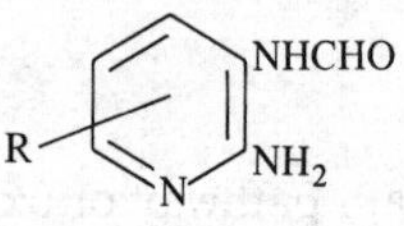 (**179**)

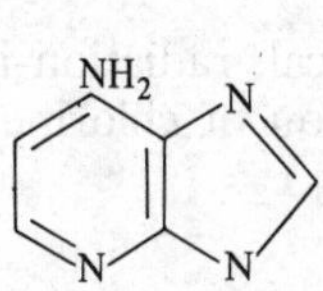

(**180**) 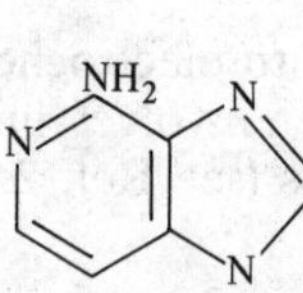 (**181**) 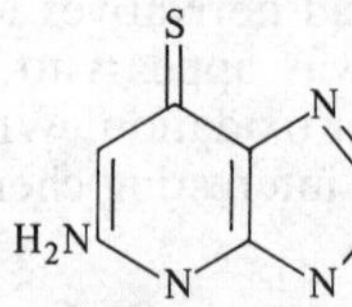 (**182**)

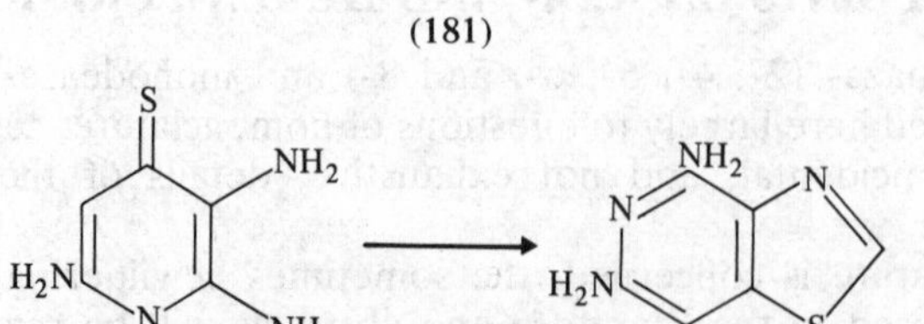

(**183**) → (**184**)

Syntheses of analogues of natural purines by these methods include that of 1-deazaguanine,[290] also prepared by dealkylation of the 7-ethoxy-analogue, and of 7-aminoimidazo[4,5-*b*]pyridine (1-deaza-adenine) (**180**) *via* reduction of the 7-nitro-derivative. The synthesis of the latter directly from 2,3,4-triaminopyridine is complicated by the possibility of an alternative ring-closure to give 4-aminoimidazo[4,5-*c*]pyridine, a 3-deazapurine (**181**). This reaction has been studied using diethoxymethyl acetate[210] and several other cyclizing agents and shown to always give a mixture of the two isomers,[211] although the 1-deaza-analogue predominates. Direct synthesis of the 7-amino-derivative is achieved with formamidine acetate and triaminopyridine. The 7-thioguanine analogue (**182**) could not be prepared in this way, as the intermediate thiopyridine (**183**) gave the non-rearranging thiazole derivative (**184**); nor could the 7-oxo-analogue be satisfactorily thiated with P_2S_5. The derivative was finally prepared by dealkylation of the 7-diphenylmethylthio-compound with trifluoroacetic acid.[212]

Use of 2-substituted-amino- (alkyl, aryl, dialkylaminoalkyl) 3-aminopyridines gives the 3-substituted imidazopyridines, whilst 3-substituted-amino-2-aminopyridines gave the corresponding 1-substituted analogues. These unambiguously substituted derivatives have been used as models in the assignment of structures to products of alkylation and glycosidation reactions (see below).

The preparation of 2-substituted (alkyl, aryl, heterocyclyl) imidazo[4,5-*b*]pyridines has usually been effected by heating the diaminopyridine with the appropriate acid, either alone at high temperature, or in the presence of polyphosphoric acid at 170–200 °C.[213] Acids used include aromatic (including polymethoxybenzoic), lower and long-chain alkanoic, arylacetic, anthranilic, trifluoracetic, heterocyclic, glycollic, and bromo- and amino-acids (β-alanine); bis-acids such as succinic[214] and stilbene-4,4′-dicarboxylic[215] give [(ly)–(my)] 2,2′-bisimidazopyridines using polyphosphoric acid.

Other acid derivatives which have been utilized include anhydrides, aromatic methyl and ethyl esters, nitriles (*e.g.* 1- and 3-isoquinolinecarbonitriles, chloroacetonitrile), and imidates [phenylacetimidates, heterocyclic carboximidates, and imidates from mandelonitriles,[216] *e.g.* (**185**)→(**186**) (Scheme 12)]. The 2-benzyl and α-hydroxybenzyl derivatives obtained in this last cyclization gave 2-benzoyl compounds on MnO_2 oxidation.[216]

Use of acid chlorides with triethylamine, or in some cases, acetic anhydride, gave 2,3-diacylaminopyridines which cyclized on pyrolysis to 2-alkyl derivatives.

OR NH CH—C OEt + NH$_2$ N NH$_2$ ⟶ OR N CH—Ph N N H

(**185**) (**186**) R = H or Ac

SCHEME 12

5-Nitrofurfural on heating with 2,3-diaminopyridine gave the 2-(5-nitrofuryl) derivative directly.

Many of these methods have given 1- and 3-alkyl derivatives with the appropriately *N*-substituted 2,3-diaminopyridines.

Other reactions which parallel those in the purine series include preparation of 2-thio-derivatives from 2,3-diaminopyridines using CS_2[24,217] (alone or in methanol, or with pyridine or KOH) or, less often, using thiophosgene and HCl or alkali, and the preparation of 2-oxo-derivatives using phosgene,[205] fusion with urea at 180 °C (many examples),[217] or reaction with carbonyldi-imidazole. This last is especially useful for 2-arylamino-3-aminopyridines.[218] In one case the 3-urethane (**187**) was cyclized by pyrolysis at 160–165 °C (ly) to give (**188**).[219]

Alternatively, reduction of a 3-pyridine substituent (such as nitro or arylazo) may be followed by cyclization with a 2-acylamido-group to give a 2-substituted imidazopyridine.

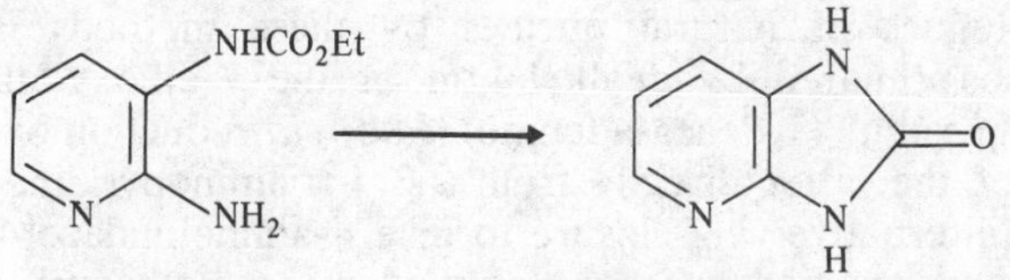

(187) (188)

This method was used for a series of herbicidal 2- (di- or tri-fluoroalkyl) derivatives,[220] *e.g.* (**189**; $R^1 = H$), and in this case selective partial reduction of the nitro-group with a palladium catalyst gave the 1-hydroxyanalogues (**189**; $R^1 = OH$) (my–hy).[221] Another method involved reduction of a 3-nitro-2-substituted aminopyridine in the presence of 98% formic acid. By this means the eritadenine analogue (**190**) was prepared, but was relatively inactive as a hypocholesteraemic agent.[222]

Miscellaneous cyclizations of 2,3-diaminopyridines include the use of tetra-alkyl pyrocarbonates to give the 2-alkoxy-derivatives (**191**),[40] and treatment of the thioureido-compound (**192**) (from the diamine and methoxycarbonyl isothiocyanate) with copper acetate in acetic acid to give (**193**), an anthelmintic.[223]

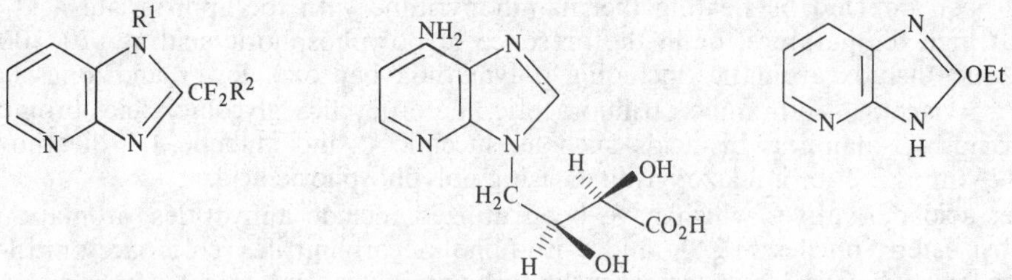

(189) (190) (191)

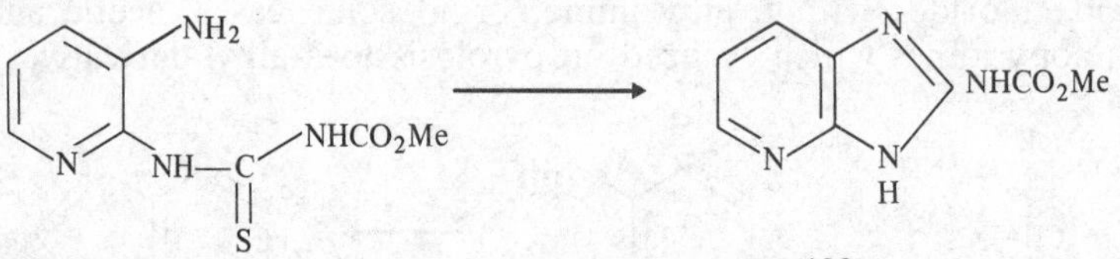

(192) (193)

Among other miscellaneous preparations of imidazo[4,5-*b*]pyridines must be mentioned the Hoffmann and Curtius degradations[224] of 2-aminopyridine-3-carboxylic amides and -carboxylic azides, respectively, which lead to 2-oxo-derivatives.

A curious reaction involves preparation of the parent compound (**196**; R = H) by a ring contraction from the azaquinazoline (**194**) on zinc dust distillation.[225] Thermal decomposition (ly) of 3-azido-2-dimethylaminopyridine (**195**) gives the 3-methyl derivative (**196**; R = Me), also obtained on performic acid oxidation of 3-amino-2-methylaminopyridine.[226] Cyclization of the *N*-chloroamidines (**197**) with ethanolic KOH (or photochemically) gives the 2-(4-thiazolyl)imidazopyridine (**199**) *via* the nitrene (**198**).[227]

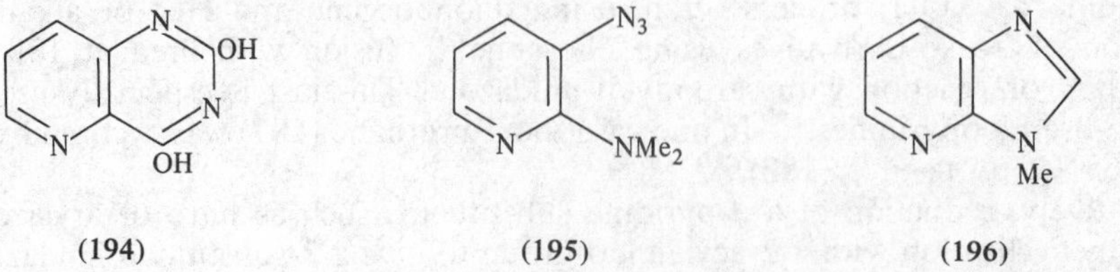

(194) (195) (196)

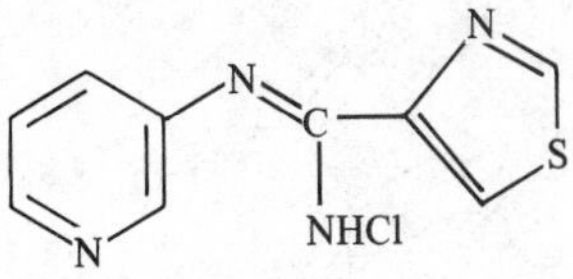

(197) 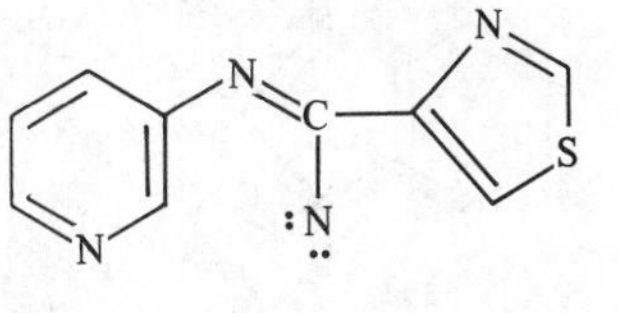 (198) 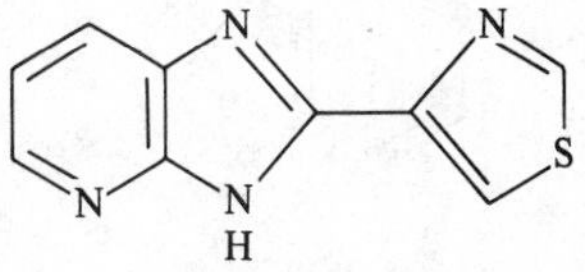 (199)

A series of interesting reactions occurs when 2,3-diaminopyridine is refluxed with ethyl acetoacetate in xylene, when the 1-isopropenylimidazopyrid-2-one (**201**) is obtained through rearrangement of the initially formed diazepinone (**200**). Reduction gives the 1-isopropyl derivative, different from the authentic 3-isopropyl isomer synthesized from 3-amino-2-isopropylaminopyridine and phosgene. When the reaction is performed with neat ethyl acetoacetate the opposite isomer (**202**) of the diazepinone (giving the authentic 3-isopropylimidazopyrid-2-one (**203**) on rearrangement and reduction) is first formed (15 min) but rearranges to (**200**) on further heating (90 min).[228,231] Similar reactions have been observed with ethyl benzoylacetate,[229] and diallyl malonic ester, which gives finally the imidazopyridyl lactone (**204**).[230] Other interesting reactions recently reported are the conversion of purine (**205**) with malonodinitrile to 5-aminoimidazo[4,5-*b*]pyridine-6-carbonitrile (**206**), *via* an initial ring-opening and recyclization of the dicyanovinyl intermediate (**207**),[232] and the conversion of the pyridine *N*-oxide (**208**) to 2,3-diphenylimidazo[4,5-*b*]pyridine on attempted reaction with *N*-phenylbenzimidoyl chloride.[233]

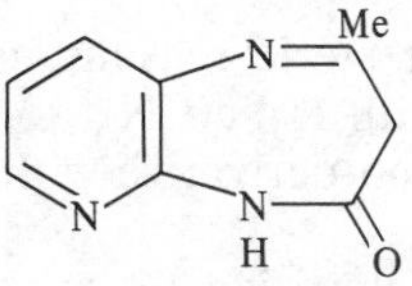

(200) 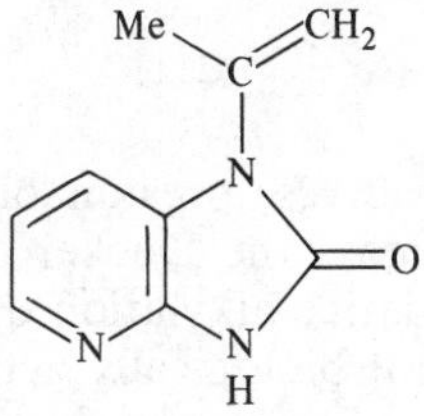 (201) 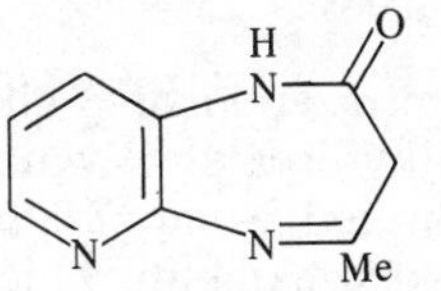 (202)

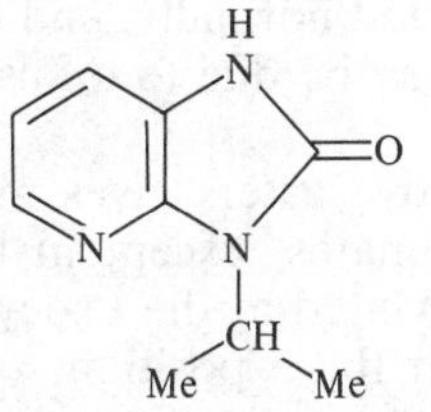

(203) 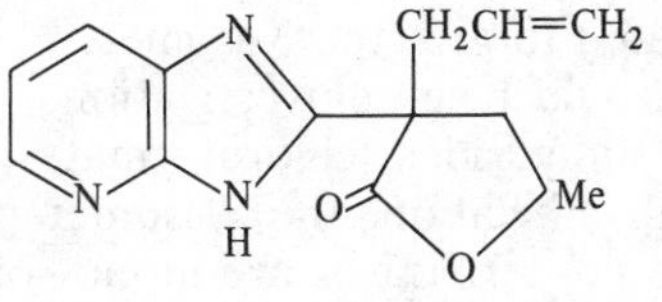 (204)

Two photochemical cyclizations of interest provide the 1-hydroxy-derivative (**209**) from 3-nitro-2-pyridyl-leucine on irradiation at 290 nm at pH 2,[234] and the tetrahydro-imidazopyrid-5-one (**211**) from the imidazole (**210**).[235]

On monoalkylation either with neat methyl iodide (150 °C), or on methylation of salts, mainly 3-alkyl derivatives are obtained, although with larger alkyl groups (*e.g.* benzyl or dialkylaminoalkyl) some 1-derivative is formed, presumably due to steric interactions. This steric effect is enhanced by 2-substituents (Me, OH), when mainly 1-substituted derivatives are formed on alkylation or Michael addition. Alkylation of the free base with excess of dimethyl sulphate in acetic acid/nitromethane at 90 °C, however, gives mainly the N^4-methyl derivative (**212**) (plus a little 3-methyl).[236] The 1,4-dialkyl derivative is also formed with the 2-oxo free base.

The 1-alkyl derivative also undergoes quaternization with methyl halides[237] at N-4, unless the presence of a nitro- or chloro-group causes deactivation of N-4, when the 1,3-salt is formed.

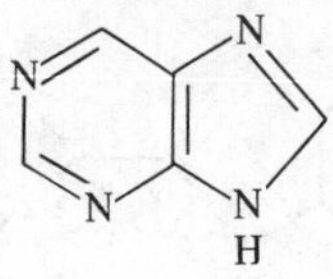

(205) 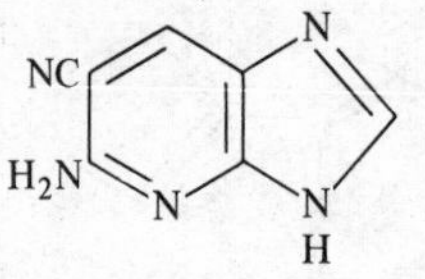 (206) 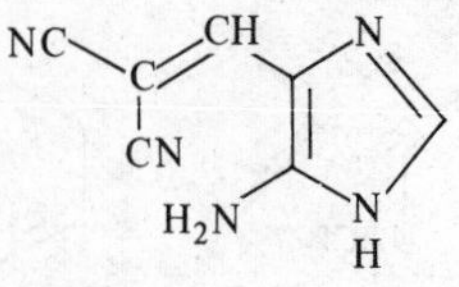 (207)

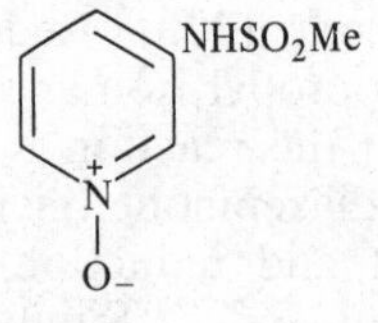

(208) 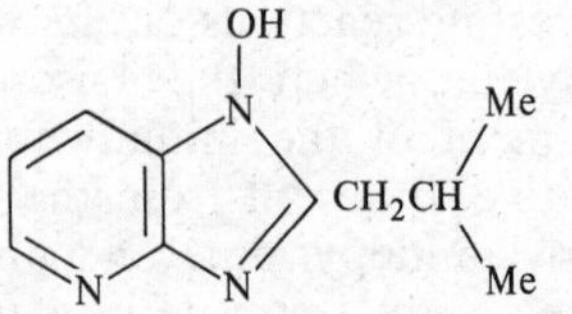 (209)

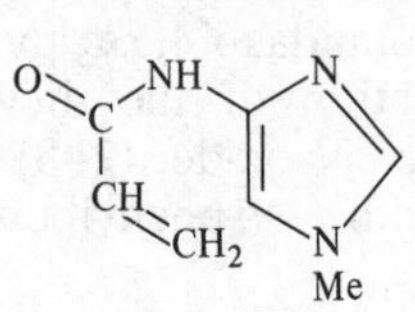

(210) 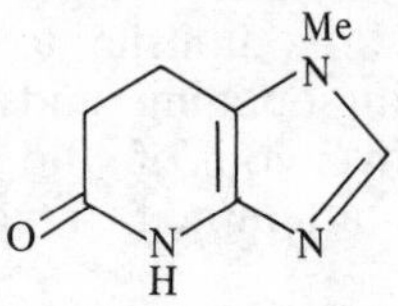 (211) 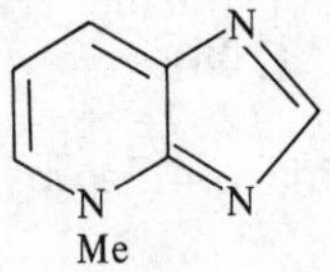 (212)

Quaternization of 3-alkyl derivatives, presumably again for steric reasons, gives the 1,3-dialkyl isomer[237] which undergoes the Decker oxidation with $K_3Fe(CN)_6$ to give the disubstituted 2-one.[238] Direct oxidative alkylation of the 7-amino-derivative with furfural gave 1-deazakinetin, which was not biologically active.

Glycosidation under various conditions mainly gives the 3-glycoside (β- plus a little α-anomer). The 6-bromo-derivative, however, was said to furnish mainly the 1-glycoside,[239] although the 6-chloro and 7-bromo reacted normally, and the 6-bromo itself was previously stated to give the 3-isomer.[240] This may be due to a false identification of the product, and could bear reinvestigation.

Acylation with anhydrides, chloroformates, or active esters gives mixtures of 1- and 3-isomers; as with alkylation, the 3-isomer predominates, except in the presence of a 2-substituent. Amino-derivatives are mono- or di-acylated at the exocyclic amino-group, as with purines. The Mannich reaction also occurs at the 3-position.

Nitration and halogenation normally occur at the 6-position, but with a 2-oxo-substituent, the 5-nitro-isomer was formed.[241]

N-Oxides are formed at the 4-position, and rearrange to ring-chloro-derivatives with phosphorus oxychloride.[242] The 1-oxides have been formed by cyclizative synthesis (see above), and are reduced catalytically, or by thionyl chloride/dimethylformamide, to the *N*-unsubstituted derivatives.

2-Alkyl groups are activated, being oxidized to aldehydes by SeO_2, and condensing with aldehydes to give 2-styryl derivatives.

Halogens in the 5- or 7-positions are reactive to nucleophilic substitution or hydrogenation, *e.g.* in a synthesis of 1-deazaisoguanine,[243] but the 6-chloro is only aminated with difficulty.[244] An attempted Chichibabin amination failed.[245] Hydrogen-exchange occurs at the 2-position *via* an ylide intermediate.[246]

1-Deazapurines do not show facile ring-opening reactions.

Studies of tautomerism show 3-H forms to predominate,[247] whilst 2-oxo- and thioxo-groups exist in that form. 2-Amino-substituents are mainly in the amine form.[248]

U.v., dipole moment, and n.m.r. measurements have been made, the latter serving for assignments of position of substitution and anomeric configuration in nucleosides.

Although a wide variety of biological activities have been reported, the principal interest seems to have been in herbicides, fungicides, anthelmintics, analgesics, anti-inflammatories, bactericides, and photographic anti-fogging agents.

Two short reviews on specialist aspects of 1-deazapurines are available.[7,249]

17.5.17 2-AZAPURINES

2-Azapurines are imidazo[4,5-*d*]-1,2,3-triazines, with numbering as shown in (**213**). Again, no member is known to occur naturally. They are weakly acidic compounds, with little or no basic properties.

In contrast to the many types of synthesis of the 1-deazapurines, the synthesis of 2-azapurines utilizes basically one reaction, the diazotization of 5-(4-)aminoimidazole derivatives with ring-closure to an adjacent 4-(5-) substituent (Scheme 13), *e.g.* (**214**) to (**215**). The stable isolated diazo-derivative (**216**) is sometimes used. Similar reactions with *vic*-triazoles give 2,8-diazapurines.

(213) (214) (215)

HNO_2, X = O

NH_4OH, X = S

(216) (217)

SCHEME 13

When the 4-substituent is the carboxamide, the 4-one (**215**; X=O) is obtained; the carboxamidine gives the 4-amino-compound, and the acetoxime the 4-methyl derivative, whilst the thioamide gives the 4-thione after rearrangement with ammonia of the first-formed thiadiazine (**217**).[250]

The reaction has often been performed on *N*-glycosylimidazoles to give 2-azapurine nucleosides, both 5- and 7-isomers; α- and β-anomers are obtained *via* suitably substituted imidazole intermediates, identification being by means of n.m.r. spectra. In these cases the cyclization is best performed at −25 °C.[251]

The use of imidazoles with 2-substituents, either preformed (2-Me, 2-OH), or inserted at an intermediate stage in the synthesis (by bromination, or nucleophilic substitution[252]), gives 6-substituted imidazotriazines.

Other *N*-substituents give rise to corresponding *N*-substituted 2-azapurine derivatives. For instance, *N*-substituted carboxamides give 3-substituted imidazotriazines, *e.g.* the formation of 3-oxides on treatment of the diazo-compound (**218**) with Na_2S_2 or hydrazine in attempted thiation or deamination.[253]

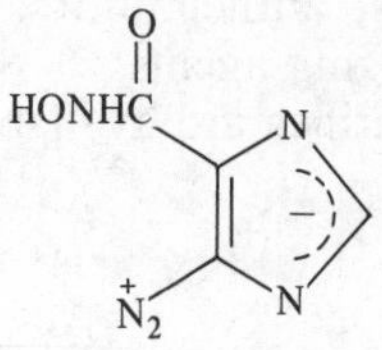

(218) 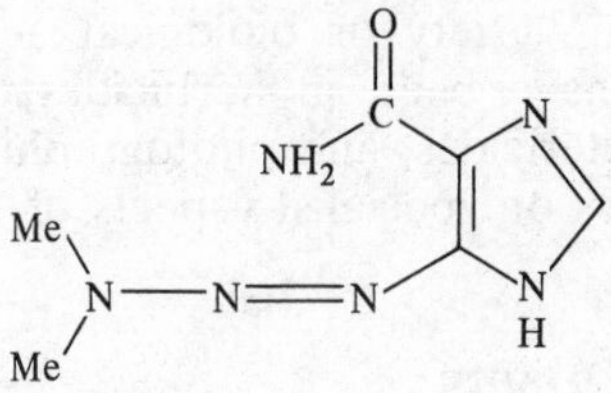 (219)

A photochemical synthesis proceeds by irradiation of 5-triazenylimidazole-4-carboxamides (**219**).[254]

Direct glycosylation of isolated 2-azapurines has not often been performed, but with the mercuric cyanide/bromo-sugar method the 4-methylthio-derivative seems to give the 7-β-nucleoside. However, these are usually prepared from imidazole intermediates.

N-Oxidation occurs mainly at the 1-position, with a small amount of another isomer, but the N^3-oxides are more usually prepared by cyclization of imidazole-4-carboxamidoximes and 4-ketoximes.[255] *N*-Amination has recently been reported at N-7 using the Krause reagent. The 7-amino-derivatives have also been prepared by cyclization of *N*-aminoimidazoles.[256]

Nucleophilic substitution occurs at the 6-position and at the 4-position with the thio- and methylthio-derivatives, *e.g.* in the synthesis of puromycin analogues. With the 4-chloro-derivatives, however, ring-opening occurs to give a cyanoimidazole, *e.g.* (**220**) to (**221**).[257] A different form of ring-opening occurs on reduction of the 6-substituted 4-ones, *e.g.* (**222**), with Raney nickel catalyst, when the corresponding 5-amino-2-substituted imidazole-4-carboxamides are formed.[252]

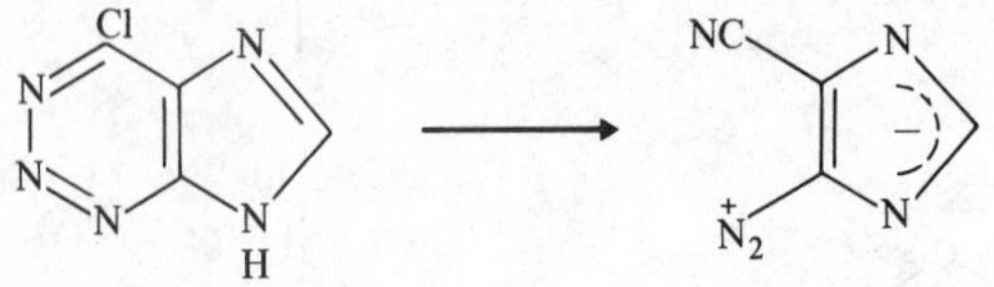

(220) (221) 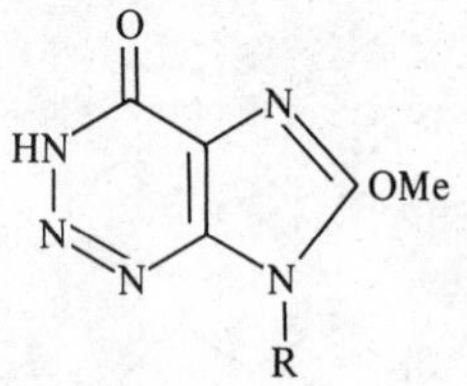 (222) R = β-D-ribofuranosyl

2-Azapurines, particulary the nucleosides, have considerable anti-tumour activity, although not as good as that found in the 8-derivatives.[7] They also have some bactericidal and other actions.

Reviews of their synthesis and biological properties are available.[7,258,259]

17.5.18 3-DEAZAPURINES

3-Deazapurines are derivatives of the imidazo[4,5-*c*]pyridine ring system, the numbering being as shown in (**223**). The pK_a's of the parent base are 6.10 and 10.88. Unlike the 1-deazapurines they are found to a limited extent in Nature, the 6-carboxylic acid (**224**; $R = CO_2H$), spinacine, being isolated from the green leaves of spinach[259a] and from the cutaneous glands and liver of some sharks and amphibians such as *Acanthias vulgaris, Leptodactylus pentadactylus, Salamandra maculosa, etc.*,[260,261] together with its decarboxylation product, spinaceamine (**224**; R = H), the structure of which has been proved by X-ray analysis. The other substantial natural occurrence of imidazo[4,5-*c*]pyridines is in the streptothricin group of antibiotics[262] (such as streptothricin A (**225**), racemycin, citromycin, *etc.*). These are 2-aminoglycosyl derivatives of a 6-hydroxy-4-oxohexahydro-imidazo[4,5-*c*]pyridine system. The full structures of streptothricins A–F and X have recently been confirmed.[263]

The earliest synthetic routes to the ring system were similar to the now supposed biogenetic pathway to spinacine. Histidine and methylal/HCl at 100 °C,[264] or, later, formalin[264a] gave (hy) spinacine, spinaceamine being similarly prepared from histamine,[265] whilst *N*-methylhistidine gave the N^5-methyl-6-carboxylic acid.[265a] With other aldehydes, including pyridoxal,[266] histamine has given 4-substituted spinaceamines,[267] (**227**; $R^2 = H$), *via* the Schiff bases (**226**), isolated under mild conditions using aromatic aldehydes.[268] With ketones, 4,4-disubstituted derivatives (**227**; R^1, R^2 = alkyl, aryl) are formed.[267]

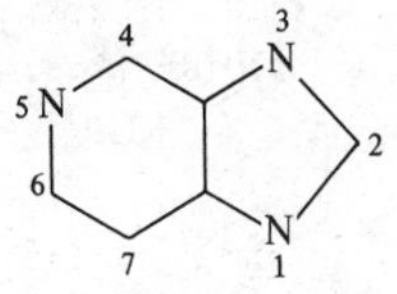

(223) 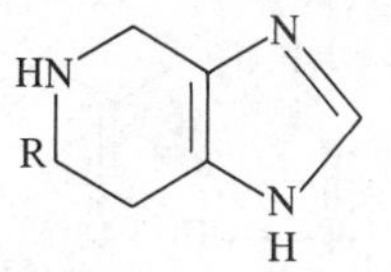 (224)

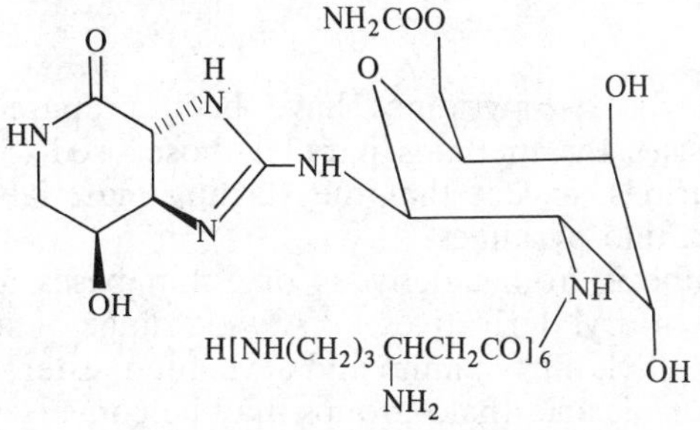

(225)

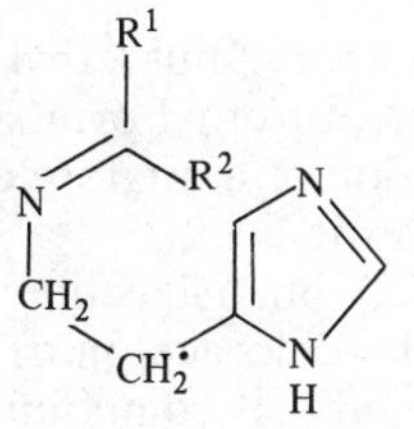

(226) 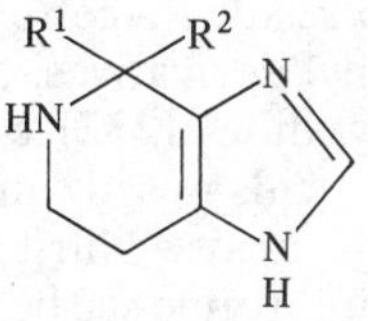 (227)

More recent syntheses involved the reaction of 4-(dimethylaminomethyleneamino-methyl)imidazole with methylal/HCl to give *N*-methylspinaceamine derivatives.[265] Spinacemines form complexes with Cu^{2+} and Ni^{2+} ions.[269] Other imidazole syntheses include the preparation of the N^5-methyl-2-oxo-derivatives (**228**) from 2-hydroxyhistamine on reductive alkylation (Pd/C) with formaldehyde/methanol, as well as the cyclization of 1-benzyl-2-methyl-5-(*N*-alkyl-*N*-dialkoxyethylaminomethyl)imidazoles with acid to give N^5-alkyl-2-methyl-3-benzyl-7-hydroxy-4,5,6,7-tetrahydroimidazo[4,5-*c*]pyridines, *e.g.* (**229**).

Finally, a recent series of papers deals with the synthesis of 3-deazapurines from 4-cyanoimidazole-5(4)-acetonitriles. The parent compound with 48% HBr gave 6-amino-4-bromo-derivatives (**230**; R = Br),[270] which gave the expected 4-thio- and seleno-derivatives with the usual reagents, whilst the corresponding carboxamide gives the anti-metabolite 6-amino-4-oxo-analogue (3-deazaguanine) (**230**; R = OH) with alkali,[271] and the 4,6-dioxo-compound is formed from the ethoxycarbonylacetamide (**231**). The use of 4-carboxamidoimidazole-5-acetonitriles with a 1-glycosyl substituent gives the 6-amino-4-hydroxy-1-*β*-D-ribofuranoside (3-deazaguanosine), whilst with the isomeric 4-methoxycarbonyl-5-acetonitrile 1- and 3-glycosides and ammonia, the 6-amino-4-oxo-1- and -3-nucleosides are respectively obtained.[272]

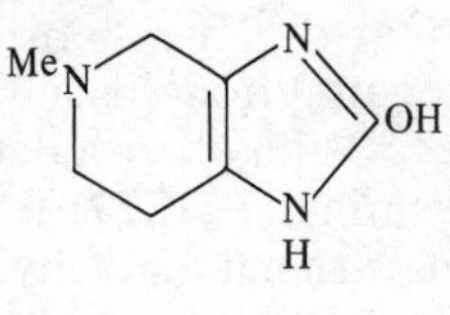

(228) 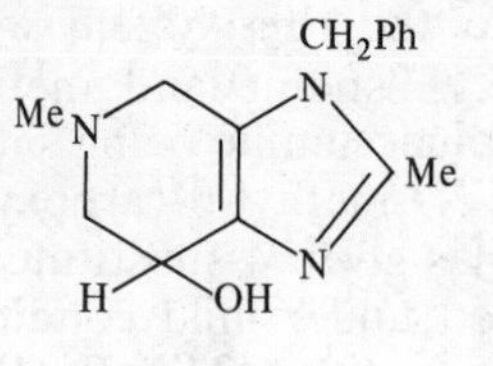 (229) 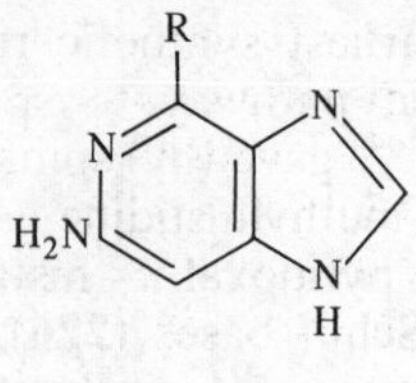 (230)

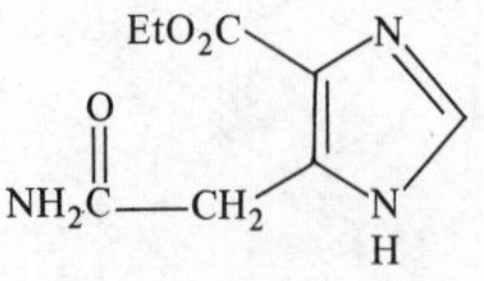

(231) 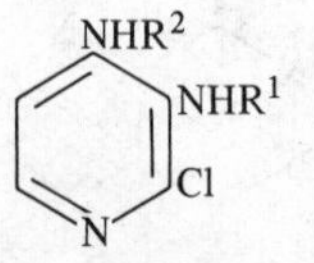 (232) a; $R^1 = H, R^2 = Ac$
b; $R^1 = Ac, R^2 = H$

The majority of imidazo[4,5-*c*]pyridines have been prepared from pyridine intermediates, and in most cases the methods parallel those used (Section 17.5.16) for the corresponding 1-deazapurines, except that the starting materials are the isomeric 3,4-diamino- or substituted-amino-pyridines.

The Wiedenhagen synthesis from aldehydes or Schiff bases and copper acetate was formerly used for 2-alkyl or aryl derivatives.[273] A wide range of such pyridines (including those with hydroxy, halo, alkyl, nitro, amino and acylamino, ester, and amide substituents) has been cyclized using formic acid (halo-groups may be converted to hydroxy), whilst 4-alkylamino-3-aminopyridines and 3-alkylamino-4-aminopyridines give 1- and 3-alkyl derivatives, respectively.

Cyclization with thioformic acid was less satisfactory, but ethyl orthoformate/acetic anhydride gives good results. Acetic anhydride alone, or ethyl orthoacetate/acetic anhydride, gives the 2-methyl derivatives, as does cyclization of a mixture of the monoacetyl-2-chloro-3,4-diaminopyridines (**232**) with polyphosphoric acid.[274]

Other carboxylic acids used include glycollic, phenylacetic, trifluoroacetic, and quinoline-2-carboxylic acids. Nitriles used include chloracetonitrile, as well as 2,2′-bipyridyl- and phenanthroline-carbonitriles, which furnish compounds useful in the colorimetric analysis of metal ions.[275] Amidines (useful for halo-derivatives), and iminoethers, particularly of phenylacetic acids and their α-hydroxy-derivatives, have been used as in Section 17.5.16, as have carbonyldi-imidazole[276] and carbon disulphide to give the 2-ones and 2-thiones, respectively.

An interesting variation uses 3,4-diaminopyridine *N*-oxide (**233**), which with formic acid gives the N^5-oxide (**234**).[277] The latter rearranges with refluxing phosphorus oxychloride (my) to give the 4-chloro-derivative (**235**) (Scheme 14), aminated with ammonia/copper sulphate at 160–170 °C to give the same 4-amino-derivative that was

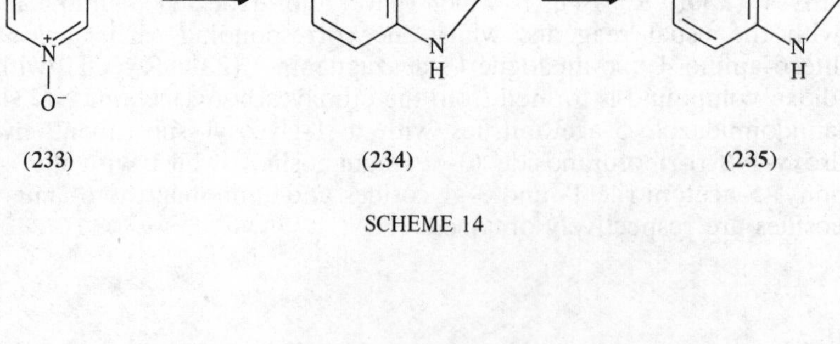

(233) $\xrightarrow{HCO_2H}$ (234) $\xrightarrow{POCl_3}$ (235)

SCHEME 14

isolated as the minor product from the mixture of isomers obtained from 2,3,4-triaminopyridine with formic acid (see Section 17.5.16).[211,277] A similar sequence of reactions has been observed with the 1- and 3-nucleoside 5-oxides,[278] but in these cases the amination is considerably more facile.

Another reaction paralleling the behaviour of the [4,5-*b*] series is that with ethyl acetoacetate in toluene or xylene, which in this case gives only one of the isomeric benzdiazepinones (**236**), leading on dry fusion[279] to the isopropenylimidazo[4,5-*c*]pyrid-2-one (**237**). The other diazepinone isomer (**238**) was prepared by the cyclization of the appropriate pyridyl aminocrotonate with base, and on dry fusion gives (**239**).

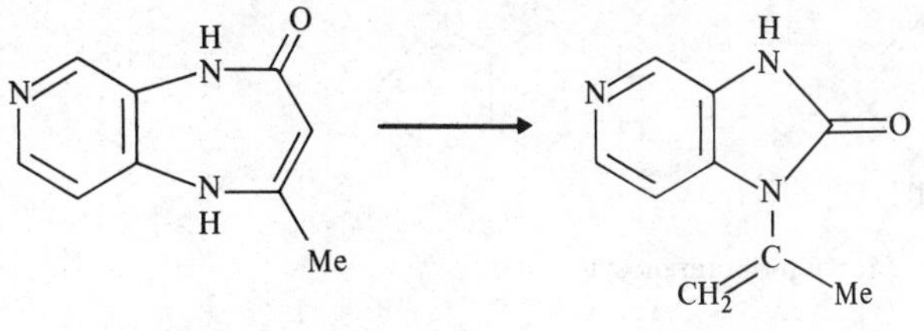

(236) (237)

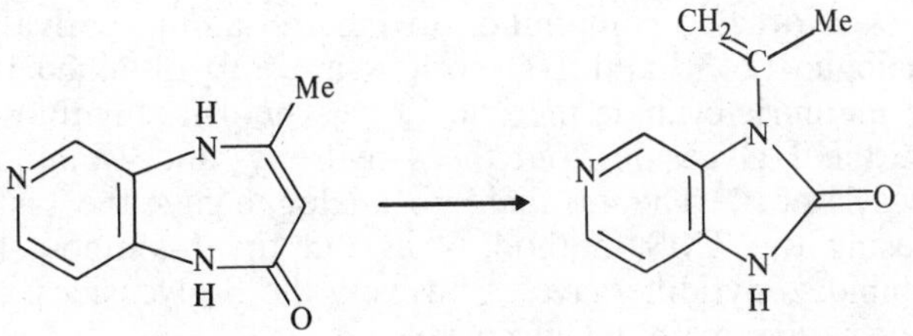

(238) (239)

Miscellaneous syntheses include the thermolytic cyclization of 3-azido-4-dimethylaminopyridine to the *N*-methyl-3-deazapurine (**240**),[226] and the formation of a 1-aminoimidazo[4,5-*c*]pyridine (**241**) on contractive rearrangement of ethyl 2-(3-aminopyridin-4-yl)hydrazinecarboxylate.[280]

Attempted alkylation of 2-methylthio- and 6-chloro-4-methylthio-imidazo[4,5-*c*]pyridines with diazomethane or methyl iodide/alkali gave a mixture of N-1, N-3, and N-5 derivatives.[281] Most 1- and 3-monomethyl derivatives have been made by cyclization of preformed methylaminopyridine derivatives, (see above). The 2-methyl analogue, however, with methyl iodide or dimethyl sulphate and alkali smoothly gave the 1,2-dimethyl product, which with excess methyl tosylate at 140 °C ultimately gave the 1,2,3,5-tetramethyl diquaternary ditosylate (**242**), whose 2-methyl group is very reactive, *e.g.* with aldehydes[282] and in the formation of polymethine dyes with triethyl orthoformate in nitrobenzene.[283] Benzylation of 2-substituted imidazo[4,5-*c*]pyridines also has been found to occur preferentially at N-3, as with the imidazo[*b*]pyridines, but some N^5-benzylation was also observed.[284]

The preparation of glycosides, on the other hand, has been much more widely studied. Imidazo[4,5-*c*]pyridine itself, as the chlormercuri-salt, was at first reported to give the 3-β-D-ribofuranoside (ly) with the protected halo-sugar,[240] but later work showed the product to be a mixture of equal parts of the 1- and 3-nucleosides,[278] the 4-chloro-analogue behaving similarly. Another report of 3-riboside formation using the mercuric cyanide/bromo-sugar/nitromethane method[285] is therefore suspect, but it has been noted elsewhere that mercury salt methods of glycosylation tend to favour unusual derivatizations.

The 7-nitrochloromercuri-salt gave an unidentified mixture of products.[240] On the other hand, the acetyl sugar fusion method, using deoxy-sugar derivatives, gives with the 4-chloro-base the 4-chloro-N^1-glycoside as the major product.[286] This was converted *via*

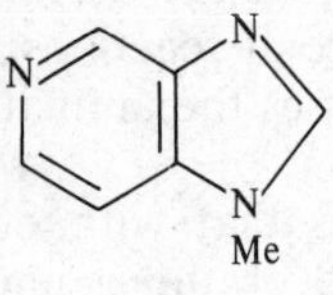

(240) 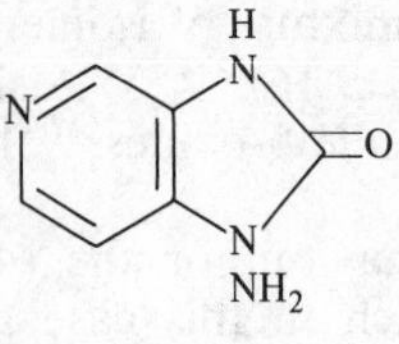 (241) 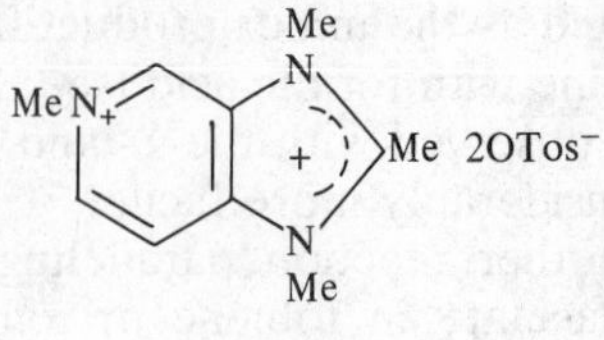 (242)

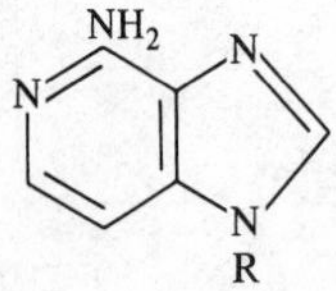

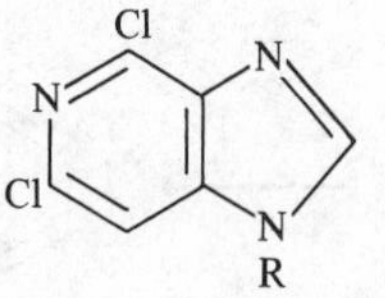

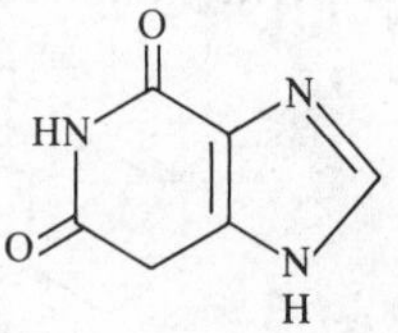

R = β-D-ribofuranosyl

(243) (244) (245)

reduction of the 4-hydrazino-compound to the 4-amino-derivative, *e.g.* the 3-deoxyadenosine analogue (**243**), and 1-glycosides are also obtained from the 4-chloro-compound with the mercuric cyanide method.[274,287] The 4-methylthio- and 4-acylamino-analogues also gave the 1-glycosides, but the 4-hydroxy- and 4-amino-derivatives could not be directly glycosylated.[288] The 4,6-dichloro-analogue gave the 1-glycoside (my), with a little 3-isomer, using the TMS method,[289] as did the 6-amino-4-bromo-derivatives, whilst the 2-methylimidazopyridine gave exclusively the 3-glycosides.[290] The 4-chloro-2-methyl bases, however, give mainly 1-glycosides.[274]

In other electrophilic substitutions the 2-unsubstituted analogue reacts with sulphur on fusion to give a good yield of 2-thioxo-derivatives, which can be desulphurized with nitric acid.[291]

Acetylation of 4-amino-derivatives takes place at the exocyclic amino-group, and *N*-oxide formation, as expected, at N-5,[277] whilst nitration of the unsubstituted or 1-methylimidazo[4,5-*c*]pyrid-2-ones is said to give the 4-nitro-derivatives, which undergo nucleophilic attack with HCl or HBr to give the 4-chloro- or 4-bromo-derivatives.[292] 2-Methyl compounds, especially with 1-alkyl, 1-benzyl, or 1-aryl substituents, are oxidized by SeO_2 to aldehydes, which can be converted to nitriles (*via* oximes) or to acrylic acid derivatives.[293] Although 4-chloro-groups are relatively unreactive to ammonia, requiring high temperature and a catalyst, glycosyl substituents facilitate the reaction (see above). Amines, such as aniline, also react more readily. The 4,6-dichloro-1-nucleoside (**244**), for example, gives the 4-amino-6-chloro-analogue at 110 °C without catalyst,[294] and several other nucleophiles (hydrazine, thiourea, NaSH, NaOEt, amines) react easily.[289] Subsequent reduction of the 6-chloro-group gives the 4-substituted imidazopyridines, including 3-deaza-adenosine.[294] The 4-chloro-1-glycoside with furfurylamine gives the 3-deazakinetin, which was rather inactive as a cytokinin.[295]

N.m.r. spectroscopy has been useful in the establishment of the structure of the histidine–pyridoxal reaction product,[296] in structural work on the streptothricin antibiotics, in the establishment of the structure and conformations of imidazopyridine nucleosides (see above), and in work on the tautomerism of the 4,6-dihydroxy-derivative (**245**), which was shown to exist in the dioxo-form.[297]

The pK_a's of a series of imidazo[4,5-*c*]pyridines with halo-, oxo-, thioxo-, and methylthio-substituents have been measured recently.[281] MO calculations have been made on the interaction energies of base pairs containing 3-deazapurine bases,[298] and the coding properties of 3-deazadenines have been discussed. Apart from the naturally-occurring antibiotics, imidazo[4,5-*c*]pyridines have been claimed to have useful analgesic, herbicidal, anti-inflammatory, anti-tumour, anti-viral, and other activities.[7]

17.5.19 4-AZAPURINES

4-Azapurines would be *sym*-triazolo[5,1-*f*]*asym*-triazines with numbering as shown in (**246**). One report, in a German patent,[300] is mentioned in the Abstract literature, but this seems to have been erroneous, due to a printing error in the original, so a genuine 4-azapurine has yet to be made.

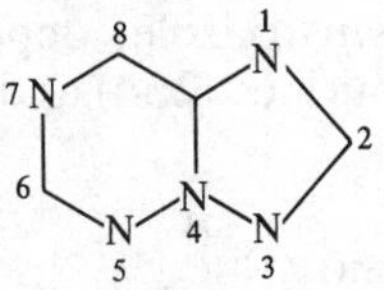

(246) 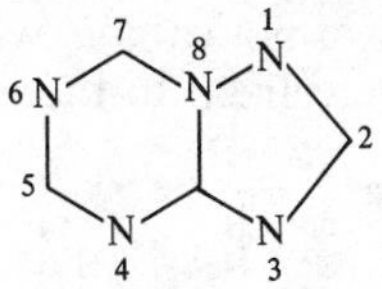 (247)

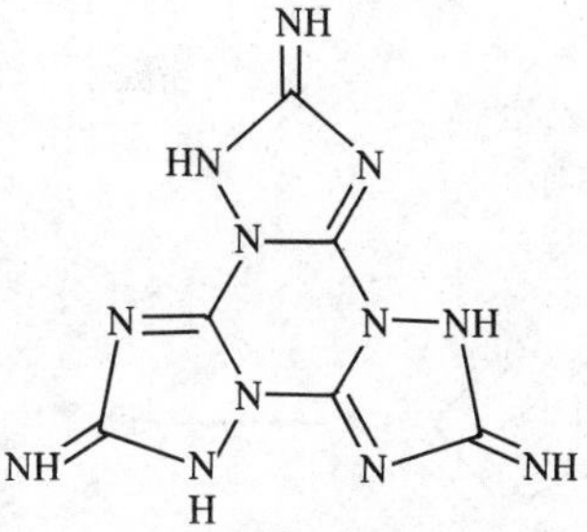

(249)

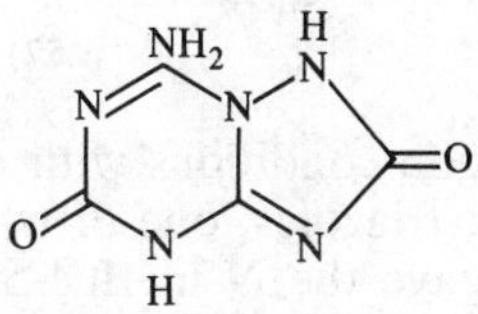

(248) 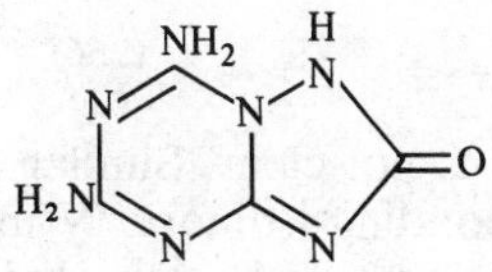 (250)

17.5.20 5-AZAPURINES

5-Azapurines are *sym*-triazolo[1,5-*a*]-*sym*-triazines, the numbering being as shown in the formula (**247**). They do not appear to occur in Nature, but have been known since 1912, when Hofmann and Ehrhart[301] obtained a derivative (**248**) by $KMnO_4$ oxidation of pyroguanazole, a tetracyclic product (**249**) from the reaction of dicyandiamide and hydrazine. The diamino-compound (**250**) was obtained using alkaline H_2O_2. No further work was done, however, until recent times when 5-aza-adenine (**251**) was obtained[302] by the reaction of the calcium cyanamide/citric acid complex ('Temposil') with *N,N'*-bis-(3-*sym*-triazolyl)formamidine (**252**) (from 3-amino-*sym*-triazole and ethyl orthoformate), *via* 3-cyanamidomethyleneamino-*sym*-triazole, and most subsequent syntheses have involved *sym*-triazole intermediates. Reaction of 3-amino-*sym*-triazole with potassium cyanate gave 5-amino-*sym*-triazole-1-carboxamide, which with ethyl orthoformate gave 5-azahypoxanthine, 7-oxo-6,7-dihydro-[6*H*]-*sym*-triazolo[1,5-*a*]-*sym*-triazine (**253**).[302] On alkylation with methyl sulphate or diazomethane this compound gave the 6-methyl-7-hydroxy-analogue, which underwent ring-opening with alkali to 3-formylamino-*sym*-triazole-1-*N*-methylcarboxamide. Final hydrolysis with HCl gave 3-amino-*sym*-triazole.[302] 5-Azaguanine was similarly prepared using urea as the cyclization agent.[303]

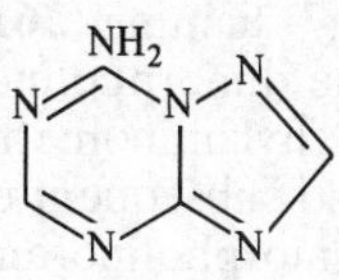

(251) 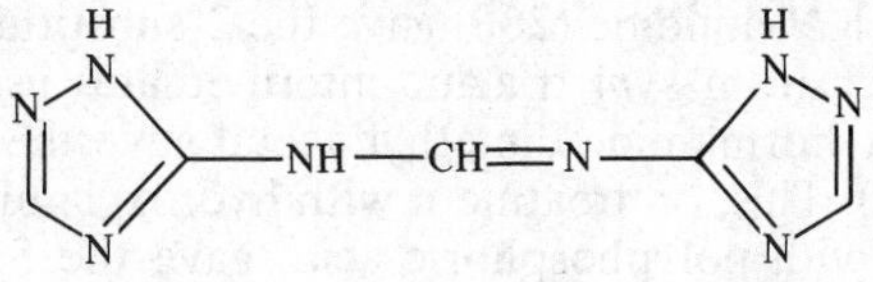 (252) 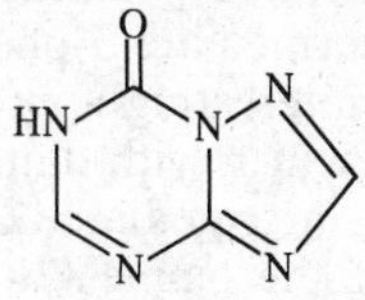 (253)

The preparation of the 5-hydroxy-7-thioxo-derivative (**254**) by aqueous alkaline cyclization (hy) of 1-ethoxycarbonylaminothioformyl-*sym*-triazole has been claimed and this reaction was later extended to 5-substituted triazoles. These 5-azapurines underwent alkylation to give the 7-methylthio-compounds, and ring-opening with HCl to give 5-substituted 3-thioureido-*sym*-triazoles.[304] Later work by American workers,[305] however, showed that different isomers, either (**255**) or (**256**), could be obtained by reaction of ethoxycarbonyl isothiocyanate with 3-amino-*sym*-triazole, depending on the conditions, and that these cyclized to the isomeric 5-azapurines (**254**) and (**257**) respectively, so

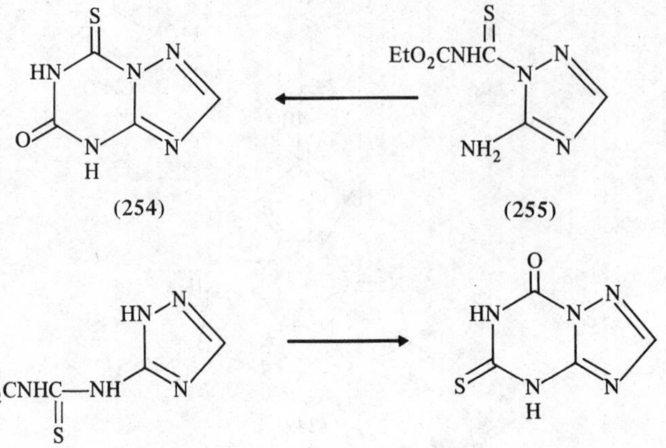

(254) (255)

(256) (257)

the earlier work is not clear. Similar dual reaction conditions with methyl isothiocyanate gave the corresponding isomeric *N*-methyl-*sym*-triazoles, one of which, on cyclization at low temperature with diethoxymethyl acetate, gave the *N*-methyl-5-azathiopurine (**258**); this on oxidation with chlorine gave the same oxo-*N*-methyl-5-azapurine as was obtained by use of KCNO followed by cyclization and methylation. Similar results had been reported by Capuano,[306] who obtained only the isomer (**257**) (derived from the more stable intermediate isomer obtained in his preparation) on heating in pyridine or *in vacuo* at 100 °C. Methylation at N-6 and alkaline or acid degradation to 5-thioureido-*sym*-triazoles was studied by both these groups of workers.

The related (3-substituted) *sym*-triazol-5-ylisothioureas, obtained from the above thioureas by methylation with alkyl halide/base, cyclized rapidly with ethyl orthoformate to the 2-substituted 5-alkylthio-analogues *e.g.* (**259**) (Scheme 15).

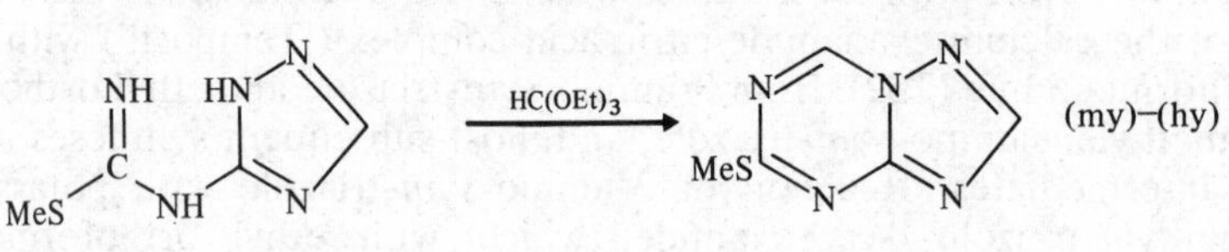

$HC(OEt)_3$ (my)–(hy)

(259)

SCHEME 15

Other approaches involved the 1-guanylation of 3-amino-*sym*-triazole with 1-guanylbenztriazole, followed by cyclization with ethyl orthoformate or 100% formic acid, to give 2-substituted 7-amino-*sym*-triazolo[1,5-*a*]-*sym*-triazines. On refluxing alone in xylene the intermediate 3-phenyl-*N*-amidine (**260**) gave the 2-substituted 5,7-diamine (**261**).[308]

An interesting approach from *sym*-triazine intermediates used the dimorpholinoamine (**262**), which with dimethylformamide dimethyl acetal gave the dimethylaminomethyleneamino-intermediate (**263**). This, on treatment with hydroxylamine and subsequent cyclization of the oxime (**264**) with polyphosphoric acid, gave the 5,7-bismorpholino-analogue (**265**).[309]

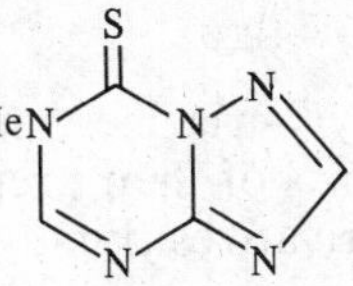

(258)

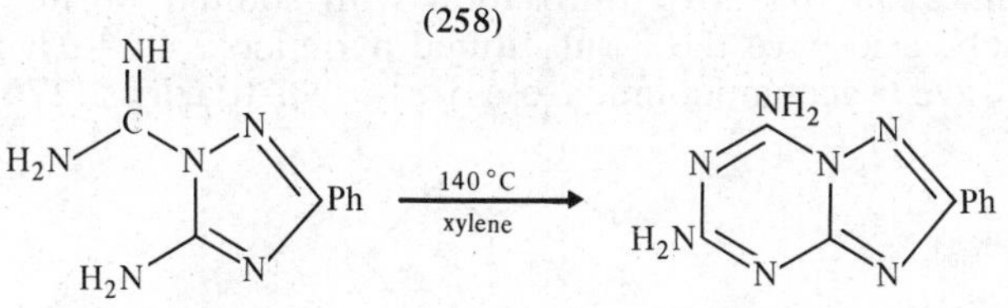

(260)

(261)

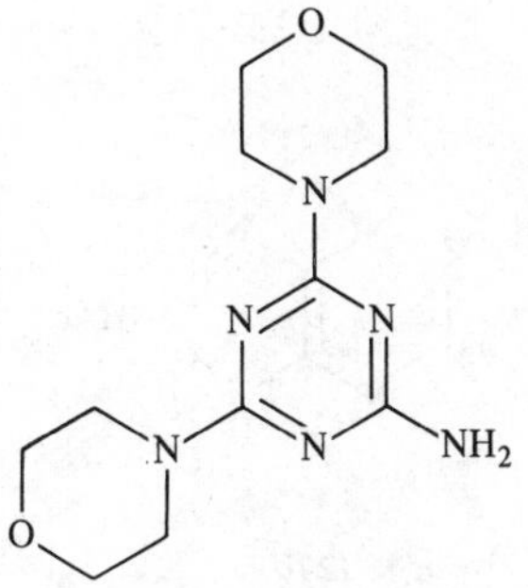

(262)

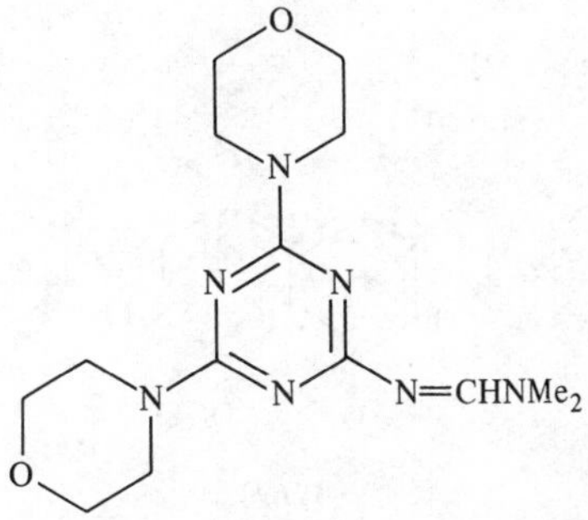

(263)

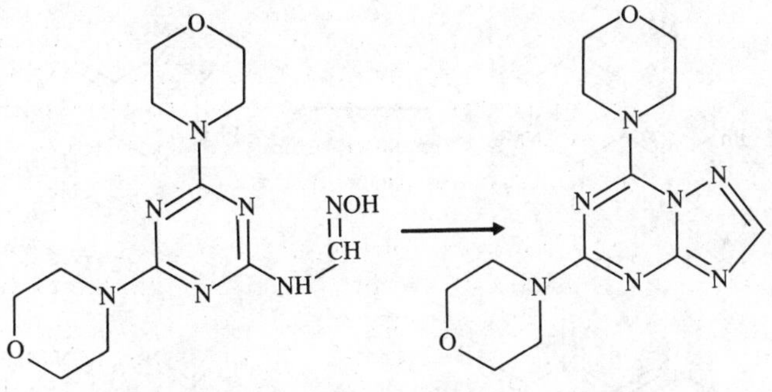

(264)

(265)

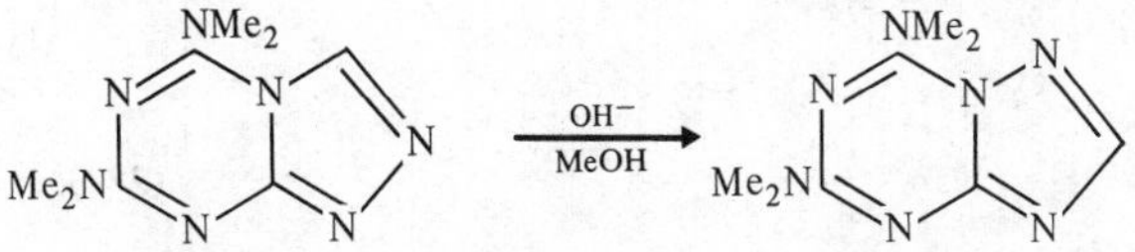

(266)

(267)

Finally, although Jacquier had predicted Dimroth rearrangement of 5-azapurines from the low electron density at the 7-position in MO calculations,[310] it appears that this is not so, as two representatives have been prepared by Dimroth rearrangements of other ring systems. Firstly, the cyclodehydration of 2-acylhydrazino-*sym*-triazines gives *sym*-triazolo[4,3-*a*]*sym*-triazines, *e.g.* (**266**), which undergo Dimroth rearrangement to the isomeric *sym*-triazolo[1,5-*a*]*sym*-triazines, *e.g.* (**267**),[311] whilst 2,6-bis(dimethylamino)-7-methylthio-*sym*-triazolo[4,3-*a*]-*sym*-triazines with piperidine (or aniline) gave 5-dimethylamino-7-piperidino-2-methylthio-*sym*-triazolo[1,5-*a*]-*sym*-triazines *via* a Dimroth-type rearrangement,[315] the structure of the product being proved by X-ray analysis.

5-Azapurines have been mentioned biologically only as cytostatics and CNS-active agents.

17.5.21 6-AZAPURINES

6-Azapurines are imidazo[4,5-*e*]*asym*-triazines, numbered as shown in (**268**). They do not occur naturally and only two reports of their preparation are extant. Both involve the ring-contraction of 7-azapteridine derivatives. Firstly, 2-amino-4-hydroxy-6-benzylidene-hydrazinopyrimidines on oxidative nitrosation with sodium nitrite and diethyl azodicarboxylate in acetic acid gave the 3-substituted pyrimido[5,4-*e*]-*asym*-triazine 4-oxides, *e.g.* (**269**), which gave 6-acetamidoimidazo[4,5-*e*]-*asym*-triazines (**270**) (Scheme 16). The

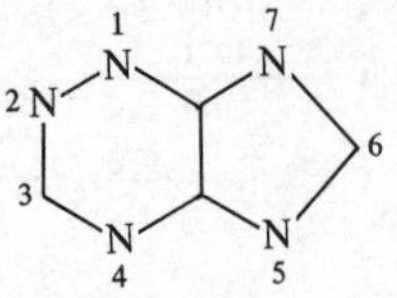

(268)

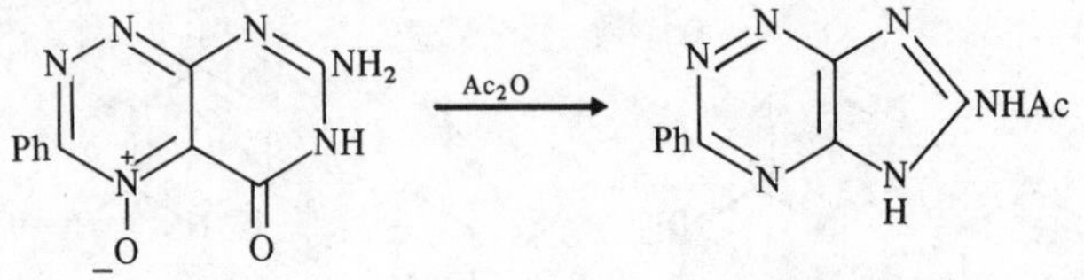

(269) (270)

SCHEME 16

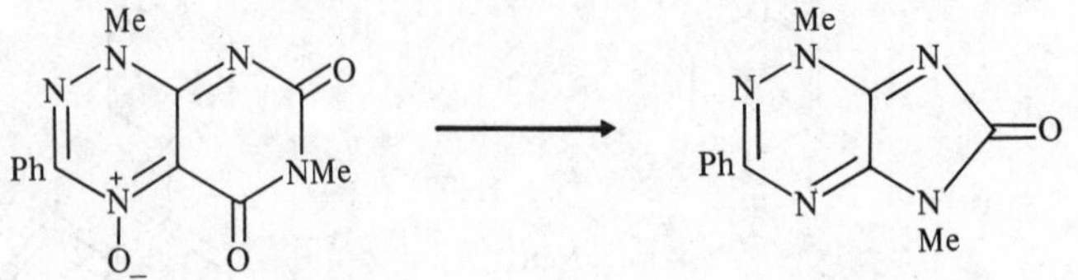

(271) (273)

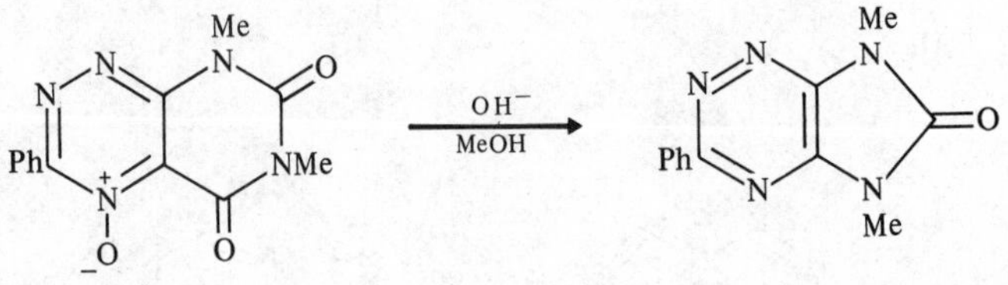

(272) (274)

naturally occurring toxoflavin (**271**) and fervenulin (**272**) 4-oxides also show the rearrangement, believed to be of the benzilic acid type, giving (**273**) and (**274**) respectively. Finally, 1,3-dimethyl-6-phenyl-7-azalumazine (**275**) on heating at 60 °C for a short time in 10% alcoholic sodium hydroxide gave 5,7-dimethyl-3-phenylimidazo[4,5-*e*]*asym*-triazin-6-one (**276**).[314]

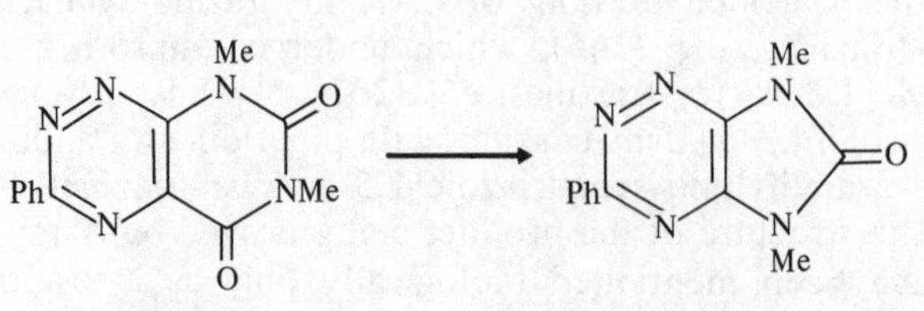

(275) (276)

17.5.22 7-DEAZAPURINES

The 7-deazapurines, pyrrolo[2,3-*d*]pyrimidines, are numbered as shown in (**277**). They are an important class of compound, their chemistry having been dominated by two classes of naturally occurring antibiotics, tubercidin (sparsomycin) (**278**) from *Strep. tubercidicus* or *S. sparsogenes*, and toyocamycin (sangivamycin) (**279a, b**) from *S. toyocaensis*, which have proved to possess considerable anti-leukaemia activity. Other naturally occurring 7-deazapurines include the *S*-adenosylmethionine analogue (**280**) and the rare ribonucleoside Q (**281**), a component of the first-position codon on *E. coli* tRNA.[315,316]

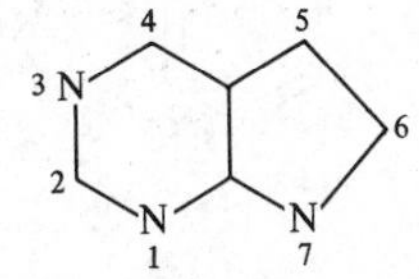

(277)

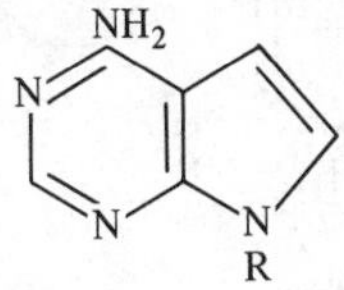

(278) R = β-D-ribofuranosyl

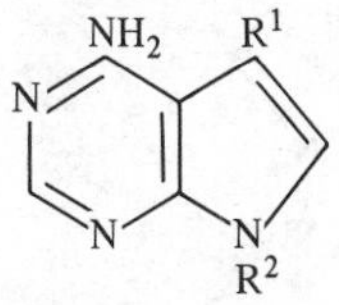

R^2 = β-D-ribofuranosyl

(279) a; R^1 = CN
b; R^1 = $CONH_2$

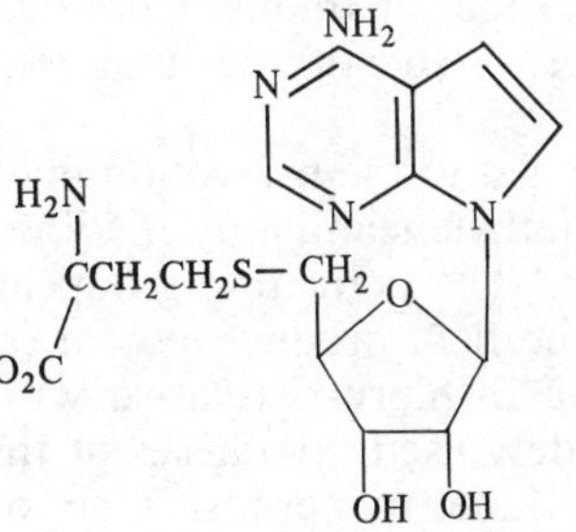

(280)

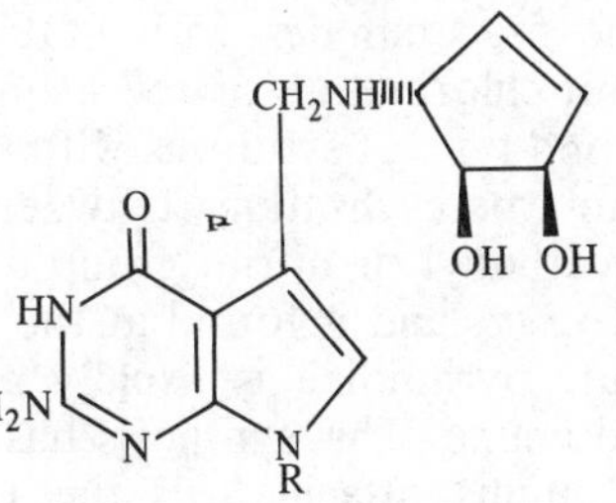

(281) R = β-D-ribofuranosyl

The synthesis of 7-deazapurines from pyrimidine intermediates can be divided conveniently into a number of types. Firstly, those in which two or three of the final 5-, 6-, or 7-atoms are initially attached to position-5 of the pyrimidine; secondly, those in which two or three of these atoms are initially attached to the 4(6)-position of the pyrimidine; and finally other miscellaneous syntheses.

The first type with two attached atoms is exemplified by the reaction in which a 2-methyl-4-methylaminopyrimidine-5-acetic acid (**282**), on attempted esterification with ethanolic HCl cyclized to the 2,7-dimethyl-pyrrolo[2,3-*d*]pyrimid-6-one (**283**), which forms a 5-isopropylidene derivative with acetone.[317] The earliest synthesis[317a] involved a similar cyclization of a 4-chloropyrimidine-5-acetic ester with ammonia. The use of 4-amino- or 4-alkylamino-5-acetaldehyde derivatives, either formed *in situ* by periodate oxidation of a dihydroxypropyl derivative[318] or in the form of the diethyl acetals (*e.g.* in a synthesis of 7-substituted analogues of tubercidin),[319] has been widespread and gives the fully aromatic derivatives, *e.g.* (**284**), with acid. The 5-acetonyl derivatives give 6-methyl derivatives, and 5-alkyl groups arise from branched-chain substituents. Chloro-, alkyl, thio-, hydroxy- and amino-substituents are compatible. Using 80% acetic acid as cyclizing agent, 7-glycosyl substituents are preserved.[320]

A variation employs the 5-(2-chloroethyl) analogues, leading to 5,6-dihydro-compounds, including 7-alkyl derivatives, which are readily dehydrogenated by Pd/C. Sometimes, 4-chloro-5-haloalkylpyrimidines are directly aminated and cyclized in a high-boiling amine. Another variant uses *in situ* $POCl_3$ chlorination and cyclization with amines of a 4-hydroxypyrimidine.[321] When hydrazine was used with 4(6)-chloro-pyrimidines, the products were 7-amino-substituted pyrrolo[2,3-*d*]pyrimid-6-ones (**285**),[322] whilst use of N^1-hydroxyuracils gave the corresponding 1-hydroxypyrrolo-pyrimidines, *e.g.* (**286**).[323]

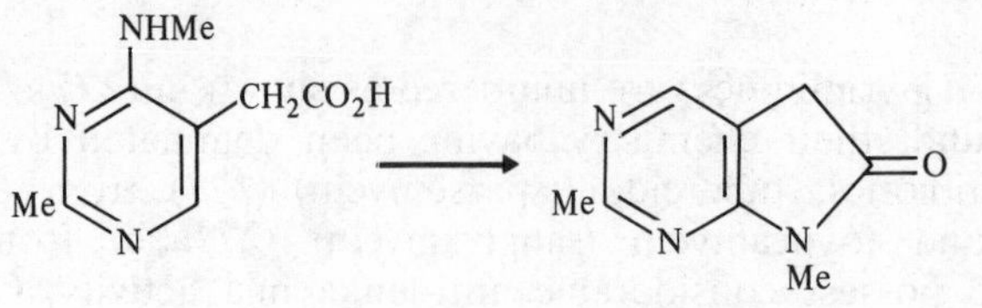

(282) (283) 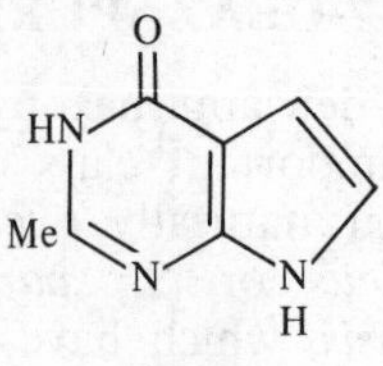 (284)

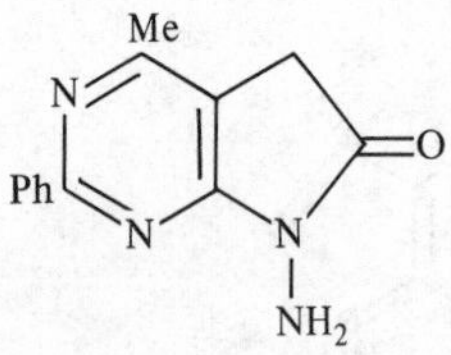

(285) 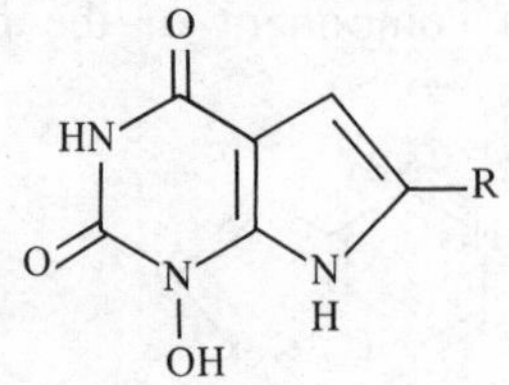 (286) 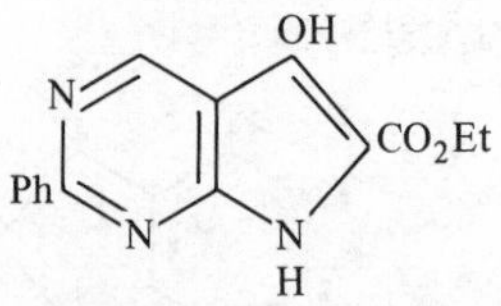 (287)

Syntheses with all three atoms initially attached to the 5-position are much rarer, the first being the cyclization of 5-β-aminoalkyl-6-hydroxypyrimidines on chlorination with such reagents as oxalyl chloride, phosphorus oxychloride, or thionyl chloride.[324] Cyclization of the 5-acetamides with $POCl_3$ gives 6-chloro-7-substituted analogues following concomitant chlorination of the 6-one.

The second type of synthesis with two atoms attached to the 6-position involves mainly Dieckmann-type cyclizations between 6-activated methyleneaminopyrimidine groups and an ethoxycarbonyl or nitrile group at the 5-position. The activating group may itself be CN, CO_2Et, *etc.* and is found at the 6-position in the final product, *e.g.* in (**287**).[325] The intermediate pyrimidine is usually prepared from a dichloro-derivative with a suitably substituted amine. This synthesis has been quite widely used. Syntheses of this type with the three atoms attached to the 6-position are quite numerous. One of the most interesting is the pyrimidine analogue of the Fischer indole synthesis, in which the 4-pyrimidylhydrazone of an aldehyde or ketone is converted into the pyrrolo[2,3-*d*]pyrimidine, *e.g.* (**288**), presumably *via* a similar mechanism.[326]

Another uses the condensation of a 6-aminopyrimidine with an α-halo-aldehyde or -ketone,[329] such as chloroacetaldehyde or phenacyl bromide,[351] sometimes in the form of its acetal[328] or dithioacetal, the products being 6(7)-alkyl- or -aryl-substituted pyrrolopyrimidines. Another possible reactant is the protected derivative of an α-dicarbonyl compound, *e.g.* glyoxylic acid hemiacetal, with which further condensation leads to a pyrimidyl-substituted product (**289**).[329] Cyclizations may be brought about thermally or by alkaline catalysis. An α-halo-ester gives the corresponding 6-one.[327] Unsaturated derivatives may also be used, such as diacylethylenes, which give, *e.g.* (**290**), in a type of Nenitzescu reaction.[330] A photochemical equivalent is seen in the cyclization on irradiation of, for example, 6-anilino-4-chloropyrimidine to the 6,7-benzo-fused pyrrolopyrimidine (**291**).[331]

Miscellaneous cyclizations include one of the earliest syntheses of the ring-system in which diethyl α-cyanosuccinate was condensed with urea and sodium ethoxide to give the 2,4,6-trihydroxy-derivative (**292**).[332] The use of the corresponding iminoether with amidines gave the 2-alkyl-4,6-dihydroxy-derivatives, whilst the diethoxyethyl cyanoacetate derivative (**293**) gave the 2,4-dihydroxy-analogue. Another is the formation of the 2-methyl-6-one on attempted conversion of 2-methyl-4-aminopyrimidine-5-acethydrazide to the azide with nitrous acid.[333]

Syntheses from pyrrole intermediates are more numerous than with the 9-deaza-isomer, owing to their greater availability. Among the earliest was a synthesis of toyocamycin aglycone from 2-thio-4-amino-3,5-dicyanopyrrole, which with trimethyl orthoformate (concomitant methylation of thio group), followed by ammonia, gave the 4-amino-5-cyano-6-methylthiopyrrolopyrimidine. Similar results were obtained on cyclization with formamide, except that the free 2-thio-analogue was obtained.

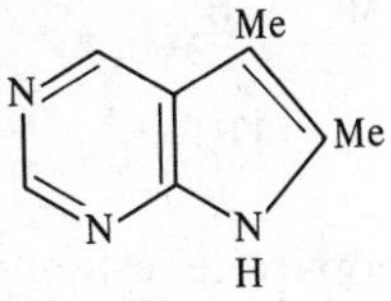

(288) 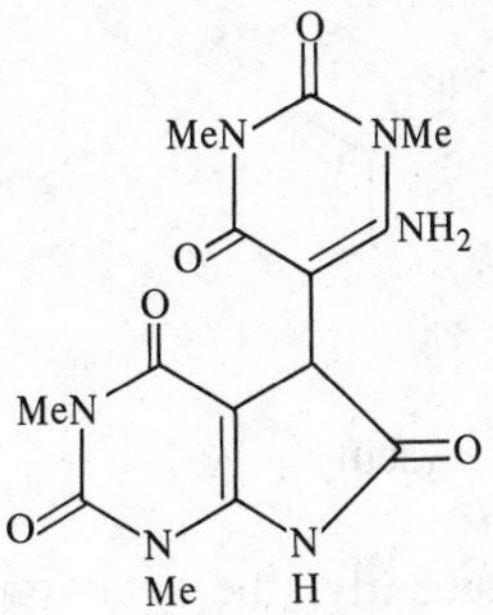 (289) 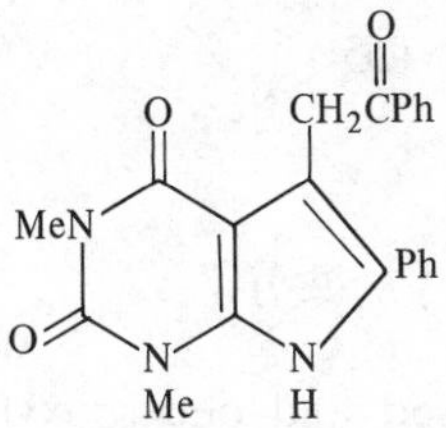 (290)

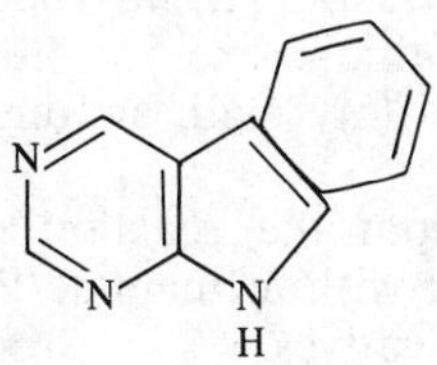

(291) 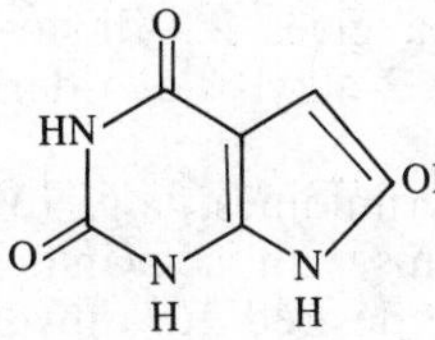 (292) 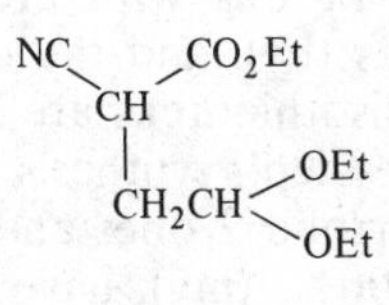 (293)

Similar aminocyanopyrrole based syntheses have furnished 5-methyl- and 5-phenyl-pyrrolopyrimidines from the appropriately substituted intermediates, although in these cases final cyclization required sodium methoxide/pyridine.[334] Desulphurization with Raney nickel furnished the desired aglycones, *e.g.* (**294**). Another synthesis of toyocamycin itself also involved a similar pyrrole intermediate, the 5-bromo-2-amino-3,4-dicyano-derivative. This gave the aminobromopyrrolopyrimidine, which after protection and glycosylation was reductively debrominated to give the antibiotic.[335]

2-Bismethylthiomethyleneamino-4-phenyl-3-cyanopyrrole with alkylamines gave the 4,7-dihydro-3-alkyl-4-amino-2-methylthio-products (**295**),[336] whilst an aminocyano-pyrrole and CS_2 gave the thiazine (**296**), converted to thiopyrrolo[2,3-*d*]pyrimidines in a variety of ways.[336] 1-Substituted 2-amino-3-cyanopyrroles give 7-substituted pyrrolo-pyrimidines on cyclization in a mixture of formic acid, formamide, and DMF.

When aminopyrrole esters are used,[336] cyclization with acetanilides gives 3-arylpyrrolo-pyrimidin-4-one-5-carboxylic esters, *e.g.* (**297**), a methaqualone metabolite, after being

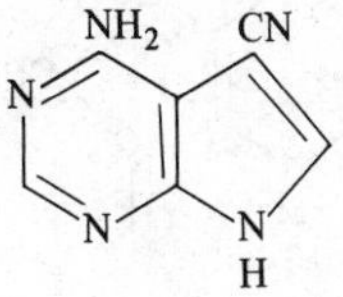

(294) 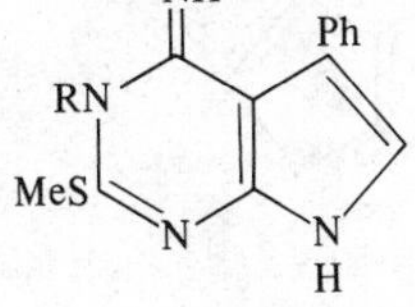 (295) 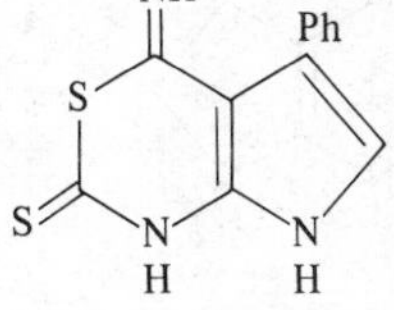 (296)

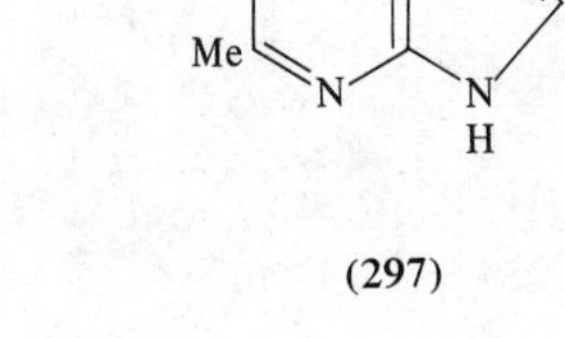

(297) 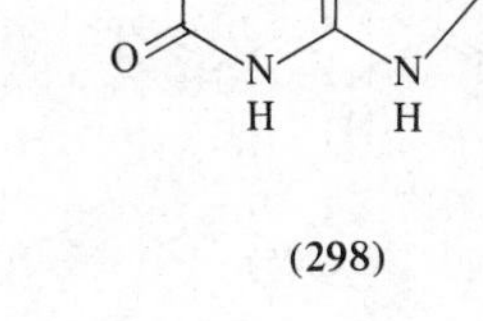 (298)

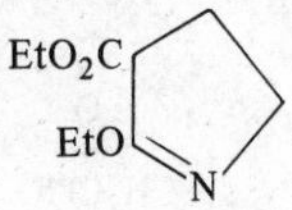

(299) 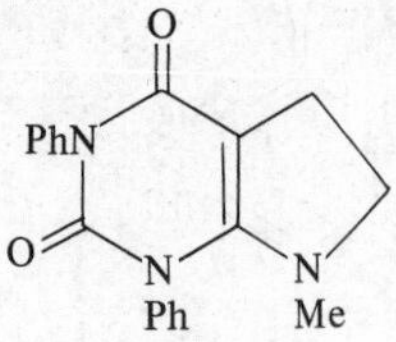 (300) 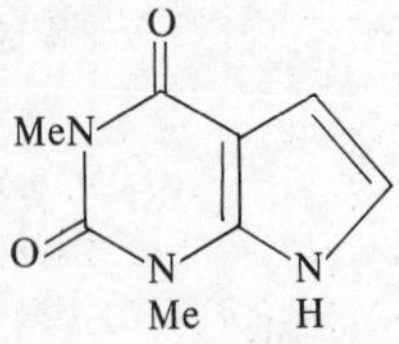 (301)

hydrolysed and decarboxylated.[337] Recently, the reaction of aminopyrrole esters with isocyanates to give, *e.g.* (**298**), after alkaline cyclization, has been described.[338]

A more flexible pyrrole-based synthesis is based on the interesting intermediate (**299**), obtained from ethyl pyrrolid-2-one-3-carboxylate with triethyloxonium fluoroborate. Reaction of this with urea or thiourea gives 2,4-dioxo- or 2-thio-4-oxo-5,6-dihydro-derivatives (hy), and the corresponding 2-alkyl-4-oxo-derivatives (my) with amidines.[339] *S*-Methylisothioureas are less successful.

A 'one-step' synthesis of pyrrolopyrimidines, *e.g.* (**300**), from the reaction of 1-methylpyrrolid-2-ones and similar lactams with isocyanates,[340] or with formamide/$POCl_3$ at 160–200 °C (my), have furnished the desired 5,6-dihydro-derivatives.

Alkylation of 4-substituted pyrrolo[2,3-*d*]pyrimidines in alkaline conditions (methyl iodide/NaOMe,[341] or NaH or DMF), or Michael addition of acrylonitrile,[342] occurs at the 7-position as expected, as does alkylation of the 7-deazatheophylline (**301**) with methyl sulphate/alkali.

4-Acetamido(diacetylamino)-5-cyano-6-bromo-[343] or 6-methylthio-compounds[344] gave 7-nucleosides with the acetyl sugar fusion method [acid or bis(*p*-nitrophenyl) phosphate catalyst], or *via* the chloromercuri-derivative, but the 4-amino- and 4-chloro-derivatives could not be glycosylated directly.[344,345] Use of the trimethylsilyl (Wittenberg) procedure gave a mixture of 1- 3-, and 7-glycosides,[346] but the use of silyl intermediates with the halo-sugar and silver oxide in acetonitrile gave a good yield of 7-nucleoside.[346] The structures were interpreted by the use of n.m.r. and u.v. spectra.

Electrophilic substitution in most 4-substituted pyrrolo[2,3-*d*]pyrimidines occurs at the 5-position, including halogenation (chlorine, iodine,[349] bromine,[347] or *N*-bromo-amides[347,348]), nitration,[347] sulphonation,[348] azo-coupling,[318] and the Mannich reaction, the products of which [*e.g.* (**302**)] behaved like the indole analogue, gramine, in undergoing a variety of nucleophilic substitution reactions to give, for example, (**303**).[349]

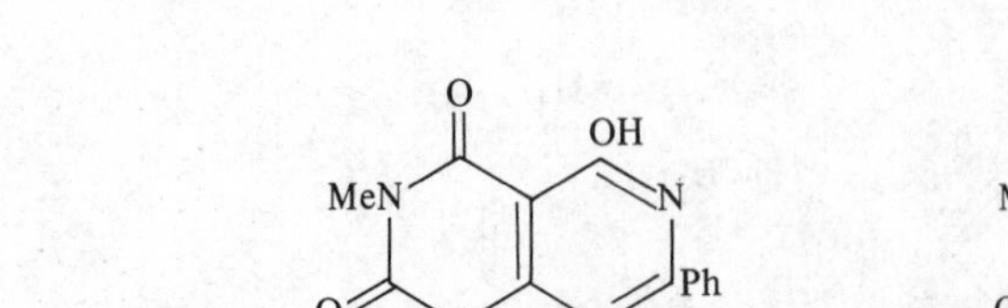

NaCN

(302) (303) 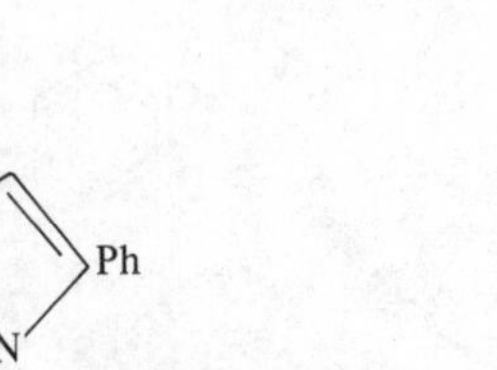 (304)

(305)  (306)

With the theophylline analogue (**301**), however, the Mannich reaction and the Vilsmeier–Haack reaction are reported to occur in the 6-position,[350] proved in the latter case by oxidation of the derived aldehyde (**304**) to the known 6-carboxylic acid.[350] This difference is postulated to be due to the increased electron density of the alkylated N-1 position, but may be due to a secondary rearrangement. On nitrosation of this type of pyrrolopyrimidine the first-formed nitroso-compound, on reduction with dithionite or bisulphite, or treatment with amines, undergoes a complex rearrangement to give pyrimido[4,5-*d*]pyrimidine derivatives, *e.g.* (**305**).[351]

In nucleophilic substitutions, oxo-groups at the 4-position are readily replaced (my–hy) by chloro-groups using $POCl_3$, and chlorination of theophylline analogues (hy) may occur, *e.g.* to (**306**). The use of $POCl_3$/amines gives 4-amino-compounds, or the 4-(*p*-dimethylaminophenyl) compound with dimethylaniline[352]. The 4-chloro-compounds thus obtained are not particularly reactive. Reaction with ammonia requires pressure and a temperature of 145–160 °C,[341] but reaction with amines proceeds at moderate temperatures (best at 125 °C with a trace of HCl catalyst)[353] to give a wide variety of 4-amino-derivatives, including analogues of tubercidin, toyocamycin, and sangivamycin, the kinetins, or chloroquine antimalarials. The kinetics of the reaction have been compared with those for 2-chloropyridine.[354] Reactions occur smoothly with a wide variety of other nucleophiles such as thiourea (to give thio-derivatives), methoxide, mercaptides (KSR/115 °C/EtOH), and hydroxylamine. Thio- and methylthio-groups are readily replaced by hydrogen using Raney nickel in ethanol, or by ammonia. Chloro- and bromo-groups at the 5- or 6-positions appear to be unreactive, but 6-thio- and methylthio-groups have also been removed using Raney nickel.[334] The *O*-acetyl or *O*-tosyl derivatives of 1-hydroxyxanthine analogues undergo remote nucleophilic substitution at the 6- (or sometimes 5-) position to give, *e.g.* (**307**) with methionine, presumably by a similar mechanism to that observed in 3-acetoxyxanthines (Section 17.5.8).[323]

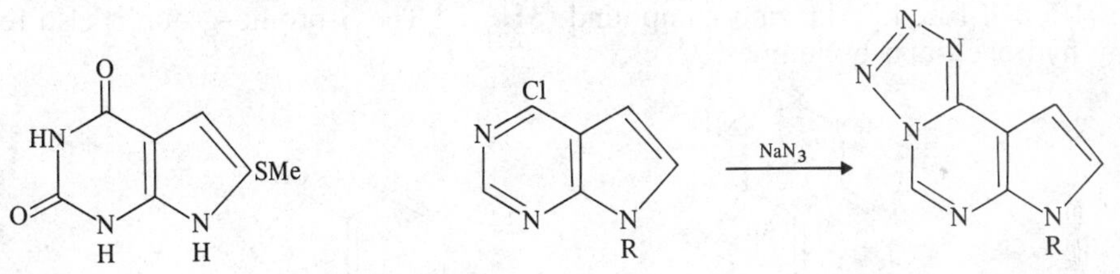

(307) (308) R = β-D-ribofuranosyl (309)

Some pyrrolo[2,3-*d*]pyrimidines (theophylline analogues) undergo facile catalytic (Pd/C) hydrogenation to dihydro-derivatives at the 5,6-positions.

Several fused cyclic derivatives similar to those seen in the purine field (Section 17.5.11) have been reported, including 2,3-thiazolo-derivatives,[355] 3,4-imidazo-compounds (tubercidin, toyocamycin, and sangivamycin analogues of wyosine)[356] and a corresponding series of products from 4-chlorosubstituted nucleosides, *e.g.* (**308**), with azide ion (hy), believed to have the analogous 3,4-tetrazolo-structure (**309**).[357]

The chemistry of tubercidin (sparsomycin) (**278**) will be discussed here for convenience of reference. The antibiotic[358] is obtained from fermentation batches using the organisms *S. sparsogenes* and *S. tubercidicus*, improved yields being given by mutant strains.[359] The biosynthesis of the aglycone has been discussed.[360] The structure as 4-aminopyrrolo[2,3-*d*]pyrimidine-7-β-D-ribofuranoside was established by degradation and spectroscopic studies, especially using u.v. and n.m.r., and confirmed first by X-ray analysis and later by total synthesis.[343]

The 4-amino-group is best converted to the 4-oxo-derivative using barium nitrite in acetic acid at 65 °C. Thiation of the lead salt with H_2S then gives the 4-thio-compound, which has anti-tumour activity.[361]

Methylation of tubercidin with methyl iodide, methyl sulphate, or methyl tosylate in DMF or DMAC gives mainly the N^3-methyl derivative (hy), plus a little N^4-methyl isomer, into which it is completely converted by Dimroth rearrangement on treatment with aqueous sodium hydroxide.[362] Alkylation with ethylene oxide, however, gives the N^4-hydroxylalkyl derivative direct. The latter also has anti-tumour activity.

Tubercidin gives a N^3-oxide with peracids and is aminated at the same position by Sheradsky's reagent, or hydroxylamine-*O*-sulphonic acid.[104]

The binding of various metal ions to tubercidin has been studied by an n.m.r. line-broadening method,[363] in which H-2 showed greater broadening than H-8, indicating that the metal was attached at the pyrimidine ring, probably at N-1, in agreement with perturbation theory.[364] In addition to its cytotoxicity and anti-tumour activity, tubercidin has been shown to have anti-bilharzial and insecticidal properties. Two useful reviews exist on the antibiotic.[365]

Toyocamycin (**279a**)[366] and its close relation sangivamycin (**279b**)[367] are isolated from cultures of *S. toyocaensis*. The biosynthesis of toyocamycin[365,368] and its relation to that of nucleoside Q[369] have been discussed. The structures were determined, as for tubercidin, by degradative[370] and spectroscopic (n.m.r.) studies and confirmed by total synthesis.[335,343]

Fusion of toyocamycin with alkali converts the 4-amino-group to an oxo-group with loss of the 5-cyano-group. More moderate hydrolysis conditions, or use of hydrogen peroxide/ammonia, cause conversion to the amide, which is sangivamycin. The 5-cyano-group is less susceptible to attack if an anion-forming group is present at the 4-position. In acid conditions the reverse is true.[371] The cyano-group in toyocamycin analogues is readily converted to the imino-ether and thence to the amidine, to the amidoxime with hydroxylamine, and to the thioamide with H_2S. 1,3-Dipolar cycloaddition at the nitrile group leads to 5-heterocyclyl derivatives.[372]

Hydrazine leads to formation of a cyclic aminopyrazolo-derivative, bridged at C-5—C-6, from 6-bromotoyocamycin,[373] whilst from the 4-chloro-analogue, methylhydrazine gives a 4/5 *peri*-fused pyridazino-compound (**310**).[374] The 6-bromo-group is also replaceable by hydrogen and by amines.[373]

MeN N NH$_2$ N N N R

(310) R = β-D-ribofuranosyl

OH N CN Ph N N Me

(311)

O N R Ph N N CN Me

(312) a; R = Cl
b; R = OMe

MeO OMe N O Ph N N Me

(313)

Electrophilic substitution of toyocamyin takes place readily at the 6-position with bromine, but other electrophilic reagents were either unreactive or caused extensive decomposition.[373]

In the case of an analogue of toyocamycin with the nitrile group at the 6-position (**311**), an interesting rearrangement occurred on treatment with thionyl chloride. This gave the

6-chloro-6-cyano-5-oxo-derivative (**312a**), which with refluxing methanol gave the 6-methoxy-6-cyano-compound (**312b**). With sodium methoxide, (**312a**) gave the 5,5-dimethoxy-6-one (**313**) *via* an oxirane intermediate, whilst (**312b**) gave an imidate, together with ring-opening to a pyrimidine derivative.[375]

N.m.r. spectroscopy has proved useful for structure details,[376] especially for cyclonucleosides.[377] In addition, aromaticity in 7-deazapurines has been studied by n.m.r.,[378] as has their tautomerism using ^{13}C n.m.r.[379]

Some semi-empirical MO calculations have been performed on the interaction energies of 7-deazapurine base pairs and the coding properties of a series of poly-7-deaza-adenines has been studied.[299] Mass spectra of 4-dimethylaminopyrrolo[2,3-*d*]pyrimidines showed considerable differences from those of the corresponding adenine derivatives, indicating that initial electron attack in these adenines may be at N-7.[380]

Reviews are available on the synthesis[381] and biological effects[7] of pyrrolo[2,3-*d*]pyrimidines, and on their 4-seleno-7-nucleosides.[382]

17.5.23 8-AZAPURINES

8-Azapurines, *vic*-triazolo[4,5-*d*]pyrimidines [numbering as shown in (**314**)], only occur in Nature in the fermentation products (vasocidin, pathocidin, *i.e.* 8-azaguanine) of a rare variety of *S. albus*,[383] but have nevertheless been the most widely studied of all the azapurines. This is principally on account of the very high experimental anti-tumour activity[384] of the 5-amino-7-hydroxy-derivative 8-azaguanine (**315**) (which has been clinically disappointing), leading to the search for other active congeners, whilst recently very high antiallergic activity has been discovered in some 8-azahypoxanthines.[385]

8-Azapurine is a weak base (pK_a 2.05, but this is probably that of the covalent hydrate, the anhydrous form being weaker see Ref. 386), and many of its oxo-derivatives are quite strong acids.[42]

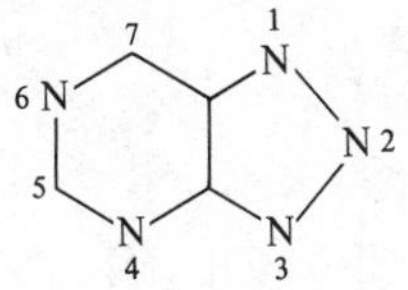

(314) 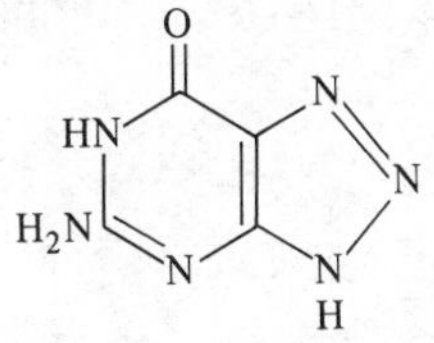 (315)

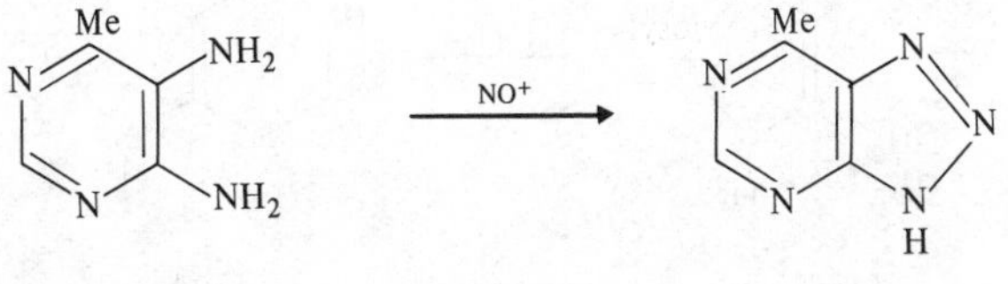

(316) $\xrightarrow{NO^+}$ (317)

The first synthesis of an 8-azapurine in 1901[387] was by the 'Traube' method by treatment of a 4,5-diaminopyrimidine derivative (**316**) with cold nitrous acid to give (**317**), and this has remained the major method of preparation ever since. A wide variety of substituent groups may be used in the remaining pyrimidine positions, leading to 5,7-disubstituted triazolopyrimidines. *N*-Substitution in the 4-position leads to the appropriately N^3-substituted triazolopyrimidines. (N^1- and N^2-Substituted derivatives must be prepared by other means, see under). Here again the substitution may be quite wide-ranging, including alkyl, aralkyl, aryl, cycloalkyl, dialkylaminoalkyl, and a variety of glycosyl and sugar-alcohol derivatives. One limitation is that the 5-amino-4-alkylamino-pyrimidines cannot be prepared by the commonly used nitrosation and reduction of

5-unsubstituted intermediates, as prior cyclization to a purine then takes place (see Section 17.5.5).

The reaction conditions usually involve use of sodium nitrite in hydrochloric or acetic acid,[358] isopentyl nitrite in ethanol or dioxan, or *in situ* reduction of a nitrosopyrimidine with a bisulphite/nitrite mixture.[389] The best conditions are those given by Roblin *et al.*,[390] or in a recent patent.[391]

The method is applicable to 1,3-dialkyluracils and has been used in a study of *N*-methylated 8-azaxanthines.[392] The use of appropriately substituted pyrimidine *N*-oxides gives the 8-azapurine *N*-oxide,[393] whilst a 1-oxide is obtained from a 5-nitro-6-*N*-methylbenzylidenehydrazinouracil on photolysis.[394] The cyclization of the 4-*p*-nitrobenzylidenehydrazino-5-aminopyrimidine (**318**) gave the N^3-amino-product (**319**), but the benzylidene protecting group could not be removed, however, without complete breakdown of the molecule.[395]

A convenient method for 2-aryl-*vic*-triazolo[4,5-*d*]pyrimidines consists in the oxidative cyclization of 4-amino-5-arylazopyrimidines (**320**) with copper sulphate in refluxing pyridine,[396] or with lead tetra-acetate in acetic acid.[397] Again the other substituents in the pyrimidine ring may be quite varied (but not thio), and the *N*-alkyluracils give very good yields.[42,398]

Me; NH_2; N; N; NH; N=CH; NO_2; NO^+; Me; N; N; N; N; N; N=CH; NO_2

(318) (319)

R^2; N=NAr; N; R^1; N; NH_2

(320)

O; C; NH_2; N; N; NH_2; N; Ph; i or ii; O; HN; N; N; N; N; Ph

(321) (322)

i, $HC(OEt)_3$, Ac_2O; ii, $HCONH_2$, reflux.

SCHEME 17

The other substantially useful methods for preparation of *vic*-triazolo[4,5-*d*]pyrimidines utilize *vic*-triazole intermediates. The methods broadly correspond to those used in the purine field with imidazoles, the first synthesis by Dornow involving the 1-phenyl-5-amino-4-carboxamide (**321**) (readily available from phenylazide and cyanacetamide) to give the 3-phenyl-7-oxotriazolopyrimidine (**322**) (Scheme 17).[399] Other aryl, benzyl, or alkyl substituents may be used in place of phenyl and give good yields, whilst glycosyl substituents have also been used. Subsequent removal of 3-benzyl groups by hydrogenation, or with sodium/liquid ammonia,[42,391,400] gives the unsubstituted derivative or the benzyl may also be thus removed before cyclization and the *N*-unsubstituted triazole cyclized in a variety of ways.[42,400] In some cases prior Dimroth rearrangement of the *vic*-triazole may produce the 3-unsubstituted 4-anilino-product.[399,401] Appropriately 1- and 2-substituted triazoles also lead to the corresponding 1- and 2-substituted triazolopyrimidines; for example, the 2-(2-benzimidazolyl) and 2-(4-styrylphenyl) derivatives have been used in the preparation of optical brightening agents. Albert has investigated the chemistry of the *vic*-triazole intermediates for the preparation of 1-, 2-, and 3-methyl analogues, and shown that cyclization can occur with the 4-carboxamide using formamide,[402] or with thiourea to give the 5-thio-7-oxo-derivative.[403] The 4-amino-5-carboxylic esters give a good yield with formamide,[399] but are relatively inaccessible. 4-*N*-Methylcarboxamides give 6-methyl-7-oxo-compounds (ly) with formamide.[404]

Attempts to use ring *N*-acyl-*vic*-triazolecarboxamides to give *N*-acyltriazolopyrimidines have been disappointing.[42]

The nitrile (**323**) was cyclized to give the 7-amino-analogue (**324**) with formamide,[399] and subsequently nitriles have been used with diethoxymethyl acetate followed by methanolic ammonia (especially useful for glycosyl triazoles),[405] with amidines,[406] and with CS_2/pyridine, potassium ethyldithiocarbonate, when 5,7-dithio-compounds are formed,[407] or isothiocyanates. The use of arylcarboxamidines with the 4-carboxamides or *N*-methylcarboxamides (which eliminate methylamine at the high temperatures required, giving the same 2-aryl-7-hydroxy-derivatives) proved disappointing.[42] The cyclization of 5-amino-*vic*-triazole-4-carboxamides themselves with various reagents has been successful.[408]

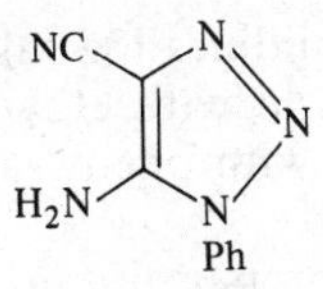

(323) 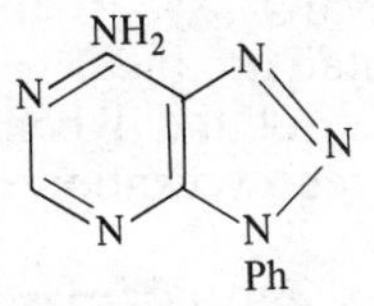 (324) 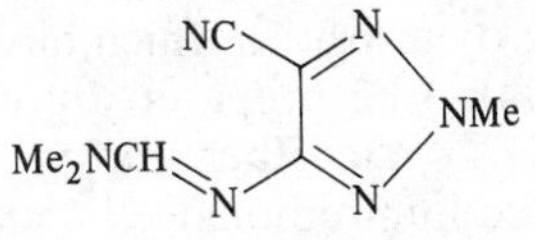 (325)

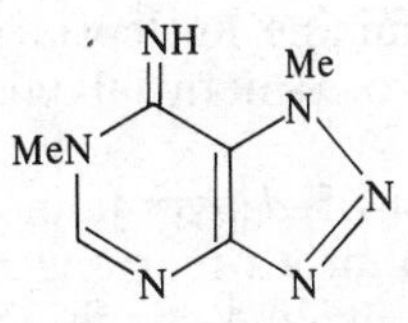

(326) 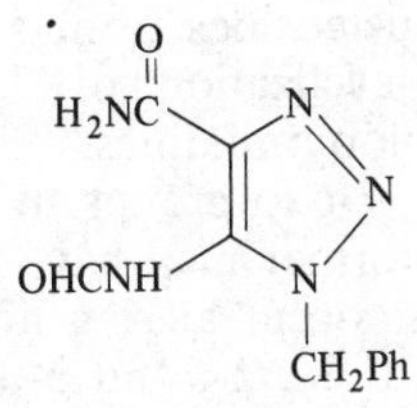 (327) 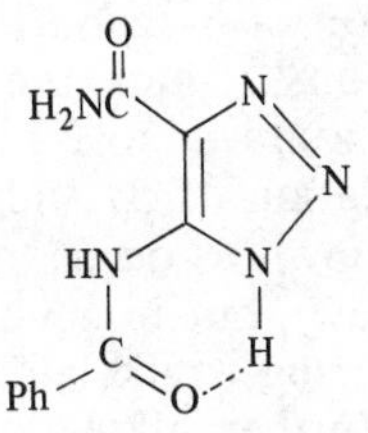 (328)

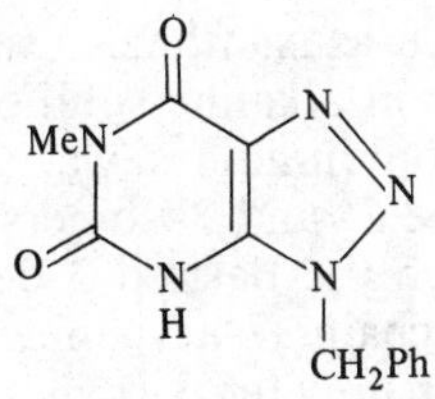

(329) 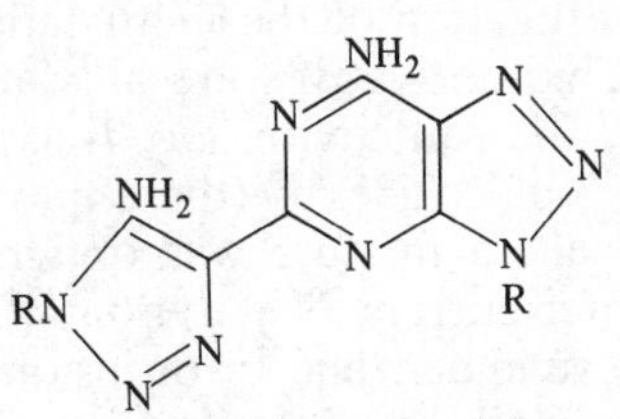 (330) a; R = CH_2Ph
b; R = Ph

Isolation of intermediates and subsequent cyclization of these has been observed in the case of the 4-dimethylaminomethyleneamino-5-cyano-*vic*-triazoles, *e.g.* (**325**) (formed from the aminocarboxamides with phosphorus oxychloride and DMF), which with ammonium acetate or amines gave the 5,7-diamino-derivatives, and with sodium hydrosulphide the 5-thio-7-amino-analogues.[409] Another isolated intermediate (from the aminonitrile with ethyl orthoformate/Ac_2O) is the 5-ethoxymethyleneamino-4-nitrile, which with alkylamines gives the 6-alkyl-7-imino compounds (hy), *e.g.* (**326**),[410] and with sodium hydrosulphide furnishes the best synthesis of the 5-thio-7-aminotriazolopyrimidines.[407] The most versatile of these intermediates, however, are undoubtedly the 5-formylamino-derivatives, *e.g.* (**327**), isolated by Albert in the form of its anhydro-dimer derivative,[411] which themselves cyclize with alkali to give 8-azahypoxanthines,[411] but on deformylative acylation and alkaline cyclization of the intermediate 5-acylaminoamides give a variety of *N*-substituted 2-alkyl- (aralkyl)[391,412] and 2-aryl-hypoxanthines.[42,391] The *N*-unsubstituted acylaminoamides only cyclize (ly) using strongly alkaline conditions as they appear to exist in the H-bonded form (**328**).[42] 2-Substituted hypoxanthines have also been prepared by cyclization of the aminoamides with a new reagent, the trithioorthoester, in the presence of an acid catalyst,[42,391] whilst use of another new reagent, phenyl chlorothioformate in pyridine or collidine, gives good yields of xanthine derivatives, *e.g.* (**329**).[42]

Recently, a synthesis was reported of the 7-unsubstituted triazolopyrimidines themselves, which had previously been mainly prepared by indirect methods, by treatment of the 5-amino-4-aldehydes with ethyl orthoformate or DMF/$POCl_3$ followed by ammoniacal ring-closure,[413] whilst the 5-amino-4-aminomethyltriazoles with orthoesters or amides give the interesting 1,6-dihydro-derivatives (see below), readily oxidized to 8-azapurines with permanganate or MnO_2.[414]

Two curiosities which have been noted are the 5-triazolyltriazolopyrimidines (**330a**) and (**330b**), one prepared by dimerization of the 1-benzyl-5-amino-4-nitrile with ethanolic KOH,[401] the other during an attempted cyclization of malonodinitrile with phenyl azide and sodium ethoxide.[412]

A comparison of the pyrimidine and *vic*-triazole methods of preparation of 8-azapurines has been made.[400]

Miscellaneous syntheses of 8-azapurines include the work of Fox[415] on the cyclization of 5-nitrouracils or 5-nitro-2-oxopyrimidines with sodium azide in DMF (in which the azide ion attacks the free 6-position, followed by cyclization and elimination of nitrite[416]), the formation of 8-azapurine-3,4-cyclonucleosides from 5-bromopyrimidine nucleosides and azide ion,[417] and the nitrosative cyclization of 1,3-dimethyl-6-α-methylalkylidenehydrazinouracils to 2-vinyl-*vic*-triazolopyrimidines.[418]

Alkylation of 5- and 7-amino- or 7-oxo-groups in *vic*-triazolo[4,5-*d*]pyrimidines has been known to occur (although the former may have arisen by Dimroth rearrangement, see below), but in general alkylations occur at ring nitrogens. In alkaline conditions the triazole ring is alkylated, a mixture of 1-, 2-, and 3-alkyl derivatives being obtained,[419] whilst further alkylation of 3-derivatives occurs at the 1-position to give quaternary salts.[42,393,420] 7-Dimethylaminotriazolopyrimidines gave mainly 2- and 3-alkyl derivatives.[421] Alkylation of the pyrimidine ring is only observed in the 4- and 6-positions of xanthine analogues when the triazole ring is first substituted.[42,393,422]

A thorough investigation of the methylation of various 8-azaxanthines was carried out by Pfleiderer.[392] Thio-derivatives are alkylated at sulphur in alkaline solution.

The mercurichloride route with, *e.g.* 7-acetamidotriazolopyrimidines, was said to give a mixture of 1- and 3-glycosides[423] (these may have been the 2- and 3-isomers, see below), and with sodium salts a mixture was obtained from which the desired 3-derivative was isolated by counter-current separation.[424] With long-chain 7-acylamino-derivatives, using molecular sieve in benzene,[425] or fusion methods,[426] mainly the 3-isomer was isolated (ly). On the other hand, the 7-nonanoylaminotriazolo[4,5-*d*]pyrimidine was said to give mainly the 2-glycoside by the mercurichloride method in dimethylacetamide,[425] as was the 7-dimethylamino-base.[421]

The 7-methylthio-analogue also gave a mixture of 2- and 3-glycosides by the fusion method,[427] whilst with molecular sieve the ratio of isomers varied with the solvent.[428] Later work showed that a small amount of 1-isomer was also formed in the fusion method.[429] The preparation of nucleosides (especially 8-azaguanosine) by fermentation, *e.g.* using *B. cereus* or *Brevibacter ammoniagenes*, has been widely reported.

Nuclear electrophilic substitutions have not been reported for 8-azapurines, but external aryl substituents are nitrated[331,430] and sulphonated[431] in the usual way. Acylations and thioacylations of 7-oxo-compounds occur mainly at the 2- and 3-positions, the proportions varying widely with the conditions and the acylating agent.[42,432] 7-Derivatives are only found in very forcing conditions.[42] The Mannich reaction of 8-azatheophyllines was said to have occurred at the 1-position, but this is not certain.[433] Japp–Klingemann type azo-coupling at a 7-methyl group was reported,[434] whilst *N*-oxidation gave the N^6-oxide, together with other *N*-oxides.[255]

In nucleophilic substitutions the 7-chloro-*vic*-triazolopyrimidines, prepared from oxo-compounds with PCl_5 or $POCl_3$, or thionyl chloride/DMF, are much more reactive than their 6-chloropurine counterparts; for instance, amination occurs at 100 °C with concentrated ammonia.[435] The kinetics of the reaction with piperidine have been compared with those theoretically calculated.[436] Other groups which have been substituted for chloro include substituted amino, oxo (also from amino with HCl or HNO_2), thio (thiourea or NaHS), alkoxy, alkylthio, hydrazino, and pyridinio (direct from $POCl_3$/pyridine on the quaternary oxo-compounds[437]). The methylthio-groups (sometimes *via* oxidation to alkylsulphonyl) and pyridinio-groups may themselves be replaced by amines, $(NH_4)_2S$, *etc.* The 3-glycosylderivatives seem to behave normally. In a biosynthetic reaction, 8-azaguanine is enzymically converted to 8-azaxanthine by guanase at an optimum pH of 6.3.[438]

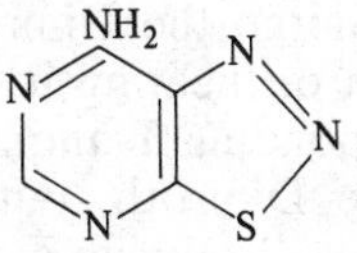

(331) 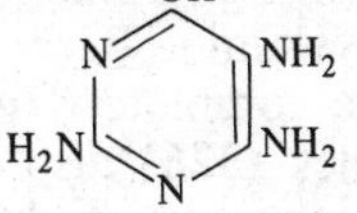 (332) 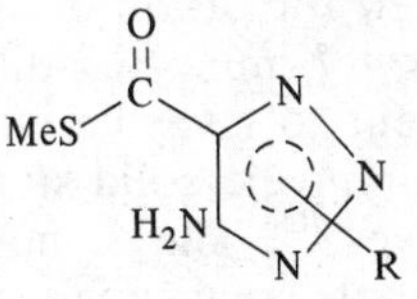 (333)

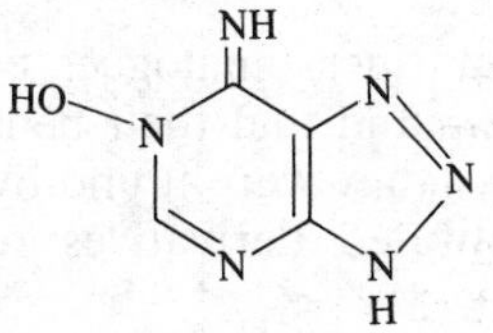

(334) 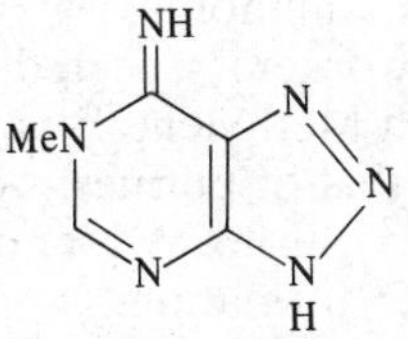 (335)

The replacement of 7-oxo-groups by thio-groups has presented some difficulties, owing to the formation of disulphide derivatives and to the occurrence of a rearrangement to 7-amino-1,2,3-thiadiazolo[5,4-*d*]pyrimidine derivatives, *e.g.* (**331**).[42,415,439] The mechanism and kinetics have been studied for a series of 3-benzyl analogues,[440] and the rearrangement shown to be favoured by electron-withdrawing groups and high temperatures. Cold alkaline conditions favour the 7-thio-form.

4-Acetoxytriazolo[4,5-*d*]pyrimidines do not undergo remote nucleophilic substitution as do the corresponding 2-C analogues (Section 17.5.8), showing only the radical reduction and hydrolysis pathways.[441] They are relatively inactive as oncogens.

8-Azapurines are reduced chemically[386] and polarographically to give 1,6-dihydro-derivatives, whilst covalent hydration[386,411] and addition of other nucleophiles (bisulphite, methanol, thiols, *etc.*) across this bond are also observed.[442]

In the degradative ring-opening of 8-azapurines, the *vic*-triazole ring is normally resistant to both acid and alkaline attack. Only with 8-azaguanine and 6 N HCl at 150 °C was degradation to a pyrimidine (**332**) observed,[443] whilst the fiercest alkaline hydrolysis (6 N NaOH at 180 °C) converted 8-azaguanine to 4-amino-*vic*-triazole.[444] The 4-amino-*vic*-triazole-carboxamide was formed from 7-oxo-*vic*-triazolo[4,5-*d*]pyrimidine with 2 N HCl at 90 °C,[443] and from the 5,7-dioxo-derivative or 8-azaguanine with ammonia at 200 °C.[445] Several 7-methylamino-derivatives were rapidly converted (by boiling N HCl) to the corresponding 4-amino-*vic*-triazole-5-carboxamidines, although the 7-amino-analogues required more severe conditions unless dioxan was used as solvent.[446] A series of 7-methylthiotriazolopyrimidines on boiling with HCl produced the methylthiocarbonyl derivatives (**333**),[404] whilst Albert has noted the great lability to hydrolysis, at room temperature with both acid and bases, of 7-unsubstituted or 7-methyltriazolopyrimidines to the corresponding 5-amino- or 5-formylamino-*vic*-triazole-4-carboxaldehydes or -4-methyl ketones, except when a 5-methyl group was present.[447]

6-Methyl-4-aryl-5,7-dioxo-derivatives boiled with 2 N NaOH gave 5-arylamino-N^4-methylcarboxamides,[422] whilst the 6-alkyl-6,7-dihydro-7-imino-analogues also gave *N*-alkylcarboxamides.[408] The 6-oxide (**334**) also gave a 4-carboxamidoxime with concentrated HCl at room temperature.[255]

The use of hydrazine hydrate was reported to give the 5-amino-4-carboxyhydrazide from the 5,7-dioxo-compounds,[448] but the 7-unsubstituted triazolopyrimidines gave 5-(aminomethyleneamino)-4-aldehyde hydrazones and azines, and corresponding oximes and methoximes are formed with hydroxylamines[449] and, for example, dicyanovinyl derivatives with malononitrile[448] and its relatives. The latter were readily recyclized, *e.g.* to 5-amino-6-cyano-*vic*-triazolo[4,5-*b*]pyridine derivatives.[449]

A ring opening to a *vic*-triazole complex (shown by X-ray analysis) was observed in attempted formation of a copper chloride complex of 8-aza-adenine.[450]

6-Alkyl-7-imino-6,7-dihydro-*vic*-triazolopyrimidines (**335**) undergo the Dimroth rearrangement to give the expected 7-alkylamino-derivatives, which on heating in refluxing DMF, or in the solid state, rearrange completely to the corresponding 7-amino-3-alkyl derivatives.[408] The 1-methyl analogues (**326**) also undergo the Dimroth, but not the subsequent, rearrangement.[410] On the other hand, 5-amino-7-ethylaminotriazolopyrimidines show the same type of rearrangement, giving 3-ethyl-5,7-diamino-derivatives.[451] Pyrolysis of 8-azapurines, unlike that of the corresponding triazolopyridines (which give cyanopyrroles) induces a profound decomposition to intractable products, HCN, and ammonia.[452]

Although early workers reported that 8-azapurines (analogues of purine bases) were not photolysed with Methylene Blue as sensitizer and that their behaviour in this respect was different from that of purines (see Ref. 453), they were found by other workers to be more reactive than purines,[454] and 7-unsubstituted derivatives readily underwent 7-oxidation and dimer formation.[455]

MO calculations on 8-azapurines have covered such topics as electronic structure, including π-electron densities, localization and delocalization energies, energies of charge-transfer complexation and base-pair interactions, u.v. bathochromic shifts on anion formation, n.m.r. proton chemical shifts, and tautomeric equilibria.

The question of tautomeric equilibria in 8-azapurines remains unsettled. Pfleiderer, from pK_a and n.m.r. studies, favoured the 1*H*-form for a series of azaxanthines,[392] but later workers have concluded that a mixture of 1*H*-, 2*H*-, and 3*H*-forms were present.[42,456] X-ray analysis indicates that in the crystal lattice the 3*H*-form predominates,[457] in agreement with MO predictions of stabilities.[458]

The ionization and protonation of triazolopyrimidines has been widely studied. Individual values for pK_a may be found in the series of papers by Albert and Pfleiderer, or in Albert's compilation of purine values,[459] together with spectroscopic (u.v., i.r., and n.m.r.) data. N.m.r. (proton and ^{13}C) has also been useful in determination of the site of attachment and conformation of glycosidic substituents.[460,469]

The mass spectra of 8-azapurine cytokinin analogues[462] have been studied.

Various charge-transfer and other complexes have been reported for 8-azapurines, including those with uracils, chloranil, metals such as copper and dinitrosyliron, and other 8-azapurine molecules.

Apart from the high anti-tumour and anti-allergic activities already mentioned, 8-azapurines have been reported, *inter alia*, as having coronary-dilating, anti-viral, anti-bacterial, fungicidal, diuretic, and xanthine oxidase inhibitory activity, and being of use in various photographic applications.

Short reviews are available on the synthesis and biological evaluation of 8-azapurine nucleosides.[7,258]

17.5.24 9-DEAZAPURINES

9-Deazapurines, pyrrolo[3,2-*d*]pyrimidines, are numbered as shown in (**336**) and are not known naturally (although at one time the antibiotic viomycin was wrongly thought to have a pyrrolo[3,2-*d*]pyrimidine structure). It is a stronger base than purine, in fact stronger even than pyrimidine, owing to resonance in the cation of the type shown in (**337**) ↔ (**338**).[463] Syntheses of 9-deazapurines from pyrimidines, by far the largest number, fall into three distinct types: those in which the final 6,7-atoms are initially both attached to a nitrogen at the 5-pyrimidine position; secondly those in which these two atoms are originally attached directly to the 4(6)-position of the pyrimidine; and finally all other miscellaneous syntheses.

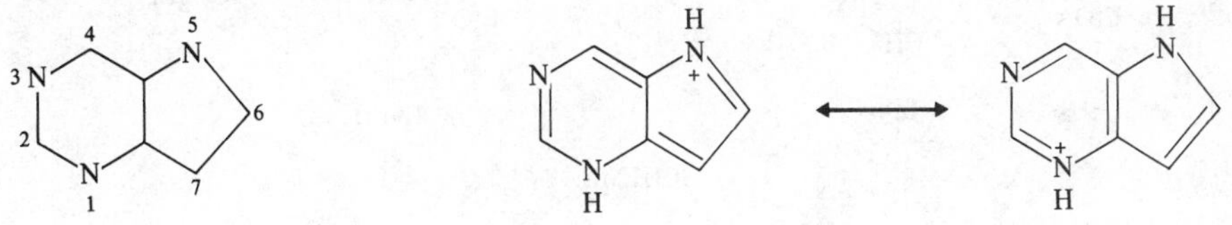

(336) (337) (338)

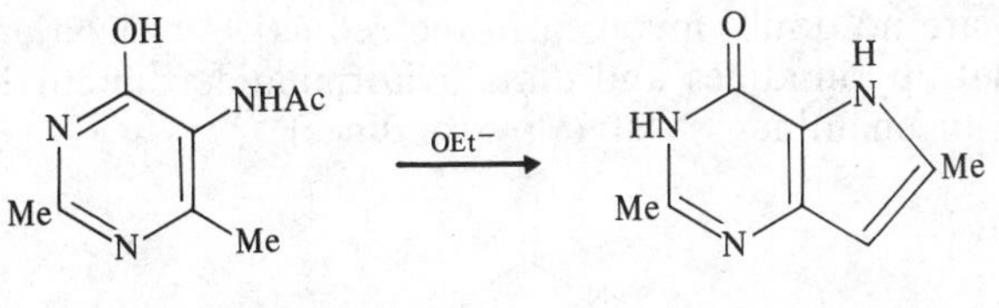

(339) (340)

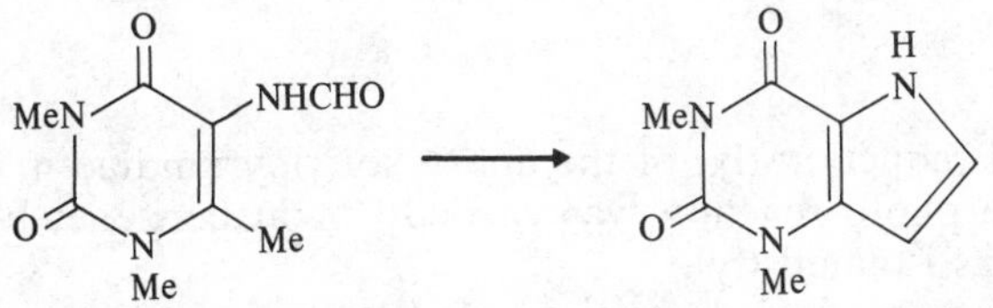

(341) (342)

There seems to be no truly genuine example of a ring-closure of the first type. The first preparations of the ring system were of the third type, the 6-methyl-5-acylamino-pyrimidines (**339**) being heated at 300 °C under nitrogen with dry sodium ethoxide to give (my) 2-alkyl or 2-unsubstituted 6-substituted 4-hydroxypyrrolo[3,2-*d*]pyrimidines, *e.g.* (**340**). Various alkyl and aryl derivatives were prepared,[464] but the reaction failed with 2-hydroxy-, 2-amino-, or 2-methylthio-pyrimidines. More recently a thermal cyclization of the formylamino-analogue (**341**) gave (**342**).[465] The reaction conditions were, of course,

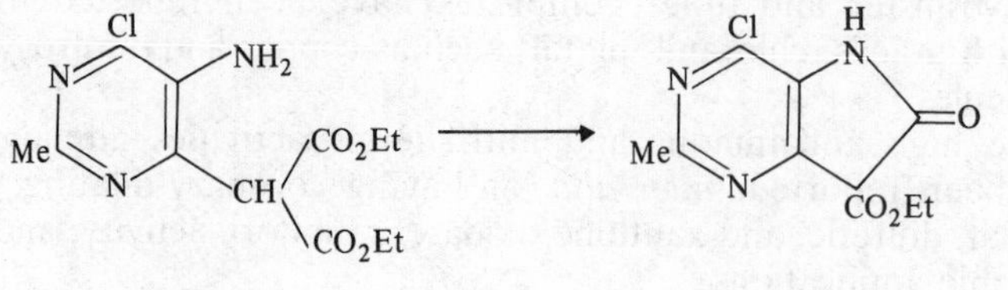

(343) (344)

rather forbidding and the route was soon superseded by one of the second type, in which a 5-amino-6-pyrimidylmalonic ester (*e.g.* **343**) was cyclized with hot aqueous sodium carbonate to give a 7-ethoxycarbonyl derivative (**344**). This route was successful with halo- and amino-substituents.[466]

A closely related synthesis which gives an isomeric 6-ethoxycarbonyl derivative uses esters, *e.g.* (**345**), prepared from 5-nitro-6-methylpyrimidines and ethyl oxalate, which on reductive cyclization give, for example, (**346**) (Scheme 18).[464] A close relative of this substitutes an acidic hydrolysis of a 5-benzylideneamino-group for the previous nitro-reduction.[467] Decarboxylations of the derived acids using copper catalysts furnished the 6- or 7-unsubstituted analogues from the above esters.[464]

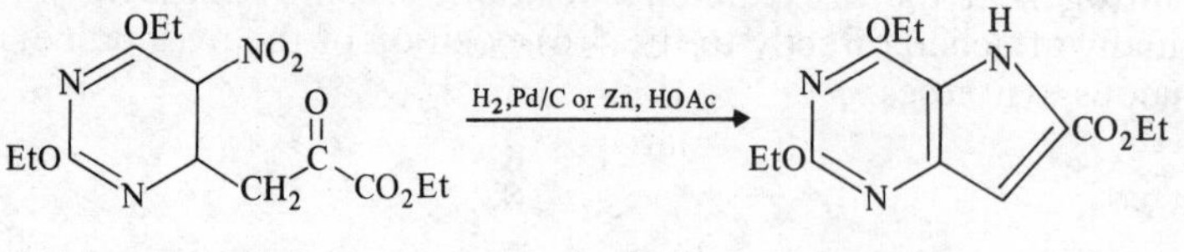

H_2,Pd/C or Zn, HOAc

(345) (346)

SCHEME 18

Further examples of the second type involve the triethyl phosphite reduction of 5-nitro-6-(*p*-substituted styryl)uracils (**348**), giving 6-aryl derivatives of type (**347**),[468] whilst similar reactions are no doubt involved in the reductive cyclizations of the intermediates from various 5-nitropyrimidines and dimethylformamide dimethyl acetal, ketene diethyl acetals,[469] (**349**), or enamines (*N*-styrylpyrrolidines).[470]

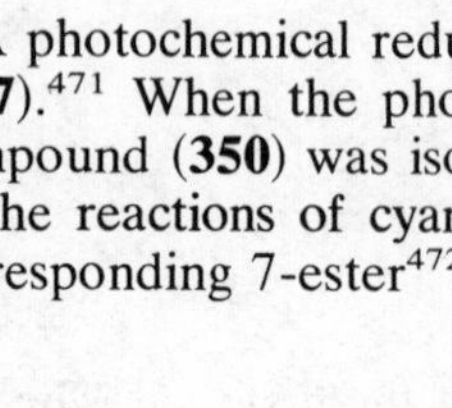

(347) 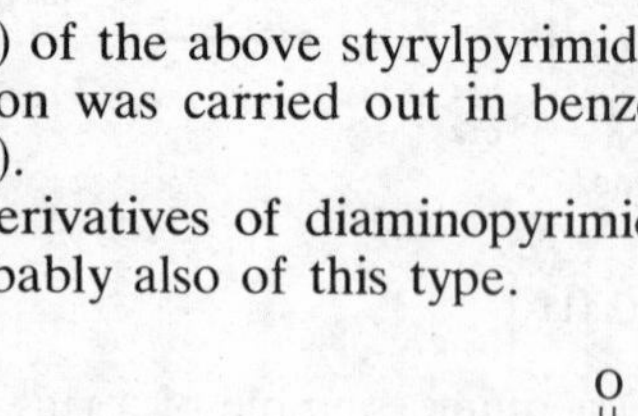 (348) 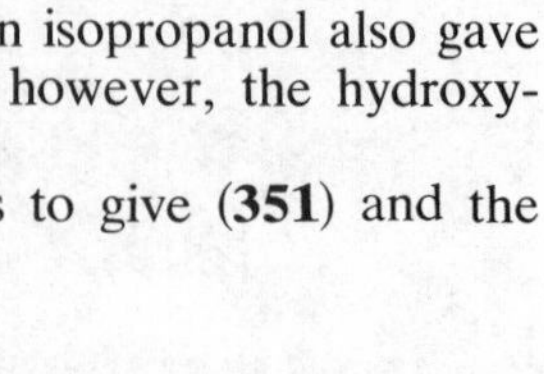 (349)

A photochemical reduction (ly) of the above styrylpyrimidine in isopropanol also gave (**347**).[471] When the photo-reaction was carried out in benzene, however, the hydroxy-compound (**350**) was isolated (ly).

The reactions of cyanovinyl derivatives of diaminopyrimidines to give (**351**) and the corresponding 7-ester[472] are probably also of this type.

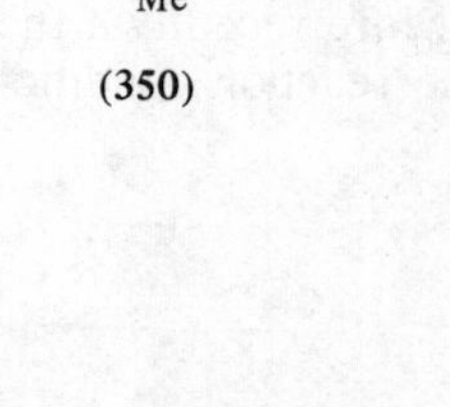

(350)  (351)

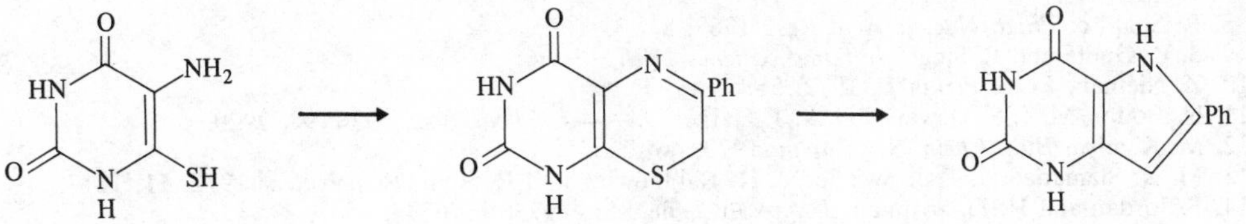

(352) (353) (354)

Other miscellaneous reactions of the third type include a cyclization of the 5-amino-6-mercaptopyrimidine-2,4-dione (**352**) to the 6-phenyl-3,4-dihydropyrrolo[3,2-*d*]pyrimid-2-one (**354**) with phenacyl bromide, *via* a pyrimidothiazine intermediate (**353**),[473] and a synthesis of 9-deaza-adenine nucleosides from furazano[3,4-*d*]pyrimidines.

The lone example of a synthesis from pyrrole intermediates is the cyclization of the aminopyrrolecarboxamide (**355**) with formic acid to give the 9-deazapurine ester (**356**).[474]

Nucleophilic substitution reactions have been used to interconvert various pyrrolo[3,2-*d*]pyrimidine derivatives. 2- and 4-chloro-groups have been reduced with H_2/Pd, or introduced by treatment of the dioxo-derivatives (from hydrolysis of the diethoxy-precursors) with phosphorus oxychloride (hy).[464] Reactions of 2- and 4-chloro-groups with ammonia[464] (requiring 200 °C), amines, thiourea, mercaptides, alkoxides, or hydrazine hydrate proceed normally and oxo-compounds (from nitrous acid treatment of the corresponding amines) are converted to thio-compounds with P_2S_5. 6-Halo-groups have also been removed by catalytic hydrogenation.

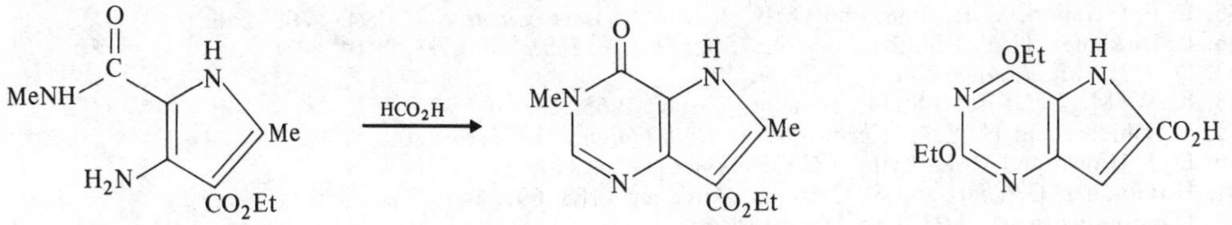

(355) (356) (357)

Alkylation of thio-groups gives the expected alkylthio-derivatives[465] (a good route to hydrazino- and amino-derivatives), but an attempted bromination of the 6-carboxylic acid (**357**) gave the 6,7-dibromo-derivatives *via* a type of Hunsdiecker reaction. The corresponding ester, however, brominated smoothly at the 7-position.[475]

The claimed activities of 9-deazapurines include antibacterial, diuretic, cardiac-stimulant, CNS-stimulant (weaker than caffeine), and cytokinin (weak) activity. A semi-empirical MO calculation has been made of the interaction energies of some 9-deazapurine base pairs.

Finally, a recent review is available on the synthesis of some pyrrolopyrimidines, including pyrrolo[3,2-*d*]pyrimidines.[381]

References

1. E. Fischer, *Ber.*, 1884, **17,** 329.
2. J. H. Lister, in 'Fused Pyrimidines. Part II. The Purines', ed. D. J. Brown, Wiley-Interscience, London, 1971.
3. 'Synthetic Procedures in Nucleic Acid Chemistry', ed. W. W. Zorbach and R. S. Tipson, Wiley-Interscience, London, 1973, vol. 2.
4. B. Pullman and A. Pullman, *Adv. Heterocyclic Chem.*, 1971, **13,** 77.
5. E. S. Golovchinskaya, *Russ. Chem. Rev.*, 1973, **42,** 413.
6. F. Bergmann, D. Lichtenberg, U. Reichman, and Z. Neiman, 'Proc. 6th Jerusalem Symposium on Quantum Chemistry and Biochemistry', ed. E. D. Bergmann and B. Pullman, Jerusalem, 1974, 397, and refs. therein.
7. 'Proc. 4th Internat. Symp. Quantum Chem. and Biochem.', ed. E. D. Bergmann and B. Pullman, Jerusalem, 1972.

8. R. Shapiro, *Progr. Nucleic Acid Res.*, 1968, **8,** 73.
9. S. P. Gupta and P. Singh, *Indian J. Chem.*, 1975, **13,** 668.
10. Z. Nieman, *J.C.S. Perkin II*, 1972, 568.
11. N. Bodor, M. J. S. Dewar, and A. J. Harget, *J. Amer. Chem. Soc.*, 1970, **92,** 2929.
12. M. Kamiya, *Bull. Chem. Soc. Japan*, 1973, **46,** 2235.
13. M. N. Stamatiadou, T. J. Swissler, J. R. Rabinowitz, and R. Rein, *Biopolymers*, 1972, **11,** 1217.
14. F. Jordan and H. D. Sostman, *J. Amer. Chem. Soc.*, 1973, **95,** 6544.
15. A. Pullman, Proc. 2nd Internat. Symp. Quantum Chem. and Biochem., Jerusalem, 1969, p. 9.
16. P. D. Lawley, in Ref. 2, p. 439.
17. J. Deutsch, Z. Neiman, and F. Bergmann, in Ref. 7, p. 402.
18. W. Traube, *Ber.*, 1900, **33,** 1371.
19. G. B. Elion, W. H. Lange, and G. H. Hitchings, *J. Amer. Chem. Soc.*, 1956, **78,** 2858.
20. G. D. Daves, C. W. Noell, R. K. Robins, H. C. Koppel, and A. G. Beaman, *J. Amer. Chem. Soc.*, 1960, **82,** 2633.
21. H. Yamasaki and T. Hashizume, *Agric. Biol. Chem.* (*Japan*), 1971, **35,** 122.
22. J. Clark and J. H. Lister, *J. Chem. Soc.*, 1961, 5048.
23. B. Gepner and L. Krebs, *Zhur. obshchei Khim.*, 1946, **16,** 179.
24. D. J. Brown and N. W. Jacobsen, *J. Chem. Soc.*, 1965, 3770.
25. S. C. J. Fu, B. J. Hargis, E. Chinoporos, and S. Malkiel, *J. Medicin. Chem.*, 1967, **10,** 109.
26. G. P. Hager, J. C. Krantz, J. C. Harmon, and R. M. Burgison, *J. Amer. Pharm. Assoc. Sci. Edn.*, 1965, **43,** 152.
27. W. Traube and co-workers, *Annalen.*, 1923, **432,** 266.
28. A. Albert, D. J. Brown and G. Cheeseman, *J. Chem. Soc.*, 1952, 1620.
29. V. M. Cherkasov and L. K. Kurilenko, *Khim. geterotsikl. Soedinenii*, 1970, 1432.
30. *Ger. Pat.* 2 058 912 (1971).
31. H. F. Ridley, R. G. W. Spickett, and G. M. Timmis, *J. Heterocyclic Chem.*, 1965, **2,** 453.
32. H. G. Mautner, *J. Org. Chem.*, 1961, **26,** 1914.
33. S. Wawzonek, *J. Org. Chem.*, 1976, **41,** 310.
34. *Brit. Pat.*, 1 201 997 (1970).
35. L. F. Cavalieri, V. E. Blair, and G. B. Brown, *J. Amer. Chem. Soc.*, 1948, **70,** 1240.
36. L. Birkhofer, H. P. Kuhlthau, and A. Ritter, *Chem. Ber.*, 1960, **93,** 2810.
37. D. J. Brown, *J. Appl. Chem.*, 1959, **9,** 203.
38. K. W. Merz and P. H. Stahl, *Arzneim.-Forsch.*, 1965, **15,** 10.
39. M. Ishidate and H. Yuki, *Chem. Pharm. Bull.* (*Japan*), 1957, **5,** 240.
40. D. J. Brown and R. K. Lynn, *J.C.S. Perkin I*, 1974, 349.
41. H. Goldner, G. Dietz and E. Carstens, *Annalen*, 1966, **691,** 142.
42. Unpublished work from these laboratories.
43. F. Yoneda, K. Ogiwara, M. Kanahori, and S. Nishigaki, *Chem. Comm.*, 1970, 1068.
44. F. Yoneda and T. Nagamatsu, *J.C.S. Perkin I*, 1976, 1547.
45. F. Yoneda, M. Higuchi, T. Matsuuma, and K. Senga, *Bull. Chem. Soc. Japan*, 1973, **46,** 1836.
46. E. C. Taylor and F. Yoneda, *J. Org. Chem.*, 1972, **37,** 4464.
47. F. Yoneda, S. Matsumoto, and M. Higuchi, *J.C.S. Chem. Comm.*, 1974, 551.
48. F. Yoneda, S. Matsumoto, and M. Higuchi, *J.C.S. Chem. Comm.*, 1975, 146.
49. K. Senga, H. Kanazawa, and S. Nishigaki, *J.C.S. Chem. Comm.*, 1976, 155.
50. H. Bredereck, F. Effenberger, and E. H. Schweizer, *Chem. Ber.*, 1962, **95,** 956.
51. R. A. Baxter and F. S. Spring, *J. Chem. Soc.*, 1945, 232.
52. L. Bauer, C. N. V. Nambury, and D. Dhawan, *J. Heterocyclic Chem.*, 1964, **1,** 275; 1965, **2,** 220.
53. A. Kolb, C. Gouyette, T. H. Dinh, and J. Igolen, *J. Heterocyclic Chem.*, 1975, **12,** 111.
54. A. H. Cook and E. Smith, *J. Chem. Soc.*, 1949, 2329.
55. M. Okutsu and A. Yamazaki, *Nucleic Acids Res.*, 1976, **3,** 231.
56. A. Edenhofer, *Helv. Chim. Acta*, 1975, **58,** 2192.
57. R. B. Meyer, D. A. Shuman, and R. K. Robins, *J. Amer. Chem. Soc.*, 1974, **96,** 4962.
58. P. C. Srivastava, G. A. Ivanovics, R. J. Rousseau, and R. K. Robins, *J. Org. Chem.*, 1975, **40,** 2920.
59. A. Kolb, T. H. Dinh, and J. Igolen, *J. Carbohydrates, Nucleosides, Nucleotides*, 1975, **2,** 37.
60. *Ger. Pat.* 2 166 506 (1974).
61. *Ger. Pat.* 2 160 674 (1972).
62. R. Manimoto, Y. Yoshioka, O. Miyashita, S. Shima, K. Imai, K. Kawazoe, and M. Honjo, *Chem. Pharm. Bull.* (*Japan*), 1975, **23,** 759.
63. G. Barnathan, T. H. Dinh, A. Kolb, and J. Igolen, *European J. Med. Chem.*, 1976, **11,** 67.
64. C. C. Duke, A. J. Liepa, J. K. Macleod, D. S. Letham, and C. W. Parker, *J.C.S. Chem. Comm.*, 1975, 964.
65. S. Kikkomaa-Shoyu, N. Yamaji, Y. Yuasa, and M. Kato, *Chem. Pharm. Bull.* (*Japan*), 1976, **24,** 1561.
66. R. J. Badger and G. B. Barlin, *J.C.S. Perkin I*, 1976, 151.
67. R. Hayatsu, M. H. Studier, L. P. Moore, and E. Anders, *Geochem. Cosmochim. Acta*, 1975, **39,** 471.
68. C. Simionescu, T. Lixandru, C. Gorea and V. Gorduza, *Compt. rend.* (*C*), 1975, **280,** 685.
69. J. H. Boyer, J. Dunn, and J. Kooi, *J.C.S. Perkin I*, 1975, 1743.
70. *U.S.S.R. Pat.* 412 196 (1974), 455 959 (1975).
71. M. Sekiya and J. Suzuki, *Chem. Pharm. Bull.* (*Japan*), 1975, **23,** 2401.

72. M. Sekiya, J. Suzuki, and Y. Terao, *Chem. Pharm. Bull.* (*Japan*), 1976, **24,** 1331.
73. T. Ichikawa, T. Kato, and T. Takenishi, *J. Heterocyclic Chem.*, 1965, **2,** 253.
74. E. C. Taylor, G. P. Beardsley, and Y. Maki, *J. Org. Chem.*, 1971, **36,** 3211.
75. C. Temple, C. L. Kussner, and J. A. Montgomery, *J. Org. Chem.*, 1971, **36,** 3502.
76. Y. Ohtsuka, *Bull. Chem. Soc.* (*Japan*), 1970, **43,** 3909.
77. Y. Ohtsuka, *Bull. Chem. Soc.* (*Japan*), 1973, **46,** 506.
78. C. Temple, R. L. McKee, and J. A. Montgomery, *J. Org. Chem.*, 1962, **27,** 1671.
79. A. Nagel and H. C. van der Plas, *Chem. Pharm. Bull.* (*Japan*), 1975, **23,** 2678.
80. W. Hutzenlaub, H. Yamamoto, G. B. Barlin, and W. Pfleiderer, *Chem. Ber.*, 1973, **106,** 3203.
81. F. Yoneda, K. Ogiwara, M. Kanahori, and S. Nikigashi, Proc. 4th Int. Symp. Chem. Biol. Pteridines 1969, 145.
82. F. Yoneda and T. Nagamatsu, *Heterocycles*, 1976, **4,** 749.
83. M. Israel, S. K. Tinter, D. H. Trites, and E. J. Modest, *J. Heterocyclic Chem.*, 1970, **7,** 1029.
84. D. W. Wiley, O. W. Webster, and E. P. Blanchard, *J. Org. Chem.*, 1976, **41,** 1889.
85. Y. Suzuki, *Bull. Chem. Soc.* (*Japan*), 1974, **47,** 898.
86. E. R. Garrett and P. J. Mehta, *J. Amer. Chem. Soc.*, 1972, **94,** 8532.
87. R. L. Erskine and T. A. Steedman, *J. Appl. Chem. Biotechnol.*, 1975, **25,** 221.
88. G. L. Gonser, H. d'A. Heck, and M. Anbar, *Anal. Biochem.*, 1976, **71,** 519.
89. Y. Suzuki, *Bull. Chem. Soc.* (*Japan*), 1974, **47,** 2077.
90. E. G. Chauser, S. G. Verenikina, and A. M. Yurkevich, *Zhur. obshchei. Khim.*, 1972, **42,** 1845.
91. T. Fujii, C. C. Wu, and T. Itaya, *Chem. Pharm. Bull.* (*Japan*), 1973, **21,** 1835.
92. J. C. Parham, N. J. M. Birdsall, T.-C. Lee, and T. J. Delia, *J. Org. Chem.*, 1971, **36,** 2635.
93. T. C. Lee, F. L. Lam, and G. B. Brown, *J. Org. Chem.*, 1975, **40,** 1547.
94. J. Guettner, M. Horn, and W. Jungstand, *Arzneim.-Forsch.*, 1971, **21,** 356.
95. G. Zvilichovsky and J. Feingers, *J.C.S. Perkin I*, 1976, 1507.
96. F. Yoneda and Y. Sakuma, *Chem. Pharm. Bull.* (*Japan*), 1973, **21,** 448.
97. A. A. Watson, *J. Org. Chem.*, 1974, **39,** 2911.
98. U. Wölcke, W. Pfleiderer, T. J. Delia, and G. B. Brown, *J. Org. Chem.*, 1969, **34,** 981.
99. G. Stoehrer, G. Salemnick, and G. B. Brown, *Biochemistry*, 1973, **12,** 5084.
100. E. C. Miller and J. A. Miller, in 'The Molecular Biology of Cancer', ed. H. Busch, Academic Press, New York, 1974, p. 377.
100a. N. J. M. Birdsall, U. Wölcke, T.-C. Lee, and G. B. Brown, *Tetrahedron*, 1971, **27,** 5961, 5969.
101. T. Fujii and T. Itaya, *Chem. Pharm. Bull.* (*Japan*), 1971, **19,** 1611.
102. T. Fujii, T. Itaya, and S. Moro, *Chem. Pharm. Bull.* (*Japan*), 1972, **20,** 958.
103. G.-F. Huang, M. Maeda, T. Okamoto, and Y. Kawazoe, *Tetrahedron*, 1975, **31,** 1363.
104. G.-F. Huang, T. Okamoto, M. Maeda, and Y. Kawazoe, *Tetrahedron Letters*, 1973, 4541.
105. D. F. Wiemer and N. J. Leonard, *J. Org. Chem.*, 1974, **39,** 3438.
106. J. A. Montgomery and C. Temple, *J. Amer. Chem. Soc.*, 1960, **82,** 4592.
107. E. Fischer, *Ber.*, 1899, **32,** 435.
108. E. S. Golovchinskaya, *Russ. Chem. Rev.*, 1974, **43,** 1089.
109. H. Vorbruggen and K. Krolikiewicz, *Annalen*, 1976, 745 and refs. therein.
110. E. Arutyunyan, V. I. Gunar, E. P. Gratcheva, and S. I. Zavyalov, *Izvest. Akad. Nauk. S.S.S.R., Ser. Khim.*, 1969, 655.
111. E. Y. Sutcliffe and R. K. Robins, *J. Org. Chem.*, 1963, **28,** 1662.
112. J. Filip and L. Bohacek, *Radioisotopy*, 1971, **12,** 949.
113. R. W. Balsiger and J. A. Montgomery, *J. Org. Chem.*, 1960, **25,** 1573.
114. L. A. Gutorov, L. A. Nikolaeva, and E. S. Golovchinskaya, *Khim.-Farm. Zhur.*, 1971, **5,** 17; *Chem. Abs.*, **76,** 126 935.
115. A. Bendich, P. J. Russell, and J. J. Fox, *J. Amer. Chem. Soc.*, 1954, **76,** 6073.
116. A. Giner-Sorolla and A. Bendich, *J. Medicin. Chem.*, 1965, **8,** 667.
117. C. Temple and J. A. Montgomery, *J. Org. Chem.*, 1966, **31,** 1417.
118. G. H. Milne and L. B. Townsend, *J. Heterocyclic Chem.*, 1971, **8,** 379.
119. A. Albert and D. J. Brown, *J. Chem. Soc.*, 1954, 2060.
120. *Ger. Pat.* 2 056 327 (1971).
121. E. C. Taylor and S. F. Martin, *J. Amer. Chem. Soc.*, 1974, **96,** 8095.
122. H. Lettré, H. Ballweg, H. Maurer, and D. Rehberger, *Naturwiss*, 1963, **250,** 224.
123. M. Ikehara and Y. Ogiso, *J. Carbohydrates, Nucleosides, Nucleotides*, 1974, **1,** 401.
124. W. A. Nasutavicus and J. Love, *J. Heterocyclic Chem.*, 1974, **11,** 77.
125. A. G. Beaman, J. F. Gerster, and R. K. Robins, *J. Org. Chem.*, 1962, **27,** 986.
126. C. L. Schmidt and L. B. Townsend, *J.C.S. Perkin I*, 1975, 1257.
127. R. J. Badger and G. B. Barlin, *J.C.S. Perkin II*, 1976, 1176.
128. E. P. Girtcheuc, Z. S. Volkova, V. I. Gunar, E. Arutyuuyan, and S. I. Zavyalov, *Izvest. Akad. Nauk. SSSR, Ser. Khim*, 1970, 420; *Chem. Abs.*, **73,** 3890.
129. R. J. Badger and G. B. Barlin, *J.C.S. Perkin II*, 1974, 1854.
130. *Jap. Pat.* 70 39 709 (1970); *Chem. Abs.*, **75,** 5954.
131. C. G. Tindall, R. K. Robins, and R. L. Tolman, *J. Org. Chem.*, 1972, **37,** 3985.

132. J. A. Montgomery and K. Hewson, *J. Amer. Chem. Soc.*, 1960, **82,** 463.
133. J. H. Lister and J. Kiburis, *J. Chem. Soc. (C)*, 1971, 1587.
134. G. B. Barlin and A. C. Young, *J. Chem. Soc. (B)*, 1971, 821.
135. J. Kiburis and J. H. Lister, *Chem. Comm.*, 1969, 381.
136. J. Kiburis and J. H. Lister, *J. Chem. Soc. (C)*, 1971, 3942.
137. J. Donat and E. Carstens, *Chem. Ber.*, 1959, **92,** 1500.
138. K. Bhushan and J. H. Lister, *Austral. J. Chem.*, 1976, **29,** 891.
139. M. Eckstein, M. Gorczyca, and A. Zejc, *Acta. Pharm. Jugoslav.*, 1972, **22,** 133; *Chem. Abs.*, **78,** 72 070.
140. F. Cramer, K. Randerath, and E. A. Schafer, *Angew. Chem.*, 1958, **70,** 571; *Biochem. Biophys. Acta*, 1963, **72,** 150.
141. A. Giner-Sorolla, E. Thom, and A. Bendich, *J. Org. Chem.*, 1964, **29,** 3209.
142. U. Langenbeck and J. E. Seegmiller, *Anal. Biochem.*, 1973, **56,** 34.
143. A. A. A. Ismail and T. A. Dakin, *J. Chromatog.*, 1975, **110,** 182.
144. N. J. Leonard and J. A. Deyrup, *J. Amer. Chem. Soc.*, 1962, **84,** 2148.
145. D. H. de Kock and H. G. Raubenheimer, *J. S. African Chem. Inst.*, 1971, **24,** 91; *Chem. Abs.*, **75,** 63 742.
146. G. E. Ustyuzhanin, V. V. Kolomeitseva, and N. S. Tikhomerova-Sidorova, *Zhur. obshchei Khim.*, 1973, **43,** 2093.
147. K. Yamauchi, M. Hayashi, and M. Kinoshita, *J. Org. Chem.*, 1975, **40,** 385.
148. T. P. Seden and R. W. Turner, *J. Heterocyclic Chem.*, 1975, **12,** 1045.
149. M. Rasmussen and J. H. S. Chan, *Austral. J. Chem.*, 1975, **28,** 1031.
150. T. Kureida and B. Witkop, *J. Org. Chem.*, 1970, **35,** 3981.
151. P. D. Lawley, D. J. Orr, and M. Jarman, *Biochem. J.*, 1975, **145,** 73.
152. P. F. Swann and P. M. Magee, *Biochem. J.*, 1971, **125,** 841.
153. F. W. Kreuger, R. Preussman, and N. Niepelt, *Biochem. Pharmacol.*, 1971, **20,** 529.
154. T. Fujii, F. Tanaka, K. Mohri, T. Itaya, and T. Saito, *Tetrahedron Letters*, 1973, 4873.
155. L. G. Marzilli, L. A. Epps, T. Sorrell, and T. J. Kistenmacher, *J. Amer. Chem. Soc.*, 1975, **97,** 3351.
156. E. C. Taylor, Y. Maki, and A. McKillop, *J. Org. Chem.*, 1969, **34,** 1170.
157. L. Birkofer, A. Ritter, and H. P. Kühlthau, *Chem. Ber.*, 1964, **97,** 934.
158. R. A. Carapellotti, Ph. D. Thesis, School of Pharmacy, State University of New York, 1976.
159. R. A. Coburn, *J. Heterocyclic Chem.*, 1971, **8,** 881.
160. R. A. Coburn, R. A. Carapellotti, and R. A. Glennon, *J. Heterocyclic Chem.*, 1973, **10,** 479.
161. M. Rahat, F. Bergmann, and I. Tamir, *J.C.S. Perkin I*, 1974, 2229.
162. Z. Neiman, F. Bergmann, D. Lichtenberg, and J. Deutsch, *J. Chem. Soc. (C)*, 1971, 1822.
163. F. Bergmann, G. W. Chen, and M. Rahat, *J.C.S. Perkin I*, 1976, 90.
164. M. Lidaks, J. Sluke, and S. Poritere, *Khim. geterotsikl. Soedinenii*, 1972, 1561.
165. K. Okumura, T. Oine, Y. Yamada, M. Tonue, T. Adachi, T. Nagura, M. Kawazu, T. Mizoguchi, and I. Inoue, *J. Org. Chem.*, 1971, **36,** 1573.
166. I. Gratcheva, A. Y. Veinberg, and G. I. Samokhvalov, *Zhur. obshchei Khim.*, 1971, **41,** 1376; *Chem. Abs.*, **75,** 88 579.
167. P. A. Lyon and C. B. Reese, *J.C.S. Perkin I*, 1974, 2645.
168. G. B. Chheda and C. I. Hong, *J. Medicin. Chem.*, 1971, **14,** 748.
169. C. I. Hong, G. B. Chheda, S. P. Dutta, A. O'Grady-Curtis, and G. L. Tritsch, *J. Medicin. Chem.*, 1973, **16,** 139.
170. W. Schulze, G. Letsch, and D. Tresselt, *Pharmazie*, 1975, **30,** 498.
171. *Ger. Pat.* 2 513 693 (1975).
172. J. P. Weidner and S. S. Block, *J. Medicin. Chem.*, 1972, **15,** 564.
173. H. Kasai and co-workers, *Biochemistry*, 1976, **15,** 898.
174. C. Nofre, A. Lefier, and A. Cier, *Compt. rend.*, 1961, **253,** 687.
175. Z. Neiman, *Israel J. Chem.*, 1971, **9,** 119.
176. J. Salomon and D. Elad, *J. Org. Chem.*, 1973, **38,** 3420.
177. N. C. Yang, L. S. Gorelic and B. Kim, *Photochem. Photobiol.*, 1971, **13,** 275.
178. J. Salomon and D. Elad, *Tetrahedron Letters*, 1971, 4783; *Photochem. Photobiol.*, 1974, **19,** 21.
179. D. Leonov and D. Elad, *J. Org. Chem.*, 1974, **39,** 1470.
180. J. Salomon and D. Elad, *J. Amer. Chem. Soc.*, 1974, **96,** 3295.
181. M. Maeda, K. Nushi, and Y. Kawazoe, *Tetrahedron*, 1974, **30,** 2677.
182. J. L. Fourrey, P. Jouin, and J. Moron, *Tetrahedron Letters*, 1974, 3005.
183. F. L. Lam and J. C. Parham, *J. Amer. Chem. Soc.*, 1975, **97,** 2839.
184. C. Helene, *Photochem. Photobiol.*, 1973, **18,** 534.
185. A. J. Varghese, *Photophysiology*, 1972, **7,** 207.
186. J. D. Engel, *Biochem. Biophys. Res. Comm.*, 1975, **64,** 581.
187. G. Grenner and H.-L. Schmidt, *Chem. Ber.*, 1977, **110,** 373.
188. T. Fujii, T. Itaya, and T. Saito, *Chem. Pharm. Bull. (Japan)*, 1975, **23,** 54.
189. G. B. Chheda and R. H. Hall, *Biochemistry*, 1966, **5,** 2082.
190. T. Itaya, T. Saito, S. Kawakatsu, and T. Fujii, *Chem. Pharm. Bull. (Japan)*, 1975, **23,** 2643.
191. A. Vineze, R. E. L. Henderson, J. J. McDonald, and N. J. Leonard, *J. Amer. Chem. Soc.*, 1973, **95,** 2677.
192. T. Fujii and T. Saito, *Chem. Pharm. Bull. (Japan)*, 1973, **21,** 1954.
193. C. R. Frihart and N. J. Leonard, *J. Amer. Chem. Soc.*, 1973, **95,** 7174; 1974, **96,** 5894.
194. B. N. Holmes and N. J. Leonard, *J. Org. Chem.*, 1976, **41,** 568.

195. H. M. Berman, R. J. Rousseau, R. W. Mancuso, G. P. Kreishman, and R. K. Robins, *Tetrahedron Letters*, 1973, 3099.
196. Kh. L. Muravich-Aleksandr, A. V. Eltsov, and I. El-Sakka, *Zhur. org. Khim.*, 1972, **8,** 1110.
197. Y. Maki, M. Suzuki, and K. Ozeki, *Tetrahedron Letters*, 1976, 1199.
198. S. M. Hecht, B. L. Adams, and J. W. Kozarich, *J. Org. Chem.*, 1976, **41,** 2303.
199. W. Pendergast, *J.C.S. Perkin I*, 1973, 2759.
200. W. Girke, *Tetrahedron Letters*, 1976, 3537.
201. W. Pendergast, *J.C.S. Perkin I*, 1975, 2240.
202. 'The Ring Index', ed. A. M. Patterson, L. T. Capell, and D. F. Walker, American Chemical Society, Washington 1959 (and Suppls.).
203. A. E. Chichibabin and A. W. Kirsanow, *Ber.*, 1927, **60,** 768.
204. V. Petrow and J. Saper, *J. Chem. Soc.*, 1948, 1389.
205. J. R. Vaughan, J. Krapcho, and J. P. English, *J. Amer. Chem. Soc.*, 1949, **71,** 1885.
206. A. Albert and C. Pedersen, *J. Chem. Soc.*, 1956, 4683.
207. F. Kögl, G. M. van der Want, and C. A. Salemink, *Rec. Trav. chim.*, 1948, **67,** 29.
208. C. A. Salemink and G. M. van der Want, *Rec. Trav. chim.*, 1949, **68,** 1013.
209. J. E. Schelling and C. A. Salemink, *Rec. Trav. chim.*, 1974, **93,** 160.
210. J. A. Montgomery and K. Hewson, *J. Medicin. Chem.*, 1965, **8,** 708.
211. K. B. de Roos and C. A. Salemink, *Rec. Trav. chim.*, 1969, **88,** 1263.
212. C. Temple, B. H. Smith, C. L. Kussner, and J. A. Montgomery, *J. Org. Chem.*, 1976, **41,** 3784.
213. D. L. Garmaise and J. Komlossy, *J. Org. Chem.*, 1964, **29,** 3403.
214. G. R. Revanker and S. Siddappa, *Monatsh.*, 1964, **95,** 242.
215. *French Pat.* 2 046 545 (1971).
216. *Swiss Pat.* 560, 594–596 (1975).
217. M. Israel and A. R. Day, *J. Org. Chem.*, 1959, **24,** 1455.
218. *U.S. Pat.* 3 719 683 (1973).
219. J. W. Clark-Lewis and M. J. Thompson, *J. Chem. Soc.*, 1957, 442.
220. *U.S. Pat.* 3 818 022 (1975).
221. *U.S. Pat.* 3 813 407 (1974).
222. A. Tensho, I. Shimizu, T. Takenawa, H. Kikuchi, T. Rokujo, and T. Kamiya, *J. Pharm. Soc. Japan*, 1974, **94,** 708.
223. *Ger. Pat.* 2 147 103 (1972).
224. A. Dornow, O. Hahmann, and E. E. Rohe, *Arch. Pharm.*, 1957, **290,** 20; 1958, **291,** 368.
225. V. Oakes, R. Pascoe, and H. N. Rydon, *J. Chem. Soc.*, 1956, 1045.
226. R. K. Smalley, *J. Chem. Soc.* (*C*), 1966, 80.
227. *U.S. Pat.* 3 325 506 (1967).
228. M. Israel, L. C. Jones, and E. J. Modest, *Tetrahedron Letters*, 1968, 4811; A. Nawojski, *Roczniki Chem.*, 1968, **42,** 1641.
229. M. Israel and L. C. Jones, *J. Heterocyclic Chem.*, 1969, **6,** 735.
230. B. Bobranski and J. Stankiewicz, *Diss. Pharm. Pharmacol.*, 1972, **24,** 301; *Chem. Abs.*, **77,** 101 547.
231. M. Israel and L. C. Jones, *J. Heterocyclic Chem.*, 1973, **10,** 201.
232. A. Albert and W. Pendergast, *J.C.S. Perkin I*, 1973, 1794.
233. R. A. Abramovitch and R. B. Rogers, *J. Org. Chem.*, 1974, **39,** 1802.
234. G. G. Aloisi, E. Bordignon, and A. Signor, *J.C.S. Perkin II*, 1972, 2218.
235. M. Ogata and H. Matsumoto, *Chem. Pharm. Bull.* (*Japan*), 1972, **20,** 2264.
236. J. Elguero, A. Fruchier, and S. Mignonac-Mondon, *Bull. Soc. chim. France*, 1972, 2916.
237. R. M. Bystrova and Y. M. Yutilov, *Khim. geterotsikl. Soedinenii*, 1973, 570.
238. Y. M. Yutilov and R. M. Bystrova, *Khim. geterotsikl. Soedinenii*, 1968, 954.
239. N. S. Miroshnichenko, O. I. Shkreblii, and A. V. Stetsenko, *Ukrain. khim. Zhur.*, 1973, **39,** 277; *Chem. Abs.*, **79,** 19 020.
240. P. C. Jain, S. K. Chatterjee, and N. Anand, *Indian J. Chem.*, 1963, **1,** 30.
241. R. M. Bystrova and Y. M. Yutilov, *Khim. geterotsikl. Soedinenii*, 1969, 378.
242. *Brit. Pat.* 1 114 199 (1968).
243. J. E. Schelling and C. A. Salemink, *Rec. Trav. chim.*, 1972, **91,** 650.
244. F. Korte, *Chem. Ber.*, 1952, **85,** 1012.
245. Y. M. Yutilov and R. M. Bystrova, *Khim. geterotsikl. Soedinenii*, 1970, 1146.
246. J. A. Elvidge, J. R. Jones, C. O'Brien, A. E. Evans, and H. C. Sheppard, *J.C.S. Perkin II*, 1973, 1889.
247. Y. M. Yutilov, N. R. Kalnitskii, and R. M. Bystrova, *Khim. geterotsikl. Soedinenii*, 1971, 1436.
248. N. G. Korzhenevskaya, E. V. Titov, Y. M. Yutilov, and R. M. Bystrova, *Khim. geterotsikl. Soedinenii*, 1975, 90.
249. S. M. Hecht, R. B. Frye, D. Werner, T. Fukui, and S. D. Hawrelak, *Biochemistry*, 1976, **15,** 1005.
250. V. I. Ofitserov, Z. V. Pushkareva, V. S. Mokrushin, and K. V. Aglitskaya, *Khim. geterotsikl. Soedinenii*, 1973, 1292.
251. R. P. Panzica and R. K. Robins, *J. Heterocyclic Chem.*, 1972, **9,** 623; M. Kawana, G. A. Ivanovics, R. J. Rousseau, and R. K. Robins, *J. Medicin. Chem.*, 1972, **15,** 841.
252. G. A. Ivanovics, R. J. Rousseau, M. Kawana, P. C. Srivastava, and R. K. Robins, *J. Org. Chem.*, 1974, **39,** 3651.

253. V. I. Ofitserov, V. S. Mokrushin, V. I. Nifonotov, N. V. Nikiforova, and L. N. Lych, *Khim. geterotsikl. Soedinenii*, 1975, 1550.
254. Y. F. Shealy, C. A. Krauth, and J. A. Montgomery, *J. Org. Chem.*, 1962, **27,** 2150.
255. M. A. Stevens, H. W. Smith, and G. B. Brown, *J. Amer. Chem. Soc.*, 1960, **82,** 3189.
256. R. N. Naylor, G. Shaw, D. V. Wilson, and D. N. Butler, *J. Chem. Soc.*, 1961, 4845.
257. V. I. Ofitserov, Z. V. Pushkareva, V. S. Mokrushin, and T. V. Rapakova, *Khim. geterotsikl. Soedinenii*, 1974, 428.
258. J. A. Montgomery, R. D. Elliott, and H. J. Thomas, *Ann. N.Y. Acad. Sci.*, 1975, **255,** 292.
259. V. S. Mokrushin, V. I. Ofitserov, T. V. Rapakova, A. G. Tsaur, and Z. V. Pushkareva, *Khim. geterotsikl. Soedinenii*, 1976, 556.
259a. A. C. Chibnall, *J. Biol. Chem.*, 1924, **61,** 303.
260. D. Ackermann and M. Mohr, *Z. Biol.*, 1936, **98,** 73.
261. G. Habermehi and H. J. Preusser, *Z. Naturforsch.* (*B*), 1970, **25,** 1451; *Z. Naturforsch.* (*C*), 1976, **31,** 118.
262. S. A. Waksman and H. B. Woodruff, *Proc. Soc. Exp. Biol. Med.*, 1942, **49,** 207; E. E. van Tamelen and co-workers, *J. Amer. Chem. Soc.*, 1961, **83,** 4295.
263. A. S. Khokhlov and K. I. Shutova, *J. Antibiot.*, 1972, **25,** 501.
264. J. Wellisch, *Biochem. Z.*, 1913, **49,** 173.
264a. A. Neuberger, *Biochem. J.*, 1944, **38,** 309.
265. A. Frankel and K. Zeimer, *Biochem. Z.*, 1920, **110,** 234.
265a. T. Vitali and G. Bertaccini, *Gazzetta*, 1964, **94,** 296.
266. D. Heyl, E. Luz, S. A. Harris, and K. Folkers, *J. Amer. Chem. Soc.*, 1948, **70,** 3429, 3669.
267. T. Vitali, F. Mossini, and G. Bertaccini, *Farmaco Ed. Sci.*, 1965, **20,** 634; 1967, **22,** 821.
268. F. B. Stocker, M. W. Fordice, J. K. Larson, and J. H. Thorstenson, *J. Org. Chem.*, 1966, **31,** 2380.
269. A. Brabianti, F. Dallavalle, E. Leporati, and G. Mori, *J.C.S. Dalton*, 1973, 323.
270. R. J. Rousseau, J. A. May, R. K. Robins, and L. B. Townsend, *J. Heterocyclic Chem.*, 1974, **11,** 233.
271. R. K. Robins and co-workers, *J. Amer. Chem. Soc.*, 1975, **97,** 2916.
272. *U.S. Pat.* 3 919 193 (1975).
273. K. Knobloch and H. Kuehne, *J. prakt. Chem.*, 1962, **17,** 199.
274. A. V. Stetsenko and N. S. Miroshnichenko, *Ukrain. khim. Zhur.*, 1973, **39,** 703; *Chem. Abs.*, **79,** 105 515.
275. F. H. Case, *J. Heterocyclic Chem.*, 1967, **4,** 157.
276. *U.S. Pat.* 3 759 933 (1973).
277. Y. Mizuno, T. Itoh, and K. Saito, *Chem. Pharm. Bull.* (*Japan*), 1964, **12,** 866.
278. Y. Mizuno, T. Itoh, and K. Saito, *J. Org. Chem.*, 1964, **29,** 2611.
279. M. Israel and L. C. Jones, *J. Heterocyclic Chem.*, 1971, **8,** 797.
280. G. C. Wright, *J. Heterocyclic Chem.*, 1976, **13,** 601.
281. G. B. Barlin, *J. Chem. Soc.* (*B*), 1966, 285.
282. N. K. Beresneva, E. R. Zakhs, L. S. Efros, and Y. M. Treger, *Khim. geterotsikl. Soedinenii*, 1972, 683.
283. A. V. Kazymov, E. P. Shchelkina, L. V. Ivanova, N. V. Monich, and A. F. Vompe, *Khim. geterotsikl. Soedinenii*, 1970, 228.
284. G. Cleve, H. Gibian, G.-A. Hoyer, D. Rahtz, E. Schroder, and G. Schulz, *Annalen*, 1971, **747,** 158.
285. *Jap. Pat.* 6 925 580 (1969); *Chem. Abs.*, **72,** 21 927.
286. R. J. Rousseau, L. B. Townsend, and R. K. Robins, *Biochemistry*, 1966, **5,** 756.
287. Y. Mizuno, S. Tazawa, and K. Kageura, *Chem. Pharm. Bull.* (*Japan*), 1968, **16,** 2011.
288. P. C. Jain and N. Anand, *Indian J. Chem.*, 1968, **6,** 616.
289. J. A. May and L. B. Townsend, *J.C.S. Perkin I*, 1975, 125.
290. N. S. Miroshnichenko, A. S. Kovalenko, and A. V. Stetsenko, *Ukrain. khim. Zhur.*, 1974, **40,** 258; *Chem. Abs.*, **81,** 4200.
291. Y. M. Yutilov and J. A. Svertilova, *Khim. geterotsikl. Soedinenii*, 1971, 428.
292. Y. M. Yutilov and J. A. Svertilova, *Khim. geterotsikl. Soedinenii*, 1973, 138; 1974, 854.
293. Y. M. Yutilov and L. I. Kovaleva, *Khim. geterotsikl. Soedinenii*, 1975, 1389.
294. J. A. May and L. B. Townsend, *J.C.S. Chem. Comm.*, 1973, 64.
295. S. Kitano, A. Nomura, Y. Mizuno, T. Okamoto, and Y. Isogai, *J. Carbohydrates, Nucleosides, Nucleotides*, 1975, **2,** 299.
296. Y. Matsushima and T. Hino, *Chem. Pharm. Bull.* (*Japan*), 1968, **16,** 2277.
297. R. K. Robins, J. K. Horner, C. V. Greco, C. W. Noell, and C. G. Beames, *J. Org. Chem.*, 1963, **28,** 3041.
298. B. L. Sukhorurov, V. L. Poltev, R. V. Polozov, and I. A. Ilicheva, *Doklady Akad. Nauk SSSR*, 1973, **208,** 443.
299. L. Hagenberg, H. G. Gassen, and H. Matthaei, *Biochem. Biophys. Res. Comm.*, 1973, **50,** 1104.
300. *Ger. Pat.* 2 349 504 (1974); *Chem. Abs.*, **81,** 19 229.
301. K. A. Hofmann and O. Ehrhart, *Ber.*, 1912, **45,** 2731.
302. E. C. Taylor and R. W. Hendess, *J. Amer. Chem. Soc.*, 1965, **87,** 1980.
303. O. Castillani and J. F. Fernandes, *Rev. Inst. Med. Trop. Sao Paulo*, 1965, **7,** 275; *Chem. Abs.*, **64,** 11 598.
304. R. Bokaldere and A. Liepina, *Khim. geterotsikl. Soedinenii*, 1973, 276.
305. T. Hirata, L. M. Twanmoh, H. B. Wood, A. Goldin, and J. S. Driscoll, *J. Heterocyclic Chem.*, 1972, **9,** 99.
306. L. Capuano and H. J. Schrepfer, *Chem. Ber.*, 1971, **104,** 3039.
307. R. Bokaldere, A. Liepina, I. Mazeika, I. S. Yankovskaya, and E. Liepins, *Khim. geterotsikl. Soedinenii*, 1973, 419.

308. R. Bokaldere and V. Grinsteins, *Khim. geterotsikl. Soedinenii*, 1970, 563.
309. S. Polanc, B. Vercek, B. Sek, B. Stanovnik, and M. Tisler, *Tetrahedron Letters*, 1973, 1677; *J. Org. Chem.*, 1974, **39,** 2143.
310. P. Guerret, R. Jacquier, and G. Maury, *J. Heterocyclic Chem.*, 1971, **8,** 643.
311. R. J. Deshpande and A. V. R. Roa, *Synthesis*, 1974, 863.
312. A. B. de Milo, J. E. Oliver, and R. D. Gilardi, *J. Heterocyclic Chem.*, 1973, **10,** 231.
313. F. Yoneda, T. Nagamura, and M. Kawamura, *J.C.S. Chem. Comm.*, 1976, 658.
314. F. Yoneda, M. Kawamura, T. Nagamatsu, K. Kuretoni, A. Hoshi, and M. Igo, *Heterocycles*, 1976, **4,** 1503.
315. H. Kasai and co-workers, *Biochemistry*, 1975, **14,** 4198.
316. T. Ohgi, T. Goto, H. Kasai, and S. Nishimura, *Tetrahedron Letters*, 1976, 367.
317. P. Nesbitt and P. Sykes, *J. Chem. Soc.*, 1954, 3057.
317a. T. B. Johnson, *J. Amer. Chem. Soc.*, 1911, **33,** 758.
318. J. Davoll, *J. Chem. Soc.*, 1960, 131.
319. J. A. Montgomery and K. Hewson, *J. Medicin. Chem.*, 1967, **10,** 665.
320. Y. Mizuno, M. Ikehara, K. A. Watanabe, and S. Suzuki, *Chem. Pharm. Bull. (Japan)*, 1963, **11,** 1091.
321. E. Bisagni, J. P. Marquet, and J. Andre-Louisfert, *Bull. Soc. chim. France*, 1969, 4338.
322. A. A. Santilli, A. C. Scotese, and D. H. Kim, *J. Heterocyclic Chem.*, 1975, **12,** 1291.
323. T. C. Lee, G. Salemnick, and G. B. Brown, *J. Org. Chem.*, 1974, **39,** 2963.
324. *Ger. Pat.* 1 100 030 (1961).
325. D. H. Kim and A. A. Santilli, *J. Heterocyclic Chem.*, 1969, **6,** 819.
326. P. A. Crooks and B. Robinson, *Canad. J. Chem.*, 1969, **47,** 2061.
327. H. Ogura, M. Sakaguchi, and K. Takeda, *Chem. Pharm. Bull. (Japan)*, 1972, **20,** 404 and refs. therein.
328. *U.S. Pat.* 3 897 433 (1975).
329. W. Pfleiderer, L. Groezinger, and F. Sagi, *Chem. Ber.*, 1966, **99,** 3524.
330. Y. Tamura, T. Sakaguchi, T. Kawasaki, and Y. Kita, *Chem. Pharm. Bull. (Japan)*, 1976, **24,** 1160.
331. *S. Afr. Pat.* 6 902 227 (1970).
332. T. B. Johnson and E. F. Kohmann, *Amer. Chem. J.*, 1913, **49,** 186.
333. J. Biggs and P. Sykes, *J. Chem. Soc.*, 1959, 1849.
334. E. C. Taylor and R. W. Hendess, *J. Amer. Chem. Soc.*, 1964, **86,** 951; 1965, **87,** 1995.
335. R. L. Tolman, R. K. Robins, and L. B. Townsend, *J. Amer. Chem. Soc.*, 1968, **90,** 524.
336. T. D. Duffy and D. G. Wibberley, *J.C.S. Perkin I*, 1974, 1921.
337. H. Poehlmann, S. Heyne, R. Kraft, and S. Pfiefer, *Pharmazie*, 1976, **31,** 28.
338. H. Wamhoff and B. Wehling, *Chem. Ber.*, 1976, **109,** 2983.
339. V. G. Granik and R. G. Gloshkov, *Khim.-Farm. Zhur.*, 1967, **1,** 16; *Chem. Abs.*, **68,** 12 941.
340. R. Richter and H. Ulrich, *J. Org. Chem.*, 1973, **38,** 2614.
341. R. A. West and L. Beauchamp, *J. Org. Chem.*, 1961, **26,** 3809.
342. R. A. West, *J. Org. Chem.*, 1963, **28,** 1991.
343. R. L. Tolman, R. K. Robins, and L. B. Townsend, *J. Amer. Chem. Soc.*, 1969, **91,** 2102; M. Bobek, R. L. Whistler, and A. Bloch, *J. Medicin. Chem.*, 1972, **15,** 168.
344. H. Iwamura and T. Hashizume, *Agric. Biol. Chem. (Japan)*, 1968, **32,** 1010.
345. Y. Mizuno, M. Ikehara, K. A. Watanabe, and S. Suzuki, *J. Org. Chem.*, 1963, **28,** 1331.
346. R. L. Tolman, G. L. Tolman, R. K. Robins, and L. B. Townsend, *J. Heterocyclic Chem.*, 1970, **7,** 799.
347. J. F. Gerster, B. C. Hinshaw, R. K. Robins, and L. B. Townsend, *J. Heterocyclic Chem.*, 1969, **6,** 207, 215.
348. K. H. Schman and L. B. Townsend, *J. Carbohydrates, Nucleosides, Nucleotides*, 1975, **2,** 177.
349. R. A. West, *J. Org. Chem.*, 1961, **26,** 4959; *Brit. Pat.* 981 458 (1965).
350. S. Senda and K. Hirota, *Chem. Pharm. Bull. (Japan)*, 1974, **22,** 2921.
351. F. Yoneda and M. Higuchi, *Chem. Pharm. Bull. (Japan)*, 1972, **20,** 2076; F. Yoneda, M. Higuchi, K. Senga, M. Kanohori, and S. Nishigaki, *ibid.*, 1973, **21,** 473.
352. K. Senga, S. Nishigaki, M. Higuchi, and F. Yoneda, *Chem. Pharm. Bull. (Japan)*, 1972, **20,** 1473.
353. *Brit. Pat.* 915 303 (1963).
354. J.-D. Bourzat and E. Bisagni, *J. Heterocyclic Chem.*, 1973, **10,** 511.
355. *Ger. Pat.* 2 042 253 (1971).
356. K. H. Schram and L. B. Townsend, *Tetrahedron Letters*, 1974, 1345.
357. K. H. Schram, S. J. Manning, and L. B. Townsend, *J. Heterocyclic Chem.*, 1975, **12,** 1021.
358. K. Anzai, G. Nakamura, and S. Suzuki, *J. Antibiotics (Japan)*, 1957, **10A,** 201.
359. S. Shirato, Y. Miyazaki, and I. Suzuki, *Hakko Kogaku Zasshi*, 1967, **45,** 60; *Chem. Abs.*, **69,** 26 024.
360. M. E. Smilson and R. J. Suhadolnik, *J. Biol. Chem.*, 1967, **242,** 2872.
361. J. E. Pike, L. Slechta, and P. F. Wiley, *J. Heterocyclic Chem.*, 1964, **1,** 159.
362. J. F. Gerster, B. Carpenter, R. K. Robins, and L. B. Townsend, *J. Medicin. Chem.*, 1967, **10,** 326.
363. N. A. Berger and G. L. Eichhorn, *Biochemistry*, 1971, **10,** 1847.
364. T. A. Glassman, G. Klopman, and C. Cooper, *Biochemistry*, 1973, **12,** 5013.
365. G. Acs and E. Reich, *Antibiotics*, 1967, **1,** 494; R. J. Suhadolnik, *ibid.*, 1967, **2,** 400, 448.
366. H. Nishimura, K. Katagiri, K. Sato, M. Mayama, and N. Shimaoka, *J. Antibiotics (Japan)*, 1956, **9A,** 60.
367. K. V. Rao and D. W. Renn, *Antimicrobial Agents and Chemotherapy*, 1963, 77.
368. R. J. Suhadolnik and T. Uematsu, *J. Biol. Chem.*, 1970, **245,** 4365.
369. Y. Kuchimo, H. Kasai, K. Nihei, and S. Nishimura, *Nucleic Acids Res.*, 1976, **3,** 393.
370. K. Ohkuma, *J. Antibiotics (Japan)*, 1960, **13A,** 361.

371. B. C. Hinshaw, J. F. Gerster, R. K. Robins, and L. B. Townsend, *J. Org. Chem.*, 1970, **35,** 236.
372. K. H. Schram and L. B. Townsend, *J. Carbohydrates, Nucleosides, Nucleotides*, 1974, **1,** 39.
373. R. L. Tolman, R. K. Robins, and L. B. Townsend, *J. Heterocyclic Chem.*, 1971, **8,** 703.
374. K. H. Schram and L. B. Townsend, *Tetrahedron Letters*, 1971, 4757.
375. D. H. Kim and A. A. Santilli, *Tetrahedron Letters*, 1971, 2441; 1972, 2301; *U.S. Pat.* 3 850 928 (1974).
376. J.-L. Imbach, J. L. Barascut, B. L. Kam, B. Rayner, C. Tamby, and C. Tapiero, *J. Heterocyclic Chem.*, 1973, **10,** 1069.
377. K. Anzai and M. Matsui, *Agric. Biol. Chem.* (*Japan*), 1973, **37,** 2431; *Bull. Chem. Soc. Japan*, 1974, **47,** 417.
378. C. W. Noell and R. K. Robins, *J. Heterocyclic Chem.*, 1964, **1,** 34.
379. M. J. Chenon, R. J. Pugmire, D. M. Grant, R. P. Panzica, and L. B. Townsend, *J. Amer. Chem. Soc.*, 1975, **97,** 4627, 4636.
380. D. L. von Minden, J. G. Liehr, M. H. Wilson, and J. A. McCloskey, *J. Org. Chem.*, 1974, **39,** 285.
381. V. Amarnath and R. Madhav, *Synthesis*, 1974, 837.
382. L. B. Townsend and G. H. Milne, *Ann. N.Y. Acad. Sci.*, 1975, **255,** 91.
383. *Jap. Pat.* 25 298 (63); *Chem. Abs.*, **60,** 7415.
384. R. C. Parks and K. C. Agarwal, *Handb. Exp. Pharmacol.*, 1975, **38,** 47.
385. K. R. H. Wooldridge and co-workers, *Nature*, 1974, **251,** 650; *J. Medicin. Chem.*, 1975, **18,** 1117.
386. A. Albert, *J. Chem. Soc.* (*B*), 1966, 427, 433.
387. S. Gabriel and J. Colman, *Ber.*, 1901, **34,** 1234.
388. I. K. W. Merz and P. H. Stahl, *Beitr. Biochem. Physiol. Naturstoffen Festschr.*, 1965, 285; *Chem. Abs.*, **65,** 3877.
389. D. Sen and P. Sengupta, *J. Indian Chem. Soc.*, 1973, **50,** 260.
390. R. O. Roblin, J. O. Lampen, J. P. English, Q. P. Cole, and J. R. Vaughan, *J. Amer. Chem. Soc.*, 1945, **67,** 290.
391. *Brit. Pat.* 1 338 235 (1973).
392. G. Nübel and W. Pfleiderer, *Chem. Ber.*, 1965, **98,** 1060.
393. R. M. Cresswell, H. K. Maurer, T. Strauss, and G. B. Brown, *J. Org. Chem.*, 1965, **30,** 408.
394. Y. Maki, M. Suzuki, K. Izuta, and S. Iwai, *Chem. Pharm. Bull.* (*Japan*), 1974, **22,** 1269.
395. E. C. Taylor, J. W. Barton, and W. D. Paudler, *J. Org. Chem.*, 1961, **26,** 4961.
396. F. R. Benson, L. W. Hartzel, and W. L. Savell, *J. Amer. Chem. Soc.*, 1950, **72,** 1816.
397. Y. Maki and E. C. Taylor, *Chem. Pharm. Bull.* (*Japan*), 1972, **20,** 605.
398. I. Jezo and Z. Voticky, *Chem. Zvesti*, 1952, **6,** 357; *Chem. Abs.*, **48,** 7019.
399. A. Dornow and J. Helberg, *Chem. Ber.*, 1960, **93,** 2001.
400. A. Albert, *J. Chem. Soc.* (*C*), 1969, 152.
401. A. Albert, *J. Chem. Soc.* (*C*), 1970, 230.
402. A. Albert and K. Tratt, *J. Chem. Soc.* (*C*), 1968, 344.
403. A. Albert and H. Taguchi, *J.C.S. Perkin I*, 1972, 449.
404. A. Albert, *J. Chem. Soc.* (*C*), 1969, 2379.
405. C. W. Smith, R. W. Sidwell, R. K. Robins, and R. L. Tolman, *J. Medicin. Chem.*, 1972, **15,** 883.
406. A. Albert, *J.C.S. Perkin I*, 1975, 345.
407. A. Albert and C. J. Lin, *J.C.S. Perkin I*, 1977, 210.
408. A. Albert, *J.C.S. Perkin I*, 1974, 2030.
409. A. Albert, *J.C.S. Perkin I*, 1972, 461.
410. A. Albert, *J.C.S. Perkin I*, 1973, 2659.
411. A. Albert, *J. Chem. Soc.* (*C*), 1968, 2076.
412. D. R. Sutherland and G. Tennant, *J. Chem. Soc.* (*C*), 1971, 706.
413. A. Albert and H. Taguchi, *J.C.S. Perkin I*, 1973, 2037.
414. A. Albert, *J.C.S. Perkin I*, 1976, 291.
415. H. V. Blank, I. Wempen, and J. J. Fox, *J. Amer. Chem. Soc.*, 1968, **90,** 7175.
416. H. V. Blank, I. Wempen, and J. J. Fox, *J. Org. Chem.*, 1970, **35,** 1131.
417. T. Sasaki, K. Minamoto, M. Kino, and T. Mizuno, *J. Org. Chem.*, 1976, **41,** 1100.
418. K. Senga, Y. Kanamori, S. Nishigaki, and F. Yoneda, *Chem. Pharm. Bull.* (*Japan*), 1976, **24,** 1917.
419. A. Albert, W. Pfleiderer, and D. Thacker, *J. Chem. Soc.* (*C*), 1969, 1084.
420. H. Bredereck and W. Baumann, *Annalen*, 1967, **701,** 143.
421. R. B. Angier and J. W. Marisco, *J. Org. Chem.*, 1960, **25,** 759.
422. M. Ridi, G. Franchi, S. Mangiavacchi, and M. P. Lombardini, *Boll. Chim. Farm.*, 1968, **107,** 401.
423. J. Davoll, *J. Chem. Soc.*, 1958, 1593.
424. *U.S. Pat.* 2 993 039 (1961).
425. J. A. Montgomery, H. J. Thomas, and S. J. Clayton, *J. Heterocyclic Chem.*, 1970, **7,** 215.
426. W. W. Lee, A. P. Martinez, G. L. Tong, and L. Goodman, *Chem. and Ind.* (*London*), 1963, 2007.
427. W. Hutzenlaub, R. L. Tolman, and R. K. Robins, *J. Medicin. Chem.*, 1972, **15,** 879.
428. J. A. Montgomery and R. D. Elliott, *J.C.S. Chem. Comm.*, 1972, 1279.
429. D. A. Baker, R. A. Harder, and R. L. Tolman, *J.C.S. Chem. Comm.*, 1974, 167.
430. *U.S. Pat.* 2 543 333 (1951).
431. *Belg. Pat.* 841 181 (1976).
432. F. F. Blicke and H. C. Godt, *J. Amer. Chem. Soc.*, 1954, **76,** 2798.

433. D. S. Bariana and C. Groundwater, *J. Heterocyclic Chem.*, 1969, **6,** 583; *J. Medicin. Chem.*, 1971, **14,** 543.
434. R. S. Karlinskaya and N. V. Khromov-Borisov, *Zhur. obshchei Khim.*, 1962, **32,** 1858; *Chem. Abs.*, **58,** 4554.
435. H. Ballweg, *Annalen*, 1962, **657,** 141.
436. B. M. Lynch, A. J. Robertson, and J. G. K. Webb, *Canad. J. Chem.*, 1969, **47,** 1129.
437. H. Bredereck and W. Baumann, *Annalen*, 1967, **701,** 157.
438. A. Roush and E. R. Norris, *Arch. Biochem.*, 1950, **29,** 124.
439. R. D. Elliott and J. A. Montgomery, *J. Medicin. Chem.*, 1976, **19,** 1186.
440. D. J. Brown and M. N. Paddon-Row, *J. Chem. Soc.* (*C*), 1967, 1856.
441. D. R. Sutherland and G. B. Brown, *J. Org. Chem.*, 1973, **38,** 1291.
442. A. Albert and W. Pendergast, *J.C.S. Perkin I*, 1972, 457.
443. Y. Hirata, K. Iwashita, and K. Teshima, *Nagaya Sangyo Kagaku*, 1957, No. 9, 83; *Chem. Abs.*, **51,** 12 074.
444. S. Yamada, T. Mizoguchi, and A. Ayada, *J. Pharm. Soc.* (*Japan*), 1957, **77,** 441; *Chem. Abs.*, **51,** 12 107.
445. L. L. Bennett and H. T. Baker, *J. Org. Chem.*, 1957, **22,** 707.
446. Y. F. Shealy and C. A. O'Dell, *J. Org. Chem.*, 1965, **30,** 2488.
447. A. Albert and C. J. Lin, *J.C.S. Perkin I*, 1977, 1819.
448. *U.S. Pat.* 2 714 110 (1955).
449. A. Albert and W. Pendergast, *J.C.S. Perkin I*, 1973, 1625.
450. L. G. Purnell, J. C. Shepherd, and D. J. Hodgson, *J. Amer. Chem. Soc.*, 1975, **97,** 2376.
451. C. Temple, B. H. Smith, and J. A. Montgomery, *J.C.S. Chem. Comm.*, 1972, 52.
452. C. Wentrup and W. D. Crow, *Tetrahedron*, 1970, **26,** 3965.
453. L. Kittler and G. Loeber, *Stud. Biophys.*, 1973, **36/37,** 5.
454. L. Kittler and H. Berg, *Photochem. Photobiol.*, 1967, **6,** 199.
455. D. McDaniel, unpublished work.
456. F. Bergmann, M. Rashi, V. Reichman, and Z. Nieman, *Israel J. Chem.*, 1970, **8,** 919.
457. D. J. Hodgson, personal communication.
458. J. Elguero, C. Marzin, A. R. Katritzky, and P. Linda, *Adv. Heterocyclic Chem.*, Suppl. 1, 1976.
459. A. Albert, in 'Physical Methods in Heterocyclic Chemistry', ed. A. R. Katritzky, Academic Press, New York, vol. 1 (1963) p. 1; vol. 3 (1971) p. 117; Ref. 3, p. 1.
460. P. Dea, G. R. Revankar, R. L. Tolman, R. K. Robins, and M. P. Schweizer, *J. Org. Chem.*, 1974, **39,** 3226.
461. H. D. Luedemann, E. Westhof, and I. Cuno, *Z. Naturforsch.* (*C*), 1976, **31,** 135.
462. S. M. Hecht, *Anal. Biochem.*, 1971, **44,** 262.
463. A. Albert, *Chem. and Ind.* (*London*), 1970, 365.
464. K. Tanaka, T. Sugawa, R. Nakamori, Y. Sanno, Y. Ando, and K. Imai, *J. Pharm. Soc. Japan*, 1955, **75,** 770; *Chem. Abs.*, **49,** 14 001; *Chem. Pharm. Bull.* (*Japan*), 1964, **12,** 1024, 1030.
465. S. Senda and K. Hirota, *Chem. Pharm. Bull.* (*Japan*), 1974, **22,** 2593.
466. F. L. Rose, *J. Chem. Soc.*, 1954, 4116.
467. W. Pfleiderer and H. Mosthaf, *Chem. Ber.*, 1957, **90,** 738.
468. E. C. Taylor and E. E. Garcia, *J. Org. Chem.*, 1965, **30,** 655.
469. J. A. Montgomery and K. Hewson, *J. Org. Chem.*, 1965, **30,** 1528.
470. J. A. Montgomery and A. G. Laseter, *J. Heterocyclic Chem.*, 1972, **9,** 1077.
471. S. Senda, K. Hirota, and M. Takahashi, *Heterocycles*, 1976, **4,** 461.
472. Y. Okamoto and T. Ueda, *Chem. Pharm. Bull.* (*Japan*), 1976, **24,** 547.
473. H. Fenner and H. Motschall, *Tetrahedron Letters*, 1971, 4185.
474. T. Murata and K. Ukawa, *Chem. Pharm. Bull.* (*Japan*), 1974, **22,** 240.
475. F. Cassidy, R. K. Olsen, and R. K. Robins, *J. Heterocyclic Chem.*, 1968, **5,** 461.

17.6
3-, 4-, and 7-Membered Aza-heterocycles

R. K. SMALLEY
University of Salford

17.6.1 INTRODUCTION

Despite their diverse structures, the heterocycles discussed herein have some common features, not least of which is their lack of aromatic character. In fact the instability of several of the ring systems is consequent on them having a destabilizing anti-aromatic[1] $4n\pi$-electron arrangement. The microcycles, particularly 2*H*-azirines, are highly useful synthons for the preparation of mono-, di-, and tri-azepines since many of their reactions are accompanied by ring-expansion. Conversely the azepines, being non-planar, non-aromatic polyenes, undergo a variety of ring-contractions to arenes, heteroarenes, and bicyclic valence-bond isomers. Ring transformations of heterocycles have been the subject of an extensive and authoritative review.[2]

In this chapter, discussion of the many hydro-, oxo-, and dioxo-derivatives, particularly of azepines, has been kept to a minimum, although leading references to their chemistry are given in the appropriate sections. The chemistry of the unsaturated systems is emphasized.

17.6.2 AZIRINES

Azirine is the name used to denote the azacyclopropene system. Two isomeric forms are possible, namely 1*H*-azirine (**1**) and 2*H*-azirine (**2**). This system of nomenclature is that favoured by the Chemical Society, Chemical Abstracts, and the Ring Index. However, confusion can arise since an alternative system, which labels (**1**) and (**2**) as 2-azirine and 1-azirine respectively, has been suggested[3] and enjoys common usage, particularly in the American literature.

17.6.2.1 1*H*-Azirines

Unlike 2*H*-azirines, whose general chemistry is well documented[4], the 1*H*-azirines are as yet uncharacterized and are known only as transient intermediates. Their instability is due not only to the large bond-angle strain inherent in any unsaturated three-membered ring, but also to the overlap of the nitrogen lone-pair electrons with the carbon–carbon double bond, a situation which gives rise to a destabilizing anti-aromatic 4π-electron system. A measure of this destabilization can be seen in the results of *ab initio* calculations,[5] which indicate that 2*H*-azirine is 169 kJ mol^{-1} more stable than the 1*H*-isomer. Some relief is possible by distorting the nitrogen atom out of the ring-plane, and later theoretical studies[6] show that the value of ΔH (heat of hydrogenation), taken as a measure of anti-aromatic destabilization, decreases from 263–277 kJ mol^{-1} to 162–199 kJ mol^{-1} on going from a planar to a non-planar 1*H*-azirine structure. The azirinyl cation (**3**) has extensive π-delocalization,[7] and may be obtainable as a stable aromatic species. Some theoretically derived molecular dimensions and physical constants for the azirines and their cation are appended to structures (**1**)–(**3**).[5–7]

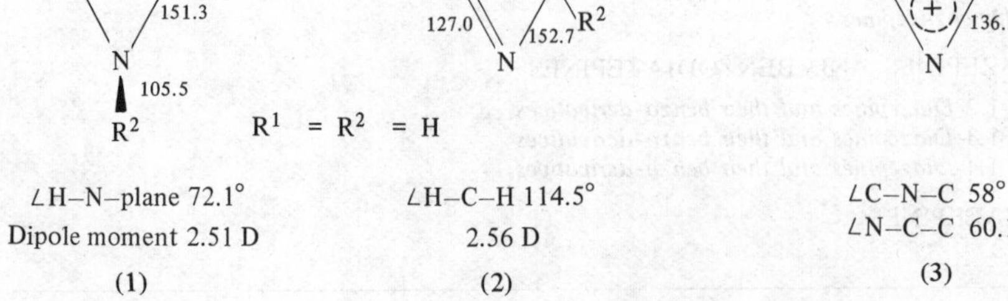

R^1 R^1 107.6 126.6 151.3 N 105.5 R^2

R^1 = R^2 = H

R^1 R^1 108.0 145.8 127.0 152.7 R^2 N

135.8 (+) 136.1 N

∠H–N–plane 72.1°
Dipole moment 2.51 D
(1)

∠H–C–H 114.5°
2.56 D
(2)

∠C–N–C 58°
∠N–C–C 60.1°
(3)

R1, N, N, R3, N, R2 (4) —i→ [R1, R2, N, R2] (5) → R1, R3, R2, N (6)

R^2 = phthalimido (N–)

a; R^1 = Me, R^3 = Me
b; R^1 = Ph, R^3 = Me

i, 400–500 °C, 0.02 mmHg.

SCHEME 1

A likely mode of 1*H*-azirine synthesis is the addition of a nitrene to an alkyne. Early attempts using ethoxycarbonylnitrene (:$\dot{N}CO_2Et$) failed.[4] However, by treating hex-3-yne with *N*-nitrenophthalimide, Rees and his co-workers[8a] were able to isolate the 2*H*-azirine (**2**; R^1 = Et, R^2 = as in Scheme 1) which most probably is formed by rearrangement of the initially produced but highly unstable 1*H*-azirine (**1**; R^1 = Et, R^2 = as in Scheme 1). The same authors have also demonstrated unequivocally the intermediacy of 1*H*-azirines in the vacuum pyrolysis of the isomeric triazoles (**4a**) and (**4b**) (Scheme 1).[8b,c] The isolation of identical mixtures of the isomeric 2*H*-azirines (**6a**) and (**6b**) (plus their pyrolysis products) suggests a common reaction intermediate, which the authors conclude to be the 1*H*-azirine (**5**). Subsequent pyrolysis studies[9] on substituted [5-^{13}C]-labelled 1,2,3-triazoles have confirmed these results.

17.6.2.2 2*H*-Azirines

2*H*-Azirines were first isolated as by-products from the synthesis of α-amino-ketones by the action of *p*-tosyl chloride on ketoximes in pyridine solution (Neber reaction: Scheme 5, path a). The highly strained nature of these intermediates rendered structures of type (**7**) doubtful, but a critical re-examination of the reaction by Cram and Hatch confirmed the 2*H*-azirine structure.[4] However, the Neber reaction and its many variations have little merit as a general synthetic method for 2*H*-azirines, although a modified Neber reaction on isobutyrophenone dimethylhydrazone methiodide has afforded 2,2-dimethyl-3-phenyl-2*H*-azirine (**6**; R^1 = Ph, $R^2 = R^3$ = Me) in 85% yield.[4]

The first general synthesis of 2*H*-azirines was developed by Smolinsky[10a] and employed the vapour-phase pyrolysis of α-vinyl azides. In contrast, the so-called terminal azides, *i.e.* β-vinyl azides, were thought[10b] to give products other than 2*H*-azirines. It was subsequently shown that, whereas β-styryl azides [*e.g.* (**8**)] decompose in boiling n-hexadecane (b.p. 287 °C) to yield mainly indoles, thermolysis at lower temperatures (100–120 °C) or, better, photolysis under nitrogen at −30 °C, permits isolation of 2*H*-azirines [*e.g.* (**9**)] in quantitative yields (Scheme 2).[11a,b]

Me, N, H ←i— Ph, N_3, H, Me (8) —ii→ Ph, Me, H, N (9)

i, n-$C_{16}H_{34}$ at 287 °C; ii light petroleum at 100–105 °C, or $h\nu$, λ = 365 nm, under N_2 at −30 °C.

SCHEME 2

$$RCH{=}CH_2 \xrightarrow{i} RCH(N_3)CH_2I \xrightarrow{ii} R{-}C(N_3){=}CH_2 \xrightarrow{iii} \left[R{-}C({:}\ddot{N}{:}){=}CH_2 \right] \longrightarrow$$ 2H-azirine (R, N)

(10)

i, ICl, NaN_3, MeCN, ii, Bu^tOK, Et_2O, 0 °C, iii, $h\nu$

SCHEME 3

The synthetic value of this method has been enhanced by the discovery[12] that addition of iodine azide (generated *in situ* by reaction of iodine monochloride and sodium azide in dimethylformamide or acetonitrile solution) to alkenes followed by dehydroiodination (Scheme 3) provides a general synthetic route to the otherwise difficultly accessible vinyl azides. Photolytic decomposition is the method of choice[3] since reaction temperatures may be kept low, so minimizing the chance of decomposing the often thermo-labile 2*H*-azirine.

A wide range of 2*H*-azirines, including the fused bicyclic system (**11**), have been prepared[3,13] and in general yields are good and by-products few. Azirine formation probably involves a vinylnitrene intermediate [*e.g.* (**10**)], although alternative routes have been considered.[3,4]

Other synthetic methods are available.[4] In particular, the thermal decomposition of oxazaphospholes (**12**) (readily prepared by 1,3-dipolar cycloaddition of nitrile oxides and alkylidenephosphoranes) is a useful route and has yielded some novel 2*H*-azirines such as the spiro-compound (**13**) (84%).

(11) (12) R_3P, O, N, R^1_2, R^2 (13) Ph, N

The photo-induced isoxazole–oxazole rearrangement (Scheme 4), which involves an isolable 2*H*-azirine intermediate, is noteworthy, not as a preparative method but for the remarkable wavelength dependence which it displays.[4]

Ph, O, N, Ph $\underset{ii}{\overset{i}{\rightleftharpoons}}$ Ph, H, COPh, N (12%) $\xrightarrow{i}$ N, Ph, Ph, O (82%)

i, $h\nu$, λ = 253.7 nm; ii, $h\nu$, λ > 300 nm

SCHEME 4

An attempt to prepare azirinones (azacyclopropenones) by a variation of the vinylazide method failed.[14] The azido-keten precursor (**14**), although trappable at −60 °C as a cycloadduct with benzylideneaniline, proved to be unstable at temperatures above −30 °C and decomposed, possibly by way of azirinone (**15**), to benzonitrile and carbon monoxide.

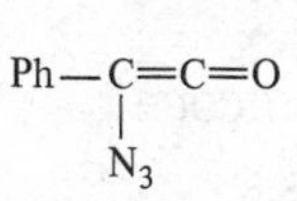

Ph—C=C=O with N_3 on C

(14)

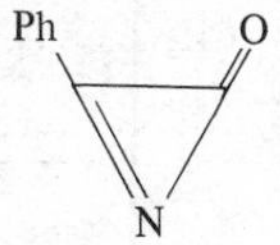

(15)

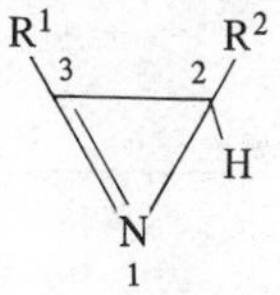

(16)

	C-2	C-3 (p.p.m.)
R^1 = Ph, R^2 = H	19.58	165.74
R^1 = Ph, R^2 = Me	22.33	172.43
R^1 = R^2 = Ph	22.66	163.80

17.6.2.3 Physical and spectroscopic properties

The high reactivity and relative ease of preparation of 2*H*-azirines make them highly desirable synthetic reagents. However, some care is needed in their use since the majority of the alkyl and aryl derivatives are evil-smelling, skin-irritating liquids of varying thermal stability.[4,10a,11b] It is for these reasons that reactions, particularly with unstable 2*H*-azirines, are often carried out by generating the azirine *in situ* from the appropriate vinyl azide.

The abnormally high stretching frequency ν(C=N) (1770–1775 cm^{-1}) of 3-alkyl-2*H*-azirines (lowered to 1740 cm^{-1} by a 3-phenyl group) is characteristic and reflects the highly strained nature of this heterocycle.[3] In contrast, 3-unsubstituted 2*H*-azirines display ν(C=N) at 1650–1660 cm^{-1}, in keeping with the value normally observed for open-chain imines.[11b] ^{1}H N.m.r spectra of several 2*H*-azirines have been tabulated,[11b,15] and reveal that the imine hydrogen, *i.e.* at C-3, is remarkably deshielded and has a chemical shift close to 0 τ.[3,11b] Some ^{13}C n.m.r. data for representative 2*H*-azirines are appended to diagram (**16**). Of note is the large $J_{C\text{-}2,H}$ coupling (186 Hz), which is taken as being indicative of the relatively high degree (*ca.* 37.3%) of *s*-character in the C—H bond, and also the chemical shift of C-2, which is advocated as being diagnostic for 2*H*-azirines.[16]

17.6.2.4 Reactions with electrophilic and nucleophilic reagents

2*H*-Azirines are non-basic, being insoluble in dilute hydrochloric acid but soluble with slow decomposition in concentrated acid. In anhydrous perchloric acid, azirines behave as ring-opened cations, *e.g.* (**17**), which with acetone or acetonitrile form cycloadducts, *e.g.* (**18**).[4]

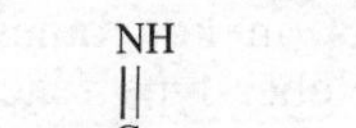

Ph—C(=NH)—$\overset{+}{C}Me_2$

(17)

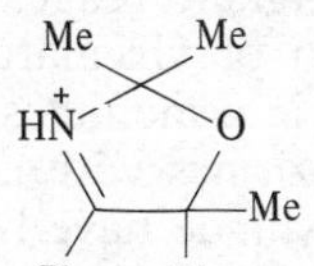

(18)

Acid-catalysed hydrolysis of 2*H*-azirines to α-amino-ketones is a useful adjunct to the Neber reaction in that it allows preparation of amino-ketones, *e.g.* (**19**) (Scheme 5, path b), not always available by the latter reaction (Scheme 5, path a). Yields of amino-ketone are much improved using the variation outlined in Scheme 5b. The ketal (**20**) is rarely isolated but is hydrolysed directly to the desired product.[3] Adaptation of the reaction to

$PhCH_2C(=NOTs)CH_3$ —(a) i→ [Ph, CH_3 azirine (7)] —(a) iii→ $PhCH(NH_2)COCH_3$

$PhCH_2C(N_3)=CH_2$ —(b) ii→ $PhCH_2C(OMe)_2CH_2NH_2$ (20) —(b) iii→ $PhCH_2COCH_2NH_2$ (19)

i, C_5H_5N; ii, $h\nu$, MeOH, NaOMe, iii, H^+.

SCHEME 5

the synthesis of α-aminoaldehydes (not available by Neber reaction) has been suggested[4] but apparently not yet exploited.

Addition of organic acids to the imine bond is followed by rapid ring cleavage of the resulting 2-acyloxyaziridine. Toluene-*p*-sulphinic acid, however, gives a stable adduct, *e.g.* (**21**; $R^1=p\text{-}MeC_6H_4SO_2$, $R^2=Me$ or Ph). Likewise the addition of acid chlorides, anhydrides, and imidoyl chlorides proceeds readily to give *N*-acyl-2-substituted aziridines, which, under mild conditions, rearrange to either larger heterocycles and/or acyclic products.[2,4] Acid chloride adducts have proved useful for preparing functionalized (*e.g.* azido and methoxy) *N*-acylaziridines and oxazolines.[17]

(21) (23)

Lithium aluminium hydride reduction of 2*H*-azirines is rapid and stereospecific, thereby providing a useful synthesis of *cis*-aziridines.[4] In contrast, catalytic reduction (Pd/C or Raney Ni) promotes cleavage of the C—N bond to give unstable imines. The stereospecific *cis*-addition of Grignard reagents to the imine bond is reminiscent of the Campbell aziridine synthesis (action of Grignard reagents on ketoximes) and there is some evidence[18] that azirines are intermediates in this Neber-type reaction.

The reactions of 2*H*-azirines with carbanions derived from acetophenone, ethyl benzoylacetate, and benzyl cyanide have been investigated.[4] In each case initial attack at the imine carbon is followed by rearrangement, generally to a five-membered heterocycle. Of particular interest is the reaction of 3-phenyl-2*H*-azirine with dimethylsulphonium methylide (Scheme 6), which constitutes the first synthesis of the 1-azabicyclo-[1,1,0]butane (**22**; $R^1=R^2=H$) ring system.[19]

Other attempts to prepare the azabicyclobutane by addition of dichlorocarbenes to 2*H*-azirines gave only *N*-vinylimines. However, at −100 °C trichloromethyl-lithium and 2-methyl-3-phenyl-2*H*-azirine yielded adduct (**21**; $R^1=CCl_3$, $R^2=Me$), which on treatment with strong base (Bu^tOK/DMSO) rearranged to dichloro-3,4-dihydro-1-azetine (**23**) (55%) *via* the azabicyclobutane (**22**; $R^1=Cl$, $R^2=Me$).[20]

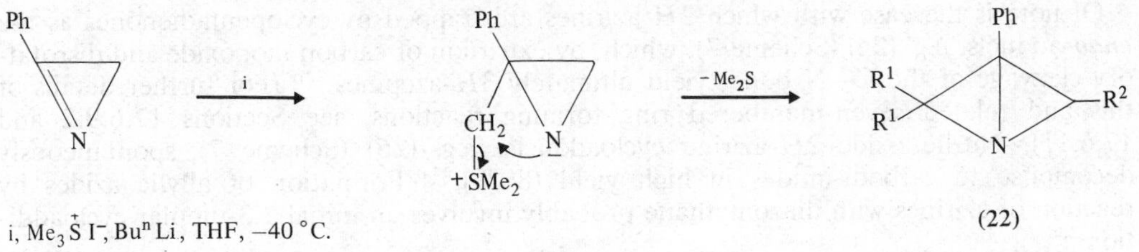

i, $Me_3\overset{+}{S}I^-$, Bu^nLi, THF, −40 °C.

(22)

SCHEME 6

17.6.2.5 Dimerization

Pyrazines and dihydropyrazines, which have been noted as products from the acid treatment of 3-aryl-[4] and 3-amino-2*H*-azirines,[21] also result from the Group VI (Mo, Cr, W) metal carbonyl[22a] or silver ion[22b] induced dimerization of 2*H*-azirines. The course of the dimerization, however, is strongly metal dependent since di-iron enneacarbonyl [$Fe_2(CO)_9$] favours 2,5-diarylpyrrole formation,[22c] whereas with rhodium(I) complexes at ambient temperatures 2-styrylindoles are obtained.[22d]

17.6.2.6 Cycloadditions

The reactivity of the azirine double bond is reflected in the readiness with which it participates as a 2π component in Diels–Alder type reactions, and to a lesser extent in 1,3-dipolar cycloadditions. Many examples are known,[23a] and some representative examples are gathered in Scheme 7.

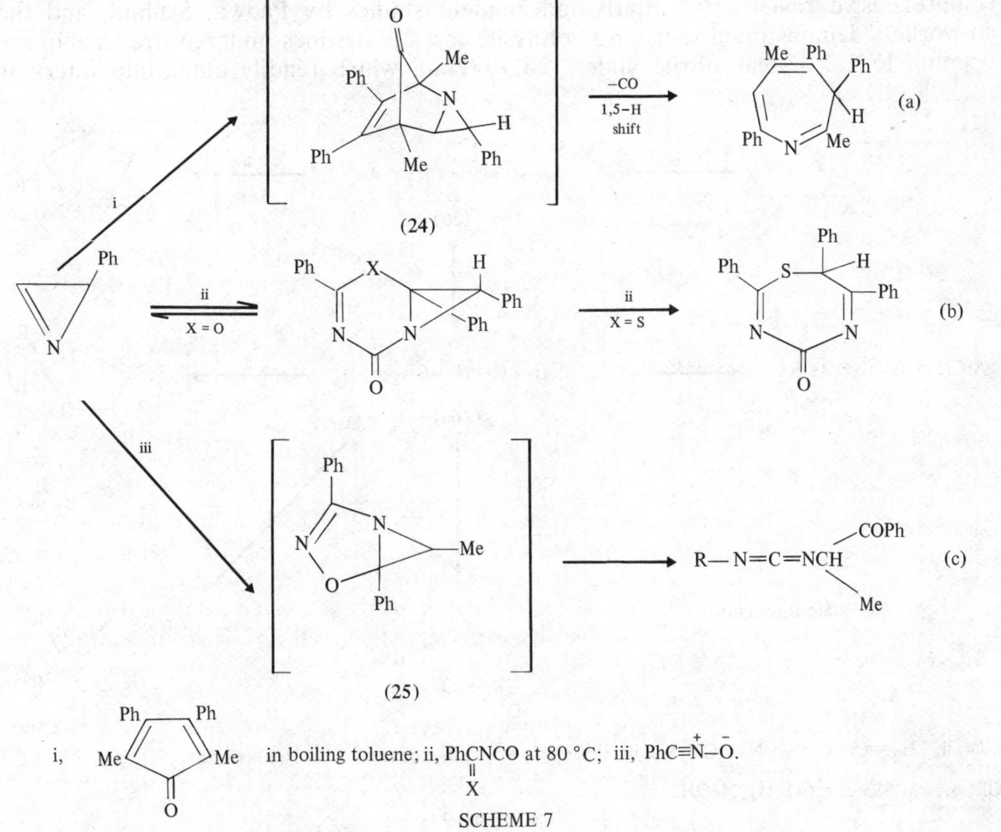

(24)

(25)

i, (Ph, Ph, Me, Me cyclopentadienone) in boiling toluene; ii, PhC(=X)NCO at 80 °C; iii, $PhC\equiv\overset{+}{N}-\overset{-}{O}$.

SCHEME 7

Of note is the ease with which 2*H*-azirines are trapped by cyclopentadienones as the *endo*-adducts, *e.g.* (**24**) (Scheme 7), which, by extrusion of carbon monoxide and disrotatory cleavage of the C—N bond, yield ultimately 3*H*-azepines.[23b] (For further details of this and related seven-membered ring forming reactions, see Sections 17.6.9.2 and 17.6.11). Nitrile oxide–2*H*-azirine cycloadducts, *e.g.* (**25**) (Scheme 7), spontaneously decompose to carbodi-imides in high yield (80%).[23a] Formation of allylic azides by reaction of azirines with diazomethane probably involves an initial 1,3-dipolar cycloaddition.[23a]

The exclusive [4+2] cycloadditions observed with the thiobenzoyl isocyanate and benzoyl isocyanate (Scheme 7; X=S and O respectively) are in direct contrast to the [2+2] polar stepwise cycloadditions undergone with benzoyl isothiocyanate (PhCONCS) (Scheme 8) and diphenylketen.[23a] Regioselective [2+2] cycloadditions between 3-aryl-2*H*-azirines and carbon disulphide have also been noted.[23a]

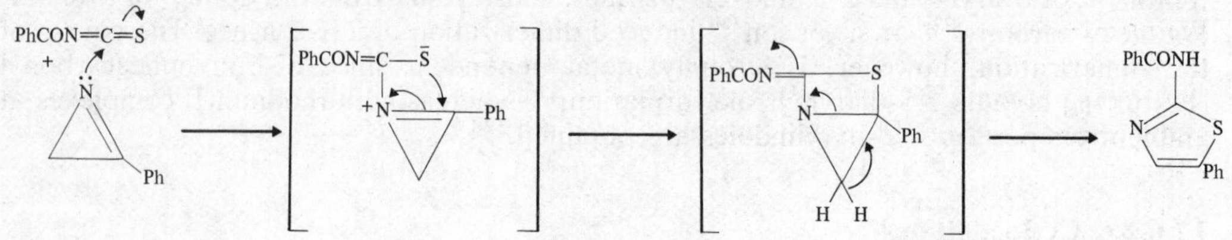

SCHEME 8

17.6.2.7 Photolytic and thermolytic reactions

The photochemistry of 2*H*-azirines has aroused much interest, and the mechanistic and synthetic aspects of this branch of azirine chemistry have been summarized in recent comprehensive reviews.[25,26] Early independent studies by Padwa, Schmid, and their co-workers demonstrated that, on photolysis, aryl-2*H*-azirines undergo irreversible ring-opening to 1,3-dipolar nitrile ylides (**26**) ↔ (**26a**), which readily enter into inter- and

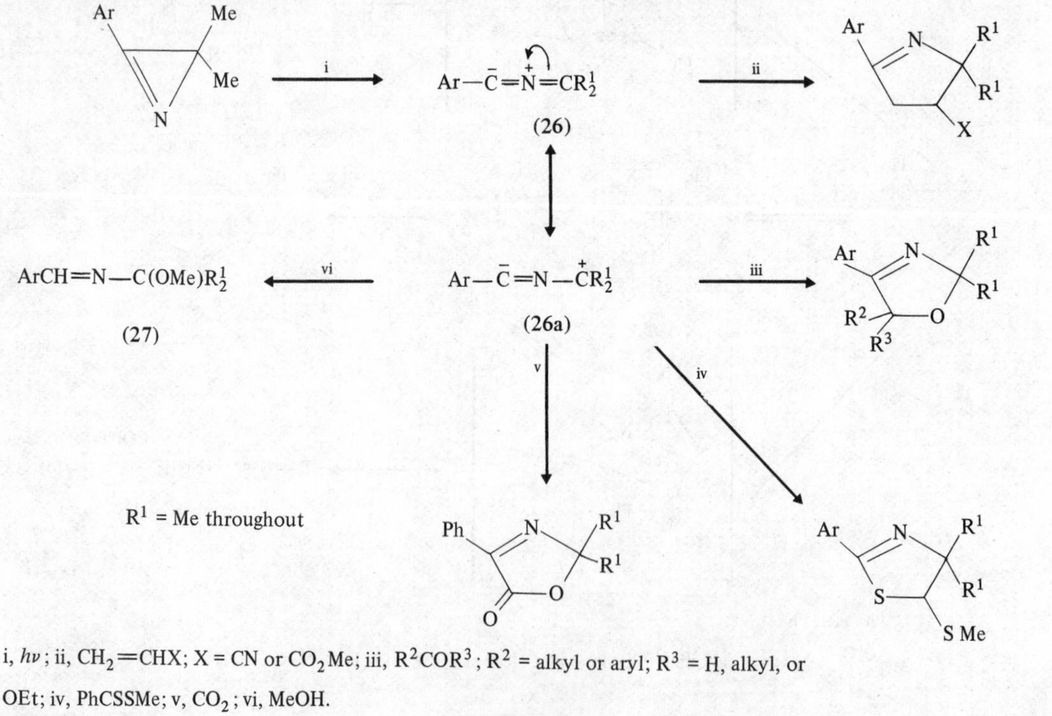

R^1 = Me throughout

i, $h\nu$; ii, $CH_2{=}CHX$; X = CN or CO_2Me; iii, R^2COR^3; R^2 = alkyl or aryl; R^3 = H, alkyl, or OEt; iv, PhCSSMe; v, CO_2; vi, MeOH.

SCHEME 9

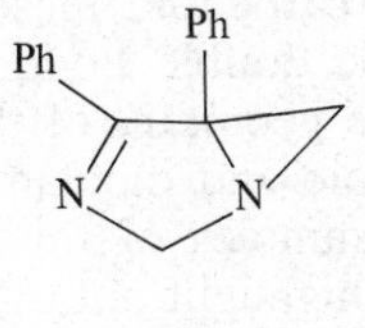

(28) 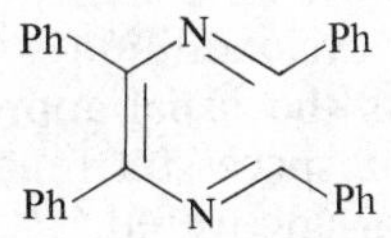 (29)

intra-molecular 1,3-dipolar cycloadditions with a wide variety of dipolarophiles, including a rare example of the dipolarophilic character of carbon dioxide[25] (see Scheme 9).

Recent interest has focused on the regio- and stereo-selectivity aspects of 1,3-dipolar cycloadditions. Frontier molecular orbital treatment[26] of 1,3-dipoles has shown that for nitrile ylides it is the HOMO (dipole) −LUMO (dipolarophile) interactions which stabilize the cycloaddition transition state. Hence a knowledge of the relative magnitudes of the coefficients of the dipole HOMO and dipolarophile LUMO enables predictions regarding the regiospecificity of cycloaddition reactions to be made. Such information for nitrile ylide (**26a**; $R^1 = Me$) has been obtained by photolysing the 2-arylazirine in methanol, whereupon only one methoxy-imine (**27**; $R^1 = Me$) was formed.[24a] On the basis of this result it was concluded that the preferred form of the dipole is mesomer (**26a**) in which the greatest electron density is at the di- rather than the tri-substituted carbon centre. The regiospecificity observed with nitrile ylide cycloadditions confirm this assignment.

In the absence of dipolarophiles, or with multiple bonds of low dipolarophilicity, the nitrile ylide (**26a**; $R^1 = H$) cycloadds to a second molecule of azirine to give the diazabicyclohexane (**28**) and not, as was first thought, a diazabicyclopentane.[24a] Low-temperature (−160 °C) photolysis of 2,3-diphenyl-2*H*-azirine yields the hexatriene (**29**).[24a,25]

A rapidly developing branch of 1,3-dipole chemistry concerns intramolecular cycloadditions, which are possible when dipole and dipolarophile are located at geometrically favoured sites in the same molecule.[24b] Many examples have been recorded[24b,27a] involving nitrile ylides derived from 2*H*-azirines (Scheme 10; path a). Of interest is the *trans*-styrylazirine (**30**; X = CHPh), which reacts as in path (a) to give 2,3-diphenylpyrrole, in contrast to its *cis*-isomer which undergoes a novel transformation to the 3*H*-2-benzazepine (**31**) (Scheme 10; path c)[24b] (see also Section 17.6.9.2).

During the course of their studies on nitrile ylide intramolecular cycloadditions, Padwa and his co-workers[24a] discovered the first example of a cycloaddition in which the nitrile

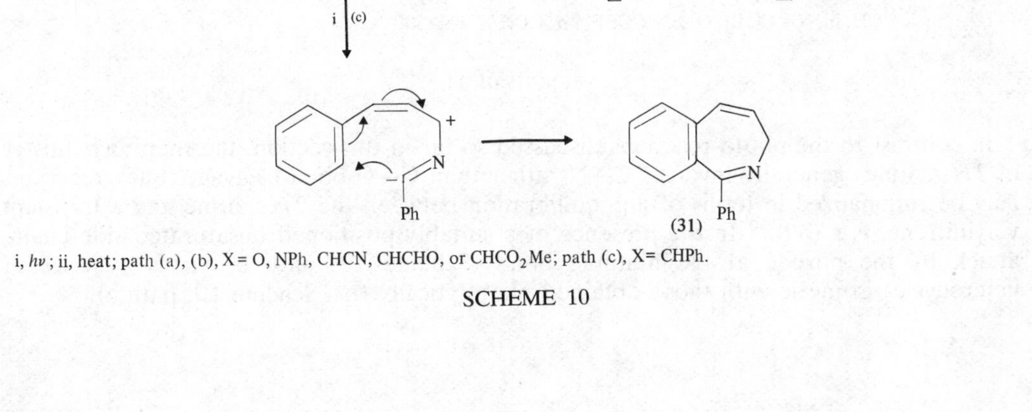

(30) (39) (31)

i, *hν*; ii, heat; path (a), (b), X = O, NPh, CHCN, CHCHO, or $CHCO_2Me$; path (c), X = CHPh.

SCHEME 10

ylide (**32**) behaves not as a 1,3-dipole but as the carbenoid mesomer (**33**), and adds intramolecularly to the favourably positioned allylic double bond, giving bicycle (**34**). Normal 1,3-dipolar cycloaddition is restricted by the geometry of the *o*-allylnitrile ylide (**32**), which prevents the usual approach of the dipole and dipolarophile in the parallel planes necessary for successful reaction. Further examples of this type of nitrile ylide behaviour have been uncovered,[27b] and the authors highlight not only the non-concerted nature of these cycloadditions but also their similarity to the provocative alternative stepwise biradical mechanism for 1,3-dipolar cycloadditions championed by Firestone[28] but rigorously refuted by Huisgen.[29]

(32) (33) (34)

Another innovation concerns the photo-transformations of 2-(hydroxymethyl)-2*H*-azirine (**35**; X = OH) and its derivatives.[24a] Photolysis of the parent alcohol generates nitrile ylide (**36**), which by proton transfer yields oxazoline (**37**) (Scheme 11; path a). However, if X is a good leaving group then smooth rearrangement takes place to the otherwise difficultly accessible but synthetically useful 1-substituted 2-azabutadienes (**38**) (Scheme 11; path b).

(35) (36) (37)

$PhC(X){=}NCH{=}CH_2$ (38)

Ph—CH=N—CH=CH₂

(40)

i, *hν*; ii, X = OH; iii, X = Cl, Br, OAc, $OCOCF_3$, or OCOAr; iv, >500 °C.

SCHEME 11

In contrast to the photo-reactions discussed so far in this section, the thermochemistry of 2*H*-azirines generally involves C—N rather than C—C bond cleavage. Such reactions may be rationalized in terms of an equilibration between the 2*H*-azirine and a transient vinylnitrene, *e.g.* (**39**).[30] In the presence of a suitably positioned unsaturated side-chain, attack by the nitrene at the multiple bond (Scheme 10; path b) yields a series of heterocycles isomeric with those obtained photolytically (*i.e.* Scheme 10; path a).

Thermal C—C bond cleavage of 2*H*-azirines is a much rarer event. However, it is observed (Scheme 11; path c) during the thermolysis of 2-phenyl-3-methyl-2*H*-azirine, giving imino-carbene (**40**) and ultimately, by proton shift, the 2-azabutadiene.[31] This reaction is very similar in nature to that observed[21] in the vapour-phase pyrolysis of 3-amino-2*H*-azirine.

17.6.3 BENZAZIRINES

Neither of the two possible isomeric benzazirines (**41**) and (**45**) have as yet been isolated, although both have been proposed as transient reaction intermediates.

1*H*-Benzazirine (**41**), like its monocyclic counterpart, suffers from anti-aromatic destabilization, which although lessened by annulation to a benzene ring is still sufficient to prevent isolation. In the early work on the formation of 2-amino-3*H*-azepines by thermolysis of aryl azides in amine solution (see Section 17.6.8.3), 1*H*-benzazirines were suspected as reaction intermediates. However, their participation was conclusively disproved by careful carbon-labelling studies.[32a] In contrast, an equally elegant and meticulous study[33a,b] of the vapour-phase (400–600 °C) chemistry of phenylnitrene, generated from various precursors, has firmly established the intermediacy of 1*H*-benzazirine in the quantitative interconversion of the imino-carbenes (**42**) and (**43**) (Scheme 12) prior to ring contraction by Wolff rearrangement to the isolable cyanocyclopentadiene (**44**).

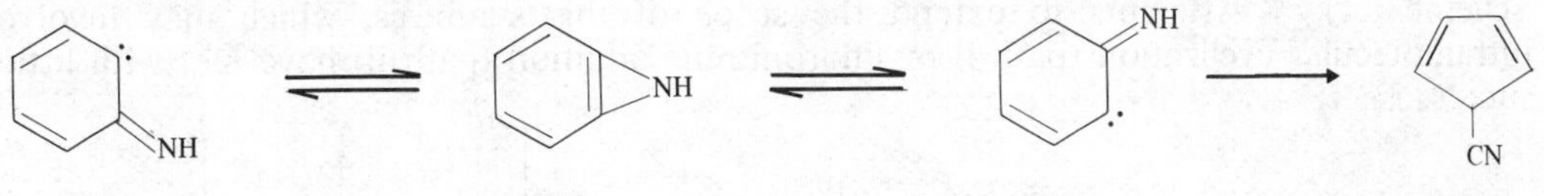

NH NH NH CN

(42) (41) (43) (44)

SCHEME 12

2*H*-Benzazirines (**45**) (also known as azanorcaradienes) are the benzo-derivatives of the well-characterized 2*H*-azirines, and as such might well be expected to be available by similar synthetic routes. However, despite sustained patient and exhaustive effort by many researchers, all attempts to isolate a 2*H*-benzazirine from either the thermolytic or photolytic decomposition of aryl azides has so far failed. Evidence for their existence is as yet purely circumstantial, although their participation in the formation of 3*H*-azepines from aryl azides and other arylnitrene precursors (see Section 17.6.8.3) is most likely.

17.6.4 DIAZIRINES

3*H*-Diazirine (**46**) and its derivatives have been well characterized, and are of interest as the cyclic isomers of the more familiar diazoalkanes. The structure of the parent heterocycle has been determined by microwave spectroscopy and its structural parameters along with those theoretically derived[5] for 1*H*-diazirine are appended to diagrams (**46**) and (**47**), respectively. 1*H*-Diazirine (**47**), like its monoaza-analogue, has anti-aromatic character and is as yet unknown. Calculations[5] show it to be 116.6 kJ mol^{-1} less stable than the 3*H*-tautomer. The diazirinyl cation, like the monoaza-cation (**3**), has extensive π-delocalization and may be preparable as a relatively stable aromatic species.[7]

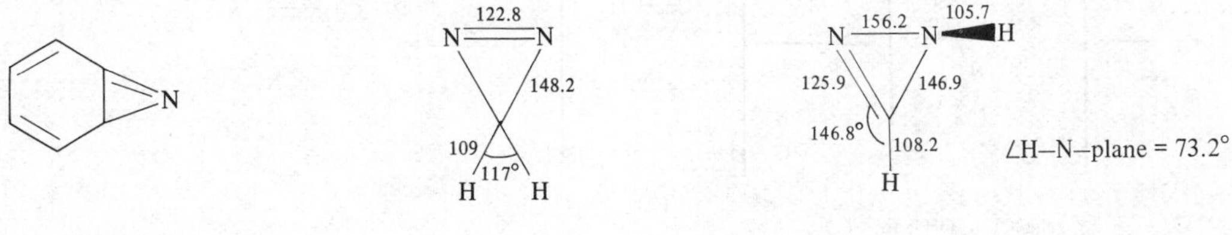

N

122.8 N N 148.2 109 117° H H

156.2 105.7 N N H 125.9 146.9 146.8° 108.2 H ∠H–N–plane = 73.2°

(45) (46) (47)

17.6.4.1 Synthesis of 3*H*-diazirines

Studies on diaziridines, *e.g.* (**48**), show that they possess considerable reducing properties and that they are readily oxidized by a variety of common reagents (*e.g.* HgO, $KMnO_4/OH^-$, Ag_2O) to diazirines.[34] This method provides the main synthetic route to a wide variety of alkyl- and aryl-3*H*-diazirines, particularly as the parent diaziridines are cheaply available from common materials (Scheme 13).

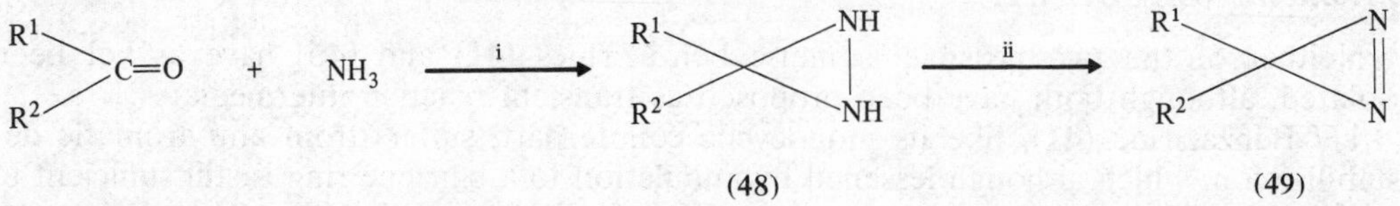

i, MeOH, Bu^{t}OCl. −40°C; ii, oxidizing agent.

SCHEME 13

A simple variation of this reaction permits preparation of 3-monoalkyl-3*H*-diazirines from aldehydes. 3*H*-Diazirine (**49**; $R^1 = R^2 = H$) can be obtained using formaldehyde, chloramine, and ammonia followed by oxidation of the resulting diaziridine with acidic dichromate solution. An alternative and superior synthesis of the parent heterocycle employs the addition of difluoramine to a formaldehyde imine (**50**; R = t-butyl or t-octyl; Scheme 14).[35a] Attempts to extend the scope of this synthesis, which may involve intramolecular cyclization (path i) or fluoronitrene addition (path ii) have met with little success.[35b]

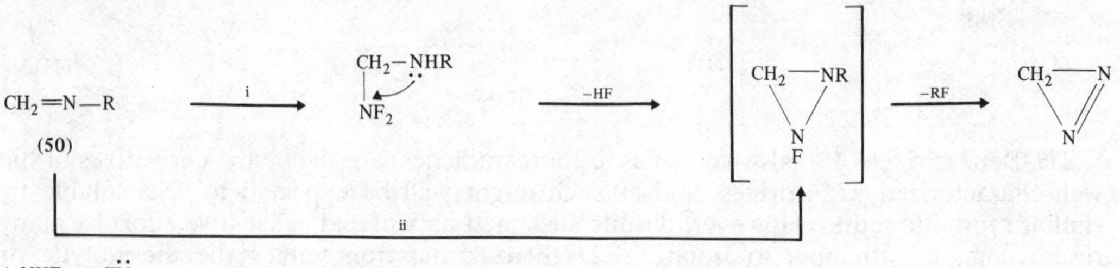

i, NHF_2; ii, FN̈.

SCHEME 14

3-Halogeno-3*H*-diazirines are accessible by halogenation of alkyl- and aryl-amidines in aqueous dimethyl sulphoxide[35c] (Scheme 15). The overall process resembles the Neber reaction (Scheme 5, path a) and is of interest in that cyclization of the nitreno-imine (**51**) generates an unstable 1*H*-diazirine, spontaneous rearrangement of which yields the diazirinium cation (**52**).

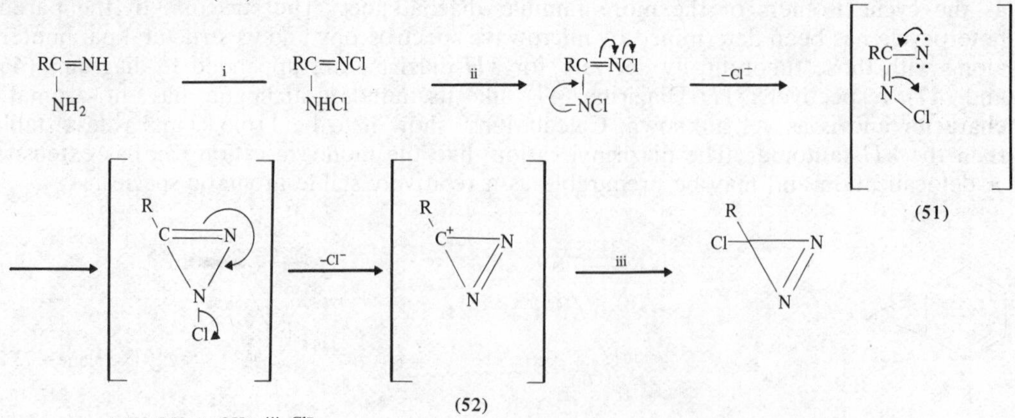

i, 2NaOCl; ii, OH^-; iii, Cl^-

SCHEME 15

17.6.4.2 Physical and spectroscopic properties

Most of the diazirines are liquids, although 3-methyl- and 3,3-dimethyl-3*H*-diazirine, like the parent diazirine, are gases at room temperature. In common with their cyclic isomers, 3*H*-diazirines are thermally unstable. The low molecular weight monoalkyl derivatives can be violently explosive and extreme caution in handling these compounds, particularly in the condensed phase, has been advocated.[34] 3,3-Difluoro-3*H*-diazirine, although explosive, is much less reactive than other diazirines.[36]

Ultraviolet,[34,35c] infrared,[35c] ^{1}H n.m.r.,[35a] ^{19}F n.m.r.,[36] and mass spectra[35a,37] of 3*H*-diazirines have been reported.

17.6.4.3 Chemical properties

Despite their highly strained structure, 3*H*-diazirines are surprisingly unreactive. They are unaffected by alkalis and, unlike the diazoalkanes, do not react with acids or other electrophilic reagents. They will, however, react with Grignard reagents at 0 °C to give 1-substituted diaziridines. Other organometallic reagents add similarly. In contrast, strong reducing agents yield diaziridines only as difficultly isolable intermediates, since further reduction is accompanied by ring-opening to ammonia and amines.

The main interest in diazirine chemistry concerns their controlled thermolytic and photolytic decomposition, which provides a useful source of carbenes.[34,38] Isomerization of diazirines prior to decomposition has been observed under both thermal[39] and photolytic conditions.[38]

17.6.5 AZETINES AND AZETES

Azacyclobutadiene is known formally as azete, the partially reduced derivatives being termed 1- and 2-azetine, respectively. All early claims to the synthesis of these systems were shown to be erroneous and have been discussed in detail.[40]

17.6.5.1 1- and 2-azetines and azetinones

The first unequivocal synthesis of a 1-azetine involved ethylation of a 3,3-disubstituted azetidin-2-one or -2-thione using triethyloxonium fluoroborate (Meerwein's reagent).[41] The method is general[42a,b] since the precursory *N*-unsubstituted azetidinones are readily available (70% yield) by [2+2] cycloaddition of chlorosulphonyl isocyanate to alkenes followed by reductive dechlorosulphonylation with, for example, sodium sulphite.[43]

An alternative and useful route to 1-azetines involves pyrolysis of cyclopropyl azides,[44,45] as illustrated in Scheme 16. Contrary to earlier reports, 1-azetines are also available, although in moderate yield (30–40%), by [2+2] photoaddition of alkenes and aryl cyanides.[46] Other methods of preparation involving cycloaddition are (a) the [2+2]

Cl R^2 Cl R^3 R^1 N$_3$ —i→ [Cl R^2 Cl R^3 R^1 N̈] → Cl Cl R^2 R^3 R^1 N

i, 105–125 °C; R^1 = Me or Ph, R^2 = R^3 = H, Me, or Ph.

SCHEME 16

photoaddition of ketene acetals to 1,3-oxazin-4-ones followed by thermal retro-Diels–Alder reaction of the cycloadduct;[47] and (b) by an unusual thermal, but photo-reversible, [3+1] cycloaddition of a nitrile ylide, *e.g.* (**26**; $R^1 = CF_3$), derived from 1,3,5-oxazaphospholes (**12**; R^2 = aryl or t-butyl; $R^1 = CF_3$) with an isocyanide (RNC; R = *p*-$NO_2C_6H_4$ or C_6H_{11}),[48] *e.g.* (**53**).

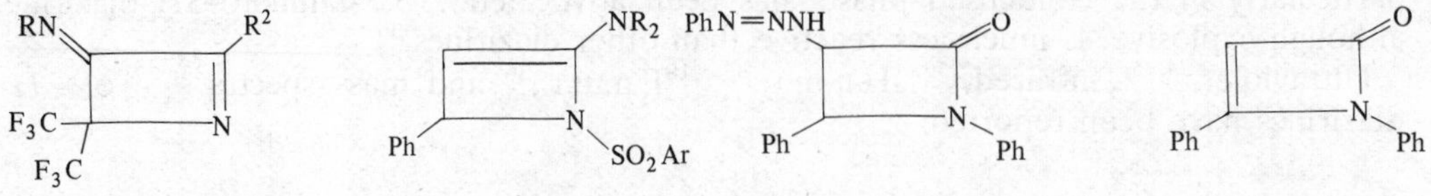

(53) (54) (55) (56)

2-Azetines are uncommon but have been prepared by the thermal [2+2] cycloaddition of *N*-sulphonylimines (PhCH=SO_2Ar) and keten *N,N*-acetals [CH_2=C(NR_2)$_2$].[49]

Azetinones have been prepared, the first authenticated example (**56**) being obtained,[50] when all other methods had failed, by cleavage of triazene (**55**) with boron trifluoride etherate. Its structure was established by spectroscopic methods and by its reduction (Pt/H_2) to the known 1,4-diphenylazetidinone. *N*-Methylazetinone (**58a**) appears to be an intermediate in the photoisomerization (Scheme 17) of the cyclopentadienone–(*N*-methyl-2-pyridone) photodimer, cycloadduct (**57a**).[51] It is, however, thermally unstable and in the presence of methanol yields a mixture of methyl *cis*- and *trans*-β-methylaminoacrylate by way of its imino-keten valence isomer (**59a**).

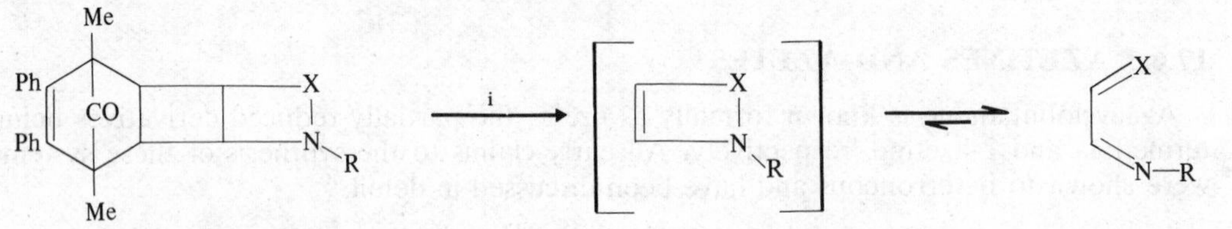

(57) a; R = Me, X = CO
b; R = CO_2Me, X = NCO_2Me
(58) (59)

i, *hν*, MeOH; –CO, –1,4-diMe-2,3-diPh-benzene.

SCHEME 17

17.6.5.2 Physical properties and spectroscopic data

The alkoxy-1-azetines are, in general, thermo-labile liquids somewhat more unstable than the corresponding thioethers, and with a tendency towards polymerization even at room temperature.[41,42a] A notable exception is 3,3-diphenyl-2-ethoxy-1-azetine, which is a stable solid, m.p. 60 °C. Unstable 2-alkoxy-derivatives have been characterized as their picrates.[41] Alkyl- and aryl-1-azetines appear to be more stable and some have been purified by vacuum sublimation.[46]

Infrared,[41,45,46,48] ^{1}H n.m.r.,[41,44,45] ^{13}C n.m.r.,[20] and mass spectral[52] data for 1-azetines are available.

17.6.5.3 Chemical properties

1-Azetines behave chemically as cyclic imines, being hydrolysed by mineral acids to β-aminoketones, and reduced by lithium aluminium hydride to azetidines. The 2-alkoxy-derivatives are lactim ethers and as such are susceptible to nucleophilic displacement of the alkoxy-group.[42b]

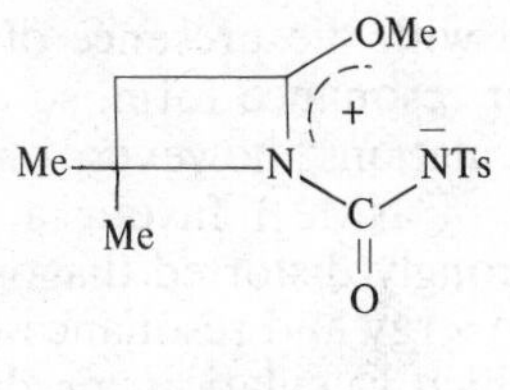

(60) 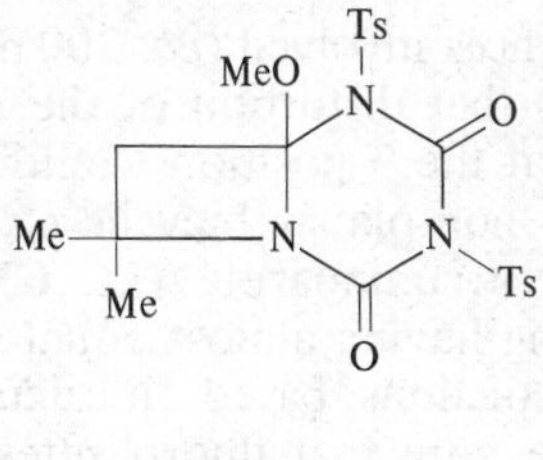 (61)

2-Alkoxyazetines with acetylenedicarboxylates[53a] or with ketens[53b] yield synthetically useful 1,4-dipolar intermediates. Of particular interest is the isolable 1,4-dipole (**60**), formed with *p*-tosylisocyanate (1 mol)[53c]; with an excess of isocyanate the 2:1 adduct (**61**) is obtained in near quantitative yield.

Azetines have been detected, albeit in low concentration (<2%), as bicyclic valence tautomers, *e.g.* (**62**), of azocines (Scheme 18).[54]

(62)

SCHEME 18

17.6.5.4 Azetes

Several abortive attempts have been made[55] to synthesize azetes, the aza-analogue of the equally elusive and anti-aromatic cyclobutadiene. Calculations indicate,[56a] however, that anti-aromatic destabilization of azete (resonance energy $-64.8\,\mathrm{kJ\,mol^{-1}}$)[57a] is considerably less than for cyclobutadiene (resonance energy $-75.2\,\mathrm{kJ\,mol^{-1}}$).[55] It has been suggested[56c] that stabilization of these systems will be enhanced not only by the presence of bulky groups but also by substituents having both electron-donating and -attracting properties, *i.e.* the so-called 'push–pull' effect. In fact, prior to this publication, Seybold and his co-workers[55] had used this principle in the synthesis (Scheme 19) of the so far solitary fully-characterized (mass, u.v., i.r., n.m.r.) example of the azete system. Azete (**63**), which condenses out of the pyrolysate at −196 °C as a red solid, can be represented by several resonance forms, including the trans-annular mesomer (**64**) (Scheme 19). It has been argued[58] that transannular interactions between atoms at opposite corners of a four-membered ring are highly stabilizing and that actual bonding may occur as the

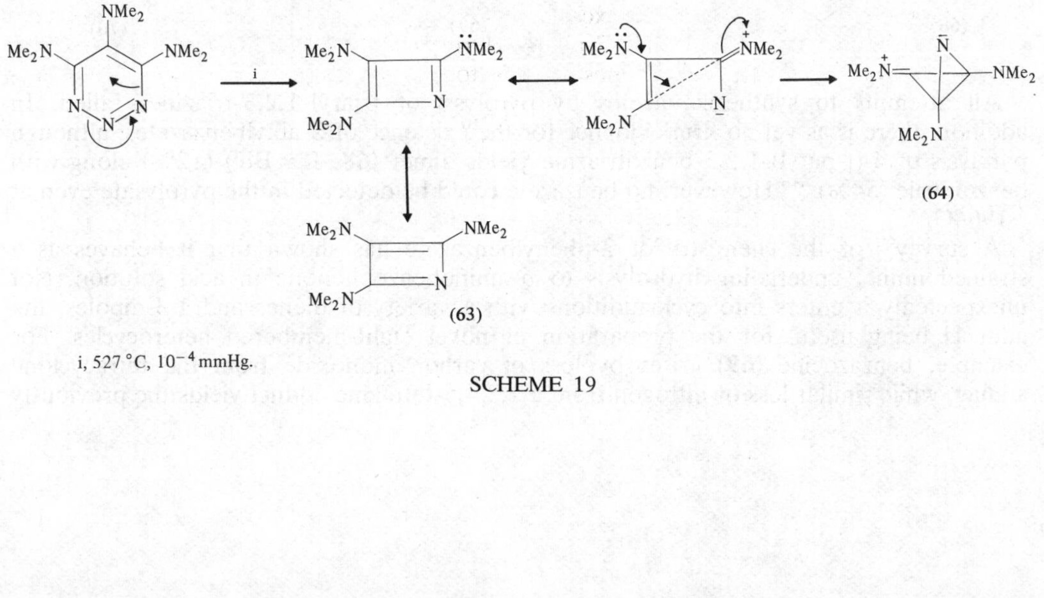

(63) (64)

i, 527 °C, 10^{-4} mmHg.

SCHEME 19

diagonal distances involved (*ca.* 200 pm) are of normal bond length order. It was expected for azete (**63**) that distortion of the molecule coupled with the presence of an electron-donor group at the 3-position would support the polar resonance form, so conferring on the azete the non-planar bicyclic structure (**64**). Calculations, however, on a simplified model of the triaminoazete (*i.e.* **63**; NH_2 in place of NMe_2) favour a planar four-membered ring having almost equal sides but with strongly distorted diagonals.

Current predictions, based on calculated resonance energy and resonance energy per π electron values, are that fluoroazetes (*e.g.* **65**), for which calculations predict a positive resonance energy, and 2,4-diaminoazetes, which lack a destabilizing 3-amino substituent, should be more stable than triamine (**63**).

17.6.6 BENZAZETES, BENZAZETINES, AND BENZAZETINONES

Derivatives of the title compounds are known and all may be prepared either by thermal and/or photolytic decomposition of the appropriately substituted benzo-1,2,3-triazine.[59] Benzazetes are the most elusive members of the series, but they have been prepared.[60] 2-Phenylbenzazete (**67a**; Ar = Ph) was obtained (60%) at −80 °C as a dark-red condensate from the flash vacuum pyrolysis (420–450 °C; 0.03 Torr) of 3-phenyl-1,2,3-benzotriazine (**66**). As expected, the anti-aromatic azete ring is somewhat stabilized by fusion to a benzene ring, stabilization being more marked in the *p*-methoxyphenyl (**67a**; Ar = *p*-$MeOC_6H_4$) and naphthalene (**67b**; Ar = Ph) derivatives. Even so, the benzazetes are very thermolabile and on warming undergo rapid dimerization to yellow dimers of the form (**68**; R = Ar). Subsequently it was found[61a] that photolysis of the 1,2,3-benzotriazines in tetrahydrofuran at −80 °C constitutes a more convenient preparative method.

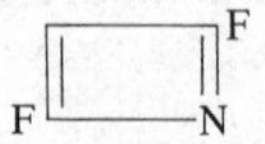

(**65**) 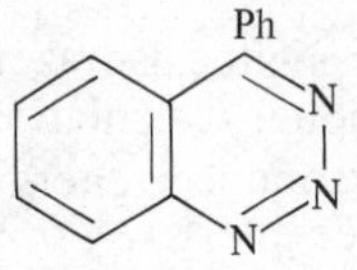 (**66**) 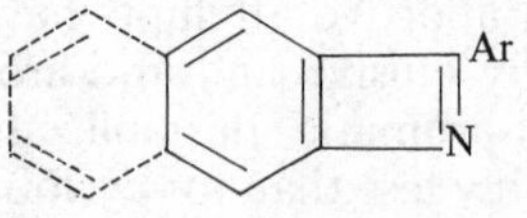 (**67**) a; benzo b; naphtho

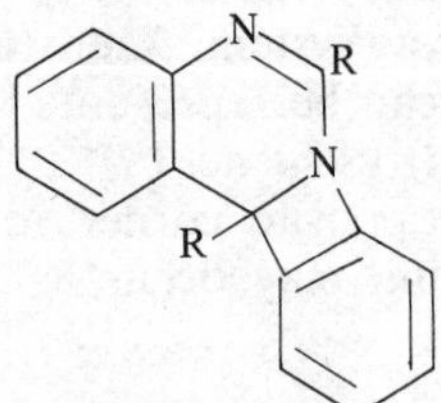

(**68**) 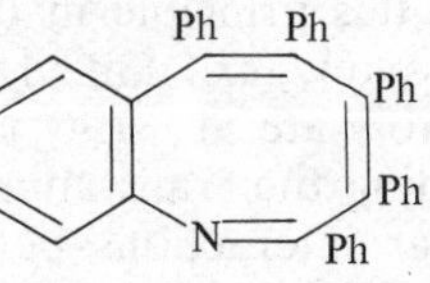 (**69**) 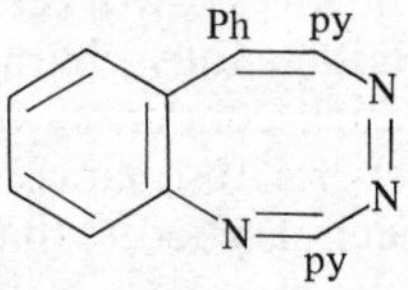 (**70**)

All attempts to synthesize azetes by pyrolysis of triaryl-1,2,3-triazines failed. In addition, there is as yet no firm evidence for the existence of 2-alkylbenzazetes, although pyrolysis of 4-(t-butyl)-1,2,3-benzotriazine yields dimer (**68**; R = Bu^t) (12%) along with benzonitrile (54%).[61b] However, no benzazete could be detected in the pyrolysate even at −196 °C.

A survey[60] of the chemistry of 2-phenylbenzazete has shown that it behaves as a strained imine, undergoing hydrolysis to *o*-aminobenzophenone in acid solution. Not unexpectedly it enters into cycloadditions with a variety of dienes and 1,3-dipoles, the adducts being useful for the preparation of novel eight-membered heterocycles. For example, benzazocine (**69**) arises by loss of carbon monoxide from the tetracyclone adduct, while similar loss of nitrogen from a 1,2,4,5-tetrazine adduct yields the previously

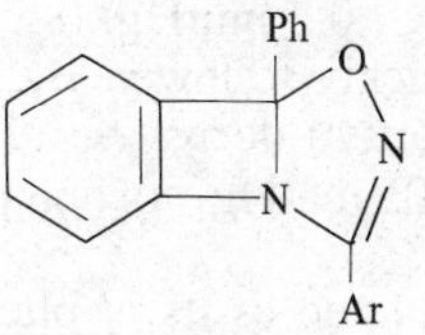

(71) 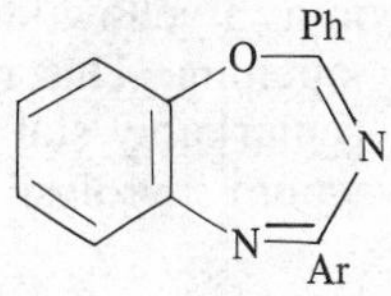 (72) 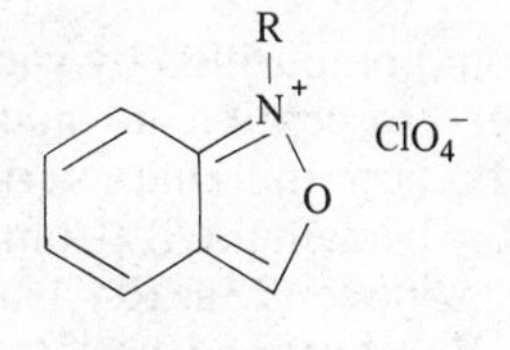 (77)

unknown 1,2,4-benzotriazocine system (**70**; py = pyridine). Also worthy of comment is the novel rearrangement undergone by the highly strained nitrile oxide adduct (**71**), which rearranges not to the expected *o*-quinonoid 2,3,5-oxadiazepine but to the 1,3,5-isomer (**72**), probably by an oxygen walk of a 3,4-epoxyquinazoline intermediate.[62]

Benzoazetinones (**74a**) were first detected as intermediates in the photolysis of 3-aryl-1,2,3-benzotriazin-4-ones (**73a**) (Scheme 20).[59] The benzazetinones were not isolated but were generally obtained as adducts of the imino-keten valence tautomer (**75a**); *e.g.* with methanol, methyl *N*-phenylanthranilate is obtained (*cf.* Scheme 17). The 3-phenyl derivative (**74a**; R = Ph) in the absence of an added nucleophile rearranges to acridone (Scheme 20), whereas on thermolysis the parent benzotriazinone (**73a**; R = H) yields benzo-3,1-oxazin-4-one (**76**; Ar = o-$NH_2C_6H_4$) *via* a [4+2] cycloaddition of the imino-keten (**75**; R = H) with itself (Scheme 20).[59]

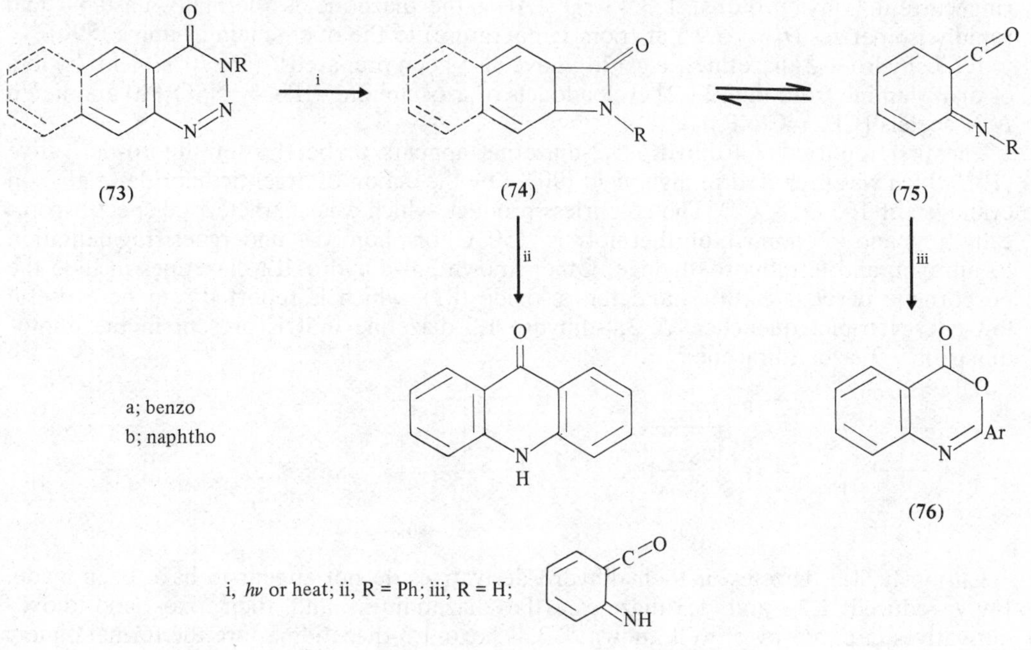

(73) (74) (75) (76)

a; benzo
b; naphtho

i, *hν* or heat; ii, R = Ph; iii, R = H;

SCHEME 20

Confirmatory evidence for the participation of benzazetinones in these reactions followed when *N*-phenylnaphth[2,3,-*b*]azetinone (**74b**; R = Ph) was isolated in high yield (93%) as a moderately stable yellow crystalline solid from the photolysis of the 3-phenylnaphthotriazinone (**73b**; R = Ph). Subsequently, *N*-aminonaphthazetinone (**74b**; R = NH_2) was isolated from the photolysis of the 3-aminonaphthotriazinone (**73b**; R = NH_2).[59]

The first isolable benzoazetinone (**74a**; R = Bu^t) was prepared by Olofson *et al.*[63] by treating 3-unsubstituted anthranilium salts, *e.g.* (**77**), with triethylamine. The *N*-methyl, *N*-ethyl, and *N*-isopropyl derivatives proved to be unstable, but the *N*-t-butyl derivative could be purified by vacuum distillation as a yellow liquid (b.p. 84 °C; 0.2 mmHg) (85%), very susceptible to nucleophilic ring opening. This report was quickly followed by the isolation and characterization of the remarkably stable 3-(1-adamantyl) derivative (**74a**; R = 1-adamantyl) from the flash vacuum pyrolysis (500 °C; 0.05 mmHg) of benzotriazinone (**73a**; R = 1-adamantyl).[59]

Benzazetine, the least-known member of this series, has been isolated as its *N*-phenyl derivative from the photolysis of the 3,4-dihydro-derivative of 3-phenylbenzotriazine (**66**). The reaction is analogous to benzazetinone formation (Scheme 20), but is thought to involve homolytic rather than heterolytic cleavage of the triazine ring.[59] An alternative synthesis is the photoisomerization of 2-cyano-1,2-dihydroquinolines.[64]

17.6.7 DIAZETES

Diazetines are known whereas the fully unsaturated diazetes, *i.e.* the 1,2- and 1,3-diazacyclobutadienes, await synthesis and characterization. Early unsuccessful attempts at their preparation have been reviewed.[40]

Of the two isomeric 1,2-diazetines, the 1,2-dihydro-derivatives are of interest in that if planar they constitute 6π aromatic systems, isoelectronic with cyclobutadiene dianion. ^{1}H N.m.r studies on the 1,2-di(ethoxycarbonyl) derivative (**58b**), prepared in a manner analogous to that used for making azetinones (Scheme 17), however, show no significant ring current (vinyl protons at 3.43 τ).[65] Also, the diazetine is thermally unstable and rapidly isomerizes ($t_{1/2} = 6.9$ h at room temperature) to the open-chain di-imine (**59b**).

1,2-Dihydro-1,2-diazetines, *e.g.* (**78**), have also been prepared[49] (66–70% yield) by loss of dialkylamine from the [2+2] cycloadducts of azosulphones (PhN=NSO$_2$Ph) and keten *N,N*-acetals [CH$_2$=C(NR$_2$)$_2$].

The first reported 3,4-dihydro-1,2-diazetine appears to be the tetrafluoro-derivative (**79**), which was prepared in high yield (90%) by the action of argentic fluoride (AgF$_2$) on cyanogen at 105–115 °C.[66] The colourless product, which was characterized spectroscopically (u.v. and ^{19}F n.m.r.), on thermolysis (150 °C) or photolysis undergoes fragmentation to nitrogen and tetrafluoroethylene. Other known 3,4-dihydro-1,2-diazetines include the norbornene derivative (**80**)[67] and the di-oxide (**81**), which is reported[68] to be a useful low-energy triplet quencher. A 3,4-dihydro-1,2-diazetine unit is present in the photodimer of 9,9′-azoanthracene.[69]

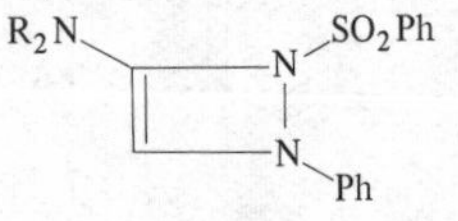

R$_2$N, SO$_2$Ph, N, N, Ph (78) F, F, F, F, N, N (79) N, N (80) Me, Me, Me, Me, N$^+$, O$^-$, N$^+$, O$^-$ (81)

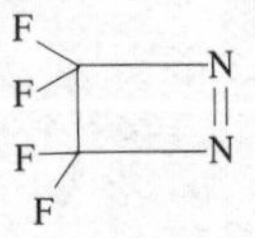

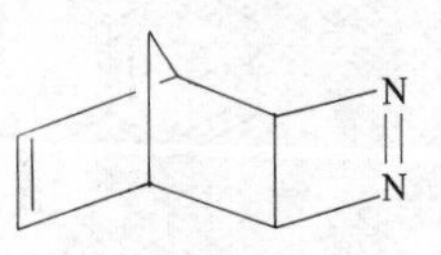

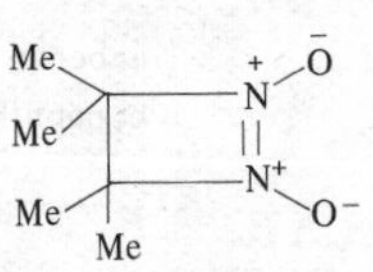

Curiously, 1,3-diazetes or their dihydro-derivatives do not appear to have been made. Fully reduced 1,2- and 1,3-diazetes, (the diazetidines) and their oxo- and dioxo-derivatives, are, however, well known;[40] 2,4-dioxo-1,3-diazetidines are the formal dimers of isocyanates.

17.6.8 AZEPINES

Four tautomeric forms, designated as 1*H*-, 2*H*-, 3*H*-, and 4*H*-azepine, may be drawn for azacycloheptatriene. None of the parent heterocycles have yet been isolated although 3*H*-azepine (**114**) has been characterized by g.l.c.–mass spectral techniques as a product from the pyrolysis of the carbinol (**113**) (Scheme 27).[70]

N-Unsubstituted 1*H*-azepines are rare since they readily tautomerize to the 3*H*-isomer, in whose preparation they are often considered to be transient intermediates. This rearrangement is particularly apparent with 2-amino- and 2-alkoxy-derivatives since stabilization of the 3*H*-azepine is then possible by amidine and imidate type mesomeric interactions. The stability of 1*H*-azepines is enhanced by electron-withdrawing substituents, especially at the 1-position, and many such derivatives have been prepared.[71] Of the remaining isomers the order of stability is 3*H*>4*H*>2*H*.[23b,72] Relatively few 4*H*- and 3*H*-azepines lacking a 2-substituent have been synthesized; even more rare are 2*H*-azepines, although their benzo-derivatives have recently been prepared.[23b]

As expected for a seven-membered ring, the azepines are non-planar, X-ray and ^{1}H n.m.r. analysis indicating that they exist predominantly in the boat conformation.[23b,71,72,73] The conformation of 1*H*-azepines has aroused much theoretical interest since a planar 1*H*-azepine would be an 8π anti-aromatic system. Theoretical calculations[57b] on 1*H*-azepine predict a resonance energy of -7.5 kJ mol^{-1} compared with a value of 0.96 kJ mol^{-1} for 3*H*-azepine. Earlier calculations on 1*H*-azepine had indicated extensive localization of electrons on nitrogen and the carbon–carbon double bonds, from which it was concluded that 1*H*-azepines should exhibit marked polyene character[71] (see Section 17.6.8.9).

17.6.8.1 Synthesis of 1*H*-azepines

The main synthetic routes to 1*H*-azepines are represented in the generalized reaction Scheme 21.

The first 1*H*-azepines was prepared[71] by ring expansion of the azanorcaradiene (**82**) formed by attack of a thermally, photolytically, or base-catalytically derived singlet ethoxycarbonylnitrene (:$\ddot{N}CO_2Et$)[74] on benzene (Scheme 21; path a). 1-Ethoxycarbonyl-1*H*-azepine (**83**; R$=CO_2Et$) was characterized by reduction (Pd/H_2) to the known

i, *hν* or heat or $PhSO_2ONHCO_2Et$, base; ii, powdered NaOMe, THF; iii, Br_2, CH_2Cl_2 at -70°C; iv, *hν*, R$=$*p*-tosyl, COMe, CO_2Me, $CONH_2$, or *p*-$NO_2C_6H_4$; E$=CO_2Me$.

SCHEME 21

N-ethoxycarbonylperhydroazepine. Subsequently the ring expansion of arenes by other thermally or photolytically generated singlet nitrenes (R$\ddot{N}$; R$=CO_2Ph$, CN, or SO_2Me[75]) has been accomplished. However, this method is limited to the preparation of 1*H*-azepines bearing electron-withdrawing substituents on nitrogen that stabilize the ring, presumably by decreasing the electron density of the reactive polyene system. Another disadvantage is that in the majority of cases nitrene attack on a monosubstituted benzene is non-regiospecific and mixtures of difficulty separable isomeric azepines are formed,[76a] although some measure of regiospecificity is observed if bulky groups, *e.g.* t-butyl, are present.[76b]

A much more versatile synthesis has been introduced by Paquette and his co-workers,[71,77] which allows the preparation of specifically substituted 1-alkoxycarbonyl- and 1-arylsulphonyl-1*H*-azepines (Scheme 21; path b). Monoaddition of iodine isocyanate to a 1,4-dihydrobenzene (available by Birch reduction), followed by methanol addition and base-catalysed cyclization of the resulting iodocarbamate (**84**), yields the fused aziridine (**85**). Low-temperature bromine addition and then dehydrohalogenation produces the azanorcaradiene (**82**), which undergoes spontaneous valence-bond isomerization to 1*H*-azepine (**83**).

4,5-Disubstituted 1*H*-azepines are available by rearrangement of the thermally labile 3-azaquadricyclanes (**87**) (Scheme 21; path c).[71,78] That the 1*H*-azepines are substituted at nitrogen by an electron-withdrawing group appears to be essential in promoting formation of the precursory [4+2] pyrrole–acetylenedicarboxylate adducts (**86**).

17.6.8.2 2*H*-Azepines

So far no monocyclic 2*H*-azepines have been isolated, although their presence as intermediates in the formation of 3*H*-azepines by cycloaddition of 2*H*-azirines and cyclopentadienones is well-established[23a,b] (see Scheme 7). By the judicious choice of dienone it is possible, however, to prepare annulated derivatives, *i.e.* 2*H*-benzazepines, and these are discussed in Section 17.6.9.2.

17.6.8.3 Synthesis of 3*H*-azepines

As with the 1*H*-azepines, the main synthetic routes to 3*H*-azepines involve ring-expansion procedures, and some general methods are summarized in Scheme 22.

Historically, the first 3*H*-azepine was obtained by Wolff from the decomposition of phenyl azide in boiling aniline. However, the actual structure of this product, which was

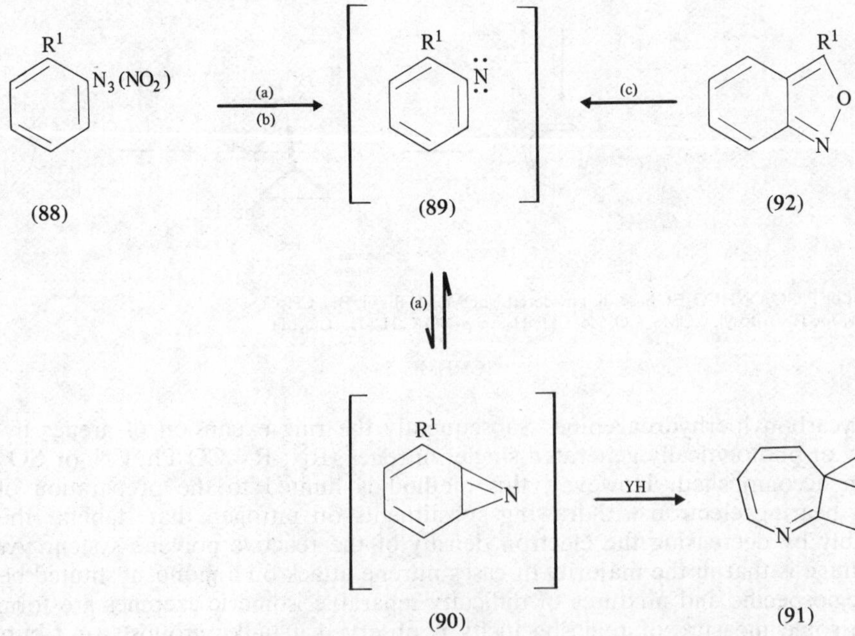

(88) (89) (92) (90) (91)

SCHEME 22

named 'Dibenzamil', remained unknown until 1955 when Huisgen and his co-workers[32a,b,c] showed it to be 2-anilino-3*H*-azepine (**91**; R^1 = H, Y = NHPh). Subsequently this reaction has become a standard method of preparing 2-amino-3*H*-azepines.

The reaction pathway as proposed by Huisgen and his co-workers involves generation of a singlet nitrene (**89**), which is held to be in dynamic equilibrium with benzazirine (**90**; Scheme 22). Addition of the amine to the imine bond followed by electrocyclic ring-opening of the substituted azanorcaradiene (**90**) is thought to give firstly a 1*H*-azepine, which subsequently rearranges to the more stable 3*H*-azepine. Originally the product was formulated as the 7-anilino-2*H*-azepine, but detailed ^{1}H n.m.r. evidence confirmed[79] the product as the 3*H*-tautomer.

As mentioned in the benzazirine section (17.6.3), unequivocal evidence for the benzazirine intermediate (**90**) is lacking, and for certain arylnitrene reactions, particularly in the gas phase, the azepinylidene moiety (**93**) has been proposed[33a,b] as a viable alternative intermediate. Evidence in favour of the proposed reaction pathway (Scheme 22; path a) comes from the work of Sundberg and his co-workers,[80] who succeeded in identifying 3-alkyl-2-diethylamino-1*H*-azepines as oxygen-sensitive, metastable intermediates in the photolysis of *o*-alkylphenyl azides in diethylamine (see also Section 17.6.8.7). Later studies[81] on the flash photolysis of aryl azides in dialkylamine solution have provided kinetic data which not only confirm the 1*H*- to 3*H*-azepine pathway, but which also strongly support the involvement of a benzazirine intermediate.

The photo- and thermo-decompositions of aryl azides in the presence of nucleophiles other than amines, as preparative routes to 3*H*-azepines, have met with only limited success. For example, photolysis of phenyl azide in liquid ammonia gives 2-amino-3*H*-azepine (**91**; R^1 = H, Y = NH_2) (25%), whereas in ethereal hydrogen sulphide only a 5% yield of 3*H*-azepin-2-thione is obtained.[79] By photolysing certain *ortho*-substituted aryl azides (**88**; R^1 = $CONHR^2$, CO_2R^3, or CN) in alcohol solution, however, practicable yields (50–80%) of 3-substituted 2-alkoxy-3*H*-azepines are obtained.[82,83a] Related is the formation of 3-acyl-2-alkoxy-3*H*-azepines and -3*H*-azepin-2-ones by the photo-induced ring-expansion of 3-substituted anthranils (**92**; R^1 = alkyl or aryl) in alcohol and aqueous alcohol solution, respectively[84a] (Scheme 22; path c), and the preparation of 3- and 7-acetyl-3*H*-azepin-2-ones by photolysis of 2,3-dimethylindazole in dilute sulphuric acid.[85] In both these reactions it is assumed that ring opening to an *o*-nitreno-ketone, *e.g.* (**89**; R^1 = acyl), followed by benzazirine formation and nucleophilic attack as in Scheme 22, is responsible for azepine production.

An alternative and major method of synthesizing 2-amino-3*H*-azepines utilizes the deoxygenation of nitroso- and nitro-arenes in the presence of a secondary amine (Scheme 22b).[86,87,88,89] Deoxygenation is accomplished by using a trivalent phosphorus compound, *e.g.* triethyl phosphite, and with nitroso-compounds proceeds under mild conditions. The more readily available nitro-compounds, however, require much harsher conditions, a factor which originally proved troublesome. However, the discovery of more powerful deoxygenating agents (Scheme 23) now enables the deoxygenation of nitro-compounds to be carried out smoothly and efficiently under relatively mild conditions.[88]

The intermediacy of a discrete nitrene species in these deoxygenations has been the subject of some debate.[89] However, supporting evidence comes from studies[90] on the deoxygenation of a series of mainly *m*-nitro-arenes in the presence of diethylamine, which reveal that the percentage composition of the resulting mixture of 4- and 6-substituted 3*H*-azepines (Scheme 23) are, within experimental error, identical to those obtained by thermolysis of the corresponding aryl azides under similar conditions. Results[87,90] suggest that when X is electron-withdrawing, the 6-substituted 3*H*-azepine (**94**) is preferred, whereas electron donors favour formation of the 4-substituted isomer (**95**) (Scheme 23). Methyl groups do not exert a critical directing influence.

Arylnitrenes and subsequently 3*H*-azepines have also been generated by decomposition of oxaziranes,[89] and lately in almost quantitative yield by thermolysis (100 °C) of *N,O*-bis(trimethylsilyl)phenylhydroxylamine [Ph($OSiMe_3$)$NSiMe_3$] in the presence of dialkylamines.[91]

(95) (94)

i, $(EtO)_2MeP$ or $Ph_2(OEt)P$; ii, R_2NH.

SCHEME 23

Other synthetic approaches to 3*H*-azepines include [4+2] cycloadditions between 2*H*-azirines and cyclopentadienones[23a,b] (Scheme 7a), and the hydrogen bromide induced cyclization of 1,4-dicyanobut-1-enes.[71] Also of use is the ring expansion of 1,2-dihydro-2-(tosyloxymethyl)pyridines in hot pyridine,[92] a reaction which resembles the ring expansion of 4-(chloromethyl)-1,4-dihydropyridines to 4*H*-azepines (see Scheme 24).

(100) (98) (99) (101) (102) (103)

(a) $R^1 = R^2 = H$, $E = CO_2Me$
(b) $R^1 = Me$, $R^2 = Ph$, $E = CN$

i, NaOMe, MeOH, cold; ii, warm EtOH; iii, NaOMe, Et_2O; iv, NaOMe, hot EtOH; v, H^+; vi, C_5H_5N; vii, Bu^tOK at 0 °C.

SCHEME 24

3*H*-Azepin-2-ones are prepared by hydrolysis of 2-alkoxy-3*H*-azepines (Scheme 26; path ii), by photo or thermal decomposition of aryl azides in wet solvents,[84a] and by base-promoted ring opening and rearrangement of spiro-quinol ethers.[93] The last reaction has synthetic potential since the spiro-ethers are easily prepared in high yield by oxidative coupling of 2,4-disubstituted phenols. Also useful is the mechanistically similar ring expansion of 2,6-dialkylphenolates by ethereal chloramine at −70 °C.[71]

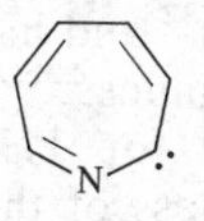

(93) 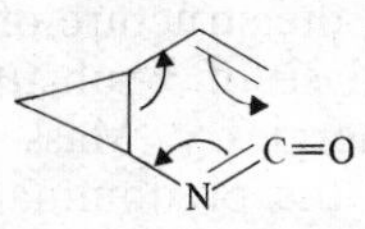 (96) 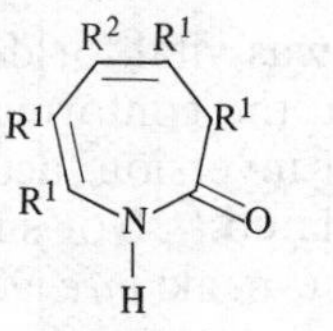 (97) 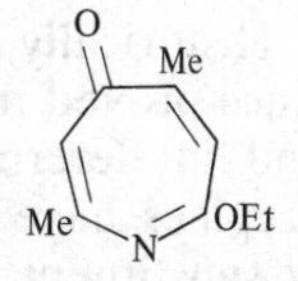 (105)

3*H*-Azepin-2-one (**97**; $R^1 = R^2 = H$) is available in 60% yield by a novel Cope-type rearrangement of the vinylcyclopropyl isocyanate (**96**).[71] Lithium aluminium hydride reduction of the azepinone yields 2,3-dihydro-1*H*-azepine as a colourless oil, which on standing rapidly polymerizes. In common with many other nitrogen heterocycles,[94] the 2-azepinones exist solely as the amide tautomers.

17.6.8.4 Synthesis of 4*H*-azepines

4*H*-Azepines are more rare than the 1*H*- and 3*H*-isomers and only a few synthetic approaches have been developed. Highly substituted 4*H*-azepines may be obtained by base-catalysed ring expansion of 4-(chloromethyl)-1,4-dihydropyridines.[71] However, the reaction is temperature and solvent sensitive, as illustrated in Scheme 24a. Azepines (**98**)–(**100**) have been isolated and characterized, but as later studies[95] on cyano-derivatives (Scheme 24b) show, the reaction is even more complex than was first thought. The azabicyclo[4,1,0]hept-4-ene (**101**) has been isolated, whereas methyleno-4*H*-azepine (**102**) has been detected only by ^{1}H n.m.r., all attempts at isolation resulting in rearrangement to the 1*H*-azepine (**103**).

[4+2] Cycloaddition of cyclopropenes to 1,2,4-triazines provides alternative access to 4*H*-azepines (Scheme 25)[72] (*cf.* Scheme 40). Base or heat treatment (190 °C) of the metastable 4*H*-azepine (**104**) brings about rearrangement to a mixture of the isomeric 3,5-diphenyl- and 4,6-diphenyl-3*H*-azepine tetraesters.

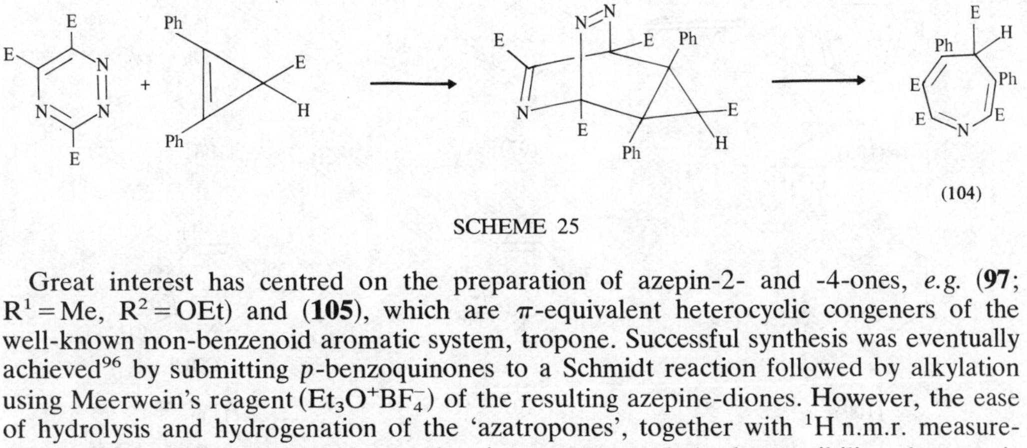

(104)

SCHEME 25

Great interest has centred on the preparation of azepin-2- and -4-ones, *e.g.* (**97**; $R^1 = Me$, $R^2 = OEt$) and (**105**), which are π-equivalent heterocyclic congeners of the well-known non-benzenoid aromatic system, tropone. Successful synthesis was eventually achieved[96] by submitting *p*-benzoquinones to a Schmidt reaction followed by alkylation using Meerwein's reagent ($Et_3O^+BF_4^-$) of the resulting azepine-diones. However, the ease of hydrolysis and hydrogenation of the 'azatropones', together with ^{1}H n.m.r. measurements which displayed no evidence of a ring current, exclude the possibility of aromatic character in these derivatives.

17.6.8.5 Physical and spectroscopic properties

The ultraviolet spectra of 1*H*-azepines have been well correlated.[71,77] ^{1}H N.m.r. data on 1*H*-,[15,71,77] 3*H*-,[15,79,80,83a,84a,87] and 4*H*-azepines[15,72,96] are available. ^{13}C N.m.r. chemical shifts[83b] (δ; CCl_4 solution) for 2-methoxy-3-methoxycarbonyl-3*H*-azepine are appended to formula (**112**; $R^1 = Me$, $R^2 = CO_2Me$).

Historically, ^1H n.m.r. was vital for determining the structure of 3*H*-azepines[82] and has since proved invaluable in the conformational analysis of 3-substituted 3*H*-azepines,[23b,71] and in determining ring-inversion activation energies.[74] Mass spectral data for 1*H*-azepines have been tabulated.[77] Not surprisingly, the predominant feature is loss of the *N*-substituent to give base-peak *m/e* 92, the azatropylium cation.

17.6.8.6 Valence isomerism

A feature of 1*H*-azepine chemistry that intrigued early research workers was the possibility of azepine–azanorcaradiene valence isomerism,[97] (**106**) ↔ (**107**), such as exists in the benzene oxide–oxepine system.[71] Paquette and his co-workers[71,77] have demonstrated that bridging the 2,7-positions by a trimethylene chain forces the 1*H*-azepine into adopting the iminobenzene or azanorcaradiene structure (**108**). The methylene chain length is critical and the tetramethyleno-derivative exists solely as the bridged azepine (**109**). Early variable-temperature (−90 to +130 °C) ^1H n.m.r. studies on other 1*H*-azepines, however, failed to detect any bicyclic isomers.[71] Later studies by Prinzbach and his co-workers[98] were more successful and provided evidence for the existence of an azepine–azanorcaradiene equilibrium in systems of type (**106**; $R^1 = H$, $R^2 =$ *p*-tosyl or Ac, $E = CO_2Me$). The iminobenzene form (**107**) has been trapped as a bis-1,3-dipolar adduct with diazomethane, and the ^1H n.m.r. spectrum ($CDCl_3$ solution) of the *N*-(*p*-tosyl) derivative shows temperature dependence between −68 and +40 °C. Subsequently the diphenyl derivative (**106**; $R^1 = Ph$, $R^2 =$ *p*-tosyl, $E = CO_2Me$) was shown to be a 97 : 3 azepine (**106**) : benzeneimine (**107**) mixture at room temperature,[99] and it was concluded that the bicyclic isomer is stabilized by electron-withdrawing groups. The equilibrium position of this type of valence isomeric system, together with the effect of electron-withdrawing substituents on the stability of 1*H*-azepines and their dimerization (Section 17.6.8.9), have been considered theoretically using extended Hückel calculations.[100]

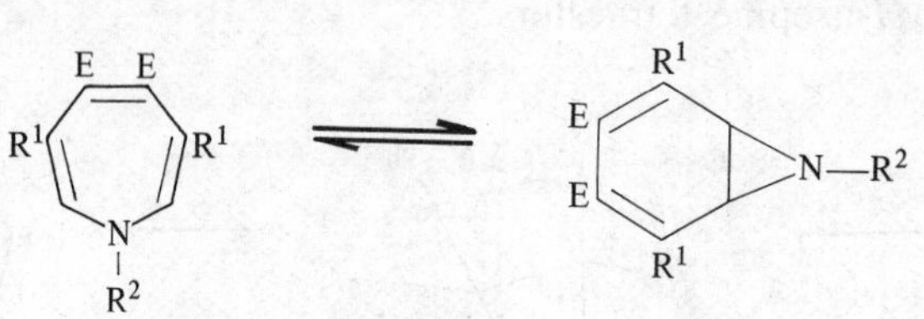

(106) (107) 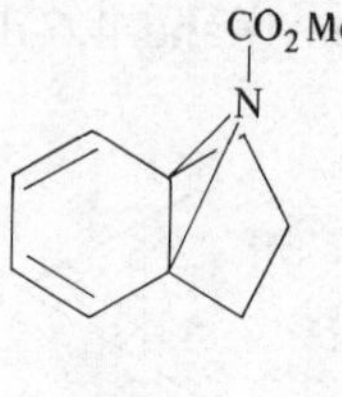 (108)

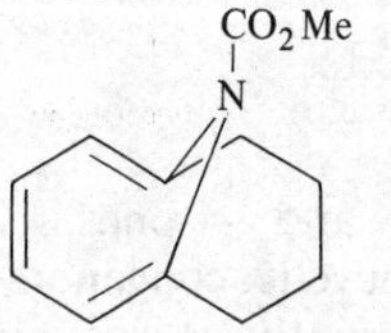

(109) 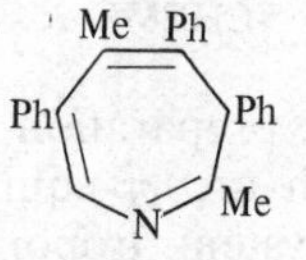 (110) 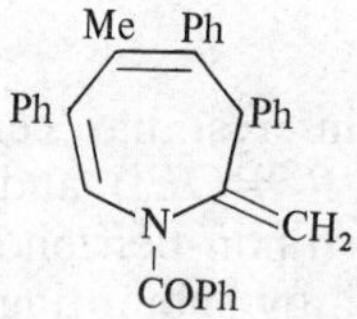 (111)

The 4*H*-azepine–azanorcaradiene valence isomerism has also come under close scrutiny,[101] the conclusion being that the 4*H*-azepine is energetically more favourable than its bicyclic isomer.

17.6.8.7 Oxidation and reduction

The early general chemistry of 1*H*-azepines[71,102a] and 2-amino-3*H*-azepines[32b,c,102a] has been well documented, and in this section only recent developments will be considered.

2-Dimethylamino-3*H*-azepine is reduced in the presence of 5% Pd/C to a mixture of 4,5-dihydro- (70%) and 4,5,6,7-tetrahydro-3*H*-azepine (30%).[103a] 3*H*-Azepin-2-ones with a platinum catalyst reduce to caprolactams, whereas with lithium aluminium hydride 1,3,5,7-tetramethyl-3*H*-azepin-2-one gives the unstable dienamine, 1,3,5,7-tetramethyl-1,2-dihydro-3*H*-azepine, in 91.8% yield.[71]

General information on the oxidation of azepines is scarce. However, recent results of Sundberg and his co-workers[80] suggest that 3-alkyl-2-dialkylamino-1*H*-azepines undergo aerial oxidation to 1*H*-azepin-2-ones (2-azatropones), whereas in certain cases oxidation in the presence of cupric ion yields the isomeric 4*H*-azepin-4-ones (4-azatropones). Ring-contraction products, *e.g.* 3-alkyl-2-alkylaminopyridines, are also produced.

17.6.8.8 Reaction with electrophilic reagents

Acid treatment of 1-alkoxycarbonyl-1*H*-azepines results in rapid aromatization to *N*-phenylurethanes.[71,104a] This ring-contraction is slower with 2-substituted 1*H*-azepines, a fact made use of in the hydrolytic separation of some isomeric 1*H*-azepines.[76c] 2-Amino-3*H*-azepines possess an amidine unit and as such are readily soluble in dilute acids, from which solution they are precipitated unchanged by base. In contrast, 4*H*-azepines, prepared by base-promoted ring expansion of dihydropyridines, undergo a retro-reaction on treatment with dilute hydrochloric acid (Scheme 24a; path v).

Acylating agents bring about ring contraction of 2-substituted 3*H*-azepines to *N*-acyl-*o*-substituted anilines.[32b,71] Atypically, the highly substituted 3*H*-azepine (**110**) on treatment with benzoyl chloride in the presence of 1,4-diazabicyclo[2,2,2]octane ('Dabco') yields the 1-benzoyl derivative (**111**).[105] Attempts to isomerize (**111**) to the conjugated 1*H*-azepine with weak base failed; strong base regenerated 3*H*-azepine (**110**). In marked contrast, glacial acetic acid caused rapid aromatization to 3,6-dimethyl-2,4,5-triphenylaniline.

Other general reactions of 2-substituted 3*H*-azepines and 3*H*-azepin-2-ones, and of 1-alkoxycarbonyl-1*H*-azepines, are summarized in Schemes 26 and 27, respectively. Notable is the reaction of 1-methoxycarbonyl-1*H*-azepine with phenyl-lithium (Scheme 27, reaction v). The structure of the 3-carbinol (**113**), which on pyrolysis yields 3*H*-azepine (**114**), has been confirmed by X-ray analysis.[73]

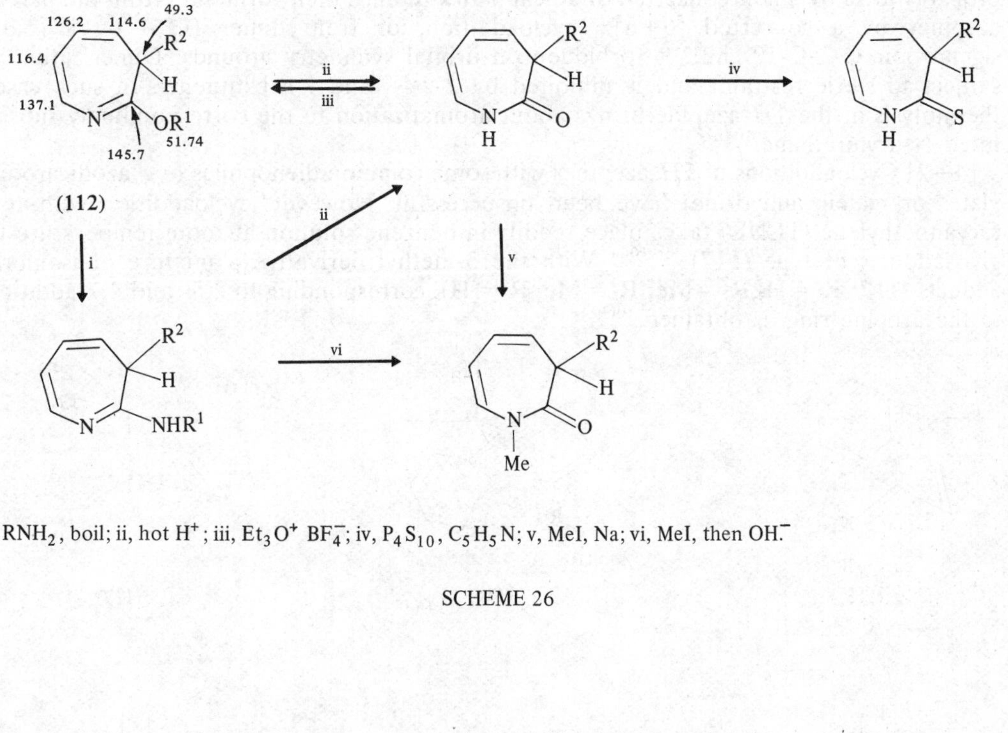

126.2 114.6 49.3 116.4 137.1 145.7 51.74 R^2 H N OR^1 (112)

i, RNH_2, boil; ii, hot H^+; iii, $Et_3O^+\ BF_4^-$; iv, P_4S_{10}, C_5H_5N; v, MeI, Na; vi, MeI, then OH^-

SCHEME 26

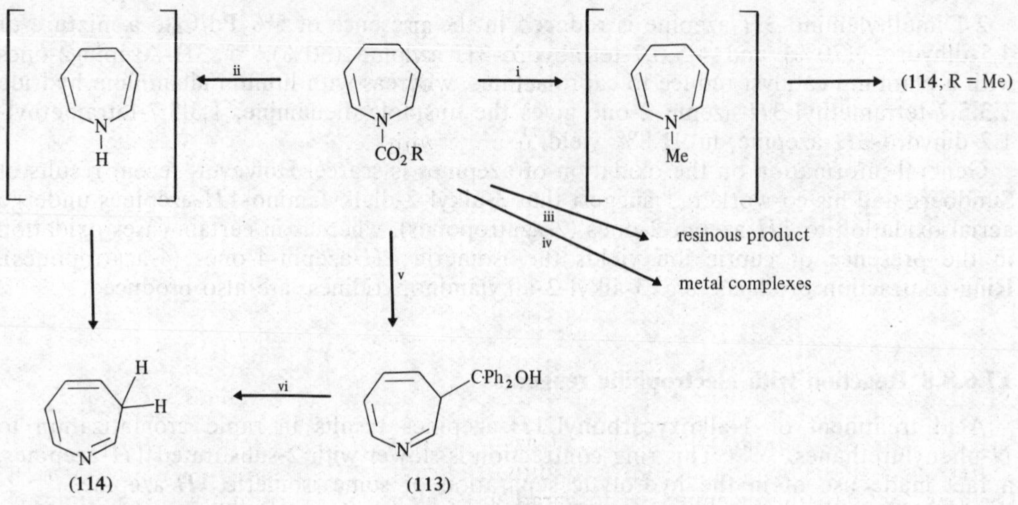

(114; R = Me)

resinous product

metal complexes

(114) (113)

i, $LiAlH_4$; ii, OH^-; iii, H^+; iv, metal carbonyls; v, PhLi; vi, g.l.c.–mass spectra.

SCHEME 27

17.6.8.9 Dimerization and cycloaddition reactions

The polyene nature of 1*H*-azepines is highlighted by the ease with which they enter into cycloaddition reactions with a wide variety of reagents. Their behaviour is somewhat unique in that not only do they display diene and dienophilic character, but they can also function as the 2π, 4π, or 6π component in the cycloaddition reaction.

Dimerization of 1*H*-azepines is a well-studied phenomenon involving a temperature-dependent cycloaddition process.[71,104b] At low temperatures (0 °C; R = Me; 130 °C; R = CO_2R or CN) a kinetically controlled, thermally allowed [6 + 4] dimerization to give *exo*-adducts (**115**) takes place, accompanied by a small amount (<10%) of symmetrical dimers (**116**). The latter are thermodynamically favoured and become the major products (83%) when the 1*H*-azepines are heated briefly at 200 °C. The symmetrical dimers probably arise by a non-concerted diradical pathway since their formation from the parent azepines by a concerted $[6+6]\pi$ cycloaddition, or from dimer (**115**) by a [1,3]-sigmatropic C-2–C-10 shift, is forbidden on orbital symmetry grounds. Dimerization is subject to steric restraint and is inhibited by 2-, 4-, and 7-substituents. In such cases thermolysis of the 1*H*-azepine brings about aromatization to the correspondingly substituted *N*-arylurethane.[104b]

[4 + 2] Cycloadditions of 1*H*-azepines with some common dienophiles (*e.g.* azodicarboxylates or maleic anhydride) have been unsuccessful. However, cycloaddition with tetracyanoethylene (TCNE) takes place readily in benzene solution at room temperature to give adducts of type (**117**).[76a,106a] With the 3-methyl derivative a mixture of isomeric adducts (**117**; R^1 = H, R^2 = Me; R^1 = Me, R^2 = H), corresponding to 2,5- and 4,7-addition to the azepine ring, is obtained.[106a]

(115) 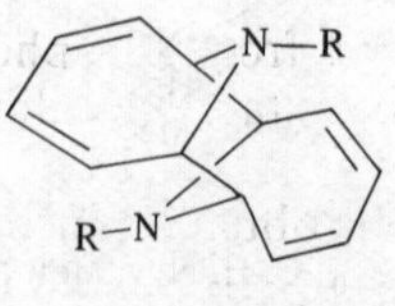 (116) 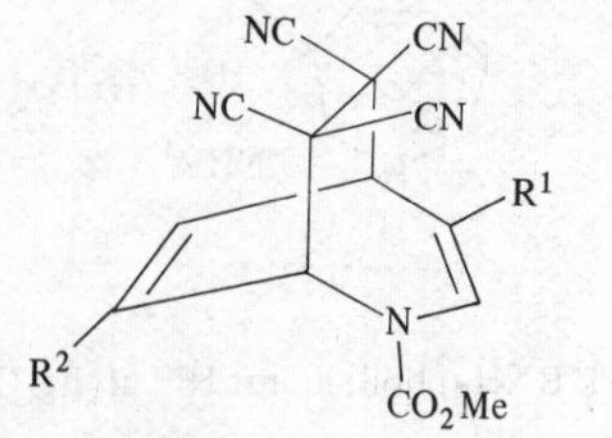 (117)

2,7-Cycloadducts, *e.g.* (**118**), which undergo rearrangement in boiling xylene to the 2,5-adducts, have been reported between 3,6-dialkyl-1*H*-azepines and TCNE.[76b]

In contrast to the above, 1*H*-azepines behave as dienophiles (4,5-double bond) towards isobenzofurans,[106a] cyclopentadienones,[106a] and 1,2,4,5-tetrazines.[107] Polar stepwise cycloadditions at the 2,7-positions have been realized with azodicarboxylates and nitrosobenzene,[108] and using phencyclone a novel non-concerted [6+2] cycloaddition at the azepine 2,3-double bond is claimed.[109]

Curiously, all attempts to coax 1*H*-azepines into functioning as 6π-systems towards other dienes in a [6+4] cycloaddition, such as operates in their low temperature dimerization, have failed.[110]

17.6.8.10 Ring contractions

Ring transformations of 1*H*- and 3*H*-azepines to substituted benzenes and pyridines under thermal, acidic, and acylating conditions have already been mentioned in previous sections, as has the azepine–azanorcaradiene valence bond isomerism. In addition, azepines undergo photo-ring contraction to bicyclic valence tautomers of the type shown in Scheme 28. These transformations, which parallel those undergone by cycloheptatrienes, are often thermally reversible and occur by an orbital symmetry-controlled disrotatory electrocyclic process.[106b,111] The methyl-substituted azepines, *e.g.* (**119**; R = Me), give a mixture of isomers (**120** and **121**; R = Me) corresponding to the two allowed electrocyclic reactions (Scheme 28; paths a and b, respectively).[106b]

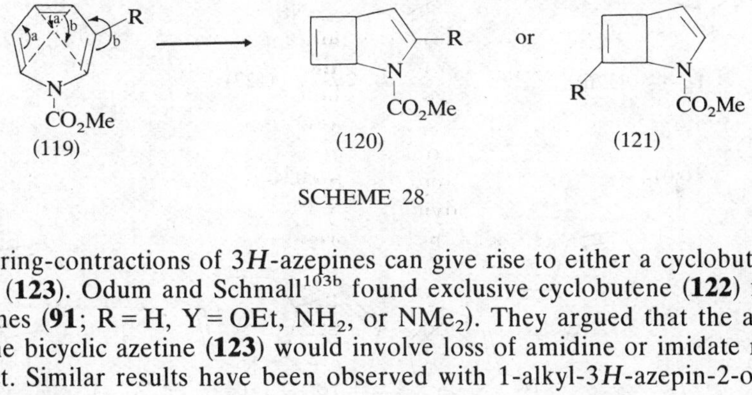

(119) (120) or (121)

SCHEME 28

Analogous ring-contractions of 3*H*-azepines can give rise to either a cyclobutene (**122**) or an azetine (**123**). Odum and Schmall[103b] found exclusive cyclobutene (**122**) formation for 3*H*-azepines (**91**; R = H, Y = OEt, NH_2, or NMe_2). They argued that the alternative pathway to the bicyclic azetine (**123**) would involve loss of amidine or imidate resonance in the product. Similar results have been observed with 1-alkyl-3*H*-azepin-2-ones.[87]

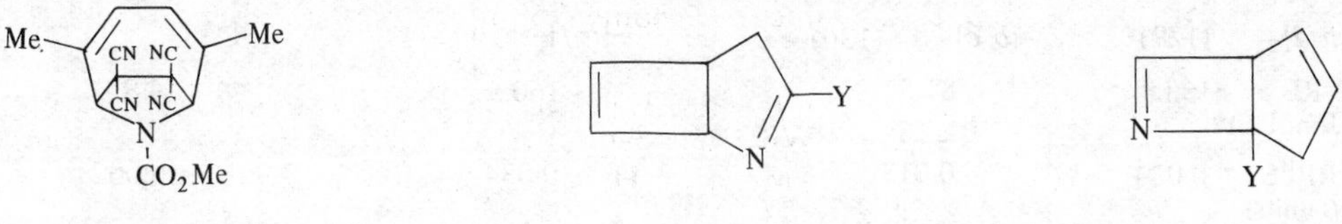

(118) (122) (123)

A bicyclic azetine has been isolated from the photosensitized (Ph_2CO or PhCOMe) ring-contraction of the otherwise photo-stable 4,5-dihydro-2-ethoxy-4,4,6-trimethyl-3*H*-azepine.[112]

2-Alkoxy-3-acylpyridines have been obtained in poor yield from the photolysis of 3-acyl-2-alkoxy-3*H*-azepines.[84b]

Ring contractions of yet a different kind are observed with 1*H*-azepines of type (**124**), which on heating yield 6-substituted fulvenes, *e.g.* (**125**), as illustrated in Scheme 29.[113]

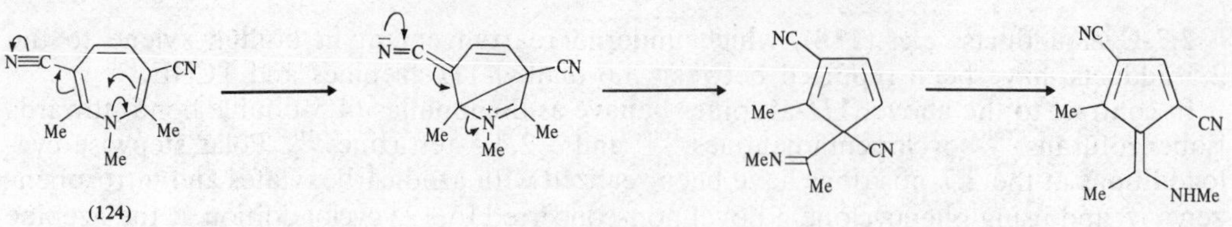

(124) (125)

SCHEME 29

17.6.9 BENZAZEPINES AND DIBENZAZEPINES

Annulation of a benzene ring to the azepine nucleus can be achieved in three different ways, and derivatives of the resulting 1- (**126**), 2- (**127**), and 3-benzazepines (**128**) are well-documented. Dibenzazepines, of which there are four isomers, are equally well known particularly as some derivatives have important pharmacological properties.[102b,114] The system of nomenclature for these compounds accompanies diagrams (**126**)–(**132**), the position of saturation, *i.e.* the 'floating hydrogen', being denoted by a locant.[102a] Also appended to structures (**126**)–(**132**) are the theoretically derived resonance energies (RE)[57b] and resonance energy per π-electron (REPE)[56b] values, which clearly indicate that the most stable isomers are those in which the carbocyclic rings retain their fully benzenoid character.

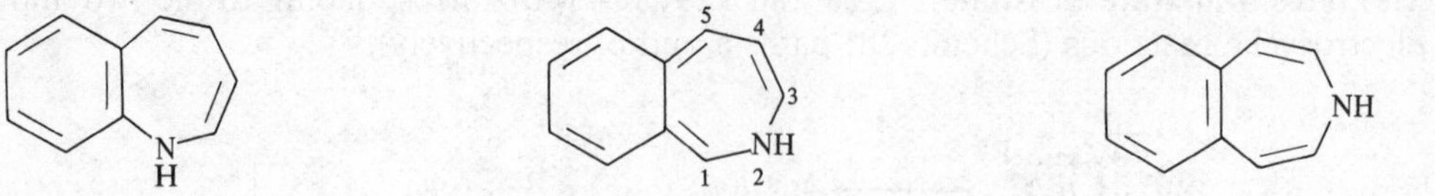

	1H-1- (126)	*2H*-2- (127)	*3H*-3- (128)
RE (kJ mol^{-1})	76.7	−3.2	73.2
REPE (β units)	0.014	−0.024	0.015

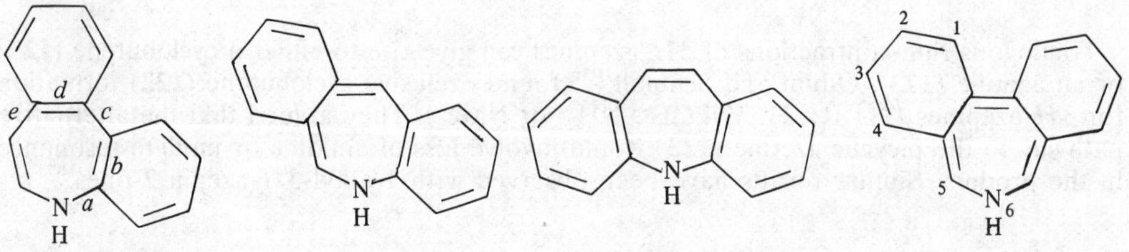

	-[*b*, *d*]- (129)	-[*b*, *e*]- (130)	-[*b*, *f*]- (131)	-[*c*, *e*]- (132)
RE (kJ mol^{-1})	158.4	81.5	160.5	4.9
REPE (β units)	0.034	0.013	0.034	0.002

17.6.9.1 1-Benzazepines

In contrast to the azepines, fully unsaturated 1-benzazepines (**126**) are little known and synthetic approaches are few. However, hydro-[115] and oxo-derivatives[115,116] are plentiful and are generally prepared by classical ring-closure methods, *e.g.* Dieckmann reaction, intramolecular Friedel–Crafts acylation, *etc.*[115] Also useful is the ring expansion of cyclic ketones and their derivatives by either the Beckmann or Schmidt (NaN_3/H_2SO_4) reactions. The former can yield either a 1- or 2-benzazepine, depending on whether an aryl

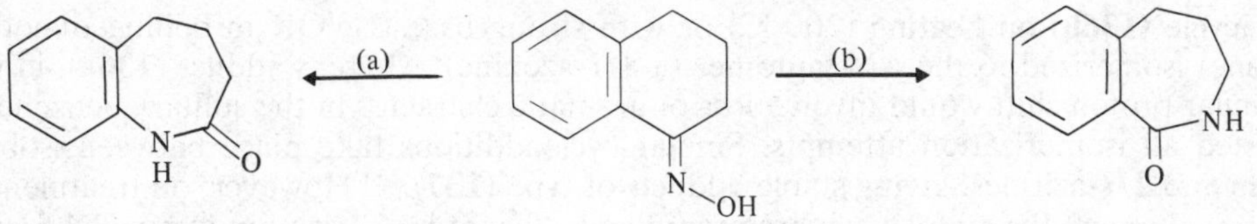

SCHEME 30

(Scheme 30; path a) or alkyl (path b) migration takes place. Comprehensive reviews are available on the syntheses of partially and fully reduced benzazepines.[102a,115]

The ring expansion procedures used for azepine synthesis (Schemes 21 and 22) have until recently met with little or no success when applied to the synthesis of benzazepines. For example, photolysis of ethyl azidoformate in the presence of naphthalene gave only the *N*-arylurethane.[74] However, preliminary studies on the ring expansion of naphthyl azides in amine solution have given encouraging yields (*ca.* 50%) of aminobenzazepines.[117]

The 3*H*-1-benzazepine (**134**) has been prepared by ring-expansion of the 2-ethoxyindole–dimethyl acetylenedicarboxylate 1,4-dipolar cycloadduct (**133**) (Scheme 31).[115] The corresponding *N*-methylindole reacts similarly to give 1-methyl-1*H*-1-benzazepine. When the reaction is carried out on 1-acetyl-3-piperidinoindole, the tricyclic intermediate corresponding to (**135**) can be isolated.

The ring expansion of quinolines by use of dibromocarbene[116] and diazomethane[115] have some merit for the synthesis of 1- and 2-benzazepines and benzazepinones.

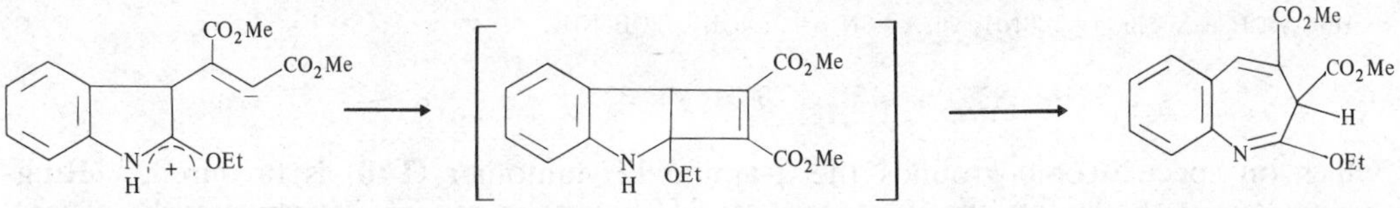

(133) (135) (134)

SCHEME 31

17.6.9.2 2-Benzazepines

2*H*-2-Benzazepine (**127**) is the least stable of the three isomers and only hydro- and oxo-derivatives are known.[115] By means of a modification of the 2*H*-azirine–cyclopentadienone 3*H*-azepine synthesis (see Section 17.6.2.6), Hassner and Anderson[23b] have been able to prepare derivatives, *e.g.* (**136a**) and (**136b**), of the more stable 3*H*-2-benzazepines. It is argued that the 2*H*-azepine which is formed initially in the cycloaddition might be prevented from rearranging to the normally more stable 3*H*-azepine if the 5,6 carbon–carbon double bond is incorporated as an integral part of a benzenoid system. Accordingly the 2*H*-azirine cycloaddition with 1,3-diphenylindan-2-one and phencyclone afforded the desired 3*H*-2-benzazepines (**136a**) and (**136b**). The

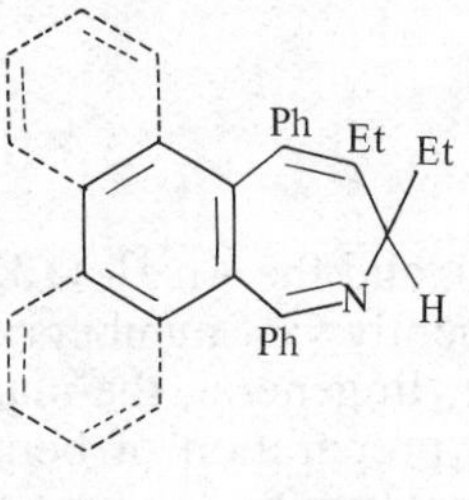

(136) a; benzo
b; phenanthro

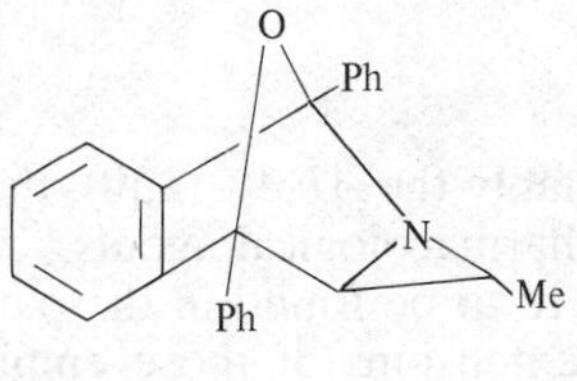

(137)

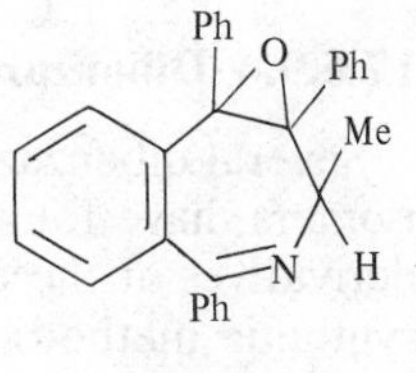

(138)

tetracycle (**136b**) on heating (200 °C) or with strong base (Bu^tOK in boiling dimethoxyethane) isomerized to the 4*H*-tautomer (a 3*H*-azepine), whereas adduct (**136a**), in which a similar proton shift would involve loss of aromatic character in the solitary benzene ring, resisted all isomerization attempts. Similar cycloadditions take place between isobenzofuran and 2*H*-azirines, giving stable adducts of type (**137**).[23a] However, on treatment with neutral alumina, the adducts undergo rapid and efficient isomerization to novel 4,5-epoxy-4,5-dihydro-3*H*-2-benzazepines, *e.g.* (**138**).

Access to 3*H*-2-benzazepines is also possible by photorearrangement of 3-phenyl-2-styryl-2*H*-azirines, as discussed in Section 17.6.2.7 (Scheme 10; path c).[24a]

Hydro-2*H*-benzazepines are available by von Braun, intramolecular Friedel–Crafts, and Bischler–Napieralski reactions on suitably mono- or *o*-di-substituted benzenes, and also by Beckmann and Schmidt rearrangements of α-tetralones and their oximes.[115]

17.6.9.3 3-Benzazepines

Few fully unsaturated 3-benzazepines (**128**) have been described. Hydrogen bromide induced cyclization of dinitrile (**139**; X = CN, $n = 1$; Scheme 32) gave a 3-benzazepine for

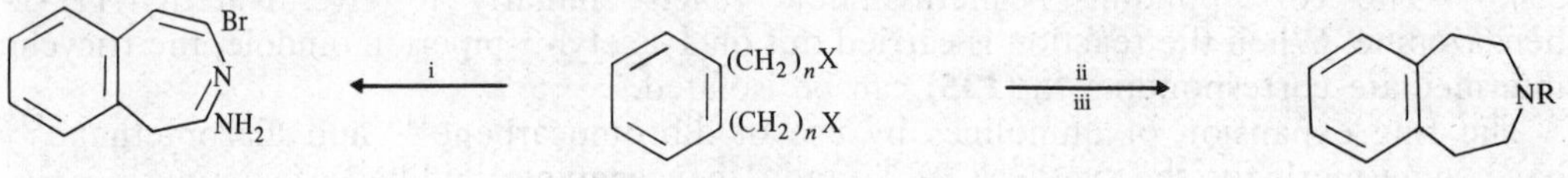

(140) (139)

i, HBr, AcOH, ii, X = Br, $n = 2$, RNH_2; iii, X = CN, $n = 2$; Ni, H_2, EtOH, NH_3.

SCHEME 32

which on spectroscopic grounds the 2-amino-1*H*-tautomer (**140**) is favoured.[71] Ring-expansion methods for the synthesis of 3*H*-benzazepines are conspicuously absent, although 4,5-dihydro-3*H*-3-benzazepines, e.g. (**141**), are available in practicable yield (50–60%) from 4-cyanoisoquinolines as outlined in Scheme 33.[118]

Routes to hydro- and oxo-derivatives of 3-benzazepines include Friedel–Crafts intramolecular acylation, ring closure of *o*-di(halogenoalkyl)benzenes with amines (Scheme 32; path ii; X = halogen, $n = 2$) and reductive cyclization of dinitriles (Scheme 32; path iii; X = CN, $n = 2$).[115]

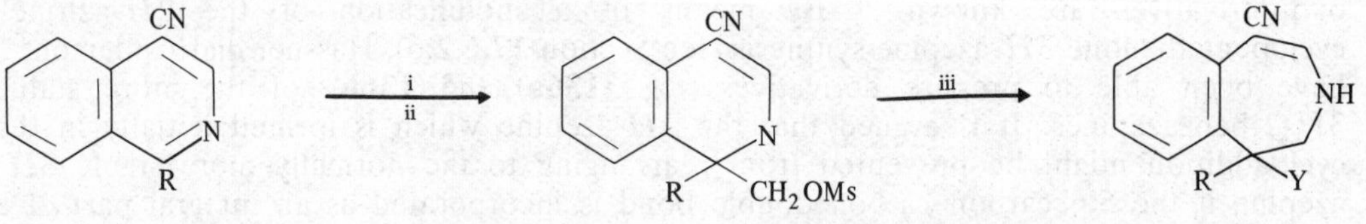

(141)

i, MeOH, $h\nu$; ii, $MeSO_2Cl$; iii, YH, Y = RO, RS, or R_2N.

SCHEME 33

17.6.9.4 Dibenzazepines

Several dibenzazepines, notably the -[*b,e*]- (**130**), the -[*b,f*]- (**131**), and the -[*c,e*]- (**132**) isomers, have found use as pharmacological agents.[102b,114] Consequently vast numbers of derivatives of these systems are to be found in the patent literature. In general, the main synthetic methods used are extensions of those employed for the preparation of benzazepines. For example, dibenzo[*b,d*]azepinones arise by intramolecular cyclization ($AlCl_3/CHCl_3$ at −70 °C) of *N*-tosyl derivatives of type (**142**).[119] Detosylation followed by

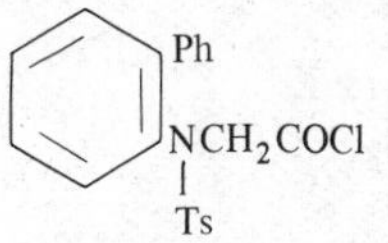

Ph
NCH$_2$COCl
Ts
(142)

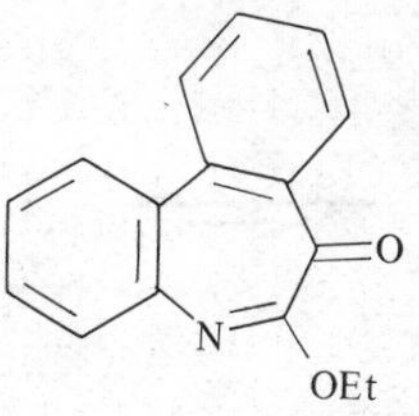

O
N
OEt
(143)

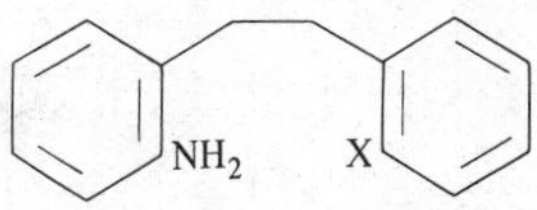

NH$_2$ X
(144)

manganese dioxide oxidation of the resulting 2-ethoxy-derivative provides a synthesis of the non-aromatic dibenzo-azatropone (**143**).

Dibenzo[*b,e*]azepines (**130**) (morphanthridines) are available by several methods, of which polyphosphoric acid cyclization of *o*-acylaminodiphenylmethanes[120] and the Schmidt reaction on anthraquinones[121] are useful examples.

The chemistry of dibenzo[*b,f*]azepines (**131**) has been reviewed in depth.[122] Main methods of preparation are by catalytic (Pd/C) dehydrogenation of the 10,11-dihydro-derivatives, which are readily obtained by acid cyclization of 2,2′-diaminobibenzyls, *e.g.* (**144**; X = NH_2), and by dehydrochlorination of the corresponding chloro-derivatives (**144**; X = Cl) using copper powder in boiling dimethylformamide. Monobromination (*N*-bromosuccinimide) of the 10,11-dihydro-compounds at either of the benzylic carbons followed by dehydrobromination is a useful alternative synthetic route.

6*H*-Dibenzo[*c,e*]azepines (**132**) are the least stable of the four isomers, since they have only one fully benzenoid carbocyclic ring. However, their 5*H*-derivatives are well known and may be prepared by cyclization of 2,2′-di(bromomethyl)biphenyls with amines,[123] by Pschorr cyclization ($NaNO_2$, $CuSO_4$, Cu powder) of *N*-(*o*-aminobenzyl)benzylamines,[124] and by photolysis of *N*-(*o*-iodobenzyl)benzylamines.[125] Demesylation (Bu^tOK/DMSO) of the 5,7-dihydro-6-mesyl derivative (**145**) gives the parent 5*H*-dibenzo[*c,e*]azepine (**146**; X = CH) in 85% yield. The authors note that (**146**) shows no tendency to rearrange to the *o*-quinonoid 6*H*-tautomer (**132**).[126]

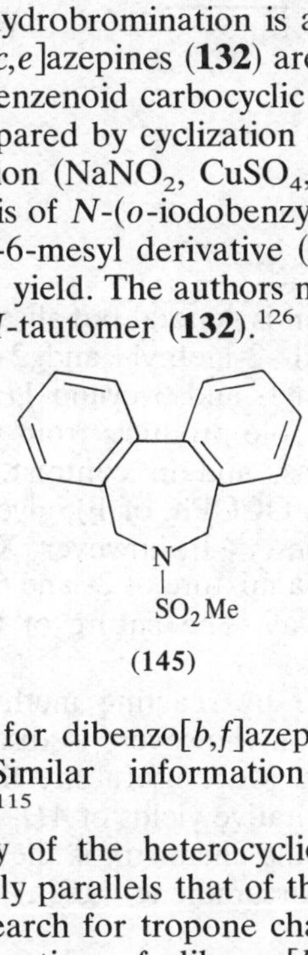

N
SO$_2$Me
(145)

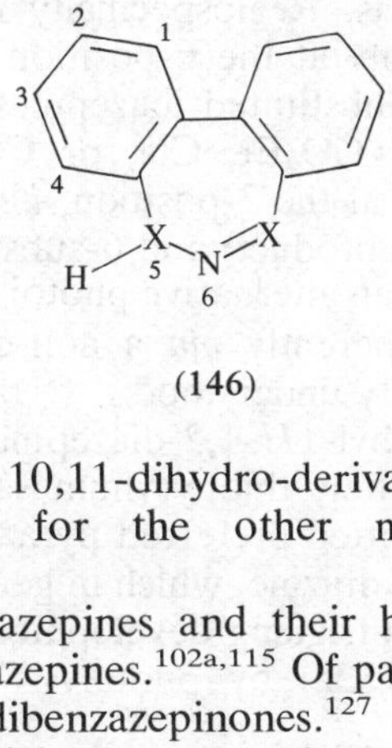

2 1 3 4 X 5 N 6 X H
(146)

Spectral data for dibenzo[*b,f*]azepines and their 10,11-dihydro-derivatives have been correlated.[122] Similar information is available for the other mono- and di-benzazepines.[15,115]

The chemistry of the heterocyclic ring in benzazepines and their hydro- and oxo-derivatives closely parallels that of the monocyclic azepines.[102a,115] Of particular note has been the vain search for tropone character in the dibenzazepinones.[127]

With the exception of dibenzo[*b,f*]azepine, which, for example, as the free base undergoes Friedel–Crafts acylation at positions-2 and -8, and as the *N*-acetyl derivative at the 10-position,[122] surprisingly little is known about the electrophilic substitution of mono- and di-benzazepines.[102a,115]

17.6.10 DIAZEPINES AND BENZODIAZEPINES

Four tautomeric diazepines are possible. However, those containing only C=C and C=N units (*i.e.* the 1*H*- and 4*H*-tautomers) are more stable than the 3*H*- and 5*H*-tautomers, which incorporate a N=N unit. In fact ^{1}H n.m.r. evidence[101] indicates that the

(147) (148)

5H-isomers (**147**) exist predominantly as the bicyclic diazanorcaradiene forms (**148**), which are estimated to be 12 kJ mol^{-1} more stable.

17.6.10.1 1,2-Diazepines and their benzo-derivatives

A wide range of 1H-1,2-diazepines are available,[128] mainly by photo-rearrangement of N-iminopyridinium ylides, *e.g.* (**149**) (Scheme 34). The reaction is presumed to involve initial electrocyclic ring closure of the ylide followed by ring expansion of the resulting 1,7-diazanorcaradiene. However, unequivocal chemical or physical evidence for a bicyclic

(151) (149) (150)

SCHEME 34

intermediate has not yet been produced. The reaction is general but sh ws some puzzling substituent effects. Regiospecificity is observed with 2-methyl- and 2-cyano-pyridines, which cyclize only at the 6-position to give 3-methyl- and 3-cyano-1H-1,2-diazepines, respectively. 4-Substituted diazepines (**150**) are the sole products from the 3-substituted ylides (**149**; X = CO_2Et, CN, or $CONH_2$), whereas, and in contrast, some electron-donating groups at the 3-position, *i.e.* (**149**; X = Me, OCOPh, or F), give rise to mixtures of the 4- (major product) and 6-substituted diazepines.[129] If, however, X is OH, NH_2, or NHCOPh, then an alternative photoisomerization to a mixture of 2- and 6-aminopyridines is preferred, apparently *via* a non-diazepine pathway. The nature of these substituent effects is not fully understood.

A few 1-methyl-1H-1,2-diazepines are available by reacting methylhydrazine with pyrylium or, better, thiopyrylium salts. The reaction, however, is disadvantaged by a competing and often preferred pyrazoline formation. More useful are the corresponding reactions with hydrazine, which in general give quantitative yields of 4H-1,2-diazepines.

A novel and intriguing development in 1,2-diazepine chemistry is the synthesis of 4H-1,2-diazepine (**153a**) by photolysis of the diazanorcaradiene (**151a**) (Scheme 35).[130]

(151) (152) (153)

a; X = CMe_2
b; X = NCH_2Ph or NC_6H_{11}

i, NH_2NH_2; ii, $h\nu$.

SCHEME 35

Careful investigation reveals that the product arises by an isopropylidene 'walk' (**151a**) → (**152a**) prior to ring expansion. Direct electrocyclic ring-opening of intermediate (**151a**) would yield the less-stable[101] N=N-containing 5*H*-diazepine.

3*H*-1,2-Diazepines have been synthesized only recently[131] from β-substituted α,β-γ,δ-unsaturated ketone *N*-tosylhydrazones by an extension of the method used for preparing benzodiazepines (Scheme 36). Apparently, unlike the 5*H*-tautomers and despite the presence of the destabilizing N=N moiety, the 3*H*-tautomers exist in the bicyclic diazepine form.

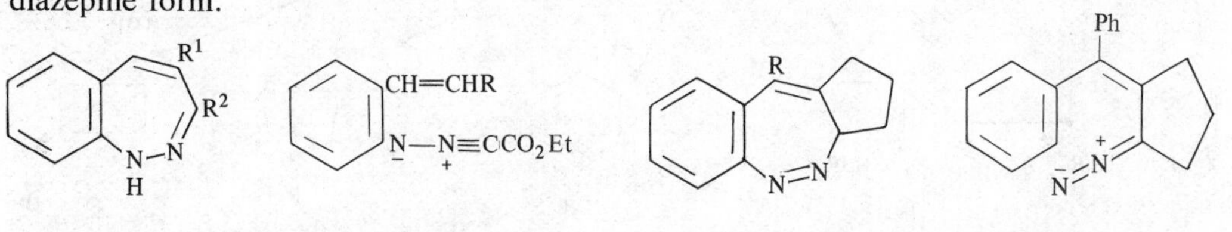

(**154**) (**155**) (**156**) (**157**)

Not unexpectedly, 1*H*-1,2-benzodiazepines (**154**) may be synthesized by photo-expansion of the dimeric *N*-iminoquinolinium ylides; actual examples, however, are few.[128] An alternative and useful preparative method is the ring closure of *o*-vinylnitrilimines, *e.g.* (**155**), a reaction which involves either an intramolecular 1,3-dipolar cycloaddition followed by ring opening of the highly strained tricyclic intermediate, or a direct 1,7-electrocyclic ring-closure and H-shift.[132] 1*H*-1,2-Benzodiazepines are the products of the acid- or base-catalysed isomerization of the 3*H*-tautomers, *e.g.* (**156**), which are obtained either by dehydrogenation of 2,3-dihydro-1*H*-1,2-benzodiazepines,[133] or by a 1,7-electrocyclic ring closure of the 8π-diazoalkene system (**157**).[134] This latter reaction proceeds *via* a 1,5-hydrogen shift of the initially formed cycloadduct. Application of this method to α-aryldiazoalkenes, *e.g.* (**158**; Scheme 36) allows the preparation of isomeric 1*H*-2,3-benzodiazepines (**159**).[134]

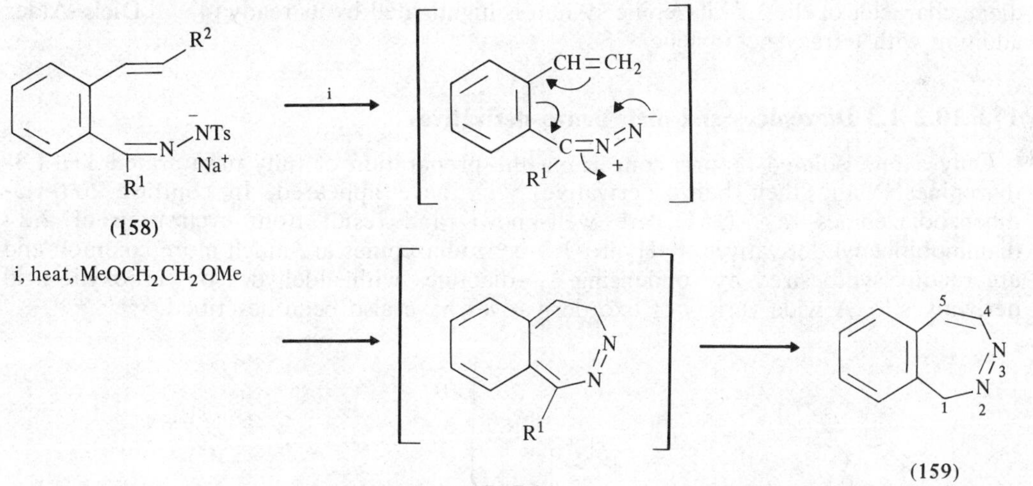

(**158**)

i, heat, MeOCH₂CH₂OMe

(**159**)

SCHEME 36

Apparently 3*H*-2,3-diazepines are unknown, whereas 5*H*-2,3-diazepines may be prepared either by thermal or base-catalysed isomerization of the 1*H*-tautomers.[128]

Dihydro- and oxo-derivatives of 2,3-diazepines are known, the latter in abundance since, like their more famous isomers the 1,4-benzodiazepinones (Section 17.6.10.3), they have also been found to exhibit tranquilizing activity.[128]

The general chemistry of 1*H*-1,2-diazepines resembles in many aspects that of 1*H*-azepines and is briefly summarized in Scheme 37. Of note are dimerization (route v), which is acid-catalysed and involves a [4+2] cycloaddition rather than the [6+4] process

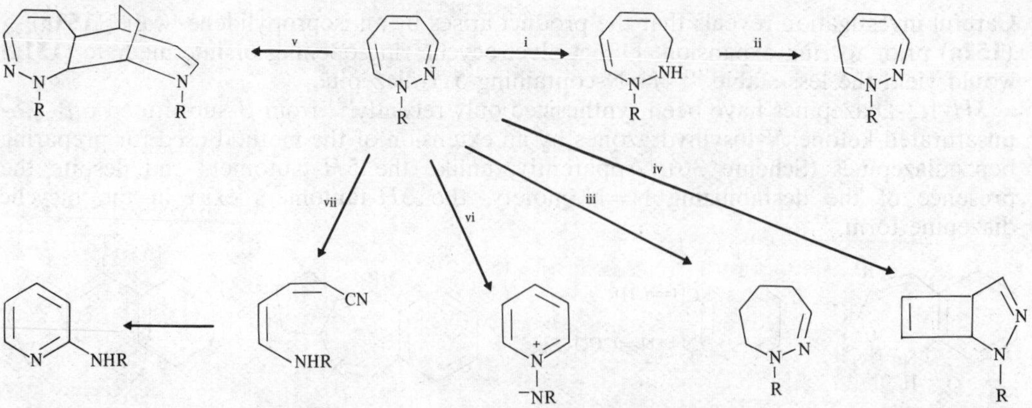

(160)

i, $NaBH_4$; ii, NaOMe, iii, H_2, Pd; iv, $h\nu$; v, HCO_2H, CF_3CO_2H, or BF_3; vi, H^+; vii, RONa, ROH.

SCHEME 37

associated with 1*H*-azepines, and the base-catalysed transformation to 2-aminopyridines (route vii) which proceeds *via* the isolable *cis,cis*-aminonitrile (**160**) and not as was first thought *via* a diazanorcaradiene.

In contrast, 4*H*-1,2-diazepines, on heating, ring-contract to pyridazines by way of sometimes isolable 3,4-diazanorcaradiene intermediates. ^{1}H N.m.r. studies show that in acid solution 1*H*-1,2-diazepines protonate at C-4, whereas 4*H*-1,2-diazepines protonate at N-1 and then collapse to give mixtures of pyrazoles and pyridines.

In addition to the reactions displayed in Scheme 37, 1*H*-1,2-diazepines undergo a variety of cycloaddition reactions, including regiospecific [2+2] addition with ketens at the 2,3-bond, and 1,3-dipolar cycloaddition with diazomethane at the 4,5-position.[128] Dienophilic activity (4,5-bond) is shown towards diphenylisobenzofuran, whereas the diene character of the 1,2-diazepine system is highlighted by its ready [4+2] Diels–Alder addition with tetracyanoethylene.

17.6.10.2 1,3-Diazepines and their benzo-derivatives

Only a few isolated reports concerning the preparation of fully unsaturated 1*H*-1,3-diazepines[136] and their benzo-derivatives[135,137] have appeared. In contrast, 5*H*-1,3-dibenzodiazepines, *e.g.* (**161**), are well-known and result from cyclization of 2,2′-diaminobiphenyl derivatives.[135] Hydro-1,3-benzodiazepines are much more common and are readily synthesized by condensing 1,4-diamines with aldehydes or carboxylic acid derivatives.[143] A wide variety of oxo-derivatives have also been described.[102a,135,137]

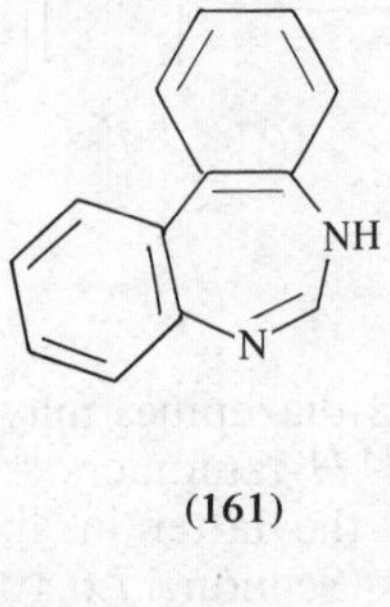

(161) 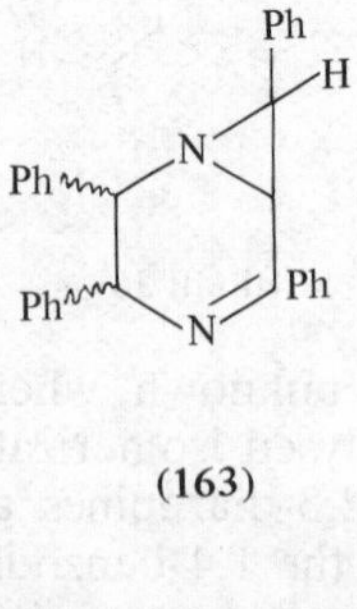 (163) 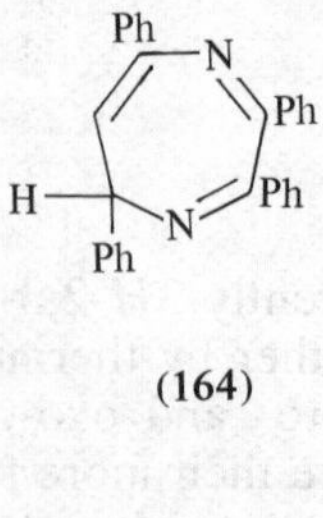 (164)

17.6.10.3 1,4-Diazepines and their benzo-derivatives

Only a few fully unsaturated 1,4-diazepines have been described. An obvious but only recently exploited route[138,139] to the 6*H*-tautomers, *e.g.* (**162**), is the condensation of

i, heat, EtOH, H_2O, AcOH.

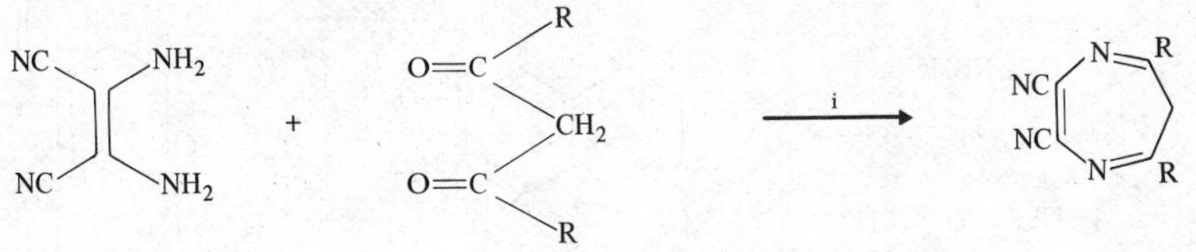

SCHEME 38

1,2-diaminoalkenes with 1,3-dicarbonyl compounds (Scheme 38). The 7*H*-1,4-diazepine (**164**) is a product (65%) from the ring expansion of diazabicycloheptene (**163**) in cold base (Bu^tOK/Bu^tOH).[140] Pyrazines are also formed.

2,3-Dihydro-1*H*-1,4-diazepines (**165a**) are much more common,[141] and are prepared by condensing 1,2-diamines with β-dicarbonyl compounds. The main interest in these hydrodiazepines stems from the 'quasi-aromatic' character displayed by their monocations, which are best represented as the delocalized mesomer (**166**). The enhanced stability of the cation is reflected in pK_a measurements, which indicate a resonance energy of 79 kJ mol^{-1} (cation) compared with a value of 42–50 kJ mol^{-1} for the neutral ring. The dihydroazepines, which exist as the conjugated form (**165a**) rather than the bis-imine form (**167a**), are themselves quite stable and show resistance to dehydrogenation, oxidation, and hydrolysis. Electrophilic substitution (deuteration, halogenation, nitration, and diazonium coupling) of the cation is facile, and, as expected, takes place at the 6-position.[141]

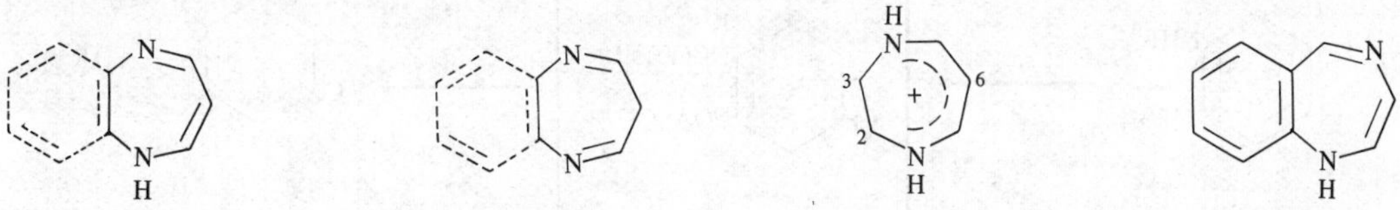

(165) (167) (166) (168)

a, monocyclic systems; b, dotted lines – bicyclic systems

Benzo-fusion to the 1,4-diazepine nucleus produces two sets of isomers, namely 1,5-benzo- (**165b**) and 1,4-benzo-diazepines (**168**). The 1,5-benzodiazepines, like their monocyclic counterparts (**165a**), are most conveniently prepared by condensing *o*-phenylenediamines with β-dicarbonyl compounds;[135,137,142] for maximum yield condensation at $pH \approx 5$ is recommended. An alternative condensation uses β-chlorovinyl ketones and *o*-phenylenediamines in alcoholic hydrogen chloride.

The 1,5-benzodiazepines differ considerably from the monocyclic 1,4-diazepines. Spectral data[142] indicate a dianil structure (**167b**) rather than the conjugated 1*H*-tautomer (**165b**). In acid solution the colourless free base is converted into the bright blue or violet monocation, for which several mesomeric forms may be drawn. The monocations are, however, much less stable than the monocyclic analogue (**166**) in that under warm acid conditions they ring-transform into benzimidazoles. Also the 1,5-benzodiazepines are much weaker bases than the 1,4-diazepines. Electrophilic substitution (Br^+) takes place at the carbocyclic ring, whereas nitration gives only tarry products.[142] Some resistance to Raney nickel and lithium aluminium hydride reduction has been noted.[142]

The 3-oxo-derivatives have aroused interest as hetero-analogues of 4,5-benzotropone. However, infrared studies [ν(C=O) 1680 cm^{-1}] and chemical instability (*e.g.* ready decarbonylation to quinoxaline) do not support a stabilized dipolar tropone-like structure. Other oxo-derivatives have been prepared.[102a,135,137]

The vast amount of literature, mostly in the form of patents, concerning the chemistry of 1,4-benzodiazepines and their oxo-derivatives derives from the discovery in the late

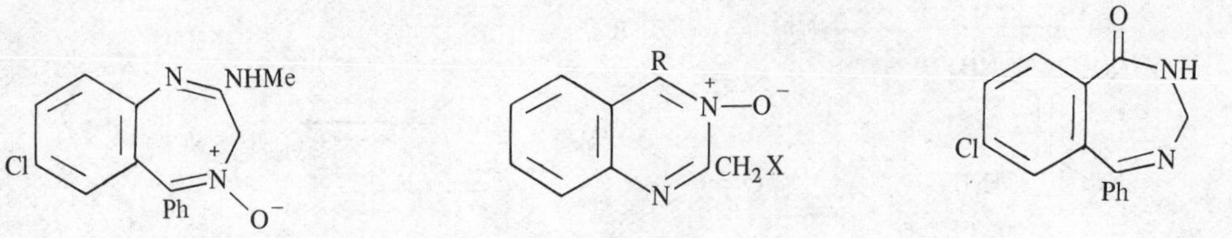

(169)

(171)

(172)

1950s of the remarkable tranquilizing, hypnotic, and central nervous system activity associated with certain members of this heterocyclic system. Of particular importance are the diazepine *N*-oxide (**169**) and the diazepinone (**170**; $R^1 = Me$, $R^2 = H$, Ar = Ph), which are marketed as chlorodiazepoxide (Librium) and diazepam (Valium), respectively.

Chlorodiazepoxide (**169**) was first synthesized by a novel amine-induced ring-expansion of quinazoline 3-oxides of type (**171**).[137,143] Apparently the *N*-oxide function is necessary for ring expansion as the deoxygenated quinazoline under the same conditions only suffers halogen replacement. The *N*-oxide also undergoes ring-expansion with sodium methoxide and sodium hydroxide to give 2-methoxy-1,4-benzodiazepines and 1,4-benzodiazepin-2-ones, respectively.

Although this method is general for benzodiazepines, the diazepam type compounds, *i.e.* the diazepinones, are most conveniently prepared by cyclization of a suitably modified *o*-aminobenzophenone.[143] A versatile general three-step synthesis is outlined in Scheme 39.

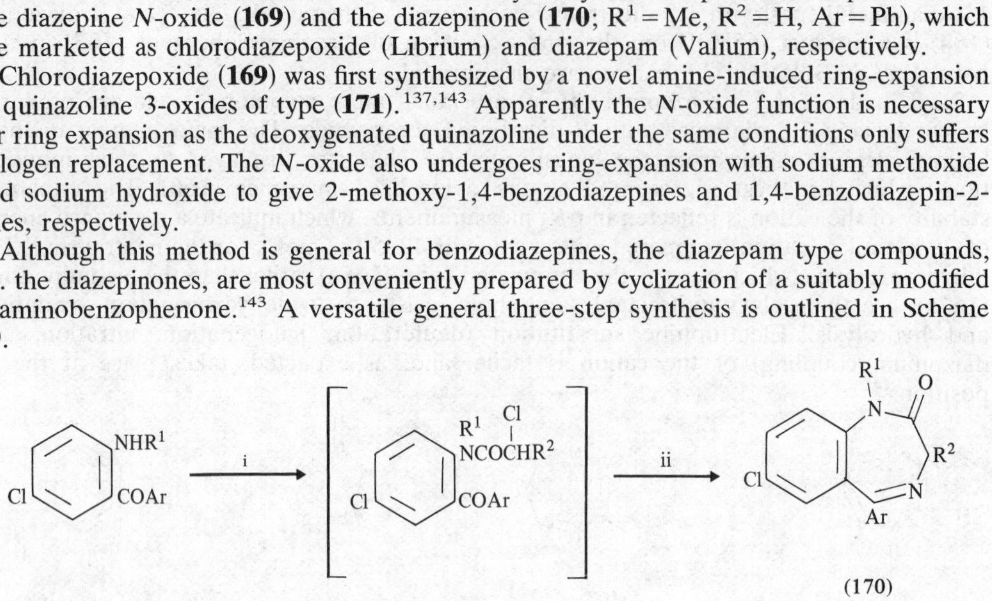

(170)

i, $ClCH(R^2)COCl$; ii, NH_3.

SCHEME 39

As is to be expected, the chemistry of these compounds has been the subject of intense scrutiny and several detailed accounts dealing with their general reactions[102a,b,143,144] and with their many and varied ring-contractions[145] are available.

The isomeric 1,4-benzodiazepin-5-ones have also been characterized and although lacking the dramatic CNS activity of diazepam and related compounds they have been found as constituents in anthrymycin-type antibiotics and other mould metabolites.[137]

The pharmacological, metabolic, and structure–activity relationship studies on benzodiazepines lie well-beyond the scope of this review and for such information specialized texts[144,146] should be consulted. However, it is worth noting that of the many successful structural changes tried and tested since benzodiazepine CNS activity was first discovered

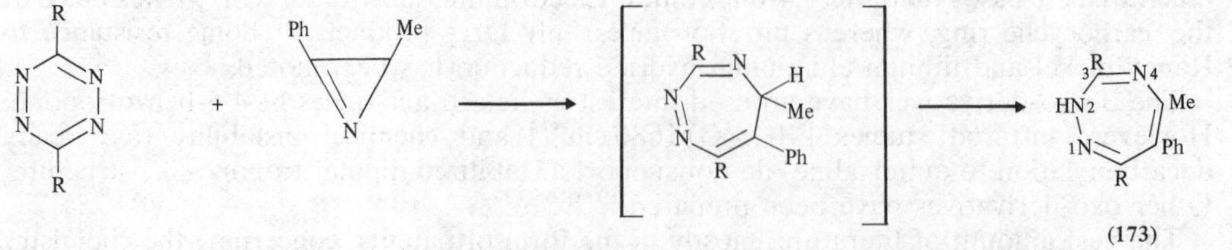

(173)

SCHEME 40

almost all have a 7-chloro and a 5-aryl substituent. This 7-substituent along with the 2-carbonyl group appears to be of paramount importance for psychopharmacological activity.

2,4-Benzodiazepin-1-one (**172**) has recently been synthesized and found to have CNS activity of diazepam-like proportions.[147]

17.6.11 TRIAZEPINES

There are four fully unsaturated isomeric triazepines, all of which exhibit many tautomeric forms. 2*H*-1,2,4-Triazepines arise by cycloaddition of 2*H*-azirines and *sym*-tetrazines[148] (Scheme 40), a reaction analogous to the synthesis of 4*H*-azepines from 1,2,4-triazines (Scheme 25).

Also of interest is the photo-ring-expansion *via* a nitrogen walk of the 3,4,7-triazanorcaradienes (**151b**) to 4-alkyl-4*H*-1,2,4-triazepines (**153b**) (Scheme 35b).[149a] Although analogous 'methylene' (see Scheme 35a) and oxygen 'walks' are well established, this reaction is claimed as the first authenticated example of a nitrogen 'walk'. Subsequent studies[149b] suggest that the reaction is thermally reversible, the resulting triazanorcaradienes undergoing fragmentation to mixtures of pyrazoles, indazoles, and pyrimidines.

1,3,4-Benzotriazepin-5-ones have recently become available by treating 1-(*o*-aminoacyl)-1-methylhydrazines with ortho-esters,[150] while the first 1,2,3-triazepine has been obtained[151] as the dibenzo-derivative (**146**; X = N) by careful neutralization (aqueous ammonia at 0 °C) of tetrazotized 2,2′-diaminobiphenyl.

1,3,5-Benzotriazepines are known mainly as their oxo-derivatives.[152]

References

1. R. Breslow, *Accounts Chem. Res.*, 1973, **6,** 393.
2. H. C. van der Plas, 'Ring Transformations of Heterocycles', Academic Press, London, 1973, vols. 1 and 2.
3. A. Hassner and F. W. Fowler, *J. Amer. Chem. Soc.*, 1968, **90,** 2869.
4. F. W. Fowler, *Adv. Heterocyclic Chem.*, 1971, **13,** 45.
5. W. A. Lathan, L. Radom, P. C. Hariharan, W. J. Hehre, and J. A. Pople, *Topics Current Chem.*, 1973, **40,** 1.
6. M. J. S. Dewar and C. A. Ramsden, *J.C.S. Chem. Comm.*, 1973, 688.
7. C. U. Pittman, Jr., A. Kress, T. B. Patterson, P. Walton, and L. D. Kispert, *J. Org. Chem.*, 1974, **39,** 373.
8. (a) D. J. Anderson, T. L. Gilchrist, G. E. Gymer, and C. W. Rees, *J.C.S. Perkin I*, 1973, 550; (b) T. L. Gilchrist, G. E. Gymer, and C. W. Rees, *ibid.*, p. 555; (c) *ibid.*, 1975, 1.
9. T. L. Gilchrist, C. W. Rees, and C. Thomas, *J.C.S. Perkin I*, 1975, 8.
10. (a) G. Smolinsky, *J. Org. Chem.*, 1962, **27,** 3557; (b) G. Smolinsky and C. A. Pryde, *ibid.*, 1968, **33,** 2411.
11. (a) K. Isomura, S. Kobayashi, and H. Taniguchi, *Tetrahedron Letters*, 1968, 3499; (b) K. Isomura, M. Okada, and H. Taniguchi, *ibid.*, 1969, 4073.
12. F. W. Fowler, A. Hassner, and L. A. Levy, *J. Amer. Chem. Soc.*, 1967, **89,** 2077.
13. G. L'Abbe, *Chem. Rev.*, 1969, **69,** 345.
14. A. Hassner, R. J. Isbister, R. B. Greenwald, J. T. Klug, and E. C. Taylor, *Tetrahedron*, 1969, **25,** 1637.
15. T. J. Batterham, 'N.m.r. of Simple Heterocycles', Wiley, New York, 1973.
16. V. Nair, *Org. Magn. Resonance*, 1974, **6,** 483.
17. A. Hassner, S. S. Burke, and J. Cheng-fan, *J. Amer. Chem. Soc.*, 1975, **97,** 4692.
18. R. Chaabouni, A. Laurent, and P. Mison, *Tetrahedron Letters*, 1973, 1343.
19. A. G. Hortmann and D. A. Robertson, *J. Amer. Chem. Soc.*, 1972. **94,** 2758.
20. A. Hassner, J. O. Currie, Jr., A. S. Steinfeld, and R. F. Atkinson, *J. Amer. Chem. Soc.*, 1973, **95,** 2982.
21. A. Demoulin, H. Gorissen, A.-M. Hesbain-Frisque, and L. Ghosez, *J. Amer. Chem. Soc.*, 1975, **97,** 4409, and references cited therein.
22. (a) H. Alper and S. Wollowitz, *J. Amer. Chem. Soc.*, 1975, **97,** 3541; H. Alper and J. E. Prickett, *J.C.S. Chem. Comm.*, 1976, (b) p. 983; (c) p. 191; (d) p. 483.
23. D. J. Anderson and A. Hassner, (a) *Synthesis*, 1975, 483; (b) *J. Org. Chem.*, 1974, **39,** 3070.
24. A. Padwa, (a) *Accounts Chem. Res.*, 1976, **9,** 371; (b) *Angew. Chem. Internat. Edn.*, 1976, **15,** 123.
25. P. Gilgen, H. Heimgartner, H. Schmid, and H.-J. Hansen, *Heterocycles*, 1977, **6,** 143.
26. K. N. Houk, *Accounts Chem. Res.*, 1975, **8,** 361.
27. (a) A. Padwa and N. Kamigata, *J. Amer. Chem. Soc.*, 1977, **99,** 1871; (b) A. Padwa and P. H. J. Carlsen, *ibid.*, p. 1514.
28. R. A. Firestone, *J. Org. Chem.*, 1972, **37,** 2181, and references cited therein.
29. R. Huisgen, *J. Org. Chem.*, 1976, **41,** 403.

30. A. Padwa, J. Smolanoff, and A. Tremper, *J. Org. Chem.*, 1976, **41,** 543.
31. L. A. Wendling and R. G. Bergman, *J. Org. Chem.*, 1976, **41,** 831.
32. (a) R. Huisgen, D. Vossius, and M. Appl, *Chem. Ber.*, 1958, **91,** 1; (b) R. Huisgen and M. Appl, *ibid.*, p. 12; (c) *ibid.*, 1959, **92,** 2961.
33. (a) C. Thetaz and C. Wentrup, *J. Amer. Chem. Soc.*, 1976, **98,** 1258; (b) C. Wentrup, *Topics Current Chem.*, 1976, **62,** 173.
34. E. Schmitz, *Adv. Heterocyclic Chem.*, 1963, **2,** 122.
35. W. H. Graham, *J. Amer. Chem. Soc.*, (a) 1962, **84,** 1063; (b) *ibid.*, 1966, **88,** 4677; (c) *ibid.*, 1965, **87,** 4396.
36. C. G. Krespan and W. J. Middleton, *Fluorine Chem. Rev.*, 1971, **5,** 57.
37. Q. N. Porter and J. Baldas, 'Mass Spectrometry of Heterocyclic Compounds', Wiley, New York, 1971.
38. H. M. Frey, *Adv. Photochem.*, 1966, **4,** 225.
39. B. M. Jennings and M. T. H. Liu, *J. Amer. Chem. Soc.*, 1976, **98,** 6416.
40. J. A. Moore, in 'Chemistry of Heterocyclic Compounds', ed. A. Weissberger, Interscience, New York, 1964, vol. 19, part II, p. 916.
41. G. Pifferi, P. Consonni, G. Pelizza, and E. Testa, *J. Heterocyclic Chem.*, 1967, **4,** 619.
42. D. Bormann (a) *Annalen*, 1969, **725,** 124; (b) *Chem. Ber.*, 1970, **103,** 1797.
43. T. Durst and M. J. O'Sullivan, *J. Org. Chem.*, 1970, **35,** 2043.
44. A. B. Levy and A. Hassner, *J. Amer. Chem. Soc.*, 1971, **93,** 2051.
45. G. Szeimies, U. Siefken, and R. Rinck, *Angew. Chem. Internat. Edn.*, 1973, **12,** 161.
46. N.-C. C. Yang, B. Kim, W. Chiang, and T. Hamada, *J.C.S. Chem. Comm.*, 1976, 729.
47. T. H. Koch, R. H. Higgins, and H. F. Schuster, *Tetrahedron Letters*, 1977, 431.
48. K. Burger, W. Thenn, and E. Muller, *Angew. Chem. Internat. Edn.*, 1973, **12,** 155, and references cited therein.
49. F. Effenberger and R. Maier, *Angew. Chem. Internat. Edn.*, 1966, **5,** 416.
50. K. R. Henery-Logan and J. V. Rodricks, *J. Amer. Chem. Soc.*, 1963, **85,** 3524.
51. G. Kretschmer and R. N. Warrener, *Tetrahedron Letters*, 1975, 1335.
52. R. G. Kostyanovsky, V. I. Markov, I. M. Gella, K. Khafizov, and V. G. Plekhanov, *Org. Mass Spectrometry*, 1972, **6,** 661.
53. D. H. Aue and D. Thomas, *J. Org. Chem.*, 1975, **40,** (a) p. 2360; (b) p. 2552; (c) p. 2356.
54. L. A. Paquette, T. Kakihana, and J. F. Kelly, *J. Org. Chem.*, 1971, **36,** 435, and references cited therein.
55. G. Seybold, U. Jersak, and R. Gompper, *Angew. Chem. Internat. Edn.*, 1973, **12,** 847, and references cited therein.
56. B. A. Hess, Jr., L. J. Schaad, and C. W. Holyoke, Jr., *Tetrahedron*, (a) 1975, **31,** 295; (b) *ibid.*, 1972, **28,** 3657; (c) B. A. Hess, Jr. and L. J. Schaad, *J. Org. Chem.*, 1976, **41,** 3058.
57. M. J. S. Dewar and N. Trinajstic, (a) *Theoret. Chim. Acta*, 1970, **17,** 235; (b) *Tetrahedron*, 1970, **26,** 4269.
58. H.-U. Wagner, *Angew. Chem. Internat. Edn.*, 1973, **12,** 848.
59. R. J. Kobylecki and A. McKillop, *Adv. Heterocyclic Chem.*, 1976, **19,** 215, and references cited therein.
60. B. M. Adger, C. W. Rees, and R. C. Storr. *J.C.S. Perkin I*, 1975, 45, and references cited therein.
61. C. W. Rees, R. C. Storr, and P. J. Whittle, (a) *J.C.S. Chem. Comm.*, 1976, 411; (b) *Tetrahedron Letters*, 1976, 4647.
62. C. W. Rees, R. Somanathan, R. C. Storr, and A. D. Woolhouse, *J.C.S. Chem. Comm.*, 1975, 740.
63. R. A. Olofson, R. K. Van der Meer, and S. Stournas, *J. Amer. Chem. Soc.*, 1971, **93,** 1543.
64. M. Ikeda, S. Matsugashita, F. Tabusa, H. Ishibashi, and Y. Tamura, *J.C.S. Chem. Comm.*, 1975, 575.
65. E. E. Nunn and R. N. Warrener, *J.C.S. Chem. Comm.*, 1972, 818.
66. H. J. Emeleus and G. L. Hurst, *J. Chem. Soc.*, 1962, 3276.
67. N. Rieber, J. Alberts, J. A. Lipsky, and D. M. Lemal, *J. Amer. Chem. Soc.*, 1969, **91,** 5668.
68. E. F. Ullmann and P. Singh, *J. Amer. Chem. Soc.*, 1972, **94,** 5077.
69. D. E. Applequist, M. A. Lintner, and R. Searle, *J. Org. Chem.*, 1968, **33,** 254.
70. G. Schaden, *Chem. Ber.*, 1973, **106,** 2084, and references cited therein.
71. L. A. Paquette, in 'Non-Benzenoid Aromatics', ed. J. P. Snyder, Academic Press, New York, 1969, vol. 1, p. 287.
72. E. Carstensen-Oeser, *Chem. Ber.*, 1972, **105,** 982, and references cited therein.
73. H. J. Lindner and B. von Gross, *Chem. Ber.*, 1972, **105,** 434; *ibid.*, 1973, **106,** 1033.
74. W. Lwowski, *Angew. Chem. Internat. Edn.*, 1967, **6,** 897.
75. R. A. Abramovitch, T. D. Bailey, T. Takaya, and V. Uma, *J. Org. Chem.*, 1974, **39,** 340.
76. J. M. Photis, *J. Heterocyclic Chem.*, (a) 1970, **7,** 1249; (b) 1971, **8,** 729; (c) *ibid.*, p. 167.
77. L. A. Paquette, D. E. Kuhla, J. H. Barrett, and R. J. Haluska, *J. Org. Chem.*, 1969, **34,** 2866.
78. H. Prinzbach, G. Kaupp, R. Fuchs, M. Joyeux, R. Kitzing, and J. Markert, *Chem. Ber.*, 1973, **106,** 3824.
79. W. E. von Doering and R. A. Odum, *Tetrahedron*, 1966, **22,** 81.
80. R. J. Sundberg, S. R. Suter, and M. Brenner, *J. Amer. Chem. Soc.*, 1972, **94,** 513.
81. B. A. DeGraff, D. W. Gillespie, and R. J. Sundberg, *J. Amer. Chem. Soc.*, 1974, **96,** 7491.
82. A. C. Mair and M. F. G. Stevens, *J. Chem. Soc.* (*C*), 1971, 2317.
83. (a) R. Purvis, R. K. Smalley, W. A. Strachan, and H. Suschitzky, *J.C.S. Perkin I*, 1978, 191; (b) *idem.*, unpublished results.
84. M. Ogata, H. Matsumoto, and H. Kano, *Tetrahedron*, 1969, **25,** (a) p. 5205; (b) p. 5217.
85. W. Heinzelmann and M. Marky, *Helv. Chim. Acta*, 1973, **56,** 1852.

86. R. J. Sundberg, B. P. Das, and R. H. Smith, Jr., *J. Amer. Chem. Soc.*, 1969, **91,** 658.
87. F. R. Atherton and R. W. Lambert, *J.C.S. Perkin I*, 1973, 1079.
88. J. I. G. Cadogan and R. K. Mackie, *Chem. Soc. Rev.*, 1974, **3,** 87.
89. J. H. Boyer, in 'Nitrenes', ed. W. Lwowski, Interscience, New York, 1970, p. 163.
90. T. de Boer, J. I. G. Cadogan, H. M. McWilliam, and A. G. Rowley, *J.C.S. Perkin II*, 1975, 554.
91. F. P. Tsui, Y. H. Chang, T. M. Vogel, and G. Zon. *J. Org. Chem.*, 1976, **41,** 3381.
92. T. J. van Bergen and R. M. Kellogg, *J. Org. Chem.*, 1971, **36,** 978.
93. H.-D. Becker and K. Gustafsson, *Tetrahedron Letters*, 1976, 1705.
94. J. Elguero, C. Marzin, A. R. Katritzky, and P. Linda, *Adv. Heterocyclic Chem.*, 1976, Suppl. 1, 554.
95. G. B. Gill, D. J. Harper, and A. W. Johnson, *J. Chem. Soc.* (*C*), 1968, 1675.
96. E. Moriconi and I. A. Maniscalco, *J. Org. Chem.*, 1972, **37,** 208, and references cited therein.
97. L. A. Paquette, *Angew. Chem. Internat. Edn.*, 1971, **10,** 11.
98. H. Prinzbach, D. Stusche, and R. Kitzing, *Angew. Chem. Internat. Edn.*, 1970, **9,** 377.
99. H. Gunther, J. B. Pawliczek, B. D. Tunggal, H. Prinzbach, and R. H. Levin, *Chem. Ber.*, 1973, **106,** 984.
100. W.-D. Stohrer, *Chem. Ber.*, 1973, **106,** 970.
101. A. Steigel, J. Saueur, D. A. Kleier, and G. Binsch, *J. Amer. Chem. Soc.*, 1972, **94,** 2770.
102. (a) J. A. Moore and E. Mitchell, in 'Heterocyclic Compounds', ed. R. C. Elderfield, Wiley, New York, 1967, vol. 9, p. 224; (b) J. Diamond, *ibid.*, p. 355.
103. (a) E. Lerner, R. A. Odum, and B. Schmall, *J.C.S. Chem. Comm.*, 1973, 327; (b) R. A. Odum and B. Schmall, *ibid.*, 1969, 1299.
104. L. A. Paquette, J. H. Barrett, and D. E. Kulha, *J. Amer. Chem. Soc.*, 1969, **91,** (a) p. 2879; (b) p. 3616.
105. D. J. Anderson, A. Hassner, and D. Y. Tong, *J. Org. Chem.*, 1974, **39,** 3076.
106. (a) L. A. Paquette, D. E. Kuhla, J. H. Barrett, and L. M. Leichter, *J. Org. Chem.*, 1969, **34,** 2888; (b) L. A. Paquette and D. E. Kuhla, *ibid.*, p. 2885.
107. G. Seitz and T. Kampchen, *Chem.-Ztg.*, 1975, **99,** 503.
108. W. S. Murphy and J. P. McCarthy, *J.C.S. Chem. Comm.*, 1970, 1120, and references cited therein.
109. T. Sasaki, K. Kanematsu, and K. Iizaka, *J. Org. Chem.*, 1976, **41,** 1105.
110. V. Snieckus and D. J. Harris, in 'Photochemistry of Heterocyclic Compounds', ed. O. Buchardt, Wiley, New York, 1976, p. 288.
111. G. Jones II and L. J. Turbini, *J. Org. Chem.*, 1976, **41,** 2362.
112. T. H. Koch and D. A. Brown, *J. Org. Chem.*, 1971, **36,** 1934.
113. M. Mahendran and A. W. Johnson, *J. Chem. Soc.* (*C*), 1971, 1237, and references cited therein.
114. 'Medicinal Chemistry', ed. A. Burger, Wiley-Interscience, New York, 3rd edn., 1970, part II.
115. S. Kasparek, *Adv. Heterocyclic Chem.*, 1974, **17,** 45.
116. M. Lennon, A. McLean, I. McWatt, and G. R. Proctor, *J.C.S. Perkin I*, 1974, 1828 and references cited therein.
117. B. Nay, E. E. V. Scriven, and H. Suschitzky, University of Salford, unpublished results.
118. M. Natsume and M. Wada, *Chem. Pharm. Bull.* (*Japan*), 1972, **20,** 1837.
119. W. C. Peaston and G. R. Proctor, *J. Chem. Soc.* (*C*), 1968, 2481.
120. F. Hunziker, F. Kunzle, and J. Schmutz, *Helv. Chim. Acta*, 1966, **49,** 1433.
121. L. H. Werner, S. Ricca, E. Mohacsi, A. Rossi, and V. P. Ayra, *J. Medicin. Chem.*, 1965, **8,** 74.
122. L. J. Kricka and A. Ledwith, *Chem. Rev.*, 1974, **74,** 101.
123. R. Kreher and H. Pawelczyk, *Z. Naturforsch.*, 1974, **29B,** 425, and references cited therein.
124. J. L. Huppatz, *Austral. J. Chem.*, 1973, **26,** 1307.
125. P. W. Jeffs, J. F. Hansen, and G. A. Brine, *J. Org. Chem.*, 1975, **40,** 2883, and references cited therein.
126. R. Kreher and W. Gerhardt, *Angew. Chem. Internat. Edn.*, 1975, **14,** 265.
127. A. Cromarty, G. R. Proctor, and M. Shabbir, *J.C.S. Perkin I*, 1972, 2012, and references cited therein.
128. M. Nastasi, *Heterocycles*, 1976, **4,** 1509.
129. J. Streith and J. L. Schuppiser, *Tetrahedron Letters*, 1976, 4859.
130. H. E. Zimmerman and W. Eberbach, *J. Amer. Chem. Soc.*, 1973, **95,** 3970.
131. C. D. Anderson, J. T. Sharp, H. R. Sood, and R. S. Strathdee, *J.C.S. Chem. Comm.*, 1975, 613, and references cited therein.
132. L. Garanti, A. Scandroglio, and G. Zecchi, *Tetrahedron Letters*, 1975, 3349.
133. J. Kurita and T. Tsuchiya, *J.C.S. Chem. Comm.*, 1974, 936.
134. J. T. Sharp, R. H. Findlay, and P. B. Thorogood, *J.C.S. Perkin I*, 1975, 102, and references cited therein.
135. F. D. Popp and A. C. Noble, *Adv. Heterocyclic Chem.*, 1967, **8,** 21.
136. F. Troxler, H. P. Weber, A. Jaunin, and H. R. Loosli, *Helv. Chim. Acta*, 1974, **57,** 750.
137. G. A. Archer and L. H. Sternbach, *Chem. Rev.*, 1968, **68,** 747.
138. F. D. Popp, *J. Heterocyclic Chem.*, 1974, **11,** 79.
139. Y. Ohtsuka, *J. Org. Chem.*, 1976, **41,** 629, and references cited therein.
140. A. Padwa, L. Gehrlein, and R. B. Kinnel, *J. Org. Chem.*, 1975, **40,** 1683.
141. D. Lloyd, H. P. Cleghorn, and D. R. Marshall, *Adv. Heterocyclic Chem.*, 1974, **17,** 1.
142. D. Lloyd and H. P. Cleghorn, *Adv. Heterocyclic Chem.*, 1974, **17,** 27.
143. L. H. Sternbach, *Angew. Chem. Internat. Edn.*, 1971, **10,** 34.
144. L. O. Randall, W. Schallek, L. H. Sternbach, and R. Y. Ning, in 'Psychopharmacological Agents', ed. M. Gordon, Academic Press, New York, 1974, vol. III, p. 175.
145. R. I. Fryer, *J. Heterocyclic Chem.*, 1972, **9,** 747.

146. 'The Benzodiazepines', ed. S. Garattini, E. Mussini, and L. O. Randall, Raven Press, New York, 1973.
147. U. Golik, *Tetrahedron Letters*, 1975, 1327.
148. M. Takahashi, N. Suzuki, and Y. Igari, *Bull. Chem. Soc.* (*Japan*), 1975, **48,** 2605, and references cited therein.
149. I. Saito, A. Yazaki, and T. Matsuura, *Tetrahedron Letters*, 1976, (a) p. 2459; (b) p. 4753.
150. R. W. Leiby and N. D. Heindel, *J. Org. Chem.*, 1976, **41,** 2736, and references cited therein.
151. S. F. Gait, M. E. Peek, C. W. Rees, and R. C. Storr, *J.C.S. Perkin I*, 1975, 19.
152. G. Dolescall, G. Hornyak, B. Agai, G. Simig, J. Fetter, and K. Lempert, *Tetrahedron*, 1976, **32,** 57.

PART 18

OXYGEN SYSTEMS

18.1

Pyrylium Salts

J. STAUNTON
University of Cambridge

18.1.1 INTRODUCTION

Pyrylium salts form one of those classes of compounds whose surprising properties make organic chemistry such a fascinating subject to study: they are organic cations which are stable in aqueous solution but which lack the nitrogen atom which is normally associated with such a property in an organic molecule. Among the first clearly characterized simple representatives of the class were the methoxypyrylium compounds prepared by methylation of the pyrone (**1**) using a powerful methylating agent such as dimethyl sulphate.[1,2] For some time the structure of the product was in dispute, but eventually the form (**2**) was abandoned in favour of the currently accepted structure (**3**).

Since that time our knowledge of pyrylium chemistry has grown enormously and several distinct types of pyrylium compound can be recognized. Those which will be discussed in this account are firstly the simple pyrylium compounds represented by (**4**); secondly, the two important benzannulated derivatives (**5**) and (**6**); thirdly, the important naturally occurring anthocyanins, which are derivatives of the flavylium salts (**7**); and finally the doubly benzannulated derivative (**8**). The currently accepted names of these compounds are shown with the structures; in the case of (**5**) and (**6**) the alternative trivial names chromylium and isochromylium are still widely used.

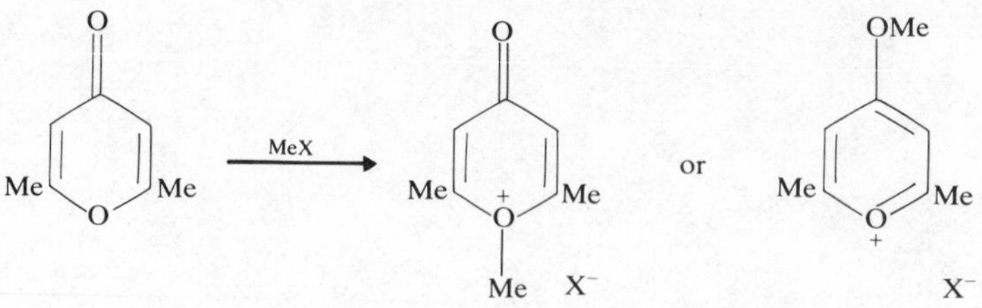

(1) MeX → (2) X⁻ or (3) X⁻

In addition to the parent pyrylium compounds, two important types of derivative deserve mention here. Firstly, the dihydro-derivatives (**9**) and (**10**); compounds of this type are frequently encountered as intermediates in the reactions of pyrylium compounds. Secondly, the corresponding carbonyl derivatives (**11**) and (**12**); these pyrones (or pyranones as they are sometimes called) have a rich chemistry of their own which will be covered in full later.

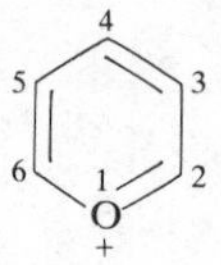

(**4**) Pyrylium 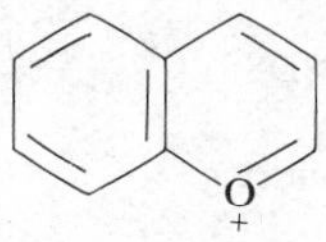 (**5**) 1-Benzopyrylium (chromylium) 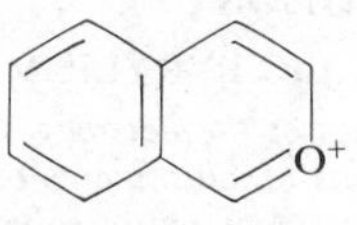 (**6**) 2-Benzopyrylium (isochromylium)

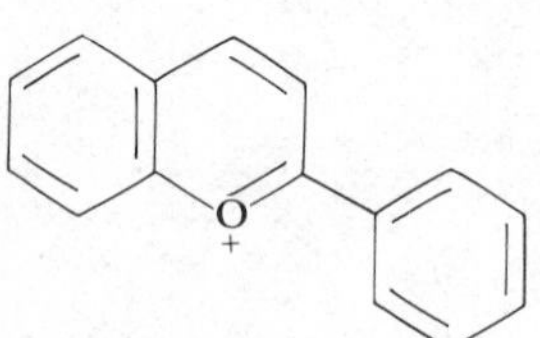

(**7**) Flavylium 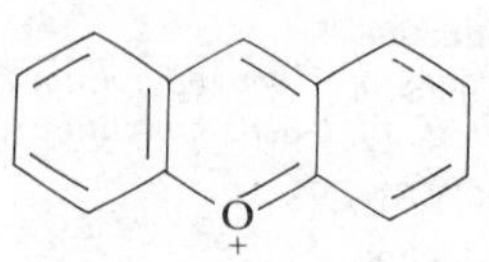 (**8**) Xanthylium

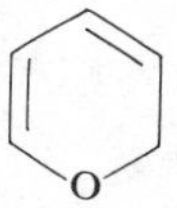

(**9**) 2*H*-Pyran 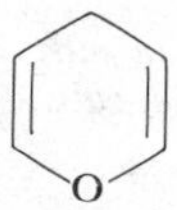 (**10**) 4*H*-Pyran

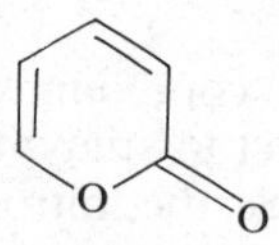

(**11**) α-Pyrone 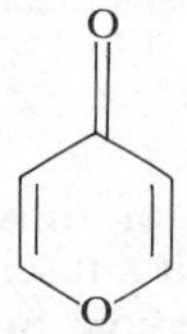 (**12**) γ-Pyrone

18.1.2 MONOCYCLIC PYRYLIUM COMPOUNDS

18.1.2.1 Properties of the heterocyclic ring

The most striking feature of pyrylium chemistry lies in the properties of the heterocyclic ring, which is remarkably stable despite the fact that it is unsaturated and bears a formal positive charge. An interesting comparison can be made with the oxonium ion (**13**), which is produced by protonation of tetrahydropyran in the generation of tetrahydropyranyl ethers. This has only a transient existence as a reactive intermediate and is rapidly attacked by the nucleophilic solvent (reaction 1). In contrast, the pyrylium ion is stable

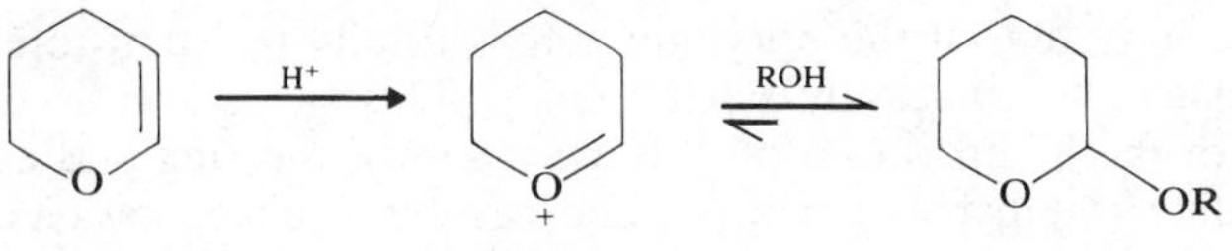

(1)

(13)

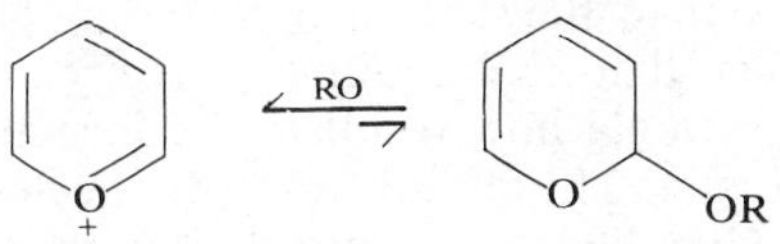

(2)

(14) (15)

compared with its adduct under the same conditions (reaction 2), and remains so as long as the pH is kept low.[3]

This extra stability can be attributed to the fact that the π-system of the pyrylium ring has six electrons and so has aromatic character equivalent to that of benzene or pyridine. Apparently the degree of stabilization is sufficient to tip the balance in the equilibrium between the cation (**14**) and the solvated derivative (**15**) towards the former. Another indication of the stability of the pyrylium cation can be found in the ready oxidation of (**16**) to (**17**) by transfer of a hydride ion to the triphenylmethane cation (reaction 3).[4,5] It

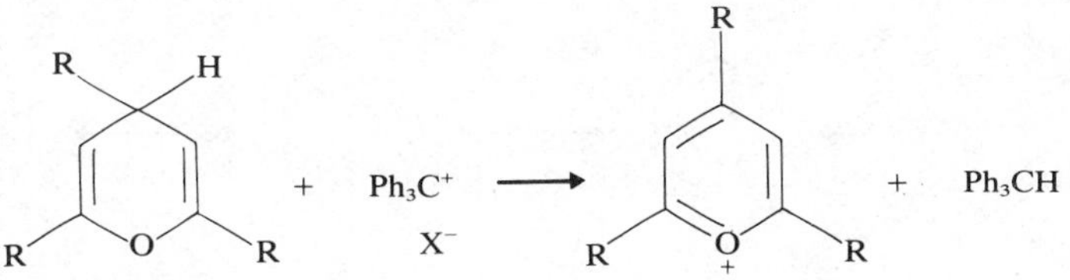

$$(16) + Ph_3C^+ X^- \longrightarrow (17)\ X^- + Ph_3CH \quad (3)$$

must be emphasized, however, that this stability is only relative and that the situation parallels that of the tropylium cation, which is also endowed with an unexpected degree of stability because of its aromatic character. In both cases we have a cation which is more stable than would be expected by comparison with structurally related compounds, but both are very susceptible to nucleophilic attack given the correct combination of reagents and conditions.

The aromatic character of the pyrylium ring is also reflected in the n.m.r. spectra of pyrylium compounds, where the ring protons show marked downfield shifts, indicating a strong ring current. For example, in the simple pyrylium compound (**18**) all the protons absorb in the region 8.5–9.6 δ.[6] The u.v. spectra of pyrylium compounds are also consistent with an aromatic system, as can be seen from the close similarity between the spectrum of the 2,4,6-trimethylpyrylium cation (**19**) and that of the *N*-methylpyridinium analogue (**20**).

Though it has aromatic character in the form of extra stability, the pyrylium ring has properties which are in complete contrast to those of benzene: whereas benzene is readily attacked by electrophiles but not by nucleophiles (unless there is a suitable substituent

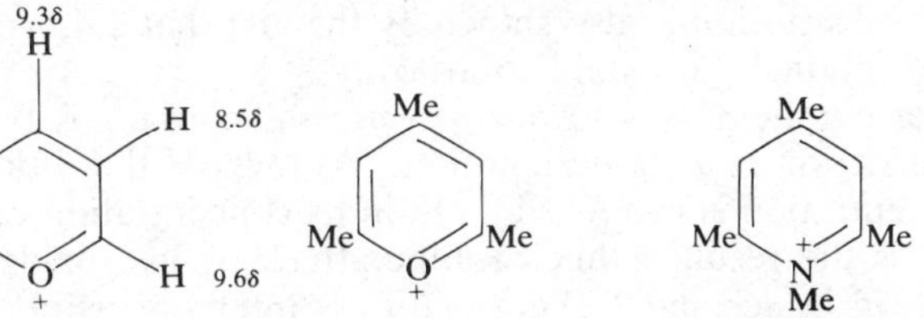

9.3δ H; H 8.5δ; H 9.6δ

(18)

λ_{max} 285, 230 nm

(19)

λ_{max} 268, 221 nm

(20)

such as nitro), the π-system of the pyrylium ring is totally resistant to electrophilic attack and is subject instead to attack by nucleophiles. This behaviour is determined by the perturbing effect of the hetero-atom on the π-system of the ring and again a parallel can be found in pyridine chemistry. Thus pyridine can be attacked by electrophilic reagents but much less readily than benzene because the hetero-atom has a greater resistance to electron deficiency than the carbon it replaces. This effect is amplified in the pyrylium system by the fact that the hetero-atom already bears a formal positive charge—electrophilic attack would now result in the formation of a highly unfavourable, doubly-charged intermediate. The trend towards nucleophilic attack is already present in pyridine, which, unlike benzene, is attacked fairly readily at the 2- and 4-positions, and it continues in the pyrylium ring where nucleophilic attack at C-2, C-4, or C-6 becomes the overriding reaction of the system. Indeed, the ease of such attack is comparable with that on quaternary pyridinium compounds, which are therefore the best models for pyrylium chemistry in the more familiar field of pyridine chemistry.

Nucleophilic attack is confined to C-2, C-4, or C-6. This is what would be expected on the basis of the various canonical forms which can be written for the pyrylium system in which the positive charge is located on carbon rather than oxygen. Attack at any one of these three positions is favourable because it leads to a neutral intermediate [*e.g.* (**21**) or (**22**)]; attack at C-3 would lead to a zwitterion (**23**) which should be less stable (though it would presumably be more stable than the analogous carbanion intermediate in the chemistry of benzene).

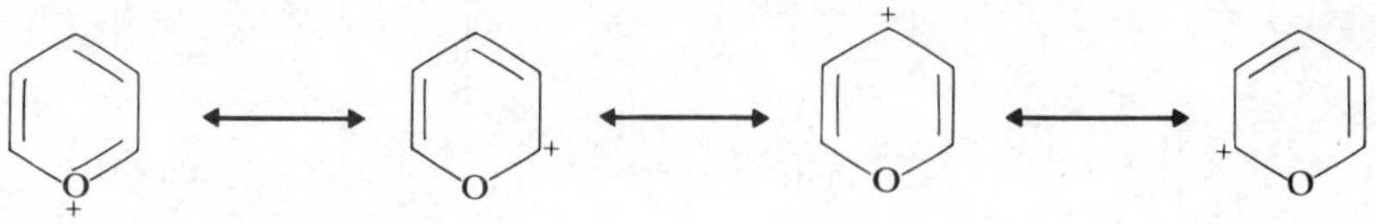

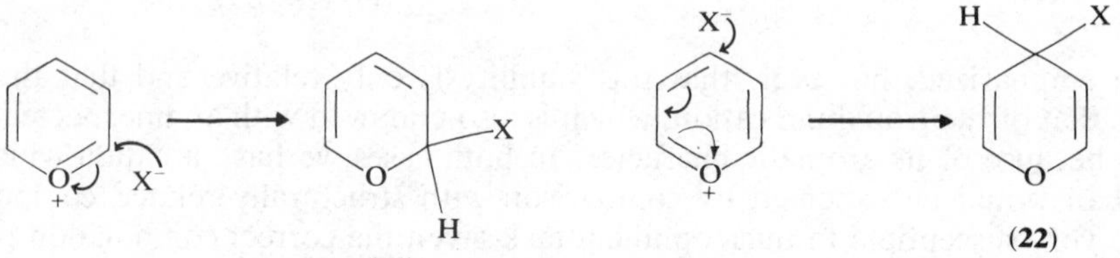

(**21**) (**22**) 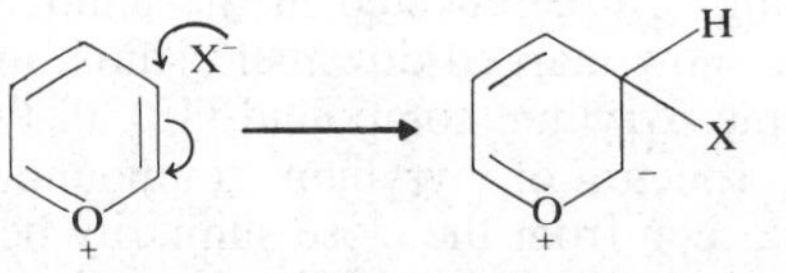 (**23**)

Of the favoured sites for attack, C-2 (or C-6) is normally preferred over C-4, possibly because the intermediate (**21**) is more conjugated than (**22**), though the behaviour of the system in this respect is markedly dependent on the substitution pattern. Thus 2,6-disubstituted pyrylium rings are attacked preferentially at C-4, the least hindered site.[7,8] The importance of substitution is also shown by the fact that 2,4,6-trisubstituted pyrylium compounds are exceptionally resistant to attack.

The behaviour of pyrylium salts towards nucleophilic attack is typified by the reaction of (**24**) with hydroxide ion in aqueous solution.[3] At low pH the cation is stable, but as the hydroxide ion concentration is increased a fall in its concentration can be detected by u.v. spectroscopy. This is the result of nucleophilic attack by hydroxide or water to give the adduct (**25**), as shown in Scheme 1. This in turn is interconvertible with the ring-opened derivative (**26**) and its tautomers. In alkaline solution these are deprotonated to form an enolate ion (**27**) and the equilibrium is pulled further to the right.

SCHEME 1

The initial pyrylium salt can be regenerated from (**26**) by strongly acidifying the solution. However, prolonged treatment with aqueous hydroxide may result in hydrolytic cleavage of the α,β-unsaturated carbonyl residue in (**26**), in which case recyclization to form the pyrylium compound is no longer possible.

The opening of the heterocyclic ring in the $2H$-pyran derivatives formed by addition of nucleophiles to pyrylium salts is a decisive feature in the chemistry of these compounds. The reaction may take place by a Cope rearrangement as shown in Scheme 1: $2H$-pyrans are known to undergo such reactions with exceptional ease.[9] However, it could take place equally well in this case by an ionic mechanism involving initial deprotonation of the adduct. Other transformations in which nucleophilic attack on a pyrylium ring is followed by ring opening are shown in Scheme 2.[10–12]

i, R_2NH; ii, $NaBH_4$; iii, PhLi; iv, NaCN.

SCHEME 2

The ring-opened products are frequently unstable under the conditions of the reaction and special care is needed to avoid further transformations. These may be relatively trivial, such as isomerization of carbon–carbon double bonds,[13] or they may be more profound as in the hydrolytic cleavage of the α,β-unsaturated ketone mentioned earlier.

The problem of over-reaction can be illustrated by the borohydride reduction (reaction 4).[14] As indicated, attack takes place mainly at C-2 to give (**28**). However, this product

(4)

can only be isolated in good yield if the reaction is carried out at low temperature; otherwise further reduction of the ring-opened form takes place and a complex mixture is the result. This difficulty probably explains many of the conflicting reports on the behaviour of pyrylium compounds.

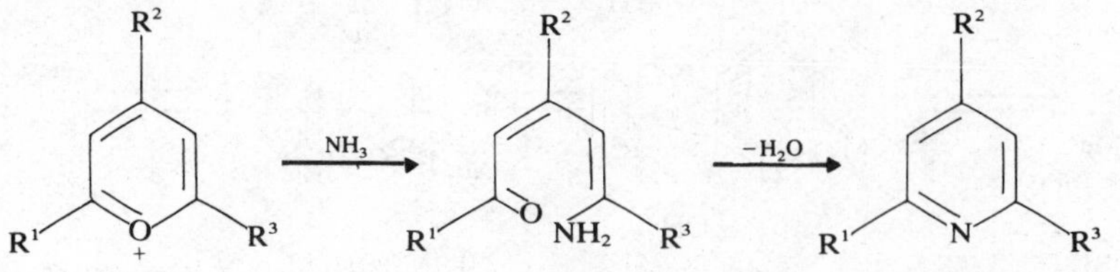

(29) (30) (31)

SCHEME 3

The further reaction of the ring-opened derivative can take a very interesting course with synthetically useful results, when recyclization takes place to form a new ring. The synthetic potential of such reactions will be discussed at length later. Here we will only consider a few key examples to illustrate the nature of the process. The ring-opened intermediate (**30**), for instance, formed after nucleophilic addition of ammonia to the pyrylium ring of (**29**), undergoes a ready recyclization to form the pyridine (**31**) as shown in Scheme 3.[15] Secondary amines generate the corresponding *N*-alkylpyridinium derivative. Similarly, the pyrylium salt can be converted to its thiapyrylium analogue (**32**) or the

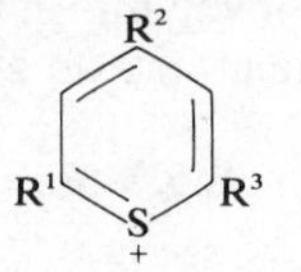

(32) 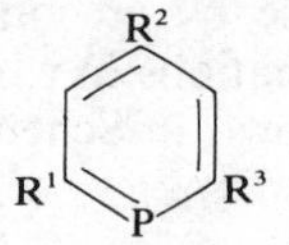 (33)

remarkable phosphorus analogue of pyridine (**33**) by treatment with H_2S or $P(CH_2OH)_3$, respectively.[16,17] Of particular importance in this connection is the conversion of pyrylium salts to benzene derivatives by reaction with a carbon nucleophile, as in the conversion of (**34**) to (**35**) in Scheme 4.[18]

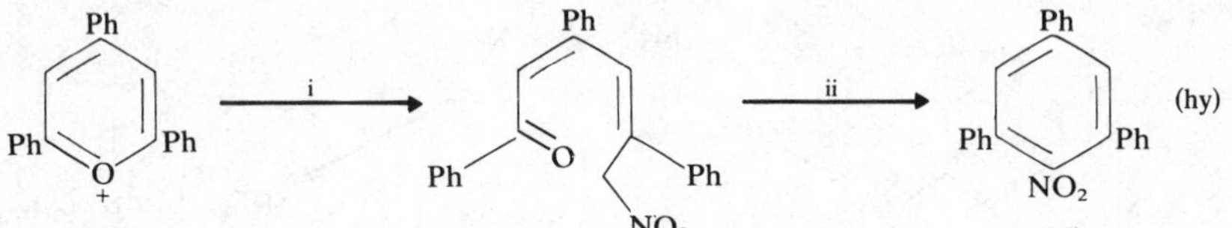

(34) (35) (hy)

i, CH_3NO_2, base; ii, base.

SCHEME 4

When a pyrylium ring carries a potential leaving group at C-2, C-4, or C-6, nucleophilic attack may result in substitution rather than ring opening.[19] For instance, a 4-methoxypyrylium salt can be converted to the ethoxy-analogue by refluxing in ethanol

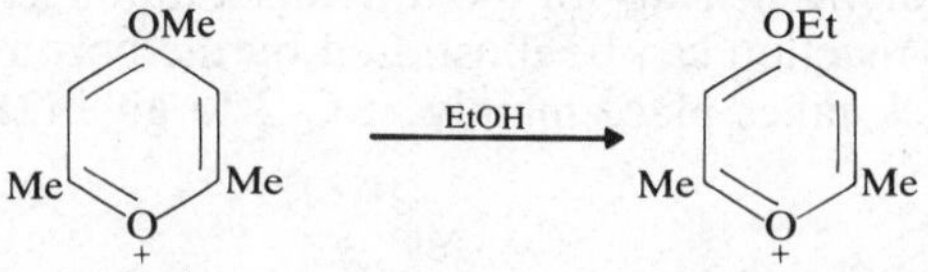

(5)

(reaction 5). The corresponding dialkylamino- and thio-derivatives, (**36**) and (**37**), are formed by attack of a secondary amine or a thiol, respectively. Equivalent substitution reactions take place at C-2.

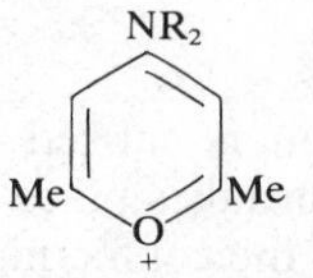

(36) 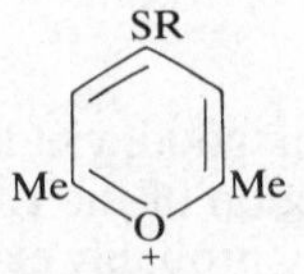 (37)

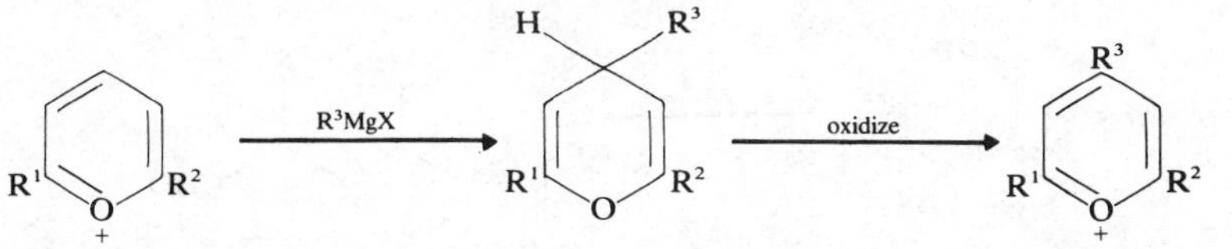

(6)

Nucleophilic attack at C-4, in contrast to C-2, usually does not lead to useful products. However, one important exception to this generalization is shown (reaction 6), where addition of a Grignard reagent at C-4 in a 2,6-disubstituted pyrylium ring is followed by oxidation of the intermediate 4*H*-pyran to a pyrylium compound.[8] A wide range of oxidizing agents can be used for this purpose, including Ph_3C^+, $FeCl_3$, or molecular oxygen. The reaction has acquired importance as a convenient and efficient route to 2,4,6-trialkyl- or -aryl-pyrylium compounds.

Before leaving the subject of nucleophilic addition to the pyrylium ring it is appropriate to comment on the exchange process shown in reaction 7.[20] At first sight this might

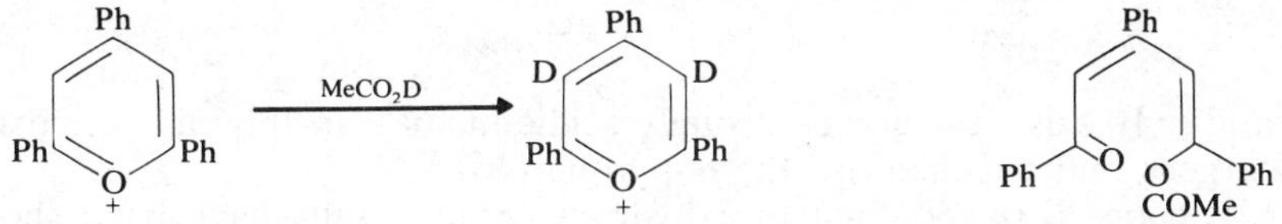

(7)

(38)

appear to involve electrophilic addition (of a deuteron) to the π-system of the heterocyclic ring. However, bearing in mind the inertness of the pyrylium cation to electrophilic attack it is clear that this explanation is unlikely, and a more reasonable interpretation is that the exchange takes place in (**38**) (formed by nucleophilic addition of acetate followed by ring cleavage), which would be in equilibrium with the pyrylium salt, and could readily exchange the relevant hydrogens for deuterium under the conditions of the reaction. A convincing demonstration of the inertness of the pyrylium ring to electrophilic attack is found in nitration (reaction 8), which takes place exclusively on the benzene rings and the pyrylium ring emerges unscathed. One point to note in this reaction is the differing site of attack in the phenyl ring attached to C-4.[21]

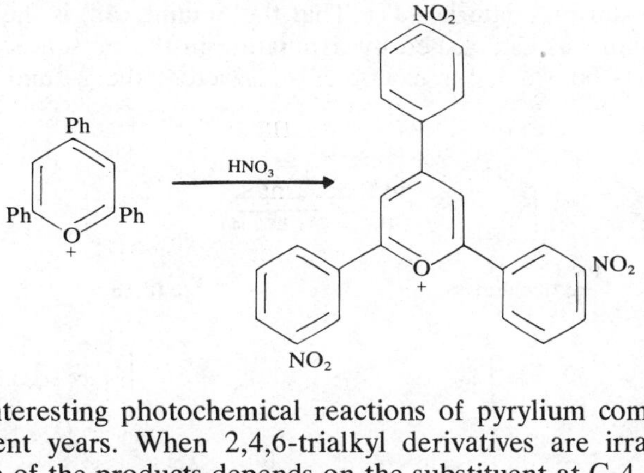

(8)

A number of interesting photochemical reactions of pyrylium compounds have been discovered in recent years. When 2,4,6-trialkyl derivatives are irradiated in aqueous solution the nature of the products depends on the substituent at C-4. Thus the 4-methyl derivative (**39**) gave the ring-opened product (**40**);[22] the ethyl analogue gave the equivalent product, (**42**), plus a mixture of isomeric cyclopentenones corresponding to (**43**);[23] when the substituent at C-4 is t-butyl a complicated mixture of products was obtained, including some in which the 4-alkyl group had been attacked.[24] These results were rationalized in terms of the intermediate formation of the corresponding oxoniabenzvalene (**44**). An equivalent intermediate is formed on irradiation of the 4-hydroxypyrylium salt (**45**) in concentrated sulphuric acid (this salt is the protonated form of a

(39) R = Me
(41) R = Et

(40) R = Me
(42) R = Et

(43)

(44)

γ-pyrone and only exists as such in strongly acidic media). In this case the oxoniabenzvalene undergoes rearrangement to the α-pyrone (46).[25,26]

(45)

(46)

Pyrylium betaines corresponding to 3-hydroxypyrylium salts have been shown to be responsible for the red colour which develops when colourless solutions of 6-oxabicyclo[3,1,0]hex-3-en-2-ones are irradiated at short wavelengths. As indicated in Scheme 5, irradiation at longer wavelengths[27] leads to the disappearance of the colour and regeneration of the starting material (47). That the betaine (48) is the red component of this photo-equilibrium was established by irradiating in the presence of a dipolarophile. Diethyl acetylenedicarboxylate, for example, adds across the 2- and 6-positions of the

(47) colourless

(48) red

(50)

(49)

SCHEME 5

pyrylium ring to give the adduct (**49**). The photo-equilibrium is unstable to prolonged irradiation because of a side reaction in which the epoxyketone is converted slowly and irreversibly to the α-pyrone (**50**). Nevertheless, it has found practical application as the basis of a photosensitive film.

18.1.2.2 Reactions of substituents on the pyrylium ring

A close parellel between the chemistry of pyrylium compounds and their pyridine analogues can also be found in the properties of substituents attached to the two rings. Thus in each case alkyl groups attached to the 2-, 4-, and 6-positions are activated towards deprotonation at the carbon adjacent to the ring. In the case of pyridine compounds a strong base is required and the product is a carbanion. In comparison, the process is greatly facilitated in pyrylium compounds and the product, a neutral methylenepyran, is often isolable when a suitable substituent such as carbonyl or nitrile is present at the site of deprotonation[28] (*e.g.* reactions 9 and 10).

NC CH CO_2Et Me O + Me —base→ NC CO_2Et Me O Me (9)

OMe Me O + CO_2Et —OH^- / H_2O→ OMe Me O $CHCO_2Et$ (10)

More usually, however, the methylenepyran is not isolable, but its existence as a reactive intermediate can be inferred from reactions which occur at the α-carbons. For example, hydrogens at these positions can be exchanged for deuterium in D_2O (reaction 11).[29] Interestingly the rate of exchange at the C-4 methyl is ten times greater than that at

CH_3 CH_3 O + CH_3 —D_2O→ CD_3 CD_3 O + CD_3 (11)

C-2. Methylenepyrans are almost certainly involved as intermediates in condensation reactions (*e.g.* reaction 12).[30] Similar reactions take place with nitroso and phenyldiazonium compounds.[31]

Ph Ph O + Me + PhCHO ⟶ Ph Ph O Ph (12)

Hydroxypyrylium salts show the same potential for tautomerism as the corresponding pyridine derivatives. Thus hydroxyl groups at C-2 or C-4 readily lose a proton to form a carbonyl group (reactions 13 and 14). The products, α- and γ-pyrones respectively, are the dominant species and the protonated forms exist only in strongly acidic media. A hydroxyl group at C-3 is also readily deprotonated, but in this case the product is a betaine rather than a carbonyl compound.

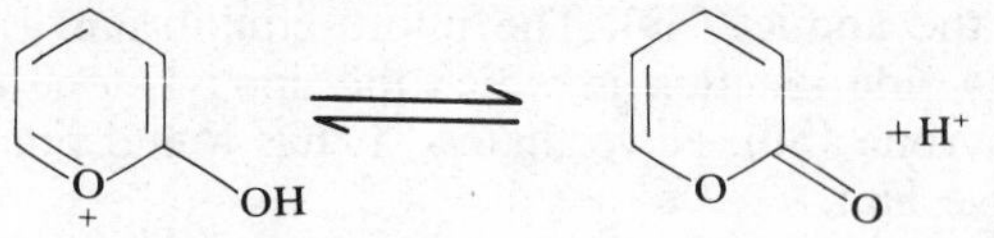

(13)

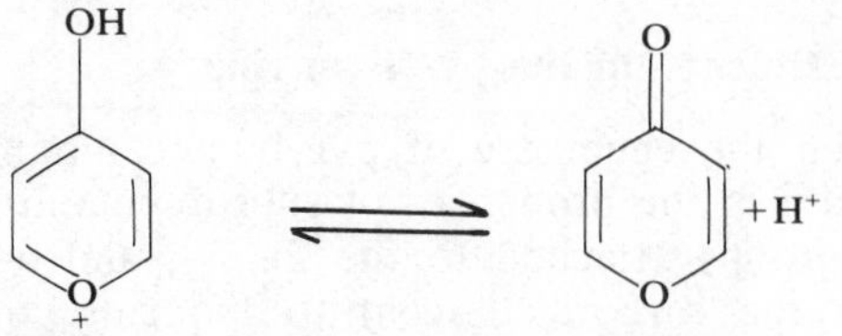

(14)

Methoxy-derivatives of pyrylium compounds are also important. As might be expected, the 2- and 4-isomers are readily hydrolysed to the corresponding pyrones.[32] This takes place in basic as well as acidic media. The base-catalysed reaction may involve nucleophilic addition of hydroxide to the ring followed by elimination of methoxide (reaction 15) or, possibly, direct S_N2 attack of hydroxide on the methyl group. The former mechanism must be followed in those reactions, mentioned earlier, in which the methoxy-group at C-2, C-4, or C-6 is replaced by an amino-group or a thiol.

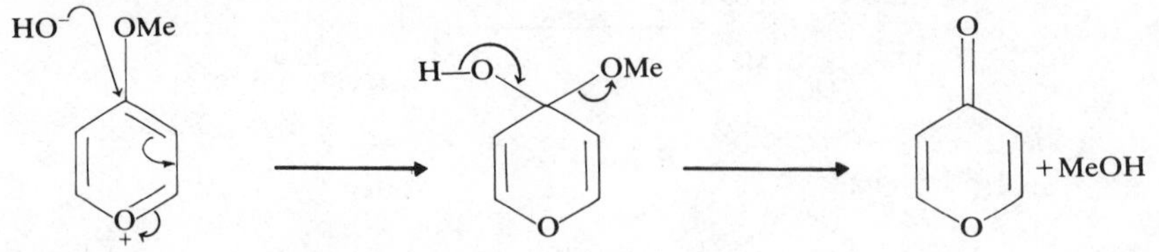

(15)

18.1.2.3 Synthesis of pyrylium salts

Pyrylium salts are prepared by cyclization of suitably functionalized five-carbon chains.[33] A large number of syntheses are based on the intermediacy of a 1,5-diketone. The intervening carbon–carbon bonds may be saturated, as in (**51**), which is the key intermediate in Scheme 6.[34] Here the 1,5-diketone is cyclized to a 4*H*-pyran which

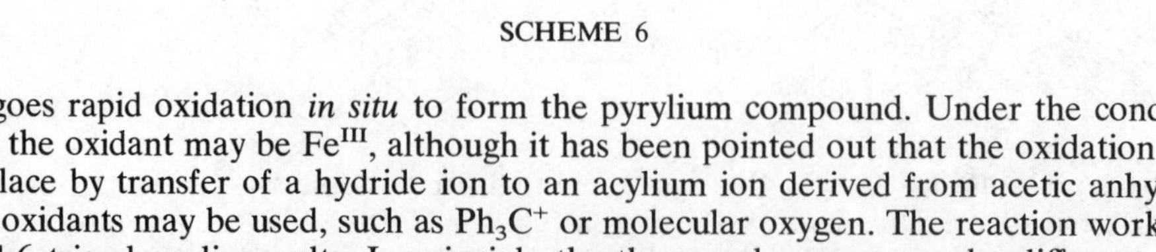

(**51**)

i, base; ii, $(MeCO)_2O$, $FeCl_3$.

SCHEME 6

undergoes rapid oxidation *in situ* to form the pyrylium compound. Under the conditions shown the oxidant may be Fe^{III}, although it has been pointed out that the oxidation could take place by transfer of a hydride ion to an acylium ion derived from acetic anhydride. Other oxidants may be used, such as Ph_3C^+ or molecular oxygen. The reaction works best for 2,4,6-triarylpyrylium salts. In principle the three aryl groups may be different.

When one of the intervening carbon–carbon bonds of the 1,5-diketone is unsaturated, the oxidation step is unnecessary. The cyclization takes place spontaneously in strong acid solution in what is effectively a reversal of the hydrolysis shown in Scheme 1.

The requisite intermediates can be prepared in a number of ways. The earliest route employed acylation of the γ-carbon of an α,β-unsaturated ketone; the resulting 1,5-diketone then cyclizes spontaneously (reaction 16). By this means it is possible to prepare pyrylium salts in which the 2- and 4-substituents are alkyl groups.[35]

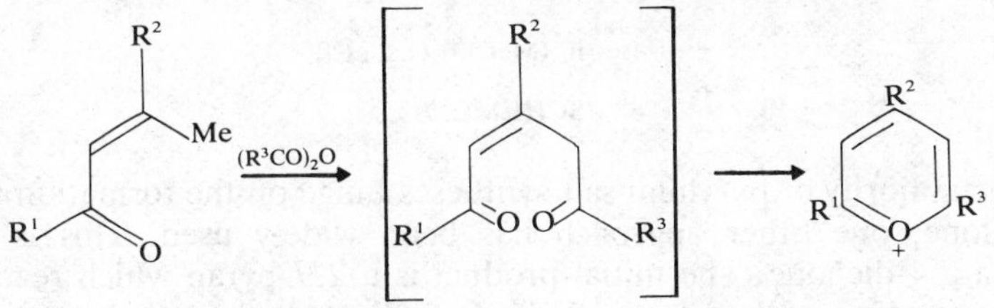

$(R^3CO)_2O$ (16)

A more recent synthesis utilizes alkene acylation (reaction 17). It is interesting that although the acylation of alkenes has been studied regularly since 1892, the formation of pyrylium compounds was overlooked until 1959 because early workers only investigated those products which dissolved in organic solvents and discarded the water-soluble

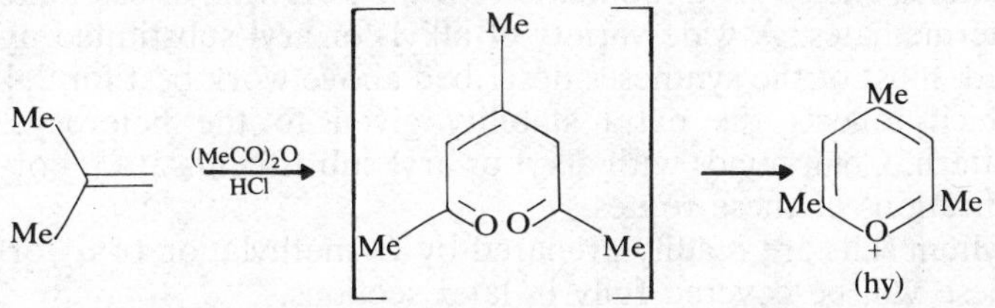

$(MeCO)_2O$, HCl (hy) (17)

pyrylium salts on work-up.[36] This method is complementary to those described above because it works best for alkyl- rather than aryl-substituted pyrylium salts.[37] By varying the structure of the alkene it is possible to prepare pyrylium salts with different substituents at C-4 or with an alkyl substituent at C-3 or C-5, though complex mixtures often result from isomerization of the alkene prior to acylation.

Condensation of a β-dicarbonyl substrate with a ketone can also be used to generate an appropriate 1,5-diketone (reaction 18). In the example shown it was not possible to detect the 4-methyl isomer, which would be formed as a result of carbon–carbon bond formation at the carbonyl adjacent to the methyl group of (**52**).[21]

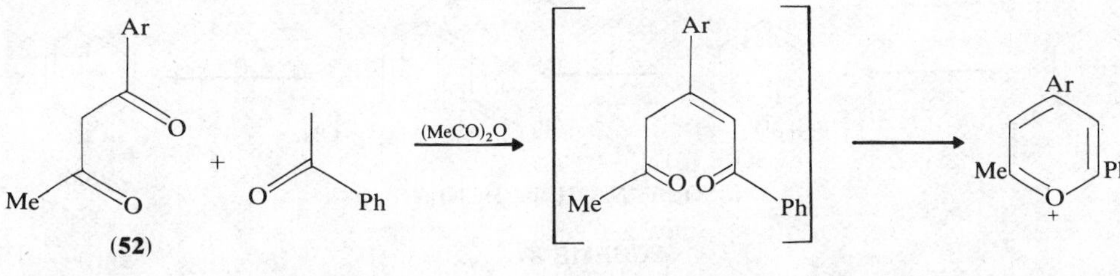

(**52**) $(MeCO)_2O$ (18)

A recently discovered variation of reaction 18 utilizes chlorovinyl ketones (reaction 19).[38] Chlorovinyl ketones are readily prepared by acylation of an alkyne and so this route can be used to prepare pyrylium salts with three different substituents. Other variations include the use of an acetylene or an enamine in place of the methyl ketone.

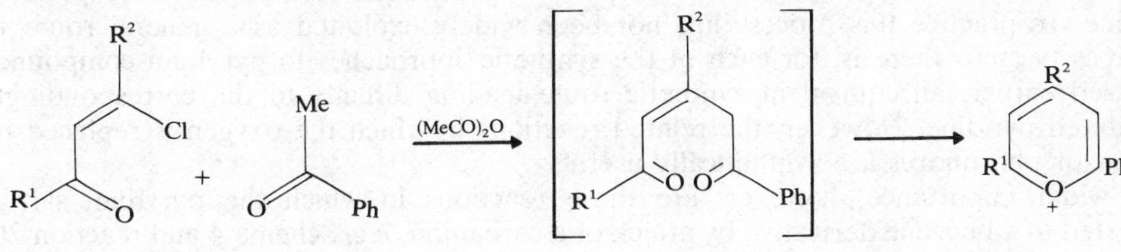

$(MeCO)_2O$ (19)

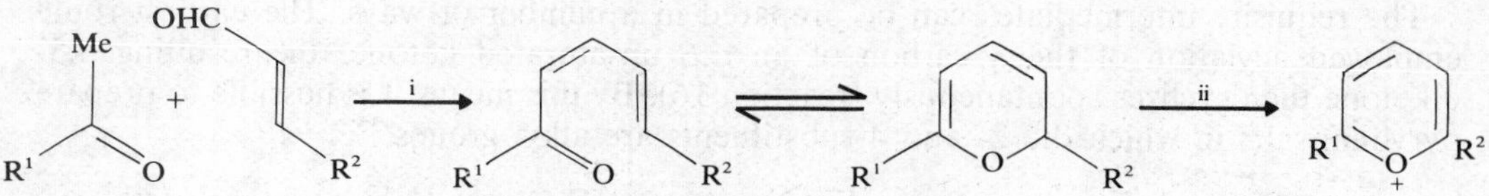

i, Base; ii, $(MeCO)_2O$, $FeCl_3$.

SCHEME 7

Although the majority of pyrylium salt syntheses hinge on the formation and cyclization of a 1,5-diketone, one other approach has been widely used. This is based on the cyclization of a 2,4-dienone. The initial product is a 2*H*-pyran which readily oxidizes *in situ* to form a pyrylium salt (Scheme 7).[39] As indicated, the requisite dienones can be prepared by condensation of a methyl ketone with an unsaturated aldehyde. The method works best when the substituents are aryl groups and it is particularly useful for the preparation of 2,6-disubstituted pyrylium compounds.

At this point it is desirable to comment on the availability of variously substituted pyrylium compounds and so lay a foundation for the following discussion of their potential as synthetic intermediates. A wide variety of alkyl- or aryl-substituted pyrylium salts can be made; indeed, most of the syntheses described above work best for 2,4,6-trisubstituted derivatives, which reflects the extra stability given to the heterocyclic ring by this substitution pattern. Compounds with alkyl or aryl substituents at C-3 or C-5 can also be made by modifications of these routes.

Methoxypyrylium salts are readily prepared by *O*-methylation of α- or γ-pyrones. The syntheses of these will be covered fully in later sections.

Finally it should be borne in mind that the substitution pattern in the primary synthetic products can be modified by some of the reactions described earlier. The nucleophilic displacement of methoxy-groups at C-2 or C-4 deserves special mention, as do the many reactions which involve condensation at the α-carbon of an alkyl group at C-2, C-4, or C-6. The ready exchange of hydrogen for deuterium at these positions has potential for the preparation of isotopically labelled compounds.

It seems appropriate, however, to close this account of the synthesis of pyrylium compounds by mentioning that the elusive parent compound (**53**) was finally synthesized in 1953 by the route shown in Scheme 8.[40] It involves a reversal of one of the synthetic transformations of pyrylium salts to be discussed below!

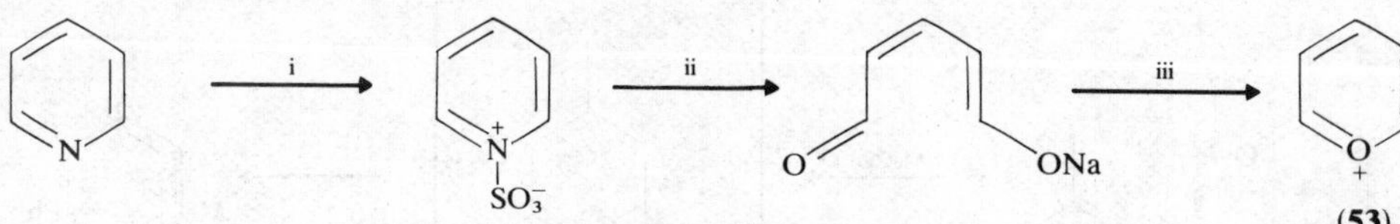

(**53**)

i, SO_3; ii, NaOH; iii, $HClO_4$.

SCHEME 8

18.1.2.4 Synthetically useful transformations of pyrylium salts

Pyrylium compounds are potentially useful synthetic intermediates. Thus Scheme 3 illustrated that the heteroatom can be replaced by nitrogen to give the corresponding pyridine. In practice this process has not been widely exploited as a general route to pyridines because there is, for each of the synthetic approaches to pyrylium compounds discussed earlier, an equivalent synthetic route leading directly to the correspondingly substituted pyridine. However, the related reactions in which the oxygen is replaced by sulphur or phosphorus are synthetically useful.

Of wider importance, however, are those reactions in which the pyrylium salt is converted to a benzene derivative by attack of a carbanion, *e.g.* Scheme 4 and reaction 20.

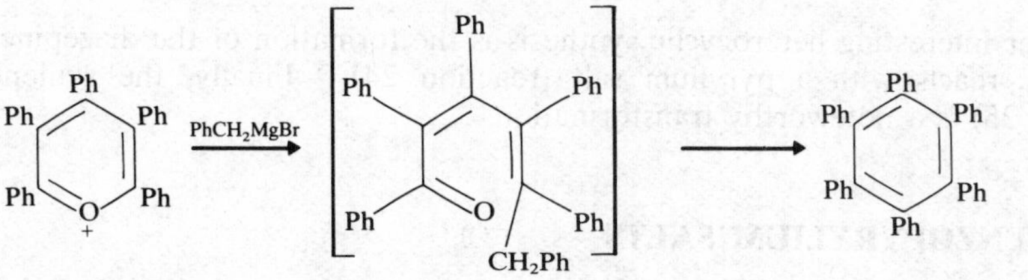

(20)

In both cases the product has a substitution pattern which would be difficult to achieve by conventional methods in which substituents are introduced on to a preformed benzene ring by electrophilic attack. Moreover, problems of isomer formation, which often seriously lower the efficiency of syntheses based on the classical approach, do not arise in a synthesis *via* a pyrylium compound.[41]

The efficiency of this type of reaction is greatly increased by the reaction of a Wittig reagent with a pyrylium compound[19,42] (reaction 21). The Wittig reagent can vary widely in structure, thereby greatly increasing the scope of the reaction.

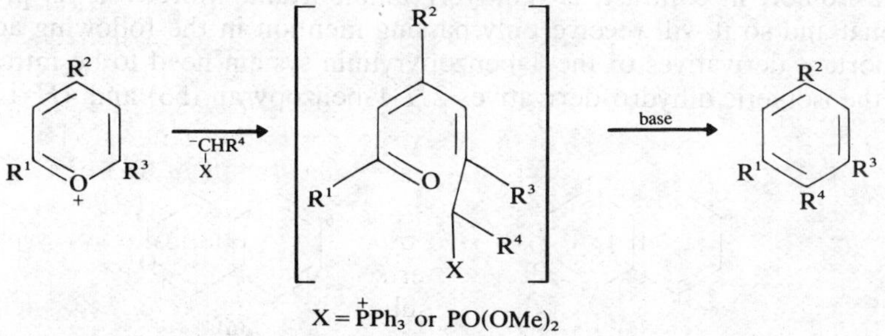

(21)

$X = \overset{+}{P}Ph_3$ or $PO(OMe)_2$

The reaction with a sulphonium ylide follows a different path leading to the formation of a furan, *e.g.* reaction 22.[43] Furans are also produced when pyrylium salts are oxidized with hydrogen peroxide (reaction 23).[44]

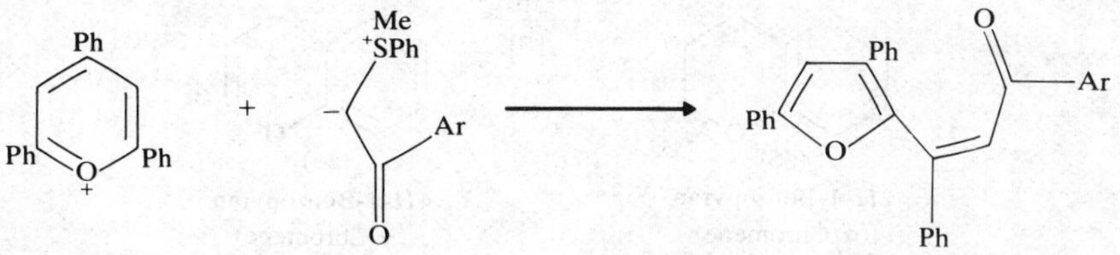

(22)

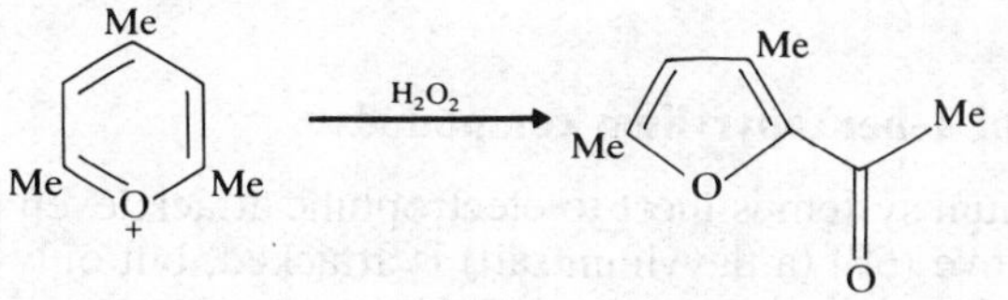

(23)

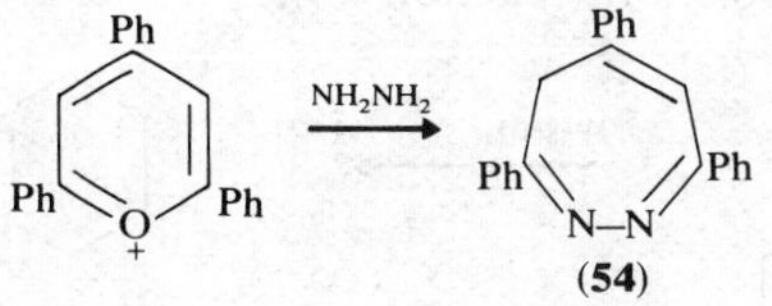

(24)

(54)

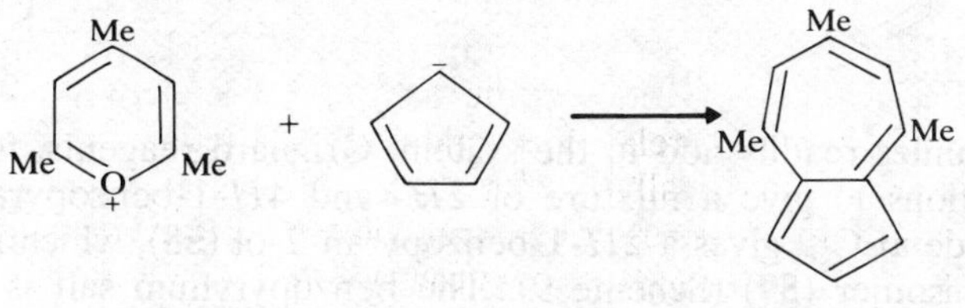

(25)

Another interesting heterocyclic synthesis is the formation of the diazepine (**54**) when hydrazine reacts with a pyrylium salt (reaction 24).[45] Finally, the azulene synthesis (reaction 25) is a noteworthy transformation.[46]

18.1.3 BENZOPYRYLIUM SALTS

18.1.3.1 Introduction

As was mentioned earlier, two isomeric benzopyrylium salts (**5**) and (**6**) are possible, depending on the site of benzannulation. These are commonly designated 1- and 2-benzopyrylium, respectively, although the alternative trivial names chromylium and isochromylium are still in common use. The 1-benzopyrylium system is the more interesting, firstly because it occurs as the basic heterocyclic system of the important plant pigments, the anthocyanins, and secondly because the properties of its heterocyclic ring show a major difference from those of simple pyrylium systems as a result of benzannulation. The 2-isomer, in contrast, is relatively unimportant. Moreover, its properties are unexceptional and so it will receive only passing mention in the following account.

Two important derivatives of the 1-benzopyrylium system need to be introduced here. These are the isomeric dihydro-derivatives 2*H*-1-benzopyran (**55**) and 4*H*-1-benzopyran (**56**).

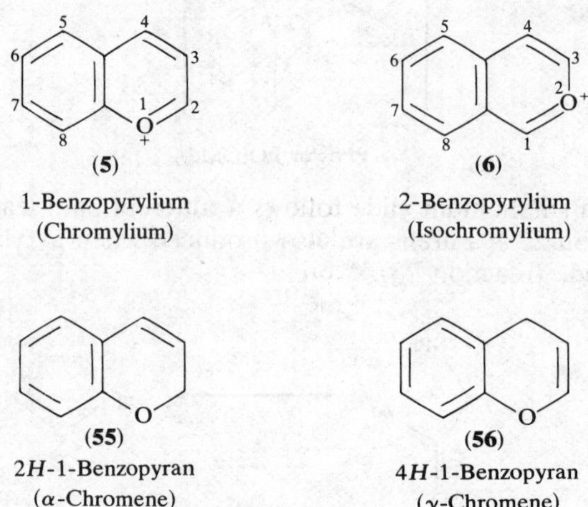

(**5**) 1-Benzopyrylium (Chromylium)

(**6**) 2-Benzopyrylium (Isochromylium)

(**55**) 2*H*-1-Benzopyran (α-Chromene)

(**56**) 4*H*-1-Benzopyran (γ-Chromene)

18.1.3.2 Reactions of 1-benzopyrylium compounds

The 1-benzopyrylium system is inert to electrophilic attack, even in the carbocyclic ring. The 2-phenyl derivative (**57**) (a flavylium salt) is attacked, but only in the substituent ring which is nitrated in the *m*-position (reaction 26).[47]

(**57**) $\xrightarrow{HNO_3}$ (26)

However, nucleophiles readily add to the system. Grignard reagents, for example, add at the 2- or 4-positions to give a mixture of 2*H*- and 4*H*-1-benzopyran derivatives.[48] Addition of hydroxide at C-2 gives a 2*H*-1-benzopyran-2-ol (**58**), which is in equilibrium with a ring-opened isomer (**59**) (Scheme 9). The benzopyrylium salt is regenerated on

(58) (59)

SCHEME 9

addition of strong acid. Prolonged treatment with alkali, however, leads to hydrolytic cleavage of the α,β-unsaturated aldehyde.

So far the reactions of the benzannulated compound have run parallel with those of simple pyrylium compounds. Where the two diverge is in their reaction with ammonia or primary amines. The typical behaviour of the 1-benzopyrylium salt is shown in Scheme 10. The reaction stops at the ring-opened derivative (**61**) and no recyclization to form the

(60)

(61)

SCHEME 10

nitrogen analogue of the pyrylium salt (in this case a quinoline) occurs.[49] The reason for this is two-fold. Firstly, attack of the nucleophile takes place at only one of the two sites adjacent to the oxygen: attack at the second site to give (**60**) is unfavourable because it would lead to temporary disruption of the π-system of the carbocyclic ring. Secondly, recyclization of (**61**) would also temporarily disrupt this π-system and so this characteristic reaction of the simple pyrylium system is not observed. These constraints are absent in the 2-benzopyrylium system, which, with ammonia, leads to the rapid formation of the corresponding isoquinoline, as shown in Scheme 11.[50]

SCHEME 11

Alkyl substituents on the 2- or 4-position are activated in the same way as those of simple pyrylium compounds (*e.g.* reaction 27).[51]

(27)

18.1.3.3 Synthesis of 1-benzopyrylium salts

The two commonly used synthetic routes to 1-benzopyrylium compounds are shown in reactions 28 and 29.[52] Both are closely related to a synthesis of simple pyrylium

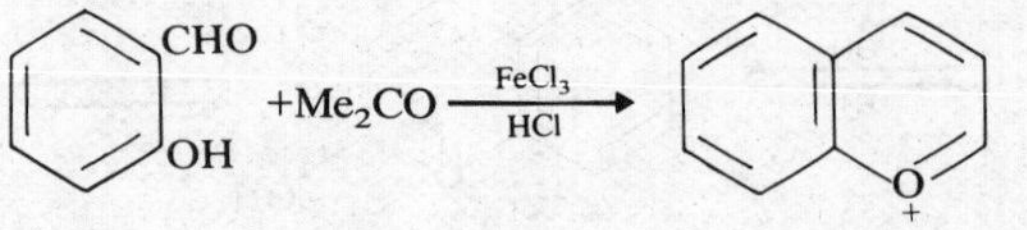

(28)

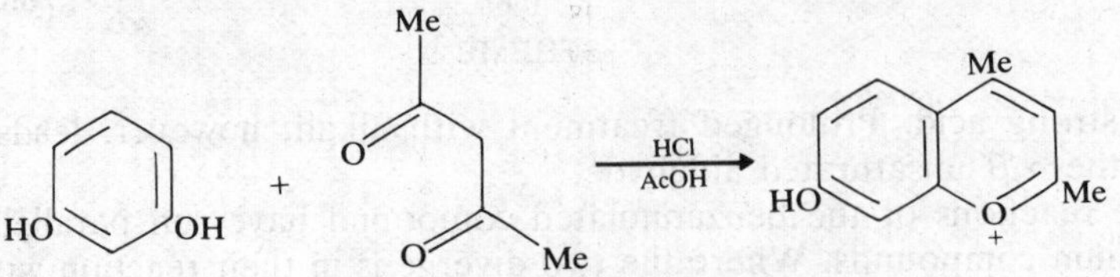

(29)

compounds described earlier (reaction 18). The latter synthesis only works satisfactorily with a highly reactive aryl ring, such as a polyphenol or a naphthol.

18.1.4 ANTHOCYANINS

18.1.4.1 Introduction

The anthocyanins are derivatives of the flavylium salt (**63**) and hence belong to the flavanoid family of natural products, the parent structure of which is the flavan (**62**). They are brightly coloured and are responsible for the brilliant reds and blues found in the flowers and fruits of the plant kingdom. Taking into account the closely related flavones which are yellow, members of this family contribute all three primary colours to Nature's palette. In some flowers, anthocyanins form 25% of the dry weight of the petals.

The oxidation level of the flavylium salt (**63**) is intermediate between that of the flavan (**62**) and the more highly oxidized flavone (**64**). Flavones will be discussed later. Here we will only be concerned with the chemistry of flavylium salts, which have distinctive properties and are therefore best treated as a separate group.

Anthocyanins are invariably highly oxygenated, as can be seen from the structure of cyanin (**65**) (which is the red colouring matter occurring in dahlias and roses), and the composite structure (**66**) which summarizes the various substitution patterns which are

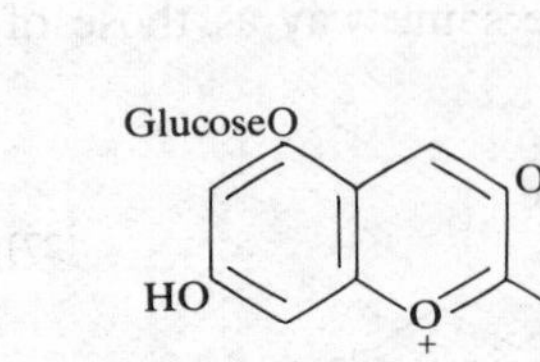

(**62**) A flavan

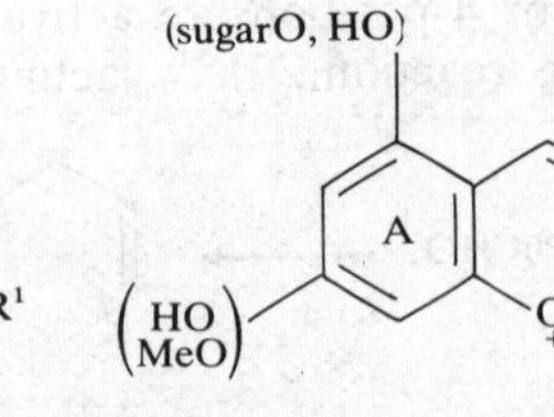

(**63**) A flavylium salt

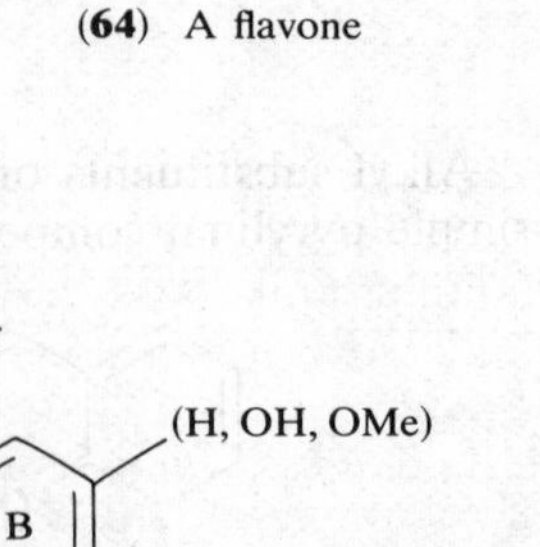

(**64**) A flavone

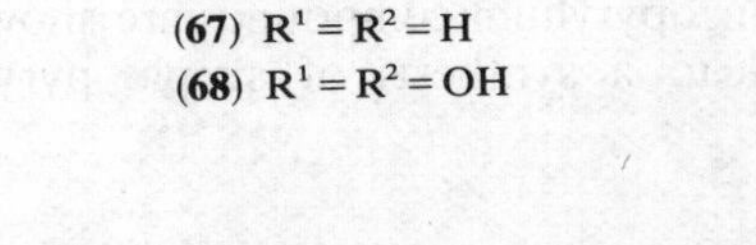

(**65**) $R^1 = H, R^2 = OH$
(**67**) $R^1 = R^2 = H$
(**68**) $R^1 = R^2 = OH$

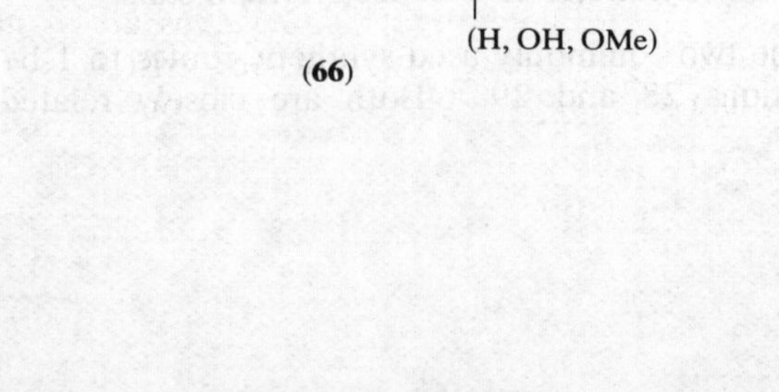

(**66**)

commonly found in Nature.[53] The differing oxygenation patterns in the two rings reflect the mixed biosynthesis of flavanoids: ring A is derived from acetate *via* the polyketide pathway and so has oxygen substituents on alternate carbons; the occurrence of adjacent oxygen substituents in ring B is characteristic of a biosynthesis from shikimate.[54] The pattern of derivatization of the hydroxyls at various positions deserves comment. Note that the C-3 hydroxyl (and sometimes that at C-5) is converted to a glycoside; the sugars most commonly used are glucose, rhamnose, or galactose. Methyl ethers are often found at C-3′ and C-5′, less frequently at C-7, but never at C-4′; the latter position always carries a free hydroxy-group which plays a vital role in the chromatic properties of the anthocyanins.

The pattern of oxygenation has a marked influence on the colour of the flavylium salt. Thus a shift in colour from red to bluish-red takes place when the number of oxygen substituents is increased from three in pelargonin (**67**) to six in delphin (**68**).

Even more striking, however, is the way the colour of a given anthocyanin is altered by changes in its environment. The effect of changes in pH has been studied on the model compound (**69**), which has the key hydroxy-group at C-4′.[55] This is exceptionally acidic (pK_a 4). In strongly acid media the cation is the dominant species, but as the pH is raised the hydroxy-group undergoes deprotonation to give the neutral derivative (**70**). This has an extended chromophore and so absorbs at a longer wavelength. As the pH is raised further the second hydroxy-group (pK_a 7.5) undergoes deprotonation to give (**71**), which absorbs at even longer wavelengths. Equivalent changes in an anthocyanin are accompanied by a change in colour from red to blue, which goes some way to explain how the same anthocyanin pigments can be responsible for the red colour of a rose and the blue of a cornflower.

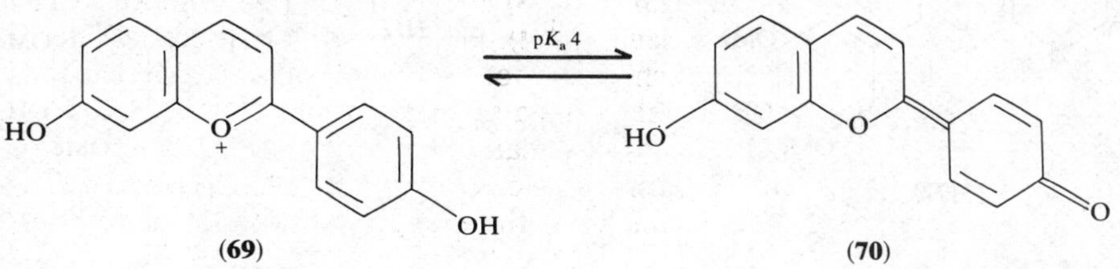

(**69**) λ_{max} 459 nm $\xrightarrow{pK_a\ 4}$ (**70**) λ_{max} 492 nm $\xrightarrow{pK_a\ 7.5}$ (**71**) λ_{max} 537 nm

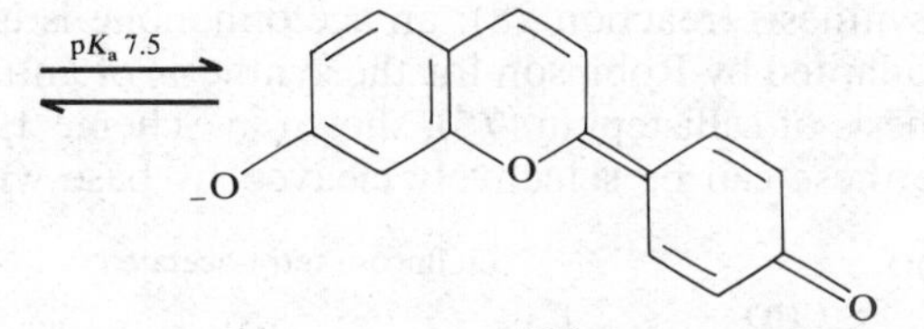

However, the pH of the sap is not the only factor which controls the colour of an anthocyanin in the petals of a flower. The presence of metal ions such as Fe^{III} can produce an equivalent change in colour when, as is so often the case, ring B possesses two adjacent free hydroxyl groups capable of chelation. Also, there is reason to believe that other cell constituents such as tannins may have a considerable influence on the colour of the pigment.

Whatever the explanation for these striking changes in colour may be, there is no denying that Nature has evolved an impressively economical system for colouring her flowers. Unfortunately the anthocyanins are unstable and short lived. While this is no serious disadvantage in the transient existence of a flower, it has precluded the use of flavylium salts in the dyeing industries of mankind.

18.1.4.2 Classical methods of structural elucidation

The main cause of instability in the flavylium system is the susceptibility of the heterocyclic ring to hydrolytic cleavage. This process was widely exploited by chemists in their structural investigations. Initially any sugar residues were cleaved by mild acid hydrolysis to give an aglycone. These derivatives were given the trivial name anthocyanidins. Malvin (**72**), for example, is converted to malvidin (**73**) (reaction 30). Further cleavage of the anthocyanidin by hydroxide gave phloroglucinol from ring A, and a substituted benzoic acid from ring B which could be identified in the usual way.[56]

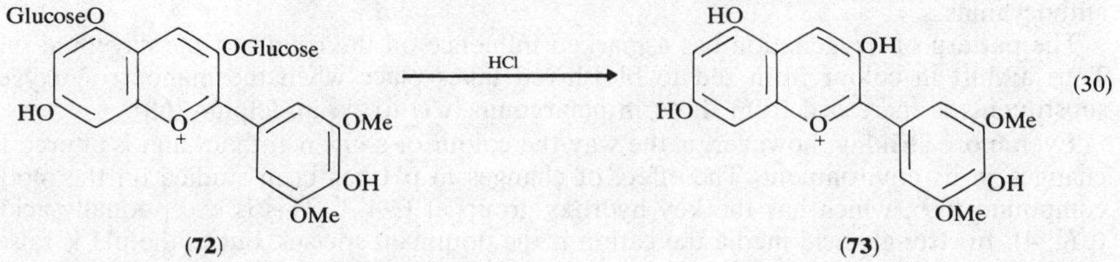

GlucoseO, OGlucose, HO, O+, OMe, OH, OMe (**72**) —HCl→ HO, OH, HO, O+, OMe, OH, OMe (**73**) (30)

Another degradation which played an essential part in structural studies is the oxidative cleavage of the heterocyclic ring with hydrogen peroxide. Malvin, for example, affords the ester (74) (reaction 31). By carrying out further degradation of the product, malvone (**74**), it was possible to elucidate the sites of the glycoside links in the starting anthocyanin.[57]

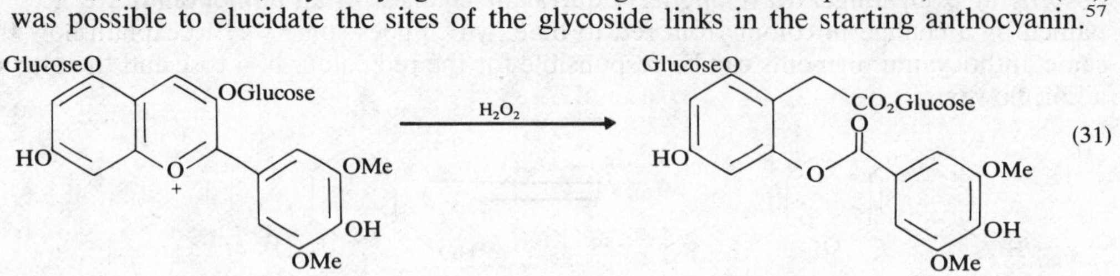

GlucoseO, OGlucose, HO, O+, OMe, OH, OMe (**72**) —H_2O_2→ GlucoseO, CO_2Glucose, HO, O, O, OMe, OH, OMe (**74**) (31)

18.1.4.3 Synthesis of anthocyanins

Flavylium salts in general are best synthesized by a modification of the benzopyrylium salt synthesis (reaction 28): an acetophenone is used in place of acetaldehyde. This route was adapted by Robinson for the synthesis of anthocyanins. His first success came with the synthesis of callistephin (**75**), shown in Scheme 12.[58] Esters are used as protecting groups since these can be selectively cleaved by base without affecting the glycosidic linkage.

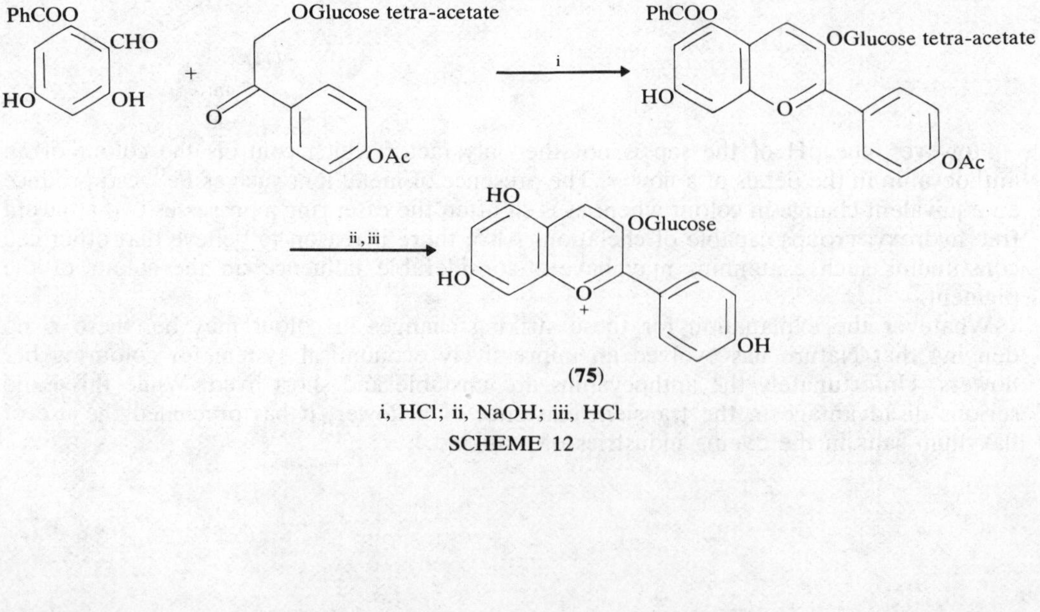

PhCOO, CHO, HO, OH + OGlucose tetra-acetate, O, OAc —i→ PhCOO, OGlucose tetra-acetate, HO, O, OAc —ii, iii→ HO, OGlucose, HO, O+, OH (**75**)

i, HCl; ii, NaOH; iii, HCl

SCHEME 12

18.1.5 XANTHYLIUM SALTS

18.1.5.1 Introduction

A xanthylium salt such as (**76**) may be regarded as a pyrylium salt in which the heterocyclic ring is doubly benzannulated. The extended conjugated system allows many canonical forms to be written, of which one is shown in (**77**). The chemical shifts of the signals in the ^{13}C n.m.r. spectra of xanthylium salts are consistent with extensive delocalization of the positive charge.[59]

Two important relatives of the xanthylium salt deserve mention. These are xanthene (**78**) (equivalent to a 4*H*-pyran) and xanthone (**79**) (equivalent to a γ-pyrone).

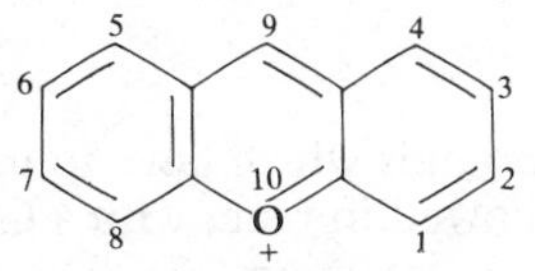

(**76**) 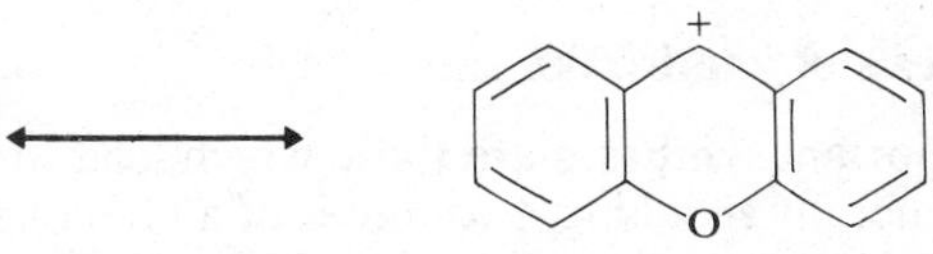 (**77**)

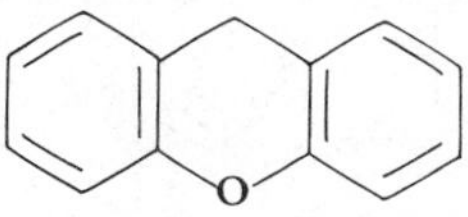

(**78**) 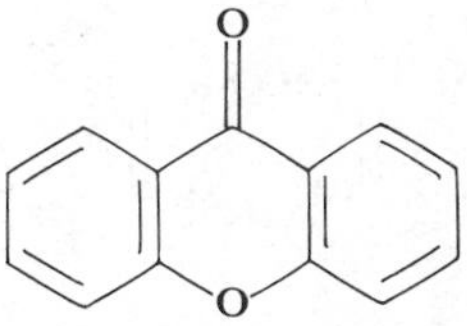 (**79**)

18.1.5.2 Properties of xanthylium salts[60]

The properties of xanthylium salts are markedly different from those of simple pyrylium salts and follow instead those of triarylmethyl carbonium (carbenium) ions: so much so that they are best thought of in terms of the carbonium ion structure (**77**) rather than the oxonium ion (**76**).

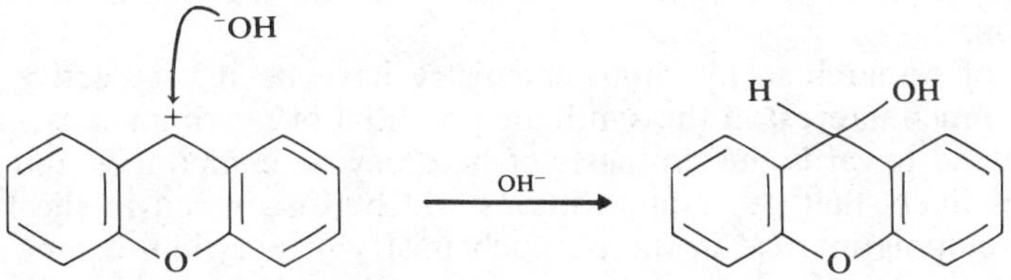

(**77**) (**80**)

Thus attack of nucleophiles such as hydroxide takes place exclusively at C-9 to give a xanthydrol (**80**). The inertness of the sites adjacent to the oxygen (equivalent to C-2 and C-6 of a simple pyrylium salt) has already been discussed in connection with benzopyrylium salts. One consequence of this lack of attack is the absence of reactions in which the heterocyclic ring undergoes ring-cleavage. Paradoxically, however, the xanthylium cation is much less stable than a simple pyrylium compound and only exists as such in the presence of strong acid; in aqueous solution the stable species is usually the solvated derivative (**80**). This reaction can be followed by changes in the visible spectrum: the cation is yellow or red depending on its substitution pattern, whereas the adduct is colourless.

The cation is made more favourable by the presence of suitable substituents in the carbocyclic rings. Particularly effective ones are dimethylamino- or hydroxy-groups.

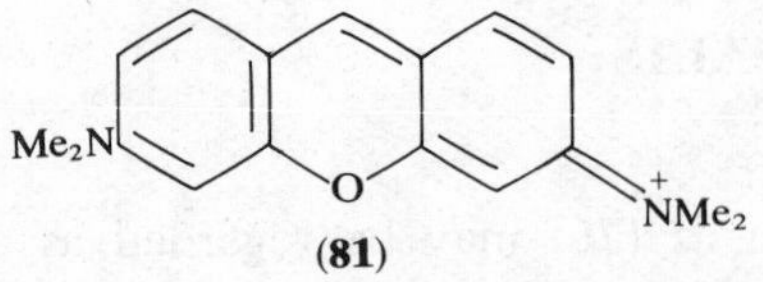

(**81**)

Certain derivatives of this type have therefore been used as dyestuffs, although they have generally been superseded in recent years. A typical example is Pyronine G (**81**).

The hydroxy-group of a xanthydrol is readily displaced by nucleophiles, presumably by an S_N1 mechanism in which the xanthylium ion is formed as an intermediate. It is also readily oxidized to give a xanthone.

18.1.5.3 Synthesis of xanthylium salts

The most important syntheses are those which lead to compounds which have found use as dyestuffs. Generally speaking, two moles of a phenol are allowed to react with a C_1 unit which may be a formate ester (reaction 32) or formaldehyde (reaction 33). The intermediate xanthene produced in the formaldehyde reaction is readily oxidized to the corresponding xanthylium salt (in this case pyronine G).

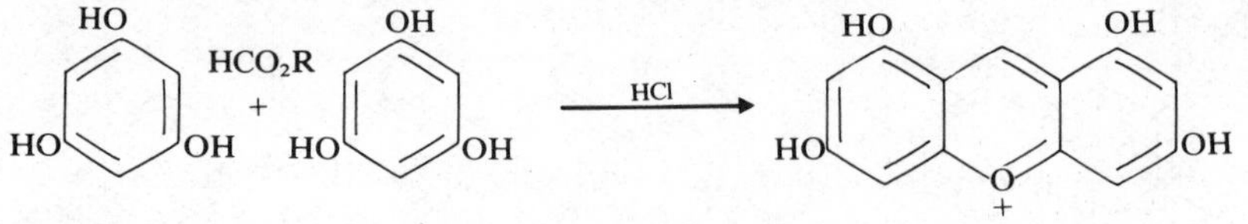

(32)

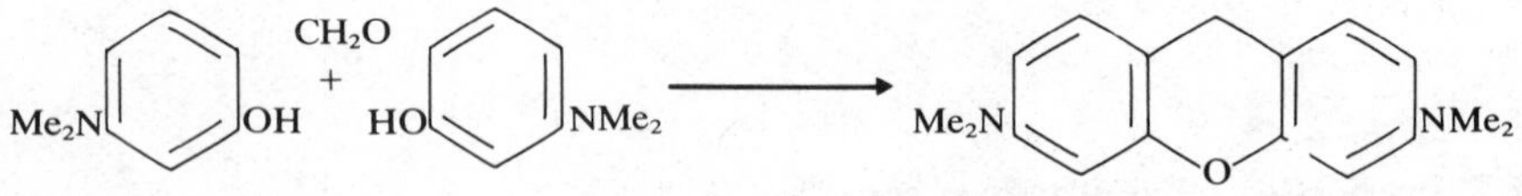

(33)

18.1.6 CONCLUSION

The characteristic reaction of pyrylium chemistry is addition of a nucleophile to the heterocyclic ring. In simple pyrylium salts attack takes place chiefly at one of the sites adjacent to oxygen and is followed by ring cleavage; this reaction is suppressed at a site of benzannulation.

Two areas of research in pyrylium chemistry have been very active in recent years. First, there is much interest in the synthetic potential of pyrylium salts; most effort so far has been directed towards the synthesis of heterocyclic compounds and benzene derivatives, but it is likely that other applications will be discovered in the future. Secondly, there is a developing interest in the photochemistry of pyrylium compounds. What little work has been done so far has produced intriguing results and there is hope for more in the future.

A major limiting factor in studies of pyrylium chemistry has been the difficulty of making compounds other than those with substituents at the 2-, 4-, and 6-positions. A worthwhile goal for future research would therefore be a search for more flexible methods of pyrylium salt synthesis.

Further reading on pyrylium compounds can be found in references 33, 61, 62, and 63.

References

1. A. Baeyer and V. Villiger, *Ber.*, 1901, **34,** 2679.
2. F. Kehrmann and A. Duttenhöfer, *Ber.*, 1906, **39,** 1299.
3. G. Schwarzenbach and K. Lutz, *Helv. Chim. Acta*, 1940, **23,** 1147.
4. I. Degani, R. Fochi, and C. Vincenzi, *Gazzetta*, 1964, **94,** 203.

5. K. Dimroth, W. Kinzbach, and M. Soyka, *Chem. Ber.*, 1966, **99,** 2351.
6. A. T. Balaban, G. R. Bedford, and A. R. Katritzky, *J. Chem. Soc.*, 1964, 1946.
7. F. Kröhnke and K. Dickoré, *Chem. Ber.*, 1959, **92,** 46.
8. K. Dimroth and K. H. Wolf, *Angew. Chem.*, 1960, **72,** 777.
9. E. N. Marvell, G. Caple, T. A. Gosink, and G. Zimmer, *J. Amer. Chem. Soc.*, 1966, **88,** 619.
10. R. Lombard and A. Kress, *Bull. Soc. chim. France*, 1960, 1528.
11. G. Köbrich and D. Wunder, *Annalen*, 1962, **654,** 131.
12. A. T. Balaban, T. H. Crawford, and R. H. Wiley, *J. Org. Chem.*, 1965, **30,** 879.
13. G. Rio and Y. Fellion, *Tetrahedron Letters*, 1962, 1213.
14. A. T. Balaban, G. Mihai, and C. D. Nenitzescu, *Tetrahedron*, 1962, **18,** 257.
15. R. Lombard and J. P. Stephan, *Bull. Soc. chim. France*, 1957, 1369.
16. R. Wizinger and P. Ulrich, *Helv. Chim. Acta*, 1956, **39,** 207.
17. L. Mark, *Angew. Chem. Internat. Edn.*, 1966, **5,** 846.
18. K. Dimroth, G. Bräuninger, and G. Neubauer, *Chem. Ber.*, 1957, **90,** 1634, 1668.
19. R. M. Anker and A. H. Cook, *J. Chem. Soc.*, 1946, 117.
20. E. Gârd, I. I. Stănoiu, F. Chiraleu, and A. T. Balaban, *Rev. Roumaine Chim.*, 1969, **14,** 247.
21. R. J. W. Le Fevre and J. Pearson, *J. Chem. Soc.*, 1933, 1197.
22. J. A. Barltrop, K. Dawes, A. C. Day, and A. J. H. Summers, *J.C.S. Chem. Comm.*, 1972, 1240.
23. J. A. Barltrop, K. Dawes, A. C. Day, S. J. Nuttall, and A. J. H. Summers, *J.C.S. Chem. Comm.*, 1973, 410.
24. J. A. Barltrop, K. Dawes, A. C. Day and A. J. H. Summers, *J. Amer. Chem. Soc.*, 1973, **95,** 2406.
25. J. W. Pavlik and J. Kwong, *J. Amer. Chem. Soc.*, 1973, **95,** 7914.
26. J. W. Pavlik and E. L. Clennan, *J. Amer. Chem. Soc.*, 1973, **95,** 1697.
27. (a) A. Schönberg, 'Preparative Organic Photochemistry', Springer-Verlag, Berlin, 1968, p. 409; (b) R. O. Kan, 'Organic Photochemistry', McGraw-Hill, New York, 1966, p. 132; (c) N. R. Bertoniere and G. W. Griffin, in 'Organic Photochemistry', ed. O. L. Chapman, Dekker, New York, 1973, vol. 3, chapter 2.
28. J. A. Van Allan, G. A. Reynolds, and D. P. Meier, *J. Org. Chem.*, 1968, **83,** 4418.
29. E. Gard, A. Vasilescu, G. D. Mateescu, and A. T. Balaban, *J. Labelled Compounds*, 1967, **3,** 196.
30. W. Schneider, *Annalen*, 1923, **432,** 297.
31. N. V. Khromov-Borisov and L. A. Gavrilova, *Zhur. obshchei Khim.*, 1962, **32,** 3211.
32. A. Baeyer, *Ber.*, 1910, **43,** 2337.
33. A. T. Balaban, W. Schroth, and G. Fischer, *Adv. Heterocyclic Chem.*, 1969, **10,** 249.
34. W. Dilthey, *J. prakt. Chem.*, 1916, **94,** 53.
35. W. Schneider and A. Sack, *Ber.*, 1923, **56,** 1786.
36. A. T. Balaban and C. D. Nenitzescu, *Annalen*, 1959, **625,** 66.
37. A. T. Balaban and C. D. Nenitzescu, *Org. Synth.*, 1964, **44,** 98.
38. G. Fischer and W. Schroth, *Z. Chem.*, 1963, **3,** 266.
39. W. Dilthey, *Ber.*, 1917, **50,** 1008.
40. F. Klages and H. Träger, *Chem. Ber.*, 1953, **86,** 1327.
41. K. Dimroth, K. H. Wolf, and H. Kroke, *Annalen*, 1964, **678**, 183.
42. G. Märkl, *Angew Chem.*, 1962, **74,** 696.
43. A. R. Katritzky, S. Q. A. Rizvi, and J. W. Suwinski, *J.C.S. Perkin I*, 1975, 2489.
44. A. T. Balaban and C. D. Nenitzescu, *Chem. Ber.*, 1960, **93,** 599.
45. A. T. Balaban, *Tetrahedron*, 1970, **26,** 739.
46. K. Hafner and H. Kaiser, *Annalen*, 1958, **618,** 140.
47. R. J. W. Le Fevre, *J. Chem. Soc.*, 1929, 2771.
48. A. Löwenbein, *Ber.*, 1924, **57,** 1517.
49. H. R. Hensel, *Annalen*, 1958, **611,** 97.
50. K. T. Potts and Sir R. Robinson, *J. Chem. Soc.*, 1955, 2675.
51. I. M. Heilbron, H. Barnes, and R. A. Morton, *J. Chem. Soc.*, 1923, **123,** 2559.
52. S. Wawzonek, in 'Heterocyclic Compounds', ed. R. C. Elderfield, Wiley, New York, 1951, vol. 2, chapter 9.
53. (a) K. Hayashi, in 'The Chemistry of Flavonoid Compounds', ed. T. A. Geissman, Pergamon Press, Oxford, 1962, chapter 9; (b) F. M. Dean, 'Naturally Occurring Oxygen Ring Compounds', Butterworths, London, 1963, chapter 13.
54. J. H. Richards and J. B. Hendrickson, 'The Biosynthesis of Steroids, Terpenes, and Acetogenins', Benjamin, New York, 1964, p. 160.
55. Von H. Kuhn and W. Sperling, *Experientia*, 1960, **16,** 237.
56. P. Karrer and R. Widmer, *Helv. Chim. Acta*, 1927, **10,** 5.
57. P. Karrer, R. Widmer, A. Helfenstein, W. Hürliman, O. Nievergelt, and P. Monsarrat-Thoms, *Helv. Chim. Acta*, 1927, **10,** 729.
58. (a) A. Robertson and R. Robinson, *J. Chem. Soc.*, 1928, 1460; (b) P. V. Nair and R. Robinson, *J. Chem. Soc.*, 1934, 1611.
59. E. Dradi and G. Gatti, *J. Amer. Chem. Soc.*, 1975, **97,** 5472.
60. S. Wawzonek, in 'Heterocyclic Compounds', ed. R. C. Elderfield, Wiley, New York, 1951, vol. 2, p. 472.
61. N. Campbell, in 'Chemistry of Carbon Compounds', ed. E. H. Rodd, Elsevier, Amsterdam, 1959, vol. 4B, chapter 8, pp. 966–983.
62. H. Perst, 'Oxonium Ions in Organic Chemistry', Verlag Chemie, Weinheim, 1971.
63. S. Wawzonek, in 'Heterocyclic Compounds', ed. R. C. Elderfield, Wiley, New York, 1951, vol. 2, p. 277.

18.2

α-Pyrones and Coumarins

J. STAUNTON
University of Cambridge

18.2.1 INTRODUCTION

Pyrones are closely related in structure to pyrylium salts in that they consist of a six-membered heterocyclic ring containing one oxygen atom and five sp^2 hybridized carbons. One of the carbons in the pyrone can be written as a carbonyl group. Two isomers are possible, (**1**) and (**2**), which will be referred to by their trivial names α- and γ-pyrone, respectively. Other names in common use are given in brackets. In each case there is an alternative canonical form which consists of a pyrylium betaine. Note that a carbonyl form cannot be written for the pyrylium betaine (**3**) in which the oxygen substituent is sited at C-3.

(**1**) (**1a**) (**2**) (**2a**) (**3**)

α-Pyrone
(2*H*-Pyrone)
(2*H*-Pyran-2-one)

γ-Pyrone
(4*H*-Pyrone)
(4*H*-Pyran-4-one)

The relationship of pyrones to pyrylium salts is stressed in Scheme 1, which shows the stepwise increase in oxidation level which takes place in passing from a 2*H*-pyran (**4**) through the pyrylium salt (**5**) and the equivalent 2*H*-pyranol (**6**) to the 2*H*-pyranone, otherwise the α-pyrone (**1**). It should be stressed, however, that this is only a formal

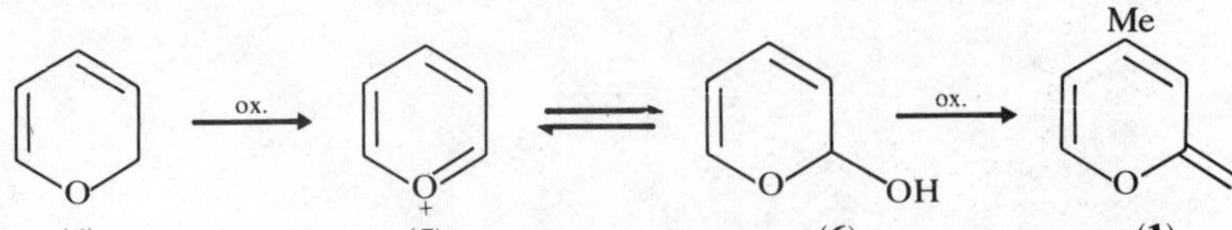

SCHEME 1

relationship and that the efficient transformation of a pyrylium compound to a pyrone cannot normally be achieved in practice.

Only α-pyrones will be covered in this chapter; γ-pyrones, which have a markedly different chemistry, are the subject of the next Chapter 18.3. Benzannulated pyrones are also of interest. Four types are possible, **(7)**, **(8)**, **(9)**, and **(10)**. The latter is a γ-pyrone and so will be treated later. Of the other three types, which are all derivatives of α-pyrone, only the first will be covered. This, the coumarin system **(7)**, deserves detailed consideration on the grounds of both its frequent occurrence in compounds of physiological interest and also its special properties, which diverge interestingly from those of simple pyrones. The isocoumarin system **(8)** occurs in many interesting compounds, but the properties of its heterocyclic ring are not sufficiently different from those of a simple pyrone to justify special treatment.[1] The third isomer **(9)** is relatively obscure.

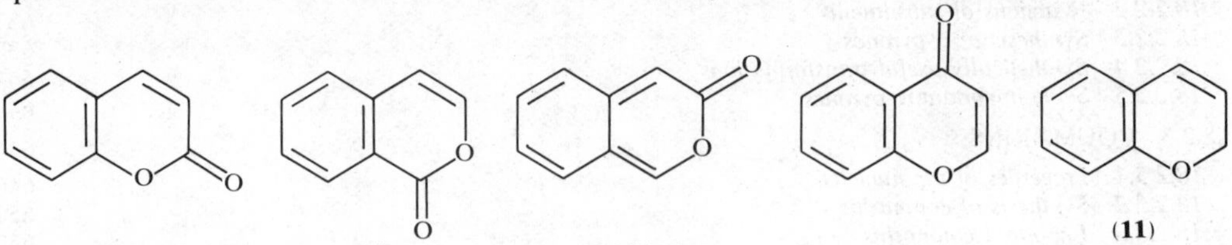

The pyran analogues of the coumarins are still commonly referred to by their trivial family name, α-chromene, as indicated in structure **(11)**. The continued use of trivial names for relatives of the coumarin family is a reflection of the importance of this class of compound.

18.2.2 α-PYRONES

18.2.2.1 Properties of the heterocyclic ring

Running through the whole of pyrone chemistry there is an intriguing question: is the heterocyclic ring aliphatic or aromatic? The potential for aromaticity is brought out in the betaine structures **(1a)** and **(2a)**, which emphasize the relationship of the pyrone system to the pyrylium system discussed in the previous chapter.

The question is an intriguing one because no clear cut answer can be given for either α- or γ-pyrones. Both types of compound show some reactions characteristic of alkenes and others characteristic of arenes. In the case of α-pyrone, however, the balance does lie heavily in favour of reactions of the former type and it can be stated at the outset that this heterocyclic ring is best thought of as an enol lactone rather than as a pyrylium betaine.

Thus the i.r. spectra of a series of pyrones of general structure **(15)** show carbonyl peaks in the range 1736–1730 cm^{-1}, which is approximately what would be expected on the basis of typical values for the undoubtedly aliphatic models **(12)**, **(13)**, and **(14)**; the expected value for the pyrylium betaine would be much lower.[2] The u.v. spectra of α-pyrones such as **(16)** are also consistent with an aliphatic system.[3]

The n.m.r. spectra of α-pyrones throw an interesting light on the electronic structure of the heterocyclic ring. Thus in the ^1H n.m.r. spectrum of **(17)** the proton resonances lie in

(12) (13) (14) (15)

ν_{max} cm^{-1} (CCl$_4$ solution): 1740–1730, 1710, 1775–1770, 1736–1730

(16) λ_{max} 295 nm (ε 5000)

(17) δ7.4 H; δ6.4 H; H δ6.4; δ7.8 H

the olefinic region and there is no evidence of a marked downfield shift to suggest the existence of a significant ring current.[4] The chemical shifts for the ring carbons of a number of α-pyrones are given in Table 1. Note the alternating pattern in going from C-2 to C-6 in the parent compound; the significant downfield shifts for C-4 and C-6 relative to C-3 and C-5 are consistent with a high degree of positive charge being localized at those positions. A second interesting point emerges from a comparison of the spectra of 6-methyl- and 4-methoxy-6-methyl-2-pyrones: the signals from both C-3 and C-5 move downfield as a consequence of the introduction of the methoxy-group at C-4, but note that the shift for C-3 (25.3 p.p.m.) is much larger than that for C-5 (3.1 p.p.m.). This result indicates that the degree of double-bond character between C-4 and C-5 is much less than that between C-4 and C-3, as would be expected of the aliphatic system.

TABLE 1

Chemical Shifts of Ring Carbons in the ^{13}C n.m.r. of α-Pyrones (p.p.m. rel. to Me_4Si)[a]

Compound	*C-2*	*C-3*	*C-4*	*C-5*	*C-6*
2-Pyrone[a]	162	116.7	144.3	106.8	153.3
2-Pyrone[b]	—	117.0	142.8	106.0	152.0
4-Methyl-[b]	161.8	113.7	156.1	109.3	151.1
3-Methoxy-[b]	158.6	145.9	112.9	106.0	142.9
5-Methyl-[b]	161.2	115.7	146.5	114.7	148.0
6-Methyl-[b]	162.0	112.6	144.1	103.4	162.9
4-Methoxy-6-methyl-[b]	162.1	87.3	171.4	100.3	164.6
4,6-Diphenyl-[b]	160.3	109.1	155.5	101.2	162.5
Coumarin[b]	160.4	116.4	143.4	(118.7)[d]	(153.8)[e]

[a] W. V. Turner and W. H. Pirkle, *J. Org. Chem.*, 1974, **39**, 1935. [b] In [$^2H_6$]acetone. [c] In $CDCl_3$. [d] C-4a. [e] C-8a.

The potential aromatic character of the α-pyrone ring is, however, manifest in certain of its reactions. First the carbonyl oxygen shows enhanced nucleophilicity so that with powerful methylating agents such as trimethyloxonium tetrafluoroborate or methyl fluorosulphonate it is methylated to give a pyrylium derivative[5–7] (reaction 1). It is interesting to note that in the n.m.r. spectrum all the proton resonances of the heterocyclic ring move

$\xrightarrow{Me_3O^+BF_4^-}$ OMe BF_4^- (1)

(18)

downfield by approximately 1 p.p.m. as a result of this change: in CD_3CN there are two multiplets at δ 7.5 and 8.6 in the spectrum of the pyrylium salt (**18**).

The aromatic potential of the α-pyrone ring is also realized in a number of electrophilic substitutions. Thus the ring can be nitrated,[5] sulphonated,[8] and halogenated.[9] In each case substitution takes place at C-3 or C-5. The course of the nitration (reaction 2) has been

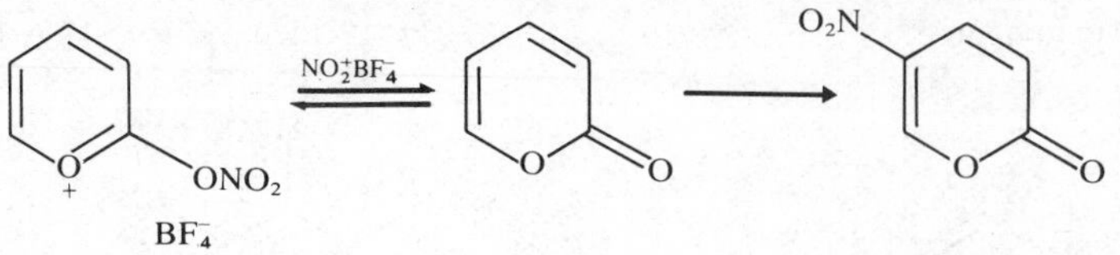

$NO_2^+BF_4^-$ ONO$_2$ BF_4^- O_2N (2)

(**19**) (**20**)

followed by n.m.r. spectroscopy. On adding the nitrating agent to the solution of the pyrone an immediate downfield shift of the proton resonances was observed. This change was attributed to formation of the *O*-nitropyrylium salt (**19**). The signals for this compound slowly disappeared and were replaced by those of the product (**20**).

The ease of electrophilic attack is increased when alkyl substituents are present on the pyrone ring. A hydroxy-group at C-4 makes the reaction particularly favourable; however, in the absence of such activation, electrophilic attack requires forcing conditions and the balance between substitution and addition is finely poised, as in bromination (reaction 3). Though the final product (**22**) corresponds to an electrophilic substitution, it is in fact formed *via* an intermediate dibromo-compound (**21**) resulting from electrophilic addition.

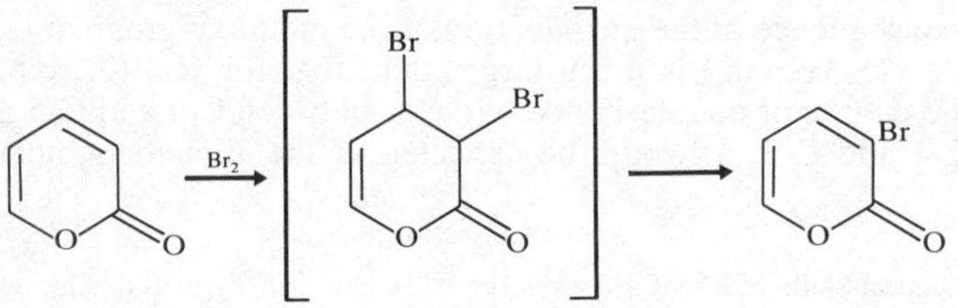

Br_2 Br Br Br (3)

(**21**) (**22**)

The resistance of the pyrone ring to electrophilic attack is in marked contrast to the ease with which it is attacked by nucleophiles. Reagents of this type add at the 2-, 4-, or 6-positions, strong nucleophiles tending to add at C-2, weaker ones at C-6.[10] Some typical reactions are given in Scheme 2.[11]

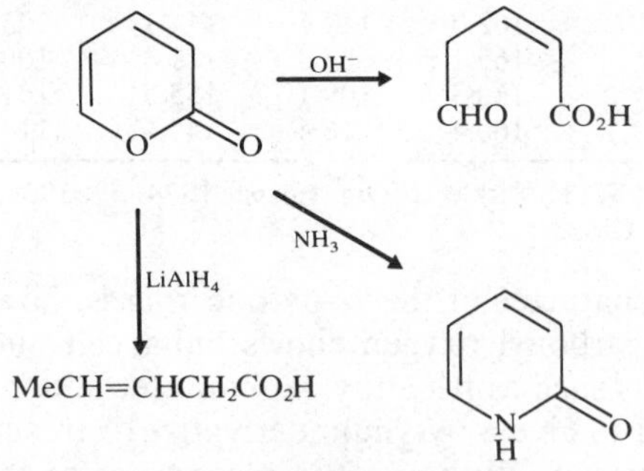

OH^- CHO CO_2H

NH_3

$LiAlH_4$

$MeCH{=}CHCH_2CO_2H$

N H O

SCHEME 2

In general, nucleophilic addition is followed by cleavage of the heterocyclic ring. Of particular importance is the reaction with ammonia in which ring cleavage is followed by recyclization to form a pyridone in a process which is reminiscent of one of the characteristic transformations of pyrylium compounds.

The interesting replacement of hydroxyl by chlorine (reaction 4) probably proceeds by nucleophilic addition of chlorine at C-4 with subsequent loss of the oxygen substituent as a phosphate derivative.[12] Once again a precedent can be found in pyrylium chemistry in the various nucleophilic displacements that take place at C-2 and C-4 of such compounds.

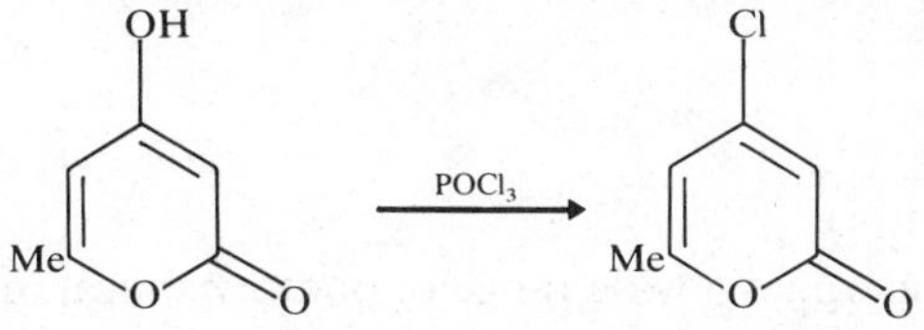

OH Me O O $\xrightarrow{POCl_3}$ Cl Me O O (4)

However, although the parallel with the chemistry of pyrylium compounds has been stressed in the treatment of these nucleophilic additions to the α-pyrone ring, the reactions can be rationalized equally well in terms of Michael additions to an unsaturated carbonyl group and so they throw no light on the key question of where the balance lies between aromatic and aliphatic character in the properties of the pyrone ring.

The ring exhibits characteristic olefinic behaviour in its susceptibility to catalytic reduction in the presence of Raney nickel. Hydrogenation (reaction 5) of **(23)** results in a mixture of valeric acid **(24)** and δ-valerolactone **(25)**.[3]

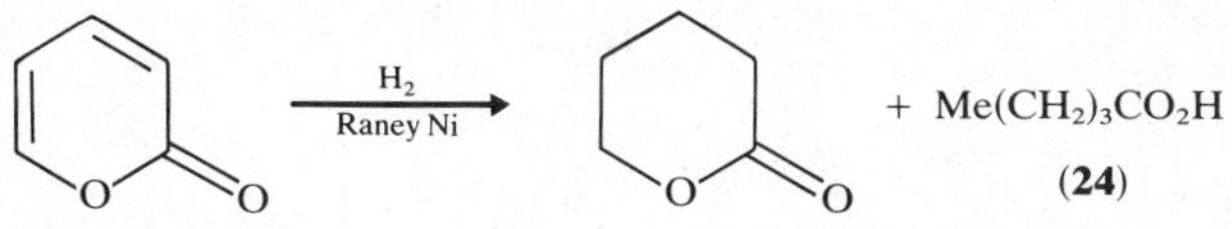

(23) $\xrightarrow[\text{Raney Ni}]{H_2}$ **(25)** + $Me(CH_2)_3CO_2H$ **(24)** (5)

Probably the most convincing demonstration of the predominantly aliphatic nature of the α-pyrone ring is to be found in its ability to serve as a diene in Diels–Alder reactions. The course of the process with maleic anhydride is shown in Scheme 3. Addition of a first molecule of dienophile leads to **(26)**, which undergoes a facile pericyclic decarboxylation to form a new diene **(27)**; this then reacts with a second molecule of dienophile to generate the final product **(28)**.[13]

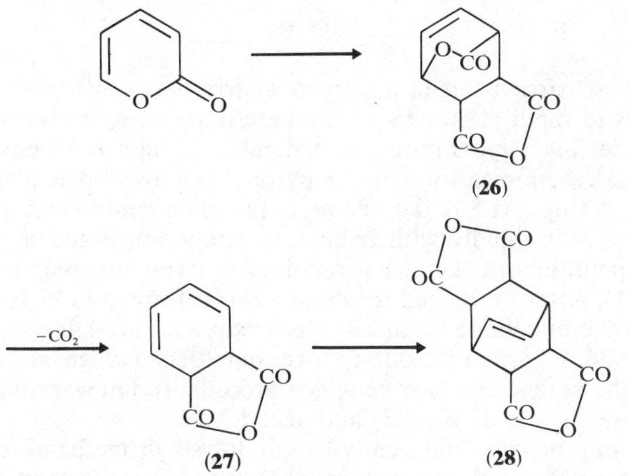

(26) $\xrightarrow{-CO_2}$ **(27)** → **(28)**

SCHEME 3

Acetylenedicarboxylate adds similarly (reaction 6), but here the product of decarboxylation is a benzene derivative **(29)** and so a second addition step does not occur. This reaction has been used as a method of determining the substitution pattern in pyrone rings.[14] It also has interesting synthetic potential.[15]

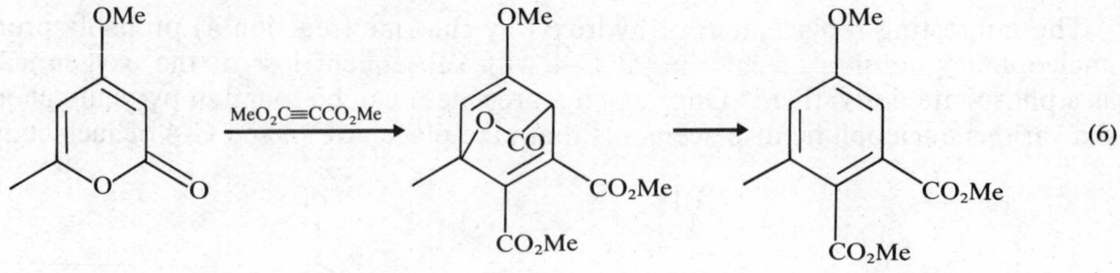

OMe $\xrightarrow{MeO_2C{\equiv}CCO_2Me}$ OMe, CO_2Me, CO_2Me $\longrightarrow$ OMe, CO_2Me, CO_2Me **(29)** (6)

The α-pyrone ring undergoes a wide range of photochemical reactions, one of which has allowed the attainment of one of the most elusive goals in organic chemistry in recent years — the synthesis of cyclobutadiene. This was achieved by photolysis of the parent α-pyrone as shown in Scheme 4. The importance of the transformation has led to its being explored in great detail.[16]

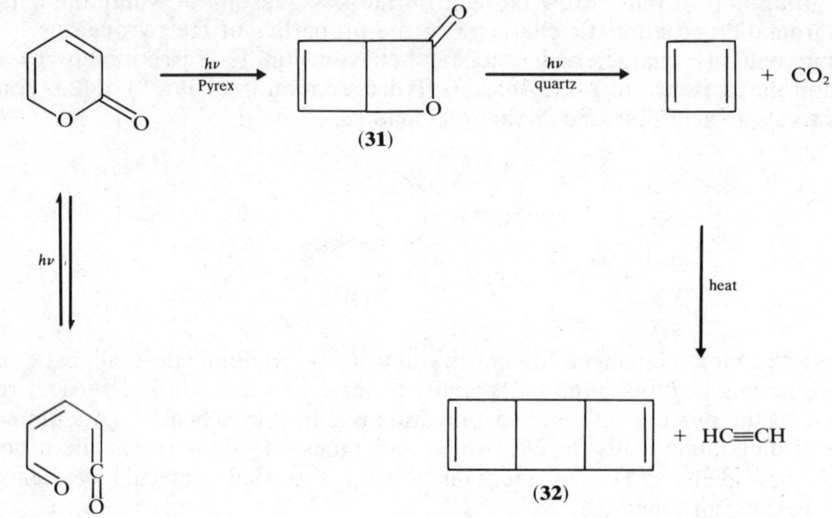

$\xrightarrow[\text{Pyrex}]{h\nu}$ **(31)** $\xrightarrow[\text{quartz}]{h\nu}$ + CO_2

$h\nu$ **(30)** heat **(32)** + HC≡CH

SCHEME 4

The reaction is best carried out in an argon matrix between 8 and 20 K. Irradiation through Pyrex leads to rapid photolysis of the heterocyclic ring to form the ketene (**30**), which exists at these low temperatures as a number of non-interconverting rotational isomers. The reverse reaction to form the α-pyrone ring also takes place and a state of photoequilibrium is set up. At 8 K the ketene is the main component of the equilibrium but its concentration falls rapidly with increasing temperature and at 77 K it is barely detectable. In competition with ketene formation, a secondary reaction takes place to form the lactone (**31**). After prolonged irradiation this is formed in high yield and it is the key intermediate in the butadiene synthesis. On irradiation through quartz it undergoes a facile photoelimination of carbon dioxide to form butadiene (which was detected spectroscopically, as were the ketene and lactone which precede it). On warming the solution the butadiene decomposes to the dimer (**32**) and acetylene.

The heterocyclic ring behaves differently on photolysis in methanol solution (reaction 7). The four isomers represented by structures (**33**) and (**34**) are formed in equal amounts. Contrary to earlier reports which suggested that they arose by solvation of the ketene (**37**), it has been shown that they are formed by acid-catalysed opening of the heterocyclic ring of the photoinduced methanol adducts (**35**) and (**36**).[17]

The photochemical behaviour of the pyrone ring is markedly dependent on its substitution pattern as well as its environment, which is evident in the very different behaviour of

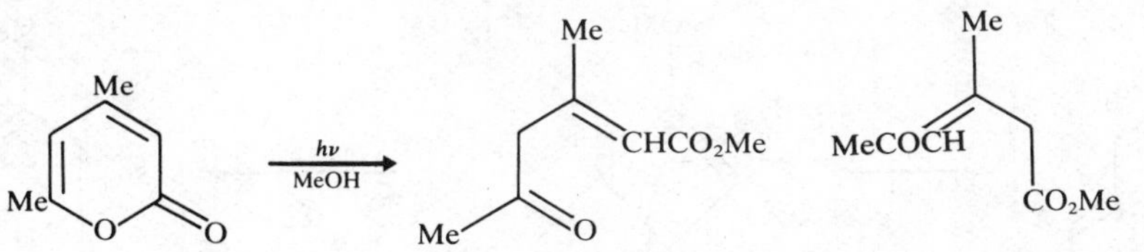

(7)

(33) (34)

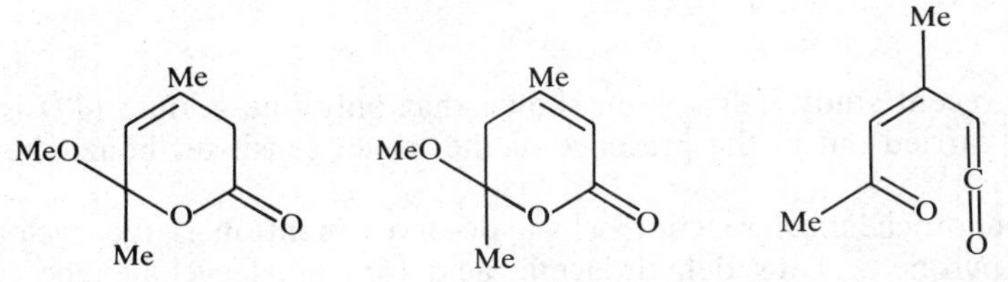

(35) (36) (37)

the 4-hydroxy-2-pyrone (**38**) and its *O*-methyl derivative (**44**).[18] The former undergoes competing ring cleavage to form (**40**) *via* the ketene (**39**), and skeletal rearrangement to give a mixture of (**42**) and (**43**) *via* the intermediacy of the β-lactone (**41**) (see Scheme 5).

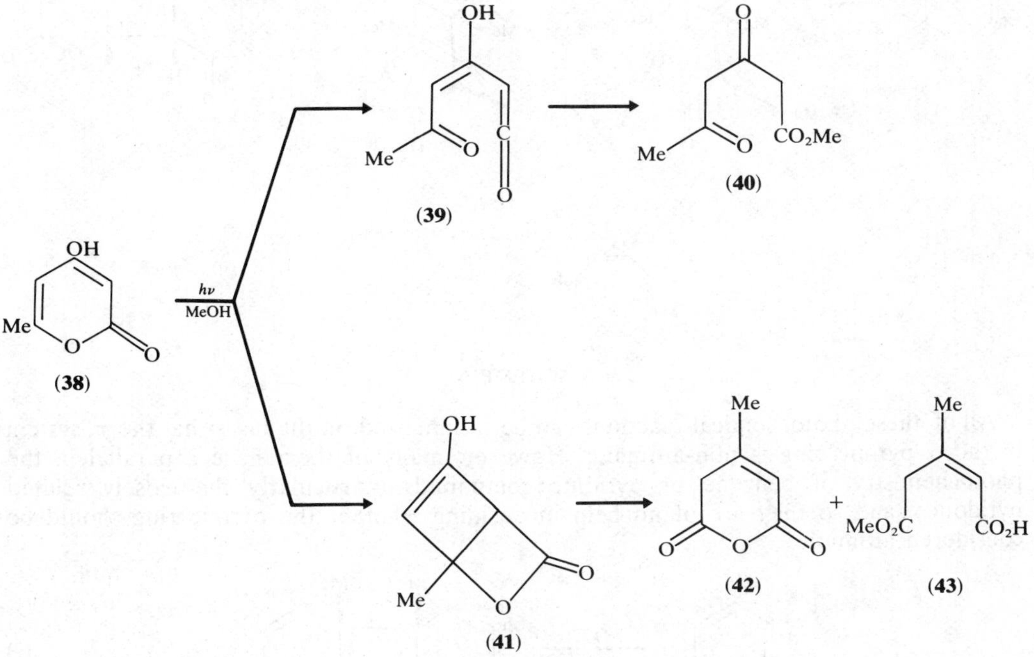

SCHEME 5

The skeletal rearrangement is the dominant process with the methoxypyrone (reaction 8); the primary photochemical process in the formation of the product (**46**) is cyclization to give (**45**). Note that the photochemical behaviour of (**38**) and (**44**) is equivalent to that of the parent pyrone shown in Scheme 4 and that the influence of the oxygen substituent is only felt in the secondary chemical transformations of the intermediate β-lactone.

Irradiation of suitably substituted α-pyrones in benzene can lead to the formation of dimers. A well-studied example is given in Scheme 6. This process was originally conceived as a potential route to substituted cyclo-octatetraenes and the goal was indeed achieved by heating two of the products as shown.[19]

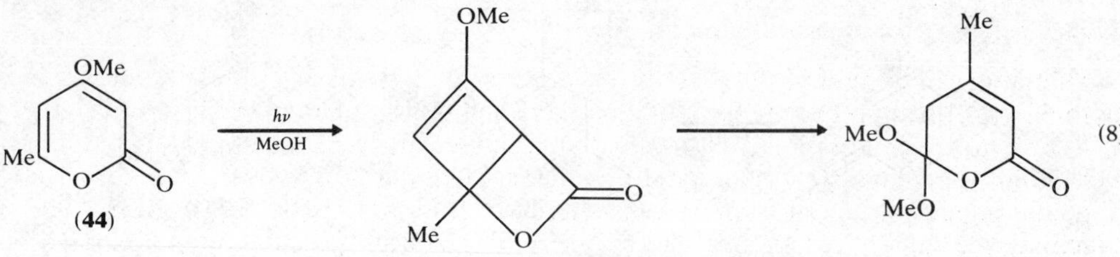

(8)

(44) (45) (46)

In a more recent study it has been shown that only one adduct (**47**) is formed when irradiation is carried out in the presence of the triplet sensitizer benzophenone (reaction 9).[20]

One final photochemical reaction which deserves mention is the cycloaddition of an alkene to a pyrone.[21] Thus dehydroacetic acid (**48**) adds cyclohexene to form a diastereomeric mixture of products having the structure (**49**) (reaction 10). Dimerization of the pyrone takes place in competition.

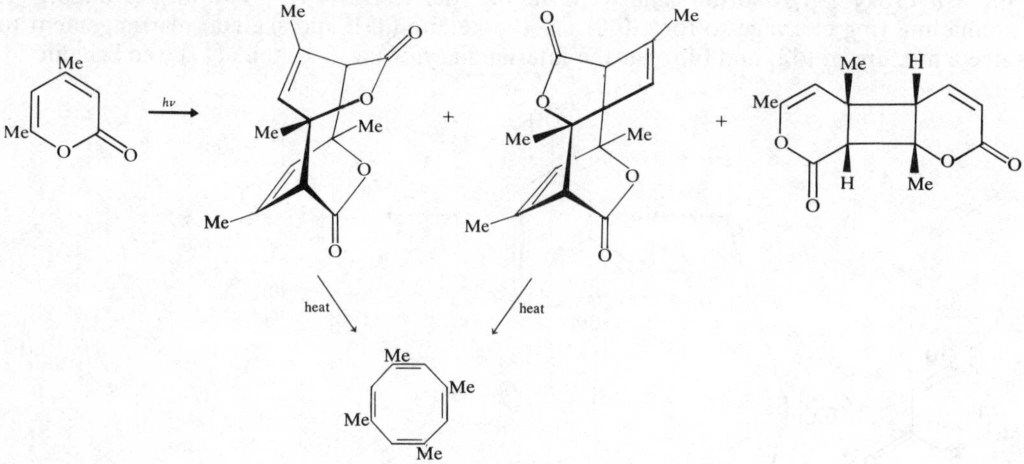

SCHEME 6

All of these photochemical reactions can be rationalized on the basis that the π-system in the α-pyrone ring is non-aromatic. However, many of them have a parallel in the photochemistry of benzene or pyridine compounds (particularly the closely related pyridones) and so they are of no help in deciding whether the pyrone ring should be considered aromatic.

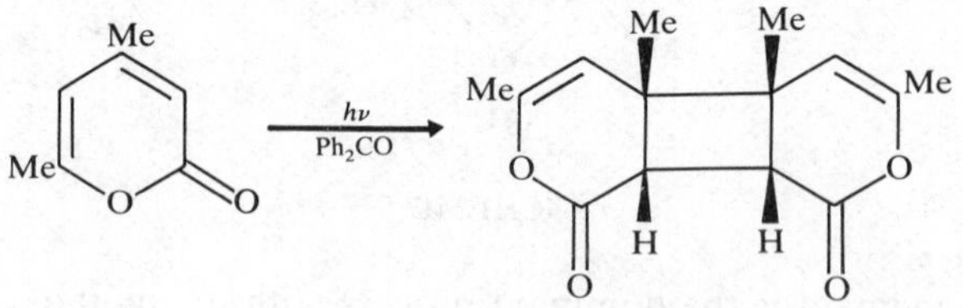

(9)

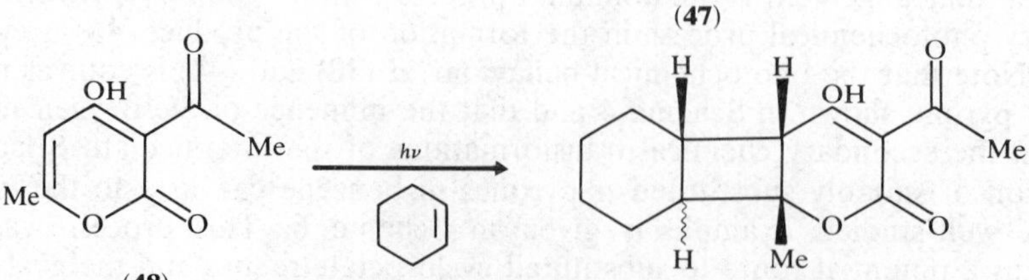

(47)

(10)

(48) (49)

18.2.2.2 Reactions of substituents

The same can be said of the various reactions which involve substituents attached to the pyrone ring. Thus precedent for the characteristic behaviour of alkyl substituents at C-2, C-4, or C-6, which are readily deprotonated at the α-carbon, can be found not only in the behaviour of analogous pyrylium and pyridinium compounds but also in the properties of aliphatic models such as crotonic esters. Two transformations which hinge on this reactivity are shown (reactions 11 and 12).[14,22] The radical bromination (reaction 13) can also be accounted for in terms of an aliphatic or aromatic system.[23]

Hydroxy-groups at C-4 (and C-6) can give rise to tautomerism (reaction 14). In this particular instance the equilibrium lies strongly in favour of the α-pyrone tautomer.[14]

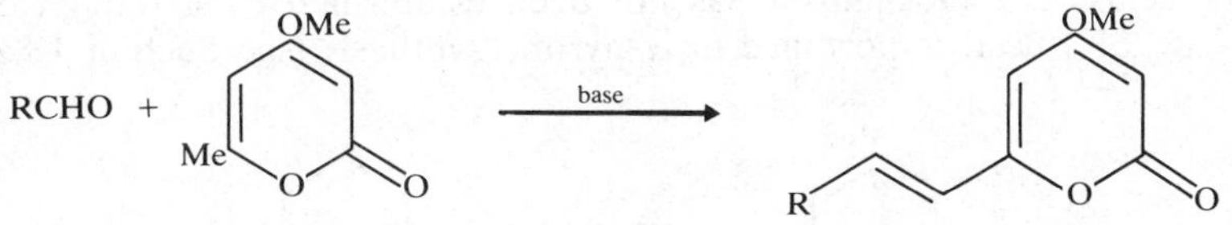

RCHO + [4-methoxy-6-methyl-2-pyrone] --base--> [6-(2-R-vinyl)-4-methoxy-2-pyrone] (11)

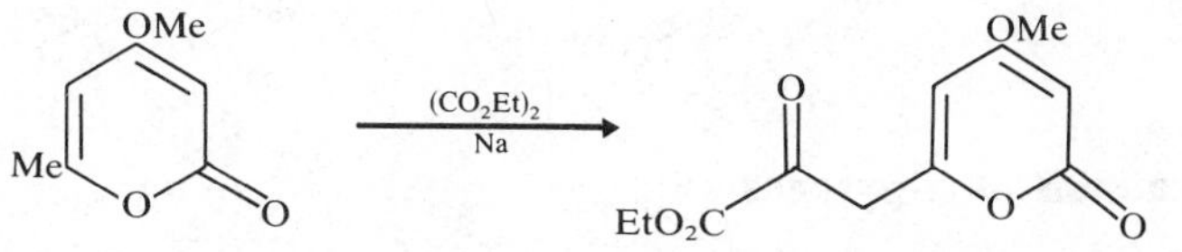

[4-methoxy-6-methyl-2-pyrone] --$(CO_2Et)_2$, Na--> EtO_2C–CO–CH_2–[4-methoxy-2-pyrone] (12)

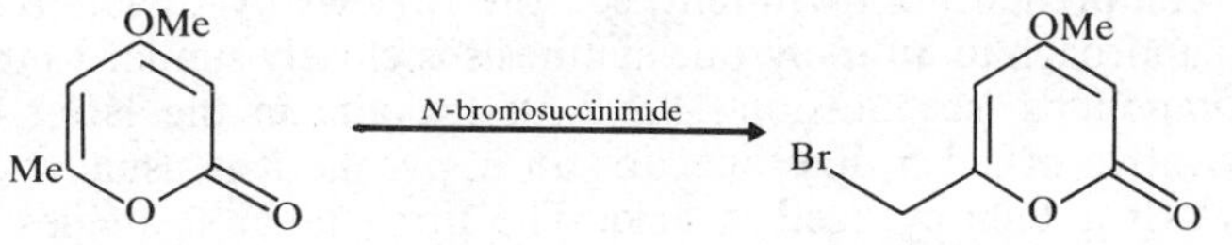

[4-methoxy-6-methyl-2-pyrone] --*N*-bromosuccinimide--> BrCH_2–[4-methoxy-2-pyrone] (13)

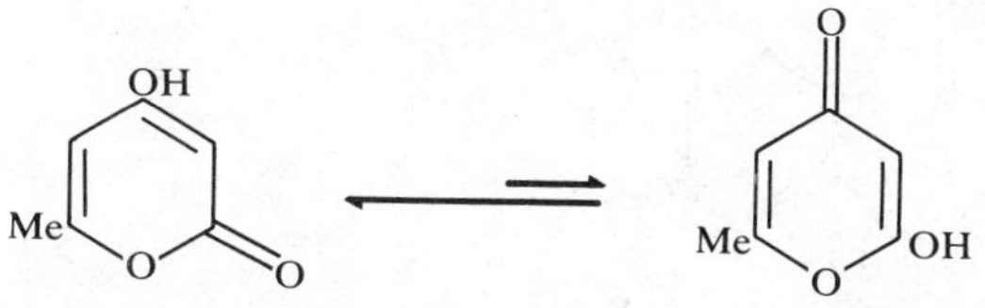

[4-hydroxy-6-methyl-2-pyrone] ⇌ [2-hydroxy-6-methyl-4-pyrone] (14)

However, the 6-phenyl analogue exists as an α-pyrone in ethanol but as the γ-pyrone in the solid state.[24] Other work has shown that the equilibrium can be displaced one way or the other according to the nature of the substitution pattern.

Hydroxy-groups at C-4 or C-6 are readily methylated by either an alkylating agent in the presence of base or by treatment with diazomethane. Interestingly, different products can be formed according to the nature of the reagent, as shown in Scheme 7.[14]

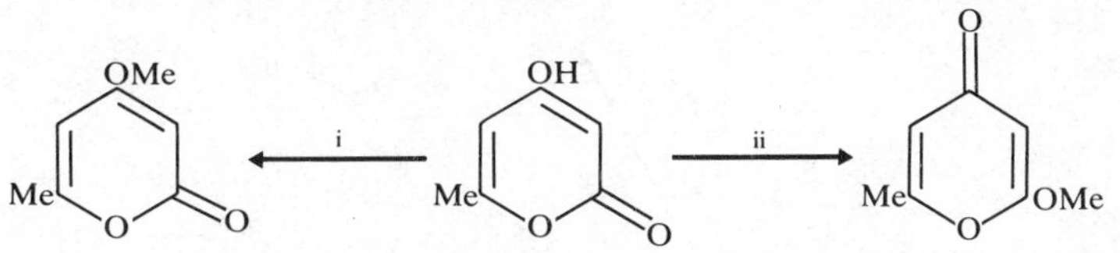

[4-methoxy-6-methyl-2-pyrone] <--i-- [4-hydroxy-6-methyl-2-pyrone] --ii--> [2-methoxy-6-methyl-4-pyrone]

i, Me_2SO_4, K_2CO_3, Me_2CO; ii, CH_2N_2.

SCHEME 7

A hydroxy-group at C-4 can be replaced by chlorine on treatment with $POCl_3$ (see Scheme 8).[12] This process is explicable on the basis of either an aromatic or aliphatic nucleus. What is surprising, however, is the ease with which the halogen is then replaced by reduction with zinc in boiling hydrochloric acid. The mechanism of this reaction (which has obvious synthetic potential) is not clear.

i, $POCl_3$; ii, Zn, HCl.

SCHEME 8

Another enigmatic reaction of this heterocyclic ring is the facile removal of acyl groups from the 3- or 5-positions of an α-pyrone ring by treatment with strong acid (reaction 15).[25] The ease of this process contrasts with the relative stability of pyridine-3- and -5-carboxylic acids. The mechanism has not been established even though deacylations such as this are of crucial importance in α-pyrone synthesis (see Section 18.2.2.3).

conc. H_2SO_4, 160 °C (15)

18.2.2.3 Synthesis of α-pyrones

In the following account, only the more widely used synthetic routes will be presented. For a more comprehensive treatment, see the reviews by Cavalieri[26] and Fried.[27]

The usual approach to an α-pyrone synthesis is closely similar to that in the synthesis of pyrylium compounds (see Section 18.1.2.3): the aim in the latter case is to produce a suitable derivative of a 1,5-diketone; for an α-pyrone it is usual to aim for a suitable 5-keto-acid which is then cyclized in acid. The three main strategies are shown in formal terms in Scheme 9.

SCHEME 9

An important synthesis which is based on strategy (a) is presented in Scheme 10. This is a key method because of the synthetic potential of 6-alkyl-4-alkoxy-2-pyrones. Note the interesting deacylation in step 2.[14,28]

$MeCOCH_2CO_2R$ →(i) [3-acetyl-4-hydroxy-6-methyl-α-pyrone] →(ii) [4-hydroxy-6-methyl-α-pyrone]

i, HCl gas; ii, aq. H_2SO_4.

SCHEME 10

Another process of this type is also given (reaction 16).[29] Note here that an acetylenic ketone is used as an alternative to a β-diketone as the electrophilic component: it has the advantage that the two terminal substituents are differentiated so that α-pyrones carrying different alkyl or aryl substituents at C-4 and C-6 can be prepared. This reaction can be adapted for the preparation of 3-alkyl derivatives of α-pyrones by using a suitably alkylated derivative of the malonic ester.

$R^2C{\equiv}CCOR^1 + CH_2(CO_2Et)_2$ →(i, base; ii, acid) [4-R^2-6-R^1-3-ethoxycarbonyl-α-pyrone] (16)

Another interesting reaction based on the same strategy is given in Scheme 11.[30] It is significant that the diazoester is at the correct oxidation level to give, after reaction with the enone, an intermediate (**50**) which can subsequently form an α-pyrone without the need for an oxidation step. Again a variety of 4,6-disubstituted pyrones are accessible using this reaction. The ester groups in the products of reaction 16 and Scheme 11 (at C-3 and C-5, respectively) can be selectively removed by treatment with acid.

[$EtO_2C(MeCO)C{=}CHPh$] + N_2CHCO_2Et → (**50**) → [5-ethoxycarbonyl-6-methyl-4-phenyl-α-pyrone]

SCHEME 11

The first synthesis based on strategy (b) is shown in Scheme 12 and it deserves particular mention as it was the classical method of preparing the parent α-pyrone.[31] The synthesis pivots on the characteristic decarbonylation of α-hydroxyacids to give an aldehyde (in this case formylacetic acid). Two molecules of this condense *in situ* to give an α-pyrone (**51**) as shown. The decarboxylation of this on heating with copper powder is not a clean reaction and so nowadays the unsubstituted α-pyrone is usually prepared by decarboxylation of the 6-carboxy-derivative (which is also readily available).[3]

[malic acid] →(i) [HO_2CCH_2CHO + $OHCCH_2CO_2H$] → (**51**) →(ii) [α-pyrone]

i, H_2SO_4; ii, heat, Cu.

SCHEME 12

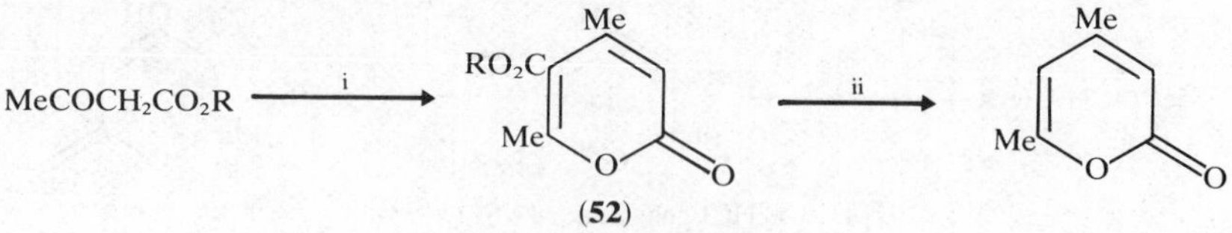

$MeCOCH_2CO_2R$ →(i) (52) →(ii)

i, HCl gas; ii, aq. H_2SO_4.

SCHEME 13

β-Ketoesters undergo a related reaction as shown in Scheme 13.[25] The mode of condensation in this acid-catalysed reaction contrasts interestingly with that which occurs in the base-catalysed condensation in Scheme 10. The product (**52**) is readily deacylated as indicated to give a 3,5-dialkylpyrone. The 3,5-dimethyl derivative shown is readily prepared by this route, which accounts for the frequency with which it has been used in photochemical investigations.

Only pyrones in which the two alkyl substituents are the same can be readily prepared by the route above. However, a recent modification in which one of the two dicarbonyl components is replaced by an acetylenic ester can be used to overcome this disadvantage (reaction 17).[32]

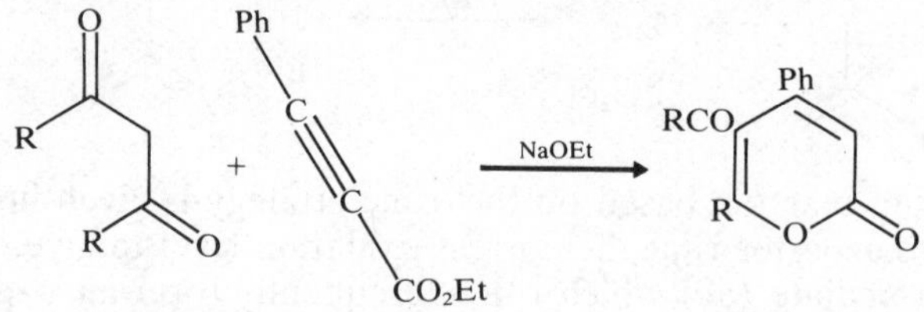

(17)

Another synthesis in which the key step involves a carbon–carbon bond formation between C-4 and C-5 of the pyrone ring (reaction 18) capitalizes on the use of the thallium salt, which provides an alternative to that shown in Scheme 10 for the preparation of the 4-hydroxy-derivative of α-pyrone.[33]

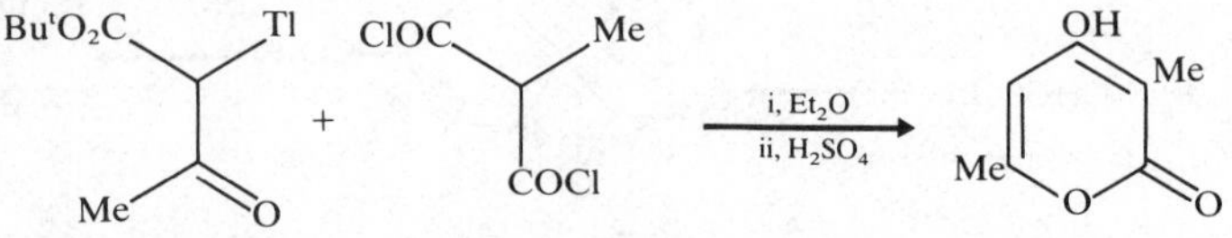

(18)

Another 4-hydroxypyrone synthesis is given in Scheme 14 and it introduces the synthetic approaches based on strategy (c) of Scheme 9. The main point to note here is the use of a dianion of the β-ketoester in order to direct reaction to the γ-position. This is a particularly efficient and flexible method of preparation.[34]

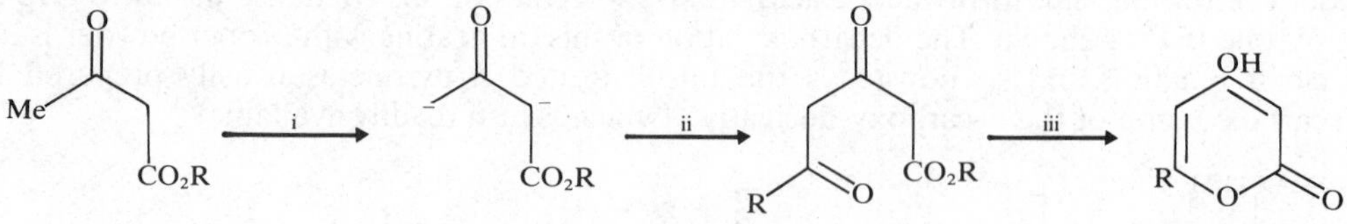

i, $LiN(Pr^i)_2$; ii, RCO_2Et; iii, CF_3CO_2H, $(CF_3CO)_2O$.

SCHEME 14

The second synthesis of this type (Scheme 15) involves base-catalysed acylation at the γ-position of an α,β-unsaturated ester. This route was originally developed for the preparation of 5-alkyl-2-pyrones (this structure occurs in an important group of natural products), but it has the potential to produce a wider range of substitution pattern.[3]

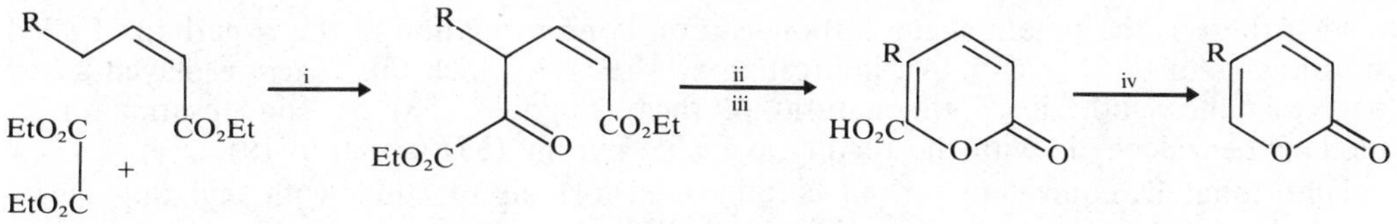

i, KOEt; ii, HCl; iii, HBr, $MeCO_2H$; iv, Cu.

SCHEME 15

The synthesis shown in Scheme 16 is closely related to that in the preceding scheme.[35] Two points deserve special comment. First, the starting material (**53**) is potentially a 6-hydroxypyrone and this tautomer is responsible for the formation of the acyl derivative (**54**). The second step presumably involves ring cleavage and recyclization to give a 5-carboxy-derivative of (**55**) which spontaneously decarboxylates under the conditions of the reaction.

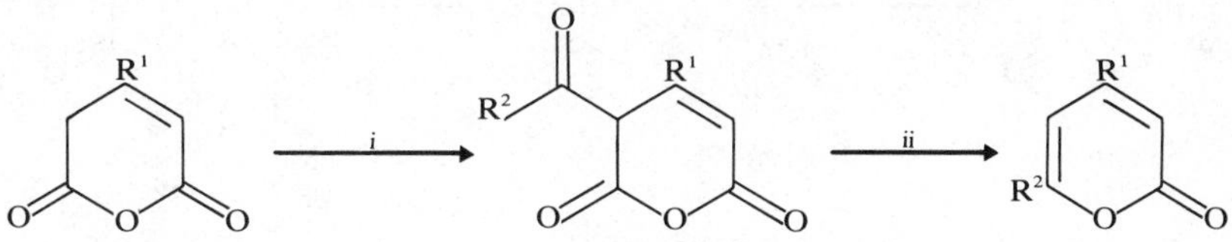

(hy)

(**53**) (**54**) (**55**)

i, R^2COCl; ii, heat.

SCHEME 16

Finally, in Scheme 17 an entirely different synthetic approach is given.[36] It is of interest because it leads to a compound such as (**56**) which is a tautomer of a 3,5-dihydroxy-derivative of α-pyrone. The fact that the 6-position of such compounds is very susceptible to electrophilic attack by aldehydes was used recently in a natural product synthesis.

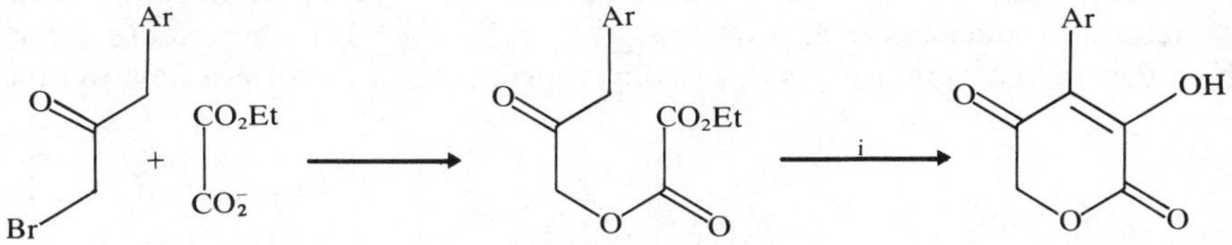

(my)

(**56**)

i, Bu^tOK, HMPT.

SCHEME 17

18.2.2.4 Synthetically useful transformations

A major part of the research in the pyrone field in recent years has been directed towards the exploitation of the synthetic potential of some of the reactions discussed earlier. Many important advances have resulted and so it is appropriate to reconsider certain aspects of pyrone chemistry with this consideration in mind.

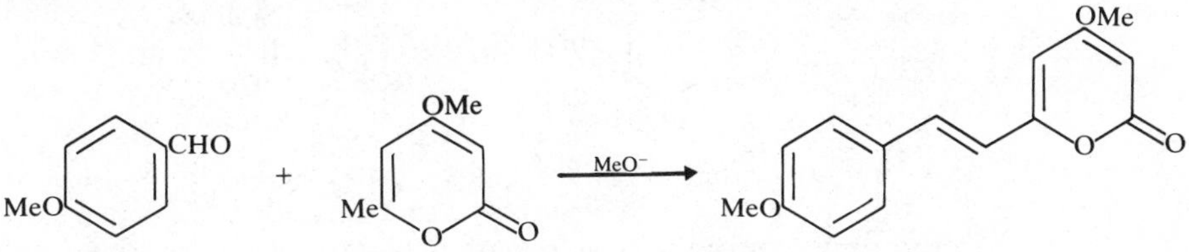

(19)

(**57**) (**58**)

First there is the potential for carbon–carbon bond formation at the α-carbon of alkyl substituents at C-4, or C-6. An interesting synthesis in which this reaction played a key part was the synthesis of the natural product yangonin (**58**) by condensation of 4-methoxybenzaldehyde with the readily available pyrone (**57**) (reaction 19).

Unfortunately, equivalent carbon–carbon bond-forming reactions with acylating agents other than aryl esters have proved less reliable. However, acylated derivatives can be prepared in high yield by the modified approach shown in Scheme 18.[23] Here the starting pyrone is converted to the ylide (**59**) in which the relevant carbon is activated by the presence of a phosphorane group. Treatment with keten leads to the formation of the allene (**60**), which was readily solvated to form the desired acylated product (**61**). This is the methyl ether of tetracetic lactone, a natural product. The route was adapted for the preparation of several other naturally occurring compounds.

OMe OMe OMe Me O O i O O ii O O Br Ph_3P (**59**)

OMe OMe O iii $CH_2{=}C{=}CH$ O O iv Me O O

(**60**) (**61**)

i, NBS; ii, Ph_3P; iii, keten; iv, 80% H_2SO_4.

SCHEME 18

α-Pyrones have also been used as the basis of several biomimetic syntheses of polyketide natural products. One approach is given in Scheme 19, where there are several points which deserve comment.[37] First, the acylation in step i to form a new pyrone ring

O O OH OH i Me O O Me O O (**62**) HO_2C O O CO_2H Me O (**63**)

ii iii

OH (my) Me OH CO_2H OH CO_2Me (ly) HO OH Me O

i, $CH_2(COCl)_2$; ii, KOH, H_2O; iii, $Mg(OMe)_2$, MeOH.

SCHEME 19

provides at once a striking example of the ease with which a 4-hydroxy-derivative of an α-pyrone can be acylated, and also an interesting method of generating an α-pyrone ring. Secondly, both pyrone rings in the pyrano-pyrone (**62**) are susceptible to hydrolytic cleavage as shown. Thirdly, the product (**63**) of ring cleavage is a β-polyketo-acid, which is closely related to the hypothetical polyketone intermediates of polyketide biosynthesis. Finally, this polyketone can cyclize to form a variety of benzene derivatives. Unfortunately a mixture of cyclization products tends to be formed, though some degree of selectivity can be achieved by careful control of reaction conditions as indicated.

This imaginative approach provided the basis of some of the earliest successful biomimetic syntheses in the polyketide area. Its basic strategy is to use a suitable pyranopyrone such as (**62**) as a masked form of a highly reactive polyketone—the heterocycle is easy to handle and yet can be conveniently converted to the desired reactive intermediate when required. The main weakness of this strategy lies in the fact that relatively drastic conditions are required to cleave the pyrone rings so that the derived polyketones do not survive but cyclize *in situ*. Consequently the conditions of the cyclization step are necessarily those required for ring cleavage rather than those which would give the optimum yield of the desired cyclization product.

Because of these limitations, alternative direct synthetic routes to polyketones have found favour in recent years. Simple α-pyrones still have a place in this field, however. Thus acylphloroglucinols can be prepared in excellent yield from the corresponding phenacylpyrones as indicated in the conversion of (**64**) to (**65**) (reaction 20).[34] This

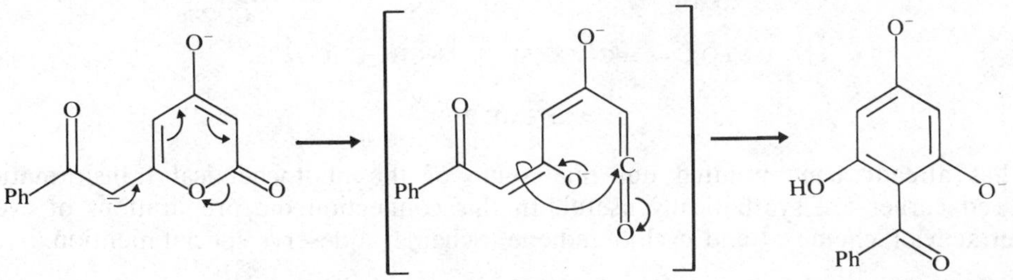

(20)

(**64**) (**65**)

process deserves mention on mechanistic as well as synthetic grounds because it has been suggested that the cleavage of the α-pyrone ring is initiated as shown by deprotonation at the α-carbon of the substituent at C-6, rather than by the usual process of addition of a nucleophile at C-2. It would be interesting to discover whether this is a general property of such carbanions. A second striking example of this type of transformation is given in reaction 21.[34]

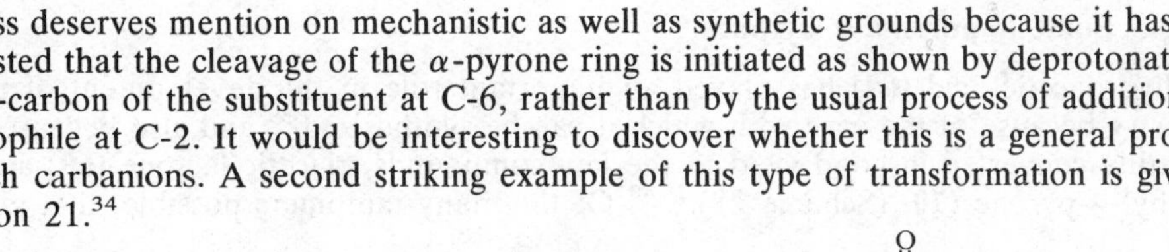

(21)

Another synthesis in which an α-pyrone is converted to a benzene derivative is shown in Scheme 20.[7] Note that in this case the overall reaction is closely related to certain of the synthetic transformations of pyrylium compounds discussed in Section 18.1.2.4.

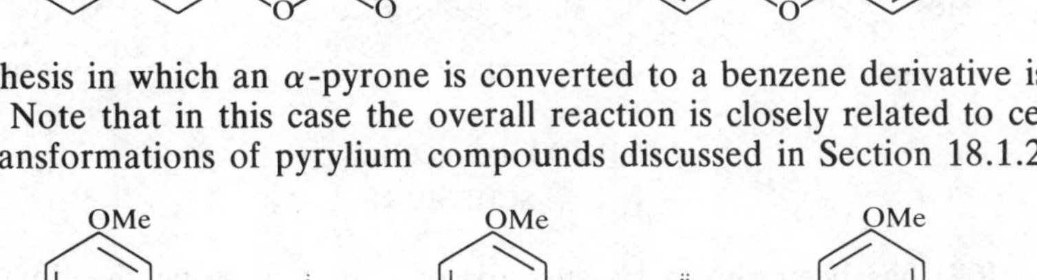

i, $MeSO_3F$; ii, $EtO_2CCH_2PO(OMe)_2$, NaH.

SCHEME 20

The aromatic compounds produced in these reactions are polyphenols or their derivatives. Many of them are natural products and the power of these biomimetic syntheses is brought home when the difficulty of synthesizing such compounds by more conventional synthetic methods is taken into account.

Other reactions of pyrones are also of synthetic potential, particularly Diels–Alder reactions. The scope of this reaction has been extended recently by the use of the readily available 3-hydroxy-2-pyrone (**66**).[38] As shown in Scheme 21, this undergoes cycloaddition in a regiospecific manner to give good yields of benzene or dihydrobenzene derivatives depending on the nature of the dienophile. These reactions are noteworthy because in them the 3-hydroxypyrone is functioning as a synthetic building block equivalent to the vinylketen (**67**).

(**67**) ≡ (**66**) —ii→ OH, CO_2Me, Me (my)

(**66**) —i→ OH, CO_2Me, CO_2Me (hy)

i, $MeO_2C—C{\equiv}C—CO_2Me$; ii, $MeCH{=}CHCO_2Me$.

SCHEME 21

It has already been pointed out that some of the photochemical transformations discussed earlier are synthetically useful. In this connection the preparations of cyclooctatetraenes (Scheme 6) and cyclobutadiene (Scheme 4) deserve special mention.

18.2.2.5 Some important α-pyrones

Dehydroacetic acid (**68**) has played an important role in the development of pyrone chemistry because of the ease with which it can be synthesized,[39] and also because it can be readily converted in good yield to the key compounds triacetic lactone (**69**) and 2,6-dimethyl-4-pyrone (**70**) (Scheme 22).[28,40] Of the many tautomers possible, that given in

(**69**) ←i— (**68**) —ii, iii→ (**70**)

i, H_2SO_4; ii, $Ba(OH)_2$; iii, H^+.

SCHEME 22

structure (**68**) has been shown by spectroscopic studies to be the dominant species in solution.[41] Apart from its academic interest the compound has importance as a fungicide. It is produced industrially by dimerization of keten.

Triacetic lactone (**69**) and tetra-acetic lactone (**71**) are of interest in biosynthetic circles because of their relationship to the polyketone acids postulated as intermediates in polyketide biosynthesis. Significantly, both compounds were produced by cultures of the mould *Penicillium stipitatum*, which had been inhibited by treatment with ethionine from

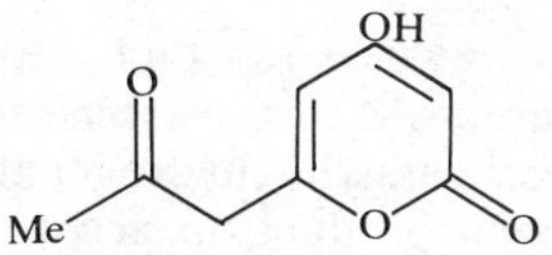

(71) 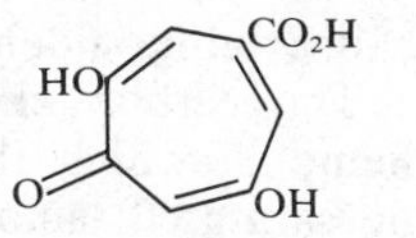 (72)

producing its normal polyketide metabolite stipitatic acid (**72**).[42] There is therefore a possibility that the lactones are intermediates on the biosynthetic pathway and that they accumulate when one of the later steps is blocked. However, an attempt to gain support for this suggestion by incorporation experiments gave negative results and it seems more likely that the true intermediates are too unstable to accumulate and that the pyrones are artefacts derived from them.

A number of steroidal α-pyrones have achieved considerable importance on account of their powerful action on the heart muscle. Typical examples are scillaren A (**73**), produced by *Scilla maritima*, and bufatolin (**74**), one of the active ingredients of the poisonous secretion found in the skin of certain toads.[27,43] Note the close structural relationship of these compounds to the more widely known steroidal butenolides typified by strophanthidin (**75**), which is the kernel of several cardioactive glycosides. It is interesting that the two different heterocyclic systems should be associated in these compounds with the same physiological activity.

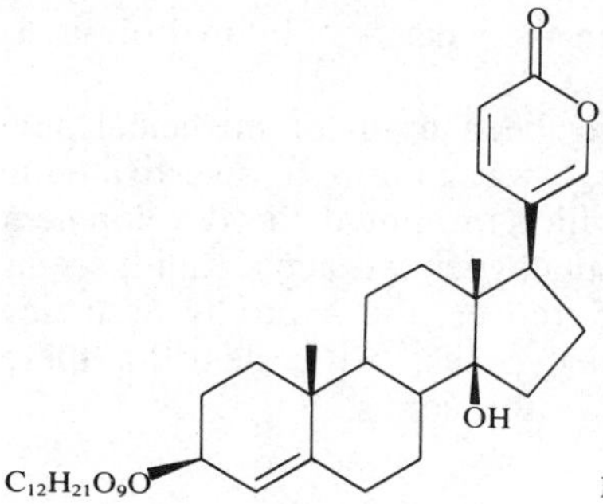

(73) 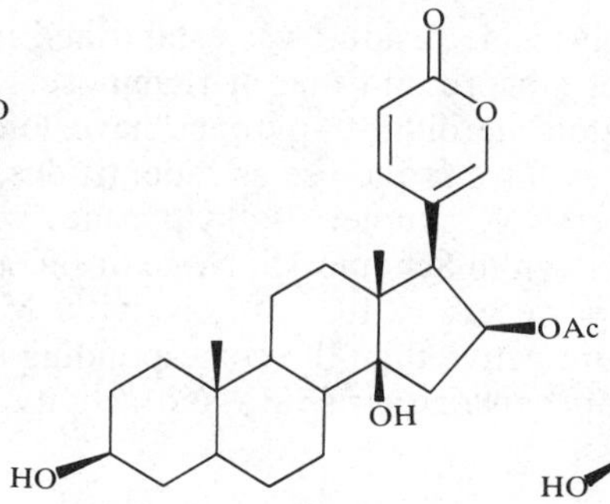 (74) 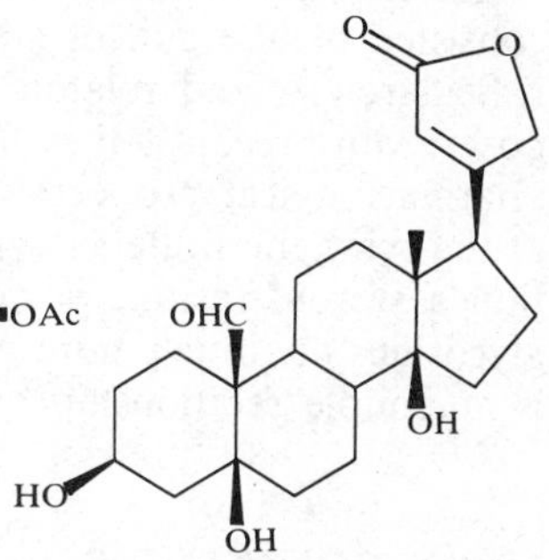 (75)

The presence of an α-pyrone ring in (**73**) was suggested by the i.r. spectrum (ν_{max} 1720 and. 1639 cm^{-1}) and the u.v. spectrum (λ_{max} 300 nm). Unfortunately, early attempts to corroborate this conclusion by chemical tests were complicated by the instability of the other substituents on the steroid nucleus. Eventually this problem was solved by vigorous treatment with acid, which led to the formation of the relatively stable triene (**76**) and which has chemical properties consistent with the presence of an α-pyrone ring (reaction 22): it dissolves slowly in alkali to give a salt from which it is regenerated by treatment

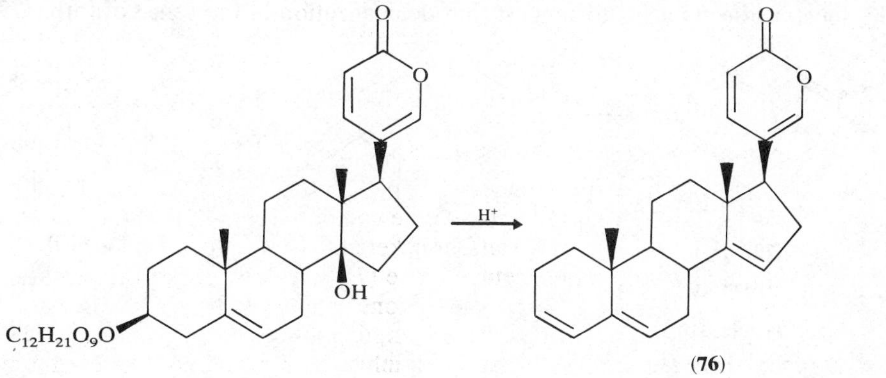

(76) (22)

with acid; the salt exhibits aldehydic properties which shows that the 6-position of the pyrone is free. The carbon skeleton of scillaren A was established by a different degradation scheme in which the first step was catalytic hydrogenation. One of the products was the saturated lactone **(77)**. A second product, in accord with the normal behaviour of an α-pyrone, was the related carboxylic acid **(78)**. The latter was converted by further steps to a known compound.

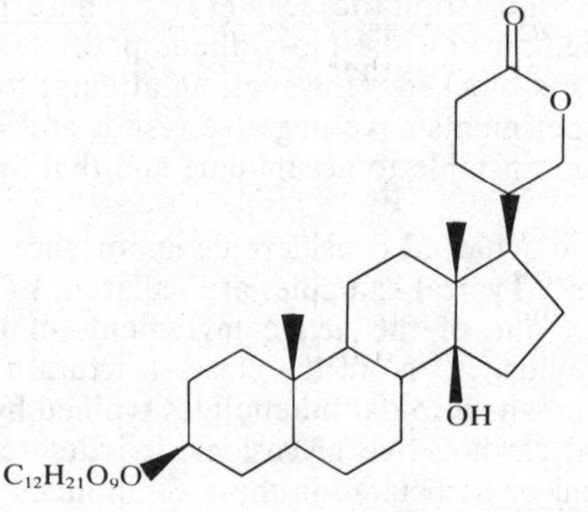

$C_{12}H_{21}O_9O$ OH **(77)**

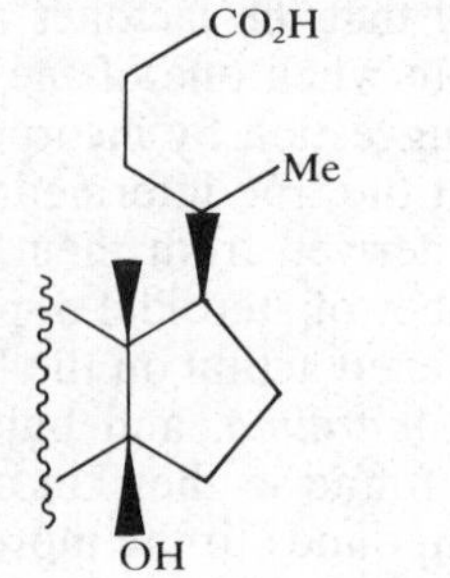

CO_2H Me OH **(78)**

The identity of the glycoside residue was established in the usual way by hydrolysis. It consisted of one unit of glucose and one of rhamnose.

Scillaren A and related steroidal α-pyrones have long been used for medicinal purposes. More recently they have found use as rodenticides. It was a desire to investigate the pharmacological properties of simpler 5-alkylpyrones which prompted the development of the synthetic route shown in Scheme 15. None of the analogues was active and it seems that a steroid residue is necessary for cardioactivity. Moreover, the naturally occurring glycosides are much more active than the corresponding aglycones, so it is clear that there is no simple relationship between structure and activity in this series.

18.2.3 COUMARINS

18.2.3.1 Properties of the nucleus

The heterocyclic ring of a coumarin poses exactly the same question as that considered in connection with the simple pyrone ring: is it aromatic or aliphatic? Once again no clear-cut answer can be given: in some respects the heterocyclic ring behaves as an aromatic system, in others as though it were aliphatic. Overall the balance once again lies strongly in favour of aliphatic properties.

As before, the essence of the problem under consideration is brought out by referring to the alternative canonical forms **(79)** and **(79a)** of coumarin. No evidence can be found in the spectra of coumarins to suggest that delocalization in the sense of form **(79a)** takes

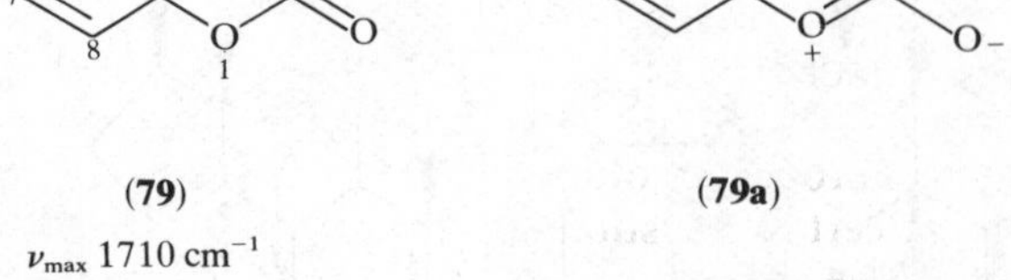

(79)

ν_{max} 1710 cm^{-1}

λ_{max} 275, 284, 310 nm

(ε 10 700, 9120, 5250)

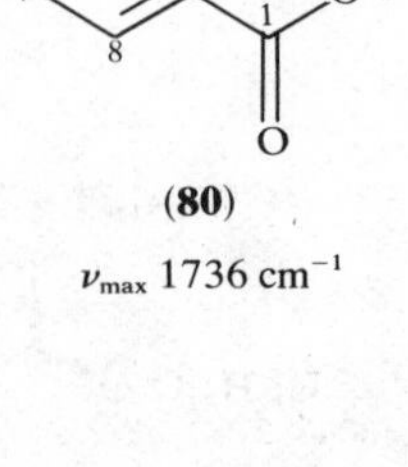

(79a)

(80)

ν_{max} 1736 cm^{-1}

place to a serious extent. The position of the carbonyl frequency in the i.r. spectrum (ν_{max} 1710 cm^{-1}) is in accord with an enol lactone. Thus it is fairly close to that of the isocoumarin (**80**) (ν_{max} 1736 cm^{-1}), but differs significantly from the value (1665 cm^{-1}) for chromone (**10**), the γ-pyrone isomer mentioned in the introduction (this comparison will be discussed more fully in Chapter 18.3)[44] Similarly, the u.v. spectrum of (**79**) can be satisfactorily accommodated in terms of an enol lactone structure.[45]

In the ^{1}H n.m.r. spectrum of coumarin the signals from the protons at C-3 and C-4 appear in the olefinic region at δ 6.45 and 7.09, respectively, with a coupling constant of 10 Hz. These values are typical of a *cis*-alkene rather than an aryl ring.[46]

Finally, the ^{13}C n.m.r. spectra of coumarins are consistent with an essentially aliphatic heterocyclic ring: the chemical shifts of C-2, C-3, and C-4 at 160.4, 116.4, and 143.4 p.p.m. respectively are remarkably close to the value for the corresponding carbons in α-pyrone (see Table 1). It has been shown by ^{13}C n.m.r. spectroscopy that coumarin exists in the protonated form in concentrated sulphuric acid.[47]

Nevertheless, coumarin does show some aromatic character in its pattern of reactivity. For example, the carbonyl oxygen can be alkylated by powerful alkylating agents to give stable pyrylium salts (reaction 23).

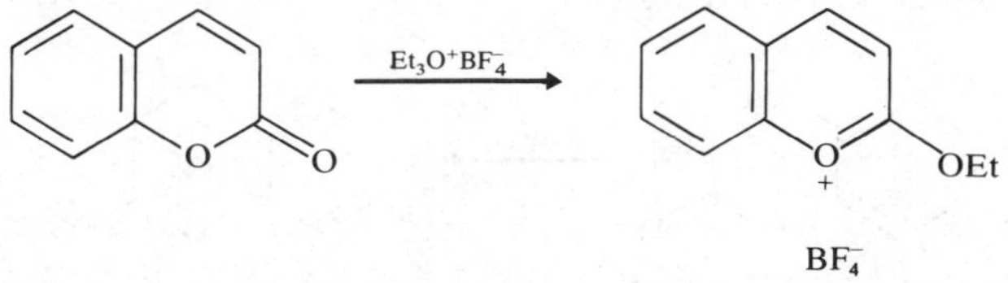

$Et_3O^+BF_4^-$ OEt BF_4^- (23)

In addition, the coumarin nucleus is susceptible to electrophilic substitution. Sulphonation takes place initially in the carbocyclic ring at C-6. Under more forcing conditions a second group can be introduced at C-3 (see Scheme 23). Nitration follows a similar pattern.[48] In contrast, substitution by bromine takes place at C-3 rather than in the carbocyclic ring. However, this reaction follows the same course as the reaction of bromine with α-pyrone: the initial product is a dibromo-adduct (**81**) which subsequently eliminates HBr to form the 3-bromocoumarin (**82**) (see Scheme 24).[49,50] A similar sequence of events may be involved in the chloromethylation at C-3 (reaction 24).

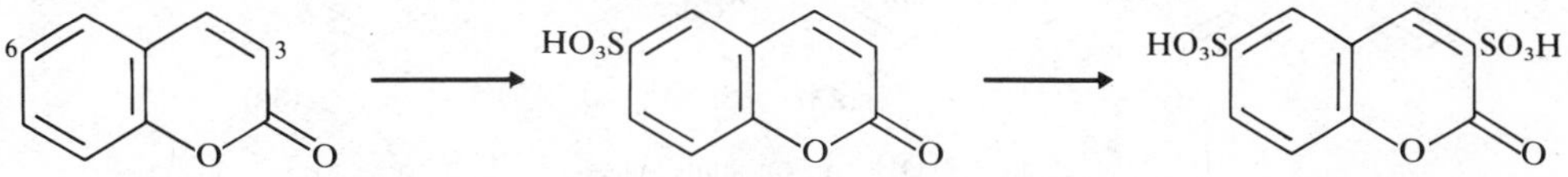

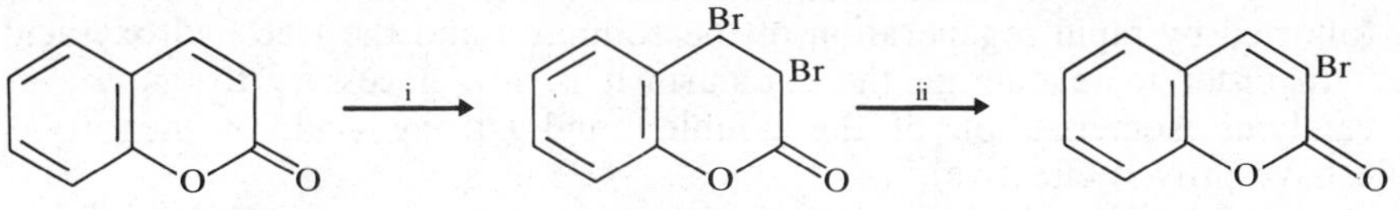

6 3 HO_3S HO_3S SO_3H

SCHEME 23

Br Br Br (**81**) (**82**)

i, Br_2; ii, heat.
SCHEME 24

As in the case of simple pyrones, the properties of the heterocyclic ring of coumarin are greatly influenced by the presence of certain substituents. Thus a hydroxy-group at C-4 enhances the electrophilic activity of the pyrone to such an extent that substitution by nitro and sulphonate as well as halogen takes place at C-3 in preference to C-6.[51] The system is also susceptible to attack by aldehydes to form bis-adducts (reaction 25).

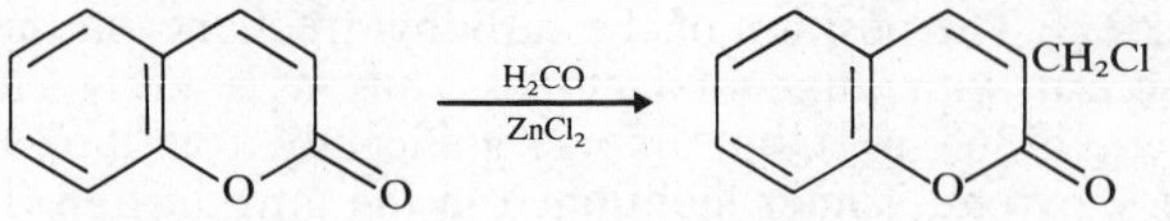

(24)

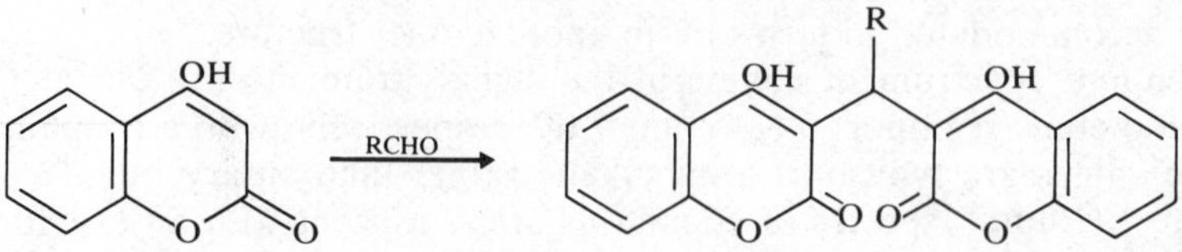

(25)

The dominant reactivity of the heterocyclic ring, however, is attack by nucleophiles rather than electrophiles and in this respect the coumarin ring follows the pattern set by simple pyrones but with one important difference: as in the case of benzopyrylium salts, reaction can take place at C-2 or C-4 but not at the position equivalent to C-6 of a simple pyrone. Generally speaking, weaker nucleophiles attack at C-4 to give an addition product, as in the reaction of bisulphite (reaction 26).[52] Other nucleophiles which behave in a similar way include CN^- and MeO^-.[53]

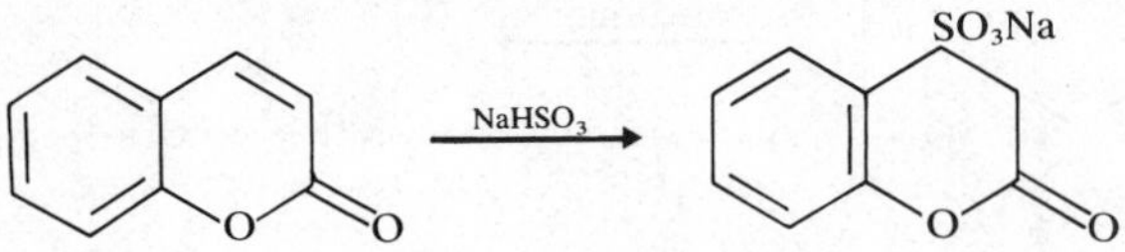

(26)

Stronger nucleophiles give products resulting from attack at C-2. This behaviour is exemplified by the reaction with water in Scheme 25.[54] Following addition of the nucleophile at C-2 the heterocyclic ring cleaves to give the dianion of a phenolic acid.

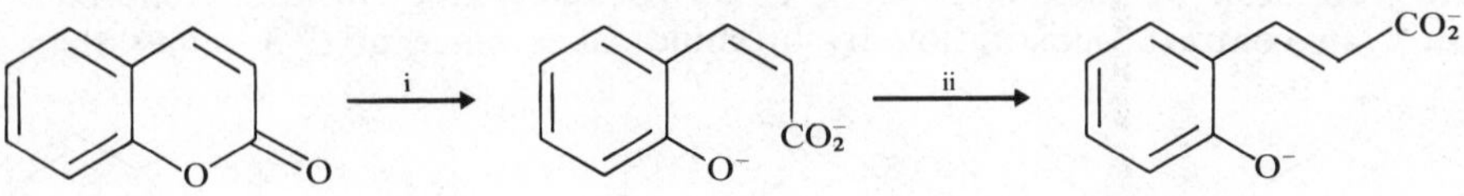

(**83**) (**84**)

i, OH^-; ii, on standing.

SCHEME 25

Initially the double bond is *cis* as shown in (**83**) (this product is known as the coumarinic acid). Acidification of the solution at this stage is followed by a rapid recyclization to reform the coumarin (this reaction is so favourable it can even be observed under weakly basic conditions). However, on prolonged contact with base the *cis*-acid slowly isomerizes to the *trans*-isomer (**84**) (known as the coumaric acid). Acidification at this stage is no longer followed by rapid regeneration of the coumarin and the free hydroxy-acid can be isolated. In order to regenerate the coumarin it is now necessary to employ a reagent which catalyses isomerization of the double bond (strong acid, or mercury salts for example have proved effective).[55]

Hydroxylamine adds at both C-2 and C-4, leading to the hydroxamic acid (**85**) (reaction 27).[56] Unlike coumarins, isocoumarins can react further with amines and, like their

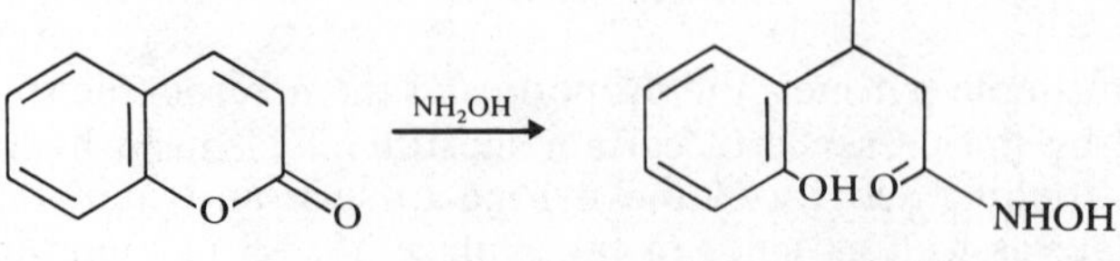

(27)

(**85**)

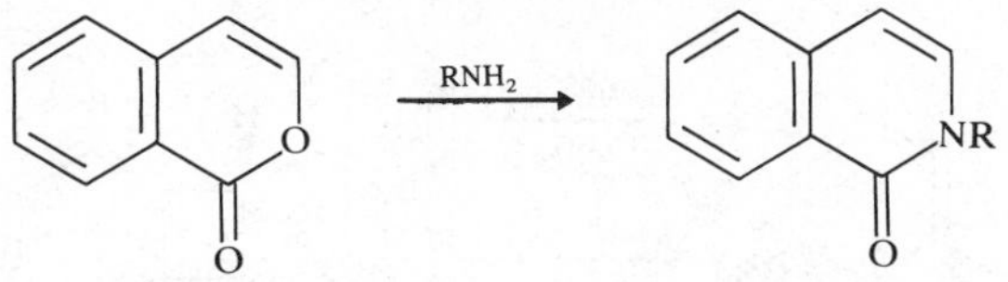

(28)

pyrylium salt analogues, may be converted to the corresponding nitrogen heterocycles (reaction 28).

Reaction of the coumarin ring with a Grignard reagent can lead to different products, depending on the site of attack (*e.g.* reaction 29[57]). Usually, however, a complex mixture is obtained and the reaction is not of practical value. The problem here is the irreversible nature of the addition, which means that the addition is subject to kinetic control. In

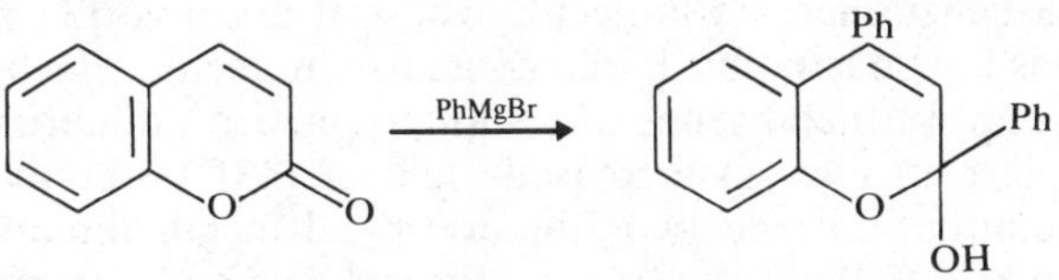

(29)

contrast, the reactions discussed previously are reversible and, being subject to thermodynamic control, they can occur to form the thermodynamically most stable product, with a greater chance of discrimination between the various types of reaction. In the attack of hydroxide, for example (Scheme 25), nucleophilic addition at C-4 may well be more rapid than that at C-2. However, the product of the latter mode of attack can ionize after ring opening and so it provides the thermodynamic trough into which the bulk of the material eventually finds its way.

It is not surprising therefore that kinetically controlled reduction of the heterocyclic ring with $LiAlH_4$ gives a mixture of products and so is not usually preparatively useful (reaction 30). Catalytic hydrogenation, on the other hand, is useful. The reaction is slow but the heterocyclic ring is reduced selectively to the 3,4-dihydro-derivative (**86**) (reaction 31).[58] This reduction is effected more conveniently by sodium amalgam.[59]

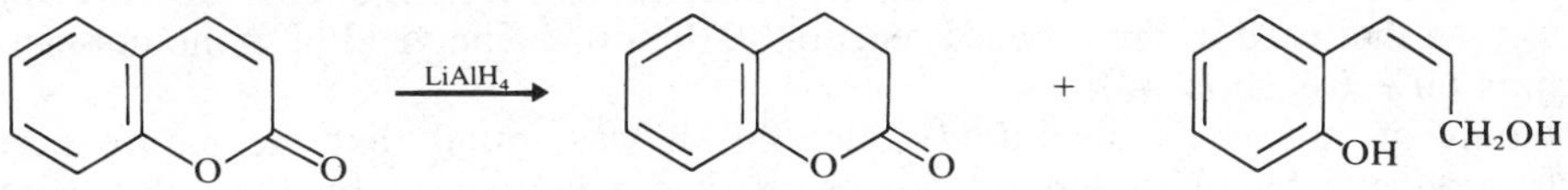

(30)

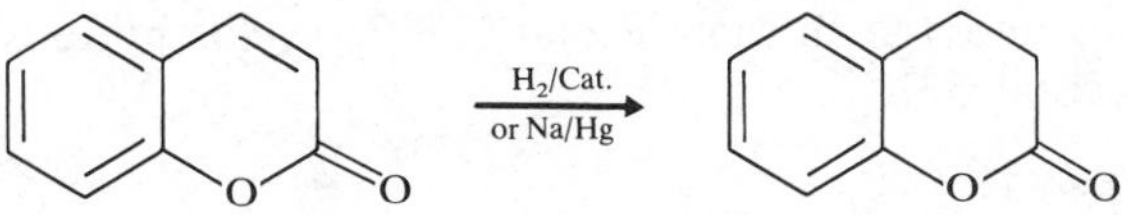

(31)

(**86**)

Benzannulation causes a profound modification of the behaviour of the heterocyclic ring in pericyclic reactions. Thus the system no longer behaves as a diene towards Diels–Alder reagents because in doing so it would have to suffer disruption of the aromatic π-system of the carbocyclic ring.

The photochemical behaviour of the heterocyclic ring is also profoundly altered and the various ring-contraction and rearrangement reactions which dominate the photochemistry of simple pyrones are not observed. Instead the dominant reaction with coumarins is one of the minor reactions of simple pyrones: [2+2] cycloaddition either with a second molecule of coumarin to form a dimer or with a different alkene to form a mixed cycloadduct.

The former process can lead to four isomeric products, (**87**)–(**90**) and all have been detected.[60] The head-to-head dimer (**87**) is formed in high yield in the presence of the triplet sensitizer benzophenone (reaction 32). The process has been thoroughly studied

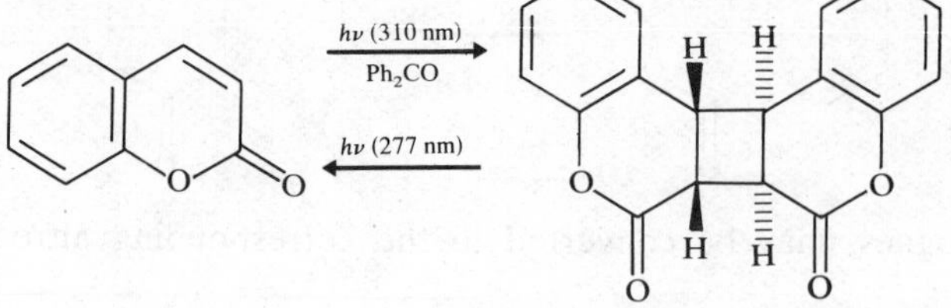

$h\nu$ (310 nm), Ph_2CO ⇌ $h\nu$ (277 nm) (32)

(87)

and has been found to follow a remarkably complicated course.[61] First, coumarin itself absorbs the radiant energy and is converted to its singlet excited state; singlet excitation is then transferred to a benzophenone molecule where it is converted to triplet excitation; the triplet excitation is now passed back to a coumarin molecule which then participates in the dimerization process. In the absence of a triplet sensitizer, dimerization *via* the singlet excited state takes place to give an increased yield of **(88)** at the expense of its stereoisomer **(87)**. Small amounts of the two isomeric head-to-tail dimers **(89)** and **(90)** are sometimes formed in addition, depending on the solvent and the nature of the excited species undergoing dimerization.

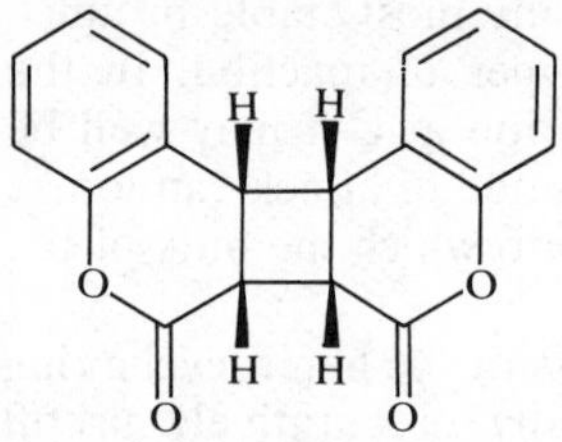

(88) 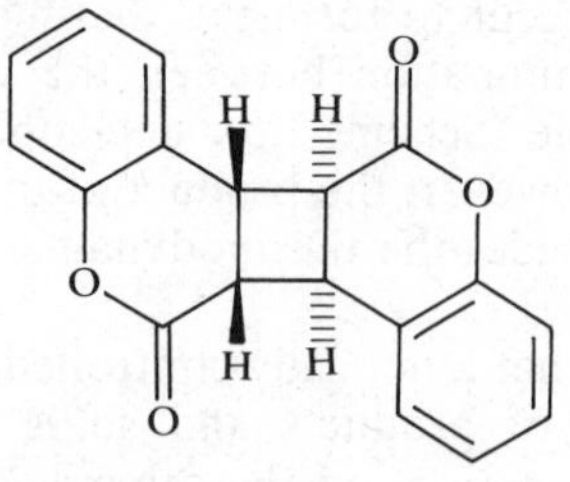 **(89)** 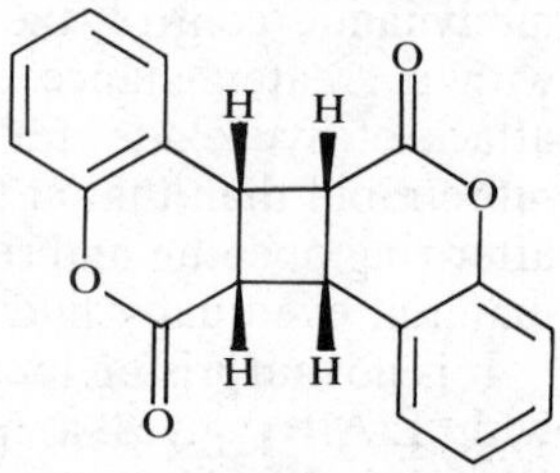 **(90)**

There is continuing interest in the photodimerization. Recently it has been shown that the process can be reversed by irradiating the dimer **(87)** with light of shorter wavelength (277 nm) than that used in the forward reaction (310 nm).[62] The yield of monomer can be as much as 60% (reaction 32).

In a study of various 6-substituted coumarins it was found that the nature of the products produced by photodimerization is markedly influenced by the substituent.[63] There is also much interest in the photochemistry of compounds such as **(91)** in which two linked coumarin rings undergo intramolecular [2+2] cycloadditions to give macrocyclic rings, as in **(92)** (reaction 33).[64]

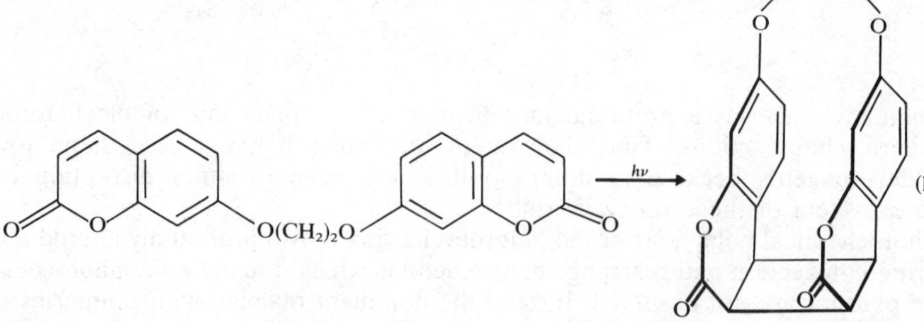

$O(CH_2)_2O$ —$h\nu$→ $(CH_2)_2$ (hy) (33)

(91) **(92)**

Such photodimerizations are not peculiar to the α-pyrone ring. Cinnamic acids undergo equivalent dimerizations to form dimeric products (the truxillic and truxinic acids) and so in these photochemical reactions the heterocyclic ring behaves in essentially the same way as its open-chain aliphatic analogues.[65]

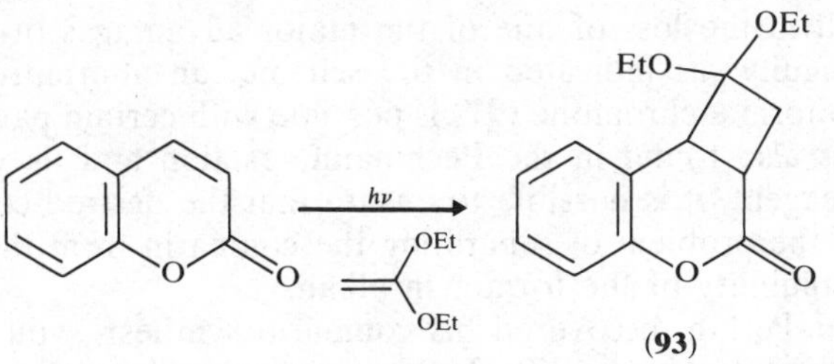

(34)

(93)

The same can be said of the [2+2] cycloadditions with added alkenes. These take place in competition with photodimerization in the presence of suitable alkenes. A typical example is the formation of **(93)** (reaction 34).[66]

18.2.3.2 Synthesis of coumarins

The history of coumarin synthesis began in the mid-nineteenth century with the discovery by Perkin of the famous synthesis which now bears his name.[67] The parent compound was obtained by heating *o*-hydroxybenzaldehyde with acetic anhydride in the presence of sodium acetate (reaction 35). The *trans*-isomer of *o*-hydroxycinnamate can be

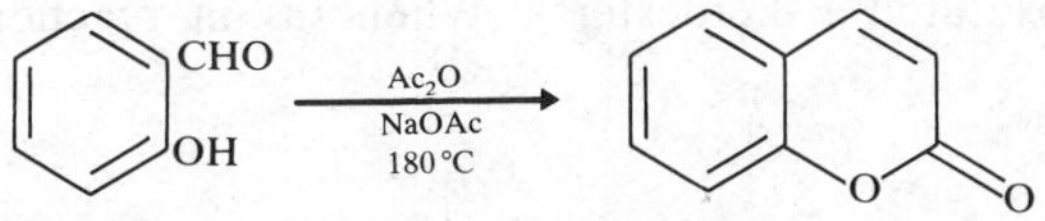

(35)

ruled out as an intermediate in the process because it does not cyclize under the conditions of the reaction. Any *trans*-cinnamic acid formed as a by-product in the reaction can be readily cyclized to the coumarin following isomerization of the *trans*-double bond by irradiation or treatment with iodine.

A variant of the Perkin reaction, in which a coumarin is formed under much milder conditions, utilizes malonic acid (reaction 36). The product **(94)** may be decarboxylated to a coumarin by one of the standard methods.

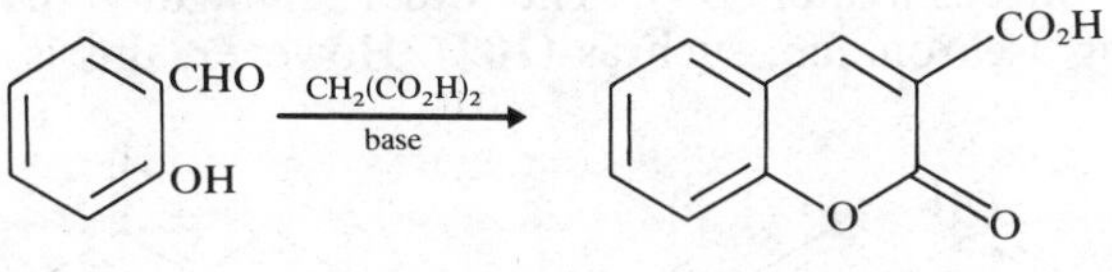

(36)

(94)

Another variant, the Kostanecki–Robinson reaction,[68] can be used to prepare 3- and 4-substituted coumarins as indicated in structure **(96)** (Scheme 26), although this extra

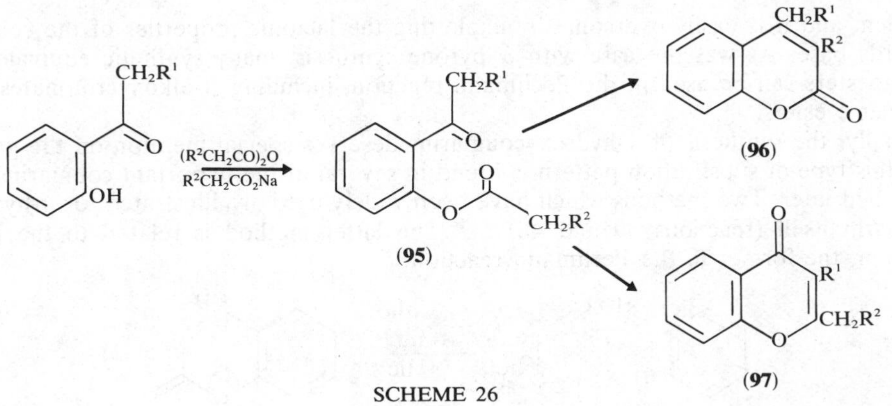

(95) **(96)** **(97)**

SCHEME 26

flexibility may result in the loss of one of the major advantages of the Perkin reaction, which is its unambiguity: as indicated in the scheme, an alternative cyclization of the intermediate (**95**) to form a chromone (**97**) is possible with certain patterns of substitution. This disadvantage is also found in the Pechmann reaction and its variants. Usually, by suitable choice of reagent, it is possible to ensure that the desired coumarin is the major product. Moreover, the problem of separating the coumarin from the chromone is made easy by the ready solubility of the former in alkali.

Fifteen years after Perkin discovered his coumarin synthesis, von Pechmann reported his alternative method (reaction 37). In the example illustrated, malic acid (**98**) is

CO_2H HO OH + $\xrightarrow{H_2SO_4}$ O O (37)

CO_2H

(**98**)

converted *in situ* to formyl acetic acid, which is the active participant in the formation of the heterocyclic ring.[69] In its more general form the synthesis employs a β-ketoester and thus gives ready access to 4-substituted coumarins (coumarins substituted at C-3 can be prepared from α-substituted β-ketoesters). When (as in reaction 38), two modes of

R O + CO_2Et HO OH $\xrightarrow{H^+}$ R HO O O (38)

(**99**) (**100**)

cyclization on to the benzene ring are available, the formation of a 7-hydroxycoumarin (**100**) is favoured over the 5-hydroxy-isomer (**101**); this follows the normal pattern of electrophilic attack on resorcinol (**99**). The other ambiguity in this reaction is the possibility of forming a chromone, such as (**102**). However, this is not usually a serious

OH R O O O HO O R

(**101**) (**102**)

problem and it is easily overcome by exploiting the lactonic properties of the coumarin towards base. As was the case with α-pyrone synthesis, many synthetic equivalents of β-ketoesters can be used in the Pechmann reaction, including β-alkoxycrotonates[70] and acetylenic esters.[71]

Finally, the synthesis of 4-hydroxycoumarins deserves special mention on the grounds that this type of substitution pattern is found in several of the important coumarins to be described later. Two methods which have been widely used are illustrated for 4-hydroxycoumarin itself (reactions 39 and 40).[72,73] The latter method is related to the Perkin reaction, the former to the Pechmann reaction.

OH + HO_2C CO_2H $\xrightarrow[ZnCl_2]{POCl_3}$ OH O O (my) (39)

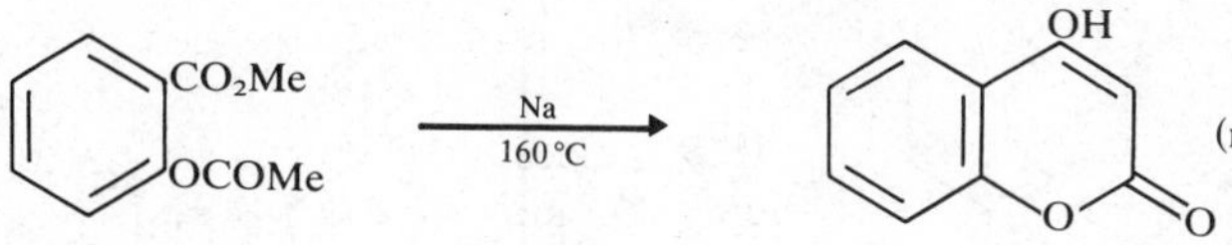

$$\text{C}_6\text{H}_4(\text{CO}_2\text{Me})(\text{OCOMe}) \xrightarrow[160\,°\text{C}]{\text{Na}} \text{4-hydroxycoumarin} \quad (\text{my}) \qquad (40)$$

18.2.3.3 Important coumarins

Coumarin itself is a natural product found in a number of plant sources, including the sweet clover and the tonka bean. It has an aromatic odour often flatteringly associated with the sweet fragrance of freshly mown hay. It was once used as a flavouring matter, particularly in chocolate, but the gradual tightening of controls on the use of food additives has led to its being abandoned on the grounds of toxicity.

Coumarin derivatives occur widely in the plant kingdom.[43,57] Usually the carbocyclic ring of the nucleus is oxygenated at C-7, less frequently at C-5, C-6, and C-8. These extra phenolic hydroxyls are sometimes derivatized as glycosides. Typical examples are umbelliferone (**103**), which occurs in many different plants, and skimmin, its β-glycoside, which is found in *Skimmia japonica* Thunb. Derivatives of umbelliferone have attracted interest as sun-burn preventatives on account of their wide spectrum of ultraviolet absorption.

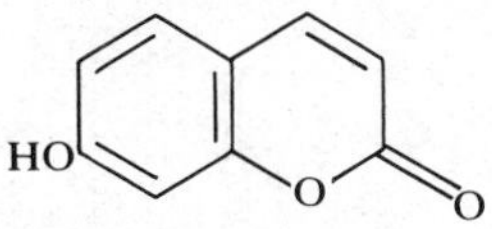

HO (**103**)

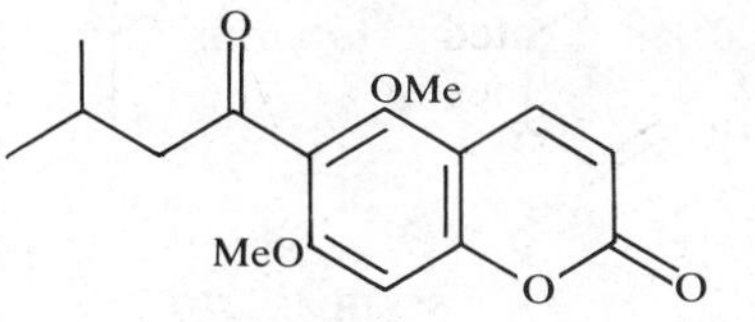

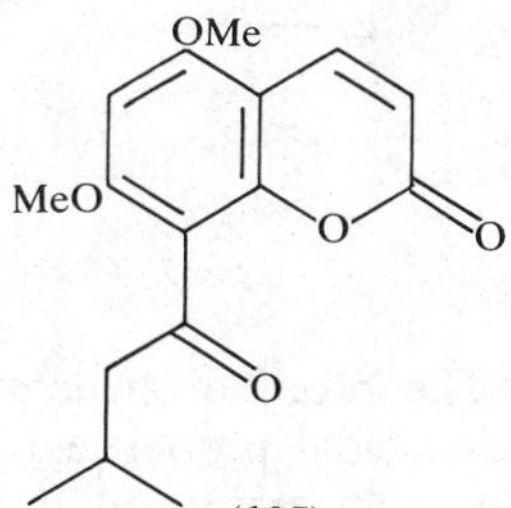

OMe, MeO, O (**104**)

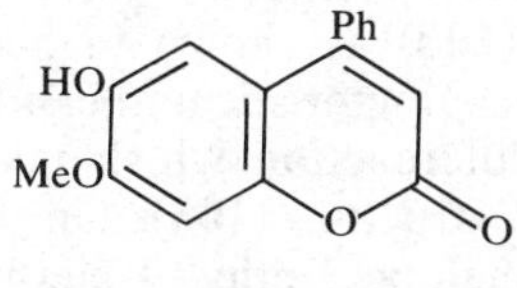

OMe, MeO, O (**105**)

Ph, HO, MeO (**106**)

OH, HO, O (**107**)

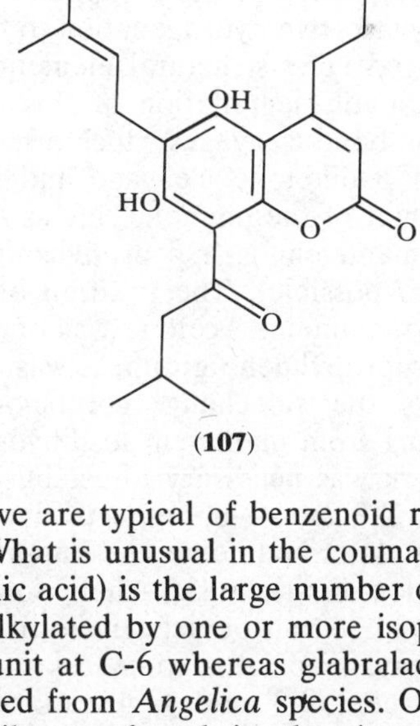

The patterns of oxygenation mentioned above are typical of benzenoid rings in C_6–C_3 units derived *via* the shikimic acid pathway. What is unusual in the coumarin family (by comparison with alkaloids derived from shikimic acid) is the large number of compounds in which the aryl ring of the C_6–C_3 units is alkylated by one or more isoprenoid units. Angelicone (**104**), for example, carries a C_5 unit at C-6 whereas glabralactone (**105**) is substituted at C-8. Both compounds are isolated from *Angelica* species. Other substitution patterns peculiar to the coumarin family are found in the 4-aryl compounds represented by dalbergin (**106**) (from the heartwood of *Dalbergia sissoo*) and the 4-alkyl derivatives such as mammein (**107**) (from the seeds of *Mammea americana* L.). Note too that (**104**), (**105**), and (**107**) have the phloroglucinol pattern of oxygenation which is unusual in a C_6–C_3 unit outside the coumarin family.

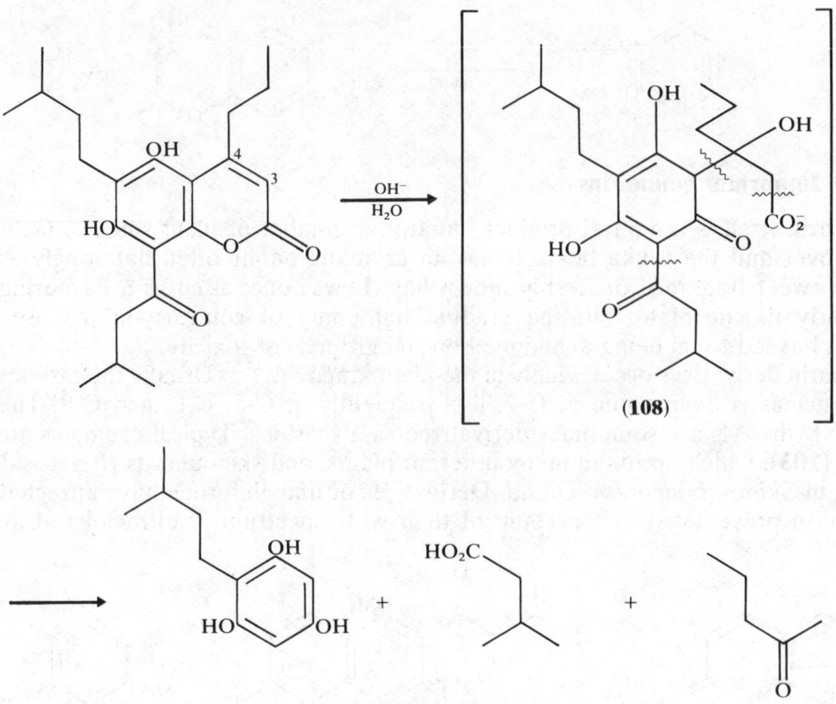

SCHEME 27

The proof of structure for mammein (which is of special interest on account of its insecticidal properties) provides an interesting case study which illustrates the methods generally employed in the coumarin field.[43] Evidence for the presence of a coumarin nucleus was to be found in the i.r. (ν_{max} 1724 cm^{-1}) and u.v. spectra (λ_{max} 295 nm), and in the lactonic properties of the compound. The presence of an isolated double bond was established by selective hydrogenation to form a dihydro-derivative in which the coumarin nucleus survived. The structural elements of the carbon skeleton were ascertained by vigorous hydrolytic degradation of this derivative (**108**) as shown in Scheme 27. The carbon–carbon bond cleavages which occur here are characteristic and can be rationalized formally as a β-diketone cleavage and a retro-aldol reaction which takes place on a ketonic tautomer of the phenolic ring as indicated in structure (**108**). On this basis it was argued that mammein is a 4-propylcoumarin (though a 3-ethyl-4-methyl substitution pattern is also possible). The position of the isolated double bond was determined by ozonolysis of mammein. Acetone was one of the products and so one of the C_5 residues contains an isopropylidene group; it was assigned to the non-ketonic residue on account of the lack of spectral change corresponding to the reduction of an α,β-unsaturated ketone in going from mammein to dihydromammein.

Further work was necessary to establish that the ketonic and non-ketonic C_5 residues were sited at C-8 and C-6, respectively, rather than the reverse. However, the experiments already described cover the important facets of coumarin chemistry illustrated in this investigation. First, the side-chain double bond is much more reactive than that of the pyrone ring and so can be selectively ozonized or reduced. Secondly, the characteristic hydrolytic cleavage of the coumarin nucleus, although it should be mentioned that cleavage between C-3 and C-4 can compete with cleavage between C-4 and the benzenoid ring. An example can be found in Perkin's original work on coumarin, which included an alkaline degradation (reaction 41). It was this observation which inspired him to try what is essentially the reverse reaction and so discover his famous synthesis.

Another important group of naturally occurring coumarins are the furocoumarins, represented here by psoralen (**109**), xanthotoxin (**110**), and angelicin (**111**).[43] Members of

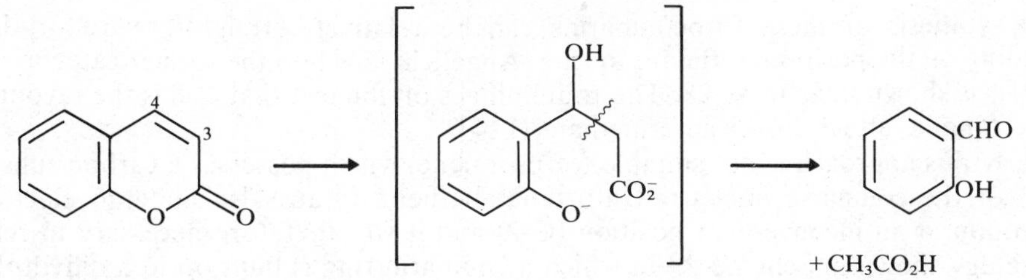

(41)

$+CH_3CO_2H$

this group have interesting physiological properties. Derivatives of angelicin, for example, are toxic to fish but not seriously so to humans, which has led to their being used (in crude form as plant extracts) in South East Asia to simplify the task of gathering fish for food. Other members of the group, including psoralen, have a photosensitizing action on the skin of man which has been exploited in medicine. Xanthotoxin, for example, is found in the seeds of *Ammi majus* L., a plant common in the Nile delta, and the powdered seeds have been eaten locally for centuries as a remedy for improper pigmentation of the skin. After exposure to sunlight the affected areas blister and then gradually assume a normal level of pigmentation.

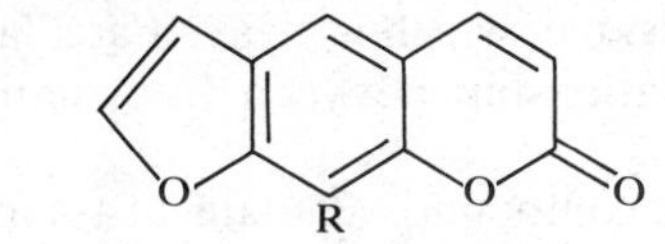

(109) R = H
(110) R = OMe

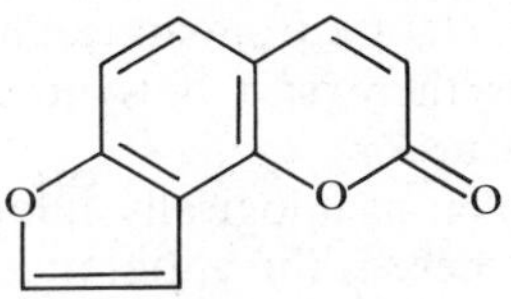

(111)

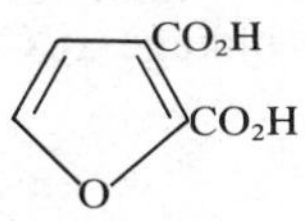

(112)

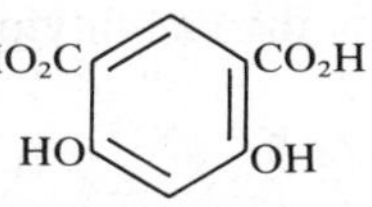

(113)

In determining the structure of these compounds it was usual to resort to oxidative as well as hydrolytic degradations. From controlled oxidation of psoralen, for example, the furan dicarboxylic acid **(112)** was isolated. This result together with the isolation of **(113)**, derived from the benzenoid ring, and the lactonic properties of the compound, leads to a unique structure for psoralen.

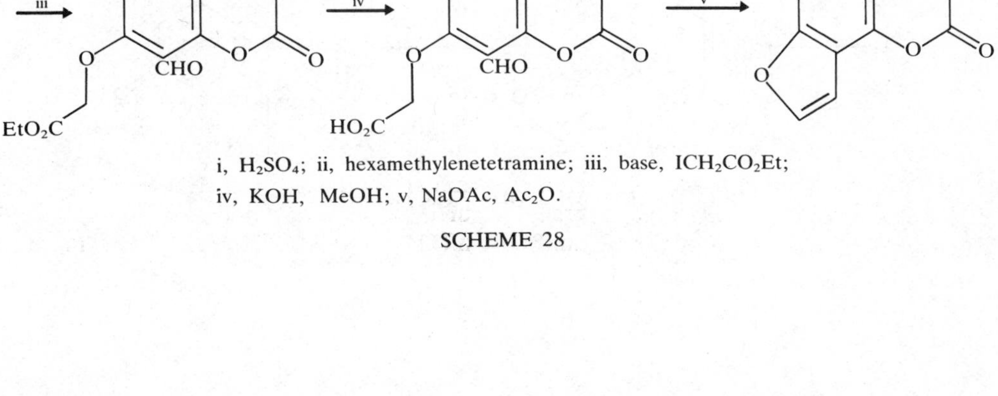

(114)

i, H_2SO_4; ii, hexamethylenetetramine; iii, base, ICH_2CO_2Et; iv, KOH, MeOH; v, NaOAc, Ac_2O.

SCHEME 28

The synthesis of these furocoumarins can be relatively straightforward or difficult, depending on the position of the furan ring. Angelicin falls into the former category and its synthesis is shown in Scheme 28. The route hinges on the fact that C-8 is the favoured site of electrophilic attack in the intermediate (**114**).[43]

Clearly this approach is not suitable for psoralene, which possesses a carbon substituent at C-6 of the coumarin nucleus. Unfortunately, benzofurans also undergo electrophilic substitution at an inconvenient position (C-2) and it was therefore necessary to resort to the strategy shown in Scheme 29 in which a coumarin ring is built on to a dihydrobenzofuran ring which is dehydrogenated in the last step.

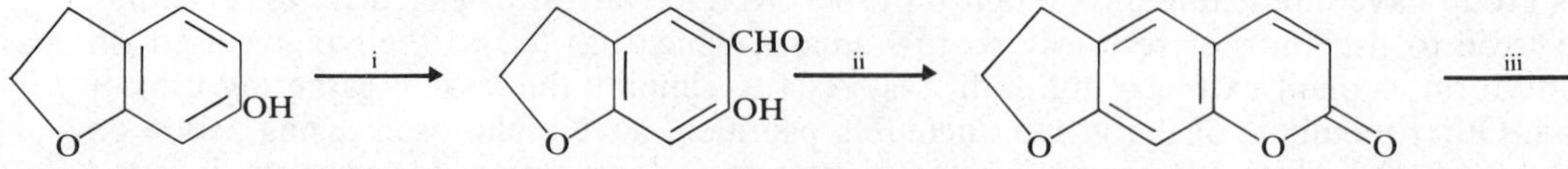

i, HCl, HCN; ii, Ac_2O, NaOAc; iii, Pd, C.

SCHEME 29

In passing it can be mentioned that the two carbons of the furan residue of furocoumarins are derived in Nature from an isopentyl residue, which is cleaved at a late stage of the biosynthesis. Thus there is a biosynthetic relationship between furocoumarins and isoprenylated coumarins.

A number of pharmacologically interesting compounds contain a 4-hydroxycoumarin residue.[43] Firstly, there is the antibiotic novobiocin (**115**), which contains an extraordinary confusion of structual residues derived from several different biosynthetic sources. The only one relevant to the present account is the coumarin. Its presence was established by the degradation shown in Scheme 30, which starts from the abbreviated structure (**115a**). The main point to note is the stability of the amino linkage at C-3: though formally part

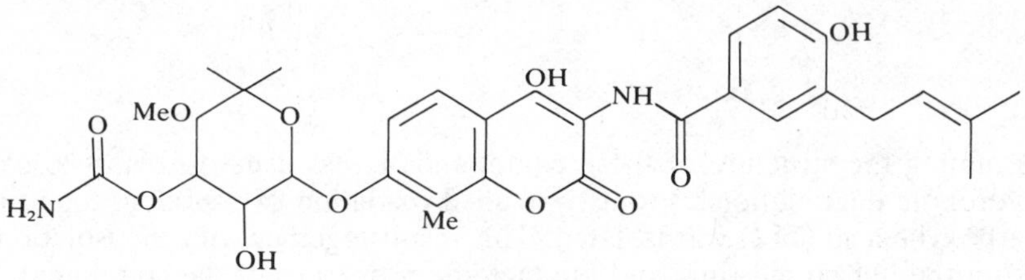

(**115**)

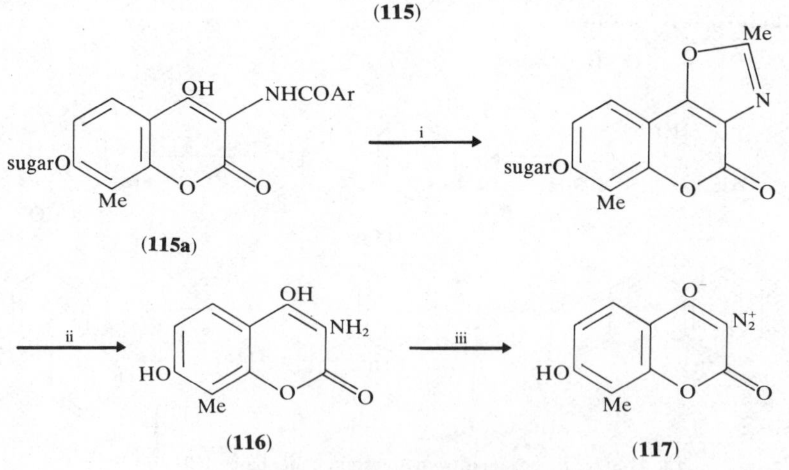

(**115a**)

(**116**)

(**117**)

i, Ac_2O; ii, OH^-, H_2O; iii, HNO_2.

SCHEME 30

of an enamine residue this survived the hydrolytic steps and thus shows in (**116**) some of the stability of an arylamine. In keeping with this, the group can be diazotized to give a stable 3-diazo-derivative (**117**).

A second important group of 4-hydroxycoumarins includes compounds which have an anticoagulant effect on blood. The first to be discovered was dicoumarol (**118**), which was identified as the cause of sweet clover disease in cattle. This condition, which is brought on by eating spoiled sweet clover, produces serious and often fatal haemorrhages in affected beasts. Dicoumarol can be readily synthesized by condensation of 4-hydroxy-coumarin with formaldehyde (reaction 42). For a while it was used in medicine to reduce

(**118**)

the blood-clotting ability of patients suffering from cardiovascular disease. It has now been superseded by other more effective 4-hydroxycoumarins. Chief among these is warfarin (**119**) (though this compound is perhaps better known for its use a rodenticide than its use in medicine).

(**119**)

18.2.4 CONCLUSION

The properties of the α-pyrone ring are essentially those of an aliphatic system and there is little evidence of aromatic character. Many of the reactions of the system have synthetic potential. This aspect of pyrone chemistry has attracted much interest in recent years and seems likely to continue to do so in the future.

The coumarin system shows relatively little modification of the properties of the heterocyclic ring as a result of benzannulation. However, given the important physiological properties shown by many of its derivatives, coumarin can be justifiably regarded as one of the more important classes of heterocyclic compound.

For further reading on α-pyrones and coumarins, see references 27, 43, 57, and 68.

References

1. S. Wawzonek, in 'Heterocyclic Compounds', ed. R. C. Elderfield, Wiley, New York, 1951, vol. 2, chapter 7.
2. A. R. Katritzky and A. P. Ambler, in 'Physical Methods in Heterocyclic Chemistry', ed. A. R. Katritzky, Academic Press, New York, 1963, vol. 2, p. 252.
3. J. Fried and R. C. Elderfield, *J. Org. Chem.*, 1941, **6,** 566.
4. W. H. Pirkle and M. Dines, *J. Heterocyclic Chem.*, 1969, **6,** 1.
5. W. H. Pirkle and M. Dines, *J. Heterocyclic Chem.*, 1969, **6,** 313.
6. S. Sib, *Tetrahedron*, 1975, **31,** 2229.
7. D. A. Griffin and J. Staunton, *J.C.S. Chem. Comm.*, 1975, 675.
8. N. P. Shusherina, N. D. Dmitrieva, and R. Y. Levina, *Zhur. obshchei Khim.*, 1961, **31,** 2794.
9. W. H. Pirkle and M. Dines, *J. Org. Chem.*, 1969, **34,** 2239.
10. G. Vogel, *Chem. and Ind.* (*London*), 1962, 968, 1829.

11. J. A. Leben, *Ber.*, 1896, **29,** 1673.
12. M. J. D. van Dam and F. Kögl, *Rec. Trav. chim.*, 1964, **83,** 39.
13. O. Diels and K. Alder, *Annalen*, 1931, **490,** 257.
14. J. D. Bu'Lock and H. G. Smith, *J. Chem. Soc.*, 1960, 502.
15. J. A. Reed, C. L. Schilling, Jr., R. F. Tarvin, T. A. Rettig, and J. K. Stille, *J. Org. Chem.*, 1969, **34,** 2188.
16. W. M. Horspool, in 'Photochemistry', ed. D. Bryce-Smith, Chemical Society, London, 1974, vol. 5, p. 385.
17. C. L. McIntosh and O. L. Chapman, *J. Amer. Chem. Soc.*, 1973, **95,** 247.
18. C. T. Bedford, J. M. Forrester, and T. Money, *Canad. J. Chem.*, 1970, **48,** 2645.
19. P. de Mayo and R. W. Yip, *Proc. Chem. Soc.*, 1964, 84.
20. M. van Meerbeck, S. Toppet, and F. C. de Schryver, *Tetrahedron Letters*, 1972, 2247.
21. H. Takeshita, R. Kikuchi, and Y. Shoji, *Bull. Chem. Soc. Japan*, 1973, **46,** 690.
22. J. L. Douglas and T. Money, *Canad. J. Chem.*, 1968, **46,** 695.
23. J. L. Bloomer, S. M. H. Zaidi, J. T. Strupczewski, C. S. Brosz, and L. A. Gudzyk, *J. Org. Chem.*, 1974, **39,** 3615.
24. D. Herbst, W. B. Mors, O. R. Gottlieb, and C. Djerassi, *J. Amer. Chem. Soc.*, 1959, **81,** 2427.
25. R. Anschütz, P. Bendix, and W. Kerp, *Annalen*, 1890, **259,** 148.
26. L. F. Cavalieri, *Chem. Rev.*, 1947, **41,** 525.
27. J. Fried, in 'Heterocyclic Compounds', ed. R. C. Elderfield, Wiley, New York, 1950, vol. 1, chapter 7.
28. F. Arndt, in 'Organic Syntheses', ed. C. F. H. Allen, Wiley, New York, 1940, p. 26.
29. R. M. Anker and A. H. Cook, *J. Chem. Soc.*, 1945, 311.
30. E. Buchner and H. Schröder, *Ber.*, 1902, **35,** 782.
31. R. H. Wiley and N. R. Smith, *Org. Synth.*, 1951, **31,** 23.
32. S. Ruhemann, *J. Chem. Soc.*, 1899, **75,** 245.
33. E. Suzuki and S. Inoue, *Synthesis*, 1975, 259.
34. T. M. Harris, C. M. Harris, and K. B. Hindley, *Fortschr. Chem. org. Naturstoffe*, 1974, 217.
35. G. R. Gogte, *Proc. Indian Acad. Sci.*, 1938, **7A,** 214.
36. H. J. Lohrisch and W. Steglich, *Tetrahedron Letters*, 1975, 2905.
37. T. Money, *Chem. Rev.*, 1970, **70,** 553.
38. E. J. Corey and A. P. Kozikowski, *Tetrahedron Letters*, 1975, 2389.
39. F. Arndt, *Org. Synth. Coll. Vol. 3*, 1955, 231.
40. L. C. King, F. J. Ozog, and J. Moffat, *J. Amer. Chem. Soc.*, 1951, **73,** 300.
41. E. E. Royals and J. C. Leffingwell, *J. Org. Chem.*, 1965, **30,** 1255.
42. R. Bentley and P. M. Zwitkowits, *J. Amer. Chem. Soc.*, 1967, **89,** 676, 681.
43. F. M. Dean, 'Naturally Occurring Oxygen Ring Compounds', Butterworths, London, 1963.
44. V. Prey, B. Kerres, and H. Berbalk, *Monatsh.*, 1960, **91,** 774.
45. B. K. Ganguly and P. Bagchi, *J. Org. Chem.*, 1956, **21,** 1415.
46. M. J. Cook, A. R. Katritzky, and P. Linda, *Adv. Heterocyclic Chem.*, 1974, **17,** 321.
47. S. A. Sojka, *J. Org. Chem.*, 1975, **40,** 1175.
48. A. Clayton, *J. Chem. Soc.*, 1910, **97,** 1388.
49. W. H. Perkin, *Annalen*, 1871, **157,** 115.
50. C. Mentzer, *Compt. rend.*, 1946, **223,** 1141.
51. P. Meunier, C. Mentzer, and A. Vinet, *Helv. Chim. Acta*, **29,** 1291.
52. F. D. Dodge, *J. Amer. Chem. Soc.*, 1930, **52,** 1724.
53. J. Bredt and J. Kallen, *Annalen*, 1896, **293,** 366.
54. E. Cingolani, *Gazzetta*, 1959, **89,** 985, 999.
55. T. R. Seshadri, *Chem. and Ind.* (*London*), 1954, 308.
56. T. Posner and R. Hess, *Ber.*, 1913, **46,** 3816.
57. N. Campbell, in 'Chemistry of Carbon Compounds', ed. E. H. Rodd, Elsevier, Amsterdam, 1959, vol. 4B, chapter 8, p. 875.
58. E. Späth and O. Pesta, *Ber.*, 1933, **66,** 754.
59. F. Wessely and F. Kallab, *Monatsh.*, 1932, **59,** 161.
60. C. H. Krauch, S. Farid, and G. O. Schenck, *Chem. Ber.*, 1966, **99,** 625.
61. G. S. Hammond, C. A. Stout, and A. A. Lamola, *J. Amer. Chem. Soc.*, 1964, **86,** 3103.
62. M. Hasegawa, Y. Suzuki, and N. Kita, *Chem. Letters*, 1972, 317.
63. D. V. Rao, H. Ulrich, F. A. Stuber, and A. A. R. Sayigh, *Chem. Ber.*, 1973, **106,** 388.
64. F. C. De Schryver, J. Put, L. Leenders, and H. Loos, *J. Amer. Chem. Soc.*, 1974, **96,** 6994.
65. M. D. Cohen, G. M. J. Schmidt, and F. I. Sonntag, *J. Chem. Soc.*, 1964, 2000.
66. J. W. Hanifin and E. Cohen, *Tetrahedron Letters*, 1966, 1419.
67. J. R. Johnson, *Org. Reactions*, 1942, **1,** 210.
68. S. Wawzonek, in 'Heterocyclic Compounds', ed. R. C. Elderfield, Wiley, New York, 1951, vol. 2, p. 176.
69. S. Sethna and R. Phadke, in *Org. Reactions*, 1953, **7,** 1.
70. O. Dann and G. Illing, *Annalen*, 1957, **605,** 158.
71. E. Fischer and O. Nouri, *Ber.*, 1917, **50,** 693.
72. V. R. Shah, J. L. Bose, and R. C. Shah, *J. Org. Chem.*, 1960, **25,** 677.
73. H. Pauly and K. Lockemann, *Ber.*, 1915, **48,** 28.

18.3

γ-Pyrones and Chromones

J. STAUNTON
University of Cambridge

18.3.1 INTRODUCTION

Many pertinent comments on the nature of pyrones and their relationship to other oxygen heterocycles were made in the introduction to Chapters 18.1 and 18.2, and so need not be repeated here. Suffice to say that this section will deal with γ-pyrone (**1**) and its important benzannulated derivative chromone (**2**). In the doubly benzannulated derivative xanthone (**3**) the characteristic properties of the pyrone ring are almost completely suppressed for reasons already discussed in connection with the related xanthylium salts. This system will therefore not be covered; its properties have been reviewed elsewhere.[1] Probably the most thoroughly investigated compounds in this area are the flavones and isoflavones, which are derivatives of 2-arylchromone (**4**) and

(**1a**) γ-Pyrone (4*H*-Pyran-4-one)

(**1b**)

(**2**) Chromone (Benzo-4*H*-pyran-4-one)

(**3**) Xanthone

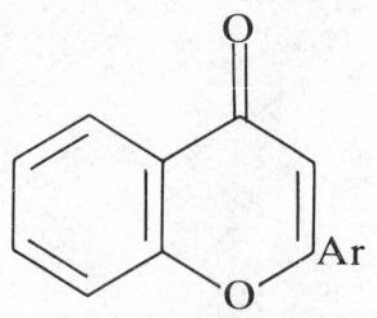

(4)
Flavone

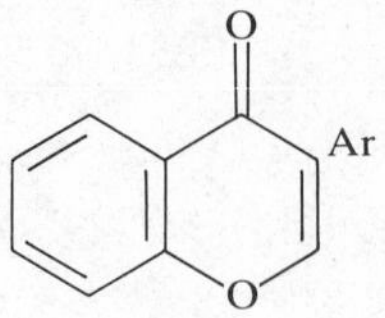

(5)
Isoflavone

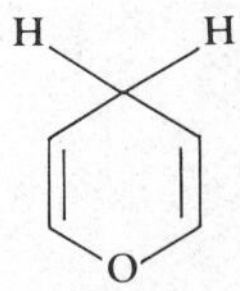

(6)
4*H*-Pyran

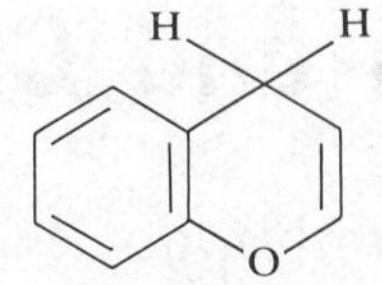

(7)
4*H*-1-Benzopyran
(4*H*-Chromene)
(γ-Chromene)

3-arylchromone (**5**), respectively. These are key members of the flavonoid family of natural products and in view of their importance they are treated separately from simple chromones.

The pyran analogues of γ-pyrone and chromone need to be introduced because they play a major role in the chemistry to be covered in this chapter. These are 4*H*-pyran (**6**) and 4*H*-1-benzopyran (**7**), respectively.

18.3.2 γ-PYRONES

18.3.2.1 Properties of the heterocyclic ring

The chemistry of the γ-pyrone ring poses the same question as that considered in connection with the α-pyrone ring: is the system best viewed as an aromatic pyrylium betaine (**1b**) or an aliphatic dienone (**1a**)? Again it can be stated that most of the properties of the heterocyclic ring can be satisfactorily explained in terms of the aliphatic system. However, some of the properties are unexpected on this basis and therefore the chemistry of γ-pyrones presented a challenging puzzle for early workers in the field. First, the carbonyl group is devoid of normal ketonic properties such as hydrazone or oxime formation (the system does not react normally with phenylhydrazine; cleavage of the heterocyclic ring occurs).[2] This lack of reaction at the carbon of the carbonyl group is compensated by exceptional reactivity at the oxygen of this group. This was first manifested in the isolation of a crystalline hydrochloride on treatment of 2,6-dimethyl-4-pyrone with hydrogen chloride.[3] At first there was dispute as to the structure of this product, some favouring the structure (**8**), in which protonation has taken place on the carbonyl oxygen, others the alternative product (**9**), in which protonation had taken place on the ether oxygen. This intriguing aspect of γ-pyrone chemistry was reinforced by the discovery that crystalline methoxonium salts can be prepared by treatment of γ-pyrones

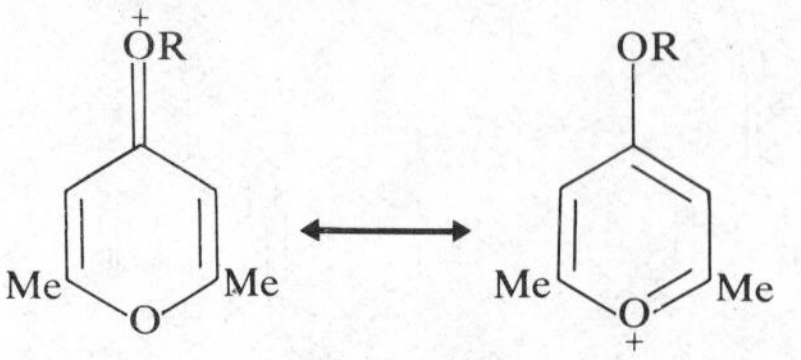

(**8**) R = H
(**10**) R = Me

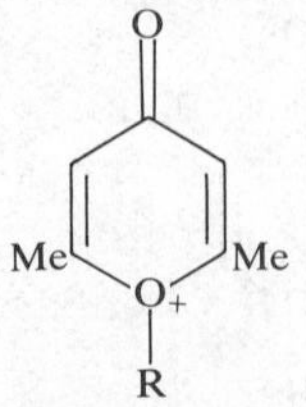

(**9**) R = H
(**11**) R = Me

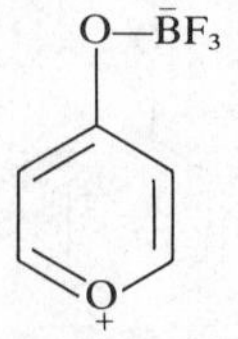

(**12**)

with dimethyl sulphate.[4,5] Again there was debate as to the structure of these salts, both (**10**) and (**11**) being considered as possible candidates.

The puzzle was resolved in 1924 by Arndt, who pointed out that the ether oxygen in γ-pyrone could interact electronically with the carbonyl group and so modify the properties of the latter.[6] This was the first exposition of the theory of mesomerism which was to have a profound influence on the thinking of theoretical chemists in the following decades, and hence γ-pyrone has an important place in the history of organic chemistry.

The fact that γ-pyrone has some of the chemical reactivity which would be expected of the pyrylium betaine structure (**1b**) gives an interesting twist to its chemistry, though it is important to keep the size of this effect in perspective. The pK_a of pyrone is only 0.1. Even in the 2,6-dimethyl derivative, where two methyl groups are ideally placed to stabilize the positive charge in the cation, the value is no more than 0.4.[7] These values are comparable with those of urea (pK_a 0.1) and *p*-nitroaniline (pK_a 1.0), compounds which in the realm of nitrogen bases rate very low in basic strength. Nevertheless, the value for γ-pyrone is remarkably high in comparison with those of other oxygen bases such as acetone (pK_a −7.2).

The basic and nucleophilic properties of the carbonyl oxygen have been much investigated since the pioneering work carried out at the turn of the century. Lewis acids as well as protic acids form oxonium salts; boron trifluoride, for instance, forms derivatives of the type shown in structure (**12**).[8] A wide variety of methylating agents can be used to prepare methoxonium salts, including dimethyl sulphate, trimethoxonium tetrafluoroborate,[9] dimethoxyphenylcarbonium fluoroborate,[10] methyl *o*-nitrobenzenesulphonate,[11] and methyl fluorosulphonate.[12] Methyl tosylate and methyl iodide are not effective, although the latter works well in association with silver tetrafluoroborate.[13] The ethyl analogue can be made using diethyl sulphate. An *O*-acetoxy-compound has been prepared by acetylation of 2,6-dimethyl-4-pyrone with 2-methyl-1,3-dioxolenium fluoroborate as acetylating agent (reaction 1).[8] It is clear from this work that γ-pyrones have a greater degree of aromaticity than α-pyrones.

O, Me, Me, O + Me, O, O⁺ ⟶ OAc, Me, Me, O, + (my) (1)

In spite of the relative ease with which it forms methoxonium salts, the γ-pyrone ring, like the α-isomer, has physical properties which are consistent with an aliphatic rather than an aromatic structure. Thus although the dipole moment (4 D) of γ-pyrone is higher than the calculated value for a non-aromatic system (1.8 D), which indicates a degree of polarization characteristic of a pyrylium betaine, the value calculated for the latter is much higher (22 D) so the degree of delocalization is relatively small.[14] Similarly, the spectral properties of the γ-pyrone ring are consistent with an aliphatic structure. The carbonyl stretching frequency in the i.r. spectrum occurs near 1650 cm^{-1}.[15] The corresponding peaks in the spectra of α-pyrones are found in the range 1730–1740 cm^{-1} and so at first sight it seems that there might be a greater degree of single-bond character in the carbonyl group of the γ-isomer than would be expected on the basis of the aliphatic structure. A better basis for comparison is, however, the carbonyl stretching frequency of cross-conjugated cyclohexadienones which are necessarily aliphatic. These also absorb at about 1650 cm^{-1} and so there is no need to invoke aromaticity in the π-system of the γ-pyrone ring in order to account for the low carbonyl stretching frequency. Moreover, there appears to be a significant degree of double-bond character in the bond between C-4 and oxygen even in the hydrochloride salts of γ-pyrones. Thus the stretching frequency assigned to this bond is usually near 1500 cm^{-1}, which is much higher than that of the C—O bond in phenols (1230–1140 cm^{-1}).[16]

The u.v. spectra of simple γ-pyrones are characterized by an intense absorption band (ε 13 000) near 250 nm.[17] Typical examples are given in formulae (**13**) and (**14**). In both cases the observed λ_{max} can be satisfactorily explained in terms of the aliphatic structure shown.

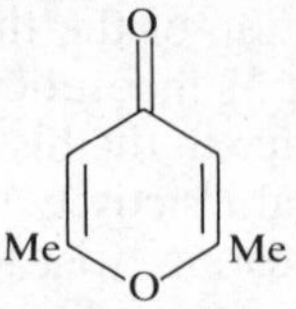

λ_{max} 247 nm (ε 14 000) (**13**)

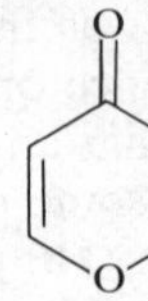

λ_{max} 245 nm (ε 13 000) (**14**)

Some evidence for an aromatic ring current can be found in the n.m.r. spectra of γ-pyrones. Relevant data from the ^{1}H n.m.r. spectra of key compounds are given in Figure 1.[18] Note that the proton resonances in the γ-pyrone (**15**) are both approximately

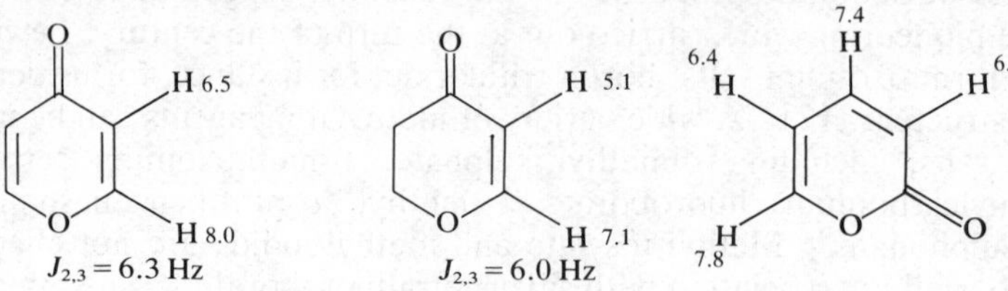

$J_{2,3} = 6.3$ Hz (**15**)

$J_{2,3} = 6.0$ Hz (**16**)

(**17**)

Chemical shifts (δ) in the ^{1}H n.m.r. spectra

FIGURE 1

1 p.p.m. down-field from the corresponding signals in the spectrum of the model compound (**16**). However, from a comparison with the equivalent signals in the spectrum of α-pyrone given in diagram (**17**), it would appear that the ring current in the two pyrones is approximately equal. Thus the signal for the proton at C-2 in γ-pyrone appears at approximately the same chemical shift as that from the proton at C-6 in the α-isomer (both carbons adjacent to the heterocyclic oxygen), and the proton attached to C-3 in the former gives a signal in essentially the same position as that from the proton at C-2 of the latter (both positions adjacent to the carbonyl group).

The same pattern emerges from a comparison of the ^{13}C n.m.r. spectra of equivalent α- and γ-pyrones (Figure 2). Once again the chemical shifts for C-6 and C-2 of the α-isomer (**18**) correspond with those of C-6 (or C-2) and C-3 (or C-5), respectively, of the γ-isomer (**19**). Note, however, that there is an interesting difference in the chemical shifts

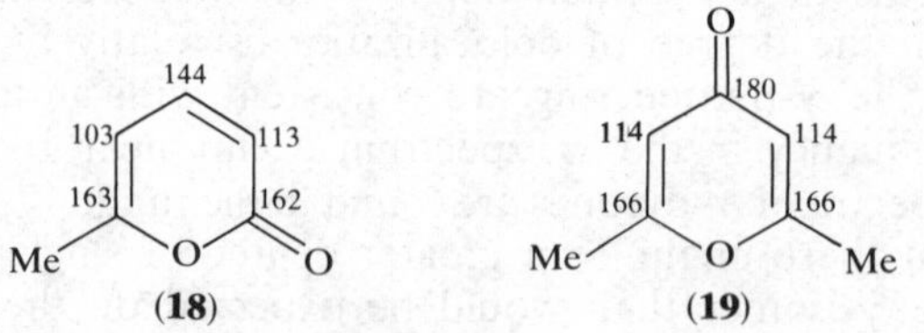

(**18**) (**19**)

Chemical shifts (rel. to Me_4Si) in the ^{13}C n.m.r. spectra of pyrones

FIGURE 2

of the carbons of the carbonyl groups. The δ value of around 180 p.p.m. in (**19**), which seems to be characteristic of γ-pyrone rings, as can be seen from Table 1, is markedly different from the typical value of about 160 p.p.m. found in the spectra of (**18**) and other α-pyrones (see Table 1 in Section 18.2.2.1). Hence ^{13}C n.m.r. spectroscopy may provide an additional method for differentiating α- from γ-pyrones in the future.

TABLE 1

^{13}C N.m.r. of γ-pyrones: chemical shifts of ring carbons (in $CDCl_3$; p.p.m. relative to Me_4Si)[a]

Compound	*C-2*	*C-3*	*C-4*
2,6-Dimethyl-4-pyrone	165.6	113.6	179.9
2,6-Diphenyl-4-pyrone	163.2	111.2	180.0
2,6-Dimethoxy-4-pyrone	152.7	120.1	178.7
2-Phenyl-4-pyrone	164.0	112.2	179.2
Chromone	155.6	113.0	177.6

[a] I. W. J. Still, N. Plavac, D. M. McKinnon, and M. S. Chauhan, *Canad. J. Chem.*, 1976, **54,** 280.

The chemical properties of the γ-pyrone ring are very similar to those of the α-pyrone ring, although the different position of the carbonyl group introduces extra flexibility in some types of reaction and limits the scope for reaction in others. In the following account, therefore, the main theme of the discussion will be a comparison of the chemical properties of the two pyrone systems. An additional consideration which needs to be borne in mind is the potential aromaticity of the heterocyclic ring. Most of the reactions to be discussed from now on, however, can be satisfactorily explained by considering the γ-pyrone ring as an α,β-unsaturated ketone. Therefore the question of aromaticity will only be raised when it helps to throw extra light on the nature of the reaction under consideration.

A γ-pyrone is susceptible to nucleophilic attack at the 2-, 4-, or 6-position of the heterocyclic ring. Attack at C-2 or C-6 is followed by cleavage of the heterocyclic ring in a process which is reminiscent of the similar behaviour of pyrylium salts. In this case the ring-opened product is either a 1,3,5-triketone or an equivalent derivative (see reactions 2, 3, and 4). The reaction with hydroxide is particularly important.[19] In suitable cases the ring-opened product can be isolated after careful acidification. Usually, however, the triketone recyclizes spontaneously to the γ-pyrone and so cannot be isolated.

The ring-opened product of attack by ammonia is not stable but undergoes recyclization to give a γ-pyridone (reaction 5).[20] This transformation provides one of the most efficient

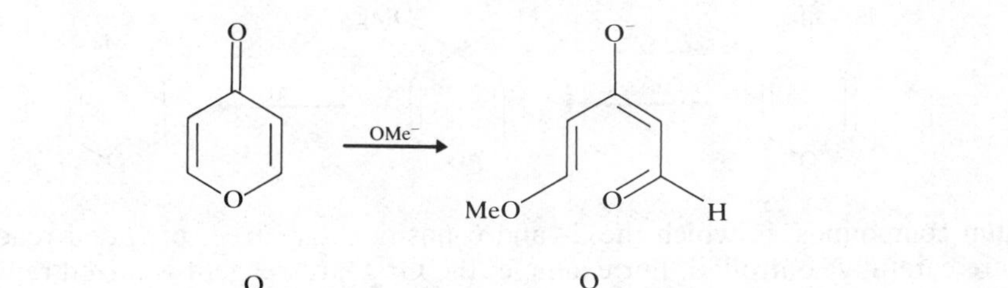

OH⁻ / H⁺ Me Me O OH Me O⁻ (2)

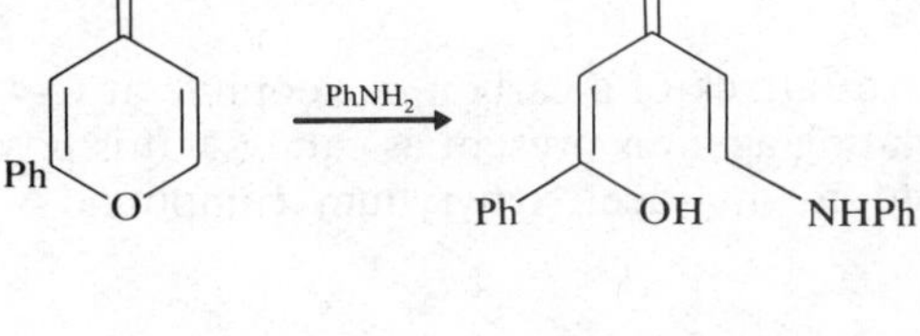

OMe⁻ MeO O H O⁻ (3)

PhNH₂ Ph O OH NHPh (4)

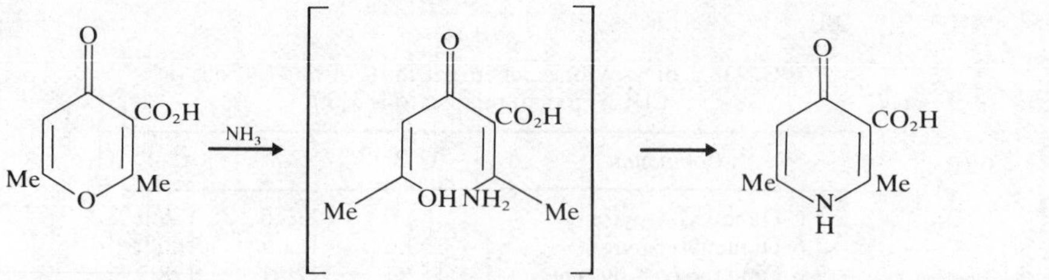

(5)

routes for the synthesis of this type of nitrogen heterocycle. The product of reaction 2 can also undergo cyclization *in situ* after deprotonation of one of the methyl groups to give a carbanion, which can attack the carbonyl group at the other end of the chain. The product is a benzene derivative, as shown in Scheme 1. It was his discovery of this process that inspired Collie to propose his theory of polyketide biosynthesis.[21] These transformations highlight the fact that γ-pyrones, like α-pyrones and pyrylium compounds, have interesting synthetic potential, although the γ-pyrone system has so far been almost completely neglected in this respect.

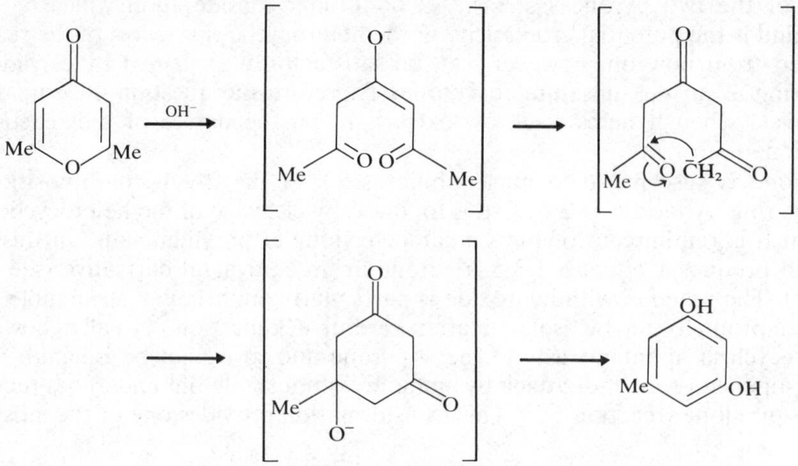

SCHEME 1

In some reactions, attack takes place at C-4 rather than C-2. Grignard reagents, for instance, react preferentially at the carbonyl group to give, initially, a 4*H*-pyran-4-ol which is converted after acidification to the corresponding 4-substituted pyrylium compound[22] (*e.g.* reaction 6).[23] This method provides a useful synthetic route to 4-substituted

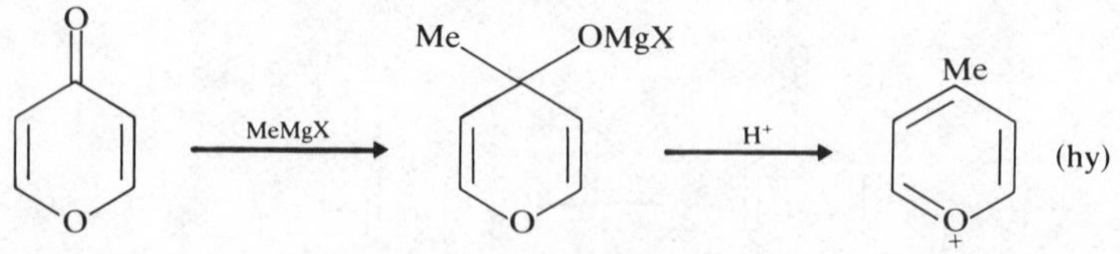

(hy) (6)

pyrylium compounds, in which the 2- and 6-positions are free, provided reaction conditions are carefully controlled. For example, the Grignard reagent is mixed rapidly with an ether solution of the pyrone, and after 1–2 min the reaction mixture is poured into ice-cold aqueous acid.

A related process utilizes attack of a carbon nucleophile at C-4 of a γ-pyrone (reaction 7).[24,25] Note that an acylating agent is present as catalyst. It is possible, therefore, that the pyrone is first converted to a 4-acetoxypyrylium compound which then undergoes a

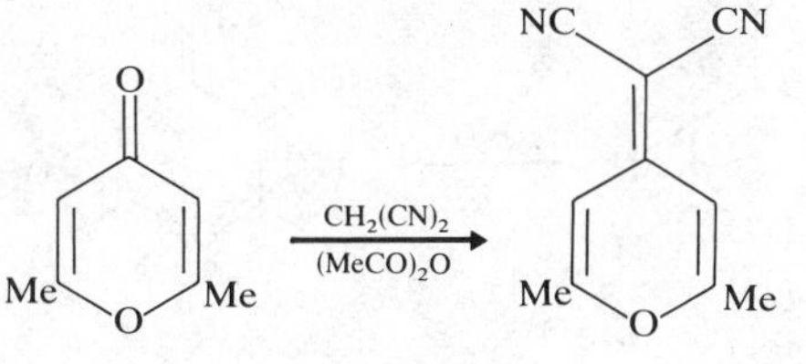

$$\text{2,6-dimethyl-}\gamma\text{-pyrone} \xrightarrow[(MeCO)_2O]{CH_2(CN)_2} \text{4-(dicyanomethylene)-2,6-dimethyl-4H-pyran} \quad (7)$$

standard nucleophilic displacement of the type shown previously (reaction 5 of Chapter 18.1). The same comments apply to the conversion of (**20**) to (**21**) (reaction 8).[26]

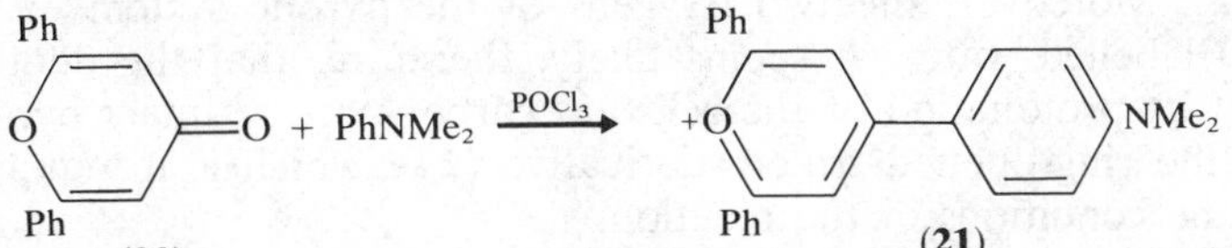

$$\text{(20)} + PhNMe_2 \xrightarrow{POCl_3} \text{(21)} \quad (8)$$

Finally, in the formation of pyrazoles such as (**23**), when phenyl hydrazine attacks a γ-pyrone (reaction 9), nucleophilic attack has occurred at all three susceptible sites of the heterocyclic ring.[27] It seems likely that attack takes place initially at C-2 followed by ring-opening to give an intermediate β-diketone (**22**); such compounds are known to form pyrazoles under the conditions of the reaction.

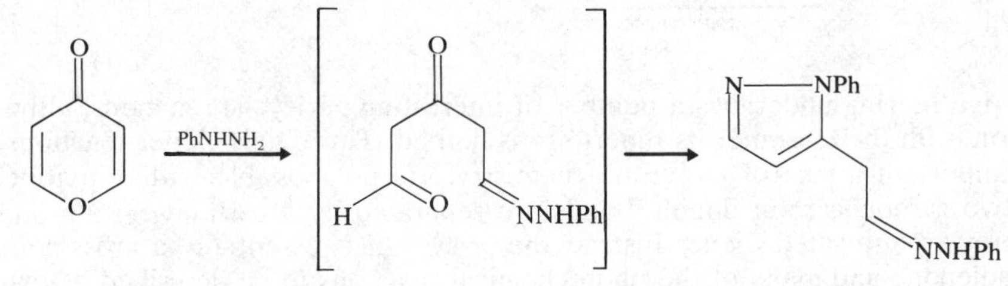

$$\gamma\text{-pyrone} \xrightarrow{PhNHNH_2} [\text{(22)}] \longrightarrow \text{(23)} \quad (9)$$

Examples of electrophilic substitution in a γ-pyrone ring are relatively few, possibly because with most reagents attack takes place preferentially at the carbonyl oxygen to give an inert pyrylium derivative. The most clear-cut example is the acylation of a γ-pyrone at the 3- and 5-positions (either one or both depending on the conditions) using an acyl halide or anhydride in the presence of trifluoroacetic acid as catalyst (reaction 10).[28] This reaction bears a superficial resemblance to the Friedel–Crafts acylation of benzene derivatives.

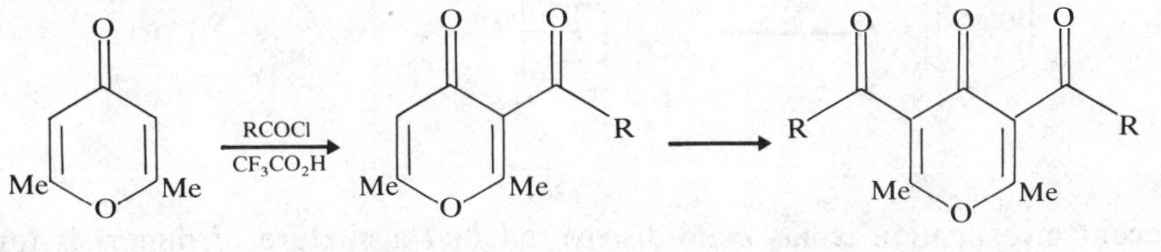

$$\text{2,6-dimethyl-}\gamma\text{-pyrone} \xrightarrow[CF_3CO_2H]{RCOCl} \text{3-acyl derivative} \longrightarrow \text{3,5-diacyl derivative} \quad (10)$$

Equivalent mono- and di-bromo-compounds can be made. Here, however, the reaction takes place by electrophilic addition to give a perbromo-derivative which is subsequently decomposed to the substitution product by heating in steam (reaction 11).[29]

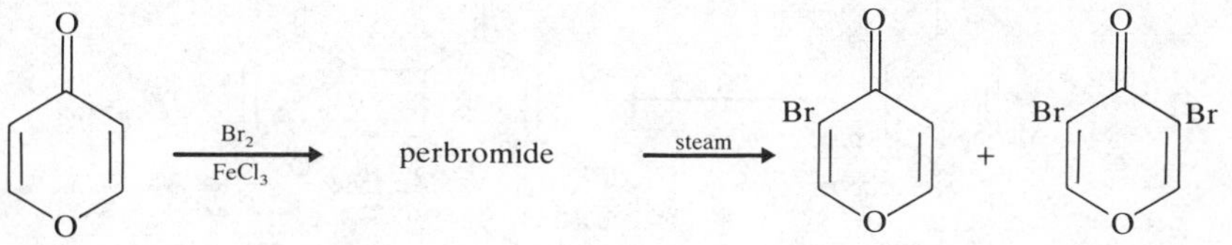

$$\gamma\text{-pyrone} \xrightarrow[FeCl_3]{Br_2} \text{perbromide} \xrightarrow{\text{steam}} \text{3-bromo-}\gamma\text{-pyrone} + \text{3,5-dibromo-}\gamma\text{-pyrone} \quad (11)$$

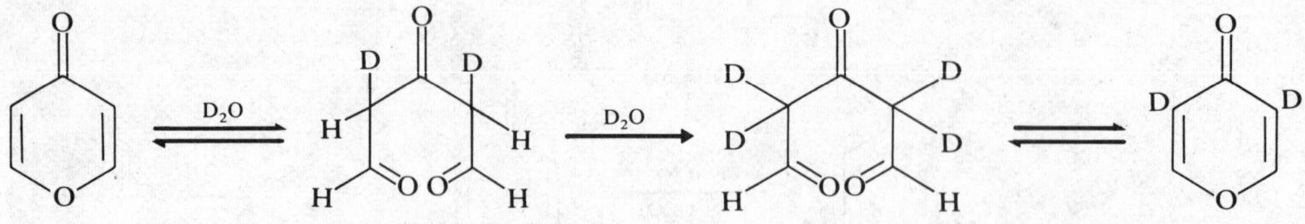

(12)

(24)

Another substitution reaction which is not as simple as it appears on the surface is deuterium exchange (reaction 12).[30] The reaction is not observed with the 2,6-dimethyl derivative (which would be expected to be more, rather than less, susceptible to electrophilic attack). Moreover, the two oxygens of the pyrone system exchange with the oxygen of ^{18}O-labelled water. It seems likely, therefore, that the deuterium exchange takes place not by protonation of the relevant carbons of the intact pyrone ring, but by enolization in the ring-opened triketo-derivative (**24**), which is in equilibrium with the pyrone under the conditions of the reaction.

The most reliable method of reducing the γ-pyrone ring is catalytic hydrogenation. The main product is usually the tetrahydropyranol, although sometimes the corresponding ketone can be isolated as well (*e.g.* reaction 13).[31] Other methods of reduction are not reliable and usually give complex mixtures when applied to γ-pyrones.

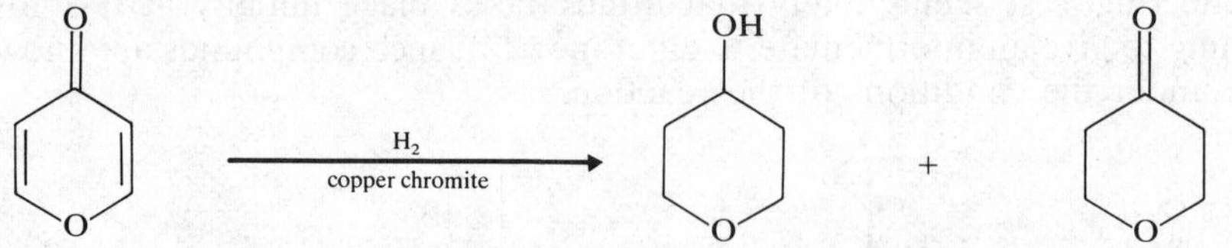

(13)

The γ-pyrone ring undergoes a number of interesting pericyclic reactions, although by comparison with the α-isomer its repertoire is limited. Thus Diels–Alder reactions, which form so important a part of α-pyrone chemistry, are not possible with γ-pyrone, given that its two carbon–carbon double bonds are separated by a carbonyı group and so no longer form a conjugated diene. Instead the π-system is essentially a cross-conjugated cyclohexadienone and many of the photochemical reactions to be described below can be rationalized on that basis.

On irradiation either in concentrated solution or in the crystalline state, 2,6-dimethyl-4-pyrone forms a high yield of dimer to which the head-to-tail cage structure (**25**) has been assigned. When the reaction is carried out in dilute solution, a small amount of the furan (**26**) is formed in addition, but the dimer is still the major product (reaction 14).[32] In a

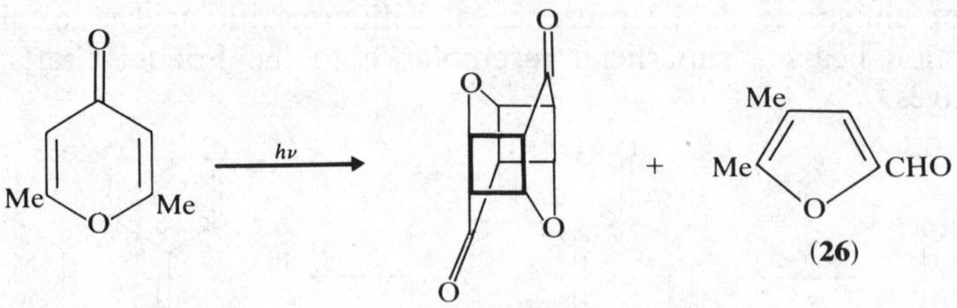

(14)

(26)

(25)

more recent investigation it has been discovered that a mixture of dimers is formed on irradiation at 254 nm and that in this case the main product is the open head-to-head dimer (**27**) (reaction 15).[33] The caged dimer (**25**) reverts to the monomer on treatment with acid.

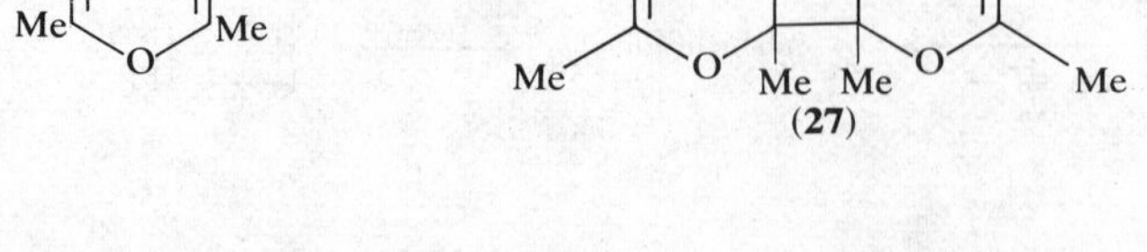

(15)

(27)

The photochemical behaviour of γ-pyrones is dramatically altered by an increase in the bulk of the substituents. Thus, with the highly substituted derivative (**28**), dimerization is suppressed and the dominant photochemical reaction becomes rearrangement to form an isomeric α-pyrone (**30**).[34] This reaction is not sensitized by acetone or acetophenone and is therefore thought to involve singlet excited states and to proceed *via* the intermediate (**29**) (reaction 16).

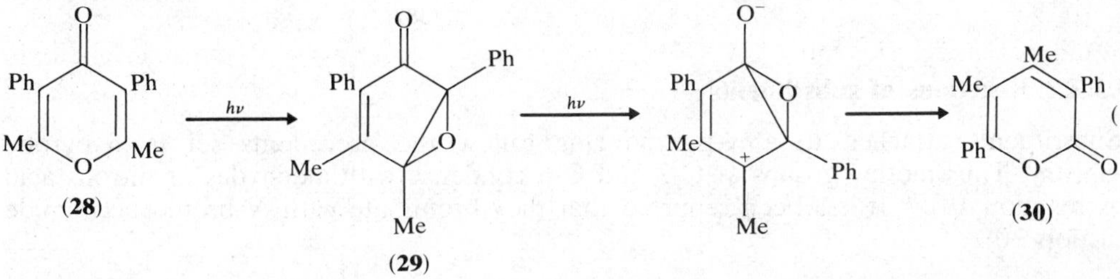

(**28**) → $h\nu$ → (**29**) → $h\nu$ → → (**30**) (16)

When a hydroxy-group is present at C-3 the rearrangement only goes part way. In reaction 17, for example, irradiation of (**31**) in methanol gave a 1:1 mixture of products (**32**) and (**33**), which are formed in the same way as the intermediate (**29**) of the previous reaction.[34]

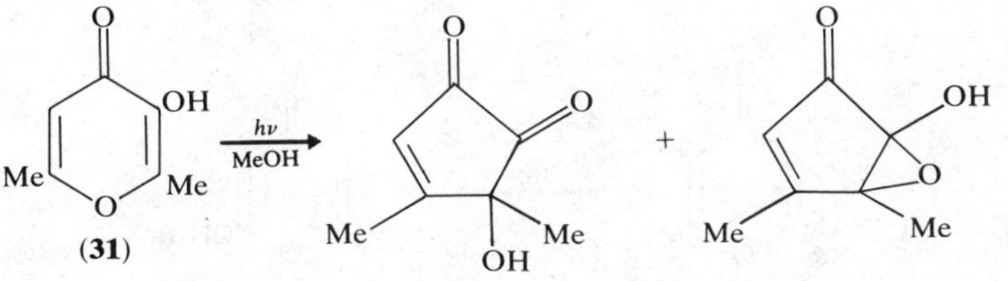

(**31**) → $h\nu$, MeOH → (**32**) + (**33**) (17)

The monomethyl analogue (**34**) undergoes an essentially equivalent photochemical reaction (Scheme 2), but the intermediate (**35**) dimerizes to form (**36**) (or an isomer). The dimer can be converted by hydrogenolysis to the cyclopentane derivative (**37**), which is of interest as one of the flavouring constituents of coffee.

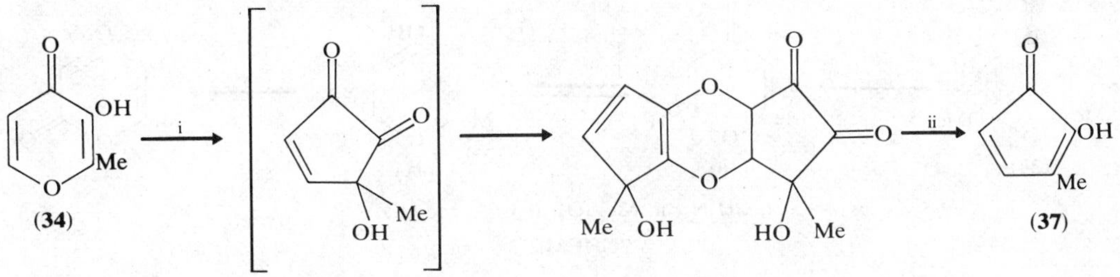

(**34**) → i → [(**35**)] → (**36**) → ii → (**37**)

i, $h\nu$; ii, H_2, catalyst.

SCHEME 2

These photochemical rearrangements of the γ-pyrone ring are closely related to certain of the photochemical reactions of pyrylium compounds which were considered in Chapter 18.1. This applies particularly to the reactions of 4-hydroxypyrylium salts, which are formed *in situ* by dissolving the appropriate γ-pyrone in strong acid.

Apart from dimerization and rearrangement, γ-pyrones undergo one other photochemical reaction of note. This is a [2+2] cycloaddition of a ketene to the carbonyl group followed by loss of carbon dioxide to give a methylenepyran such as (**38**) (reaction 18).[35] In this reaction, as in all the other photochemical reactions described above, the γ-pyrone ring is following the pattern set by simpler aliphatic models.

(18)

(**38**)

18.3.2.2 Reactions of substituents

Substituents attached to a γ-pyrone ring follow the precedents set in α-pyrone chemistry. Thus methyl groups at C-2 and C-6 condense with aldehydes or nitrous acid (*e.g.* reaction 19).[36] It has been reported that they brominate with *N*-bromosuccinimide (reaction 20).[37]

(19)

(20)

When a hydroxy-group is sited at C-2 or C-6, as in (**40**), there is the possibility of tautomerization to give a 4-hydroxy-2-pyrone (**41**). Although the α-pyrone is normally the main component of the equilibrium (see Section 18.2.2.2), it is possible to prepare either a 2-methoxy-4-pyrone (**39**) or a 4-methoxy-2-pyrone (**42**) according to the choice of reaction conditions (Scheme 3). The further example of this type of tautomerism

(**39**) (**40**) (**41**) (**42**)

i, CH_2N_2; ii, Me_2SO_4, K_2CO_3, Me_2CO.

SCHEME 3

(reaction 21) is worthy of special note because, in this instance, the dominant species is neither the γ- nor the α-pyrone but a monoenol such as (**43**).[38] This observation strikingly demonstrates the point which has been made repeatedly in the treatment of both α- and γ-pyrones: although both heterocycles are potentially aromatic, the degree of stabilization of the π-system is insignificant in comparison with that found in benzene and its derivatives.

(21)

(**43**)

18.3.2.3 Synthesis of γ-pyrones

Most of the important syntheses of γ-pyrones have as their basis the cyclodehydration of a 1,3,5-triketone (reaction 22). The process corresponds to a reversal of the hydrolysis

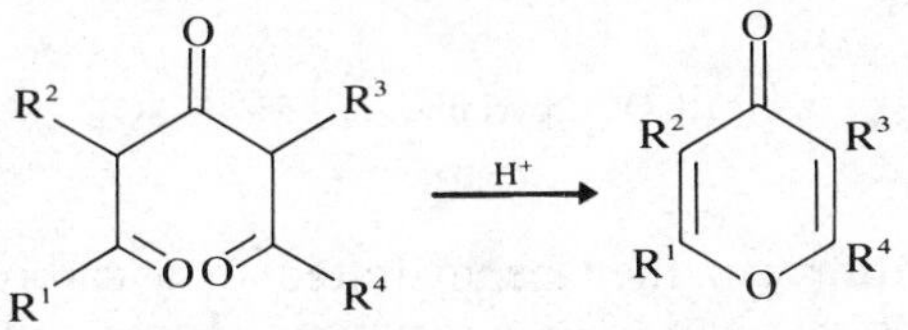

(22)

discussed earlier (reaction 2) and is so facile it normally takes place spontaneously. The limiting factor until very recently was the relative inaccessibility of 1,3,5-triketones and so for many years γ-pyrones were made by special reactions of limited scope. The first of these, which leads to the parent pyrone, is given in Scheme 4.[39] The lack of flexibility of this approach needs no comment.

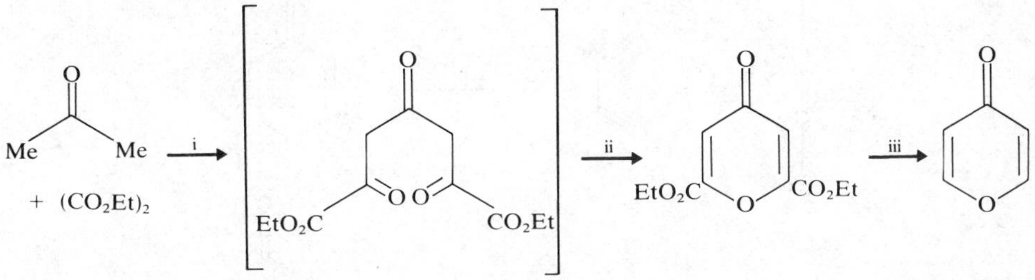

i, NaOEt; ii, conc. HCl; iii, Cu, heat.

SCHEME 4

2,6-Dimethyl-4-pyrone is readily prepared by the remarkable condensation shown (reaction 23).[40] Again the process is limited in scope but it provides a very convenient

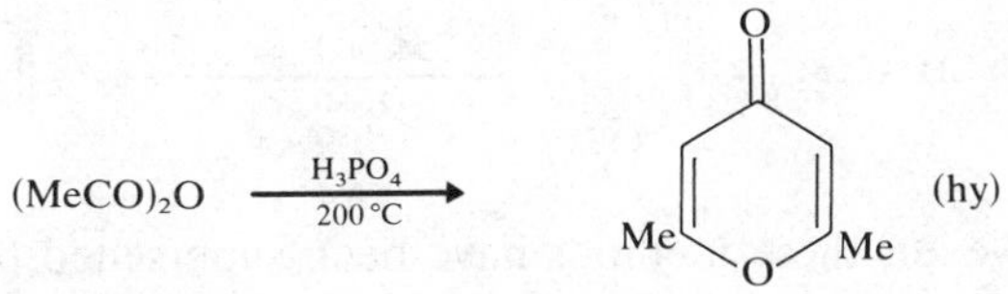

$$(MeCO)_2O \xrightarrow[200\,°C]{H_3PO_4} \text{(hy)} \quad (23)$$

source of symmetrical derivatives of γ-pyrone. The synthesis given in Scheme 5 also leads to 2,6-disubstituted γ-pyrones.[41]

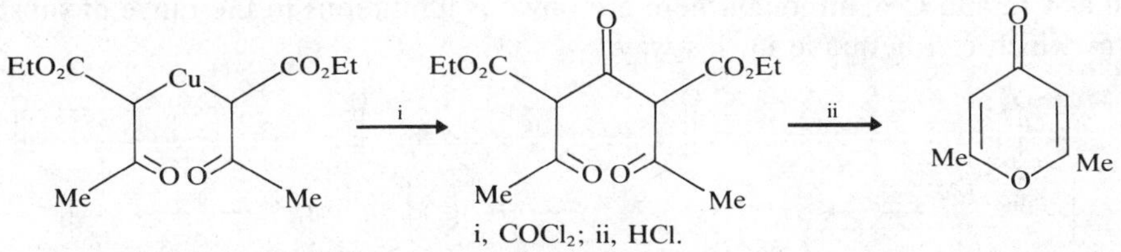

i, $COCl_2$; ii, HCl.

SCHEME 5

These approaches to γ-pyrones only work efficiently when the two substituents are identical. γ-Pyrones with two different substituents at C-2 and C-6 can be made, however, by the synthesis given in Scheme 6.[42,43] In the example illustrated one of the substituents is methyl but this can be varied: the requisite α-pyrones can be prepared by one of the methods given in Chapter 18.2.

A number of attempts have been made to circumvent the inaccessibility of β-triketones by synthesizing an intermediate in which one of the carbonyl groups is replaced by an

i, RCOCl, pyridine; ii, 85% H_2SO_4.

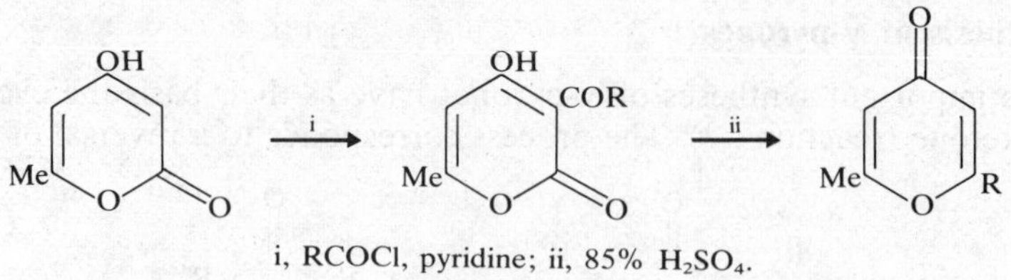

SCHEME 6

acetylenic group. The latter can then be hydrated to a carbonyl group by one of the standard methods prior to cyclization to a γ-pyrone. Three routes based on this strategy are given in reactions 24,[44] 25,[45] and 26.[46] The starting materials are readily prepared by the standard methods of acetylene chemistry. The scope of these approaches has not been fully explored, but clearly the first two could be used in principle to prepare a wide range of γ-pyrones substituted at C-2 and C-6 with different groups.

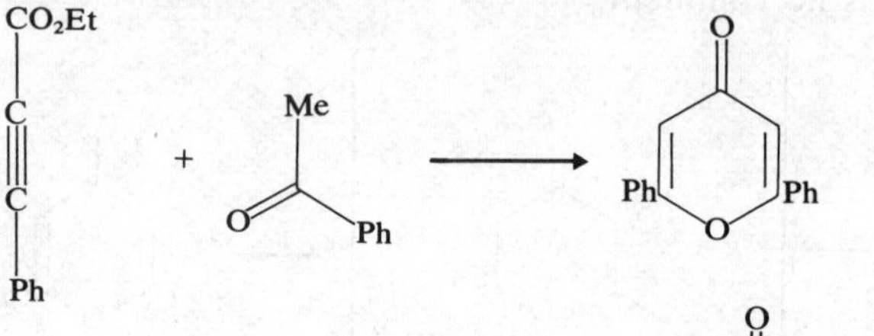

(24)

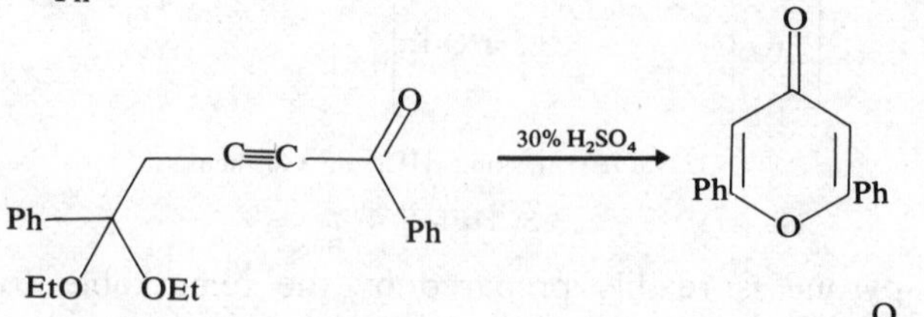

30% H_2SO_4

(25)

MeOCH=CH—C≡C—CH(OEt)₂ → (MeOH/H_2O, $HgSO_4$, H_2SO_4)

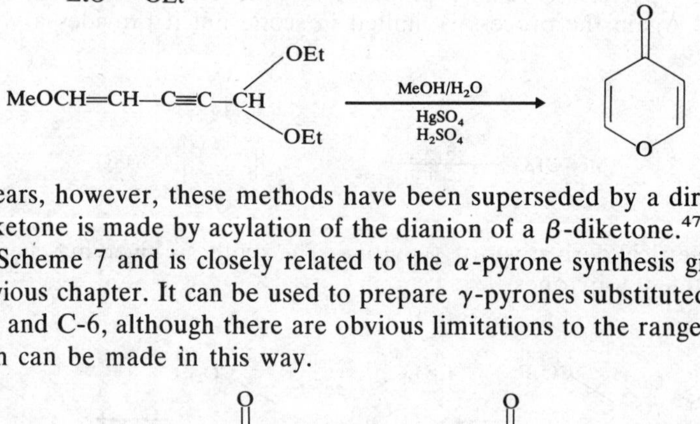

(26)

In recent years, however, these methods have been superseded by a direct synthesis in which a β-triketone is made by acylation of the dianion of a β-diketone.[47] The method is illustrated in Scheme 7 and is closely related to the α-pyrone synthesis given in Scheme 14 of the previous chapter. It can be used to prepare γ-pyrones substituted at C-3 or C-5 as well as C-2 and C-6, although there are obvious limitations to the range of substitution patterns which can be made in this way.

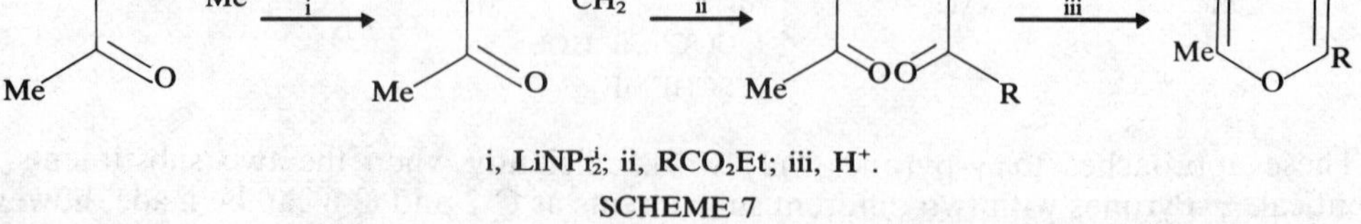

i, $LiNPr^i_2$; ii, RCO_2Et; iii, H^+.

SCHEME 7

18.3.2.4 Important γ-pyrones[48–50]

A few simple γ-pyrones have been reported as natural products. One of them is maltol (**44**), which has been isolated from pine needles and larch bark.[51] It is of special interest, particularly to brewers, as one of the flavouring agents produced when barley is roasted in the production of malt.

Two other naturally occurring γ-pyrones are chelidonic acid (**45**) and meconic acid (**46**). The first is produced by *Chelidonium magus* along with isoquinoline and protoberberine alkaloids;[52] the second is produced by the opium poppy, *Papaver somniferum*, together with the notorious opium alkaloids, which are also members of the benzylisoquinoline family.[53] It is not clear whether this association of γ-pyrones with benzylisoquinoline alkaloids is fortuitous or physiologically significant.

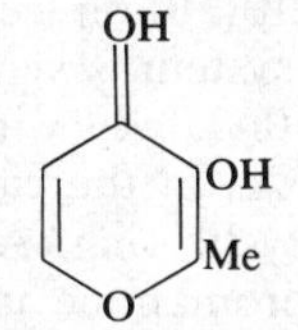

(**44**) 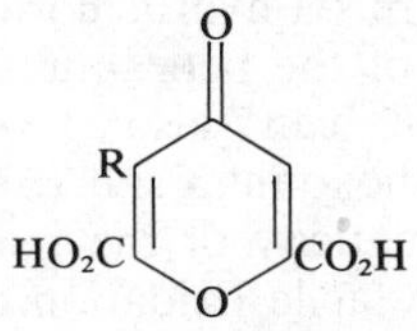 (**45**) R = H (**46**) R = OH (**47**)

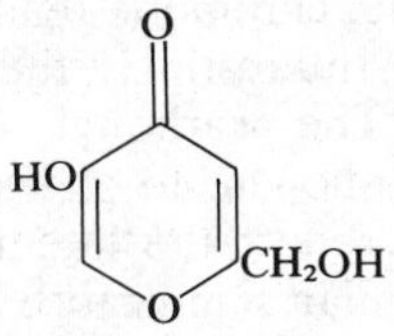

Kojic acid (**47**) is an intriguing compound produced when various moulds (mainly *Aspergillus* species) are grown on glucose or xylose.[54] Under optimum conditions as much as 60% of the sugar consumed may be converted to kojic acid. The reason for this transformation is not clear, although kojic acid has been reported to show weak antibiotic activity.

The chemistry of kojic acid is of special interest because it is a 3-hydroxypyrone. The hydroxy-group in this position is unusually acidic (pK_a 7.9), which explains why the compound is named as an acid. The acidic hydroxyl is preferentially methylated by dimethyl sulphate in the presence of alkali; its methyl ether can be hydrolysed by base.

3-Hydroxypyrones cannot be synthesized readily by any of the routes given earlier. Kojic acid diacetate (**49**), however, has been synthesized from (**48**) (a straightforward derivative of glucose) *via* the route given in Scheme 8.[55]

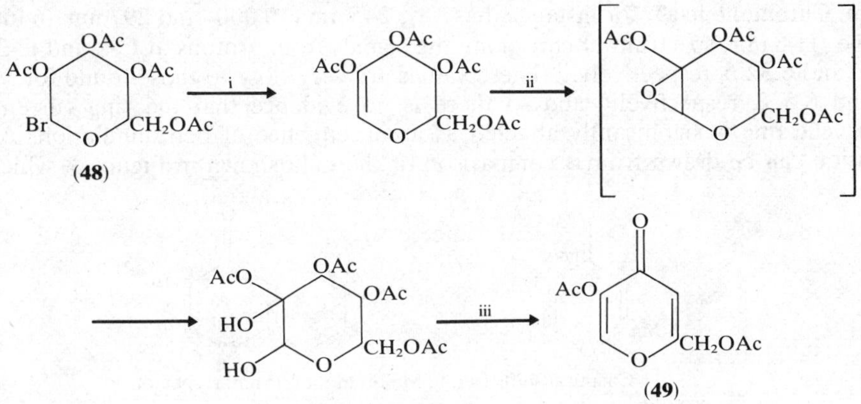

(**48**) (**49**)

i, Et_2NH; ii, $PhCO_3H$; iii, $MeCO_2Na$, $(MeCO)_2O$.

SCHEME 8

18.3.3 CHROMONES

18.3.3.1 Properties of the nucleus

For the most part the properties of the heterocyclic ring of the chromone nucleus follow the precedents set by γ-pyrone although, as might be expected on the basis of coumarin and benzopyrylium chemistry, some reactions of the system are modified as a result of benzannulation.

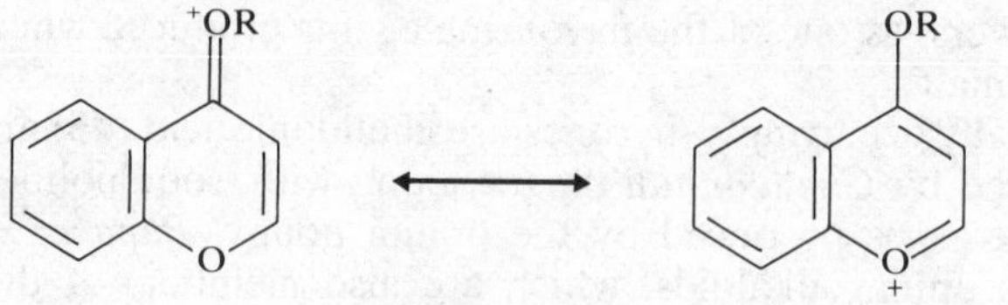

(50) R = H
(51) R = Me

Like γ-pyrone, chromone readily forms a hydrochloride salt **(50)** on treatment with dry HCl in ether.[56] Interestingly the pK_a of the benzannulated system is significantly higher (2.0 *vs.* 0.1).[57] The carbonyl oxygen can also be methylated with methyl *O*-nitrobenzenesulphonate to give a methoxonium salt **(51)**.[11] As in the case of γ-pyrone, these reactions reflect the latent aromaticity of the heterocyclic ring. Another reaction which has a similar significance is the facile oxidation of chromanone to chromones on treatment with relatively mild oxidizing agents such as the triphenylmethyl carbonium ion (reaction 27).[58]

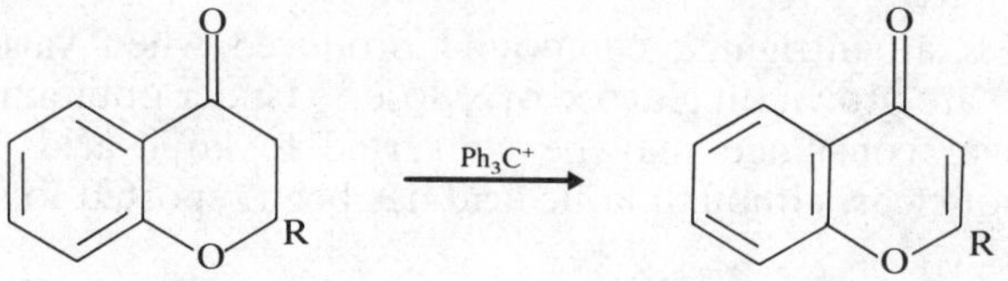

Ph_3C^+ (27)

The spectroscopic properties of chromones, like those of γ-pyrone, are best interpreted in terms of an aliphatic π-system in the heterocyclic ring. First, the carbonyl stretching frequency in the i.r. spectrum of chromone occurs at 1660 cm^{-1}.[59] This is somewhat higher than that of γ-pyrone (ν_{max} 1650 cm^{-1}) but is much lower than that of coumarin (ν_{max} 1710 cm^{-1}). Hence i.r. spectroscopy can be used to distinguish chromones from coumarins.

The u.v. spectra of chromones are characterized by two strong peaks at about 225 and 290 nm. Chromone itself, for instance, has λ_{max} 245 nm (10 000) and 297 nm (6460).[60]

In the ^1H n.m.r. spectrum of chromone the signals from protons at C-2 and C-3 appear at 7.88 and 6.32 δ, respectively.[18] These values are very close to those found for γ-pyrone (8.0 and 6.5 δ, respectively) and so there is no evidence that the ring current in the heterocyclic ring is significantly affected as a consequence of benzannulation. A similar conclusion can be drawn from a comparison of the data shown in Figure 3, which relate

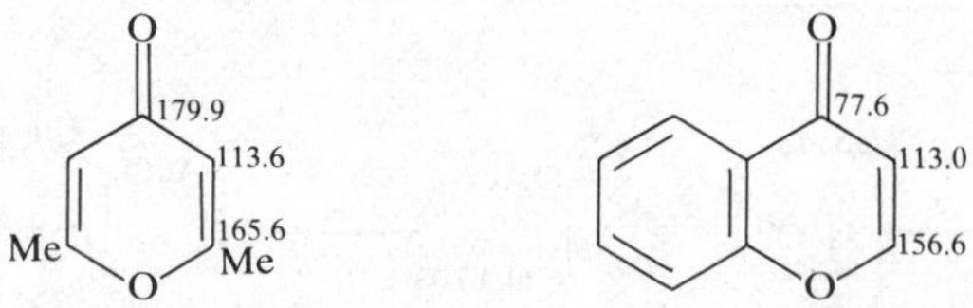

179.9 113.6 165.6 Me Me

177.6 113.0 156.6

Chemical shifts (rel. to Me_4Si) in the ^{13}C n.m.r. spectra

FIGURE 3

to the ^{13}C n.m.r. spectra of chromone and a γ-pyrone (the downfield shift of 10 p.p.m. given by the signal from C-2 in 2,6-dimethyl-4-pyrone can be accounted for by the replacement of a hydrogen by methyl).

The behaviour of the heterocyclic rings towards nucleophiles follows a predictable pattern: C-2 is susceptible to nucleophilic addition, as in simple γ-pyrones, but C-6 being a site of benzannulation is not. Following addition of hydroxide, the heterocyclic ring undergoes a ring-opening to give a salt of a β-diketone (reaction 28). This cyclizes to reform the chromone ring on acidification and so the free diketone cannot usually be isolated.[61] Prolonged vigorous treatment with alkali can lead to carbon–carbon bond cleavage of the β-diketone residue to give a mixture of products which may be mainly

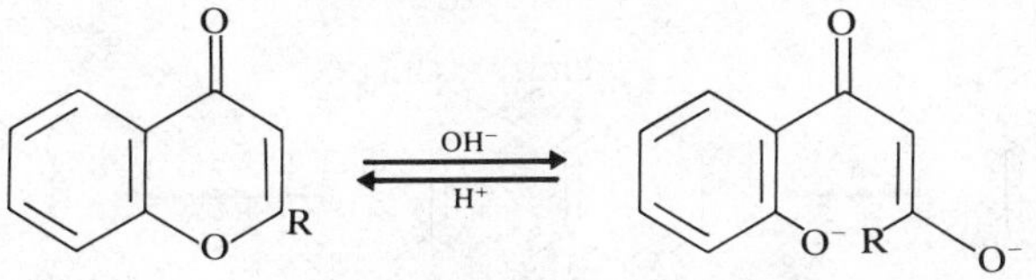

(28)

(**52**) plus (**53**), or (**54**) plus (**55**), depending on the favoured site of cleavage (reaction 29).[62] This type of hydrolytic cleavage has been widely used for the elucidation of structure in the chromone field.

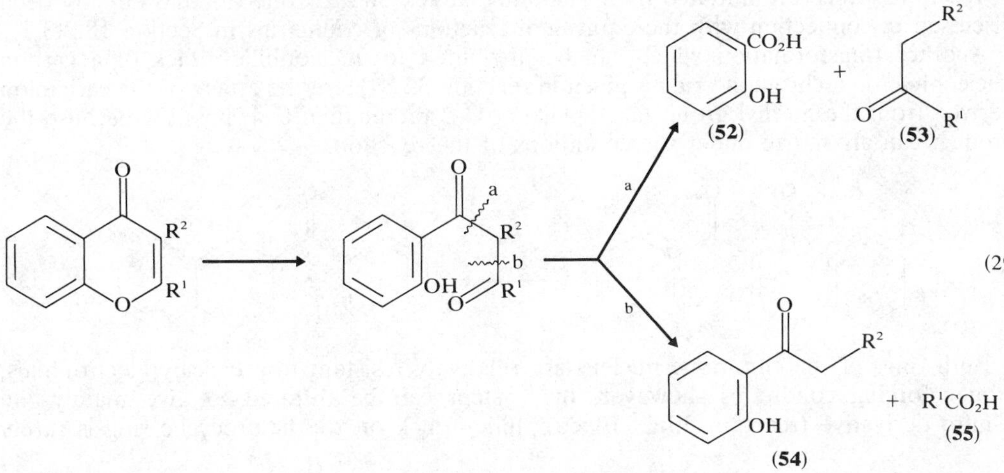

(29)

The reaction with amines only goes as far as the ring-opened stage (**56**), as shown in reaction 30.[63] As was the case in benzopyrylium and coumarin chemistry, ring-opened products of this type cannot recyclize to form the nitrogen analogue of the starting oxygen heterocycle because of the resistance of the benzene ring to nucleophilic attack.

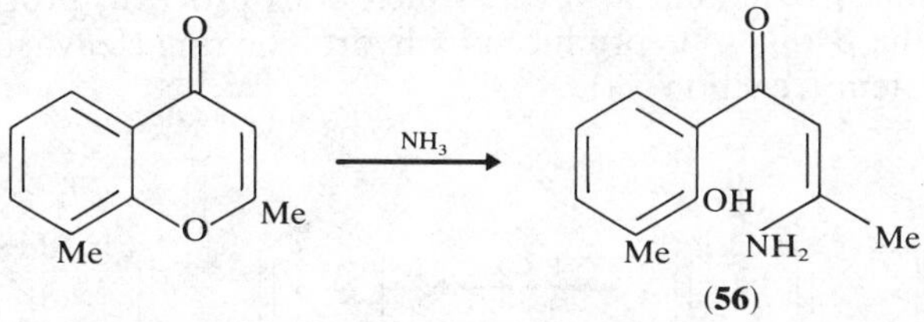

(30)

In the reaction with hydrazine and its derivatives, ring-opening is followed by cyclization to give a pyrazole (**57**).[64] In this transformation (reaction 31) the chromone follows the behaviour of simple γ-pyrones, but the reaction goes further with the latter systems by attack of a second molecule of the nucleophile at the position equivalent to the site of the phenolic group in (**57**).

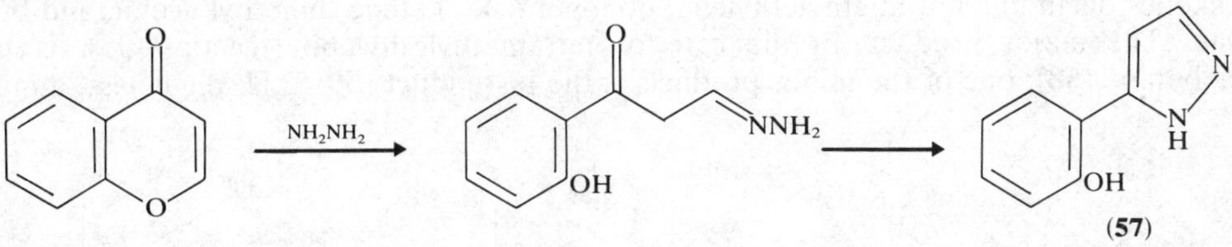

(31)

Products of attack at C-4 rather than C-2 can sometimes be formed. In Scheme 9, for example, a Grignard reagent adds at the carbonyl group.[65] This reaction is of course subject to kinetic control. The significance of this factor in determining the overall course

i, PhMgX; ii, H^+, H_2O.

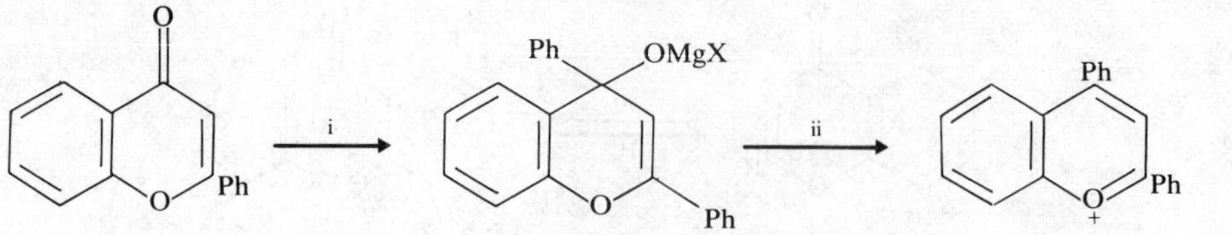

SCHEME 9

of reactions which are initiated by nucleophilic attack on a pyrone ring has already been discussed in connection with the equivalent reactions of coumarins in Section 18.2.3.1.

Another transformation which can be attributed to nucleophilic attack of a carbon nucleophile on a chromone ring is given in reaction 32.[66] Here the attack of the carbanion derived from the methyl group takes place at C-2 rather than C-4, possibly because the product can aromatize under the conditions of the reaction.

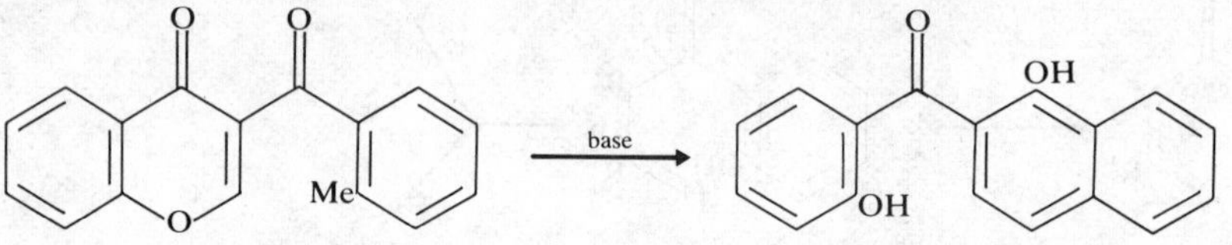

(32)

Both rings of the chromone nucleus are relatively resistant to attack by electrophiles. Under forcing conditions, however, the system can be nitrated to give mainly the 6-nitro-derivative (reaction 33).[67] Electrophilic attack on the heterocyclic ring is rarely

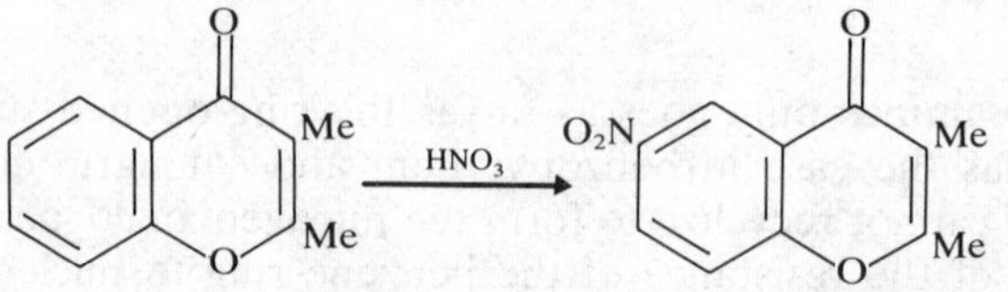

(33)

observed. The Mannich reaction, however, which does proceed, probably involves electrophilic attack on the β-diketone produced by hydrolytic ring cleavage rather than on the intact chromone system (reaction 34).

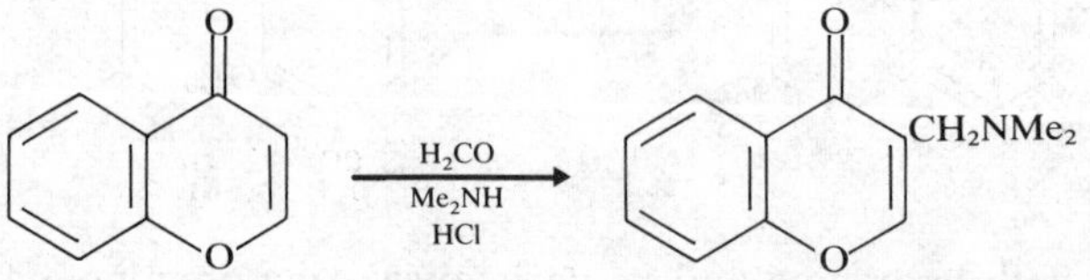

(34)

In its photochemical behaviour the chromone nucleus generally follows the pattern set by γ-pyrone. In one important respect, however, it diverges markedly from the simple pyrone and also from coumarin: dimer formation does not seem to be a dominant reaction. Instead the system undergoes efficient unsensitized photoaddition of a wide range of alkenes, including tetramethylethylene, cyclopentene, ketene dimethyl acetal, and but-2-yne. The major product, in the case of tetramethylethylene (reaction 35), is the cyclobutane (**58**); one of the minor products is the bis-adduct (**59**).[68] Nevertheless, dimers

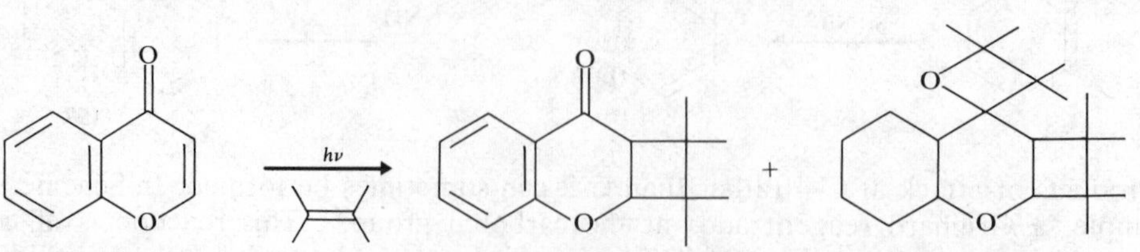

(**58**) (**59**) (35)

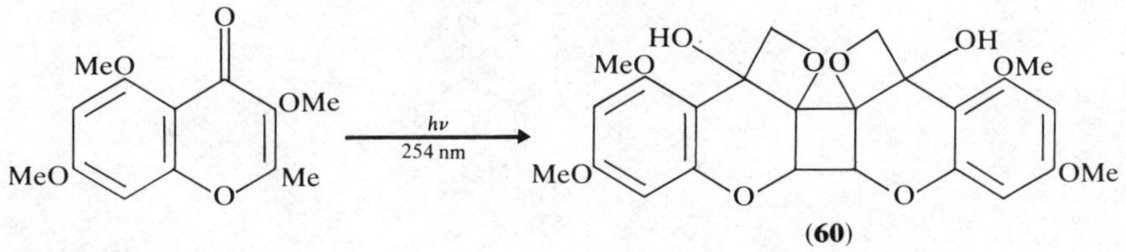

(36)

(60)

can be formed from chromones under suitable conditions (reaction 36).[69] The oxetan rings in (**60**) are thought to arise from secondary photochemical reactions following the formation of a simple dimer. Needless to say, benzannulation of the heterocyclic ring suppresses the formation of caged dimers, equivalent to those which are the characteristic products of irradiation of simple γ-pyrones.

The rearrangement of the 2-aryl-3-hydroxychromone (**61**) to the diketone (**62**) has a direct precedent in the photochemistry of simple γ-pyrones (see reaction 17). In the case of the corresponding enol ether (**63**) the rearrangement is prevented and so cyclization to form an additional pyran ring as in (**64**) becomes the dominant reaction.[70]

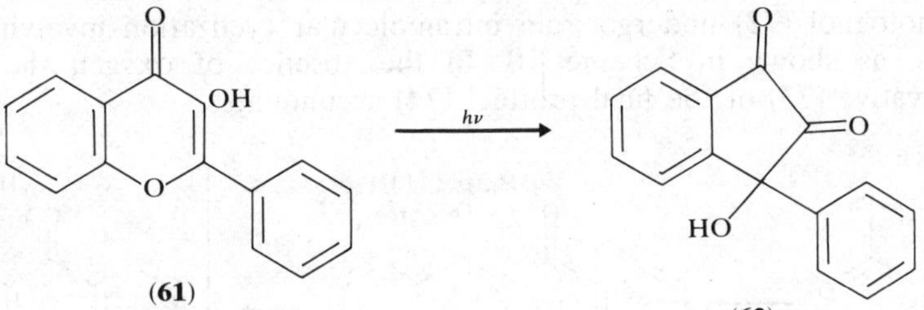

(**61**) (**62**)

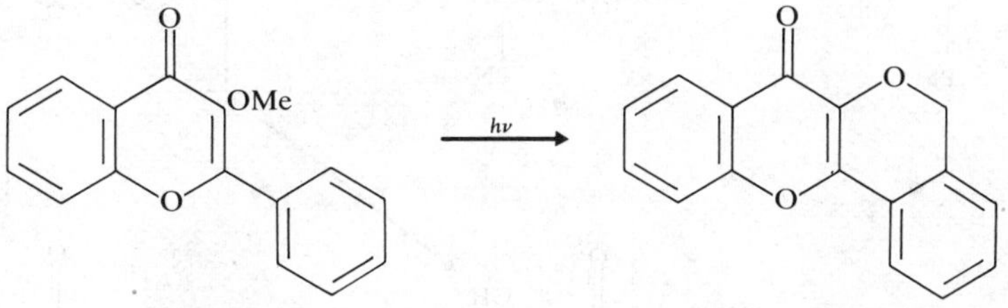

(**63**) (**64**)

Suitably substituted chromones undergo an important photochromic reaction (*e.g.* reaction 37). Here 3-acyl-2-alkylchromone (**65**) is converted to a bright orange enol (**66**) on irradiation. Reversion to the colourless chromone can take place photochemically or by solvent-mediated proton transfer.

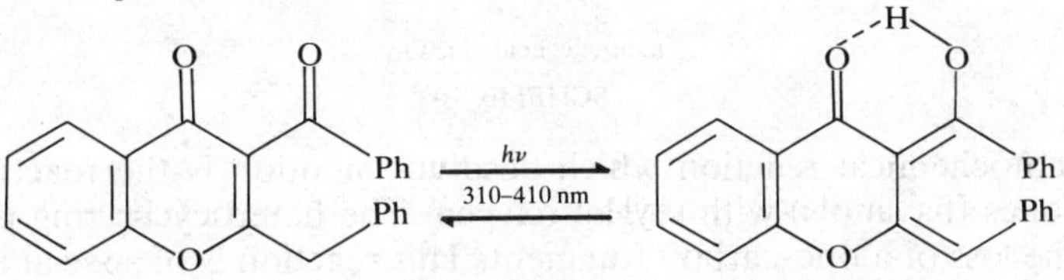

(37)

(**65**) (**66**)

A precedent for this photoenolization can be found in the photochemistry of 1-methylbenzophenones.[71] However, in the chromone series the process is markedly dependent on the substitution pattern. Thus the chromone (**67**) does not form an enol on irradiation whereas its isomer (**68**), in which the potentially acidic methyl group is attached to the carbocyclic rather than the heterocyclic ring, does so readily.[66] The photoenol (**69**) so formed reacts further to generate the tetracyclic product (**70**) by cycloaddition on to the chromone system.

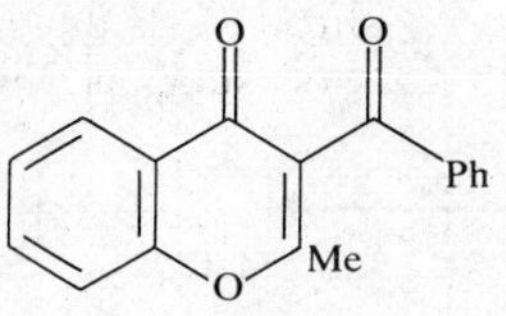

(67)

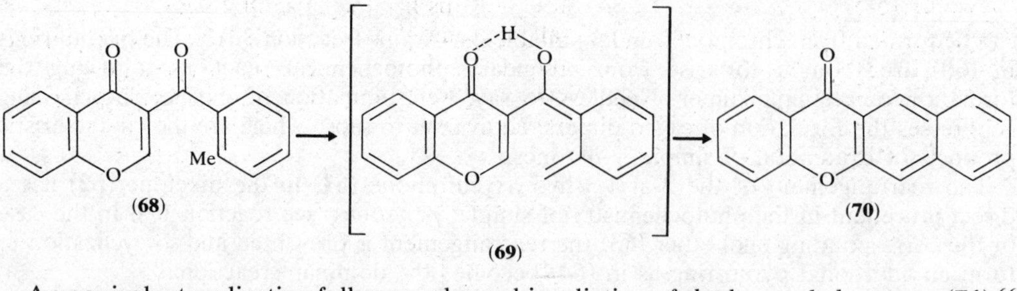

(68) (69) (70)

An equivalent cyclization follows prolonged irradiation of the benzoyl chromone (**71**).[66] Here the photoenol (**72**) undergoes an intramolecular cyclization involving one of the phenyl rings, as shown in Scheme 10. In the absence of oxygen the intermediate dihydro-derivative (**73**) of the final product (**74**) accumulates.

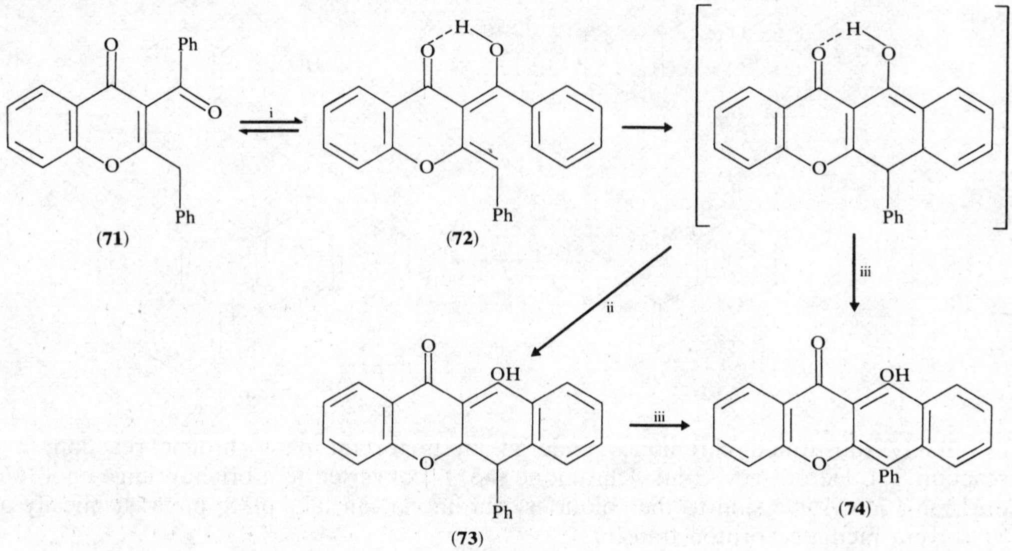

(71) (72) (73) (74)

i, $h\nu$; ii, heat; iii, O_2.

SCHEME 10

One final photochemical reaction which deserves mention is the reaction of 2-aryl-3-hydroxychromones (flavanols) with singlet oxygen. The heterocyclic ring is cleaved (reaction 38) with the loss of a one-carbon fragment. This reaction is of special interest because it is closely related to a biological degradation of naturally occurring chromones.[72]

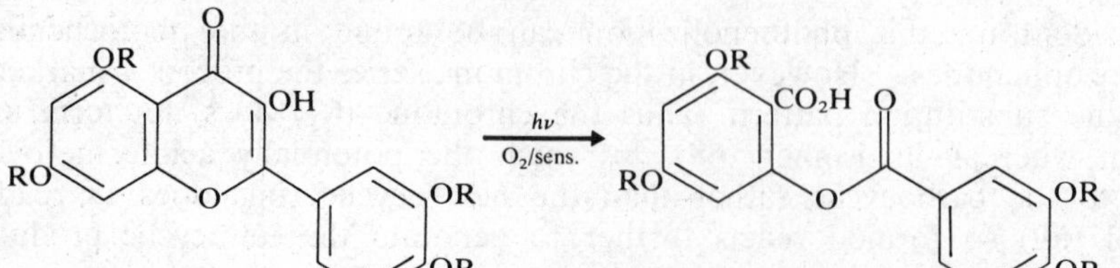

$$\xrightarrow[O_2/\text{sens.}]{h\nu} \qquad (38)$$

Many reports on the reduction of chromones have appeared. Complex mixtures are usually formed both with hydride donors, such as sodium borohydride, or on catalytic hydrogenation.[73]

18.3.3.2 Reactions of substituents

Two useful reactions deserve attention under this heading. The first is the formation of condensation products by reaction of 2-alkylchromones with aldehydes.[74] Reaction takes place at the α-carbon; alkyl groups at position-3 do not react (reaction 39).

The second reaction of note is the selective removal of an acyl substituent at C-3 of the chromone nucleus by controlled hydrolysis (reaction 40).[75]

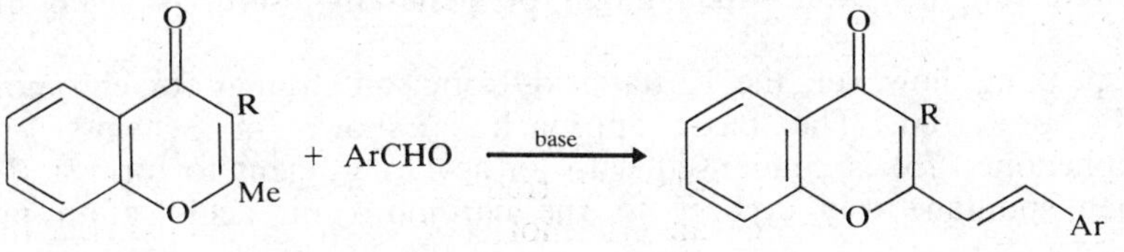

(39)

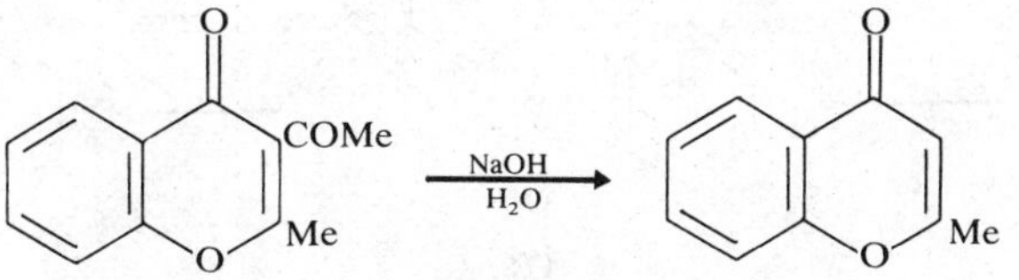

(40)

18.3.3.3 Synthesis of chromones

Two widely used methods of chromone synthesis will be discussed here. The first is the Simonis reaction in which a β-ketoester is condensed with a phenol. The condensation can take place either to give a chromone or a coumarin (reaction 41).[76] The formation of

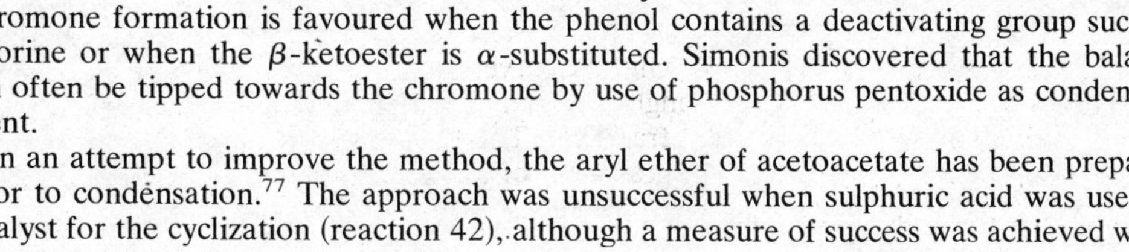

(41)

coumarins (the Pechmann reaction) has already been discussed in Section 18.2.3.2. Chromone formation is favoured when the phenol contains a deactivating group such as chlorine or when the β-ketoester is α-substituted. Simonis discovered that the balance can often be tipped towards the chromone by use of phosphorus pentoxide as condensing agent.

In an attempt to improve the method, the aryl ether of acetoacetate has been prepared prior to condensation.[77] The approach was unsuccessful when sulphuric acid was used as catalyst for the cyclization (reaction 42), although a measure of success was achieved when

(only) (42)

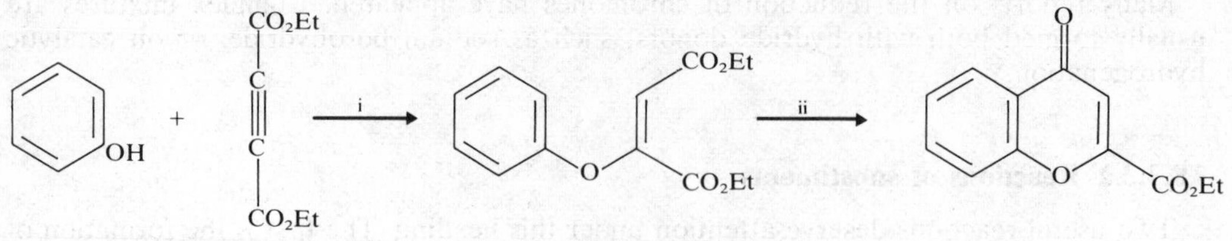

i, base; ii, MeCOCl, H_2SO_4.

SCHEME 11

the ester was hydrolysed to the free carboxylic acid followed by cyclization to the chromone using a Friedel–Crafts reaction.

Another interesting variation which might be generally useful is given in Scheme 11.[78,79]

Generally speaking, however, the Kostanecki–Robinson strategy for chromone synthesis is to be preferred. The basic approach is shown in Scheme 12: an *o*-hydroxyacetophenone (**75**) is condensed with an acylating agent to form a β-diketone (**76**); this then spontaneously cyclizes to the chromone on acidification in what is

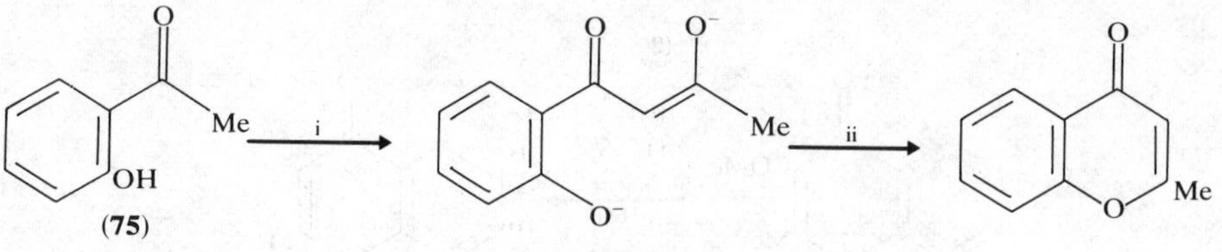

i, Na, CH_3CO_2Et; ii, H^+.

SCHEME 12

effectively a reverse of chromone hydrolysis.[80,81] The strategy still has a potential ambiguity, in principle at least, in that the acyl group might form a carbanion which could then attack the ketonic carbonyl, leading ultimately to the formation of a coumarin. In practice the protons on the carbon adjacent to the ketone are more acidic and so chromone formation is the favoured outcome.

A useful variation which is often more convenient in practice is shown in Scheme 13. The key step here is the Baker–Venkatamaran rearrangement of (**77**) to (**78**).[82,83] A possibly ambiguous outcome is avoided because the hydrogens next to the ketonic carbonyl are more acidic than those adjacent to the ester carbonyl.

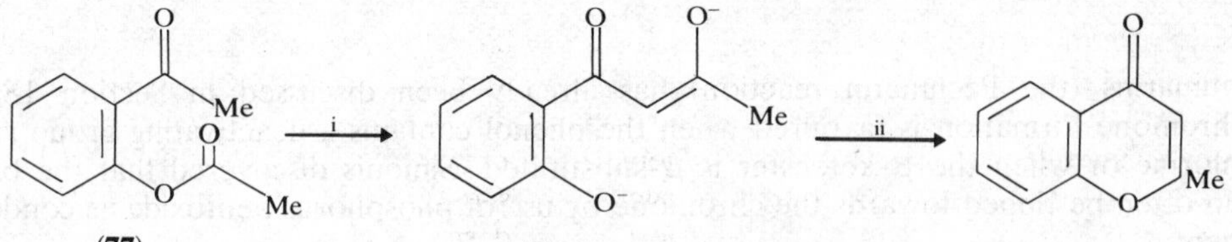

i, base; ii, H^+.

SCHEME 13

The Kostanecki–Robinson method of chromone synthesis can also be adapted for the preparation of 3-alkyl- or 3-alkoxy-chromones.[81,84] These are made by condensing the appropriate substituted acetophenone of general structure (**79**) with an ester, as indicated in Scheme 14.

Another important class of compound which can be made by a variation of the Kostanecki–Robinson synthesis is represented by the 3-acylchromone (**81**).[66] In reaction

(79) R^1 = alkyl or Oalkyl

i, NaOEt, R^2CO_2Et; ii, H^+.

SCHEME 14

42a, two different cyclizations of the intermediate triketone (**80**) are possible, leading to either a 3-acetyl- or a 3-aroyl-chromone. The latter is the major product, presumably because the acetyl group is more reactive in the cyclization step.

(80) (81) (42a)

Finally, methyl ethers have been used in place of the free phenol in the Kostanecki–Robinson chromone synthesis (Scheme 15). After formation of the β-diketone (**82**), the ether link is cleaved using hydriodic acid. The chromone forms spontaneously under the conditions of the reaction. It is not clear whether this variation offers any advantage over the more direct methods described above.

(82)

i, NaOEt, $MeCO_2Et$; ii, HI.

SCHEME 15

18.3.3.4 Naturally occurring chromones

Only 2-alkylchromones will be covered under this heading. The naturally occurring 2-aryl- and 3-aryl-chromones will be covered separately later.

Naturally occurring 2-alkylchromones form a relatively small group compared with the much more numerous and wide-spread coumarin family. Interestingly, a similar range of structural types, including prenylated derivatives and furan derivatives, are known.[85,86]

One of the simplest chromones is eugenin (**83**) a constituent of the wild clove *Eugenia caryophyllata* L. Thunb.[87] Its basic structure was established by demethylation to the known 5,7-dihydroxy-2-methylchromone (**84**), and so all that remains is to decide the position of the *O*-methyl group. This was placed at C-7 rather than C-5 on the basis of

(83) (84)

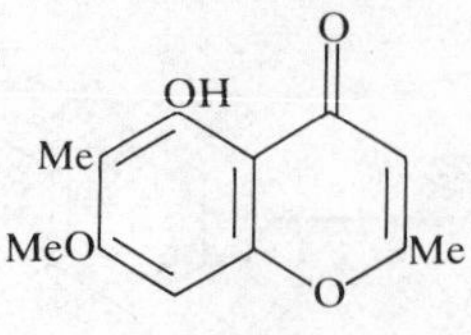

(**85**) 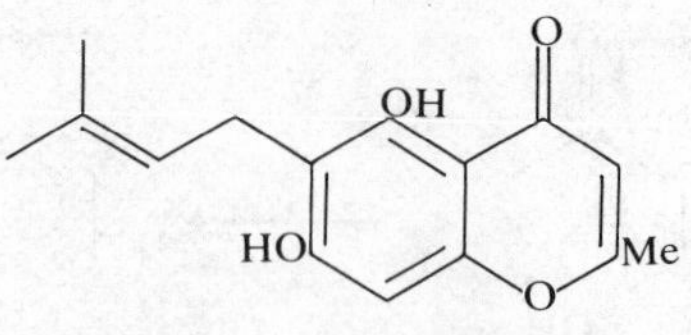 (**86**)

two pieces of evidence. First, eugenin gives the intense purple colour with alcoholic ferric chloride, which is characteristic of compounds in which a strong chelate can be formed. Secondly, remethylation of (**84**) under mild conditions generates a monomethyl ether which is identical with eugenin. Under these conditions a non-chelated hydroxy-group (in this case that at C-7) is known to react more rapidly than one which is chelated (that at C-5).

A number of chromones have been isolated in which the carbon skeleton has been methylated or prenylated. Two examples are eugenitin (**85**),[88] which occurs with eugenin in the wild clove, and peucenin (**86**),[89] which is found in the rhizome of *Imperatoria ostruthium* (it would seem that the latter plant has a hyperactive isoprenoid metabolism for it produces prenylated coumarins as well as chromones).

Khellin (**87**) is an interesting furochromone found in the fruits and seeds of *Ammi visnaga.*[90] Extracts of this plant have been used in medical practice in countries bordering the Eastern Mediterranean. Khellin, which is an important constituent of these extracts, induces muscular relaxation and has therefore been used to alleviate the symptoms of bronchial asthma. This therapeutic property is found in a number of simpler 2-methylchromones.

The determination of the structure of khellin was far from straightforward and makes an interesting story. On hydrolytic cleavage with hot alkali it gives acetic acid and khellinone (**88**). The presence of the furan ring was established by oxidation of the latter to furan-2,3-dicarboxylic acid (Scheme 16). At first these results were interpreted in terms

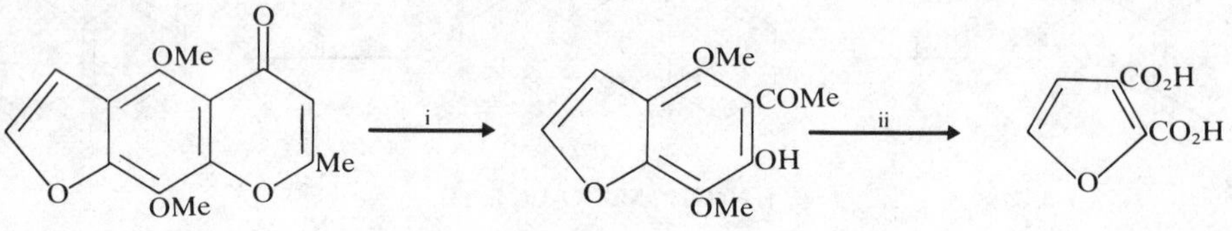

(**87**) (**88**)

i, OH^-, H_2O; ii, O_3.

SCHEME 16

of a 4-methylcoumarin structure (which would give the same products on hydrolysis), but the structure was revised to a 2-methylchromone when it was discovered that khellin forms a condensation product with the aromatic aldehyde piperonal (*cf.* reaction 39).

The problem now was to discover how the two heterocyclic rings and the two methoxy-groups are placed on the central carbocyclic ring. That this was no easy task can be seen from the major rearrangement which took place when khellin was demethylated under standard conditions with hydriodic acid (reaction 43). Fortunately it was possible to

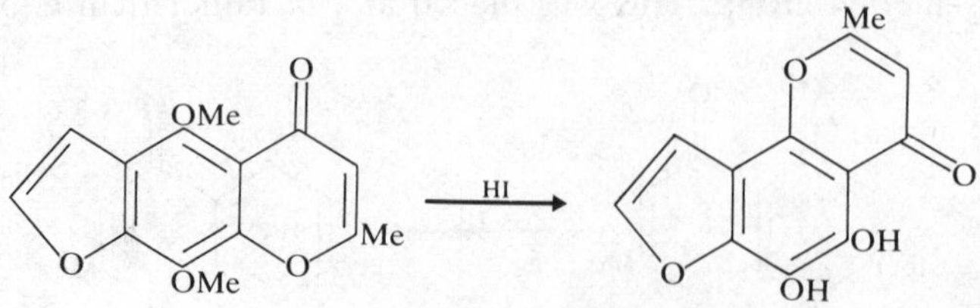

(43)

(**89**)

establish that the product (**89**) had rearranged by the fact that on remethylation it gave an isomer of khellin rather than khellin itself.[91] This type of rearrangement is commonly observed with 5-hydroxychromones under these conditions and is known as the Wessely–Moser rearrangement. It takes place by cleavage of the heterocyclic ring to form a 1,3-diketone in which the dicarbonyl residue is flanked on both sides by a free hydroxy-group and so can recyclize in either one of two ways to give the original chromone or an isomer.

Eventually the pattern of substitution of the carbocyclic ring was established unambiguously by degradation of khellinone (**87**) by standard reactions *via* the sequence (**88**), (**90**), (**91**), (**92**), (**93**), and thence to the known compound (**94**) [the furan ring was cleaved by ozonolysis in going from (**90**) to (**91**)].

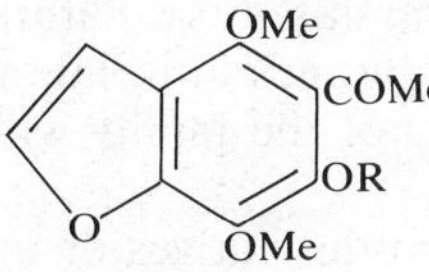

(**88**) R = H
(**90**) R = OMe

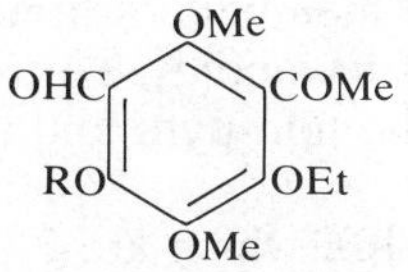

(**91**) R = H
(**92**) R = Et

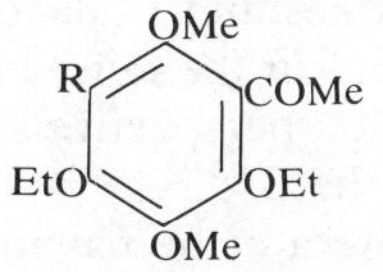

(**93**) R = CO_2H
(**94**) R = H

In view of its pharmacological interest, khellin has attracted the interest of several groups of synthetic chemists. Of the various routes which have been devised for its synthesis, that shown in Scheme 17 is probably of most interest in the context of

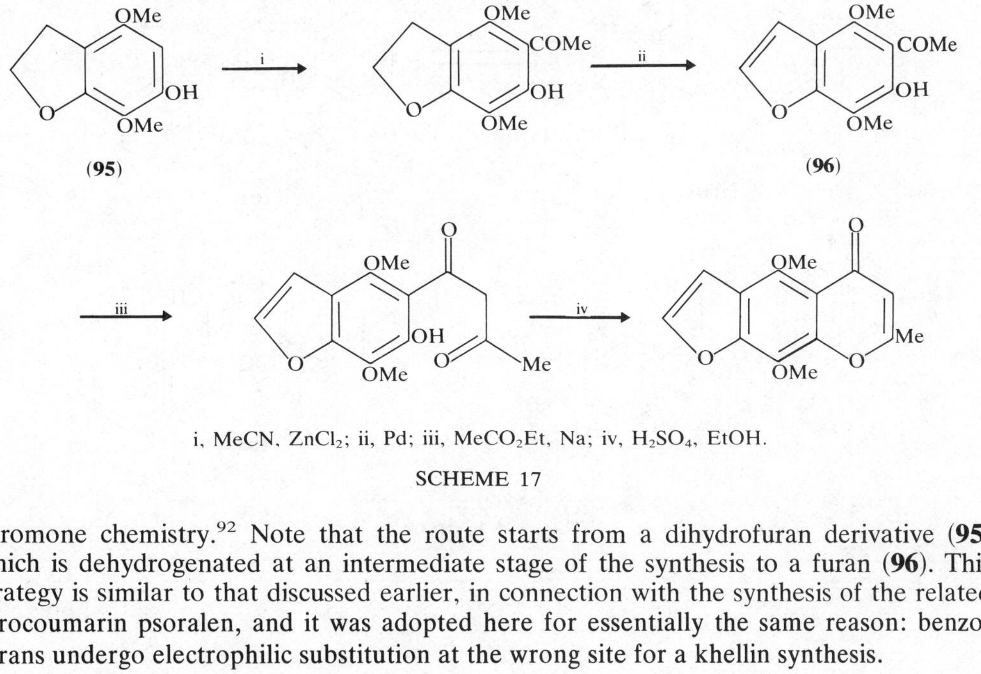

(**95**)

(**96**)

i, MeCN, $ZnCl_2$; ii, Pd; iii, $MeCO_2Et$, Na; iv, H_2SO_4, EtOH.

SCHEME 17

chromone chemistry.[92] Note that the route starts from a dihydrofuran derivative (**95**) which is dehydrogenated at an intermediate stage of the synthesis to a furan (**96**). This strategy is similar to that discussed earlier, in connection with the synthesis of the related furocoumarin psoralen, and it was adopted here for essentially the same reason: benzofurans undergo electrophilic substitution at the wrong site for a khellin synthesis.

18.3.4 FLAVONES AND ISOFLAVONES

18.3.4.1 The flavanoid family

The compounds of primary concern in this section are the important naturally occurring aryl chromones related to flavone (**97**) and isoflavone (**98**). 3-Hydroxyflavone (**99**) and its derivatives are recognized as a separate subgroup called the flavonols.

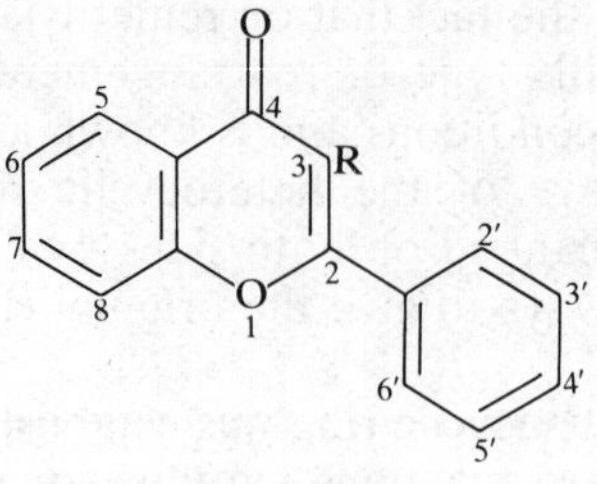

(97) R = H
(99) R = OH

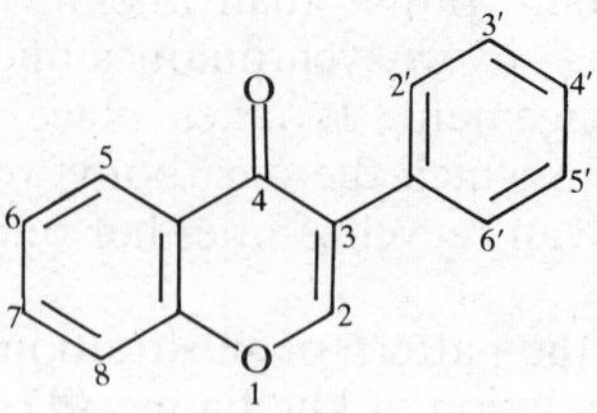

(98)

The flavones and isoflavones are members of a much wider family of natural products which rejoice in the family name of flavanoids. To treat the pyrones in isolation would be to lose a vital perspective and so the non-pyronoid members of the family will be briefly introduced here.[93,94]

First there are the flavanones (**100**), which are 2,3-dihydro-derivatives of the flavones. These are readily interconvertible with the isomeric chalcones (**101**). Next come the anthocyanins represented by (**102**). These possess a benzopyrylium nucleus and have already been discussed in Chapter 18.1. Finally the catechin group is characterized by a tetrahydropyranol ring as in (**103**).

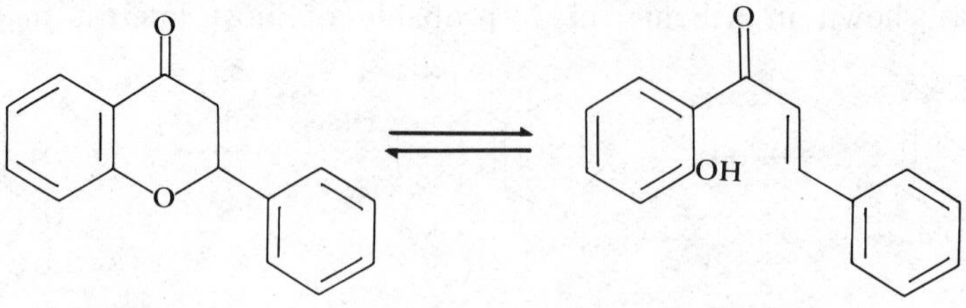

(100)

(101)

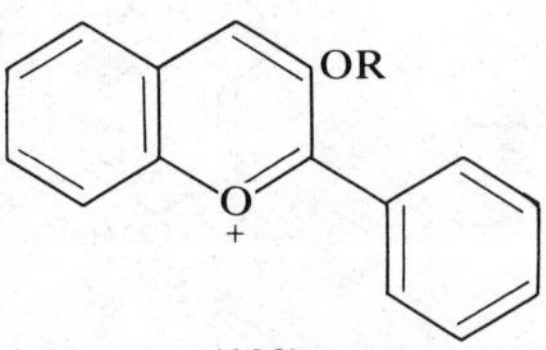

(102)

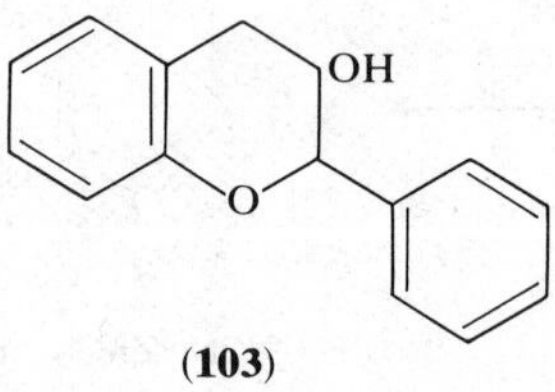

(103)

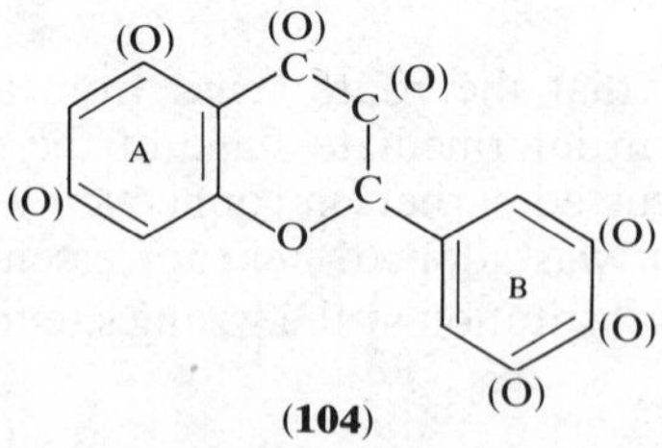

(104)

With the notable exception of isoflavones, all of these systems have the same basic skeleton, which is indicated in diagram (**104**). The key feature which distinguishes one structural type from another is the oxidation level of the various carbons in the heterocyclic ring. At one extreme the flavones and flavonols are in a high state of oxidation; the catechins at the other end are highly reduced. In between lie the flavanones and anthocyanins.

The flavanoids are found almost exclusively in the plant kingdom and there is no established record of their being produced in animals, as opposed to being taken in as part of a vegetarian diet. Many members of the family are highly coloured and as a consequence play a vital role in the ecology of plants by making the flowers and fruits attractive to the bees and birds, respectively. Others, however, are colourless and in any case flavanoids occur in all parts of plants so they may be useful in other respects also. It may be significant in this connection that although flavanoids occur widely in higher plants and in ferns, they are absent from lower forms of plant life such as lichens, mosses, and algae. There is no record of their being produced in fungi or bacteria either.

In spite of the strong family resemblance in the form of a common carbon skeleton (isoflavones excepted) the number of individuals is very great. The source of the variation within the different subgroups is, as indicated in diagram (**104**), the incidence of additional oxygen functions at key sites in the carbon skeleton. The characteristic patterns of oxygenation and their biosynthetic significance has already been noted in the treatment of anthocyanins: ring A is usually hydroxylated on alternate carbons (C-5 and C-7) whereas ring B is almost invariably hydroxylated on adjacent carbons (primarily at C-4′, secondarily at C-3′ and C-5′).[95]

This pattern of hydroxylation reflects the differing biosynthetic origins of the two rings, as indicated in the biosynthetic pathway shown in Scheme 18.[96] A fuller treatment of this

3 MeCO₂H + (cinnamic acid, CO₂H) → (polyketide, COSR) → (chalcone, HO, HO, OH) → (**105**) (HO, HO, 2, 3, B, O) → Flavanoids

SCHEME 18

aspect of flavanoid chemistry can be found elsewhere. Suffice to say here that the carbon skeleton is generated by condensation of a unit of cinnamate with three acetate units in a polyketide biosynthesis. Cyclization of the polyketone chain gives rise to ring A and so its three oxygen functions come from the carbonyl groups of acetate. In contrast, the oxygen functions of ring B are introduced by oxidation at the expense of molecular oxygen (one or more of these hydroxyls may be introduced prior to formation of the heterocycle). The isoflavone skeleton is generated from (**105**) by migration of ring B from C-2 to C-3 at a subsequent stage of the biosynthesis.

Superimposed on the hydroxylation pattern at a later stage in the biosynthesis is a further source of structure variation in the form of *O*-alkylation (methyl or methylenedioxy) and glycoside formation. The sugars most frequently found in association with the flavanoids are glucose and rhamnose. The standard patterns of derivatization found in the anthocyanin group have already been mentioned. The situation in the case of the flavones and isoflavones is much more variable and no clear trend can be discerned in the pattern of derivatization.

18.3.4.2 Determination of structure

The classical method of structure determination for flavones and isoflavones is alkaline hydrolysis. The reaction proceeds *via* an intermediate β-dicarbonyl compound which can break down in a number of ways. The general pattern for flavones is given in Scheme 19

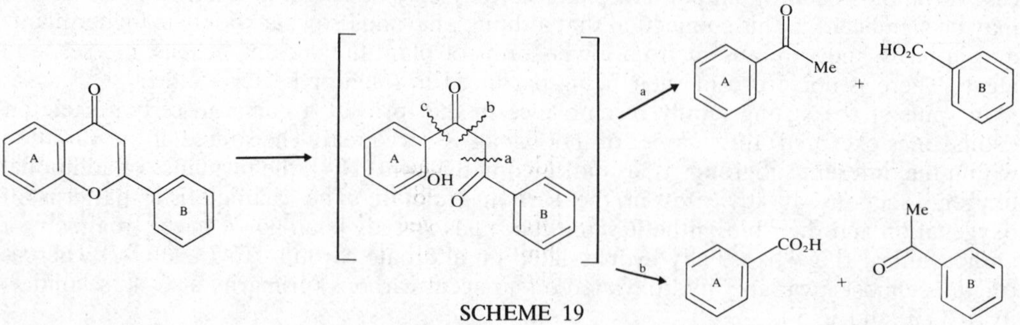

SCHEME 19

and for isoflavones in Scheme 20. In the latter case cleavage at site (a) is favoured over site (b). When, as is so often the case, ring A is a phloroglucinol derivative, cleavage at site (c) may also be observed in both systems. This is found, for example, in the degradation of chrysin (**106**) (reaction 44).[97]

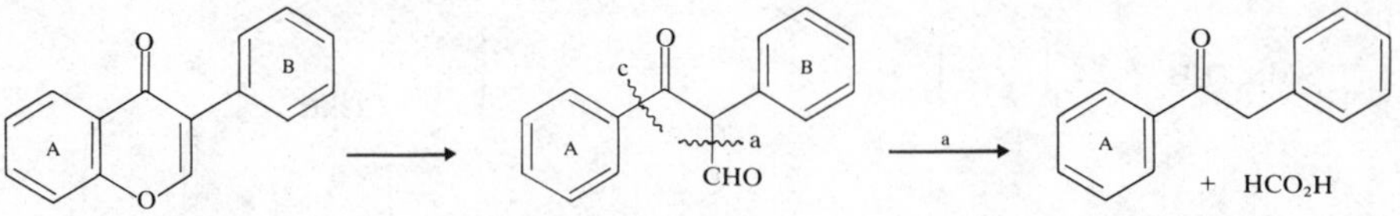

SCHEME 20

Usually the yield in these degradations is improved by methylating the free hydroxyls prior to degradation of the pyrone. If methoxy-groups are already present, ethylation can be used instead, so that the positions of free hydroxyls in the flavone or isoflavone can be inferred from the positions of the ethoxy-groups in the degradation products.

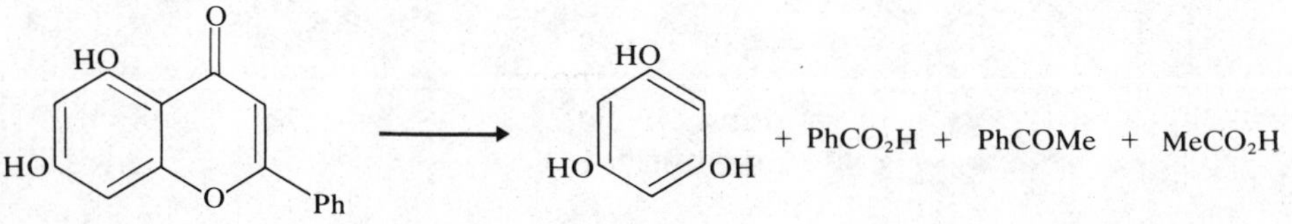

$$+ PhCO_2H + PhCOMe + MeCO_2H \quad (44)$$

(**106**)

Needless to say, there has been a considerable effort to replace these tedious and destructive chemical methods of structure determination by physical methods based on spectroscopy. Unfortunately i.r. spectroscopy has not proved useful because the carbonyl stretching frequency in flavones and isoflavones almost always occurs near 1650 cm^{-1} and its position is not sensitive to the substitution pattern, except that when a hydroxy-group is present at C-3 the peak is sometimes found at a lower frequency.[59]

U.v. spectroscopy is more useful.[98] In flavone itself the spectrum shows two peaks at 250 and 297 nm. The latter peak shifts to longer wavelengths as hydroxy-groups are introduced into either ring A or ring B. This trend is apparent in the spectra given in Figure 4.

N.m.r. spectroscopy is even more useful for structure determination.[99] The definitive assignment of the signals in the ^1H n.m.r. spectrum is usually not easy, although the signal from a proton at C-3 will be a singlet and so can be distinguished from others; this feature

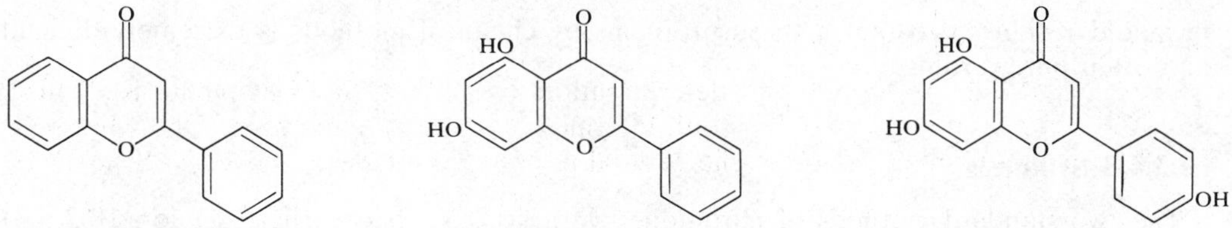

λ_{max} 250, 297 nm 270, 330 nm 265, 340 nm

U.v. spectra of flavones

FIGURE 4

can be used to distinguish flavones from flavonols. The power of the n.m.r. method has been greatly improved by using lanthanide shift reagents to assist in making the assignments. The signals from protons at C-3, C-6, and C-8, for example, are markedly different in the degree to which their chemical shift is changed on addition of Eu(fod)$_3$.[100]

The most promising spectroscopic technique of all is ^{13}C n.m.r. Spectral data for two flavones are given in Figure 5.[101] With the interesting exception of C-3′, C-5′, and C-6,

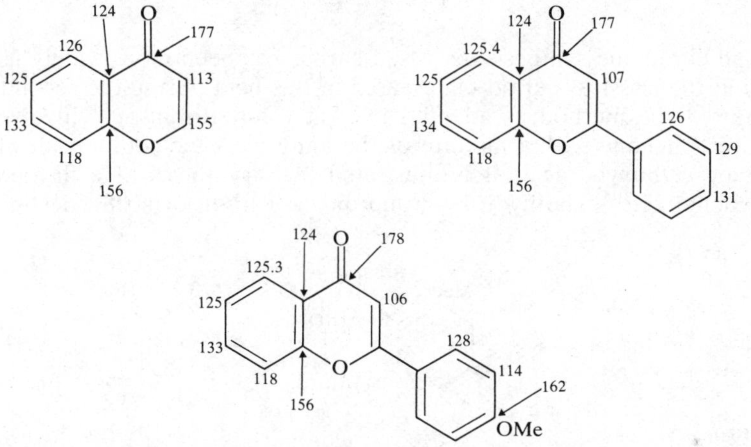

Chemical shifts (rel. to Me_4Si) in the ^{13}C n.m.r. spectra

FIGURE 5

calculated values of chemical shift based on simple models are in good agreement with observed values. The chemical shift for C-4 (the carbonyl) is remarkably constant at about 178 p.p.m. for a wide range of substitution patterns. If the trends established in preliminary studies hold good for a wider range of compounds, ^{13}C n.m.r. will play a major part in structure elucidation in the flavone field from now on.

Armed with these new techniques, chemists will be much better placed to tackle the structures of bis-flavones such as (**107**), which are produced by oxidative coupling of two

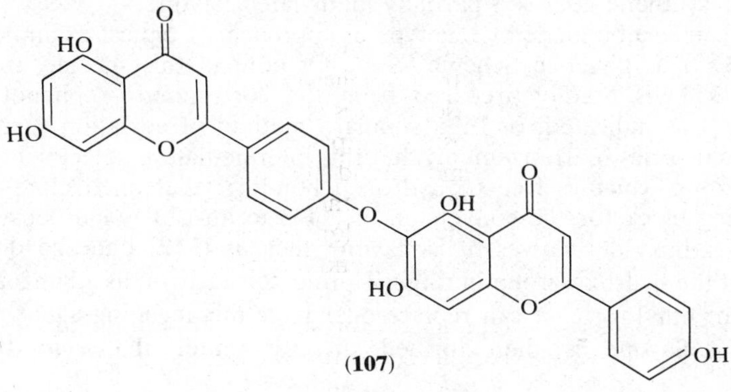

(**107**)

flavanoid residues. To tackle these problems by chemical methods is extremely difficult and often unrewarding.

18.3.4.3 Synthesis

The two standard methods of chromone synthesis described earlier (Section 18.3.3.3) have been widely used for the preparation of flavones. Thus a polyhydroxyphenol condenses with an aroylacetic ester (reaction 45). Ideally the reaction is carried out without a catalyst by heating the mixture under reduced pressure so that ethanol and water are removed as they form. The alternative condensation to form a coumarin does not take place under these conditions.[102]

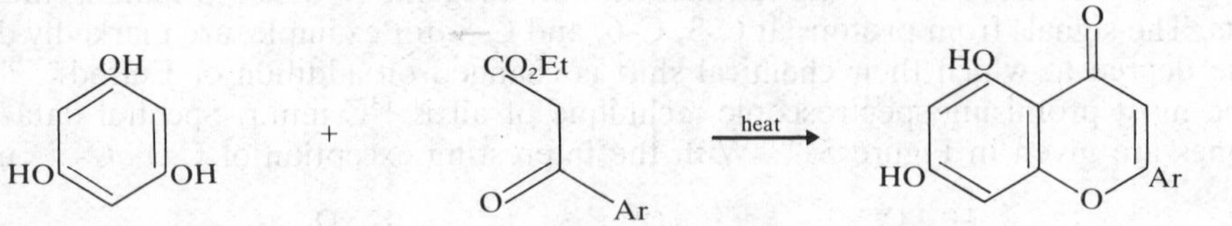

(45)

The second chromone synthesis discussed earlier has been more widely used, however, particularly in the classical period of research in this field during the second two decades of this century. The method as adapted by Allan and Robinson[81] involves heating an *o*-hydroxyacetophenone with a mixture of the anhydride and sodium salt of the appropriate aromatic carboxylic acid. It is illustrated by a synthesis of a flavonol, herbacetin (**109**), in Scheme 21.[103] The hydroxy-group on the carbon of (**108**), destined to become

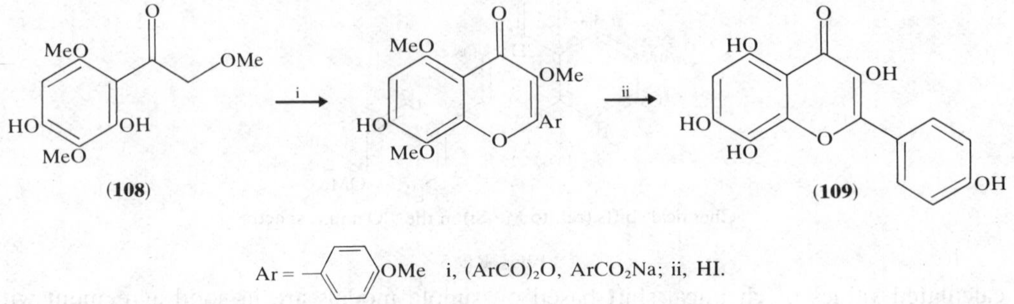

(**108**) (**109**)

Ar = C_6H_4-OMe i, $(ArCO)_2O$, $ArCO_2Na$; ii, HI.

SCHEME 21

C-3 of the final product, is protected by methylation. This is usually a necessary precaution, as is the protection of any hydroxy-groups in the anhydride component. On the other hand, it is not usually essential to protect the phenol hydroxyls in the *o*-hydroxyacetophenone. Benzyl, isopropyl, and methoxymethyl protecting groups can be used when the synthetic goal is a partially methylated flavone.

Isoflavones can also be made by this type of approach. A typical example, the synthesis of prunetin (**113**), is given in Scheme 22.[104] Of note in this scheme are that the key intermediate (**111**) is readily prepared from the corresponding phenol (**110**) by the Hoesch reaction as indicated, or by a standard method of acylation such as a Friedel–Crafts reaction or Fries rearrangement; that this intermediate is selectively methylated on the hydroxy-group which is free of hydrogen bonding; that an oxalate derivative is a suitable building block for the conversion of (**111**) to an isoflavone because of the ease with which 2-carboxy-derivatives of isoflavone such as (**112**) undergo decarboxylation; and finally that the hydroxy-group in ring B is protected early on as a benzoate ester. Ethyl orthoformate or ethyl formate can replace oxalate as the acylating agent with advantage because the isoflavone is then formed directly rather than *via* the 2-carboxy-derivative.[105,106]

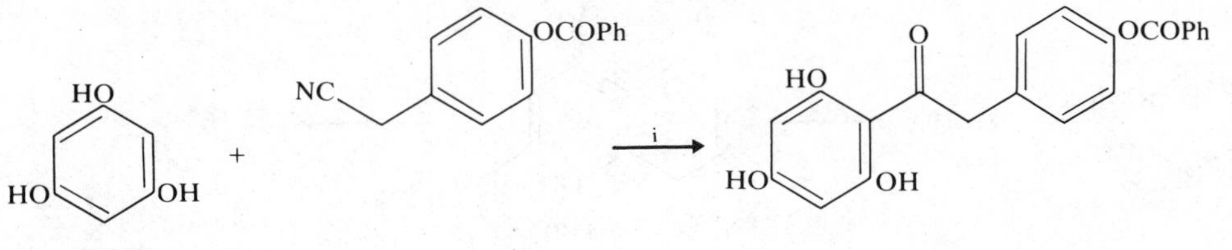

(110) (111)

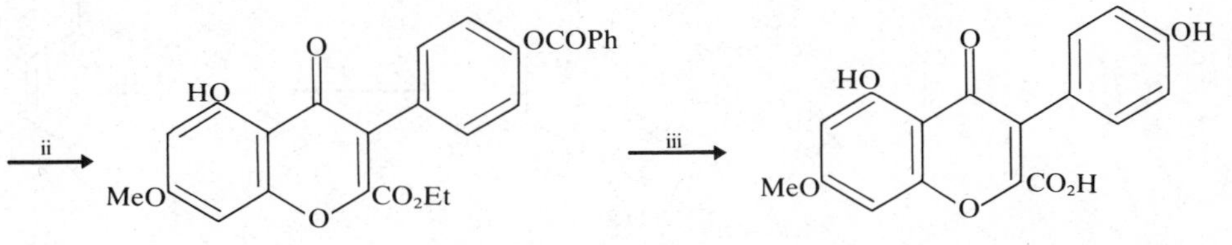

(112)

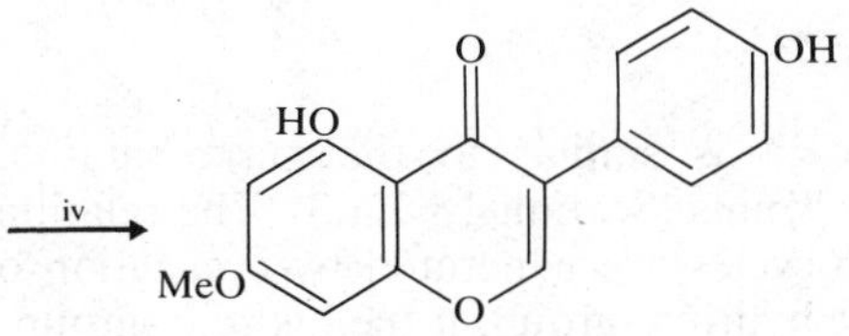

(113)

i, $ZnCl_2$; ii, $ClOCCO_2Et$, pyridine; iii, NaOH; iv, heat.

SCHEME 22

Flavones can be prepared from 4-hydroxycoumarins *via* the sequence of reactions shown in Scheme 23. The acylation of (**114**) to form (**115**) is usually effected in two steps *via* the *O*-acyl derivative and a Fries rearrangement. Thereafter the synthesis runs parallel to one of the classical γ-pyrone syntheses given in Section 18.3.2.3: the α-pyrone ring of (**115**) is hydrolytically cleaved to give a diketo-acid, which decarboxylates to the β-diketone precursor of (**116**).[107,108]

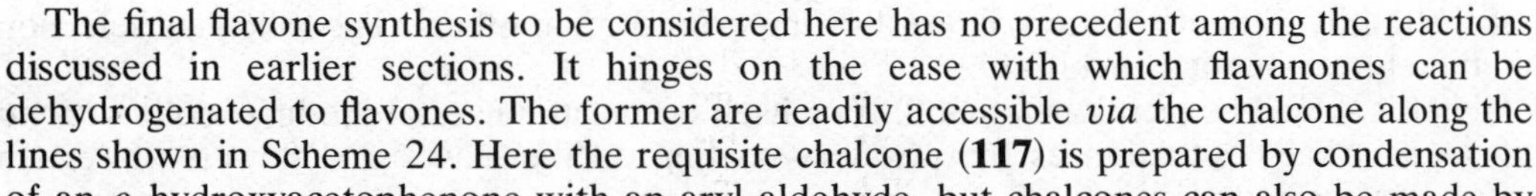

(114) (115) (116)

i, ArCOCl, $POCl_3$; ii, HCl, EtOH.

SCHEME 23

The final flavone synthesis to be considered here has no precedent among the reactions discussed in earlier sections. It hinges on the ease with which flavanones can be dehydrogenated to flavones. The former are readily accessible *via* the chalcone along the lines shown in Scheme 24. Here the requisite chalcone (**117**) is prepared by condensation of an *o*-hydroxyacetophenone with an aryl aldehyde, but chalcones can also be made by *C*-acylation of a phenol with the appropriate cinnamate either directly or *via* a Fries rearrangement.[109] Cyclization to the flavanone (**118**) is spontaneous. The key dehydrogenation in the final step can be effected by a variety of reagents, including selenium dioxide or triphenylmethyl carbonium ion.[58] If nitrous acid (or a nitrosating agent) is used as oxidizing agent in the last step, the product is a flavonol rather than the flavone, so this is an exceptionally flexible method of flavanoid synthesis.[110]

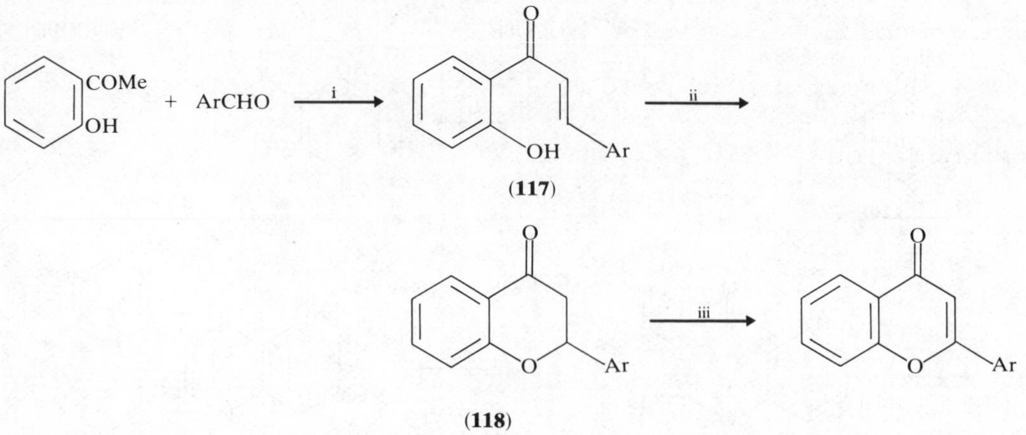

(117)

(118)

i, base; ii, H^+; iii, SeO_2 or Ph_3C^+.

SCHEME 24

The flavanone synthesis given in Scheme 24 is similar to the major method of synthesizing the pyrylium nucleus of the anthocyanins (Section 18.1.4.3). The relationship is illustrated in Scheme 25, where both heterocycles are generated by cyclization of an α,β-unsaturated ketone residue with a phenolic hydroxy-group; if the hydroxy-group is so

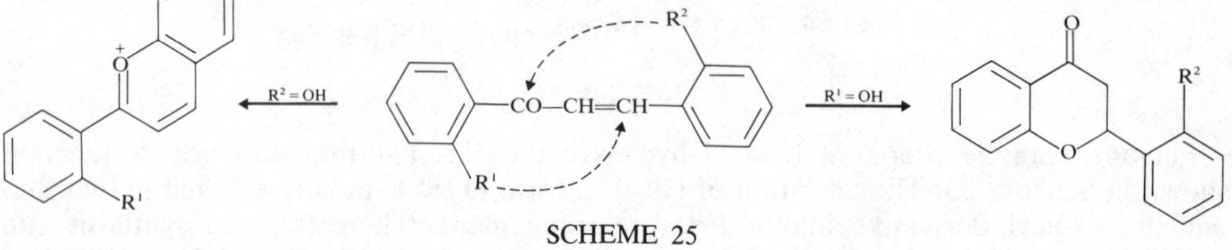

SCHEME 25

placed that it attacks the carbonyl group, the product is a pyrylium salt, whereas if it is in the other aryl ring the result is Michael addition to give a flavanone. Thus the position of the free *o*-hydroxy-group is critical and this is a key factor in most of the synthetic approaches to flavanoids. Fortunately for synthetic chemists, ring B in flavanoid natural products is derived from cinnamic acid rather than acetate and so only rarely does it carry a hydroxy-group in a position which would give rise to ambiguity, *i.e.* adjacent to the heterocyclic ring [see structure (**104**) and Scheme 18].

18.3.4.4 Typical flavones and isoflavones

Flavones are so numerous and widespread it seems invidious to select certain members of the family as being of more interest or importance than others. The following selection is therefore essentially arbitrary.

First it should be mentioned that flavone itself occurs as a natural product and it is found in a remarkable situation as the powdery coating on the leaves and flowers of certain species of primula.[111,112]

Quercetin (**119**) deserves a mention if only because it is the most widely occurring of all flavone derivatives. It is found, for example, in lemons and onions as well as many different flowering plants.[81] The 3-rhamnoside (**120**), called quercetrin, is one of the constituents of a yellow vegetable dye isolated from the bark of *Quercus velutina*, a North American oak.[113] Another flavone which has played an important role in dyeing for over 2000 years is luteolin (**121**), which is a colour constituent of weld, the dyestuff extracted

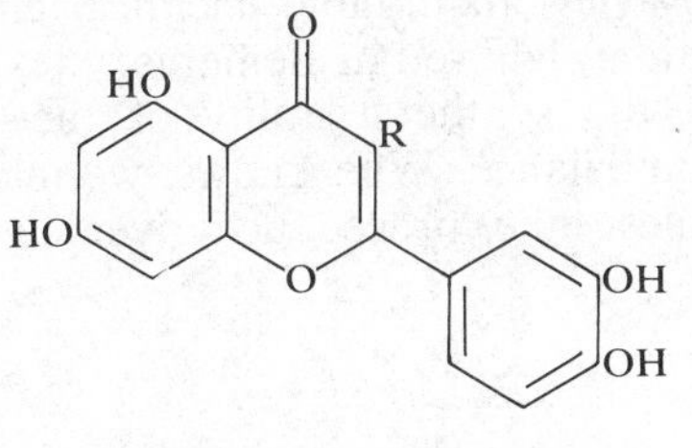

(**119**) R = OH
(**120**) R = O—rhamnoside
(**121**) R = H

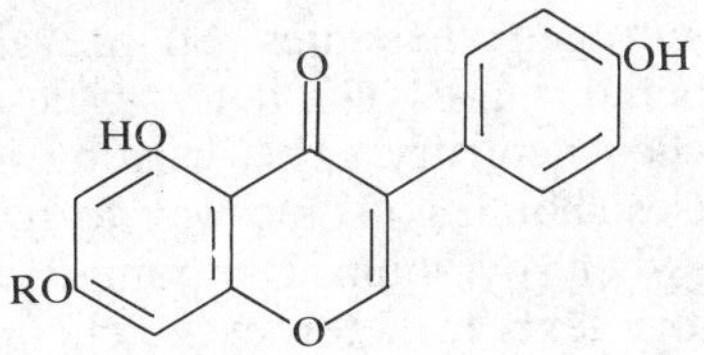

(**122**) R = Me
(**123**) R = H

from *Reseda luteola*.[114] The successful use of these flavones as dyes hinges on the fact that they are relatively stable; the related anthocyanins are more brilliant and offer a wider range of colour, but are too unstable to be exploited in this way.

Compared with the flavones the isoflavones are relatively rare. The first to be identified (in 1910) was prunetin (**122**) from the bark of a Prunus species.[104] Some members of the group possess interesting pharmacological properties. Genistein (**123**), for example, is mildly oestrogenic;[115] others are poisonous to fish. Isoflavones occur in some foodstuffs, although their concentration is never large enough to cause problems under normal conditions. It is fortunate indeed that it is not they but their innocuous cousins the flavones which occur so widely, otherwise the range of plants suitable for human consumption would be greatly restricted.

18.3.5 CONCLUSION

The properties of the γ-pyrone ring are essentially aliphatic, even though it gives reactions characteristic of an aromatic pyrylium betaine more readily than α-pyrone. Monocyclic γ-pyrones were, until recently, relatively inaccessible because of the difficulty of preparing 1,3,5-triketones. This deficiency has now been remedied. The heterocycle is equivalent to a β-triketone and so has interesting potential in synthesis, but this aspect of its chemistry has not so far been given due attention.

Chromones are readily prepared *via* phenolic β-diketones using long established methods. The properties of the heterocyclic ring are not greatly modified as a consequence of benzannulation.

Relatively few simple γ-pyrones occur in Nature, but this deficiency is more than compensated by the enormous number of naturally occurring chromones in the flavanoid family. Taking them into account the γ-pyrone ring must be acknowledged as one of the most important heterocycles in natural product chemistry.

Further reading on the topics covered in this section can be found in references 116, 117, 118, and 119.

A few closing remarks of a more general nature concerning the position of pyrylium compounds, and α- and γ-pyrones, in the hierarchy of heterocyclic compounds will help to set their chemistry in perspective. In the opening to Chapter 18.1 the relationship of the pyrylium cation to pyridine and its pyridinium derivatives was discussed. Similarly, α-pyrones can be compared with α-pyridones and γ-pyrones with γ-pyridones. Another comparison of this type which is frequently made is that between the chemistry of the pyran heterocycles and their sulphur analogues. Unfortunately, while such comparisons can be helpful, they can also lead to a confusing network of cross-relationships in which the chemistry of individual heterocycles is lost from view. It was to avoid this risk that the comparative approach was largely ignored in the present treatment. The reader can now undertake such an exercise by turning to the relevant chapters of this work.

This present method of treatment also avoids the second pitfall of a comparative approach. All too often the impression is given that pyran heterocycles are inferior in status to their nitrogen analogues. So prevalent is this view in organic chemical circles that six-membered ring-oxygen heterocycles are sometimes ignored in elementary textbooks of heterocyclic chemistry. The current interest in the synthetic utility of these oxygen heterocycles should go some way to redress the imbalance. Who knows, we may yet see the day when pyrylium compounds and pyrones take precedence over pyridine in introductory texts to heterocyclic chemistry!

References

1. S. Wawzonek, in 'Heterocyclic Compounds', ed. R. C. Elderfield, Wiley, New York, 1951, vol. 2, chapter 12.
2. F. Arndt, G. T. O. Merlin, and J. R. Partington, *J. Chem. Soc.*, 1935, 602.
3. J. N. Collie and T. Tickle, *J. Chem. Soc.*, 1899, **75,** 710.
4. A. Baeyer and V. Villiger, *Ber.*, 1901, **34,** 2679.
5. K. Hafner and H. Kaiser, *Annalen*, 1958, **618,** 140.
6. F. Arndt, E. Scholz, and P. Nachtwey, *Ber.*, 1924, **57,** 1903.
7. D. D. Perrin, 'Dissociation Constants of Organic Bases in Aqueous Solution', Butterworths, London, 1965.
8. H. Meerwein, K. Bodenbenner, P. Borner, F. Kunert, and K. Wunderlich, *Annalen*, 1960, **632,** 38.
9. H. Meerwein, G. Hinz, P. Hofmann, E. Kroning, and E. Pfeil, *J. prakt. Chem.*, 1937, **147,** 257.
10. K. Dimroth and P. Heinrich, *Angew. Chem. Internat. Edn.*, 1966, **5,** 676.
11. A. I. Tolmachev and V. P. Sribnaya, *Zhur. obshchei Khim.*, 1965, **35,** 316.
12. D. A. Griffin and J. Staunton, *J.C.S. Chem. Comm.*, 1975, 675.
13. H. Meerwein and K. Wunderlich, *Angew. Chem.*, 1957, **69,** 481.
14. (a) E. C. E. Hunter and J. R. Partington, *J. Chem. Soc.*, 1933, 87; (b) G. G. Le Fevre and R. J. W. Le Fevre, *ibid.*, 1937, 1088.
15. N. J. Leonard and D. Chroudhury, *J. Amer. Chem. Soc.*, 1957, **79,** 156.
16. D. Cook, *Canad. J. Chem.*, 1963, **41,** 505.
17. S. Aronoff, *J. Org. Chem.*, 1940, **5,** 561.
18. M. J. Cook, A. R. Katritzky, and P. Linda, *Adv. Heterocyclic Chem.*, 1974, **17,** 321.
19. J. N. Collie and N. T. M. Wilsmore, *J. Chem. Soc.*, 1896, **69,** 293.
20. C. F. Rassweiler and R. Adams, *J. Amer. Chem. Soc.*, 1924, **46,** 2758.
21. J. N. Collie, *J. Chem. Soc.*, 1907, 1806.
22. A. T. Balaban, W. Schroth, and G. Fischer, in *Adv. Heterocyclic Chem.*, 1969, **10,** 261.
23. G. Köbrich, *Annalen*, 1961, **648,** 114.
24. L. L. Woods, *J. Amer. Chem. Soc.*, 1958, **80,** 1440.
25. J. A. VanAllan, G. A. Reynolds, and D. P. Maier, *J. Org. Chem.*, 1968, **33,** 4418.
26. R. Wizinger, A. Grüne, and E. Jacobi, *Helv. Chim. Acta*, 1956, **39,** 1.
27. C. Ainsworth and R. G. Jones, *J. Amer. Chem. Soc.*, 1954, **76,** 3172.
28. L. L. Woods, *J. Org. Chem.*, 1959, **24,** 1804; 1962, **27,** 696.
29. (a) F. Feist and E. Baum, *Ber.*, 1905, **38,** 3562; (b) F. Feist, *ibid.*, 1907, **40,** 3647.
30. P. Beak and G. A. Carls, *J. Org. Chem.*, 1964, **29,** 2678.
31. R. Mozingo and H. Adkins, *J. Amer. Chem. Soc.*, 1938, **60,** 669.
32. P. Yates and M. J. Jorgensen, *J. Amer. Chem. Soc.*, 1963, **85,** 2956.
33. N. Ishibe, M. Sunami and M. Odani, *J. Amer. Chem. Soc.*, 1973, **95,** 463.
34. M. Shiozaki and T. Hiraoka, *Tetrahedron Letters*, 1972, 4655.
35. H. Staudinger and N. Kon, *Annalen*, 1911, **384,** 38, 129.
36. A. A. Boon, K. J. McKenzie, and J. Trotter, *Proc. Chem. Soc.*, 1914, **30,** 206.
37. Buu-Hof and J. Lecocq, *Compt. rend.*, 1946, **222,** 1441.
38. F. J. Leeper and J. Staunton, unpublished work.
39. R. Cornubert and P. Robinet, *Bull. Soc. chim. France*, 1933, **53,** 565.
40. E. B. Mullock and H. Suschitzky, *J. Chem. Soc. (C)*, 1967, 828.
41. M. Conrad and M. Guthzeit, *Ber.*, 1886, **19,** 19.
42. F. Kögl and C. A. Salemink, *Rec. Trav. chim.*, 1955, **74,** 221.
43. S. Iguchi and K. Hisatsune, *J. Pharm. Soc. Japan*, 1957, **77,** 94.
44. (a) S. Ruhemann, *J. Chem. Soc.*, 1908, **93,** 431; (b) G. Barger and W. W. Starling, *ibid.*, 1915, **107,** 418.
45. F. Gaudemar-Bardone and M. Gaudemar, *Bull. Soc. chim. France*, 1966, 3033.
46. A. Dornow and F. Ische, *Angew. Chem.*, 1955, **67,** 653.
47. T. M. Harris, C. M. Harris, and K. B. Hindley, *Fortschr. Chem. org. Naturstoffe*, 1974, **31,** 217.
48. F. M. Dean, 'Naturally Occurring Oxygen Ring Compounds', Butterworths, London, 1963, pp. 107–113.
49. J. Fried, in 'Heterocyclic Compounds', ed. R. C. Elderfield, Wiley, New York, 1950, vol. I, pp 382–389.
50. N. Campbell in 'Chemistry of Carbon Compounds', ed. E. H. Rodd, Elsevier, Amsterdam, 1959, vol. 4B, pp. 828–835.
51. O. E. May, A. J. Moyer, P. A. Wells, and H. T. Herrick, *J. Amer. Chem. Soc.*, 1931, **53,** 774.
52. Dr. Probst, *Annalen*, 1839, **29,** 113.

53. J. P. Wibaut and R. J. C. Kleipool, *Rec. Trav. chim.*, 1947, **66,** 24.
54. T. Yabuta, *J. Chem. Soc.*, 1924, **125,** 575.
55. M. Stacey and L. M. Turton, *J. Chem. Soc.*, 1946, 661.
56. G. Wittig, *Annalen*, 1925, **446,** 157.
57. A. I. Tolmachev, L. M. Shulezhko and A. A. Kisilenko, *Zhur. obshchei Khim.*, 1965, **35,** 1707.
58. A. Schönberg and G. Schütz, *Chem. Ber.*, 1960, **93,** 1466.
59. J. H. Looker and W. W. Hanneman, *J. Org. Chem.*, 1962, **27,** 381.
60. N. Campbell, in 'Chemistry of Carbon Compounds', ed. E. H. Rodd, Elsevier, Amsterdam, 1959, vol. 4B, p. 897.
61. G. Wittig, *Ber.*, 1924, **57,** 88.
62. H. Simonis, *Ber.*, 1917, **50,** 779.
63. G. Wittig and H. Blumenthal, *Ber.*, 1927, **60,** 1085.
64. W. Baker, J. B. Harborne, and W. D. Ollis, *J. Chem. Soc.*, 1952, 1303.
65. I. M. Heilbron, R. N. Heslop, and F. Irving, *J. Chem. Soc.*, 1933, 430.
66. P. G. Sammes and T. W. Wallace, *J.C.S. Perkin I*, 1975, 1845.
67. E. Petschek and H. Simonis, *Ber.*, 1913, **46,** 2014.
68. J. W. Hanifin and E. Cohen, *Tetrahedron Letters*, 1966, 5421.
69. S. C. Gupta and S. K. Mukerjee, *Tetrahedron Letters*, 1973, 5073.
70. T. Matsuura, T. Takemoto, and R. Nakashima, *Tetrahedron*, 1973, **29,** 3337.
71. W. A. Henderson and E. F. Ullman, *J. Amer. Chem. Soc.*, 1965, **87,** 5424.
72. T. Matsuura, H. Matsushima, and R. Nakashima, *Tetrahedron*, 1970, **26,** 435.
73. S. Wawzonek, in 'Heterocyclic Compounds', ed. R. C. Elderfield, Wiley, New York, 1951, vol. 2, p. 256.
74. I. M. Heilbron, H. Barnes, and R. A. Morton, *J. Chem. Soc.*, 1923, **123,** 2559.
75. J. Allan and R. Robinson, *J. Chem. Soc.*, 1924, **125,** 2192.
76. S. Sethna and R. Phadke in 'Organic Reactions', ed. R. Adams, Wiley, New York, 1953, vol. VII, chapter 1.
77. O. Dann and G. Illing, *Annalen*, 1957, **605,** 146.
78. S. Ruhemann and H. E. Stapleton, *J. Chem. Soc.*, 1900, **77,** 1179.
79. E. Gottesmann *Ber.*, 1933, **66,** 1168.
80. T. A. Geissman, *J. Amer. Chem. Soc.*, 1949, **71,** 1498.
81. J. Allan and R. Robinson, *J. Chem. Soc.*, 1926, 2334.
82. W. Baker, *J. Chem. Soc.*, 1933, 1381.
83. H. S. Mahal and K. Venkataraman, *J. Chem. Soc.*, 1934, 1767.
84. (a) H. M. Lynch, T. M. O'Toole, and T. S. Wheeler, *J. Chem. Soc.*, 1952, 2063; (b) W. D. Ollis and D. Weight, *ibid.*, 1952, 3826.
85. F. M. Dean, 'Naturally Occurring Oxygen Ring Compounds', Butterworths, London, 1963, pp. 251–265.
86. N. Campbell, in 'Chemistry of Carbon Compounds', ed. E. H. Rodd, Elsevier, Amsterdam, 1959, vol. 4B, pp. 898–903.
87. Th. M. Meyer and H. Schmid, *Helv. Chim. Acta*, 1948, **31,** 1603.
88. H. Schmid, *Helv. Chim. Acta*, 1949, **32,** 813.
89. A. Bolleter, K. Eiter, and H. Schmid, *Helv. Chim. Acta*, 1951, **34,** 186.
90. E. Spath and W. Gruber, *Ber.*, 1938, **71,** 106.
91. J. R. Clarke and A. Robertson, *J. Chem. Soc.*, 1949, 302.
92. R. A. Baxter, G. R. Ramage, and J. A. Timson, *J. Chem. Soc.*, 1949, S30.
93. F. M. Dean, 'Naturally Occurring Oxygen Ring Compounds', Butterworths, London, 1963, pp. 280–449.
94. T. A. Geissman, 'The Chemistry of Flavanoid Compounds', Pergamon Press, Oxford, 1962.
95. T. A. Geissman, in 'Comprehensive Biochemistry', ed. M. Florkin and E. M. Stotz, Elsevier, Amsterdam, 1963, vol. 9, pp. 213–265.
96. J. H. Richards and J. B. Hendrickson, 'The Biosynthesis of Steroids, Terpenes and Acetogenins', Benjamin, New York, 1964, pp. 160–164.
97. S. von Kostanecki, *Ber.*, 1893, **26,** 2901.
98. F. M. Dean, 'Naturally Occurring Oxygen Ring Compounds', Butterworths, London, 1963, pp. 286–287.
99. A. Grouiller, *Bull. Soc. chim. France*, 1966, 2405.
100. M. Okigawa, N. U. Khan, N. Kawano, and W. Rahman, *J.C.S. Perkin I*, 1975, 1563.
101. C. A. Kingsbury and J. A. Looker, *J. Org. Chem.*, 1975, **40,** 1120.
102. R. Teoule, G. Grenier, H. Pacheco, and J. Chopin, *Bull. Soc. chim. France*, 1961, 546.
103. L. J. Goldsworthy and R. Robinson, *J. Chem. Soc.*, 1938, 56.
104. W. Baker, J. Chadderton, J. B. Harborne, and W. D. Ollis, *J. Chem. Soc.*, 1953, 1852.
105. V. R. Sathe and K. Venkataraman, *Current Sci.*, 1949, **18,** 373.
106. A. Schönberg and A. Sina, *J. Amer. Chem. Soc.*, 1950, **72,** 3396.
107. K. Veres, J. Jonas, and A. Horeni, *Coll. Trav. chim. Tchecosl.*, 1959, **24,** 3471.
108. C. Mentzer, J. Chopin, and M. Mercier, *Compt. rend.*, 1956, **242,** 1034.
109. V. T. Ramakrishnan and J. Kagan, *J. Org. Chem.*, 1970, **35,** 2901.
110. S. von Kostanecki and V. Lampe, *Ber.*, 1904, **37,** 773, 1402.
111. P. Karrer and G. Schwab, *Helv. Chim. Acta*, 1941, **24,** 297.
112. W. Baker, N. C. Brown and J. A. Scott, *J. Chem. Soc.*, 1939, 1922.
113. C. H. Ice and S. H. Wender, *J. Amer. Chem. Soc.*, 1952, **74,** 4606.
114. W. A. Hutchins and T. S. Wheeler, *J. Chem. Soc.*, 1939, 91.

115. R. B. Bradbury and D. E. White, *J. Chem. Soc.*, 1953, 871.
116. A. H. Corwin, in 'Heterocyclic Compounds', ed. R. C. Elderfield, Wiley, New York, 1950, vol. I, chapter 6, p. 277.
117. S. Wawzonek, in 'Heterocyclic Compounds', ed. R. C. Elderfield, Wiley, New York, 1951, vol. II, chapter 8, p. 229.
118. F. M. Dean, 'Naturally Occurring Oxygen Ring Compounds', Butterworths, London, 1963.
119. N. Campbell, in 'Chemistry of Carbon Compounds', ed. E. H. Rodd, Elsevier, Amsterdam, 1959, vol. 4B, pp 810–961.

18.4

Furans

M. V. SARGENT
University of Western Australia
AND T. M. CRESP
University College, London

18.4.1 INTRODUCTION

Furan is numbered as in (**1**). Its first derivative was reported in 1780 but the most active period in furan research was between *ca.* 1930 and 1945 since furan-2-carboxaldehyde then became readily available by acid hydrolysis of carbohydrate-containing agricultural waste. This period culminated in the publication of an authoritative and comprehensive text.[1] Some aspects of the more recent literature have also been reviewed.[2]

18.4.2 AROMATICITY

Aromaticity is concerned with the ground-state properties of fully conjugated cyclic systems. A number of methods have been used to disclose information on the aromaticity of furan, particularly in relation to benzene, thiophen, and pyrrole.

In the valence-bond approach furan is regarded as a resonance hybrid of the canonicals (**1**)–(**5**). The direction of the molecular dipole (0.72 D), contrary to textbook fallacy, is such that the oxygen is negative so that it withdraws electrons inductively from the ring. The situation is the same in thiophen, but reversed in pyrrole.[3] Many MO calculations have been made for furan, but the results are widely divergent.[4] Values for the resonance energy of furan obtained thermochemically fall in the range 66–96 kJ mol^{-1}.[5] The bond angles and lengths for thiophen, pyrrole, and furan have been determined by microwave spectroscopy. Comparison of the ratios of the 2,3 to 3,4 bond lengths has been used as a criterion of aromaticity, but the validity of this approach has been questioned.

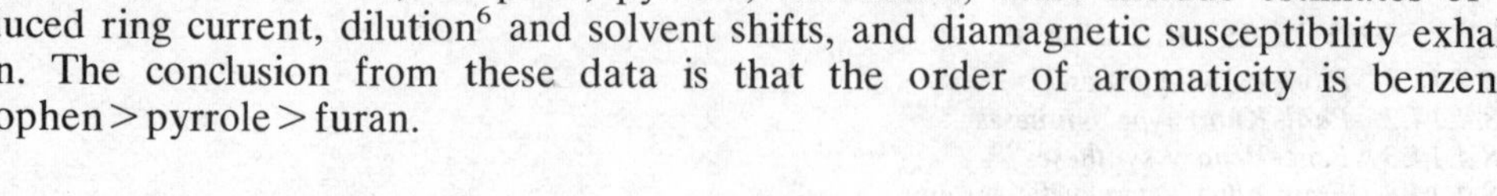

4, 3(β), 5, 2(α), O 1 ↔ O+ ↔ O+ ↔ O+ ↔ O+

(**1**) (**2**) (**3**) (**4**) (**5**)

A number of methods based on n.m.r. spectroscopy[5] have been used to compare the aromaticities of benzene, thiophen, pyrrole, and furan; these include estimates of the induced ring current, dilution[6] and solvent shifts, and diamagnetic susceptibility exhaltation. The conclusion from these data is that the order of aromaticity is benzene > thiophen > pyrrole > furan.

18.4.3 PHYSICAL PROPERTIES

The infrared, ^{1}H n.m.r., electronic, and mass spectra of furan and its derivatives have been fully reviewed.[7] ^{13}C N.m.r. spectral assignments for furan and a number of its derivatives have been made.[8] 2-Carbonyl substituted furans exhibit rotational isomerism. This usually results in the appearance of two i.r. carbonyl bands, but in the case of furan-2-carboxaldehydes these are due to Fermi resonance.[9]

18.4.4 ELECTROPHILIC SUBSTITUTION

18.4.4.1 General remarks

Furan exhibits great reactivity towards electrophiles. Furans substituted with electron-releasing groups are unstable to mineral acids and protonation at an α-position occurs,[10] and this may be followed by ring-opening (Section 18.4.9) or polymerization; furans with electron-withdrawing substituents are more stable to acids. The conditions under which electrophilic substitutions are carried out thus require to be carefully controlled. The mechanism of reaction is often not so simple as the electrophilic substitution characteristic of benzenoid compounds, and may be complicated by 1,2- and 1,4-addition to the diene system.

The reactivity order of the five-membered heterocyclic systems has been quantified by kinetic investigations.[4] The relative rates of electrophilic substitution at the 2-position have been determined for a number of reactions (Table 1). Owing to complications

TABLE 1

Relative Rates of Electrophilic Substitution at the 2-Position of [five-membered ring with X]

X	*Bromination* (25 °C)	*Trifluoroacetylation* (75 °C)	*Acetylation* (25 °C)	*Formylation* (30 °C)
O	1	1	1	1
NH	4.9×10^{6}	3.8×10^{5}	—	—
S	8.3×10^{-3}	7.1×10^{-3}	8.4×10^{-2}	9.7×10^{-2}
Se	4.0×10^{-1}	5.2×10^{-2}	1.9×10^{-1}	3.6×10^{-1}
Te	—	3.3×10^{-1}	6.3×10^{-1}	3.5

with the parent systems, the data for bromination were determined with the 2-methoxycarbonyl compounds. Comparison of these data with the known rates of bromination of thiophen and benzene gives the following partial rate factors for the 2-position relative to a benzene position: pyrrole 3×10^{18}; furan 6×10^{11}; thiophen 5×10^{9}; selenophen 2.4×10^{11}. It has been suggested that the greater reactivity of pyrrole compared with furan is due to the greater electronegativity of oxygen, and the greater reactivity of furan over thiophen is due to the greater conjugative electron donation of first-row elements.[11] It would appear that ring electron-density is not the only factor determining reactivity towards electrophiles, since furan and thiophen derivatives also display greater reactivity towards nucleophiles than their benzene analogues. Factors such as ground-state aromaticity and position of the transition state along the reaction coordinate are undoubtedly important.

The 2-position of the furan nucleus is much more reactive towards electrophiles than the 3-position and this can be rationalized in terms of the greater number of resonance canonical forms for the Wheland intermediate for the 2-position. The 2:3 ratios for furan, pyrrole, and thiophen have been determined for acetylation with acetyl trifluoroacetate in 1,2-dichloroethane at 75 °C: furan, 6000; thiophen, 71.4; pyrrole, 6.[12,13] The reason for the large differences between these heterocycles is little understood.

Linear free energy relationships have been used in two ways to study the reactivity of furan. In the 'extended selectivity treatment' the logarithms of the partial rate factors for both 2- and 3-substitution are separately plotted against the reaction constant (ρ) for the analogous reactions of benzene derivatives. As well as electrophilic substitutions such as bromination, chlorination, acetylation, trifluoroacetylation, protodesilylation, and protodemercuration, side-chain carbonium ion reactions have been included in the plots.[4,13] The slope of the lines gives a substituent constant (σ^+) for the replacement of CH=CH in benzene by O. The reactivity of thiophen has been similarly treated (Table 2). These data indicate that the 3-position of furan is least reactive, but more reactive than a benzene position.

TABLE 2
Substituent Constants

Compound	σ_2^+	σ_3^+
Thiophen	−0.79	−0.52
Furan	−0.93	−0.44

The other way of applying linear free energy relationships is to study the effect of substituents on the rate of electrophilic substitution. A Hammett plot has been made of the rates of trifluoroacetylation at the 5-position of seven 2-substituted furans against the benzene σ_p^+ substituent constants.[14] The furan ring appears to be more sensitive to substituent effects than thiophen since the ρ values obtained for this reaction are −10.7 and −7.4, respectively. A similar treatment has been applied to the solvolysis of 5-substituted 1-(2-furyl)ethanol derivatives and a ρ value of −8 was obtained.[15] Theoretically derived parameters have been used for correlations of the solvolysis rates of substituted 3-furyl derivatives.[16]

Empirical correlations can readily be made on the directing effect of substituents from the large body of experimental data available.[2,4] Furans substituted in the 2-position with a $+I$ or a $-I,+M$ substituent always undergo electrophilic substitution at the 5-position. For a $-I,-M$ substituent at the 2-position the mesomeric effect will deactivate the 5-position, but this is generally overwhelmed by the powerful orientational effect of the heteroatom and 5-substitution ensues. Lewis acids enhance the $-M$ effect of the substituent in 2-carbonyl substituted furans and electrophilic substitution gives appreciable amounts of the 2,4-isomer.[17] For a $+I$ substituent at the 3-position the 2-position is more activated by electronic effects, but the 5-position is unhindered and mixtures of 2- and 5-substituted isomers are usually obtained. The proportions depend on the size of the 3-substituent and the nature of the electrophile. Few data are available on $-I,+M$ substituents at the 3-position, but the 2-position is favoured on electronic grounds. For a $-I,+M$ substituent at the 3-position the effects of the heteroatom and the substituent are in alliance and direct the electrophile to the 5-position.

For disubstituted compounds the position of attack may be deduced from the directing effects of the heteroatom and the substituents. In some cases of 2,5-disubstituted compounds, replacement of a substituent such as a carbonyl group may occur in preference to attack at the 3- or 4-position.

18.4.4.2 Halogenation

Early work on direct bromination of furan gave ill-characterized products,[1] but reaction of furan with dioxan dibromide at −5 °C gives 2-bromofuran in good yield.[18] Bromination of furan in methanol at −10 °C gives 2,5-dihydro-2,5-dimethoxyfuran (47%), which suggests a 1,4-addition of bromine to the diene system. By an n.m.r. study of the bromination of furan in carbon disulphide at −50 °C, evidence has been obtained for the formation of *cis*- and *trans*-2,5-dibromo-2,5-dihydrofuran, and *trans*-2,3-dibromo-2,3-dihydrofuran.[19] These products, on warming to room temperature, give 2-bromofuran, which strongly suggests that addition of bromine is followed by elimination of hydrogen bromide, but does not preclude elimination of bromine and subsequent electrophilic substitution. Bromination of furans substituted with electron-withdrawing groups at the 2-position proceed by an electrophilic substitution mechanism and generally yield the 5-bromo-compounds. Treatment of 2-carbonyl substituted furans with one mol. equiv. of bromine in the presence of an excess of aluminium chloride gives starting material, 5-bromo- (**6**), and 4,5-dibromo-compounds (**7**).[20,21] In the presence of solvents, Br to Cl exchange occurs and compounds (**8**) and (**9**) also result. Bromination of 3-acetylfuran (**10**) in the presence of an excess of aluminium chloride gives the 5-substitution products (**11**) and (**12**) (Scheme 1).[22]

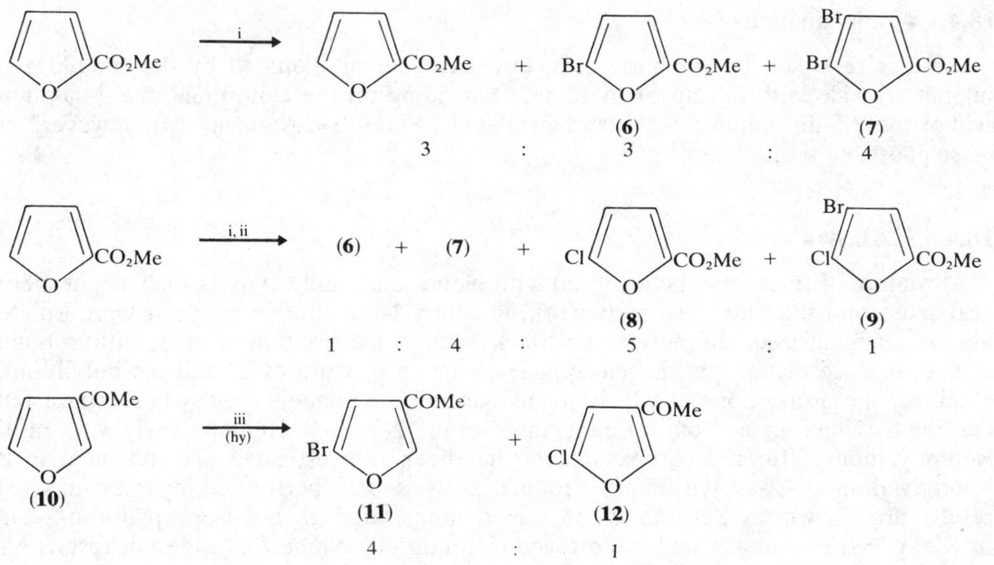

i, 1 mol Br_2, 2.5 mol $AlCl_3$; ii, 12 mol $ClCH_2CH_2Cl$; iii, 1.15 mol Br_2, 2.5 mol $AlCl_3$, CH_2Cl_2, heat.

SCHEME 1

Gradual treatment of furan with 1.6 mol. equiv. of chlorine in dichloromethane at −40 °C gives a mixture consisting of 2-chloro- (64%), 2,5-dichloro- (29%), and 2,3,5-trichlorofuran (7%). The yield of the 2-chloro-compound may be enhanced by using less chlorine. With 3 mol. equiv. or more of chlorine, tetrachlorofuran and the addition product (**13**) are obtained as well.[1] 2-Furanmethanol has been converted to the trichloro derivative (72%) by chlorination at −30 °C.[2] The chlorination of methyl furan-2-carboxylate obeys second-order kinetics under conditions where the 5-chloro-compound is the only product.[4]

18.4.4.3 Nitration

Nitration of furans with a mixture of fuming nitric acid and acetic anhydride at low temperature proceeds largely by an addition–elimination mechanism. In certain cases the addition products have been isolated and they may require treatment with a base such as pyridine to effect elimination. An addition product has been isolated from the reaction of furan at −5 °C with acetic anhydride and nitric acid and assigned structure (**14**).[1] The initial products from the nitration of furan-2-carboxaldehyde and its diacetate are (**15**)–(**18**), so that both 1,4- and 1,2-addition to the diene system occur.[23] Compounds (**17**) and (**18**) are reported to arise by direct electrophilic substitution. Analogous addition products are also isolated from the nitration of methyl furan-2-carboxylate.[24] A 14% yield of 2-nitrofuran is obtained by treatment of furan with nitronium tetrafluoroborate.[25]

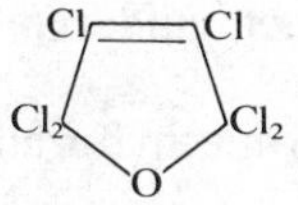

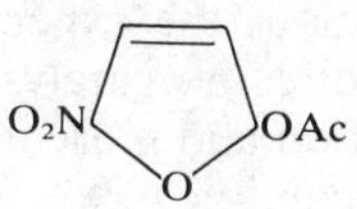

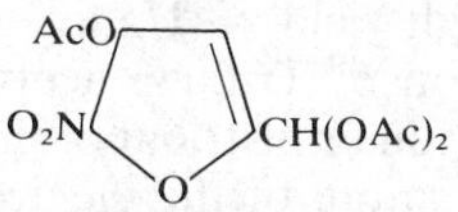

18.4.4.4 Sulphonation

Furan is resinified by sulphuric acid, but it can be sulphonated by the complexes of sulphur trioxide with dioxan or pyridine. Depending on the conditions, the 2-sulphonic acid or the 2,5-disulphonic acid may be isolated.[2] Furan-2-carboxylic acid, however, may be sulphonated with oleum.[1]

18.4.4.5 Alkylation

Alkylation of furan may be achieved with olefins and mild catalysts such as phosphoric acid or boron trifluoride. The early work, in which 2-substitution products were isolated, has been summarized.[2] In more recent work, such as the reaction of furan with isobutene in the presence of phosphoric acid on kieselguhr, a mixture of 2- and 3-t-butylfuran is obtained, the proportions of which are temperature dependent. It has been suggested[17] that the 3-isomer arises from rearrangement of the 2-isomer. Gilman's early work on the isopropylation of furan-2-carboxaldehyde has been reinvestigated and extended to the isopropylation of 2-acetylfuran.[26] Product analysis was performed by g.l.c./m.s. The results are shown in Scheme 2. It was demonstrated that 5-isopropylfuran-2-carboxaldehyde is unchanged under the reaction conditions. Somewhat different results have

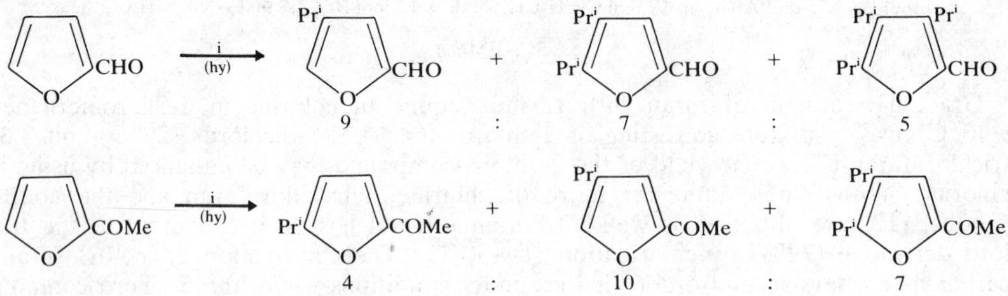

i, 1.2 mol $AlCl_3$, 1 mol Pr^iCl, CS_2, 25 °C.

SCHEME 2

been obtained with the isopropylation of methyl furan-2-carboxylate.[27] It was found that the 5-isomer rearranges to the 4-isomer under the reaction conditions. The variation in results obtained by different workers with this type of reaction may be in part attributed to differences in the aluminium chloride used.

The chloromethylation of furans has been reviewed.[28]

18.4.4.6 Acylation and formylation

Formylation is best carried out by the Vilsmeier–Haack method at low temperature. Acylations may be performed with acyl anhydrides and catalysts such as boron trifluoride etherate or phosphoric acid.[2] Acetylation with acetyl *p*-toluenesulphonate gives particularly high yields.[29] Under other conditions the expected products undergo further transformations.[30] Trifluoroacetylation requires no catalyst.[4]

The isomer distribution in the acylation and mercuration of 3-methylfuran is dependent on the nature of the electrophile.[31]

18.4.4.7 Mercuration

Alkyl-substituted furans readily give chloromercuri-derivatives on treatment with mercuric chloride and sodium acetate in aqueous ethanol. The α-positions, if vacant, are always attacked in preference to β-positions which, however, are attacked if the furan is 2,5-disubstituted.[32] Furans substituted with electron-withdrawing groups are difficult to mercurate, but a 2-carboxylic acid may be replaced: thus treatment of sodium 5-nitrofuran-2-carboxylate with aqueous mercuric chloride at 150–160 °C yields 5-nitro-2-furylmercuric chloride.[1] Treatment of furan-2-carboxylic acid with aqueous mercuric acetate yields a salt which on pyrolysis affords predominantly the 3-mercurated derivative, presumably *via* a cyclic transition state.[33]

18.4.4.8 Hydrogen exchange

The rates for acid-catalysed protodedeuteration, protodetritiation, deuteriodeprotonation, and deuteriodetritiation for the 5-position of 2-methylfuran have been determined. The isotope effects and activation parameters thus obtained are typical of an $A\text{–}S_E2$ mechanism.[34]

18.4.5 NUCLEOPHILIC SUBSTITUTION

From the few mechanistic studies that have been reported it appears that the reaction follows a normal course and no evidence has been advanced for the intermediacy of furynes. 2-Bromo- and 2-chloro-furan react with piperidine about ten times faster than the corresponding halobenzenes.[35] The rates of substitution for 5-substituted 2-nitrofurans and methyl furan-2-carboxylates are also higher than for similar benzene derivatives.[35,36] The ^{1}H n.m.r. spectra of the Meisenheimer complexes obtained on treatment of 2-nitrofuran with nucleophiles have been reported.[37] Examples of typical nucleophilic substitution reactions are shown (Scheme 3).

OHC–(furan)–S–(furan)–CHO ←ii (hy)— Br–(furan)–CHO —i (hy)→ I–(furan)–CHO

PhS–(furan)–CHO ←iv (my)— O_2N–(furan)–CHO —iii (my)→ N_3–(furan)–CHO

Br–(furan)–NO_2 —v (hy)→ Et_2N–(furan)–NO_2

i, NaI, AcOH, heat; ii, Na_2S, H_2O, 100 °C; iii, NaN_3, Me_2SO, 20 °C; iv, PhSNa, MeOH, 20 °C; v, Et_2NH.

18.4.6 RADICAL REACTIONS

Furan undergoes both substitution and addition on reaction with radicals. Early work,[38] of a preparative nature, was confined to arylation usually using the Gomberg reaction. 2-Substituted furans invariably gave 2,5-disubstituted products. Both dibenzoyl peroxide[39] and phenylazotriphenylmethane[40] give addition products with furan (Scheme 4). In

i, $PhCO_2$—O_2CPh; ii, PhN=$NCPh_3$.

SCHEME 4

the case of phenylazotriphenylmethane this has been attributed to the high stationary-state concentration of the fairly stable triphenylmethyl radical.

Competition experiments with aryl radicals have been carried out for furan, thiophen, and benzene.[41] The aryl radicals were generated by the decomposition of *N*-nitroso-acetanilide, the aprotic diazotization of anilines, or under Gomberg conditions. Furan undergoes arylation solely at the 2-position, whereas thiophen gives some 3-arylation products. Furan was found to be 11.6 times more reactive than benzene and 4.5 times more reactive than thiophen. Partial rate factors with respect to benzene were determined for the 2-position of furan and *m*- and *p*-substituted aryl radicals. A Hammett plot of these against the σ^+ values gave a good correlation and a ρ value of +0.46. Thiophen similarly gave a ρ value of +0.225.

Reaction of 2,5-dimethylfuran with *N*-nitrosoacetanilide gives 2-benzyl-5-methylfuran and no 3-phenyl-2,5-dimethylfuran. This reaction is thought to involve a π-complex between the diazonium cation and the furan, rather than phenyl radicals.[42] When furan is allowed to react with *p*-bromophenylthio-radicals generated by treatment of the thiol with Fenton's reagent, 2-monosubstituted and 2,3-disubstituted products are obtained.[43] When the thio-radical is generated by other means, only 2-substitution occurs. A plausible mechanism was advanced. In the reaction of furan with t-butyl hydroperoxide and Fe^{2+} ions in methanol, addition products are obtained,[44] but reaction of alkylfurans with diacetyl peroxide is reported to give only substitution products.[45] The radicals produced by addition of HO· and H· at the α-position of substituted furans have been studied by e.s.r. spectroscopy.[46]

18.4.7 REDUCTION

Catalytic hydrogenation and hydrogenolysis of furans have been studied under a wide range of conditions, since the outcome of the reactions is dependent on such variables as catalyst, temperature, pressure, and solvent.[1,2]

The alkali metal/liquid ammonia reduction of furans has also been studied.[47] Treatment of furan-2-carboxylic acid with lithium and ammonia at −78 °C followed by addition of ammonium chloride gives 2,5-dihydrofuran-2-carboxylic acid (80%). Reductive alkylation of furan-2-carboxylic acid similarly gives 2-alkyl-2,5-dihydrofuran-2-carboxylic acids (68–95%). 5-Alkylfuran-2-carboxylic acids on reduction with lithium and methanol in ammonia give *cis*- and *trans*-5-alkyl-2,5-dihydrofuran-2-carboxylic acids (71–85%).

Furan-3-carboxylic acid also undergoes reduction with sodium and liquid ammonia. With propan-2-ol as the proton source, 2,3-dihydrofuran-3-carboxylic acid results, but with methanol or ethanol as proton source, addition to the double bond occurs during work-up. With no proton source, ring opening and further reduction occurs.

18.4.8 OXIDATION

The furan nucleus is particularly sensitive to oxidation, and most reactions involve 1,4-addition of the oxidant to the diene system. Addition to furans of singlet oxygen, usually produced by photosensitization, gives transannular peroxides which undergo solvolysis or rearrangement.[48] The peroxide (**19**) obtained from furan is thermally unstable, but can be isolated at −100 °C. When the photo-oxidation reaction is conducted in methanol the lactone (**20**) is obtained. With 2,5-dimethylfuran in methanol, the hydroperoxide (**21**) can be isolated, and can readily be converted to 1,2-diacetylethylene. With furan-2-carboxaldehyde in ethanol the lactone (**22**) is obtained by rearrangement of the intermediate hydroperoxide with loss of formic acid. More complex products are obtained with the cyclophane (**23**).[49]

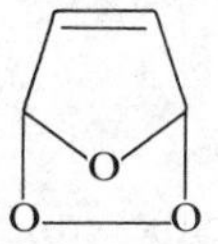

(**19**)

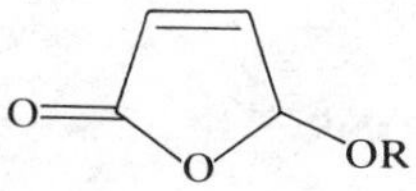

(**20**) R = Me
(**22**) R = Et

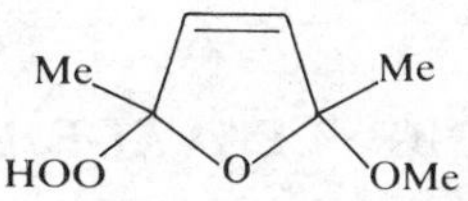

(**21**)

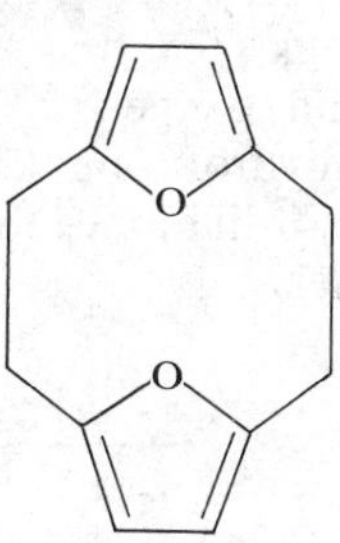

(**23**)

The anodic oxidation of furans may be carried out under various conditions.[50] The mechanism (Scheme 5) involves the formation of a radical cation which is captured by a nucleophile, yielding a radical which on further oxidation yields a cation, which is in turn captured.

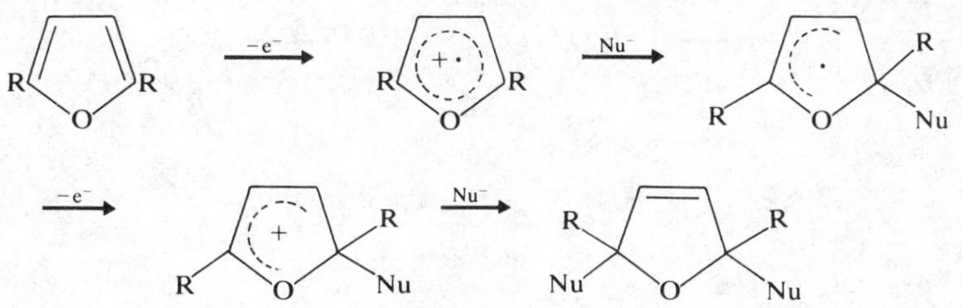

SCHEME 5

Electrochemical oxidation carried out in methanol yields *cis*- and *trans*-2,5-dihydro-2,5-dimethoxyfurans.[50,51] Treatment of furans with bromine and methanolic sodium acetate yields the same products. In the electrolysis reaction the intermediate cation may be intercepted by other nucleophiles.[52] Electrochemical acetoxylation of furan in acetic acid gives *cis*- and *trans*-2,5-diacetoxy-2,5-dihydrofuran (12%), but alkylfurans give side-chain acetoxylated products. Lead tetra-acetate in acetic acid gives similar results.[51]

Treatment of furan with hydrogen peroxide at low temperature in the presence of acid affords the peroxide (**24**), which may be semi-hydrogenated to malealdehyde.[53] The reaction with 2,5-dimethylfuran is more complicated and yields (**25**)–(**27**); a mechanism has been proposed.[54]

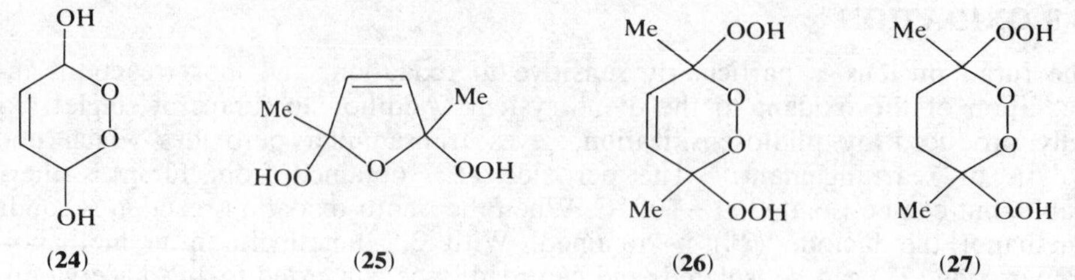
(24) (25) (26) (27)

The ozonolysis of furans may be achieved at low temperature.[2] Furans are also susceptible to oxidation by permanganate or nitric acid, the products being ketones or acids.

18.4.9 RING CLEAVAGE REACTIONS

Alkylfurans may be readily ring-opened to 1,4-diketones under acidic conditions, the reverse of the Paal–Knorr synthesis (Section 18.4.14.2), but unless precautions are taken yields may be low owing to polymerization.[55] Furans with electron-withdrawing substituents are generally resistant to attack. The first step of the reaction involves protonation at an α-position rather than on oxygen.[56] 2,5-Dialkoxy-2,5-dihydrofurans (**28**) (Section 18.4.8), or the derived tetrahydrofurans (**29**) obtained on catalytic hydrogenation, may also serve as precursors to 1,4-diketones (Scheme 6). These reactions have been extensively exploited in organic synthesis.

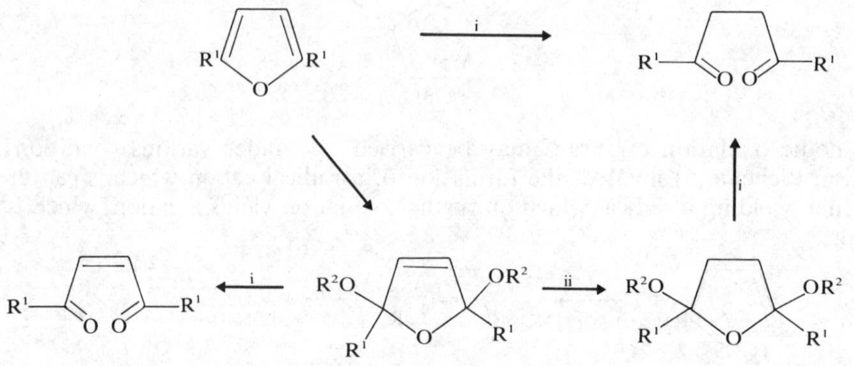
(28) (29)

i, H_3O^+; ii, H_2, $Pd/CaCO_3$.

SCHEME 6

cis-Jasmone (**32**) is synthesized by aldol condensation of the 1,4-diketone (**31**) obtained by ring-opening of the furan (**30**) (Scheme 7).[57] It has been found that ring-opening under ketal-forming conditions often gives higher yields; thus the bis-ketal (**34**), a key intermediate in a biomimetic synthesis of (±)-16,17-dehydroprogesterone, was obtained from the furan (**33**) (Scheme 7).[58] The dihydrofuran (**35**) on hydrolysis and reduction yields *Z*-3-methyl-2-penten-1,4-diol (**36**) (Scheme 8).[59]

2-Furanmethanols (**37**) are cleaved to derivatives of laevulinic ester (**38**) by hot methanolic hydrogen chloride. A mechanism involving the carbonium ion (**39**) has been proposed. Under similar conditions $\alpha\beta$-unsaturated carbonyl compounds (**40**) yield keto-esters (**41**). This reaction, known as the Marckwald rearrangement, is probably mechanistically similar to that of 2-furanmethanols.[1] It has been used in a synthesis of equilenin; thus the furan (**42**) on treatment with hot ethanolic hydrogen chloride yields the diketo-ester (**43**) (55%).[60]

(30) → i → (31)

→ ii → (32) (my)

Br (33) → iii (hy) → (34) Br

i, H_2SO_4, AcOH; ii, EtOH, NaOH; iii, $HOCH_2CH_2OH$, H^+.

SCHEME 7

H, MeO, Me, OMe, Me (35) → i → [CHO, Me, Me, O] → ii (hy) → Me, Me, OH, HO (36)

i, H_2O; ii, $NaBH_4$.

SCHEME 8

CR_2OH (37) MeO_2C, O, CHR_2 (38) +, −, CR_2 (39)

CH=CHCOR (40) MeO_2C, O, CH_2CH_2COR (41)

MeO, COCH=CH (42) EtO_2C, O, O, MeO (43)

Furans, by virtue of their ease of ring opening, have also been converted into a variety of other heterocycles.[61] When a mixture of furan (or 2-methylfuran) and hydrogen sulphide is passed over an alumina catalyst activated with potassium tungsten phosphate, high yields of the corresponding thiophens are obtained. Tetrahydrofurans on passing over an alumina catalyst with ammonia or a primary amine yield pyrrolidines or *N*-substituted pyrrolidines. 2,5-Dialkoxy-2,5-dihydrofurans similarly yield *N*-alkylpyrroles. This transformation can also be achieved by heating the dihydrofuran with a primary amine in acetic acid.

Substituted furan-2-carboxaldehydes and 2-acylfurans on treatment with ammonia or ammonium salts yield 3-hydroxypyridines. Nucleophilic attack of ammonia is thought to occur at the 5-position and ring opening of the intermediate (**44**) yields the amino-compound (**45**) and thence the 3-hydroxypyridine (**46**). In another synthesis of 3-hydroxypyridines an acylamine (**47**) is subjected to electrolytic methoxylation to yield (**48**). Hydrolytic ring fission followed by reclosure of the intermediate in the alternative sense then affords the 3-hydroxypyridine (**46**) (Scheme 9). This type of route has been used for a synthesis of pyridoxine.

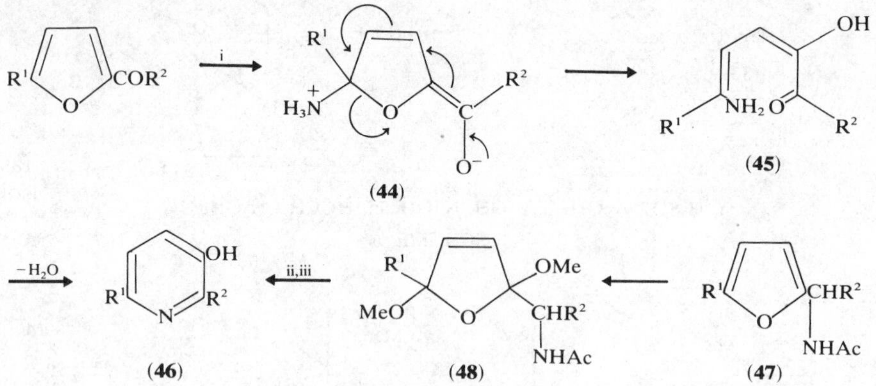

(**44**) (**45**) (**46**) (**48**) (**47**)

i, NH_3, ii, OH^-; iii, H^+.

SCHEME 9

18.4.10 CYCLOADDITION REACTIONS

18.4.10.1 Thermal reactions

Unlike pyrroles and thiophens, furans readily undergo $[_\pi 4_s + _\pi 2_s]$ cycloaddition—the classical Diels–Alder reaction.[2,62] The *exo*-adducts are thermodynamically more stable than the *endo*-adducts, but under mild conditions *endo*-adducts can sometimes be isolated. In the addition of furan to maleic anhydride at 25 °C, twice as much *exo*-isomer (**49**) as *endo*-isomer (**50**) is initially formed and the latter reverts to addends.[63] With maleimide the *endo*-adduct (**54**) results at 25 °C, but the *exo*-adduct (**53**) at 90 °C; the

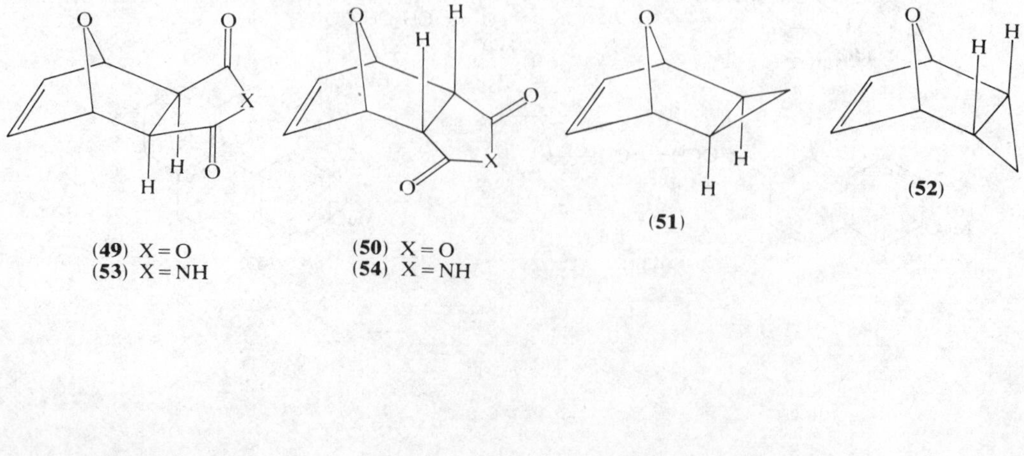

(**49**) X=O
(**53**) X=NH

(**50**) X=O
(**54**) X=NH

(**51**)

(**52**)

endo-adduct also yields the *exo*-adduct on heating. Furan reacts with cyclopropene at room temperature to yield a 1:1 mixture of *exo*- (**51**) and *endo*-adducts (**52**).[64] Acid-catalysed rearrangement of the *endo*-adduct yields 2,4-cycloheptadienone. With halocyclopropenes, *e.g.* (**53**), only the *endo*-adducts are formed and these may undergo electrocyclic ring opening and stereospecific 1,2-halogen migration (Scheme 10).[65] With

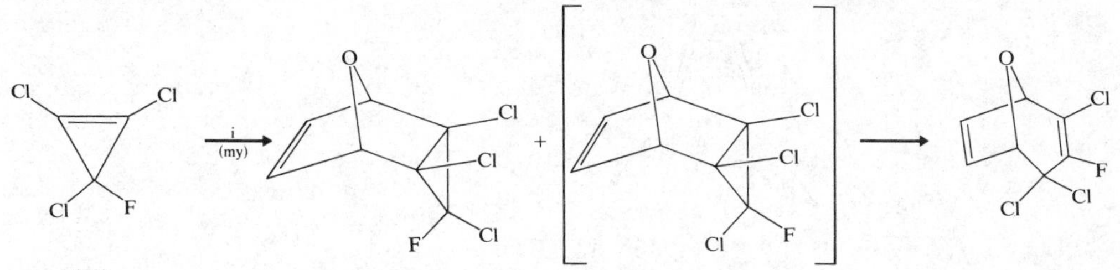

(**53**) 1.5 1

i, furan, CCl_4, heat.

SCHEME 10

dichlorovinylene carbonate in boiling dichlorobenzene the adducts (**54**) and (**55**) are obtained in the ratio 1:2. The adducts on base hydrolysis yield the 1,2-diketone (**56**).[66] Reaction of furan with maleic acid in water occurs slowly and the adducts readily revert thermally to the addends. Initially the *endo*-isomer (**57**) is formed four times faster than the *exo*-isomer (**58**), but after ten days the ratio is 1:1.[63] Both isomers may be isolated.[67]

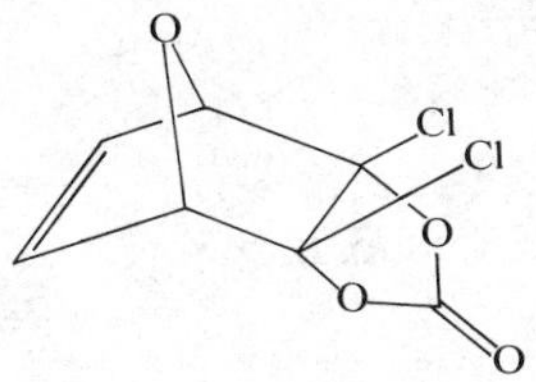

(**54**) 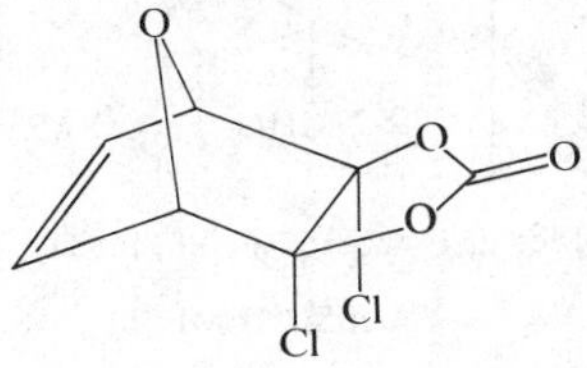 (**55**) 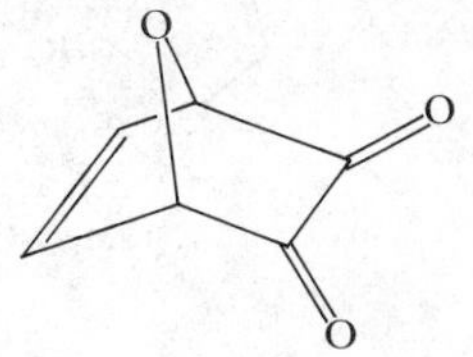 (**56**)

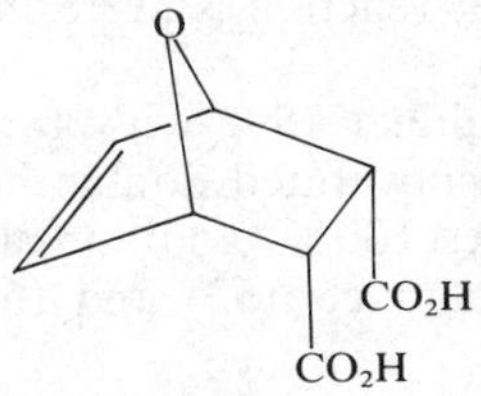

(**57**) 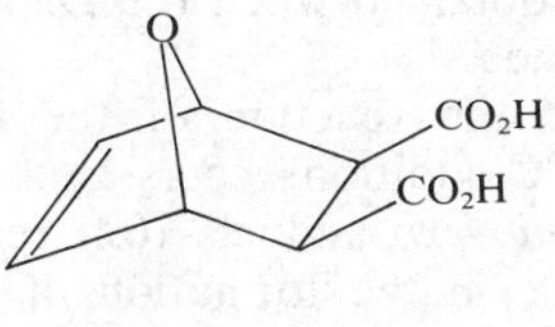 (**58**)

Ethylenic dienophiles with only one electron-withdrawing group react only slowly with furans at room temperature to afford low-to-moderate yields of both *endo*- and *exo*-adducts. At higher temperatures the reverse reaction, as well as polymerization of the dienophile, generally occur. By use of very high pressures it is possible to overcome these difficulties and good yields may be obtained.[68]

Furans generally react with acetylenic dienophiles under mild conditions, affording good yields of oxanorbornadienes. Oxanorbornadienes have found a variety of uses in organic synthesis. The unsubstituted double bond may be hydrogenated and the product on thermolysis undergoes retro-Diels–Alder reaction; this provides a convenient synthesis of 3,4-disubstituted furans (Scheme 11). Conversion to 3,4-disubstituted furans has also been achieved by treating (**60**) with 3,5-di-(2′-pyridyl)-*sym*-tetrazine (Scheme 12).[69]

Ultraviolet irradiation of the norbornadienes yields oxaquadricyclanes, which on thermolysis afford the benzene oxide–oxepin system.[70] On treatment with aluminium

i, $MeO_2C{\equiv}CCO_2Me$; ii, $F_3CC{\equiv}CCF_3$; iii, Pt, H_2; iv, 400 °C; v, Ni(R), H_2.

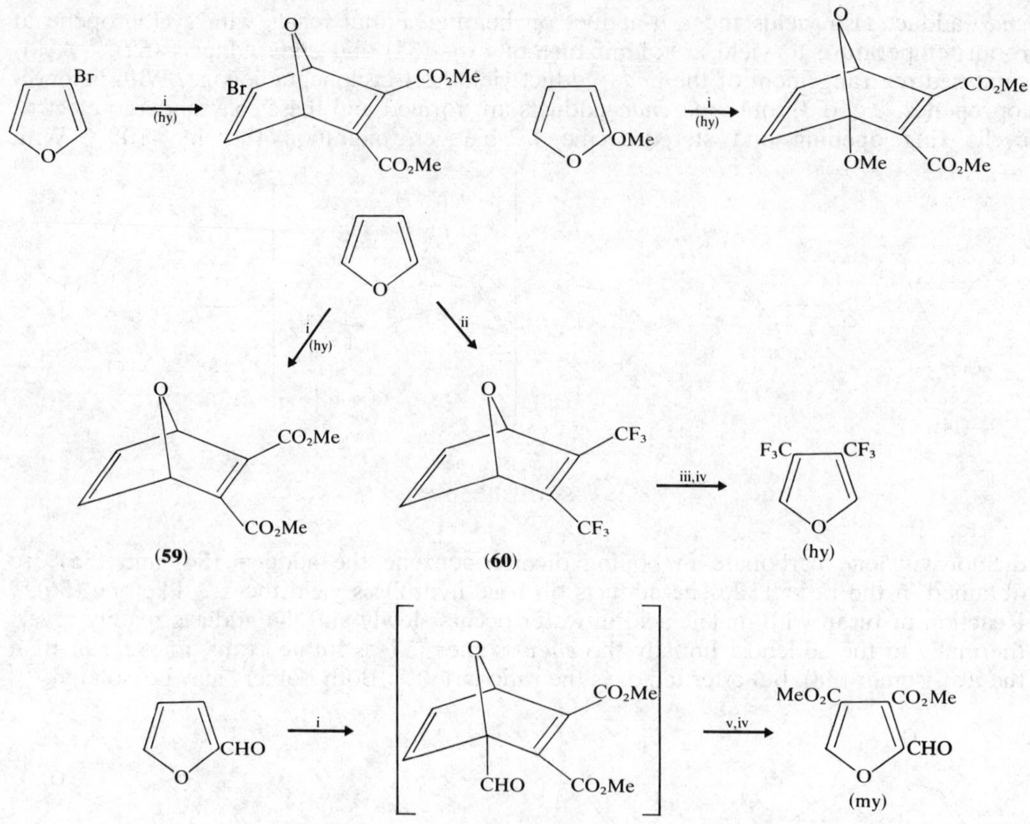

SCHEME 11

chloride[71] or trifluoroacetic acid[72] the oxanorbornadienes rearrange to phenols. Comparable results may be obtained when the Diels–Alder reaction is performed in the presence of aluminium chloride.

The outcome of the reaction of the monoadduct (**59**) with furan is temperature dependent. At 50 °C addition occurs at the tetrasubstituted double bond and the *endo*-*endo*- (**61**) and *endo*-*exo*-adducts (**62**) result. At 100 °C addition to the disubstituted double bond results in the formation of the diadducts (**63**) and (**64**) as well as four

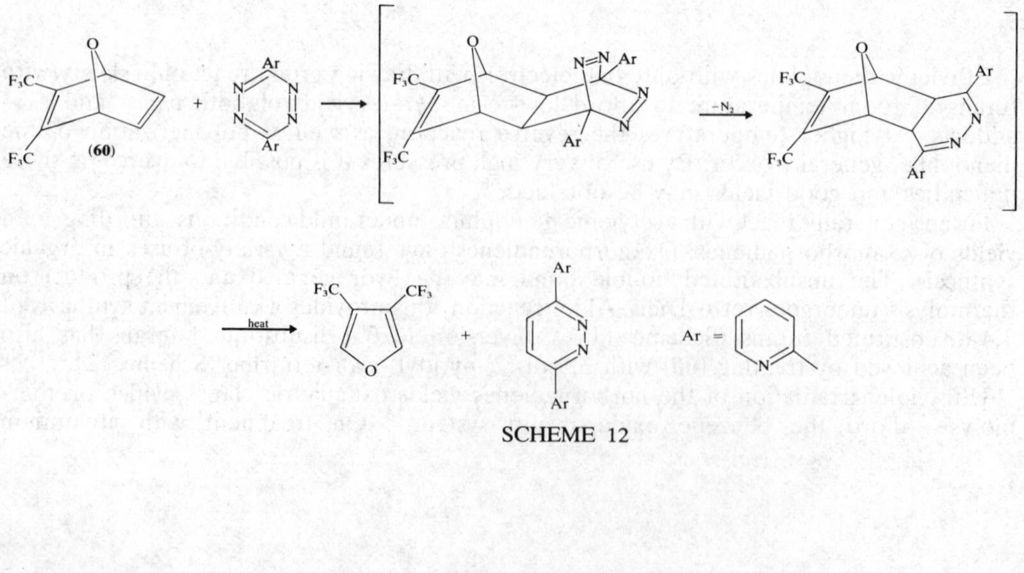

SCHEME 12

triadducts, *e.g.* (**65**). Furan fails to react with ethyl propiolate at 25 °C, but at 130 °C the adduct (**66**) (9%) is obtained. 2,5-Dimethylfuran, however, reacts with ethyl propiolate at 95–120 °C to give initially the 1:1 adduct (**68**), which undergoes Diels–Alder reaction with itself, or the diene, [2+2] addition with the dienophile, or retro-Diels–Alder reaction with loss of acetylene. The product (**69**) from homo-Diels–Alder reaction of (**68**) with itself undergoes retrogression, yielding (**67**). The ratio of products is dependent on the reaction conditions.[73] The cyclophane (**23**) reacts with dimethyl acetylenedicarboxylate to yield an internal adduct which dissociates on heating.[74]

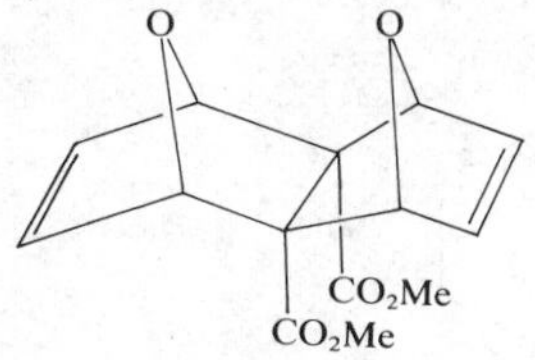

(**61**) 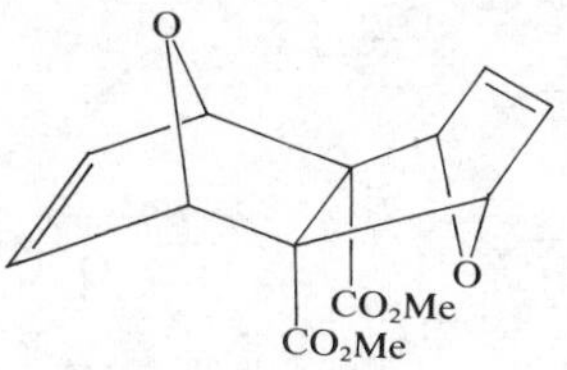 (**62**) 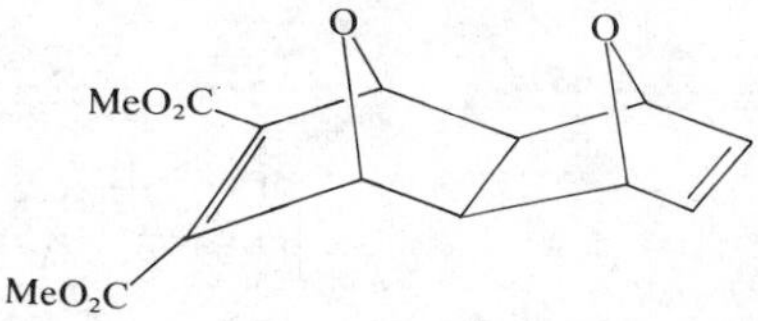 (**63**)

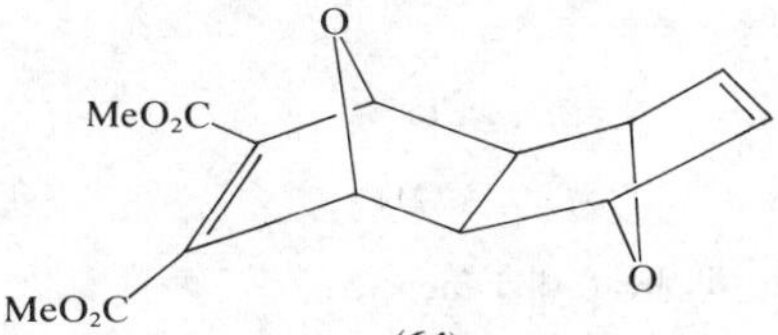

(**64**) 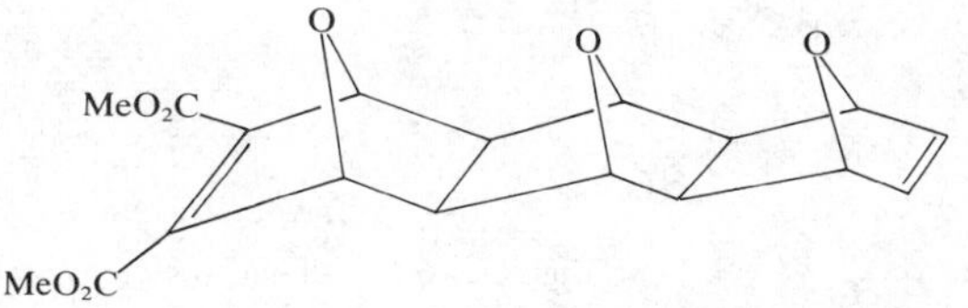 (**65**)

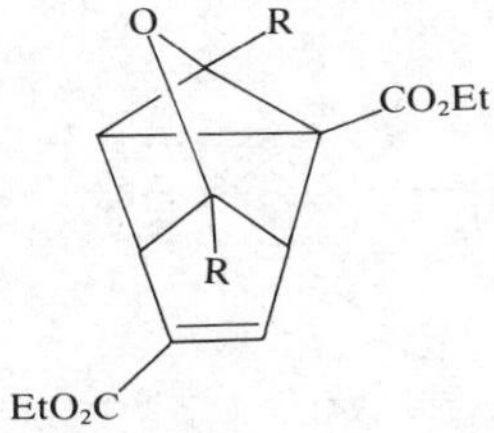

(**66**) R = H
(**67**) R = Me

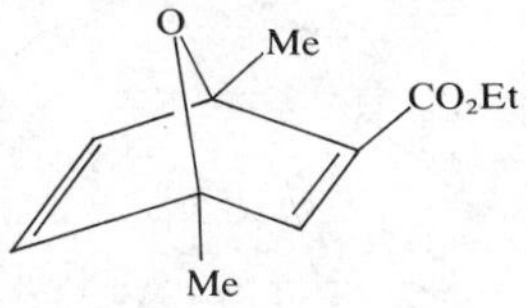

(**68**) 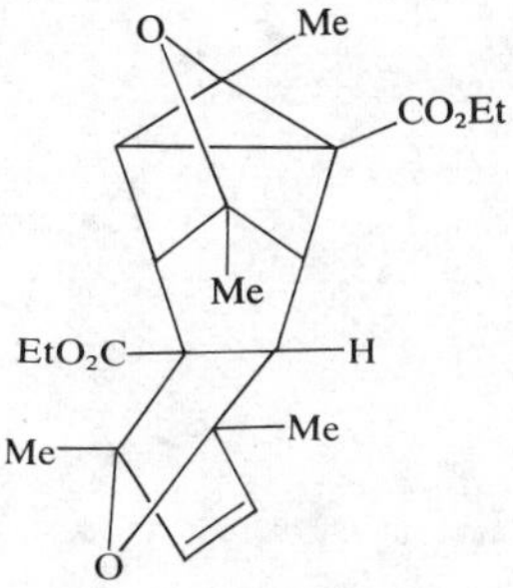 (**69**)

Examples of furans reacting as dienophiles are rare. The diterpene (**70**) on heating undergoes a [1,5]-sigmatropic shift and lactonization, yielding (**71**), which, because of the favourable stereochemistry, affords (**72**) by an intramolecular Diels–Alder reaction (Scheme 13).[75] Reaction of furan-2-carboxaldehyde with butadiene gives mono- and bis-adducts.[2]

Furan, in keeping with its aromaticity and in contrast to 2,3- and 2,5-dihydrofurans, behaves only as a weak dipolarophile. Generation of benzonitrile oxide in presence of a large excess of furan yields (91%) the adducts (**73**) and (**74**) in the ratio 97:3 by $[_{\pi}4_s + _{\pi}2_s]$ addition to a double bond of the furan ring.[76]

Arynes may be trapped by furan or alkyl furans in $[_{\pi}4_s + _{\pi}2_s]$ reactions.[2] This has been used in a synthesis of benzo[*c*]furan-4,7-quinone (**75**) (Scheme 14).[77] The cyclophane (**23**) gives both mono- and bis-adducts on reaction with benzyne.[78]

Strained bridgehead olefins may also be trapped with furans.[79]

Methyl cyclopropanone and 2,2-dimethylcyclopropanone (**76**) react *via* their tautomeric oxyallyl cations with furan and alkylfurans to yield $[_{\pi}4 + _{\pi}2]$ adducts (Scheme 15).[80] With furan-2-carboxaldehyde and (**76**), only one $[_{\pi}4 + _{\pi}2]$ adduct (**77**) results. The dioxolan (**78**), resulting from a 1,3-dipolar addition, is also formed. Oxyallyl cations may be generated by debromination of α,α'-dibromoketones with a zinc–copper couple or with $Fe_2(CO)_9$ and the cations undergo cycloaddition with furan.[81,82]

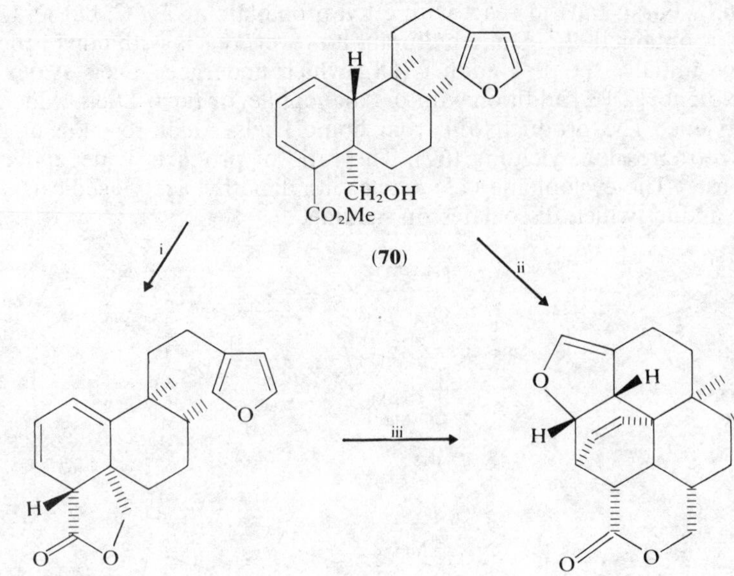

(**70**)

(**71**)

(**72**)

i, heat, diglyme, 6 h; ii, heat, diglyme, 48 h; iii, heat, diglyme, 32 h.

SCHEME 13

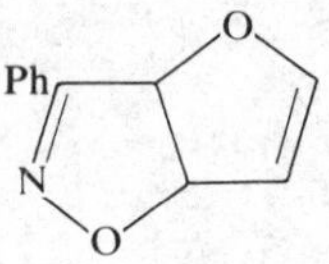

(**73**)

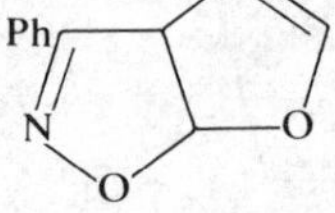

(**74**)

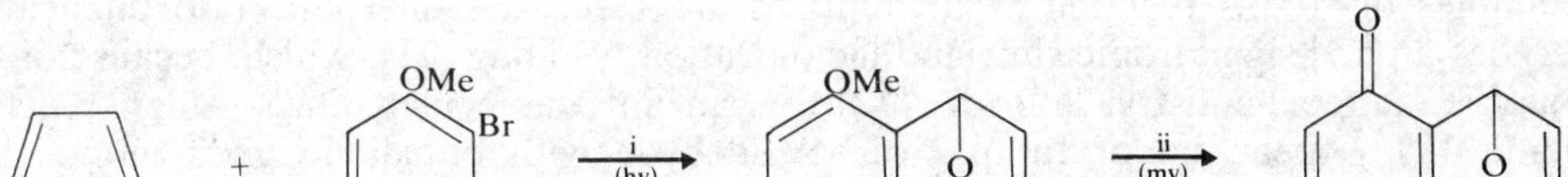

(**75**)

i, $NaNH_2$, THF, heat; ii, AgO, dioxan, HNO_3; iii, 3,6-di-(2′-pyridyl)-*sym*-tetrazine.

SCHEME 14

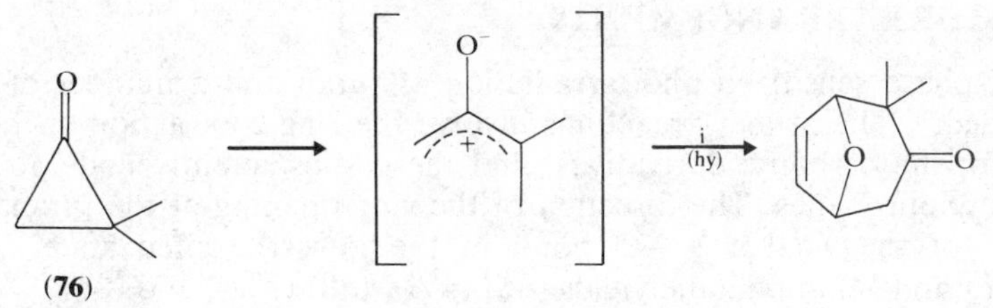

O O⁻ (hy) O O
(76)

i, furan, CH_2Cl_2, 25 °C.
SCHEME 15

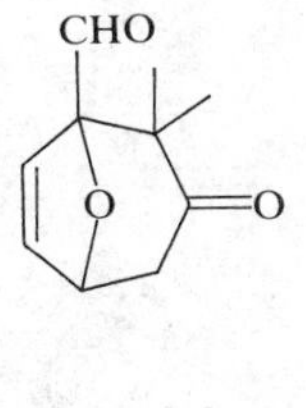

CHO O O O O
(77) 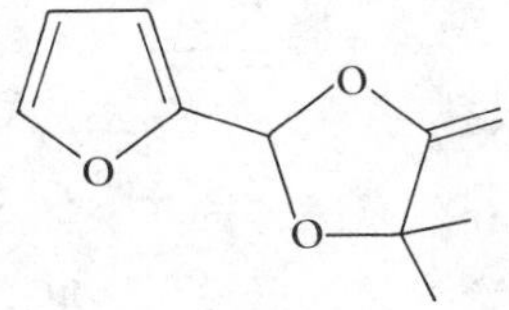 (78)

18.4.10.2 Photochemical reactions

Furan undergoes both $[_{\pi}4+_{\pi}4]$ and $[_{\pi}2+_{\pi}2]$ photoaddition reactions. Irradiation of a mixture of benzene and furan gives largely the $[_{\pi}4+_{\pi}4]$ adduct (**79**), which undergoes a Cope rearrangement to afford (**82**), and a $[_{\pi}2+_{\pi}2]$ photoaddition to give (**83**). Minor adducts (**80**) and (**81**) are also formed (Scheme 16). The proportions of the products are markedly dependent on the light source, the reactant concentration, and the isolation procedure.[83]

Furans also undergo $[_{\pi}4+_{\pi}4]$ photoadditions with naphthalenes and anthracenes.[84] Aldehydes and ketones undergo $[_{\pi}2+_{\pi}2]$ photoaddition to furan, yielding oxetanes.[85]

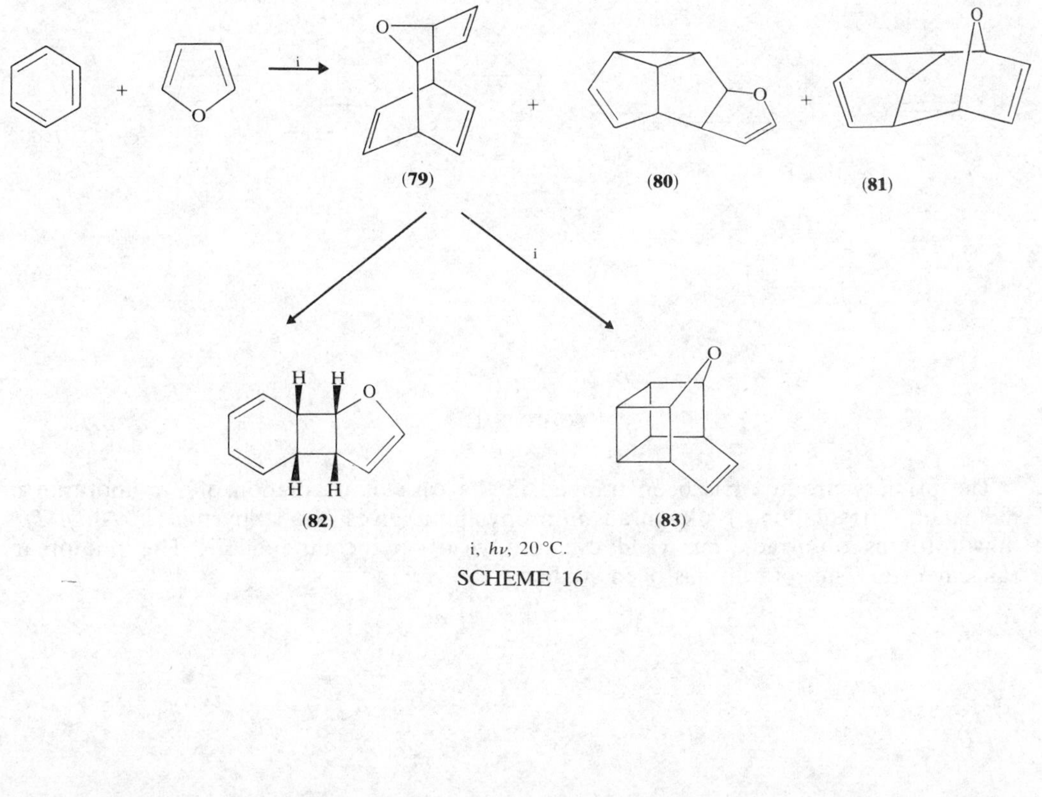

(79) (80) (81)

(82) (83)

i, $h\nu$, 20 °C.
SCHEME 16

18.4.11 PHOTOREARRANGEMENTS

The vapour-phase sensitized photoirradiation of furan and a number of methylfurans has been studied.[86] The primary reactions involve the ring contraction of the substrate to methylcyclopropene carbonyl derivatives, and these subsequently undergo decarbonylation yielding cyclopropenes. The direction of the ring opening of the furans is controlled by the fission of the weaker C—O bond in the lowest excited state. Irradiation of 2,5-di-t-butylfuran (**84**) in pentane yields (**85**) (4%), (**88**) (9%), and (**86**) (9%). The initial product is (**85**), which can ring-open by photolysis of the 2,3- or 1,3-bonds, hence yielding (**88**), or (**86**) by H-atom migration of the intermediate diradical (**87**). In the case of 2,3,5-tri-t-butylfuran (**89**) the products are (**91**) (90%) and (**90**) (5%), which arise by alternative modes of ring opening (Scheme 17).[87]

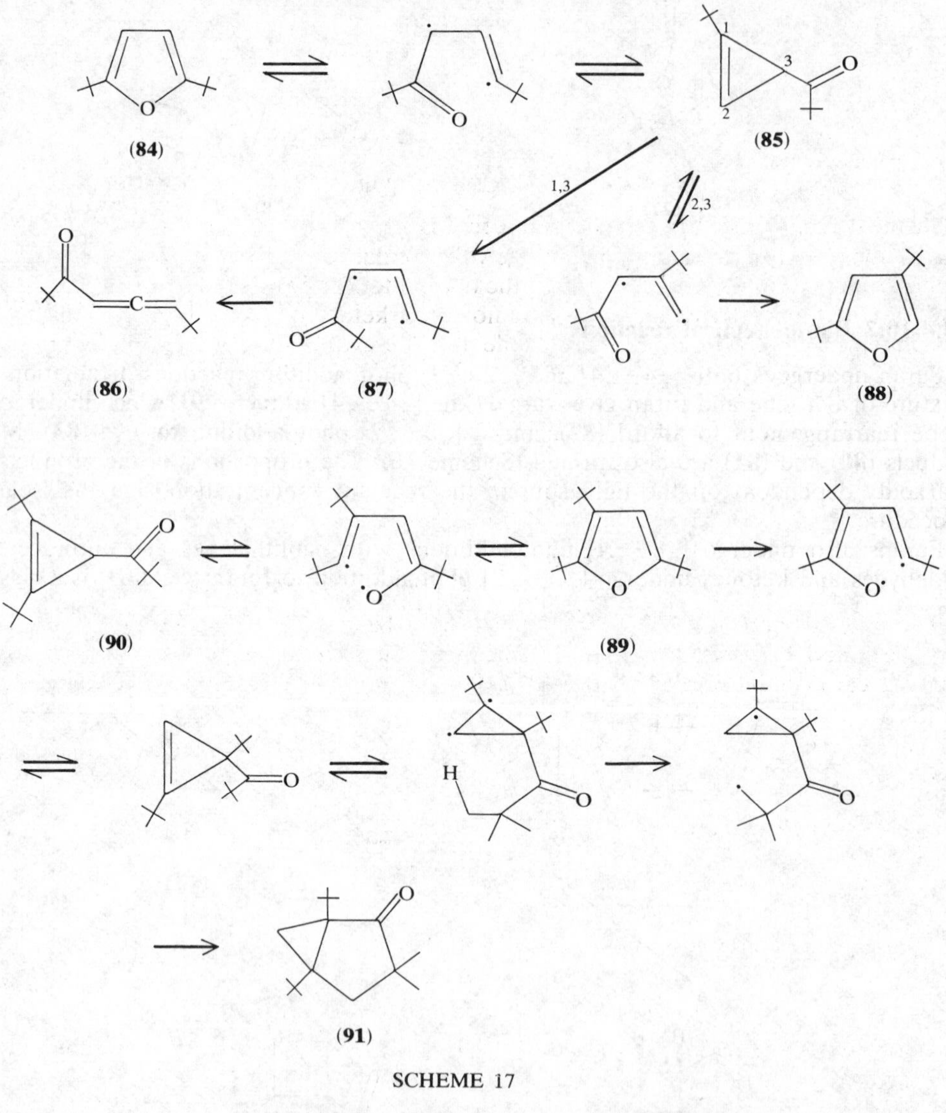

(**84**) (**85**) 1,3 2,3 (**86**) (**87**) (**88**) (**90**) (**89**) (**91**)

SCHEME 17

The primary product has been trapped in the case of irradiation of 2-cyanofuran in methanol.[88] Irradiation of alkylfurans in propylamine gives *N*-alkylpyrroles.[89] Alkyl-2,3-dihydrofurans on irradiation yield cyclopropylcarbonyl compounds.[90] The photorearrangement of 2-nitrofuran has been studied.[91]

18.4.12 CARBENES AND NITRENES

Carbene yields the expected product (**92**) with furan.[92] Ultraviolet irradiation of a solution of methyl diazoacetate in furan yields the carbene adduct (**93**), which on brief heating to 160 °C undergoes valence tautomerism to the *cis,trans*-muconaldehydate (**94**)

(**92**) (**93**) (**94**)

i, CH_2N_2, Cu_2Br_2; ii, N_2CHCO_2Me, $h\nu$; iii, 160 °C.

SCHEME 18

(Scheme 18).[93] A number of substituted furans on thermal reaction with ethyl diazoacetate in the presence of copper(II) sulphate yield similar dienes.[94] The reaction with vinylcarbenes follows a similar course.[95] When the diazoketone (**95**) is heated in cyclohexane in the presence of copper(II) sulphate, an intramolecular ketocarbene addition takes place; again the initial product (**96**) is thermally labile and the cyclopentenone (**97**) results.[96]

(**95**) (**96**) (**97**)

Examples of both 2- and 3-furylcarbenes have been generated. Short-path pyrolysis of the diazomethane (**98**), generated *in situ* from the sodium salt of the tosylhydrazone of furan-2-carboxaldehyde, yields the *cis*- (**100**) and *trans*-aldehydes (**101**) by fragmentation

(**98**) (**99**) (**100**) (**101**)

i, heat; ii, cyclo-octane, 140 °C; iii, $PhCH{=}CH_2$, 140 °C.

SCHEME 19

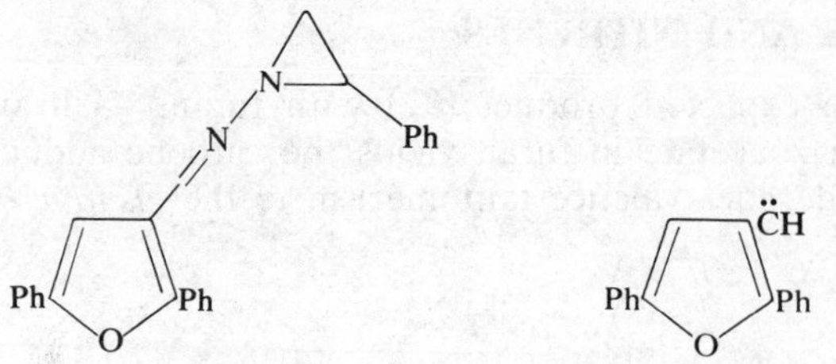

(**102**) (**103**)

of the intermediate 2-furylcarbene (**99**). The initial product is the *cis*-isomer (**100**). The carbene undergoes an insertion reaction when the sodium salt of the tosylhydrazone is thermolysed in cyclo-octane, or an addition reaction with styrene under similar conditions (Scheme 19).[97] The reaction has been extended to substituted furans. The 3-furylcarbene (**103**) is generated by pyrolysis or photolysis of the aziridinylimine (**102**). It does not fragment and its reactions are similar to those of phenylcarbene.[98]

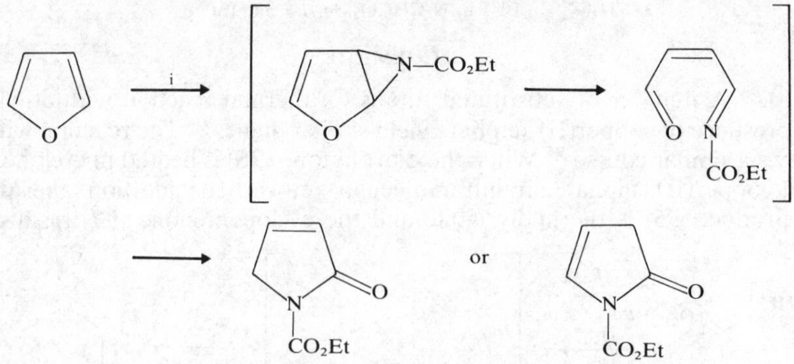

(**104**) or (**105**)

i, N_3CO_2Et, 130 °C.

SCHEME 20

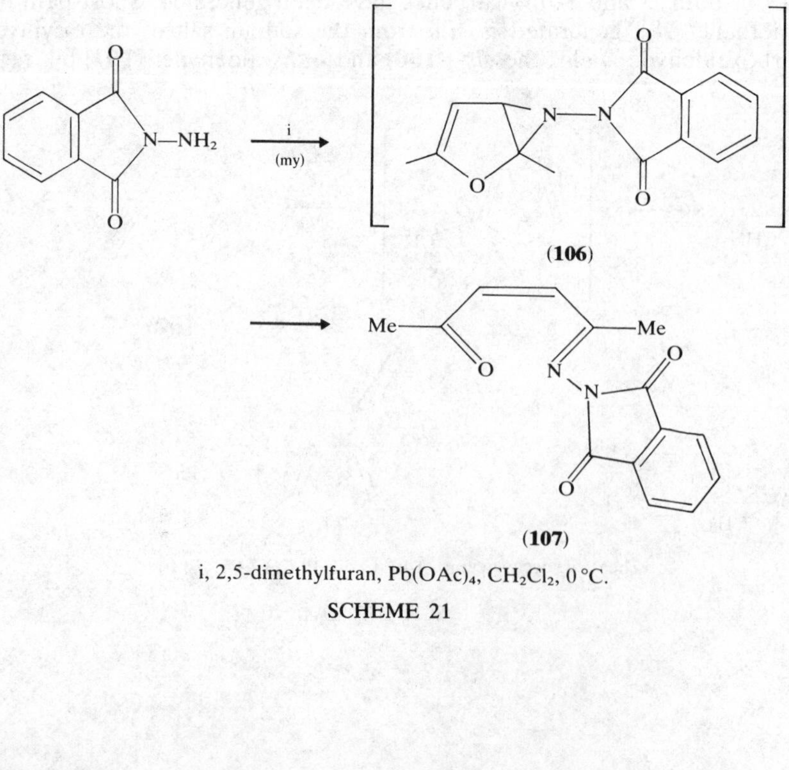

(**106**)

(**107**)

i, 2,5-dimethylfuran, $Pb(OAc)_4$, CH_2Cl_2, 0 °C.

SCHEME 21

Few reactions of furans with nitrenes have been reported.[99] Thermolysis of ethyl azidoformate in furan is reported to give (**104**) or (**105**) (Scheme 20). Generation of phthalimidonitrene in the presence of 2,5-dimethylfuran yields the 2-phthaloylhydrazone of *cis*-hex-3-ene-2,5-dione (**107**). The nitrene presumably undergoes 1,2-addition to the diene system, yielding the aziridine (**106**), and thence (**107**) by C—O or C—N bond cleavage (Scheme 21). 2-Substituted furans give similar products which arise by 4,5-addition of the nitrene to the furan.

18.4.13 ORGANOMETALLIC COMPOUNDS

The preparation of mercurichlorides has been mentioned (18.4.4.7). 3-Chloromercurifuran is a valuable intermediate for the preparation of 3-substituted furans, since on treatment with iodine and potassium iodide it yields 3-iodofuran (75%).[33,100] Treatment of the mercurichloride derivatives with aqueous sodium thiosulphate yields difurylmercury compounds.[1] These may be converted to the lithio-derivatives on treatment with lithium as in a synthesis of rosefuran (**108**) (Scheme 22).[101]

Me, CO_2Na —i→ Me, HgCl —ii→ Me, Hg, Me —iii→ [Me, Li] —iv→ Me (my) (**108**)

i, $HgCl_2$; ii, $Na_2S_2O_3$; iii, Li, Et_2O; iv, 1-bromo-3-methylbut-2-ene.

SCHEME 22

Grignard reagents in the furan series are difficult to prepare and their use has been superseded by the more tractable lithio-compounds. Furan readily undergoes hydrogen–lithium interconversion at the 2-position on treatment with butyl-lithium.[102] Halogen–lithium interconversion may also be achieved with butyl-lithium. Exclusive hydrogen–lithium interconversion for halofurans is achieved with lithium di-isopropylamide at −70 °C, and this occurs at a vacant α-position.[103] The lithio-derivatives undergo the usual carbonation, formylation, and alkylation reactions with the appropriate reagents; on treatment with copper(II) chloride[104] they undergo coupling reactions (Scheme 23).

Treatment of lithiofurans with copper(I) bromide yields furylcopper derivatives.[105]

18.4.14 SYNTHESIS OF FURANS

18.4.14.1 From carbohydrates

The most valuable source of furans is furan-2-carboxaldehyde since it can be converted into furan by decarbonylation, into furan-2-carboxylic acid by oxidation, or into furan-2-carboxylic acid and 2-furanmethanol by Cannizzaro reaction. Furan-2-carboxaldehyde is available from such materials as oat husks and maize cobs, which are relatively rich in 'pentosans'. These on acid treatment are degraded to pentoses (**109a**), which are converted, again by acid treatment, into furan-2-carboxaldehyde (**112a**). Hexoses (**109b**) under similar conditions are converted into 5-(hydroxymethyl)furan-2-carboxaldehyde (**112b**). These reactions have been studied by isotopic labelling techniques.[106] The aldose (**109**), or 2-ketose, is initially converted to the 1,2-enediol (**110**) in a reversible reaction.

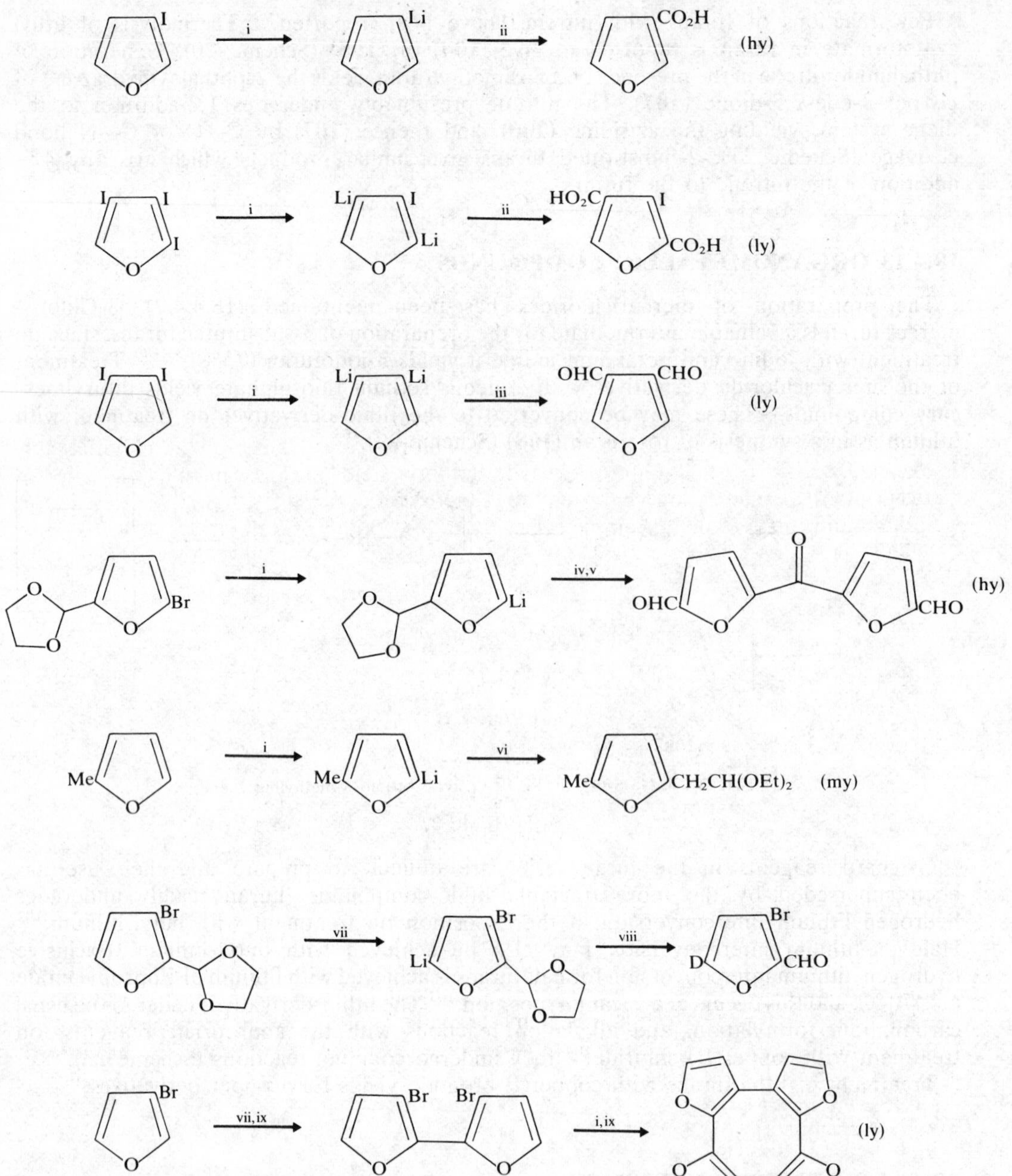

i, Bu^nLi; ii, CO_2; iii, Me_2NCHO; iv, Me_2NCO_2Et; v, H^+; vi, $BrCH_2CH(OEt)_2$; vii, $LiN(CHMe_2)_2$; viii, D_3O^+; ix, $CuCl_2$.

SCHEME 23

The enediol (**110**) suffers consecutive dehydrations to give (**111**), which then undergoes acid-catalysed cyclization. This last step is the same as the Paal–Knorr reaction (Section 18.4.14.2). Sucrose is converted to (**112b**) with iodine in hot *N,N*-dimethylformamide,[107] or into (**112c**) with dry hydrogen chloride in carbon tetrachloride.

Monosaccharides react with a variety of 1,3-dicarbonyl compounds in the presence of zinc chloride in ethanolic or aqueous solution. Thus D-glucose and D-mannose with ethyl

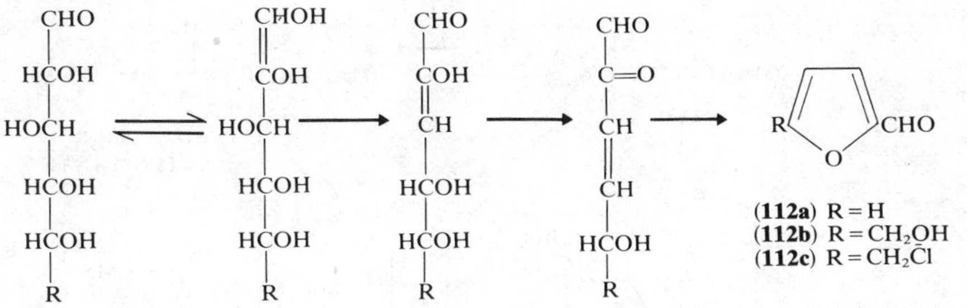

CHO, HCOH, HOCH, HCOH, HCOH, R ⇌ CHOH, COH, HOCH, HCOH, HCOH, R → CHO, COH, CH, HCOH, HCOH, R → CHO, C=O, CH, CH, HCOH, R → R-furan-CHO

(**109a**) R = H
(**109b**) R = CH_2OH
(**110**)
(**111**)
(**112a**) R = H
(**112b**) R = CH_2OH
(**112c**) R = CH_2Cl

acetoacetate yield (**113**) (20%), and D-fructose similarly yields (**114**) (7%).[108] The tetrahydroxybutyl side-chains are oxidized to aldehyde with lead tetra-acetate, and the aldehydes are converted to acids with alkaline silver oxide. Lower sugars, α-hydroxy-aldehydes and ketones, and glyoxal[109] undergo similar reactions. Thus D-glycerose with ethyl acetoacetate yields (**115**), and glycolaldehyde similarly yields (**116**). A possible mechanism involves aldol condensation between the methylene group of the 1,3-dicarbonyl compound and the carbonyl group of the sugar followed by dehydration and ring closure.

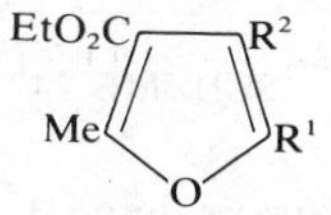

EtO_2C, R^2, Me, R^1, O

(**113**) R^1 = —$(CHOH)_3CH_2OH$, R^2 = H
(**114**) R^1 = H, R^2 = —$(CHOH)_3CH_2OH$
(**115**) R^1 = CH_2OH, R^2 = H
(**116**) R^1 = R^2 = H

18.4.14.2 Paal–Knorr type syntheses

Acidic reagents bring about the cyclization of 1,4-diketones, presumably *via* the mono-enol.[110] The requisite dicarbonyl compounds may be obtained by standard methods[111] including the *C*-alkylation of β-ketoesters with α-haloketones. Suitable intermediates may also be prepared by hydroformylation of the acetals of $\alpha\beta$-unsaturated aldehydes.[112] β-Chloroallyl ketones, which are readily available by alkylation of enamines or enolate anions, behave as masked 1,4-diketones, and on treatment with acid afford furans.[113]

Appropriately substituted oxiranes, by taking advantage of the acid-catalysed oxirane–ketone rearrangement, also behave as masked 1,4-diketones. Thus Darzens reaction of methyl chloroacetate with (**117**) yields the glycidic ester (**118**), which on heating in the presence of a trace of hydrogen chloride yields the furan (**119**). The reaction of the oxirane (**120**) with acid may take a similar course (Scheme 24).

The enolate anions derived from β-ketoesters undergo Michael addition to 1-nitroalkenes; the resultant nitronic acids yield 1,4-diketones on Nef reaction and subsequent ring-closure affords furans.[2]

18.4.14.3 Feist–Benary syntheses

Condensation of an α-chloroaldehyde or an α-chloroketone with a β-ketoester, usually in the presence of aqueous base, provides a synthesis of furans bearing an ester substituent at a β-position. The reaction is thought to be of the aldol type, and the

(MeO)$_2$CHCH$_2$COMe (**117**) →(i) (MeO)$_2$CHCH$_2$–C(Me)(epoxide)–CO$_2$Me (**118**) →(ii) [(MeO)$_2$CH–CH$_2$–CH(Me)–CO–CO$_2$Me] → 3-methylfuran-2-carboxylate methyl ester (**119**) (my)

(**120**) →(iii) (my) 3-methylfuran

i, ClCH$_2$CO$_2$Me, NaOMe; ii, H$^+$, 150 °C; iii, H$_2$SO$_4$.

SCHEME 24

intermediate 3-hydroxy-2,3-dihydrofurans have been isolated in certain cases (Scheme 25).[114] The condensation of ethyl bromopyruvate and sodium oxaloacetate is mechanistically similar.[111]

18.4.14.4 From other heterocyclic systems

Pyrylium salts (see Chapter 18.1) on treatment with aqueous acidic hydrogen peroxide suffer oxidation and ring-contraction, yielding furans.[61] 2-Pyrones may be converted to furans: thus 3-bromo-2-pyrone-5-carboxylic acid and 3-bromo-4,6-dimethyl-2-pyrone-5-carboxylic acid on treatment with base yield furan-2,4-dicarboxylic acid and 2,4-dimethylfuran-3-carboxylic acid, respectively.

1,2-Oxathiin 2,2-dioxides (δ-sultones) are readily prepared by treating $\alpha\beta$- or $\beta\gamma$-unsaturated ketones with a cold mixture of acetic anhydride and concentrated sulphuric

i, Na$_2$CO$_3$, H$_2$O, 0–8 °C; ii, PhH, MeC$_6$H$_4$SO$_3$H, heat.

SCHEME 25

acid. On pyrolysis under basic conditions they lose sulphur dioxide and afford furans. This method is useful for the preparation of alkyl- and alkyl/aryl-substituted furans.

Oxazoles undergo $[_{\pi}4_s + _{\pi}2_s]$ reaction with acetylenic dienophiles and the intermediate adducts suffer chelotropic loss of a nitrile, thus yielding furans.[115] This method provides a convenient route to furan-3-carboxylic acid and 3-acetylfuran (Scheme 26).

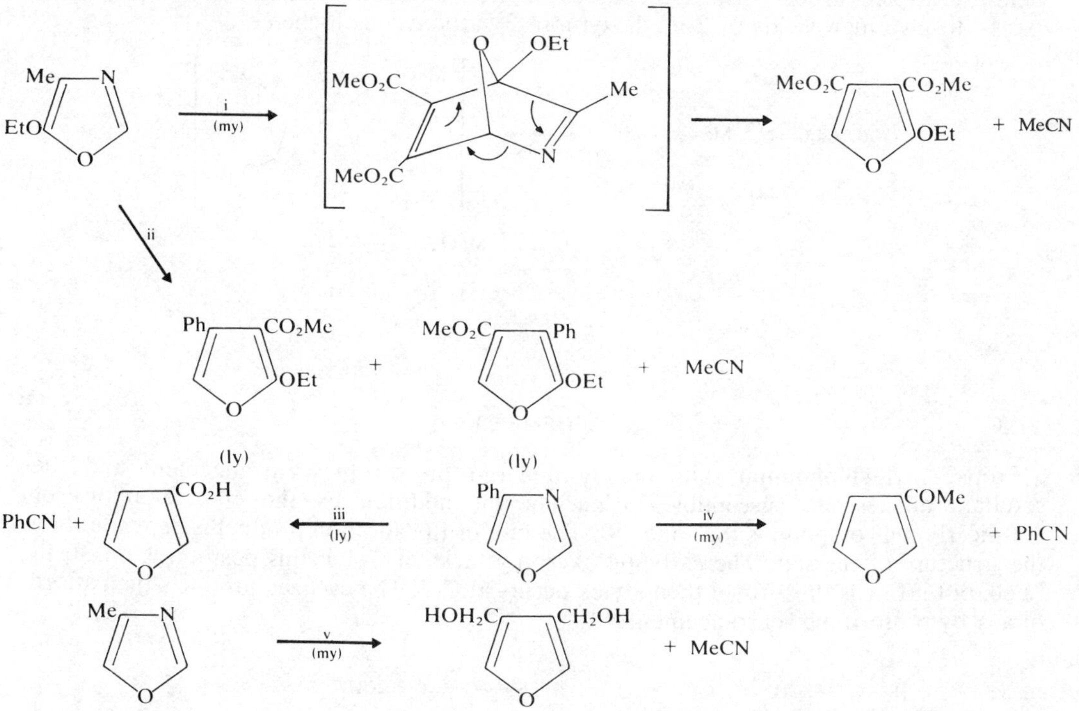

i, $MeO_2CC{\equiv}CCO_2Me$, Et_2O, 25 °C; ii, $PhC{\equiv}CCO_2Me$, PhMe, heat, $BF_3 \cdot Et_2O$; iii, $HC{\equiv}CCO_2H$, 115 °C; iv, $HC{\equiv}CCO_2Me$, 115 °C; v, $HOCH_2C{\equiv}CCH_2OH$, 170 °C.

SCHEME 26

2(5*H*)-Furanones (Section 18.4.20) (butenolides) are readily reduced to furans with di-isobutylaluminium hydride. Butenolides may be functionalized by reaction with diazoalkanes and pyrolysis of the pyrazolines so formed. Reduction, which may be controlled to preserve an ester substituent, then affords furans.[116] Tetrahydro-3-furanones may be alkylated in the 3-position on treatment with lithium di-isopropylamide and an alkyl halide. The products on reaction with phenyl selenylchloride and lithium di-isopropylamide, and subsequent oxidation, yield 3-alkyl-2(5*H*)-furanones, which on reduction afford furans.[117]

The tosylhydrazones of tetrahydro-2-furanones on Bamford–Stevens reaction afford dihydrofurans, which may be oxidized to furans with 2,3-dichloro-5,6-dicyano-1,4-benzoquinone (Scheme 27).[118] 3-Acyl-2(5*H*)-furanones, readily available by reaction of acyloins with diketen,[119] on acyl–lactone rearrangement,[120] yield furans.

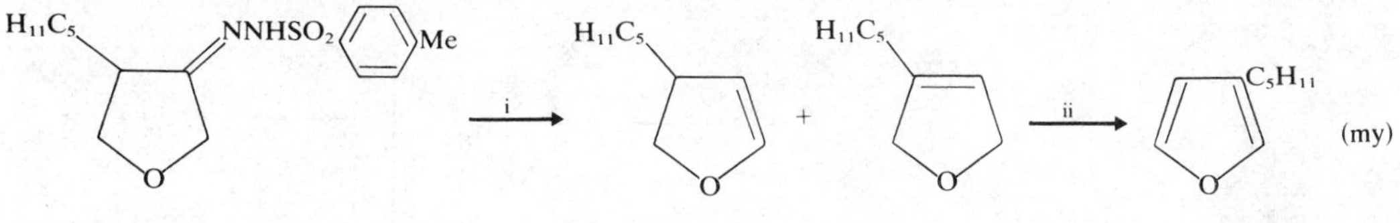

i, $(HOCH_2CH_2)_2O$, Na; ii, DDQ, PhH, 25 °C.

SCHEME 27

18.4.14.5 From acetylenes

Reaction of alkoxycarbonyl- and keto-carbenes with alkynes produces 3-carbonyl substituted cyclopropenes, generally in poor yield. These compounds on metal-catalysed ring-opening yield furans.[121] Photosensitized decomposition of diazomalonate (**121**) yields the triplet carbene, which undergoes stepwise addition to alkynes, when present in excess, to give high yields of 2-methoxyfuran-3-carboxylates (Scheme 28).[122]

$(MeO_2C)_2C{=}N_2 + Me{-}{\equiv}{-}Me$ (**121**) $\xrightarrow{i}$ [diradical intermediate: MeO_2C, Me, MeO, Me] → 2-methoxy-3-methoxycarbonyl-4,5-dimethylfuran (MeO_2C, Me, MeO, Me)

i, Ph_2CO, *hν*.

SCHEME 28

Prop-2-ynyl sulphonium salts readily undergo prototropic rearrangement and the resultant allenes are susceptible to nucleophilic addition by the enolate anions of 1,3-dicarbonyl compounds (Scheme 29). The fate of the addition product is dependent on the structure of the salt. The carbonyl oxygen attacks at C-1 if this position is unsubstituted, but if C-1 is substituted then attack occurs at C-3. The cyclized products then afford furans by prototropic rearrangement.[123]

$Br^- Me_2\overset{+}{S}{-}\overset{1}{C}H(Me){-}\overset{2}{C}{\equiv}\overset{3}{C}H$ + $MeCOCH_2CO_2Et$ $\xrightarrow{i}$ [$Me_2\overset{+}{S}{-}C(Me){=}C{=}CH_2$, $MeCO\bar{C}HCO_2Et$] → $Me_2\overset{+}{S}{-}CHMe$, CH_2, EtO_2C, Me → [MeHC, CO_2Et, Me dihydrofuran] → Et, CO_2Et, Me furan

$Me_2\overset{+}{S}{-}\overset{1}{C}H_2{-}\overset{2}{C}{\equiv}\overset{3}{C}H$ + $MeCOCH_2COMe$ $\xrightarrow{i}$ [$Me_2\overset{+}{S}{-}CH{=}C{=}CH_2$, $MeCO\bar{C}HCOMe$] → $Me_2\overset{+}{S}{-}CH_2{-}C{=}CH_2$, COMe, Me → [$CH_2$, COMe, Me dihydrofuran] → Me, COMe, Me furan

i, NaOEt, EtOH, heat.

SCHEME 29

trans-Enynols on treatment with base undergo cyclization to furans, presumably *via* the *cis*-compounds;[124] the same reaction may be achieved with aqueous acidic mercuric sulphate.[125]

2-Alkyl-2-ethynyloxiranes, *e.g.* (**122**), on treatment with aqueous acidic mercuric sulphate yield 3-alkylfurans. The presumed intermediates are the glycols, and nucleophilic attack occurs at the terminal carbon atom.[126] Similarly, acetylenic glycols or their derivatives, *e.g.* (**123**), on base treatment yield dihydrofurans, which may readily be converted to furans (Scheme 30).[127]

On acid or base treatment, 4-ethynylketones cyclize to furans *via* their enol tautomers.

(**122**) —i→ [HC≡C–C(Me)(OH)–CH₂OH] → 3-methylfuran (my)

HC≡C–CH(OEt)–CH(OH)Ph (**123**) —ii→ 3-ethoxy-2-phenyl-dihydrofuran —iii→ 2-phenylfuran (hy)

i, $HgSO_4$, H^+; ii, NaOH, EtOH; iii, H^+.

SCHEME 30

18.4.14.6 From ylides

Ketones (**124**) may be formylated in the α-position and the resultant hydroxymethylene derivatives are readily converted into the corresponding thioenol ethers (**125**). The keto-functions of these compounds, on reaction with dimethylsulphonium methylide, are converted into oxirans (**126**), which can undergo cationic transformation to dihydrofurans (**127**), and thence afford 3,4-disubstituted furans (**128**) (Scheme 31). This method has been used for the synthesis of the terpenoid furans perillene and dendrolasin.[128] The enol ethers of 1,3-dicarbonyl compounds undergo a similar transformation with dimethylsulphonium methylide.[129]

Dimethylsulphonium acylylides add to dimethyl acetylenedicarboxylate, yielding furan-3,4-dicarboxylic esters.[130] Sulphonium diacylmethylides also yield furans with acetylenedicarboxylic ester.[131]

$R^1COCH_2R^2$ (**124**) —i,ii→ $R^1COCR^2{=}CHSBu^n$ (**125**)

—iii→ (**126**) —iv→ [(**127**)] → (**128**) (ly–my)

i, NaOMe, HCO_2Et; ii, Bu^nSH, PhH, $MeC_6H_4SO_3H$, heat; iii, Bu^nLi, $Me_3S^+BF_4^-$; iv, $HgSO_4$, H_2O.

SCHEME 31

The anions derived from acyloins undergo nucleophilic addition to β-ethoxyvinylphosphonium salts (**129**). The resultant ylide (**130**) then yields a dihydrofuran (**131**) by intramolecular Wittig reaction. The dihydrofurans are readily converted to furans (**132**) by elimination of ethanol (Scheme 32).[132] With symmetrical acyloins only one product is obtained by this method. Unsymmetrical acyloins may give two products because of their ready tautomerism under basic conditions.

PhCHCOPh(OH) —i→ [Ph—CH(O⁻)—COPh + $EtOCH{=}CHR\overset{+}{P}Ph_3$ Br^- (**129**) ⇌ PhCO $^-\overset{+}{C}RPPh_3$ / Ph—CH—O—CH—OEt (**130**)] → (**131**) —ii→ (**132**) (hy)

(**131**) (**132**) R = H, Me

i, NaH, $MeO(CH_2)_2OMe$, 25 °C; ii, PhH, $MeC_6H_4SO_3H$, heat.

SCHEME 32

18.4.14.7 Miscellaneous methods

Catalytic oxidative dehydrogenation of 1,3-alkadienes and alkenes yields furans.[133]

Sodium succinate in hot acetic anhydride will react with β-dicarbonyl compounds to yield furans.[134] The reaction is thought to involve an aldol condensation between succinic anhydride and the 1,3-dicarbonyl compound, followed by an internal Michael addition.

Glycidaldehyde, readily available by epoxidation of acrolein, will undergo base-catalysed condensation with 1,3-dicarbonyl compounds to yield furans.[135]

18.4.15 ALKYLFURANS

Alkylfurans are readily available by ring-synthesis (Section 18.4.14), by the decarboxylation of alkylfurancarboxylic acids, or by the reaction of furan carbonyl compounds with Grignard reagents and subsequent steps. The alkylation of furans is also useful (Section 18.4.4.5). Wolff–Kishner reduction of furan aldehydes and ketones has also been used. The Wolff–Kishner reduction of furan-2-carboxaldehyde, however, gives 2,5-dihydro-5-methylenefuran, as well as 2-methylfuran.[136] In the t-butylation of certain t-butylfurans, similar non-aromatic products are also formed. Thus with 5-t-butyl-2,3-dimethylfuran (**133**), minimization of steric interference occurs by formation of the 2,5-dihydrofuran (**134**). t-Butylation of 3-t-butyl-2,5-dimethylfuran (**135**) does not yield the 3,4-disubstituted product, but instead the product (**136**) arises from substitution at the lesser hindered 5-position. On the other hand, the t-butylation of 2,5-di-t-butylfuran (**84**) yields 2,3,5-tri-t-butylfuran (**89**) (Scheme 33).[137] The best preparation of 2,5-dimethylfuran (94%) is the distillation of hexane-2,5-dione from an acidic ion-exchange resin.[138] Alkylfurans are particularly important as synthons for 1,4-diketones (Section 18.4.9).

The furan ring is common in many terpenoid compounds; its synthesis is achieved *via* acetylenes as in bilabone (**137**),[139] reduction of butenolides as in actractylon (**138**),[140] or

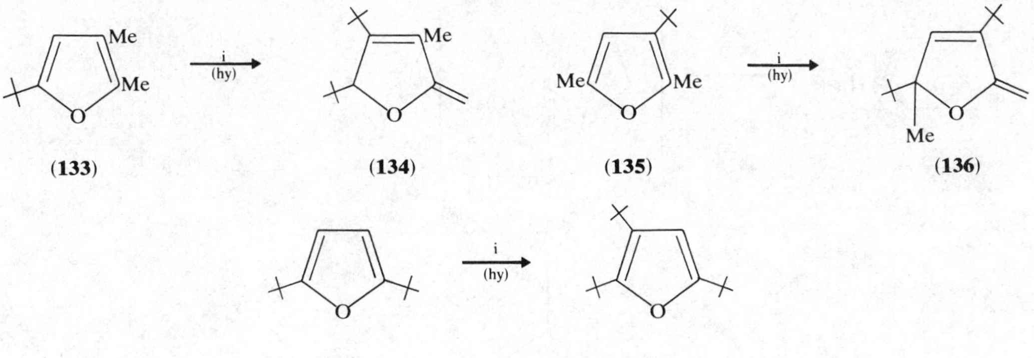

(133) (134) (135) (136)

(84) (89)

i, Bu^{t}Cl, $AlCl_3$.

SCHEME 33

by use of a furyl-lithium as in the case of the degraded limonoid fraxinellone (**139**)[141] (Scheme 34).

The chemistry of side-chain-substituted furans is conventional, but the reaction of 2-(chloromethyl)furan with nucleophiles is worthy of comment. With aqueous cyanide the reaction appears to proceed through the intimate ion pair (**140**). Cyanide ion may attack to give the 'normal' product (**141**) or the 'abnormal' product (**143**), by aromatization of the intermediate (**142**), which may be isolated under certain conditions. Spectroscopic evidence for the intermediacy of (**142**) has been obtained and the effect on the product

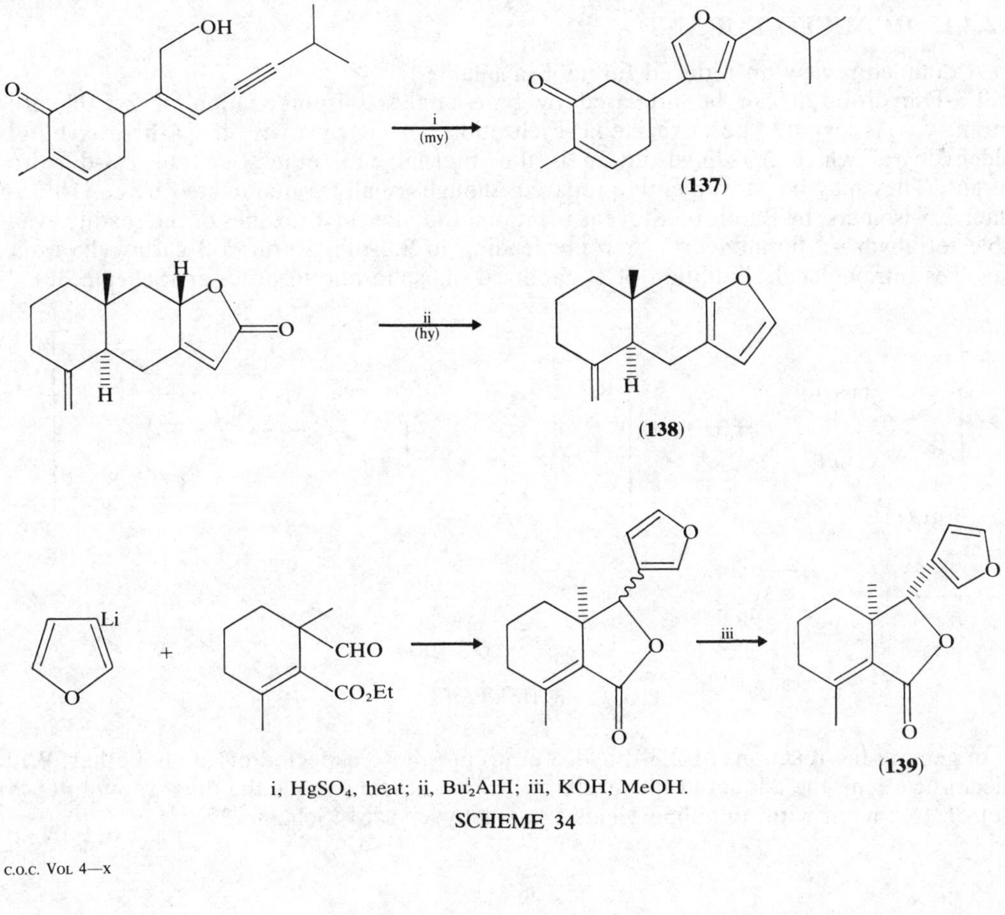

(137)

(138)

(139)

i, $HgSO_4$, heat; ii, Bu^i_2AlH; iii, KOH, MeOH.

SCHEME 34

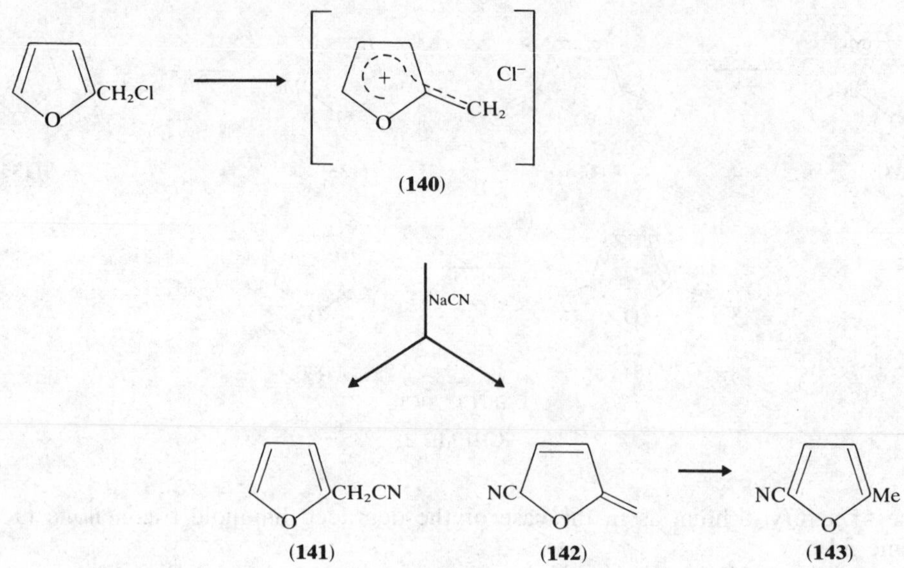

(140)

(141) (142) (143)

SCHEME 35

ratio of different substituents has been investigated (Scheme 35).[142] 2,5-Bis(chloromethyl)furan similarly gives a mixture of the normal product and 2-chloromethyl-2,5-dihydro-5-methylenefuran-2-carbonitrile.[143]

18.4.16 REDUCED FURANS

A detailed review on reduced furans is available.[144]

2,3-Dihydrofuran can be prepared by base-catalysed isomerization of 2,5-dihydrofuran.[145] As would be expected, cyclization of derivatives of 4-hydroxybutyraldehyde, of which 2,3-dihydrofuran is the internal enol ether, lead to 2,3-dihydrofurans. They may be conveniently prepared, though usually contaminated by *ca.* 10% of their 2,5-isomers, by Bamford–Stevens reaction of the tosylhydrazones of the readily available tetrahydro-3-furanones.[146] A route leading to 2,3-dihydrofuran-3-carboxylic esters involves intramolecular addition of a stabilized phosphorane to an ester (Scheme 36).[147]

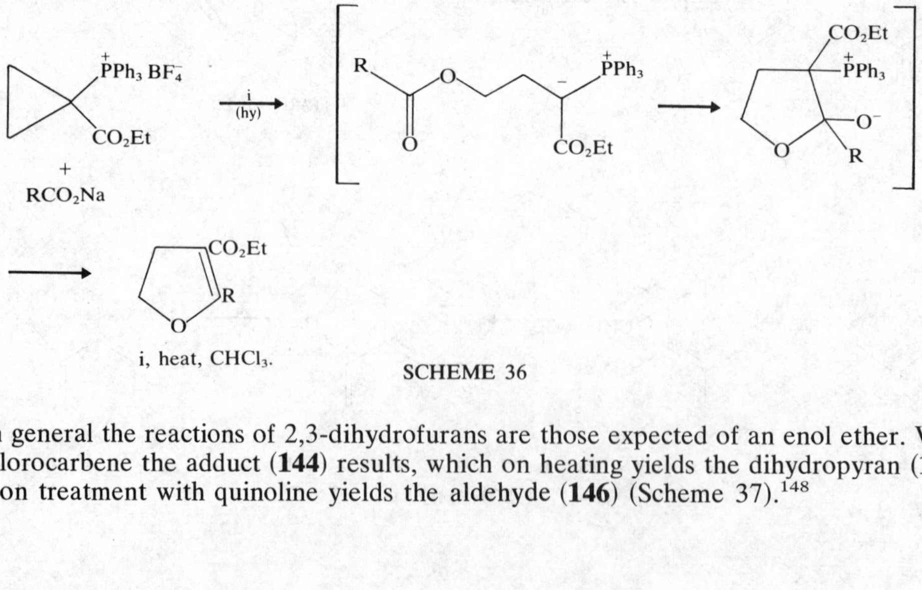

i, heat, $CHCl_3$.

SCHEME 36

In general the reactions of 2,3-dihydrofurans are those expected of an enol ether. With dichlorocarbene the adduct (**144**) results, which on heating yields the dihydropyran (**145**) but on treatment with quinoline yields the aldehyde (**146**) (Scheme 37).[148]

(144)

(145) (146)

i, $:CCl_2$; ii, heat; iii, quinoline, heat.

SCHEME 37

Acid-catalysed dehydration of *cis*-but-2-ene-1,4-diol affords 2,5-dihydrofuran.[149] The preparation of 2,5-dihydrofurans by 2,5-addition across the furan ring (Section 18.4.8), and the use of the products in synthesis (Section 18.4.9), have been discussed. The intramolecular Wittig reaction of appropriately constituted ylides (**147**) provides a general synthesis of 2,5-dihydrofurans, but often in low yield (Scheme 38).[150]

(**147**)

i, NaH; ii, $CH_2{=}CH\overset{+}{P}Ph_3\ Br^-$

SCHEME 38

2,5-Dihydrofuran is more stable to acid than its 2,3-isomer, and it behaves as a dienophile in Diels–Alder reactions with such dienes as butadiene and anthracene.[149] It can be vinylcyclopropanated by reaction with vinyldiazomethane.[151]

The end groups of certain carotenoids contain the 2,5-dihydrofuran ring, but these compounds may be artifacts.[152] The synthesis of the 2,5-dihydrofurans (**149**) and (**150**) which occur in Japanese hops has been accomplished by the oxidative cyclization of the trihydroxydiene (**148**) (Scheme 39).[153] Similar spirans occur in *Santalina* species.[154] The

(**148**) (**149**)

(**150**)

i, quinoline-poisoned 5% Pd–$BaSO_4$, H_2; ii, MnO_2; iii, Pt, H_2.

SCHEME 39

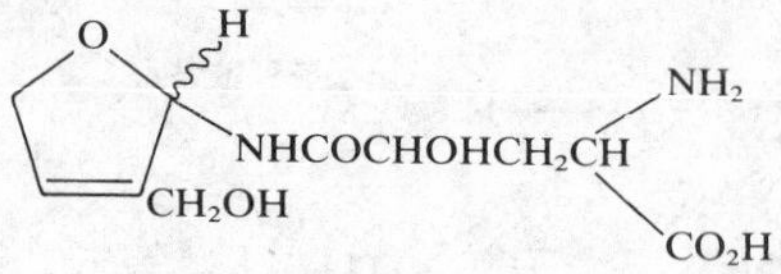

(**151**)

unusual amino-acid amide oxypinnatine (**151**) occurs in a *Staphylea* species; on acidic hydrolysis it affords 3-furanmethanol.[155]

Tetrahydrofuran is manufactured on a large scale by catalytic reduction of furan and by dehydration of butane-1,4-diol (from butyne-1,4-diol). Base treatment of 4-haloalkan-1-ols, and acidic treatment of alk-3-en-1-ols or alk-4-en-1-ols, also yields tetrahydrofurans.[144] Tetrahydrofuran is a commonly used solvent, having the advantage of being a relatively inert ether which is miscible with both water and organic solvents.

Streptomyces spp. produce a family of macrotetrolide antibiotics termed nactins (**152**), which are based on a 32-membered ring containing four tetrahydrofuran units. These compounds transport potassium ions across cell membranes and have been shown to markedly influence mitochondrial respiration and phosphorylation. Extensive research has culminated in the total synthesis of nonactin (**153**).[156]

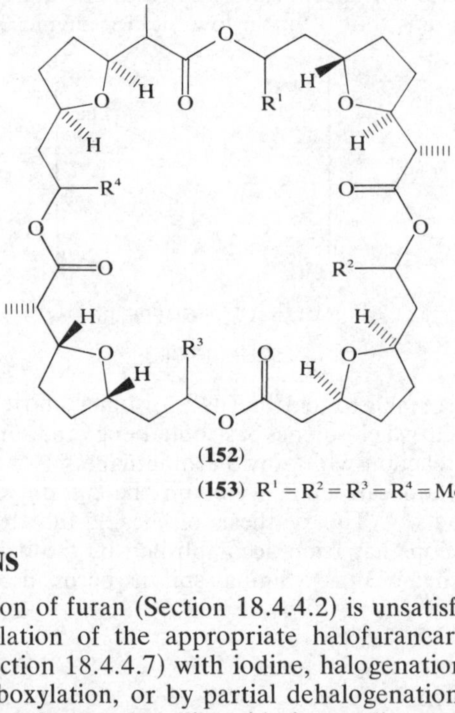

(**152**)

(**153**) $R^1 = R^2 = R^3 = R^4 = Me$

18.4.17 HALOFURANS

The direct halogenation of furan (Section 18.4.4.2) is unsatisfactory and halofurans are prepared by decarboxylation of the appropriate halofurancarboxylic acid, reaction of chloromercurifurans (Section 18.4.4.7) with iodine, halogenation of furancarboxylic acids with concomitant decarboxylation, or by partial dehalogenation of halofurans.

Decarboxylation of halofurancarboxylic acids is usually carried out with copper and quinoline at 150–230 °C and the product is often removed immediately by distillation. Heating chloromercurifurans with iodine and potassium iodide in water yields the iodofurans. Thus 3,4-bis(chloromercuri)-2,5-dimethylfuran yields 3,4-di-iodo-2,5-dimethylfuran (41%), and tetrakis(chloromercuri)furan yields tetraiodofuran (67%).[1] Boiling sodium furan-2,5-dicarboxylate with potassium iodide and iodine in water yields 2,5-di-iodofuran. 2,5-Dibromofuran (78%) may be prepared from sodium 5-bromofuran-2-carboxylate by heating in water with bromine and potassium bromide.[157]

Tetraiodofuran may be converted to 3,4-di-iodofuran (48%), or 2,3,4-tri-iodofuran (85%), depending on the proportions of reagents, by brief entrainment with bromoethane and magnesium and subsequent hydrolysis. 2,3-Dibromofuran and 3,4-di-iodofuran on reaction with 1 mol. equiv. of ethyl-lithium and subsequent hydrolysis are converted into 3-bromofuran (57%) and 3-iodofuran (41%), respectively.[158]

Tetrafluorofuran has been synthesized *via* reaction of tetrahydrofuran with cobalt(II) fluoride at 100–120 °C.[159]

The halofurans are usually unstable to light and air. The 3-halofurans are useful for obtaining other 3-substituted furans *via* the lithio-derivatives or by nucleophilic substitutions. Thus 3-iodofuran may be converted to 3-methoxyfuran (65%) with sodium methoxide and copper(II) oxide in methanol, to 3-cyanofuran (48%) with copper(I) cyanide in quinoline, or *via* the lithio-compound to furan-3-carboxaldehyde (39%).[33] The halogen atoms in 2-halofurans substituted in the 5-position with an electron-withdrawing group are readily displaced by nucleophiles.

18.4.18 AMINOFURANS

Few simple aminofurans have been characterized and there is a corresponding lack of spectroscopic information on the position of the imine–amine tautomeric equilibrium. Their behaviour, however, is in no sense that of aromatic amines. Most classical synthetic approaches to 2-aminofuran have failed.[1] Ethyl 2-furylcarbamate and *N*-2-furylacetamide both undergo loss of ammonia on acidic or basic hydrolysis. The former compound is obtained from ethyl 5-nitrofuran-2-carboxylate, or by treatment of 2-furylisocyanate, available by Curtius reaction of the azide of furan-2-carboxylic acid, with methylmagnesium iodide. However, hydrazinolysis of *N*-(2-furyl)phthalimide, obtained (30%) from phthalimide and 2,5-dihydro-2,5-dimethoxyfuran, gives 2-aminofuran, which was not isolated but its presence was inferred by g.c.–m.s. and ^{1}H n.m.r. spectroscopy. The n.m.r. spectrum is claimed to reveal the absence of imino-tautomers.[160]

2-Dialkylaminofurans (**156**)[161] are prepared from the tertiary amides of β-aroylpropionic acids (**154**), which on treatment with acetic anhydride and perchloric acid yield 2,3-dihydrofuran-2-furylideneammonium salts (**155**); on deprotonation the salts yield the stable dialkylaminofurans (**156**). The chemistry of these compounds is typical of that of enamines. Protonation occurs on carbon and the iminium salts (**155**) are obtained. The furans are stable to aqueous base, but with dilute acid hydrolysis to the butenolide (**157**) occurs. The furans fail to give well-defined products with electrophiles, but the salt (**158**) condenses with benzaldehyde to afford the benzylidene salt (**160**). This reaction is thought to occur by proton exchange between the salt and benzaldehyde. The resultant free base (**159**) then undergoes electrophilic attack by the cation Ph$\overset{+}{C}$HOH and dehydration of the product then gives (**160**). The furan (**159**) gives coupling products (**161**) with diazonium salts. Unlike simple enamines the furan (**159**) does not function as a dipolarophile, but takes part in Diels–Alder reactions with maleic anhydride and *N*-phenylmaleimide. The initial adducts undergo rearrangement to the dienols (**162**), which are readily dehydrated to the aromatic compounds (**163**) (Scheme 40).

2-Aminofurans substituted with electron-withdrawing groups are stable compounds and exist as the amino-tautomers, but their reactions are not characteristic of aromatic amines. Compound (**164**) may be nitrated and brominated. The nitro-compound (**165**), thus obtained, undergoes acidic hydrolysis to the amine (**166**) and reduction to (**167**). The last-mentioned compound on diazotization gives a 2-naphthol coupling product, but no other reactions characteristic of diazonium salts.[1] 5-Nitrofuran-2-carboxaldehyde and some of its condensation products may be catalytically hydrogenated to the 5-amino-compounds.[162]

A stable furan-2-diazonium fluoroborate has been prepared which gives some reactions characteristic of an aromatic species.[163]

2-Amino-3-cyanofurans (**169**)[164] are obtained by base-catalysed condensation of acyloins (**168**) with malondinitrile (Scheme 41). The aminofuran (**174**), however, is unique and undergoes spontaneous Diels–Alder dimerization to (**175**).[165] The aminofurans (**169**),

i, $HClO_4$, Ac_2O; ii, Et_3N; iii, $HClO_4$; iv, 2 M HCl aq.; v, PhCHO, AcOH, 100 °C; vi, $Ar\overset{+}{N}_2\,BF_4^-$, Et_3N; vii, maleic anhydride or *N*-phenylmaleimide.

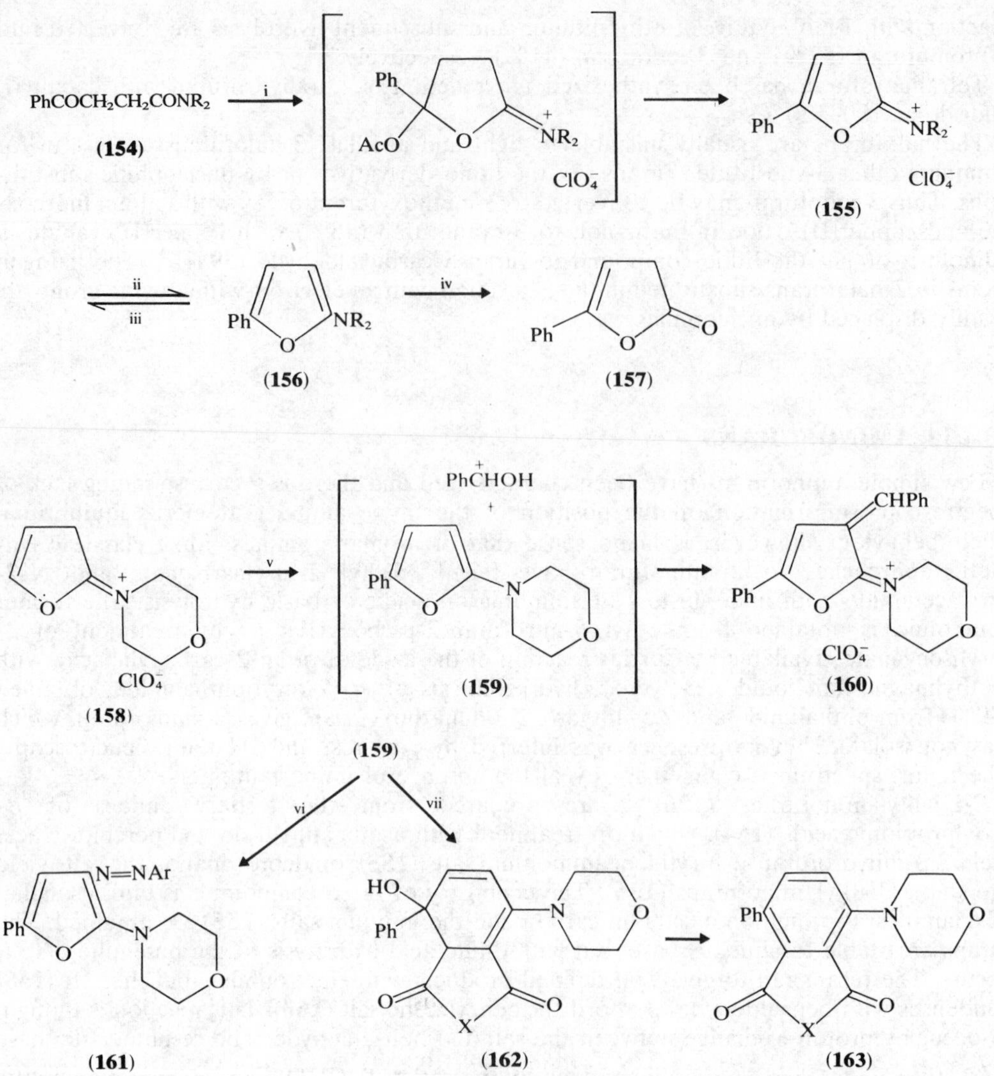

SCHEME 40

as expected of enamines, yield (**170**) on protonation.[166] On acidic hydrolysis they yield butenolides (**171**), and they condense with benzaldehyde to give Schiff bases (**172**). With maleic anhydride they undergo Diels–Alder reaction, but the initial adducts lose water and yield phthalic anhydrides (**173**).

Two *C*-alkyl-3-aminofurans are known; they both resinify very rapidly in air.[1] They are sensitive to hot acid and alkali and ammonia is liberated. With base, products characteristic of the cleavage of 2,5-dimethyl-3(2*H*)-furanone are obtained (Section 18.4.20). The aminofurans react with nitrous acid and the products couple with 2-naphthol, but they give no other reactions characteristic of diazonium salts. It has been suggested that they exist as the imino-tautomers, but there is no spectroscopic information available on this point.

(**164**) $R^1 = Ac$, $R^2 = H$
(**165**) $R^1 = Ac$, $R^2 = NO_2$
(**166**) $R^1 = H$, $R^2 = NO_2$
(**167**) $R^1 = Ac$, $R^2 = NH_2$

R[1]—C(=O); R[2]—CHOH (**168**) —i (my-hy)→ (**169**) —ii→ (**170**) NH$_2$ CF$_3$CO$_2^-$

(**169**) —iii→ (**171**); —iv→ (**172**) N=CHPh; —v→ (**173**)

(**171**) (**172**) (**173**)

$R^1 = R^2 =$ Me or Ph; $R^1,R^2 =$ —$(CH_2)_4$—

[(**174**)] → (**175**)

i, $CH_2(CN)_2$, MeOH, KOH or Et_3NH, <40 °C; ii, CF_3CO_2H; iii, HCl, EtOH, heat; iv, PhCHO, AcOH, EtOH; v, maleic anhydride.

SCHEME 41

18.4.19 NITROFURANS

The nitration of furans has been discussed previously (Section 18.4.4.3). Carboxylic acid groups are often replaced. Thus 5-bromofuran-2-carboxylic acid on nitration yields 5-bromo-2-nitrofuran, which may be readily converted to 5-iodo-2-nitrofuran, and thence to 2,5-dinitrofuran.[167]

Nitrofurans, in contrast to alkylfurans, are usually acid-stable but they resinify in aqueous alkali. Thus 5-nitrofuran-2-carboxaldehyde may be converted into the nitro-acid by treatment with *aqua regia*. The halogen atoms in 2-halo-5-nitrofurans can readily be displaced by nucleophiles.[168] A nitro-group in compounds such as 5-nitrofuran-2-carboxaldehyde and 2,5-dinitrofuran may also be displaced by nucleophiles.[169]

Some nitrofurans exhibit antimicrobial properties and have been used as chemotherapeutic agents.[162,170]

18.4.20 HYDROXYFURANS

2-Hydroxyfuran (**176**) is formally the enolic form of 2(3*H*)-furanone ($\Delta^{\beta,\gamma}$-butenolide) (**177**), but it and 2(5*H*)-furanone ($\Delta^{\alpha,\beta}$-butenolide) (**178**) show little evidence of enolic behaviour. The chemistry of the butenolides has been comprehensively reviewed.[171] 2-Methoxyfuran and 2-acetoxyfuran are, however, well-known compounds. 2-Methoxyfuran (11%) and 2-acetoxyfuran (81%) are prepared by heating the appropriate 2,5-disubstituted 2,5-dihydrofuran with naphthalene-2-sulphonic acid at high temperature.[50] 2-Alkoxyfurans are also obtained by reaction of 5-bromofuran-2-carboxylates with sodium alkoxides, followed by hydrolysis and decarboxylation.[172] The 2-alkoxyfurans are unstable and undergo polymerization under nitrogen at 0 °C. Both the 2-alkoxy- and 2-acetoxyfurans undergo normal Diels–Alder reactions. Acid treatment of 2-alkoxy- and 2-acetoxyfurans yields derivatives of 2-formylpropionic acid; thus 2-methoxyfuran with dry methanol and hydrogen chloride yields (**179**). With halogens or lead tetra-acetate, 2-acetoxyfuran yields 5-halo- or 5-acetoxy-2(5*H*)-furanones, which are readily hydrolysed to *Z*-3-formylacrylic acid.

2(5*H*)-Furanones are the stable butenolide isomers since they are conjugated. They may be prepared from 2(3*H*)-furanones by the action of bases. The 2(5*H*)-furanones are widespread in Nature and examples are found in the sesquiterpenes,[173] and in the lichen pulvinic acid derivatives.[174] 4-Hydroxy-2(5*H*)-furanones are known as tetronic acids; ascorbic acid belongs to this group. The α-methylene-γ-butyrolactone unit (**180**) is found in a number of biologically active terpenes. Synthetic methods in this area have been reviewed.[175] One important class of naturally occurring 2(5*H*)-furanones is the cardiac active steroidal lactones, cardenolides, which contain the system (**181**), and many of the synthetic routes to the 2(5*H*)-furanones have been developed in this connection.[176] Thus

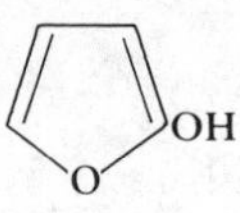

(**176**) 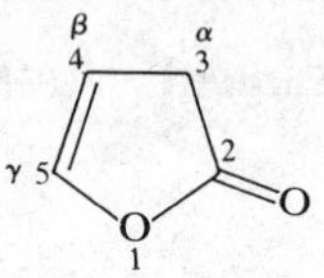 (**177**) 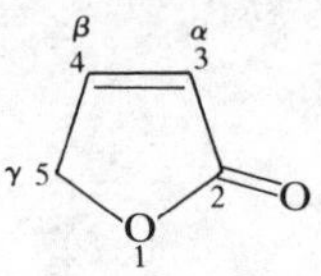 (**178**)

$(MeO)_2CHCH_2CO_2Me$

(**179**) (**180**) 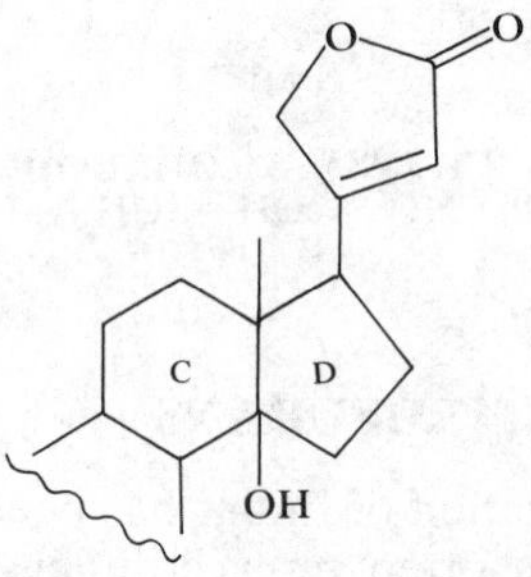 (**181**)

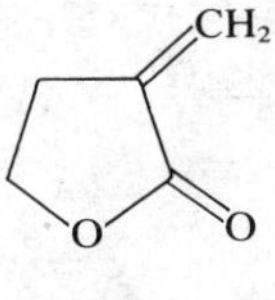

the ketone (**182**) on reaction with lithium ethoxyacetylide gives the carbinol (**183**), which undergoes acid-catalysed rearrangement to the $\alpha\beta$-unsaturated ester (**184**). Allylic oxidation of (**184**) with selenium dioxide under mild conditions then affords digitoxigenin acetate (**185**) (Scheme 42). The cyanohydrins of 3-hydroxyaldehydes or ketones on acid hydrolysis afford hydroxy-lactones, which readily undergo dehydration. Partial hydrogenation of γ-hydroxy-$\alpha\beta$-acetylenic acids gives 2(5*H*)-furanones directly.

2(3*H*)-Furanones may be prepared from enolizable γ-ketoacids, *e.g.* laevulinic acid, by slow distillation or by treatment with acetic anhydride. β,γ-Dibromoacids on boiling with water or aqueous sodium carbonate afford 2(3*H*)-furanones by solvolysis and dehydrobromination. Treatment of 2-lithiofurans with ethyl borate and hydrolysis of the resultant boronic ester also yields 2(3*H*)-furanones.[177]

The butenolides, in contrast to their thiophen analogues, show little evidence of enolic behaviour, although α-angelica lactone (**186**) is reported to liberate *ca.* 0.3 mol. equiv. of methane on Zerewitinoff determination. They fail to undergo *O*-alkylation; thus treatment of α-angelica lactone (**186**) with methyl iodide under phase transfer conditions gives the *C*-alkylation products (**187**) and (**188**), and β-angelica lactone (**189**) in low yield.[178] Other aspects of their chemistry have been reviewed.[171]

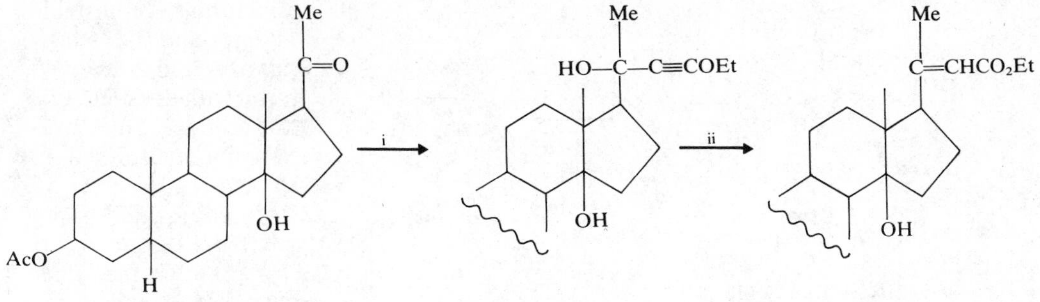

(182) (183) (184)

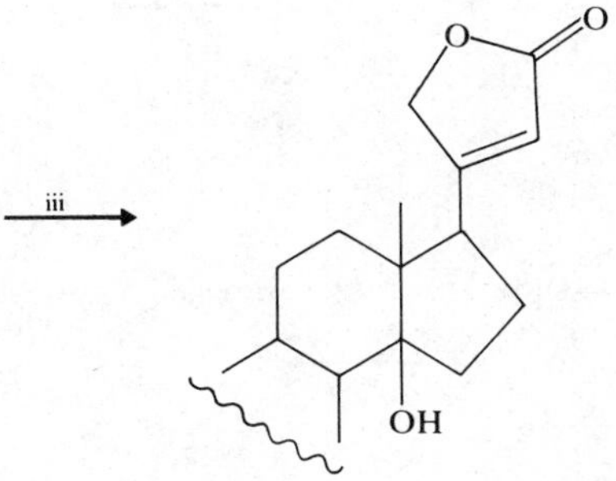

(185)

i, LiC≡COEt, THF; ii, 2% H_2SO_4, aq. MeOH; iii, SeO_2.

SCHEME 42

3-Hydroxyfuran (**190**) is formally the enolic form of 3(2*H*)-furanone (**191**). Compounds (**192**) with a carbonyl group at the 2-position behave as genuine 3-hydroxyfurans, but without this substituent the compounds behave as vinylogous lactones. Isomaltol, 3-hydroxy-2-acetylfuran (**194**), is available by hydrolysis of its β-D-galactopyranoside (**193**), which is obtained by treatment of lactose with salts of secondary amines (Scheme 43). Isomaltol (**194**) is an acidic enol (pK_a 5.7) which gives a deep red colour with iron(III) chloride. On treatment with diazomethane it affords the methyl ether (**195**). This can be degraded to 3-methoxyfuran (**198**) by base treatment of the pyridinium salt (**196**) and subsequent decarboxylation of the acid (**197**). 3-Methoxyfuran (**198**) affords the Diels–Alder adduct (**199**) with maleic anhydride and may be hydrolysed to (**191**) under mild

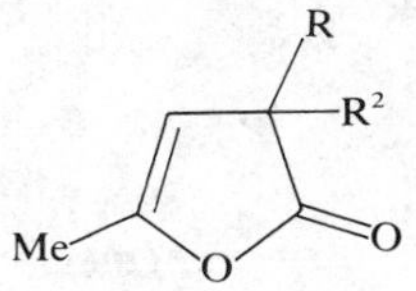

(**186**) $R^1 = R^2 = H$
(**187**) $R^1 = R^2 = Me$
(**188**) $R^1 = Me$, $R^2 = H$

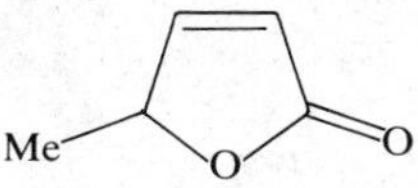

(**189**) 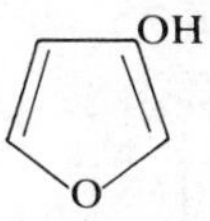 (**190**)

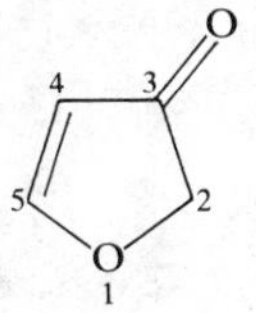

(**191**) 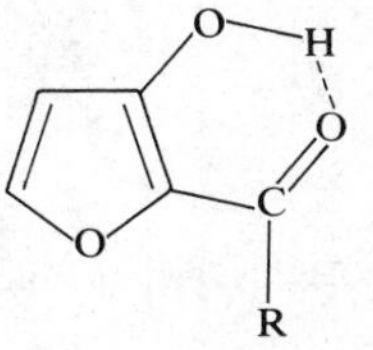 (**192**)

(193) → i → (194) → ii (my) → (195)

→ iii (my) → (196) → iv (ly) → (197)

→ v (my) → (198) → vi (my) → (199)

(198) → vii (ly) → (191)

i, H^+; ii, CH_2N_2; iii, C_5H_5N, I_2, 100 °C; iv, OH^-; v, CuO, quinoline, heat; vi, maleic anhydride, Et_2O, 5 °C; vii, Dowex 50 (H^+), H_2O, heat.

SCHEME 43

(200) → i → (201) → ii → (202)

→ iii, iv → (203) → vi → (205) → vii → (206)

(203) → v → (204)

i, NaH; ii, CH_2N_2; iii, OH^-; iv, Cu, quinoline, heat; v, $SnCl_4$, Ac_2O; vi, maleic anhydride; vii, NaOEt.

SCHEME 44

conditions.[2] 3-Methoxyfuran is also available from 3-iodofuran (Section 18.4.17). Reaction of methyl oxalate with the glycollic ester (**200**) yields the 3,4-dihydroxyfuran (**201**) (Scheme 44).[179] This compound can readily be *O*-alkylated, and the derived methyl ether (**202**) on hydrolysis and decarboxylation affords 3,4-dimethoxyfuran (**203**). This undergoes Friedel–Crafts reaction with acetic anhydride and tin(IV) chloride to afford (**204**).[1] Compound (**203**) also reacts exothermically in the Diels–Alder reaction with such dienophiles as maleic anhydride and benzoquinone. The adduct (**205**) affords the phthalic anhydride (**206**) on base treatment.[2]

2,5-Dimethyl-3(2*H*)-furanone (**209**) has been synthesized by a Curtius reaction on the ester (**207**), and acid hydrolysis of the intermediate urethane (**208**) (Scheme 45). This type

(**207**) (**208**) (**209**)

i, H_2NNH_2; ii, HOAc, $NaNO_2$; iii, Pr^iOH; iv, H^+, heat.

SCHEME 45

of route has been used for the synthesis of muscarine and its steroisomers.[180] Acylation of the ethoxymagnesio-derivatives of 1,3-dicarbonyl compounds with α-chloroacid chlorides produces 4-carbonyl substituted 3(2*H*)-furanones.[181]

The 3(2*H*)-furanones, since they are vinylogous lactones, do not give such ketonic derivatives as semicarbazones. Spectroscopic evidence indicates that they exist exclusively in the keto-form. They may be converted to 3-acetoxyfurans on reaction with sodium acetate and acetic anhydride, but they undergo *C*-alkylation with methyl iodide and strong base. They are generally stable to acid, merely becoming protonated. The 4-alkoxy-3(2*H*)-furanones, however, are easily converted to tetronic acids by mild acid or base treatment or even by leaving in water. 3(2*H*)-Furanones are degraded by aqueous base, which attacks in a Michael sense. 2,5-Dimethyl-3(2*H*)-furanone (**209**) gives acetate ion and acetoin, whereas compounds with an ester group at the 4-position, *e.g.* (**210**), yield tetronic acids (Scheme 46).

(**209**)

$MeCOCH_2COCHMe(O^-)$ → $MeCO_2^-$ + $MeCOCHOHMe$

(**210**)

i, OH^-; ii, 10% aq. Na_2CO_3.

SCHEME 46

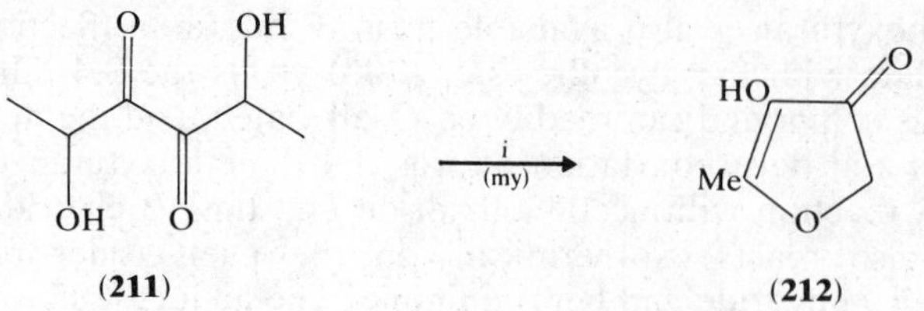

(211) (212)

i, $(CO_2H)_2$, heat, or $NaHCO_3$, C_5H_{12}, heat.

SCHEME 47

Only a few natural 3(2*H*)-furanones are known.[173] Furaneol (**212**), present in strawberries and pineapples, is obtained by acid or base treatment of the dihydroxydiketone (**211**) (Scheme 47).[182]

The tautomerism of hydroxyfurans has been reviewed.[183]

18.4.21 ALDEHYDES AND KETONES

Furan-2-carboxaldehyde (furfural) is used in the vegetable oil, petroleum, plastics, and rubber industries, as well as being a versatile starting material for furan syntheses. In particular, its ability to extract unsaturated hydrocarbons from hydrocarbon mixtures is important in the purification of petroleum and vegetable oils. As a consequence of its industrial significance and ready availability, the chemistry of furfural has been extensively studied. Electrophilic substitution occurs more readily than for benzaldehyde and the electrophile enters the 5-position except in the presence of aluminium chloride. It behaves like an aromatic aldehyde in most condensation reactions. Aqueous acids generally induce polymerization and many oxidants induce ring cleavage.[1] Autoxidation yields resins.[184] With a primary arylamine in acidic solution, ring-opening affords a salt (**213**) which is cyclized to a pyridinium salt (**215**) with acid and to a cyclopentenone (**214**) with base (Scheme 48). Cyclopentenones can be formed directly by the reaction of furan-2-carboxaldehydes with arylamines in the absence of acid.[185]

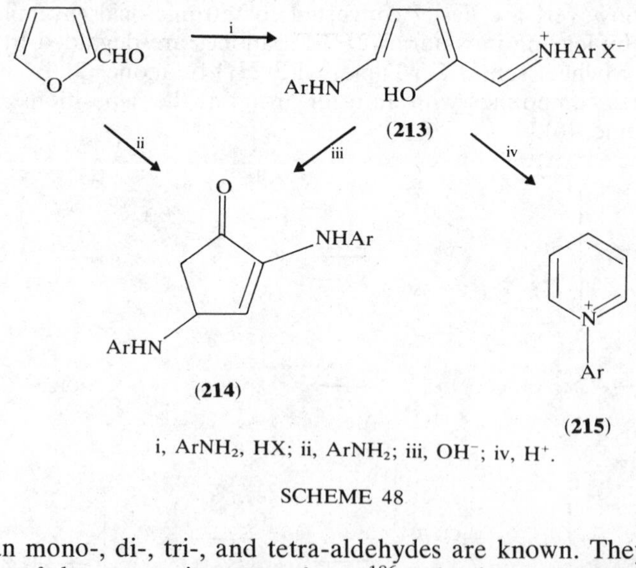

(213) (214) (215)

i, $ArNH_2$, HX; ii, $ArNH_2$; iii, OH^-; iv, H^+.

SCHEME 48

All nine furan mono-, di-, tri-, and tetra-aldehydes are known. They can be prepared by the reduction of the appropriate cyanofuran,[186] reduction and subsequent oxidation of the appropriate ester,[187] or from halofurans by lithiation and treatment of the lithiofurans with *N,N*-dimethylformamide.[188] Furan-2,5-dicarboxaldehyde is conveniently prepared from sucrose.[189] Formylation of alkylfurans by the Vilsmeir–Haack reaction is also a convenient synthesis of furan aldehydes.

2-Furylketones can be prepared by Friedel–Crafts acylation of furans (Section 18.4.4.6) or from derivatives of the furan-2-carboxylic acids such as the nitriles, amides, and chlorides by standard reactions. These last-mentioned methods are also applicable to 3-furyl ketones. The reactions of simple furyl ketones in general parallel those of the aldehydes. Interconversion between the acetyl methyl group and the 2-alkyl substituent of 2-alkyl-3-acetylfurans occurs at high temperature and a radical mechanism has been proposed.[190] 2-Acetylfuran forms enamines (**216**), which on heating yield *o*-dialkylaminophenols (**217**) (Scheme 49).[191]

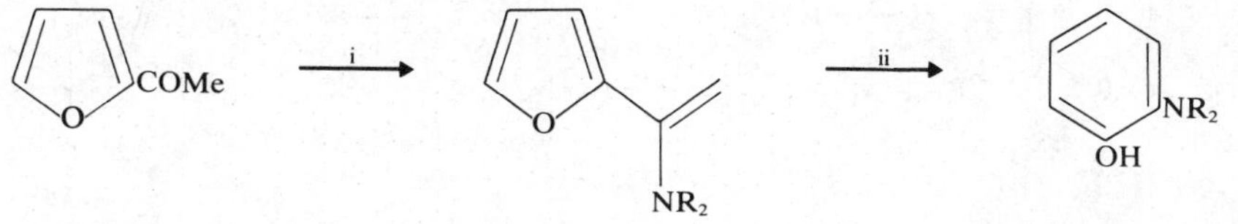

COMe (**216**) NR_2 (**217**) NR_2 OH

i, NHR_2, AcOH; ii, heat.

SCHEME 49

18.4.22 CARBOXYLIC ACIDS

Furan carboxylic acids are usually prepared by ring synthesis using the Feist–Benary or Paal–Knorr methods. Reaction of lithiofurans with carbon dioxide is also useful. Furan-2-carboxylic acid (pK_a 3.6) is a much stronger acid than its 3-isomer (pK_a 4.5) because of the inductive effect of the oxygen atom. Similarly, the 2-carboxylic acids are much more readily decarboxylated than the 3-carboxylic acids, a striking example being the stepwise preparation of furan-3-carboxylic acid from the tetracarboxylic acid.[192] A more convenient synthesis of this compound is from the commercially available ethyl furan-3,4-dicarboxylate (Scheme 50).[193]

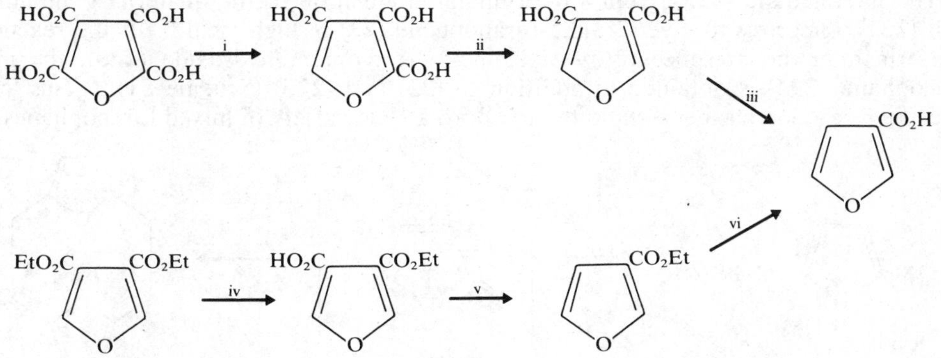

i, 270 °C; ii, 280 °C; iii, Cu, quinoline, heat; iv, NaOH, EtOH, 0 °C; v, Cu, heat; vi, NaOH, EtOH, heat.

SCHEME 50

18.4.23 SYSTEMS WITH MORE THAN ONE FURAN NUCLEUS

Ullmann reaction of 2-halofurans gives symmetrical 2,2′-bifuryls, while the unsymmetrical compounds are prepared by Feist–Benary synthesis of the second ring.[194] Copper(II) chloride induces coupling of lithiofurans to 2,2′-bifuryls (Section 18.4.13). 3,3′-Bifuryl is prepared by reaction of tetrahydro-3-furanone with 3-lithiofuran and subsequent dehydrogenation of the product.[195]

Under acidic conditions furan reacts with aldehydes and ketones to produce difurylmethanes and higher homologues. With aldehydes the difurylmethane (**218**) is usually the major product,[196] whilst with ketones tetramerization occurs and results in the formation of quaterenes (**219**).[197] The condensation of furan with acetone to give the quaterene (**220**) shows a marked 'template effect' in the presence of metal salts. With lithium perchlorate as the template agent the yield of (**220**) is raised to 40–45%.[198] A number of furan-containing crown ethers have been prepared.[199]

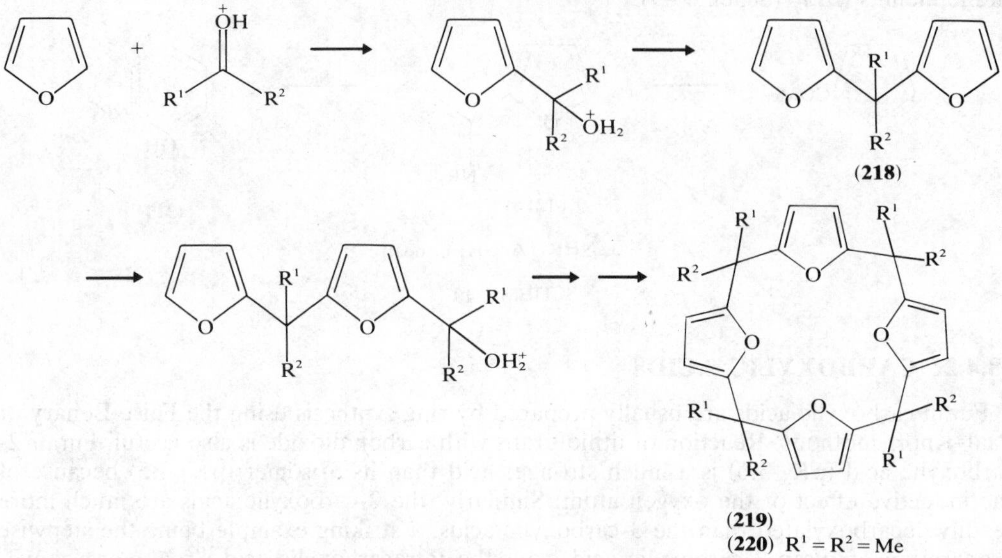

(**218**)

(**219**)
(**220**) $R^1 = R^2 = Me$

Furilic acid, originally claimed to have been isolated by treatment of furil with base, appears to be unstable under the reaction conditions.[200]

The intermediate (**222**), from 1,6-Hofmann elimination of the quaternary ammonium salt (**221**), dimerizes to give [2,2]2,5-furanophane (**23**) in high yield.[201] If the reaction is performed in the presence of *p*-xylyltrimethylammonium hydroxide (**223**), the mixed cyclophane (**224**) is obtained in addition to (**23**) and (**225**) (Scheme 51).[74] This 'cross-breeding' reaction has been exploited to afford a wide variety of mixed furanophanes, and

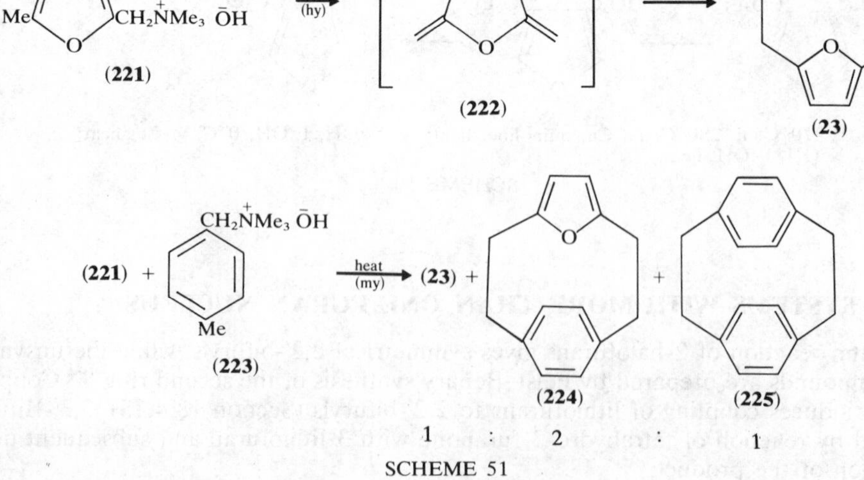

(**221**) (**222**) (**23**)

(**221**) + (**223**) heat (my) → (**23**) + (**224**) + (**225**)

1 : 2 : 1

SCHEME 51

their chemistry has been studied largely in relation to conformational effects and to the properties resultant from the forced proximity of π-electron systems.[202]

A number of cyclic conjugated systems containing the furan ring have been studied. The diatropic [18]annulene analogues (**226**) and (**227**) are synthesized by a route based on a Perkin reaction.[203] The self-Wittig reaction of the phosphobium salt (**228**) also yields (**226**) and higher vinylogues such as (**229**) (Scheme 52).[204] Porphin analogues containing furan rings have also been synthesized.[205]

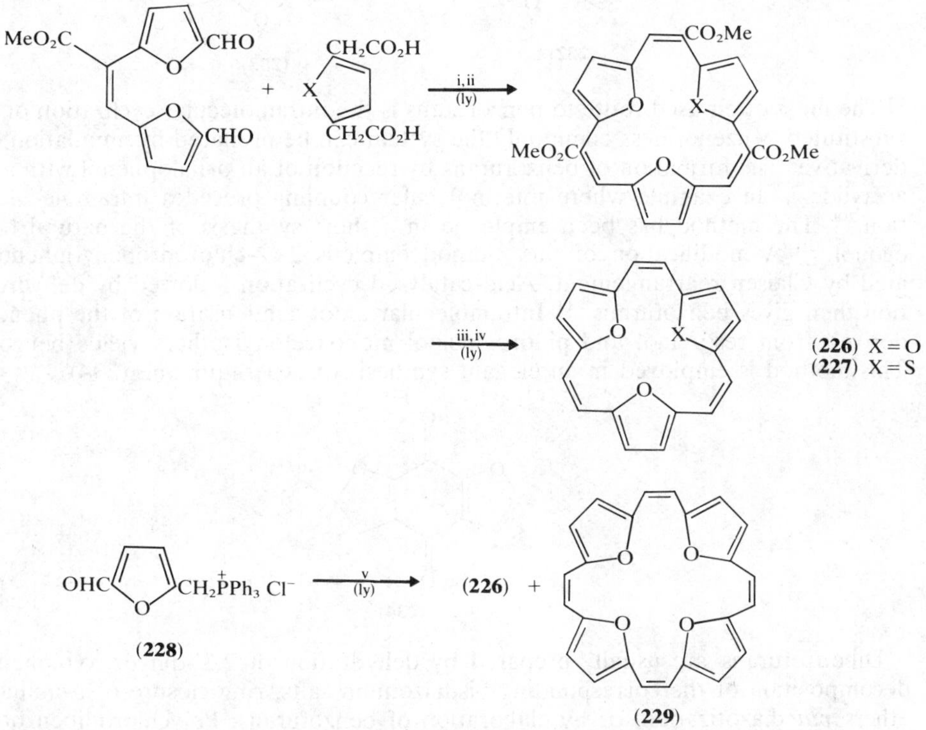

(**226**) X = O
(**227**) X = S

(**228**)

(**229**)

i, Ac_2O, Et_3N, ii, MeOH, HCl; iii, KOH, aq. EtOH; iv, $Cu_2Cr_2O_4$, quinoline, heat; v, LiOEt, Me_2NCHO.

SCHEME 52

A number of annulenones containing the furan ring have also been synthesized.[206] Thus the paratropic [17]annulenones (**230**) and (**231**) are obtained by Wittig reaction (Scheme 53); higher members of the series have also been synthesized.

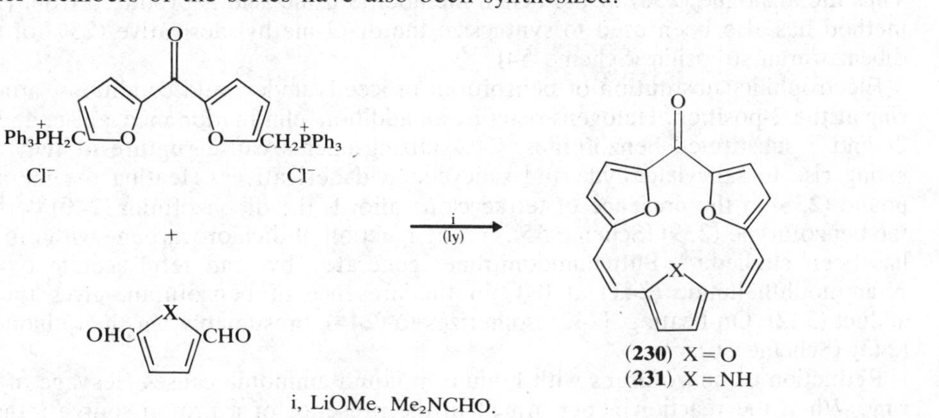

(**230**) X = O
(**231**) X = NH

i, LiOMe, Me_2NCHO.

SCHEME 53

18.4.24 BENZO[*b*]FURAN AND DIBENZOFURAN

The chemistry of benzo[*b*]furan, commonly known as benzofuran (**232**), has been comprehensively reviewed,[207] but that of dibenzofuran (**233**) has received scant treatment.[208]

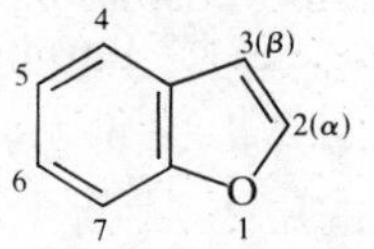

(**232**) 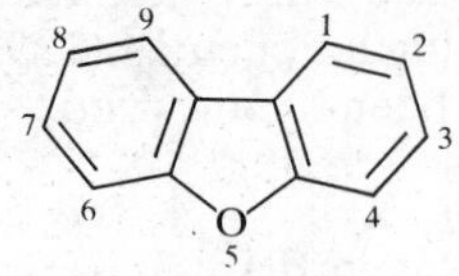 (**233**)

The most often used route to benzofurans is the intramolecular cyclization of a suitably substituted benzene. Less commonly the system can be prepared by annulation of a furan derivative. The formation of benzofurans by reaction of an *o*-halophenol with a copper(I) acetylide is an example where intermolecular coupling precedes intramolecular cyclization.[209] The method has been employed in a short synthesis of the natural benzofuran egonol.[210] A modification of this method employs 2-(2-chloropropenyl)phenols generated by Claisen rearrangement. Acid-catalysed cyclization followed by dehydrochlorination then gives benzofurans.[211] Intramolecular aldol condensation of the phenoxide ions derived from resorcinol and phloroglucinol monoacetonyl ethers yields benzofurans.[212] This method is employed in an elegant synthesis of cyperaquinone (**234**).

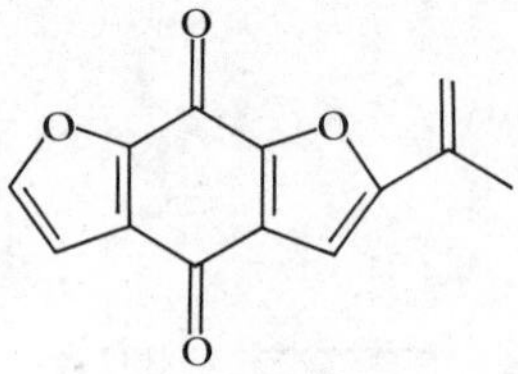

(**234**)

Dibenzofurans are usually prepared by dehydration of 2,2′-dihydroxybiphenyls,[213,214] decomposition of the corresponding bisdiazonium salts, ring closure of *o*-aminodiphenyl ethers *via* diazotization, or by elaboration of benzofurans. Polychlorodibenzofurans are extremely toxic trace atmospheric contaminants. To aid identification and to study structure–toxicity relationships their preparation by diazotization of chloro-*o*-phenoxyanilines and by photocyclization of chlorodiphenyl ethers has been described.[215] The preparation of substituted dibenzofurans by photocyclization of diphenyl ethers has been investigated systematically.[216] 2-Vinylbenzofurans undergo $[_{\pi}4_s + _{\pi}2_s]$ addition with dienophiles, providing a versatile route to dibenzofurans and 3,9b-dihydrobenzofurans. Thus the analogue (**236**) of the lichen metabolite usnic acid is produced from (**235**). This method has also been used to synthesize the di-*O*-methyl derivative (**237**) of the lichen dibenzofuran strepsilin (Scheme 54).[217]

Electrophilic substitution of benzofuran proceeds under mild conditions, attack occurring at the 2-position. Halogens react by an addition–elimination mechanism and yield the 2- and 3-substituted benzofurans.[218] Oxidizing agents cause rupture of the furan ring, giving rise to salicylaldehyde and salicyclic acid derivatives. Heating the mercury compound (**238**) in the presence of tetracyclone affords the dibenzofuran (**240**) derived from the benzofuryne (**239**) (Scheme 55).[219] The reaction of dichlorocarbene with dibenzofuran has been studied.[220] Phthalimidonitrene, generated by lead tetra-acetate oxidation of *N*-aminophthalimide (**241**) at 0 °C in the presence of benzofuran, gives the unstable adduct (**242**). On heating, (**242**) isomerizes to (**244**), presumably *via* the quinone methide (**243**) (Scheme 56).[221]

Reduction of benzofurans with lithium in liquid ammonia causes cleavage of the furan ring. When the reaction is performed in the presence of a proton source (ethanol), the furan ring is reduced.[222] Dibenzofuran is reduced by sodium in liquid ammonia to

(235) (236)

(237)

i, $MeO_2CC{\equiv}CCO_2Me$, heat; ii, $BrCH_2CHO$, K_2CO_3, Me_2NCHO; iii, $Ph_3P{=}CHOMe$; iv, *N*-bromosuccinimide; v, $LiAlH_4$; vi, $Na_2Cr_2O_7$, H^+.

SCHEME 54

(238) (239)

(240)

i, $HgBr_2$; ii, tetraphenylcyclopentadienone.

SCHEME 55

(241) (242)

(243) (244)

I, $Pb(OAc)_4$, 0 °C; ii, PhMe, heat.

SCHEME 56

1,4-dihydrodibenzofuran.[223] With alkali metals in ethereal solvents, cleavage of the furan ring occurs *via* radical anion formation. Carboxylation of the intermediate from treatment of dibenzofuran with lithium yields 3,4-benzocoumarin.[224]

18.4.25 BENZO[*c*]FURAN

Benzo[*c*]furan, or isobenzofuran, numbered as in (**248**), is iso-π-electronic with naphthalene and might be assumed to be a stable molecule. It may also be expected to be extremely reactive since addition of dienophiles across the furan diene system generates a benzene ring. The 1,3-diarylbenzo[*c*]furans are stable molecules, in contrast to the parent system. They may be prepared by reaction of aryl Grignard reagents with 3-arylphthalides or with 2-formylbenzoyl chloride and subsequent dehydration.[225] Another convenient route to 1,3-diphenylisobenzofurans (**248**) involves the dehydration of 1,2-dibenzoylcyclohexa-1,4-dienes (**247**) available by addition of butadienes (**245**) to dibenzoylacetylene (**246**) (Scheme 57).[226] 2-(Bromomethyl)benzophenone on heating loses hydrogen bromide to afford 1-phenylisobenzofuran, which is detected by conducting the experiment in the presence of dienophiles.[227] The carbene generated by decomposition of

(245) (246) (247) (248) (ly–my)

R^1,R^2 = H, Me, Ph

i, heat; ii, $MeC_6H_4SO_3H$, PhH, heat.

SCHEME 57

methyl *o*-diazomethylbenzoate undergoes intramolecular carbene–carbonyl reaction to give 1-methoxyisobenzofuran, which can be trapped with dienophiles or alcohols.[228]

1,3-Diphenylisobenzofuran behaves as a diene in cycloaddition reactions, *e.g.* with cyclopropenone,[229] 3-methyl-2-phenyl-1-azirine,[230] oxygen,[231] and benzocyclobutadiene.[232]

The adduct (**250**), from α-pyrone and naphthalene oxide (**249**), on heating in the presence of (**249**) affords the adducts (**251**) and (**252**), in addition to benzene and carbon dioxide.[233] That the products arise by chelotropic loss of carbon dioxide from (**250**) followed by retro-Diels–Alder reaction of (**253**) to isobenzofuran (**254**), which is trapped by the dienophile (**249**), is shown by the isolation of both (**253**) and (**254**).[234] Vacuum pyrolysis of the adduct (**255**) also affords isobenzofuran (**254**) (Scheme 58).[235] The driving

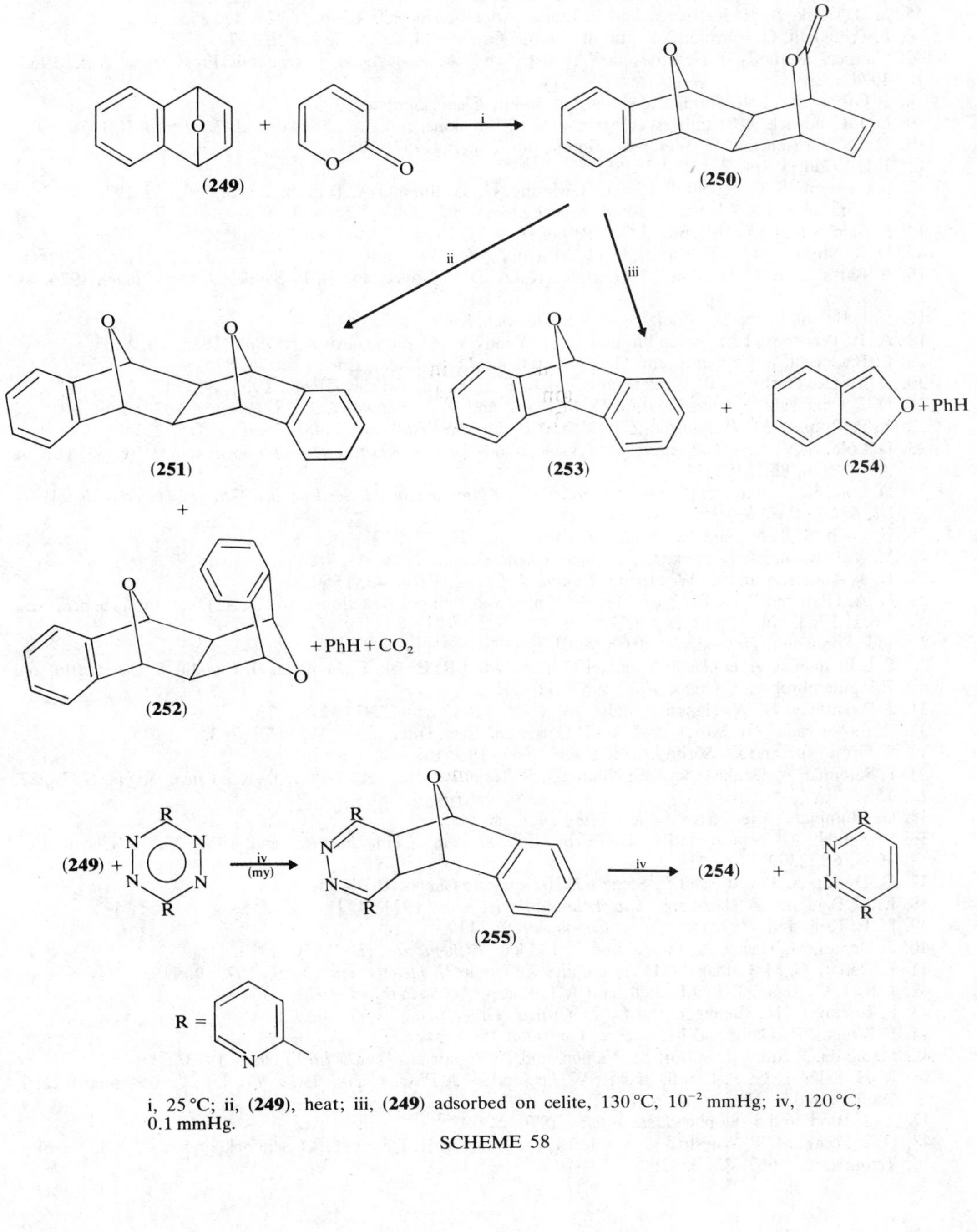

i, 25 °C; ii, (**249**), heat; iii, (**249**) adsorbed on celite, 130 °C, 10^{-2} mmHg; iv, 120 °C, 0.1 mmHg.

SCHEME 58

force in the retro-Diels–Alder reaction in both these preparations of (**254**) is the extrusion of an aromatic species. That this expediency is unnecessary has been elegantly shown by the vacuum pyrolysis of (**249**), which affords ethylene and isobenzofuran in very high yield.[236] Isobenzofuran is a low melting solid which is stable in dilute solution in the absence of oxygen.

References

1. A. P. Dunlop and F. N. Peters, 'The Furans', Reinhold, New York, 1953.
2. P. Bosshard and C. H. Eugster, *Adv. Heterocyclic Chem.*, 1966, **7,** 377.
3. T. J. Barton, R. W. Roth, and J. G. Verkade, *J. Amer. Chem. Soc.*, 1972, **94,** 8854.
4. G. Marino, *Adv. Heterocyclic Chem.*, 1971, **13,** 235.
5. M. J. Cook, A. R. Katritzky, and P. Linda, *Adv. Heterocyclic Chem.*, 1974, **17,** 255.
6. F. Fringuelli, G. Marino, A. Taticchi, and G. Grandolini, *J.C.S. Perkin II*, 1974, 332.
7. 'Physical Methods in Heterocyclic Chemistry', ed. A. R. Katritzky, Academic Press, New York, 1963–1974.
8. S. Gronowitz, I. Johnson, and A.-B. Hornfeldt, *Chem. Scripta*, 1975, **7,** 211.
9. D. J. Chadwick, J. Chambers, R. Macrae, G. D. Meakins, and R. L. Snowden, *J.C.S. Perkin II*, 1976, 597.
10. U. E. Wiersum and H. Wynberg, *Tetrahedron Letters*, 1967, 2951.
11. J. H. Ridd in Ref. 7, vol. 4, p. 55.
12. S. Clementi, F. Fringuelli, P. Linda, G. Marino, G. Savelli, and A. Taticchi, *J.C.S. Perkin II*, 1973, 2097.
13. G. Ciranni and S. Clementi, *Tetrahedron Letters*, 1971, 3833.
14. S. Clementi and G. Marino, *J.C.S. Perkin II*, 1972, 71.
15. D. S. Noyce and G. V. Kaiser, *J. Org. Chem.*, 1969, **34,** 1008.
16. B. Bartman, E. C. Gordon, M. Gonzalez-Kutas, D. S. Noyce, and B. B. Sandel, *J. Org. Chem.*, 1976, **41,** 776.
17. L. I. Belen'kii, *Izvest. Akad. Nauk S.S.S.R., Ser. Khim.*, 1975, 344.
18. A. P. Terent'ev, L. I. Belen'kii, and L. A. Yanovskaya, *Zhur. obshchei Khim.*, 1955, **24,** 1265.
19. E. Baciocchi, S. Clementi, and G. V. Sebastiani, *J.C.S. Chem. Comm.*, 1975, 875.
20. Ja. L. Gol'dfarb, Ju. B. Volkenstein, and L. I. Belen'kii, *Angew. Chem.*, 1968, **80,** 547.
21. D. J. Chadwick, J. Chambers, G. D. Meakins, and R. L. Snowden, *J.C.S. Perkin I*, 1973, 1766.
22. B.-P. Roques, M.-C. Fournie-Zaluski, and R. Oberlin, *Bull. Soc. chim. France*, 1975, 2334.
23. D. Lola, K. Venters, E. Liepins, M. Trusule, and S. Hillers, *Khim. geterotsikl. Soedinenii*, 1976, 601 (*Chem. Abs.*, 1976, **85,** 77957).
24. D. Lola, K. Venters, E. Liepins, and S. Hillers. *Khim. geterotsikl. Soedinenii*, 1975, 883 (*Chem. Abs.*, 1976, **84,** 4749).
25. G. Olah, S. Kuhn, and A. Mlinko, *J. Chem. Soc.*, 1956, 4257.
26. M. Valenta and I. Koubek, *Coll. Czech. Chem. Comm.*, 1976, **41,** 78.
27. H. J. Anderson and C. W. Huang, *Canad. J. Chem.*, 1970, **48,** 1550.
28. G. A. Olah and W. S. Tolgyesi, 'Friedel Crafts and Related Reactions', ed. G. A. Olah, Interscience, New York, 1964, vol. 2, part 2, p. 671.
29. S. I. Pennanen, *Heterocycles*, 1976, **4,** 1021.
30. S. I. Pennanen, *Acta Chem. Scand.*, 1972, **26,** 2907; R. Ercoli, E. Mantica, G. Claudia, S. Chiozzotto, and E. Santambrogio, *J. Org. Chem.*, 1967, **32,** 2917.
31. J. P. Kutney, H. W. Hanssen, and G. V. Nair, *Tetrahedron*, 1971, **27,** 3323.
32. P. E. Verkade, Th. Morel, and H. G. Gerritsen, *Rec. Trav. chim.*, 1955, **74,** 763.
33. S. Gronowitz and G. Sörlin, *Arkiv Kemi*, 1962, **19,** 515.
34. P. Salomaa, A. Kankaanperä, E. Nikander, K. Kaipainen, and R. Aaltonen, *Acta Chem. Scand.*, 1973, **27,** 153.
35. G. Illuminati, *Adv. Heterocyclic Chem.*, 1964, **3,** 285.
36. D. Spinelli, G. Guanti, and C. Dell'Erba, *Boll. Sci. Fac. Chim. Ind. Bologna*, 1967, **25,** 71 (*Chem. Abs.*, 1968, **68,** 2389).
37. G. Doddi, A. Poretti, and F. Stegel, *J. Heterocyclic Chem.*, 1974, **11,** 97.
38. K. C. Bass and P. Nabasing, *Adv. Free Radical Chem.*, 1971, **4,** 1.
39. K. E. Kolb and W. Black, *Chem. Comm.*, 1969, 1119.
40. L. Benati, M. Tiecco, A. Tundo, and F. Taddei, *J. Chem. Soc.* (*B*), 1970, 1443.
41. L. Benati, C. M. Camaggi, M. Tiecco, and A. Tundo, *J. Heterocyclic Chem.*, 1972, **9,** 919.
42. J. I. G. Cadogan, J. R. Mitchell, and J. T. Sharp, *J.C.S. Perkin I*, 1972, 1304.
43. L. Benati, C. M. Camaggi, and G. Zanardi, *J. Org. Chem.*, 1975, **40,** 966.
44. F. Minisci, R. Galli, and M. Cecere, *Gazzetta*, 1964, **94,** 67.
45. M. Janda, J. Srogl, I. Stibor, M. Nemec, and P. Vopatrna, *Tetrahedron Letters*, 1973, 637.
46. R. H. Schuler, G. P. Laroff, and R. W. Fessenden, *J. Phys. Chem.*, 1973, **77,** 456; C. L. Greenstock, I. Dunlop, and P. Neta, *ibid.*, p. 1187.
47. A. J. Birch and J. Slobbe, *Heterocycles*, 1976, **5,** 905.
48. C. S. Foote, M. T. Wuestoff, S. Wexler, I. G. Burstain, R. Denny, G. O. Schenck, and K.-H. Schulte-Elte, *Tetrahedron*, 1967, **23,** 2583.

49. H. H. Wasserman and R. Kitzing, *Tetrahedron Letters*, 1969, 5315; T. J. Katz, V. Balogh, and J. Schulman, *J. Amer. Chem. Soc.*, 1968, **90,** 734.
50. N. L. Weinberg and H. R. Weinberg, *Chem. Rev.*, 1968, **68,** 449; N. Elming, *Adv. Org. Chem.*, 1960, **2,** 67.
51. A. J. Baggaley and R. Brettle, *J. Chem. Soc. (C)*, 1968, 969.
52. K. Yoshida and T. Fueno, *J. Org. Chem.*, 1971, **36,** 1523; J. Froborg, G. Magnusson, and S. Thoren, *ibid.*, 1975, **40,** 122.
53. N. A. Milas, R. L. Peeler, Jr., and O. L. Mageli, *J. Amer. Chem. Soc.*, 1954, **76,** 2322.
54. D. Seebach, *Chem. Ber.*, 1963, **96,** 2712.
55. A. I. Meyers, 'Heterocycles in Organic Synthesis', Wiley-Interscience, New York, 1974, p. 222.
56. E. J. Stamhuis, W. Drenth, and H. van den Berg, *Rec. Trav. chim.*, 1964, **83,** 167.
57. G. Büchi and H. Wüest, *J. Org. Chem.*, 1966, **31,** 977.
58. W. S. Johnson, M. B. Gravestock, and B. E. McCarry, *J. Amer. Chem. Soc.*, 1971, **93,** 4332.
59. K. C. Rice and J. R. Dyer, *J. Heterocyclic Chem.*, 1975, **12,** 1325.
60. A. J. Birch and G. S. R. Subba Rao, *Austral. J. Chem.*, 1970, **23,** 547.
61. H. C. van der Plas, 'Ring Transformations of Heterocycles', Academic Press, London, 1973.
62. A. S. Onishchenko, 'Diene Synthesis', Israel Program for Scientific Translations, Jerusalem, 1964.
63. F. A. L. Anet, *Tetrahedron Letters*, 1962, 1219.
64. R. W. La Rochelle and B. M. Trost, *Chem. Comm.*, 1970, 1353.
65. D. C. F. Law and S. W. Tobey, *J. Amer. Chem. Soc.*, 1968, **90,** 2376.
66. H.-D. Scharf, P. Friedrich, and A. Linckens, *Synthesis*, 1976, 256.
67. T. A. Eggelte, H. de Koning, and H. O. Huisman, *Tetrahedron*, 1973, **29,** 2491.
68. W. G. Dauben and H. O. Krabbenhoft, *J. Amer. Chem. Soc.*, 1976, **98,** 1992.
69. W. S. Wilson and R. N. Warrener, *J.C.S. Chem. Comm.*, 1972, 211.
70. H. Prinzbach, D. Stusche, M. Breuninger, and J. Markert, *Chem. Ber.*, 1976, **109,** 2823.
71. A. W. McCulloch and A. G. McInnes, *Canad. J. Chem.*, 1974, **52,** 143.
72. P. Vogel, B. Willhalm, and H. Prinzbach, *Helv. Chim. Acta*, 1969, **52,** 584.
73. A. W. McCulloch, D. G. Smith, and A. G. McInnes, *Canad. J. Chem.*, 1973, **51,** 4125; 1974, **52,** 1013; A. W. McCulloch and A. G. McInnes, *ibid.*, 1975, **53,** 1496.
74. D. J. Cram and G. R. Knox, *J. Amer. Chem. Soc.*, 1961, **83,** 2204.
75. E. L. Ghisalberti, P. R. Jefferies, and T. G. Payne, *Tetrahedron*, 1974, **30,** 3099.
76. P. Caramella, G. Cellerino, A. C. Coda, A. G. Invernizzi, P. Grünanger, K. N. Houk, and F. M. Albini, *J. Org. Chem.*, 1976, **41,** 3349.
77. G. M. L. Cragg, R. G. F. Giles, and G. H. P. Roos, *J.C.S. Perkin I*, 1975, 1339.
78. L. A. Kapicak and M. A. Battiste, *J.C.S. Chem. Comm.*, 1973, 930.
79. A. H. Alberts, J. Strating, and H. Wynberg, *Tetrahedron Letters*, 1973, 3047.
80. N. J. Turro, S. S. Edelson, J. R. Williams, T. R. Darling, and W. B. Hammond, *J. Amer. Chem. Soc.*, 1969, **91,** 2283.
81. R. Noyori, S. Makino, and H. Takaya, *Tetrahedron Letters*, 1973, 1745; S. Ito, H. Ohtani, and S. Amiya, *ibid.*, 1973, 1737.
82. J. G. Vinter and H. M. R. Hoffmann, *J. Amer. Chem. Soc.*, 1973, **95,** 3051; 1974, **96,** 5466.
83. J. C. Berridge, D. Bryce-Smith, A. Gilbert, and T. S. Cantrell, *J.C.S. Chem. Comm.*, 1975, 611.
84. K. Mizuno, C. Pac, and H. Sakurai, *J.C.S. Perkin I*, 1974, 2360.
85. K. Shima and H. Sakurai, *Bull. Chem. Soc. Japan*, 1966, **39,** 1806; J. Leitich, *Tetrahedron Letters*, 1967, 1937; G. R. Evanega and E. B. Whipple, *ibid.*, 1967, 2163.
86. H. Hiraoka, *J. Phys. Chem.*, 1970, **74,** 574; S. Boué and R. Srinivasan, *J. Amer. Chem. Soc.*, 1970, **92,** 1824.
87. E. E. van Tamelen and T. H. Whitesides, *J. Amer. Chem. Soc.*, 1971, **93,** 6129.
88. H. Hiraoka, *Tetrahedron*, 1973, **29,** 2955.
89. A. Couture, A. Devallee, and A. Lablache-Combier, *Tetrahedron*, 1975, **31,** 785.
90. P. Scribe, C. Nouet, and J. Wiemann, *Tetrahedron Letters*, 1970, 4375.
91. R. Hunt and S. T. Reid, *J.C.S. Perkin I*, 1972, 2527.
92. E. Müller, H. Kessler, H. Fricke, and H. Suhr, *Tetrahedron Letters*, 1963, 1047.
93. G. O. Schenck and R. Steinmetz, *Annalen*, 1963, **668,** 19.
94. O. M. Nefedov, V. M. Shostakovskii, M. Ya. Samoilova, and M. I. Kravchenko, *Izvest. Akad. Nauk S.S.S.R., Ser. Khim.*, 1971, 1590; 1972, 2342.
95. M. Franck-Neumann and C. Buchecker, *Tetrahedron Letters*, 1973, 2875; *Angew. Chem.*, 1970, **82,** 549.
96. M. N. Nwaji and O. S. Onyiriuka, *Tetrahedron Letters*, 1974, 2255.
97. R. V. Hoffman and H. Shechter, *J. Amer. Chem. Soc.*, 1971, **93,** 5940.
98. T. L. Gilchrist and D. P. J. Pearson, *J.C.S. Perkin I*, 1976, 1257.
99. D. W. Jones, *J.C.S. Perkin I*, 1972, 2728.
100. R. A. Bell, M. B. Gravestock, and V. Y. Taguchi, *Canad. J. Chem.*, 1972, **50,** 3749.
101. G. Büchi, E. Sz. Kovats, P. Enggist, and G. Uhde, *J. Org. Chem.*, 1968, **33,** 1227.
102. V. Ramanathan and R. Levine, *J. Org. Chem.*, 1962, **27,** 1216.
103. G. M. Davies and P. S. Davies, *Tetrahedron Letters*, 1972, 3507.
104. T. Kauffmann, B. Greving, J. König, A. Mitschker, and A. Woltermann, *Angew. Chem.*, 1975, **87,** 745.
105. C. Ullenius, *Acta Chem. Scand.*, 1972, **26,** 3383; D. W. Knight and G. Pattenden, *J.C.S. Perkin I*, 1975, 641.

106. M. S. Feather, D. W. Harris, and S. B. Nichols, *J. Org. Chem.*, 1972, **37,** 1606.
107. C. H. Fawcett, *et al.*, *J. Chem. Soc.* (*C*), 1968, 2455.
108. F. García González, *Adv. Carbohydrate Chem.*, 1956, **11,** 97.
109. M. Selim-Dorgans, M. Selim, and H. Gault, *Compt. rend.*, 1957, **244,** 1047.
110. L. D. Krasnoslobodskaya and Ya. L. Gol'dfarb, *Russ. Chem. Rev.*, 1969, **38,** 389.
111. E. C. Kornfeld and R. G. Jones, *J. Org. Chem.*, 1954, **19,** 1671.
112. C. Botteghi, L. Lardicci, and R. Menicagli, *J. Org. Chem.*, 1973, **38,** 2361.
113. E. J. Nienhouse, R. M. Irwin, and G. R. Finni, *J. Amer. Chem. Soc.*, 1967, **89,** 4557.
114. E. Bisagni and C. Rivalle, *Bull. Soc. chim. France*, 1974, 519.
115. R. Lakhan and B. Ternai, *Adv. Heterocyclic Chem.*, 1974, **17,** 99.
116. S. W. Pelletier, Z. Djarmati, S. D. Lajšić, I. V. Mićović, and D. T. C. Yang, *Tetrahedron*, 1975, **31,** 1659.
117. P. A. Grieco, C. S. Pogonowski, and S. Burke, *J. Org. Chem.*, 1975, **40,** 542.
118. M. A. Gianturco and P. Friedel, *Canad. J. Chem.*, 1966, **44,** 1083.
119. R. N. Lacey, *J. Chem. Soc.*, 1954, 816, 822.
120. F. Korte and K. H. Büchel, *Angew. Chem.*, 1959, **71,** 709; F. Korte, R. Heinz, and D. Scharf, *Chem. Ber.*, 1961, **94,** 825.
121. M. I. Komendantov, I. N. Domnin, and E. V. Bulucheva, *Tetrahedron*, 1975, **31,** 2495.
122. M. E. Hendrick, *J. Amer. Chem. Soc.*, 1971, **93,** 6337.
123. J. W. Batty, P. D. Howes, and C. J. M. Stirling, *J.C.S. Perkin I*, 1973, 65.
124. P. H. M. Schreurs, W. G. Galesloot, and L. Brandsma, *Rec. Trav. chim.*, 1975, **94,** 70.
125. R. Fuks and H. G. Viehe, 'Chemistry of Acetylenes', ed. H. G. Viehe, Dekker, New York, 1969, p. 425.
126. D. Miller, *J. Chem. Soc.* (*C*), 1969, 12.
127. S. Holand, F. Mercier, N. Le Goff, and R. Epsztein, *Bull. Soc. chim. France*, 1972, 4357.
128. M. E. Garst and T. A. Spencer, *J. Amer. Chem. Soc.*, 1973, **95,** 250.
129. T. M. Harris, C. M. Harris, and J. C. Cleary, *Tetrahedron Letters*, 1968, 1427.
130. M. Higo and T. Mukaiyama, *Tetrahedron Letters*, 1970, 2565.
131. T. Takaku, Y. Hayasi, and H. Nozaki, *Tetrahedron Letters*, 1969, 2053.
132. M. E. Garst and T. A. Spencer, *J. Org. Chem.*, 1974, **39,** 584.
133. *U.S. Pat.*, 39712763 (*Chem. Abs.*, 1976, **84,** 4807); *U.S. Pat.*, 3894056 (*Chem. Abs.*, 1975, **83,** 163984).
134. G. H. Suverkropp and H. Wynberg, *Chem. Ind.* (*London*), 1963, 1803.
135. P. H. Williams, G. B. Payne, W. J. Sullivan, and P. R. Van Ess, *J. Amer. Chem. Soc.*, 1960, **82,** 4883.
136. H. L. Rice, *J. Amer. Chem. Soc.*, 1952, **74,** 3193.
137. H. Wynberg and U. E. Wiersum, *Tetrahedron Letters*, 1975, 3619.
138. L. T. Scott and J. O. Naples, *Synthesis*, 1973, 209.
139. G. Büchi and H. Wüest, *J. Org. Chem.*, 1969, **34,** 857.
140. H. Minato and T. Nagasaki, *J. Chem. Soc.* (*C*), 1966, 1866.
141. Y. Fukuyama, T. Tokoroyama, and T. Kubota, *Tetrahedron Letters*, 1973, 4869.
142. S. Divald, M. C. Chun, and M. M. Jouillié, *J. Org. Chem.*, 1976, **41,** 2835.
143. K. Yu. Novitskii, Yu. K. Yur'ev, and V. N. Zhingareva, *Zhur. obshchei Khim.*, 1962, **32,** 3303.
144. 'Methoden der Organischen Chemie (Houben-Weyl),' Thieme, Stuttgart, 1965, vol. 6/3.
145. R. Paul, H. Fluchaire, and G. Collardeau, *Bull. Soc. chim. France*, 1950, 668.
146. M. A. Gianturco, P. Friedel, and V. Flanagan, *Tetrahedron Letters*, 1965, 1847.
147. W. G. Dauben and D. J. Hart, *Tetrahedron Letters*, 1975, 4353.
148. J. C. Anderson, D. G. Lindsay, and C. B. Reese, *Tetrahedron*, 1964, **20,** 2091.
149. W. Reppe, *Annalen*, 1955, **596,** 80; N. O. Brace, *J. Amer. Chem. Soc.*, 1955, **77,** 4157.
150. E. E. Schweizer and J. G. Liehr, *J. Org. Chem.*, 1968, **33,** 583.
151. R. G. Salomon, M. F. Salomon, and T. R. Heyne, *J. Org. Chem.*, 1975, **40,** 756.
152. B. C. L. Weedon, in 'Carotenoids', ed. O. Isler, Birkhäuser, Basel, 1971, p. 42.
153. A. W. Burgstahler and G. N. Widiger, *J. Org. Chem.*, 1973, **38,** 3652.
154. F. Bohlmann and Ch. Zdero, *Chem. Ber.*, 1973, **106,** 845.
155. M. D. Grove, D. Weisleder, and M. E. Daxenbichler, *Tetrahedron*, 1973, **29,** 2715.
156. U. Schmidt, J. Gombos, E. Haslinger, and H. Zak, *Chem. Ber.*, 1976, **109,** 2628.
157. Z. N. Nazarova, *Zhur. org. Khim.*, 1974, **10,** 1341.
158. M.-C. Zaluski, M. Robba, and M. Bonhomme, *Bull. Soc. chim. France*, 1970, 1838.
159. J. Burdon, J. C. Tatlow, and D. F. Thomas, *Chem. Comm.*, 1966, 48.
160. J. Reisch and H. Labitzke, *Arch. Pharm.*, 1975, **308,** 713.
161. G. V. Boyd and K. Heatherington, *J.C.S. Perkin I*, 1973, 2523.
162. F. Ebenito, *Heterocycles*, 1974, **2,** 391.
163. W. Ried and W. Bodenstedt, *Annalen*, 1963, **667,** 96.
164. K. Gewald, *Chem. Ber.*, 1966, **99,** 1002.
165. J. L. Isidor, M. S. Brookhart, and R. L. McKee, *J. Org. Chem.*, 1973, **38,** 612.
166. C. T. Wie, S. Sunder, and C. D. Blanton, Jr., *Tetrahedron Letters*, 1968, 4605.
167. Z. N. Nazarova and V. N. Novikov, *Zhur. obshchei Khim.*, 1961, **31,** 263.
168. Z. N. Nazarova, V. N. Novikov, and V. Ts. Bukhaeva, *Zhur. obshchei Khim.*, 1964, **34,** 705.
169. T. Severin and H. Kullmer, *Chem. Ber.*, 1973, **106,** 1688.
170. K. Miura and H. K. Reckendorf, *Progr. Medicin. Chem.*, 1967, **5,** 320.
171. Y. S. Rao, *Chem. Rev.*, 1976, **76,** 625.

172. D. G. Manly and E. D. Amstutz, *J. Org. Chem.*, 1956, **21,** 516.
173. F. M. Dean, 'Naturally Occurring Oxygen Ring Compounds', Butterworths, London, 1963.
174. C. F. Culberson, 'Chemical and Botanical Guide to Lichen Products', University of North Carolina Press, Chapel Hill, 1969.
175. P. A. Grieco, *Synthesis*, 1975, 67.
176. F. Sondheimer, *Chem. Britain*, 1965, **1,** 454.
177. A.-B. Hörnfeldt, *Arkiv Kemi*, 1969, **29,** 229.
178. B. Cederlund, A. Jesperson, and A.-B. Hörnfeldt, *Acta Chem. Scand.*, 1971, **25,** 3656.
179. D. W. Henry and R. M. Silverstein, *J. Org. Chem.*, 1966, **31,** 2391.
180. S. Wilkinson, *Quart. Rev.*, 1961, **15,** 153.
181. S. Gelin and A. Galliand, *Compt. rend.*, 1972, **275C,** 897; T. P. C. Mulholland, R. Foster, and D. B. Haydock, *J.C.S. Perkin I*, 1972, 1225.
182. G. Büchi and E. Demole, *J. Org. Chem.*, 1973, **38,** 123; L. Re, B. Maurer, and G. Ohloff, *Helv. Chim. Acta*, 1973, **56,** 1882.
183. J. Elguero, C. Marzin, A. R. Katritzky, and P. Linda, *Adv. Heterocyclic Chem.*, 1976, Suppl. 1, 214.
184. D. A. Isăcescu, I. Gavăt, C. Stoicescu, and M. Sterescu, *Bull. Soc. chim. France*, 1967, 2171.
185. K. G. Lewis and C. E. Mulquiney, *Austral. J. Chem.*, 1970, **23,** 2315.
186. M.-C. Zaluski, M. Robba, and M. Bonhomme, *Bull. Soc. chim. France*, 1970, 1445.
187. T. M. Cresp and F. Sondheimer, *J. Amer. Chem. Soc.*, 1977, **99,** 194.
188. M.-C. Zaluski, M. Robba, and M. Bonhomme, *Bull. Soc. chim. France*, 1970, 1838.
189. G. Drechsler and K. Kopperschlaeger, *East Ger. Pat.*, 26 542 (*Chem. Abs.*, 1964, **61,** 4315).
190. M. Tada and T. Takahashi, *Tetrahedron Letters*, 1973, 3999.
191. L. Birkofer and G. Daum, *Chem. Ber.*, 1962, **95,** 183.
192. T. Reichstein, A. Grüssner, K. Schindler, and E. Hardmeier, *Helv. Chim. Acta*, 1933, **16,** 276.
193. M. R. Boyd, T. M. Harris, and B. J. Wilson, *Synthesis*, 1971, 545.
194. R. Grigg, J. A. Knight, and M. V. Sargent, *J. Chem. Soc.* (*C*), 1966, 976.
195. H. Wynberg and J. W. van Reijendam, *Rec. Trav. chim.*, 1967, **86,** 381.
196. V. G. Glukhovstev and S. K. Zakharova, *Khim. geterotsikl. Soedinenii*, 1968, 947.
197. W. H. Brown and W. N. French, *Canad. J. Chem.*, 1958, **36,** 537.
198. M. Chastrette and F. Chastrette, *J.C.S. Chem. Comm.*, 1973, 534; A. J. Rest, S. A. Smith, and I. D. Tyler, *Inorg. Chim. Acta*, 1976, **16,** L1.
199. D. N. Reinhoudt, R. T. Gray, C. J. Smit, and M. I. Veenstra, *Tetrahedron*, 1976, **32,** 1161.
200. S. I. Pennanen, *Acta Chem. Scand.*, 1972, **26,** 1280.
201. H. E. Winberg, F. S. Fawcett, W. E. Mochel, and C. W. Theobald, *J. Amer. Chem. Soc.*, 1960, **82,** 1428.
202. M. Nakazaki, K. Yamamoto, and S. Tanaka, *J. Org. Chem.*, 1976, **41,** 4081; C. Wong and W. W. Paudler, *ibid.*, 1974, **39,** 2570; S. Mizogami, N. Osaka, and T. Otsubo, *Tetrahedron Letters*, 1974, 799; H. Wynberg and R. Helder, *ibid.*, 1971, 4317.
203. P. J. Garratt and M. V. Sargent, *Adv. Org. Chem.*, 1969, **6,** 50.
204. J. A. Elix, *Austral. J. Chem.*, 1969, **22,** 1951.
205. M. J. Broadhurst, R. Grigg, and A. W. Johnson, *J. Chem. Soc.* (*C*), 1971, 3681.
206. M. V. Sargent and T. M. Cresp, *Fortschr. Chem. Forsch.*, 1975, **57,** 111.
207. P. Cagniant and D. Cagniant, *Adv. Heterocyclic Chem.*, 1975, **18,** 337; A. Mustafa, 'Benzofurans', Wiley-Interscience, New York, 1974.
208. W. E. Parham, in 'Heterocyclic Compounds', ed. R. C. Elderfield, Wiley, New York, 1951, vol. 2, p. 123.
209. C. E. Castro, R. Havlen, V. K. Honwad, A. Malte, and S. Moje, *J. Amer. Chem. Soc.*, 1969, **91,** 6464.
210. F. G. Schreiber and R. Stevenson, *J.C.S. Perkin I*, 1976, 1515.
211. W. K. Anderson, E. J. La Voie, and J. C. Bottaro, *J.C.S. Perkin I*, 1976, 1.
212. J. K. MacLeod, B. R. Worth, and R. J. Wells, *Tetrahedron Letters*, 1972, 241.
213. B. G. Pring and N. E. Stjernström, *Acta Chem. Scand.*, 1968, **22,** 681.
214. N. H. Anderson, W. D. Ollis, J. G. Underwood, and R. M. Scrowston, *J. Chem. Soc.* (*C*), 1969, 2403.
215. A. P. Gray, V. M. Dipinto, and I. J. Solomon, *J. Org. Chem.*, 1976, **41,** 2428; A. Norstrom, K. Andersson, and C. Rappe, *Chemosphere*, 1976, 21.
216. J. A. Elix and D. P. Murphy, *Austral. J. Chem.*, 1975, **28,** 1559; K.-P. Zeller and H. Petersen, *Synthesis*, 1975, 532.
217. J. A. Elix and D. Tronson, *Austral. J. Chem.*, 1973, **26,** 1093; J. D. Brewer and J. A. Elix, *Tetrahedron Letters*, 1969, 4139.
218. E. Baciocchi, S. Clementi, and G. V. Sebastiani, *J.C.S. Perkin II*, 1976, 266.
219. G. Wittig, *Angew. Chem.*, 1962, **74,** 479.
220. W. E. Parham, C. G. Fritz, R. W. Soeder, and R. M. Dodson, *J. Org. Chem.*, 1963, **28,** 577.
221. D. W. Jones, *J.C.S. Perkin I*, 1972, 225.
222. S. D. Darling and K. D. Wills, *J. Org. Chem.*, 1967, **32,** 2794.
223. H. Gilman and C. W. Bradley, *J. Amer. Chem. Soc.*, 1938, **60,** 2333.
224. A. G. Evans, P. B. Roberts, and B. J. Tabner, *J. Chem. Soc.* (*B*), 1966, 269; H. Gilman and J. J. Dietrich, *J. Org. Chem.*, 1957, **22,** 851.
225. R. C. Elderfield, in 'Heterocyclic Compounds', ed. R. C. Elderfield, Wiley, New York, 1951, vol. 2, p. 68.
226. J. D. White, M. E. Mann, H. D. Kirshenbaum, and A. Mitra, *J. Org. Chem.*, 1971, **36,** 1048.
227. R. Faragher and T. L. Gilchrist, *J.C.S. Perkin I*, 1976, 336.

228. M. Hamaguchi and T. Ibata, *Chem. Letters*, 1976, 287.
229. M. Odo, R. Breslow, and J. Pecoraro, *Tetrahedron Letters*, 1972, 4419.
230. V. Nair, *J. Org. Chem.*, 1972, **37,** 2508.
231. F. Nahavandi, F. Razmara, and M. P. Stevens, *Tetrahedron Letters*, 1973, 301.
232. M. P. Cava and A. C. Hsu, *J. Amer. Chem. Soc.*, 1972, **94,** 6441.
233. L. F. Fieser and M. J. Haddadin, *Canad. J. Chem.*, 1965, **43,** 1599.
234. D. Wege, *Tetrahedron Letters*, 1971, 2337.
235. R. N. Warrener, *J. Amer. Chem. Soc.*, 1971, **93,** 2346.
236. U. E. Wiersum and W. J. Mijs, *J.C.S. Chem. Comm.*, 1972, 347.

18.5
Mixed Oxygen Heterocyclic Systems

D. M. HARRISON
New University of Ulster, Coleraine

In this chapter the chemistry of the major mixed polyoxa systems is surveyed. The majority of the examples discussed have been chosen, inevitably, from the field of natural products chemistry. Comprehensive reviews by Whalley[1] and Mustafa[2] cover a similar range of material. An excellent summary by Dean of the chemistry of the naturally occurring oxygen heterocycles[3] and a useful survey of synthetic methods by the same author[4] are also relevant to this chapter.

18.5.1 FUROCOUMARINS

18.5.1.1 Introduction

The furocoumarins[1,2] constitute an important group of natural products isolated, mainly, from higher plants of the Rutaceae, Umbelliferae, Leguminosae, and Moraceae families. The vast majority of these compounds are derivatives of the linear furocoumarin

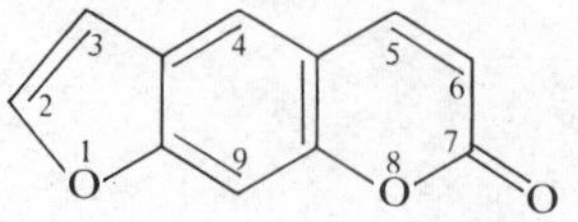

(1)

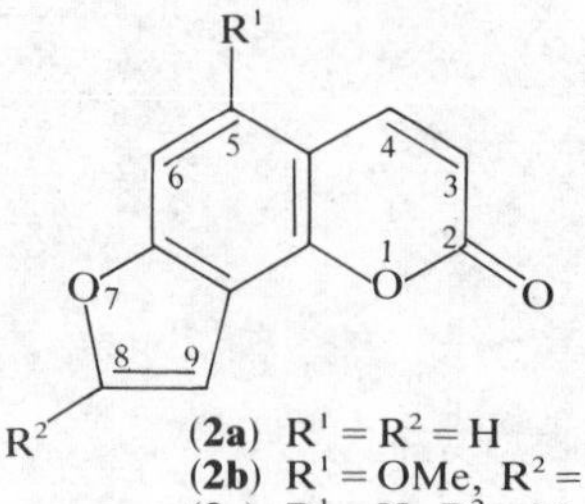

(2a) $R^1 = R^2 = H$
(2b) $R^1 = OMe$, $R^2 = H$
(2c) $R^1 = H$, $R^2 = Me_2CH$

psoralen (**1**) or of its angular isomer angelicin (**2a**). This account is concerned mainly with the chemistry of these two typical skeleta, notwithstanding the existence of other, mainly synthetic, types.

Several systems of nomenclature are in use. In the IUPAC system psoralen and angelicin are numbered as shown in (**1**) and (**2**) respectively. More commonly the numbering illustrated in diagrams (**3**) and (**4**) is used. The latter system is generally more convenient and will be adopted in the following discussion.

18.5.1.2 Furocoumarins: physical and chemical properties

Furocoumarins are generally colourless crystalline solids which retain the properties of the coumarin system. Thus they exhibit fluorescence under u.v. light and are insoluble in aqueous acid. Furocoumarins dissolve readily in warm aqueous alkali owing to opening of the lactone ring; subsequent neutralization with aqueous acid usually brings about smooth relactonization, although skeletal rearrangement can sometimes occur, as discussed below. Furocoumarins develop a deep violet colouration on treatment with alkali in the presence of 8-amino-5-hydroxyfuro[3′,2′:6,7]chromone and may thus be readily distinguished from furochromones, which do not show this colour test.[5]

Spectroscopic studies play an important role in the structure determination of natural furocoumarins. N.m.r. spectroscopy is particularly important in identifying substituents; lanthanide shift reagents may be used to aid the determination of the substitution pattern.[6] The long-range coupling between protons on the furan and homocyclic rings, which is sometimes observed, may be of diagnostic value.[7] Strong molecular ions are observed in the mass spectra of simple furocoumarins. The major fragmentation pathways usually involve loss of CO from the pyrone ring, preceded or followed by loss of a methyl radical in the case of methoxy-derivatives. The furan ring remains intact until late in the fragmentation pathway.[8,9]

Furocoumarins which lack substituents in the furan ring are degraded to furan-2,3-dicarboxylic acid on treatment with alkaline hydrogen peroxide.[10] In contrast, chromic oxide treatment,[11] or ozonolysis,[10] usually destroys the furan ring first and in favourable cases a hydroxycoumarin containing a vicinal formyl group may be isolated. The products of reduction similarly depend on the reagent used. Catalytic hydrogenation reduces the furan ring to yield the 4′,5′-dihydro-derivative, while lithium aluminium hydride cleaves the pyrone ring *via* reduction of the carbonyl group.[10]

Linear furocoumarins exhibit a strong tendency to rearrange to their thermodynamically more stable angular isomers whenever possible. For example, demethylation of bergapten (**3a**) with hydriodic acid followed by remethylation gives the angular isomer (**4**); this can arise only by acid-catalysed opening of the furan ring of the intermediate phenol (**3b**) and recyclization involving the C-5 hydroxyl.[12] Base-catalysed opening of the pyrone ring followed by relactonization can also supply the opportunity for isomerization of furocoumarins when a 5-hydroxy-substituent is present. Thus, although diazomethane treatment of bergaptol (**3b**) gives the expected product (**3a**), treatment of the former with alkaline dimethyl sulphate yields the isomeric methyl ether (**2b**).[13] The synthetic furocoumarin (**3c**) similarly undergoes skeletal rearrangement, on treatment with alkaline

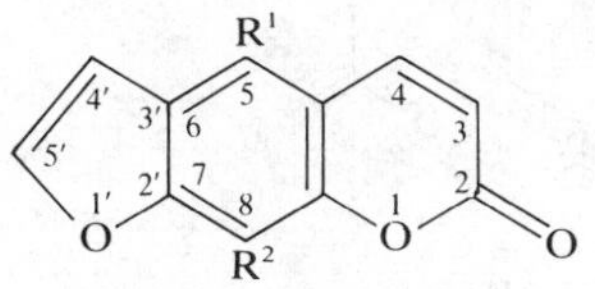

(**3a**) $R^1 = OMe$, $R^2 = H$
(**3b**) $R^1 = OH$, $R^2 = H$
(**3c**) $R^1 = CH_2OH$, $R^2 = OMe$

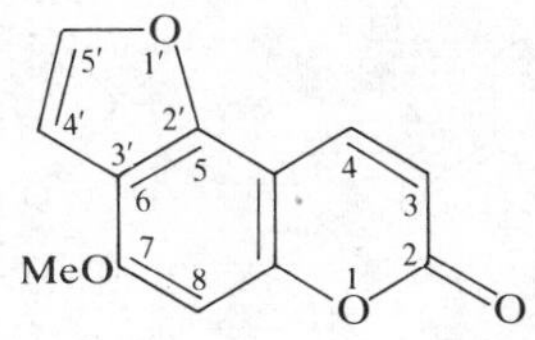

(**4**)

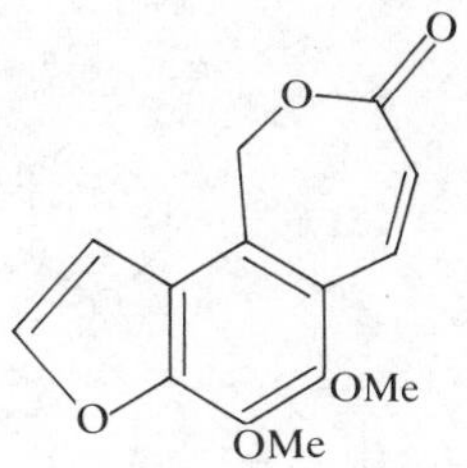

(**5**)

dimethyl sulphate, to form (**5**).[14] Many naturally occurring furocoumarins contain dimethylallyl (3-methyl-but-2-enyl) or dimethylallyloxy-substituents, or epoxides or diols derived from these groups. One such compound is the epoxide byakangelicol (**6**), which undergoes rearrangement to dioxene (**7a**) on treatment with aqueous alkali. Treatment of (**6**) with dimethylamine similarly gives amide (**7b**) and the product (**7c**) of Michael addition on to the latter.[15] Hydrazine reacts with simple furocoumarins by nucleophilic attack on the carbonyl group, followed by opening of the pyrone ring to yield acid hydrazides;[10] thiols, in contrast, react at C-4 to form Michael addition products.[16]

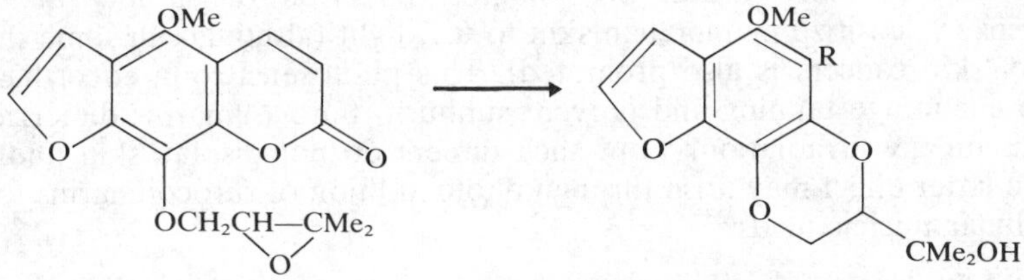

(**6**)

(**7a**) $R = CH{=}CHCO_2H$
(**7b**) $R = CH{=}CHCONMe_2$
(**7c**) $R = CH(NMe_2)CH_2CONMe_2$

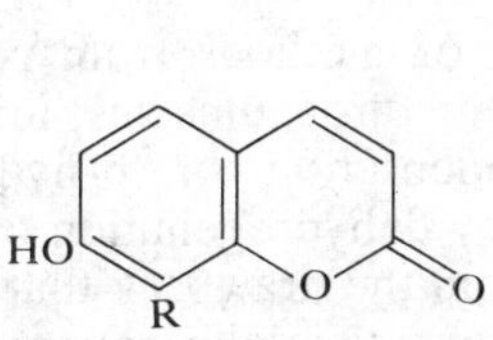

(**8a**) $R = CH_2CHPh_2$
(**8b**) $R = CHPhCH_2Ph$

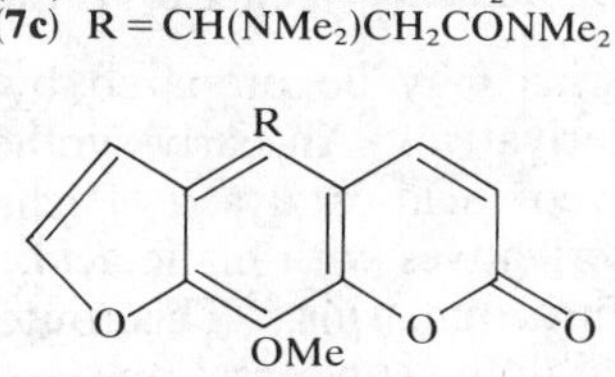

(**9a**) $R = H$
(**9b**) $R = NO_2$
(**9c**) $R = NH_2$
(**9d**) $R = Br$

Furocoumarins which are not substituted in the furan ring react with benzene in the presence of aluminium chloride to give hydroxycoumarins containing a vicinal diphenylethyl substituent, such as (**8a**) derived from angelicin (**2a**).[17] The structures of these products were incorrectly assigned (*cf.* **8b**) in the earlier literature. No systematic studies on the electrophilic substitution of furocoumarins have been reported. The methoxyfurocoumarins bergapten (**3a**) and xanthotoxin (**9a**) undergo electrophilic substitution in the homocyclic ring. For example, xanthotoxin forms a nitro-derivative (**9b**) which may be reduced to the amine (**9c**). A number of derivatives have been prepared from the diazonium salt derived from this amine,[10] while chromic acid oxidation of the amine gives quinone (**10**) in low yield.[18] Xanthotoxin similarly gives a monobromo-derivative (**9d**), although in the presence of excess bromine addition to the 4′(5′) double bond also occurs.[10] In contrast 4,5′,8-trimethylpsoralen (**11a**) reacts with electrophilic reagents exclusively at the C-4′ position. The 4′-bromo-derivative (**11b**) is further brominated with difficulty at C-3, while the 4′-nitro-derivative undergoes further substitution in

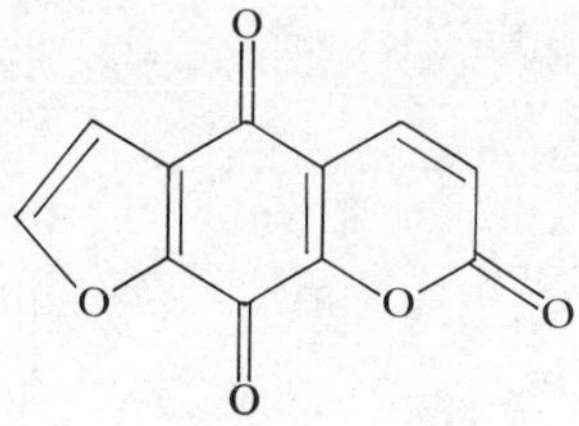

(10)

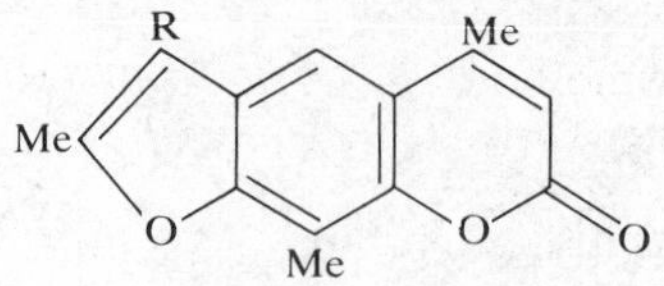

(11a) R = H
(11b) R = Br
(11c) R = NO_2

the homocyclic ring.[19] 4′,5′-Dihydrofurocoumarins undergo electrophilic substitution at C-3;[19,20] the interesting ring contraction indicated (reaction 1) occurs on treatment of their 3-bromo-derivatives with hot aqueous alkali.[20]

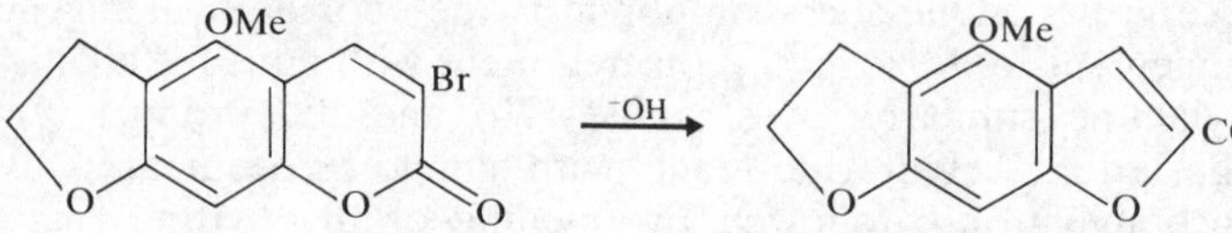

$$\xrightarrow{^{-}OH} \quad CO_2H \qquad (1)$$

Furocoumarins display considerable toxicity towards cold-blooded animals, especially fresh-water fish. A more useful physiological effect is the ability of some linear furocoumarins to sensitize mammalian skin to u.v. light (although, in some instances, the formation of skin cancers is also promoted). This photosensitizing effect has been used clinically to encourage tanning and prevent sunburn. Furocoumarins dimerize across the 3,4-positions on u.v. irradiation,[21] but such dimers do not display skin photosensitizing activity. The latter effect may arise through photoaddition of furocoumarins to pyrimidine bases of cellular nucleic acids.[22]

18.5.1.3 Furocoumarins: general synthetic methods

Furocoumarins may be prepared by a variety of methods from coumarins or from benzofuran derivatives. An early synthesis of linear furocoumarins, initially explored by Späth, involved acid-catalysed Pechmann condensation of 6-hydroxy-2,3-dihydrobenzofuran derivatives with malic acid, followed by dehydrogenation of the intermediate 4′,5′-dihydrofurocoumarins.[23] This route suffers from the drawback that isomeric products may be formed in the condensation step. Furthermore, the dehydrogenation reaction gives variable and frequently low yields; the use of unprotected 6-hydroxybenzofuran derivatives in the Pechmann condensation is precluded by the acid-sensitivity of the furan ring. In principle any β-ketoacid derivative may be condensed with the hydroxydihydrobenzofuran; a preparation of psoralen (**1**) by this route is illustrated in Scheme 1.[24] Variations include Lewis acid catalysed condensation of acrylonitrile with dihydrobenzofuran (**12a**) to give 3,4,4′,5′-tetrahydropsoralen, which may be dehydrogenated to psoralen with palladium–carbon.[25]

(12a) R^1 = Ac, R^2 = H
(12b) R^1 = R^2 = H
(12c) R^1 = H, R^2 = CHO

(13a) R^1 = CO_2H, R^2 = H
(13b) R^1 = H, R^2 = CO_2H
(13c) R^1 = R^2 = H

(12a) $\xrightarrow{\text{i (ly)}}$ **(13a)** $\xrightarrow{\text{ii (my)}}$ **(1)**

i, $EtO_2CCH_2COCO_2Et$, H_2SO_4; ii, Pd–C, Ph_2O, reflux.

SCHEME 1

Addition of the α-pyrone ring to a hydroxybenzofuran derivative which contains a vicinal formyl substituent, by acetic anhydride–sodium acetate treatment, or any other conventional method, gives a product of clearly defined constitution. The requisite benzofuran derivatives are, however, not readily prepared from 6-hydroxybenzofuran owing to the reactivity of the furan ring. This problem may be overcome by use of a removable 2-carboxylic acid substituent or by the use of a hydroxy-2,3-dihydrobenzofuran. The latter solution is exemplified by the synthesis of psoralen described by Foster *et al.*[26] (Scheme 2).

(**12b**) —i (hy)→ (**12c**) —ii (my–hy)→ (**13b**) —iii (ly)→ (**13c**) —iv (ly)→ (**1**)

i, HCN, $ZnCl_2$, HCl; ii, $NCCH_2CO_2H$; iii, Cu, quinoline, heat; iv, Pd–C, heat.

SCHEME 2

4-Hydroxyfurocoumarins may be prepared from acetylbenzofuran derivatives which contain a vicinal hydroxy-group, as exemplified in reaction 2 by the preparation of (**15a**).[5] The starting benzofuran (**14a**) is readily available by degradation of the furochromone

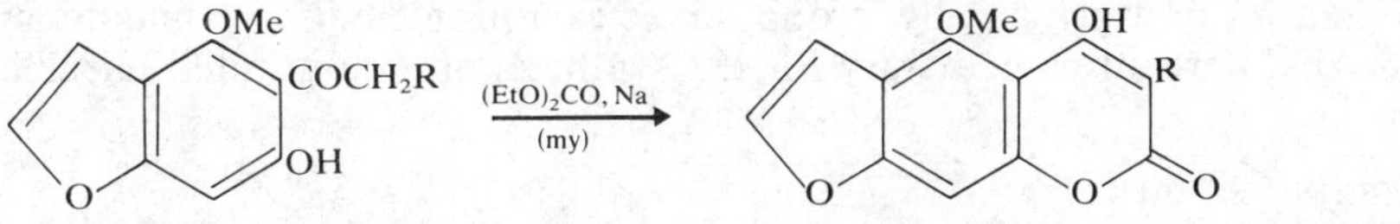

OMe, $COCH_2R$, OH —$(EtO)_2CO$, Na (my)→ OMe, OH, R, O, O, O (2)

(**14a**) R = H
(**14b**) R = OMe

(**15a**) R = H
(**15b**) R = OMe

visnagin. The unusual trimethoxyfurocoumarin halfordin (**16**) has been similarly prepared from (**14b**) by diazomethane treatment of the condensation product (**15b**).[27] 4-Hydroxyfurocoumarin (**15a**) reacts with formaldehyde to give the dimeric product (**17**).[5] 3-Substituted 4-methylfurocoumarins have been prepared *via* reaction of 5-acetyl-6-hydroxybenzofurans with β-ketoacids and their derivatives.[28]

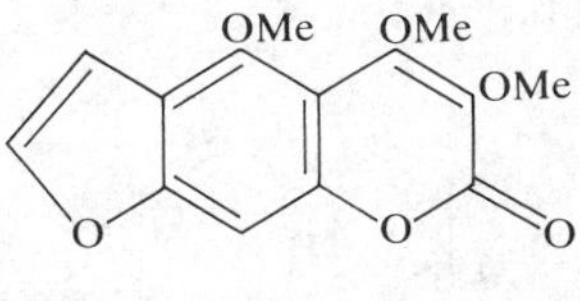

OMe OMe OMe (**16**)

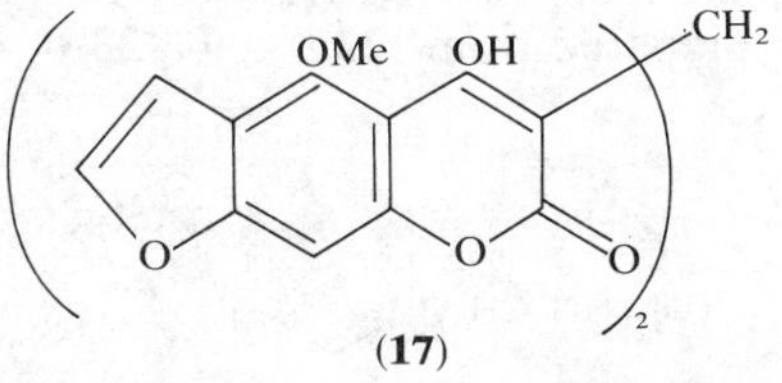

OMe OH CH_2 (...)$_2$ (**17**)

Angular furocoumarins are readily prepared from formylcoumarins containing a vicinal hydroxy-substituent. Either Dieckman cyclization of ester[29] (**19**) or Perkin cyclization of the derived carboxylic acid[30] may be used (Scheme 3). The requisite 8-formyl derivative (**18**) is available in moderate yield by mild formylation of 7-hydroxycoumarin with

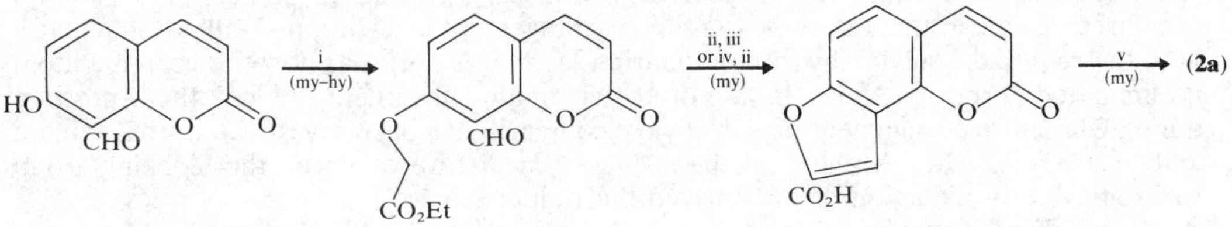

HO, CHO (**18**) —i (my–hy)→ CHO, CO_2Et (**19**) —ii, iii or iv, ii (my)→ CO_2H (**20**) —v (my)→ (**2a**)

i, $BrCH_2CO_2Et$, K_2CO_3; ii, KOH; iii, NaOAc, Ac_2O; iv, NaOEt; v, Cu, quinoline, heat.

SCHEME 3

hexamethylenetetramine in acetic acid.[30] A variation on this route involves base-catalysed condensation of diethyl bromomalonate with (**18**) to give diester (**21**) which, after phosphoric oxide treatment followed by base-catalysed hydrolysis, gives the original carboxylic acid (**20**).[31] Linear furocoumarins which lack substitution at C-8 are not readily available by these routes since 7-hydroxycoumarin is formylated most easily at C-8. 4′-Substituted furocoumarins may be prepared as in Scheme 3, utilizing instead acylcoumarins containing a vicinal hydroxy-substituent derived from Fries rearrangement of acyloxycoumarins.[25,32]

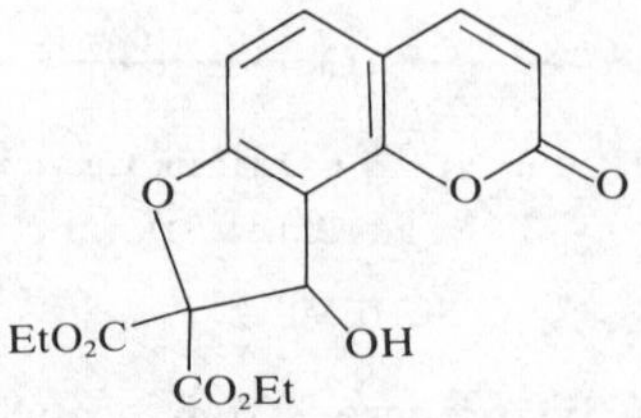

EtO_2C OH CO_2Et

(**21**)

Seshadri and co-workers have developed an interesting and versatile synthesis of furocoumarins in which an allyl group serves as a protected formylmethyl group (Scheme 4).[33] 6-Allyl derivatives necessary for the synthesis of 8-substituted linear furocoumarins

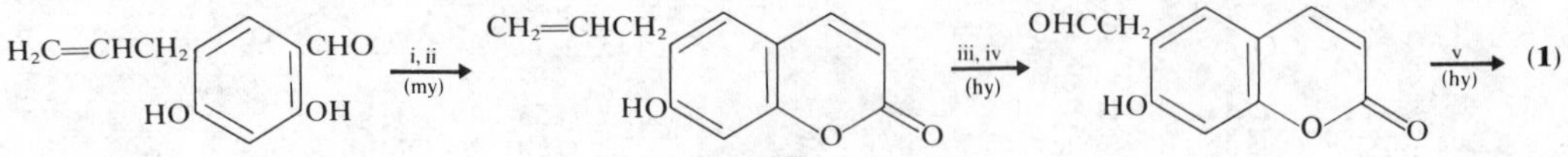

$H_2C{=}CHCH_2$ CHO HO OH $\xrightarrow[\text{(my)}]{\text{i, ii}}$ $CH_2{=}CHCH_2$ HO (**22**) $\xrightarrow[\text{(hy)}]{\text{iii, iv}}$ $OHCCH_2$ HO $\xrightarrow[\text{(hy)}]{\text{v}}$ (**1**)

i, NaOAc, Ac_2O; ii, KOH; iii, O_3; iv, H_2, Pd; v, polyphosphoric acid.

SCHEME 4

are conveniently prepared by Claisen rearrangement of the appropriate 7-allyloxycoumarin.[34] However, 7-allyloxycoumarins which lack a substituent at C-8 yield exclusively the 8-allyl isomer (reaction 3). Angelicin (**2a**) and its 3,4-dimethoxy-derivative have been prepared from (**23a**) and (**23b**), respectively.[35]

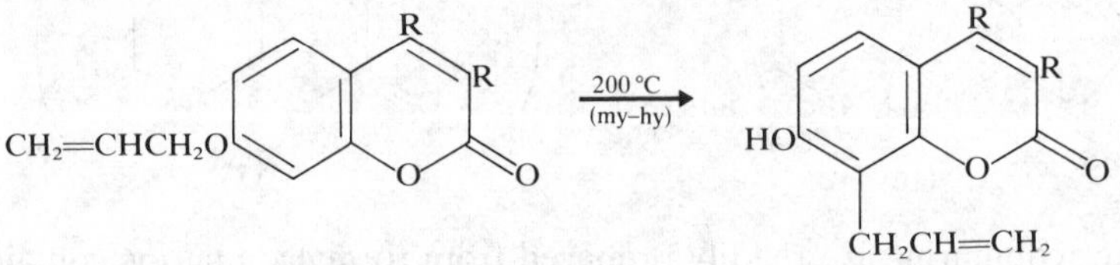

$CH_2{=}CHCH_2O$ R R $\xrightarrow[\text{(my–hy)}]{200\,°C}$ HO R R $CH_2CH{=}CH_2$ (3)

(**23a**) R = H
(**23b**) R = OMe

Trivedi and co-workers have reported an unusual synthesis of psoralen (**1**) from 2-bromoresorcinol; in the key step, Claisen rearrangement of 8-bromo-7-allyloxycoumarin gave the required 6-allyl-7-hydroxycoumarin (**22**).[36] It is not clear how the bromine atom is eliminated in this reaction. The use of an acetamido-substituent to block the 8-position during Claisen rearrangement of a 7-allyloxycoumarin has been investigated by Kaufman and co-workers. After synthesis of the required linear furocoumarin, the blocking group was removed by reduction of the derived diazonium salt.[37]

5′-Methylfurocoumarins may be prepared from appropriate hydroxyallylcoumarins as exemplified in Scheme 5.[38] Alternatively, these coumarin derivatives may be cyclized with concentrated sulphuric acid and the resultant 5′-methyl-4′,5′-dihydrofurocoumarins dehydrogenated over palladium black.[39] Linear furocoumarins substituted at C-4′ are prepared in excellent yield by addition of α-chloromethyl ketones to 7-hydroxycoumarin

(23a) → i, ii (hy) → [AcO, Br, Br] → iii (my) → [O, Me]

i, Ac_2O, NaOAc; ii, Br_2; iii, KOH, EtOH.

SCHEME 5

in the presence of potassium carbonate (Scheme 6). The phenolate anion (**24**) is most probably an intermediate in this reaction.[40]

HO → i (hy) → (**24**) [R, $CH{=}CHCO_2^-$] → [R]

i, $RCOCH_2Cl$, K_2CO_3.

SCHEME 6

Furo[3′,2′:3,4]coumarins may be prepared from 3-allyl-4-hydroxy- or 3-acyl-4-hydroxy-coumarins by routes already discussed. An interesting alternative is exemplified in Scheme 7. The starting bromopyrone (**25**) was prepared by Pechmann condensation of 4-hydroxy-7-methoxy-8-methylcoumarin with ethyl acetoacetate followed by bromination of the pyronocoumarin product.[39]

(**25**) [Br, Me, MeO, Me] → i (my) → [Me, CO_2H, MeO, OH, Me] → ii (my) → [Me, MeO, Me]

i, Na_2CO_3, aqueous, reflux; ii, HCl, EtOH, reflux.

SCHEME 7

18.5.1.4 Dihydrofurocoumarins and related natural products

Linear and angular hydroxyisopropyldihydrofurocoumarins such as (**26a**) and (**27**) co-occur in Nature with the simpler furocoumarins and most probably are biosynthetic precursors of the latter.[41] Marmesin (**26a**), the enantiomer of nodakenetin, occurs both in

(**26a**) R = OH
(**26b**) R = H; 4′,5′-dehydro

(**27**)

(**28**) [CHO, OH, OH]

the free state and as a glycoside. Phosphorus pentoxide treatment of the aglycone yields the anhydro-derivative (**26b**). Marmesin undergoes electrophilic substitution at C-3,[42] as described earlier for 4′,5′-dihydrofurocoumarins. (+)-Marmesin has been shown to possess the 5′-*S* configuration by its synthesis from the optically pure dihydrobenzofuran derivative (**28**).[43] The angular compound dihydrooroselol (or columbianetin) (**27**) is found in Nature as its tiglate ester and as a glucoside. Destruction of the pyrone ring by ozonolysis and correlation of the product with tubaic acid, a degradation product of rotenone (Section 18.5.5), established that dihydro-oroselol has the *S*-configuration depicted.[44] Racemic hydroxyisopropyldihydrofurocoumarins may be prepared by peracid treatment of appropriate dimethylallyl-substituted hydroxycoumarins, as illustrated in Scheme 8. In the presence of a trace quantity of mineral acid the putative epoxide

(**27**) ← i (hy) — ii (hy) → (**29**)

i, *m*-$ClC_6H_4CO_3H$, ether; ii, *m*-$ClC_6H_4CO_3H$, $CHCl_3$, trace HCl.

SCHEME 8

intermediate undergoes an alternative mode of cyclization to give the pyranocoumarin lomatin (**29**).[45,46] The latter undergoes facile acid-catalysed dehydration and rearrangement to give the isopropylfurocoumarin (**2c**).[47] The esters athamantin (**30a**) and archangelicin (**30b**) are of interest owing to the remarkable transformations which occur on treatment with sodium methoxide (Scheme 9). On heating the reaction mixture a molecule of methanol is eliminated from (**31**) and the methyl ester of the coumaric acid is formed together with the coumarin derivative, among other products.[48]

(**30a**) R = $COCH_2CHMe_2$
(**30b**) R = COCMe=CHMe

(**31**)

i, MeONa, MeOH, 5 °C.

SCHEME 9

The aflatoxins are a group of highly toxic and carcinogenic mold metabolites isolated mainly from *Aspergillus* and *Penicillium* species. Aflatoxins B_1 (**32**) and G_1 (**33**) are typical members of this group of bisfurocoumarins.[49] Büchi and Weinreb have reported a particularly elegant total synthesis of aflatoxin B_1, the final step of which utilizes a new coumarin synthesis (reaction 4). Zinc or magnesium salts catalyse this step, presumably owing to formation of a chelated β-diketone intermediate (**34**). Aflatoxin G_1 was similarly

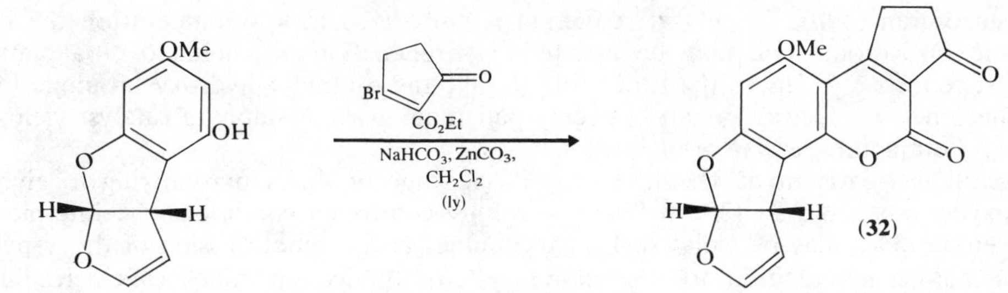

CO_2Et, $NaHCO_3$, $ZnCO_3$, CH_2Cl_2 (ly) (4)

(32)

prepared by use of the requisite bromopyranone.[50] The aflatoxins are structurally and biosynthetically related to a group of fungal bisfuroxanthones typified by sterigmatocystin (**35**).[49]

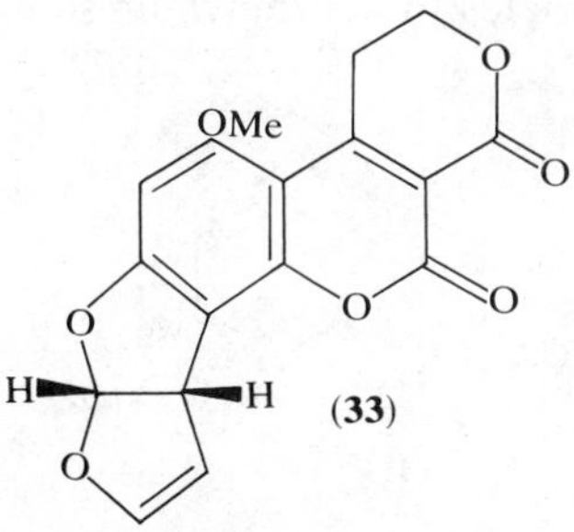

(**33**) 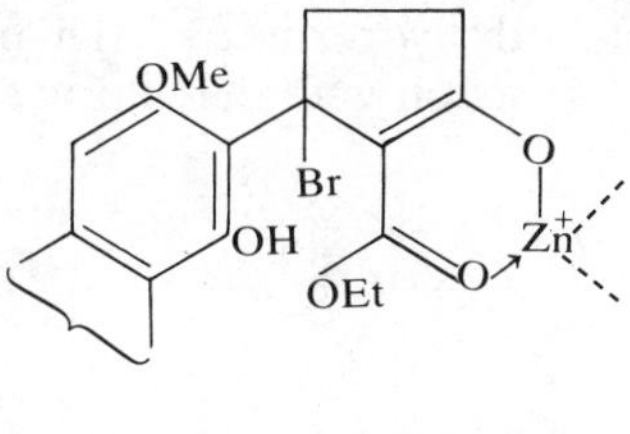 (**34**) 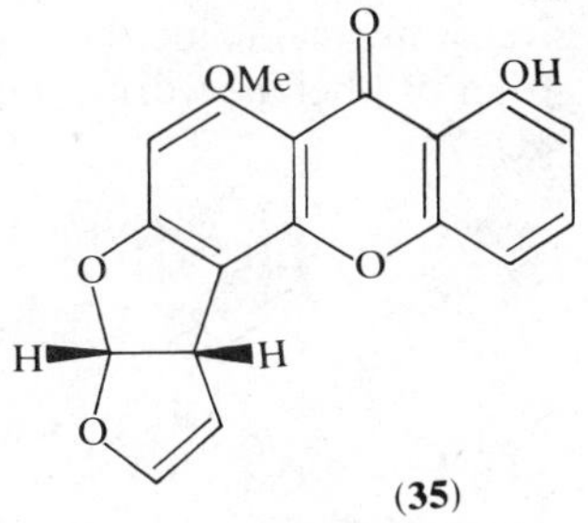 (**35**)

18.5.2 FUROCHROMONES

18.5.2.1 Introduction

The furochromones[1,2] are a relatively small group of natural products typified by the 2-methyl derivatives khellin (**36a**) and visnagin (**36b**), and the furoflavone karanjin (**37**). Some furochromones display interesting physiological activity; khellin in particular is important as an antispasmodic agent and as a coronary vasodilator.[51] The 2-methylfurochromones may be recognized by the deep-violet colour which they give on treatment with aqueous alkali.[52]

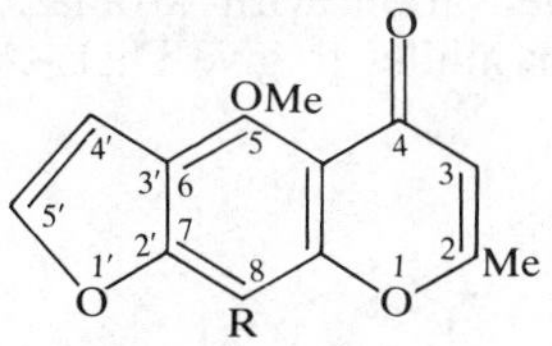

(**36a**) R = OMe
(**36b**) R = H

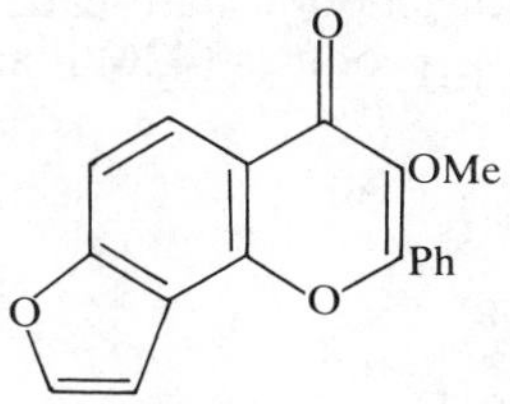

(**37**)

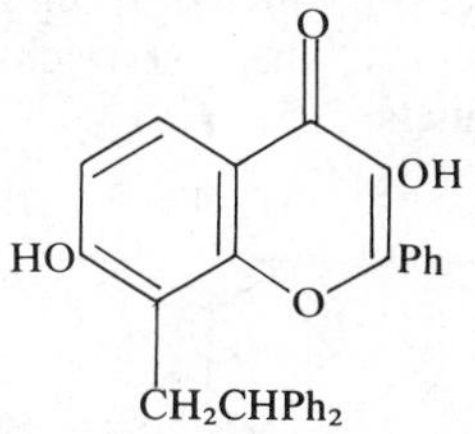

CH_2CHPh_2

(**38**)

18.5.2.2 Furochromones: chemical properties

The γ-pyrone ring of the furochromones is weakly basic, as expected, and khellin forms an orange crystalline perchlorate salt on treatment with perchloric acid.[2] The furan ring is cleaved by aluminium chloride in benzene. Karanjin (**37**) gives the flavone derivative (**38**) under these conditions.[17] Simple furochromones are ultimately degraded to furan-2,3-dicarboxylic acid by treatment with alkaline hydrogen peroxide; the intermediate hydroxycarboxylic acid (**39a**) may be isolated in good yield when khellin is oxidized under these conditions. In contrast, chromic acid cleaves the furan ring to give the hydroxyformylchromone (**40a**).[11] Lithium aluminium hydride reduction of khellin yields the acylbenzofuran (**39b**). When this reduction is performed at low temperature the intermediate furochromanone may be isolated. Hydrogenation of khellin over a platinum catalyst reductively cleaves the furan ring to give the 6-ethyl-7-hydroxychromone (**40b**). On the other hand, hydrogenation over a palladium–barium sulphate catalyst yields the 2,3,4′,5′-tetrahydro-derivative of khellin.[53]

Hot alkali treatment of khellin results in cleavage of the γ-pyrone ring to give the hydroxyacetylbenzofuran (**39c**).[5] Under carefully controlled conditions the intermediate β-diketone (**39d**) may be isolated. Primary amines also promote fission of the γ-pyrone ring *via* Michael addition to the enone system to give enamines (**39e**). Khellin is reconstituted upon mild acid treatment of these enamines or the β-diketone (**39d**).[54] Under normal conditions condensation of hydrazine with khellin yields pyrazole (**39f**) or its tautomer; hydroxylamine similarly forms an isoxazole. However, when these condensations are performed in strictly anhydrous solvents, khellin forms a normal oxime, and gives a normal azine on reaction with hydrazine.[55] Grignard reagents add to the carbonyl group of khellin; work-up of the reaction with perchloric acid yields furochromylium salts (**41**).[56]

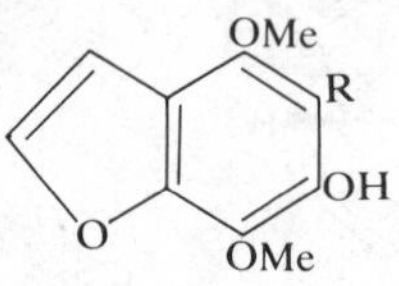

(**39a**) R = CO_2H
(**39b**) R = $CO(CH_2)_2CH_3$
(**39c**) R = COMe
(**39d**) R = $COCH_2COMe$
(**39e**) R = COCH=CMeNHR
(**39f**) R = [pyrazolyl, Me, HN—N]

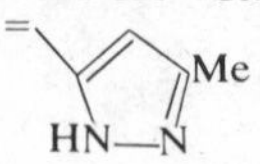

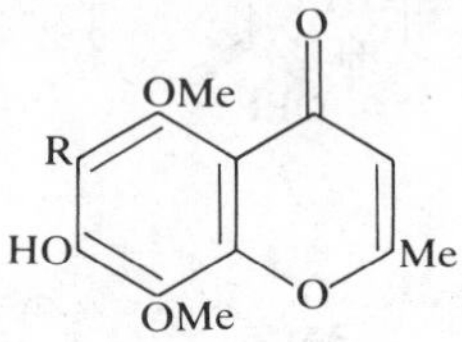

(**40a**) R = CHO
(**40b**) R = Et

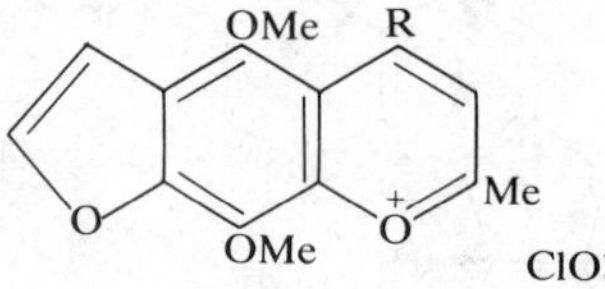

(**41**) [ClO_4^-]

The dimethoxyfurochromone (**42a**) reacts with formaldehyde in the presence of dimethylamine hydrochloride to give the expected Mannich condensation product (**42b**). Under similar conditions the 2-methylfurochromones condense with three equivalents of formaldehyde; khellin gives the tetra-alkylammonium salt (**43**). The ‘expected’ product (**42c**) is formed when the β-diketone (**39d**) is submitted to the conditions of the Mannich reaction.[57] Khellin undergoes smooth base-catalysed condensation with aromatic aldehydes. The 2-styrylchromone products (**42d**) react with dienophiles to give Diels–Alder adducts.[58]

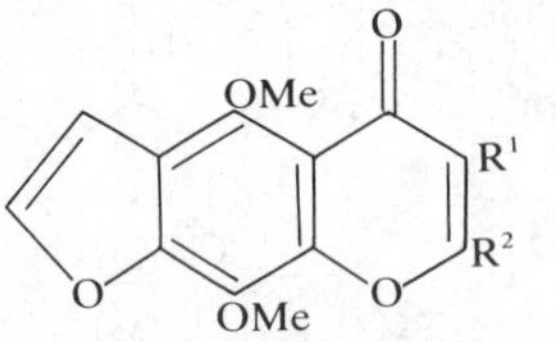

(**42a**) $R^1 = R^2 = H$
(**42b**) $R^1 = CH_2NMe_2$, $R^2 = H$
(**42c**) $R^1 = CH_2NMe_2$, $R^2 = Me$
(**42d**) $R^1 = H$, $R^2 = CH{=}CHAr$

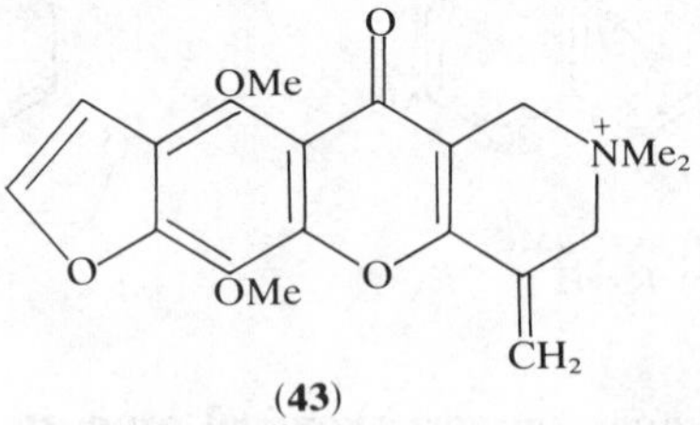

(**43**)

Refluxing hydrochloric acid smoothly demethylates the 5-methoxyfurochromone visnagin (**36b**). Similar treatment of khellin results in partial hydrolysis to the nor-derivative (**44**).[54] Such 5-hydroxy-derivatives are insoluble in aqueous alkali and display considerable stability towards normal methylation procedures owing to hydrogen bonding between the hydroxy-group and the pyrone carbonyl group. Selective demethylation of the 5-methoxy-substituent of khellin may also be achieved with thiophenols plus piperidine[16] or with potassium iodide in formic acid.[54] Complete demethylation of khellin with refluxing concentrated hydriodic acid results in skeletal rearrangement to the angular furochromone (**45**). This is an example of the Wesely–Moser rearrangement, in which the γ-pyrone ring opens and recyclizes at the alternative site to form the thermodynamically more stable product. Similar hydriodic acid treatment of (**46**) yields the same rearranged furochromone (**45**); in this example both heterocyclic rings have migrated.[59] Complete demethylation of methoxyfurochromones, such as khellin, may be achieved without rearrangement by treatment with magnesium iodide.[60]

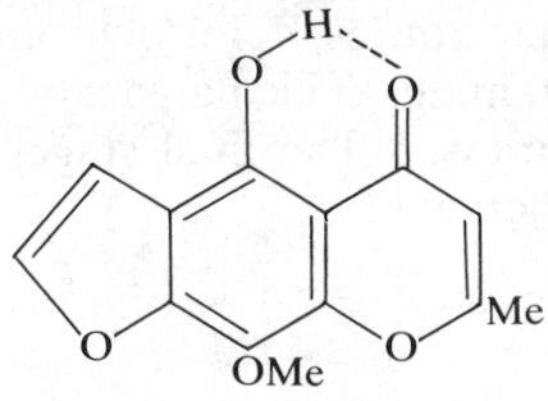

(**44**) 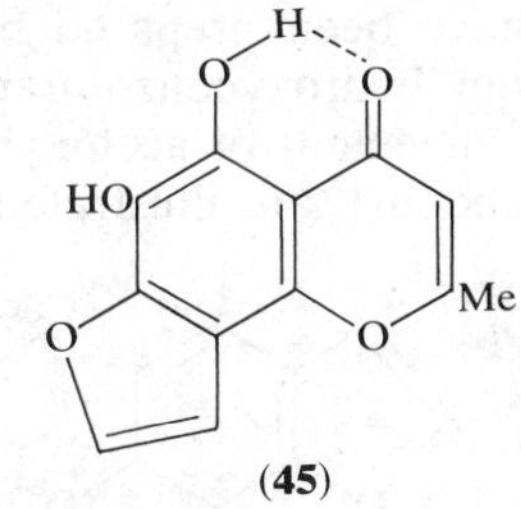 (**45**) 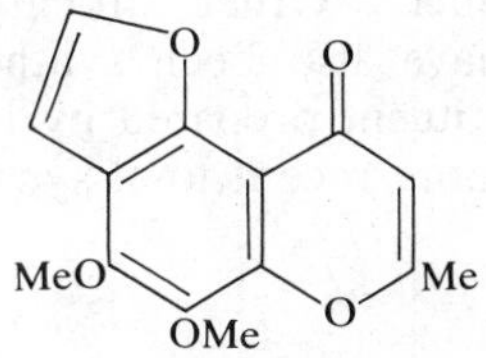 (**46**)

18.5.2.3 Furochromones: synthetic methods

The khellin degradation product khellinone (**39c**) has been utilized as the starting point for the preparation of a wide range of 2-substituted 5,8-dimethoxyfurochromones. Base-catalysed condensation of khellinone with aromatic aldehydes followed by selenium dioxide treatment of the resultant chalcones (**47**) yields the 2-arylfurochromones (**50**).[61] Alternatively, treatment of these chalcones with triethyl orthoformate and perchloric acid gives the furochromylium perchlorates (**48**), which may be hydrolysed to the furochromone system in excellent yield.[62] Khellin analogues are also available from the condensation of khellinone with appropriate carboxylate esters, followed by acid-catalysed cyclization of the resultant β-diketones (**49**)[63] (Scheme 10). The total synthesis of khellin has been achieved *via* the latter route. Khellinone was prepared either by Friedel–Crafts acetylation of the protected 6-hydroxybenzofuran (**51a**) followed by saponification and thermal decarboxylation, or by Hoesch reaction of dihydrobenzofuran

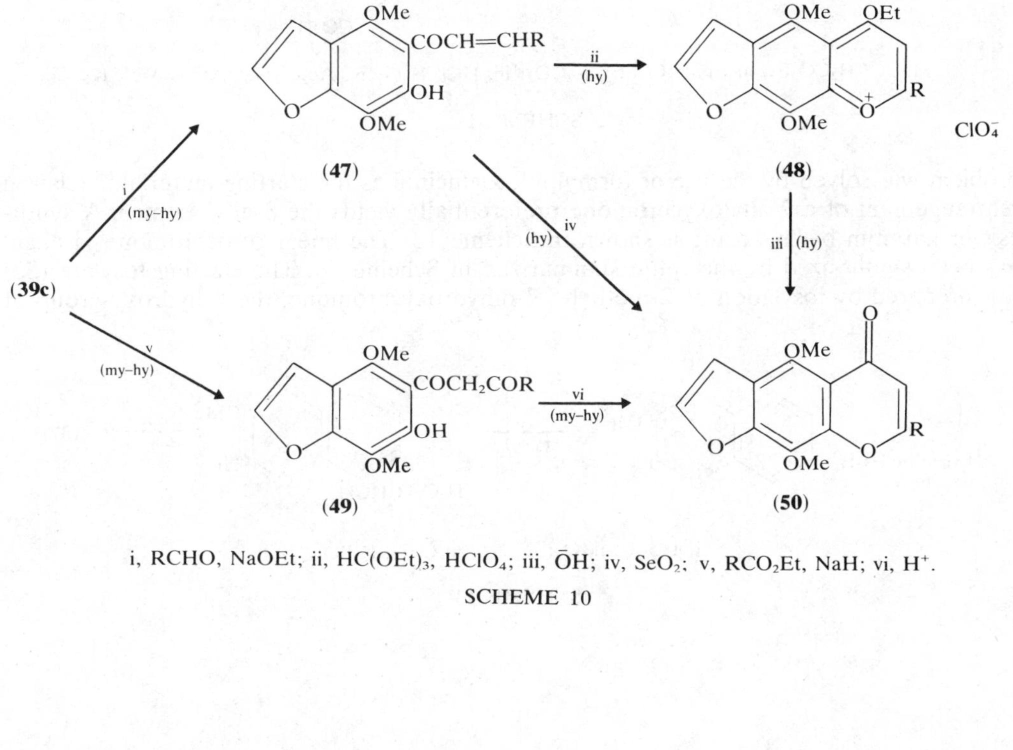

i, RCHO, NaOEt; ii, $HC(OEt)_3$, $HClO_4$; iii, $\bar{O}H$; iv, SeO_2; v, RCO_2Et, NaH; vi, H^+.

SCHEME 10

OMe R OH O OMe

(**51a**) $R = CO_2Me$
(**51b**) $R = H$

OMe OH O OMe

(**52**)

(**52**) with acetonitrile and zinc chloride, with subsequent dehydrogenation of the product.[64] Other furochromones have been prepared by similar routes.[65] Furochromones have also been synthesized from hydroxybenzofurans containing a vicinal acetyl substituent prepared by Fries rearrangement of acetoxybenzofurans.[66] The final stages of a more recent total synthesis of khellin[67] are illustrated in Scheme 11.

(**51b**) —i, ii (my)→ [OMe, CO_2H, O, OMe, O, Me] —iii (ly)→ (**36a**)

i, $MeCCl{=}CHCO_2Me$, K_2CO_3; ii, NaOH; iii, MeCOCl, H_2SO_4.

SCHEME 11

Furochromones may also be prepared from suitable hydroxychromones, containing a vicinal formyl or allyl substituent, using routes discussed above for the synthesis of furocoumarins. Introduction of a formyl substituent into a hydroxychromone is often difficult to achieve in good yield; in the total synthesis of visnagin shown in Scheme 12 this

[OH, OHC, HO, OH] —i→ [H, O, O, OHC, HO, O, Me] —ii, iii→ [H, O, O, OHC, O, CO_2H, O, Me] —iv, iii, v→ (**36b**)

i, $MeCOCH_2CO_2Et$; ii, $BrCH_2CO_2Et$, K_2CO_3; iii, HCl, H_2O; iv, Ac_2O, NaOAc; v, MeI, K_2CO_3.

SCHEME 12

problem was solved by the use of formylphloroglucinol as the starting material.[68] Claisen rearrangement of a 7-allyloxychromone preferentially yields the 8-allyl isomer. A synthesis of karanjin by this route is shown in Scheme 13. The linear furochromone visnagin has been synthesized by the route summarized in Scheme 14. The starting tosylate (**53**) was prepared by tosylation of 2-methyl-5,7-dihydroxychromone, the 5-hydroxy-group of

[O, OMe, $H_2C{=}CHCH_2O$, O, Ph] —i (my–hy)→ [O, OMe, HO, $H_2C{=}CHCH_2$, O, Ph] —ii, iii, iv (my–hy)→ (**37**)

i, 195 °C; ii, O_3; iii, H_2, Pd/C; iv, H_3PO_4.

SCHEME 13

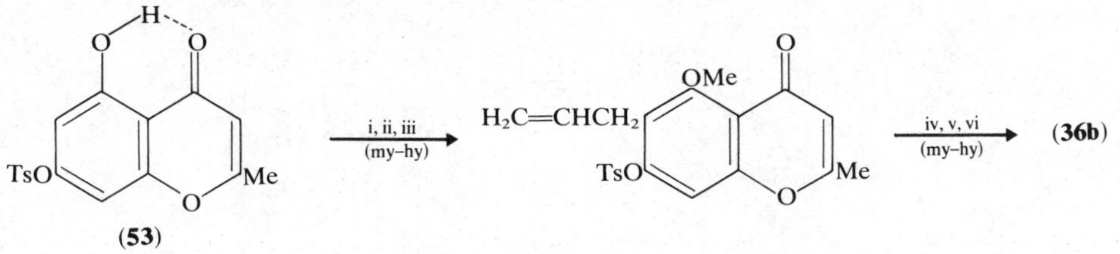

(53)

i, $CH_2{=}CHCH_2Br$, K_2CO_3; ii, 195 °C; iii, MeI, K_2CO_3; iv, NaOH; v, O_3; vi, polyphosphoric acid.

SCHEME 14

which is protected by intramolecular hydrogen bonding.[69] A recent example of the use of hydroxyallyl derivatives in the preparation of furochromones is provided by the synthesis of the furochromone (**46**).[59]

18.5.3 CHROMENOPYRONE SYSTEMS

The 2,2-dimethylchromene system is widespread among natural products. The present discussion is limited to a brief survey of the chemistry of the derived chromenopyrone system. The simplest examples of this system are the linear coumarin xanthyletin (**54a**) and its angular isomer seselin (**55**).[1] Among the many more complex derivatives known may be cited the flavone pongachromene (**56**)[70] and the isoflavone robustic acid (**54b**).[71]

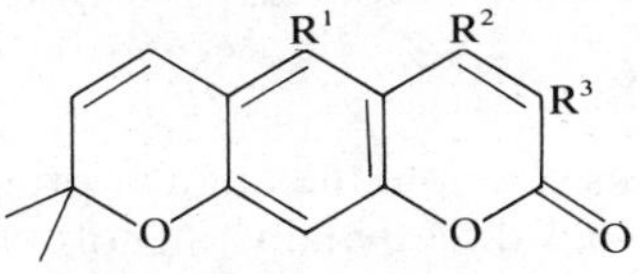

(**54a**) $R^1 = R^2 = R^3 = H$
(**54b**) $R^1 = OMe$, $R^2 = OH$, $R^3 = p\text{-}MeOC_6H_4$
(**54c**) $R^1 = OMe$, $R^2 = R^3 = H$

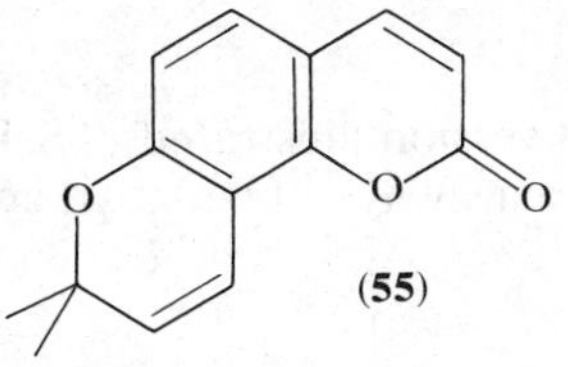

(**55**)

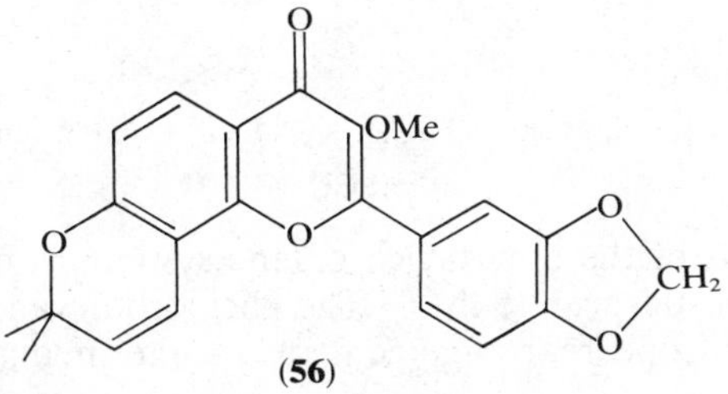

(**56**)

Catalytic hydrogenation of the typical chromenopyrone xanthyletin yields the corresponding chromanopyrone. The pyran ring of the former is cleaved by permanganate oxidation or by ozonolysis; with care, 6-formyl-7-hydroxycoumarin may be isolated from the latter reaction. Osmium tetroxide treatment converts the pyran double bond to the *cis*-glycol. Treatment of xanthyletin with strong alkali yields acetone, among other products. This result appears to be quite general, although at first sight surprising, and is best rationalized by the route shown in Scheme 15 in which acetone arises ultimately by retro-aldol cleavage of 3-methylbut-2-enal. The dimethylchromenopyrones suffer cleavage of the pyrone ring on treatment with alkaline hydrogen peroxide; hydroxychromenecarboxylic acids may be isolated from this reaction.

Good synthetic routes for the elaboration of the chromenopyrone system are all recent discoveries.[4] Pongachromene (**56**) has been prepared in an overall yield of 0.1% by the

(54a) $\xrightarrow[\text{(ly)}]{\text{i}}$ ... CH=CHCO$_2^-$ ⇌ ... CH=CHCO$_2^-$ ⇌ ... CH=CHCO$_2^-$ ⟶ ... CH=CHCO$_2^-$ ⟶ ... ⟶ acetone

i, conc. alkali.

SCHEME 15

route shown in Scheme 16; attempted direct addition of the dimethylpyran ring to the requisite flavone failed altogether.[70] The method used here for elaboration of the pyran ring was first introduced by Späth in the late 1930s and is now essentially replaced by a

COCH$_2$OMe, HO, OH $\xrightarrow{\text{i (ly)}}$ COCH$_2$OMe, OH $\xrightarrow{\text{ii (ly)}}$ (56)

i, $Me_2C(OH)C{\equiv}CH$, $ZnCl_2$; ii, piperonylic anhydride, potassium piperonylate.

SCHEME 16

more recent version illustrated in Scheme 17. In this example the angular product, seselin, is formed exclusively.[72] Oxidative cyclization of *o*-3,3-dimethylallylphenols constitutes an

HO ... $\xrightarrow{\text{i (ly)}}$ (57) $\xrightarrow{\text{ii (ly)}}$ (55)

i, $Me_2CClC{\equiv}CH$, K_2CO_3; ii, Et_2NPh, reflux.

SCHEME 17

good alternative synthesis of the dimethylchromene system, as exemplified in Scheme 18 by the synthesis of robustic acid (**54b**).[71] The dichlorodicyanobenzoquinone mediated cyclization must involve Cope rearrangement of an intermediate *o*-quinonemethide. In

(58) OMe, OMe, HO, OH $\xrightarrow[\text{(my–hy)}]{\text{i}}$ OMe, OMe, OH $\xrightarrow[\text{(hy)}]{\text{ii}}$ (54b)

i, [Cl, Cl, CN, CN benzoquinone]; ii, $ClCO_2Et$, K_2CO_3.

SCHEME 18

this example the necessary starting material (**58**) was prepared together with the alternative isomer by boron trifluoride catalysed addition of 2-methylbut-3-en-2-ol to the parent phenol;[71] a more lengthy but usually better route to such *o*-dimethylallylphenols involves partial reduction of aryl propargyl ethers of type (**57**) to the corresponding aryl allyl ethers followed by Claisen rearrangement of the latter.[73]

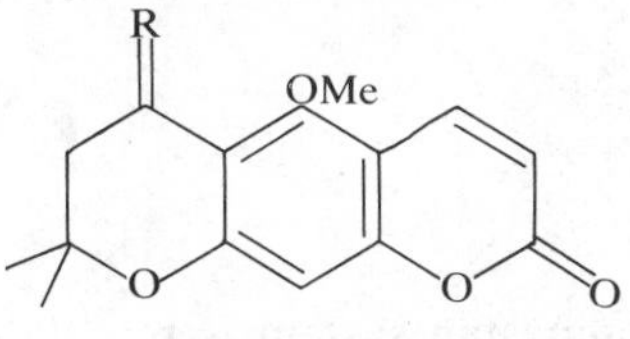

(**59a**) R = O
(**59b**) R = H, OH

Chromanone derivative (**59a**) has been converted into xanthoxyletin (**54c**) by sodium borohydride reduction to the alcohol (**59b**) and mild acid-catalysed dehydration of the latter.[74] The dehydration may be accomplished by a variety of methods, including sublimation of the required product from sodium hydrogen sulphate. The requisite chromanone system is available through Lewis acid catalysed condensation of the acid chloride of 3-methylbut-2-enoic acid with the appropriate phenol or its methyl ether.[1,4]

18.5.4 COUMARONOCOUMARINS, COUMARANOCHROMANS, AND RELATED SYSTEMS

18.5.4.1 Introduction

The coumaronocoumarins and the coumaranochromans are isoflavanoids isolated mainly from plants of the Papilionaceae sub-family of the Leguminosae.[1,2,75] The structures of these two groups of compounds are based on the 'coumestan' (**60**) and 'pterocarpan' (**61**) skeleta, respectively. The natural coumestans are found mainly in ladino clover and alfalfa. Interest in these compounds is associated at least in part with the oestrogenic activity displayed by coumestrol (**62a**) and the synthetic analogue (**62b**).[76] The related pterocarpans are traditionally associated with the 'redwood', *Pterocarpus* genus, although it is now clear that they are widely distributed in the Leguminosae family. Typical members of this group are pterocarpin (**63a**) and homopterocarpin (**64**). Some pterocarpan derivatives,[77] notably pisatin (**63b**) and phaseollin[78] (**65**) are phytoalexins, *i.e.* antifungal principles produced by the host plant in response to fungal infection.

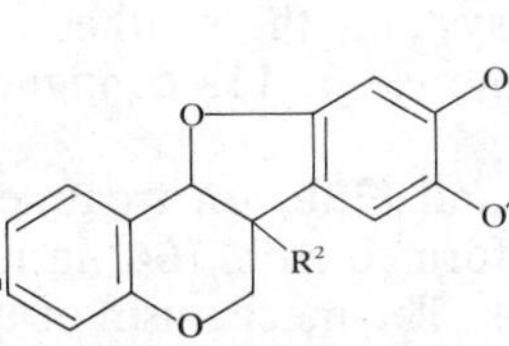

(**60**)

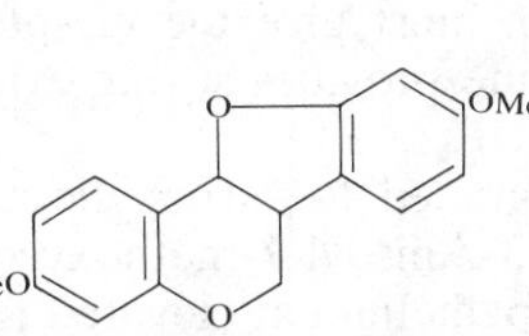

(**61**)

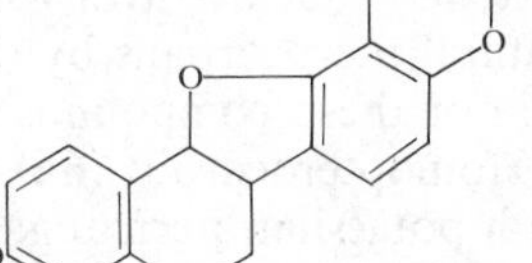

(**62a**) $R^1 = OH$, $R^2 = H$
(**62b**) $R^1 = R^2 = OH$
(**62c**) $R^1 = OMe$, $R^2 = H$

(**63a**) $R^1 = OMe$, $R^2 = H$
(**63b**) $R^1 = OMe$, $R^2 = OH$
(**63c**) $R^1 =$ O-glucosyl, $R^2 = H$

(**64**)

(**65**)

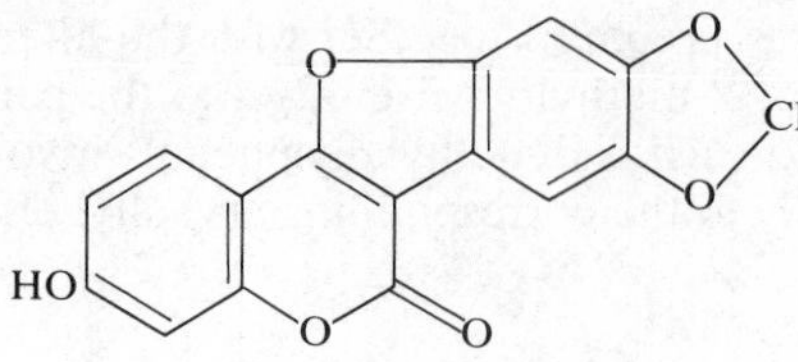

(66)

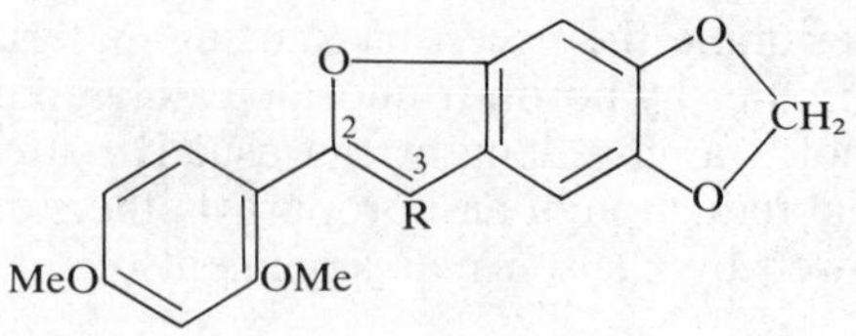

(67a) $R=CO_2Me$
(67b) $R=H$

18.5.4.2 Chemical properties

The α-pyrone ring of the coumestan system is readily opened by aqueous alkali. For example, treatment of medicagol (**66**) with alkaline dimethyl sulphate yields the benzofurancarboxylate ester (**67a**). Thermal decarboxylation of the derived carboxylic acid gives the benzofuran derivative (**67b**).[79] Further degradation may be accomplished by oxidative cleavage of the benzofuran 2,3-double bond and hydrolysis of the resulting ester. This degradative sequence has been utilized in the structure determination of several of the natural coumestans.[2]

Lithium aluminium hydride treatment of the coumestan system results in reductive cleavage of the α-pyrone ring, as illustrated in Scheme 19. Benzofuran derivative (**68**) is

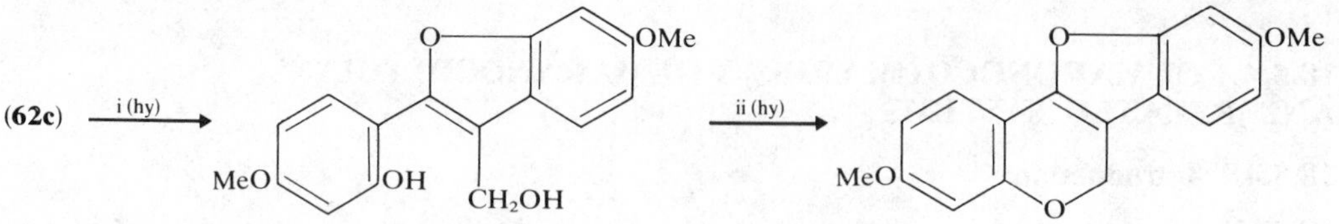

(**62c**) —i (hy)→ (**68**) —ii (hy)→ (**69**)

i, $LiAlH_4$; ii, ethylene glycol, reflux.

SCHEME 19

thermally cyclized to the coumaronochromene system (**69**),[80] which may be formed directly from the starting coumestan by diborane treatment.[81] The coumaronochromene (**69**) may be reoxidized to the starting coumestan (**62c**) in low yield. Treatment of (**69**) with t-butyl perchlorate yields the chromylium salt (**70**), while hydrogenolysis yields the 3-arylchroman (**71**), which is also formed on hydrogenolysis of homopterocarpin (**64**),

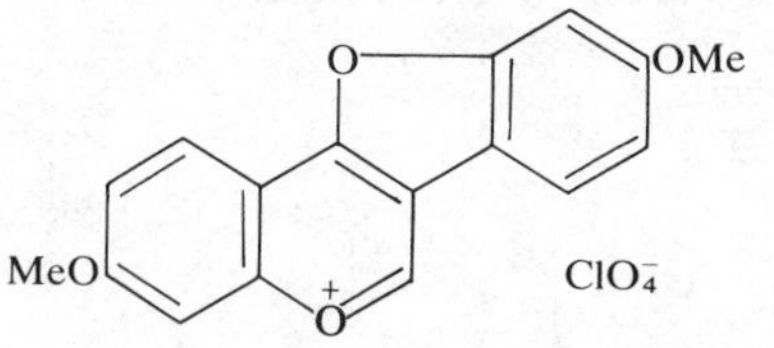

(70) 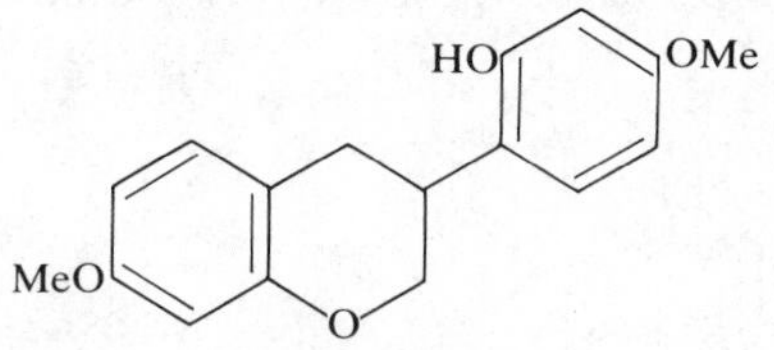 **(71)**

thus correllating the two series.[80] Catalytic reduction of coumaronochromenes such as (**69**) may also yield the pterocarpan nucleus, and constitutes one synthetic route to this system.[82] Such a hydrogenation must give the *cis*-pterocarpan system; the synthesis of natural pterocarpans by this route provides strong evidence for the *cis*-6a,11a configuration of these compounds.

Homopterocarpin (**64**) is degraded to carboxylic acids (**72a**) and (**72b**) on treatment with potassium permanganate. 3-Anisoyl-7-methoxycoumarin is formed from (**64**) in low yield on dehydrogenation over palladium at elevated temperatures. The mechanism of the deep-seated rearrangement which is involved is not clear.[80] Mild acid treatment of pterocarpin gives the chromene derivative (**73a**). Racemic pisatin (**63b**) may be prepared

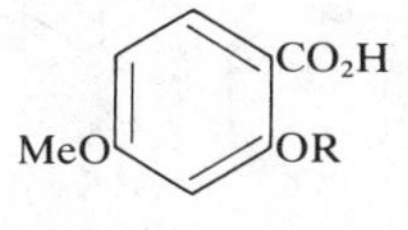

(**72a**) R = H
(**72b**) $R = CH_2CO_2H$

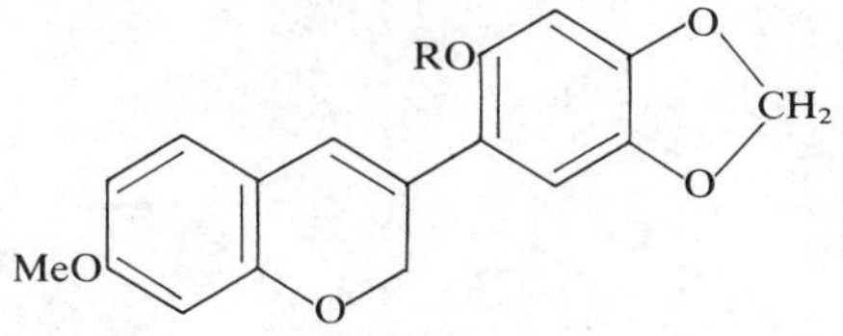

(**73a**) R = H
(**73b**) R = Ac

by osmium tetroxide treatment of the derived acetate (**73b**), followed by base-catalysed hydrolysis of the osmate ester so formed.[83] Pisatin and other 6a-hydroxypterocarpans readily form the 6a,12a-anhydro-derivatives on treatment with acids.

The absolute stereochemistry of the pterocarpan glycoside trifolirhizin (**63c**) has been determined directly by hydrogenolysis of the tetra-acetate to give a 3-arylchroman which retains the 6a-chiral centre, followed by ozonolysis of the latter which gave (2*S*)-2-hydroxymethylbutanedioic acid; hence the glycoside (**63c**) has the 6a*R*-configuration and, owing to the known *cis* relationship, must also have the 11a*R*-configuration.[84] It has been suggested that the natural pterocarpans which display a negative optical rotation have the 6a*R*,12a*R*-configuration;[84] o.r.d. studies are more useful in establishing the absolute stereochemistry of these compounds.[85]

18.5.4.3 Synthetic methods

Coumestans may be synthesized by a variety of routes, the first of which is illustrated in Scheme 20. The dimethoxydeoxybenzoin derivative, formed by the Hoesch reaction depicted, is converted to the trimethoxy-derivative (**74**). The remaining phenolic hydroxyl

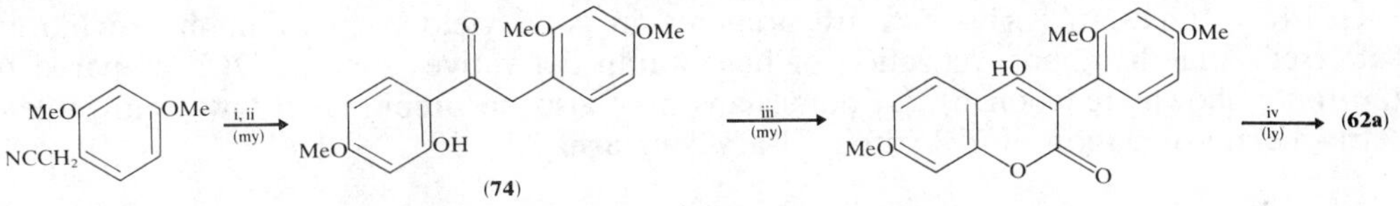

(**74**)

i, HO–C₆H₄–OH, HCl; ii, CH_2N_2; iii, $(MeO)_2CO$, Na; iv, HI, AcOH.

SCHEME 20

is protected from methylation by virtue of an intramolecular hydrogen bond to the carbonyl group. Conversion of (**74**) to a 4-hydroxy-3-arylcoumarin is followed by hydrogen iodide catalysed demethylation concomitant with cyclization to coumestrol. This route is improved by condensation of (**74**) with methyl chloroformate and demethylation–cyclization of the resulting 4-hydroxycoumarin.[86] The necessary deoxybenzoin derivatives may also be prepared by alkaline hydrolysis of suitably constituted isoflavones.[87] An alternative synthesis of coumestrol is illustrated in Scheme 21;[88] the brevity of this route adequately compensates for the somewhat low overall yields obtained.[88,89]

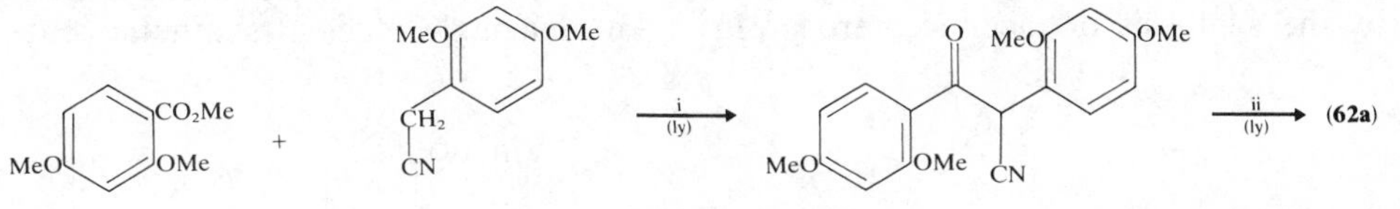

i, NaH; ii, py.HCl.

SCHEME 21

The method of choice for the synthesis of 8,9-dihydroxycoumestans is illustrated by the preparation of wedelolactone (reaction 5). The mechanism of this remarkably efficient reaction must involve oxidation of catechol to *o*-benzoquinone, addition of the latter to

(5)

the nucleophilic C-3 of the 4-hydroxycoumarin (**75**), reoxidation to a substituted *o*-benzoquinone, and ring-closure by internal Michael addition of the 4-hydroxy-group to the enone of the quinone moiety.[90] Other oxidants such as potassium periodate or electrochemical oxidation may be used.[91] The ease of selective methylation of 4,5,7-trihydroxycoumarin with alkaline dimethyl sulphate to yield 7-methoxycoumarin (**75**) is another example of the protection afforded a phenolic hydroxyl by intramolecular hydrogen bonding. Another excellent synthesis of coumestan derivatives relies on the ring contraction which occurs on hydrogen peroxide oxidation of 2-arylchromylium salts (Scheme 22). The requisite starting material is readily available by standard methods.[92]

i, H_2O_2; ii, HCl.

SCHEME 22

8-Hydroxycoumestan derivatives are prepared in good yield by pyridine hydrochloride catalysed demethylation–cyclization of benzofuran derivatives such as (**76**), prepared by the route shown (reaction 6).[93] Coumestans may also be prepared in low yield by lead tetra-acetate oxidation of 2′-hydroxy-3-arylcoumarins.[94]

(6)

Synthetic routes to the pterocarpan system are less well developed. One route, namely catalytic reduction of 6a,11a-dehydropterocarpans derived from coumestan derivatives, has already been mentioned above. An alternative route involves sodium borohydride reduction of an appropriate 2′-hydroxyisoflavone; cyclization of the intermediate hydroxychroman is performed under very mild conditions. This route is illustrated in Scheme 23 by the synthesis of racemic pterocarpin.[95] An amazingly efficient synthetic route to

i, $NaBH_4$; ii, AcOH.

SCHEME 23

pterocarpin is illustrated in equation 7. The requisite chloromercuriphenol is readily available through the action of mercuric acetate, followed by sodium chloride, on 3,4-(methylenedioxy)phenol.[96]

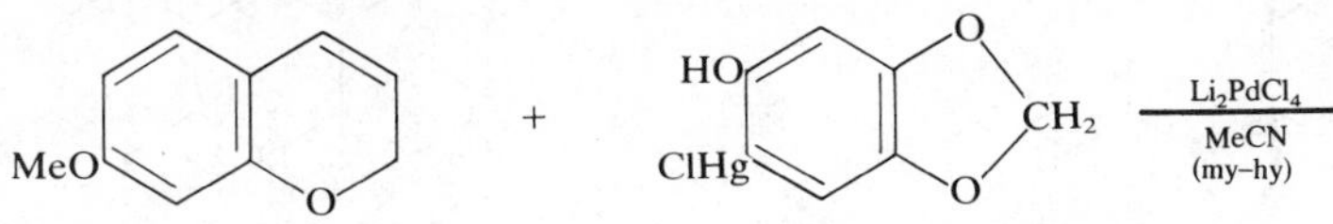

MeO ... + HO ... ClHg ... CH_2 —(Li_2PdCl_4, MeCN (my–hy))→ **(63a)** (7)

18.5.4.4 Coumaronochromones

These systems have not been extensively studied, probably owing to the dearth of natural products of this type. The unique isoflavanoid lisetin (**77**) shows a carbonyl stretching frequency at 1693 cm^{-1} in the infrared; in contrast, the isomeric coumestan system shows ν_{max} in the region 1700–1740 cm^{-1}. Alkali treatment of the tri-*O*-methyl

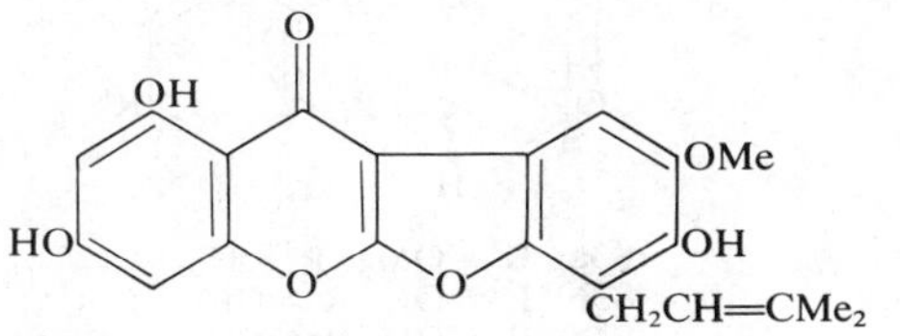

(**77**) 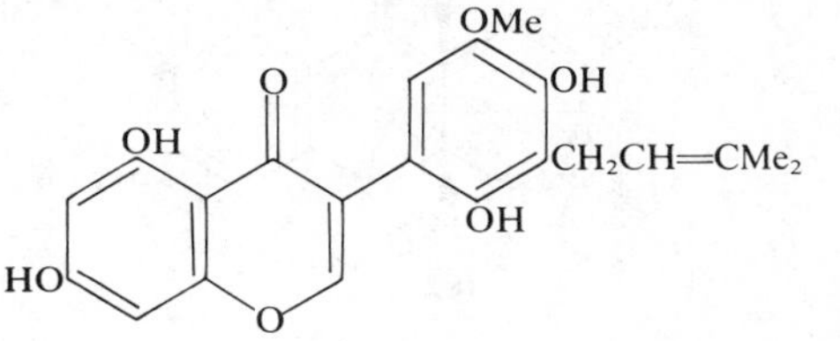 (**78**)

derivative of lisetin gave the derived 3-aryl-4-hydroxy-5,7-dimethoxycoumarin *via* conjugate addition of hydroxide to the pyrone ring. Lisetin has been synthesized by potassium ferricyanide oxidation of isoflavone (**78**).[97] The isomeric coumaronochromone skeleton has been prepared by base-catalysed cyclization of keto-ester (**79**) followed by acid-catalysed ring closure of the resulting coumaran-3-one (reaction 8).[98]

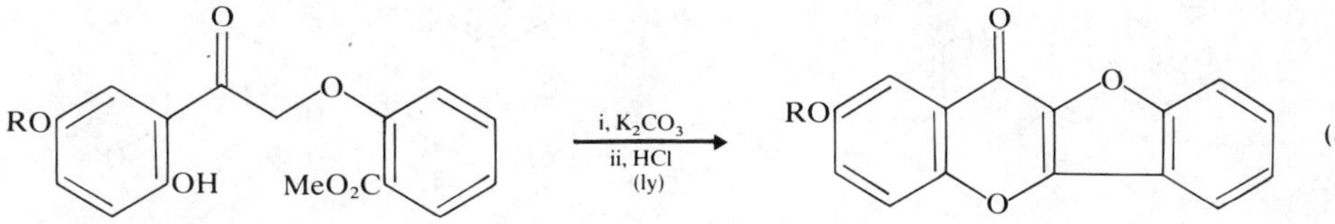

(**79**) —(i, K_2CO_3; ii, HCl (ly))→ (8)

18.5.5 CHROMANOCHROMANONES: THE ROTENOIDS

18.5.5.1 Introduction

The rotenoids are a group of complex isoflavanoids based on the chromanochromanone system. The structural variation found in these natural products is well illustrated by the typical 'angular' rotenoids rotenone (**80**), α-toxicarol (**81**), and the 'linear' pachyrrhizone (**82**). The simplest natural rotenoid known is the tetracyclic compound munduserone (**83a**). Rotenone and its analogues are found in the roots of a number of species of the Papilionacea sub-family of the Leguminosae; the most important sources are the genera *Derris, Lonchocarpus, Tephrosia,* and *Mundelea.* Rotenoids cause paralysis in fish and display highly insecticidal properties.[99] Commercial preparations containing rotenoids are in common use for insect control.

Most of the following discussion on the rotenoid group will be illustrated through the chemistry of the typical member rotenone (**80**). Although rotenone is one of the more complex of these compounds, it was the first to be extensively studied. The correct structure of rotenone was advanced in 1932 by several independent workers, but not until

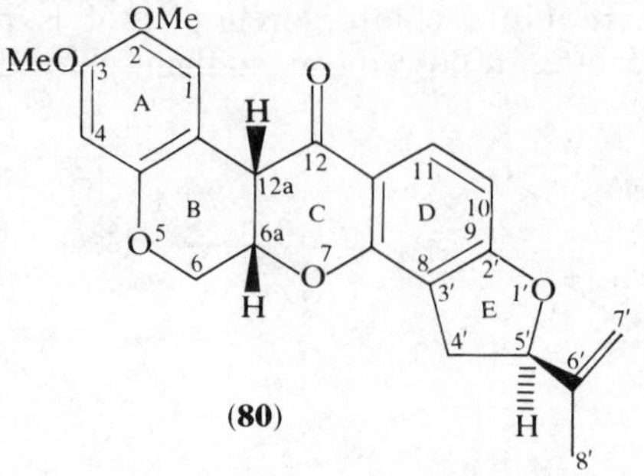

(80)

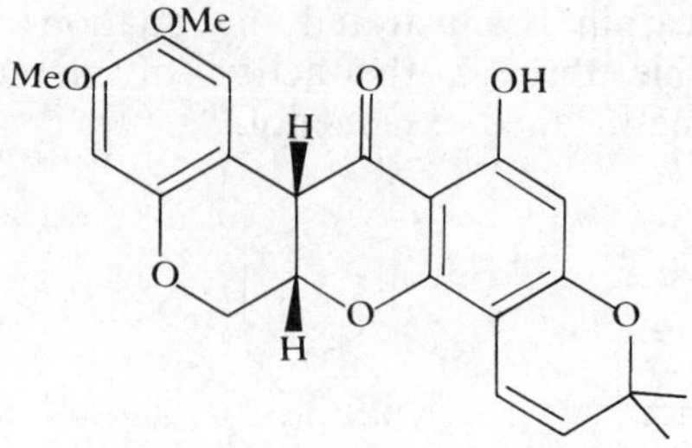

(81)

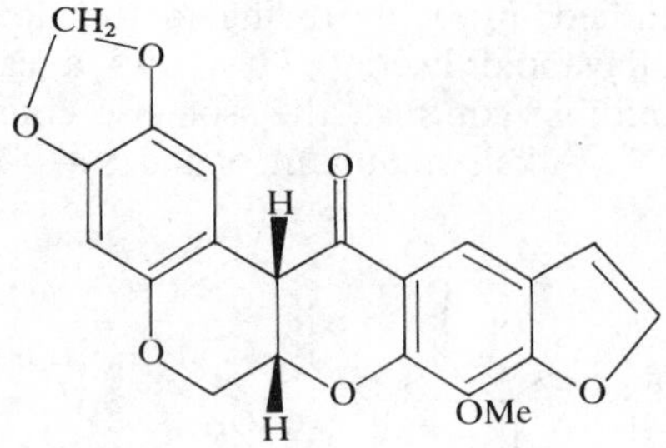

(82)

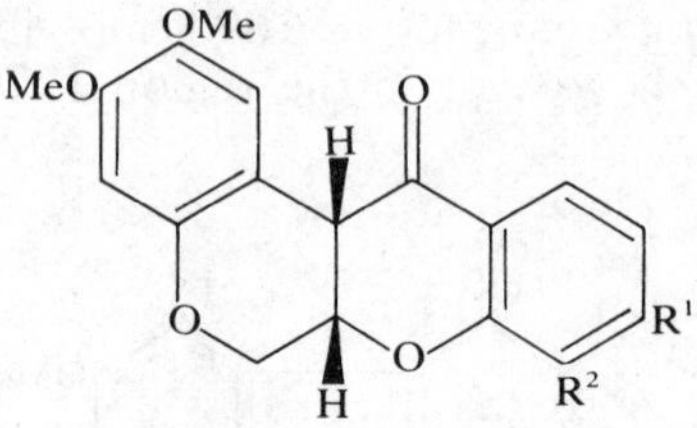

(83a) $R^1 = OMe$, $R^2 = H$
(83b) $R^1 = OH$, $R^2 = CH_2CH{=}CMe_2$
(83c) $R^1 = OH$, $R^2 = CH_2CH_2CHMe_2$

1961 was the stereochemical detail defined.[100] The early literature has been reviewed extensively.[1,2,75,101] A semisystematic nomenclature for the rotenoids based on the 'rotoxen' skeleton (**84**) has been proposed.[100]

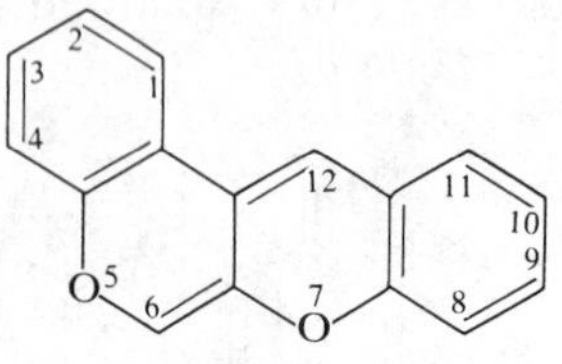

(84)

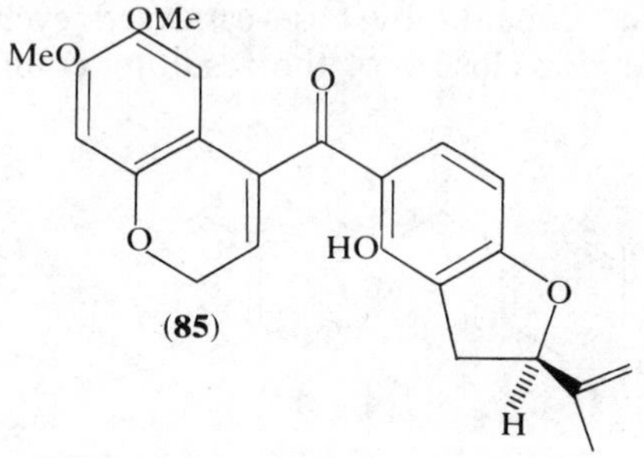

(85)

18.5.5.2 Chemical properties

Rotenone is a colourless crystalline material which exhibits polymorphism. Rapid equilibration with its diastereoisomer, 6a,12a-epirotenone, occurs on mild base treatment (reaction 9) through the transient formation of the Δ^3-chromene (**85**). The equilibrium (**80**) ⇌ (**85**) is reminiscent of the flavanone–chalcone equilibrium, but dehydrorotenol (**85**)

(80) 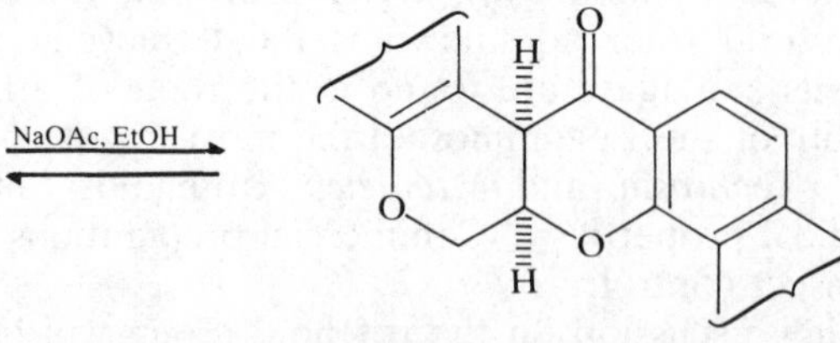 $\xrightarrow{\text{NaOAc, EtOH}}$ (9)

lacks the conjugative stabilization available to the chalcones and is isolable only with difficulty.[102] 'Angular' 11-hydroxyrotenoids such as α-toxicarol (**81**) undergo facile base-catalysed equilibration with their racemic 'linear' analogues owing to similar opening of ring-c followed by ring closure with the 11-hydroxy-group.[102,103] Hydroxylamine reacts with alkaline solutions of rotenone to give an isoxazoline, *via* the oxime of dehydrorotenol (**85**). Rotenone forms a normal oxime at neutral pHs.[102]

The stable diastereoisomeric Δ^2-chromenes (**86**) may be isolated from strongly alkaline solutions of rotenone. Recyclization to rotenone occurs on treatment of (**86**) with wet sodium carbonate, owing to equilibration with Δ^3-chromene (**85**).[102] An interesting skeletal rearrangement occurs on treatment of (**86**) with sulphuric acid in methanol (reaction 10), following acid-catalysed opening of the pyran ring.[104]

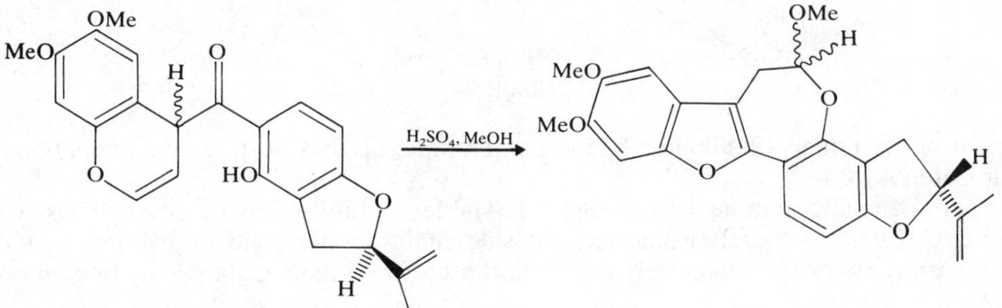

OMe MeO H O HO O H (**86**) $\xrightarrow{H_2SO_4,\ MeOH}$ OMe H MeO MeO O O H O (10)

Rotenone readily undergoes dehydrogenation on treatment with manganese dioxide. The carbonyl group of the product, 6a,12a-dehydrorotenone (**87a**), is unreactive; ring opening occurs with aqueous potassium hydroxide to form derrisic acid (**88a**) *via* conjugative addition of hydroxide and ring fission of the β-diketone intermediate (**89a**). Further

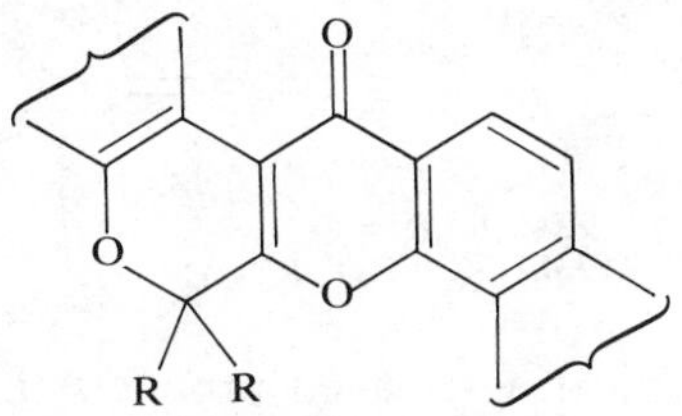

(**87a**) $R_2 = H_2$
(**87b**) $R_2 = O$

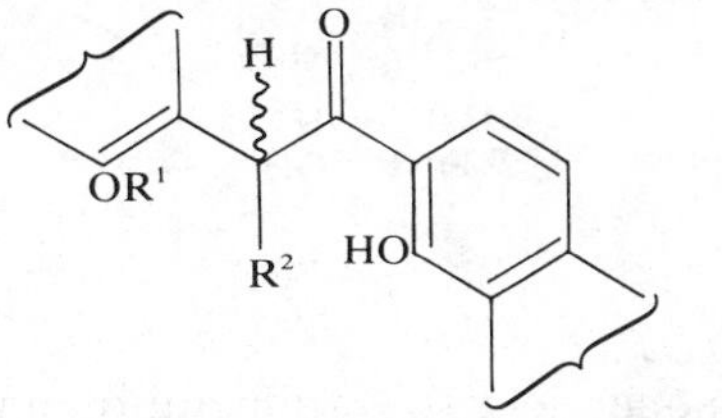

(**88a**) $R^1 = CH_2CO_2H$, $R^2 = H$
(**88b**) $R^1 = R^2 = H$
(**88c**) $R^1 = H$, $R^2 = COCO_2H$

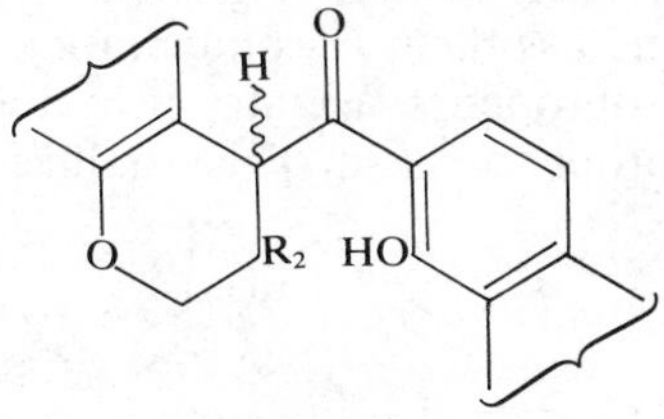

(**89a**) $R_2 = O$
(**89b**) $R_2 = H_2$

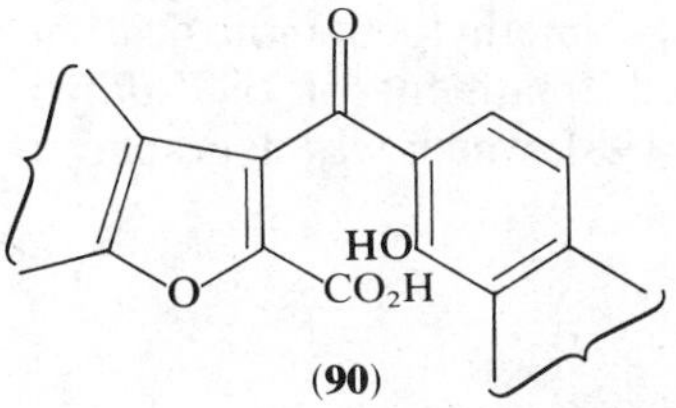

(**90**)

oxidation of dehydrorotenone, preferably with lead tetra-acetate, yields the yellow α-pyrone rotenonone (**87b**). Mild alkali treatment of the latter gives benzofuran (**90**), while more vigorous alkali treatment yields oxalic acid and derritol (**88b**). Both routes must involve the intermediate formation of β-diketone (**88c**).[102] Rotenonone has recently been isolated as a natural product.[105]

Rotenone is degraded by zinc and alkali to rotenol (**89b**) and derritol (**88b**), as illustrated in Scheme 24. The formation of derritol occurs through elimination of acetaldehyde from intermediate (**91**) by a retro-aldol reaction. Rotenol (**89b**) is oxidized to 6a,12a-dehydrorotenone (**87a**) and spiro-ketone (**92a**) by manganese dioxide; zinc–alkali reduction of (**92a**) regenerates rotenol.[102] Hydroxyspiro-ketone (**92b**) is formed on

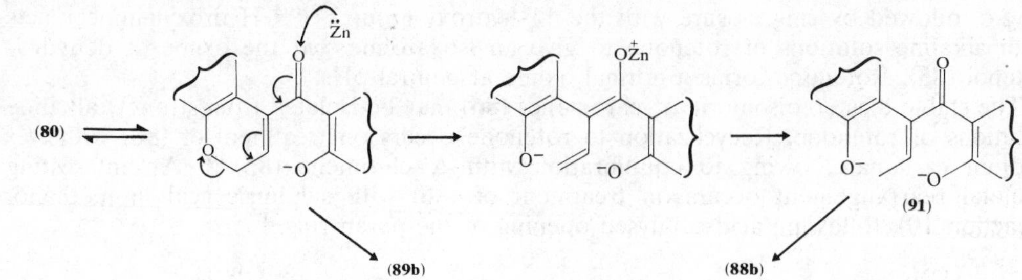

Zinc–alkali reduction of rotenone.

SCHEME 24

treatment of rotenone with alkaline hydrogen peroxide, *via* base-catalysed epoxidation of dehydrorotenol (**85**).[106]

The chromanochromanone system itself is stable to the action of acids. Thus acid treatment of rotenone merely isomerizes the side-chain double-bond to give isorotenone (**93a**), the properties of which closely resemble those of rotenone. Catalytic hydrogenation

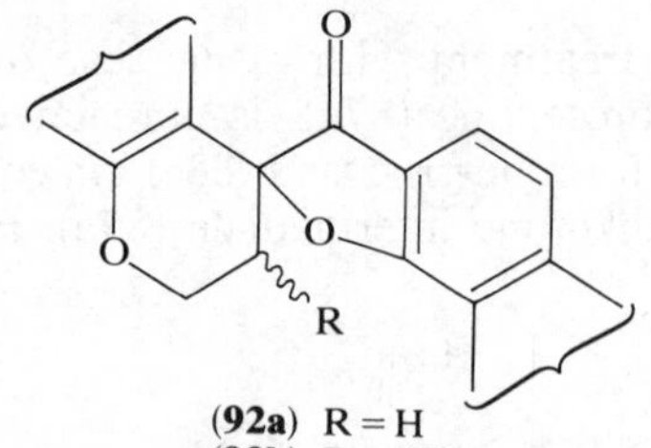

(**92a**) R = H
(**92b**) R = OH

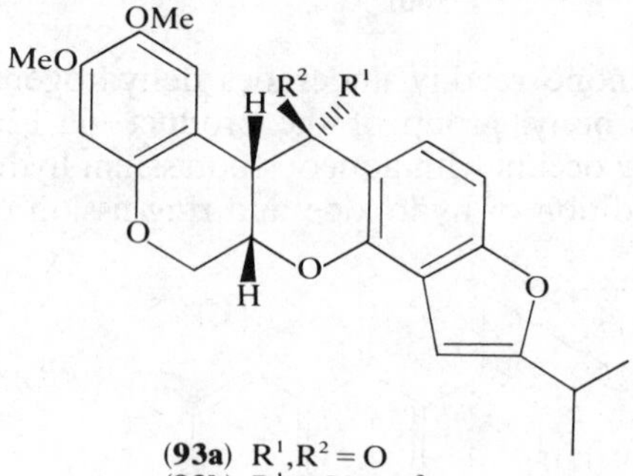

(**93a**) R^1,R^2 = O
(**93b**) R^1 = OH, R^2 = H

of rotenone yields 6′,7′-dihydrorotenone, or the furan ring-cleaved products (**83b**) and (**83c**), depending on the conditions. Reduction of the carbonyl group is also possible.[2]

The stereochemistry of rotenone was determined by Büchi *et al.* in 1961.[100] Prolonged alkaline treatment of rotenone gave tubaic acid (**94a**) which was degraded to (*R*)-3-hydroxy-2-methylpentanoic acid; hence natural rotenone has the 5′*R*-configuration. Acid-catalysed treatment of 6′,7′-dihydrorotenone with isopropenyl acetate gave the enol acetate (**95**), which was degraded to *R*-2,3-dihydroxypropanoic acid. This established the

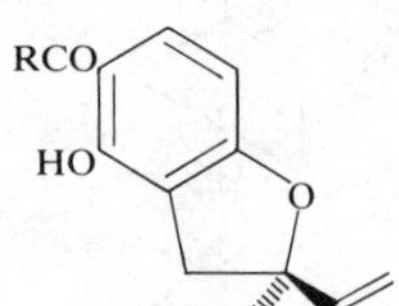

(**94a**) R = OH
(**94b**) R = Cl

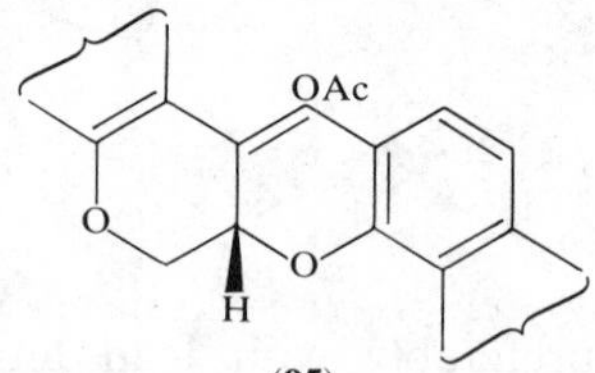

(**95**)

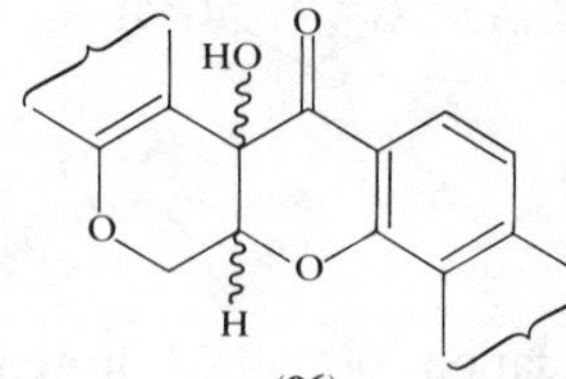

(**96**)

6a*S*-configuration for rotenone. The configuration at the third chiral centre (12a) in rotenone (**80**) was determined by an elegant demonstration that rings B and C are *cis*-fused. That the B/C ring junction has the thermodynamically stable configuration had long been known, since base treatment of rotenone yielded only one pair of diastereoisomers, rotenone and epirotenone, while isorotenone similarly yielded a single racemate, (±)-isorotenone. Sodium borohydride reduction of isorotenone gave the acid-labile alcohol (**93b**) in high yield. The i.r. spectrum of this product showed the existence of an intramolecular hydrogen bond. Study of molecular models revealed that only the B/C *cis*-fused alcohol (**93b**) could form an intramolecular hydrogen bond between the hydroxy-group and the 5-oxygen atom. Hence *cis*-fusion of the B/C ring junction was established, and the final chiral centre, 12a, of rotenone has the *S* configuration.[100] Single-crystal X-ray diffraction studies on 8′-bromorotenone have confirmed the relative and absolute stereochemistry of rotenone (**80**).[107]

O.r.d. studies have proved useful in correlating the absolute stereochemistry of other rotenoids with rotenone; the natural rotenoids all show a positive Cotton effect and possess the same absolute stereochemistry at 6a and 12a.[103,108] Proton n.m.r. spectroscopy has also proved useful in establishing the *cis* B/C ring fusion of the natural rotenoids. The C-1 proton in a *cis* B/C ring-fused rotenoid resonates some 1 p.p.m. upfield from that in the *trans*-fused isomer, owing to the large deshielding of this proton by the carbonyl group in the latter isomer.[109]

Isorotenone is oxidized by air in an alkaline medium to give two racemic 12a-hydroxy-derivatives (**96**), which differ only in the stereochemistry of the B/C ring fusion. These isorotenolones are stable to manganese dioxide or alkali treatment but undergo acid-catalysed dehydration to give 6a,12a-dehydroisorotenone. Their structures have been investigated by Crombie and Godin.[106] A number of 12a-hydroxy- and 12a-methoxy-rotenoids have been isolated from natural sources.[105,110]

18.5.5.3 Synthesis

Rotenoids are readily synthesized by the route depicted in Scheme 25. The Hoesch reaction between benzyl cyanides (**97**) and the requisite phenol component generally gives acceptable yields of the deoxybenzoin esters (**98**), but if either component contains an acid-sensitive group protection is necessary. The ester product (**98**) is hydrolysed and the carboxylic acid cyclized with acetic anhydride–sodium acetate.[111,112] Better yields are obtained in this step if the carboxylic acid is cyclized with dicyclohexylcarbodi-imide[113,114] or the esters (**98**) treated with sodium ethoxide[114] or potassium carbonate.[115] The 6a,12a-dehydrorotenoid product (**99**) is finally reduced with sodium borohydride and the *cis*-dihydrobenzylic alcohol product is reoxidized to the required rotenoid.[116] This sequence

CN
CO_2Et
(**97**)
\+
HO R
i (ly–my)
O
HO
CO_2Et
(**98**)

ii, iii (ly–my)
O O O
(**99**)
iv, v (ly–my)
H O H O O

i, HCl, $ZnCl_2$; ii, KOH; iii, Ac_2O, NaOAc; iv, $NaBH_4$; v, Me_2CO, $Al(OPr^i)_3$.

SCHEME 25

of reactions has been utilized in the total synthesis of several rotenoids, including rotenone, since the dehydrorotenoid intermediates (**99**) and their deoxybenzoin carboxylic acid precursors are readily available as relay compounds through degradation procedures discussed above. The formal total synthesis of rotenone claimed by Miyano and co-workers[111] suffers from two major drawbacks, namely the exceedingly low overall yield of synthetic racemic derrisic acid (**88a**) and the non-resolution of this synthetic material. A more satisfactory total synthesis of rotenone utilizes dehydrorotenone prepared through condensation of enamine (**100**) with tubaic acid chloride (**94b**) (reaction 11).[117]

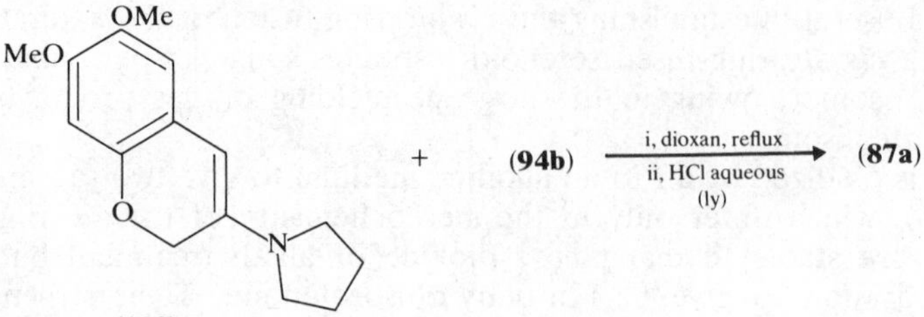

(**100**) + (**94b**) —(i, dioxan, reflux; ii, HCl aqueous (ly))→ (**87a**) (11)

Suitably constituted natural or synthetic isoflavones provide an alternative source of deoxybenzoin carboxylic acids.[118–120] In the total synthesis of munduserone (**83a**) described by Seshadri and co-workers, isoflavone (**101a**) was selectively demethylated in high yield with aluminium trichloride, presumably *via* a chelated intermediate such as (**102**). The phenol product (**101b**) was converted quantitatively to the ester (**101c**), which underwent base-catalysed hydrolysis to give the required deoxybenzoin derivative (**103**). The latter was converted to munduserone by the route already discussed.[118]

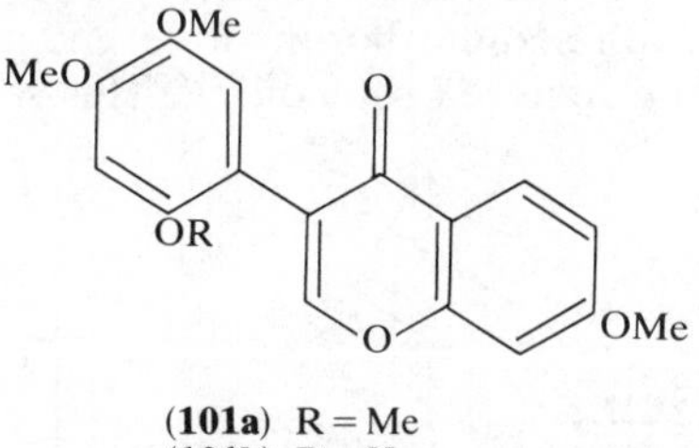

(**101a**) R = Me
(**101b**) R = H
(**101c**) R = CH_2CO_2Me

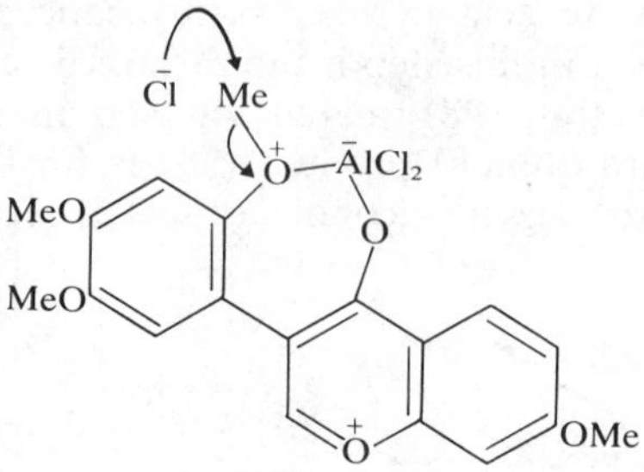

(**102**)

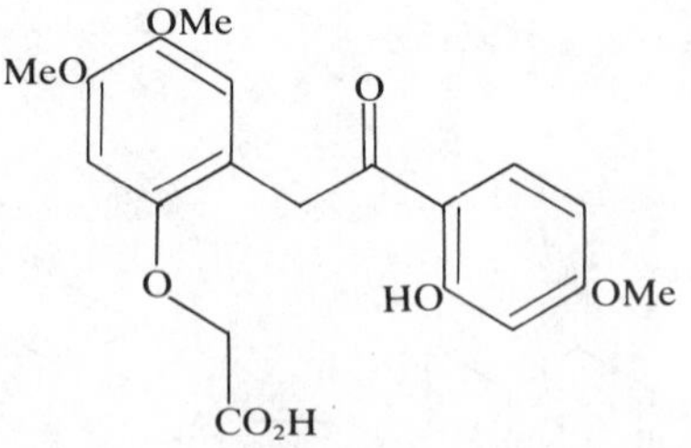

(**103**)

Several synthetic routes have been reported which by-pass the dehydrorotenoid system and hence avoid the reduction–oxidation sequence of Scheme 25. Thus pyrolysis of acetylenic ketone (**104**) gives munduserone directly,[121] while heating of deoxybenzoin aldehyde (**105**) in pyridine yields isorotenone.[122] The most interesting route is illustrated

OMe MeO O O HO OMe (**104**)

OMe MeO O O HO O CHO CHMe₂ (**105**)

by the synthesis of isorotenone in Scheme 26. Conjugate addition of one equivalent of dimethylsulphoxonium methylide to isoflavone (**106**) followed by pyrone ring-cleavage and recyclization gives vinyl coumaranone (**107**). The latter is isomeric with isorotenone, which is formed on heating in pyridine, through the intermediacy of quinone methide (**108**) and 6a,12a-dehydroisorotenol.[123]

(**106**) (**107**) (**108**) (**93a**)

i, $Me_2\overset{+}{S}O\overset{-}{C}H_2$ (1 equivalent); ii, pyridine, reflux.

SCHEME 26

References

1. W. B. Whalley, in 'Heterocyclic Compounds', ed. R. C. Elderfield, Wiley, New York, 1961, vol. 7, p. 1.
2. A. Mustafa, in 'Heterocyclic Compounds', ed. A. Weissberger, Wiley-Interscience, London, 1967, vol. 23, p. 1.
3. F. M. Dean, 'Naturally Occurring Oxygen Ring Compounds', Butterworths, London, 1963, p. 1.
4. F. M. Dean, in 'The Total Synthesis of Natural Products', ed. J. ApSimon, Wiley-Interscience, New York, 1973, vol. 1, p. 467.
5. A. Schönberg, N. Badran, and N. A. Starkowsky, *J. Amer. Chem. Soc.*, 1955, **77,** 5438.
6. A. I. Gray, R. D. Waigh, and P. G. Waterman, *J.C.S. Chem. Comm.*, 1974, 632.
7. M. W. Jarvis and A. G. Moritz, *Austral. J. Chem.*, 1968, **21,** 2445.
8. C. S. Barnes and J. L. Occolowitz, *Austral. J. Chem.*, 1964, **17,** 975.
9. J. K. MacLeod and M. Nakayama, *Org. Mass Spectrometry*, 1972, **6,** 293.
10. M. E. Brokke and B. E. Christensen, *J. Org. Chem.*, 1958, **23,** 589.
11. A. Schönberg, N. Badran, and N. A. Starkowsky, *J. Amer. Chem. Soc.*, 1955, **77,** 1019.
12. S. Bhanu, T. R. Seshadri, and S. K. Mukerjee, *Indian J. Chem.*, 1975, **13,** 15 (*Chem. Abs.*, 1975, **83,** 28 132).
13. E. Späth and L. Socias, *Ber.*, 1934, **67,** 59.
14. A. F. Aboulezz, A. A. El-Attar, and M. A. El-Sockary, *Acta Chim. Acad. Sci. Hung.*, 1973, **77,** 205 (*Chem. Abs.*, 1973, **79,** 92 054).
15. M. Murayama, H. Murai, K. Sempuku, and T. Suminokura, *Chem. Pharm. Bull. (Japan)*, 1970, **18,** 2453 (*Chem. Abs.*, 1971, **74,** 53 677).

16. W. Asker, A. F. A. M. Shalaby, and S. M. A. Zayed, *J. Org. Chem.*, 1958, **23,** 1781.
17. B. S. Joshi, K. D. Dabholkar, and D. H. Gawad, *Indian J. Chem.*, 1972, **10,** 567 (*Chem. Abs.*, 1973, **78,** 29 547).
18. M. E. Brokke and B. E. Christensen, *J. Org. Chem.*, 1959, **24,** 523.
19. K. D. Kaufman, L. R. Worden, E. T. Lode, M. K. Strong, and N. C. Reitz, *J. Org. Chem.*, 1970, **35,** 157.
20. E. A. Abu-Mustafa and E. A. M. Elkhrisy, *J. Heterocyclic Chem.*, 1974, **11,** 1119.
21. C. H. Krauch and S. Farid, *Chem. Ber.*, 1967, **100,** 1685.
22. L. Musajo and G. Rodighiero, *Photophysiology*, 1972, **7,** 115.
23. E. C. Horning and D. E. Reisner, *J. Amer. Chem. Soc.*, 1948, **70,** 3619.
24. R. C. Esse and B. E. Christensen, *J. Org. Chem.*, 1960, **25,** 1565.
25. D. K. Chatterjee and K. Sen, *J. Indian Chem. Soc.*, 1971, **48,** 387 (*Chem. Abs.*, 1971, **75,** 76 641).
26. R. T. Foster, A. Robertson, and A. Bushra, *J. Chem. Soc.*, 1948, 2254.
27. S. Bhanu, T. Saroja, T. R. Seshadri, and S. K. Mukerjee, *Indian J. Chem.*, 1971, **9,** 380 (*Chem. Abs.*, 1971, **75,** 48 943).
28. L. P. Vettori, G. Auzzi, P. Papini, and F. Bruni, *Farmaco Ed. Sci.*, 1975, **30,** 754 (*Chem. Abs.*, 1976, **84,** 43 896).
29. C. Antonello, *Gazzetta*, 1958, **88,** 430.
30. R. M. Naik and V. M. Thakor, *J. Org. Chem.*, 1957, **22,** 1696.
31. Y. Kawase, M. Nakayama, and H. Tamatsukuri, *Bull. Chem. Soc. Japan*, 1962, **35,** 149.
32. K. N. H. Pardanani, M. G. Parekh, and K. N. Trivedi, *J. Indian Chem. Soc.*, 1969, **46,** 1014 (*Chem. Abs.*, 1970, **72,** 78 918).
33. T. R. Seshadri and M. S. Sood, *Indian J. Chem.*, 1963, **1,** 291 (*Chem. Abs.*, 1963, **59,** 12 773).
34. V. K. Ahluwalia, T. R. Seshadri, and P. Venkateswarlu, *Indian J. Chem.*, 1971, **9,** 194 (*Chem. Abs.*, 1971, **75,** 5753).
35. V. K. Ahluwalia and C. Prakash, *Indian J. Chem.*, 1975, **13,** 791 (*Chem. Abs.*, 1975, **83,** 164 039).
36. N. H. Pardanani and K. N. Trivedi, *Austral. J. Chem.*, 1972, **25,** 1537.
37. K. D. Kaufman, W. E. Russey, and L. R. Worden, *J. Org. Chem.*, 1962, **27,** 875.
38. K. D. Kaufman, *J. Org. Chem.*, 1961, **26,** 117.
39. V. N. Dholakia and K. N. Trivedi, *J. Indian Chem. Soc.*, 1973, **50,** 813 (*Chem. Abs.*, 1974, **81,** 152 056).
40. J. K. MacLeod and B. R. Worth, *Tetrahedron Letters*, 1972, 237.
41. H. G. Floss, *Recent Adv. Phytochem.*, 1972, **4,** 143.
42. E. A. Abu-Mustafa and M. B. E. Fayez, *J. Org. Chem.*, 1961, **26,** 161.
43. I. Harada, Y. Hirose, and M. Nakazaki, *Tetrahedron Letters*, 1968, 5463.
44. B. E. Nielsen and J. Lemmich, *Acta Chem. Scand.*, 1964, **18,** 2111.
45. R. D. H. Murray, M. Sutcliffe, and P. H. McCabe, *Tetrahedron*, 1971, **27,** 4901.
46. W. Steck, *Canad. J. Chem.*, 1971, **49,** 2297.
47. V. S. Kamat, G. K. Trivedi, and S. C. Bhattacharyya, *J.C.S. Perkin I*, 1976, 1857.
48. B. E. Nielsen and J. Lemmich, *Acta Chem. Scand.*, 1964, **18,** 932.
49. J. C. Roberts, in 'Progress in the Chemistry of Organic Natural Products', ed. W. Herz, H. Grisebach, and G. W. Kirby, Springer-Verlag, Vienna, 1974, vol. 31, p. 119.
50. G. Büchi and S. M. Weinreb, *J. Amer. Chem. Soc.*, 1971, **93,** 746.
51. E. E. Galal, A. Kandil, M. Abdel Latif, and T. Khedr, *J. Drug. Res.*, 1975, **7,** 145.
52. M. M. Sidky and M. R. Mahran, *J. Amer. Chem. Soc.*, 1962, **84,** 4112.
53. O. Dann and G. Volz, *Annalen*, 1965, **685,** 167.
54. A. Mustafa, M. M. Sidky, and M. R. Mahran, *Annalen*, 1967, **704,** 182.
55. R. Beugelmans and C. Morin, *Tetrahedron Letters*, 1976, 2145.
56. G. N. Dorofeenko, Yu. A. Zhdanov, and T. G. Soroka, *Zhur. obshchei Khim.*, 1967, **37,** 743 (*Chem. Abs.*, 1967, **67,** 43 694).
57. F. Eiden and U. Rehse, *Chem. Ber.*, 1974, **107,** 1057.
58. A. Schönberg, A. Mustafa, and G. Aziz, *J. Amer. Chem. Soc.*, 1954, **76,** 4576.
59. S. Raychaudhuri, T. R. Seshadri, and S. K. Mukerjee, *Indian J. Chem.*, 1972, **10,** 1125 (*Chem. Abs.*, 1973, **78,** 124 474).
60. A. Schönberg and G. Aziz, *J. Amer. Chem. Soc.*, 1953, **75,** 3265.
61. A. Schönberg, N. Badran, and N. A. Starkowsky, *J. Amer. Chem. Soc.*, 1955, **77,** 5390.
62. G. N. Dorofeenko, V. V. Tkachenko, and V. V. Mezheritskii, *Khim. geterotsikl. Soedinenii*, 1973, 1020 (*Chem. Abs.*, 1973, **79,** 126 348).
63. A. Schönberg, N. Badran, and N. A. Starkowsky, *J. Amer. Chem. Soc.*, 1955, **77,** 5439.
64. T. A. Geissman and T. G. Halsall, *J. Amer. Chem. Soc.*, 1951, **73,** 1280.
65. P. K. Ramachandran, A. T. Tefteller, G. O. Paulson, T. Cheng, C. T. Lin, and W. J. Horton, *J. Org. Chem.*, 1963, **28,** 398.
66. M. Nanbu, S. Yamaguchi, Y. Sugimasa, T. Miyaura, and Y. Kawase, *Bull. Chem. Soc. Japan*, 1975, **48,** 3423.
67. O. Dann and H. G. Zeller, *Chem. Ber.*, 1960, **93,** 2829.
68. M. M. Badawi and M. B. E. Fayez, *Tetrahedron*, 1965, **21,** 2925.
69. R. Anjena, S. K. Mukerjee, and T. R. Seshadri, *Tetrahedron*, 1958, **3,** 230.
70. S. K. Mukerjee, S. C. Sarkar, and T. R. Seshadri, *Tetrahedron*, 1969, **25,** 1063.
71. A. C. Jain and S. M. Jain, *Tetrahedron*, 1973, **29,** 2803.

72. J. Hlubucek, E. Ritchie, and W. C. Taylor, *Austral. J. Chem.*, 1971, **24,** 2347.
73. R. D. H. Murray, M. M. Ballantyne, and K. P. Mathai, *Tetrahedron*, 1971, **27,** 1247.
74. A. K. Ganguly, B. S. Joshi, V. N. Kamat, and A. H. Manmade, *Tetrahedron*, 1967, **23,** 4777.
75. E. Wong, in 'Progress in the Chemistry of Organic Natural Products', ed. W. Herz, H. Grisebach, and A. I. Scott, Springer-Verlag, Vienna, 1970, vol. 28, p. 1.
76. A. L. Livingstone, E. M. Bickoff, R. E. Lundin, and L. Jurd, *Tetrahedron*, 1964, **20,** 1963.
77. R. L. Lyne, L. J. Mulheirn, and D. P. Leworthy, *J.C.S. Chem. Comm.*, 1976, 497.
78. D. R. Perrin, C. P. Whittle, and T. J. Batterham, *Tetrahedron Letters*, 1972, 1673.
79. A. L. Livingstone, S. C. Witt, R. E. Lundin, and E. M. Bickoff, *J. Org. Chem.*, 1965, **30,** 2353.
80. W. J. Bowyer, J. N. Chatterjea, S. P. Dhoubhadel, B. O. Handford, and W. B. Whalley, *J. Chem. Soc.*, 1964, 4212.
81. D. Bouwer, C. Van der M. Brink, J. P. Engelbrecht, and G. J. H. Rall, *J.S. Afr. Chem. Inst.*, 1968, **21,** 159 (*Chem. Abs.*, 1969, **70,** 77 837).
82. K. Fukui, M. Nakayama, and K. Tsuzuki, *Bull. Chem. Soc. Japan*, 1969, **42,** 2395.
83. A. J. Birch, B. Moore, S. K. Mukerjee, and C. W. L. Bevan, *Tetrahedron Letters*, 1962, 673.
84. S. Ito, Y. Fujise, and A. Mori, *Chem. Comm.*, 1965, 595.
85. A. Pelter and P. I. Amenechi, *J. Chem. Soc.* (*C*), 1969, 887.
86. O. H. Emerson and E. M. Bickoff, *J. Amer. Chem. Soc.*, 1958, **80,** 4381.
87. M. Nakayama, T. Harano, and K. Fukui, *Experientia*, 1971, **27,** 361.
88. Y. Kawase, *Bull. Chem. Soc. Japan*, 1959, **32,** 690.
89. J. N. Chatterjea, K. D. Banerji, and N. Prasad, *Chem. Ber.*, 1963, **96,** 2356.
90. H. Wanzlick, R. Gritzky, and H. Heidepreim, *Chem. Ber.*, 1963, **96,** 305.
91. Z. Grujić, I. Tabaković, and M. Trkovnik, *Tetrahedron Letters*, 1976, 4823.
92. L. Jurd, *J. Org. Chem.*, 1964, **29,** 3036.
93. H. L. McPherson and B. W. Ponder, *J. Heterocyclic Chem.*, 1976, **13,** 909.
94. K. Kurosawa and K. Nogami, *Bull. Chem. Soc. Japan*, 1976, **49,** 1955.
95. M. Uchiyama and M. Matsui, *Agric. Biol. Chem.* (*Japan*), 1967, **31,** 1490 (*Chem. Abs.*, 1968, **69,** 18 974).
96. H. Horino and N. Inoue, *J.C.S. Chem. Comm.*, 1976, 500.
97. C. P. Falshaw, W. D. Ollis, J. A. Moore, and K. Magnus, *Tetrahedron*, 1966, Suppl. 7, 333.
98. R. Bryant, *J. Chem. Soc.*, 1965, 5140.
99. L. Feinstein and M. Jacobson, *Fortschr. Chem. org. Naturstoffe*, 1953, **10,** 423.
100. G. Büchi, L. Crombie, P. J. Godin, J. S. Kaltenbronn, K. S. Siddalingaiah, and D. A. Whiting, *J. Chem. Soc.*, 1961, 2843.
101. L. Crombie, *Fortschr. Chem. org. Naturstoffe*, 1963, **21,** 275.
102. L. Crombie, P. J. Godin, D. A. Whiting, and K. S. Siddalingaiah, *J. Chem. Soc.*, 1961, 2876.
103. L. Crombie and R. Peace, *J. Chem. Soc.*, 1961, 5445.
104. D. J. Adam, L. Crombie, and D. A. Whiting, *J. Chem. Soc.* (*C*), 1966, 550.
105. M. E. Oberholzer, G. J. H. Rall, and D. G. Roux, *Phytochemistry*, 1976, **15,** 1283.
106. L. Crombie and P. J. Godin, *J. Chem. Soc.*, 1961, 2861.
107. M. J. Begley, L. Crombie, and D. A. Whiting, *J.C.S. Chem. Comm.*, 1975, 850.
108. C. Djerassi, W. D. Ollis, and R. C. Russell, *J. Chem. Soc.*, 1961, 1448.
109. D. E. Carlson, D. Weisleder, and W. H. Tallent, *Tetrahedron*, 1973, **29,** 2731.
110. P. N. Sarma, G. Srimannarayana, and N. V. S. Rao, *Indian J. Chem.*, 1976, **14B,** 152 (*Chem. Abs.*, 1976, **85,** 74 926.
111. M. Miyano, A. Kobayashi, and M. Matsui, *Agric. Biol. Chem.* (*Japan*), 1961, **25,** 673 (*Chem. Abs.*, 1962, **56,** 1415).
112. M. Uchiyama and H. Shimotori, *Agric. Biol. Chem.* (*Japan*), 1973, **37,** 1227 (*Chem. Abs.*, 1973, **79,** 53 210).
113. M. Miyano, *J. Amer. Chem. Soc.*, 1965, **87,** 3962.
114. N. Nakatani and M. Matsui, *Agric. Biol. Chem.* (*Japan*), 1968, **32,** 769 (*Chem. Abs.*, 1968, **69,** 77 135).
115. N. Nakatani, H. Ohta, and M. Matsui, *Agric. Biol. Chem.* (*Japan*), 1973, **36,** 2433 (*Chem. Abs.*, 1973, **78,** 111 169).
116. M. Miyano and M. Matsui, *Bull. Agric. Chem. Soc. Japan*, 1958, **22,** 128 (*Chem. Abs.*, 1959, **53,** 345).
117. M. Miyano, *J. Amer. Chem. Soc.*, 1965, **87,** 3958.
118. V. Chandrashekar, M. Krishnamurti, and T. R. Seshadri, *Tetrahedron*, 1967, **23,** 2505.
119. T. Harano, *Bull. Chem. Soc. Japan*, 1970, **43,** 1560.
120. M. Krishnamurti, Y. R. Sambhy, and T. R. Seshadri, *Indian J. Chem.*, 1971, **9,** 297 (*Chem. Abs.*, 1971, **75,** 48 944).
121. H. Omokawa and K. Yamashita, *Agric. Biol. Chem.* (*Japan*), 1973, **37,** 1717 (*Chem. Abs.*, 1973, **79,** 78 639).
122. D. Carson, M. W. Cass, L. Crombie, and D. A. Whiting, *J.C.S. Chem. Comm.*, 1975, 802.
123. L. Crombie, P. W. Freeman, and D. A. Whiting, *J.C.S. Perkin I*, 1973, 1277.

18.6
Other Oxygen Heterocycles

D. M. HARRISON
New University of Ulster, Coleraine

18.6.1 OXIREN

This chapter surveys the chemistry of the known oxygen heterocycles of type (**1**), other than furan. The simplest member of the series, oxiren (**2**), remains unknown. Oxiren, a 4π-electron heterocycle, is expected to be anti-aromatic; results of molecular orbital calculations suggest that oxiren is some 50 kJ mol^{-1} higher in energy than its isomer, formylcarbene.[1] Nevertheless, oxiren derivatives have been postulated as reaction intermediates in the peracid oxidation of acetylenes (ethynes) and in the photochemical Wolff rearrangement of α-diazocarbonyl compounds. The photolyses of keten[2] and of α-diazoacetaldehyde[3] have been studied using ^{13}C labelling. In each case a symmetrical intermediate, presumably oxiren, was detected in minor reaction pathways.

O
(CH=CH)$_n$
(**1**)

O
CH=CH
(**2**)

18.6.2 OXEPIN–BENZENE OXIDE AND RELATED SYSTEMS

18.6.2.1 Introduction

The first synthesis of oxepin (**3**) was reported in 1964 by Vogel *et al.*[4] It was subsequently demonstrated that oxepin exists in rapid equilibrium with its valence tautomer, benzene oxide (**4**) (reaction 1). Indeed, most of the chemistry of this equilibrium mixture is that of the oxide component. This discussion will principally survey the chemistry of the oxepins, their arene oxide valence tautomers, and the benzoxepins. The

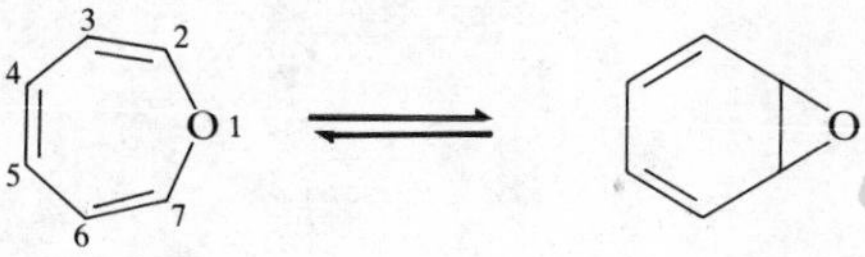

(3) ⇌ (4) (1)

stable arene oxides which do not equilibrate with their formal oxepin valence tautomers will be discussed only briefly.

The chemistry of the oxepins, benzoxepins, and their hydro-derivatives has been comprehensively reviewed.[5–7]

18.6.2.2 General synthetic methods

Oxepin was first prepared by treatment of dibromoepoxide (**5**) with the mild dehydrohalogenating agent 1,5-diazabicyclo[4,3,0]non-5-ene or with sodium methoxide in ether.[6] Substituted oxepins may be prepared similarly from appropriate cyclohexa-1,4-diene derivatives[6,8,9] (*cf.* Scheme 1). Important variations on this route are exemplified by the

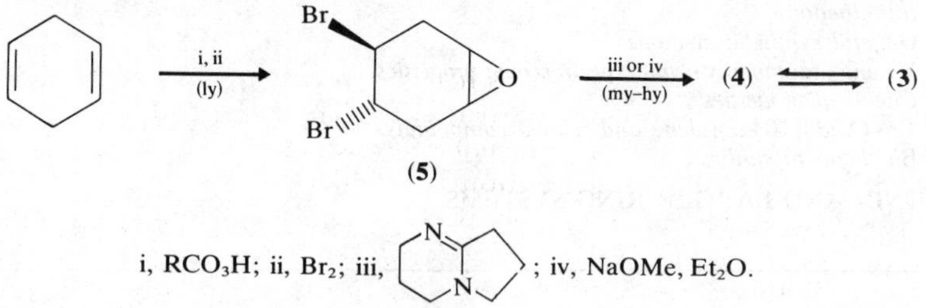

i, ii (ly) → (**5**) → iii or iv (my–hy) → (**4**) ⇌ (**3**)

i, RCO_3H; ii, Br_2; iii, [1,5-diazabicyclo[4,3,0]non-5-ene]; iv, NaOMe, Et_2O.

SCHEME 1

dehydrobromination of allylic bromide (**6**) and the debromination of dibromide (**7**) to yield oxepin derivatives[6] (reactions 2 and 3).

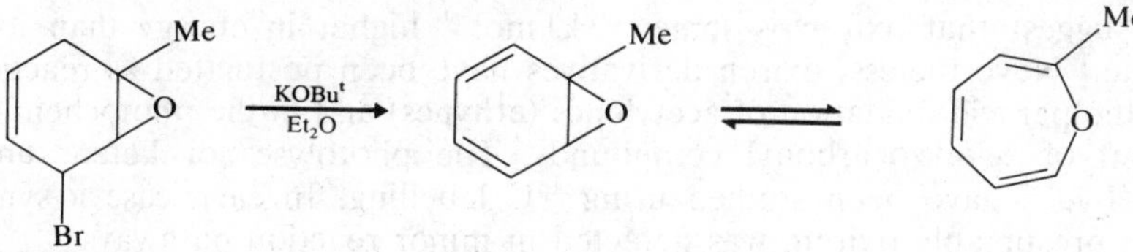

(**6**) → $KOBu^t$, Et_2O → ⇌ (2)

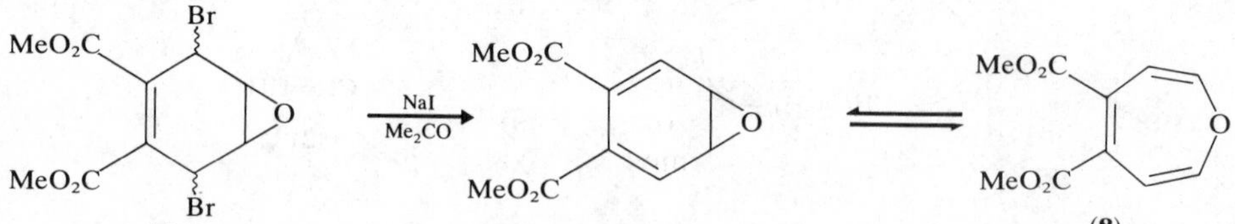

(**7**) → NaI, Me_2CO → ⇌ (**8**) (3)

Prinzbach *et al.* have prepared oxepin by thermally induced rearrangement of 7-oxaquadricyclane.[10] Substituted oxepins such as diester (**8**) may be prepared in a similar manner from oxaquadricyclane derivatives obtained by photolysis of Diels–Alder adducts of furans with acetylene derivatives[11] (Scheme 2). Transition metal catalysed isomerization of 7-oxaquadricyclane derivatives, in strictly aprotic media, yields oxepins isomeric with the products of thermal rearrangement.[12] 3-Benzoxepin (**9**) has been prepared by photolysis of the Diels–Alder adduct (**10**) of furan with benzyne. The oxaquadricyclane

i, $MeO_2CC{\equiv}CCO_2Me$, heat; ii, $h\nu$; iii, heat.

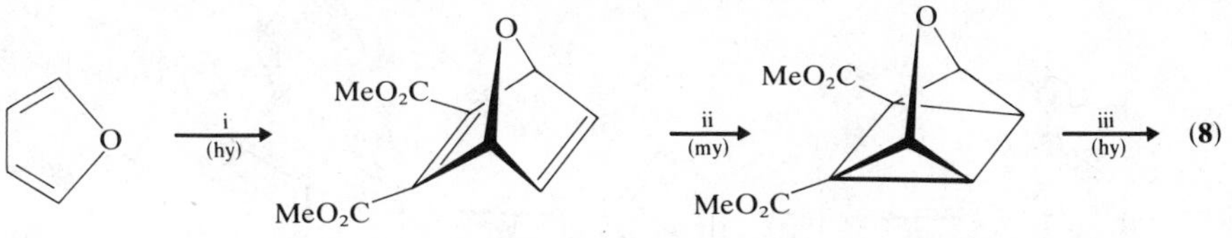

SCHEME 2

intermediate has not been isolated. The brevity of this synthetic route makes it attractive for the preparation of substituted 3-benzoxepins despite low yields (*ca.* 5%) in the photolysis step.[13]

van Tamelen and Carty have described the preparation of oxepin by thermal or photochemical isomerization of the epoxide (**11**) of 'Dewar benzene'.[14] Highly substituted

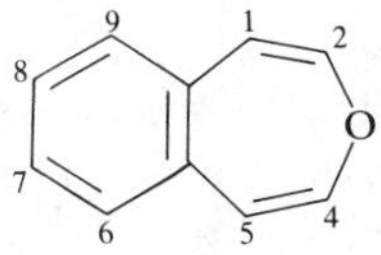

(**9**) 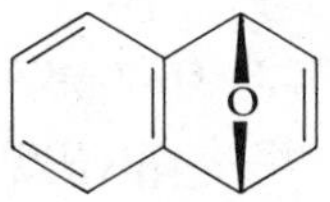 (**10**) 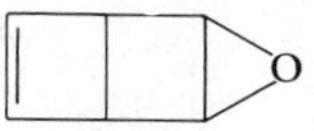 (**11**)

and sterically crowded oxepin derivatives have been prepared by acid-catalysed dehydration of cyclohexa-2,5-diene-1,4-diol derivatives.[15] The extreme lability of simple oxepin derivatives towards acid treatment precludes the use of this route as a general synthetic method. A novel synthesis of 3-substituted oxepins, whose generality has yet to be established, involves the elimination of carbon dioxide from a β-lactone in the final step.[16]

Schweizer *et al.* have described the preparation of 1-benzoxepin from salicylaldehyde (Scheme 3). In the key step, base treatment of phosphonium salt (**12**) induced an

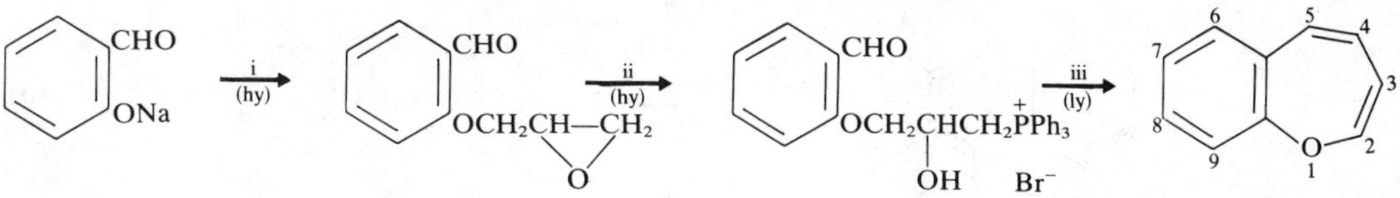

(**12**) (**13**)

i, $ClCH_2CH{-}CH_2$ (epoxide); ii, $Ph_3P{\cdot}HBr$; iii, NaOEt, EtOH.

SCHEME 3

intramolecular Wittig reaction together with dehydration, giving 1-benzoxepin (**13**) in 5% yield.[17] A particularly interesting route to 1-benzoxepin derivatives is exemplified in Scheme 4.[18] 3-Benzoxepin (**9**) and its dicarboxylic acid derivative (**14**) may be prepared by means of the double Wittig[19] or double Stobbe[20] reactions shown (Scheme 5). A large number of derivatives of 2,3:6,7-dibenzoxepin (**15**) have been prepared by a variety of methods.[21] The parent compound is formed from xanthylium perchlorate on ring expansion with diazomethane[22] (reaction 4).

Arene oxides of type (**16**), derived from polycyclic aromatic hydrocarbons, may be synthesized by the route shown in Scheme 6. The allylic bromination step usually gives low yields and even in the synthesis of naphthalene 1,2-oxide from 1,2-dihydronaphthalene the overall yield is less than 20%.[23] *N*-Bromosuccinimide treatment of bromo-esters of type (**17**), followed by base treatment of the resultant allylic bromides (**18**), gives arene oxides (**16**) in higher overall yield.[24] Arene oxides of type (**19**), the '*K*-region' oxides of polycyclic aromatic compounds, may be prepared in a variety of ways,[25,26] including, in favourable circumstances, direct peracid oxidation of polycyclic hydrocarbons.[27]

i, piperidine, H^+; ii, $MeO_2CC{\equiv}CCO_2Me$, Et_2O; iii, 100 °C.

SCHEME 4

i, $O(CH_2\overset{+}{P}Ph_3)_2$, MeONa; ii, $O(CH_2CO_2Et)_2$, $KOBu^t$.

SCHEME 5

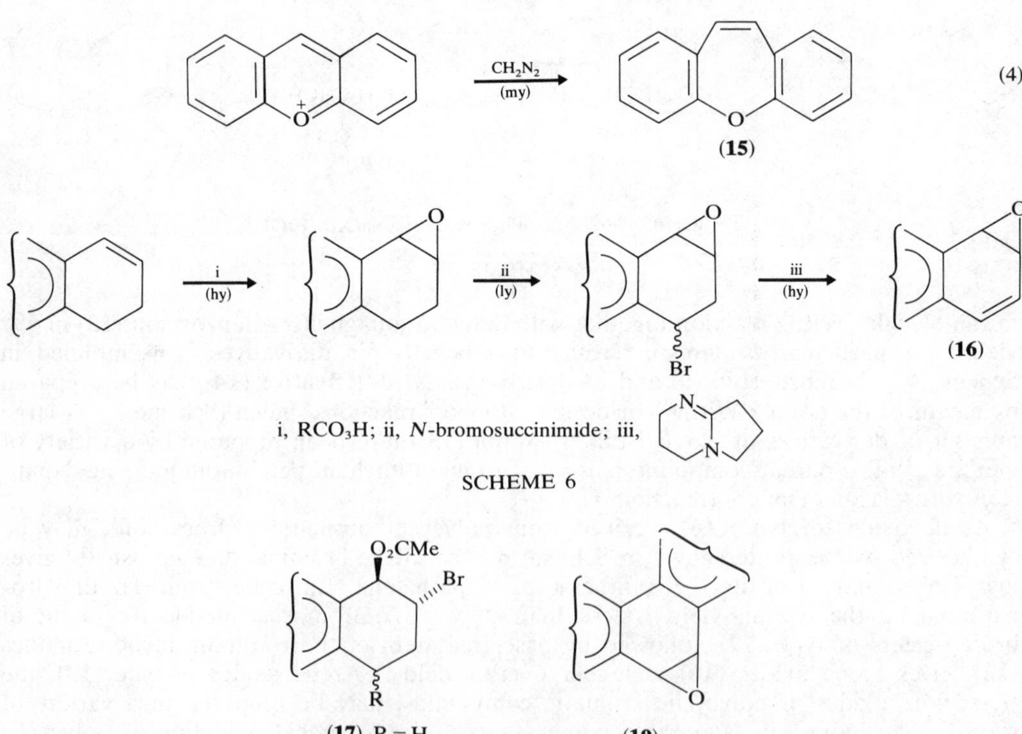

i, RCO_3H; ii, *N*-bromosuccinimide; iii,

SCHEME 6

18.6.2.3 Valence tautomerism and spectroscopic properties

Spectroscopic studies have played a crucial role in the investigation of the oxepin–benzene oxide equilibrium[5,6] (equation 1). The u.v. spectrum of the orange valence tautomeric mixture in iso-octane solution shows λ_{max} 271 nm, assigned to oxide (**4**) with a shoulder at 305 nm due to oxepin (**3**). The spectrum is strongly solvent dependent, owing to the solvent dependence of the equilibrium composition. Polar solvents favour the oxide valence tautomer and the composition of the oxepin–benzene oxide system varies from 70% oxepin in hydrocarbon solvents to 90% oxide in aqueous methanol. The proton n.m.r. spectrum of oxepin–benzene oxide is temperature dependent; a time-averaged spectrum of the two valence tautomers is observed at room temperature.[6] The ^{13}C n.m.r. spectrum of oxepin has been reported.[28]

Simple monosubstituted oxepins in general exist in rapid equilibrium with comparable concentrations of their oxide valence tautomers.[6] An important exception is 2-acetyloxepin in which the oxepin form is stabilized by conjugation. 2,7-Dimethyloxepin, a yellow-orange liquid, also exists overwhelmingly in the oxepin form, and shows a broad band at 297 nm in its u.v. spectrum. The corresponding oxide valence tautomer is presumably destabilized by the 1,2-eclipsing interaction between the methyl substituents. Annulated oxide (**20**) exists in 1:1 equilibrium with the relatively unstrained oxepin (**21**) and the system possesses the orange colour typical of the oxepins. In sharp contrast, indane 8,9-oxide (**22**), which does not equilibrate with the highly strained oxepin (**23**), is a colourless compound whose u.v. spectrum (λ_{max} 258 nm) is typical for a cyclohexa-1,3-diene derivative.[6]

O (CH$_2$)$_n$ O (CH$_2$)$_n$

(**20**) $n=3$ (**21**) $n=3$
(**22**) $n=1$ (**23**) $n=1$

Naphthalene 1,2-oxide does not equilibrate with its formal oxepin valence tautomer, presumably because conversion of (**24**) to (**25**) would involve loss of aromaticity. Hence it has been possible to prepare chiral naphthalene oxide in a state of high optical purity.[29] Similarly, 1-benzoxepin and 3-benzoxepin show no evidence of equilibration with their arene oxide valence tautomers.

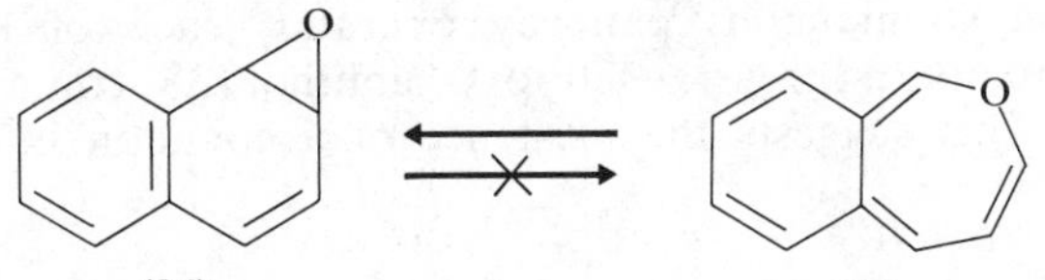

(**24**) (**25**)

18.6.2.4 Chemical properties

Oxepin and its derivatives undergo facile thermal and acid-catalysed rearrangement to phenols *via* their arene oxide valence tautomers.[6] In aqueous solutions these aromatization reactions follow the rate law:

$$k_{observed} = k_0 + k_1\, a_{H^+}$$

where k_0 and k_1 are the rate constants for the spontaneous (thermal) and acid-catalysed rearrangements, respectively.[30] The use of suitable isotopic labelling has revealed that these reactions are generally accompanied by 1,2-hydride migration, as exemplified by the isomerization of deuterium-labelled 4-methyloxepin (**26**) to *p*-cresol (**29**). The acid-catalysed pathway is illustrated (Scheme 7); the spontaneous pathway must involve the conjugate bases of (**27**) and (**28**).[8,30] The extent of deuterium retention in *p*-cresol (**29**)

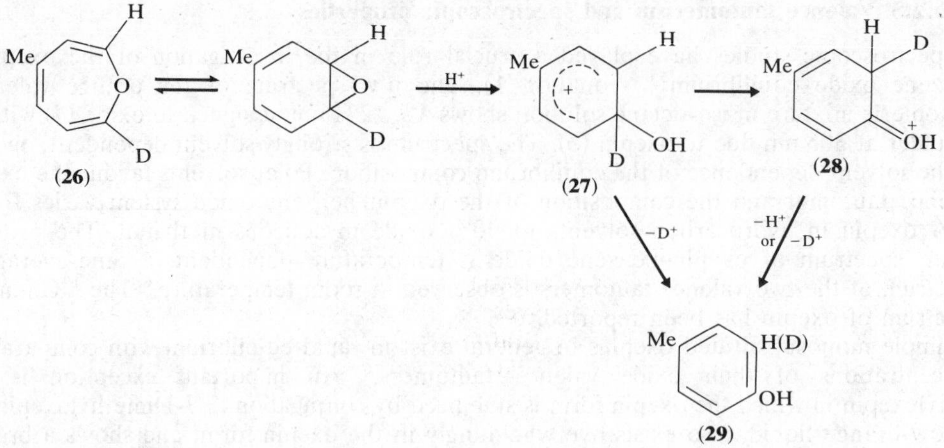

SCHEME 7

varies from 37% at low pH to 75% at pH 7.[8] The corresponding aromatization reaction of 2,5-dimethyloxepin *via* arene oxide (**30**) proceeds, in part, by way of a 1,2-methyl migration (reaction 5). The product (**31**) of methyl migration predominates at high pH, while the isomeric phenol (**32**) is the major product of the reaction in strongly acid conditions.[9,31]

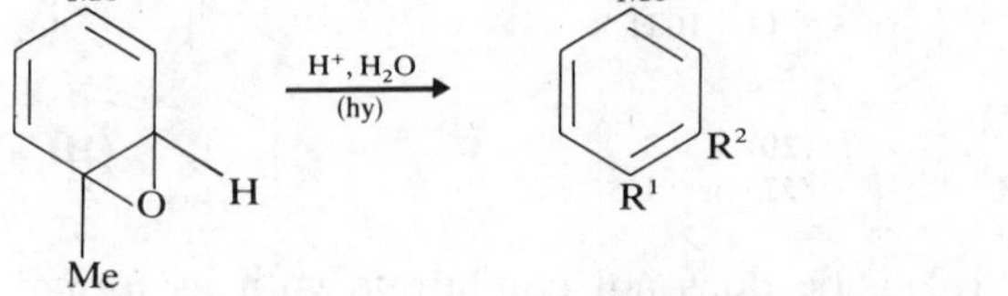

(**30**) → (H^+, H_2O (hy)) (**31**) $R^1 = OH$, $R^2 = Me$; (**32**) $R^1 = Me$, $R^2 = OH$ (5)

Naphthalene 1,2-oxide also undergoes aromatization in aqueous solution by acid-catalysed and spontaneous pathways,[30] each of which involves a 1,2-hydride migration. At pH 7, when only the spontaneous pathway operates, each of the two isomerically deuterated oxides (**33**) and (**34**) rearranged to 1-naphthol (**35**) containing 80% deuterium at C-2 (reaction 6). This suggests the existence of a common intermediate (**36**) from

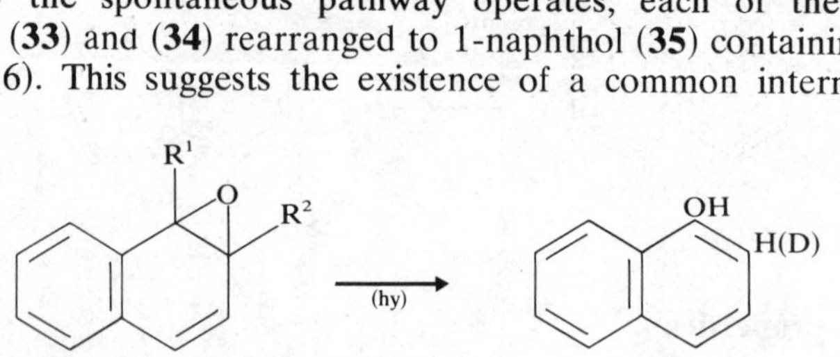

(**33**) $R^1 = H$, $R^2 = D$; (**34**) $R^1 = D$, $R^2 = H$ → (hy) (**35**) (6)

which loss of H or D depends solely on the deuterium isotope effect.[32] The 1,2- and 3,4-oxides of phenanthrene readily isomerize to phenols and resemble naphthalene 1,2-oxide in their behaviour. Phenanthrene 9,10-oxide (**37**), a '*K*-region' oxide, is more thermally stable since there is less gain in resonance energy on aromatization. In aqueous acid solution oxide (**37**) isomerizes to 9-phenanthrol, while at high pH slow hydration leads to *trans*-9,10-dihydroxy-9,10-dihydrophenanthrene.[33] The kinetics and mechanisms of these isomerization reactions have been discussed in detail by Bruice and Bruice.[34] The 1,2-hydride (or alkyl) migrations which accompany isomerization of oxepin–arene oxide systems to phenols were first observed in a biochemical context (see Section 18.6.2.6).

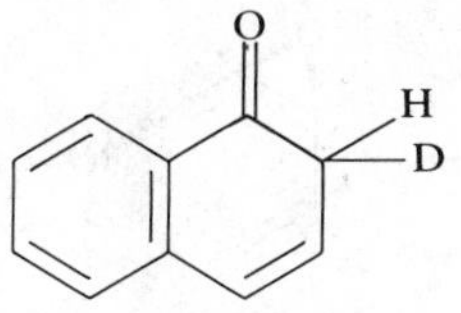

(36)

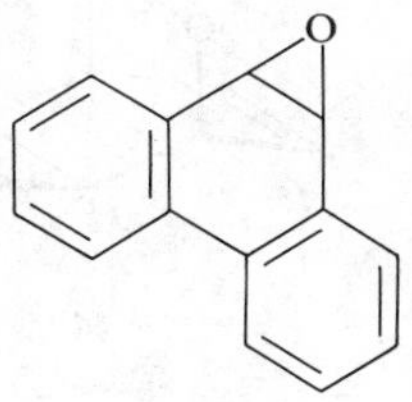

(37)

1-Benzoxepin (**13**) undergoes isomerization to 1-naphthol in poor yield on treatment with protic or Lewis acids.[35] Derivatives of 1-benzoxepin rearrange thermally to 1-naphthol derivatives at moderate temperatures.[18] 3-Benzoxepin undergoes ring contraction to indene derivative (**38**) in strong acid solutions,[19] but 3-benzoxepin-2,4-dicarboxylic acid (**14**) is surprisingly stable to acids and will survive dissolution in concentrated mineral acids.[20] Decarboxylation occurs at 300 °C to yield 2-naphthol-3-carboxylic acid, presumably through the intermediacy of the oxide valence tautomer.[36]

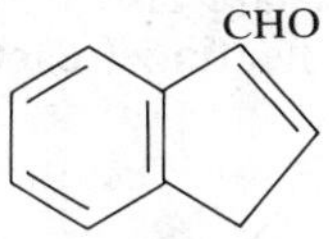

(38)

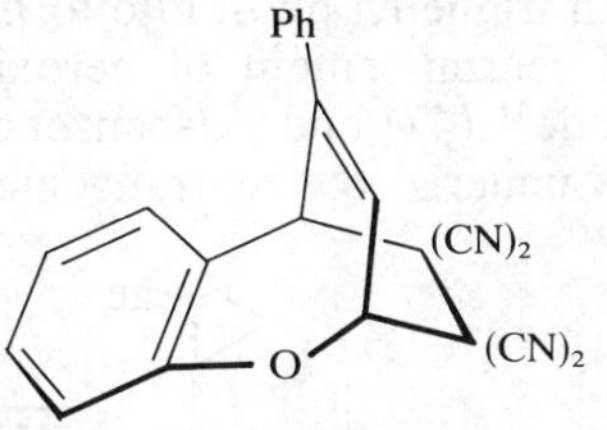

(39)

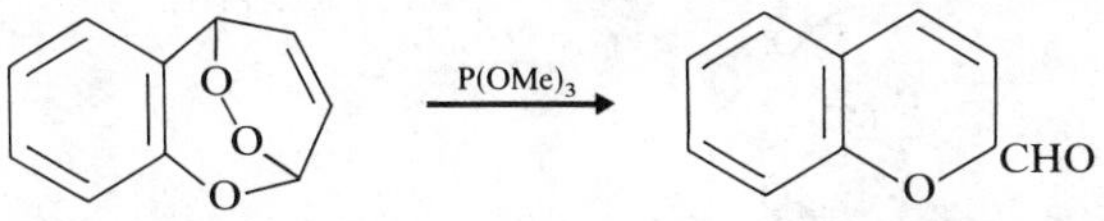

$\xrightarrow{P(OMe)_3}$

(40) (41)

4-Phenyl-1-benzoxepin gives Diels–Alder adduct (**39**) in quantitative yield on reaction with tetracyanoethylene.[37] The parent benzoxepin reacts with singlet oxygen to give peroxide (**40**). The latter undergoes an interesting ring contraction to benzpyran (**41**) on treatment with trimethyl phosphite.[38] Oxepin and its simple derivatives react with dienophiles through their arene oxide valence tautomers exclusively. Thus oxepin reacts rapidly with maleic anhydride at room temperature to yield adduct (**42**),[6] while treatment

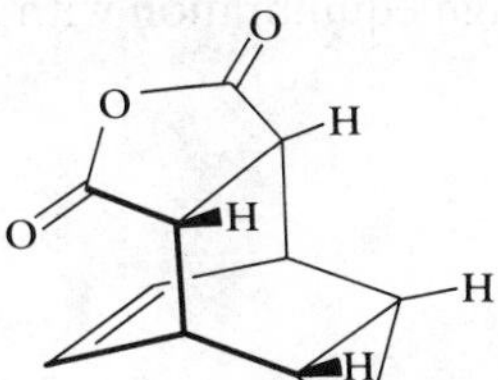

(42) 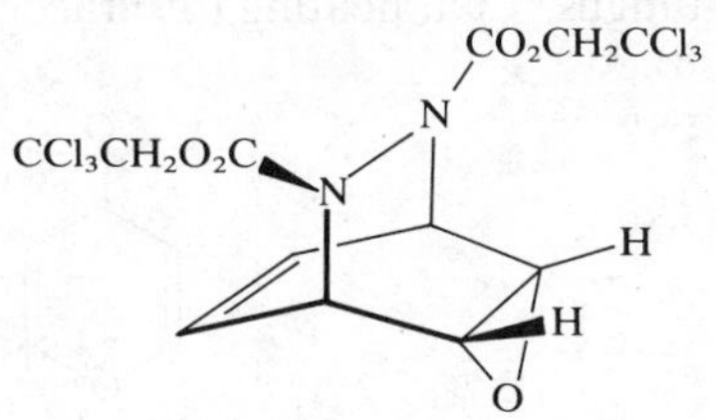 (43)

of oxepin with bis(trichloroethyl) azodicarboxylate gives (**43**) in quantitative yield.[39] The latter adduct was utilized as the starting point for an elegant synthesis of oxepin 4,5-oxide, which undergoes the degenerate Cope rearrangement depicted (reaction 7).

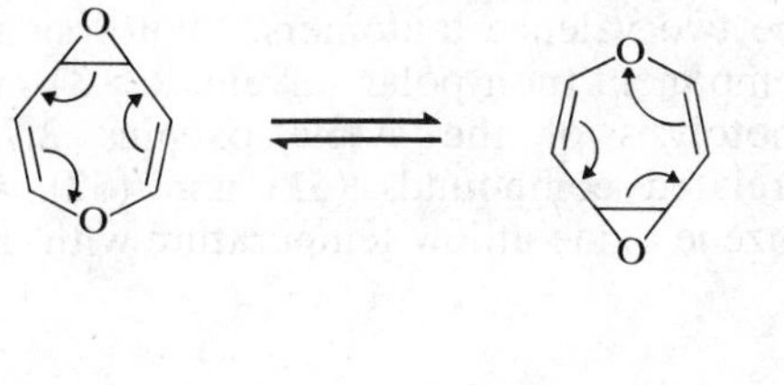

(7)

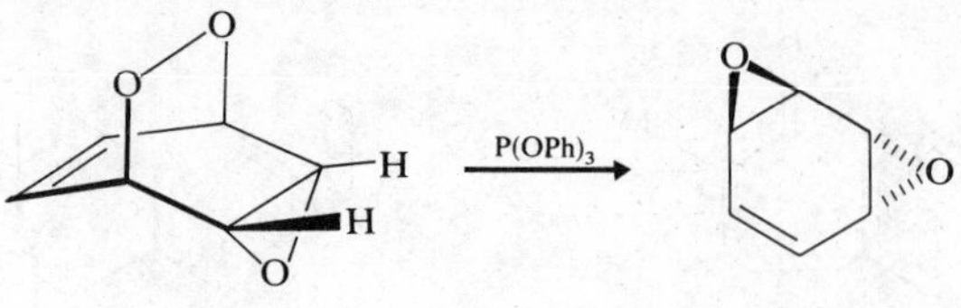

(44) (45)

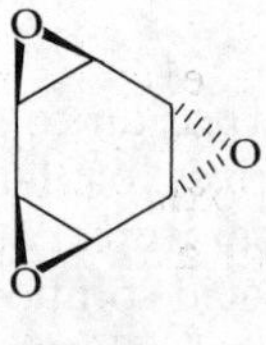

(46)

Singlet oxygen adds to oxepin–benzene oxide to give peroxide (**44**), which may be deoxygenated with triphenyl phosphite to *trans*-benzene dioxide (**45**). Epoxidation of the latter or thermal rearrangement of peroxide (**44**) gives *trans*-benzene trioxide (**46**).[40] *cis*-Benzene dioxide[41] (**47**) and *cis*-benzene trioxide[42] (**48**) are also known. These interesting compounds undergo the rearrangements shown, on heating (reactions 8 and 9). A

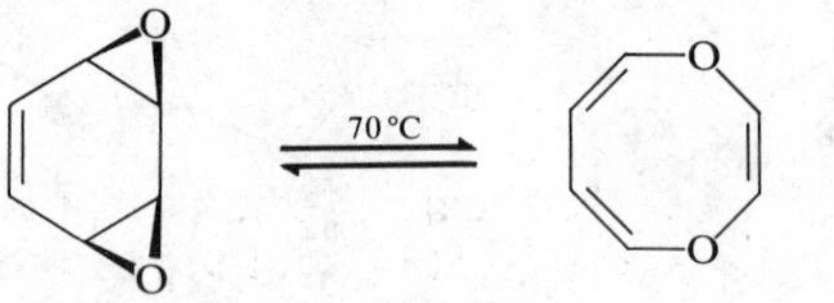

(8)

(47)

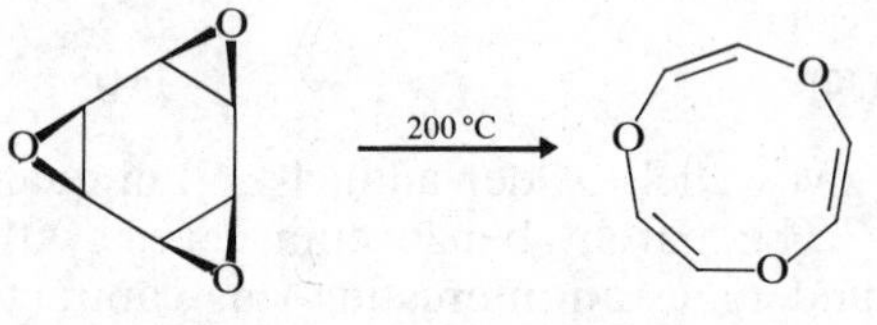

(9)

(48)

natural product (**49**) containing the *cis*-benzene dioxide moiety has been isolated from an unidentified fungus.[43] On heating (**49**) racemizes through equilibration with a 1,4-dioxocin (*cf.* reaction 8).

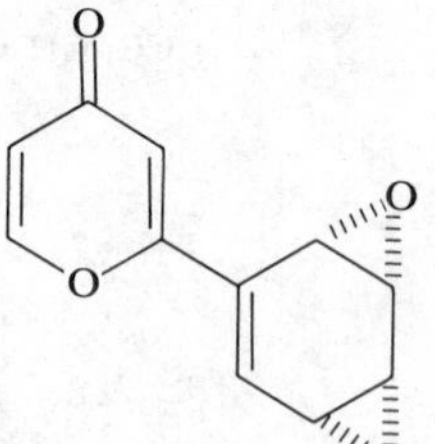

(49)

Study of the photolysis of the oxepin–benzene oxide system is complicated by the rapid equilibration between the two valence tautomers. Photolysis under conditions chosen to excite only the oxepin component (non-polar solvent, $\lambda > 310$ nm) gave *cis*-bicycle (**50**) in quantitative yield.[44] Photolysis of the stable oxepins 2,7-dimethyloxepin[45] and 1-benzoxepin[6] gave the related compounds (**51**) and (**52**), respectively. Alternatively, irradiation of oxepin–benzene oxide at low temperature with short wavelength u.v. light,[44]

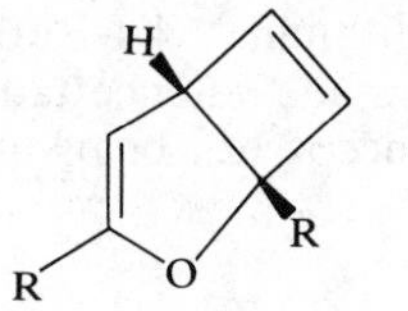

H R R O (50) R=H (51) R=Me

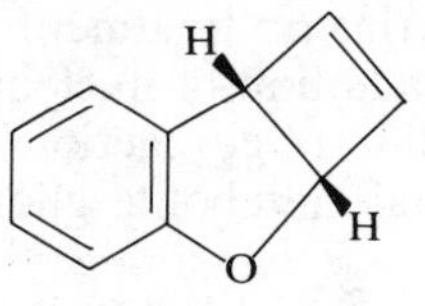

H H O (52)

conditions expected to favour selective excitation of the oxide valence tautomer, gave phenol as the major product, presumably formed *via* cyclohexa-2,4-dienone. On photolysis in acetone, deuterium labelled benzene oxide (**53**) equilibrated with its isomer (**54**) *via* an 'oxygen walk' mechanism (Scheme 8) at a rate comparable with that of isomerization to

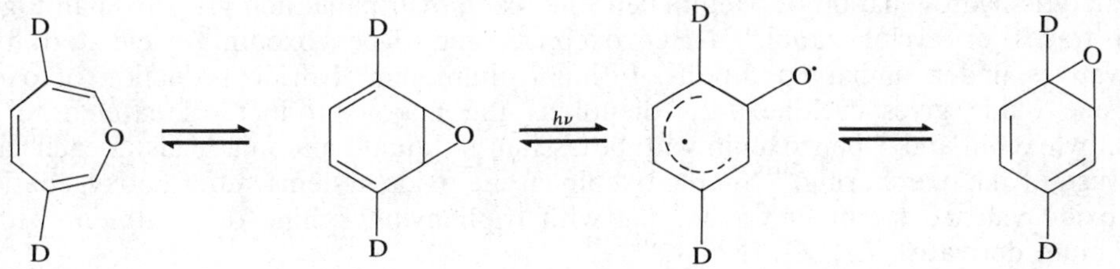

(53) (54)

SCHEME 8

phenol. A similar 'oxygen walk' may explain the photoisomerization of phenanthrene 9,10-oxide (**37**) to 2,3:4,5-dibenzoxepin (**55**) and 9-phenanthrol.[46]

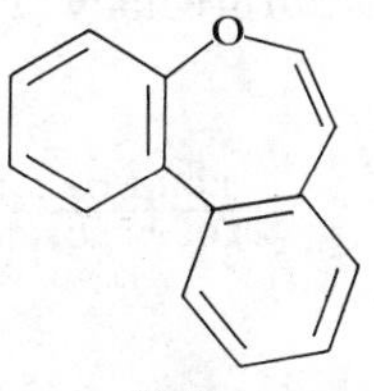

(55)

Nucleophilic addition to oxepin–benzene oxide has been studied in depth by Jeffrey *et al.*[47] Soft, polarizable nucleophiles, such as thiolate and azide anions, readily yield *trans*-addition compounds on reaction with the oxide valence tautomer (*cf.* reaction 10).

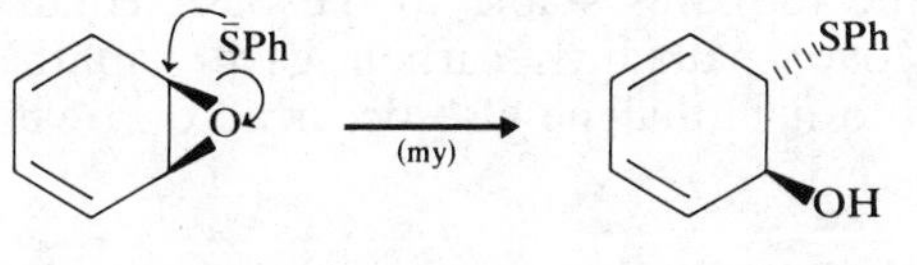

(10)

Hard nucleophiles such as methoxide add slowly, while the amide anion in liquid ammonia does not react with benzene oxide. Methyl-lithium and dimethylmagnesium yield *cis*- and *trans*-addition compounds, the former products arising from 1,6-addition (reaction 11). Polarizable nucleophiles react with naphthalene 1,2-oxide by *trans*-addition at C-2. Nucleophiles are unreactive towards benzoxepins, although 3-benzoxepin undergoes

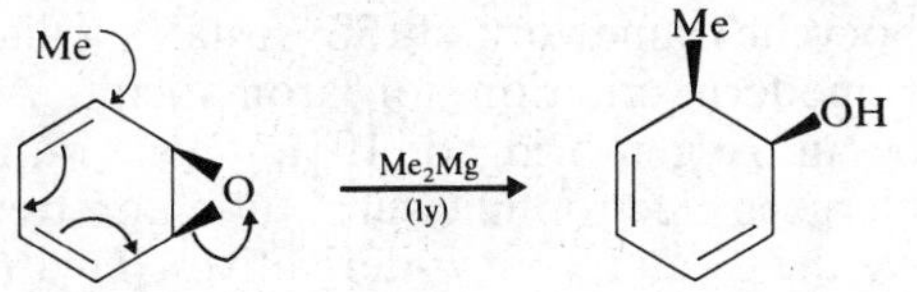

(11)

fission of the oxepin ring on treatment with methyl-lithium.[47] 4-t-Butyloxycarbonyloxepin sustains facile *trans*-addition of nucleophiles to the oxide valence tautomer (**56**) aided by initial Michael addition (*e.g.* reaction 12). This concept has been utilized in an elegant synthesis of the fungal metabolite gliotoxin.[48]

CO_2Bu^t (**56**) $\xrightarrow[\text{(ly)}]{HO^-}$ product with CO_2Bu^t, OH, OH (12)

Catalytic hydrogenation of oxepin–benzene oxide over palladium gives oxepan together with traces of cyclohexanol.[6] 1-Benzoxepin[35] and 3-benzoxepin[19] yield tetrahydro-derivatives under similar conditions. Lithium aluminium hydride reduction of oxepin–benzene oxide gives cyclohexa-2,4-dienol as the major product.[6] Treatment of 2,7-dimethyloxepin and 3-benzoxepin with potassium in liquid ammonia leads in each case to cleavage of the oxepin ring.[49] Simple oxepin–arene oxide systems suffer deoxygenation of the oxide valence tautomer on heating with triphenylphosphine or treatment with the chromium derivative $Cr(CO)_3(NH_3)_3$.[6]

18.6.2.5 1,6-Oxido[10]annulene and related compounds

Dehydrobromination of tetrabromoepoxide (**57**) gives 1,6-oxido[10]annulene as a yellow solid, in 50–60% yield[6,35] (reaction 13). The oxide (**59**) is a presumed intermediate in this transformation. Annulene (**58**) is surprisingly stable to heat, light, and oxygen but is

(**57**) $\xrightarrow[\text{or KOBu}^t\text{, Et}_2\text{O, }-10\,°\text{C (my)}]{\text{KOH, EtOH, 55 °C}}$ (**58**) (13)

highly sensitive to acid conditions. Smooth isomerization to 1-benzoxepin (**13**) occurs during chromatography on acid-washed alumina or silica gel,[6] while 1-naphthol is the main product of treatment with boron trifluoride.[35] 1-Benzoxepin was isolated as a by-product in the dehydrobromination of (**57**) at 55 °C and must have arisen directly from oxide (**59**) since annulene (**58**) was stable to the basic conditions involved.[35] Catalytic hydrogenation of (**58**) over palladium–carbon gave naphthalene as the first-formed product. Deoxygenation to naphthalene also occurred on treatment with triphenyl phosphine or $Cr(CO)_3(NH_3)_3$.[6,35]

(**59**) (**60**)

The 10π-electron carbocyclic framework of (**58**) sustains a diamagnetic ring current and hence satisfies a major modern criterion for aromaticity. Accordingly, (**58**) is more properly considered to be an oxygen-bridged [10]annulene rather than an oxepin derivative. Compound (**58**) undergoes electrophilic substitution on treatment with cupric nitrate in acetic anhydride to give the 2- and 3-nitro-derivatives. Treatment of (**58**) with bromine led to the formation of addition compounds, while other attempts at electrophilic substitution resulted only in the isolation of naphthalene derivatives.[6,35]

1,6:8,13-bisoxido[14]annulene (**60**) has also been prepared. This carmine-red compound also shows aromatic properties, is insensitive to air, and undergoes electrophilic substitution.[50]

18.6.2.6 Biochemical studies

Aromatic hydroxylation is an important step in the biosynthesis of a wide range of natural products. Such hydroxylation reactions are catalysed by mixed-function oxygenases and normally occur through the intermediate formation of oxepin–arene oxide systems. Isomerization of these arene oxide intermediates occurs spontaneously and involves a 1,2-hydride migration from the site of ultimate hydroxylation (*cf.* Scheme 7). This 1,2-hydride migration has been called the N.I.H. shift after the National Institutes of Health where the phenomenon was first observed in the mid 1960s.[51] The N.I.H. shift has been reviewed in detail[52] and will be discussed further in Chapter 30.3.

Oxidation by non-specific microsomal mixed-function oxygenases is important in the metabolism of foreign, ingested aromatic hydrocarbons in mammalian liver tissue. The arene oxides formed by such oxidations can suffer a variety of fates (Scheme 9), among

(**61**) $R^2 = CO(CH_2)_2CH(NH_2)CO_2H$, $R^3 = NHCH_2CO_2H$

(**62**) $R^2 = Ac$, $R^3 = OH$

Metabolic transformations of aromatic hydrocarbons *via* the intermediacy of arene oxides.

SCHEME 9

which are enzyme-catalysed hydration to *trans*-dihydrodiol derivatives and spontaneous or enzyme-catalysed reaction with cellular glutathione to yield premercapturic acids (**61**), which are subsequently metabolized to mercapturic acids (**62**). These processes result in detoxification of the starting aromatic hydrocarbons. However, it has long been known that some aromatic compounds are highly toxic towards mammalian tissue. The bone marrow toxicity of benzene, hepatotoxicity of halobenzenes, and the mutagenicity and carcinogenicity of polycyclic hydrocarbons such as benz[*a*]anthracene and dibenz[*a,h*]anthracene may be cited. It now seems likely that arene oxides derived from these compounds are the true cytotoxic materials, and that they act by reaction with nucleophilic groups in essential cellular components, especially proteins and nucleic acids. These aspects of the chemistry of the arene oxides have been extensively reviewed.[52–55]

18.6.3 OXONIN AND LARGER RING SYSTEMS

Oxonin (**64**) was prepared for the first time in 1969 by photolysis of cyclo-octatetraene monoepoxide (reaction 14). Oxonin proved to be very labile and could be isolated only at

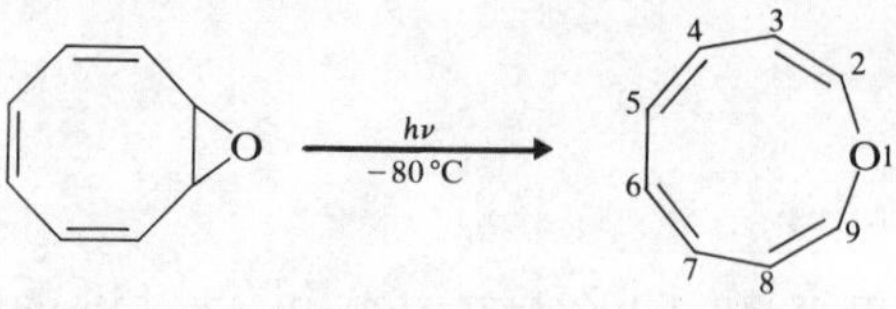

(14)

(63) **(64)**

low temperatures. Thermally allowed disrotatory cyclization to form *cis*-bicycle (**65**) occurred rapidly at 30 °C, while the only product of photolysis of oxonin at 0 °C was epoxide (**63**).[56] Oxonin reacted with 4-phenyl-1,2,4-triazoline-3,5-dione, even at low temperatures, to give adduct (**66**). The latter was most probably formed *via* thermally allowed $_{\pi}8_s + _{\pi}2_s$ cycloaddition followed by disrotatory cyclization of the initial product.[57] Oxonin is formally a 10π-electron heterocycle and might have been expected to display aromatic properties. In the event, both the proton and ^{13}C n.m.r. spectra[58] of oxonin may be interpreted in terms of a non-planar conformation, and the heterocycle does not sustain a diamagnetic ring current. In contrast, the related heterocycle azonin exists in a planar conformation and displays aromatic properties. The chemistry of oxonin and related heteronins has been reviewed.[59,60]

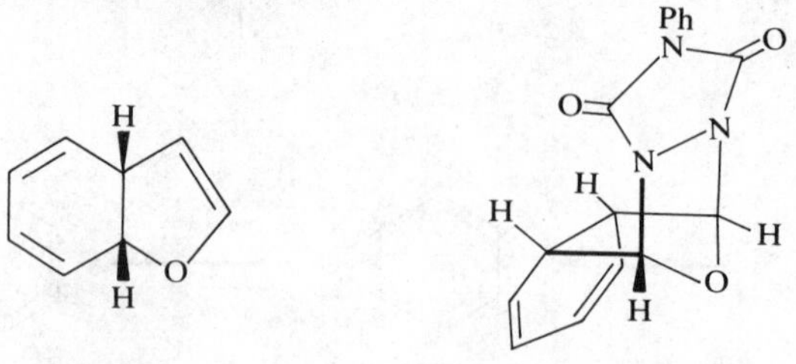

(65) **(66)**

4,5:6,7-Dibenzoxonin (**67**) was prepared in 1968 in 6% yield by means of the double Wittig reaction shown (reaction 15). Catalytic hydrogenation of (**67**) gave the tetrahydro-derivative (**68**), while acid treatment brought about ring contraction to aldehyde (**69**).[61]

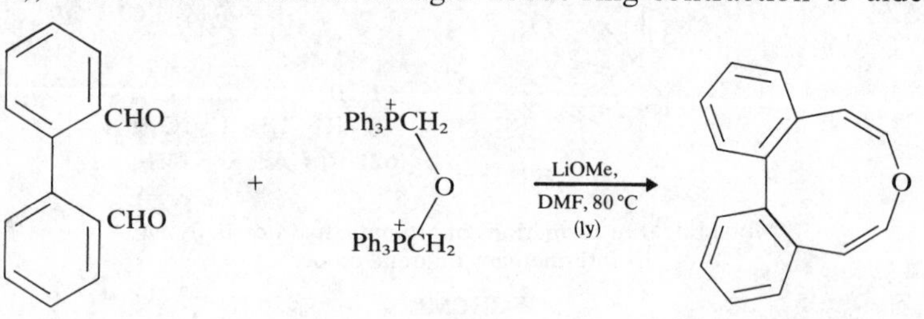

(15)

(67)

No example of a monobenzoxonin has yet been reported, although the pyridazino-analogue (**70**) has been prepared from cyclo-octatetraene epoxide.[62]

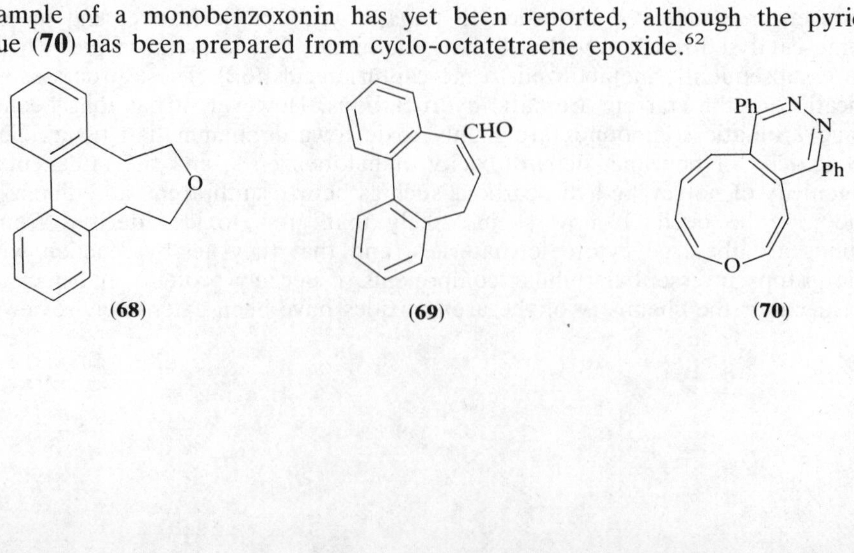

(68) **(69)** **(70)**

Oxygen heterocycles larger than oxonin may readily accommodate one or more *trans* double bonds and hence geometrical isomers are possible. Two isomeric oxa[13]annulenes (**71**) have been prepared in low yield by the photolytic route shown (reaction 16). These

$hν$, −75 °C (ly) → $(CH{=}CH)_6$ (16)

(**71**)

two coloured compounds, whose configurations are not reliably known, each gave oxacyclotridecane on catalytic hydrogenation.[63]

Three isomeric oxa[17]annulenes have been reported, one of which was prepared in 40% yield by the photolytic route shown (reaction 17). Each of these isomers is a

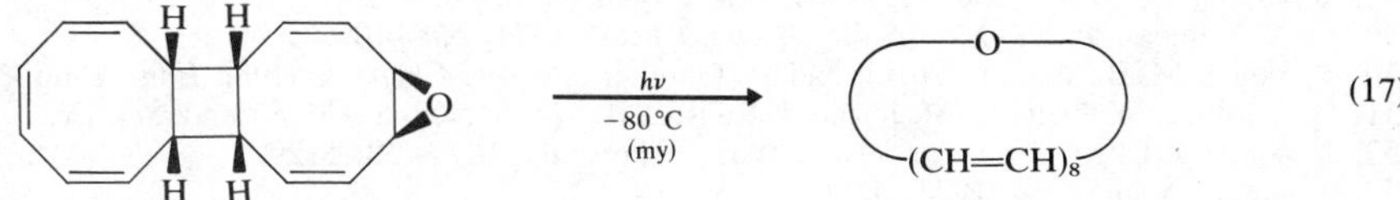

thermally stable, oxygen-sensitive, red crystalline compound, which yields oxacycloheptadecane on catalytic hydrogenation. The configurations of the three isomers have not been determined.[64] Neither the oxa[13]annulenes nor the oxa[17]annulenes show evidence of a diamagnetic ring current despite their formal $(4n+2)\pi$-electron constitutions.

Analogues of oxa[13]annulene and oxa[15]annulene containing two triple bonds in the heterocyclic ring have been prepared.[65,66]

References

1. O. P. Strausz, R. K. Gosavi, A. S. Denes, and I. G. Csizmadia, *J. Amer. Chem. Soc.*, 1976, **98,** 4784.
2. R. L. Russell and F. S. Rowland, *J. Amer. Chem. Soc.*, 1970, **92,** 7508.
3. K. P. Zeller, *Tetrahedron Letters*, 1977, 707.
4. E. Vogel, R. Schubart, and W. A. Böll, *Angew. Chem. Internat. Edn.*, 1964, **3,** 510.
5. A. Rosowsky, in 'Heterocyclic Chemistry', ed. A. Rosowsky, Wiley-Interscience, New York, 1972, vol. 26, p. 1.
6. E. Vogel and H. Günther, *Angew. Chem. Internat. Edn.*, 1967, **6,** 385.
7. D. M. Jerina, H. Yagi, and J. W. Daly, *Heterocycles*, 1973, **1,** 267.
8. D. M. Jerina, J. W. Daly, and B. Witkop, *J. Amer. Chem. Soc.*, 1968, **90,** 6523.
9. E. A. Fehnel, *J. Amer. Chem. Soc.*, 1972, **94,** 3961.
10. H. Prinzbach and H. Babsch, *Angew. Chem. Internat. Edn.*, 1975, **14,** 753.
11. H. Prinzbach, M. Arguëlles, and E. Druckrey, *Angew. Chem. Internat. Edn.*, 1966, **5,** 1039.
12. R. Roulet, J. Wenger, M. Hardy, and P. Vogel, *Tetrahedron Letters*, 1974, 1479.
13. G. R. Ziegler, *J. Amer. Chem. Soc.*, 1969, **91,** 446.
14. E. E. van Tamelen and D. Carty, *J. Amer. Chem. Soc.*, 1971, **93,** 6102.
15. A. Rieker, G. Henes, and S. Berger, *Chem. Ber.*, 1975, **108,** 3700.
16. G. W. Holbert, L. B. Weiss, and B. Ganem, *Tetrahedron Letters*, 1976, 4435.
17. E. E. Schweizer, M. S. El-Bakoush, K. K. Light, and K. H. Oberle, *J. Org. Chem.*, 1968, **33,** 2590.
18. D. N. Reinhoudt and C. G. Kouwenhoven, *Tetrahedron Letters*, 1972, 5203.
19. K. Dimroth, G. Pohl, and H. Follmann, *Chem. Ber.*, 1966, **99,** 634.
20. K. Dimroth and H. Freyschlag, *Chem. Ber.*, 1957, **90,** 1623.
21. A. Rosowsky, in Ref. 5, pp. 154–176.
22. H. W. Whitlock, *Tetrahedron Letters*, 1961, 593.
23. E. Vogel and F. G. Klärner, *Angew. Chem. Internat. Edn.*, 1968, **7,** 374.
24. H. Yagi and D. M. Jerina, *J. Amer. Chem. Soc.*, 1975, **97,** 3185.
25. S. H. Goh and R. G. Harvey, *J. Amer. Chem. Soc.*, 1973, **95,** 242.
26. P. Dansette and D. M. Jerina, *J. Amer. Chem. Soc.*, 1974, **96,** 1224.
27. K. Ishikawa, H. C. Charles, and G. W. Griffin, *Tetrahedron Letters*, 1977, 427.
28. H. Günther and G. Jikeli, *Chem. Ber.*, 1973, **106,** 1863.
29. M. N. Akhtar and D. R. Boyd, *J.C.S. Chem. Comm.*, 1975, 916.
30. G. J. Kasperek and T. C. Bruice, *J. Amer. Chem. Soc.*, 1972, **94,** 198.

31. G. J. Kasperek, T. C. Bruice, H. Yagi, N. Kaubisch, and D. M. Jerina, *J. Amer. Chem. Soc.*, 1972, **94,** 7876.
32. D. R. Boyd, J. W. Daly, and D. M. Jerina, *Biochemistry*, 1972, **11,** 1961.
33. P. Y. Bruice, T. C. Bruice, H. G. Selander, H. Yagi, and D. M. Jerina, *J. Amer. Chem. Soc.*, 1974, **96,** 6814.
34. T. C. Bruice and P. Y. Bruice, *Accounts Chem. Res.*, 1976, **9,** 378.
35. A. Shani and F. Sondheimer, *J. Amer. Chem. Soc.*, 1967, **89,** 6310.
36. R. Huisgen, E. Laschtuvka, I. Ugi, and A. Kammermeier, *Annalen*, 1960, **630,** 128.
37. H. Hofmann and P. Hofmann, *Annalen*, 1975, 1571.
38. J. E. Baldwin and O. W. Lever, *J.C.S. Chem. Comm.*, 1973, 344.
39. W. H. Rastetter, *J. Amer. Chem. Soc.*, 1976, **98,** 6350.
40. C. H. Foster and G. A. Berchtold, *J. Org. Chem.*, 1975, **40,** 3743.
41. E. Vogel, H. J. Altenbach, and D. Cremer. *Angew. Chem. Internat. Edn.*, 1972, **11,** 935.
42. E. Vogel, H. J. Altenbach, and C. D. Sommerfeld, *Angew. Chem. Internat. Edn.*, 1972, **11,** 939.
43. D. B. Borders and J. E. Lancaster, *J. Org. Chem.*, 1974, **39,** 435.
44. J. M. Holovka and P. D. Gardner, *J. Amer. Chem. Soc.*, 1967, **89,** 6390.
45. L. A. Paquette and J. H. Barrett, *J. Amer. Chem. Soc.*, 1966, **88,** 1718.
46. D. M. Jerina, B. Witkop, C. L. McIntosh, and O. L. Chapman, *J. Amer. Chem. Soc.*, 1974, **96,** 5578.
47. A. M. Jeffrey, H. J. C. Yeh, D. M. Jerina, R. M. DeMarinis, C. H. Foster, D. E. Piccolo, and G. A. Berchtold, *J. Amer. Chem. Soc.*, 1974, **96,** 6929.
48. T. Fukuyama and Y. Kishi, *J. Amer. Chem. Soc.*, 1976, **98,** 6723.
49. L. A. Paquette and T. McCreadie, *J. Org. Chem.*, 1971, **36,** 1402.
50. E. Vogel, M. Biskup, A. Vogel, and H. Günther, *Angew. Chem. Internat. Edn.*, 1966, **5,** 734.
51. G. Guroff, J. W. Daly, D. M. Jerina, J. Renson, B. Witkop, and S. Udenfriend, *Science*, 1967, **157,** 1524.
52. J. W. Daly, D. M. Jerina, and B. Witkop, *Experientia*, 1972, **28,** 1129.
53. F. Oesch, *Xenobiotica*, 1973, **3,** 305.
54. D. M. Jerina and J. W. Daly, *Science*, 1974, **185,** 573.
55. P. Sims and P. L. Grover, *Adv. Cancer Res.*, 1974, **20,** 165.
56. A. G. Anastassiou and R. P. Cellura, *Chem. Comm.*, 1969, 1521.
57. A. G. Anastassiou and R. P. Cellura, *Chem. Comm.*, 1970, 484.
58. A. G. Anastassiou and E. Reichmanis, *J. Amer. Chem. Soc.*, 1976, **98,** 8266.
59. A. G. Anastassiou, *Accounts Chem. Res.*, 1972, **5,** 281.
60. A. G. Anastassiou, *Pure Appl. Chem.*, 1975, **44,** 691.
61. A. P. Bindra, J. A. Elix, P. J. Garratt, and R. H. Mitchell, *J. Amer. Chem. Soc.*, 1968, **90,** 7372.
62. A. G. Anastassiou and S. J. Girgenti, *Angew. Chem. Internat. Edn.*, 1975, **14,** 814.
63. W. Henne, G. Plinke, and G. Schröder, *Chem. Ber.*, 1975, **108,** 3753.
64. G. Schröder, G. Plinke, and J. F. M. Oth, *Angew. Chem. Internat. Edn.*, 1972, **84,** 72.
65. J. Ojima, A. Kimura, and M. Ikeguchi, *Bull. Chem. Soc. Japan*, 1976, **49,** 3709.
66. R. L. Wife, P. J. Beeby, and F. Sondheimer, *J. Amer. Chem. Soc.*, 1975, **97,** 641.

PART 19

SULPHUR AND OTHER HETEROATOM SYSTEMS

19.1

Thiophens

O. METH-COHN
University of Salford

Good reviews covering all aspects of the chemistry of thiophen are available and are referred to where appropriate throughout this survey. Where no reference is cited, then the reader should consult the major reviews on thiophen chemistry, in particular those by Gronowitz,[1–3] the Specialist Periodical Reports[2,3] being an on-going series. Earlier literature is well collated in Hartough's monographs.[4,5]

19.1.1 INTRODUCTION

The well-known, accidental discovery of thiophen (38 years after the unwitting preparation of the first thiophen derivative 'thionessal', or tetraphenylthiophen, by Laurent) by Victor Meyer in 1882 as a contaminant in coal-tar benzene, initiated an extensive and fascinating area of exploration that is still under active investigation. Following Meyer's pioneering work involving 106 papers published during five years, Steinkopf and his co-workers firmly established the classical concepts of thiophen chemistry as a close heterocyclic analogue of the parent of aromatic systems, benzene, his work being summarized in a thorough monograph.[6] However, thiophen remained a laboratory speciality until the middle 1940's, when its availability in tonnage quantities from butane and sulphur was demonstrated at the Socony-Vacuum Laboratories in the U.S.A. The impetus this gave to research culminated, in 1952, in Hartough's 'bible' of thiophen chemistry,[4] still valuable today. However, at the time of his writing, little commercial use for thiophen existed. Since that time, major centres of research all over the world have discovered new and often unique characteristics of the thiophen family that have fulfilled Hartough's prediction that 'Organic chemists who in the future will be the greatest contributors to thiophene chemistry will . . . proceed on the theory that the chemistry of thiophene and benzene is to be compared about as closely as a zoologist would compare a tortoise and the boa constrictor; they are in the same class but of widely separated species.' Excellent reviews of recent developments include those by Gronowitz,[1–3] Hurd,[7] and the useful flow-charts of Clark.[8] The chemistry of fused thiophens,[5] benzothiophen,[9] and dibenzothiophen[10] have also been covered and continue to be monitored in Gronowitz's surveys.[2,3]

Today, thiophen research is undergoing a new boom as a result of new and cheaper industrial routes to it and its homologues, and the discovery of important applications, particularly in the drug and dyestuff fields.

19.1.2 NOMENCLATURE AND NUMBERING

Although the nomenclature of thiophen is uncomplicated and generally agreed upon (Scheme 1), the spelling without the terminal 'e' is common in Great Britain and Germany, while the 'e' is generally added in North America, the remainder of Europe, and the USSR. Trivial names such as thianaphthene (for benzothiophen) are used, but less commonly in recent years.

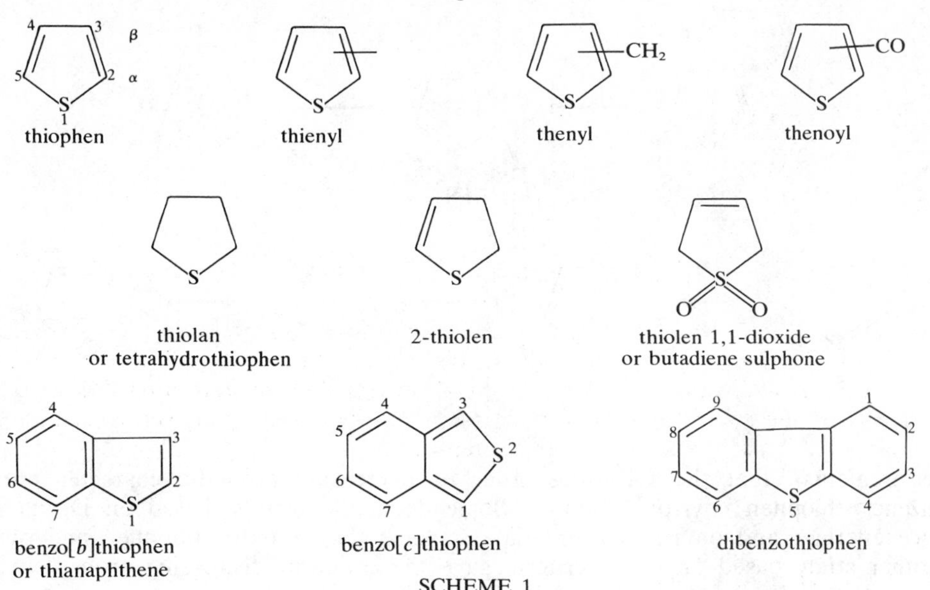

SCHEME 1

19.1.3 STRUCTURE OF THIOPHEN

The extent to which sulphur *d*-orbitals participate in the structure and reactivity of thiophens had long been a bone of contention. Early pioneering work by Schomaker and Pauling[11] on the geometry of thiophen (by electron diffraction) has been largely corroborated by recent work[12] and refined by microwave studies to give dimensions, median values of which are shown in (**1**). The latter method gives highly precise data.[13] The

(**1**)

results underline the surprisingly high double-bond character of the C—S bond, not evident from the usual Kekulé structure. X-ray data have supplied useful molecular dimensions of numerous thiophen derivatives,[14] while comparisons of bond parameters from microwave data for a wide variety of homo- and hetero-cyclic analogues of thiophen have also been collated.[15] Recent SCF-MO, PPP, and CNDO non-empirical treatments and *ab initio* calculations all seem to support the contention that the ground-state properties of thiophen require only a minor involvement of *d*-orbitals, if at all.[16] From a valence bond viewpoint (Scheme 2), mesomers such as (**2**)–(**4**) are important while others, *e.g.* (**5**)–(**8**), are of little weight in ground-state terms. Indeed, Pauling's approximate quantifying of the relative significance of structures (**2**:**3**+**4**:others) as 70:30:10 may be an overestimate of the role of the sulphur decet and 'long-bonded' thiophens (*e.g.* **8**). The effect of substituents is discussed later (p. 796).

The resonance energy and aromaticity of thiophen is another area of considerable debate. A recent review concludes, on the basis of virtually all known criteria (twelve

(2) (3) (4)

(5) (6) (7) (8)

SCHEME 2

measurements), that the following order of decreasing aromatic character prevails: benzene > thiophen > pyrrole > furan.[17] Some debatable results, based on Dewar resonance energies and on ring-current data, reverse the order of thiophen and pyrrole. Another study based on seven criteria gave the order of decreasing aromaticity, *viz.* benzene > thiophen > selenophen > tellurophen > furan.[18]

19.1.4 PHYSICAL PROPERTIES AND SPECTROSCOPY

19.1.4.1 Physical properties

Some of the significant physical properties of thiophen are collected in Table 1.

TABLE 1
Physical Properties of Thiophen

Property	*Value*
Freezing point (°C)	−38.30
Boiling point (°C) at 760 mmHg	84.16
Flash point (°C)	−6.7
Density ($g\,ml^{-1}$): d_4^0	1.0873
Refractive index (n_D^{25})	1.52572
Critical constants	
temp. (°C)	307.2
pressure ($N\,m^{-2}$)	56.9×10^5
density ($g\,ml^{-1}$)	0.385
Heat of vaporization at 84.12 °C ($kJ\,mol^{-1}$)	32.483
Heat of formation, ΔH ($kJ\,mol^{-1}$)	−82.13
Heat of combustion[a] at 760 mmHg and 20 °C ($kJ\,mol^{-1}$)	2806.7
Dielectric constant	2.76 ± 2%
Dipole moment ($\mu\times10^{18}$)	0.53–0.63

[a] Liquid C_4H_4S to liquid water and gaseous CO_2 and H_2S.

19.1.4.2 Ultraviolet spectroscopy

Although thiophen in the vapour phase gives three distinct absorptions at 240.5, 232.6 and 220.4 nm, the solution spectra, more commonly utilized by organic chemists, show only a broad band at about 220–250 nm (log $\varepsilon\approx3.9$) and two very low intensity bands at 313 and 318 nm. Thus the study of substituent effects on the ultraviolet absorption has been more difficult than in other related systems. This broad absorption is an overlap of two or three transitions of similar intensity, which with suitable substitution in the thiophen ring, may be separated. Thus $-I$, $-M$ substituents in the 2-position show two close, high-intensity absorptions which are progressively shifted to longer wavelengths with increasing efficiency of conjugation, the extinction coefficient increasing in the same order as in the benzene series ($NO_2 > CHO > COCH_3 > COOH > CN > SO_2CH_3$). Similar, but smaller, shifts are noted for these substituents in the 3-position (only one band being observed), suggesting that substituents conjugate better at the 2- than the 3-position.

With $+M$ substituents (*e.g.* alkyl, halogen, OMe, and SMe) the effect of the substituent is reversed in that 2-substituents give a single band while 3-isomers exhibit two peaks, both of lesser extinction.

In general, the thiophen analogues of a benzenoid compound show a bathochromic shift, making thiophen derivatives of considerable commercial interest as, for example, dyes (*e.g.* the red benzenoid azo-dyes having blue-green thiophen isosteres[19]) and optical brightening agents.

19.1.4.3 Infrared spectroscopy

Infrared absorption is relatively little used in distinguishing between differently substituted thiophens compared with its application in the benzene series, largely perhaps because of the greater ease of interpretation of thiophen n.m.r. spectra. Nevertheless, significant absorptions at 1537–1509, 1444–1402, and 1365–1339 cm^{-1} characterize ring stretching in 2-substituted thiophens, with in-plane bending signals at 1086–1077 and 1053–1031 cm^{-1}, and out-of-plane deformations at 938–905 and 863–841, and a ring breathing absorption at 839–790 cm^{-1}. The 3-isomers reveal a diagnostic absorption at 760–780 cm^{-1}.

The most commonly utilized absorption, the C=O stretching frequency, appears at lower positions in the 2-isomer relative to the 3-isomer, indicative of the greater single-bond character of the former as a result of more efficient conjugation. Thus, in chloroform solution, thiophen-2- and -3-carboxylic acids absorb at 1679 and 1690 cm^{-1}, respectively, and the corresponding aldehydes at 1673 and 1691 cm^{-1}. The appearance of carbonyl doublets in many aldehydes and ketones has been shown to derive from the presence of rotamers, and intensity data have been used to quantify the conformational preference.[20]

19.1.4.4 Nuclear magnetic resonance spectroscopy

Proton n.m.r. spectroscopy is valuable for structure elucidation of thiophens.[21] Chemical shifts are well documented and the characteristic coupling constants are in the range $J_{2,3} = 4.90–5.80$, $J_{3,4} = 3.45–4.35$, $J_{2,4} = 1.25–1.70$, $J_{2,5} = 3.20–3.65$ Hz. Furthermore, useful and definitive long-range coupling occurs between substituents (*e.g.* methyl, alkyl, formyl, aldoxime, thiol, alkynyl) and ring protons, enabling easy assignment of more complex and fused thiophens. For example, this coupling has been utilized to demonstrate the preferred conformation of 2-formylthiophen (**9**), while 3-hydroxythiophen-2-aldehyde exists in the intramolecularly hydrogen bonded form (**10**) in carbon tetrachloride but in the alternative rotamer (**11**) in acetone solution, the long-range coupling always following an extended W pathway. The magnitude of coupling constants in substituted thiophens increases with increasing electronegativity of the substituent and, with 2-substituted cases, in a linear manner.

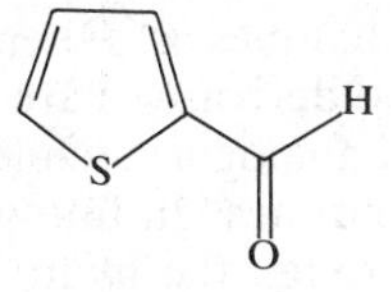

(**9**) 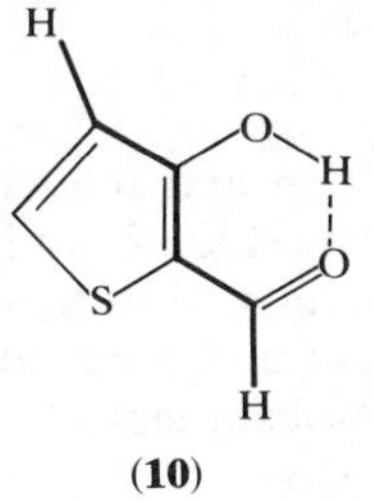 (**10**) 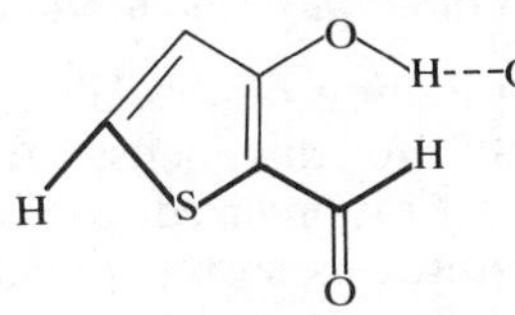 (**11**)

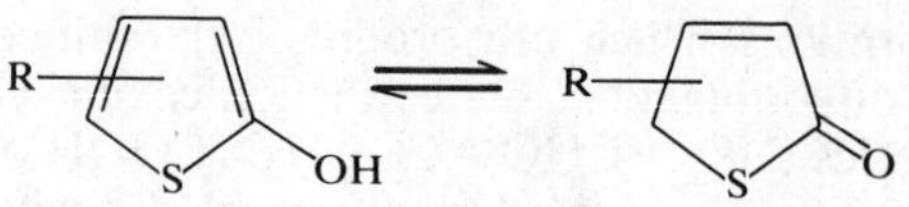

(12) (13)

^{1}H n.m.r. spectroscopy has been used in the study of potentially tautomeric thiophens, the hydroxy-derivatives in particular showing considerable diversity, with the tautomers (**12** and **13**) being easily assigned (see p. 820).

^{13}C n.m.r. spectra of thiophens are also documented,[21] and n.q.r. resonance frequencies of a number of thiophens have been collated.[22]

19.1.4.5 Mass spectrometry

Simple thiophens show characteristic fragmentation pathways (Scheme 3). Extensive use of D and ^{13}C labelling indicates that rapid randomization of thiophenic CH's precedes fragmentation, probably *via* a 'Ladenburg thiophen' intermediate, in a manner related to the photoisomerization of 2- to 3-substituted thiophens (see p. 818). The mass spectra of thiophen derivatives and fused thiophens have been usefully reviewed.[23]

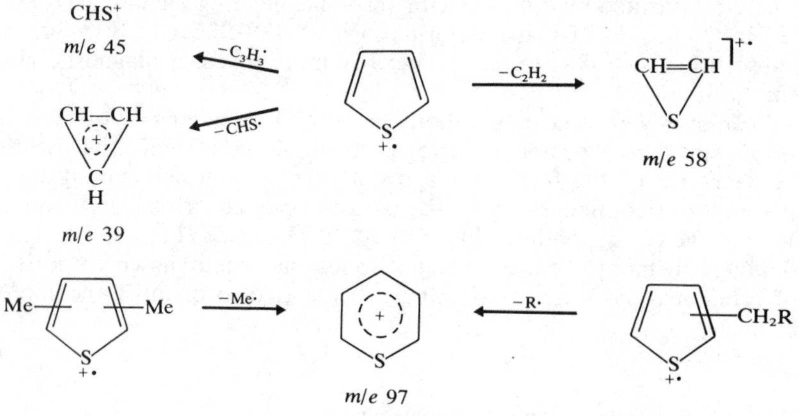

CHS^+ *m/e* 45 ← $-C_3H_3^{\cdot}$; ← $-CHS^{\cdot}$ CH—CH / C / H *m/e* 39; $-C_2H_2$ → CH=CH / S $]^{+\cdot}$ *m/e* 58

Me— —Me (S +•) → $-Me^{\cdot}$ → *m/e* 97 ← $-R^{\cdot}$ ← —CH_2R (S +•)

SCHEME 3

19.1.4.6 Electron spin resonance spectroscopy

The study of e.s.r. spectra of thiophens[24] has largely been a research field in its own right rather than a tool for structural elucidation. However, confirmation of *p*- rather than *d*-orbital participation in thiophen radical anions, the quantitative study of acyl thiophen rotamers (as radical anions) and restricted rotation in thenyl radicals, and a comparison of the stability of thienyl and phenyl substituted radicals represent typical problems conveniently studied by this technique.

19.1.4.7 Other physical methods

The application of the Hammett equation and related kinetic techniques is generally very effective for thiophens,[25] though σ-values for 2- and 3-substituted derivatives are not always easily rationalized. ρ-Values are similar to those of benzenoid analogues, while 4- and 5-substituents correlate well with *meta* and *para* positions in benzenes. In line with MO calculation data, the transmission of substituent effects occurs through the hydrocarbon rather than sulphur part of the molecule.

Useful summaries of the application of photoelectron spectroscopy to heterocycles in general, and thiophens in particular,[26] and of polarography, have appeared.[27]

19.1.5 ELECTROPHILIC REACTIONS OF THIOPHENS*

19.1.5.1 Reactivity of thiophens

Some generalizations regarding the reactivity of thiophen are apparent. (i) Thiophen is much more reactive than benzene towards electrophilic substitution (differences of 600–100 000 times are quoted for different cases — see Table 2), though the majority of

TABLE 2

Relative Rates of Electrophilic Substitution of Thiophens and Benzenes

Reaction	*Rate (thiophen)* / *Rate (benzene)*	*Rate (α-thienyl)* / *Rate (β-thienyl)*	*Ref.*
H/D exchange (CF_3COOH)	—	3400	a
H/D exchange (aq. H_2SO_4)	—	1045	b
H/T exchange (aq. H_2SO_4)	~600	955	c
Desilylation	5000	43.5	d
Nitration	850	6.2	e
Formylation (Vilsmeyer)	—	>1000	f
Bromination (Br_2)	—	490	f
Chlorination (Cl_2)	—	100	f
Acetylation ($Ac_2O/SnCl_4$)	—	200	f
Acetoxymercuration	97 000	—	g

[a] E. N. Zvyagintseva, T. A. Yukushina, and A. I. Shatenshtein, *J. Gen. Chem.* (*U.S.S.R.*), 1968, **38,** 1933. [b] B. Östman and S. Olsson, *Arkiv Kemi,* 1960, **15,** 275. [c] K. Halvarson and L. Melander, *Arkiv Kemi,* 1956, **8,** 29. [d] F. B. Deans and C. Eaborn, *J. Chem. Soc.*, 1959, 2303. [e] B. Östman, *Arkiv Kemi,* 1962, **19,** 499. [f] S. Clementi, P. Linda, and G. Marino, *J. Chem. Soc.* (*B*), 1968, 397. [g] R. Motoyama, S. Nishimura, E. Imoto, Y. Murakami, K. Hari, and J. Ogawa, *Nippon Kagaku Zasshi,* 1957, **78,** 962 (*Chem. Abs.*, 1960, **54,** 14 224).

reactions follow the same mechanistic pathway as those of benzene. Indeed, benzene is often conveniently used as a solvent. The reactivity order pyrrole > furan > selenophen > thiophen > benzene applies in electrophilic substitution and yields of products are generally high. (ii) α-Substitution is generally observed, often exclusively, even causing expulsion or rearrangement of a substituent (*e.g.* Br, COOH, or alkyl). The α-selectivity decreases with increasing vigour of the reagent (nitration of thiophen gives up to 15% of the 3-isomer in addition to 2-nitrothiophen) or the conditions (increasing the temperature during acetylation of thiophen from 25 to 75 °C decreases the α/β ratio from 200 to 82) (see Table 2). (iii) Steric effects are less severe in thiophens than in benzene analogues because of a more favourable geometry. (iv) The presence of a substituent has a much greater influence at the α-position in thiophen than in benzene. Thus 2,5-dimethylthiophen efficiently competes with thiophen and even 2-methylthiophen during acetylation, while 2-alkoxy- and 2-alkylthio-thiophens give considerable amounts of 3-substitution as well as the expected 5-isomer (see Table 3). Furthermore, deactivating groups strongly resist *ortho*-substitution, even when the remaining sites are blocked. (v) Reaction conditions cause a greater variation in the nature and course of an electrophilic substitution in the thiophen than in the benzene series. These factors allow a much wider range of reagents and conditions to be considered in thiophen chemistry, but make specific β-substituted derivatives difficult of access.

* An excellent comparative review[28] supplements the coverage by Gronowitz[1–3] on this subject.

TABLE 3

Products from the Electrophilic Substitution of 2-Substituted Thiophens ($\pm I,+M$ Substituents)

Substituent	*Reaction and/or reagent*	*Substn. position (ratio)* 3	5	*Ref.*
Me	nitration	30	70	a
	chlorination, SO_2Cl_2		only	a
	bromination, Br_2		mostly	a
	bromination, NBS		only	a
	H exchange		only	b
	alkylation, $Bu^tCl/FeCl_3$		only	a
	acylation		only	a
	thallation		only	c
Bu^t	acylation		only	a
	nitration		only	a
Cyclopropyl	deuteration		only	d
	halogenation		only	d
	formylation		only	d
Ph	bromination	4	96	e
	formylation	0.1	99.9	e
	acetylation	1	99	e
NHAc	chlorosulphonation		only	a
	formylation		only	a
OMe	bromination, NBS		only	a
	bromination, Br_2		mostly	a
	formylation		only	a
	acetylation		mostly	f
	nitration	40	60	a
SMe	formylation		only	a
	iodination		only	a
	acetylation		91	f
	nitration		only	a
Cl, Br, I	sulphonation	9	only	a
	bromination		only	a
	formylation		only	a
	nitration		only	a

[a] See Ref. 1, Section IV-B-2, p. 47. [b] E. N. Zvyagintseva, T. A. Yukushina, and A. I. Shatenshtein, *J. Gen. Chem. (U.S.S.R.)*, 1968, **38,** 1933. [c] A. McKillop, D. Bromley, and E. C. Taylor, *J. Org. Chem.*, 1972, **37,** 88. [d] J. Fabian, S. Scheithauer, and R. Mayer, *J. prakt. Chem.*, 1969, **311,** 45. [e] N. Gjøs and S. Gronowitz, *Acta Chem. Scand.*, 1970, **24,** 99. [f] S. Clementi, P. Linda, and G. Marino, *J. Chem. Soc. (B)*, 1970, 1153.

19.1.5.2 Directing effects

Ideally, one should consider the transition state energies leading to differently substituted thiophens in order to decide orientation. However, Wheland intermediates (or σ-complexes) (Scheme 4) are reasonable models of the transition state with thiophens and give a good understanding of the nature of electrophilic substitution. It is immediately evident that there is greater delocalization in the σ-complex, leading to α-substitution (equation 1) than to β-substitution (equation 2), accounting for the dominance of the former mode of reaction.[29]

An exactly parallel situation prevails when substituted thiophens undergo electrophilic substitution (Scheme 5), but superimposed upon the delocalized σ-complex is the stabilizing or destabilizing influence of the substituent. Thus for 2-substituents, of the four possible sites for attack (excluding sulphur — see p. 803), α-attack (equation 3) will

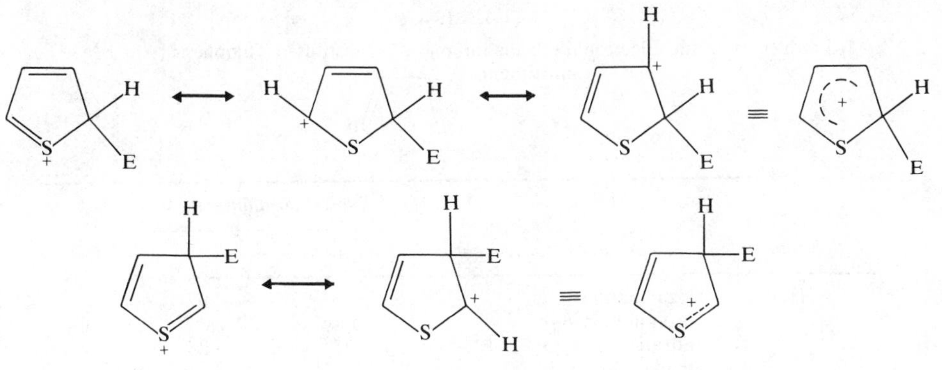

(1)

(2)

SCHEME 4

clearly be favoured for $\pm I,+M$ substituents (R = Me, Cl, Br, OMe, NHAc, *etc.*) since the substituent further stabilizes the positive charge by extending the mesomerism (Table 3). This situation is not possible for attack at the 4-position (equation 4) and is less efficient but prevalent at the 3-position (equation 5). However, *ipso*-attack (equation 6) allows extensive delocalization, making it an attractive alternative to β-substitution. With $-I,-M$ substituents this conjugation of the substituent with the charged ring (equations 3 and 5) results in destabilization of the σ-complex, making 4-substitution (equation 4) a more significant possibility. The more powerful this electronic effect (either of the substituents or the reaction conditions — see, for example, the swamping catalyst effect, Table 4) the greater the degree of 4-substitution (see Table 4).

With 3-substituted thiophens (Scheme 6) the same kind of arguments suggest that $\pm I,+M$ substituents should favour 2-substitution since only in this position does the

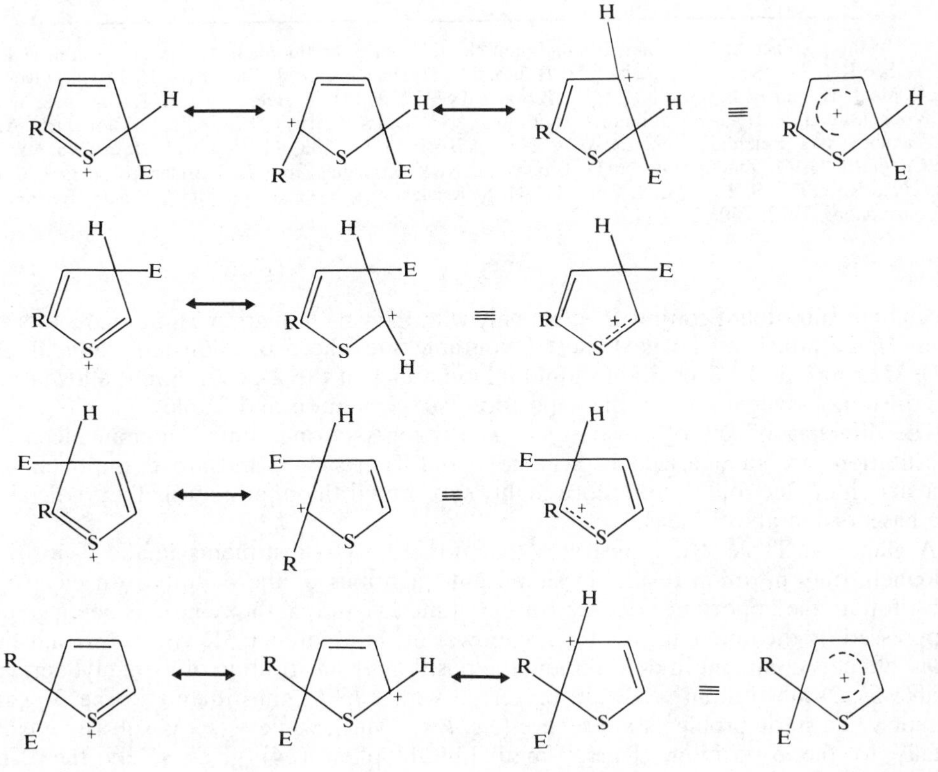

(3)

(4)

(5)

(6)

SCHEME 5

TABLE 4

Products from the Electrophilic Substitution of 2-Substituted Thiophens ($-I,-M$ Substituents)

Substituent	*Reaction and/or reagent*	*Substn. position (ratio)* 4	5	*Ref.*
CHO	bromination	3	97	a
	bromination*	only		a
	nitration	75	25	b
	thallation		only	c
Ac	chloromethylation†	50	50	d
	acetylation†		only	c
	nitration	75	25	b
	isopropylation*	99	1	a
	bromination*	mostly		a
	acetylation*	90	10	e
	chloromethylation*	mostly		d
CN	nitration	43	57	b
COOH	nitration, HNO_3/AcOH	31	69	b
	nitration, HNO_3/H_2SO_4	56	40	f
NO_2	nitration, HNO_3/AcOH	85	15	c
	nitration, HNO_3/H_2SO_4	63	37	f
SO_3H	nitration		mostly	a
SO_2Cl	nitration	75	25	a
	chlorination	mostly		a
	chlorsulphonation	mostly		a
$\overset{+}{N}Me_3$	nitration		only	g
$\overset{+}{S}Me_2$		60	40	h

* In the presence of 2–3 moles of aluminium chloride per mole thiophen. † Very low overall yield.
[a] See Ref. 30. [b] See Ref. 1, Section IV-B-3, p. 52. [c] D. J. Chadwick, J. Chambers, H. E. Hargraves, G. D. Meakins, and R. L. Snowden, *J.C.S. Perkin I*, 1973, 2327. [d] L. I. Belen'kii, I. B. Karmanova, Yu. B. Vol'kenshtein, and Ya. L. Gol'dfarb, *Bull. Acad. Sci. U.S.S.R.*, 1971, 878. [e] Ya. L. Gol'dfarb, A. P. Yakubov, L. I. Belen'kii, *Doklady Akad. Nauk S.S.S.R.*, 1969, **185,** 91. [f] See Ref. 28, p. 296. [g] Ya. L. Gol'dfarb, G. M. Zhidomirov, N. D. Chuvylkin, N. S. Ksenzhek, and L. I. Belen'kii, *J. Org. Chem. (U.S.S.R.)*, 1973, **9,** 153. [h] L. I. Belen'kii, N. S. Ksenzhek, and Ya. L. Gol'dfarb, *Chem. Heterocyclic Compounds*, 1972, 280.

incumbent substituent conjugate effectively with the ring (equation 7). Steric factors might cause some attack at the next most favourable site, the 5-position (equation 8). Again $-I,-M$ groups in the 3-position should inhibit attack at the 2-position, but allow attack at the 5- and, less significantly, the 4-position (see Scheme 6 and Table 5).

The directing effects of a variety of substituents during some common electrophilic substitutions are summarized in Scheme 7 and Tables 3–5, and are exemplified below. The action of electrophiles on more highly substituted thiophens is usually predictable on the basis of the above ideas.

A glance at Table 3 demonstrates that, for $\pm I,+M$ substituents in the 2-position of thiophen, only nitration results in significant quantities of the 3-isomer on electrophilic substitution, the important *ortho*-activation of methyl and methoxy-groups being generally suppressed by the dominant α-directing power of the thiophen. However, this substituent *ortho*-effect is dominant in determining the position of substitution of 3-thienyl derivatives (Table 5). 2-Substitution is always preferred with $\pm I,+M$ substituents in the 3-position, except when steric problems are severe (*e.g.* R$=$CMe$_3$), while $-I, -M$ substituents direct totally to the 5-position. Thus 3-methylthiothiophen (**14**) gives solely the 2-nitro-derivative, while the corresponding 3-dimethylthienyl sulphonium salts (**15**) give only the

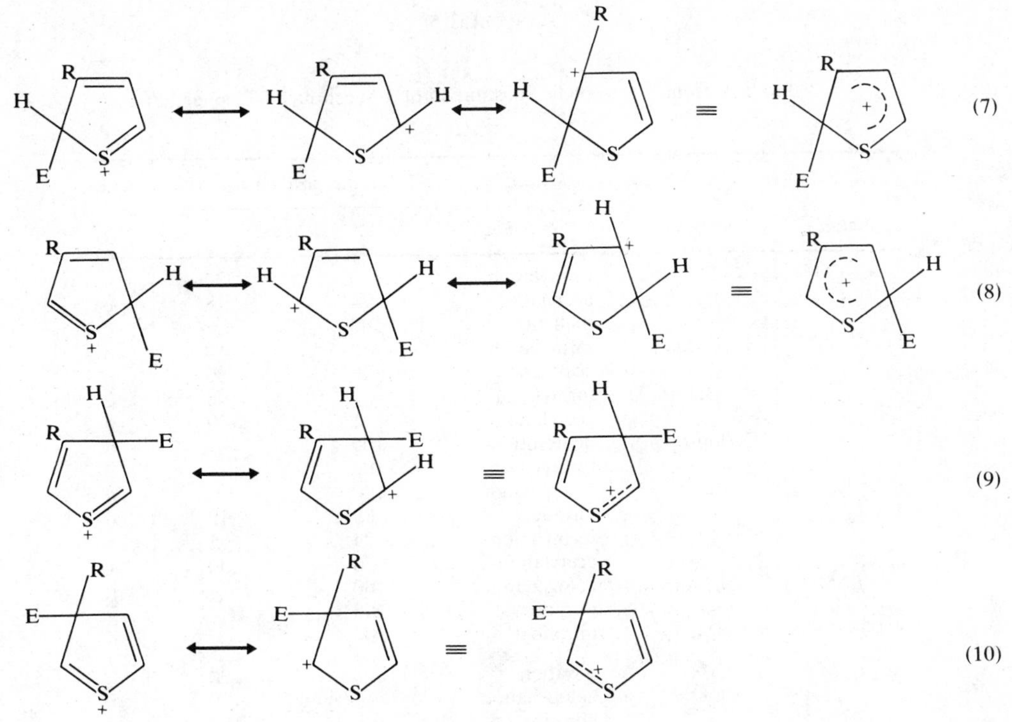

(7)

(8)

(9)

(10)

SCHEME 6

5-isomer. α-Electron-withdrawing substituents ($-I,-M$) (Table 4) give variable proportions of 4- and 5-substituted derivatives on electrophilic attack.

Interesting effects derive from the use of a large excess of a Lewis acid catalyst (particularly aluminium chloride—the swamping catalyst effect[30]). By complexing a 2-carbonyl group (*e.g.* as in **16**), increased deactivation at the 3- and 5-positions results and further substitution gives largely or exclusively the 2,4-disubstituted product (see Table 4). By this means, Gol'dfarb and his co-workers have successfully alkylated, halogenated, acylated, and chloromethylated various 2-acylthiophens to give otherwise difficultly accessible products. Curiously, the chloroalkanes often used as solvents in Friedel–Crafts

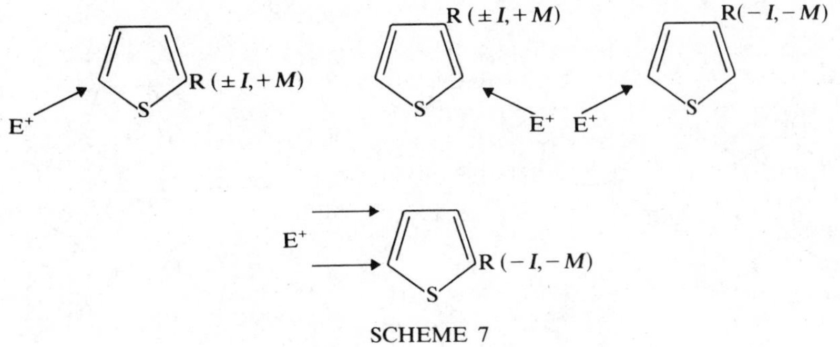

SCHEME 7

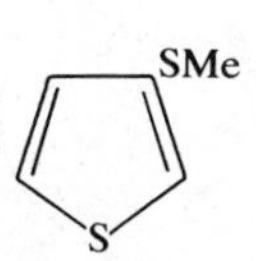

(14) 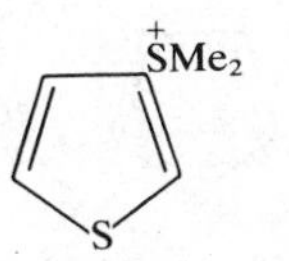 **(15)**

TABLE 5

Products from Electrophilic Substitution of 3-Substituted Thiophens

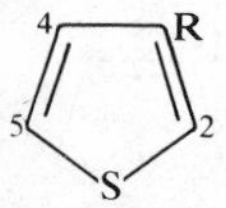

Nature	Substituent	Reaction	Substn. position (ratio) 2	5	Ref.
$+I,+M$	Me	formylation	80	20	a
		acetylation	74	18	b
		thallation	100		c
	$CHMe_2$	formylation	50	50	d
		acetylation	40	60	b
	Bu^t	formylation		100	e
		acetylation		100	b
	Cyclopropyl	deuteration	100		d
		formylation	100		d
		halogenation	100		d
		nitration	88	12	d
	Ph	formylation	94	6	f
		acetylation	70	30	f
$-I,+M$	NHAc	halogenation	100		g
		nitration	100		g
	SMe	nitration	100		h
$-I,-M$	$\overset{+}{S}Me_2$	nitration		100	h
	COOH	halogenation		100	i
		nitration		100	i
	CHO	nitration		100	i
	Ac	acetylation*		100	j

* In the presence of 2–3 moles aluminium chloride per mole thiophen.
[a] R. M. Kellogg and J. Buter, *J. Org. Chem.*, 1971, **36,** 2236. [b] See Ref. 28, p. 294. [c] See Table 3, Ref. c. [d] See Table 3, Ref. d. [e] M. Sy, N. P. Buu-Hoi and N. D. Xuong, *J. Chem. Soc.*, 1955, 21. [f] See Table 3, Ref. e. [g] E. Campaigne and P. A. Monroe, *J. Amer. Chem. Soc.*, 1954, **76,** 2447. [h] Ya. L. Gol'dfarb, N. S. Ksenzhek, and L. I. Belen'kii, *Bull. Acad. Sci. U.S.S.R.*, 1973, 1124. [i] See Ref. 1, p. 55. [j] See Table 4, Ref. e.

reactions (methylene chloride, chloroform, and carbon tetrachloride) alkylate 2-acetylthiophen under these conditions to give bis- and tris-(2-acetyl-4-thienyl)methanes (**17**, **18**, and **19**, respectively[31]).

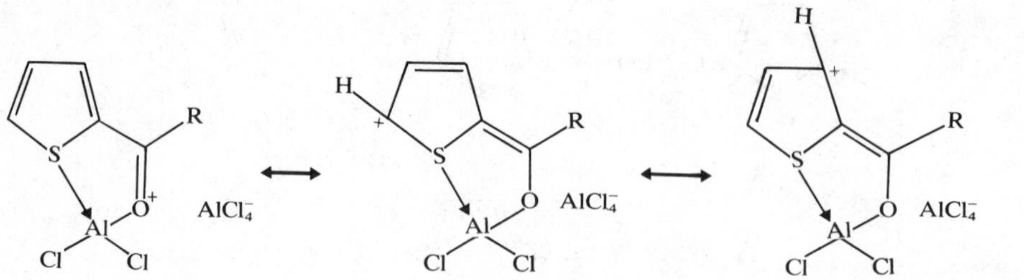

(**16a**) (**16b**) (**16c**)

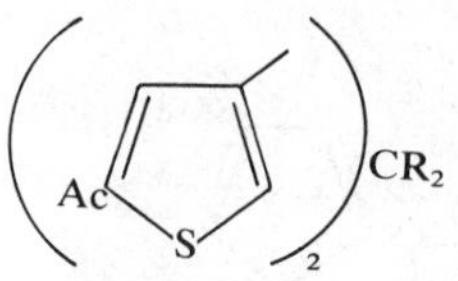

(**17**) R = H
(**18**) R = Cl

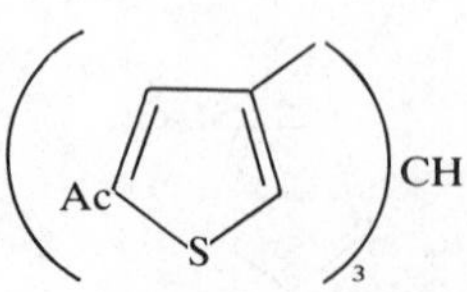

(**19**)

TABLE 6

Synthetic Methods for the Preparation of Some Useful Thiophens by Electrophilic Substitution of Thiophen

Conditions	*Product* (% *Yield*)	*Comment*	*Ref.*
PhN(Me)CHO/$POCl_3$	2-Thiophenaldehyde (71–74)	For reduction to 2-Me	a
$Me_2NCHO/POCl_3$	2-Thiophenaldehyde (67)	(78%) see Ref. 6	b
Ac_2O/H_3PO_4	2-Acetylthiophen (74–79)*	Also with $(EtCO)_2O$ and $(PrCO)_2O$ (85–95%), and for 2-chloro-5-methylthiophen (65–70%) and 3- and 4-methyl-2-acetylthiophen	c
$RCOCl/SnCl_4$	2-Acetylthiophen (79–83)	Better method for higher acyl homologues	d
$RCOCl/AlCl_3$	3-Acetyl-2,5-dichlorothiophen (73); 2-benzoylthiophen (88–89)	Better method for deactivated thiophens and aroyl halides	e
CH_2O/HCl	2-Chloromethylthiophen (40–41)		f
$CH_2O/HCl/ZnCl_2$	Di-(2-thienyl)methane (55)		g
R_2CO/aq. H_2SO_4	Di-(2-thienyl)alkanes (39–79)		h
CH_3CHO/HCl then pyridine	2-Vinylthiophen (50–55)	5-Bromo-derivative (35%)	i
$Me_2C{=}CH_2/H_2SO_4$	2- and 3-, and di-t-butyl-thiophen		j
Cl_2	2-Chlorothiophen and 2,5-dichlorothiophen		h
Br_2/AcOH	2-Bromo- and 2,5-dibromo-thiophen		k
Br_2/AcOH	2,3,5-Tribromothiophen (75–85)	Source of 3-bromothiophen (89–90%)	l
Br_2/AcOH	3-Bromothiophen→2,3-dibromothiophen		m
Br_2	Thenoic acid→4,5-dibromothenoic acid (90–92)		n
NBS/$CHCl_3$/AcOH	2-Bromo-3-methylthiophen	Good for monohalogenation	o
I_2/HgO	2-Iodothiophen (72–75)		p
I_2/HNO_3	2-Iodothiophen (68–72)		q
I_2/HgO/AcOH	Tetraiodothiophen (68–77)		f
HNO_3/Ac_2O	2-Nitrothiophen (70–85)	→2-Aminothiophen salt	t
$Cu(NO_3)_2$/AcOH	2-Nitrothiophen (~85)		s
$ClSO_3H$	Thiophen-2-sulphonyl chloride (72)		t
SO_3/pyridine	Thiophen-2-sulphonic acid (86)		u

* Treatment of product with halogen and sodium hydroxide gives 2-thenoic acid (86–95%) (see Ref. e).

[a] A. W. Weston and R. J. Michaels, *Org. Synth., Coll. Vol. IV*, 1963, 915. [b] E. Campaigne and W. L. Archer, *J. Amer. Chem. Soc.*, 1953, **75,** 989 (reduction: W. J. King, and F. F. Nord, *J. Org. Chem.*, 1949, **14,** 638). [c] A. I. Kosak and H. D. Hartough, *Org. Synth., Coll. Vol. III*, 1953, 14; also Ref. 4, p. 503–504. [d] J. R. Johnson and G. E. May, *Org. Synth., Coll. Vol. II*, 1943, 8. [e] See Ref. 4, p. 505; also W. Minnis, *Org. Synth., Coll. Vol. II*, 1943, 520. [f] See Ref. 4, p. 501; and K. B. Wiberg and H. F. McShane, *Org. Synth., Coll. Vol. III*, 1955, 197. [g] See Ref. 129. [h] See Ref. 33. [i] See Ref. 127. [j] See Ref. 4, p. 497. [k] See Ref. 4, p. 498. [l] S. Gronowitz and T. Raznikiewicz, *Org. Synth., Coll. Vol. V*, 1973, 149. [m] S.-O. Lawesson, *Arkiv Kemi*, 1957, **11,** 373. [n] See Ref. 4, p. 509. [o] See Ref. 35. [p] W. Minnis, *Org. Synth., Coll. Vol. II*, 1943, 357. [q] H. Y. Lew and C. R. Noller, *Org. Synth., Coll. Vol. IV*, 1953, 545. [r] V. S. Babasinian, *Org. Synth., Coll. Vol. II*, 1943, 466; see also Ref. 4, p. 512. [s] N. I. Putokhin, *Sbornik Nauch. Trudov, Kuibyshev Ind. Inst. im. V. V. Kuibysheva*, 1955, 271 (*Chem. Abs.*, 1957, **51,** 16 419). [t] See Ref. 4, p. 513; and J. Cymerman-Craig, G. N. Vaughan, and W. K. Warburton, *J. Chem. Soc.*, 1956, 4114. [u] A. P. Terent'ev and G. M. Kadatskii, *Zhur. org. Khim.*, 1952, **22,** 153 (*Chem. Abs.*, 1952, **46,** 11 178).

19.1.5.3 Synthetic applications

From a synthetic viewpoint, electrophilic substitution of thiophens is very important. Conditions for some typical cases are recorded in Table 6. Alkylation of thiophen is not normally an effective reaction, but many alkyl derivatives are available by reduction of the acyl derivative by means of the Huang–Minlon (hydrazine/KOH) or Clemmensen (Zn/Hg/HCl) reduction, which are both highly effective in the thiophen series.[32] Di-2-thienylmethanes derive from treatment of thiophens with formaldehyde and hydrochloric acid in the presence of zinc chloride, and analogous alkanes, *e.g.* (**20**), by the action of a ketone and 70% sulphuric acid.[33] 2,5-Dimethylthiophen gives a cyclic trimer (**21**) and tetramer under these conditions.[34] Halogenation of thiophens can be carried out in a stepwise manner to the perhalo-derivative without problems, although some addition products are also found as by-products. However, mono-halogenation is sometimes awkward using elemental halogens (Cl_2 or Br_2), this problem being overcome by use of kinetically controlled methods such as the use of *N*-bromosuccinimide in acetic acid/chloroform mixture[35] or *N*-chlorobenzotriazole,[36] or by way of the thallated thiophen (see Table 3). An excellent procedure for 3-bromothiophen, a vital intermediate for the synthesis of 3-substituted thiophens (see p. 809), depends upon the selective α-debromination of 2,3,5-tribromothiophen with zinc and acetic acid (see p. 806 for another route to 3-bromothiophen). By the same debromination method, 3,4-dibromothiophen, 3-methyl-4-bromothiophen, and 2-methyl-3-bromothiophen are conveniently prepared from tetrabromothiophen, 3-methyl-2,4,5-tribromothiophen, and 2-methyl-3,5-dibromothiophen, respectively. Other methods for α-dehalogenation include the use of sodium in liquid ammonia, zinc, sodium or aluminium amalgam, copper and hypophosphorous acid, sodium telluride (good for Cl), and sodium alkoxide in dimethyl sulphoxide.[37] 3-Nitrothiophen can be obtained by steam distillation of 4-nitrothiophen-2-sulphonyl chloride (see Table 4), derived by nitration of thiophen-2-sulphonyl chloride.

The lability of alkyl groups under certain conditions has already been referred to (p. 795). Some remarkable examples are recorded below (reactions 11–13). Decarboxylation is especially noted during nitrations. Thus the method of choice for making 2,4-dinitro-5-bromothiophen involves nitration of 5-bromothiophen-2-carboxylic acid with mixed sulphuric and nitric acids; the ester nitrates normally.[43]

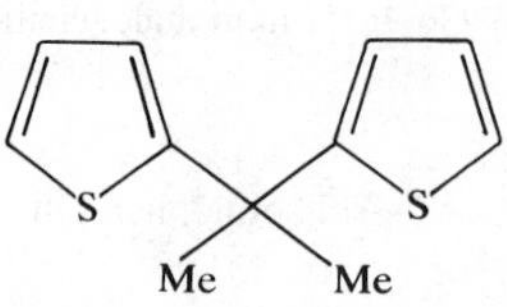

(20)

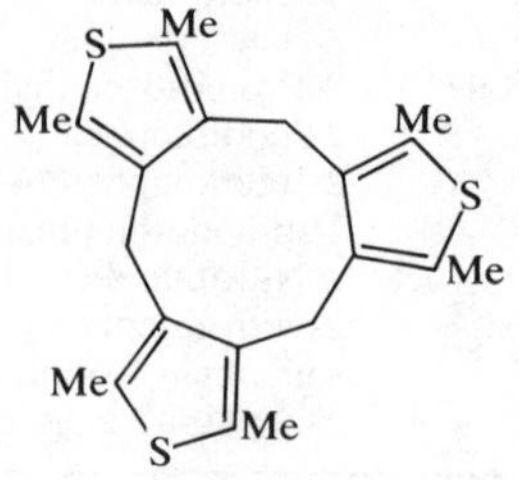

(21)

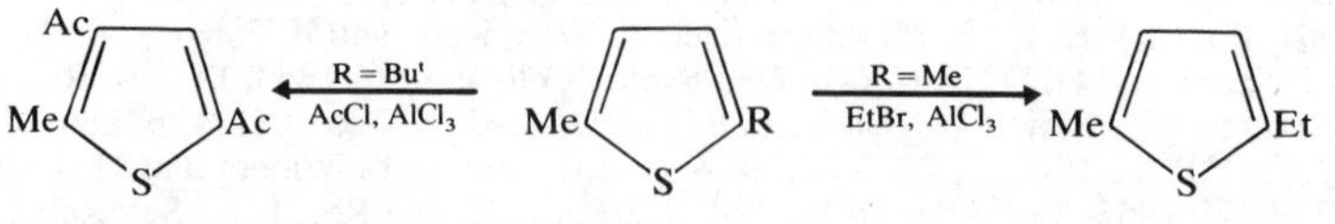

R = Bu^t, AcCl, $AlCl_3$ (Ref. 39); R = Me, EtBr, $AlCl_3$ (Ref. 38) (11)

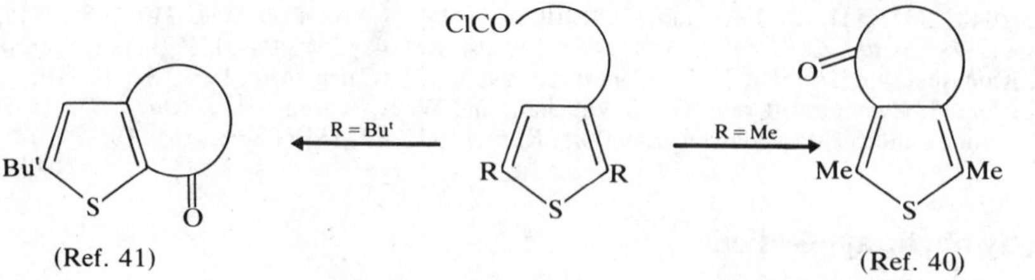

R = Bu^t (Ref. 41); R = Me (Ref. 40) (12)

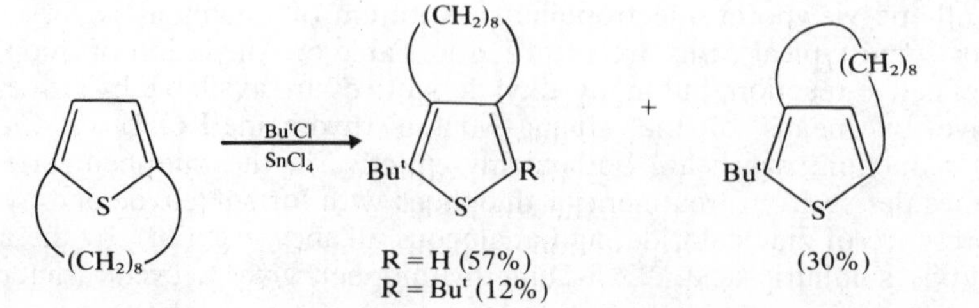

Bu^{t}Cl, $SnCl_4$: R = H (57%), R = Bu^t (12%) + (30%) (Ref. 42) (13)

19.1.5.4 Electrophilic ring-closures

Intramolecular acylation (*e.g.* **22**→**23** and **24**→**25**) has been widely used in the thiophen series, the method of choice often being use of polyphosphoric acid on the free acid in place of the usual Friedel–Crafts techniques. Cyclization always prefers to take place at an α-position, but can be forced to attack the β-position if chlorines, methyl groups, or other alkyl groups occupy the α-sites Bromo and t-butyl groups are particularly prone to elimination if used as blocking groups. Unlike the benzene series, annulation of a five-membered ring proceeds with difficulty, especially at the 2,3-positions, indicating the greater angular requirements of thiophen.[44] Medium sized rings also are difficult to annulate. Long-chain ω-(2-thienyl)alkanoic acids (**22**; X = OH; $n > 9$) cyclize at the 4-position if the 5-position is blocked (to give **26**), and with $n = 13$–17 cyclize at the 5-position to yield the interesting macrocycles (**27**). By desulphurization (see p. 816) of these and related systems, Gol'dfarb's group made a number of muscone and civetone perfumery analogues. With lower members (**22**, $n = 5$–9) the cyclic dimers (**28**) were isolated.[32]

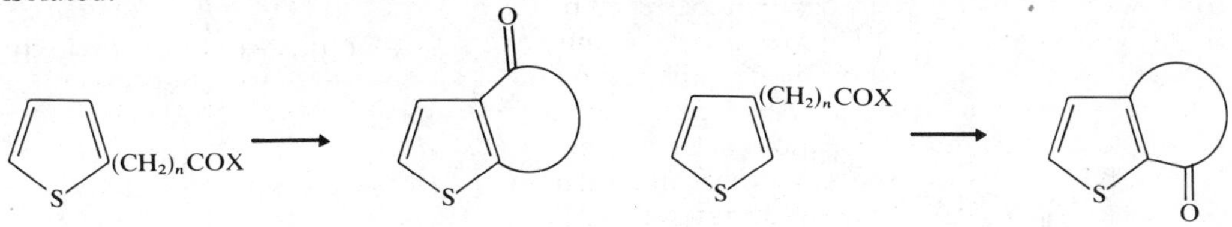

$(CH_2)_nCOX$ → (22) (23) $(CH_2)_nCOX$ → (24) (25)

$(CH_2)_{10 \text{ or } 11}$ Me (26) $(CH_2)_n$ (27) $(CH_2)_n$ CO $(CH_2)_n$ (28)

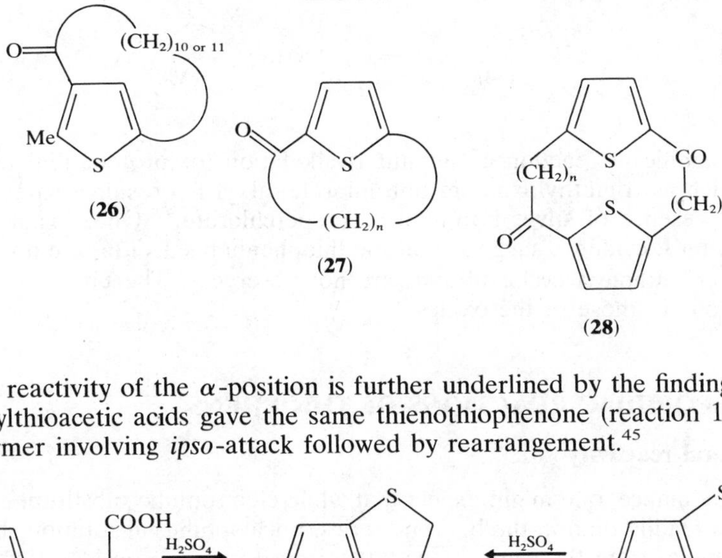

The great reactivity of the α-position is further underlined by the finding that both 2- and 3-thienylthioacetic acids gave the same thienothiophenone (reaction 14) on cyclization, the former involving *ipso*-attack followed by rearrangement.[45]

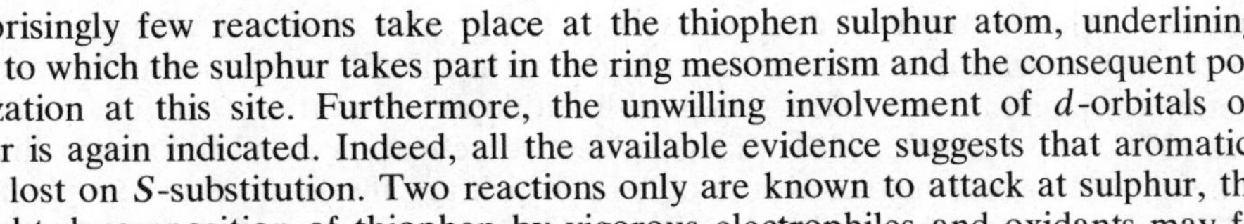

COOH CH_2 H_2SO_4 → ← H_2SO_4 COOH (14)

19.1.5.5 Reactions at sulphur

Surprisingly few reactions take place at the thiophen sulphur atom, underlining the extent to which the sulphur takes part in the ring mesomerism and the consequent positive polarization at this site. Furthermore, the unwilling involvement of d-orbitals of the sulphur is again indicated. Indeed, all the available evidence suggests that aromaticity is totally lost on S-substitution. Two reactions only are known to attack at sulphur, though no doubt decomposition of thiophen by vigorous electrophiles and oxidants may follow this course. The first is the oxidation with sources of hydroxonium ions (OH^+) such as peroxides, peracids, and hypochlorites. All attempts to isolate thiophen sulphoxide (**29**) by such oxidation have failed since its formation was rapidly followed by dimerization and further oxidation to (**30**). Similarly, the sulphone (**31**) is highly reactive, but isolable. Both oxides behave efficiently as both dienes and dienophiles in Diels–Alder cycloadditions. Ring substituents confer stability on both the sulphoxide and sulphone and 2,5-dimethyl, 2,5-di-butyl, 3,4-dichloro-, tetraphenyl-, benzo-, and dibenzo-derivatives, for example, may be isolated as stable species. The tetrahedral nature of the sulphur is supported by the non-equivalence of the methyl groups[46] in the proton n.m.r. spectrum of (**32**), from which the inversion activation energy was found to be 61.95 kJ mol^{-1}. The diene character of thiophen sulphone has been elegantly harnessed by, for example, the synthesis of juglone (**33**)[47] and azulene (**35**)[48] by interaction with benzoquinone and the fulvene (**34**), respectively. In both cases, cycloaddition was rapidly followed by cheletropic elimination of sulphur dioxide.

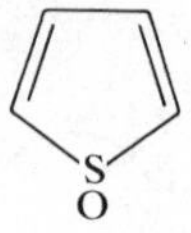

SO S O O₂ O₂ $C_5H_{11}Me_2C$ $CMe_2C_5H_{11}$

(**29**) 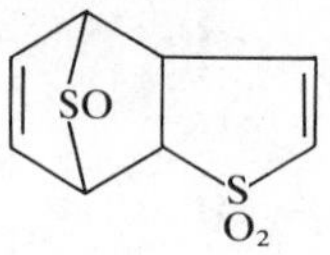 (**30**) 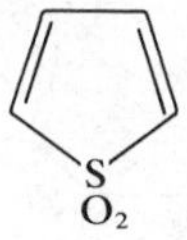 (**31**) 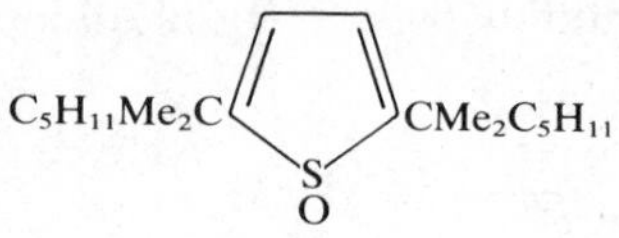 (**32**)

OH O O NMe_2 (**31**) S+ R

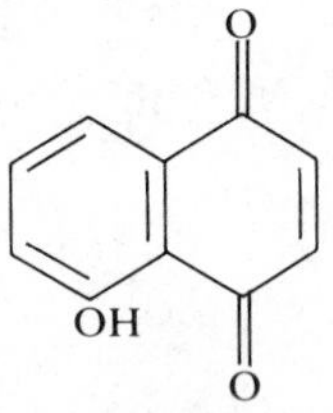

(**33**) 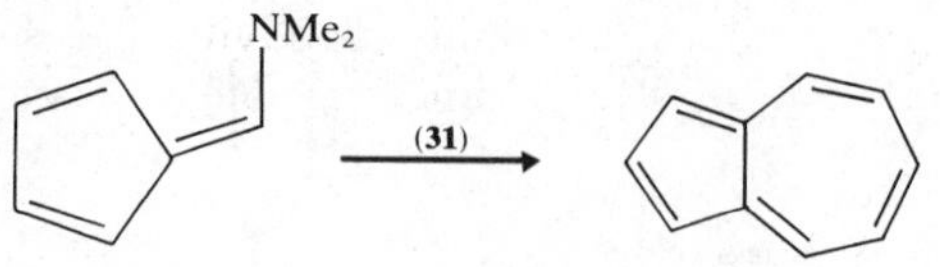 (**34**) (**35**) 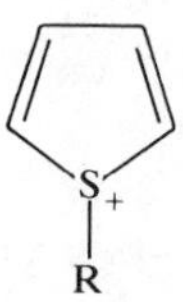 (**36**)

The other significant reaction at sulphur is alkylation (or protonation) using a 'hard' electrophile such as trimethyloxonium borofluoride, alkyl fluorosulphonates, or an alkyl halide in the presence of silver borofluoride or perchlorate.[49] Once again the spectroscopic and chemical evidence suggests that the thiophenium salts (**36**) are non-planar with dienic character (although cycloadditions are not observed). The ultraviolet spectra are closely analogous to those of the oxides.

19.1.6 NUCLEOPHILIC REACTIONS OF THIOPHENS

19.1.6.1 General reactivity

At first acquaintance, one might expect that while electrophilic substitution of thiophen proceeds more readily than in the benzene series, nucleophilic substitution should be less effective. This is far from the truth. Indeed, theoretical studies, consideration of Wheland intermediates, and experimental data indicate the fact that a similar increase in reactivity ($>1000\times$) applies to both types of substitution, as seen in Table 7. Furthermore, unlike the benzenes, and in the case of nitro-halothiophens in particular, every positional combination of the nitro- and halo-substituents strongly activates the halide towards nucleophilic substitution. This is understandable when Wheland intermediates are considered (*e.g.* Scheme 8) which demonstrates (i) the greater efficiency with which a nitro-group can delocalize the charge in the σ-complex in thiophens relative to benzenes; (ii) the better conjugation of a nitro-group with a thiophen resulting from charged species such as (**37**) compared with (**38**); (iii) the effective activation in all halo-nitro-thiophens with the exception of 4-halo-2-nitrothiophens, which are the least reactive and show the most limited scope for charge delocalization. (They are still, however, almost as reactive

TABLE 7
Relative Rates of Reaction of Bromonitrothiophens and Bromonitrobenzenes with Piperidine at 25 °C[a]

Compound	*Relative rate*	*Compound*	*Relative Rate*
m-Bromonitrobenzene	1	4-Bromo-2-nitrothiophen	1360
		4-Nitro-2-bromothiophen	very fast
p-Bromonitrobenzene	185	5-Bromo-2-nitrothiophen	28 400
o-Bromonitrobenzene	1620	2-Bromo-3-nitrothiophen	632 000
		2-Nitro-3-bromothiophen	2 500 000
		3-Bromo-4-nitrothiophen	very fast

[a] Taken from Ref. 1, p. 69.

as *o*-nitrohalobenzenes). Although most of the work in the thiophen series on nucleophilic substitution involves halo-nitro-derivatives, the above principles seem of general application. Thus, 3-bromothiophen-2-aldehyde reacts well with mild nucleophiles such as sodium azide in dimethyl sulphoxide solution, giving 3-azidothiophen-2-aldehyde, a useful precursor to 3-aminothiophen-2-aldehyde and thence thieno[3,2-*b*]pyridines.[50]

This greater reactivity shows itself in some remarkable ways. Thus, 3,4,5-trinitrobromothiophen can only be recrystallized from concentrated nitric acid, so great is its reactivity with all protic solvents.[51] Other nitrothiophens show substitution of nitro-groups, ciné substitution, or ring-opening, as illustrated below (Scheme 9).

In the light of this enhanced nucleophilic reactivity, it is not surprising that thiophens make excellent candidates for Meisenheimer complex formation.[54] Thus the equilibrium

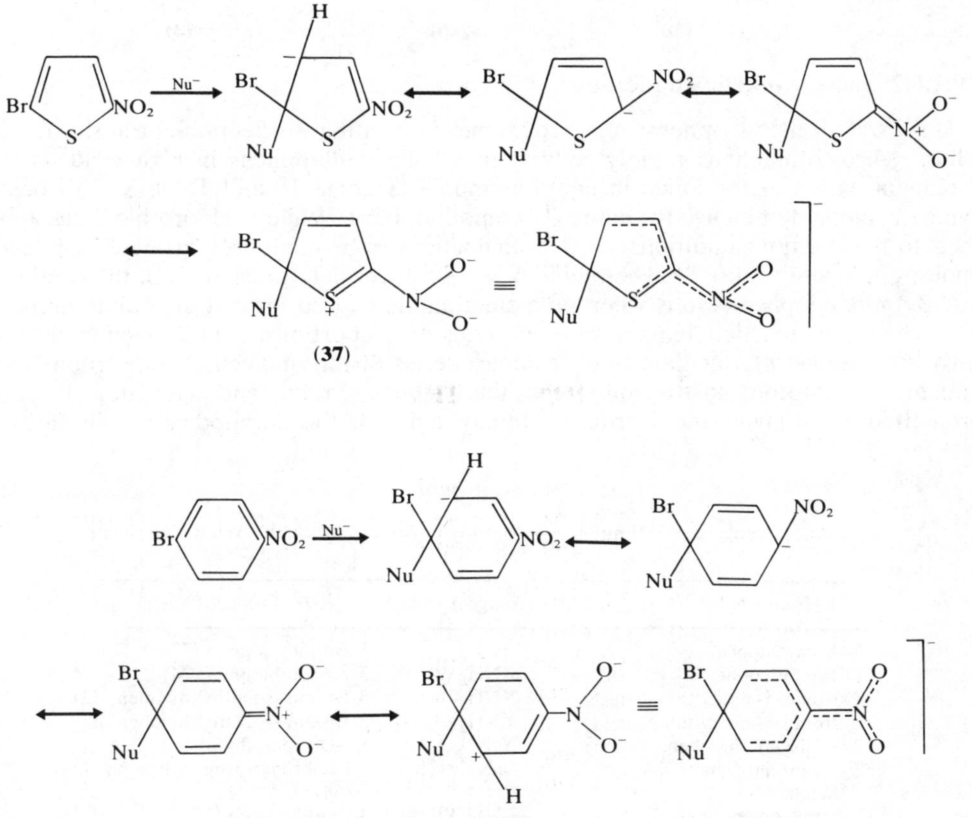

(37)

(38)

SCHEME 8

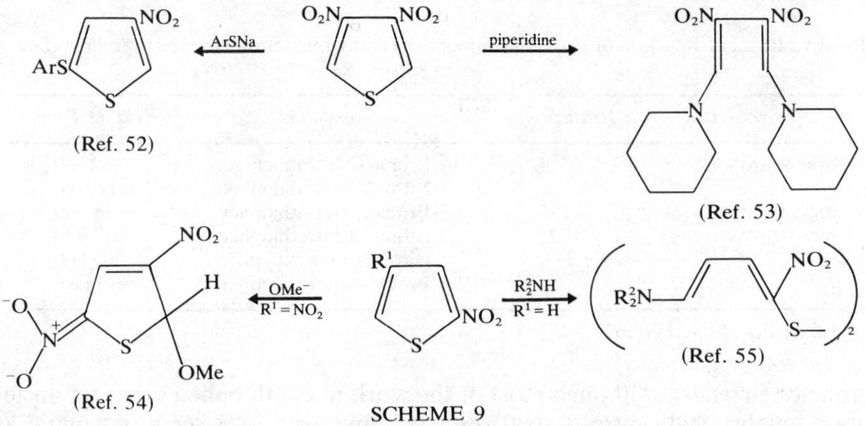

(Ref. 52)

(Ref. 53)

(Ref. 54)

(Ref. 55)

SCHEME 9

constant for formation of the complex (**39**) from 2-methoxy-3,5-dinitrothiophen is greater than that for 2,4,6-trinitroanisole under comparable conditions.[56] The mononitromethoxythiophens (**40**) and (**41**) are also efficiently transformed into Meisenheimer complexes by the action of methoxide at the positions shown.

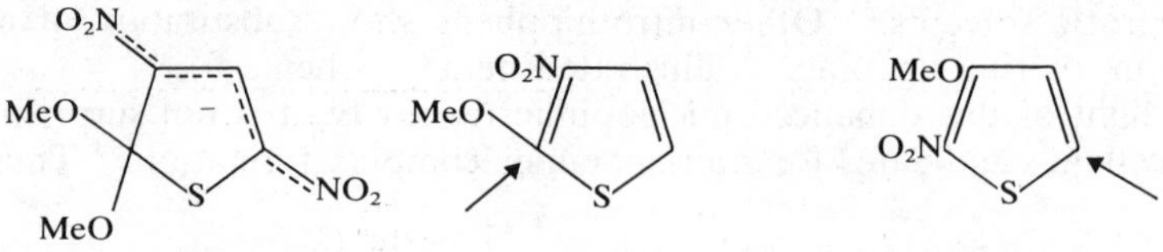

(**39**) (**40**) (**41**)

19.1.6.2 Some synthetic applications

Unactivated halothiophens undergo some fascinating nucleophilic transformations. Thus, α-bromothiophens rapidly convert into β-bromothiophens in high yield on brief treatment with a metal amide in liquid ammonia (Scheme 10 and Table 8), this being a synthetic method of choice for many β-bromothiophens. While α-chlorothiophens appear inert to the reaction conditions, α-iodo-analogues can be made to rearrange by judicious choice of reagent (Table 8). The reaction is very susceptible to changes in the conditions, and 3-aminothiophen results when potassium amide is used instead of sodium amide, or with change in reaction temperatures or reagent proportions.[57] The reaction does not involve an aryne intermediate but a complex series of intermolecular transbrominations, initiated by proton abstraction from the thiophen ring, and mediated by polybromothiophens, giving the 3-bromo-2-thienyl anion as the thermodynamically favoured

TABLE 8

Rearrangements of α-Halothiophens with Metal Amides in Liquid Ammonia (20 min, $-33\,°C$)[a]

α-Halothiophen	*Reagent (mol)*	*Product (%)*
2-Bromothiophen	$NaNH_2$ (2)	3-Bromothiophen (73)
2-Bromothiophen	KNH_2 (2)	3-Aminothiophen (74)
2-Bromo-5-methylthiophen	KNH_2 (6)	3-Bromo-5-methylthiophen (72)
2-Bromo-3-methylthiophen	KNH_2 (3)	3-Bromo-4-methylthiophen (67)
2,5-Dibromothiophen	$NaNH_2$ (4)	3,4-Dibromothiophen (73)
2,4-Dibromo-5-methylthiophen	$NaNH_2$ (3)	3,4-Dibromo-5-methylthiophen (64)
2-Iodothiophen	KNMePh (6)	3-Iodothiophen (77)

[a] See Ref. 53.

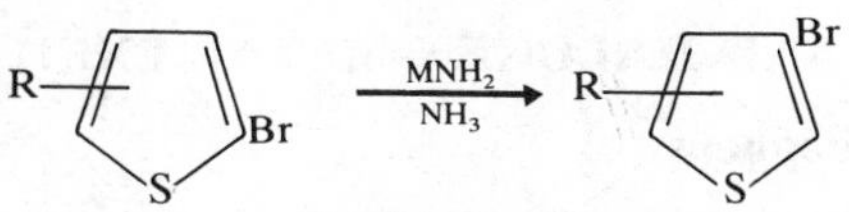

SCHEME 10

ion. Amination appears to proceed from a polybromothiophen rather than from 3-bromothiophen.

Synthetically valuable nucleophilic substitutions occur in copper-mediated reactions of halothiophens, *e.g.* Table 9. Especially worthy of note is the ready access now available to β-chlorothiophens from the corresponding bromo-derivatives, all kinds of which readily exchange halide with cuprous chloride in hot DMF. The classical Hurtley reaction, though grossly and unjustly maligned in a recent publication,[58] is an efficient reaction in the thiophen series, whereby 3-bromo-4-, 3-bromo-2-, and 2-bromo-3-thenoic acids[36] are readily reacted with a variety of active methylene compounds such as ethyl acetoacetate, benzoylacetonitrile, acetylacetone, *etc.* (Table 9). Numerous natural acetylenes have been made by interaction of copper acetylides with iodothiophens, as shown in Table 9.

TABLE 9

Some Copper-catalysed Reactions of Halothiophens

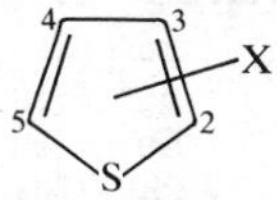

Halogen	*Method*	*Product* (%)	*Ref.*
2-I-5-Me	CuCN, pyridine, Δ	2-CN-5-Me (73)	a
All four di-Br_2's	CuCN, quinoline, Δ	di-CN (46–71)	b
3-Br	CuO, KI, MeOH, Δ	3-OMe (81)	c
2-Br	$CuSBu^n$, quinoline	2-SBu^n (92)	d
2-I	Cu, K_2CO_3 ArSH, pyridine, Δ	2-SAr (55–75)	e
2-I	Cu, K_2CO_3, $PhNO_2$, PhNHAc, Δ	2-NMePh (34)	f
2-I	Cu propynide, pyridine, Δ	2-C≡C—Me	g
2-I	Cu, K_2CO_3, $I(CF_2)_3$ I, pyridine, Δ	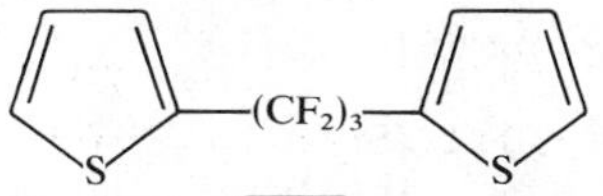 —$(CF_2)_3$— (25)	h
2- and 3-I	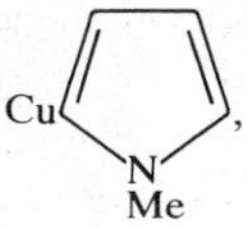 Cu, pyridine, Δ	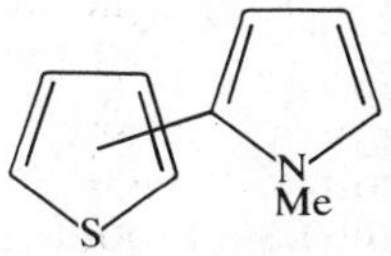	i
All mono, di, tri, and tetra Br's	Cu_2Cl_2, DMF, Δ	All corresponding chloro-compounds: mono (95–98); di (80–95); tri (60–85); tetra (30)	j
3-Br-4-CO_2H	NaOEt, EtOH, Cu, $AcCH_2CO_2Et$	3-CH_2CO_2Et-4-CO_2H (95)	k
3-Br-2-CO_2H	NaOEt, EtOH, Cu, $AcCH_2CO_2Et$	3-CH_2CO_2Et-2-CO_2H (89)	k

[a] A. Vecchi and G. Melone, *J. Org. Chem.*, 1957, **22,** 1636. [b] See Ref. 77. [c] S. Gronowitz, *Arkiv Kemi*, 1958, **12,** 239. [d] R. Adams and A. Ferretti, *J. Amer. Chem. Soc.*, 1959, **81,** 4927. [e] M. Rajsner, J. Metysova, and M. Protiva, *Coll. Czech. Chem. Comm.*, 1970, **35,** 378. [f] R. R. Estes and P. Panzera, *J. Amer. Chem. Soc.*, 1952, **74,** 853. [g] R. F. Curtis and J. A. Taylor, *J. Chem. Soc.* (*C*), 1969, 1813. [h] V. C. R. McLoughlin and J. Thrower, *Tetrahedron*, 1969, **25,** 5921. [i] N. Gjøs and S. Gronowitz, *Acta Chem. Scand.*, 1971, **25,** 2596. [j] S. Conde, C. Corral, R. Madronero, A. S. Alvarez-Insua, *Synthesis*, 1976, 412. [k] D. E. Ames and O. Ribiero, *J.C.S. Perkin I*, 1975, 1390.

The easily available iodonium salts (**42**) undergo nucleophilic substitution, *e.g.* with sodium nitrite the difficultly accessible 3-nitrothiophen (**43**) is produced.[59]

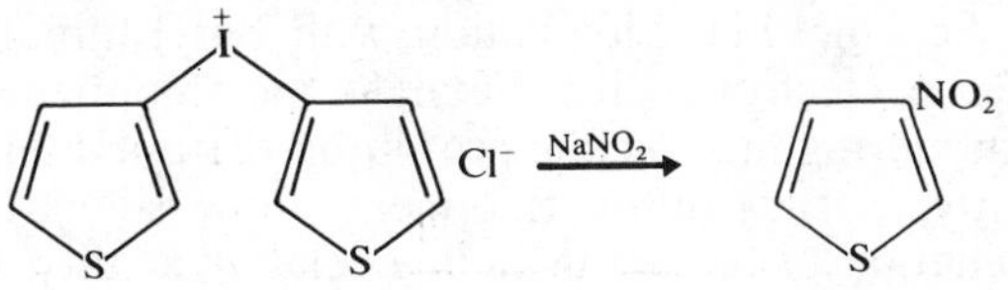

Cl^- $\xrightarrow{NaNO_2}$

(**42**) (**43**)

19.1.7 METALLATION AND HALOGEN–METAL EXCHANGE

19.1.7.1 Metallation of thiophens

The Gilman reaction is one of the most versatile and synthetically useful of processes, especially with thiophens. Metallation of thiophens is best conducted with an organolithium derivative in ether solution, commonly n-butyl-lithium, but organomagnesium, organosodium, and organomercury derivatives are occasionally utilized. The metallation is rapid and essentially quantitative at ambient temperatures, being highly regiospecific in most cases, giving an α-metallated thiophen (Table 10). Unlike other

TABLE 10
Metallation of Thiophen Derivatives

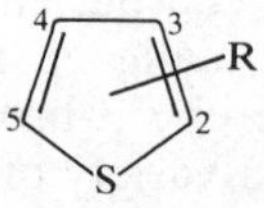

Substituent	*Metallating agent (solvent)*	*Position metallated, proportion* (%)	*Total yield* (%)	*Ref.*
H	Bu^nLi (Et_2O)	2	87, 100	a
H	Bu^nLi (Et_3N)	2	100	a
2-Me	Bu^nLi (Et_2O)	5	75	a
3-Me	PhLi (Et_2O)	2 (21), 5 (79)	87	a
3-Me*	Bu^nLi (Et_2O)	2 (20), 5 (80)	79	b
3-Et*	Bu^nLi (Et_2O)	2 (15), 5 (85)	75	b
3-Pr^i*	Bu^nLi (Et_2O)	2 (5), 5 (95)	69	b
3-Bu^t*	Bu^nLi (Et_2O)	2 (1), 5 (99)	73	b
3-Ph	Bu^nLi (Et_2O)	2 (45), 5 (55)	72	a
3-CH_2Ph	Bu^nLi (Et_2O)	2 (12), 5 (88)	80	c
2-CH_2NMe_2	Bu^nLi (THF)	5	50	a
3-CH_2NMe_2	Bu^nLi (Et_2O)	2	80	a
2-CH_2OMe		5	good	d
3-CH_2OMe		2	good	d
3-$CH(OEt)_2$	Bu^nLi (Et_2O)	2	16	e
3-$CONMe_2$		2	41	d
2-SO_2Bu^t	Bu^nLi (Et_2O)	both 3 and 5	59–68	f
3-SO_2Bu^t	Bu^nLi (Et_2O)	both 2 and 5	36	f
2-OMe	Bu^nLi (Et_2O)	5	67	a
3-OMe	Bu^nLi (Et_2O)	2	86	a
3-OBu^t	Bu^nLi (Et_2O)	2		a
2-SMe	Bu^nLi (Et_2O)	5	87	a
3-SMe	Bu^nLi (Et_2O)	2	70	a
2-Br	$LiNPr^i_2$ (Et_2O) (−70 °C)	5	good	g
3-Br	Bu^nLi (Et_2O)	2†		a
3-CN	3-thienyl-lithium (Et_2O)	2	80	a
3-NO_2	$LiNPr^i_2$ (Et_2O, HMPA)	2	good	g

* Addition of TMEDA gave 93–100% 5- and very little 2-lithio-derivative. † Formed indirectly by halogen–metal exchange and metallation processes (see p. 809).

[a] See B. J. Wakefield, 'The Chemistry of Organolithium Compounds', Pergamon Press, Oxford, 1974, p. 46. [b] See Ref. 71. [c] D. W. H. MacDowell and A. T. Jeffries, *J. Org. Chem.*, 1972, **37,** 514. [d] D. W. Slocum and P. L. Gierer, *J. Org. Chem.*, 1976, **41,** 3668. [e] P. Pirson, A. Schonne, and L. Christiaens, *Bull. Soc. chim. belges*, 1970, **79,** 575. [f] F. M. Stoyanovich and B. P. Fedorov, *Chem. Heterocyclic Compounds*, 1967, **3,** 650. [g] G. M. Davies and P. S. Davies, *Tetrahedron Letters*, 1972, 3507.

nucleophilic substitution reactions, no 2-substituents (except α-pyridyl, Scheme 11) have any effect on this positional selectivity, but only on the rate of metallation, as indicated by competitive reactions (Scheme 11). Metallation with butyl-lithium is brought about by initial coordination of the electropositive metal to the thiophenic sulphur, followed by abstraction of the neighbouring most acidic proton by the butyl anion and substitution by the metal. Consequently, β-substituted thiophens show selectivity during metallation determined by a combination of factors including relative acidity, coordination capability of the β-substituent, and steric crowding. Thus 3-alkylthiophens show an increasing

R^1	R^2	*Ref.*
Me	Me	60a
Me	OMe	60b
Me	SMe	60c

(Ref. 61) (Ref. 60c) (Ref. 62) (Ref. 63) $+PhC{\equiv}CH$ (Ref. 64) Bu^nLi/Et_2O (brief) Bu^nLi/THF or $LiN(Pr^i)_2$ (Ref. 65)

Sites of attack in the competitive metallation of thiophens

SCHEME 11

proportion of 5-substitution with increasing bulk of the 3-substituent, a process which is enhanced by use of a more sterically demanding reagent such as butyl-lithium/tetramethylethylenediamine (TMEDA) (Table 10). However, if the 3-substituent bears a heteroatom capable of coordinating a 2-lithio-group, solely 2-metallation is observed. A sulphone group in the 2- or 3-position is especially activating, and dimetallation is easily achieved. 2,5-Disubstitution allows metallation to take place at the β-position, again the ease and position of metallation being dictated by coordination and acidity factors (Scheme 11). Unlike the benzenoid series, side-chain metallation is relatively uncommon with thiophens, but metallation of acidic side-chains such as SO_2Me and $C{\equiv}CH$ can compete with α-metallation (Scheme 11). Thiophens are more prone to metallation than furan or pyrrole analogues and functional groups such as aldehydes and ketones can be protected by acetal formation.

Metallation may be enhanced by addition of coordinating amines such as TMEDA or by use of powerful, non-nucleophilic bases such as lithium di-isopropylamide, whereby even bromo- and nitro-thiophens are preferentially metallated at $-70\,°C$.

19.1.7.2 Halogen–metal exchange

Of equal applicability and value in syntheses is the halogen–metal exchange reaction wherein nucleophilic attack occurs at the halogen. While fluoro- and chloro-thiophens are generally inert, bromo- and iodo-derivatives react almost instantaneously with n-butyllithium at $-70\,°C$ or even $-100\,°C$ in ether solution, giving an equilibrium mixture almost completely in favour of the thienyl-lithium (Scheme 12). Where alternatives are possible, at $-70\,°C$ halogen–metal exchange is always preferred to metallation, and α-halogens exchange before β-, while, at suitable temperatures, iodo–lithium interconversion takes place in preference to bromo–metal exchange (Scheme 12). One, two, or more halogens may be selectively exchanged at one time. If the lithium atom does not take up the most acidic position, the reaction mixture, which is stable at $-70\,°C$, undergoes a complex

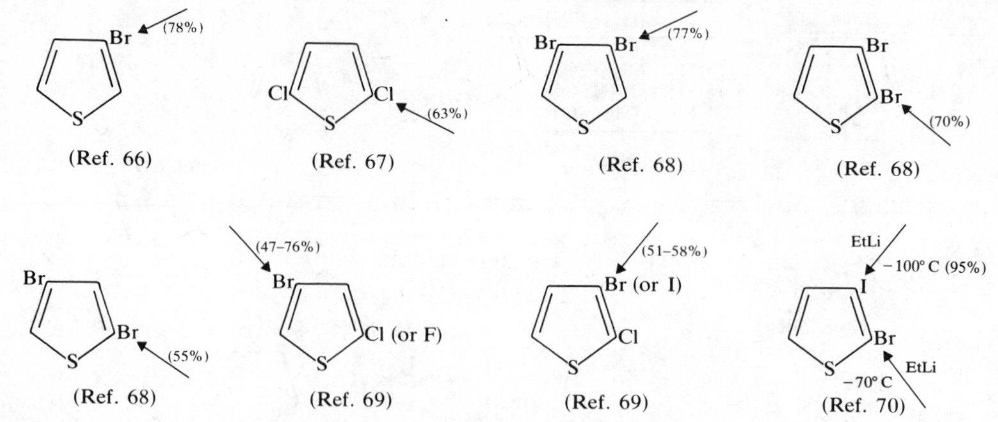

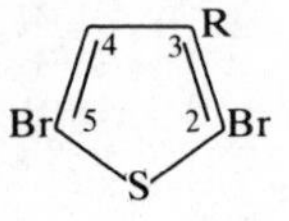

(Ref. 66) (Ref. 67) (Ref. 68) (Ref. 68)

(Ref. 68) (Ref. 69) (Ref. 69) (Ref. 70)

R	2	5	*Total* (%)
H	—	—	81
Me	19	81	88
Et	23	77	81
Pr^i	28	72	70
Bu^n	57	43	80
Br	100	0	75

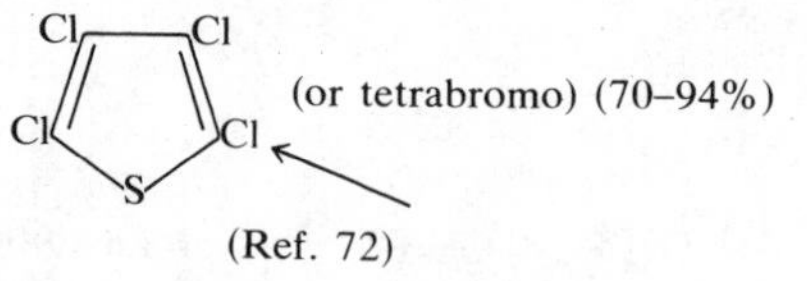

(Ref. 71) (or tetrabromo) (70–94%) (Ref. 72)

Position of halogen–metal exchange in the thiophens (yield %) with Bu^nLi in ether

SCHEME 12

series of metallation/halogen–metal exchange equilibria on warming to ambient temperature, whereby a mixture of products is formed with lithium at the most acidic centre (*e.g.* Scheme 13). [This rearrangement is related to the ciné substitution of bromothiophens with metal amides (p. 806)]. Thus 2,4-dibromothiophen at ambient temperature yields a mixture of 3,5-dibromo- (**44**) and 3-bromo-2-thienyl-lithium (**45**). The process may thus be viewed as nucleophilic attack at an electropositive metal. Thus, while 2-chlorothiophen metallates at the free 5-position, 2,5-dichloro- and tetrachloro-thiophen both undergo halogen–metal exchange at ambient temperature, reflecting the considerable $-I$ effect of the extra halogens. Furthermore, since steric crowding decreases upon exchange of Br for Li, the reverse ratio of 2- to 5-attack is observed compared with the metallation process (compare Scheme 12 with Table 10). *o*-Halophenyl-lithium derivatives, even at $-100\,°C$, eliminate lithium halide to give benzyne. However, the corresponding thiophens show no

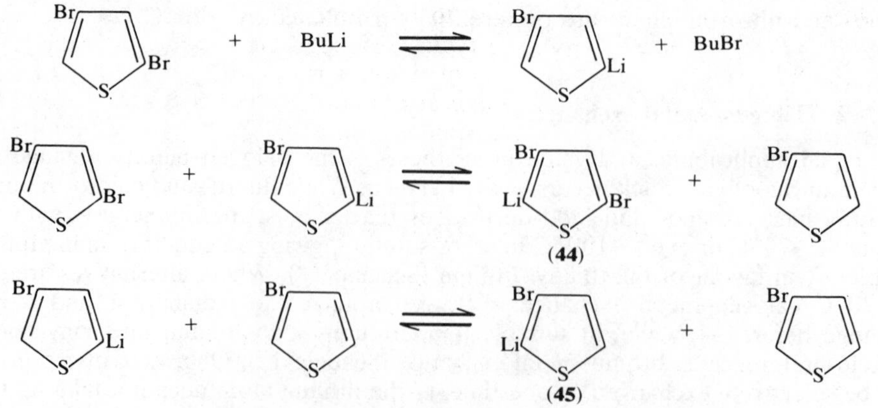

(**44**)

(**45**)

SCHEME 13

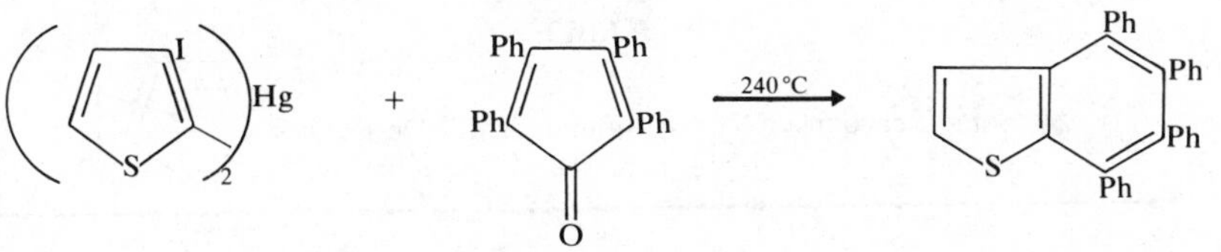

(15)

such tendency, even the iodothienylmercury derivative giving little evidence of thiophyne at 240 °C (reaction 15).

3-Thienyl-lithium derivatives undergo a reversible ring-opening in refluxing ether to give a mercaptovinylacetylene (Scheme 14), which may be trapped with the butyl bromide

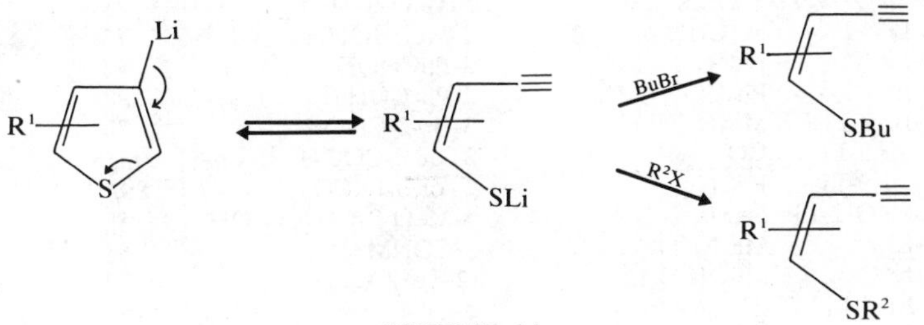

SCHEME 14

derived from the butyl-lithium or with an added electrophile.[73] The analogous ring-opening of an α-lithiothiophen in the presence of an excess of butyl-lithium and hexamethylphosphoric triamide (HMPT) has been demonstrated (Scheme 15).[74]

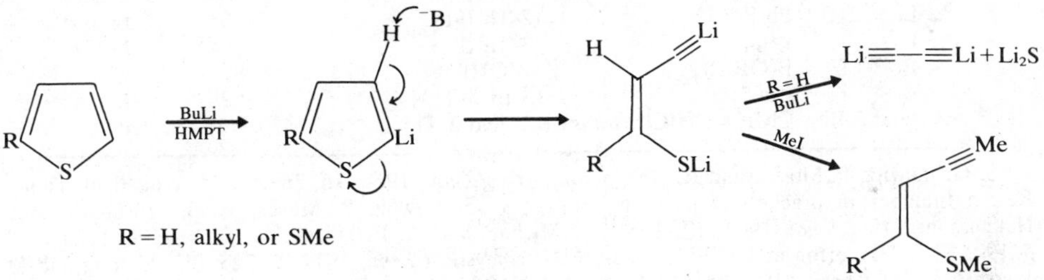

R = H, alkyl, or SMe

SCHEME 15

19.1.7.3 Some synthetic applications

The metallated thiophens are amongst the most versatile of synthetic intermediates, allowing a wide variety of substituents to be introduced regiospecifically at the α- or β-position into a thiophen ring. Some typical examples are recorded in Table 11. Yields are generally high and a great diversity of reagents is available. Two similar substituents may be introduced simultaneously using a dilithio-derivative,[75] while sequential disubstitution of dissimilar groups may be introduced in a 'one-pot' manner.[76,77]

19.1.8 RADICAL REACTIONS OF THIOPHENS

While ionic reactions are by far the most important and useful reactions in the thiophen series, the high α-preference for reactions of thiophens with radical reagents (as predicted by theoretical considerations) should make this approach more attractive. Potentially attractive routes to derivatives which are difficult of access by conventional means may be conceived using radicals. However, relatively little work has been reported in this area.

TABLE 11

Some Reactions of Thienyl-lithium (ThLi) Derivatives

Substituent	*Reactant*	*Product (substituent)*	*Yield* (%)	*Ref.*
2- or 3-Li	D_2O	2- or 3-D	high	a
2- or 3-Li	T_2O	2- or 3-T	high	b(164)
2-Li	EtBr	2-Et	61	b(146)
2-Li	Me_2SO_4	2-Me	65	b(159)
5-MeO-2-Li	TsOMe	5-MeO-3-Me	78	b(159)
2-Li	MeCHO	2-MeCHOH	73	b(130)
2-Li	PhCHO	2-PhCHOH	54	b(130)
2-Li	Ph_2CO	2-Ph_2COH	84	b(130)
3-Br-2-Li	Me_2NCHO	3-Br-2-CHO	70	c
2- or 3-Li	CO_2	2- or 3-COOH	high	b(124)
2- or 3-Li	PhCN	2- or 3-PhCO	89/75	b(117)
5-MeO-2-Li	oxiran	5-MeO-2-CH_2CH_2OH	50	b(200)
2-Li	Me_2NCOCl	2-CONMe_2	62	d
2-Li	Me_2NCO_2Et	(2-Th)$_2$CO	92	d
3-Br-4-Li	Cl_2	3-Br-4-Cl	55	b(194)
2-Li	Cl_3CCN	2-Cl	40	e
2-Li	$FClO_3$	2-F	49	b(218)
5-EtS-2-Li	S_8 then EtI	2,5-(SEt)$_2$	71	b(193)
3-Li	SO_2	3-SO_2H	83	b(222)
3-Li	(4-Br-3-Th-S-)$_2$	4-Br-3-Th-S-3-Th	80	f
2-Li	Ph_3SiCl	2-SiPh_3	63	b(261)
2-Li	$SnCl_4$	(2-Th)$_4$Sn	66	b(266)
2-Li	Pb($Pr^i CO_2$)$_4$	(2-Th)$_4$Pb	53	b(269)
3-Li	PBr_3	(3-Th)$_3$P	50	b(271)
4-Br-2-Li	B(OBu^n)$_3$	-B(OH)$_2$	80	g
3-Br-4-Li	$CuCl_2$	(3-Br-4-Th)$_2$	52	h
2- and 3-Li	ClCH=CHCl$_2$ (*trans*)	(2- and 3-Th)$_2I^+$	69/61	i

[a] S. Gronowitz, P. Moses, and R. Håkansson, *Arkiv Kemi*, 1960, **16,** 267. [b] B. J. Wakefield, Table 10, Ref. a (number in brackets is page in book). [c] S. Gronowitz, P. Moses, A. B. Hörnfeldt, and R. Håkansson, *Arkiv Kemi*, 1961, **17,** 165. [d] U. Michael and A. B. Hörnfeldt, *Tetrahedron Letters*, 1970, 5219. [e] F. H. Pinkerton and S. F. Thames, *J. Heterocyclic Chem.*, 1972, **9,** 725. [f] F. de Jong and M. J. Janssen, *J. Org. Chem.*, 1971, **36,** 1998. [g] A. B. Hörnfeldt and S. Gronowitz, *Arkiv Kemi*, 1964, **21,** 239. [h] S. Gronowitz, *Acta Chem. Scand.*, 1961, **15,** 1393. [i] See Ref. 59.

The Gomberg–Bachmann reaction offers one of the best means of making arylthiophens.[78,79] Though yields are moderate (20–50%), the starting materials are readily available and the method is simple. Optimum yields require use of a large excess of the thiophen and aprotic conditions are usually preferred. The small amount of β-arylated isomer (5–15%) is usually easily removed, and numerous heteroaryl groups have been introduced in this way. The related Meerwein reaction gives similar results, and in both cases substituent effects are small.[80] However, some unexpected results have been noted with strongly electrophilic diazonium salts such as 2,4-dinitrobenzenediazonium chloride,[81] which arylates thiophen, 2- and 3-methylthiophen even at 0 °C but gives azo-coupled derivatives with 2-phenyl-, 2-t-butyl- and 2,4-dimethyl-thiophen (Scheme 16). With polymethylthiophens, attack at an α-methyl group is observed, giving the hydrazones (**46**) in up to 93% yield. The mechanisms of these reactions are still unclear.

Most other arylating agents show a 90–95% preference for α- over β-substitution, except for phenylazotriphenylmethane, which gives a 4 : 1 ratio of α- to β-phenylated product.

Thiophens show only a slight increase in reactivity towards radicals compared with benzene. Thus figures of 1–2 times reactivity with aryl radicals and up to 9 times for alkyl radicals have been reported.[79] Partial rate factors of 7.25 and 0.5 for the phenylation at α- and β-thiophenic positions respectively is noted.

$R = Ph, CMe_3$, or 2,4-Me_2

$R = H$, 3-Me, or 3,4-Me_2 (**46**)

SCHEME 16

Attempted phenylation of thiophen with benzoyl peroxide (Scheme 17) preferentially yields 2,2′-bithienyl (**47**) by an addition–dimerization–elimination process.[82] Again, in contrast to the benzenoid analogue, thenoyl peroxide gives products primarily derived from thenoyloxyl radicals, decomposition in hot benzene in the presence of cupric ions giving phenyl thenoate in high yield and suggesting some kind of stabilized radical intermediate (Scheme 17).[83]

(**47**)

SCHEME 17

Another potential means of arylation involves irradiation of an aryl iodide in an aromatic solvent. Thus, 3-iodopyridine[84a] and 5-iodopyrimidines[84b] on photolysis in thiophen give reasonable yields of primarily the 2-thienyl derivatives (Scheme 18). Similarly, iodothiophens give the corresponding thienylbenzenes on irradiation in a benzenoid solvent.[85] However, since 2-arylthiophens undergo a photo-rearrangement (see p. 818) on prolonged irradiation, this method should be used with caution.

High-temperature bromination, probably a homolytic process, is claimed to give a low yield of 3-bromothiophen, while alkylthiophens undergo side-chain bromination under

X = CH, 37%
X = N, 58%

X = CH, 5%
X = N, 6%

SCHEME 18

radical conditions (see p. 825).[86] Thiophen may be α-aminated with dialkylamino-radicals, though details have not yet been published.[82]

19.1.9 ADDITION REACTIONS OF THIOPHENS

Until recently it was commonly believed that while thiophen showed considerably increased reactivity over benzene, dienic reactions were rare, unlike the case of furan. However, it is now evident that reactions such as the Diels–Alder cycloaddition and related processes are often readily brought about, if appropriate reagents or substituents are used. Surprisingly, it has been realised from the earliest days of thiophen chemistry that addition of halogens is a common process, which can compete with substitution in suitable cases. Nevertheless, developments in the addition reactions of thiophens have only occurred in the last few years.

19.1.9.1 Addition during electrophilic reactions

Some addition always accompanies the action of halogens on thiophens and the removal of these unstable, undesired by-products is usually accomplished by the action of hot alkali. However, only the chloro-addition derivatives have been isolated, the use of low temperatures favouring their formation. Thus tetra- (**48**) and penta-chlorothiolane (**49**) have been isolated from the low-temperature chlorination of thiophen and 2-chlorothiophen, respectively, while perchlorination of thiophen or 2,5-dichlorothiophen yields hexachlorothiolan (**50**) in 80% yield.[87] When molar amounts of iodine are used as a catalyst, octachlorothiophen (**51**) is formed in 70% yield, while iodine in small amounts produces the hexachlorothiolen (**52**) in 80% yield.[87] Presumably similar types of

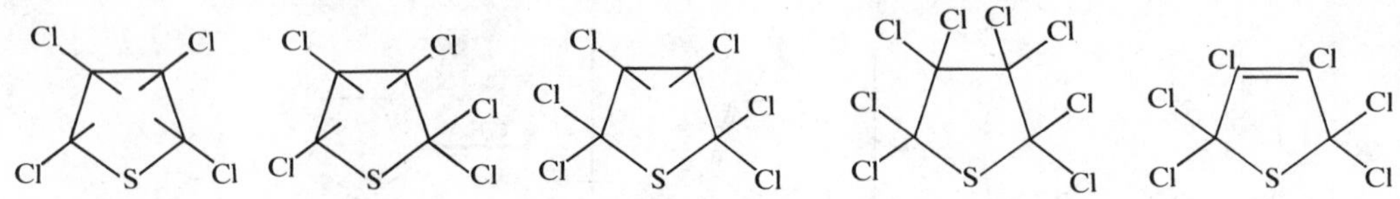

(**48**) (**49**) (**50**) (**51**) (**52**)

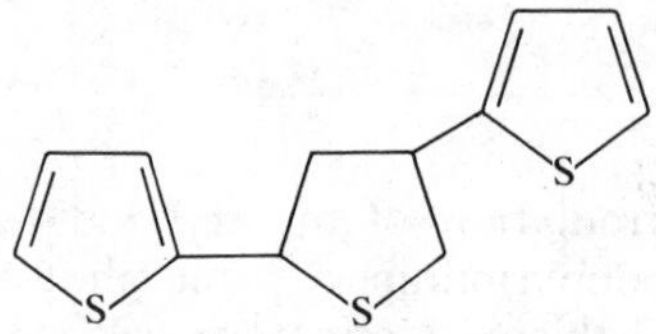

(**53**) 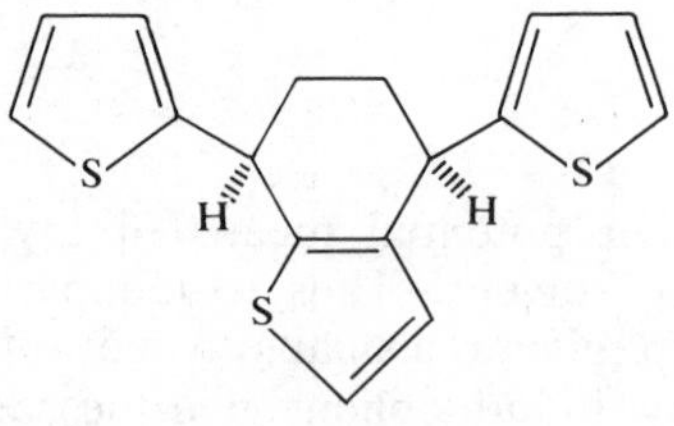 (**54**)

addition compounds accompany other electrophilic substitution reactions and probably account for the loss of yield and discoloured by-products sometimes noted. Thus maleic acid has been isolated from nitration of thiophen, while the action of phosphoric acid on thiophen, giving a 'trimer' and a 'pentamer', involves addition giving (**53**) and (**54**), respectively. The 'pentamer' is actually a tetramer, whose formation involves sulphur extrusion, as shown by X-ray data.[88]

19.1.9.2 Cycloadditions

Diels–Alder cycloadditions are now well documented,[89,90] although not thoroughly investigated. Acetylenic dienophiles are particularly effective and the reaction generally

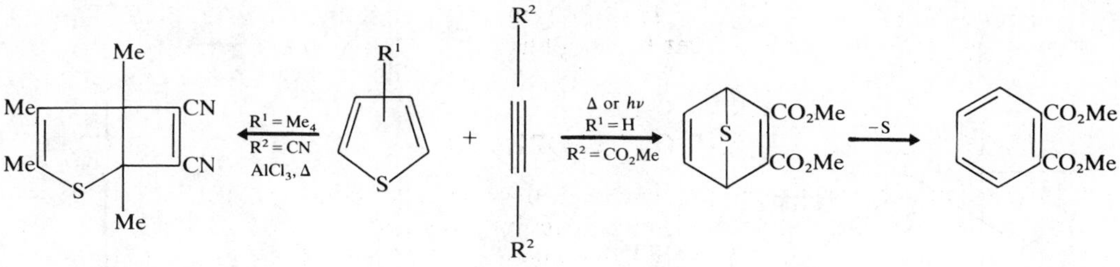

SCHEME 19

involves cheletropic removal of sulphur from the initially formed but unstable intermediate, giving a benzene derivative (Scheme 19). In some cases, particularly those involving the enamine-like-3-aminothiophens, [2+2] cycloaddition is the preferred mode of addition (*e.g.* Schemes 19[89] and 20[90]), each product in the subsequent fascinating

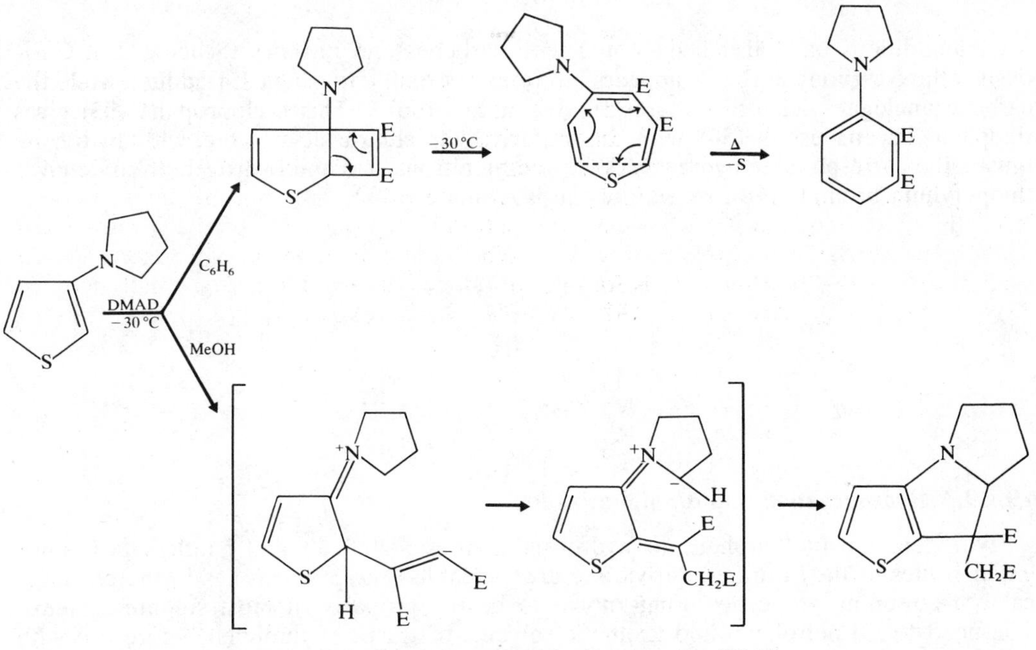

DMAD = dimethyl acetylenedicarboxylate; E = CO_2Me

SCHEME 20

sequence of pericyclic processes being detectable. The delicate balance between polar and non-polar intermediates is revealed by the change of course of the cycloadditions in different solvents (Scheme 20). Thiophen itself behaves as a $_{\pi}2$ system in its reaction with tetracyanoethylene oxide (a potential 4π, 1,3-dipolar carbonyl ylide with considerable electrophilic character), giving a tetrahydrofuranothiophen (Scheme 21).[91] Other interesting photo-initiated cycloadditions are dealt with in the chapter on photochemistry of thiophens (p. 818).

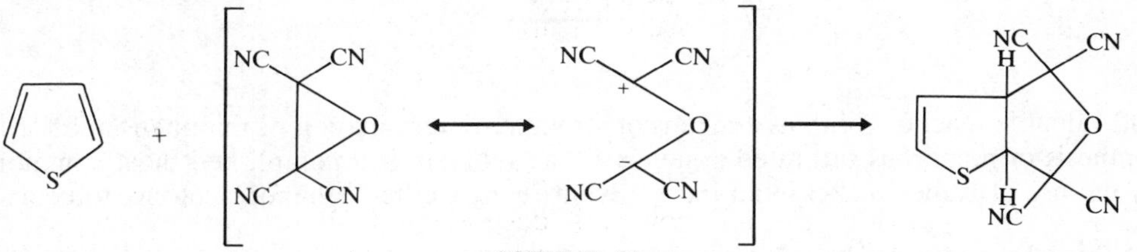

SCHEME 21

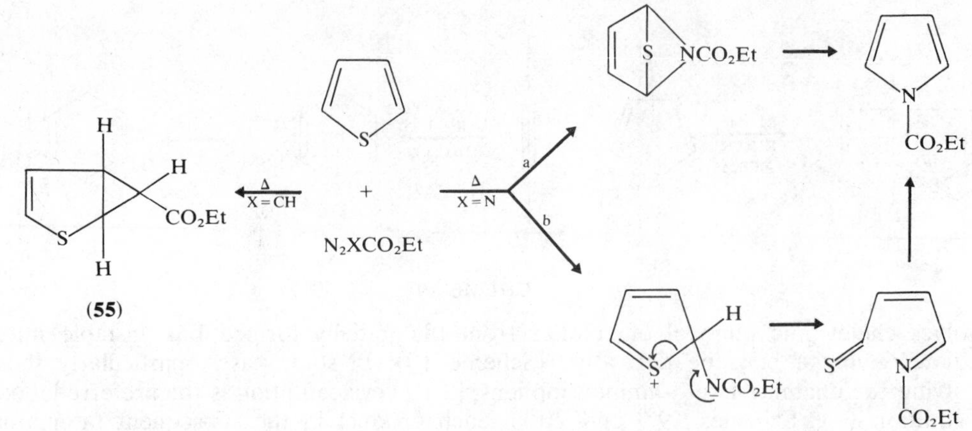

SCHEME 22

Cycloadditions have also been noted with carbenes and nitrenes (Scheme 22). Curiously, ethoxycarbonylnitrene[92] appears* to react *via* path a to give a 1,4-adduct, while the carbene analogue yields the interesting 1,2-adduct (**55**).[93] This cyclopropane (**55**) gives thiophen-3-acetic esters (**56**) with methanolic hydrochloric acid,[93] but with hydrazine (giving the corresponding hydrazide) followed by nitrous acid and hydrogen bromide gives thiopyrylium bromide (**58**), by way of the isocyanate (**57**).[94]

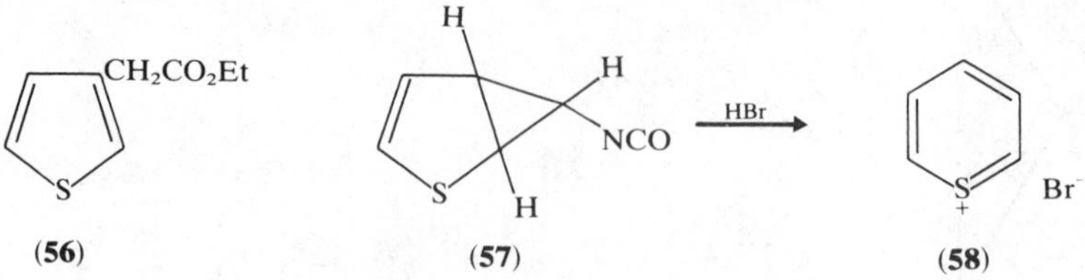

19.1.9.3 Hydrogenation and desulphurization

Hydrogenation of thiophens to hydrothiophens is a difficult and fruitless task since easier routes to the saturated derivatives are available. The problem lies in the powerful catalyst-poisoning properties, long-known to be the property of most sulphur systems. The need to rid petroleum and aromatic solvents of traces of thiophen before use with catalysts (*e.g.* in cracking of petroleum, as organic reaction solvents, *etc.*) was recognized by Victor Meyer. Thus palladium–charcoal will catalyse the hydrogenation of thiophen to tetrahydrothiophen in 70% yield – so long as 2 parts of catalyst are used for each part of thiophen![95] Certain transition-metal sulphides are reported to effect the hydrogenation of thiophen under high temperature and pressure conditions (*e.g.* molybdenum sulphide or rhenium sulphide at 250 °C, 100 atm). Curiously, Raney nickel in acetic anhydride is not poisoned and effectively reduces nitrothiophens to acetylaminothiophens.[96] However, the hydrogenative desulphurization of thiophens to paraffins (reaction 16) is a most efficient

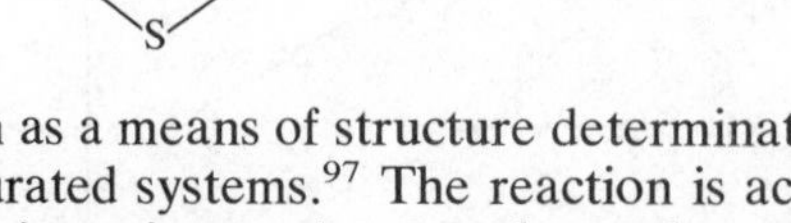

(16)

and valuable reaction, both as a means of structure determination of thiophens and for the synthesis of numerous saturated systems.[97] The reaction is accomplished most commonly by the use of Raney nickel in an inert solvent such as ether, ethanol, acetone, toluene, or

* The author prefers the alternative mechanism, path b.

ammonium hydroxide, but an alternative method, especially useful for alkali-soluble compounds, involves use of the Raney alloy and aqueous sodium hydroxide. All olefinic and acetylenic bonds are reduced, as are nitro, oximino, and aldehyde functions. Ketones are reduced with active catalysts but may be retained by use of lower temperatures and by use of acetone as a solvent. Carbonyl groups may be protected as acetals. Some typical applications include the synthesis of mono- and di-functional alkanoic acids (Scheme 23),

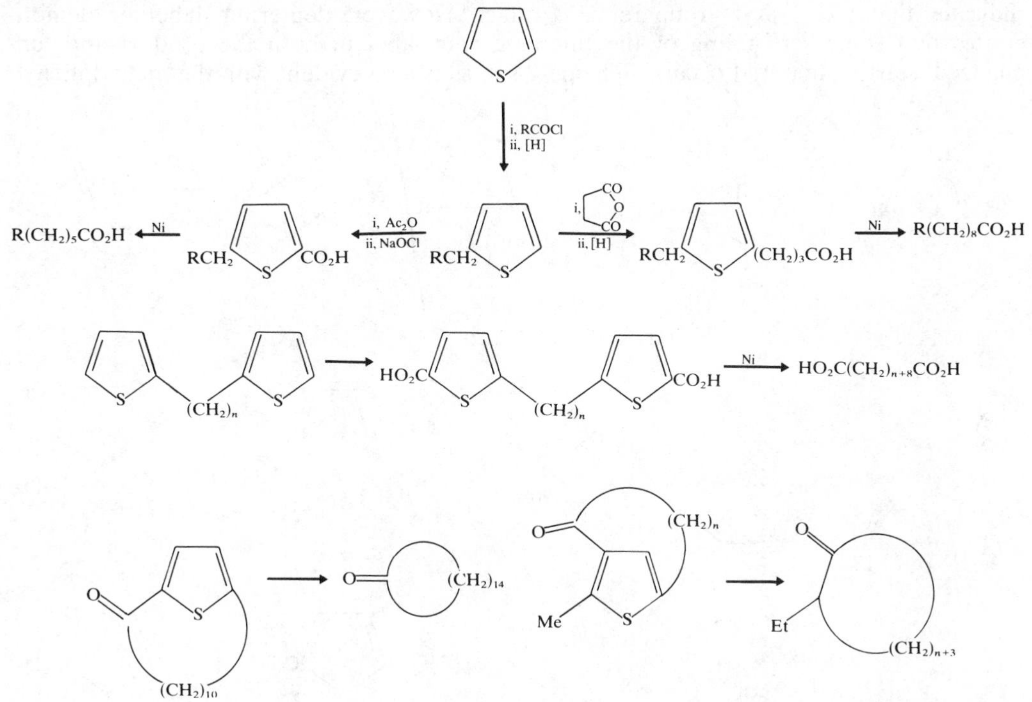

SCHEME 23

whereby thiophen is used as a chain-extending group (≡4 carbons). By use of dithienylalkanes, even more effective chain-extension may be obtained (Scheme 23), while appropriate substitution allows the synthesis of amino-acids.[98] ^{13}C and ^{14}C labelled fatty acids and partially fluorinated ketones have been synthesized, as have macrocyclic ketones[32,99] such as exaltone, civetone, and muscone, important perfumery ingredients. Monocyclic borazaropyridines (*e.g.* **59**) are readily obtained by building up from a simple thiophen followed by desulphurization.[76] Functionalization of adamantanes by way of a thienyl derivative (all four possible thienyladamantanes have been made) such as (**60**) is reported,[100] and the correlation of chiral thiophens with acyclic analogues is readily performed by a desulphurization method. A chiral thiophen was neatly employed to demonstrate the virtually zero rotation of a chiral paraffin.[101]

The Birch reduction of thiophens offers an entry to the rare dihydrothiophens.[102] Despite low yields the reaction is amenable to large-scale operation and the mixture of 2,3- and 2,5-dihydro-derivatives is readily separated by distillation. The products have been thoroughly characterized as their sulphones, sulphoxides, and various addition products.

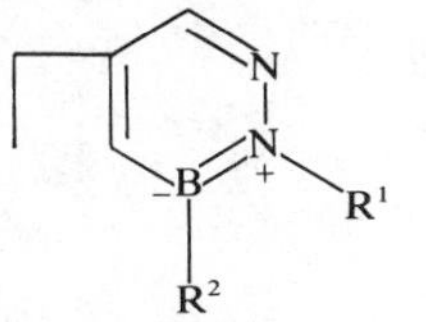

(**59**) 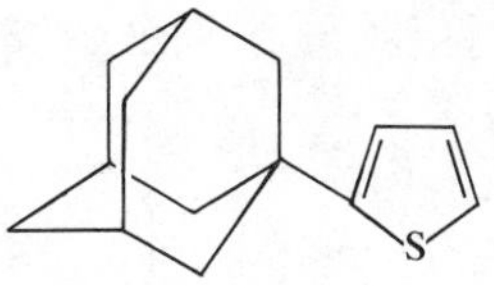 (**60**)

19.1.10 PHOTOCHEMISTRY OF THIOPHENS

As with other aromatic systems, some fascinating chemistry has derived from a study of the ultraviolet irradiation of the thiophens.[103] The most studied area is the isomerization of 2-substituted thiophens to their 3-isomers (Scheme 24a).[101] Although much is known about this reaction, the mechanism is still not fully clear, and several competing processes may well be involved. Thus arylthiophens rearrange most efficiently and ^{14}C labelling indicates that the *C*-aryl group remains intact. However, deuterium labelling demonstrates that some scrambling of the thiophen ring label in both the product and 'unchanged' starting material occurs (Scheme 24b), as is also evident with the methylphenylthiophens. It is clear that valence isomerization precedes rearrangement and of the numerous mechanisms postulated, intermediates such as (**61**), (**62**), and (**63**) may be considered. Thus a bicyclic system (**64**) has been isolated in one case,[104] while other workers have trapped the apparent cyclopropenethioaldehyde by photolysis in an aliphatic

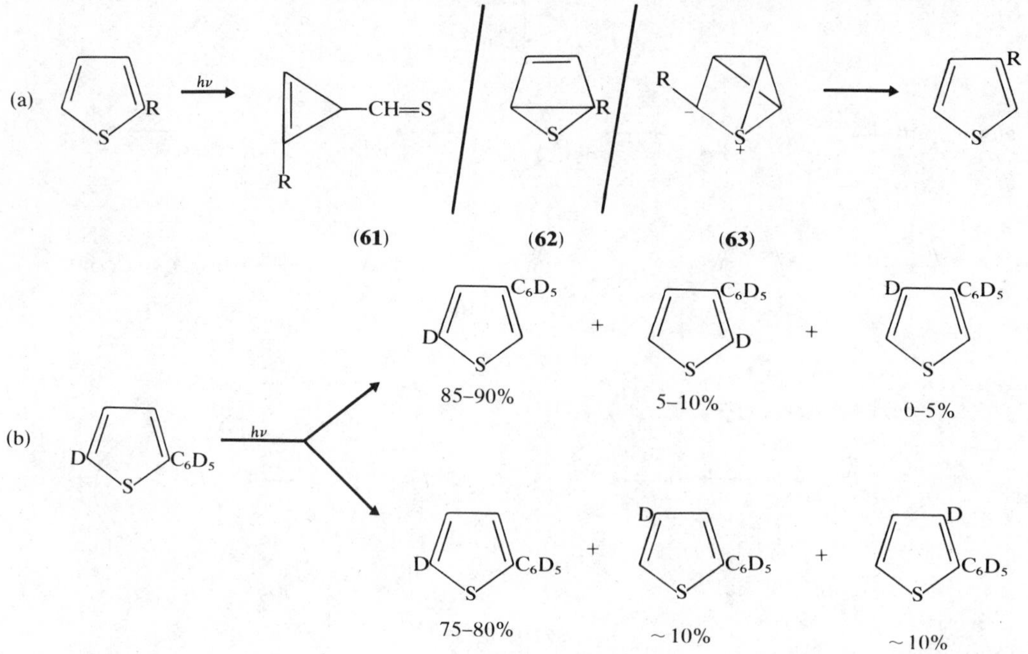

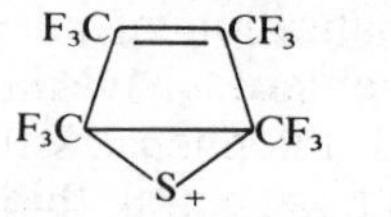

(a) hν; —CH=S; R; (**61**) (**62**) (**63**)

(b) hν; D; C_6D_5; S; 85–90%; 5–10%; 0–5%; 75–80%; ~10%; ~10%

SCHEME 24

F_3C; CF_3; S_+; (**64**); N; alk; (**65**)

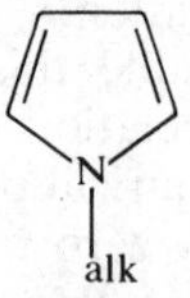

primary amine, to give low yields (5–8%) of *N*-alkylpyrroles (*e.g.* **65**) derived from the corresponding thiophen.[105] *S*-Alkylthiophenium salts similarly rearrange on photolysis (reaction 17).

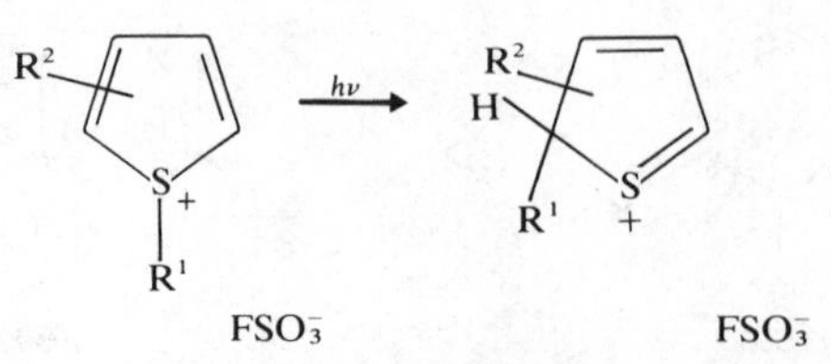

R^2; R^1; H; S^+; hν; FSO_3^-; FSO_3^- (17)

Photo-cycloaddition reactions are also well documented in the thiophen series. Thus photo-generated singlet oxygen does not attack thiophen (unlike furan and pyrrole), but reacts with alkylthiophens to give products explicable either by 1,4-addition (Scheme 25) or *S*-peroxide formation.[106] Alkenes, alkynes, and ketones have been utilized in photo-cycloaddition to thiophens, some examples of which, involving both [4+2] (see p. 815) and [2+2] reactions, are given in Scheme 25.[107–109]

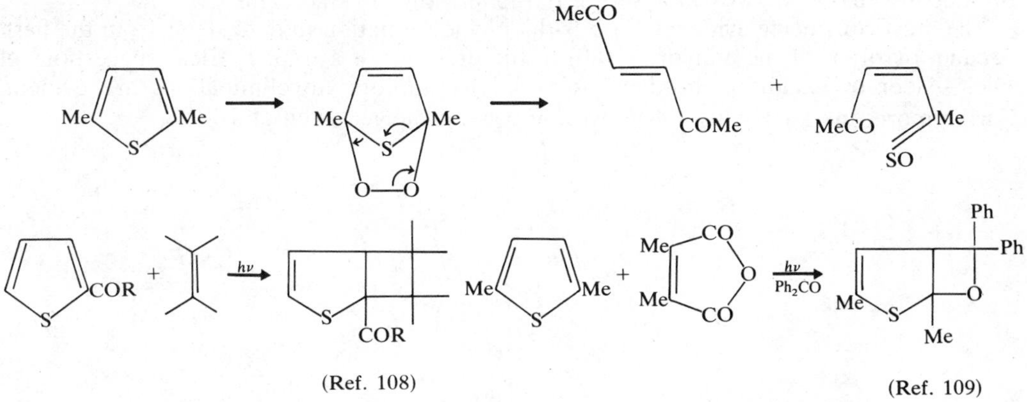

(Ref. 108) (Ref. 109)

SCHEME 25

19.1.11 DEHYDROTHIOPHENS

The first aryne, postulated 75 years ago as a reactive intermediate, was 2,3-dehydrobenzofuran. Ironically, however, until very recently the formation of five-membered hetarynes has been shown to be very questionable, even though apparently reasonable evidence was produced.[110] In contrast to benzyne and related six-membered arynes, the *ortho*-halothienylmetal derivatives (**66**) are remarkably stable and are useful intermediates

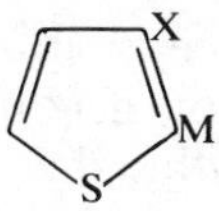

X = Cl, Br, or I

M = Li, MgX, Na, or Zn

(**66**)

in synthesis (see p. 810). Similarly, the 3,4-iodolithium derivative (**67**) prefers to disproportionate and then further react, rather than form the 3,4-thiophyne (Scheme 26). That ring strain is not the only cause of the unwillingness to form a hetaryne is underlined by the observation that cyclopentyne can be made by eliminations related to those above.

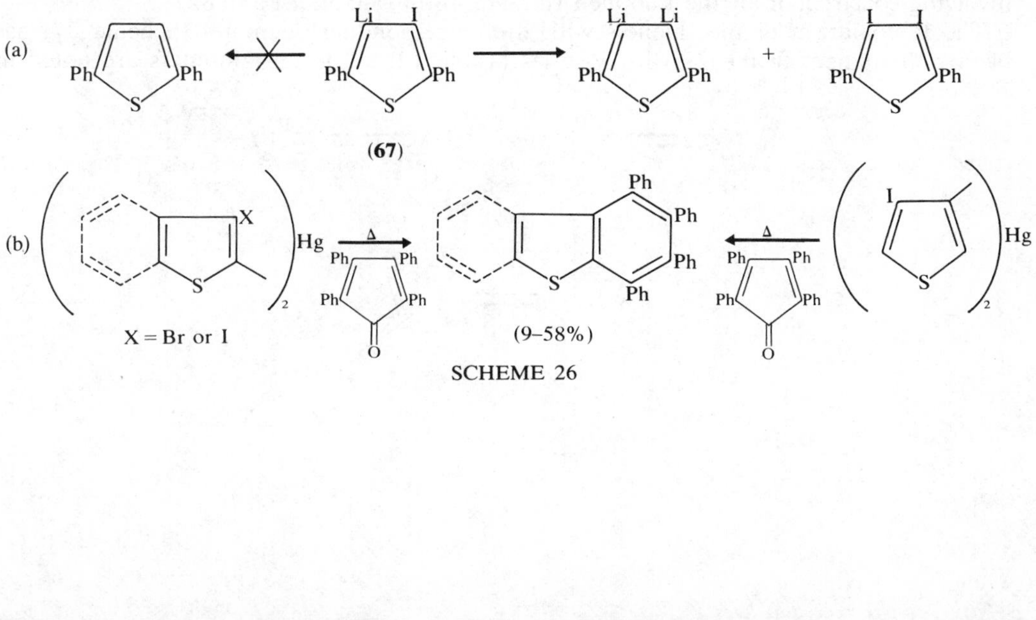

(a)

(**67**)

(b)

X = Br or I

(9–58%)

SCHEME 26

It would appear that potential thiophyne precursors are able to be diverted to follow alternative, less energetically demanding, pathways not open to the cyclopentene equivalents. The apparent indications of thiophyne shown in Scheme 26 are discounted by the observation that the precursors have been shown to yield 3-iodothiophen and 3-bromobenzothiophen on pyrolysis and that these simple halothiophens react efficiently with tetracyclone to give the apparent heterаryne-derived product in good yield. Thus, 3-bromobenzothiophen gives 77% of the tetraphenyldibenzothiophen.

The most convincing evidence for 2,3-thiophyne formation to date stems from the flash vacuum pyrolysis of the anhydride (**68**) in the presence of a trap.[111] Clear suggestions of Diels–Alder cycloaddition products followed by appropriate eliminations are evident. Furthermore, the analogous 3,4-anhydride was unchanged even at 700 °C.

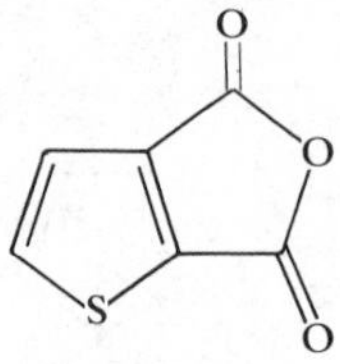

(**68**)

19.1.12 TAUTOMERISM OF THIOPHENS[112]

In striking contrast to benzenoid chemistry, in which phenols, anilines, and thiophenols occupy a vital role as synthetic intermediates, the hydroxy-, amino-, and mercapto-thiophens are often unstable, not of easy access, and tautomerically divergent. Indeed, 2-hydroxythiophen was not isolated until 1950. Despite this unpromising summary, with suitable substitution, stable and useful derivatives now abound and the chemistry of these systems is well understood.

19.1.12.1 Hydroxythiophens

2-Methyl-5-hydroxythiophen was the first reported thienol, prepared by the Paal cyclization (see p. 829) of 4-oxopentanoic acid in 1886. Other work shortly after reported the synthesis of 2-phenyl-3-hydroxythiophen and in 1912 Friedländer noted the tautomeric character of both the α- and β-hydroxy-derivatives, by virtue of their reactivity, both as enols (*e.g.* *O*-acylation) and keto-derivatives, (*e.g.* condensation with aldehydes). General methods to their synthesis have since been elaborated, including the reaction of thienyl Grignard reagents with oxygen in the presence of isopropyl magnesium bromide, thienyl-lithium derivatives with hydroperoxides, acid-catalysed dealkylation of t-butyl thienyl ethers,[113] and, finally, the cleavage of thienylboronic acids (easily obtained from thienyl-lithium derivatives and a trialkylborate) with hydrogen peroxide[114]. Routes involving construction of the thiophen ring are also available (see p. 827).

The tautomerism of the thienols with the corresponding thienones (Scheme 27) has been well studied. In the 2-hydroxy-series (Table 12), all three tautomers are noted in

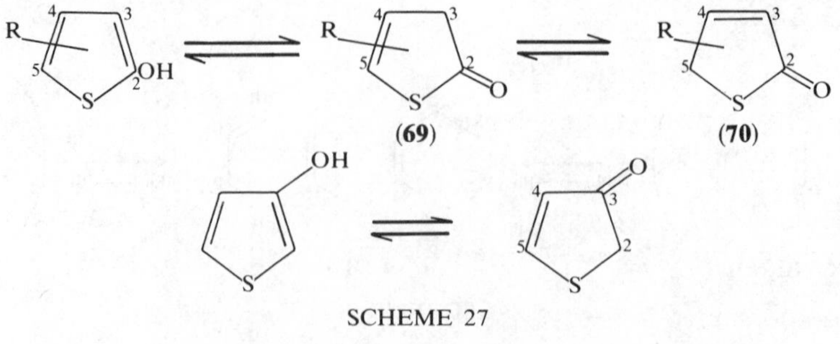

(**69**) (**70**)

SCHEME 27

TABLE 12

The Tautomerism of 2-Hydroxythiophens

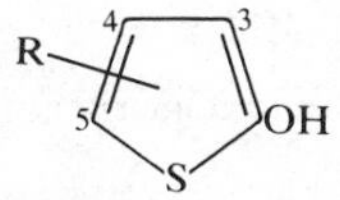

R	*Conditions*	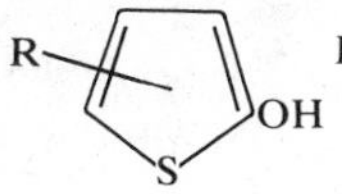 2-hydroxythiophen (%)	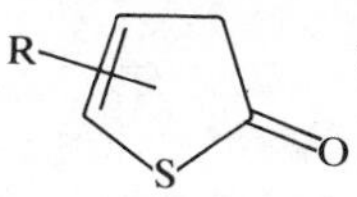 Δ⁴-thiolenone (%)	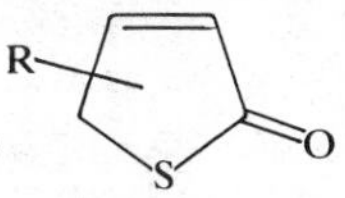 Δ³-thiolenone (%)	*Ref.*
H	CCl_4	—	—	~100	a
3-Me	$CDCl_3$	—	—	~100	b
4-Me	$CDCl_3$	—	—	~100	b
5-Me	CCl_4	—	20	80	c
3,5-Me	$CDCl_3$	—	<2%	mostly	d
5-Et	CCl_4	—	15	85	e
5-Pr^n	CCl_2	—	18	82	e
5-Pr^i	C_6H_{12}	—	11.5	88.5	e
5-Bu^n	CCl_4	—	18	82	e
5-Bu^t	CCl_4	—	8	92	e
5-$CH_2CH=CH_2$	CCl_4	—	15	85	c
5-CH_2Ph	CCl_4	—	30	70	c
5-Ph	CCl_4/solid	—	~100	—	e
5-Ph	MeOH	—	~100	—	e
3-Ac	$CCl_4/CDCl_3$/neat	~100	—	—	f
3-CO_2Et	$CCl_4/CDCl_3$/neat	~100	—	—	f
5-CO_2Et	CCl_4	85	15	—	c
5-SMe	CCl_4	—	15	85	c
3-OMe	C_6H_{12}/neat	—	—	~100	g
5-Cl	CCl_4	—	2	98	h
3-Br	C_6H_{12}/neat	—	—	~100	g
4-Br	C_6H_{12}/neat	—	—	~100	g
5-Br	CCl_4	—	—	~100	h

[a] S. Gronowitz and R. A. Hoffman, *Arkiv Kemi*, 1960, **15,** 499. [b] A. B. Hörnfeldt and S. Gronowitz, *Acta Chem. Scand.*, 1962, **16,** 789. [c] H. J. Jakobsen, E. H. Larsen, and S.-O. Lawesson, *Tetrahedron*, 1963, **19,** 1867. [d] E. B. Pedersen and S.-O. Lawesson, *Tetrahedron*, 1971, **27,** 3861. [e] A. B. Hörnfeldt, *Arkiv Kemi*, 1964, **22,** 211. [f] H. J. Jakobsen and S.-O. Lawesson, *Tetrahedron*, 1967, **23,** 871. [g] A. B. Hörnfeldt and S. Gronowitz, *Arkiv Kemi*, 1963, **21,** 239. [h] H. J. Jakobsen, *Tetrahedron*, 1967, **23,** 3737; A. B. Hörnfeldt, *Arkiv Kemi*, 1968, **29,** 427.

appropriate derivatives, and in general the non-aromatic, γ-thiolactones are preferred, the 3-thiolenone (**70**) being more common than the 4-thiolenone form (**69**). Polar solvents slightly favour the Δ^4-derivatives, as do conjugating and hyperconjugating substituents in the 5-position. Electron-withdrawing substituents, especially when located in the 3-position and thus allowing intramolecular hydrogen bonding, favour the hydroxy-form. Potential 5-vinyl-2-hydroxythiophens (**71**) tend to exist solely as the thiolenones (**72**).

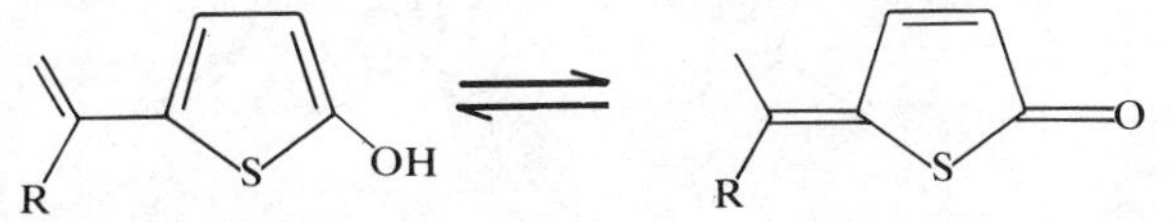

R = H, Me, or Ph

(71) **(72)**

With the 3-hydroxythiophens, only two tautomers are feasible (Scheme 27) and in general the hydroxy-form is more favoured than in the isomeric series (Table 13). Conjugating and hyperconjugating groups in the 5-position favour the oxo-tautomer, while the same groups in the 2-position render the hydroxy-form more stable. Electron-withdrawing groups in any position shift the equilibria virtually completely towards the enol tautomer, as do hydrogen-bond accepting solvents (*e.g.* acetone), in contrast to hydrogen-bond donating solvents (*e.g.* nitromethane or chloroform).

TABLE 13

The Tautomerism of 3-Hydroxythiophens

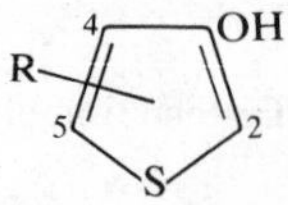

R	*Conditions*	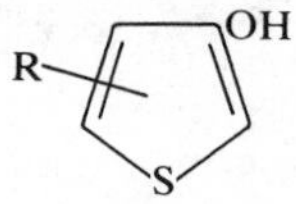 (%)	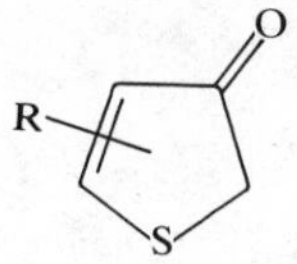 (%)	*Ref.*
H	neat	both forms present		a
2-Me	CS_2	80	20	a
2-Bu^t	CS_2	55	45	a
2,5-Me_2	CS_2	32	68	a
2-Me-5-Et	CS_2	35	65	b
2-Me-5-Pr^i	CS_2	39	61	b
2-Me-5-Bu^t	CS_2	43	57	b
2-Et-5-Me	CS_2	29	71	b
2-Bu^t-5-Me	CS_2	26	74	b
2,5-Bu^t_2	CS_2/neat	0	100	c
2-Ph-5-Me	$CDCl_3$*	100	0	d
2-CHO	Me_2CO	100	0	e
4-CHO	Me_2CO	100	0	e
5-CO_2H	Me_2CO	100	0	e
2-CO_2H	Me_2CO	100	0	e
2-COMe	solid	100	0	f
5-Me-2,4-$(CN)_2$	solid	100	0	g
2,5-$(CHO)_2$	solid	100	0	h

* After 1 month standing, solution contained 16% oxo-form.

[a] A.-B. Hörnfeldt, *Acta Chem. Scand.*, 1965, **19**, 1249. [b] R. Lantz and A.-B. Hörnfeldt, *Chem. Scripta*, 1972, **2,** 9. [c] A.-B. Hörnfeldt, *Acta Chem. Scand.*, 1972, **26,** 31. [d] B. Hedegaard, J. Z. Mortensen, and S.-O. Lawesson, *Tetrahedron*, 1971, **27,** 3853. [e] S. Gronowitz and A. Bugge, *Acta Chem. Scand.*, 1966, **20,** 261. [f] H. J. Jakobsen and S.-O. Lawesson, *Tetrahedron*, 1965, **21,** 3331. [g] K. Harke and F. Meissner, *Tetrahedron*, 1972, **28,** 875. [h] C. Pauliner, J. Morel, D. Semard and P. Pastour, *Bull. Soc. chim. France*, 1973, 2434.

Apart from the 2,5-isomer, all the potential dihydroxythiophens prefer to exist as the mono-oxo-tautomer (Scheme 28). Thus the 2,3-analogue exists as the 3-hydroxythiol-3-en-2-one in all phases and with alkyl or acyl substituents in the 4- or 5-position. Surprisingly, the 4-bromo-derivative is reported to be largely the diketo-tautomer in carbon tetrachloride solution. Intramolecular hydrogen bonding, as with simpler hydroxythiophens, will tend to favour the aromatic structure, as with 3,4-dihydroxy-2,5-thiophendicarboxylates.

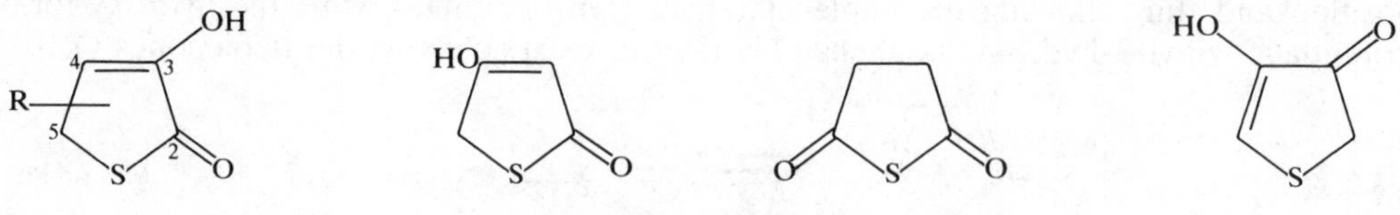

SCHEME 28

Why do hydroxythiophens forego the aromatic stabilization and prefer to exist as thiolactone tautomers? It would appear that the resonance energy associated with an S—CO—C=C unit is greater than that of an S—C=C—CO group and that both are significant in comparison with the corresponding hydroxythiophen resonance energies. The conjugating efficiency is clearly heightened in the non-aromatic γ-thiolactones compared with their acyclic counterparts by the strictures of a five-membered ring. Furthermore, further conjugation or hyperconjugation of the thiolactone tends to emphasize the stability of the non-aromatic tautomer.

19.1.12.2 Mercapto-, amino-, and alkyl-thiophens

Tautomerism in the mercapto- and amino-thiophens is much less significant than with the hydroxy-derivatives. Thus, the constraints of hydrogen bonding (as in **73**⇌**74**) or another tautomerizable group (as in **75**) seems to be required to force the thione tautomer to show up, the imino-group being even less prone to occur.

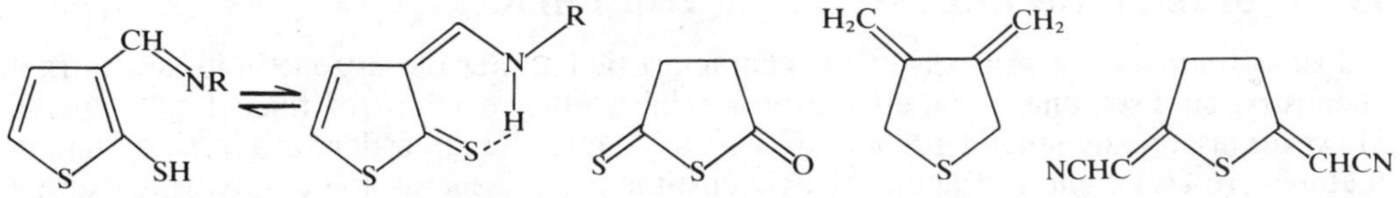

(73) (74) (75) (76) (77)

Significantly, several alkylene tautomers of thiophens (*e.g.* **76**[115] and **77**[116]) are known, but no study of the tautomerism of the systems has been reported to date. The bis-cyanomethylene (**77**) was obtained in both geometrical isomers from the corresponding 2,5-diketothiophen by way of a Wittig reaction (Scheme 29), which yielded only the thiophen-2,5-diacetic ester with the analogous phosphorane ester.[116]

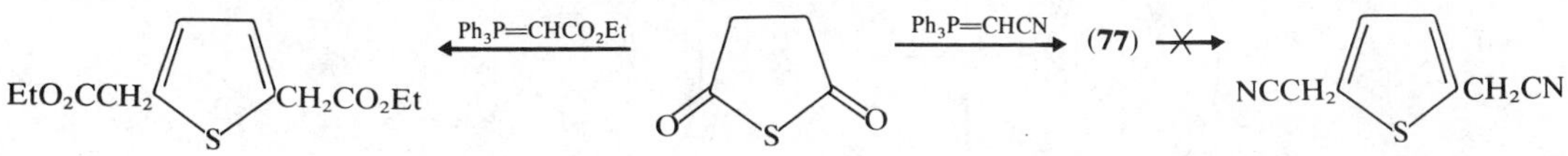

EtO_2CCH_2 ... CH_2CO_2Et ← $Ph_3P{=}CHCO_2Et$ — → $Ph_3P{=}CHCN$ (77) ↛ $NCCH_2$ ... CH_2CN (78)

SCHEME 29

19.1.12.3 Reactions of the tautomeric thiophens

The hydroxy-, amino-, and mercapto-derivatives are all prone to decomposition on storage, particularly in air. The hydroxy-compounds undergo both *O*- and *C*-alkylation with alkylating agents, dimethyl sulphate[117] and the use of thallium[118] or (better) quaternary ammonium salts[119] of the thienols favouring the formation of *O*-alkyl derivatives. Oxidative coupling of the hydroxy-compounds with potassium ferricyanide has been observed[120].

Thiophenthiols are best obtained by treatment of thienyl-lithium or Grignard derivatives with sulphur.[121] Alternatively, reduction of sulphonyl chlorides or cleavage of thienyl alkyl sulphides with sodium in liquid ammonia has been used. *S*-Alkylation and -arylation is readily performed by treatment of the thiolate salt with an alkyl or activated aryl halide, or with an aryl diazonium salt.

The aminothiophens may be isolated by reduction of their nitro-analogues, when they are best stored as their salts (especially their stannichlorides) or as their acyl derivatives. Surprisingly, Raney nickel in acetic anhydride effectively accomplishes reduction of the nitro-group and acetylation, without any desulphurization and with minimal catalyst poisoning.[96] The Hofmann rearrangement of the thenamides is only effective with the 3-isomers, and the Beckmann rearrangement of ketone oximes is of limited value in the acyclic series as a source of acylaminothiophens since mixtures are produced. Better results are reported with annulated ketone oximes. The ready electrophilic substitution of the thiophen ring results in self-coupling and consequent polymer formation during the normal diazotization of aminothiophens. However, high yields of diazonium salts and their derivatives are reported when the stannic chloride salts are diazotized, allowing formation even of bis-diazonium salts.[122]

It should again be emphasised that when the hydroxy-, mercapto-, or amino-groups are attached to a thiophen ring containing an electron-withdrawing group ($-M$), the push–pull effect of the two substituents makes for a stable, normal system with chemistry more akin to the benzenoid analogues. Thus, for example, the Gewald reaction (p. 833) which allows synthesis of numerous aminothiophens, yields stable crystalline intermediates of great value in the synthesis of important thiophens for pharmaceutical application, especially thieno-fused nitrogen heterocycles.

19.1.13 SIDE-CHAIN REACTIVITY OF THIOPHENS

Just as thiophen shows many of the characteristic features of benzene in its substitution chemistry, so also some parallels are noticeable in the reactions of thienyl substituents. However, as already noted with amino-, hydroxy-, and mercapto-derivatives, some unique features are also conferred upon thienyl substituents. In general, the thienyl ring exerts a $-I$ effect on attached groups, but has the remarkable facility to stabilize both positive and negative charges on a substituent much more effectively than the phenyl group. This is seen, for example, in the higher migratory aptitude of thienyl over phenyl (1000 times, twice that of *p*-methoxyphenyl) in the acid-catalysed pinacol-type rearrangement (Scheme 30a)[123] and tautomerism (Scheme 30b; 40 times that of the phenyl analogue), which is

(a) Ph, C, OH, 2 → H^+ → Ph, CCOPh, 2

(b) CHCH=CHR, OH → H^+ → CH=CHCHR, OH

SCHEME 30

due to the electron-donor ability of the thiophen ring,[124] and in the base-catalysed racemization of mandelic acid (**79**; Ar = Ph) which is 33 times slower than with its thienyl analogue (**79**; Ar = 2-Th), showing the electron-acceptor nature of the thienyl group.[125]

$ArCH(OH)CO_2H$
(79)

19.1.13.1 Thenyl derivatives

In the case of thenyl halides, this increased reactivity over the benzyl analogue is revealed by some surprising reactions. The halides are readily prepared either by haloalkylation of a suitable thiophen (Scheme 31) (allowing either mono- or di-alkylation

chloroalkylation → CH_2Cl 40% (Ref. 126), 50–55% (Ref. 127), $ClCH_2$ CH_2Cl 79% (Ref. 128), and 50–55% (Ref. 129)

Cl, Cl → $MeOCH_2Cl$ / $SnCl_4$ → $ClCH_2$, CH_2Cl, Cl, Cl 75% (Ref. 130)

MeO_2C → $MeOCH_2Cl$ / $ZnCl_2$ → CH_2Cl, MeO_2C, CH_2CL 85% (Ref. 131)

SCHEME 31

Me — NBS, CCl_4, Δ or Br_2, AZBN, $h\nu$, Δ, CCl_4 → CH_2Br (**80**) NBS 79% Br_2 47%; → Br_2, $h\nu$, Δ, AZBN → $CHBr_2$ (overall 58%) → OH^- → CHO 72%; (80) → $(CH_2)_6N_4$ → CHO

Et — Br_2, $h\nu$, Δ, AZBN → Br 70% → [O] → Ac; → vinyl

SCHEME 32

or further reaction of the intermediate), or by side-chain halogenation of an alkyl thiophen. The latter method offers a useful entry to otherwise difficultly accessible 3-substituted thiophens and may be accomplished using either *N*-bromosuccinimide[132] or bromine under homolytic conditions (Scheme 32).[133] Since the alkylthiophens are now available in large supply at reasonable prices, these routes once again become attractive alternatives to the metallation method (see p. 808) for β-substituted thiophens in particular.

Some reactions of the highly lachrymatory thenyl halides are summarized in Scheme 33, which demonstrates the remarkable reactivity and diversity of these underrated intermediates. The Grignard derivative makes available the otherwise inaccessible 2-methyl-3-thienyl series,[137] while several types of annulation derive from the bis-halomethyl derivatives. The cyclizations of 3,4-bis(halomethyl) derivatives are subject to steric limitations with bulky α-substituents, causing dimeric products to form.[130, 138–141] Some other typical useful applications of thenyl and thenal halides are summarized in Scheme 34.

Thenyl cyanides are not only of importance as intermediates to thienylacetic acids (of great value in synthetic penicillin analogues and related pharmaceutical preparations), but as 'active' methylene systems, capable of ready condensation with esters, aldehydes, carbonates, and alkyl halides. Cyclization of the *o*-bis(cyanomethyl) derivatives are also effective means of annulation.[142, 146, 147]

19.1.13.2 Thenoyl derivatives

A primary source of acylbenzenes is by oxidation of alkylbenzenes. However, in the thiophen series this approach is generally disappointing, owing to the sensitivity of the

70% (Ref. 134)

NaCN aq. acetone

29% 16%

CO_2 HCHO

49%

$(CH_2)_6N_4$ Mg ether

74% (Ref. 135) 98% (Ref. 137)

i, $ClCO_2Et$ ii, NaOH

72%

PPh_3 AcCl

RCHO base

(Ref. 136) 39% 34%

SCHEME 33

ring to oxidation. Some surprising exceptions include the conversion of 2- or 3-methylthiophen to the corresponding carboxylic acids in high yield with aqueous dichromate (Reaction 18).[148] The reaction will tolerate halogen substituents if ammonium dichromate is used, giving up to 55% yield.[36] Another promising method, applicable to a variety of alkylthiophens, is catalytic aerial oxidation using a cobalt salt.[149] Probably the most versatile oxidative approach to thenoic acids is the hypohalite oxidation of acetylthiophens, which invariably proceeds in high yield.[150]

The reactivity of the thenoyl group is very similar to that of the benzoyl series. Thus thenaldehydes undergo condensation reactions with aliphatic aldehydes, methyl ketones, cyclic ketones, benzyl halides, aliphatic nitro-compounds, and active methylenes and methyl heterocycles. Formation of Schiff bases, acetals, aminals, azlactones and all the usual derivatives such as oximes, hydrazones, and semicarbazones proceed normally. The Cannizzaro and benzoin condensation and the usual oxidations and reductions are

HBr

(Ref. 143)

Δ

(Ref. 144)

NaI

(Ref. 145)

SCHEME 34

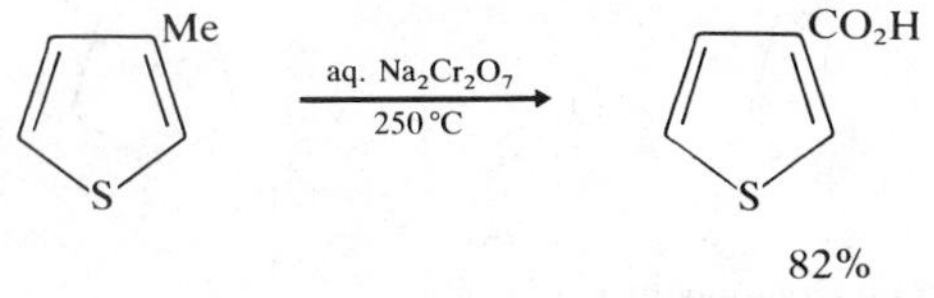

$$\text{2-methylthiophene} \xrightarrow[250\,^{\circ}\text{C}]{\text{aq. } Na_2Cr_2O_7} \text{thiophene-2-carboxylic acid } (CO_2H) \quad 82\% \qquad (18)$$

unexceptional, while the Huang–Minlon adaption of the Wolff–Kishner, and the Clemmensen reduction of acyl- to alkyl-thiophens, are particularly suited to the thiophens. Reactions such as the Fischer indole synthesis, the Leuckart, Pfitzinger, Hunsdieker, Beckmann, Hofmann, Curtius, and Reformatsky reactions are all successful in the thiophen series, and the Willgerodt conversion of acetylthiophens to thienylacetic acids is effective. The Wittig reaction has proved particularly useful in the synthesis of thiophen-derived helicenes. Trifluoroacetoacetylthiophen (**81**) has proved an especially effective complexing agent.

$COCH_2COCF_3$ (2-thienyl)

(**81**)

19.1.14 PREPARATIVE METHODS

In order to systematize the vast array of routes to the thiophen ring system, the methods have been classified according to the number of components (one to five) utilized in construction of the ring. Excluding those based on modification of an existing hydrothiophen (briefly reviewed at the end of the section), 19 combinations are possible, summarized in Scheme 35, the solid diagrams being those used to date. Clearly, the sub-division is somewhat artificial in that, for example, one-component methods will often feature as intermediates arising from a multi-component pathway. Furthermore, little virtue exists in trying to fill in gaps using, say, four-component methods. However, a route using the 2f principle would prove very worthwhile for deriving the difficultly accessible 3-substituted thiophens.

The survey below follows the most significant methods in the order of Scheme 35, and is illustrative rather than comprehensive. Early methods (up to about 1946) have been comprehensively reviewed[151] and more recent literature features in the general reviews of thiophen.[1–3, 7]

19.1.14.1 One-component pathways

All types of one-component ring-closures have been explored, the first [1a] invariably involving the attack of a thiol derivative on an endo- or exo-cyclic multiple bond, as exemplified in Scheme 36. The last method has particularly wide application. Several methods involving ring contraction by way of ring-opening–ring-closure sequences or by cheletropic removal of sulphur are also formally within this framework (Scheme 37). The conversion of 1,4-dithiins to thiophens (route 1c) proceeds both photochemically and thermally (Scheme 37) and is of general application for substituted thiophens. The 1,2-dithiins[161] and thiophens are found, together with polyacetylenes, in various plants.

Another promising approach to thiophens makes use of the masked carbonyl reactivity of oxathiolium salts (Scheme 38), whereby treatment with an active-methylene compound and base yields thiophens by either route 1b (*e.g.* X=Y=CN; X=CO_2Et, Y=CN; X=Y=CO_2Et) or route 1c (X=H, Y=NO_2).

Other useful examples of the route 1c approach involve the readily available β-mercaptocrotonic ester derivatives (**82**) and the fascinating synthesis of the elusive and unstable thienocyclobutadiene (**83**).

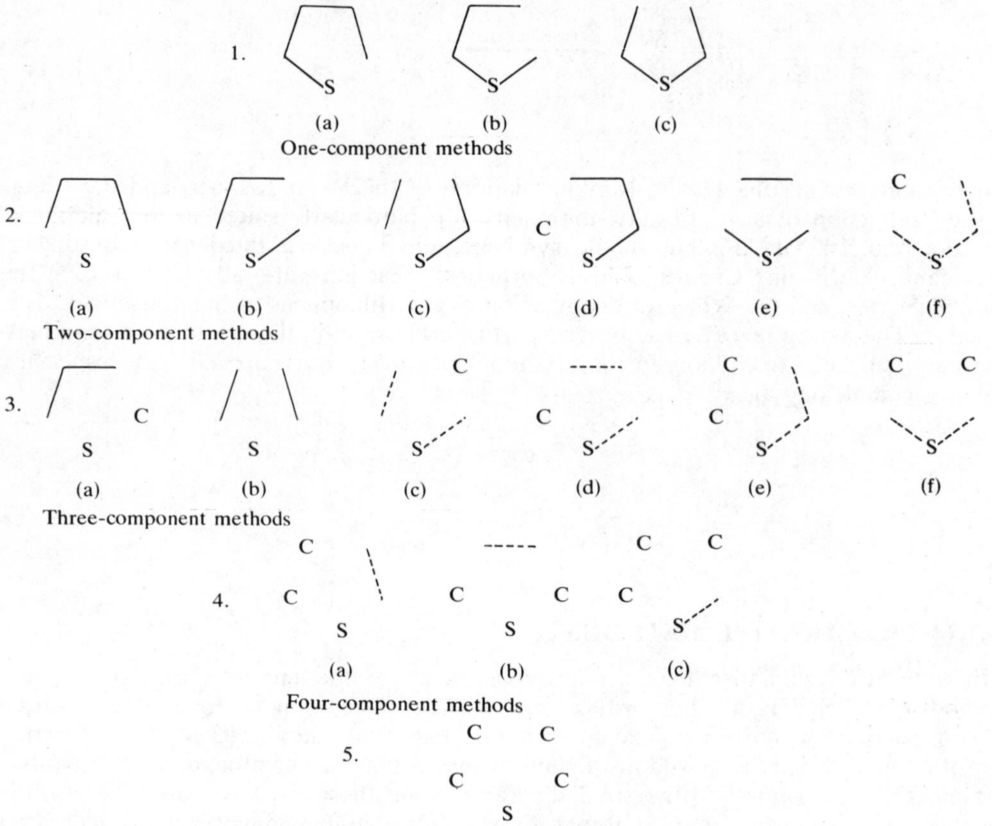

Five-component methods
(solid figures indicate those utilized in the literature)

SCHEME 35

HO_2C, SH, Ph $\xrightarrow{Cl_2,\ CCl_4}$ HO_2C, Ph (Ref. 152)

PhCH$_2$S, C≡N $\xrightarrow{HCl}$ $NH_2 \cdot HCl$ (Ref. 153)

MeO_2C $\xrightarrow{Me_2SO,\ 8\ h,\ 125\ °C}$ CO_2Me, MeO_2C, $\overset{+}{S}Me_2$, O^- $\xrightarrow{SOCl_2}$ MeO_2C, CO_2Me 91% (Ref. 154)

R, R $\xrightarrow{KHSO_4,\ 150–200\ °C}$ R, R $\longrightarrow$ R, R (Ref. 155)

R $\xrightarrow{H^+\ or\ B^-}$ R $\longrightarrow$ R, Me (Ref. 156)

SCHEME 36 1a routes

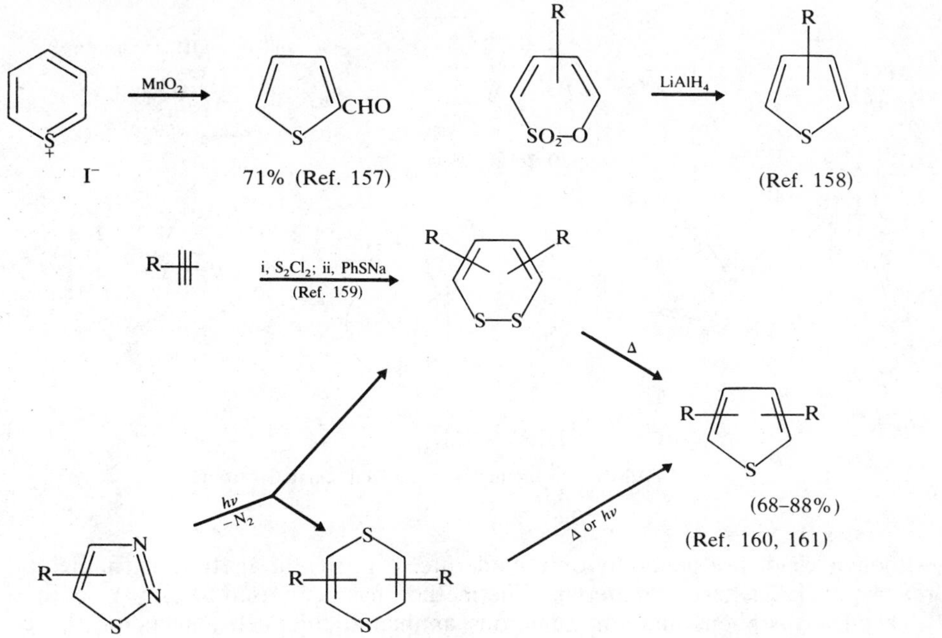
MnO2
CHO
71% (Ref. 157)
LiAlH4
(Ref. 158)
i, S2Cl2; ii, PhSNa
(Ref. 159)
hν
−N2
Δ
Δ or hν
(68–88%)
(Ref. 160, 161)

SCHEME 37 1b and 1c routes

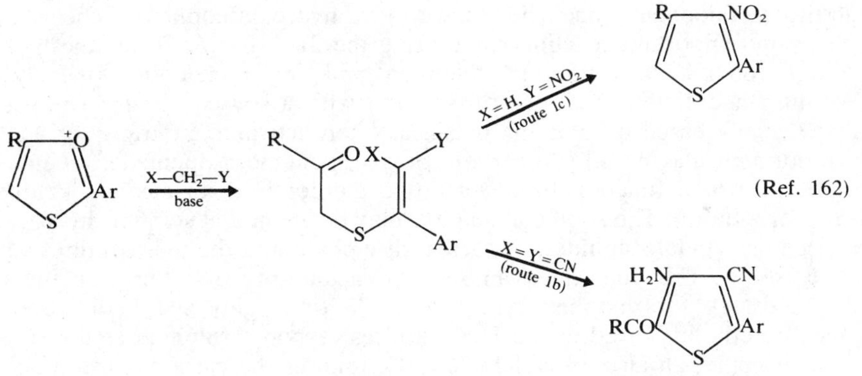
X—CH2—Y
base
X=H, Y=NO2
(route 1c)
X=Y=CN
(route 1b)
(Ref. 162)

SCHEME 38

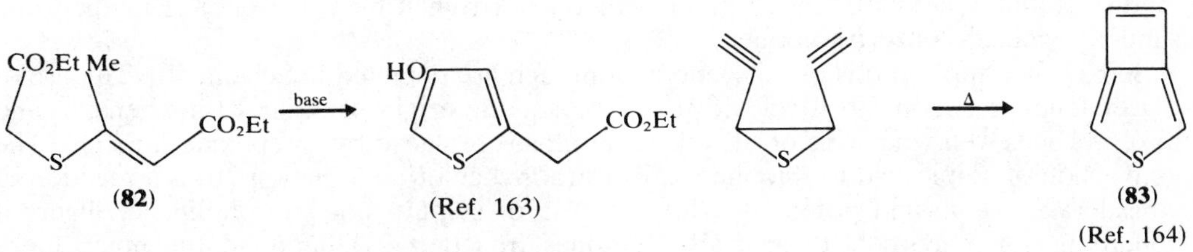
base
(82)
(Ref. 163)
Δ
(83)
(Ref. 164)

19.1.14.2 Two-component pathways

These are probably the most important and most common means for thiophen synthesis, encompassing both the vital industrial methods and such routes as the Paal and Hinsberg reactions.

The Paal reaction (Scheme 39), discovered in 1885 and involving the action of a phosphorus sulphide on a 1,4-dioxobutane derivative, is still a useful reaction today.[165]

SCHEME 39 Examples of the Paal reaction — route 2a

Although yields are generally only moderate to good, the method is simple and utilizes readily available starting materials. The method has been used to prepare both S- and α-C-labelled thiophens and for numerous annulated thiophens, including the cyclobutathiophens (**84**) and (**85**),[166] the cyclophanes (**86**) and (**87**),[167] and several 'non-classical' thiophens such as the thieno[*c*]pyrrole (**88**).[168] The carbonyl group may be an aldehyde, ketone, or acid function, use of acid salts being preferable for the synthesis of α-unsubstituted thiophens since acids tend to give hydroxythiophens (Scheme 39).

The second important reaction route using the 2a principle forms the basis of all the important commercial routes to thiophen and its homologues, namely the high-temperature interaction of a four-carbon unit with a source of sulphur over a suitable catalyst (usually based on alumina or silica, often activated by transition metal oxides). The carbon unit may be a hydrocarbon (butane, pentane, isopentane, butene, butadiene, isoprene, *etc.*) or a functionalized derivative thereof (*e.g.* an alcohol, ketone, epoxide, chloride, or a furan). The source of sulphur may be elemental sulphur, hydrogen sulphide, sulphur oxides, metal sulphides, or carbon disulphide and the temperatures vary between 180 and 700 °C. Thus the American Socony-Vacuum process[169] uses butane and sulphur as the feedstock for thiophen synthesis, while the highly successful Croda Synthetic Chemicals method,[170] used in the U.K., utilizes carbon disulphide as the sulphur source and, for example, an alcohol, aldehyde, or ketone as the carbon source. The method is highly adaptable giving, for example, thiophen (from n-butanol or crotonaldehyde), 2-methyl- (from n-pentanol), 3-methyl- (from 3-methylbutan-1-ol), 2,4-dimethyl- (from methyl isobutyl ketone), and 3-phenyl-thiophen (from s-butylbenzene). Ethylbenzene similarly yielded benzo[*b*]thiophen.

Some other applications of this general approach are depicted in Scheme 40. The use of diacetylenes (reaction 19) is related to the biogenetic origin of natural thiophens,[173] and also proceeds well when one of the alkyne groups is replaced by an epoxide function. The conversion of adipic acid to thiophen-2,5-dicarbonyl chloride (reaction 20) is a reaction of considerable value and potential, while the use of sulphur and morpholine (Willgerodt conditions) in reaction with aryl alkyl ketones (reaction 21) heralded the much more versatile Gewald reaction (see p. 833). The classical Benary reaction has numerous variations and is valuable for certain hydroxythiophen syntheses.[177]

The 2b approach to thiophens is also useful and invariably incorporates the use of a thioglycolic ester (Scheme 41) with a three-carbon unit. The Fiesselman synthesis of thiophen-2-carboxylates (**89**)[179] makes use of a wide variety of substrates: esters yield 3-hydroxythiophens; aldehydes or their acetals give 3-unsubstituted systems; while ketones and nitriles give 3-alkyl- (or 3-aryl-) and 3-amino-thiophens, respectively, all in

Ar[1], Ar[2] —S, 200–250 °C→ 15–60% (Ref. 171)

Cl, Cl, Cl, Cl, Cl, Cl —S, Δ, $-SCl_2$→ Cl, Cl, Cl, Cl ~100% (Ref. 172)

R^1, R^2 —H_2S, alkali→ 81–89% (Ref. 173)

R^1,R^2 = Ar, alkyl, $(C{\equiv}C)_n$Ar, 2-Th, 3-py

(19)

HO_2C, CO_2H —$SOCl_2$→ ClCO, COCl (63%) (Ref. 175)

(20)

Ph, O, O or Ph, O —S, morpholine→ Ph, N, O 37% (Ref. 176)

(21)

SCHEME 40 Other routes of type 2a

COMe + $HSCH_2CO_2Et$ → OH, Me, CO_2Et, S —i, PPA; ii, $(PhS)_2$→ Me, CO_2Et, S 70% (Ref. 178)

CO_2Me, R + $HSCH_2CO_2Me$ —aq. NaOH→ CO_2Me, R, S, CO_2Me → OH, R, CO_2Me, S (**89**)

also with: Br, CO_2Me, MeO_2C, Br , COR, RCO , CHO, R, Cl , Br, COMe, Ph , Cl, CN, Me, Cl (Ref. 179)

SCHEME 41 2b routes

(Ref. 180; see also 181 and 182)

(22)

(Ref. 183)

(23)

(Ref. 184)

SCHEME 42 2c and 2d routes

high yield. The correct oxidation level is attained by incorporation of an alkyne, carbonyl, halovinyl, or α,β-dihaloalkyl group in the three-carbon substrate.

Methods using the 2c and 2d mode of reaction are known but are less common and not of wide application, some examples being collated in Scheme 42. The second class of reactions (reactions 22 and 23) depends upon the availability of α,β-unsaturated thiones or their equivalents, some of which are readily generated from, for example, isothiocyanates, from the action of carbon disulphide or active-methylene derivatives, or from easily obtained monothio-β-diketones.

R^1,R^2 = H, alkyl, aryl, or OMe

(**90**)

good yields

(Ref. 185, 186)

(**91**) (**92**) (**93**)

SCHEME 43 The Hinsberg reaction — route 2e

An important source of β-substituted thiophens is the Hinsberg reaction, virtually the only significant use of the 2e method (Scheme 43). The reaction has been shown to give the half-ester (**90**) in a Stobbe manner;[186] the yields are usually high and *o*-quinones are particularly effective. A related method makes use of a bis-ylide in place of the thiodiglycolate ester, enabling isolation of the highly strained cyclobutathiophens (**91**), (**92**), and (**93**).[187]

19.1.14.3 Three-component pathways

Only one of the six possible three-component reactions has been extensively explored but this route [3b], the symmetrical assembly of sulphur and two two-carbon fragments, is very valuable. Indeed, one of the earliest industrial routes to thiophen involved the action of sulphur or a sulphide on acetylene.[188] Several useful syntheses of otherwise hard-won 2,4-disubstituted thiophens depend upon this approach. Thus, 2,4-diphenylthiophen is formed when acetophenone is treated with sulphur and ammonia.[189] Similarly, sulphur and acrylonitrile react at 160–200 °C to give 2,4-dicyanothiophen.[190] The action of hot sulphur on dimethyl maleate or fumarate[191] or on stilbene[192] gives tetramethoxycarbonyl- and tetraphenyl-thiophen, respectively. However, the most useful reaction in this class is undoubtedly the Gewald method for aminothiophen synthesis,[193] whereby an activated acetonitrile (such as ethyl cyanoacetate, cyanoacetone, benzoylacetonitrile, *etc.*), an aldehyde or ketone (containing the $COCH_2$ group), sulphur, and an amine such as morpholine are heated in ethanol solution (reaction 24). Yields are high and the products are stable, highly crystalline solids. An alternative approach employs an α-mercaptoaldehyde or ketone, giving the same products in high yields (reaction 24). The

$$R^1\text{-C(=O)-CH} + R^2\text{CH}_2\text{C}\equiv\text{N} \xrightarrow[\text{EtOH, }\Delta]{\text{morpholine, S}_8} \text{2-amino-thiophen }(R^1, R^2, \text{NH}_2) \xleftarrow[\text{EtOH, }\Delta]{\text{morpholine}} R^1\text{-C(=O)-CH-SH} + R^2\text{CH}_2\text{C}\equiv\text{N} \quad (24)$$

value of these methods is evident from the large number of papers reporting such syntheses. The products form excellent substrates for annulation of all kinds of rings, particularly nitrogen heterocycles, thereby making the derivatives of great pharmaceutical value.

19.1.14.4 Five-component pathways

Finally, the methods involving a five-component reaction, which are surprisingly effective for the synthesis of tetra-arylthiophens, need mention. Historically this was the first method employed for the synthesis of a thiophen, when Laurent, in 1844, obtained tetraphenylthiophen from polymeric thiobenzaldehyde by pyrolysis. Almost any Ph—C precursor, on heating with sulphur, yields tetraphenylthiophen.[194] Thus toluene, benzyl chloride, alcohol, sulphide, or disulphide, phenylacetic acid, and various derivatives of benzoic acid give the same product, which is clearly the thermodynamically favoured end product since it is unchanged even on distillation at 460 °C!

19.1.14.5 Hydrothiophen pathways

Numerous routes based upon a preformed hydrothiophen ring are reported,[151] but are largely outside of the scope of this chapter. The most useful aspect of this approach lies in the synthesis of 3-substituted thiophens from easily derived 3-ketotetrahydrothiophen, exemplified in Scheme 44. A recent patent describing routes to important intermediates,

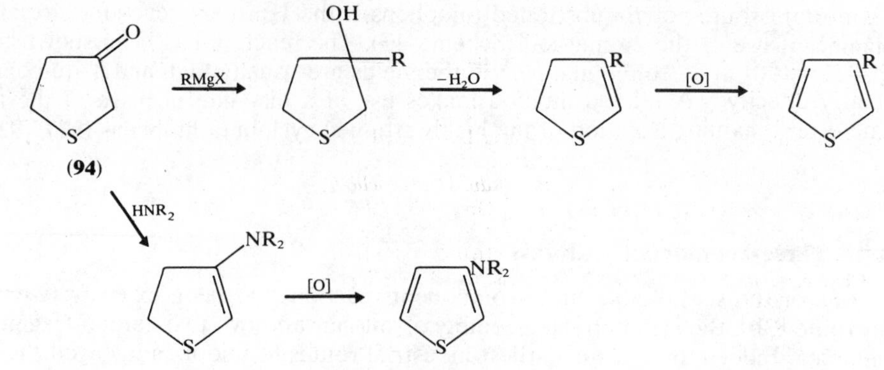

SCHEME 44

such as 3-thienylacetic acid for incorporation into penicillins, utilizes the same ketone (**94**) which was catalytically aromatized after condensation with cyanoacetic acid, giving 3-thienylacetic acid.[195]

19.1.15 NATURALLY OCCURRING THIOPHENS

Higher plants, particularly of the Compositae family, which contain polyacetylenes are frequently also found to contain related thiophens, which are probably derived by addition of hydrogen sulphide or its biogenetic equivalent to a conjugated diacetylene (see p. 831).[173] Micro-organisms such as the fungi Basidiomycetes also show the co-occurrence of polyacetylenes and thiophens.[173] As well as α-substituted thiophens, simple derivatives of di- and ter-thienyls (including terthienyl itself) are also found naturally, some having useful nematocidal properties. Thus, typically, the leaves of the plant *Wedelia forsteriana* Endl. contain an interrelated series of compounds (Scheme 45), as well as further functionalized derivatives.[173]

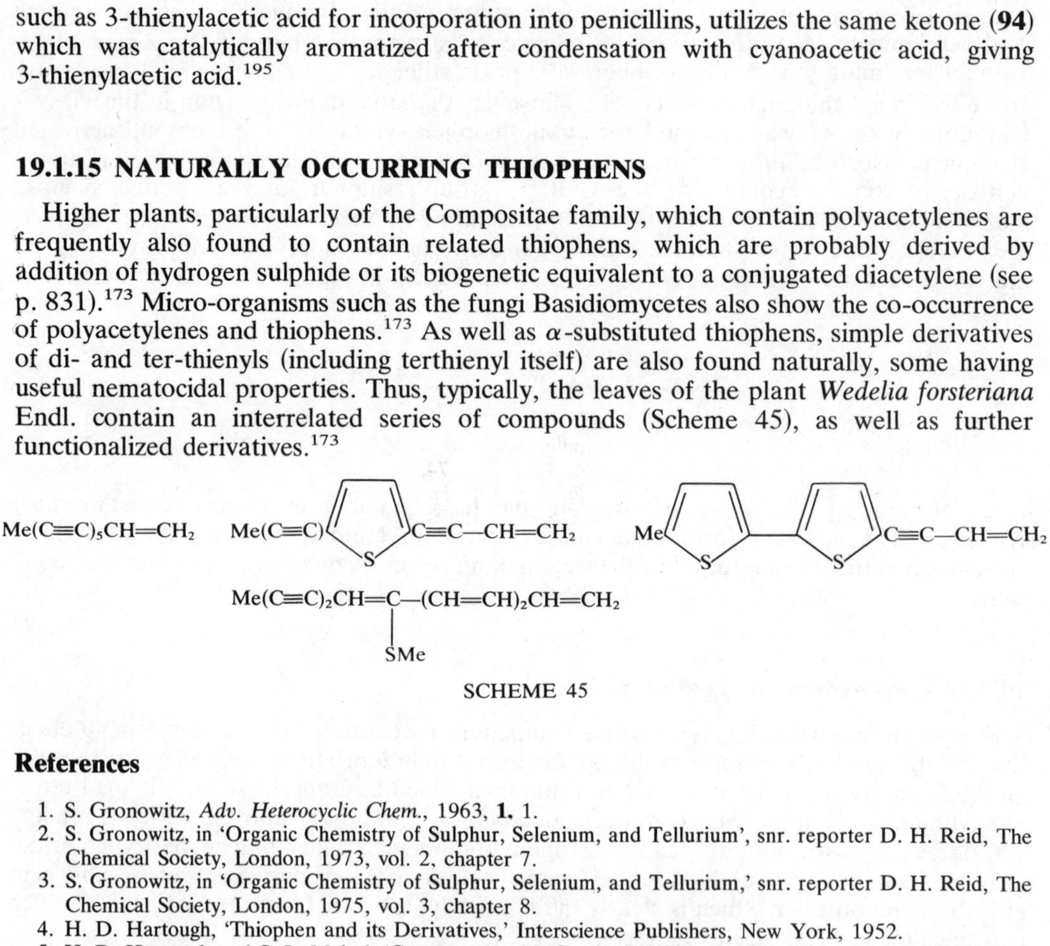

$Me(C{\equiv}C)_5CH{=}CH_2$

$Me(C{\equiv}C)_2CH{=}C(SMe){-}(CH{=}CH)_2CH{=}CH_2$

SCHEME 45

References

1. S. Gronowitz, *Adv. Heterocyclic Chem.*, 1963, **1.** 1.
2. S. Gronowitz, in 'Organic Chemistry of Sulphur, Selenium, and Tellurium', snr. reporter D. H. Reid, The Chemical Society, London, 1973, vol. 2, chapter 7.
3. S. Gronowitz, in 'Organic Chemistry of Sulphur, Selenium, and Tellurium,' snr. reporter D. H. Reid, The Chemical Society, London, 1975, vol. 3, chapter 8.
4. H. D. Hartough, 'Thiophen and its Derivatives,' Interscience Publishers, New York, 1952.
5. H. D. Hartough and S. L. Meisel, 'Compounds with Condensed Thiopen Rings,' Interscience Publishers, New York, 1954.
6. W. Steinkopf, 'Die Chemie des Thiophens,' T. Steinkopff, Leipzig, 1941.
7. C. D. Hurd, 'Thiophene Chemistry 1962–1968,' *Quart. Reports Sulphur Chem.*, 1969, **4,** 75.
8. N. R. Clark, *Manuf. Chemist,* 1972, **43,** 3.
9. B. Iddon and R. M. Scrowston, *Adv. Heterocyclic Chem.*, 1970, **11,** 177.
10. J. Ashby and C. C. Cook, *Adv. Heterocyclic Chem.*, 1974, **74,** 351.
11. V. Schomaker and L. Pauling, *J. Amer. Chem. Soc.*, 1939, **61,** 1769.
12. W. R. Harshbarger and S. H. Bauer, *Acta Cryst.*, 1970, **B26,** 1010.
13. B. Bak, D. Christensen, L. H. Nygaard, and J. Rastrup-Andersen, *J. Mol. Spectroscopy*, 1961, **7,** 58.

14. P. J. Wheatley, in 'Physical Methods in Heterocyclic Chemistry,' ed. A. R. Katritzky, Academic Press, New York, 1972, vol. 5.
15. J. Sheridan, in 'Physical Methods in Heterocyclic Chemistry', ed. A. R. Katritzky, Academic Press, New York, 1974, vol. 6, chapter 2.
16. R. Zahradník, *Adv. Heterocyclic Chem.*, 1965, **5,** 1; see also D. T. Clark, in 'Organic Compounds of Sulphur, Selenium, and Tellurium,' snr. reporter D. H. Reid, The Chemical Society, London, 1970, vol. 1. chapter 1.
17. M. J. Cook, A. R. Katritzky, and P. Linda, *Adv. Heterocyclic Chem.*, 1974, **17,** 255.
18. F. Fringuelli, G. Marino, A. Taticchi, and G. Grandolino, *J.C.S. Perkin II*, 1974, 332.
19. *E.g.* D. B. Baird, A. R. Costello, B. R. Fishwick, R. D. McClelland, and P. Smith, *Brit. Pat.* 1 394 367/1975 and 1 394 368/1975 (*Chem. Abs.*, 1975, **83,** 116 943 and 116 944).
20. D. J. Chadwick, J. Chambers, G. D. Meakins, and R. L. Snowden, *J.C.S. Perkin II*, 1972, 1959; 1974, 1181; 1975, 604; 1976, 1.
21. T. J. Batterham, 'Nmr Spectra of Simple Heterocycles,' Wiley, New York, 1973, p. 429.
22. I. P. Biryukov, M. G. Voronkov, and I. A. Safin, 'Tables of Nuclear Quadrupole Resonance Frequencies,' Israel Program for Scientific Translations, Jerusalem, 1969; E. A. C. Lucken, in 'Physical Methods in Heterocyclic Chemistry,' ed. A. R. Katritzky, Academic Press, New York, 1971, vol. 4, chapter 2.
23. G. Spiteller, *Adv. Heterocyclic Chem.*, 1966, **7,** 312; Q. N. Porter and J. Baldas, 'Mass Spectrometry of Heterocyclic Compounds,' Wiley, New York, 1971, p. 243.
24. B. C. Gilbert and M. Trenwith, in 'Physical Methods in Heterocyclic Chemistry,' ed. A. R. Katritzky, Academic Press, New York, 1974, vol. 6, chapter 3.
25. P. Tomasik and C. D. Johnson, *Adv. Heterocyclic Chem.*, 1976, **20,** 34.
26. E. Heilbronner, J. P. Maier, and E. Hasselbach, in 'Physical Methods in Heterocyclic Chemistry,' ed. A. R. Katritzky, Academic Press, New York, 1974, vol. 6, chapter 1.
27. J. Volke, in 'Physical Methods in Heterocyclic Chemistry,' ed. A. R. Katritzky, Academic Press, New York, 1963, vol. 1, chapter 6.
28. G. Marino, *Adv. Heterocyclic Chem.*, 1971, **13,** 235.
29. M. J. S. Dewar, 'The Electronic Theory of Organic Chemistry,' Clarendon Press, Oxford, 1949, p. 188.
30. Ya. L. Gol'dfarb, Yu. B. Volkenstein, and L. I. Belen'kii, *Angew. Chem. Internat. Edn.*, 1968, **7,** 519.
31. A. P. Yakubov, Yu. K. Sudarushkin, L. I. Belen'kii, and Ya. L. Gol'dfarb, *J. Org. Chem.* (*U.S.S.R.*), 1973, **9,** 1549.
32. *E.g.* see Ya. L. Gol'dfarb, S. Z. Taits, and L. I. Belen'kii, *Tetrahedron*, 1963, **19,** 1851.
33. J. W. Schick and D. J. Crowley, *J. Amer. Chem. Soc.*, 1951, **73,** 1377.
34. O. Meth-Cohn, *Tetrahedron Letters*, 1973, 91.
35. R. M. Kellogg, A. P. Schaaf, E. T. Harper, and H. Wynberg, *J. Org. Chem.*, 1968, **33,** 2902.
36. O. Meth-Cohn, unpublished observations.
37. J. M. Barker, I. G. C. Coutts, and P. R. Huddleston, *J.C.S. Chem. Comm.*, 1972, 615.
38. N. Messina and E. V. Brown, *J. Amer. Chem. Soc.*, 1952, **74,** 920.
39. P. Cagniant, A. Reisse, and D. Cagniant, *Bull. Soc. chim. France*, 1969, 991.
40. P. Cagniant, G. Merle, and D. Cagniant, *Bull. Soc. chim. France*, 1970, 322.
41. G. Muraro, D. Cagniant, and P. Cagniant, *Compt. rend.* (*C*), 1972, **274,** 201.
42. R. Helder and H. Wynberg, *Tetrahedron*, 1975, **31,** 2551.
43. R. Motoyama, K. Sato, and E. Imoto, *Nippon Kagaka Zasshi*, 1957, **78,** 779 (*Chem. Abs.* 1960, **54,** 22 559).
44. *E.g.* see O. Meth-Cohn and S. Gronowitz, *Acta Chem. Scand.*; 1966, **20,** 1577.
45. S. Gronowitz and P. Moses, *Acta Chem. Scand.*, 1962, **16,** 155.
46. W. L. Mock, *J. Amer. Chem. Soc.*, 1970, **92,** 7610.
47. K. Torsell, *Acta Chem. Scand.* (*B*), 1976, **30,** 353.
48. D. Copland, D. Leaver, and W. B. Menzies, *Tetrahedron Letters*, 1977, 639.
49. R. M. Acheson and D. R. Harrison, *J. Chem. Soc.* (*C*), 1970, 1764; R. F. Heldweg and H. Hogeveer, *Tetrahedron Letters*, 1974, 75.
50. S. Gronowitz, C. Westerlund, and A-B. Hörnfeldt, *Acta Chem. Scand.* (*B*), 1975, **29,** 233, 244.
51. A. H. Blatt, N. Gross, and C. W. Tristram, *J. Org. Chem.*, 1957, **22,** 1588.
52. C. Dell'Erba, D. Spinelli, and G. Leandri, *Gazetta*, 1969, **99,** 535.
53. M. G. Reinecke, H. W. Adickes, and Ch. Pyun, *J. Org. Chem.*, 1971, **36,** 3820.
54. G. Doddi, G. Illuminati, and F. Stegel, *Tetrahedron Letters*, 1973, 3221, and C. Paulmier, M. P. Simmonin, A. P. Chatrousse, and F. Terrier, *Tetrahedron Letters*, 1973, 1123.
55. G. Guanti, C. Dell'Erba, and S. Thea, *J.C.S. Perkin I*, 1974, 2357.
56. G. Doddi, G. Illuminati, and F. Stegel, *J. Org. Chem.*, 1971, **36,** 1918.
57. H. C. van der Plas, D. A. de Bie, G. Geurtsen, M. G. Reinecke, and H. W. Adickes, *Rec. Trav. chim.*, 1974, **93,** 33.
58. A. Bruggink and A. McKillop, *Tetrahedron*, 1975, **31,** 2607.
59. S. Gronowitz and B. Holm, *Chem. Scripta*, 1972, **2,** 245.
60 (a) J. A. Clarke, S. McNamara, and O. Meth-Cohn, *Tetrahedron Letters*, 1974, 2373; (b) J. Sice, *J. Amer. Chem. Soc.*, 1953, **75,** 3697; (c) Ya. L. Gol'dfarb, M. A. Kalik, and M. L. Kirmalova, *J. Gen. Chem.* (*U.S.S.R.*), 1959, **29,** 3592.
61. S. Gronowitz, *Arkiv Kemi*, 1954, **7,** 361.

62. D. A. Shirley and J. Goan, *J. Organometallic Chem.*, 1964, **2,** 304.
63. T. B. Patrick, J. M. Disher, and W. J. Probst, *J. Org. Chem.*, 1972, **37,** 4467.
64. Ya. L. Gol'dfarb and Ya. L. Danyushevskii, *Bull. Acad. Sci. U.S.S.R.*, 1963, 483.
65. T. Kauffman and A. Mitschker, *Tetrahedron Letters*, 1973, 4039.
66. P. Moses and S. Gronowitz, *Arkiv Kemi*, 1962, **18,** 119.
67. G. B. Backman and L. W. Heisey, *J. Amer. Chem. Soc.*, 1948, **70,** 2378.
68. S. Gronowitz, A.-B. Hörnfeldt, and R. Håkensson, *Arkiv Kemi*, 1961, **17,** 165.
69. S. Gronowitz and B. Holm, *Acta Chem. Scand.* (*B*), 1976, **30,** 505.
70. S. Gronowitz and B. Holm, *Acta Chem. Scand.*, 1969, **23,** 2207.
71. S. Gronowitz, B. Cederlund, and A.-B. Hörnfeldt, *Chem. Scripta*, 1974, **5,** 217.
72. I. Haiduc and H. Gilman, *Rev. Roumaine Chim.*, 1971, **16,** 305; B. Decroix, J. Morel, C. Paulmier, and P. Pastour, *Bull. Soc. chim. France*, 1972, 1848.
73. S. Gronowitz and T. Frejd, *Acta Chem. Scand.*, 1970, **24,** 2656; *ibid.* (*B*), 1976, **30,** 485 and papers cited therein.
74. R. Grafing and L. Brandsma, *Rec. Trav. chim.*, 1976, **95,** 264.
75. C. Paulmier, J. Morel, and P. Pastour, *Bull. Soc. chim. France*, 1969, 2511.
76. U. Michael and S. Gronowitz, *Acta Chem. Scand.*, 1968, **22,** 1353.
77. S. Gronowitz and A. Bugge, *Acta Chem. Scand.*, 1965, **19,** 1271.
78. M. Gomberg and W. E. Bachmann, *J. Amer. Chem. Soc.*, 1924, **46,** 2339.
79. G. Vernin, J. Metzger, and C. Parkanyi, *J. Org. Chem.*, 1975, **40,** 3183.
80. R. Frimm, L. Fisera, and J. Kovac, *Coll. Czech. Chem. Comm.*, 1973, **38,** 1809.
81. M. Bartle, S. T. Gore, R. K. Mackie, and J. M. Tedder, *J.C.S. Perkin I*, 1976, 1636; S. T. Gore, R. K. Mackie, and J. M. Tedder, *ibid.*, 1976, 1639.
82. F. Minisci and O. Porta, *Adv. Heterocyclic Chem.*, 1974, **16,** 123.
83. S. Hashimoto, W. Hoike, and M. Oyama, *J. Chem. Soc. Japan, Chem. Ind. Chem.*, 1972, 375; 1973, 1139.
84 (a) H.-S. Ryang and H. Sakurai, *J.C.S. Chem. Comm.*, 1972, 594; (b) D. W. Allen, D. J. Buckland, B. G. Hutley, A. C. Oades, and J. B. Turner, *J.C.S. Perkin I*, 1977, 621.
85. L. Benati and M. Tiecco, *Boll. Sci. Fac. Chim. Ind. Bologna, Suppl.*, 1966, **24,** 225.
86. *E.g.* see S. Gronowitz, and T. Raznikiewicz, *Org. Synth., Coll. Vol. V*, p. 149.
87. H. L. Coonradt and H. D. Hartough, *J. Amer. Chem. Soc.*, 1948, **70,** 1158.
88. R. F. Curtis, D. M. Jones, G. Ferguson, D. M. Hawley, J. G. Sime, K. K. Cheung, and G. Germain, *Chem. Comm.*, 1969, 165; D. M. Hawley and G. Ferguson, *J. Chem. Soc.* (*B*), 1971, 843; R. F. Curtis, D. M. Jones, and W. A. Thomas, *J. Chem. Soc.* (*C*), 1971, 234.
89. D. N. Rheinhoudt and C. G. Kouwenhoven, *Tetrahedron*, 1974, **30,** 2093.
90. D. N. Rheinhoudt, W. P. Trompenaars, and J. Geevers, *Tetrahedron Letters*, 1976, 4777, and references cited therein.
91. S. Gronowitz and B. Uppström, *Acta Chem. Scand.*, 1975, **29,** 441.
92. K. Hafner and W. Kaiser, *Tetrahedron Letters*, 1964, 2185.
93. G. O. Schenk and R. Steinmetz, *Angew. Chem.*, 1958, **70,** 504.
94. R. Pettit, *Tetrahedron Letters*, 1960, 11.
95. R. Mozingo, S. A. Harris, D. E. Wolf, C. E. Hoffhine, N. R. Easton, and K. Folkers, *J. Amer. Chem. Soc.*, 1945, **67,** 2092.
96. Ya. L. Gol'dfarb, M. M. Polonskaya, B. P. Fabrichnyi, and I. F. Shalavina, *Proc. Acad. Sci.* (*U.S.S.R.*), 1959, **126,** 331.
97. See *e.g.* A. I. Meyers, 'Heterocycles in Organic Synthesis', Wiley-Interscience, New York, 1974, pp. 15, 228, 243, 266.
98. Ya, L. Gol'dfarb, B. P. Fabrichnyi, and I. F. Shalavina, *Tetrahedron*, 1962, **18,** 21.
99. O. A. Kalinovskii, S. Z. Taits, and Ya. L. Gol'dfarb, *Bull. Acad. Sci. U.S.S.R.*, 1970, 2189.
100. W. Hoek, J. Strating, and H. Wynberg, *Rec. Trav. chim.*, 1966, **85,** 1054.
101. H. Wynberg, *Accounts Chem. Res.*, 1971, **4,** 65.
102. S. F. Birch and D. T. McAllan, *J. Chem. Soc.*, 1951, 2556 and 3411.
103. For a general review see 'Photochemistry of Heterocycles', ed. O. Buchardt, Wiley-Interscience, New York, 1976, pp. 145, 483, 564.
104. H. A. Wiebe, S. Braslavsky, and J. Heicklen, *Canad. J. Chem.*, 1972, **50,** 2721.
105. A. Couture, A. Delevallee, A. Lablache-Combier, and C. Parkanyi, *Tetrahedron*, 1975, **31,** 785 and Refs. cited therein.
106. C. N. Skold and R. H. Schlessinger, *Tetrahedron Letters*, 1970, 791; H. H. Wasserman and W. Strehlow, *ibid.*, 1970, 795.
107. H. J. Kuhn and K. Gollnick, *Tetrahedron Letters*, 1972, 1909.
108. T. S. Cantrell, *J. Org. Chem.*, 1974, **39,** 2242.
109. C. Rivas, M. Velez, and O. Crescente, *Chem. Comm.*, 1970, 1474.
110. For a general review see R. W. Hoffmann, 'Dehydrobenzene and Cycloalkynes,' Academic Press, New York, 1967, p. 290.
111. M. G. Reinecke and J. G. Newsom, *J. Amer. Chem. Soc.*, 1976, **98,** 3021.
112. For an excellent comparative review see J. Elguero, C. Marzin, A. R. Katritzky, and P. Linda, 'The Tautomerism of Heterocycles', Academic Press, New York, 1976.
113. C. Frisell and S.-O. Lawesson, *Org. Synth., Coll. Vol. V*, 642.
114. A.-B. Hörnfeldt and S. Gronowitz, *Acta Chem. Scand.*, 1962, **16,** 789.

115. S. Sadeh and Y. Gaoni, *Tetrahedron Letters*, 1973, 2365.
116. W. Flitsch, J. Schwieger, and U. Strunk, *Annalen*, 1975, 1967; see G. M. Badger, J. A. Elix, and G. E. Lewis, *Austral. J. Chem.*, 1965, **18,** 70, for the 2,5-bis(cyanomethyl) tautomer.
117. B. Cederlund and A.-B. Hörnfeldt, *Acta Chem. Scand.*, 1971, **25,** 3324, 3546.
118. E. B. Pedersen and S.-O. Lawesson, *Tetrahedron*, 1971, **27,** 3861.
119. B. Cederlund, Å. Jesperson, and A.-B. Hörnfeldt, *Acta Chem. Scand.*, 1971, **25,** 3656.
120. A.-B. Hörnfeldt and P.-O. Sundberg, *Acta Chem. Scand.*, 1972, **26,** 31.
121. S. Gronowitz and R. Håkansson, *Arkiv Kemi.*, 1960, **16,** 309.
122. N. I. Putohkim and V. I. Yakovlev, *Doklady Akad. Nauk S.S.S.R.*, 1954, **98,** 89 (*Chem. Abs.*, 1955, **49,** 12 431).
123. M. R. Kegelman and E. V. Brown, *J. Amer. Chem. Soc.*, 1953, **75,** 5961; 1954, **76,** 2711.
124. E. A. Braude and J. S. Fawcett, *J. Chem. Soc.*, 1952, 4158.
125. S. Gronowitz, *Arkiv Kemi.*, 1958, **13,** 231.
126. K. B. Wiberg and H. F. McShane, *Org. Synth.*, *Coll. Vol. III*, 197.
127. W. S. Emerson and T. M. Patrick, *Org. Synth.*, *Coll. Vol. IV*, 980.
128. J. M. Griffing and L. F. Salisbury, *J. Amer. Chem. Soc.*, 1948, **70,** 3416.
129. Ya. L. Gol'dfarb and Ya. L. Danyushevskii, *Bull. Acad. Sci. U.S.S.R.*, 1956, 1395.
130. D. J. Zwanenburg and H. Wynberg, *J. Org. Chem.*, 1969, **34,** 333.
131. D. J. Zwanenburg and H. Wynberg, *J. Org. Chem.*, 1966, **31,** 3363.
132. E. Campaigne and B. F. Tullar, *Org. Synth.*, *Coll. Vol. IV*, 921; see also S. Gronowitz and T. Frejd, *Synth. Comm.*, 1976, **6,** 475.
133. J. A. Clarke and O. Meth-Cohn, *Tetrahedron Letters*, 1975, 4705.
134. K. Pettersson, *Acta Chem. Scand.*, 1950, **4,** 395.
135. S. J. Angyal, D. R. Penman, and G. P. Warwick, *J. Chem. Soc.*, 1953, 1742.
136. Ref. 3, p. 431.
137. R. Gaertner, *J. Amer. Chem. Soc.*, 1951, **73,** 3934.
138. J. Feijen and H. Wynberg, *Rec. Trav. chim.*, 1970, **89,** 639.
139. M. Peyrot, D. Villessot, and Y. Lepage, *Compt. rend.* (*C*), 1976, **282,** 607.
140. Ya. L. Gol'dfarb and M. S. Kondakova, *Bull. Acad. Sci. U.S.S.R.*, 1956, 1235.
141. D. J. Zwanenburg and H. Wynberg, *J. Org. Chem.*, 1969, **34,** 340.
142. R. Helmers, *J. prakt. Chem.*, 1972, **314,** 334.
143. M. Neeman, E. Krakauer, and Y. Shorr, *J. Amer. Chem. Soc.*, 1957, **79,** 4380.
144. H. E. Winberg, F. S. Fawcett, W. E. Mochel, and C. W. Theobald, *J. Amer. Chem. Soc.*, 1960, **82,** 1428.
145. B. E. Ayres, S. W. Longworth, and J. F. W. McOmie, *Tetrahedron*, 1975, **31,** 1755.
146. R. Helmers, *Annalen*, 1973, 890.
147. C. Hoogzand, J. Nielsen, and E. H. Braye, *Chem. Comm.*, 1971, 1520.
148. L. Friedman, D. L. Fishel, and H. Shechter, *J. Org. Chem.*, 1965, **30,** 1453.
149. P. A. Konstantinov, T. V. Schredrinskaya, I. V. Zakharov, and M. N. Volkov, *J. Org. Chem.* (*U.S.S.R.*), 1972, **8,** 2639.
150. See *e.g.* H. D. Hartough and L. G. Conley, *J. Amer. Chem. Soc.*, 1947, **69,** 3096.
151. D. E. Wolf and K. Folkers, *Org. Reactions*, 1951, **6,** 410.
152. P. M. Chakrabarti, N. P. Chapman, and K. Clarke, *Tetrahedron*, 1969, **25,** 2781.
153. D. L. Eck and G. W. Stacy, *J. Heterocyclic Chem.*, 1969, **6,** 147.
154. E. Winterfeldt and H. Dillinger, *Chem. Ber.*, 1966, **99,** 1558.
155. H. Boelens and L. Brandsma, *Rec. Trav. chim.*, 1972, **91,** 141.
156. J. Meyer, P. Vermeer, H. J. T. Bos, and L. Brandsma, *Rec. Trav. chim.*, 1974, **93,** 26 and previous related papers cited therein.
157. I. Degani, R. Fochi, and C. Vincenzi, *Gazzetta*, 1967, **97,** 397.
158. W. Treibs, *Annalen*, 1960, **630,** 120.
159. W. Reid and W. Ochs, *Chem. Ber.*, 1972, **105,** 1093.
160. K.-P. Zeller, H. Meier, and E. Müller, *Annalen*, 1972, **766,** 32.
161. Ref. 1, p. 31, and Ref. 7, p. 88.
162. H. Hartmann, *Z. Chem.*, 1971, **11,** 421; H. Hartmann, H. Schäfer, and K. Gewald, *J. prakt. Chem.*, 1973, **315,** 497; K. Hirai and T. Ishiba, *Chem. and Pharm. Bull.* (*Japan*), 1972, **20,** 2384.
163. N. K. Chakrabarty and S. K. Mitra, *J. Chem. Soc.*, 1940, 1385.
164. K. P. C. Vollhardt and R. G. Bergman, *J. Amer. Chem. Soc.*, 1973, **95,** 7538.
165. See Ref. 151, Ref. 4 and R. Phillips, *Org. Synth.*, *Coll. Vol. II*, 578.
166. P. J. Garratt and S. B. Neoh, *J. Org. Chem.*, 1975, **40,** 970.
167. H. Nozaki, T. Koyama, and T. Mori, *Tetrahedron*, 1969, **25,** 5357; S. Bradamante, R. Fusco, A. Marchesini, and G. Pagani, *Tetrahedron Letters*, 1970, 11.
168. M. P. Cava and M. V. Lakshmikantham, *Accounts Chem. Res.*, 1975, **8,** 139.
169. See Ref. 4, p. 55.
170. N. R. Clark and W. E. Webster, *Brit. Pat.* 1 345 203/1974.
171. See Ref. 1, p. 25.
172. E. J. Geering, *J. Org. Chem.*, 1959, **24,** 1128.
173. See Ref. 7, p. 83, and F. Bohlmann, T. Burkhardt, and C. Zdero, 'Naturally Occurring Acetylenes', Academic Press, London and New York, 1973.

175. S. Nakagawa, J. Okurama, F. Sakai, H. Hoshi, and T. Naito, *Tetrahedron Letters*, 1970, 3719.
176. F. Bottino and G. Purello, *Gazzetta*, 1965, **95,** 1062.
177. See Ref. 4, p. 405.
178. B. Tilak and S. S. Gupte, *Indian J. Chem.*, 1969, **7,** 9.
179. See Ref. 1, p. 29.
180. L. Henriksen and H. Autrup, *Acta Chem. Scand.*, 1970, **24,** 2629.
181. R. P. Napier and C. Chu, *Internat. J. Sulfur Chem.* (*A*), 1971, **1,** 62.
182. J. M. McIntosh and F. P. Seguin, *Canad. J. Chem.*, 1975, **53,** 3526.
183. R. Gompper, E. Kutter, and W. Toepfl, *Annalen*, 1962, **659,** 90.
184. See Ref. 2 (Refs. 1, 6 and 7) and J.-C. Meslin, *Compt. rend.* (*C*), 1973, **277,** 1391.
185. See Ref. 4, pp. 291, 401 and Ref. 151, p. 435 and C. J. Chadwick, J. Chambers, G. D. Meakins, and R. L. Snowden, *J.C.S. Perkin I*, 1972, 2079.
186. H. Wynberg and H. J. Kooreman, *J. Amer. Chem. Soc.*, 1965, **87,** 1739.
187. P. J. Garratt and D. N. Nicolaides, *J. Org. Chem.*, 1974, **39,** 2222; see also Ref. 145 and Refs. cited therein.
188. See Ref. 4, p. 49.
189. F. Asinger, M. Thiel, P. Puechel, F. Haaf, and W. Schaeffer, *Annalen*, 1962, **660,** 85.
190. A. Bantjes, *Canad. Pat.* 672 716/1963 (*Chem. Abs.*, 1963, **60,** 7999).
191. H. Hopff and J. von der Crone, *Chimia* (*Switz.*), 1959, **13,** 107.
192. L. Fortina, *Ann. Chim.* (*Italy*), 1959, **49,** 2047; 1960, **50,** 445.
193. K. Gewald, *Chem. Ber.*, 1965, **98,** 3571; K. Gewald, E. Schinke, and H. Boettcher, *ibid.*, 1966, **99,** 94; also see Ref. 3, p. 401.
194. See Ref. 4, p. 473.
195. G. P. Jansen, *Brit. Pat.* 1 359 992/1974.

19.2
Other Sulphur Systems

N. D. A. WALSHE
University of Salford

19.2.1 THIIRENS AND RELATED COMPOUNDS

19.2.1 Thiiren

Thiiren (**1**) belongs to the class of 4π-electron anti-aromatic heterocycles, isoelectronic with cyclobutadiene. Unsurprisingly, therefore, thiirens have only been observed as transient intermediates, *e.g.* by addition of singlet sulphur (from flash photolysis of carbonyl sulphide) to a triple bond.[1] The C—S bond of thiiren (**1**), from flash photolysis of 1,2,3-thiadiazole, adds to a triple bond to give thiophens.[2] Attempts to isolate thiiren as a metal complex led only to complexes of the isomeric α-thiocarbonyl carbene (**2**).[3] However, a benzothiiren is proposed as an intermediate in the formation of a dithiin from *o*-bromobenzenethiolates.[4]

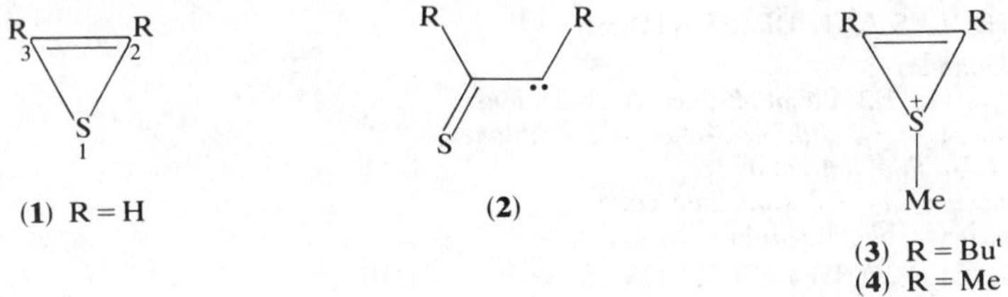

(**1**) R = H

(**2**)

(**3**) R = Bu[t]
(**4**) R = Me

19.2.1.2 Thiirenium salts

These derivatives, and thiiren 1-oxides and 1,1-dioxides, no longer have a free lone pair on sulphur for effective conjugation, and thus resemble 2π-systems (cyclopropenes) in their enhanced stability over the parent thiirens. *S*-Alkylthiirenium salts have been held responsible for the ease of solvolysis of certain vinylsulphonates.[4] Their control of the *trans*-stereochemistry of the addition of sulphenyl halides to acetylenes has been proved by the isolation of (**3**) from methanesulphenyl chloride and di-t-butylacetylene.[5] This is stable to its melting point; in contrast, the dimethyl analogue (**4**) decomposes rapidly above −40 °C.

19.2.1.3 Thiiren 1-oxides

The 2,3-diphenyl derivative (**5**) is obtained from α,α'-dibromodibenzyl sulphoxide (**6**) and triethylamine.[6] It is a potent dipolarophile, giving 3,4,5-triphenylpyrazole with phenyldiazomethane (Scheme 1). Nucleophiles (Grignard reagents, hydroxylamine, *etc*) give ring-opened, often sulphur-free, products, *e.g.* (**7**)–(**10**). Aerial oxidation would appear responsible for the formation of dioxime (**8**), and also for the formation of benzil on thermolysis of (**5**) in air; photolysis of (**5**) gives tolan.

Ph, Ph, Ph, SPh (9); Ph—C≡C—Ph; Br, Br, Ph, S, O, Ph (6); $(PhCH_2C{=}N{-})_2$, Ph (10); NOH, Ph, Ph + dioxime (7) (8); Ph, Ph, S, O (5); Ph, Ph, Ph, N, N, H

i, Et_3N; ii, $PhCHN_2$; iii, $h\nu$; iv, N_2H_4; v, PhMgBr; vi, NH_2OH.

SCHEME 1

19.2.1.4 Thiiren 1,1-dioxides

Ab initio CNDO/2 calculations predict[7] a decrease in anti-aromaticity on passing from thiiren to the 1,1-dioxide, attributed to a stabilizing π(carbon)–π(oxygen) interaction, which makes the 1,1-dioxides the most stable members of this elusive thiiren family.

Synthesis of thiiren dioxides hinges on the intermediate 2-bromothiiran (**11**), made either (a) by base-catalysed ring-closure[8] of dibromosulphones (**12**) and (**13**), or (b) by carbene addition to an α-bromosulphene;[9] dehydrobromination of (**11**) then gives (**14**) (Scheme 2).

$R^1CHBrSO_2CHBrR^2$ (**12**); $R^1CBr_2SO_2CH_2R^2$ (**13**) → R^1, R^2, Br, SO_2 (**11**) → R^1, R^2, SO_2 (**14**)

(**15**) $R^1 = R^2 = Ph$

SCHEME 2

The reactions of 2,3-diphenylthiiren 1,1-dioxide (**15**) mostly depend on the intermediacy of the highly stabilized sulphonyl anion (**16**), formed by addition of relatively soft nucleophiles to the α,β-unsaturated sulphone moiety. On quenching (**16**) with a hard electrophile (a proton), the resulting thiiran 1,1-dioxide extrudes sulphur dioxide. Thus, for example, the nitrogen nucleophiles hydroxylamine and dimethylamine attack at carbon to yield the oxime (**7**) and enamine (**17**; $R^1 = Me_2N$, $R^2 = H$), respectively.[8,10] Addition of phenyldiazomethane gives 3,4,5-triphenylpyrazole.[8] The adducts from enamines and ynamines,[4] which presumably are formed by a polar process, do not lose sulphur dioxide, but give the divinyl sulphone (**18**) and the thiophen 1,1-dioxide (**19**), respectively. Pyridinium ylides[10] give adducts, *e.g.* (**20**), probably *via* intramolecular quenching of a dipolar intermediate (Scheme 3).

The ambident nature of the ylide (**16**; X = CN) is shown by quenching with the soft electrophile iodomethane, which gives the sulphur-alkylated product (**17**; $R^1 = CN$, $R^2 = SO_2Me$).

Hard nucleophiles (*e.g.* OH^-, MeO^-) attack the sulphur atom of (**15**) to give the sulphonic acid derivatives (**17**; $R^1 = SO_3H$ or SO_3Me, $R^2 = H$). Kinetic analysis[12] of this unexpectedly fast reaction is offered as evidence against *d*-orbital stabilization of (**15**). Less conventional than the above nucleophilic reactions is the formation, *via* (**16**), of the vinylogous ylides (**21**; X = N or P) from (**15**) and tertiary nitrogen or phosphorus nucleophiles.[11] The phosphorus compound is mesomeric with the interesting conjugated phosphorane-sulphene (**22**). Thermolysis of (**15**) expels sulphur dioxide to give tolan.

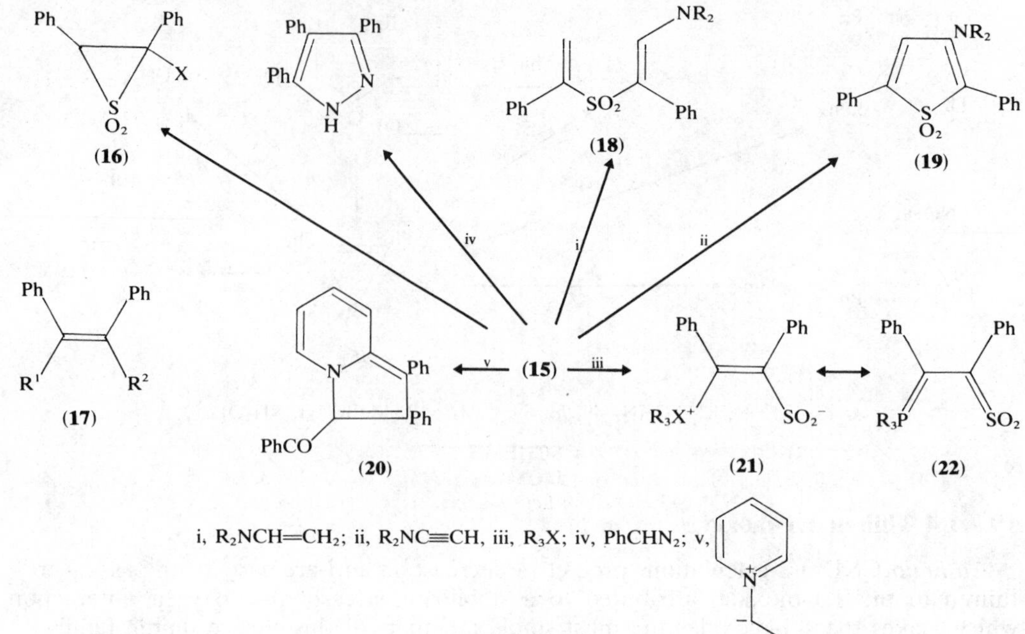

(16) (17) (18) (19) (20) (21) (22)

i, $R_2NCH{=}CH_2$; ii, $R_2NC{\equiv}CH$, iii, R_3X; iv, $PhCHN_2$; v, (N-phenacylpyridinium)

SCHEME 3

19.2.2 THIETS AND RELATED COMPOUNDS

19.2.2.1 Thiet 1,1-dioxides

These sulphones (**23**) are isoelectronic with cyclobutene, while thiets (**24**) are more akin to cyclopentadiene, and are thermodynamically less stable than their oxides. Most

(**23**) (**24**)

syntheses of thiet 1,1-dioxides (**25**) involve a regiospecific 'head-to-tail' cycloaddition of a sulphene with an electron-rich olefin, then modification of the thietan (**26**) thus formed. If acetylenes are used, thiet dioxides are obtained directly (Table 1 and Scheme 4).

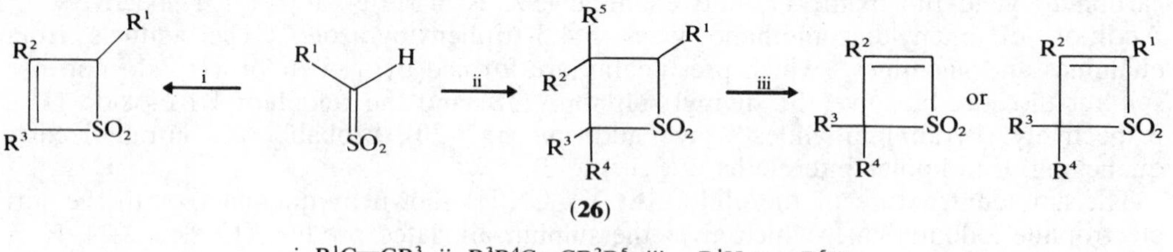

(**26**)

i, $R^1C{\equiv}CR^3$; ii, $R^3R^4C{=}CR^2R^5$; iii, $-R^1H$ or $-R^5H$.

SCHEME 4

Other routes include the first synthesis of the parent compound (**23**) from epichlorohydrin.[13] A 3-bromo-derivative has been made by free radical bromination/dehydrobromination of thietan 1,1-dioxide.

Benzo[*b*]thiet 1,1-dioxide (**28**) has been obtained from (**27**), the key step being a cycloaddition with butadiene.

TABLE 1

Synthesis of Thiet 1,1-Dioxides (**25**)

R^1	R^2	R^3	R^4	R^5	*Method of elimination*	*Ref.*
H	Me_2N	alkyl	alkyl	EtO	*E*2 elimination of EtOH	a
H	C_4H_8ON	H	H	C_4H_8ON	Cope	a
H	H	Me	Me	Me_2N	Hofmann	b
Ph	H	Me	Me	C_4H_8ON	Cope	c
H	H	Me	H	Ph(Me)CHNMe	Hofmann; optically active product	d
Ph	H	Ph	H	Et_2N	Hofmann	e
H	—CH=CH—$(CH_2)_4$—		H	Me_2N	Hofmann	f
H	—$(CH_2)_5$—		H	Me_2N	Hofmann	g
Br	C_4H_8OH	alkyl	alkyl	H	*E*2 elimination of HBr	h
Ph	NEt_2	Me	—	—	Made from ynamine	i
Ph	NEt_2	CN	—	—		

[a] R. H. Hasek, P. G. Gott, R. H. Mee, and J. C. Martin, *J. Org. Chem.*, 1963, **28,** 2496. [b] W. Truce, J. Norell, J. Richman, and J. Walsh, *Tetrahedron Letters*, 1963, 1677. [c] V. Drozd and V. Sergeichuk, *J. Org. Chem.* (*U.S.S.R.*), 1975, **11,** 1301. [d] L. Paquette, J. Freeman, and P. Houser, *J. Org. Chem.*, 1969, **34,** 2901. [e] D. Eckroth and G. Love, *J. Org. Chem.*, 1969, **34,** 1136. [f] L. Paquette and R. Begland, *J. Org. Chem.*, 1969, **34,** 2896. [g] D. C. Dittmer and F. Davis, *J. Org. Chem.*, 1966. **29,** 3131. [h] P. Del Buttero, S. Maiorano, and M. Trautluft, *J. C. S. Perkin I*, 1974, 1411. [i] M. Kuchni and H. Linde, *J. Org. Chem.*, 1972, **37,** 1846.

The chemistry of thiet 1,1-dioxides centres on the array of cycloadditions undergone by the strained double bond (Scheme 5).[13b,14] Though the double bond is sufficiently electron poor to give a Michael adduct with cyanide anion and nitromethane,[15] it reacts with tetraphenylcyclopentadienone[14b] to give a [4+2] adduct (**29**), as it also does with a typical

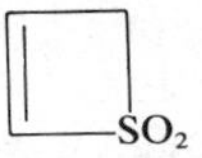

SO_2 SO_2

(**27**) 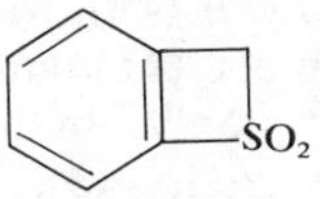 (**28**)

dienamine to give (**30**).[14a] It is also a powerful, if non-regioselective, dipolarophile, forming (**31**) and its regioisomer with diphenyldiazomethane.[14c] Woodward–Hoffmann rules notwithstanding, it also forms [2+2] adducts with enamines to give (**32**) and ynamines to give (**33**), from which the ketone (**34**) is isolated.[14a]

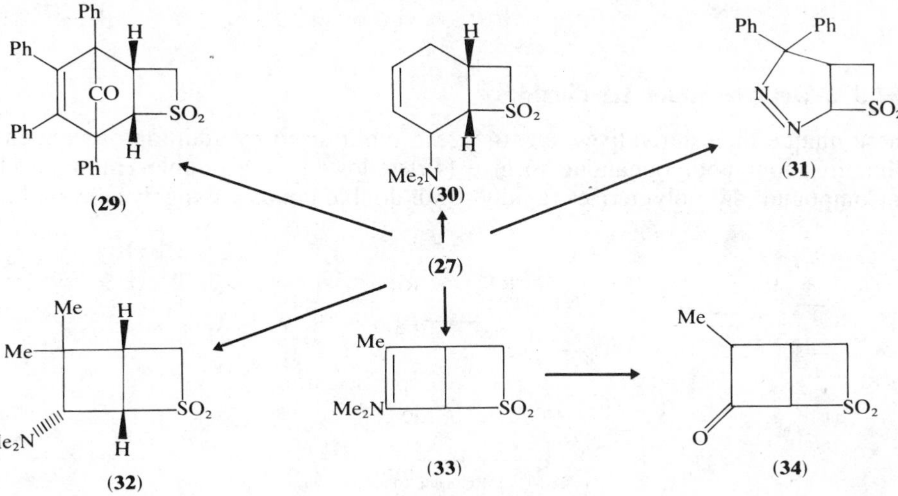

Ph H Ph CO SO_2 Ph H Ph (**29**)

H SO_2 H Me_2N (**30**)

Ph Ph N N SO_2 (**31**)

(**27**)

Me H Me SO_2 Me_2N H (**32**)

Me Me_2N SO_2 (**33**)

Me SO_2 O (**34**)

SCHEME 5

The second class of reactions involving thiet dioxides concerns the (possibly concerted) ring cleavage[15] to a vinylsulphene (**35**), which can be trapped (a) intermolecularly with phenol to yield the allyl sulphonate (**37**) or (b) intramolecularly as the oxathiolene 1-oxide (**36**). The benzothiet dioxide (**29**) similarly gives a fused analogue of (**36**). Oddly, a vinylsulphene is also thought to intervene in the photochemical conversion[16] of thiet 1,1-dioxides to give, in contrast to the thermal process, an α,β-unsaturated ketone and a cyclohexadiene (**38**) (Scheme 6).

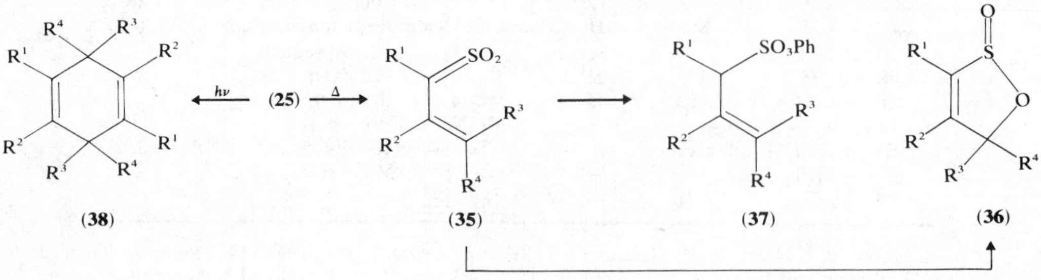

(**38**) (**35**) (**37**) (**36**)

SCHEME 6

19.2.2.2 Thiets

These are usually made[17] by $LiAlH_4$ reduction of the sulphonyl group of the adducts (**26**) to a sulphide, followed by a Hofmann or Cope elimination. Benzo[*b*]thiet has been prepared in 45% yield by thermolysis of benzo[*b*]thiophen 1,1-dioxide.[18]

As with cyclopentadiene, thiet should form a stable, 6π monoanion. However, its preparation is thwarted by the probable equilibrium of the thiet–enethione (**39**) valence tautomers, since even weakly nucleophilic bases cause ring cleavage.[17] Attack at sulphur by n-butyl-lithium yields sulphides (**40**), whilst trityl anion and 2,4-dinitrophenylhydrazine attack at carbon to give thiols (**41**) and hydrazones (**42**), respectively; mixtures of double-bond isomers are commonly obtained.[12]

Benzothiets thermally dimerize to (**43**),[18,19] although it would be rash to assume that an enethione such as (**39**) is the intermediate.

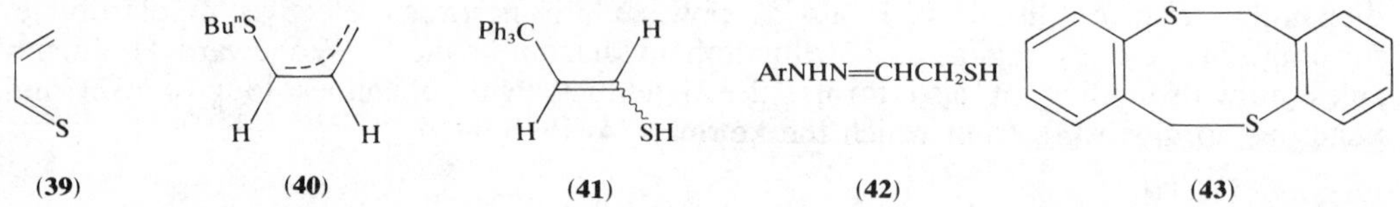

ArNHN=CHCH$_2$SH

(**39**) (**40**) (**41**) (**42**) (**43**)

19.2.2.3 2-Methylenethiet 1,1-dioxides[20]

These unique thiet derivatives, *e.g.* (**44**), are synthesized by addition of sulphenes to a β-(dimethylaminomethyl)enamine to give (**45**) followed by a double Hofmann elimination. Compound (**44**) polymerizes rapidly. Both double bonds are highly activated towards

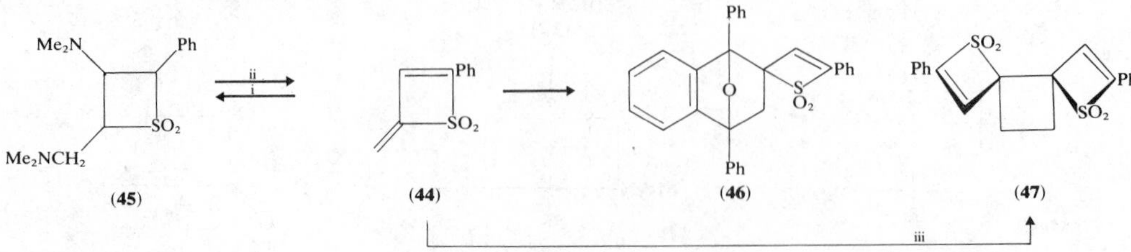

(**45**) (**44**) (**46**) (**47**)

i, Me_2NH; ii, Hoffmann elimination; iii, $h\nu$.

SCHEME 7

Michael addition and precursor (**45**) is regenerated by reaction with two molecules of amine. The exo-double bond is particularly reactive, being selectively hydrogenated. 2,5-Diphenylbenzoisofuran gives the spiro-adduct (**46**), whilst photolysis yields, with impressive regiospecificity, the unusual cyclobutene dimer (**47**) (Scheme 7).

19.2.2.4 Thietium salts

These possess a stability comparable with that of thiet 1,1-dioxides. The cation (**48**) is obtained as a stable, solid intermediate[21] in the rearrangement of (**49**) to (**50**).

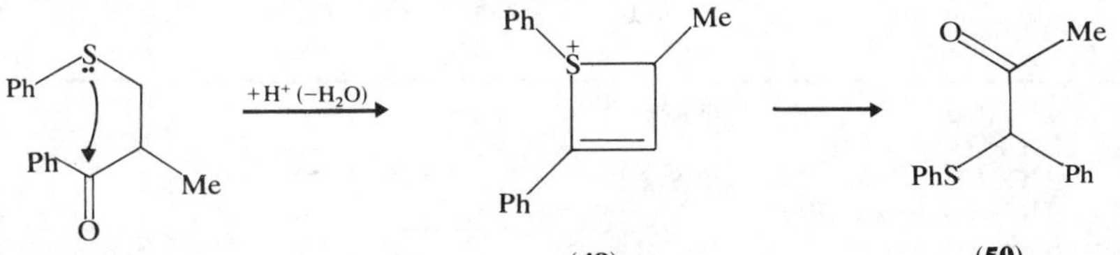

$+H^+ (-H_2O)$

(**49**) (**48**) (**50**)

19.2.2.5 Dithiets

Photochemical extrusion of carbon monoxide from a 1,3-dithiole-2-one gives compound (**51**), indistinguishable from its valence tautomer, the 1,2-dithione (**52**).[22] However, a genuine dithiet (**53**), locked in a rigid steroidal cage, has been obtained from a fused dithian by photochemical expulsion of ethylene.[23]

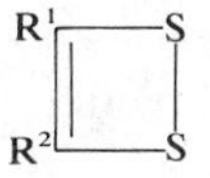

(**51**) 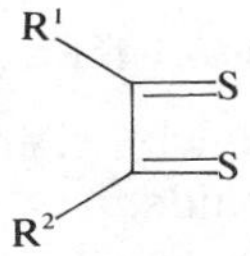 (**52**) 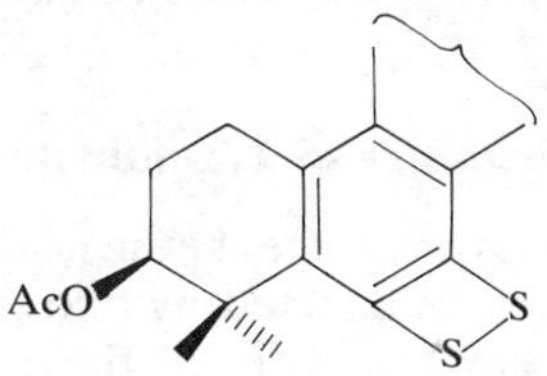 (**53**)

19.2.3 1,2-DITHIOLES AND DERIVATIVES

19.2.3.1 1,2-Dithioles[24]

The parent compound (**54**) is unknown. All its simple derivatives are destabilized by strain in the weak S—S bond, and by the non-aromatic, cycloheptatriene-like, electronic structure; they are discussed in Section 19.2.3.6. However, in contrast, a huge number of stable 1,2-dithiole-3-ones (**55**) and especially -3-thiones (**56**) are known. These, and their 3-dithiolylidene derivatives (**57**) and (**58**), are planar, mesomeric with the aromatic, 6π dipolar forms (**59**)–(**61**). These probably contribute appreciably to the ground-state energy of the system; the thiones (**56**) possess both an unusually long C=S bond, and a high dipole moment.[24]

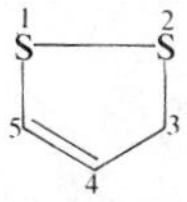

(**54**)

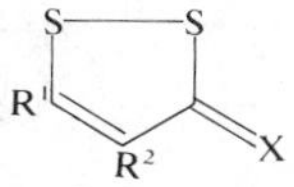

(**55**) X = O
(**56**) X = S
(**57**) $X = CR_2$
(**58**) X = NR

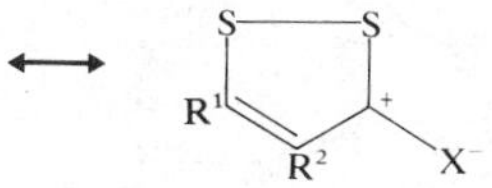

(**59**) 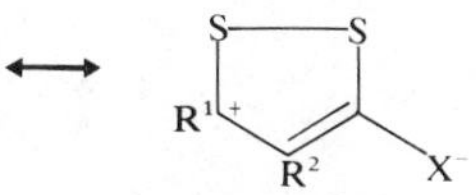 (**60**) 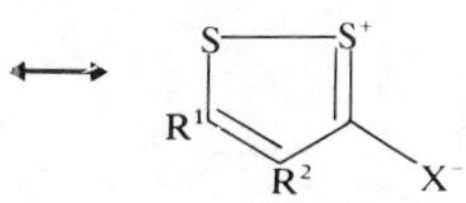 (**61**)

19.2.3.2 Synthesis of 1,2-dithiole-3-ones and -3-thiones

Many methods are based on *ad hoc* procedures. Heating various three-carbon units with sulphur or phosphorus sulphide ('sulphuration') is a popular, often high-yield, reaction (Table 2). The use of 1,1-dithiolates from carbon disulphide and enolate anions is a recent specific method.

N.B. Misleading trivial names, whose use is to be disparaged, include 'Baumann–Fromm disulphides' for (**55**), and 'trithiones' for (**56**).

TABLE 2

Syntheses of 1,2-Dithiole-3-ones and -3-thiones

Method	*Starting material*	*Sulphur source*	*Product* (**55**) *or* (**56**) R^1	R^2	X	*Yields* (%)	*Ref.*
(1)	R^1CHMe_2	S_8/base	Aryl	H	S	40–80*	a
(2)	$R^1CH_2(R^2)C{=}CH_2$	S_8	R	Aryl	S	25–60*	b
(3)	$R^1COCH(R^2)Me$	S_8/P_4S_{10}	Aryl	R	S	20–35	c
(4)	$R^1CH{=}C(R^2)CO_2Et$	S	R	R	O	40–65	d
(5)	β-ketoesters	P_4S_{10}/S	R	R	S	40–50	e
(6)	$R^1COCH_2R^2$	CS_2/base; P_4S_{10}	R	R	S	50–90	f

* Yield based on sulphur; excess of hydrocarbon used.

[a] R. Wegler, E. Kuhle, and W. Schafer, in 'Newer Methods of Preparative Organic Chemistry', ed. W. Forest, Vol. III, Academic Press, New York, 1964. [b] J. P. Brown, *J. Chem. Soc.* (*C*), 1968, 1077. [c] Ref. 24, p. 369. [d] M. Cadec, C. Trebaul, and J. Teste, *Bull. Soc. chim. France*, 1968, 2964. [e] C. Trebaul and J. Teste, *Bull. Soc. chim. France*, 1969, 2456. [f] Ref. 24, p. 384; L. Henriksen, *Chem. Comm.*, 1969, 1408; R. Coutourier, D. Paquer, and A. Thullier, *Compt. rend.* (*C*), 1970, **70**, 1878; E. Takesima, M. Yokoyama, N. Fukada, and M. Akano, *J. Org. Chem.*, 1970, **35**, 2438.

19.2.3.3 Chemistry of 1,2-dithiole-3-ones and -3-thiones

The thiones give the ketones on reaction with mercuric acetate, while the reverse transformation is effected by phosphorus sulphide.[25]

The complex chemistry of these compounds acting as electrophiles is governed by (a) the hard–soft character of both the nucleophile and the position of attack on the dithiole ring; and (b) the thermodynamic stabilities of the products. The electron-deficient sites of (**56**), in decreasing order of hardness, are positions 3, 5, 2, 1, giving us the possibilities of attack (i) at hard carbon-3 by a hard nucleophile, to give ring-cleavage and loss of sulphur; (ii) at soft sulphur-2 by a soft nucleophile, causing ring-opening; (iii) at carbon-5, usually by a soft nucleophile, leading to (α) ring-opening or (β) ejection of a leaving

SCHEME 8

TABLE 3
1,2-Dithiole-3-ylidene Compounds as Electrophiles

Type	*Dithiole* (**55**)–(**58**) R^1	R^2	X	*Conditions*	*Product*
(i)	Aryl	Cl	O	MeO^-, MeI	$Ar(SMe)C{=}C(Cl)CO_2Me$[a]
(i)	Et_2N	Cl	O	MeO^-, H^+	Tetrasubstituted 1,2-dithiin
(ii)	Et_2N	Aryl	O	PhMgBr	$Et_2N(SH)C{=}C(Ar)COSPh$[c]
(ii)	Aryl	H	NAr	RMgX	$ArCSCH{=}C(SR)Ar$[d]
(ii)	Cl	Aryl	O	RMgX	A dithietan[e] ($R^2 = Ph$)
		Cl			A 1,3-dithiole[e] ($R^2 = Cl$)
(iii)	Cl	R	O, S, NR	RS^-, R_2NH	(**55**, **56**, or **58**; $R^1 = RS$, R_2NH)[f]
(iii)	Ph	H	S	Acenaphthenone	Acenaphthylene-fused 2*H*-thiin-2-thione[g]
(iv)	Aryl	Aryl	S	$Y{-}\bar{C}H{-}Z$	(**57**; X = C(YZ); R^1, R^2 = aryl)[h]
(iv)	Aryl	H	S	NH_2OH	(**58**; X = NOH, R^1 = aryl)[i]

[a] F. Boberg, *Annalen*, 1967, **683,** 132. [b] F. Boberg, H. Niemann, and K. Kirchoff, *Annalen*, 1969, **728,** 32. [c] F. Boberg and R. Schardt, *Annalen*, 1969, **728,** 44; 1970, **734,** 173. [d] F. Boberg and W. Gentzkow, *Annalen*, 1973, 247, 256; D. McKinnon and M. Hassan, *Canad. J. Chem.*, 1973, **51,** 3081. [e] F. Boberg, M. Ghoudikian, and M. Khorgami, *Annalen*, 1974, 1261. [f] J. Bader, *Helv. Chim. Acta*, 1968, **51,** 1409. [g] K. N'Guyen, R. Pinel, and Y. Mollier, *Bull. Soc. chim. France*, 1973, 3334. [h] Ref. 25. [i] A. Grandin and J. Vialle, *Bull. Soc. chim. France*, 1967, 1851.

group, if present, from carbon-5; (iv) at hard carbon-3 by a soft nucleophile, giving stable products. Scheme 8 and Table 3 exemplify these processes.

The nucleophilicity of the exo-X atom of (**56**) and (**57**) is due to the formation of a (formally) aromatic dithiolium cation (**62**), with such electrophiles as a proton,[24] iodomethane (Section 19.2.3.4), carbenes, and nitrenes (**63**). The latter complex decomposes thermally, by sulphur extrusion from a thiaziridine, to the imine (**64**).[26]

(**62**) (**63**) (**64**) (**65**)

Ring substitution by electrophiles (*e.g.* chlorine) probably results from addition/elimination, or from dithiolium intermediates.[27]

As expected, thione (**65**) has an acidic 5-methyl group, the anion of which condenses with benzaldehyde to give a styryl derivative,[25] with carbon disulphide to give a trithiapentalene precursor, and with DMF diethyl acetal to give a 5-dimethylaminovinyl derivative.[28]

The potential use of 1,2-dithiole-3-thiones as 1,3-dipoles is clear from mesomer (**61**), the general reaction (Scheme 9) with acetylenic dipolarophiles[29] giving a 1,3-dithiolylidene derivative (**66**). The imino-derivatives (**58**) in this reaction give thiazoles.[29]

(**66**) (**67**)

SCHEME 9

The α,β-unsaturated thione moiety in (**66**) is a voracious Diels–Alder diene,[30] giving spiro-2:1 adducts of type (**67**) with an excess of dipolarophile. Common dipolarophiles used include electron-deficient acetylenes,[31] keten, and phenylacetylene, which reacts anomalously, in the absence of an acid catalyst, to give a trithiapentalene.[32] Products (**66**) with a thial group are thermally converted to the coupled olefins (**68**) by sulphur extrusion.[31a]

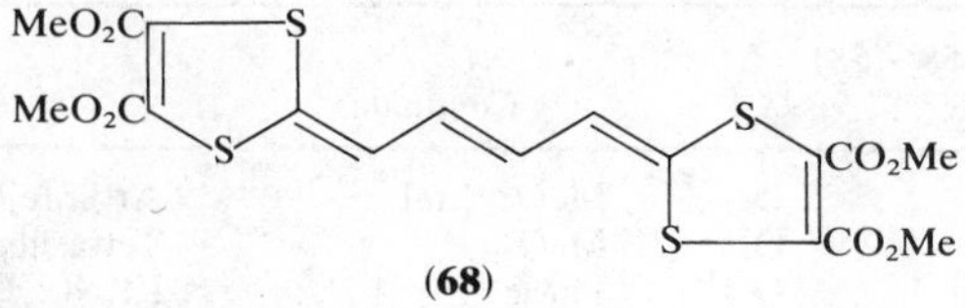

(**68**)

19.2.3.4 The 1,2-dithiolium cation (69)

This aromatic, 6π-electron system, related to 1,2-dithiole as is the tropylium cation to cycloheptatriene, is both planar and symmetrical about an axis from C-4 to the mid-point of the S—S bond, as emphasized by structure (**69**).[33] However, as canonical forms (**71**) and (**72**), and MO calculations both imply, carbon atoms 3 and 5 are more electron-deficient than carbon-4.[34] The terms 'dithiolylium' and 'dithiolium' are both used in the literature; although the former is more systematic, the latter is more euphonious and is more widely used.

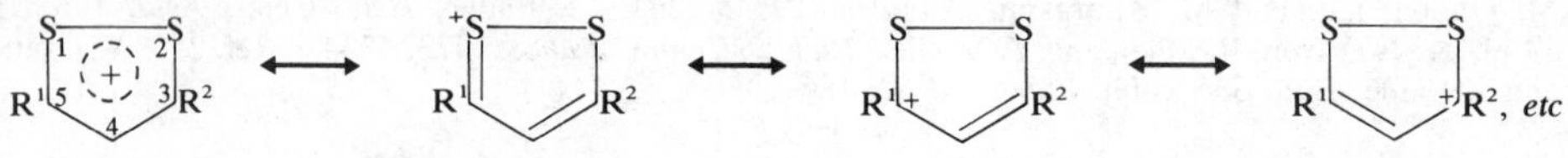

(**69**) ⟷ (**70**) ⟷ (**71**) ⟷ (**72**), *etc*

(**86**) $R^1 = R^2 = MeCONH$

19.2.3.5 Synthesis of the 1,2-dithiolium cation

(i) 1,2-Dithiole derivatives may be transformed, *e.g.* the thiones (**57**) may be oxidized with peracid to give 3-unsubstituted cations;[35] the 3-ylidene derivatives (**56**)–(**58**) may be X-alkylated.[33,34] (ii) Methods using acyclic precursors include the unique formation of 3,4,5-trichloro-1,2-dithiolium chloride from perchloropropane and sulphur in high yield.[36] Cyclization of 1,3-dicarbonyl compounds with hydrogen disulphide (H_2S_2) or diacetyl disulphide is a general method, as is iodine oxidation of 1,3-dithiones, often generated *in situ* from 1,3-diketones and H_2S and acid.[37]

19.2.3.6 Chemistry of the 1,2-dithiolium cation

As expected, the cation behaves as a powerful electrophile, as is foreshadowed by its great sensitivity to aqueous bases. In all cases, nucleophilic attack gives a 3-substituted 1,2-dithiole (**73**) as an intermediate, occasionally isolable[34] and referred to as a leuco-compound. The possible fates of the usually elusive (**73**) are classified as: (a) X = H, nucleophile has no labile proton attached to its attacking centre. The leuco-compound is oxidized by air, or more often, by hydride capture from an unreacted cation to give the more stable, substituted cation (**74**). The dithiole (**54**) formed concomitantly decomposes to unknown products, so yields do not exceed 50%. (b) X = H, nucleophile has a labile proton. Ring opening can now occur, and the intermediate (**75**) may cyclize if it still retains a nucleophilic centre, *e.g.* with ammonia, isothiazoles are obtained,[38] or acyclic products are formed by loss of sulphur. Less commonly, oxidation of (**73**) by route (b′) to a dithiolylidene compound (**76**) occurs. (c) X = leaving group (Cl, OR, SR), nucleophile has no labile proton. X is displaced, and the leuco-compound reverts to cation (**74**). A competing process (c′) lowers yields by ring fission to (**77**). (d) X = leaving group, nucleophile has labile proton. X is displaced and a dithiolylidene derivative (**76**) is formed. Normally, if X is a leaving group, routes (c) and (d) are preferred; however, probably owing to strain relief, the fused derivative (**78**) prefers to ring-open by route (b). Scheme 10 and Table 4 provide examples.

SCHEME 10

As an electrophile, the 1,2-dithiolium cation can accept electrons to form nucleophilic dithiolyl radicals.[39] These may couple to form the bis-dithiole (**79**), whilst those derived from a 3-chloro-cation give the *cis*- and *trans*-isomers of the tetrathiafulvalene (**80**).

Complementing the electrophilic activity in the 3-position is the strong acidity of methyl(ene) groups at this site.[34] The dithiafulvene (**81**), resulting from proton abstraction from (**82**), is nucleophilic, like ketones (**55**) and thiones (**56**). Its basicity and nucleophilicity are (a) very great when R^1, R^2 = alkyl, *etc.*, owing to the driving force provided by aromatization; and (b) small if R^1, R^2 are $-M$ groups, which counteract this driving force. Thus, aromaticity is achieved by attack on unreacted cation, or by adding an electrophile,

TABLE 4
1,2-Dithiolium Cations as Electrophiles

Type	*Cation* (**69**) R^1	R^2	R^3	*Nucleophile*	*Products*	*Refs.*
(a)	Aryl	H	H	$C_6H_5NMe_2$	(**69**; R^1 = Aryl, R^2 = 4-$Me_2NC_6H_4$)	a
(a)	R	R	R	MeO^-	3,3,4,5-Tetrasubstituted 1,2-dithioles	a
(b)	R	R	H	NH_3	Isothiazoles	b
(b)	SBr	R	H	RNH_2	Isothiazole-5-thiones	c
(b)	R	R	R	RNHNHR	Pyrazoles or pyrazolium cations	d
(b)	H	Ar^1	H	Ar^2NH_2	$Ar^1\overset{+}{N}H{=}CHCH{=}NAr^2$	e
(b)	Ar	H	Ar	RNH_2	Ar(RNH)C═CHCSAr	b,f
(b)	NR_2	H	R	$\bar{C}H_2NO_2$	A 2-nitro-3-dialkylaminothiophen	g
(b′)	H	R	R	Y—C̄H—Z	(**58**)	a
(c)	Cl	R	R	AcSH R_2NH	(**69**; R^1 = AcS, R_2N)	h
(d)	Cl	Cl	Cl	H_2O	(**59**; R^1 = R^2 = Cl)	i
(d)	R	R	SMe	Y—C̄H—Z RNH_2	(**57**) and (**58**) respectively	a,e

[a] Refs. 33, 34. [b] Ref. 38; S. Coen, J.-C. Poite, and J.-P. Roggero, *Bull. Soc. chim. France*, 1975, 611. [c] F. Boberg and W. Gentzkow, *J. prakt. Chem.*, 1973, **315,** 965, 970. [d] E. Klingsberg, *J. Org. Chem.*, 1963, **28,** 529. [e] C. Paulmier, Y. Mollier, and N. Lozac'h, *Bull. Soc. chim. France*, 1965, 2463. [f] D. Leaver, D. McKinnon, and W. Robertson, *J. Chem. Soc.*, 1965, 32; G. Le Coustumer and Y. Mollier, *Bull. Soc. chim. France*, 1973, 3349. [g] B. Bartho, J. Faust, R. Pohl, and R. Mayer, *J. prakt. Chem.*, 1976, **318,** 221. [h] F. Boberg, M. Koepke, and G.-J. Wentrup, *Synthesis*, 1975, 525. [i] Ref. 36.

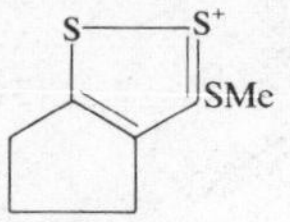

(78) 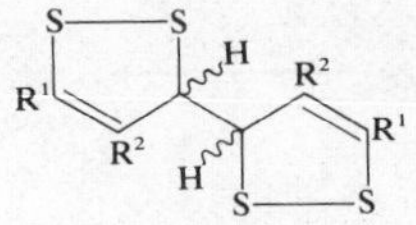 (79) 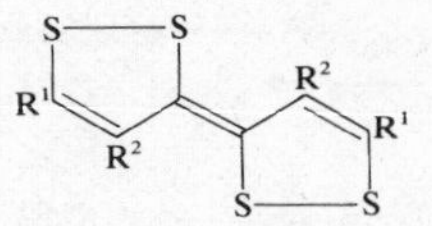 (80) 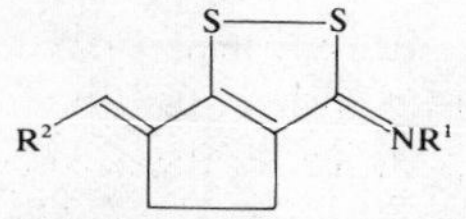 (83)

e.g. an aldehyde, giving typical Claisen–Schmidt condensation products.[33,34] The formation of cyanine dyes (**83**) from (**78**) is a very versatile process of great technical importance.[34]

Similarly, the 3-hydroxy- and 3-alkoxy-1,2-dithiolium cations, and their sulphur analogues (**62**), readily regenerate the parent 3-ylidene derivatives (**55**) and (**56**) by deprotonation or dealkylation with weak bases. Such salts are often efficient alkylating agents.[40] The arylogue (**84**) of 3-hydroxy-1,2-dithiolium salts is readily deprotonated[41] to give the quinoid species (**85**). In the nitrogen series, the 3,5-diacetylamino-1,2-dithiolium cation (**86**) possesses very acidic NH protons, removal of one of which gives a very stable imino-base.[42]

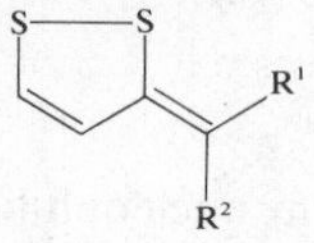

(81) 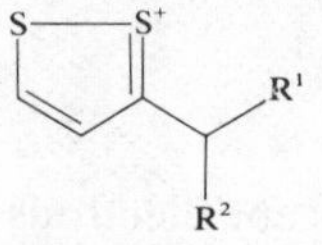 (82) 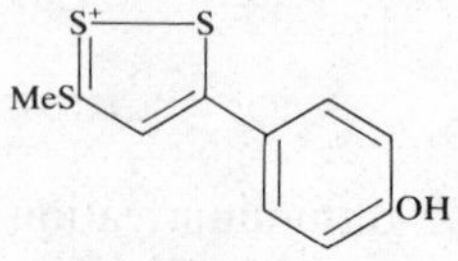 (84) 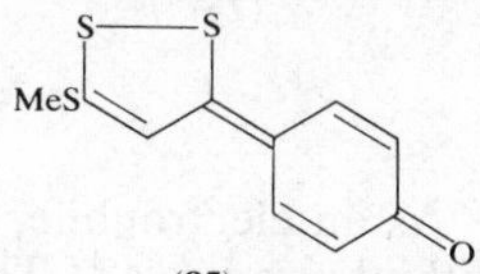 (85)

19.2.4 1,3-DITHIOLES AND DERIVATIVES

19.2.4.1 1,3-Dithioles (87)[43]

Unlike 1,2-dithioles, these do not contain a weak S—S bond and are much more stable. Syntheses of (**87**) and the 2-ylidene compounds (**88**) usually utilize a C—C—S, S—C—S, or S—C—C—S synthon or a 1,3-ditholium cation and a nucleophile. Such methods are classified in Table 5.

The known chemistry of 1,3-dithioles is unremarkable, and is briefly summarized below. The older literature has been ably reviewed.[43]

TABLE 5
Preparations of 1,3-Dithioles and Derivatives

Type	*Synthon*	*Other reagents*	*Product*	R^1	R^2	R^3	R^4	X	*Ref.*
C—C—S	PhC≡CS⁻	PhNCS	(**88**)	Ph	H	—	—	PhN, S	a
S—C—S	$Me_2NCS_2^-$	$ClCH_2CHO$	(**88**)	H	H	—	—	S	b
	$Pr^iOCS_2^-$	α-Bromoketone	(**88**)	R	R	—	—	O	c
	$MeOCS_2^-$	RCS_2Me/CH_2N_2	(**88**)	R	H	—	—	S	d
	$Bu_3\overset{+}{P}CS_2^-$	RC≡CR/ArCHO	(**88**)	R	R	—	—	CHAr	e
	$(NaS)_2C{=}C(CN)_2$	Benzoquinone	(**88**)	Benzo		—	—	$C(CN)_2$	f
	1,3-Dithiolan-2-thione	$EtO_2CC{\equiv}CCO_2Et$	(**88**)	H	H	CO_2Et	CO_2Et	S	g
	$R_2C(SH)_2$	α-Bromoketone	(**87**)	R	R	R	R	—	h
S—C—C—S	NaS(CN)C=C(CN)SNa	$CXCl_2$	(**88**)	CN	CN	—	—	O, S	c,i
	ArCSCSAr	CH_2N_2	(**87**)	Ar	Ar	H	H	—	j
	1,2-Dimercaptobenzene	ArCHO	(**87**)	Benzo		Ar	H	—	c

[a] H. Spies, K. Gewald, and R. Mayer, *J. prakt. Chem.*, 1971, **313,** 804. [b] Y. Ueno, A. Nakayama, and M. Okawara, *Synthesis,* 1975, 277. [c] A. Bhattacharva and A. Hortmann, *J. Org. Chem.*, 1974, **39,** 95. [d] J. M. Beiner, D. Lecadet, D. Paquer, A. Thuillier, and J. Vialle, *Bull. Soc. chim. France,* 1973, 1979. [e] H. Hartzler, *J. Amer. Chem. Soc.*, 1971, **93,** 4961. [f] K. Klemm and B. Geiger, *Annalen,* 1969, **726,** 103. [g] C. Pittman, Jr., M. Narita, and Y. Liang, *J. Org. Chem.*, 1976, **41,** 2855. [h] E. Campaigne and F. Haaf, *J. Heterocyclic Chem.*, 1964, **1,** 163. [i] Ref. 43. [j] P. de Mayo and W. Kusters, *J. Amer. Chem. Soc.*, 1974, **96,** 3502.

Simple 1,3-dithioles, *e.g.* (**89**), can be oxidized to stable 1,1,3-trioxides and 1,1,3,3-tetroxides, unlike 1,2-dithioles. The 2-thiones readily undergo nucleophilic attack at position-4 (hydroxylamine gives glyoxime), or position-2 (phenylhydrazine gives a hydrazone).[44] The exo-sulphur atom is readily methylated, giving a 1,3-dithiolium cation (Section 19.2.4.3).

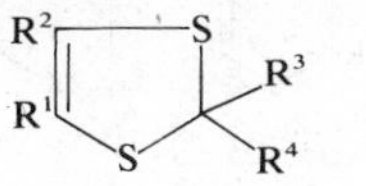

(**87**)
(**89**) R^3,R^4 = $PhCH_2$

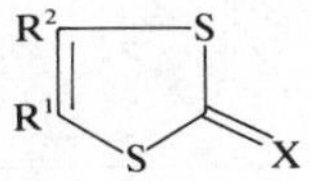

(**88**) 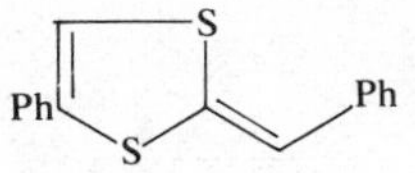 (**90**) 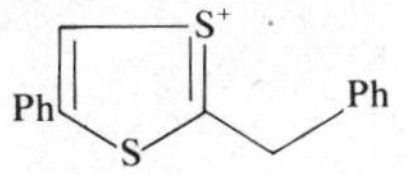 (**91**)

The benzylidene compound (**90**) is protonated reversibly to give the cation (**91**). It is surprisingly easily reduced to the 1,3-dithiole (**92**) by zinc and acetic acid. Oxidation gives the tetroxide (**93**) which undergoes a double Michael addition, followed by ring cleavage, with methoxide ion,[45] to give (**94**).

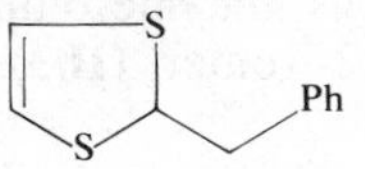

(**92**) 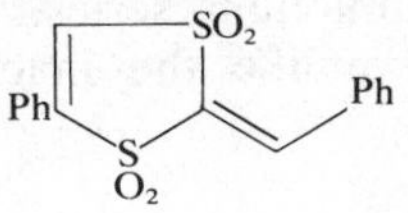 (**93**) 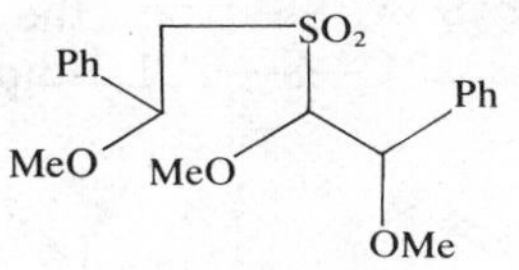 (**94**) 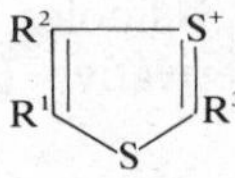 (**95**)

19.2.4.2 The 1,3-dithiolium cation (95)

Synthetic methods discussed under 1,2-dithiolium salts are often applicable here. For instance, 1,3-dithiole-2-thiones may be oxidized with peracid,[35] or *S*-alkylated;[35] 1,3-dithioles may undergo hydride abstraction or oxidation;[34] or dithioesters RCS_2CH_2X may be cyclized to 4-hydroxy-cations ($X = CO_2H$) or 4-amino-cations (X = CN).[46]

19.2.4.3 Chemistry of the 1,3-dithiolium cation

Although, in principle, the cation appears to be a 6π-electron aromatic system, it seems that a three-centre, 4π-system is spread over the S—C—S bonds, position-2 being most electropositive, effectively insulated from the Δ^4-double bond. Physical properties relevant to the problem of bonding in the cation are discussed in two reviews.[33,34]

Thus, even more than the 1,2-analogue, the 1,3-dithiolium cation is a potent electrophile. Nucleophilic attack invariably occurs at position-2, and, lacking a weak S—S bond, the leuco-1,3-dithiole (**96**) cannot easily ring open and is often isolable.[47] As Scheme 11 indicates, leaving groups at position-2 are then displaced, regenerating cation (**97**) by route (a); or else the 2-ylidene derivative (**98**) is found by route (a′). A proton at position-2 may be removed by oxidation or hydride transfer to unreacted cation to generate the same products as by routes (b) and (b′). Nucleophiles used include: (i) alcohols,[34] amines,[47] azide ion,[48] borohydrides,[49] and cyclopentadienide anion[34] to generate (**96**); (ii) amines, olefins, cinnamic acids, and electron-rich arenes[34] to give (**97**); and (iii) pyrroles[50] and active-methylene compounds[34] to form (**98**).

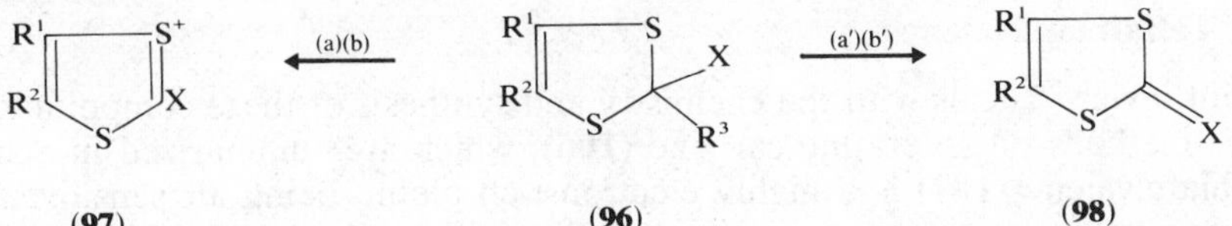

(**97**) ← (a)(b) — (**96**) — (a′)(b′) → (**98**)
(**99**) X = CH_2

SCHEME 11

As a corollary of this electrophilic activity, protons on carbon atoms (or heteroatoms) at position-2 are acidic; the neutral compound (**99**) is strongly nucleophilic, as demonstrated by its behaviour towards electrophiles,[34,51] as set out in Scheme 12. Not surprisingly, 4-methyl groups are not acidic.

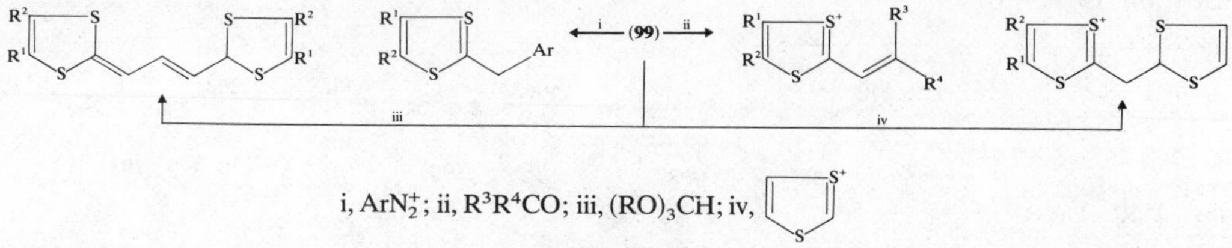

i, ArN_2^+; ii, R^3R^4CO; iii, $(RO)_3CH$; iv,

SCHEME 12

1,3-Dithiole-2-ones or -2-thiones are reversibly protonated to 2-hydroxy- or 2-mercapto-1,3-dithiolium cations.[33] Deprotonation of the phenol (**100**) gives the quinoid compound (**101**).[34] As a counterpart to the 1,3-dipole activity of 1,2-dithiole-3-thiones, 1,3-dipoles of the S—C—S type, the 1,3-dithiolium series, offers us the mesoionic derivative (**102**) as a C—S—C 1,3-dipole,[52] unlike the inactive 1,2-isomer (**103**).[54]

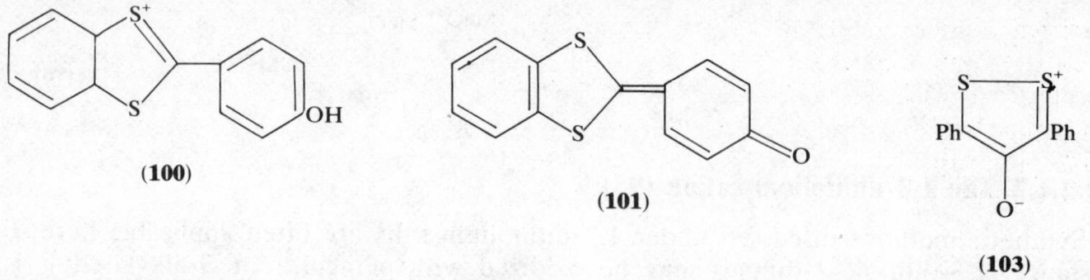

(**100**) (**101**) (**103**)

Compound (**102**) gives *cis,endo*-adducts (**104**) with typical dipolarophiles (maleic anhydride, *etc.*). If acetylenes are used, the adducts expel carbonyl sulphide to give thiophens (**105**) (Scheme 13).[53] Photolysis of (**102**; $R^1=R^2=Ph$) gives tolan,[55] *via* a proposed thiiren intermediate.

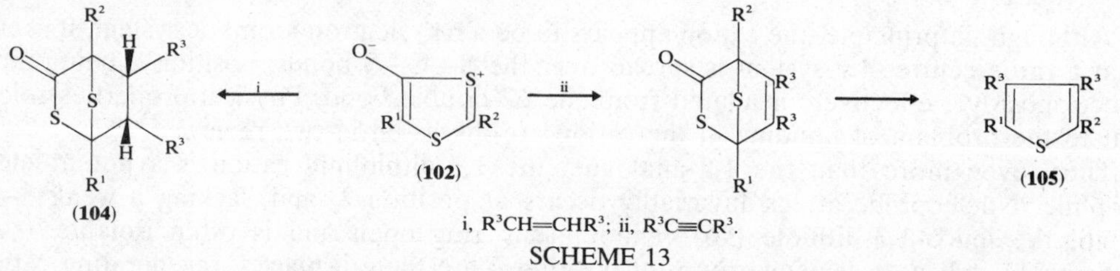

(**104**) (**102**) (**105**)

i, $R^3CH{=}CHR^3$; ii, $R^3C{\equiv}CR^3$.

SCHEME 13

The strongly electrophilic 2-position of the 1,3-dithiolium cation (**95**) bears an acidic proton, removal of which gives the carbene (**106**). This has a number of ylide-like mesomers, which are responsible for its longevity, thus permitting a 'dimerization' to tetrathiafulvalenes (**107**).[56]

19.2.4.4 Tetrathiafulvalenes

A recent review[56] deals with the chemistry and synthesis of these compounds, including numerous methods of generating carbene (**106**), which are summarized in Scheme 14.

Tetrathiafulvalene (**107**) is a highly electron-rich olefin, being air-sensitive, and easily oxidized to a radical cation or a dication. It is metallated by butyl-lithium,[57] providing a route to numerous derivatives. However, its most significant property is its ability to form stable 1:1 charge-transfer complexes with electron-deficient π acceptors, such as

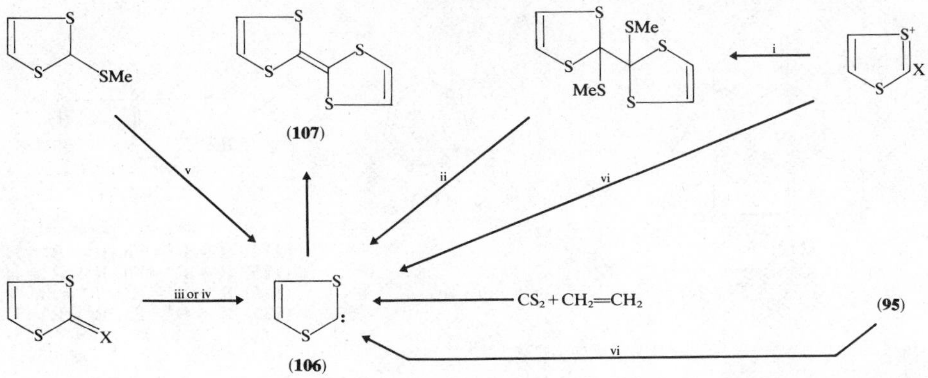

(107)

(106)

(95)

i, electrolysis; ii, Δ; iii, X = S, Ph_3P; iv, X = NNHTs, Δ; v, base; vi, X = H, base.

SCHEME 14

tetracyanoquinodimethane (TCNQ). Such complexes have a very high conductivity[58] even at room temperature, and have thus been christened 'organic metals'. The 1:1 complex of (**107**) and TCNQ has a maximum conductivity of 1.47×10^4 ohm^{-1} cm^{-1} at 58 K, and 652 ohm^{-1} cm^{-1} at 25 °C (*cf.* conductivity of copper: 5.6×10^5 ohm^{-1} cm^{-1} at 25 °C; pure insulator, 10^{-12} ohm^{-1} cm^{-1}). Very good conductivity is also recorded for the 1 : 1 complex of the phenyl-bridged compound (**108**) and TCNQ.[59]

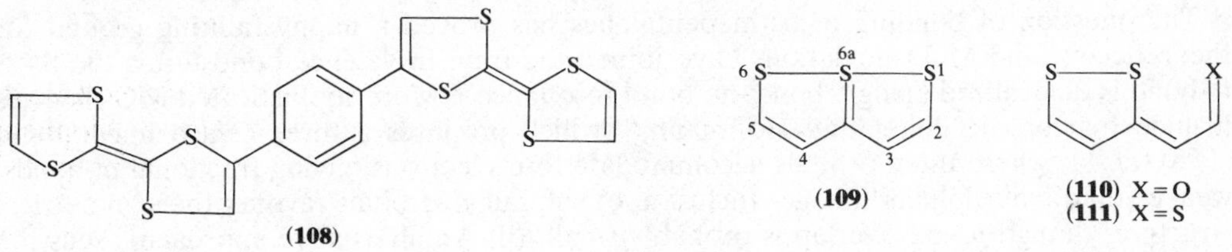

(108) (109) (110) X = O (111) X = S

19.2.5 TRITHIAPENTALENES AND RELATED COMPOUNDS

19.2.5.1 Nomenclature

This unique structural unit (**109**) has been named 'thio- or 'thia-thiophthen', or '1,6,6aS^{IV}-trithiapentalene', of which the last is unambiguous, euphonious, and stresses the unique character of the 6a sulphur atom. The fixed structures (**110**) and (**111**) may be called '1,2-dithiol-3-ylidene ketone' or -thione', respectively.[60]

19.2.5.2 Synthesis

(a) Ketones (**110**) are converted to trithiapentalenes with phosphorus sulphide; their synthesis is reviewed elsewhere.[60] (b) Direct syntheses of (**109**) are classified into (i) modification of 1,2-dithioles; (ii) total ring syntheses. The former include reaction of (**69**; R^1 = Me) with dithioesters or thioformamides,[61] while 2-mono- and 2,5-di-alkylthio-trithiapentalenes are respectively formed from (**69**; R^1 = SMe) and dithioesters, and (**56**; R^1 = Me) and CS_2.[62] The latter type use 1,3,5-triones,[63] 4*H*-thiin-4-thiones, and sulphide ion,[64] or ketones and CS_2.[65] The planar compound (**112**), C_9S_9, is formed by sulphuration,[66] and is hopefully the forerunner of the class of 'sulphocarbons'.

(112)

(113)

(114) $R^1 = R^3 = Ph$, $R^2 = R^4 = H$
(115) $R^2 = R^3 = Ph$, $R^1 = R^4 = H$
(116) $R^1 = R^2 = R^3 = R^4 = Ph$
(117) $R^1 = R^4 = Ph$, $R^2 = R^3 = H$

19.2.5.3 Physical properties; bonding

Although previously formulated as 1,2-dithiepin-5-thiones (**113**), and as thiones (**111**),[60] it is now accepted that trithiapentalenes are best represented as a planar, symmetrical structure (**109**), with the sulphur atoms collinear, and a formally tetracovalent 6a sulphur. This explains the chemical and magnetic equivalence of positions-2 and -5, and -3 and -4.[67] Explanations invoking a rapid interchange of thione tautomers (**111**) are refuted by X-ray diffraction data, *e.g.* the S—S distance (236 pm) is much shorter than the sum of the covalent radii (~340 pm), proving the existence of a true S—S bond.[68] In non-symmetrical cases, *e.g.* (**114**), the two S—S distances are not equal, as is also the case for the sterically crowded 3,4-diphenyl derivative (**115**), although the tetraphenyl compound (**116**) is symmetrical. The 2,5-diphenyl derivative (**117**) is symmetrical in solution, although not in the solid state.[69]

The question of bonding in trithiapentalenes has proved a happy hunting ground for theoreticians, and MO calculations have joined the fray. In valence-bond terms the S—S σ-bond is delocalized ('single bond–no bond resonance'). More sophisticated MO theories include that of (i) Gleiter and Hoffmann,[70] which proposes a three-centre linear bond (**118**) (*cf.* I_3^-) where three orbitals accommodate four electrons, giving fractional σ-bonds, weaker than a disulphide linkage. Inclusion of sulphur d-orbitals favours the symmetrical structure, though p_π–p_π overlap is probably small. (ii) An alternative approach[71] sees the 6a-sulphur using three singly occupied p^2d hybrids to form the σ-framework. The orthogonal, singly occupied p_z-orbital then overlaps with the remaining p_z-orbitals to give a bicyclic 10π aromatic system (**119**) (one electron from each carbon, two each from sulphur atoms 1 and 6). This model provides bond lengths and strengths in good agreement with experiment.

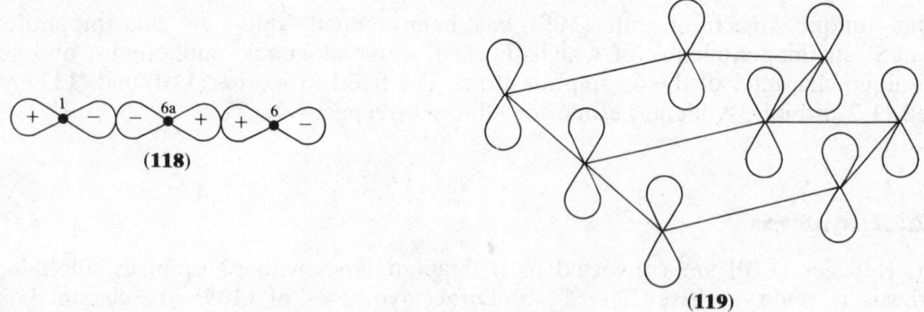

(118) **(119)**

Calculations predict that the less electronegative of the two sulphur atoms 1 and 6 will form the stronger, *i.e.* shorter, bond with the 6a-sulphur.[72]

Numerous isosteres of trithiapentalenes have been made. Not all of these possess a linear three-centre bond, *e.g.* 1,3-dithiole (**66**) does not. The orbital energies of the component atoms appear to be critical.

19.2.5.4 Chemical properties

These have been rather overshadowed by the intense interest in the physical properties of these molecules.

Electrophilic bromination and formylation[73] occurs in positions-3 and -4, as predicted by charge-density calculations.[74] Nitration, nitrosation, or reactions with diazonium cation rearranges trithiapentalenes[73] to their 1,2-oxaza-, or 1,2-diaza-isosteres, respectively (Scheme 15). Alkylating agents[75] attack at sulphur 1, forming 1,2-dithiolium derivatives (**120**).

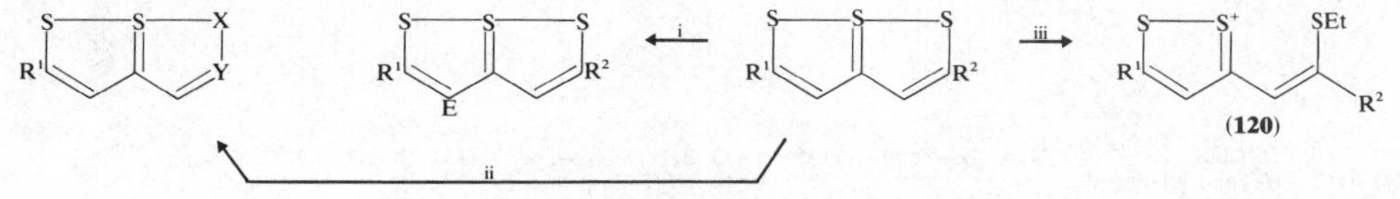

i, $E^+ = Br^+$, CHO^+; ii, $(X—Y)^+ = NO^+$, ArN_2^+; iii, EtI.

SCHEME 15

Nucleophilic attack of hydrazine, and HS^- ion, both at position-2, gives (a) a hydrazone of ketone (**110**), and (b) 4*H*-thiin-4-thione, respectively.[76] Leaving groups, *e.g.* SMe, at position-2 are readily displaced.[60]

Methyl(ene) groups in the 2-position of trithiapentalenes are acidic,[60] and give styryl derivatives, *e.g.* (**121**), with benzaldehyde and base. If the anion (**122**) is capable of rotation about the C-2—C-3 axis, rearrangement by attack of the soft carbanion at sulphur gives a thiophen[73] (**123**) after alkylation (Scheme 16).

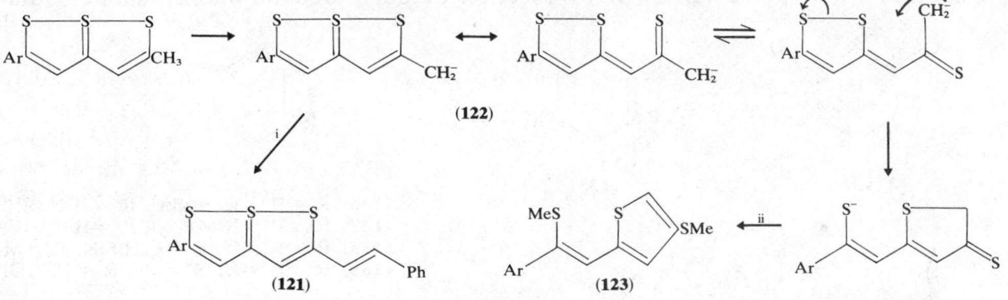

i, PhCHO; ii, MeI.

SCHEME 16

19.2.5.5 Extended systems[77]

These may be formally generated from trithiapentalenes[60] as shown in Scheme 17, by successive addition of a 1,2-dithiol-3-ylidene-methyl group to the 2-position; only one of

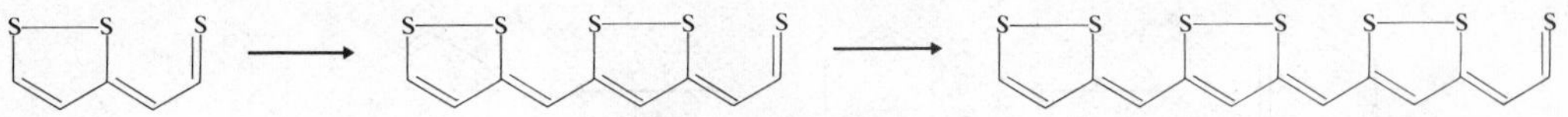

SCHEME 17

the plethora of mesomers is shown. The compound (**124**) has been obtained[78] from trithiapentalene (**125**), where rearrangement of the intermediate anion is prevented by ring fusion (Scheme 18). X-ray analysis of compound (**124**; $R^1 = R^2 = Bu^t$) shows that it is symmetrical with five collinear sulphur atoms in a five-centre bond.

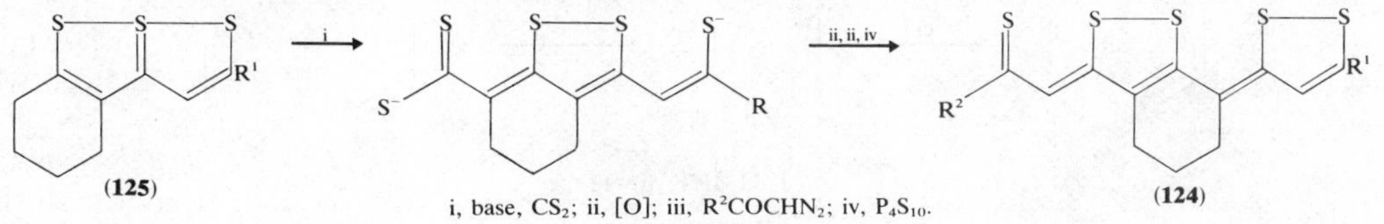

i, base, CS_2; ii, [O]; iii, R^2COCHN_2; iv, P_4S_{10}.

SCHEME 18

19.2.6 THIINS AND RELATED COMPOUNDS

19.2.6.1 Nomenclature

The preferred, systematic names for compounds (**126**) and (**128**) are 2*H*-thiin and 4*H*-thiin, respectively, rather than the universally used 'thiopyran' or 'thiapyran' system. To avoid ambiguity, the terminology of Scheme 19 will be adopted.

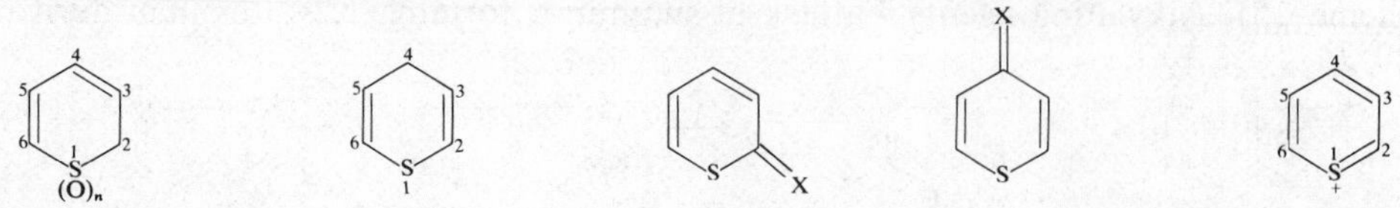

(**126**) $n = 0$, 2*H*-Thiin
(**127**) $n = 2$, 2*H*-Thiin 1,1-dioxide
(**128**) 4*H*-Thiin
(**129**) X = O, 2*H*-Thiin-2-one
(**130**) X = S, 2*H*-Thiin-2-thione
(**131**) X = O, 4*H*-Thiin-4-one
(**132**) X = S, 4*H*-Thiin-4-thione
(**133**) Thiinylium cation

SCHEME 19

19.2.6.2 Synthesis

The chemistries of compounds (**126**)–(**133**) are so intimately inter-twined that we often find, for example, that a 'reaction' of a thiinylium salt is a 'synthesis' of a thiin. Independent syntheses of all types of compound will therefore be described first.

(i) Synthesis of 2*H*-thiins is achieved by cyclization of penta-2,4-dienals with H_2S;[79] or by elaboration[80] of dione (**134**); or by reaction of benzaldehyde and ethanolic sodium sulphide[81] to give (**135**). Scheme 20 summarizes some general, unambiguous syntheses,[82] *e.g.* method (a), where a series of sigmatropic shifts culminates in a Claisen rearrangement.[83]

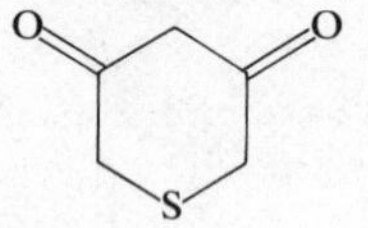

(**134**)

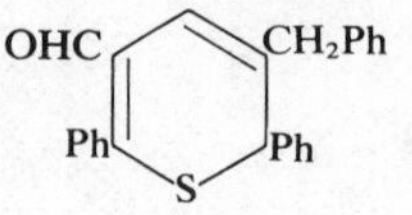

(**135**)

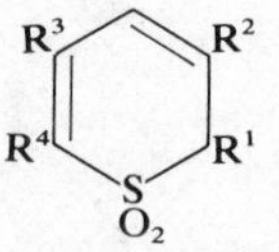

(**136**) $R^1 = R^2 = R^4 = \text{alkyl}$, $R^3 = H$
(**137**) $R^1 = R^3 = R^4 = \text{alkyl}$, $R^2 = H$
(**138**) $R^1 = R^4 = H$, $R^2 = C_4H_8ON$, $R^3 = Me$
(**139**) $R^1 = R^4 = H$, $R^2 = Me$, $R^3 = C_4H_8ON$

(ii) 4*H*-Thiins are easily prepared (Table 6) by cyclization of 1,5-diketones with either H_2S/HCl in methanol or phosphorus sulphide.

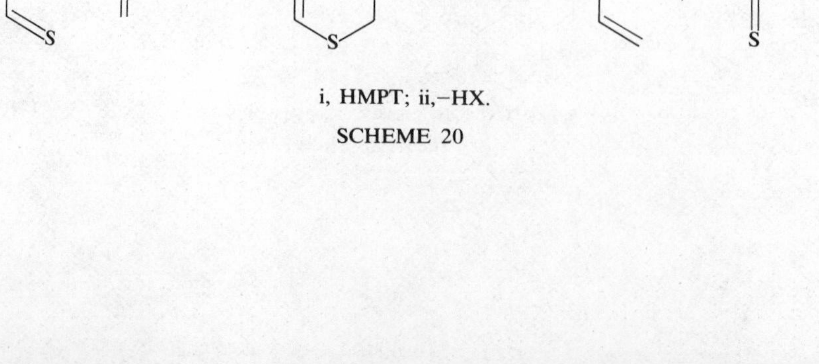

i, HMPT; ii, –HX.

SCHEME 20

TABLE 6
Synthesis of Thiins and Thiinylium Cations

Substituents in products					4H-Thiins		Thiinylium cation	
1	2	3	4	5	Yield (%)	Ref.	Yield (%)	Ref.
H	H	H	H	H	20	a	~20	g
Ph	Me	Ph	H	Ph	95	b	64	h
H	H	Me	H	H	low	c	—	c
Ph	H	Ph	H	Ph	75	d	70–90	d,i
—$(CH_2)_4$—		H	—$(CH_2)_4$—		65	e	40	j
—$(CH_2)_3$—		H	—$(CH_2)_3$—		89	f	58	f
Cl	Ph	H	H	Ph	—	—	85	k

[a] J. Strating and E. Molenaar, *Org. Prep. Procedures*, 1969, **1**, 21. [b] V. Kharchenko, V. Kleimenova, and A. Yakoreva, *Chem. Heterocyclic Compounds*, 1970, **6**, 834. [c] Ref. 80. [d] V. Kharchenko, S. Klimenko, A. Plaksina, and A. Yakoreva, *J. Org. Chem. (U. S. S. R.)*, 1966, **2**, 1116. [e] V. Kharchenko, T. Krupina, S. Klimenko, and A. Rassudova, *Chem. Heterocyclic Compounds*, 1972, **8**, 1081. [f] V. Kharchenko, M. Martem'yanova, N. Zaitseva, and M. Kuramshin, *J. Org. Chem. (U. S. S. R.)*, 1976, **12**, 1766. [g] R. Pettit, *Tetrahedron Letters*, 1960, 11. [h] V. Kharchenko, M. Stankevitch, N. Kupranets, A. Yakoreva, V. Kleimenova, and S. Klimenko, *J. Org. Chem. (U.S.S.R.)*, 1970, **8**, 197. [i] Ref. 106. [j] V. Kharchenko, V. Kleimenova, N. Kupranets, N. Polikarpova, and A. Yakoreva, *J. Org. Chem. (U. S. S. R.)*, 1968, **4**, 1984. [k] G. Laban and R. Mayer, *Z. Chem.*, 1967, **7**, 227.

(iii) Thiin 1,1-dioxides are made by direct oxidation only of polysubstituted thiins.[84] Elaboration of suitable tetrahydrothiins furnishes 2*H*-thiin 1,1-dioxide (**127**) and its 3-phenyl derivative.[84] Base-catalysed ring-closure of dialkylaminovinyl allyl sulphones[85] provides a mixture of isomers (**136**) and (**137**). Reaction of morpholine and a dipropargyl sulphone[86] gives a 5:1 mixture of sulphones (**138**) and (**139**).

(iv) If the 1,5-diketone cyclization of (ii) above is conducted in acetic acid, a thiinylium cation is obtained by disproportionation of a 4*H*-thiin (Section 19.2.6.4). Diels–Alder reaction of thiophosgene and a diene conveniently gives a 2-chlorothiinylium salt. Pyrylium salts and sulphide ion, or ring expansion of thiophen, are other methods (Scheme 21 and Table 6).

Cl Cl
iii
iv
X = H or Cl
i, ii
v

i, N_2CHCO_2Et, ii, N_2H_4, HNO_2; iii, Na_2S; iv, Δ; v, H_2S, HCl; or P_4S_{10}.

SCHEME 21

(v) The single-step condensations of Scheme 22 lead to 2*H*-thiin-2-ones (**129**) or -2-thiones (**130**), exemplified in Table 7 (see also Table 3). The 4*H*-thiin-4-ones (**131**) and -4-thiones (**132**) are made (a) from suitable tetrahydro-derivatives,[87] (b) from diynyl ketones,[88] or (c) from trithiapentalenes (Section 19.2.5.4). 2,6-Dimercapto-4*H*-thiin-4-ones are easily made from ketones and carbon disulphide[89] in high yield.

For the interconversion of ketones and thiones, see Section 19.2.3.3.

19.2.6.3 Chemistry of thiins and thiin 1,1-dioxides[84]

(i) Simple 2*H*- and 4*H*-thiins are notoriously unstable, decomposing above −30 °C in air, in contrast to the stable polysubstituted derivatives. Of the two isomers, the 2*H*-thiins appear to be more stable thermodynamically. One theory[90] suggests that 2*H*- but not 4*H*-thiins can homoconjugate *via* sulphur *d*-orbitals. The two isomeric 2,4,6-triphenylthiins give the same monoanion, which gives the (probably) less stable 4*H*-thiin

SCHEME 22

TABLE 7
Synthesis of 2*H*-Thiin-2-ones and -2-thiones

Method (*Scheme* 22)	R^1	R^2	R^3	R^4	X	*Yield* (%)	*Ref.*
(a)	COPh	H	H	Ph	S	20	a
(a)	H	Ph	H	Ph	S	37	b
(b)	Me	H	Me	H	S	55	c
(c)	—$(CH_2)_4$—		CN	NH_2	S	85	d
(d)	Ph	H	H	Me	S	—	e
(e)	CN	Ph	H	Ph	O	33	e
(f)	Ph	H	H	Ph	O	53	e
(g)	Ph	OH	CO_2Et	Ph	O	70	f

[a] J.-P. Pradère, A. Guenec, G. Duguay, J.-P. Guemas, and H. Quiniou, *Compt. rend.* (*C*), 1969, **269,** 929. [b] H. Behringer and A. Grimm, *Annalen*, 1964, **682,** 188. [c] R. Mayer, G. Laban, and M. Wirth, *Annalen*, 1967, **703,** 140. [d] K. Gewald, *J. prakt. Chem.*, 1966, **31,** 205. [e] S. Scheithauer and R. Mayer, *Z. Chem.*, 1969, **9,** 59. [f] Ref. 118.

on methylation.[91] 4*H*-Thiins are isomerized by acids to their 2*H*-isomers, if necessary by migration of a 4-substituent to position-2 by an intermolecular anion transfer.[92] This is related to the rearrangement[93] of 4*H*-thiin (**140**) in acid to the naphthalene (**141**). Dichlorocarbene inserts into the weak 4-C—H bond of 4*H*-thiin,[84] as well as giving the expected cyclopropane derivative.

(**140**) (**141**) (**142**) (**143**)

Free-radical halogenation of 2,4,6-trisubstituted thiins gives 3-mono- and 3,5-di-bromo-derivatives,[94] while bromine adds to 4*H*-thiin[84] to form a tetrabromide.

(ii) 4*H*-Thiin 1,1-dioxide is apparently unstable and attempted preparation gives the 2*H*-isomer.[95] Non-symmetrical 2*H*-thiin 1,1-dioxides, *e.g.* (**142**) and (**143**), are easily equilibrated and most syntheses produce mixtures.[80,96]

The sulphone (**142**) gives a Diels–Alder dimer (**144**).[96] 1,3-Dipoles attack the $\Delta^{3,4}$-double bond regiospecifically, *e.g.* diazomethane gives pyrazoline (**145**). Soft nucleophiles give adducts[95,97] (**146**) by Michael addition and rearrangement (Scheme 23).

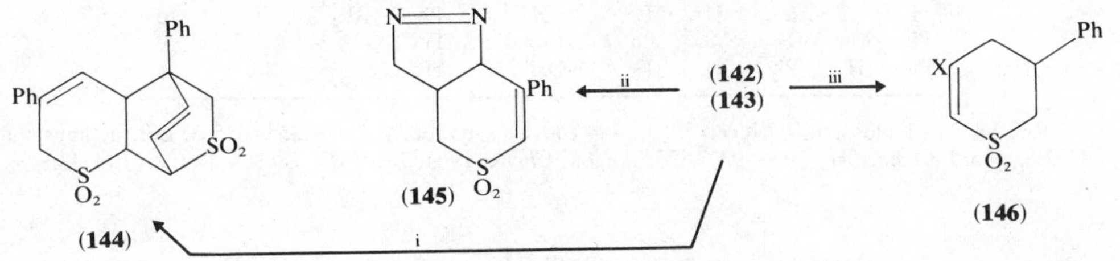

i, Δ; ii, CH_2N_2; iii, $X^- = OMe^-$, SPh^-; H^+.

SCHEME 23

Removal of a very acidic 2-proton from sulphone (**127**) gives an anion,[98] with a trigonal 2-carbon, which has typical carbanionoid properties. Its ring protons are shifted downfield in the n.m.r. spectrum relative to those of sulphone (**127**), and it may be a homoaromatic 6π-system (**147**). The derivative (**148**) possesses considerable dipolar character (**149**), and may be protonated or electrophilically substituted.[99] The aromaticity of anion (**147**) is explained either by homosulphonyl conjugation or by sulphur *d*-orbital participation.

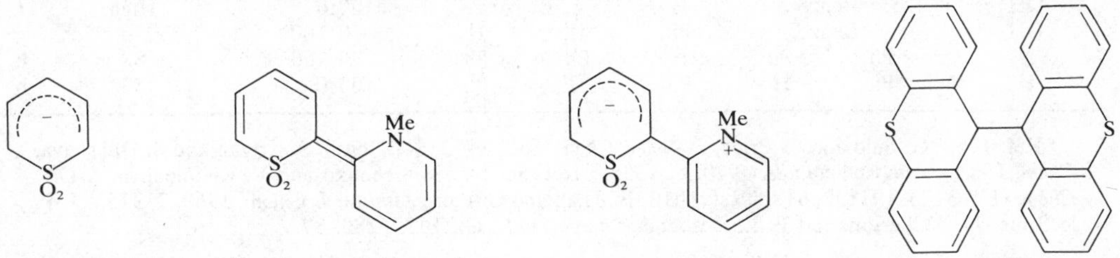

(**147**) (**148**) (**149**) (**150**)

19.2.6.4 Interconversions of thiins and thiinylium cations

(i) There is a powerful driving force urging both the thiins (**126**) and (**128**) towards the same, planar, symmetrical aromatic 6π thiinylium cation (**133**) (*cf.* cycloheptatriene and tropylium cation). Hydride abstraction from thiins is thus very favourable, whilst elimination of a leaving group from the 2(4)-position of a 2(4)*H*-thiin, generated *in situ*, is even easier. Examples are collected in Table 8.

Protonated thiins act as their own hydride acceptors,[84] and thus disproportionate in acids[100] to cation (**133**) and a tetrahydrothiin in the ratio 2:1.

(ii) Conversely, thiinylium cations are electrophilic and add nucleophiles to give 2*H*- and 4*H*-thiins (Table 9: also Section 19.2.6.5). Bulky nucleophiles choose less hindered sites for attack, while hard nucleophiles prefer position-2 to position-4. Soft organolithium reagents give thiabenzenes (Section 19.2.7.2).

Radicals generated by one-electron transfer to thiinylium cations[101] give coupled 4*H*-thiins, *e.g.* (**150**).

TABLE 8

Hydride Abstraction from Thiins

Substituents in cation product 2	3	4	5	6	*Type of thiin*	*Reagent*	*Ref.*
H	H	H	H	H	4*H*	Cl_2; pyrylium cation	a
Me	H	H	H	H	2*H*?	$Ph_3C^+ ClO_4^-$	a
Benzo		H	Benzo		4*H*	Br_2	b
Benzo		H	Benzo		3-OH-4*H*	H_2SO_4	c
Benzo		H	H	H	2*H*	SO_2Cl_2	d
H	Benzo		H	H	2*H*	$Ph_3C^+ BF_4^-$	d
Ph	Me	H	Me	Ph	2*H* or 4*H*	$FeCl_3$	e
Ph	H	Ph	H	H	5-OH-2*H*	$HClO_4$	f

[a] Ref. 84. [b] E. Fields and S. Meyerson, *J. Org. Chem.*, 1965, **30,** 937. [c] Ref. 126. [d] A. Luttringhaus and N. Engelhard, *Chem. Ber.*, 1960, **93,** 1525; *Angew. Chem.*, 1961, **73,** 218. [e] Table 6, Ref. b. [f] Ref. 113.

TABLE 9

Nucleophilic Additions to Thiinylium Salts

Nucleophile	*Substituent positions in starting cation* 2	3	4	5	6	*Product ratio* 4*H*:2*H*	*Yield* (%)	*Ref.*
ArMgX	Benzo		H	Benzo		100:0	50–90	a
RMgX	Alkyl Aryl	Benzo		H	H	0:100	40–90	a
RMgX	Ph	H	Ph	H	Ph	3:1	72	b
H_2O	Ph	H	H	H	Ph	100:0	40	c
OH^-, MeO^-	H	H	H	H	H	0:100	40	d
RMgX	Ph	Ph	H	Ph	Ph	100:0	80	e
OH^-	Benzo		H	Benzo		100:0	High	f
OH^-	Benzo		Ph	H	H	0:100	—	g
N_3^-	Ph	Ph	H	Ph	Ph	0:100	88	h
H^-	Ph	H	Ph	H	Ph	100:0	53	b

[a] Ref. 126. [b] G. Suld and C. Price, *J. Amer. Chem. Soc.*, 1962, **84,** 2090. [c] S. Krivun and S. Dul'skaya, *Chem. Heterocyclic Compounds*, 1970, **6,** 1355. [d] Ref. 86. [e] V. Kharchenko and V. Kleimenova, *J. Org. Chem.* (*U.S.S.R.*), 1971, **7,** 618. [f] Ref. 101. [g] B. Tilak and G. Panse, *Indian J. Chem.*, 1969, **7,** 315. [h] J.-P. le Roux, J.-C. Cherton, and P.-L. Desbene, *Compt. rend.* (*C*), 1975, **280,** 37.

i, X^-; ii, RX; iii, X^-.

SCHEME 24

19.2.6.5 Thiinylium cation chemistry

These cations (i) are electrophilic and (ii) activate 2- and 4-methyl substituents, as predicted by MO calculations.[102]

The leuco-thiin obtained by nucleophilic addition to a cation may either be stable and isolable (Section 19.2.6.4); be oxidized, or expel a leaving group to give an ylidene derivative (Section 19.2.6.7) or a new cation;[103,104] or ring-open. Ring-opened products of type (a) (Scheme 24) are obtained from 1-phenylpyrazol-5-one and 4-methoxythiinylium cation,[105] whilst attack of dimethylamine on **(133)** gives type (b) compounds.[104] The formation of pyridinium salts with primary amines,[106] or benzenoid compounds, *e.g.* with malononitrile,[107] exemplifies type (c). Attack of hydrazines may give 1,2-diazepins.[108]

The thiinylium betaine **(151)**, a 1,3-dipole, dimerizes to **(152)** and its *anti*-isomer.[109]

Syntheses of numerous dyestuffs exploit the acidity of 4-methyl substituted thiinylium cations.[110]

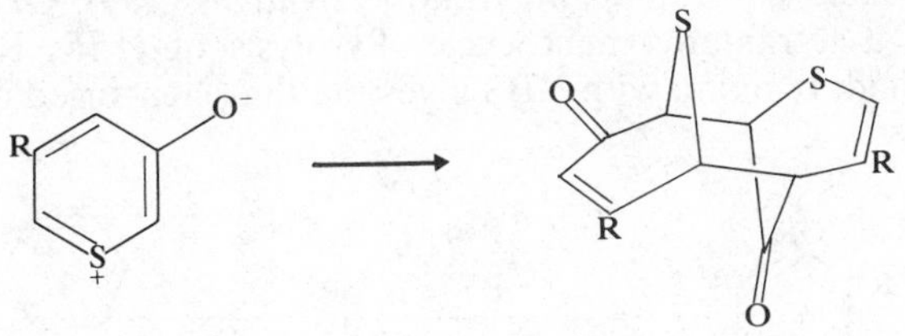

(151) **(152)**

19.2.6.6 Interconversions of thiinylium and thiinylidene derivatives

Nucleophilic addition may convert the cation into 2*H*- or 4*H*-thiinylidene derivatives[111] (Scheme 25), while the reverse transformation is effected (a) by electrophilic attack[89,112] at the *exo*-atom X of **(129)**–**(132)**, (b) by peracid oxidation[113] of a 2*H*-thione, or (c) by Reformatsky reaction[114] of a 4*H*-thione.

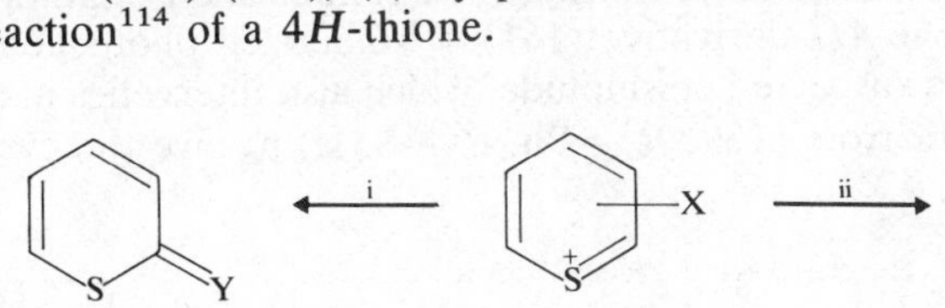

X = 2-Cl, Y = O, S, NR
X = 2-SMe, Y = NR, CR_2

X = 4-H, 4-Cl, Z = CR_2
X = 4-OMe, Z = S

i, H_2Y; ii, H_2Z.

SCHEME 25

19.2.6.7 Chemistry of thiinylidene compounds

A valuable collection of mainly physical data of these compounds is available.[88]

A vast number of active-methylene compounds give condensation products with the 4-ones and 4-thiones,[88,115] which also form oximes, hydrazones, *etc.*,[87] as does 4,6-diphenyl-2*H*-thiin-2-thione.[112] With primary amines in hydroxylamine, 2-ones and 2-thiones usually give pyrid-2-one and -2-thione derivatives; a similar ring-opening has been observed in the 4*H*-series.[88,112] Episulphides are formed from 4-thiones and diazoalkanes, giving an ylidene derivative by sulphur extrusion.[116] Pyrolysis of thiones **(130)** and **(132)** gives sulphur extrusion with formation of compounds **(153)** and **(154)**, respectively.[88]

4-Acyloxy-2*H*-thiin-2-ones are rearranged to the *C*-acylated **(155)** by titanium tetrachloride.[117]

The 2,6-methyl groups of the 4*H*-derivative **(156)** are acidic,[110] while the methyl group of 4-hydroxy-6-methyl-2*H*-thiin-2-one condenses with benzaldehyde in base to give a styryl derivative.[118] Nucleophiles displace chloride ion from 3-chloro-4*H*-thiin-4-ones.[87]

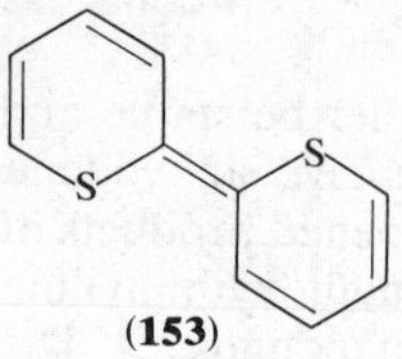

(153) 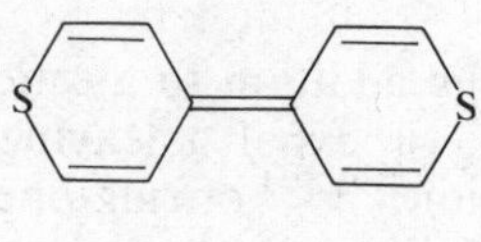 (154) 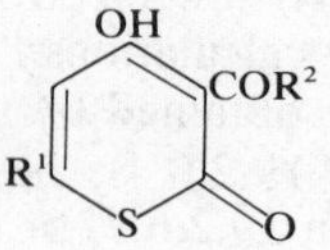 (155) 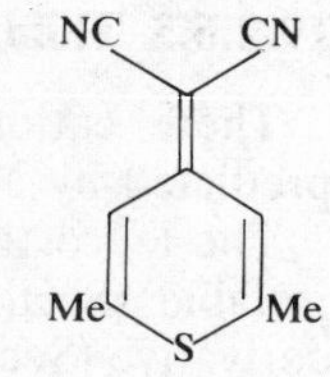 (156)

Photolysis of the 4*H*-4-ones gives, in the presence of an acetylene, thiepins (Section 19.2.9.2) or, on their own, cyclobutane dimers or cage dimers.[119]

2,6-Dimercapto-4*H*-thiin-4-ones (**157**) are correctly formulated[120] as the tautomer (**158**). They are alkylated[89] to the bis-alkylthio-derivatives (**159**). Oxidation of thiol (**158**; $R^1 = CO_2Me$) gives a tetrameric macrocycle. Pyrolysis of (**158**; $R^1 = Ph$) also gives a macrocycle in low yield; reaction with NBS gives the thiophen-fused derivative (**160**).[120]

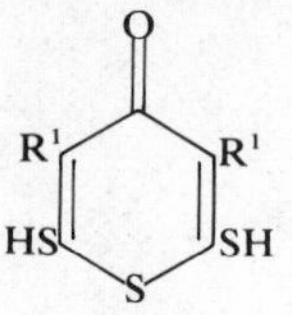

(157) 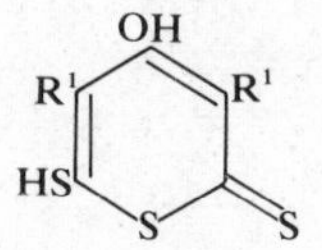 (158) 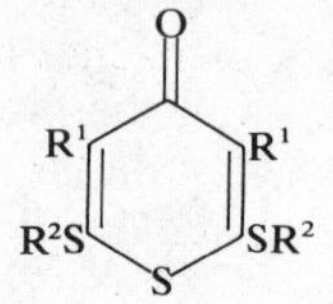 (159) 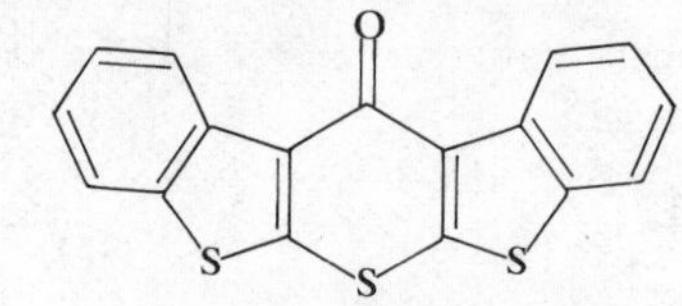 (160)

The *S*-alkylated (**159**) is oxidized to the 2,6-bis(alkylsulphonyl) compound.[88] Sulphide (**159**) is converted by phosphorus sulphide to the 4-thione, which rearranges thermally to the 2*H*-thione by intermolecular alkyl transfer.[89] A similar alkyl migration from oxygen to sulphur is known.[121] The 4*H*-derivative (**161**) is isomerized photochemically[122] to the 2*H*-compound (**162**) *via* a suspected episulphide, which also intervenes in the photochemical extrusion[123] of sulphur from (**159**; $R^1 = Ph$, $R^2 = SMe$) to give the cyclopentadienone (**163**).

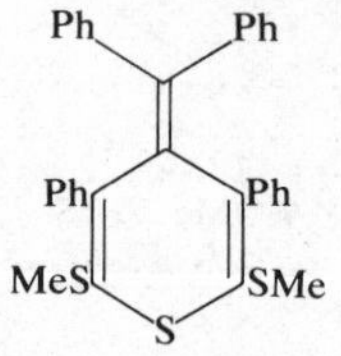

(161) 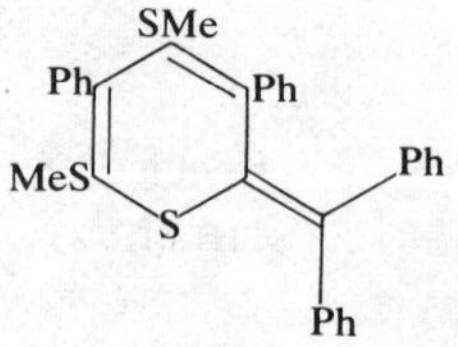 (162) 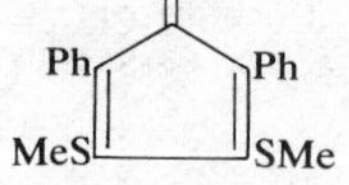 (163)

19.2.6.8 Other thiin derivatives

These include thiochromenes and thiochromones, recently reviewed,[124] and a recent publication gives leading references to thiocoumarins.[125]

19.2.7 THIABENZENES

19.2.7.1 A cautionary note

'Thiabenzene' describes the neutral structural unit (**164**). Almost the entire literature on thiabenzenes — especially 'stable thiabenzenes' — published before Mislow's elegant work[126] of 1975, is a welter of confusion and error, and should be viewed with circumspection.

19.2.7.2 Synthesis[126]

These elusive compounds are obtained as highly coloured solutions, stable only at low temperatures either by deprotonation of a sulphonium salt (**165**) or by *S*-addition of organolithium reagents to thiinylium salts.

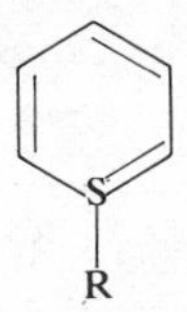

(**164**) 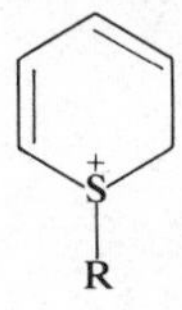 (**165**) 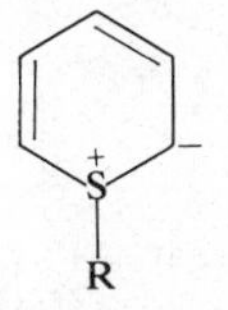 (**166**) 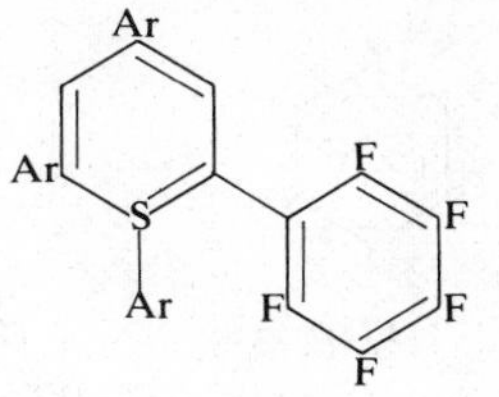 (**167**)

19.2.7.3 Chemistry of thiabenzenes[126]

The ylide structure (**166**), rather than the tetracovalent sulphur structure (**164**), rationalizes these observations. (i) Thiabenzenes are stabilized by $-M,-I$ 2,6-substituents and by polar solvents. (ii) The protons on the more negative 2-carbon are shifted upfield in the n.m.r. spectrum. (iii) The sulphur atom has a stable, tetrahedral configuration, since thiabenzenes are obtainable optically active. (iv) Thiabenzenes decompose, *via* an intramolecular Stevens rearrangement, to a mixture of 2*H*- and 4*H*-thiins, by migration to positions-2 or -4 of the R group originally bonded to sulphur.

19.2.7.4 'Stable thiabenzenes'[126]

With the sole exception of (**167**), all thiabenzenes are thermally unstable. The 'stable' products from reaction of aryl-lithiums and thiinylium cations are amorphous oligomers, formed by proton abstraction.

The thione[127] (**168**) may exist as the 'stable thiabenzene' (**169**).

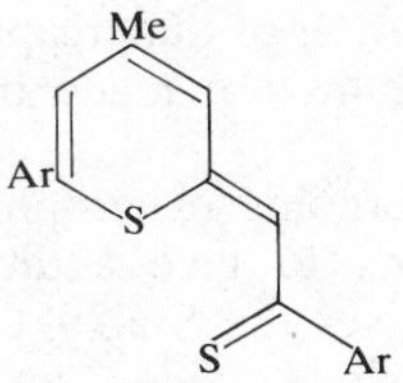

(**168**) 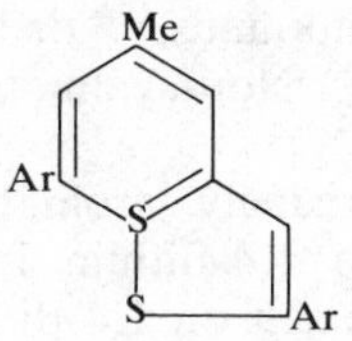 (**169**) 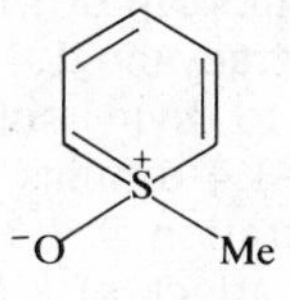 (**170**)

19.2.7.5 Thiabenzene 1-oxides (170)

Substituted derivatives are made from dimethylsulphoxonium methylide and various enones or ynones.[128] Like thiabenzenes, the 1-oxides behave as cyclic ylides, but are much more stable. The carbanionoid character of position-2 is disclosed (i) by the upfield shift of the 2,6-protons, and (ii) by its basicity, being exchangeable in acids. The *S*-methyl group is acidic, and may be alkylated.[128]

19.2.8 DITHIINS

19.2.8.1 1,2-Dithiins (171)[129]

Oxidation of all *cis*-butadiene-1,4-dithiolates (from bis-addition of benzyl thiol to 1,3-diynes) is a general synthesis of 3,6-disubstituted 1,2-dithiins.[130] The role of the dithione valence tautomer (**172**) in the chemistry of these nominally anti-aromatic, 8π-molecules is probably small.[131] The S—S bond of 1,2-dithiins is nonetheless weak, being cleaved by oxidation, nucleophilic attack (Scheme 26), or thermal sulphur extrusion to a thiophen.

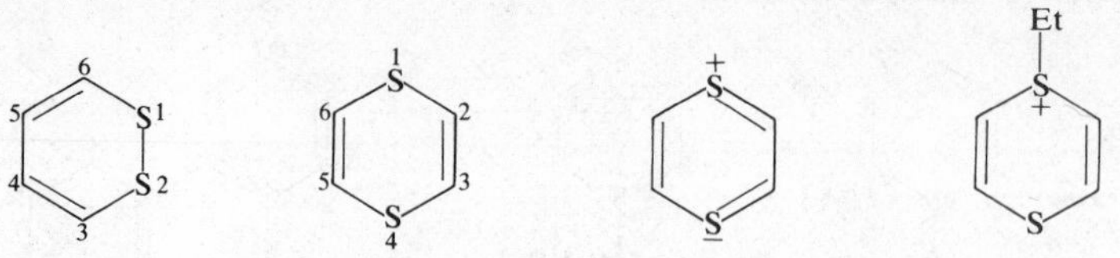

(**171**) (**173**) (**174**) (**175**)

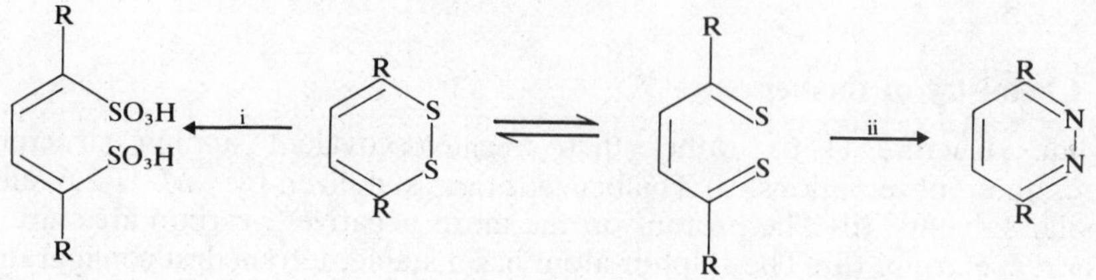

(**172**)

i, [O]; ii, N_2H_4.

SCHEME 26

19.2.8.2 1,4-Dithiin (173)

These compounds lack a weak S—S bond, and are thus more stable than the 1,2-isomer. It is folded, in an easily inverted 'book' conformation, about the S—S axis,[132] implying *d*-orbital contributions (**174**) to the resonance energy of 54 kJ mol^{-1}.[133]

Many syntheses[132,133] of 1,4-dithiins give symmetrical products, *e.g.* dehydration of the dimers of α-mercaptocarbonyl compounds,[134] or the oxidation of dimercaptomaleonitrile to 2,3,5,6-tetracyano-1,4-dithiin.[133] Non-symmetrical dithiins are made by addition of sulphide ion to diynyl sulphides.[135]

2,5-Diaryl-1,4-dithiins are surprisingly 'aromatic', undergoing electrophilic halogenation and nitration,[134,136] although 1,4-dithiin itself tends to give addition products. Electrophilic attack of a Meerwein salt on 1,4-dithiin[137] gives the *S*-alkyl cation (**175**), a powerful alkylating agent. 1,4-Dithiins form rather labile 1-oxides, and very stable 1,1-dioxides and 1,1,4,4-tetroxides.[132] Oxidation removes the electrons occupying the very high energy HOMO of 2,3,5,6-tetraphenyl-1,4-dithiin,[138] giving an aromatic 6π-electron dication (**176**).

Nucleophilic attack of n-butyl-lithium on the soft sulphur atom of 2,5-diphenyl-1,4-dithiin gives sulphide (**177**) after methylation.[132] Leaving groups (*e.g.* MeS) are activated by vicinal nitro-groups. Amines or cyanide ion[133] ring-cleave the very electrophilic tetranitrile (**178**).

The double bonds of electron-deficient 1,4-dithiins, *e.g.* tetraester (**179**), are powerful dienophiles.[133] Photolysis of 2,5-diphenyl-1,4-dithiin gives a cyclobutane dimer and a cage dimer.[139]

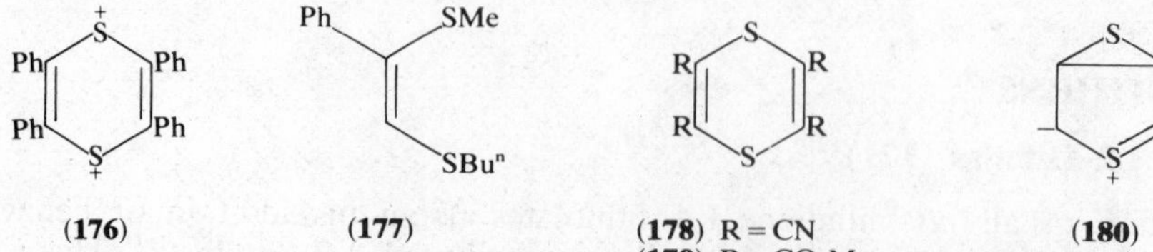

(**176**) (**177**) (**178**) R = CN (**179**) R = CO_2Me (**180**)

Thiophens are formed by extrusion of sulphur from 1,4-dithiins[132,133] either (i) thermally or (ii) more easily, oxidatively, *via* the monosulphoxide. $-M$ Substituents assist in case (i) only, and ensure the retention of the most electron-rich sulphur, *i.e.* that next to the substituent. Mechanistically, the thermal reaction is a $_{\pi}4_s + _{\pi}2_s$ disrotatory process,[140] where the dipole (**180**) undergoes cheleotropic elimination to the thiophen. The oxidative pathway is perhaps a bimolecular redox process.[132]

19.2.8.3 Benzo[*b*]-1,4-dithiin (181)

This compound is more 'aromatic' than 1,4-dithiin, but has similar reactions (oxidation, electrophilic substitution, sulphur extrusion).[132]

19.2.8.4 Dibenzo-1,4-dithiins (thianthrenes)[132] (182)

The latter is an accepted, widely-used trivial name. Owing to the difficulty of unambiguous synthesis, many structural assignments in thianthrene literature await rigorous proof.

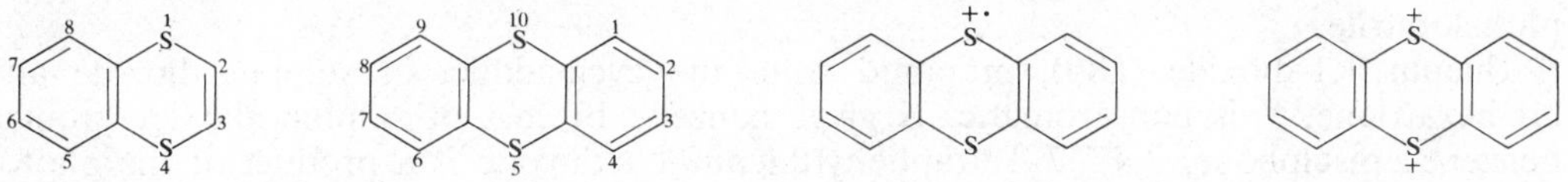

(**181**) (**182**) (**183**) (**184**)

Syntheses[132] of thianthrenes include the photolysis of benzothia-diazines, the treatment of a thiophenol with oleum, and the reaction of arenes with disulphur dichloride and aluminium chloride, which is convenient, though not always unambiguous. Thianthrene is a stable solid, structurally similar to 1,4-dithiin.

Controlled oxidation[132] of thianthrene gives all the various mono-, di-, tri-, and tetra-oxides, except the 5,5-dioxide, made by reduction of the 5,5,10-trioxide. The 5,10-dioxide exists as two stereoisomers, with 'diequatorial' or 'equatorial–axial' S—O bonds.

Thianthrene and Lewis acids (*e.g.* H_2SO_4) give a blue paramagnetic complex ('halochromism'), which contains the radical cation (**183**), formed by removal of one electron from the high-energy HOMO.[141] This radical equilibrates with a small concentration of thianthrene and the highly electrophilic aromatic dication (**184**). This complex is readily hydrolysed to an equimolar mixture of thianthrene and its 5-oxide. It also attacks arenes to give 5-arylthianthrenium salts and gives sulphonium salts with ketones, and sulphilimines with amines.[142]

Electrophilic attack (*e.g.* acylation[143]) on the aromatic rings[132] gives the 2- or 2,7-isomers. Metallation occurs in position-1; the 5,5-dioxide is bis-metallated at positions-1 and -9. Hydroxy- or amino-thianthrenes behave like phenols or anilines.[132]

19.2.8.5 1,3-Dithiins

These are reviewed elsewhere.[144]

19.2.9 THIEPINS[145]

19.2.9.1 General characteristics[145]

Thiepin (**185**), an isostere of cyclo-octatetraene, is non-planar and anti-aromatic. An electrocyclic ring closure to benzene episulphide (**186**), followed by sulphur extrusion, is a general reaction (Scheme 27) of virtually all thiepins. Thus, thiepins are stabilized by (a)

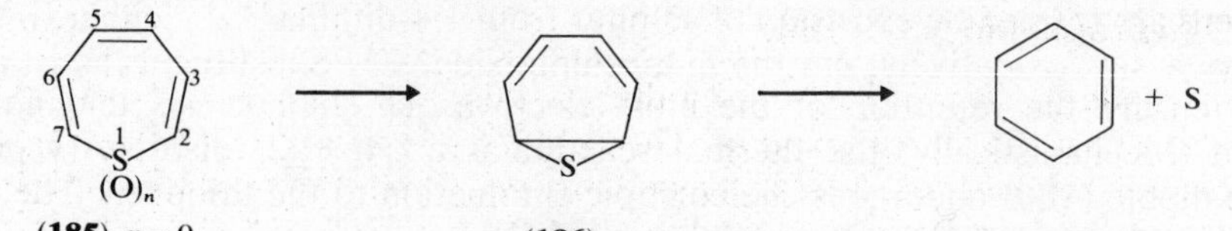

(185) $n=0$
(189) $n=2$

(186)

SCHEME 27

decreasing the π-electron density, by $-M$ substituents or by oxidation to a sulphone; or (b) by benzo-fusion, benzo[*b*]thiepins being particularly stable, as formation of the episulphide disrupts the benzene ring.

19.2.9.2 Non-fused thiepins

Most attempted syntheses give benzene derivatives. However, tetramethylthiophen and dicyanoacetylene give thiepin (**187**) *via* the 2+2 adduct (**188**).[146] It forms a stable sulphone and, on warming, gives an isomer of adduct (**188**). Pyrolysis gives tetramethylphthalonitrile.

Thiepin 1,1-dioxide (**189**), prepared from the cycloadduct of sulphur dioxide and *cis*-hexatriene,[147] is non-aromatic. It gives benzene by loss of sulphur dioxide from a benzene episulphone. 2,4,5,7-Tetraphenylthiepin 1,1-dioxide is a product of the photolysis of 2,6-diphenyl-4*H*-thiin-4-one and tolan.[148]

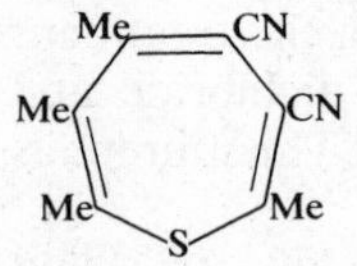

(187) 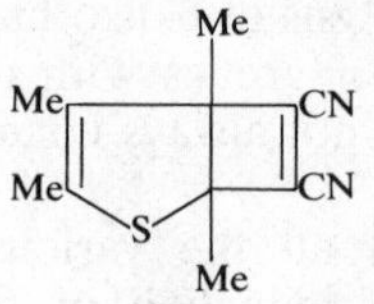 (188) 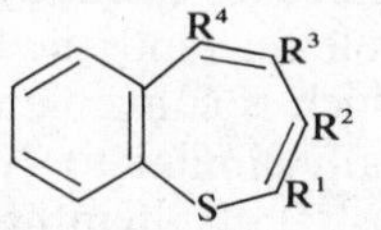 (190) 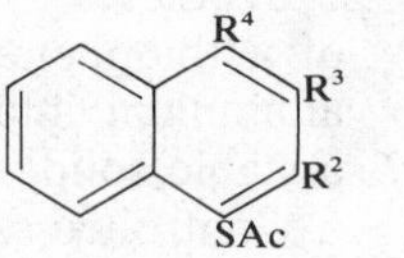 (191)

19.2.9.3 Benzo[*b*]thiepins (190)

These are prepared (a) by ring expansion of (i) a benzothiin by a carbene,[145] or (ii) a benzo[*b*]thiophen by an acetylene;[149] (b) by elaboration of a hydrogenated benzothiepin.[150] Thiepins (190) give naphthalenes on thermolysis; if $R^1=H$, heating with acetic anhydride retains the sulphur as thiolacetate (**191**). Sulphide (**190**) is oxidized to the stable sulphone, distillable without decomposition. There is little conjugation of the sulphonyl group with the double bonds.[145]

19.2.9.4 Benzo[*c*]thiepins[145]

The diacid (**192**) is made from phthalaldehyde, and readily extrudes sulphur, although its diester is stable. The monoester (**193**), unstable above −40 °C, is obtained by a novel ring-expansion of 2-thianaphthalenium salts.[151] Benzo[*c*]thiepin 1,1-dioxide is less stable than its [*b*]-fused isomer, and expels sulphur dioxide to give naphthalene.

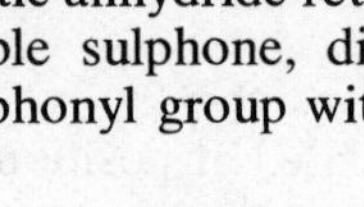

(192) $R^1=R^2=CO_2H$
(193) $R^1=H$, $R^2=CO_2Et$

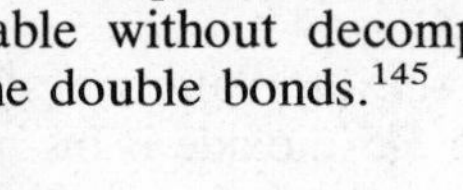

(194) 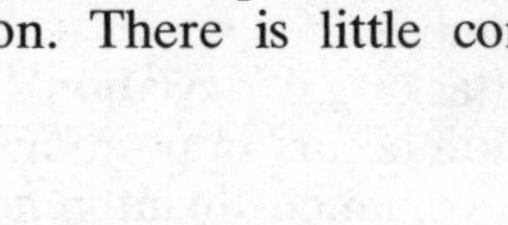 (195) 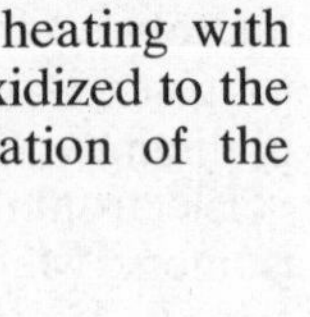 (196)

19.2.9.5 Dibenzo[*b,f*]thiepins (194)

These are made by ring expansion of a dibenzothiin by a carbonium ion mechanism, or by elaboration of a dihydro-derivative. They are very stable yellow solids, only forming phenanthrenes when pyrolysed with copper bronze. They resemble *cis*-stilbene in electronic structure. The parent thiepin (**194**) is oxidized to a sulphone, which is converted by bromination/dehydrobromination to the strained acetylene (**195**); this undergoes various nucleophilic additions, and cycloadditions, *e.g.* with diphenylbenzoisofuran.[145]

The very rare dibenzo[*b,d*]thiepins (**196**) are now available by ring expansion of dibenzothiinylium salts with the lithium salt of ethyl diazoacetate.[151] For a discussion of tribenzothiepins, the standard monograph[145] should be consulted.

References

1. O. Strausz, J. Font, E. Dedio, P. Kebarle, and H. Gunning, *J. Amer. Chem. Soc.*, 1967, **89,** 4805.
2. G. Schrauzer and H. Kisch, *J. Amer. Chem. Soc.*, 1973, **95,** 2501.
3. J. Cadogan, J. Sharp, and M. Trattles, *J. C. S. Chem. Comm.*, 1974, 900.
4. V. Caló, G. Scorrano, and G. Modena, *J. Org. Chem.*, 1969, **34,** 2020.
5. G. Capozzi, V. Luccini, G. Modena, and P. Scrimin, *Tetrahedron Letters*, 1977, 911.
6. H.-W. Chen, Ph.D. Thesis, University of Massachusetts, 1972; *Diss. Abs. Internat.* (*B*), 1972, **33,** 2525.
7. D. Clark, *Internat. J. Sulfur Chem.* (*C*), 1972, **7,** 11.
8. (a) L. Carpino, L. McAdams, R. Rynbrandt, and J. Spiewak, *J. Amer. Chem. Soc.*, 1971, **93,** 476; (b) J. Philips, J. Swisher, D. Haidukewych, and O. Morales, *Chem. Comm.*, 1971, 22.
9. M. Rosen and G. Bonet, *J. Org. Chem.*, 1974, **39,** 3805.
10. Y. Hayasi, H. Nakamura, and H. Nozaki, *Bull. Chem. Soc. Japan*, 1973, **46,** 667.
11. B. Jarvis, W. Tong, and H. Ammon, *J. Org. Chem.*, 1975, **40,** 3189.
12. F. Bordwell and S. Crooks, *J. Amer. Chem. Soc.*, 1969, **91,** 2084.
13. D. Dittmer and T. Nelsen, *J. Org. Chem.*, 1976, **41,** 3044 and references therein.
14. (a) L. Paquette, R. Houser, and M. Rosen, *J. Org. Chem.*, 1970, **35,** 905; (b) D. Dittmer, K. Ikura, J. Balquist, and N. Takashina, *ibid.*, 1972, **37,** 225 and references therein; (c) J. McCaskie, T. Nelsen, and D. Dittmer, *ibid.*, 1973, **38,** 3963.
15. J. Coates, *Diss. Abs. Internat.* (*B*), 1972, **33,** 2537.
16. R. Langendries and F. de Schryver, *Tetrahedron Letters*, 1972, 4781.
17. D. Dittmer, P. Chang, F. Davis, M. Iwanami, I. Stamos, and K. Takahashi, *J. Org. Chem.*, 1972, **37,** 1111, 1116.
18. W. van Tilbourg and R. Plomp, *J. C. S. Chem. Comm.*, 1977, 130.
19. O. Chapman and C. McIntosh, *J. Amer. Chem. Soc.*, 1970, **92,** 7001.
20. L. Paquette and M. Rosen, *J. Org. Chem.*, 1968, **33,** 3027.
21. R. Devdhar, V. Gogte, and B. Tilak, *Tetrahedron Letters*, 1974, 3911.
22. B. Saville and M. Steer, *J. C. S. Chem. Comm.*, 1972, 616.
23. R. Boar, D. Hawkins, J. McGhie, and D. H. R. Barton, *J. C. S. Perkin I*, 1977, 515.
24. D. Breslow and H. Skolnik, 'Multi-sulfur and Sulfur–oxygen Five- and Six-Membered Heterocycles', Interscience, New York, 1966, pp. 347–404.
25. N. Lozac'h, in 'The Chemistry of Organic Sulphur Compounds', vol. 2, ed. N. Kharasch and C. Meyers, Pergamon Press, Oxford, 1966, p. 257.
26. S. Tamagaki and S. Oae, *Tetrahedron Letters*, 1972, 1159.
27. Ref. 24, pp. 390.
28. C. Métayer, G. Duguay, and H. Quiniou, *Bull. Soc. chim. France*, 1974, 163.
29. J. Buchshriber, D. McKinnon, and M. Ahmed, *Canad. J. Chem.*, 1970, **48,** 1991.
30. D. Easton and D. Leaver, *Chem. Comm.*, 1965, 585.
31. (a) J. Buchshriber and D. McKinnon, *Canad. J. Chem.*, 1971, **49,** 3299; (b) H. Davy and J. Vialle, *Bull. Soc. chim. France*, 1975, 1435.
32. H. Behringer and R. Wiedenmann, *Tetrahedron Letters*, 1965, 3705.
33. E. Campaigne and R. Hamilton, *Quart. Reports Sulfur Chem.*, 1970, **5,** 275.
34. H. Prinzbach and E. Futterer, *Adv. Heterocyclic Chem.*, 1966, **7,** 39.
35. E. Klingsberg, *J. Amer. Chem. Soc.*, 1961, **83,** 2934.
36. F. Boberg, *Annalen*, 1964, **679,** 107.
37. J.-P. Guemas and H. Quiniou, *Bull. Soc. chim. France*, 1973, 592; A. Hendrickson and R. Martin, *J. Org. Chem.*, 1973, **38,** 2548.
38. R. Olofson, J. Landesberg, R. Berry, D. Leaver, W. Robertson, and D. McKinnon, *Tetrahedron*, 1966, **22,** 2119.
39. H. Behringer and E. Meinetsberger, *Tetrahedron Letters*, 1975, 3473.
40. Y. Mollier and N. Lozac'h, *Bull. Soc. chim. France*, 1961, 614.
41. R. Mayer and H. Hartmann, *Chem. Ber.*, 1964, **97,** 1886.

42. K. Jensen, H. Baccaro, and O. Buchardt, *Acta Chem. Scand.*, 1963, **17,** 163.
43. Ref. 24, pp. 546–551, 557–565.
44. F. Runge, Z. El-Hewehi, H. Renner, and E. Taeger, *J. prakt. Chem.*, 1960, **283,** 284.
45. W. Kirmse and L. Horner, *Annalen*, 1958, **614,** 4.
46. K. Potts and V. Singh, *Chem. Comm.*, 1969, 569.
47. A. Takamizawa and K. Hirai, *Chem. and Pharm. Bull.* (*Japan*), 1969, **17,** 1931.
48. E. Fanghanel, *J. prakt. Chem.*, 1976, **318,** 127.
49. F. Wudl, M. Kaplan, E. Hufnagel, and E. Southwick, *J. Org. Chem.*, 1974, **39,** 3608.
50. R. Weiss and P. Gompper, *Tetrahedron Letters*, 1970, 481.
51. L. Nivorozhkin, N. Loseva, and V. Minkin, *Chem. Hererocyclic Compounds*, 1972, **8,** 288.
52. H. Gotthardt and B. Christl, *Tetrahedron Letters*, 1968, 4743, 4747, 4751.
53. H. Gotthardt, M. Weisshuhn, and B. Christl, *Chem. Ber.*, 1976, **109,** 740, 753.
54. D. Barillier, P. Rioult, and J. Vialle, *Bull. Soc. chim. France*, 1973, 3031.
55. H. Kato, M. Kawamura, T. Shiba, and M. Ohta, *Chem. Comm.*, 1970, 959.
56. M. Narita and C. Pittman, Jr., *Synthesis*, 1976, 489.
57. D. Green, *J. C. S. Chem. Comm.*, 1977, 161.
58. T. Phillips, T. Kistenmacher, J. Ferraris, and D. Cowan, *J. C. S. Chem. Comm.*, 1973, 471.
59. M. Kaplan, R. Haddon, and F. Wudl, *J. C. S. Chem. Comm.*, 1977, 388.
60. N. Lozac'h, *Adv. Heterocyclic Chem.*, 1971, **13,** 162.
61. C. Métayer, G. Duguay, and H. Quiniou, *Bull. Soc. chim. France*, 1972, 4576; J. Dingwall, S. McKenzie, and D. Reid, *J. Chem. Soc.* (*C*), 1968, 2543.
62. F. Clesse and H. Quiniou, *Compt. rend*, (*C*), 1969, **268,** 637; C. Bouillon and J. Vialle, *Bull. Soc. chim. France*, 1968, 4560.
63. M. Stavaux and N. Lozac'h, *Bull. Soc. chim. France*, 1967, 2082.
64. J. Dingwall, D. Reid, and J. Symon, *Chem. Comm.*, 1969, 466.
65. A. Thullier and J. Vialle, *Bull. Soc. chim. France*, 1962, 2182, 2194.
66. J. Brown and T. Gay, *J. C. S. Perkin I*, 1974, 866.
67. D. Reid and J. Symon, *Chem. Comm.*, 1969, 1314.
68. L. Hansen and A. Hordvik, *Acta Chem. Scand.*, 1973, **27,** 411.
69. S. Johnson, M. Newton, I. Paul, R. Beer, and D. Cartwright, *Chem. Comm.*, 1967, 1170; F. Leung and S. Nyburg, *ibid.*, 1969, 137; see also Ref. 68 and references therein.
70. R. Gleiter and R. Hoffmann, *Tetrahedron*, 1968, **24,** 5899.
71. R. Johnstone and S. Ward, *Theor. Chim. Acta*, 1969, **14,** 420.
72. A. Hordvik, E. Sletten, and J. Sletten, *Acta Chem. Scand.*, 1969, **23,** 1852.
73. R. Christie, A. Ingram, D. Reid, and R. Webster, *J. C. S. Perkin I*, 1974, 772.
74. D. Clark and D. Kilcast, *Tetrahedron*, 1971, **27,** 4367.
75. H. Behringer and J. Falkenberg, *Chem. Ber.*, 1969, **102,** 1580.
76. D. Reid and R. Webster, *J. C. S. Perkin I*, 1972, 1447.
77. M. Stavaux and N. Lozac'h, *Bull. Soc. chim. France*, 1967, 3557.
78. M. Stavaux, *Bull. Soc. chim. France*, 1971, 4418, 4426, 4429; M. Stavaux and N. Lozac'h, *ibid.*, 4419, 4423.
79. A. Chechak, M. Stern, and C. Robeson, *J. Org. Chem.*, 1964, **29,** 187.
80. J. Degani, R. Fochi, and C. Vincenzi, *Gazzetta*, 1967, **97,** 397.
81. S. Cremer and A. Subbaratoram, *Chem. Comm.*, 1967, 33.
82. J. Boerma, N. Nilsson, and A. Senning, *Tetrahedron*, 1974, **30,** 2735; J. Pradere, Y. N'Guessan, and H. Quiniou, *ibid.*, 1975, **31,** 3059; R. Kalish, A. Smith, and E. Smutny, *Tetrahedron Letters*, 1971, 2241.
83. R. Van Der Welle and L. Brandsma, *Rec. Trav. chim.*, 1973, **92,** 667; J.-P. Pradere and H. Quiniou, *Compt. rend.* (*C*), 1972, **275,** 677.
84. V. Kharchenko, S. Chalaya, and T. Konovalova, *Chem. Heterocyclic Compounds*, 1974, **9,** 1003.
85. L. Skattebøl, B. Boulette, and S. Solomon, *J. Org. Chem.*, 1967, **33,** 548; S. Bradamante, S. Maiorana, and G. Pagani, *J. C. S. Perkin I*, 1972, 282.
86. I. El-Sayid El-Kholy and F. Kamal Rafla, *J. Chem. Soc.* (*C*), 1969, 315.
87. R. Mayer and W. Broy, *Adv. Heterocyclic Chem.*, 1967, **8,** 219.
88. Ref. 87, p. 240; see also Ref. 110.
89. H. Teague and W. Tucker, *J. Org. Chem.*, 1967, **32,** 3140, 3144.
90. T. Parasaran and C. Price, *J. Org. Chem.*, 1964, **29,** 946.
91. R. Schmidt and V. Burkert, *Tetrahedron Letters*, 1973, 4355.
92. V. Kharchenko and A. Rassudova, *J. Org. Chem.* (*U.S.S.R.*), 1973, **9,** 2190.
93. K. Dimroth, K. Wolf, and H. Kroke, *Annalen*, 1964, **678,** 183, 202.
94. U. Eisner and T. Krishnamurthy, *J. Org. Chem.*, 1972, **37,** 150.
95. E. Molenaar and J. Strating, *Rec. Trav. chim.*, 1967, **86,** 436, 1047.
96. G. Pagani, *Gazzetta*, 1967, **97,** 1518.
97. S. Bradamante, A. Mangia, S. Maiorana, and G. Pagani, *Tetrahedron Letters*, 1969, 2971.
98. S. Bradamante, A. Mangia, and G. Pagani, *Tetrahedron Letters*, 1970, 3381; *J. Chem. Soc.* (*B*), 1971, 545.
99. G. Gaviraghi and G. Pagani, *J. C. S. Perkin II*, 1973, 50.
100. B. Tilak, R. Mitra, and Z. Muljiani, *Tetrahedron*, 1969, **25,** 1939; E. de Waard, W. Vloon, and H. Huisman, *Chem. Comm.*, 1970, 841.

101. C. Price, M. Siskin, and C. Miao, *J. Org. Chem.*, 1971, **36,** 794.
102. S. Yoneda, T. Sugimoto, and Z. Yoshida, *Tetrahedron*, 1973, **29,** 2009.
103. S. Baranov, A. Burzak, and S. Krivun, *Khim. geterotsikl. Soedinenii*, 1971, **7,** 14, 279.
104. S. Yoneda, T. Sugimoto, O. Tanaka, Y. Moriya, and Z. Yoshida, *Tetrahedron*, 1975, **31,** 2669.
105. A. Tolmachev, *J. Gen. Chem.* (*U.S.S.R.*), 1963, **33,** 1813.
106. R. Wizinger and P. Ulrich, *Helv. Chim. Acta*, 1956, **39,** 207.
107. G. Reynolds and J. Van Allen, *J. Heterocyclic Chem.*, 1971, **8,** 301.
108. D. Harris, G. Kan, V. Snieckus, and E. Klingsberg, *Canad. J. Chem.*, 1974, **52,** 2798.
109. S. Baklien, P. Groth, and K. Undheim, *J. C. S. Perkin I*, 1975, 2099.
110. A. Tolmachev and V. Sribnaya, *Chem. Heterocyclic Compounds*, 1966, **2,** 128.
111. B. Eistert, A. Schmitt, and Th. Arackal, *Chem. Ber.*, 1976, **109,** 1549; E. Brown, D. Leaver, and D. McKinnon, *J. Chem. Soc.* (*C*), 1970, 1202; H. Hartmann, *J. prakt. Chem.*, 1971, **313,** 1113.
112. J. Faust, G. Speier, and R. Mayer, *J. prakt. Chem.*, 1969, **311,** 61, and references therein.
113. D. McKinnon, *Canad. J. Chem.*, 1970, **48,** 3388.
114. S. Krivun, A. Buryak, and S. Baranov, *Chem. Heterocyclic Compounds*, 1973, **9,** 1191.
115. G. Seitz and H. Lehmann, *Arch. Pharm.*, 1974, **307,** 853; *Annalen*, 1975, 331.
116. D. Lloyd and F. Wasson, *J. Chem. Soc.* (*C*), 1966, 1086.
117. O. Caputo, L. Cattel, F. Viola, and G. Biglino, *Ann. Chim.* (*Italy*), 1975, **65,** 635.
118. F. Splinter and H. Arold, *J. prakt. Chem.*, 1968, **310,** 142.
119. N. Sugiyama, Y. Sato, and C. Kashima, *Bull. Chem. Soc. Japan*, 1970, **43,** 3205; N. Ishibe and M. Odani, *J. Org. Chem.*, 1971, **36,** 4132.
120. A. Schönberg and R. von Ardenne, *Chem. Ber.*, 1966, **99,** 3316, 3327.
121. H. Teague and W. Tucker, *J. Org. Chem.*, 1970, **35,** 1968.
122. N. Ishibe and M. Tamura, *J. C. S. Chem. Comm.*, 1974, 48.
123. N. Ishibe and M. Odani, *Chem. Comm.*, 1971, 702.
124. S. Schneller, *Adv. Heterocyclic Chem.*, 1975, **18,** 79.
125. A. Ruivet and M. Renson, *Bull. Soc. chim. belges*, 1969, **78,** 449, 459.
126. B. Maryanoff, J. Stackhouse, G. Senkler, Jr., and K. Mislow, *J. Amer. Chem. Soc.*, 1975, **97,** 2718.
127. M. Bard and G. Duguay, *Compt. rend.* (*C*), 1972, **275,** 905.
128. A. Hortmann and R. Harris, *J. Amer. Chem. Soc.*, 1971, **93,** 2471.
129. U. Eisner and T. Krishnamurthy, *Internat. J. Sulfur Chem.* (*B*), 1972, **7,** 101.
130. Refs. 6 and 7 in Ref. 129; M. Bard, J. Meslin, and H. Quiniou, *J. C. S. Chem. Comm.*, 1973, 672; W. Reid and W. Ochs, *Chem. Ber.*, 1977, **105,** 1093.
131. R. Borsdorf, H. Hofmann, H. Kohler, M. Scholz, and J. Fabian, *Tetrahedron*, 1970, **26,** 3227.
132. Ref. 24, p. 1112.
133. C. Buess, *Mechanisms Reactions Sulfur Compounds*, 1968, **3,** 15.
134. C. Buess, V. Brandt, R. Srivastava, and W. Carper, *J. Heterocyclic Chem.*, 1972, **9,** 887.
135. A. Zilverschoon, J. Meijer, P. Vermeer, and L. Brandsma, *Rec. Trav. chim.*, 1974, **94,** 163.
136. M. Oki and K. Kobayashi, *Bull. Chem. Soc. Japan*, 1973, **46,** 687.
137. W. Schroth and M. Hassfield, *Z. Chem.*, 1970, **10,** 296.
138. W. Schroth, R. Borsdorf, R. Herzschuh, and J. Seidler, *Z. Chem.*, 1970, **10,** 147.
139. K. Kobayashi and T. Ohi, *Chem. Letters*, 1973, 645.
140. R. Grigg, R. Hayes, and J. Jackson, *Chem. Comm.*, 1969, 1167.
141. J. Silber and H. Shine, *J. Org. Chem.*, 1971, **36,** 2923.
142. B. Bandlish, A. Padilla, and H. Shine, *J. Org. Chem.*, 1975, **40,** 2590; B. Bandlish, S. Mani, and H. Shine, *ibid.*, 1977, **42,** 1538.
143. J. Servoin-Sidione, M. Montaigne-Lépine, and G. Saint-Ruf, *Bull. Soc. chim. France*, 1973, 1460.
144. U. Eisner and T. Krishnamurthy, *Tetrahedron*, 1971, **27,** 5753.
145. A. Rosowsky, 'Seven-Membered Heterocyclic Compounds Containing Oxygen and Sulphur', Wiley-Interscience, New York, 1972, p. 573.
146. H. Wynberg and R. Helder, *Tetrahedron Letters*, 1972, 3647.
147. W. Mock, *J. Amer. Chem. Soc.*, 1967, **89,** 1281.
148. N. Ishibe, K. Hashimoto, and M. Sunami, *J. Org. Chem.*, 1974, **39,** 103.
149. D. Reinhoudt and C. Kouwenhoven, *Tetrahedron*, 1974, **30,** 2431.
150. V. Traynelis, Y. Yoshikawa, J. Sih, L. Miller, and J. Livingston, Jr., *J. Org. Chem.*, 1973, **38,** 3978.
151. K. Nakasuji, K. Kawamura, T. Ishihara, and I. Murata, *Angew. Chem. Internat. Ed.*, 1976, **15,** 611.

19.3

Systems with Heteroatoms Other Than Nitrogen, Oxygen, or Sulphur

R. LIVINGSTONE
Huddersfield Polytechnic

19.3.1 NOMENCLATURE

In the main this chapter will be concerned with the chemistry of five- and six-membered heterocycles containing one heteroatom other than nitrogen, oxygen, and sulphur. The heteroatoms will be some of those found in Groups IIIB, IVB, VB, VIB, and VIIB of the Periodic Table, and since the preceding chapters describe heterocycles containing one sulphur atom, it is appropriate that this chapter should commence with the chemistry of heterocycles containing other atoms found in Group VIB, *i.e.* selenium and tellurium. There follows a description of heterocycles containing one heteroatom from Groups in the order VB, IVB, IIIB, and VIIB.

Heterocycles containing one atom from these groups were in some cases originally named by combining the appropriate prefix from Table 1(a) with the name of the related carbocyclic compound. Now they are generally named in one of two ways, (i) by combining the appropriate prefix from Table 1(a) (eliding 'a' where necessary) with a stem from Table 2, or (ii) by combining the appropriate prefix from Table 1(b) with the name

TABLE 1

Nomenclature

Element		*Prefix* (*b*)
Selenium	Selena	Seleno
Tellurium	Tellura	Telluro
Phosphorus	Phospha[a]	
Arsenic	Arsa[a]	
Antimony	Stiba[a]	
Bismuth	Bisma	
Silicon	Sila	
Germanium	Germa	
Tin	Stanna	
Lead	Plumba	

[a] 'Phospha' is replaced by 'phosphor' when immediately followed by 'in' or 'ine'; similarly, 'arsa' is replaced by 'arsen' and 'stiba' by 'antimon'.

TABLE 2

Number of members in the ring	*Unsaturated ring*	*Saturated ring*
3	-iren[c]	-iran[b,c]
4	-et[c]	-etan[c]
5	-ole	-olan[c]
6	-in[a]	-an[b,c]
7	-epin	-epan[c]
8	-ocin	-ocan[c]
9	-onin	-onan[c]
10	-ecin	-ecan[c]

[a] See ref. a, Table 1. [b] Not applicable to silicon, germanium, tin and lead. In this case, 'perhydro' is prefixed to the name of the corresponding unsaturated compound. [c] Only under U.K. Chemical Society rules; IUPAC rules add a terminal 'e' to all these endings, giving -irene, *etc.*

of the related heterocycle containing an oxygen atom. The latter method is generally used for six-membered heterocycles containing a selenium or a tellurium atom which may or may not be fused to an aromatic ring system. In the case of five-membered heterocycles containing a selenium or tellurium atom, the appropriate prefix from Table 1(b) replaces 'thio' in the name of the related compound containing one sulphur atom, *e.g.* 'thiophen' and 'benzo[*b*]thiophen' become 'selenophen' and 'benzo[*b*]selenophen', respectively.

Names of compounds according to the IUPAC Commission on the Nomenclature of Organic Chemistry will be included in their description.

19.3.2 HETEROCYCLES CONTAINING ONE SELENIUM[1] ATOM

The names in general use for heterocycles containing one selenium atom are given in Table 3. Those for saturated compounds follow the definitive IUPAC nomenclature,[2,3] whereas trivial names related to those of the corresponding sulphur compounds are used for selenium heterocycles containing a five- and six-membered unsaturated ring, which may or may not be fused to an aromatic ring. This is permissible according to Section B-2 of the IUPAC rules.

TABLE 3

Names of Heterocycles containing One Selenium Atom

Number of members in ring including one Se atom	*Name of heterocycle*	
	Saturated	*Unsaturated*
3	Seleniran	Seleniren
4	Selenetan	Selenet
5	Selenolan	Selenole or Selenophen
6	Selenan	2*H*-Selenin 2*H*-Selenopyran 4*H*-Selenin 4*H*-Selenopyran
7	Selenepan	Selenepin
Number of members in heterocyclic ring fused on to one benzene ring		
(a) Se atom adjacent to C_6H_6 ring		
5	2,3-Dihydrobenzo[*b*]-selenophen	Benzo[*b*]selenophen
6	3,4-Dihydro-2*H*-benzo[*b*]-selenopyran or 3,4-Dihydro-2*H*-1-benzo-selenopyran	2*H*-Benzo[*b*]selenopyran or 2*H*-1-Benzoselenopyran 4*H*-Benzo[*b*]selenopyran or 4*H*-1-Benzoselenopyran
(b) Se atom not adjacent to C_6H_6		
5	1,3-Dihydrobenzo[*c*]-selenophen	Benzo[*c*]selenophen
6	3,4-Dihydro-1*H*-benzo[*c*]-selenopyran or 3,4-Dihydro-1*H*-2-benzo-selenopyran	1*H*-Benzo[*c*]selenopyran or 1*H*-2-Benzoselenopyran
Number of members in heterocyclic ring fused on to two C_6H_6 rings		
(a) Linear		
5		Dibenzoselenophen
6		Selenoxanthen or Dibenzo[*b,e*]selenopyran
(b) Angular		
6		6*H*-Dibenzo[*b,d*]selenopyran

19.3.2.1 Three- and four-membered rings containing one selenium atom

With the saturated rings it is found that the stability of the selenacycloalkanes increases with increasing ring size. The selenirans, *e.g.* (**1**), have never been isolated, but they have been observed as transient intermediates in the reaction between selenium and olefins.[4]

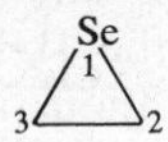

(**1**)

Some seleniranium salts, *e.g.* (**2**), and selenirenium salts, *e.g.* (**3**), have been prepared by treating the appropriate alkenes or alkynes with areneselenyl hexafluorophosphate or hexafluoroantimonates (reactions 1 and 2).[5]

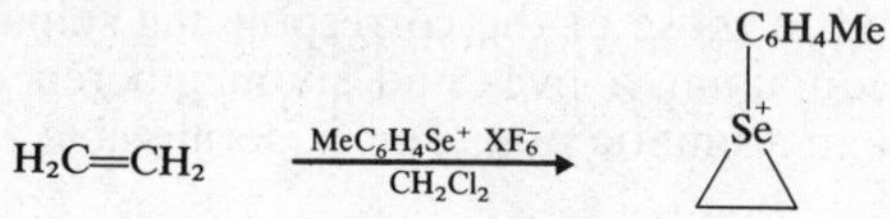

$$H_2C{=}CH_2 \xrightarrow[CH_2Cl_2]{MeC_6H_4Se^+\ XF_6^-} \text{(2)}\ XF_6^- \quad (X = P \text{ or } Sb) \qquad (1)$$

$$MeC{\equiv}CMe \xrightarrow[CH_2Cl_2]{PhSbF_6} \text{(3)}\ SbF_6^- \qquad (2)$$

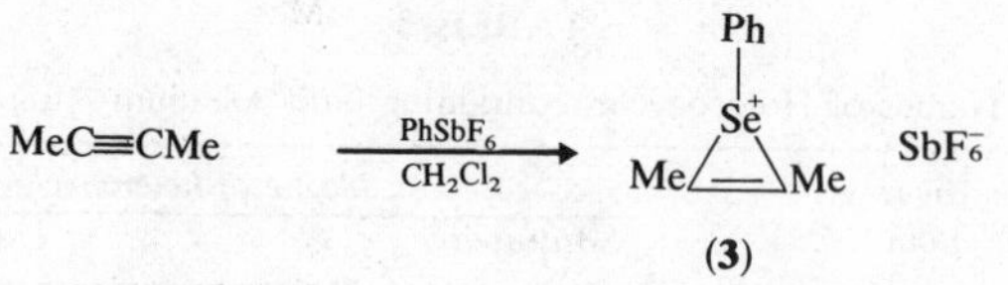

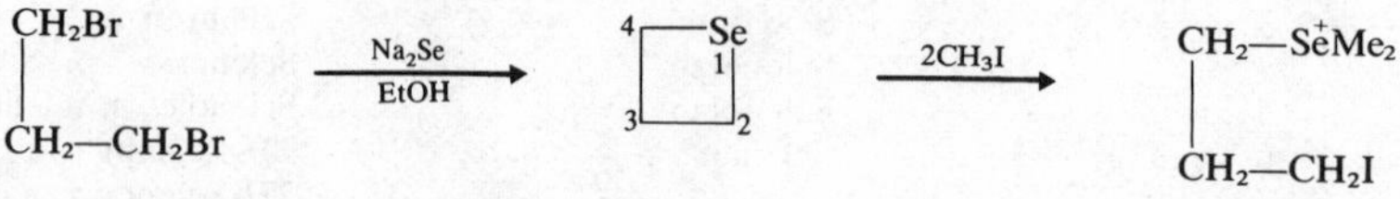

$$CH_2Br\text{–}CH_2\text{–}CH_2Br \xrightarrow[EtOH]{Na_2Se} \text{(4)} \xrightarrow{2CH_3I} CH_2\text{–}\overset{+}{Se}Me_2\ I^- \ / \ CH_2\text{–}CH_2I \ \text{(5)} \qquad (3)$$

Selenetan (**4**), a colourless oil having an exceedingly powerful and penetrating smell, has been obtained in low yield from the reaction between 1,3-dibromopropane and an alkali selenide (reaction 3).[6] It is stable when stored under cool, dark conditions, but polymerizes easily on distillation or by the action of mineral acids. Selenetan reacts with an alcoholic solution of iodine to give 1,1-di-iodoselenetan and an amorphous violet polymer, and with alcoholic mercuric chloride a white crystalline complex separates, which on heating decomposes to yield mercuric selenide and 1,3-dichloropropane, and on treatment with sodium hydroxide regenerates selenetan (**4**). The non-planarity of the selenetan ring and the occurrence of rapid inversion has been confirmed by n.m.r. spectroscopy.[7] The selenetan ring may be opened by reaction with methyl iodide to afford 3-iodopropyldimethylselenonium iodide (**5**).

The 3,3-disubstituted selenetans, like the analogous sulphur compounds, are considerably more stable than selenetan. 3,3-Dimethylselenetan (**6**) may be obtained by reacting 1,3-dibromo-2,2-dimethylpropane with potassium selenide.[6] Similarly to selenetan, the 3,3-disubstituted compounds form 1,1-di-iodo-derivatives and addition compounds with

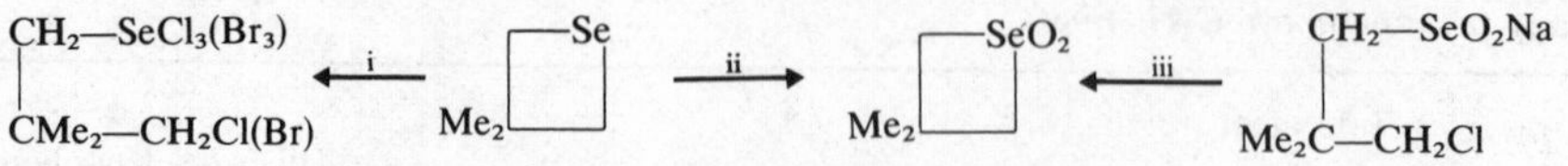

$$CH_2\text{–}SeCl_3(Br_3) / CMe_2\text{–}CH_2Cl(Br) \xleftarrow{i} \text{(6)} \xrightarrow{ii} \text{(7)} \xleftarrow{iii} CH_2\text{–}SeO_2Na / Me_2C\text{–}CH_2Cl \ \text{(8)}$$

i, $2Cl_2$ or $2Br_2$; ii, H_2O_2; iii, sealed tube, 85 °C.

SCHEME 1

mercuric chloride. Ring opening occurs when they are treated with methyl iodide, chlorine, or bromine. Oxidation of 3,3-dimethylselenetan (**6**) using hydrogen peroxide gives the 1,1-dioxide (**7**), also obtained by the cyclization of sodium 3-chloro-2,2-dimethylpropane-1-seleninate (**8**) (Scheme 1).

19.3.2.2 Five-membered ring containing one selenium atom

(i) Selenophens[1,8] *and their synthesis*

2,5-Dimethylselenophen (**10**), the first selenophen to be described, was obtained by heating acetonylacetone (**9**) with phosphorus pentaselenide[9] (reaction 4). Selenophen (**11**), b.p. 110 °C, is prepared by passing acetylene over selenium at 400 °C (reaction 5),[10] and 2,5-disubstituted selenophens (**13**) are obtained by the addition of hydrogen selenide to the appropriate diyne in the presence of either silver or cuprous ion,[11] or by the condensation of ketoacetylenic esters with hydrogen selenide (reaction 6).[12] Selenophen (**11**) is also formed when furan (**12**) is heated with hydrogen selenide in the presence of magnesium oxide (reaction 5).[13]

H_2C—CH_2 / MeCO COMe (**9**) —P_2Se_5, Δ→ Me–selenophen–Me (**10**) (4)

CH≡CH + CH≡CH —H_2Se, 450 °C→ selenophen (**11**) ←H_2Se, MgO, Δ— furan (**12**) (5)

$R^2C≡C–C≡CR^1$ —H_2Se→ (**13**) (R^2, R^1, Se) ←R^1 = Me— R^3O_2C, R^2, Me, Se ←H_2Se— R^3O_2CCH—CH_2 / R^2CO C≡CH (6)

2-Substituted 4-methylselenophens (**15**) may be prepared by the reaction between the necessary acetylenic epoxide (**14**) and hydrogen selenide (reaction 7),[14] while 3,4-diphenylselenophen (**17**) is obtained following the hydrolysis and decarboxylation of the products formed by the Hinsberg condensation of dimethyl selenodiacetate (**16**) with benzil (reaction 8).[15] 2,4-Diarylselenophens (**18**) may be obtained by fusing anils of aliphatic aromatic ketones with selenium at 240 °C (reaction 9).[16]

Me, C—C≡CR, O, CH_2 (**14**) —H_2Se→ [Me—C(OH)(CH_2SeH)—C=CR] → [Me—C(=HC–SeH)—C=CR] → Me, R, Se (**15**) (7)

2,3,4,5-Tetraphenylselenophen (**20**) is prepared from 1,4-dilithio-1,2,3,4-tetraphenylbutadiene (**19**) and diselenium dibromide or from 1,4-di-iodo-1,2,3,4-tetraphenylbutadiene and lithium selenide,[17] and also when 2,3,4,5-tetraphenylcyclopentadien-1-one (**22**) is fused with selenium (Scheme 2).[18] 2-Methyl-3,4,5-trisubstituted

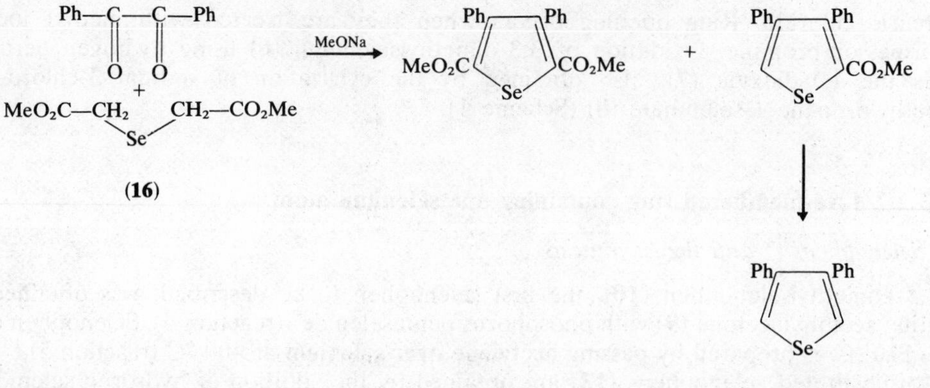

(**16**) (8)

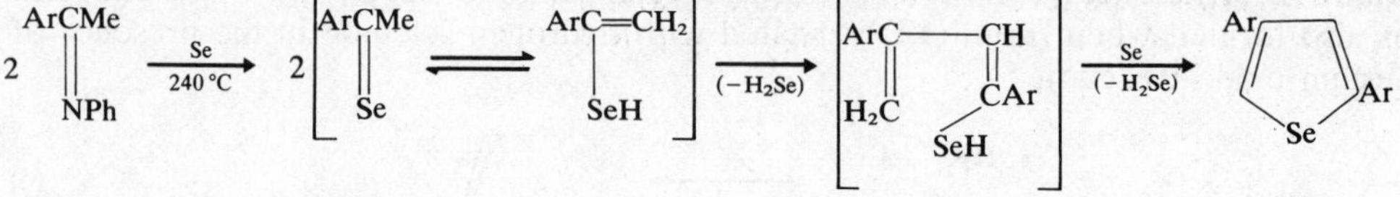

(**17**)

(9)

(**18**)

selenophens (**24**) may be obtained by the reaction of the indicated type of unsaturated ketone (**23**) with hydrogen selenide and hydrogen chloride[12] (reaction 10).

(*ii*) *Properties of selenophens*

Selenophen is little affected under moderate conditions by acid, alkali, or oxidizing or reducing reagents, and gives a greenish-blue indophenine reaction on treatment with

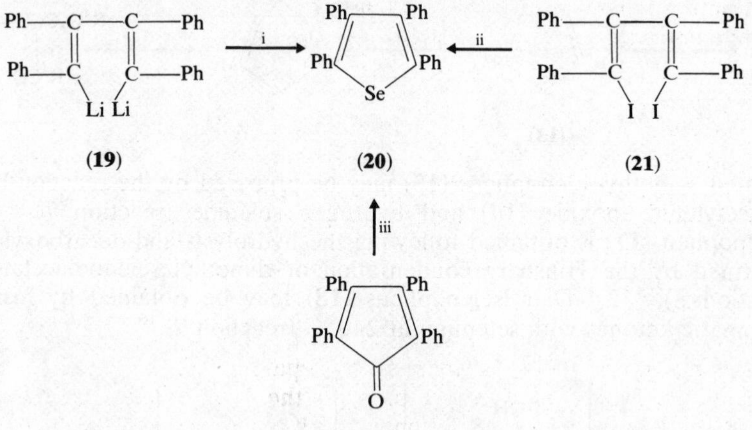

(**19**) (**20**) (**21**)

(**22**)

i, Se_2Br_2; ii, Li_2Se; iii, Se.

SCHEME 2

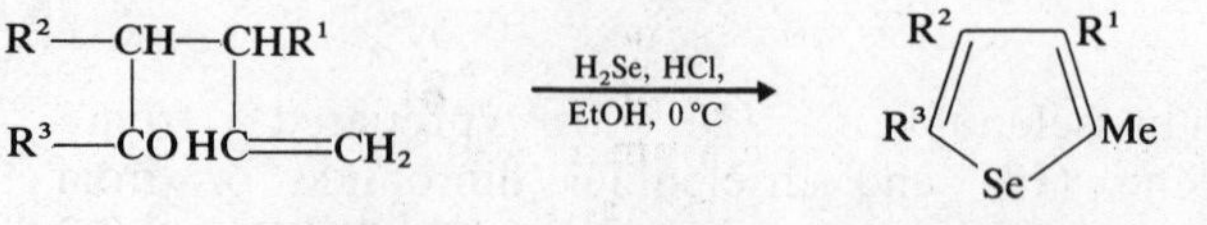

(10)

(**23**) (**24**)

(25)

isatin in concentrated sulphuric acid. The bond lengths and angles of the selenophen molecule (**25**) have been determined from microwave spectral data[19] by the Kraitchman–Costain method[20] and compared with those obtained for furan[21] and thiophen[22] (Table 4). Some workers[23,24] believe that the evidence provided by X-ray analysis indicates that the selenium atom in selenophen-2-carboxylic acid is about 6 pm above the plane of the ring.

(*iii*) *Halogenation, nitration, and sulphonation*

Electrophilic substitution occurs more readily in selenophen than in benzene and, as in the cases of furan and thiophen, more easily in the α- than the β-positions. This is best understood in terms of resonance stabilization or delocalization of positive charge in the intermediate cation, which is greater in the intermediate (**26**), derived by α-addition, than in (**27**), derived by β-addition and in which the C-4 — C-5 double bond cannot be mesomerically involved in the delocalization of the positive charge. With selenophen there are no clear-cut orientation rules for electrophilic substitution, but regardless of what substituent may be in the 2-position, further electrophilic substitution occurs chiefly at the 5-position. Kinetic studies have shown that selenophen is more susceptible to electrophilic and nucleophilic substitution reactions than is thiophen.

(26) **(27)**

Direct halogenation of selenophen occurs at the 2-position and yields 2-bromo- and 2-chloro-selenophen, but the iodoselenophens are prepared either by reaction with iodine in the presence of yellow mercuric oxide, or from the mercurated selenophen, which is readily formed by treatment of the selenophen with a mercuric salt, *e.g.* mercuric acetate in dilute acetic acid.[25] Depending on the reaction conditions, further halogenation occurs at the 5-position to give 2,5-dibromo- and 2,5-dichloro-selenophen.[10] 2-Chloroselenophen is also obtained when chlorination is carried out using sulphuryl chloride.[26] The 3-halogenoselenophens cannot be obtained by direct methods; thus 3-bromoselenophen is prepared by the debromination of 2,3,5-tribromoselenophen using zinc and acetic acid,[27] and 3-iodoselenophen by treating tetraiodoselenophen with aluminium amalgam.[28] Tetrachloroselenophen (**28**) may be obtained by the reaction between hexachlorobutadiene and selenium at 250 °C (reaction 11).[29]

Selenophen, on nitration with nitric acid in acetic anhydride, yields mainly 2-nitroselenophen and a small amount of 3-nitroselenophen,[30] which may also be obtained

TABLE 4

Bond Lengths and Angles of Furan, Thiophen, and Selenophen

Compound		*Bond length* (pm)					*Bond angle*				
		X—C-2	C-2—C-3	C-3—C-4	C-2—H	C-3—H	X—C-2—C-3	C-5—X—C-2	C-4—C-3—C-2	X—C-2—H	C-4—C-3—H
Furan	X = O	136.21	136.09	143.09	107.50	107.68	110°41′	106°33′	106°3′	115°55′	127°57′
Thiophen	X = S	171.40	136.96	142.32	107.76	108.05	111°28′	92°10′	112°27′	119°51′	124°16′
Selenophen	X = Se	185.47	136.95	143.32	107.00	107.92	111°34′	87°46′	114°33′	121°44′	122°52′

$ClC{=}CCl_2$ / $ClC{=}CCl_2$ (hexachlorobutadiene) $\xrightarrow[250\,°C]{Se}$ tetrachloroselenophen **(28)** (11)

by the decarboxylation of 4-nitroselenophen-2-carboxylic acid.[31] Similarly, decarboxylation of 5-nitroselenophen-2-carboxylic acid affords 2-nitroselenophen. Nitration thus occurs preferentially at the 2-position, but if this position is blocked by an electron-accepting group, *e.g.* nitro, formyl, or acetyl, then nitration takes place mainly at the 4-position with some at the 5-position.[32] In the case of the 2-acetyl derivative a third product, 2,4-dinitroselenophen, is also obtained.[31]

Sulphonation of selenophen using sulphuric acid[33] or pyridine–sulphur trioxide[34] yields the 2-sulphonic acid, and selenophen-2-carbaldehyde with dioxan–sulphur trioxide gives selenophen-2-carbaldehyde-5-sulphonic acid. Treatment of selenophen-2-carboxylic acid with oleum affords the 5-sulphonic acid derivative along with some of the 4-isomer. By reaction with fuming nitric acid the sulphonic acid group is readily replaced by a nitro-group.[35]

(*iv*) *Formylation, acylation, and chloromethylation*

Selenophen-2-carbaldehyde (**29**) is obtained by the formylation of selenophen using dimethylformamide in the presence of phosphoryl chloride (reaction 12).[36] Alkylselenophens behave similarly, *e.g.* 2-methyl- and 3-methyl-selenophen give 2-methylselenophen-5-carbaldehyde and 3-methylselenophen-2-carbaldehyde, respectively.[37]

selenophen $\xrightarrow[POCl_3]{DMF}$ selenophen-2-CHO **(29)** (12)

3-iodoselenophen $\xrightarrow{PhLi}$ 3-lithioselenophen $\xrightarrow{DMF}$ selenophen-3-CHO **(30)** (13)

3-(CH_2Br)selenophen $\xrightarrow[\text{ii, 50\% AcOH}]{\text{i, } C_6H_{12}N_4,\ CHCl_3}$ selenophen-3-CHO **(30)** $\xleftarrow{LiAlH_4}$ selenophen-3-CN (14)

The 3-carbaldehyde (**30**) is prepared by treating 3-selenienyl-lithium with dimethylformamide (reaction 13),[28] and also by the Sommelet reaction from 3-bromomethylselenophen, and from selenophen-3-carbonitrile (reaction 14).[38]

Selenophen-2-carbaldehyde behaves as an aromatic aldehyde. It reacts readily with ammonia and aromatic amines, and gives the Perkin reaction to yield (β-(2-selenienyl)-acrylic acid (**31**), also obtained by the Knoevenagel condensation with malonic acid (reaction 15).[36] 2-Selenienyl alkyl carbinols may be obtained by the Grignard reaction between the 2-carbaldehyde (**29**) and an alkylmagnesium halide.[39]

selenophen-2-CHO $\longrightarrow$ selenophen-2-$CH{=}CHCO_2H$ **(31)** (15)

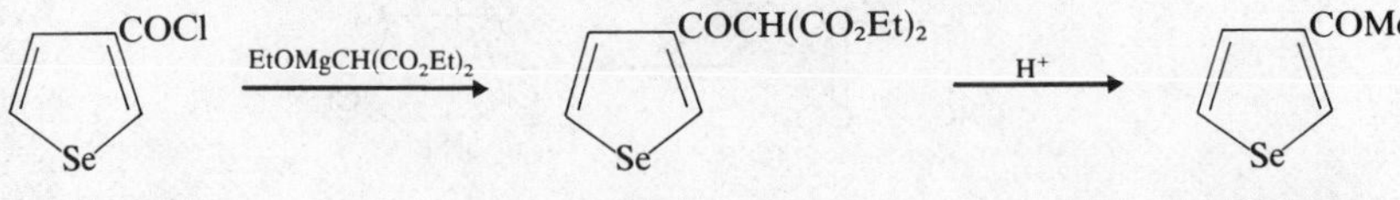

(32) COCl, Se —$EtOMgCH(CO_2Et)_2$→ (33) $COCH(CO_2Et)_2$, Se —H^+→ (34) COMe, Se (16)

Treatment of selenophen with acyl chlorides under Friedel–Crafts conditions results in acylation at the 2-position.[33,40] The same products are formed by using acid anhydrides in the presence of phosphoric acid[41] and tetra-acyloxysilanes in the presence of stannic chloride.[42] Similarly to 3-substituted derivatives mentioned previously, the 3-acylselenophen can only be obtained by an indirect method, *e.g.* 3-acetylselenophen (**34**) is prepared by reacting selenophen-2-carboxylic acid chloride (**32**) with diethyl ethoxymagnesium malonate and then treating the product (**33**) with acid (reaction 16).[43]

If hydrogen chloride is passed into a mixture of selenophen and formalin in ethylene dichloride, chloromethylation occurs as expected at the 2-position to give 2-chloromethylselenophen (**35**) (reaction 17) and a small amount of 2,5-bischloromethyselenophen.[44]

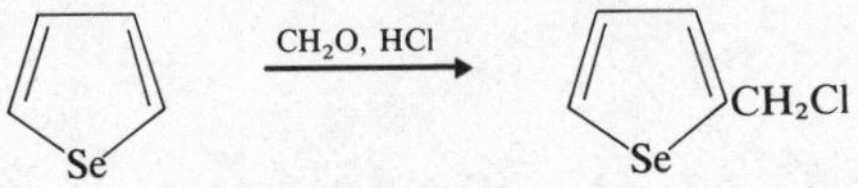

Se —CH_2O, HCl→ (35) CH_2Cl, Se (17)

(*v*) *Other derivatives and reactions of selenophen*

Metallation occurs at the 2-position to give 2-selenienyl-lithium when selenophen is treated with butyl-lithium.[43] The same result can be obtained by reacting 2-iodoselenophen with phenyl-lithium.[45] Similar treatment of 3-bromo- or 3-iodoselenophen affords 3-selenienyl-lithium. The metallated compounds may be used to obtain the related carbaldehydes or carboxylic acids (**36**)[46] and (**37**)[43,47] (reactions 18 and 19). Both these acids may be prepared by alternative methods, *e.g.* selenophen-2-carboxylic acid is formed by the oxidation of either the 2-carbaldehyde or the 2-acetyl derivative with either alkaline peroxide[39] or potassium permanganate, respectively. The 2-carboxylic acid is also obtained by the carbonation of 2-selenienylmagnesium iodide,[48] and the 3-acid (**37**) by hydrolysis of the 3-cyano-compound (**38**) formed from the related 3-bromo-derivative (reaction 20).[43] 2-Chloromethylselenophen (**39**) behaves as a typical allylic halide, *e.g.* reaction 21.[49]

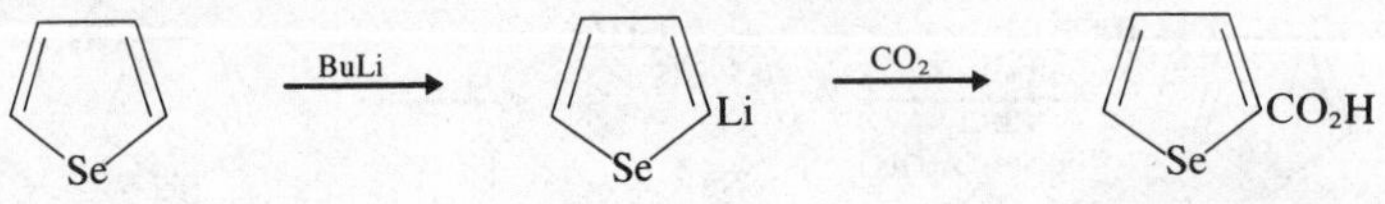

Se —BuLi→ Li, Se —CO_2→ (36) CO_2H, Se (18)

Br, Se —PhLi→ Li, Se —CO_2→ (37) CO_2H, Se (19)

Br, Se —CuCN, C_5H_5N→ (38) CN, Se —HCl→ (37) CO_2H, Se (20)

(39) CH_2Cl, Se —$NaCH(CO_2Et)_2$→ $CH_2CH(CO_2Et)_2$, Se —aq. KOH, H^+→ $CH_2CH_2CO_2H$, Se (21)

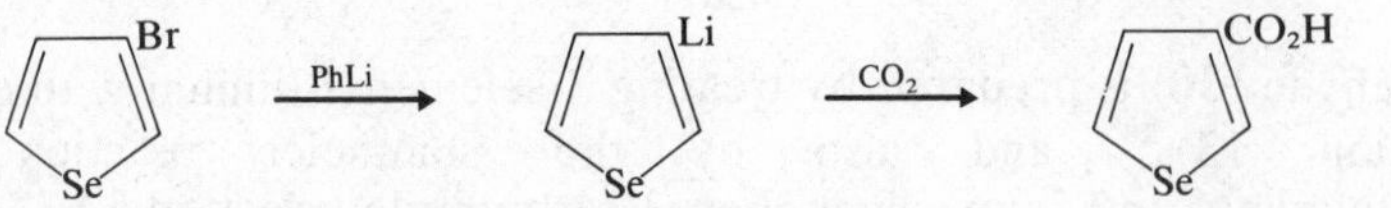

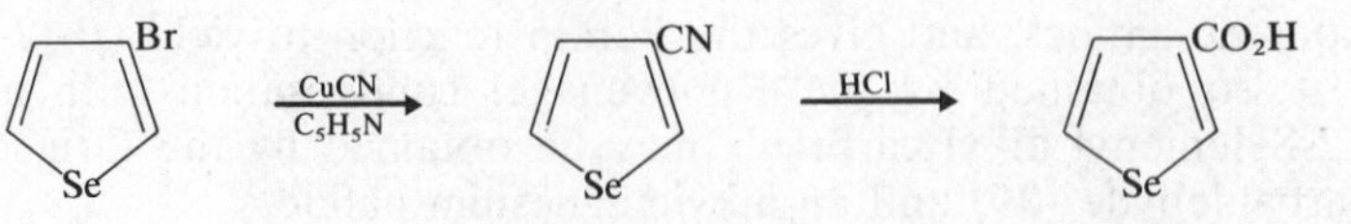

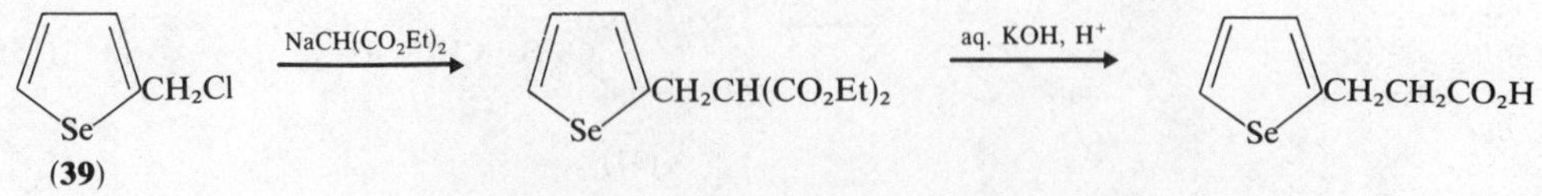

Some mono-, di-, and tetra-deuterioselenophens[50] are obtained by iodine–deuterium exchange when the corresponding iodoselenophens are heated with deuterium oxide, zinc filings, and acetic anhydride. 2-Deuterio- (**40**), 2,5-dideuterio-, and 2,3,4,5-tetradeuterio-selenophens are prepared by this method (reaction 22), but 3-deuterioselenophen (**41**) is obtained by treating 3-selenienyl-lithium, formed from 3-bromoselenophen, with deuterium oxide (reaction 23).

Zn, Ac_2O, D_2O (22)

(**40**)

EtLi, −70 °C; Ac_2O, D_2O, −70 °C (23)

(**41**)

The rate constant for deuterium exchange of the α-H of selenophen with a solution of lithium t-butoxide in dimethyl sulphoxide at 25 °C is $1.5\times10^{-4}\,l\,mol^{-1}\,s^{-1}$, whereas the rate constant of the α-H exchange for α-deuteriothiophen under the same conditions is $1.0\times10^{-4}\,l\,mol^{-1}\,s^{-1}$, which shows that the mobility of the α-H in selenophen is enhanced only very slightly compared with thiophen.[51]

Biselenienyl (**42**) is formed when 2-iodoselenophen is treated with activated copper[52] (reaction 24).

Cu (24)

(**42**)

(*vi*) *Di- and tetra-hydroselenophens*

When selenophens (**43**; R = Me, CHO, Ac, SMe, OBu^t) of the type illustrated are distilled in the presence of toluene-*p*-sulphonic acid, they give a mixture of 5-substituted 2,3-dihydro- and 5-substituted 2,5-dihydro-selenophen-2-ones (**44**) and (**45**)[53] (reaction 25). If compound (**43**; R = OBu^t) is treated with acid, the resulting product is 2,5-dihydroselenophen-2,5-dione (**46**), the selenium homologue of maleic anhydride.

H^+ (25)

(**43**) (**44**) (**45**)

(**46**)

Selenolan (tetrahydroselenophen) (**48**), b.p. 135 °C, was originally prepared by reacting 1,4-dibromobutane (**47**) with sodium selenide in water under hydrogen[54] (reaction 26). A

Na_2Se, H_2O, H_2, 80 °C (26)

(**47**) (**48**)

better method, which eliminates the use of metallic sodium in the preparation of the sodium selenide, and the atmosphere of hydrogen, is to react the dibromo-compound (**47**) with a mixture of selenium powder, sodium formaldehyde sulphoxylate (Rongalite), sodium hydroxide, and water,[55] which mixture behaves as sodium selenide (reactions 27–29). It may also be obtained by passing tetrahydrofuran over alumina at 400 °C in a stream of hydrogen selenide.[56]

$$HOCH_2SO_2^- + Se + 2OH^- \longrightarrow HOCH_2SO_3^- + Se^{2-} + H_2O \quad (27)$$

$$Br(CH_2)_4Br + Se^{2-} \longrightarrow Br(CH_2)_4Se^- + Br^- \quad (28)$$

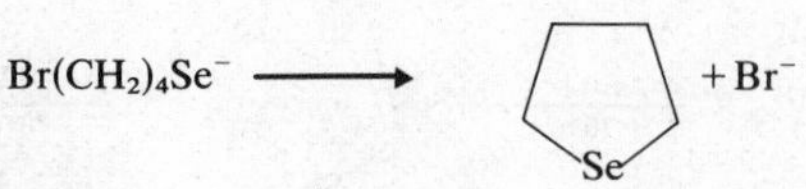

$$Br(CH_2)_4Se^- \longrightarrow \text{(48)} + Br^- \quad (29)$$

(**48**)

Selenophen reacts with chlorine under a variety of conditions to give a number of chlorinated tetrahydroselenophens, *e.g.* selenophen with an excess of chlorine affords 2,2,3,4,5,5-hexachloroselenolan (**49**), and in carbon disulphide at −15 °C with chlorine yields 2,3,4,5-tetrachloroselenolan (**50**).[10] The reaction between selenophen and bromine in carbon disulphide gives 1,1,2,2,5,5-hexabromoselenolan (**51**), which on treatment with

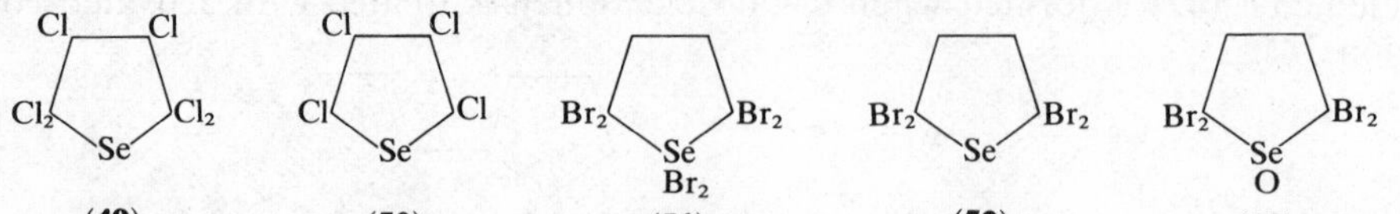

(**49**) (**50**) (**51**) (**52**) (**53**)

water yields a mixture of 2,2,5,5-tetrabromoselenolan (**52**) and 2,2,5,5-tetrabromoselenolan 1-oxide (**53**). 2,2,3,3,4,4,5,5-Octafluoroselenolan (**54**) is obtained directly, from the reaction of tetrafluoroethylene with selenium in the presence of iodine at 250 °C under pressure (reaction 30).[57]

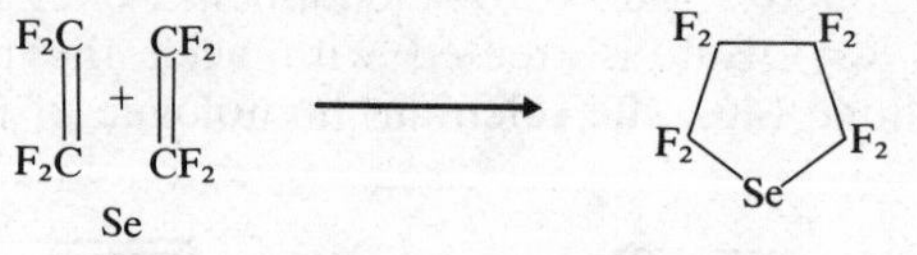

$$2\,F_2C{=}CF_2 + Se \longrightarrow \text{(54)} \quad (30)$$

(**54**)

$$HO_2C\text{-}CHBr\text{-}CH_2\text{-}CH_2\text{-}CHBr\text{-}CO_2H \xrightarrow[Me_2CO]{K_2Se_2 \text{ or } K_2Se} \text{(55)} \quad (31)$$

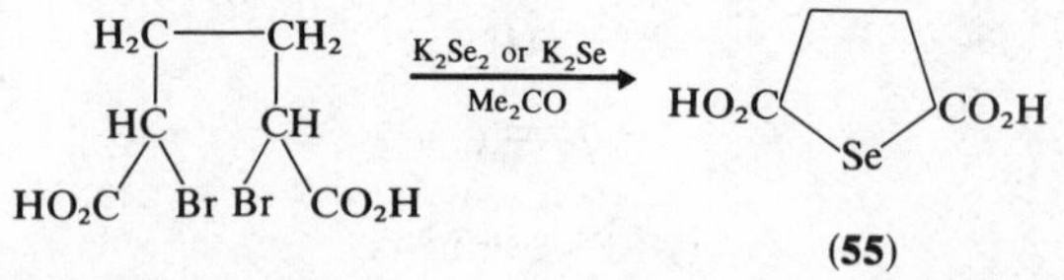

(**55**)

Selenolan-2,5-dicarboxylic acid (**55**), which may exist in either the *cis*- or *trans*-form, is obtained by reacting the appropriate α,α'-dibromoadipic acid with either potassium diselenide or monoselenide (reaction 31).[58] The *cis*-compound is obtained from *meso*-α,α'-dibromoadipic acid and the *trans*-isomer from racemic α,α'-dibromoadipic acid. The latter acid has been resolved by making use of either its brucine or quinine salt.[59]

A drastic change is encountered on comparison of the mass spectra of selenolan (**48**) and tetrahydrothiophen. In the latter the base peak corresponds to the elimination of 28 mass units, but with selenolan it corresponds to the loss of SeH, with the charged species being $C_4H_7^+$ (($m/e = 55$).[61] Also present in the spectrum are peaks at $m/e = 107$ and $m/e = 94$ (Scheme 3).

(48)

$m/e = 55$

$m/e = 107$

$m/e = 94$

SCHEME 3

1,6-Di-iodo-3,4-bisiodomethylhexane (**56**) on boiling with sodium selenide in absolute alcohol under nitrogen affords 3,3′-biselenolan (octahydro-3,3′-biselenophen (reaction 32).[60]

(56) $\xrightarrow[\text{EtOH}]{Na_2Se}$ (57) (32)

19.3.2.3 Five-membered rings containing one selenium atom and fused to a benzene ring

(*i*) *Benzo*[*b*]*selenophens and their synthesis*

Benzo[*b*]selenophens are obtained in general by methods related to those used for the preparation of furans and thiophens. Benzo[*b*]selenophen (**60**), m.p. 50–51 °C, is prepared by the cyclization of *ω*-dimethoxyethyl phenyl selenide (**59**), formed by the reaction (under hydrogen) between benzeneselenol (**58**) and bromoacetal in the presence of sodium ethoxide (reaction 33).[62] It is also isolated, along with selenophen, following the

(58) + $BrCH_2CH(OMe)_2$ $\xrightarrow[H_2]{\text{Na, EtOH}}$ (59) $\xrightarrow[H_3PO_4]{P_2O_5}$ (60) (33)

passage of acetylene over heated elemental selenium,[63] or by the passage of styrene over a heated chromium–aluminium oxide catalyst in the presence of selenium dioxide vapour.[64]

α-(2-Carboxyphenylseleno) propionic acid (**61**) on heating with anhydrous potassium acetate and acetic anhydride at 115 °C, and then at 135–140 °C after the subsidence of the evolution of carbon dioxide, affords an enol acetate, which on hydrolysis with potassium hydroxide, followed by acidification, yields 2-methyl-2,3-dihydro-3-oxobenzo[*b*]selenophen (**62**). The parent compound 2,3-dihydro-3-oxobenzo[*b*]selenophen (**62**; Me = H) may be obtained from (2-carboxyphenyl)acetic acid. Absorption and n.m.r. spectral data show that in neutral and acidic solvents compound (**62**) is mainly present as the ketone.[65] By comparison, the sulphur analogue exists, both in neutral and acidic solvents, in the enol form.[66] The ketone (**62**) on reduction with sodium amalgam in ethanol gives 2-methylbenzo[*b*]selenophen (**63**) (reaction 34).[67]

(61) —i, KOAc, Ac_2O; ii, KOH, H^+→ (62) —Na/Hg, EtOH→ (63) (34)

2-Acetylphenylselenacetic acid (**64**) with alkali gives 3-methylbenzo[*b*]selenophen-2-carboxylic acid (**65**), which may be decarboxylated to yield 3-methylbenzo[*b*]selenophen (**66**) (reaction 35).[68]

(64) —i, 30% KOH; ii, H^+→ (65) —Cu bronze, quinoline→ (66) (35)

(ii) Properties of benzo[b]selenophen and the preparation of some of its derivatives

The acetylation of benzo[*b*]selenophen (see reaction 36) affords 2-acetylbenzo[*b*]selenophen (**67**),[64] rather than the 3-isomer. Direct acetylation of 3-methylbenzo[*b*]selenophen (**66**) gives a moderate yield of the 2-acetyl derivative (**65**) (reaction 37), but with 2-methylbenzo[*b*]selenophen (**63**) only a very poor yield of the 3-acetyl derivative (**69**) is obtained. A moderate yield of (**69**) is achieved by treating 2-methylbenzo[*b*]selenophen with mercuric acetate followed by acetyl chloride (reaction 38).[69]

(60) —AcOH, $SiCl_4$, $SnCl_4$, C_6H_6→ (67) (36)

(66) —AcCl, CS_2, $SnCl_4$→ (68) (37)

(63) —i, $Hg(OAc)_2$; ii, AcCl→ (69) (38)

(70) —SeO_2, C_5H_5N→ (71) (39)

Neither benzo[*b*]selenophen-2-carbaldehyde nor -3-carbaldehyde (**73**) has so far been obtained by direct formylation of benzo[*b*]selenophen. The 2-carbaldehyde (**71**) is formed when selenochromene (2*H*-benzo[*b*]selenopyran) (**70**) is oxidized with selenium dioxide in boiling pyridine (reaction 39),[70] or when the hydrolysis product from selenochromylium perchlorate (see Section 19.3.2.6) is oxidized with manganese dioxide. Other related 2-carbaldehydes may be prepared by these methods. Benzo[*b*]selenophen-3-carbaldehyde

(66) $\xrightarrow[\text{Bz}_2\text{O}_2,\ \text{C}_6\text{H}_6]{\text{NBS,}}$ (72) $\xrightarrow[\text{ii, 50\% AcOH}]{\text{i, C}_6\text{H}_{12}\text{N}_4,\ \text{CHCl}_3}$ (73) (40)

(66) **(72)** **(73)**

(**73**) is prepared from 3-methylbenzo[*b*]selenophen (**66**), which is first converted to 3-bromomethylbenzo[*b*]selenophen (**72**) by reaction with *N*-bromosuccinimide. The bromo-compound (**72**) is converted into the aldehyde (**73**) by Sommelet reaction (equation 40).[68]

Benzo[*b*]selenophen on bromination yields a mixture of products, from which is isolated 3-bromo- (**74**), 2,3-dibromo- (**75**), and 2,3,6-tribromo-benzo[*b*]selenophen (**76**), and 1,3-dibromobenzo[*b*]selenienyl bromide (**77**). A mixture of bromo-derivatives (**74**)

(74) **(75)** **(76)** **(77)**

and (**75**) is obtained by reacting phenylacetylene with selenium tetrabromide. Bromination of either 2-methyl- (**63**) or 3-methyl-benzo[*b*]selenophen (**66**) results in electrophilic substitution taking place at the vacant 3- or 2-position to give the respective 3-bromo- (**78**) and 2-bromo- (**79**) derivatives[69] (reaction 41). Treatment of 2-bromo-3-methylbenzo[*b*]selenophen (**79**) with lithium, followed by carbonation, yields 3-methylbenzo[*b*]selenophen-2-carboxylic acid (**80**), also obtained by treating 3-methylbenzo[*b*]selenophen (**66**) with butyl-lithium and dry ice (reaction 42). 2-Methylbenzo[*b*]selenophen-3-carboxylic acid (**81**) is formed by pouring the Grignard reagent prepared from 3-bromo-2-methylbenzo[*b*]selenophen (**78**) and magnesium in tetrahydrofuran on to dry ice (reaction 43).

(63) → (78)   (66) → (79) (41)

(63) **(78)** **(66)** **(79)**

(79) $\xrightarrow{\text{i, Li, ii, CO}_2}$ (80) $\xleftarrow[\text{ii, CO}_2]{\text{i, BuLi,}}$ (66) (42)

(79) **(80)** **(66)**

(78) $\xrightarrow[\text{ii, CO}_2]{\text{i, Mg, THF,}}$ (81) (43)

(78) **(81)**

2-Bromo-3-methylbenzo[*b*]selenophen (**79**), like 3-methylbenzo[*b*]selenophen (**66**), may be converted to 2-bromobenzo[*b*]selenophen-3-carbaldehyde (**82**) and thence to benzo[*b*]selenophen-2,3-dicarboxylic acid (**84**) by oxidation of the intermediate (**83**) using silver oxide (Scheme 4). Similarly, benzo[*b*]selenophen-2-carbaldehyde-3-carboxylic acid (**85**) may be obtained and converted to the dicarboxylic acid (**84**).

(79) (82) (83) (84) (85)

i, EtOH, $HC(OEt)_3$, $KHSO_4$; ii, BuLi, CO_2; iii, Ag_2O.

SCHEME 4

2-Methyl-2,3-dihydrobenzo[*b*]selenophen (**87**) is formed when allyl phenyl selenide (**86**) is boiled in excess of quinoline under dry nitrogen (reaction 44). No other products

$$\text{PhSeCH}_2\text{CH=CH}_2\ (\mathbf{86}) \xrightarrow[\text{N}_2]{\text{quinoline,}} (\mathbf{87}) \qquad (44)$$

are obtained, *e.g.* 2-allylselenophenol or selenochroman, and no cyclization occurs when the selenide (**86**) is heated alone at a temperature >200 °C. The cyclization mechanism appears, therefore, to be different from the classical Claisen rearrangement.[71]

(88) (89)

2,3-Dihydro-3-oxobenzo[*b*]selenophen on oxidation with potassium ferricyanide in alkali gives a mixture of *cis*- and *trans*-selenoindigo, (**88**) and (**89**).[72] It is reported that 2,3-dihydro-5-methyl-3-oxobenzo[*b*]selenophen when treated under the same conditions gives the *trans*-selenoindigo.[73]

–Se; $m/e = 102$; $-C_2H_2$; $m/e = 76$; –H; $m/e = 75$

–CSe; $m/e = 90$; –H; $m/e = 89$; $-C_2H_2$; $m/e = 63$

SCHEME 5

U.v. spectral data of benzoselenophen have been reported.[74] The mass spectrometry of benzo[*b*]selenophen and some of its derivatives shows that whilst the fragmentation of these compounds follows the pattern of the corresponding benzo[*b*]thiophens, the benzo-[*b*]selenophens are relatively more fragile, selenium being easily split off (Scheme 5).[75]

19.3.2.4 Five-membered rings containing one selenium atom and fused to two benzene rings

(i) Synthesis of dibenzoselenophens (diphenylene selenides)

Selenanthrene (**90**) on heating with copper-bronze under nitrogen gives dibenzo-selenophen (**91**), m.p. 78 °C, also obtained from the diazonium salt formed from 2-aminodiphenyl selenide (**92**)[76] (reaction 45). Treatment of bis-2-biphenyl diselenide (**93**) with bromine also yields dibenzoselenophen (reaction 46);[77] it is also obtained either by the cyclodehydration of diphenyl selenoxide (**94**) using sodamide[78] (reaction 47), or by heating biphenylene (**95**) with selenium, when it is formed as an insertion product[79] (reaction 48).

(90) —230 °C→ (91) ←$NaNO_2$, H_2SO_4— (92) (45)

(93) —Br_2→ [2-biphenylyl-SeBr] → (91) (46)

(94) —$NaNH_2$→ (91) (47)

(95) —Se, Δ→ (91) (48)

(ii) Properties of dibenzoselenophen and the preparation of some of its derivatives

Nitration of dibenzoselenophen occurs mainly in the 2-position to give 2-nitrodibenzoselenophen (**96**), which on reduction yields 2-aminobenzoselenophen (**97**).[80] A second product formed during the nitration has been shown to be 4-nitrodibenzoselenophen (**98**)[81] (reaction 49).

3-Nitrodibenzoselenophen 5-oxide (**100**) is obtained by nitrating dibenzoselenophen 5-oxide (**99**) with a mixture of nitric acid, sulphuric acid, and acetic acid at 0–10 °C. On reduction it affords 3-aminodibenzoselenophen (**101**) (reaction 50).[82]

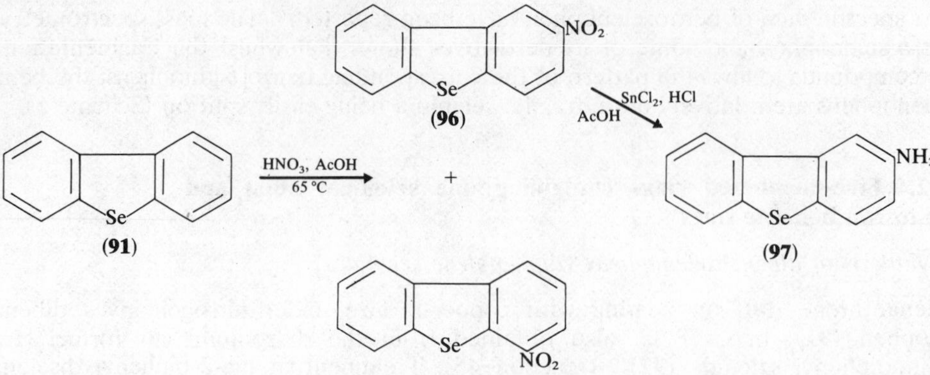

(96) (91) (97) (98) (49)

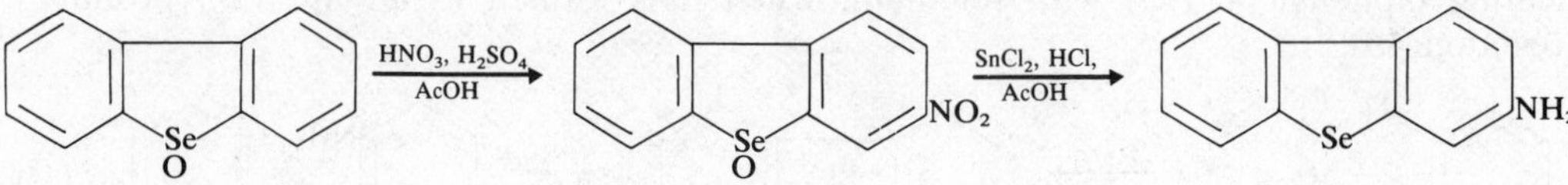

(99) (100) (101) (50)

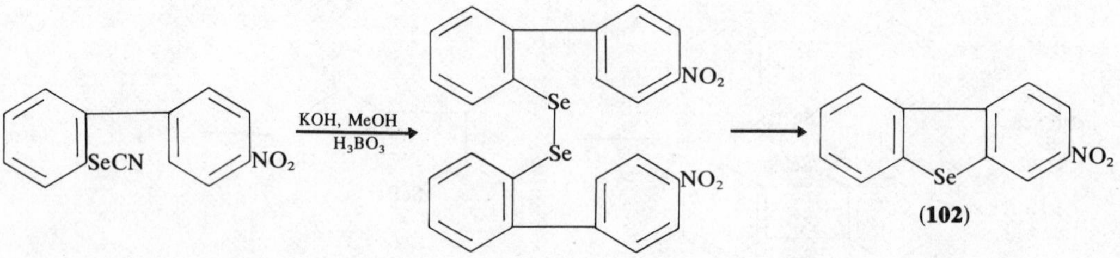

(102) (51)

These results are similar to those obtained on direct nitration of dibenzothiophen and its 5-oxide. 3-Nitrodibenzoselenophen (**102**) has been synthesized (reaction 51)[83] (see also reaction 46). In a similar way, 2,7- and 3,7-dinitrodibenzoselenophen, (**103**) and (**104**), may be prepared from 2-selenocyanato-5,4′- and 2-selenocyanato-4,4′-dinitrobiphenyl.

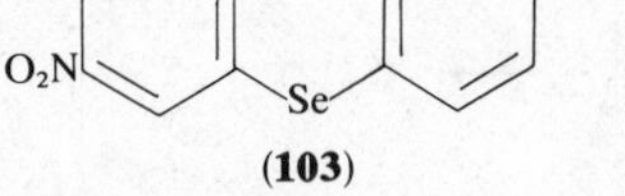

(**103**)

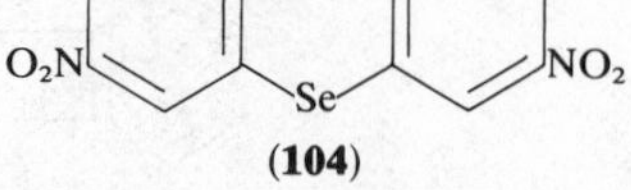

(**104**)

2-Iododibenzoselenophen (**105**) is obtained by diazotization of 2-aminodibenzoselenophen (**97**), then treating the resulting solution with potassium iodide[84] (reaction 52).

The direct bromination of bis-2-biphenyl diselenide (**93**)[77] leads to a mixture of dibenzoselenophen dibromide (**106**) and 2-bromodibenzoselenophen dibromide (**107**)

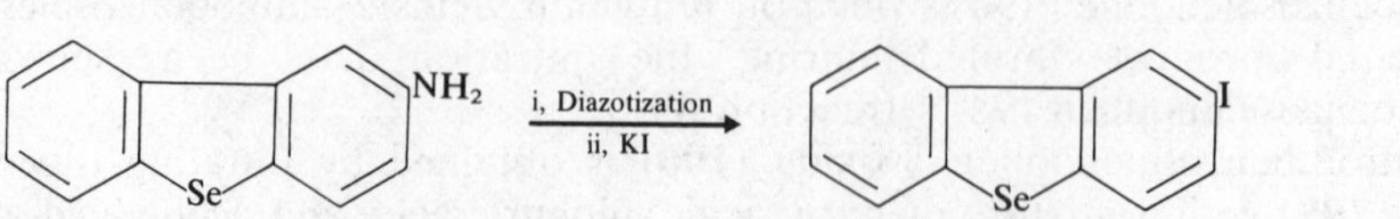

(97) (105) (52)

(reaction 53). It is found that addition of bromine to a solution of 2-selenocyanatobiphenyl (**108**) gives a good yield of the former dibromide (**106**).[85] Treatment of the dibromide (**106**) with 3% aqueous sodium hydroxide affords dibenzoselenophen 5-oxide (**99**) (reaction 54), also obtained by the oxidation of dibenzoselenophen using 40% peracetic acid in acetic acid.[77]

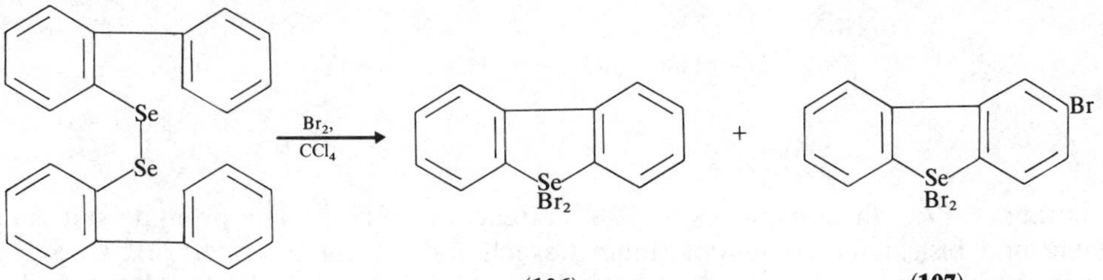

(53)

(**93**) (**106**) (**107**)

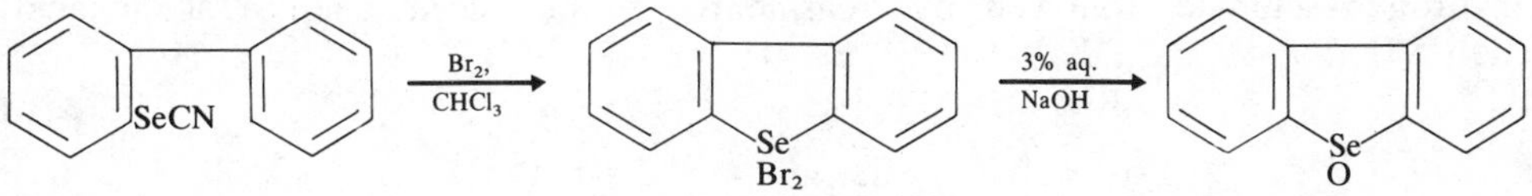

(54)

(**108**) (**106**) (**99**)

When selenium and 1,2-di-iodotetrafluorobenzene (**109**) are heated in a sealed tube at 375 °C, besides the expected selenanthrene (**111**), some octafluorodibenzoselenophen (**110**) is also formed by the occurrence of ring coupling[86] (reaction 55).

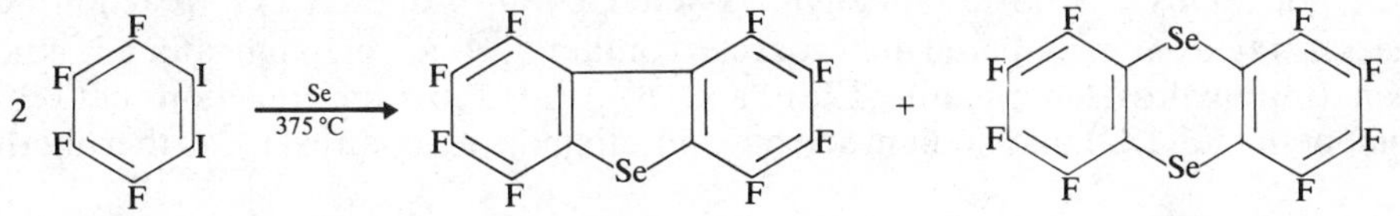

(55)

(**109**) (**110**) (**111**)

The reaction between dibenzoselenophen (**91**) and butyl-lithium results in metallation, occurring most probably at the 4-position to give dibenzoselenienyl-lithium (**112**). The lithium compound on treatment with iodine, dimethyl sulphate, or carbon dioxide affords 4-iodo- (**113**; R = I), or 4-methyl-dibenzoselenophen (**113**; R = Me), or the 4-carboxylic acid (**113**; R = CO_2H), respectively[87] (reaction 56).

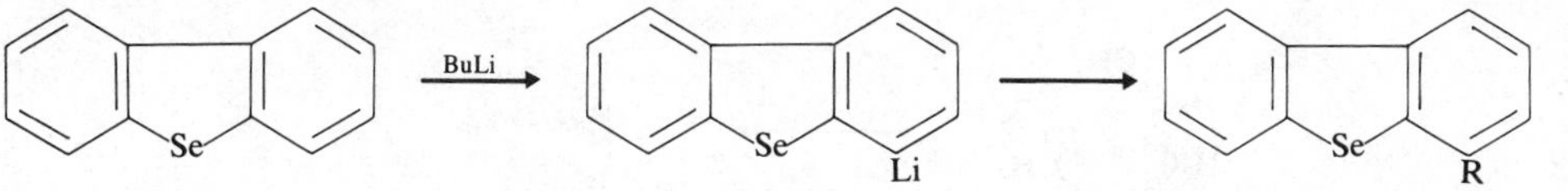

(56)

(**91**) (**112**) (**113**)

19.3.2.5 Six-membered ring containing one selenium atom

(i) Selenopyrans

4*H*-Selenopyran (**116**), an unstable liquid, is prepared from glutaraldehyde (**114**), which, on reacting in methylene dichoride with dry hydrogen chloride and hydrogen selenide, is converted to 2,6-dichloroselenan (2,6-dichlorotetrahydroselenopyran) (**115**). Dehydrochlorination of the dichloro-derivative (**112**) with diethylaniline results in the formation of 4*H*-selenopyran[88] (reaction 57).

Treatment of 4*H*-selenopyran (**116**) with a suitable oxidizing agent such as triphenylmethyl perchlorate results in the formation of selenopyrylium perchlorate (seleninium

(114) $\xrightarrow[\text{CH}_2\text{Cl}_2,\ -10\text{ to }0\,°\text{C}]{\text{H}_2\text{Se, HCl,}}$ (115) $\xrightarrow[130\text{–}170\,°\text{C}]{\text{PhNEt}_2}$ (116) (57)

(116) + Ph_3C^+ ClO_4^- → Ph_3CH + (117) ClO_4^- (58)

perchlorate) (**117**), which explodes at 228 °C (reaction 58).[88,89] The bromide salt is also unstable and bis(selenopyrylium)platinum hexachloride decomposes at 210 °C. Selenoprylium perchlorates (**119**; R^1 = Ph, 4-MeC_6H_4, 4-$MeOC_6H_4$; R^2 = H, Me, Ph) have been prepared by the cyclization of the appropriate 1,5-diketones (**118**) with hydrogen chloride and hydrogen selenide, followed by treatment with perchloric acid in acetic acid[90] (reaction 59).

(118) $\xrightarrow[\text{ii, HClO}_4,\ \text{AcOH}]{\text{i, H}_2\text{Se, HCl,}}$ (119) ClO_4^- (59)

The photolysis of the di-iron carbonyl complex (**120**) in the presence of potassium polyselenide yields 2,3,5,6-tetraphenyl-4*H*-selenopyran-4-one (**121**)[91] (reaction 60).

The saturated six-membered heterocycle containing one selenium atom is known as selenan (tetrahydroselenopyran) (**123**), and is formed by the reaction between 1,5-dibromopentane (**122**) and sodium selenide in ethanol (reaction 61).[92] Other methods of

(120) $\xrightarrow[\text{K}_2\text{Se}_x]{h\nu}$ (121) (60)

(122) → (123) (61)

preparing selenan include reacting the dibromopentane (**122**) with a mixture of selenium powder, sodium formaldehyde sulphoxylate (Rongalite), sodium hydroxide, and water ('sodium selenide'),[55] passing oxan (**124**) and hydrogen selenide over alumina at 400 °C,[56] and by heating what originally was thought to be 1,2-diselenepan (**125**),[92] but which is now

(124) (125) (126)

regarded as a low molecular weight polymer (poly[(diseleno)pentamethylene]) of this monomer.[93] Similarly, poly[(diseleno)hexamethylene] on heating affords 2-methylselenan (**126**).[94]

(*ii*) *Selenans* (*tetrahydroselenopyrans*)

Selenan (**123**) reacts readily with chlorine, bromine, or iodine to form the corresponding selenan 1,1-dihalide,[92] and it has been shown that the 1,1-dibromide exists as a trigonal bipyramid and the 1,1-di-iodide as a molecular complex.[95] Reaction with methyl iodide affords selenan 1-methiodide (1-methylselenanium iodide), and with one molar proportion of 30% hydrogen peroxide in a sealed tube, or two molar proportions of hydrogen peroxide and some heating, yields the 1-oxide or the 1,1-dioxide, respectively. The axial preference of the 1-oxide and the proton in 1-protonated selenan has been demonstrated by n.m.r. spectroscopy. Selenan 1-oxide is found to have a larger proportion of axial isomer (84%) than thian 1-oxide (62%) because the longer carbon–selenium bond produces more attractive 1,3- and 1,5-interactions. For the same reason, selenan 1-methiodide (**127**) appears to be almost entirely axial, whereas thian 1-methiodide (**128**) is almost entirely equatorial.[96]

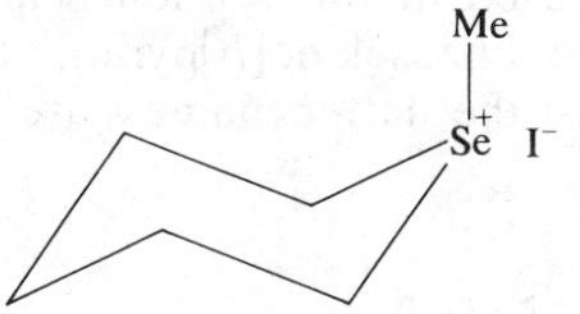

Me
$\overset{+}{Se}$ I^-
$\overset{+}{S}$ I^-
Me

(**127**)  (**128**)

The reaction between α,α'-dibromopimelic acid (**129**) and sodium selenide gives selenan-2,6-dicarboxylic acid (**130**)[97] (reaction 62). The *cis*-isomer is thus obtained from *meso*-dibromopimelic acid and the (±)-*trans*-isomer from the (±)-form of the acid.

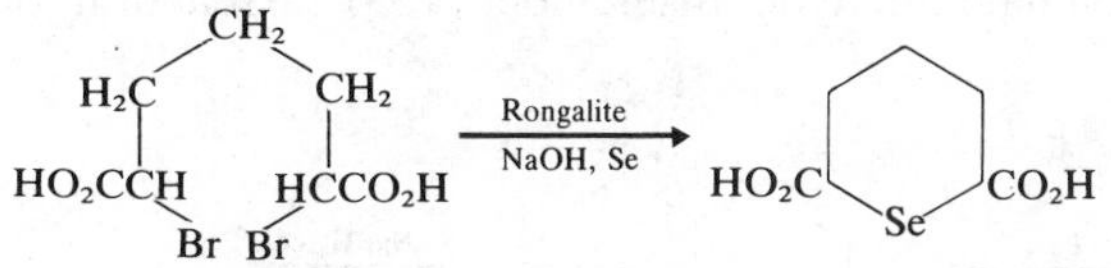

CH_2 H_2C CH_2 HO_2CCH $HCCO_2H$ Br Br —Rongalite / NaOH, Se→ HO_2C Se CO_2H (62)

(**129**) (**130**)

The latter may be resolved by means of its quinine or brucine salt. Diethyl 2,6-dimethyl-4-oxoselenan-3,5-dicarboxylate (**132**) is formed when diethyl acetonedicarboxylate (**131**), acetaldehyde, and hydrogen selenide are condensed together in ethanol in the presence of 1,4-diazabicyclo[2,2,2]octane (DBO) (reaction 63). It has been shown by the application of n.m.r. spectroscopy that the two methyl groups in compound (**132**) are in the *cis* configuration to each other, as are the two ethoxycarbonyl groups.[98] It is also found that the chair form is the preferred conformation with the substituent at C-2, C-3, C-5, and

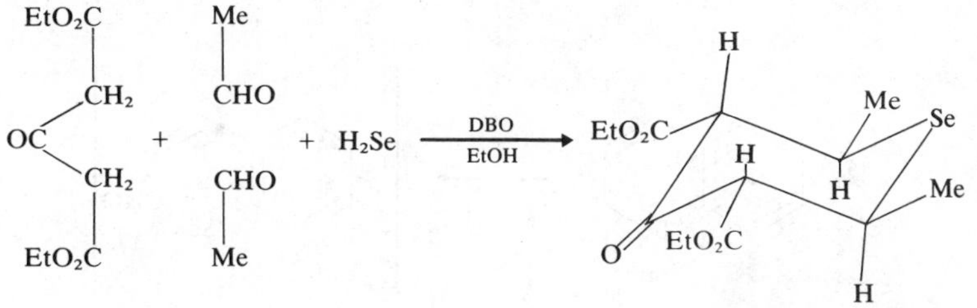

EtO_2C CH_2 OC CH_2 EtO_2C + Me CHO CHO Me + H_2Se —DBO / EtOH→ H EtO_2C H Me Se H Me O EtO_2C H (63)

(**131**) (**132**)

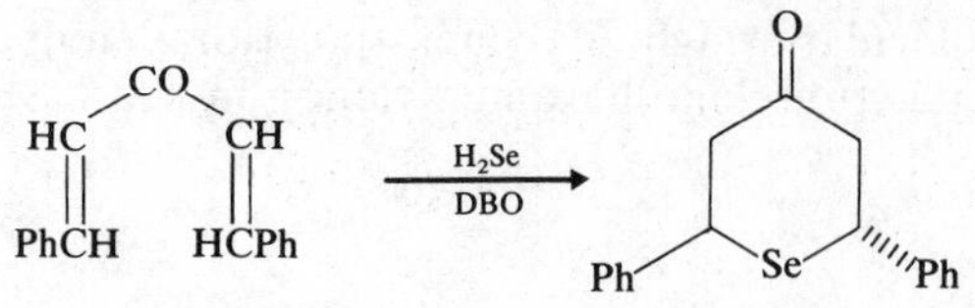

(64)

(133) (134)

C-6 equatorial. Starting with the unsaturated ketone 1,5-diphenyl-3-oxopenta-1,4-diene (**133**) and reacting it with hydrogen selenide in the presence of 1,4-diazabicyclo[2,2,2]octane gives *trans*-2,6-diphenylselenan-4-one (**134**)[99] (reaction 64).

19.3.2.6 Six-membered rings containing one selenium atom fused to a benzene ring

The names of these compounds are derived from those given to the related oxygen heterocycles, thus 2*H*-selenochromen (**138**) comes from 2*H*-chromen and selenochroman (**154**) from chroman. These trivial names are still acceptable according to IUPAC rules and since they are still in general use they will be used in this section. The systematic names are 2*H*-benzoseleno[*b*]pyran and dihydrobenzoseleno[*b*]pyran, and 2*H*-1-benzoselenopyran and dihydro-1-benzoselenopyran; the latter names are used more frequently.

(*i*) *Selenochromen, selenochroman, and some derivatives*

2*H*-Selenochromen (**138**) is formed on dehydrating selenochroman-4-ol (**137**), obtained by reducing selenochroman-4-one (**136**), which is prepared by the cyclization of β-phenylselenopropionic acid (**135**)[100,101] (reaction 65).

2*H*-Selenochromen (**138**) on oxidation with chromium trioxide–pyridine yields a small amount of 2,2′-biscinnamaldehyde diselenide (**139**)[69] (reaction 66), and treatment in

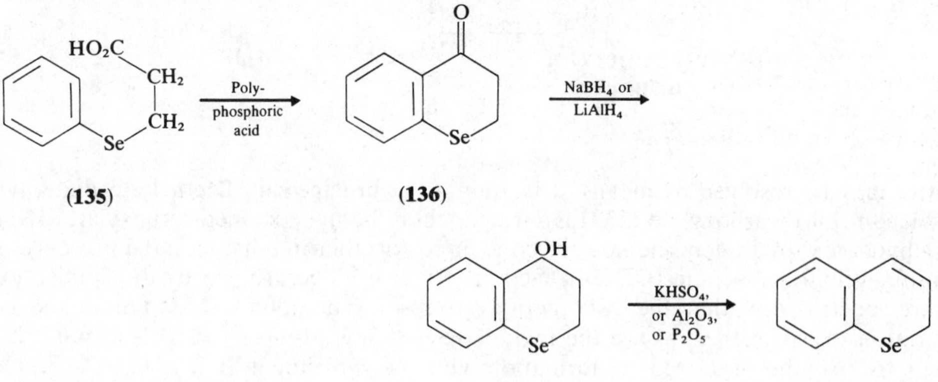

(135) (136)

(137) (138) (65)

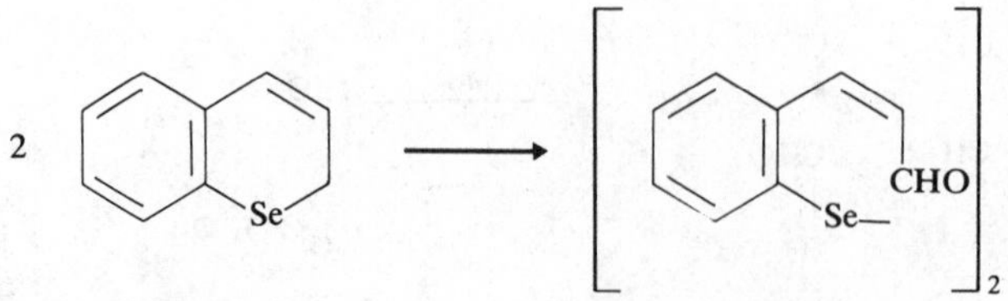

(138) (139) (66)

acetonitrile with triphenylmethyl perchlorate gives selenochromylium perchlorate (benzoselenopyrylium or benzoseleninium perchlorate) (**140**),[100] which is oxidized by manganese dioxide in either boiling chloroform or acetonitrile to afford benzo[*b*]selenophen-2-carbaldehyde (**141**)[69,102] (reaction 67).

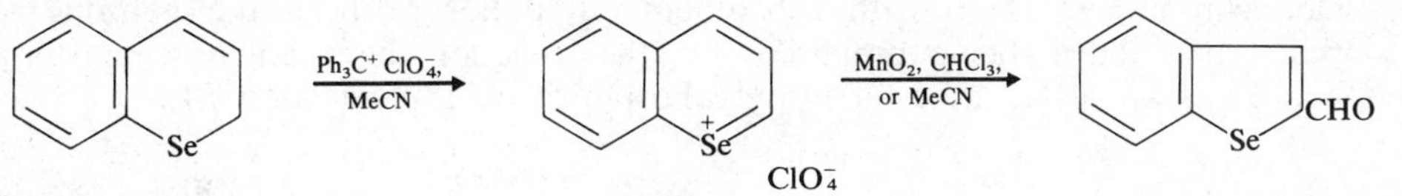
Se
$Ph_3C^+ ClO_4^-$,
MeCN
Se
ClO_4^-
MnO_2, $CHCl_3$,
or MeCN
Se
CHO

(67)

(**138**) (**140**) (**141**)

Selenochromylium perchlorate (**140**) is attacked by nucleophiles, *e.g.* thiophenol and selenophenol, at the 2-position to give the respective 2-substituted selenochromen (**142**; R = PhS or PhSe)[103] (reaction 68). The perchlorate (**140**) reacts with phenylmagnesium bromide to give a mixture of 2-phenyl-2*H*-selenochromen (**143**) and 4-phenyl-4*H*-selenochromen (**144**) (reaction 69). It is found that 2- or 4-arylselenochromylium cation reacts with nucleophiles only in the unsubstituted reactive position, independently of the nature of the nucleophile, and that electron-donor or electron-withdrawing groups on the aryl substituent do not affect the reaction.[104]

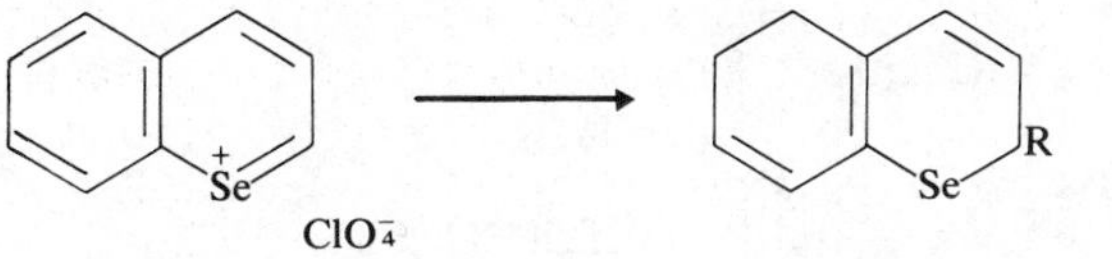
Se
ClO_4^-
Se
R

(68)

(**140**) (**142**)

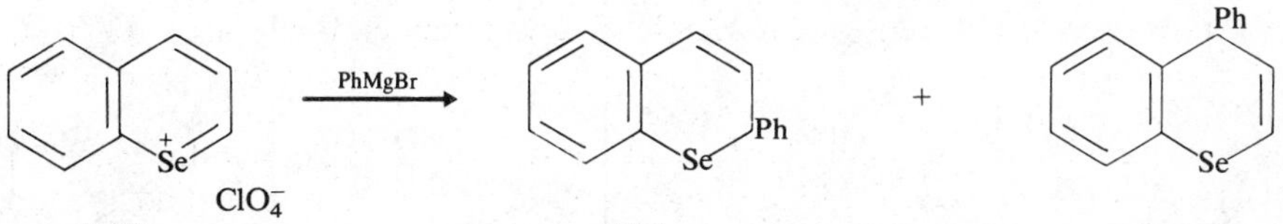
Se
ClO_4^-
PhMgBr
Se
Ph
+
Ph
Se

(69)

(**140**) (**143**) (**144**)

Selenocoumarins, including selenocoumarin (**146**), are prepared by the cyclization of the appropriate 2-(methylseleno)cinnamic acid (**145**) using either phosphoryl chloride or aluminium chloride in carbon disulphide, or by the cyclization of a selenocinnamic acid chloride (**147**) in pyridine[105,106] (reaction 70). They are also formed by the selective oxidation of selenochromens (**138**) with chromium trioxide in pyridine[105] (reaction 71). 4-Hydroxyselenocoumarin (**148**) may be prepared from malonyl dichloride and selenophenol (reaction 72).[107] Some 3-alkyl- and 3-aryl-4-hydroxyselenocoumarins have been obtained by the reaction between selenophenols and a substituted malonic acid in polyphosphoric acid.[108]

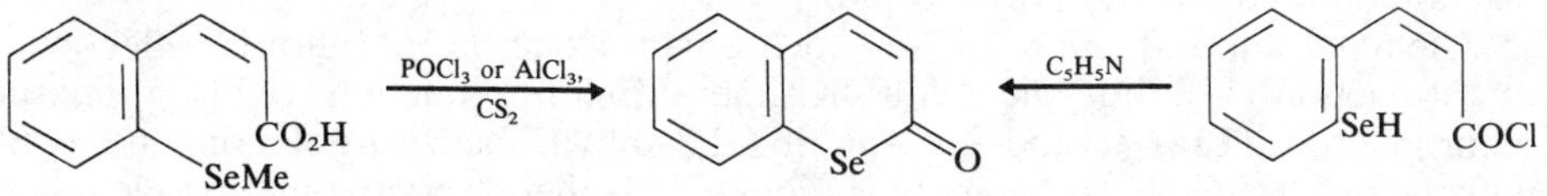
CO_2H
SeMe
$POCl_3$ or $AlCl_3$,
CS_2
Se
O
C_5H_5N
SeH
COCl

(70)

(**145**) (**146**) (**147**)

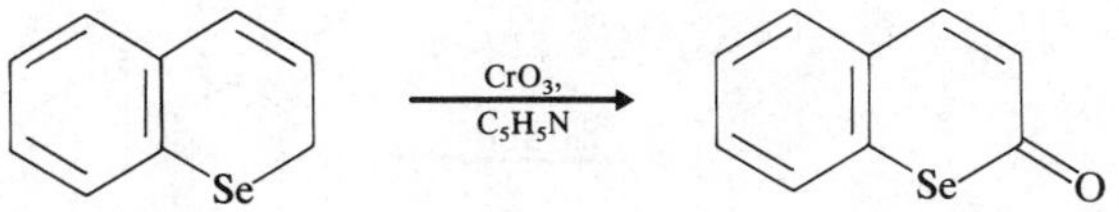
Se
CrO_3,
C_5H_5N
Se
O

(**138**) (**146**)

The dehydrogenation of selenochroman-4-one (**136**) to give selenochromone (**149**) may be achieved by use of either triphenylmethyl perchlorate[100] or chloranil (reaction 73).[101] Alternatively, selenochromone may be prepared by the oxidation of selenochromen (**138**) using selenium dioxide.[69] Some selenochromones and selenoflavones (2-phenylselenochromones) (**150**) with substituents in either or both the benzene and heterocyclic rings have been prepared by the reaction between an appropriate selenophenol and an acylacetic ester in polyphosphoric acid[109] (reaction 74).

$CH_2(COCl)_2$ + 2PhSeH —Δ→ $CH_2(COSePh)_2$ —i, $AlCl_3$, 170 °C; ii, dil. HCl→ (**148**) (72)

(**136**) —$-H_2$→ (**149**) (73)

SeH + $EtO_2CCH{=}C(OH)Ph$ —Polyphosphoric acid→ (**150**) (74)

(**151**) → (**152**) → [ ] → (**153**) (75)

An interesting reaction which leads to the formation of 2-methylselenochromone (**153**) is that between selenophenol and diketen (**151**). The initial reaction is carried out in the presence of toluene-*p*-sulphonic acid and the resulting product (**152**) undergoes rearrangement and cyclization on treatment with polyphosphoric acid[110] (reaction 75). The reaction may be completed without the isolation of the intermediate product by adding the reactants directly to the polyphosphoric acid.

The Clemmensen reduction of selenochroman-4-one (**136**) affords selenochroman (**154**)[101] (reaction 76), but the Wolff–Kishner–Minlon reduction yields a mixture of selenochroman (**154**) and 2-methyl-2,3-dihydrobenzo[*b*]selenophen (**87**).[111] Selenoflavanone (**156**) is prepared by boiling 2-(β-phenylacryloyl)methylselenobenzene (**155**) in a saturated solution of hydrogen bromide in acetic acid (reaction 77).[112]

(**136**) —Zn, HCl→ (**154**) (76)

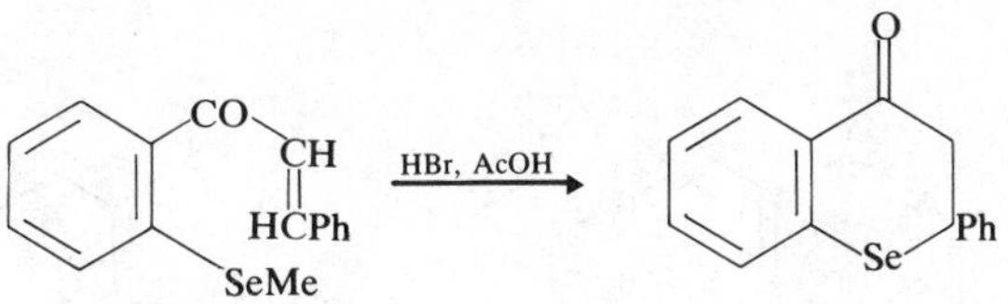

(77)

(155) **(156)**

(*ii*) *Isoselenochromen and some derivatives*

Isoselenochroman-4-one (**159**) is obtained when 2-carboxybenzylselenoacetic acid (**157**) and potassium acetate are boiled in acetic anhydride and the resulting mixture of (**159**) and its enol acetate (**158**) is hydrolysed with 10% alcoholic potassium hydroxide. The isoselenochromanone on reduction affords isoselenochroman-4-ol (**160**), readily dehydrated to yield isoselenochromen (**161**), which may be converted to isoselenochromylium perchlorate (**162**) by treatment with sulphuryl chloride and then perchloric acid (Scheme 6).[113]

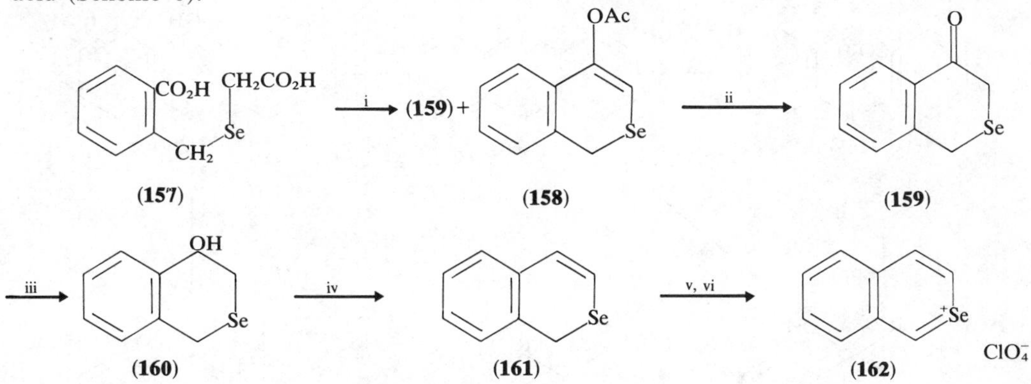

(157) **(158)** **(159)**

(160) **(161)** **(162)**

i, KOAc, Ac_2O; ii, 10% KOH in EtOH; iii, $NaBH_4$, MeOH; iv, $KHSO_4$; v, SO_2Cl_2; vi, 70% $HClO_4$.

SCHEME 6

19.3.2.7 Six-membered ring containing one selenium atom and fused to two benzene rings

The names given to the linear compounds formed by the fusion of the heterocyclic ring and two benzene rings are derived from the related oxygen heterocycle xanthen. The parent compound of this class of heterocycles thus becomes selenoxanthen (**163**). The names of the angular compounds are derived from selenopyran, *e.g.* 6*H*-dibenzo[*b*, *d*]selenopyran (**164**).

(*i*) *Selenoxanthen, selenoxanthone, and related compounds*

Methods of preparation used for this series of compounds are similar to those involved in the related oxygen series. A key starting material is thus selenoxanthone (selenoxanthen-9-one) (**166**) prepared by cyclizing 2-(phenylseleno)benzoic acid (**165**).[114] Selenoxanthone (**166**), on reduction with sodium tetrahydridoborate, yields selenoxanthydrol (**167**), which on treatment with perchloric acid at −40 °C gives selenoxanthylium perchlorate (**168**). Subsequent reduction using lithium tetrahydridoaluminate affords selenoxanthen (**163**). Heating selenoxanthone with red phosphorus and fuming hydriodic acid gives selenoxanthen in almost quantitative yield.[115] Reaction between the salt (**168**)

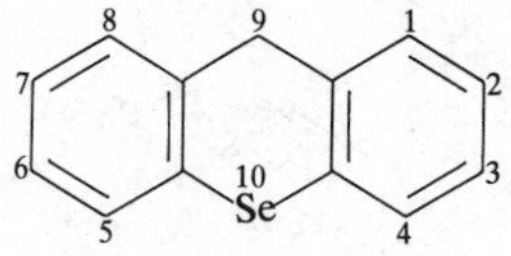

(163)

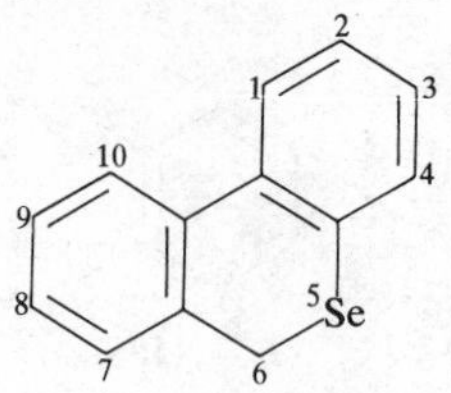

(164)

and a Grignard reagent, *e.g.* methyl- or phenyl-magnesium halide, gives the 9-substituted selenoxanthen (**169**; R = Me or Ph).[116] The same compound (**169**) may be obtained by first reacting selenoxanthone (**166**) with the appropriate Grignard reagent, then treating the resulting selenoxanthydrol (**170**) with perchloric acid to form the selenoxanthylium perchlorate (**171**), and finally completing the preparation by reducing the salt. If the salt (**171**) is treated with a Grignard reagent a 9,9-disubstituted selenoxanthen (**172**) is obtained (Scheme 7).

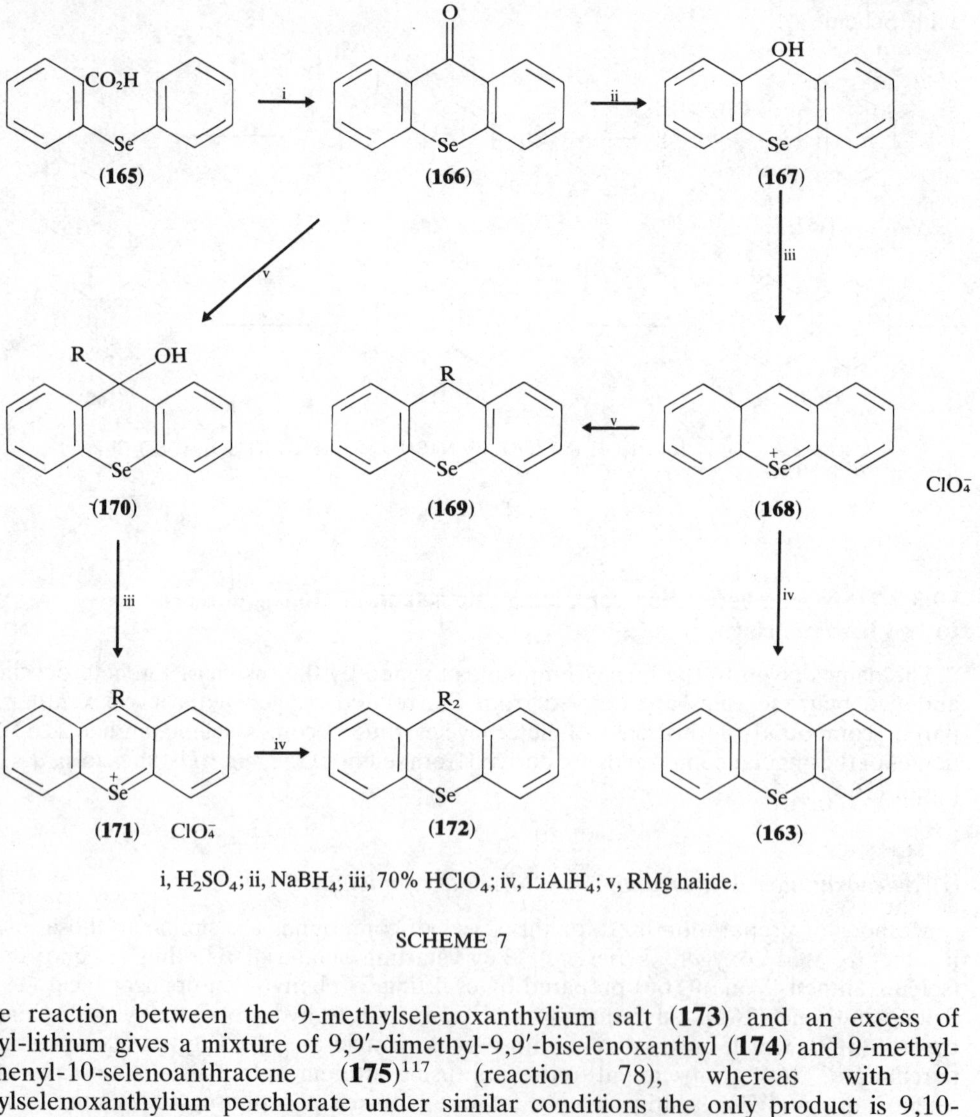

(165) (166) (167) (170) (169) (168) (171) ClO_4^- (172) (163)

i, H_2SO_4; ii, $NaBH_4$; iii, 70% $HClO_4$; iv, $LiAlH_4$; v, RMg halide.

SCHEME 7

The reaction between the 9-methylselenoxanthylium salt (**173**) and an excess of phenyl-lithium gives a mixture of 9,9′-dimethyl-9,9′-biselenoxanthyl (**174**) and 9-methyl-10-phenyl-10-selenoanthracene (**175**)[117] (reaction 78), whereas with 9-phenylselenoxanthylium perchlorate under similar conditions the only product is 9,10-diphenyl-10-selenoanthracene. The selenoanthracene derivatives are very stable to light,

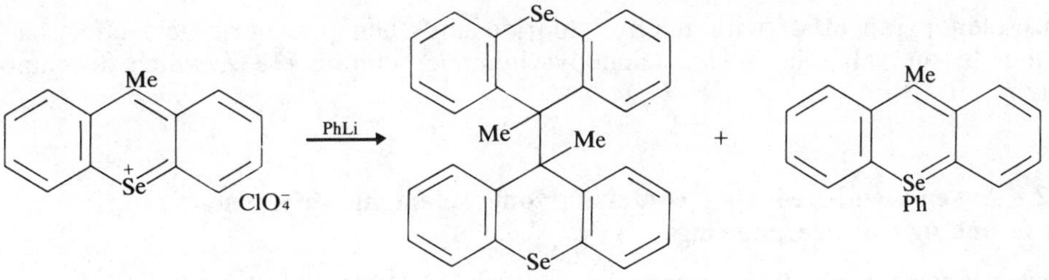

(173) (174) (175) (78)

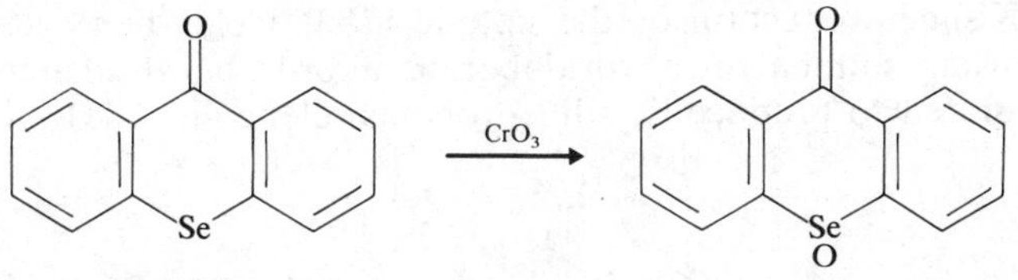

(166) (176) (79)

heat, and oxygen, although oxidation of selenoxanthone (**166**) with chromium trioxide gives selenoxanthone 10-oxide (**176**) (reaction 79).[118]

Selenoxanthone (**166**) gives a mass spectrum of which one of the prominent features is the decarbonylation species, *i.e.* the dibenzo[*b,d*]selenophen ion (**177**). The other principal mode of fragmentation results in the formation of the fluorenone ion (**178**) and its known decomposition product, biphenylene (**179**), (Scheme 8).[75]

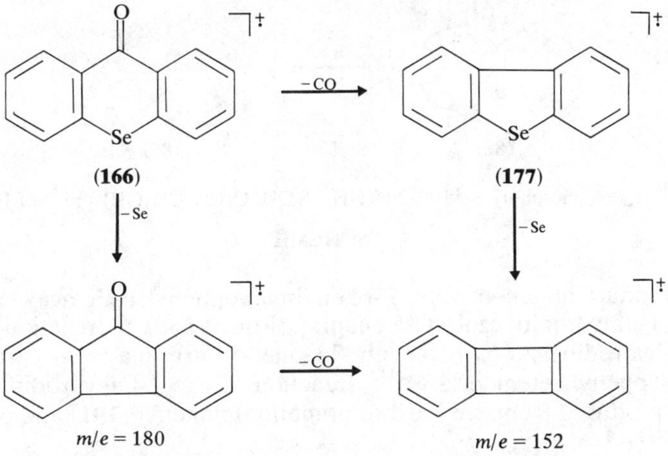

(166) (177)

$m/e = 180$ (178) $m/e = 152$ (179)

SCHEME 8

(*ii*) 6*H*-Dibenzo[*b,d*]selenopyran

The angular compound 6*H*-dibenzo[*b,d*]selenopyran (**164**) may be obtained from 2-aminobenzylphenyl selenide (**180**) by the Pschorr reaction, which involves diazotization of the amino-group and subsequent ring closure[119] (reaction 80). Treatment of the

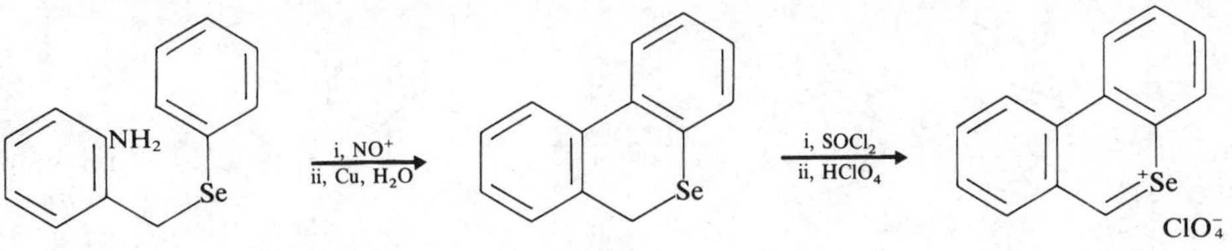

(180) (164) (181) (80)

dibenzoselenopyran (**164**) with thionyl chloride and then perchloric acid effects a net oxidation to the salt dibenzo[*b,d*]selenopyrylium perchlorate (**181**), which decomposes rapidly in air.

19.3.2.8 Seven-membered ring containing one selenium atom and fused to one or two benzene rings

Benzoselenepin-5-one (homoselenochroman-5-one) (**183**), like the related selenochroman, is prepared by the intramolecular condensation of 4-phenylselenobutyric acid (**182**).[101,120] Wolff–Kishner reduction of the ketone (**183**) yields benzoselenepan (**184**), whereas reduction using sodium tetrahydridoborate affords benzoselenepan-5-ol (**185**) which may be dehydrated to produce 2,3-dihydrobenzoselenepin (**186**) (Scheme 9).[121]

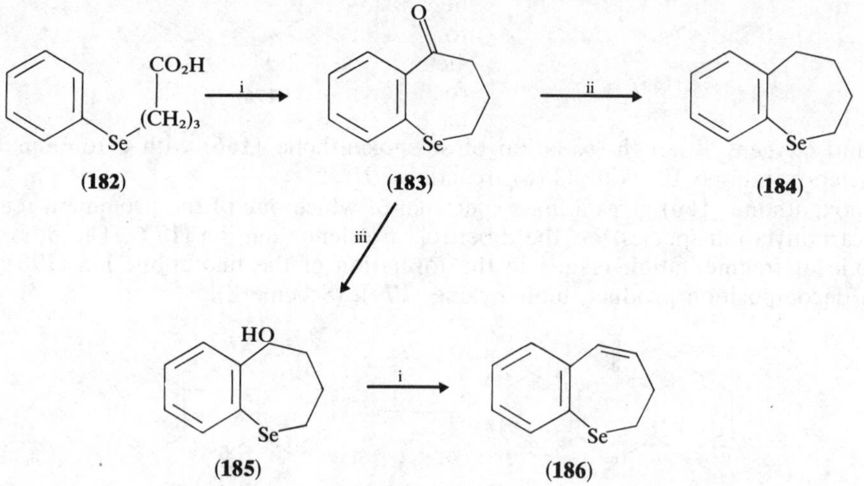

(**182**) (**183**) (**184**) (**185**) (**186**)

i, Polyphosphoric acid; ii, NH_2NH_2, KOH, $O(CH_2CH_2OH)_2$; iii, $NaBH_4$.

SCHEME 9

The main product obtained when 2-phenylselenophenylacetic acid (**187**) is cyclized as above is 10,11-dihydrodibenzo[*b,f*]selenepin-10-one (**188**)[120] (reaction 81). In a similar way, 6,11-dihydrodibenzo[*b,e*]selenepin-11-one (**190**) may be prepared from 2-carboxybenzyl phenyl selenide (**189**)[122] (reaction 82). 5,7-Dihydrodibenzo[*c,e*]selenepin (**192**) may be produced from 2,2′-bis(bromomethyl)biphenyl (**191**) and potassium selenide (reaction 83).[123]

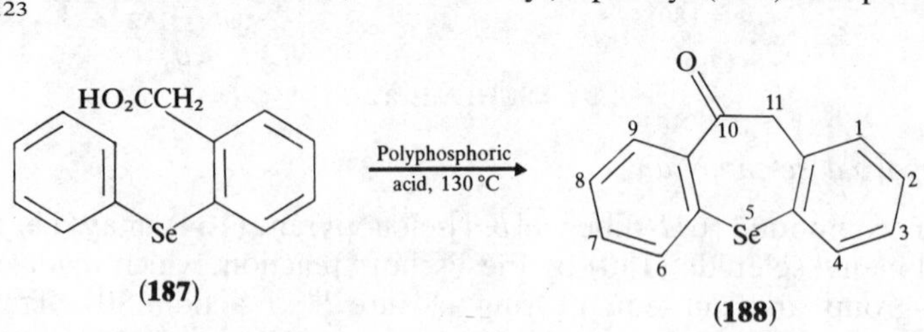

(**187**) —Polyphosphoric acid, 130 °C→ (**188**) (81)

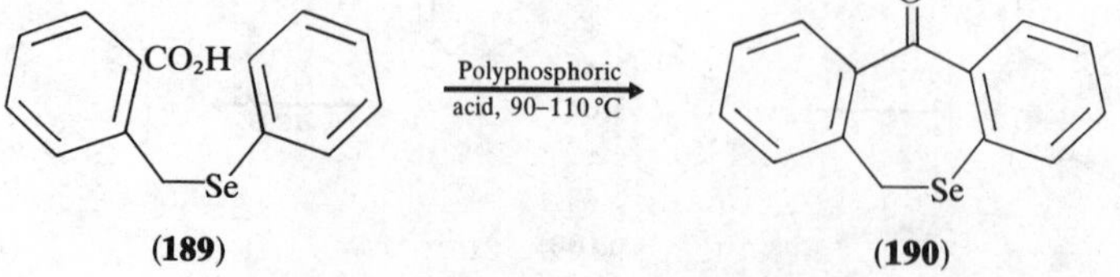

(**189**) —Polyphosphoric acid, 90–110 °C→ (**190**) (82)

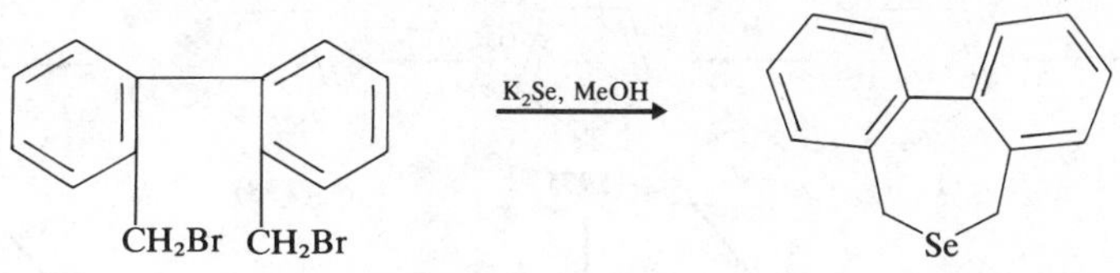

$$\xrightarrow{K_2Se,\ MeOH} \quad (83)$$

CH_2Br CH_2Br Se

(191) **(192)**

19.3.3 HETEROCYCLES CONTAINING ONE TELLURIUM ATOM

The name of the parent compound of the five-membered ring containing one tellurium atom is obtained by replacing thio in the name of the analogous sulphur compound by telluro; thus 'thiophen' becomes 'tellurophen' (**193**). In the case of compounds containing a six-membered ring with one tellurium atom the name is formed by combining the prefix, telluro, with the name of the related heterocycle containing one oxygen atom, *e.g.* 4*H*-telluropyran (**194**) and 2-phenyltellurochroman-4-one (telluroflavanone) (**195**).

Various cyclic tellurium compounds have been prepared in order to examine the possibility of their use as bactericidal or bacteriosteric agents.

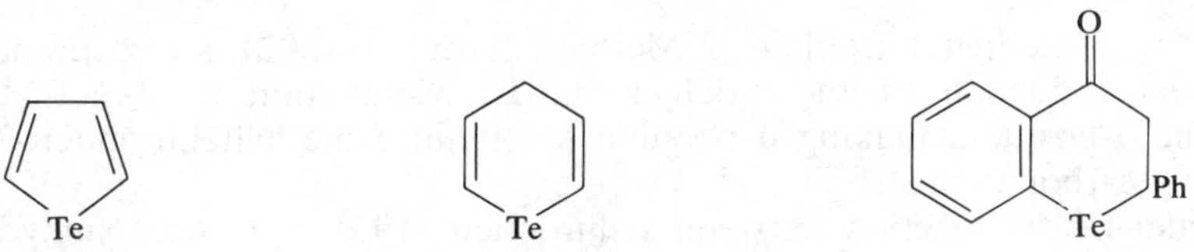

Te Te O Ph Te

(193) **(194)** **(195)**

19.3.3.1 Five-membered ring containing one tellurium atom

Tellurophen (**193**) is prepared from butadiyne (**196**) and sodium telluride (reaction 84), but to obtain any product great care must be taken[124] and the following procedures observed. (i) Moisture and oxygen must be rigorously excluded; (ii) butadiyne is very readily oxidized and polymerized and therefore must be used directly after preparation by bubbling into a methanolic solution of sodium telluride; (iii) anhydrous and iron-free liquid ammonia is necessary for the preparation of sodium telluride; and (iv) commercial metallic grey tellurium must be used, for with amorphous or partly oxidized tellurium, no reaction occurs.

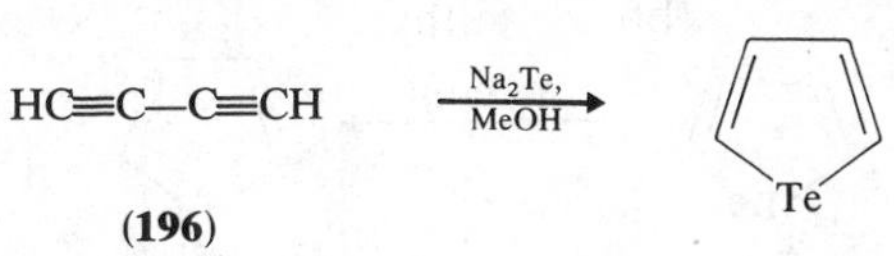

$$HC{\equiv}C{-}C{\equiv}CH \xrightarrow[MeOH]{Na_2Te,} \quad (84)$$

Te

(196) **(193)**

Originally a few derivatives of tellurophen were synthesized directly from suitable substituted butadiynes, but this approach had only limited application[17,125] and subsequently a number of derivatives have been prepared by the application of classical substitution reactions to tellurophen.[124] Metallation of tellurophen (**193**) with n-butyllithium followed by carbonation gives tellurophen-2-carboxylic acid (**198**), which may be converted to the methyl ester (**199**) and then to 2-hydroxymethyltellurophen (**200**); the latter is also obtained by reduction of tellurophen-2-carbaldehyde (**201**), prepared by the metallation of tellurophen followed by treatment with *N*-methylformanilide (Scheme 10).

Tellurophen-2-yl-lithium (**197**) on alkylation with dimethyl sulphate yields a mixture of 2-methyltellurophen (**202**), unchanged tellurophen (**193**), and a trace of a compound

i, Bu^nLi, C_6H_{14}, Et_2O; ii, CO_2, Et_2O; iii, CH_2N_2, Et_2O; iv, $LiAlH_4$, Et_2O; v, HCONMePh, Et_2O; vi, $LiAlH_4$, Et_2O; vii, Me_2SO_4, Et_2O; viii, NH_2NH_2, $HOCH_2CH_2OH$, KOH; ix, Bu^nLi, C_6H_{14}, Et_2O, CO_2; x, Ac_2O, $SnCl_4$.

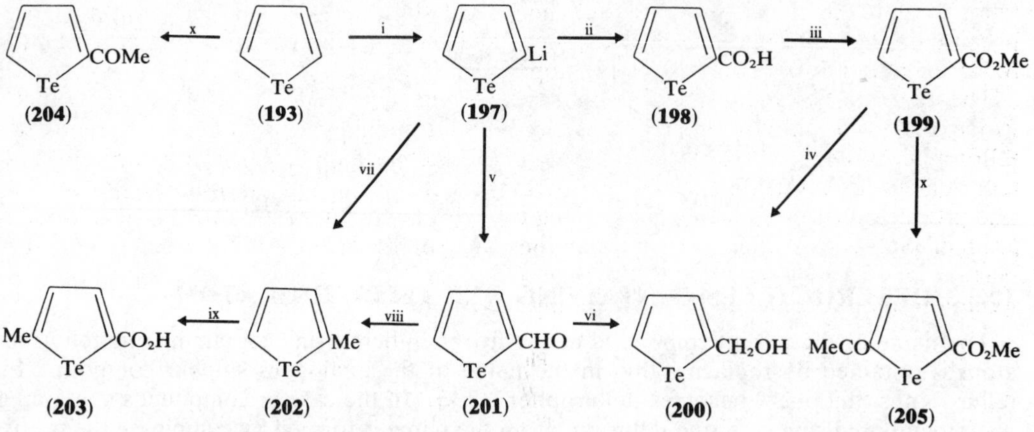

SCHEME 10

believed to be 3-methyltellurophen. 2-Methyltellurophen (**202**) is also prepared by the Wolff–Kishner reduction of the aldehyde (**201**). Metallation of 2-methyltellurophen occurs at the 5-position, making it possible to obtain 5-methyltellurophen-2-carboxylic acid (**203**) by carbonation.

The Friedel–Crafts reaction between tellurophen (**193**) and acetic anhydride in the presence of stannic chloride affords 2-acetyltellurophen (**204**), and the acetylation of methyl tellurophen-2-carboxylate (**199**) gives methyl 5-acetyltellurophen-2-carboxylate (**205**).

The above reactions show that tellurophen has a typical heteroaromatic five-membered ring, in that it undergoes electrophilic and nucleophilic substitutions, and is more reactive at the 2- than at the 3-position, but the presence of the tellurium atom induces significant differences from furan, thiophen, and selenophen. Tellurophen is very sensitive to mineral acids and with halogens gives 1,1-dihalogeno-derivatives, *e.g.* 1,1-dibromotellurophen (**206**) (reaction 85). Nitration of tellurophen using nitric acid–acetic acid under the usual conditions does not give the desired products and attempts to convert the acetyl group of methyl 5-acetyltellurophen-2-carboxylate (**205**) into a carboxylic acid group *via* the haloform reaction is unsuccessful. The alkylation and arylation of tellurophen-2-yllithium with ethyl bromide and bromobenzene respectively gave unsatisfactory results.

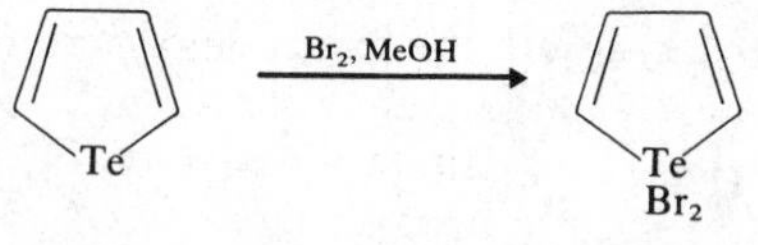

(**193**) $\xrightarrow{Br_2,\ MeOH}$ (**206**) (85)

The n.m.r. spectrum of tellurophen shows two multiplets at τ 1.13 and 2.22; the low-field multiplet is assigned to the 2- and 5-protons and the latter to the 3- and 4-protons by analogy with the spectra of furan, thiophen, and selenophen.[126] The spectra of substituted tellurophens can be readily analysed, owing to the large differences in chemical shifts of the α- and β-protons, and the known ring-proton coupling constants for substituted tellurophens are not greatly influenced by substituents. The coupling constants $J_{2,3}$ and $J_{3,4}$ increase from furan to tellurophen, in agreement both with the decrease in electronegativity of the heteroatom and with the change of angle between the bonds to the protons and the carbon–carbon bond. An electron-withdrawing substituent at the C-2 position shifts the ring-proton signal to low field, whereas an electron-donating substituent tends to shift the signal to high field. The largest shift is shown by the 3-proton signal, because of electronic and anisotropic effects.[124] These observations are in agreement with those for derivatives of furan[127] and thiophen.[128]

The reaction between 1,4-dilithiotetraphenylbutadiene (**207**) and tellurium tetrachloride, or 1,4-di-iodotetraphenylbutadiene (**209**) and lithium telluride, yields tetraphenyltellurophen (**208**)[17,125] (reaction 86). Shaking hexachlorobutadiene (**210**) with powdered tellurium yields tetrachlorotellurophen (**211**); the tellurium tetrachloride produced as a side product must be removed by washing out with hydrochloric acid before distilling the product. Subsequent chlorination yields hexachlorotellurophen (**212**) (reaction 87). Reduction of either hexachlorotellurophen (**212**) or 1,1-dibromotellurophen (**206**) with sodium bisulphite yields the corresponding tellurophens (**211**) and (**193**).

(**207**) $\xrightarrow{TeCl_4}$ (**208**) $\xleftarrow{Li_2Te}$ (**209**) (86)

(**210**) $\xrightarrow[250\,°C]{Te}$ (**211**) $\xrightarrow{Cl_2}$ (**212**) (87)

(**213**) $\xrightarrow{Na_2Te}$ (**214**) $\xrightarrow{Br_2}$ (**215**) (88)

The related saturated five-membered ring compound tetrahydrotellurophen is generally called tellurolan (**214**) and is prepared by the reaction between 1,4-dibromobutane (**213**) and sodium telluride,[129] made *in situ* by the reaction of tellurium with a hot aqueous solution of sodium formaldehyde sulphoxylate (Rongalite) in an inert atmosphere (reaction 88). Tellurolan on treatment with bromine gives 1,1-dibromotellurolan (**215**); the 1,1-dichloro-derivative may be similarly obtained.

In the mass spectrum of tellurolan the base peak occurs at *m/e* 55 ($C_4H_7^+$) and is somewhat more intense than is found for selenolan (p. 882). No $M-CH_3$ ion is present in the spectrum, while Te^+ and TeH^+ constitute more of the total ionization than is observed for the corresponding ions of other atoms of Group VI of the Periodic Table. A weak group of peaks representing a $M-28$ ion is also present in the spectrum, and a second series of peaks between *m/e* 136 and *m/e* 144 ($M-42$) originates from the loss of cyclopropane; a mechanism analogous to that used in the formation of species (*m/e* 94) in selenolan (p. 882) can be invoked in this instance.[61]

3,3′-Bitellurolan (octahydro-3,3′-bitellurophen) (**216**) is prepared from 1,6-di-iodo-3,4-bisiodomethylhexane (**56**) (p. 883) and sodium telluride.[60]

(**216**)

19.3.3.2 Five-membered ring containing one tellurium atom and fused to one or two benzene rings

3-Hydroxybenzo[*b*]tellurophen (**220**) is obtained by first treating diazotized anthranilic acid (**217**) with sodium telluride at 0 °C in a current of hydrogen and then heating the solution to boiling point; acidification affords bis-(2-carboxyphenyl)telluride (**218**), which on reduction with hot aqueous potassium hydroxide and zinc, followed by reaction with sodium chloroacetate and acidification, yields (2-carboxyphenyl)telluroglycolic acid (**219**). Ring closure of (**219**) is effected by boiling with acetic anhydride and sodium acetate. Removal of the acetic anhydride and hydrolysis of the product with boiling aqueous sodium hydroxide solution, followed by acidification, gives 3-hydroxybenzo[*b*]tellurophen (**220**) and carbon dioxide (Scheme 11).[130]

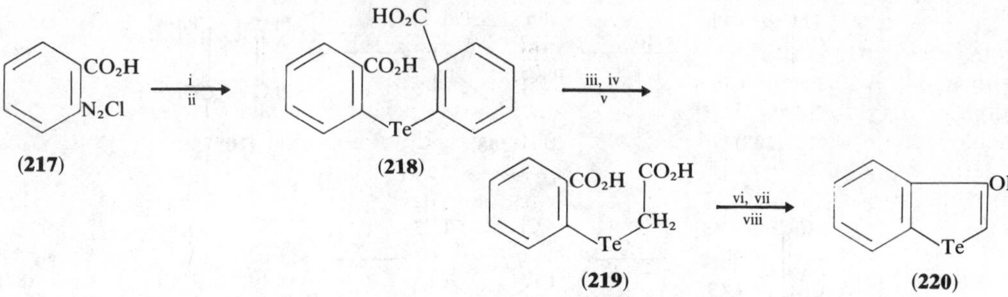

i, Na_2Te, H_2O, Na_2CO_3, H_2; ii, HCl; iii, 10% KOH, Zn; iv, $ClCH_2CO_2Na$; v, H_2SO_4; vi, NaOAc, Ac_2O; vii, 10% NaOH; viii, HCl.

SCHEME 11

2-Chloro-1,2-diphenylvinyltellurotrichloride (**221**) on boiling in 1,2,4-trichlorobenzene yields 2-phenyl-1,1,3-trichlorobenzo[*b*]tellurophen (**222**), which may be converted to 3-chloro-2-phenylbenzo[*b*]tellurophen (**223**) (reaction 89).[131]

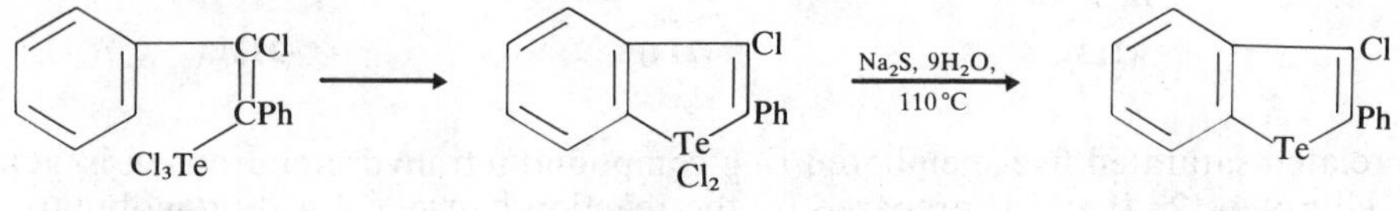

(89)

Benzo[*b*]tellurophen (**226**) has been prepared by the decarboxylation with copper in quinoline of benzo[*b*]tellurophen-2-carboxylic acid (**225**), obtained by the cyclization of 2-(carboxymethyltelluro)benzaldehyde (**224**) using acetic anhydride in pyridine[132] (reaction 90). Its mass spectrum has been reported,[133] and is remarkably simple. The base peak

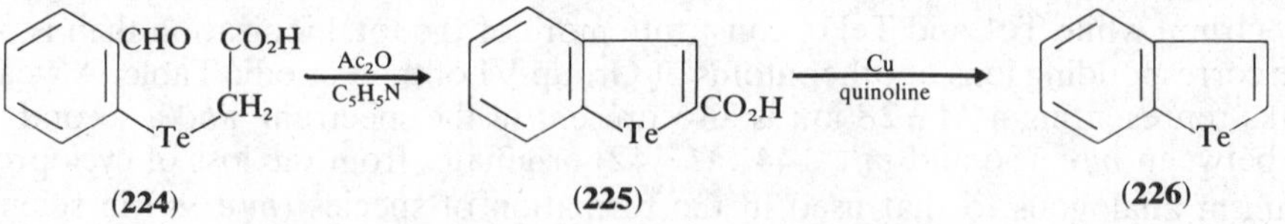

(90)

corresponds to a detellurated species, probably benzocyclobutadiene (**227**), which loses ethylene to give the benzyne species (**228**) and its dehydrogenation product (**229**). Another recognizable detellurated species is the ion (**231**) which, on grounds of what had been observed previously in the benzo[*b*]thiophen series, could be considered as arising from the bicyclic ion (**230**) by loss of ethylene. Direct extrusion of ethylene from (**226**), leading to the bicyclic species (**232**), is distinctly recognizable by a series of small peaks corresponding to the most frequent isotopes of tellurium (especially $m/e = 206$ and $m/e = 204$, corresponding to ^{130}Te and ^{128}Te, respectively).

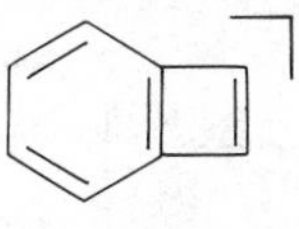

m/e 102 (**227**)   *m/e* 76 (**228**)   *m/e* 75 (**229**)

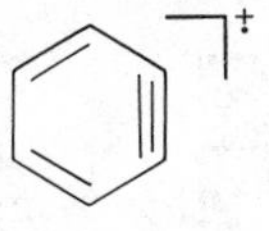

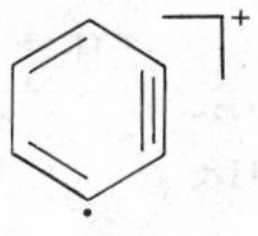

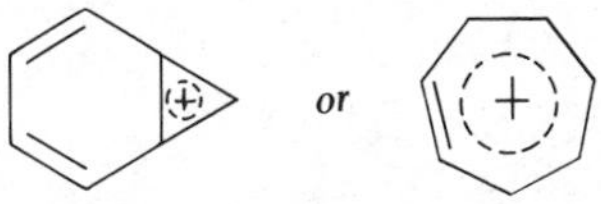

or

m/e 89 (**230**)   *m/e* 63 (**231**)   Te (**232**)

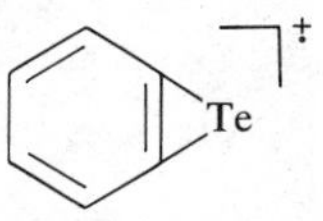

Dibenzotellurophen is formed when thianthrene 5,5,10,10-tetroxide (**233**) is heated with finely powered tellurium in an atmosphere of carbon dioxide until the evolution of sulphur dioxide stops[134] (reaction 91). It may also be obtained by treating 5,5-dichlorodibenzotellurophen (**236**) with potassium dithionite; the former is prepared by reacting biphenyl (**235**) with tellurium tetrachloride (reaction 92).[135] Treatment of dibenzotellurophen with nitric acid affords 2-nitrobenzotellurophen.[134]

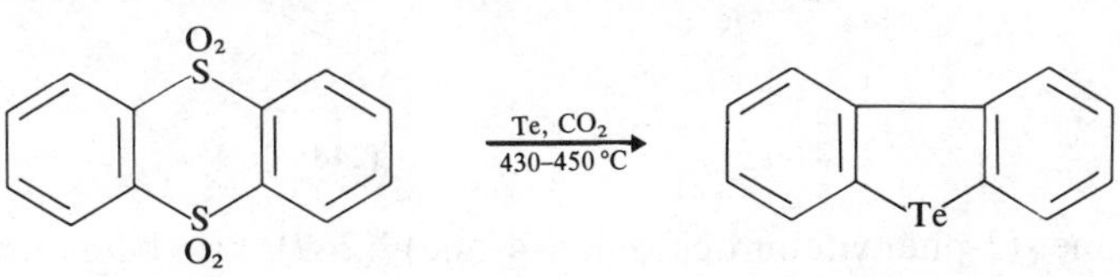

O_2 S S O_2 (**233**) → Te, CO_2, 430–450 °C → Te (**234**) (91)

(**235**) → $TeCl_4$, 140–165 °C → Te Cl_2 (**236**) → $K_2S_2O_4$ → (**234**) (92)

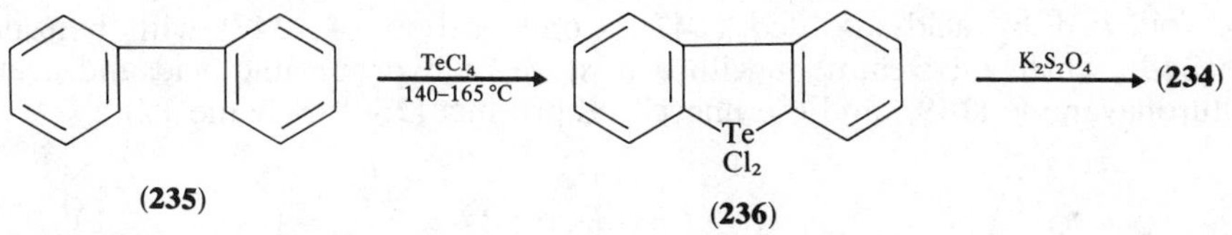

19.3.3.3 Six-membered ring containing one tellurium atom

A small amount of telluran (tetrahydrotelluropyran) (**241**) was obtained originally as one of the products of the reaction between a metallic telluride, preferably the aluminium compound, and a 1,5-dihalogenopentane (**237**; X = Cl, Br, or I). The dibromide (**238**; X = Br) is, however, more readily prepared by adding the calculated amount of bromine to a mixture of telluran and 1,5-dibromopentane produced by the thermal dissociation, under reduced pressure, of the two complex telluronium bromides (**239**) and (**240**) also formed during the above reaction (reaction 93).[136]

A simpler and more satisfactory synthesis of 1,1-dihalotelluran (**238**) has, however, been achieved by treating 1,5-dihalopentane (**237**) with sodium telluride (p. 901) and then with either bromine or chlorine[129] (reaction 94). The 1,1-di-iodo-derivative is obtained by heating together amorphous tellurium and 1,5-di-iodopentane at 150 °C. Telluran oxidizes rapidly in air and when treated with hydrogen peroxide in warm methanol yields the 1,1-dioxide (**242**) (reaction 95).[136]

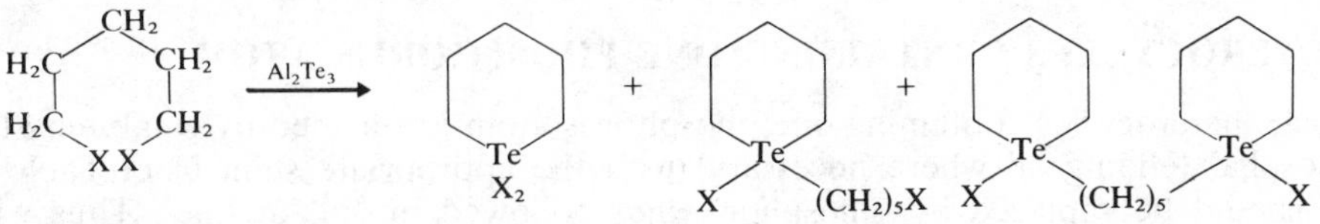

CH_2 H_2C CH_2 H_2C CH_2 X X (**237**) → Al_2Te_3 → Te X_2 (**238**) + Te X $(CH_2)_5X$ (**239**) + X Te $(CH_2)_5$ Te X (**240**) (93)

(94)

(237) $\xrightarrow{Na_2Te}$ **(241)** $\xrightarrow{Cl_2 \text{ or } Br_2}$ **(238)**

(95)

(241) $\xrightarrow[MeOH]{H_2O_2}$ **(242)**

The reaction between acetylacetone (**243**) and tellurium tetrachloride affords 1,1-dichlorotelluran-3,5-dione (**244**) (reaction 96). The structure of (**229**) is supported by its n.m.r. spectrum, which shows two peaks of relative intensity 2 : 1.[137]

(96)

(243) $\xrightarrow{TeCl_4}$ **(244)**

Telluroflavanone (2-phenyltellurochroman-4-one) (**249**) has been obtained from 2-(methyltelluro)acetophenone (**247**), prepared by reacting the dioxolan (**245**) with lithium and then with tellurium and methyl iodide to give the salt (**246**), which on treatment with pyridine followed by acid afforded (**247**). Condensation of (**247**) with benzaldehyde yielded (**248**), which on treatment with a mixture of hydrobromic acid and acetic acid gave telluroflavanone (**249**), and the uncyclized product (**250**) (Scheme 12).[138]

(245) $\xrightarrow{i, ii}$ **(246)** $\xrightarrow{iii, iv}$ **(247)** $\xrightarrow{v}$ **(248)** $\xrightarrow{vi}$ **(249)** + **(250)**

i, Li; ii, Te, MeI; iii, C_5H_5N; iv, H_2SO_4; v, PhCHO; vi, HBr, AcOH.

SCHEME 12

19.3.4 HETEROCYCLES CONTAINING ONE PHOSPHORUS ATOM

Monocyclic heterocycles containing one phosphorus atom are named by combining the prefix 'phospha' (eliding 'a' where necessary) with the appropriate stem from Table 2. 'Phospha' should be replaced by 'phosphor' when followed by 'in' or 'ine'. Thus with monocyclic heterocycles containing one phosphorus atom the names in general use are those derived from the related heterocycles containing one nitrogen atom. Even with the heterocyclic ring fused on to a benzene ring the name of the compound is derived from that of the related compound containing a nitrogen atom in place of the phosphorus atom, *e.g.* indole and quinoline become phosphindole and phosphinoline, respectively.

19.3.4.1 Three- and four-membered rings containing one phosphorus atom

Phosphiran (phosphacyclopropane) (**251**), b.p. 36.5 °C, has been obtained by the reaction between 1,2-dichloroethane and phosphinide ion in anhydrous ammonia. The reaction did not follow the nucleophilic substitution pathway usually observed with simple alkyl halides, but yielded phosphiran, phosphine, and ethylene (reaction 97). A tentative mechanism for the ring closure is illustrated by reactions 98, 99, and 100.[139]

$$ClCH_2CH_2Cl \xrightarrow[NH_3]{2NaPH_2} \underset{(\mathbf{251})}{\text{(HP ring)}} + PH_3 + CH_2{=}CH_2 \quad (97)$$

$$ClCH_2CH_2Cl + PH_2^- \longrightarrow ClCH_2CH_2PH + Cl^- \quad (98)$$

$$ClCH_2CH_2PH_2 + PH_2^- \longrightarrow ClCH_2CH_2PH + PH_3 \quad (99)$$

$$ClCH_2CH_2\bar{P}H \longrightarrow \underset{(\mathbf{251})}{\text{HP}(CH_2{-}CH_2)} + Cl^- \quad (100)$$

$$\underset{(\mathbf{252})}{BrCH_2CH_2PBr_2} \xrightarrow{LiAlH_4} \underset{(\mathbf{251})}{\text{(HP ring)}} + \underset{(\mathbf{253})}{EtPH_2} + CH_2{=}CH_2 + PH_3 \quad (101)$$

The reduction of 2-bromoethylphosphonous dibromide (**252**) with lithium tetrahydridoaluminate in diethylene glycol diethyl ether gives phosphiran (**251**) along with ethylphosphine (**253**), ethylene, and phosphine (reaction 101). Phosphiran is conveniently separated from ethylphosphine (**253**) by gas–liquid chromatography and has been spectrally characterized (i.r., n.m.r., and mass spectra). It is thermally unstable in the liquid phase and decomposes at 25 °C to give a viscous non-volatile liquid plus ethylphosphine and ethylene. Phosphiran is also rapidly decomposed below room temperature by hydrogen chloride, but it is not rapidly decomposed by methanol at room temperature. The ^{1}H n.m.r. spectrum of the solution indicates that the hydrogen on the phosphorus is not undergoing rapid exchange with solvent hydrogen atoms, and is in contrast to the behaviour of methylphosphine, which does undergo a rapid exchange reaction in methanol solution at room temperature. This may indicate that phosphiran is a weaker base than methylphosphine and that it is not capable of accepting a proton from the weak acid methanol. It has been previously observed that aziridine is an unusually weak base (pK_a = 8.04) compared with acyclic analogues, *e.g.* dimethylamine (pK_a = 10.77), and this may also hold for the analogous phosphorus bases.

The addition of 1,5-diazabicyclo[4,3,0]non-5-ene to a solution of bis-(α-bromobenzyl)phenylphosphine oxide (**254**) in benzene results in the immediate formation of 1,2,3-triphenylphosphiren oxide (**255**)[140] (reaction 102). 1,2,3-Triphenylphosphiren oxide (**255**) gives a mass spectrum showing a parent ion at *m/e* 302 and diphenylacetylene, *m/e* 178. On pyrolysis it yields diphenylacetylene (**256**) and on addition of aqueous sodium hydroxide it affords the 1,2-diphenylvinylphosphinic acid (**257**), which on pyrolysis gives *cis*- and *trans*-stilbene (**258**) (3 : 1) as the only volatile products (reaction 103). The triphenylphosphiren oxide (**255**) is a member of a new class of potentially aromatic phosphacyclopropenes.

$(PhCHBr)_2P(O)Ph$ (**254**) $\xrightarrow{-HBr}$ [intermediate] $\xrightarrow{-HBr}$ (**255**) (102)

$PhC{\equiv}CPh$ (**256**) $\xleftarrow[10^{-5}-10^{-6}\ Torr]{120\,°C}$ (**255**) $\xrightarrow{NaOH}$ (**257**) $\xrightarrow{\Delta}$ $PhCH{=}CHPh$ (**258**) (103)

A phosphetan, 1-chloro-2,2,3,4,4-pentamethylphosphetan 1-oxide (1,1,2,3,3-pentamethyltrimethylenephosphinic acid chloride) (**260**) has been obtained by the addition of 2,4,4-trimethylpent-2-ene (**259**) to a cold suspension of equimolar amounts of phosphorus trichloride and aluminium chloride in methylene chloride, followed by partial hydrolysis with water. Further hydrolysis gives a chlorine-free acidic material, 1-hydroxy-2,2,3,4,4-pentamethylphosphetan 1-oxide (1,1,2,3,3-pentamethyltrimethylenephosphinic acid) (**261**) (Scheme 13).[141]

(**259**) $+ PCl_3 \xrightarrow{-Cl^-}$ … $\xrightarrow{H_2O}$ … $\xrightarrow{-HCl}$ (**260**) $\xrightarrow{H_2O}$ (**261**)

SCHEME 13

The most important features of the suggested mechanism for the formation of the phosphetan (**260**) are illustrated. Initial attack at the 2-position rather than the 3-position of the olefin is suggested for steric reasons, the bulky reagent being precluded from attack at the neopentyl-type 3-position. Shift of a methyl group from C-4 to C-3 then occurs, converting the initially formed secondary carbonium ion into the more stable tertiary carbonium ion. Then follows cyclization of the rearranged ion in preference to the loss of a proton, and finally the resulting cyclic cation reacts with a solvent anion to form a neutral species, which loses hydrogen chloride to give the phosphetan (**260**). This four-membered ring is stable to several powerful reagents such as hot concentrated nitric acid, hot concentrated sulphuric acid, and boiling concentrated aqueous sodium hydroxide. The reaction of compound (**260**) with t-butyl-lithium at 0 °C, followed by trichlorosilane reduction of the resulting oxide (**262**), affords 1-t-butyl-2,2,3,4,4-pentamethylphosphetan (**263**; $R = Bu^t$) (reaction 104).[142]

Treatment of 2,4,4-trimethylpent-2-ene (**259**) with the preformed complex from phenylphosphonous dichloride and aluminium chloride yields 2,2,3,4,4-pentamethyl-1-phenylphosphetan 1-oxide (**264**). Although it is possible to prepare oxide (**264**) from the

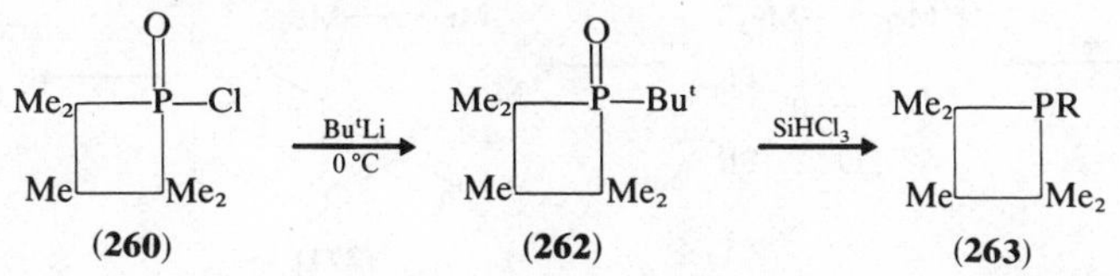

(104)

acid chloride (**260**) by treatment with phenyl-lithium, the yield is low owing to the subsequent, rapid reaction of (**264**) with phenyl-lithium to give ring opening (see reaction 108).

Inversion of configuration at the phosphorus of the phosphetans (**263**) has been studied by n.m.r. techniques and it has been found that the methylphosphetan (**263**; R = Me) did not invert at 162 °C for 4 days, while the t-butyl- and phenyl-phosphetans (**263**; R = Bu^t or Ph) inverted remarkably rapidly in view of the increased strain expected in the four-membered ring in the transition state (reaction 105). Hence, a clear steric factor seems to

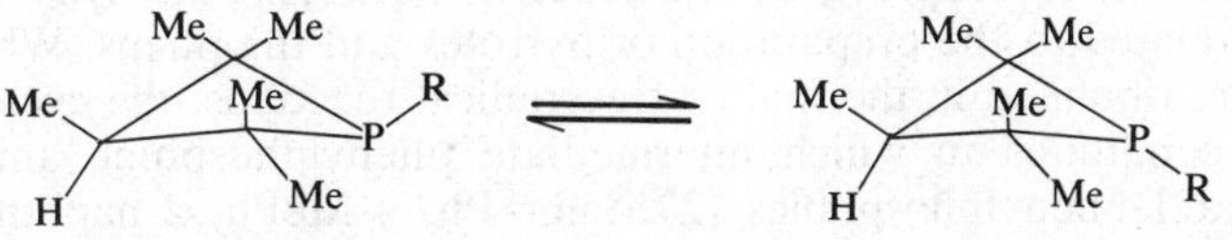

(105)

be operative in the t-butyl derivative owing to non-bonded interactions with the adjacent methyl groups in the ring. In the case of the 1-phenyl derivative, the possibility of either p_π–p_π or d_π–p_π overlap on the transition state may serve to lower the energy barrier for interconversion of the isomers.[143]

Reduction of 2,2,3,4,4-pentamethyl-1-phenylphosphetan 1-oxide (**264**) with lithium tetrahydridoaluminate and quaternization with methylene di-iodide yields the phosphetanium salt (**265**), which on alkaline hydrolysis affords the five-membered ring oxide (**267**). Presumably in the intermediate (**266**) from the hydrolysis of the salt (**265**) the four-membered ring is constrained to occupy apical-equatorial positions, thus blocking the expected migration, which must occur from an apical position. Thus migration of the apical CMe_2 occurs to the methylene group with expulsion of iodide ion (reaction 106).[144]

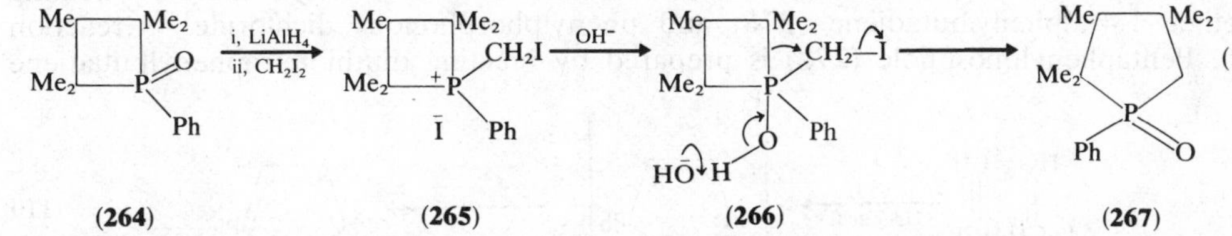

(106)

The alkaline hydrolysis of 1,2,2,3,4,4-hexamethyl-1-phenylphosphetanium iodide (**268**) conforms to the above general pattern, and since the phenyl group cannot be lost, migration of the apical CMe_2 occurs to the phenyl group, giving a cyclohexadienyl anion (**269**), which protonates to give the non-conjugated isomer (**270**) (reaction 107). An

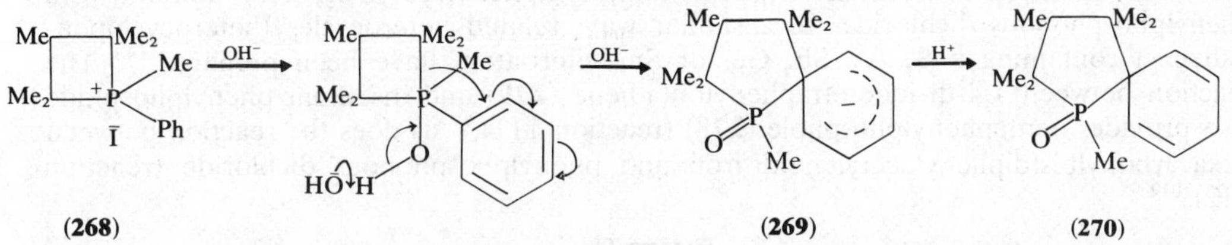

(107)

intermediate analogous to (**269**) is probably involved in the action of phenyl-lithium on 2,2,3,4,4-pentamethyl-l-phenylphosphetan 1-oxide (**264**), but here the cyclohexadienyl anion (**271**) cannot be protonated and, therefore, the five-membered ring opens to give an anion (**272**), which when quenched with methyl iodide yields an open-chain phosphine oxide (**273**) (reaction 108).[145]

(108)

(264) (271) (272)

$\xrightarrow{\text{MeI}}$ PhC(Me$_2$)CH(Me)C(Me$_2$)—P(=O)(Me)Ph

(273)

19.3.4.2 Five-membered ring containing one phosphorus atom

The five-membered heterocycles containing a phosphorus atom are analogous to the pyrroles. Phospholes are prepared by the reaction between butadiynes and phenylphosphine, which is related to the preparation of pyrroles and thiophens. Whereas only traces of phospholes are obtained in thermal or free-radical reactions, the cycloaddition may be effected under conditions in which intermediate phenylphosphine anions are formed. 2,5-Disubstituted 1-phenylphospholes (**275**; R = Ph, 4-MePh, 2-naphthyl), for instance, are obtained by the reaction between bis(hydroxymethyl)phenylphosphine and butadiynes (**274**) in boiling pyridine, or from phenylphosphine itself, which adds smoothly to butadiynes (**274**; R = Me, 4-BrPh) in benzene even at room temperature in the presence of a catalytic amount of phenyl-lithium (reaction 109).[146]

RC≡C—C≡CR (**274**) $\xrightarrow[\text{or PhPH}_2\text{, C}_6\text{H}_6\text{, PhLi}]{\text{either PhP(CH}_2\text{OH)}_2\text{, C}_5\text{H}_5\text{N}}$ (**275**) (109)

1,2,5-Triphenyphosphole (**275**; R = Ph), m.p. 187–189 °C, is also formed by heating together 1,4-diphenylbutadiene (**276**) and phenylphosphonous dichloride[147] (reaction 110). Pentaphenylphosphole (**278**) is prepared by reacting dilithiotetraphenylbutadiene

(**276**) $\xrightarrow[\text{214–230 °C}]{\text{PhPCl}_2}$ [intermediate] $\longrightarrow$ (**275**) (110)

(**277**), obtained by causing the dimerization of diphenylacetylene with lithium, with phenylphosphonous dichloride. In a similar way, related heterocycles (heterocyclopentadienes) containing a S, As, Sb. Ge, or Sn heteroatom have been prepared.[148] The reaction between 1,4-di-iodotetraphenylbutadiene (**279**) and disodium phenylphosphide also provides pentaphenylphosphole (**278**) (reaction 111),[17] as does the reaction between hexacarbonylbis(diphenylacetylene)di-iron and phenylphosphonous dichloride (reaction 112).[149]

(**277**) $\xrightarrow{\text{PhPCl}_2}$ (**278**) $\xleftarrow{\text{PhPNa}_2}$ (**279**) (111)

$$Fe_2(CO)_6(PhC{\equiv}CPh)_2 \xrightarrow[140\,°C]{PhPCl_2} \textbf{(278)} \qquad (112)$$

1,2,5-Triphenylphosphole (**275**) forms an oxide (**280**), sulphide (**281**), and selenide (**282**), although the sulphide appears to dimerize. With bromine, only the *P*-dibromide (**283**) is formed, without substitution of the ring, and quaternization occurs smoothly with methyl iodide to give the methiodide (**284**), but proceeds more slowly with ethyl bromoacetate to give the phosphonium salt (**285**), which on treatment with aqueous sodium hydroxide affords the phosphorane (**286**). This phosphorane does *not* undergo the Wittig reaction on prolonged boiling with cyclohexanone in benzene (Scheme 14).[147] Hydrogenation of the triphenylphosphole oxide (**280**) gives 1,2,5-triphenylphospholan 1-oxide (1,2,5-triphenylphosphacyclopentane 1-oxide) (**287**). The phosphole cannot be hydrogenated, presumably because the tervalent phosphorus poisons the catalyst.

i, H_2O_2; ii, S, $C_6H_4Me_2$; iii, Se, C_6H_6; iv, Br_2, CCl_4; v, MeI; vi, $BrCH_2CO_2Et$, C_6H_6; vii, NaOH, H_2O; viii, H_2, Pt.

SCHEME 14

1,2,5-Triphenylphosphole absorbs in the u. v. region at a much longer wavelength than does 1,2,5-triphenylpyrrole, which suggests that the phosphole has less aromatic character. This is also supported by chemical evidence, which indicates that the heterocyclic ring in the phosphole (**275**) has little or no aromatic character, but its n.m.r. spectrum might be construed as indicating aromaticity. Here all the protons resonate as a complex multiplet between 2.35 and 2.9 τ, but in the oxide, in which the phosphorus lone-pair is not available to the ring, this multiplet is broadened and one peak of the doublet expected from the two hetero-ring protons is separated from the main band and appears at 3.15 τ. Again it is not a characteristic diene, for it fails to react with either maleic anhydride or acrylonitrile in benzene at 80 °C, conditions under which its oxide forms normal adducts.

However, at 150–200 °C, both the phosphole and its oxide undergo Diels–Alder reactions with maleic anhydride, with elimination of the PPh bridge and aromatization to 3,6-diphenylphthalic anhydride. 1,2,5-Triphenylphosphole (**259**) reacts with dimethyl acetylenedicarboxylate to give a stable 1 : 2 yellow adduct, which has been shown to be the tricycloallylidenephosphorane (**288**).[150]

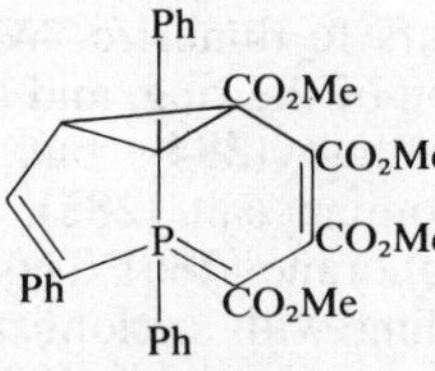

(288)

1,3-Dienes (**289**) on treatment with phosphorus trichloride or tribromide in the presence of copper stearate (a polymerization inhibitor) give 1,1,1-trihalogeno-3-phospholens (**290**), which on reaction with triethylamine and ethanol afford 1-ethoxyphospholen 1-oxides (**291**).[151] It is found that phosphorus tribromide reacts more energetically with 1,3-dienes than does the trichloride, which tends to yield a lot of tar. The position of the double bond is determined by means of n.m.r. spectral data, and it is found that this depends not only on the method of synthesis but also on the nature of the diene and on the degree of symmetry of the arrangement of substituents on the ring.[152] The oxide (**291**) may be converted to the very reactive 1-ethoxyphosphole 1-oxide (**292**) by the route indicated (Scheme 15).[153] The ester dimerizes too readily to allow its isolation, but it has been identified by its u.v. spectrum and by trapping it as a Diels–Alder adduct (**293**) with cyclopentadiene (Scheme 15).

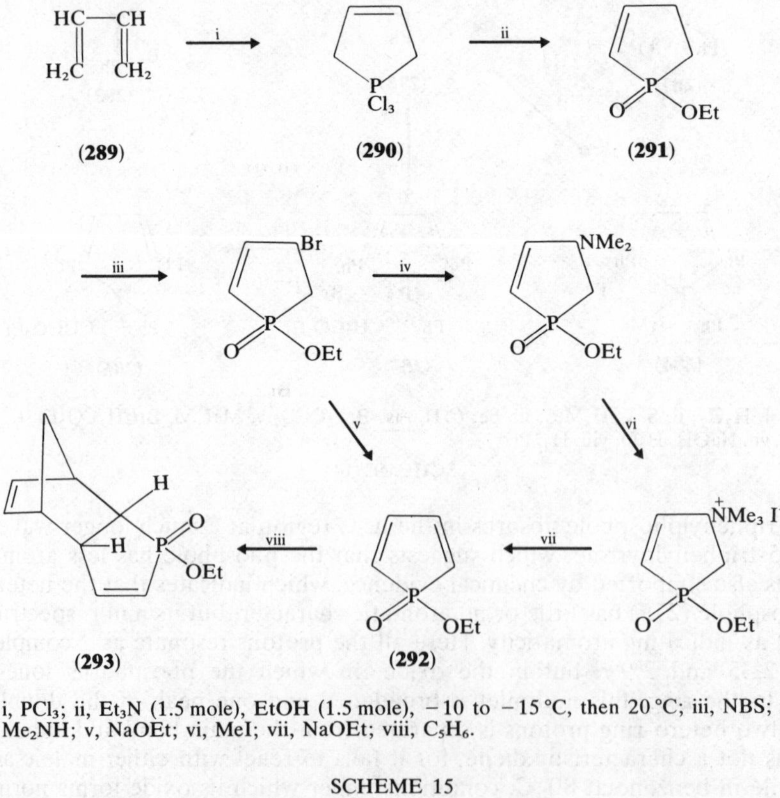

(289) **(290)** **(291)**

(293) **(292)**

i, PCl_3; ii, Et_3N (1.5 mole), EtOH (1.5 mole), −10 to −15 °C, then 20 °C; iii, NBS; iv, Me_2NH; v, NaOEt; vi, MeI; vii, NaOEt; viii, C_5H_6.

SCHEME 15

O=P(OEt) ring

(294)

1-Alkoxy-3-phospholen 1-oxides (**294**) are obtained from 1,1,1-tribromo-3-phospholens (**290**; Cl = Br) and the appropriate alcohol.[154] It is found that heating butadiene (**289**) with butyl dichlorophosphite, in the presence of copper stearate in an autoclave, affords 1-chloro-3-phospholen 1-oxide (**295**), whereas if the reaction is carried out under similar conditions, but with phosphorus trichloride also present, then a mixture of the 2- and 3-phospholen is obtained from which the 1-chloro-2-phospholen 1-oxide (**296**) is separated by fractional distillation (reaction 113).[155]

$$\text{(289)} \xrightarrow[\text{PCl}_3,\ 120\,^\circ\text{C}]{\text{BuOPCl}_2} \text{(295)} + \text{(296)} \qquad (113)$$

1-Phenyl-3-phospholen 1-oxide (**297**) is obtained by treating with water the product formed by the 1,4-cycloaddition of butadiene (**289**) to phenyldibromophosphine. Reduction of the oxide (**297**) using phenylsilane yields 1-phenyl-3-phospholen (**298**) (reaction 114).[156]

$$\text{(289)} \xrightarrow[\text{ii, H}_2\text{O}]{\text{i, PhPBr}_2} \text{(297)} \xrightarrow{\text{PhSiH}_3} \text{(298)} \qquad (114)$$

It is found that the base-catalysed cycloaddition of diacetylene to phenylphosphine does not afford 1-phenylphosphole (**301**), but gives an intermediate 1-phenyl-3-phospholen 1-oxide (**297**) from which the desired compound may be obtained. The oxide (**297**) adds on bromine smoothly to give 3,4-dibromo-1-phenylphospholan 1-oxide (**299**), which on reduction followed by dehydrobromination with 1,5-diazabicyclo[5,4,0]undec-5-ene (DBU) yields 3,4-dibromo-1-phenylphospholan (**300**) and hence 1-phenylphosphole (**301**) (reaction 115).

$$\text{(297)} \xrightarrow{\text{Br}_2,\ \text{CHCl}_3} \text{(299)} \xrightarrow{\text{PhSiH}_3} \text{(300)} \xrightarrow{\text{DBU}} \text{(301)} \qquad (115)$$

3,4-Dibromo-1-phenylphospholan 1-oxide (**299**) on dehydrobromination using potassium t-butoxide in DMF does not give the expected 1-phenylphosphole 1-oxide, but the Diels–Alder adduct (**302**), also obtained when 1-phenylphosphole (**301**) is treated with hydrogen peroxide in acetone (reaction 116). The analogous tricyclo-sulphide (**302**; O = S) is formed when 1-phenylphosphole is boiled with sulphur in benzene. The reaction between the dibromo-compound (**300**) and t-butyl-lithium gives 1-phenyl-3-phospholen (**298**) (reaction 117).

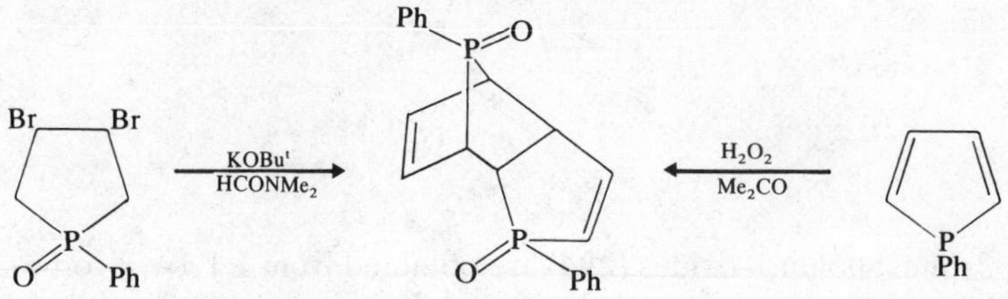

(116)

(299) (302) (301)

(117)

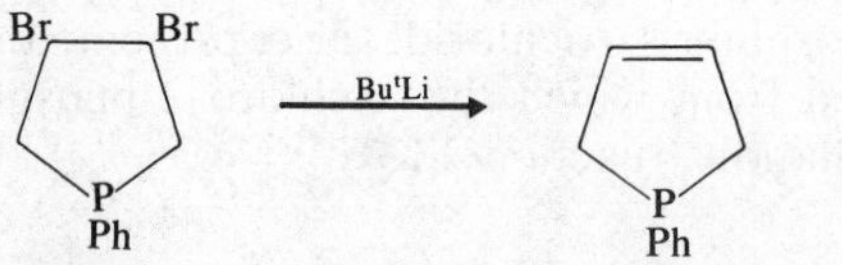

(300) (298)

The five-membered heterocycle phospholan (phospholidine, phosphacyclopentane) (**305**) containing a saturated ring has been obtained by heating 1-dimethylaminophospholan-boron trihydride (**304**), formed by the reaction between the aminocyclophosphine (**303**) and diborane (reaction 118).[157]

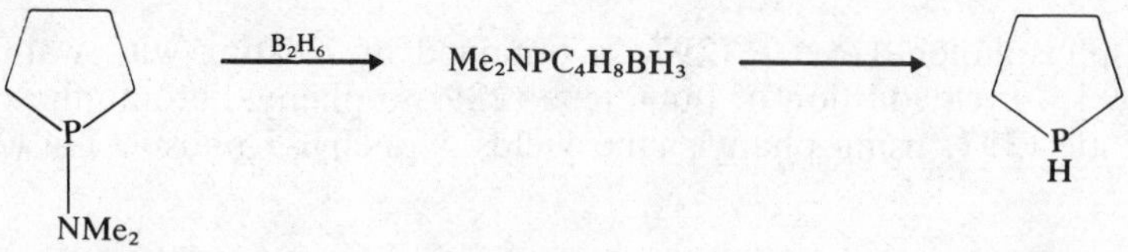

$$\xrightarrow{B_2H_6} Me_2NPC_4H_8BH_3 \longrightarrow \qquad (118)$$

(303) (304) (305)

19.3.4.3 Five-membered ring containing one phosphorus atom and fused to one or two benzene rings

Phosphindolines may be synthesized and converted to phosphindoles by bromination followed by dehydrobromination. 2-Phenylethylphosphonous dichloride (**306**) on heating at 170 °C with zince chloride, followed by hydrolysis using hydrochloric acid and oxidation, affords 1-hydroxyphosphindoline 1-oxide (**307**) (reaction 119). Bromination of 1-methoxyphosphindoline 1-oxide (**308**) with *N*-bromosuccinimide and subsequent dehydrobromination yields 1-methoxyphosphindole (**309**) (reaction 120).[158]

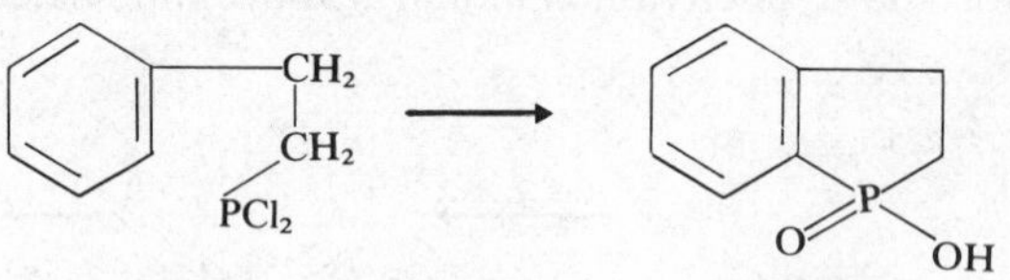

(119)

(306) (307)

(120)

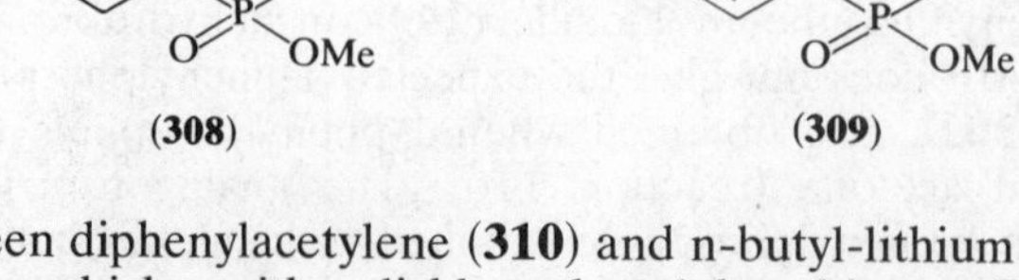

(308) (309)

The reaction between diphenylacetylene (**310**) and n-butyl-lithium produces a dilithium intermediate (**311**), which with dichlorophenylphosphine affords 3-n-butyl-1,2-diphenylphosphindole (**312**) (reaction 121)[159].

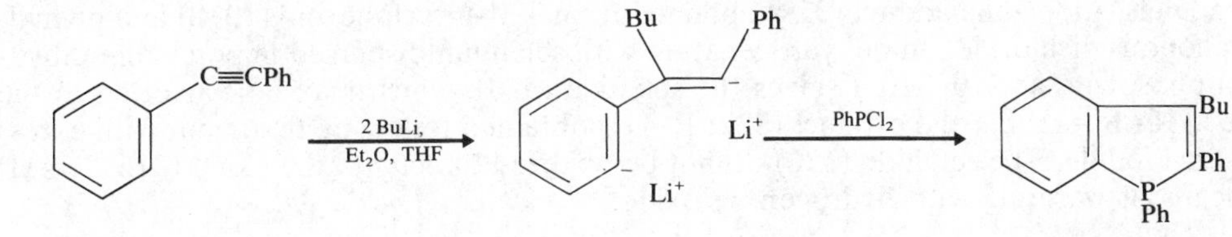

(310) (311) (312) (121)

5-Phenyldibenzophosphole 5-oxides (9-phenyl-9-phosphafluorene 9-oxides) (**316**) are prepared by the routes shown in Scheme 16. The biphenyl(phenyl)phosphinic acids (**314**) are obtained from the corresponding 2-aminobiphenyls (**313**) either by Doak and Freedman's diazonium fluoroborate method, *i.e.* the aminobiphenyl is converted into the diazonium fluoroborate, and the dry salt is decomposed in ethyl acetate in the presence of phenylphosphonous dichloride and copper bronze, or through the reaction between the Grignard reagent from the iodide (**315**) and phenylphosphonous dichloride.[160] Cyclization of the phosphinic acids (**314**) presents considerable difficulty, *e.g.* the use of aluminium chloride as catalyst in the normal Friedel–Crafts reaction on the chloride of the acid (**314**; R = H) fails. The cyclization is effected by the use of a large excess of phosphorus pentachloride in nitrobenzene at 160–180 °C. 3-Carboxy-5-phenyldibenzophosphole 5-oxide (**316**; R = CO_2H) has been resolved by separation of the diastereoisomeric amides obtained from (+)- and (−)-1-phenylethylamine. Reduction of the (+)-oxide, with lithium tetrahydridoaluminate, gives 3-hydroxymethyl-5-phenyldibenzophosphole (**317**; R = CH_2OH), with complete loss of optical activity.

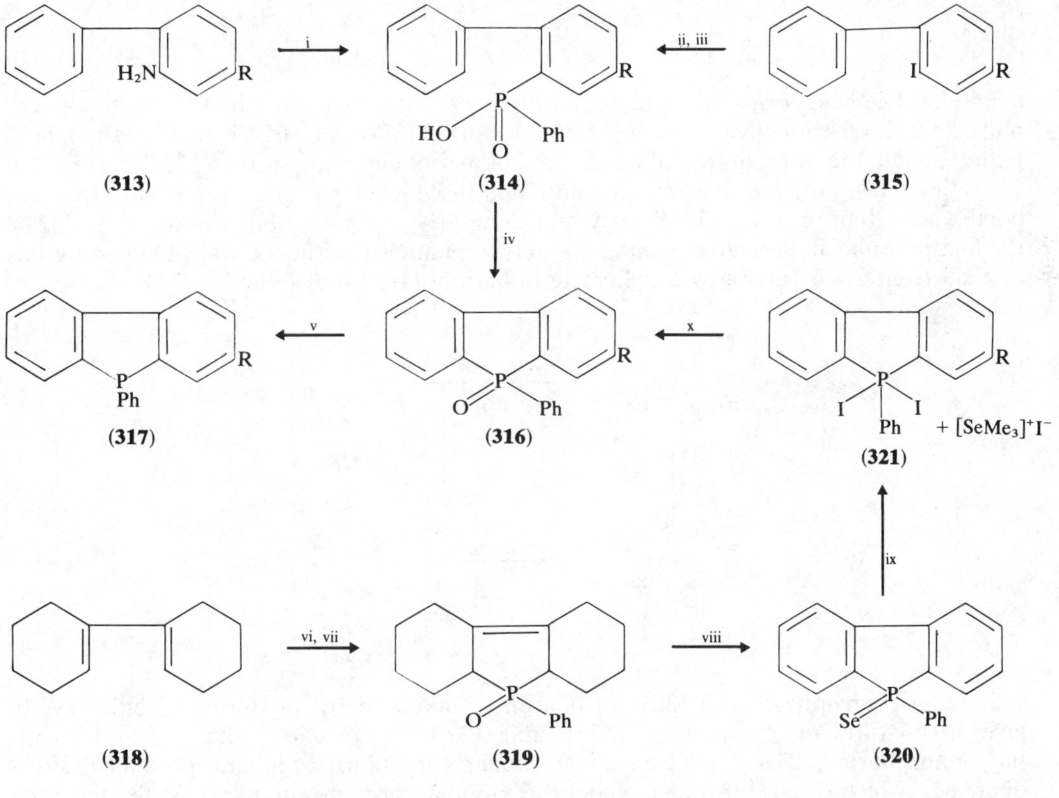

(313) (314) (315)

(317) (316) (321) + $[SeMe_3]^+I^-$

(318) (319) (320)

i, Diazonium fluoroborate, $PhPCl_2$, EtOAc; ii, Mg, Et_2O; iii, $PhPCl_2$, then H_2O; iv, PCl_5, $PhNO_2$, then H_2O; v, $LiAlH_4$, Bu_2O, C_6H_6; vi, $PhPCl_2$; vii, H_2O_2; viii, Se, KH_2PO_4; ix, MeI; x, NaOH, H_2O.

SCHEME 16

The Diels–Alder-like adduct (**319**) obtained from 1,1′-bicyclohexenyl (**318**) and phenylphosphonous dichloride, on dehydrogenation with selenium dispersed in potassium dihydrogen phosphate at 270–370 °C gives the selenide (**320**), which may be converted to the oxide (**316**; R = H) *via* the product (**321**; R = H) obtained following treatment with excess of methyl iodide. The selenide (**320**) cannot be oxidized directly to the oxide (**316**; R = H) by means of warming with hydrogen peroxide.[147]

19.3.4.4 Six-membered ring containing one phosphorus atom

The generally accepted name for unsaturated six-membered heterocycles containing one phosphorus atom is phosphorin because of its relationship to pyridine, but because of its relationship to benzene the name phosphabenzene is also used. Among the new heteroaromatic compounds the chemistry of substituted phosphorins has been investigated particularly thoroughly.

Phosphorin (**323**) has been prepared by a one-step synthesis by treating 1,4-dihydro-1,1-dibutylstannabenzene (**322**) with phosphorus tribromide, which gave phosphorin hydrobromide and dibutyltin dibromide (reaction 122). Phosphorin can be freed from its salt by adding 1,5-diazabicyclo[5,4,0]undec-5-ene.[161] It is a colourless volatile liquid with a characteristic phosphine odour and is air- and moisture-sensitive, but appears to be quite stable for extended periods under an inert atmosphere.

Sn Bu$_2$ (**322**) —i, PBr_3; ii, DBU→ P (**323**) (122)

Previously, phosphorins were prepared from pyrylium salts and tris(hydroxymethyl)phosphine, *e.g.* 2,4,6-triphenylpyrylium tetrafluoroborate (**324**) and tris(hydroxymethyl)phosphine boiled together in pyridine afford 2,4,6-triphenylphosphorin (**325**)[162] (reaction 123), thus giving for the first time a compound in which a phosphorus–carbon $(3p–2p)\pi$ bond was stabilized in a Hückel π-system. Similarly, 2,4,6-tri(t-butyl)phosphorin (**327**), the first phosphorin derivative containing only aliphatic substituents, was obtained by this method from 2,4,6-tri(t-butyl)pyrylium tetrafluoroborate (**306**) (reaction 124).[163]

Ph, Ph, Ph, O^+, BF_4^- (**324**) —$P(CH_2OH)_3$, C_5H_5N→ Ph, Ph, Ph, P (**325**) (123)

Bu^t, Bu^t, Bu^t, O^+, BF_4^- (**326**) —$P(CH_2OH)_3$, C_5H_5N→ Bu^t, Bu^t, Bu^t, P (**327**) (124)

2,4,6-Triphenylphosphorin (**325**) in benzene is oxidized by air during several days to give a mixture of compounds (**328**) and (**329**)[164] (reaction 125). If 2,4,6-tri(t-butyl)phosphorin (**327**) is oxidized under similar conditions, a dimeric product (**330**) is obtained, which is also formed when the original phosphorin (**327**) is treated with bromine water, hydrogen peroxide in acetone, or a mixture of nitric acid and sulphuric acid at 0 °C. The dimer with diazomethane in ether–methanol affords a methoxy-derivative (**331**) (reaction 126). It has now been found that the reaction between

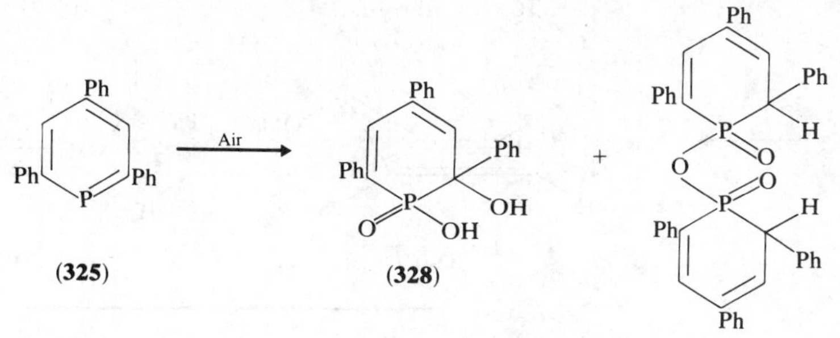

(125)

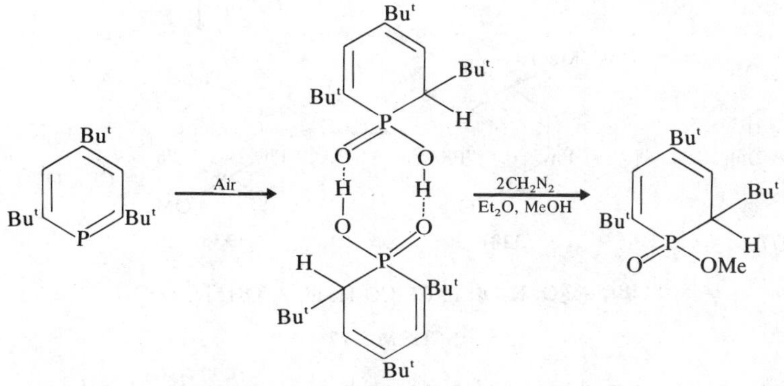

(126)

2,4,6-triphenylphosphorin and an equivalent amount of bromine or chlorine in the presence of light gives the corresponding 1,1-dihalogeno-derivative, from which the halide ion may be displaced by nucleophiles.[165]

2,4,6-Triaryl-substituted phosphorins (**332**) are oxidized to stable radical-cations by, *inter alia*, mercury(II) acetate; in the presence of alcohols or phenols the reaction proceeds with separation of mercury to form 1,1-dialkoxy- or 1,1-diaryloxy-phosphorins (**333**; R = alkyl or aryl). The reaction probably proceeds *via* the cation-radical stage, which can be detected by e.s.r. spectroscopy (reaction 127).[166] The above type of oxidation converts a 'classical' to a 'non-classical' phosphorin.

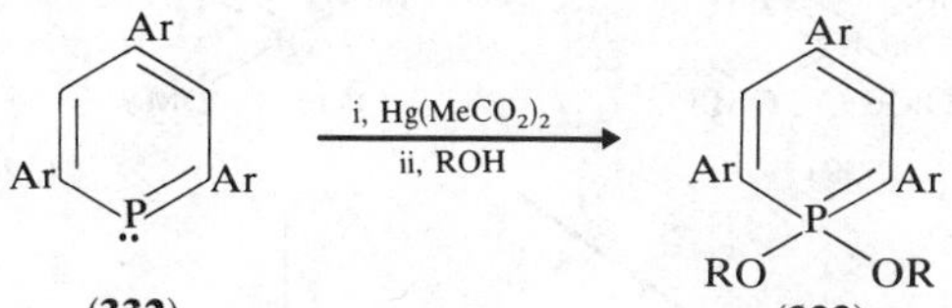

(127)

Compound (**333**; Ar = Ph, R = Me) on treatment with lithium bromide, acetic anhydride, and hydrogen peroxide in dimethoxyethane yields isomers of 4-hydroxy-1-methoxy-2,4,6-triphenyl-1-phosphacyclohexa-2,5-dienone, (**334**) and (**335**), both of which give the same deep blue salt (**336**) with trifluoroacetic acid. The salt (**336**) yields the phosphacyclohexadienones (**337**) and (**338**) when treated with methanol, and with aqueous acetic acid regenerates the isomers (**334**) and (**335**) (see Scheme 17).

Hydride-ion abstraction from 1,1-dimethoxy-2,6-diphenyl-4-methylphosphorin (**339**) with triphenylmethyl tetrafluoroborate gives the tetrafluoroborate (**340**), and the ambidentate character of the carbonium ion is shown by its reactions with nucleophiles.[167] Reduction using sodium tetrahydridoborate gives the original starting material (**339**), and cyanide and thiocyanate ions similarly add to yield the respective 4-substituted methyl derivative (**341**), whereas reaction with halide ions affords the methylene derivative (**342**).

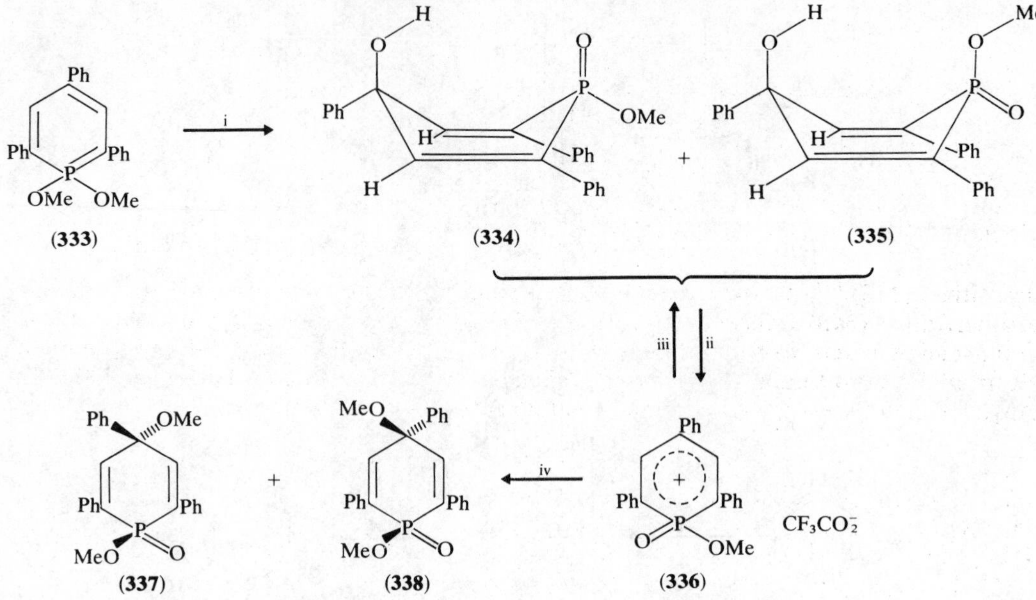

i, LiBr, Ac_2O, H_2O_2; ii, CF_3CO_2H; iii, AcOH; iv, MeOH.

SCHEME 17

Reaction with aqueous sodium hydrogen carbonate gives the bisphosphorinylmethane (**343**) (Scheme 18).

It has been shown that derivatives like (**340**) can be electrophilically substituted even in acidic media, without any evidence of a phosphonium salt as intermediate. The tetrafluoroborate derivative (**340**), in acetonitrile containing a little water, first produces the alcohol (**344**), which immediately undergoes electrophilic substitution (losing a formaldehyde cation) by reaction with (**340**) still present in the solution to give the methane

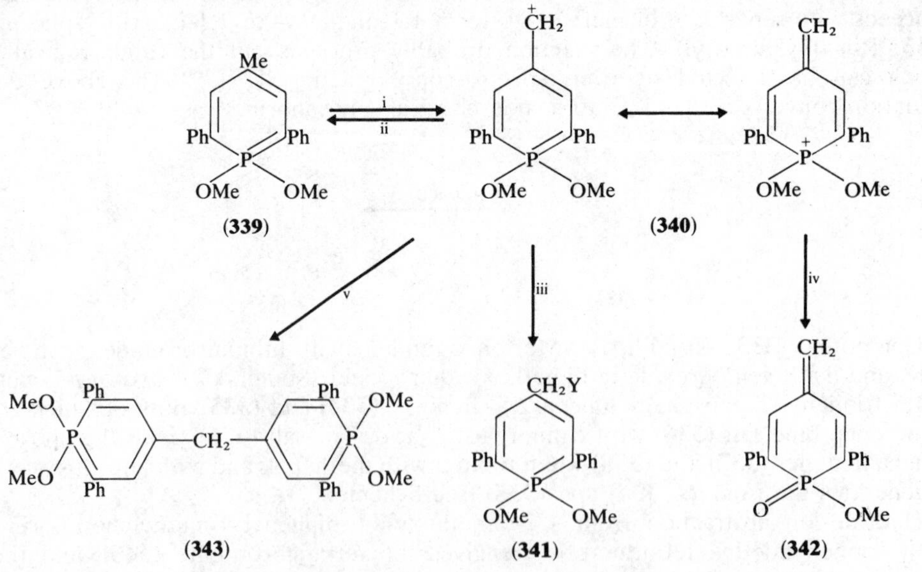

i, $Ph_3C^+BF_4^-$; ii, $NaBH_4$; iii, NaY (Y = CN or SCN); iv, NaX (X = Cl, Br, or I); v, $NaHCO_3$.

SCHEME 18

(128)

derivative (**343**),[168] or by reaction with aryldiazonium tetrafluoroborate to give the azo-derivative (**345**) (reaction 128). Compound (**343**) also contains a good electrofugal group and it reacts with diazonium salts to yield the azo-derivative (**345**). In another electrophilic substitution, (**343**) may be alkylated by triethyloxonium salts to give the 4-ethyl-1,1-dimethoxy-2,6-diphenylphosphorin (**346**).

Electrophilic substitution with removal of the cation of formaldehyde or benzaldehyde instead of a proton has long been known in the benzene series, *e.g.* it has been used to synthesize aldehydes starting from suitably *C*-substituted derivatives of *N,N*-dimethylaniline. Comparison of the reactions illustrated above with those of the corresponding derivatives of *N,N*-dimethylaniline illustrates the typical 'aromatic' electron-rich character of 1,1-dimethoxyphosphorin.

2,4,6-Triphenylphosphorin (**325**) on treatment with phenyl-lithium, followed by reaction with alkyl halide in tetrahydrofuran, gives the 1-alkyl-1,2,4,6-tetraphenylphosphorin (**348**; R = alkyl), which contains a pentavalent phosphorus. If the latter reaction is carried out in benzene then the resulting product, 2-alkyl-1,2-dihydro-1,2,4,6-tetraphenylphosphorin (**349**; R = alkyl), contains a trivalent phosphorus.[169] The above reactions appear to go *via* the intermediate anion (**347**) (reaction 129).

(129)

2,4,6-Triphenylphosphorin (**325**) possesses only slight diene reactivity, *e.g.* it reacts neither with diethyl acetylenedicarboxylate nor with maleic anhydride. However, along with other 2,4,6-triaryl- or 2,4,6-trialkyl-phosphorins it reacts with hexafluoro-but-2-yne at 100 °C to give the adduct (**350**)[170] (reaction 130), whereas the analogous reaction with

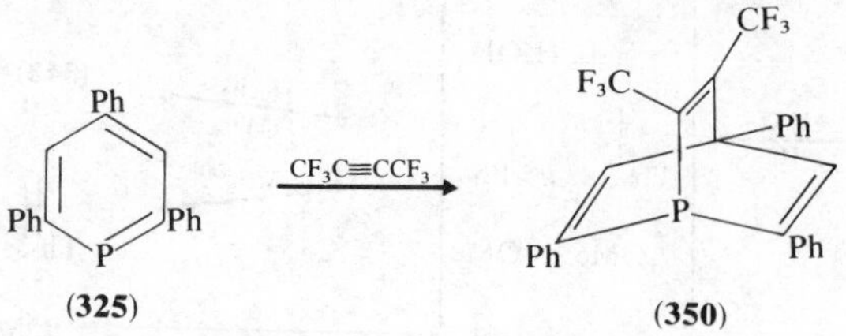

(130)

(325) (350)

aromatic hydrocarbons requires a much higher temperature, *e.g.* with benzene a temperature of approximately 200 °C.

The saturated six-membered heterocycle containing a phosphorus atom is called phosphan (**354**), but other names such as phosphacyclohexane, phosphorinan, and pentamethylenephosphine have also been used. It is the phosphorus analogue of piperidine. The initial step in the synthesis of phosphan involves a Michaelis–Arbuzov reaction between 1,5-dibromopentane (**122**) and tri-n-butyl phosphite to give di-n-butyl 5-bromopentylphosphonate (**351**), which is cyclized through a Grignard-type reaction to 1-n-butoxyphosphorane 1-oxide (**352**). The ester (**352**) is hydrolysed to the phosphinic acid (**353**), and then conveniently reduced to phosphan (**354**) by means of diphenylsilane (reaction 131).[171]

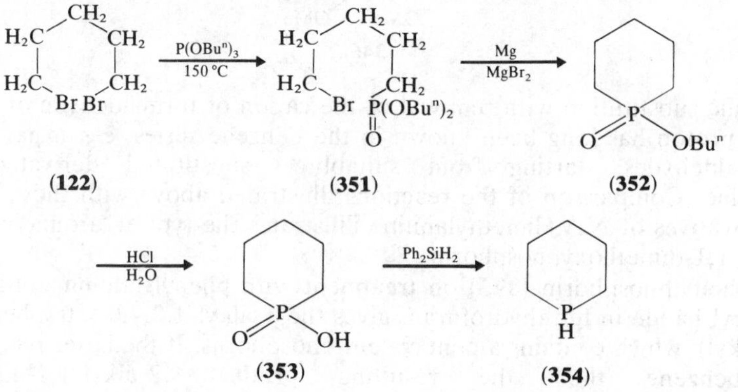

(122) (351) (352)

(131)

(353) (354)

In the n.m.r. spectrum of phosphan (**354**) the proton on the phosphorus appears as a broad triplet of triplets, which is consistent only with an axial placement, and no evidence for the presence of an equatorial conformer is obtained. The n.m.r. spectrum of phosphan 1-sulphide (**355**), prepared by direct reaction of phosphan (**354**) with an equimolar amount of elemental sulphur, leads to similar conclusions. For the sulphide (**355**), this preference is probably due to the greater steric requirements of the sulphide substituent.

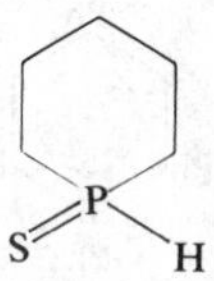

(355)

With phosphan (**354**) the geminal 'substituent' to the proton is a non-bonding pair of electrons. The axial preference of the proton in this case is more likely a result of the attractive properties of the axial substituent than any repulsive properties of the lone pair.

When an equivalent amount of methyl iodide is mixed with phosphan (**354**) at −78 °C and then allowed to warm, a reaction occurs producing white crystals of phosphan methiodide (**356**) in quantitative yield (reaction 132). From its n.m.r. spectrum no conclusions can be reached concerning the location of the proton on the phosphorus because of second-order spectral complications.

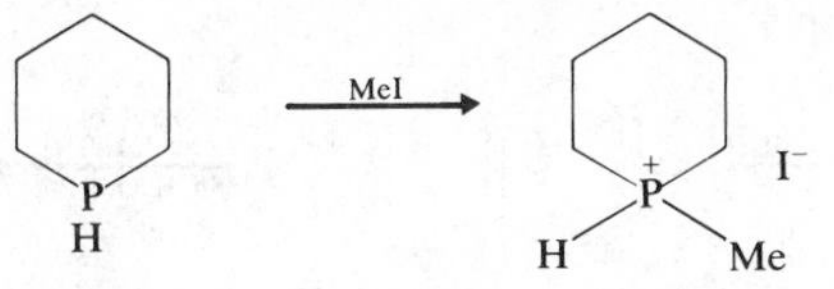

(132)

(354) (356)

1-Phenylphosphan (**354**; H replaced by Ph) may be prepared by reacting phenyldichlorophosphine with the di-Grignard reagent formed from 1,5-dibromopentane or with the related dilithio-compound. The phosphans are liquids, slowly oxidized in air, sufficiently basic to form salts with acids, and able to form quaternary salts.

19.3.4.5 Six-membered ring containing one phosphorus atom and fused to one or two benzene rings

Names for heterocycles in this class of compounds are generally derived from either the analogous quinoline and isoquinoline or naphthalene compounds, thus giving phosphinoline and isophosphinoline or 1-phosphanaphthalene and 2-phosphanaphthalene.

Methods which were used for the preparation of related arsenic heterocycles were not practicable for obtaining 1,2,3,4-tetrahydrophosphinolines and 1,2,3,4-tetrahydroisophosphinolines. Their synthesis is based essentially on the fact that many quaternary ethylphosphonium chlorides undergo smooth thermal decomposition, an ethyl group breaking off as ethylene with the formation of the tertiary phosphine hydrochloride.[172] Routes to 1-ethyl-1,2,3,4-tetrahydrophosphinoline (**357**) and 2-ethyl-1,2,3,4-tetrahydroisophosphinoline (**358**) are indicated in Scheme 19. The latter compound is both more basic and more readily oxidized than the former.

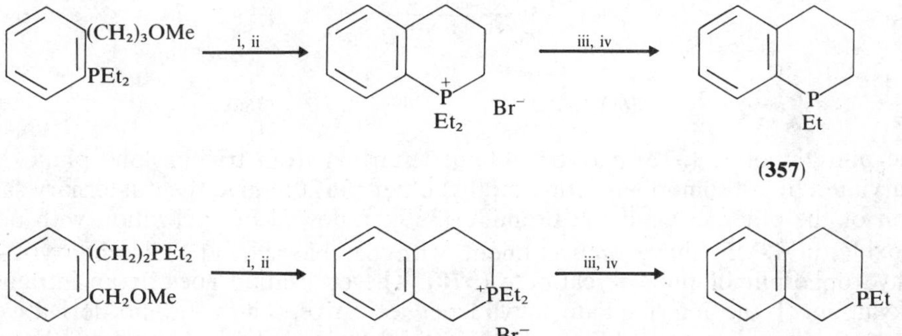

(357)

(358)

i, HBr, AcOH; ii, $NaHCO_3$; iii, prepare picrate and then treat with HCl to obtain chloride; iv, 350–370 °C, 20 mmHg.

SCHEME 19

A convenient method for obtaining tetrahydroisophosphinolinium and tetrahydrophosphinolinium hexafluorophosphates has been developed. It involves the treatment of a variety of tertiary phosphines (all containing an aryl group and/or an arylmethyl group) with allylic-type halides to give phosphonium salts containing a β-alkenyl substituent, which are subsequently ring-closed using polyphosphoric acid to yield the required heterocycle. 2,2-Diphenyl-4-methyl-1,2,3,4-tetrahydroisophosphinolinium hexafluorophosphate (**361**) is obtained in such a manner from allylbenzyldiphenylphosphonium bromide (**360**), prepared when benzyldiphenylphosphine (**359**) is quaternized with allyl bromide. Addition of a saturated aqueous solution of potassium hexafluorophosphate following cyclization results in the precipitation of the required salt (**361**) (reaction 133).[173]

PPh (**359**) — $CH_2{=}CHCH_2Br$, C_6H_6, Δ → $^{+}PPh_2$ Br^- (**360**) — i, Polyphosphoric acid, 160 °C; ii, H_2O, KPF_6 → Me $^{+}PPh_2$ PF_6^- (**361**) (133)

Surprisingly, this synthetic procedure is also applicable to the preparation of the isomeric phosphinoline derivative. For instance, it is possible to react triphenylphosphine (**362**) with 1-bromobut-2-ene, to obtain but-2-enyltriphenylphosphonium bromide (**363**), which, following ring closure, affords 1,1-diphenyl-4-methyl-1,2,3,4-tetrahydrophosphinolinium hexafluorophosphate (**364**) (reaction 134). Compound (**365**; R = H or Me) undergoes ring closure to produce the seven-membered ring heterocycle (**366**; R = H or Me), and not the six-membered heterocycle (reaction 135).

PPh_2 (**362**) — $MeCH{=}CHCH_2Br$, MePh, Δ → Me $\overset{+}{P}$ Ph_2 Br^- (**363**) — i, Polyphosphoric acid, 160 °C; ii, H_2O, KPF_6 → Me $\overset{+}{P}$ Ph_2 PF_6^- (**364**) (134)

Me R $^{+}PPh_2$ Br^- (**365**) — i, Polyphosphoric acid, 160 °C; ii, H_2O, KPF_6 → Me R $^{+}PPh_2$ PF_6^- (**366**) (135)

The phosphinoline (**373**) may be obtained starting from tribenzylphosphine, which is first arylated by 2-bromophenylethyl methyl ether (**367**) to give the quaternary salt (**368**). Fission of the ether to yield the bromide (**369**), followed by cyclization with potassium t-butoxide in DMF gives, on treatment with perchloric acid, 1,1-dibenzyl-2-phenyltetrahydrophosphinolinium perchlorate (**370**). The cyclization goes by an intramolecular *C*-alkylation of the intermediate alkylidene derivative. The 4-bromo-derivative (**371**), obtained by treating the perchlorate (**370**) with *N*-bromosuccinimide, on heating in 100% phosphoric acid dehydrobrominates and simultaneously isomerizes to give the salt (**372**), which is deprotonated by dilute sodium hydroxide solution to yield 1,1-dibenzyl-2-phenylphosphinoline (**373**) (Scheme 20).[174]

The phosphinoline (**373**) with methyl iodide forms a phosphonium salt, deprotonation of which by aqueous sodium hydroxide gives 1,1-dibenzyl-4-methyl-2-phenylphosphinoline (**374**). Treatment of (**373**) with benzoyl chloride leads directly to the 4-benzoyl derivative. The phosphonium salt first formed is immediately deprotonated by the more strongly basic phosphinoline (**373**) (transylidation). Thermolysis of the phosphinoline (**373**) at 230–250 °C results in the migration of a benzyl group to position-4, giving 1,4-dibenzyl-1,4-dihydro-2-phenylphosphinoline (**375**). In the case of 1,1-dibenzyl-2,4,6-triphenylphosphorin (**376**), heating at 220 °C gives 2,4,6-triphenylphosphorin (**325**) and bibenzyl (reaction 136). It would therefore appear that the 1,4-dihydrophosphinolines, as tertiary phosphans, are energetically favoured over the phosphinolines.

1,1-Dibenzyl-2-phenyltetrahydrophosphinolinium tetrafluoroborate (**370**; BF_4^- for ClO_4^-) is hydrolysed by alkali to give 1-benzyl-2-phenyltetrahydrophosphinoline 1-oxide

(367) (368) (369) (370) (371) (372) (373)

i, $P(CH_2Ph)_3$; ii, HBr, AcOH; iii, $KOBu^t$, DMF; iv, $HClO_4$; v, NBS; vi, 100% H_3PO_4; vii, dil. NaOH.

SCHEME 20

(374) (375) (376) (325) (136)

(370) (377) (137)

(377) (378) (379) (380) (138)

(**377**) (reaction 137). The oxide (**377**) on treatment with *N*-bromosuccinimide affords the 4-bromo-derivative (**378**), which on dehydrobromination, followed by reduction by trichlorosilane in boiling benzene, gives 1-benzyl-2-phenyl-1,2-dihydrophosphinoline (**379**). On heating, the phosphan (**379**) yields 2-phenylphosphinoline (**380**), the first phosphinoline to be synthesized (reaction 138).[175]

Dibenzo[*b,e*]phosphorin (9-phosphanthracene) (**383**) may be synthesized *via* the di-Grignard compound (**381**) from bis-(2-bromophenyl)methane, which on reacting with dichlorodiethylaminophosphine in THF, followed by isolation with cyclohexane and treatment with gaseous hydrogen chloride, affords 5-chloro-5,10-dihydrodibenzo[*b,e*]phosphorin (**382**) (reaction 139). An alternative method of obtaining

CH2 MgBr MgBr —(i, Et_2NPCl_2, THF, −80 °C; ii, HCl, C_6H_{12})→ P Cl —(DBN)→ P (139)

(**381**) (**382**) (**383**)

the chloro-compound (**382**) involves reacting the Grignard reagent (**384**) from (2-bromophenyl)phenylmethane with chlorobis(diethylamino)phosphine and isolating the product (**385**) by the method described above. The 2-benzylphenyldichlorophosphine (**385**) is cyclized by boiling with aluminium chloride in carbon disulphide (reaction 140).

CH2 MgBr —(i, $(Et_2N)_2PCl$; ii, HCl, C_6H_{12})→ CH2 PCl_2 —($AlCl_3$, CS_2)→ (**382**) (140)

(**384**) (**385**)

The chloro-compound (**382**) on dehydrochlorination in degassed toluene, using 1,5-diazabicyclo[4,3,0]non-5-ene, gives a solution containing dibenzo[*b,e*]phosphorin (**383**), which remains stable for several days (reaction 139).

Surprisingly, its u.v. spectrum resembles that of anthracene more closely than that of acridine. It has not been possible to isolate dibenzo[*b,e*]phosphorin (**383**) and its u.v. spectrum disappears immediately on the admission of atmospheric oxygen, very rapidly on addition of sodium hydroxide or anhydrous hydrogen chloride, and slowly on addition of dilute acid.[176] The instability of dibenzo[*b,e*]phosphorin (**383**) corresponds to the reactivity of positions-9 and -10 of anthracene, which results from the π-electron density. A stable compound may be obtained by the introduction of a phenyl group in position-10, when the reactivity is reduced by steric shielding. 10-Phenyldibenzo[*b,e*]phosphorin may be isolated as a crystalline solid, but it is extremely air-sensitive.[177] The chloro-compound (**382**) is oxidized in sodium hydroxide solution by hydrogen peroxide to give 5-hydroxy-5,10-dihydrodibenzo[*b,e*]phosphorin 5-oxide (**386**) (reaction 141).

P Cl —(H_2O_2, NaOH)→ P O OH (141)

(**382**) (**386**)

Dibenzo[*b,d*]phosphorin (9-phosphaphenanthrene) (**391**) has been obtained from 5,6-dihydro-5-hydroxydibenzo[*b,d*]phosphorin 5-oxide (**388**), prepared by the cyclodehydration of 2-phenylbenzylphosphonic acid (**387**). The oxide (**388**) on reduction with diphenylsilane yields 5,6-dihydrodibenzo[*b,d*]phosphorin (**389**), which on reacting with carbonyl dichloride gives 5-chloro-5,6-dihydrodibenzo[*b,d*]phosphorin (**390**). Dehydrochlorination of the chloro-compound (**390**) using 1,5-diazabicyclo[5,4,0]undec-5-ene in toluene gives a solution of dibenzo[*b,d*]phosphorin (**391**)[178] (Scheme 21). In contrast to dibenzo[*b,e*]phosphorin (**383**), the shape of its u.v. spectrum resembles that of the nitrogen analogue more than that of the carbon analogue. It could not be isolated in a pure form and its stability appears comparable with that of dibenzo[*b,e*]phosphorin (**383**).

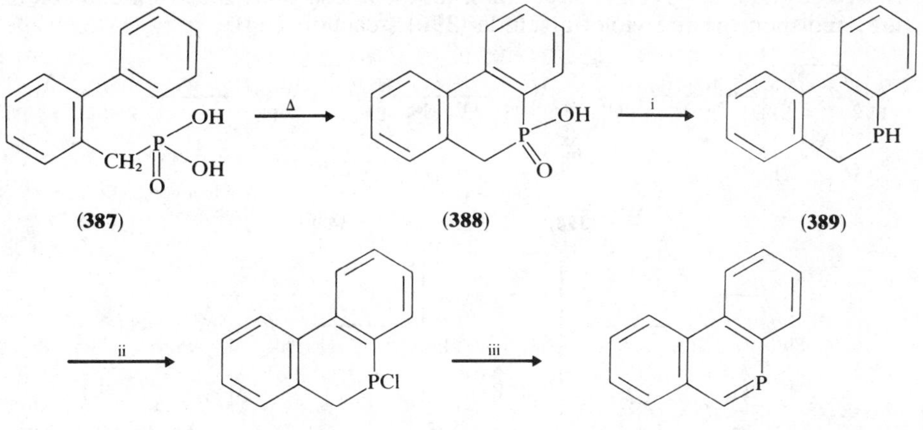

(**387**) (**388**) (**389**)

(**390**) (**391**)

i, Ph_2SiH_2; ii, $COCl_2$, CH_2Cl_2, N_2; iii, DBU, MePh.

SCHEME 21

A comparative study of the photoelectron spectra of 9-phenylphenanthrene and 6-phenyldibenzo[*b,d*]phosphorin indicates a similarity in the electronic effects produced by an sp^2 hybridized carbon atom substituted by hydrogen and an sp^2 phosphorus atom.[179]

19.3.5 HETEROCYCLES CONTAINING ONE ARSENIC, ANTIMONY, OR BISMUTH ATOM

Heterocycles containing either one arsenic or antimony atom are named in a similar fashion to those containing one phosphorus atom. Thus to obtain the names of related compounds, 'phosphor' is replaced by either 'arsen' or 'antimon', and 'phospha' by either 'arsa' or 'stiba' (eliding the 'a' where necessary).

19.3.5.1 Five-membered ring containing either one arsenic or antimony atom

The five-membered analogue of pyrrole with an arsenic atom in place of the nitrogen atom is called arsole and 1-phenyl-2,5-disubstituted arsoles (**393**; R = Ph, 4-MeC_6H_4, 4-ClC_6H_4, α-$C_{10}H_7$, Me) are readily prepared by base-catalysed cycloaddition of phenylarsine to the appropriate buta-1,3-diyne (**392**) (reaction 142).[180] It is found that the chemistry of the arsoles resembles that of pyrrole rather than the phospholes.

$$RC{\equiv}C{-}C{\equiv}CR \xrightarrow[C_6H_6,\ BuLi]{PhAsH_2} \text{(393)} \qquad (142)$$

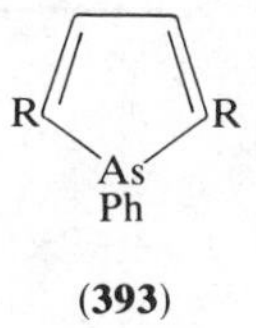

(**392**) (**393**)

Oxidation of 1,2,5-triphenylarsole (**394**) with hydrogen peroxide affords the 1-oxide (**395**) (reaction 143), but the sulphide and quaternary salts cannot be obtained. The triphenylarsole (**394**) on treatment with potassium in boiling dimethoxyethane yields 1-potassio-2,5-diphenylarsole (**396**), which reacts with alkyl halides to give 1-alkyl-2,5-diphenylarsole (**397**; R = Me, Et, CH_2OMe) (reaction 143). The reduction with either lithium or potassium in dimethoxyethane goes by a stepwise addition of two electrons. First is formed the blue-green radical anion (**398**), whose e.s.r. spectrum exhibits no fine structure, and then the red-violet arsenide (**396**) (reaction 144).

(**395**) ← [H_2O_2, Me_2CO] (**394**) → [K, $(CH_2OMe)_2$] (**396**) → [RX] (**397**) (143)

X = halogen

(**394**) → [M] (**398**) M^+ → [M] (**396**) + PhM (144)

M = Li, K

The radical anion (**398**) reacts with methylene chloride and benzal chloride to yield not only the monochloroalkylarsoles (**397**; R = CH_2Cl, CHClPh), but also the dinuclear arsoles (**399**) and (**400**) by a double nucleophilic substitution. Both the radical anion (**398**) and the anion of the lithium or potassium derivative (**396**) react with trityl perchlorate to give the biarsolyl (**401**)[181] (reaction 145), also formed by the oxidation of the radical ion in ethanol, by the oxidation of 2,5-diphenyl-1-methoxymethylarsole using ferric nitrate, and by the reaction between 1-lithio- and 1-chloro-2,5-diphenylarsole. The biarsolyl (**401**) is smoothly cleaved by treatment with alkali metal to give the alkali arsenide (**396**).

(**399**) (**400**)

(**398**) M^+ → [$Ph_3C^+ClO_4^-$] (**401**) ← [$Ph_3C^+ClO_4^-$] (**396**) (145)

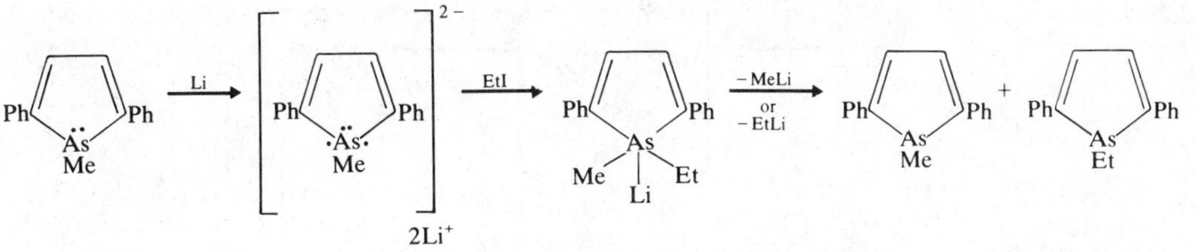

(146)

(402) (403) (404) (402) (405)

Unlike triphenylarsole (**397**), 1-methyl-2,5-diphenylarsole (**402**) produces a stable dianion (**403**) on treatment with two equivalents of alkali metal, *e.g.* lithium. The dianion reacts with ethyl iodide, probably *via* the penta-coordinate species (**404**), to give a mixture of 1-methyl- (**402**) and 1-ethyl-2,5-diphenylarsole (**405**) (reaction 146).

Both 1,2,5-triphenyl- (**394**) and 1-phenyl-2,5-dimethyl-arsole (**393**; R = Me) are unable to form quaternary salts and sulphides, and the oxides are formed only in moderate yields along with arsenic-free decomposition products on oxidation with hydrogen peroxide. The arsoles (**397**; R = Me, Ph), however, react smoothly with elemental chlorine to yield the stable 1,1-dichloro-derivatives (**406**), and with t-butyl hypochlorite to give the 1-chloro-1-t-butoxy-derivatives (**407**) (reaction 147). The dichloro-derivatives are thermolabile, but are surprisingly stable to water and alcohols. The benzyl derivative (**406**; R = $PhCH_2$) decomposes at room temperature to give 1-chloro-2,5-diphenylarsole (**408**) (reaction 148).

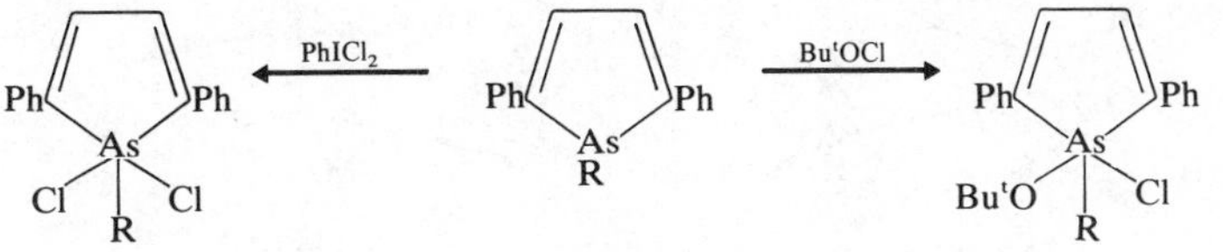

(147)

(406) (397) (407)

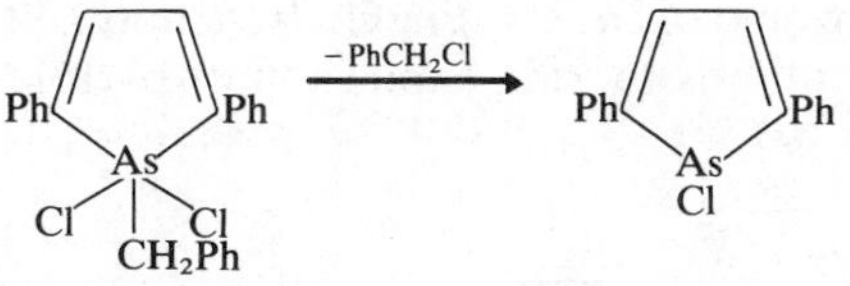

(148)

(406) (408)

Treatment of the dichloro-derivatives (**406**) with one equivalent of sodium methoxide affords the unstable 1-chloro-methoxyarsoles (**409**), which spontaneously lose methyl chloride to give the respective oxides (**410**) (reaction 149). The 1-chloro-1-t-butoxy-derivatives (**407**; R = Ph, $PhCH_2$) do not decompose to the corresponding arsole oxides,

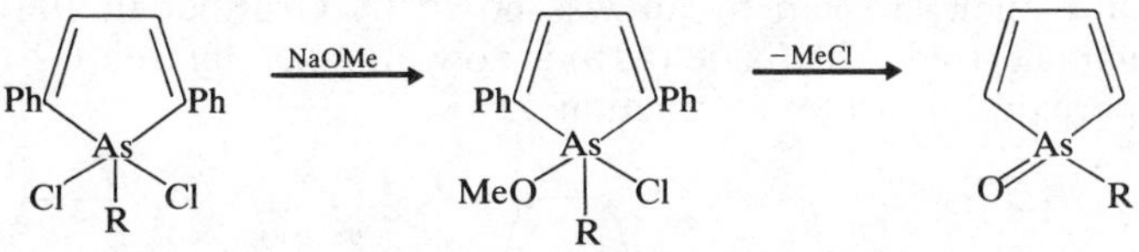

(149)

(406) (409) (410)

but they disproportionate in solution at room temperature to yield the respective dichloro-derivative (**406**). In the case when the dichloro-derivative (**406**; R = $PhCH_2$) is formed, the reaction proceeds further to give 1-chloroarsole (**408**).[182] 1-Chloro-1-t-butoxy-1,2,5-triphenylarsole (**407**; R = Ph) reacts with sodium methoxide to yield directly the oxide (**411**) (reaction 150).

The above reactions indicate that the reaction between the 1-chloro-1-t-butoxy-derivatives (**407**; R^1 = Me, Ph, $C_6H_4NMe_2$) and organolithium compounds should give

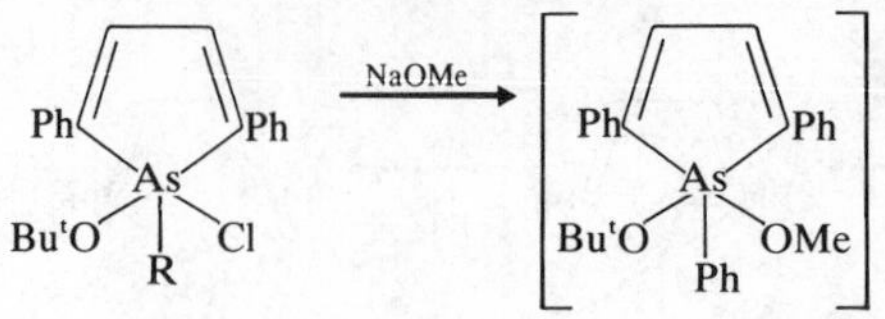

(150)

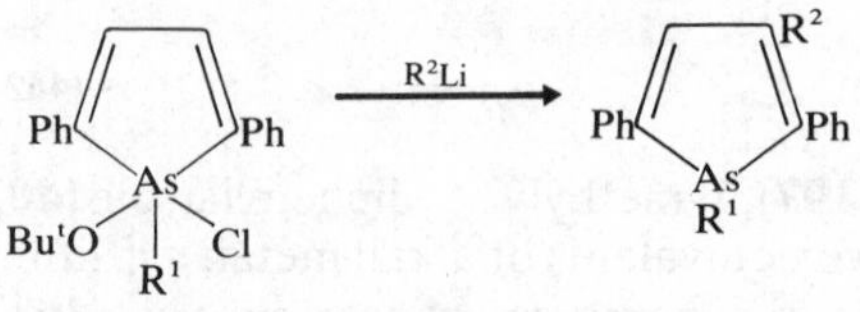

(407) (411)

(151)

(407) (412)

penta-coordinated arsoles. In fact the products are 3-substituted arsoles (**412**) (reaction 151). Their formation can be rationalized mechanistically as S_N2' substitution and subsequent aromatization by elimination of t-butanol.

Arsoles (**393**; R = Ph, Me) react with dimethyl acetylenedicarboxylate to afford 3,6-disubstituted phthalates *via* the adduct (**413**). The 2,5-dimethyl-1-phenylarsole (**393**; R = Me) on treatment with tetracyanoethylene yields a stable adduct (**414**).[180]

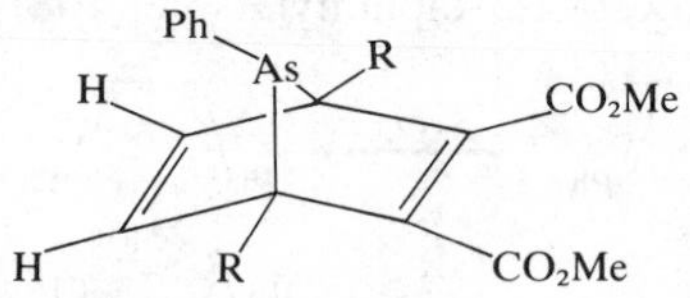

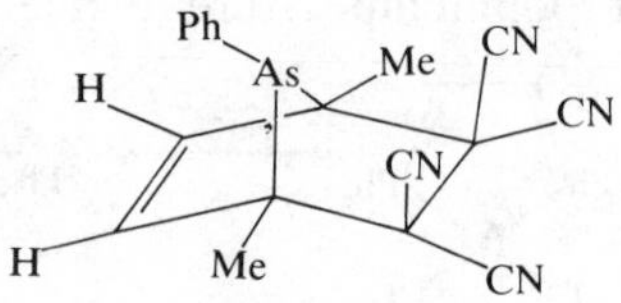

(413) (414)

2,5-Diphenyl-1-t-butylarsole (**415**) reacts with phenylchlorocarbene or dichlorocarbene to give an arsenin (arsabenzene) (**416**; R = Ph, Cl) (reaction 152). This reaction is analogous to the classical pyridine synthesis by ring expansion of pyrroles.[183]

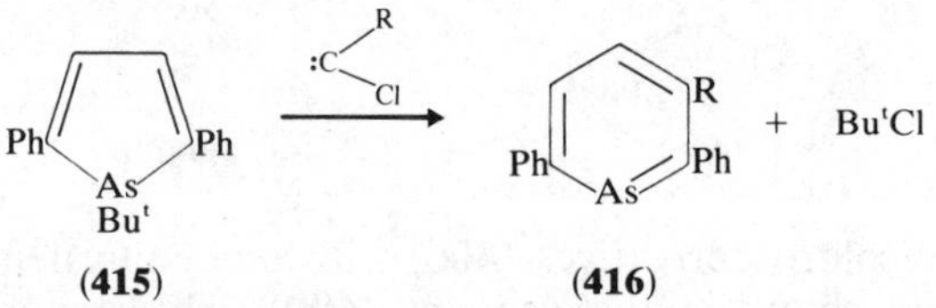

(152)

(415) (416)

1-Phenylarsolan (1-phenylarsolidine, 1-phenylarsacyclopentane) (**417**) is prepared by reacting the di-Grignard reagent from 1,4-dibromobutane with phenyldichloroarsine.[184] The oxidation of 1-phenylarsolan by air was noted by Grüttner in 1916, but the oxide (**418**) was not characterized. The oxide (**418**) is now obtained by treating the arsolan (**417**) with hydrogen peroxide in acetone (reaction 153).

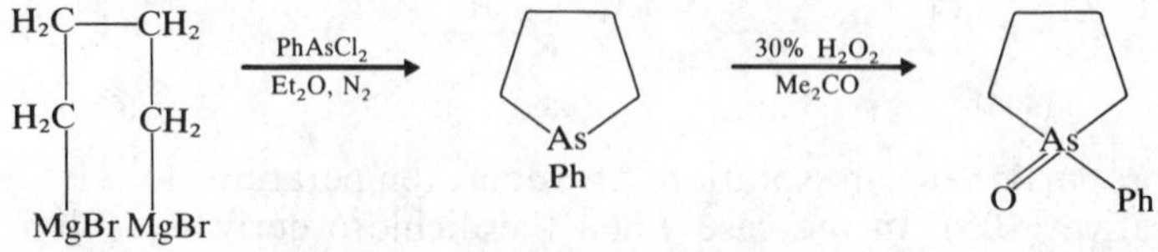

(153)

(417) (418)

The reaction between 1,4-dichlorobutane and lithium dicyclohexylarsine in dioxan gives 4-chlorobutyldicyclohexylarsine (**419**), which cyclizes on standing at room temperature for several weeks, or on being heated, to give an arsonium salt (**420**) (reaction 154).[185]

$$\begin{array}{c} H_2C\text{—}CH_2 \\ H_2CCl \quad CH_2Cl \end{array} \xrightarrow[\text{dioxan}]{LiAs(C_6H_{11})_2} Cl(CH_2)_4As(C_6H_{11})_2 \xrightarrow{170\,°C} \overset{+}{As}(C_6H_{11})_2\ Cl^- \qquad (154)$$

(419) (420)

The analogue of pyrrole containing an antimony atom in place of the nitrogen is called stibole. Pentaphenylstibole (**421**) is prepared by reacting 1,4-dilithiotetraphenylbutadiene (**277**) with phenyldichlorostibine. Oxidation of pentaphenylstibole solutions by air gives the 1-oxide (**422**)[17] (reaction 155), without any rupture of the carbon–antimony bond.

$$(277) \xrightarrow{PhSbCl_2} (421) \xrightarrow{\text{air}} (422) \qquad (155)$$

(277) (421) (422)

19.3.5.2 Five-membered ring containing either one arsenic or antimony atom and fused to one or two benzene rings

Compounds containing a five-membered ring with one heteroatom and fused to a benzene ring derive their names from the analogous compounds containing a nitrogen atom. The first member of the arsindoline series, 1-methylarsindoline (**426**), was obtained from dimethyl(2-phenylethyl)arsine (**423**), prepared by reacting 2-phenylethylmagnesium bromide with iododimethylarsine. The arsine (**423**) combined directly with chlorine to give the arsine dichloride (**424**), which when heated yielded methyl chloride and the chloroarsine (**425**). Cyclization of the chloroarsine (**425**) gave the methylarsindoline (**426**) (reaction 156).

$$(423) \xrightarrow{Cl_2} (424) \xrightarrow[-MeCl]{\Delta} (425) \xrightarrow[CS_2]{AlCl_3} (426) \qquad (156)$$

(423) (424) (425) (426)

Later, 1-phenylarsindoline (**429**) was prepared by a much shorter route, wherein 2-phenylethyl bromide is added to dichlorophenylarsine in boiling aqueous sodium hydroxide to furnish phenyl(2-phenylethyl)arsinic acid (**427**). The acid (**427**) is readily cyclized by sulphuric acid to 1-phenylarsindoline 1-oxide (**428**), which without isolation is reduced by sulphur dioxide to 1-phenylarsindoline (**429**) (reaction 157).[186]

$$\xrightarrow[NaOH,\ H_2O]{PhAsCl_2} (427) \xrightarrow[100\,°C]{H_2SO_4} (428) \xrightarrow[H_2O]{HCl,\ SO_2} (429) \qquad (157)$$

(427) (428) (429)

1-Phenylarsindoline (**429**) is *not* dehydrogenated to 1-phenylarsindole by boiling with tetrachloro-*o*-benzoquinone in xylene, or with palladized charcoal in ethylene glycol. The arsindoline (**429**) is also largely unaffected when it is boiled in benzene containing *N*-bromosuccinimide and benzoyl peroxide, but it readily combines with cold methyl iodide to give 1-phenylarsindoline methiodide.

2-Phenyl- and 2-methyl-isoarsindoline (**431**; R = Ph, Me) are formed when 1,2-di(bromomethyl)benzene (**430**) in the presence of sodium is reacted with phenyl- and methyl-dichloroarsine, respectively (reaction 158).[187] 2-Methylisoarsindoline (**431**; R = Me) combines with 1,2-di(bromomethyl)benzene to give the quaternary salt (**432**), which on heating undergoes cyclization and loss of methyl bromide to yield bisisoarsindolinium bromide (**433**). The phenyl compound (**431**; R = Ph) when treated with nitric acid is

CH_2Br / CH_2Br (**430**) $\xrightarrow[\text{Na, Et}_2\text{O}]{\text{RAsCl}_2}$ AsR (**431**) (158)

readily oxidized to the corresponding 1-hydroxy-1-nitrate, which with alkali gives the 1,1-dihydroxy-derivative. The latter on dehydration forms 2-phenylisoarsindoline oxide (**434**).[187]

Me, +As, $CH_2C_6H_4CH_2Br$, Br^- (**432**) As$^+$ Br^- (**433**) Ph, As, O (**434**)

5-Hydroxydibenzarsole 5-oxide (2,2′-biphenylenearsinic acid) (**436**) is obtained by warming biphenyl-2-arsinic acid (**435**), formed from diazotized 2-aminobiphenyl and sodium arsenite in alkali solution, with concentrated sulphuric acid. The 5-oxide (**436**) may be converted to the 5-iodo-derivative and thence on treatment with methylmagnesium iodide to 5-methyldibenzarsole (9-methyl-9-arsafluorene) (**437**), which on heating with methyl iodide affords the dimethylarsonium iodide (**438**) (reaction 159).[188]

$AsO(OH)_2$ (**435**) $\xrightarrow{H_2SO_4}$ As, O, OH (**436**) $\xrightarrow[\text{ii, MeMgI}]{\text{i, 10\% H}_2\text{SO}_4\text{, SO}_2\text{, KI,}}$ As Me (**437**) (159)

$\xrightarrow[\text{MeI}]{\Delta}$ $\overset{+}{\text{As}}$ Me_2 I^- (**438**)

2.2′-Biphenylenebis(dimethylarsine) (**439**) undergoes diquaternization with 1,2-dibromoethane and 1,3-dibromopropane to form cyclic diquaternary dibromides (**440**; $n = 2$ or 3). These salts on heating undergo ring contraction with the formation of 5-methyldibenzarsole (**437**)[189] (reaction 160). 5-Methyldibenzarsole (**437**) is also prepared, as is 5-phenyldibenzostibole (9-phenyl-9-stibiafluorene) (**442**), by the interaction of 2,2′-dilithiobiphenyl (**441**) and the appropriate dihalogeno-arsine or -stibine (reaction 161).[190]

Me_2As $AsMe_2$ (**439**) $\xrightarrow[\text{Sealed tube, 100 °C}]{\text{Br(CH}_2)_n\text{Br,}}$ Me_2As^+ $^+AsMe_2$ $(CH_2)_n$ (**440**) $\xrightarrow{\Delta}$ As Me (**437**) (160)

(437) (441) (442) (161)

19.3.5.3 Six-membered ring containing one arsenic, antimony, or bismuth atom

The six-membered ring containing one arsenic atom and analogous to pyridine and benzene is called arsenin, but the name arsabenzene is also used. In the case of the related compound containing an antimony atom the name stibabenzene is commonly used, but according to IUPAC rules the name antimonin is also applicable.

Arsenic trichloride and 1,4-dihydro-1,1-dibutylstannabenzene (**322**) (p. 914) react together exothermally in THF to give arsenin hydrochloride and dibutyltin dichloride. The arsenin (**443**) is freed from its salt by treatment with 1,5-diazabicyclo[4,3,0]non-5-ene[161] (reaction 162). It is a liquid with a characteristic onion-like odour; it is very air-sensitive and brief exposure turns it bright red. Both its u.v. and ^{1}H n.m.r. spectra indicate aromatic character.

(322) → i, $AsCl_3$; ii, DBN → (443) (162)

Some substituted arsenins may be obtained by ring-expansion brought about by the reaction between substituted arsoles (**415**) and carbenes (p. 926). A general method for the synthesis of triaryl or more highly substituted arsenins is by an intramolecular carbene insertion reaction of certain arsoles. The lithium arsenide (**444**) (see reaction 143, Li for K) reacts with sodium phenyldichloroacetate to yield the potential carbene precursor (**445**), which decomposes a little above its melting point to give a good yield of 2,3,6-triphenylarsenin (**447**), a stable colourless crystalline solid.[191] The product (**447**) results from an intramolecular carbene insertion, probably *via* the arsacyclopropane of the bicyclic betaine (**446**) (reaction 163). Non-symmetrical substituted arsoles should give two isomeric arsenins by insertion at position-2 or -5. The sodium arsolyldichloroacetate obtained from the lithium arsenide (**444**) and sodium trichloroacetate on heating at 170 °C degrades in an undefined manner.

(444) → $PhCCl_2CO_2Na$ → (445) → Δ, −NaCl, −CO_2 → [carbene] → (446) → (447) (163)

4-Substituted arsenins have been obtained from 4-substituted 4-methoxy-1,4-dihydro-1,1-dibutylstannabenzenes (**449**; R = Ph, C_6H_{11}, Bu), prepared by the reaction between pentadiynes (**448**) and di-n-butyltin dihydride in boiling methylcyclohexane containing azoisobutyrodinitrile. The 4-methoxy-1,4-dihydrostannins (**449**; R = Ph, C_6H_{11}, Bu^t) react with arsenic trichloride in boiling THF to form directly the 4-substituted arsenins (**452**). The 1,4-dihydroarsenins (**450**) that occur as intermediates can be detected by n.m.r. spectroscopy if care is taken, but they decompose spontaneously, perhaps by way of the 1-chloro-1-methoxyarsenin (**451**)[192] (reaction 164). The 4-substituted arsenins (**452**), with the exception of the 4-phenylarsenin (**452**; R = Ph), which is a crystalline solid, are colourless, distillable, air-sensitive oils. 4-Cyclohexylphosphorin is also prepared in a similar manner from 4-cyclohexyl-4-methoxy-1,4-dihydro-1-dibutylstannabenzene and phosphorus tribromide in the presence of triphenylphosphan.

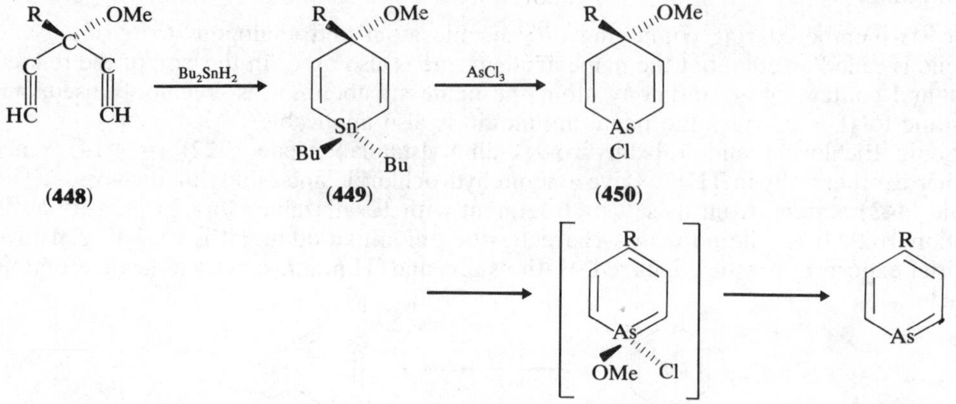

(**448**) —Bu_2SnH_2→ (**449**) —$AsCl_3$→ (**450**) → [(**451**)] → (**452**) (164)

The arsenins react with nucleophiles, *e.g.* with phenyl-lithium to give dihydroarsenins (p. 917) and with dienophiles to form adducts, *e.g.* 2,3,6-triphenylarsenin (**447**) with diethyl acetylenedicarboxylate gives adduct (**453**) (reaction 165). It also forms an adduct with hexafluorobutyne,[193] and with arynes. Aryl-substituted arsenins, therefore, add to less-reactive dienophiles than do aryl-substituted phosphorins.

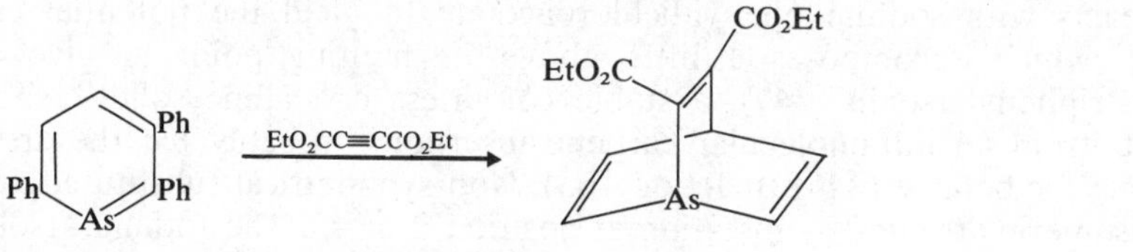

(**447**) —$EtO_2CC{\equiv}CCO_2Et$→ (**453**) (165)

The name in general use for the antimony analogue of phosphorin and arsenin is stibabenzene (**455**), and it is obtained in a similar way from 1,4-dihydro-1,1-dibutylstannabenzene (**322**) (pp. 914 and 929), *via* the chloro-derivative (**454**) (reaction 166). It is an extremely labile compound and rapidly polymerizes to an intractable brown tar at −80 °C, but it can be conveniently handled in solution under an inert atmosphere.

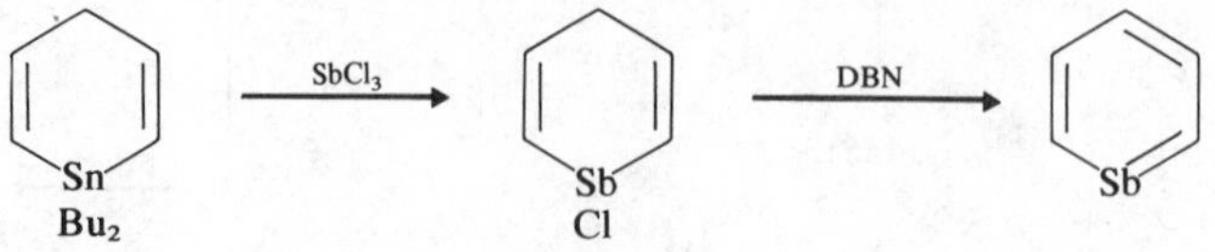

(**322**) —$SbCl_3$→ (**454**) —DBN→ (**455**) (166)

The ^{1}H n.m.r. spectrum of stibabenzene shows the characteristic pattern of the Group VB heteroaromatics. The α-protons occur as a doublet at extremely low field while the β- and γ-protons are in the normal aromatic region. The chemical shift of the α-protons is τ 1.9 for pyridine, 1.4 for phosphorin, 0.7 for arsenin, and −0.7 for stibabenzene. This progressive shift to lower field can probably be associated with magnetic anistropy changes as the heteroatom size increases. This effect should be attenuated for the more remote β- and γ-protons and, indeed, chemical-shift values of these protons do not vary greatly for the above compounds.[194]

Attempts have been made to obtain bismabenzene (bismin) (**457**) by use of the reaction between 1,4-dihydro-1,1-dibutylstannabenzene (**322**) and bismuth trichloride. Although a solid 1,4-dihydro-1-chlorobismabenzene (**456**) could be isolated, subsequent treatment with 1,5-diazabicyclo[4,3,0]non-5-ene caused an exothermic loss of hydrogen chloride with the production of polymeric material.

On the other hand, addition of hexafluorobutyne at low temperature after precipitation of the hydrochloride (**456**) gives an adduct (**458**) (reaction 167). This reaction has been cited as evidence for formation of a $(6p–2p)\pi$-bond in the transient bismabenzene![195]

Sn Bu₂ (**322**) —$BiCl_3$→ Bi Cl (**456**) —DBN→ [Bi (**457**)] —$CF_3C{\equiv}CCF_3$→ (**458**) (167)

19.3.5.4 Six-membered ring containing either one arsenic or antimony atom and fused to one or two benzene rings

The names of compounds in the group of heterocycles containing one benzene ring are derived from the related nitrogen heterocycles, *e.g.* arsinoline and isoarsinoline. 1-Methyl-1,2,3,4-tetrahydroarsinoline (**460**) is prepared by the ready cyclization of methyl(3-phenylpropyl)chloroarsine (**459**) by aluminium chloride[196] (reaction 168). It readily forms a dichloride (**461**), which when heated is smoothly decomposed to give 1-chloroarsinoline (**462**) (reaction 169). With methyl iodide it gives a methiodide, which undergoes partial ring fission in alcoholic solutions, and thus exists in equilibrium with 2-(3-iodopropyl)phenyldimethylarsine.

(**459**) $CH_2CH_2CH_2$ ClAsMe —$AlCl_3$→ (**460**) As Me (168)

(**460**) —Cl_2, CCl_4→ (**461**) As Cl Me Cl —Δ, −MeCl→ (**462**) As Cl (169)

(**463**) CH_2CH_2Br, CH_2Br —$RAsCl_2$, Na, EtOAc, Et_2O→ (**464**) AsR (170)

The Michaelis condensation of 2-(2-bromoethyl)benzyl bromide (**463**) with phenyl- and methyl-dichloroarsine in the presence of metallic sodium and ethyl acetate affords 2-phenyl- and 2-methyl-1,2,3,4-tetrahydroisoarsinoline (**464**; R = Ph, Me) (reaction 170). The methyl derivative (**464**; R = Me) is more susceptible to atmospheric oxidation than the phenyl analogue, but it forms a crystalline methiodide. Like the 2-methyltetrahydroarsinoline (**460**) it readily forms a dichloro-derivative, which on heating loses methyl chloride to yield 2-chloro-1,2,3,4-tetrahydroisoarsinoline (**464**; R = Cl).[197] These is no appreciable change on boiling this compound (**464**; R = Cl) with pyridine, no hydrogen chloride being removed to yield 3,4-dihydroisoarsinoline. The reaction between dilithium phenylarsenide and bromomethyl-2-(2-bromoethyl)benzene gives 2-phenyl-1,2,3,4-tetrahydroisoarsinoline (**464**; R = Ph).[198]

The linear compounds containing a six-membered ring with an arsenic or antimony atom and fused to two benzene rings are generally called dibenzo[*b,e*]arsenins (9-arsanthracene) and stibanthracene, respectively. The former has its name derived from pyridine and the latter from the related carbocyclic compound, anthracene.

9,10-Dihydro-9,9-dimethyl-9-stannanthracene (**465**) and arsenic trichloride on heating together afford a good yield of 5-chloro-5,10-dihydrodibenzo[*b,e*]arsenin (**466**) and dimethyltin dichloride. Treatment of the chloro-derivative (**466**) with auxiliary bases such as 1,5-diazabicyclo[5,4,0]undec-5-ene or triethylamine yields dibenzo[*b,e*]arsenin (**467**) (reaction 171), which is only stable in solution and polymerizes on attempted isolation. It decomposes in solution in the presence of oxygen or hydrogen chloride and forms a Diels–Alder adduct (**468**) with maleic anhydride. The elimination of hydrogen chloride from the chloro-compound (**466**) can be verified by mass spectrometry and the u.v. spectrum of the dibenzoarsenin (**467**)[199] corresponds with that of dibenzophosphorin (**383**). Compound (**467**) gave the first example showing that it was possible to stabilize an arsenic–carbon $(4p\text{–}2p)\pi$-bond in a Hückel aromatic compound. The stability of dibenzoarsenin is increased by the introduction of a substituent, *e.g.* a phenyl group, at position-10.

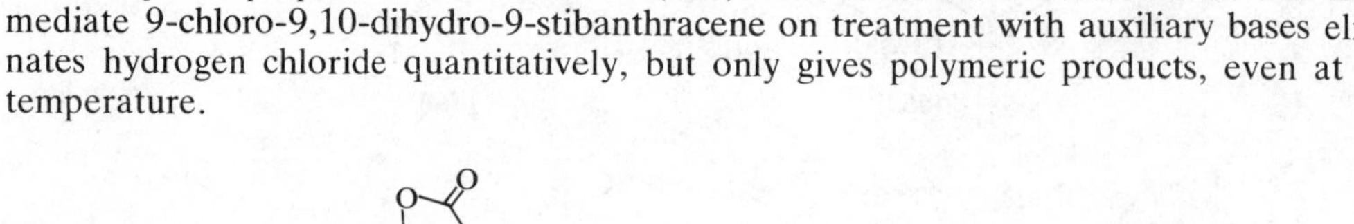

Sn Me$_2$ (**465**) —$AsCl_3$, 120 °C→ As Cl (**466**) —DBU→ As (**467**) (171)

Attempts to prepare 9-stibanthracene (**469**) have so far been unsuccessful. The intermediate 9-chloro-9,10-dihydro-9-stibanthracene on treatment with auxiliary bases eliminates hydrogen chloride quantitatively, but only gives polymeric products, even at low temperature.

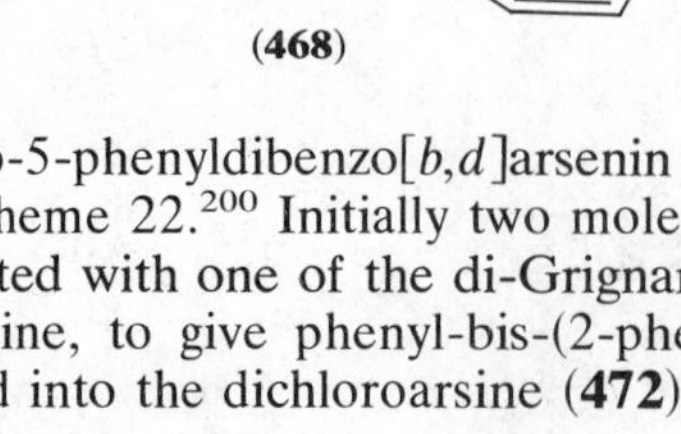

(**468**) 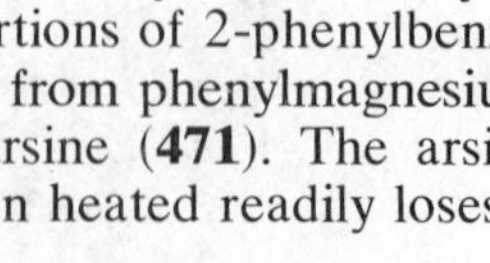 (**469**)

5,6-Dihydro-5-phenyldibenzo[*b,d*]arsenin (**474**) may be synthesized by the route outlined in Scheme 22.[200] Initially two molecular proportions of 2-phenylbenzyl bromide (**470**) are treated with one of the di-Grignard reagents, from phenylmagnesium bromide and phenylarsine, to give phenyl-bis-(2-phenylbenzyl)arsine (**471**). The arsine (**471**) is next converted into the dichloroarsine (**472**), which when heated readily loses one of the

2 (470) $\xrightarrow{i}$ (471) $\xrightarrow{ii}$ (472) $\xrightarrow{iii}$ (473) $\xrightarrow{iv}$ (474)

i, $PhAs(MgBr)_2$, C_6H_6; ii, Cl_2, CCl_4; iii, Δ; iv, $AlCl_3$, CS_2.

SCHEME 22

benzyl groups to give phenyl-2-phenylbenzylchloroarsine (**473**). Cyclization of this product (**473**) with aluminium trichloride in carbon disulphide affords 5,6-dihydro-5-phenyldibenzo[*b*,*d*]arsenin (**474**). It gives a crystalline methiodide, methopicrate, and dichloropalladium and dibromopalladium derivatives.

19.3.6 HETEROCYCLES CONTAINING ONE SILICON ATOM

Heterocycles containing one silicon atom are generally named by combining the prefix sila with the name of the related carbocyclic compound. They are also named by combining the prefix sila (eliding 'a' where necessary) with a stem from Table 2.

19.3.6.1 Four-membered rings containing one silicon atom

1-Chloro-1-methylsilacyclobutane (**476**) is prepared by boiling 3-chloropropylmethyldichlorosilane (**475**) in ether with magnesium activated by iodine vapour. The reaction between the 1-chloro-compound (**476**) and methylmagnesium iodide gives 1,1-dimethylsilacyclobutane (**477**)[201] (reaction 172) and reduction with lithium tetrahydridoaluminate affords 1-methylsilacyclobutane (**478**).

$$H_2CCl(CH_2)_2SiCl_2Me\ (\mathbf{475}) \xrightarrow[Et_2O]{Mg} (\mathbf{476}) \xrightarrow[Et_2O]{MeMgI} (\mathbf{477}) \quad (172)$$

The thermal decomposition of 1,1-dimethylsilacyclobutane (**477**) between 400 and 460 °C gives only ethylene and 1,1,3,3-tetramethyl-1,3-disilacyclobutane (**480**) as products[202] (reaction 173). A number of chemical reactions give support for the idea of the formation of an intermediate (**479**) possessing a silicon–carbon double bond. In fact the last step of the decomposition itself is an example of the very common 1,2-cycloaddition reactions of unsaturated compounds. This reaction occurs more readily when there are suitable (electron-withdrawing) substituents attached to the double bond. Perhaps the longer length of a silicon–carbon bond does not allow sufficient overlap of the *p*-electrons

(478) cyclobutyl–SiHMe; (477) cyclobutyl–$SiMe_2$ $\xrightleftharpoons{400\text{–}600\,°C}$ [$Me_2Si{=}CH_2$] (479) + $CH_2{=}CH_2$ ⟶ (480) Me_2Si–$SiMe_2$ ring (173)

$$2[Me_2Si{=}CH_2]\ (479) \xrightarrow{H_2O} 2Me_3SiOH\ (481) \rightleftharpoons Me_3SiOSiMe_3 + H_2O\ (482)$$

$$2[Me_2Si{=}CH_2]\ (479) \xrightarrow{NH_3} 2Me_3SiNH_2 \longrightarrow Me_3SiNHSiMe_3\ (483) \qquad (174)$$

of these atoms to form a very stable double bond. Therefore cyclodimerization occurs extremely easily compared with any carbon–carbon double bond compounds. Also the reaction of the intermediate (**479**) with water or ammonia leads to products (**481**) and (**482**), and (**483**) (reaction 174), which may be easily explained on the assumption of a silicon–carbon double bond. The above two reactions are faster than cyclodimerization.

19.3.6.2 Five-membered rings containing one silicon atom

The best way of incorporating a silicon atom into an unsaturated ring system, *e.g.* silacyclopentadiene (silole) (**498**), is *via* the saturated dichloro-derivative 1,1-dichlorosilacyclopentane (1,1-dichlorosilolan) (**494**), which is obtained by reacting silicon tetrachloride with the Grignard reagent (**493**) prepared from 1,4-dibromobutane[203] (reaction 175). Treatment of the dichloro-compound (**494**) with two equivalents of sulphuryl chloride and a catalytic amount of benzoyl peroxide yields a mixture of a monochlorinated product (**495**) and a dichlorinated product (**496**), separable by fractional distillation. Thermal dehydrochlorination of the dichloro-derivative (**496**) under nitrogen yields 1,1-dichlorosilacyclopentadiene (**497**), which on reduction with lithium tetrahydridoaluminate in diglyme affords silacyclopentadiene (**498**) (reaction 176).[204]

$$BrMgCH_2CH_2CH_2CH_2MgBr\ (493) \xrightarrow{SiCl_4} \text{1,1-dichlorosilacyclopentane } (494) \qquad (175)$$

$$(494) \xrightarrow[Bz_2O_2]{SO_2Cl_2} \underset{12\%}{(495)} + \underset{48\%}{(496)};\quad (496) \xrightarrow[N_2]{550\,°C} (497) \xrightarrow{LiAlH_4} (498) \qquad (176)$$

1,1-Dichlorosilacyclopentadiene (**497**), like practically all chlorosilanes, undergoes facile displacement of its halogens by nucleophilic reagents. It reacts with phenylmagnesium bromide to form 1,1-diphenylsilacyclopentadiene (**499**) (reaction 177). Silacyclopentadiene (**498**) has been found to react directly with potassium to give hydrogen and

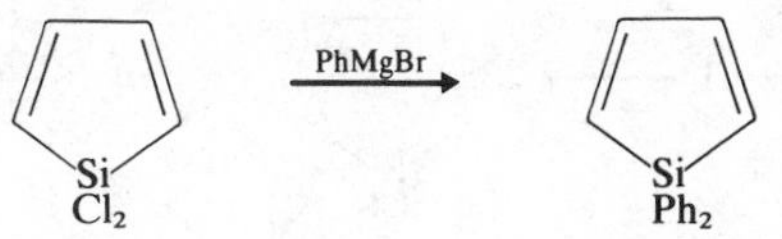

(177)

(497) (499)

(178)

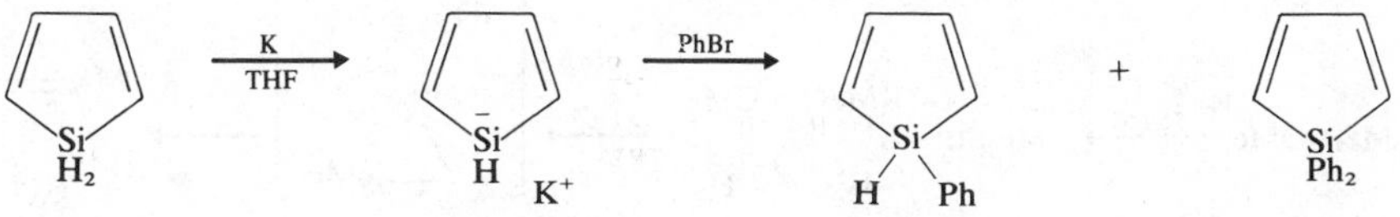

(498) (500) (501) (499)

potassium silacyclopentadienide (**500**) (reaction 178). Comparison of the n.m.r. spectra of silacyclopentadiene (**498**) and potassium silacyclopentadienide (**500**) provides evidence that it is the silanic hydrogens which are replaced by the metal in this reaction. Significantly, divinylsilane, the open-chain analogue of silacyclopentadiene, does not react with potassium at any appreciable rate under comparable conditions. This suggests that resonance stabilization of the silacyclopentadienide anion is providing a driving force for this reaction. Potassium silacyclopentadienide reacts with bromobenzene to yield a mixture of 1-phenyl- (**501**) and 1,1-diphenyl-silacyclopentadiene (**499**)[205] (reaction 178). Potassium silacyclopentadienide is coloured in THF solutions and possesses a spectrum in the visible region, which is quite similar to that of potassium cyclopentadienide.

1,1-Dimethyl-2,5-diphenyl-1-silacyclopentadiene (**502**), when reduced with sodium or potassium in dimethoxyethane, gives the radical anion (**503**), detected by e.s.r. However, further reaction with potassium leads to a decrease in the e.s.r. signal, which eventually disappears. The sequence of observations is in accord with the formation of the radical anion (**503**) which is further reduced to the dianion (**504**)[206] (reaction 179). The silacyclopentadiene (**502**) is obtained by the dehydrobromination of *trans*-2,5-dibromo-1,1-dimethyl-2,5-diphenyl-1-silacyclopentane. The reaction between 1,4-dilithiotetraphenylbutadiene (**505**) and diphenyldichlorosilane gives hexaphenylsilacyclopentadiene (**506**) (reaction 180).[17]

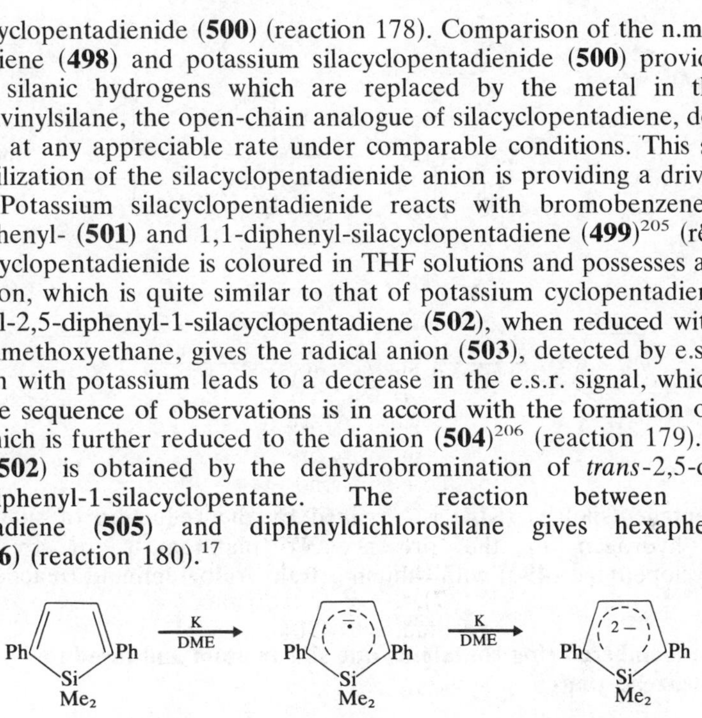

(179)

(502) (503) (504)

(180)

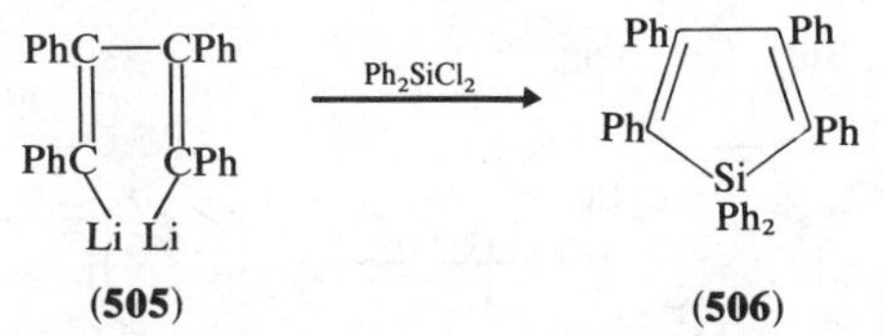

(505) (506)

A vapour-phase chromatogram of the monochlorinated product (**495**) indicates that it is a 1:9 mixture of the α- and β-isomers. Dehydrochlorination of (**495**) by heating with quinoline gives a mixture of 1,1-dichlorosilacyclopent-2-ene (**507**) and 1,1-dichlorosilacyclopent-3-ene (**508**)[204] (reaction 181). 1,1,3,4-Tetramethylsilacyclopent-3-ene (**512**) has been made by heating together 2,3-dimethylbutadiene (**511**) and dimethylsilylene (**510**), which is generated by either the pyrolysis of *sym*-dimethoxytetramethyldisilane (**509**) (reaction 182), or the 7-silanorbornadiene (**513**) (reaction 183).[207]

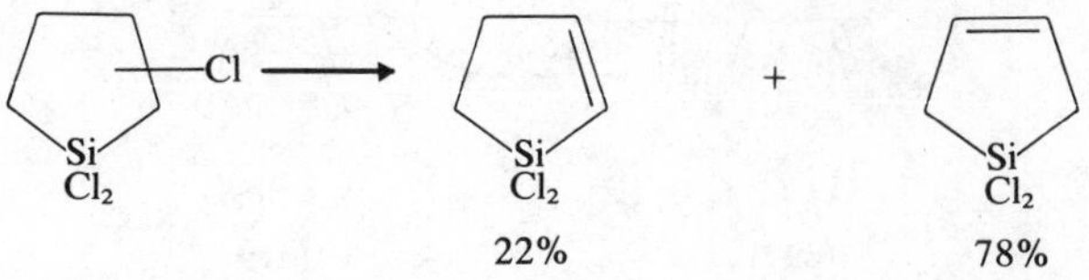

(181)

22% 78%

(495) (507) (508)

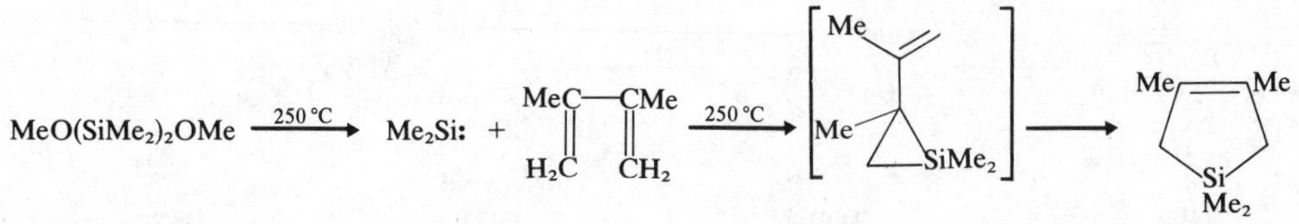

$$MeO(SiMe_2)_2OMe \xrightarrow{250\,°C} Me_2Si: + MeC(=CH_2)-C(=CH_2)Me \xrightarrow{250\,°C} [\ ] \longrightarrow$$ (182)

(509) (510) (511) (512)

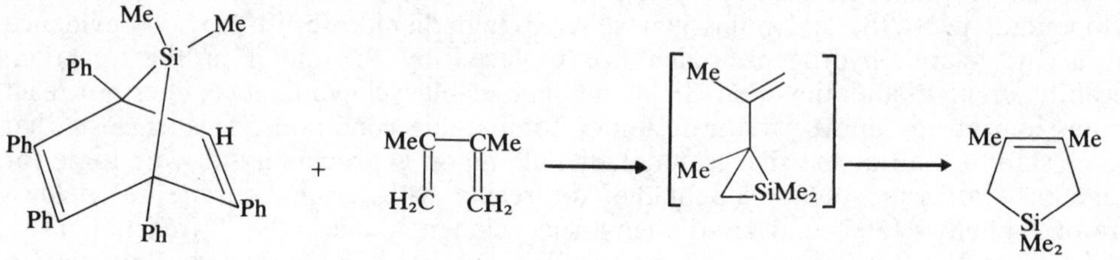

(183)

(513) (511) (512)

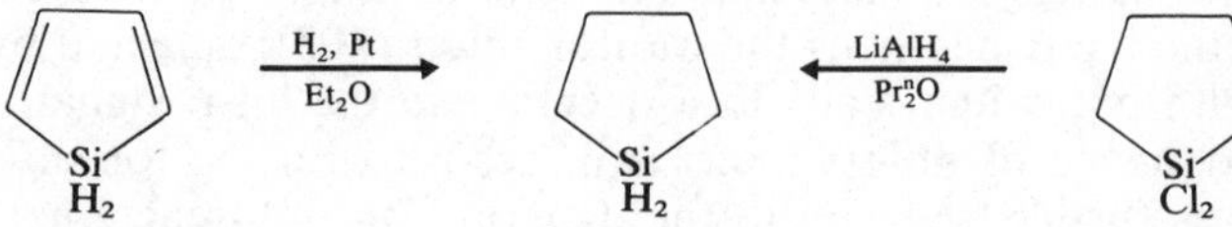

(184)

(498) (514) (494)

Silacyclopentane (silolan) (**514**) is obtained by the reduction of silacyclopentadiene (**498**) with hydrogen in the presence of platinum,[204] or by treating 1,1-dichlorosilacyclopentane (**494**) with lithium tetrahydridoaluminate (reaction 184).[203]

19.3.6.3 Five-membered ring containing one silicon atom and fused to one or two benzene rings

The reaction between the dilithium intermediate (p. 912) and dichlorodiphenylsilane yields 3-n-butyl-1,1,2-triphenyl-1-silaindene (**515**) (reaction 185).[159]

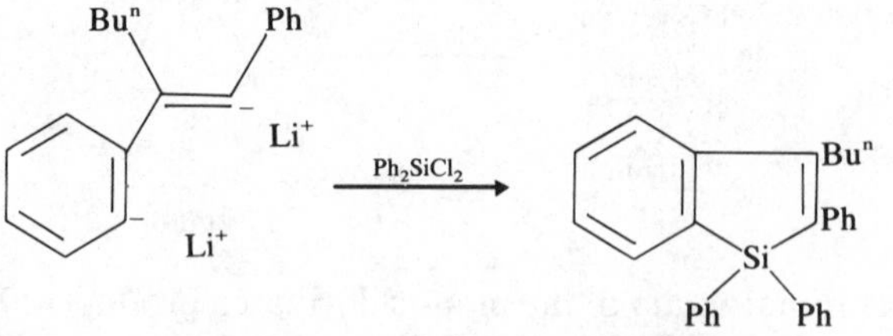

(185)

(310) (515)

1,1-Diphenyl-1-silaindan (**517**) has been prepared by ring closure of 2-(2-chlorophenyl)ethyldiphenylsilane (**516**), using sodium metal in toluene[208] (reaction 186). The required intermediate silane (**516**) is obtained by converting 1-chloro-2-(2-bromoethyl)benzene to the corresponding Grignard reagent and reacting it with chlorodiphenylsilane.

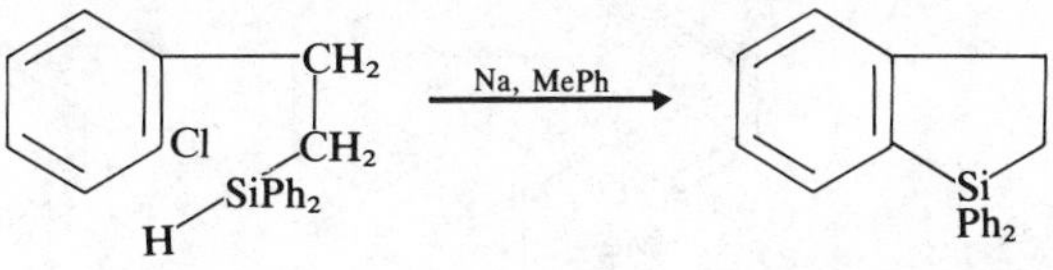

(186)

(516) (517)

2-Silaindan (**520**) has been obtained by treating with lithium tetrahydridoaluminate 2,2-dichloro-2-silaindan (**519**), formed when benzylchloromethyldichlorosilane (**518**) is boiled with aluminium chloride in benzene[209] (reaction 187). The dichloro-2-silaindan (**519**) on reacting with methylmagnesium bromide gives 2,2-dimethyl-2-silaindan (**521**), and on heating with acetic anhydride it affords the 2,2-diacetoxy-derivative (**522**) (reaction 188).

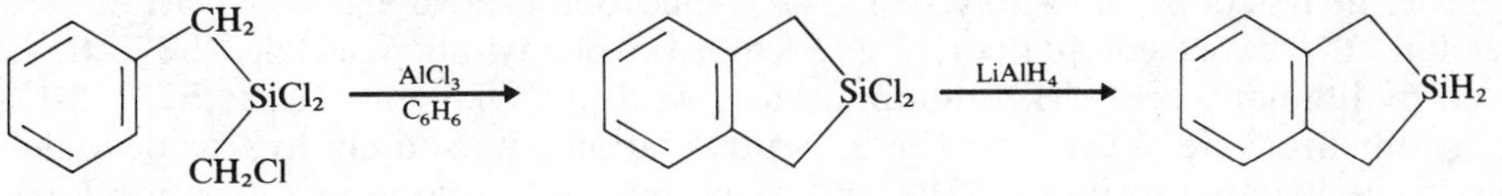

(187)

(518) (519) (520)

(188)

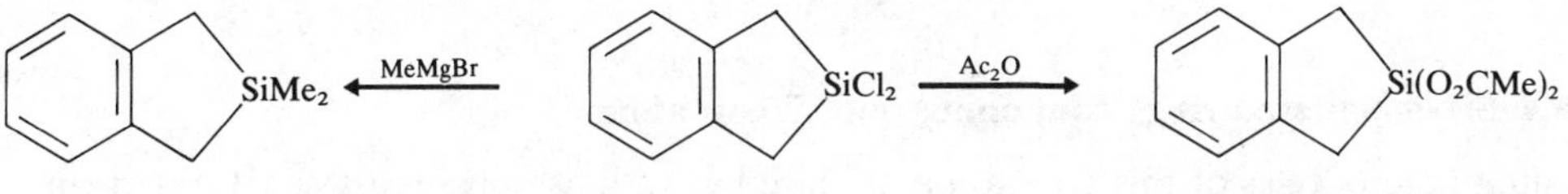

(521) (519) (522)

The first dibenzosilole (silafluorene) to be synthesized was 5,5-diphenyldibenzosilole (**524**), which was obtained by treating 2,2′-biphenylenedilithium (**523**) with dichlorodiphenylsilane[210] (reaction 189). Methyl- and phenyl-trichlorosilane react with the dilithium compound (**523**) to give an excellent yield of 5-chloro-5-methyl- and 5-chloro-5-phenyl-dibenzosilole (**525**; R = Me or Ph) respectively, along with a small amount of the related 5,5′-spirobi(dibenzosilole) (**526**) (reaction 190). 5-Chloro-5-methyldibenzosilole (**525**; R = Me) reacts with 2,2′-biphenylenedilithium (**523**) to give 5,5′-spirobi(dibenzosilole) (**526**) and 5,5-dimethyldibenzosilole (**527**)[211] (reaction 191). Furthermore, compound (**526**) is the only one isolated from the reaction between 2,2′-biphenylenedilithium (**523**) and hexachlorodisilane.

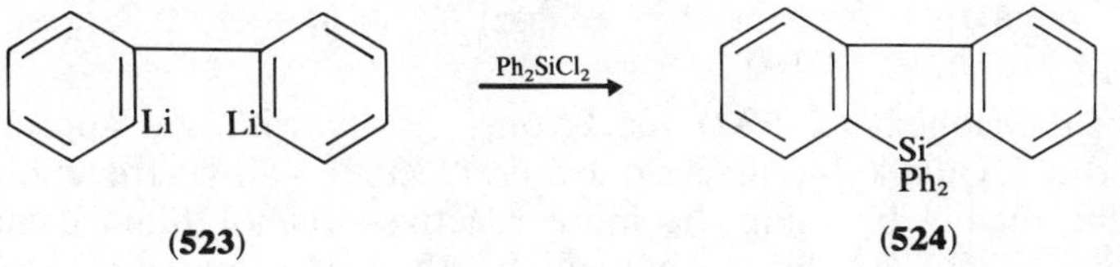

(189)

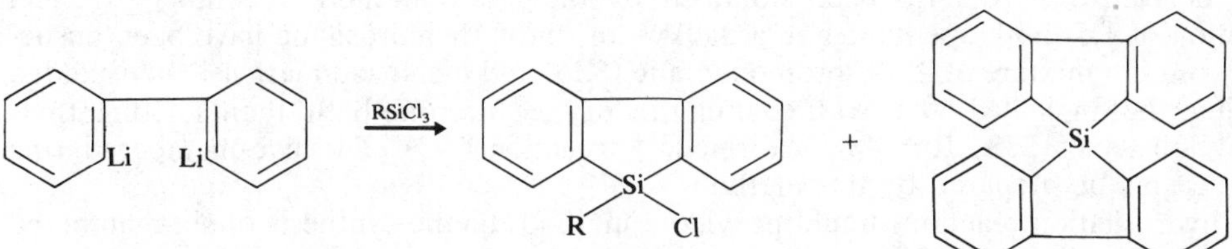

(523) (524)

(190)

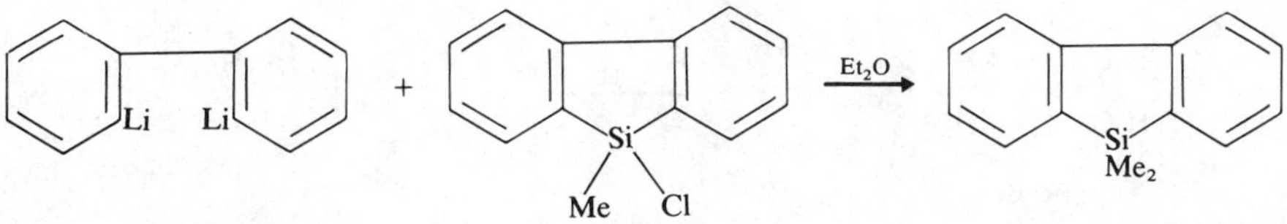

(523) (525) (526)

(191)

(523) (525) (527)

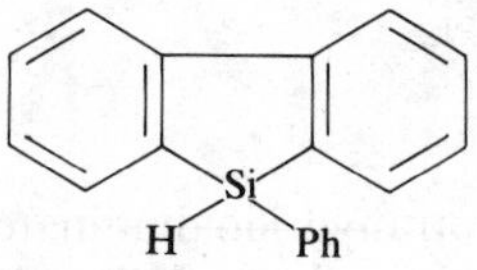

(528) 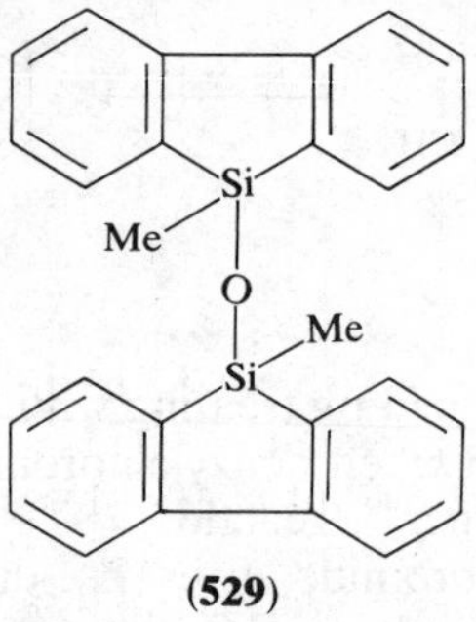 (529) 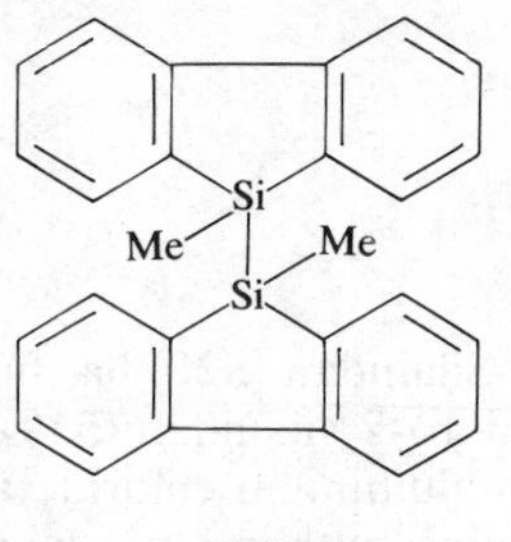 (530)

In general, reactions of 5-alkyl- or 5-aryl-chlorodibenzosiloles with various reagents proceed in the expected manner, *e.g.* 5-chloro-5-phenyldibenzosilole (**525**; R = Ph) is reduced by lithium tetrahydridoaluminate to give 5-phenyldibenzosilole (**528**). 5-Chloro-5-methyldibenzosilole (**525**; R = Me) on neutralization with sodium hydroxide yields 5,5′-oxybis-(5-methyldibenzosilole) (**529**), and on boiling with sodium in xylene it affords 5,5′-dimethyl-5,5′-bi(dibenzosilole) (**530**), which contains a silicon–silicon bond.

19.3.6.4 Six-membered rings containing one silicon atom

The first heterocycles of this type to be prepared were those possessing a saturated ring, *e.g.* 1,1-dichloro-1-silacyclohexane (**532**) was obtained by reacting the Grignard reagent (**531**) formed from 1,5-dibromopentane with silicon tetrachloride[212] (reaction 192). The same dichlorosilacyclohexane has been obtained by passing 1,5-dichloropentane over a heated mixture of silicon and copper (9 : 1).[213]

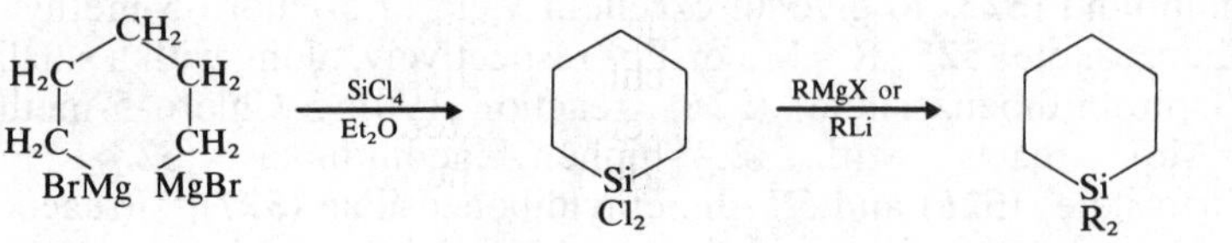

(192)

(R = alkyl, X = halogen)

(531) (532) (533)

1,1-Dichlorosilacyclohexane (**532**) reacts only slowly with the appropriate Grignard reagent to yield a 1,1-dialkyl-1-silacyclohexane (**533**; R = alkyl) (reaction 192), whereas alkylation occurs readily by using the more reactive organolithium compounds.[203] 1,1-Diaryl derivatives have also been obtained by the latter method. Attempts to effect cyclization of 1,5-dilithiopentane and a dialkyl- or diaryl-dichlorosilane have been unsuccessful, but if a mixture of 1,5-dibromopentane (**534**) and dichlorodimethylsilane is added to lithium shot in chilled ether with cooling in a dry-ice–acetone bath, then 1,1-dimethyl-1-silacyclohexane (**533**; R = Me) is formed[214] (reaction 193). The five-membered ring analogue may be prepared by this method.

The hydrosilation reaction should provide a method for the synthesis of six-membered silicon heterocycles, but with an appropriate substituted silane the ring-closure reaction

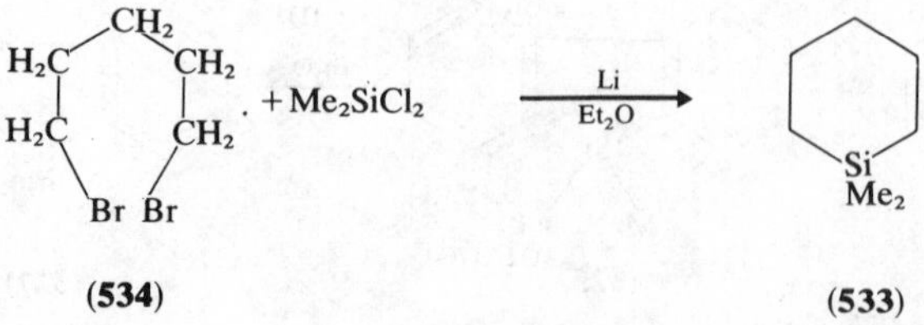

$$+ Me_2SiCl_2 \xrightarrow[Et_2O]{Li} \quad (193)$$

(534) (533)

(537) (535) $\xrightarrow{H_2PtCl_6}$ (536) (194)

results in a silacyclopentane rather than the expected silacyclohexane. 5-(Dimethylsilyl)-1-hexene (**535**), for instance, upon treatment with chloroplatinic acid, gives equal amounts of *cis*- and *trans*-1,1,2,5-tetramethylsilacyclopentane (**536**) (reaction 194), and only a trace of 1,1,2-trimethylsilacyclohexane (**537**).[215]

The reduction of 1,1-dichlorosilacyclohexane (**532**) with lithium tetrahydridoaluminate yields silacyclohexane (**538**) (reaction 195). Attempts to dehydrogenate silacyclohexane by heating it with platinum resulted in 'cracking' and failed to produce an aromatic silane. Silacyclohexane, like the corresponding straight-chain compounds, is subject to alkaline hydrolysis, and it is found that the relative reactivity of cyclic organosilanes towards alkali-catalysed hydrolysis is a function of ring size. The order of reactivity is 5 > 7 > straight-chain > 6, where the number given is the number of atoms in the ring.[216]

(532) $\xrightarrow{LiAlH_4}$ (538) (195)

1,1-Dimethylsilacyclohexane and 1,1-dimethylsilacyclopentane undergo chlorination both with chlorine and sulphuryl chloride, in the presence or absence of free radical catalysts, to yield predominantly the 1-chloromethyl derivative. The addition of free-radical catalysts does not substantially alter the extent of methyl chlorination, but does result in increased attack on the ring carbon β to the silicon and decreased attack on the ring carbon α to the silicon.[217]

CCl_2 insertion into the β-C—H linkage of sila- and stanna-cyclohexanes occurs exclusively, *e.g.* when 1,1-dimethylsilacyclohexane (**533**) and phenyl(bromodichloromethyl)-mercury are boiled in benzene the sole product is 3-dichloromethyl-1,1-dimethylsilacyclohexane (**539**) (reaction 196).[218]

(533) $\xrightarrow{PhHgCCl_2Br}$ (539) (196)

4,4-Dimethyl-4-silacyclohexanone (**542**) has been obtained by the cyclization of dimethylbis-(2-methoxycarbonylethyl)silane (**540**) using a modified Dieckmann reaction. This is of interest since a carbon–silicon bond is considerably longer than a carbon–carbon single bond. For this reason, preparing a six-membered ring containing silicon may be like preparing an all-carbon seven-membered ring system. To achieve reasonable yields, the enolate anion formed in the Dieckmann cyclization must be trapped by rapid addition of trimethylchlorosilane to yield the corresponding silyl enol ether (**541**). This prevents the reverse reaction, which occurs if quenching of the anion is slow (*e.g.* by addition of water). The silyl enol ether (**541**) is boiled with aqueous methanolic hydrogen chloride to effect hydrolysis of the trimethylsilyl protecting group and decarboxylation of the β-keto-acid[219] (reaction 197). The mass spectral fragmentation of 4,4-dimethylsilacyclohexanone (**542**) is indicated in Scheme 23.

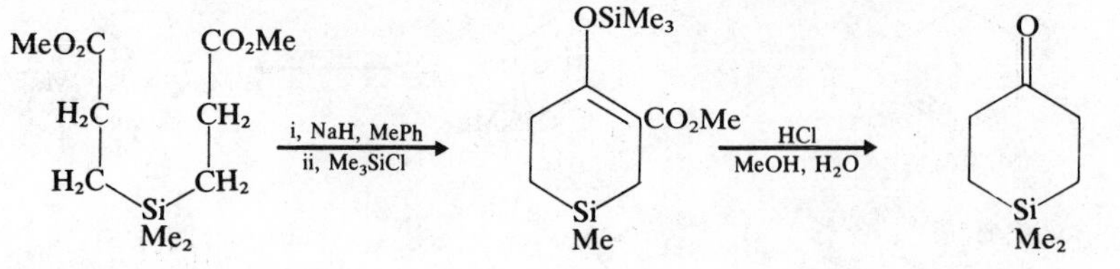

(197)

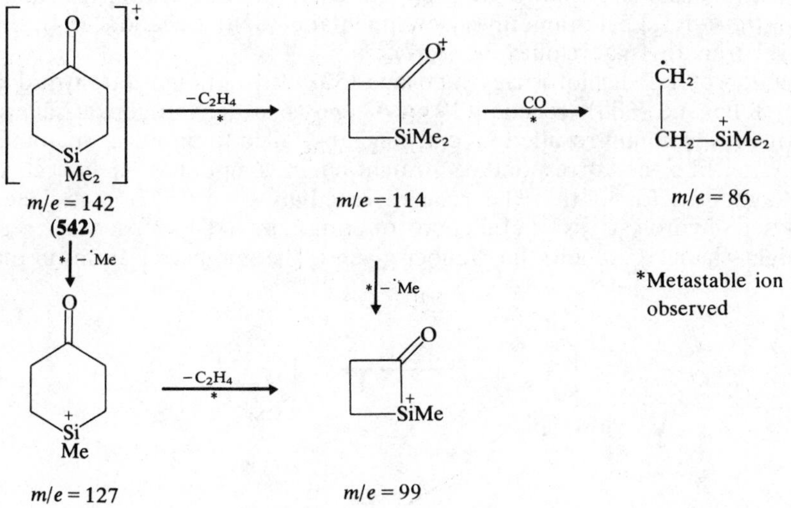

SCHEME 23

*Metastable ion observed

The selenium dioxide oxidation of 4,4-dimethylsilacyclohexanone (**542**) in boiling t-butanol affords 4,4-dimethylsilacyclohexadien-1-one (**543**) (reaction 198).[220]

The cyclic silicon olefin 1,1-dichlorosilacyclohex-2-ene (**545**) is prepared by the ring closure of 5-chloro-1-(trichlorosilyl)pent-1-ene (**544**) in ether in the presence of magnesium (reaction 199). The ring closure proceeds smoothly when the original olefin contains over 75% of the *cis*-isomer.[221] Treatment of the dichloro-compound (**545**) with phenyl-lithium yields 1,1-diphenylsilacyclohex-2-ene (**546**), which on hydroboration and oxidation gives predominantly the product (**547**) where the hydroxy-group is attached to the carbon adjacent to the silicon atom. Oxidation of this product with chromium trioxide affords mainly cleavage products, while oxidation using dicyclohexylcarbodi-imide with pyridinium trifluoroacetate in dimethyl sulphoxide gives 1,1-diphenylsilacyclohexan-2-one (**548**) (reaction 200).[222]

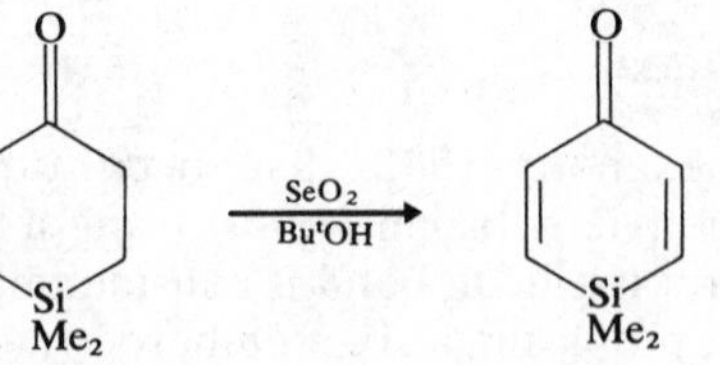

(198)

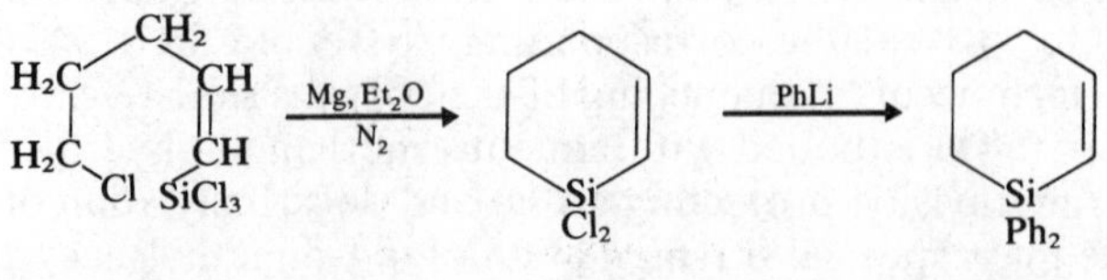

(199)

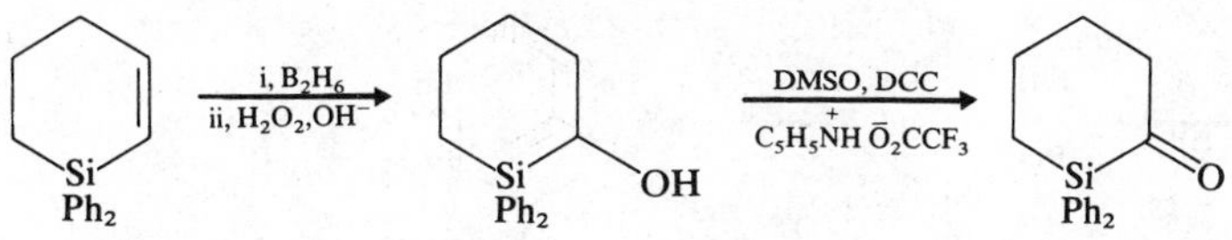

(200)

(546) (547) (548)

1,1-Dimethylsilacyclohex-2-ene (**549**) reacts with dibromocarbene, generated by the thermolysis of phenyl(tribromomethyl)mercury to give 7,7-dibromo-2,2-dimethyl-2-silanorcarane (**550**), which on treatment with tri-n-butyltin hydride yields either *endo*-(**551**) and *exo*-7-bromo-2,2-dimethyl-2-silanorcarane (**552**), or 2,2-dimethyl-2-silanorcarane (**553**), depending on the conditions of the reaction[223] (reaction 201). Treatment of the dibromo-compound (**550**) with silver ion or pyrolysis with quinoline releases hydrogen bromide and decomposes the ring system.

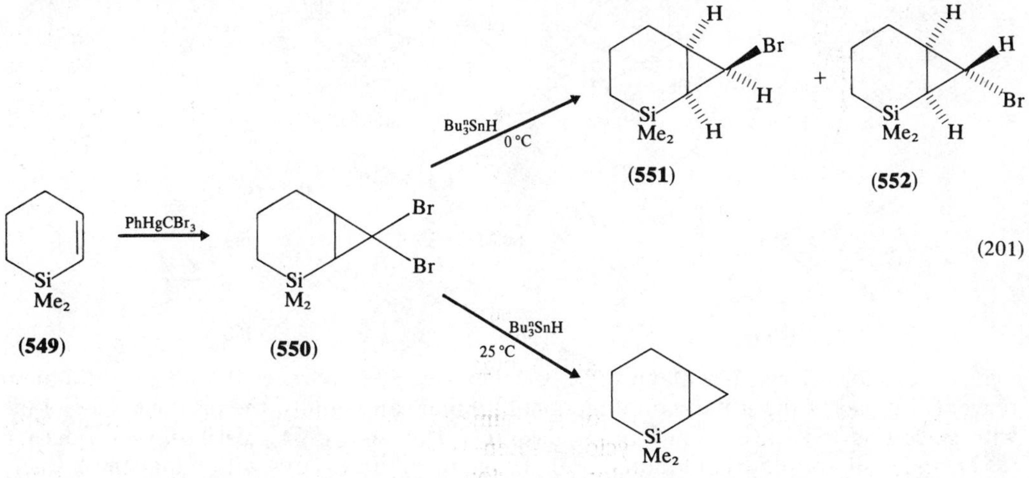

(201)

(549) (550) (551) (552) (553)

The dichlorosilacyclohexene (**545**) has been converted into 1,1-dichlorosilacyclohexadiene (**555**) by allylic bromination, followed by dehydrobromination. Bromination is readily accomplished using *N*-bromosuccinimide, when the main product is 4-bromo-1,1-dichlorosilacyclohex-2-ene (**554**). Treatment of this compound with quinoline and ferric chloride to effect dehydrobromination leads only to partial or total destruction of starting material. Pyrolysis, however, gives as main product the required 1,1-dichlorosilacyclohexadiene (**555**) along with small amounts of 1,1-dichlorosilacyclohex-2- and -3-ene (reaction 202).[221]

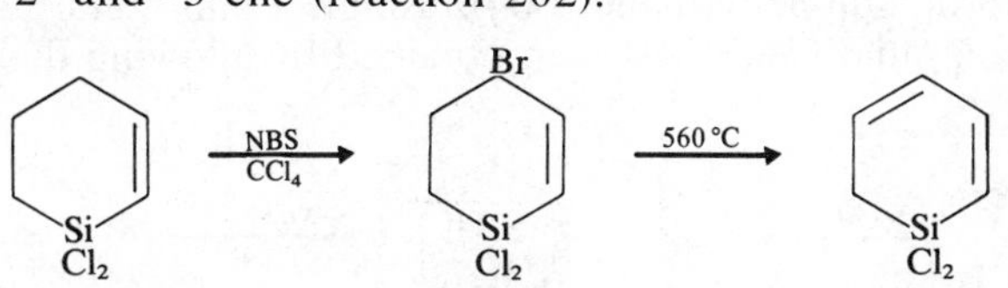

(202)

(545) (554) (555)

A silacyclohexadiene (**556**) is formed when 1,4-dilithiotetraphenylbutadiene (**505**) is reacted with chloro(chloromethyl)dimethylsilane. With butyl-lithium it gives a red anion (**557**), which with D_2O and Me_3SiCl affords the corresponding 2-substituted silacyclohexadiene (**558**; R = D or Me_3Si) (reaction 203).[224] Attempts to obtain a silabenzene have been unsuccessful, *e.g.* although hydrogen chloride can be eliminated quantitatively from 1-chloro-1,2,3,4,5-pentaphenylsilacyclohexadiene (**559**), no expected silabenzene (**560**) can be detected.

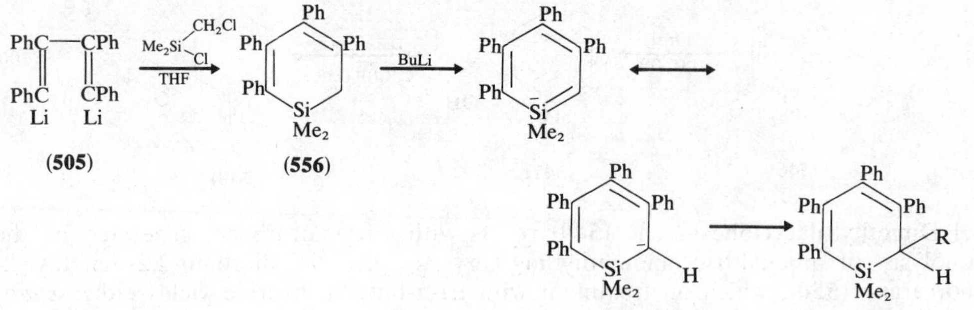

(505) (556) (557) (558) (203)

(559) (560)

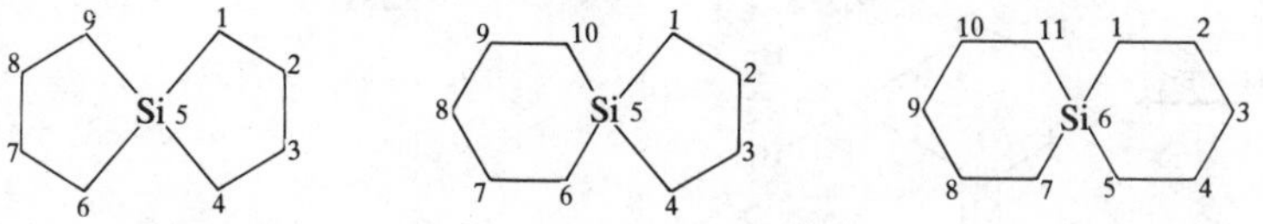

(561) (562) (563)

If 1,1-dichlorosilacyclo-pentane (**494**) or -hexane (**532**) is reacted with alkyldilithium reagents instead of the usual monofunctional lithium compounds, the product is a spirane with silicon as the spiro-atom, *e.g.* 5-silaspiro[4,4]nonane (**561**), 5-silaspiro[4,5]decane (**562**), and 6-silaspiro[5,5]undecane (**563**). They are colourless oils with faint camphoraceous odours and, compared with typical organosilicon compounds, they possess unusually high densities, refractive indices, and melting and boiling points.[203]

19.3.6.5 Six-membered ring containing one silicon atom and fused to one or two benzene rings

3,4-Dihydro-1,1-diphenyl-2*H*-1-silanaphthalene (2,3-benzo-1,1-diphenyl-1-silacyclohex-2-ene) (**565**) has been prepared by the ring closure of the trisubstituted silane 3-(2-chlorophenyl)propyldiphenylsilane (**564**), using sodium metal in toluene[208] (reaction 204). The same compound (**565**) has been obtained by allowing the di-Grignard reagent

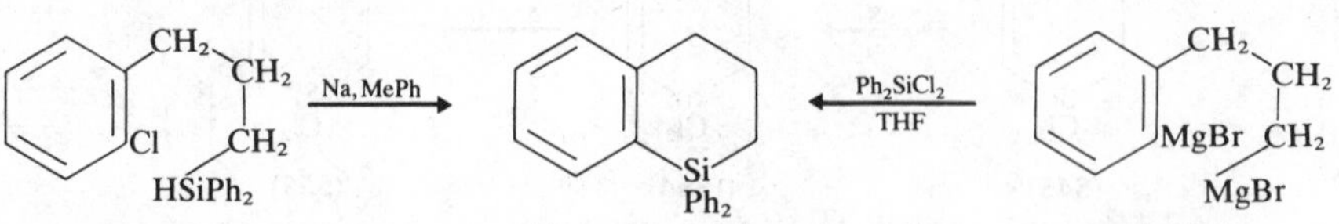

(564) (565) (566) (204)

of 1-bromo-2-(3-bromopropyl)benzene (**566**) to react with dichlorodiphenylsilane[225] (reaction 204). The actual composition of the di-Grignard reagent (**566**) is unknown. In view of the fact that magnesium bromide may be precipitated from the Grignard solutions by the addition of dioxan, and because of the proposal that Grignard reagents may better be represented by the complex $RMg{\cdot}MgX_2$, it may have the structure (**567**).

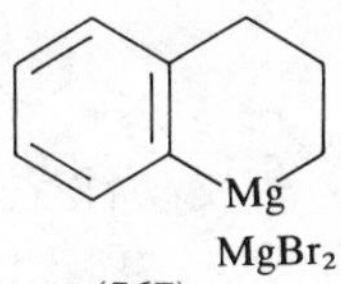

Mg
$MgBr_2$
(567)

Previous attempts to prepare 3,4-dihydro-1,1-diphenyl-2*H*-1-silanaphthalene (**565**) by the cyclization of either 3-triphenylsilylpropan-1-ol with boron trifluoride, or 3-bromopropyltriphenylsilane with ferric chloride, failed. Apparently the Lewis acids effected rupture of the phenyl–silicon bonds as none of the desired cyclic silane could be isolated.

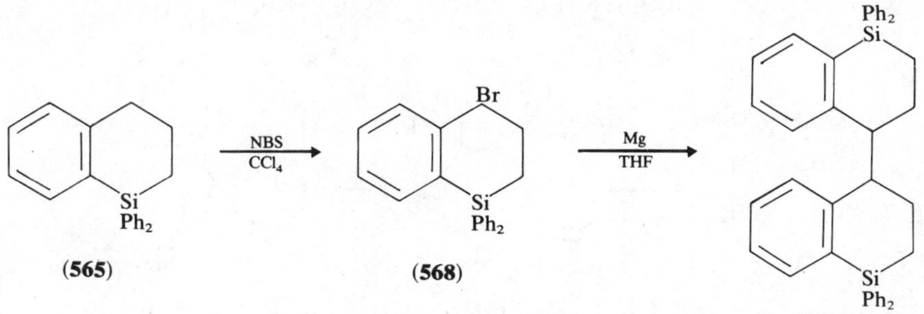

(**565**) —NBS, CCl_4→ (**568**) —Mg, THF→ (**569**) (205)

Treatment of the 3,4-dihydrodiphenyl-2*H*-silanaphthalene (**565**) with *N*-bromosuccinimide in the presence of benzoyl peroxide gives the 4-bromo-derivative (**568**), which reacts with magnesium to afford 4,4′-bi-(3,4-dihydro-1,1-diphenyl-2*H*-1-silanaphthyl) (**569**), probably by a Wurtz-type coupling reaction enhanced by tetrahydrofuran (reaction 205). Hydrolysis and oxidation of compound (**568**) with silver acetate gives ketone (**570**).

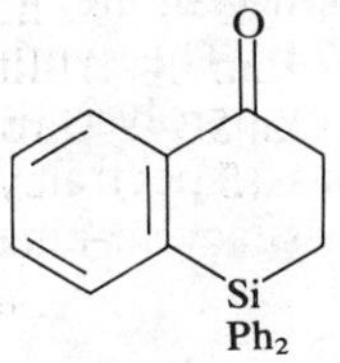

(570)

9,10-Dihydro-9-silanthracenes (**572**, R^1 = Me, Ph, Cl; R^2 = Me, Cl) have been prepared by reacting the di-Grignard reagent (**571**) formed from bis-(2-chlorophenyl)methane with the appropriate di- or tri-chlorosilane (reaction 206).[226]

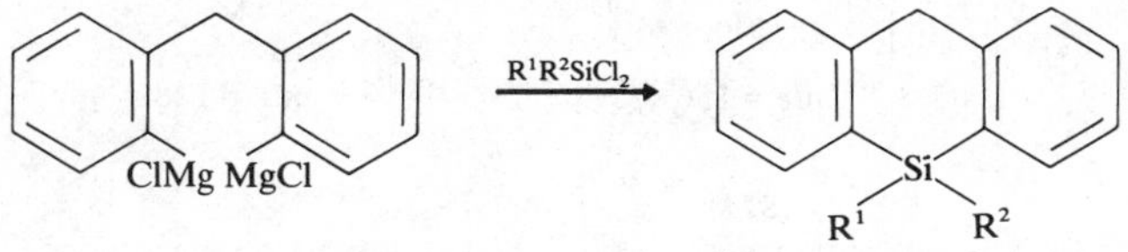

(**571**) —$R^1R^2SiCl_2$→ (**572**) (206)

As yet, 9,10-disubstituted silanthracenes have not been obtained from 9,10-disubstituted 9-chloro-9,10-dihydro-9-silanthracenes by the elimination of hydrogen chloride. Intermolecular substitution leading to a dimeric or polymeric product occurs.[183]

19.3.6.6 Large rings containing one silicon atom

Synthetic routes have been devised which provide convenient entry into organofunctional seven-, eight-, and nine-membered silicon ring systems. The route to 4,4-dimethyl-4-silacycloheptanone (**574**) is given in Scheme 24. The acyloin-type ring closure is

i, Me_2SiCl_2, C_5H_5N; ii, Na, MePh, Me_3SiCl; iii, EtOH, aq. HCl; iv, $KMnO_4$, aq. NaOH; v, CH_2N_2; vi, Na, MePh; vii, $HS(CH_2)_3SH$, $ZnCl_2$, HCl; viii, $CdCO_3$, $HgCl_2$, Me_2CO.

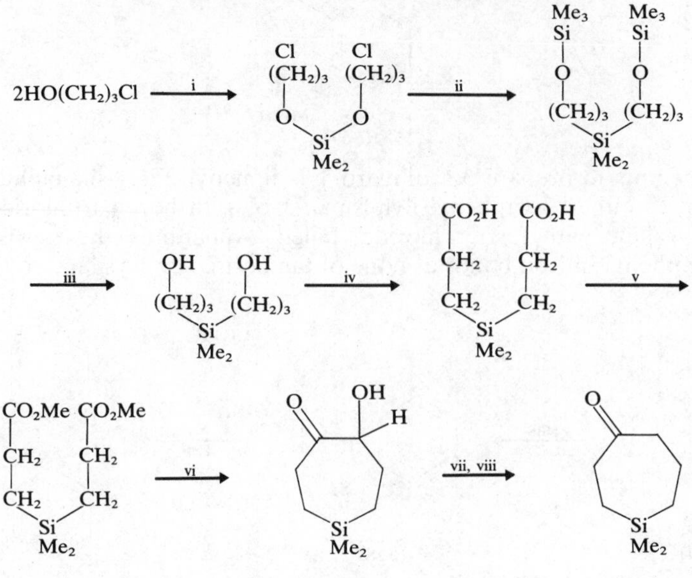

SCHEME 24

employed successfully to prepare 4,4-dimethyl-7-hydroxy-4-silacycloheptanone (**573**) and 5,5-dimethyl-9-hydroxy-5-silacyclononanone. The former is reduced successfully to 4,4-dimethyl-4-silacycloheptanone (**574**). The synthesis of 5,5-dimethyl-5-silacyclo-octanone is also achieved, but in low yield, by pyrolysis of the appropriate thorium salt to effect the required cyclization.[227] The mass spectral fragmentation of 4,4-dimethylsilacyclo-heptanone (**574**) and 5,5-dimethylsilacyclo-octanone (**575**) are indicated in Schemes 25 and 26, respectively.

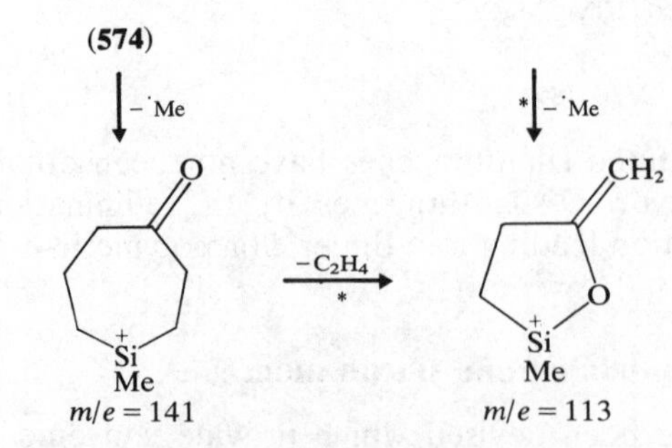

*Metastable ion observed

SCHEME 25

m/e = 170 (575) → $-C_2H_4$ * → m/e = 142 → $-C_2H_4$ → m/e = 114

↓ * $-\dot{M}e$ (each)

m/e = 155 → $-C_2H_4$ * → m/e = 127 → $-C_2H_4$ * → m/e = 99

*Metastable ion observed

SCHEME 26

19.3.7 HETEROCYCLES CONTAINING ONE GERMANIUM, TIN, OR LEAD ATOM

The names in general use for heterocycles containing one germanium, tin, or lead atom are obtained by combining the prefix 'germa', 'stanna', or 'plumba' with the name of the related carbocyclic compound. In some literature, use is made of the IUPAC names derived by combining a prefix from Table 1 with a stem from Table 2.

19.3.7.1 Four-, five-, and six-membered rings containing one germanium, tin, or lead atom

The reaction of sodium with a dilute equimolar mixture of di-n-butyldichlorogermane and 1,3-dichloropropane in xylene gives a poor yield of 1,1-dibutylgermacyclobutane (**577**), but a high yield of the desired product is obtained by treating dibutylchloro(3-chloropropyl)germane (**576**) with sodium in boiling xylene (reaction 207). 1,1-Diethylgermacyclobutane is prepared by boiling diethylchloro(3-chloropropyl)germane with liquid sodium–potassium eutectic alloy in toluene.[228]

$$CH_2{-}GeClBu_2 / CH_2{-}CH_2Cl \ \textbf{(576)} \xrightarrow{\text{Na, MePh}} \text{GeBu}_2 \text{ ring } \textbf{(577)} \qquad (207)$$

The highly strained ring in germacyclobutanes makes them very reactive, and as a consequence the ring is opened by a large variety of reagents, including alcoholic silver nitrate, mercuric chloride, sulphuryl chloride, lithium tetrahydridoaluminate, germanium tetrachloride, bromine, hydrogen halides, and other protonic acids; some examples are illustrated in reactions (208) and (209). They reduce alcoholic silver nitrate in the cold, and mercuric chloride on warming. Dichlorocarbene insertion into the germanium–carbon bond of 1,1-diethylgermacyclobutane (**578**) results in ring expansion and the formation of 2,2-dichloro-1,1-diethylgermacyclopentane (**579**) (reaction 210).[229]

$$Et_2Ge(CH_2)_3Br(\text{–}Br) \xleftarrow{Br_2} \text{(cyclo-}(CH_2)_3GeEt_2) \xrightarrow{HBr} Et_2GePr(\text{–}Br) \quad (208)$$

$$Et_2Ge(CH_2)_3GeCl_3(\text{–}Cl) \xleftarrow{GeCl_4} \text{(cyclo-}(CH_2)_3GeEt_2) \xrightarrow{H_2SO_4} Et_2GeOSO_3H(\text{–}Pr) \quad (209)$$

$$\textbf{(578)} \xrightarrow[C_6H_6]{Ph(BrCl_2C)Hg} \textbf{(579)} \quad (210)$$

The saturated five-membered heterocycle germacyclopentane (**581**) is prepared by treating its 1,1-dichloro-derivative (**580**) with lithium tetrahydridoaluminate. Originally the dichloro-derivative is formed by reacting the Grignard reagent (**493**) from 1,4-dibromobutane with germanium tetrachloride (reaction 211). Beside the desired compound the reaction also produces the spiro-compound 5-germaspiro[4,4]nonane (**583**).[230]

$$BrMgCH_2CH_2CH_2CH_2MgBr\ \textbf{(493)} \xrightarrow[N_2,\ Et_2O]{GeCl_4} GeCl_2\ \textbf{(580)} \xrightarrow{LiAlH_4} GeH_2\ \textbf{(581)} \xrightarrow{O_2} GeO\ \textbf{(582)} \quad (211)$$

(583)

Germacyclopentane (**581**) on shaking with oxygen is rapidly converted into germacyclopentane 1-oxide (**582**), also obtained by shaking an ethereal solution of dibromogermacyclopentane with the theoretical amount of aqueous sodium hydroxide. The germacyclopentane ring is stable to the action of acids and bases, with the exception of sulphuric acid, which causes ring opening (reaction 212).

$$\text{(germacyclopentane-}GeBu_2) \xrightarrow{H_2SO_4} Bu_2GeOSO_3H(\text{–}Bu) \quad (212)$$

Germacyclopentane (**581**) on heating under nitrogen in the dark with either s-butyl iodide or bromide yields the corresponding 1,1-dihalogenogermacyclopentane (**584**; X = I, Br) (reaction 213). 1-Iodogermacyclopentane is obtained by cooling germacyclopentane to 0 °C and treating it with iodine under nitrogen. The di-iodo-derivative (**584**; X = I) on boiling with silver chloride in heptane yields 1,1-dichlorogermacyclopentane. The dibromo-derivative (**584**; X = Br) reacts with ethyl- and phenyl-magnesium bromide to give 1,1-diethyl- and 1,1-diphenyl-germacyclopentane (**585**; R = Et, Ph), respectively (reaction 213). When 1-ethyl-1-phenylgermacyclopentane (**586**) is treated with bromine,

$$GeH_2\ \textbf{(581)} \xrightarrow[N_2]{Bu^sX} GeX_2\ \textbf{(584)} \xrightarrow{RMgBr} GeR_2\ \textbf{(585)} \quad (213)$$

cleavage of the Ph—Ge bond occurs to give 1-bromo-1-ethylgermacyclopentane (**587**) (reaction 214); also 1-bromo-1-phenylgermacyclopentane is obtained by reacting the 1,1-diphenyl derivative (**585**; R = Ph) with bromine in ethyl bromide. Treatment of the 1-bromo-1-phenyl compound with lithium tetrahydridoaluminate affords 1-phenylgermacyclopentane (**588**).

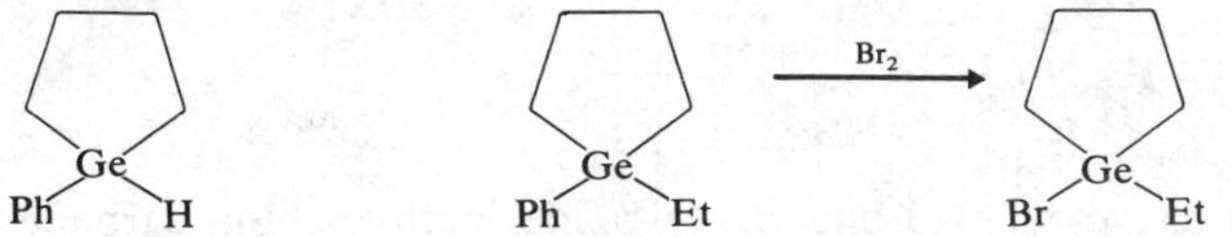

(214)

Germacyclopent-3-enes are prepared by reacting germanium di-iodide with conjugated dienes, *e.g.* germanium di-iodide with 2,3-dimethylbutadiene (**511**) affords 1,1-di-iodo-3,4-dimethylgermacyclopent-3-ene (**589**),[231] a low-melting solid which decomposes on distillation (reaction 215). The di-iodo-derivative (**589**) on boiling with freshly prepared

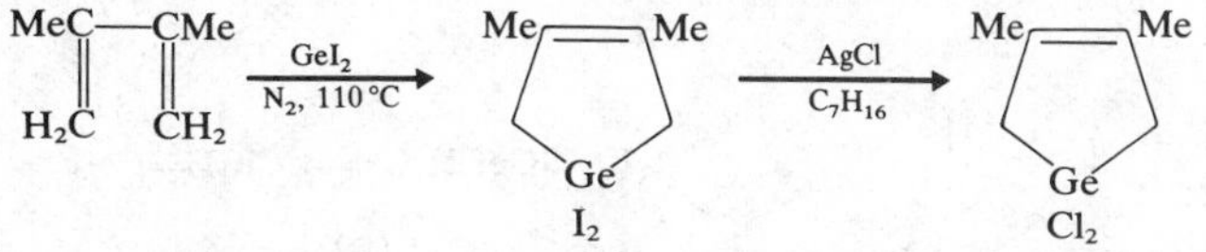

(215)

silver chloride in heptane gives the 1,1-dichloro-derivative (**590**). Reduction of the di-iodo-compound (**589**) with lithium tetrahydridoaluminate gives 3,4-dimethylgermacyclopent-3-ene (**591**), and reaction with alkyl- or aryl-magnesium halides yields the corresponding 1,1-dialkyl or 1,1-diaryl derivative, *e.g.* 1,1-diethyl-3,4-dimethylgermacyclopent-3-ene (**592**) (reaction 216).

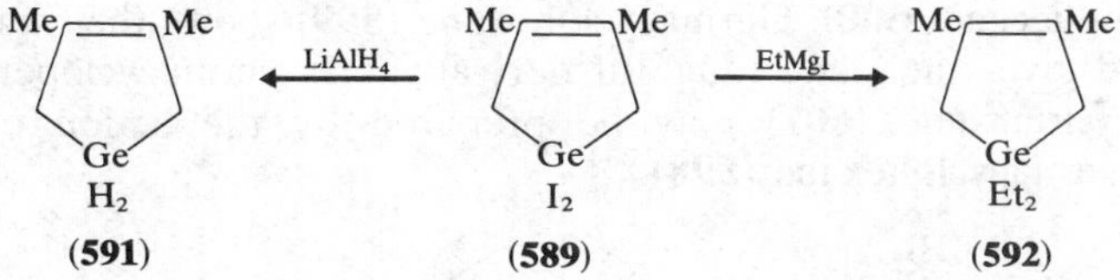

(216)

Although the germacyclopent-3-enes appear to be moderately stable to substitution, alkylation, arylation, and reduction, they are not stable to halogenation and hydrolysis. Treatment of the 1,1-diethyl derivative (**592**) with bromine in ethyl bromide at −80 °C gives dibromodiethylgermane. With 4-nitroperbenzoic acid it forms an epoxide, which with lithium tetrahydridoaluminate yields 1,1-diethyl-3,4-dimethylgermacyclopentan-3-ol (**593**) (reaction 217), also obtained from compound (**592**) by hydroboration and subsequent treatment of the product with hydrogen peroxide in sodium hydroxide solution. Alk-3-enylgermanium compounds, $Et_2GeXCH_2CH(Me)C(Me){=}CH_2$ (X = Cl, Br, OEt, or OAc), may be prepared by reacting the germacyclopentan-3-ol (**593**) with dehydrating agents; phosphoryl chloride, for example, produces the chloride.[232]

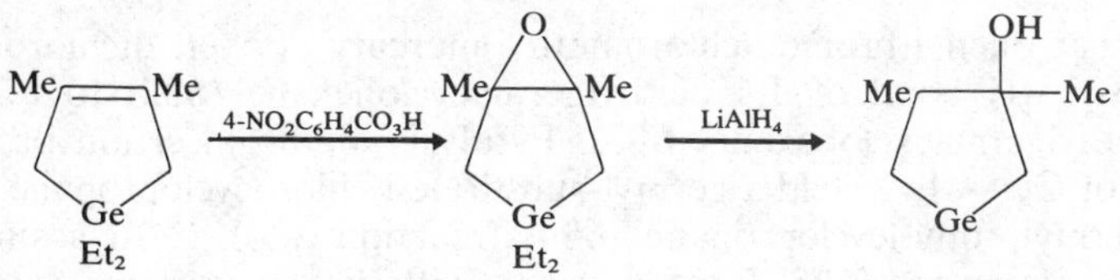

(217)

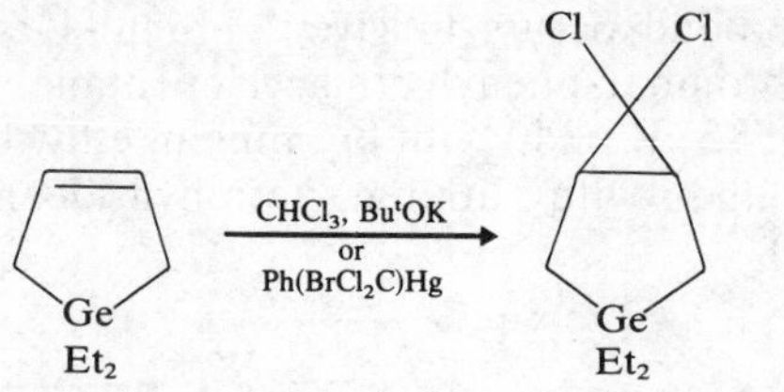

$$\text{(594)} \xrightarrow[\text{Ph(BrCl}_2\text{C)Hg}]{\text{CHCl}_3,\ \text{Bu}^t\text{OK or}} \text{(595)} \qquad (218)$$

1,1-Diethylgermacyclopent-3-ene (**594**) reacts with dichlorocarbene to form the expected adduct, 3,3-diethyl-6,6-dichloro-3-germabicyclo[3,1,0]hexane (**595**) (reaction 218), which may be isolated as a pure substance. Although adducts are also formed when there are methyl substituents on the double bond, these substituents tend to decrease the stability of the adduct.[233] They decompose to give either a 1,1-diethyl-4-chloro-1-germacyclohexa-2,4-diene (**596**) or a diethylchlorogermyl-substituted pentadiene (*e.g.* **597**). The 3,3-diethyl-6,6-difluoro-3-germabicyclo[3,1,0]hexane structure is much more stable than the analogous system containing chloro-groups.

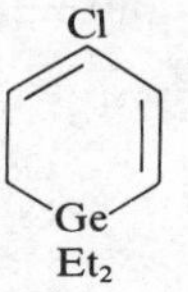

(**596**)

$Et_2ClGeCH_2CH{=}CClCH{=}CH_2$

(**597**)

The saturated six-membered heterocycles containing one germanium, tin, or lead atom are prepared in ways similar to those used for obtaining germacyclopentanes. The reaction between the di-Grignard reagent (**531**) from 1,5-dibromopentane and germanium tetrachloride gives 1,1-dichlorogermacyclohexane (**598**), which on treatment with lithium tetrahydridoaluminate affords germacyclohexane (**599**) (reaction 219).[230] The original Grignard reaction also results in the formation of a second product, namely 6-germaspiro[5,5]undecane (**600**). Germacyclohexane (**599**) shows the same type of chemical reactions and gives the same range of derivatives as germacyclopentane (**581**). 1,1-Diphenylgermacyclohexane (**601**) may be prepared by the action of phenyl-lithium on 1,1-dichlorogermacyclohexane (**598**).[234]

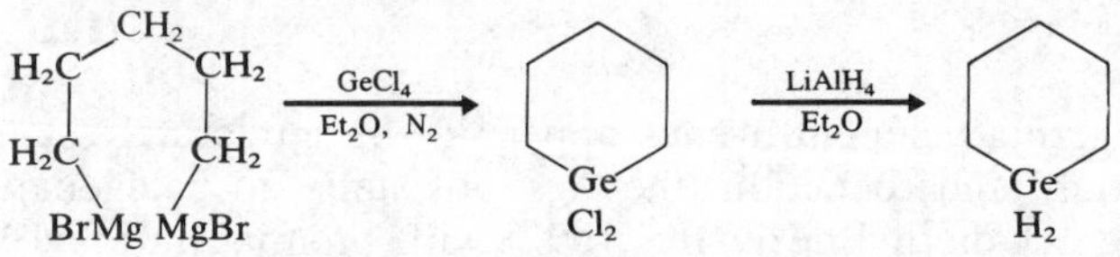

$$\text{(531)} \xrightarrow[\text{Et}_2\text{O, N}_2]{\text{GeCl}_4} \text{(598)} \xrightarrow[\text{Et}_2\text{O}]{\text{LiAlH}_4} \text{(599)} \qquad (219)$$

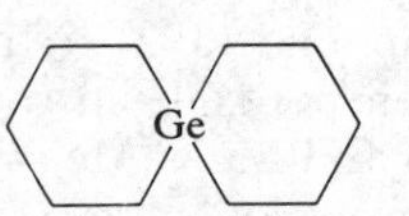

(**600**)  (**601**)

It is found that phenyl(bromodichloromethyl)mercury-derived dichlorocarbene inserts itself into a β-C—H bond of 1,1-diethylgermacyclohexane (**602**) to form 3-dichloromethyl-1,1-diethylgermacyclohexane (**603**). Pyrolysis of the insertion product results in γ-elimination of Ge—Cl to yield a germyl-substituted chlorocyclopropane, 1-chloro-2-(3-diethylchlorogermylpropyl)cyclopropane (**604**) (reaction 220).[235] In a similar way, 1,1-dimethylstannacyclohexane (**605**; R = Me) reacts with dichlorocarbene to give 3-dichloromethyl-1,1-dimethylstannacyclohexane.[218]

$$\text{(602)}\ \text{1,1-Et}_2\text{-germacyclohexane} \xrightarrow{Ph(BrCl_2C)Hg} \text{(603)}\ \text{CHCl}_2\text{-substituted Et}_2\text{Ge ring} \xrightarrow{\Delta} Et_2Ge(Cl)(CH_2)_3\text{-}C_3H_4Cl\ \text{(604)} \qquad (220)$$

The reaction between the di-Grignard reagent (**531**) from 1,5-dibromopentane and the appropriate dialkyl-tin or -lead di-iodide furnishes the desired 1,1-dialkylstannacyclohexane (**605**)[236] or plumbacyclohexane (**606**)[237] (reaction 221). These compounds are stable in the absence of air, but the tin compounds, like the open-chain tin alkyls, show a pronounced tendency to resinify by atmospheric oxidation. 1,1-Diethylplumbacyclohexane (**606**; R = Et) resinifies readily, inflames when exposed to air on a filter paper, and reduces silver nitrate immediately. With both the tin and lead heterocycles the ring is opened immediately by bromine (reaction 222). The tin-containing spiro-compound 6-stannaspiro[5,5]undecane (**607**) is prepared by the reaction between stannic chloride and a small excess of the di-Grignard reagent from 1,5-dibromopentane. The spiro-compound undergoes ring cleavage when treated with bromine, hydrogen bromide, and iodine (reaction 223).[238]

$$\text{(605)}\ R_2Sn(CH_2)_5 \xleftarrow{R_2SnI_2} BrMg(CH_2)_5MgBr\ \text{(531)} \xrightarrow{R_2PbI_2} R_2Pb(CH_2)_5\ \text{(606)} \qquad (221)$$

$$Et_2Sn(CH_2)_5 \xrightarrow{Br_2} Et_2Sn(Br)(CH_2)_4CH_2Br \qquad (222)$$

$$[XCH_2(CH_2)_4]_2SnX_2 \xleftarrow[80\text{–}85\,°C]{X_2} \text{Sn spiro}\ \text{(607)} \xrightarrow[80\text{–}85\,°C]{HBr} [Me(CH_2)_4]_2SnBr_2 \qquad (223)$$

(X = Br, I)

Both 1,1-diphenyl-stannacyclohexane (**608**) and -plumbacyclohexane (**609**) are obtained by reacting 1,5-dilithiopentane with diphenyl-tin or -lead dichloride, respectively[234] (reaction 224). Bromination of 1,1-diphenylstannacyclohexane (**608**) using either potassium bromide–potassium bromate in carbon tetrachloride or anhydrous hydrogen bromide gives 1,1-dibromostannacyclohexane (**610**). The by-product in the later reaction is benzene. Iodination with iodine in carbon tetrachloride removes only one phenyl group to

$$\text{(608)}\ Ph_2Sn(CH_2)_5 \xleftarrow{Ph_2SnCl_2} Li(CH_2)_5Li \xrightarrow{Ph_2PbCl_2} Ph_2Pb(CH_2)_5\ \text{(609)} \qquad (224)$$

furnish 1-iodo-1-phenylstannacyclohexane (**611**) (reaction 225). Since the organotin dihalides are, for the most part, oils and difficult to distil in small quantities, it is found convenient to prepare crystalline complexes through coordination with either 2,2′-bipyridine or 1,10-phenanthroline.

(225)

(610) (608) (611)

Stannohydration of diethynylmethane (**612**) with dibutyltin hydride in boiling heptane followed by heating at 200 °C gives 1,4-dihydro-1,1-dibutylstannabenzene (**322**), which on reacting with phenylboron dibromide yields 1,4-dihydro-1-phenylborabenzene (**613**). Deprotonation of the borabenzene derivation (**613**) by lithium triphenylmethide in THF or more conveniently using t-butyl-lithium in pentane–THF affords the 1-phenylborabenzene anion (**614**) (reaction 226).[239] Treating the above solutions with an excess of acetic acid gives penta-1,4–diene, *cis*-penta-1,3-diene, and benzene. The presence of anion (**614**) is established by examining the products of deuteration by [$^2H_1$]acetic acid. N.m.r. spectral data suggests significant aromatic properties for the borabenzene anion (**614**).

(226)

(612) (322) (613) (614)

19.3.7.2 Five- and six-membered rings containing either one germanium or tin atom and fused to two benzene rings

5,5-Diphenyldibenzogermole (**615**) is prepared by heating a mixture of 2,2′-dilithiobiphenyl and dichlorodiphenylgermane[240] (reaction 227). The 9,9-dialkyl derivatives of 9,10-dihydro-9-germanthracene (**616**) and 9,10-dihydrostannanthracene (**617**) are obtained by reacting 2-benzylphenylmagnesium chloride with dialkyldichloro-germane and -stannane, respectively[226] (reaction 228). In a similar manner, diaryl, alkyl-chloro-, aryl-chloro-, and dichloro-derivatives may be prepared by using the appropriate substituted germane or stannane.

(227)

(523) (615)

(228)

(616) (617)

19.3.8 HETEROCYCLES CONTAINING ONE ALUMINIUM, BORON, BROMINE, OR IODINE ATOM

The interaction between triphenylaluminium and diphenylacetylene in a melt at 200 °C results in the evolution of benzene and the formation of the benzaluminole (**621**) as a yellow solid. Its formation can be explained in the light of the properties of certain

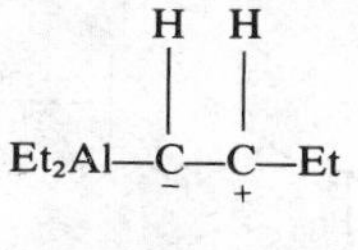

$$Et_2Al-\overset{H}{\underset{-}{C}}-\overset{H}{\underset{+}{C}}-Et$$

(618)

α,β-unsaturated aluminium compounds, *e.g.* physical data support the formulation (**618**) for the π-bond in diethylbut-1-enylaluminium. This suggests that the cyclization of the α,β-unsaturated system in (**619**) can be ascribed to the π-electron withdrawal from the vinylogous β-position (*ortho*) by the available aluminium orbital. Such π-electron polarization would enhance both the 'acidic' character of the *ortho* proton of the benzene ring and the nucleophilic character of the attacking phenyl group (**620**) (reaction 229).[241]

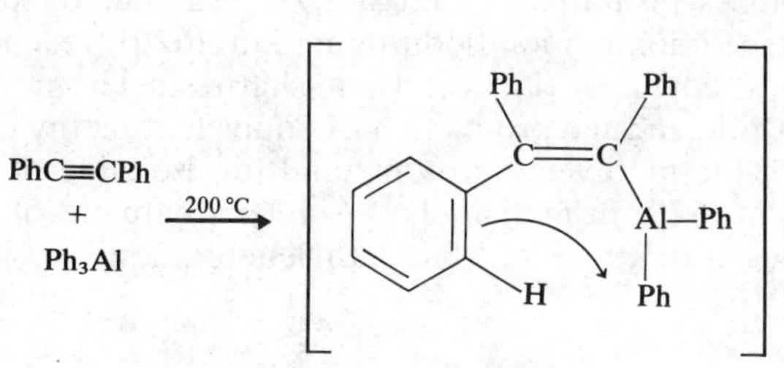

PhC≡CPh + Ph₃Al —200 °C→ [(**619**)] → [(**620**)] → (**621**) + C_6H_6 (229)

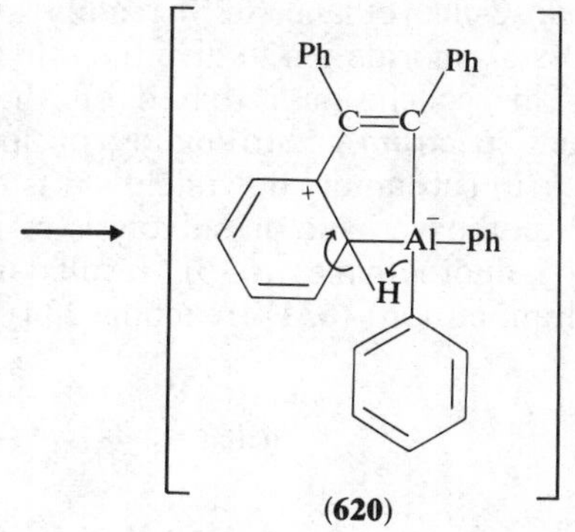

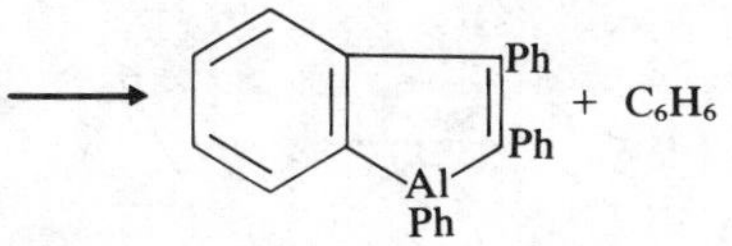

The validity of the postulated cyclization is supported by the fact that 2-biphenylyldiphenylaluminium (**622**), prepared from 2-biphenylyl-lithium and diphenylaluminium chloride, on heating at 200 °C evolves benzene and affords 5-phenyldibenzaluminole (**623**) (reaction 230).

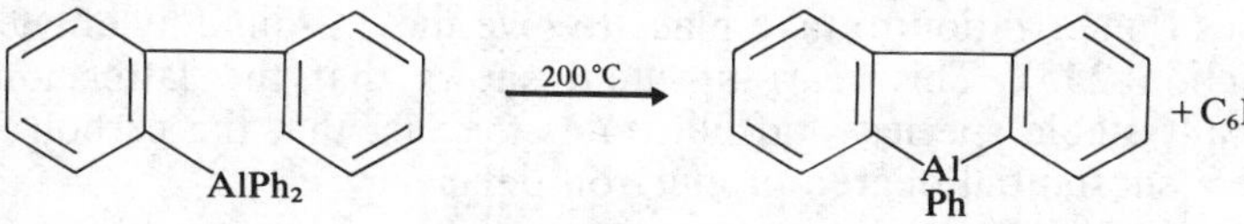

(**622**) —200 °C→ (**623**) + C_6H_6 (230)

2-Biphenyldialkylboranes (**624**; *e.g.* R = Et), prepared from 2-biphenylyl-lithium and dialkylchloroboranes, split out alkane at 180–200 °C to give 5-alkyldibenzoboroles (**625**) (9-alkyl-9-borafluorene)[242] (reaction 231). The reaction between 9,10-dihydro-9,9-dimethyl-9-stannanthracene (**617**; R = Me) and phenyldichloroborane gives 9,10-dihydro-9-phenyl-9-boranthracene (**626**). Metallation of the boranthracene (**626**) with t-butyl-lithium in THF–pentane at −78 °C yields the 9-phenyl-9-boranthracene anion, which can be isolated as the solvent-separated ion pair (**627**) (reaction 232).[243]

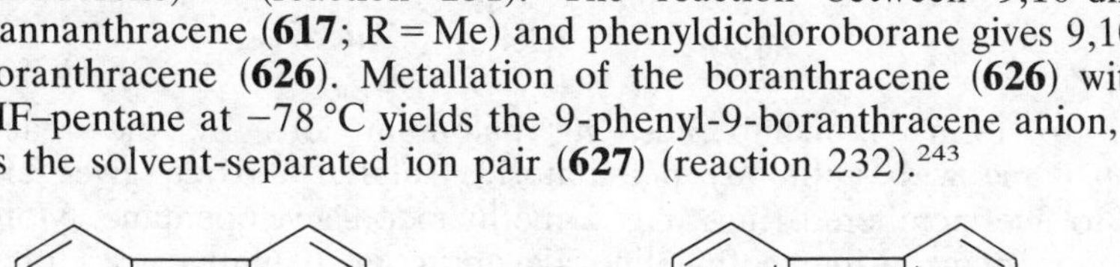

(**624**) —180–200 °C→ (**625**) + RH (231)

(617) $\xrightarrow{PhBCl_2}$ (626) $\xrightarrow[THF]{Bu^tLi}$ (627) $\overset{+}{Li}$.3.5THF (232)

The structure of the borarene (**627**) is confirmed by its n.m.r. and u.v. spectra. All borarenes so far reported are extremely sensitive to atmospheric oxygen and moisture, and the anionic aromatic compounds (**627**) and (**614**) react as resonance structures having a carbanionic centre at C-10, C-2, or C-4 and a tervalent Lewis-acid boron atom. Hence electrophilic and nucleophilic reactions are possible.

1-Iodo-2-fluorethane (**628**) readily ionizes in antimony pentafluoride–sulphur dioxide with loss of fluoride ion to give the ethyleneiodonium ion (iodiranium ion) (**629**) (reaction 233). This conclusion is based on the following evidence: (i) a sharp singlet in the ^{1}H n.m.r. spectrum, indicating loss of fluoride and the production of equivalent methylene groups; (ii) substantial downfield shifts of the methylene protons; and (iii) isolation of 1-iodo-2-methoxyethane on solvolysis of ion (**629**) in methanol.[244] Similar treatment of 1-bromo-2-fluoroethane (**630**) results in formation of the ethylenebromonium ion (bromiranium ion) (**631**) (reaction 234).

ICH_2CH_2F (**628**) $\xrightarrow[-60\,°C]{SbF_5\text{–}SO_2}$ (**629**) SbF_6^- (233)

$BrCH_2CH_2F$ (**630**) $\xrightarrow[-60\,°C]{SbF_6\text{–}SO_2}$ (**631**) SbF_6^- (234)

Starting from *erythro*- or *threo*-2-bromo-3-fluorobutane, the same mixture of *cis*- and *trans*-dimethylethylenebromonium ions is obtained, but from the corresponding iodides, stereospecific formation of the *trans*- and *cis*-1,2-dimethylethyleneiodonium ions, respectively, is observed. The latter ions are stereospecifically solvolysed in methanol to give the related methyl ethers. Warming the isomeric 2,3-dimethylethylenebromonium ion (**632**) to −40° C causes isomerization to take place to give the 2,2-dimethylethylenebromonium ion (**633**) (reaction 235). This rearrangement shows that the latter ion is the more thermodynamically stable species, and illustrates the fact that the carbon atoms in these ions must have a substantial degree of electron deficiency.

(**632**) SbF_6^- $\xrightarrow[-40\,°C]{SbF_5\text{–}SO_2}$ (**633**) SbF_6^- (235)

Attempts to form six-membered ring halonium ions by the reaction of 1,5-dihalogenopentane with antimony pentafluoride–sulphur dioxide gives exclusive rearrangement to five-membered rings, *e.g.* 2-methyliodoniacyclopentane. Monomethylation of 1,5-di-iodopentane using methyl fluoride–antimony pentafluoride, followed by cyclization, gives the six-membered ring iodoniacyclohexane ion (pentamethyleneiodonium ion) (**634**)[245] (reaction 236) containing less than 6% rearranged ion. 1,5-Dibromopentane yields the six-membered bromonia-cyclohexane ion (pentamethylenebromonium ion) (**635**), contaminated with 27–45% rearranged ion (**636**) (reaction 237).

$[H_2C(CH_2)_3CH_2\text{–I}^{+}\text{Me}]\ SbF_6^- \xrightarrow[SO_2]{MeF-SbF_5}$ (634) $+ MeI$ (236)

$[H_2C(CH_2)_3CH_2\text{–Br}^{+}\text{BrMe}]\ SbF_6^- \xrightarrow[SO_2]{MeF-SbF_6}$ (635) SbF_6^- + (636) SbF_6^- $+ MeBr$ (237)

(637) $\xrightarrow[Ac_2O]{CH_3CO_3H}$ (638) $\xrightarrow[\text{ii, } H_2O, \text{ NaCl}]{\text{i, } H_2SO_4}$ (639) Cl^- (238)

10*H*-Dibenz[*b,e*]iodininium chloride (9-iodonia-10*H*-anthracene chloride) (**639**) is made in high yield from 2-iodosobiphenyl (**638**), obtained by oxidizing 2-iodobiphenyl (**637**) with peracetic acid. The addition of concentrated sulphuric acid to the iodoso-compound (**638**) in acetic anhydride solution brings about a facile ring closure to give the iodonium salt (**639**) (reaction 238).[246]

References

1. D. L. Klayman and W. H. H. Gunther, 'Organic Selenium Compounds: Their Chemistry and Biology', Wiley-Interscience, New York, 1973.
2. IUPAC Nomenclature of Organic Chemistry, Section A (Hydrocarbons) and Section B (Fundamental Heterocyclic Systems), Butterworth, London, 1958; 3rd edn., 1971.
3. Extracts of the IUPAC Rules (Sections A and B) 'Handbook of Chemistry and Physics', Section C, 55th edn., Chemical Rubber Co., Cleveland, Ohio, 1974–1975.
4. A. B. Callear and W. J. R. Tyerman, *Proc. Chem. Soc.*, 1964, 296; *Trans. Faraday Soc.*, 1965, **61,** 2395; 1966, **66,** 371; W. J. R. Tyerman, W. B. O'Callaghan, P. Kebarle, O. P. Strausz, and H. E. Gunning, *J. Amer. Chem. Soc.*, 1966, **88,** 4277.
5. G. H. Schmid and D. G. Garratt, *Tetrahedron Letters*, 1975, 3991.
6. G. T. Morgan and F. H. Burstall, *J. Chem. Soc.*, 1930, 1497; H. J. Backer and H. J. Winter, *Rec. Trav. chim.*, 1937, **56,** 492.
7. W. B. Moniz, *J. Phys. Chem.*, 1969, **73,** 1124.
8. N. N. Magdesieva, *Adv. Heterocyclic Chem.*, 1970, **12,** 1.
9. C. Paal, *Ber.*, 1885, **18,** 2255.
10. H. V. A. Briscoe and J. B. Peel, *J. Chem. Soc.*, 1928, 1741; H. V. A. Briscoe, J. B. Peel, and P. L. Robinson, *ibid.*, p. 2628; H. Suginome and S. Umezawa, *Bull. Chem. Soc. Japan*, 1936, **11,** 157.
11. R. F. Curtis, S. N. Hasnain, and J. A. Taylor, *Chem. Comm.*, 1968, 365.
12. K. Schulte, J. Reisch, and D. Bergenthal, *Chem. Ber.*, 1968, **101,** 1540.
13. Yu. K. Yur'ev, *Zhur. obschei Khim.*, 1946, **16,** 851.
14. F. Ya. Perveev, N. I. Kudryashova, and D. N. Glebovskii, *Zhur. obshchei Khim.*, 1956, **26,** 3331.
15. H. J. Backer and W. Stevens, *Rec. Trav. chim.*, 1940, **59,** 423.
16. M. T. Bogert and P. P. Herrera, *J. Amer. Chem. Soc.*, 1923, **45,** 238; M. T. Bogert and C. N. Anderson, *ibid.*, 1926, **48,** 223; P. Demerzeman, N. P. Buu-Hoï, R. Royer, and A. Cheutin, *J. Chem. Soc.*, 1954, 2720.

17. E. H. Braye, W. Hübel, and J. Caplier, *J. Amer. Chem. Soc.*, 1961, **83,** 4406.
18. W. Dilthey, *Ger. Pat.* 628 954/1936 (*Chem. Abs.*, 1936, **30,** 6009).
19. N. M. Pozdeev, O. B. Akulinin, A. A. Shapkin, and N. N. Magdesieva, *Doklady Akad. Nauk S.S.S.R.*, 1969, **185,** 384.
20. J. Kraitchman, *Amer. J. Phys.*, 1953, **21,** 17; C. C. Costain, *J. Chem. Phys.*, 1958, **29,** 864.
21. B. Bak, D. Christensen, W. B. Dixon, L. Hansen-Nygaard, J. Rastrup-Andersen, and M. Schottlander, *J. Mol. Spectroscopy*, 1962, **9,** 124.
22. B. Bak, D. Christensen, L. Hansen-Nygaard, and J. Rastrup-Andersen, *J. Mol. Spectroscopy*, 1961, **7,** 58.
23. M. Nardelli and G. Fava, *Gazzetta*, 1958, **88,** 229.
24. N. Mario, F. Giovanna, and G. Giulia, *Acta Cryst.*, 1962, **15,** 737.
25. H. Briscoe and J. Peel, *J. Chem. Soc.*, 1929, 2589.
26. Yu. K. Yur'ev, N. N. Magdesieva, and A. T. Monakhova, *Zhur. obshchei Khim.*, 1960, **30,** 2726.
27. H. Gerding, G. Milazzo, and H. H. K. Rossmark, *Rec. Trav. chim.*, 1953, **72,** 957.
28. C. Paulmier and P. Pastour, *Compt. rend.* (*C*), 1967, **265,** 926.
29. W. Mack, *Angew. Chem.*, 1965, **77,** 260.
30. Yu. K. Yur'ev, E. L. Zaitseva, and G. G. Rosantsev, *Zhur. obshchei Khim.*, 1960, **30,** 2207.
31. Yu. K. Yur'ev and E. L. Zaitseva, *Zhur. obshchei Khim.*, 1960, **30,** 859.
32. Yu. K. Yur'ev and E. L. Zaitseva, *Zhur. obshchei Khim.*, 1958, **28,** 2164.
33. S. Umezawa, *Bull. Soc. Chem. Japan*, 1936, **11,** 775.
34. E. G. Kataev and A. E. Zimkin, *Uch. Zap. Kazansk. Gos. Univ.*, 1957, **117,** 174; Yu. K. Yur'ev and N. K. Sadovaya, *Zhur. obshchei Khim.*, 1964, **34,** 1803.
35. Yu. K. Yur'ev and N. K. Sadovaya, *Zhur. obshchei Khim.*, 1964, **34,** 2190.
36. Yu. K. Yur'ev and N. N. Mezentsova, *Zhur. obshchei Khim.*, 1957, **27,** 179.
37. Yu. K. Yur'ev, N. N. Mezentsova, and E. V. Vas'kovskii, *Zhur. obshchei Khim.*, 1957, **27,** 3155; Yu. K. Yur'ev, N. K. Sadovaya and M. A. Gal'bershtam, *ibid.*, 1958, **28,** 620, 622; 1959, **29,** 1917.
38. Yu. K. Yur'ev, N. N. Magdesieva, and A. T. Monakhova, *Khim. geterotsikl. Soedinenii*, 1968, 649.
39. Yu. K. Yur'ev, N. N. Mezentsova, and V. E. Vas'kovskii, *Zhur. obshchei Khim.*, 1958, **28,** 3262.
40. N. P. Buu-Hoï, P. Demerseman, and R. Royer, *Compt. rend.*, 1953, **237,** 397.
41. E. G. Kataev and M. V. Palkina, *Chem. Abs.*, 1956, **50,** 937; 1958, **52,** 3762.
42. Yu. K. Yur'ev and G. B. Yelyakov, *Doklady Akad. Nauk S.S.S.R.*, 1955, **102,** 763; Yu. K. Yur'ev and N. K. Sadovaya, *Zhur. obshchei Khim.*, 1956, **26,** 930; Yu. K. Yur'ev, N. K. Sadovaya, and V. V. Titov, *ibid.*, 1958, **28,** 3036.
43. Yu. K. Yur'ev, N. K. Sadovaya, and E. A. Grekova, *Zhur. obshchei Khim.*, 1964, **34,** 847.
44. Yu. K. Yur'ev, N. K. Sadovaya, and M. A. Gal'bershtam, *Zhur. obshchei Khim.*, 1962, **32,** 259.
45. Yu. K. Yur'ev and N. K. Sadovaya, *Zhur. obshchei Khim.*, 1956, **26,** 3154.
46. Yu. K. Yur'ev and N. K. Sadovaya, *Zhur. obshchei Khim.*, 1964, **34,** 1803.
47. C. Paulmier, J. Morel, P. Pastour, and D. Semard, *Bull. Soc. chim. France*, 1969, 2511.
48. Yu. K. Yur'ev and N. K. Sadovaya, *Zhur. obshchei Khim.*, 1958, **28,** 2162.
49. Yu. K. Yur'ev and M. A. Gal'bershtam, *Zhur. obshchei Khim.*, 1962, **32,** 3249.
50. Yu. K. Yur'ev, N. N. Magdesieva and L. Ya. Petrova, *Khim. geterotsikl. Soedinenii*, 1966, 910.
51. A. I. Shatenshtein, I. O. Shapiro, Yu. I. Ranneva, N. N. Magdesieva, and Yu. K. Yur'ev, *Reaktsionnaya Sposobnost Org. Soedinenii, Tartusk. Gos. Univ.*, 1964, **1,** 236 (*Chem. Abs.*, 1965, **62,** 8460).
52. L. Chierici, C. Dell'Erba, A. Guareschi, and D. Spinelli, *Ricerca sci.*, (*A*), 1965, **8,** 1537.
53. J. Morel, C. Paulmier, D. Semard, and P. Pastour, *Compt. rend.* (*C*), 1970, **270,** 825.
54. G. T. Morgan and F. H. Burstall, *J. Chem. Soc.*, 1929, 1096.
55. J. D. McCullough and A. Lefohn, *Inorg. Chem.*, 1966, **5,** 150.
56. Yu. K. Yur'ev, *J. Gen. Chem.* (*U.S.S.R.*), 1946, **16,** 851.
57. C. G. Krespan and C. M. Langkammerer, *J. Org. Chem.*, 1962, **27,** 3584.
58. A. Fredga, *J. prakt. Chem.*, 1930, **127,** 103.
59. A. Fredga, *J. prakt. Chem.*, 1931, **130,** 180.
60. E. Buchta and K. Greiner, *Chem. Ber.*, 1961, **94,** 1311.
61. A. M. Duffield, H. Budzikiewicz, and C. Djerassi, *J. Amer. Chem. Soc.*, 1965, **87,** 2920.
62. R. B. Mitra, K. Rabindran, and B. D. Tilak, *Current Sci.*, 1954, **23,** 263.
63. E. P. Mazza and L. Solazzo, *Chem. Abs.*, 1929, **23,** 2417.
64. Yu. K. Yur'ev, N. N. Mezentsova, T. A. Melent'eva, and E. G. Treshchova, *Zhur. obshchei Khim.*, 1957, **27,** 2260.
65. A. I. Kiss and B. R. Muth, *Chem. Abs.*, 1959, **53,** 3876; G. Grandolini, A. Ricci, N. P. Buu-Hoï, and F. Perin, *J. Heterocyclic Chem.*, 1968, **5,** 133.
66. B. R. Muth and A. I. Kiss, *J. Org. Chem.*, 1956, **21,** 576.
67. N. P. Buu-Hoï, V. Bellavita, A. Ricci, and G. Grandolini, *Bull. Soc. chim. France*, 1965, 2658.
68. L. Christiaens and M. Renson, *Bull. Soc. chim. belges*, 1968, **77,** 153.
69. L. Christiaens, R. Dufour, and M. Renson, *Bull. Soc. chim. belges*, 1970, **79,** 143.
70. A. Ruwet, J. Meessen, and M. Renson, *Bull. Soc. chim. belges*, 1969, **78,** 459.
71. E. G. Kataev, G. A. Chmutova, A. A. Musina and A. P. Anatas'eva, *Zhur. org. Khim.*, 1967, **3,** 597; G. A. Chmutova, *Chem. Abs.*, 1969, **70** 87 444.
72. D. L. Ross, J. Blane, and F. Matticoli, *J. Amer. Chem. Soc.*, 1970, **92,** 5750.
73. J. Gosselck, *Chem. Ber.*, 1958, **91,** 2345.

74. M. R. Padhye and J. C. Patel, *J. Sci. Ind. Res., India*, 1956, **15B,** 171; A. I. Kiss and B. R. Muth, *Acta Chim. Acad. Sci. Hung.*, 1957, **11,** 57.
75. N. P. Buu-Hoï, M. Mangane, M. Renson, and L. Christiaens, *J. Chem. Soc.* (*B*), 1969, 971.
76. N. M. Cullinane, A. G. Rees, and C. A. J. Plummer, *J. Chem. Soc.*, 1939, 151.
77. J. D. McCullough, T. W. Campbell, and E. S. Gould, *J. Amer. Chem. Soc.*, 1950, **72,** 5753.
78. C. Courtot and A. Monytamedi, *Compt. rend.*, 1934, **199,** 531.
79. J. Gaidis, *J. Org. Chem.*, 1970, **35,** 2811.
80. E. Sawicki and F. E. Ray, *J. Amer. Chem. Soc.*, 1952, **74,** 4120.
81. G. E. Wiseman and E. S. Gould, *J. Amer. Chem. Soc.*, 1955, **77,** 1061.
82. E. Sawicki and F. E. Ray, *J. Org. Chem.*, 1953, **18,** 946.
83. A. Magelli and R. Passerini, *Chem. Abs.*, 1958, **52,** 10 049.
84. N. Marziano, *Ricerca sci.*, 1960, **30,** 743.
85. L. C. Chierici and R. Passerini, *J. Chem. Soc.*, 1954, 3249.
86. S. C. Cohen, M. L. N. Reddy, and A. G. Massey, *Chem. Comm.*, 1967, 451.
87. W. J. Burlant and E. S. Gould, *J. Amer. Chem. Soc.*, 1954, **76,** 5775.
88. I. Degani, R. Fochi, and C. Vincenzi, *Gazzetta*, 1964, **94,** 203.
89. I. Degani, R. Fochi, and C. Vincenzi, *Boll. sci. Fac. Chim. ind. Bologna*, 1965, **23,** 21.
90. M. A. Kudinova, S. V. Krivun, and A. I. Tolmachev, *Khim. geterotsikl. Soedinenii*, 1973, 857.
91. K. W. Hubel and E. H. Braye, *U.S. Pat.* 3 280 017/1966.
92. G. T. Morgan and F. H. Burstall, *J. Chem. Soc.*, 1929, 2197.
93. G. Bergson, *Arkiv Kemi*, 1958, **13,** 11; J. R. Brown, G. P. Gillman, and M. H. George, *J. Polymer Sci., Part A-1*, 1967, **5,** 903.
94. G. T. Morgan and F. H. Burstall, *J. Chem. Soc.*, 1931, 173.
95. J. B. Lambert, D. H. Johnson, R. G. Keske, and C. E. Mixan, *J. Amer. Chem. Soc.*, 1972, **94,** 8172.
96. J. B. Lambert, C. E. Mixan, and D. H. Johnson, *Tetrahedron Letters*, 1972, 4355; *J. Amer. Chem. Soc.*, 1973, **95,** 4634.
97. A. Fredga and K. Styrman, *Arkiv Kemi*, 1959, **14,** 461.
98. W. Hänsel and R. Haller, *Naturwiss.*, 1968, **55,** 83; *Arch. Pharm.*, 1969, **302,** 147.
99. W. Ziriakus, W. Hänsel, and R. Haller, *Arch. Pharm.*, 1971, **304,** 681.
100. I. Degani, R. Fochi, and C. Vincenzi, *Gazzetta*, 1964, **94,** 451.
101. M. Renson, *Bull. Soc. chim. belges*, 1964, **73,** 483.
102. I. Degani and R. Fochi, *Ann. chim.* (*Rome*), 1968, **58,** 251.
103. A. Tadino, L. Christiaens, and M. Renson, *Bull. Soc. Roy. Sci. Liege*, 1973, **42,** 146.
104. A. Tadino, L. Christiaens, and M. Renson, *Bull. Soc. Roy. Sci. Liege*, 1973, **42,** 129.
105. A. Ruwet and M. Renson, *Bull. Soc. chim. belges*, 1968, **77,** 465.
106. A. Ruwet and M. Renson, *Bull. Soc. chim. belges*, 1969, **78,** 449.
107. E. Ziegler and E. Nölken, *Monatsh.*, 1958, **89,** 737.
108. A. Ruwet, C. Draguet, and M. Renson, *Bull. Soc. chim. belges*, 1970, **79,** 539.
109. F. Bossert, *Angew. Chem.*, 1965, **77,** 913; A. Ruwet and M. Renson, *Bull. Soc. chim. belges*, 1966, **75,** 260.
110. A. Ruwet, D. Janne, and M. Renson, *Bull. Soc. chim. belges*, 1970, **79,** 81.
111. N. Bellinger, D. Cagniant, and P. Cagniant, *Tetrahedron Letters*, 1971, 49.
112. J. Gosseck and E. Wolters, *Chem. Ber.*, 1962, **95,** 1237.
113. M. Renson and P. Pirson, *Bull. Soc. chim. belges*, 1966, **75,** 456.
114. K. Šindelář, E. Svátek, J. Metyšová, J. Meytš, and M. Protiva, *Coll. Czech. Chem. Comm.*, 1969, **34,** 3792.
115. B. R. Muth, *Chem. Ber.*, 1960, **93,** 283.
116. M. Renson and L. Christiaens, *Bull. Soc. chim. belges*, 1970, **79,** 511.
117. M. Hori, T. Kataoka, and C-F. Hsu, *Chem. Pharm. Bull.* (*Japan*), 1974, 15.
118. R. Lesser and R. Weiss, *Ber.*, 1924, **57B,** 1077.
119. I. Degani, R. Fochi, and G. Spunta, *Boll. sci. Fac. Chim. ind. Bologna*, 1965, **23,** 165.
120. K. Šindelář, J. Metyšová, and M. Protiva, *Coll. Czech. Chem. Comm.*, 1969, **34,** 3801.
121. N. Bellinger, P. Cagniant, and M. Renson, *Compt. rend.* (*C*), 1969, **269,** 532.
122. K. Šindelář, J. Metyšová, E. Svátek, and M. Protiva, *Coll. Czech. Chem. Comm.*, 1969, **34,** 2122.
123. W. E. Truce and D. D. Emrick, *J. Amer. Chem. Soc.*, 1956, **78,** 6130.
124. F. Fringuelli and A. Taticchi, *J. C. S. Perkin I*, 1972, 199.
125. W. Mack, *Angew. Chem.*, 1965, **77,** 260; *Angew. Chem. Internat. Edn.*, 1966, **5,** 986.
126. J. M. Read, Jr., C. T. Mathis, and J. H. Goldstein, *Spectrochim. Acta*, 1965, **21,** 85.
127. R. J. Abraham and H. J. Bernstein, *Canad. J. Chem.*, 1961, **39,** 905.
128. Y. Pascal, J. P. Morizur, and J. Wiemann, *Bull. Soc. chim. France*, 1965, 2211; J. P. Morizur and Y. Pascal, *ibid.*, 1966, 2296; A. R. Katritzky, 'Physical Methods in Heterocyclic Chemistry,' vol. II, Academic Press, New York, 1963.
129. W. V. Farrar and J. M. Gulland, *J. Chem. Soc.*, 1945, 11.
130. E. P. Mazza and E. Melchionna, *Chem. Abs.*, 1929, **23,** 2955.
131. I. D. Sadekov and V. I. Minkin, *Khim. geterotsikl. Soedinenii*, 1971, **7,** 138.
132. J. L. Piette and M. Renson, *Bull. Soc. chim. belges*, 1971, **80,** 521.
133. N. P. Buu-Hoï, M. Mangane, M. Renson, and J. L. Piette, *J. Heterocyclic Chem.*, 1970, **7,** 219.
134. R. Passerini and G. Purrello, *Ann. chim.* (*Rome*), 1958, **48,** 738.

135. C. Courtot and M. G. Bastani, *Compt. rend.*, 1936, **203,** 197.
136. G. T. Morgan and H. Burgess, *J. Chem. Soc.*, 1928, 321.
137. D. H. Dewar, J. E. Fergusson, P. R. Hentschel, C. J. Wilkins, and P. P. Williams, *J. Chem. Soc.*, 1964, 688.
138. J. L. Piette and M. Renson, *Bull. Soc. chim. belges*, 1971, **80,** 669.
139. R. I. Wagner, L. D. Freeman, H. Goldwhite and D. G. Rowsell, *J. Amer. Chem. Soc.*, 1967, **89,** 1102.
140. E. W. Koos, J. P. Van der Kooi, E. E. Green, and J. K. Stille, *J. C. S. Chem. Comm.*, 1972, 1085.
141. J. J. McBride, Jr., E. Jungermann, J. V. Killheffer, and R. J. Clutter, *J. Org. Chem.*, 1962, **27,** 1833.
142. S. E. Cremer, R. J. Chorvat, C. H. Chang, and D. W. Davis, *Tetrahedron Letters*, 1968, 5799.
143. S. E. Cremer and R. J. Chorvat, *J. Org. Chem.*, 1967, **32,** 4066.
144. S. E. Fishwick, J. Flint, W. Hawes, and S. Trippett, *Chem. Comm.*, 1967, 1113.
145. S. E. Cremer and R. J. Chorvat, *Tetrahedron Letters*, 1968, 413.
146. G. Märkl and R. Potthast, *Angew. Chem. Internat. Edn.*, 1967, **6,** 86.
147. I. G. M. Campbell, R. C. Cookson, M. B. Hocking, and A. N. Hughes, *J. Chem. Soc.*, 1965, 2184.
148. F. C. Leavitt, T. A. Manuel, F. Johnson, L. U. Matternas, and D. S. Lehman, *J. Amer. Chem. Soc.*, 1960, **82,** 5099.
149. E. H. Braye and W. Hübel, *Chem. and Ind.* (*London*), 1959, 1250.
150. N. E. Waite and J. C. Tebby, *J. Chem. Soc.* (*C*), 1970, 386.
151. U. Hasserodt, K. Hunger, and F. Korte, *Tetrahedron*, 1963, **19,** 1563.
152. B. A. Arbuzov, Yu. Yu. Samitov, A. O. Vizel, and T. V. Zykova, *Doklady Akad. Nauk S.S.S.R.*, 1964, **159,** 1062.
153. D. A. Usher, and F. H. Westheimer, *J. Amer. Chem. Soc.*, 1964, **86,** 4732.
154. U. Hasserodt, K. Hunger, and F. Korte, *Tetrahedron*, 1964, **20,** 1593.
155. B. A. Arbuzov, A. O. Vizel, Yu. Yu. Samitov, and Yu. F. Tarenko, *Izvest. Akad. Nauk S.S.S.R., Ser. khim.*, 1967, 672; B. A. Arbuzov and A. O. Vizel, *ibid.*, 1969, 460.
156. G. Märkl and R. Potthast, *Tetrahedron Letters*, 1968, 1755.
157. A. B. Burg and P. J. Slota, Jr., *J. Amer. Chem. Soc.*, 1960, **82,** 2148.
158. D. J. Collins, L. E. Rowley, and J. M. Swan, *Austral. J. Chem.*, 1974, **27,** 831.
159. M. R. Rausch and L. P. Klemann, *J. Amer. Chem. Soc.*, 1967, **89,** 5732.
160. I. G. M. Campbell and J. K. Way, *J. Chem. Soc.*, 1961, 2133.
161. A. J. Ashe, *J. Amer. Chem. Soc.*, 1971, **93,** 3293.
162. G Märkl, *Angew. Chem.*, 1966, **78,** 907.
163. K. Dimroth and W. Mack, *Angew. Chem. Internat. Edn.*, 1968, **7,** 460.
164. K. Dimroth, K. Vogel, W. Mack, and U. Schoeler, *Angew. Chem. Internat. Edn.*, 1968, **7,** 371.
165. H. Kanter and K. Dimroth, *Angew. Chem. Internat. Edn.*, 1972, **11,** 1090.
166. K. Dimroth and W. Städe, *Angew. Chem. Internat. Edn.*, 1968, **7,** 881.
167. W. Schäfer and K. Dimroth, *Tetrahedron Letters*, 1972, 843.
168. W. Schäfer and K. Dimroth, *Angew. Chem. Internat. Edn.*, 1973, **12,** 753.
169. G. Märkl and A. Merz, *Tetrahedron Letters*, 1968, 3611.
170. G. Märkl and F. Lieb, *Angew. Chem. Internat. Edn.*, 1968, **7,** 733.
171. L. B. Lambert and W. L. Oliver, Jr., *Tetrahedron*, 1971, **27,** 4245.
172. M. H. Beeby and F. G. Mann, *J. Chem. Soc.*, 1951, 411.
173. G. A. Dilbeck, D. L. Morris, and K. D. Berlin, *J. Org. Chem.*, 1975, **40,** 1150.
174. G. Märkl and K.-H. Heier, *Angew. Chem. Internat. Edn.*, 1972, **11,** 1016.
175. G. Märkl and K.-H. Heier, *Angew. Chem. Internat. Edn.*, 1972, **11,** 1017.
176. P. de Koe and F. Bickelhaupt, *Angew. Chem. Internat. Edn.*, 1967, **6,** 567.
177. H. Vermeer and F. Bickelhaupt, *Angew. Chem. Internat. Edn.*, 1968, **7,** 889.
178. P. de Koe, R. van Veen, and F. Bickelhaupt, *Angew. Chem. Internat. Edn.*, 1968, **7,** 465.
179. W. Schäfer, A. Schweig, F. Bickelhaupt, and H. Vermeer, *Rec. Trav. chim.*, 1974, **93,** 17.
180. G. Märkl and H. Hauptmann, *Tetrahedron Letters*, 1968, 3257.
181. G. Märkl and H. Hauptmann, *Angew. Chem. Internat. Edn.*, 1972, **11,** 439.
182. G. Märkl and H. Hauptmann, *Angew. Chem. Internat. Edn.*, 1972, **11,** 441.
183. P. Jutzi, *Angew. Chem. Internat. Edn.*, 1975, **14,** 232.
184. J. J. Monagle, *J. Org. Chem.*, 1962, **27,** 3851.
185. A. Tzschach and W. Fischer, *Z. Chem.*, 1967, **7,** 196.
186. E. R. H. Jones and F. G. Mann, *J. Chem. Soc.*, 1958, 1719.
187. D. R. Lyon, F. G. Mann, and G. H. Cookson, *J. Chem. Soc.*, 1947, 662.
188. J. A. Aeschlimann, N. D. Lees, N. P. McCleland and G. N. Nicklin, *J. Chem. Soc.*, 1925, **127,** 66.
189. H. Heaney, D. M. Heinekey, F. G. Mann, and I. T. Millar, *J. Chem. Soc.*, 1958, 3838.
190. D. M. Heinekey and I. T. Millar, *J. Chem. Soc.*, 1959, 3101.
191. G. Märkl, H. Hauptmann, and J. Advena, *Angew. Chem. Internat. Edn.*, 1972, **11,** 441.
192. G. Märkl and F. Kneidl, *Angew. Chem. Internat. Edn.*, 1973, **12,** 931.
193. G. Märkl, J. Advena, and H. Hauptmann, *Tetrahedron Letters*, 1972, 3961.
194. A. J. Ashe, *J. Amer. Chem. Soc.*, 1971, **93,** 6691.
195. A. J. Ashe and M. D. Gordon, *J. Amer. Chem. Soc.*, 1972, **94,** 7596.
196. G. J. Burrows and E. E. Turner, *J. Chem. Soc.*, 1921, **119,** 430.
197. F. G. Holliman and F. G. Mann, *J. Chem. Soc.*, 1943, 547.
198. D. A. Thornton, *J. S. Afr. Chem. Inst.*, 1964, **17,** 71.

199. P. Jutzi and K. Deuchert, *Angew. Chem. Internat. Edn.*, 1969, **8,** 991; H. Vermeer and F. Bickelhaupt, *ibid.*, p. 992.
200. G. H. Cookson and F. G. Mann, *J. Chem. Soc.*, 1949, 2888.
201. N. S. Nametkin, V. M. Vdovin, and P. L. Grinberg, *Doklady Akad. Nauk S.S.S.R.*, 1963, **150,** 799.
202. M. C. Flowers and L. E. Gusel'nikov, *J. Chem. Soc.* (*B*), 1968, 419.
203. R. West, *J. Amer. Chem. Soc.*, 1954, **76,** 6012.
204. R. A. Benkeser, R. F. Grossman, and G. M. Stanton, *J. Amer. Chem. Soc.*, 1962, **84,** 4732.
205. R. A. Benkeser, R. F. Grossman, and G. M. Stanton, *J. Amer. Chem. Soc.*, 1962, **84,** 4727.
206. E. G. Janzen, J. B. Pickett, and W. H. Atwell, *J. Organometallic Chem.*, 1967, **10,** P6; W. H. Atwell, D. R. Weyenberg, and H. Gilman, *J. Org. Chem.*, 1967, **32,** 885.
207. W. H. Atwell and D. R. Weyenberg, *J. Amer. Chem. Soc.*, 1968, **90,** 3438.
208. H. Gilman and O. L. Marrs, *Chem. and Ind.* (*London*), 1961, 208.
209. N. S. Nametkin, V. M. Vdovin, E. Sh. Finkel'shtein, T. N. Arkhipov, and V. D. Oppengeim, *Doklady Akad. Nauk S.S.S.R.*, 1964, **154,** 383.
210. H. Gilman and R. D. Gorsich, *J. Amer. Chem. Soc.*, 1958, **80,** 1883.
211. H. Gilman and R. D. Gorsich, *J. Amer. Chem. Soc.*, 1958, **80,** 3243.
212. A. Bygdén, *Ber.*, 1915, **48,** 1236; G. Gruttner and M. Wiernik, *ibid.*, p. 1474.
213. N. S. Nametkin, V. M. Vdovin, E. D. Babich, and T. N. Arkhipova, *Doklady Akad. Nauk S.S.S.R.*, 1966, **171,** 1345.
214. R. Fessenden and M. D. Coon, *J. Org. Chem.*, 1961, **26,** 2530.
215. R. J. Fessenden and W. D. Kray, *J. Org. Chem.*, 1973, **38,** 87.
216. R. West, *J. Amer. Chem. Soc.*, 1954, **76,** 6015.
217. R. Fessenden and F. J. Freenor, *J. Org. Chem.*, 1961, **26,** 2003.
218. D. Seyferth and S. S. Washburne, *J. Organometallic Chem.*, 1966, **5,** 389.
219. W. P. Weber, R. A. Felix, A. K. Willard, and H. G. Boettger, *J. Org. Chem.*, 1971, **36,** 4060.
220. W. P. Weber and R. Laine, *Tetrahedron Letters*, 1970, 4169.
221. R. Benkeser and R. F. Cunico, *J. Organometallic Chem.*, 1965, **4,** 284.
222. A. G. Brook and J. B. Pierce, *J. Org. Chem.*, 1965, **30,** 2566.
223. E. Rosenberg and J. J. Zuckerman, *J. Organometallic Chem.*, 1971, **33,** 321.
224. G. Märkl and P. L. Merz, *Tetrahedron Letters*, 1971, 1303.
225. H. Gilman and O. L. Marrs, *J. Org. Chem.*, 1965, **30,** 324.
226. P. Jutzi, *Chem. Ber.*, 1971, **104,** 1455.
227. R. A. Benkeser and R. F. Cunico, *J. Org. Chem.*, 1967, **32,** 395.
228. P. Mazerolles, M. Lesbre, and J. Dubac, *Compt. rend.*, 1965, **260,** 2255.
229. D. Seyferth, S. S. Washburne, T. F. Julia, P. Mazerolles, and J. Dubac, *J. Organometallic Chem.*, 1969, **16,** 503.
230. P. Mazerolles, *Bull. Soc. chim. France*, 1962, 1907.
231. P. Mazerolles and G. Manuel, *Bull. Soc. chim. France*, 1966, 327.
232. P. Mazerolles and G. Manuel, *Compt. rend.* (*C*), 1968, **267,** 1158.
233. D. Seyferth, T. F. Julia, D. C. Mueller, P. Mazerolles, G. Manuel, and F. Thoumas, *J. Amer. Chem. Soc.*, 1970, **92,** 657.
234. F. J. Bajer and H. W. Post, *J. Org. Chem.*, 1962, **27,** 1422.
235. D. Seyferth, H.-M. Shih, P. Mazerolles, M. Lesbre, and M. Joanny, *J. Organometallic Chem.*, 1971, **29,** 371.
236. R. Schwarz and W. Reinhardt, *Ber.*, 1932, **65,** 1743.
237. G. Grüttner and E. Krause, *Ber.*, 1916, **49,** 2666.
238. F. J. Bajer and H. W. Post, *J. Organometallic Chem.*, 1968, **11,** 187.
239. A. J. Ashe and P. Shu, *J. Amer. Chem. Soc.*, 1971, **93,** 1804.
240. I. M. Gverdtsiteli, T. P. Doksopulo, M. M. Menteshashvili, and I. I. Abhazavo, *Chem. Abs.*, 1960, **64,** 11 239.
241. J. J. Eisch and W. C. Kaska, *J. Amer. Chem. Soc.*, 1962, **84,** 1501.
242. R. Koester and G. Benedikt, *Angew. Chem.*, 1963, **75,** 419.
243. P. Jutzi, *Angew. Chem. Internat. Edn.*, 1972, **11,** 53.
244. G. A. Olah, J. M. Bollinger, and J. Brinich, *J. Amer. Chem. Soc.*, 1968, **90,** 2587.
245. P. E. Peterson, B. R. Bonazza, and P. M. Henrichs, *J. Amer. Chem. Soc.*, 1973, **95,** 2222.
246. J. D. Edwards, Jr., and J. J. Cashaw, *J. Amer. Chem. Soc.*, 1956, **78,** 3821.

PART 20

MIXED HETEROATOM SYSTEMS

20.1
Five-membered Ring Systems

M. M. CAMPBELL
Heriot-Watt University, Edinburgh

20.1.1 ONE NITROGEN, ONE HETEROATOM: OXAZOLES AND THEIR BENZO-FUSED ANALOGUES

Oxazoles, benzoxazoles and related systems have been the subject of major reviews[1–5] and texts.[6] Important aspects of their chemistry will therefore be summarized and recent developments discussed. Oxazole (**1**) and benzoxazole (**2**) are numbered as shown.

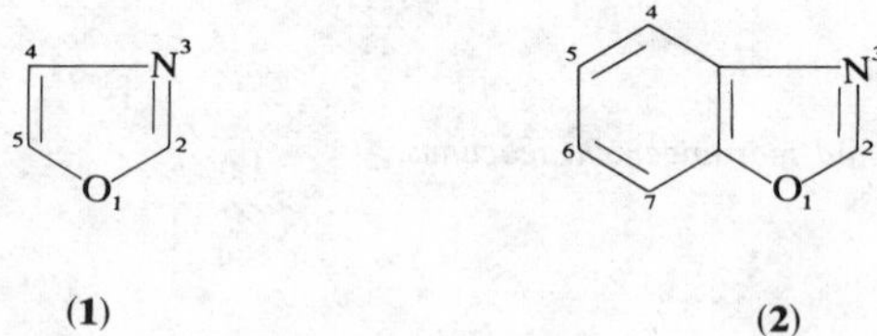

(**1**) (**2**)

The azoles are planar molecules with conjugated π-electron sextets in the cyclic system, as in the parent molecule, oxazole (**1**). Their chemical properties are aromatic in character, as will be seen from their reactions with, for example, certain electrophiles and nucleophiles. The lone-pair of electrons on nitrogen, which is coplanar with the heterocyclic ring and therefore not involved in delocalization, confers weakly basic properties as in pyridine. Many of the salts are hydrolysed by water, and stable complexes are formed with heavy-metal ions, which are often useful for isolating azoles. Associated with the aromaticity is a degree of stability, but when the oxazoles and benzoxazoles are quaternized the resulting azolium species are significantly activated towards nucleophilic attack.

20.1.1.1 Physical and spectroscopic properties

Oxazole is a liquid (b.p. 69 °C). The physical properties of it and related azoles have been summarized.[6b] The physical and structural properties of oxazoles, including X-ray crystallographic and bond length values, theoretical studies, and dipole moments have been reviewed in detail.[4]

Diagnostically-useful spectroscopic characteristics have been reviewed,[4] including n.m.r. chemical shifts, aromatic solvent-induced chemical shifts, J values and ^{14}N spectroscopy.[4,7] The u.v. spectra are similar to the related thiazoles. Oxazole (**1**) absorbs at 205 nm (log ε 3.59). Whereas the monophenyl derivatives absorb in the 245–270 nm region, alkyl substituents do not affect the parent chromophore so significantly.[4] In the i.r. spectrum a strong band at 1555–1585 cm^{-1} is attributed to an —N═C—O— ring-stretching mode.[4] The mass spectral fragmentation properties have been extensively studied[8] with the aid of deuterium-labelling techniques and high-resolution mass spectrometry.[4] In certain cases interesting correlations with photochemical behaviour were observed.[8]

20.1.1.2 Electrophilic substitution

Oxazoles do not readily undergo electrophilic substitution reactions, although many are stable to acidic conditions. The presence of the pyridine-like nitrogen in the azole ring considerably deactivates the system. A consideration of the transition states in electrophilic substitution at C-2, C-4, and C-5 (equations 1–3) indicates that the C-2 intermediate (equation 1) is highly unfavourable.

C-2 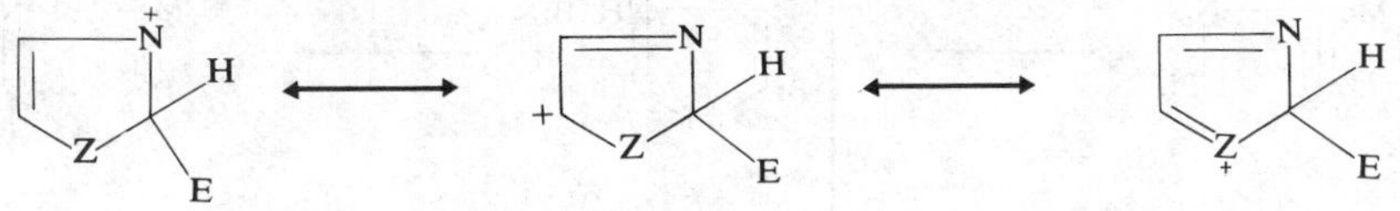 (1)

C-4 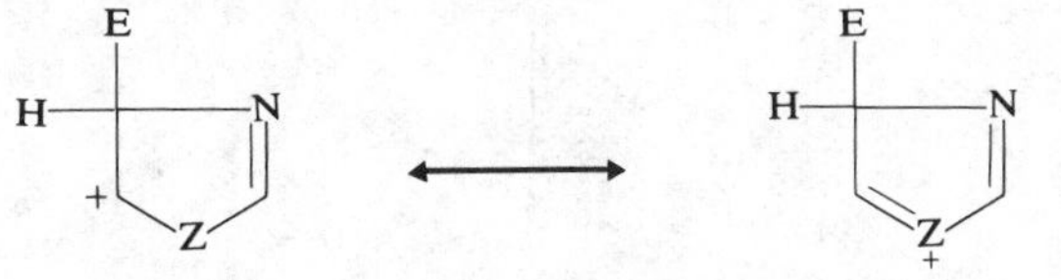 (2)

C-5 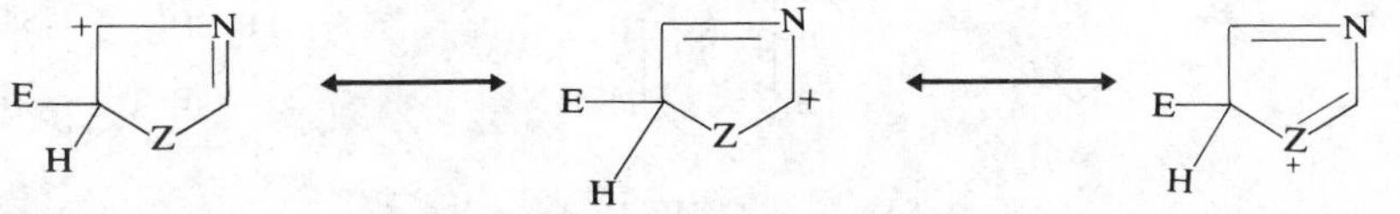 (3)

Electrophilic attack at C-4 and C-5 may occur when the ring is activated by $+M$ substituents such as amino- and hydroxy-groups. The 2-dimethylamino-oxazole (**3**), for example, affords the 5-nitro-product (**4**) (reaction 4). Examples of bromination using either bromine or N-bromosuccinimide are known in which substitution occurs at the

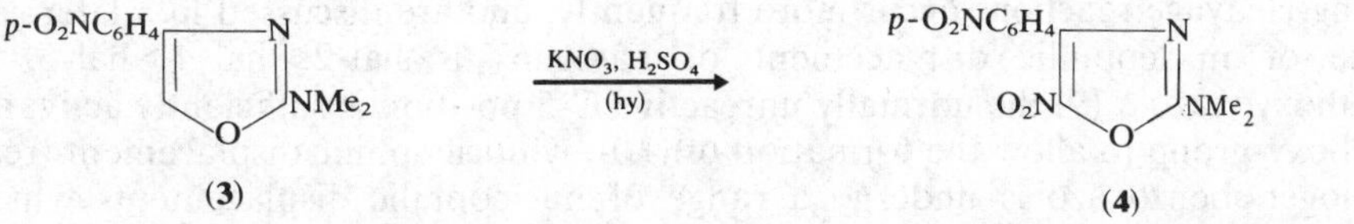

p-$O_2NC_6H_4$ — NMe_2 (**3**) $\xrightarrow{KNO_3,\ H_2SO_4\ (hy)}$ p-$O_2NC_6H_4$ — O_2N — NMe_2 (**4**) (4)

vacant 4- or 5-position in disubstituted oxazoles (Scheme 1).[6c] An intramolecular electrophilic attack is presumably involved in the Eschweiler–Clarke conversion of (**5**) into (**6**),[4] and the transformation of (**7**) into (**8**)[4] (reactions 5 and 6). It should be noted that electrophilic substitution of aryl-substituted oxazoles may occur in the aryl ring; phenyloxazoles, for example, give p-nitrophenyloxazoles.[4]

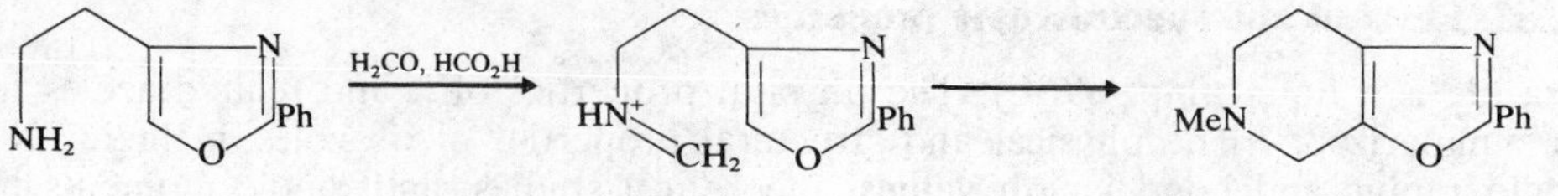

(5) $\xrightarrow{H_2CO,\ HCO_2H}$ → (6) (5)

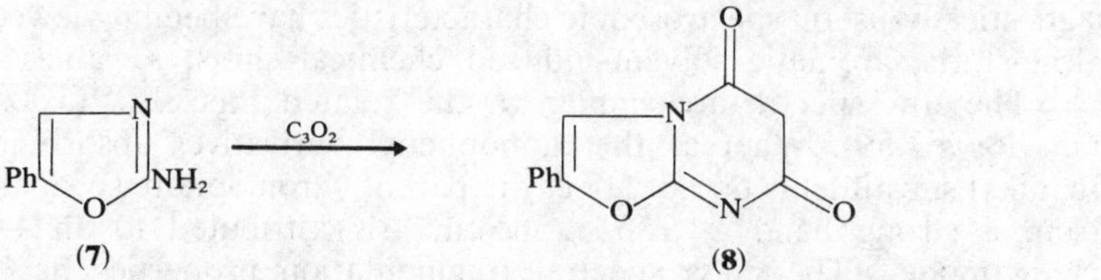

(7) $\xrightarrow{C_3O_2}$ (8) (6)

Benzoxazoles tend to react mainly at C-6 in electrophilic substitutions and to a lesser extent at C-5. Nitration of (**2**) affords the 6-nitro-product. Phenyl substituents may preferentially undergo electrophilic substitution at the *p*-position.[6b]

Electrophilic attack on the ring nitrogen leading to quaternized oxazoles will be covered separately.

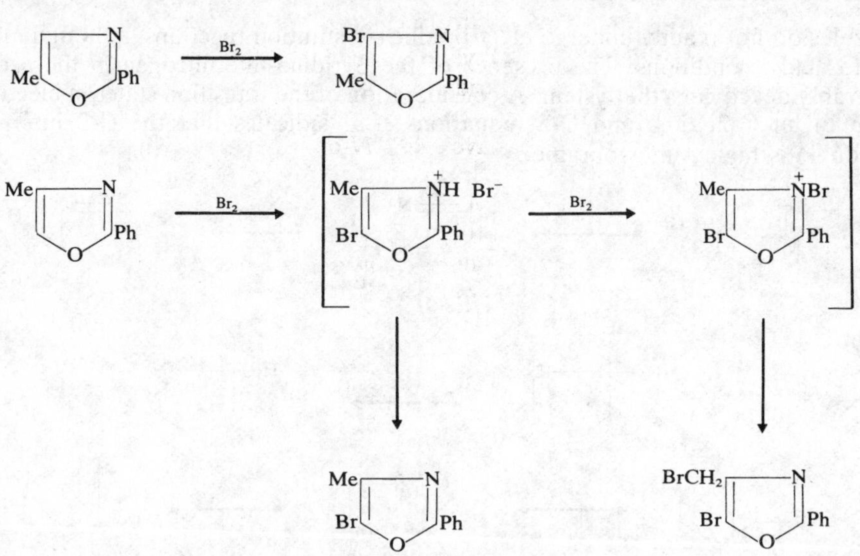

SCHEME 1

20.1.1.3 Nucleophilic substitution

Nucleophilic substitution reactions on the oxazole ring are relatively uncommon. (Ring-cleavage reactions occur more frequently, and are discussed in a later section.) The ease of nucleophilic displacement of halogens is hal-2>hal-4>hal-5.[4] In the 5-methoxyoxazole (**9**) the normally unreactive C-5 position is sufficiently activated by the 4-carboxy-group to allow the formation of (**10**) by nucleophilic displacement (reaction 7). 2-Halogenobenzoxazoles undergo a range of nucleophilic displacements which are summarized in Scheme 2.[3]

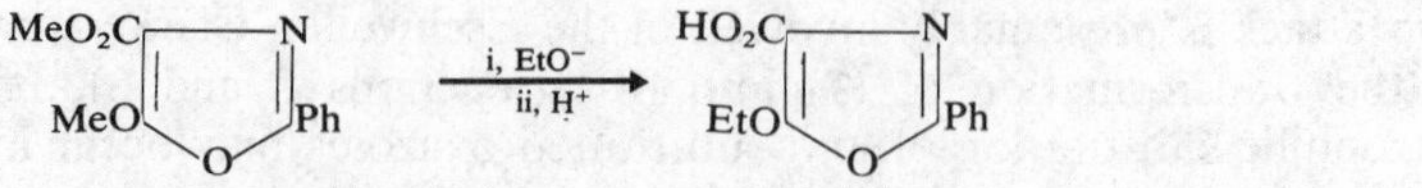

(9) $\xrightarrow{\text{i, EtO}^-;\ \text{ii, H}^+}$ (10) (7)

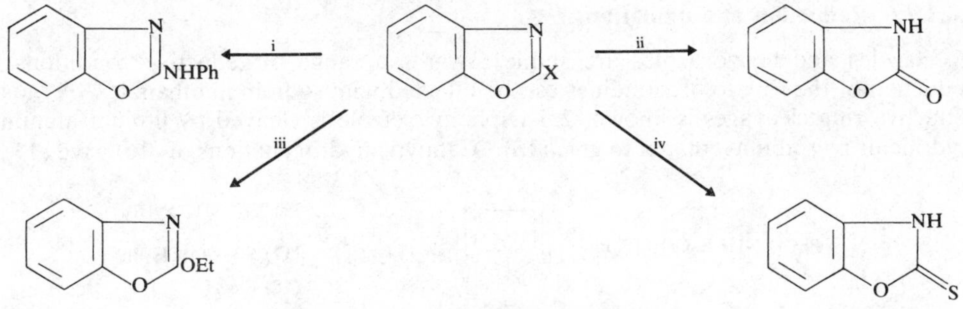

i, $PhNH_2$; ii, H_2O; iii, EtOH, KOH; iv, NaSH.

SCHEME 2

20.1.1.4 Quaternization

Oxazoles are weak bases, as is shown by the instability of their *N*-protonated salts.[4] 4-Methyloxazole ($pK_{BH^+} = -1.07$) is approximately 100 times less basic than 4-methylimidazole. A 2-methyl substituent increases the pK_a more than a 4-methyl substituent, presumably because of hyperconjugative effects. A comparison of oxazole basicity relative to pyridines emerged from alkylation studies of 2-pyridyloxazoles (which give pyridinium adducts) and 4- and 5-pyridyloxazoles (which give diquaternary salts, the pyridine being alkylated first).

A range of azoles and benzazoles have been *N*-alkylated using dimethyl sulphate or methyl iodide in dimethyl sulphoxide.[9] Difficulty was experienced in alkylating compounds of type (**11**) with either dimethyl sulphate or benzyl bromide, but quaternization was achieved with triethyloxonium fluoroborate, giving (**12**).[4,10] Benzoxazole (**2**) quaternizes to give the methiodide, but under more vigorous conditions may suffer ring cleavage (Section 20.1.1.10).

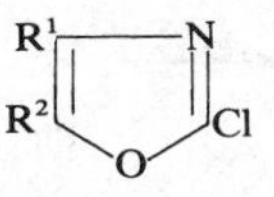

(**11**) 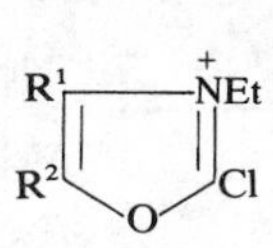 (**12**)

20.1.1.5 Proton exchange

Oxazole undergoes instantaneous proton exchange at C-2 and slower proton exchange at C-5 when treated with [$^2H_6$]dimethyl sulphoxide and sodium methoxide[4] (reaction 8). Following Breslow's important work on the ionization of the 2-proton in thiazolium cations (which is critical for the mechanism of thiamine action),[11] the acidities and exchange rates of the labile 2- and 5-protons in oxazolium species and their free bases have been

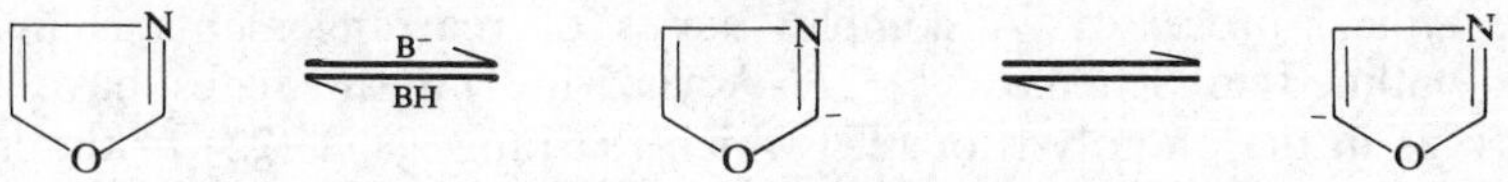

(8)

studied.[4] The relative rates of H/D exchange at C-2 in (**13**) and (**14**), for example, are $10^{5.5}$ and $10^{3.5}$, respectively, relative to the corresponding imidazolium species, illustrating that the more electronegative oxygen enhances acidity at C-2.[4,6c] The unusually high ^{13}C–H coupling constants for the oxazolium compounds reflect the high acidity of 2-H and 5-H.

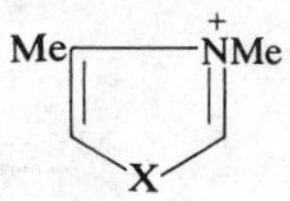

(**13**) X = O
(**14**) X = S

20.1.1.6 Reduction and oxidation

Oxazoles and benzoxazoles are stable towards a range of reductive conditions, but reduction of the ring to oxazolidines can be effected with sodium in ethanol.[4,6c] A range of reductive ring cleavages is known. 2,5-Diphenyloxazole is cleaved by lithium aluminium hydride or by sodium–ethanol to give (**15**).[4] Catalytic hydrogenations of (**16**) gave (**17**) and

$PhCH_2NHCH_2CH(OH)Ph$ (**15**)

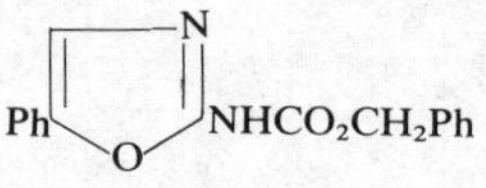

(**16**) [oxazole: Ph, $NHCO_2CH_2Ph$]

$Ph(CH_2)_2NHCONHR$
(**17**) R = CO_2CH_2Ph
(**18**) R = H

(**18**), but Raney nickel was ineffective.[4] Reductive ring cleavage was inhibited by base, but enhanced by acid. Other reductive ring cleavages, including the electrochemical reduction of (**19**) (reaction 9), have been reviewed.[4]

The oxazole ring is not normally stable to oxidative conditions, cold permanganate, chromic acid, sodium hydrobromite, fuming nitric acid, and ozone cleaving most oxazole derivatives.[4] Oxazoles are normally stable to hydrogen peroxide, but 4-(2-oxalyl)pyridine with H_2O_2–HOAc gives the pyridine *N*-oxide.[4] Photosensitized singlet oxygen addition to (**20**) gives the ring-cleaved product (**21**) (reaction 10). Labelling studies confirm that a 1,4- rather then a 1,2-cycloaddition pathway is involved. These ring-cleavage reactions have been investigated in a range of oxazoles, and may be of practical utility in, for example, the synthesis of mixed triamides.[13]

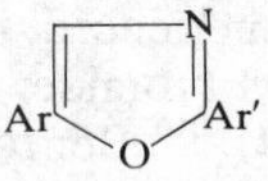

(**19**) [2-Ar′-5-Ar-oxazole] + $6e^- + 6H^+$ ⟶ $Ar(CH_2)_2NHCH(OH)Ar'$ (9)

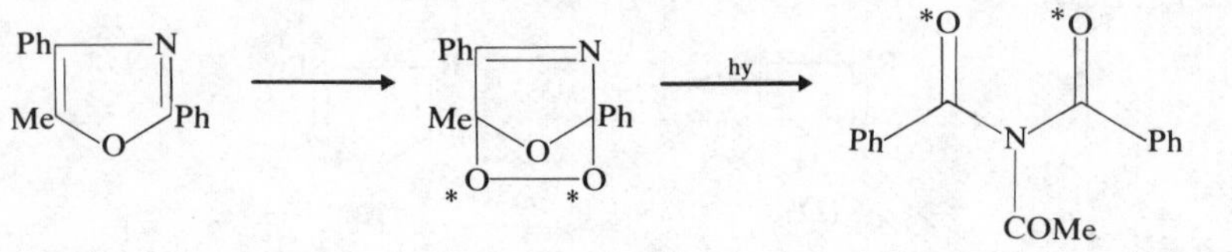

(**20**) [4-Ph-5-Me-2-Ph-oxazole] ⟶ [endoperoxide, O*–O*] $\xrightarrow{h\nu}$ (**21**) PhC(=*O)–N(COMe)–C(=*O)Ph (10)

20.1.1.7 Photochemistry

A recent text reviews the photochemistry of heterocycles, including the oxazoles.[14] Photoisomerization of the oxazole ring and photochemical interconversions with isoxazoles are typical processes. A comples series of rearrangements is involved in the photolyses outlined in Scheme 3.[14] 2*H*-Acylazirine intermediates have been isolated (Scheme 4), as in the photolysis of (**22**) which rearranges *via* (**23**) to the isoxazole (**24**). The isomeric oxazole (**27**) may result from intermediates such as (**25**) and (**26**). Other related studies further clarify this area.[15] 2-Aroylbenzoxazoles undergo photo-Fries rearrangements.[14] The first report of photo-alkylative addition to the —O—C═N— group in a heterocycle (phenanthro[9,10-*d*]oxazole) has appeared, giving in low yields the 2-alkylated products (**28**).[16]

20.1.1.8 Properties of substituents

The hydroxyoxazoles are capable of tautomerism and four ketonic forms are possible, the 2-oxazolin-5-one, 3-oxazolin-5-ones, 4-oxazolin-2-ones, and 2-oxazolin-4-ones.

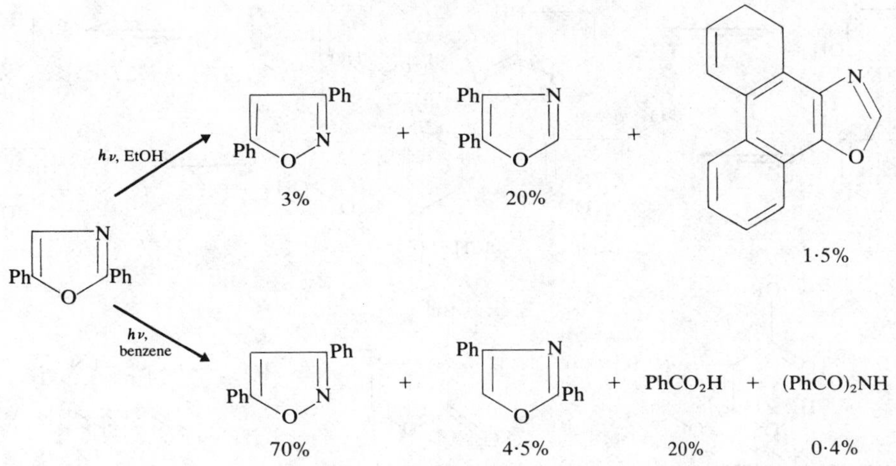

SCHEME 3

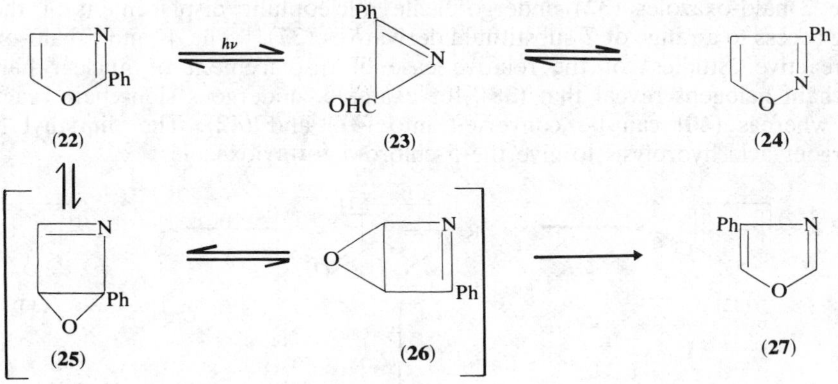

(22) (23) (24) (25) (26) (27)

SCHEME 4

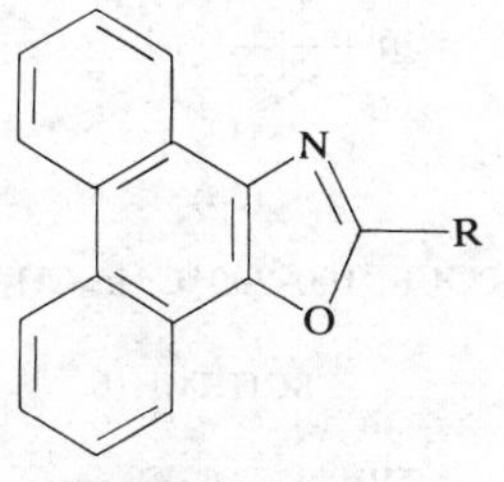

(28) R = Me, Pr^i, *etc.*

These compounds have been reviewed.[17] In general, the 2-hydroxyazoles exist predominantly in the 2-keto form (Scheme 5). Hydroxy-forms can, however, be stabilized by intramolecular hydrogen-bonding, as in (**29**).[6b] Spectroscopic data often reveal the presence of the NH and C=O groups in the oxo-forms. In the case (**30**) the compound is isolable as an optically active form.[18] The oxazolone (**31**) was shown to exist as such by ^{1}H n.m.r. spectroscopy.

Typical transformations of 2-hydroxyoxazoles are summarized in Scheme 6.[6b] The phosphorus oxychloride route to (**32**) affords synthetically-useful intermediates which cannot be made by electrophilic chlorination because of the relative lack of reactivity at C-2. Products (**33**) and (**34**) illustrate differences in reactivity of the tautomeric forms of the 2-hydroxyazoles. Benzoxazolone is *N*-alkylated by ethyl iodide and by diazomethane, affording (**35**) and (**36**).[3,18]

SCHEME 5

(29) (30) (31)

The 2-halo-oxazoles (**32**) undergo facile nucleophilic displacement of the halogen, giving access to a range of 2-substituted derivatives (**37**).[6b] The 4- and 5-halo-oxazoles are less reactive. Studies[4] of the relative ease of displacement of nuclear halogens and side-chain halogens reveal that (**38**), for example, undergoes side-chain reaction giving (**39**), whereas (**40**) can be converted into (**41**) and (**42**). The dimethyl acetal (**41**) undergoes acid hydrolysis to give the 5-chloro-4-formyloxazole.

(32) (33) (34)

i, $POCl_3$, 100–200 °C; ii, CH_2N_2; iii, MeI

SCHEME 6

The amino-oxazoles exist as the amine tautomers. For example, structures (**43**) and (**44**) have been established by u.v. spectroscopy.[18] They protonate on the ring nitrogen, and 2-aminobenzoxazole reacts with methyl iodide at 100 °C to give the N^3-alkylated product (**45**).[6d]

2-Mercapto-oxazoles exist predominantly in the thiono-form (**46**), as shown spectroscopically.[18] Similarly, 2-mercaptobenzoxazoles exist as (**47**). Alkylation of (**47**) can occur at sulphur.[2]

C-2 Alkyl groups in oxazoles and benzoxazoles are reactive, and readily afford a resonance-stabilized anion. 2-Methyloxazoles and 2-methylbenzoxazole react with, for example, benzaldehyde–zinc chloride to give the condensation products (**48**) and (**49**), respectively.[6b,c,d] 2-Benzylbenzoxazole is reactive enough to couple with diazonium salts, and will react at the 2-methylene group with aldehydes, nitroso-compounds, and amyl nitrate.[3] The intermediate hydroxy-compound in the condensation of 2-methylbenzoxazole with aryl aldehydes is isolable when NaOH is used in a polar solvent such as

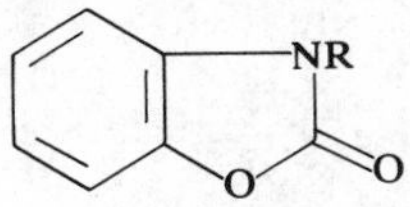

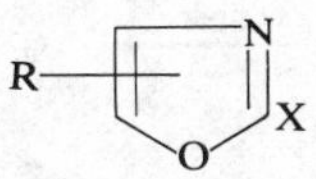

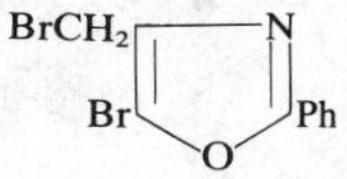

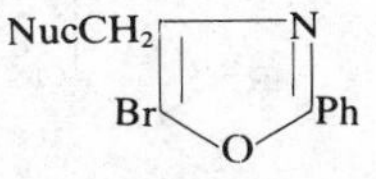

(**35**) R = Et
(**36**) R = Me

(**37**) R = OR^1, SR^1, NR^1_2

(**38**)

(**39**) Nuc = H_2NCSNH, PhNH, OEt

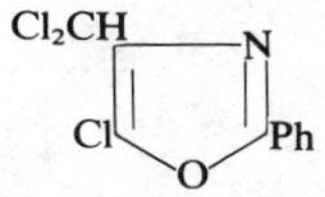

(**40**)

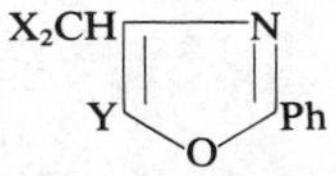

(**41**) X = OMe, Y = Cl
(**42**) X = Y = piperidino

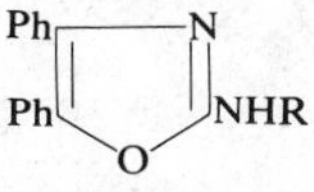

(**43**)

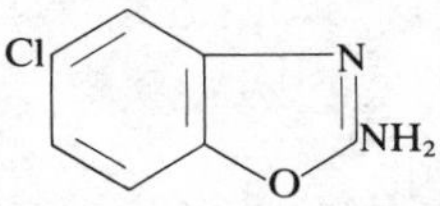

(**44**)

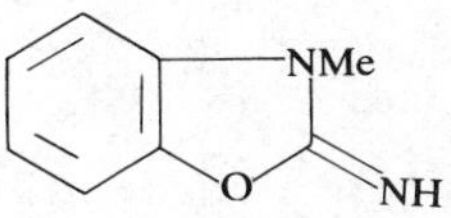

(**45**)

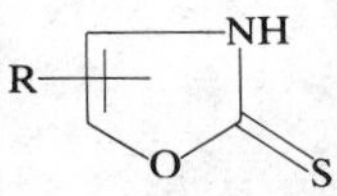

(**46**)

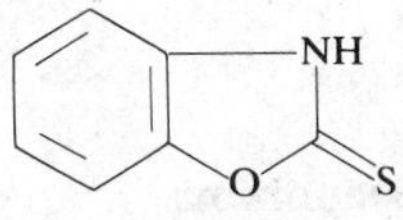

(**47**)

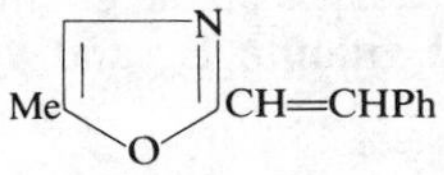

(**48**)

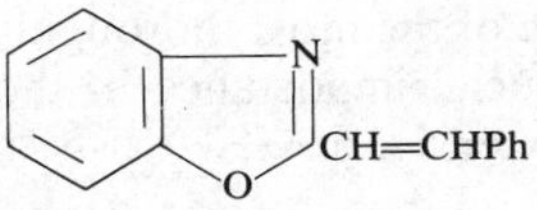

(**49**)

hexamethylphosphoric triamide.[19] Acyl substituents at C-2 may undergo Grignard reactions, as in the reaction of, for example, (**50**).[3] Other conventional C-2 functional group transformations which are possible because of the stability of the oxazole ring include the examples in Scheme 7.[4] Alkyl groups at C-4 and C-5 in oxazoles are not as reactive as C-2 alkyl groups. Alkyl groups are often oxidizable to carboxylic acids without disruption of the oxazole nucleus.

Oxazolium derivatives are particuarly prone to nucleophilic displacement of 2-halo-groups (reaction 11).[4] Interesting sulphur extrusions from intermediate thiirans occur in

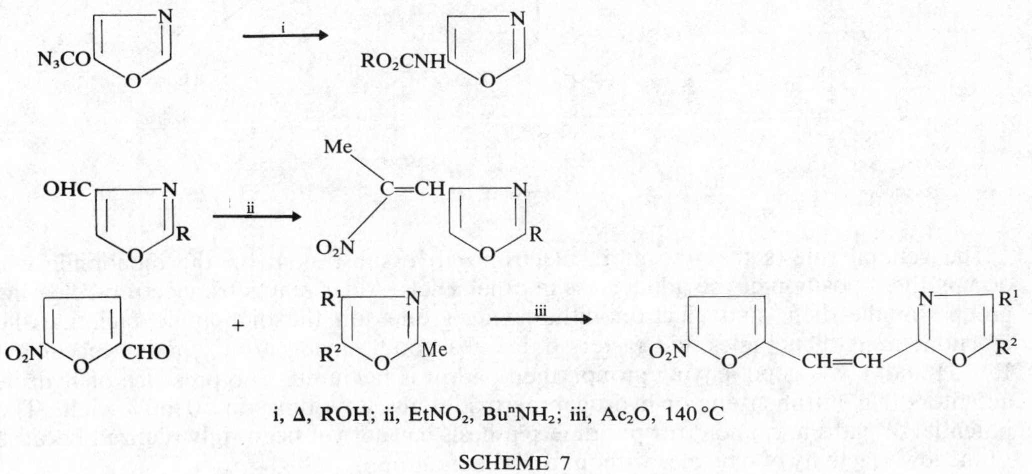

i, Δ, ROH; ii, $EtNO_2$, Bu^nNH_2; iii, Ac_2O, 140 °C

SCHEME 7

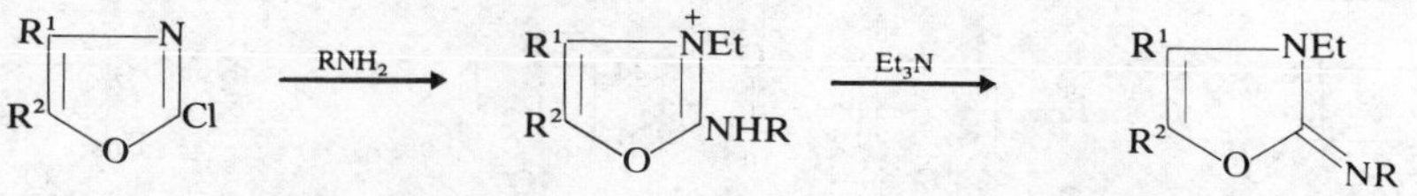

(11)

the treatment of (**51**) with sodium hydride–DMSO, giving (**52**), or with sodium hydroxide–aryl aldehyde, giving (**53**).[20] In 2-methyloxazolium salts such as (**54**) the anhydro-base (**55**) may be isolable, following proton abstraction.[6b]

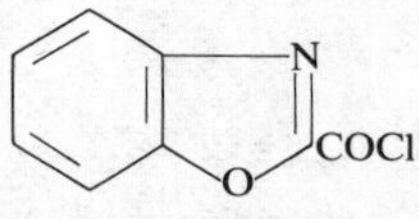

PhCOCH=CHAr

(**50**) 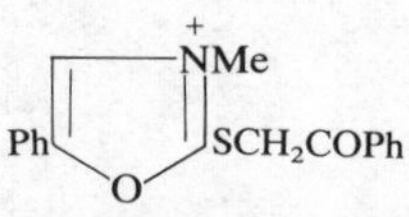 (**51**) 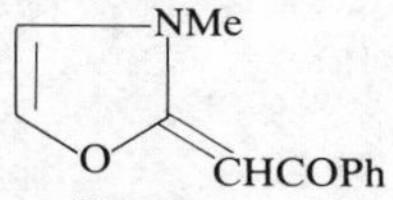 (**52**) (**53**) (**54**) 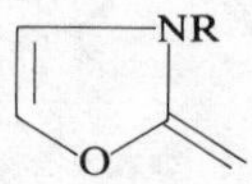 (**55**)

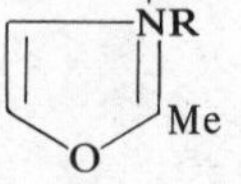

20.1.1.9 Cycloadditions

The Diels–Alder reaction of oxazoles leading to pyridines, furans, and bicyclo[2,2,1]-compounds is one of the most thoroughly explored areas in oxazole chemistry (e.g. Scheme 8), and is of especial importance in the synthesis of vitamin B_6 and various pyridoxine analogues. An extensive review exists.[4]

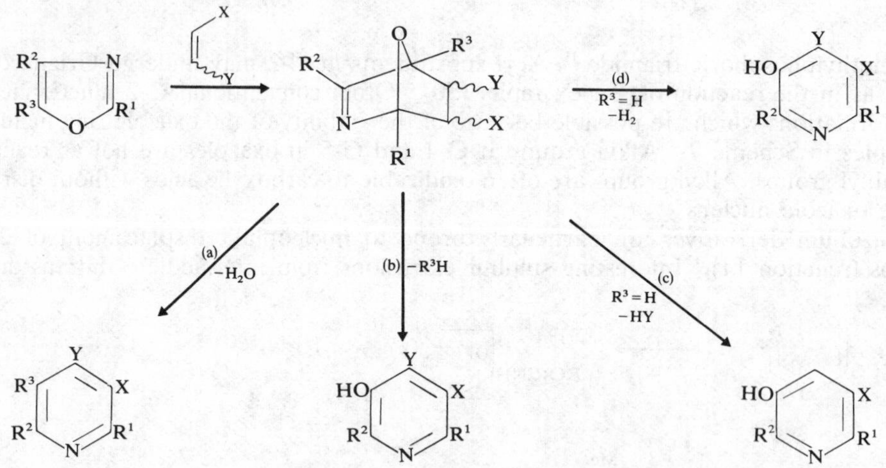

SCHEME 8

The general rule is that the more electronegative substituent on the dienophile will occupy the 4-position of the adduct. As in other Diels–Alder reactions, electron-releasing groups on the diene and electron-withdrawing groups on the dienophile facilitate the reaction; *trans*-dienophiles add faster. If R^3 is a good leaving group, path b occurs; if $R^3 = H$ and Y is a good leaving group, then path c is favoured. The presence of hydride acceptors (*e.g.* nitrobenzene or hydrogen peroxide) gives pyridines in 50–60% yield. The potential of paths a, c, and d for pyridine synthesis has not yet been fully realized because of the low reactivity of oxazoles without a 5-alkoxy-group.

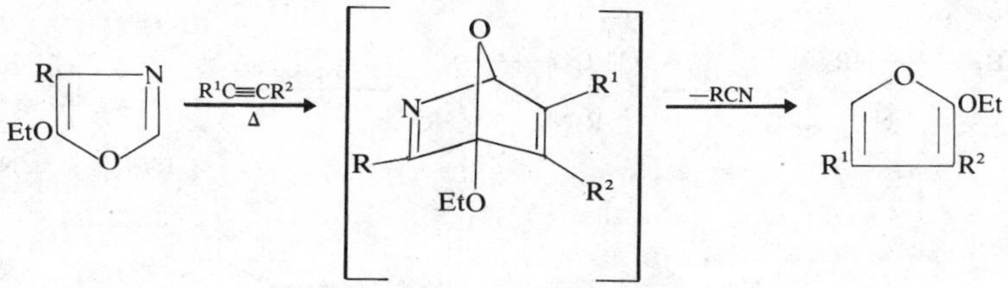

(12)

Acetylenes may lead to furan products in a regioselective cycloaddition in which the carbon of the dienophile bearing the more electronegative substituents will add to C-5 (reaction 12).[4]

Benzoxazole reacts with two moles of diphenylketen in a [2+2+2] cycloaddition involving the C═N double bond, affording an oxazinone-fused benzoxazole.[21]

20.1.1.10 Ring-cleavage reactions

Oxazoles and benzoxazole are resistant to alkaline hydrolysis, but are readily cleaved by acids, probably because of *N*-protonation which enhances the susceptibility of the ring towards nucleophilic attack. Although many oxazoles and benzoxazoles are soluble in acids without cleavage, they can be made to give acylaminoketones (reaction 13).[6b] 5-Alkoxy- and 5-amino-oxazoles tend to be readily cleaved by dilute mineral acids; the 2-amino-oxazoles are much more stable.[4] Oxazoles are cleaved by hot strong acids and 2,5-dinitrophenylhydrazine (reaction 14).[6c] Benzoxazole hydrolysis is relatively easy, 2-methylbenzoxazole giving *o*-acetamidophenol in hot water, although the reaction is more rapid in dilute acid.[3]

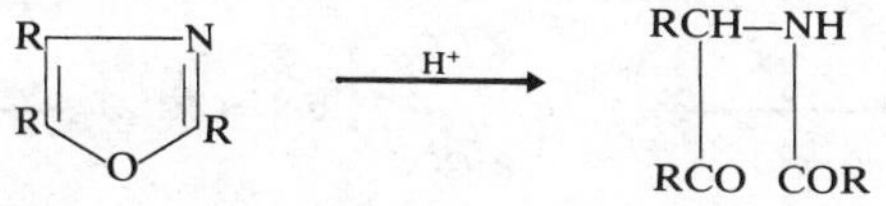

(13)

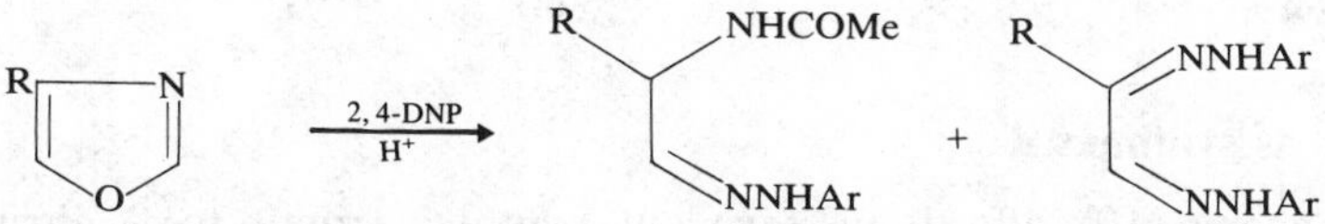

(14)

Amines cleave oxazoles, giving imidazoles or pyrimidines (Scheme 9).[4] Further examples are quoted in reaction 15.[22] Attacks by carbanions and by H_2S are depicted in reactions 16 and 17.[4]

Quaternary salts are hydrolysed much more readily, as shown by the hydrolysis of (**56**) which gives (**57**),[3] although 2-methylbenzoxazole can be cleaved by methoxide at 120 °C to give *o*-aminophenol. Quaternization of benzoxazole with methyl iodide at 120 °C leads to (**58**).[3]

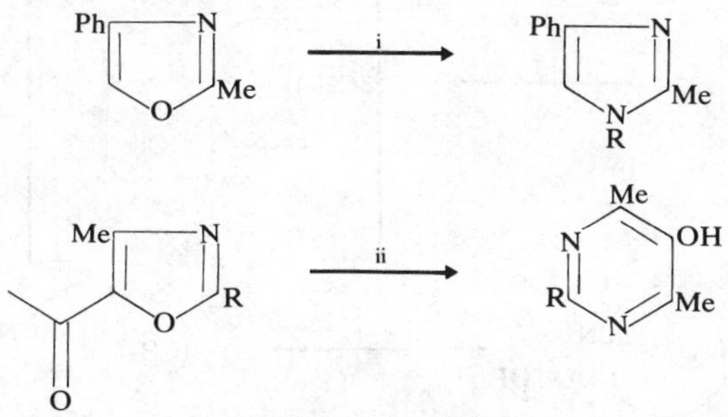

i, RNH_2, Δ; ii, NH_3, Δ

SCHEME 9

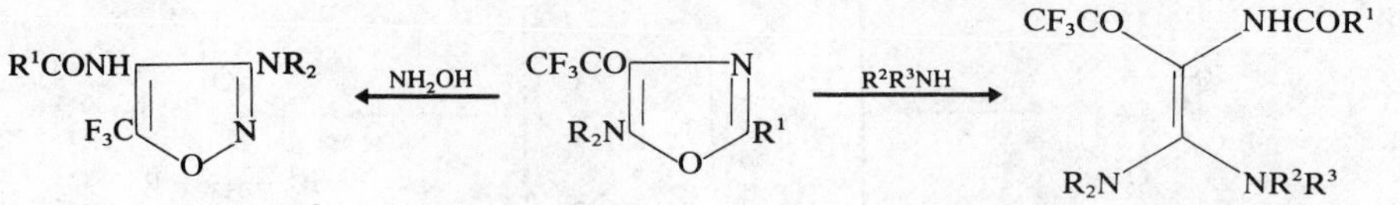

(15)

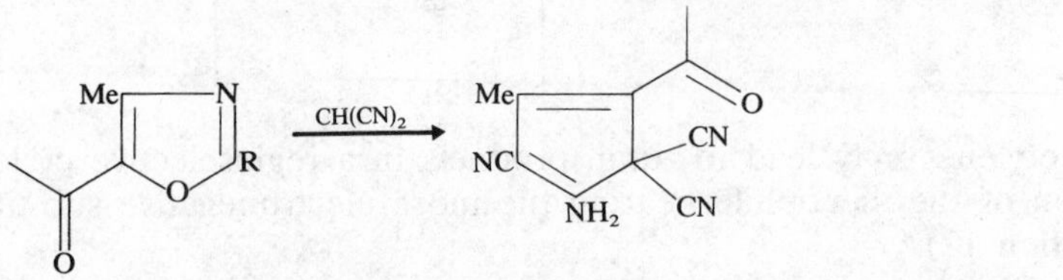

(16)

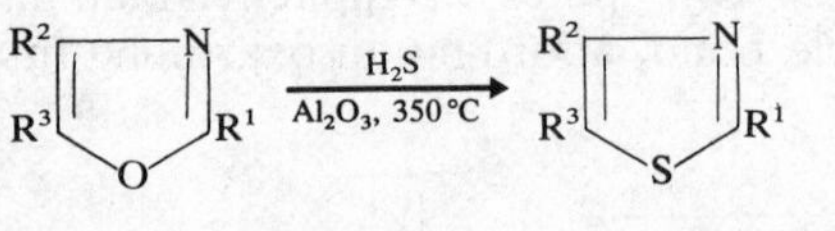

(17)

4-Acyloxazoles such as (**59**) on thermolysis undergo the Cornforth rearrangement (reaction 18) to give, *via* the intermediate (**60**), the isomerized oxazoles (**61**). This reaction, which has found widespread use in the synthesis of alanines, has been further studied mechanistically and the intermediate (**62**) postulated (first-order reaction, little solvent effect).[23]

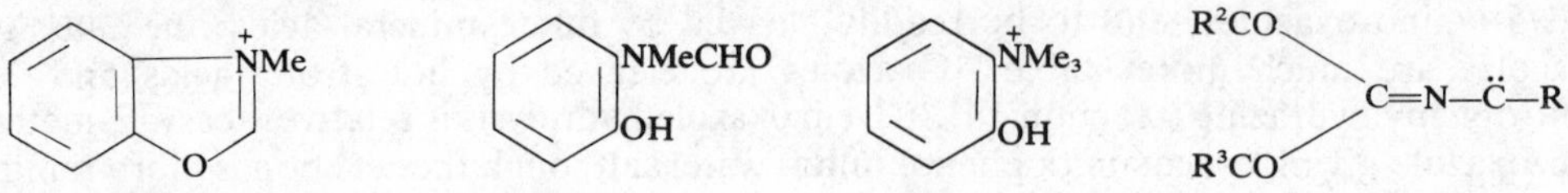

(**56**) (**57**) (**58**) (**62**)

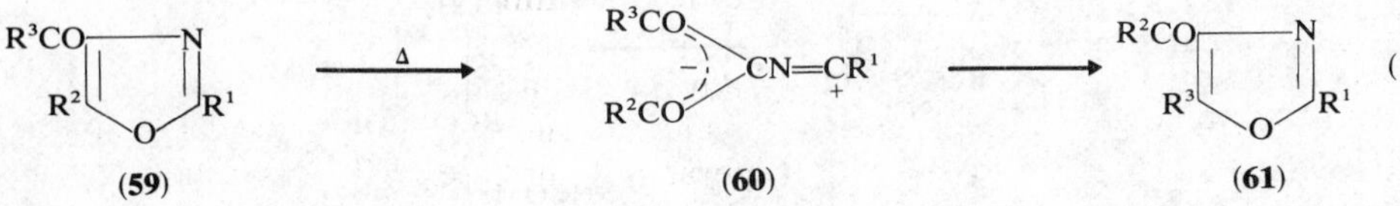

(18)

(**59**) (**60**) (**61**)

20.1.1.11 Use as synthons

Oxazoles (Scheme 10) afford an excellent synthetic route to ω-nitrilo-acids.[13] 4-Carboxyoxazoles (**63**) (Scheme 11) lead to β-aminoketones.[4]

The reaction of α-amino-acids with acid chlorides giving oxazolones and thence elaborated amino-acids is a well-established synthesis, and modifications have been reviewed.[13] Oxazoline-2-ones also offer a useful alternative to the Gabriel synthesis (reaction 19), an advantage being that in the last stage no acidic or basic hydrolysis is required, thus avoiding racemization.[13]

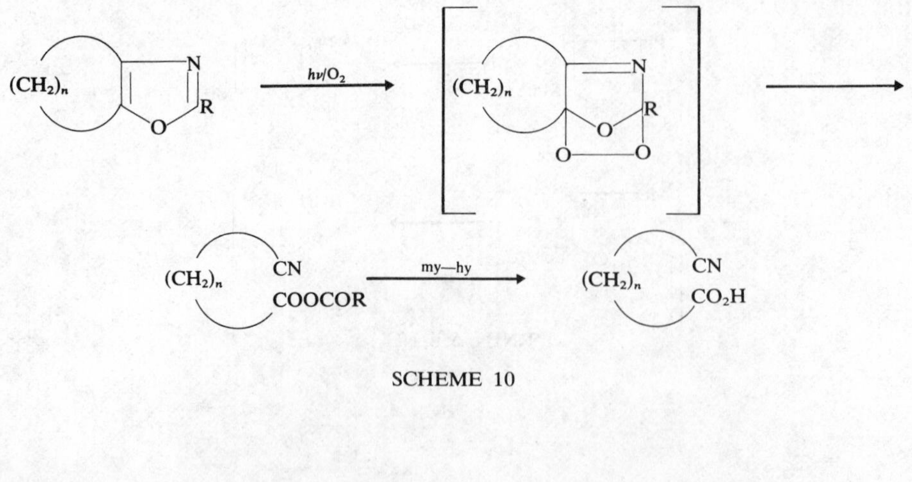

SCHEME 10

R^1O_2C, R-oxazole (63) —i→ $RCOCH_2NH_2$; (63) —ii→ $RCOCH(NH_2)CO_2R^1$ —iii→ $RCOCH_2NH_2$

i, 6N HCl, 90–95 °C; ii, 3N HCl, 50 °C; iii, 6N HCl, 90–95 °C

SCHEME 11

Ph, Ph-oxazolone (NH) —B^-, RX→ Ph, Ph-oxazolone (NR) —ox. or red.→ RNH_2 (19)

20.1.1.12 Preparations

Syntheses, which have been reviewed,[4,5] will be briefly summarized with pertinent examples and more recent modifications.

α-Haloketones and α-hydroxyketones have been widely used, reacting, for example, with formamide to give 4- and 5-substituted oxazoles, but often in low yield.[4] This approach, using ureas and thiocyanate-derived reagents, gives 2-amino- and 2-mercapto-derivatives. In an improved procedure for 2-amino-oxazoles, α-hydroxyketones are reacted with a cyanamide.[24]

The use of isocyanides overcomes disadvantages associated with obtaining appropriate α-substituted ketones. α-Metallated isocyanides condense with carboxylic acid derivatives to give 2-unsubstituted oxazoles in moderate to high yield.[4] The commercially-available tosylmethyl isocyanide ('TOSMIC') and derived reagents add the CH—N=CH fragment to unsaturated substrates (*e.g.* aldehydes) to give oxazoles in good yield.[25] Related reactions have been summarized.[4,26]

A group of syntheses which utilize amino-, imino-, and amido-nitriles give a wide range of 5-substituted oxazoles from reactions with aldehydes.[4] Nitrilium salts condense with α-haloketones in high yields (reaction 20).[4]

$$R^1—C\equiv\overset{+}{N}—\overset{-}{S}nCl_4 + PhCOCHClR \longrightarrow \text{oxazole (4-R, 5-Ph, 2-}R^1\text{)} \quad (20)$$

1,3-Dipolar cycloaddition reactions have been employed, carbonylcarbenes affording low yields of oxazoles (reaction 21)[4,27] and nitrilium methylides also giving 1,3-cycloadditions (reaction 22).[4] Carbonylnitrenes offer a third variant on 1,3-cycloadditions (reaction 23).[4] A reaction which might be classified as a 1,4-cycloaddition involved the addition of a carbenoid to an α-iminoketone.[28]

$$RCO—C(=N_2)—R^1 \xrightarrow[\text{metal cat.}]{\Delta,\ R^2CN} \left[R—C(—O^-)=\overset{+}{C}R^1 \right] \longrightarrow \text{oxazole (4-}R^1\text{, 5-R, 2-}R^2\text{)} \quad (21)$$

$$\text{R—C(Cl)=NCH}_2\text{R}^1 \xrightarrow[\text{R}^2\text{CHO}]{\text{Et}_3\text{N}} \left[\text{R—C}\equiv\overset{+}{\text{N}}\text{—}\overset{-}{\text{C}}\text{HR}^1\right] \longrightarrow \text{oxazoline} \xrightarrow{[\text{O}]} \text{oxazole} \quad (22)$$

$$\text{RCON}_3 \xrightarrow[\text{or } \Delta]{\text{PPh}_3} [\text{R—C(=N}^+\text{)O}^-] \xrightarrow{\text{R}^1\text{C}\equiv\text{CR}^1} \text{oxazole} \quad (23)$$

R = alkyl, aryl, alkoxy; R^2 = H or substit.

Oxazoles may also be synthesized by the acid-catalysed reaction of propargyl derivatives with amides or nitriles, as in the synthesis of (**64**) (reaction 24).[4]

The classical Robinson–Gabriel synthesis involving α-acylaminoketones has been improved by using new cyclodehydrating agents such as anhydrous HF, phosgene–triethylamine, and others[4] (reaction 25). The mechanism has been elucidated by ^{18}O-labelling.[29]

$$\text{HC}\equiv\text{C—CH(OH)—Me} + \text{MeCONH}_2 \xrightarrow[\text{H}_2\text{SO}_4,\ h\nu]{\text{g.HOAc,}} \text{(64)} \quad (24)$$

(**64**)

(25)

In modifications of existing reactions of α-acylaminocarbonyl compounds, enamides, enamines, and imides may be converted into 2,4,5-trisubstituted oxazoles. This group of reactions involves compounds or intermediates of general structure (**65**), which can be cyclodehydrated.[4] Benzoylalanine in a related reaction with PCl_5 gives 5-chloro-4-chloromethyl-2-phenyloxazolone.[4]

$$\text{RXC=C(R}^1\text{)NHY}$$

(**65**)

Oxazolin-5-ones (**66**) (from the cyclodehydration of α-amino-acids and acetic anhydride) under kinetically controlled conditions give (reaction 26) 5-acyloxy-derivatives (**67**).[4] 2-Oxonitrones with acetic anhydride give oxazoles,[4] and aroyl isocyanates with carbonyl-stabilized sulphonium ylides give good yields of 4-hydroxyoxazole derivatives.[4] Acylaziridines may also be thermally rearranged into trisubstituted oxazoles in high yield.[4]

$$\text{(66)} \xrightarrow{\text{R}^3\text{COX}} \text{(67)} \quad (26)$$

(**66**) (**67**)

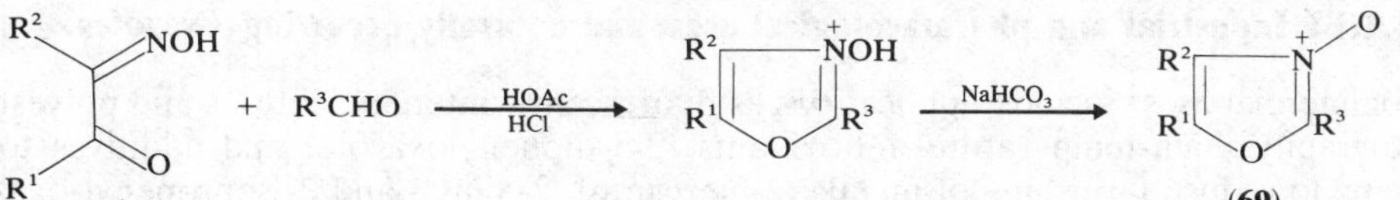

(27)

Oxazole *N*-oxides (**69**) can be prepared (reaction 27) from compounds (**68**), but are unstable; the parent molecule has not been synthesized.[4] Reduction of (**69**) gives oxazoles.

Benzoxazoles are readily prepared from *o*-aminophenols and carboxylic acid derivatives, as summarized in Scheme 12. Heating is often required to cyclize an intermediate, and yields range from moderate to excellent. Another group of syntheses utilize the

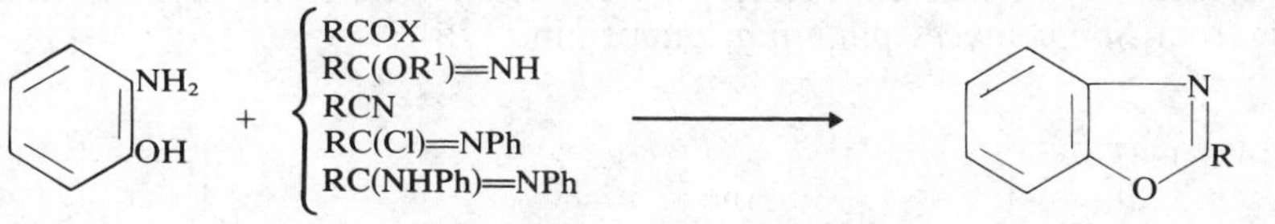

RCOX
RC(OR¹)=NH
RCN
RC(Cl)=NPh
RC(NHPh)=NPh

X = OH, OCOR, Cl, NH_2, NHPh

SCHEME 12

photochemical or oxidative ring-closure of imines (**70**), giving 2-arylbenzoxazoles.[30] *o*-Nitroaroyloxybenzenes upon reduction with Sn–HCl give benzoxazoles.[2] *o*-Hydroxyphenylketoximes (**71**) react with Pcl_5 or P_2O_5 to give benzoxazoles.[3] (This method has been extended to the synthesis of perfluorobenzoxazoles.[31]) A new synthesis involves the reaction of *o*-aminophenols with allenic and acetylenic nitriles, leading in high yield to a range of benzoxazoles.[32] Preparations of 2-oxo-, -thio-, and -amino-benzoxazoles include the thermolysis of *N*-(*o*-hydroxyphenyl)urethanes, the reaction of *o*-aminophenol with CS_2, and the mercuric oxide oxidation of *N*-(*o*-hydroxyphenol)-thiourea. These and other preparations of aryloxazoles have been summarized.[3,5]

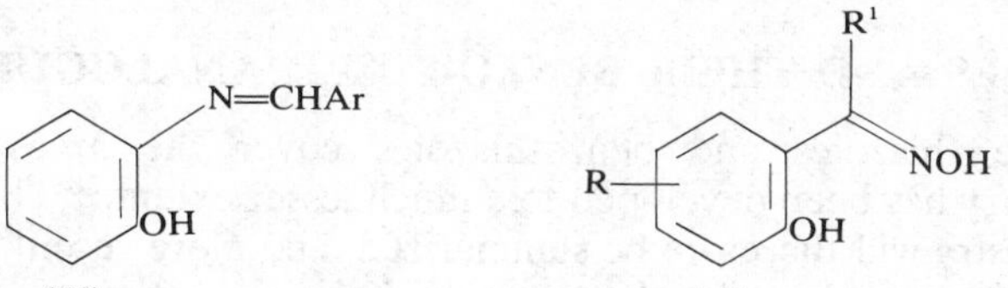

(**70**) (**71**)

20.1.1.13 Metallation

Oxazoles give adducts with mercuric acetate (reaction 28) which can be used in the preparation of halo-derivatives.[4] n-Butyl-lithium will react with halo-oxazoles (*e.g.* **72**) to give lithio-derivatives which can be quenched with D_2O (deuteration of non-halogenated oxazoles occurs mainly at C-2) (reaction 29).[4]

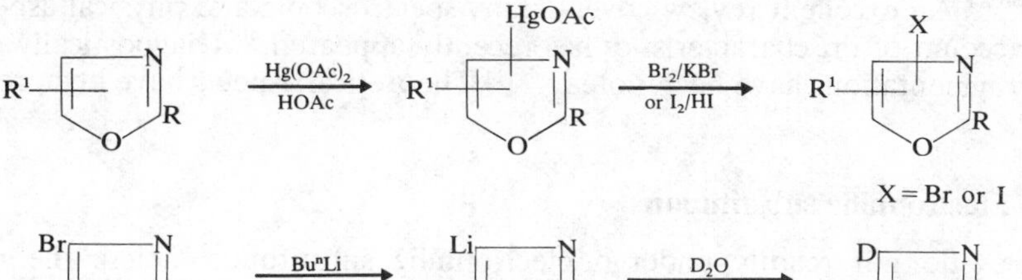

(28)

X = Br or I

(29)

(**72**)

20.1.1.14 Industrial and pharmacological uses, and naturally occurring oxazoles

Commercial uses[4] include applications as fluorescent whiteners (cottons and polyesters), photography, high-temperature antioxidants (2,5-diphenyloxazole) and additives to detergents to reduce tarnishing of metals (2-mercapto). 2-Vinyl- and 2-isopropenyl-oxazoles have been polymerized and also used as copolymers. An important aspect is the use of certain aryloxazoles as liquid scintillators. The plastic properties are better than those of terphenyls.

'Zoxazolamine' (2-amino-5-chlorobenzoxazole) is a relaxant and sedative.[33] Other oxazoles[4] have been investigated as analgesics, anti-inflammatories, antibacterials, antibiotics, and antivirals. Hypnotic and anti-hypertensive oxazoles have also appeared.[4]

Naturally-occurring oxazoles[4] include the alkaloids (**73**), *N*-methylhalfordium chloride (**74**), and, from *Streptomyces pimprina,* pimprinine (**75**).

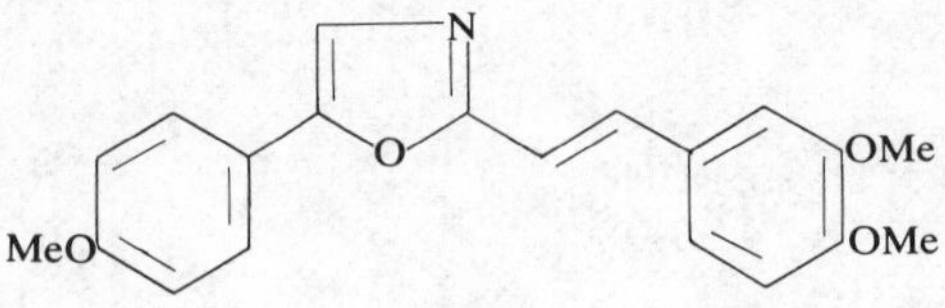

MeO OMe OMe N O

(73)

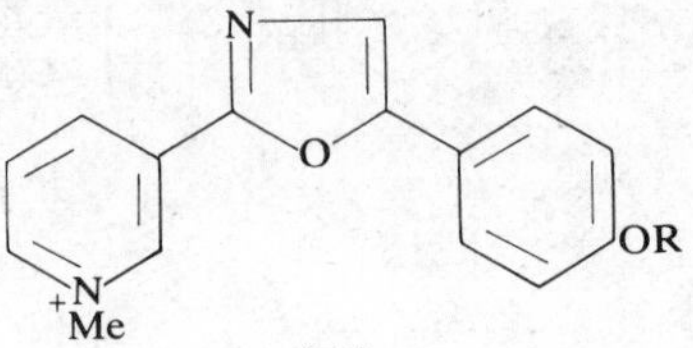

N O OR +N Me

(74)

NH O Me N

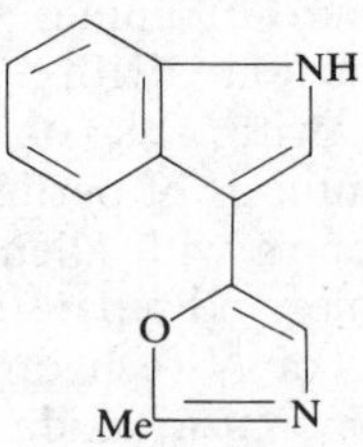

(75)

20.1.2 THIAZOLES AND THEIR BENZO-FUSED ANALOGUES

Reviews of the thiazoles and benzothiazoles cover the areas in depth.[3,6,7,34–37] Selenazole chemistry has been developed to a much lesser extent.[38] This enormous area of heterocyclic chemistry will therefore be summarized, and more recent advances discussed. The thiazoles and selenazoles, like the oxazoles, have aromatic properties and are weak bases. and afford a similar range of reactions. This section will concentrate on the chemistry of thiazoles. Some examples from selenazole chemistry will be included to illustrate similarities.

20.1.2.1 Physical and spectroscopic properties

Representative physical and spectroscopic properties of thiazoles have been summarized.[6b,34] An excellent review covers many spectroscopic and physical aspects[39] and a detailed account of i.r. characteristics has recently appeared.[40] Diagnostically useful mass spectral fragmentations have been noted.[34b,41] Theoretical aspects have been reviewed.[34]

20.1.2.2 Electrophilic substitution

Thiazoles do not readily undergo electrophilic substitution unless the molecule is suitably activated (*e.g.* by an electron-donating 2-substituent such as amino or hydroxy).[42] Thiazole (**76**) will undergo sulphonation and nitration to give the 5-derivatives, whereas bromine in chloroform gives a perbromide.[6c] In strong acid the reaction may occur on the thiazolium cation.[6c] More recently, reagents such as nitronium tetrafluoroborate have been employed,[43] and a series of HNO_3 and H_2SO_4 reactions on alkylthiazoles has been performed, and product distribution tabulated.[44] Benzothiazole on nitration gives 6-nitrobenzothiazole (**77**). Arylthiazoles are preferentially nitrated in the aromatic ring; the 2-phenyl substituent, for example, undergoes *o*-nitration, and 2-benzylthiazole gives *m*- and *p*-isomers. The heterocyclic ring exerted a strong deactivating influence and discrepancies with theoretical predictions were observed in these reactions.[45] Theoretical calculations on electrophilic attack have been reviewed.[34c] Selenazoles react similarly in the cases which have been examined, affording 5-nitro, 5-sulphonic, and 5-halo-derivatives. Gatterman and Friedel–Crafts reactions were unsuccessful.[5b] The *N*-amination using *O*-mesylhydroxylamine of the thiazole (**78**), giving (**79**),[46] is formally an electrophilic attack. *N*-Alkylations will be included in the section on quaternization.

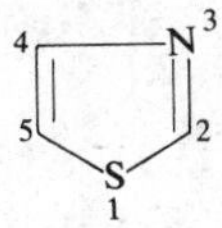

(**76**) 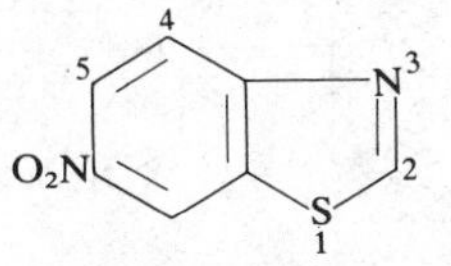 (**77**) 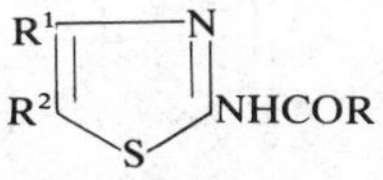 (**78**) 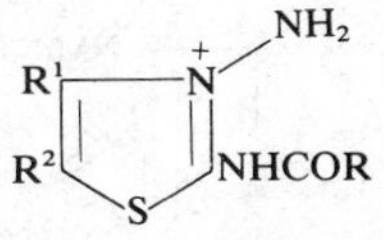 (**79**)

20.1.2.3 Nucleophilic substitution

Nucleophilic substitutions are more numerous than in oxazoles. An example of nucleophilic substitution occurring at C-2 with hydride transfer has been cited[6d] (reaction 30), but C-2 halogen substituents are readily displaced by hydroxide, alkoxide, hydrosulphide, sulphide, and a wide range of amino-groups.[4,6b] The 2-halo-group is more readily displaced than that in chlorobenzene, but 4- and 5-halo-substituents are not reactive unless activated by electron-withdrawing substituents in the α- or γ-positions of the thiazole. In an interesting study of the nucleophilic substitution of 4- and 5-halo-groups, the possibility of an elimination–addition mechanism (*via* a hetaryne) was rejected.[47]

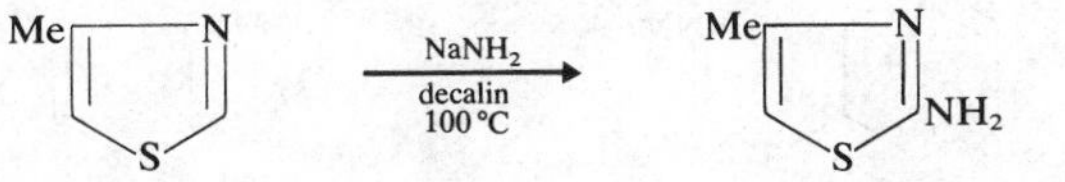

$\xrightarrow[\text{decalin, 100 °C}]{NaNH_2}$ (30)

The 2-nitro-group in 2-nitrothiazole has been displaced by fluoride, giving 2-fluorothiazole; the latter is difficult to make by other means.[42] 2-Methylsulphonylbenzothiazole reacts with alkoxides, cyanide, and carbanions to give 2-substituted products.[48]

Benzothiazolium compounds are much more susceptible to nucleophilic attack, as shown by the reactions in Scheme 13[49] and by the general nucleophilic reaction in equation 31.[50] A nucleophilic addition reaction is depicted in reaction 32.[50]

20.1.2.4 Quaternization

Thiazoles, benzothiazoles, and selenazoles can be quaternized, giving for example (**80**) and (**81**).[5b,34a] One of the most important areas of research in the chemistry of azolium species is concerned with thiamine chemistry because of the catalytic action of this coenzyme in life processes. The first step (reaction 33) in the action of this quaternized thiazole (**82**) is generation of the ylide (**83**), which subsequently adds to the ketonic carbonyl of a pyruvate unit to give (**84**).[6d,11] Extensive studies of this system emphasize the particular importance of sulphur in this system (compared with imidazolium analogues). Leading papers have been briefly summarized.[51] In relevant studies the decarboxylation of thiazolium carboxylates has been investigated, paralleling work on oxazolium and imidazolium analogues, in order to investigate the effect of sulphur.[52]

i, H_2S; ii, NH_2OH; iii, NH_3; iv, NaOH.

SCHEME 13

(31)

(32)

(**80**) (**81**)

(**82**) (**83**) (**84**)

(33)

Thiazolium, benzothiazolium and selenium analogues have been the study of an extensive theoretical investigation covering spectroscopic, polarographic reduction potential, and reactivity parameters.[53]

20.1.2.5 Proton exchange

Oxazoles, thiazoles, and selenazoles undergo H–D exchange in neutral CH_3OD. The rate of exchange falls to zero in acid and the reaction is therefore not an electrophilic

pH>1 BuLi pH 7 pH>11 pH 7

SCHEME 14

displacement involving initial addition of H^+. Under strongly basic conditions these azoles exchange by direct H^+ abstraction at C-2 and C-5, (*e.g.* Scheme 14).[6d] Later studies[51,54] have shown that with deuteroxide, or with alkoxide in deuteroalcohols, deprotonation at C-2 and C-5 occurs at similar rates. H-4 does not exchange even under forcing conditions. The mechanisms involving C-2 anions and thiazole ylides were investigated at different pH values. Studies of tritium exchange at C-2, relevant to thiamine chemistry, are also summarized.[54] The use of phase transfer methods in effecting H–D exchange was reported for the first time in heterocyclic chemistry.[55] 5-Ethylthiazole, for example, was subjected to isotopic exchange using tetrabutylammonium bromide. Base-catalysed H–D exchange at C-2 in 5- and 6-substituted benzothiazoles has been examined, and information obtained about the transmittance of substituent effects through N and S of the thiazole part of the nucleus.[56] Methoxide reversibly abstracts H-2 from 6-nitrobenzothiazole, although in dimethyl sulphoxide nucleophilic addition to C-2 occurs, giving a Meisenheimer-like adduct which can ring-open.

20.1.2.6 Radical reactions

The addition of aryl radicals to thiazole has been reviewed and product distributions analysed in detail.[39] Alkyl radicals have also been added to thiazoles and benzothiazoles. This reaction is of interest because of the nucleophilic character of a wide range of these radicals, and comparisons thus exist with electrophilic reactions.[49] For example, benzothiazole reacts with 1,4-dioxan radicals to give (**85**) in 40% yield.[49,57] The reactivity of phenyl radicals (from benzoyl peroxide thermolysis) with 2-substituted benzothiazoles is in the order C-4>C-7>C-5>C-6, in agreement with theoretical predictions. The product distribution in the reaction of a range of substituted thiazoles with cyclohexyl radicals is summarized for (**86**), (**87**), and (**88**).[58,59] Related reactions have been reviewed.[5] Acyl radicals also react with these heterocycles, adding highly selectively in the formation of (**89**).[49] This efficient trapping process is in fact of potential use for demonstrating the presence of acyl radicals. A range of homolytic *ipso* displacements of 2-substituents such as acyl, halo, and other leaving groups by adamantyl and acyl radicals has been reported.[49] Free radical bromination of, for example, 4-dibromomethyl-2-phenylthiazole, gives the 5-bromo-compound.[42] Homolytic amidation of benzothiazole gives the 2-formamido- and 2-amino-compounds in satisfactory yields.[50]

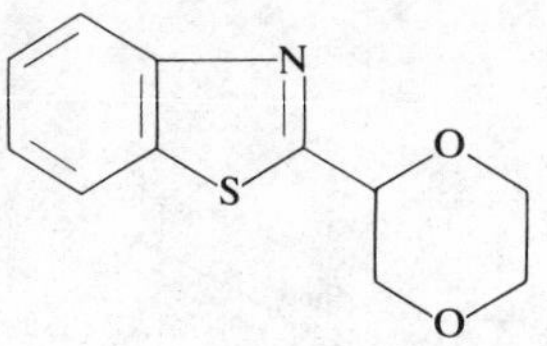

(85)

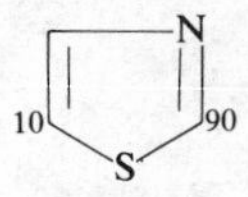

(86)

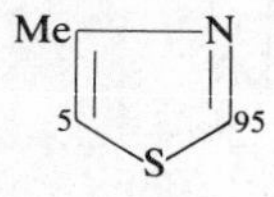

(87)

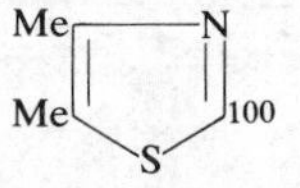

(88)

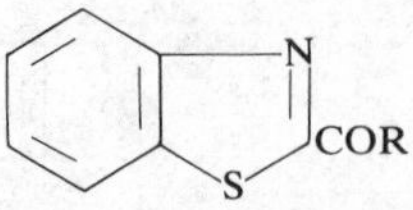

(89)

2-Thiazolyl radicals have been prepared by aprotic diazotization of the 2-amine, from 2-thiazolylhydrazones, and from 2-carbonyl peroxides, and they insert into aryl substrates.[34] All three thiazolyl radicals (*i.e.* at C-2, C-4, and C-5) can be formed by the last method and it has been shown that the C-5 radical is of relatively low reactivity.

20.1.2.7 Reduction and oxidation

Thiazole is resistant to many systems, although active Raney nickel can be used in desulphurization reactions.[6d] Benzothiazole is similarly resistant, but lithium aluminium hydride will give *o*-mercaptomethylaniline and 2,2-bisbenzothiazole gives the reduced product (**90**).[60] Benzothiazolium salts have been reduced polarographically to benzothiazolidines.[61]

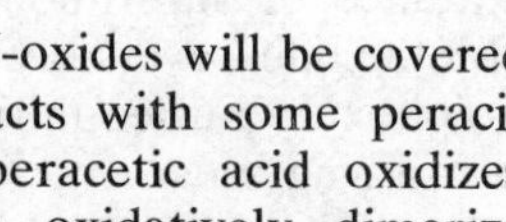

(**90**)

Oxidations leading to thiazole *N*-oxides will be covered separately (Section 20.1.2.14), but 2-(2-pyridyl)benzothiazole reacts with some peracids with initial cleavage of the pyridine ring, although trichloroperacetic acid oxidizes the thiazole ring nitrogen.[50] Certain 2-hydroxyselenazoles are oxidatively dimerized at the 5-position to give quinonoid-type dyes.[62] Unusual aromatic ring oxidative processes occur during Cu^{II}-catalysed oxygen treatment of 6-hydroxybenzothiazole with secondary amines, 4-amino-6,7-dioxo-products being formed.[49]

20.1.2.8 Photochemistry

As in the oxazoles, isomerization and interconversion with isothiazoles[63] are the predominant photochemical processes.[14] Several mechanistic pathways have been considered[64] for the transformation of (**91**) into (**92**) and (**93**). Diradical intermediates have been suggested, as have been valence isomers such as (**94**) and (**95**).

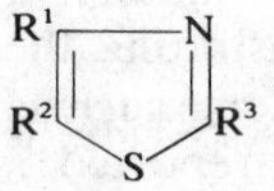

(**91**) 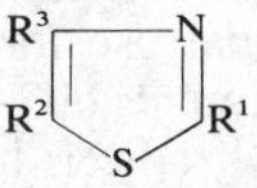 (**92**) 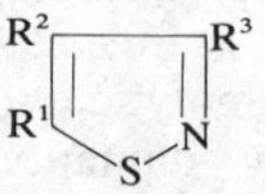 (**93**) 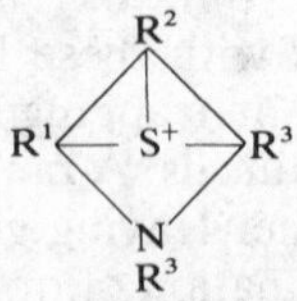 (**94**) 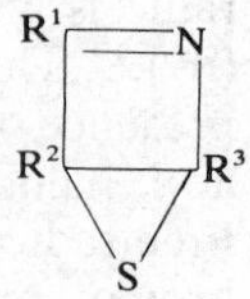 (**95**)

20.1.2.9 Reactions of substituents

The 2-hydroxythiazoles and 2-hydroxybenzothiazoles have long been known to exist in the 2-keto tautomeric form. Studies on 2,4-disubstituted 5-hydroxythiazoles have shown that of four possible forms (**96**)–(**99**) in solution, structure (**96**) predominates in chloroform when $R^1 = Ph$ and $R^2 = Pr^i$. Enolization to varying degrees occurs in more polar solvents.[65] Complementary studies showed that 2-alkoxy-analogues exist in forms (**96**) and (**97**).[66] The 2-alkyl- and 2-aryl-5-ones enolized readily and underwent acylation at oxygen, whereas the 2-alkoxy-5-ones did not enolize readily, and required strong base to effect acylation at oxygen. 2-Hydroxybenzothiazole (benzothiazol-2-one) alkylates at nitrogen, with, for example, alkyl halides in sodium ethoxide. 2-Aryloxybenzothiazoles (**100**) and (**101**) undergo the Smiles and Fries rearrangements, giving (**102**) and (**103**) respectively.[67]

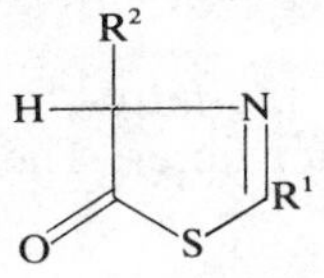

(**96**) 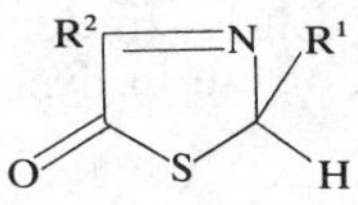 (**97**) 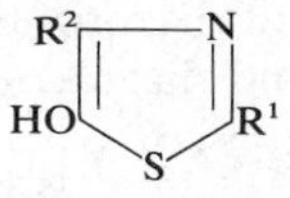 (**98**) 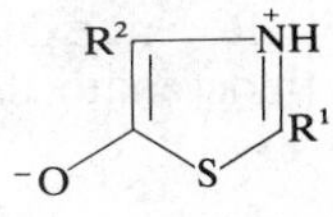 (**99**)

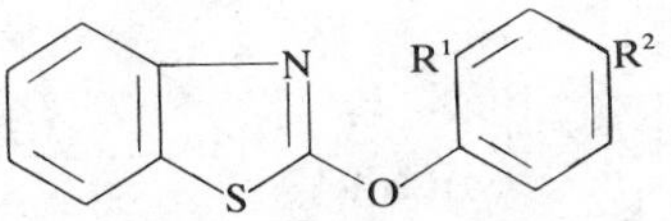

(**100**) $R^1 = NH_2$, $R^2 = H$
(**101**) $R^1 = H$

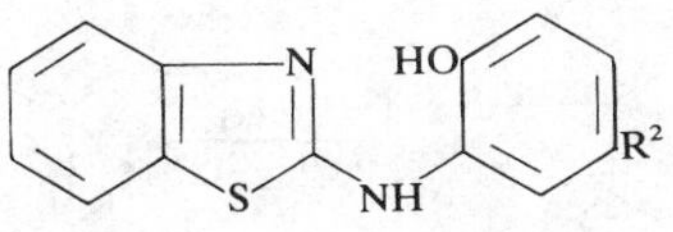

(**102**) $R^2 = H$

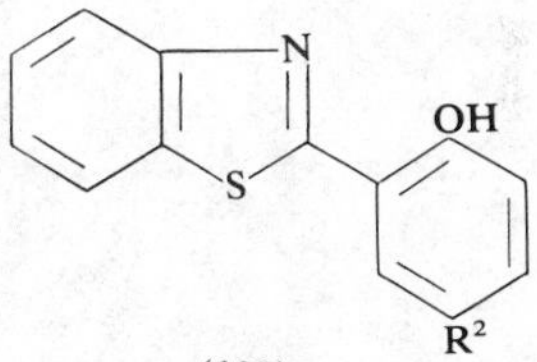

(**103**)

2-Mercaptothiazoles have been extensively investigated,[34] and recent reactions include that with diazomethane to give 2-methylthiothiazole. The variations in *N*-methyl to *S*-methyl ratios from diazomethane reactions with (**104**) have been determined for a range of substituents. Aminomethylation occurred at nitrogen in the ambident 2-mercaptobenzothiazole anion, as did Michael addition of methyl vinyl ketone.[63] In contrast, the sodium mercaptide with 1-bromo-3-chloropropane gave the adduct (**105**).[68] Important uses of 2-mercaptobenzothiazole in trapping transient sulphenic acid intermediates in penicillin chemistry have been reviewed.[69] *S*-2-Benzothiazole thiocarbonates on pyrolysis give, by S_Ni' and S_Ni processes, the products (**106**) and (**107**) in good yields.[68]

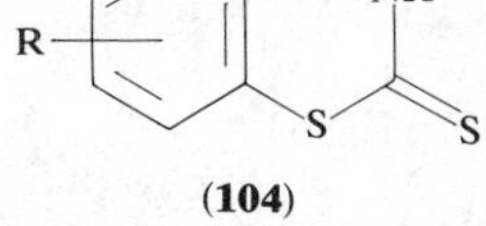

(**104**) 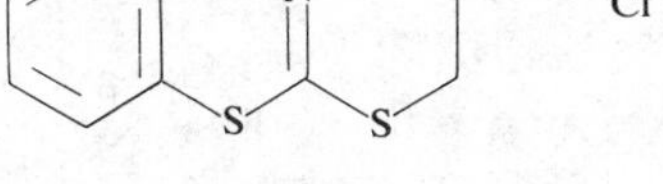 (**105**) 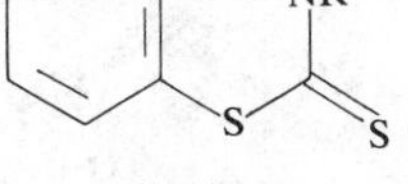 (**106**) 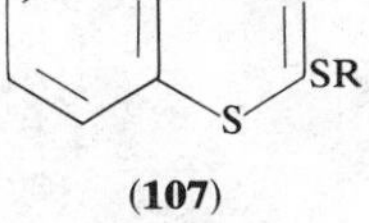 (**107**)

The 2-amino–2-imino equilibrium of 2-aminothiazoles has been discussed.[6b,18] Aminobenzothiazoles exist largely in the 2-amino-form,[49] as does 2-hydrazinobenzothiazole. Two distinct tautomeric forms (**108**) and (**109**) have been isolated and interconverted under carefully controlled conditions, and 2-amidino-geometrical isomers have been separated.[49]

Alkylation occurs on the ring nitrogen of 2-aminothiazoles in many reactions, except when an intermediate *N*-anion such as (**110**) is formed, giving (**111**).[6b] 2-Amino-4-ethylthiazole can be alkylated with secondary alcohols and acid to give (**112**).[70,71] t-Butyl alcohol under these conditions gave 2-amino-4-ethyl-5-t-butylthiazole by electrophilic attack. Acylation of 2-aminothiazole gives the 2-acetamido-compound, although certain acylating systems may give both amino and azomethine *N*-acylation.[71] Aminoalkylation of

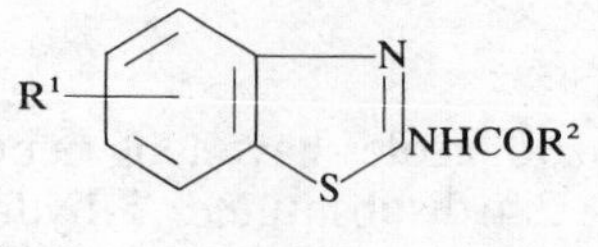

(108) 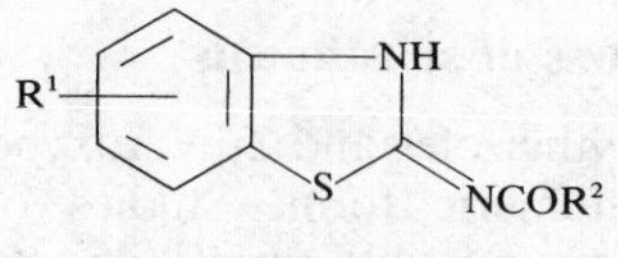 **(109)**

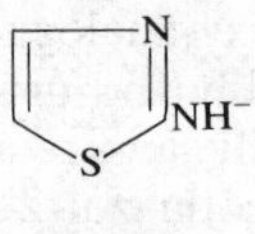

(110) 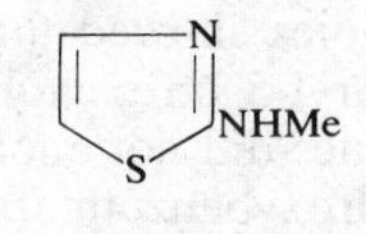 **(111)** 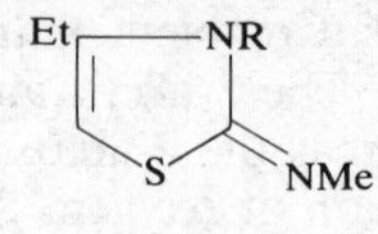 **(112)**

the 2-amino-group gave products as in reactions 34 and 35.[50,63,70] 2-Aminothiazoles and 2-aminobenzothiazoles form Schiff bases, which may be reduced to give 2-alkylamino-derivatives.[50,63,70] An interesting modification is shown in reaction 36.[50]

Recent examples[42,54] of diazotization have given 2-halo- (including fluoro-), 2-nitro-, 2-thiocyanato-thiazoles, and diazo-coupled products of use in dyeing polyester fibres. The

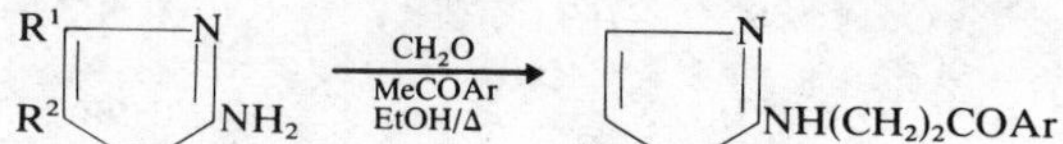

$$\xrightarrow[\text{EtOH}/\Delta]{\text{CH}_2\text{O},\ \text{MeCOAr}} \quad \text{NH(CH}_2)_2\text{COAr} \tag{34}$$

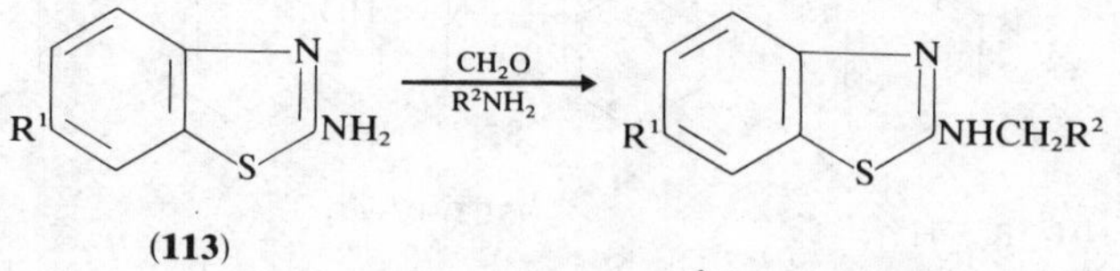

$$\xrightarrow{\text{CH}_2\text{O},\ \text{R}^2\text{NH}_2} \quad \text{NHCH}_2\text{R}^2 \tag{35}$$

(113)

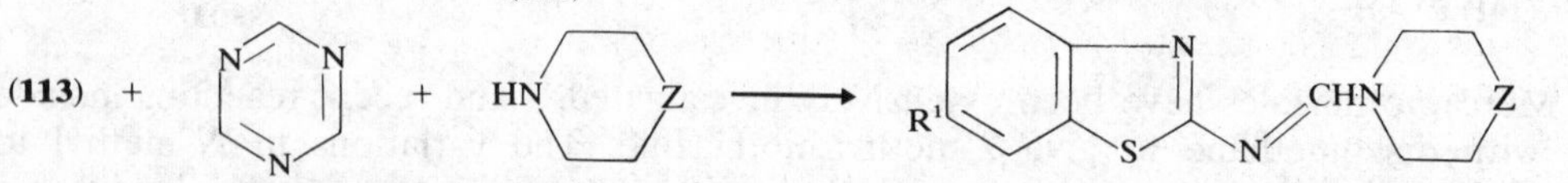

$$\textbf{(113)} + \ldots + \text{HN} \longrightarrow \ldots \text{CHN} \tag{36}$$

diazotization of (**114**) in furan, unlike that of anthranilic acid, gave (**115**), and the possible intermediacy of a 4,5-dehydro-2-methylthiazole was discounted.

Reactions of thiazole alkyl groups are important. The 2-methyl group is reactive in condensation processes, particularly when the thiazole is quaternized. The 4- and 5-methyl groups are relatively inert. For example, (**116**) gives (**117**), and (**118**) reacts with anhydrous zinc chloride and benzaldehyde to give (**119**).[42] In a related sequence,[54] lithium

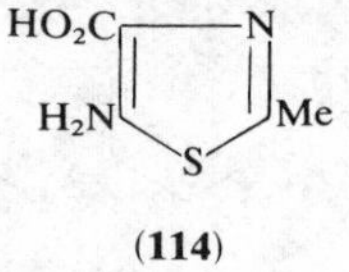

(114) 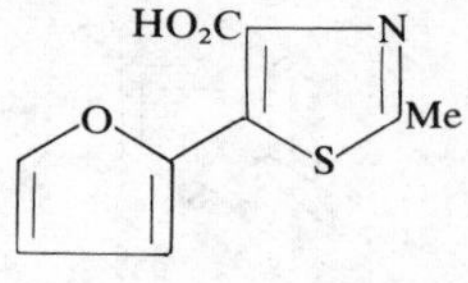 **(115)**

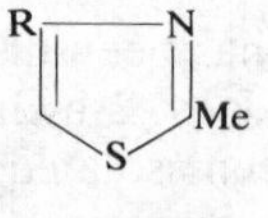

(116) R = NO_2
(118) R = Me

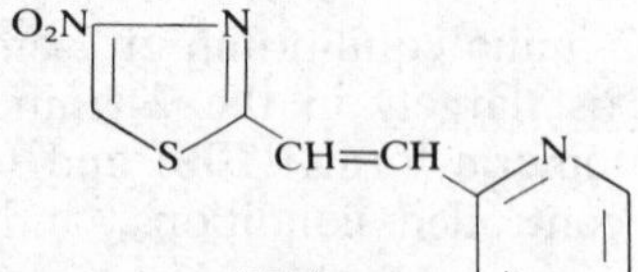

(117)

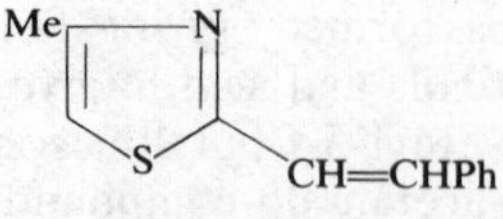

(119) 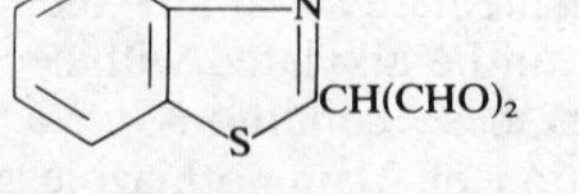 **(120)**

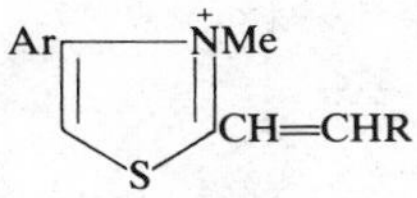

(121) 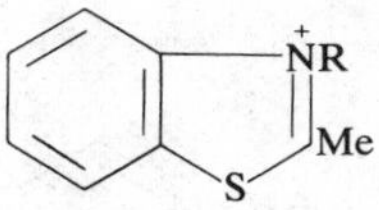 (122) R = Et (124) R = Me (123) 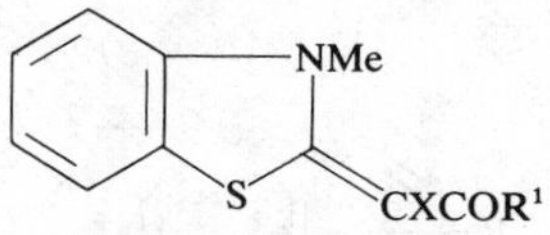 (125) X = H (126) X = COR[1]

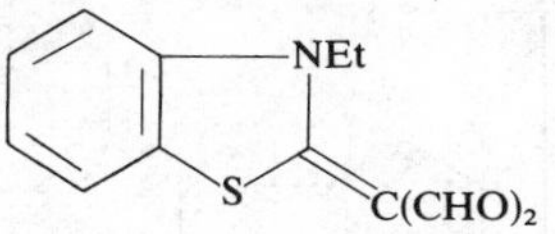

amide catalyses the condensation of ketones with the 2-methyl group, and not the 4-methyl groups. 2-Methylbenzothiazole reacts in a Vilsmeier type of process to give the bis-formylated product (**120**),[72] as does the corresponding benzoselenazole. Quaternization facilitates these condensation reactions, leading to products such as (**121**).[42] Acylation of (**122**) and its seleno-analogue with dimethylformamide gives (**123**) and its seleno-analogue,[72] and acyl chloride reactions of (**124**) give (**125**) and (**126**).[49]

Displacement of halogen from the α-position of thiazole side-chains is illustrated by the conversion of (**127**) into the substituted alanine (**128**) by a sequence involving sodio-acetamidomalonic diethyl ester.[42] The Sommelet reaction (hexamethylenetetramine) of (**129**) gives the 2-aryl-5-formylthiazole.[51]

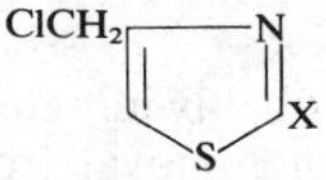

(127) X = NMe_2 (129) X = aryl (128)

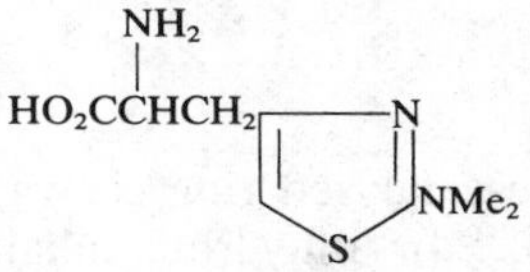

Other reactions of alkyl substituents include the formation and use of thiazolium *N*-ylides (**130**)[73] and the sulphur-catalysed condensation of *o*-aminophenol or phenylenediamine with 2-methylbenzothiazole to give the products (**131**) and (**132**).[51] In a new reaction it has been shown that 2-methylbenzothiazole will react with sodamide and an alkyl nitrate in liquid ammonia to give the 2-(nitromethyl)benzothiazole.[74]

Benzothiazole-2-carbaldehyde with primary amines readily forms Schiff bases, which are particularly useful for transamination reactions.[75]

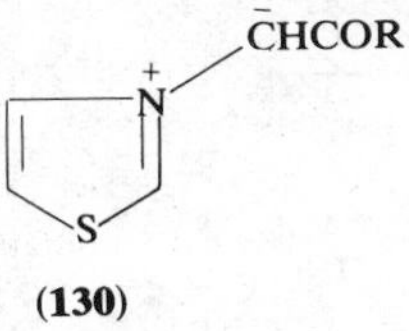

(130) 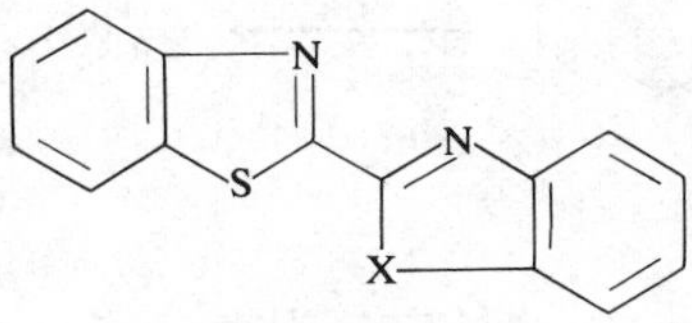 (131) X = O (132) X = NH

20.1.2.10 Cycloadditions

Benzothiazoles react with dimethyl butynedioate in aqueous methanol as shown in Scheme 15.[76] Diphenylketen reacts as it does with benzoxazole, to give a tricyclic product from [2+2+2] cycloaddition.[21]

20.1.2.11 Ring-cleavage and rearrangement reactions

The more recent literature is concerned mainly with thiazolium species, although a base-catalysed rearrangement of 2,5-diaminothiazoles (**133**) to imidazoles (**134**) has been described,[42] and photo-oxidative cleavage processes which operate by a [2+4] cycloaddition mode in polar solvent, and a [2+2] reaction in non-polar solvents, also occur.[51]

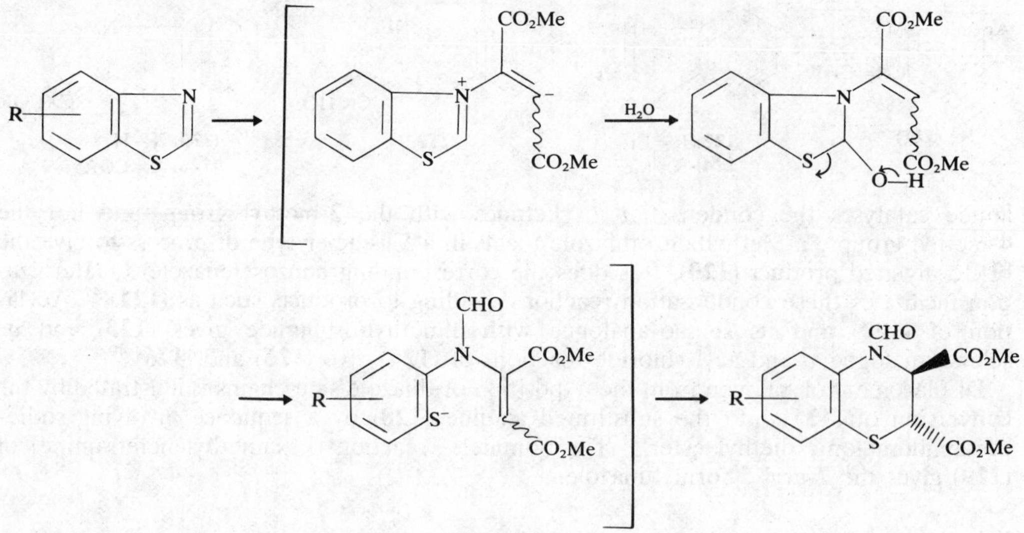

SCHEME 15

Thiazolium base-catalysed ring cleavage is probably initiated by nucleophilic attack at C-2 (reaction 37).[51] In anhydrous alkali, proton removal from the alkyl group on the quaternary nitrogen may lead to ylide formation and dimerization.[51] The thiazolium salt (**135**) (reaction 38) can be ring-expanded to the thiazine (**136**).[51] This general reaction has been extended to thiamine, and to benzothiazoliums such as (**137**), which can be converted to (**138**) (reaction 39).

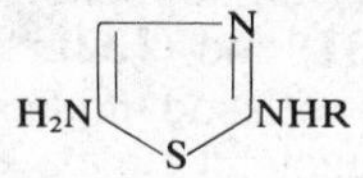

(**133**) 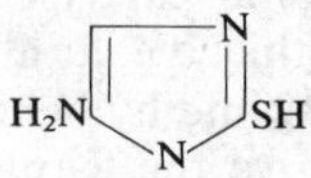 (**134**)

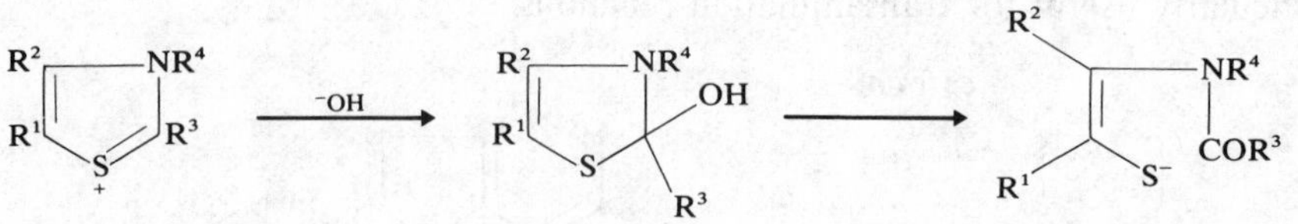

(37)

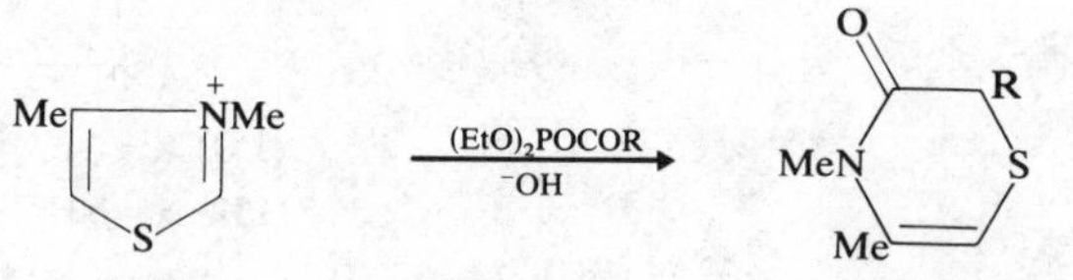

(**135**) (**136**) (38)

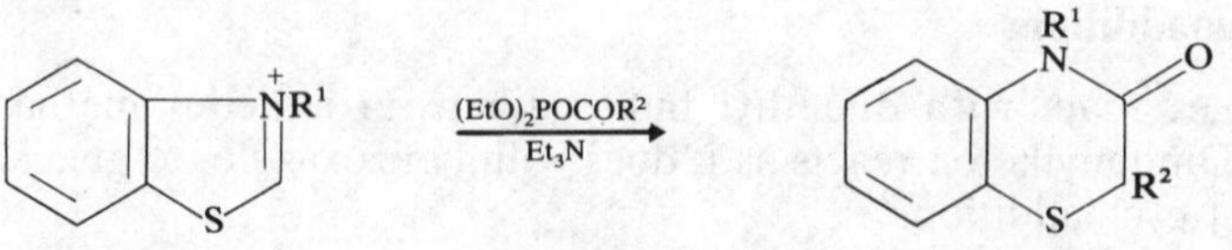

(**137**) (**138**) (39)

20.1.2.12 Use as synthons

2-Methylthiazoles are useful as synthons for aldehydes (reaction 40). For example, 3-phenylpropanal is conveniently prepared in this manner, as are many others.[77] Benzothiazoles are similarly useful, affording aldehydes and ketones from amines (Scheme 16)

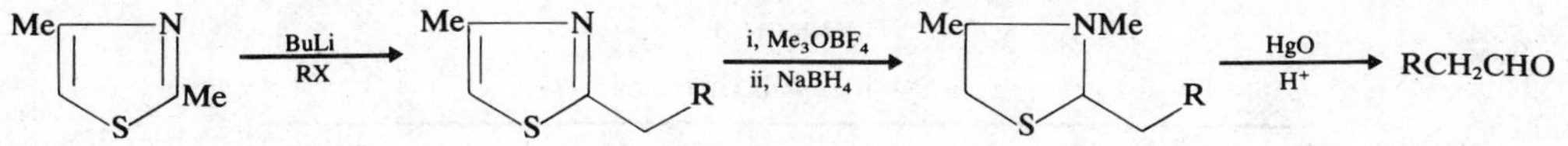

$$\text{2,4-dimethylthiazole} \xrightarrow[\text{RX}]{\text{BuLi}} \text{2-(RCH}_2\text{)-4-methylthiazole} \xrightarrow[\text{ii, NaBH}_4]{\text{i, Me}_3\text{OBF}_4} \text{3,4-dimethyl-2-(RCH}_2\text{)thiazoline} \xrightarrow[\text{H}^+]{\text{HgO}} \text{RCH}_2\text{CHO} \quad (40)$$

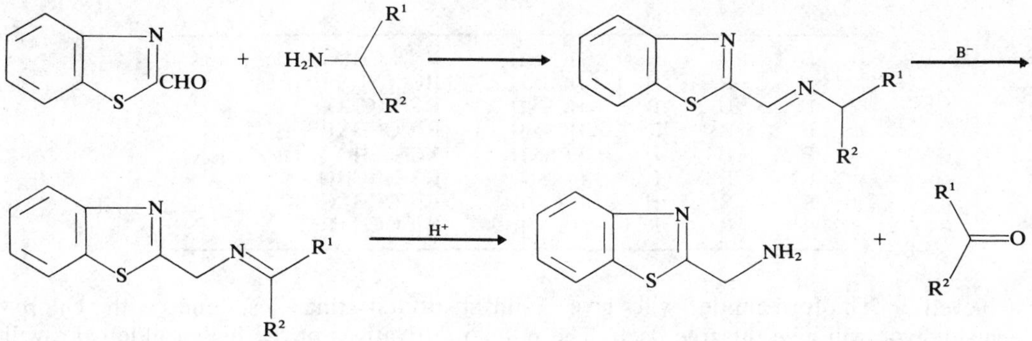

SCHEME 16

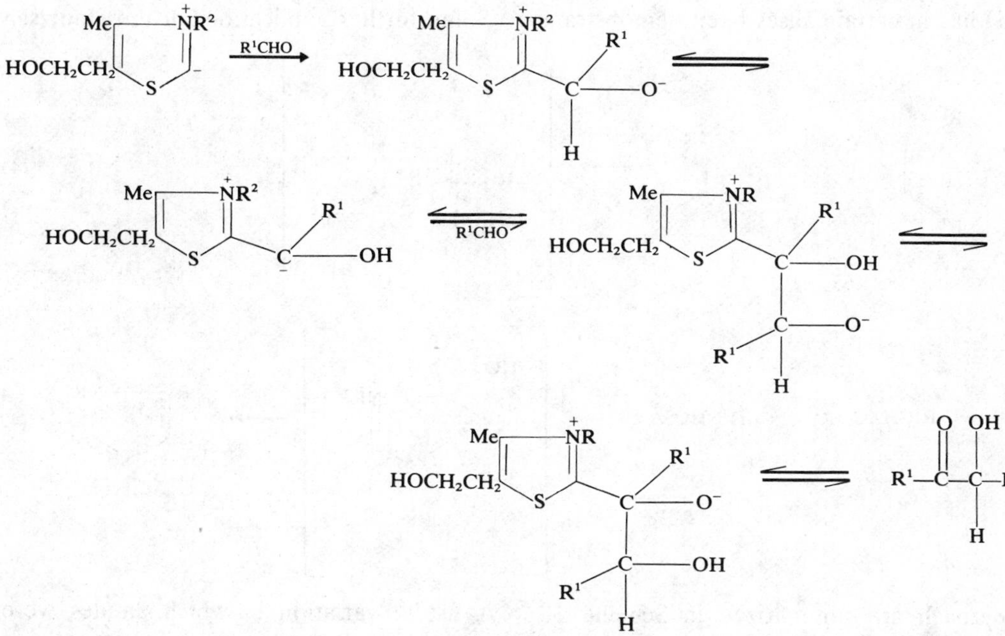

SCHEME 17

in good to excellent yields. Thiazolium salts may also be used to form acyloins and benzoins (Scheme 17).[78]

20.1.2.13 Preparations

A synthetic taxonomy which classifies the myriad syntheses of thiazoles and benzothiazoles has evolved,[34] and is based upon the nature of the skeleton of the components which join to form the ring system. This approach will be followed in this section. The literature contains several hundred relevant syntheses, and only representative examples are included.

C—C+S—C—N routes are exemplified by the old-established and extremely versatile Hantzsch synthesis. In the general case (reaction 41) an α-haloketone is reacted with an appropriate thioamide;[3] Table 1 illustrates some of the variations which can thus be

TABLE 1

Syntheses of Thiazoles

Thiazole substituent 2	4	5	*Thioamide*	*Halogeno-compound*
H	H	H	$HCSNH_2$	XCH_2CHO and derivatives
H	R	H	$HCSNH_2$	$RCOCH_2X$
H	H	R	$HCSNH_2$	RCHXCHO
H	R^1	R^2	$HCSNH_2$	$R^1COCHXR^2$
R	H	H	$RCSNH_2$	XCH_2CHO and derivatives
R^1	H	R^2	R^1CSNH_2	$R^2CHXCHO$
R^1	R^2	H	R^1CSNH_2	R^2COCH_2X
R^1	R^2	R^3	R^1CSNH_2	$R^2COCHXR^3$

achieved.[79] Thioformamide will give 2-unsubstituted thiazoles, and with chloroacetaldehyde will give thiazole itself. The α-halo-derivatives of the higher aldehydes will give 5-unsubstituted thiazoles. Thioformamide will react with, for example, chloroacetone, to give 4-methylthiazoles. An intermediate 4-hydroxythiazoline (reaction 41) has in certain cases been demonstrated.[71] Some further applications of the Hantzsch

(41)

$R^1CHXCOR^2$ + $R^3NHCSAr$ → (42)

approach are summarized in Scheme 18.[42] A useful variation in which amides were treated with P_4S_{10} and condensed *in situ* with α-halogenocarbonyls has been reported.[51] Thiazolium compounds can be synthesized by condensing secondary thioamides with α-halogenoketones (reaction 42).[42]

$R^1C(NH_2)=S$ + $BrCH_2COCO_2R^2$ →

+ $ClCH(CO_2R^2)COCO_2R^2$ →

SCHEME 18

Thioureas react similarly, giving 2-aminothiazoles.[79,80] Dithiobiurets give products such as (**139**).[81] A wide range of amino-substituted thiazoles may be prepared by modified Hantzsch syntheses employing α-diazo-ketones,[42] and 2,4-diaminothiazoles have been synthesized from reactions of α-phenylsulphonylacetonitriles with thiourea.[80] 2,5-Diaminothiazoles were made from the thioureas (**140**).[80] 2-Alkoxythiazoles result from the reactions of thiourethanes[79] and 2-mercaptothiazoles are best prepared from ammonium dithiocarbamate with α-halocarbonyls.[79]

$NHCSNH_2$

(**139**)

RNHCSNHCHRCN

(**140**) R = alkyl
(**144**) R = H

Selenazoles are also prepared by the Hantzsch synthesis from selenoureas, a wide range being known although difficulties may arise because of the instability of selenoamides and some of the products.[18] Selenoamides can, however, be generated *in situ* by the reaction of nitriles with hydroselenide anion, and thus reacted with α-halogenocarbonyls.[18]

Other useful syntheses of the C—C+S—C—N type which are summarized within Ref. 34 include the reaction of aminothiocyanogen with a range of ketones possessing an α-methylene group (leading to 2-aminothiazoles), the reaction of α-acylphosphoranes with dithiocyanogen (also giving 2-aminothiazoles, or 2-mercaptothiazoles), and the 1,3-cycloaddition of acetylenes to 'Hector's base' (reaction 43).[82]

ArN, NAr, HN, S, NH —XC≡CY→ [ArNH, Ar N, H N, X, N, S, Y] → X, N, Y, S, NHAr (43)

C—C—N+C—S syntheses are illustrated by the condensation of thioacetic acids with α-amino-acids (reaction 44),[42] and by the reaction of enamines (**141**) giving either 2-hydroxythiazoles (**142**) or 2-aminothiazoles (**143**) depending upon reaction conditions.[42] In the reaction of aminoacetonitrile with alkyl isothiocyanates, intermediates (**144**) have been demonstrated in the formation of the 2,5-diaminothiazoles (**145**).[71]

$$H_3\overset{+}{N}CHR^1CO_2^- \xrightarrow{MeCOSH} \text{(R}^1\text{, N, MeCOS, S, Me)} \quad (44)$$

C—C—N—C—S type syntheses are represented by the well-known[6b] cyclization of α-thioacylaminoketones to thiazoles. This synthesis parallels that of thiophen from 1,4-dicarbonyls. Examples include the preparation of (**148**) and (**149**) in good yield from (**146**) and (**147**).[79] The recent synthesis in high yield of 2-amino-derivatives (**150**) (reaction 45) may also be included in this category.[83]

Me, O, H_2N, R

(**141**)

Me, N, MeCO, X, S

(**142**) X = OH
(**143**) X = NH_2

N, H_2N, NHR, S

(**145**)

$PhCOCHR^2NHCSR^3$

(**146**) $R^2 = R^3 = H$
(**147**) $R^2 = R^3 = Me$

R^2, N, R^1, R^3, S

(**148**) $R^1 = Ph$, $R^2 = R^3 = H$
(**149**) $R^1 = Ph$, $R^2 = R^3 = Me$

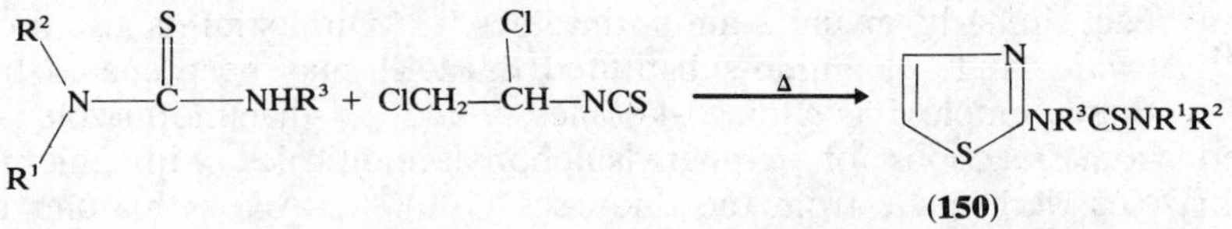

$$R^1R^2N{-}C({=}S){-}NHR^3 + ClCH_2{-}CH(Cl){-}NCS \xrightarrow{\Delta} \textbf{(150)}\ \text{2-}NR^3CSNR^1R^2\text{-thiazole} \qquad (45)$$

$$EtCOCHMeSH + R^1R^2C{=}NCH_2CO_2Et \longrightarrow \text{(thiazolidine, Et, Me, } NCH_2CO_2Et, R^1, R^2) \xrightarrow{\Delta} \text{(4-Et-5-Me-2-}R^2\text{-thiazole)} \qquad (46)$$

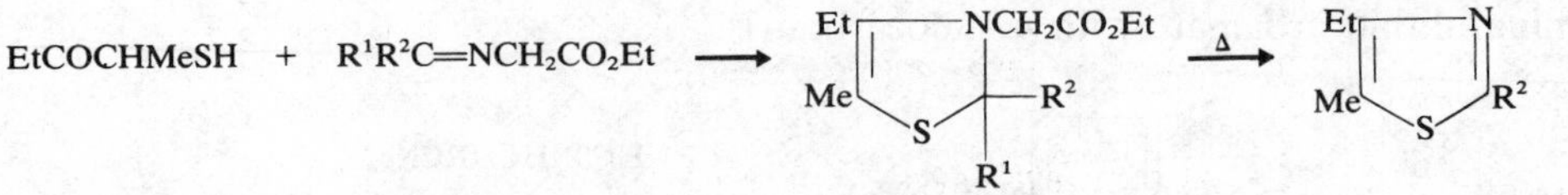

A C—C—S+N—C route is depicted in reaction 46.[42]

N—C—C—S+C approaches involve the formation of intermediates (**151**) by the reactions of an enamine with sulphur and then cyanamide, and subsequent uncatalysed cyclization to the thiazoles (**152**).[51]

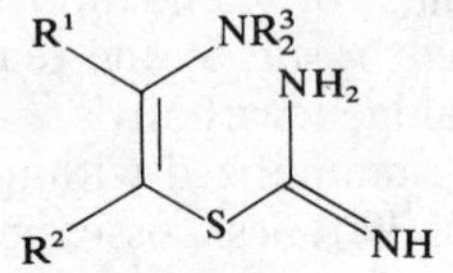

(**151**) 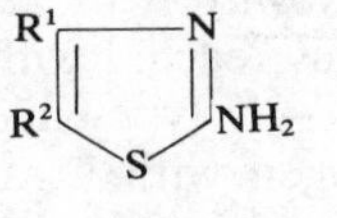 (**152**) 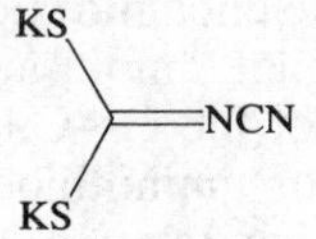 (**153**) 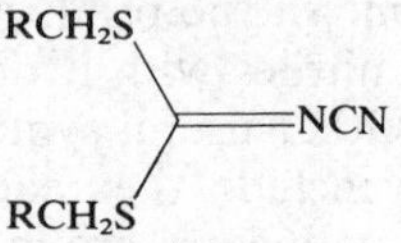 (**154**)

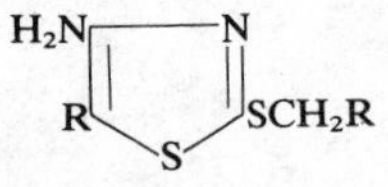

(**155**) 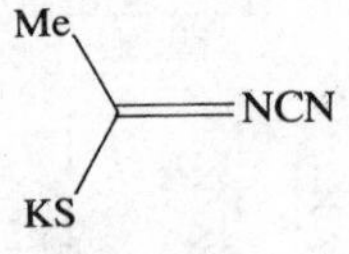 (**156**) (**157**) $ClCH_2CSOEt$ (**158**)

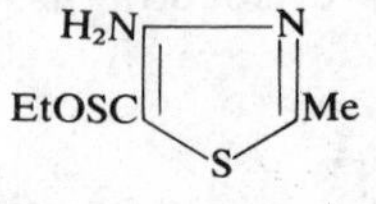

C—N—C—S+C approaches utilize condensations of the readily available dithio-salts (**153**) with alkyl halides, and base-catalysed cyclization of the products (**154**) to thiazoles such as (**155**).[71] In a related procedure the salt (**156**) was condensed with (**157**), which is a bifunctional thioacylating reagent of considerable potential use in heterocyclic synthesis, giving (**158**).[51] 5-Aroyl-2-arythiazoles are accessible as shown in reaction 47.[84]

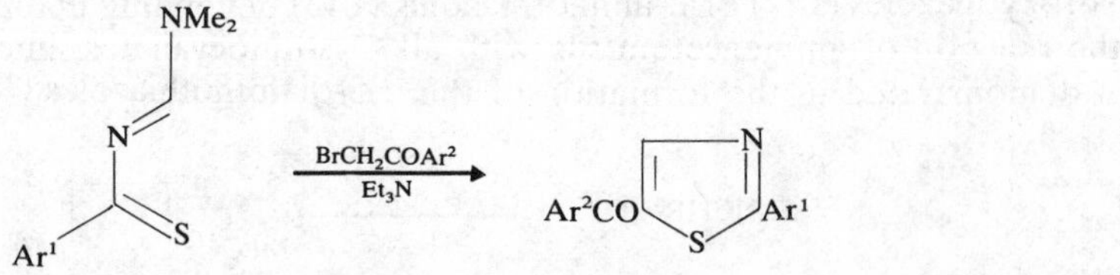

$$Ar^1C({=}S){-}N{=}CH{-}NMe_2 \xrightarrow[Et_3N]{BrCH_2COAr^2} \text{5-}Ar^2CO\text{-2-}Ar^1\text{-thiazole} \qquad (47)$$

N—C—S—C—C cyclizations are known, and generally involve the nucleophilic condensation of an imine group with a γ-carbonyl, examples being shown in Scheme 19.[51,79,80,85]

C—N—C+C—S routes include the cycloaddition of nitrile ylides with thiono-compounds[27,71] (reaction 48) and the condensation of α-metallated isocyanides with thionoesters (Scheme 20).[86] The latter route, which complements the important work of Oldenziel and van Leusen (*loc. cit.*) affords a useful general route to 4,5-disubstituted thiazoles.

C—S—C—N cyclizations find an example in the conversion of 1,3-oxathiolans (reaction 49) into 4-aminothiazoles.[71]

C—C—S—C+N routes are illustrated by a new thiazole synthesis in which thioacetic acid is reacted with an α-haloketone (reaction 50) and then the product condensed with ammonium acetate.[87]

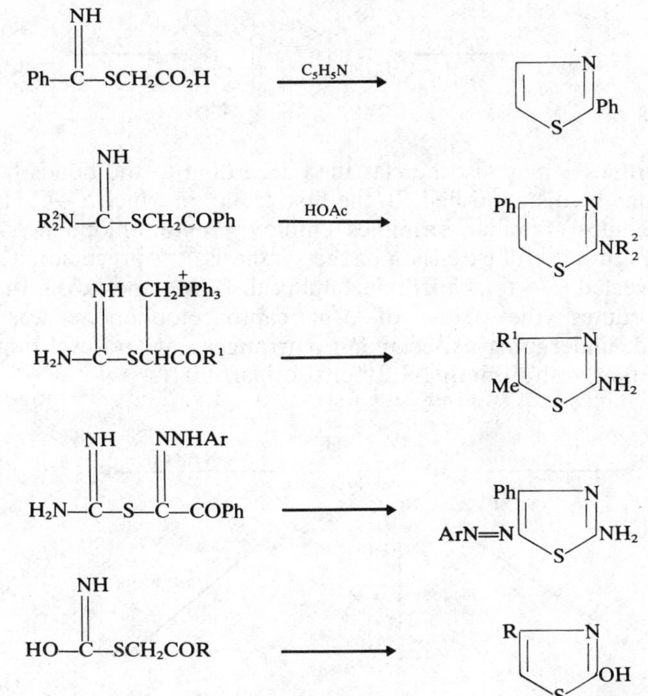

SCHEME 19

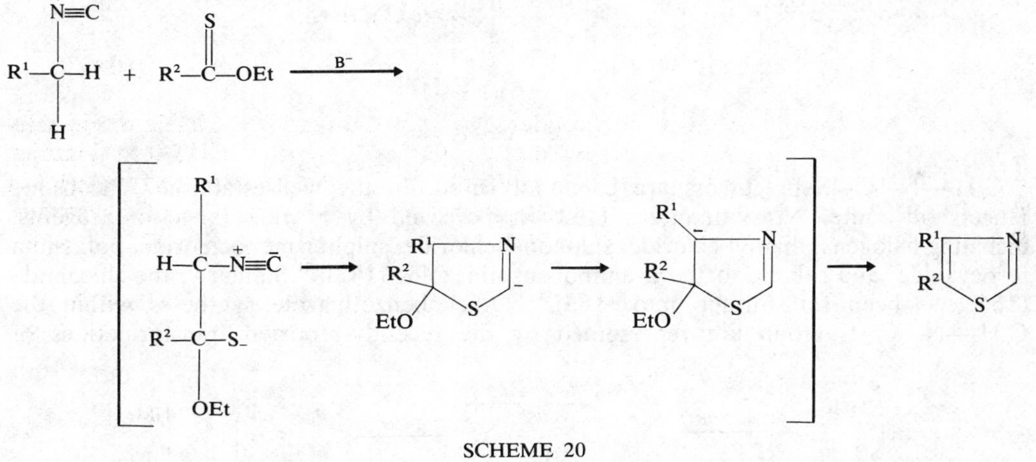

SCHEME 20

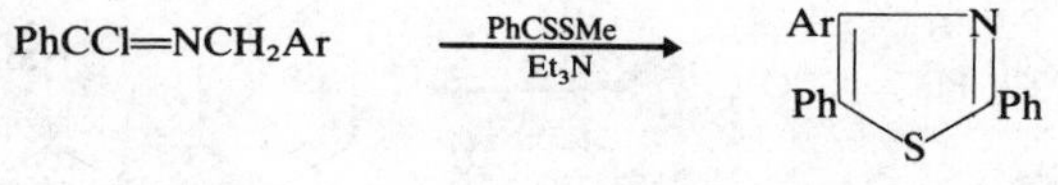

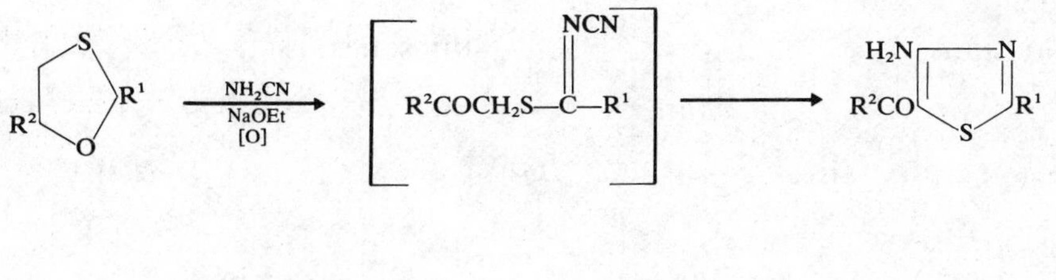

(49)

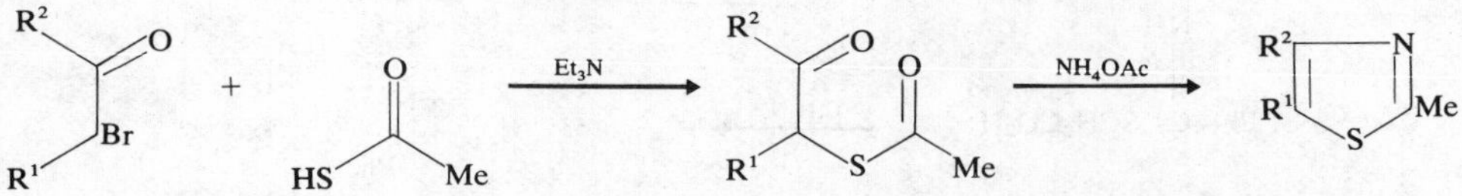

(50)

Benzothiazole syntheses may also be classified according to the bonds being formed,[34] and several types may be distinguished. In the first group, in which S—C_6H_4—N+C units are condensed, the most familiar examples employ *o*-aminothiophenols, as shown in Scheme 21.[6c,49,50,61] In a useful extension of the synthesis,[49,50] precursors (**159**) (reactions 51 and 52) are converted into the benzothiazolium salts (**160**) and (**161**). In a variation on the *o*-aminothiol routes, the oxime of *o*-mercaptoacetophenone was treated with polyphosphoric acid, undergoing a Beckmann rearrangement and cyclizing to 2-methylbenzothiazole (together with 3-methyl-1,2-benzisothiazole).[88]

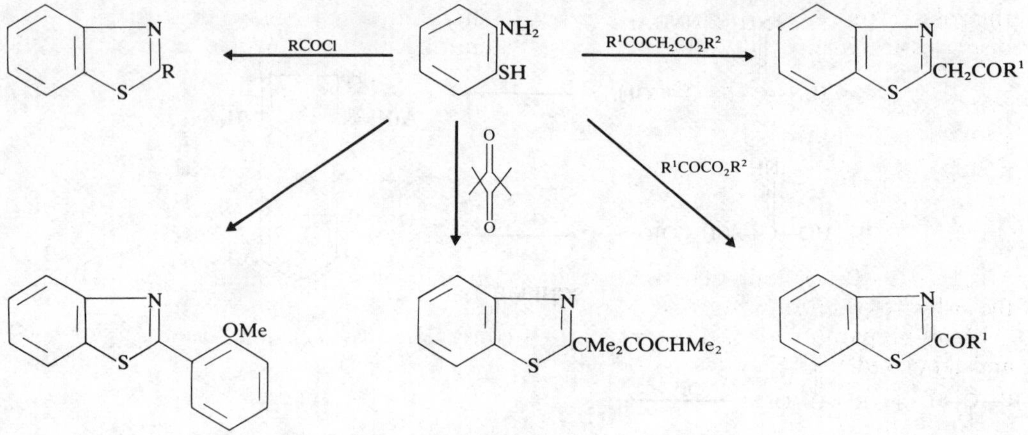

SCHEME 21

C_6H_5—N—C—S precursors are commonly used. In the well-established Jacobson–Hugershoff route, *N*-arylthioureas (**162**) are cyclized by a wide range of reagents, including halogens, thionyl chloride, sulphonyl chloride, sulphur monochloride, potassium ferricyanide, and others, to the 2-aminobenzothiazoles (**163**).[80] Similarly, the thioamide (**164**) has been transformed into (**165**).[31] Other benzothiazole syntheses within the C_6H_5—N—C—S group are represented by the recently-reported transformations of

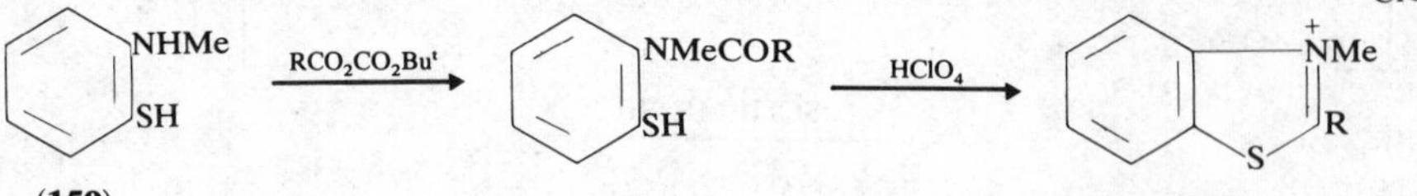

(51)

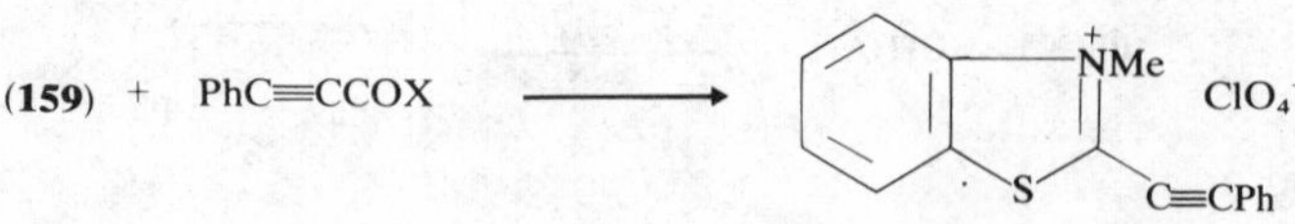

(52)

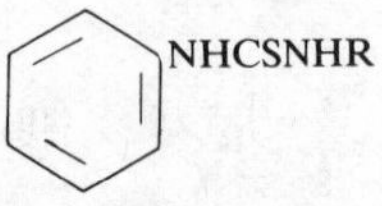

(**162**) 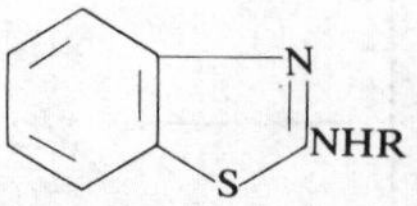 (**163**) 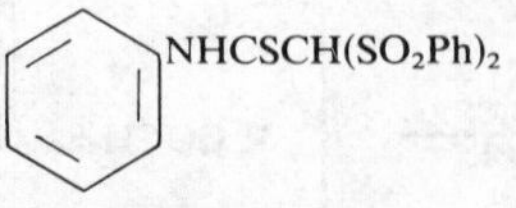 (**164**) 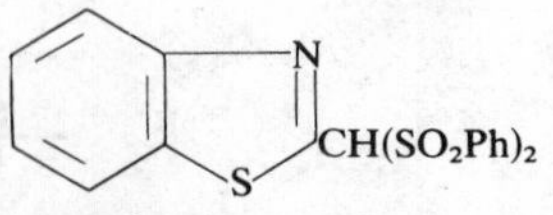 (**165**)

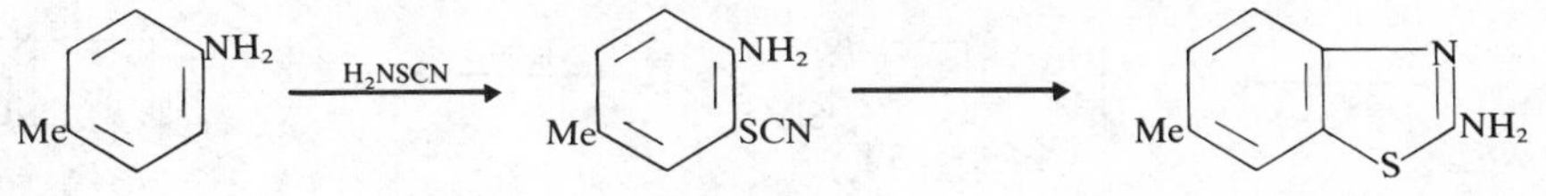

(53)

(ly)

precursors such as benzaldehye, thiobenzanilide, and benzyl benzoate with sulphur–hexamethylphosphoric triamide–aniline into 2-phenylbenzothiazole in good yield.[89]

N—C_6H_4—S—C syntheses have employed dithiocyanogen and aminothiocyanogen to introduce the thiocyanato-group into an aromatic nucleus, giving benzothiazoles (reaction 53).[50] Another example within this category is the reaction of (**166**) with methyl ketones, giving the 2-acylbenzothiazoles (**167**).[50] Possibly the major development, though, is a convenient and general synthesis which parallels the first two types of benzothiazole synthesis and which involves the reaction of an *o*-halothiobenzanilide (**168**) with a base such as methoxide, in an amide solvent, giving in good yields the substituted benzothiazoles (**169**).[90] The mechanism involves an intermediate thioimidate species which displaces the halogen in a nucleophilic reaction, and does not involve an aryne.

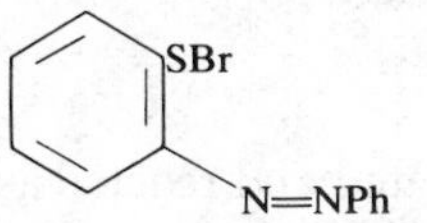

(**166**) 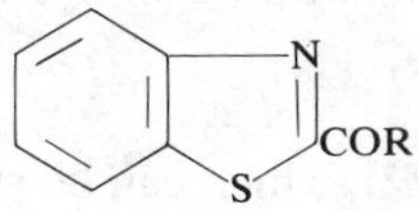 (**167**) 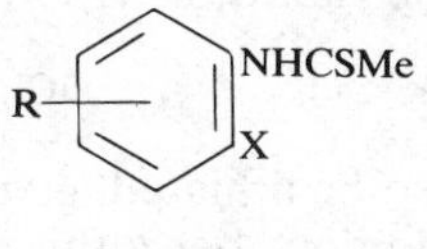 (**168**) 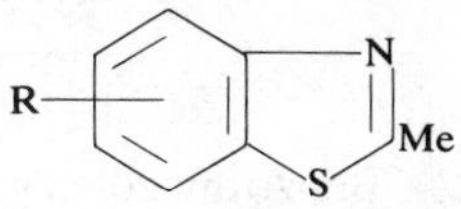 (**169**)

C_6H_5—S—C—N approaches are illustrated in reaction 54, and afford a general route to the 6-hydroxybenzothiazoles in excellent yields.[91]

Many preparations of benzothiazoles by conventional procedures have been reviewed and tabulated.[49,50,61]

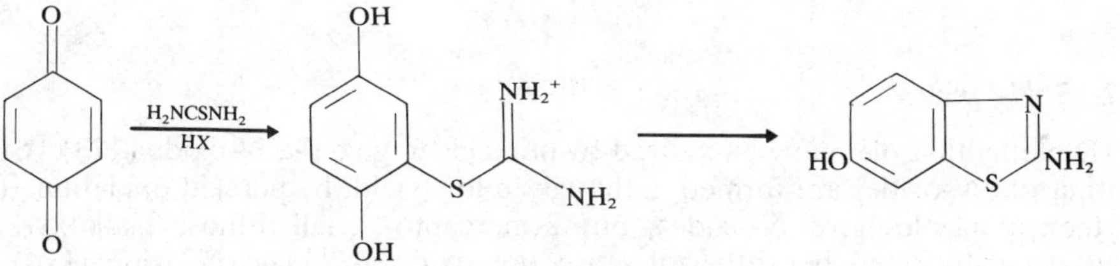

(54)

20.1.2.14 Metallation

The electrophilic mercuration of thiazole with mercuric acetate gives a derivative (**170**) which may be readily transformed into 1,3,4-tri-iodothiazole.[6c] 2,4-Dimethylthiazole gives the 5-mercuriacetyl derivative, and 4,5-dimethyl-2-phenylthiazole is mercurated in the phenyl nucleus.[51]

4,5-Disubstituted thiazoles react with, for example, phenyl-lithium, to give the 2-lithio-product.[6b] 2-Halo- and 5-halo-thiazoles are lithiated by butyl-lithium (the latter less readily) to give 2- and 5-lithio-derivatives which can be reacted with, for example, carbon dioxide, aldehydes or ketones, affording useful syntheses of substituted thiazoles.[6c,42] 2-Methylthiazoles afford lithio-derivatives (**171**) which are stable at low temperature and which rearrange and dimerize at higher temperatures.[71]

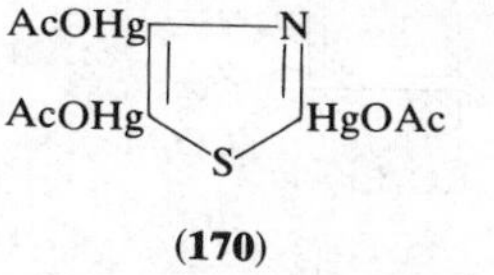

(**170**) 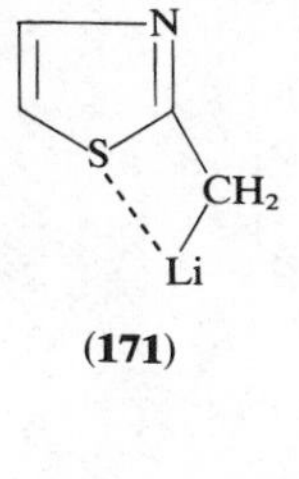 (**171**) 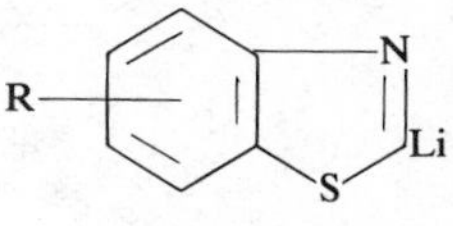 (**172**)

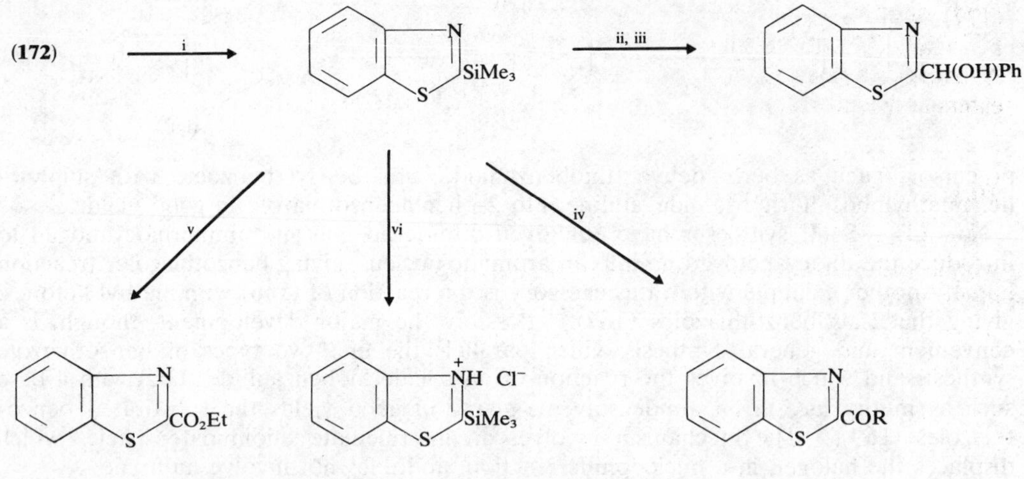

(**172**) →(i) 2-SiMe$_3$-benzothiazole →(ii, iii) 2-CH(OH)Ph-benzothiazole

i, Me_3SiCl; ii, PhCHO; iii, H^+; iv, RCOCl; v, EtOCOCl; vi, HCl, ether.

SCHEME 22

Benzothiazoles may also be lithiated, giving (**172**), which can be silylated and reacted as in Scheme 22.[49,50] In addition, benzothiazole will react as in reaction 55 *via* a copper derivative.[92]

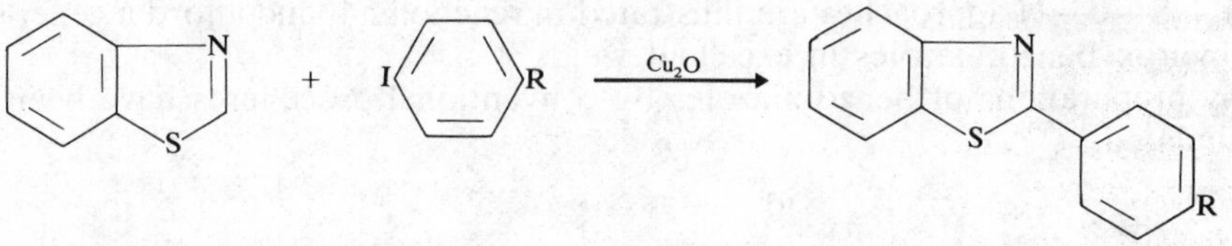

$$\text{benzothiazole} + \text{I-C}_6\text{H}_4\text{-R} \xrightarrow{Cu_2O} \text{2-(R-C}_6\text{H}_4\text{)-benzothiazole} \quad (55)$$

20.1.2.15 *N*-Oxides

2,5-Dimethylthiazole may be oxidized by peracids to give the *N*-oxide (**173**).[4] Similarly, benzothiazole *N*-oxides are formed, although in low yield, by peracid oxidation. Thus, 2-methylbenzothiazoles give *N*-oxides, but 2-mercapto-, 2-alkylthio-, 2-alkoxy-, and 2-carboalkoxy-substituted benzothiazoles are not oxidized.[61] The *N*-oxide (**174**) can be transformed in a series of reactions into the 2-ethoxycarbonyl *N*-oxide, which can be hydrolysed and decarboxylated, giving the parent *N*-oxide (**175**) which cannot be made by direct oxidation.[49] The *N*-oxides undergo 1,3-dipolar cycloadditions with benzyne, dimethyl butynedioate, phenyl isocyanate and tetracyanoethylene.[49,50] *N*-Oxide (**175**), for example, reacts with benzyne to give (**176**), and the 2-ethoxycarbonyl *N*-oxide gives

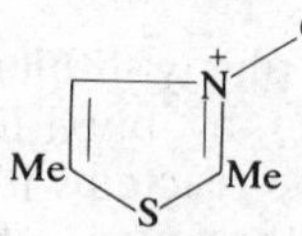

(**173**)

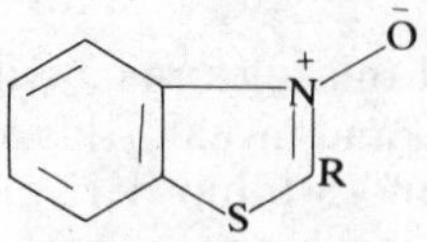

(**174**) R = Me
(**175**) R = H

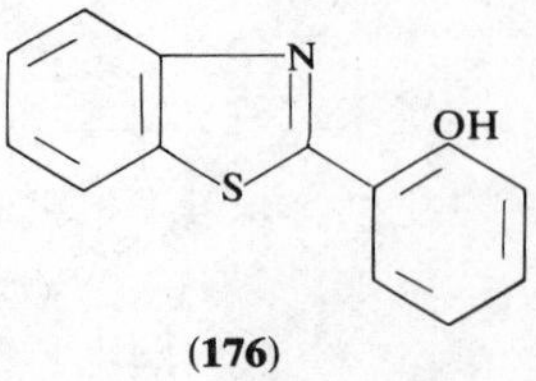

(**176**)

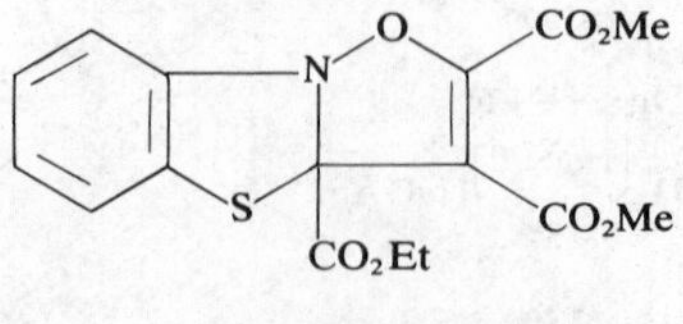

(**177**)

(**177**) with dimethyl butynedioate. Reduction of the *N*-oxides to the benzothiazoles may be achieved with trialkyl phosphites.[49,50] In an interesting synthesis, benzothiazole *N*-oxides may be prepared by treatment of *o*-chloronitrobenzenes with α-mercaptoacetic esters and triethylamine.[50]

20.1.2.16 Metal complexes

Well-characterized metal complexes of thiazoles and benzothiazoles have been obtained with a wide range of transition and other metal ligands,[49,50,61] 'TAHN' (**178**) complexes in alkali with various metals giving a sharp end-point, and forming pink-violet chelates.

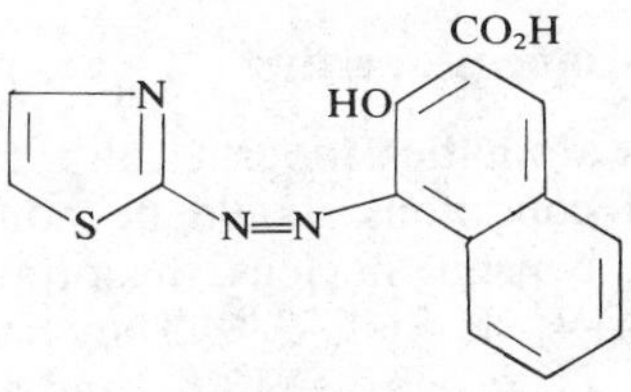

(178)

20.1.2.17 Pharmaceutical and industrial uses

An enormous amount of research into potentially pharmacologically active thiazoles and benzothiazoles has been reported, and among the many products which have emerged are antibiotics such as sulfathiazole and a host of related compounds.[34] Diuretics,[33] antihistamines,[33] anthelmintics,[33] mitodepressives,[51] mitostatics,[51] antiparasitics,[42] anti-inflammatories,[42] and antivirals[51] have been reported.

Recent industrial uses encompass vulcanization accelerators and antioxidants,[51] photochromics,[42,50] and dyestuffs.[42,49,50] A wide range of polymeric applications has been reported.[42,50,51]

20.1.2.18 Natural occurrence

The most important naturally occurring thiazole is thiamine (vitamin B_1) (**179**).[6b,11] Other systems include a new amino-acid from the fungus *Xerocumus subtomemtosus*,[42] and aeruginoic acid (**180**) from *Pseudomonas aeruginosa*.[51] The role of the benzothiazole nucleus in the bioluminescence of fireflies is important,[93] luciferin (LH_2) (**181**) being very similar to the chromophore responsible for emission. Luminescence involves intermediate dioxetans which decompose to give electronically-excited species which chemiluminesce. Another recent area of interest has involved the benzothiazole constituents of the phaemelanins.[49,50]

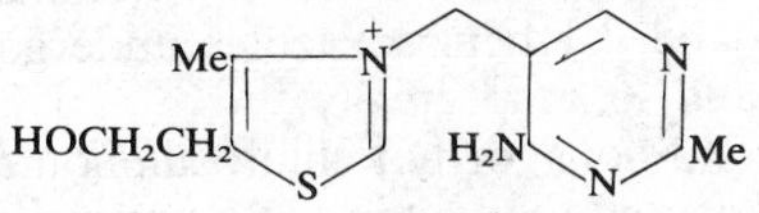

(179) 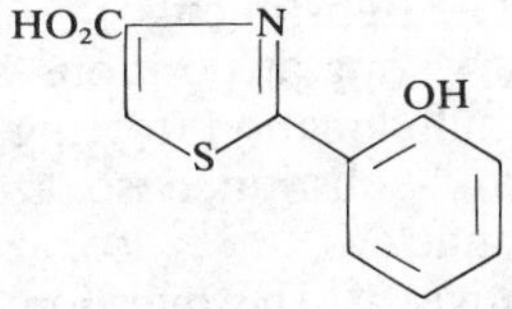 **(180)** 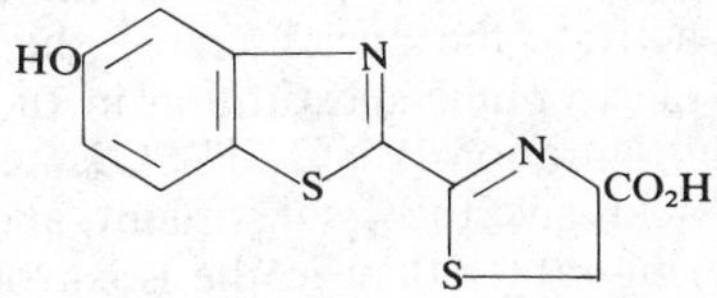 **(181)**

20.1.3 ISOXAZOLES AND THEIR BENZO-FUSED ANALOGUES

Isoxazole (**182**) is numbered as shown. Two classes of benzo-fused analogues exist, the 1,2-benzisoxazoles (**183**), sometimes termed indoxazenes, and the 2,1-benzisoxazoles

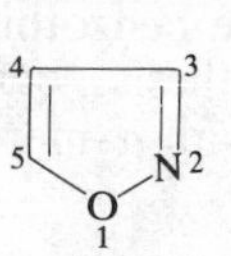

(182) 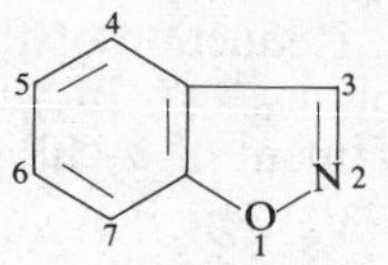 (183) 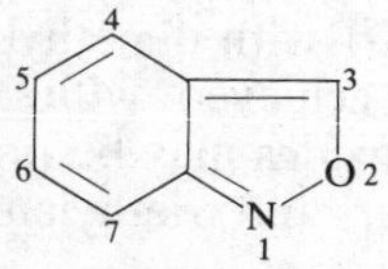 (184)

(**184**), often referred to as anthranils. Isoxazole[3,5,6,94,95] and benzisoxazole reviews[95–97] have been supplemented by more recent articles.[5] These sources, and recent publications, will be summarized.

20.1.3.1 Physical and spectroscopic properties

In general, isoxazoles exhibit aromatic properties although the lability of the ring at the N—O bond under a range of conditions should be noted. The isoxazole nucleus is a weaker electron donor than the benzene nucleus, this property decreasing at the isoxazole carbon atoms in the order C-4>C-5~C-3.[94] They are weak bases, like pyrazoles (isoxazole, pK_b 12.0; pyrazole, 11.5; pyrrole, 13.6), and are not dissolved by dilute acids although gaseous HCl will give a crystalline salt in some cases.[3]

Spectroscopic properties of isoxazoles have been reviewed.[94,95] In the u.v. spectrum, characteristic aromatic shifts in absorption maxima depend upon the number and position of substituents. C-4 Methyl substitution leads to a marked shift, whereas a C-3 methyl has little effect. The i.r. spectra have been extensively studied by Katritzky *et al.*, and have been summarized.[94]

Physical, theoretical and spectroscopic properties of the 1,2-benzisoxazoles have been reviewed.[97] Similar data for 2,1-benzisoxazoles have been summarized[97] and the weakly basic properties of these heteroaromatic 10π-electron species noted. Simple salts of 2,1-benzisoxazoles are not known, although they are protonated in acid solution.[97] They will form crystalline complexes with mercuric chloride.[95,97]

20.1.3.2 Electrophilic substitution

Isoxazoles undergo electrophilic substitution more readily than pyridine, and less readily than other five-membered ring heterocycles with one heteroatom. Theoretical predictions, substantiated by experiment, are (a) because of the electron-withdrawing effect of nitrogen, electrophilic substitution will be less ready than for benzene; (b) electrophilic substitution should occur β to nitrogen as in pyridine and other azoles; and (c) the electron-donating oxygen should facilitate such reactions in isoxazoles, compared with benzene.[98] Known reactions include nitration, sulphonation, halogenation, chloroalkylation, hydroxymethylation, and mercuration. Reactions occur readily at C-4, corresponding to pyridine, and frequently fail if C-4 is substituted.[6c] Yields may be poor to moderate, but increase with electron-donating 3- and 5-substituents (C-5 substituents activate the nucleus more than C-3 substituents[1]). 1,2- and 2,1-benzisoxazoles undergo electrophilic substitution in the aryl ring at positions-5 and -6, respectively.[97]

Nitration (HNO_3–H_2SO_4) and sulphonation (forcing conditions, or hot chlorosulphonic acid) give the 4-substituent, and aryl-substituted isoxazoles tend to react in the *homocyclic* ring rather than in the isoxazole nucleus.[6b,95,97] Isoxazole itself nitrates only in low yield and does not give dinitro-derivatives.[94] The question of the mechanism of electrophilic attack on *N*-protonated isoxazoles or on the free base has been raised.[6c,99] Benzisoxazoles undergo electrophilic substitution exclusively in the aryl ring, giving, for example, 5-nitro-1,2-benzisoxazole. Friedel–Crafts conditions may lead to rupture of 1,2-benzisoxazole rings.[95]

Halogenation may give rise to coordination compounds, 3-phenylisoxazole for example giving a perbromide.[97] 3-, 5-, and 3,5-dimethylisoxazoles give dihalides which on heating or exposure to light give 4-haloisoxazoles.[94,95] A good general method for iodination or bromination is to use iodine or bromine with concentrated nitric acid.[94]

Chloroalkylation (CH_2O–HCl–$ZnCl_2$) gives the 4-chloromethyl product, the reaction being facilitated by 5-phenyl or 3,5-dimethyl substitution.[94] (Pyridines do not readily undergo chloromethylation.)

20.1.3.3 Nucleophilic reactions

Many nucleophilic reactions result in ring cleavage (see Section 20.1.3.10). Direct substitution of H-3 and H-5 is uncommon because of the lability of the ring towards nucleophilic reagents,[94] but 5-halo- and 5-methoxy-groups are replaceable, as shown by the formation of (**186**) from (**185**). 5-Halo-groups are comparable in reactivity with acyl

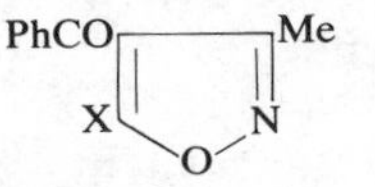

PhCO Me X N O

(**185**) X = Cl
(**186**) X = R_2N or OH

halides.[95] C-3 Substituents are more resistant to displacement, but halides may be replaced by alkoxides using, for example, hot concentrated methoxide.[95] C-4 Substituents tend to be inert to nucleophilic reaction, although the substitution by halide of a 4-diazonium group may be viewed as an S_N1 type of reaction.[94]

3-Substituents on the 1,2-benzisoxazole ring are stable to acids and alkali.[95] 3-Unsubstituted 1,2-benzisoxazoles undergo ring cleavage (Section 20.1.3.10).

20.1.3.4 Quaternization

Alkyl iodides or sulphates will give crystalline salts (**187**), which are readily cleaved by nucleophiles.[6c] Forcing conditions (methyl iodide, heat, sealed tube) are sometimes necessary. 1,2-Benzisoxazoles are quaternized by dialkyl sulphates, benzyl halides, and trialkyloxonium fluoroborates to give a range of salts.[97] 2,1-Benzisoxazoles are also quaternized, the salts being very readily ring-opened.[97]

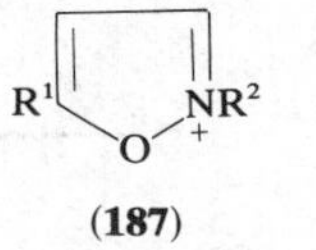

R¹ NR² O +

(**187**)

R¹ R² · N O

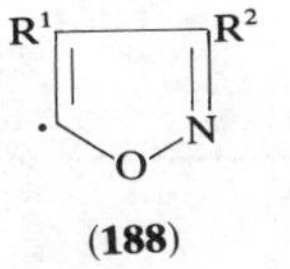

(**188**)

20.1.3.5 Isoxazole radicals

5-Aminoisoxazoles are diazotized by isopentyl nitrite to give a 5-diazo-product which, following decomposition to the 5-yl radical (**188**),[100] may be converted (EtOH–Cu) to the 5-hydro-derivative, or trapped (I_2) as the 5-iodo-derivative. Reaction with aromatics gives the 5-aryl derivatives.

20.1.3.6 Reduction and oxidation

Reduction (Na–EtOH, Raney nickel, $SnCl_2$–HCl, HI–P, N_2H_4) normally leads to ring cleavage of isoxazoles and benzisoxazoles[94,95,97] (see Section 20.1.3.10).

The isoxazole ring systems are stable to many oxidizing reagents, but generally not in alkali (although stable examples are known).[94] Thus 3-acetyl-5-methylisoxazole is oxidized by nitric acid to 3-carboxy-5-methylisoxazole, and 3-chloro-5-hydroxymethylisoxazole gives the 5-carboxy-3-chloro-derivative with alkaline permanganate. Isoxazoles are stable to peracids, but can be ozonolysed with ring cleavage to give α-keto-oxime esters.[94]

2,1-Benzisoxazoles are readily attacked both by oxidizing and reducing reagents.[95] The parent molecule (**184**) gives *o*-nitrobenzaldehyde with acidic permanganate, and *o*-aminobenzaldehyde with ferrous sulphate–ammonia.[95] 3-Methyl-2,1-benzisoxazole when heated in air gives (**189**), which on further oxidation affords dimeric indigo derivatives.

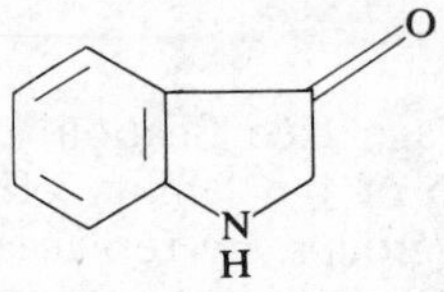

(189) 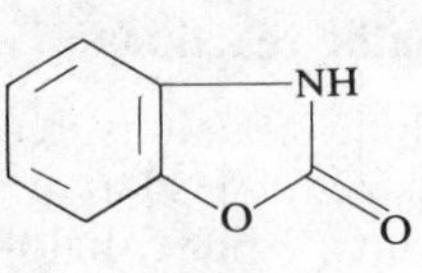 **(190)**

20.1.3.7 Photochemistry

The extensive photochemical rearrangements of isoxazoles have been thoroughly documented.[14,101,102] In many cases, azirine intermediates have been identified in the isoxazole rearrangements and interconversions with oxazoles (Scheme 23). The photo-rearrangement of 3,5-diphenylisoxazole to 2,5-diphenyloxazole has been examined as a function of varying wavelengths, and the mechanistic aspects fully discussed.[14] 3-Hydroxyisoxazole gives 2(3*H*)-oxazolone, and 3-hydroxybenzisoxazole gives the corresponding product (**190**).

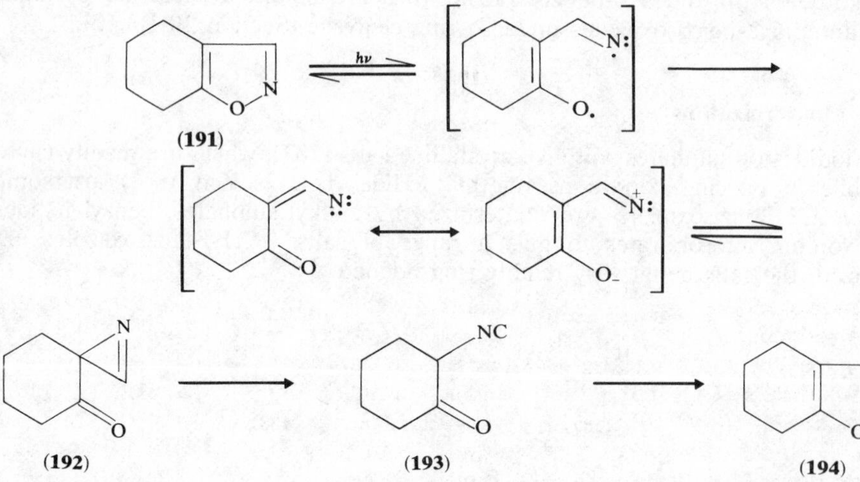

(191)

(192) **(193)** **(194)**

SCHEME 23

The isoxazole (**191**) (Scheme 23) is tranformed into the oxazole (**194**) through intermediates (**192**) and (**193**) as is shown, in quantitative yield, and the related rearrangement of the 3-methyl analogue has been studied.[103] Similarly, 1,2-benzisoxazole rearranges to benzoxazole *via* an intermediate *o*-hydroxybenzisonitrile (**195**).[103] 2,1-Benzisoxazoles

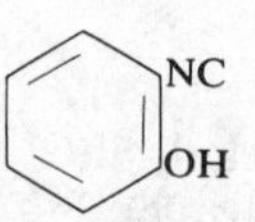

(195)

(**196**) give ring-expanded products (**197**) (reaction 56), and 3-aryl derivatives can also give substituted acridones and 2-arylbenzoxazoles,[14,104] and other rearrangements.[101,102]

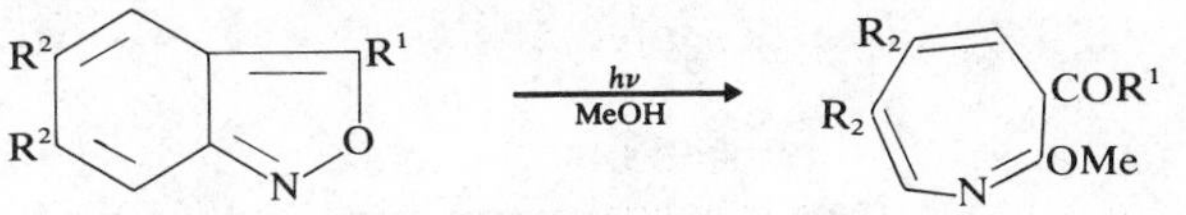

(56)

(196) (197)

20.1.3.8 Properties of substituents

3-Hydroxyisoxazole (**198**) exists mainly in the hydroxy-form, whereas 5-hydroxyisoxazole exists in the keto-form (**199**), although substituents and solvents can influence the equilibrium.[6b,6c] The latter compounds may exist in the hydroxy-form when strongly electron-withdrawing substituents are present.[6c] 5-Hydroxyisoxazoles (isoxazolin-5-ones) dissolve in sodium carbonate to give the hydroxy-salts and react with aldehydes or ketones to give C-4 condensation products such as (**200**).[6c] Self condensation may also occur. Studies of the tautomerism and protonation constants of 4-hydroxyisoxazoles have also been reported recently, and the zwitterionic form (**201**) shown to predominate.[99] 3-Hydroxy-1,2-benzisoxazoles (**202**) are readily ring-opened to give hydroxamic acids, exemplifying the lability of the N—O bond in these systems which will be discussed in more detail in Section 20.1.3.10. 3-Hydroxy-2,1-benzisoxazoles react with ethyl chloroformate and phenyl isocyanate to give *N*-substituted 3-keto-products, and with nitrous acid to give anthranilic acid.

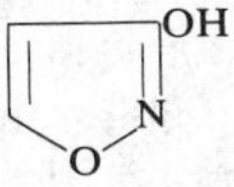

(198) 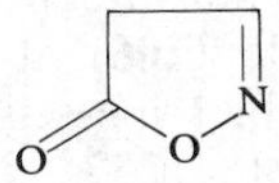 (199) 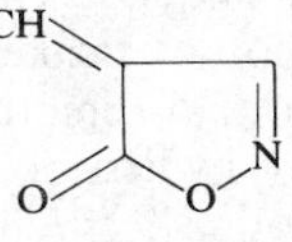 (200)

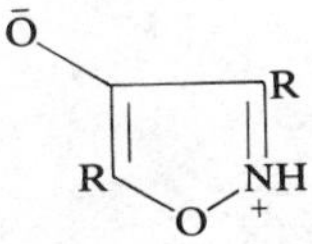

(201) 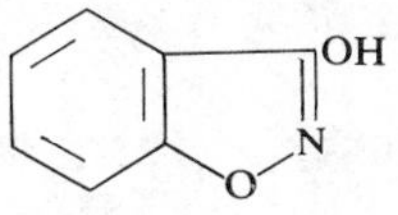 (202)

5-Aminoisoxazoles (**203**) are tautomeric with the 5-imines.[95] 4-Aminoisoxazoles, obtained by reduction of 4-nitro-precursors, exhibit properties similar to aromatic amines in that they may be diazotized (only when the isoxazole is 3,5-disubstituted) to give the 4-diazoisoxazole, which undergoes standard coupling and displacement reactions. 3-Methyl and 5-methyl substituted 4-diazonium species are very unstable, and extrude nitrogen.[95] 3-Aminoisoxazoles are prepared from the 3-azidoisoxazoles, and also give diazonium derivatives.[95] 3-Amino-1,2-benzisoxazole is stable to acids and bases, and can be transformed into 3-diazonium derivatives (**204**).[95,97]

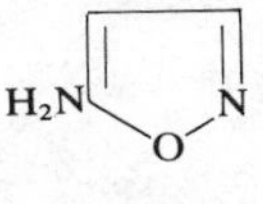

(203) 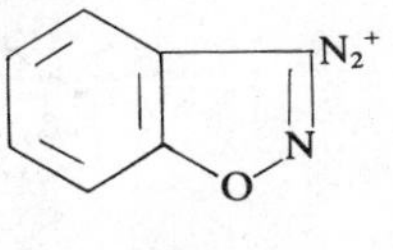 (204)

Isoxazole-3-carboxylic acids decarboxylate on thermolysis to give β-ketonitriles (reaction 57). 4-Acyl or 4-alkoxycarbonyl substituents facilitate this process to the extent that decarboxylation at C-3 may occur during hydrolysis of a 4-ester. It should be noted that crystalline isoxazole-3-carboxylic acids decarboxylate by a different mechanism involving an *N*-protonated carboxylate zwitterion. 3-Acyl- and 3-carboxy-1,2-benzisoxazoles decompose on hydrolysis to give the parent 1,2-benzisoxazole and thence an *o*-hydroxynitrile.[95] The 3-carboxy-group can be converted into the 3-amino-group by a

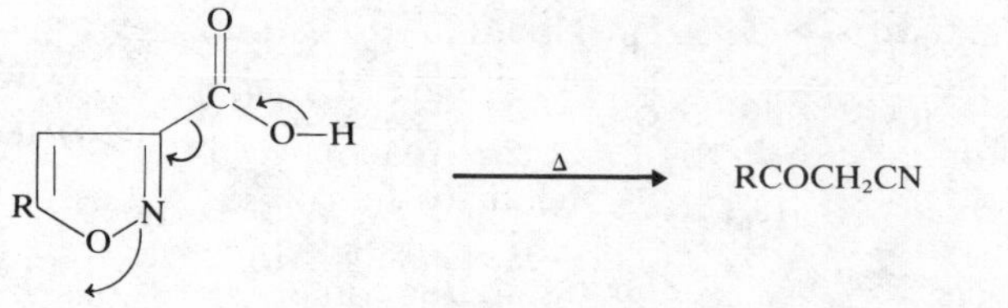

RCOCH$_2$CN (57)

Curtius process, and then by diazotization into a 3-hydroxy-group.[95] Few isoxazole-sulphonic acids are known, but (**205**) is readily hydrolysed by prolonged heating in aqueous solution giving isoxazole (**182**).

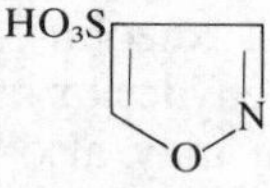

(**205**) 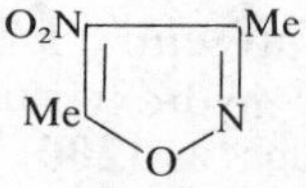 (**206**) 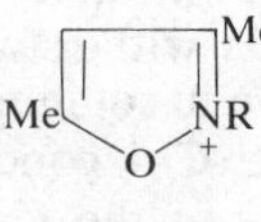 (**207**)

The reactivity of methyl groups on the isoxazole nucleus towards homolytic halogenation decreases in the order C-4>C-5>C-3, illustrating the inhibitory effect of the N-heteroatom on radical substitution at a 3-methyl group and, to a lesser extent, a 5-methyl group.[94] (In the absence of a 4-substituent, 4-halogenation may occur by an electrophilic reaction.[94]) The side-chain halides (*e.g.* 4-chloromethyl) are similar to benzyl halides.[94] The formation of carbanions from methyl substituents is shown by the Michael and aldol reactions of 3,5-dimethyl-4-nitroisoxazole (**206**) at the C-5 methyl. This reaction is catalysed by hydroxide (although reversible ring-scission is also possible) and by bases such as picoline.[94,95] The quaternized azoles (**207**) give condensation reactions at the 3- and the 5-methyl groups under mild conditions.[94]

20.1.3.9 Cycloadditions

Isoxazoles react with diphenylcyclopropenone to give 4-pyridones (**208**).[105] 2,1-Benzisoxazoles undergo Diels–Alder reactions, *N*-phenylmaleimide, for example, giving (**209**) which reacts with hydroxide to give the quinoline (**210**),[94] acetylenes also leading directly to quinolines, and benzyne giving (**211**).[104] Cycloaddition with active methylene compounds such as malononitrile gives in very high yields a whole range of quinoline *N*-oxides such as (**212**).[104]

(**208**) (**209**) 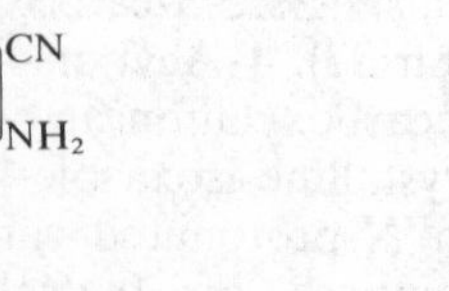 (**210**)

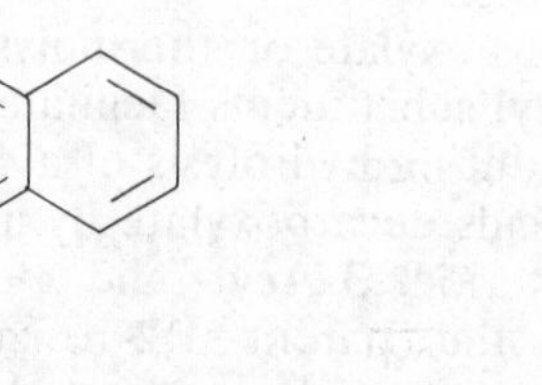

(**211**) (**212**)

20.1.3.10 Ring cleavage and rearrangement

Nucleophilic deprotonation at C-3 and C-5 occurs irreversibly in isoxazoles (but reversibly in isothiazoles), leading to fission (reactions 58 and 59). This lability towards nucleophiles distinguishes isoxazole reactivity in comparison with pyridine and other azoles.[101,102] Invariably the N—O bond is broken. 5-Aminoisoxazole is cleaved by methoxide to give α-cyanoacetamide, but the presence of electron-withdrawing substituents sharply increases the ring-lability (reaction 60),[101,102] as does quaternization (reaction 61).[101,102] The presence of a 3-acyl group in a 3,4,5-trisubstituted isoxazole may afford an alternative ring-cleavage mode (reaction 62).[101,102]

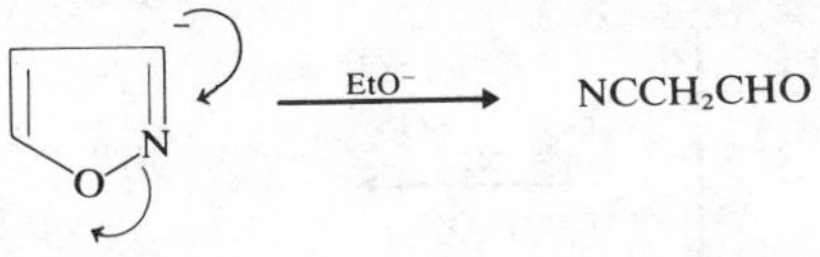

$$\xrightarrow{EtO^-} NCCH_2CHO \quad (58)$$

$$\xrightarrow{EtO^-} \longrightarrow \longrightarrow RCN + CH_3CO_2Et \quad (59)$$

$$\xrightarrow{PhNH_2} PhN{=}CHCH(NO_2)CN \quad (60)$$

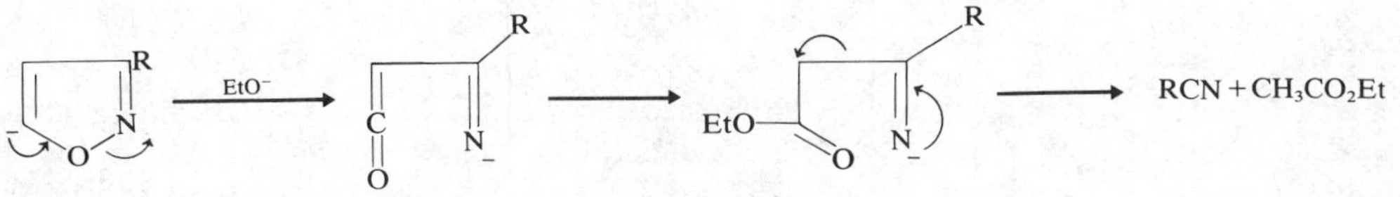

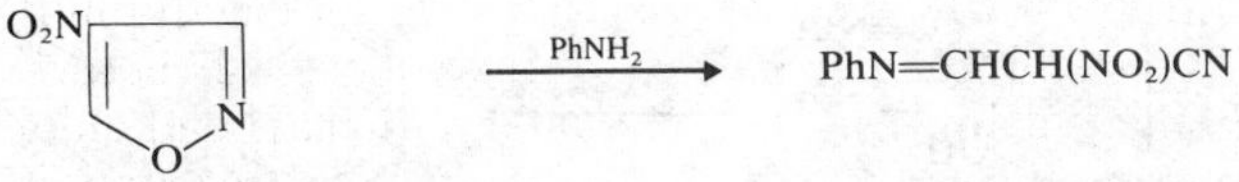

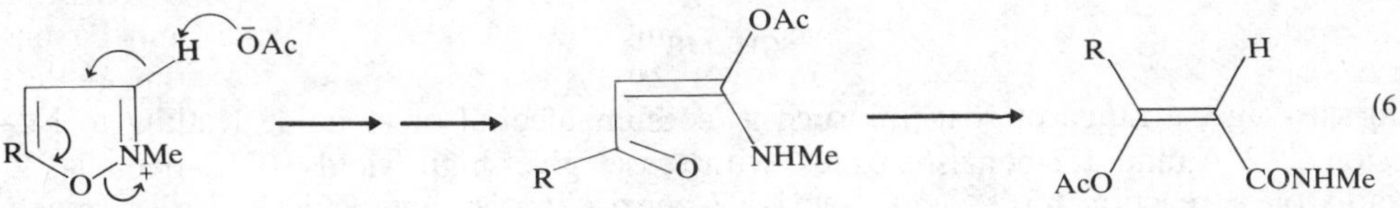

(61)

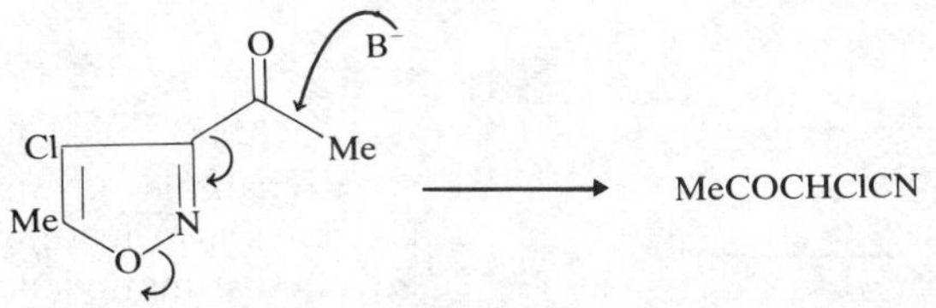

$$\longrightarrow MeCOCHClCN \quad (62)$$

Weaker nucleophiles may cleave the isoxazole nucleus, hydrazines, for example, reacting with 4-nitroisoxazoles (**213**) to give pyrazoles[101,102] (*via* reaction of the hydrazine with an intermediate α-ketonitrile).

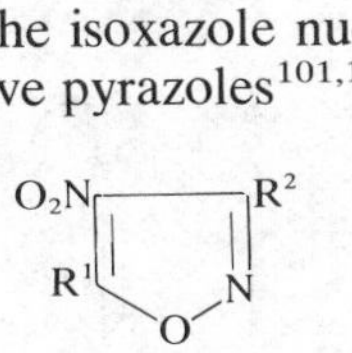

(**213**)

Reductive conditions often lead to isoxazole ring cleavage, sodium in alcohol, for example, leading to β-iminoketones (reaction 63);[95] mercury amalgam and Raney nickel reductions also give products from N—O bond scission. Such reactions involve reductive cleavage with, in the first step, addition of one or two electrons. Reducing reagents which act by other mechanisms in general do not destroy the ring (*e.g.* lithium aluminium hydride),[94] and zinc–ethanol and zinc–acetic acid systems may preserve the isoxazole nucleus. Catalytic hydrogenation often ruptures the ring.

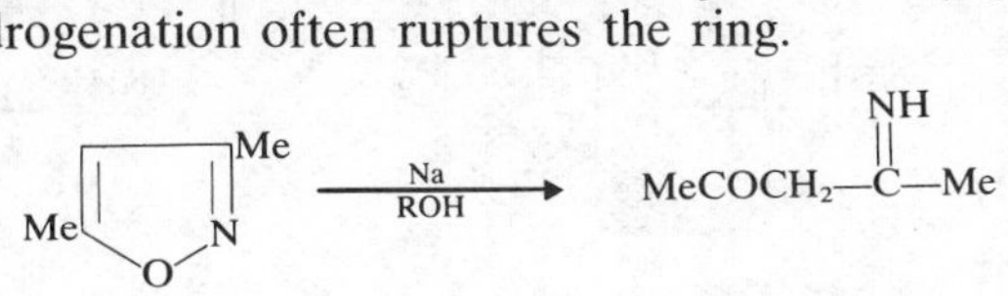

$$\xrightarrow[ROH]{Na} MeCOCH_2{-}C({=}NH){-}Me \quad (63)$$

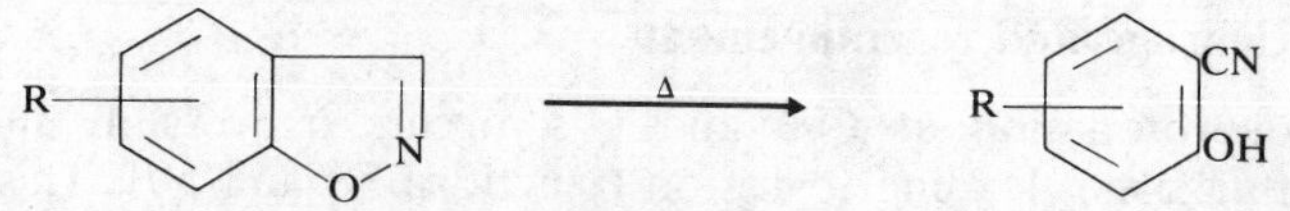

(64)

1,2-Benzisoxazoles undergo ring fission thermally. Salicylonitriles may be produced from them at the melting point (reaction 64) as a result of base-catalysed abstraction of H-3 (again giving salicylonitriles) or when treated with proton and Lewis acids.[97] Quaternized 1,2-benzisoxazoles are particularly vulnerable[97] (*e.g.* Scheme 24). Reduction also leads to

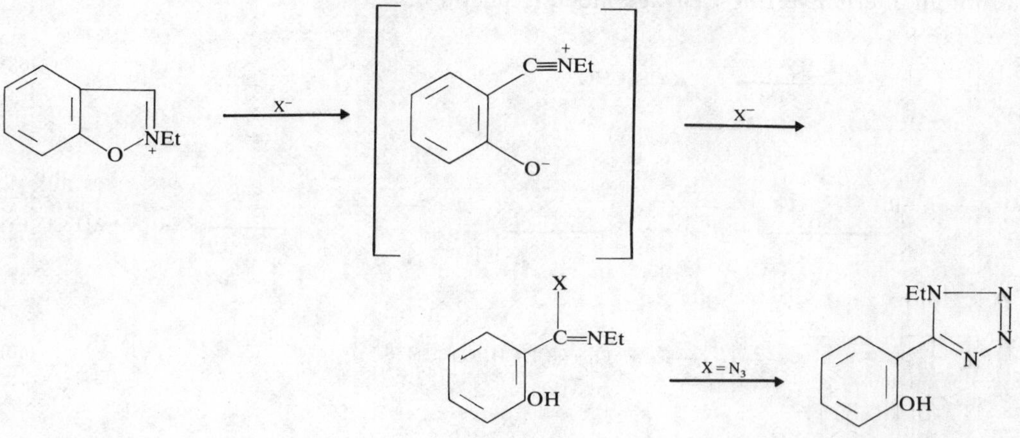

SCHEME 24

ring cleavage, a range of reagents such as sodium–alcohol or di-imide leading to N—O fission.[97] 3-Amido-1,2-benzisoxazoles with base give high yields of 3-phenyl-1,2,4-oxadiazoles (reaction 65)[104] and 3-acyl-1,2-benzisoxazoles give salicylonitriles (reaction 66).[97]

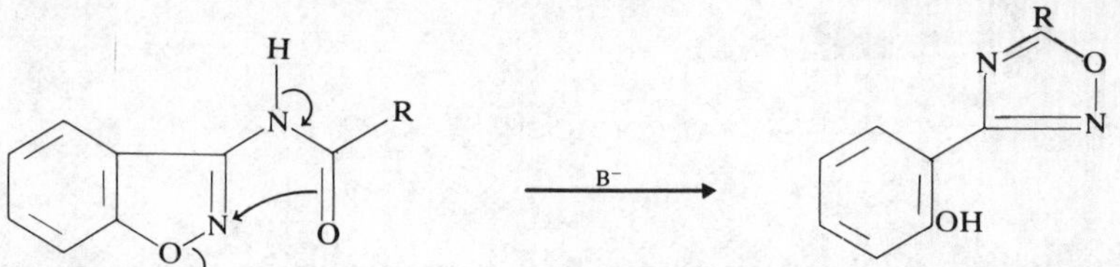

(65)

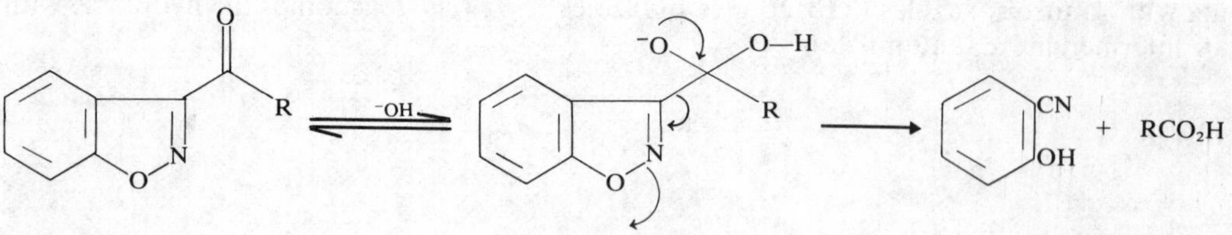

(66)

A recent interesting use of a 1,2-benzisoxazole is as a new protecting group for peptide synthesis, which withstands most conventional manipulations but is readily removed by hydrogenolysis and other methods, including aprotic base treatment (reaction 67).[106]

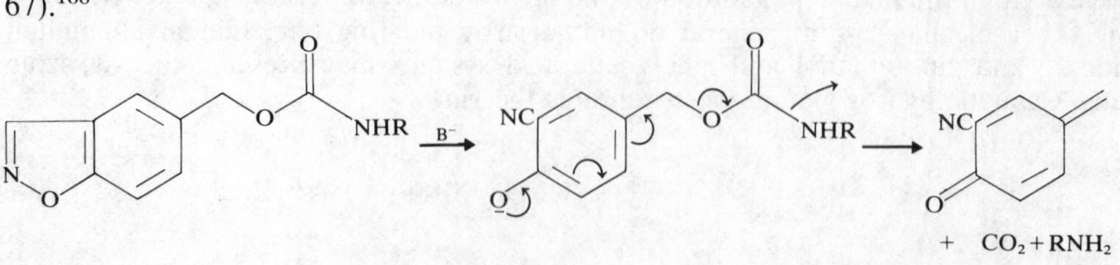

(67)

(68)

2,1-Benzisoxazoles are characteristically readily cleaved by a wide variety of reagents. Bases react with the parent molecule to give anthranilic acid *via* a 3-anion (reaction 68),[97] although ammonia attacks C-3 in a nucleophilic manner, leading to a trimeric product.[105] The 3-methyl derivative under more forcing basic conditions gives *o*-aminoacetophenone.[97] Quaternization with dimethyl sulphate leads to a mixture of methylated *o*-aminobenzaldehydes.[97] Reduction with a wide range of reagents gives *o*-amino-products and oxidation may lead to azo-coupled products or to *o*-nitroso- and *o*-nitro-products.[97] Nitrous acid leads readily to *o*-diazonium salts.[97] A large number of transformations of 2,1-benzisoxazoles into heterocyclic systems, including 2,1,3-benzoxadiazoles, 3,1-benzoxazines, benzimidazoles, acridones, indoles, and quinolines, have been reviewed.[97,104]

20.1.3.11 Use as synthons

Extensive uses of isoxazoles as synthetic precursors have been reported. For example, 3,5-disubstituted isoxazoles (Scheme 25) have been converted as shown into α,β-unsaturated ketones.[21,107]

$R^1CH_2COCH{=}C(NH_2)Me \xrightarrow{ii} R^1CH_2COCH{=}C(NHCOR^2)Me \xrightarrow{iii} R^1CH_2CH(OH)CH{=}C(NHCOR^2)Me \xrightarrow{iv} R^1CH_2CH{=}CHCOMe$

i, H_2, Pt; ii, R^2COCl; iii, $NaBH_4$; iv, H^+.

SCHEME 25

Isoxazole rings when utilized as in Scheme 26[13] can be thought of as electrophilic cyanide (CN^+) equivalents, affording a useful preparation of α-cyanoketones. A particularly useful synthetic precursor is the chloromethylisoxazole (**214**), made from 3,5-dimethylisoxazole by treatment with formaldehyde–HCl.[20] This isoxazole is in effect a four-carbon synthon, a general example being shown in Scheme 27.[20] Intermediate (**215**) is subjected to reduction of the labile isoxazole N—O bond, and is in effect a masked

i, HCO_2Et, EtO^-; ii, Me_2CHOH, H^+; iii, NH_2OH; iv, MeO^-.

SCHEME 26

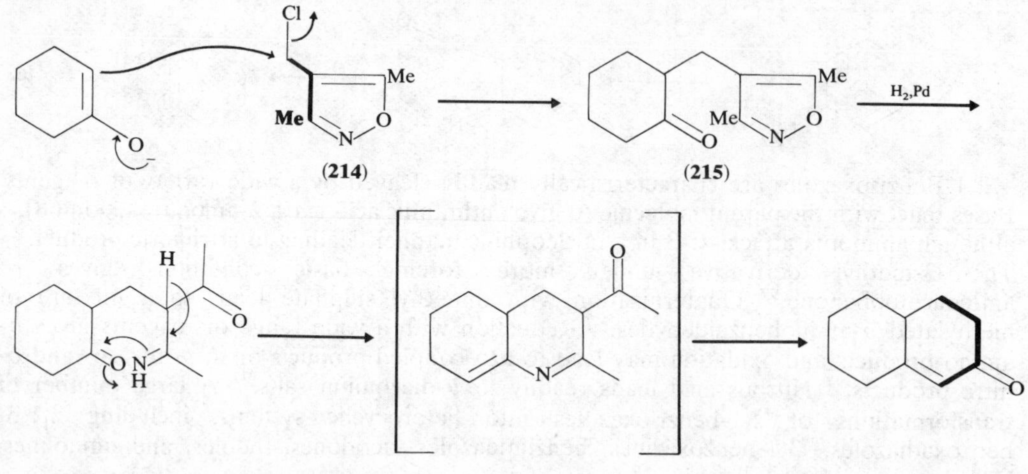

(214) (215)

SCHEME 27

triketone, leading to a ring-annulated product. Rings A and B of a steroid 4-en-3-one have been made by this approach, the isoxazole ring being maintained intact through a variety of chemical manipulations. In a synthetic variant the intermediate (**215**) was quaternized, giving (**216**). Base treatment gave a 5-methyl anion which underwent a series of reactions leading to (**217**).[13] Other uses of isoxazoles in ring annulation have been summarized.[13]

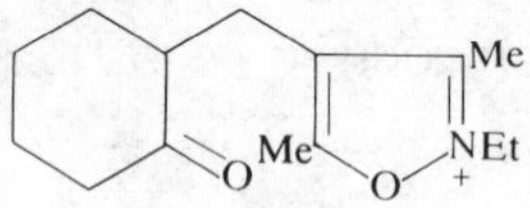

(216) 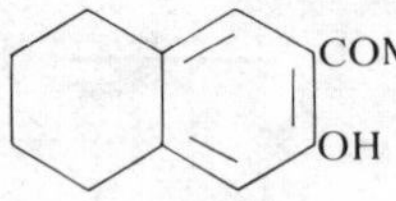 (217) 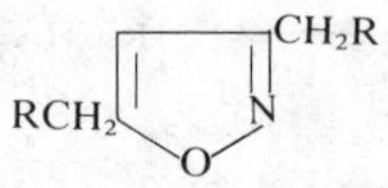 (218) RCH_2COCH_2COR (219)

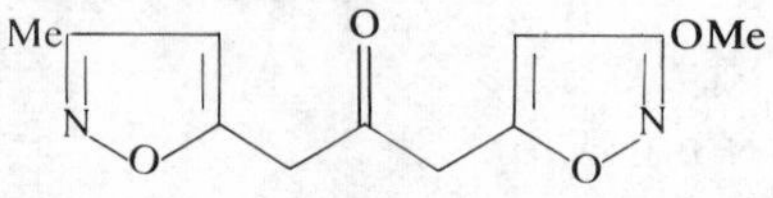

(220) 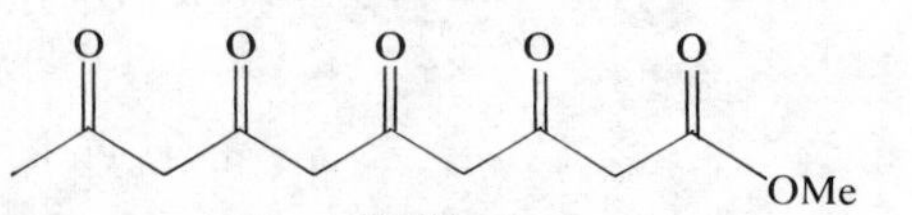 (221)

Isoxazoles have also been used as synthetic equivalents of β-dicarbonyls, as shown by the reductive transformation of (**218**) into (**219**) and of (**220**) into (**221**).[13]

The isoxazolium salts (**222**) (Scheme 28) find application in peptide synthesis, leading to high yields of coupled peptides with less than 2% racemization.[13] Benzisoxazolium salts have similarly been exploited, giving optically active protected polypeptides in high yield (Scheme 29).[13]

The synthetic transformations of isoxazoles into other heterocycles are important, as shown by the conversion of a range of substituted isoxazoles into triazoles, tetrazoles, and oxadiazoles, the general case (X,Y = C or N, Z = O or NH) being depicted in reaction 69.[108] For a specific, high-yield example, the conversion of (**223**) into (**224**) (heat or base) can be quoted.[109] In a general synthesis of 3-acylpyridines (**226**), the 4-substituted isoxazoles (**225**) were reductively cleaved and recyclized.[110] Isoxazoles have also been transformed into a wide range of heterocycles,[104] including diazines, oxazines, 1,2,4-triazines, 1,3,5-triazines, pyrroles, and seven- and eight-membered heterocyclic rings, and have been employed in the preparation of intermediates for corrin syntheses. Isoxazolium salts have been used in ring transformations, (**227**) with hydroxide giving (**228**), and (**229**) with azide ion and then heat giving (**230**).[104] The synthetically useful transformation of 3-aryl-2,1-benzisoxazoles into acridones can be depicted as in Scheme 30.[104]

$\bar{O}_3S$ (222) + HO_2CHR^1NHZ $\xrightarrow{i}$ [O-acyl enol, CHR^1NHZ, NHEt] $\xrightarrow{ii}$ HO_3S-aryl ketoamide (NHEt) + $ZNHCHR^1CONHCHR^2COX$

(**222**)

i, Et_3N; ii, H_2NCHR^2COX (X = OR, NHR)

SCHEME 28

H, B^-, NEt, O, $\xrightarrow{i}$ NEt, O $\xrightarrow{ii}$ NHEt, OSerZ, O ⇌ NEt, O, OH, SerZ ⟶ CONHEt, OSerZ $\xrightarrow{iii}$ $ZSerNHCH_2CO_2CH_2Ph$

i, B^-; ii, HOSerZ (Ser = serine); iii, $H_2NCH_2CO_2CH_2Ph$.

SCHEME 29

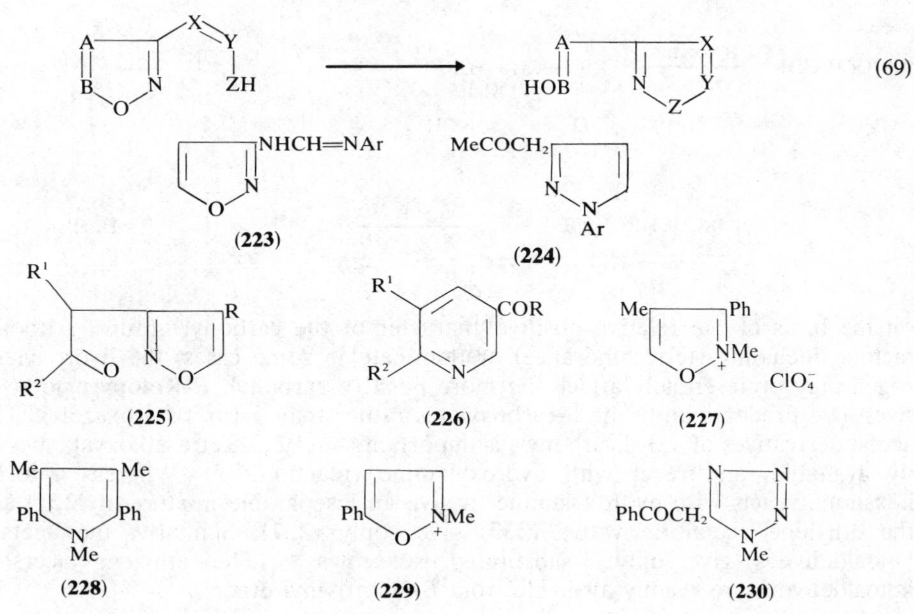

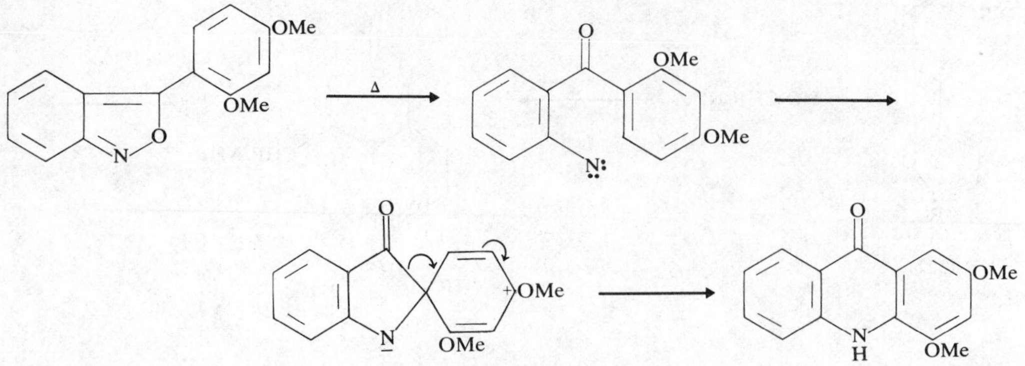

SCHEME 30

20.1.3.12 Preparation

Two main routes exist involving C—C—C+N—O and C—C+C—N—O approaches, although there are many variations within these types and there are several other synthetic methods.

C—C—C+N—O routes mostly involve condensation reactions of 1,3-dicarbonyl compounds with hydroxylamines (reaction 70). Isomeric isoxazoles may result, but the initial step in the condensation, involving nucleophilic attack by the hydroxylamine nitrogen on a carbonyl group, occurs in the order aldehyde > ketone > ester. From an extensive list of preparations[95] the sodium salt of acetoacetaldehyde is noted to react with hydroxylamine hydrochloride to give both 3- and 5-methylisoxazole, depending upon the H^+ concentration. Stable, stereoisomeric mono-oximes are often observed in these condensations (reaction 71). The direction of enolization may be a governing factor in the formation of isomeric isoxazoles. With highly enolic diketones the principal product can be predicted

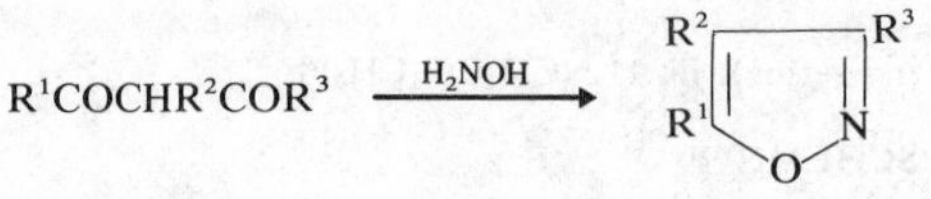

$$R^1COCHR^2COR^3 \xrightarrow{H_2NOH} \text{isoxazole} \quad R^{1,2,3} = \text{H, alkyl, aryl, CN, } CO_2R\text{, } etc. \tag{70}$$

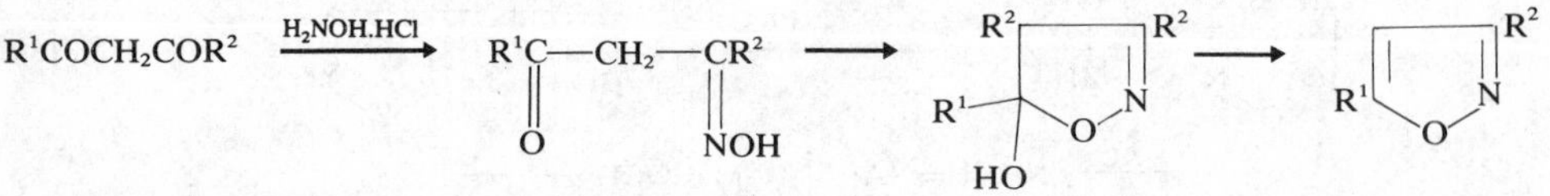

$$R^1COCH_2COR^2 \xrightarrow{H_2NOH.HCl} R^1C(=O)—CH_2—CR^2(=NOH) \longrightarrow \longrightarrow \tag{71}$$

$$(EtO)_2CHCHRCH(OEt)_2 \xrightarrow{H_2NOH} \quad R = \text{H, alkyl} \tag{72}$$

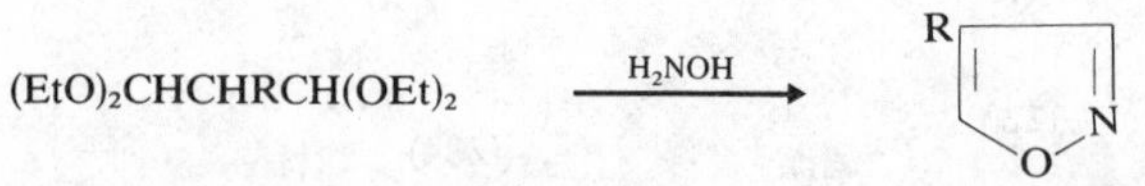

on the basis of the relative positive character of the carbonyl groups. Although other factors (including steric hindrance) predominate in some cases, the hydroxylamine nitrogen may preferentially attack the more positive carbonyl. β-Ketopyruvic acid derivatives give predominantly the 3-carboxylates rather than 5-carboxyisoxazoles. The use of acetal derivatives of 1,3-dicarbonyls is important.[94] 1,1,3,3-Tetra-alkoxyalkanes are readily available, and react with hydroxylamine (reaction 72). Whereas 2-formylcyclohexanone reacts with hydroxylamine to give an inseparable mixture of (**231**) and (**232**), the ethylene acetal derivative (**233**) gives only (**231**). Similarly, the acetals of β-ketoaldehydes give only 3-substituted isoxazoles.[94] (The ethylene acetals of β-ketoaldehydes are readily available from β-chlorovinyl ethers.)

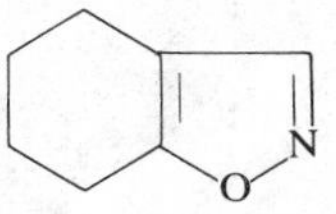

(231) 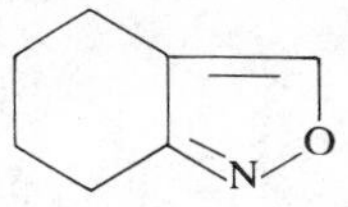 **(232)** 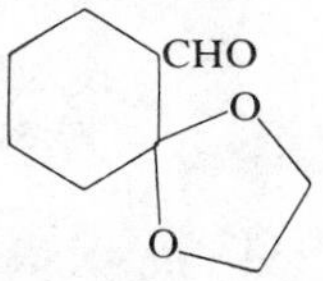 **(233)**

3- and 5-hydroxyisoxazolones may be prepared by the reaction of hydroxylamine with β-ketoesters.[95] The 4-hydroxy-derivatives arise from nitrous acid treatment of acetonedicarboxylic acid esters (reaction 73) in a process which formally may be regarded as a C—C—C+N—O reaction.[95]

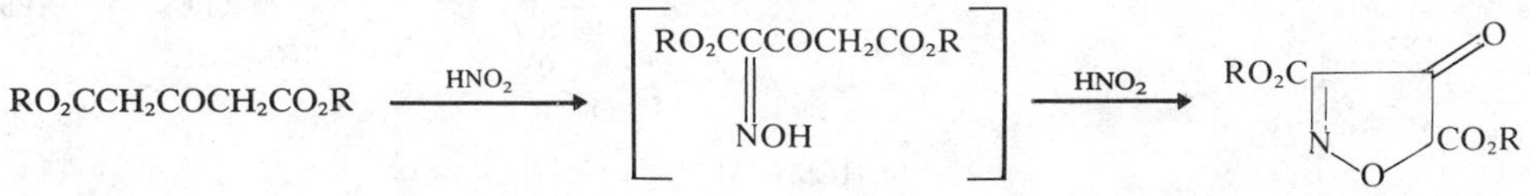

$$RO_2CCH_2COCH_2CO_2R \xrightarrow{HNO_2} [RO_2CC(=NOH)COCH_2CO_2R] \xrightarrow{HNO_2} \text{isoxazolone} \quad (73)$$

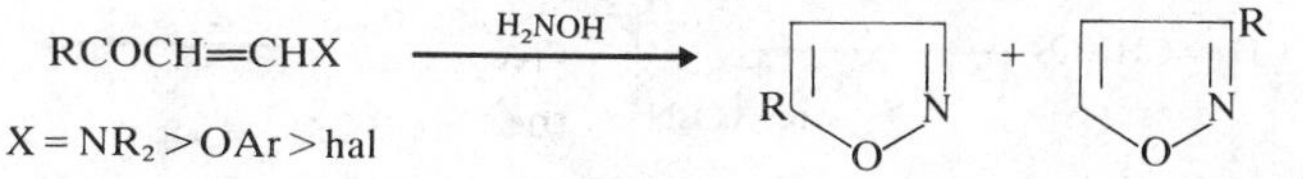

$$RCOCH{=}CHX \xrightarrow{H_2NOH} \text{5-R-isoxazole} + \text{3-R-isoxazole} \quad (74)$$

$X = NR_2 > OAr > hal$

Other variants of the C—C—C+N—O route include the reactions of β-halo-enones with hydroxylamine (reaction 74)[95] and the reaction of diethyl acetals such as (**234**), which give 4-alkylisoxazoles.[3] α-Acetylenic ketones and aldehydes and their ethylene acetal derivatives also give isoxazoles. The hydroxylamine may add to the acetylenic bond first,

$$BrCH{=}CRCH(OEt)_2$$

(234)

depending upon reaction conditions, or add to the carbonyl group, thus leading to one or both isomers.[95] α,β-Dihaloketones and α,β-epoxyketones give 3,5-disubstituted isoxazoles, and α,β-unsaturated aldehydes may react with HNO_3 to give 4-nitroisoxazoles.[95] Propargylic halides react with sodium nitrite to give 3-nitroisoxazoles.[26]

C—N—O+C—C reactions are becoming increasingly important in the synthesis of isoxazoles, and mostly involve nitrile oxide 1,3-cycloadditions (reaction 75). A large

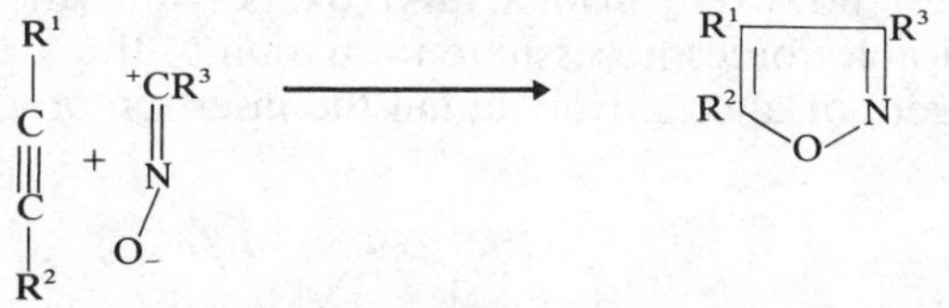

$$R^1C{\equiv}CR^2 + {}^+CR^3{=}N{-}O^- \longrightarrow \text{3-}R^3\text{-4-}R^1\text{-5-}R^2\text{-isoxazole} \quad (75)$$

number of preparations of nitrile oxides, which often have to be generated *in situ*, are available.[111] In general, electron-withdrawing substituents on the acetylene favour the process. Among the many preparations of nitrile oxides are base-catalysed 1,3-eliminations from hydroxamoyl halides (*e.g.* Scheme 31).[27] Not only do nitrile oxides react with acetylenes,[28] but also with enolizable compounds such as 1,3-dicarbonyls to give 4-acylisoxazoles, and β-ketonitriles to give 5-aminoisoxazoles.[95,97] Nitroalkanes, nitrolic acids, furoxans, and other sources afford nitrile oxides and thence isoxazoles.[26,95111] Recent variations involve cycloadditions to nitroalkenes (reaction 76) and to nitroacetonitrile (reaction 77).[112]

C—C—C—N—O routes are exemplified by the synthesis of 3,5-diarylisoxazoles from the dilithio-precursor (**235**) (reaction 78)[31] and by the related synthesis of 4-acylisoxazoles[113] (reaction 79). α,β-Unsaturated ketoximes also undergo bis(triphenylphosphine)palladium dichloride-catalysed cyclizations to isoxazoles in high yield.[141]

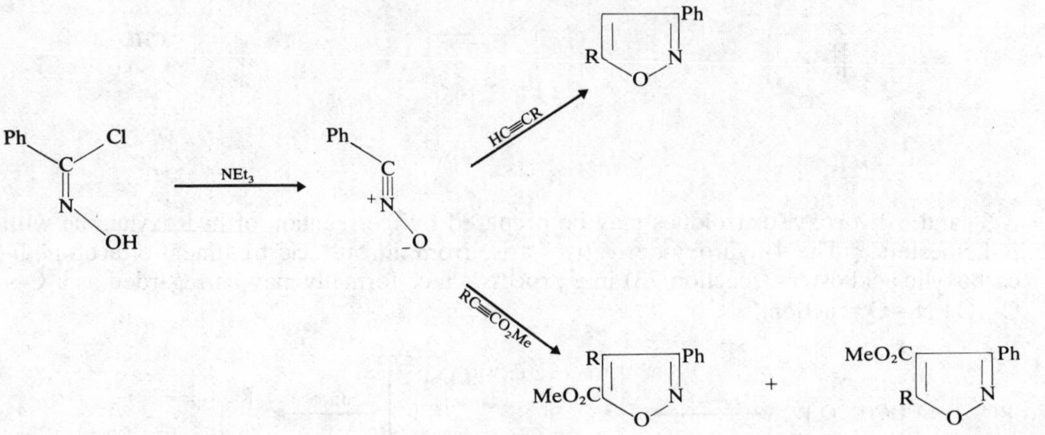

SCHEME 31

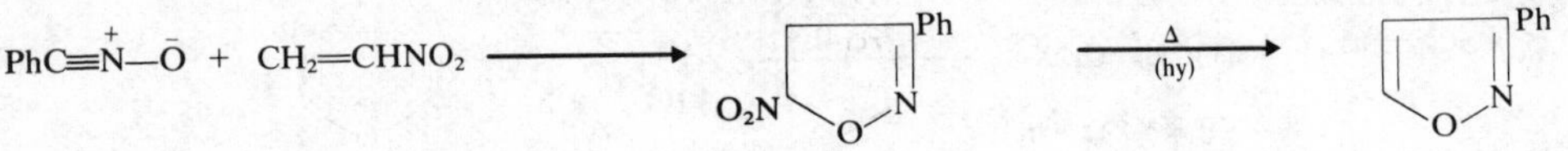

$$PhC{\equiv}\overset{+}{N}{-}\bar{O} + CH_2{=}CHNO_2 \longrightarrow \text{(3-phenyl-5-nitro-2-isoxazoline)} \xrightarrow[(hy)]{\Delta} \text{(3-phenylisoxazole)} \quad (76)$$

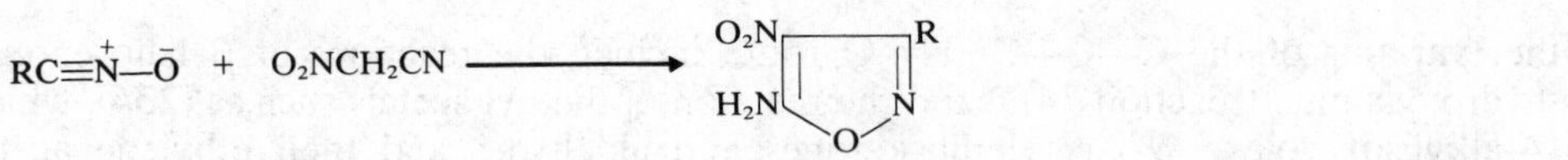

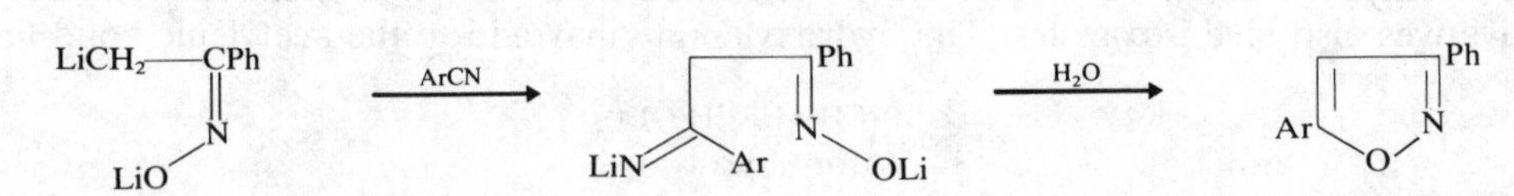

$$RC{\equiv}\overset{+}{N}{-}\bar{O} + O_2NCH_2CN \longrightarrow \text{(5-amino-4-nitro-3-R-isoxazole)} \quad (77)$$

$$LiCH_2C(Ph){=}NOLi \ (\mathbf{235}) \xrightarrow{ArCN} \xrightarrow{H_2O} \text{(5-aryl-3-phenylisoxazole)} \quad (78)$$

$$(\mathbf{235}) \xrightarrow[\text{ii, HCl}]{\text{i, Ar'CO}_2\text{Me}} \text{(4-Ar'CO-5-Ar'-3-phenylisoxazole)} \quad (79)$$

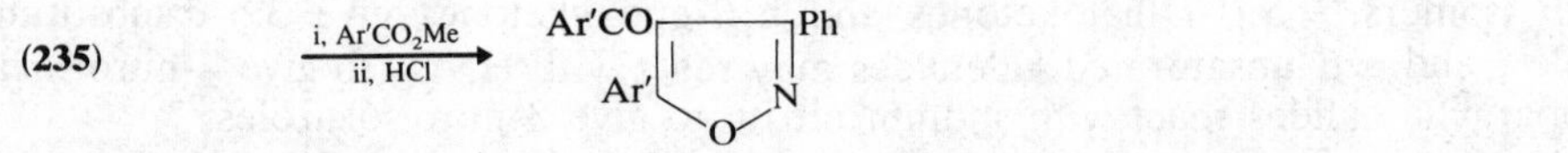

The syntheses of 1,2-benzisoxazoles can involve condensations of fragments (Scheme 32). The major routes, however, involve aryl oximes and the elimination of an *o*-substituent. Less applicable routes necessitate formation of the 1,2-bond with elimination of a group from nitrogen, or alternatively entail the insertion of a nitrogen. Other routes are even less general.

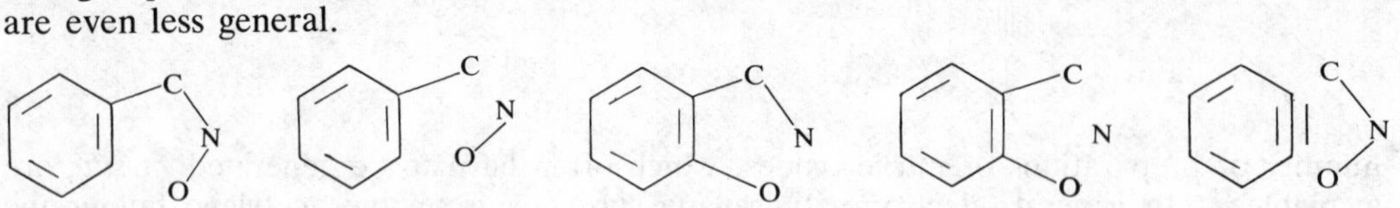

SCHEME 32

o-Substituted *syn*-oximes react more readily (reaction 80) than the *anti*-isomers, alkali catalysing expulsion of the substituent ($X=F$, Cl, Br, I, NO_2, NH_2, OH, OR, N_2^+).[95,97] When $R^2=H$, acyl, or carboxy, decomposition of the benzisoxazole ring may occur. In

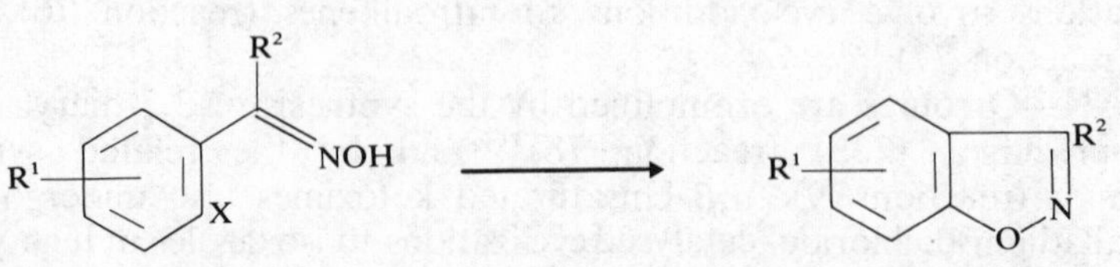

$$\text{(}R^1\text{-substituted }o\text{-X-aryl ketoxime, }C(R^2){=}NOH\text{)} \longrightarrow \text{(}R^1\text{-3-}R^2\text{-1,2-benzisoxazole)} \quad (80)$$

certain cases, Beckmann rearrangement of the oxime may compete, leading to benzoxazoles in addition to the 1,2-benzisoxazoles.[97,115] *o*-Hydroxyoxime acetates are readily converted to benzisoxazoles in good yield by refluxing in acetic anhydride.[116] 3-Hydroxy-1,2-benzisoxazoles are accessible by thionyl chloride–triethylamine treatment of salicylhydroxamic acid, and the 3-amino-derivative may be obtained by heating (**236**).[97] The

NH_2

$NOCO_2Et$

OH

(**236**)

parent 1,2-benzisoxazole is accessible by several methods, but treatment of salicylaldehyde with hydroxylamine-*O*-sulphonic acid is particularly facile.[117] 1,3-Dipolar cycloadditions are not common in the preparation of 1,2-benzisoxazoles, but benzyne will react with nitrile oxides to give the 3-phenyl substituted products.[97]

2,1-Benzisoxazoles are prepared by two major routes, *i.e.* cyclization of fragments (A) and (B) (Scheme 33) and secondly by introduction of C-3 (C). The most important routes

(A) (B) (C)

SCHEME 33

involve the reduction of *o*-nitro- and *o*-nitroso-ketones and -aldehydes (reaction 81), the oxidation of *o*-aminoaldehydes and ketones (reaction 82), and the pyrolysis of *o*-azides (reaction 83).[95,97] Isatin derivatives (**237**) can be converted by hydroxide into 3-carboxy-2,1-benzisoxazoles,[97] and a range of *o*-substituted *N*-phenylhydroxylamines has also

COR, NO_2 —reduction→ [R, O, NHOH ⇌ R, OH, NOH] → R, O, N (81)

COR, NH_2 —oxidation→ R, O, N (82)

COR, N_3 —Δ→ R, O, N (83)

been used.[97] *o*-Nitrobenzyl alcohols such as (**238**) give 3-carboxy-derivatives.[97] Several syntheses use nucleophilic attack by active methylene groups at the *ortho* position of nitrobenzenes, with substitution of either the *o*-H or *o*-halide.[31] Other useful 2,1-benzisoxazole syntheses from *o*-nitroarenes have been reviewed.[118]

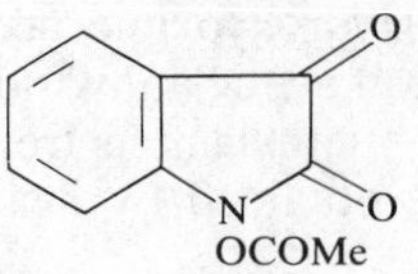

CO₂H

OCOMe

(**237**) 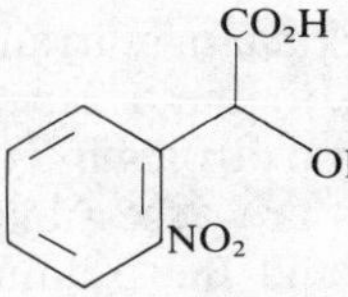 (**238**)

20.1.3.13 Metallation

Mercuration of isoxazoles may proceed by preliminary coordination, which facilitates subsequent electrophilic attack,[94] paralleling related pyridine and other azole reactions. For example, 3,5-disubstituted isoxazoles react with mercuric acetate to give the C-4 derivative (**239**) in very high yield, and thence the 4-halo-derivatives. Isoxazole itself, however, is oxidized, giving mercurous salts.

Isoxazole Grignard derivatives are known.[94] 4-Iodo-derivatives with magnesium give Grignard reagents, which react conventionally. Cleavage reactions occur in some 3-unsubstituted derivatives.

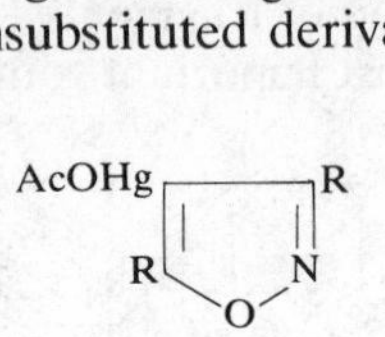

(**239**) 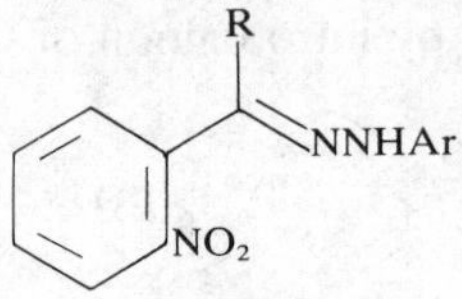 (**240**) 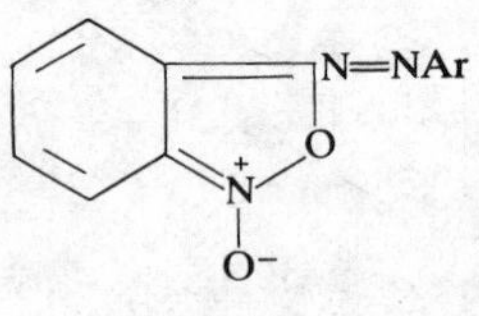 (**241**)

20.1.3.14 *N*-Oxides

Aryl hydrazones of *o*-nitrobenzaldehydes (**240**) are converted by bromine in sodium acetate into the *N*-oxides (**241**) in high yield.[118] Other isolated reports of benzisoxazole *N*-oxides have appeared.[97,119]

20.1.3.15 Pharmacological and industrial applications, and naturally occurring isoxazoles

Many industrial uses of isoxazoles and their benzo-fused analogues have been reported.[97] Pharmacologically-useful isoxazoles[33] include the antibiotic sulfisoxazoles (**242**), oxacillin (**243**), cloxacillin (**244**), and dicloxacillin (**245**), an antileprous compound (**246**), the monoamine oxidase inhibitor, Isocarboxazid (**247**) (used in psychotherapy),[94] isoxazole-fused anabolic steroids,[94] the central nervous system active isoxazole agarin (**248**) (from *Amanita muscaria*),[120] and the important isoxazole derivative cycloserine (**249**), an antituberculosis drug[94] which is recently finding other pharmaceutical uses.

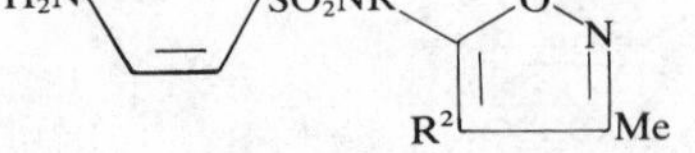

(**242**; $R^1 = H$ or Ac; $R^2 = H$ or Me)

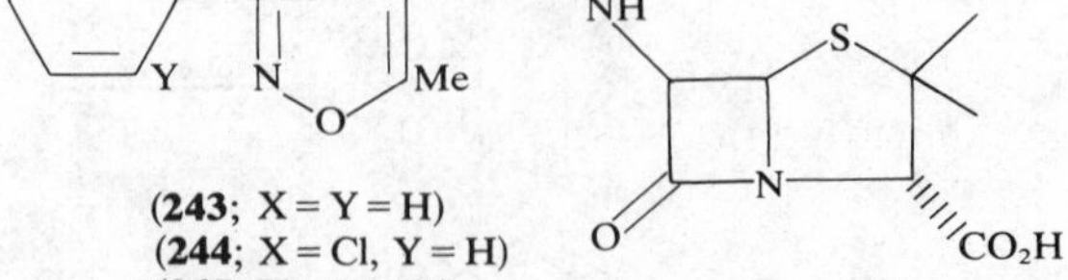

(**243**; X = Y = H)
(**244**; X = Cl, Y = H)
(**245**; X = Y = Cl)

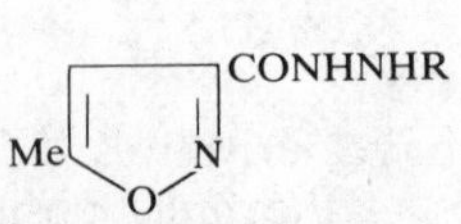

(**246**; $R = CH_2Ph$)
(**247**; R = H)

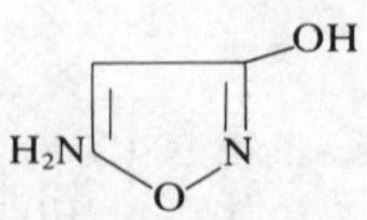

(**248**) 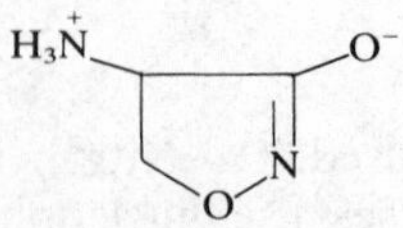 (**249**)

20.1.4 ISOTHIAZOLES AND THEIR BENZO-FUSED ANALOGUES

Isothiazoles and the two isomeric classes of benzisothiazoles are numbered in the same manner as the isoxazole analogues. Comprehensive reviews of these compounds have appeared.[121–125]

20.1.4.1 Physical and spectroscopic properties

They are weakly basic, exhibit certain aromatic properties, and, like their oxygen analogues, undergo ready fission of the S—N bond in some reactions. Isothiazole (**250**) is a colourless liquid with a pyridine-like odour, but is more toxic than pyridine. 1,2-Benzisothiazole (**251**) is a pale yellow solid which is soluble in concentrated acids, as in the 2,1-isomer (**252**). The physical and spectroscopic properties have been documented together with theoretical studies.[63,123,125,126] Isothiazole exhibits in the i.r. spectrum, in common with many other five-membered heterocycles, three bands of medium intensity between 1600 and 1300 cm^{-1} (1510, 1400, 1340 cm^{-1}). The band at 1510 cm^{-1} is displaced to higher wavenumber by electron-donating substituents such as hydroxyl, and the band at 1400 cm^{-1} is insensitive to substitution.[123,125] In the n.m.r. spectrum of (**250**),

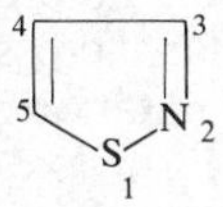

(**250**) 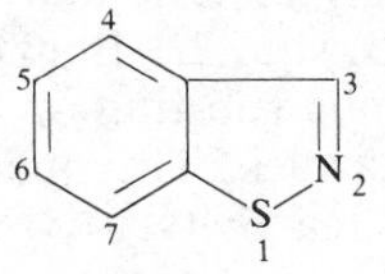 (**251**) 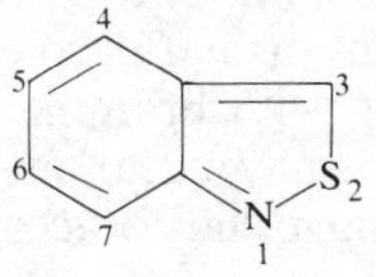 (**252**)

H-3 appears at δ8.54, whereas in (**251**) and (**252**) H-3 appears at δ8.73 and 9.06, respectively.[125,126] H-4 and H-5 in (**250**) appear at δ7.26 and 8.72. Coupling constants have been assigned. The u.v. spectra have also been studied[125,126] and mass spectral fragmentation pathways elucidated (Scheme 34). 3-Phenylisothiazole mass spectrometry has been extensively studied.[126]

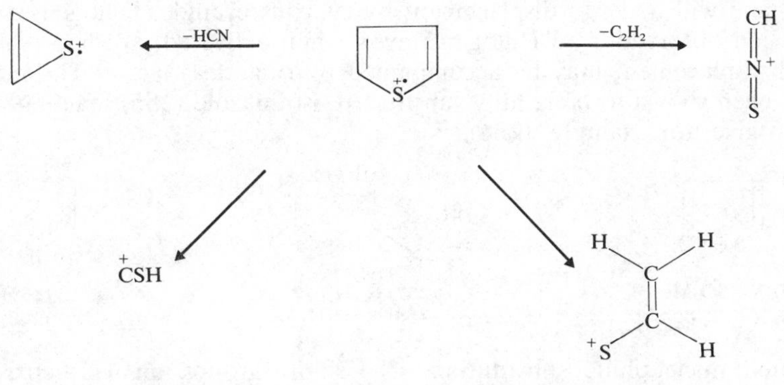

SCHEME 34

20.1.4.2 Electrophilic attack

The S—N bond is ambiphilic—electrophilic attack can occur at nitrogen or nucleophilic attack can occur at sulphur. C-3 tends to be relatively inert, but C-4 in isothiazoles is susceptible to electrophilic attack (and C-5 to nucleophilic attack).[124] Isothiazole is almost completely protonated in strong acids, but concentrated nitric–sulphuric acid gives a high yield of 4-nitroisothiazole. High yields are also obtained from

3- or 5-alkyl, or 5-bromo- or 5-acetamido-isoxazoles. It should be noted that 3-phenylisoxazole nitrates at the *meta*-position of the phenyl ring, and 4-phenylisoxazole gives *o*- and *p*-nitrophenylisoxazoles. Halogenation occurs in a variety of conditions, although in bromination, for example, poor yields of 4-bromo-derivatives are obtained because of perbromoisothiazole complex formation. Electron-releasing 3- or 5-substituents (amino, hydroxy and alkoxy) facilitate 4-halogenation, although 5-carboxy-3-methylisothiazole (**253**) gives a high yield of the 4-halogeno-derivative. 4-Methylisothiazole on chlorination gives a mixture of the 4-chloromethyl derivative and 3-chloroisothiazole.[122] The conversion of 4-diazonium derivatives to 4-haloisothiazoles has been explained as an electrophilic substitution.[124] 4-Hydroxy-3-phenylisothiazole reacts with chlorine to give the 5-chloro-product.

Me, HO_2C, N, S

(**253**)

(**254**)

2,1-Benzisothiazole gives the 5- and 7-bromo-derivatives (which differ in basicity and are separable).[63,126] Bromine–silver sulphate is an effective source of bromonium ion in this reaction. Nitration occurs mainly at C-7, and Vilsmeier–Haack and Friedel–Crafts reactions, as in thiazoles, are unsuccessful. 2,1-Benzisothiazole is a stable heteroaromatic, reacting with bromine by electrophilic substitution and not by addition across C-4 and C-5, as may occur in 2,1-benzisoxazoles. The quadricovalent representation (**254**) has been suggested to account for such reactivity.[28]

20.1.4.3 Nucleophilic substitution

Some nucleophilic reagents can cleave the ring system in isothiazoles and benzisothiazoles and their quaternary derivatives. Isothiazole 5-substituents such as halogens undergo ready nucleophilic substitution[63,123,124,126] by cyanide, hydroxide, alkoxide, and a range of related nucleophiles. For example, 4,5-dibromoisothiazole reacts with alkoxides to give 4-bromo-5-alkoxyisothiazoles, illustrating the relative susceptibilities to nucleophilic attack of C-4 and C-5. 4-Halogen substituents are more benzene-like in their reactivity, and will undergo displacement by cuprous cyanide. Halogen-exchange at C-4 has also been observed.[125] 3-Halogens, even when activated, are less reactive still, and attempted displacement may be accompanied by ring cleavage.[125] The methylsulphonyl group has been shown to be readily substituted, isothiazole (**255**) reacting with amines or alcohols to give, for example, (**256**).[125]

(**255**) X = SO_2Me
(**256**) X = OEt

(**257**) X = Cl
(**258**) X = NH_2

(**259**) X = N_2
(**260**) X = CN

Attempted nucleophilic substitution of 3-chloro-groups on 1,2-benzisothiazoles by species such as cyanide or malonate anions leads to ring cleavage (Section 20.1.4.10), although in certain cases when the nitrogen is quaternized, as in (**257**), displacement can be effected giving, for example, (**258**).[125] A nucleophilic substitution reaction of the 3-diazonium 2,1-benzisothiazole (**259**) led to the 3-cyano-derivative (**260**).

20.1.4.4 Quaternization

Adducts are obtained with strong acids and with Lewis acids. Isothiazoles decompose on attempted alkylation with methyl iodide and heat, although they react slowly at lower temperatures. The more powerful reagents methyl toluene-*p*-sulphonate, dimethyl sulphate, and triethyloxonium fluoroborate will give quaternary salts (**261**), which have been reconverted to isothiazoles by distillation or by treatment with ammonia.[125] 3-Hydroxyisothiazole gives both *O*- and *N*-alkylated products, but 3-hydroxy-5-phenylisothiazole gives (**262**) with diazomethane, and dimethyl sulphate gives salts such as (**263**).[63,126] Diazomethane reacts with 5-hydroxyisothiazoles to give the 5-methoxy-derivative.[63,126]

1,2-Benzisothiazoles do not readily quaternize, although 2,1-benzisothiazole readily forms stable salts such as (**264**),[125] and trialkyl orthoformates give (**265**).

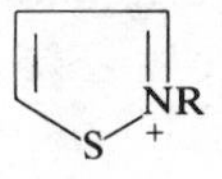

(**261**) 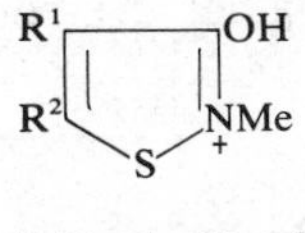 (**262**) $R^1 = H$, $R^2 = Ph$ (**263**) $R^1 = R^2 = Ph$ (**264**) 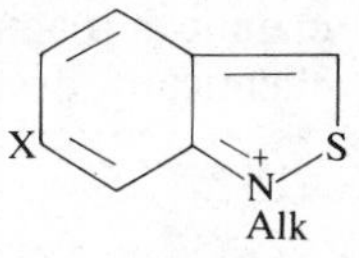 (**265**)

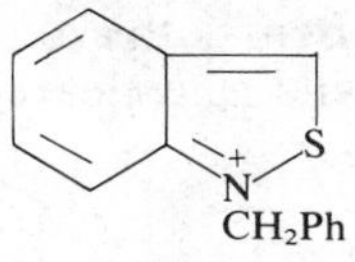

20.1.4.5 Proton exchange

Isothiazoles undergo reversible deprotonation, H-5 exchanging readily under mildly basic conditions, H-3 more slowly, and H-4 not exchanging readily. The isothiazolium counterparts exchange H-5 10^3–10^5 times more rapidly (reaction 84). The reversible deprotonation contrasts interestingly with the irreversible proton abstraction and ring cleavage of isoxazoles.[124]

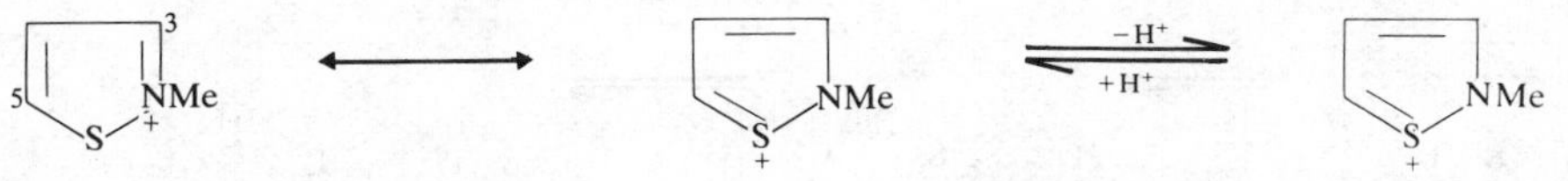

(84)

Proton exchange in isothiazole ring methyls follows the same pattern, 5-methyl groups undergoing more rapid deprotonation than 3-methyl groups and the 4-methyl being reluctant to exchange its protons.

20.1.4.6 Radical reactions

Photolysis of 5-iodoisothiazoles gives the 5-radical, which inserts in low to moderate yield into substituted benzenes, the isomeric ratio of *meta* to *ortho* and *para* substitution being lower than that of phenyl radicals reacting with the same substrates.[63,126] Decomposition of diazonium salts also leads to 5-isothiazolyl radicals.[127,128] 4-Isothiazolyl radicals insert in a similar manner, and they also resemble the phenyl radical.[63,126]

Phenyl radicals generated by thermal decomposition of dibenzoyl peroxide insert into isothiazole at C-3, C-4, and C-5 in the ratio 47:9:44. Both 3- and 4-methylisothiazole give 5-phenyl adducts, and 5-methylisothiazole gives the 3-phenyl adduct (**266**).[127,128]

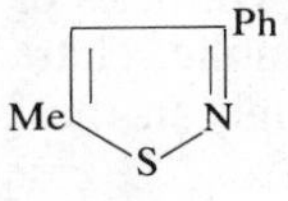

(**266**) 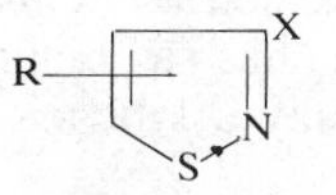 (**267**) X = Me (**268**) $X = CO_2H$

20.1.4.7 Reduction and oxidation

The isothiazole ring is generally stable to a range of oxidizing and reducing systems, including nitric acid–sulphuric acid, potassium permanganate, and chromic acid. For example, (**267**) reacts with chromium trioxide–sulphuric acid to give (**268**).

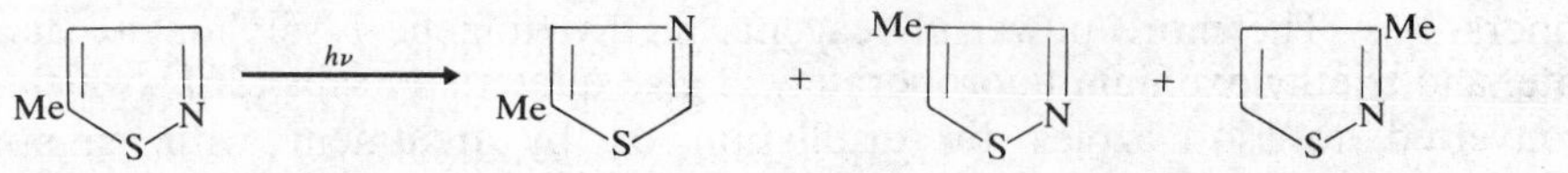

(85)

20.1.4.8 Photochemistry

The principal photoreaction is isomerism (*e.g.* reaction 85), the isomeric yields increasing with solvent polarity, supporting the zwitterionic types of intermediate proposed for oxazole and thiazole photolyses.[14] 4-Phenylisothiazole is photostable in certain photochemical reactions.[14] 1,2-Benzisothiazole gives (**269**), but 2,1-benzisothiazole is photochemically desulphurized and ring cleaved.[14]

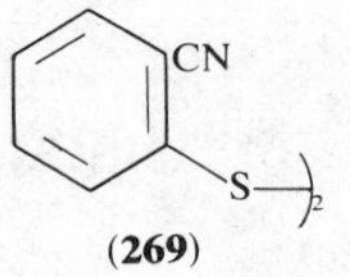

(**269**)

20.1.4.9 Properties of ring substituents

3-Hydroxyisothiazole exists as such in non-polar solvents, and as the 3-one in solvents of high dielectric constant, as shown mainly by u.v. spectroscopy.[63,126] In the 5-hydroxy equilibrium series (reaction 86) the ketonic forms predominate in the solid state, but make a small contribution in solution. 4-Bromo-5-hydroxy-3-methylisothiazole possibly exists

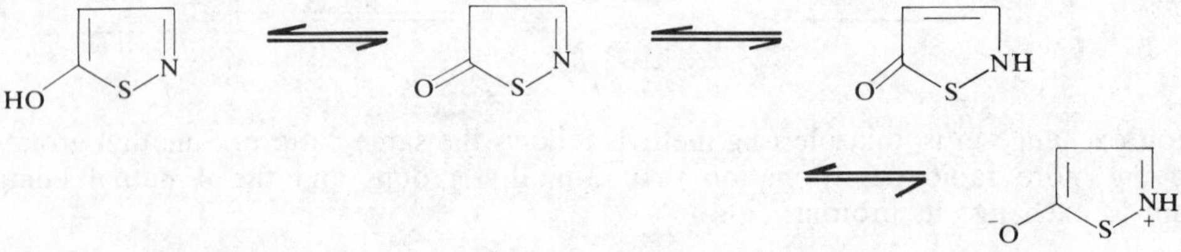

(86)

mainly in the NH form, with contributions from the zwitterionic form.[124] 3-Hydroxy-5-phenylisothiazole exists in the hydroxy-form even in methanol.[124] Methylation of 3-hydroxyisothiazole with diazomethane gives the *N*-methylisothiazolin-3-one, but this reaction depends on the rate of interconversion of the hydroxy- and keto-forms and the relative rate of reaction of these forms.[123] Acylation of 3-hydroxyisothiazole has been intensively studied.[129] Acyl chlorides and triethylamine in aprotic solvents give, in a kinetically controlled process, the 3-acetoxy-product (reaction 87). On standing, an intermolecular rearrangement (as shown by mixed migration experiments) gives the *N*-acylated product, depending upon the steric bulk of the acyl group. 3-Hydroxyisothiazole is attacked by cyanide ion at pH 0–5.5 (reaction 88) to give *cis*-3-thiocyanoacrylamide, but a second reaction pathway for the equilibrium, involving an amide N-anion, was also revealed.[130] (This is the reverse of a synthesis of 3-hydroxyisothiazole.) 3-Hydroxyisothiazole has a higher degree of phenolic character than 5-hydroxyisothiazole.[6c] 3-Hydroxy-2,1-benzisothiazole exists mainly in the ketonic form.[68,126]

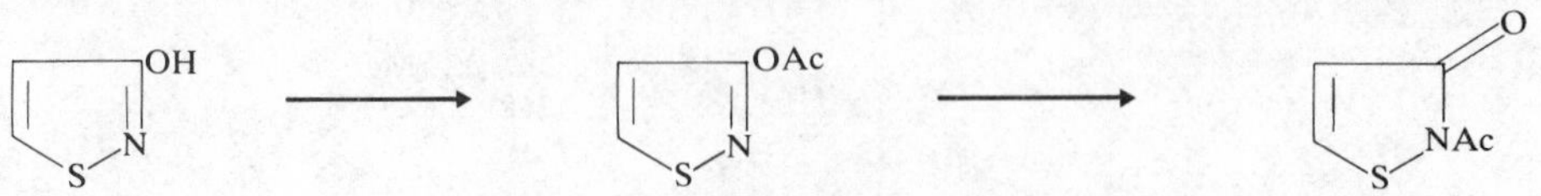

(87)

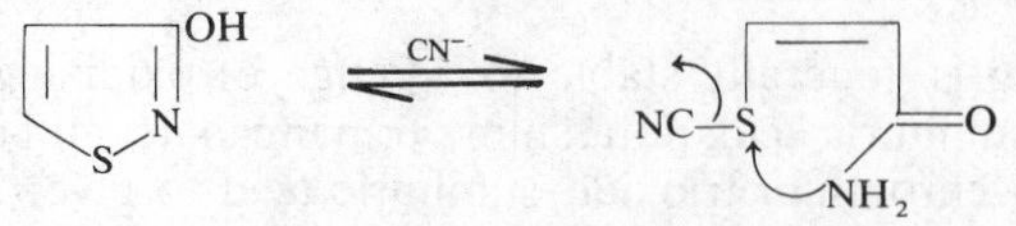

(88)

Many 3- and 5-thiols and thioethers have been made by direct synthesis and by nucleophilic displacement reactions.[124] Diazotizations of 5-aminoisothiazoles have also led to 5-thiols. The sodium salts of 5-thiols are readily alkylated, and thus converted into 5-sulphinyl and 5-sulphonyl derivatives. 1,2-Benzisothiazole-3-thiols, which are formally capable of tautomerism, all substitute at N on alkylation.

3-, 4-, and 5-aminoisothiazoles may be diazotized.[63,126] Certain 3- and 4-aminoisothiazoles give diazonium salts in solution and also give isolable solid tetrafluoroborates. 5-Aminoisothiazoles react in a similar manner, but an electron-withdrawing group at C-4 leads to stable nitrosoamines. The 3-amino-group exists in the amino-form, and is less basic than the 5-amino-group.[123] (The relative pK_a values for 3-, 4-, and 5-amino-groups are 2.49, 3.58, and 2.70, respectively, *i.e.* they behave as weak amines.[124]) Some 3-amino-1,2-benzisothiazoles are oxidized by nitric acid or by hydrogen peroxide to give the oxides (**270**) and (**271**), respectively.[63,126] On acylation, they give products such as (**272**). On alkylation, the quaternary salts (**273**) are formed. They exist in tautomeric equilibrium with the 3-imino-isomers. Diazotization of 3-amino-1,2-benzisothiazole leads to formation of (**269**).[63,126] 3-Amino-2,1-benzisothiazoles on diazotization give the diazonium salts, which can be used in the synthesis of azo-dyes. Acylation of 3-amino-2,1-benzisothiazole leads the bis-acyl derivative (**274**).

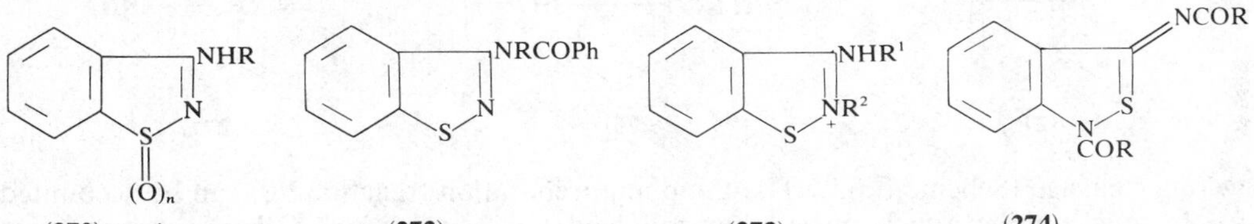

NHR N S $(O)_n$ NRCOPh N S NHR^1 NR^2 S + NCOR S N COR

(**270**) $n = 1$
(**271**) $n = 2$ (**272**) (**273**) (**274**)

5-Methylisothiazole affords an anion at the reactive 5-methyl group, which although not reactive enough to condense with benzaldehyde, will condense with *m*-nitrobenzaldehyde.[122] 4-Nitro-substitution increases the reactivity of the 5-methyl group. The 3- and 4-methyl groups are much less reactive. Side-chain halogenation occurs readily.[124]

3- and 5-carboxyisothiazoles are readily decarboxylated, whereas the 4-carboxy-derivative is more stable. The carboxylic acids may be obtained by oxidation of methyl groups.[124] 3- and 4-aldehyde groups are accessible by a range of reactions, including hydrolysis of dihalomethyl derivatives, and from Reissert compounds derived from the carboxylic acids. 5-Aldehydes can be made from the 5-nitrile, the 5-lithio, and the 5-alkoxycarbonyl derivatives.[124,125] Many syntheses of 5-acylisothiazoles have been reported.[124,125]

20.1.4.10 Ring cleavage and rearrangement

Isothiazoles are normally stable to a wide range of conditions. 5-Lithioisothiazoles, for example, can be prepared and are useful synthetic intermediates (see Section 20.1.4.12), but 3-methyl- and 3,5-dimethyl-isothiazole are ring-cleaved by butyl-lithium (reaction 89).[125] 4-Diazonium derivatives may be converted into 1,2,3-thiadiazoles by nucleophilic

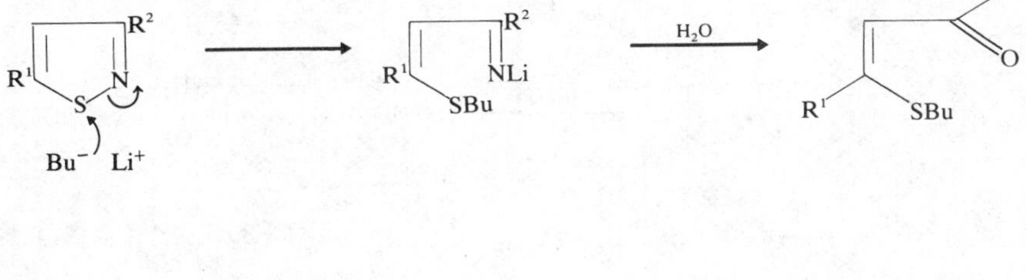

R^1 S N R^2 Bu^- Li^+ ⟶ R^1 SBu NLi R^2 $\xrightarrow{H_2O}$ R^1 SBu R^2 O (89)

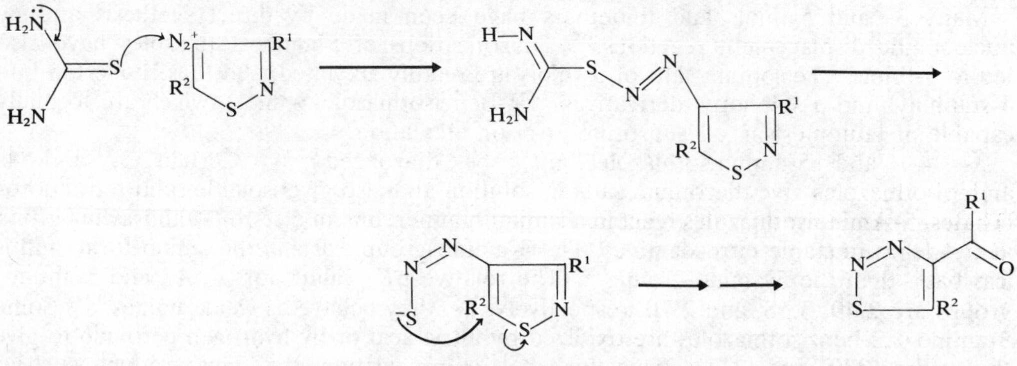

SCHEME 35

attack of thiourea on the diazonium group (Scheme 35).[63,124,126] The 5-diazonium derivative, however, gives the disulphide (**275**). The 3,5-dithiol (**276**) on standing in ethanol or acid is transformed into (**277**).[124] Quaternary derivatives are more susceptible

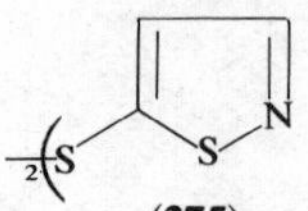

(**275**) 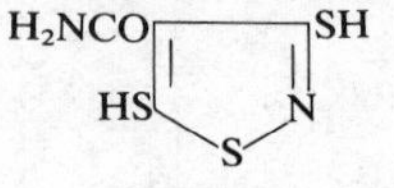 (**276**) 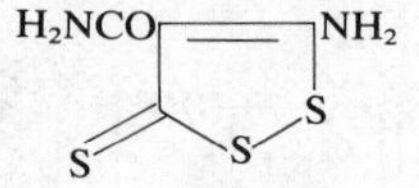 (**277**)

to ring cleavage (Scheme 36).[5,8] The thiophen preparation (reaction 90) can be accounted for by nucleophilic attack on sulphur followed by recyclization. Other rearrangement products are known, and are explained by initial attack at C-5 rather than at C-3.[63,126]

3-Chloro-1,2-benzisothiazoles and their quaternary salts undergo ready nucleophilic attack, usually with ring cleavage, as shown by the reactions with carbanions (*e.g.* reaction 91).[131] Sodium cyanide gives (**269**), by nucleophilic attack at sulphur,[101] although 3-alkoxy- and 3-amino-1,2-benzisothiazoles may be obtained by nucleophilic substitution.

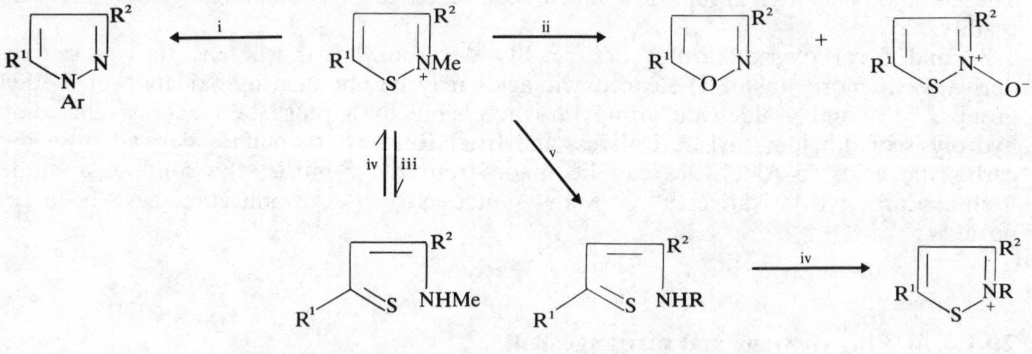

i, $ArNHNH_2$; ii, NH_2OH; iii, PhSH; iv, I_2; v, RNH_2.

SCHEME 36

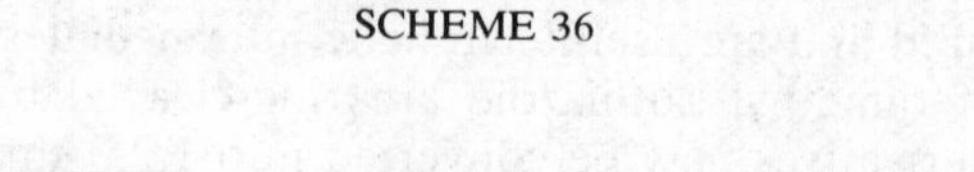

$PhCOCH_2CO_2^-$ (90)

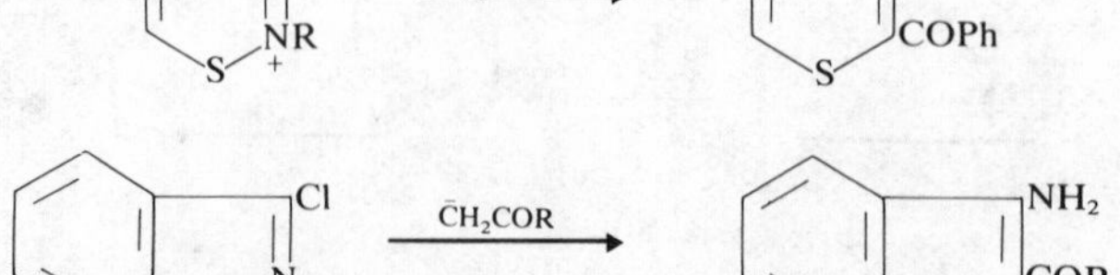

$\bar{C}H_2COR$ (91)

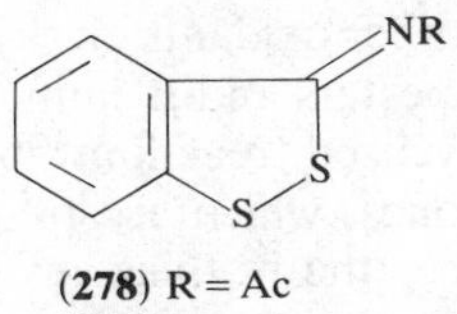

(278) R = Ac
(279) R = alkyl

(280)

Thioacetic acid gives (**278**), and reacts with alkyl 1,2-benzisothiazolium salts to give (**279**). Butyl-lithium reacts as it does with some isothiazoles, and gives (**280**).[63,126] Other ring-cleavage reactions are known.[63,126]

2,1-Benzisothiazole is stable to hot aqueous alkali, unlike 2,1-benzisoxazole which is rapidly cleaved by cold dilute alkali.[125] Hydrazine will effect nucleophilic cleavage however (reaction 92). Quaternary salts are more reactive, the salt (**281**) reacting with hydrochloric acid to give (**282**), and (**283**) reacting with amines (R^2NH_2) to give (**284**).[94]

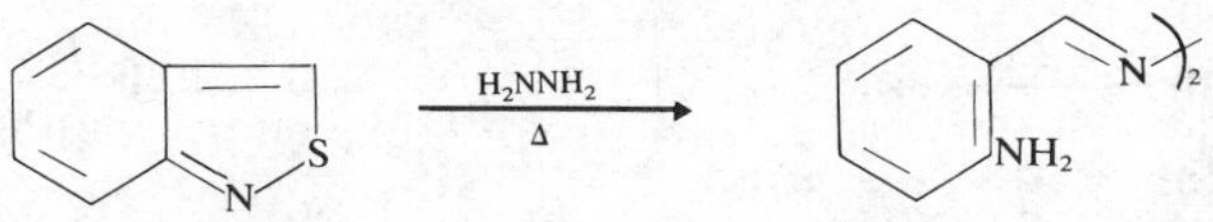

H_2NNH_2, Δ (92)

Attempted Diels–Alder reactions on 2,1-benzisothiazoles also led to ring cleavage, as shown by the reaction of (**285**) with maleic anhydride to give (**286**) *via* a bicyclic intermediate (reaction 93).[63,126]

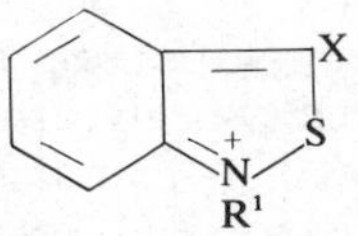

(281) R^1 = alkyl, X = NHMe
(283) R^1 = alkyl, X = Ph

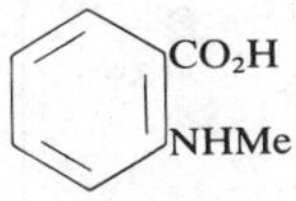

(282)

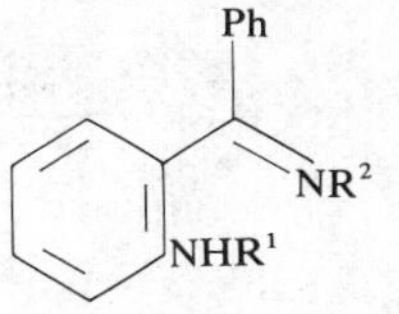

(284)

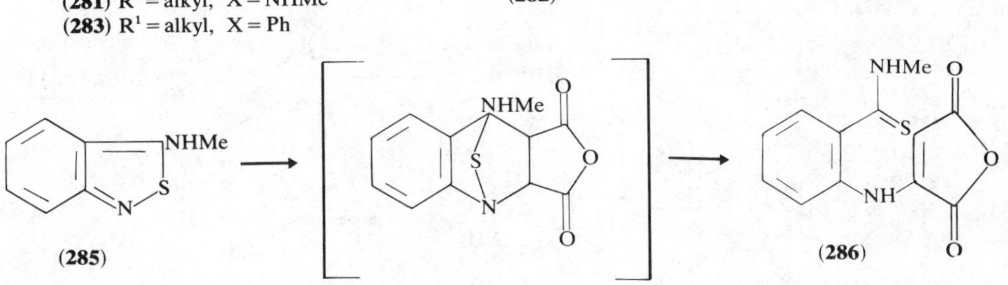

(93)

(285)

(286)

20.1.4.11 Use as synthons

Possibly the best known synthetic use of isothiazoles was in Woodward's synthesis of colchicine, in which a key step involved the desulphurization and acetylation of (**287**) to give (**288**).[124]

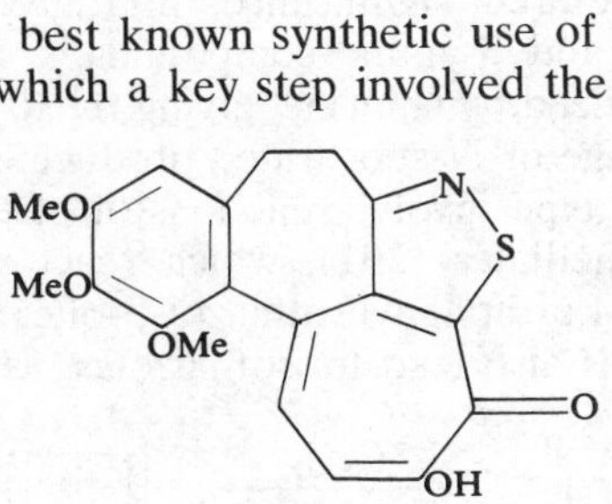

(287)

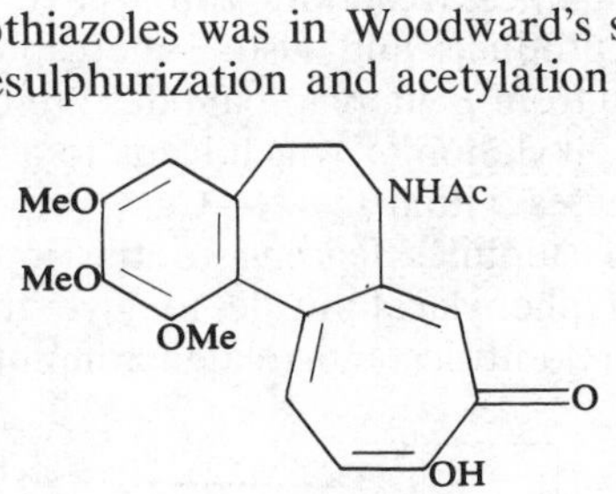

(288)

20.1.4.12 Preparations

The most important group of syntheses involve cyclization of N—C—C—C—S units. Illustrative examples only can be quoted. The oxidative cyclization of β-thionoimines using peracids, halogens, quinones, sulphur, and other oxidants gives isothiazoles (reactions 94 and 95).[123,124] In a modification,[122,132] β-iminoesters and β-iminonitriles are converted into N—C—C—C—S intermediates, which are cyclized (reactions 96 and 97). In a related synthesis, oxazoles are reduced to enamino-ketones, which are treated with phosphorus pentasulphide and chloranil to effect cyclization to the isothiazoles (reaction 98).[124] This

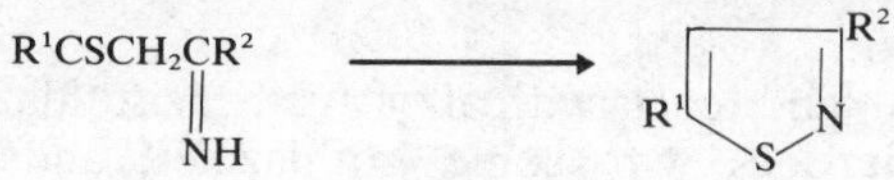

(94)

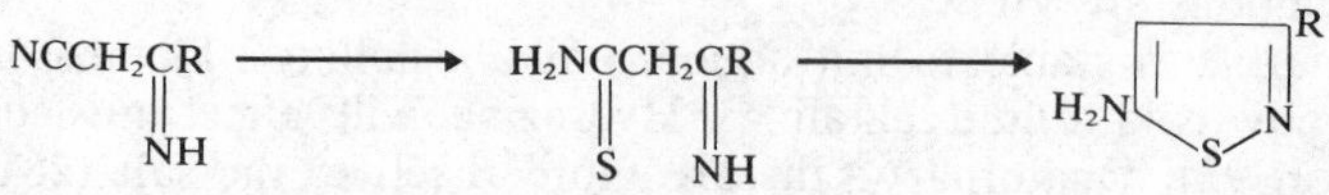

(95)

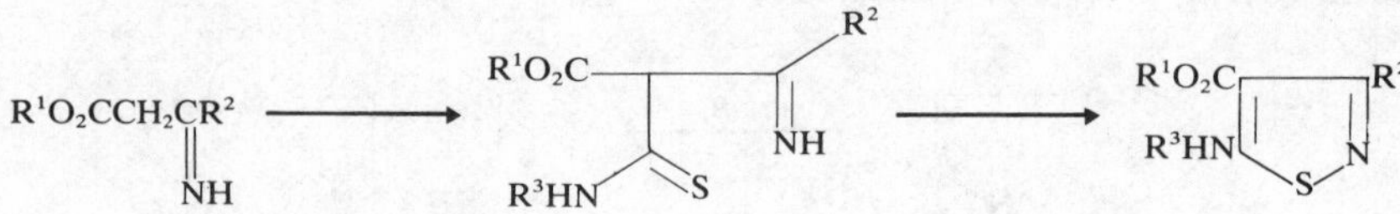

(96)

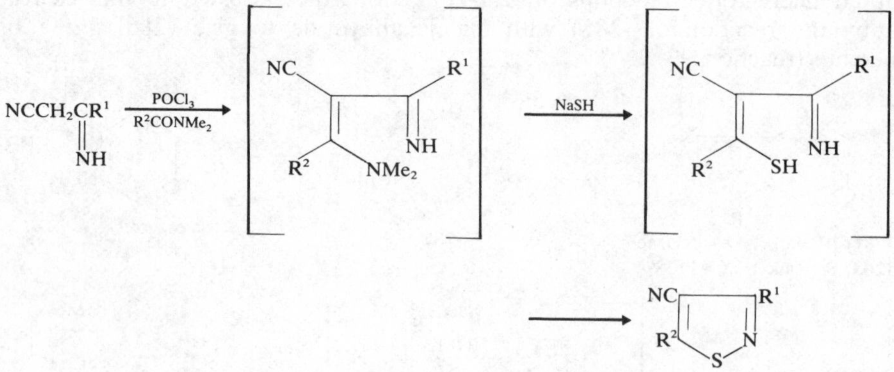

(97)

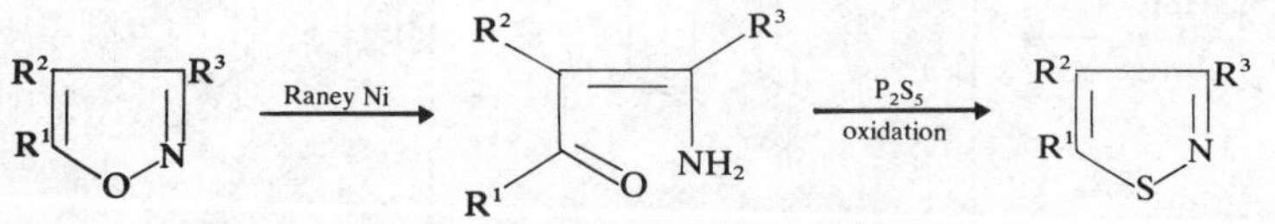

(98)

useful route gives a range of isothiazoles.[6c] The conversion of 1,2-dithiolium salts with ammonia into isothiazoles and their salts (reaction 99), and the reaction of the dithiolium salts with amines (reaction 100), have acquired significance, and give di- and tri-substituted products and also 3-mercapto- and 3-amino-compounds.[123–125] The disulphides (**289**) from β-mercaptoamides are cyclized by chlorine, giving the 4-isothiazolin-3-ones (**290**), alkylation of which leads to a range of *N*-substituted products in good yields. Novel syntheses within the N—C—C—C—S type involve reactions of phenylacetonitrile with dialkyl sulphites leading to the intermediates (**291**), which react with a further molecule of phenylacetonitrile to give the 4,5-diphenylisothiazol-3-ones (**292**).[31,63] β-Mercaptoacrylonitriles and related compounds also lead to isothiazoles, as shown by the

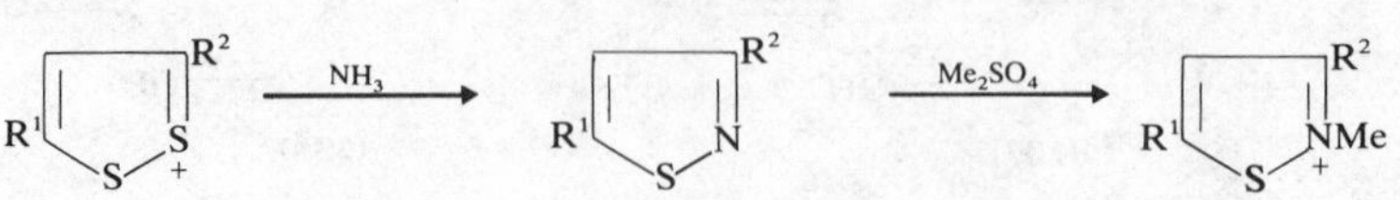

(99)

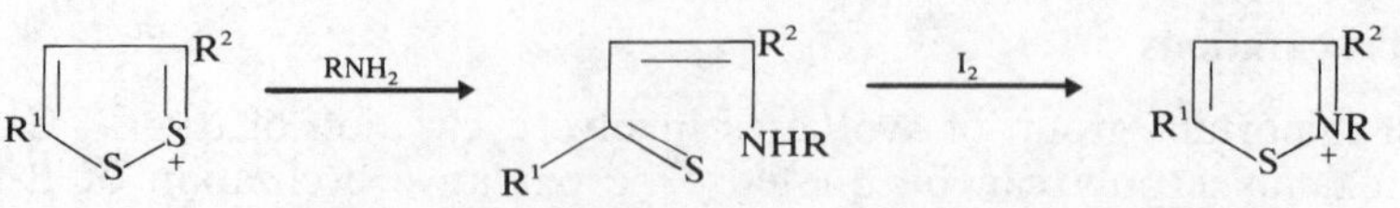

(100)

$H_2NCOCH_2CH_2S\text{)}_2$ 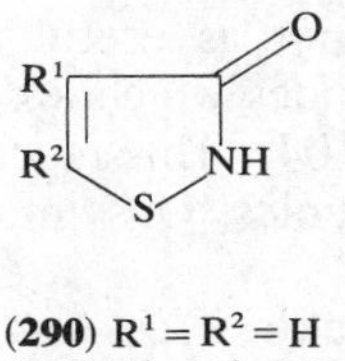 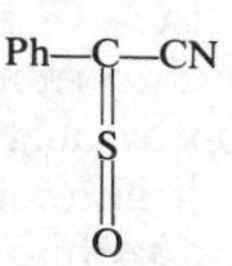 **(289)**

(290) $R^1 = R^2 = H$
(292) $R^1 = R^2 = Ph$

(291)

transformation of (**293**) into (**294**) (reaction 101). In a general example (reaction 102), a possible mechanism[124] is shown. Several routes from acetylenes exist, including the reaction of acylacetylenes with thiosulphate to give intermediates (**295**), which are cyclized by ammonia to give 3-substituted isothiazoles (**296**), and related preparations

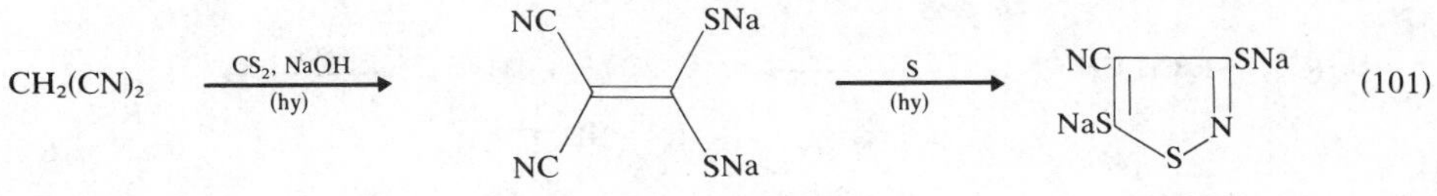

$CH_2(CN)_2$ → (CS_2, NaOH (hy)) → (**293**) → (S (hy)) → (**294**) (101)

(102)

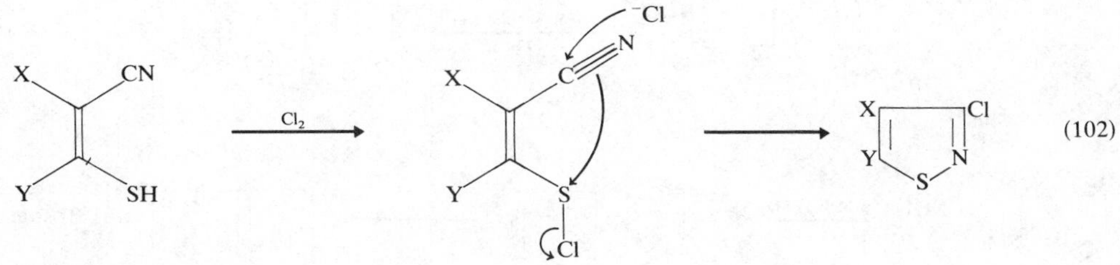

from propynal involving ring closure with potassium thiohydroxylamine-*S*-sulphonate or sodium thiocyanate (the former reagent giving the parent isothiazole in 90% yield).[124] Routes to isoselenazoles involving related N—C—C—C—Se cyclizations have been described.[123]

RCOCH═CHSSO$_3$Na **(295)** **(296)** R—C(NH$_2$)═CHCOX **(297)** X = R^1 or OR^2 **(298)**

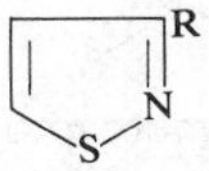

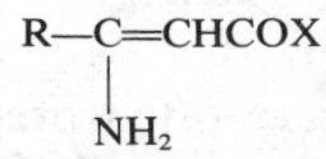

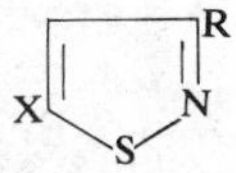

N—C—C + C—S routes are based on thiophosgene and trichloromethylsulphenyl chlorides, which afford the C—S unit. For example, thiophosgene and enamino-ketones or -esters (**297**) give the isothiazoles (**298**),[124,133] whereas trichloromethylsulphenyl chloride and base give 5-hydroxyisothiazoles (reaction 103).[63]

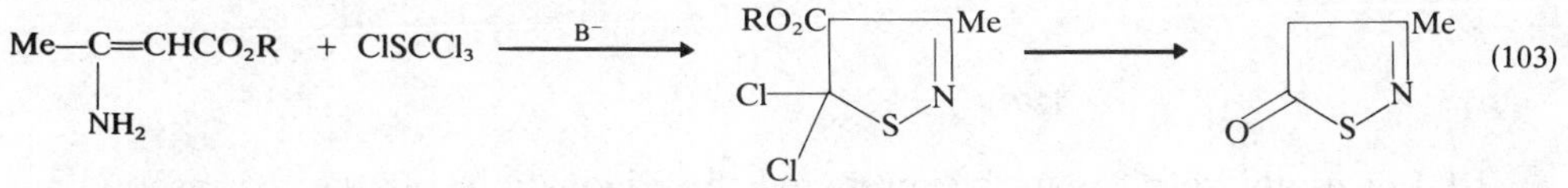

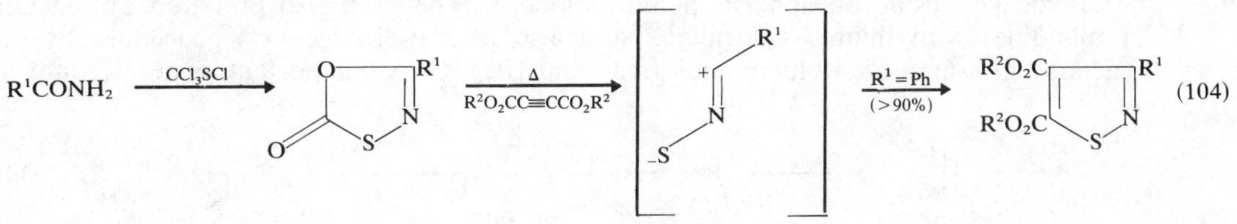

Me—C(NH$_2$)═CHCO$_2$R + ClSCCl$_3$ → (B^-) → (103)

R^1CONH_2 → (CCl_3SCl) → (Δ, $R^2O_2CC{\equiv}CCO_2R^2$) → ($R^1 = Ph$ (>90%)) (104)

C—N—S+C—C routes, although not as useful as some of the more classical approaches, utilize transient nitrile sulphides which are trapped in a 1,3-dipolar cycloaddition reaction by acetylenes (reaction 104). This reaction parallels the much more widely applicable nitrile oxide route to isoxazoles. Unsymmetrically substituted acetylenes may give isomeric isothiazole mixtures.[126,134]

C—C—C—N+S approaches (the Dakin–West synthesis) are illustrated by the reaction of α-aminoketones with thionyl chloride or sulphur monochloride to give 4-hydroxyisothiazoles.[126] Recent syntheses within this classification include the reaction of enaminonitriles with these reagents (reaction 105) to give 4-cyano- (and thus 4-carboxy-) substituted isothiazoles, although often in poor yield.[126]

Finally, the simple isothiazole syntheses from Hubenett should be mentioned, pyrolysis of olefins (*e.g.* propene) with sulphur dioxide and ammonia over a catalyst (*e.g.* alumina) leading to each of the monomethyl- and dimethyl-isothiazoles.[122,123]

$R^1CN + R^2CH_2CH_2CN \longrightarrow R^1C(NH_2){=}C(CN)CH_2R^2 \xrightarrow[\text{or } SOCl_2]{SCl_2}$ 4-cyano-isothiazole (NC, R^1, R^2) (105)

o-HS-C$_6$H$_4$-C(R)=NOH $\xrightarrow[130\,^\circ C]{\text{polyphosphoric acid}}$ 3-R-1,2-benzisothiazole (106)

o-MeS-C$_6$H$_4$-C(R)=NONa $\xrightarrow{p\text{-}MeC_6H_4SO_2Cl}$ S-methyl-1,2-benzisothiazolium (S$^+$–Me) $\longrightarrow$ 3-R-1,2-benzisothiazole + TosMe (107)

o-BrS-C$_6$H$_4$-C(R)=O $\xrightarrow{NH_3}$ 3-R-1,2-benzisothiazole (108)

1,2-Benzisothiazoles are often prepared from *o*-mercaptobenzaldoximes and ketones (reaction 106), and from their derivatives (reactions 107 and 108).[125] Another good method involves the oxidative cyclization of the aminothiols (**299**) with reagents such as iodine–potassium iodide–sodium hydroxide, or with potassium ferricyanide–sodium hydroxide,[125] giving a range of substituted products. Thioisatin (**300**) is converted by ammonia–hydrogen peroxide into the 3-carboxamido-1,2-benzisothiazole.[125]

(**299**) R-C$_6$H$_3$(CH$_2$NH$_2$)(SH) (**300**)

2,1-Benzisothiazoles may be prepared by thionation of the corresponding 2,1-benzisoxazoles, using phosphorus pentasulphide.[125] They are also prepared by reaction of toluidines with thionyl chloride,[135] and also in a useful general procedure by the reaction of *o*-amino-α-toluenethiol (reaction 109).[125] An important general route to

R-C$_6$H$_3$(Me)(NH$_2$) $\xrightarrow{SOCl_2}$ R-2,1-benzisothiazole $\longleftarrow$ R-C$_6$H$_3$(CH$_2$SH)(NH$_2$) (109)

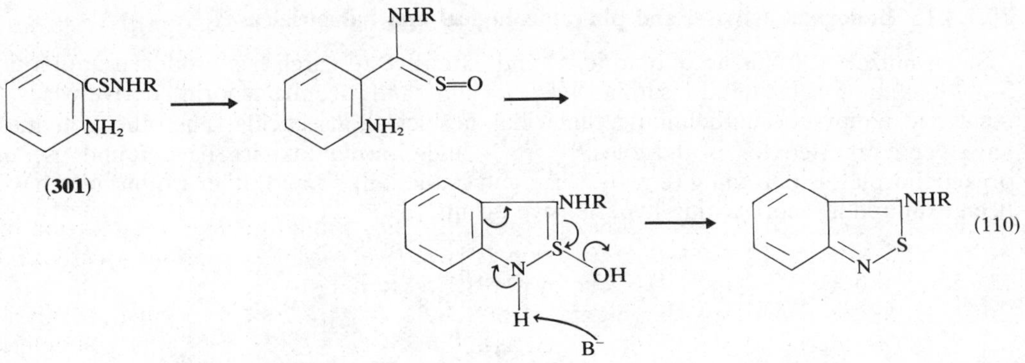

(301)

(110)

3-amino-2,1-benzisothiazoles is the hydrogen peroxide oxidation of the *o*-aminothiobenzamides (**301**), probably *via* intermediate sulphines (reaction 110).[125] Reductive routes from *o*-nitro-α-toluenethiol, *o*-nitrothiobenzamides, and related compounds are known.[125] The first synthesis of 3-hydroxy-2,1-benzisothiazole (**303**) utilized the reaction of isatoic anhydride (**302**) with metal hydrosulphides.[20]

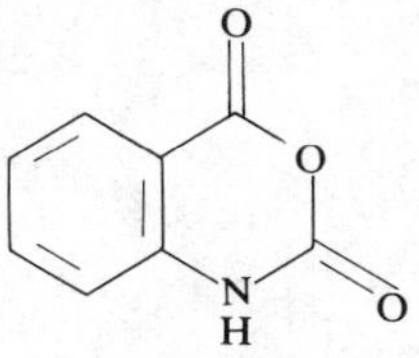

(302) 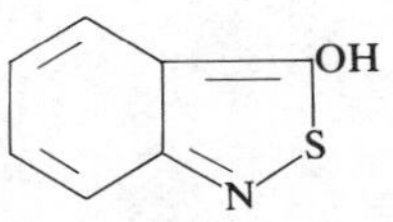 (303)

20.1.4.13 Metallation

Lithiation of isothiazoles occurs preferentially at C-5 when the site is available,[123,133] although the possible nucleophilic attack on sulphur by alkyl 'anion', as in the case of butyl-lithium, should be borne in mind. Thus, 4-methylisothiazole gives mainly the 5-lithio-derivative, with some lithiation at C-3 (which leads to ring cleavage). The 5-lithio-derivatives are readily halogenated, alkylated, carboxylated, and thus converted into a range of derivatives. 4-Nitroisothiazoles do not give 5-lithio-4-nitroisothiazoles. 5-Halogenoisothiazoles give 5-lithio-derivatives, but examples of 4-halogenoisothiazoles reacting in a similar manner are rare.[124] It should be noted that 4-halogenoisothiazoles may give the 4-halogeno-5-lithioisothiazoles when treated with butyl-lithium.

An example of acetoxymercuration is the reaction of 4-nitroisothiazole with mercuric acetate–acetic acid, giving the 5-acetoxymercury derivative (quantitative), which reacts with bromine to give the 5-halogenoisothiazole.[124]

20.1.4.14 *N*-Oxides

The quaternary salt (**304**) reacts with hydroxylamine to give the *N*-oxide hydrochloride (**305**).[123,126] This salt is of limited stability, but is water soluble. It is also accessible from peracid treatment of 3-amino-1,2-benzisoxazoles, and resembles pyridine *N*-oxides in the reduction with phosphorus trichloride to the deoxy ring system. In an interesting rearrangement, (**305**) in formic acid gives (**306**).

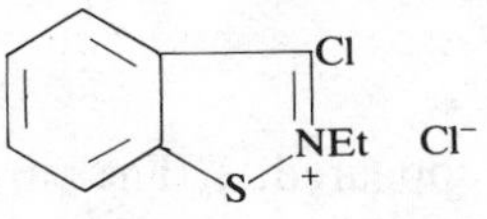

(304) 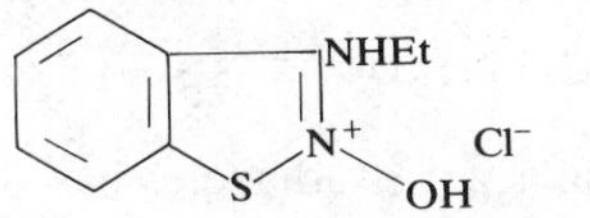 (305) 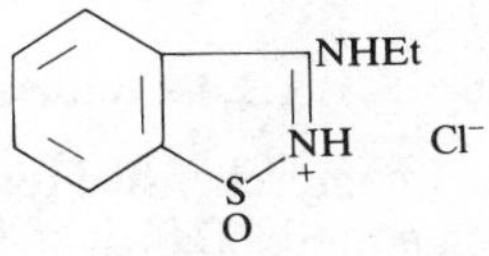 (306)

20.1.4.15 Biological activity, and pharmacological and industrial uses

Sulfasomizole (**307**) is an antibiotic,[123] and extensive research has produced antibiotic 3-, 4-, and 5-substituted isothiazolyl-penicillin and -cephalosporin derivatives.[125] Analgesic, antipyretic, anthelmintic, fungicidal, herbicidal, insecticidal, and other activities have been reported for isothiazoles.[123,125] 3- and 5-acetaldoximes have found use in organophosphorus poisoning (*e.g.* by Sarin and Paraoxon),[126] and other isothiazoles have been explored as short-lasting hypertensive agents.

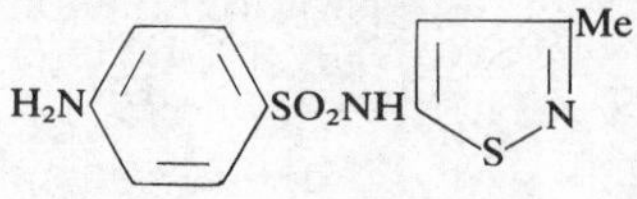

(307)

1,2-Benzisoxazoles have been widely studied in anaesthesia, and as selective phytocides.[126] Several 1,2-benzisothiazole-3-acetic acids have been investigated as plant hormones. Many other uses have been summarized.[125] Possibly the best-known derivative of 1,2-benzisoxazoles is 3-oxo-2,3-dihydrobenz[*d*]isothiazole 1,1-dioxide, saccharin (**308**).[136]

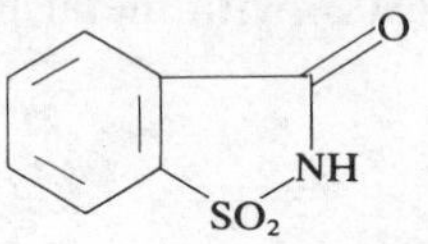

(308)

20.1.5 TWO NITROGENS, ONE HETEROATOM: THE OXADIAZOLES

Four oxadiazoles, 1,2,3- (**309**), 1,2,4- (**310**), 1,3,4- (**311**), and 1,2,5-oxadiazoles (**312**) are possible. Ring systems of type (**310**) have been termed azoximes, and the 1,2,5-oxadiazoles (**312**) are commonly called furoxans. The possible benzo-fused analogues are 1,2,3-benzoxadiazole (**313**) and 2,1,3-benzoxadiazole (**314**), known as benzofurazan. Arguments concerning the existence of (**309**) and (**313**) have been summarized,[95] but the firmly established 1,2,3-oxadiazole ring systems are of the sydnone type. The oxadiazoles will be treated systematically, and a major group of derivatives, the 1,2,5-oxadiazole *N*-oxides (**315**) (the furazans) and the related benzofurazans (**316**), will be discussed separately.

(309) **(310)** **(311)** **(312)**

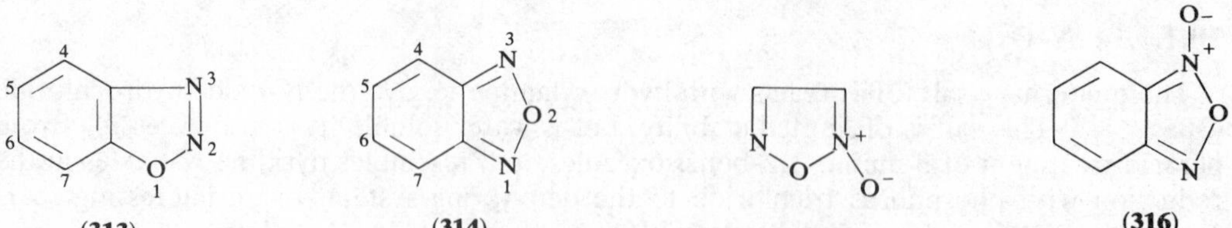

(313) **(314)** **(315)** **(316)**

20.1.5.1 1,2,4-Oxadiazoles[5,6c,95,137,138]

A comprehensive review of 1,2,4-oxadiazoles has very recently appeared.[138] The parent compound (**310**) is a relatively unstable liquid which will form complexes with heavy-metal ions.[138] The 3- and 5-derivatives are more stable,[6c] and some 3,5-disubstituted

1,2,4-oxadiazoles will dissolve in concentrated acids without reaction.[95,138] They may also be heat stable. Theoretical treatments of the system have been covered,[138] and the resonance energy has been noted as similar to that of benzene.[95]

The physical and spectroscopic properties have been reviewed,[138] the u.v. spectra of 3- and 5-phenyl-1,2,4-oxadiazoles, for example (238 and 250 nm respectively), and the 3,5-diphenyl derivative (245 nm), suggesting that the ring can be regarded as a conjugated diene. In the n.m.r. spectrum, H-3 and H-5 appear downfield from benzene (δ 8.7 and 8.2 for **310**). A recent mass spectral study showed that predominant fission processes in 3,5-disubstituted 1,2,4-oxadiazoles involved cleavage of (a) O-1—N-2 and N-4—C-5 bonds and (b) C-3—N-4 and O-1—C-5 bonds. The principal cleavage in a series of aryl derivatives involved type (b), *i.e.* a retro 1,3-cycloaddition.[138,139] Resemblances have been noted between the mass spectral and the thermal decompositions of diphenyl-1,2,4-oxadiazole.[138]

Reductions to oxadiazolines and oxadiazolidines are apparently not possible. Many reducing reagents effect ring cleavage. Zinc–hydrochloric acid gives benzonitrile, and catalytic reduction often leads to N—O bond cleavage (reaction 111). Lithium aluminium hydride causes C—O bond cleavage (reaction 112).[138]

$$\text{5-amino-3-phenyl-1,2,4-oxadiazole} \xrightarrow{\text{Pd/C}} Ph{-}C({=}NCONH_2){-}NH_2 \qquad (111)$$

$$\text{3-}R^1\text{-5-}R^2\text{-1,2,4-oxadiazole} \xrightarrow{\text{LiAlH}_4} R^1{-}C({=}NOH){-}NHCH_2R^2 \qquad (112)$$

Electrophilic reactions are difficult, particularly with the parent compound, because of instability in aqueous acids and bases. Many other substituted 1,2,4-oxadiazoles, because of their apparently low aromatic character,[138] are also reluctant to undergo electrophilic reactions. Mercuration of 3-methyl-1,2,4-oxadiazole, however, gives (**317**), which reacts with halogens to give the 5-halo-product (**318**).[6c] Bromine by itself does not give any reaction.[95] Nitration of 3- or 5-aryl-1,2,4-oxadiazoles occurs, not in the azole ring, but in the aryl substituents.[138]

(**317**) X = HgCl
(**318**) X = hal

Nucleophilic reactions may lead to ring cleavage[6c,140] (reactions 113 and 114). However, many examples of nucleophilic substitution at C-3 and C-5 are known. Chloride, for example, is displaced by amines, hydroxide, and alkoxides.[138] A comparison of the relative susceptibility of C-5 and C-3 to nucleophilic substitution emerges from the more ready displacement of the trichloromethyl group from C-5 by hydroxide.[135] The relative ease of exchange of hydrogen in oxadiazole rings, and in methyl substituent groups, has been estimated.[138] The 5-methyl group in 5-methyl-3-phenyl-1,2,4-oxadiazole exchanges deuterium faster than 5-H exchanges in 3-phenyl-1,2,4-oxadiazole.

$$\text{1,2,4-oxadiazole} \xrightarrow{^-OH} CN^- + NCO^- \qquad (113)$$

$$\text{3-R-1,2,4-oxadiazole} \xrightarrow{^-OH} RCN + NCO^- \qquad (114)$$

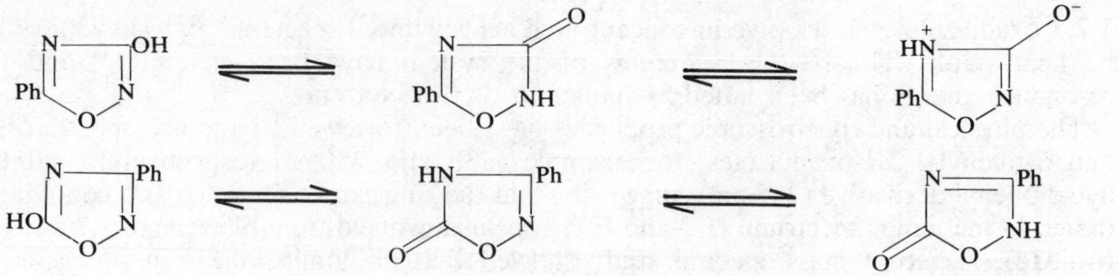

SCHEME 37

Many transformations of functional groups are comparable with the chemistry of aromatic compounds, but there may be significant differences between C-3 and C-5 substituents. The position of equilibrium of 3- and 5-hydroxy-substituents differ (Scheme 37), the hydroxy-form predominating in the former and the keto-form in the latter, although solvent differences affect the equilibria. Explanations have been summarized.[138] Diazomethane gives *O*- and N^4-methyl derivatives, whereas other methylating agents may give *N*-methylation.[138] The tautomeric situation is simpler in the 5-amino-derivatives, which exist as the amines (as in the isoxazoles). Alkoxycarbonyl groups are stable, but the 3-carboxyl group is prone to decarboxylation. Lithiation of 5-methyl-3-phenyl-1,2,4-oxadiazole gave the expected lithio-derivative of the methyl group, which reacted normally with, for example, carbon dioxide. The 3-methyl-5-phenyl derivative underwent nucleophilic attack at C-5 when treated with n-butyl-lithium (reaction 115).[138] The 5-methyl group will undergo zinc chloride catalysed condensation with aldehydes.

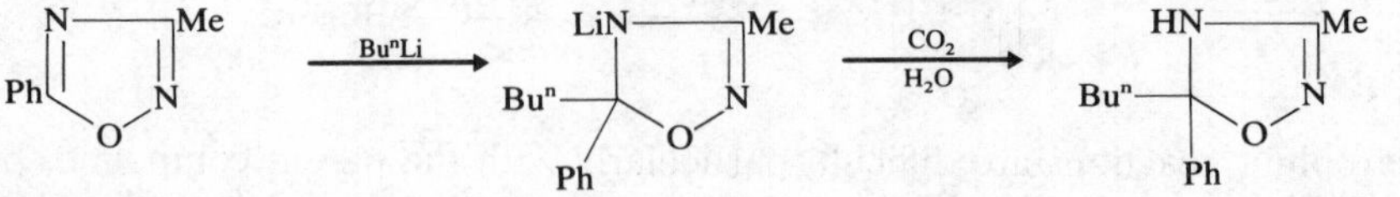

(115)

Ring-cleavage reactions are common.[138] For example, the 3-substituted 1,2,4-oxadiazole (**319**) is reversible transformed by base into the 1,2-benzisoxazole (**320**). Photochemical decomposition of 3,5-diphenyl-1,2,4-oxadiazole in ether gave *N*-benzoylbenzamidine, *via* cleavage at the N—O bond (ether furnishing the additional

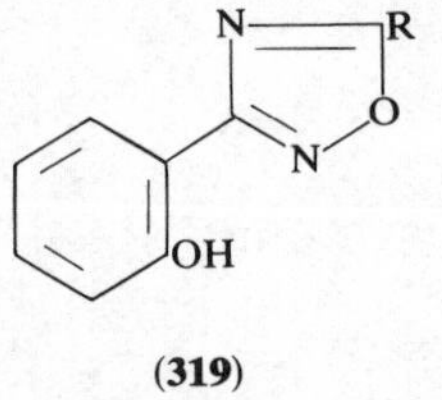

(**319**) 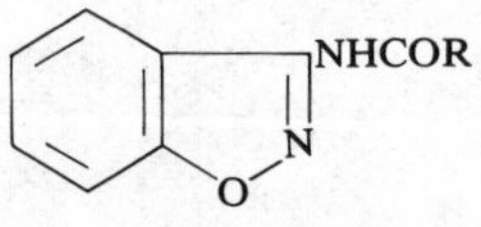 (**320**)

hydrogens). A photochemical cleavage and rearrangement may also occur (reaction 116), probably following initial N—O bond cleavage.[14] A further reductive ring cleavage is illustrated in reaction 117.[135] A series of transformations of the 1,2,4-oxadiazole ring of the types shown in reaction 69 exists, leading to 1,2,3- and 1,2,4-triazoles, 1,2,5-oxadiazoles, imidazoles, benzisoxazoles, and 1,2,4-thiadiazoles.[138]

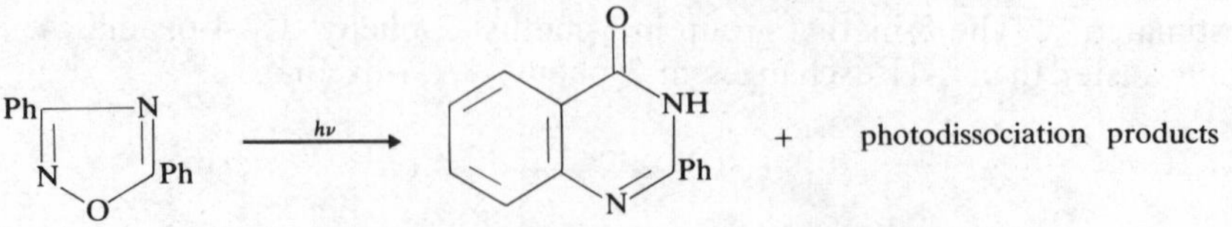

+ photodissociation products (116)

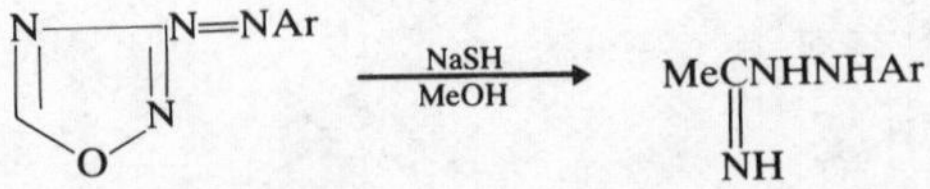

MeCNHNHAr (117)

$$R^1\text{—}C(NH_2)\text{=NOH} + \begin{cases} R^2COCl \\ R^2CO_2COR^1 \end{cases} \longrightarrow R^1\text{—}C(\text{=NOCOR}^2)\text{—}NH_2 \longrightarrow \text{5-}R^2\text{-3-}R^1\text{-1,2,4-oxadiazole} \quad (118)$$

Amidoximes, available from the chloro-oximes or from nitriles and hydroxylamine hydrochloride, are key precursors for the synthesis of a range of 1,2,4-oxadiazoles (reaction 118).[95,138] Reactions of amides with amidoximes also give good yields. Symmetrically disubstituted 1,2,4-oxadiazoles (**321**) can be made by heating the appropriately substituted amidoxime with carboxylic acids. Variations on this process (reaction 119) give 5-acyloxy-,[95,135] 5-mercapto- (reaction 120) and 5-amino-1,2,4-oxadiazoles (reaction 121). Reaction 122 illustrates a route to 3-amino-derivatives, and a further approach, recently patented, uses the cyclization of (**322**) to (**323**) by potassium hydroxide.[141] A 3-alkoxy-synthesis is shown in reaction 123 and a 3-alkylthio-5-amino-synthesis in reaction 124.[142] Alkyl- and aryl-guanidines may be converted in high yield into the disubstituted products (reaction 125).[139] 1,2-Dioximes can be converted by a variety of methods into disubstituted 1,2,4-oxadiazoles.[95]

(**321**) 3,5-R,R-1,2,4-oxadiazole (**322**) $ArCH_2CONHC(NH_2)\text{=NOH}$ (**323**) 3-NH_2-5-$ArCH_2$-1,2,4-oxadiazole

$$Ph\text{—}C(NH_2)\text{=NOH} + COCl\text{—}CO_2Et \longrightarrow Ph(NH_2)C\text{=NOCOCO}_2Et \longrightarrow \text{3-Ph-5-EtOCO-1,2,4-oxadiazole} \quad (119)$$

$$Ar\text{—}C(NH_2)\text{=NOH} \xrightarrow{\text{i, }CSCl_2\text{; ii, NaOH}} \text{3-Ar-5-HS-1,2,4-oxadiazole} \quad (120)$$

$$R^1\text{—}C(OR^2)\text{=NCN} \xrightarrow{H_2NOH} R^1\text{—}C(NHOH)\text{=NCN} \longrightarrow \text{3-}R^1\text{-5-}H_2N\text{-1,2,4-oxadiazole} \quad (121)$$

$$HONH\text{—}C(\text{=NOH})\text{—}NH_2 \longrightarrow AcONH\text{—}C(\text{=NOAc})\text{—}NH_2 \longrightarrow \text{3-NHAc-5-Me-1,2,4-oxadiazole} \quad (122)$$

$$PhCOCl \xrightarrow{\text{i, KSCN; ii, MeOH; iii, MeI}} PhCON\text{=}C(SMe)(OMe) \xrightarrow{H_2NOH} \text{3-OMe-5-Ph-1,2,4-oxadiazole} \quad (123)$$

$$NCN\text{=}C(SK)_2 \xrightarrow{MeI} NCN\text{=}C(SMe)_2 \xrightarrow{NH_2OH} \text{3-SMe-5-}H_2N\text{-1,2,4-oxadiazole} \quad (124)$$

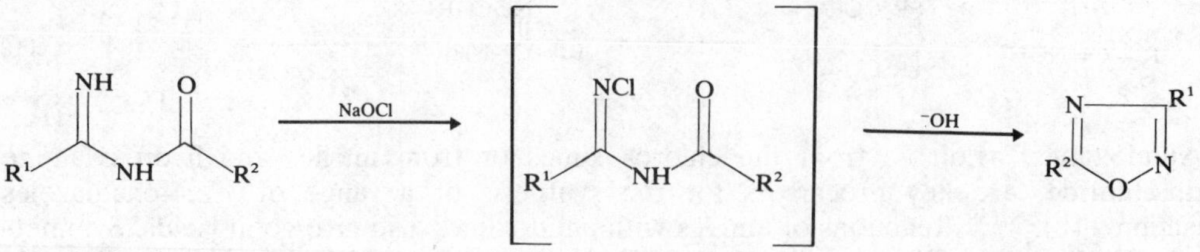

(125)

Nitrile oxides afford 1,3-cycloaddition routes to 1,2,4-oxadiazoles.[138] They will react with simple nitriles,[27] and tetracyanoethylene gives a high yield of (**324**) and thence 3-substituted 1,2,4-oxadiazoles (**325**).[143] When nitrile oxides are heated in acid or base they may condense to give the symmetrically disubstituted *N*-oxides (**326**).[6c] In an alternative route to (**326**; R = Ph) the ammonium salt of benzonitrosolic acid [PhC(NO)=NOH] is treated with HCl.[95] The *N*-oxides can be deoxygenated with PCl_5 or with zinc–acetic acid.

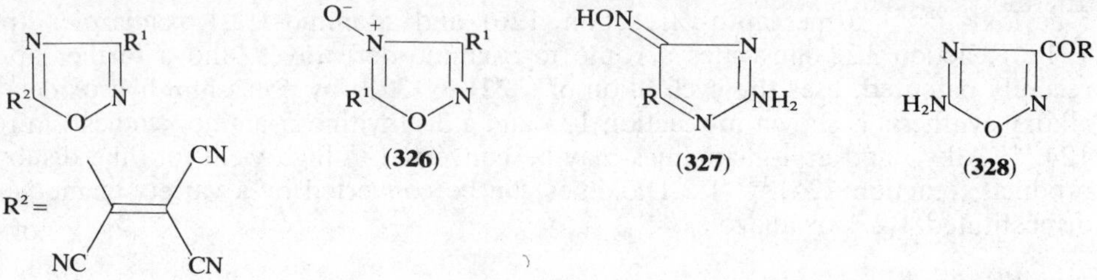

(**324**) R^2 = $C(CN)=C(CN)_2$

(**325**) $R^2 = H$

(**326**) (**327**) (**328**)

Other routes to 1,2,4-oxadiazoles include the acid-catalysed rearrangement of nitrosoimidazoles (**327**) into (**328**) by a process related to a known pyrrole–isoxazole interconversion.[144] Many other routes to these compounds have been reviewed.[138]

1,2,4-Oxadiazoles find extensive use[138] in chemotherapy and in agricultural chemistry. Industrial uses encompass textile and polymer technology, and they are also used as photosensitizers. Among the many recent patents are uses as optical brighteners, in common with the 1,3,4-isomers and 1,3,4-thiadiazole analogues.[145]

20.1.5.2 1,3,4-Oxadiazoles[6c,95,144,146,147]

Although aromatic type properties are evident in this group, the compounds are chemically fairly labile. The parent compound (**311**) is a liquid, b.p. 150 °C. Substituted 1,3,4-oxadiazoles have u.v. characteristics similar to benzenoids.[147] The spectroscopic and physical properties have been reviewed.[146,147]

Electrophilic introduction of functional groups (for example nitro or sulphuric acid groups) into the nucleus is unusual. Electrophilic substitution will, however, occur in aryl substituents. Halogenation is also difficult, but 2,5-diaryl-1,3,4-oxadiazoles afford complexes with halogens.[146] A range of alkylation and acylation reactions of hydroxy-, thio-, and amino-1,3,4-oxadiazoles is known at the ring nitrogen (*e.g.* reactions 126 and 127) and at the potentially tautomeric functional group (reaction 128).[147]

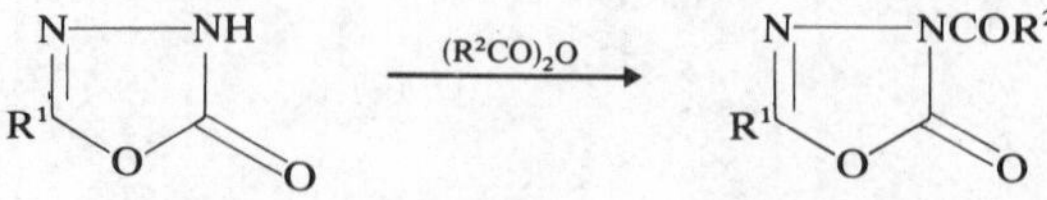

$\xrightarrow{(R^2CO)_2O}$ (126)

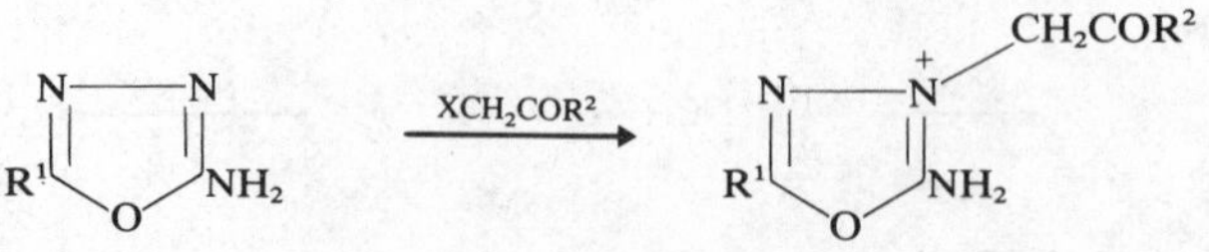

$\xrightarrow{XCH_2COR^2}$ (127)

$$\text{2-}R^1\text{-5-}NH_2\text{-1,3,4-oxadiazole} \xrightarrow{R^2NCO} \text{2-}R^1\text{-5-}NHCONHR^2\text{-1,3,4-oxadiazole} \qquad (128)$$

Nucleophilic substitution reactions are uncommon, and the known examples, including displacement of a methylsulphonyl group by ammonia, proceed in low yield.[147] Many nucleophilic reagents will cause ring cleavage, depending upon the electron density at C-2 and C-5. The acid and base hydrolysis products are shown in reaction 129. In general, aryl-substituted 1,3,4-oxadiazoles are less sensitive to acid than the alkyl-substituted derivatives. Mineral acids, and concentrated nitric and sulphuric acids, do not attack the oxadiazole ring of 2,5-diphenyl-1,3,4-oxadiazole, but 2-phenyl-5-alkoxycarbonyl derivatives afford ring-cleavage products. 2-Alkyl-5-aryl derivatives are readily cleaved by mineral acids to give the diacid hydrazides. Acid hydrolysis of 1,3,4-oxadiazoles probably occurs by initial protonation at nitrogen, followed by nucleophilic attack. 2,5-Bisperfluoroalkyl-1,3,4-oxadiazoles, which are extremely resistant to thermolysis, readily undergo fission, as in reaction 129, with acid or base, and with amines give 1,2,4-triazole products. The action of nucleophilic agents on 2-amino-1,3,4-oxadiazoles also leads to ring-cleaved products, which often cyclize to triazoles (reaction 130). Other related nucleophilic cleavage and recyclization processes have been reviewed.[147] When the 1,3,4-oxadiazole is quaternized at a ring nitrogen, ring cleavage takes place under very mild hydrolytic conditions to give 1-acylamino-4-arylimidazolin-2-ones. 5-Hydroxy- and 5-thio-derivatives are also ring-cleaved. 5-Ethoxy-2-phenyl-1,3,4-oxadiazole is cleaved by aniline (reaction 131).[147] An interesting ring-cleavage of methyl-1,3,4-oxadiazolin-5-one in aqueous acetone possibly occurs as shown in reaction 132. The hydrolytic stability of 5-hydroxy-derivatives was particularly relevant in studies of the metabolism of drugs which incorporate this structure.[147]

$$\text{2-}R^1\text{-5-}R^2\text{-1,3,4-oxadiazole} \xrightarrow[\text{or } ^-OH]{H^+} R^1CONHNHCOR^2 \xrightarrow[\text{or } ^-OH]{H^+} H_2NNH_2 + R^1CO_2H + R^2CO_2H \qquad (129)$$

$$\text{2-}H_2N\text{-5-R-1,3,4-oxadiazole} \xrightarrow{H_2O} H_2NCONHNHCOR \longrightarrow \text{5-R-1,2,4-triazol-3(2H,4H)-one} \qquad (130)$$

$$\text{2-Ph-5-OEt-1,3,4-oxadiazole} \xrightarrow{PhNH_2.HCl} [\text{2-Ph-5-OEt-5-}\overset{+}{N}H_2Ph\text{-oxadiazoline (NH)}] \longrightarrow PhCONHNHCONHPh \qquad (131)$$

$$\text{Me-oxadiazolinone (NH)} \rightleftharpoons \text{Me-oxadiazolone} \xrightarrow{Me_2CO} \text{Me-oxadiazolinone-N-}CMe_2OH \longrightarrow MeCONHN{=}CMe_2 \qquad (132)$$

Other aspects[146] of amino- and hydroxy-substituted derivatives (which may exist in the imino- or keto-forms) include the ready diazotization of the former and thus the introduction of nuclear substituents, the alkylation and acylation of the ring or exocyclic nitrogen groups, the reduction of *N*-nitroso-derivatives (**329**) to hydrazino-derivatives, and the thermolysis of (**329**), giving oxadiazolyl radicals which insert into benzene.[59]

$$\text{(329): 2-R-5-NHNO-1,3,4-oxadiazole}$$

(**329**)

Syntheses of 1,3,4-oxadiazoles often utilize intramolecular condensation of mono- and di-acid hydrazides, acylsemicarbazides and related compounds[95,144,146,147] (reaction 133). Thus, symmetrical 2,5-dialkyl, -diaryl, unsymmetrical, and unsubstituted oxadiazoles have been prepared. In a related process, diacylhydrazides have been converted by $POCl_3$ into

$$R^1CONHNHCOR^2 \xrightarrow[\text{conditions, } e.g.\ P_2O_5]{\text{dehydrating}} \text{2-}R^1\text{-5-}R^2\text{-1,3,4-oxadiazole} \quad (133)$$

the bisimidoyl chlorides (**330**), which on treatment with water give the oxadiazoles.[146] 2-Amino- and 2,5-diamino-1,3,4-oxadiazoles are formed, often in high yield, by heating 1-acylsemicarbazides and substituted hydrazodicarboxamides with $POCl_3$ or with $SOCl_2$.[147]

$$Ar{-}C(Cl){=}N{-}N{=}C(Cl){-}Ar$$

(**330**)

Oxidative cyclizations also lead to 1,3,4-oxadiazoles, lead oxide, halogens, and hypohalites being used to transform the semicarbazones of aldehydes and of α-keto-acids into a wide range of 5-substituted 2-amino-derivatives (reaction 134).[31,95,147]

$$RCH{=}NNHCONH_2 \xrightarrow{[O]} \text{2-R-5-}NH_2\text{-1,3,4-oxadiazole} \quad (134)$$

One-carbon fragments may also be introduced into acid hydrazides, giving 1,3,4-oxadiazoles.[147] Ortho-esters give intermediates (reaction 135) which cyclize with elimination of alcohol. The parent oxadiazole (**311**) may be prepared this way, and is a liquid, b.p. 150 °C. A general method for the preparation of 2,5-disubstituted derivatives is the reaction of acid hydrazides with imido-esters (reaction 136).[147] Other sources of the one-carbon fragment include phosgene, thiophosgene, and isocyanide dichlorides. Carbon disulphide gives 5-thiones. Acid hydrazides will also react with cyanogen halides to give 5-substituted 2-amino-1,3,4-oxadiazoles. This method, when extended to 4-arylsemicarbazides (reaction 137), gives 2,5-diamino-products.

$$RCONHNH_2 \xrightarrow[\Delta]{(EtO)_3CH} \left[RCONHN{=}CHOEt \right] \longrightarrow \text{2-R-1,3,4-oxadiazole} \quad (135)$$

$$R^1CONHNH_2 \xrightarrow{R^2{-}C(OR^2){=}\overset{+}{N}H_2\ Cl^-} \text{2-}R^1\text{-5-}R^2\text{-1,3,4-oxadiazole} \quad (136)$$

$$ArNHCONHNH_2 \xrightarrow{BrCN} \text{2-ArNH-5-}NH_2\text{-1,3,4-oxadiazole} \quad (137)$$

Other heterocyclic ring systems can be transformed into 1,3,4-oxadiazoles. 2-Substituted hydantoins with hypobromite give the 5-keto-derivatives,[147] tetrazoles react with acid chlorides or anhydrides to give high yields of 2,5-disubstituted products

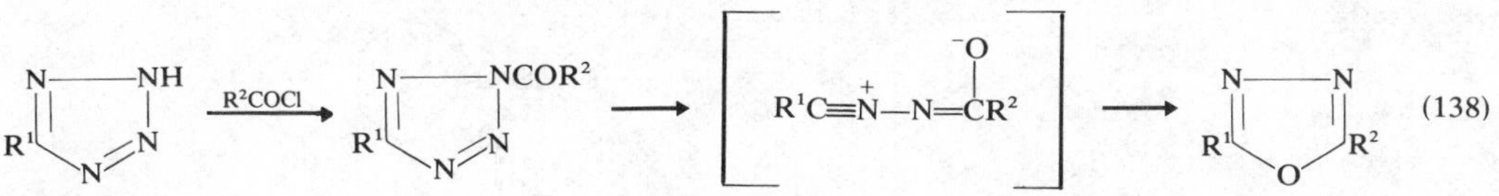

$$\text{(138)}$$

(reaction 138), and they also react with isocyanates to give 5-amino-1,3,4-oxadiazoles.[146,147]

There has been a significant increase in the use of 1,3,4-oxadiazoles in recent years in diverse areas, including drug synthesis, scintillation materials, and dyestuffs, and patents therefore comprise a large fraction of the literature. These applications have been summarized.[147]

20.1.5.3 1,2,5-Oxadiazoles (furazans) and benzo-2,1,3-oxadiazoles (benzofurazans)

1,2,5-Oxadiazoles (**312**) and their benzo-fused derivatives (**314**) appear to be relatively stable towards a range of conditions. Benzofurazan (**314**), for example, is stable to heat, aqueous acid, and alkali, and dissolves in concentrated sulphuric acid without reaction. Nucleophilic displacement reactions of 4-halogeno- and 5-halogeno-substituents in benzofurazans (**314**) have been investigated,[47] and it has been shown, for example, that one mole of methoxide ion will displace the 4-halogen from 4-halogeno-7-nitrobenzofurazans, whereas two moles of reagent gives a Meisenheimer-type adduct. The 3,4-disubstituted furazans are stable to alkali,[6c] but the parent molecule (**312**) is cleaved (reaction 139). This type of cleavage parallels the behaviour of isoxazoles and 1,2,4-oxadiazoles towards nucleophilic reagents, when initial abstraction of an oxime-type α-proton triggers the

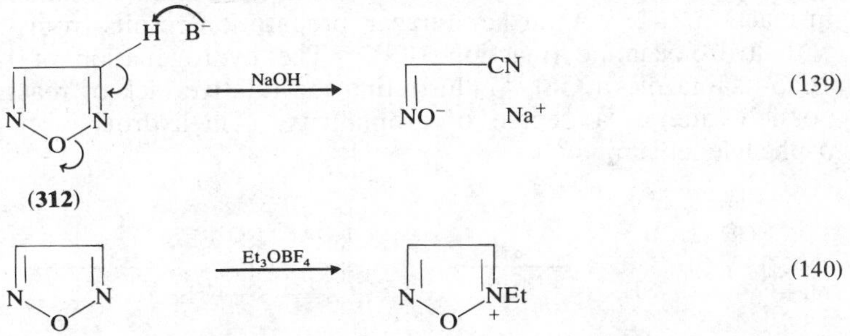

(139)

(312)

(140)

decomposition. The formation of quaternary salts (reaction 140) represents an electrophilic attack on the furazan, but the facile hydrolysis of these salts by water is a nucleophilic process. A further reaction which may be viewed as nucleophilic is the trialkylphosphine deoxygenation of benzofurazan (**314**) to give the muconitrile (**331**).[101]

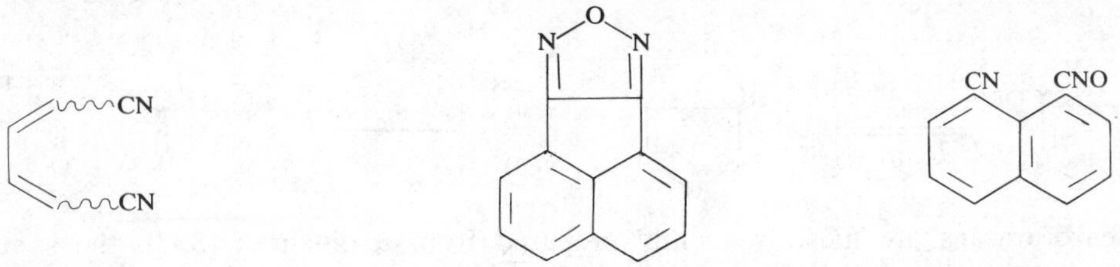

(331) **(332)** **(333)**

Thermal and photochemical processes lead to ring fission. For example, thermolysis of (**332**) leads to the nitrile oxide (**333**).[135] In a related photochemical reaction (**314**) gives, *via* the nitrile oxide (**334**), the product (**335**),[101] and 3,4-diphenyl-1,2,5-oxadiazole gives benzonitrile and benzonitrile oxide. (It is not certain if the latter product is in fact a

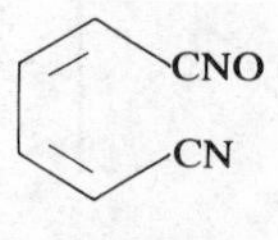

(334) 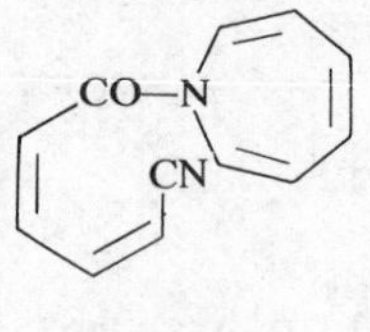 (335)

ground-state intermediate.[14]) When (**314**) is irradiated in the presence of triethyl phosphite the intermediate (**334**) is deoxygenated, again leading to muconitrile. When acenaphtho[1,2-*c*]furazan is thermolysed, the resulting nitrile oxide can be trapped with phenylacetylene, giving an isoxazole adduct.[27]

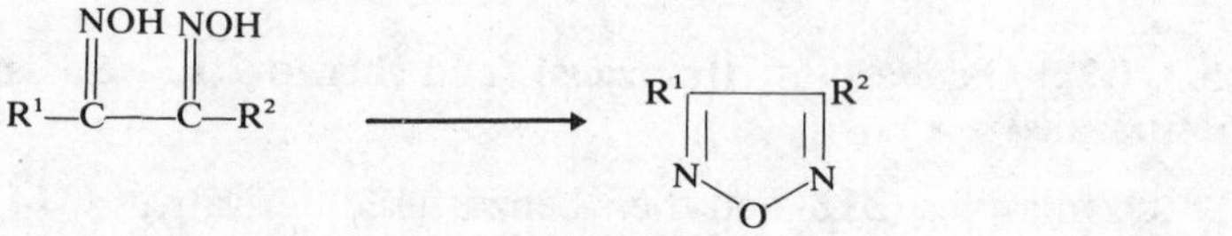

(141)

Furazans are most commonly prepared from α-dioximes (reaction 141) by heating with base or with acid anhydrides. For example, aqueous ammonia or succinic anhydride give high yields of 3,4-dimethyl-1,2,5-oxadiazole.[95] The *amphi*-forms are most easily dehydrated. The parent compound (**312**), which is a liquid, b.p. 98 °C, displaying only end absorption in the u.v., was most conveniently prepared[148] by the high-temperature dehydration of glyoxime in a mildly acid medium from which furazan could be continuously removed. Its spectroscopic properties were discussed in detail (n.m.r., δ 8.19; $J_{^{13}C\text{-}H}$ 199 Hz; mass spectrum $C_2H_2N^+$, $HCNO^+$, HCN^+, NO^+). A dubious preparation of 3-hydroxy-1,2,5-oxadiazole is known[95] and 3-amino-1,2,5-oxadiazole is prepared as shown in reaction 142.[95] A further furazan preparation results from the reaction of isoxazoles with hydroxylamine (reaction 143).[95] The hydrogenation of benzofurazan affords the 1,2,5-oxadiazoles (**336**),[149] illustrating the relative lack of reactivity of the furazan ring, possibly due to a degree of aromaticity. (Tin–hydrochloric acid will, however, give *o*-phenylenediamine.[95])

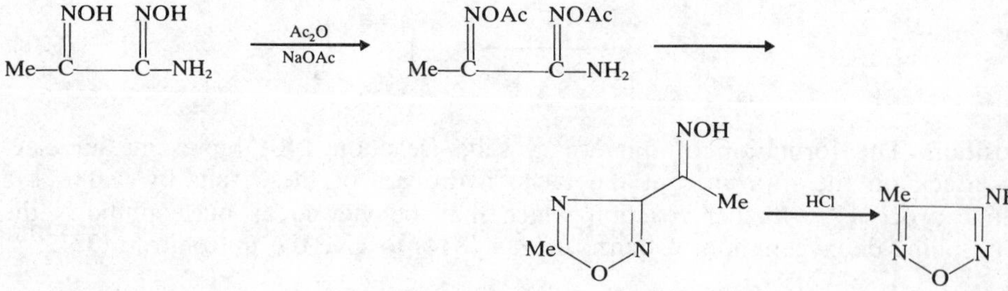

(142)

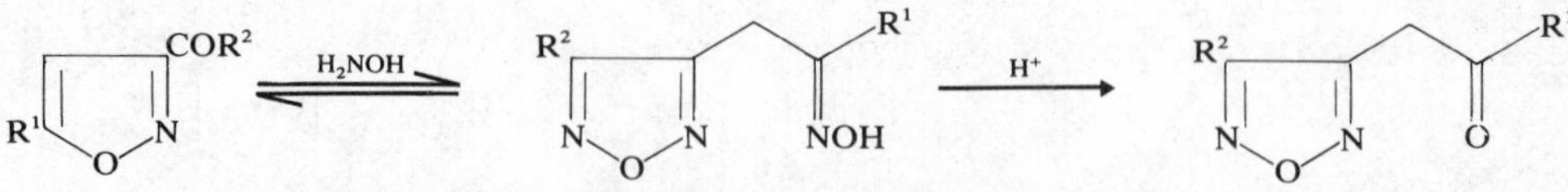

(143)

Benzofurazans are most frequently prepared from α-dioximes (**337**), the synthetic methods paralleling those of furazans.[95] Alternatively, the *o*-nitrosobenzamide (**338**) may be cyclized by hypochlorite and, in a recent and useful synthesis, the thermolysis of methyl *N*-(*o*-nitroaryl)carbamates gives moderate yields (Scheme 38), the major advantages of the method being its directness and the ready availability of starting materials.[118] Another good synthesis is from the reduction of benzofuroxans by alkaline hydroxylamine or trialkyl phosphite.[118]

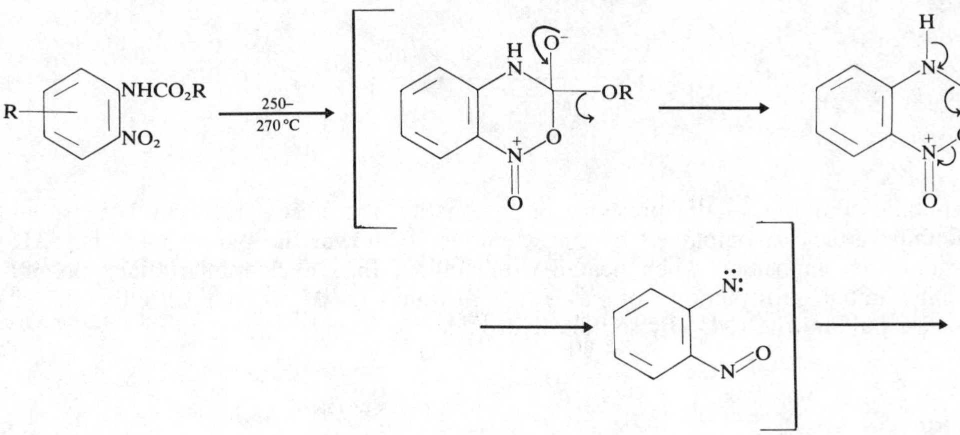

SCHEME 38

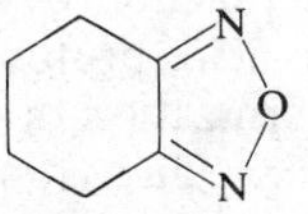

20.1.5.4 1,2,5-Oxadiazole *N*-oxides (furoxans) and benzo-2,1,3-oxadiazole *N*-oxides (benzofuroxans)

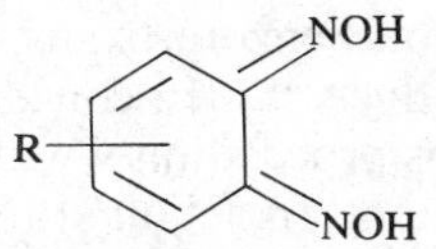

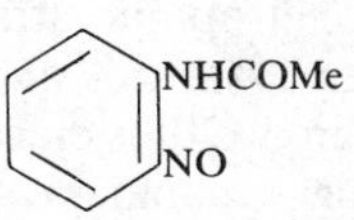

Furoxan (**315**)[119] and benzofuroxan (**316**)[119] ring systems have received much attention in the recent literature, particularly because of the intriguing isomerizations and ring interconversions of the latter group. Reviews cover the earlier developments in the area.[95,150] The physical and spectroscopic properties of the benzofuroxans have been summarized (i.r., 1630, 1600, 1545, 1500 cm^{-1}, one or more bands occasionally absent; u.v., 360 and 370 nm, similar to benzofurazan; and n.m.r.).[95,150] Benzofuroxan is a pale yellow crystalline solid, m.p. 72 °C.

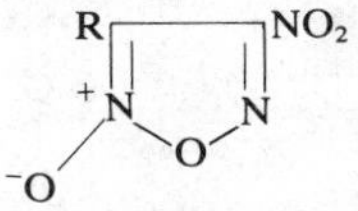

Aromatic behaviour may be inferred from nucleophilic substitution reactions of these rings. A recent example of substitution in furoxans involved displacement of the nitro-group in (**339**) by a range of nucleophiles.[151] The halobenzofuroxans undergo ready displacement of chloride and bromide from the homocyclic ring providing there is activation by a nitro-group.[150] Alkoxides, amines, and thiols, for example, act as nucleophiles. Reaction 144 illustrates a nucleophilic displacement followed by rearrangement.

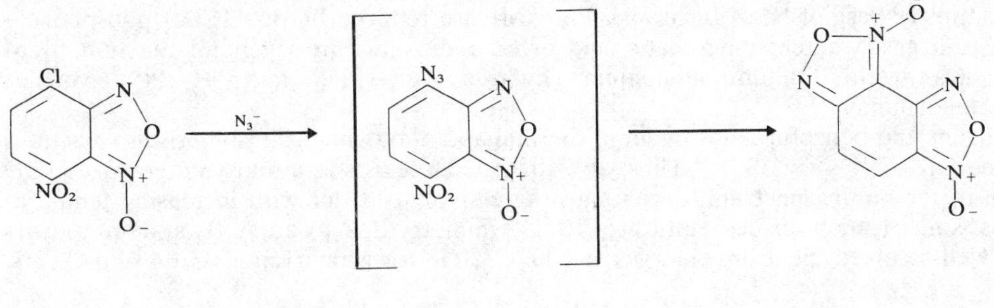

(144)

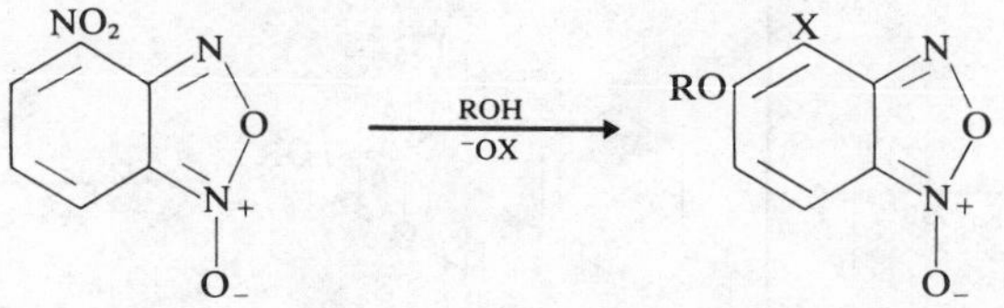

(145)

Nucleophilic additions in the presence of hypohalite also occur (reaction 145),[150] and nucleophilic attack accompanied by ring cleavage is known, the parent molecule (**316**) undergoing decomposition when heated with aniline. In a nucleophilic attack possibly involving initial abstraction of a ring proton, (**340**) is converted into 2-nitromalonamidonitrile (**341**) by sodium hydroxide.[102]

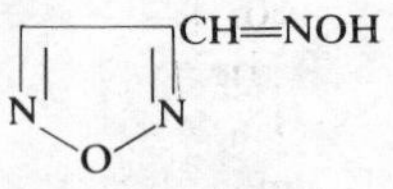

$NCCHNO_2CONH_2$

(**340**) (**341**)

Electrophilic attack, particularly on benzofuroxans, is shown by the reaction of (**316**) to give 4-nitro- and 4,6-dinitro-products.[150] (In concentrated sulphuric acid, protonation occurs.) Other electrophilic reactions proceed only with difficulty, quaternization of (**316**) being accompanied by rearrangement (reaction 146).[150] A related product (**342**) is obtained by the reaction of (**316**) with formaldehyde and hydroxide.[152] The addition of bromine gives the tetrabromo-adduct (**343**).[150]

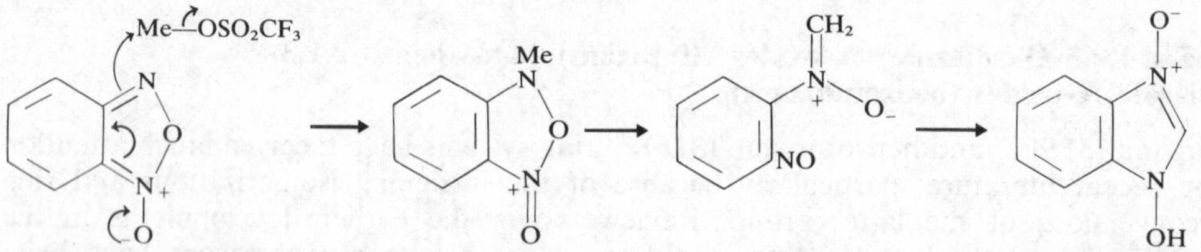

(146)

Peracid oxidation of benzofuroxans in some cases gives *o*-dinitrobenzenes, following oxidative cleavage of the oxadiazole ring.[150] The resistance of the furoxan ring to oxidation by alkaline permanganate is shown by the preferential oxidation of the aryl group in (**344**) to a carboxylic acid.

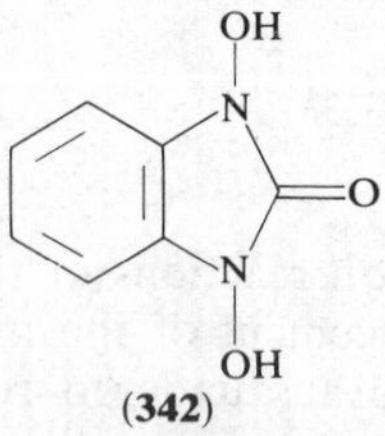

(**342**) 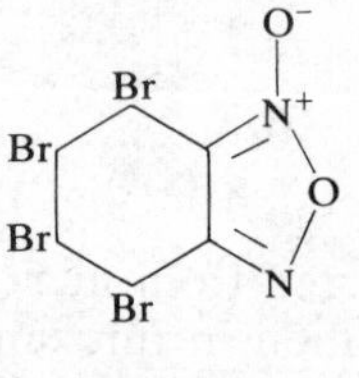 (**343**) 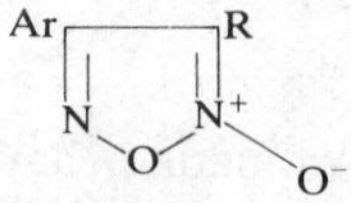 (**344**)

Reduction of furoxans and benzofuroxans to furazans and benzofurazans may be achieved with PCl_5 (as in the case of pyridine *N*-oxides).[6c,150] Alternatively, phosphines and phosphites may be used, or hydroxylamine–alkali (which proceeds *via* an *o*-quinone dioxime in the case of benzofuroxans). Furoxans are reduced by tin–HCl or phosphorus–HI to furazans, whereas zinc–acetic acid gives α-dioximes in a general reaction. Benzofuroxans with lithium aluminium hydride, tin–HCl, and H_2–Pt give *o*-phenylenediamines.[95,150]

Furoxans and benzofuroxans undergo thermal and photochemical isomerisms (reactions 147 and 148).[14,118,142,150,153,154] Thus, at −40 °C, (**316**) exists as an unsymmetrical ABCD system in the n.m.r. spectrum, becoming progressively broader with increasing temperature as isomerization ensues, until at −10 °C symmetry due to an A_2B_2 system appears and a well-resolved spectrum emerges at 100 °C. ΔG^* for isomerism is 63 ± 4 kJ mol^{-1}. A

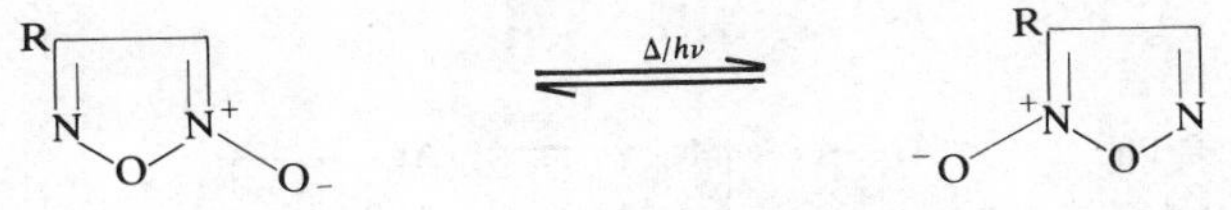

(147)

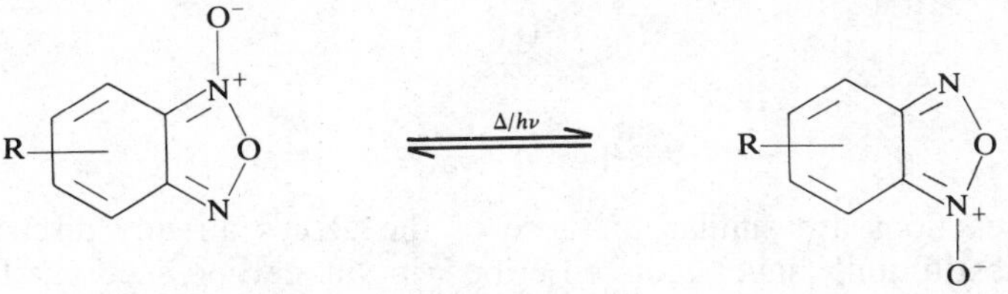

(148)

wide range of related systems has been examined, pyrido[2,3-*c*]furoxan, for example, existing almost entirely as (**345**) at −50 °C, and benzo-fusion as in (**346**) slowing the isomerization ($\Delta G^* \sim 84\ \mathrm{kJ\ mol^{-1}}$). An interesting and related isomerization occurs with adjacent nitro-groups, as in (**347**) which is thermally interconverted with (**348**), being

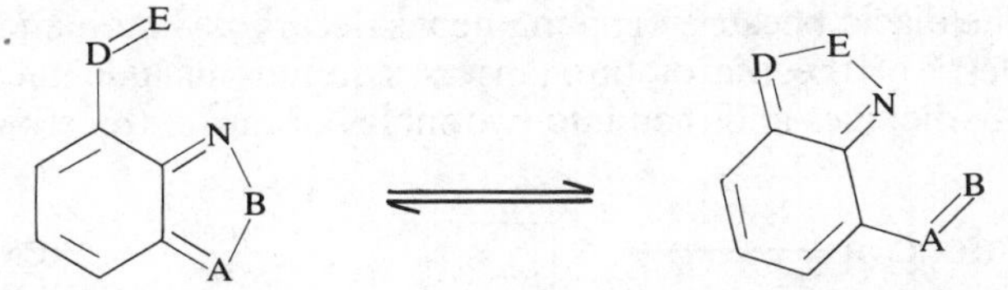

(149)

formally an interconversion of the general type shown in reaction 149.[154] Other examples of this type occur in benzofuroxans when D=E (reaction 149) is a carbonyl group, leading to 2,1-benzisoxazoles,[118] and when D=E is an imino-group.

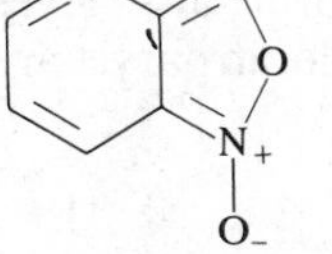

(**345**) 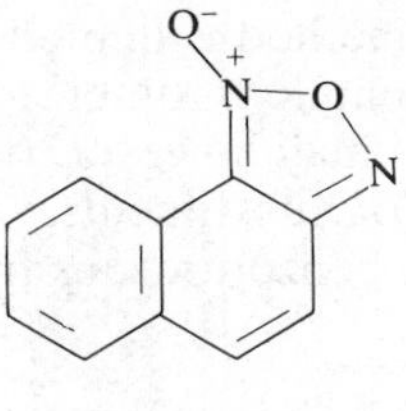 (**346**) 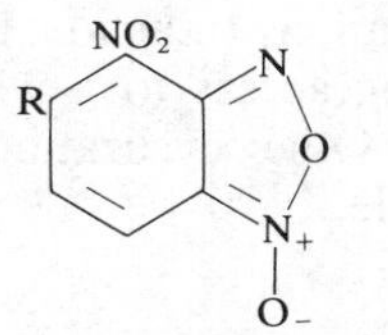 (**347**) 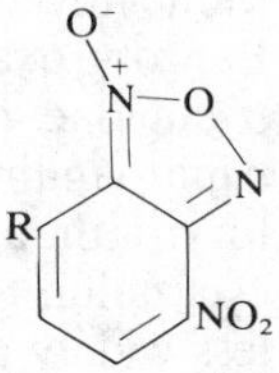 (**348**)

Other ring-cleavage reactions result from the thermolysis of (**349**), giving the intermediate (**350**) which can be trapped by phenylacetylene.[155] In the thermolysis of (**351**) a benzoyl nitrile oxide decomposition product was not trapped by phenylacetylene, but instead the furoxan behaved as a nitrone, giving the 1,3-dipolar cycloaddition product (**352**).[27] Benzofuroxans, in the 'Beirut reaction' (which has been extensively patented), add a variety of enamines and enolate anions to give in moderate yields quinoxaline 1,4-dioxides[131,150] *e.g.* (**353**). A large range of related transformations into previously unknown or difficulty accessible heterocycles are known.[156] Further ring interconversions occur when furoxans (**354**) are transformed by hydroxide into compounds (**355**).[95]

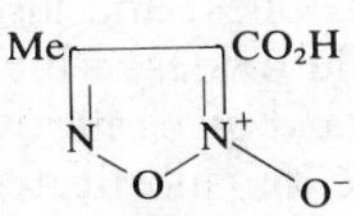

(**349**) 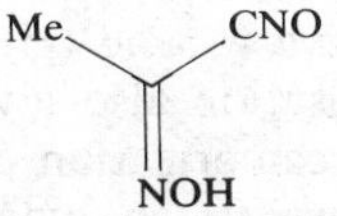 (**350**) 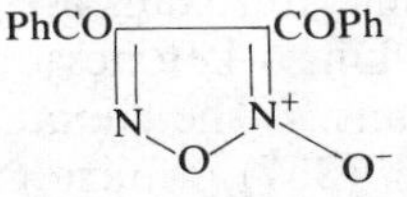 (**351**) 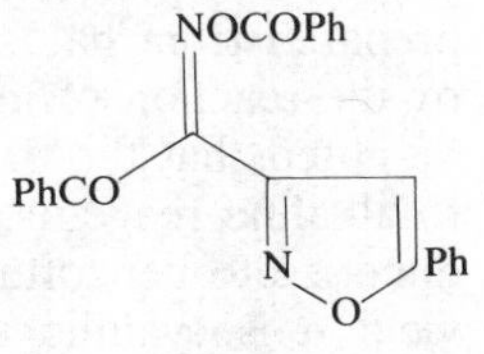 (**352**)

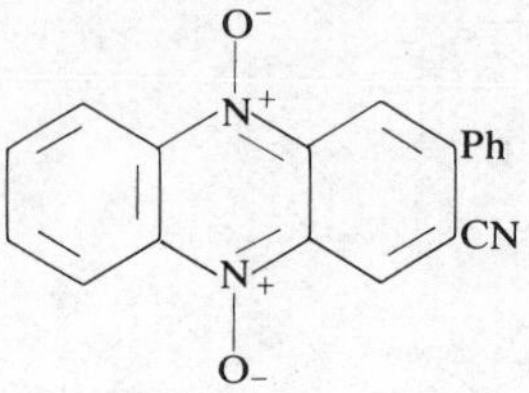

(353) 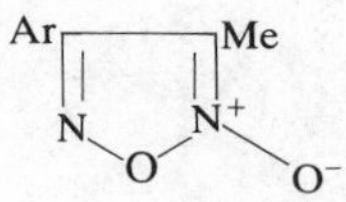 (354) 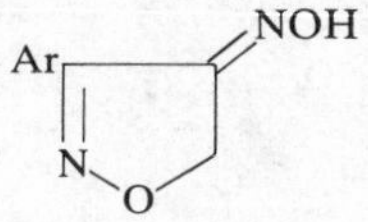 (355) 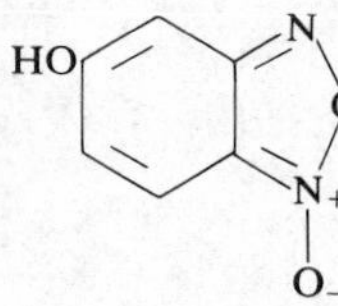 (356)

Substituent reactions are similar to those of the azoles already discussed, and some nucleophilic substitutions have been described in an earlier section. The 5-hydroxy-derivative, which is difficult to prepare by other methods, is obtained from the 5-acetate and probably exists largely in the hydroxy-form.[150] It is a fairly strong acid (pK 6.15, *cf.* 5-hydroxybenzofuroxan (**35**), 7.28). The 5-amino-derivative is obtained by a standard sequence starting from the 5-carboxy-compound, and is unstable. The 5-aldehyde has been prepared, and the 4-acetyl compound rearranges under the conditions used for its preparation, giving anthranil. The 4,5-bis(ethoxycarbonyl) derivative cannot be hydrolysed to the diacid because of spontaneous decarboxylation and decomposition.[95]

Specific examples[13] of the use of furoxans as synthons include the transformation of α-diketones into α-dioxines and then into two moles of nitrile (reaction 150).

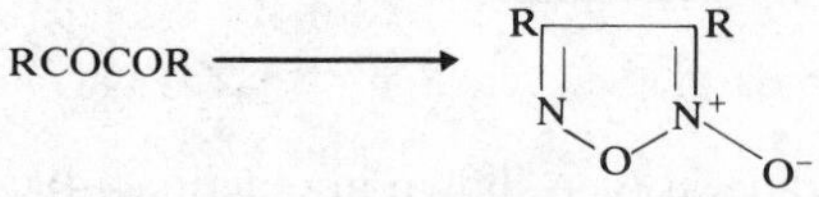

RCOCOR → [furoxan] $\xrightarrow[270\,°C]{P(OPh)_3}$ 2RCN (150)

The most common synthetic precursors of furoxans are α-dioximes, which are oxidized by a range of reagents such as hypochlorite, nitrogen dioxide, or alkaline ferricyanide. Monoalkyl derivatives may be unstable under synthetic conditions involving alkali. Benzofuroxans are similarly prepared, but the method is limited by the availability of the o-quinone or its dioxime. Thus, the most common routes to benzofuroxans involve o-nitroaniline cyclizations (reaction 151), which may however, occasionally be unsatisfactory, although giving access to a wide range of substituents.[27] Hypochlorite also gives cyclization to (**316**). Other cyclization routes to benzofuroxans from o-azidonitroaryls are less widely applicable.[18,150]

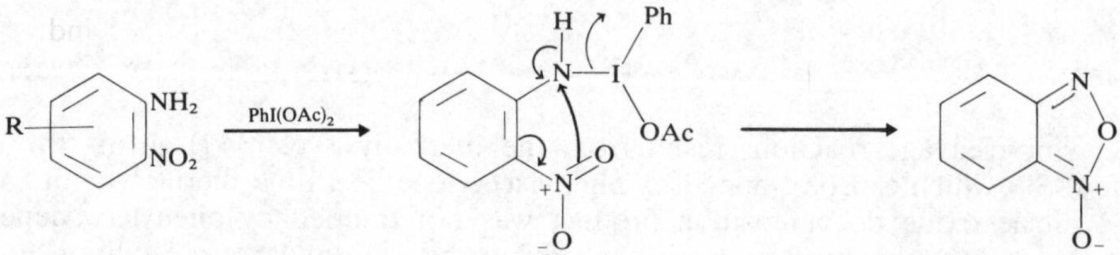

$\xrightarrow{PhI(OAc)_2}$ → (151)

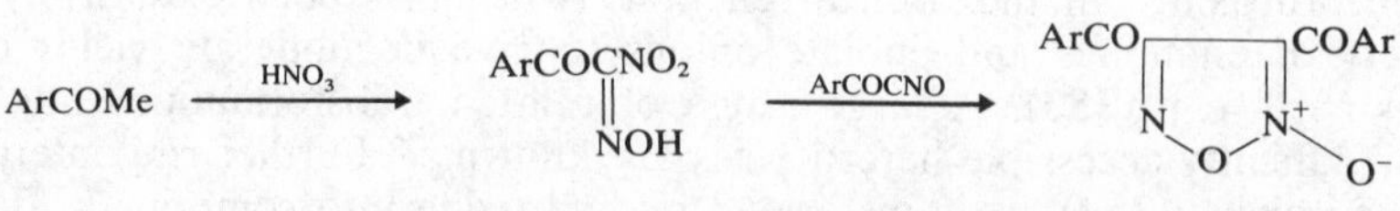

ArCOMe $\xrightarrow{HNO_3}$ ArCOCNO$_2$ (=NOH) $\xrightarrow{ArCOCNO}$ [3,4-bis(ArCO)furoxan] (152)

The dimerization of nitrile oxides also gives benzofuroxans. The nitrile oxides may be prepared from haloximes by treatment with, for example, triethylamine, or alternatively by the reaction of methyl ketones with fuming nitric acid (reaction 152).[6c,27,95] It should be noted that the 3,4-diacylfuroxans are sensitive to acid (giving 1,4-diketones) and also to alkaline reagents. α-Diazo-ketones on nitrosation also give acyl nitrile oxides, which dimerize to benzofuroxans.[27] The spontaneous dimerization of fulminic acid gives in low yield α-isocyanilic acid (**357**), which gives rise to an extensive range of substituted furoxans, including (**358**).[95]

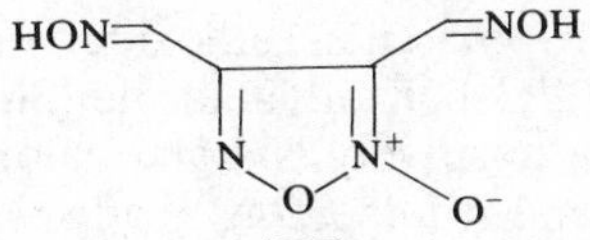

HON= =NOH (357) HO$_2$C NH$_2$ (358)

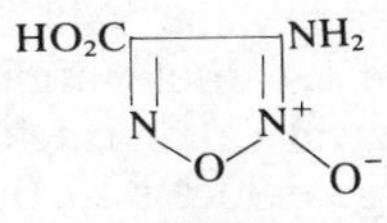

The extensive uses of benzofuroxans have been summarized[150] and of recent significance is their possible anti-leukaemic activity, which is said to be due to their ability to form σ-complexes of the Meisenheimer type.[157]

20.1.6 TWO NITROGENS, ONE HETEROATOM: THIADIAZOLES

The thiadiazole systems comprise the 1,2,3-thiadiazoles (**359**) and their benzo-fused analogues, the benzo-1,2,3-thiadiazoles (**360**), the 1,2,4-thiadiazoles (**361**), the 1,3,4-thiadiazoles (**362**), the 1,2,5-thiadiazoles (**363**) and their benzo-fused analogues the benzo-2,1,3-thiadiazoles (**364**). Some examples of analogous selenadiazole chemistry will be included to illustrate comparisons and differences in reactivity.

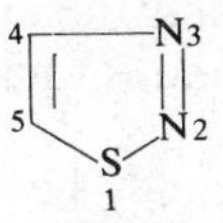

(**359**) 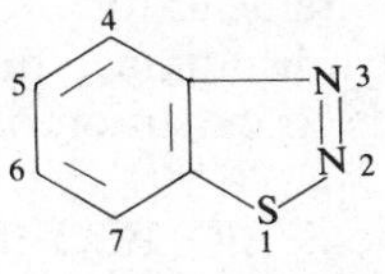 (**360**) 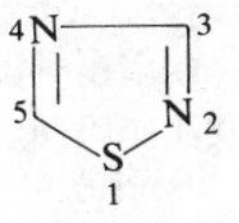 (**361**) 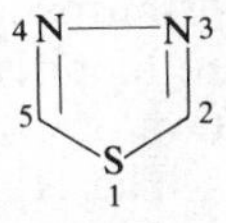 (**362**)

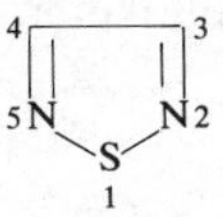

(**363**) 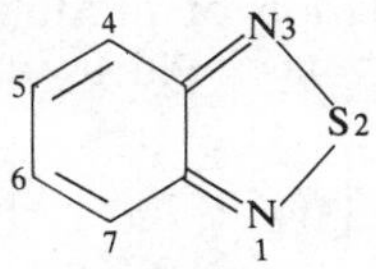 (**364**)

20.1.6.1 1,2,3-Thiadiazoles and benzo-1,2,3-thiadiazoles

Unlike the 1,2,3-oxadiazole systems, the 1,2,3-thiadiazoles are well known and extensively documented.[5,32,121,158] Physical properties have been briefly summarized[6c] and more recently there have been extensive studies of the mass spectral fragmentation, which proceeds by nitrogen extrusion and formation of a thiirenium ion.[159,160,161]

Nucleophilic reagents may cause ring cleavage, following initial abstraction of a proton (reaction 153). Similar behaviour is exhibited by 1,2,3-selenadiazoles (reaction 154), the

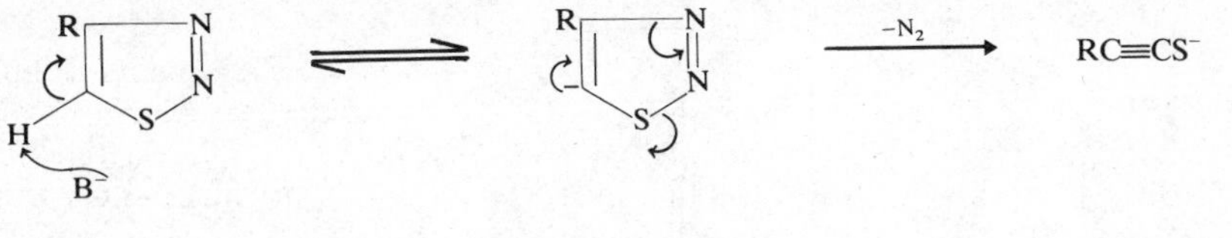

$$\xrightarrow{-N_2} RC{\equiv}CS^- \quad (153)$$

$$\xrightarrow{B^-} R{-}C{\equiv}C{-}Se^- \xrightarrow{H^+} [RHC{=}C{=}Se] \longrightarrow \quad (154)$$

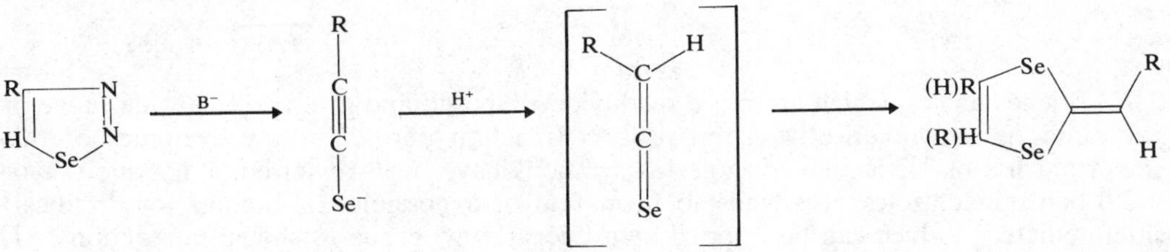

intermediate acetylene salt being isolable in dioxan–potassium ethoxide, or reacting further to give 1,4-diselenafulvenes.[161] The 6-chloro-1,2,3-benzothiadiazole (**365**) may be written in several canonical forms in which there is a positively charged chlorine and a negative charge delocalized through the ring system, and it undergoes ready nucleophilic displacement of chloride by a range of nucleophiles in aqueous DMSO.[161] The 5-chloro-derivative (366) suffers ring-cleavage, giving the disulphide (**367**). Examples of electrophilic reactions exist, the quaternization of 1,2,3-benzothiadiazoles giving exclusively N-3 derivatives such as (**368**), as was shown by an unequivocal synthesis of the product.

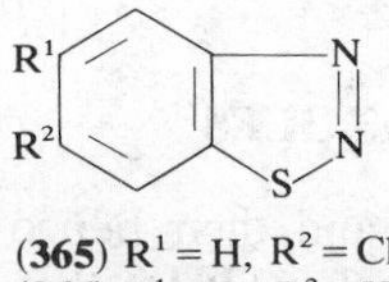

(**365**) $R^1 = H$, $R^2 = Cl$
(**366**) $R^1 = Cl$, $R^2 = H$

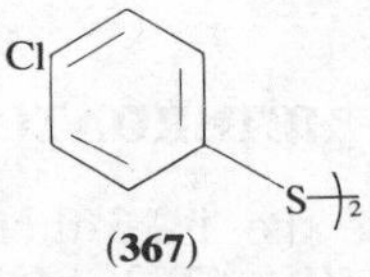

(**367**)

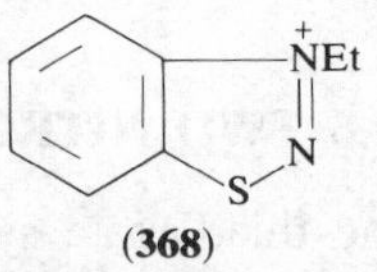

(**368**)

The comparative rates of quaternization of a range of thia- and selena-diazoles was in the order 1,2,3-benzoselenadiazole > 2,1,3-benzoselenadiazole > 1,2,3-benzothiadiazole > 2,1,3-benzothiazole, following closely the pK_a values of their first ionization constants.[161] Radicals react with 1,2,3-benzothiadiazoles, phenyl radical, for example (generated from *N*-nitrosoacetanilide) giving a mixture of diphenyl disulphide, thianthrene, and dibenzothiophen.[127,161] The phenylthiyl radical also leads to extrusion of nitrogen, and formation of a range of sulphur heterocycles.[162]

The thermal and photochemical fragmentations of 1,2,3-thiadiazoles and benzothiadiazoles and their selenium analogues have recently been extensively investigated and employed in organic synthesis. Thermolysis of 1,2,3-thiadiazoles results in extrusion of nitrogen and formation of products arising from an intermediate biradical (reaction 155).[102] This pyrolysis has been used recently to give thioketen intermediates (by Wolff

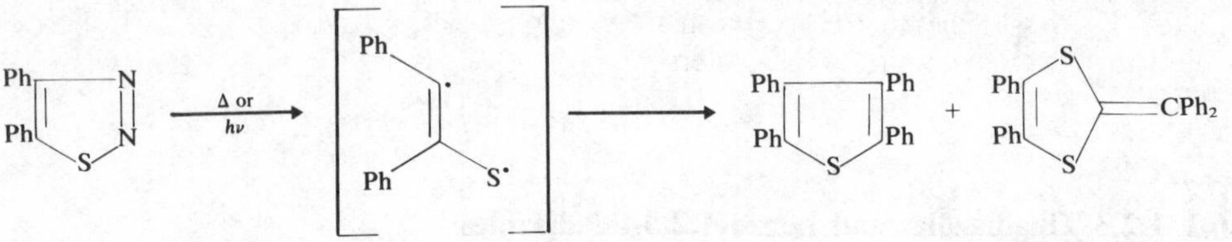

(155)

rearrangement of a possible thioacyl carbene intermediate, **369**), which are particularly useful for the synthesis of thiocarboxylic acid *O*-esters.[163] A flash thermolysis technique has been employed for the routine preparation of the difficultly accessible thioketens (**370**; $R^1,R^2 = H$, alkyl, or aryl).[164] An interesting aspect of the thermal decomposition of 1,2,3-thiadiazoles and their selenium analogues is the trapping of intermediates by di-iron nonacarbonyl as the complex (**371**).[102,161] Differences emerge in the thermolysis of

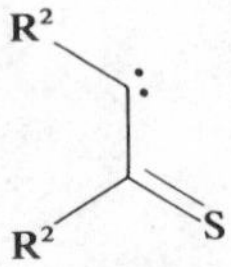

(**369**)

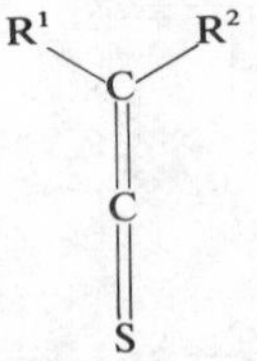

(**370**)

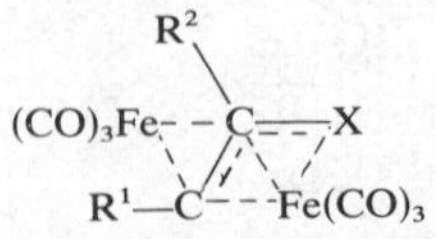

(**371**) X = S or Se

1,2,3-selenadiazoles, which are more readily decomposed and give rise to a wide range of acetylenes in a synthetically useful process (reaction 156).[160] Many examples of this important use of 1,2,3-selenadiazoles as synthons have been reviewed.[13] Thermolysis of 1,2,3-benzothiadiazoles also leads to formation of a possible 1,3-dipolar (or diradical) intermediate,[161] which can be trapped by a thiocarbonyl group as shown in reaction 157.

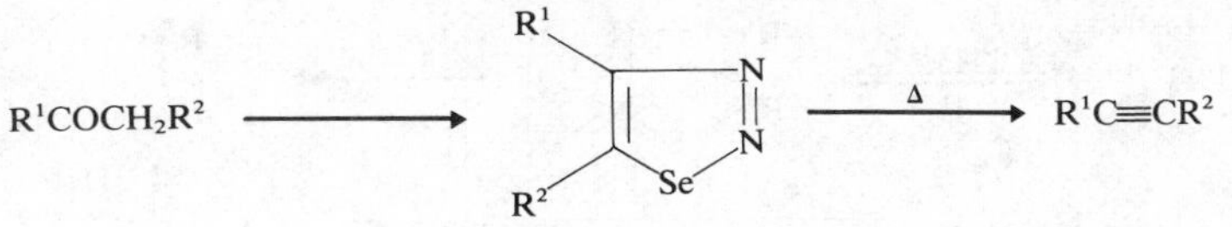

$R^1COCH_2R^2 \longrightarrow$ [4-R^1-5-R^2-1,2,3-selenadiazole] $\xrightarrow{\Delta} R^1C{\equiv}CR^2$ (156)

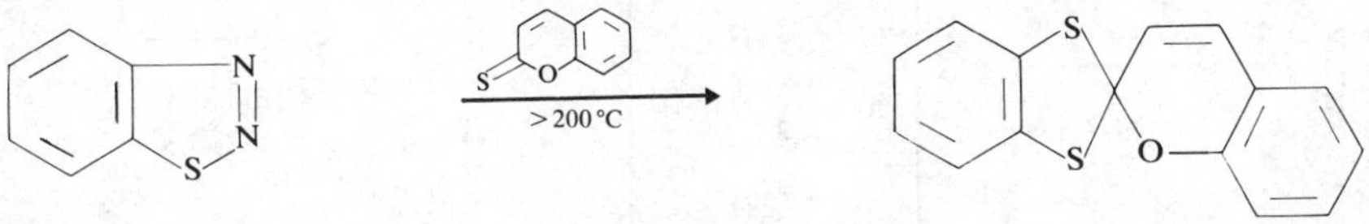

[1,2,3-benzothiadiazole] $\xrightarrow[>200\,°C]{\text{chromene-2-thione}}$ [spiro product] (157)

A paper discusses the possible thermal and photochemical intermediates which give rise to (**372**) (Scheme 39), and from a study of solvent and substituent effects favours diradical intermediates rather than dipolar or benzothiiren species.[165] Trapping of the intermediate occurred with phenylacetylene, giving each of the possible benzothiophen adducts. Photochemical investigations of the nitrogen extrusion from 1,2,3-thiadiazoles have also implicated diradical, dipolar, thiiren,[166,167] and thioketen intermediates in the formation of sulphur heterocycles.[14,27,160,161] Photolysis at 77 K allowed e.s.r. investigation of

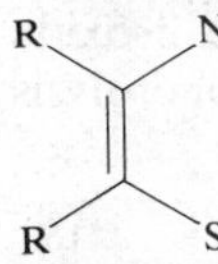

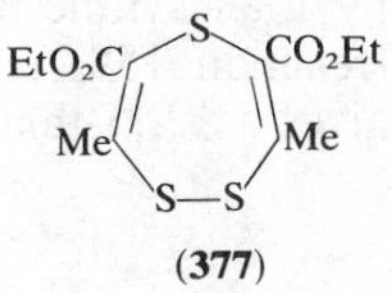

(**372**) 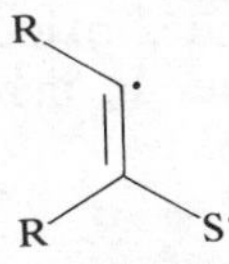 (**373**) 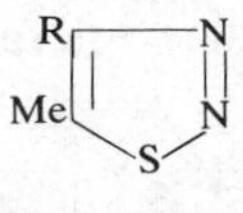 (**374**) R = PhCO (**375**) (**377**)

(**376**) R = EtO_2C

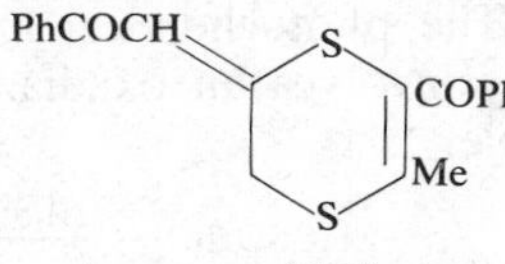

thiyl radical species (**372**) or (**373**).[14] Photolysis of (**374**) in benzene gave the adduct (**375**), and (**376**) in the gas phase gave (**377**). Selenadiazoles undergo related photofragmentations.[101,166] The photochemical production of thioacylcarbenes from 1,2,3-thiadiazoles has been used in the cross-linking of polymers.[168]

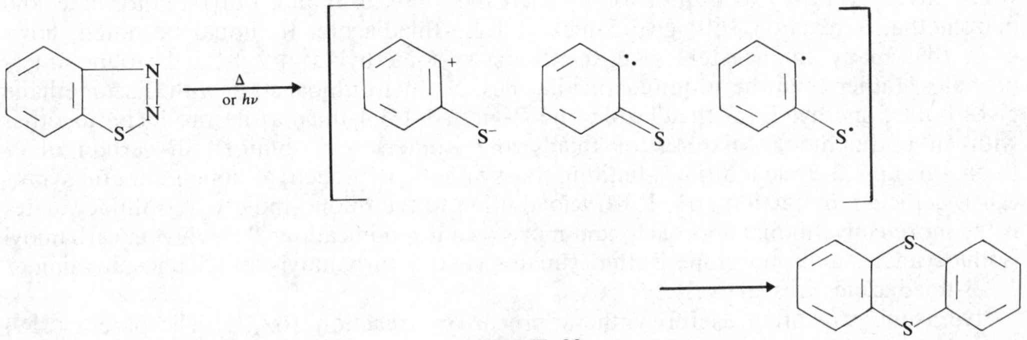

SCHEME 39

Ring substituents on 1,2,3-thiadiazoles and 1,2,3-benzothiadiazoles undergo conventional reactions, paralleling many of those in other aromatic systems.[160,161] For example, the methyl group in (**378**) can be halogenated and thus transformed by nucleophilic substitution reactions into a hydroxymethyl group, which may be oxidized to an aldehyde. These reactions illustrate the stability of the thiadiazole nucleus allow access to a range of derivatives.[168]

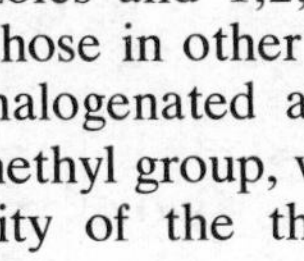

(**378**)

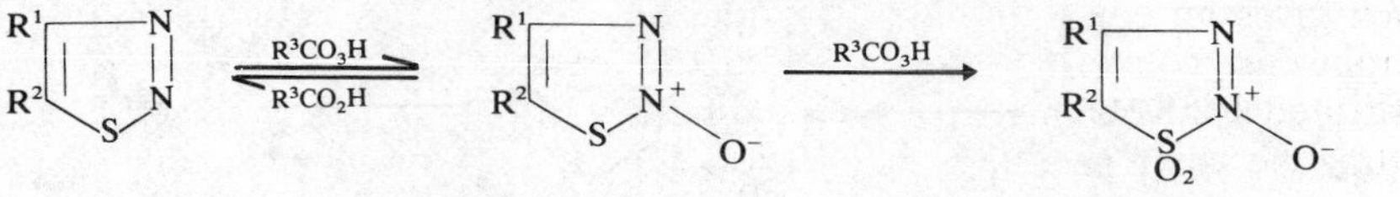

(158)

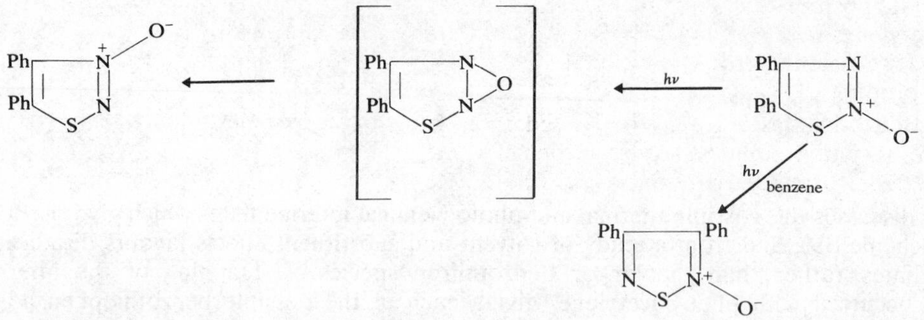

(159)

1,2,3-Thiadiazole *N*-oxides, formed reversibly by peracid treatment (reaction 158), are oxidized at N-2, which bears the higher charge density (in accord with theoretical calculations),[161] further oxidation giving a sulphone. Such sulphones have been prepared by total synthesis.[170] The photochemical rearrangement of the 2-oxide to the 3-oxide (reaction 159) possibly involves an oxadiaziridine intermediate. In benzene, photolysis gives a 1,2,5-thiadiazole.[14]

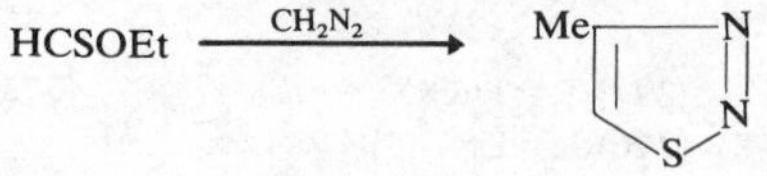

HCSOEt $\xrightarrow{CH_2N_2}$ (160)

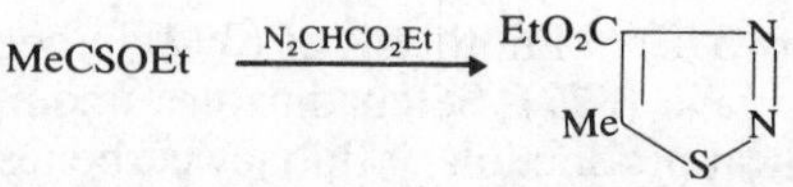

MeCSOEt $\xrightarrow{N_2CHCO_2Et}$ (161)

1,2,3-Thiadiazoles are often synthesized by modifications of Pechman's synthesis,[160,161] which involves 1,3-cycloaddition of diazoalkanes. For example, ethyl thioformate and diazomethane (reaction 160) give 5-methyl-1,2,3-thiadiazole. It should be noted, however, that many thionoesters give the 5-alkoxy-5-methyl-Δ^2-1,2,3-thiadiazoline intermediates, rather than the required thiadiazoles. Methyl dithioacetate with diazomethane gives both 5-methyl-1,2,3-thiadiazole and 2-methyl-1,3,4-thiadiazole products (together with an episulphide). Mixtures of thiadiazole isomers are obtained in certain other thiono-group 1,3-cycloadditions, limiting this synthetic approach, although a useful synthesis is depicted in reaction 161. 1,3-Cycloaddition to the thiono-moiety of isothiocyanates is the more conventional approach, and more recent modifications[160] involving carbamoyl isothiocyanates and phosphine isothiocyanates yield 5-carbamoyl- and 5-phosphinamido-1,2,3-thiadiazoles, respectively.

Hydrazones are often useful synthetic precursors (reaction 162).[161] The parent (**359**) was prepared from 4-methyl-1,2,3-thiadiazole by permanganate oxidation followed by thermal decarboxylation of the 4-carboxylic acid.[6c] An extension of this hydrazone

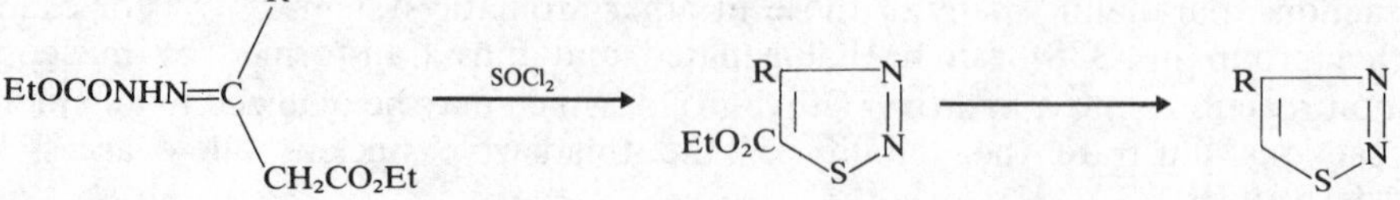

(162)

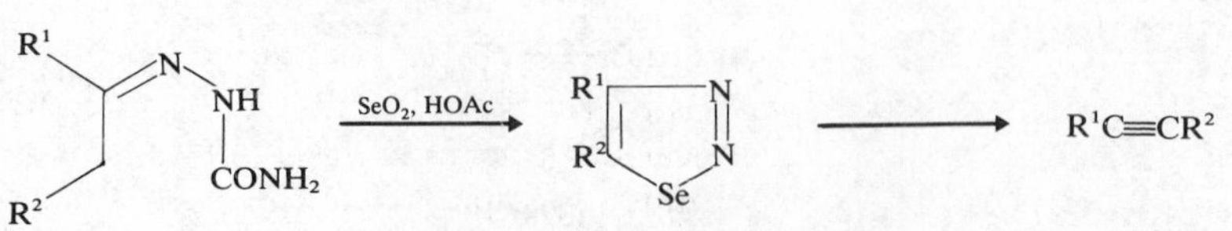

(163)

approach gives an entry to 1,2,3-selenadiazoles (reaction 163).[160] Substituted aldehydes and ketones are converted into the hydrazone, giving a wide range of 4-, 5-, mono-, and di-substituted 1,2,3-selenadiazoles. The direction of ring-closure when both α-positions of the hydrazone are available for oxidation depends on the acidity of the α-protons (the more acidic positions being oxidized). A convenient synthesis of unsymmetrically substituted acetylenes is thus obtained. Further developments, using hydrazones (**379**), gave the bicyclic 1,2,3-selenadiazoles which on thermolysis led to cyclic alkynes (**380**), together with by-products.[161]

1,2,3-Benzothiadiazoles are synthesized from *o*-mercaptoanilines, using nitrous acid, or, more recently, sodium nitrite–acetic acid.[161,171] Benzothiazathiolium salts (readily available from primary arylamines) are ring-opened by base to give *o*-mercaptoanilines and thus 1,2,3-benzothiadiazoles. This one-stage preparation gives variable yields.[161]

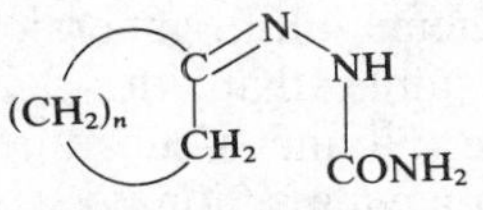

(**379**) 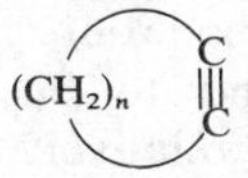 (**380**) 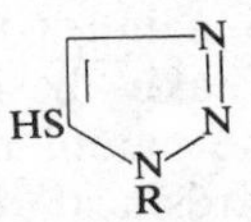 (**381**) 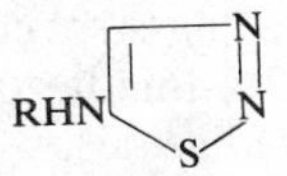 (**382**)

Ring transformations of other heterocycles may give 1,2,3-thiadiazoles, (**381**) for example, existing in equilibrium with (**382**).[161] Other rearrangements are shown in reactions 164 and 165.[161] 6-Mercapto-8-azapurines exist in equilibrium with 1,2,3-thiadiazolo[5,4-*d*]pyrimidines (reaction 166).[168]

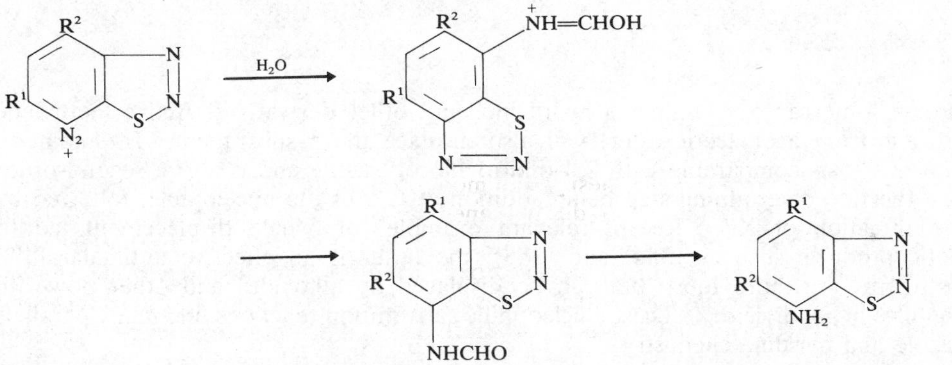

(164)

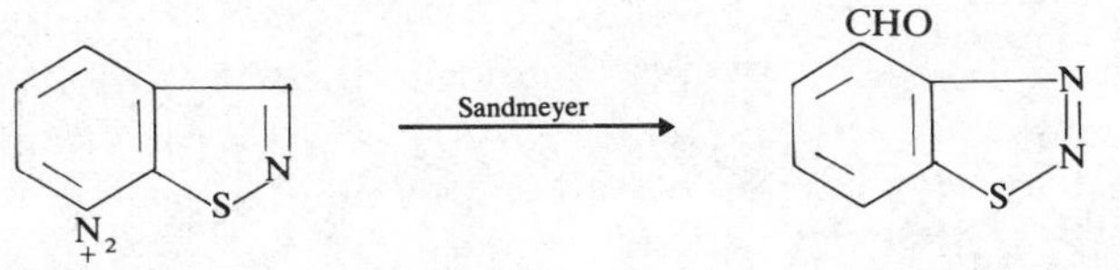

(165)

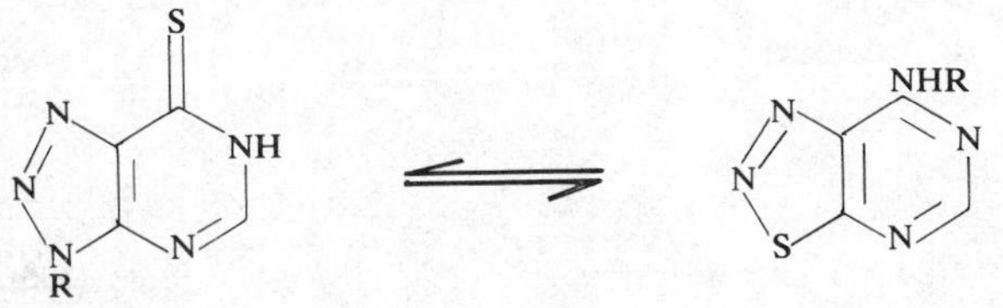

(166)

20.1.6.2 1,2,4-Thiadiazoles

This class of aromatic thiadiazoles has been reviewed,[5,5c,32,172] the parent (**361**) being a volatile liquid, b.p. 121 °C. The physical and spectroscopic properties have been summarized (the u.v. spectrum of the parent, for example, showing λ_{max} 229 nm, log ε 3.7). In

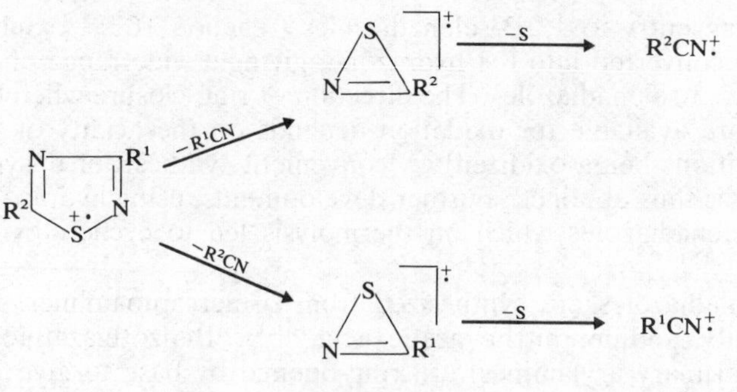

SCHEME 40

the mass spectrum[161] the main fragmentation modes are as in Scheme 40. In general, the 1,2,4-thiadiazoles are weakly basic, and tend to be more stable than the 1,2,4-oxadiazoles. Although the parent molecule is relatively reactive, 3- and 5-substituents stabilize the nucleus towards acid, alkali, oxidizing, and reducing agents.[172] Heavy-metal salts give addition compounds, and acids give salts.[172]

According to theoretical calculations, the 5-position in free 1,2,4-thiadiazoles should be the most reactive in nucleophilic substitution reactions in any of the isomeric thiadiazoles. Numerous displacements of 5-substituents are known,[168,172] (**383**; X = halogen) giving

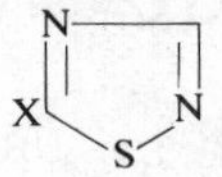

(**383**)

5-hydroxy, 5-mercapto, 5-amino, 5-hydrazino, and other derivatives. Active methylene anions (*e.g.* from acetoacetic ester) will also displace the 5-substituents. In summary, reaction at C-5 is comparable with 2,4-dinitrochlorobenzene and is also a second-order process, the rate-determining step being addition at C-5 of the nucleophile, followed by rapid elimination of X.[172] Recent relevant examples of 5-halo displacement led to antibiotic products such as (**384**) and (**385**); the latter is an effective antimalarial.[161] 3-Substituents are more inert, but can be displaced by alkoxides and other powerful nucleophiles in good yield.[172] These nucleophilic substitution reactions find many parallels in thiazole and pyridine chemistry.

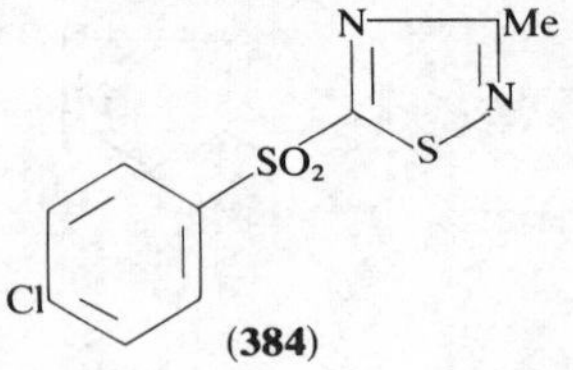

(**384**)

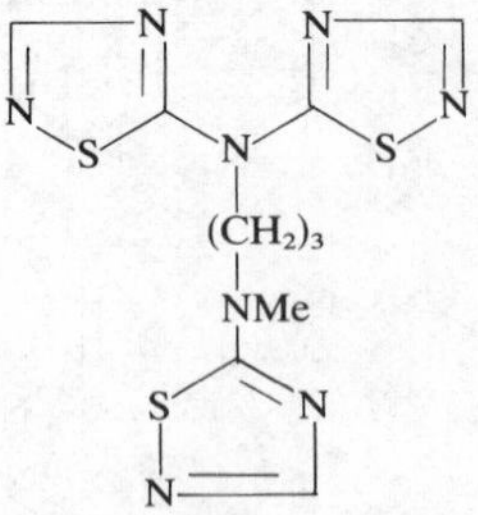

(**385**)

Electrophilic reactions are more unusual, but the quaternization of 1,2,4-thiadiazoles gives salts, 3-phenyl-5-methylthio-1,2,4-thiadiazole giving (**386**) with dimethyl sulphate.[172] 5-Amino-derivatives are acylated at the 5-position, and methylated at N-4 to give (**387**), which on standing rearranges to (**388**). Tritylation occurs at the 5-amine.[172] Diquaternary salts have been reported.[161] Acylation of the 3-hydroxy-group gives 3-acyloxy-products.

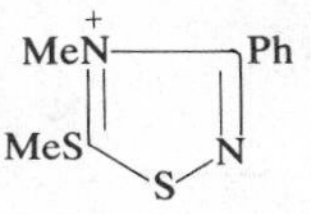

(386) 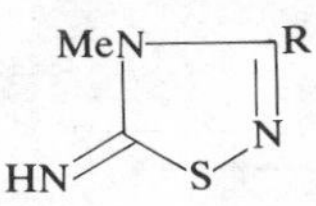 (387) 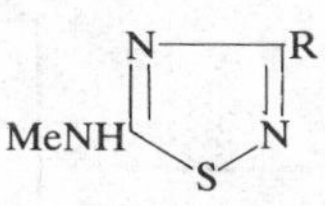 (388)

Although many 1,2,4-thiadiazoles are resistant to a range of oxidizing and reducing systems, Raney nickel gives ring-opened products, 3-hydroxy-5-phenylamino-1,2,4-thiadiazole being converted to 1-phenyl-2-thiobiuret.[172]

Methyl groups on the 1,2,4-thiadiazole nucleus can be preferentially reacted with butyl-lithium, the 5-methyl group giving a 5-lithio-derivative which can be carboxylated with CO_2 in good yield. Other thiadiazoles, including 3-methyl-1,2,3- and 3-methyl-1,2,5-, are ring-cleaved under these conditions, although 2-methyl-1,3,4-thiadiazoles may be lithiated and transformed into thiadiazolylacetic acids.[160] Other methods of making carboxylic acid derivatives of 1,2,4-thiadiazoles have been summarized.[172]

Amino-substituted 1,2,4-thiadiazoles account for a large fraction of the literature for these ring systems. 5-Amino-groups are readily diazotized (nitrosonium fluoroborate[172]). The 5-diazonium products are also made from 5-nitrosoamino and 5-diazoether precursors.[172] These 5-diazonium salts are among the most reactive known, probably because of the uniquely electron-deficient properties of C-5 of the heterocycle. Their reaction with alcohols (reaction 167) exemplifies the strongly electropositive nature of this functional grouping. Similar adducts resulting from nucleophilic attack at the terminal nitrogen by amines and thiols are known. Numerous conventional diazonium reactions at C-5 are also known, and coupling with phenols gives azo-dyes. Certain monoazo-dyes are of particular importance in the dyeing of polymers. There is spectroscopic evidence that 5-amino-derivatives exist as such, and they have been shown to be weak bases (pK_a 0.1). Zinc–HCl leads to reductive ring-cleavage.[172] 3-Amino-1,2,4-thiadiazoles may also be diazotized, are weak bases (pK_a 1.4), and are more resistant to reduction, although zinc–HCl gives ring cleavage.

(167)

5-Hydroxy-1,2,4-thiadiazoles are distinctly acidic, the 3-ethyl-5-hydroxy-derivative being more acidic than nitrophenol but less than 2,4-dinitrophenol. The 3-hydroxy-derivatives also have some phenolic character (ferric chloride test),[172] but give no ketonic derivatives. Spectroscopic evidence points to a 3-keto-structure.[18] 3- and 5-mercapto-1,2,4-thiadiazoles and their derivatives are well known,[172] 5-mercapto-derivatives being readily oxidized to disulphides. Peracid or aqueous chlorine oxidation of 3-alkylthio-derivatives occurs at the exocyclic sulphur, giving sulphoxides or sulphones.[172] Chlorine in acetic acid gives the 3-chloro-derivative, presumably *via* a chlorosulphonium intermediate.

Syntheses of 1,2,4-thiadiazoles may be categorized in terms of the arrays of atoms involved in the bond-forming process.[34,172] In the method involving N—C—S+C—N units, thioamides and thioureas are oxidized, a wide variety of oxidants including halogens, sulphur monochloride, hydrogen peroxide, ozone, and others, giving *S*-oxides (reaction 168) which show a remarkable tendency to form 1,2,4-thiadiazoles.[172] This

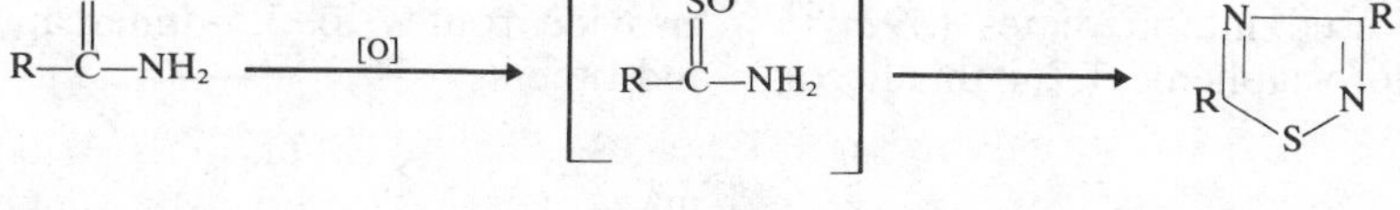

(168)

method affords 3,5-dialkyl-, -diaryl-, -dialkoxy-, and -diacylamino-1,2,4-thiadiazoles. The intermediate *S*-oxides (thioperimidic acids) may be stable at low temperatures, and condense with other thioamides to give products with dissimilar 3- and 5-substituents. A possible mechanism for these condensations is shown in Scheme 41.[172]

SCHEME 41

N—C—N+C—S syntheses utilize the condensation of amidine, amidoxime, isothiourea, guanidine, and other N—C—N fragments, with carbon disulphide, dithiocyanogen, thiocyanates, isothiocyanate esters, or halogenated methanesulphenyl chlorides (*i.e.* C—S fragments). These versatile synthetic methods are illustrated by reactions 169 and 170,[172] and lead to a wide range of 3- and 5-substituted amino, thio, alkoxy, alkyl, aryl, and hetaryl 1,2,4-thiadiazoles. The parent 1,2,4-thiadiazole has been prepared by this method.[172] Many related syntheses, particularly of 4-amino-derivatives, have been reported.[160,161,173]

$$H_2N{-}C({=}NOH){-}Ar + CS_2 \longrightarrow \text{3-aryl-5-mercapto-1,2,4-thiadiazole} \quad (169)$$

$$R{-}C({=}NH){-}NH_2 + (CNS)_2 \longrightarrow \text{5-amino-3-R-1,2,4-thiadiazole} \quad (170)$$

$$RCN + RCSNH_2 \xrightleftharpoons{HCl} R{-}C({=}NH){-}S{-}C({=}NH)R \longrightarrow R{-}C({=}NH){-}NH{-}C({=}S){-}R\ (\mathbf{389}) \xrightarrow{[O]} \text{3,5-di-R-1,2,4-thiadiazole} \quad (171)$$

N—C—N—C—S cyclization, particularly of the amidinothiono-grouping, continues to be the most versatile route into 1,2,4-thiadiazoles. 3,5-Disubstituted products may be formed in good yield as shown in reaction 171.[172] Intermediates (**389**) (reaction 171, $R=NH_2$) may also be oxidatively cyclized to 3,5-diamino-derivatives. Other variants give 3- and 5-amino-1,2,4-thiadiazoles. 3-Alkoxy- and 3-thio-products are obtained from precursors (**390**; $X=OR^1$ or SR^1), whereas (**391**) gives a 3-hydroxy-1,2,4-thiadiazole.[172] The widely used intermediates (**392**)[160,161] provide routes to 3,5-dimercapto- and 3-chloro-5-chlorosulphenyl-1,2,4-thiadiazoles, and others.

RNHC(=S)NHC(X)=NH (**390**) RNHC(=S)NHC(=S)NH₂ (**391**) $(^-S)_2C{=}NCN$ (**392**)

A synthetic route to thiadiazoles involving C—N—S + CN fragments which is mechanistically interesting, although of very limited applicability, is the 1,3-cycloaddition of nitrile sulphides to nitrile groups.[143]

Analytical, industrial, pharmacological, and other uses of 1,2,4-thiadiazoles have been reported.[168,172]

20.1.6.3 1,3,4-Thiadiazoles

Most of the published work on the four thiadiazoles has been on the 1,3,4-thiadiazoles. Physical, theoretical, and spectroscopic properties have been reviewed.[6c,160,161,168] According to microwave spectroscopy, the π-electron delocalization decreases in the order 1,2,5-thiadiazole > thiophen > 1,3,4-thiadiazole > 1,2,5-oxadiazole. U.v. spectral features have been investigated, the parent absorbing at 229 nm, and substituents with lone pairs giving rise to bathochromic shifts as in other oxa- and thia-diazoles. Of particular note is the C—Se—C bond angle (81.8°) in 1,3,4-selenadiazole, which is the smallest known angle in a planar five-membered ring.

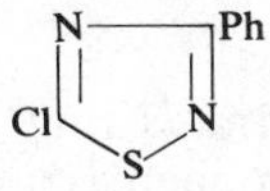

(393) 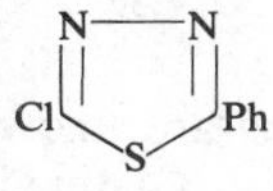 **(394)** 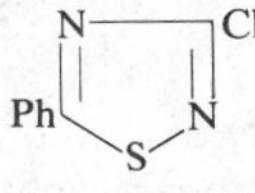 **(395)**

The majority of chemical reactions of the 1,3,4-thiadiazoles involve nucleophiles. Nucleophilic substitution of good leaving groups is a facile reaction, the relative rate constants for the thiadiazoles (**393**), (**394**), and (**395**) being 7000 : 64 : 1.[174] Scheme 42 depicts a series of nucleophilic substitutions which indicate a dependence of the leaving group potential upon the cation associated with the anionic nucleophile.[160] Hydrazine will

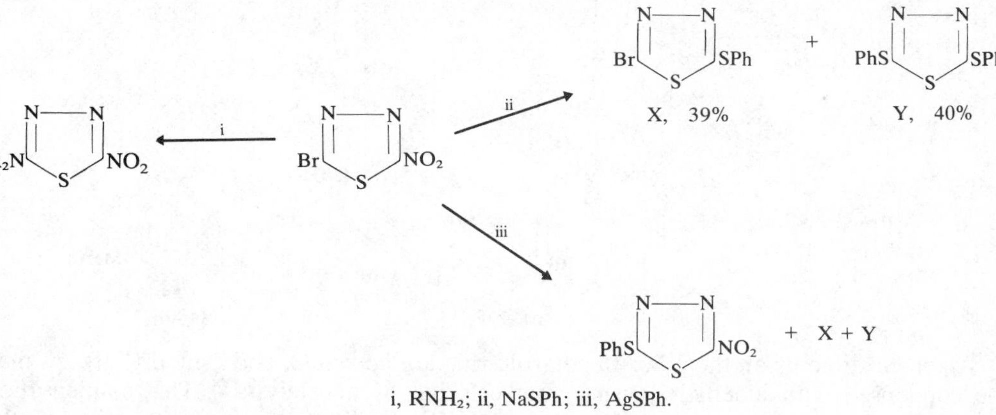

X, 39% Y, 40%

+ X + Y

i, RNH_2; ii, NaSPh; iii, AgSPh.

SCHEME 42

also displace halogens[175] and alkylsulphonyl groups,[172] giving derivatives of use, for example, in the synthesis of thiadiazolo-fused bicyclic analogues of purines.[175] 1,3,4-Thiadiazoles are vulnerable to attack by strong bases, leading to ring cleavage (reaction 172), this behaviour paralleling that of other azoles (particularly the 1,3,4-oxadiazoles).[140,161] The quaternized 1,3,4-thiadiazoles (**396**) afford a 2-anion which will attack acyl groups such as dialkylbenzoylphosphonates, giving by rearrangement ring-expanded dihydrothiadiazine products (**397**).[161] Nucleophilic attack by hydroxylamine on 1,3,4-thiadiazoles (reaction 173) leads in some cases to 2-amino-products in good yield.[160] The susceptibility to attack by nucleophiles is also demonstrated by a group of amine reactions which lead to triazolinthiones, *e.g.* reaction 174.[174]

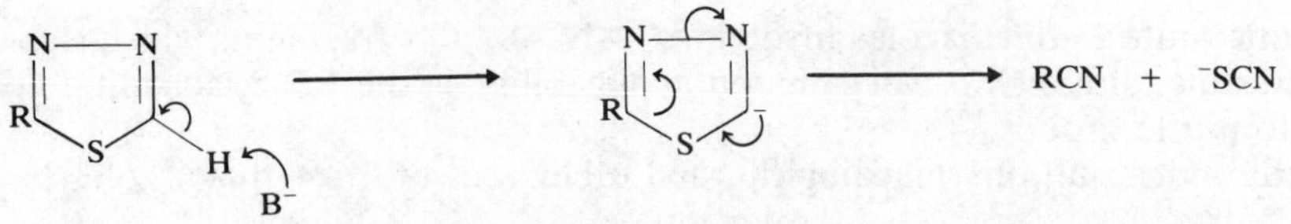

RCN + $^{-}$SCN (172)

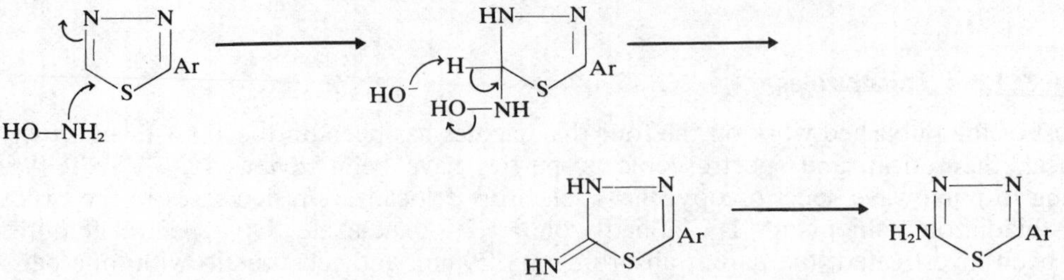

(173)

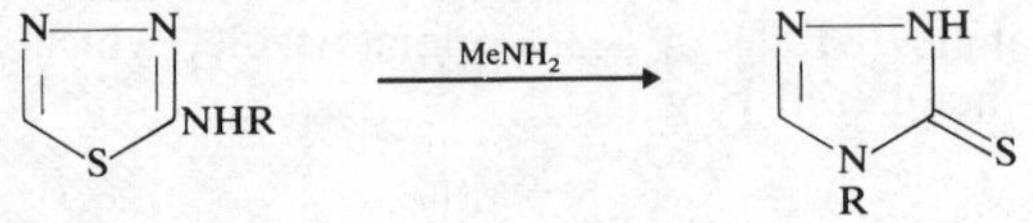

MeNH$_2$ (174)

No electrophilic substitution reactions are known in the unsubstituted ring. Nitration has not been achieved, but 2-phenyl-1,3,4-thiadiazole gives nitrophenyl products.[174] Alkylation can be achieved on ring nitrogen (even when an exocyclic 2-amino group is available)[174] and, as in the case of 1,2,4-thiadiazoles, diquaternary salts (**398**) can be made, using strong alkylating reagents such as trialkyloxonium salts.[161] Mechanistic and kinetic studies have been reported.[70,176] Differences in alkylation at ring nitrogen or at a 2-acetamido nitrogen were observed with different alkylating systems, methyl bromide–methoxide giving (**399**) and methyl iodide–butoxide giving (**400**).[174] Related studies have been performed on 1,3,4-selenadiazoles.[161]

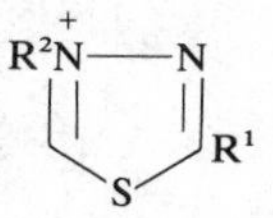

(**396**) 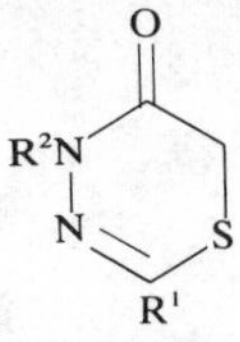 (**397**)

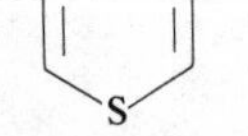

(**398**) 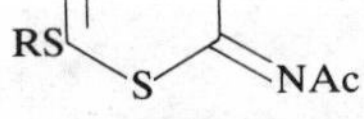 (**399**) 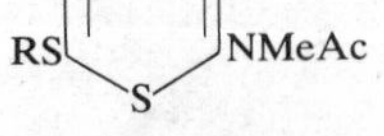 (**400**)

Alkyl substituents on the 1,3,4-thiadiazole ring are activated, and 2-methyl groups may be condensed with aldehydes, using zinc chloride as a catalyst.[174] This condensation reaction, which is common in the azole group, is facilitated by quaternization of the thiadiazole. Lithiation of the 2-methyl group leads to ring opening at or below room temperature, in a reaction type common to 2-lithiomethyl derivatives of thiazoles, 1,3,4-thiadiazoles, and 1,3,4-oxadiazoles (reaction 175).[140]

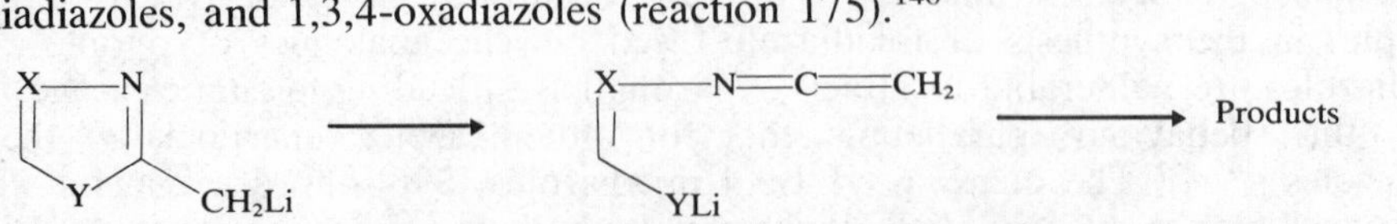

Products (175)

Amino substituents are important in 1,3,4-thiadiazole chemistry. 2-Amino-derivatives probably exist in the amino-form,[174] and they are readily diazotized, the 2-diazonium products showing remarkably strong coupling activity (as also observed for 1,2,4-thiadiazoles) in that they will even couple with *m*-xylene and mesitylene in high yields.[174]

Sodium cobaltinitrite–copper sulphide transforms the diazonium chloride into 2-nitro-1,3,4-thiadiazole.[174] Stable primary nitrosamines have been formed by diazotization, and on thermolysis in benzene they give 2-phenyl-1,3,4-thiadiazoles by a possible Gomberg–Bachman reaction, involving a thiadiazolyl radical.[161] 2,5-Diamino-1,3,4-thiadiazoles undergo a well-known isomerization (reaction 176), giving 1,3,4-triazoles.[168] Related ring rearrangements occur when 2-imino-5-thiono-1,3,4-thiazolidines react with amines.[161]

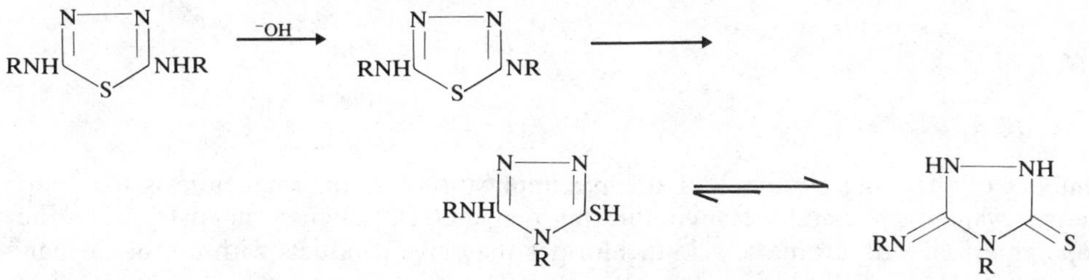

(176)

Hydroxy-1,3,4-thiadiazoles exist mainly in the keto-form in the solid state,[18,174] and 2-thiols exist mainly as the 2-thiones. The thiol–thiono equilibrium in 2,5-dithiols has been investigated spectroscopically, and a thiono-thiol structure suggested.[161] Oxidation of alkylthio-substituents gives the sulphones, without affecting the ring.[174]

Aldehyde groups at C-2 may exist as stable hydrates, reflecting the electron-attracting nature of the ring. 2-Carboxylates are rather unstable and prone to decarboxylate, although an electron-donating 5-substituent increases their stability.[174]

Syntheses of 1,3,4-thiadiazoles have been comprehensively reviewed[172,174] and only representative examples will therefore be quoted. The majority of syntheses involve cyclizations of thiosemicarbazides or compounds with the basic S—C—N—N—C—S structural unit. Dehydrative cyclization of acylthiosemicarbazides is a standard method for preparing 1,3,4-thiadiazoles. Phosphoric, sulphuric, and sulphonic acids are some of the dehydrating reagents used (reaction 177).[158,160] The reaction of diacylhydrazides with P_2S_5

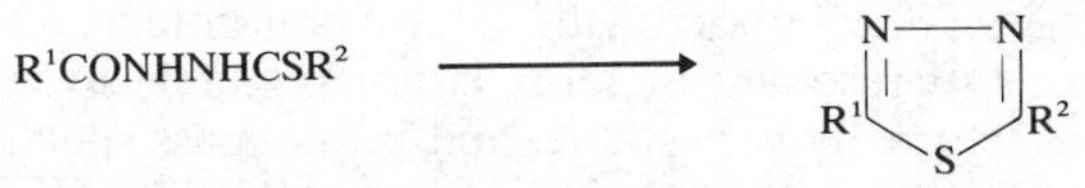

$R^1CONHNHCSR^2 \longrightarrow$ $R^1 = $ amino, alkyl, aryl (177)

gives 2,5-disubstituted 1,3,4-thiadiazoles directly. Aminoguanidines (**401**) are cyclized to derivatives of 2,5-diamino-1,3,4-thiadiazoles by HCl or acetyl chloride. Carbono- and thiocarbono-hydrazides (**402**, X = O or S) undergo an addition–cyclization with aroyl isothiocyanates, giving moderate yields of 2-aroylamido-5-hydroxy-1,3,4-thiadiazoles. Adducts of thiocarbonohydrazides and carbodi-imides also lead to 1,3,4-thiadiazoles. Dithiocarbazates (**403**), formed from (thio)benzhydrazides and CS_2, lead to mercapto-derivatives, as does the reaction of benzimidrazones (**404**) with CS_2. Dithiocarbazates will also react with cyanogen bromide to give 2-amino-5-mercapto-derivatives in an extension of a well-known synthesis of 1,3,4-thiadiazoles. The parent (**362**) is made conveniently by treating (**405**) with H_2S. Many other related intramolecular cyclizations are known,[158,160] including reactions of thiosemicarbazides with ortho-esters to give monosubstituted products. Oxadiazolium salts (**406**) on treatment with sodium sulphide, and then acetic anhydride–perchloric acid, give the 1,3,4-thiadiazolium analogues (**407**) in high yield. 1,3,4-Oxadiazoles may in some cases be directly transformed into 1,3,4-thiadiazoles by P_2S_5. Syntheses of 1,3,4-selenadiazoles based upon some of the cyclization reactions cited above, using for example, acylselenocarbazides, are known.[160] Miscellaneous preparations involve the reaction of diazomethane with thiobenzoyl chloride to give 2-phenyl-1,3,4-thiadiazole,[158] the reaction of aldehydes with hydrazine and H_2S to give (**408**), which when heated with sulphur in piperidine gives high yields of 2,5-derivatives,[174] and others recently reviewed.[161]

RCONHCSNHNHC(=NH)NHR (401)

$H_2NNHC(=X)NHNH_2$ (402)

RC(=X)NHNHCS_2K (403) X = O or S; (404) X = NH

$Me_2NCH{=}N{-}N{=}CHNMe_2$ (405)

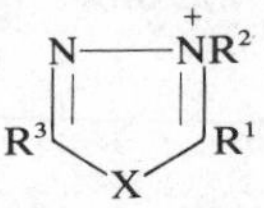

(406) X = O
(407) X = S

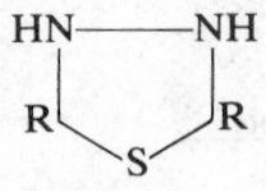

(408)

Members of this ring system find diverse applications as pharmaceuticals, oxidation inhibitors, cyanine dyes, and metal complexing reagents. In polymer chemistry when they are copolymerized with aromatic acid dichlorides they give products with useful mechanical properties.[160]

20.1.6.4 1,2,5-Thiadiazoles and 2,1,3-benzothiadiazoles[5,6c,34,158,177]

Recent investigations of theoretical, physical, and spectroscopic aspects have been summarized.[160,161] These weakly basic compounds are protonated in 96% H_2SO_4,[177] give sulphate and perchlorate salts, and heavy-metal ion complexes. The dipole moments of 2,1,3-benzoxa/thia/selenadiazoles indicate that mesomeric charge-transfer increases regularly from O to S to Se, implying that 2,1,3-benzoxadiazoles assume an *o*-quinonoid structure and that the sulphur and selenium analogues show increasing aromatic character.[160,161] Precise X-ray crystallographic data have been reported for acenaphtho[1,2-*c*]-[1,2,5]thiadiazole.[161] In the u.v. spectrum, the parent (**363**) absorbs at 253 nm (ε 7800), methyl, amino, and carboxylic acid substituents causing bathochromic shifts.[177] Halogeno-1,2,5-thiadiazoles have been subjected to detailed i.r. and Raman analysis, and assignments for all the fundamental vibrations made.[161] In the n.m.r. spectrum, 1,2,5-thiadiazole exhibits a singlet at δ 8.70, compared with 1,2,5-oxadiazole at δ 8.64, but the aromaticity as expressed in terms of resonance energy is not relatable to chemical shift values.[177] These compounds are thermally very stable, and in the mass spectrum of (**363**) the molecular ion is the base peak, fragment ions being $HCNS^+$, HCN^+ and S^+. 1,2,5-Thiadiazole is a liquid, b.p. 94 °C, which is susceptible to oxidation, peracetic acid, for example, giving ring cleavage. 1,2,5-Thiadiazoles are relatively acid stable, and certain derivatives are decomposed by base. Photochemically the 1,2,5-thiadiazoles give alkyl/aryl nitriles and sulphur.[13]

Nucleophilic substitution reactions can be performed on halogeno- and arylsulphonyl-1,2,5-thiadiazoles (reactions 178 and 179).[177,178] Metal amides turned out to be unnecessarily powerful nucleophiles, and led to ring cleavage caused by attack on sulphur similar to that observed for dialkyl- and diaryl-1,2,5-thiadiazoles with alkyl-lithiums of Grignard reagents. It was observed that the 1,1-dioxide-3,4-dichloro-derivatives behaved like acyl halides in their reactions with nucleophiles.[177]

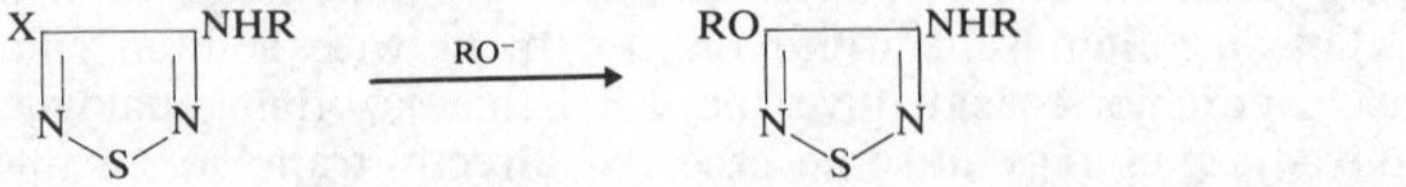

X, NHR-thiadiazole $\xrightarrow{RO^-}$ RO, NHR-thiadiazole (178)

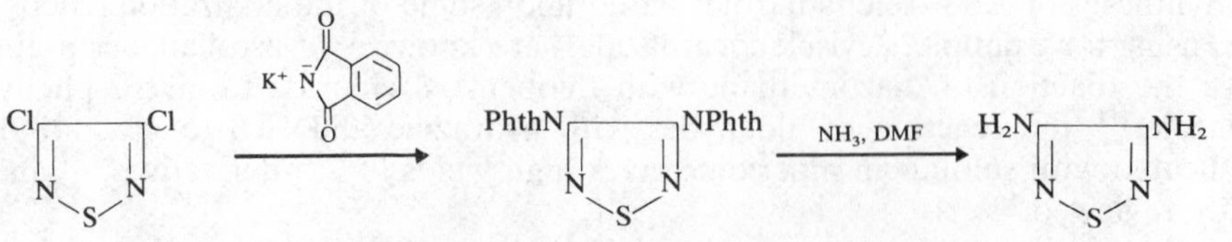

Cl, Cl-thiadiazole $\xrightarrow{K^+\ \text{phthalimide}^-}$ PhthN, NPhth-thiadiazole $\xrightarrow{NH_3,\ DMF}$ H_2N, NH_2-thiadiazole (179)

1,2,5-Thiadiazoles are relatively inert towards electrophilic substitution in attempted halogenation, nitration, and Friedel–Crafts reactions.[177] However, electrophilic deuteration can be achieved in low yield with D_3PO_4 at 250 °C. Some examples of electrophilic substitution of monosubstituted 1,2,5-thiadiazoles with activating groups (such as amino and methyl) are known, leading to halogeno-derivatives.[177] 2,1,3-Benzothiadiazoles reacted with bromine in a melt (iron catalyst) to give 4,5,6,7-tetrabromo-adducts, and 4-bromo-2,1,3-thiadiazole can be sulphonated at C-7.[161] Condensed 1,2,5-thiadiazoles such as (**409**) are nitrated at C-6 and C-9[15,86] and halogenated at C-4 and C-7,[160,161] although other isomers are obtained under differing conditions. Oxidation of (**409**) with chromic acid gave (**410**).

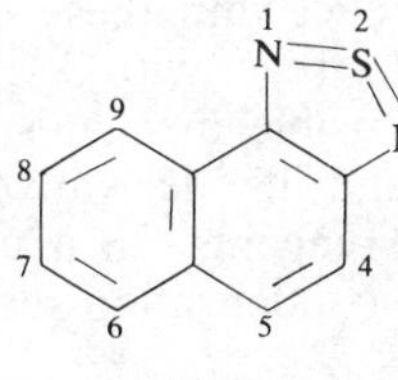

(**409**) 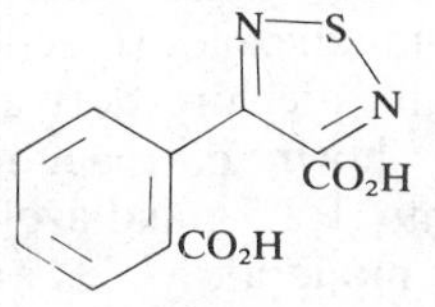 (**410**)

Oxidation of (**364**) with ozone or permanganate gave (**411**), which was decarboxylated at 200 °C to give the parent compound (**363**)[6c] or in hot nitrobenzene to give the monocarboxylic acid.[160] Reduction of 2,1,3-benzothiadiazoles by $SnCl_2$–HCl leads to reductive cleavage and formation of *o*-phenylenediamines.[140,160]

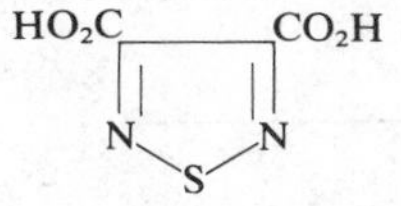

(**411**)

Alkyl substituents exhibit reactivity due to the electron-withdrawing heterocyclic ring which (a) destabilizes carbonium ions, (b) stabilizes carbanions, (c) represses S_N1 reactivity, and (d) enhances S_N2 reactivity.[177] Thus, they are very reactive in S_N2 substitution reactions, and condensation reactions can be achieved. The electron-withdrawing effect of the ring is further exemplified in the high acidity of carboxylic acid substituents. The dicarboxylic acid has pK_{a_1} 1.59 and pK_{a_2} 4.14, whereas the monocarboxylic acid has pK_a 2.47 (*o*-nitrophenol, pK_a 2.18). The halogen in acyl halides (**412**) is very readily displaced, ethoxymagnesiomalonate, for example, giving (**413**) which can be hydrolysed by acid to (**414**).[177] Unexpected reactivity due to the heterocylic ring is shown by the reaction of dialkylcadmium with (**412**) to give the alkyl ketone. Sodium borohydride or lithium borohydride reduce (**412**) to the primary alcohol. Amino-substituents are weakly basic, and are acylated, but give abnormal diazotization resulting from attack of unreacted amine on a nitrosoamine or highly electropositive diazonium group, giving (**415**). Spectroscopically it has been shown that protonation occurs on the ring nitrogen, as in 2- and 4-aminopyridine.[177] Diazotization of (**416**) gives a decomposition product (**417**), which reacts with anthracene to give 9-nitroanthracene and (**418**).[101] Hydroxy-substituents are markedly acidic (pK_a 5.10, similar to acetic acid).[177] It is interesting that an ethoxy group lowers the pK_a by 0.7 units, indicating a predominant $-I$ effect. The i.r. spectrum indicates the hydroxy-form, and alkylation occurs on oxygen. 3-Mercapto-1,2,5-thiadiazole probably exists in the thiono-form.[161]

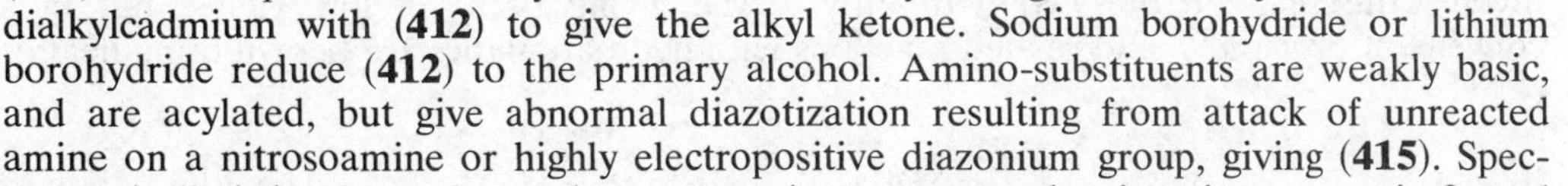

(**412**) (**413**) (**414**) (**415**)

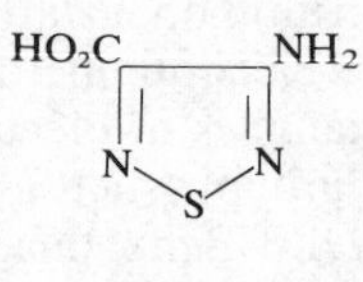

(416) 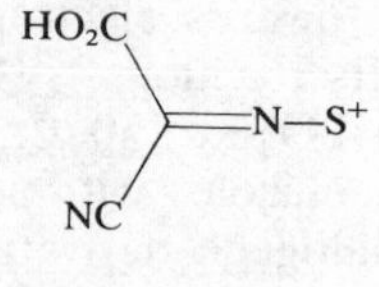 (417) 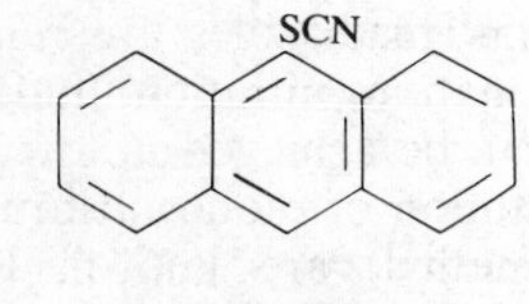 (418)

There is relatively little work on the synthesis of 1,2,5-thiadiazoles, the bulk of publications being concerned with fused ring systems. α-Diamines afford 1,2,5-thiadiazoles when treated with S_4N_4, S_2Cl_2 or $SOCl_2$. 1,2,5-Selenadiazoles are made with Se_2Cl_2.[160,177] α-Amino-acid amides and amidines also react with these reagents to give hydroxy-1,2,5-thiadiazoles (reaction 180). α-Aminonitriles react with S_2Cl_2 or Cl_2 to give monochloro- and dichloro-derivatives, respectively, a possible mechanism for the latter process being shown in reaction 181.[177] Cyanogen and its derivatives can also be transformed into 1,2,5-thiadiazoles by certain of these reagents. In a unique synthesis (reaction 182), moderate yields are obtained.[117] Related condensation reactions with, for

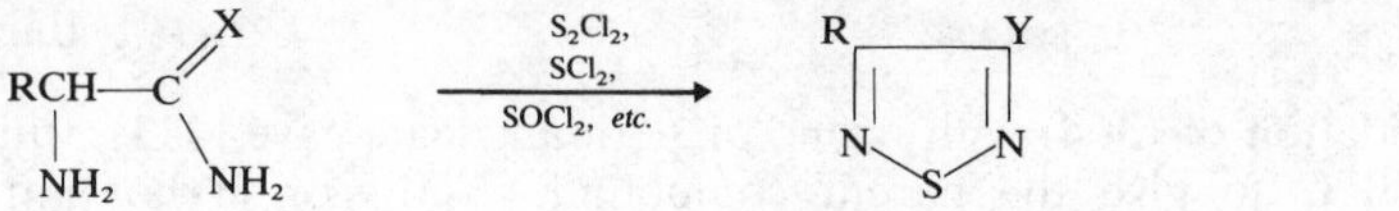

S_2Cl_2, SCl_2, $SOCl_2$, *etc.* (180)

X = O or NH Y = OH or NH_2

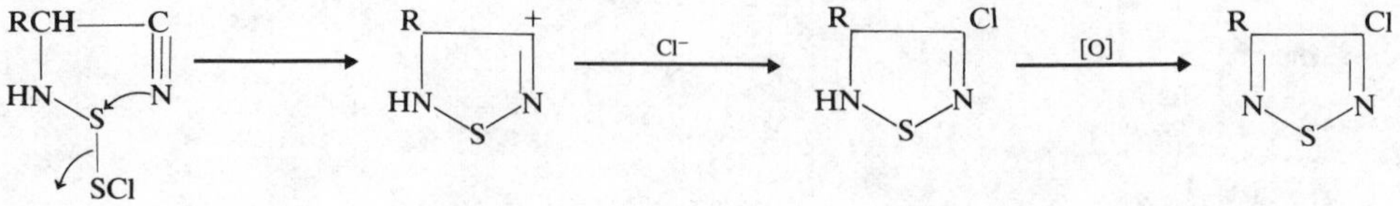

(181)

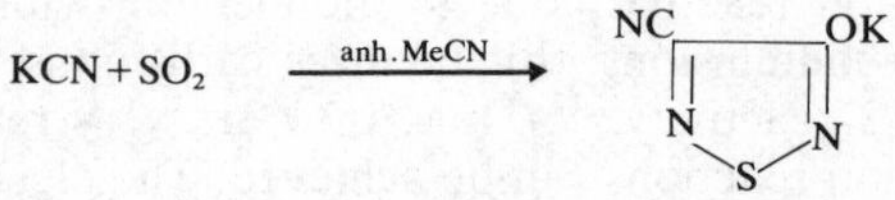

$KCN + SO_2$ $\xrightarrow{\text{anh. MeCN}}$ (182)

example, $SOCl_2$ and *o*-diamines (**419**) give rise to 2,1,3-benzothiadiazoles.[6b,6c] Selenium dioxide will give the 2,1,3-benzoselenadiazoles. In a re-investigation of earlier reports it was shown that *o*-dioximes (**420**) are converted by SCl_2 into (**364**) and the *N*-oxide (**421**).[160] Similar products were formed from (**422**). Other routes to 1,2,5-thiadiazoles involve the reaction of N_4S_4 with ethyl aryls, giving (**423**) in low yields, but in a convenient process,[177] and related reactions with hydroaromatics are known using heat or

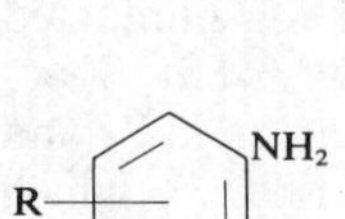

(419) 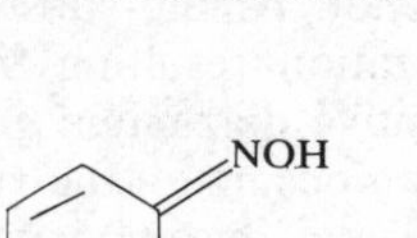 (420) 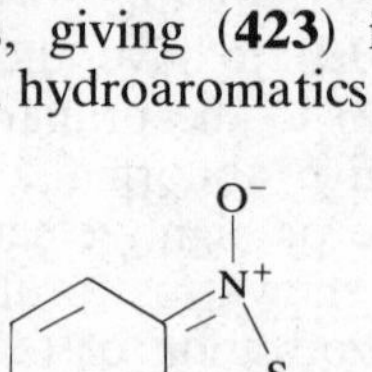 (421) 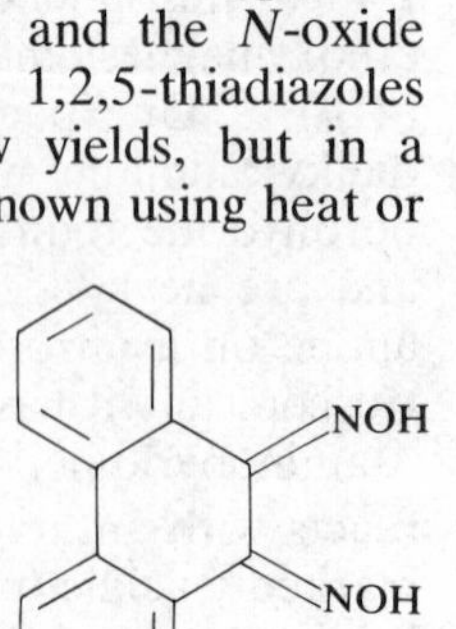 (422)

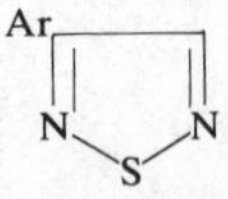

(423) 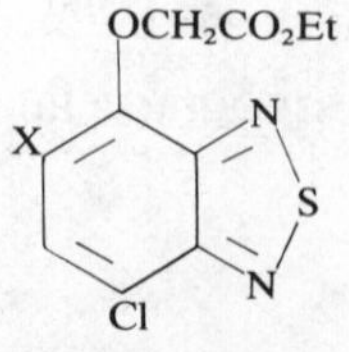 (424) X = Cl or Me

photolysis. Recently, trithiazyl trichloride ($S_3N_3Cl_3$) was shown to react with acenaphthylene and stilbene to give fused 1,2,5-thiadiazoles.[179] A different mode of reaction was, however, encountered with less-reactive olefins, which gave isothiazolo-fused heterocycles.

Uses of 1,2,5-thiadiazoles have been summarized,[34,177] and among potentially important compounds recently reported were derivatives (**424**), which are powerful and selective plant-growth regulators.

References

1. R. H. Wiley, *Chem. Rev.*, 1945, **37,** 401.
2. J. W. Cornforth, in 'The Chemistry of Penicillins', Princeton University Press, Princeton, New Jersey, 1949.
3. 'The Chemistry of Heterocyclic Compounds', ed. R. C. Elderfield, Wiley, New York, 1956, vol. 5.
4. I. J. Turchi and M. J. S. Dewar, *Chem. Rev.*, 1975, **75,** 389.
5. 'Aromatic and Heteroaromatic Chemistry' (Specialist Periodical Reports), The Chemical Society, London (a) 1973, vol 1; (b) 1974, vol. 2; (c) 1975, vol. 3.
6. See, for example, (a) L. A. Paquette, 'Principles of Modern Heterocyclic Chemistry', N. A. Benjamin, New York, 1968; (b) A. R. Katritzky and J. M. Lagowski, 'The Principles of Heterocyclic Chemistry', Chapman and Hall, London, 1967; (c) M. H. Palmer, 'The Structure and Reactions of Heterocyclic Compounds', E. Arnold, London, 1967; (d) J. A. Joule and G. F. Smith, 'Heterocyclic Chemistry', Van Nostrand Reinhold, London, 1972.
7. 'Physical Methods in Heterocyclic Chemistry', ed. A. R. Katritzky, 1971, vol. 4, p. 187.
8. J. H. Bowie, P. F. Donaghue, H. J. Rodda, R. G. Cooks, and D. H. Williams, *Org. Mass Spectrometry*, 1968, **1,** 13; H. Nakata, H. Sakurai, H. Yoshizumi, and A. Tatematsu, *ibid.*, 199.
9. L. W. Deady, *Austral. J. Chem.*, 1973, **26,** 1949.
10. R. Gompper and F. Effenberger, *Chem. Ber.*, 1959, **92,** 1928; V. M. Dziomko and I. V. Ivashchenko, *Zhur. org. Khim.*, 1973, **9,** 2191.
11. R. Breslow, *J. Amer. Chem. Soc.*, 1958, **80,** 3719.
12. H. H. Wasserman, F. J. Vinick, and Y. C. Chang, *J. Amer. Chem. Soc.*, 1972, **94,** 7180.
13. A. I. Meyers, 'Heterocycles in Organic Synthesis', Wiley-Interscience, New York, 1974.
14. O. Buchardt, 'Photochemistry of Heterocyclic Compounds', Wiley-Interscience, New York, 1976.
15. See, for example, A. Padwa, J. Smolanoff, and A. Tremper, *Tetrahedron Letters*, 1974, 29.
16. M. Maeda, Y. Iwase, and M. Kojima, *J. Heterocyclic Chem.*, 1976, **13,** 221.
17. R. Filler, *Adv. Heterocyclic Chem.*, 1965, **4,** 75.
18. A. R. Katritzky and J. M. Lagowski, *Adv. Heterocyclic Chem.*, 1963, **2,** 3.
19. V. Dryanska and C. Ivanov, *Synthesis*, 1976, 37.
20. Y. Ueno and M. Okawara, *Bull. Chem. Soc. Japan*, 1972, **45,** 1797.
21. M. J. Haddadin and A. Hassner, *J. Org. Chem.*, 1973, **38,** 2650.
22. D. Clerin, J.-P. Fleury, and H. Fritz, *J. Heterocyclic Chem.*, 1976, **13,** 825.
23. M. J. S. Dewar, P. A. Spanninger, and I. J. Turchi, *J.C.S. Chem. Comm.*, 1973, 925.
24. A. F. Cockerill, A. Deacon, R. G. Harrison, D. J. Osborne, D. M. Prime, W. J. Ross, A. Todd, and J. P. Verge, *Synthesis*, 1976, 591.
25. See, for example, H. A. Houwing, J. Wildeman, and A. M. van Leusen, *Tetrahedron Letters*, 1976, 143, and references cited therein.
26. P. A. Lowe, in Ref. 5b, chapter 3.
27. G. V. Boyd, in Ref. 5b, chapter 4.
28. G. V. Boyd, in Ref. 5c, chapter 4.
29. H. H. Wasserman and F. J. Vinick, *J. Org. Chem.*, 1973, **38,** 2407.
30. K. H. Grellmann and E. Tauer, *J. Amer. Chem. Soc.*, 1973, **95,** 3104.
31. A. W. Somerville, in Ref. 5b, chapter 2.
32. Z. T. Fomum, P. D. Landor, S. R. Landor, and G. B. Mpango, *Tetrahedron Letters*, 1975, 1101.
33. C. O. Wilson, O. Gisvold, and R. F. Doerge, 'Textbook of Medicinal and Pharmaceutical Chemistry', Pitman, London, 1966.
34. 'Sulphur, Selenium and Tellurium' (Specialist Periodical Reports), The Chemical Society, London (a) vol. 1, 1970; (b) vol. 2, 1973; (c) vol. 3, 1975.
35. R. H. Wiley, D. C. England, and L. C. Behr, 'Organic Reactions', Wiley, New York, 1951, vol. VI, pp. 367–409.
36. T. S. Griffin, T. S. Woods, and D. L. Klayman, *Adv. Heterocyclic Chem.*, 1975, **18,** 99.
37. B. Prijs, 'Kartothek der Thiazolverbindungen', 4 vols., Karger, Basel and New York, 1952.
38. See, for example, Ref. 18 and, E. Bulka, in 'Organic Selenium Compounds', ed. D. L. Klayman and W. W. H. Gunther, Wiley-Interscience, New York, 1973, pp. 449–496; L. G. S. Brooker, J. A. Ford, and E. J. van Lare, *ibid.*, p. 507.
39. J. Metzger, *Z. Chem.*, 1969, 99.
40. J. C. Panizzi, G. Davidovics, R. Guglielmetti, G. Mille, J. Metzger, and J. Chouteau, *Canad. J. Chem.*, 1971, **49,** 956.

41. R. G. Buttery, L. C. Ling, and R. E. Lundin, *J. Agric. Food Chem.*, 1973, **21,** 488.
42. F. Kurtzer, in Ref. 34a, chapter 13.
43. G. Asato, *J. Org. Chem.*, 1968, **33,** 2544.
44. H. J. M. Dou, G. Vernin, and J. Metzger, *J. Heterocyclic Chem.*, 1969, **6,** 575.
45. M. Baule, R. Vivaldi, J. C. Poite, H. J. M. Dou, G. Vernin, and J. Metzger, *Bull. Soc. chim. France*, 1972, 2679.
46. Y. Tamura, H. Hayashi, J.-H. Kim, and M. Ikeda, *J. Heterocyclic Chem.*, 1973, **10,** 947.
47. G. B. Barlin, in Ref. 5a, chapter 8.
48. J. Bourdais, D. Abenheim, B. Sabourault, and A. Lorre, *J. Heterocyclic Chem.*, 1976, **13,** 491.
49. F. Kurtzer, in Ref. 34b, chapter 14; see also M. Fiorentino, L. Testaferri, M. Tiecco, and L. Troisi, *J.C.S. Chem. Comm.*, 1977, 317 and references cited therein.
50. F. Kurtzer, in Ref. 34c, chapter 14.
51. F. Kurtzer, in Ref. 34b, chapter 13.
52. P. Haacke, L. P. Bausher, and J. P. McNeal, *J. Amer. Chem. Soc.*, 1971, **93,** 7045.
53. J. Fabian, in Ref. 34c, chapter 17.
54. D. S. Kemp and J. T. O'Brien, *J. Amer. Chem. Soc.*, 1970, **92,** 2554.
55. W. J. Spillane, H. J.-M. Dou and J. Metzger, *Tetrahedron Letters*, 1976, 2269.
56. O. Attanasi, G. Bartoli, and P. E. Todesco, *J. Heterocyclic Chem.*, 1976, **13,** 1021; G. Bartoli, M. Lelli, F. Ciminale, and O. Attanasi, *J.C.S. Perkin II*, 1977, 20, and references to σ-anionic species cited therein.
57. F. Minisci, *Synthesis*, 1973, 1.
58. M. Baule, G. Vernin, H. J.-M. Dou, and J. Metzger, *Bull. Soc. chim. France*, 1971, 2083.
59. S. R. Challand, in Ref. 5a, chapter 9.
60. S. Hunig, D. Scheutzow, H. Schaf, and H. Quast, *Annalen*, 1972, **765,** 110.
61. F. Kurtzer, in Ref. 34a, chapter 14.
62. G. V. Boyd, in Ref. 5b, chapter 10.
63. F. Kurtzer, in Ref. 34c, chapter 12.
64. See, for example, Ref. 14, and E. E. van Tamelen and T. H. Whitesides, *J. Amer. Chem. Soc.*, 1971, **93,** 6129; R. M. Kellogg, *Tetrahedron Letters*, 1972, 1429; G. Vernin, C. Riou, H. J.-M. Dou, L. Boncasse, J. Metzger, and G. Loridan, *Bull. Soc. chim. France*, 1973, 1743; M. Maeda and M. Kojima, *Tetrahedron Letters*, 1973, 3523.
65. E. Glotter and M. D. Bachi, *Israel J. Chem.*, 1970, **8,** 633.
66. J. H. Davies, R. H. Davis, and R. A. G. Carrington, *J.C.S. Perkin I*, 1972, 1983.
67. T. Nagai, Y. Fukushima, T. K. Kuroda, H. Shimizu, S. Sekiguchi, and K. Matsui, *Bull. Soc. chim. France*, 1973, 2600.
68. C. H. Cheu, *J. Heterocyclic Chem.*, 1976, **13,** 1079.
69. P. G. Sammes, *Chem. Rev.*, 1976, **76,** 113.
70. E. F. V. Scriven, in Ref. 5b, chapter 7.
71. F. Kurtzer, in Ref. 34c, chapter 13.
72. J. Ciernik, *Coll. Czech. Chem. Comm.*, 1972, **37,** 2273.
73. K. T. Potts, D. R. Choudhury and T. R. Westby, *J. Org. Chem.*, 1976, **41,** 186.
74. H. Feuer and J. P. Lawrence, *J. Org. Chem.*, 1972, **37,** 3662.
75. V. Calo, L. Lopez, and P. E. Todesco, *J.C.S. Perkin I*, 1972, 1652.
76. A. McKillop, T. S. B. Sayer, and G. C. A. Bellinger, *J. Org. Chem.*, 1976, **41,** 1328.
77. See, for example, L. J. Altman and S. L. Richheimer, *Tetrahedron Letters*, 1971, 4709; A. I. Meyers, R. Munavu, and J. Durandetta, *ibid*, 1972, 3929.
78. H. Stetter, R. Y. Ramsch, and H. Kuhlmann, *Synthesis*, 1976, 733.
79. R. H. Wiley, in 'Organic Reactions', Wiley, New York, 1951.
80. T. S. Griffin, T. S. Woods, and D. L. Klayman, *Adv. Heterocyclic Chem.*, 1975, **18,** 109.
81. I. Iwataki, *Bull. Chem. Soc. Japan*, 1972, **45,** 3218.
82. K.-Y. Akiba, M. Ochiumi, T. Tsuchiya, and N. Inamoto, *Tetrahedron Letters*, 1975, 459.
83. R. Lantzsch and D. Arlt, *Synthesis*, 1975, 675.
84. J.-C. Meslin and H. Quiniou, *Synthesis*, 1974, 298.
85. A. S. Shawali and A. O. Abdelhamid, *J. Heterocyclic Chem.*, 1976, **13,** 45, and references cited therein.
86. G. D. Hartman and L. M. Weinstock, *Synthesis*, 1976, 681.
87. P. Dubs and R. Stuessi, *Synthesis*, 1976, 696.
88. K. Clarke, C. G. Hughes, and R. M. Scrowston, *J.C.S. Perkin I*, 1973, 356.
89. J. Perregaard and S.-O. Lawesson, *Acta Chem. Scand.*, 1975, **B29,** 604.
90. M. J. Spitulnik, *Synthesis*, 1976, 730.
91. P. T. S. Lau and T. E. Gompf, *J. Org. Chem.*, 1970, **35,** 4103.
92. J. Chodowska-Palicka and M. Nilsson, *Synthesis*, 1974, 128.
93. For a recent summary, see Ref. 34c, pp. 638–641; see also J. Jung, C.-An Chin and P.-S. Song, *J. Amer. Chem. Soc*, 1976, **98,** 3949 and references cited therein.
94. N. K. Kochetkov and S. D. Sokolov, *Adv. Heterocyclic Chem.*, 1963, **2,** 365.
95. 'Five and Six-Membered Compounds with Nitrogen and Oxygen', in 'Chemistry of Heterocyclic Compounds', ed. A. Weissberger, Interscience, New York, 1962.
96. R. A. Barnes, in 'Heterocyclic Compounds', ed. R. C. Elderfield, Wiley, New York, 1967.
97. K.-H. Wunsch and A. J. Boulton, *Adv. Heterocyclic Chem.*, 1967, **8,** 277.

98. C. W. Bird and G. W. H. Cheeseman, in Ref. 5b, chapter 6.
99. E. F. V. Scriven, in Ref. 5a, chapter 7.
100. G. Vernin, C. Siv, S. Treppendahl, and J. Metzger, *Helv. Chim. Acta*, 1976, **59,** 1705.
101. T. L. Gilchrist, in Ref. 5a, chapter 11.
102. T. L. Gilchrist, in Ref. 5b, chapter 11.
103. J. P. Ferris and R. W. Trimmer, *J. Org. Chem.*, 1976, **41,** 13.
104. T. Nishiwaki, *Synthesis*, 1975, 20.
105. P. A. Lowe, in Ref. 5c, chapter 3.
106. D. S. Kemp and C. F. Hoyng, *Tetrahedron Letters*, 1975, 4625.
107. C. Kashima, *J. Org. Chem.*, 1975, **40,** 526.
108. A. J. Boulton, A. R. Katritzky, and A. Majid Hamid, *J. Chem. Soc. (C)*, 1967, 2005.
109. M. Rucchia, N. Vivona, G. Cusmano, and G. Macaluso, *J.C.S. Perkin I*, 1977, 589.
110. G. Stork, M. Ohashi, H. Kamachi, and H. Kakisawa, *J. Org. Chem.*, 1971, **36,** 2784.
111. C. Grundmann and P. Grunager, 'The Nitrile Oxides', Springer-Verlag, **Berlin, 1971.**
112. V. Dal Piaz, S. Pinzauti, and P. Lacrimini, *Synthesis*, 1975, 664; G. A. Shvekhgeimer, A. Baranski, and M. Grzegozek, *ibid*, 1976, 612.
113. R. M. Sandifer, L. M. Shaffer, W. M. Hollinger, D. C. Reames, and C. F. Beame, *J. Heterocyclic Chem.*, 1976, **13,** 607.
114. A. W. Somerville, in Ref. 5c, chapter 2.
115. P. D. Magnus, in Ref. 5c, chapter 12.
116. D. H. R. Barton, D. J. Collins, B. Halpern, and Q. N. Porter, *J. Chem. Soc. (C)*, 1971, 2166.
117. D. S. Kemp and R. B. Woodward, *Tetrahedron*, 1965, **21,** 3019.
118. P. N. Preston and G. Tennant, *Chem. Rev.*, 1972, **72,** 627.
119. A. R. Katritzky and J. M. Lagowski, 'Chemistry of the Heterocyclic N-Oxides', Academic Press, London and New York, 1971.
120. B. Loev, *J. Medicin. Chem.*, 1970, 738.
121. L. L. Bambas, in 'The Chemistry of Heterocyclic Compounds', ed. A. Weissberger, Wiley Interscience, New York, 1952, vol. 4.
122. F. Hubenett, F. H. Flock, W. Hansel, H. Heinze and Hd. Hofmann, *Angew. Chem. Internat. Edn.*, 1963, **2,** 714.
123. R. Slack and K. R. H. Wooldridge, *Adv. Heterocyclic Chem.*, 1965, **4,** 107.
124. K. R. H. Wooldridge, *Adv. Heterocyclic Chem.*, 1972, **14,** 1.
125. M. Davis, *Adv. Heterocyclic Chem.*, 1972, **14,** 43.
126. F. Kurtzer, in Ref. 34b, chapter 12.
127. S. R. Challand, in Ref. 5b, chapter 9.
128. S. R. Challand, in Ref. 5c, chapter 9.
129. A. W. K. Chan and W. D. Crow, *Austral. J. Chem.*, 1968, **21,** 2967.
130. W. D. Crow and I. Gosney, *Austral. J. Chem.*, 1967, **20,** 2729.
131. A. J. Boulton, in Ref. 5a, chapter 5.
132. J. Goerdeler and W. Mittler, *Chem. Ber.*, 1963, **96,** 944.
133. F. Kurtzer, in Ref. 34a, chapter 12.
134. J. E. Franz and H. K. Pearl, *J. Org. Chem.*, 1976, **41,** 1296.
135. M. Davis, E. Harnfeld, and T. Paproth, *Org. Prep. Proced. Internat.*, 1973, **5,** 197.
136. A. Lawson and R. B. Tinkler, *Chem. Rev.*, 1970, **70,** 593.
137. F. Eloy, *Fortschr. Chem. Forsch.*, 1963, **4,** 807.
138. L. B. Clapp, *Adv. Heterocyclic Chem.*, 1976, **20,** 65.
139. A. Selva, L. F. Zerilli, B. Cavalleri, and G. G. Gallo, *Org. Mass Spectrometry*, 1974, **9,** 558.
140. T. L. Gilchrist, in Ref. 5c, chapter 11.
141. N. Götz and B. Zeeh, *Synthesis*, 1976, 268.
142. L. S. Wittenbrook, *J. Heterocyclic Chem.*, 1975, **12,** 37.
143. J. E. Franz, R. K. Howe, and H. K. Pearl, *J. Org. Chem.*, 1976, **41,** 620.
144. G. Blankenstein and K. Möckel, *Z. Chem.*, 1962, **2,** 69.
145. A. Dorlars, C.-W. Schellhammer, and J. Schroeder, *Angew. Chem. Internat. Edn.*, 1975, **14,** 665.
146. E. P. Nesynov and A. P. Grekov, *Russ. Chem. Rev.*, 1964, 508.
147. A. Hetzheim and K. Möckel, *Adv. Heterocyclic Chem.*, 1966, **7,** 183.
148. R. A. Olofson and J. S. Michelman, *J. Org. Chem.*, 1965, **30,** 1854.
149. J. W. Barton, in Ref. 5b, chapter 12.
150. A. J. Boulton and P. B. Ghosh, *Adv. Heterocyclic Chem.*, 1969, **10,** 1.
151. A. Gasco, V. Mortarini, G. Rua, and A. Serafino, *J. Heterocyclic Chem.*, 1973, **10,** 587 and references cited within.
152. F. Seng and K. Ley, *Angew. Chem. Internat. Edn.*, 1972, **11,** 1009.
153. A. J. Boulton, in Ref. 5b, chapter 5.
154. G. Calzaferri, R. Gleiter, K.-H. Knauer, H.-D. Martin, and E. Schmidt, *Angew. Chem. Internat. Edn.*, 1974, **13,** 86.
155. A. Gasco, V. Mortarini, R. Calvino, and A. Serafino, *Tetrahedron Letters*, 1974, 627.
156. K. Ley amd F. Seng, *Synthesis*, 1975, 415.
157. P. B. Ghosh, B. Ternai, and M. W. Whitehouse, *J. Medicin. Chem.*, 1972, **22,** 123.

158. (a) F. Kurtzer and L. E. A. Godfrey, *Angew. Chem. Internat. Edn.*, 1963, **2,** 467; (b) W. A. Sherman, in 'Heterocyclic Compounds', ed. R. C. Elderfield, Wiley, New York, 1961, vol. 7, p. 541.
159. K.-P. Zeller, H. Meier, and E. Muller, *Tetrahedron*, 1972, **28,** 1353.
160. F. Kurtzer, in Ref. 32b, chapter 15.
161. F. Kurtzer, in Ref. 32c, chapter 15.
162. L. Benati, P. C. Montevecchi, A. Tundo, and G. Zanardi, *J. Org. Chem.*, 1976, **41,** 1331.
163. H. Buhl, B. Seitz, and H. Meier, *Tetrahedron*, 1977, **33,** 449.
164. G. Seybold and C. Heibl, *Angew. Chem. Internat. Edn.*, 1975, **14,** 248.
165. L. Benati, P. C. Montevecchi, and G. Zanardi, *J. Org. Chem.*, 1977, **42,** 575.
166. J. Laureni, A. Krantz, and R. A. Hajdu, *J. Amer. Chem. Soc.*, 1976, **98,** 7872.
167. T. L. Gilchrist, P. G. Mente, and C. W. Rees, *J.C.S. Perkin I*, 1972, 2165.
168. F. Kurtzer, in Ref. 34a, chapter 16.
169. A. Shafiee, *J. Heterocyclic Chem.*, 1976, **13,** 301.
170. H. Meier, G. Trickes, and H. P. Braun, *Tetrahedron Letters*, 1976, 171.
171. S. Hunig, *Annalen*, 1970, **738,** 192.
172. F. Kurtzer, *Adv. Heterocyclic Chem.*, 1965, **5,** 119.
173. K. Leverenz, *Angew. Chem. Internat. Edn.*, 1973, **12,** 237.
174. J. Sandström, *Adv. Heterocyclic Chem.*, 1968, **9,** 165.
175. A. Shafiee, I. Lalezari, and M. Mirrashed, *J. Heterocyclic Chem.*, 1976, **1,** 117.
176. E. F. V. Scriven, in Ref. 5c, chapter 7.
177. L. M. Weinstock and P. I. Pollack, *Adv. Heterocyclic Chem.*, 1968, **9,** 107.
178. A. P. Komin and M. Carmack, *J. Heterocyclic Chem.*, 1976, **13,** 13.
179. D. H. R. Barton and W. A. Bubb, *J.C.S. Perkin I*, 1977, 916.

20.2
Six-membered Ring Systems

J. K. LANDQUIST
formerly of ICI Pharmaceuticals Division, Alderley Park, Cheshire

The collective index to *Chemical Abstracts*, volumes 66–75, records over 250 six-membered ring systems having two or more different heteroatoms, and a large number of bicyclic or polycyclic fused or spiro-compounds derived from these primary ring systems. In most of these the heteroatoms, together with adjacent carbon atoms, form functional groups similar to those found in acyclic compounds and their chemistry can be inferred from that already known for the open-chain analogues. The compounds that show special chemical or physical properties dependent on their cyclic structure are generally unsaturated or aromatic in character, and these compounds provide the main subject matter of this chapter. Cyclic esters and amides of the oxyacids of phosphorus and arsenic, and cyclic silicon, boron, or organometallic compounds, are not discussed here.

Tautomerism of the type CO—NH $\rightleftharpoons$ C(OH)=N is frequently encountered in mixed nitrogen heterocycles. Many authors have chosen to name such compounds as hydroxy-compounds (the nomenclature is simpler), but this does not imply that such a compound usually exists as the hydroxy-tautomer and physical evidence, *e.g.* spectroscopic data, should be sought to elucidate such questions.

20.2.1 COMPOUNDS WITH TWO OR MORE ATOMS OF THE GROUP OXYGEN, SULPHUR, SELENIUM, TELLURIUM[1-3]

The most saturated of the compounds containing one oxygen and one sulphur atom are the 1,2-, 1,3-, and 1,4-oxathians, and the most unsaturated ones are the 1,2-, 1,3-, and 1,4-oxathiins. The 1,2- and 1,4-oxathiins contain two double bonds whereas the 1,3-oxathiins contain only one, and for this reason they are sometimes called 1,3-oxathiens.

Compounds of the 1,2-series are known only as their 2-oxides (cyclic sulphinates) and 2,2-dioxides (cyclic sulphonates, sultones). Like other sulphonic esters the 2,2-dioxides alkylate nucleophiles and give 4-sulphobutyl derivatives. The unsaturated sultone 3,6-dihydro-4,5-dimethyl-1,2-oxathiin 2,2-dioxide (**1**), made by the action of sulphur trioxide–dioxan on 2,3-dimethylbutadiene, is hydrolysed 1000 times faster than 1,2-oxathian 2,2-dioxide.

1,2-Oxathiin 2,2-dioxides are obtained by the action of sulphuric acid and acetic anhydride on α,β- or β,γ-unsaturated ketones which have carbon chains branching at the β-position, mesityl oxide giving the 4,6-dimethyl derivative (**2**). Bromination of this compound gives a monobromo-derivative, described as the 5-bromo-compound, and then an adduct with another molecule of bromine. Treatment of (**2**) with primary amines at 140–200 °C gives 1,2-thiazine 1,1-dioxides. Pyrolysis of 1,2-oxathiin 2,2-dioxides causes ring-contraction with loss of sulphur dioxide, and good yields of furans have been claimed.

4-Methyl-1,2-benzoxathiin 2,2-dioxide (**3**; R = H) and 4-methyl-3-phenyl-1,2-benzoxathiin 2,2-dioxide, (**3**; R = Ph) are made by heating the methanesulphonate and benzylsulphonate of 2-hydroxyacetophenone with potassium hydroxide in pyridine, a reaction reminiscent of the Camps quinoline synthesis. The 3,4-double bond can be reduced by boiling hydriodic acid. The 3-phenyl derivative is resistant to hydrolysis by alkali.

The 3,4-dihydro-2,1-benzoxathiin 1,1-dioxides (**4**; R = Me or H) are obtained by treating *O*-methyl- or *O*-acetyl-eugenol with sulphuric acid. 3,4-Dihydro-2,1-benzoxathiin-3-one 1-oxide (**5**) is made by diazotization of *o*-aminophenylacetic acid and treatment of the diazonium salt with sulphur dioxide and copper. The cyclic anhydride of 8-sulpho-1-naphthoic acid (**6**) is an example of a polycyclic 1,2-oxathiin derivative.

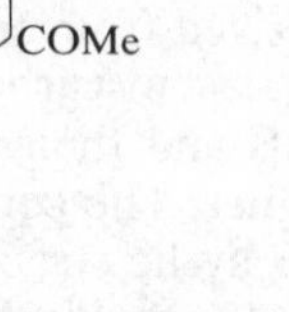

1 O, 2 SO_2, 3, 4, 5, 6, Me, Me
(**1**)

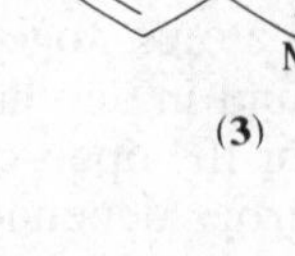

O, SO_2, Me, Me
(**2**)

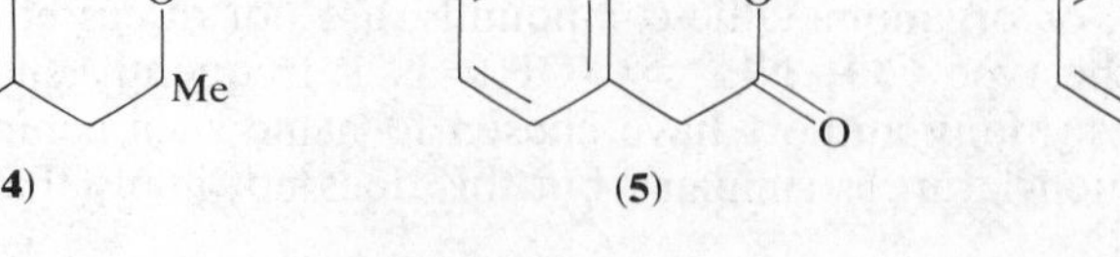

OSO_2CH_2R, COMe → O, SO_2, R, Me
(**3**)

O_2, S, O, RO, MeO, Me
(**4**)

O, S, O, O
(**5**)

O_2S, O, O
(**6**)

1,3-Oxathians are *O,S*-acetals obtained, for example, by reaction of ketones with 3-mercaptopropanol derivatives and an acid catalyst. Formation of 1,3-oxathians may be used to protect carbonyl groups, *e.g.* in steroid syntheses, and the heterocyclic ring may subsequently be removed by desulphurization with Raney nickel.

o-Mercaptobenzoic acid reacts with aldehyde precursors, *e.g.* vinyl acetate or benzylidene diacetate, to give 3,1-benzoxathiin-4-ones (**7**; R = Me or Ph), while reaction of its lead salt with phosgene gives 2,4-dioxo-3,1-benzoxathiin, which decomposes at 100–120 °C to give a polymeric thioester. 2,2-Diphenyl-3,1-benzoxathiin-4-one (**8**) is made by reaction of diazotized anthranilic acid with thiobenzophenone. Two complex 1,3-oxathian derivatives, 1,3,5,7-tetramethyl-2,6-dioxa-9,10-dithia-adamantane (**9**) and 1,3,5,7-tetramethyl-2,10-dioxa-6,9-dithia-adamantane (**10**) are obtained by the action of hydrogen sulphide on pentane-2,4-dione in concentrated hydrochloric acid.[4]

1,4-Oxathian is obtained by dehydrating 2,2′-dihydroxydiethyl sulphide or by the reaction of 2,2′-dihalogenodiethyl ethers with potassium sulphide. It is a liquid and like 1,4-dioxan it may be used as a solvent in Grignard reactions. It behaves as a dialkyl sulphide, forming complexes with heavy-metal salts and addition compounds with bromine, iodine, and alkyl iodides. Its addition compound with sulphur trioxide has been recommended as a sulphonating and sulphating agent. It is readily oxidized to a sulphoxide and a sulphone, the latter being a powerful solvent when used in admixture with other organic solvents.

2-Methoxy-1,4-oxathian (**11**) is obtained by treatment of 1-hydroxy-5,5-dimethoxy-3-thiapentane with hydrogen chloride in methanol, or by acid-catalysed addition of

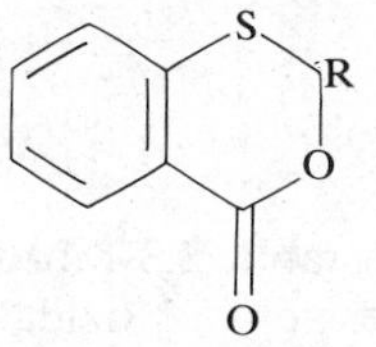

(**7**)

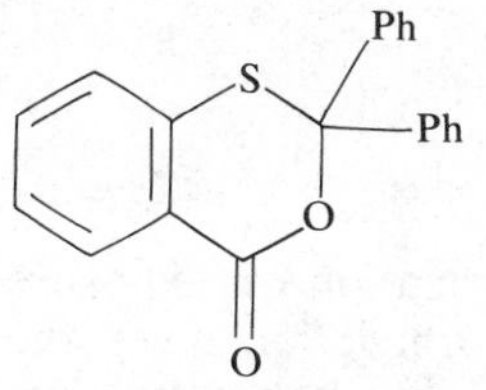

(**8**)

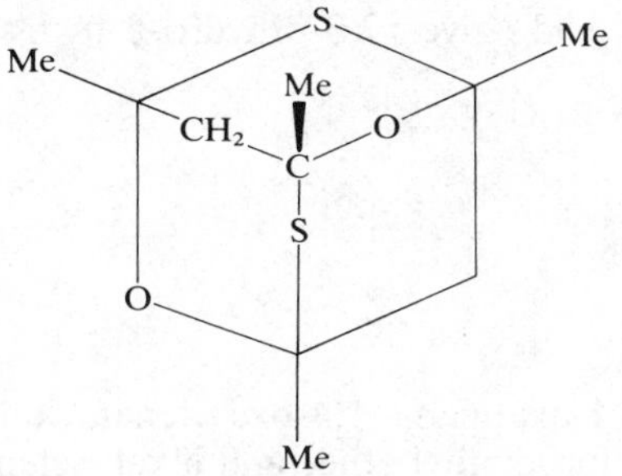

(**9**)

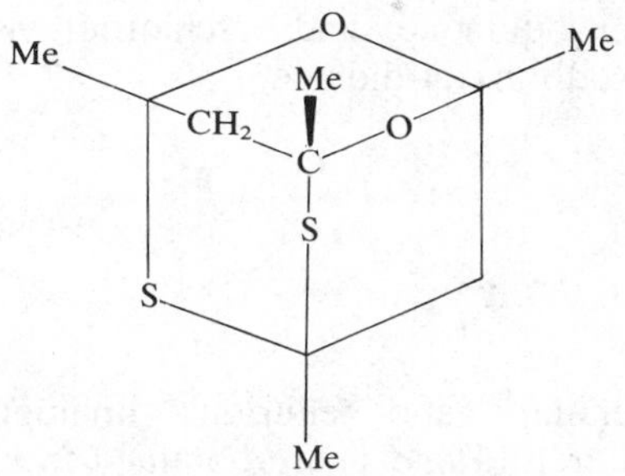

(**10**)

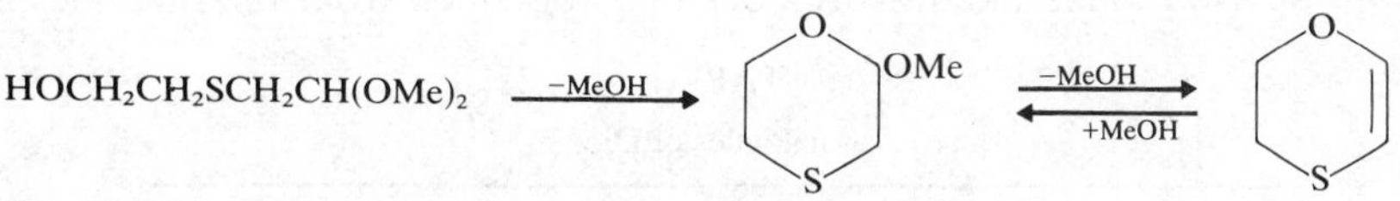

$HOCH_2CH_2SCH_2CH(OMe)_2 \xrightarrow{-MeOH}$ (**11**) $\underset{+MeOH}{\overset{-MeOH}{\rightleftharpoons}}$ (**12**)

Ph,Ph-SO_2 + $Me_2\overset{+}{S}$—$\overset{-}{C}$HCOPh ⟶ (**13**)

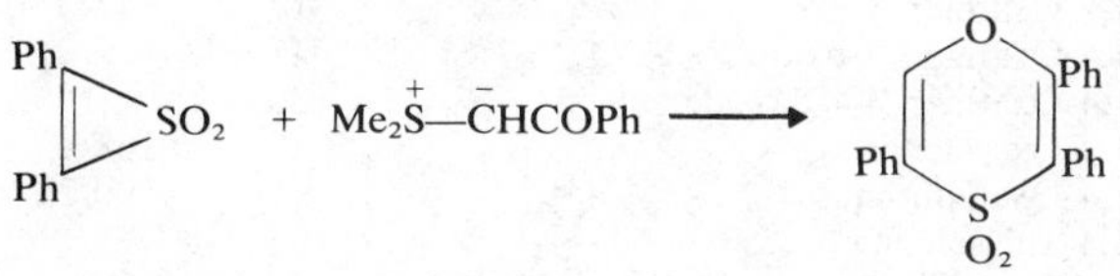

methanol to 5,6-dihydro-1,4-oxathiin. When heated over phosphorus pentoxide it gives 5,6-dihydro-1,4-oxathiin (**12**).[5] This is a vinyl ether and readily undergoes addition or fission at the vinyl ether bond. 5,6-Dihydro-2,3-diphenyl-1,4-oxathiin is made by condensation of desyl chloride with 2-mercaptoethanol and acid-catalysed ring closure of the product. 2,6-Diethoxy-1,4-oxathian 4,4-dioxide can be prepared by ozonolysis of 2,5-dihydrothiophen 1,1-dioxide in the presence of ethanol.[6]

1,4-Oxathiins are almost unknown, but reaction of diphenylthiiren dioxide with dimethylsulphonium phenacylide gives 2,3,5-triphenyl-1,4-oxathiin 4,4-dioxide (**13**)[7] and some related sulphonamides have been obtained from benzoylsulphene and diphenylcarbodi-imide.

2-Ethoxybenzo-1,4-oxathian (**14**) has been made from 2-hydroxythiophenol and bromoacetal.[8] It does not readily lose ethanol over phosphorus pentoxide, but the corresponding 2-acetoxy-compound, when heated at 450–460 °C, gives a good yield of benzo-1,4-oxathiin (**15**) (Scheme 1). This compound readily polymerizes, *e.g.* through

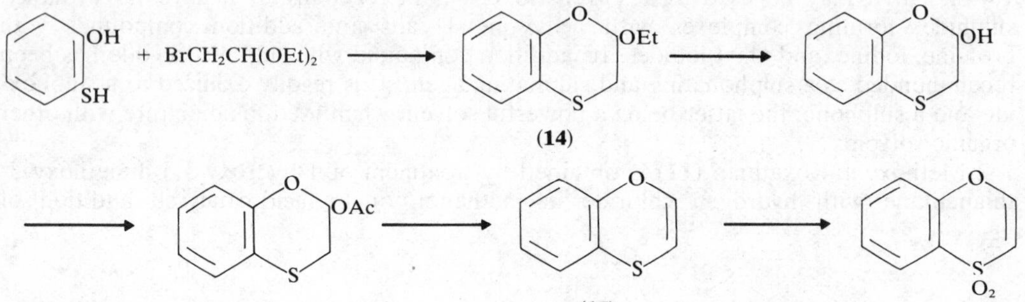

OH, SH + $BrCH_2CH(OEt)_2$ → (**14**) (OEt) → OH → OAc → (**15**) → O_2

SCHEME 1

peroxide formation on exposure to air and it gives an unstable 2,3-dibromo-derivative. Vilsmeier formylation gives the 2-formyl derivative, and peracid oxidation attacks at sulphur to give benzo-1,4-oxathiin 4,4-dioxide. Addition of mercaptoacetic acid to 1,4-benzoquinone gives 2,3-dihydro-6-hydroxy-1,4-benzoxathiin-2-one (**16**). Similarly, 1,4-benzoquinone and 2-chloroethylsulphinic acid give 2,3-dihydro-6-hydroxy-1,4-benzoxathiin 4,4-dioxide.

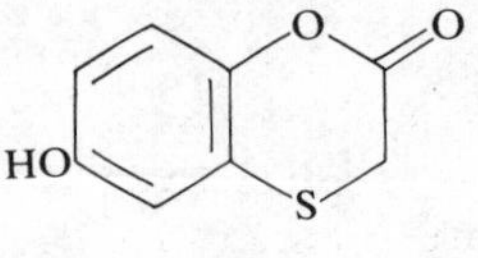

(**16**)

Selenium[2a] and tellurium[3] analogues of 1,4-oxathian, 1,4-oxaselenan, and 1,4-oxatelluran (Table 1) are obtained from 2,2′-dichlorodiethyl ether and alkali selenides or tellurides. They resemble 1,4-oxathian in giving dihalides and methiodides, but their oxidation products are alkaline dihydroxides (also obtained from the dihalides and silver

TABLE 1
Analogues of Dioxan

	M.p. (°C)	*B.p.* (°C)
1,4-Oxathian	−17	147–150
1,4-Oxathian 4-oxide	45	
1,4-Oxathian 4,4-dioxide	130	
1,4-Oxaselenan		168
1,4-Oxatelluran	6	90/21 mm
1,4-Thiaselenan	107	
1,4-Thiatelluran	69.5	

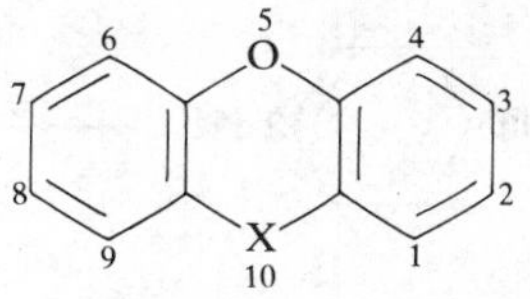

5 O, 6, 4, 7, 3, 8, 2, 9, 1, X 10

(17)

oxide) and they give hydroxynitrates when oxidized with nitric acid. 1,4-Oxatelluran is oxidized on exposure to air. 1,4-Thiaselenan and 1,4-thiatelluran, made in an analogous manner, also form halogen adducts and methiodides.

Phenoxathiin (dibenzo-1,4-oxathiin) (**17**; X = S)[9] (Table 2) is obtained by heating diphenyl ether with sulphur and aluminium chloride, or by heating phenoxatellurin (**17**; X = Te) with sulphur. Phenoxathiin derivatives have been made by treating *o*-phenoxybenzenesulphinic acids with acetic anhydride–sulphuric acid and by condensing

TABLE 2
Phenoxathiin and Analogous Compounds

	M.p. (°C)	*B.p.* (°C)
Phenoxathiin	58	311
Phenoxathiin 10-oxide	158–159	
Phenoxathiin 10,10-dioxide	147–148	
Phenoxaselenin	88	
Phenoxatellurin	78–79	
10,10-Dichlorophenoxatellurin	265	

o-hydroxythiophenol with *o*-chloronitrobenzenes having other electron-withdrawing substituents. Mauthner[10] stated that picryl chloride gave 2,4-dinitrophenoxathiin, and subsequently claimed that 4-chloro-3,5-dinitrobenzoic acid gave 1-nitrophenoxathiin-3-carboxylic acid, which was degraded to phenoxathiin-3-carboxylic acid, m.p. 223–224°C, and to phenoxathiin (the presently accepted numbering is shown, although other systems have been used in the past). The structure of this carboxylic acid was confirmed by the unambiguous synthesis of phenoxathiin-2-carboxylic acid, m.p. 260–262 °C (from phenoxathiin-2-magnesium bromide), and it is therefore likely that Mauthner's dinitro-compound is 1,3-dinitrophenoxathiin.

Phenoxathiin is a colourless substance with a geranial odour. It is reported to have insecticidal, antibacterial, and anthelmintic activity. It is isomorphous with phenothiazine, phenoxaselenin, and phenoxatellurin and the molecule is folded about the axis of the heteroatoms. The dipole moment is 1.09 D (benzene and hexane) or 0.92 D (benzene). The heterocyclic ring is cleaved by 85% potassium hydroxide at 195 °C to give 2,2′-dihydroxydiphenyl sulphide, and by sodium in liquid ammonia, giving *o*-phenoxythiophenol. Heating under reflux with diphenylsilane causes replacement of the sulphur atom by a diphenylsilicon group. It is readily oxidized to the 10-oxide and the 10,10-dioxide, the latter being resistant to reduction whereas the monoxide is readily reduced to phenoxathiin. 2,8-Dibromophenoxathiin 10,10-dioxide was made by cyclization of 4,4′-dibromo-2-chlorosulphonyldiphenyl ether with aluminium chloride. 2,8-Dimethylphenoxathiin 10-oxide was similarly obtained from 2,2′-dihydroxy-5,5′-dimethyldiphenyl sulphoxide by treatment with sulphuric acid. Electrophilic substitution of phenoxathiin (bromination, sulphonation, Friedel–Crafts acylation) occurs in the 2- and 8-positions. Nitration first effects oxidation to the dioxide and then substitution to give 2-nitrophenoxathiin 10,10-dioxide. Metallation of phenoxathiin with butyl-lithium goes in the 4-position, but the monoxide and dioxide are attacked at positions-1 and -9.

Phenoxaselenin[2b] (**17**; X = Se) is obtained by heating phenoxatellurin with selenium. The product is a molecular compound of phenoxaselenin and phenoxatellurin and is separated by bromination and selective reduction of the phenoxaselenin 10,10-dibromide

SCHEME 2

with acetone. 2-Carboxyphenoxaselenin 10-oxide was made by cyclization of 4′-carboxyphenoxyphenyl seleninic acid. Unlike phenoxathiin, phenoxaselenin forms easily reduced adducts with chlorine and bromine. It gives an unstable diacetate and a monoxide that loses oxygen at its melting point and decomposes hydrogen peroxide.

Phenoxatellurin (**17**; X = Te) is obtained by reduction (*e.g.* with bisulphite) of 10,10-dichlorophenoxatellurin (**18**), which is made by heating diphenyl ether with tellurium tetrachloride (Scheme 2). For the unambiguous synthesis of substituted phenoxatellurins the intermediate *o*-phenoxyphenyltellurium trichloride may be made from a substituted 2-phenoxyaniline by diazotization, conversion into a 2-phenoxyphenylmercurichloride, and treatment of this with tellurium tetrachloride. The chlorine atoms of 10,10-dichlorophenoxatellurin are readily exchanged for bromine, iodine, or hydrogen sulphate anions. Nitration of phenoxatellurin is preceded by oxidation to the 10,10-dinitrate and then occurs in the 2- and 8-positions, with a minor amount of nitration at the 4-position. Phenoxatellurins are very stable to strong acids, but potassium hydroxide abstracts the tellurium atom and regenerates a diphenyl ether.

Phenoxathiin, phenoxaselenin, and phenoxatellurin exhibit halochromism in concentrated sulphuric acid. This is accompanied by reduction of the sulphuric acid and the violet to red colours are attributed to radical cations formed by one-electron oxidation. Similar behaviour is shown by phenoxazine. Phenoxatellurin forms molecular compounds with substances of similar structure, and charge-transfer complexes with 1,3,5-trinitrobenzene, picric acid, and picryl chloride. Deeply coloured substances analogous to quinhydrones are formed by solid-phase reaction (without solvent) between phenoxatellurin 10,10-dinitrate (or other salt) and base metals or phenoxatellurin, phenoxaselenin, or phenoxathiin.[11]

20.2.2 COMPOUNDS CONTAINING OXYGEN AND NITROGEN IN THE RING

20.2.2.1 Oxazines[12]

There are three series of compounds having one oxygen and one nitrogen atom in a six-membered ring, derived from the parent compounds 1,2-, 1,3-, and 1,4-oxazine. Because of the divalency of oxygen the parent ring systems contain only two double bonds and there is isomerism in each series, depending on the position of the 'extra' hydrogen atom which may be located in a methylene group or an imino-group. Following the IUPAC recommendations on nomenclature, the position of this hydrogen atom (or substituent replacing it) is shown as in examples (**19**)–(**22**), which also show the numbering of the ring systems. In the older literature the locant of the hydrogen atom was often given after those of the heteroatoms, and (**19**) was described as 1,2,4*H*-oxazine or 1,2,4-oxazine.

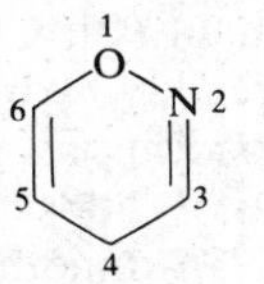

(19) 4*H*-1,2-Oxazine 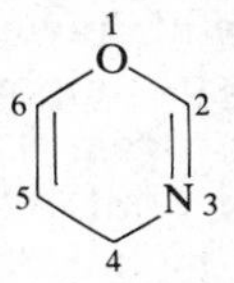 (20) 4*H*-1,3-Oxazine 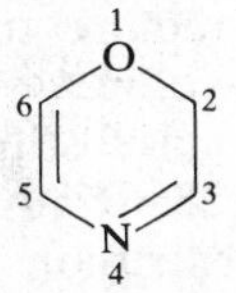 (21) 2*H*-1,4-Oxazine 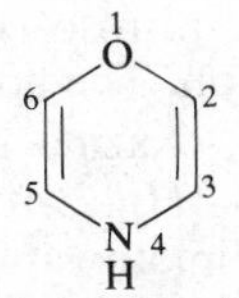 (22) 4*H*-1,4-Oxazine

1,2-Oxazines are cyclic derivatives of hydroxylamine, and this fact is reflected in a major group of preparative methods involving cyclization of oximes or reaction of hydroxylamine with unsaturated 1,4-dicarbonyl compounds or α,β-unsaturated-γ-oxo-carboxylic acids. Thus, mucobromic acid with hydroxylamine affords 4,5-dibromo-6-oxo-6*H*-1,2-oxazine (**23**). Mucochloric acid and other formylacrylic acids give oximes that may be dehydrated with sulphuric acid. The oxazines thus made (*e.g.* **23**) rearrange to the corresponding maleimides when heated (reaction 1) and to ammonium maleates upon

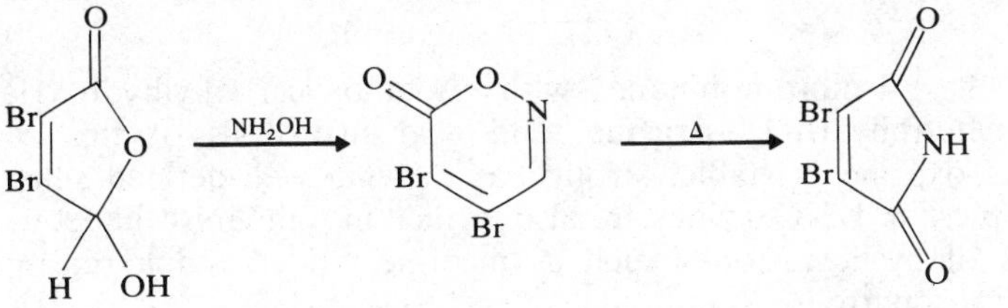

NH_2OH → (**23**) Δ → (1)

hydrolysis. In a similar way, 2-methyl-4-oxopent-2-enoic acid and 2-benzoyl-1-phenyl-acrylic acid afford 3,5-dimethyl-6-oxo-6*H*-1,2-oxazine and 6-oxo-3,5-diphenyl-6*H*-1,2-oxazine, respectively. Reaction of *cis*-phenyldibenzoylethylene with hydroxylamine gives 6-hydroxy-3,5,6-triphenyl-6*H*-1,2-oxazine, and acid-catalysed cyclization of 1,2-dioxo-4-oximino-1,3,4-triphenylbutane (**24**) in methanol gives 5-hydroxy-6-methoxy-3,4,6-triphenyl-6*H*-1,2-oxazine (**25**). This is hydrolysed by hot water, giving back (**24**), whilst with phenylmagnesium bromide it gives 5-hydroxy-3,4,6,6-tetraphenyl-6*H*-1,2-oxazine (**26**), which undergoes *N*-alkylation to give, for example, 2-ethyl-5,6-dihydro-5-oxo-3,4,6,6-tetraphenyl-2*H*-1,2-oxazine (**27**) (reaction 2). 3,6-Diphenyl-6*H*-1,2-oxazine is made by heating 3-nitro-1,4-diphenylbut-3-en-1-one with hydroxylamine in pyridine–ethanol. Acid hydrolysis of β-carbamylcinnamohydroxamic acids affords 5-aryl-3-hydroxy-6-oxo-6*H*-1,2-oxazines.

The second class of preparative methods involves Diels–Alder condensations of nitroso-compounds with dienes (reaction 3).[13] Condensation occurs readily with aromatic nitroso-compounds and with nitrososulphinyl compounds, but nitrosamines do not react and

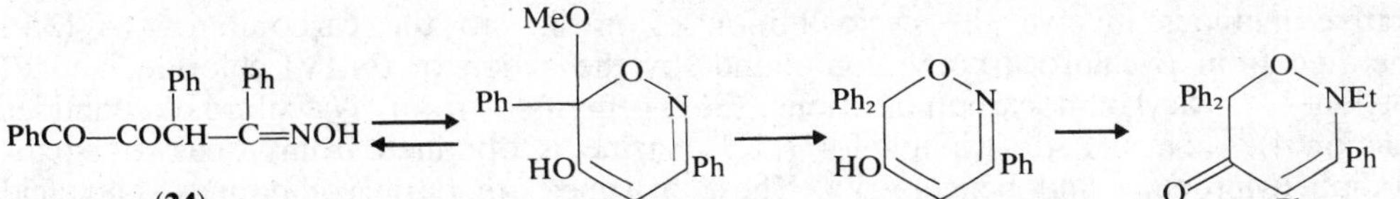

PhCO—COCH(Ph)—C(Ph)=NOH (24) ⇌ (25) → (26) → (27) (2)

RNO + diene → (28) (3)

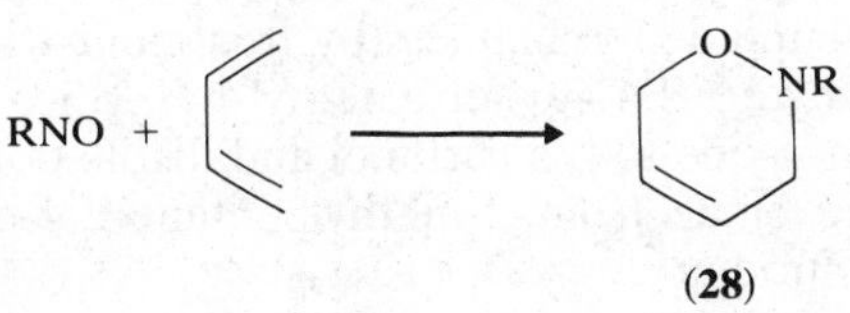

aliphatic nitroso-compounds require an electron-withdrawing substituent (Cl, CN) in the α-position. The adducts of α-chloroaliphatic nitroso-compounds are readily hydrolysed to give 1,2-oxazine derivatives unsubstituted on nitrogen. These cycloadditions give 3,6-dihydro-2*H*-1,2-oxazines (**28**; R = Ar, $ArSO_2$), but tetracyclone adds to 4-nitrosodimethylaniline with extrusion of carbon monoxide, giving 2-*p*-dimethylaminophenyl-3,4,5,6-tetraphenyl-2*H*-1,2-oxazine (**29**). Catalytic or borohydride reduction of the 3,6-dihydro-compounds gives tetrahydro-1,2-oxazines.

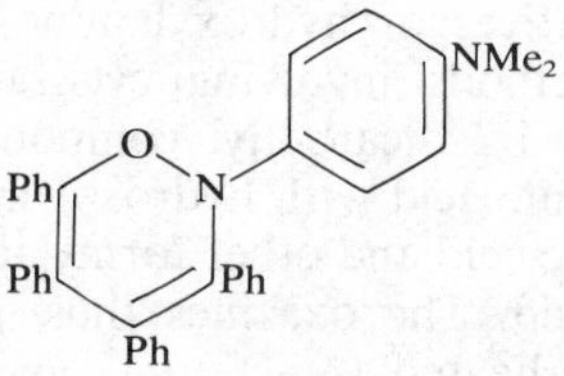

(**29**)

Reaction of 1,4-dibromobutane with *N*-ethoxycarbonylhydroxylamine gives 2-ethoxycarbonyltetrahydro-1,2-oxazine, and acid hydrolysis of this compound affords tetrahydro-1,2-oxazine, a stable, strong base, giving well-defined salts, *e.g.* the hydrochloride and picrate. 1,2-Oxazines are also formed in some miscellaneous rearrangements, *e.g.* pyrolysis of cyclic nitrones such as nicotine oxide, and a rearrangement of 2,4-dimethyl-5-nitrosopyrrole.

In general, the 1,2-oxazines are reducing agents, *e.g.* towards permanganate. Catalytic hydrogenation affords tetrahydro-derivatives, sometimes with replacement of a halogen substituent, but reduction with zinc and acetic acid or zinc and sodium hydroxide solution cleaves the O—N bond with formation of aminobutanol derivatives.

Formation and cleavage of a 1,2-oxazine ring may be used in the synthesis of γ-lactones or aldehydes from olefins with addition of two-carbon units.[14,15] The oxazines are obtained by the silver-ion catalysed addition of an α-chloronitrone (probably as a vinylnitrosonium ion) to the olefin to give a dihydro-1,2-oxazinium salt (**30**). This reacts with cyanide ion to give a 3-cyanotetrahydro-1,2-oxazine (**31**); treatment of this with base eliminates hydrogen cyanide to form a lactone imide (**32**), which gives the γ-lactone (**33**) on acid hydrolysis. With alkali the oxazinium salt (**30**) gives a 5,6-dihydro-2*H*-1,2-oxazine (**34**) which suffers cleavage of the O—N bond and the original olefinic bond on heating, yielding an aldehyde on subsequent hydrolysis (Scheme 3).

4*H*-1,3-Oxazines[16] are made by the elimination of water from β-acylamino-aldehydes, -ketones, or -esters. 2,6-Diphenyl-4*H*-1,3-oxazine is obtained from β-benzamidopropiophenone and phosphorus pentachloride, 6-ethoxy-2-phenyl-4*H*-1,3-oxazine from ethyl β-benzamidopropionate and phosphorus pentoxide, and 2-phenyl-4*H*-1,3-oxazine from *N*-(3,3-diethoxypropyl)benzamide with aqueous oxalic acid at 25 °C. Newer preparative methods involve the cycloaddition of nitriles to the carbonium ions (**35**), generated from β-chlorocarbonyl compounds by the action of tin(IV) chloride, and of acetylenes to acylaminocarbonium ions (**36**) generated from *N*-(chloroalkyl)amides (reaction 4).[17] Thus, 2,4,6-triphenyl-4*H*-1,3-oxazine is obtained from 3-chloro-1-oxo-1,3-diphenylpropane and benzonitrile. These oxazines are basic compounds, and acid hydrolysis cleaves the 2,3-bond to give a salt of a 3-aminoenol ester which rearranges to a β-acylaminocarbonyl compound upon liberation of the base. A methyl group on C-2 can be converted into a styryl group by condensation with benzaldehyde in acetic anhydride.

Dihydro-1,3-oxazines are made by several related processes from γ-acylamino-alcohols (by dehydration) or γ-acylaminoalkyl halides (by treatment with base), as exemplified by the preparation of 4,5-dihydro-2,4,4,6-tetramethyl-6*H*-1,3-oxazine by reduction of 2-acetamido-2-methylpentan-4-one to the carbinol and distillation (with dehydration) of the latter. This oxazine is also made from 2-methylpentane-2,4-diol by the Ritter reaction with acetonitrile and sulphuric acid.

i, $AgBF_4$; ii, CN^-; iii, base; iv, H^+; v, K_2CO_3; vi, Δ.

SCHEME 3

(35) (36) (4)

β-Acylamido- or ureido-acids of the general formula $R^1NHCR^2R^3CHR^4CO_2H$ are cyclized by acetic anhydride to yield 6-oxodihydro-1,3-oxazines. For such condensations to succeed it is usually necessary for the carbon atom next to the nitrogen atom to have one or two substituents. Diketen reacts with monosubstituted *S*-methylisothioureas giving dihydro-1,3-oxazine derivatives, *e.g.* 2-amino-2,3-dihydro-6-methyl-2-methylthio-4-oxo-3-phenyl-4*H*-1,3-oxazine (**37**), which are converted into 2-methylthiopyrimidines by acid. With *N,N′*-disubstituted *S*-methylisothioureas or the corresponding carbodi-imides, diketen gives 2-imino-1,3-oxazines, *e.g.* 3-ethyl-2-ethylimino-2,3-dihydro-6-methyl-4-oxo-4*H*-1,3-oxazine (**38**), weak bases that are hydrolysed by acid to dioxo-1,3-oxazines, *e.g.* 3-ethyl-2,3-dihydro-6-methyl-2,4-dioxo-4*H*-1,3-oxazine. These readily react with amines to give pyrimidines.

Dihydro-1,3-oxazines have also been made by Diels–Alder additions of electron-deficient *N*-acylimines with electron-rich olefins such as enamines.[17] Some less general

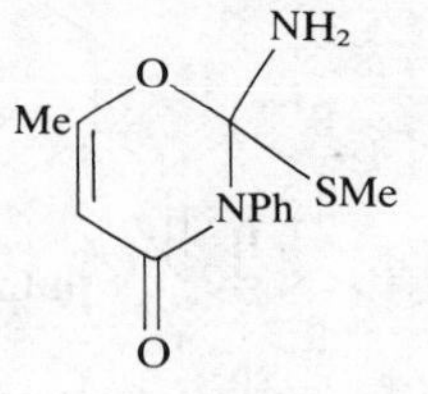

(37) 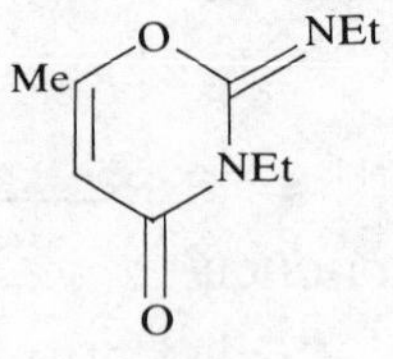 (38)

methods for the preparation of dihydro-1,3-oxazines include the reaction of sodium hypochlorite with maleimide to give 2,3-dihydro-2,6-dioxo-6*H*-1,3-oxazine, reaction of phenyl isocyanate with malonic ester giving 2,3-dihydro-2,4-dioxo-3-phenyl-6-phenylamino-4*H*-1,3-oxazine, and hydrolysis of 1,2,3,4-tetrahydro-5-methoxycarbonyl-6-methyl-1,2,3,4-tetraphenylpyrimidine to give 1,2-dihydro-5-methoxycarbonyl-6-methyl-2,3,4-triphenyl-4*H*-1,3-oxazine.

Tetrahydro-1,3-oxazines are obtained by the reaction of primary or secondary γ-aminoalcohols with aldehydes or equivalent reagents (acetylene under pressure or vinyl ethers with mercuric acetate as catalyst). When the amino-group is primary some uncyclized Schiff's base may be present in the reaction product. Ketones usually give Schiff's bases with γ-amino-alcohols, but in some instances, *e.g.* with cyclohexanone, 1,3-oxazines are formed. Condensation of olefin $R^1R^2C{=}CH_2$ (if R = H the double bond must be activated by an aryl group or a conjugated double bond) with formaldehyde and ammonium chloride or primary amine hydrochlorides affords tetrahydro-1,3-oxazines (**39**) (reaction 5). Methanolysis of these with methanol and aqueous hydrochloric acid removes

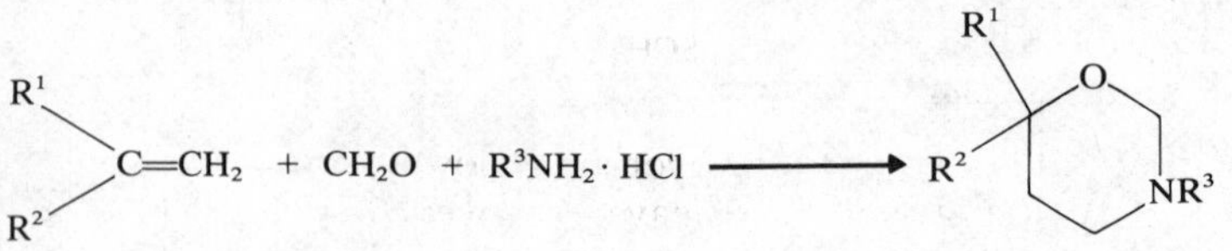

$$R^1R^2C{=}CH_2 + CH_2O + R^3NH_2\cdot HCl \longrightarrow \textbf{(39)} \qquad (5)$$

(39)

C-2 as methylal and affords a preparative route to γ-amino-alcohols.[18] Primary nitroparaffins react with formaldehyde and amines to give tetrahydro-5-nitro-1,3-oxazines (**40**). Fission of these, *e.g.* with aqueous hydrochloric acid, affords 3-amino-2-nitropropanols.[19]

(40)

4*H*-1,3-Oxazines are oxidized by 1,4-benzoquinone through the formation of adducts that are unstable to water, and break down to β-acylaminovinylcarbonyl compounds (**41**). These may be cyclized by anhydrous perchloric acid to 1,3-oxazinium perchlorates (3-azapyrylium salts) (**42**), which may also be prepared directly from 4*H*-1,3-oxazines by oxidation with trityl perchlorate. Typical compounds are 2,4,6-triphenyl-1,3-oxazinium perchlorate, and the 4,6-dimethyl-2-phenyl and 6-ethoxy-4-methyl-2-phenyl derivatives. They are very readily hydrolysed to the β-acylaminovinylcarbonyl compounds (**41**), and are very reactive towards nucleophiles through attack at C-6 and subsequent rearrangement. Ammonia gives pyrimidines, and malononitrile (base-catalysed) gives pyridine derivatives through open-chain intermediates (Scheme 4).[20]

The chemical properties of dihydro-1,3-oxazines, particularly the readily accessible 2-alkyl-5,6-dihydro-4,4,6-trimethyl-4*H*-1,3-oxazides, have been widely exploited in recent methods for preparing carbonyl compounds and carboxylic acids.[15] Reduction of the

i, benzoquinone; ii, H_2O; iii, $HClO_4$; iv, NH_3; v, $CH_2(CN)_2$.

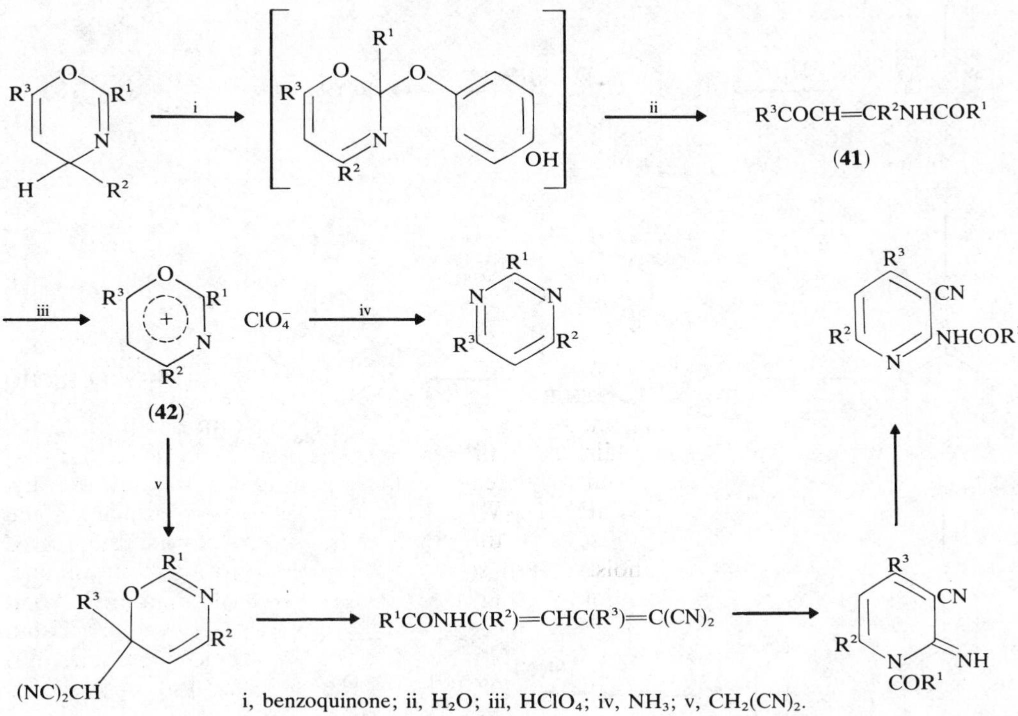

SCHEME 4

dihydro-compounds with sodium borohydride at low temperatures gives high yields of tetrahydro-oxazines which are readily cleaved to give aldehydes, *e.g.* with aqueous oxalic acid. The 2-alkyl group gives a carbanion when it is treated with butyl-lithium at $-78\,°C$, and this is highly reactive towards electrophiles (alkyl halides, carbonyl compounds, epoxides), giving substituted alkyl derivatives that may be reduced and hydrolysed (Scheme 5). The dihydro-1,3-oxazine system is not reactive towards Grignard reagents, but *N*-alkylation, *e.g.* with methyl iodide, affords iminium salts, *e.g.* 5,6-dihydro-2,4,4,6-tetramethyl-4*H*-1,3-oxazine methiodide (**43**), which react with Grignard reagents to give 2,2-disubstituted tetrahydro-oxazines that are hydrolysed to yield ketones. The methiodide (**43**) when treated with sodium hydride is converted into the keten *N*,*O*-acetal (**44**), which may be alkylated with alkyl halides to generate substituted oxazinium salts for

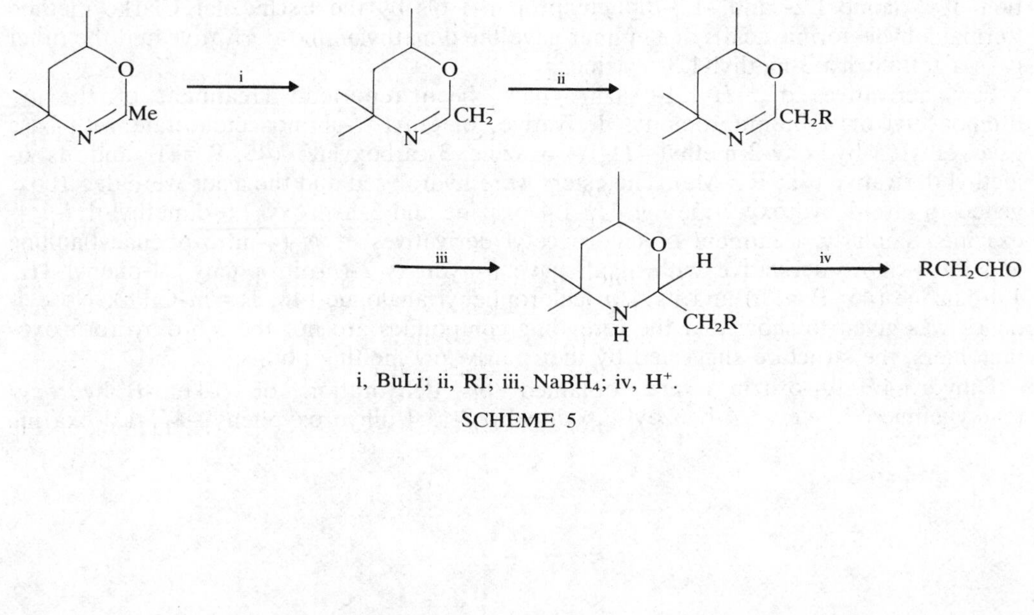

i, BuLi; ii, RI; iii, $NaBH_4$; iv, H^+.

SCHEME 5

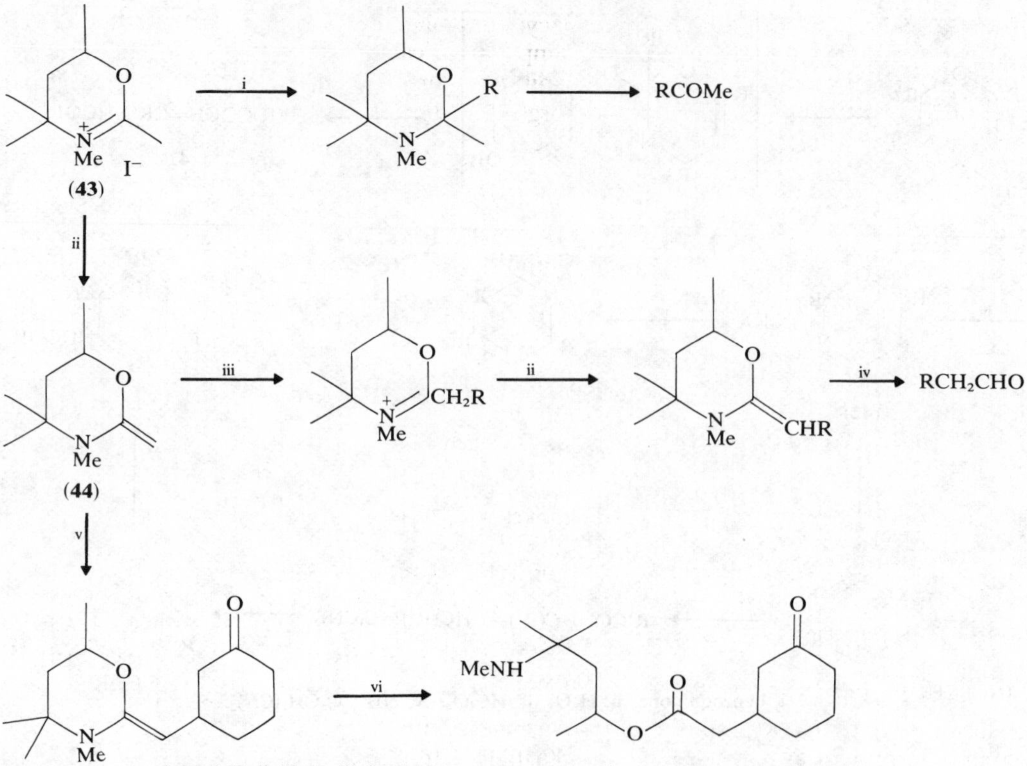

i, RMgX; ii, NaH; iii, RHal; iv, $NaBH_4$, then H^+; v, cyclohexenone; vi, H_2O.

SCHEME 6

conversion into aldehydes, or may be added to electrophilic olefins to give 2-allylidene-3-methyl-tetrahydro-1,3-oxazines that are rapidly hydrolysed by water to aminoalkyl esters of new carboxylic acids (Scheme 6). Other syntheses depend on the addition of organolithium or Grignard reagents to 2-(substituted vinyl)-dihydro-oxazines. This cleaves the oxazine ring, giving a keten-imine that undergoes further reaction with the organometallic reagent. Unsubstituted vinyl compounds are polymerized under these conditions.

The configuration of certain alkaloids and their derivatives having amino and alcohol groups separated by three carbon atoms has been established by the formation of dihydro-1,3-oxazines by reaction with a suitable carbonyl compound, *e.g.* 3-nitrobenzaldehyde. Another application of 1,3-oxazine formation to stereochemical studies involved methylation of 3-amino-1,2- and -1,3-diphenylpropan-1-ols by the Eschweiler–Clarke method (formaldehyde–formic acid); one epimer gave the dimethylamino-derivative and the other gave a tetrahydro-3-methyl-1,3-oxazine.[21]

Few derivatives of 4*H*-1,4-oxazine have been reported. Treatment of the *N*-bromoacetyl or α-bromopropionyl derivatives of ethyl α-aminoacetoacetate with base gave ethyl 5-hydroxy-2-methyl-4*H*-1,4-oxazine 3-carboxylate (**45**; R = H) and its 6-methyl derivative (**45**; R = Me). The esters were hydrolysed and the acids were decarboxylated to give 5-hydroxy-2-methyl-4*H*-1,4-oxazine and 5-hydroxy-2,6-dimethyl-4*H*-1,4-oxazine. Similarly, treatment of chloroacetyl derivatives of *N*-(4-nitro-phenacyl)aniline and its 3-chloro-derivative with alkali gave 5-hydroxy-2-(*p*-nitrophenyl)-4-phenyl-4*H*-1,4-oxazine (**46**; R = Ph) and its 4-*m*-chlorophenyl analogue (**46**; R = *m*-ClPh). No evidence was given to show that the foregoing compounds are not the 5,6-dihydro-5-oxo-tautomers, the structure suggested by their fairly low melting points.

Dihydro-4*H*-1,4-oxazines are obtained by dehydration of β-keto-β'-hydroxy-dialkylamines, *e.g.* 4-benzoyl-5,6-dihydro-2-(3,4-dihydroxyphenyl)-4*H*-1,4-oxazine from *N*-benzoyl-*N*-(3,4-dihydroxyphenacyl)ethanolamine. Similarly, dehydration of *N*-(2-hydroxyalkyl) derivatives of aminocamphor to 'camphenemorpholines' has been described. Phenacylethanolamines (**47**) exhibit tautomerism to the cyclic tetrahydro-1,4-oxazine forms (**48**), and these will undergo dehydration with acid catalysts to give 2,3-dihydro-4*H*-1,4-oxazines (**49**).

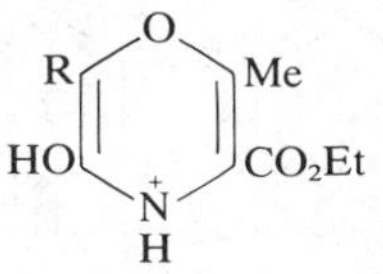

(**45**) 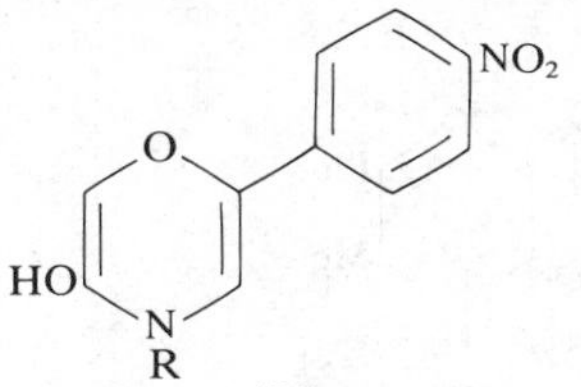 (**46**)

PhCOCHR^{1}NR^{2}CH$_2$CH$_2$OH

(**47**) (**48**) 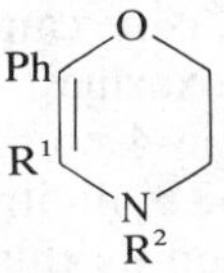 (**49**)

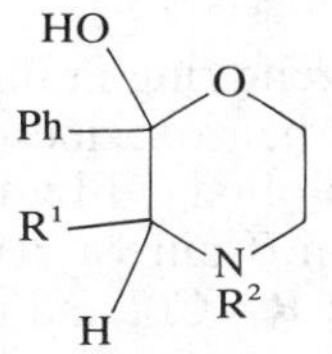

The most important monocyclic 1,4-oxazines are the tetrahydro-derivatives. The parent compound, morpholine, b.p. 128 °C, is a fairly strong base (pK_a 8.7, less than that of piperidine) and a powerful solvent with many industrial applications and uses in synthesis. Its chemical properties are those of a typical aliphatic secondary amine. The dipole moment (1.48 D in benzene) indicates that it has the chair conformation. Morpholine is made by dehydration of diethanolamine, *e.g.* by sulphuric acid, or by reaction of 2,2′-dichlorodiethyl ether with ammonia. *C*- or *N*-substituted morpholines are made similarly, and dehydration of di-(2-hydroxyalkyl)anilines affords *N*-phenylmorpholines from which the aryl group may be removed by nitrosation and alkaline hydrolysis.

The 2-oxo- and 3-oxo-derivatives of morpholine are lactones and lactams, respectively, and their syntheses and chemical properties follow from this. 3-Oxomorpholine is obtained in 23% yield by heating 2-aminoethoxyacetic acid in toluene, but is better made from ethyl chloroacetate and sodium 2-aminoethoxide. Reaction of glycine with ethylene oxide gave 2-oxomorpholine, a liquid unstable to hydrolysis and to oxidation. 3,5-Dioxomorpholine and 3,5-dioxo-4-phenylmorpholine are imides of diglycollic acid. The isomeric 2,6-dioxomorpholine (the anhydride of iminodiacetic acid) has not been recorded, but *N*-substituted compounds, *e.g.* 2,6-dioxo-4-phenylmorpholine, are known.

Several morpholine derivatives have been used as medicinal agents. 6-Morpholino-4,4-diphenylheptan-3-one (**50**) gives a hydrochloride which is the analgesic phenodoxone (Heptalgin). The diguanide 'ABOB' (**51**), obtained by reaction of morpholine with cyanoguanidine, enjoyed a short vogue as a prophylactic against influenza; its efficacy was not established. 3-Methyl-2-phenylmorpholine (**52**) is the anorexiant phenmetrazine (Preludin). It is made by dehydrating *N*-(2-hydroxyethyl)norephedrine, *e.g.* by heating with toluene-*p*-sulphonic acid in xylene, and is the *trans*-isomer. The *cis*-isomer is made stereospecifically by reaction of ethyl chloroacetate with the sodio-derivative of norephedrine and reduction of the resulting 3-methyl-5-oxo-2-phenylmorpholine with lithium aluminium hydride; the *trans*-isomer is made similarly from nor-ψ-ephedrine. In a related synthesis, 3-*o*-ethoxyphenoxy-2-hydroxypropylamine was chloroacetylated and the product cyclized to a 1,4-oxazinone with sodium methoxide. Reduction of the carbonyl group with lithium aluminium hydride gave 2-*o*-ethoxyphenoxymethylmorpholine (**53**). The hydrochloride is the antidepressant drug viloxazine hydrochloride (Vivalan).

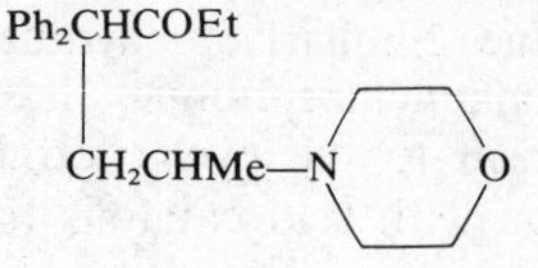

(50)

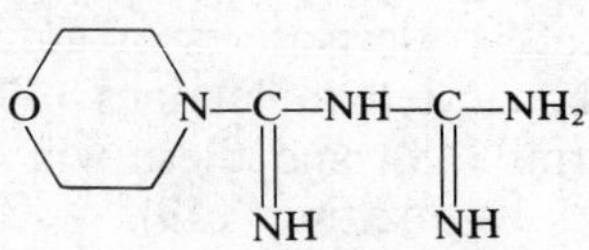

(51)

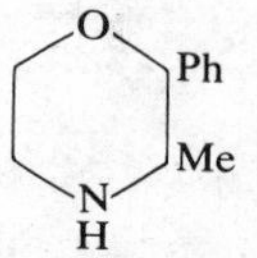

(52)

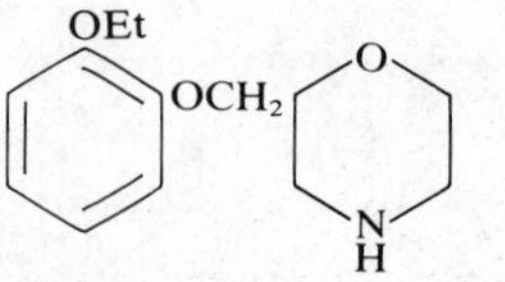

(53)

20.2.2.2 Benzoxazines

1,2-Oxazine can fuse with a benzene ring in three different ways to give 1,2-, 2,1-, and 2,3-benzoxazines. Few 1,2- or 2,1-benzoxazine derivatives are known. 3-Nitro-4*H*-1,2-benzoxazin-4-one (**54**; $R = NO_2$) is obtained by the action of nitrous gases on cinnamaldehyde or of dinitrogen tetroxide on β-nitrostyrene. Dilute hydrochloric acid replaces the nitro-group by chlorine, giving (**54**; R = Cl), and alkaline hydrolysis gives salicyclic acid.[22]

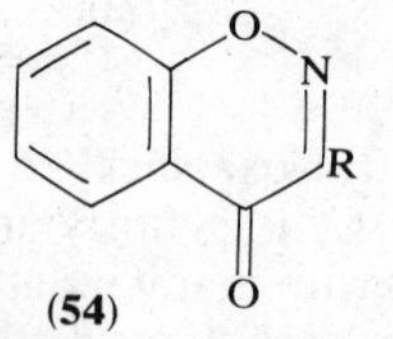

(54)

When *o*-hydroxyphenylglyoxal 2-phenylhydrazone is heated with acetic anhydride, the N—N bond undergoes fission to acetanilide and both *o*-acetoxybenzoyl cyanide and 4*H*-1,2-benzoxazin-4-one, which is acylated by a molecule of the acyl cyanide, the end-product of the reaction being 3-(*o*-acetoxybenzoyl)-4*H*-1,2-benzoxazin-4-one (**55**) (Scheme 7). 7-Nitro-3-phenyl-4*H*-1,2-benzoxazin-4-one (**56**) is obtained by heating 2,4-dinitrobenzil monoxime (α-isomer) in ethanol; displacement of nitrite occurs. Similar cyclizations of polyfluorobenzil monoximes are reported.

No 2,1-benzoxazines have been described, but some dihydro-derivatives, *e.g.* 3,4-dihydro-4,4,6,8-tetramethyl-5,7-dinitro-1*H*-2,1-benzoxazine, were obtained by photolysis of *o*-nitro-t-butylbenzenes. It is claimed that transient 3,4-dihydro-1*H*-2,1-benzoxazin-1-oxyl radicals are obtained by decomposition of *o*-nitroaryldiazonium ions in the presence of olefins and titanium(III) ions.

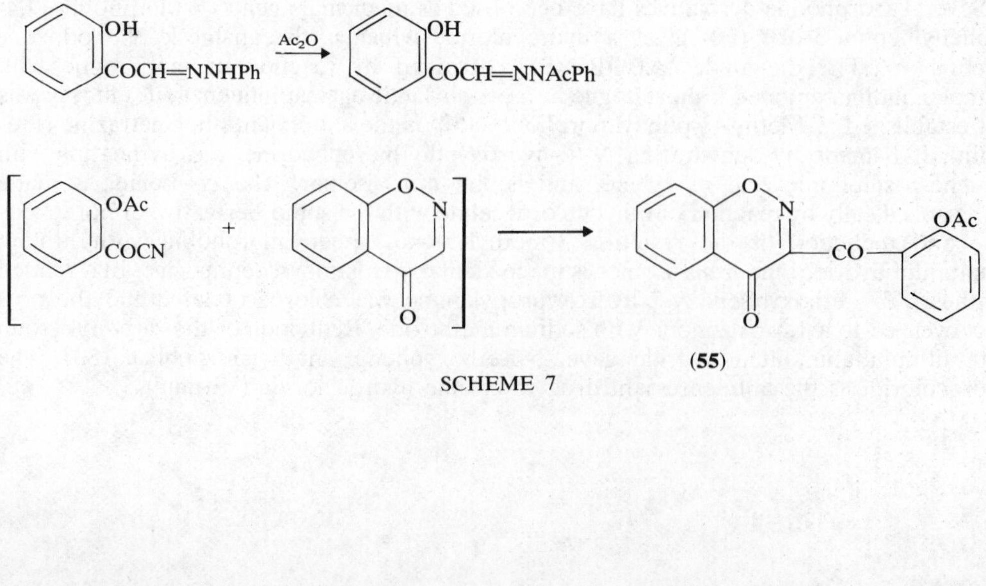

(55)

SCHEME 7

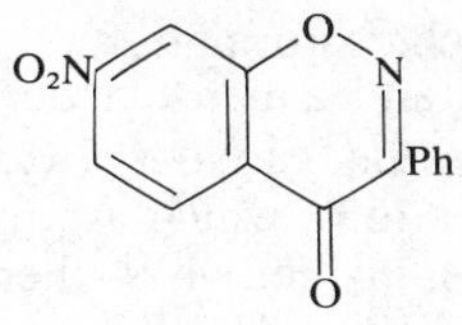

(56)

The monoxime of phthalaldehyde gives a stable tautomer, 1-hydroxy-1*H*-2,3-benzoxazine (**57**). As a hemiacetal this reacts with alcohols to give *O*-alkyl derivatives, *e.g.* the 1-methoxy-compound, and is oxidized by permanganate to 1-oxo-1*H*-2,3-benzoxazine. The latter compound, when heated, is converted into phthalimide *via* *o*-cyanobenzoic acid. Other 1-oxo-compounds ('oxime anhydrides') are obtained from *o*-carboxyphenyl ketones and hydroxylamine, *e.g.* 4-methyl-1-oxo-1*H*-2,3-benzoxazine from *o*-carboxy- or *o*-cyano-acetophenone. Phthalic anhydride reacts with *N*-alkylhydroxylamines to give 3-alkyl-3,4-dihydro-1,4-dioxo-1*H*-2,3-benzoxazines (**58**).

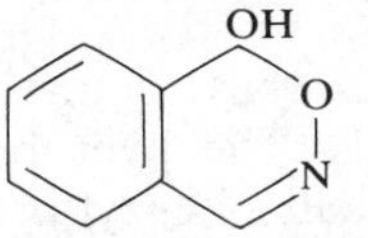

(57)

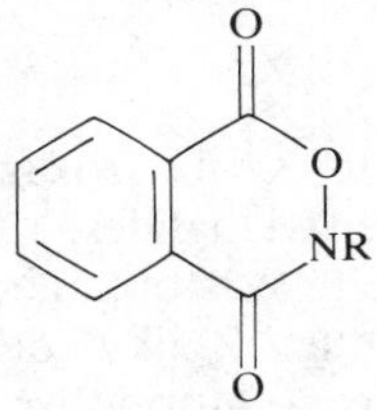

(58)

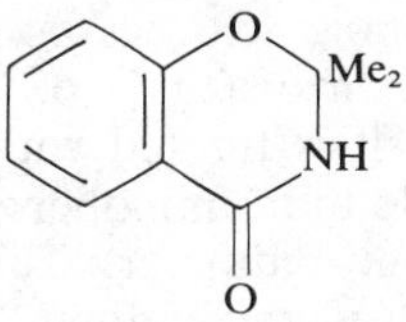

(59)

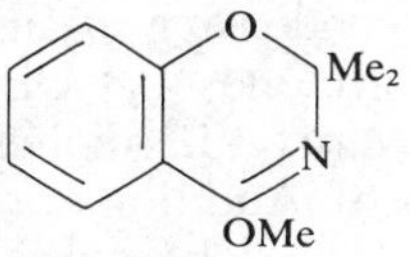

(60)

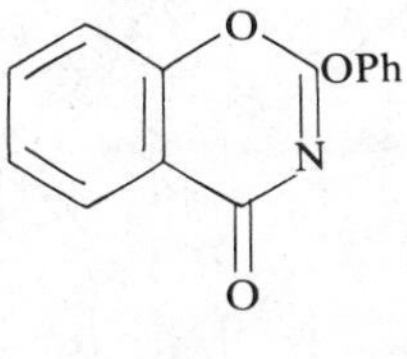

(61)

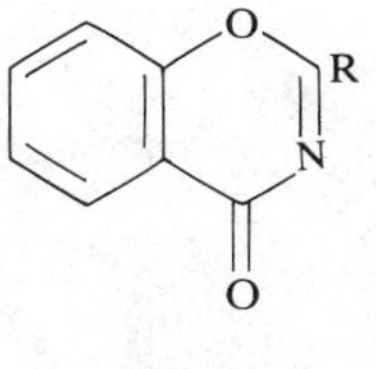

(62)

1,3-Oxazine forms two series of benzo-derivatives. Most preparative routes for 1,3-benzoxazines involve the action of carbonyl compounds on *o*-hydroxybenzylamines or salicylamides. Thus, acid-catalysed reaction of salicylamide with benzaldehyde or with acetone gives 4-hydroxy-2-phenyl-2*H*-1,3-benzoxazine or 4-hydroxy-2,2-dimethyl-2*H*-1,3-benzoxazine [as the 3,4-dihydro-4-oxo-tautomer (**59**)]. The latter can be methylated to give 4-methoxy-2,2-dimethyl-2*H*-1,3-benzoxazine (**60**). 4-Oxo-2-phenoxy-4*H*-1,3-benzoxazine (**61**) is made by base-catalysed cyclization of phenyl *o*-cyanatobenzoate; the phenoxy-group is readily displaced by nucleophiles. Dehydration of *O*- (or *N*-) benzoylsalicylamide gives 2-phenyl-4*H*-1,3-benzoxazin-4-one (**62**; R = Ph), which is also obtained by reaction of phenyl salicylate with *N*-phenylbenzamidine. Its chemistry parallels that of the flavones, but acids hydrolyse it to *N*-benzoylsalicylamide. Since 1,3-benzoxazin-4-ones are derivatives of salicylamide, many have been investigated as anti-inflammatory agents, and chlorthenoxazine (**62**; R = CH_2CH_2Cl) has been used in osteoarthritis. 2,3-Dihydro-2,4-dioxo-4*H*-1,3-benzoxazine is obtained by reaction of salicylamide with phosgene, ethyl chloroformate, urea, or diaryl carbonates. It is acidic and gives an *N*-methyl derivative and a benzoyl derivative. 3-Aryl derivatives of the dioxo-compound are obtained by the cautious hydrolysis of 3-aryl-2-arylimino-2,3-dihydro-4-oxo-4*H*-1,3-benzoxazines (**63**), which are made by reaction of salicylic esters with *S*-methyl-*N*,*N*′-diarylthioureas. *S*-Methyl-*N*-phenylthiourea with phenyl salicylate gives 2,3-dihydro-2-phenylimino-4-oxo-4*H*-1,3-benzoxazine. 2,3-Dihydro-2-oxo-4*H*-1,3-benzoxazine is obtained from the decomposition of *o*-hydroxyphenylacetyl azide. *p*-Substituted phenols react with formaldehyde and primary alkylamines or aralkylamines to give dihydro-1,3-benzoxazines, *e.g.* 3-benzyl-2,3-dihydro-6-methyl-4*H*-1,3-benzoxazine. These are stable to alkali, but acid hydrolysis gives *o*-hydroxybenzylamines.

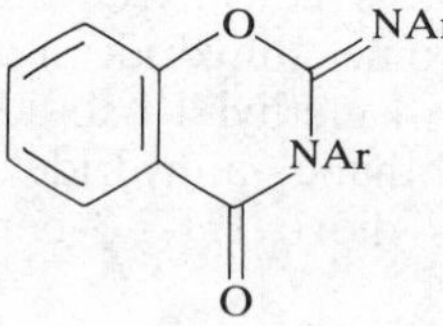

(63) 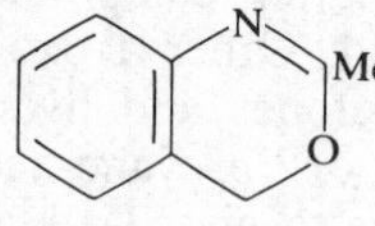 **(64)**

Synthetic routes to 3,1-benzoxazines often use as starting materials *o*-aminobenzyl alcohols, *o*-aminobenzyl halides, or anthranilic acids. The hydrobromide of 2-methyl-4*H*-3,1-benzoxazine (**64**) is obtained by warming *o*-aminobenzyl bromide hydrobromide with acetic acid. Other 2-alkyl derivatives are made by cyclizing *o*-acylaminobenzyl alcohols with hydrogen chloride or sulphuric acid. They are bases, stable to alkali, and with acids they give salts that are readily hydrolysed by water to regenerate the *o*-acylaminobenzyl alcohols. The dihydro-derivatives are more stable to hydrolysis. They are obtained by condensing *o*-aminobenzyl alcohols with carbonyl compounds, *e.g.* formaldehyde or acetone. 2,4-Diphenyl-4*H*-3,1-benzoxazine (**65**) undergoes a ring-contraction to 3-hydroxy-2,3-diphenylindole when it is treated with sodamide in liquid ammonia (reaction 6). Addition of carbon disulphide to *o*-aminobenzyl alcohol in ethanol gives 2-mercapto-4*H*-3,1-benzoxazine or its thiono-tautomer. This is a stable solid, soluble in alkali. Cyclization of *o*-hydroxymethyldiarylureas affords 2-arylimino-4*H*-3,1-benzoxazines.

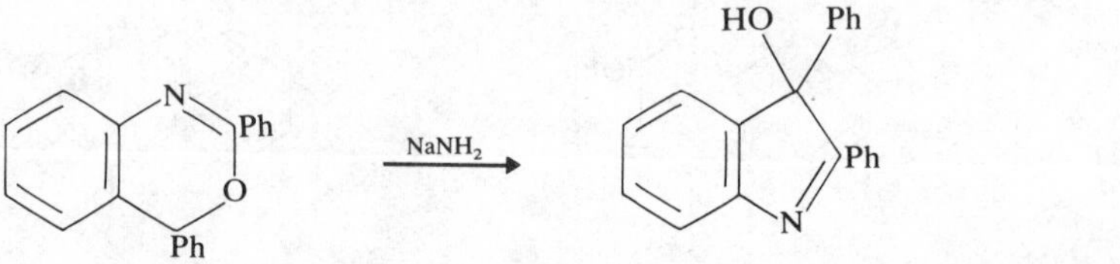

(65) $\xrightarrow{NaNH_2}$ (6)

N-Acylanthranilic acids are cyclized by acetic anhydride to 4-oxo-4*H*-3,1-benzoxazines (acylanthranils) (**66**), which are also obtained by the action of acyl halides on anthranils. 4-Oxo-4*H*-3,1-benzoxazine (formanthranil), the 2-methyl derivative (acetylanthranil), and other 2-alkyl derivatives are very susceptible to hydrolysis or to nucleophilic attack. With amines they give quinazolones or *o*-acylaminoanthranilamides, and they behave as acid anhydrides in Friedel–Crafts reactions. With Grignard reagents they give *o*-acylaminophenyl ketones or carbinols. 2-Phenyl-4*H*-3,1-benzoxazin-4-one (**66**; R = Ph),

(66)

from dehydration of *N*-benzoylanthranilic acid, is also obtained by photochemical rearrangement of 3-oxo-2-phenylindole 1-oxide, and is less reactive than the 2-alkyl compounds. It is used as a herbicide (Bentranil).

The Schiff's bases from aldehydes and anthranilic acids may be transformed into their tautomers, the 1,2-dihydro-4*H*-3,1-benzoxazin-4-ones (**67**). In some cases, *e.g.* formalide (**67**; R = H), the cyclic form is the stable reaction product. In other cases, *e.g.* (**67**; R = Ph), the cyclic form may be stabilized by acylation with acetic anhydride or phenyl isocyanate to give, for example, 1-acetyl- or 1-phenylcarbamyl-1,2-dihydro-4-oxo-2-phenyl-4*H*-3,1-benzoxazine.

2,4-Dioxodihydro-3,1-benzoxazines (isatoic anhydrides) are obtained by reaction of anthranilic acids with phosgene or ethyl chloroformate. The parent compound (**68**) is also made by oxidation of isatin or indigo with chromic acid. Isatoic anhydrides are weak acids which may be recovered unchanged after brief solution in cold, dilute alkali; under more vigorous treatment with alkali, however, they are rapidly hydrolysed. Nucleophiles such as ammonia, amines, or alcohols attack either of the two carbonyl groups with ring-opening to give anthranilamides, *o*-ureidobenzoic acids, quinazoline-2,4-diones, and polyanthranilic acids.

Derivatives of 1,4-benzoxazine are obtained by reduction of *o*-nitrophenoxymethyl ketones with tin(II) chloride. More vigorous reduction conditions, *e.g.* tin and hydrochloric acid or catalytic hydrogenation, give dihydro-derivatives (phenomorpholines). 3-Methyl-2*H*-1,4-benzoxazine (**69**; R = Me) is an oil that is unstable in air. The 3-phenyl compound (**69**; R = Ph), a solid, is a weak base that is precipitated from solutions of its salts on dilution. The phenomorpholines, *e.g.* 3-methylphenomorpholine or 4-methylphenomorpholine, are typical aromatic secondary or tertiary amines.

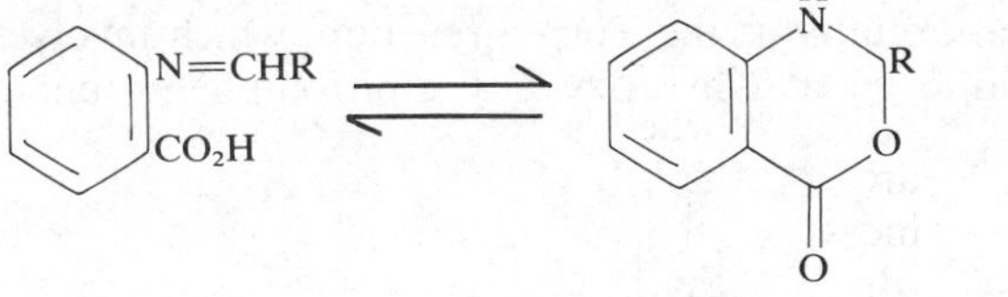

(67) 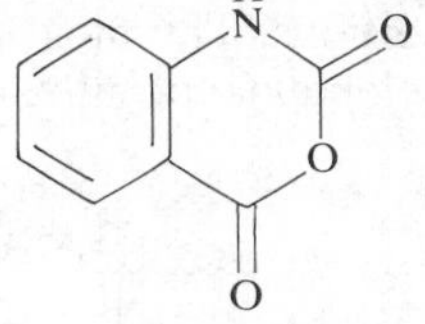 **(68)**

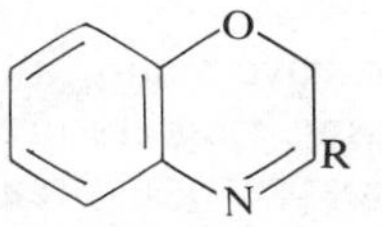

(69) 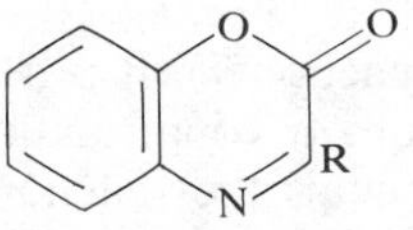 **(70)**

Condensation of α-ketoesters, *e.g.* ethyl pyruvate, ethyl phenylpyruvate, or ethyl ethoxalylacetate, with *o*-aminophenols gives 2-oxo-2*H*-1,4-benzoxazines (**70**; R = Me, $PhCH_2$, CO_2Et). Reaction of chloroacetic acid with *o*-methylaminophenol gives 3,4-dihydro-4-methyl-2-oxo-2*H*-1,4-benzoxazine, but *o*-hydroxyphenylglycine does not give a similar lactone. 3,4-Dihydro-3-oxo-2*H*-1,4-benzoxazines (3-phenomorpholones) are obtained by reaction of α-haloacyl halides with *o*-aminophenols, and also by the lactamization (usually spontaneous) of *o*-aminoaryloxyacetic acids. 3-Phenomorpholone is a tautomer of 3-hydroxy-4*H*-1,4-benzoxazine. Methylation of its sodium salt gives 4-methyl-3-phenomorpholone, but the silver salt may be alkylated to give 3-alkoxy-4*H*-1,4-benzoxazines, which are typical iminoethers, readily hydrolysed by acids and reacting with amines to give amidines (3-alkylamino- or arylamino-2*H*-1,4-benzoxazines). Dihydro-1,4-benzoxazine-2,3-dione is obtained from *o*-aminophenol and oxalyl chloride; with ethyl oxalate, *o*-aminophenol gives 2,2′-dihydroxyoxanilide, but some substituted *o*-aminophenols will react with the ester to give dihydrobenzoxazinediones.

20.2.2.3 Phenoxazines (dibenzo-1,4-oxazines)[23]

The largest and most important part of oxazine chemistry is that of the phenoxazines. The system of numbering now employed (IUPAC, *Chemical Abstracts*, and Ring Index) is shown in structure (**71**), but the numbering of (**72**) has been widely used in German literature and two other systems have also been employed. Care is therefore necessary in interpreting older nomenclature. To meet the requirements of valency, most phenoxazines have added hydrogen or its equivalent at position-10 or position-3, or are phenoxazonium salts. A most important group are the 3-oxo-3*H*-phenoxazines (phenoxazones)[24] and their 3-imino-derivatives (phenoxazimes).

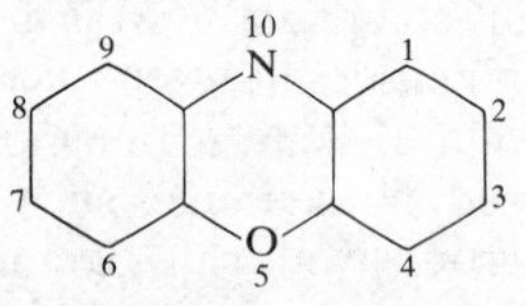

(**71**) IUPAC numbering 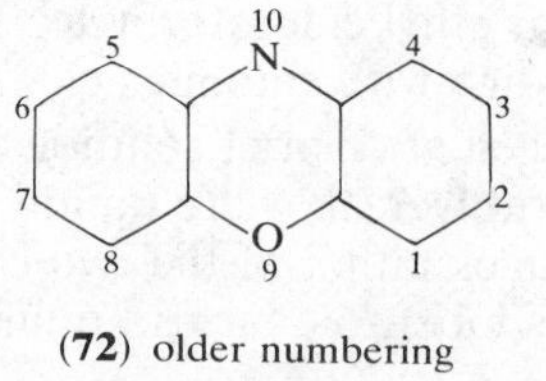 (**72**) older numbering 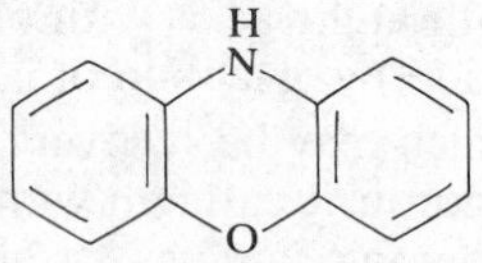 (**73**)

Phenoxazines are usually prepared by condensation reactions of *o*-aminophenols, *o*-hydroxydiphenylamines, or *o*-aminodiphenyl ethers. *o*-Aminophenol undergoes self-condensation to give a poor yield of phenoxazine (**73**) on heating its hydrochloride, or better by heating it with iodine at 270°C (30–35% yield). Phenoxazine is also obtained by heating *o*-aminophenol with catechol. It has been claimed that catechol acts only as a source of hydrogen ions, but this is not consistent with the fact that unsymmetrically substituted phenoxazines are obtained by heating substituted catechols with *o*-aminophenol. 1,2-Dichloro-3,5-dinitrobenzene condenses with *o*-aminophenol to give 1,3-dinitrophenoxazine. *o*-Hydroxydiphenylamines may undergo ring closure under various conditions. The most general example is the Turpin reaction, which involves the loss of the elements of nitrous acid from a 2-hydroxy-2′,4′-dinitrodiphenylamine (**74**) on

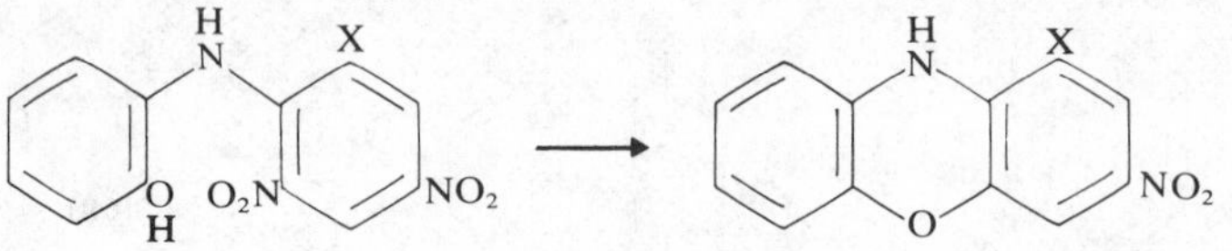

(**74**) (7)

heating with a base (reaction 7). It is desirable to have a substituent in the 6′-position, to bring the molecule into a conformation favouring ring-closure, and alkylation of the amino-nitrogen atom helps the reaction. The most effective reaction conditions involve heating the sodium salt of the hydroxydiphenylamine in dimethyl sulphoxide. Some 2-amino-2′,4′-dinitrodiphenyl ethers will also undergo the same reaction with a base, the ring-closure being preceded by a rearrangement (like the Smiles rearrangement) to the isomeric 2-hydroxydiphenylamine. Some other ring-closures of diphenyl ethers are shown in Table 3.

Phenoxazine is a colourless solid giving solutions that fluoresce violet or violet-red. The molecule is slightly folded along its shorter axis (dipole moment 1.93 D in benzene) and the ^{1}H n.m.r. spectrum is considered to indicate some flexibility in the non-planar

TABLE 3

Cyclization of Diphenyl Ethers to Phenoxazines

Substituents in diphenyl ether	*Reaction conditions*
2-Amino-2′-hydroxy-	Heat the hydrochloride
2,2′-diamino-	Heat the hydrochloride
2-amino-2′-bromo-	K_2CO_3, Cu, DMF
2-nitro-2′-methyl-	Heat with iron(II) oxalate

heterocyclic ring. Its aromaticity is less than that of phenazine. It does not form a hydrochloride, and *N*-alkylation is difficult, requiring the use of sodamide to generate the anion for attack by alkyl halides or epoxides. In contrast to this, cyanoethylation by means of acrylonitrile and benzyltrimethylammonium methoxide proceeds readily, giving 10-(β-cyanoethyl)phenoxazine. Arylation may be achieved with aryl iodides or aryl bromides under Ullmann conditions (heating with copper and potassium carbonate). Phenoxazine is readily *N*-acylated by acid anhydrides or acid chlorides. With phosgene it gives 10-chlorocarbonylphenoxazine (**75**), which is stable to water but reacts slowly with methanol to give 10-methoxycarbonylphenoxazine, and with dialkylaminoalkanols it gives esters (**76**) that are converted into 10-dialkylaminoalkylphenoxazines (**77**) with loss of carbon dioxide when they are heated to *ca.* 220 °C (reaction 8).

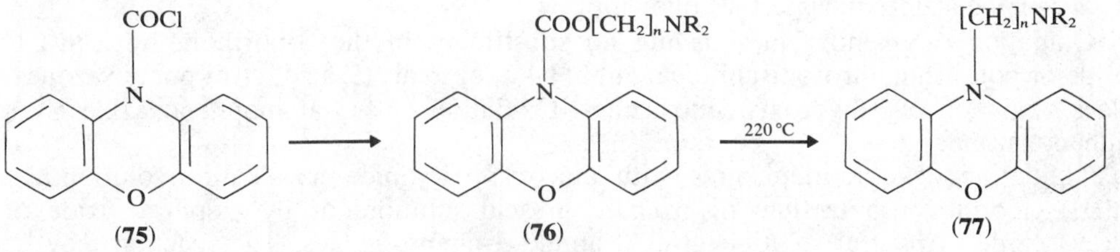

COCl — N — O (**75**) → $COO[CH_2]_nNR_2$ — N — O (**76**) —220 °C→ $[CH_2]_nNR_2$ — N — O (**77**) (8)

Phenoxazine is readily oxidized in one-electron steps to give firstly a radical cation and then a phenoxazonium ion (reaction 9). The radical cation is coloured, showing strong absorption at 530–535 nm, and its formation in a solution of phenoxazine in sulphuric acid on keeping, or on addition of hydrogen peroxide, may be demonstrated by e.s.r. spectroscopy. In neutral or alkaline solution the radical cation loses a proton to give a neutral radical that decays to a phenoxazine polymer by the formation of 3–10 bonds. At lower pH the oxidation leads to a 3-phenoxazone. The potassium salt of phenoxazine oxidizes to give two dimers, 3,10 and 1,10.

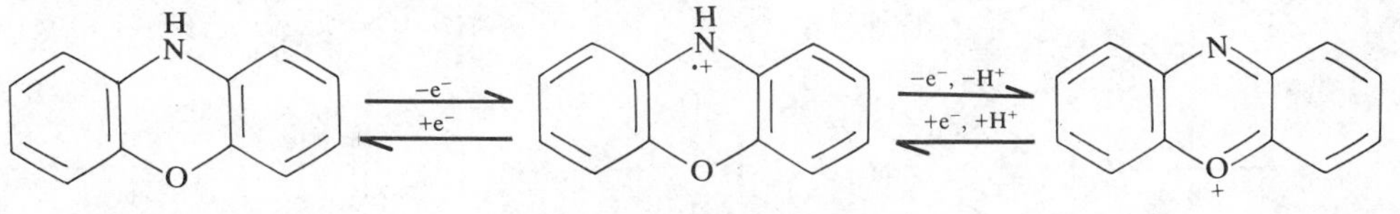

H N O $\underset{+e^-}{\overset{-e^-}{\rightleftharpoons}}$ H N •+ O $\underset{+e^-,\ +H^+}{\overset{-e^-,\ -H^+}{\rightleftharpoons}}$ N O + (9)

Oxidation of phenoxazine

Electrophilic substitution of phenoxazine takes place readily, the 3- and 7-positions being first attacked. Bromination of phenoxazine in benzene gives the 3-bromo- and 3,7-dibromo-derivatives with much polymeric and oxidation products. *N*-Acetylphenoxazine brominates cleanly to give 10-acetyl-1,3,7,9-tetrabromophenoxazine. Nitration of phenoxazine is violent, and gives the 1,3,7,9-tetranitro-compound. Nitration of *N*-acetylphenoxazine gives 10-acetyl-3-nitrophenoxazine or 10-acetyl-3,7-dinitrophenoxazine. Friedel–Crafts acetylation of 10-acetylphenoxazine gives, after hydrolysis, 2-acetylphenoxazine or 2,8-diacetylphenoxazine because the 3- and 7-positions are rendered deficient in electrons by complexing of the Lewis acid with the 10-acetyl group. 10-Ethylphenoxazine is acetylated in positions-3 and -7, and in the Vilsmeier–Haack reaction it gives 10-ethyl-3-formylphenoxazine. 10-Acetyl-2,5-dimethylphenoxazine is oxidized to 10-acetylphenoxazine-2,8-dicarboxylic acid by permanganate in pyridine, and ethylphenoxazines may be oxidized to methyl ketones. Other carboxylic acids are made by oxidation of acetyl substituents. Metallation of phenoxazine and its 10-alkyl and 10-aryl derivatives with butyl-lithium is slow, and subsequent carbonation gives the 4-carboxylic acids.

Some 1,4,6,8-tetrahydro-1,4,6,8-tetraoxophenoxazines, *e.g.* (**78**; R^1 = tridecyl, R^2 = *p*-dimethylaminophenyl), have been made by reaction of *p*-nitrosodimethylaniline with 3-alkyl-2,5-dihydroxy-1,4-benzoquinones, *e.g.* rapanone (**79**; R^1 = tridecyl).

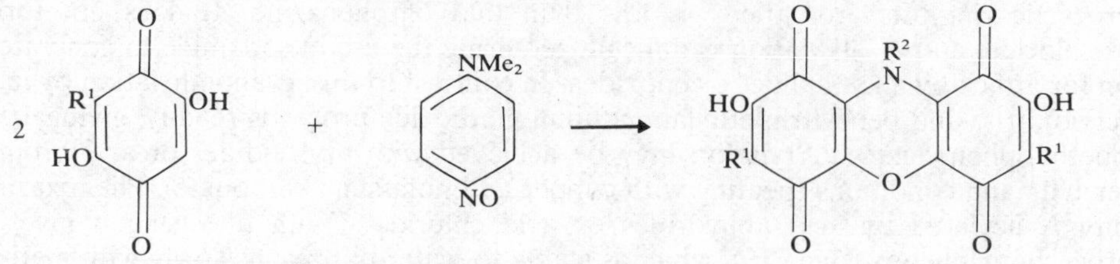

(10)

(79) **(78)**

Phenoxazone (*cf.* **80**) and its imino-analogue (*cf.* **81**) are the chromophores of a number of synthetic dyestuffs and natural colouring matters. Preparative routes differing from the routes to phenoxazines include the following.

(i) Oxidation of phenoxazines having no substituent in the 3-position, by iron(III) chloride or potassium nitrosodisulphonate at pH 4.5, or in acetic acid, gives phenoxazones, as does oxidation of 3-hydroxyphenoxazines. Oxidation of a 3-aminophenoxazine gives the phenoxazime.

(ii) Oxidation of *o*-aminophenols with air or ferricyanide in alkaline solution, by iron(III) chloride or potassium dichromate in acid solution, or by mercuric oxide or lead(IV) oxide in neutral solution gives 2-amino-3-phenoxazones. The reaction may be explained as an oxidation of the *o*-aminophenol to an *o*-quinone-imine followed by 1,4-addition of the amino-group of another molecule of *o*-aminophenol to the C=C—C=N system, further oxidation to the quinonoid state, 1,4-addition of the hydroxy-group to the C=C—C=O system, and final oxidation of the 3-hydroxyphenoxazine to give (**80**). The alternative addition of the amino-group to the less-reactive C=C—C=O system in the *o*-quinoneimine, leading to the isomer (**81**), is also observed (Scheme 8).

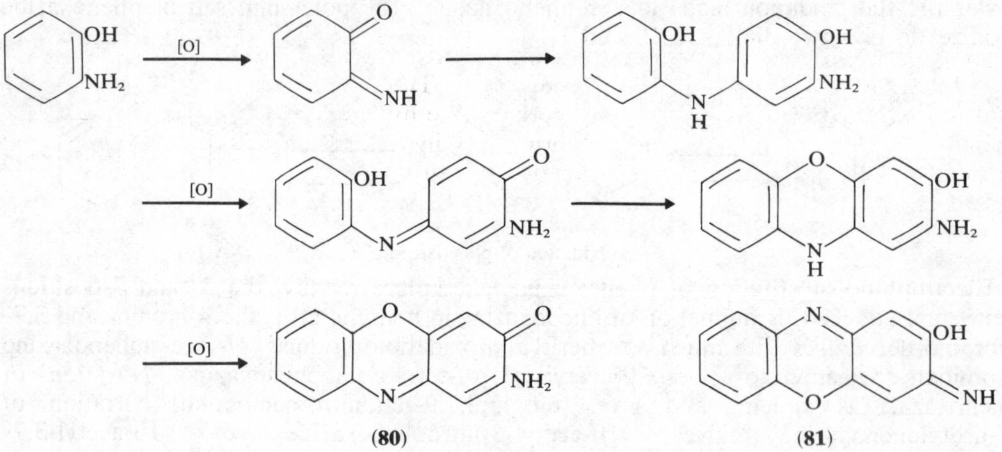

(80) **(81)**

SCHEME 8

The imino-group is readily removed from the hydroxyphenoxazime by hydrolysis, and the corresponding hydroxyphenoxazone may be isolated. This oxidative condensation is prevented by substituents (CN, NO_2, CO_2H, CO_2R) *para* to the amino-group, but phenoxazone formation occurs with *para*-substituents (halogen, OMe, NMe_2, Ac) that can be eliminated during the reaction.

(iii) Oxidative condensation of *o*-aminophenols with catechols. This requires preferential oxidation of the catechol to an *o*-quinone, and depends on the relative redox potentials of the two possible *o*-quinonoid products. The oxidation may be enzymic (tyrosinase and air) or chemical (ferricyanide). The products are 2-hydroxy-3-phenoxazones, which are comparable in acidity with carboxylic acids.

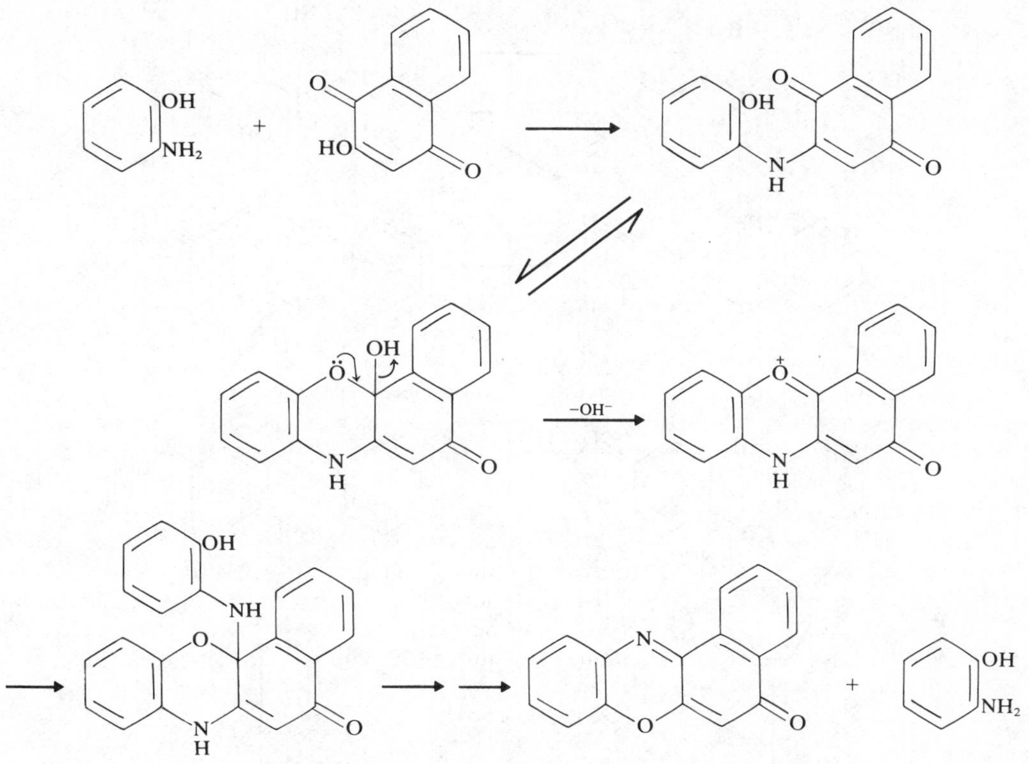

SCHEME 9

(iv) Condensation of *o*-aminophenols with hydroxyquinones has been widely applied, particularly with 2-hydroxy-1,4-naphthoquinones. The reaction mechanism[25] is complex, and the major pathway involves replacement of the hydroxy-group of the quinone by the amino-group of the aminophenol to form an *o*-hydroxyanilinoquinone (which may be isolated in some instances), and reaction of the hydroxy-group with the adjacent carbonyl group to form a hemiacetal. This may then lose a hydroxyl ion, generating an oxonium ion which is attacked by the amino-group of a second molecule of *o*-aminophenol. Finally, cyclization occurs with expulsion of the original molecule of *o*-aminophenol (Scheme 9). With hydroxybenzoquinones the same mechanism can operate, but alternative pathways are also possible. For example, the carbonium ion formed from the hemiacetal is electron-deficient at position-3 and may add *o*-aminophenol at this position with subsequent cyclization and oxidation to a triphendioxazine (Scheme 10). Alternatively, if there is a hydroxy- or alkoxy-group in the *o*-aminophenol *para* to the amino-group, or if a 2,5-dihydroxybenzoquinone is used, the carbonium ion may rearrange directly to a hydroxyphenoxazone in which the carbonyl group is now different from the one in the precursor[26] (Scheme 11). These reactions may be further complicated by oxidation of the *o*-aminophenol by the quinone.

(v) Condensation of 2-amino-5-nitrophenol with chloranil gives 1,2,4-trichloro-7-nitrophenoxaz-3-one, which may be reduced with sodium dithionite to the aminodihydro-compound; oxidation of this gives 7-amino-1,2,4-trichlorophenoxaz-3-one. 2-Amino-5-chlorophenol has similarly been condensed with bromanil, and 3-chloro-2,6-dihydroxybenzoquinone with *o*-aminophenol in acid solution gives 1-chloro-2-hydroxyphenoxaz-3-one. It is not known whether these reactions also involve the addition of *o*-aminophenol in the reverse direction with subsequent displacement.

(vi) 2,4,4′-Trihydroxydiphenylamines, which may be made by reaction of *p*-aminophenols with 1,2,4-trihydroxybenzene or aminoresorcinol, may be oxidized in acid solution to give hydroxyphenoxazones.

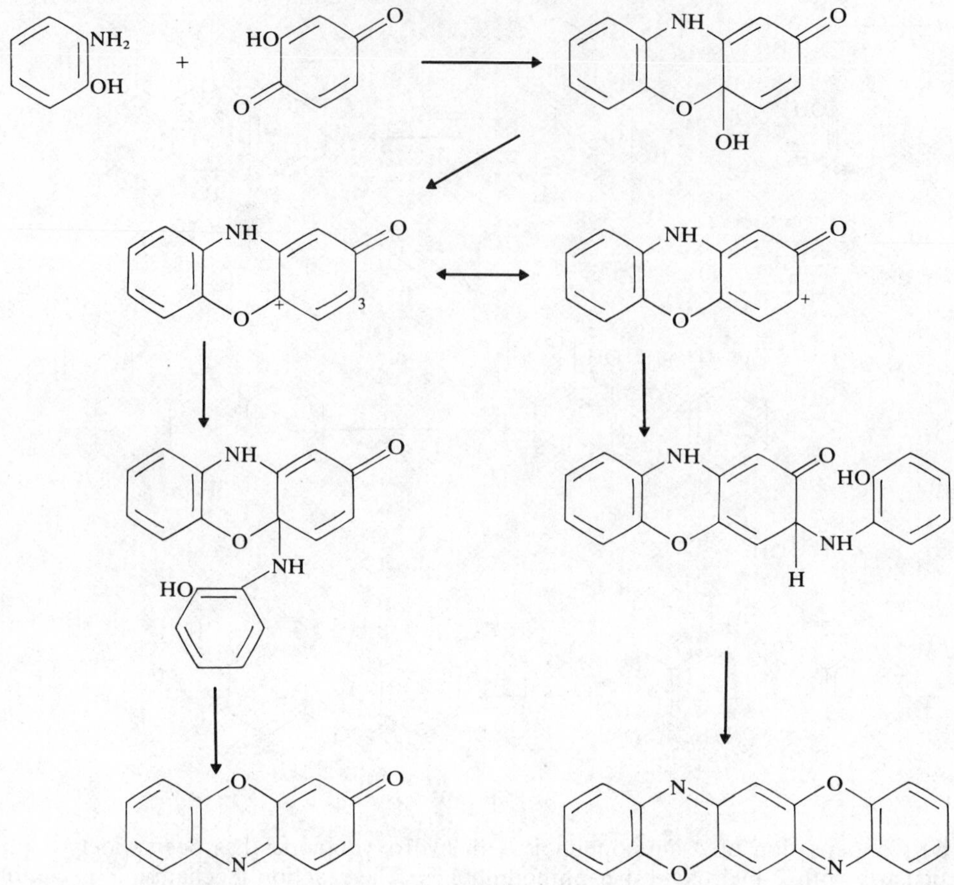

SCHEME 10

10-Methyl-2,3-dioxophenoxazine (**82**), a red solid, is obtained by oxidation of *o*-methylaminophenol sulphate with potassium ferricyanide. Concentrated sodium hydroxide solution decomposes it to *o*-methylaminophenol and 2,5-dihydroxybenzoquinone. Oxidation of phenoxazine with iron(III) chloride in the presence of aniline hydrochloride affords 3*H*-3-phenyliminophenoxazine (**83**). The hydrochloride of this compound reacts further with aniline, and atmospheric oxidation of the adduct gives 3,7-dianilinophenoxazonium chloride (**84**). Further examples of substitution by amines are found among the oxazine dyestuffs.

Treatment of a solution of resorcinol in ether with fuming nitric acid gives 7-hydroxyphenoxazone (resorufin) (**85**) and its *N*-oxide (resazurin) (**86**). Both compounds

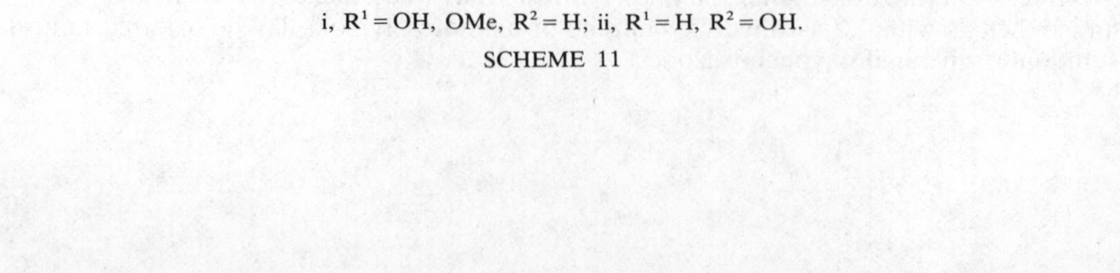

i, $R^1 = OH, OMe, R^2 = H$; ii, $R^1 = H, R^2 = OH$.

SCHEME 11

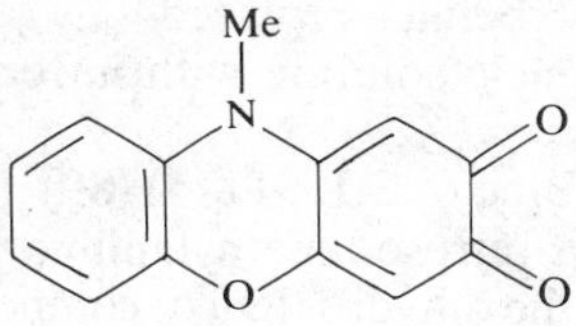

(82)

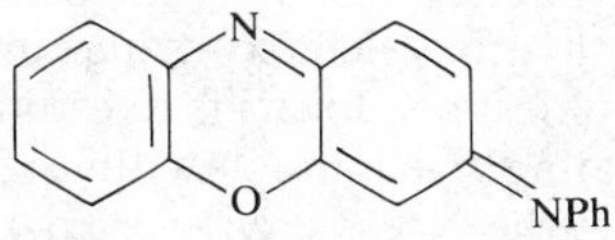

(83)

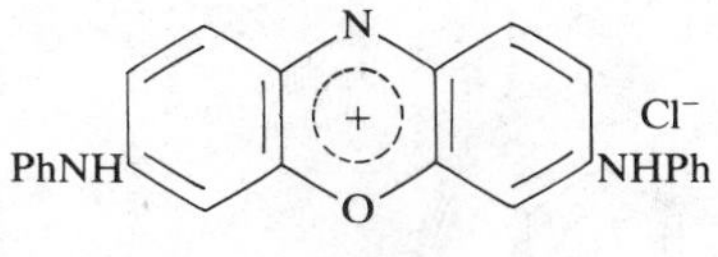

(84)

can be reduced to 3,7-dihydroxyphenoxazine (dihydroresorufin), which is readily reconverted into resorufin by atmospheric oxidation. Oxidation of resorufin to resazurin is difficult, but may be achieved with hydrogen peroxide. It is likely, therefore, that resazurin is a primary product in the foregoing synthesis. Resorufin is also obtained by various condensations, *e.g.* of *p*-nitrosophenol or quinone monochloroimide with resorcinol. Resorufin dissolves in alkali to a red solution with a red fluorescence. Resazurin dissolves in ethanol with a red fluorescence, and its alkaline solution is blue. Both compounds give monoacetyl derivatives. Benzoquinone dichloroimide condenses with resorcinol to give 7-aminophenoxazone, which is insoluble in aqueous alkali but gives red solutions with acids. Its alkaline alcoholic solution fluoresces orange-red.

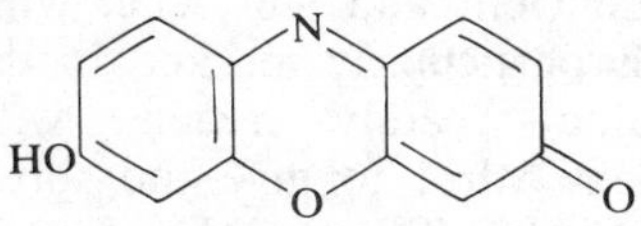

(85)

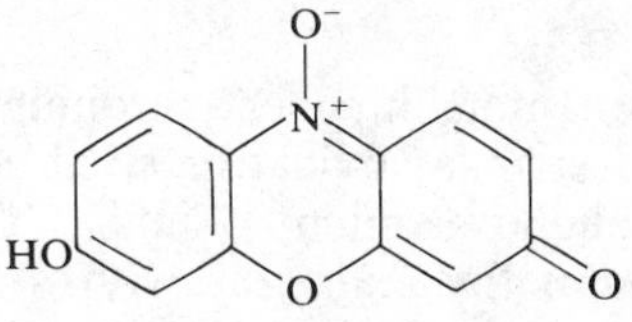

(86)

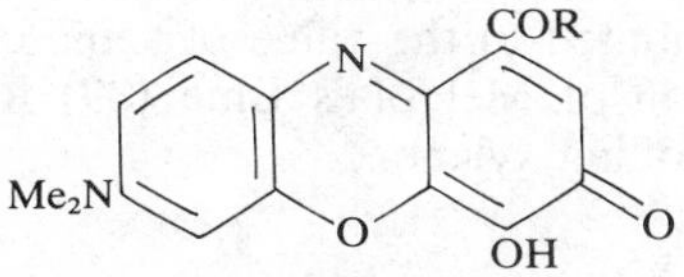

(87)

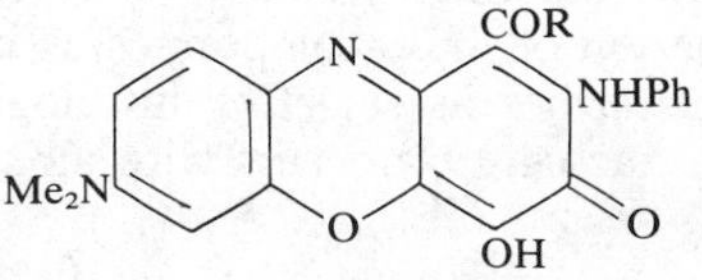

(88)

Oxazine dyes[27,28] are made by condensation of *p*-nitrosodialkylanilines with resorcinol, *m*-aminophenols, pyrogallolsulphonic acid, or naphthols, or of 5-dialkylamino-2-nitrosophenols with the same phenols or naphthols or with 1-naphthylamines. Condensation of gallic acid and its derivatives with *p*-nitrosodialkylanilines affords phenoxazones known as the gallocyanine dyes. Gallocyanine (Mordant Blue 10, C.I. 51030) (**87**; R = OH) dyes chrome-mordanted wool blue-violet. Its methyl ester (Prune, Mordant Violet 54, C.I. 51040) (**87**; R = OMe), amide (Gallamine Blue, Mordant Blue 45, C.I. 51050) (**87**; R = NH_2), anilide (Gallanil Violet), and other derivatives have been used as dyestuffs. When these latter compounds (ester, amide, *etc.*) are heated with aniline, preferably in a stream of air to effect oxidation, an aniline residue is introduced into the 2-position (**88**). With Gallocyanine the reaction is accompanied by decarboxylation and it has been supposed that in this case the anilino group enters the 1-position. Gallocyanine also reacts with aniline at lower temperatures without decarboxylation, and the product loses carbon dioxide when heated to 100°C to give a substance which was thought to be 2-anilino-7-dimethylamino-4-hydroxyphenoxaz-3-one. The sulphonic acids made by sulphonating these two anilinophenoxazones have been used as dyestuffs. Gallocyanine reacts with resorcinol in alkaline solution, the carboxy-group being replaced by a *m*-hydroxyphenoxy-group, and the ester and amide react in acid solution with introduction of a *m*-hydroxyphenoxy-group in position-2.

The oldest phenoxazine dyestuff, Meldola's Blue (Basic Blue 6, C.I. 51175) (**89**), is made by treating 2-naphthol with a 50% molar excess of *p*-nitrosodimethylaniline hydrochloride, the excess serving as an oxidizing agent to oxidize the dihydro (leuco) compound to the phenoxazonium salt (Scheme 12). Unsubstituted positions *para* to the heterocyclic

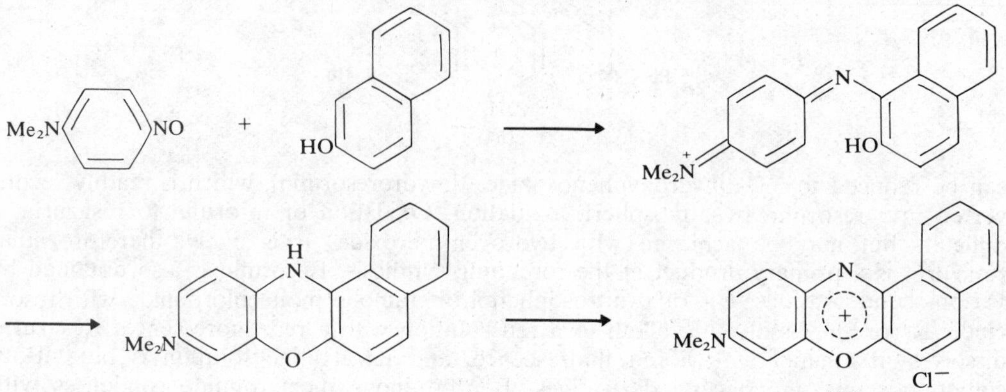

(**89**)

SCHEME 12

nitrogen atom in phenoxazonium salts are electron-deficient and will react with nucleophiles under oxidative conditions. This reactivity is particularly marked in the 1,2-benzophenoxazonium salts. Thus, Meldola's Blue readily reacts with *p*-aminodimethylaniline or with 4,4′-bisdimethylaminobenzhydrol to give the substituted phenoxazonium salts New Blue B (Basic Blue 10, C.I. 51190) (**90**) and New Fast Blue F (C.I. 51200) (**91**). Amino- and substituted amino-groups *para* to the ring nitrogen are readily removed by acid or alkaline hydrolysis with formation of phenoxazones, and the formation of these compounds as impurities has contributed to the value of some of the phenoxazines as selective histological stains. Oxidation of Meldola's Blue (**89**) to the phenoxazone (**92**) occurs with air and dilute acid in boiling xylene.

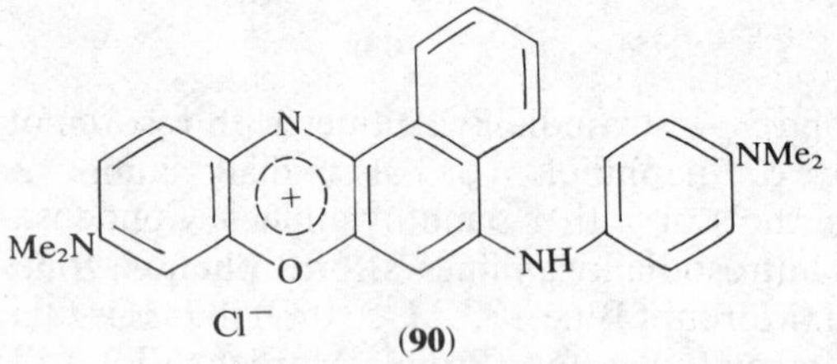

(**90**) 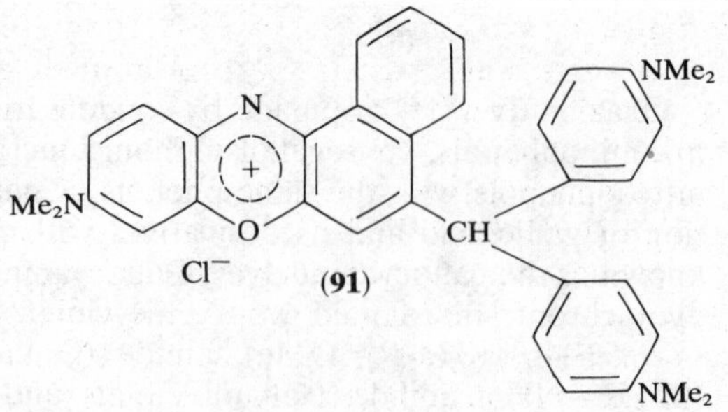 (**91**)

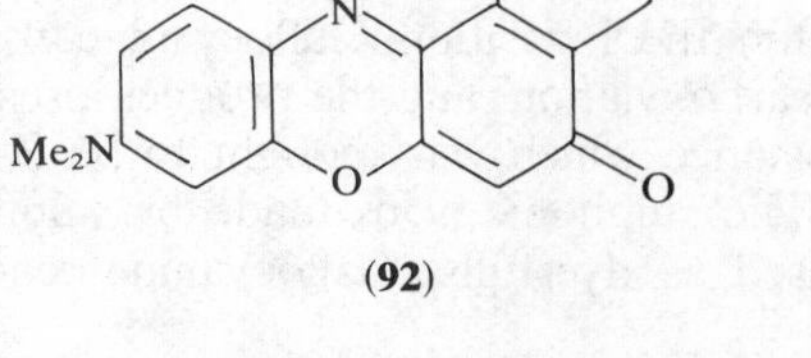

(**92**) 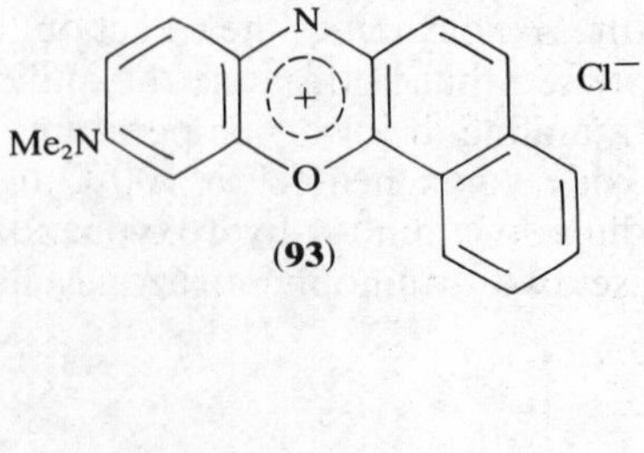 (**93**)

1,2-Naphthaquinones react with *o*-aminophenols to give 3,4-benzophenoxazines in which the amino-group has condensed with the carbonyl group in the 2-position. Thus, 1,2-naphthaquinone and 2-amino-5-dimethylaminophenol give the isomer (**93**) of Meldola's Blue. Yet another preparative method for phenoxazones is exemplified in the preparation of Alizarin Green G (C.I. 51405) (**94**) by condensation of 1,2-naphthaquinone-4-sulphonic acid with 1-amino-2-naphthol-6-sulphonic acid; in this case the amino-group displaces the labile sulphonic acid group in the naphthaquinone (reaction 11).

O O SO$_3$H + NH$_2$ OH HSO$_3$ → HSO$_3$ N O OH O (11)

(**94**)

The oxazine dyes are basic or chrome-mordant dyes that are of value in calico printing because they are destroyed by oxidizing agents, *e.g.* chlorate, and can be used for discharge effects. Their use in dyeing textiles has declined, but some are of value as vital stains in microscopy, examples being Brilliant Cresyl Blue (C.I. 51010) (**95**) and Nile Blue 2B (C.I. 51185) (**96**). The latter and some analogues have been investigated as tumour inhibitors. They are very active against tuberculosis in small rodents, but are too toxic for use in human medicine. Oxazine dyes and phenoxazones are also useful as analytical reagents, *e.g.* specific precipitants for ions and as redox indicators.

Me R^1_2N N R^2 + O NH_2 Cl^-

$R^1 = Et$, $R^2 = NH_2$ or $R^1 = Me$, $R^2 = H$

(**95**)

Et_2N N + O $NHCH_2Ph$ Cl^-

(**96**)

The phenoxazone chromophore has been found in a number of mould metabolites and other natural products. The actinomycins[29,30] are a group of closely related antibiotics produced by *Streptomyces* species. They are very toxic, but some of them are effective chemotherapeutic agents in certain neoplastic diseases. Actinomycins complex with double-stranded helical DNA, apparently by hydrogen bonding to the guanine residues, with inhibition of RNA synthesis, and they are valuable tools in molecular biology. The actinomycins are chromopeptides formed by combination of 2-amino-4,6-dimethylphenoxaz-3-one-1,9-dicarboxylic acid (actinocin) (**97**; R = OH) with two pentapeptide lactones, and the actinomycins containing different peptide chains are designated as actinomycins D, C, C_3, *etc.* Actinomycin was synthesized by ferricyanide oxidation of 3-hydroxy-4-methylanthranilic acid, and the total synthesis of a number of actinomycins (**97**; Pep = pentapeptide) was achieved by oxidation of peptides containing this anthranilic acid. In these condensations the desired 2-aminophenoxaz-3-ones were obtained in high yield (90%), but ferric chloride oxidation of the isomeric 3-amino-4-methylsalicylic acid (as its methyl ester) gave both the 2-amino-3-phenoxazone (13%) and its isomer, a 2-hydroxyphenoxazime (53%), which was isolated as the hydroxyphenoxazone (**98**) after hydrolysis (Scheme 13).

The synthesis of analogues and derivatives of the actinomycins has illustrated other features of phenoxazine chemistry. Actinomycin C_2 was reduced to the dihydro-compound (**99**), which condensed with pyruvic acid to give a 1,4-oxazinophenoxazine

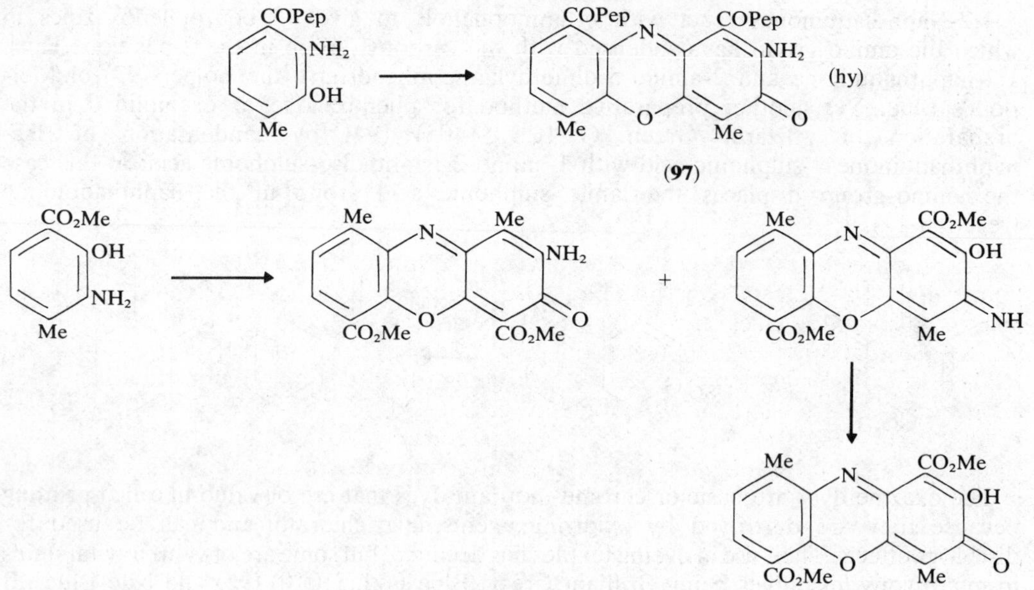

(97)

SCHEME 13

(98)

(**100**). This could be nitrated in the 7-position of the original phenoxazine ring, or oxidized to generate a carbonyl group at this position, and after cleavage of the pyruvic acid, 7-nitroactinomycin C_2 (**101**) or 7-hydroxyactinomycin C_2 (**102**) was obtained (Scheme 14). Oxidation of methyl 4-t-butyl-3-hydroxyanthranilate tended to stop at the quinoneimine stage (**103**) (reaction 12), and phenoxazone formation was wholly prevented by attachment of a peptide residue to the carboxy-group.

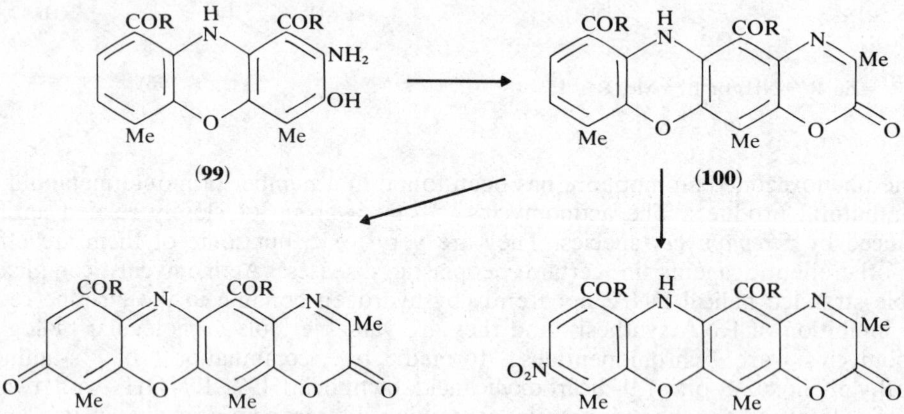

(99) (100)

(102) (101)

SCHEME 14

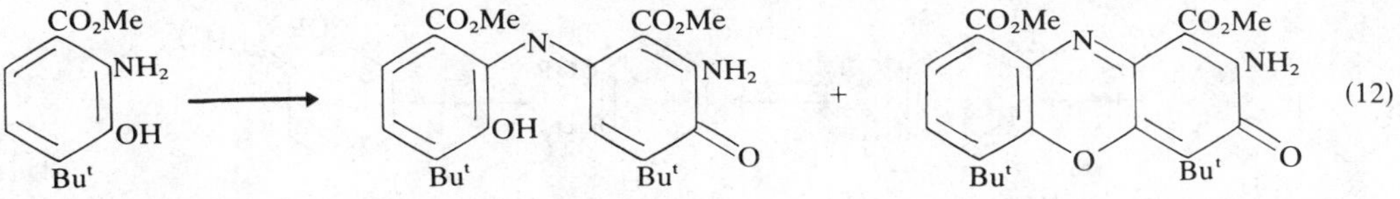

(12)

(103)

2-Aminophenoxaz-3-one has been isolated as a mould metabolite and named questiomycin A. Its *N*-acetyl derivative has also been found. A number of enzymes from plants, mammalian tissues, and *Streptomyces* species catalyse the formation of questiomycin A by oxidation of *o*-aminophenol.

Three coloured metabolites from certain wood-rotting fungi have been identified as 2-aminophenoxaz-3-ones, cinnabarin or polystictin (**104**; $R^1 = CH_2OH$, $R^2 = CO_2H$), cinnabaric acid (**104**; $R^1 = R^2 = CO_2H$), and tramesanguin (**104**; $R^1 = CO_2H$, $R^2 = CHO$). Structure determination was assisted by fission of the heterocyclic ring by mild alkaline hydrolysis, giving derivatives of 2,5-dihydroxybenzoquinone, and by permanganate oxidation of cinnabarin to give benzoxazolone-5-carboxylic acid (reaction 13). Oxidation of 2-aminophenoxaz-3-one with alkaline hydrogen peroxide gives benzoxazolone, a reaction that has been applied in the degradation of actinomycins.

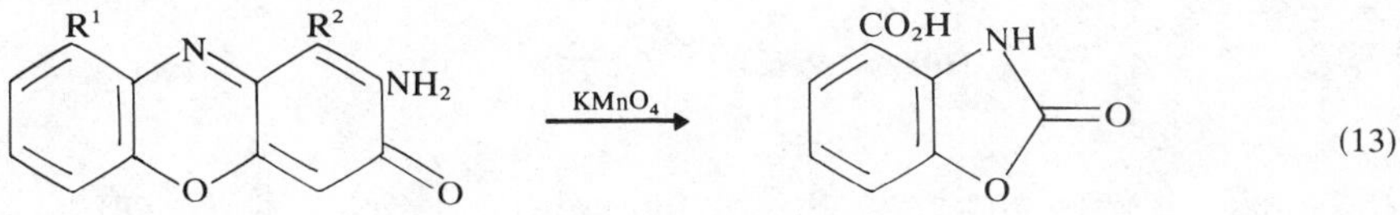

(13)

(104)

Ommochromes are a group of acidic pigments found in the eyes and skins of arthropods and cephalopods and in some butterfly wings.[31] They are derived from tryptophan metabolism. Xanthommatin is the pyridophenoxaz-3-one (**105**), which arises from tryptophan by enzymic fission to kynurenine, hydroxylation of this compound to the *o*-aminophenol hydroxykynurenine (**106**), and oxidation of the latter. This oxidation has been carried out in the laboratory with ferricyanide at pH 7.0. The 2-aminophenoxaz-2-one first produced loses ammonia with formation of the pyridone ring, and further oxidation gives xanthommatin (Scheme 15). A total synthesis that confirmed the structure (**105**) involved the preparation of 2-carboxy-4,6-dihydroxyquinoline-5,8-quinone (**107**) by a conventional synthesis from 2,4,5-trimethoxyaniline and ethyl oxaloacetate and condensation of this with (**106**) in acetic acid. Xanthommatin is a yellow substance that can be reduced reversibly to a red dihydro-derivative (**108**). The latter may be stabilized by protection of the hydroxy-group formed in the reduction, and the ommochromes rhodommatin and ommatin D are respectively the β-*O*-glucoside and the sulphate ester of dihydroxanthommatin.

Orcein is a violet colouring matter isolated from certain lichens by the action of air and ammonia. It has been separated into 14 components which are complex phenoxazones and phenoxazimes of the general structures (**109**) and (**110**). They are produced by autoxidation of orcinol with amination, condensation of quinonoid products to give phenoxazones, and nucleophilic attack by orcinol on quinonoid intermediates. This recalls the reaction of resorcinol with the Gallocyanine dyestuffs and suggests a re-examination of the structures of the products. Because of restricted rotation of the pendant phenyl groups in the orceins, numerous stereoisomers are formed.[32] Litmus consists of similar, more complex compounds obtained by adding lime, potassium carbonate, and gypsum to the autoxidation mixture. Autoxidation of resorcinol in ammoniacal solution also gives a blue phenoxazone.

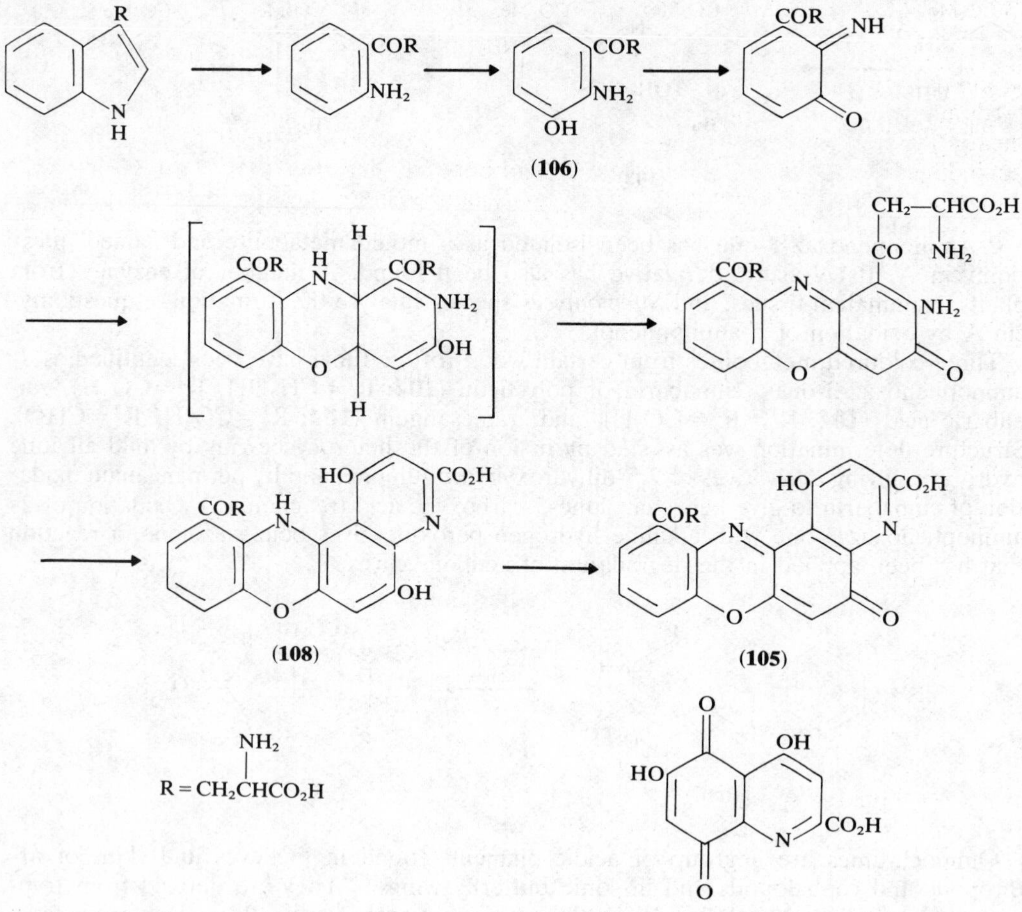

(**106**)

(**108**)

(**105**)

$R = CH_2CH(NH_2)CO_2H$

(**107**)

SCHEME 15

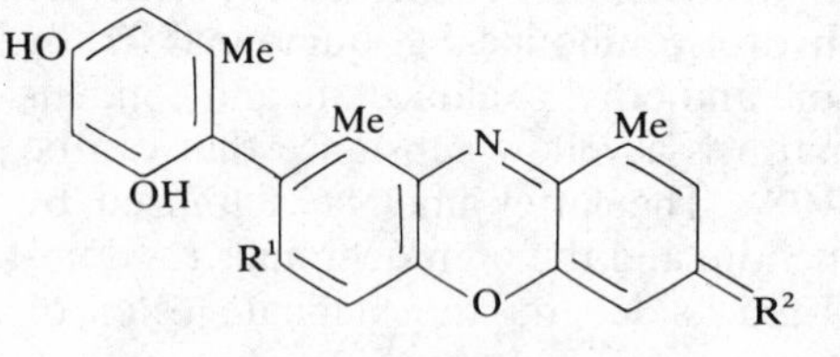

(**109**)

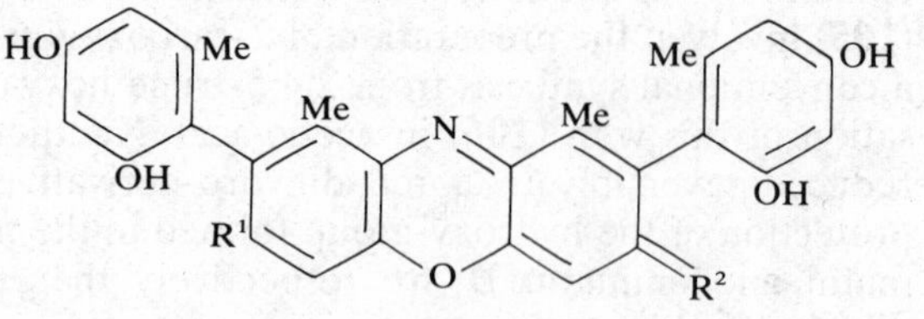

(**110**)

$R^1 = OH$ or NH_2, $R^2 = O$ or NH

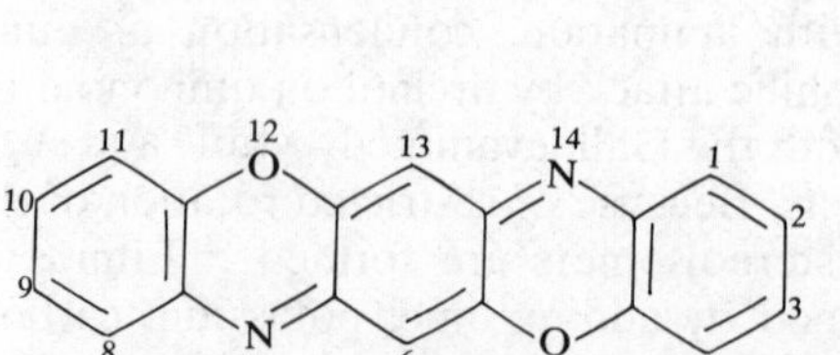

(**111**)

20.2.2.4 Triphendioxazines[27,28]

Triphendioxazine (**111**) is an inert aromatic compound which is produced by oxidizing *o*-aminophenol under many conditions: with air in acid solution, with hydrogen peroxide, with aminoazobenzene, and with nitrosobenzene in acetic acid (my). It is also formed by heating *o*-aminophenol with 4,6-diaminoresorcinol sulphate, by condensing it with 2-aminophenoxaz-3-one or 2,5-dihydroxybenzoquinone, and by fusing *o*-nitrophenol with potassium hydroxide and a little water. It crystallizes in needles described as red, violet, or steel blue, subliming at 240–250 °C. Hydriodic acid at 170 °C degrades it to *o*-aminophenol, but otherwise it is stable to concentrated acid and alkali and it can be sulphonated. Nitration in acetic acid gives a mononitro-derivative. Phenylhydrazine reduces it to 7,14-dihydrotriphendioxazine, which is readily reoxidized to triphendioxazine but may be stabilized as the 7,14-diacetyl derivative. The same preparative methods will give substituted triphendioxazines, *e.g.* 2-amino-5-dimethylaminophenol and 2,5-dihydroxybenzoquinone condense to give 3,10-bisdimethylaminotriphendioxazine, but reaction of nitrosobenzene with substituted *o*-aminophenols gives mainly *o*-hydroxyazobenzenes. 2,9-Dinitrotriphendioxazine has been obtained from 2-amino-4-nitrophenol both with nitrosobenzene and with 2,5-dihydroxybenzoquinone.

The other important preparative method is the oxidative cyclization of 2,5-diarylaminobenzoquinones. These are usually made from chloranil, which will react with two equivalents of an arylamine to give a 2,5-diarylamino-3,6-dichlorobenzoquinone (reaction 14). Stepwise reaction of 2,3,5-trichloro-6-methoxybenzoquinone with two

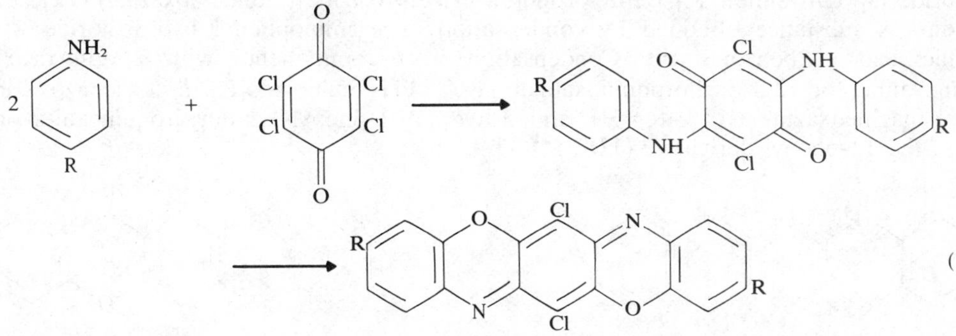

(14)

(**112**)

different arylamines gives unsymmetrical diarylaminodichlorobenzoquinones. Cyclization is effected by sulphuric acid, or by benzoyl chloride in nitrobenzene. If an *o*-alkoxyarylamine has been used, the alkoxy-groups participate in the ring closure (with dealkylation) and no oxidative step is then involved. The products of cyclization are 6,13-dichlorotriphendioxazines (**112**) and those from simple arylamines are orange or red in colour. The products derived from chloranil and polynuclear arylamines or aminodiphenylamines are blue or violet and are valuable pigments which may be sulphonated to give direct cotton dyestuffs. They have high light-fastness, which reflects the chemical stability of the chromophore. An example is Sirius Light Blue FF2GL (Direct Blue 106, C.I. 51300) (**113**) from 4-aminodiphenylamine. Triphendioxazines in the solid state show dimorphism, which can affect their colour. The chromophoric system does not require auxochromic groups for the generation of intense absorption bands. These compounds are at the oxidation level of phenoxazones, and both reduced and oxidized forms are less stable.

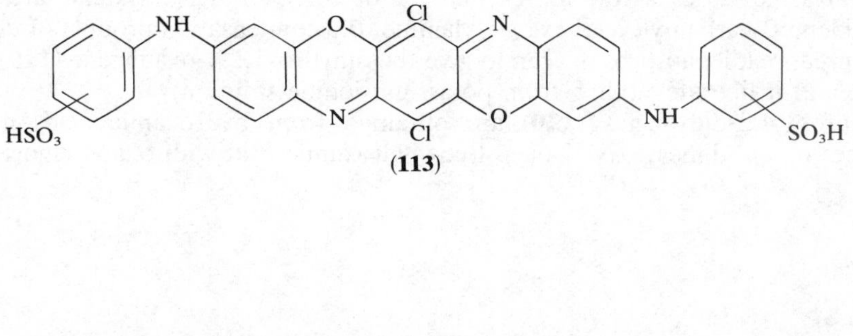

(**113**)

Methylation of triphendioxazine with dimethyl sulphate gives a poor yield of a 7-methyltriphendioxazinium salt (**114**). A better synthesis is by reaction of *o*-aminophenol hydrochloride with 10-methyl-2,3-dioxophenoxazine (**82**) in acetic acid at room temperature (reaction 15). The compound is isolated as its nitrate, which is violet, and it decomposes to triphendioxazine when its aqueous solution is warmed. A synthetic method

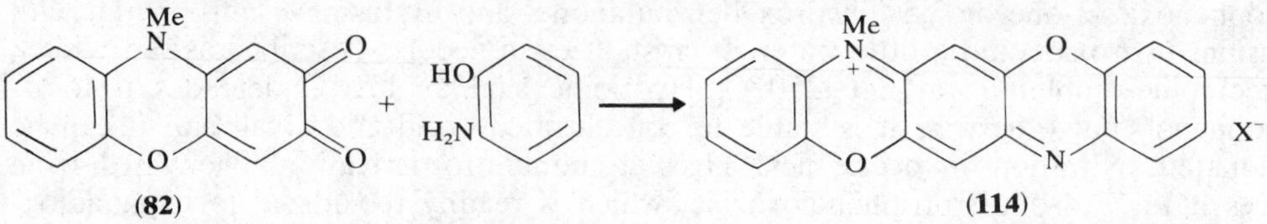

(**82**) + HO/H_2N → (**114**) X^- (15)

that has been less widely used employs the condensation of chloranil or bromanil with a 2-aminophenol having an electron-withdrawing substituent (Cl or NO_2) in the 4-position to give a 1,2,4-trihalogenophenoxaz-3-one with a chloro- or nitro-substituent at the 8-position. This will react with a further molecule of a 2-aminophenol to give a 6,13-dichloro- or -dibromo-triphendioxazine.[33]

Other fusion products of 1,4-benzoxazine with heterocyclic systems are known, *e.g.* the quinoxalo[2,3-*b*]phenoxazines (**115**) and (**116**), obtained by reaction of *o*-phenylenediamine derivatives with 2-hydroxyphenoxazones or 10-alkyl-2,3-dioxophenoxazines, or by reaction of *o*-aminophenol with 3-amino-5-phenylphenazonium chloride (aposafranine). 1,4-Benzoxazino[3,2-*b*]benzoxazine (diphendioxazine) (**117**) is a colourless substance obtained by condensation of *o*-aminophenol hydrochloride with oxalic acid in benzoic acid. Condensation of *o*-aminophenol with 2,3-dihydroxyquinoxaline or 2,3-dichloroquinoxaline gives 11*H*-quinoxalo[2,3-*b*][1,4]benzoxazine (diphenazineoxazine) (**118**; R = H), and 3-hydroxy-1-phenyl-1,2-dihydroquinoxal-2-one gives the 11-phenyl derivative (**118**; R = Ph).

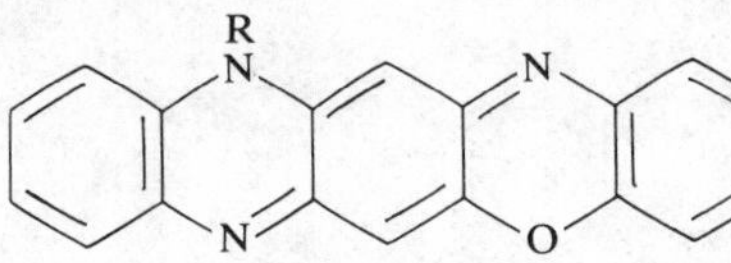

(**115**) 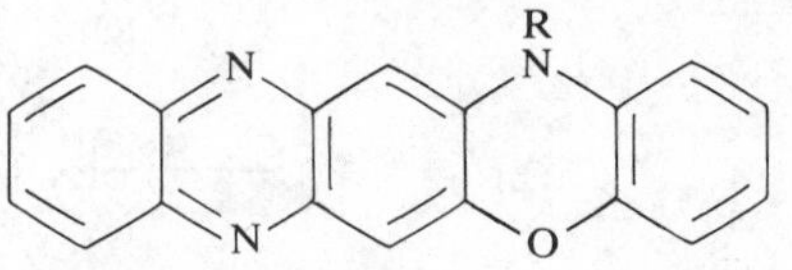 (**116**)

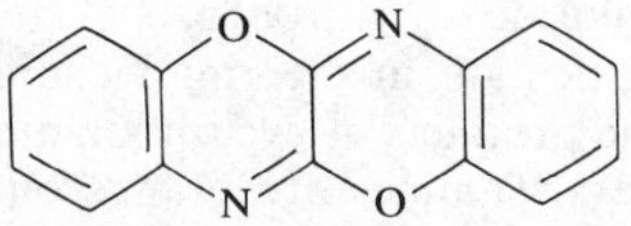

(**117**) 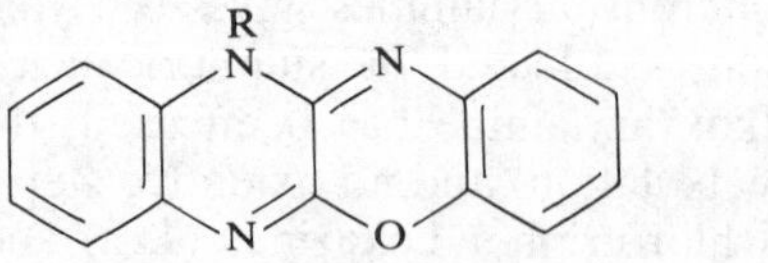 (**118**)

20.2.2.5 Dioxazines and oxadiazines

Compounds containing two oxygen atoms and one nitrogen atom (dioxazines) may theoretically be derived from six parent ring systems, and similarly there are six parent ring systems for compounds containing two nitrogen atoms and one oxygen atom (oxadiazines), neglecting isomerism due to extra hydrogen atoms (Scheme 16).

In the dioxazine series a few representatives of three of these systems are known. Cyclohexylidene-2-carbamylcyclohexan-1-ylamine, the condensation product of cyclohexanone and urea, readily absorbs oxygen to give the dihydro-1,2,4-oxadiazine (**119**). This is a peroxide, and it liberates iodine from potassium iodide solution.

5,6-Dihydro-1,4,2-dioxazines (**120**) are obtained from hydroxamic acids and *vic*-dihaloalkanes or *vic*-dimesyloxyalkanes. Benzhydroxamic acid with oxalyl chloride gives

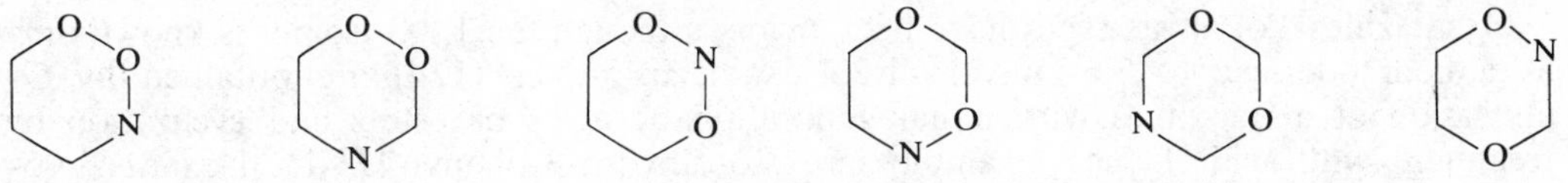

1,2,3- 1,2,4- 1,3,2- 1,3,4- 1,3,5- 1,4,2-

Dioxazines

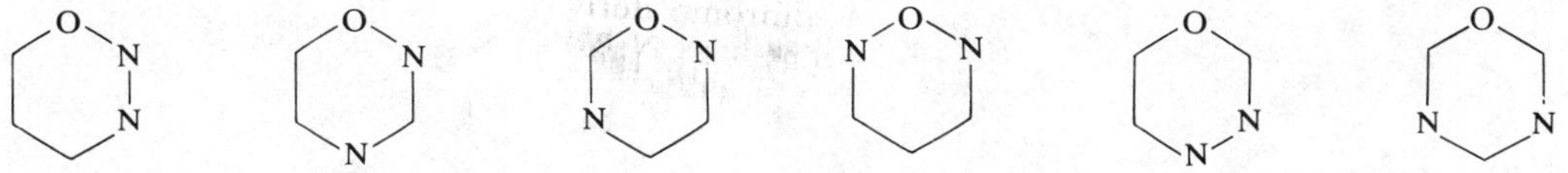

1,2,3- 1,2,4- 1,2,5- 1,2,6- 1,3,4- 1,3,5-

Oxadiazines

SCHEME 16

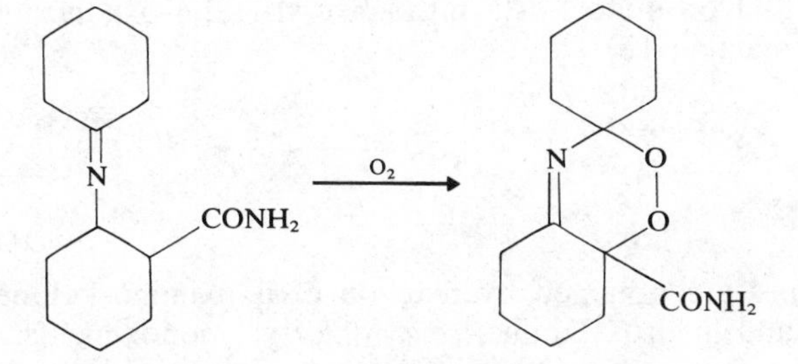

O_2 → CONH$_2$

(**119**)

5,6-dihydro-3-phenyl-1,4,2-dioxazine-5,6-dione (**121**) (reaction 16). Tetrahydro-1,4,2-dioxazines have been obtained from *O*-(2-hydroxyethyl)-*N*-methylhydroxylamine with formaldehyde or acetone.

Unstable products formulated as 5,6-dihydro-1,3,5-dioxazines have been obtained from amines by reaction with formaldehyde or acetaldehyde, *e.g.* 5,6-dihydro-5-ethyl-1,3,5-dioxazine, which is characterized as its picrate. Reaction of sulphonamides with formaldehyde (as trioxin) in a strongly acid medium gives a mixture of *N*-sulphonyl-5,6-dihydro-1,3,5-dioxazine (**122**; R = Me, Ph), *N,N'*-disulphonyltetrahydro-1,3,5-oxadiazine (**123**), and trisulphonylhexahydro-1,3,5-triazine (reaction 17). Treatment of 2-dimethylaminoethylamine with formaldehyde similarly gives a mixture of *N*-substituted dioxazine, oxadiazine, and triazine. Cyanamide reacts with hexafluoroacetone, giving 6-amino-2,2,4,4-tetrakis(trifluoromethyl)-1,3,5-dioxazine.

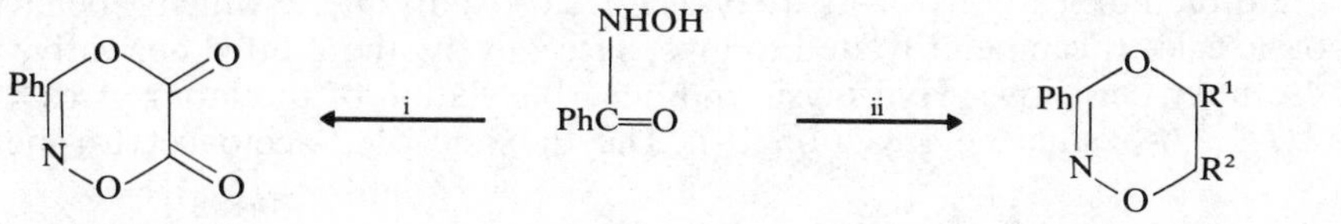

(**121**) ←i— PhC(=O)NHOH —ii→ (**120**) (16)

i, Oxalyl chloride; ii, $R^1CHXCHXR^2$.

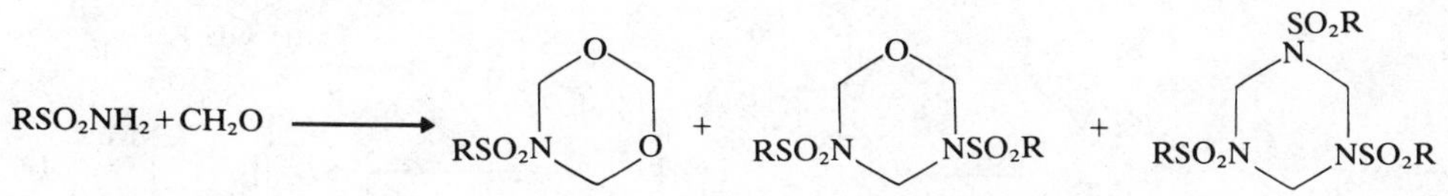

$RSO_2NH_2 + CH_2O$ → (**122**) + (**123**) + (17)

Oxadiazines[34] of all six types have been made, although the 1,2,3-isomer is known only as a benzo-derivative. 5,6-Dihydro-1,2,4-oxadiazin-5-ones (**124**) are obtained by *O*-alkylation of amidoximes with α-halogeno-aliphatic acids or esters and cyclization by treatment with acid. Typical examples are 5,6-dihydro-3-phenyl-4*H*-1,2,4-oxadiazin-5-one (**124**; R = Ph) and the 6,6-dimethyl-3-phenyl derivative. They are acidic compounds which give benzonitrile upon oxidation with permanganate. The ring-closure is reversed by drastic hydrolysis with conc. hydrochloric acid at 130 °C. 3,5-Dioxotetrahydro-1,2,4-oxadiazine (6-oxadihydrouracil) (**125**) is obtained by cyclizing

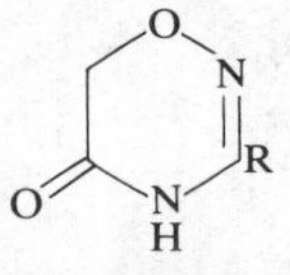

(**124**) 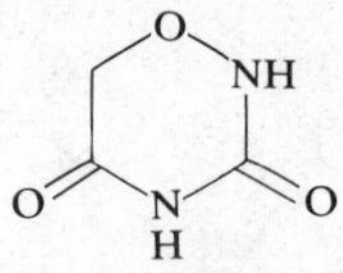 (**125**)

ethyl ureido-oxyacetate with sodium hydroxide. It inhibits the growth of some micro-organisms, and some 2-substituted analogues have been patented as plant growth regulators. 5-Amino-3-(5-nitro-2-furyl)-6*H*-1,2,4-oxadiazine (**126**) is made by ring-enlargement of 5-chloromethyl-3-(5-nitro-2-furyl)-1,2,4-oxadiazole by treatment with liquid ammonia (reaction 18).

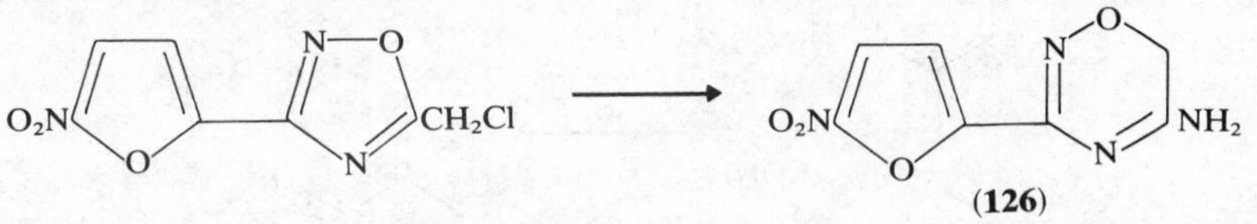

(**126**) (18)

1,2,5-Oxadiazines are obtained by reaction of α-oximino-ketones with aldoximes or with aldehydes and hydroxylamine, *e.g.* diacetyl monoxime gives 4-hydroxy-3,4,6-trimethyl-4*H*-1,2,5-oxadiazine (**127**) with acetoxime and the 3,4-dimethyl-6-phenyl analogue with benzaldehyde and hydroxylamine. These are stable bases with fairly high melting points, and their formation involves an oxidation of the aldehyde function. Acetaldehyde reacts with *syn*-oximes of ω-aminoacetophenones to give 3-aryl-5,6-dihydro-6-methyl-4*H*-1,2,5-oxadiazines (**128**).

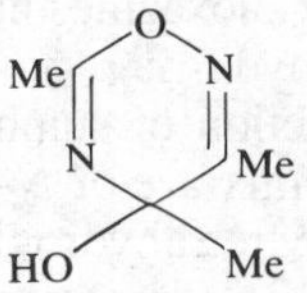

(**127**) 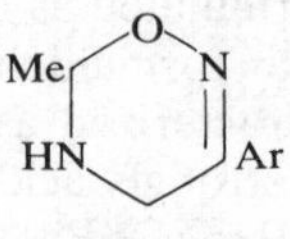 (**128**)

Treatment of ethyl 2-oxoglutarate with fuming nitric acid gives a yellow ester formulated as a 1,2,6-oxadiazine *N*-oxide (**129**). Reduction of this ester with sodium bisulphite gives 3,6-diethoxycarbonyl-4-hydroxy-2*H*-1,2,6-oxadiazine, which behaves as a monobasic acid. It can be ethylated to give, successively, the 2-ethyl derivative and the 2-ethyl-4-ethoxy-derivative. Hydrolysis and decarboxylation of the latter gives 4-ethoxy-2-ethyl-2*H*-1,2,6-oxadiazine (reaction 19). The dioxime of 2-acetoacetyl-4-methylphenol

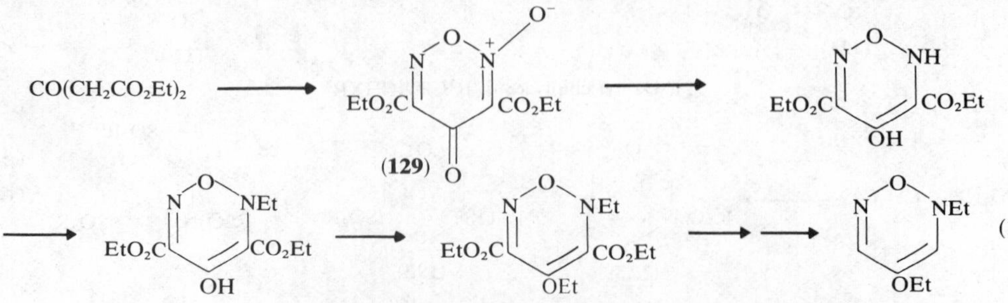

$CO(CH_2CO_2Et)_2$ → (**129**) → ... (19)

when treated with dilute sodium hydroxide gives two isomeric 3-aryl-5-methyl-1,2,6-oxadiazines (2*H* and 6*H*) which differ in their acid strength.

1,3,4-Oxadiazines are made from acylhydrazines and α-chloroacyl chlorides, *e.g.* N^2-acetylphenylhydrazine with chloroacetyl chloride gives N^2-acetyl-N^1-chloroacetylphenylhydrazine, which is cyclized by potassium carbonate to give 5,6-dihydro-2-methyl-4-phenyl-4*H*-1,3,4-oxadiazin-5-one (**130**). This is stable to water but is hydrolysed by potassium hydroxide or by acid. The 2,4-diphenyl compound is more stable to alkali. Oxalyl chloride reacts with N^2-acetylphenylhydrazine, giving 5,6-dihydro-2-methyl-4-phenyl-4*H*-1,3,4-oxadiazine-5,6-dione (**131**), (reaction 20), a yellow compound which is

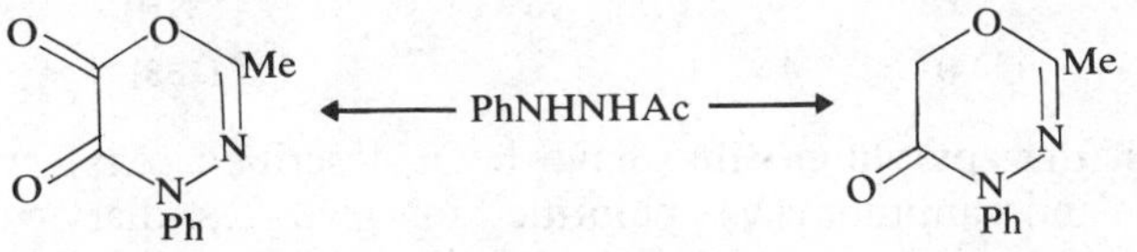

(**131**) ⟵ PhNHNHAc ⟶ (**130**) (20)

slowly hydrolysed by water and reacts with ethanol to give N^2-acetyl-N^1-ethoxalylphenylhydrazine. N^1,N^2-Dibenzoylhydrazine and oxalyl chloride give 4-benzoyl-5,6-dihydro-2-phenyl-4*H*-1,3,4-oxadiazine-5,6-dione. Treatment of N^2-(3,4,5-trimethoxybenzoyl)-N^1-methyl-N^1-2-hydroxy-2-methylpropylhydrazine with sulphuric acid gives 5,6-dihydro-2-(3,4,5-trimethoxyphenyl)-4,6,6-trimethyl-4*H*-1,3,4-oxadiazine.

Benzoyl isocyanate reacts with benzoyl chloride and aqueous pyridine to give 2,6-diphenyl-4*H*-1,3,5-oxadiazin-4-one. Mild pyrolysis of α-acetamidocinnamic azides gives 4-arylidene-2,3-dihydro-6-methyl-4*H*-1,3,5-oxadiazin-2-ones, which are hydrolysed by acid to give arylacetic acids. This has been recommended as a preparative method, starting with the aromatic aldehyde (Scheme 17). Trimerization of methyl isocyanate in

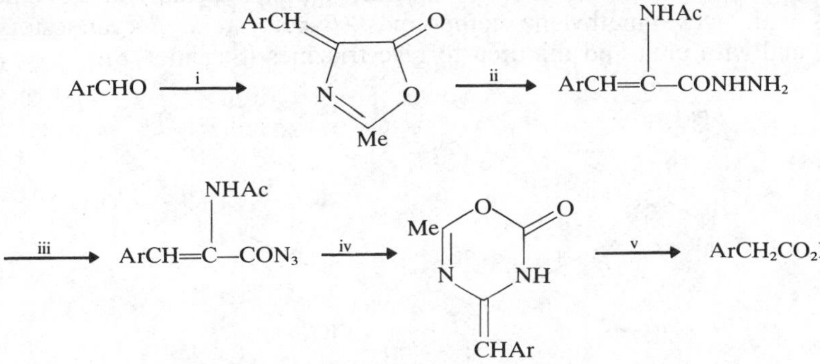

ArCHO $\xrightarrow{i}$ [oxazolone] $\xrightarrow{ii}$ ArCH=C(NHAc)—CONHNH$_2$ $\xrightarrow{iii}$ ArCH=C(NHAc)—CON$_3$ $\xrightarrow{iv}$ [oxadiazinone, =CHAr] $\xrightarrow{v}$ ArCH$_2$CO$_2$H

i, 2-methylisoxazolone; ii, N_2H_4; iii, HNO_2; iv, 80 °C; v, H^+.

SCHEME 17

the presence of triethylphosphine affords 3,5-dimethyl-2-methyliminotetrahydro-1,3,5-oxadiazine-4,6-dione (**132**) or 3,5-dimethyltetrahydro-1,3,5-oxadiazine-2,4,6-trione (**133**) (reaction 21). Numerous 3,5-bisalkoxymethylenetetrahydro-1,3,5-oxadiazines have been made by the reaction of urea with formaldehyde, barium hydroxide, and alcohols. 3,5-Dinitrotetrahydro-1,3,5-oxadiazine (oxyhexogen, cyclonite oxide) is one of the nitration products of hexamethylenetetramine.

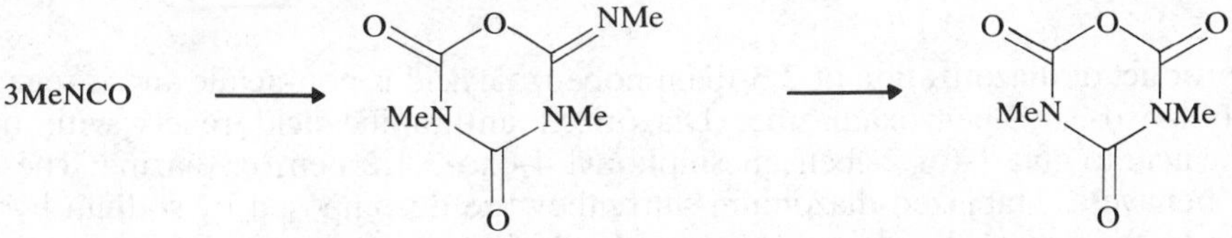

3MeNCO ⟶ (**132**) ⟶ (**133**) (21)

Guanylchloroformamidine hydrochlorides (made by heating dialkylcyanamide hydrochlorides) react with sodium bicarbonate, giving 2,4-bis(dialkylamino)-6*H*-1,3,5-oxadiazin-6-ones (**134**). These react with nucleophiles as if the 1–6 bond had undergone cleavage to give an isocyanate. They are alkylated by triethyloxonium fluoroborate to 2,4-bis(dialkylamino)-6-ethoxy-1,3,5-oxadiazinium fluoroborates (**135**).[35] Several other 1,3,5-

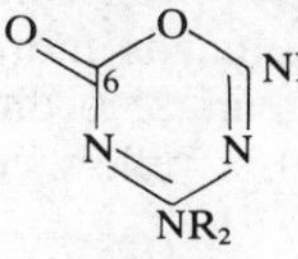

(134) 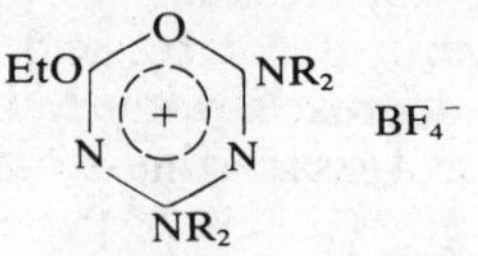 **(135)**

oxadiazinium salts (diazapyrylium salts) have been described. Aryl cyanates react with benzoyl chloride and antimony(V) chloride to give 2,4-diaryloxy-6-phenyl-1,3,5-oxadiazinium hexachloroantimonates (**136**) (reaction 22). These react with ammonia to

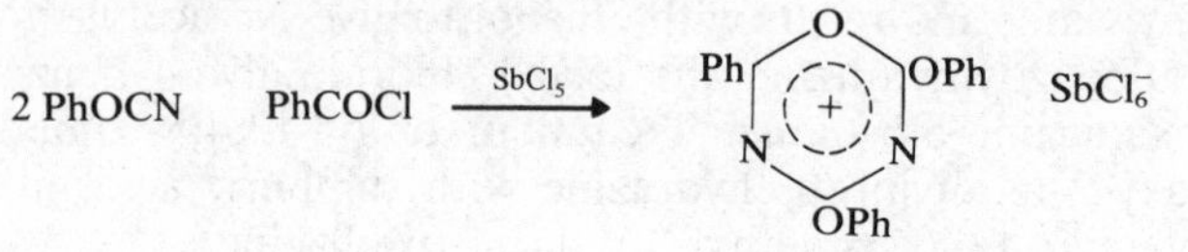

$$2\,\text{PhOCN} \quad \text{PhCOCl} \xrightarrow{\text{SbCl}_5} \textbf{(136)}\ \text{SbCl}_6^- \qquad (22)$$

give ring-opened products and small amounts of 1,3,5-triazine derivatives.[36] Dimethylcyanamide with benzoyl chloride at 150 °C gives 2,4-bis(dimethylamino)-6-phenyl-1,3,5-oxadiazinium chloride. 2,4,6-Triphenyl-1,3,5-oxadiazinium salts are obtained by treating benzonitrile and benzoyl chloride with Lewis acids, *e.g.* tin(IV) chloride.[37] The perchlorate (**137**) reacts with active methylene compounds (β-ketoesters, β-cyanoesters) to give pyrimidines, and with urea and thiourea to give triazines (Scheme 18).

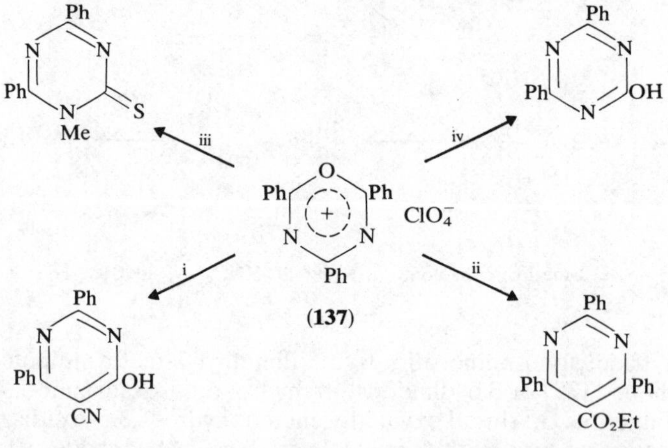

(137)

i, $CH_2(CN)CO_2Et$; ii, $PhCOCH_2CO_2Et$; iii, $MeNHCSNH_2$; iv, NH_2CONH_2.

SCHEME 18

The product of diazotization of 2,3-diaminobenzoic acid is non-acidic and is regarded as 8-amino-4-oxo-3,1,2-benzoxadiazine. Diazotized anthranilic acid reacts with benzene sulphinic acid to give 1-(or 2-)benzenesulphonyl-4-oxo-3,1,2-benzoxadiazine. These compounds behave as stabilized diazonium salts; they are decomposed by sodium hydroxide to generate the parent diazonium compound, which will couple with aromatic amines or phenols to give azo-compounds.

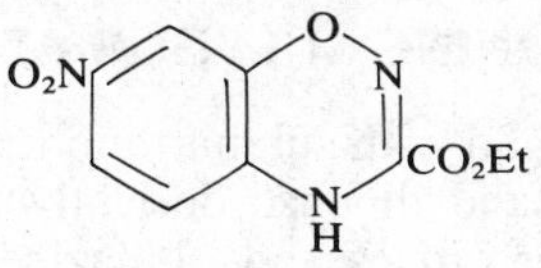

(138)

Ethyl anilino-oximinoacetate when treated with nitrous acid gives 3-ethoxycarbonyl-7-nitro-4*H*-1,2,4-benzoxadiazine (**138**) from which 7-nitro-4*H*-1,2,4-benzoxadiazine is obtained by hydrolysis and decarboxylation. Other 7-nitro-4*H*-1,2,4-benzoxadiazines (**139**) are obtained from 2,4-dinitrophenyl ethers of amidoximes by treatment of an ethanolic solution with potassium hydroxide (reaction 23). Nitrile oxides react with *N*-aryl-*S*,*S*-dimethylsulphimines to give 2*H*-1,2,4-benzoxadiazines (reaction 24), *e.g.* 3-*p*-tolyl-2*H*-1,2,4-benzoxadiazine (**140**; R = *p*-tolyl) and 7-chloro-3-ethoxycarbonyl-2*H*-1,2,4-benzoxadiazine, which is also made by oxidation of ethyl *p*-chloroanilinooximinoacetate with lead tetra-acetate. These benzoxadiazines are converted into benzoxazoles when they are heated at 80–140 °C.

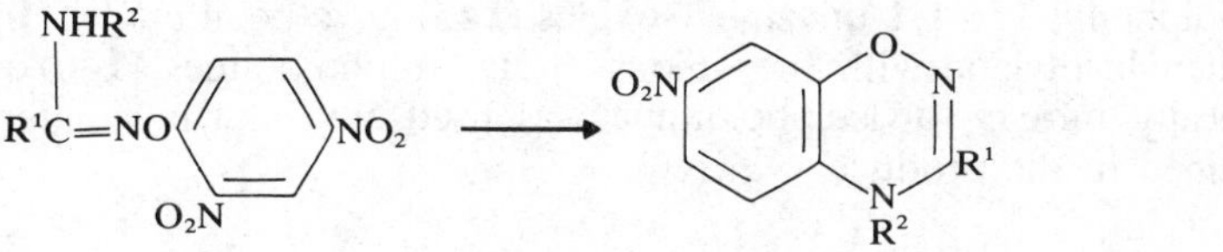

(23)

(139)

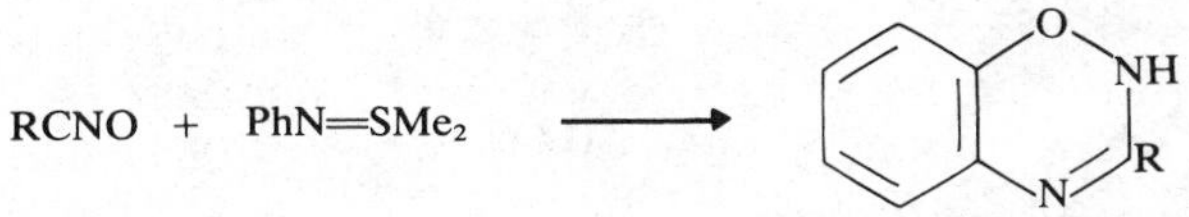

RCNO + PhN=SMe$_2$ ⟶ (24)

(140)

4*H*-1,3,4-Benzoxadiazines (**141**) are obtained from N^1-acetyl-N^2-aroyl-2,4-dihalogenophenylhydrazines by treatment with base. The reaction goes most readily with fluoro-derivatives. Another route to compounds of this series employs the reaction of aliphatic diazo-compounds with diazotized *o*-aminophenols. The initial reaction product is

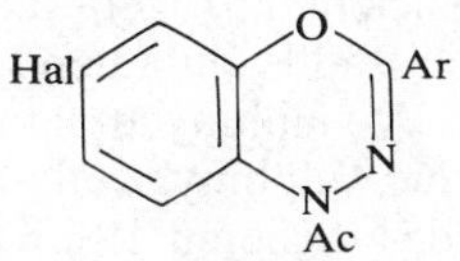

(141)

a 2*H*-1,3,4-benzoxadiazine (**142**), which may be trapped by addition of diphenylketen across the 3,4-double bond, but rearrangement to the 4*H*-1,3,4-benzoxadiazine (**143**) occurs under the influence of bases (reaction 25).

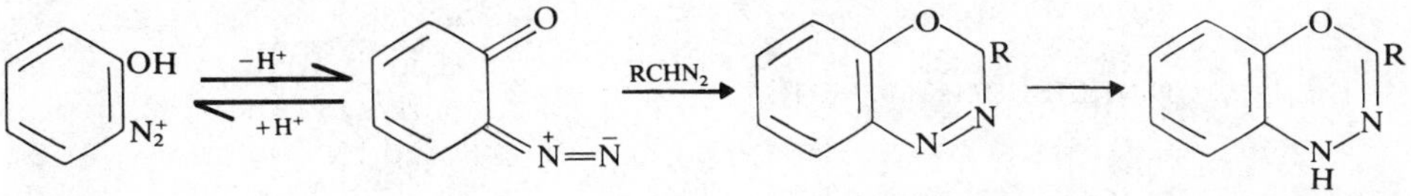

(25)

(142) (143)

20.2.3 COMPOUNDS CONTAINING SULPHUR AND NITROGEN

The principal types of compound with sulphur and nitrogen in a six-membered ring are the thiazines and their benzo- and dibenzo-derivatives with two heteroatoms, and the thiadiazines and their benzo-derivatives and dithiazines with three heteroatoms. A few series of compounds are known with more heteroatoms.

20.2.3.1 Thiazines[38,39]

The 1,2-, 1,3-, and 1,4-thiazines show many analogies with the oxazines, and the three fundamental ring systems are numbered similarly. The scope of their chemistry is broadened considerably by the ability of sulphur to exhibit higher valency than 2 and by the facility with which it is extruded from some cyclic structures.

1,2-Thiazines having two double bonds in the ring are known only in *S*-oxidized forms. The 1,1-dioxides (unsaturated sultams), *e.g.* (**144**; R = H, Me, Ph), are obtained by the action of ammonia or primary amines on 1,2-oxathiin 2,2-dioxides (reaction 26). When heated they lose sulphur dioxide to give good yields of pyrroles. Chiral 1-alkyl (or 1-aryl) 3,5-diphenyl-1*H*-1,2-thiazine 1-oxides (**145**) have been made[40] by reaction of 1-benzoyl-2-phenylacetylene with the anions of the sulphoximines (**146**) (reaction 27). When the optically pure *p*-tolylsulphoximine was used as a starting material, optical activity was retained in the product.

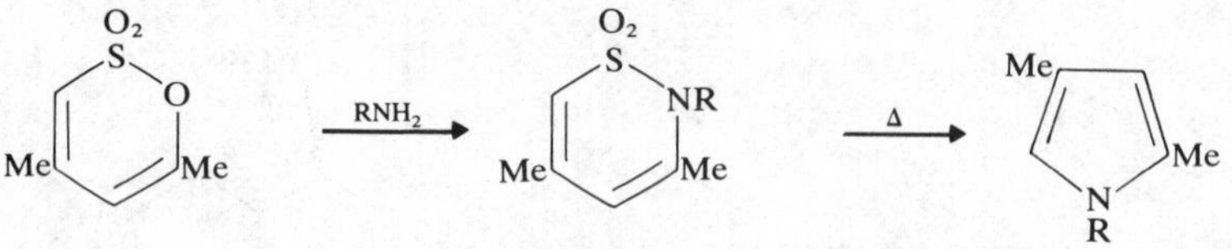

O_2 S O Me Me —RNH_2→ O_2 S NR Me Me —Δ→ Me Me N R (26)

(**144**)

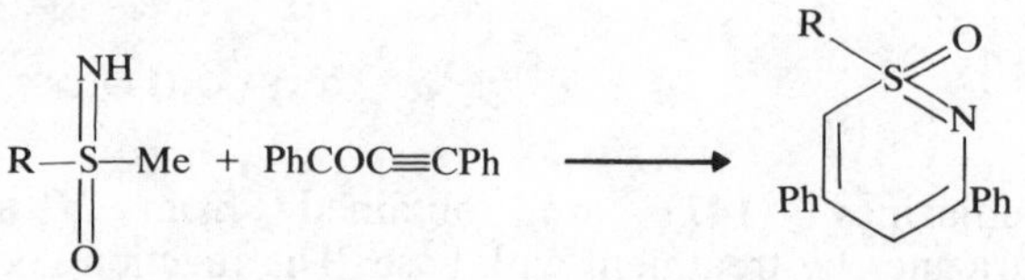

R—S(=NH)(=O)—Me + PhCOC≡CPh ——→ R S O N Ph Ph (27)

(**146**) R = Me, *p*-tolyl (**145**)

3,6-Dihydro-2*H*-1,2-thiazines are made by Diels–Alder cycloadditions of dienes with compounds containing electron-deficient nitrogen–sulphur double bonds. The transient intermediate *N*-thioaniline (PhN=S) adds butadiene to give 3,6-dihydro-2-phenyl-2*H*-1,2-thiazine. Other dienophiles are *N*-sulphinylarylamines, sulphinylsulphonamides (very reactive), sulphinylcarboxamides, and sulphinylurethanes, all of which give 2-substituted 3,6-dihydro-2*H*-1,2-thiazine 1-oxides (Scheme 19). These are mostly solids with melting

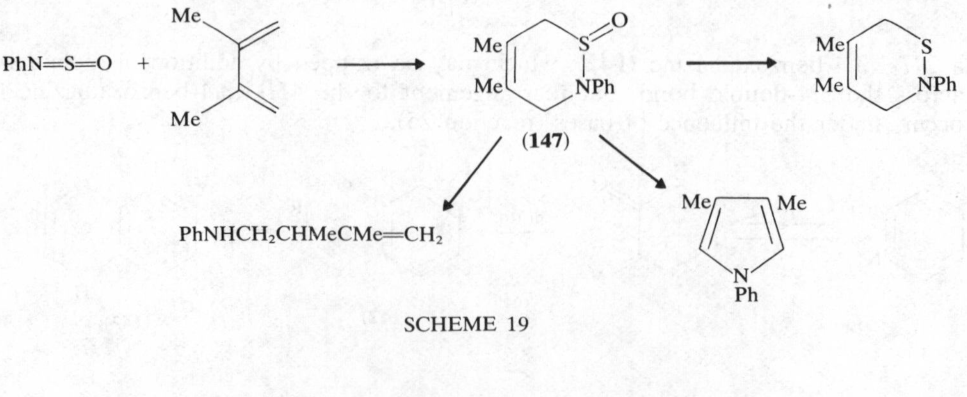

PhN=S=O + Me Me ——→ Me Me S O NPh (**147**) ——→ Me Me S NPh

PhNHCH$_2$CHMeCMe=CH$_2$ Me Me N Ph

SCHEME 19

points around 100 °C. Acid or alkaline hydrolysis cleaves the 1,2- and 1,6-bonds with elimination of sulphur dioxide. The 4,5-dimethyl-2-phenyl derivative (**147**), for example, gives an unsaturated alkylaniline on acid hydrolysis and 3,4-dimethyl-1-phenylpyrrole with alkali; reduction with lithium aluminium hydride gives 3,6-dihydro-4,5-dimethyl-2-phenyl-2*H*-1,2-thiazine. Since compounds unsubstituted on nitrogen cannot be obtained by hydrolysis of *N*-acyl derivatives, they have been made by addition of dienes to *N*-sulphinyl-2,2,2-trichloroethylurethane and elimination of the trichloroethoxycarbonyl group from the resulting thiazine by treatment with zinc dust. The diene addition has also been done with bis-*N*-sulphonyl sulphur di-imides ($RSO_2N{=}S{=}NSO_2R$) to give 3,6-dihydro-2-sulphonyl-1-sulphonimido-2*H*-1,2-thiazines.

Tetrahydro-1,2-thiazine 1,1-dioxides (δ-sultams) are made by reduction of the unsaturated 1,2-thiazine 1,1-dioxides or by reaction of ammonia or amines with 1,2-oxathian 2,2-dioxides (δ-sultones). Some of them have anticonvulsant properties, and the 2-*p*-sulphonamidophenyl derivative (Ospolot), made from sulphanilamide and 4-chlorobutanesulphonyl chloride, has been used in the treatment of epilepsy. Photochemical sulphochlorination of n-butylamine hydrochloride with chlorine and sulphur dioxide gives the 3- and 4-chlorosulphonyl derivatives, which are cyclized by alkali to 5-methyltetrahydro-1,2-thiazole 1,1-dioxide and tetrahydro-1,2-thiazine 1,1-dioxide.

1,3-Thiazines have been extensively studied. They are claimed to have utility as fungicides, pharmaceuticals, and vulcanizing agents, and the discovery of the cephalosporin antibiotics stimulated interest in the 6*H*-1,3-thiazine series. The 1,3-thiazine system is usually stable, and sulphur cannot be abstracted by lead salts, but electron-withdrawing groups make the ring susceptible to cleavage by nucleophilic attack. Compounds may be classified according to the position of the extra hydrogen atom in the parent structure.

2*H*-1,3-Thiazines are the least studied. Addition of thiourea to phenyl ethinyl ketone in acid solution gives a benzoylvinylisothiouronium salt which cyclizes to 2-imino-4-phenyl-2*H*-1,3-thiazine (**148**); this easily rearranges under basic conditions, giving 1,2-dihydro-4-phenylpyrimidine-2-thione (reaction 28). 5,6-Dihydro-2*H*-1,3-thiazines are obtained by the action of carbonyl compounds and ammonia on β-mercaptoketones,[41] *e.g.* the adduct from mesityl oxide and hydrogen sulphide reacts with acetone and ammonia to give 5,6-dihydro-2,2,4,6,6-pentamethyl-2*H*-1,3-thiazine (**149**), a liquid with an odour like camphor. The group in the 2-position in this compound may be exchanged for the residue of another carbonyl compound, *e.g.* heating with cyclohexanone and oxalic acid hydrate gives the spiro-compound (**150**) (reaction 29). Reduction with lithium aluminium hydride cleaves

PhCOCH=CHS—C(=NH·HCl)—NH₂ ⟶ (**148**) ⟶ (28)

Me₂C(SH)—CH₂COMe ⟶ (**149**) —cyclohexanone, H⁺⟶ (**150**) (29)

the ring to give a γ-mercaptoamine. Some 2-alkylimino- or 2-arylimino-3,4-dihydro-4,4,6-trimethyl-2*H*-1,3-thiazines have been obtained by rearrangement of 1-alkyl (or 1-aryl) 1,2,3,4-tetrahydro-4,4,6-trimethylpyrimidine-2-thiones in conc. hydrochloric acid.

Heating 2-acetamido-2-methylpentan-4-one with phosphorus pentasulphide affords 2,4,4,6-tetramethyl-4*H*-1,3-thiazine, the methiodide of which has been described as an intermediate for cyanine dyestuffs. Addition of carbon disulphide to 2-amino-2-methylpentan-4-one gives a dithiocarbamic acid, which may be dehydrated to give 2-mercapto-4,4,6-trimethyl-4*H*-1,3-thiazine. Condensation of thioamides and aldehydes with phenylacetylenes, catalysed by boron trifluoride, gives 4*H*-1,3-thiazines and small amounts of the isomeric 6*H*-1,3-thiazines (reaction 30), *e.g.* thiobenzamide, benzaldehyde, and phenylacetylene give 2,4,6-triphenyl-4*H*-1,3-thiazine (**151**; $R^1 = R^2 = Ph$, $R^3 = H$) and 2,4,5-triphenyl-4*H*-1,3-thiazine (**152**; $R^1 = R^2 = Ph$, $R^3 = H$). The latter compound is isomerized to 2,4,5-triphenyl-6*H*-1,3-thiazine by treatment with triethylamine in ether. Most other 4*H*-1,3-thiazines are 4-oxo-derivatives. 2-Chloro-4-oxo-4*H*-1,3-thiazine is made by treating β-thiocyanatoacryloyl chloride with hydrogen chloride, and 2-ethylthio-5-methyl-4-oxo-4*H*-1,3-thiazine is made from ethyl dithiocarbamate and 2-methoxy-1-methylacryloyl chloride. Carbon disulphide condenses with cyanoacetamide to give the ethylene (**153**), the mono-*S*-methyl derivative of which condenses with aromatic carboxylic acids in polyphosphoric ester to give good yields of 2-aryl-5-carbamoyl-6-methylthio-4*H*-1,3-thiazin-4-ones (**154**) (reaction 31); aliphatic acids

$R^1CSNH_2 + R^2CHO + PhC{\equiv}CR^3 \longrightarrow$ (**151**) + (**152**) (30)

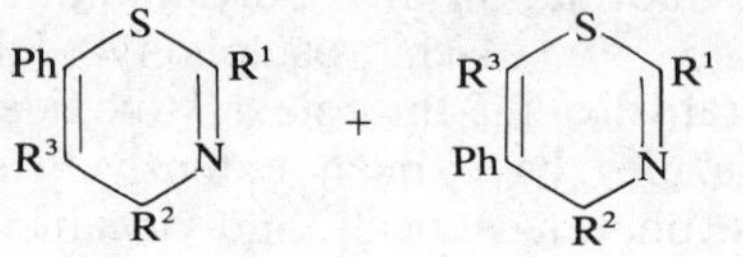

(**153**) $(HS)_2C{=}C(CN)CONH_2 \longrightarrow (MeS)(HS)C{=}C(CN)CONH_2 \longrightarrow$ (**154**) (31)

give small yields of analogous thiazine 1-oxides. Ethyl *N*-cyanoacetyldithiocarbamate reacts with ethyl orthoformate to give 5-cyano-2-ethylthio-4-oxo-4*H*-1,3-thiazine (**155**), which can undergo cleavage by amines through attack at the 6-position. The acyclic compounds thus generated are cyclized to 2-thiouracils by sodium hydroxide (Scheme 20).

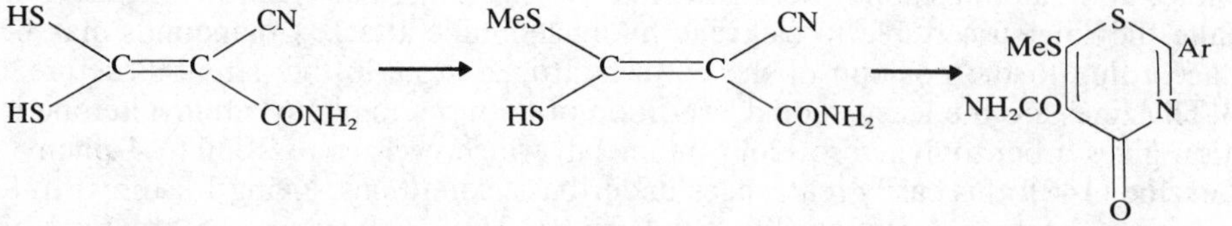

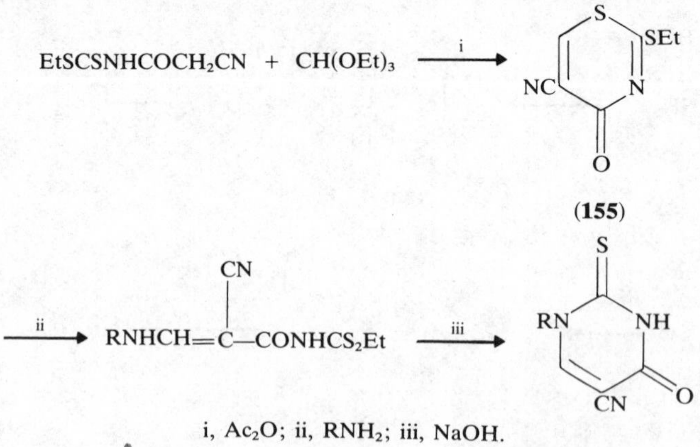

$EtSCSNHCOCH_2CN + CH(OEt)_3 \xrightarrow{i}$ (**155**)

$\xrightarrow{ii} RNHCH{=}C(CN){-}CONHCS_2Et \xrightarrow{iii}$

i, Ac_2O; ii, RNH_2; iii, NaOH.

SCHEME 20

5,6-Dihydro-4*H*-1,3-thiazines, 'penthiazolines', are made by reaction of thioamides with trimethylene dihalides or γ-halogenoamines, of nitriles with γ-mercaptoamines, or of γ-acylaminopropanols with phosphorus pentasulphide. The simpler compounds are basic oils or low-melting solids. The 2-(2-thienylethyl) derivative is a powerful anthelmintic.

The Ritter reaction with a nitrile, 3-hydroxy-3-methylbutane-1-thiol and sulphuric acid gives 2-substituted 5,6-dihydro-4,4-dimethyl-4*H*-1,3-thiazines. Condensation of vinyl ketones with thioamides under neutral, basic, or acid conditions gives 4-hydroxy-derivatives, *e.g.* 5,6-dihydro-4-hydroxy-4-methyl-2-phenyl-4*H*-1,3-thiazine (**156**) from thiobenzamide and methyl vinyl ketone. Attempts to dehydrate (**156**) were unsuccessful. 2-Amino-5,6-dihydro-4*H*-1,3-thiazines are obtained by treating γ-hydroxy- or γ-halogeno-propylamines with isothiocyanates, or by cyclization of γ-amino-propylisothiouronium bromides (this occurs spontaneously in an acetate buffer). Treatment of N^1-t-butyl-N^2-3-hydroxypropylthiourea with conc. hydrochloric acid gives 2-amino-5,6-dihydro-4*H*-1,3-thiazine, which reacts as the imino-tautomer (**157**); acylation gives *N,N′*-diacyl derivatives from which the endocyclic acyl group is readily removed by hydrolysis to give, for example, the 2-acetamido-derivative. Alkylation with alkyl halides gives 3-alkyl-2-iminotetrahydro-1,3-thiazines, and with bifunctional reagents bicyclic compounds are formed by reaction with both nitrogen atoms, *e.g.* ethyl malonate gives 2,3,7,8-tetrahydro-4*H*,6*H*-pyrimido[2,1-*b*]-1,3-thiazine-6,8-dione (**158**). 5,6-Dihydro-2-(2,6-dimethylanilino)-4*H*-1,3-thiazine (xylazine, Rompun) is used as a veterinary sedative, analgesic, and muscle relaxant. 5,6-Dihydro-2-mercapto-4*H*-1,3-thiazine (or the thione tautomer), which is made from 3-mercaptopropylamine and carbon disulphide, is a fungicide.

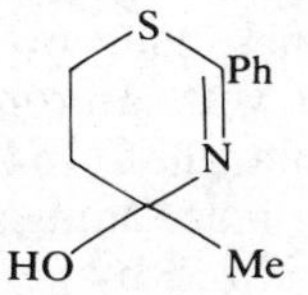

S Ph N HO Me

(156)

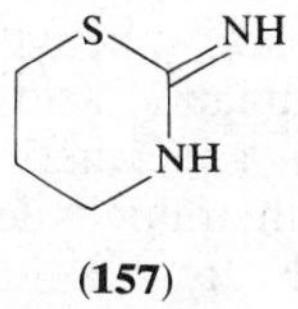

S NH NH

(157)

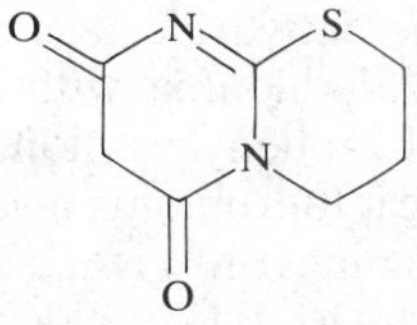

O N S N O

(158)

6*H*-1,3-Thiazines are made by condensation of thioamides with α,β-unsaturated ketones in the presence of perchloric acid, *e.g.* thiobenzamide and phenyl styryl ketone give 2,4,6-triphenyl-6*H*-1,3-thiazine perchlorate. 6*H*-1,3-Thiazine-4-carboxylic esters (**159**) have been made by base-catalysed condensation of 1-chloropropan-2-ones with diethyl α-thioformamidophosphonoacetates (reaction 32) and have been used in the total synthesis

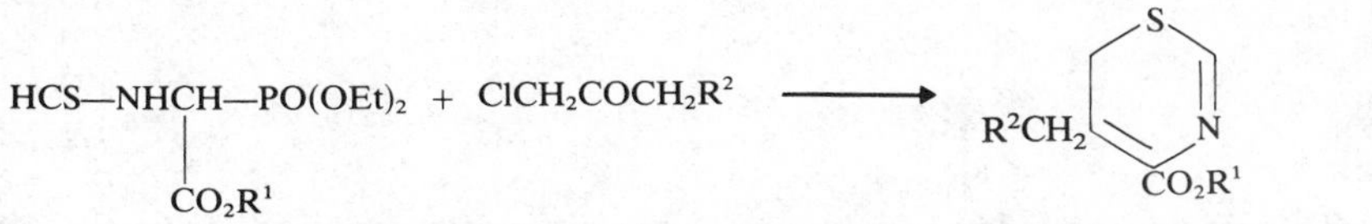

$$\text{HCS—NHCH(CO}_2\text{R}^1\text{)—PO(OEt)}_2 + \text{ClCH}_2\text{COCH}_2\text{R}^2 \longrightarrow \text{(159)} \quad (32)$$

S N R^2CH_2 CO_2R^1

(159)

of cephalosporin analogues. In most cephalosporin syntheses, however, the thiazine ring has been built up on a β-lactam structure.[42] Condensation of thiourea with α,β-unsaturated ketones in hydrochloric acid gives 2-amino-6*H*-1,3-thiazines (or their imino-tautomers), *e.g.* mesityl oxide gives 2-amino-4,6,6-trimethyl-6*H*-1,3-thiazine. Reaction of 1-ethoxymethyl-2-methoxyacrylonitrile with thiourea or phenylthiourea gives 5-cyano-2,3-dihydro-2-oxo-6*H*-1,3-thiazine, with loss of ammonia or aniline. Thioamides condense with malonyl chloride, carbon suboxide, or with substituted malonic acids and phosphoryl chloride to give high yields of 4-hydroxy-6-oxo-6*H*-1,3-thiazine derivatives. These are readily substituted in the 5-position by electrophilic reagents such as formaldehyde or diazonium salts. 4-Amino-6-oxo-2-phenyl-6*H*-1,3-thiazine is obtained in poor yield from thiobenzamide, cyanoacetic acid, and phosphoryl chloride. Condensation of thioamides with 2,2-dichlorovinyl ketones gives 6-thioxo-6*H*-1,3-thiazines (**160**; X = S) (reaction 33), which are alkylated to give 6-alkylthiothiazinium salts (**161**; Y = SR) or are converted into the corresponding 6-oxo-compounds (**160**; X = O) which react with oxalyl chloride to give 6-chloro-1,3-thiazinium salts (**161**; Y = Cl).

$$R^1COCR^2{=}CCl_2 + 2R^3CSNH_2 \longrightarrow \textbf{(160)} \longrightarrow \textbf{(161)}\ Cl^- \qquad (33)$$

(160) **(161)**

2,3-Dihydro-4-oxo-2-thioxo-4*H*-1,3-thiazines (**162**; X = S) are obtained by addition of dithiocarbamic acids to propiolic acid.[43] Methylation of (**162**; X = S, R = H) gives 2-methylthio-4-oxo-4*H*-1,3-thiazine (**163**), which is hydrolysed by acid to the dioxo-compound (**162**; X = O, R = H). This reacts with ammonia to give uracil. Sulphurization of (**162**; X = S, R = H) with phosphorus pentasulphide in xylene gives the corresponding 2,4-dithioxo-1,3-thiazine, but in pyridine the reaction gives 1,2-dithiol-3-thione. Nucleophiles (alkali, hydroxylamine, hydrazine) cleave the 3,4-bond of 5,6-dihydro-4-oxo-2-thioxo-compounds to give derivatives of carboxyethyldithiocarbamates. Thioureas condense with acetylenic esters or nitriles to give 1,3-thiazines, *e.g.* 2,3-dihydro-4-imino-3-methyl-2-methylimino-4*H*-1,3-thiazine from cyanoacetylene and *N,N′*-dimethylthiourea, but in many instances the primary adducts do not cyclize but decompose with transfer of sulphur to the acetylene moiety. Derivatives of 2,4-dioxotetrahydro-1,3-thiazine have been made by reaction of sodium thiocyanate, thiocarbamates, dithiocarbamates, or thioureas with propiolactone or β-halogenopropionic acids, cyclization being effected by thionyl chloride, acetic anhydride, *etc.* 5-Ethyl-2,4-dioxo-6-phenyltetrahydro-1,3-thiazine, made by reaction of thiourea with *trans*-α-ethylcinnamic acid in sulphuric acid or with 2-bromo-1-ethyl-2-phenylpropionic acid, is an anaesthetic and sedative, Dolitrone (**164**). 2-Mercaptopropionic acid condenses with araldehydes and ammonia or primary amines (not arylamines) to give 2-aryl-4-oxotetrahydro-1,3-thiazines, which are oxidized by permanganate to 1,1-dioxides. These compounds are muscle relaxants and depressants of the central nervous system. 2-*p*-Chlorophenyl-3-methyl-4-oxotetrahydro-1,3-thiazine 1,1-dioxide (**165**) is the tranquilizer chlormazanone.

(162) **(163)** **(164)** **(165)**

1,3-Thiazinium salts (**166**) analogous to pyrylium salts are made from the corresponding 1,3-oxazinium salts (**42**) by addition of hydrogen sulphide to give β-acylaminovinyl thioketones and dehydration of these with perchloric acid (reaction 34). The products, *e.g.* (**166**; $R^1 = R^2 = R^3 = Ph$), react readily with oxygen, sulphur, or carbon nucleophiles with cleavage of the 5,6-bond to give thioacyl derivatives of vinylamines. Reduction with

$$\textbf{(42)} \longrightarrow R^3CS{-}CH{=}C(R^2){-}NHCOR^1 \longrightarrow \textbf{(166)}\ ClO_4^- \qquad (34)$$

(42) **(166)**

sodium borohydride affords 2*H*-, 4*H*-, and 6*H*-1,3-thiazines, which may be converted into 1,3-thiazine anions (**167**) by proton abstraction by butyl-lithium (reaction 35). The anions are energy-rich 8π-electron systems which readily extrude sulphur to form pyrrole anions.[44]

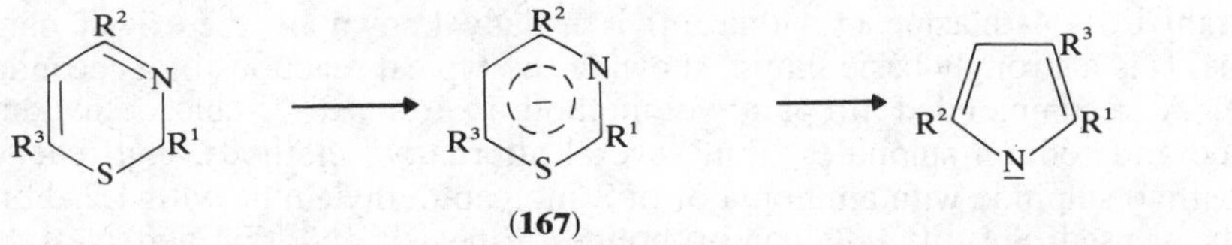

(35)

(167)

$2H$-1,4-Thiazine (**168**) is made in poor yield by reduction of thiodiglycollimide (3,5-dioxotetrahydro-1,4-thiazine) with aluminium powder at 450 °C. It is a weakly basic oil and is not attacked by the usual acylating agents. Aryl derivatives, *e.g.* 3,5-diphenyl-$2H$-1,4-thiazine (**169**; Ar = Ph), are obtained by reaction of diphenacyl sulphides with ammonia. They undergo oxidative coupling to 2,2′-dithiazinyl derivatives (**170**) when treated with aromatic nitro-compounds or quinones. Catalytic reduction converts them into 2-methylthiazolines, probably by cleavage of the 1–2 bond to a methyl and a thiol group, the latter then adding to the imine double bond. Reaction of diphenacyl sulphones with ammonia or urea in acetic acid gives 3,5-diaryl-$4H$-thiazine 1,1-dioxides (**171**). 3-Methyl-5-phenyl-$4H$-1,4-thiazine 1,1-dioxide, made in a similar manner, extrudes sulphur dioxide on pyrolysis, giving 2-methyl-5-phenylpyrrole. $4H$-1,4-Thiazine 1,1-dioxide is prepared by reaction of ammonium chloride with 2,6-diethoxy-1,4-oxathian 4,4-dioxide. Treatment of thiodiglycollic dinitrile with hydrogen bromide and then with acetic anhydride gives 3-amino-5-bromo-$2H$-1,4-thiazine hydrobromide (**172**).

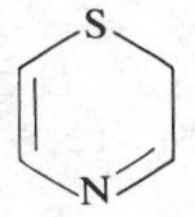

(**168**) 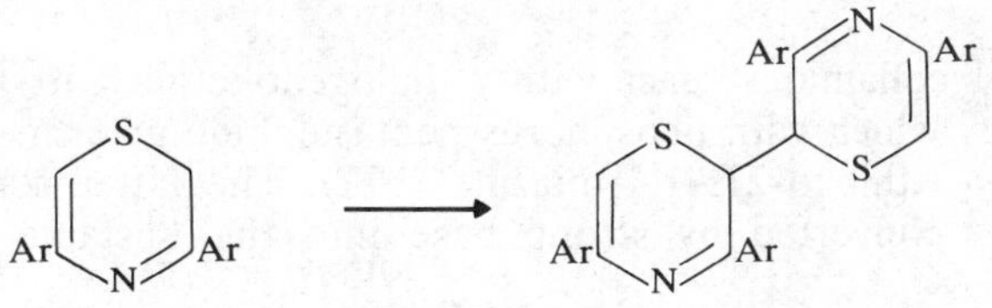 (**169**) (**170**)

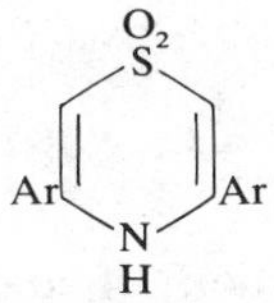

(**171**) 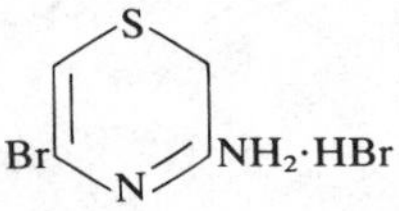 (**172**)

5,6-Dihydro-$4H$-1,4-thiazines are obtained by reaction of aziridine with α-mercaptoketones or with aldehydes or ketones and sulphur, *e.g.* reaction (36). In the latter case, mixtures of $4H$- and $2H$-1,4-thiazines may be obtained, *e.g.* 2-methylbutan-3-one gives 5,6-dihydro-3-isopropyl-$4H$-1,4-thiazine (**173**) and 5,6-dihydro-2,2,3-trimethyl-$2H$-1,4-thiazine (**174**). The 5,6-dihydro-$4H$-1,4-thiazines are basic and they can be acylated. Because of their enamine character they are unstable. They are reduced by formic acid to give N-formyltetrahydro-1,4-thiazines, and treatment with hydrogen sulphide converts them into thiazolines (**175**), from which they are regenerated by oxidation with sulphur (reaction 37).[45]

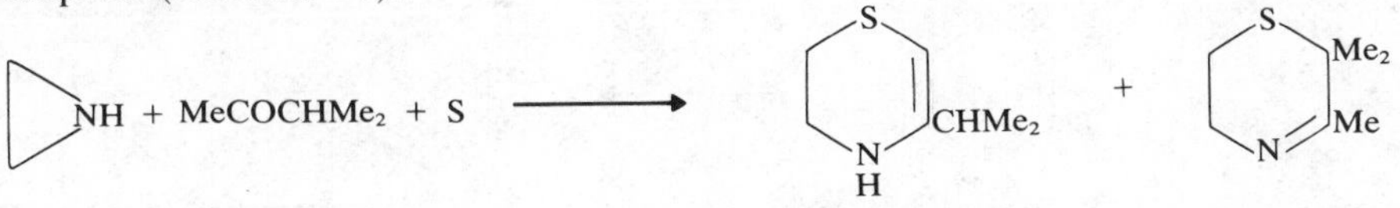

NH + MeCOCHMe$_2$ + S → (36)

(**173**) (**174**)

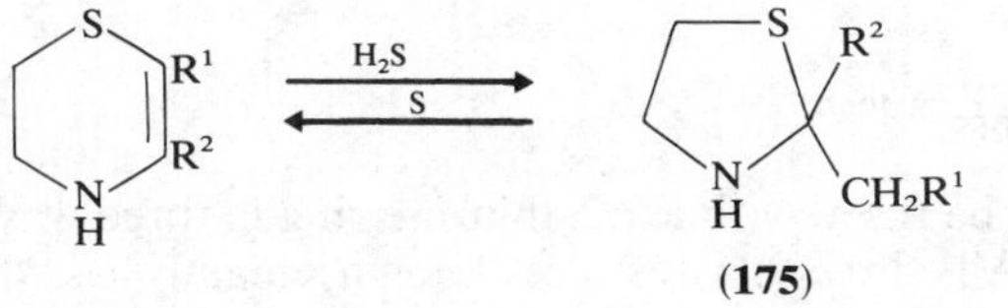

(37)

(**175**)

Tetrahydro-1,4-thiazine (1,4-thiazan) is usually known by the trivial name thiomorpholine. It is a strongly basic liquid, showing the typical reactions of a secondary aliphatic amine. A recommended preparative method is from di-(2-chloroethyl)amine hydrochloride and sodium sulphide,[46] but several alternative methods, *e.g.* reaction of di-(2-chloroethyl) sulphide with ammonia or of 2-mercaptoethylamine with 1,2-dibromoethane, have been used. Substituted thiomorpholines, especially *N*-alkyl and *N*-aryl derivatives, are made similarly. Di-(2-chloroethyl) sulphoxide and sulphone react with ammonia or primary amines to give thiomorpholine 1-oxides and 1,1-dioxides. The latter compounds are also made by the addition of amines, *e.g.* methylamine or aniline, to divinyl sulphone, and thiomorpholine 1,1-dioxide is obtained by heating di-(2-hydroxyethyl) sulphone or 1,4-oxathian 4,4-dioxide with aqueous ammonia at 150 °C. The amino-acid 3-carboxytetrahydro-1,4-thiazine 1-oxide (chondrine) (**176**; R = H) has been isolated from certain seaweeds, and its 5-methyl derivative (cycloalliin) (**176**; R = Me) is found in onions.

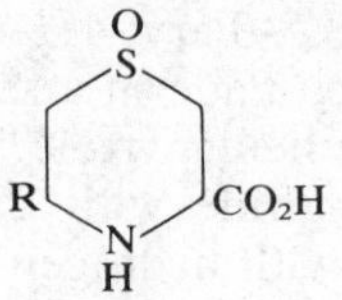

O S R CO_2H N H

(**176**)

Thioglycollamides react with α-halogenoketones to give 3,4-dihydro-3-oxo-2*H*-1,4-thiazines, which with phosphorus pentasulphide give the 3-thiones, *e.g.* 3,4-dihydro-4,5-dimethyl-3-thioxo-2*H*-1,4-thiazine (**177**). The latter compound gives an *S*-methiodide which is converted by strong base into the unstable thiazine (**178**) (reaction 38).[47]

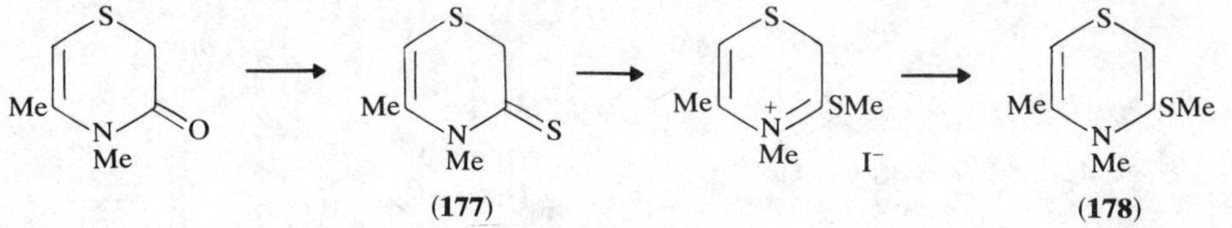

S Me N Me O → S Me N Me S → S Me $\overset{+}{N}$ Me SMe I^- → S Me N Me SMe (38)

(**177**) (**178**)

3-Oxothiomorpholines are lactams made by reaction of α-halogenoesters with 2-mercaptoethylamine or of aziridine with thioglycollic esters. Oxidation with hydrogen peroxide in acetic acid converts them into 1-oxides and 1,1-dioxides. The latter compounds have a reactive methylene group in the 2-position and will condense with aldehydes and ketones. Alkylation of 3-oxothiomorpholine with triethyloxonium fluoroborate gives 3-ethoxy-5,6-dihydro-2*H*-1,4-thiazine.

Benzonitrile reacts with dimethylsulphoxonium methylide to give a small yield of a thiazine for which the ylide structure (**179**) is proposed. It is unaffected by peracetic acid, and this excludes the isomeric 4-methyl-1,4-thiazine 1-oxide structure that would also agree with the spectroscopic data.

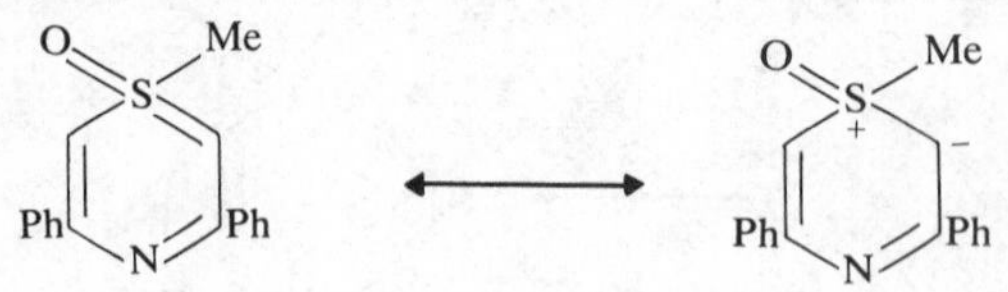

O Me S Ph N Ph ⟷ O Me $\overset{+}{S}$ $^-$ Ph N Ph

(**179**)

20.2.3.2 Benzothiazines[38,39]

A benzene ring can be fused with a 1,2-thiazine ring in three ways, giving 1,2-, 2,1-, and 2,3-benzothiazines. All three classes are known, mainly as their *S,S*-dioxides. The

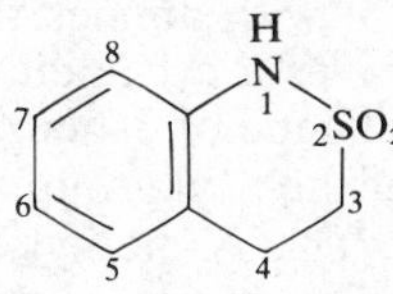

(180) 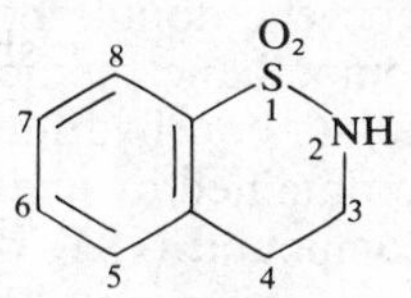 (181)

dihydro-compounds 3,4-dihydro-1*H*-2,1-benzothiazine 2,2-dioxide (**180**) and 3,4-dihydro-2*H*-1,2-benzothiazine 1,1-dioxide (**181**) are cyclic sulphonamides made by dehydrating *o*-aminophenylethane sulphonic acid and *o*-sulphophenylethylamine. They may be *N*-alkylated and undergo electrophilic substitution, *e.g.* nitration, in the benzene nucleus. The 7-sulphonamido-derivative of (**181**) is obtained by treating β-chloroethylbenzene-2,4-disulphonyl chloride with ammonia.

3-Acyl-3,4-dihydro-4-oxo-2*H*-1,2-benzothiazine 1,1-dioxide derivatives (**182**) are made by *N*-alkylating saccharin with an α-bromoketone and treating the product with a base. The reactions of these compounds are shown in Scheme 21. They have a strong

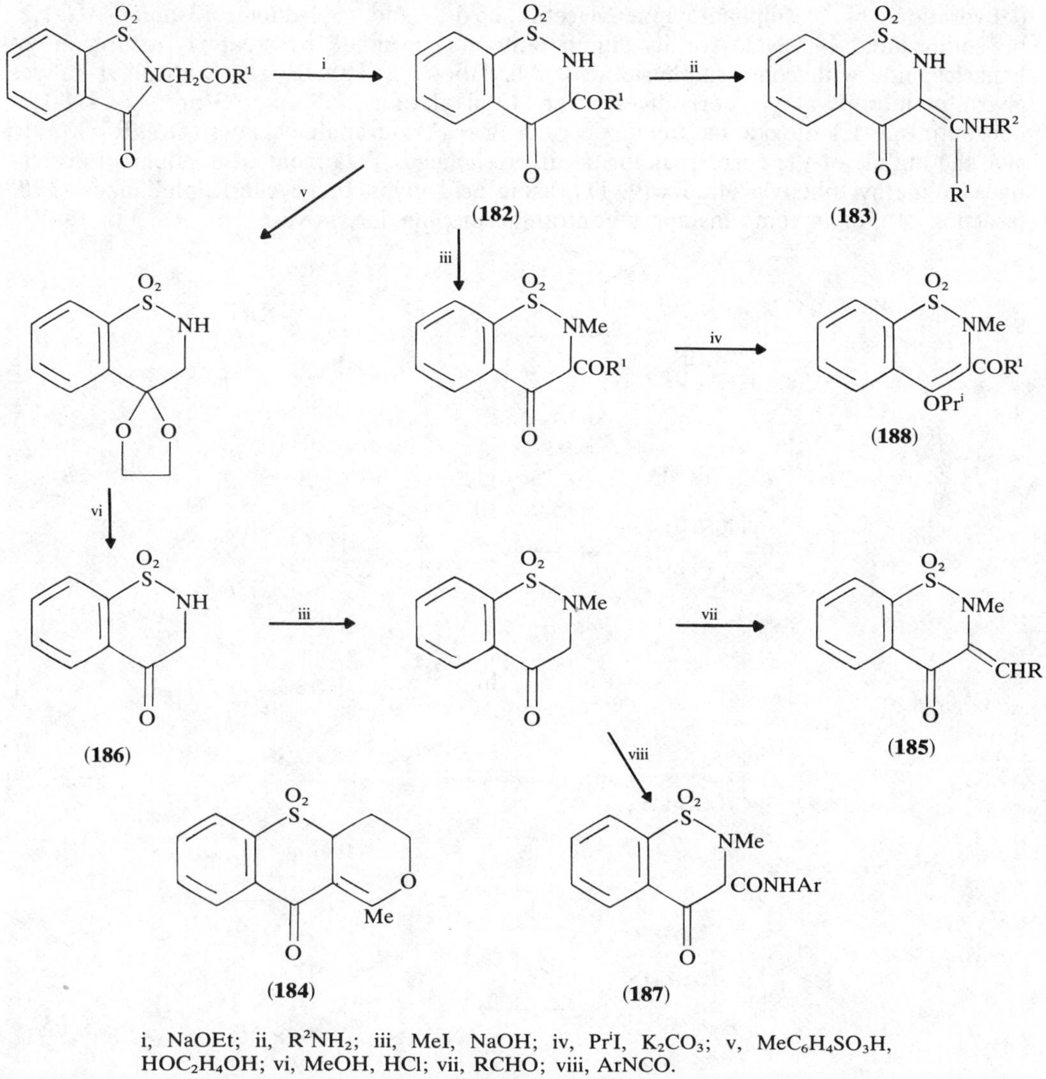

i, NaOEt; ii, R^2NH_2; iii, MeI, NaOH; iv, Pr^iI, K_2CO_3; v, $MeC_6H_4SO_3H$, HOC_2H_4OH; vi, MeOH, HCl; vii, RCHO; viii, ArNCO.

SCHEME 21

tendency to form exocyclic double bonds at position-3: with amines they give enamines (**183**), with 1,2-dibromoethane the acetyl compound gives a cyclic enol ether (**184**), and on reduction with sodium borohydride they give 3-alkylidene or 3-benzylidene derivatives (**185**). These are also obtained by hydrolytic removal of the acyl group and reaction of the 3,4-dihydro-4-oxo-compound (**186**) with an aldehyde. The *N*-methyl derivative of (**186**) reacts with aryl isocyanates to give arylamides (**187**). These compounds, particularly the 2-thiazolyl compound (Sudoxicam), have strong anti-inflammatory activity. They are more readily made by ring-enlargement of *N*-methoxycarbonylmethylsaccharin to give 3,4-dihydro-3-methoxycarbonyl-4-oxo-2*H*-1,2-benzothiazine 1,1-dioxide, which is *N*-methylated and then condensed with the appropriate arylamine. Methylation of (**182**) gives only the 2-methyl derivative, but this will give an enol acetate and the isopropyl enol ether (**188**). It is probable that the enolizable compounds of the foregoing series exist as the 4-hydroxy-tautomers since they give colour reactions with iron(III) chloride.[48]

Decomposition of *o*-diazoacetylbenzenesulphonamides in formic acid also gives 3,4-dihydro-4-oxo-2*H*-1,2-benzothiazine 1,1-dioxides, but thermal decomposition in boiling chlorobenzene gives both these and the 3-oxo-isomers (**189**) (reaction 39), and photochemical decomposition gives only the 3-oxo-compounds. These are also made by dehydration of *o*-sulphamoylphenylacetic acids, and 3,4-dihydro-3-imino-2*H*-1,2-benzothiazine 1,1-dioxide (or its amino tautomer) is made by treating *o*-sulphamoylbenzylcyanide with conc. sulphuric acid. Alkylation of (**189**; R = H) with alkyl halides gives mainly *N*-alkyl derivatives, but *O*-alkylation to give 3-propoxy-4*H*-1,2-benzothiazine 1,1-dioxide has been reported. Phenyl isocyanate acylates (**189**; R = Me) to give the anilide of the corresponding 4-carboxylic acid. Treatment of *o*-sulphinyl derivatives of methyl phenylacetate with hydrazoic acid gives the cyclic sulphilimines (**190**) (reaction 40), or in some instances benzothiadiazepine derivatives.

SO_2NHR, $COCHN_2$ → ; SO_2NHR, CH_2CO_2H → (**189**) (39)

SOR, CH_2CO_2Me + HN_3 → (**190**) (40)

$NR^2SO_2CH_2R^1$, COR^2 → (**191**) (41)

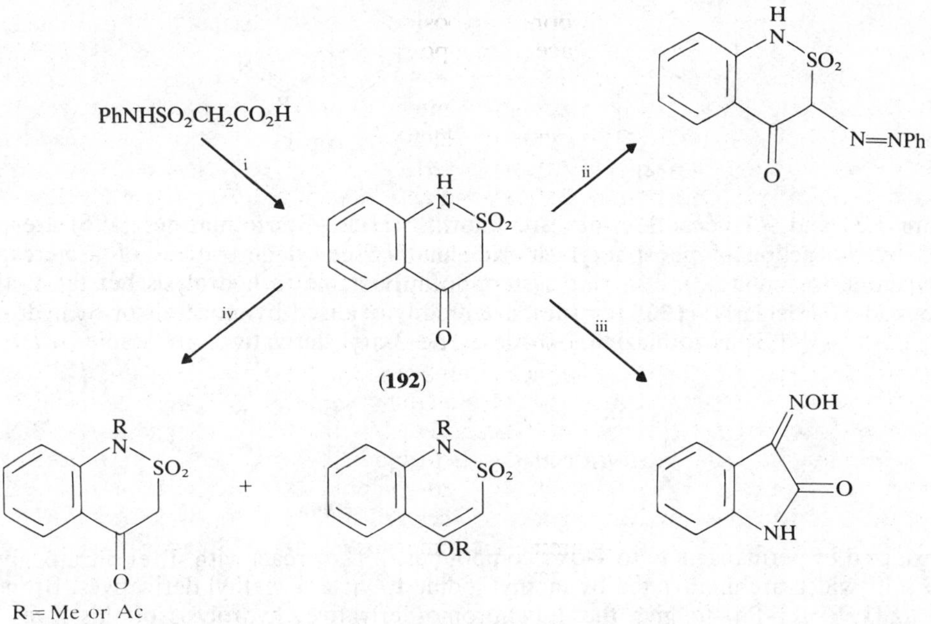

$PhNHSO_2CH_2CO_2H$

(192)

R = Me or Ac

i, Polyphosphoric acid; ii, PhN_2^+; iii, $C_5H_{11}ONO$; iv, Me_2SO_4 or Ac_2O.

SCHEME 22

1*H*-2,1-Benzothiazine 2,2-dioxide derivatives **(191)** are made by the base-catalysed cyclization of *o*-methylsulphonamido- or *o*-benzylsulphonamido-phenyl ketones (reaction 41). Phenylsulphamoylacetic acid cyclizes in polyphosphoric acid to give 3,4-dihydro-4-oxo-1*H*-2,1-benzothiazine 2,2-dioxide **(192)**, the reactions of which are shown in Scheme 22.[49] It fails to react with hydroxylamine, benzaldehyde, phosphoryl chloride, or primary or secondary amines, but is very reactive in other respects. Bamford–Stevens reduction, or borohydride reduction, converts it into 1*H*-2,1-benzothiazine 2,2-dioxide (sulphostyril), from which the *N*-methyl derivative has been made. The 3,4-benzo-derivative, the sultam of 2-aminobiphenyl-2-sulphonic acid, is both a 2,1- and a 1,2-benzothiazine dioxide; its polychloro-derivatives have been claimed as pesticides. Methyl *N*-methyl-*N*-methanesulphonylanthranilate is cyclized by sodium hydride in high yield to 3,4-dihydro-1-methyl-4-oxo-1*H*-2,1-benzothiazine 2,2-dioxide **(193)**, which is acylated by aryl isocyanates to give arylamides of 3-carboxy-4-hydroxy-1-methyl-1*H*-2,1-benzothiazine 2,2-dioxide (reaction 42). These are fairly strong enolic acids.

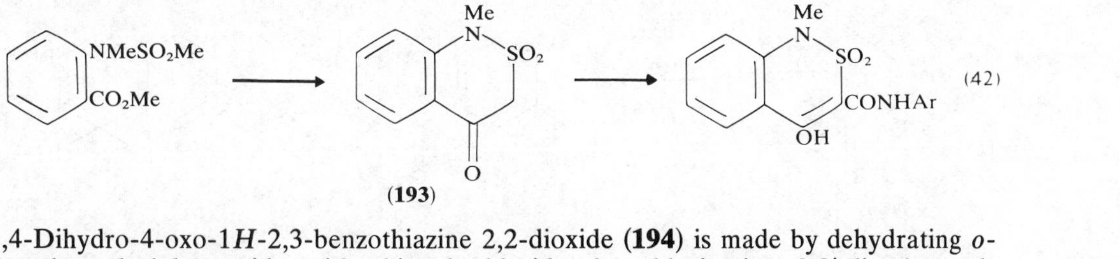

(42)

(193)

3,4-Dihydro-4-oxo-1*H*-2,3-benzothiazine 2,2-dioxide **(194)** is made by dehydrating *o*-carboxybenzylsulphonamide with thionyl chloride, by chlorinating 2,2′-dicarbamoyldibenzyl disulphide, or by hydrolysis of the 4-imino-derivative. The latter is obtained by treating *o*-cyanobenzylsulphonamide with sulphuric acid. Alkylating agents usually convert **(194)** into *N*-alkyl derivatives (*e.g.* *N*-methyl and *N*-ethyl), but *O*-alkylation (*e.g.* to the isopropoxy-compound) has been reported. It may be nitrated, substitution occurring in the benzene nucleus. Reaction with phosphoryl chloride gives 4-chloro-1*H*-2,3-benzothiazine 2,2-dioxide (**195**; R = Cl), from which the 4-alkoxy-compounds (**195**; R = *O*-alkyl) are obtained by nucleophilic substitution.

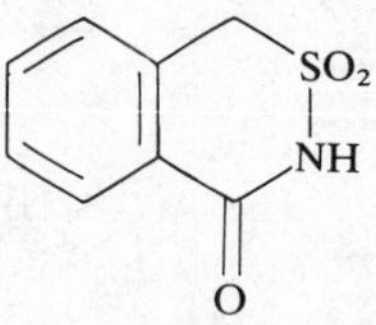

(194)

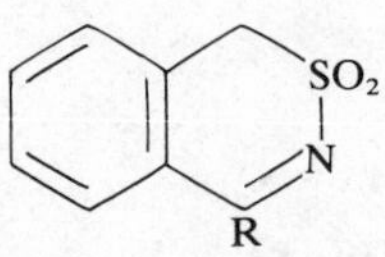

(195)

Both 1,3- and 3,1-benzothiazines are known. 4*H*-1,3-Benzothiazines (**196**) are prepared by the action of phosphoryl chloride on *N,S*-diacyl derivatives of 2-mercaptobenzylamine (reaction 43). The ring system is fairly stable to hydrolysis but the 2-alkyl compounds, particularly (**196**; R = Me), are readily oxidized by moist air or by hydrogen peroxide to 4*H*-1,3-benzothiazine 1-oxides. The 2-aryl derivatives are stable to air but

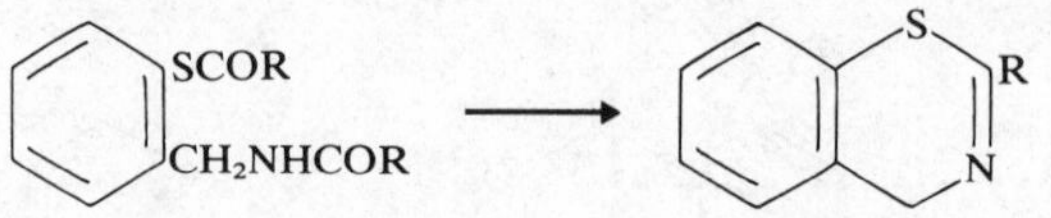

(43)

(196)

are oxidized by permanganate to 4-oxo-compounds.[50] They react with silver nitrate, giving silver salts which are methylated by methyl iodide to yield 4-methyl derivatives. Bromine adds to (**196**; R = Ph) to give the 1,1-dibromo-derivative; hydrolysis of this gives the 1-oxide. 3,4-Dialkoxythiophenols condense with *N*-hydroxymethylbenzamides to give *N*-dialkoxyphenylthiomethylbenzamides; these are rearranged and dehydrated by phosphoryl chloride, giving 6,7-dialkoxy-2-aryl-4*H*-1,3-benzothiazines. Some 2-substituted 4,4-diethyl-4*H*-1,3-benzothiazines have been made from 3-σ-mercaptophenylpentan-3-ol and nitriles by the Ritter reaction.

4-Oxo-4*H*-1,3-benzothiazines are obtained from *o*-mercaptobenzamide by acylation and ring-closure with hydrogen chloride, or from *o*-mercaptobenzoic acid and nitriles by the action of hydrochloric acid. Examples are 2-phenyl-4*H*-1,3-benzothiazin-4-one and 2-carboxy-4*H*-1,3-benzothiazin-4-one. The latter compound was said to be readily decarboxylated to give 4*H*-1,3-benzothiazin-4-one, m.p. 138 °C, but another report describes the unsubstituted compound as an unstable and transient intermediate which may be trapped as an adduct with a diene and is obtained by thermal decomposition of 2-ethoxy-2,3-dihydro-4*H*-1,3-benzothiazin-4-one (made from *o*-mercaptobenzamide and ethyl orthoformate). 2-Aryl-4*H*-1,3-benzothiazin-4-ones (**197**; R = aryl) react with Grignard reagents both at the 2- and 4-positions, giving 2-disubstituted 2,3-dihydro-4*H*-1,3-benzothiazin-4-ones and 4-substituted 2-aryl-4-hydroxy-4*H*-1,3-benzothiazines (reaction 44).

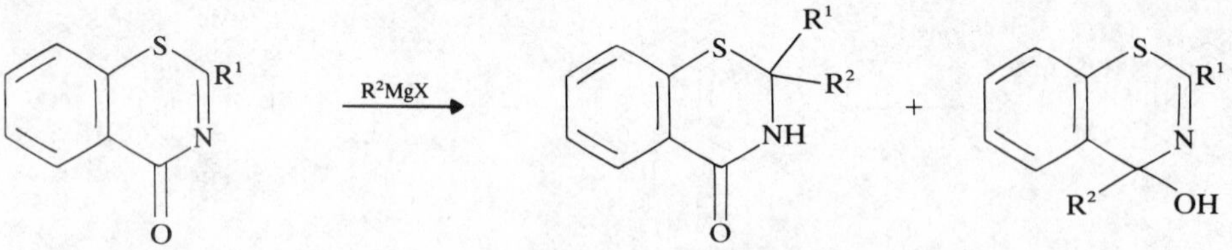

(44)

(197)

Condensation of phenyl cyanate with methyl *o*-mercaptobenzoate gives the 2-phenoxy-compound (**197**; R = OPh). 2-Chloro-4*H*-1,3-benzothiazin-4-one is obtained by treatment of *o*-thiocyanatobenzoic acid with phosphorus pentachloride, and the selenium analogue is made similarly. The chlorine atom may be replaced by nucleophiles such as amines and carbanions and the products tautomerize, if possible, to derivatives of 2,3-dihydro-4-oxo-4*H*-1,3-benzothiazine, *e.g.* (**198**) from reaction with ammonia and (**199**) from ethyl malonate (reaction 45). Cyanogen reacts with *o*-mercaptobenzoic acid, giving 2,2′-bi-(4-oxo-4*H*-1,3-benzothiazinyl) (**200**), but with *o*-mercaptobenzhydrazide it gives

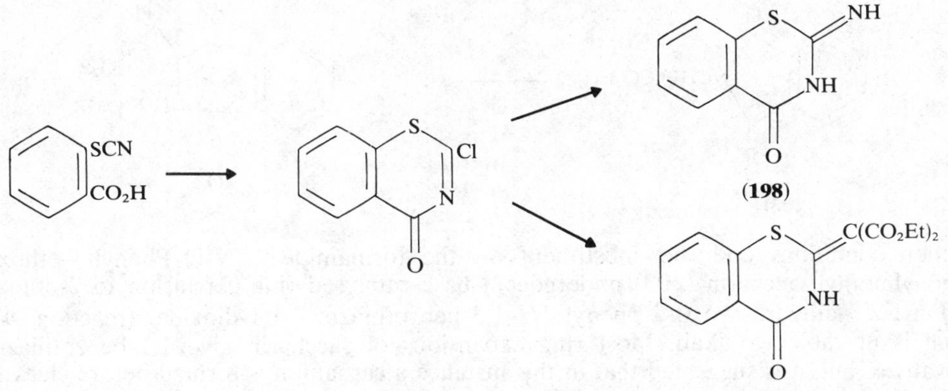

(45)

(**198**)

(**199**)

3-amino-2,3-dihydro-2-imino-4*H*-1,3-benzothiazin-4-one (**201**), with liberation of hydrogen cyanide.

2,4-Dioxodihydro-1,3-benzothiazine is made from *o*-mercaptobenzoic acid by the successive action of ethyl chloroformate and ammonia, or by heating it with urea. Similar reactions with thiourea, cyanamide, cyanoguanidine, or potassium thiocyanate give a variety of related thioxo- and imino-derivatives. The 2-imino-4-oxo-compound dissolves

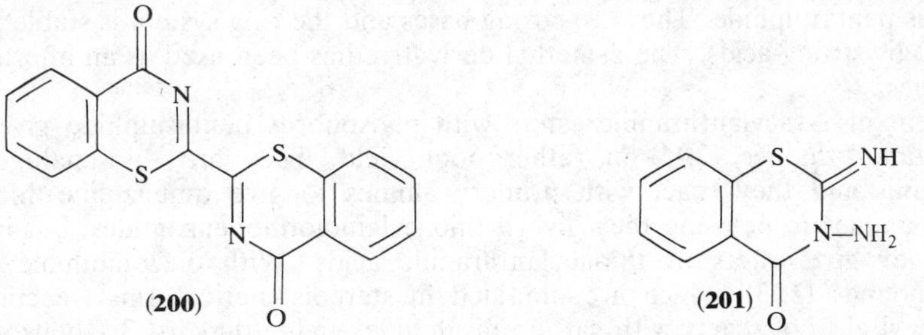

(**200**) (**201**)

in sodium hydroxide solution through a reversible cleavage to the dianion of *o*-mercaptobenzoylcyanamide, and it reacts with aniline to give 2-phenylimino-4-oxodihydro-1,3-benzothiazine. 4-Oxo-2-thioxodihydro-1,3-benzothiazine acylates aniline, giving *o*-mercaptobenzanilide, and with phosphorus pentasulphide it is thionated to the 2,4-dithioxo-derivative. These and related compounds may be made by ring-enlargement of benzisothiazole derivatives. When benzo-1,2-thiazol-3-one is heated with aqueous–alcoholic potassium cyanide, it gives 2-imino-4-oxodihydro-1,3-benzothiazine. 3-Chlorobenzisothiazolium chlorides (**202**) condense with formamides in pyridine at 35 °C or in *o*-dichlorobenzene at 100 °C, giving 2-imino-4-oxo- and/or 4-imino-2-oxo-dihydro-1,3-benzothiazines (reaction 46), the direction taken by the reaction depending on the

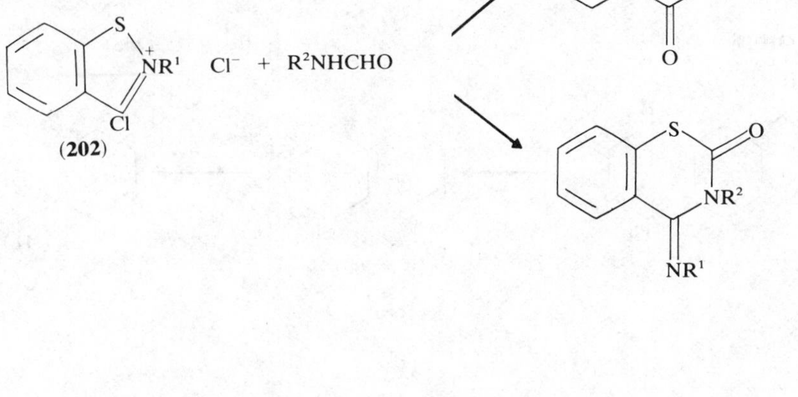

$Cl^- + R^2NHCHO$ (46)

(**202**)

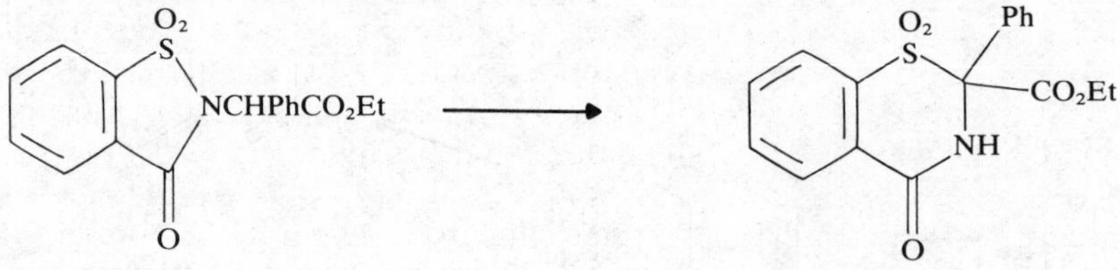

(47)

(203)

reaction conditions and the substituent on the formamide.[51] *N*-(1-Phenyl-1-ethoxycarbonylmethyl)saccharin (**203**) undergoes a base-catalysed ring expansion to 2-ethoxycarbonyl-2,3-dihydro-4-oxo-2-phenyl-4*H*-1,3-benzothiazine 1,1-dioxide (reaction 47), which is unstable to alkali. Most ring expansions of saccharin give 1,2-benzothiazine derivatives, and it is suggested that in this instance a carbanion is formed before cleavage of the benzisothiazole ring occurs and that it attacks the electrophilic SO_2 group. The structure of the product was established by *N*-methylation, hydrolysis, and decarboxylation to give 2,3-dihydro-3-methyl-4-oxo-2-phenyl-4*H*-1,3-benzothiazine 1,1-dioxide, which was identical with a sample made by condensation of benzaldehyde with *N*-methyl-*o*-mercaptobenzamide and oxidation of the resulting dihydro-3-methyl-4-oxo-2-phenyl-1,3-benzothiazine.

4*H*-3,1-Benzothiazine and its homologues are obtained by condensation of thioamides with *o*-aminobenzyl chloride or by treatment of *o*-acylaminobenzyl halides or esters with phosphorus pentasulphide. They are strong bases and the ring system is stable, but it may be cleaved by strong acids. The 2-methyl derivative has been used as an intermediate for cyanine dyes.

Treatment of *N*-acylanthranilic esters with phosphorus pentasulphide gives 4*H*-3,1-benzothiazine-4-thiones (**204**) in rather poor yield. They are thio-analogues of the acylanthranils and they react with primary amines to give quinazoline-thiones. With secondary amines in benzene they give *o*-thioacylaminothiobenzamides, but in alcoholic solution they give esters of thioacylanthranilic acids. With diazomethane they yield spiro-compounds (**205**), which are obtained in stereoisomeric forms (reaction 48). *o*-Aminobenzyl alcohol reacts with carbon disulphide, giving dihydro-3,1-benzothiazine-2-thione. This will react with arylamines with replacement of either sulphur atom, giving a 3-aryltetrahydroquinazoline-2-thione and a 2-aryliminodihydro-3,1-benzothiazine. The 4-phenyl derivative, made similarly from *o*-aminobenzhydrol, affords 2-alkylthio-4-phenyl-4*H*-3,1-benzothiazines (**206**) on alkylation, and when these are treated with sodamide they extrude sulphur to give 2-alkylthio-3-phenylindoles (reaction 49).

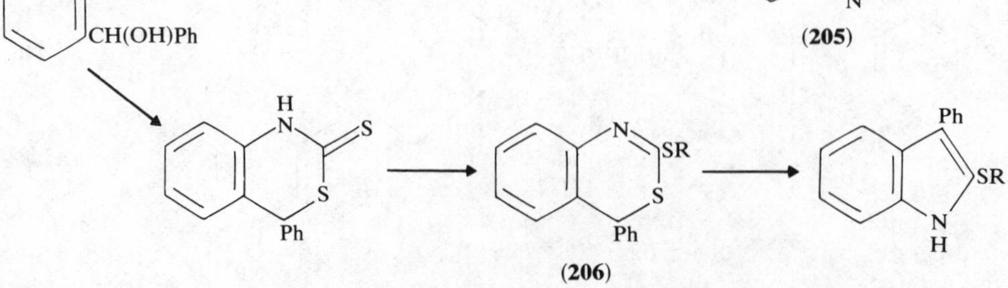

(48)

(204)

(205)

(49)

(206)

Condensation of *o*-aminobenzhydrol with thiourea in conc. hydrobromic acid gives 2-amino-4-phenyl-4*H*-3,1-benzothiazine or its imino-tautomer. 1-Alkyl- and 1-aryl-dihydro-3,1-benzothiazine-4-thiones are made by thionation of the corresponding 3,1-benzoxazin-4-ones with phosphorus pentasulphide. They are low-melting solids and the 1-methyl compound is unstable and is oxidized on exposure to air. Permanganate oxidation converts them into 1-substituted dihydro-3,1-benzothiazin-4-ones.

Thiophosgene reacts with anthranilic acid to give 2-thioxo-3,1-benzoxazin-4-one, and this rearranges to 2,4-dioxodihydro-3,1-benzothiazine when heated. 2,4-Dithioxodihydro-3,1-benzothiazines are minor products from the reaction of carbon disulphide with thioanthranilamides.

2-Aminothiophenol reacts with phenacyl bromide to give a sparingly soluble benzo-1,4-thiazine hydrobromide. The base, m.p. 48 °C, liberated from this hydrobromide by sodium hydroxide contains 60–80% of 3-phenyl-2*H*-1,4-benzothiazine (**207**) and 40–20% of 3-phenyl-4*H*-1,4-benzothiazine (**208**).[52] The base is unstable in air, and undergoes oxidation to the 2,2′-dimer. When acetylated in alkaline solution it is cleaved to give

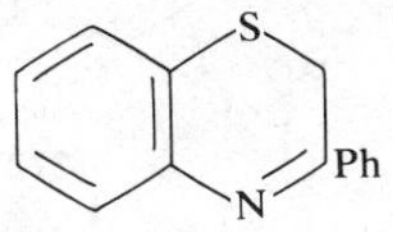

S Ph N

(**207**)

S Ph N H

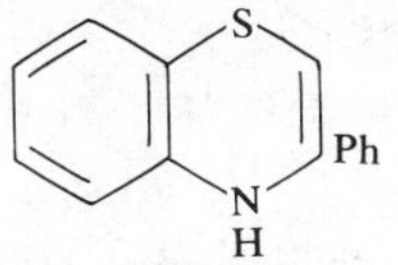

(**208**)

2,2′-diacetamidodiphenyl disulphide. Reduction of (**207**) by lithium aluminium hydride gives 3-phenyldihydrobenzo-1,4-thiazine, but catalytic reduction cleaves the 1,2-bond and the product is 2-methyl-2-phenylbenzothiazole. 2-Methylaminothiophenol and α-halogenoketones give 4-methyl-4*H*-1,4-benzothiazines, *e.g.* 4-methyl-3-phenyl- and 2,3,4-trimethyl-4*H*-1,4-benzothiazine. The latter compound is protonated by perchloric acid, giving 2,3,4-trimethyl-2*H*-1,4-benzothiazinium perchlorate, which has been used as an intermediate for cyanine dyestuffs. Instead of halogenoketones, ketones with a methylene group next to the carbonyl group, including cyclic and steroidal ketones, may be heated with 2,2′-diaminodiphenyl disulphide to give 2,3-disubstituted 2*H*-1,4-benzothiazines. Reaction of 2-aminothiophenol with nitromethyl ketones ($RCOCH_2NO_2$) gives 2-oximino-2*H*-1,4-benzothiazines. 1,4-Benzothiazines are also obtainable from benzothiazolium salts by ring fission and expansion (reaction 50).[53] Dilute alkali cleaves

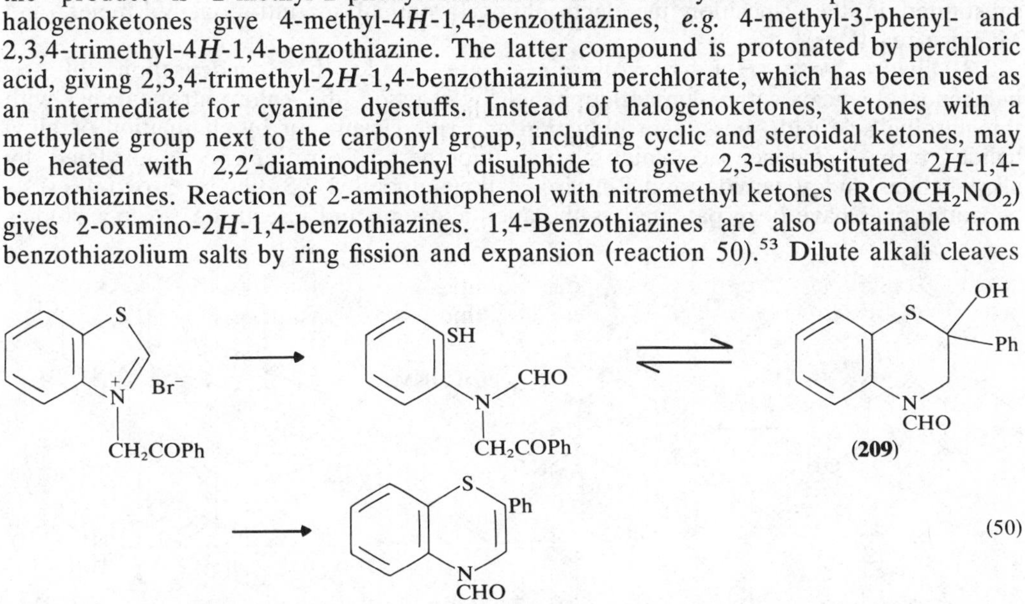

S $\overset{+}{N}$ Br^- CH_2COPh ⟶ SH CHO N CH_2COPh ⇌ OH S Ph N CHO (**209**)

⟶ S Ph N CHO (50)

phenacylbenzothiazolium bromide to 2-(*N*-formyl-*N*-phenacyl)aminothiophenol, which is in equilibrium with its hemithioacetal, 4-formyl-2,3-dihydro-2-hydroxy-2-phenyl-4*H*-1,4-benzothiazine (**209**). This is dehydrated by perchloric acid to give 4-formyl-2-phenyl-4*H*-1,4-benzothiazine. Treatment of benzothiazolium salts with phenacyl bromide and sodium hydroxide gives 2-phenacylthioformanilides, which are cyclized by methanolic sodium hydroxide to 4-substituted 2-benzoyl-4*H*-1,4-benzothiazines (**210**) (reaction 51). Alkyl 4*H*-1,4-benzothiazine-2-carboxylates are obtained from 2-amino- or 2-methylamino-thiophenol and alkyl 1-chloro-2-oxopropionates. With dimethyl sulphate they give 1-methyl-1,4-benzothiazinium salts (**211**).

(51)

(210)

Ozonolysis of allyl *o*-nitrophenyl sulphone followed by reduction of the nitro-group gives 4*H*-1,4-benzothiazine 1,1-dioxide (**212**). Substituted derivatives, *e.g.* the 3-phenyl compound, are made by reaction of sodium *o*-nitrophenylsulphinate with α-bromoketones and catalytic reduction of the resulting sulphones. The reduction may be

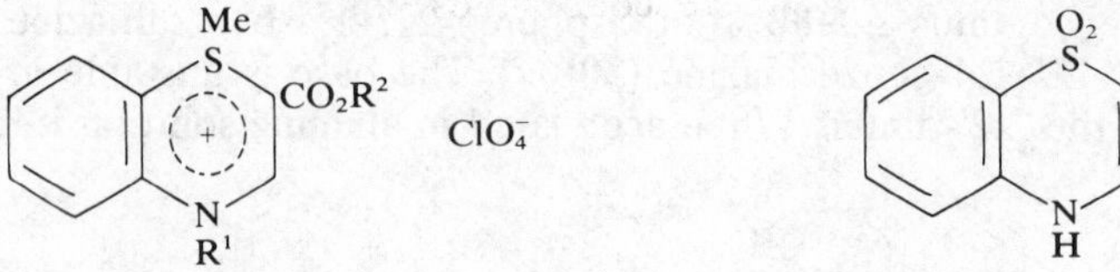

(211) (212)

stopped at the hydroxylamine stage giving, for example, 4-hydroxy-3-phenyl-4*H*-1,4-benzothiazine 1,1-dioxide. This compound gives an *O*-benzoate, 4-benzoyloxy-3-phenyl-4*H*-1,4-benzothiazine 1,1-dioxide, which is isomerized to 2-benzoyloxy-3-phenyl-4*H*-1,4-benzothiazine 1,1-dioxide when it is heated in pyridine with pyridine hydrochloride.[54] The 4*H*-1,4-benzothiazine 1,1-dioxides are very resistant to cold mineral acids and are substituted in the 2-position by electrophiles (bromination, nitration, Vilsmeier, and Mannich reactions).

2,3-Dihydro-3-oxo-4*H*-1,4-benzothiazines are made by condensation of α-halogenoacetic acids with 2-aminothiophenol, by reaction of 2-chloronitrobenzenes with thioglycollic acid with subsequent reduction and ring closure, or by elimination of alkyl halides from 2-(halogenoacetamido)phenyl thioethers. They can be *N*-alkylated by alkyl halides and sodamide, and 2-aryl-4-dialkylaminoalkyl-2,3-dihydro-3-oxo-4*H*-1,4-benzothiazines have been patented as therapeutic agents and insecticides. N.m.r. studies

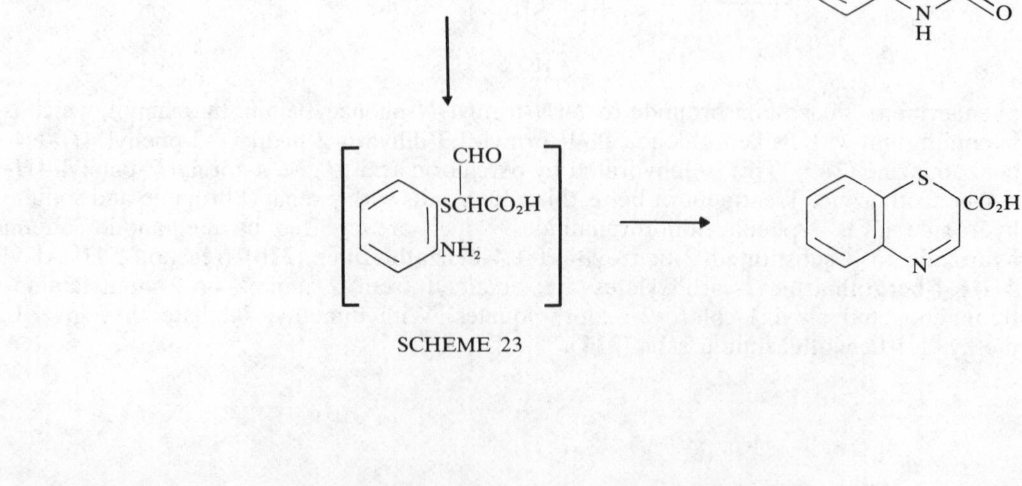

(213)

SCHEME 23

have shown that there is no enolic tautomer in the 3-oxodihydro-1,4-benzothiazines and their 1,1-dioxides. A Vilsmeier reaction on the unsubstituted 3-oxo-compound affords 3-chloro-2-dimethylaminomethylene-2*H*-1,4-benzothiazine (**213**), and under varying conditions of hydrolysis this gives 3-chloro-2-formyl-2*H*-1,4-benzothiazine, 2-formyl-2,3-dihydro-3-oxo-4*H*-1,4-benzothiazine, or 2*H*-1,4-benzothiazine-2-carboxylic acid (Scheme 23). 2,3-Dihydro-4-hydroxy-3-oxo-4*H*-1,4-benzothiazines and their 1,1-dioxides, cyclic hydroxamic acids, are obtained by reducing alkyl *o*-nitrophenylthioacetates or -sulphonylacetates with sodium borohydride and a palladium/carbon catalyst. The 3-oxodihydro-1,4-benzothiazine 1,1-dioxides cannot be obtained by oxidation of 3-oxodihydro-1,4-benzothiazines, but they are made by reduction of 2-nitrophenylsulphonylacetic acids. 3-Thioxodihydro-1,4-benzothiazine 1,1-dioxide is made by thionation of the 3-oxo-compound with phosphorus pentasulphide. Because of the multiplicity of nucleophilic centres in these compounds, alkylating agents give a variety of products (Scheme 24).[55]

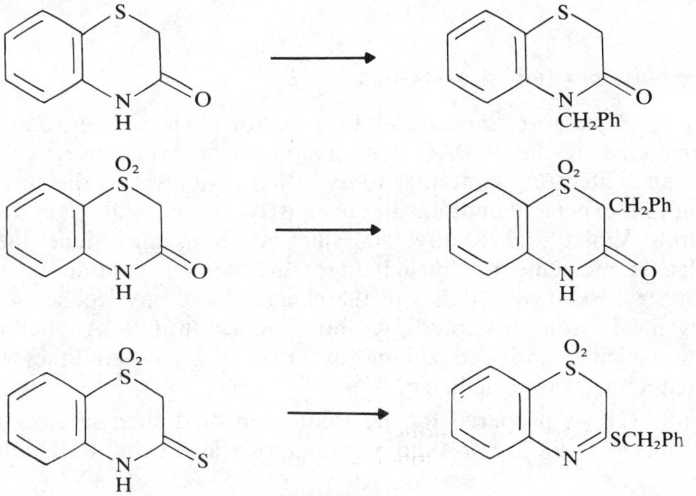

S N H O → S N CH_2Ph O

O_2 S N H O → O_2 S CH_2Ph N H O

O_2 S N H S → O_2 S N SCH_2Ph

SCHEME 24

Dihydro-1,4-benzothiazine (phenothiomorpholine) is a low-melting secondary amine obtained by treatment of 2-aminothiophenol with 1,2-dibromoethane or ethylene oxide. It is notable for the ready cleavage of the 3,4-bond, *e.g.* by the action of nitrous acid or by acylation in alkaline solution. 2-Substituted derivatives, *e.g.* the 2-phenyl compound, are obtained by reduction of the 3-oxodihydro-1,4-benzothiazines with lithium aluminium hydride. 5-Hydroxy-7-methyldihydro-1,4-benzothiazine is obtained by addition of 2-mercaptoethylamine to the *o*-quinone produced by ferricyanide oxidation of 4-methylcatechol.

The trichosiderins, a group of red and yellow pigments of hair and feathers, are $\Delta^{2,2'}$-bi-(2*H*-1,4-benzothiazines). It is postulated that these compounds, *e.g.* trichosiderin C (**214**), arise from oxidative coupling of 3,4-dihydroxyphenylalanine and cysteine. The formation of a 1,4-benzothiazine derivative from oxidation of 4-methylcatechol and cysteine, and the oxidation of this to a 2,2′-dimer have been demonstrated. The parent ring system (**215**) of trichosiderins has been synthesized from 2-aminothiophenol and mucochloric

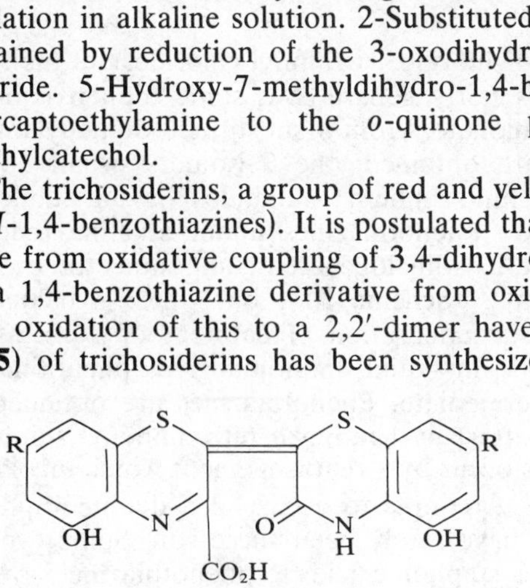

R S N OH CO_2H S O N H R OH

(**214**) $R = CH_2CH(NH_2)CO_2H$

S N S O N H

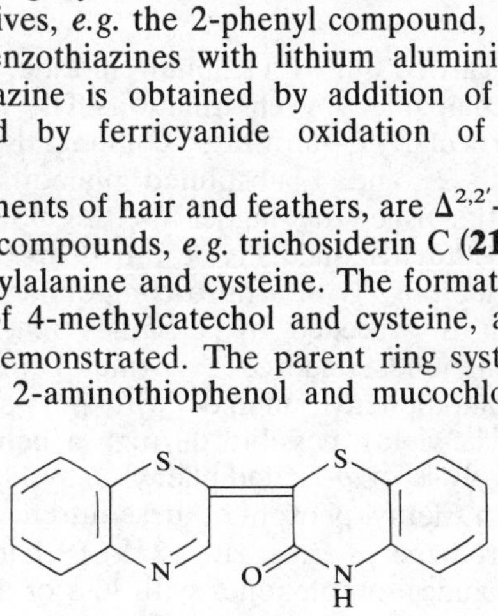

(**215**)

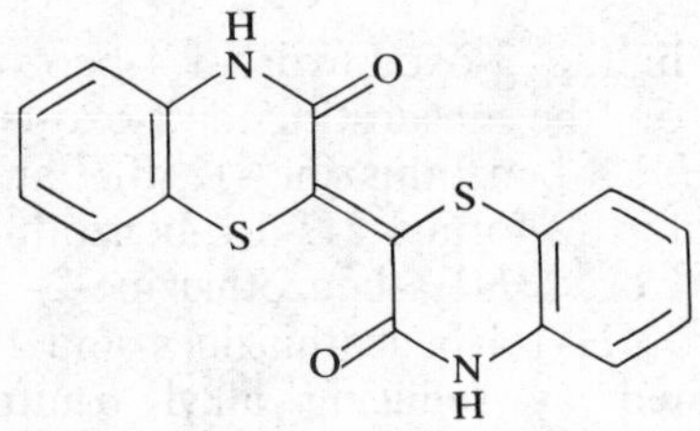

(216)

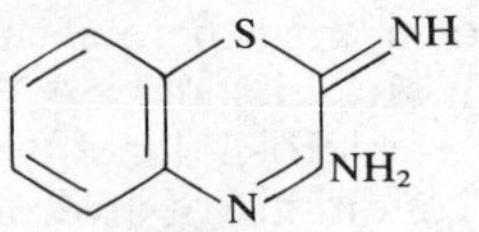

(217)

acid (2,3-dichloro-2,5-dihydro-5-hydroxyfuran-2-one), and related indigoid colouring matters, both the *cis*-isomer and the more stable *trans*-isomer (**216**), are obtained from 2-aminothiophenol and dichloromaleic anhydride.[56] Other indigoid compounds have been made by conventional methods, *e.g.* by condensation of 4-methyl-4*H*-1,4-benzothiazine with isatin α-chloride. 3-Amino-2-imino-2*H*-1,4-benzothiazine (**217**) is obtained from 2-aminothiophenol and oxaldi-imidic acid.

20.2.3.3 Phenothiazines (dibenzo-1,4-thiazines)[57,58]

The numbering of phenothiazines resembles that of phenoxazines, and the Ring Index system recommended by the IUPAC commission on nomenclature is used here. Much German and French literature continues to use other systems, and this has led to errors in abstracting and indexing. Phenothiazine chemistry began with the discovery of the dyestuffs Lauth's Violet and Methylene Blue in 1876, and since the early use of Methylene Blue in medicine by Ehrlich there has been a continuing association with medicinal chemistry. Extensive studies of the chemical and physicochemical properties of phenothiazines have been prompted by the practical utility of phenothiazine as an anthelmintic, insecticide, and antioxidant for lubricating oils and the wide use of 10-substituted phenothiazines in medicine.[59]

Phenothiazine (**218**) is prepared by the thionation of diphenylamine by heating with sulphur, preferably with iodine or aluminium chloride as a catalyst. Thionation may also

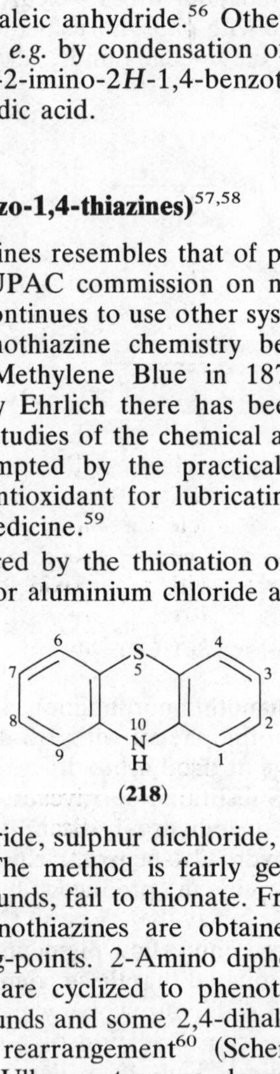

(218)

be carried out with sulphur chloride, sulphur dichloride, or thionyl chloride, but then it is accompanied by chlorination. The method is fairly general but some diphenylamines, particularly 2-substituted compounds, fail to thionate. From 3-substituted diphenylamines both 2- and 4-substituted phenothiazines are obtained, the 2-isomers usually being distinguished by higher melting-points. 2-Amino diphenyl sulphides having labile 2′-substituents (halogen or nitro) are cyclized to phenothiazines under alkaline reaction conditions. With *o*-nitro-compounds and some 2,4-dihalogenated compounds the cyclization is preceded by a Smiles rearrangement[60] (Scheme 25), but other *o*-halogeno-compounds undergo a normal Ullmann-type condensation. Treatment of 2-amino-2′-bromodiphenyl sulphide with potassium amide in liquid ammonia gives phenothiazine (35% yield), possibly through a benzyne intermediate. Phenothiazines are obtained by pyrolysis of *o*-azidodiphenyl sulphides, or (better) by heating *o*-nitrodiphenyl sulphides with triethyl phosphite; these nitrene reactions occur by a rearrangement which involves a spirodiene intermediate (**219**) (Scheme 26).[61] Attempts to cyclize 2,2′-diaminodiphenyl sulphides or sulphones with loss of ammonia have rarely been successful. Some cyclizations of diphenylamine-2-sulphinic acids and -2-sulphonic acids to phenothiazine 5-oxides and 5,5-dioxides have been reported.

SCHEME 25

Phenothiazine may be detected by colour reactions with oxidising agents — green with iron(III) chloride and red with acidified hydrogen peroxide. As a secondary amine it is readily alkylated or acylated to give, for example, 10-methylphenothiazine or 10-acetylphenothiazine, and with iodobenzene under Ullmann conditions it gives 10-phenylphenothiazine. Dipole moments of phenothiazine and its 10-methyl and 10-phenyl derivatives show that in solution the molecules have a dihedral angle of about 150° with the 10-substituent equatorial to the heterocyclic ring. Phenothiazine is a good electron donor and forms charge-transfer complexes with a number of acceptors. The formation of stable semiquinones or cation radicals is characteristic of phenothiazines.

Oxidation of phenothiazine by hydrogen peroxide or potassium permanganate gives phenothiazine 5-oxide or 5,5-dioxide; with other oxidizing agents *C*-oxidation occurs at positions-3 and -7. The phenothiazine 5-oxides are reactive compounds that may readily be reduced to phenothiazines, but the 5,5-dioxides are much more stable and reduction has only been achieved polarographically. The oxidation may be carried out in successive one-electron steps electrochemically or by chemical means such as sulphuric acid (*cf.* phenoxazine), iron(III), or cerium(IV). The first oxidation step gives an orange-yellow radical cation, which may become deprotonated to give a phenothiazinyl radical; this decays fairly quickly. The second oxidation step gives a phenazathionium cation which is protonated in anhydrous acid solution to a green dication. Loss of a hydroxyl radical from protonated phenothiazine 5-oxide will also generate the radical cation (Scheme 27). The phenazathionium ion can rearrange to a mesomeric form having the positive charge at the 3-position and attack by nucleophiles may then occur. Hydration and rearrangement gives 3-hydroxyphenothiazine, attack by amines gives 3-amino-derivatives, and chloride ion

(219)

SCHEME 26

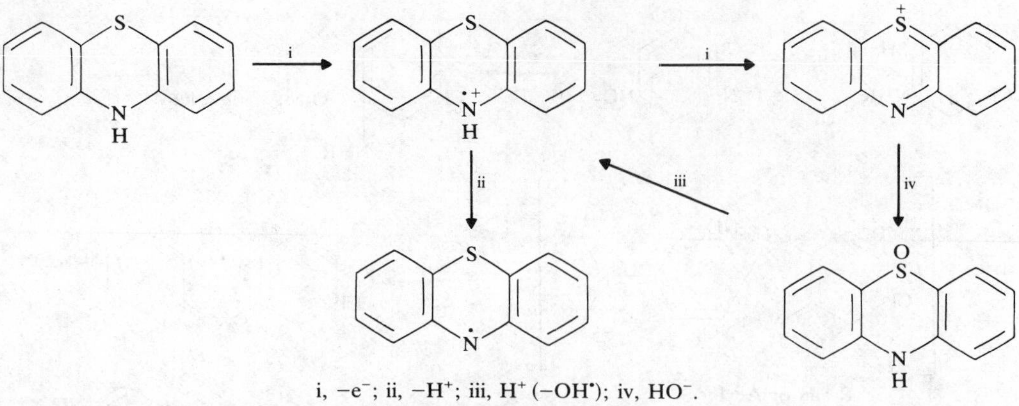

i, $-e^-$; ii, $-H^+$; iii, H^+ ($-OH^•$); iv, HO^-.

SCHEME 27

gives 3-chlorophenothiazine. By oxidation of phenothiazine with iron(III) chloride in the presence of sodium *p*-tolylsulphinate, sodium nitrite, or thiourea, the 3-*p*-tolylsulphonyl, 3-nitro, and (after hydrolysis of the thiouronium salt) 3-mercapto-derivatives are obtained. Attack by compounds with reactive methylene groups, *e.g.* indane-1,3-dione or dimedone, gives quinonoid colouring matters, *e.g.* (**220**).[62] Treatment of phenazathionium perchlorate with methylmagnesium chloride gives 5-methyl-5*H*-phenothiazine (**221**). The radical from a one-electron oxidation with iodine or with acetic anhydride in DMSO dimerizes to 3,10′-biphenothiazinyl, which can undergo further oxidation to fairly stable free radicals.

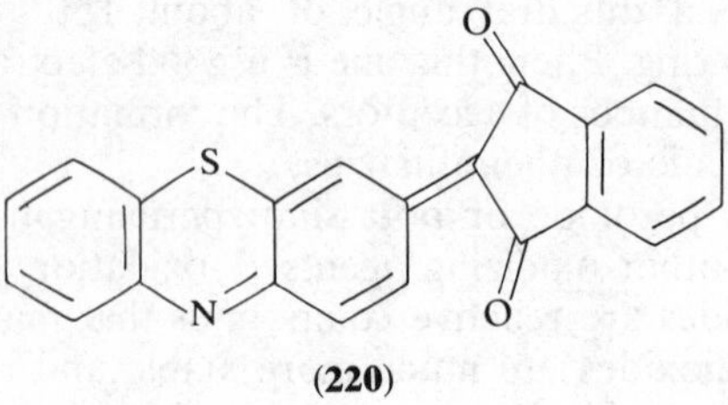

(**220**) 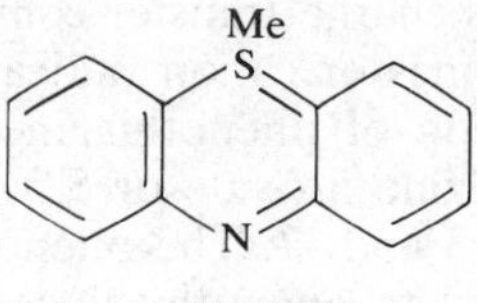 (**221**)

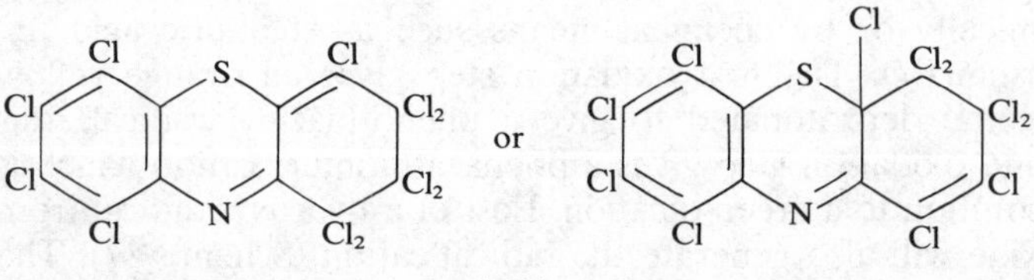

(**222**) 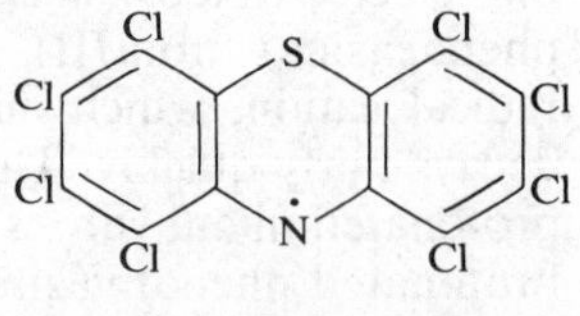 (**223**)

Electrophilic substitution of phenothiazines is frequently complicated by oxidation. Thus, nitration with nitric acid gives 3-nitro- and 3,7-dinitro-phenothiazine 5-oxide, but nitration with nitrous acid gives 3,7-dinitrophenothiazine. The action of bromine gives phenazathionium perbromide, which is unstable and gives ring-brominated compounds (3,7-, 1,3,7-, and 1,3,7,9-) on keeping or on heating in acetic acid. Bromination in nitrobenzene gives 1,2,3,7,8,9-hexabromophenothiazine. Chlorination in acetic acid gives 3,7-dichloro- and 1,3,7,9-tetrachloro-phenothiazine, but in nitrobenzene chlorination can be continued until eleven chlorine atoms have been introduced with loss of aromaticity of one ring. The product (**222**) when heated at 180 °C loses three atoms of chlorine, giving 30% of the very stable free radical (**223**) and 70% of 10,10′-bi(octachlorophenothiazinyl). Thionyl chloride will chlorinate phenothiazines, and so converts phenothiazinecarboxylic acids into chlorophenothiazine acid chlorides. Friedel–Crafts acylation of 10-alkylphenothiazines gives 3-acyl or 3,7-diacyl derivatives, but in 10-acylphenothiazines the electron availability from the nitrogen atom is reduced and the reaction is controlled by the sulphur atom, giving substitution at positions-2 and -8. With unsubstituted phenothiazine the reaction is complicated by *N*-acylation, and besides 2,10-diacyl derivatives some products of uncertain orientation have been isolated. Vilsmeier–Haack formylation of 10-alkylphenothiazines gives 3-formyl derivatives, but 10-acylphenothiazines are deacylated and converted into 10-formyl derivatives. Phenothiazine may be sulphonated with chloro-sulphonic acid. Alkenes catalysed by boron trifluoride alkylate phenothiazine at positions-3 and -7. Metallation of phenothiazine with butyl-lithium gives the 1,10-dilithium derivative, which is converted into phenothiazine-1-carboxylic acid on carbonation. 10-Alkylphenothiazines when treated with butyl-lithium give 4-substituted compounds, but carbonation gives both 1- and 4-carboxylic acids.

3-Hydroxyphenothiazine, obtained by thionation of 4-hydroxydiphenylamine, is readily oxidized to the red compound 3*H*-phenothiazin-3-one (phenothiazone) (**224**), and is regenerated from its oxidation product by reduction, *e.g.* with zinc and acetic acid. One-electron oxidation of the hydroxy-compound or reduction of phenothiazone gives a stable violet semiquinone (**225**). This is also obtained from phenothiazone in strong acid solution because the *O*-protonated phenothiazonium ion is unstable and decays through loss of a hydroxyl radical (Scheme 28). The relationship between these compounds resembles that between phenothiazine, its radical cation, phenazathionium ions and the 5-oxide. The two oxidation series are linked because hydrolysis of phenazathionium ions can give 3-hydroxyphenothiazine. Phenothiazones, particularly those with more positive redox potentials (*e.g.* nitro- and 1-halogeno-phenothiazones) add hydrochloric or hydrobromic acid across the quinoneimine ring, giving 2- and 4-halogenated 3-hydroxy-phenothiazines. Oxidation of phenothiazone, or prolonged oxidation of phenothiazine, gives the red-brown 7-hydroxy-3-phenothiazone (thionol), which gives colourless 3,7-dihydroxyphenothiazine on reduction. A semiquinone has been obtained from thionol at low pH.

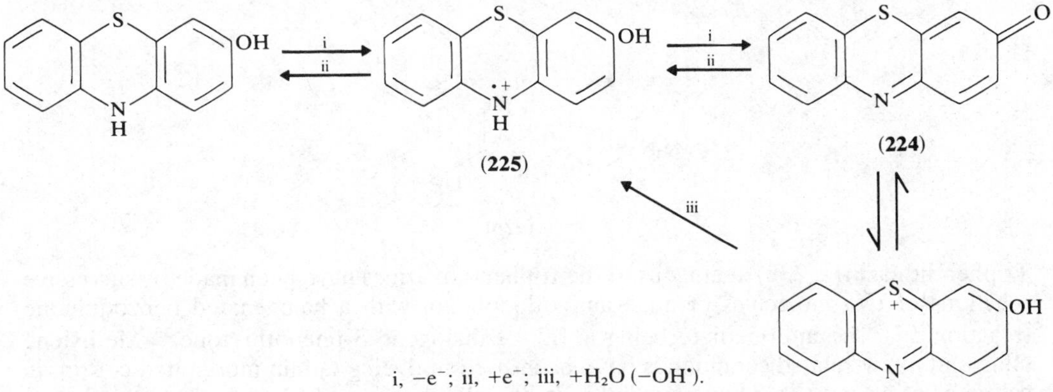

(225) **(224)**

i, $-e^-$; ii, $+e^-$; iii, $+H_2O$ ($-OH^\cdot$).

SCHEME 28

3-Aminophenothiazine, made by thionation of 4-aminodiphenylamine or by reduction of 3-nitrophenothiazine 5-oxide, is oxidized (reaction 52) by iron(III) chloride to give 3-phenothiazime (**226**), a red substance which gives phenothiazone on alkaline hydrolysis. Similarly, 3,7-diaminophenothiazine is oxidized to 7-aminophenothiazime (thionin), the hydrochloride of which is Lauth's Violet. This dyestuff, and Methylene Blue, were originally obtained by oxidation of *p*-phenylenediamine (or 4-aminodimethylaniline) with iron(III) chloride in the presence of hydrogen sulphide. Methylene Blue (Basic blue 9, C.I.

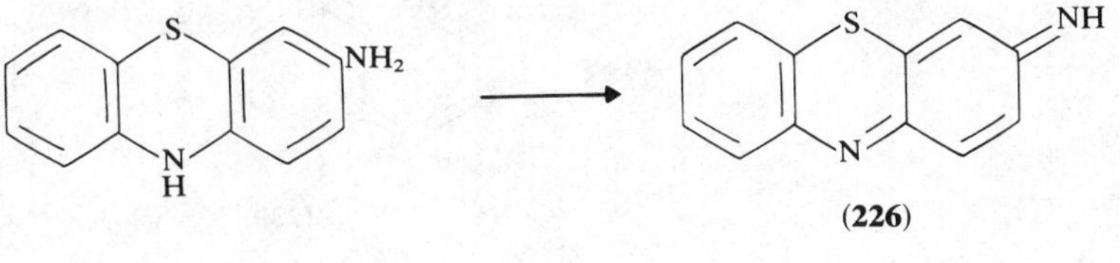

(52)

(226)

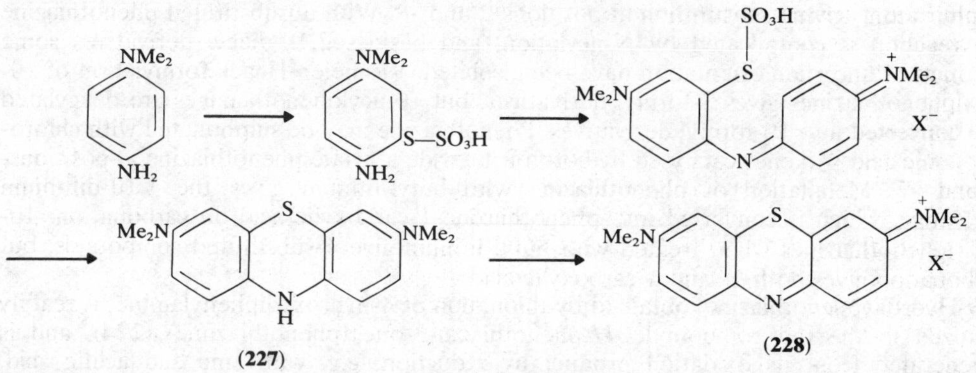

(227)

(228)

SCHEME 29

52015) was subsequently made by addition of thiosulphate ion across the quinonoid system generated by oxidation of 4-aminodimethylaniline, oxidation of the product together with dimethylaniline to give an indaminethiosulphonic acid, and cyclization of this in acid solution, giving leuco Methylene Blue (**227**). The leuco-compound may be isolated at this stage, but oxidation is usually continued to give the dyestuff (**228**), which is isolated as its double salt with zinc chloride (Scheme 29). Methylene Blue has also been made by heating 3,7-dibromophenazathionium bromide with dimethylamine. The related dyestuff Brilliant Alizarin Blue 3R (**229**) is made by condensing 4-aminodimethylaniline-3-thiosulphonic acid with 1,2-naphthaquinone-6-sulphonic acid.[27,28] Certain of the sulphur colours, dyestuffs made by heating diphenylamines, indophenols, or other intermediates with sodium polysulphide, are believed to be complex phenothiazine derivatives.

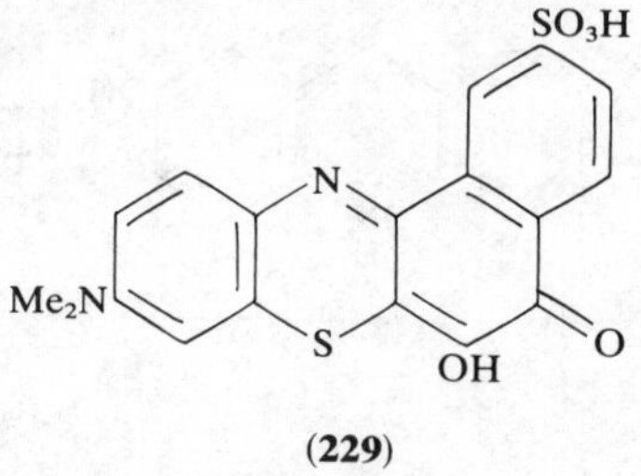

(229)

Triphendithiazines (**230**) analogous to the triphendioxazines have been made by successive reactions of two molecules of an *o*-aminothiophenol with a halogenated benzoquinone (reaction 53), the intermediate being a 1,2,4-trihalogeno-3-phenothiazone.[33] Methylene Blue and a few related compounds have been used in dyeing tannin mordanted cotton, in calico printing, and in hectograph inks. It has also been used in biological stains, as a

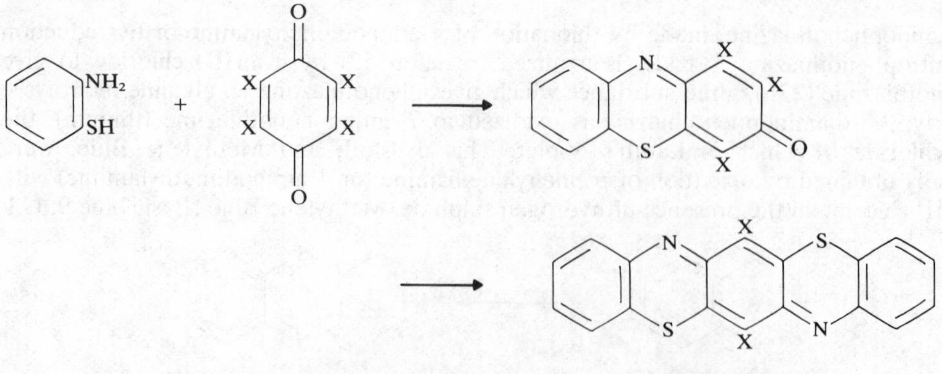

(53)

(**230**) X = halogen

redox indicator, and as a mild antiseptic. It is nitrated in the 4-position, giving Methylene Green. Further nitration gives an olive-green compound, throught to be oxidized to the 5,5-dioxide and to have a nitramine group (from loss of a methyl group), and finally a compound formulated as 3,7-di(methylnitramino)-2,4,6,8-tetranitrophenothiazine 5,5-dioxide is obtained. This is hydrolysed by alkali to 3,7-dihydroxy-2,4,6,8-tetranitrophenothiazine 5,5-dioxide.

The widespread use of 10-substituted phenothiazines in medicine followed the introduction of 10-(2-dimethylaminopropyl)phenothiazine (promethazine) (**231**) as an antihistaminic. This compound is prepared by alkylating phenothiazine with sodamide and 1-chloro-2-dimethylaminopropane or 2-chloro-1-dimethylaminopropane, the latter compound undergoing rearrangement through an aziridinium salt under the alkaline reaction conditions. It may also be made by reaction of phenothiazinyl-10-magnesium chloride with a chloroalkylamine, by reaction of dimethylamine with 10-(2-chloropropyl)-phenothiazine, or by thionating 2-dimethylamino-1-diphenylaminopropane with sulphur and iodine. The diethylamino-analogue, ethopropazine, is used for treatment of parkinsonism. Attempts to enhance the sedative action found with promethazine led to a series of compounds with profound depressant effects on the central nervous system. These are characterized by having a three-carbon chain from the side-chain basic group to the ring and a substituent such as Cl, CF_3, SO_2NMe_2, SMe, or OMe in the 2-position. The most widely studied example is chlorpromazine (**232**), but greater potency is found in analogous compounds with piperazine groups in the side-chain. These compounds are widely used as antipsychotic agents. Alternative ways of introducing the basic side-chain have been investigated. Reaction of phosgene with phenothiazine gives 10-chlorocarbonyl-phenothiazine (**233**), which gives esters with amino-alcohols, and these esters eliminate carbon dioxide to give the aminoalkylated phenothiazine when they are heated with

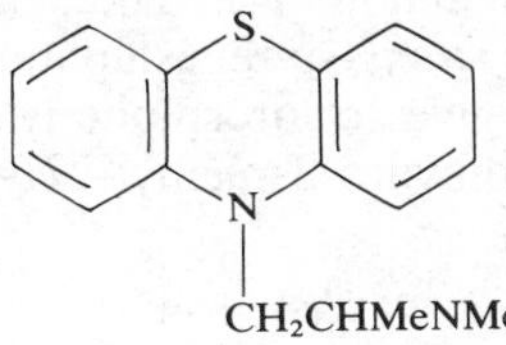

$CH_2CHMeNMe_2$

(**231**)

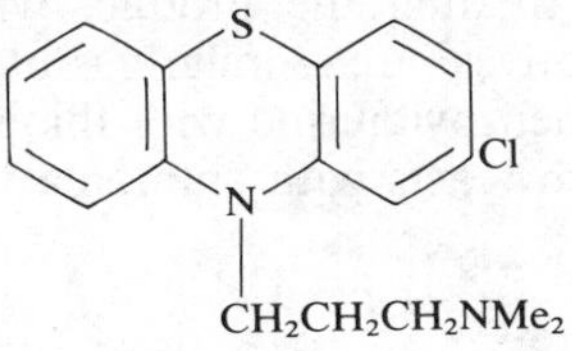

$CH_2CH_2CH_2NMe_2$

(**232**)

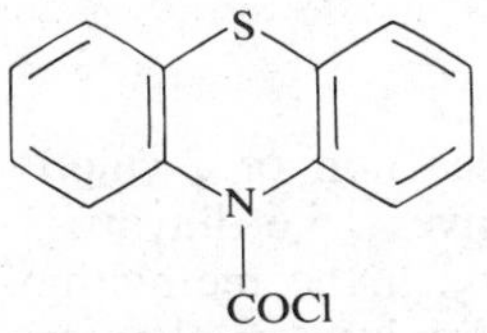

COCl

(**233**)

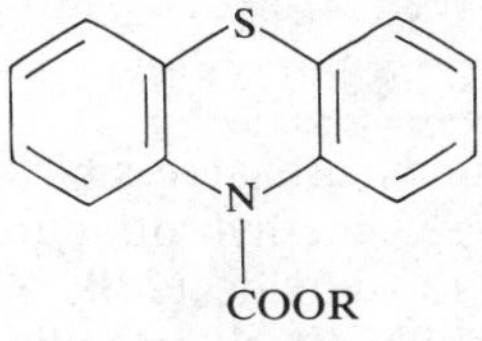

COOR

(**234**)

$COCH_2CH_2NEt_2$

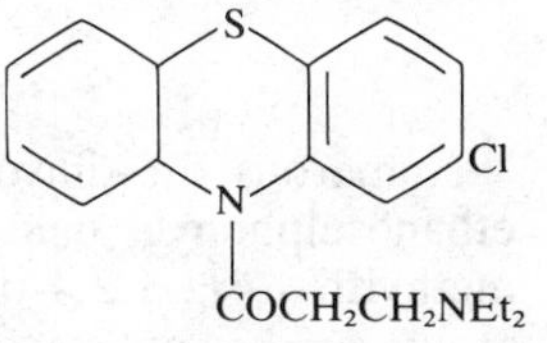

(**235**)

copper powder under reduced pressure. This permits the introduction of alkali-sensitive groups. The corresponding thiolesters are more stable but they will eliminate carbonyl sulphide when heated more strongly. The ester (**234**; $R = C_2H_4OC_2H_4NMe_2$), dimethoxanate, is useful as an antitussive agent. Phenothiazines may be acylated with halogenated acids and the halogen may then be replaced by a substituted amino-group. The acyl compound (**235**), chloracizine, has anti-depressant and coronary-dilating activity. Treatment of these acyl derivatives with lithium aluminium hydride causes cleavage to phenothiazine, but the isomeric amidocarbonylalkylphenothiazines can be reduced to aminoalkylphenothiazines. Metabolic degradation of phenothiazines involves oxidation to the 5-oxide, hydroxylation at positions-3 and -7 and, in the case of 10-substituted compounds, *N*-oxidation and dealkylation of the basic side-chain.

20.2.3.4 Thiadiazines[34] and dithiazines

These compounds have a sulphur atom and two nitrogen atoms, or two sulphur atoms and a nitrogen atom, in a six-membered ring. In both these series six isomeric ring systems are possible, and representatives of all six kinds of thiadiazine are known. The thiadiazine *S,S*-dioxides have been widely investigated.[63]

Some derivatives of 2*H*-1,2,3-thiadiazine (**236**) have been obtained by treating arylhydrazones of 2-thiocyanatocycloalkene-1-aldehydes with 70% perchloric acid. They give stable, colourless or yellow salts; the free bases are more deeply coloured. Application of the same reagent to thiocyanatoacetone 2,4-dinitrophenylhydrazone, however, gives 2-imino-5-methyl-3-(2,4-dinitrophenyl)-4*H*-1,3,4-thiadiazine. The cyclic sulphonylhydrazide tetrahydro-3-phenyl-1,2,3-thiadiazine 1,1-dioxide (**237**) is obtained in poor yield by treating N^1-phenyl-N^2-3-chloropropylsulphonylhydrazine with base.

It is claimed that 4*H*-1,2,4-thiadiazine derivatives (**238**) are produced by treating thiamine and related thiazolium compounds with hydroxylamine in strong alkali. They are said to be thiamine antagonists. Apart from this the 1,2,4-series is represented only by 1,1-dioxides. 5-Aryl derivatives of 4*H*-1,2,4-thiadiazine 1,1-dioxide (**239**; R = Me, Ph) are obtained by acylating acetamidine or benzamidine with α-bromostyrene-α-sulphonyl chloride and treating the product with alkali (reaction 54). The corresponding 5,6-dihydro-derivatives are similarly obtained from *trans*-styrenesulphonylamidines. The treatment of benzoyltaurine with thionyl chloride gives chlorosulphonylethylbenzimidoyl chloride, which reacts with ammonia to give 5,6-dihydro-3-phenyl-2*H*-1,2,4-thiadiazine 1,1-dioxide.

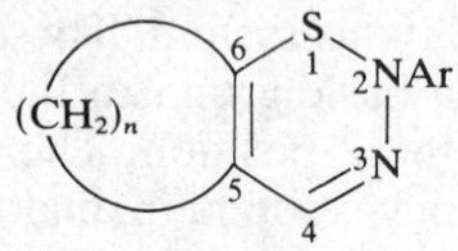

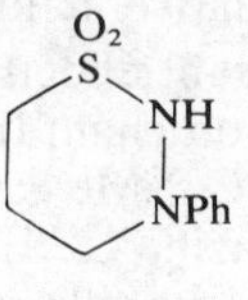

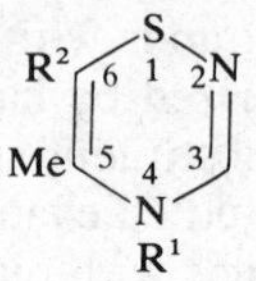

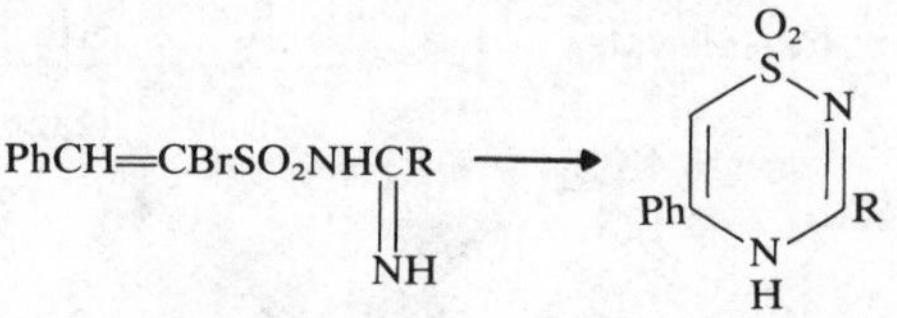

S, 1, 2 NAr, 3 N, 4, 5, 6, $(CH_2)_n$

(236)

O_2 S, NH, NPh

(237)

S, R^2, 6, 1, 2 N, 3, 4, 5, Me, N, R^1

(238)

$PhCH{=}CBrSO_2NHC(=NH)R \longrightarrow$ **(239)** (ring: O_2S, N, R, N–H, Ph) (54)

Numerous 5,6-dihydro-compounds are accessible through the reaction of 2-chloroethanesulphonyl chloride with *S*-methylisothiourea.[64] This gives 5,6-dihydro-3-methylthio-4*H*-1,2,4-thiadiazine 1,1-dioxide (**240**; R^1 = SMe) in which the methylthio-group can be replaced by nucleophilic attack by sodium thiolates, alkoxides, or phenates, or by amines. Chlorination of (**240**; R^1 = SMe) under anhydrous conditions gives the 3-chloro-compound (**240**; R^1 = Cl), which is more reactive towards nucleophiles. Alkylation or acylation of (**240**; $R^1 = SR^2$, OR^2, or Ph) takes place at the 4-position, but 3-amino-derivatives react at the exocyclic nitrogen atom. When the group R^1 is displaceable (OR^2, SR^2, NH_2), reaction with isocyanates or isothiocyanates R^3NCO or R^3NCS gives triazinothiadiazines (**241**; X = O or S). 5,6-Dihydro-3-methylthio-5-phenyl-4*H*-1,2,4-thiadiazine is obtained by cyclization of styrenesulphonylthiourea and methylation of the product. The methylthio-group can be replaced by chlorine and by nucleophiles. Chlorination of (**240**; R^1 = SMe) under aqueous conditions gives 3-oxotetrahydro-1,2,4-thiadiazine (**242**; R^1 = H), which is also made by heating 2-ureidoethanesulphonamide in pyridine. 2,4-Disubstituted derivatives of (**242**) are obtained by oxidation of 2-iminothiazolines with potassium chlorate and hydrochloric acid. They are readily hydrolysed to derivatives of taurine, and the ring of (**242**; R^1 = H) is opened by acetic anhydride giving a triacetyl derivative of taurine amide. 3,5-Dioxotetrahydro-1,2,4-thiadiazine 1,1-dioxide (**243**) is made from chlorosulphonylacetyl chloride through the diphenyl ester and the diamide. This is converted into carbamoylmethylsulphonylurea, which is cyclized by heating in dry pyridine. An alternative route employs ethyl sulphamoylacetate, which is converted into the sulphonylurea and this is cyclized with sodium ethoxide. The product is an analogue of barbituric acid, which it resembles in being high-melting, sparingly soluble, and strongly acidic (pK_a 2.7). The methylene group is reactive and may be nitrosated or coupled with aryldiazonium compounds.

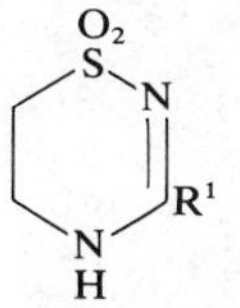

(**240**) 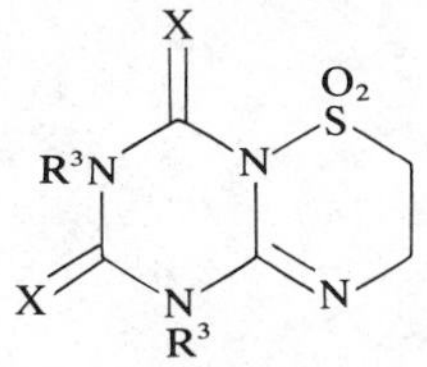 (**241**) 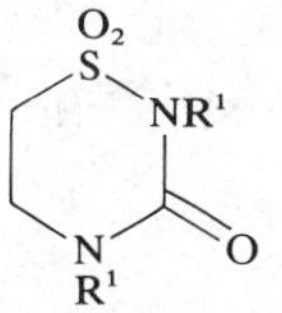 (**242**) 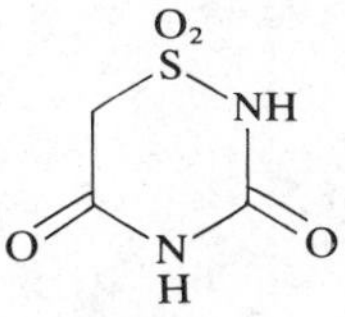 (**243**) 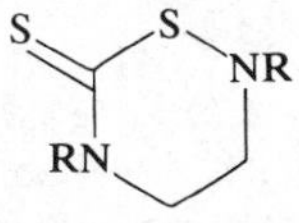 (**244**)

1,2,5-Thiadiazines are almost unknown, but a few tetrahydro-1,2,5-thiadiazine-6-thiones (**244**) have been made by oxidation of the aminoethyldithiocarbamic acids obtained from *N,N'*-dialkylethylenediamines and carbon disulphide. They are described as insecticides, fungicides, and vulcanizing agents for rubber.

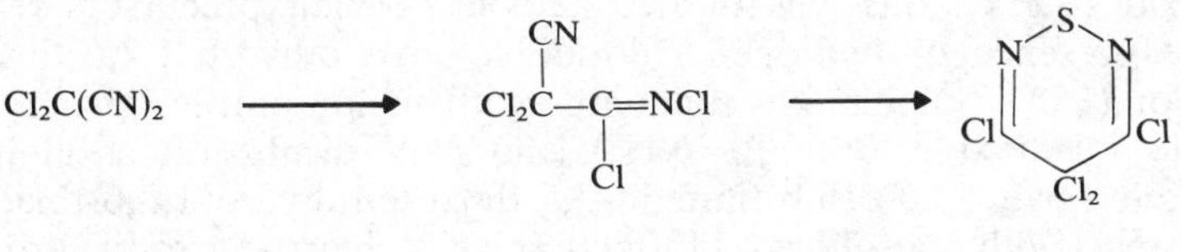

$Cl_2C(CN)_2$ ⟶ $Cl_2C(CN)$—C(Cl)=NCl ⟶ (**245**) (55)

The chemistry of 1,2,6-thiadiazines and their 1,1-dioxides is fairly extensive and varied. Treatment of dichloromalononitrile with sulphur dichloride, or chlorination and treatment of the resulting *N*,2,2-trichlorocyanacetimidoyl chloride with sulphur gives 3,4,4,5-tetrachloro-4*H*-1,2,6-thiadiazine (**245**) (reaction 55). This is unstable, undergoing ring-fission in moist air, but it reacts with formic acid to give a high yield of 3,5-dichloro-4*H*-1,2,6-thiadiazin-4-one (**246**), a much more stable compound in which the chlorine atoms may be replaced successively by reaction with ammonia, primary or secondary amines, or alcoholate or thiolate anions, replacement of the second chlorine atom requiring fairly elevated temperatures.[65] Treatment of di(*p*-toluenesulphonyl)sulphur di-imide with 1,3-diaminopropane gives 4,5-dihydro-3*H*-1,2,6-thiadiazine (**247**), a liquid that is fairly stable to water but which is hydrolysed to the parent diamine by acid. Di-t-butylsulphur di-imide reacts with substituted malonyl chlorides of the general structure $R_2CHCH(COCl)_2$ to give derivatives of 2,6-di-t-butyl-3,5-dioxotetrahydro-1,2,6-thiadiazine (**248**), and *S,S*-dialkylsulphur di-imides react with disubstituted malonyl chlorides, giving 1,1-dialkyl-1,1,4,5-tetrahydro-3,5-dioxo-3*H*-1,2,6-thiadiazines (**249**). 2,6-Dialkyltetrahydro-1,2,6-thiadizines and the corresponding 1-oxides and 1,1-dioxides, shown to have insecticidal activity, are made by treatment of 1,3-di(alkylamino)-propanes with sulphur dichloride, thionyl chloride, or sulphuryl chloride.

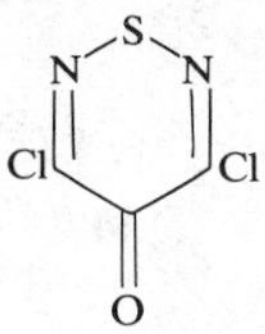

(**246**)  (**247**)

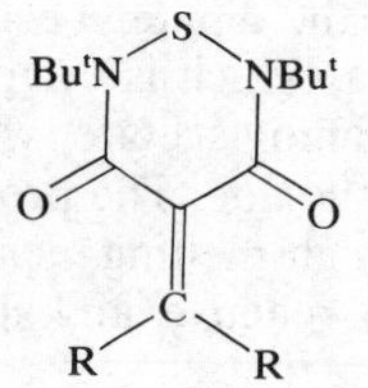

(248) 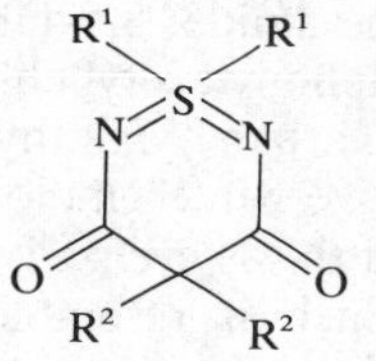 (249) 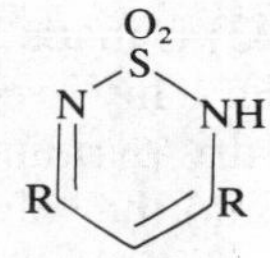 (250)

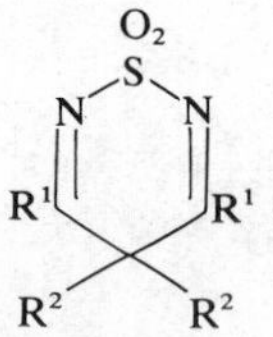

(251) 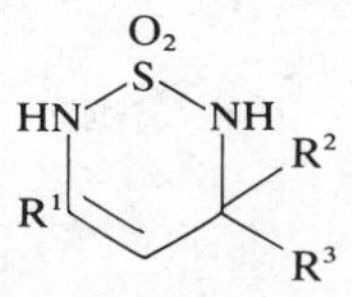 (252) 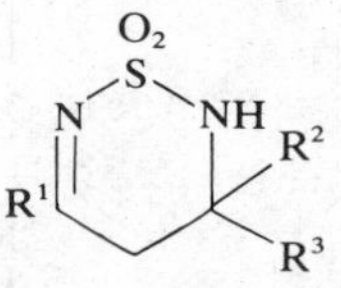 (253)

Many 1,2,6-thiadiazine 1,1-dioxides have been made from sulphamide. Acid-catalysed condensation with 1,3-dicarbonyl compounds gives 2*H*-1,2,6-thiadiazine 1,1-dioxides (**250**), which are acidic compounds that may be alkylated on N-2 but undergo electrophilic substitution (halogenation, nitrosation, azo-coupling, Vilsmeier, and Mannich reactions) at the 4-position.[66] 2,2-Disubstituted 1,3-diketones with sulphamide give 4*H*-1,2,6-thiadiazine dioxides (**251**). α,β-Unsaturated ketones or their precursors react with sulphamide in the presence of hydrogen chloride to give dihydro-1,2,6-thiadiazine 1,1-dioxides (**252**) or (**253**). Sulphamide condenses with malononitrile, giving 3,5-diamino-1,2,6-thiadiazine 1,1-dioxide (**250**; $R=NH_2$), and *N,N'*-disubstituted sulphamides with malonyl chloride give 2,6-disubstituted 3,5-dioxotetrahydro-1,2,6-thiadiazine 1,1-dioxides. These react with phosphoryl chloride to give 3-chloro-5-oxodihydro-derivatives.

1,3,4-Thiadiazines are obtained by the reaction of α-halogenocarbonyl compounds with thiosemicarbazides or dithiocarbazinates. In the reaction of thiosemicarbazide with α-halogenoketones, either the 2-amino-6*H*-1,3,4-thiadiazine (**254**) or a 3-amino-2-iminothiazoline (**255**) may be produced. If the reaction is carried out in dilute hydrochloric acid, the first product is the thiosemicarbazone of the halogenoketone. When this is warmed in ethanolic solution it cyclizes to the thiadiazine (**254**), but if the latter is heated with conc. hydrochloric acid it is isomerized to the thiazoline (**255**). Similar compounds substituted on nitrogen atoms are obtained from alkyl- or aryl-thiosemicarbazides. 2-Amino-5-phenyl-6*H*-1,3,4-thiadiazine (**254**; R = Ph) is also obtained by the action of hydrazine on phenyl thiocyanomethyl ketone. *p*-Nitrobenzaldehyde condenses with (**254**; R = Ph) with ring-fission and loss of sulphur, the product being a chalcone semicarbazone, but from the 2-dimethylamino-analogue a 6-*p*-nitrobenzylidene derivative may be isolated. This affords evidence of the 6*H*-structure, but the products from 4-substituted thiosemicarbazides must be the 4*H*-1,3,4-thiadiazines. 2-Amino-6*H*-1,3,4-thiadiazines are destabilized by electron-withdrawing substituents, *e.g.* aryl or alkoxycarbonyl groups, in the 6-position and they readily extrude sulphur to give pyrazole derivatives. 2-Methylamino-5-phenyl-6*H*-1,3,4-thiadiazine (**256**; $R^1=H$, $R^2=NHMe$) is converted into 3-methylamino-5-phenylpyrazole (**257**; $R^1=H$, $R^2=NHMe$) when it is heated in acetic acid. The dimethylamino-compound (**256**; $R^1=H$, $R^2=NMe_2$) is more stable, but cleaves in a different way with acetic anhydride, giving the 4-acetylthiopyrazole derivative (**258**).[67,68]

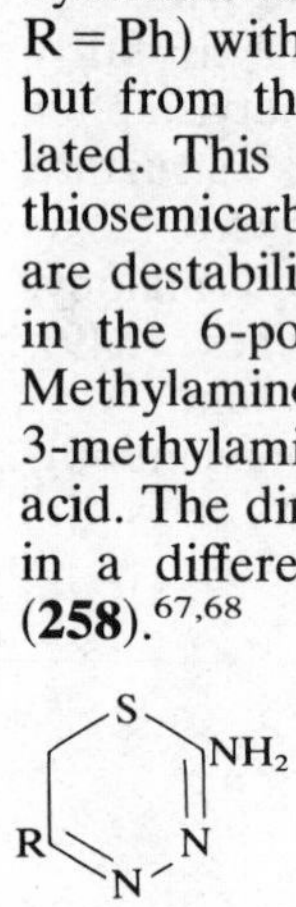

(254) 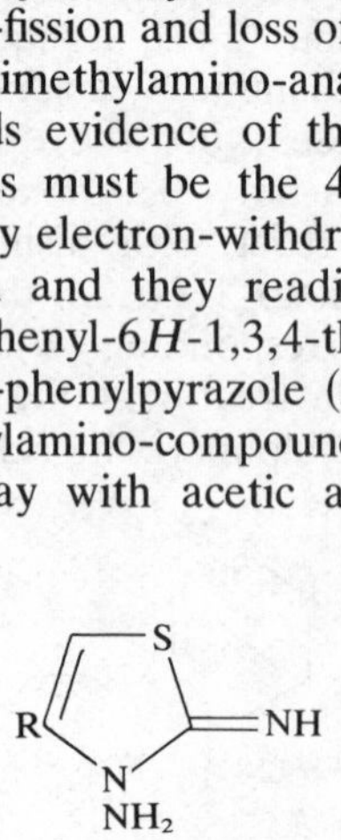 (255) 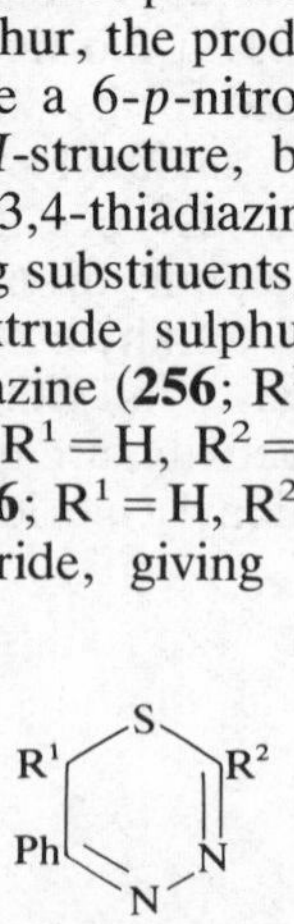 (256) 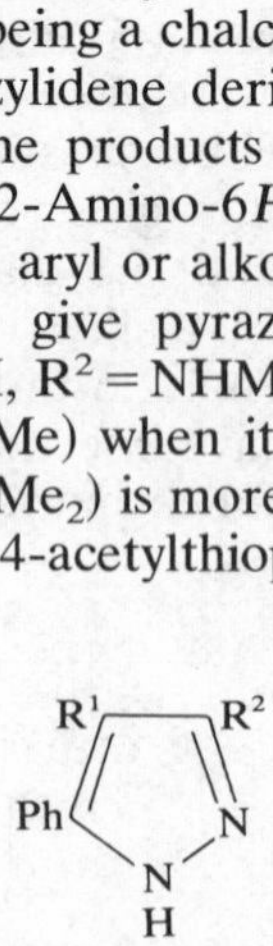 (257) 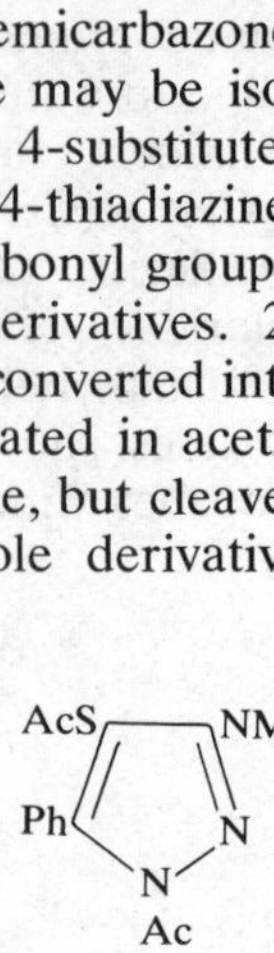 (258)

Methyl dithiocarbazinate and α-halogenoketones condense in an alkaline medium to give 2-methylthio-6H-1,3,4-thiadiazines, *e.g.* (**256**; $R^2 = SMe$), which also extrude sulphur when they are warmed in ethanol. The 6-substituted compounds lose elemental sulphur and give 3-methylthio-4,5-disubstituted pyrazoles (**257**; $R^2 = SMe$), but (**256**; $R^1 = H$, $R^2 = SMe$) is converted into 4-mercapto-3-methylthio-5-phenylpyrazole. Sulphur extrusion from 2-dimethylamino- or 2-methylmercapto-6H-1,3,4-thiadiazines can also be brought about by sodium ethoxide in boiling ethanol, or at low temperatures by generating the anions with butyl-lithium. These anions are unstable and very readily contract to pyrazoles or mercaptopyrazoles.[69]

α-Halogenoesters react with thiosemicarbazides or dithiocarbazinates, giving 5,6-dihydro-5-oxo-1,3,4-thiadiazines, *e.g.* the 2-amino-derivative from ethyl chloroacetate and thiosemicarbazide. This is soluble in alkali and may tautomerize to 2-amino-5-hydroxy-6H-1,3,4-thiadiazine. Chloroacetamide and thiosemicarbazide give 2,5-diamino-6H-1,3,4-thiadiazine. In a similar manner, thiobenzoic hydrazide and bromoacetic acid give 5,6-dihydro-5-oxo-2-phenyl-1,3,4-thiadiazine. Further ways of synthesizing the 1,3,4-thiadiazine ring are exemplified by the cyclization of the ethoxycarbonylmethyl ester of benzoyldithiocarbazinic acid with ammonia to give 6-carbamoyl-2-mercapto-5-phenyl-6H-1,3,4-thiadiazine (**259**) (reaction 56); by the reaction of 2,5-dihydroxy-1,4-dithian

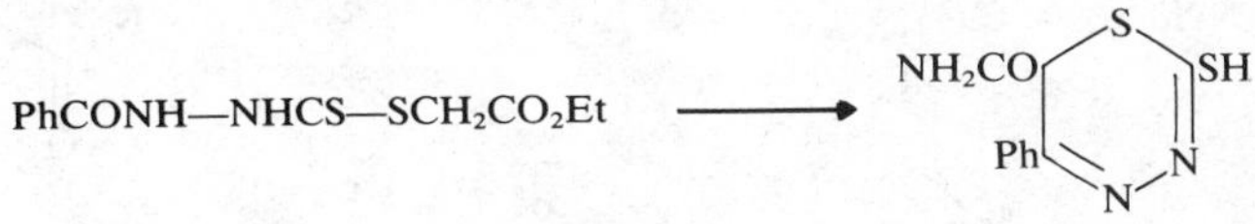

PhCONH—NHCS—SCH_2CO_2Et → (**259**) (56)

with semicarbazide, giving 3-carbamoyl-2,3-dihydro-2-mercaptomethyl-6H-1,3,4-thiadiazine (**260**); and by the reaction of hydrazine or phenylhydrazine with the 2-immonio-1,3-oxathiole (**261**), giving 5-phenyl-2-piperidino- or 4,5-diphenyl-2-piperidino-4H-1,3,4-thiadiazine (**262**; R = H or Ph). 5,6-Dihydro-1,3,4-thiadiazines are obtained by cyclization and thionation of N^1-acyl-N^2-methyl-N^2-(β-hydroxyalkyl)hydrazines with phosphorus pentasulphide, and by condensation of nitriles, imidates, cyanogen bromide, or ethyl orthoformate with 1-methyl-1-(β-mercaptopropyl)hydrazine. The latter compound reacts with benzaldehyde to give a tetrahydro-1,3,4-thiadiazine derivative.

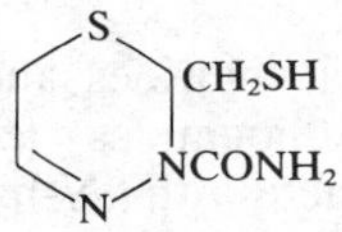

(**260**)

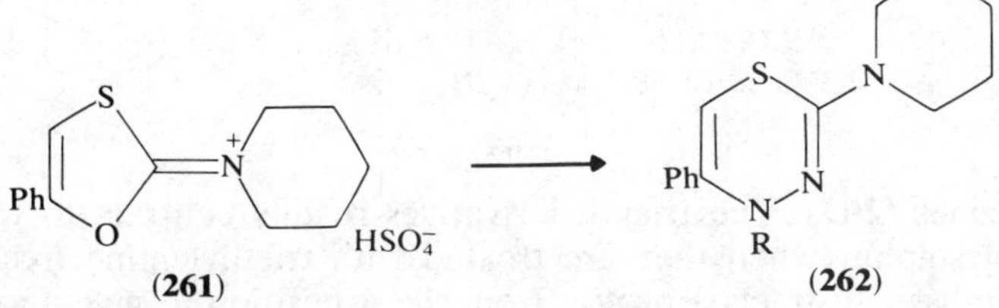

(**261**) → (**262**)

The unstable tetrahydro-4,6-dimethyl-1,3,5-thiadiazine-2-thione (carbothialdine) (**263**) was first prepared from acetaldehyde–ammonia and carbon disulphide in 1848. The isomeric 3,5-dimethyl derivative, dazomet (**264**), is representative of a large number of 1,3,5-thiadiazines that are made from amines, formaldehyde (or other aldehyde), and carbon disulphide. Acid hydrolysis of (**264**) liberates carbon disulphide, but in the soil it breaks down to release methyl isothiocyanate, and it is upon this that its use as a fungicide, nematocide, and herbicide depends. Related compounds have been described

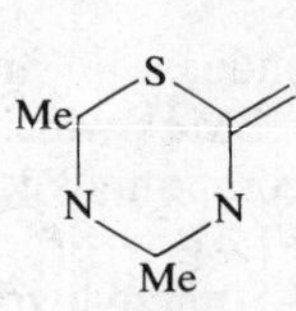

(263) 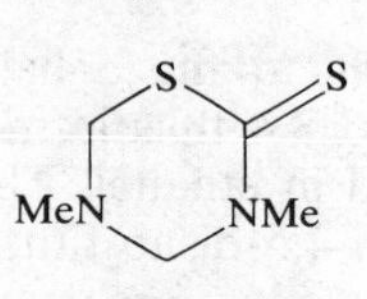 (264)

as vulcanizing agents for synthetic rubber. Methyleneimines react with formaldehyde and hydrogen sulphide under various conditions to give thiazetidines, dihydro-1,3,5-dithiazines, or tetrahydro-1,3,5-thiadiazines.

Thiobenzoyl isocyanate undergoes 1,4-addition with phenyl isothiocyanate, carbodi-imides, and azomethines to give 4-oxo-1,3,5-thiadiazine derivatives, and it also dimerizes on standing, giving the 3-thioacyl derivative (**265**). When this is heated it loses carbonyl sulphide and gives 4-oxo-2,6-diphenyl-1,3,5-thiadiazine (Scheme 30). Similar reactions have been carried out with dialkylcarbamyl isothiocyanates, dialkylthiocarbamyl isocyanates, and dialkylthiocarbamyl isothiocyanates.[70,71]

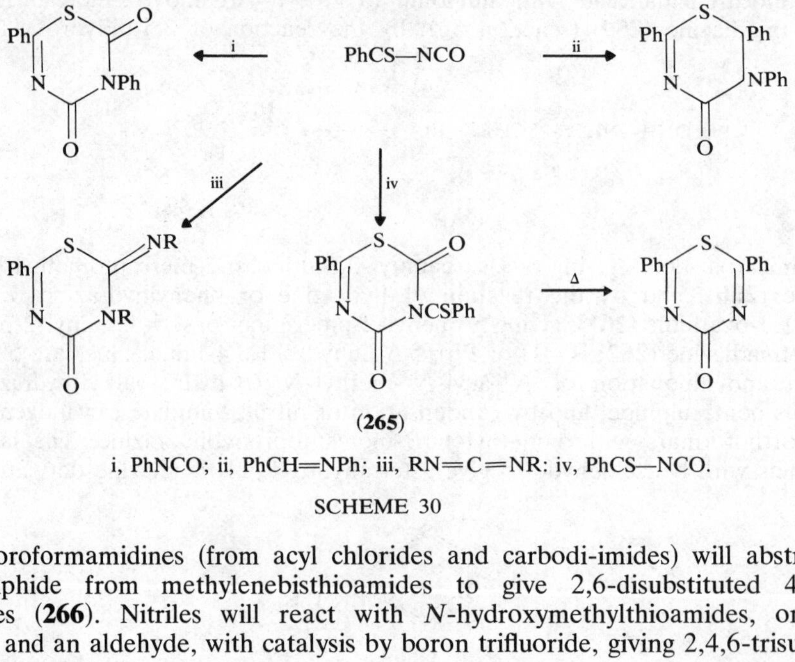

(**265**)

i, PhNCO; ii, PhCH=NPh; iii, RN=C=NR; iv, PhCS—NCO.

SCHEME 30

Acylchloroformamidines (from acyl chlorides and carbodi-imides) will abstract hydrogen sulphide from methylenebisthioamides to give 2,6-disubstituted 4*H*-1,3,5-thiadiazines (**266**). Nitriles will react with *N*-hydroxymethylthioamides, or with a thioamide and an aldehyde, with catalysis by boron trifluoride, giving 2,4,6-trisubstituted

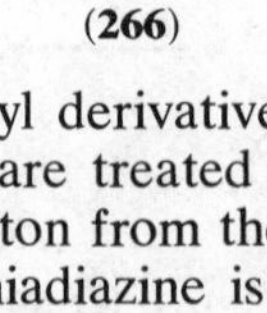

(**266**)

4*H*-1,3,5-thiadiazines (**267**). The triaryl derivatives readily contract to triarylimidazoles with extrusion of sulphur when they are treated with triethylamine (reaction 57). It is suggested that the base abstracts a proton from the 4-position to give an unstable anion. 2,6-Diamino-4,4-dimethyl-4*H*-1,3,5-thiadiazine is obtained by condensation of acetone with thiodiformamidine.

$$\mathrm{ArCN + RCHO + Ar'CSNH_2} \longrightarrow (\mathbf{267}) \longrightarrow \text{imidazole} \quad (57)$$

(**267**)

Substituted amidinothioureas of the type (**268**) react with thiophosgene, giving 2,4-diamino-6*H*-1,3,5-thiadiazine-6-thiones (**269**), which can be *S*-methylated with methyl iodide to give 2,6-diamino-6-methylthio-1,3,5-thiadiazinium salts (**270**) (reaction 58). Primary and secondary amines readily displace methanethiol from these salts, giving derivatives of 2,4,6-triamino-1,3,5-thiadiazine.[72]

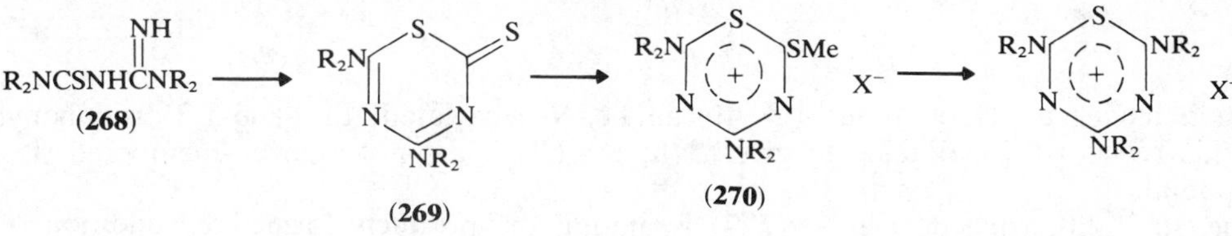

$R_2NCSNHC(=NH)NR_2$ (**268**) → (**269**) → (**270**) X^- → X^- (58)

Representatives of five of the six possible dithiazine systems have been prepared, the oldest being 5,6-dihydro-2,4,6-trimethyl-1,3,5-dithiazine, thialdine (**271**), which was made in 1847 by the action of hydrogen sulphide on acetaldehyde–ammonia. Thialdine has been identified among the substances responsible for the flavour of beef broth, and other 1,3,5-dithiazines have been proposed as flavour enchancement agents.

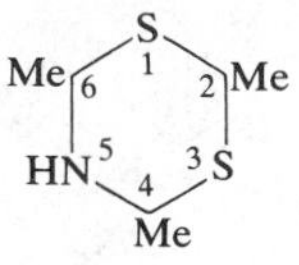

(**271**) 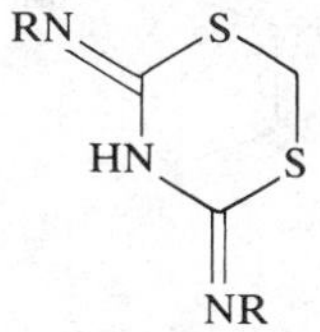 (**273**) 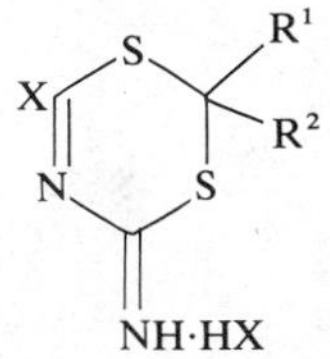 (**274**)

Thialdine is a weak base (pK_a 2.85 in aqueous ethanol). It is oxidized to ethanedisulphonic acid by aqueous permanganate and is hydrolysed by dilute sulphuric acid, giving trithioacetaldehyde. Many other dihydro-1,3,5-dithiazines have been made by the reaction of hydrogen sulphide and primary amines with aldehydes, both aliphatic and aromatic, and have been claimed to have use as pesticides or as rubber vulcanizing agents. Tri- and tetra-phenyl derivatives (**272**) are obtained by the photochemical addition of thiobenzophenone to benzaldimines (reaction 59).

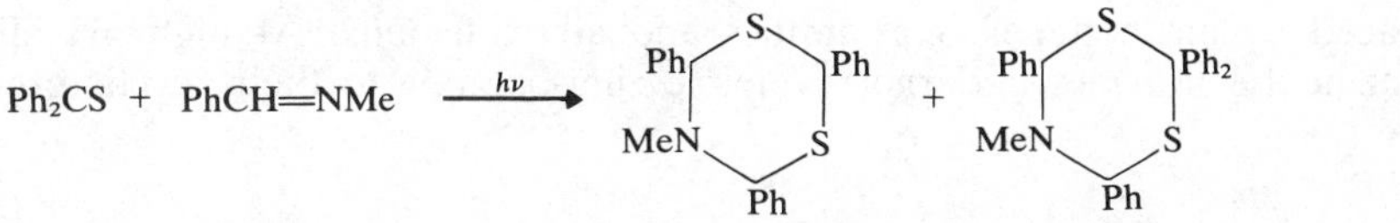

$Ph_2CS + PhCH{=}NMe \xrightarrow{h\nu}$ + (59)

(**272**)

4,6,-Di-iminodihydro-1,3,5-dithiazine (**273**; R = H) and its 2-phenyl derivative are obtained from thiourea and methylene dihalides or benzal chloride, the first products of the reaction (methylenedi-isothiouronium salts) being unstable and cyclizing with loss of ammonia. Methylene iodide or bromide reacts with phenylthiourea or with 1,5-diphenyldithiobiuret to give (**273**; R = Ph), which is more stable than the unsubstituted compound. 6-Halogeno-4-imino-1,3,5-dithiazine salts (**274**), useful as pesticides, are made by treating bisthiocyanates $R^1R^2C(SCN)_2$ with hydrogen bromide or hydrogen iodide in an inert solvent.

3-Anilino-5,6-dihydro-1,2,4-dithiazine (**275**) is obtained by the action of hydrochloric acid on 2-phenylthioureidoethyl thiosulphate.

1,3,2-Dithiazines are known only as the 1,1,3,3-tetroxides, which are the imides of propane-1,3-disulphonic acid. They are made by treating 3-chloropropanesulphonamide with ammonium sulphite and cyclization of the resulting 3-sulphopropanesulphonamide with phosphoryl chloride. The parent compound (**276**; R = H) is strongly acidic, a 1%

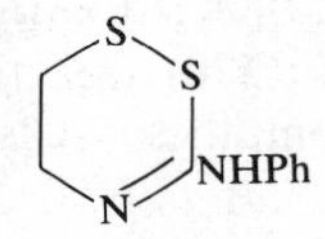

(275) 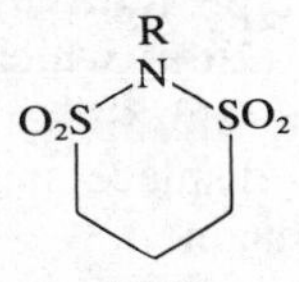 (276) 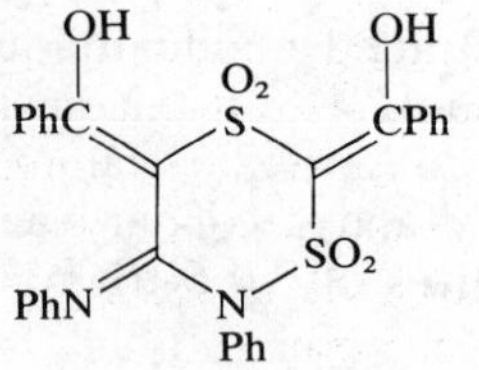 (277)

solution having a pH of about 1.5. It can be *N*-benzylated. Propane-1,3-disulphonyl chloride reacts with hydrazine to give (**276**; $R=NH_2$) and not a seven-membered ring compound.

The 1,5,2-dithiazine derivative (**277**) is among the products formed by addition of benzoylsulphene to diphenylcarbodi-imide.

2-Methylthio-1,3-dithiolium salts react with azide to give unstable intermediates which lose nitrogen at room temperature, giving 3-methylthio-1,4,2-dithiazines (**278**). These are weakly basic compounds that decompose at 100–160 °C with extrusion of sulphur and formation of 3-methylthioisothiazoles (Scheme 31). Photochemical decomposition causes

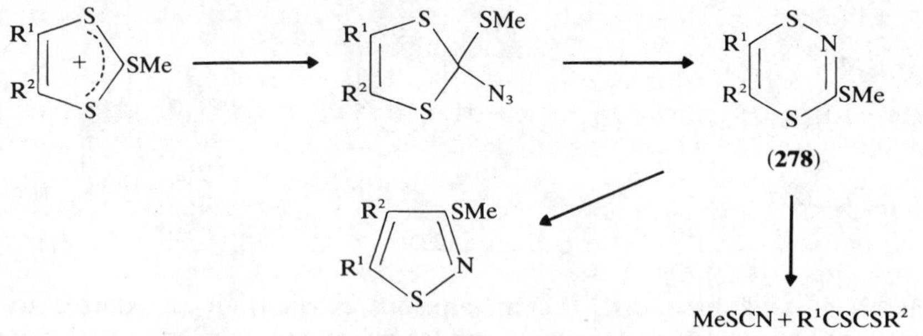

(278)

$MeSCN + R^1CSCSR^2$

SCHEME 31

elimination of methyl isothiocyanate and leaves a 1,2-dithione.[73] Vinylsulphonamide and styrylsulphonamides undergo base-catalysed cycloadditions with carbon disulphide or isothiocyanates to give 3-mercapto- or 3-amino-5,6-dihydro-1,4,2-dithiazine 1,1-dioxides (**279**; X = SH or NHR^2) (reaction 60). The mercapto-group may be methylated and the methylthio-group may be replaced by chlorine (by chlorination); the chlorine can be displaced by nucleophiles, *e.g.* amines and alkoxide ions. At high pH all these 1,4,2-dithiazine 1,1-dioxides undergo reversible ring-opening to their acyclic precursors.[74]

$$R^1CH{=}CHSO_2NH_2 \longrightarrow R^1CH{=}CHSO_2NHCSX \rightleftharpoons \text{(279)} \qquad (60)$$

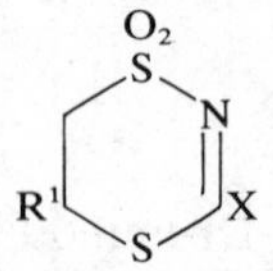

(279)

20.2.3.5 Benzothiadiazines[34]

Five of the six thiadiazine systems can give benzo-derivatives, and examples of all five types are known.

The ring closure of *o*-aminobenzenesulphonamide with formic acid to give benzo-1,2,4-thiadiazine 1,1-dioxide (**280**) has been known since 1902, and the preparation of the 3-methyl derivative by heating 2-acetamidobenzenesulphonamide was described in the following year. *o*-Aminobenzenesulphonamides were later found to react with formaldehyde in alkaline solution, giving 3,4-dihydro-2*H*-1,2,4-thiadiazine 1,1-dioxides, and the reaction was extended to other aldehydes with the use of an acid catalyst. During an investigation of benzene-1,3-disulphonamides with diuretic activity it was found that the formyl derivative of 6-amino-4-chlorobenzene-1,3-disulphonamide (**281**) has only a transitory existence and it rapidly cyclizes to the 1,2,4-benzodithiazine 1,1-dioxide chlorothiazide (**282**), a most potent diuretic agent with low toxicity. The publication of this discovery in 1957 stimulated the investigation of several hundred similar compounds.[75] The 3,4-dihydro-compound (hydrochlorothiazide) obtained by reduction of (**282**) with sodium borohydride or by treatment of (**281**) with formaldehyde proved to be more active, and further improvements were achieved by replacing the chlorine atom by a trifluoromethyl group (flumethazide and hydroflumethazide) and by variations in the substituent at the 3-position. These compounds are slowly hydrolysed in

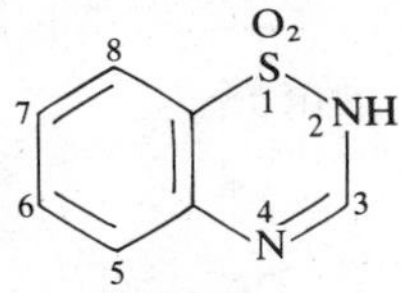

(**280**) 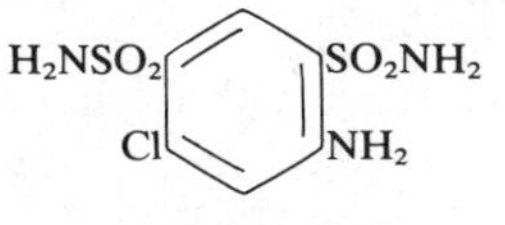 (**281**) 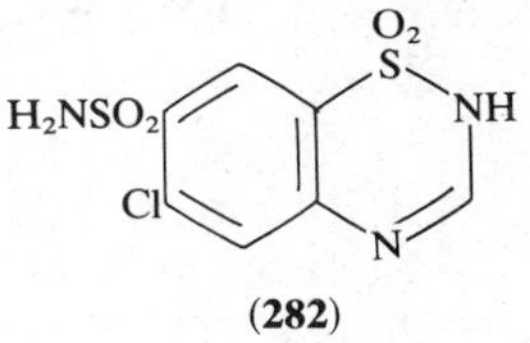 (**282**)

aqueous solution (acid, neutral, or alkaline) with cleavage of the 2,3-bond, the dihydro-compounds being more stable than chlorothiazide and the 3-substituted compounds generally less stable. Hydrochlorothiazide is stable to catalytic hydrogenation or to reduction by lithium aluminium hydride, but the 2,3-bond can be cleaved by sodium borohydride in methanol or by trimethylamine-borane in acetic acid. High diuretic activity is found only in compounds having a sulphamyl group in the 7-position, but many 1,2,4-benzothiadiazine 1,1-dioxides have anti-hypertensive activity and 7-chloro-3-methyl-1,2,4-benzothiadiazine 1,1-dioxide (diazoxide) is used in the treatment of hypertension. These compounds may be made by reaction of orthoesters with *o*-aminobenzenesulphonamides, or by base-catalysed ring-closure of *o*-aminobenzenesulphonylamidines. They are also obtained by a rearrangement of 4-alkyl-3-aryl-2-oxo-1,2,3,5-oxathiadiazoles (**283**) on heating in toluene (reaction 61). Reaction of *o*-aminobenzenesulphonamides with anhydrides of dibasic acids (*e.g.* succinic or phthalic acid) gives tricyclic compounds incorporating the benzo-1,2,4-thiadiazine system, *e.g.* (**284**) from succinic anhydride and 2-amino-5-chloro-4-methylbenzenesulphonamide.

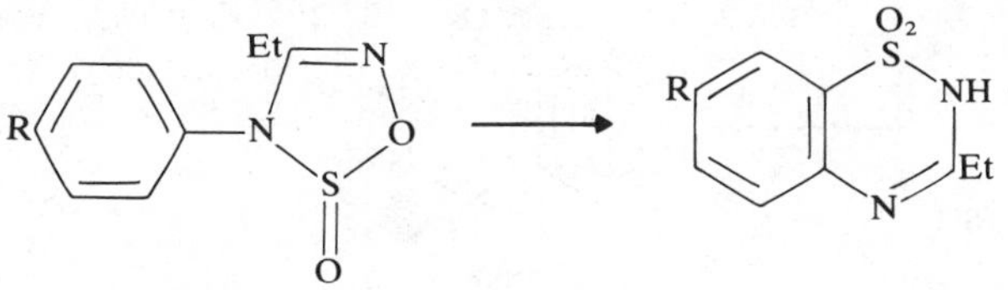

(**283**) (61)

Treatment of 4-ureidotoluene-3-sulphonamide with phosphoryl chloride gives 3-chloro-6-methyl-1,2,4-benzothiadiazine 1,1-dioxide (**285**), an acidic compound in which the chlorine atom is relatively inert. The 3,4-dihydro-3-oxo-compounds are more conveniently made by heating *o*-aminobenzenesulphonamides with urea or by reaction with

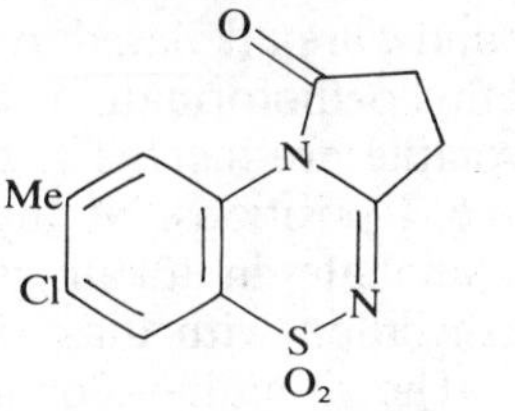

(**284**) 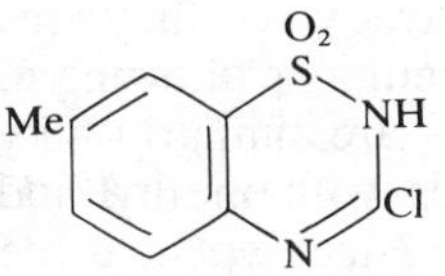 (**285**)

phosgene; they are also obtained by the action of phenyl isocyanate on *o*-aminobenzenesulphonamides. Alkyl and aryl isothiocyanates, on the other hand, react with *o*-aminobenzenesulphonamides to give 3-alkylimino- or 3-arylimino-3,4-dihydro-1,2,4-benzothiadiazine 1,1-dioxides (**286**) (reaction 62). The unsubstituted 3-imino-3,4-dihydro-compound (or 3-amino-compound) is obtained by fusion of *o*-aminobenzenesulphonamide with guanidine carbonate or by alkaline hydrolysis of the pyridinium betaine (**287**), which is made by treating 3,4-dihydro-3-oxo-1,2,4-benzothiadiazine with arenesulphonyl chlorides in pyridine (reaction 63). The 3-imino-compounds are weakly

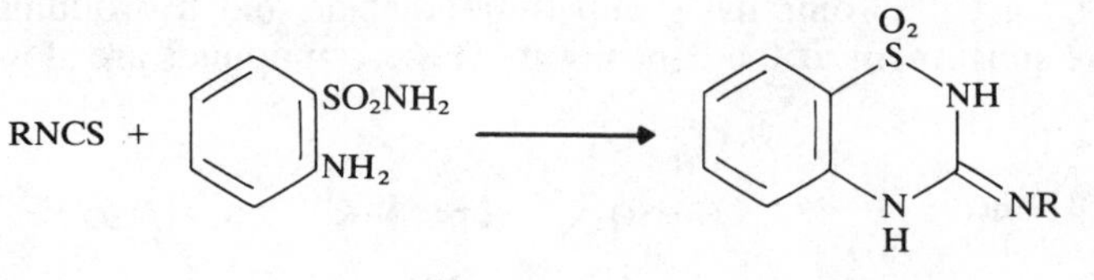

RNCS + [o-$H_2NC_6H_4SO_2NH_2$] → (**286**) (62)

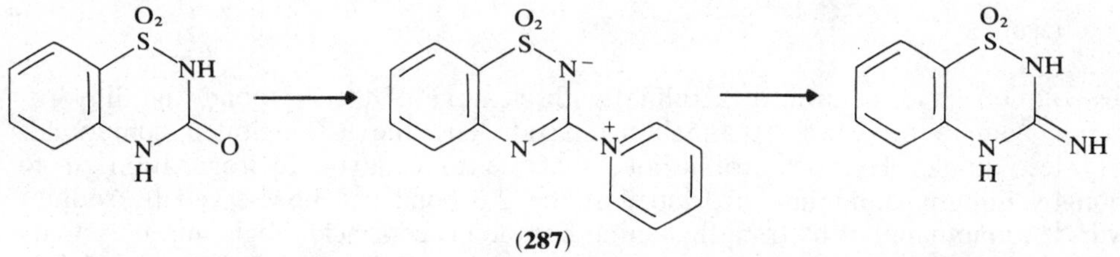

→ (**287**) → (63)

acidic (pK_a 9–11) and are stable to hydrolysis by acids or bases. Treatment of *o*-aminobenzenesulphonamide with thiophosgene gives 3,4-dihydro-1,2,4-benzothiadiazine-3-thione 1,1-dioxide, which is a fungicide. The tautomeric structure 3-mercapto-1,2,4-benzothiadiazine 1,1-dioxide was preferred by Raffa *et al.*,[76] who obtained it by fusion of *o*-aminobenzenesulphonamide with thiourea and showed that methylation with diazomethane gives the 2-methyl-3-methylthio-derivative (**288**) with a smaller proportion of the 4-methyl-3-methylthio-compound (**289**) (reaction 64). Methylation with methyl iodide in sodium bicarbonate solution gives 3-methylthio-1,2,4-benzothiadiazine 1,1-dioxide, from which 3-amino-derivatives are made by reaction with primary or secondary amines or hydrazine.

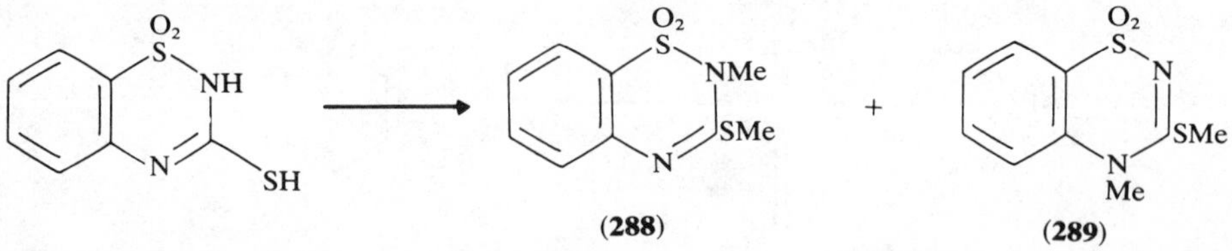

→ (**288**) + (**289**) (64)

The tautomerism of the 1,2,4-benzothiadiazine 1,1-dioxides has been investigated both by spectroscopy and by alkylation experiments. Molecular orbital calculations favour the 4*H*-structure and the u.v. spectra support this conclusion for solutions in ethanol but suggest that the 2*H*-tautomer is present in aqueous alkali. Methylation of 3-methyl-1,2,4-benzothiadiazine 1,1-dioxide gives the 3,4-dimethyl derivative, but methylation of hydrochlorothiazide gives the 2-methyl compound. 1,2,4-Benzothiadiazine 1,1-dioxides of the chlorothiazide type, but not the 3,4-dihydro-compounds, are alkylated to give both 2- and 4-alkyl derivatives when they are heated with trimethyl orthoformate or triethyl orthoformate under conditions allowing distillation of the volatile products. The corresponding 3-oxo-derivatives are similarly alkylated at the 2- and 4-positions. Methylation of the 3-oxo-compounds with methyl iodide or dimethyl sulphate in alkali gives isomeric 3-methoxy- and 2-methyl-3-oxo-derivatives, but methylation with diazomethane gives 4-methyl-3-methoxy and 2,4-dimethyl derivatives. The 3-amino- or 3-iminodihydro-derivatives appear to have the amino structure (from u.v. spectroscopic evidence) and they are alkylated by methyl iodide–potassium carbonate to give the 4-methyl derivative; further methylation occurs on the exocyclic nitrogen atom to give methylamino- and dimethylamino-derivatives. Diazomethane, however, converts the 3-amino-compound into both 2- and 4-methyl derivatives.

3-Alkoxy-1,2,4-benzothiadiazine 1,1-dioxides are isomerized to 2-alkyl-3,4-dihydro-3-oxo-1,2,4-benzothiadiazine 1,1-dioxides when they are heated. If the alkyl group is α,β-unsaturated, it undergoes inversion as in the Claisen rearrangement.

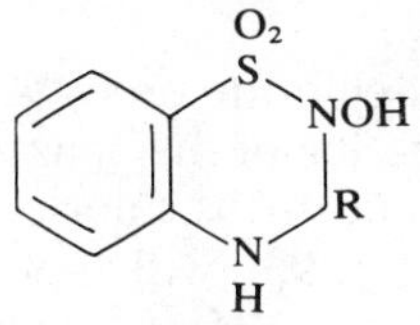

O_2 S NOH R N H

(290)

Treatment of *o*-aminobenzenesulphonylhydroxylamine with aldehydes or with phosgene gives 2-hydroxy-3-alkyl (or 3-oxo) 3,4-dihydro-1,2,4-benzothiadiazine 1,1-dioxides (**290**) and not the isomeric seven-membered ring compounds (benzoxathiadiazepines). 3-Chloromethyl-3,4-dihydro-1,2,4-benzothiadiazine 1,1-dioxides undergo ring-enlargement to give benzothiadiazepine derivatives (**291**) when treated with sodium methoxide (reaction 65).

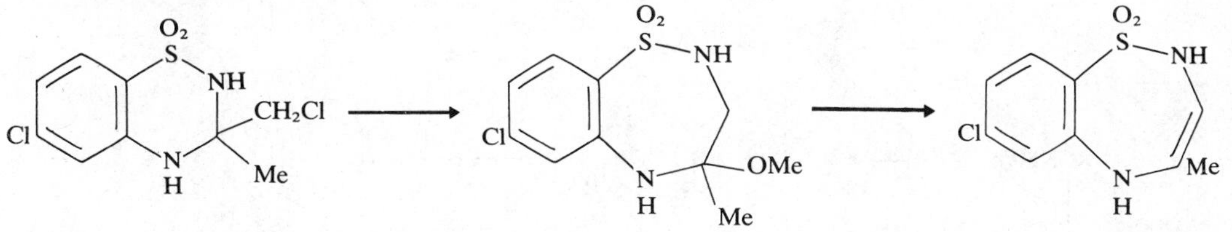

O_2 S NH CH_2Cl Cl N H Me → O_2 S NH Cl N H OMe Me → O_2 S NH Cl N H Me (65)

(291)

N-Arylamidines react with *N*-sulphinylsulphonamides (**292**; $R^2 = Me$, Ph, *p*-tolyl), giving 1-sulphonylimido-2*H*-1,2,4-benzothiadiazines (**293**), which undergo acid hydrolysis to give the 1-oxides (**294**). Thionyl chloride reacts with *N*-phenylbenzamidine, giving 7-chloro-3-phenyl-2*H*-1,2,4-benzothiadiazine 1-oxide (**294**; $R^1 = Cl$) as well as 7-chloro-3-phenyl-2*H*-1,2,4-benzothiadiazine (**295**) (Scheme 32). The latter compound is obtained in

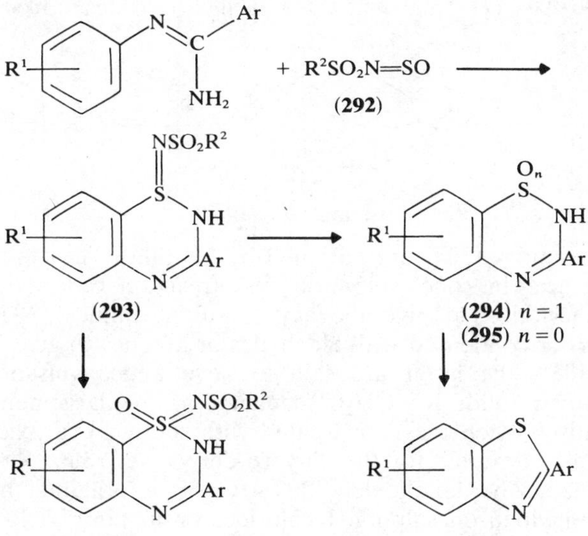

R^1 N C Ar NH_2 + $R^2SO_2N{=}SO$ (**292**) →

NSO_2R^2 S NH R^1 Ar N (**293**) → O_n S NH R^1 Ar N (**294**) $n = 1$ (**295**) $n = 0$

O S NSO_2R^2 NH R^1 Ar N (**297**)

S R^1 Ar N (**296**)

SCHEME 32

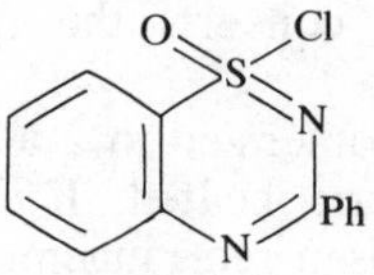

(298) 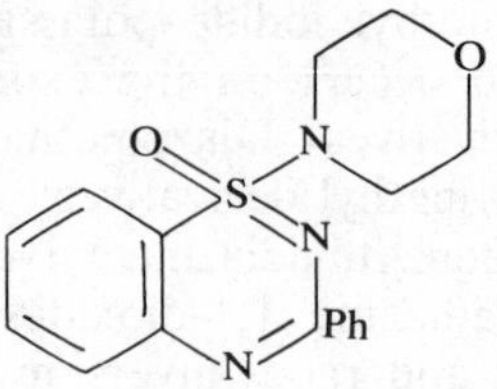 (299)

better yield by the action of thionyl chloride on (**294**; $R^1 = H$), reduction of the sulphinyl group being accompanied by chlorination of the benzene ring. Reduction of (**294**) with zinc and acetic acid or tin(II) chloride and hydrochloric acid causes ring-contraction to give benzothiazoles (**296**). Oxidation of (**293**) with permanganate gives the corresponding sulphoximine (**297**).[77] Chlorination of (**294**; $R^1 = H$, Ar = Ph) gives 1-chloro-3-phenyl-1*H*-1,2,4-benzothiadiazine 1-oxide (**298**), which may be hydrolysed to 3-phenyl-2*H*-1,2,4-benzothiadiazine 1,1-dioxide or treated with morpholine to give the 1-morpholino-compound (**299**). Treatment of *N*-phenylamidines (**300**; R = CCl_3 or 2,4-$Cl_2C_6H_3$) with sulphur dichloride gives 1,7-dichloro-1*H*-1,2,4-benzothiadiazines (**301**), from which the 1-morpholino-derivatives are obtained by reaction with morpholine (reaction 66).

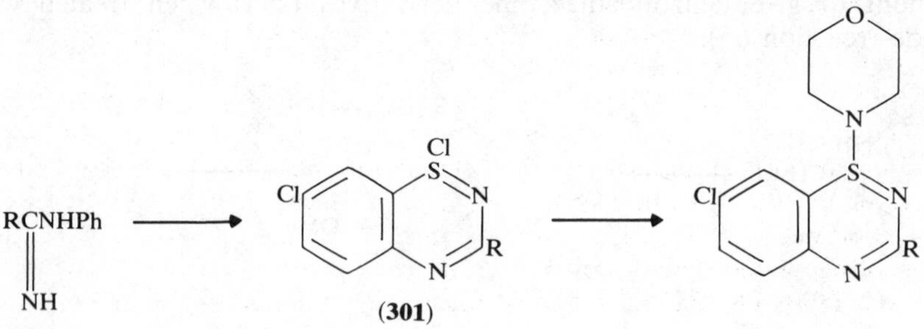

(66)

(300) (301)

o-Alkylthiophenylureas and *o*-arylthiophenylureas are cyclized by bromine and sodium methoxide in anhydrous methanol to 1-alkyl (or 1-aryl) 3,4-dihydro-3-oxo-1*H*-1,2,4-benzothiadiazines (**302**), which are extensively hydrolysed to acyclic sulphoxides by moist solvents but may be oxidized by permanganate in acetonitrile to give stable 1-alkyl (or 1-aryl) 3,4-dihydro-3-oxo-1*H*-1,2,4-benzothiadiazine 1-oxides (**303**) (reaction 67).[78]

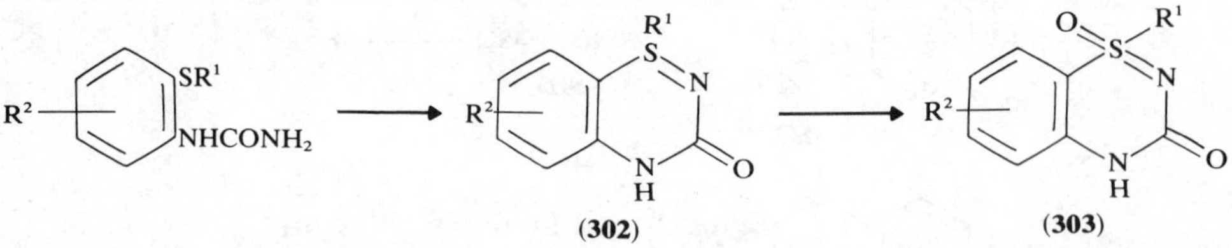

(67)

(302) (303)

Similar cyclic sulphoximides (**304**) are obtained by treating *o*-acylaminophenyl sulphoxides with hydrazoic acid in conc. sulphuric acid (reaction 68); the reaction is often accompanied by deacylation to give the acyclic sulphoximines (**305**), which may be acylated to give (**304**) or condensed with aldehydes or ketones to give 3,4-dihydro-1,2,4-benzothiadiazines (**306**); the latter are said to have anticonvulsant activity. The 3-chloromethyl derivatives (**304**; R = CH_2Cl) undergo ring-enlargement to 1,2,5-benzothiadiazepine derivatives when they are treated with sodium hydroxide solution, and a ring-enlargement is also reported for the dihydro-derivatives (**306**; R = CH_2Cl).[79]

2*H*-1,2,3-Benzothiadiazine 1,1-dioxide (**307**; R = H) is obtained by the reaction of hydrazine with *o*-formylbenzenesulphonyl chloride, or in poor yield by cyclization of *o*-sulphobenzaldehyde hydrazone with phosphorus pentachloride and phosphoryl chloride. It may be pyrolysed to give small amounts of the sultine (**308**; R = H) and

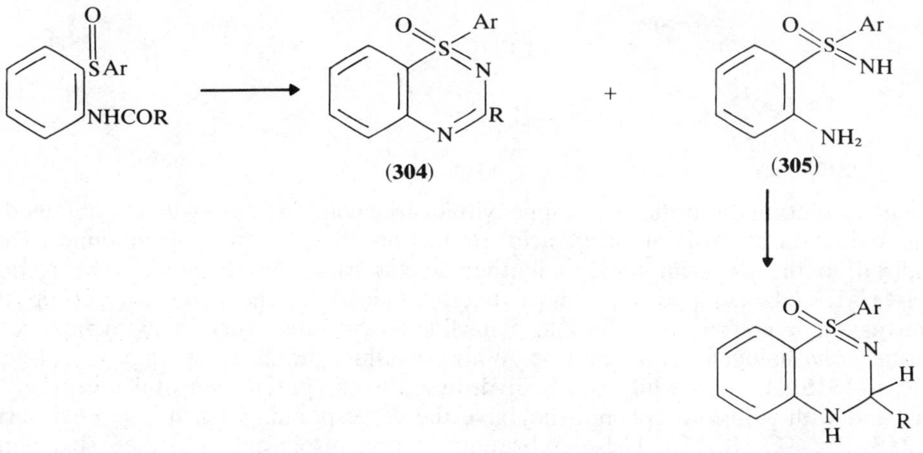

(304) (305) (306) (68)

chlorinated to give 3-chloro-3*H*-2,1-benzoxathiol 1-oxide (**309**). 4-Aryl derivatives (**307**; R = Ar) are made from hydrazine and 2-chlorosulphonylbenzophenones. These compounds are acidic and can be alkylated or acylated at the 2-position. Some 2-alkyl, 2-alkylcarbamyl, and 2-arylcarbamyl derivatives of (**307**; R = H) have been described as antibacterials and fungicides. Hydrogenation gives 3,4-dihydro-derivatives. The 3,4-dihydro-4-oxo-derivative is made by reaction of hydrazine with isopropyl 2-chlorosulphonylbenzoate. Treatment of 2-cyanobenzenesulphonyl chloride with hydrazine gives 4-hydrazino-2*H*-1,2,3-benzothiadiazine 1,1-dioxide (**307**; R = $NHNH_2$), which shows typical reactions of the hydrazine group. Oxidation of the isopropylidene derivative of 7-ethoxy-4-hydrazino-2*H*-1,2,3-benzothiadiazine 1,1-dioxide gives the stable 4-imino-4*H*-1,2,3-benzothiadiazine derivative (**310**). The related ring system naphtho[1,8-*de*]-1,2,3-thiadiazine 1,1-dioxide (**311**) is made by diazotizing 8-aminonaphthalene-1-sulphinic acid; on photolysis this compound loses nitrogen to give naphtho[1,8-*bc*]thietan 1,1-dioxide. Hydrazobenzene reacts with 2-chlorosulphonylbenzoyl chloride in the presence of triethylamine, giving 3,4-dihydro-4-oxo-2,3-diphenyl-2*H*-1,2,3-benzothiadiazine 1,1-dioxide (**312**).

2-Phenylazo-4*H*-1,3,4-benzothiadiazine (**313**) is a purple compound formed by atmospheric oxidation of dithizone (**314**) in acid solution. It is conveniently made by

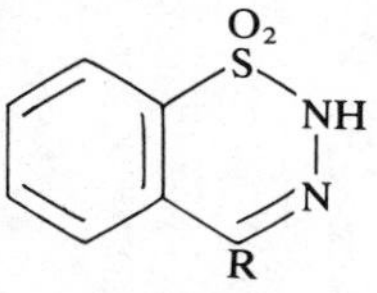

(307) 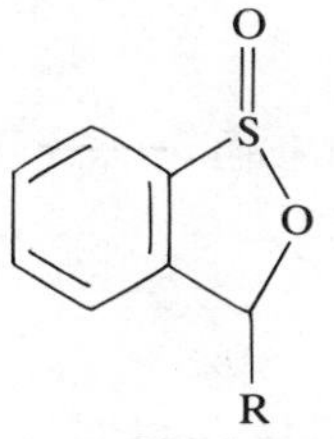 (308) 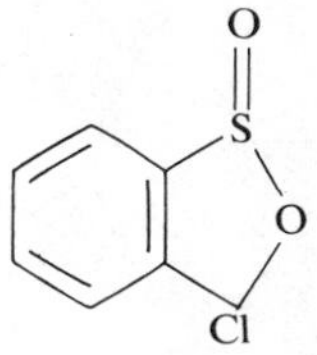 (309)

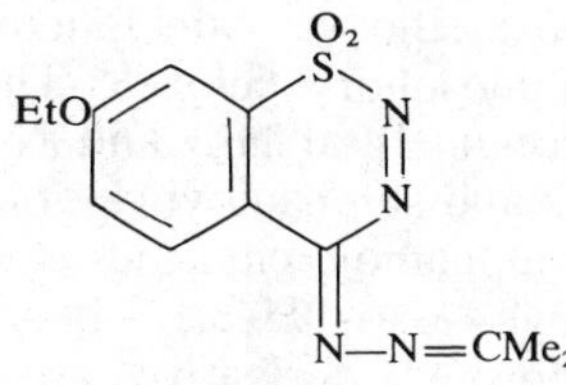

(310) 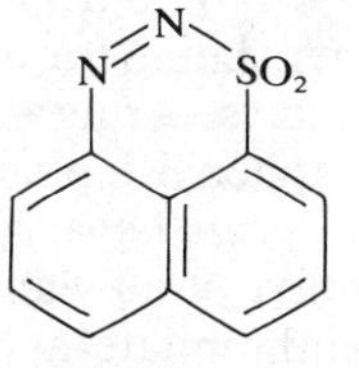 (311) 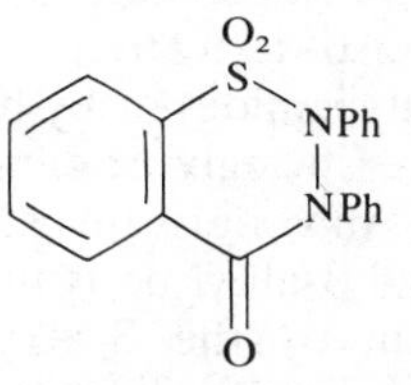 (312)

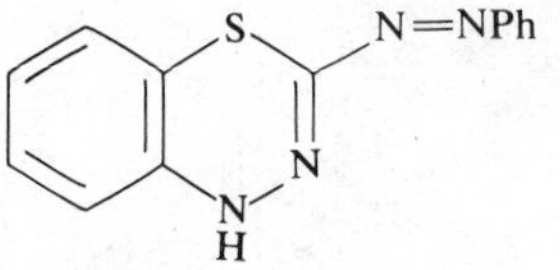

(313) 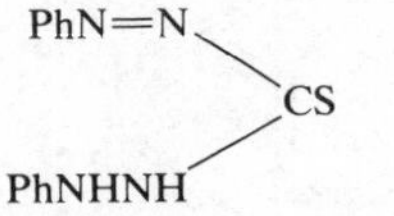 (314) 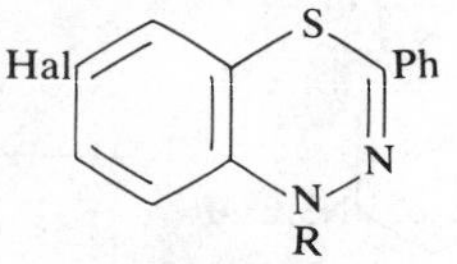 (315)

refluxing a solution of dithizone, diphenylthiocarbazide, or the sydnone obtained by alkaline oxidation of (**314**), in acetic acid. Reduction gives a dihydro-compound (readily re-oxidized to the azo-compound) and then a colourless amine. Some 7-halogeno-2-phenyl-4*H*-1,3,4-benzothiadiazines have been obtained by the action of certain thiol derivatives on α-chloro- or α-bromo-benzylidene 2,4-dihalogenophenylhydrazones in which the *ortho* halogen is not chlorine. With potassium dithioacetate they give 4-acetyl derivatives (**315**; R = Ac), which can be hydrolysed to the parent benzothiadiazines (**315**; R = H), and with *p*-nitrothiophenol they give the corresponding 4-*p*-nitrophenyl derivatives (**315**; R = 4-$O_2NC_6H_4$). These cyclizations involve displacement of the α-halogen by the thiol and migration of the acetyl or *p*-nitrophenyl group; mechanisms have been suggested.[80] A variation of this synthesis of rather wider application is the cyclization of *N*-acetyl-*N*-aryl-*N*′-thioaroylhydrazines by treatment with triethylamine. The aryl group must be 2,4-disubstituted with suitable electron-withdrawing substituents. An earlier report that 2-phenyl-4*H*-1,3,4-benzothiadiazine is obtained by heating *N*′-benzoyl-N^2-phenylhydrazine with sulphur is erroneous; the product is 2-phenylbenzothiazole. The name preferred in the Ring Index for these compounds is 1*H*-benzo-4,1,2-thiadiazine, and this nomenclature is used in some publications.

1*H*-2,1,3-Benzothiadiazine 2,2-dioxides (**316**) are obtained by heating *o*-aminoaryl ketones with sulphamide (reaction 69). They may be alkylated at position-1, and hydrogenation converts them into 3,4-dihydro-derivatives (**317**). The latter compounds can also be made by heating *o*-aminobenzylamines with sulphamide in pyridine (reaction 70), or

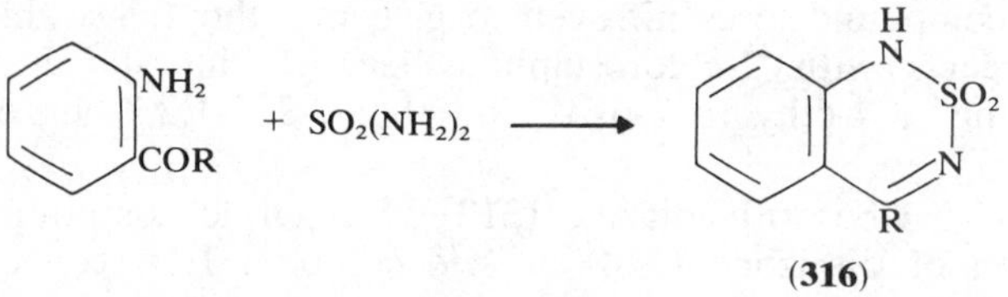

$$+\ SO_2(NH_2)_2 \longrightarrow \quad (69)$$

(316)

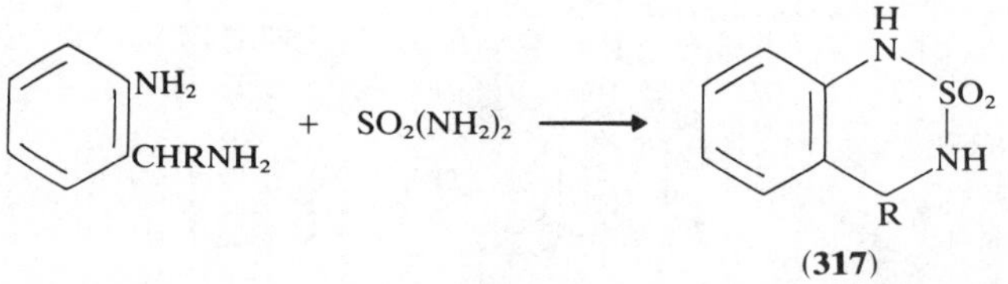

$$+\ SO_2(NH_2)_2 \longrightarrow \quad (70)$$

(317)

by treating them with sulphuryl chloride (which also effects nuclear chlorination). The dihydro-compounds are oxidized to (**316**) by permanganate. Their disodium salts may be alkylated with 1,2-dibromoethane, 1,3-dibromopropane, or *o*-xylylene dibromide to give 1,3-bridged derivatives, two stereoisomers being formed from (**317**; R = Ph). 3,4-Dihydro-4-oxo-derivatives are made by base-catalysed ring closure of *o*-methoxycarbonylphenylsulphamides or by heating an *o*-carboxyphenylsulphamide with phosphoryl chloride. They are strongly acidic substances and they can be alkylated successively at N-3 and N-1. 1-Alkyl derivatives are made from methyl *N*-alkylanthranilates and sulphamoyl chloride, and 3-alkyl derivatives from methyl anthranilate and *N*-alkylsulphamoyl chlorides. Certain of the 3-alkyl compounds, *e.g.* 3,4-dihydro-3-isopropyl-4-oxo-1*H*-2,1,3-benzothiadiazine 2,2-dioxide (bentazon) (**318**) and the 3-(1-methylpropyl) derivative, are of importance as herbicides. 1-Halogenoanthraquinones condense with sulphamide to give

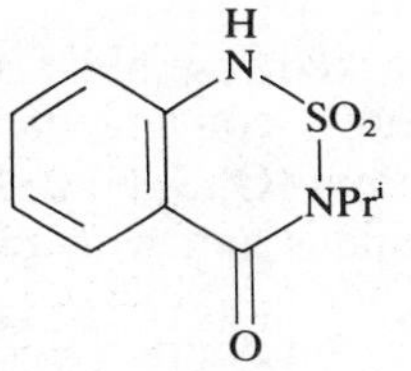

(318) 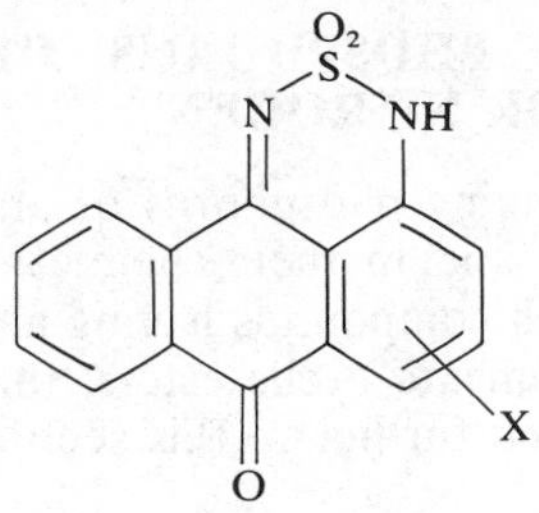 (319) 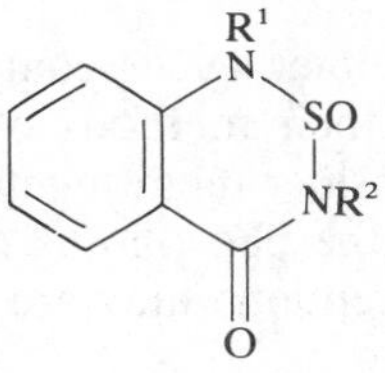 (320)

2,1,3-benzothiadiazine derivatives (**319**) that are useful as dyestuffs and intermediates. 2-Alkylamino-*N*-alkyl (or aralkyl) benzamides having an electron-withdrawing substituent in the 5-position react with thionyl chloride to give 1,3-disubstituted 3,4-dihydro-4-oxo-1*H*-2,1,3-benzothiadiazine 2-oxides (**320**). The unsubstituted compounds (**320**; R^1 and $R^2 = H$) and the analogous 4-thiones are made by the action of *N*-sulphinyl-*p*-toluenesulphonamide on anthranilamide or 2-aminothiobenzamide. Oxidation of (**320**; $R^1 = R^2 = H$) with hydrogen peroxide at 20 °C gives the 2,2-dioxide, but at 100 °C it is converted into indazolone with extrusion of sulphur dioxide. *o*-Aminobenzamide reacts with *N,N'*-bis(toluenesulphonyl)sulphur di-imide, giving 1,3-didehydro-4-oxo-2,1,3-benzothiadiazine (**321**), which is thermally stable but very sensitive to moisture, controlled hydrolysis giving (**320**; $R^1 = R^2 = H$).[81] Naphtho[1,8-*cd*]-1,2,6-thiadiazine (**322**) is obtained by treating 1,8-diaminonaphthalene with *N*-sulphinylaniline or with sulphur dioxide in the presence of triethylamine. With a weaker base (pyridine) or no base, the cyclic sulphinamide (**323**) is obtained. This is unstable and is readily hydrolysed to 1,8-diaminonaphthalene.

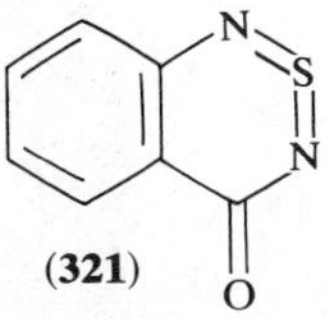

(321) 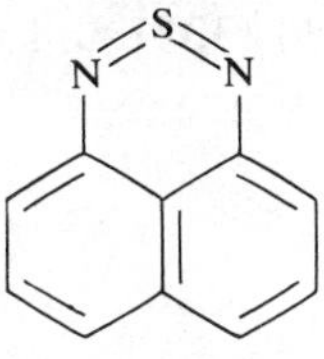 (322) 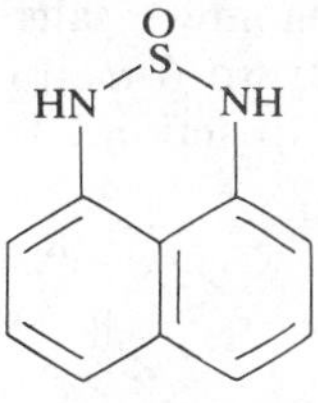 (323)

3-Anilino-1*H*-2,1,4-benzothiadiazine (**324**; R = Ph) is obtained by oxidation of *N'*-(2-aminophenyl)-N^2-phenylthiourea with iron(III) chloride. Similar compounds having carbamate substituents (**324**; $R = CO_2Me$) are made by oxidation of N^1-(2-aminophenyl)-N^2-methoxycarbonylthioureas, *e.g.* with *N*-chlorosuccinimide, or by reduction of the corresponding 2-nitrophenylthioureas with sodium dithionite. They readily suffer a ring-contraction to give the corresponding benzimidazoles, and this reaction is accelerated by acid or by triphenylphosphine. Oxidation with *m*-chloroperoxybenzoic acid gives the corresponding benzothiadiazine 2-oxides, which also decompose readily to give benzimidazoles.[82] The 6-oxo-6*H*-2,1,4-benzothiadiazine derivative (**325**) is obtained by reaction of methyl isothiocyanatoformate with 2-amino-4-hydroxyazobenzene.

4-Oxo-3,1,2-benzothiadiazine (**326**) is obtained by diazotization of thioanthranilic acid in a non-aqueous medium. It is more stable than a diazonium salt, but when heated or photolysed it loses nitrogen, and at higher temperatures carbonyl sulphide, to give 3*H*-1,2-benzodithiol-3-one and products indicating the probable formation of benzyne.

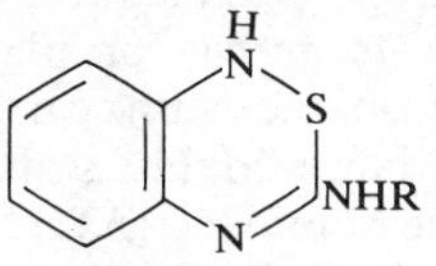

(324) 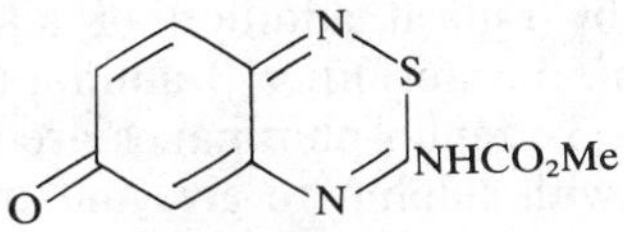 (325) 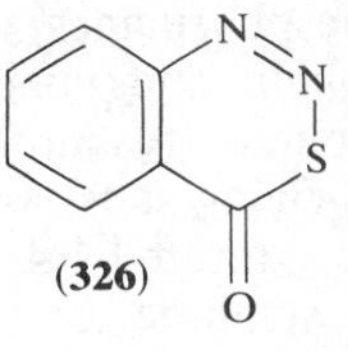 (326)

20.2.4 RINGS CONTAINING PHOSPHORUS OR ARSENIC AND OXYGEN, SULPHUR, OR NITROGEN[83]

Heterocyclic compounds containing phosphorus or arsenic show marked resemblances in their methods of preparation and in their chemical properties and is convenient to consider them together. Most such compounds having another heteroatom (O, S, N) *ortho* to the phosphorus or arsenic atom are cyclic esters, thioesters, or amides and with two exceptions they are not considered further in this section.

20.2.4.1 Monocyclic compounds

The unsaturated compounds with two double bonds in the ring are called oxaphosphorin, thiaphosphorin, oxarsenin, and thiarsenin, and the saturated compounds are oxaphosphorinan, oxarsenan, *etc.* Azaphosphorine and azarsenine have three double bonds in the ring. The naming of some compounds presents difficulty and substitutive nomenclature is sometimes used.

Phosphonodithioic anhydrides (**327**) undergo cycloaddition with dienes at 100 °C, giving 1,2-thiaphosphorin sulphides (**328**) (reaction 71). When these are treated with strong bases in aprotic solvents (*e.g.* sodium hydride in diglyme) they undergo cleavage of the 1–6 bond to give dienephosphinodithioic acids. By treatment of phenylphosphonodithioic acid dichloride with magnesium in the presence of dimethylbutadiene, the transient intermediate (**329**) is trapped (reaction 72) to give 4,5-dimethyl-2-phenyl-3*H*,6*H*-1,2-thiaphosphorin (**330**), which is attacked by oxygen or by sulphur and is isolated as the sulphide (**328**; $R^1 = Ph$, $R^2 = Me$) or the corresponding oxide. Tributyl-4-oxo-1-pentylphosphonium salts when treated with potassium t-butoxide give 2,2,2-tributyl-2,2,3,4-tetrahydro-6-methyl-1,2-oxaphosphorin (**331**) and its siomer with the double bond exocyclic (reaction 73).

(**327**) + diene → (**328**) (71)

$PhP(S)Cl_2$ → [PhP=S̈] (**329**) → (**330**) (72)

$Bu_3\overset{+}{P}[CH_2]_3COMe$ → (**331**) (73)

Few 1,3-compounds have been made. Acid-catalysed condensation of ketones with 3-hydroxypropylphosphines gives 1,3-oxaphosphorinans. The nitrogen analogues, 1,3-azaphosphorinanes, are made by radical addition of allylamine to butyl- or phenylphosphine and condensation of the resulting 3-aminopropylphosphine with carbonyl compounds (reaction 74). The 1,3-azaphosphorinanes are usually basic liquids, stable to acids, giving salts, and reacting with sulphur to give the crystalline sulphides (**332**).

The saturated 1,4-compounds were first prepared by reaction of 2,2′-dihalogenodiethyl ether, sulphide, or amines with the magnesium bromide compounds (**333**) obtained by

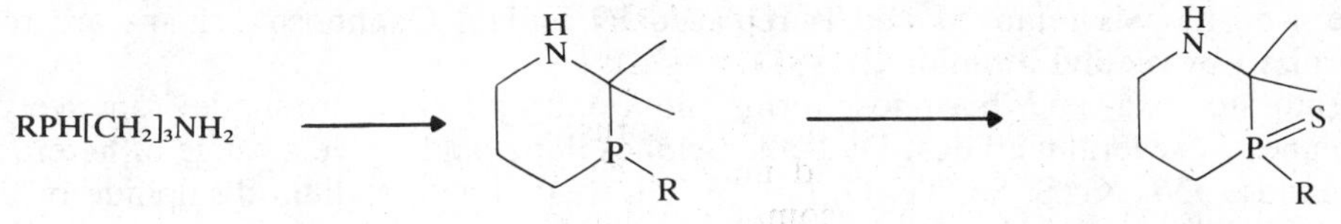

$$RPH[CH_2]_3NH_2 \longrightarrow \text{(2,2-dimethyl-1-R-1,3-azaphosphorinane)} \longrightarrow \text{(332)} \quad (74)$$

$$PhX(MgBr)_2 + (ClCH_2CH_2)_2Y \longrightarrow PhX\langle\ \rangle Y \quad (75)$$

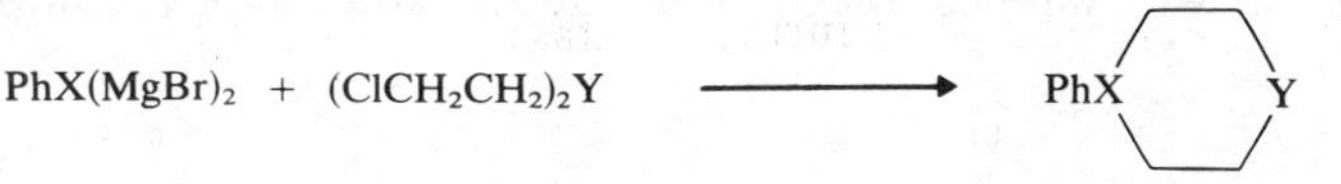

(**333**) X = P, As; Y = O, S, NR

treating phenyldichlorophosphine or phenyldichloroarsine with Grignard reagents (reaction 75). Phenyldichlorophosphine reacts with vinylmagnesium bromide to give phenyldivinylphosphine, which will combine with primary amines to yield 1,4-azaphosphorinanes. Phosphine and alkylphosphines will add to divinyl ether when irradiated or activated by a free-radical initiator, giving oxaphosphorinans. The sodium salt of phenylphosphine, prepared in liquid ammonia, may be alkylated with 2-aminoethyl chloride. A second alkylation gives di-(2-aminoethyl)phenylphosphine, which cyclizes with loss of ammonia when it is distilled. If the second alkylation is performed with chloroacetamide the product is 2-oxo-4-phenyl-1,4-azaphosphorinane. 1,4-Oxaphosphorinans have been made by dehydrating di-(2-hydroxyalkyl)phenylphosphines over alumina at 300 °C. The phosphorus or arsenic in these compounds is readily oxidized to a higher valency state. Thus hydrogen peroxide in acetone oxidizes 4-phenyl-1,4-oxarsenan to the 4-oxide (**334**; X = O), which is converted into the 4-sulphide (**334**; X = S) by hydrogen sulphide and into an ionic hydroxy-chloride by hydrochloric acid (reaction 76). Sulphur dioxide reduces the hydroxy-chloride to 4-phenyl-1,4-oxarsenan. The phosphorus compounds are oxidized to 4-oxides by hydrogen peroxide and to 4-sulphides by sulphur in boiling benzene. Oxaphosphorinan oxides and azaphosphorinane oxides and sulphides have been made directly from phenyldivinylphosphine oxide (or sulphide) by treatment with sodium hydroxide or amines. Diallyldiphenyl-arsonium and -phosphonium salts are isomerized in basic medium to dipropenyl compounds which undergo nucleophilic attack by dilute sodium hydroxide, sodium hydrogen sulphide, or primary amines to give the cyclic arsonium or phosphonium salts (**335**). Reduction of these (electrolytic, or by lithium aluminium hydride) removes one phenyl group to give the trivalent arsenic or phosphorus compound (reaction 77).[84]

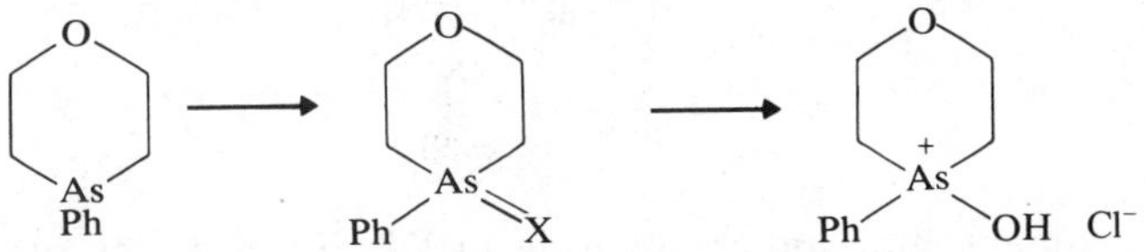

(76) [oxarsenan ring, As–Ph] → (**334**) [As(=X)Ph] → [$\overset{+}{As}$(Ph)OH Cl^-]

$$Ph_2\overset{+}{X}(CH_2CH{=}CH_2)_2 \longrightarrow Ph_2\overset{+}{X}(CH{=}CHMe)_2 \longrightarrow \text{(335)} \longrightarrow \text{(X–Ph ring)} \quad (77)$$

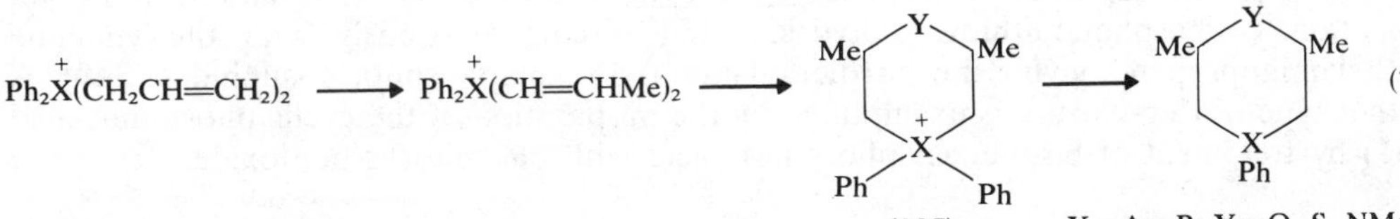

(**335**) X = As, P; Y = O, S, NMe

1,4-Diphenyl-1,4-azaphosphorinane is a weak base that is stable to air and to constant boiling hydriodic acid. It forms complexes with palladium salts. It gives a mono- and a di-hydrochloride and a monomethiodide quaternized on phosphorus. A dimethiodide is obtained with difficulty. 1,4-Diphenyl-1,4-azarsenane gives only a monomethiodide (quaternized on arsenic), and when it is heated with conc. hydriodic acid the phenyl group at the 4-position is removed and is replaced by iodine. Oxaphosphorinans are readily quaternized by methyl bromide.

Unsaturated oxa- and thia-phosphorins and azaphosphorines are made from acetylenic phosphines by several methods. Di-1-alkynylphosphine oxides give a range of heterocyclic compounds (**336**; X = S, Se, Te, O, NR^3) when treated with sodium disulphide (which is safer than sodium sulphide with intermediates sensitive to strong bases), sodium selenide, sodium telluride, or with amines (reaction 78). In the latter case the initial product is an

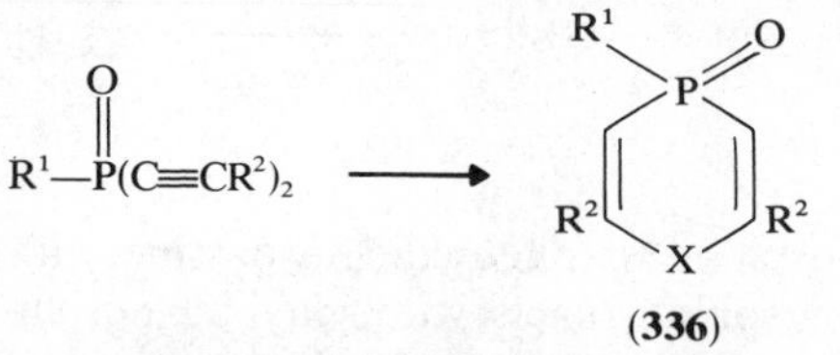

$$R^1{-}P(O)(C{\equiv}CR^2)_2 \longrightarrow \textbf{(336)} \quad (78)$$

(336)

enamine which may cyclize to (**336**; X = NR^3), but in most cases it is hydrolysed (in solution or during chromatography on basic alumina) to give an enol that cyclizes to (**336**; X = O).[85] Alternatively, phenyldi-(1-alkynyl)phosphine oxides will give bis-enamines with butylamine; acid hydrolysis of these gives phenyldi-(2-oxoalkyl)phosphine oxides which are converted into (**336**; X = NH) when heated with ammonium carbonate. When tertiary phosphines with one or more 1-alkynyl groups are treated with enolizable α-halogenoketones, the products are 4,4-disubstituted 4-phosphoniapyran salts (**337**). These are non-aromatic, non-planar compounds with the charge located on the phosphorus atom (reaction 79). Hydrolysis with sodium hydroxide removes a hydrocarbon residue, *e.g.* benzene, from the phosphorus and gives an oxaphosphorin oxide (**336**; X = O). Acid hydrolysis of (**337**; $R^1 = R^2 = R^3 = R^5 = Ph$, $R^4 = H$) gives a diphenyldiphenacylphosphonium salt, which reacts with ammonium acetate in acetic acid to give the cyclic phosphonium salt (**338**) (reaction 80). This is non-aromatic, but removal of hydrogen

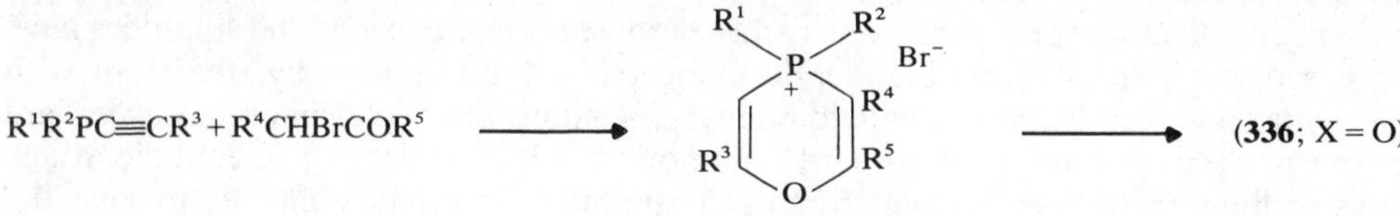

$$R^1R^2PC{\equiv}CR^3 + R^4CHBrCOR^5 \longrightarrow \textbf{(337)}\ Br^- \longrightarrow \textbf{(336; X = O)} \quad (79)$$

(337)

$$Ph_2\overset{+}{P}(CH_2COPh)_2\ Br^- \longrightarrow \textbf{(338)}\ Br^- \rightleftharpoons \textbf{(339)} \quad (80)$$

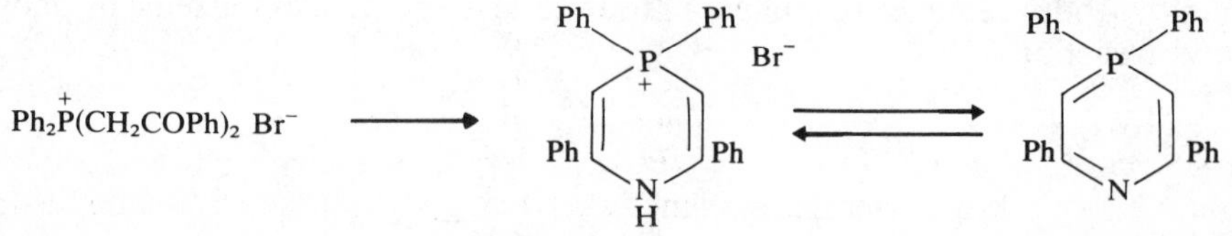

(338) **(339)**

bromide by treatment with sodium hydride in DMSO gives the aromatic base 2,4,4,6-tetraphenylphosphapyridine (**339**). This is reconverted into the non-aromatic form by salt formation.[86] Phosphonodithioic anhydrides (**327**) react vigorously with the ynamine diethylaminopropyne, giving 2,6-bis(diethylamino)-1,4-thiaphosphorin sulphides (**340**). A further type of ring-closure is exemplified by the preparation of the cyclic phosphinic acid (**341**) by treatment of bisphenacylphosphinic acid with phosphorus pentoxide.

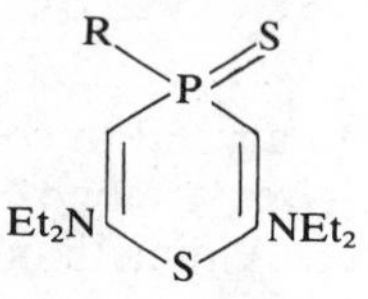

(340) 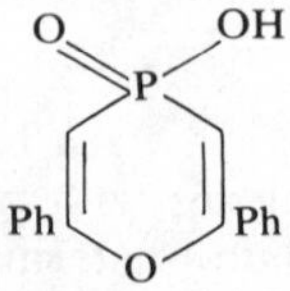 **(341)** 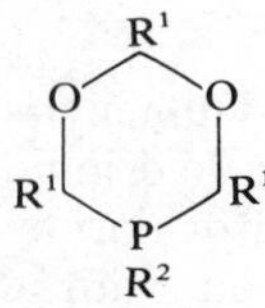 **(342)**

1,3-Dioxa-5-phosphacyclohexanes (**342**) are obtained by nucleophilic addition of phosphine to α-branched aldehydes or by reaction of aromatic aldehydes with phosphine or arylphosphines. They show considerable stability to hydrolysis by acid and to autoxidation, probably because of the steric constraints imposed by the cyclic structure.

20.2.4.2 Dibenzo-derivatives

The reaction of arsenic trichloride with diphenylamine and with diphenyl ether to give tricyclic compounds was studied during World War I because the strongly irritant and sternutatory properties of the products gave them potential as war gases. The reaction of phosphorus trichloride with diphenylamine had been described in 1890, but the structure then proposed for the product was not confirmed until 1938. The phosphorus and arsenic analogues of phenoxazine, phenothiazine, and phenazine are known as phenoxaphosphine (**343**; X = P), phenothiaphosphine (**344**; X = P), phenophosphazine (**345**; X = P), phenoxarsine (**343**; X = As), phenothiarsine (**344**; X = As), and phenarsazine (**345**; X = As). The system of numbering now accepted is shown. In the older literature the heteroatoms are numbered 9 and 10 or 1 and 6, and the numbering in the benzene rings is different. All these compounds form 10-hydroxy-10-oxides which are acids known as

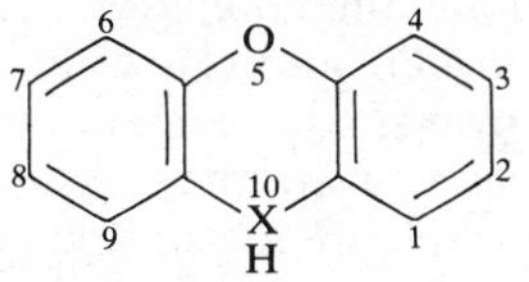

(343) 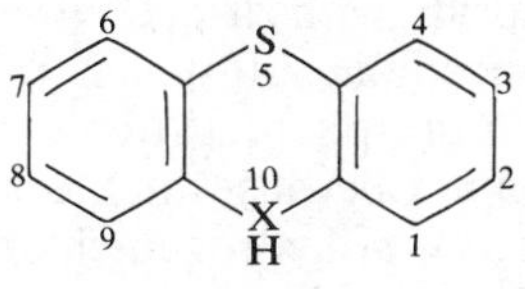 **(344)** 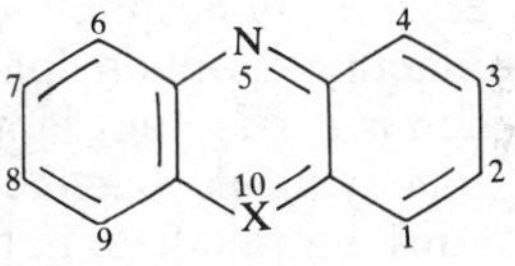 **(345)**

phenoxaphosphinic acid, phenothiaphosphinic acid, phenophosphazinic acid, phenoxarsinic acid, and phenarsazinic or phenazarsinic acid. The main preparative routes to the six series of compounds differ in details but are broadly similar.

Arsenic trichloride reacts with diphenylamine on heating, *e.g.* in *o*-dichlorobenzene, giving 10-chloro-5,10-dihydrophenarsazine (Adamsite) (**346**; X = As). Arsenic tribromide gives the corresponding 10-bromo-compound. The reaction with diphenyl ether requires catalysis by a little aluminium chloride and occurs at reflux temperature (170–200 °C), giving 10-chlorophenoxarsine. This is also obtained from diphenyl ether, arsenious oxide, and aluminium chloride at 175–260 °C. Phosphorus trichloride reacts thermally with diphenylamine, evolution of hydrogen chloride beginning at 70 °C. The initial product, diphenylphosphoramidous dichloride, Ph_2NPCl_2, undergoes rearrangement and condensation at 200–220 °C to give a compound assigned as 10-chloro-5,10-dihydrophenophosphazine (**346**; X = P) together with the spiro-compound (**347**). The chloro-compound (**346**; X = P) has not been isolated but has been hydrolysed by water to give 5,10-dihydrophenophosphazine 10-oxide (**348**), which was at first thought to be 10-hydroxy-5,10-dihydrophenophosphazine. From X-ray crystallographic studies and n.m.r. evidence it

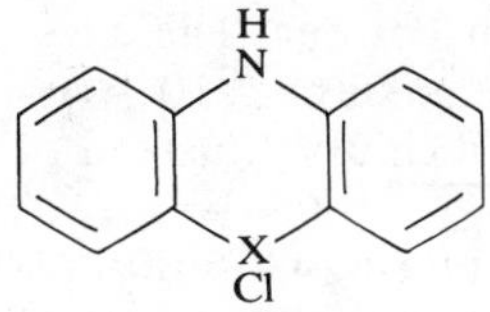

(346) 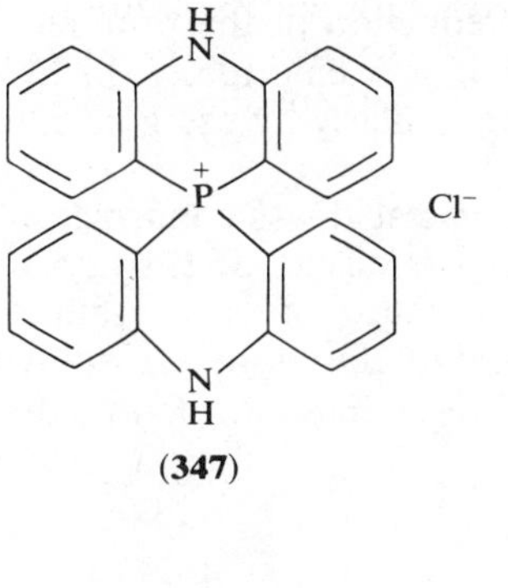 **(347)** 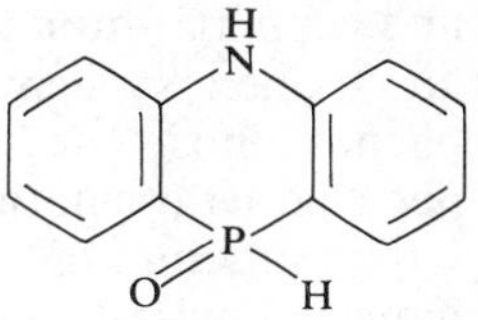 **(348)**

appears that the heterocyclic rings of (**347**) have aromatic character.[87] Diphenyl ether is substituted mainly in the 4-position when it is heated with phosphorus trichloride and aluminium chloride, but a small amount of a phenoxaphosphine derivative may be isolated if a large quantity of aluminium chloride is used. Formation of phenoxaphosphine derivatives occurs readily if the *para* positions in the diaryl ether are protected by substitution (*e.g.* by methyl or chlorine) or by a bulky *meta* substituent (*e.g.* bromine). The 10-chlorophenoxaphosphines undergo hydrolysis during the work-up of the reaction mixture, and the products isolated are phenoxaphosphine 10-oxides (**349**) or their oxidation products the phenoxaphosphinic acids (**350**). In a similar way, diaryl sulphides react with phosphorus trichloride–aluminium chloride, giving 10-chlorophenothiaphosphines or, after hydrolysis and oxidation, phenothiaphosphinic acids (**351**).

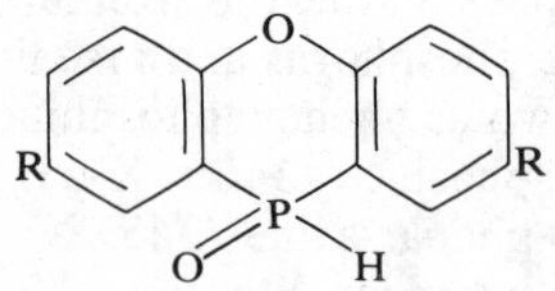

(**349**) 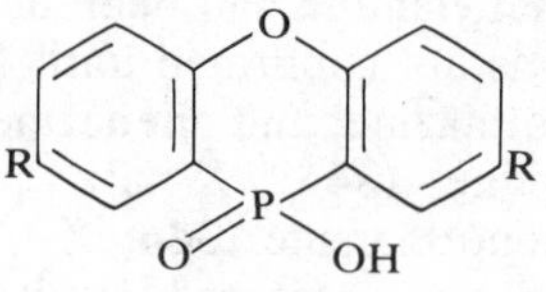 (**350**) 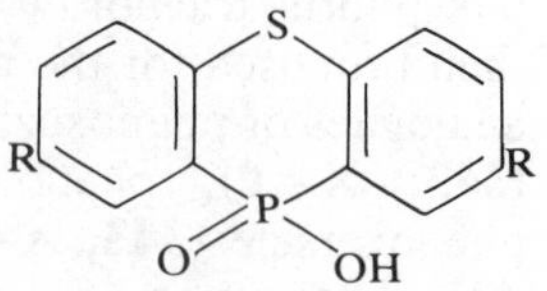 (**351**)

Methylphosphonous dichloride and phenylphosphonous dichloride react with diaryl sulphides and aluminium chloride, giving 10-methyl- and 10-phenyl-phenothiaphosphines; phenylphosphonous dichloride with di-*p*-tolyl ether and aluminium chloride gives 2,8-dimethyl-10-phenylphenoxaphosphine 10-oxide, oxidation occurring even when the reaction is carried out in an inert atmosphere. Attempts to perform similar condensations with phenyldichloroarsine have failed because of the ready fission of arsenic trichloride from this intermediate in the presence of hydrogen chloride. Thus the products of reaction with diphenylamine are 10-chloro-5,10-dihydrophenarsazine and benzene. Reaction of phenylphosphonous dichloride with 2,2′-dilithiated derivatives of diphenyl ether, diphenyl sulphide, diphenyl sulphone, or *N*-alkyldiphenylamines (or corresponding Grignard reagents) gives 10-phenyl derivatives of phenoxaphosphine, phenothiaphosphine, phenothiaphosphine 5,5-dioxide, or 5-alkylphenophosphazines.

Unambiguous syntheses of substituted compounds are possible through arylarsonic acids. Diphenylamine-2-arsonic acids are made by reaction of 2-bromophenylarsonic acid with arylamines or by condensation of 2-aminophenylarsonic acid with reactive aryl halides, *e.g.* chloronitrobenzenes, under Ullmann conditions. Diphenylamine-2-arsonic acid is cyclized to phenazarsinic acid when it is boiled with hydrochloric acid. Alternatively, the arsonic acid group may be reduced with sulphur dioxide in hydrochloric acid to give a dichloroarsine substituent which then condenses to give a 10-chloro-5,10-dihydrophenarsazine (Scheme 33). 2-Aminodiphenyl ether and 2-aminodiphenyl sulphide are converted into 2-arsonic acids by diazotization and treatment with sodium arsenite (Bart reaction). *o*-Phenoxyphenylarsonic acids are cyclized to phenoxarsinic acids by sulphuric acid and to 10-chlorophenoxarsines by reduction with sulphur dioxide–hydrochloric acid. *o*-Phenylthiophenylarsonic acid is not cyclized by sulphuric acid but gives a small yield of 10-chlorophenothiarsine by the reduction method. Attempts to cyclize *o*-phenoxyphenylphosphonic acid by dehydration failed, but 10-chlorophenoxaphosphines may be prepared from *o*-phenoxyphenyldiazonium fluoroborates by reaction with phosphorus trichloride and copper(I) bromide and reduction of the product with aluminium powder. This gives an *o*-phenoxyphenylphosphonous chloride, which cyclizes spontaneously. 10-Arylphenoxaphosphines are made similarly by using an arylphosphonous dichloride instead of phosphorus trichloride.

The structure of phenazarsinic acid was confirmed by its synthesis by ring-closure of 2-aminophenyl-2-bromophenylarsinic acid (**352**) by an Ullmann condensation. This method has not been widely used because the preparation of intermediates is tedious. The reduction of phenazarsinic acid to 10-chloro-5,10-dihydrophenarsazine by sulphur dioxide in hydrochloric acid and a little iodine afforded further confirmation of the structures of these compounds.

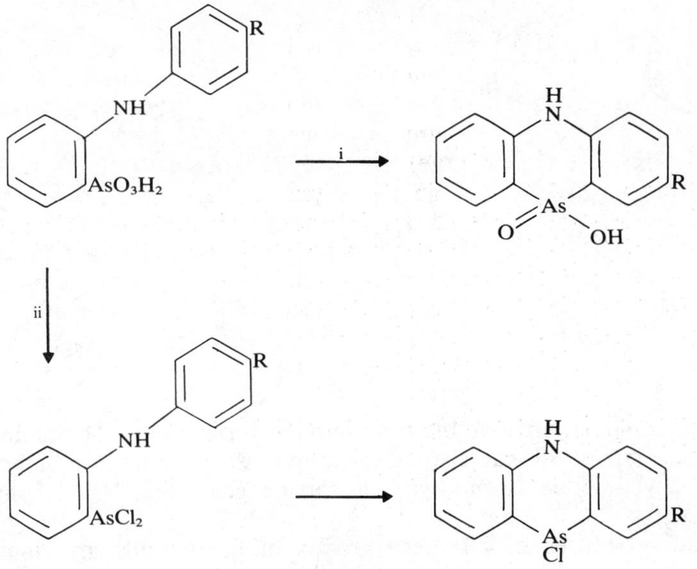

i, HCl; ii, SO_2, HCl.

SCHEME 33

10-Chloro-5,10-dihydrophenophosphazine 10-sulphides (**353**) are obtained by heating diphenylamines with thiophosphoryl chloride at 180–200 °C. They are acid chlorides of phenophosphazinic thioic acids and may be hydrolysed to the thioic acids or converted into amides (10-amino-compounds), esters, or thioesters.

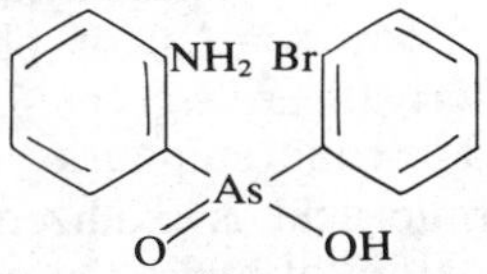

(**352**) 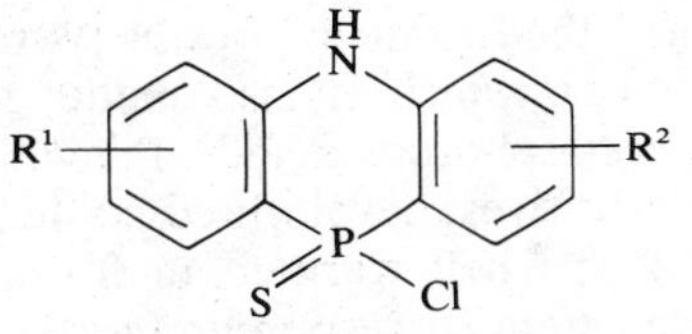 (**353**)

10-Alkyl and 10-aryl derivatives of phenoxarsine and 5,10-dihydrophenarsazine are obtained by the action of Grignard reagents on the 10-chloro-compounds. The application of this method to other members of the group has not been reported, but the 10-substituted phosphorus compounds are available through other routes. The 10-alkyl and 10-aryl derivatives may be oxidized by hydrogen peroxide to give 10-oxides (sometimes hydrates or dihydroxy-compounds), and they are quaternized on the arsenic or phosphorus atoms by alkyl halides. Quaternary compounds are also obtained by the action of Grignard reagents on 10-alkyl-5,10-dihydrophenarsazine 10-oxides. 10-Methyl- and 10-phenyl-5,10-dihydrophenarsazine react with chlorine to give dichlorides which are converted into 10-chloro-5,10-dihydrophenarsazine on heating.

Attempts to synthesize 5-methyl-5,10-dihydrophenarsazine derivatives or 5-methylphenazarsinic acid have failed, but 10-alkyl- and 10-aryl-5,10-dihydrophenarsazines can be acylated or nitrosated on nitrogen. 5,10-Dihydrophenophosphazine 10-oxide is acylated on nitrogen by phenyl isocyanate. Most derivatives having substituents in the benzene rings have been made from intermediates already substituted, but carboxylic acids have been made by oxidation of methyl groups in phenoxaphosphinic acids, phenoxarsines, and phenoxarsinic acids, and in phenarsazines. Phenazarsinic acid and 10-chloro-5,10-dihydrophenarsazine can be nitrated to give 2- and 4-nitro-compounds and 2,8-dinitro-derivatives. The 2-nitro- and 2,8-dinitro-phenazarsinic acids give disodium

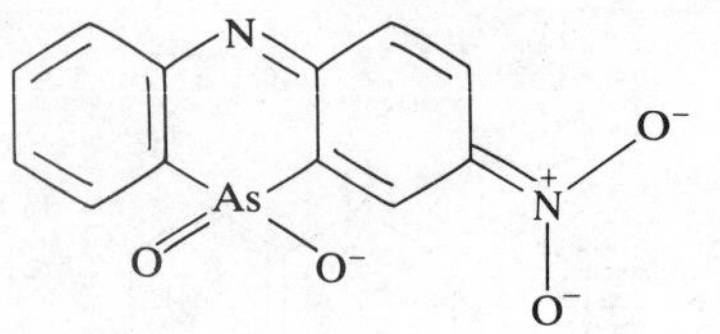

(354)

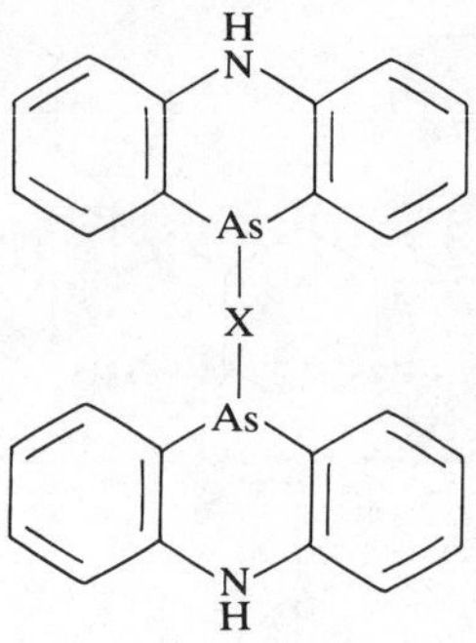

(355)

salts with intense colours, attributable to quinonoid forms (**354**). The amines obtained by reduction of these nitro-acids can also be oxidized to deeply coloured quinonoid products. Friedel–Crafts acylation of 10-methylphenoxarsine gives the 2-acetyl and 2,8-diacetyl derivatives.

The chemical properties of all these groups of compounds are dominated by the chemical properties of the arsenic or phosphorus atom. The main outlines of phenarsazine chemistry were already established in 1923.[88] 10-Chloro-5,10-dihydrophenarsazine (**346**; X = As) reacts with sodium methoxide to give the 10-methoxy-compound, and with alkali to give the 10-oxide (**355**; X = O); this reacts with hydrogen sulphide to give the sulphide (**355**; X = S) and with hydrochloric acid to regenerate (**346**; X = As). Reaction of the oxide with other acids gives 10-acyl derivatives, *e.g.* the 10-acetyl and 10-formyl compounds. Oxidation of (**346**; X = As) with hydrogen peroxide or with Chloramine T gives phenazarsinic acid. The claim that the parent heterocyclic compound, phenarsazine, is obtained by heating the 10-methoxy-5,10-dihydro-derivative (with expulsion of methanol) is erroneous; the product of this pyrolysis is 10,10′-(5*H*,5′*H*-biphenarsazine) (**355**; X = direct bond), which is also made by reduction of (**346**; X = As) or phenazarsinic acid by hypophosphorous acid.[89] Dihydrophenarsazine derivatives, *e.g.* the 10-chloro or 10-formyl derivatives, develop red colours when dissolved in conc. sulphuric acid or other strong acids, or when warmed in formic acid. The formic acid is oxidized to carbon dioxide to an extent corresponding to the addition of one atom of hydrogen per molecule of dihydrophenarsazine derivative, and the coloured species, which increases the electrical conductivity of the solution, is probably a radical cation. It is readily re-oxidized to the oxide (**355**; X = O) by air. Prolonged heating of the dihydrophenarsazine derivative with formic acid causes decomposition to diphenylamine and arsenic. Bromination of 5,10-dihydrophenarsazines causes cleavage of the heterocyclic ring with removal of arsenic; treatment of the dihydrophenarsazines with dry hydrogen chloride also eliminates arsenic, the 10-chloro-compound giving arsenic trichloride and the 10-alkyl and 10-aryl compounds giving alkyl- or aryl-phosphonous dichlorides. This has been recommended as a preparative method for the latter compounds.

10-Chlorophenoxarsine gives a similar oxide to (**355**; X = O) when it is treated with alkali, and its chlorine atom is rapidly exchanged for bromine, iodine, or thiocyanate when its solution is treated with potassium bromide, iodide, or thiocyanate. The oxide is converted into di(phenoxarsine-10-yl) sulphide by hydrogen sulphide and is reduced by hypophosphorous acid to 10,10′-biphenoxarsine.

The hydrolysis of 10-chloro-5,10-dihydrophenophosphazine takes a different course; the 10-hydroxy-compound which may first be formed isomerizes to a secondary phosphine oxide (**348**) which is soluble only in very polar solvents. It is oxidized by sulphur in acetic acid to 10-hydroxy-5,10-dihydrophenophosphazine 10-sulphide (thiophenophosphazinic acid) (**356**; X = S), and by iodine or chloramine T in DMF to 10,10′-(5*H*,5*H*′-biphenophosphazine) 10,10′-dioxide (**357**). Oxidation by aeration of a hot solution gives phenophosphazinic acid (**358**; R = OH), a fairly strong acid that can be

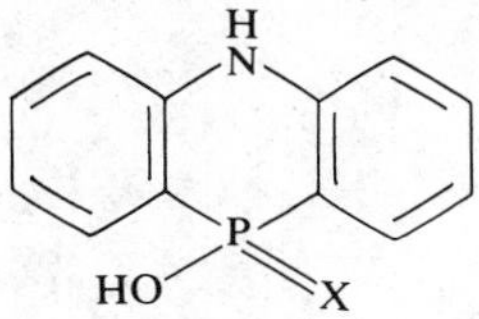

(356) 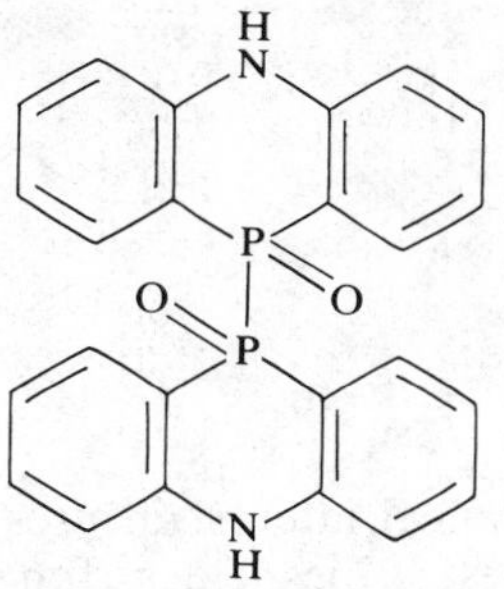 (357) 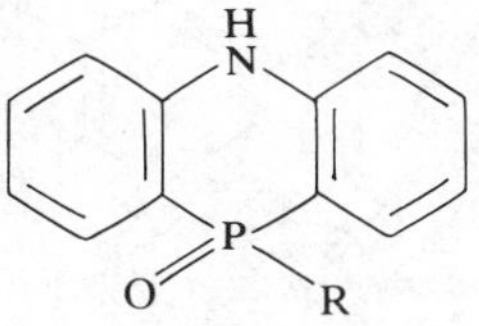 (358)

acylated on its hydroxy-group. It is converted into its acid chloride (10-chloro-5,10-dihydrophenophosphazine 10-oxide) (**358**; R = Cl) by thionyl chloride or phosphorus pentachloride and this gives esters (**358**; R = *O*-alkyl) and amides (**358**; R = NR^1R^2) in the usual way.

The phenoxaphosphine 10-oxides (**349**) produced by hydrolysis of 10-chloropheno-xaphosphines are very readily converted into phenoxaphosphinic acids (**350**). The 2,8-dimethyl compound undergoes autoxidation during its isolation from the reaction mixture, and the 2,8-dichloro- and 2,8-difluoro-compounds dissolve in sodium hydroxide or sodium methoxide solution with evolution of hydrogen and formation of (**350**; R = Cl or F).[90] Treatment of the acids with thionyl chloride gives 10-chlorophenoxaphosphine 10-oxides, which react with aryl Grignard reagents to give 10-arylphenoxaphosphine 10-oxides. These are reduced to 10-arylphenoxaphosphines by trichlorosilane. 2,8-Dimethylphenoxaphosphinic acid (**350**; R = Me) is reduced to the secondary phosphine, 2,8-dimethylphenoxaphosphine, by diphenylsilane. The methiodides of 10-arylphenoxa-phosphines undergo cleavage by alkali to give *o*-phenoxyphenylphosphine oxides (**359**) (reaction 81).

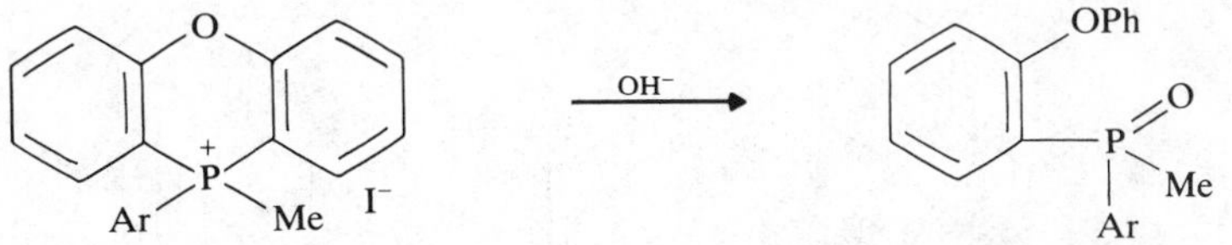

$\xrightarrow{OH^-}$ (81)

(359)

To account for the yellow colour of 10-chloro-5,10-dihydrophenarsazine and related compounds (*e.g.* the bromo-compound and sulphate), alternative structures (**360**) or (**361**) have been proposed having the proton of the heterocyclic ring on a pentavalent arsenic atom. Strong evidence for the accepted structure (**346**; X = As) was provided by X-ray crystallography, which showed that the molecule is slightly folded along the N—As axis (dihedral angle 169°) and the As—Cl bond which is out of plane has a length approximating to that in diphenylphosphonous chloride. Evidence of molecular asymmetry was provided by the preparation of two (+)-bromocamphorsulphonates with different optical rotations from the benzo-derivative (**362**). In the phenoxarsine series the dihedral angle is more acute, and several compounds (**363**; R = Me, Et, Ph) have been resolved into enantiomers possessing considerable optical stability. Racemization occurs through reactions affecting the atomic radius of arsenic, *e.g.* quaternization with methyl iodide or oxidation of (**363**; R = Ph) to the 10-oxide. The latter compound, however, has been resolved.

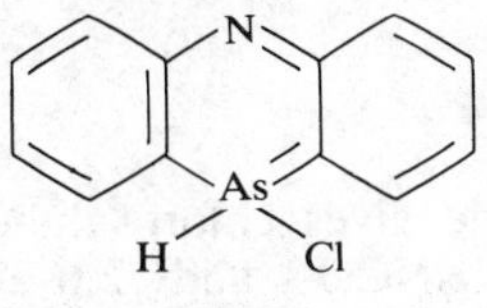

(360) 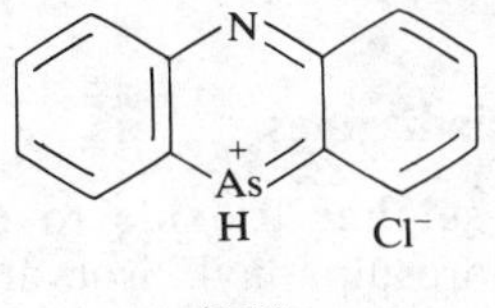 (361)

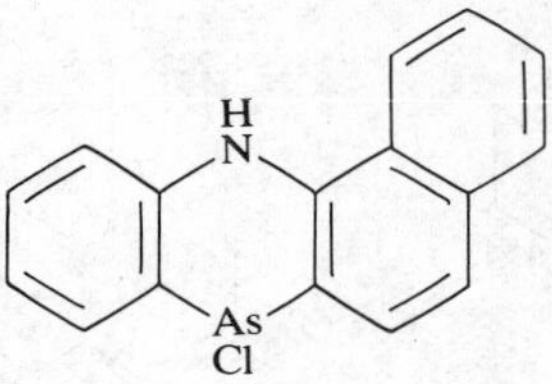

(362) 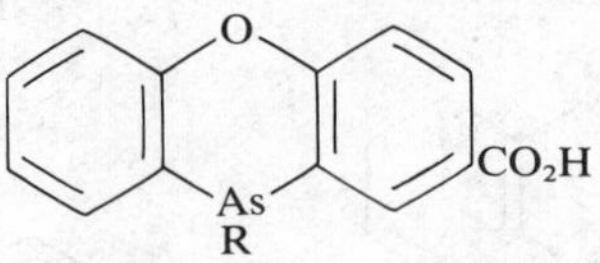 (363)

There has been continuing technical interest in uses for the phenarsazine derivatives and the less irritant phenoxarsines as pesticides, fungicides, herbicides, stabilizers for poly(vinyl chloride), and particularly as antifouling agents for marine paints. For these purposes many derivatives have been made from the 10-chloro-compounds by replacement of the chlorine by other halogens, carboxylate radicals, and aliphatic, aromatic, and heterocyclic thiols.

A few derivatives of 5,6-dihydrodibenz[*c,e*]-1,2-azaphosphorine (**364**) are known. Phosphorus trichloride reacts with 2-aminobiphenyl or substituted 2-aminobiphenyls to give *N*-(2-biphenylyl)phosphoramidous dichlorides, which cyclize at higher temperatures to 6-chloro-5,6-dihydrodibenzo[*c,e*]-1,2-azaphosphorines (**364**; R = Cl). These are very sensitive to moisture, hydrolysing to the 6-oxides (**365**; R = H). With Grignard reagents they give the 6-methyl and 6-aryl derivatives (**364**; R = Me or Ar). Reports on the stability of these compounds in air differ, but in solution they are readily oxidized to the 6-oxides (**365**; R = Me or Ar) which are stable to dilute alkali but are cleaved by acids to 2′-amino-2-biphenylylarylphosphinic acids. Addition of methyl iodide to (**364**; R = Me, Ph) occurs at the phosphorus atom. Resolution of (**365**; R = *p*-$Me_2NC_6H_4$) was accomplished with camphorsulphonic acid, and reduction of the enantiomeric oxides with lithium aluminium hydride gave optically active azaphosphorines, but resolution of racemic (**364**; R = *p*-$Me_2NC_6H_4$) was not possible because of autoxidation in solution.

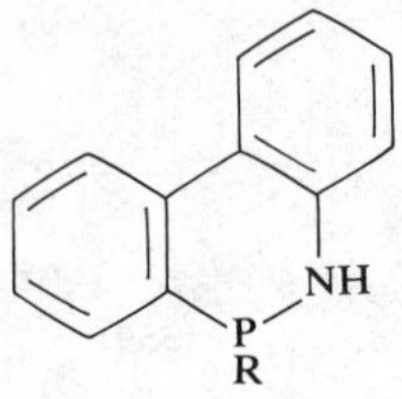

(364) 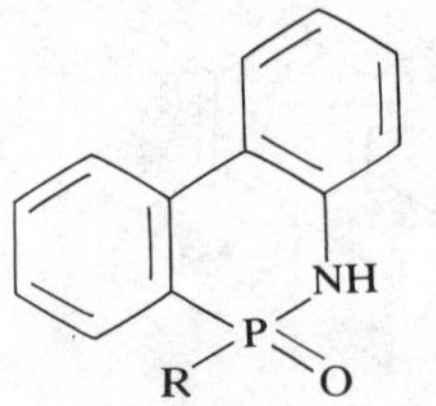 (365)

20.2.5 MISCELLANEOUS: COMPOUNDS WITH THREE OR MORE HETEROATOMS

Apart from cyclic esters, thioesters, and amides of the oxyacids of phosphorus and arsenic, the numerous six-membered ring systems with three or more heteroatoms usually contain nitrogen, sulphur, and oxygen. Only a limited systematic treatment of such ring systems is possible at present, but it is possible to distinguish families that have arisen from the use of new reagents, *e.g.* chlorosulphonyl isocyanate, and others that come from synthetic methods analogous to those used in making more familiar heterocyclic systems, *e.g.* benzothiadiazines.

20.2.5.1 Oxathiadiazines

Addition of sulphur trioxide to cyanogen chloride gives chlorosulphonyl isocyanate (**366**), pyrochlorosulphonyl isocyanate, $ClSO_2OSO_2NCO$, and 2,6-dichloro-1,4,3,5-oxathiadiazine 4,4-dioxide (**367**; R = Cl). This is a stable solid that gives sulphamide upon hydrolysis with water and is decomposed into cyanogen chloride and **(366)** when it is heated.[91] Aryl nitriles react with sulphur trioxide, giving 1,3,2,4,5-dioxadithiazine 2,2,4,4-tetroxides **(368)**, which react further with nitriles to give 1,2,3,5-oxathiadiazine 2,2-dioxides **(369)** and 1,4,3,5-oxathiadiazine 4,4-dioxides (**367**; R = Ar). These may be distinguished by their hydrolysis products, *N*-acylbenzamides from **(369)** and *N*-aroylsulphamides from (**367**; R = Ar). Aryl cyanates react with sulphur trioxide, giving (**369**; R = *O*-aryl), and polyhalogenated acetonitriles give (**367**; R = CCl_3, *etc.*).

$ClSO_2NCO$ **(366)** **(367)** 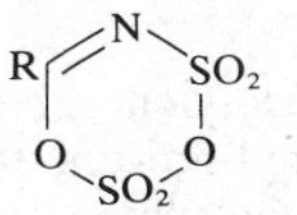 **(368)** 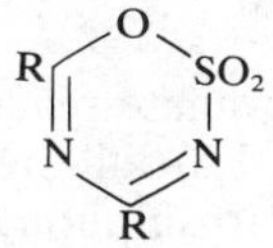 **(369)** 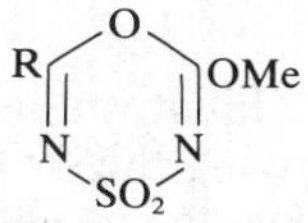 **(370)**

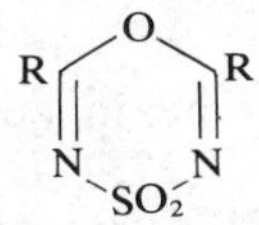

The sodium salt of methoxycarbonylsulphamoyl chloride reacts with acetonitrile or with dimethylcyanamide to give the methoxy-derivatives (**370**; R = Me) and (**370**; R = NMe_2). These compounds decompose on heating to give methyl *N*-sulphonylurethane ($MeO_2CN{=}SO_2$).[92]

Alkyl isocyanates react with **(366)** in the presence of Lewis acids (*e.g.* tin(IV) chloride) to give carbamyl chlorides (**371**; X = Cl) derived from 2-amino-1,4,3,5-oxathiadiazin-6-one 4,4-dioxides. These are converted into the corresponding amines **(372)** by treatment with formamide, and into the corresponding urethanes (**371**; X = OR) by controlled treatment with an alcohol and pyridine. They react with alcohols to give 5-alkoxy-2,3-dihydro-4*H*-1,2,4,6-thiatriazin-3-one 1,1-dioxides **(373)**.

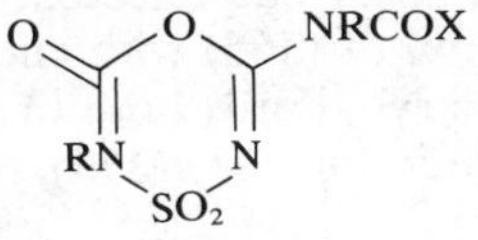

(371) 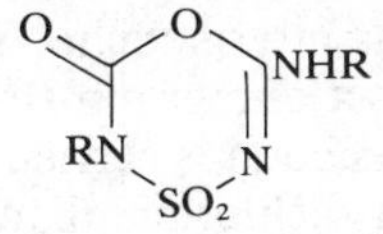 **(372)** 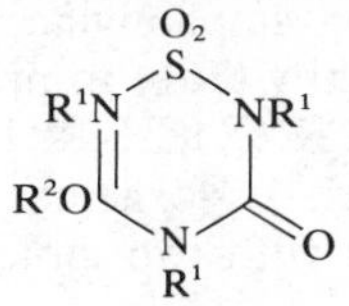 **(373)**

20.2.5.2 Oxathiazines

Acetylenes, *e.g.* hex-3-yne or 1-ethoxybut-1-yne, react with **(366)** to give 1:1 adducts which are 6-chloro-1,2,3-oxathiazine 2,2-dioxides (**374**; $R^1 = R^2 =$ Et, or $R^1 =$ OEt, $R^2 =$ Et). It is postulated that the reaction follows the pathway shown in Scheme 34. The chlorine atom may be replaced by reaction with thiophenol and pyridine, giving the 6-phenylthio analogue of **(374)**, but most nucleophiles cause degradation of the ring, *e.g.* aqueous sodium bicarbonate gives a ketone, methanol, and a β-ketoester, and sodium methoxide a bis-ester of a β-amino(*N*-sulphonic acid)acrylic acid. The 3,4-double bond can be reduced by lithium aluminium hydride, and alkylation of the nitrogen by methyl iodide is then possible.[93]

$R^1C{\equiv}CR^2$ + **(366)** ⟶

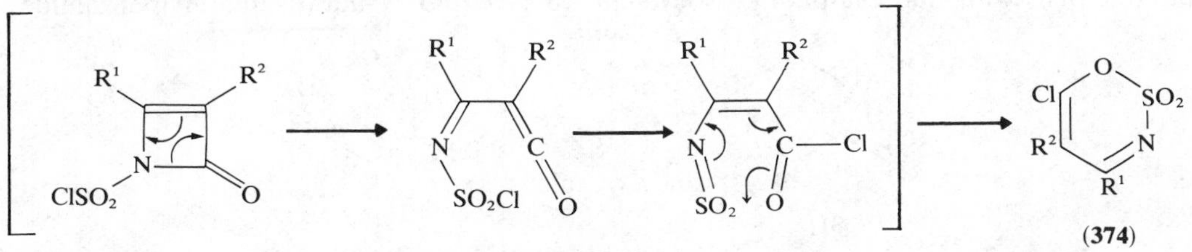

(374)

SCHEME 34

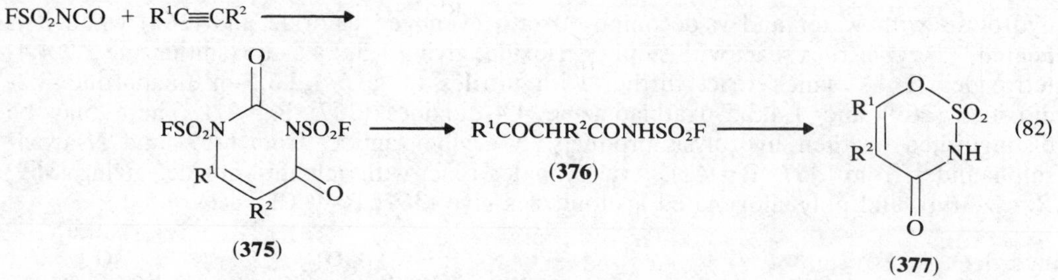

$$FSO_2NCO + R^1C{\equiv}CR^2 \longrightarrow$$

(375) $\longrightarrow$ $R^1COCHR^2CONHSO_2F$ (376) $\longrightarrow$ (377) (82)

Fluorosulphonyl isocyanate reacts with acetylenes, *e.g.* but-2-yne, in a different manner, giving the 2:1 adduct 1,3-difluorosulphonyl-5,6-dimethyluracil (**375**). Hydrolysis of this compound gives *N*-fluorosulphonyl-α-methylacetoacetamide (**376**; $R^1 = R^2 = Me$), which is readily cyclized to 5,6-dimethyl-3*H*-1,2,3-oxathiazin-4-one 2,2-dioxide (**377**; $R^1 = R^2 =$ Me) (reaction 82). Related compounds, particularly the 6-methyl derivative (**377**; $R^1 =$ Me, $R^2 = H$), are of interest as synthetic sweetening agents.[94] They are strong monobasic acids and form stable neutral salts with alkali metals. These are stable to hydrolysis, but (**377**; $R^1 = Me$, $R^2 = H$) is hydrolysed to acetone by boiling water. Methylation of these compounds by dimethyl sulphate gives *N*-methyl derivatives, but diazomethane gives both *N*-methyl and *O*-methyl compounds. Reaction with phosphorus pentachloride gives 4-chloro-1,2,3-oxathiadiazine 2,2-dioxides (**378**), in which the chlorine atom can be replaced by nucleophiles. Several alternative routes to the *N*-fluorosulphonyl-β-oxocarboxamides (**376**) have been devised, including the reaction of fluorosulphonyl isocyanate with ketones, β-diketones, and β-oxocarboxylic acids and esters, and other leaving groups have been used. Aryloxysulphonyl isocyanates will acylate ketones to give *N*-aryloxysulphonylacetoacetamides that will cyclize to (**377**), and t-butyl acetoacetate is acylated by (**366**) to provide a similar intermediate. The reaction of (**366**) with enolizable ketones $R^1COCH_2R^2$ in which at least one group R^1 or R^2 is aryl affords both (**377**) and 2,4-dioxo-1,3-oxazines (**379**) if the reaction is carried out in ether, but only (**379**), through reaction with two moles of (**366**), in dichloromethane.

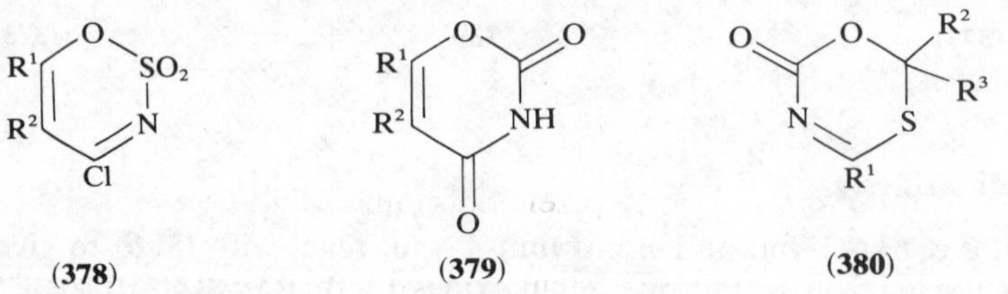

(**378**) (**379**) (**380**)

The condensation of thiocarbonyl isocyanates R^1CSNCO ($R^1 = EtO$ or Ph) with carbonyl compounds R^2R^3CO (chloral, benzaldehyde, acetone) gives 1,3,5-oxa-thiazin-6-ones (**380**).

20.2.5.3 Benzoxathiazines

1,2,3-Benzoxathiazine 2,2-dioxides (**381**; R = Me, Ph) are obtained by fusion of *o*-hydroxyaryl ketones with sulphamide at 130–180 °C (*cf.* the preparation of 2,1,3-benzothiadiazine 2,2-dioxides). The 4-oxo-derivative (**382**) is made by reaction of *o*-benzyloxybenzoic acid with fluorosulphonyl isocyanate to give the *N*-fluorosulphonylbenzamide,

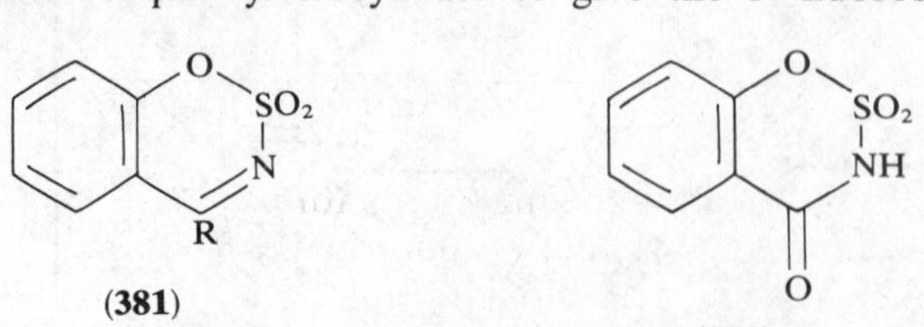

(**381**) (**382**)

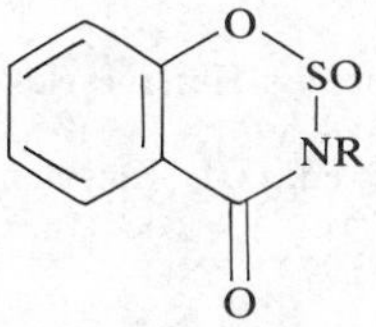

(383) 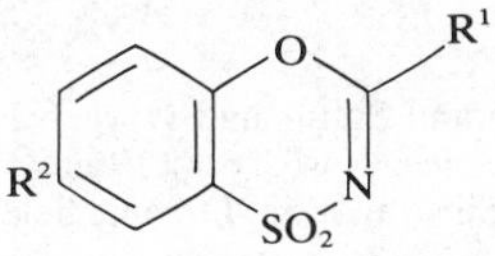 (384) 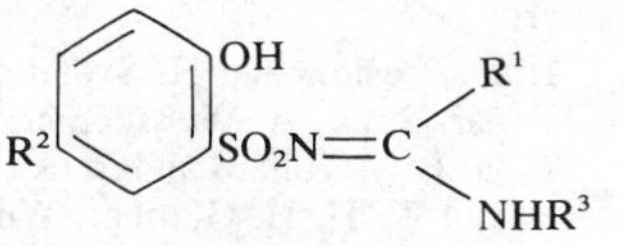 (385)

hydrogenolysis of the benzyl group, and cyclization by sodium hydroxide. This compound has a sweet and bitter taste. 4-Oxo-1,2,3-benzoxathiazine 2-oxides (**383**; R = H or aryl) are obtained by reaction of salicylamide or salicylanilides with thionyl chloride. They are readily hydrolysed by acid or base with regeneration of the original amide.

1,4,3-Benzoxathiazine 4,4-dioxides (**384**) are made by the cyclization of *N*-acyl-*o*-hydroxybenzenesulphonamides with phosphoryl chloride. They are readily hydrolysed to the parent *N*-acylsulphonamides, and with amines they give *o*-hydroxybenzenesulphonylamidines (**385**).

20.2.5.4 Benzodithiadiazines and benzothiatriazines

o-Aminobenzenesulphonamide reacts with *N*-sulphinyltoluene-*p*-sulphonamide and with *N,N'*-bis(toluene-*p*-sulphonyl)sulphur di-imide to give 1,3,2,4-benzodithiadiazine 1,1,3-trioxide (**386**) and 2,4-didehydro-1,3,2,4-benzodithiadiazine 1,1-dioxide (**387**) (*cf.* the preparation of analogous benzothiadiazines from anthranilamide).[81]

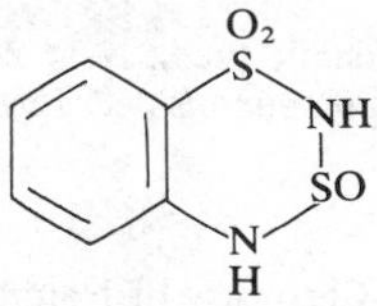

(386) 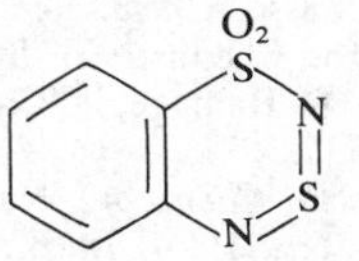 (387)

The diazonium salt from 2-aminobenzenesulphonanilide gives 2-phenyl-1,2,3,4-benzothiatriazine 1,1-dioxide (**388**) when it is treated with sodium acetate. The diazonium salt is regenerated by dilute acids, and treatment of (**388**) with sodium hydroxide solution and precipitated copper gives the dibenzothiazine dioxide (**389**). Photolysis gives the benzothiazete (**390**), which decomposes thermally to phenothiazine 5,5-dioxide and photochemically to carbazole.[95] The use of (**388**) and analogous 2-alkyl-1,2,3,4-benzothiatriazine

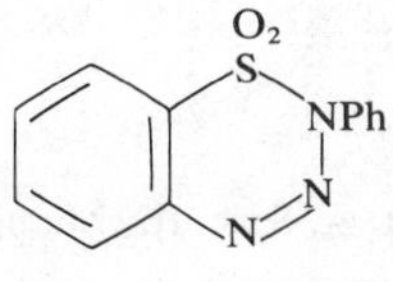

(388) 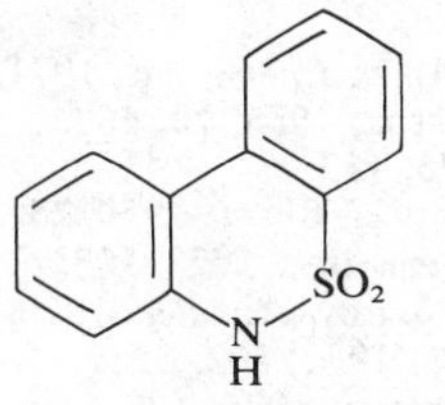 (389) 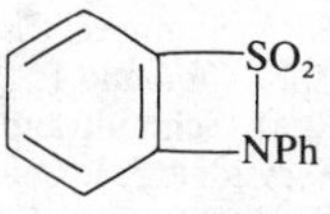 (390)

1,1-dioxides (prepared similarly) as blowing agents for cellular rubber and plastics, and as polymerization catalysts for acrylonitrile, *etc.*, has been claimed in patents. The preparation of benzo-1,2,3,4-thiatriazine 1,1-dioxides analogous to the chlorothiazide (benzothiadiazine) diuretics has also been described.

References

1. D. Breslow and H. Skolnik, 'Multi-Sulfur and Sulfur and Oxygen Five- and Six-Membered Heterocycles' part 2, ed. A. Weissberger, Wiley-Interscience, New York, 1966.
2. (a) L. Mortillaro and M. Russo, (b) R. B. Silverman, in 'Organic Selenium Compounds', ed. D. L. Klayman and W. H. H. Gunther, Wiley, New York, 1973.
3. K. J. Irgolic, 'The Organic Chemistry of Tellurium', Gordon and Breach, New York, 1974.
4. A. Brändström, *Arkiv Kemi*, 1951, **3,** 41 (*Chem. Abs.* 1951, **45,** 7574).
5. W. E. Parham, I. Gordon, and J. D. Swalen, *J. Amer. Chem. Soc.*, 1952, **74,** 1824; W. E. Parham, J. Heberling, and H. Wynberg, *ibid.*, 1955, **77,** 1169.
6. W. E. Noland and R. E. DeMaster, *Org. Synth.*, **52,** 135
7. Y. Hayasi, H. Nakamura, and H. Nozaki, *Bull. Chem. Soc. Japan*, 1973, **46,** 667.
8. W. E. Parham and J. D. Jones, *J. Amer. Chem. Soc.*, 1954, **76,** 1068.
9. C. L. Deasy, *Chem. Rev.*, 1943, **32,** 173.
10. F. Mauthner, *Ber.*, 1905, **38,** 1411.
11. H. D. K. Drew, *J. Chem. Soc.*, 1926, 3054.
12. R. L. McKee, in 'Five- and Six-Membered Compounds with Nitrogen and Oxygen', ed. R. H. Wiley, Wiley-Interscience, New York, 1962.
13. J. Hamer, '1,4-Cycloaddition Reactions', Academic Press, New York, 1967.
14. T. K. Das Gupta, D. Felix, U. M. Kempe, and A. Eschenmoser, *Helv. Chim. Acta*, 1972, **55,** 2198; P. Gygax, T. K. Das Gupta, and A. Eschenmoser, *ibid.*, 2205.
15. A. I. Meyers, 'Heterocyclics in Organic Synthesis', Wiley, New York, 1974.
16. Z. Eckstein and T. Urbański, *Adv. Heterocyclic Chem.*, 1963, **2,** 311.
17. R. R. Schmidt, *Synthesis*, 1972, 333.
18. S. L. Meisel, J. J. Dickert, and H. D. Hartough, *J. Amer. Chem. Soc.*, 1956, **78,** 4782.
19. T. Urbański, *Synthesis*, 1974, 613.
20. R. R. Schmidt, D. Schwille, and U. Sommer, *Annalen*, 1969, **723,** 111.
21. G. Drefahl and H. H. Hörhold, *Chem. Ber.*, 1961, **94,** 1657.
22. H. G. Viehe, V. Jäger, and F. Compernolle, *Angew. Chem. Internat. Edn.*, 1969, **8,** 979.
23. M. Ionescu and H. Mantsch, *Adv. Heterocyclic Chem.*, 1976, **8,** 83.
24. W. Schäfer, *Progr. Org. Chem.*, 1964, **6,** 135.
25. A. Butenandt, E. Biekert, and W. Schäfer, *Annalen*, 1960, **632,** 143.
26. H. Beecken and H. Musso, *Chem. Ber.*, 1961, **94,** 601.
27. K. Venkataraman, 'The Chemistry of Synthetic Dyes', Academic Press, New York, 1952.
28. H. T. Howard and G. R. Ramage, in 'Chemistry of Carbon Compounds', ed. E. H. Rodd, Elsevier, London, 1960, vol. IVc, p. 1535.
29. H. Brockmann, *Fortschr. Chem. org. Naturstoffe*, 1960, **18,** 1.
30. H. Hollstein, *Chem. Rev.*, 1974, **74,** 625.
31. A. Butenandt and W. Schäfer, in 'Recent Progress in the Chemistry of Natural and Synthetic Colouring Matters', ed. T. S. Gore, B. S. Joshi, S. V. Sunthankar, and B. D. Tilak, Academic Press, New York, 1962, p. 13.
32. H. Beecken, E. M. Gottschalk, U. von Gizycki, H. Krämer, D. Maassen, H.-G. Matthies, H. Musso, C. Rathjen, and U. J. Záhorszky, *Angew. Chem.*, 1961, **73,** 665.
33. R. L. Mital and S. K. Jain, *J. Chem. Soc. (C)*, 1971, 1875.
34. G. W. Stacy, in 'Heterocyclic Compounds', ed. R. C. Elderfield, Wiley, New York, 1961, vol. 7, p. 797.
35. R. Gompper and F. Towae, *Synthesis*, 1975, 522.
36. D. Martin and A. Weise, *Chem. Ber.*, 1967, **100,** 3736.
37. R. R. Schmidt, *Chem. Ber.*, 1965, **98,** 334.
38. R. C. Elderfield and E. E. Harris, in 'Heterocyclic Compounds', ed. R. C. Elderfield, Wiley, New York, 1957, vol. 6, p. 601.
39. G. R. Ramage, E. H. Rodd, and J. K. Landquist, in 'Chemistry of Carbon Compounds', vol. IVc, ed. E. H. Rodd, Elsevier, London, 1960.
40. T. R. Williams and D. J. Cram, *J. Amer. Chem. Soc.*, 1971, **93,** 7333.
41. F. Asinger and M. Thiel, *Angew. Chem.*, 1958, **70,** 667.
42. P. G. Sammes, *Chem. Rev.*, 1976, **76,** 113.
43. E. N. Cain and R. N. Warrener, *Austral. J. Chem.*, 1970, **23,** 51.
44. R. R. Schmidt and M. Dimmler, *Chem. Ber.*, 1975, **108,** 6.
45. F. Asinger, H. Offermanns, and D. Neuray, *Annalen*, 1970, **739,** 32; F. Asinger, A. Saus, H. Offermanns, D. Neuray, and K. H. Lim, *Monatsh.*, 1971, **102,** 321.
46. H. Hückstädt and K. H. Mayer, *Synthesis*, 1970, 183.
47. C. R. Johnson and C. B. Thanawalla, *J. Heterocyclic Chem.*, 1969, **6,** 247.
48. C. R. Rasmussen, *J. Org. Chem.*, 1974, **39,** 1554.
49. B. Loev and K. M. Snader, *J. Heterocyclic Chem.*, 1967, **4,** 403.
50. D. Bourgoin-Legay and R. Boudet, *Bull. Soc. chim. France*, 1969, 2524.
51. H. Böshagen, W. Geiger, H. Hulpke, and C. Wünsche, *Chem. Ber.*, 1971, **104,** 3757.
52. M. Wilhelm and P. Schmidt, *J. Heterocyclic Chem.*, 1969, **6,** 635.
53. W. Friedrich, F. Kröhnke, and P. Schiller, *Chem. Ber.*, 1965, **98,** 3804.

54. G. Pagani, *Gazzetta*, 1967, **97,** 1804.
55. R. N. Prasad, *J. Medicin. Chem.*, 1969, **12,** 290.
56. B. L. Kaul, *Helv. Chim. Acta*, 1974, **57,** 2664.
57. S. P. Massie, *Chem. Rev.*, 1954, **54,** 797.
58. C. Bodea and I. Silberg, *Adv. Heterocyclic Chem.*, 1968, **9,** 321.
59. E. Schenker and H. Herbst, *Progr. Drug Res.*, 1963, **5,** 269.
60. W. E. Truce, E. M. Kreider, and W. W. Brand, *Org. Reactions*, 1970, **18,** 99.
61. J. I. G. Cadogan, S. Kulik, C. Thomson, and M. J. Todd, *J. Chem. Soc. (C)*, 1970, 2437.
62. J. Daneke and H. W. Wanzlick, *Annalen*, 1970, **740,** 52.
63. A. Lawson and R. B. Tinkler, *Chem. Rev.*, 1970, **70,** 593.
64. A. Etienne, A. Le Barre, and J. P. Giorgetti, *Bull. Soc. chim. France*, 1973, 985, 2361; 1974, 1395.
65. J. Geevers and W. P. Trompen, *Rec. Trav. chim.*, 1974, **93,** 270.
66. G. Pagani, *J. C. S. Perkin I*, 1974, 2050.
67. H. Beyer, *Z. Chem.*, 1969, **9,** 361.
68. H. Beyer, H. Honeck, and L. Reichelt, *Annalen*, 1970, **741,** 45.
69. R. R. Schmidt and H. Huth, *Tetrahedron Letters*, 1975, 33.
70. J. Goerdeler and D. Wobig, *Annalen*, 1970, **731,** 120.
71. J. Goerdeler and H. Lüdke, *Chem. Ber.*, 1970, **103,** 3393.
72. J. E. Oliver and A. B. De Milo, *J. Heterocyclic Chem.*, 1971, **8,** 1087.
73. E. Fanghänel, *J. prakt. Chem.*, 1976, **318,** 127.
74. K. Hasegawa and S. Hirooka, *Bull. Chem. Soc. Japan*, 1972, **45,** 1567.
75. E. Schlittler, G. De Stevens, and L. Werner, *Angew. Chem. Internat. Edn.*, 1962, **1,** 235.
76. L. Raffa, M. Di Bella, M. Melegari, and G. Vampa, *Il Farmaco Ed. Sci.*, 1962, **17,** 320, 331.
77. G. Kresze, C. Seyfried, and A. Trede, *Annalen*, 1968, **715,** 223.
78. A. W. Wagner and G. Reinöhl, *Annalen*, 1964, **675,** 189.
79. E. Cohnen and J. Mahnke, *Chem. Ber.*, 1972, **105,** 757; 1973, **106,** 3368.
80. A. J. Elliott, P. D. Callaghan, M. S. Gibson, and S. T. Nemeth, *Canad. J. Chem.*, 1975, **53,** 1484.
81. H. Grill and G. Kresze, *Annalen*, 1971, **749,** 171.
82. I.C.I. Ltd., *Brit. Pat.* 1 350 277/1974 (*Chem. Abs.*, 1974, **81,** 25 669).
83. F. G. Mann, 'The Heterocyclic Derivatives of Phosphorus, Arsenic, Antimony, and Bismuth', 2nd edn, Wiley-Interscience, New York, 1970.
84. S. Samaan, *Tetrahedron Letters*, 1974, 3927.
85. A. Naaktgeboren, J. Meijer, P. Vermeer, and L. Brandsma, *Rec. Trav. chim.*, 1975, **94,** 92.
86. M. H. Mebazaa and H. Simalty, *Tetrahedron Letters*, 1972, 4363.
87. R. N. Jenkins and L. D. Freedman, *J. Org. Chem.*, 1975, **40,** 766.
88. H. Wieland and W. Rheinheimer, *Annalen*, 1923, **423,** 1.
89. H. Vermeer, R. Lourens, and F. Bickelhaupt, *Tetrahedron*, 1975, **31,** 2529.
90. J. Granoth, A. Kalir, Z. Pelah, and E. D. Bergmann, *Tetrahedron*, 1970, **26,** 813.
91. R. Graf, *Chem Ber.*, 1956, **89,** 1071.
92. E. M. Burgess and W. M. Williams, *J. Org. Chem.*, 1973, **38,** 1249.
93. E. J. Moriconi and Y. Shimakawa, *J. Org. Chem.*, 1972, **37,** 196.
94. K. Clauss and H. Jensen, *Angew. Chem. Internat. Edn.*, 1973, **12,** 869.
95. M. S. Ao and E. M. Burgess, *J. Amer. Chem. Soc.*, 1971, **93,** 5298.

20.3
Other Mixed Heteroatom Systems

J. K. LANDQUIST
formerly of ICI Pharmaceuticals Division, Alderley Park, Cheshire

The majority of cyclic compounds that have distinctive 'heterocyclic' properties have five- or six-membered rings. Compounds with smaller rings are of interest because they are strained systems and often display high reactivity. With larger ring sizes, heterocyclic compounds become increasingly like acyclic compounds with similar functional groups, but effects of proximity and steric constraints can still be found in seven- and eight-membered rings and the much larger rings of cyclic peptides and depsipeptides sometimes show interesting trans-annular interactions.

20.3.1 THREE-MEMBERED RINGS

Three-membered ring structures were proposed for a number of compounds in the older literature without supporting evidence. For example, the adducts of triethylphosphine with diphenylketen and with carbon disulphide were formulated as the oxaphosphiran (**1**) and the thiaphosphiran (**2**), and the adduct of triethylamine with sulphur trioxide was thought to be the oxathiazidine (**3**), but dipolar, acyclic structures could obviously be considered for such compounds. Well authenticated examples of three-membered rings with mixed heteroatoms have been described in more recent years. The systems containing one atom of carbon are oxaziridines (**4**) and thiaziridines (**5**). Some compounds without a carbon atom in the ring clearly belong to organic chemistry rather than inorganic chemistry. Such compounds include oxadiaziridines (**6**) and thiadiaziridines (**7**).

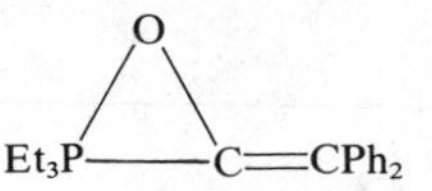

O; Et_3P—C=CPh_2 **(1)** S; Et_3P—C=S **(2)** O; Et_3N—SO_2 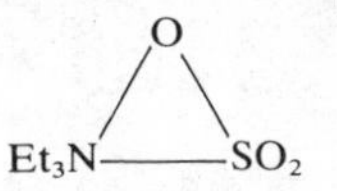 **(3)**

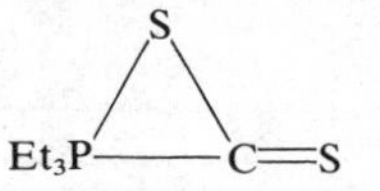

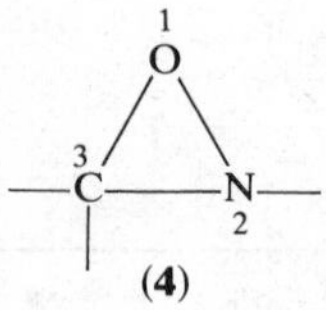

(4) 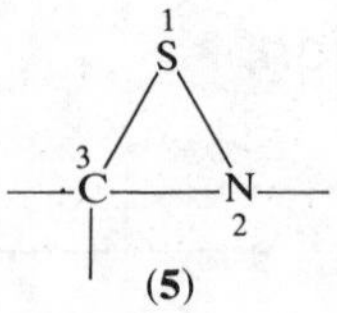 **(5)** 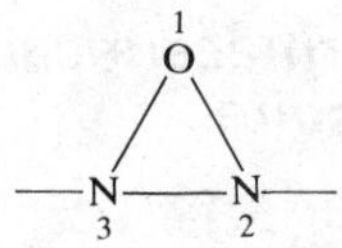 **(6)** 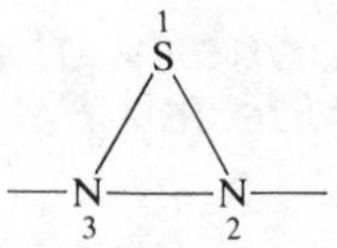 **(7)**

20.3.1.1 Oxaziridines (oxaziranes)[1–3]

The isolation of oxaziridines dates from 1952, although in the older literature the structure was sometimes ascribed to nitrones. The first, and most general, method of preparation was the oxidation of imines (Schiff's bases) with organic peracids, *e.g.* anhydrous peracetic acid, peroxyphthalic acid, or *m*-chloroperoxybenzoic acid. A convenient alternative is the use of hydrogen peroxide in the presence of a nitrile (reaction 1); this can be done in aqueous or aqueous-alcoholic solution. Addition of 95% hydrogen peroxide to Schiff's bases in anhydrous ether gives hydroperoxyamines, *e.g.* (**8**), which are dehydrated to oxaziridines by heating at 80 °C in an inert solvent (reaction 2). Oxaziridines have been made by oxidation of imines from aldehydes and ketones, of cyclic imines (*e.g.* 1-pyrrolines), and of iminoethers, the latter compounds giving 3-alkoxyoxaziridines. Oxidation has also been performed with ozone, but the yields are poor and the method is limited in its application. Ozonolysis of olefins in the presence of primary amines or ammonia at low temperatures (−20 °C) gives oxaziridines *via* hydroperoxyamines. The scale of these preparations is limited by the explosion risk.

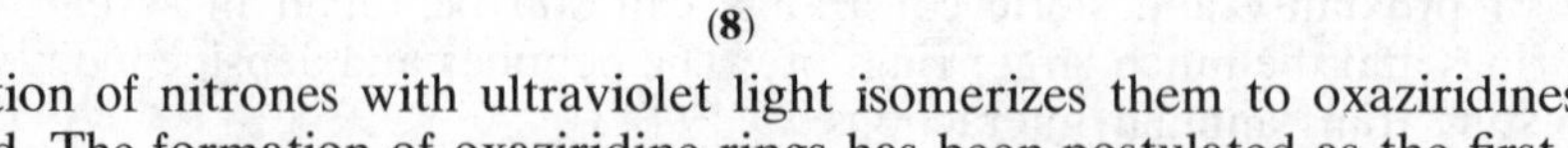

$$R^1R^2C{=}NR^3 + H_2O_2 + R^4CN \longrightarrow R^1R^2\overset{O}{C}{-}NR^3 + R^4CONH_2 \quad (1)$$

$$(\text{cyclohexylidene})=N-C_6H_{11} \longrightarrow C_6H_{10}(OOH)NH-C_6H_{11}\ (\mathbf{8}) \longrightarrow \text{spiro-oxaziridine } N-C_6H_{11} \quad (2)$$

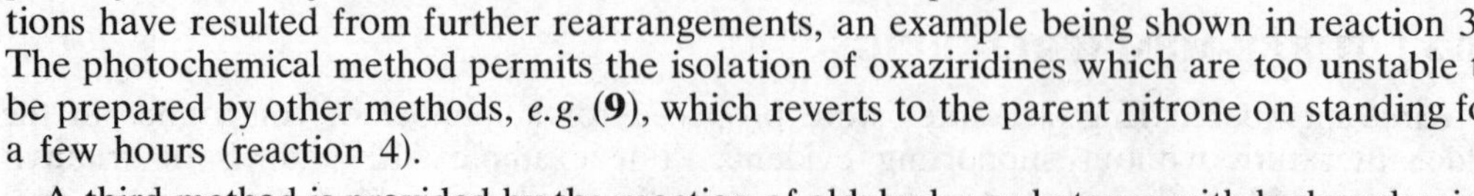

Irradiation of nitrones with ultraviolet light isomerizes them to oxaziridines in fair to good yield. The formation of oxaziridine rings has been postulated as the first step in the photolysis of many heteroaromatic *N*-oxides, but the products isolated from such reactions have resulted from further rearrangements, an example being shown in reaction 3.[4] The photochemical method permits the isolation of oxaziridines which are too unstable to be prepared by other methods, *e.g.* (**9**), which reverts to the parent nitrone on standing for a few hours (reaction 4).

A third method is provided by the reaction of aldehydes or ketones with hydroxylamine *O*-sulphonic acids in alkaline solution (reaction 5). *N*-Chloromethylamine has also been employed. This method will provide oxaziridines unsubstituted on the nitrogen atoms, *e.g.*

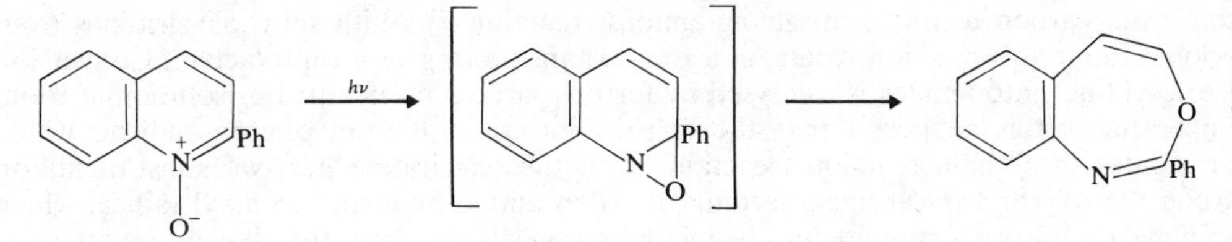

(3)

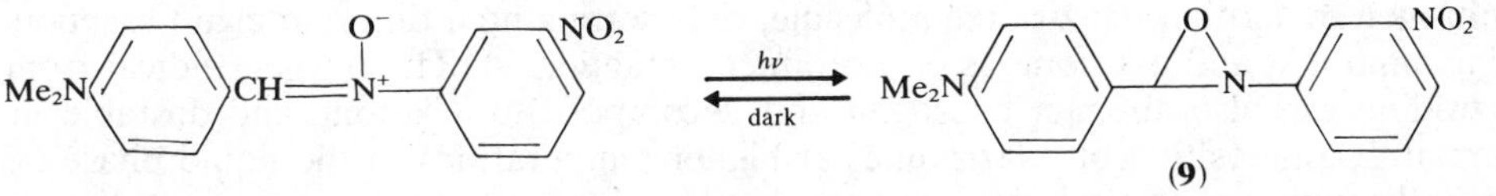

(4)

(9)

(**10**). Such compounds are usually unstable, but they are converted into much more stable 2-acyloxaziridines (**11**) by treatment with acid chlorides, acid anhydrides, phenyl isocyanate, or cyanic acid. 3,3-Diphenyloxaziridine and 3,3-di-t-butyloxaziridine have been made by oxidation of the corresponding imines, and have been further oxidized with lead dioxide to give long-lived oxaziridinyl radicals (**12**; R = Ph or Bu^t).

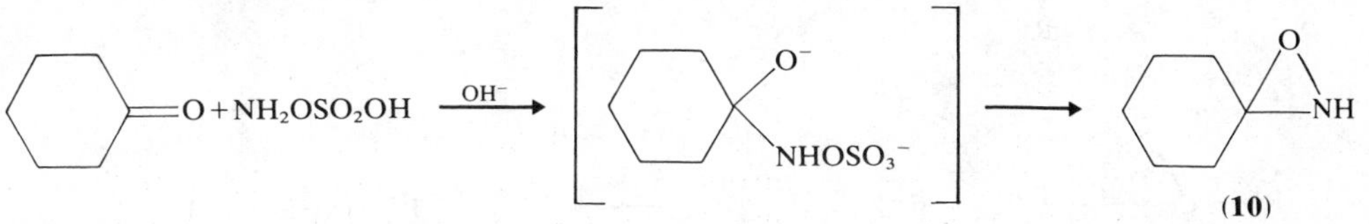

(5)

(10)

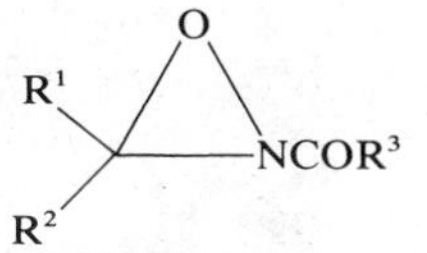

(11) 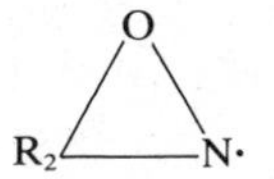 **(12)**

Photolysis of ethyl azidoformate in acetone gives 2-ethoxycarbonyl-3,3-dimethyloxaziridine (**13**) by a nitrene addition (reaction 6), but in cyclohexanone the reaction takes a different course and no oxaziridine is formed.

$N_3CO_2Et + Me_2CO \longrightarrow$ (6)

(13)

Oxaziridines which are unsymmetrically substituted at carbon have an asymmetric carbon atom, and the partial resolution of 2-butyl-3-isobutyl-3-methyloxaziridine by a stereoselective oxidation in the presence of brucine provided evidence for the cyclic structure of these compounds. Oxaziridines with chirality due only to asymmetry at the tercovalent nitrogen atom have been made by oxidation of imines with optically active peracids, *e.g.* (+)-peroxycamphoric acid. Peracid oxidation of imines derived from optically active amines has given high optical yields of diastereomers, showing chirality both at nitrogen and in the carbon chain of the parent amine.

Oxaziridines are active oxygen compounds which liberate iodine from potassium iodide and can be estimated iodometrically, and although many can be purified by distillation it is suggested that they should be treated with the same precautions as organic peroxides. They are reduced to the parent imines by tertiary phosphines and the reaction with brucine, mentioned previously, gave brucine *N*-oxide. The ring is unstable, and oxaziridines revert to nitrones on standing or on heating, but stability is usually enhanced by t-alkyl substituents on the nitrogen atom. Pyrolysis at 200–300 °C in the vapour phase breaks the oxygen–nitrogen bond and causes migration of an alkyl group or hydrogen atom from carbon to nitrogen, giving amides (reaction 7). With spiro-oxaziridines from cyclohexanone this reaction results in a ring-expansion to give a caprolactam. Conversion of oxaziridines into amides is catalysed by ferrous salts, even in aqueous solution at room temperature. It is supposed that the ferrous ion generates an oxygen radical, which participates in a chain reaction (reaction 8). If the oxaziridine has two substituents on carbon the oxygen radical must decompose to an amide by losing an alkyl radical which can dimerize, disproportionate, or abstract a hydrogen atom from the alkyl group attached to nitrogen in another oxaziridine molecule, thus setting up a different chain reaction and giving ammonia and a ketone as end products (reaction 9). The oxygen radical from the oxaziridine can also abstract hydrogen and decompose to a ketone and unstable imine. Thermal decomposition of oxaziridines at higher temperatures in the liquid phase occurs largely through such radical chain-reactions, giving ketones, ammonia, and small amounts of amide.

$$R^1C(R^2)(O)NR^3 \longrightarrow R^1CONR^2R^3 + R^2CONR^1R^3 \quad (7)$$

$$R^1C(H)(O)N{-}R^2 + Fe^{2+} + H^+ \longrightarrow R^1CH(O\cdot)NHR^2 + Fe^{3+}$$

$$R^1CH(O\cdot)NHR^2 + R^1C(H)(O)N{-}R^2 \longrightarrow R^1CONHR^2 + R^1CH(O\cdot)NHR^2 \quad (8)$$

$$R^1_2C(O)NCH_2R^2 + R^{1\cdot} \longrightarrow \left[R^1_2C(O)N\dot{C}HR^2\right] + R^1H$$

$$\longrightarrow R^1_2C(O\cdot)N{=}CHR^2 \xrightarrow{R^1H} R^1_2CO + HN{\equiv}CHR^2 + R^{1\cdot} \quad (9)$$

Acid hydrolysis of oxaziridines can take two courses. 3-Aryl derivatives which can form a resonance-stabilized benzylcarbonium ion (**14**) by cleavage of the carbon–oxygen bond are hydrolysed to aromatic aldehydes and hydroxylamines (reaction 10). This has been suggested as a preparative route for converting amines, *e.g.* amino-acids, into the corresponding hydroxylamines. 3-Alkyloxaziridines, on the other hand, undergo fission of the oxygen–nitrogen bond with migration of hydrogen or an alkyl group from the 2-substituent on to nitrogen. Further hydrolysis then affords ketones and ammonia (or an amine). Amines may thus be converted into aldehydes or ketones by condensation with acetone, oxidation to the oxaziridine, and acid hydrolysis (reaction 11). Alkaline hydrolysis of oxaziridines with an α-hydrogen atom in the 2-substituent gives carbonyl compounds and a quantitative yield of ammonia, but oxaziridines with tertiary alkyl substituents on the nitrogen atom are resistant to alkali. 3-Alkoxyoxaziridines (**15**) are

$$\text{Ph—}\overset{\text{O}}{\text{C}}(\text{H})\text{—N—R} \longrightarrow \text{Ph}\overset{+}{\text{C}}\text{H—N(OH)—R} \longrightarrow \text{PhCHO} + \text{RNHOH} \quad (10)$$

(14)

$$\text{R}_2\text{CHNH}_2 \longrightarrow \text{Me}_2\text{C=NCHR}_2 \longrightarrow \text{Me}_2\text{C(O)N—CHR}_2 \longrightarrow \text{Me}_2\text{C}(\overset{+}{\text{O}}\text{H})\text{N—CR}_2\text{H}$$

$$\longrightarrow \text{R}_2\overset{+}{\text{C}}\text{—NHCMe}_2\text{OH} \longrightarrow \text{R}_2\text{CO} + \text{NH}_3 + \text{Me}_2\text{CO} \quad (11)$$

$$\text{Bu}^t(\text{MeO})\text{C=N—Bu}^t \longrightarrow \text{Bu}^t(\text{MeO})\text{C(O)N—Bu}^t \xrightarrow{\text{H}_3\text{O}^+} \text{Bu}^t\text{CO}_2\text{Me} + \text{Bu}^t\text{NHOH} \quad (12)$$

(15)

stable to aqueous alkali but are hydrolysed by acid, giving hydroxylamines and esters (reaction 12).

Reduction of oxaziridines with lithium aluminium hydride usually gives the parent imine, but 2-t-octyloxaziridine is reduced to t-octylmethylamine. Reduction to amines can also be done with hydrogen and a platinum catalyst. Treatment of oxaziridines with sulphur nucleophiles (thiourea, potassium thiocyanate, potassium methyl xanthate, or triphenylphosphine sulphide) under conditions expected to give thiaziridines leads to the formation of the parent imines and sulphur.

Most other reactions of oxaziridines involve fission of the molecule and the combination of only one fragment with the reagent. Oxidation, *e.g.* with peracetic acid, affords a good preparative route to nitrosoalkanes, which are isolated as their *trans*-dimers (**16**). Reaction of oxaziridines with carbon disulphide gives isothiocyanates in quantitative yield, possibly through a thiaziridine intermediate (**17**) (Scheme 1). Heterocumulenes such as diphenylketen and diphenylcarbodi-imide give products in which only the nitrogen atom and 2-substituent of the oxaziridine are incorporated. Isocyanates, however, react with

$$\text{R}^1_2\text{C(O)NR}^2 \longrightarrow \text{R}^1_2\text{CO} + \text{R}^2\text{NO} \longrightarrow \text{R}^2\overset{+}{\text{N}}(\text{O}^-)\text{=}\overset{+}{\text{N}}(\text{O}^-)\text{R}^2$$

(16)

$$\downarrow$$

$$\left[\text{R}^2\text{N—C(=S)S}\right] \longrightarrow \text{R}^2\text{NCS}$$

(17)

SCHEME 1

oxaziridines to give oxadiazolidinones (**18**). Phenyl isothiocyanate reacts with 3-phenyloxaziridines at 100 °C, giving benzaldehyde and a carbodi-imide (**19**) or its decomposition products; at 90 °C thiadiazolinethiones (**20**) and (**21**) are obtained (reaction 13).

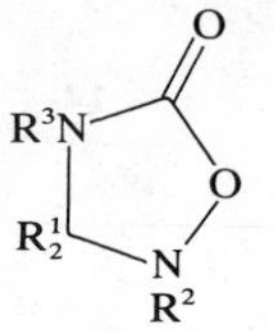

(**18**)

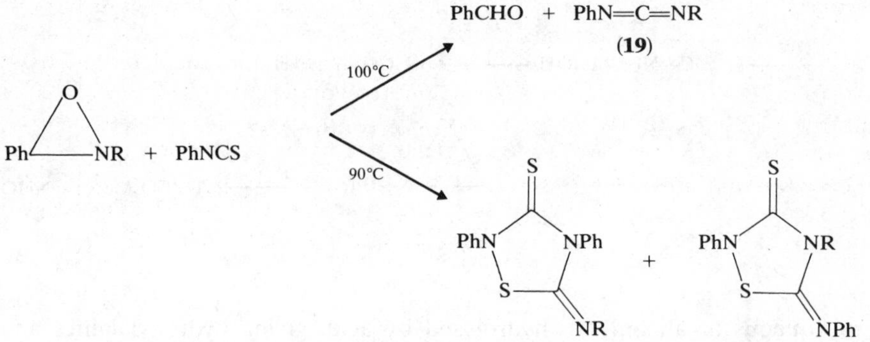

PhCHO + PhN=C=NR (**19**)

Ph–oxaziridine–NR + PhNCS (100 °C; 90 °C) (13)

(**20**) (**21**)

2-Acyloxaziridines undergo ring enlargement to 1,3,4-dioxazoles (**22**) when they are heated. They are powerful aminating agents and will convert amines into acylhydrazines, *e.g.* (**23**), and hydrazines into triazanes (**24**) or triazanium betaines (**25**) from (11; $R^1 = Ph$, $R^2 = H$, $R^3 = NHR^1$).[5]

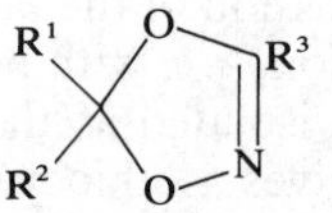

(**22**)

$R^2NHNHCONHR^1$ (**23**)

$NH_2N(R^2)NHCONHR^1$ (**24**)

$NH_2\text{—}\overset{+}{N}(Me)_2\text{—}\bar{N}\text{—}CONHR^1$ (**25**)

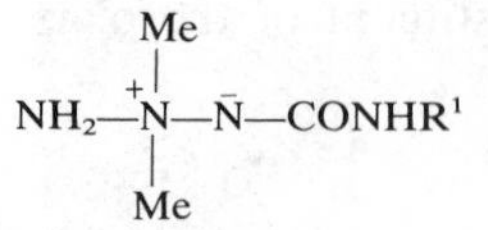

20.3.1.2 Oxadiaziridines

The oxadiaziridine ring system is little known, the first authentic example being reported in 1967, although 2,3-diaryloxadiaziridines have been postulated as intermediates in the Wallach rearrangement of azoxybenzenes to 4-hydroxyazobenzenes in strong acid. The oxadiaziridines that have been characterized are 2,3-dialkyl compounds (**26**) and are obtained by photoisomerization of aliphatic azoxy-compounds (reaction 14). They lack the ultraviolet absorption of the azoxy-compounds. They are thermally less stable than the oxaziridines and they readily revert to mixtures of the *cis*- and *trans*-forms of the parent azoxy-compounds. They are unreactive towards dipolarophiles and are stable to water. They oxidize iodide ion, and the rate of ring-opening is increased by acids.

Reduction of (**26**; R = Bu^n) with lithium aluminium hydride at 20 °C gives azobutane, but the t-butyl analogue is not affected by this reagent under the same conditions. Methyllithium reduces (**26**; R = Bu^n) to azobutane and butyraldehyde butylhydrazone.

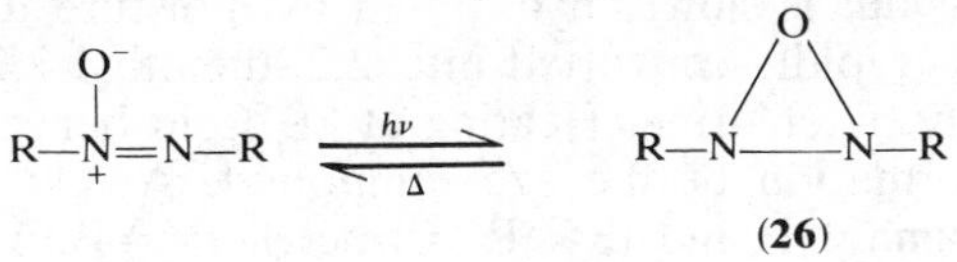

$$R{-}\overset{O^-}{\underset{+}{N}}{=}N{-}R \underset{\Delta}{\overset{h\nu}{\rightleftharpoons}} R{-}N{\overset{O}{\triangle}}N{-}R \quad (\mathbf{26}) \tag{14}$$

20.3.1.3 Thiaziridines

The transient existence of thiaziridines has been postulated to explain the course of a number of photochemical reactions, the deoxygenation of oxaziridines by sulphur nucleophiles, and other reactions, but no simple thiaziridine has been isolated. Thermal decomposition at 60 °C of 4-alkyl-5-arylsulphonyliminothiatriazolines (**27**) (the adducts from alkyl azides and arylsulphonyl isothiocyanates) gives transient intermediates that may be thiaziridine imines (**28**) or the dipolar compounds (**29**) (Scheme 2). They may be captured by phosphorus ylides, heterocumulenes (isocyanates, isothiocyanates, ketens, or carbodi-imides), or enamines, giving products with five-membered rings, *e.g.* (**30**), but in the absence of such reagents they decompose to give arylsulphonylcarbodiimides (**31**).[6]

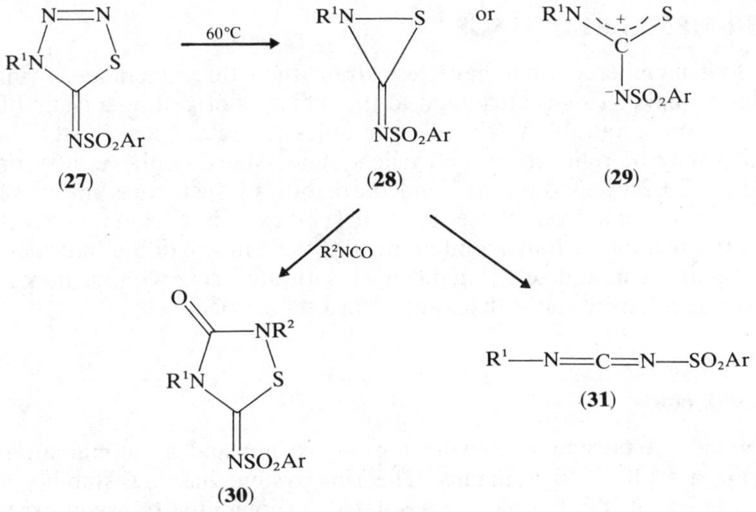

$$(\mathbf{27}) \xrightarrow{60\,^{\circ}C} (\mathbf{28}) \text{ or } (\mathbf{29}); \quad (\mathbf{28}) \xrightarrow{R^2NCO} (\mathbf{30}); \quad (\mathbf{28}) \rightarrow R^1{-}N{=}C{=}N{-}SO_2Ar \ (\mathbf{31})$$

SCHEME 2

2,3-Di-t-butylthiaziridine 1,1-dioxide (**32**) has been made by the reaction of 2,2-dimethyldiazopropane with the *N*-sulphonylamine (**33**), prepared *in situ* from t-butylsulphamoyl chloride and triethylamine (reaction 15). It forms colourless crystals and decomposes quantitatively at 60–80 °C into sulphur dioxide and the azomethine $Bu^tCH{=}NBu^t$. Titration with perchloric acid in acetic acid gives the perchlorate of this azomethine.[7]

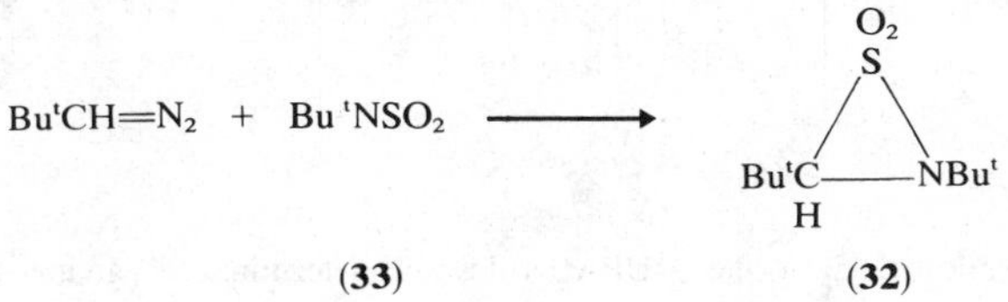

$$Bu^tCH{=}N_2 + \underset{(\mathbf{33})}{Bu^tNSO_2} \longrightarrow \underset{(\mathbf{32})}{Bu^tCH{\overset{SO_2}{\triangle}}NBu^t} \tag{15}$$

20.3.1.4 Thiadiaziridines

Oxidation of the sodium salt of *N,N′*-di-t-butylsulphamide with t-butyl hypochlorite gives 2,3-di-t-butylthiadiaziridine 1,1-dioxide (**34**; $R^1 = R^2 = Bu^t$). This is fairly stable to dilute acids and alkalis, but is slowly hydrolysed by moisture to 1,2-di-t-butylhydrazine hydrogen sulphate. It is rapidly converted into 2,2′-dimethyl-2,2′-azopropane by t-butyl hypochlorite, and slowly by chlorine. Heating at 80 °C in benzene causes elimination of sulphur dioxide and formation of the azo-compound. A few related compounds (**34**; $R^1 = R^2 =$ t-octyl or adamantyl) and (**34**; $R^1 =$ t-octyl, $R^2 = Bu^t$) have been made by the same process. The compounds with the larger alkyl substituents have much higher thermal stability. The formation of a Diels–Alder addition product from (**34**; $R^1 = R^2 = Bu^t$) and 1,3-diphenylisobenzofuran was reported. The diadamantyl compound oxidizes potassium iodide quantitatively with regeneration of the sulphamide. Other reducing agents (thiophenol, hydrogen–palladium, and Grignard reagents) regenerate the parent sulphamide, but with lithium aluminium hydride both the azo-compound and sulphamide are obtained, and phenyl-lithium abstracts sulphur dioxide to give the azo-compound.

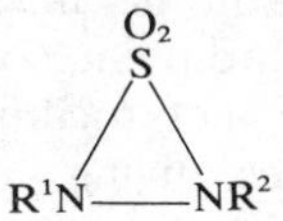

O_2 S, R^1N—NR^2

(34)

20.3.2 FOUR-MEMBERED RINGS

Although four-membered rings have less strain than three-membered rings, they are usually harder to make. This is attributed to the difficulty of getting a chain of four atoms into a conformation favourable to ring-closure unless there are special steric or electronic constraints that may be relieved in the cyclic system. Many syntheses of four-membered rings depend on 2+2 cycloadditions,[8] and the fission of such rings into 2+2 fragments (often the opposite pairs from those used in synthesis) is a common feature of their chemistry. In the naming of four-membered rings the suffix -etidine indicates a saturated ring containing nitrogen, and -etan* indicates a saturated ring without nitrogen. Unsaturated rings are denoted by the suffixes -ete and -et,* respectively.

20.3.2.1 Oxathietans

The compounds at present known having an oxygen and a sulphur atom in a four-membered ring are all 1,2-oxathietans. The ring system has low stability and no 1,2-oxathietan with divalent sulphur has been isolated, although the transient existence of (**35**) has been suggested to account for the photolytic breakdown of 2,3-dibenzoyl-2,3-diphenylthiiran 1-oxide (reaction 16). Some very unstable 1,2-oxathietan 2-oxides (β-sultines) have been generated by the action of *N*-chlorosuccinimide on β-hydroxyalkyl

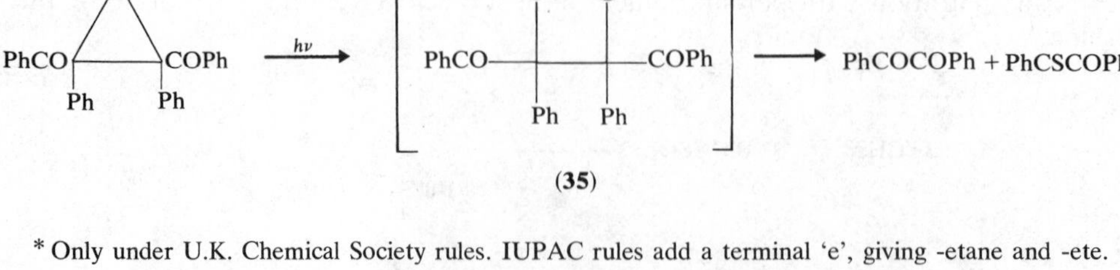

PhCO(Ph)C—C(Ph)COPh (S=O bridge) $\xrightarrow{h\nu}$ [PhCO(Ph)C—C(Ph)COPh, S—O ring] $\longrightarrow$ PhCOCOPh + PhCSCOPh (16)

(35)

* Only under U.K. Chemical Society rules. IUPAC rules add a terminal 'e', giving -etane and -ete.

t-butyl sulphoxides, and a more stable compound, 3,3-dimethyl-4,4-diphenyl-1,2-oxathietan 2-oxide (**36**), is obtained by treating the sulphoxide (**37**) with sulphuryl chloride at −70 °C.[9] This may be kept for some days at room temperature, but at 30 °C it decomposes into sulphur dioxide and 1,1-dimethyl-2,2-diphenylethylene (reaction 17).

$$Ph_2C(OH)—CMe_2SOBu^t \text{ (37)} \longrightarrow \text{(36)} \longrightarrow Ph_2C{=}CMe_2 \qquad (17)$$

(**37**) (**36**)

There has been considerable investigation of 1,2-oxathietan 2,2-dioxides (β-sultones), which are made by the addition of olefins to sulphur trioxide, the reactivity of the latter compound being controlled by using it as its adduct with dioxan or pyridine. The 4-phenyl derivative obtained from styrene is very labile, but polyfluoro-derivatives made from highly fluorinated olefins are much more stable.[10] The addition of sulphur trioxide to *cis*- and *trans*-but-2-enes is stereospecific and gives *cis*- and *trans*-3,4-dimethyl-1,2-oxathietan 2,2-dioxides. This suggests that the reaction is a cycloaddition, but it is generally supposed to go through a polar, open-chain intermediate, since the reaction can also give 1:2 and 2:1 addition products (1,2-oxathian 2,2-dioxides and 1,3-dioxa-2,4-dithian 2,2,4,4-tetroxides). Thus dodec-1-ene and dodec-2-ene react with sulphur trioxide–dioxan to give (**38**; $R^1 = C_{10}H_{21}$, $R^2 = H$ or $R^1 = C_9H_{19}$, $R^2 = Me$) and (**39**; $R^1 = C_{10}H_{21}$, $R^2 = H$ or $R^1 = C_9H_{19}$, $R^2 = Me$). These compounds differ in their reaction with aniline (reactions 18 and 19), (**38**) undergoing fission of the 1,4-bond and giving β-anilinoalkanesulphonic acids (**40**; $R^1 = C_{10}H_{21}$, $R^2 = H$ or $R^1 = C_9H_{19}$, $R^2 = Me$), whereas a sulphur–oxygen bond in (**39**) is broken, giving the β-hydroxyalkanesulphonanilides (**41**; $R^1 = C_{10}H_{21}$, $R^2 = H$ or $R^1 = C_9H_{19}$, $R^2 = Me$).

$$\text{(38)} \longrightarrow R^1CH(NHPh)CHR^2SO_3H \text{ (40)} \qquad (18)$$

$$\text{(39)} \longrightarrow R^1CH(OH)CHR^2SO_2NHPh \text{ (41)} \qquad (19)$$

A second method for preparing β-sultones is the addition of a sulphene (prepared *in situ* from an alkanesulphonyl chloride and triethylamine) to a perhalogenated carbonyl compound, *e.g.* chloral or penta- or hexa-halogenoacetones (reaction 20). The reaction fails to give β-sultones if there is steric hindrance at the carbonyl group, *e.g.* with bromal, hexachloroacetone, or α,α,α-trifluoroacetophenone.[11] The sultone (**42**; $R^1 = R^2 = H$, $R^3 = CCl_3$) was hydrolysed to sodium 2-hydroxy-3,3,3-trichloropropanesulphonate, which was also synthesized by a different method. Aminolysis of (**42**; $R^1 = R^2 = H$, $R^3 = CCl_3$), *e.g.*

$$R^1CH{=}SO_2 + R^2COR^3 \longrightarrow \text{(42)} \qquad (20)$$

$R^1 = H, Me, Br$

(**42**)

with ammonia or morpholine, was also said to open the 1,2-bond, giving a hydroxypropanesulphonamide. This reaction site differs from the one in non-halogenated compounds, *e.g.* (**38**; R^1 and R^2 = alkyl or H) and 4-phenyl-1,2-oxathietan 2,2-dioxide (**38**; R^1 = Ph, R^2 = H), in which nucleophiles such as aniline, pyridine, or methanol attack the carbon atom in the 4-position.

Ring opening of the perfluoro- and chlorofluoro-β-sultones takes place at the 1,2-bond, but is complicated by the very ready expulsion of a fluoride or chloride ion from the β-carbon atom; this ion may subsequently attack the positive centre formed in the ring cleavage. Thus triethylamine will isomerize the sultone (**43**) catalytically to difluoro(fluorosulphonyl)acetyl fluoride (**44**) (reaction 21), and with other nucleophiles (water, alcohols, amines, thiocyanate ion) this rearrangement is followed by a reaction of the acyl fluoride with the nucleophile. The same isomerization of (**43**) to (**44**) is also produced by prolonged heating. 3,4,4-Trifluoro-3-trifluoromethyl-1,2-oxathietan 2,2-dioxide is isomerized to the fluorosulphonylacyl fluoride by heating under pressure for 1 hour at 150–160 °C. Thermal isomerization of (**38**; R^1 = Ph, R^2 = H) goes very readily and gives styrene-β-sulphonic acid.

F_2 F_2 / O—SO_2 (**43**) → [CF_2—CF_2 / O^- $\overset{+}{SO_2NEt_3}$] → $FC(=O)CF_2SO_2\overset{+}{N}Et_3$ F^- → $FC(=O)CF_2SO_2F$ (**44**) (21)

Nitrosyl chloride opens the 2,3-bond of (**43**), giving nitrosyl fluoride and CF_2=$CFOSO_2Cl$. Ethylene reacts with (**43**) to give ethylene-1,1-bis(sulphonic acid) and tetrafluoroethylene. It is probable that the labile ethylene β-sultone is an intermediate in this reaction.

Several useful applications have been suggested for the polyfluorosultones, *e.g.* as stabilizers for liquid sulphur trioxide (because of their reactivity with moisture) and as intermediates for polymers and ion-exchange resins. The trichloromethyl compound (**42**; R^1 = R^2 = H, R^3 = CCl_3) is claimed to be a catalyst in acylations of alcohols and amines.

20.3.2.2 Oxazetidines

There are two series of compounds with one oxygen and one nitrogen atom in a four-membered ring: the 1,2- and 1,3-oxazetidines. 1,2-Oxazetidines can be obtained by the cycloaddition of nitroso-compounds to olefins, but the published examples fall into rather special classes and it is not known whether this is a general reaction. The addition of nitrosobenzene to diphenylketen, first described in 1911, gives the 1,2-oxazetidinones (**45**) and (**46**), but the second isomer is unstable and loses carbon dioxide to give benzophenone anil, which reacts further with the diphenylketen. The ratio of isomers is little affected by the polarity of the reaction medium, but 4-dimethylaminonitrosobenzene gives the unstable isomer whereas other 4-substituted nitrosobenzenes (Me, MeO, CO_2Me) give good yields of the stable isomers.[12] Condensation of diphenylchloroacetyl chloride with phenylhydroxylamine provides an alternative synthesis of (**45**), and similar oxazetidines are obtained from alkyl- and cycloalkyl-hydroxylamines. Catalytic reduction of (**45**) gives the anilide of benzilic acid, but the *N*-alkyl compounds are cleaved to

Ph_2 (4) — (3) C=O / O (1) — NPh (2) (**45**)

Ph_2 — C=O / PhN—O (**46**)

diphenylacetylhydroxylamines. Thermal decomposition of (**45**) gives benzophenone and phenyl isocyanate, probably through a nitrenium intermediate (**47**). Decomposition in methanol gives the methoxy-compounds (**48**) and (**49**) which could be derived from such an intermediate (reaction 22). Diphenylketen *N*-*p*-tolylimide reacts with nitrosobenzene, giving the *p*-tolylimide of (**45**) and no isomer related to (**46**). Thermal decomposition of this imide gives benzophenone and phenyl-*p*-tolylcarbodi-imide.

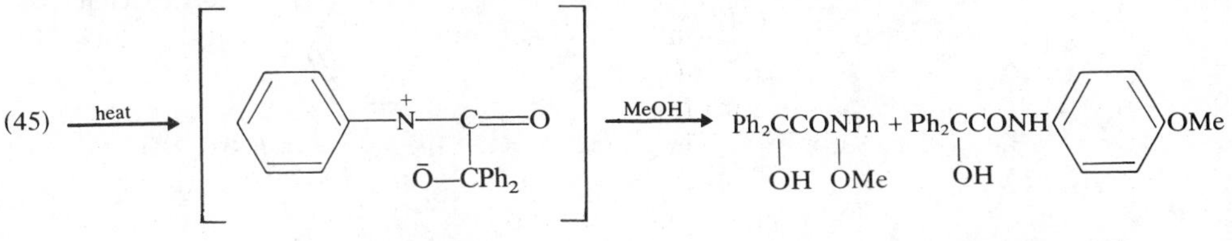

(45) —heat→ [C_6H_5–$\overset{+}{N}$–C=O / $\bar{O}$–CPh_2] —MeOH→ $Ph_2C(OH)CONPh(OMe)$ + $Ph_2C(OH)CONH$–C_6H_4–OMe (22)

(**47**) (**49**)

Numerous polyfluoro-oxazetidines have been made by addition of highly fluorinated alkenes to trifluoronitrosomethane, or to nitrosyl fluoride. In the latter case the primary adduct is a nitroso-fluorocarbon, and this undergoes cycloaddition with a further molecule of the alkene. Oxazetidine formation is prevented, however, if there is steric hindrance through branching of the olefin chain.[13] 2-Chloro-2-ethoxy-1,1,2-trifluoronitrosoethane adds to trifluorohalogenoethylenes to give oxazetidines (**50**) (reaction 23), from which the *N*-substituent may be removed by hydrolysis with sulphuric acid to yield (**51**; X = F, Cl, Br) as pungent, low-boiling, unstable compounds. Chlorination of (**51**; X = F) gives the *N*-chloro-derivative. Photolysis or pyrolysis of perfluoro-oxazetidines, *e.g.* (**52**), cleaves the ring at the 1,2- and 3,4-bonds.

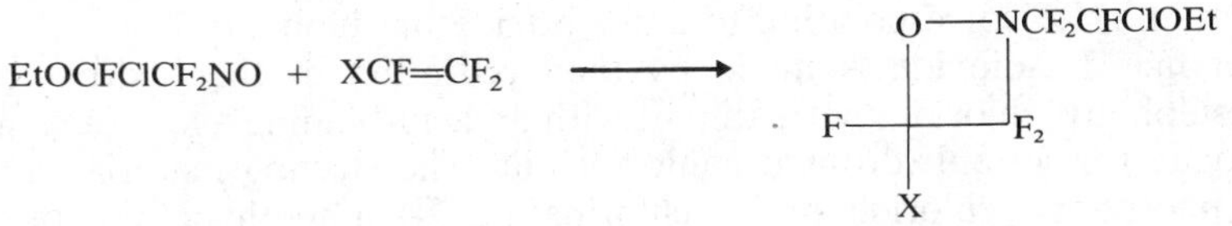

$EtOCFClCF_2NO$ + $XCF{=}CF_2$ ⟶ O–$NCF_2CFClOEt$ / F, X–C–CF_2 (23)

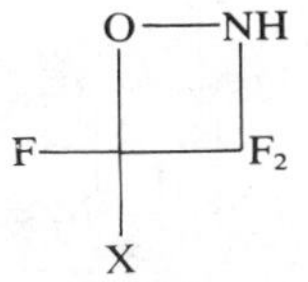

(**50**)

O–NH / F, X–C–CF_2 (**51**)

O–NCF_3 / F_2C–CF_2 (**52**)

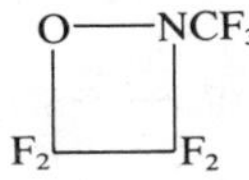

Nitrosobenzene adds to tetramethoxyethylene to give 3,3,4,4-tetramethoxy-2-phenyl-oxazetidine. 2-Ethoxycarbonyl-1,2-oxazetidine has been made from *N*-hydroxyurethane by converting it into an *O*-(2-bromoethyl) derivative and heating the sodium salt of this in DMF.

1,3-Oxazetidines (*e.g.* **53**) have been made in low to moderate yield by addition of electron-rich isocyanates (methyl, butyl) to ketones with electronegative substituents, *e.g.* hexafluoroacetone and chlorofluoro-analogues. They are claimed to be catalysts for lactam polymerization. Phenyl isocyanate reacts with tetraoxan and boron trifluoride, giving 2-oxo-3-phenyl-1,3-oxazetidine (**54**), which gives polymers with trioxan. The 1,3-oxazetidines (**55**; R = NO_2 or Ac) are among the products of reaction of keten with *N*-isopropylideneanilines.

Derivatives of the unsaturated 1,2-oxazete ring system (**56**) and some 1,2-oxazete *N*-oxides (**57**) have been made from nitroso- and nitro-alkenes in which bulky t-butyl groups produce steric overcrowding in the acyclic molecules. They readily decompose to give di-t-butyl ketone.[14]

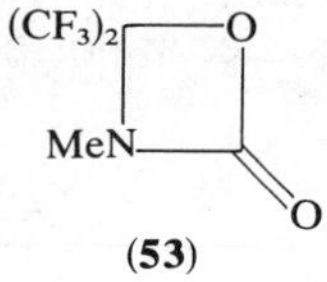

(53) 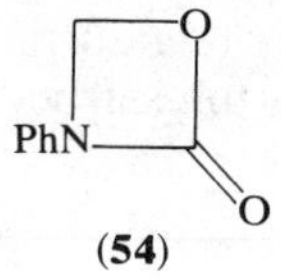 (54) 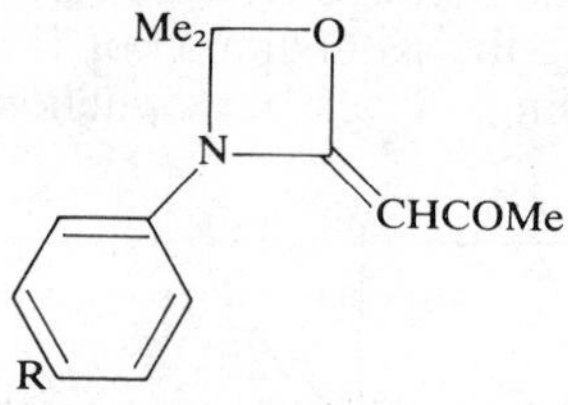 (55)

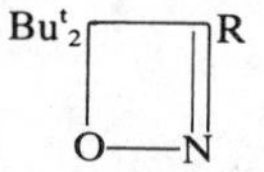

(56) 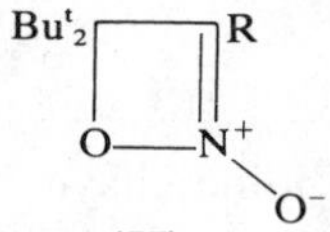 (57)

20.3.2.3 Thiazetidines

Compounds with one sulphur and one nitrogen atom in a saturated four-membered ring are 1,2- and 1,3-thiazetidines. A few unsaturated 1,2- and 1,3-thiazetes are also known. Preparative methods include 2+2 cycloadditions and cyclization of open chains of four atoms. In the 1,2-thiazetidines and thiazetes that are known at present, the sulphur atom is in an oxidation state corresponding to a sulphoxide or sulphone.

1,2-Thiazetidine 1,1-dioxide is made by the cyclization of the hydrochloride of 2-aminoethanesulphonyl chloride by treating it with an acid-binding agent (*e.g.* ammonia or sodium carbonate) in an anhydrous organic solvent. The starting material is made from taurine and phosphoryl chloride or by chlorinating 2-aminoethanethiol or sodium 2-aminoethylthiosulphate in an aqueous medium. The thiazetidine is slowly hydrolysed by water to taurine.

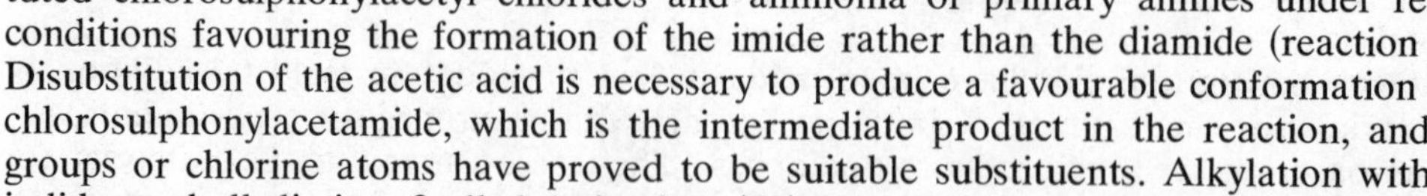

(58) (59) (60) (24)

4,4-Disubstituted 3-oxo-1,2-thiazetidine 1,1-dioxides (**58**) are obtained from disubstituted chlorosulphonylacetyl chlorides and ammonia or primary amines under reaction conditions favouring the formation of the imide rather than the diamide (reaction 24).[15] Disubstitution of the acetic acid is necessary to produce a favourable conformation of the chlorosulphonylacetamide, which is the intermediate product in the reaction, and alkyl groups or chlorine atoms have proved to be suitable substituents. Alkylation with alkyl iodides and alkali gives 2-alkyl derivatives (**59**), but with alkyl iodides and silver oxide 3-alkoxy-1,2-thiazete 1,1-dioxides (**60**) are obtained. Alkylation with diazomethane gives (**59**) and (**60**) in the ratio 4:1. Cleavage of the ring of (**58**) occurs fairly readily through acid or alkaline hydrolysis and breaks the C—N or the N—S bond. Hydrazinolysis of (**58**; $R^1 = Me$) gives 1-sulphoisobutyric hydrazide, but reduction with lithium aluminium hydride breaks the C—N bond to yield 1-hydroxymethyl-1-methylethanesulphonamide. The

i, H_2O_2; ii, NH_3; iii, H_2O; iv, N_2H_4.

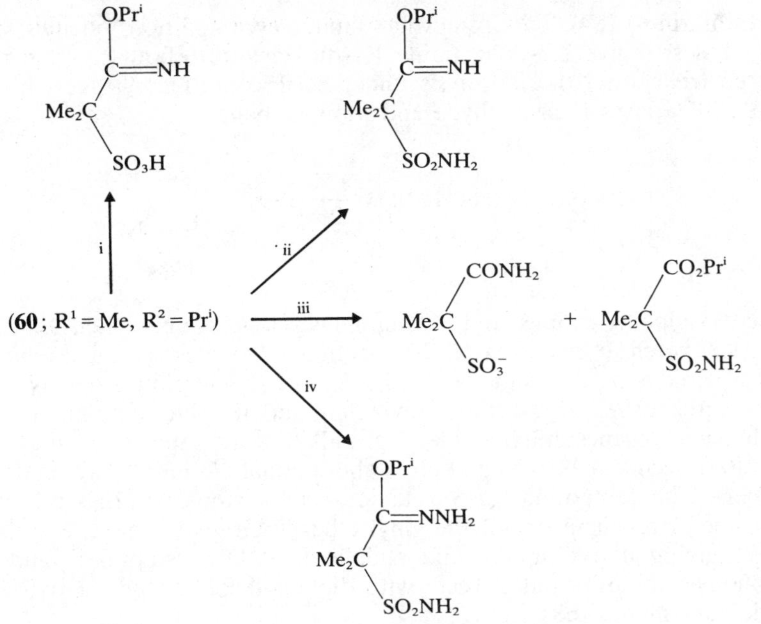

SCHEME 3

3-alkoxythiazetes, *e.g.* (**60**; $R^1 = Me$, $R^2 = Pr^i$), are cyclic imino-ethers and this is reflected in their pattern of ring-opening (Scheme 3).

1,2-Thiazetidine 1,1-dioxides are also obtained by addition of *N*-sulphonylamines (generated from sulphamoyl chlorides and triethylamine) to nucleophilic olefins.[16] *N*-Sulphonylethylamine reacts in this manner with keten acetals and with enamines, giving, for example, the pyrrolidino-compound (**61**), which is very easily hydrolysed to α-(ethylsulphamoyl)isobutyraldehyde (reaction 25). The more reactive *N*-sulphonylbenzamide

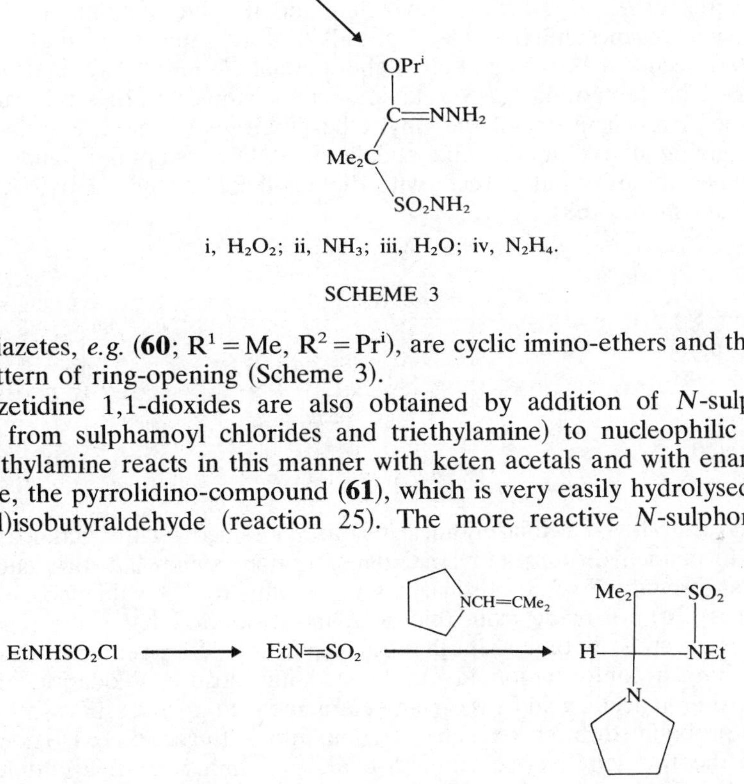

(25)

(**61**)

will add to vinyl ethers to give, for example, (**62**). The reactivity of *N*-sulphonylamines generated with triethylamine is diminished by complexing with the amine, and a more strongly electrophilic reagent is obtained by treating methoxycarbonylsulphamoyl chloride with sodium hydride in THF. The resulting methyl *N*-sulphonylurethane reacts with moderately nucleophilic olefines to give both 2+2 adducts (thiazetidines) (**63**) and 2+4

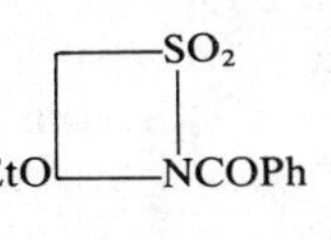

(**62**) 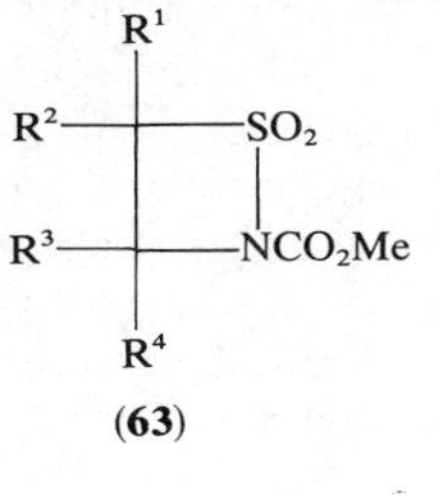 (**63**) 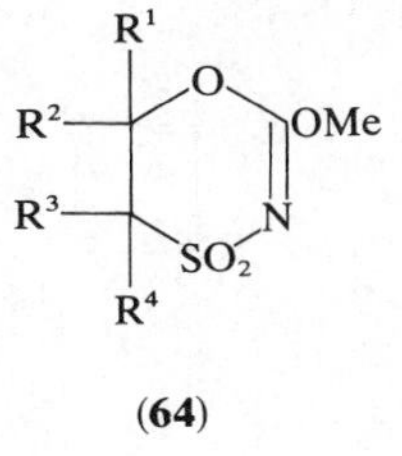 (**64**)

adducts (oxathiazines) (**64**). In a somewhat similar cycloaddition, phenylsulphene reacts with Schiff's bases to give 1,2-thiazetidine 1,1-dioxides (**65**), both *cis*- and *trans*-isomers being isolated (reaction 26). Hydrolysis under basic conditions cleaves the N—S bond; pyrolysis at 220 °C gives benzaldehyde and *trans*-stilbene.[17]

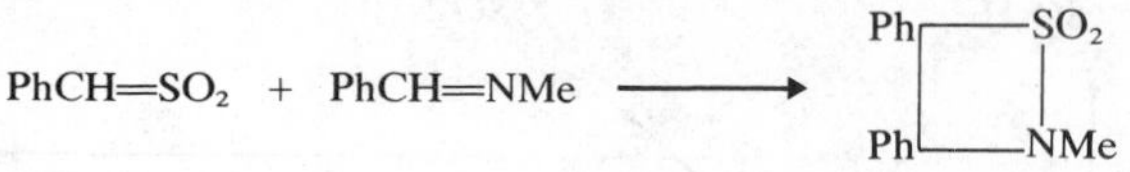

$$PhCH{=}SO_2 + PhCH{=}NMe \longrightarrow \text{(65)} \quad (26)$$

(**65**)

Aromatic *N*-sulphinylamines and *N*-sulphinylcyclohexylamine add to diphenylketen and biphenyleneketen to give 3-oxo-1,2-thiazetidine 1-oxides (**66**). These are reduced or hydrolysed to amides by hydriodic acid, with loss of the sulphur atom. *N*-Sulphinylaryl-sulphonamides react similarly with diphenylketen, and also with enol ethers. The addition of *N*-sulphinyl-*p*-toluenesulphonamide with alkyl vinyl ethers to give the alkoxy-compounds (**67**) occurs at −30 °C and gives about equal amounts of *cis* and *trans* (RO to S=O) isomers. The compounds dissociate into the starting materials on warming, with retention of the stereochemistry of the vinyl ether.[18] Grignard reagents cleave the N—S bond of (**67**), giving alkoxysulphonamido-sulphoxides. Diarylsulphurdi-imides, and di-*p*-toluenesulphonylsulphurdi-imides, react with diphenylketen to give 3-oxo-4,4-diphenyl-1,2-thiazetidine-1-imides (**68**).

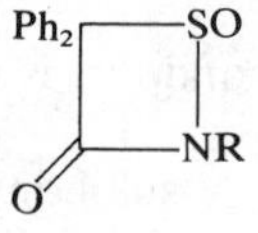

(**66**) 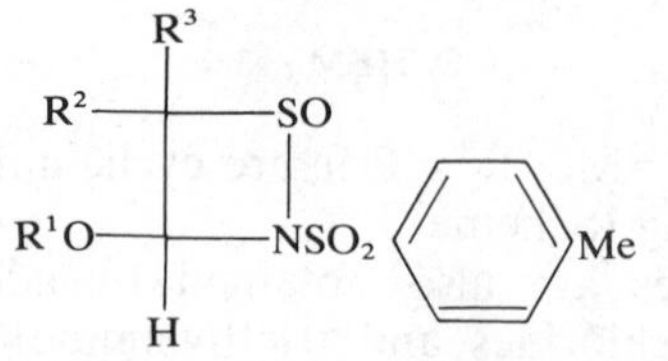 (**67**) 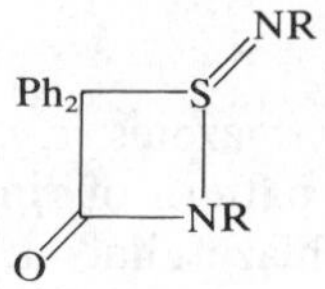 (**68**)

2-Alkyl- and 2-aralkyl-1,3-thiazetidines (**69**) are obtained by the action of hydrogen sulphide and formaldehyde on methyleneimines;[19] under somewhat different conditions these reagents give dihydro-1,3,5-dithiazines or tetrahydro-1,3,5-thiadiazines. 2-Imino-1,3-thiazetidines (**70**) are made from thioureas and di-iodomethane, the reaction being facilitated if the group R^2 in the parent thiourea can form a hydrogen bond with the group RNH, establishing a conformation favourable to ring closure. Oxidation of (**70**) with hydrogen peroxide in acetic acid causes ring enlargement to give the 1,2,4-thiadiazolidine 1-oxide (**71**), probably through oxygen insertion into a thiazetidine 1-oxide and rearrangement of the transient oxathiazoline *S*-oxide.[20] 2-Imino-1,3-thiazetidines (**70**) are also made by treatment of isothiouronium salts from thioureas and chloromethyl methyl sulphoxide with sulphuric acid, methylsulphinyl serving as a leaving group for the transfer of the methylene group to the ring. In a few instances carbamoyl or toluenesulphonyl isothiocyanates add to Schiff's bases to give 4-aryl-2-imino-1,3-thiazetidines.[21]

Diphenylthiourea reacts with phosgene to give diphenylchloroformamidinium chloride and a little 3-phenyl-2-phenylimino-1,3-thiazetidin-4-one, the preparation of which

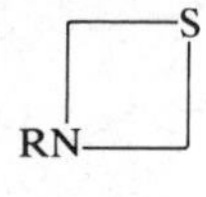

(**69**) 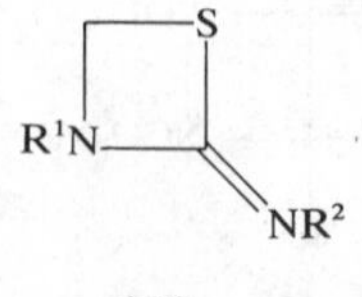 (**70**) 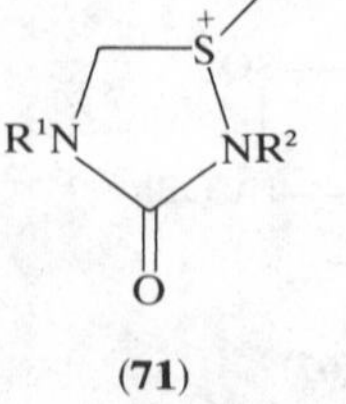 (**71**)

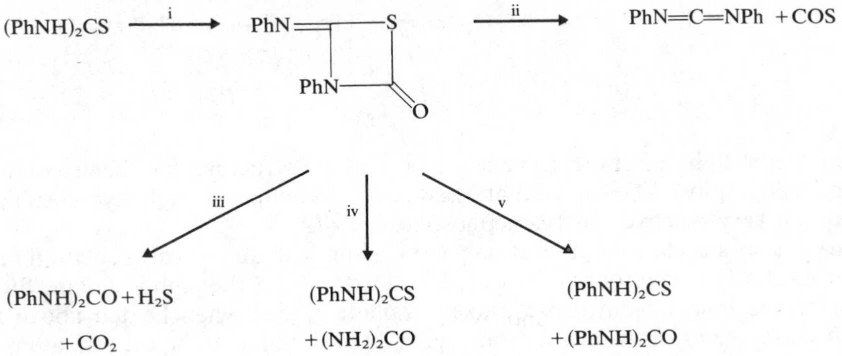

(PhNH)$_2$CS →(i) PhN=C(–S–)(–NPh–C=O) →(ii) PhN=C=NPh + COS

iii: (PhNH)$_2$CO + H_2S + CO_2

iv: (PhNH)$_2$CS + $(NH_2)_2CO$

v: (PhNH)$_2$CS + (PhNH)$_2$CO

i, $COCl_2$; ii, heat; iii, H_2O; iv, NH_3; v, $PhNH_2$.

SCHEME 4

was first reported in 1881! The reactions of this thiazetidinone are shown in Scheme 4. 1-Alkyl- or 1-aryl-sulphonyl-3-arylthioureas react similarly with phosgene, giving 2-alkyl- or 2-aryl-sulphonylimido-3-aryl-1,3-thiazetidin-4-ones, which yield sulphonylarylcarbodi-imides on pyrolysis, but analogous 1,3-thiazetidines are not obtained from 1-alkyl- or 1-aryl-sulphonyl-3-alkylthioureas.[22] Electron-deficient isothiocyanates add to carbodi-imides, giving 2,4-di-imino-1,3-thiazetidines [**72**; $R^1 = MeC_6H_4SO_2$ or $Ph_2P(S)$]. 3,5-Bis(dialkylamino)phenols react with isocyanatoformates and isothiocyanatoformates to give quinonemethide derivatives (**73**; X = O or S), which lose alcohols when heated above their melting points, giving 1,3-oxazetidine-2-ones (**74**; X = O) or 1,3-thiazetidin-2-ones (**74**; X = S).

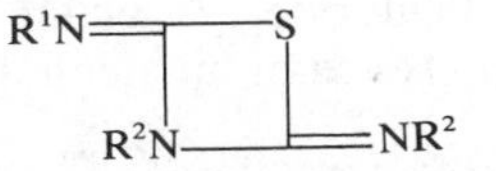

R^1N=, S, R^2N, =NR^2

(**72**)

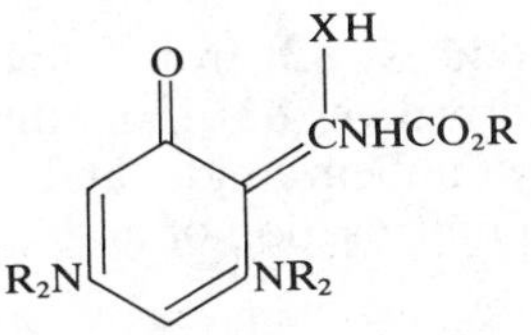

O, XH, $CNHCO_2R$, R_2N, NR_2

(**73**)

X, O, NH, O, R_2N, NR_2

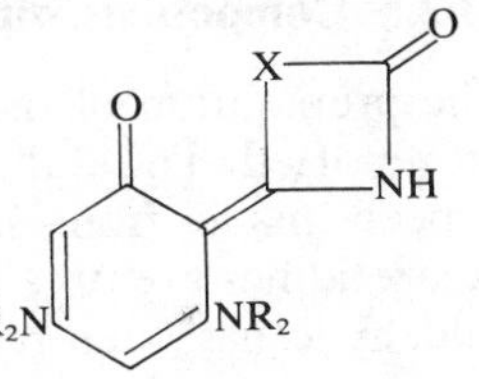

(**74**)

2*H*-1,3-Thiazetes have been obtained from hexafluoroacetone and thioamides. The initial condensation, effected by phosphoryl chloride and pyridine, gives 2,2,6,6-tetrakis(trifluoromethyl)-6*H*-1,3,5-oxathiazines (**75**), which undergo a retro-Diels–Alder reaction and cyclization on thermolysis at 120–140 °C. The 1,3-thiazetes (**76**) are stable at room temperature, but at high temperatures they are in equilibrium with the open-chain *N*-(perfluoroisopropylidene)thiocarboxamides (**77**).[23]

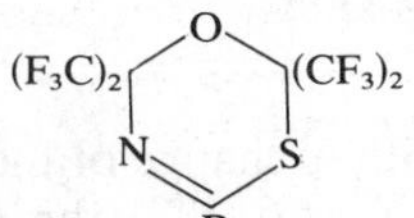

O, $(F_3C)_2$, $(CF_3)_2$, N, S, R

(**75**)

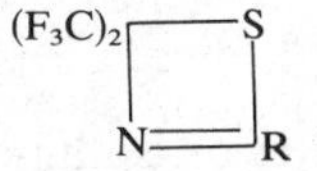

$(F_3C)_2$, S, N, R

(**76**)

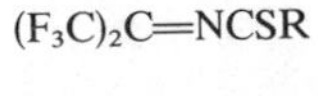

$(F_3C)_2C$=NCSR

(**77**)

20.3.2.4 Oxaphosphetans

It is generally supposed that in the Wittig reaction between a phosphorane and a carbonyl compound the initial adduct has a betaine structure and that this eliminates phosphine oxide through an unstable four-membered ring intermediate (an oxaphosphetan). The stable oxaphosphetan (**78**) has been obtained from hexafluoroacetone and

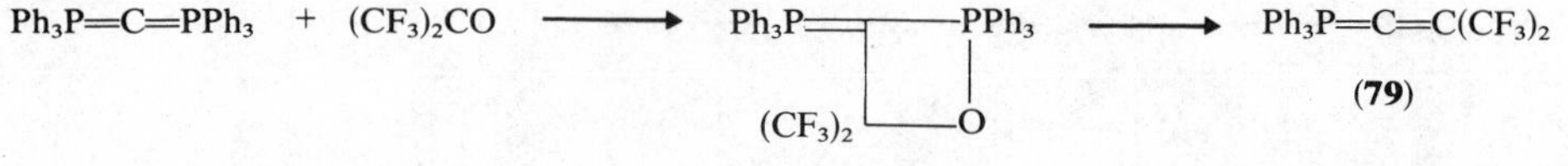

$$Ph_3P{=}C{=}PPh_3 + (CF_3)_2CO \longrightarrow \underset{(\mathbf{78})}{\text{4-membered ring: } Ph_3P{=}C{-}PPh_3,\ (CF_3)_2C{-}O} \longrightarrow \underset{(\mathbf{79})}{Ph_3P{=}C{=}C(CF_3)_2} \quad (27)$$

hexaphenylcarbodiphosphorane (reaction 27), and its structure has been confirmed by X-ray crystallography. When heated above 110 °C it eliminates triphenylphosphine oxide and gives the very reactive vinylidenephosphorane (**79**).[24]

Tertiary phosphines having at least one alkyl group with an α-hydrogen atom react with hexafluoroacetone to give derivatives of 2,2-dihydro-1,3,2-dioxaphospholan (**80**). These are transformed into 2,2-dihydro-1,2-oxaphosphetans (**81**) when heated above 80 °C, a hydrogen atom being transferred from the alkylphosphine to a carbon atom from a carbonyl group. These compounds decompose to give olefins at *ca.* 120°C (reaction 28). The phosphetan from diethylphenylphosphine was obtained as two diastereomers at phosphorus.[25]

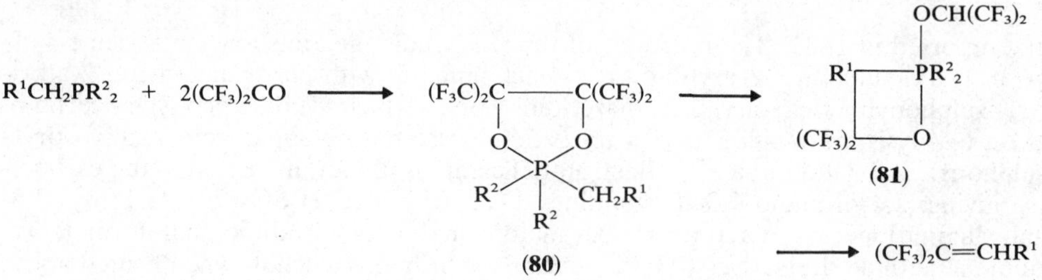

$$R^1CH_2PR^2_2 + 2(CF_3)_2CO \longrightarrow \underset{(\mathbf{80})}{(F_3C)_2C{-}C(CF_3)_2,\ O{-}P(R^2)_2(CH_2R^1){-}O} \longrightarrow \underset{(\mathbf{81})}{R^1C{-}PR^2_2(OCH(CF_3)_2),\ (CF_3)_2C{-}O} \longrightarrow (CF_3)_2C{=}CHR^1 \quad (28)$$

20.3.2.5 Compounds with one carbon atom in the ring

The preparation of oxathiazetidines (**82**) from sulphinylamines and arylaldehydes has been reported. The dithiazete derivative (**83**), 3-methyl-4-phenyldithiazetinium chloride, has been made from *N*-methylthiobenzamide and sulphur dichloride. A claim that oxadiazetidines are obtained from nitrosation of amines in the presence of an alcohol and an aldehyde has been refuted.

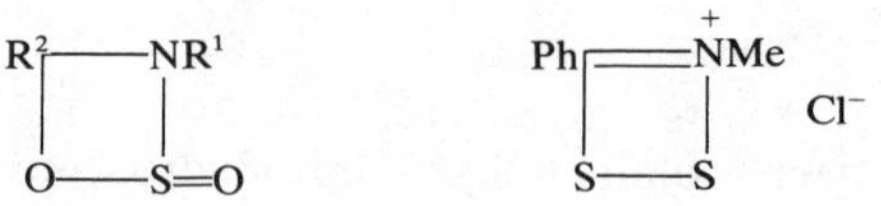

$$\underset{(\mathbf{82})}{R^2C{-}NR^1,\ O{-}S{=}O} \qquad \underset{(\mathbf{83})}{PhC{=}\overset{+}{N}Me,\ S{-}S\quad Cl^-}$$

$$R^1NSO + R^2NCX \rightleftharpoons \left[R^1N{-}S{=}O,\ X{=}C{-}NR^2\right] \rightleftharpoons R^2NSO + R^1NCX \quad (29)$$

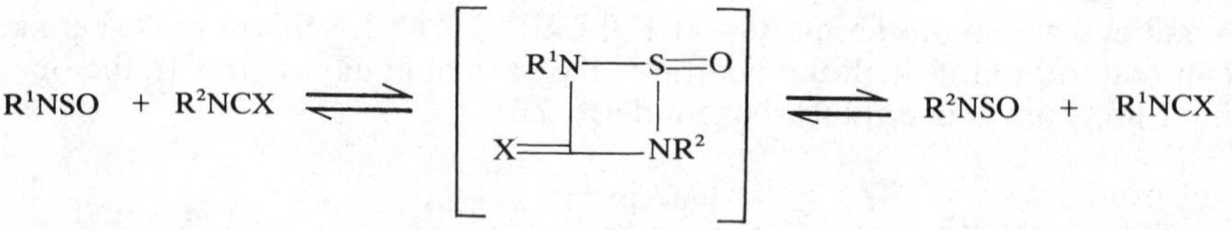

Sulphinylamines react with isocyanates or isothiocyanates with exchange of the carbonyl (or thiocarbonyl) groups for sulphinyl groups (reaction 29; X = O or S) through intermediate cycloadducts which are too unstable to isolate. With carbodi-imides or ketenimines, sulphinylsulphonamides give 3-imino- or 3-ylidene-1,2,4-thiadiazetidine 1-oxides (**84**) and (**85**), which will undergo the exchange reaction (29) on thermolysis and

$$\underset{(\mathbf{84})}{R^1SO_2N{-}S{=}O,\ R^2N{=}C{-}NR^2} \qquad \underset{(\mathbf{85})}{R^1SO_2N{-}S{=}O,\ R^3R^4C{=}C{-}NR^2}$$

are readily hydrolysed to guanidines or amidines. The reaction of *N*-sulphinylacylamides with carbodi-imides gives mixtures of 3-imino-4-acyl-1,2,4-thiadiazetidine 1-oxides (**86**) and the 1,4-cycloadducts, 6-imino-1,4,3,5-oxathiadiazines (**87**).[26]

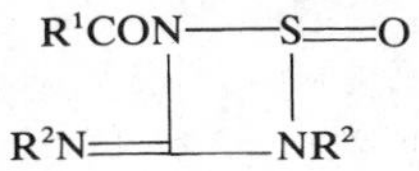

(**86**)

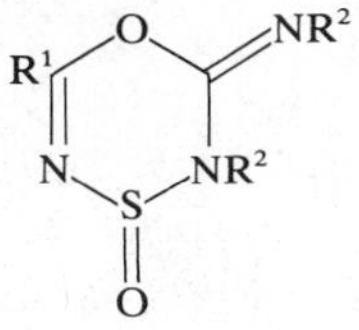

(**87**)

20.3.3 SEVEN-MEMBERED RINGS

Seven-membered ring compounds are named by combining prefixes defining the heteroatoms with the stem -epine for unsaturated compounds containing nitrogen, -epin for unsaturated compounds with no nitrogen, or -epan* for saturated compounds without nitrogen. Many of these compounds are synthesized by the union of two bifunctional reagents through familiar reactions such as alkylation or acylation of amino, hydroxy, or thiol groups, a typical example being 1,4,5-oxadithiepan (**88**), which is made from 2,2′-dichlorodiethyl ether and sodium disulphide. (**88**) is of interest for giving rubber-like polymers after fission of the disulphide link, *e.g.* by the action of sodium methoxide. Other syntheses of seven-membered rings involve some ring enlargements that have some unusual features. The most widely investigated classes are benzo- and dibenzo-derivatives, made as analogues of compounds with valuable pharmacological properties.

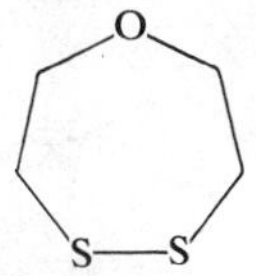

(**88**)

20.3.3.1 Oxazepines and benzo-derivatives

There are three isomeric systems having one oxygen and one nitrogen atom in the ring, the 1,2-, 1,3-, and 1,4-oxazepines, and these can give rise to ten isomeric benzoxazepines and six dibenzoxazepines. In the literature there are inconsistencies in the nomenclature of these compounds.

Perhydro-1,2-oxazepine is made by alkylation of *N*-ethoxycarbonylhydroxylamine with 1,5-dibromopentane and removal of the ethoxycarbonyl group by acid hydrolysis. Similar processes give 2,3- and 3,2-benzoxazepines. 7-Aryl-2-methyl derivatives (**89**) are produced by thermal rearrangement of 2-aryl-1-methylpiperidine *N*-oxides, *e.g.* in dimethylacetamide at 170 °C (reaction 30). They are basic oils, giving crystalline picrates. 5-Oxoperhydro-1,2-oxazepines (**90**; R = H or Me) are obtained by the addition of hydroxylamine or *N*-methylhydroxylamine to phorone. 2,3,3,7,7-Pentamethyl-5-oxoperhydro-1,2-oxazepine (**90**; R = Me) is made by oxidation of 1,2,2,6,6-pentamethylpiperid-4-one with hydrogen peroxide.[27]

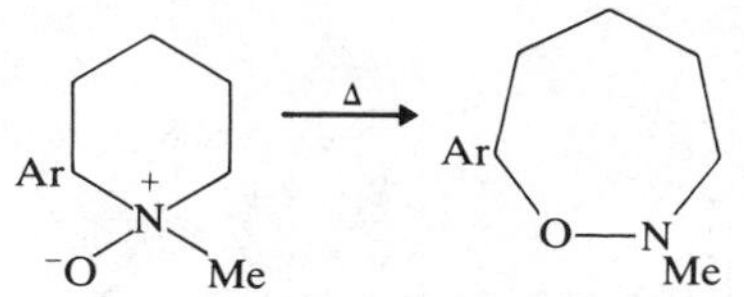

(30)

(**89**)

*-epane under IUPAC nomenclature.

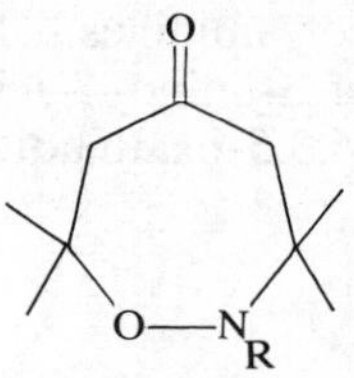

(90)

1,3-Oxazepines are obtained by photolysis of pyridine *N*-oxides if electron-withdrawing groups are present to stabilize the products. 2,6-Dicyanopyridine 1-oxide gives 2,4-dicyano-1,3-oxazepine, and tetra- and penta-phenylpyridine 1-oxides also give 1,3-oxazepines. Thermolysis of 2,4,5,7-tetraphenyl-1,3-oxazepine (**91**) causes rearrangement to 3-benzoyl-2,4,5-triphenylpyrrole (reaction 31), and this compound, *N*-benzoyl-2,4,5-triphenylpyrrole, and other products are formed by acid hydrolysis.[28] Irradiation of 2,4,6-triphenylpyridine *N*-oxide gives both 2,4,6-triphenyl-1,3-oxazepine and 2-benzoyl-3,5-diphenylpyrrole. Another photochemical reaction giving a 1,3-oxazepine is the rearrangement of 4-phenyl-2,3-oxazabicyclo[3,2,0]hepta-3,6-diene (**92**) to 2-phenyl-1,3-oxazepine (**93**), probably through an unstable 1,2-oxazepine. Acid hydrolysis of (**93**) gives 1-benzoylpyrrole and 3-hydroxy-2-phenylpyridine (reaction 32). 1,3-Oxazepines are also obtained by ring-enlargement of 1,3-oxazinium salts (azapyrylium salts), *e.g.* (**94**), by reaction with aliphatic diazo-compounds (reaction 33). Tetra- and penta-phenyl-1,3-oxazepines (**95**; R = H or Ph) are obtained from decomposition of azidopyrans, which are made by addition of azide to pyrylium salts (reaction 34).

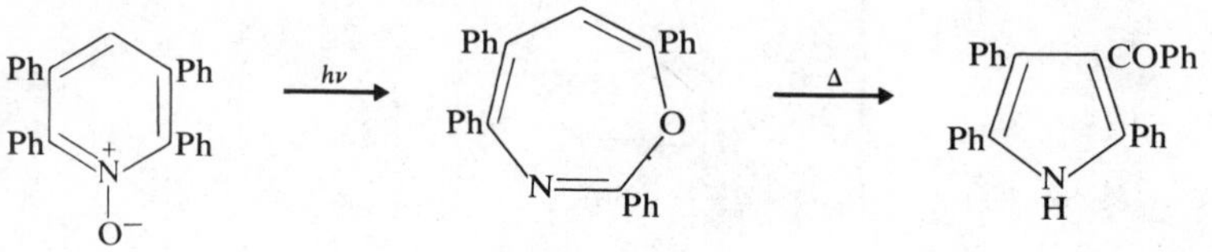

(31)

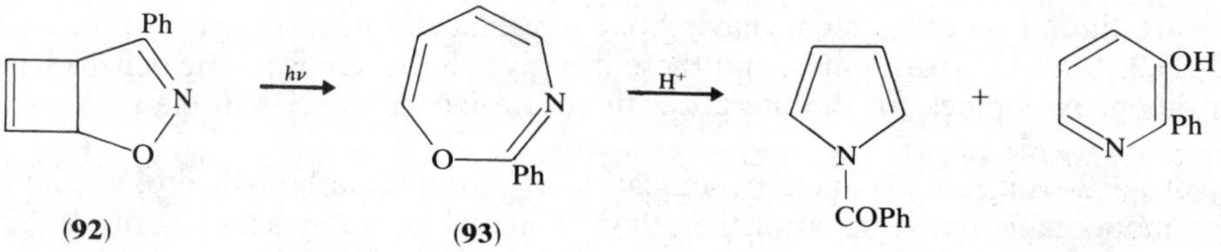

(91)

(32)

(92) **(93)**

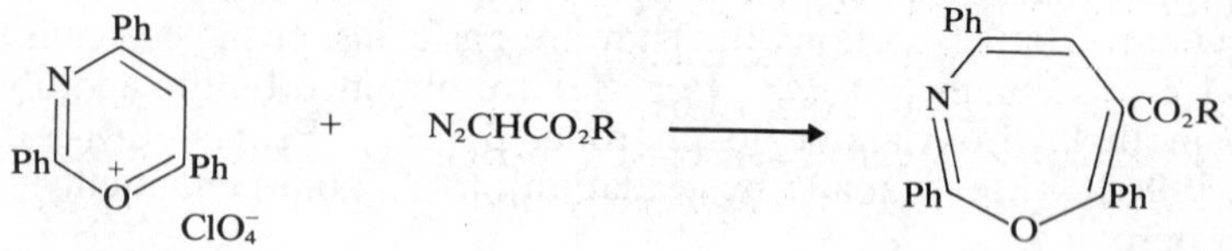

(33)

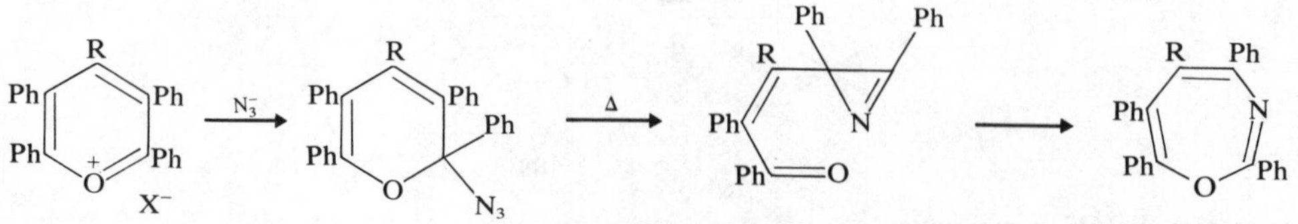

(94)

(34)

(95)

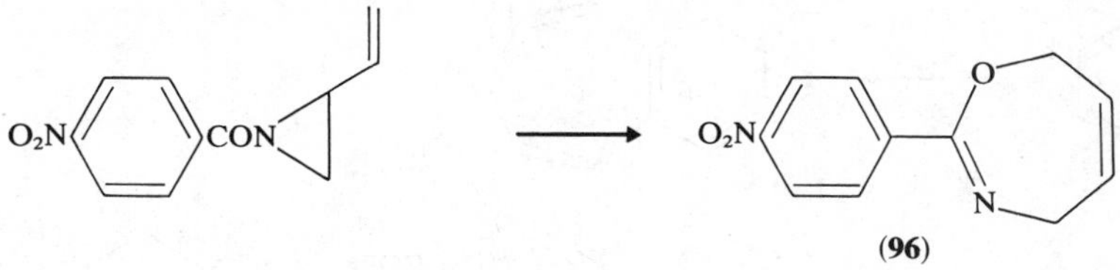

(35)

(96)

N-p-Nitrobenzoyl-2-vinylaziridine is rearranged to a dihydro-1,3-oxazepine (**96**) when it is heated in toluene (reaction 35).[29] The imino-ether ring is readily cleaved by bases, giving *cis-p*-nitrobenzamidobutadiene, and by acids to give 1-*p*-nitrobenzamidobut-2-en-4-ol. Ring-contraction to oxazolidine derivatives is caused by several reagents, *e.g.* iodide or bromide ion.

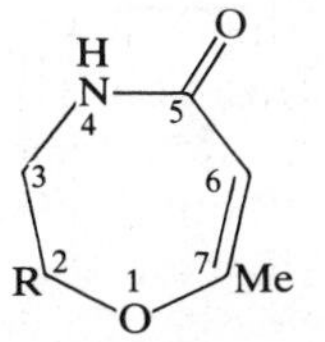

(97)

2,3-Dihydro-1,4-oxazepin-5-ones (**97**) are enol ethers made by dehydrating *N*-(2-hydroxyalkyl)acetoacetamides with thionyl chloride. They give *N*-acetyl derivatives, are brominated at the 6-position, and can be reduced to 5,6-dihydro-compounds. These are also obtainable from 2,6-dialkyl-4-oximinopyrans by the Beckmann rearrangement with polyphosphoric acid.

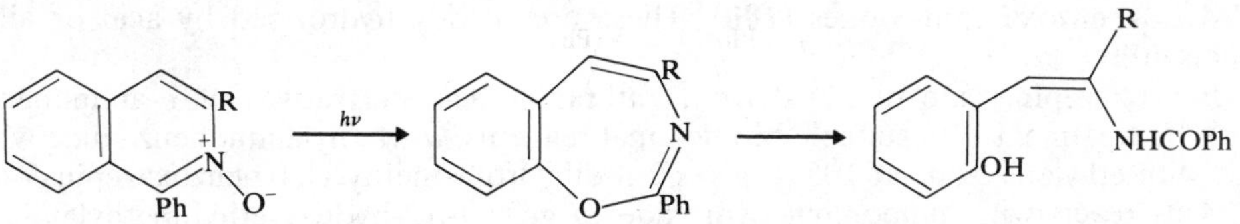

(36)

1,3-Benzoxazepines are obtained by photolysis of isoquinoline *N*-oxides in aprotic solvents. Cleavage of the imino-ether function occurs readily in aqueous ethanol (reaction 36). Irradiation of quinoline *N*-oxides in aprotic solvents gives 3,1-benzoxazepines if the quinoline has a cyano-group in the 2-position or an aryl group in the 2- or 3-position, but the photochemistry of other quinoline *N*-oxides is complex. Carbostyrils are frequently produced in protic solvents. The 3,1-benzoxazepines are imino-ethers of enols and are very readily attacked by water, giving indoles or 2-hydroxy-2,3-dihydroindole derivatives (reactions 37 and 38).[30] 2-Cyano-5-methyl-3,1-benzoxazepine (from 2-cyano-4-methylquinoline oxide) reacts with dialkylamines to give 2-amino-3-dialkylamino-4-methylquinolines. It is thought that the initial photochemical reaction with heteroaromatic *N*-oxides is the formation of an oxaziridine ring; various mechanisms have been proposed to account for the further rearrangements. 1,3-Benzoxazepines are found among the products of photolysis of some t-butylbenzoquinones (**98**) in the presence of nitriles (reaction 39).

2-Aminomethylbenzyl alcohols react with aldehydes (with acid catalysis) to give 1,3,4,5-tetrahydro-2,4-benzoxazepines (**99**), which are stable to dilute alkali but are

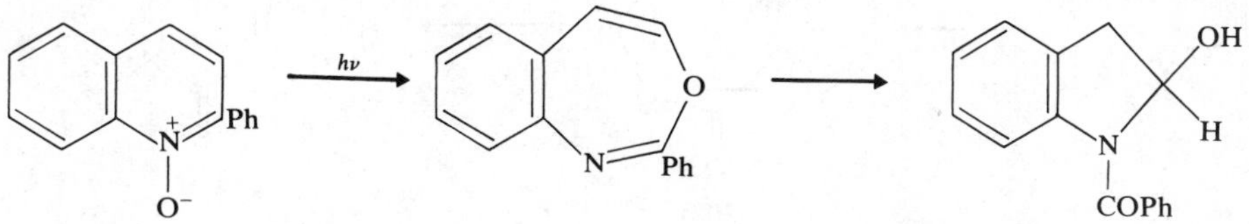

(37)

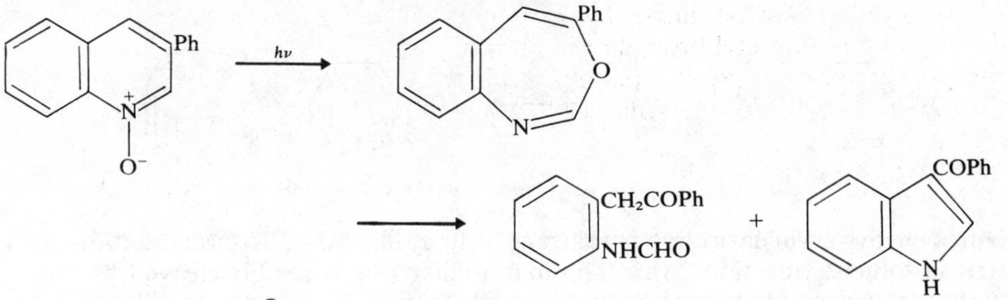

(38)

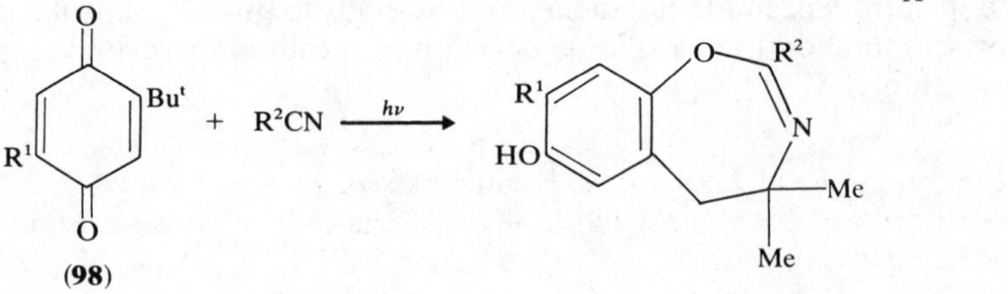

(39)

(98)

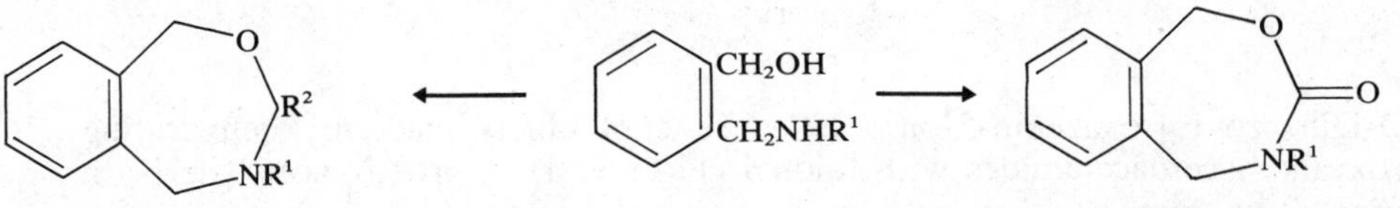

(40)

(99) **(100)**

rapidly hydrolysed by acids. With phosgene the aminomethylbenzyl alcohols give 1,5-dihydro-2,4-benzoxazepin-3-ones (**100**). These are readily hydrolysed by acid or alkali (reaction 40).

4,1-Benzoxazepines are obtained from anthranilic acid derivatives or *o*-aminobenzyl alcohols by treatment with suitable bifunctional reagents. *o*-Methylaminobenzamide when treated with ethylene oxide at 105 °C gives 2,3-dihydro-1-methyl-4,1-benzoxazepin-5-one (**101**). This reacts with ammonium hydroxide to give *o*-[(2-hydroxyethyl)methylamino]-benzamide, which is reconverted into (**101**) by heating (reaction 41).[31] *N*-α-Halogenoacyl derivatives of anthranilic acids are cyclized to 2,5-dioxo-derivatives of 4,1-benzoxazepine (**102**) by heating in DMF. These are converted into quinazolines by reaction with ammonia (reaction 42). Similarly, the *N*-chloroacetyl derivatives of *o*-aminobenzhydrols are cyclized (by sodium ethoxide) to 3,5-dihydro-2-oxo-4,1-benzoxazepines, *e.g.* (**103**).[32]

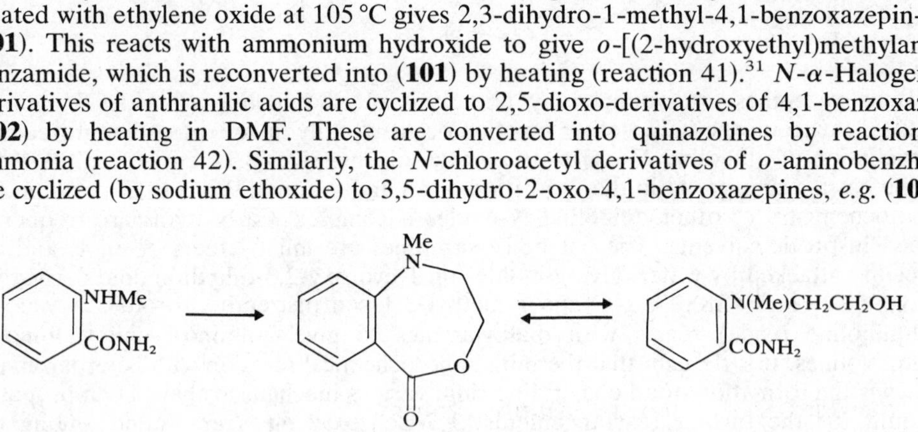

(41)

(101)

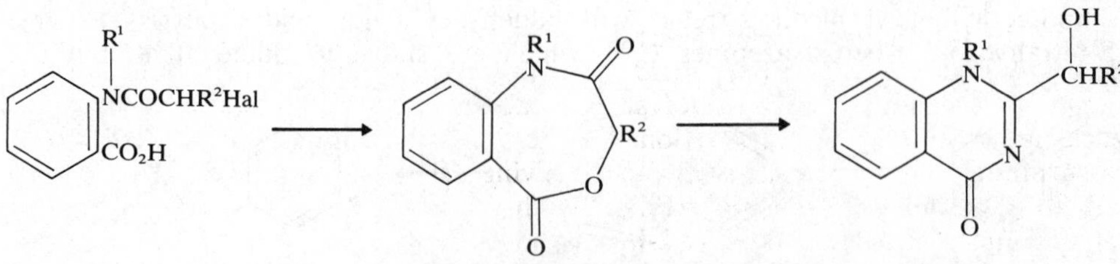

(42)

(102)

1,4-Benzoxazepines may be made by a number of similar processes. *o*-Hydroxybenzylamines with bromoacetyl bromide give 4,5-dihydro-3-oxo-1,4-benzoxazepines, *e.g.* (**104**),[33] and the isomeric 2-oxo-compounds (**105**) are made by Mannich reactions with phenols, formaldehyde, and α-amino-acids, ring-closure being effected by thionyl

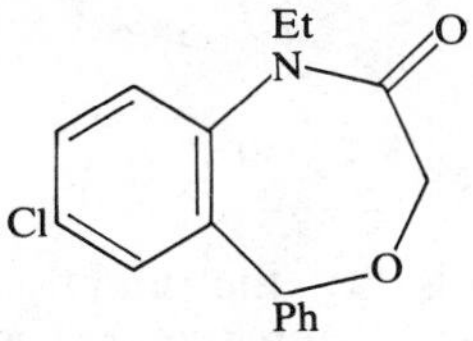

(**103**) 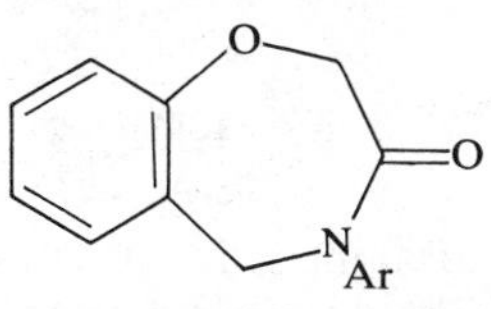 (**104**) 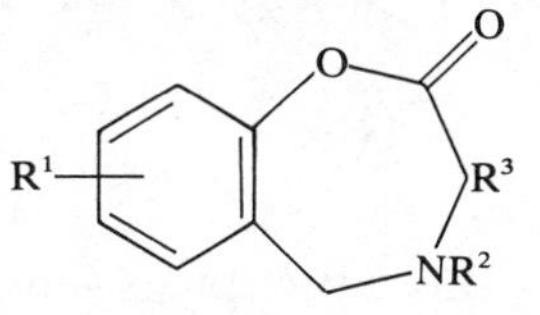 (**105**)

chloride. *o*-(2-Chloroethoxy)acetophenone reacts with amines to give 4-alkyl-2,3,4,5-tetrahydro-5-methylene-1,4-benzoxazepines (**106**), provided that there is no steric hindrance; reduction of the double bond of these enamines gives stable 5-methyltetrahydro-1,4-benzoxazepines (reaction 43). *O*-Phenacylsalicylamide is dehydrated (with acid catalysis) to give 3-phenyl-1,4-benzoxazepin-5-one (**107**), which can be *N*-alkylated, or converted by triethyloxonium fluoroborate into 5-ethoxy-3-phenyl-1,4-benzoxazepine.

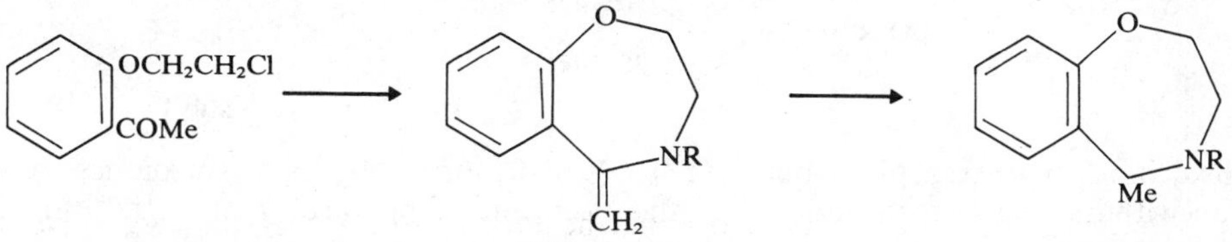

(**106**) (43)

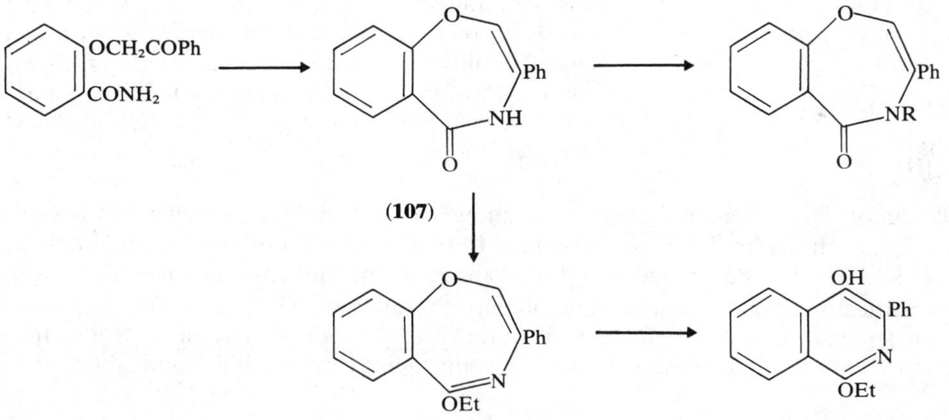

(**107**) (44)

This is converted into 1-ethoxy-4-hydroxy-3-phenylisoquinoline when it is boiled in methanol (reaction 44). Treatment of (**107**) with sodamide also gives hydroxyisoquinolines.[34]

2,3-Dihydro-1,4-benzoxazepin-5-one (**108**; $R^1 = R^2 = H$) is made by the Beckmann rearrangement of chromanone oxime with polyphosphoric acid, and also by the action of sodium azide and conc. sulphuric acid on chromanone. The Schmidt reaction on chromanones, flavanones, and isoflavanones also gives the isomeric 4-oxo-1,5-benzoxazepine derivatives (**109**) (reaction 45); the proportion of the latter is increased in 3- or 5-substituted chromanones, isoflavanone giving a 40% yield of (**108**; $R^1 = H, R^2 = Ph$) and 20% of (**109**; $R^1 = H$, $R^2 = Ph$).[35] With conc. sulphuric acid, (**108**; $R^1 = Ph$, $R^2 = H$) undergoes cleavage at the 1,2-bond and rearrangement to give 2-*o*-hydroxyphenyl-5-phenyloxazoline and 1,2,3,4-tetrahydro-1,8-dihydroxy-3-phenylisoquinoline.

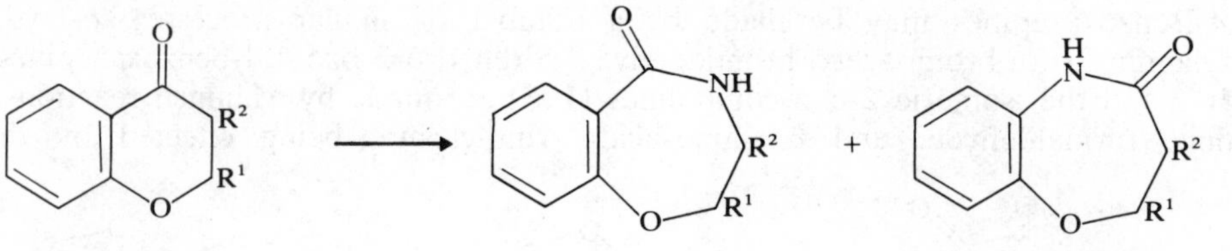

(45)

(108) **(109)**

1,5-Benzoxazepines are also obtained from *o*-aminophenols by the addition of three-carbon chains. Thus 2,4-bis(alkoxycarbonylcyanomethylene)-1,3-dithietans react with two molecules of *o*-aminophenol (reaction 46) to give 2-imino-4-mercapto-1,5-benzoxazepines (**110**);[36] the sodium salt of *N*-methanesulphonyl-*o*-aminophenol is alkylated by epichlorohydrin to give 2,3,4,5-tetrahydro-3-hydroxy-5-methanesulphonyl-1,5-benzoxazepine (**111**); and 2-*o*-aminophenoxypropionic acid is cyclized by heating to give

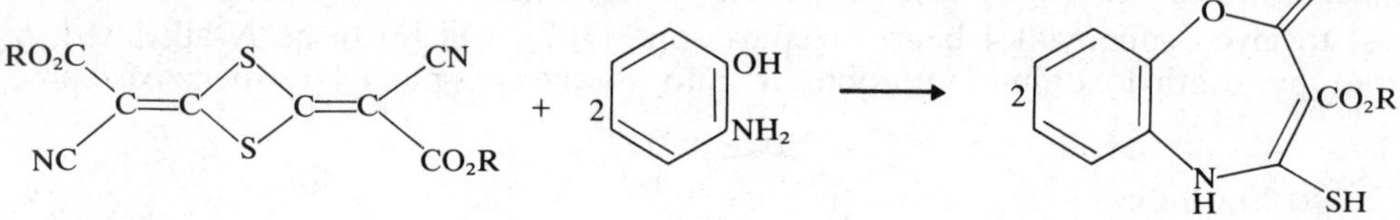

(46)

(110)

2,3-dihydro-1,5-benzoxazepin-4-one (**112**). *N*-(3-Chloropropyl)benzoxazolones having electron-withdrawing substitutents in the benzene ring are rearranged to 1,5-benzoxazepines (**113**) on treatment with pyrrolidine.

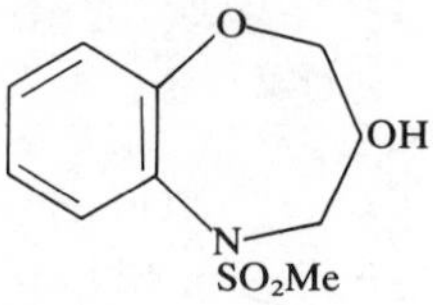

(111) 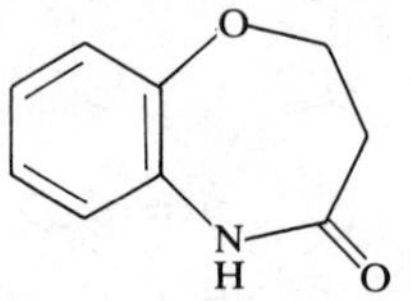 **(112)** 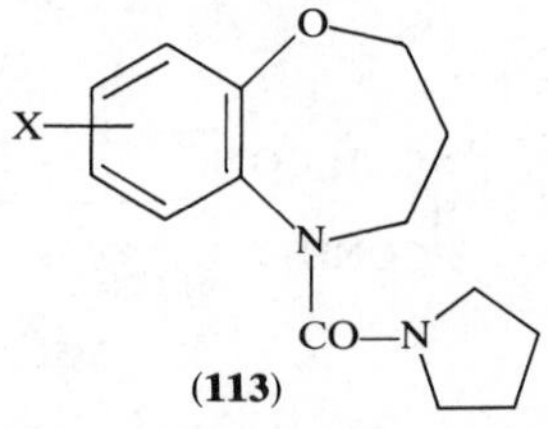 **(113)**

Reduction of chromanone oxime or isoflavanone oxime with lithium aluminium hydride affords 2,3,4,5-tetrahydro-1,5-benzoxazepine (**114**; $R^1 = R^2 = H$) or the 3-phenyl derivative (**114**; $R^1 = H$, $R^2 = Ph$) (reaction 47). Flavanone oxime and chromanone oximes with bulky substituents at the 2-position give mainly the amines (**115**). This ring-expansion has been applied to the production of tetrahydro-1,5-benzothiazepines from thiochromanone oximes and tetrahydro-1,5-benzodiazepines from dihydroquinolone oximes.[37,38]

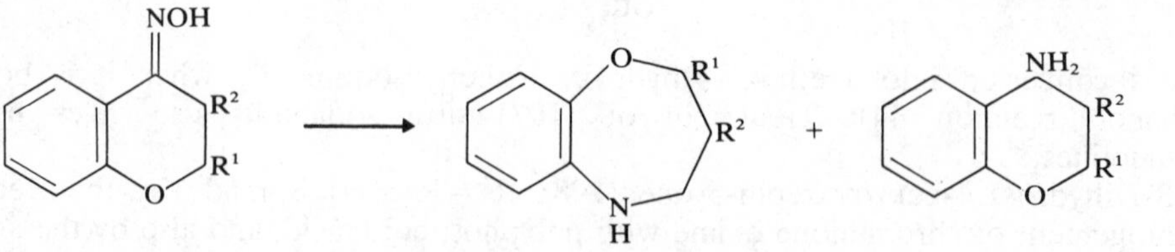

(47)

(114) **(115)**

Dibenzo-derivatives of 1,4-oxazepine have been studied extensively in attempts to find new medicinal agents, particularly neuroleptics. The compounds of interest for this purpose usually have tertiary amino-groups either at the 11-position, *i.e.* the 5-position of the 1,4-oxazepine ring, or in a side-chain introduced by *N*-aminoalkylating the parent dibenzoxazepine.

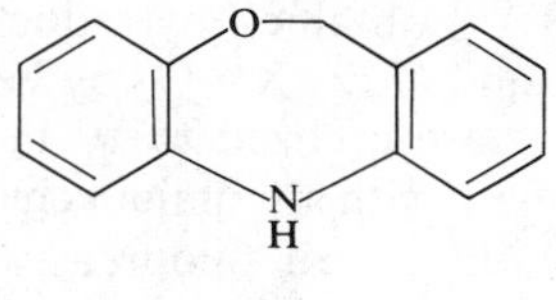

(116)

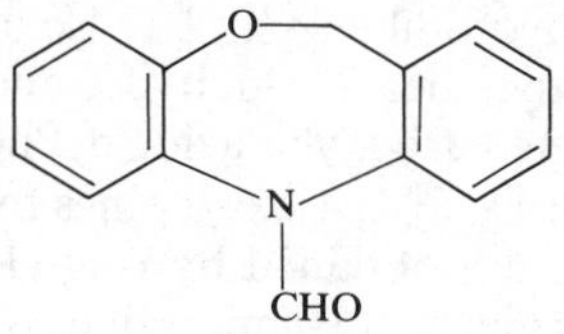

(117)

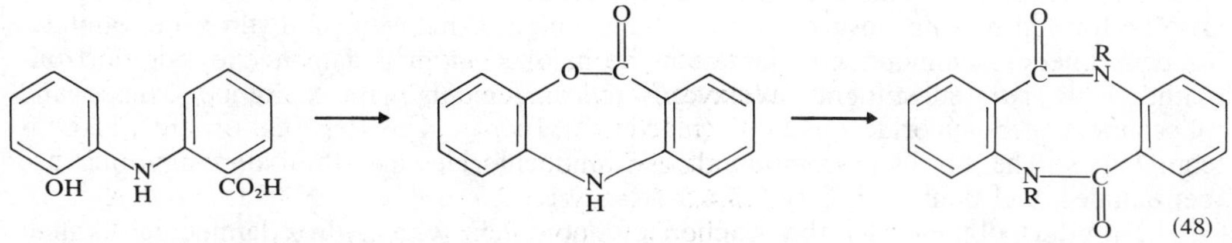

(118) (119) (48)

Derivatives of dibenz[*b*, *e*][1, 4]oxazepine (**116**), *e.g.* the formyl derivative (**117**), are made by Ullmann condensations of *o*-(2-bromobenzyloxy)formanilides or by lactonization of 2-hydroxy-2′-carboxydiphenylamines. In the latter reactions, thionyl chloride, or (better) dicyclohexylcarbodi-imide, is used as a dehydrating agent.[39] The lactone 5,11-dihydro-11-oxo-dibenz[*b*, *e*][1, 4]oxazepine (**118**) is rather unstable, and with sodium hydride it gives the dibenzodiazocine (**119**; R = *o*-hydroxyphenyl) (reaction 48). The anion of (**116**), generated with sodium hydride, also gives the dimer (**120**) and polymeric material.

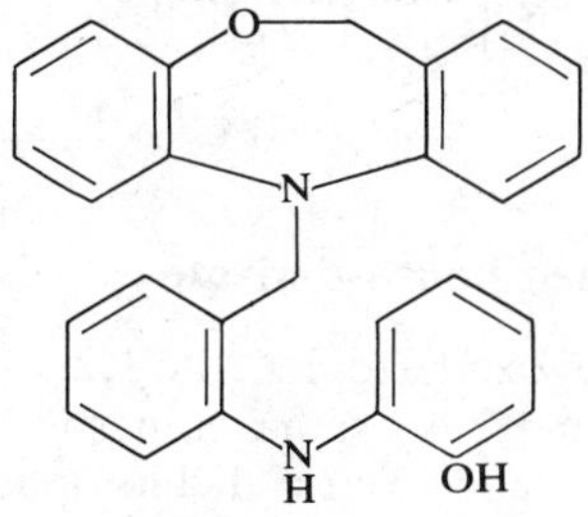

(120)

Photolysis of acridine *N*-oxide in methanol or ethanol gives 11-methoxy- or 11-ethoxy-5,11-dihydrodibenz[*b*, *e*][1, 4]oxazepine (**121**; R = Me or Et), which are acetals and are readily hydrolysed to 2-hydroxydiphenylamine-2′-aldehyde.[40]

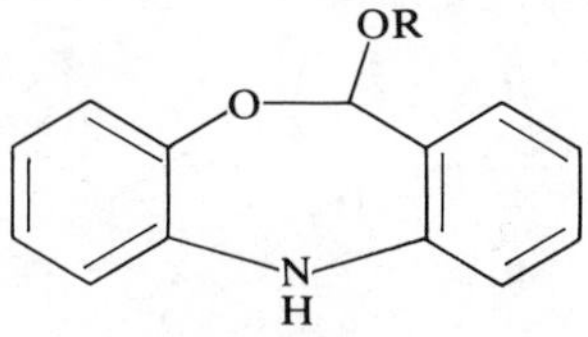

(121)

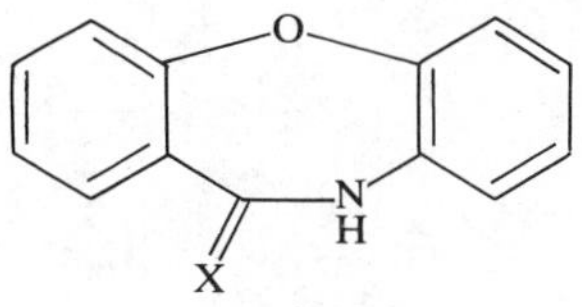

(122)

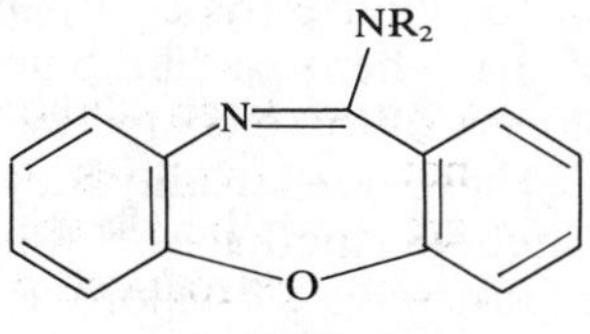

(123)

10,11-Dihydrodibenz[*b*, *f*][1, 4]oxazepine (**122**; X = H_2) is obtained by reduction of 2-aminodiphenylether-2′-aldehyde, and the 11-oxo-compound (**122**; X = O) by heating the corresponding carboxylic acid. *o*-Phenoxyaryl isocyanates are cyclized to 10,11-dihydro-11-oxodibenz[*b*, *f*][1, 4]oxazepines by heating or by treatment with aluminium chloride, but better yields are obtained by ring-closure of *o*-phenoxyarylureas or thioureas with phosphoryl chloride to 11-aminodibenz[*b*,*f*][1,4]oxazepines (**123**), which can be hydrolysed to the oxo-compounds by acid. Polyphosphoric acid cyclizes *o*-phenoxyaryl-thioureas or -thiocarbamates to 11-thioxo-compounds, *e.g.* (**122**; X = S). The seven-membered ring can also be formed by ring closure of the sodium salts of 2-halogeno-2′-hydroxybenzanilides or benzylidene-*o*-aminophenols in which the halogen atom is activated by an electron-withdrawing *para*-substituent. Beckmann rearrangements with xanthone oximes and phosphorus pentachloride give 11-chlorodibenz[*b*, *f*][1, 4]oxazepines or the 11-oxo-compounds. The 11-chloro-compounds are imidochlorides and the oxo-compounds are benzamides, and their chemistry follows from this.

The product obtained by the reaction of fluorescein with hydroxylamine in alkaline solution is the dibenz[*b*, *f*][1, 4]oxazepine (**124**). Decarboxylation and methylation with diazomethane give 3,7-dimethoxy-11-phenyldibenz[*b*, *f*][1, 4]oxazepine, which has also been synthesized from 2-hydroxy-4-methoxybenzophenone and 3-bromo-4-nitroanisole by condensation to give the diaryl ether, reduction of the nitro-group, and cyclization.[41]

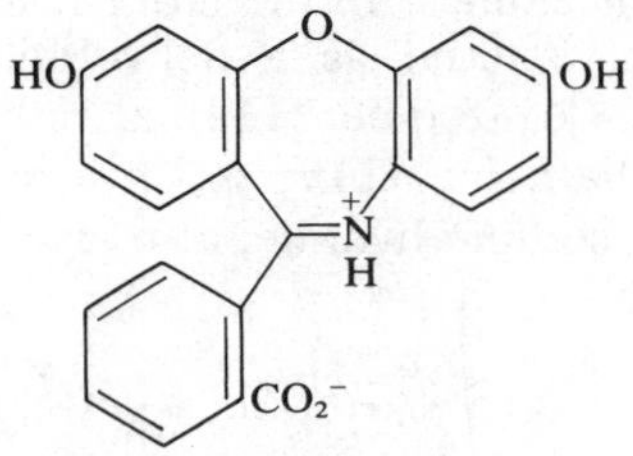

(124)

20.3.3.2 Oxadiazepines and their benzo-derivatives

Of the nine possible isomers of oxadiazepine, the 1,2,4- and 1,4,5-compounds are the best known. Examples are the 3-oxo-1,2,4-oxadiazepines (**125**), made by addition of hydroxyurea to aryl vinyl ketones,[42] and 3-(2,6-dichlorophenyl)-dihydro-1,2,4-oxadiazepine, made from 2,6-dichlorobenzamidoxime and 1,3-dibromopropane. This has herbicidal activity owing to release of 2,6-dichlorobenzonitrile in the soil. Hexahydro-1,4,5-oxadiazepines (**126**) are obtained by alkylation of 1,2-dialkylhydrazines with 2,2′-dichlorodiethyl ether.

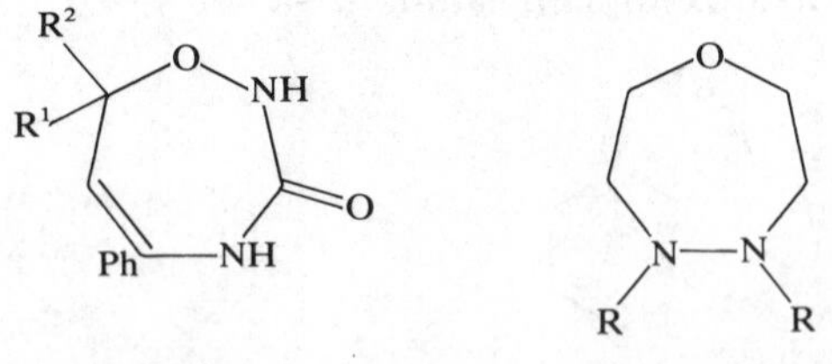

(125) **(126)**

The Beckmann rearrangement of oximes of *o*-acylaminoaryl ketones gives products that were formerly thought to be 3,1,4-benzoxadiazepines (**127**) but are now known to be quinazoline 3-oxides (**128**). Photolysis of these compounds causes ring enlargement to give 1,3,5-benzoxadiazepines (**129**). These are readily degraded to benzoxazoles and *o*-acylaminophenols by boiling aqueous methanol.[43] Phosgene reacts with 2-amino-5-chlorobenzophenone α-oxime to give 7-chloro-1,2-dihydro-2-oxo-5-phenyl-3,1,4-benzoxadiazepine (**130**), but it converts the β-oxime into 6-chloro-4-phenylquinazol-2-one

3-oxide. The 2-methylamino-3,1,4-benzoxadiazepine derivative (**127**; R = NHMe) is made by treating 5-chloro-2-methylthioureidobenzophenone oxime with mercuric oxide. It is rearranged photolytically to 5-chloro-1-methylcarbamyl-3-phenylindazole, and hydrochloric acid converts it into 6-chloro-2-methylamino-3-phenylquinazol-4-one.[44] Photolysis of 2-arylquinoxaline *N*-oxides gives 3,1,5-benzoxadiazepines (**131**; R = H or Ph), which react with water to give diacyl *o*-phenylenediamines.

2-Oxo-2,3-dihydro-1,4,5-benzoxadiazepines (**132**; X = O) are obtained by addition of ketens to *o*-quinonediazides. They will add a further molecule of keten across the 4,5-double bond, and on heating they dissociate into the starting materials.[45] Ketenimines react similarly in inert solvents, giving the 2-imino-compounds, *e.g.* (**132**; X = NC_6H_4Me), but in alcoholic solvents they give 2-alkoxy-2-arylamino-2,3-dihydro-1,4,5-benzoxadiazepines.

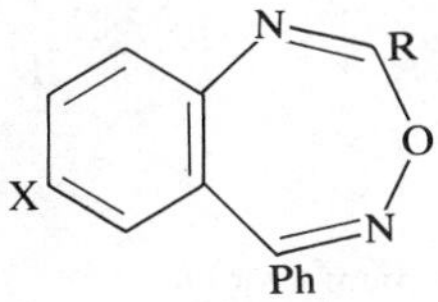

(**127**) X = H or Cl

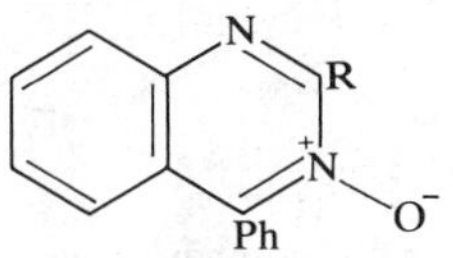

(**128**)

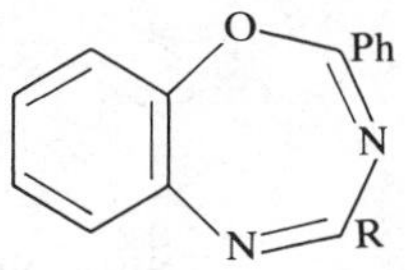

(**129**)

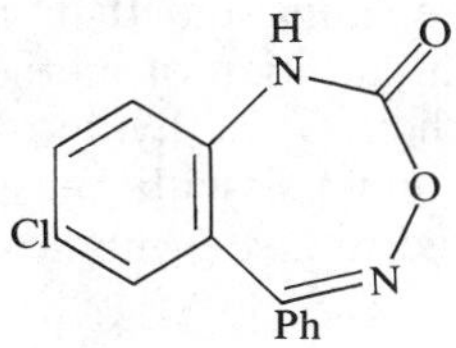

(**130**)

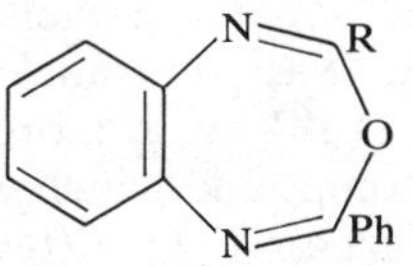

(**131**)

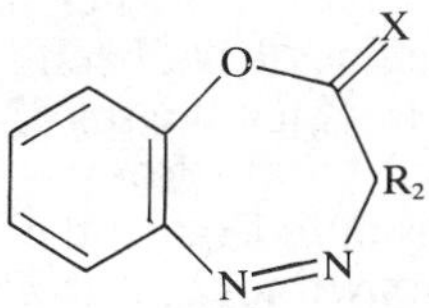

(**132**)

20.3.3.3 Thiazepines and benzo-derivatives[46]

Thiazepines resemble oxazepines in the number of possible isomers, in many of their preparative routes, and in the interest of many of them as actual or potential therapeutic agents. Among the preparative methods common to both series are the ring-enlargement of cyclic ketones by the Schmidt reaction or the Beckmann rearrangement of their oximes, and the lithium aluminium hydride reduction of thiochromanone oxime, which gives 2,3,4,5-tetrahydro-1,5-benzothiazepine (34% yield) as well as 4-aminothiochromanone (24% yield). The Schmidt reaction with 1-thiochroman-4-one gives both 2,3-dihydro-1,4- and -1,5-benzothiazepinones, but benzo-halogenated compounds and the *S*-oxides and *S,S*-dioxides usually give only dihydro-1,5-benzothiazepinones.[47] The Beckmann rearrangement of 1-thiochroman-4-one oxime gives a high yield of 2,3-dihydro-1,5-benzothiazepinone (Scheme 5). 1,4-Thiopyrone oxime gives 2,3,5,6-tetrahydro-1,4-thiazepin-5-one on rearrangement, and thian-3-one *syn*-oxime gives 4,5,6,7-tetrahydro-1,4-thiazepin-3-one (**133**) in good yield, while the *anti*-oxime of thian-3-one gives a low yield (27%) of 2,3,6,7-tetrahydro-1,3-thiazepin-4-one (**134**) (reaction 49).[48]

1,4-Thiazepines are among the products of rearrangement of penicillins. Thus, methyl 6-β-phthalimidopenicillanate is converted into 2,2-dimethyl-(3*S*)-methoxycarbonyl-7-oxo-6-phthalimido-2,3,4,7-tetrahydro-1,4-thiazepine (**135**) by triethylamine (reaction 50).[49] There has been considerable interest in the isomeric 5-oxo-1,4-thiazepine derivatives, *e.g.* 3-carboxy-2,2-dimethyl-5-oxo-6-phenylacetamidoperhydro-1,4-thiazepine (**136**), two diastereomers of which are obtained by condensation of D-penicillamine with methyl α-phenylacetamidoacrylate (reaction 51).[50] It has been postulated that such cyclic

i, HN_3; ii, NH_2OH; iii, polyphosphoric acid; iv, lithium aluminium hydride.

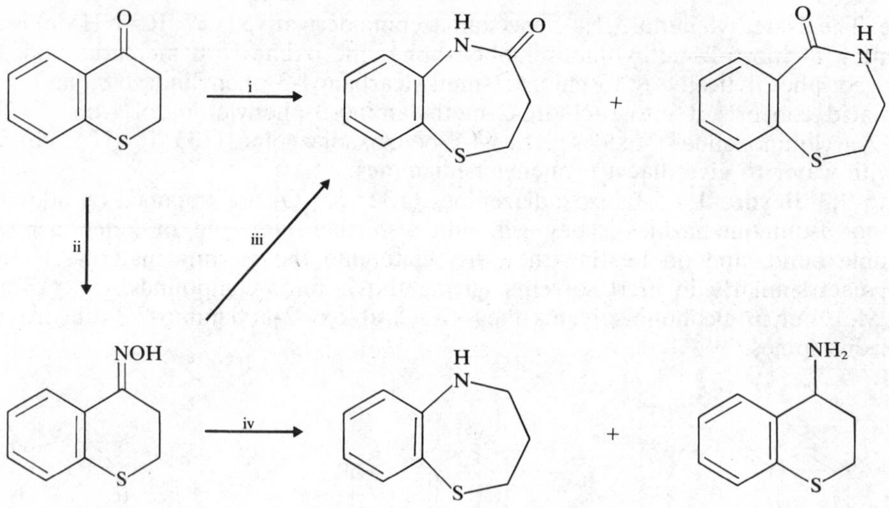

SCHEME 5

peptides are the biosynthetic precursors of penicillins, but attempts to cyclize such thiazepines to penicillins have failed. Simpler analogues of penicillins and their related thiazepines have been investigated, *e.g.* the β-lactam (**137**) from addition of methoxyketen to 2-methylthio-2-thiazoline was rearranged to 6-methoxy-7-methylthio-5-oxo-2,3,4,5-tetrahydro-1,4-thiazepine (**138**) by trifluoroacetic acid, and this with phosphorus pentasulphide gave the 5-thioxo-compound which was methylated to give 6-methoxy-5,7-bismethylthio-2,3-dihydro-1,4-thiazepine (**139**) (reaction 52).

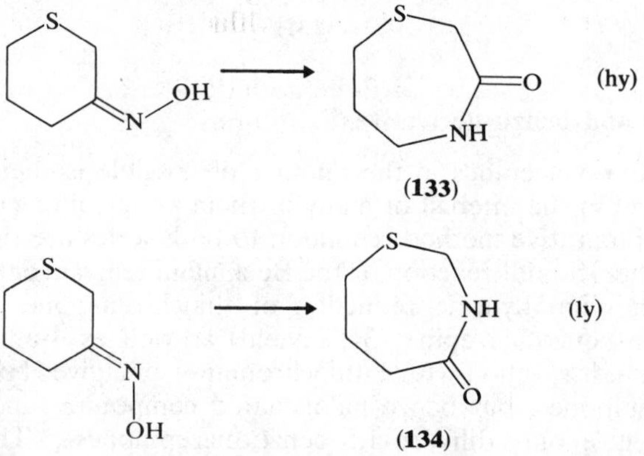

(hy)

(**133**)

(49)

(ly)

(**134**)

1,4-Thiazepine derivatives are obtained by reaction of 2-mercaptoethylamine with α,β-unsaturated esters and ketones, but not with acetylenic ketones, in which the triple bond reacts preferentially with the amino-group to give β-(2-mercaptoethylamino)vinyl ketones.

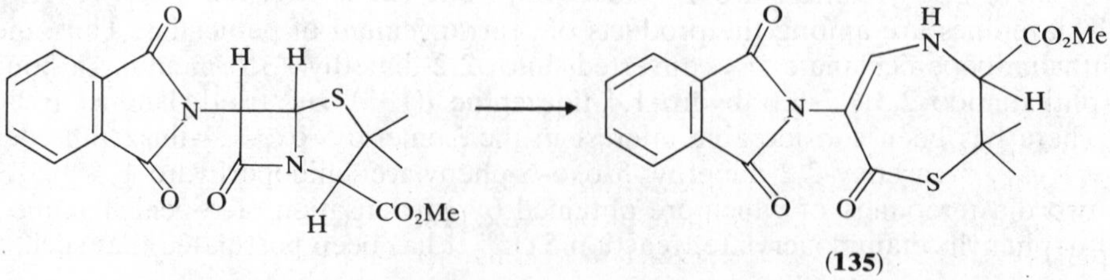

(50)

(**135**)

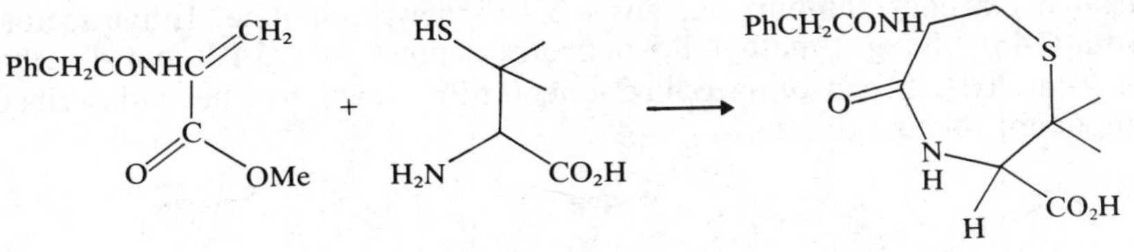

(51)

(136)

1,3-Thiazepines are obtained by the reaction of thioureas or thioamides with 1,4-dibromoalkanes (*o*-xylylene dibromide gives 2,4-benzothiazepines) or 3-bromobutyryl bromide.[51] 2-Vinylaziridine reacts with phenyl isothiocyanate to give the 4,7-dihydro-1,3-thiazepine (**140**; R = PhNH) and with carboxymethyl 4-chlorodithiobenzoate to yield (**140**; R = ClC_6H_4).

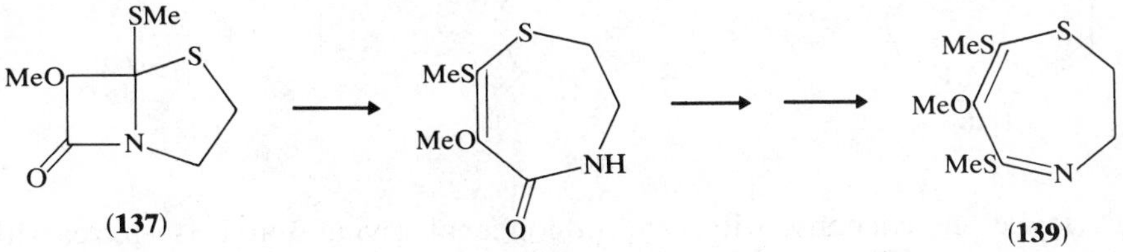

(52)

(137) **(138)** **(139)**

4,1-Benzothiazepin-2-ones are obtained from thioglycollic acid by condensation with *o*-aminobenzhydrol in hydrochloric acid, giving (**141**; R = Ph), or by reaction with *o*-nitrobenzyl bromide followed by reduction of the nitro-group and cyclization to (**141**; R = H).

Methods for the preparation of 1,4-benzothiazepines include cyclization of *o*-(carboxymethylthio)benzamides with thionyl chloride, giving 3,5-dioxo-compounds (**142**), treatment of 2-mercaptobenzoic acid with aziridine to give 2,3-dihydro-1,4-benzothiazepin-5-one (**143**), and cyclization of *N*-(*p*-chlorophenylthioethyl)benzamide with phosphoryl chloride, giving (**144**).

1,5-Benzothiazepines are made from 2-aminothiophenol (**145**) by addition of three-carbon-atom chains. Heating (**145**) with α,β-unsaturated acids, *e.g.* cinnamic, crotonic, or atropic acid, gives 2,3-dihydro-1,5-benzothiazepin-4-ones, *e.g.* (**146**). *N*-Aminoalkylation

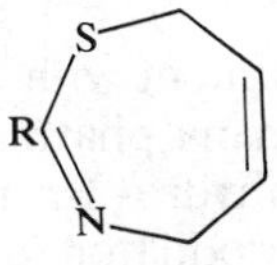

(140) 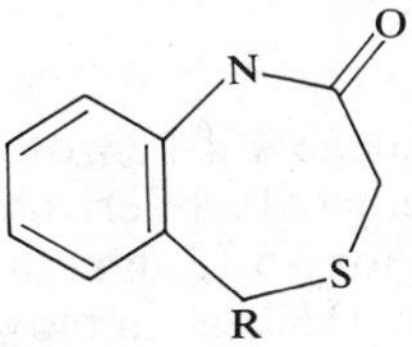 **(141)**

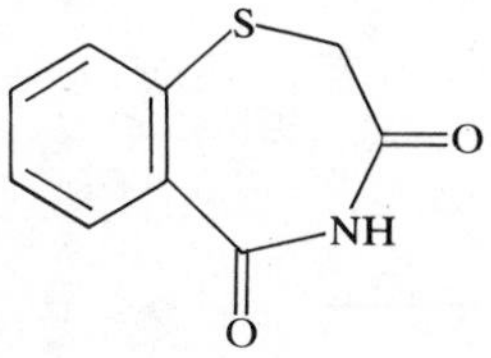

(142) 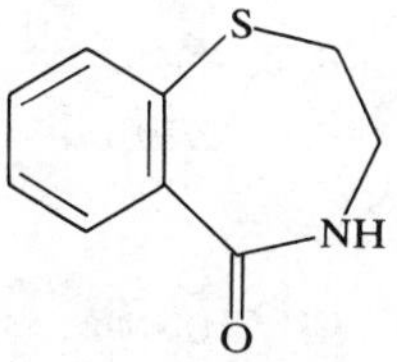 **(143)** 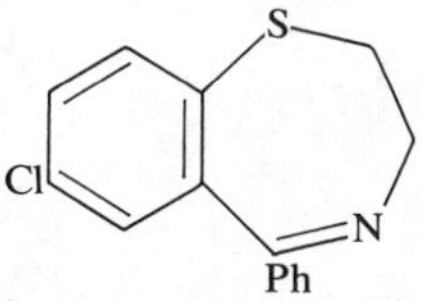 **(144)**

of (**146**) gives compounds with interesting pharmacological properties, the 2-dimethylaminoethyl derivative (**147**) being the antidepressant drug thiazesim.[52] The hydrochloride of (**145**) condenses with unsaturated ketones, *e.g.* mesityl oxide, giving 2,3-dihydro-1,5-benzothiazepines, and with 1,3-diphenylpropynone (**145**) gives an adduct which cyclizes to 2,4-diphenyl-1,5-benzothiazepine (**148**) in boiling methanol. When (**148**) is heated at

65 °C *in vacuo* it extrudes sulphur and gives 2,4-diphenylquinoline. Ethyl acetoacetate condenses with (**145**), giving 2-methyl-1,5-benzothiazepin-4-one (**149**), but diketen gives the isomeric 4-methyl-1,5-benzothiazepin-2-one (**150**), which has been described as a coupling component for azo-dyestuffs.

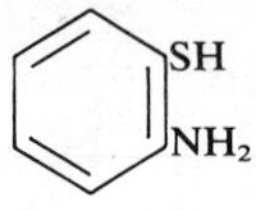

SH NH$_2$ (**145**) (**146**) CH$_2$CH$_2$NMe$_2$ (**147**)

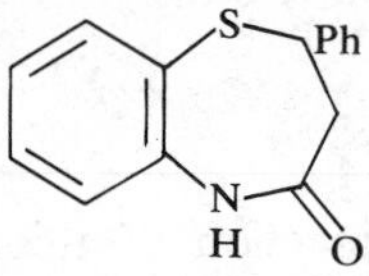

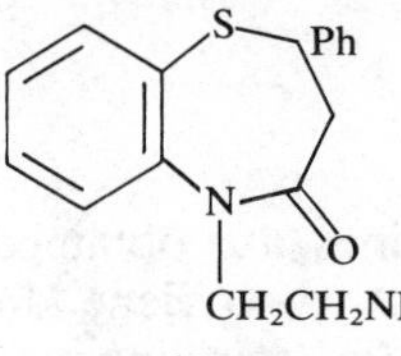

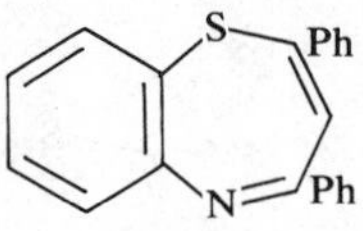

(**148**) 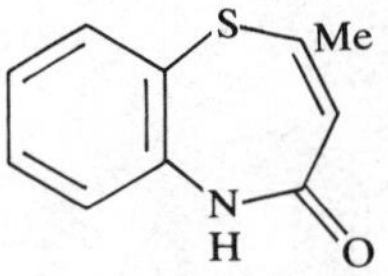 (**149**) 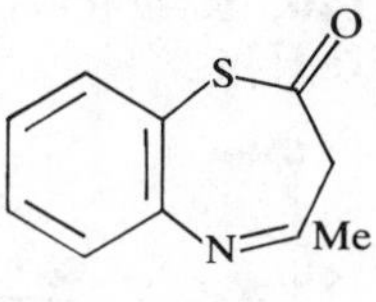 (**150**)

Arylglycidyl esters condense with *o*-nitrothiophenol, giving 3-aryl-3-*o*-nitroarylthio-2-hydroxypropionates (**151**) which are reduced and cyclized (*e.g.* by hydrolysis to the acid and heating the latter) to 2-aryl-2,3-dihydro-3-hydroxy-1,5-benzothiazepin-4-ones (**152**) (reaction 53). These are also obtained directly from the glycidic esters and (**145**). They are

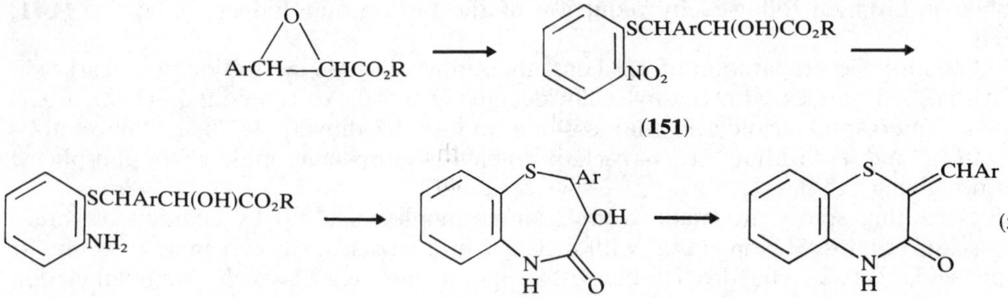

ArCH—CHCO$_2$R → SCHArCH(OH)CO$_2$R NO$_2$ (**151**) →

SCHArCH(OH)CO$_2$R NH$_2$ → (**152**) → CHAr (53)

rearranged to 2-arylidene-1,4-benzothiazin-3-ones by treatment with thionyl chloride and pyridine.[53] The *N*-aminoalkyl derivatives of (**152**) have many pharmacological activities, and *cis*-(+)-3-acetoxy-5-(2-dimethylaminoethyl)-2,3-dihydro-2-*p*-methoxyphenyl-1,5-benzothiazepin-4-one (**153**) is employed as a coronary vasodilator (dilitazem).

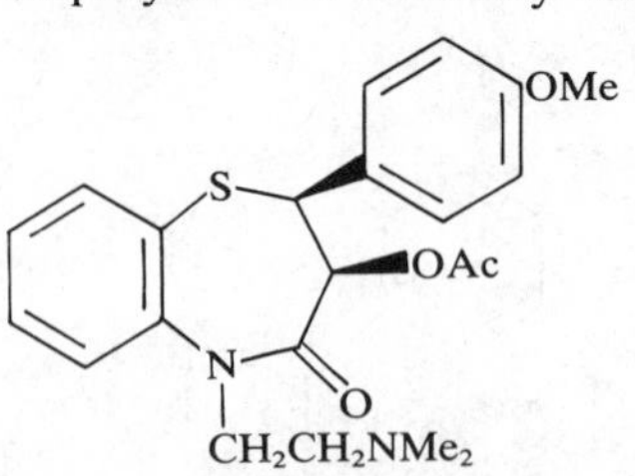

OMe OAc CH$_2$CH$_2$NMe$_2$ (**153**)

Reaction of 2-phenyl-1,5-benzothiazepin-4-one with phosphorus pentasulphide gives the 4-thione (**154**), which extrudes sulphur when treated with bases to give 4-phenylquinoline-2-thione. The *S*-methyl derivative also extrudes sulphur to give 2-methylthio-4-phenylquinoline, but 2,3-dihydro-4-methylthio-2-phenyl-1,5-benzothiazepine (**155**) loses methanethiol and gives 2-styrylbenzothiazole (Scheme 6).

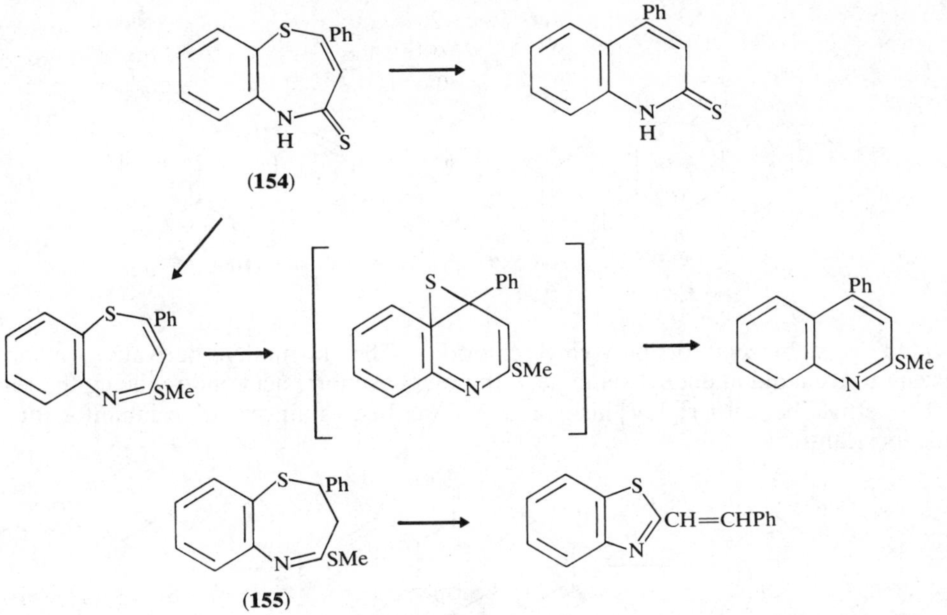

(**154**)

(**155**)

SCHEME 6

Dibenzo[*b*, *e*][1, 4]thiazepines are obtained by Ullmann-type condensation of 2-bromobenzyl 2′-aminophenyl sulphides, and thermolysis of 2-azido-2′,6′-dimethyldiphenyl sulphide gives the 5,11-dihydro-4-methyl derivative (**156**). To explain the latter reaction, a rearrangement through a spirodiene intermediate is postulated (Scheme 7).[54]

A few dibenzo[*c*, *f*][1, 2]thiazepines have been made by Friedel–Crafts cyclization of the acid chlorides of *o*-carboxybenzenesulphon-*N*-alkylanilides. The 11-oxo-compounds (**157**) thus made have been reduced to carbinols with sodium borohydride, and aminoalkyl ethers of these have been made and tested for pharmacological activity.

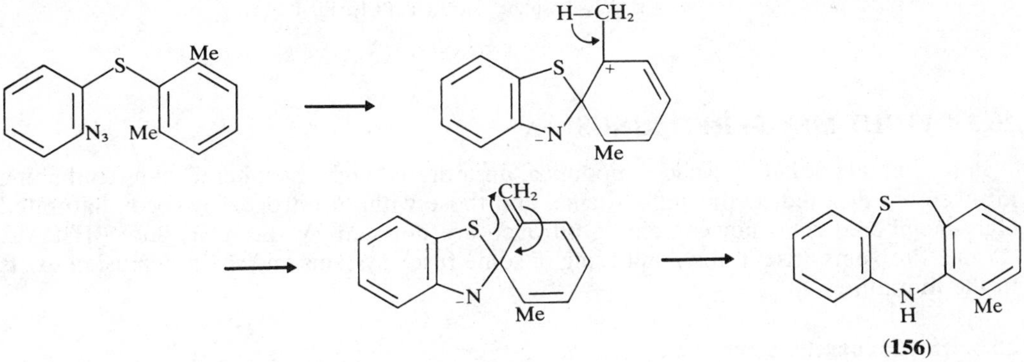

(**156**)

SCHEME 7

Dibenzo[*b*, *f*][1, 4]thiazepines are obtainable by the Beckmann rearrangement of thioxanthone oxime. With polyphosphoric acid the 10,11-dihydro-11-oxo-compound (**158**) is formed, but phosphorus pentachloride gives 11-chlorodibenzo[*b*, *f*][1, 4]-thiazepine (**159**), which will react with amines to give 11-amino-derivatives. The latter compounds are also obtained by cyclization of *o*-ureidodiphenyl sulphides with phosphoryl chloride (Scheme 8). The 11-(*N*-methylpiperazine) derivative (**160**) is the tranquilizer clothiapine. Reduction of (**158**) with lithium aluminium hydride gives 10,11-dihydro-dibenzo[*b*, *f*][1, 4]thiazepine. *N*-Acyl and *N*-alkyl derivatives of this have been

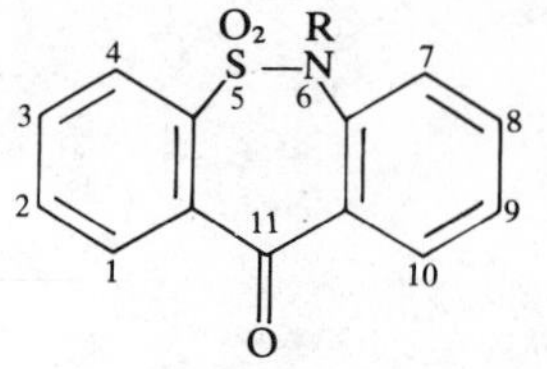

(157)

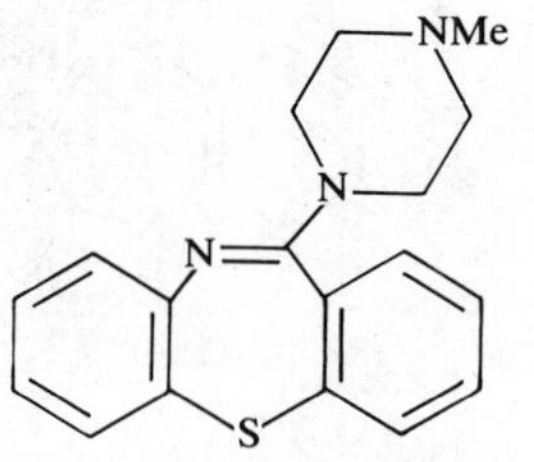

(160)

patented as stabilizers for poly(vinyl chloride). The 11-methyl derivative, made by cyclizing *o*-acetamidodiphenyl sulphide with polyphosphoric acid and reducing the resulting 11-methyldibenzo[*b*, *f*][1, 4]thiazepine, is described as an oxidation inhibitor for synthetic lubricants.

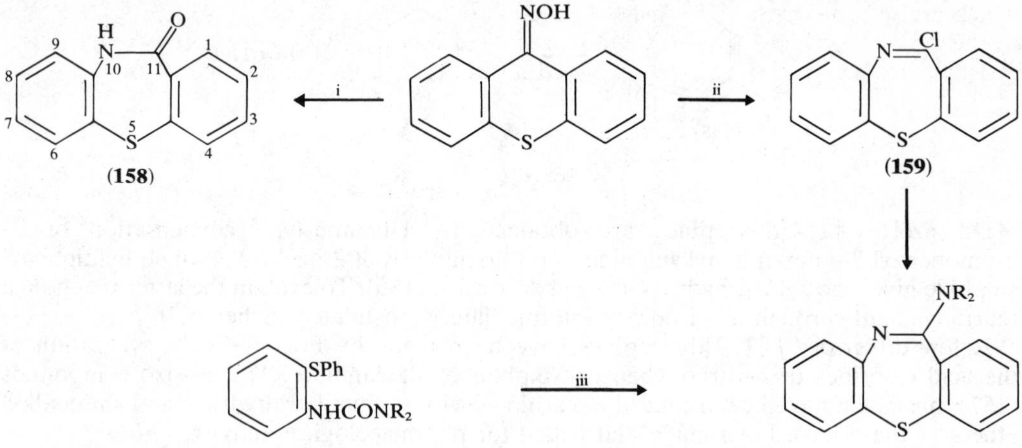

(158)

(159)

i, polyphosphoric acid; ii, PCl_5; iii, $POCl_3$.

SCHEME 8

20.3.4 EIGHT-MEMBERED RINGS

In the naming of heterocyclic compounds, unsaturated eight-membered rings containing nitrogen are denoted by the suffix -ocine, and those without nitrogen by -ocin. Saturated eight-membered rings not containing nitrogen are denoted by -ocan or, under IUPAC, -ocane. Problems arise in the numbering of some fused systems and some confusion exists in the literature.

20.3.4.1 Monocyclic compounds

There has been relatively little systematic study of eight-membered ring compounds, and far fewer such compounds are known than in the five-, six-, and seven-membered series. This may reflect the fact that conditions are less favourable for cyclizing chains of eight or more atoms, but the three modes of synthesis — ring expansion, ring closure, and addition of smaller units — have all been exemplified.

2-Methyl-8-phenylhexahydro-1,2-oxazocine (**161**) is produced by thermal rearrangement of 1-methyl-2-phenylhexahydroazepine 1-oxide. It is reduced and cleaved to 6-methylamino-1-phenylhexan-1-ol by zinc and hydrochloric acid. Reaction of di-(3-chloro-3-phenyl)propyl ether with methylamine is reported to give 5-methyl-4,6-diphenylhexahydro-1,5-oxazocine (**162**). Condensation of maleic anhydride with ethanolamides

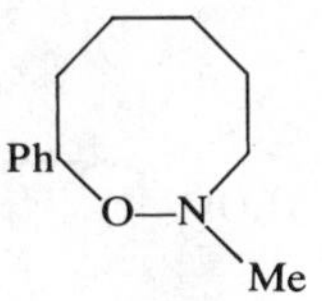

(161) 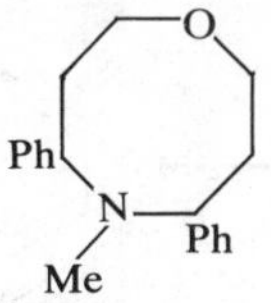 (162)

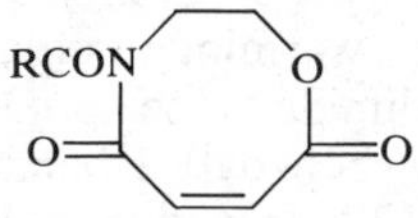

(163) 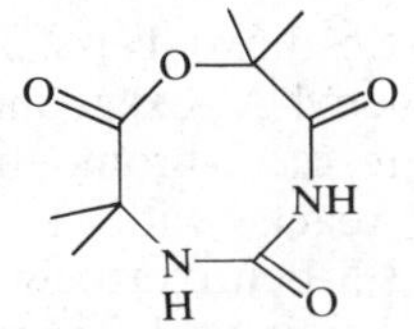 (164)

of long-chain acids gives cyclic ester-imides, possibly 1,4-oxazocine derivatives (**163**), which are intermediates for polymers. Oxadiazocine derivatives, *e.g.* (**164**), are obtained by heating tetramethylglycollide with urea or thiourea. They are said to give crease-resistant finishes to textiles, and to afford antibacterial *N*-halogeno-derivatives.

20.3.4.2 Benzoxazocines

Application of the Schmidt reaction to 3,4-dihydro-2*H*-1-benzoxepin-5(6*H*)-one (**165**) gives 3,4-dihydro-2*H*-1,6-benzoxazocin-5-one (**166**) (also described as 2,3,4,5-tetrahydro-6*H*-benzo[*b*]-1,4-oxazocin-5-one) and 2,3,4,5-tetrahydro-6*H*-1,5-benzoxazocin-6-one (**167**) in the ratio 11 : 1 if the reaction is done in sulphuric–acetic acid, and 7 : 3 in trichloracetic acid.[55,56] Beckmann rearrangement of the oxime of (**165**) also gives (**166**) and (**167**) (reaction 54).

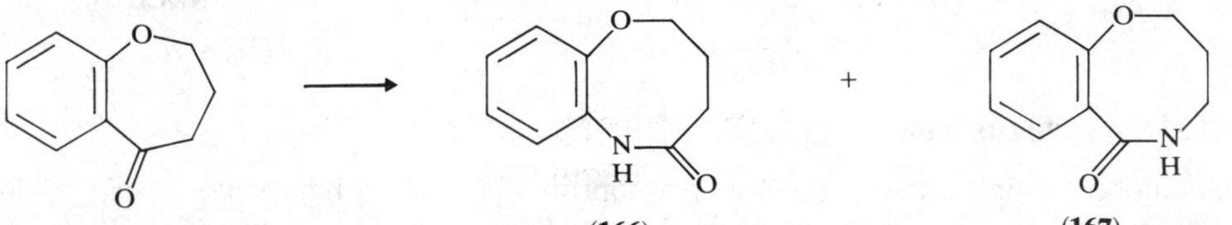

(165) → (166) + (167) (54)

The glycidyl ethers from *o*-hydroxy-aromatic ketones react with ammonia, giving aminoalcohols which slowly cyclize to 3,4-dihydro-3-hydroxy-1,5-benzoxazocines (**168**) (reaction 55). The 6-methyl derivative (**168**; R = Me) is a blocker of β-adrenergic receptors. It condenses with arylaldehydes to give styryl derivatives, and the ring is readily cleaved by acid or by acetic anhydride.[57]

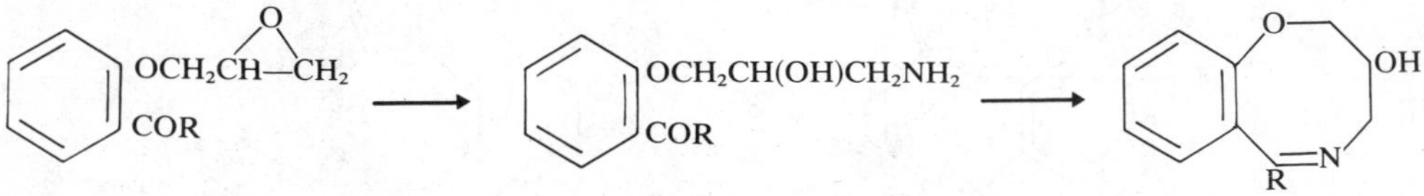

$OCH_2CH(O)CH_2$, COR → $OCH_2CH(OH)CH_2NH_2$, COR → (168) (55)

The *N*-methyl-*N*-2-hydroxyethylamide of *o*-benzoylbenzoic acid is reduced to the aminomethylbenzhydrol (**169**) by lithium aluminium hydride (reaction 56), and this cyclizes when heated in benzene with toluene-*p*-sulphonic acid, giving 3,4,5,6-tetrahydro-5-methyl-1-phenyl-1*H*-2,5-benzoxazocine (nefopam) (**170**). This is described variously as an anti-depressant, skeletal-muscle relaxant, and non-narcotic analgesic. Some *N*-aminoalkyl analogues are said to be CNS depressants.

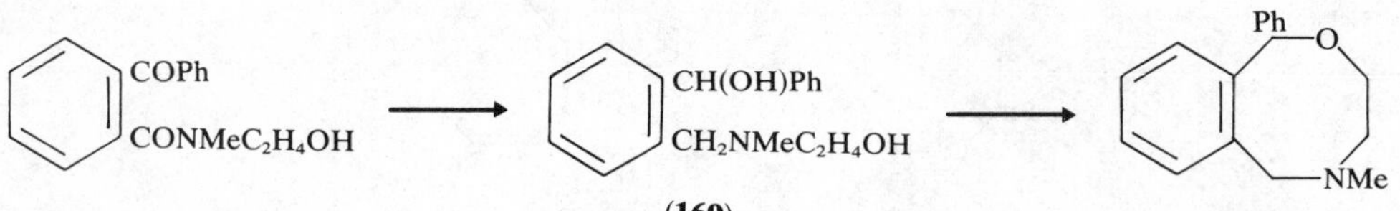

(56)

(169) (170)

3,4,5,6-Tetrahydro-1,6-dioxo-1*H*-2,5-benzoxazocine (**171**) is obtained by rearrangement of 2-*o*-carboxyphenyl-Δ^2-oxazoline (**172**) by warming in ethanol (reaction 57); (**172**) is made by treatment of 2-bromoethylphthalimide with potassium hydroxide. 1,2-Dibromopropionitrile reacts with 1-(*o*-hydroxybenzoyl)-2-benzylhydrazine, giving 5-benzylamino-2-imino-1,5-benzoxazocin-6-one (**173**) (reaction 58). This is hydrolysed by hydrochloric acid to salicylic acid, benzylhydrazinoacrylic acid, and ammonium chloride.

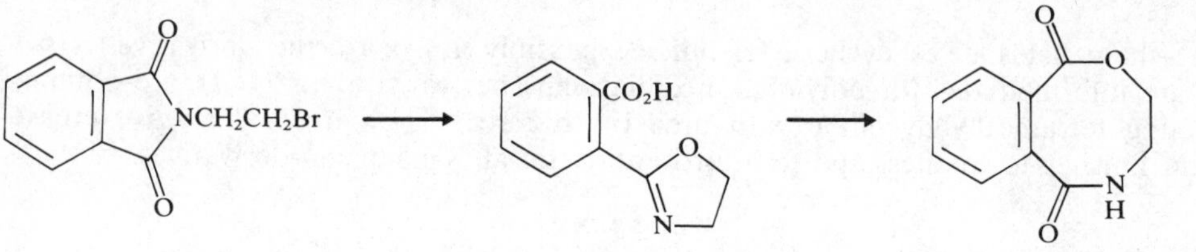

(57)

(172) (171)

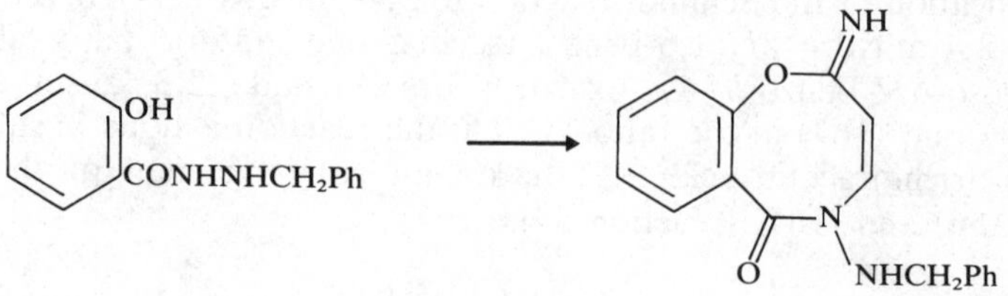

(58)

(173)

20.3.4.3 Benzothiazocines

o-Aminobenzhydrols react with 2-mercaptopropionic acid in hydrochloric acid, giving 3,4-dihydro-6-phenyl-5,1-benzothiazocin-2-ones (**174**).[58] Similar compounds without the 6-phenyl group are made from *o*-nitrobenzyl halides by reaction with 2-mercaptopropionic acid, reduction, and cyclization. The Schmidt reaction with 3,4-dihydro-7-chloro-2*H*-1-benzothiepin-5-one gives 3,4-dihydro-8-chloro-2*H*-1,6-benzothiazocin-5(6*H*)-one (**175**). 3,4-Dihydro-2-phenyl-2*H*-1,6-benzothiazocin-5(6*H*)-one is made by condensation of ethyl 4-chloro-4-phenylbutyrate with 2-aminothiophenol, hydrolysis to the acid, and ring-closure with thionyl chloride.

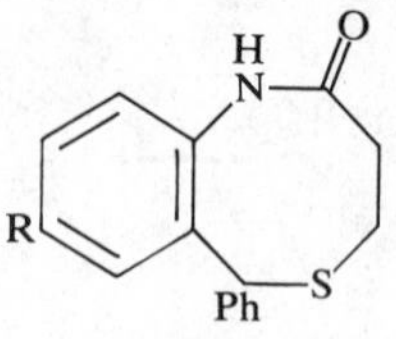

(174) 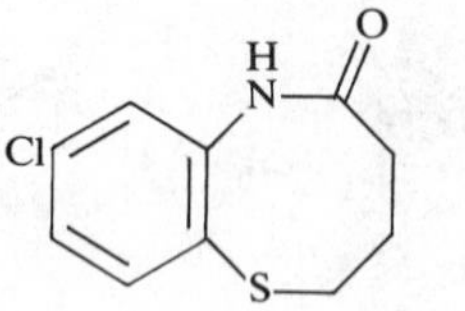 (175)

20.3.4.4 Dibenzoxazocines and dibenzothiazocines

Ring expansion of the dibenzoxepinone (**176**)[59] and the dibenzothiazepinone (**177**) has been achieved by the Schmidt reaction or the Beckmann rearrangement of their oximes (reactions 59 and 60). 2,2′-Bis(bromomethyl)diphenyl ether or sulphide react with amines

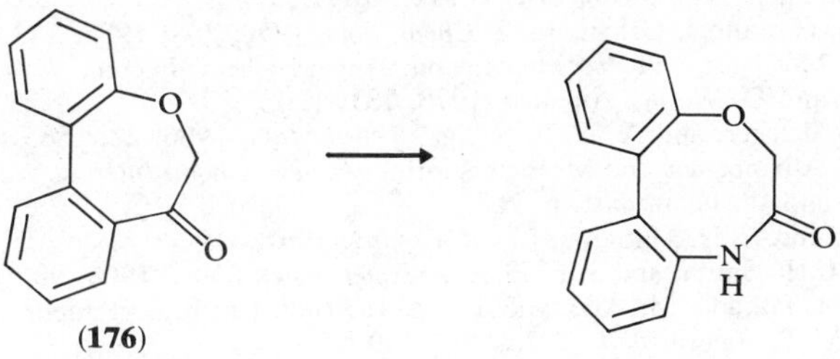

(176) (59)

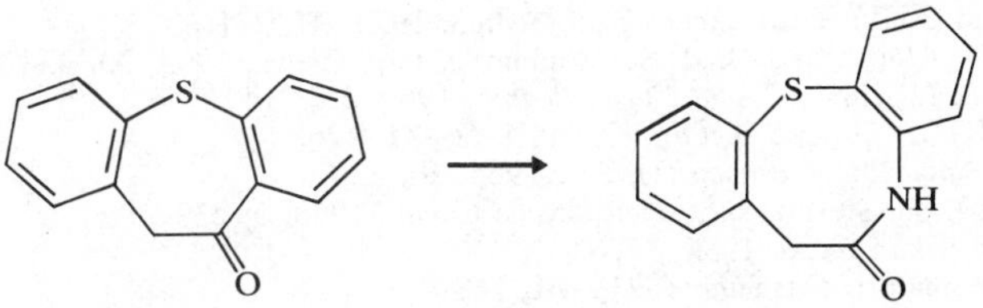

(177) (60)

to give 6,7-dihydro-5*H*-dibenz[*b*, *g*][1, 5]-oxazocines (and -thiazocines) (**178**), together with dimeric compounds containing a 16-membered ring. 6*H*-Dibenzo[*b*, *f*][1, 4]-oxazocine-11(12*H*)-one (**179**; R=H, X=O, Y=O) is made by dehydrating *o*-carboxybenzyl-*o*′-aminophenyl ether. The related thiazocinone (**179**; R=Ph, X=O, Y=S) is prepared by treating *o*-aminothiophenol with 3-phenylphthalide and cyclizing the adduct with thionyl chloride. The reduction product (**179**; R=Ph, $X=H_2$, Y=S) is rearranged during alkylation with dimethylaminoethyl chloride to 1-phenyl-2-*o*-dimethylaminoethylthiophenylisoindoline (**180**).

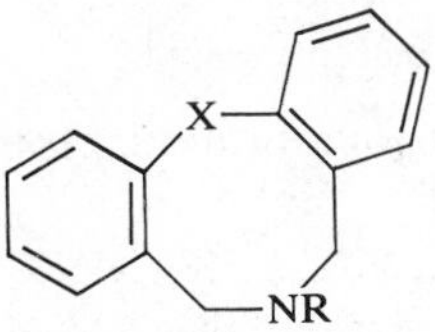

(**178**) X=O or S 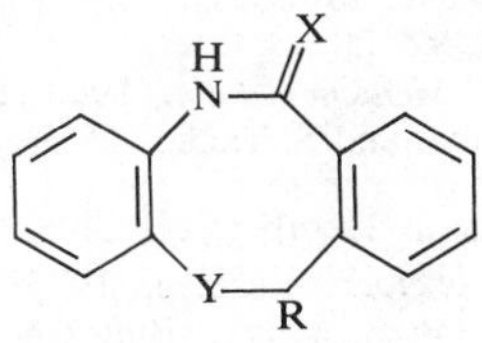 (**179**) 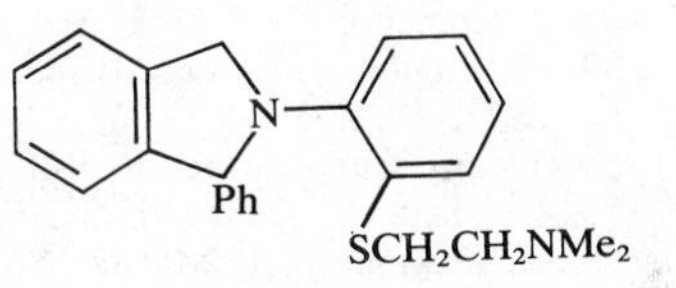 (**180**)

References

1. W. D. Emmons, in 'Heterocyclic Compounds with Three and Four Membered Rings,' ed. A. Weissberger, Wiley-Interscience, New York, 1964, pp. 624, 978.
2. R. Livingstone, in 'Rodd's Chemistry of Carbon Compounds', 2nd edn, ed. S. Coffey, Elsevier, London, 1973, vol, IVa, p. 40.
3. E. Schmitz, *Adv. Heterocyclic Chem.*, 1963, **2,** 83.
4. G. G. Spence, E. C. Taylor, and O. Buchardt, *Chem. Rev.*, 1970, **70,** 231.
5. E. Schmitz, S. Schramm, and H. Simon, *Angew. Chem. Internat. Edn.*, 1966, **5,** 578.
6. G. L'Abbe, *Bull. Soc. chim. France*, 1975, 1127.
7. H. Quast and F. Kees, *Angew. Chem. Internat. Edn.*, 1974, **13,** 742.
8. L. L. Muller and J. Hamer, '1,2-Cycloaddition Reactions,' Wiley-Interscience, New York, 1967.
9. T. Durst and B. P. Gimbarzevsky, *J.C.S. Chem. Comm.*, 1975, 724.
10. J. L. Knunyants and G. A. Sokolski, *Angew. Chem. Internat. Edn.*, 1972, **11,** 583.
11. W. E. Truce and L. K. Lin, *Chem. and Ind.* (*London*), 1969, 457.
12. R. C. Kerber and M. C. Cann, *J. Org. Chem.*, 1974, **39,** 2552.
13. S. Andreades, *J. Org. Chem.*, 1962, **27,** 4163.
14. K. Weiser and A. Berndt, *Angew. Chem. Internat. Edn.*, 1975, **14,** 69, 70.
15. B. J. R. Nicolaus, L. Bellasio, and E. Testa, *Helv. Chim. Acta*, 1962, **45,** 717; 1963, **46,** 450.
16. G. M. Atkins, Jr. and E. M. Burgess, *J. Amer. Chem. Soc.*, 1972, **94,** 6135.
17. T. Hiraoka and T. Kobayashi, *Bull. Chem. Soc. Japan*, 1975, **48,** 480.

18. W. Wucherpfennig, *Tetrahedron Letters*, 1971, 1891.
19. E. R. Braithwaite and J. Graymore, *J. Chem. Soc.*, 1950, 208; 1953, 143.
20. W. Ried, O. Mösinger, and W. Schuckmann, *Angew. Chem. Internat. Edn.*, 1976, **15,** 103.
21. J. Goerdeler and D. Wobig, *Annalen*, 1970, **731,** 120.
22. H. Ulrich, B. Tucker, and A. A. R. Sayigh, *Tetrahedron*, 1966, **22,** 1565.
23. K. Burger, J. Albanbauer, and M. Eggersdorfer, *Angew. Chem. Internat. Edn.*, 1975, **14,** 766; K. Burger, J. Albanbauer, and W. Foag, *ibid.*, p. 767.
24. G. H. Birum and C. N. Mathews, *Chem. Comm.*, 1967, 137.
25. F. Ramirez, C. P. Smith, and J. F. Pilot, *J. Amer. Chem. Soc.*, 1968, **90,** 6726.
26. T. Minami, M. Fukuda, M. Abe, and T. Agawa, *Bull. Chem. Soc. Japan*, 1973, **46,** 2156; T. Minami, F. Takimoto, and T. Agawa, *ibid.*, 1975, **48,** 3259.
27. K. C. Rice and U. Weiss, *Tetrahedron Letters*, 1973, 1615.
28. C. L. Pedersen and O. Buchardt, *Acta Chem. Scand.*, 1973, **27,** 271.
29. P. G. Mente, H. W. Heine, and G. R. Scharoubim, *J. Org. Chem.*, 1968, **33,** 4547.
30. O. Buchardt and P. L. Kumler, *Acta Chem. Scand.*, 1969, **23,** 2149.
31. A. A. Santilli and T. S. Osdene, *J. Org. Chem.*, 1966, **31,** 4268.
32. E. Testa and L. Fontanella, *Farmaco Ed. Sci.*, 1965, **20,** 323.
33. M. E. Derieg and L. H. Sternbach, *J. Heterocyclic Chem.*, 1966, **3,** 237.
34. K. Schenker, *Helv. Chim. Acta*, 1968, **51,** 413.
35. D. Misiti and V. Rimatori, *Gazzetta*, 1971, **101,** 167.
36. K. Peseke, *J. prakt. Chem.*, 1975, **317,** 648.
37. S. Ito, *Bull. Chem. Soc. Japan*, 1970, **43,** 1824.
38. N. V. Dudykina and V. A. Zagorevski, *Chem. Abs.*, 1966, **65,** 683.
39. S. Raines, C. A. Kovacs, S. Goldstein, and E. P. Palopoli, *J. Medicin. Chem.*, 1968, **11,** 895.
40. H. Mantsch, V. Zanker, W. Seiffert, and G. Prell, *Annalen*, 1969, **723,** 95.
41. H. Lund, P. Lunde, and F. Kaufmann, *Acta Chem. Scand.*, 1966, **20,** 1631.
42. R. Jacquier, J.-L. Olive, C. Petrus, and F. Petrus, *Tetrahedron Letters*, 1975, 2979.
43. C. Kaneko and S. Yamada, *Tetrahedron Letters*, 1967, 5233.
44. W. Metlesics, G. Silverman, and L. H. Sternbach, *Monatsh.*, 1967, **98,** 633.
45. W. Ried and R. Dietrich, *Annalen*, 1963, **666,** 113, 135.
46. K.-H. Wünsch and A. Ehlers, *Z. Chem.*, 1970, **10,** 361.
47. I. W. J. Still, M. T. Thomas, and A. M. Clish, *Canad. J. Chem.*, 1975, **53,** 276.
48. C. A. Grob and J. Ide, *Helv. Chim. Acta*, 1974, **57,** 2562.
49. O. K. J. Kovacs, B. Edström, and B. Sjöberg, *Acta Chem. Scand.*, 1973, **27,** 677.
50. N. J. Leonard and R. Y. Ning, *J. Org. Chem.*, 1966, **31,** 3928.
51. D. N. Reinhoudt, *Rec. Trav. chim.*, 1973, **92,** 20.
52. J. Krapcho, C. F. Turk, and J. J. Piala, *J. Medicin. Chem.*, 1968, **11,** 361.
53. H. Kugita, H. Inoue, M. Ikezaki, M. Konda, and S. Takeo, *Chem. Pharm. Bull.* (*Japan*), 1970, **18,** 2284; 1971, **19,** 595.
54. J. I. G. Cadogan and S. Kulik, *Chem. Comm.*, 1970, 233.
55. D. Misiti, V. Rimatori, and F. Gatta, *J. Heterocyclic Chem.*, 1973, **10,** 689.
56. H. Kawamoto, T. Matsuo, S. Morosawa, and A. Yokoo, *Bull. Chem. Soc. Japan*, 1973, **46,** 3898.
57. B. Basil, E. J. C. Coffee, D. C. Gell, D. R. Maxwell, D. J. Sheffield, and K. R. H. Wooldridge, *J. Medicin. Chem.*, 1970, **13,** 403.
58. J. Klosa, *J. prakt. Chem.*, 1967, **36,** 5.
59. T. A. Harrow, C. E. Harrison, and W. R. N. Williamson, *J. Chem. Soc.* (*C*), 1971, 2098.

20.4
Meso-ionic Heterocycles

C. A. RAMSDEN
May and Baker Ltd., Dagenham

20.4.1 DEFINITION

The concept of meso-ionic molecules was introduced in 1949 by Baker and Ollis,[1a] who recognized that *N*-phenylsydnone (**2**; $R^1 = Ph$, $R^2 = H$), the cyclodehydration product of *N*-nitroso-*N*-phenylglycine (**1**; $R^1 = Ph$, $R^2 = H$), has a structure which can only be represented as a resonance hybrid of many dipolar and tetrapolar canonical forms, *e.g.* (**3**)–(**5**). In order to describe this type of heterocycle they coined the term 'meso-ionic' (*meso*meric + *ionic*) and it was recognized that in addition to the sydnones (**2**) other meso-ionic compounds had been reported. These included the 1,3,4-thiadiazolium-2-thiolates (**6**) prepared by Busch (1895),* the 1,2,3,4-tetrazolium-5-thiolate (dehydrodithizone) (**7**) prepared by Emil Fischer (1882), and the pigment Besthorn's Red (**8**) (1904). The possible existence of other meso-ionic systems was also predicted and subsequently many new systems have been prepared. The chemistry of meso-ionic compounds has been the subject of several comprehensive reviews;[2–9] these include an account of the historical development of the subject.[2] In a survey covering the literature up to the end of 1975, Ollis and Ramsden[9] have recommended the following modification of the original definition of a meso-ionic compound:

> '*a compound may be appropriately called meso-ionic if it is a five-membered heterocycle which cannot be represented satisfactorily by any one covalent or polar structure and possesses a sextet of electrons in association with the five atoms comprising the ring.*'

In accord with the above definition, meso-ionic molecules are represented by formulae of the general type (**9**), in which a, b, c, d, e, and f refer to atoms or groups derived from suitably substituted carbon or heteroatoms. The restrictions imposed on the atoms or groups a–f in structure (**9**), the number of possible mesomeric structural variations (see Section 20.4.2), and the relationship of meso-ionic compounds (**9**) to other heterocyclic systems having the same σ-skeleton (**10**) have been discussed in a review.[9] In this context, it is generally accepted that five-membered heterocyclic *N*-oxides (and related *N*-imines and ylides) exemplified by compounds (**11**) and (**12**) are *satisfactorily* represented by a single dipolar structure. Compounds of this type are not therefore described as meso-ionic. Originally it was proposed that meso-ionic compounds should be represented by structures of the type (**13**), but this vague description is no longer favoured and the symbolism (**9**) is preferred. The large full circle and the positive sign in structure (**9**) symbolize a cyclic delocalization of the π-electrons of the meso-ionic ring in association with a partial positive charge; the exocyclic group (f) is associated with a corresponding partial negative charge. This polarization of meso-ionic compounds in which the ring

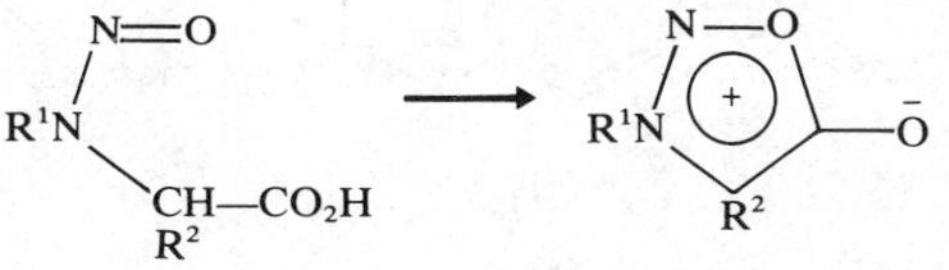

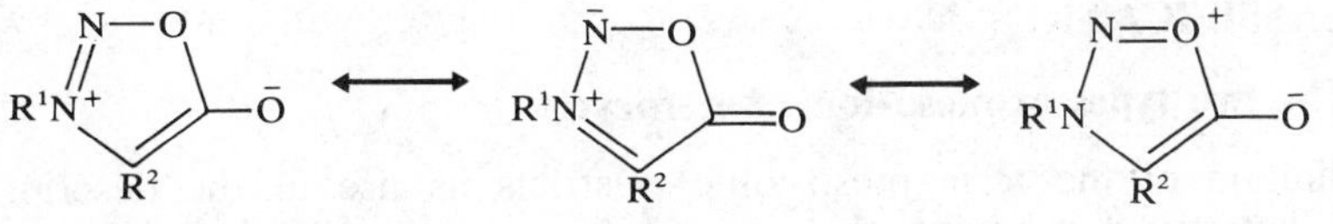

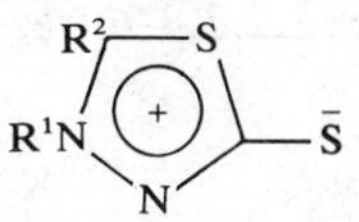

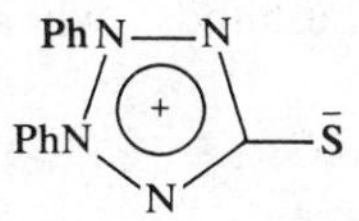

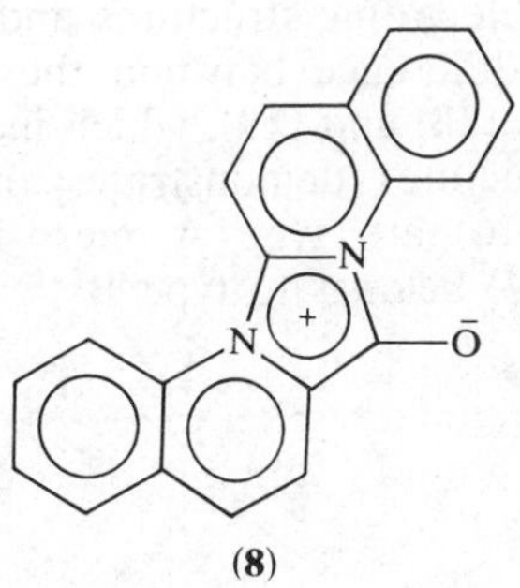

* The dipolar structure was recognized by Schönberg in 1938.[1b]

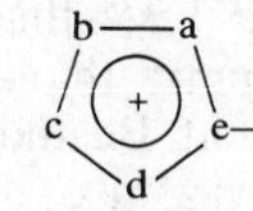

(9) 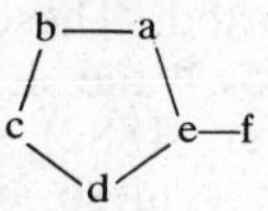 (10) 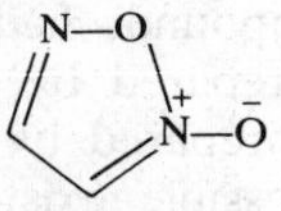 (11)

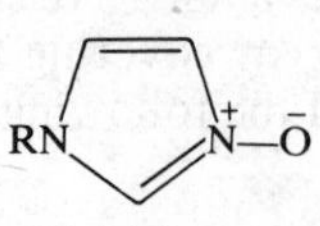

(12) 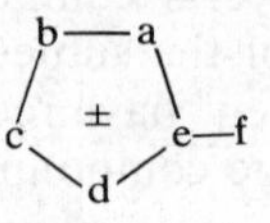 (13) 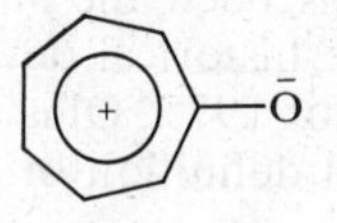 (14)

tends towards a structure associated with a sextet of electrons has led to their description as aromatic and analogies with the structure of tropone (**14**) have been drawn. Some meso-ionic molecules do have properties which are associated with classical aromaticity, but because of the diverse properties of this large group of heterocycles it no longer seems useful to describe them as aromatic.

The original definition of meso-ionic embraced the possibility of six-membered meso-ionic compounds. In order to retain the usefulness of the term, the modified definition[9] has been specifically restricted to five-membered heterocycles of type (**9**). Accordingly, six-membered dipolar heterocycles exemplified by the pyridinium-3-olates (**15**)[10,11] are best described as mesomeric betaines. Similarly, polycyclic systems such as the triazaphenalenes (**16**)[11] and the pyrazolo[3,4-*d*]pyrazoles (**17**)[12] should also be described as mesomeric betaines and meso-ionic compounds can be regarded as a specific type of five-membered mesomeric betaine.

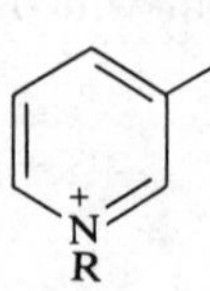

(15) 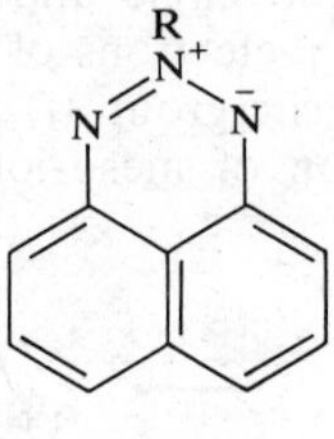 (16) 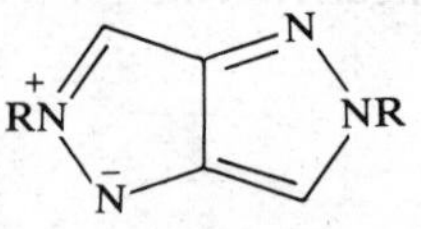 (17)

20.4.2 CLASSIFICATION

20.4.2.1 The two types of meso-ionic heterocycle

The definition of the term meso-ionic[9] restricts its use to the description of five-membered heterocycles having the general formula (**9**) (Section 20.4.1). Meso-ionic compounds (**9**) can be further subdivided into two distinct types of molecule having different electronic structures and which are conveniently described as type A and type B.[9] The difference between these two classes can be appreciated by considering the structures (**18**) and (**19**), which indicate the origin of the eight π-electrons. Inspection of these structures demonstrates that the sydnones (**2**) and the 1,3,4-thiadiazolium-2-thiolates (**6**) are type A meso-ionic compounds, whereas the 1,2,3,4-tetrazolium-5-thiolate (**7**) belongs to type B.

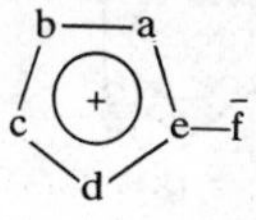

(9) 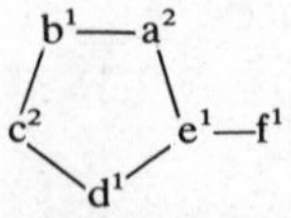 (18) type A 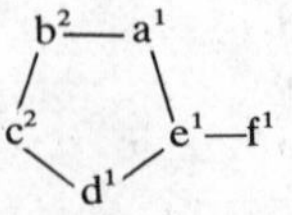 (19) type B

20.4.2.2 Type A meso-ionic compounds

If the atoms or groups a, b, c, d, e, and f in the general formula (**9**) of the type A meso-ionic compounds are chosen from suitably substituted carbon, nitrogen, oxygen, or sulphur atoms, then 144 different meso-ionic structural types are possible. Derivatives of 48 of these systems have been reported.[9] In addition, two examples of type A meso-ionic heterocycles containing selenium are known.

Many type A meso-ionic compounds participate in 1,3-dipolar cycloaddition reactions and in this connection canonical forms of the general type (**22**) clearly make an important contribution to the electronic structure. Structure (**22**) can be regarded as being formed by union of a 1,3-dipole (**20**) and a heterocumulene (**21**), and this association of type A meso-ionic compounds with a 1,3-dipolar fragment provides a convenient and useful method of subdividing this class.[9] Most of the 50 known type A systems are associated with the nitrogen-containing 1,3-dipolar fragments (**23**), (**24**), and (**25**). In addition, meso-ionic systems derived from the chalcogen-containing 1,3-dipoles (**26**; X = O, S, Se) and the thiocarbonyl imines (**27**) are also known.

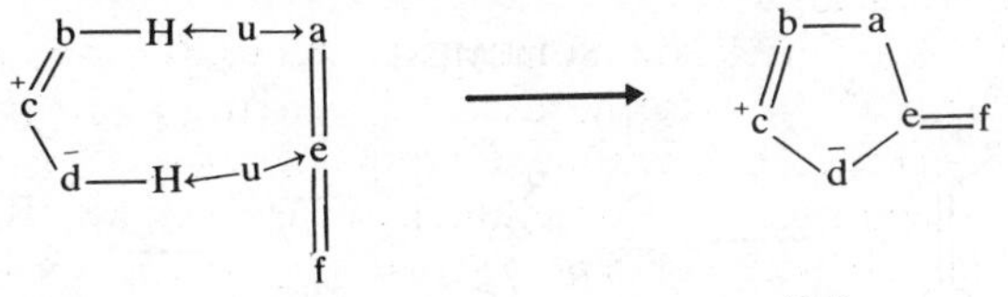

1,3-Dipolar cycloaddition reactions of type A systems with a wide variety of 1,3-dipolarophiles are known.[13] The modes of reaction with alkynes, alkenes, and heterocumulenes are generalized in Schemes 1, 2, and 3. Addition of alkynes (Scheme 1) initially gives a bicyclic adduct (**28**) which is rarely isolated but which fragments, giving the heterocyclic system (**30**) and a heterocumulene (a=e=f). This provides a valuable synthetic route to many systems of this general type (**30**) (see Section 20.4.22). In a few cases an alternative fragmentation of the adducts (**28**; c = S, Se) occurs, giving a six-membered product (**29**). Addition of alkenes (Scheme 2) also gives bicyclic adducts (**31**) and in many reactions this product (**31**) is stable and can be isolated. With some systems, however, the adduct (**31**) undergoes a retro-1,3-dipolar cycloaddition, giving the transient 1,3-dipole (**32**) which then either rearranges to a stable isomer by proton transfer (**32** to **34**; R^1 = H) or, in the presence of excess 1,3-dipolarophile, gives the adduct (**33**). With heterocumulenes (Scheme 3), four isomeric bicycloadducts are possible (*e.g.* **35** and **36**). A number of adducts of this type have been isolated, but often the loss of the alternative heterocumulene fragment (a=e=f) occurs to give a new type A meso-ionic system (*i.e.* **37** or **38**). This provides a useful route to several meso-ionic systems.

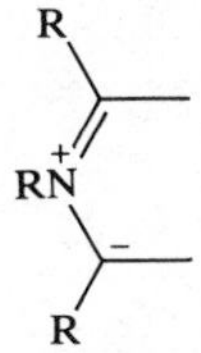

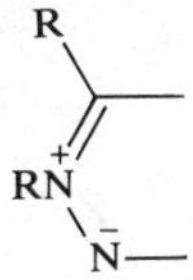

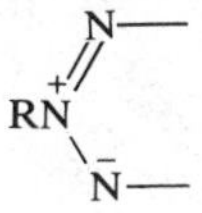

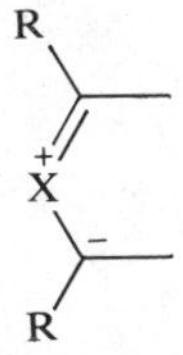

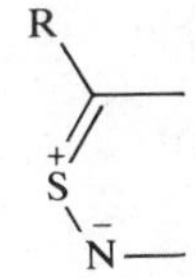

These modes of 1,3-dipolar cycloaddition of type A meso-ionic compounds (Schemes 1, 2, and 3) are exemplified by three reactions of the 1,3-thiazolium-5-olate (**39**) (Scheme 4).[14] The reactivity of type A systems towards 1,3-dipolarophiles decreases as the number of nitrogen atoms in the 1,3-dipolar fragment increases, *i.e.* decreasing activity along the series (**23**), (**24**), and (**25**). Thus, the 1,3-oxazolium-5-olates (münchnones) (**40**) are more reactive than the sydnones (**41**), whereas the 1,2,3,4-oxatriazolium-5-olates (**42**) show little, if any, 1,3-dipolar reactivity.

Some cycloaddition reactions of type A systems (**9**) suggest that a small concentration of the valence tautomer (**43**) may be in equilibrium with the meso-ionic isomer (**9**).[7] The

(9) (28) (29) (30)

SCHEME 1

(9) (31) (32) (33) (34)

SCHEME 2

(9) (35) and/or (36) (37) (38)

SCHEME 3

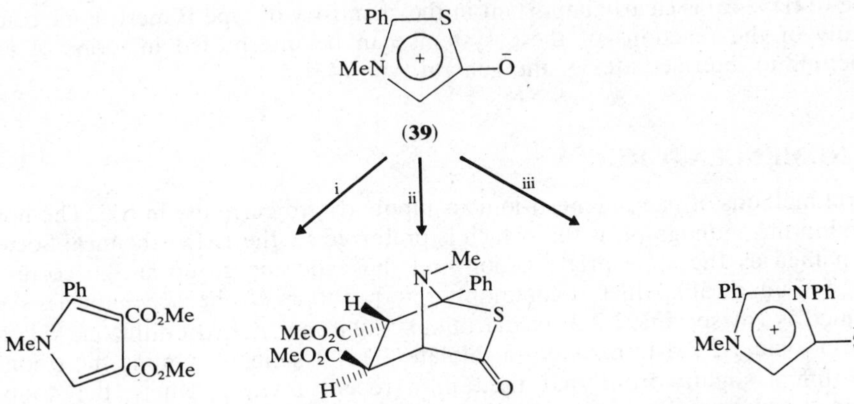

(39)

i, $MeO_2CC{\equiv}CCO_2Me$; ii, $MeO_2CCH{=}CHCO_2Me$; iii, PhNCS.

SCHEME 4

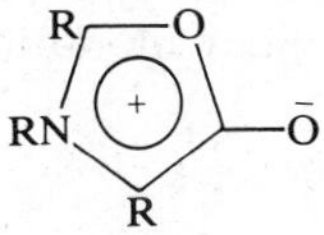

(40) 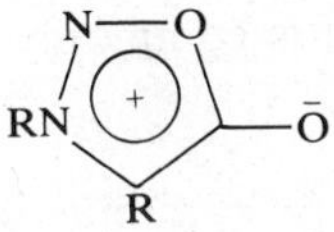 (41) 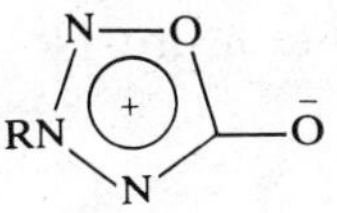 (42)

(9) (43)

valence tautomers (**43**) may also be generated in some photochemical transformations of type A compounds.

20.4.2.3 Type B meso-ionic compounds

84 different type B meso-ionic systems (**44**) are possible if the atoms or groups a–f are selected from suitably substituted carbon, nitrogen, oxygen, and sulphur atoms. So far, representatives of eight of these systems have been reported.[9] Some of these 84 possibilities can be expected to be more stable as the valence tautomer (**45**). For example, although dehydrodithizone (**46**) is stable with respect to the acyclic isomer (**47**), the triketones (**49**) are more stable than the meso-ionic isomers (**48**). Valence tautomerism of

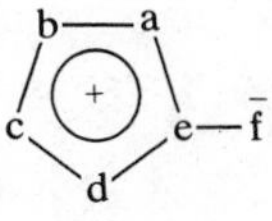

(44) 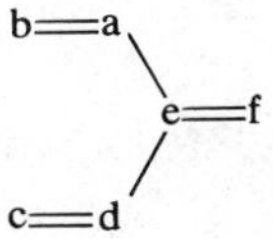 (45) 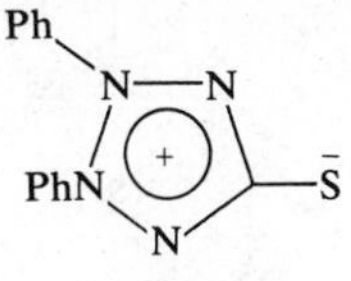 (46)

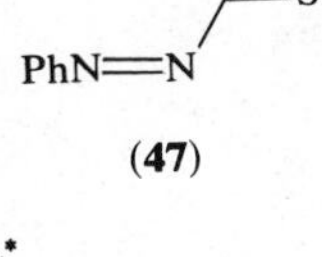

(47) 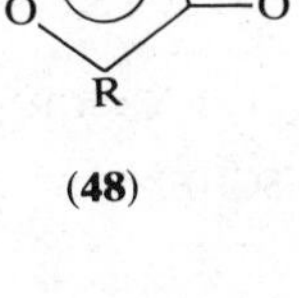 (48) 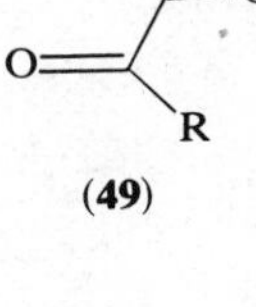 (49)

the type (**44**)⇌(**45**) is clearly important in the chemistry of type B meso-ionic compounds and many of the reactions of these systems can be interpreted in terms of an initial ring-opening to intermediates of the general type (**45**).

20.4.3 NOMENCLATURE

Several methods of naming meso-ionic compounds are currently in use. The nomenclature used in this monograph is that which is preferred by the U.K. Chemical Society. The ring is named as the appropriate cation and the exocyclic group as the corresponding anion. Compound (**50**), which is commonly referred to as *N*-phenylsydnone, is systematically named as a meso-ionic 1,2,3-oxadiazolium-5-olate and dehydrodithizone (**51**) is named as a meso-ionic 1,2,3,4-tetrazolium-5-thiolate. This system of naming meso-ionic compounds differs slightly from that used in a recent review,[9] where the double-bond character of the exocyclic bond is emphasized. For example, *N*-phenylsydnone (**50**) would be named as a meso-ionic 1,2,3-oxadiazole-5-one.

An alternative method which is often used is to name the meso-ionic system as an anhydride. *N*-Phenylsydnone (**50**) is then named anhydro-5-hydroxy-3-phenyl-1,2,3-oxadiazolium hydroxide. This system is commonly used in the older literature.

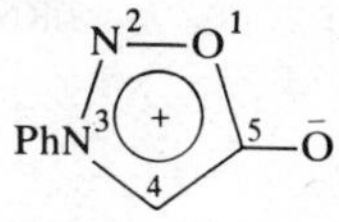

(**50**) 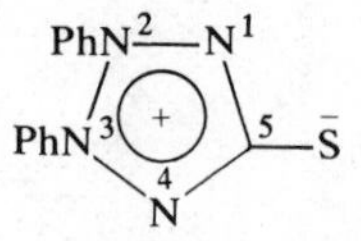 (**51**)

20.4.4 MESO-IONIC OXAZOLES OF TYPE A

20.4.4.1 1,3-Oxazolium-4-olates (54)

Derivatives of this meso-ionic system (**54**) have been prepared (my) by thermal decomposition of the diazo-derivatives (**52**) using cupric acetylacetonate.[15,16] The mechanism of this reaction is assumed to involve an intermediate carbene (**53**), which cyclizes to the red crystalline meso-ionic product (**54**) (Scheme 5).

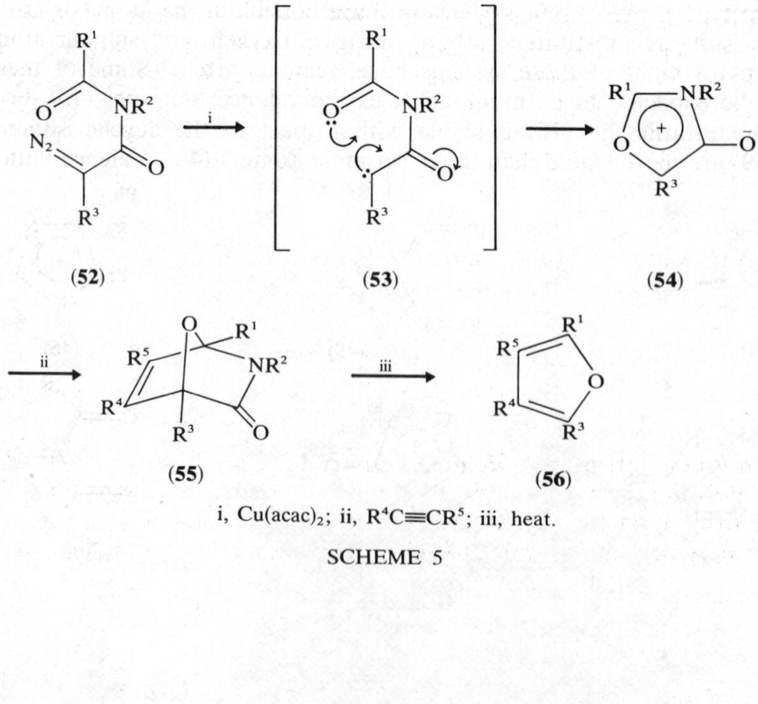

(**52**) (**53**) (**54**)

(**55**) (**56**)

i, $Cu(acac)_2$; ii, $R^4C{\equiv}CR^5$; iii, heat.

SCHEME 5

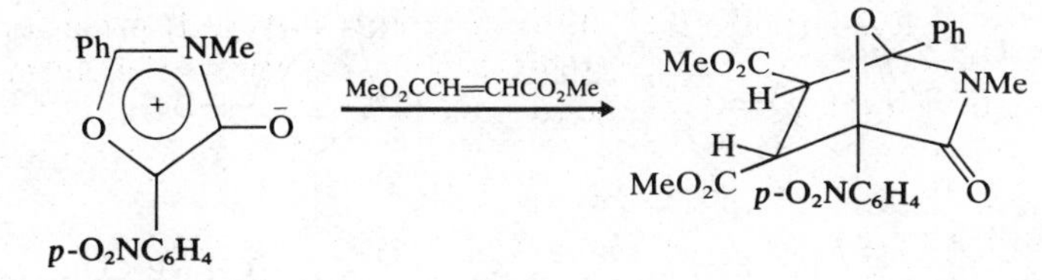

SCHEME 6

The 1,3-oxazolium-4-olates (**54**) are reactive 1,3-dipoles. With acetylenes both the primary 1,3-dipolar cycloadduct (**55**) and the substituted furan (**56**) are isolated. Addition of dimethyl fumarate to the derivative (**57**) gives the adduct (**58**) in quantitative yield (Scheme 6).[15,16]

20.4.4.2 1,3-Oxazolium-4-aminides (59)

Although the salts (**60**) are prepared by treatment of the nitriles (**61**) with acid or acyl perchlorates, conversion of the salts (**60**) into the free meso-ionic systems (**59**) has not been successful.[17]

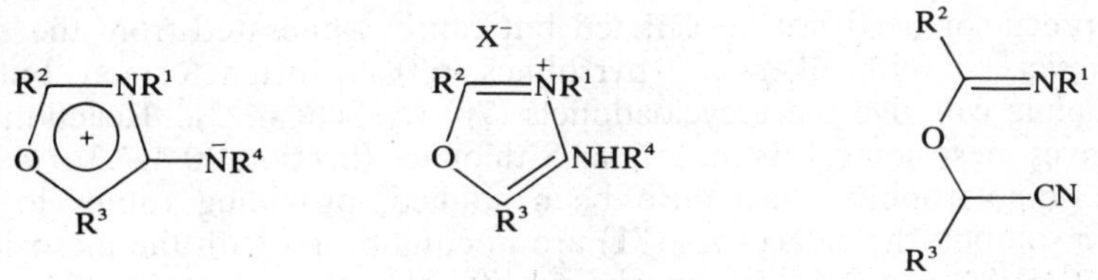

20.4.4.3 1,3-Oxazolium-5-olates (münchnones) (63)

The meso-ionic 1,3-oxazolium-5-olates (**63**) are commonly referred to as münchnones (*cf.* sydnones, Section 20.4.10.1) and aspects of their chemistry have been subjects of review articles.[7,9,18] Their preparation (hy) involves the cyclodehydration of α-acylamino-acids (**62**), usually using acetic anhydride at temperatures not exceeding 60 °C.[7] At higher temperatures (100 °C) the meso-ionic product (**63**) reacts with acetic anhydride to give the methyl ketones (**64**) (Scheme 7), suggesting that meso-ionic 1,3-oxazolium-5-olates (**63**) are intermediates in the Dakin–West reaction of amino-acids [$RCH(NH_2)CO_2H \leftarrow$ **62** $\rightarrow$ **63** $\rightarrow$ **64**].[19] Derivatives of the bicyclic system (**67**) have been prepared by a similar route (my).[20]

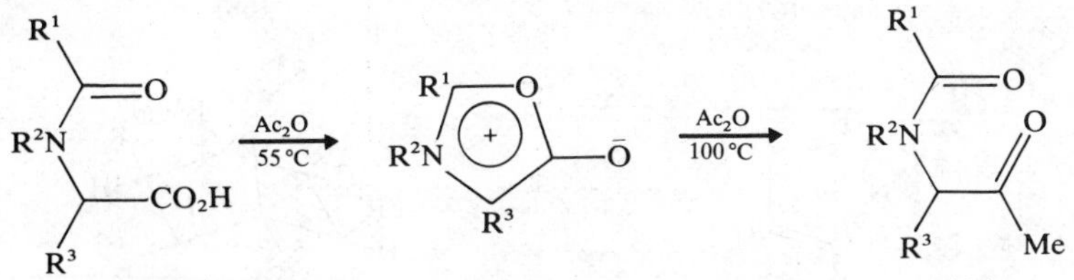

SCHEME 7

The trisubstituted derivatives (**63**) are obtained as yellow crystals which are extremely sensitive to moisture,[7] whereas derivatives unsubstituted at position-4 (**63**; $R^3 = H$) are too unstable to be isolated.[18] Perchlorates of these unsubstituted derivatives (**65**) can be prepared from α-acylamino-acids (**62**; $R^3 = H$) if a mixture of acetic anhydride and

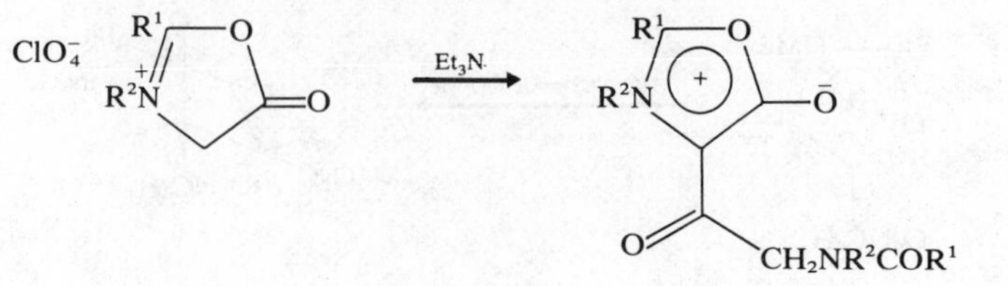

(65) **(66)**

SCHEME 8

perchloric acid is used as cyclodehydrating agent.[21] Treatment of solutions of these perchlorates (**65**) with base generates the free meso-ionic compounds (**63**; $R^3 = H$), which must be used immediately otherwise dimerization to the substitution product (**66**) occurs (Scheme 8).[21]

Münchnones (**63**) are highly susceptible to nucleophilic attack by water, alcohols, and primary amines, giving the corresponding α-acylamino-acid derivatives (**68**; Nu = OH, OR, or NHR).[7] The unsubstituted derivatives (**63**; $R^3 = H$) undergo electrophilic substitution at position-4, but the substituted product (**63**; $R^3 \neq H$) can only be isolated if the substituent is strongly electron withdrawing (*e.g.* **63**; $R^3 = CF_3CO$, ArN_2).[18,21]

1,3-Oxazolium-5-olates (**63**) are highly reactive 1,3-dipoles and their cycloadditions have been studied extensively.[7,9] Addition of alkynes gives pyrroles (*cf.* Scheme 1); the meso-ionic precursor need not be isolated but can be generated from the α-acylamino-acid (**62**) *in situ*.[7,22] With alkenes, 2-pyrrolines (**69**) are often formed, but excess of a 1,3-dipolarophile can give the bicycloadducts (**70**) (*cf.* Scheme 2).[7] Reaction with carbon disulphide gives meso-ionic 1,3-thiazolium-5-thiolates (Section 20.4.7.3).[7] A wide variety of other 1,3-dipolarophiles have also been studied, providing routes to many novel products.[9] In solution the azlactones (**71**) are in equilibrium with the meso-ionic tautomers (**72**), and in this context it is relevant to note that the azlactones (**71**) participate in 1,3-dipolar cycloadditions with alkynes and alkenes.[23]

Some cycloaddition reactions of meso-ionic 1,3-oxazolium-5-olates (*e.g.* **73**) suggest that a small equilibrium concentration of the valence tautomeric keten (*e.g.* **74**) exists in solution. With carbodi-imides, 2,4-diphenyl-3-methyl-1,3-oxazolium-5-olate (**73**) gives iminoazetidones (**75**) (Scheme 9) (my).[24]

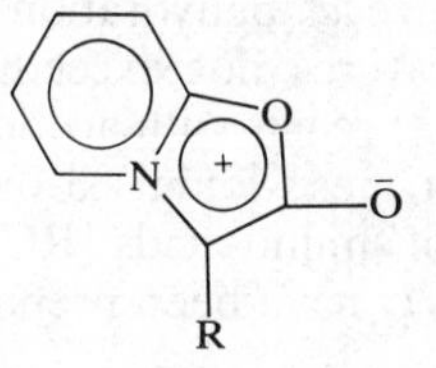

(67) 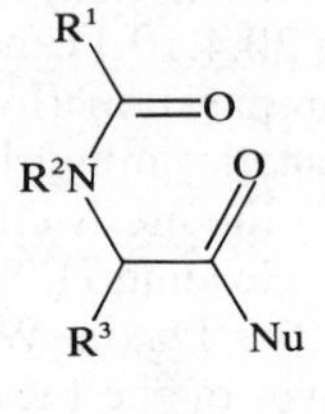 **(68)**

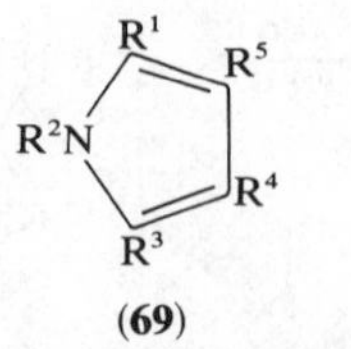

(69) 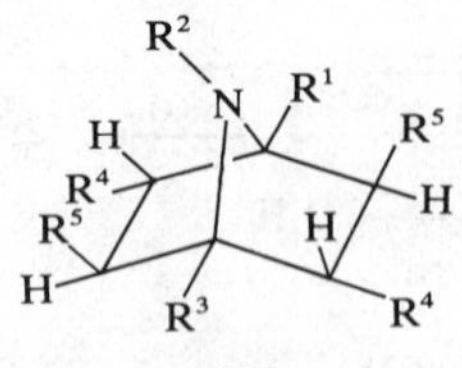 **(70)**

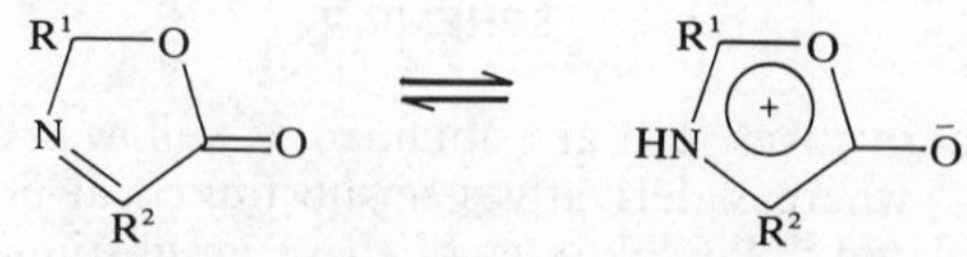

(71) **(72)**

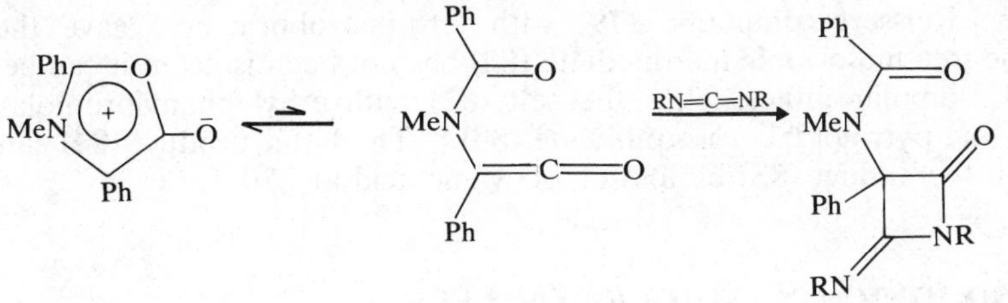

(73) (74) (75)

SCHEME 9

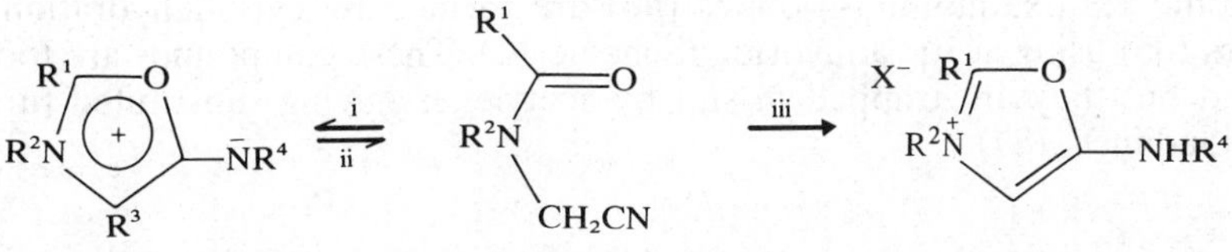

(76) (77) (78)

i, $(CF_3CO)_2O$; ii, H^+, H_2O; iii, HCl or $RCO^+ClO_4^-$.

SCHEME 10

20.4.4.4 1,3-Oxazolium-5-aminides (76)

Treatment of the nitrile derivatives (**77**) with trifluoroacetic anhydride gives the crystalline but relatively unstable meso-ionic 1,3-oxazolium-5-aminides (**76**; $R^3 = H$, $R^4 = COCF_3$), which revert to the precursor (**77**) upon mild acid hydrolysis (Scheme 10).[25] With hydrogen chloride, acetyl perchlorate, or benzoyl perchlorate the nitriles (**77**) give the 1,3-oxazolium salts (**78**; $R^4 = H$, MeCO, PhCO).[26] With dimethyl acetylenedicarboxylate the derivatives (**76**; $R^3 = H$, $R^4 = COCF_3$) undergo 1,3-dipolar cycloaddition reactions, giving the corresponding pyrrole together with elimination of trifluoroacetyl isocyanate.[25]

The preparation of aldehydes from Reissert compounds by acid hydrolysis has been shown to involve a meso-ionic 1,3-oxazolium-5-aminide intermediate (Scheme 11).[27] For

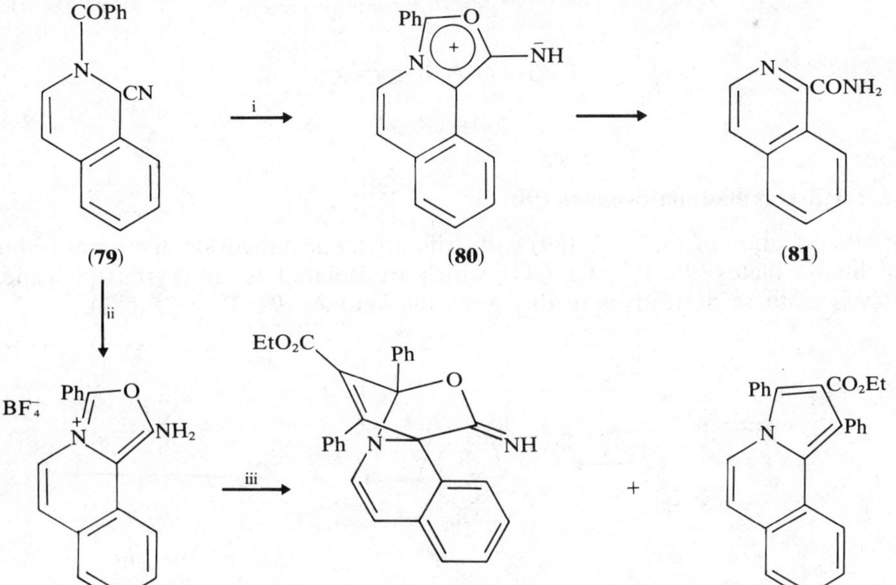

(79) (80) (81)

(82) (83) (84)

i, H^+, H_2O; ii, HBF_4; iii, $PhC{\equiv}CCO_2Et$.

SCHEME 11

example, the Reissert compound (**79**) with tetrafluoroboric acid gave the salt (**82**). Although the free meso-ionic intermediate (**80**) has not been isolated it can be trapped by acetylenic 1,3-dipolarophiles. Thus the salt (**82**) with ethyl phenylpropiolate gave the adduct (**83**) and pyrrolo[2,1-*a*]isoquinoline (**84**).[27] The latter product (**84**) can be formed directly from the adduct (**83**) by loss of isocyanic acid at 220 °C.

20.4.5 MESO-IONIC OXATHIOLES OF TYPE A

20.4.5.1 1,3-Oxathiolium-4-olates (86)

Meso-ionic 1,3-oxathiolium-4-olates (**86**) are formed by cyclodehydration of the acid derivatives (**85**) using acetic anhydride (Scheme 12). These compounds are too unstable to be isolated but they are trapped *in situ* by acetylenes, giving substituted furans (**88**) *via* the bicycloadducts (**87**).[28]

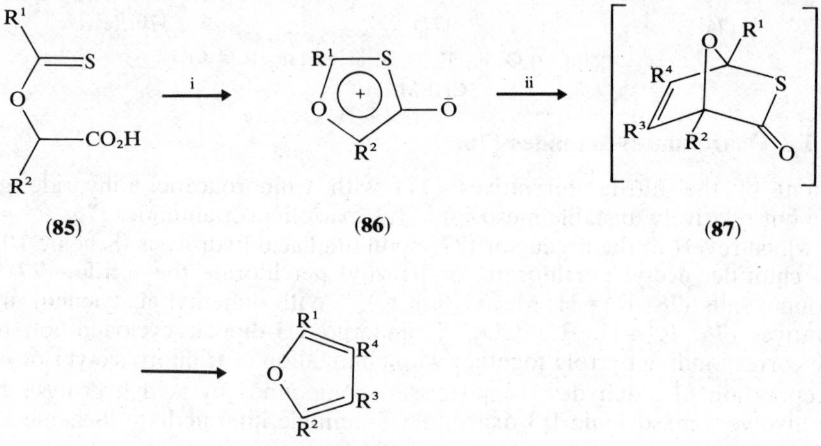

(**85**) (**86**) (**87**) (**88**)

i, $(CH_3CO)_2O$; ii, $R^3C{\equiv}CR^4$.

SCHEME 12

20.4.5.2 1,3-Oxathiolium-5-olates (90)

Cyclodehydration of the acids (**89**) with trifluoroacetic anhydride gives meso-ionic 1,3-oxathiolium-5-olates (**90**; $R^2 = CF_3CO$), which are isolated as red crystals (Scheme 13).[29] Hydrolysis of these derivatives readily gives the ketones (**91**; $R^2 = CF_3CO$).

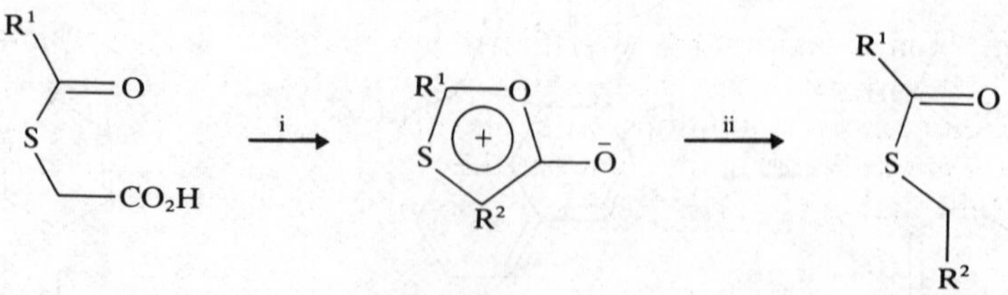

(**89**) (**90**) (**91**)

i, $(CF_3CO)_2O$; ii, H_2O.

SCHEME 13

20.4.6 MESO-IONIC DIAZOLES OF TYPE A

20.4.6.1 1,3-Diazolium-4-olates (93)

Methods of preparing meso-ionic 1,3-diazolium-4-olates (**93**) are shown in Scheme 14. Good general routes are the cyclization of the amino-amides (**92**) using triethyl orthoformate or *N*-phenylbenzimidoyl chloride,[30] or anhydro-acylation of *N*-(*N*-phenylbenzimidoyl)amino-acids (**94**) to (**93**; $R^4 = COR$).[31] Other preparations include: (i) 1,3-dipolar cycloaddition of phenyl isocyanate to meso-ionic 1,3-oxazolium-5-olates (**95**; $R^2 = R^4 = Ph$, $R^3 = Me$),[32] and (ii) treatment of benzoyl formic acid anil (**96**; $R^1 = R^2 = Ph$) with trifluoroacetic anhydride and pyridine followed by hydrolysis;[33] the intermediate trifluoroacetate salt can be isolated prior to hydrolysis.

(92) (94) (93) (95) (96)

i, EtO_3CH or PhC(Cl)=NPh; ii, $(R^4CO)_2O$; iii, PhNCO; iv, $(CF_3CO)_2O$, py; v, H_2O.

SCHEME 14

The 1,3-diazolium-4-olates (**93**) are surprisingly stable, being apparently unreactive towards acid, alkali, or amines. The tetraphenyl derivative (**93**; $R^1 = R^2 = R^3 = R^4 = Ph$) undergoes 1,3-dipolar cycloadditions with alkynes to give the anticipated pyrroles.[33] With dimethyl fumarate or dimethyl maleate the bicyclic 1,3-dipolar cycloadducts (**98**) are not isolated but undergo elimination of methanol, giving the bicyclo-product (**99**) (Scheme 15).[33]

20.4.6.2 1,3-Diazolium-4-aminides (102)

Representatives of this system are prepared by reaction of the nitrile (**100**; $R^1 = R^2 =$ Ph, $R^3 = Me$) with acid chlorides and treatment of the intermediate salts (**101**; X = Cl)

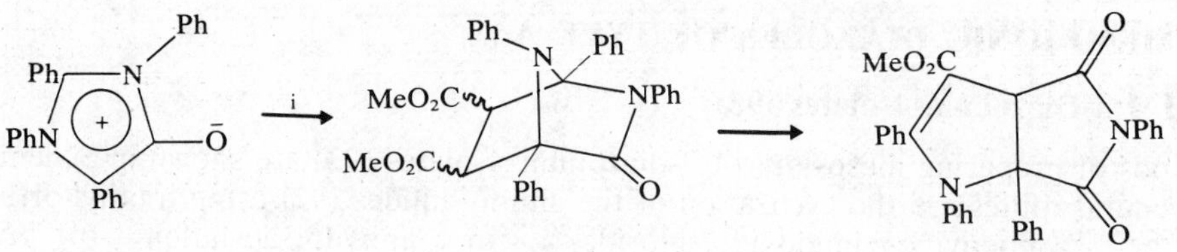

(97) (98) (99)

i, $MeO_2CCH{=}CHCO_2Me$ (*cis* or *trans*).

SCHEME 15

with aqueous sodium hydrogen carbonate (Scheme 16).[34] These heterocycles are yellow crystalline solids. Treatment of amidines ($R^3N{=}CR^2{-}NHR^1$) with halogenoacetonitriles (XCH_2CN) gives the diazolium salts (**101**; $R^4=H$), but for these derivatives reaction with base does not give the free meso-ionic system (**102**; $R^4=H$);[35] the bromide (**101**; $R^1=R^2=R^3=Ph$, $R^4=H$, $X=Br$) with sodium hydroxide gives the nitrile (**100**; $R^1=R^2=R^3=Ph$). The 4-amino-1,3-diazolium salts (**101**; $R^4=H$) with acetic anhydride give the corresponding *N*-acetyl derivatives (**101**; $R^4=COMe$).[35]

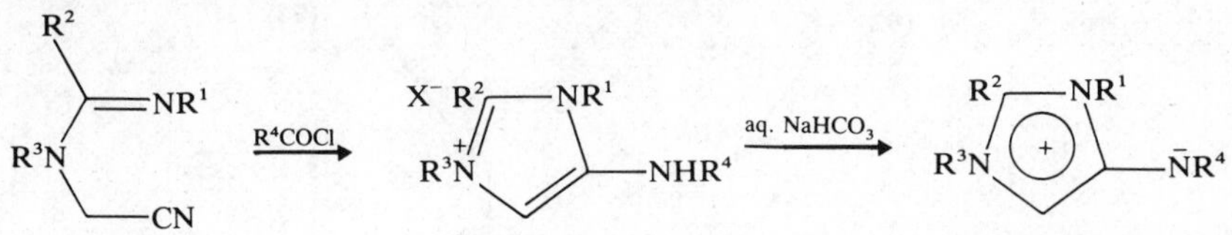

(100) (101) (102)

SCHEME 16

With dimethyl acetylenedicarboxylate the 1,3-diazolium-4-aminide (**104**) rapidly gives the pyrrole (**103**) (Scheme 17). Using diethyl azodicarboxylate the bicycloadduct (**105**) is obtained: a similar 1 : 1 adduct is formed with dimethyl maleate.[34]

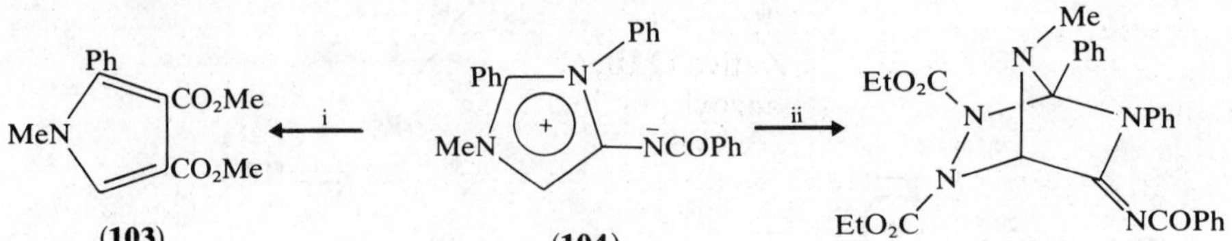

(103) (104) (105)

i, $MeO_2CC{\equiv}CCO_2Me$; ii, $EtO_2CN{=}NCO_2Et$.

SCHEME 17

20.4.6.3 1,3-Diazolium-4-thiolates (107)

This system is prepared by 1,3-dipolar cycloaddition of phenyl isothiocyanate to meso-ionic 1,3-oxazolium-5-olates (**106**; $X=O$)[36] or 1,3-thiazolium-5-olates (**106**; $X=S$).[37] A representative (**107**; $R^1=R^3=Me$, $R^2=R^4=Ph$) has also been prepared (hy) by treatment of the thiazolium methiodide (**108**) with methylamine (Scheme 18). The stable meso-ionic systems (**107**) form salts with methyl iodide.

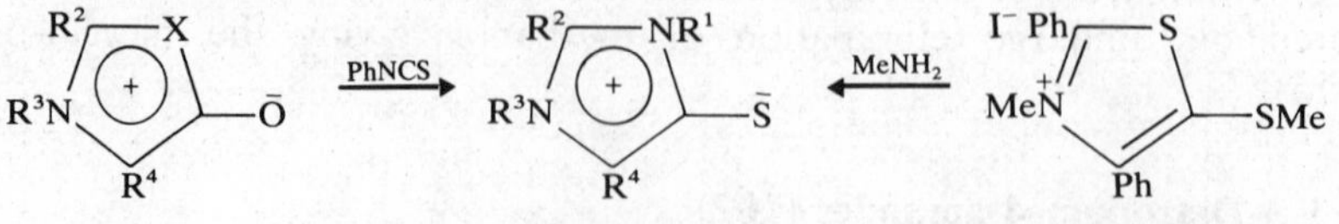

(106) (107) (108)

SCHEME 18

20.4.7 MESO-IONIC THIAZOLES OF TYPE A

20.4.7.1 1,3-Thiazolium-4-olates (110)

Cyclodehydration of the acids (**109**) using acetic anhydride–triethylamine at room temperature gives meso-ionic 1,3-thiazolium-4-olates (**110**) (my) (Scheme 19). A similar sequence using (2-pyridylthio)acetic acid gives bicyclic derivatives. Mild conditions are essential for the preparation of 5-unsubstituted derivatives (**110**; $R^3 = H$), otherwise further reaction of the meso-ionic product occurs.[38] An alternative preparative route involves treatment of 5-phenylrhodanines (*e.g.* **114**) with alkyl iodides and sodium ethoxide (Scheme 20) (hy).[39]

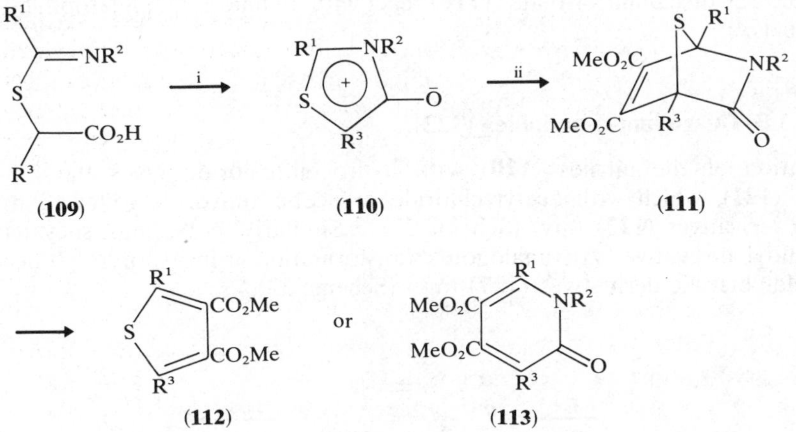

(**109**) (**110**) (**111**)

(**112**) or (**113**)

i, Ac_2O, Et_3N; ii, $MeO_2CC{\equiv}CCO_2Me$.

SCHEME 19

The meso-ionic compounds (**110**; $R^3 = H$) are sensitive to moisture. Sodium borohydride reduction of the triphenyl derivative (**110**; $R^1 = R^2 = R^3 = Ph$) gives the corresponding thiazolium-4-one, and acetyl or benzoyl perchlorate give thiazolium perchlorates. Upon irradiation the derivative (**115**) rearranges to the isomeric 1,3-thiazol-2-one (**116**) (ly) (Scheme 20).

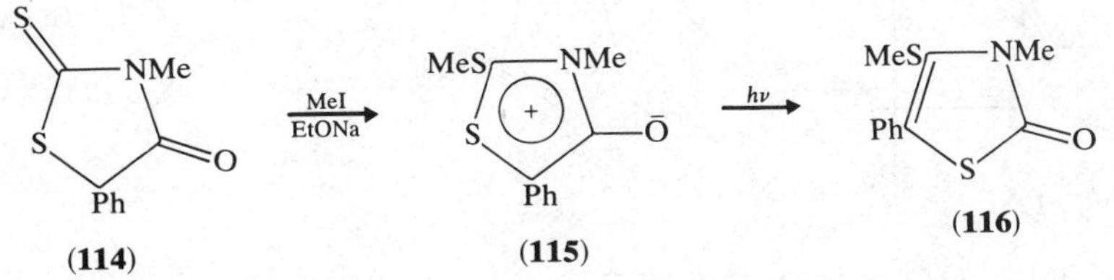

(**114**) (**115**) (**116**)

SCHEME 20

The mode of reaction of 1,3-thiazolium-4-olates (**110**) with acetylenes is dependent upon the substitution pattern. The triphenyl derivative (**110**; $R^1 = R^2 = R^3 = Ph$) with dimethyl acetylenedicarboxylate undergoes a conventional 1,3-dipolar cycloaddition with elimination of phenyl isocyanate, giving the thiophen (**112**; $R^1 = R^3 = Ph$) (hy) (Scheme 19). In contrast, the 5-unsubstituted derivate (**110**; $R^1 = R^2 = Ph$, $R^3 = H$) with dimethyl acetylenedicarboxylate gives the pyridone (**113**; $R^1 = R^2 = Ph$, $R^3 = H$) (hy) (Scheme 19) by elimination of sulphur from the bicyclic intermediate (**111**).[38] Using benzyne as 1,3-dipolarophile the adduct (**118**) is isolated (hy). Thermal decomposition of

(117) (118) (119)

SCHEME 21

compound (**118**) gives diphenylbenzo[*c*]thiophen (**117**) (my) (Scheme 21), whereas photochemical fragmentation gives 1,2,4-triphenyl-3-isoquinolone (**119**) (my) (Scheme 21).[40] Meso-ionic 1,3-thiazolium-4-olates (**110**) react with olefinic 1,3-dipolarophiles, giving 1 : 1 cycloadducts.[41]

20.4.7.2 1,3-Thiazolium-4-aminides (122)

Cyclization of the nitriles (**120**) with hydrogen chloride gives the 1,3-thiazolium chlorides (**121**), which with acetyl chloride or acetic anhydride followed by base give *N*-acetyl derivatives (**122**) (my) (Scheme 22).[42] Similarly, potassium isocyanate gives an *N*-carbamoyl derivative. An analogous transformation using (2-pyridylthio)acetonitrile (**126**) yields bicyclic derivatives (**127**) (my) (Scheme 23).[43]

(120) (121) (122)

(123) (124) (125)

i, HCl; ii, AcCl or Ac_2O; iii, $NaHCO_3$; iv, $h\nu$; v, $PhNH_2$; vi, $MeO_2CC{\equiv}CCO_2Me$.

SCHEME 22

The diphenyl-1,3-thiazolium-4-aminide (**122**; $R^1 = R^2 = Ph$, $R^3 = H$) undergoes a 1,3-dipolar cycloaddition reaction with dimethyl acetylenedicarboxylate to give the thiophen (**125**; $R^1 = Ph$, $R^3 = H$) and *N*-benzoyl-*N'*-phenylcarbodi-imide (Scheme 22).[44]

Base treatment of the salts (**121**) does not yield unsubstituted meso-ionic derivatives. At room temperature the chlorides (**121**; $R^1 = Ph$ or p-$MeOC_6H_4$, $R^2 = Ph$, $R^3 = H$) with aniline give 4-anilino-1,3-thiazoles (**124**; $R^1 = Ph$ or p-$MeOC_6H_4$, $R^3 = H$) (my) (Scheme 22).[45] Photolysis of the chlorides (**121**) in aqueous solution yields the disulphides (**123**) (ly) (Scheme 22) and keto-amidines ($R^1COCHR^3C{=}NHNHR^2$) (hy).[46]

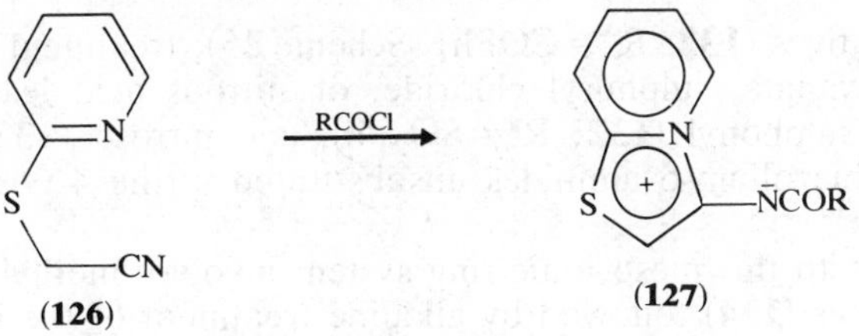

(126) → RCOCl → **(127)**

SCHEME 23

20.4.7.3 1,3-Thiazolium-5-olates (129)

The meso-ionic compounds (**129**; $R^3 = H$) are best prepared by treatment of *N*-thiobenzoylglycines (**128**) with acid anhydrides or with acid chlorides in pyridine at room temperature (hy) (Scheme 24).[47,48] At higher temperatures 4-acyl derivatives (**129**; $R^3 =$ COR) are obtained.[48] Similar electrophilic substitutions at the 4-position have led to bromo, iodo, mercurichloro, and mercuric acetosy-derivatives (**129**; $R^3 =$ Br, I, HgCl, or HgOCOMe).[48] An alternative preparative route to 1,3-thiazolium-5-olates (**129**) involves 1,3-dipolar cycloaddition (ly) of carbon oxysulphide to meso-ionic 1,3-oxazolium-5-olates (**63**) (Section 20.4.4.3).[36,49]

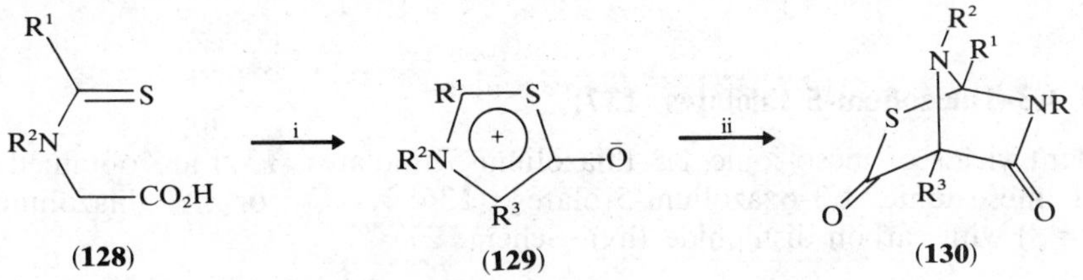

(128) → i → **(129)** → ii → **(130)**

i, $(RCO)_2O$ or RCOCl, py; ii, RNCO.

SCHEME 24

The meso-ionic 1,3-thiazolium-5-olates (**129**) are stable, pale yellow crystalline compounds. The 4-acyl derivatives (**129**; $R^3 =$ COR) are very stable, being unreactive towards 1,3-dipolarophiles, but the 4-unsubstituted derivatives (**129**; $R^3 = H$) undergo a wide range of cycloadditions. Acetylenic dipolarophiles yield pyrroles (hy); the meso-ionic precursor (**129**) can be conveniently generated *in situ*. Structurally interesting 1:1 adducts are formed with olefinic dipolarophiles or heterocumulenes. Notably, activated isocyanates (*i.e.* toluene-*p*-sulphonyl and benzoyl isocyanates) give the cycloadducts (**130**) at room temperature (hy).[14]

20.4.7.4 1,3-Thiazolium-5-aminides (132)

With benzoyl chloride followed by aqueous alkali the thiobenzamidoaminoacetonitriles (**131**; $R^1 = Ph$) yield *N*-benzoyl-1,3-thiazolium-5-aminides (**132**; $R^1 = Ph$, $R^4 = COPh$) (hy) (Scheme 25). Alternatively, the nitriles (**131**) with hydrogen chloride give 5-amino-1,3-thiazolium chlorides (**133**) (hy), which are similarly converted to the meso-ionic

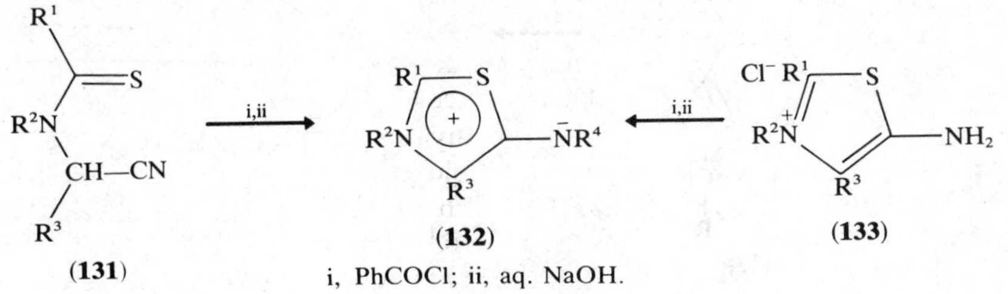

(131) → i,ii → **(132)** ← i,ii ← **(133)**

i, PhCOCl; ii, aq. NaOH.

SCHEME 25

N-benzoyl derivatives (**132**; $R^4 = COPh$) (Scheme 25); treatment of the derivatives (**133**) with phenyl isocyanate, sulphuryl chloride, or nitrous acid leads to benzamido (**132**; $R^4 = CONHPh$), sulphonyl (**132**; $R^4 = SO_2Ph$), and nitroso (**132**; $R^4 = NO$) derivatives. Meso-ionic 1,3-thiazolium-5-aminides unsubstituted in the 4-position (**132**; $R^3 = H$) can be brominated.[30,50]

A second route to this meso-ionic ring system involves methylation of 5-acylamino-4-thiazoline-2-thiones (**134**) followed by alkaline treatment of the intermediate iodide (hy) (Scheme 26).[51]

(**134**) → (**135**)

i, MeI; ii, aq. NaOH.

SCHEME 26

20.4.7.5 1,3-Thiazolium-5-thiolates (137)

Excellent yields of meso-ionic 1,3-thiazolium-5-thiolates (**137**) are obtained by treatment of meso-ionic 1,3-oxazolium-5-olates (**136**; X = O) or 1,3-thiazolium-5-olates (**136**; X = S) with carbon disulphide (hy) (Scheme 27).[36,52]

(**136**) → (**137**)

SCHEME 27

In a closely related reaction, tricyclic derivatives (**139**) are prepared by addition of carbon disulphide to the azomethine ylides (**138**) (Scheme 28); the structure of the *p*-bromophenyl derivative (**139**; R = *p*-BrC_6H_4) has been confirmed by X-ray crystallography.[53] The bicyclic derivative (**141**) has been prepared by reaction of *N*-thiobenzoylproline (**140**) with warm thioacetic acid (Scheme 29).[52]

The meso-ionic derivatives (**137**) are stable crystalline compounds which form salts with methyl iodide.

(**138**) → (**139**)

SCHEME 28

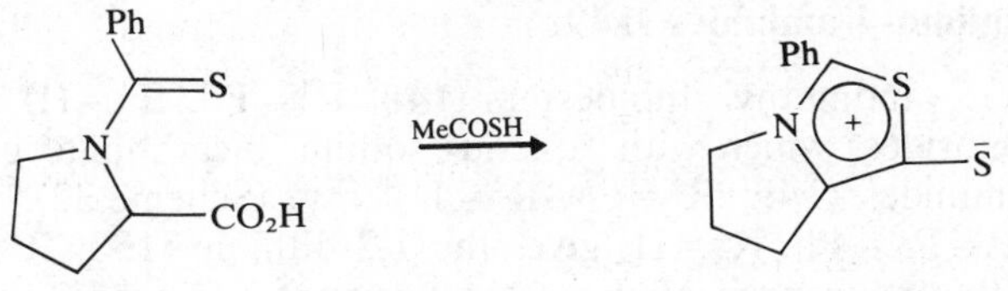

(140) (141)

SCHEME 29

20.4.8 MESO-IONIC SELENAZOLES OF TYPE A

20.4.8.1 1,3-Selenazolium-4-olates (143)

Cyclodehydration of the α-selenoacid (**142**; $R^1 = R^2 = R^3 = Ph$) using a mixture of acetic anhydride and triethylamine at room temperature gives the meso-ionic 1,3-selenazolium-4-olate (**143**; $R^1 = R^2 = R^3 = Ph$) (hy) (Scheme 30), which reacts slowly with dimethyl acetylenedicarboxylate to give the pyridone (**144**; $R^1 = R^2 = R^3 = Ph$) (my) and selenium (hy).[54]

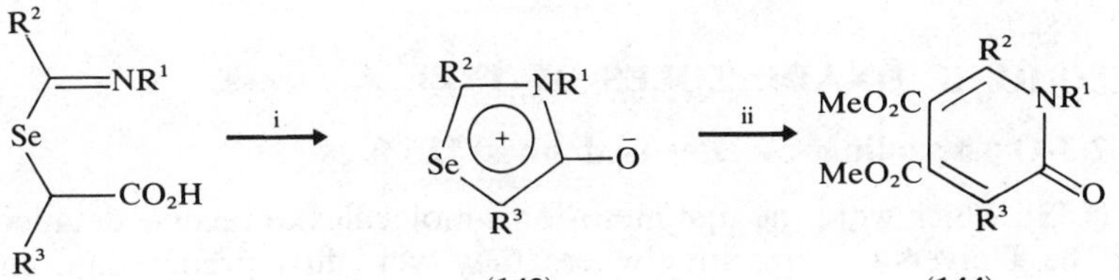

(142) (143) (144)

i, Ac_2O, Et_3N; ii, $MeO_2CC{\equiv}CCO_2Me$.

SCHEME 30

20.4.9 MESO-IONIC DITHIOLES OF TYPE A

20.4.9.1 1,3-Dithiolium-4-olates (146)

The highly coloured derivatives (**146**) are made by cyclodehydration of thioglycolic acid derivatives (**145**) using acetic anhydride–triethylamine at 0–10 °C (hy) (Scheme 31).[55,56] These molecules (**146**) participate in 1,3-dipolar cycloaddition reactions, giving thiophens with alkynes and the bicyclic 1:1 adducts (**147**) (Scheme 31) with alkenes.[56,57] Photolysis of the diphenyl derivative (**146**; $R^1 = R^2 = Ph$) gives tetraphenyldithiin.[58]

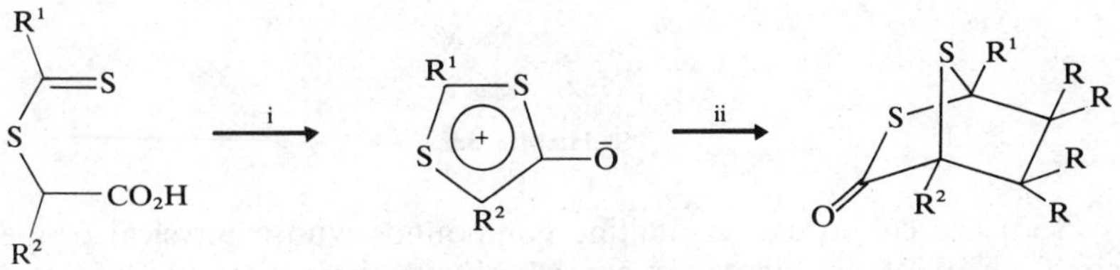

(145) (146) (147)

i, Ac_2O, Et_3N; ii, $R_2C{=}CR_2$.

SCHEME 31

20.4.9.2 1,3-Dithiolium-4-aminides (149)

Treatment of the cyanomethyl dithioesters (**148**; $R^1 = Ph$, $R^2 = H$) with acid chlorides gives dithiolium chlorides, which with aqueous sodium bicarbonate give red meso-ionic 1,3-dithiolium-4-aminides (**149**; $R^1 = Ph$, $R^2 = H$) (my) (Scheme 32).[59] Photolysis of the derivative (**149**; $R^1 = R^3 = Ph$, $R^2 = H$) gives the 1,2-dithiole (**150**; $R^1 = R^3 = Ph$, $R^2 = H$) (hy) (Scheme 32); the mechanism of this rearrangement has been discussed.[60]

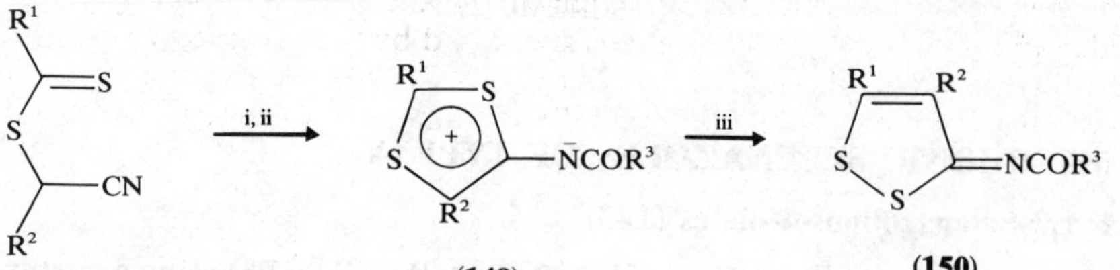

(148) (149) (150)

i, RCOCl; ii, aq. $NaHCO_3$; iii, $h\nu$.

SCHEME 32

20.4.10 MESO-IONIC OXADIAZOLES OF TYPE A

20.4.10.1 1,2,3-Oxadiazolium-5-olates (sydnones) (153)

Sydnones (**153**), which were the first meso-ionic molecules to receive detailed study, are named after the University of Sydney where they were first prepared. Comprehensive reviews of their chemistry have been published.[2,4,5,8]

In spite of extensive studies of the sydnone ring, only one preparative route is available; this method involves the cyclodehydration of *N*-nitroso-α-amino-acids (**151**) using acetic anhydride (Scheme 33) (my–hy). This route, which has not been significantly improved since it was discovered by Earl and Mackney in 1935, has yielded a wide variety of derivatives.[4] The preparation is successful for the unsubstituted derivatives (**153**; $R^2 = H$) but not for the derivatives (**153**; $R^1 = H$). In some cases the cyclization step is slow and the reaction takes place considerably more rapidly using trifluoroacetic anhydride as cyclodehydrating agent. The mechanism of the reaction (**151**) → (**153**) (Scheme 33) involves a mixed anhydride intermediate (**152**).[1]

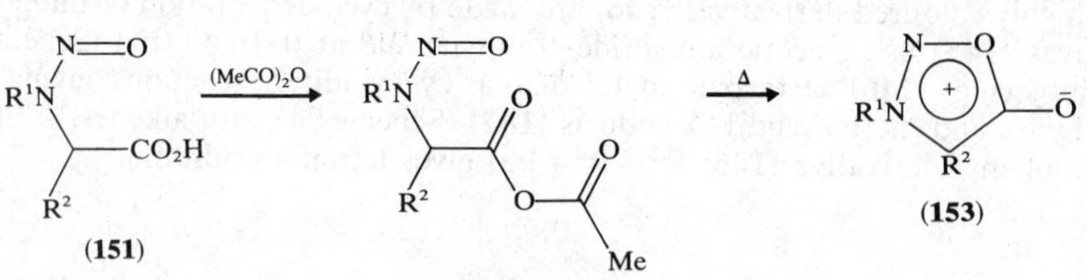

(151) (152) (153)

SCHEME 33

Sydnones (**153**) are colourless, crystalline compounds whose physical properties have been studied. Their dipole moments are consistent with their meso-ionic formulation (*e.g.* **153**; $R^1 = Ph$, $R^2 = H$, $\mu = 6.5$ D).[2] In the i.r. spectra the position of the carbonyl band varies over the region 1720–1770 cm^{-1}.[2,4,8] Mass spectra have been recorded; in unsubstituted derivatives (**153**; $R^2 = H$), progressive loss of NO, CO, and HCN is observed.[61] The crystal structures of two sydnones have been determined; the geometry is essentially the same in both molecules and consistent with the planar oxadiazolium structure.[9,62] ESCA spectra of *N*-phenyl- and *N*-methyl-sydnones (**153**; R^1 = Ph or Me, R^2 = H) demonstrate considerable difference between the formal charges on the two nitrogen atoms.[9] Interest in the bonding of sydnones (**153**) has led to a large number of theoretical studies, which have been discussed in a review.[9]

Acid hydrolysis yields alkyl- or aryl-hydrazines (my), a carboxylic acid, and carbon dioxide (Scheme 34).[2] This is a useful route to hydrazines which are inaccessible using other methods (*e.g.* 1-hydrazinoadamantane).[63] A mechanism involving an intermediate hydrazide (**154**) has been proposed and is supported by the observation that *N*-formyl-*N*-phenylhydrazine (**154**; R^1 = Ph, R^2 = H) can be isolated in the acid hydrolysis of *N*-phenylsydnone (**153**; R^1 = Ph, R^2 = H).[2,5] Base-catalysed hydrolysis leads to regeneration of *N*-nitroso-α-amino-acids (**151**) (hy).[2]

(153) —H^+→ —H_2O, −CO_2→ —→ (154) —H_2O→ R^1NHNH_2 + R^2CO_2H

SCHEME 34

N.m.r. studies of *N*-phenylsydnone (**153**; R^1 = Ph, R^2 = H) in 'super acid' solution (FSO_3–SbF_5) show that protonation occurs on the exocyclic oxygen atom.[64] Treatment of *N*-phenylsydnone with triethyloxonium tetrafluoroborate results in *O*-alkylation, giving the salt (**155**).[65]

(155) [BF_4^-, PhN, OEt]

The photochemical transformations of sydnones have been discussed,[9] but the scope and mechanism of these reactions is not yet clear. Irradiation of *N*-phenylsydnone (**153**; R^1 = Ph, R^2 = H) gives 4-phenyl-Δ^2-1,3,4-oxadiazolium-5-one (**157**; R^1 = Ph, R^2 = H);[66] the mechanism may involve a 1*H*-diazirine intermediate (**156**) (Scheme 35).

(153) —→ —−CO_2→ (156) —→ $R^1\bar{N}$–$\overset{+}{N}{\equiv}CR^2$ —CO_2→ (157)

SCHEME 35

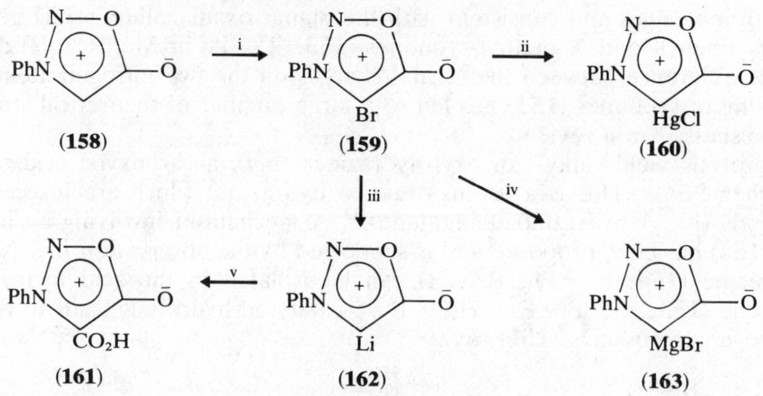

i, Br_2; ii, $HgCl_2$, NaOAc; iii, Bu^nLi; iv, Mg; v, CO_2.

SCHEME 36

The sydnones can be regarded as aromatic in that 4-unsubstituted derivatives (**153**; $R^2=H$) undergo electrophilic substitution. Thus *N*-phenylsydnone (**158**) with bromine or *N*-bromosuccinimide gives the 4-bromo-derivative (**159**) (hy).[2,4,8] Similarly, a hydrogen atom in the 4-position can be replaced by chloro, iodo, nitro, sulphonyl, deutero, acyl, and formyl groups.[4,8]

Organometallic derivatives of sydnones are known (*e.g.* Scheme 36). Treatment of 3-phenyl-4-bromosydnone (**159**) with mercuric chloride and sodium acetate gives the 4-chloromercuri-derivative (**160**).[67] 4-Lithiosydnones (*e.g.* **162**) are made by reaction of either the unsubstituted sydnone[68] or the bromo-derivative[69] with n-butyl-lithium. These derivatives (*e.g.* **162**) are useful intermediates for the preparation of other sydnone molecules. Reaction with carbon dioxide gives the 4-carboxylic acid (*e.g.* **161**), which can be converted to the ester, acid chloride, or amide. In contrast to the lithium derivatives the Grignard reagents (*e.g.* **163**),[70] prepared in the usual way, do not react with carbon dioxide. The reaction of 3-phenyl-4-lithiosydnone (**162**) with a number of acid chlorides and non-metallic chlorides has been reported; reaction with ketones gives the anticipated tertiary alcohols.

1,3-Dipolar cycloaddition reactions of sydnones have received detailed attention.[7,8] Reaction with acetylene dipolarophiles gives pyrazoles (hy) and carbon dioxide. With methyl acetylenecarboxylate in xylene at 120 °C, 4-methyl-3-phenylsydnone (**164**) gives a

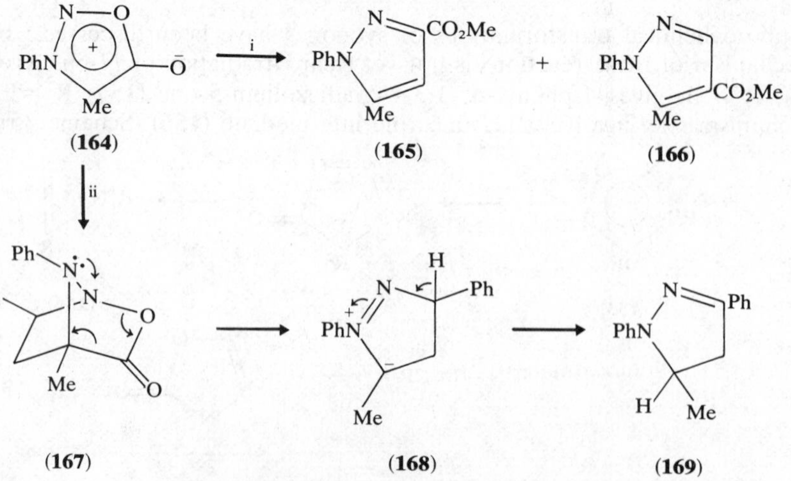

i, HC≡CCO$_2$Me; ii, PhCH=CH$_2$.

SCHEME 37

mixture of the isomeric pyrazoles (**165**) (55%) and (**166**) (29%) (Scheme 37);[7] symmetrical alkynes give a single product. This reaction provides a useful route to pyrazoles and, conveniently, the sydnone can be generated *in situ*.[71]

With olefinic dipolarophiles sydnones give 2-pyrazolines (hy). Thus the 4-methyl-3-phenyl derivative (**164**) gives 5-methyl-1,3-diphenyl-2-pyrazoline (**169**) (Scheme 37) when reacted with styrene at 135 °C.[7] The mechanism of the transformation (**164**) → (**169**) involves the bicyclic adduct (**167**) and its decomposition product (**168**) (Scheme 37). Other 1,3-dipolarophiles including isocyanates, 1,4-quinones, benzyne, and nitriles have been used.[8]

20.4.10.2 1,2,3-Oxadiazolium-5-aminides (sydnone imines) (170)

Review articles covering the chemistry of the sydnone imines (**170**) and their salts (**171**) have been published.[2,4,5,8]

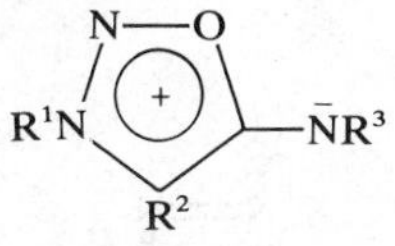

(**170**) 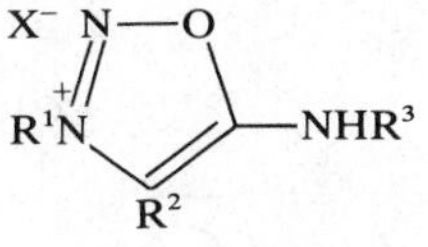 (**171**) 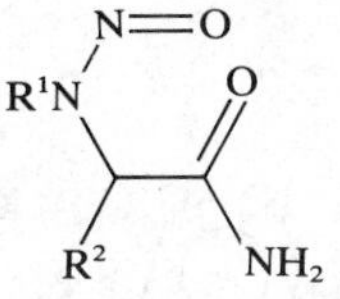 (**172**)

Treatment of *N*-nitroso-α-aminoacetonitriles (**173**) with acid gives 5-amino-1,2,3-oxadiazolium salts (**174**) (Scheme 38) (hy).[72] Attempts to convert these salts (**174**) to the free meso-ionic 1,2,3-oxadiazolium-5-aminides (**170**; $R^3 = H$) were not successful; treatment with sodium bicarbonate in ether solution regenerates the nitrile (**173**), whereas cold aqueous alkali gives the amides (**172**).[72]

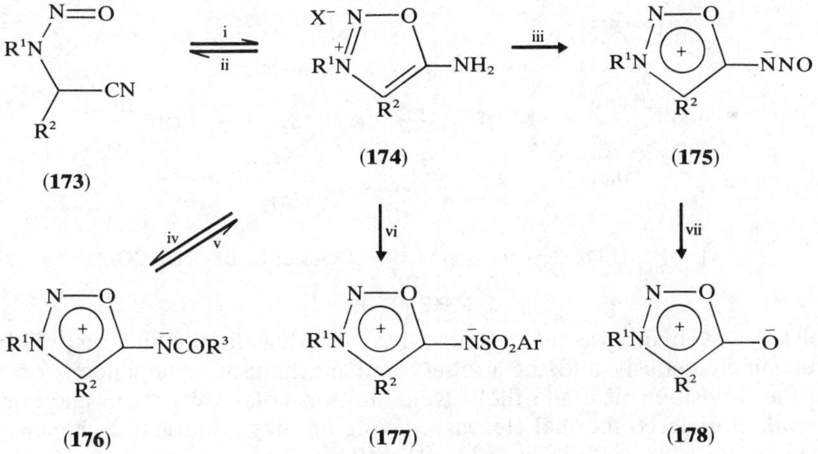

(**173**) (**174**) (**175**) (**176**) (**177**) (**178**)

i, mineral acid (HX); ii, aq. $NaHCO_3$, Et_2O; iii, HONO; iv, R^3COCl, py; v, H^+, H_2O; vi, $ArSO_2Cl$, py; vii, Δ.

SCHEME 38

Stable sydnone imines (**170**) are obtained by substitution of the exocyclic nitrogen atom. With acetyl chloride and pyridine, 5-amino-1,2,3-oxadiazolium chlorides (**174**; X = Cl) give the free *N*-acetylsydnone imines (**176**; $R^3 = Me$) (my).[73] Similarly, *N*-benzoyl derivatives (**176**; $R^3 = Ph$) are formed and benzenesulphonyl chloride gives the *N*-benzenesulphonyl derivatives (**177**; Ar = Ph) (my).[73] Other stable systems include *N*-carbomethoxy (**170**; $R^3 = CO_2Me$), *N*-phenylcarbamyl (**170**; $R^3 = CONHPh$), and *N*-carbamyl (**170**; $R^3 = CONH_2$).[8] The 1,2,3-oxadiazolium nitrates (**174**; $X = NO_3$) are dehydrated by acetic anhydride or concentrated sulphuric acid to give *N*-nitro-derivatives (**170**; $R^3 = NO_2$).[72] Nitrosation of the chlorides (**174**; X = Cl) gives *N*-nitroso-derivatives (**175**) (Scheme 38).[74]

The *N*-acylsydnone imines are stable crystalline compounds whose structures are supported by their spectroscopic properties.[4,8] They form salts with acids, protonation taking place at the exocyclic nitrogen atom; with alkyl halides quaternization occurs. Hot aqueous mineral acids hydrolyse the *N*-acyl derivatives to 5-amino-1,2,3-oxadiazolium salts, (**176**) → (**174**) (Scheme 38).[73,75] With aqueous alkali the *N*-acyl derivatives usually form open-chain amides (**172**) or the corresponding acid. However, *N*-acetyl-3-phenylsydnone imine (**179**) undergoes a skeletal rearrangement to 1-phenyl-4-hydroxy-1,2,3-triazole (**180**) in aqueous alkali (Scheme 39).[76]

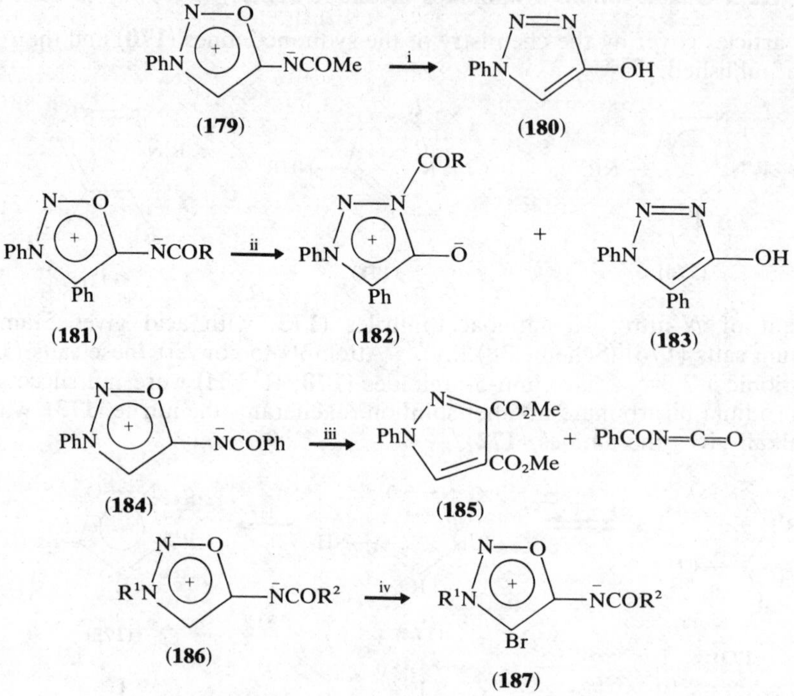

(**179**) → i → (**180**)

(**181**) → ii → (**182**) + (**183**)

(**184**) → iii → (**185**) + PhCON=C=O

(**186**) → iv → (**187**)

i, OH^-, H_2O; ii, $h\nu$, iii, $MeO_2CC{\equiv}CCO_2Me$; iv, Br_2, $NaHCO_3$.

SCHEME 39

Thermolysis and hydrolysis of 5-amino-1,2,3-oxadiazolium chlorides (**174**; X = Cl) often occur simultaneously and the product and mechanism is dependent on the substituent in the 3-position (R^1) and the solvent. In non-polar solvents, 3-alkyl derivatives (**174**; R^1 = alkyl) undergo thermal cleavage, giving an alkyl chloride, 2-chloroacetamide, and *N*-alkyl-*N*-nitrosoglycinonitrile (**173**; R^1 = alkyl). The same products are formed in aqueous solutions containing a high concentration of chloride ion, where dissociation of the ion pair (**174**; X = Cl) is suppressed. In aqueous solution hydrolytic cleavage gives the alkyl alcohol and 2-hydroxyacetamide.[77] The 3-benzyl derivative (**174**; $R^1 = PhCH_2$, R^2 = H, X = Cl) gives similar thermolysis products, but the mechanism has been shown to differ from that of the simple alkyl derivatives.[77,78] Hydrolysis of 3-aryl-5-amino-1,2,3-oxadiazolium chlorides (**174**; R^1 = Ar, X = Cl) with boiling 6N hydrochloric acid gives the corresponding phenol and chloroacetic acid.[76] Pyrolysis of *N*-nitrososydnone imines (**175**) in DMF gives the sydnones (**178**) (Scheme 38) (hy), with elimination of nitrogen.[74]

Irradiation of the 3,4-diphenyl-*N*-acylsydnone imines (**181**; R = Me, Ph) gives the isomeric meso-ionic 1,2,3-triazolium-5-olates (**182**; R = Me, Ph) (Scheme 39) (ly) and 3,4-diphenyl-5-hydroxy-1,2,3-triazole (**183**) (my).[79]

N-Acylsydnone imines (**176**) participate in 1,3-dipolar cycloaddition reactions, but little attention has been paid to this aspect of their chemistry. *N*-Benzoyl-3-phenylsydnone imine (**184**) with dimethyl acetylenedicarboxylate gives 1-phenylpyrazole-3,4-dicarboxylic acid dimethyl ester (**185**) and benzoyl isocyanate (Scheme 39).[44]

Treatment of 4-unsubstituted *N*-acetyl and *N*-carbomethoxysydnone imines (**186**; $R^2 = Me$ or OMe) with bromine in ether, in the presence of sodium bicarbonate, gives the 4-bromo-derivatives (**187**) (Scheme 39).[80]

20.4.10.3 1,3,4-Oxadiazolium-2-olates (isosydnones) (189)

This meso-ionic system is prepared by condensation of *N*-acylhydrazines (**188**),[81] or their hydrochlorides,[82] with phosgene (Scheme 40) (my). Bicyclic derivatives are similarly obtained from *N*-aminopyrid-2-ones.[83] Spectroscopic properties of isosydnones have been recorded:[82] the i.r. spectra show a carbonyl absorption at $1760\,cm^{-1}$; mass spectra are characterized by a molecular ion and the fragment ion $R^1—C{\equiv}O^+$. Dipole moments (*e.g.* **189**; $R^1 = R^2 = Ph$, $\mu = 7.8$ D) have been measured.[82]

i, $COCl_2$; ii, H_2S, Et_3N; iii, ROH; iv, RNH_2; v, $PhC{\equiv}CCO_2Et$

SCHEME 40

Hydrolysis gives *N*-acylhydrazines (**188**) and the urea derivatives $(R^1COR^2NNH)_2CO$.[81] In mechanistically similar reactions, alcoholysis gives urethanes (**191**) and aminolysis gives semicarbazides (**192**) (Scheme 40).[81] With hydrogen sulphide and tertiary base, isosydnones (**189**) give *N*-thioacylhydrazines (**190**) (my), which often cannot be obtained by alternative routes (*e.g.* **190**; $R^1 = R^2 = Ph$).[84]

Like sydnones (**153**), the isosydnones (**189**) participate in 1,3-dipolar cycloaddition reactions (*e.g.* **189** → **193**) (Scheme 40), but the rate of addition is relatively slow and these reactions are not of synthetic importance.[84]

20.4.10.4 1,3,4-Oxadiazolium-2-aminides (195)

N-Acylhydrazines (**188**) condense with aryl isocyanide dichlorides ($ArN{=}CCl_2$), giving 1,3,4-oxadiazolium chlorides (**194**) (ly) which are converted to the free meso-ionic system (**195**) (ly) by treatment with diazomethane (Scheme 41).[85] A similar procedure using acyl isocyanide dichlorides yields *N*-acyl derivatives (**195**; $R^3 = SO_2R$ or COR) (my–hy).[86]

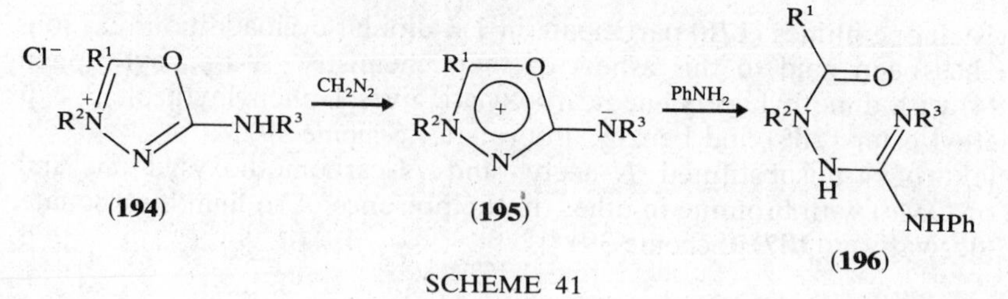

SCHEME 41

The yellow crystalline N-aryl derivatives (**195**; $R^3 = Ar$) regenerate the salts (**194**) when treated with hydrogen chloride.[85] The derivative (**195**; $R^1 = R^3 = Ph$, $R^2 = Me$) with aniline gives the guanidine (**196**; $R^1 = R^3 = Ph$, $R^2 = Me$) (Scheme 41).[85] Reduction of the derivative (**195**; $R^1 = p\text{-}Me_2NC_6H_4$, $R^2 = Ph$, $R^3 = SO_2Ph$) with Raney nickel gives p-dimethylaminobenzanilide ($p\text{-}Me_2NC_6H_4CONHPh$) (hy), whereas compound (**195**; $R^1 = p\text{-}Me_2NC_6H_4$, $R^2 = Me$, $R^3 = SO_2C_6H_4Me\text{-}p$) with sodium borohydride gives 1-methyl-1-(p-dimethylaminobenzyl)-4-tosyl semicarbazide ($p\text{-}Me_2NC_6H_4CH_2NMeNHCO\text{-}NHSO_2C_6H_4Me\text{-}p$) (hy).[86]

20.4.10.5 1,3,4-Oxadiazolium-2-thiolates (198)

Three general methods for preparing derivatives of this meso-ionic system from N-acylhydrazines (**197**) are available. Reaction with thiophosgene in the presence of potassium carbonate or with carbon disulphide and diethylcarbodi-imide gives good yields (55–95%) of the meso-ionic 1,3,4-oxadiazolium-2-thiolates (**198**).[87] Alternatively, the N-acylhydrazines (**197**) can be converted to the dithiocarbamic acid salts (**199**), which with phosphoryl chloride and triethylamine give the meso-ionic derivatives (**198**) (Scheme 42) (ly).[88]

These yellow crystalline compounds form salts with methyl iodide and triethyloxonium tetrafluoroborate. With secondary amines, ring opening to thiosemicarbazides (**200**) occurs, but primary amines give meso-ionic 1,2,4-triazolium-2-thiolates (**201**), presumably

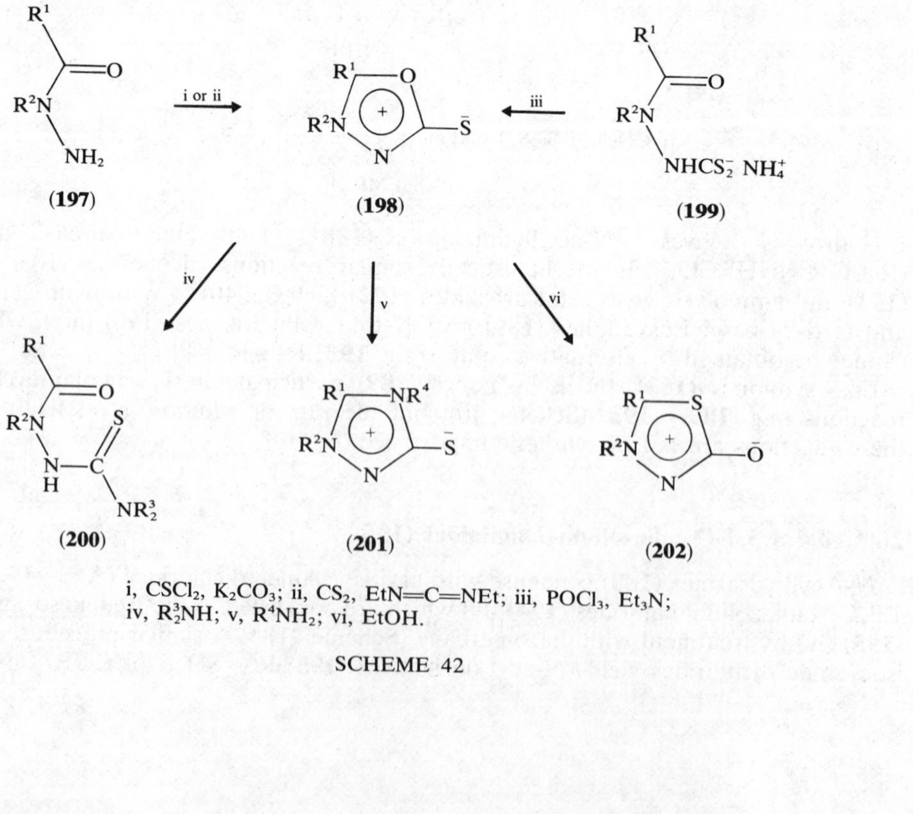

i, $CSCl_2$, K_2CO_3; ii, CS_2, EtN=C=NEt; iii, $POCl_3$, Et_3N; iv, R^3_2NH; v, R^4NH_2; vi, EtOH.

SCHEME 42

via an intermediate thiosemicarbazide.[84,87] In hot ethanol, rearrangement to isomeric meso-ionic 1,3,4-thiadiazolium-2-olates (**202**) (Scheme 42) has been demonstrated.[84]

20.4.10.6 1,3,4-Oxadiazolium-2-methylides (205)

The synthesis of derivatives of this system (**205**; X = CN or CO_2Me) (hy) has been achieved by reaction of *N*-acylhydrazines (**203**) with 3,3-dichloroacrylonitriles (**204**; X = CN or CO_2Me) in the presence of base (Scheme 43). The products (**205**) are greenish-yellow crystalline compounds.[89]

R¹C(=O)N(R²)NH₂ (**203**) + $Cl_2C{=}C(X)CN$ (**204**) —[K_2CO_3 or Et_3N]→ (**205**)

SCHEME 43

20.4.11 MESO-IONIC OXATHIAZOLES OF TYPE A

20.4.11.1 1,3,2-Oxathiazolium-5-olates (207)

Aryl derivatives of meso-ionic 1,3,2-oxathiazolium-5-olates (**207**; R^1 = Ar) are prepared (hy) by treatment of α-mercaptocarboxylic acids (**206**) with ethyl nitrite, dicyclohexylcarbodi-imide, and a catalytic amount of sulphuric acid (Scheme 44).[90] An intermediate *S*-nitroso-acid ($ONSCHArCO_2H$) is presumably formed, but not isolated. The dipole moment ($\mu = 4.5$ D), i.r. spectrum ($\nu_{CO} = 1700\ \text{cm}^{-1}$), and X-ray crystal structure of the phenyl derivative (**207**; R^1 = Ph) support the meso-ionic structure.[90,91]

Treatment with ethanolic ammonia is reported to yield the acyclic sulphonamide derivatives (**209**) and (**210**) (Scheme 44); excess of alkylamines containing a trace of water give the salts (**211**).[92]

(**206**) —i→ (**207**) —ii→ (**208**)

(**207**) —iii→ $H_2NSO_2CH_2R^1$ (**209**) + $H_2NSO_2CH(R^1)CO_2H$ (**210**)

(**207**) —iv→ $R^2N{=}S(NH{-}O)(CH_2R^1)$ $CO_2^-\ H_3\overset{+}{N}R^2$ (**211**)

i, EtONO, $C_6H_{11}N{=}C{=}NC_6H_{11}$, H^+; ii, $R^2C{\equiv}CR^3$; iii, NH_4OH, EtOH; iv, R^2NH_2, H_2O.

SCHEME 44

1,3,2-Oxathiazolium-5-olates (**207**) are associated with a thiocarbonyl imine 1,3-dipolar fragment ($R\bar{C}H{-}\overset{+}{S}{=}NH$) and accordingly they participate in cycloaddition reactions with subsequent elimination of carbon dioxide. Reaction with alkynes provides a convenient synthesis of isothiazoles (**208**) (Scheme 44) (my). Typically, dimethyl acetylenedicarboxylate and the phenyl derivative (**207**; $R^1 = Ph$) give compound (**208**; $R^1 = Ph$, $R^2 = R^3 = CO_2Me$) (56%).[90]

Photolysis of 4-phenyl-1,3,2-oxathiazolium-5-olate (**212**) (Scheme 45) has been studied. Irradiation at 405 nm in ether–dichloromethane solution gives benzonitrile, sulphur, and carbon dioxide (hy). The mechanistic sequence (**212**) → (**216**) → (**217**) → (**218**) → (**219**) has been postulated.[90] In ethanol, photolysis gives ethyl phenylglyoxylate (**215**) (65%); spectroscopic evidence for the intermediate (**213**) has been presented.[93]

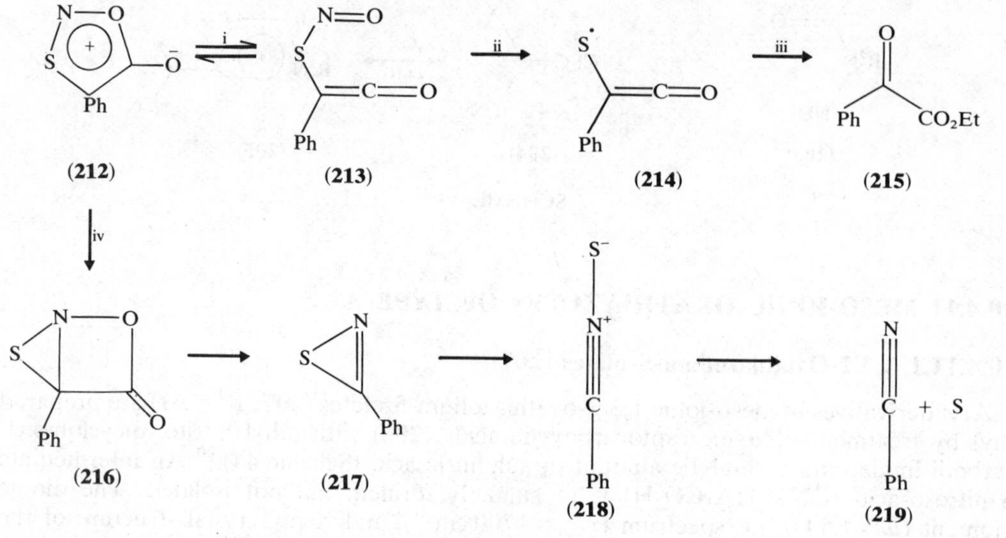

i, $h\nu$ ($\lambda > 300$ nm); ii, $h\nu$; iii, EtOH, O_2; iv, $h\nu$ (λ405 nm).

SCHEME 45

20.4.12 MESO-IONIC TRIAZOLES OF TYPE A

20.4.12.1 1,2,3-Triazolium-4-olates (224)

A number of methods for preparing derivatives of this meso-ionic system are available.[9] The most generally useful routes are shown in Scheme 46. Cyclization of arylazoaminoacetates (**220**) using thionyl chloride and pyridine gives the meso-ionic triazoles (**224**; $R^2 = H$) (ly) directly.[94] Alternatively, methylation of 4-hydroxy-1,2,3-triazoles (**223**) with diazomethane gives a mixture of the meso-ionic derivative (**224**) (ly–my) and the isomeric systems (**226**) (my) and (**228**) (ly).[95] Conveniently the 4-methoxy-1,2,3-triazoles (**226**) are converted to the meso-ionic system (**224**) (hy) by treatment with methyl iodide. This transformation (**226**) → (**224**) (Scheme 46) proceeds *via* 1,2,3-triazolium iodides (**227**), which are not isolated.[95] Two methods of converting 1,2,3-triazolium tosylates (**222**) to meso-ionic 1,2,3-triazolium-4-olates (**224**) have been reported. Treatment with *N*-bromoacetamide gives the bromo-derivatives (**221**), which with aqueous alkali give the meso-ionic products (**224**) (ly).[96] In another sequence the tosylate (**222**; $R^1 = Me$) with sodium hydride gives the ylide (**225**; $R^1 = Me$), which with oxygen and cupric chloride catalyst gives the meso-ionic 1,2,3-triazolium-4-olate (**224**; $R^1 = Me$, $R^2 = H$).[97]

The colourless crystalline 1,2,3-triazolium-4-olates (**224**) are stable compounds showing a carbonyl absorption at 1650 cm^{-1}. With hydrogen chloride they give triazolium

(220) (221) (222) (223) (224) (225) (226) (227) (228)

i, $SOCl_2$, py; ii, aq. NaOH; iii, MeCONHBr; iv, NaH; v, CH_2N_2; vi, O_2, $CuCl_2$; vii, MeI.

SCHEME 46

chlorides (**233**; R = H, X = Cl) and with triethyloxonium tetrafluoroborate *O*-alkylation occurs, giving the salts (**233**; R = Et, X = BF_4).[94] 1,3-Dipolar cycloaddition reactions have been demonstrated. The derivative (**230**; Ar = p-MeC_6H_4) with dimethyl acetylenedicarboxylate gives the pyrazole (**232**) (my) *via* a bicyclic adduct (**229**) (Scheme 47). Diethyl azodicarboxylate gives the stable adduct (**231**) (hy).[94] The 5-unsubstituted 1,2,3-triazolium-5-olates, *e.g.* (**230**), undergo electrophilic substitution: bromine in acetic acid gives 5-bromo-derivatives, *e.g.* (**230**) → (**234**).[94]

(229) (230) (231) (232) (233) (234)

i, $MeO_2CC{\equiv}CCO_2Me$; ii, $EtO_2CN{=}NCO_2Et$; iii, HCl or $Et_3O^+ BF_4^-$; iv, Br, AcOH.

SCHEME 47

20.4.12.2 1,2,3-Triazolium-4-aminides (235)

The preparation of these compounds (**235**) resembles that of the sydnone imines (**170**) (Section 20.4.10.2) and like the sydnone imines only *N*-acyl derivatives are sufficiently stable to be isolated. *N*-Methylaminoacetonitrile ($MeNHCH_2CN$) with aryldiazonium salts gives *N*-methyl-*N'*-arylazoaminoacetonitriles (**236**; $R^2 = Me$), which with hydrogen chloride give the 1,2,3-triazolium chlorides (**237**) (Scheme 48). Treatment of the chlorides (**237**) with acetic anhydride gives the *N*-acyl derivatives (**238**), which are alternatively prepared by cyclization of the nitriles (**236**) using acetyl chloride. These *N*-acyl salts (**238**) with ammonium hydroxide give *N*-acyl-1,2,3-triazolium-4-aminides (**239**), which are colourless and crystalline;[94] their chemistry has not been examined.

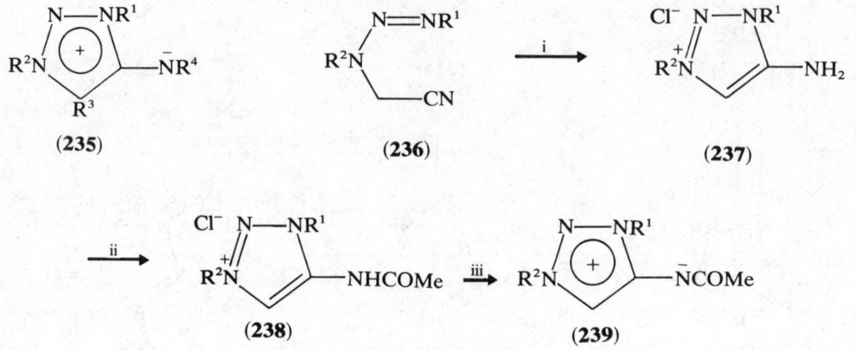

(235) (236) (237) (238) (239)

i, HCl; ii, Ac_2O; iii, NH_4OH.

SCHEME 48

20.4.12.3 1,2,3-Triazolium-4-thiolates (241)

Treatment of 4-halo-1,2,3-triazolium tosylates (**240**; X = Cl or Br) with sodium sulphide in DMF gives meso-ionic 1,2,3-triazolium-4-thiolates (**241**) (my) (Scheme 49) as colourless or pale-cream crystals.[98,99] Another route to this system (**241**) (my) involves methylation of 4- or 5-methylmercapto-1,2,3-triazoles (**243** or **245**) using methyl tosylate, followed by demethylation of the intermediate 1,2,3-triazolium tosylates (**244**) by hot piperidine.[100]

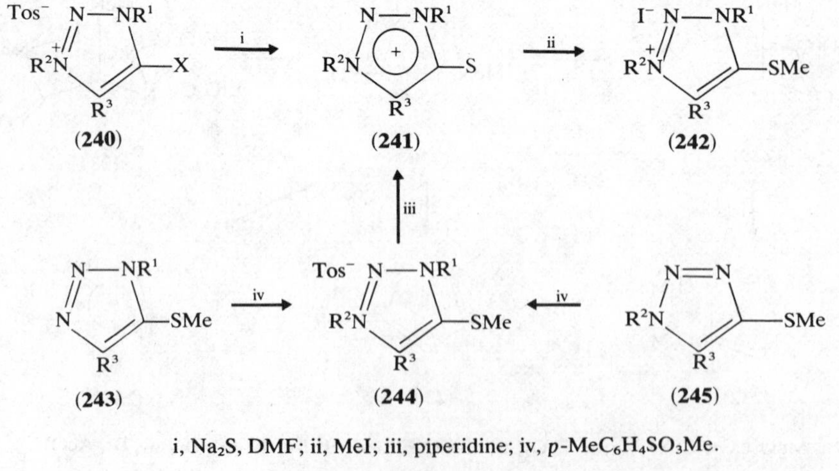

(240) (241) (242) (243) (244) (245)

i, Na_2S, DMF; ii, MeI; iii, piperidine; iv, p-$MeC_6H_4SO_3Me$.

SCHEME 49

(246) **(247)** **(248)**

(249) **(250)** **(251)**

SCHEME 50

With methyl iodide, the 1,2,3-triazolium-4-thiolates (**241**) give methiodides (**242**) (Scheme 49). With benzoyl chloride and pyridine the derivative (**246**) gives an *S*-benzoyl-1,2,3-triazolium chloride which is not isolated but dealkylates, giving a mixture of the 1,2,3-triazoles (**247** and **248**) (Scheme 50).[99] Pyrolysis of the derivative (**246**) at 180 °C results in isomerization to the 1,2,3-triazole (**249**) (hy). The 1,2,3-triazoles (**250**) and (**251**) (ly) are also produced and are presumably formed by an intermolecular transfer of alkyl groups.[98,99]

20.4.12.4 1,2,4-Triazolium-3-olates (253)

The stable colourless meso-ionic derivatives (**253**) are easily prepared and have been studied in detail. Two widely applicable preparative methods are available. Condensation of *N*-amino-amidines (**252**) with phosgene is a good route (Scheme 51) (hy) if the amidine (**252**) is available.[101] Alternatively, base-catalysed cyclization of *N*-aroylsemicarbazides (**254**) gives (**253**) in good yield and has the advantage that the precursor (**254**) is readily

(252) **(253)** **(254)**

(255) **(256)** **(257)**

i, $COCl_2$; ii, NaOEt; iii, R^3NCO; iv, MeI, base or CH_2N_2; v, OH^-, H_2O.

SCHEME 51

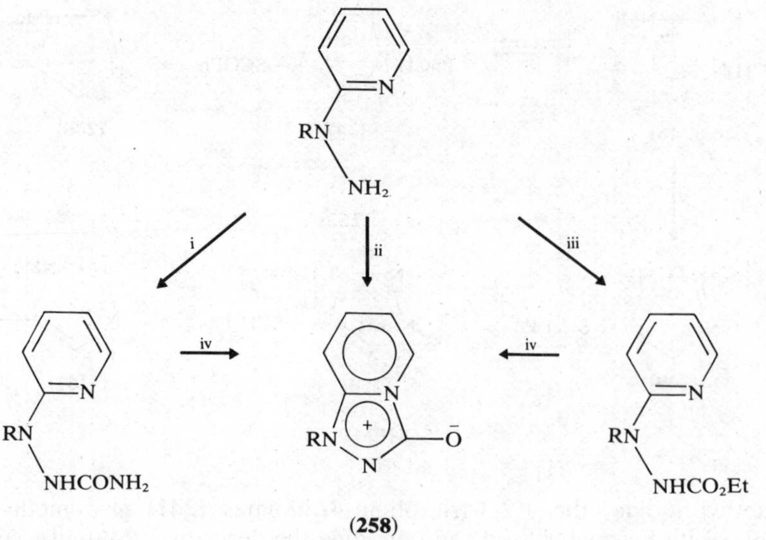

(258)

i, HCl, KCNO; ii, $COCl_2$ or $(NH_2)_2CO$; iii, $ClCO_2Et$; iv, Δ.

SCHEME 52

prepared from an *N*-aroylhydrazine ($R^2CONR^1NH_2$) and the appropriate isocyanate (R^3NCO).[85] Similar routes have been employed to prepare bicyclic derivatives (**258**) (Scheme 52).[102,103]

Other preparative routes which have been employed include: (i) 1,3-dipolar cycloaddition of isocyanates to sydnones (**255**) (Scheme 51) (hy);[104] (ii) methylation of 3-hydroxy-1,2,4-triazoles (**256**) using methyl iodide or diazomethane (ly);[105] (iii) alkaline hydrolysis of 3-methylthio-1,2,4-triazolium iodides (**257**) (hy).[106]

The meso-ionic derivatives (**253**) are associated with a carbonyl absorption at 1660 cm^{-1};[101] their mass spectra are characterized by a molecular ion and the fragment ion $R^2C{\equiv}\overset{+}{N}R^3$.[107] Reduction with lithium aluminium hydride gives 1,2,4-triazolidin-3-ones (**259**);[85] treatment with triethyloxonium tetrafluoroborate gives the salts (**260**).[108] The 1,2,4-triazolium-3-olates (**253**) are not reactive 1,3-dipoles. The only reported cycloaddition is that of the diphenyl derivative (**253**; $R^1 = R^3 = Ph$, $R^2 = H$) with benzyne, giving 2-phenylindazole (**261**) (ly).[40] Photolysis of the same diphenyl derivative in methanol–dichloromethane gives methyl phenylcarbamate (25%), benzimidazole (18%), and azobenzene (7%),[60] but the mechanism of this reaction is still open to speculation.[9]

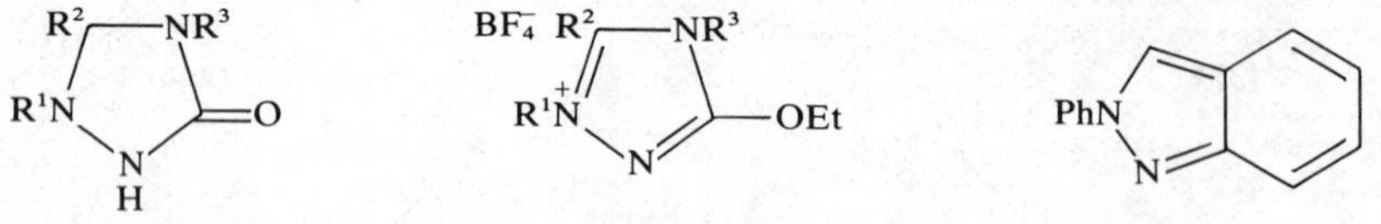

(259) (260) (261)

20.4.12.5 1,2,4-Triazolium-3-aminides (263)

Reaction of the substituted guanidines (**262**) with acid chlorides gives the meso-ionic derivatives (**263**) (Scheme 53) (my).[109,110] A similar route involves condensation of the guanidines with aldehydes and oxidation of the resulting 1,2,4-triazolidines (**265**) (my).[109,110] Stable derivatives are obtained when the exocyclic nitrogen substituent (R^4) is an aryl group, but the above routes have the disadvantage that no control over the position of the substituents R^3 and R^4 can be exercised. A specific preparative route (my) i, R^2COCl; ii, $R^4N{=}CCl_2$; iii, R^2CHO; iv, Fe^{3+}; v, $LiAlH_4$; vi, HX; vii, MeI.

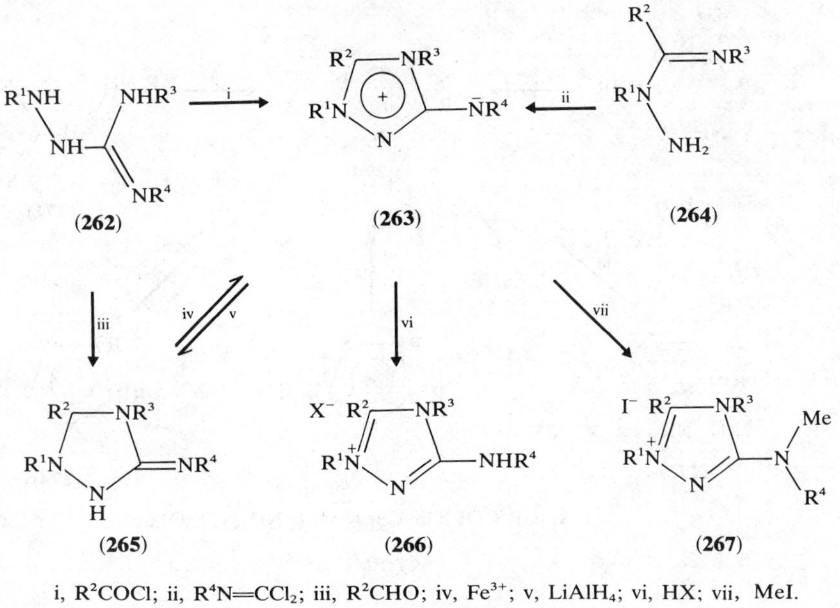

SCHEME 53

involves reaction of *N*-amino-amidines (**264**) with aryl isocyanide dichlorides ($ArN{=}CCl_2$).[111,112]

The 1,2,4-triazolium-3-aminides (**263**) are obtained as stable yellow crystals. The derivative (**263**; $R^1 = R^3 = R^4 = Ph$, $R^2 = H$) is commonly known as nitron and has been used as an analytical reagent since it forms a very insoluble nitrate (**266**; $R^1 = R^3 = R^4 =$ Ph, $R^2 = H$, $X = NO_3$).[113] In general the meso-ionic derivatives (**263**) form salts (**266**) with mineral acids and triazolium iodides (**267**) with methyl iodide. Reduction with lithium aluminium hydride gives 1,2,4-triazolidines (**265**).[111]

In hot ethanol, equilibration of the isomeric systems (**263**) and (**268**) (Scheme 54) is observed.[111] The position of this equilibrium depends on the nature of the aryl substituents R^3 and R^4. The *p*-chlorophenyl derivative (**263**; $R^1 = Me$, $R^2 = R^4 = Ph$, $R^3 = p\text{-}ClC_6H_4$) in ethanol at reflux temperature undergoes a quantitative transformation to the isomer (**268**; $R^1 = Me$, $R^2 = R^4 = Ph$, $R^3 = p\text{-}ClC_6H_4$). Under the same conditions each of the *p*-tolyl derivatives (**263** and **268**; $R^1 = Me$, $R^2 = R^4 = Ph$, $R^3 = p\text{-}MeC_6H_4$) gives a mixture (*ca.* 1 : 1) of the two isomers.[111] The isomeric species (**263**) and (**268**) are distinguished by mass spectrometry: the fragmentations (**263**; $M^{\cdot +} \rightarrow R^2C{\equiv}\overset{+}{N}R^3$) and (**268**; $M^{\cdot +} \rightarrow R^2C{\equiv}\overset{+}{N}R^4$) are unambiguous.[107]

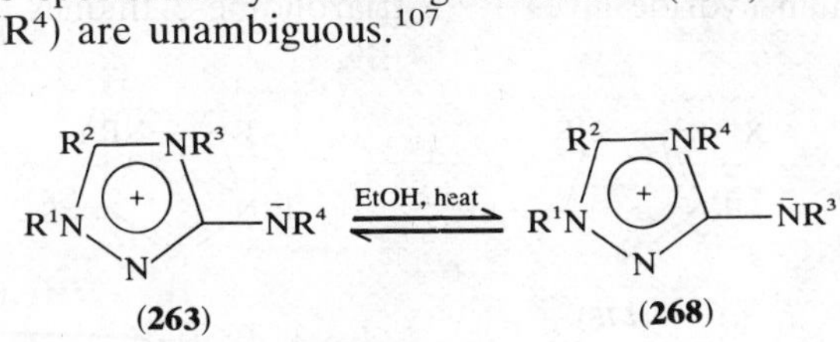

EtOH, heat

SCHEME 54

20.4.12.6 1,2,4-Triazolium-3-thiolates (270)

Representatives of this well known meso-ionic system (**270**) can be prepared by a variety of routes (Scheme 55). The methods of synthetic importance are: (i) thermal

i, Δ; ii, R^2COCl; iii, $CSCl_2$; iv, R^3NH_2; v, EtOH.

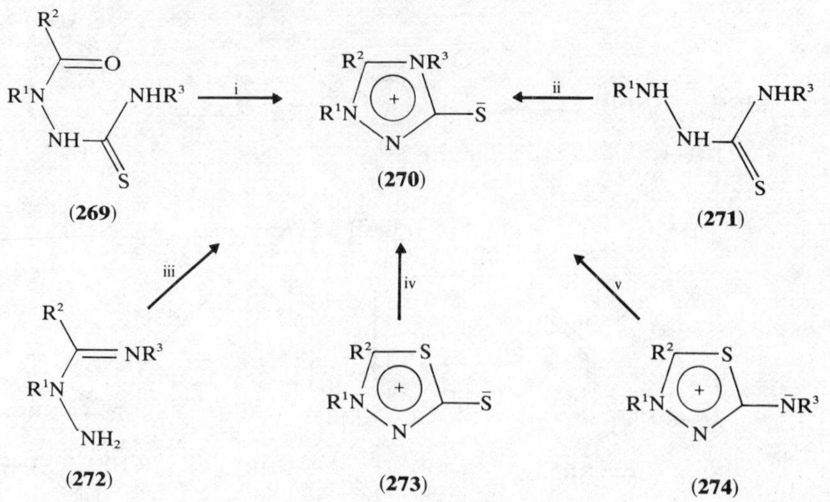

SCHEME 55

cyclization of *N*-acylthiosemicarbazides (**269**) (my);[114] (ii) anhydro-acylation of 1,4-disubstituted thiosemicarbazides (**271**) using acid chlorides (my);[101,114] (iii) condensation of *N*-aminoamidines (**272**) with thiophosgene (my);[101,114] and (iv) reaction of meso-ionic 1,3,4-thiadiazolium-2-thiolates (**273**) (Section 20.4.13.5) (or their methiodides) with primary amines (my).[2,8]

Formation of the 1,2,4-triazolium-3-thiolates (**270**) by methods (i), (ii), and (iv) occurs by cyclization of intermediate thiosemicarbazide derivatives. These routes could in principle lead to the isomeric meso-ionic 1,3,4-thiadiazolium-2-aminides (**274**) (Section 20.4.13.4), but none of this isomeric species (**174**) is formed. The meso-ionic thiadiazoles (**274**) are in fact thermodynamically and kinetically unstable with respect to the triazoles (**270**) and the rearrangement (**274**)→(**270**) readily occurs in ethanol solution (Scheme 55) (hy).[115]

Derivatives of the system (**270**) are very stable. The dipole moment of compound (**270**; $R^1 = Me$, $R^2 = R^3 = Ph$) ($\mu = 9.1$ D) supports the meso-ionic structure.[115] The mass spectra of these species (**270**) show a molecular ion and the fragment ion $R^2C{\equiv}\overset{+}{N}R^1$, which readily distinguished them from the isomeric molecules (**274**) ($M^{\cdot+} \rightarrow R^2C{\equiv}\overset{+}{S}$).[107] Treatment with methyl iodide gives methiodides (**275**; $R^4 = Me$, $X = I$)[114] and, similarly, hydrochlorides (**275**; $R^4 = H$, $X = Cl$) are formed using hydrogen chloride.[101] Reduction with lithium aluminium hydride gives 1,2,3-triazolidine-3-thiones (**276**).[115]

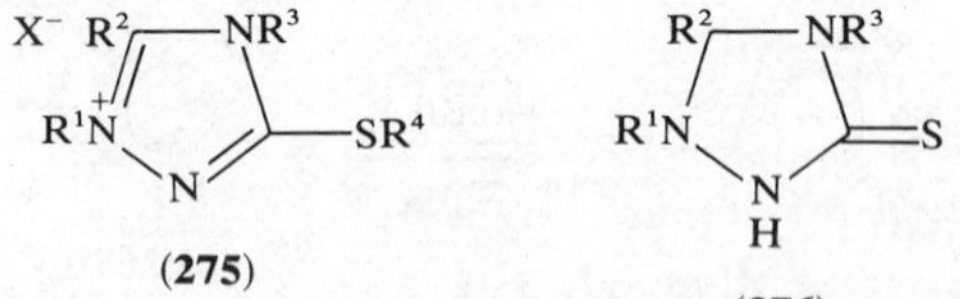

The *N*-amino-derivative (**278**), prepared from the salt (**277**) and hydrazine hydrate (Scheme 56), is transformed to the disulphide (**279**) by nitrous acid.[116] Condensation of this compound (**278**) with *p*-chlorobenzaldehyde gives the benzylidene derivative (**281**), which is also obtained by rearrangement of the meso-ionic thiadiazole (**280**) in ammoniacal ethanol solution (ly).[112]

i, $N_2H_4 \cdot H_2O$; ii, HONO; iii, EtOH, NH_4OH; iv, p-ClC_6H_4CHO.

SCHEME 56

20.4.12.7 1,2,4-Triazolium-3-methylides (283)

3-Methylmercapto-1,2,4-triazolium iodides (**282**) react with malononitrile in the presence of potassium carbonate to give the meso-ionic derivatives (**283**) (Scheme 57) (ly). These colourless-to-yellow crystalline compounds show great stability. Considerably better yields are obtained by treatment of *N*-amino-amidines (**284**) with 3,3-bis(methylthio)-acrylonitriles [$(MeS)_2C{=}C(X)CN$; X = CN or CO_2Me] (my–hy).[117]

i, $CH_2(CN)_2$, K_2CO_3; ii, $(MeS)_2C{=}C(X)CN$

SCHEME 57

20.4.12.8 1,2,4-Triazolium-3-selenates (285)

The preparation of derivatives of this meso-ionic system (**285**) has been achieved by reaction of *N*-acylhydrazines [$R^1N(R^2CO)NH_2$] with phenyl selenoisocyanate (PhN=C=Se) in the presence of tertiary base (my).[118]

20.4.13 MESO-IONIC THIADIAZOLES OF TYPE A

20.4.13.1 1,2,3-Thiadiazolium-4-olates (287)

The meso-ionic compounds (**287**) are prepared (my–hy) by cyclodehydration of arylazomercaptoacetic acids (**286**) (Scheme 58) using acetic anhydride and pyridine.[119] Diaryl (**287**; $R^1 = R^2 = Ar$) and monoaryl (**287**; $R^1 = Ar$, $R^2 = H$) derivatives, obtained as

(**286**) (**287**)

SCHEME 58

stable yellow solids, have been made by this route.[119] The chemistry of 5-unsubstituted derivatives (**292**) has received detailed study (Scheme 59). Reduction with zinc in sulphuric acid (5N) gives an arylhydrazine (hy) and thioglycolic acid (hy).[119] Treatment with hydrogen chloride gives thiadiazolium chlorides (**291**) (my)[119] and Meerwein alkylation gives tetrafluoroborates (**293**) (hy).[108] Electrophilic substitution yields 5-nitro-derivatives (**288**) (my) and 5-bromo-derivatives (**289**) (my). With diazonium salts elimination of nitrogen occurs, giving the diaryl derivatives (**290**) (ly); with benzylamine the amide $PhCH_2NHCOCSNHCH_2Ph$ (my) is formed.[119]

(**288**) (**289**) (**290**) (**291**) (**292**) (**293**)

i, HNO_3, H_2SO_4; ii, Br_2, AcOH, AcO^-Na^+; iii, $Ar'N_2^+$; iv, HCl; v, $Et_3O^+BF_4^-$.

SCHEME 59

20.4.13.2 1,2,4-Thiadiazolium-3-aminides (294)

It is reported that the thiadiazolium iodide (**295**) with silver oxide gives the meso-ionic derivative (**294**; $R^1 = Ph$, $R^2 = Me$, $R^3 = H$).[120]

(**294**) (**295**)

20.4.13.3 1,3,4-Thiadiazolium-2-olates (297)

Condensation of *N*-thioacylhydrazines (**296**) with phosgene[84] or methyl chlorocarbonate[121] (Scheme 60) (my) provides a good route to derivatives of this meso-ionic system (**297**), which are obtained as stable, crystalline solids (ν_{max} 1650 cm^{-1}; C=O).[84] Their mass spectra[107] show the fragmentation $M^{\cdot+} \rightarrow R^2C\equiv\overset{+}{S}$, which distinguishes them from the isomeric species (**298**) (Section 20.4.10.5) from which they are formed by rearrangement in hot ethanol (**298** → **297**; R^1 = Me, R^2 = Ph) (hy).[84] Dipole moments have been measured (**297**; $R^1 = R^2$ = Ph; μ = 7.75 D).

(296) (297) (298) (299) (300) (301)

i, $COCl_2$ or COCl(OMe); ii, EtOH; iii, Et_3O^+ BF_4^-; iv, $PhNH_2$; v, *hν*.

SCHEME 60

Alkylation with triethyloxonium tetrafluoroborate gives the salts (**299**) (hy);[108] treatment with aniline results in ring opening to the semicarbazides (**300**) (my).[121] Photolysis of compound (**297**; R^1 = Me, R^2 = Ph) in acetonitrile gives *N*-methylthiobenzamide (**301**; R^1 = Me, R^2 = Ph) (Scheme 60) (ly) [122]

20.4.13.4 1,3,4-Thiadiazolium-2-aminides (304)

Two methods have been used to prepare derivatives of this system from *N*-thioacylhydrazines (**302**). Reaction with isocyanide dichlorides ($R^3N{=}CCl_2$) gives the salts (**303**; X = Cl) (my), which with base yield the free meso-ionic system (**304**) (hy) (Scheme 61).[86,115,123] In an alternative procedure the hydrazines (**302**) with isothiocyanates

(302) (303) (304)

i, $R^3N{=}CCl_2$; ii, R^3NCS, HX; iii, base.

SCHEME 61

i, PhNCO; ii, PhNCS; iii, HCl, HNO_3 or MeI; iv, EtOH; v, Δ.

SCHEME 62

and an acid catalyst also give the 1,3,4-thiadiazolium salts (**303**) (hy).[86,123] It is significant that if the acid catalyst is omitted the product is a meso-ionic 1,2,4-triazolium-3-thiolate (**270**) (Section 20.4.12.6). These two methods yield 1,3,4-thiadiazolium-2-aminides (**304**) in which the exocyclic nitrogen substituent is an aryl group (**304**; $R^3 = Ar$),[115] an aroyl group (**304**; $R^3 = COAr$),[86] or a sulphonyl group (**304**; $R^3 = SO_2Ar$).[86]

The structure of these derivatives has been confirmed by mass spectrometry ($M^{\cdot +} \rightarrow R^2—C{\equiv}S^+$),[107] dipole moments studies (**306**; $\mu = 6.7$ D),[115] and ESCA spectroscopy.[124] Rearrangement of the derivative (**306**) to the isomeric 1,2,4-triazolium-3-thiolate (**310**) (hy) (Scheme 62)[115] takes place under the following conditions: (i) heating in ethanol solution; (ii) treatment with phenyl isothiocyanate in benzene solution at room temperature; (iii) pyrolysis at 120 °C. It is proposed that the intermediate betaine (**309**) is involved in method (i). Method (ii) almost certainly involves the 1,3-dipolar cycloadduct (**307**); evidence for this mechanism (Scheme 62) is provided by the isolation of the adduct (**305**) when phenyl isocyanate is used.[115]

Meso-ionic 1,3,4-thiadiazolium-2-aminides (**304**) form thiadiazolium salts (*e.g.* **308**) when treated with methyl iodide or dilute mineral acids. Lithium aluminium hydride reduction of the derivative (**306**) gives N^1-benzyl-N^1-methyl-N^3-phenylthiosemicarbazide [$PhCH_2N(Me)NHCSNHPh$].[115]

20.4.13.5 1,3,4-Thiadiazolium-2-thiolates (312)

Compounds of this type have been known for many years[2,8,9] and a large number of derivatives have been reported. Important preparative routes are shown in Scheme 63. The salts (**311**) with acid chlorides give the thiadiazoles (**312**) directly (ly);[125] using sodium dithioformate, derivatives unsubstituted in position-5 (**312**; $R^2 = H$) (ly) are obtained.[125] The same salts (**311**) with aldehydes give thiadiazolidine-2-thiones (**314**) (my), which are converted to the meso-ionic system (**312**) by oxidation and thermal disproportionation of intermediate disulphides (**315** → **312** + **314**) (hy).[126] These routes starting from the salts (**311**) are not suitable for preparing 4-alkyl derivatives (**312**; $R^1 = $ alkyl). A more generally

i, R^2COCl or $HCS_2^- Na^+$; ii, $CSCl_2$; iii, R^2CHO; iv, Δ; v, CS_2; vi, $FeCl_3$ or I_2.

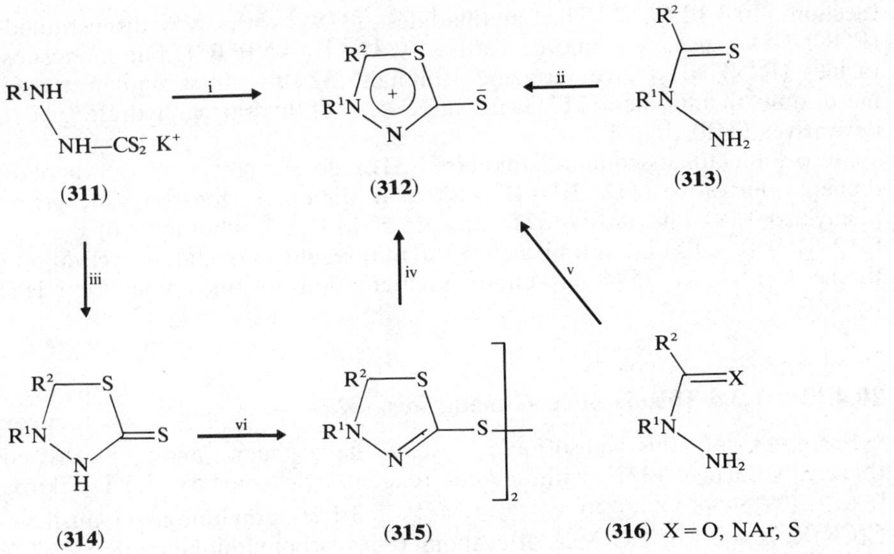

SCHEME 63

applicable method involves condensation of *N*-thioacylhydrazines (**313**) with thiophosgene (my).[127] Alternatively, the hydrazine derivatives (**316**) with carbon disulphide also give these meso-ionic compounds (**312**) (hy).[123,127]

The bright yellow compounds (**312**) (ν_{max} 1320 cm^{-1}; C═S) are very stable. Their mass spectra ($M^{\cdot+} \rightarrow R^2C\overset{+}{\equiv}S$),[107] dipole moments (**312**; $R^1 = R^2 = Ph$; $\mu = 9.25$ D),[128] u.v. spectra,[129] and ESCA spectra[130] have been measured. The ESCA spectra are in accord with a betaine structure.

The 1,3,4-thiadiazolium-2-thiolates (**312**) form salts (**319**) with methyl iodide (my).[126] Treatment of either the meso-ionic compounds (**312**) or their methiodides (**319**) with primary amines gives meso-ionic 1,2,4-triazolium-3-thiolates (**317**) (Scheme 64) (hy)

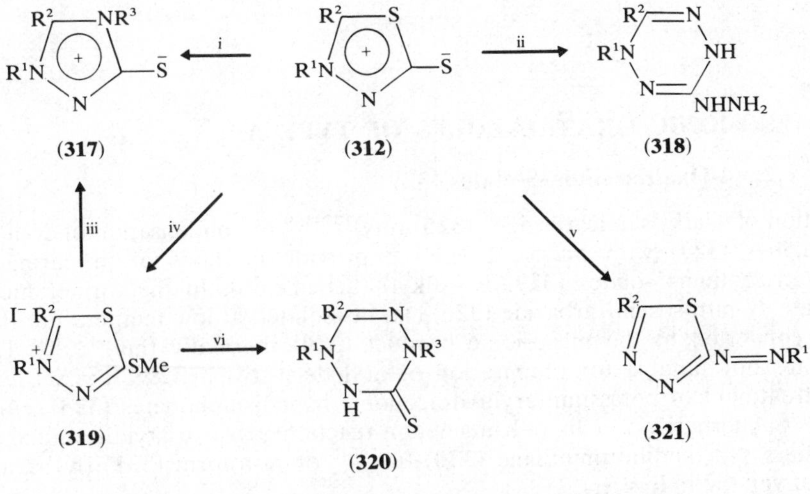

i, R^3NH_2 where R^3 = alkyl or aryl; ii, $N_2H_4 \cdot H_2O$; iii, R^3NH_2 where R^3 = alkyl, aryl, or NR_2; iv, MeI; v, $MeO_2CN{=}NCO_2Me$; vi, R^3NHNH_2.

SCHEME 64

(Section 20.4.12.6).[126] The methiodides (**319**) with *N,N*-disubstituted hydrazines ($R^4R^5NNH_2$) give *N*-amino-derivatives (**317**; $R^3 = NR^4R^5$), but monosubstituted hydrazines (R^4NHNH_2) give tetrazine-3-thiones (**320**).[131] In a similar transformation the meso-ionic thiadiazoles (**312**) with an excess of hydrazine hydrate give the tetrazine derivatives (**318**) (hy).[132]

Meso-ionic thiadiazolium-2-thiolates (**312**) do not react as 1,3-dipolarophiles. The diphenyl derivative (**312**; $R^1 = R^2 = Ph$) with dimethyl azocarboxylate gives 2-phenyl-5-phenylazo-1,3,4-thiadiazole (**321**; $R^1 = R^2 = Ph$) (ly).[133] Photolysis of the same derivative (**312**; $R^1 = R^2 = Ph$) in acetonitrile (> 300 nm) results in oxidative cyclization to a phenanthrene (ly).[134] At 253.7 nm photo-fragmentation to thioamides (R^1NHCSR^2) occurs (ly–my).[122]

20.4.13.6 1,3,4-Thiadiazolium-2-methylides (322)

Examples of this meso-ionic system have been made by reaction of *N*-thioacylhydrazines (**323**) with several reagents (Scheme 65). 3,3-Dichloroacrylonitriles [$Cl_2C{=}C(CN)X$; X = CN or CO_2Me], 3,3-bis(methylthio)acrylonitriles [$(MeS)_2C{=}C(CN)X$; X = CN or CO_2Me], or cyanomethoxycarbonylthioketen [$S{=}C{=}C(CN)CO_2Me$] yield the yellow-to-orange cyano-derivatives (**322**) (my).[89] In another study, *N*-thiobenzoyl-*N*-methylhydrazine (**323**; $R^1 = Me$, $R^2 = Ph$) reacted with phenylthioacetylthioglycolic acid ($PhCH_2CSSCH_2CO_2H$), giving the red crystalline meso-ionic thiadiazolium-2-methylide (**324**) (my).[135] The structure of this product (**324**) has been confirmed by X-ray crystallography.[136]

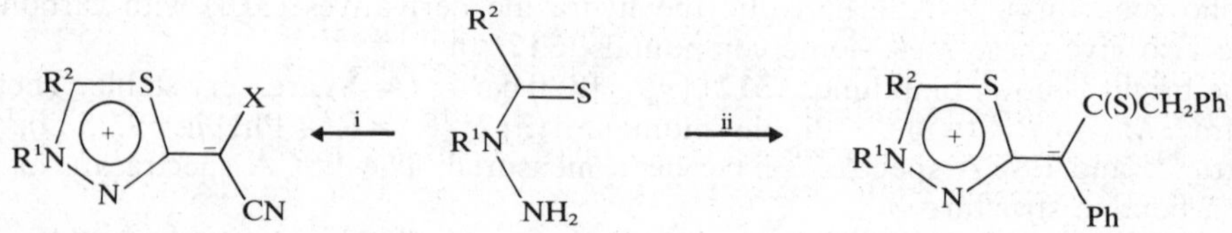

(**322**) X = CN, CO_2Me (**323**) (**324**)

i, $Cl_2C{=}C(CN)X$, $(MeS)_2C{=}C(CN)X$, or SCC(CN)X; ii, $PhCH_2C(S)SCH_2CO_2H$.

SCHEME 65

20.4.14 MESO-IONIC OXATRIAZOLES OF TYPE A

20.4.14.1 1,2,3,4-Oxatriazolium-5-olates (329)

Nitrosation of 1-alkylsemicarbazides (**325**) (my)[137–139] or condensation of *N*-nitroso-*N*-alkylhydrazines (**327**) with phosgene (my)[140] provide methods of preparing 3-alkyl-1,2,3,4-oxatriazolium-5-olates (**329**; R = alkyl) (Scheme 66). In the former method, an intermediate *N*-nitrososemicarbazide (**326**) can be isolated at low temperature (hy),[138,139] and then converted to the meso-ionic product (**329**) thermally (my).[138,139] These approaches are not suitable for preparation of aryl derivatives (**329**; R = Ar), which are made by treatment of potassium arylhydrazonomethanedisulphonates (**328**) with nitrous acid (my).[141] Alternatively, a Japp–Klingemann reaction between aryldiazonium salts and the potassium salt of dinitromethane (**330**) (my)[142] or nitroform (**331**) (ly)[139] also gives aryl derivatives (**329**; R = Ar).

The colourless crystalline compounds (**329**) are characterized by a carbonyl absorption in the region 1790–1800 cm^{-1}.[137,139] The dipole moment of the phenyl derivative (**329**; R = Ph; $\mu = 6.14$ D)[9] supports the meso-ionic structure.

RNH—NHCONH$_2$ (**325**) $\xrightarrow{\text{i}}$ RN(N=O)—NHCONH$_2$ (**326**); RN(N=O)—NH$_2$ (**327**)

(**328**) RNH—N=C($SO_3^-K^+$)$_2$ $\xrightarrow{\text{i}}$ (**329**) $\xleftarrow{\text{iv}}$ $K^+\bar{C}H(NO_2)_2$ (**330**) or $HC(NO_2)_3$ (**331**)

i, HONO; ii, Δ; iii, $COCl_2$; iv, $RN_2^+X^-$(R = Ar).

SCHEME 66

Acid hydrolysis of the 3-phenyl derivative (**329**; R = Ph) gives phenyl azide. The 3-cyclohexyl derivative (**329**; R = C_6H_{11}) is resistant to hydrolysis by dilute mineral acids, but concentrated sulphuric acid gives cyclohexanol (60%).[139] The mechanism of hydrolysis of the isopropyl derivative (**329**; R = Pr^i) to isopropanol or isopropyl chloride has been discussed.[143] Photolysis of the 3-cyclohexyl compound (**329**; R = C_6H_{11}) and subsequent hydrolysis gives cyclohexanone (27%);[144] mechanistic proposals for this reaction have been put forward.[9,144]

Meso-ionic 1,2,3,4-oxatriazolium-5-olates (**329**) do not react as 1,3-dipoles. The 3-cyclohexyl derivative (**329**; R = C_6H_{11}) upon heating with diphenylacetylene gives 1-cyclohexyl-4,5-diphenyltriazole (37%) and not the anticipated isomer.[145] Presumably the meso-ionic compound dissociates and rearranges to the azide prior to cycloaddition.

20.4.14.2 1,2,3,4-Oxatriazolium-5-aminides (334)

Whereas nitrosation of alkylsemicarbazides gives meso-ionic 1,2,3,4-oxatriazolium-5-olates (**329**) (Section 20.4.14.1), nitrosation of thiosemicarbazides (**332**) gives the oxatriazolium salts (**333**) (my) (Scheme 67).[146,147] When the exocyclic substituent (R^2) is an aryl group, treatment with base readily gives the bright-red crystalline meso-ionic compounds (**334**; R^2 = Ar) (hy).[146,147] The physical properties of these compounds, (ν_{max} 1675–1680 cm^{-1}; C=N) (**334**; $R^1 = R^2$ = Ph, μ = 5.4 D), are consistent with the oxatriazolium structure.[9]

R^1NH—NHCSNHR^2 (**332**) $\xrightarrow{\text{HONO}}$ (**333**) $\xrightarrow{\text{NH}_3}$ (**334**)

SCHEME 67

Acid hydrolysis of the diphenyl derivative (**334**; $R^1 = R^2$ = Ph) gives 3-phenyl-1,2,3,4-oxatriazolium-5-olate (**329**; R = Ph) (ly) (Section 20.4.14.1)[147] and treatment with hydrogen sulphide regenerates 1,4-diphenylthiosemicarbazide.[147] In hot aqueous ethanolic alkali the diaryl derivatives (**334**; $R^1 = R^2$ = Ar) undergo rearrangement to meso-ionic

1,2,3,4-tetrazolium-5-olates (**344**) (Scheme 70) (Section 20.4.15.1).[146,147] Other rearrangements of meso-ionic compounds to more stable meso-ionic isomers are known and possible mechanisms have been discussed[9] (See also **342** → **343**; Scheme 69).

The unsubstituted 1,2,3,4-oxatriazolium salts (**336**) are alternatively prepared by acid-catalysed cyclization of 1-substituted 1-nitroso-2-cyanohydrazines [$R^1N(NO)NHCN$], prepared *in situ* from a monosubstituted hydrazine (**335**) and BrCN followed by nitrosation (Scheme 68) (my).[148] The cyclohexyl derivative (**336**; $R^1 = C_6H_{11}$) has been made by acid-catalysed cyclization of the guanidine (**337**) (my).[149]

R^1NHNH_2 (**335**) —i, ii, iii→ (**336**) ←iv— (**337**)
(**336**) —v→ (**338**); (**336**) ⇌ vi/vii (**339**); (**336**) —viii→ (**340**)
(**338**) ←ii— (**339**) —viii→ (**340**)

i, BrCN; ii, HONO; iii, HCl, MeOH; iv, conc. HCl; v, aq. $NaNO_2$; vi, aq. $NaHCO_3$ at 0 °C; vii, HCl; viii, R^2COCl, py.

SCHEME 68

Treatment of the salts (**336**) with cold aqueous sodium bicarbonate is reported to give the unsubstituted meso-ionic derivatives (**339**) (hy) as yellow crystals, which are reconverted to the salts (**336**) by acid treatment.[148] With acyl chlorides in pyridine the meso-ionic compounds (**339**) form *N*-acyl derivatives (**340**) (Scheme 68), which are also formed directly from the salts (**336**) (hy). Similarly, the salts (**336**) with aqueous sodium nitrite, or the free base (**339**) with nitrous acid, give *N*-nitroso-derivatives (**338**).[148]

$RNHNHCS_2^- \overset{+}{H_3N}NHR$ (**341**) —i→ (**342**) —ii→ (**343**)

i, $NaNO_2$, dil. HCl at 0 °C; ii, EtOH, NH_4OH.

SCHEME 69

20.4.14.3 1,2,3,4-Oxatriazolium-5-thiolates (342)

Aryl derivatives of this meso-ionic system are obtained by nitrosation of the dithiocarbazinic acid salts (**341**) (my) (Scheme 69).[150] The structures of these crystalline compounds (ν_{max} 1365–1375 cm^{-1}; C=S) are supported by their mass spectra ($M^{\cdot+} \rightarrow M^{\cdot+} - NO$) and dipole moments (**342**; R = Ph; $\mu = 6.8$ D).[150]

Acid hydrolysis of the phenyl derivative (**342**; R = Ph) gives phenol, whereas alkaline hydrolysis gives phenyl azide. Treatment with triethyloxonium tetrafluoroborate results in *S*-alkylation, giving the corresponding 1,2,3,4-oxatriazolium tetrafluoroborate. In hot ethanolic ammonia, rearrangement to isomeric meso-ionic 1,2,3,4-thiatriazolium-5-olates (**343**) (my) (Scheme 69) occurs, providing a preparative route to this previously unknown meso-ionic system (**343**) (Section 20.4.16.1).[150]

20.4.15 MESO-IONIC TETRAZOLES OF TYPE A

20.4.15.1 1,2,3,4-Tetrazolium-5-olates (344)

Only diaryl derivatives of this meso-ionic system are well characterized. A convenient preparative route involves base-catalysed rearrangement of meso-ionic 1,2,3,4-oxatriazolium-5-aminides (**334** → **344**) (hy) (Scheme 70).[146,147] In an alternative procedure, treatment of arylhydrazonomethanedisulphonates (**345**) with diazonium salts also gives these meso-ionic derivatives (my),[141] but this route is not particularly suited for mixed aryl derivatives (**344**; $R^1 \neq R^2$) since aryl group exchange occurs. Similar preparations (ly) have been achieved by treating bisalkylsulphonylmethanes $[CH_2(SO_2R)_2]$[142,151] and dinitromethane $[CH_2(NO_2)_2]$[142] with diazonium salts. Nitrosation of *p*-ethoxyphenylurea (dulcin) gives the di-*p*-ethoxyphenyl derivative (**344**; $R^1 = R^2 = p\text{-}EtOC_6H_4$), but the scope of this preparative method has not been explored.[152] The diphenyl derivative has been made (my) by thermolysis of the isomeric type B meso-ionic 2,3-diphenyl-1,2,3,4-tetrazolium-5-olate (Section 20.4.21.1).

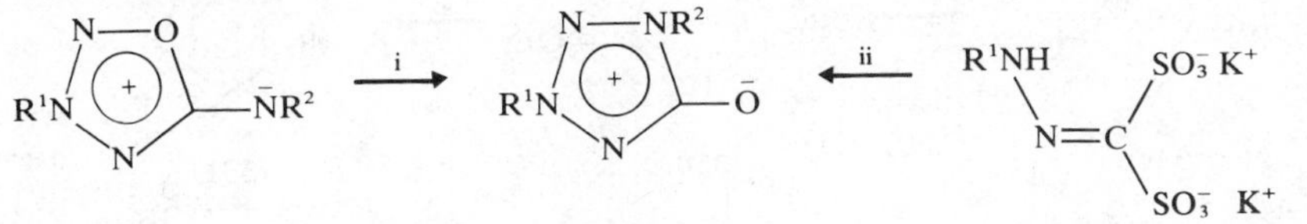

(334) (344) (345)

i, aq. NaOH, EtOH; ii, $R^2N_2^+ Cl^-$.

SCHEME 70

The colourless crystalline tetrazolium-5-olates (**344**) (ν_{max} 1700 cm^{-1}; C=O) (**344**; $R^1 = R^2 = Ph$; $\mu = 5.4$ D) are quite distinct from the bright-red isomers (**334**). Chemically these compounds (**344**; R^1 and $R^2 = Ar$) are very stable; treatment with triethyloxonium tetrafluoroborate gives 5-ethoxy-1,2,3,4-tetrazolium tetrafluoroborates (**350**) (hy).

20.4.15.2 1,2,3,4-Tetrazolium-5-aminides (347)

Only dialkyl derivatives (**347**; $R^1 = R^2 = $ alkyl) of this meso-ionic system have been reported. Their preparation involves alkylation of 5-amino-1- or -2-methyltetrazoles (**348** and **346**) using dimethyl sulphate (my) or methyl benzenesulphonate (hy), respectively (Scheme 71).[153–155] In addition to the unsubstituted derivatives (**347**; $R^1 = R^2 = Me$, $R^3 = H$) which readily form a hydrochloride,[153,154] derivatives in which the exocyclic nitrogen substituent (R^3) is MeCO, $p\text{-}MeC_6H_4SO_2$, or $p\text{-}ClC_6H_4SO_2$ have also been obtained as colourless crystalline compounds.[155]

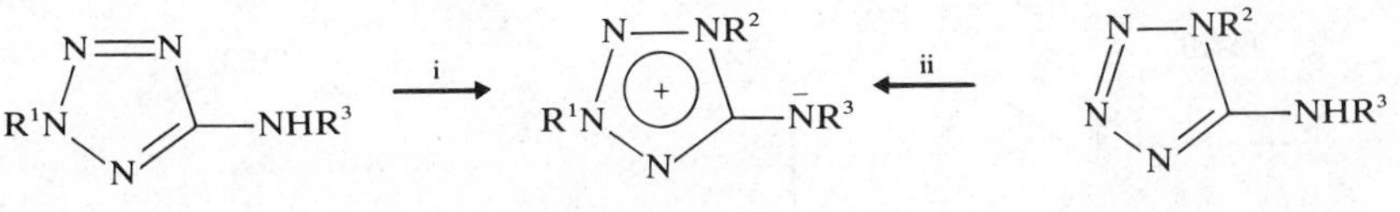

(346) (347) (348)

i, $PhSO_3Me$; ii, Me_2SO_4.

SCHEME 71

Acid hydrolysis of the acyl derivative (**347**; $R^1 = R^2 = Me$, $R^3 = COMe$) affords the hydrochloride of the derivative (**347**; $R^1 = R^2 = Me$, $R^3 = H$).[154] Alkaline hydrolysis of the sulphonyl derivatives (**347**; $R^1 = R^2 = Me$, $R^3 = p\text{-}MeC_6H_4SO_2$ or $p\text{-}ClC_6H_4SO_2$) gives the corresponding sulphonamides as the only isolated products.[155] The same derivatives are demethylated by piperidine or hydrazine hydrate, giving 5-amino-1-methyltetrazole derivatives (**348**).[155]

20.4.15.3 1,2,3,4-Tetrazolium-5-thiolates (349)

Treatment of the tetrazolium tetrafluoroborates (**350**) (see Section 20.4.15.1) with sodium sulphide in hot dimethylformamide gives the yellow crystalline meso-ionic 1,2,3,4-tetrazolium-5-thiolates (**349**) (my) (Scheme 72). The chemistry of these species (ν_{max} 1365–1370 cm^{-1}, C═S) has not been investigated but the dipole moment of the diphenyl derivative (**349**; $R^1 = R^2 = Ph$, $\mu = 6.5$ D) supports the meso-ionic formulation.[156] Compound (**349**; $R^1 = R^2 = Ph$) is also formed (ly) by thermolysis of dehydrodithizone at 150 °C (Section 20.4.21.3).

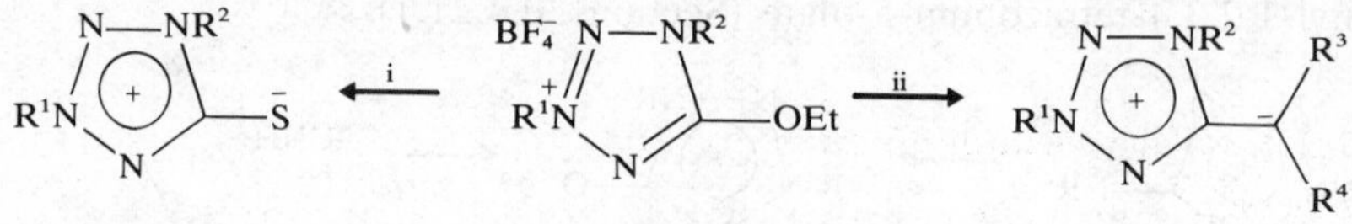

(349) (350) (351)

i, Na_2S, DMF; ii, $CH_2(CN)_2$, Et_3N.

SCHEME 72

20.4.15.4 1,2,3,4-Tetrazolium-5-methylides (351)

Dicyanomethylide derivatives (**351**; $R^3 = R^4 = CN$) of this system are made by heating the tetrazolium tetrafluoroborates (**350**) with malononitrile and triethylamine in acetonitrile solution (my) (Scheme 72). The yellow crystalline diphenyl derivative (**351**; $R^1 = R^2 = Ph$) has a dipole moment of 9.5 D.[156]

20.4.16 MESO-IONIC THIATRIAZOLES OF TYPE A

20.4.16.1 1.2.3.4-Thiatriazolium-5-olates (343)

Compounds of this type are prepared by rearrangement of meso-ionic 1,2,3,4-oxatriazolium-5-thiolates (**342** → **343**) (Scheme 69) (Section 20.4.14.3) in hot ethanolic ammonia solution (my).[150] The structure of these colourless crystalline derivatives (**343**) is

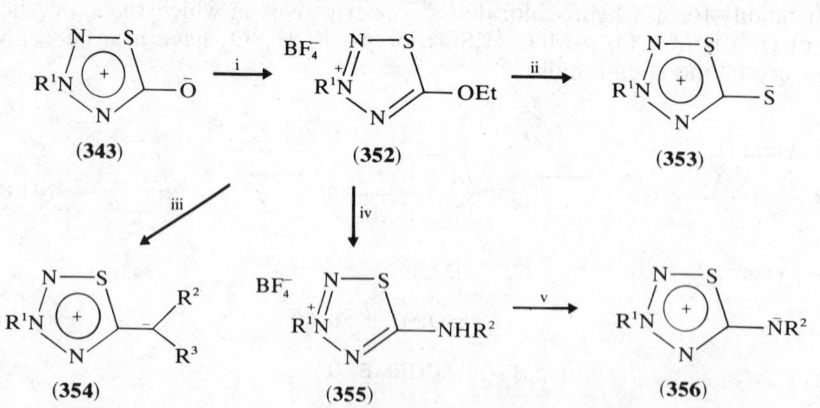

(343) (352) (353)

(354) (355) (356)

i, $Et_3O^+BF_4^-$; ii, Na_2S, aq. DMF; iii, $CH_2(CN)_2$, Et_3N; iv, R^2NH_2; v, aq. NaOH.

SCHEME 73

supported by their i.r. spectra (ν_{max} 1675–1680 cm^{-1}; C═O), mass spectra ($M^{\cdot+} \rightarrow M^{\cdot+} - NS^{\cdot}$) and dipole moments (**343**; $R^1 = Ph$; $\mu = 4.4$ D). The phenyl derivative is stable to acid, but alkaline hydrolysis gives phenyl azide. Alkylation using triethyloxonium tetrafluoroborate in methylene chloride gives 1,2,3,4-thiatriazolium tetrafluoroborates (**352**), which are useful intermediates (Scheme 73) for the preparation of other meso-ionic thiatriazolium derivatives (Sections 20.4.16.2–4).[150]

20.4.16.2 1,2,3,4-Thiatriazolium-5-aminides (356)

The 5-ethoxy-1,2,3,4-thiatriazolium tetrafluoroborates (**352**) with aromatic amines in hot benzene solution give 5-arylamino-1,2,3,4-thiatriazolium tetrafluoroborates (**355**) (Scheme 73) (my), which with sodium hydroxide in warm ethanol afford meso-ionic 1,2,3,4-thiatriazolium-5-aminides (**356**) (my). The chemistry of this system has not been investigated.[156]

20.4.16.3 1,2,3,4-Thiatriazolium-5-thiolates (353)

These orange crystalline compounds are made by reaction of the 1,2,3,4-thiatriazolium salts (**352**) with sodium sulphide in aqueous DMF (Scheme 73) (hy). The chemistry of these derivatives (**353**) has not been examined but their meso-ionic structure is supported by mass spectrometry ($M^{\cdot+} \rightarrow M^{\cdot+} - NS^{\cdot}$).[150]

20.4.16.4 1,2,3,4-Thiatriazolium-5-methylides (354)

With malononitrile and triethylamine in acetonitrile solution the 1,2,3,4-thiatriazolium tetrafluoroborates (**352**) give yellow crystalline meso-ionic 1,2,3,4-thiatriazolium-5-biscyanomethylides (**354**; $R^2 = R^3 = CN$). The dipole moment of the phenyl derivative (**354**; $R^1 = Ph$, $R^2 = R^3 = CN$; $\mu = 8.8$ D) supports the meso-ionic formulation.[156]

20.4.17 MESO-IONIC OXAZOLES OF TYPE B

20.4.17.1 1,2-Oxazolium-4-olates (359)

Derivatives of this system have not been isolated. The oxazolium perchlorates (**358**; $X = ClO_4$) are formed by methylation of the appropriate isoxazoles (**357**) (Scheme 74), and it is possible that a yellow colouration observed when these salts (**358**) are dissolved in aqueous alkali is due to the free meso-ionic system (**359**).[157]

(**357**) Y = OAc (**358**) Y = OAc (**359**) Z = O

(**360**) Y = NHTs (**361**) Y = NHTs (**362**) Z = NTs

i, $MeOSO_2F$; ii, $HClO_4$; iii, OH^-, H_2O or Et_3N.

SCHEME 74

20.4.17.2 1,2-Oxazolium-4-aminides (362)

Derivatives of this meso-ionic system have not been isolated. Methylation of the isoxazoles (**360**) gives the salts (**361**; $X = ClO_4$) (Scheme 74), but treatment with base did not afford an isolable meso-ionic product.[157]

20.4.18 MESO-IONIC DIAZOLES OF TYPE B

20.4.18.1 1,2-Diazolium-4-olates (364)

Crystalline derivatives of this meso-ionic system have been prepared but their chemistry has not received detailed study and in some cases it is not clear if the product is a monohydrate of the meso-ionic heterocycle (**364**) or a diazolium hydroxide (**367**; X= OH).[9] In general the 1,2-diazolium-4-olates (**364**) are obtained by methylation of 4-hydroxypyrazoles (**366**) (hy), followed by treatment of the intermediate 1,2-diazolium salts (**367**) with aqueous base (Scheme 75) (hy).[158] A convenient method of preparing the 1,2-dimethyl derivative (**364**; $R^1 = R^2 = Me$, $R^3 = R^4 = Ph$) involves condensation of *N,N'*-dimethylhydrazine dihydrochloride with 2-acetoxy-1,3-diphenylpropane-1,3-dione (**363**; $R^3 = R^4 = Ph$), followed by treatment with aqueous sodium hydroxide (my) (Scheme 75).[158]

This meso-ionic system is characterized by an i.r. absorption in the region 1530–1550 cm^{-1} which has been assigned to the exocyclic carbonyl group.[159] Pyrolysis of the derivative (**364**; $R^1 = R^2 = Me$, $R^3 = R^4 = Ph$) gives the 4-hydroxypyrazole (**366**; $R^2 = Me$, $R^3 = R^4 = Ph$) (23%) together with 3,5-diphenyl-1-methyl-4-methoxypyrazole (**365**; $R^1 = R^2 = Me$, $R^3 = R^4 = Ph$) (2%).[157] With dimethyl acetylenedicarboxylate the same derivative gives the pyrazole [**365**; $R^2 = Me$, $R^3 = R^4 = Ph$, $R^1 = C(CO_2Me){=}CH(CO_2Me)$].[157]

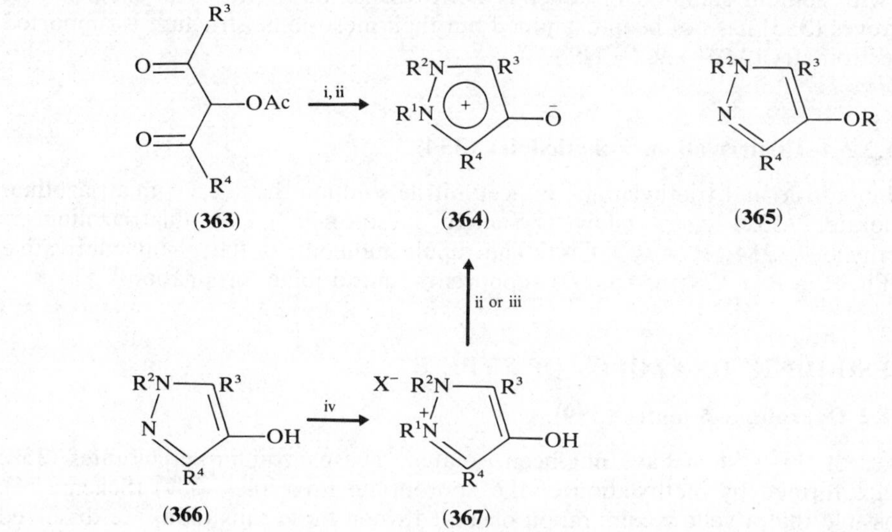

(363) (364) (365)

(366) (367)

i, MeNHNHMe·2HCl; ii, aq. NaOH; iii, Ag_2O, H_2O; iv, Me_2SO_4 or MeI.

SCHEME 75

20.4.18.2 1,2-Diazolium-4-aminides (370)

Examples of this meso-ionic system in which the exocyclic nitrogen atom is stabilized by sulphonyl groups have been prepared by methylation of the pyrazoles (**368**) using methyl fluorosulphonate (hy) and subsequent treatment of the diazolium perchlorates (**369**; $R^1 = Me$) with aqueous potassium hydroxide (Scheme 76).[157] This method gives the derivatives (**370**; $R^4 = Me$ or p-MeC_6H_4) as stable colourless solids (my–hy). Perchloric acid regenerates the diazolium perchlorates (**369**); treatment with methyl fluorosulphonate results in methylation of the exocyclic nitrogen atom, giving diazolium salts which have been characterized as perchlorates.[157]

In boiling benzonitrile the tetramethyl derivatives (**370**; $R^1 = R^2 = R^3 = R^5 = Me$, $R^4 =$ Me or p-MeC_6H_4) rearrange to isomeric pyrazoles (**372**) (my) (Scheme 76). The derivative

(368) (369) (370)

(371) (372)

i, $MeOSO_2F$; ii, $HClO_4$; iii, aq. KOH: iv, $MeO_2CC{\equiv}CCO_2Me$; v, Δ.

SCHEME 76

(**370**; $R^1 = R^2 = Me$, $R^3 = R^5 = Ph$, $R^4 = p\text{-}MeC_6H_4$) appears to be thermally stable, but with dimethyl acetylenedicarboxylate gives a 25% yield of the Michael adduct (**371**; $R^2 = Me$, $R^3 = R^5 = Ph$).[157]

20.4.19 MESO-IONIC THIAZOLES OF TYPE B

20.4.19.1 1,2-Thiazolium-4-aminides (375)

A single example of this system is known. Thus, the isothiazole (**373**; $R^2 = H$, $R^3 = p\text{-}MeC_6H_4$, $R^4 = Me$) is readily converted to the perchlorate (**374**; $R^1 = R^4 = Me$, $R^2 = H$, $R^3 = p\text{-}MeC_6H_4$) (Scheme 77), which with aqueous alkali gives the meso-ionic 1,2-thiazole (**375**; $R^1 = R^4 = Me$, $R^2 = H$, $R^3 = p\text{-}MeC_6H_4$) (40%) as a pale yellow solid.[157]

(373) (374) (375)

i, $MeOSO_2F$; ii, $HClO_4$; iii, aq. KOH.

SCHEME 77

20.4.20 MESO-IONIC DITHIOLES OF TYPE B

20.4.20.1 1,2-Dithiolium-4-olates (381)

Several closely related methods have been reported for preparing diaryl derivatives of this meso-ionic system (**381**; R^1 and R^2 = aryl). Treatment of 1,3-diarylpropanetriones (**376**) with hydrogen sulphide and hydrogen chloride gives 1,2-dithiolium salts (**377**),[160] which are also prepared from the propane-1,3-dione esters (**378**) using hydrogen disulphide or tetraphosphorus decasulphide (Scheme 78).[161] The salts (**377**; X = Cl or ClO_4) are deprotonated by tertiary base, giving meso-ionic 1,2-dithiolium-4-olates (**381**) (my).[160,161] A similar sequence starting from propane-1,3-dione esters (**379** → **382** → **381**) also gives the meso-ionic product (ly–my).[161] In an alternative route, tetrabromo-1,3-diphenylacetone (**380**) combines with potassium ethylxanthate to give the diphenyl compound (**381**; $R^1 = R^2 = Ph$) (my).[162]

(**376**) (**377**) (**378**) R = COMe, COPh (**379**) R = Me, Et

(**380**) (**381**) (**382**) R = Me, Et

i, HCl, H_2S; ii, HCl, H_2S_2; iii, P_4S_{10}; iv, $HClO_4$; v, Et_3N or py; vi, $EtOCS_2^- K^+$; vii, py.

SCHEME 78

The 3,5-diaryl-1,2-dithiolium-4-olates (**381**) are obtained as high-melting crystalline solids; an absorption at 1495 cm^{-1} in their i.r. spectra has been attributed to the exocyclic carbonyl group.[160–162] A number of reactions of the diphenyl derivative (**381**; $R^1 = R^2 =$ Ph) have been investigated (Scheme 79).[162] With acetic anhydride and perchloric acid the

(**383**) (**384**) (**385**) (**386**) (**387**) (**388**)

i, Ac_2O, $HClO_4$; ii, Raney Ni; iii, active MnO_2; iv, $NH_2NH_2 \cdot H_2O$; v, $PhNHNH_2$; vi, Δ.

SCHEME 79

salt (**383**) is formed (hy). Reduction by Raney nickel followed by treatment with activated manganese dioxide gives dibenzyl ketone (**385**) (hy). Hydrazine hydrate gives the pyrazole (**386**) (hy), whereas phenylhydrazine gives the acyclic hydrazone (**387**) (hy). Thermolysis (20 min at 275 °C) gives tetraphenyl-*p*-benzoquinone (my) and sulphur (Scheme 79).[162]

20.4.21 MESO-IONIC TETRAZOLES OF TYPE B

20.4.21.1 1,2,3,4-Tetrazolium-5-olates (391)

Oxidation of 1,5-diaryl-3-nitroformazans (**389**) or 1,5-diaryl-3-chloroformazans (**390**) with pentyl nitrite in acid solution gives colourless crystalline meso-ionic tetrazolium-5-olates (**391**) (hy) (Scheme 80).[163] The diphenyl compound (**391**; $R^1 = R^2 = Ph$) is also made by oxidation of diphenylcarbazide (**392**; $R^1 = R^2 = Ph$) or diphenylcarbazone (PhN=NC(O)NHNHPh).[164]

$R^2N{=}N$, $R^1NH{-}N$, X → $R^2N{-}N$, R^1N, N, $\bar{O}$ ⇌ (ii / iii) $R^2NH{-}NH$, =O, $R^1NH{-}NH$

(**389**) $X = NO_2$
(**390**) $X = Cl$

(**391**)

(**392**)

i, $C_5H_{11}ONO$, HCl; ii, MnO_2; iii, $(NH_4)_2S$.

SCHEME 80

Only diaryl derivatives of this system (**391**) have been prepared and little is known about their chemistry. The u.v. spectrum of the diphenyl compound lends support to the meso-ionic formulation,[165] but other spectral properties have not been reported. Compound (**391**; $R^1 = R^2 = Ph$) forms salts with acids and is reduced to the carbazide (**392**; $R^1 = R^2 = Ph$) by ammonium sulphide.[166] Thermolysis at 150 °C gives the type A isomer (**344**; $R^1 = R^2 = Ph$) (my).[167]

20.4.21.2 1,2,3,4-Tetrazolium-5-aminides (393)

Two examples of meso-ionic tetrazolium-5-aminides have been reported: both are prepared from the tetrazolium chloride (**395**) (Scheme 81).[9,166] With nitrous acid the orange crystalline *N*-nitroso-derivative (**396**) is formed (my). This product (**396**) rearranges to the *N*-oxide (**398**) in hot ethanol and the mechanism of this transformation (**396**) → (**398**) may well involve the valence tautomer (**394**); *cf.* thermal rearrangement of dehydrodithizone (**405**) (Section 20.4.21.3). Treatment of compound (**395**) with toluene-*p*-sulphonyl chloride followed by aqueous alkali gives the colourless derivative (**397**; $Ar = p\text{-}MeC_6H_4SO_2$) (my), but a similar procedure using benzoyl chloride gives a formazan and not the meso-ionic product. No other information about this meso-ionic system (**393**) is available.

20.4.21.3 1,2,3,4-Tetrazolium-5-thiolates (401)

The only known example of this system is the diphenyl derivative (**401**; $R^1 = R^2 = Ph$), which is referred to as dehydrodithizone.

Dehydrodithizone (**401**; $R^1 = R^2 = Ph$) is prepared as red crystals (hy) by oxidation of dithizone (**399**; $R^1 = R^2 = Ph$) using manganese dioxide (Scheme 82).[168] Oxidation using selenium dioxide in concentrated hydrochloric acid or iodine in water gives the formazan

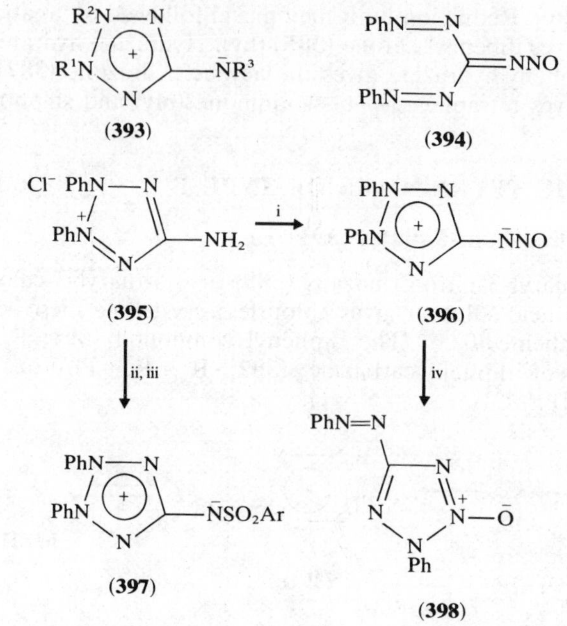

(393) (394)

(395) (396)

(397) (398)

i, HONO; ii, p-$MeC_6H_4SO_2Cl$; iii, aq. NaOH; iv, Δ.

SCHEME 81

disulphide (**400**), which disproportionates to dithizone (**399**; $R^1 = R^2 = Ph$) and dehydrodithizone (**401**; $R^1 = R^2 = Ph$) on standing at room temperature.[169] The structure of this meso-ionic compound (**401**; $R^1 = R^2 = Ph$) has been confirmed by X-ray crystallography.[170] Treatment with alkyl halides results in *S*-alkylation, giving tetrazolium salts.[171] Thermolysis in sand at 150 °C results in a rearrangement (ly) to the isomeric type A meso-ionic 1,2,3,4-tetrazolium-5-thiolate (**349**; $R^1 = R^2 = Ph$) (Section 20.4.15.3).[167]

Many of the reactions of dehydrodithizone (**405**) can be interpreted in terms of an initial ring-opening to the valence tautomer (**406**) (Scheme 83). In hot acetic acid the reddish-bronze isomer (**402**) is formed (my), presumably by an electrocyclic ring-closure and hydrogen transfer.[171,172] Treatment with electron-rich olefins gives products (my) which can be regarded as Diels–Alder adducts of the species (**406**).[173] Typically, 1,2,5,6-tetrahydro-1-methyl-4-morpholinopyridine gives the red crystalline 1,3,4-thiadiazine

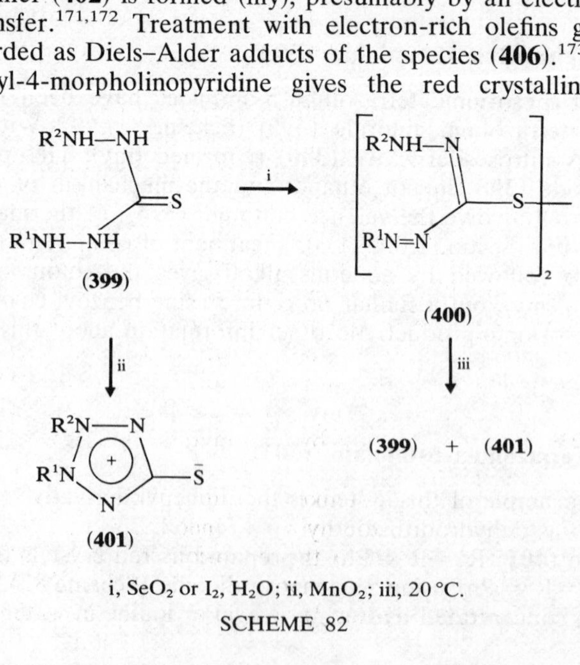

(399) (400)

(399) + (401)

(401)

i, SeO_2 or I_2, H_2O; ii, MnO_2; iii, 20 °C.

SCHEME 82

(**402**) (**403**) (**404**)

(**405**) (**406**)

(**407**) (**408**) (**409**)

i, hot $MeCO_2H$; ii, $Fe_2(CO)_{10}$; iii, 2-morpholinoindene; iv, $MeC{\equiv}CNEt_2$; v, 1,2,5,6-tetrahydro-1-methyl-4-morpholinopyridine.

SCHEME 83

(**409**), whose structure has been determined by an X-ray study.[173] With electron-rich acetylenes, *e.g.* 1-diethylaminopropyne, the product is a pyrazole derivative (my) (*e.g.* **407**), which is best regarded as being formed by elimination of sulphur from the antiaromatic primary adduct (**408**).[173] An unusual variation of the reaction of dehydrodithizone (**405**) with enamines is the behaviour with indene derivatives. In this case the products (ly–my), exemplified by that formed from 2-morpholinoindene, have spiro-structures (*e.g.* **404**).[173] Iron pentacarbonyl and dehydrodithizone (**405**) give the 1,3,4-thiadiazol-2-one (**403**) (my), presumably by carbonyl insertion into the valence tautomer (**406**).[174]

With electron-deficient alkynes and alkenes a different type of cycloaddition to the dehydrodithizone skeleton occurs. Dimethyl acetylenedicarboxylate gives the adduct (**410**) (Scheme 84), presumably by addition to the meso-ionic structure (**411**) rather than the valence tautomer.[173,175] Tetracyanoethylene gives compound (**412**)[175] and tetraphenylcyclopentadienone gives a similar adduct.[173] A sound rationale, based on frontier-orbital

(413) → (414) → (415)

TABLE 1
The Preparation of Five-membered Heteroaromatic Systems (**415**) from Type A Meso-ionic Heterocycles (**414**) and Acetylenes ($R^2C{\equiv}CR^3$)[a]

Heteroaromatic ring system	*Precursor* (**413** or **414**)					*Cyclo-dehydrating agent*[b]	*Product* (**415**)					*Yield* (%)
		X	Y	Z	R^1		Y	Z	R^1	R^2	R^3	
Pyrroles	(**413**)	O	PhC	MeN	Ph	Ac_2O	PhC	MeN	Ph	CO_2Me	CO_2Me	93[c]
	(**413**)	O	PhC	MeN	Ph	Ac_2O	PhC	MeN	Ph	CO_2Me	Ph	87[c]
	(**414**)	O	PhC	MeN	Ph	—	PhC	MeN	Ph	H	H	96[c]
	(**413**)	O	HC	PhN	H	Ac_2O	HC	PhN	H	CO_2Me	CO_2Me	91[c]
	(**413**)	O	MeC	PhN	Me	Ac_2O	MeC	PhN	Me	CO_2Me	CO_2Me	80[c]
	(**413**)	O	PhC	MeN	H	Ac_2O	PhC	MeN	H	PhCO	PhCO	65[d]
Furans	(**413**)	S	piperidino-C	O	Ph	Ac_2O	piperidino-C	O	Ph	CO_2Me	CO_2Me	37[e]
	(**413**)	S	piperidino-C	O	Ph	DCCI	piperidino-C	O	Ph	CO_2Me	CO_2Me	67[e]
	(**413**)	S	pyrrolidino-C	O	Ph	DCCI	pyrrolidino-C	O	Ph	H	CO_2Me	55[e]
	(**413**)	S	pyrrolidino-C	O	Ph	Ac_2O	pyrrolidino-C	O	Ph	CO_2Me	CO_2Me	60[e]
	(**413**)	S	morpholino-C	O	Ph	Ac_2O	morpholino-C	O	Ph	CO_2Me	CO_2Me	49[e]
	(**413**)	S	MeSC	O	Ph	DCCI	MeSC	O	Ph	CO_2Me	CO_2Me	80[e]
Thiophens	(**414**)	S	PhC	S	Ph	—	PhC	S	Ph	Ph	H	91[f]
	(**414**)	S	PhC	S	Ph	—	PhC	S	Ph	PhCO	Ph	78[f]
	(**414**)	S	PhC	S	Ph	—	PhC	S	Ph	Ph	CO_2Me	82[f]
	(**414**)	S	PhC	S	Me	—	PhC	S	Me	CO_2Me	CO_2Me	67[f]
	(**414**)	S	PhC	S	Ph	—	PhC	S	Ph	CO_2Me	CO_2Me	99[f]
	(**414**)	S	PhC	S	H	—	PhC	S	H	CO_2Me	CO_2Me	40[g]
	(**414**)	S	*p*-$MeOC_6H_4C$	S	H	—	*p*-$MeOC_6H_4C$	S	H	CO_2Me	CO_2Me	47[g]
	(**414**)	PhN	PhC	S	Ph	—	PhC	S	Ph	PhCO	PhCO	42[d,h]
	(**414**)	PhN	PhC	S	Ph	—	PhC	S	Ph	CO_2Me	CO_2Me	90[h]
	(**414**)	PhN	PhC	S	Ph	—	PhC	S	Ph	CF_3	CF_3	90[h]
	(**414**)	PhN	PhC	S	Ph	—	PhC	S	Ph	CN	CN	95[h]

Pyrazoles	(**414**) O	N	PhN	H	—	N	PhN	H	H	H	75[i]
	(**414**) O	N	PhN	H	—	N	PhN	H	H	CH_2OH	66[i]
	(**414**) O	N	PhN	H	—	N	PhN	H	H	Ph	79[i]
	(**414**) O	N	$PhCH_2N$	H	—	N	PhN	H	H	Ph	74[i]
	(**414**) O	N	PhN	Ph	—	N	PhN	Ph	Ph	Ph	98[i]
	(**414**) O	N	PhN	Me	—	N	PhN	Me	Ph	Ph	96[i]
	(**414**) O	N	3-pyridyl-N	H	—	N	3-pyridyl-N	H	H	Ph	92[i]
	(**414**) O	N	PhN	H	—	{N	PhN	H	H	CO_2Me	70[i]
						N	PhN	H	CO_2Me	H	22[i]
	(**414**) O	N	PhN	H	—	N	PhN	H	CO_2Me	CO_2Me	92[i]
	(**414**) O	N	$PhCH_2N$	H	—	N	$PhCH_2N$	H	CO_2Me	CO_2Me	93[i]
	(**414**) O	N	PhN	Me	—	N	PhN	Me	CO_2Me	CO_2Me	100[i]
	(**414**) O	N	PhN	H	—	N	PhN	H	MeCO	Ph	99[i]
	(**413**) O	N	PhN	H	Ac_2O	N	PhN	H	PhCO	PhCO	63[j]
	(**414**) O	N	PhN	H	—	N	PhN	H	PhCO	PhCO	85[j]
	(**414**) O	N	NH_2N	H	—	N	NH_2N	H	H	Ph	60[k]
Isothiazoles	(**414**) O	N	S	Ph	—	N	S	Ph	CO_2Me	CO_2Me	56[l]
	(**414**) O	N	S	p-MeC_6H_4	—	N	S	p-MeC_6H_4	CO_2Me	CO_2Me	63[l]
	(**414**) O	N	S	p-BrC_6H_4	—	N	S	p-BrC_6H_4	CO_2Me	CO_2Me	57[l]
	(**414**) O	N	S	Ph	—	{N	S	Ph	Ph	H	43[l]
						N	S	Ph	H	Ph	1[l]
	(**414**) O	N	S	Ph	—	N	S	Ph	PhCO	PhCO	18[m]

[a] The meso-ionic precursor (**414**) may be isolated or generated *in situ* from the acid (**413**). [b]Only given for systems where the meso-ionic intermediate is generated *in situ* in the presence of the appropriate acetylene, [c] R. Huisgen, H. Gotthardt, H. O. Bayer, and F. C. Schaefer, *Chem. Ber.*, 1970, **103,** 2611. [d] K. T. Potts and D. McKeough, *J. Amer. Chem. Soc.*, 1974, **96,** 4268, [e] H. Gotthardt, M. C. Weisshuhn, and K. Dorhofer, *Angew. Chem. Internat. Edn.*, 1975, **14,** 422. [f] H. Gotthardt and B. Christl, *Tetrahedron Letters*, 1968, 4747. [g] K. T. Potts and U. P. Singh, *Chem. Comm.*, 1969, 569. [h] K. T. Potts, E. Houghton, and U. P. Singh, *J. Org. Chem.*, 1974, **39,** 3627. [i] R. Huisgen, H. Gotthardt, and R. Grashey, *Chem. Ber.*, 1968, **101,** 536. [j] K. T. Potts and D. McKeough, *J. Amer. Chem. Soc.*, 1974, **96,** 4276. [k] S. Kishimoto, S. Noguchi, and K. Masuda, *Chem. Pharm. Bull. (Japan)*, 1976, **24,** 3001. [l] H. Gotthardt, *Chem. Ber.*, 1972, **105,** 196. [m] H. Gotthardt and F. Reiter, *Tetrahedron Letters*, 1976, 2163.

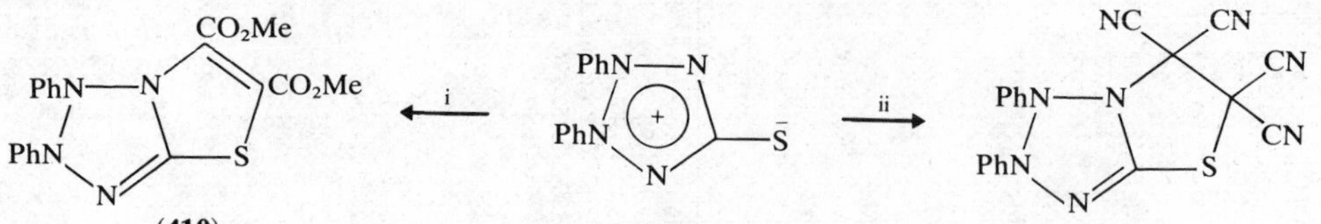

(410) **(411)** **(412)**

i, $MeO_2CC{\equiv}CCO_2Me$; ii, $(NC)_2C{=}C(CN)_2$.

SCHEME 84

theory, has been proposed to account for the differing behaviour of dehydrodithizone with electron-deficient and electron-rich dipolarophiles.[173]

20.4.22 SYNTHESIS OF FIVE-MEMBERED HETEROCYCLES FROM TYPE A MESO-IONIC HETEROCYCLES

The 1,3-dipolar cycloaddition of alkynes to meso-ionic compounds giving five-membered heterosystems *via* bicyclic adducts (see Scheme 1, Section 20.4.2.2) has proved to be a valuable synthetic reaction. The method is particularly advantageous when the liberated heterocumulene is carbon dioxide (*i.e.* **414**; X = O), but other leaving groups

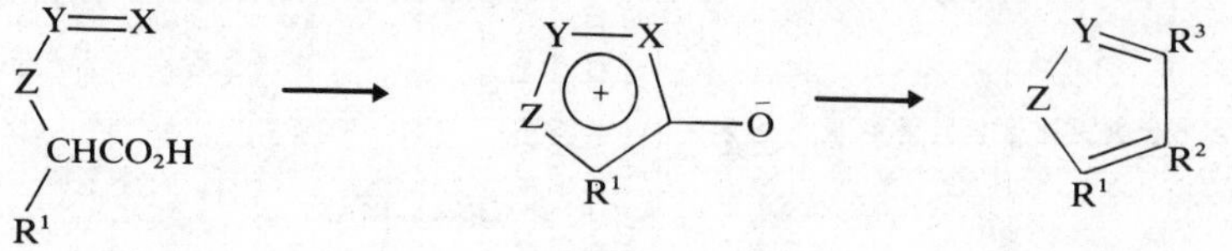

(413) **(414)** **(415)**

have also been highly successful. Many novel derivatives of pyrroles, furans, thiophens, pyrazoles, and isothiazoles have been made in high yield by this route (**414** → **415**; Table 1). Another attractive facet of the method is that the meso-ionic substrate (**414**) need not be isolated but can be generated *in situ* from its acyclic precursor (**413**) in the presence of the acetylene.

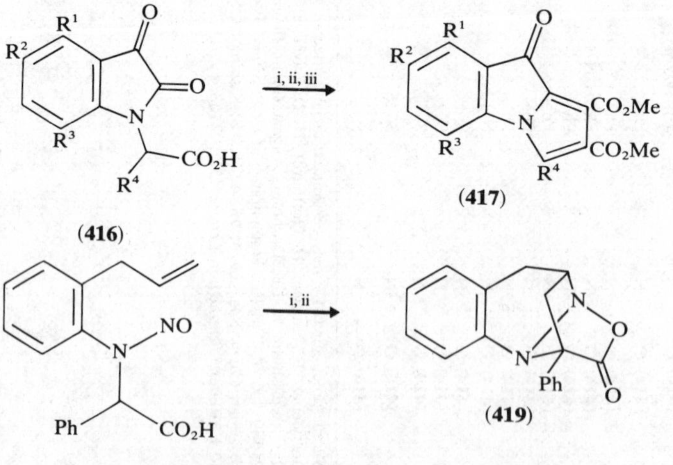

(416) **(417)**

(418) **(419)**

i, Ac_2O; ii, Δ; iii, $MeO_2CC{\equiv}CCO_2Me$.

SCHEME 85

(**414**) (**420**) (**421**)

i, PhCOC≡CCOPh; ii, P_4S_{10}.

SCHEME 86

In the preparation of pyrazoles or isothiazoles using unsymmetrical acetylenes, isomeric products are formed. In practice, although the reaction is seldom regiospecific, one isomer usually predominates. Examples of the value of meso-ionic synthons for making fairly complex heterosystems are the preparation of the tricyclic pyrrole (**417**) from the münchnone precursor (**416**)[176] and the intramolecular adduct (**419**) from the sydnone precursor (**418**) (Scheme 85).[177] The dibenzoyl heterosystems (**420**) are particularly useful for the preparation of the bicyclic mesomeric betaines (**421**) (Scheme 86).[12]

20.4.23 BIOLOGICAL ACTIVITY OF MESO-IONIC HETEROCYCLES

A wide spectrum of biological activity has been claimed for a variety of type A meso-ionic compounds. This aspect of meso-ionic chemistry has been the subject of a review[5] and the main claims for biological activity have also been summarized in a recent monograph.[9] Particular attention has been directed towards the sydnones (**153**; Section 20.4.10.1) and the sydnone imines (**170**; Section 20.4.10.2). In particular, their properties as antibacterial, analgesic, anti-inflammatory, antimalarial, and insecticidal agents have been of interest. Two sydnone imine derivatives, sydnofen and sydnocarb,[9] have undergone clinical trials as antidepressants.

The following claims for activity have also been made: 1,3-diazolium-4-olates (**98**) as anthelmintics; 1,2,3-triazolium-4-olates (**224**) as herbicides; 1,2,4-triazolium-3-olates (**253**) as analgesics and anti-inflammatory agents; 1,3,4-thiadiazolium-2-aminides (**304**) as sedatives; 1,3,4-thiadiazolium-2-thiolates (**312**) as antibacterial agents; 1,2,3,4-oxatriazolium-5-olates (**329**) and 1,2,3,4-oxatriazolium-5-aminides (**334**) as hypotensive agents.

No biological activity of type B meso-ionic molecules has been claimed.

References

1. (a) W. Baker, W. D. Ollis, and V. D. Poole, *J. Chem. Soc.*, 1949, 307; 1950, 1542; (b) A. Schönberg, *ibid.*, 1938, 824.
2. W. Baker and W. D. Ollis, *Quart. Rev.*, 1957, **11,** 15.
3. Y. Noël, *Bull. Soc. chim. France*, 1964, 173.
4. F. H. C. Stewart, *Chem. Rev.*, 1964, **64,** 129.
5. L. B. Kier and E. B. Roche, *J. Pharm. Sci.*, 1967, **56,** 149.
6. N. Sucia, *Stud. Cercet. Chim.*, 1968, **16,** 117.
7. R. Huisgen, *Chem. Soc. Spec. Publ.*, 1967, No. 21, p. 51.
8. M. Ohta and H. Kato, in 'Nonbenzenoid Aromatics,' ed. J. P. Snyder, Academic Press, New York, 1969, pp. 117–248.
9. W. D. Ollis and C. A. Ramsden, *Adv. Heterocyclic Chem.*, 1976, **19,** 1.
10. N. Dennis, A. R. Katritzky, and Y. Takeuchi, *Angew. Chem. Internat. Edn.*, 1976, **15,** 1.
11. C. A. Ramsden, *J. C. S. Chem. Comm.*, 1977, 109.
12. C. A. Ramsden, *Tetrahedron*, 1977, **33,** 3203.
13. R. Huisgen, *Rev. Real Acad. Cienc. Exactas, Fis. Natur. Madrid*, 1971, **65,** 293.
14. K. T. Potts, J. Baum, E. Houghton, D. N. Roy, and U. P. Singh, *J. Org. Chem.*, 1974, **39,** 3619.
15. M. Hamaguchi and T. Ibata, *Tetrahedron Letters*, 1974, 4475.
16. T. Ibata, M. Hamaguchi, and H. Kiyohara, *Chem. Letters*, 1975, 21.

17. A. Chinone, S. Sato, T. Mase, and M. Ohta, *Bull. Chem. Soc. Japan*, 1969, **42,** 2310.
18. G. V. Boyd, in 'Aromaticity, Pseudo-aromaticity, Anti-aromaticity,' ed. E. D. Bergmann and B. Pullman, Israel Acad. Sci. Humanities, Jerusalem, 1971, p. 166.
19. R. Knorr and R. Huisgen, *Chem. Ber.*, 1970, **103,** 2598.
20. A. Lawson and D. H. Miles, *J. Chem. Soc.*, 1959, 2865; 1960, 1945.
21. G. V. Boyd and P. H. Wright, *J. C. S. Perkin I*, 1972, 909, 914.
22. K. T. Potts and D. McKeough, *J. Amer. Chem. Soc.*, 1973, **95,** 2749.
23. R. Huisgen, H. Gotthardt, and H. O. Bayer, *Chem. Ber.*, 1970, **103,** 2368.
24. R. Huisgen, E. Funke, F. C. Schaefer, and R. Knorr, *Angew. Chem. Internat. Edn.*, 1967, **6,** 367.
25. D. Clerin, B. Meyer, J. P. Fleury, and H. Fritz, *Tetrahedron*, 1976, **32,** 1055.
26. S. Sato, T. Mase, and M. Ohta, *Bull. Chem. Soc. Japan*, 1968, **41,** 2218.
27. W. E. McEwen, I. C. Mineo, Y. H. Shen and G. Y. Han, *Tetrahedron Letters*, 1968, 5157.
28. H. Gotthardt, M. C. Weisshuhn, and K. Dörhöfer, *Angew. Chem. Internat. Edn.*, 1975, **14,** 422.
29. K. T. Potts, J. Kane, E. Carnahan, and U. P. Singh, *J. C. S. Chem. Comm.*, 1975, 417.
30. R. Grashey, E. Jänchen, and J. Litzke, *Chem. Z.*, 1973, **97,** 657.
31. A. Lawson and D. H. Miles, *J. Chem. Soc.*, 1959, 2865.
32. T. Shiba and H. Kato, *Bull. Chem. Soc. Japan*, 1970, **43,** 3941.
33. G. Singh and P. S. Pande, *Tetrahedron Letters*, 1974, 2169.
34. K. T. Potts and S. Husain, *J. Org. Chem.*, 1971, **36,** 3368.
35. A. Chinone, S. Sato, and M. Ohta, *Bull. Chem. Soc. Japan*, 1971, **44,** 826.
36. R. Huisgen, E. Funke, F. C. Schaefer, H. Gotthardt, and E. Brunn, *Tetrahedron Letters*, 1967, 1809.
37. K. T. Potts and D. N. Roy, *Chem. Comm.*, 1968, 1062.
38. K. T. Potts, E. Houghton, and U. P. Singh, *J. Org. Chem.*, 1974, **39,** 3627.
39. S. Abrahamsson, A. Westerdahl, G. Isaksson, and J. Sandström, *Acta Chem. Scand.*, 1967, **21,** 442.
40. S. Nakazawa, T. Kiyosawa, K. Hirakawa, and H. Kato, *J. C. S. Chem. Comm.*, 1974, 621.
41. K. T. Potts, J. Baum, and E. Houghton, *J. Org. Chem.*, 1974, **39,** 3631.
42. M. Ohta, K. Yoshida, and S. Sato, *Bull. Chem. Soc. Japan*, 1966, **39,** 1269.
43. H. Kato and M. Ohta, *Bull. Chem. Soc. Japan*, 1966, **39,** 1253.
44. K. T. Potts, S. Husain, and S. Husain, *Chem. Comm.*, 1970, 1360.
45. S. Sato and M. Ohta, *Bull. Chem. Soc. Japan*, 1968, **41,** 2801.
46. A. Chinone, Y. Huseya, and M. Ohta, *Bull. Chem. Soc. Japan*, 1970, **43,** 2650.
47. A. Lawson and C. E. Searle, *J. Chem. Soc.*, 1957, 1556.
48. M. Ohta and C. Shin, *Bull. Chem. Soc. Japan*, 1965, **38,** 704.
49. E. Funke, R. Huisgen, and F. C. Schaefer, *Chem. Ber.*, 1971, **104,** 1550.
50. T. Shiba and H. Kato, *Bull. Chem. Soc. Japan*, 1971, **44,** 1864.
51. T. Shiba and H. Kato, *Bull. Chem. Soc. Japan*, 1973, **46,** 964.
52. G. C. Barrett, A. R. Khokhar, and J. R. Chapman, *Chem. Comm.*, 1969, 818.
53. J. E. Baldwin, M. C. McDaniel, M. G. Newton, and I. C. Paul, *Tetrahedron Letters*, 1966, 4239.
54. M. P. Cava and L. E. Saris, *J. C. S. Chem. Comm.*, 1975, 617.
55. H. Gotthardt and B. Christl, *Tetrahedron Letters*, 1968, 4743; H. Gotthardt, M. C. Weisshuhn and B. Christl, *Chem. Ber.*, 1976, **109,** 740.
56. K. T. Potts and U. P. Singh, *Chem. Comm.*, 1969, 569; K. T. Potts, D. R. Choudhury, A. J. Elliott, and U. P. Singh, *J. Org. Chem.*, 1976, **41,** 1724.
57. H. Gotthardt and B. Christl, *Tetrahedron Letters*, 1968, 4747, 4751.
58. H. Kato, M. Kawamura, T. Shiba, and M. Ohta, *Chem. Comm.*, 1970, 959.
59. M. Ohta and M. Sugiyama, *Bull. Chem. Soc. Japan*, 1963, **36,** 1437.
60. H. Kato, T. Shiba, and Y. Miki, *J. C. S. Chem. Comm.*, 1972, 498.
61. J. H. Bowie, R. A. Eade, and J. C. Earl. *Austral. J. Chem.*, 1968, **21,** 1665.
62. H. Bärnighausen, F. Jellinek, J. Munnik, and A. Vos, *Acta Cryst.*, 1963, **16,** 471; W. E. Thiessen and H. Hope, *J. Amer. Chem. Soc.*, 1967, **89,** 5977.
63. H. U. Daeniker, *Helv. Chim. Acta*, 1976, **50,** 2008.
64. G. A. Olah, D. P. Kelly, and N. Sucia, *J. Amer. Chem. Soc.*, 1970, **92,** 3133.
65 K. T. Potts, E. Houghton, and S. Husain, *Chem. Comm.*, 1970, 1025.
66. C. H. Krauch, J. Kuhls, and H.-J. Piek, *Tetrahedron Letters*, 1966, 4043.
67. K. Nakahara and M. Ohta, *J. Chem. Soc. Japan*, 1956, **77,** 1306.
68. C. V. Greco, M. Pesce, and J. M. Franco, *J. Heterocyclic Chem.*, 1966, **3,** 391.
69. H. Kato and M. Ohta, *Bull. Chem. Soc. Japan*, 1959, **32,** 282.
70. M. Ohta and H. Kato, *J. Chem. Soc. Japan*, 1957, **78,** 1653.
71. K. T. Potts and U. P. Singh, *Chem. Comm.*, 1969, 66.
72. P. Brookes and J. Walker, *J. Chem. Soc.*, 1957, 4409.
73. H. U. Daeniker and J. Druey, *Helv. Chim. Acta*, 1962, **45,** 2441, 2462.
74. L. E. Kholodov and V. G. Yashunskii, *Proc. Acad. Sci. (U.S.S.R.)*, 1968, **179,** 248.
75. H. U. Daeniker and J. Druey, *Helv. Chim. Acta*, 1963, **46,** 805.
76. H. U. Daeniker and J. Druey, *Helv. Chim. Acta*, 1962, **45,** 2426.
77. L. E. Kholodov and V. G. Yashunskii, *J. Gen. Chem. (U.S.S.R.)*, 1965, **35,** 1555.
78. H. U. Daeniker, *Helv. Chim. Acta*, 1964, **47,** 2421.
79. A. Chinone and M. Ohta, *Chem. Letters*, 1972, 969.

80. V. G. Yashunskii and V. S. Fedorovich, *J. Gen. Chem. (U.S.S.R.)*, 1964, **34,** 3112.
81. M. Hashimoto and M. Ohta, *Bull. Chem. Soc. Japan*, 1961, **34,** 668.
82. A. R. McCarthy, W. D. Ollis, A. N. M. Barnes, L. E. Sutton, and C. Ainsworth, *J. Chem. Soc. (B)*, 1969, 1185.
83. K. Hoegerle, *Helv. Chim. Acta*, 1958, **41,** 548.
84. A. R. McCarthy, W. D. Ollis, and C. A. Ramsden, *J. C. S. Perkin I*, 1974, 627.
85. W. D. Ollis and C. A. Ramsden, *J. C. S. Perkin I*, 1974, 642.
86. R. Grashey, R. Hamprecht, N. Keramaris, and M. Baumann, *Tetrahedron Letters*, 1972, 2943.
87. R. Grashey, N. Keramaris, and M. Baumann, *Tetrahedron Letters*, 1970, 5087.
88 A. Ya. Lazaris, *J. Org. Chem. (U.S.S.R.)*, 1967, **3,** 1856; 1968, **4,** 1786.
89. R. Grashey, M. Baumann, and R. Hamprecht, *Tetrahedron Letters*, 1970, 5083.
90. H. Gotthardt, *Chem. Ber.*, 1972, **105,** 188, 196.
91. G. D. Andreetti, G. Bocelli, L. Cavalca, and P. Sgarabotto, *Gazzetta*, 1972, **102,** 23.
92. A. Alemagna and T. Bacchetti, *Chim. Ind. (Milan)*, 1972, **54,** 1105.
93. A. Holm, N. Harrit, and N. H. Toubro, *Tetrahedron*, 1976, **32,** 2559.
94. K. T. Potts and S. Husain, *J. Org. Chem.*, 1970, **35,** 3451; 1972, **37,** 2049.
95. M. Begtrup and C. Pedersen, *Acta Chem. Scand.*, 1967, **21,** 633.
96. M. Begtrup, *Acta Chem. Scand.*, 1971, **25,** 249.
97. M. Begtrup, *J. C. S. Chem. Comm.*, 1975, 334.
98. M. Begtrup, *Tetrahedron Letters*, 1971, 1577.
99. M. Begtrup, *Acta Chem. Scand.*, 1972, **26,** 1243.
100. M. Begtrup, *Acta Chem. Scand.*, 1971, **25,** 3500.
101. K. T. Potts, S. K. Roy, and D. P. Jones, *J. Heterocyclic Chem.*, 1965, **2,** 105.
102. G. Palazzo and L. Baiocchi, *Ann. Chim. (Rome)*, 1965, **55,** 935.
103. K. T. Potts, S. K. Roy, S. W. Schneller, and R. M. Huseby, *J. Org. Chem.*, 1968, **33,** 2559.
104. H. Kato, S. Sato, and M. Ohta, *Tetrahedron Letters*, 1967, 4261.
105. S. Kubota and M. Uda, *Chem. Pharm. Bull. (Japan)*, 1976, **24,** 1336.
106. G. Doleschall, *Tetrahedron*, 1976, **32,** 2549.
107. W. D. Ollis and C. A. Ramsden, *J. C. S. Perkin I*, 1974, 645.
108. K. T. Potts, E. Houghton, and S. Husain, *Chem. Comm.*, 1970, 1025.
109. M. Busch, *Ber.*, 1905, **38,** 856; M. Busch and G. Mehrtens, *ibid.*, 1905, **38,** 4049.
110. M. Busch, H. Brandt, and G. Blume, *J. prakt. Chem.*, 1906, **74,** 533.
111. W. D. Ollis and C. A. Ramsden, *J. C. S. Perkin I*, 1974, 638.
112. E. Cawkill, W. D. Ollis, C. A. Ramsden, and G. P. Rowson, *J. C. S. Chem. Comm.*, 1976, 439.
113. A. I. Vogel, 'A Textbook of Quantitative Inorganic Analysis,' Longmans, London, 3rd edn., 1962, p. 131.
114. K. T. Potts, S. K. Roy, and D. P. Jones, *J. Org. Chem.*, 1967, **32,** 2245.
115. W. D. Ollis and C. A. Ramsden, *J. C. S. Perkin I*, 1974, 633.
116. W. D. Ollis and G. P. Rowson, *J. C. S. Chem. Comm.*, 1976, 440.
117. R. Grashey and M. Baumann, *Tetrahedron Letters*, 1972, 2947; *Angew. Chem. Internat. Edn.*, 1969, **8,** 133.
118. A. Y. Lazaris and A. N. Egorochkin, *Izvest. Akad. Nauk S.S.S.R., Ser. khim.*, 1976, 1191 (*Chem. Abs.*, 1976, **85,** 108 584n).
119. G. F. Duffin and J. D. Kendall, *J. Chem. Soc.*, 1956, 3189.
120. J. Goerdeler and W. Roth, *Chem. Ber.*, 1963, **96,** 534.
121. R. Grashey, M. Baumann, and W.-D. Lubos, *Tetrahedron Letters*, 1968, 5877.
122. R. M. Moriarty and R. Mukherjee, *Tetrahedron Letters*, 1969, 4627.
123. R. Grashey, M. Baumann, and R. Hamprecht, *Tetrahedron Letters*, 1972, 2939.
124. P. Thieme, M. Patsch, and H. König, *Annalen*, 1972, **764,** 94.
125. W. Baker, W. D. Ollis, A. Phillips, and T. Strawford, *J. Chem. Soc.*, 1951, 289.
126. M. Ohta, H. Kato, and T. Kaneko, *Bull. Chem. Soc. Japan*, 1967, **40,** 579.
127. R. Grashey, M. Baumann, and W.-D. Lubos, *Tetrahedron Letters*, 1968, 5881.
128. C. W. Atkin, A. N. M. Barnes, P. G. Edgerley, and L. E. Sutton, *J. Chem. Soc. (B)*, 1969, 1194.
129. P. B. Talukdar and S. K. Sengupta, *J. Indian Chem. Soc.*, 1970, **47,** 49; A. M. Kiwan and H. M. N. H. Irving, *J. Chem. Soc.* (*B*), 1971, 898; A. M. Kiwan and H. M. Marafie, *J. Heterocyclic Chem.*, 1976, **13,** 1273.
130. M. Patsch and P. Thieme, *Angew. Chem. Internat. Edn.*, 1971, **10,** 569.
131. R. Grashey, C. Knorn, and M. Weidner, *Chem. Z.*, 1973, **97,** 565.
132. A. Y. Lazaris, S. M. Shmuilovich, and A. N. Egorochkin, *Khim. geterotsikl. Soedinenii*, 1976, 713.
133. W. L. Mosby and M. L. Vega, *Chem. Comm.*, 1971, 837.
134. R. M. Moriarty, J. M. Kliegman, and R. B. Desai, *Chem. Comm.*, 1967, 1255.
135. R. M. Moriarty, Ŕ. Mukherjee, J. L. Flippen, and J. Karle, *Chem. Comm.*, 1971, 1436.
136. J. L. Flippen, *Acta Cryst.*, 1972, **B28,** 2749.
137. J. H. Boyer and F. C. Canter, *J. Amer. Chem. Soc.*, 1955, **77,** 1280.
138. T. L. Thomas, M. Fedorchuk, B. V. Shetty, and F. E. Anderson, *J. Medicin. Chem.*, 1970, **13,** 196.
139. J. H. Boyer and J. A. Hernandez, *J. Amer. Chem. Soc.*, 1956, **78,** 5124.
140. M. Hashimoto and M. Ohta, *Bull. Chem. Soc. Japan*, 1962, **35,** 766.
141. W. V. Farrar, *J. Chem. Soc.*, 1964, 906.
142. S. Hünig and O. Boes, *Annalen*, 1953, **579,** 28.

143. E. A. Isukul, R. Ranson and J. G. Tillett, *J. Org. Chem.*, 1976, **41,** 3040.
144. H. Kato, T. Shiba, H. Yoshida, and S. Fujimori, *Chem. Comm.*, 1970, 1591.
145. R. Huisgen, H. Gotthardt, and R. Grashey, *Chem. Ber.*, 1968, **101,** 536.
146. M. Busch and W. Schmidt, *Ber.*, 1929, **62,** 1449.
147. C. Christopherson and S. Treppendahl, *Acta Chem. Scand.*, 1971, **25,** 625.
148. K. Masuda, T. Kamiya, and K. Kashiwa, *Chem. Pharm. Bull. (Japan)*, 1971, **19,** 559.
149. W. G. Finnegan and R. A. Henry, *J. Org. Chem.*, 1965, **30,** 567.
150. R. N. Hanley, W. D. Ollis, and C. A. Ramsden, *J. C. S. Chem. Comm.*, 1976, 306.
151. H. J. Backer, *Rec. Trav. chim.*, 1951, **70,** 733.
152. Z. Tamura, Y. Iitaka, H. Tanabe, and S. Uchiyama, *Chem. Pharm. Bull. (Japan)*, 1970, **18,** 2359.
153. J. H. Bryden, R. A. Henry, W. G. Finnegan, R. H. Boschan, W. S. McEwan, and R. W. Van Dolah, *J. Amer. Chem. Soc.*, 1953, **75,** 4863.
154. R. A. Henry, W. G. Finnegan, and E. Lieber, *J. Amer. Chem. Soc.*, 1954, **76,** 2894.
155. V. P. Shchipanov, Y. N. Sheinker, and I. Y. Postovskii, *J. Org. Chem. (U.S.S.R.)*, 1966, **2,** 342; V. P. Shchipanov, *ibid.*, 1966, **2,** 347, 1471.
156. R. N. Hanley, W. D. Ollis, and C. A. Ramsden, *J. C. S. Chem. Comm.*, 1976, 307.
157. G. V. Boyd and T. Norris, *J. C. S. Perkin I*, 1974, 1028.
158. M. J. Nye and W.-P. Tang, *Tetrahedron*, 1972, **28,** 455.
159. M. J. Nye, M. J. O'Hare, and W.-P. Tang, *J. C. S. Chem. Comm.*, 1973, 402.
160. A. Chinone, K. Inouye, and M. Ohta, *Bull. Chem. Soc. Japan*, 1972, **45,** 213.
161. D. Barillier, P. Rioult, and J. Vialle, *Bull. Soc. chim. France*, 1976, 444.
162. A. Schönberg and E. Frese, *Chem. Ber.*, 1970, **103,** 3885.
163. R. G. Dubenko and P. S. Pel'kis, *J. Gen. Chem.* (*U.S.S.R.*), 1959, **29,** 200; M. O. Lozinskii and P. S. Pel'kis, *ibid.*, 1963, **33,** 106.
164. P. Cazeneuve, *Bull. Soc. chim. France*, 1901, **25,** 315.
165. P. B. Talukdar, S. K. Sengupta, A. K. Datta, and A. Chakravorty, *Indian J. Chem.*, 1973, **11,** 611.
166. E. Bamberger, R. Padova, and E. Ormerod, *Annalen*, 1926, **446**, 260.
167. P. N. Preston, K. K. Tiwari, K. Turnbull, and T. J. King, *J. C. S. Chem. Comm.*, 1976, 343.
168. E. Fischer and E. Besthorn, *Annalen*, 1882, **212**, 316.
169. R. S. Ramakrishna and H. M. N. H. Irving, *Chem. Comm.*, 1969, 1356.
170. Y. Kushi and Q. Fernando, *J. Amer. Chem. Soc.*, 1970, **92,** 1965.
171. J. W. Ogilvie and A. H. Corwin, *J. Amer. Chem. Soc.*, 1961, **83,** 5023.
172. W. S. McDonald, H. M. N. H. Irving, G. Raper, and D. C. Rupainwar, *Chem. Comm.*, 1969, 392.
173. G. V. Boyd, T. Norris, and P. F. Lindley, *J. C. S. Perkin I*, 1976, 1673.
174. P. N. Preston, N. J. Robinson, K. Turnbull, and T. J. King, *J. C. S. Chem. Comm.*, 1974, 998.
175. P. Rajagopalan and P. Penev, *Chem. Comm.*, 1971, 490.
176. W. K. Anderson and P. F. Corey, *J. Org. Chem.*, 1977, **42,** 559.
177. H. Meier, H. Heimgartner, and H. Schmid, *Helv. Chim. Acta*, 1977, **60,** 1087.